Shooting free throws

Is it better to shoot a free throw overhand or underhand? Mathematics can be used to improve an athlete's performance. Learn how parabolas, the angle of release, and the velocity of the basketball all play a role in determining whether the ball goes through the hoop. (See the Chapter 3 opener on page 169 to learn more.)

Modeling movement of weather

How do meteorologists know where a cold front will be tomorrow? How do they know that one city will be hit by a blizzard, and a city 100 miles away will get only flurries? Scientists model weather systems and make predictions by translating and transforming graphs on a weather map. (Learn more in Section 3.5 on pages 222 and 238.)

Waiting in line

Have you ever noticed how a slight increase in the number of cars exiting a parking garage or trying to get through a construction zone can make your wait much longer? These long lines of waiting cars are subject to a nonlinear effect. To make predictions about traffic congestion, highway engineers often use rational functions. (See the application and Example 2 in Section 4.6 on page 309 to learn more.)

4th edition

PRECALCULUS

with Modeling & Visualization

A RIGHT TRIANGLE APPROACH

Gary K. Rockswold

Minnesota State University, Mankato

Addison-Wesley

Boston   San Francisco   New York
London   Toronto   Sydney   Tokyo   Singapore   Madrid
Mexico City   Munich   Paris   Cape Town   Hong Kong   Montreal

Executive Editor Anne Kelly
Executive Project Manager Christine O'Brien
Senior Project Manager Chere Bemelmans
Assistant Editor Leah Goldberg
Senior Managing Editor Karen Wernholm
Senior Production Supervisor Kathleen A. Manley
Art Direction Beth Paquin
Senior Media Producer Ceci Fleming
Associate Media Producer Nathaniel Koven
Software Development Mary Durnwald, TestGen, and Rebecca Williams, MathXL
Senior Marketing Manager Katherine Greig
Marketing Assistant Katherine Minton
Senior Author Support/Technology Specialist Joe Vetere
Senior Manufacturing Buyer Carol Melville
Production Coordination Kathy Diamond
Text Design, Composition, and Illustrations Nesbitt Graphics, Inc.
Cover Designer Nancy Goulet, studio;wink

Cover photo: Jim Wehtje/Photographer's Choice/Getty Images

Photo credits appear on page P-1.

Library of Congress Cataloging-in-Publication Data
Rockswold, Gary K.
   Precalculus with modeling & visualization / by Gary Rockswold — 4th ed.
     p. cm.
   Includes bibliographical references and index.
   ISBN 0-321-56800-1 (student ed. : alk. paper)

   1. Functions. 2. Graph Theory. I. Title.
QA331.3.R63 2010
512'.13—dc22   2008039185

4 5 6 7 8 9 10—CRK—12 11 10

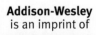

Addison-Wesley
is an imprint of

www.pearsonhighered.com   ISBN-10: 0-321-56800-1
                                     ISBN-13: 978-0-321-56800-7

In memory of my Uncle Jerry,

who passed before his time

Contents

3 Quadratic Functions and Equations *169*

4 | More Nonlinear Functions and Equations   *248*

7 | Trigonometric Identities and Equations   *577*

8 Further Topics in Trigonometry *646*

9 Systems of Equations and Inequalities *730*

10 Conic Sections   *832*

11 Further Topics in Algebra *873*

R Reference: Basic Concepts from Algebra and Geometry *R-1*

Preface

This textbook offers an innovative approach that consistently links mathematical concepts to real-world applications by moving from the concrete to the abstract. It demonstrates the relevance of mathematics and answers the question, "When will I ever need to know this?" This text provides a comprehensive curriculum with the balance and flexibility necessary for today's college mathematics courses. The early introduction of functions and graphs allows the instructor to use applications and visualization to present mathematical topics. Real data, graphs, and tables play an important role in the course, giving meaning to the numbers and equations that students encounter. This approach increases students' interest and motivation and their likelihood of success.

The vast majority of students who study mathematics in college will not become professional mathematicians. Because the importance of mathematics and its applications is accelerating rapidly, however, essentially all of these students will use mathematics during their lifetimes to a much greater extent than they might anticipate. Mathematics courses must prepare students with a variety of skills and the understanding needed to be productive and informed members of society.

Approach

Instructors are free to strike their own balance of skills, rule of four, applications, modeling, and technology. With a flexible approach to the rule of four (verbal, graphical, numerical, and symbolic methods), instructors can easily emphasize one rule more than another to meet their students' needs. This approach also extends to modeling and applications. The use of technology, which helps students visualize mathematical concepts, is *optional* and not a requirement for students to benefit from this approach. Nevertheless, the text still provides a strong option for instructors who wish to implement graphing calculator technology. The text contains numerous applications, including models of real-world data with functions and problem-solving strategies. It is not necessary for an instructor to discuss any particular application; rather, an instructor has the option to choose from a wide variety of topics.

The concept of a function is the unifying theme of the text. It is common for students to examine a type of function and its graph and then use this knowledge to solve associated equations and inequalities. For example, students apply their knowledge of quadratic functions and parabolas to solve quadratic equations and inequalities. This visual approach complements the traditional symbolic approach to solving equations and inequalities and allows students to solve a problem by using more than one method. Functions and their graphs are frequently used to solve applications and model real-world data, and students are often asked to interpret their results. Mathematical skills also play an important role in this text. Numerous exercises have been included so that students can practice their skills.

When students arrive in a college-level mathematics class, they often lack a full mastery of intermediate algebra. Rather than reviewing all of the necessary intermediate algebra skills in the first chapter in hopes that "saying it louder and faster" will help students remember it better, this book integrates them seamlessly throughout the early chapters. In addition, review notes and geometry notes appearing in the margins refer students "just in time" to extra help found in Chapter R: Basic Concepts from Algebra and Geometry.

Content Changes to the Fourth Edition

The fourth edition contains several important changes, which are the result of the many comments and suggestions made by instructors, students, and reviewers. This text contains eleven chapters plus a review chapter at the end of the text. The breadth and depth of several topics have been increased. Some highlights of this revision include the following:

- Hundreds of examples and exercises have been added and revised.
- Several new chapter and section openers have been written using current data to increase student interest.
- Arrows have been added to graphs whenever appropriate.
- Calculator graphs have a new, smooth look, and equation labels have been added to make these graphs easier to read.
- Greater use has been made of headings and subheadings to add clarity to the text.
- Each exercise set has been carefully revised to ensure that there are sufficient exercises of various types for each example and mathematical concept. Exercise sets are carefully graded with several levels of difficulty.
- Real-world data have been updated or added to make the text more current.
- Chapter 1 has been expanded from four sections to five. Increasing and decreasing functions, average rates of change, and difference quotients are now discussed in Section 1.5. Circles now appear in Section 1.2 instead of Chapter R.
- In Chapter 2, piecewise-linear functions are now discussed earlier, in Section 2.1.
- Complex numbers have been moved to Chapter 3 as a new section.
- Chapter 4 has been expanded and reorganized from seven to eight sections, making it easier to cover one section per class. Division of polynomials and real zeros of polynomial functions are now in separate sections. Coverage of several topics, such as rational functions, has been revised and enhanced.
- In Chapter 5, the change of base formula is now presented in Section 5.4 and additional modeling has been included in Section 5.7.
- In Chapter 6 a unit circle approach that uses the wrapping function has been included.
- Chapter 7 has more explanation and exercises related to solving trigonometric equations.
- In Chaper 8 additional explanation of the ambiguous case for law of sines has been added. There also is more explanation of position vectors, components, and direction angles.
- Chapter 9 has been reorganized so that systems of equations are discussed in Section 9.1 and systems of inequalities and linear programming are discussed in Section 9.2. Substitution and elimination are both presented in Section 9.1.
- Chapter R includes additional problems involving factoring and rational expressions.
- Discussion of the following important topics has been added or enhanced:

Order of operations	Graphing rational functions by hand
Rational functions and expressions	Concavity
Factoring	Vectors
Interpolation and extrapolation	Wrapping function
Modeling real data	

Features

- **Chapter and Section Introductions**
  Many college algebra students have little or no understanding of mathematics beyond basic computation. To motivate students, chapter and section introductions explain some of the reasons for studying mathematics. (See pages 1, 76, 146, and 249.)

■ **Now Try**

This feature occurs after each example. It suggests a similar exercise students can work to see if they understand the concept presented in the example. (See pages 5, 115, and 189.)

NEW!

■ **Getting Started**

This new feature occurs in selected examples that require multistep solutions. Getting Started helps students develop an overall problem-solving strategy before they begin writing a detailed solution. (See pages 6, 98, and 225.)

■ **Algebra and Geometry Review Notes**

Throughout the text, Algebra and Geometry Review Notes, located in the margins, direct students "just in time" to Chapter R, where important topics in algebra and geometry are reviewed. Instructors can use this chapter for extra review or refer students to it as needed. This feature *frees* instructors from having to frequently review material from intermediate algebra and geometry. (See pages 124 and 174.)

■ **Calculator Help Notes**

The Calculator Help Notes in the margins direct students "just in time" to Appendix A: Using the Graphing Calculator. This appendix shows students the keystrokes necessary to complete specific examples from the text. This feature *frees* instructors from having to teach the specifics of the graphing calculator and gives students a convenient reference written specifically for this text. (See pages 6, 22, and 119.)

■ **Class Discussion**

This feature, included in most sections, poses a question that can be used for either classroom discussion or homework. (See pages 33, 50, and 253.)

■ **Making Connections**

This feature, which occurs throughout the text, shows students how concepts covered previously are related to new concepts being presented. (See pages 31, 32, 226, and 286.)

■ **Putting It All Together**

This helpful feature at the end of each section summarizes techniques and reinforces the mathematical concepts presented in the section. It is given in an easy-to-follow grid. (See pages 125–126, 348–349, and 394–396.)

■ **Checking Basic Concepts**

This feature, included after every two sections, provides a small set of exercises that can be used for review. These exercises require about 15 or 20 minutes to complete and can be used for collaborative learning exercises if time permits. (See pages 113, 204, and 276–277.)

■ **Exercise Sets**

The exercise sets are the heart of any mathematics text, and this text includes a large variety of instructive exercises. Each set of exercises covers skill building, mathematical concepts, and applications. Graphical interpretation and tables of data are often used to extend students' understanding of mathematical concepts. The exercise sets are graded carefully and categorized according to topic, making it easy for an instructor to select appropriate assignments. (See pages 107–113 and 200–204.)

■ **Chapter Summaries**

Chapter summaries are presented in an easy-to-read grid. They allow students to quickly review key concepts from the chapter. (See pages 240–243 and 354–358.)

■ **Chapter Review Exercises**

This exercise set contains both skill-building and applied exercises. These exercises stress different techniques for solving problems and provide students with the review necessary to pass a chapter test. (See pages 71–74 and 358–361).

- **Extended and Discovery Exercises**
  Extended and Discovery Exercises occur at the end of selected sections and at the end of every chapter. These exercises are usually more complex and challenging than the rest of the exercises and often require extension of a topic presented or exploration of a new topic. They can be used for either collaborative learning or extra homework assignments. (See pages 74–75, 246–247, and 362.)
- **Cumulative Review Exercises**
  These comprehensive exercise sets, which usually occur after every two chapters, allow students to review previous material. (See pages 165–168 and 363–366.)

Instructor Supplements

ANNOTATED INSTRUCTOR'S EDITION

- A special edition of the text
- Includes new Teaching Examples, an extra set of examples for instructors to present in class, doubling the number of examples available for instructors
- Includes new Teaching Tips, helpful ideas about presenting topics or teaching from the text
- Includes all of the answers to the exercise sets, usually right on the page where the exercise appears

ISBN: 0-321-56509-6 / 978-0-321-56509-9

INSTRUCTOR'S SOLUTIONS MANUAL

- By David Atwood and Terry Krieger, *Rochester Community and Technical College*
- Provides complete solutions to all text exercises, excluding Writing about Mathematics

ISBN: 0-321-57710-8 / 978-0-321-57710-8

INSTRUCTOR'S TESTING MANUAL (DOWNLOAD ONLY)

- By Vincent McGarry, *Austin Community College*
- Provides prepared tests for each chapter of the text, as well as answers
- Available for download at www.pearsonhighered.com/irc
- An access code is required.

TESTGEN™ (DOWNLOAD ONLY)

- Enables instructors to build, edit, print, and administer tests using a computerized bank of questions that cover all the objectives of the text
- Using algorithmically based questions, allows instructors to create multiple but equivalent versions of the same question or test with the click of a button

- Lets instructors modify test bank questions or add new questions
- Provides printable or online tests
- Available for download from Pearson Education's online catalog (www.pearsonhighered.com/irc)
- An access code is required.

INSIDER'S GUIDE

- Includes resources to help faculty with course preparation and classroom management
- Provides helpful teaching tips correlated to the sections of text, as well as general teaching advice

ISBN: 0-321-57717-5 / 978-0-321-57717-7

POWERPOINT PRESENTATION (DOWNLOAD ONLY)

- Classroom presentation software correlated specifically to this textbook sequence
- Available for download within MyMathLab or at www.pearsonhighered.com/irc
- An access code is required.

PEARSON MATH ADJUNCT SUPPORT CENTER

The **Pearson Math Adjunct Support Center** (www.pearsontutorservices.com/math-adjunct.html) is staffed by qualified instructors with more than 50 years of combined experience at both the community college and the university level. Assistance is provided for faculty in the following areas:

- Suggested syllabus consultation
- Tips on using materials packed with your book
- Book-specific content assistance
- Teaching suggestions, including advice on classroom strategies

Student Supplements

STUDENT'S SOLUTIONS MANUAL

- By David Atwood and Terry Krieger, *Rochester Community and Technical College*
- Provides complete solutions to all odd-numbered text exercises, excluding Writing about Mathematics and Extended and Discovery Exercises

ISBN: 0-321-57709-4 / 978-0-321-57709-2

VIDEOS ON DVD WITH OPTIONAL SUBTITLES

- DVD format enables students to watch the videos at home or on campus

- Affordable and portable for students
- Ideal for distance learning or supplemental instruction
- Include optional subtitles in English and Spanish

ISBN: 0-321-57715-9 / 978-0-321-57715-3

A REVIEW OF ALGEBRA

- By Heidi Howard, *Florida Community College at Jacksonville*
- Provides additional support for those students needing further algebra review

ISBN: 0-201-77347-3 / 978-0-201-77347-7

Technology Resources

MathXL®

MathXL is a powerful online homework, tutorial, and assessment system that accompanies Pearson Education's textbooks in mathematics or statistics. With MathXL, instructors can create, edit, and assign online homework and tests, using algorithmically generated exercises correlated at the objective level to the textbook. They can also create and assign their own online exercises and import TestGen tests for added flexibility. All student work is tracked in MathXL's online gradebook. Students can take chapter tests in MathXL and receive personalized study plans based on their test results. The study plan diagnoses weaknesses and links students directly to tutorial exercises for the objectives they need to study and retest. Students can also access supplemental animations and video clips directly from selected exercises. MathXL is available to qualified adopters. For more information, visit our Web site at www.mathxl.com or contact your sales representative.

MyMathLab®

MyMathLab is a series of text-specific, easily customizable online courses for Pearson Education's textbooks in mathematics and statistics. Powered by CourseCompass™ (our online teaching and learning environment) and MathXL® (our online homework, tutorial, and assessment system), MyMathLab gives you the tools you need to deliver all or a portion of your course online, whether your students are in a lab setting or working from home. MyMathLab provides a rich and flexible set of course materials, featuring free-response exercises that are algorithmically generated for unlimited practice and mastery. Students can also use online tools, such as video lectures, animations, and a multimedia textbook, to independently improve their understanding and performance. Instructors can use MyMathLab's homework and test managers to select and assign online exercises correlated directly to the textbook, and they can also create and assign their own online exercises and import TestGen tests for added flexibility. MyMathLab's online gradebook—designed specifically for mathematics and statistics—automatically tracks students' homework and test results and gives the instructor control over how to calculate final grades. Instructors can also add offline (paper-and-pencil) grades to the gradebook. MyMathLab also includes access to the **Pearson Tutor Center** (www.pearsontutorservices.com). The Tutor Center is staffed by qualified mathematics instructors who provide textbook-specific tutoring for students via toll-free phone, fax, email, and interactive Web sessions. MyMathLab is available to qualified adopters. For more information, visit our Web site at www.mymathlab.com or contact your sales representative.

MathXL® Tutorials on CD ISBN 0-321-57707-8 / 978-0-321-57707-8

This interactive tutorial CD-ROM provides algorithmically generated practice exercises that are correlated at the objective level to the exercises in the textbook. Every practice exercise is accompanied by an example and a guided solution designed to involve students in the solution process. Selected exercises may also include a video clip to help students visualize concepts. The software provides helpful feedback for incorrect answers and can generate printed summaries of students' progress.

InterAct Math Tutorial Web Site: www.interactmath.com

Get practice and tutorial help online! This interactive tutorial website provides algorithmically generated practice exercises that correlate directly to the exercises in the textbook. Students can retry an exercise as many times as they like, with new values each time, for unlimited practice and mastery. Every exercise is accompanied by an interactive guided solution that provides helpful feedback for incorrect answers, and students can also view a worked-out sample problem that steps them through an exercise similar to the one they're working on.

Acknowledgments

Many individuals contributed to the development of this textbook. I would like to thank the following reviewers, whose comments and suggestions were invaluable in preparing this and previous editions of the text.

Randy Baker	*Fairmont State College*
Warren Bernard	*Santa Fe Community College*
Connie Buller	*Metropolitan Community College*
Beth Burns	*Bowling Green State University*
Rose Cavin	*Chipola Junior College*
Sarah Cook	*Washburn University*
Amy Daniel	*University of New Orleans*
Ann Darke	*Bowling Green State University*
William Fox	*Francis Marion University*
John C. Hake	*St. Louis Community College at Florissant Valley*
Johanna G. Halsey	*Dutchess Community College*
Robert Hanes	*Middlesex County College*
Libby Higgins	*Greenville Technical College*
Gloria Hitchcock	*Georgia Perimeter College*
Mike Huff	*Austin Community College, Cypress Creek Campus*
Peter Jarvis	*Georgia College and State University*
Tuesday J. Johnson	*New Mexico State University*
Susan Jones	*Nashville State Technical Institute*

Cheryl Kane	*University of Nebraska*
Mike Keller	*St. Johns River Community College*
Tom Kelly	*Metropolitan State College of Denver*
Mary Kilbride	*Augustana College*
Nancy Kolakowsky	*Paradise Valley Community College*
Marlene Kovaly	*Florida Community College, Kent Campus*
Michael Lloyd	*Henderson State University*
Judith Maggiore	*Holyoke Community College*
Anne Magnuson	*Wake Technical Community College*
Mary Martin	*Middle Tennessee State University*
Nancy Mauldin	*College of Charleston*
Joe May	*North Hennepin Community College*
Nolan Mitchell	*Chemeketa Community College*
Canda Mueller-Engheta	*University of Kansas*
Rebecca Muller	*Southeastern Louisiana University*
Stephen Nicoloff	*Paradise Valley Community College*
Donna Norman	*Jefferson Community College*
Wing Park	*College of Lake County*
Nancy Pevey	*Pellissippi State Community College*
David Platt	*Front Range Community College, Larimer*
Joan Raines	*Middle Tennessee State University*
Laura Ralston	*Floyd College at North Metro Tech*
Jolene Rhodes	*Valencia Community College*
Judith D. Smalling	*St. Petersburg Junior College*
Lorna TenEyck	*Chemeketa Community College*
Vanessa White	*Southern University of Shreveport*
Christine Wise	*University of Louisiana at Lafayette*

A special thank you is due Terry Krieger at Rochester Community and Technical College for his exceptional work with the manuscript and answers. I would also like to thank Paul Lorczak, Christine Nguyen, Marcia Nermoe, Kathleen Pellissier, and David Atwood at Rochester Community and Technical College and Janis Cimperman at St. Cloud State University for their superb work with proofreading and accuracy checking. Thanks also to Jessica Rockswold for preparing the art manuscript and proofreading the text.

Without the excellent cooperation from the professional staff at Addison-Wesley, this project would have been impossible. They are, without a doubt, the best. Thanks go to Greg Tobin for his support of this project. Particular recognition is due Anne Kelly and Chere Bemelmans, who gave advice, support, assistance, and encouragement. The outstanding contributions of Kathy Manley, Leah Goldberg, Barbara Atkinson, Beth Paquin, Katherine Greig, Bonnie Gill, and Joe Vetere are much appreciated. The outstanding work of Kathy Diamond was instrumental to the success of this project.

Thanks go to Wendy Rockswold, who gave invaluable assistance and encouragement throughout the project.

A special thank you goes to the many students and instructors who used the first three editions. Their suggestions were insightful. Please feel free to contact me at *gary.rockswold@mnsu.edu* or Department of Mathematics, Minnesota State University, Mankato, MN 56001 with your comments. Your opinion is important.

Gary Rockswold

Introduction to Functions and Graphs

Have you ever thought about how we "live by the numbers"? Money, digital televisions, speed limits, grade point averages, gas mileages, and temperatures are all based on numbers. When we are told what our weight, blood pressure, body mass index, and cholesterol level are, these numbers can even affect how we feel about ourselves. Numbers permeate our society.

People are concerned about our environment and how it is changing. Do cars and their carbon dioxide emissions contribute to global warming? Conventional cars are inherently inefficient because they burn gasoline when they are not moving. Hybrid vehicles may be a viable option, but no doubt numbers will be used to make a decision. Rates of change, consumption, efficiency, and pollution levels are all described by numbers.

Numbers are part of mathematics, but mathematics is *much more* than numbers. Mathematics also includes techniques to analyze these numbers and to guide our decisions about the future. Mathematics is used not only in science and technology; it is also used to describe almost every facet of life, including consumer behavior and the Internet.

In this chapter we discuss numbers and how functions are used to perform computations with these numbers. Understanding numbers and mathematical concepts is essential to understanding and dealing with the many changes that will occur in our lifetimes. Mathematics makes life easier!

Source: Andrew Frank, "Plug-in Hybrid Vehicles for a Sustainable Future," *American Scientist,* March–April, 2007.

The essence of mathematics is not to make simple things complicated, but to make complicated things simple.
—Stanley Gudder

1.1 Numbers, Data, and Problem Solving

- **Recognize common sets of numbers**
- **Evaluate expressions by applying the order of operations**
- **Learn scientific notation and use it in applications**
- **Apply problem-solving strategies**

Introduction

Because society is becoming more complex and diverse, our need for mathematics is increasing dramatically each year. Numbers are essential to our everyday lives. For example, the iPhone is 4.5 inches in height, 2.4 inches in width, and 0.46 inch in thickness. It has an 8-gigabyte flash drive, a 2-megapixel camera, and 480-by-320-pixel screen resolution, and it can operate at temperatures between 32° and 95°F.   (**Source:** Apple Corporation.)

Mathematics not only provides numbers to describe new products, but also gives us problem-solving strategies. This section discusses basic sets of numbers and introduces some essential problem-solving strategies.

Sets of Numbers

One important set of numbers is the set of **natural numbers**. This set comprises the *counting numbers* $N = \{1, 2, 3, 4, \ldots\}$. Natural numbers can be used when data are positive and not presented in fractional parts.

The **integers** $I = \{\ldots, -3, -2, -1, 0, 1, 2, 3, \ldots\}$ are a set of numbers that contains the natural numbers, their additive inverses (negatives), and 0. Historically, negative numbers were not readily accepted. Today, however, when a person overdraws a personal checking account for the first time, negative numbers quickly take on meaning. There is a significant difference between a positive and a negative balance.

A **rational number** can be expressed as the *ratio* of two integers $\frac{p}{q}$, where $q \neq 0$. Rational numbers include the integers. Examples of rational numbers are

$$\frac{2}{1}, \frac{1}{3}, -\frac{1}{4}, \frac{-50}{2}, \frac{22}{7}, 0, \sqrt{25}, \text{ and } 1.2.$$

Note that 0 and 1.2 are both rational numbers. They can be represented by the fractions $\frac{0}{1}$ and $\frac{12}{10}$. Because two fractions that look different can be equivalent, rational numbers have more than one form. A rational number can always be expressed in a decimal form that either *repeats* or *terminates*. For example, $\frac{2}{3} = 0.\overline{6}$, a repeating decimal, and $\frac{1}{4} = 0.25$, a terminating decimal. The overbar indicates that $0.\overline{6} = 0.6666666\ldots$.

Real numbers can be represented by decimal numbers. Since every rational number has a decimal form, real numbers include rational numbers. However, some real numbers cannot be expressed as a ratio of two integers. These numbers are called **irrational numbers**. The numbers $\sqrt{2}$, $\sqrt{15}$, and π are examples of irrational numbers. They can be represented by nonrepeating, nonterminating decimals. Note that for any positive integer a, if $\sqrt{a}$ is not an integer, then $\sqrt{a}$ is an irrational number.

Real numbers include both rational and irrational numbers and can be *approximated* by a terminating decimal. Examples of real numbers include

$$2, -10, -131.3337, \frac{1}{3} = 0.\overline{3}, -\sqrt{5} \approx -2.2361, \text{ and } \sqrt{11} \approx 3.3166.$$

NOTE The symbol $\approx$ means **approximately equal**. This symbol is used in place of an equals sign whenever two unequal quantities are close in value. For example, $\frac{1}{4} = 0.25$, whereas $\frac{1}{3} \approx 0.3333$.

EXAMPLE 1 Classifying numbers

Classify each real number as one or more of the following: natural number, integer, rational number, or irrational number.

$$5, -1.2, \frac{13}{7}, -\sqrt{7}, -12, \sqrt{16}$$

SOLUTION

5: natural number, integer, and rational number

−1.2: rational number

$\frac{13}{7}$: rational number

$-\sqrt{7}$: irrational number

−12: integer and rational number

$\sqrt{16} = 4$: natural number, integer, and rational number   *Now Try Exercise 7* ◀

Order of Operations

Does $6 - 3 \cdot 2$ equal 0 or 6? Does -5^2 equal 25 or −25? Figure 1.1 correctly shows that $6 - 3 \cdot 2 = 0$ and that $-5^2 = -25$. Because multiplication is performed before subtraction, $6 - 3 \cdot 2 = 0$. Similarly, because exponents are evaluated before performing negation, $-5^2 = -25$. It is essential that algebraic expressions be evaluated consistently, so the following rules have been established.

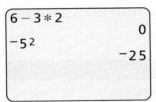

Figure 1.1

Order of Operations

Using the following order of operations, first perform all calculations within parentheses, square roots, and absolute value bars and above and below fraction bars. Then use the same order of operations to perform any remaining calculations.

1. Evaluate all exponents. Then do any negation *after* evaluating exponents.
2. Do all multiplication and division from *left to right*.
3. Do all addition and subtraction from *left to right*.

EXAMPLE 2 Evaluating arithmetic expressions

Evaluate each expression by hand.

(a) $3(1 - 5)^2 - 4^2$ **(b)** $\dfrac{10 - 6}{5 - 3} - 4 - |7 - 2|$

SOLUTION

(a) $3(1 - 5)^2 - 4^2 = 3(-4)^2 - 4^2$ **(b)** $\dfrac{10 - 6}{5 - 3} - 4 - |7 - 2| = \dfrac{4}{2} - 4 - |5|$

$\qquad\qquad\qquad\quad = 3(16) - 16 \qquad\qquad\qquad\qquad\qquad\qquad\quad = 2 - 4 - 5$

$\qquad\qquad\qquad\quad = 48 - 16 \qquad\qquad\qquad\qquad\qquad\qquad\qquad = -2 - 5$

$\qquad\qquad\qquad\quad = 32 \qquad\qquad\qquad\qquad\qquad\qquad\qquad\qquad = -7$

NOTE $(-4)^2 = (-4)(-4) = 16$ and $-4^2 = -(4)(4) = -16$.

Now Try Exercises 19 and 21 ◀

Scientific Notation

Numbers that are large or small in absolute value are often expressed in scientific notation. Table 1.1 lists examples of numbers in **standard (decimal) form** and in **scientific notation**.

Table 1.1

Standard Form	Scientific Notation	Application
93,000,000 mi	9.3×10^7 mi	Distance to the sun
13,517	1.3517×10^4	Radio stations in 2005
9,000,000,000	9×10^9	Estimated world population in 2050
0.00000538 sec	5.38×10^{-6} sec	Time for light to travel 1 mile
0.000005 cm	5×10^{-6} cm	Size of a typical virus

To write 0.00000538 in scientific notation, start by moving the decimal point to the right of the first nonzero digit, 5, to obtain 5.38. Since the decimal point was moved six places to the *right*, the exponent of 10 is -6. Thus, $0.00000538 = 5.38 \times 10^{-6}$. When the decimal point is moved to the *left*, the exponent of 10 is positive, rather than negative. Here is a formal definition of scientific notation.

Calculator Help

To display numbers in scientific notation, see Appendix A (page AP-2).

> ## Scientific Notation
>
> A real number r is in **scientific notation** when r is written as $c \times 10^n$, where $1 \le |c| < 10$ and n is an integer.

An Application   Nanotechnology involves extremely small electrical circuits. Someday this technology may use the movement of the human body to power tiny devices such as pacemakers. The next example demonstrates how scientific notation appears in the description of this new technology.

EXAMPLE 3   Analyzing the energy produced by your body

Nanotechnology is a technology of the very small: on the order of one billionth of a meter. Researchers are looking to power tiny devices with energy generated by the human body. (**Source:** Z. Wang, "Self-Powered Nanotech," *Scientific American*, January 2008.)
(a) Write one billionth in scientific notation.
(b) While typing, a person's fingers generate about 2.2×10^{-3} watt of electrical energy. Write this number in standard (decimal) form.

SOLUTION
(a) One billionth can be written as $\frac{1}{1,000,000,000} = \frac{1}{10^9} = 1 \times 10^{-9}$.
(b) Move the decimal point in 2.2 three places to the left: $2.2 \times 10^{-3} = 0.0022$.

Now Try Exercise 83 ◀

The next two examples illustrate how to evaluate expressions involving scientific notation.

EXAMPLE 4   Evaluating expressions by hand

Evaluate each expression. Write your result in scientific notation and standard form.

(a) $(3 \times 10^3)(2 \times 10^4)$ **(b)** $(5 \times 10^{-3})(6 \times 10^5)$ **(c)** $\dfrac{4.6 \times 10^{-1}}{2 \times 10^2}$

SOLUTION

(a)

$$\begin{aligned}
(3 \times 10^3)(2 \times 10^4) &= 3 \times 2 \times 10^3 \times 10^4 && \textit{Commutative property} \\
&= 6 \times 10^{3+4} && \textit{Add exponents.} \\
&= 6 \times 10^7 && \textit{Scientific notation} \\
&= 60{,}000{,}000 && \textit{Standard form}
\end{aligned}$$

Algebra Review
To review exponents, see Chapter R (page R-7).

(b)

$$\begin{aligned}
(5 \times 10^{-3})(6 \times 10^5) &= 5 \times 6 \times 10^{-3} \times 10^5 && \textit{Commutative property} \\
&= 30 \times 10^2 && \textit{Add exponents.} \\
&= 3 \times 10^3 && \textit{Scientific notation} \\
&= 3000 && \textit{Standard form}
\end{aligned}$$

(c)

$$\begin{aligned}
\frac{4.6 \times 10^{-1}}{2 \times 10^2} &= \frac{4.6}{2} \times \frac{10^{-1}}{10^2} && \textit{Multiplication of fractions} \\
&= 2.3 \times 10^{-1-2} && \textit{Subtract exponents.} \\
&= 2.3 \times 10^{-3} && \textit{Scientific notation} \\
&= 0.0023 && \textit{Standard form}
\end{aligned}$$

Now Try Exercises 53, 55, and 57 ◀

Calculators   Calculators often use E to express powers of 10. For example, 4.2×10^{-3} might be displayed as 4.2E–3. On some calculators, numbers can be entered in scientific notation with the (EE) key, which you can find by pressing (2nd) (,).

EXAMPLE 5   Computing in scientific notation with a calculator

Approximate each expression. Write your answer in scientific notation.

(a) $\left(\dfrac{6 \times 10^3}{4 \times 10^6}\right)(1.2 \times 10^2)$ **(b)** $\sqrt{4500\pi}\left(\dfrac{103 + 450}{0.233}\right)^3$

SOLUTION

(a) The given expression is entered in two ways in Figure 1.2. Note that in both cases

$$\left(\frac{6 \times 10^3}{4 \times 10^6}\right)(1.2 \times 10^2) = 0.18 = 1.8 \times 10^{-1}.$$

(b) Be sure to insert parentheses around 4500π and around the numerator, $103 + 450$, in the ratio. From Figure 1.3 we can see that the result is approximately 1.59×10^{12}.

Calculator Help
To enter numbers in scientific notation, see Appendix A (page AP-2).

```
(6*10^3)/(4*10^6
)*(1.2*10^2)
              .18
(6E3)/(4E6)*(1.2
E2)
              .18
```

Figure 1.2

```
√(4500π)*((103+4
50)/.233)^3
    1.58960355E12
```

Figure 1.3

Now Try Exercises 61 and 63 ◀

EXAMPLE 6 Computing with a calculator

Use a calculator to evaluate each expression. Round answers to the nearest thousandth.

(a) $\sqrt[3]{131}$ **(b)** $\pi^3 + 1.2^2$ **(c)** $\dfrac{1 + \sqrt{2}}{3.7 + 9.8}$ **(d)** $|\sqrt{3} - 6|$

SOLUTION

Algebra Review
To review cube roots, see Chapter R (page R-40).

(a) On some calculators the cube root can be found by using the MATH menu. If your calculator does not have a cube root key, enter 131^(1/3). From the first two lines in Figure 1.4, we see that $\sqrt[3]{131} \approx 5.079$.

(b) Do *not* use 3.14 for the value of π. Instead, use the built-in key to obtain a more accurate value of π. From the bottom two lines in Figure 1.4, $\pi^3 + 1.2^2 \approx 32.446$.

(c) When evaluating this expression be sure to include parentheses around the numerator and around the denominator. Most calculators have a special square root key that can be used to evaluate $\sqrt{2}$. From the first three lines in Figure 1.5, $\dfrac{1 + \sqrt{2}}{3.7 + 9.8} \approx 0.179$.

(d) The absolute value can be found on some calculators by using the MATH NUM menus. From the bottom two lines in Figure 1.5, $|\sqrt{3} - 6| \approx 4.268$.

Calculator Help
To enter expressions such as $\sqrt[3]{131}$, $\sqrt{2}$, π, and $|\sqrt{3} - 6|$, see Appendix A (page AP-2).

```
³√(131)
         5.078753078
π^3+1.2²
         32.44627668
```

Figure 1.4

```
(1+√(2))/(3.7+9.
8)
         .1788306342
abs(√(3)−6)
         4.267949192
```

Figure 1.5

Now Try Exercises 67, 69, 71, and 73 ◀

Problem Solving

Many problem-solving strategies are used in algebra. However, in this subsection we focus on two important strategies that are used frequently: making a sketch and applying one or more formulas. These strategies are illustrated in the next three examples.

EXAMPLE 7 Finding the speed of Earth

Earth travels around the sun in an approximately circular orbit with an average radius of 93 million miles. If Earth takes 1 year, or about 365 days, to complete one orbit, estimate the orbital speed of Earth in miles per hour.

SOLUTION

Getting Started Speed S equals distance D divided by time T, $S = \frac{D}{T}$. We need to find the number of miles Earth travels in 1 year and then divide it by the number of hours in 1 year. ▶

Geometry Review
To find the circumference of a circle, see Chapter R (page R-2).

Distance Traveled A sketch of Earth orbiting the sun is shown in Figure 1.6. In 1 year Earth travels the circumference of a circle with a radius of 93 million miles. The circumference of a circle is $2\pi r$, where r is the radius, so the distance D is

$$D = 2\pi r = 2\pi(93{,}000{,}000) \approx 584{,}300{,}000 \text{ miles.}$$

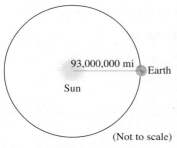

93,000,000 mi

Earth

Sun

(Not to scale)

Figure 1.6 Earth's Orbit

Hours in 1 Year The number of hours H in 1 year, or 365 days, equals

$$H = 365 \times 24 = \mathbf{8760} \text{ hours.}$$

Speed of Earth $S = \dfrac{D}{H} = \dfrac{584{,}300{,}000}{8760} \approx 66{,}700$ miles per hour.

Now Try Exercise 85 ◀

Many times in geometry we evaluate formulas to determine quantities, such as perimeter, area, and volume. In the next example we use a formula to determine the number of fluid ounces in a soda can.

EXAMPLE 8 Finding the volume of a soda can

The volume V of the cylindrical soda can in Figure 1.7 is given by $V = \pi r^2 h$, where r is its radius and h is its height.
(a) If $r = 1.4$ inches and $h = 5$ inches, find the volume of the can in cubic inches.
(b) Could this can hold 16 fluid ounces? (*Hint:* 1 cubic inch equals 0.55 fluid ounce.)

Figure 1.7 A Soda Can

SOLUTION
(a) $V = \pi r^2 h = \pi(1.4)^2(5) = 9.8\pi \approx 30.8$ cubic inches.
(b) To find the number of fluid ounces, multiply the number of cubic inches by 0.55.

$$30.8 \times 0.55 = 16.94$$

Yes, the can could hold 16 fluid ounces.

Now Try Exercise 93 ◀

Measuring the thickness of a very thin layer of material can be difficult to do directly. For example, it would be difficult to measure the thickness of a sheet of aluminum foil or a coat of paint with a ruler. However, it can be done indirectly using the following formula.

$$\text{Thickness} = \frac{\text{Volume}}{\text{Area}}$$

That is, the thickness of a thin layer equals the volume of the substance divided by the area that it covers. For example, if a volume of 1 cubic inch of paint is spread over an area of 100 square inches, then the thickness of the paint equals $\frac{1}{100}$ inch. This formula is illustrated in the next example.

EXAMPLE 9 Calculating the thickness of aluminum foil

A rectangular sheet of aluminum foil is 15 centimeters by 35 centimeters and weighs 5.4 grams. If 1 cubic centimeter of aluminum weighs 2.7 grams, find the thickness of the aluminum foil.   (**Source:** U. Haber-Schaim, *Introductory Physical Science.*)

SOLUTION
Getting Started Start by making a sketch of a rectangular sheet of aluminum, as shown in Figure 1.8. To complete this problem we need to find the volume V of the aluminum foil and its area A. Then we can determine the thickness T by using the formula $T = \dfrac{V}{A}$. ▶

T

35 cm

15 cm

(Not to scale)

Figure 1.8 Aluminum Foil

NOTE For the rectangular box shape shown in Figure 1.8 on the previous page,

$$\text{Volume} = \underbrace{\text{Length} \times \text{Width}}_{\text{Area}} \times \textbf{Thickness}.$$

It follows that **Thickness** $= \frac{\text{Volume}}{\text{Area}}$.

Geometry Review
To find the area of a rectangle, see Chapter R (page R-1). To find the volume of a box, see Chapter R (page R-3).

Volume   Because the aluminum foil weighs 5.4 grams and each 2.7 grams equals 1 cubic centimeter, the volume of the aluminum foil is

$$\frac{5.4}{2.7} = 2 \text{ cubic centimeters.} \qquad \textit{Divide weight by density.}$$

Area   The aluminum foil is rectangular with an area of $15 \times 35 = 525$ square centimeters.

Thickness   The thickness of **2** cubic centimeters of aluminum foil with an area of **525** square centimeters is

$$\textbf{Thickness} = \frac{\text{Volume}}{\text{Area}} = \frac{2}{525} \approx 0.0038 \text{ centimeter.}$$

Now Try Exercise 89 ◀

1.1 Putting It All Together

Numbers play a central role in our society. Without numbers, data could be described qualitatively but not quantitatively. For example, we could say that the day seems hot but would not be able to give an actual number for the temperature. Problem-solving strategies are used in almost every facet of our lives, providing the procedures needed to systematically complete tasks and perform computations.

The following table summarizes some of the concepts in this section.

Concept	Comments	Examples
Natural numbers	Sometimes referred to as the *counting numbers*	$1, 2, 3, 4, 5, \ldots$
Integers	Include the natural numbers, their opposites, and 0	$\ldots, -2, -1, 0, 1, 2, \ldots$
Rational numbers	Include integers; all fractions $\frac{p}{q}$, where p and q are integers with $q \neq 0$; all repeating and all terminating decimals	$\frac{1}{2}, -3, \frac{128}{6}, -0.335, 0,$ $0.25 = \frac{1}{4}, 0.\overline{33} = \frac{1}{3}$
Irrational numbers	Can be written as nonrepeating, nonterminating decimals; cannot be a rational number; if a square root of a positive integer is not an integer, it is an irrational number.	$\pi, \sqrt{2}, -\sqrt{5}, \sqrt[3]{7}, \pi^4$

Concept	Comments	Examples
Real numbers	Any number that can be expressed in standard (decimal) form Include the rational numbers and irrational numbers	$\pi, \sqrt{7}, -\dfrac{4}{7}, 0, -10, 1.237$ $0.\overline{6} = \dfrac{2}{3}, 1000, \sqrt{15}, -\sqrt{5}$
Order of operations	Using the following order of operations, first perform all calculations within parentheses, square roots, and absolute value bars and above and below fraction bars. Then perform any remaining calculations. 1. Evaluate all exponents. Then do any negation *after* evaluating exponents. 2. Do all multiplication and division from *left to right.* 3. Do all addition and subtraction from *left to right.*	$\begin{aligned}-4^2 - 12 \div 2 - 2 &= -16 - 12 \div 2 - 2\\ &= -16 - 6 - 2\\ &= -22 - 2\\ &= -24\end{aligned}$ $\begin{aligned}\dfrac{2 + 4^2}{3 - 3 \cdot 5} &= \dfrac{2 + 16}{3 - 15}\\[2mm] &= \dfrac{18}{-12}\\[2mm] &= -\dfrac{3}{2}\end{aligned}$
Scientific notation	A number in the form $c \times 10^n$, where $1 \le |c| < 10$ and n is an integer Used to represent numbers that are large or small in absolute value	$3.12 \times 10^4 = 31{,}200$ $-1.4521 \times 10^{-2} = -0.014521$ $5 \times 10^9 = 5{,}000{,}000{,}000$ $1.5987 \times 10^{-6} = 0.0000015987$

1.1 Exercises

Classifying Numbers

Exercises 1–6: Classify the number as one or more of the following: natural number, integer, rational number, or real number.

1. $\frac{21}{24}$ (Fraction of people in the United States completing at least 4 years of high school)

2. 20,082 (Average cost in dollars of tuition and fees at a private college in 2004)

3. 7.5 (Average number of gallons of water used each minute while taking a shower)

4. 25.8 (Nielsen rating of the TV show *Grey's Anatomy* the week of February 12–18, 2007)

5. $90\sqrt{2}$ (Distance in feet from home plate to second base on a baseball field)

6. -71 (Wind chill when the temperature is $-30°F$ and the wind speed is 40 mph)

Exercises 7–10: Classify each number as one or more of the following: natural number, integer, rational number, or irrational number.

7. $\pi, -3, \frac{2}{9}, \sqrt{9}, 1.\overline{3}, -\sqrt{2}$

8. $\frac{3}{1}, -\frac{5}{8}, \sqrt{7}, 0.\overline{45}, 0, 5.6 \times 10^3$

9. $\sqrt{13}, \frac{1}{3}, 5.1 \times 10^{-6}, -2.33, 0.\overline{7}, -\sqrt{4}$

10. $-103, \frac{21}{25}, \sqrt{100}, -\frac{5.7}{10}, \frac{2}{9}, -1.457, \sqrt{3}$

Exercises 11–16: For the measured quantity, state the set of numbers that most appropriately describes it. Choose from the natural numbers, integers, and rational numbers. Explain your answer.

11. Shoe sizes

12. Populations of states

13. Gallons of gasoline

14. Speed limits

15. Temperatures in a winter weather forecast in Montana

16. Numbers of compact disc sales

Order of Operations

Exercises 17–28: Evaluate by hand.

17. $|5 - 8 \cdot 7|$

18. $-2(16 - 3 \cdot 5) \div 2$

19. $-6^2 - 3(2 - 4)^4$

20. $(4 - 5)^2 - 3^2 - 3\sqrt{9}$

21. $\sqrt{9 - 5} - \dfrac{8 - 4}{4 - 2}$

22. $\dfrac{6 - 4^2 \div 2^3}{3 - 4}$

23. $\sqrt{13^2 - 12^2}$

24. $\dfrac{13 - \sqrt{9 + 16}}{|5 - 7|^2}$

25. $\dfrac{4 + 9}{2 + 3} - \dfrac{-3^2 \cdot 3}{5}$

26. $10 \div 2 \div \dfrac{5 + 10}{5}$

27. $-5^2 - 20 \div 4 - 2$

28. $5 - (-4)^3 - (4)^3$

Scientific Notation

Exercises 29–40: Write the number in scientific notation.

29. 184,800 (New lung cancer cases reported in 2005)

30. 29,285,000 (People worldwide living with HIV)

31. 0.04361 (Proportion of U.S. deaths attributed to accidents in 2004)

32. 0.62 (Number of miles in 1 kilometer)

33. 2450

34. 105.6

35. 0.56

36. −0.00456

37. −0.0087

38. 1,250,000

39. 206.8

40. 0.00007

Exercises 41–52: Write the number in standard form.

41. 1×10^{-6} (Wavelength in meters of visible light)

42. 9.11×10^{-31} (Weight in kilograms of an electron)

43. 2×10^8 (Years required for the sun to orbit our galaxy)

44. 9×10^{12} (Federal debt in dollars in 2007)

45. 1.567×10^2

46. -5.68×10^{-1}

47. 5×10^5

48. 3.5×10^3

49. 0.045×10^5

50. -5.4×10^{-5}

51. 67×10^3

52. 0.0032×10^{-1}

Exercises 53–60: Evaluate the expression by hand. Write your result in scientific notation and standard form.

53. $(4 \times 10^3)(2 \times 10^5)$

54. $(3 \times 10^1)(3 \times 10^4)$

55. $(5 \times 10^2)(7 \times 10^{-4})$

56. $(8 \times 10^{-3})(7 \times 10^1)$

57. $\dfrac{6.3 \times 10^{-2}}{3 \times 10^1}$

58. $\dfrac{8.2 \times 10^2}{2 \times 10^{-2}}$

59. $\dfrac{4 \times 10^{-3}}{8 \times 10^{-1}}$

60. $\dfrac{2.4 \times 10^{-5}}{4.8 \times 10^{-7}}$

Exercises 61–66: Use a calculator to approximate the expression. Write your result in scientific notation.

61. $\dfrac{8.947 \times 10^7}{0.00095}(4.5 \times 10^8)$

62. $(9.87 \times 10^6)(34 \times 10^{11})$

63. $\left(\dfrac{101 + 23}{0.42}\right)^2 + \sqrt{3.4 \times 10^{-2}}$

64. $\sqrt[3]{(2.5 \times 10^{-8}) + 10^{-7}}$

65. $(8.5 \times 10^{-5})(-9.5 \times 10^7)^2$

66. $\sqrt{\pi(4.56 \times 10^4) + (3.1 \times 10^{-2})}$

Exercises 67–76: Use a calculator to evaluate the expression. Round your result to the nearest thousandth.

67. $\sqrt[3]{192}$

68. $\sqrt{(32 + \pi^3)}$

69. $|\pi - 3.2|$

70. $\dfrac{1.72 - 5.98}{35.6 + 1.02}$

71. $\dfrac{0.3 + 1.5}{5.5 - 1.2}$

72. $3.2(1.1)^2 - 4(1.1) + 2$

73. $\dfrac{1.5^3}{\sqrt{2 + \pi} - 5}$

74. $4.3^2 - \dfrac{5}{17}$

75. $15 + \dfrac{4 + \sqrt{3}}{7}$

76. $\dfrac{5 + \sqrt{5}}{2}$

Applications

Exercises 77–80: Percent Change   If an amount changes from A to a new amount B, then the percent change is $\dfrac{B - A}{A} \times 100$.

Calculate the percent change for the given A and B. Round your answer to the nearest tenth of a percent when appropriate.

77. $A = \$8, B = \$13$ **78.** $A = \$0.90, B = \$13.47$

79. $A = 1.4, B = 0.85$ **80.** $A = 1256, B = 1195$

81. *Percent Change*   Suppose that tuition is initially $100 per credit and increases by 6% from the first year to the second year. What is the cost of tuition the second year? Now suppose that tuition decreases by 6% from the second to the third year. Is tuition equal to $100 per credit the third year? Explain.

82. *Tuition Increases*   From 1976 to 2004, average annual tuition and fees at public colleges and universities increased from $433 to $5132. Calculate the percent change over this time period.

83. *Nanotechnology*   (Refer to Example 3.) During inhalation, the typical body generates 0.14 watt of electrical power, which could be used to power tiny electrical circuits. Write this number in scientific notation.   (**Source:** *Scientific American,* January 2008.)

84. *Movement of the Pacific Plate*   The Pacific plate (the floor of the Pacific Ocean) near Hawaii is moving at about 0.000071 kilometer per year. This is about the speed at which a fingernail grows. Use scientific notation to determine how many kilometers the Pacific plate travels in one million years.

85. *Orbital Speed*   (Refer to Example 7.) The planet Mars travels around the sun in a nearly circular orbit with a radius of 141 million miles. If it takes 1.88 years for Mars to complete one orbit, estimate the orbital speed of Mars in miles per hour.

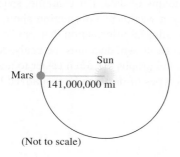

(Not to scale)

86. *Size of the Milky Way*   The speed of light is about 186,000 miles per second. The Milky Way galaxy has an approximate diameter of 6×10^{17} miles. Estimate, to the nearest thousand, the number of years it takes for light to travel across the Milky Way.   (**Source:** C. Ronan, *The Natural History of the Universe.*)

87. *Federal Debt*   The amount of federal debt changed dramatically during the 30 years from 1970 to 2000. (**Sources:** Department of the Treasury, Bureau of the Census.)

(a) In 1970 the population of the United States was 203,000,000 and the federal debt was $370 billion. Find the debt per person.

(b) In 2000 the population of the United States was approximately 281,000,000 and the federal debt was $5.54 trillion. Find the debt per person.

88. *Discharge of Water*   The Amazon River discharges water into the Atlantic Ocean at an average rate of 4,200,000 cubic feet per second, the highest rate of any river in the world. Is this more or less than 1 cubic mile of water per day? Explain your calculations.   (**Source:** *The Guinness Book of Records 1993.*)

89. *Thickness of an Oil Film*   (Refer to Example 9.) A drop of oil measuring 0.12 cubic centimeter is spilled onto a lake. The oil spreads out in a circular shape having a *diameter* of 23 centimeters. Approximate the thickness of the oil film.

90. *Thickness of Gold Foil*   (Refer to Example 9.) A flat, rectangular sheet of gold foil measures 20 centimeters by 30 centimeters and has a mass of 23.16 grams. If 1 cubic centimeter of gold has a mass of 19.3 grams, find the thickness of the gold foil.   (**Source:** U. Haber-Schaim, *Introductory Physical Science.*)

91. *Analyzing Debt*   A 1-inch-high stack of $100 bills contains about 250 bills. In 2000 the federal debt was approximately 5.54 trillion dollars.

(a) If the entire federal debt were converted into a stack of $100 bills, how many feet high would it be?

(b) The distance between Los Angeles and New York is approximately 2500 miles. Could this stack of $100 bills reach between these two cities?

92. *Volume of a Cone* The volume V of a cone is given by $V = \frac{1}{3}\pi r^2 h$, where r is its radius and h is its height. Find V when $r = 4$ inches and $h = 1$ foot. Round your answer to the nearest hundredth.

93. *Size of a Soda Can* (Refer to Example 8.) The volume V of a cylindrical soda can is given by $V = \pi r^2 h$, where r is its radius and h is its height.
   (a) If $r = 1.3$ inches and $h = 4.4$ inches, find the volume of the can in cubic inches.

   (b) Could this can hold 12 fluid ounces? (*Hint:* 1 cubic inch equals about 0.55 fluid ounce.)

94. *Volume of a Sphere* The volume of a sphere is given by $V = \frac{4}{3}\pi r^3$, where r is the radius of the sphere. Calculate the volume if the radius is 3 feet. Approximate your answer to the nearest tenth.

95. *Thickness of Cement* (Refer to Example 9.) A 100-foot-long sidewalk is 5 feet wide. If 125 cubic feet of cement are evenly poured to form the sidewalk, find the thickness of the sidewalk.

96. *Depth of a Lake* (Refer to Example 9.) A lake covers 2.5×10^7 square feet and contains 7.5×10^8 cubic feet of water. Find the average depth of the lake.

Writing about Mathematics

97. Describe some basic sets of numbers that are used in mathematics.

98. Suppose that a positive number a is written in scientific notation as $a = b \times 10^n$, where n is an integer and $1 \le b \le 10$. Explain what n indicates about the size of a.

EXTENDED AND DISCOVERY EXERCISE

1. If you have access to a scale that weighs in grams, find the thickness of regular and heavy-duty aluminum foil. Is heavy-duty foil worth the price difference? (*Hint:* Each 2.7 grams of aluminum equals 1 cubic centimeter.)

1.2 Visualizing and Graphing Data

- Analyze one-variable data
- Find the domain and range of a relation
- Graph in the *xy*-plane
- Calculate distance
- Find the midpoint
- Learn the standard equation of a circle
- Learn to graph equations with a calculator (optional)

Introduction

Technology is giving us access to huge amounts of data. For example, space telescopes, such as the Hubble telescope, are providing a wealth of information about the universe. The challenge is to convert the data into meaningful information that can be used to solve important problems. Before conclusions can be drawn, data must be analyzed. A powerful tool in this step is visualization, as pictures and graphs are often easier to understand than words. This section discusses how different types of data can be visualized by using various mathematical techniques.

One-Variable Data

Data often occur in the form of a list. A list of test scores without names is an example; the only variable is the score. Data of this type are referred to as **one-variable data**. If the values in a list are unique, they can be represented visually on a number line.

Means and medians can be found for one-variable data sets. To calculate the **mean** (or **average**) of a set of n numbers, we add the n numbers and then divide the sum by n. The **median** is equal to the value that is located in the middle of a *sorted* list. If there is an odd number of data items, the median is the middle data item. If there is an even number of data items, the median is the average of the two middle items.

EXAMPLE 1 Analyzing a list of data

Table 1.2 lists the monthly average temperatures in degrees Fahrenheit at Mould Bay, Canada.

Table 1.2 Monthly Average Temperatures at Mould Bay, Canada

Temperature (°F)	−27	−31	−26	−9	12	32	39	36	21	1	−17	−24

Source: A. Miller and J. Thompson, *Elements of Meteorology.*

(a) Plot these temperatures on a number line.
(b) Find the maximum and minimum temperatures.
(c) Determine the mean of these 12 temperatures.
(d) Find the median and interpret the result.

SOLUTION
(a) In Figure 1.9 the numbers in Table 1.2 are plotted on a number line.

Figure 1.9 Monthly Average Temperatures

CLASS DISCUSSION

In Example 1(c), the mean of the temperatures is approximately 0.6°F. Interpret this temperature. Explain your reasoning.

(b) The maximum temperature of 39°F is plotted farthest to the right in Figure 1.9. Similarly, the minimum temperature of −31°F is plotted farthest to the left.
(c) The sum of the 12 temperatures in Table 1.2 equals 7. The mean, or average, of these temperatures is $\frac{7}{12} \approx 0.6°F$.
(d) Because there is an even number of data items, the median is the average of the middle two values. From the number line we see that the middle two values are −9°F and 1°F. Thus the median is $\frac{-9 + 1}{2} = -4°F$. This result means that half the months have an average temperature that is greater than −4°F and half the months have an average temperature that is below −4°F. ***Now Try Exercises 1 and 5*** ◄

Two-Variable Data

Sometimes a relationship exists between two lists of data. Table 1.3 lists the monthly average precipitation in inches for Portland, Oregon. In this table, 1 corresponds to January, 2 to February, and so on, until 12 represents December. We show the relationship between a month and its average precipitation by combining the two lists so that corresponding months and precipitations are visually paired.

Table 1.3 Average Precipitation for Portland, Oregon

Month	1	2	3	4	5	6	7	8	9	10	11	12
Precipitation (inches)	6.2	3.9	3.6	2.3	2.0	1.5	0.5	1.1	1.6	3.1	5.2	6.4

If x is the month and y is the precipitation, then the **ordered pair** (x, y) represents the average amount of precipitation y during month x. For example, the ordered pair $(5, 2.0)$ indicates that the average precipitation in May is 2.0 inches, whereas the ordered pair $(2, 3.9)$ indicates that the average precipitation in February is 3.9 inches. *Order is important in an ordered pair.*

Since the data in Table 1.3 involve two variables, the month and precipitation, we refer to them as **two-variable data**. It is important to realize that a relation established by two-variable data is between two lists rather than within a single list. January is not related to August, and 6.2 inches of precipitation is not associated with 1.1 inches of precipitation. Instead, January is paired with 6.2 inches, and August is paired with 1.1 inches. We now define the mathematical concept of a relation.

Relation

A **relation** is a set of ordered pairs.

If we denote the ordered pairs in a relation by (x, y), then the set of all x-values is called the **domain** of the relation and the set of all y-values is called the **range**. The relation shown in Table 1.3 has domain

$$D = \{1, 2, 3, 4, 5, 6, 7, 8, 9, 10, 11, 12\} \qquad \text{x-values}$$

and range

$$R = \{0.5, 1.1, 1.5, 1.6, 2.0, 2.3, 3.1, 3.6, 3.9, 5.2, 6.2, 6.4\}. \qquad \text{y-values}$$

EXAMPLE 2   Finding the domain and range of a relation

A physics class measured the time y that it takes for an object to fall x feet, as shown in Table 1.4. The object was dropped twice from each height.

Table 1.4   Falling Object

x (feet)	20	20	40	40
y (seconds)	1.2	1.1	1.5	1.6

(a) Express the data as a relation S.
(b) Find the domain and range of S.

SOLUTION
(a) A relation is a set of ordered pairs, so we can write

$$S = \{(20, 1.2), (20, 1.1), (40, 1.5), (40, 1.6)\}.$$

(b) The domain is the set of x-values of the ordered pairs, or $D = \{20, 40\}$. The range is the set of y-values of the ordered pairs, or $R = \{1.1, 1.2, 1.5, 1.6\}$.

Now Try Exercise 47

To visualize a relation, we often use the **Cartesian (rectangular) coordinate plane**, or ***xy*-plane**. The horizontal axis is the ***x*-axis** and the vertical axis is the ***y*-axis**. The axes

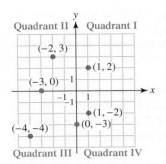

Figure 1.10   The *xy*-plane

intersect at the **origin** and determine four regions called **quadrants**, numbered I, II, III, and IV, counterclockwise, as shown in Figure 1.10. We can plot the ordered pair (*x*, *y*) using the *x*- and *y*-axes. The point (1, 2) is located in quadrant I, (−2, 3) in quadrant II, (−4, −4) in quadrant III, and (1, −2) in quadrant IV. A point lying on a coordinate axis does not belong to any quadrant. The point (−3, 0) is located on the *x*-axis, whereas the point (0, −3) lies on the *y*-axis.

The term **scatterplot** is given to a graph in the *xy*-plane where distinct points are plotted. Figure 1.10 is an example of a scatterplot.

EXAMPLE 3   Graphing a relation

Complete the following for the relation

$$S = \{(5, 10), (5, -5), (-10, 10), (0, 15), (-15, -10)\}.$$

(a) Find the domain and range of the relation.
(b) Determine the maximum and minimum of the *x*-values and then of the *y*-values.
(c) Label appropriate scales on the *xy*-axes.
(d) Plot the relation.

SOLUTION

(a) The elements of the domain correspond to the first number in each ordered pair. Thus

$$D = \{-15, -10, 0, 5\}.$$

Similarly, the elements of the range correspond to the second number in each ordered pair. Thus

$$R = \{-10, -5, 10, 15\}.$$

(b) *x*-minimum: −15; *x*-maximum: 5; *y*-minimum: −10; *y*-maximum: 15
(c) An appropriate scale for both the *x*-axis and the *y*-axis might be −20 to 20, with each tick mark representing a distance of 5. This scale is shown in Figure 1.11.

Figure 1.11

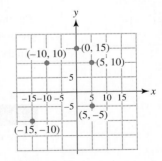

Figure 1.12

(d) The points (5, 10), (5, −5), (−10, 10), (0, 15), and (−15, −10) have been plotted in Figure 1.12.   *Now Try Exercise 51* ◀

Sometimes it is helpful to connect consecutive data points in a scatterplot with straight-line segments. This type of graph, which visually emphasizes changes in the data, is called a **line graph**.

EXAMPLE 4   Making a scatterplot and a line graph

Use Table 1.3 on page 13 to make a scatterplot of average monthly precipitation in Portland, Oregon. Then make a line graph.

SOLUTION   Use the x-axis for the months and the y-axis for the precipitation amounts. To make a scatterplot, simply graph the ordered pairs (1, 6.2), (2, 3.9), (3, 3.6), (4, 2.3), (5, 2.0), (6, 1.5), (7, 0.5), (8, 1.1), (9, 1.6), (10, 3.1), (11, 5.2), and (12, 6.4) in the xy-plane, as shown in Figure 1.13. Then connect consecutive data points to make a line graph, as shown in Figure 1.14.

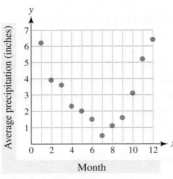

Figure 1.13   Monthly Average Precipitation

Figure 1.14   A Line Graph

Now Try Exercise 57 ◀

Geometry Review
To review the Pythagorean theorem, see Chapter R (page R-2).

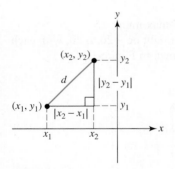

Figure 1.15

The Distance Formula

In the xy-plane, the length of a line segment with endpoints (x_1, y_1) and (x_2, y_2) can be calculated by using the **Pythagorean theorem**. See Figure 1.15.

The lengths of the legs of the right triangle are $|x_2 - x_1|$ and $|y_2 - y_1|$. The distance d is the hypotenuse of the right triangle. Applying the Pythagorean theorem to this triangle gives $d^2 = (x_2 - x_1)^2 + (y_2 - y_1)^2$. Because distance is nonnegative, we can solve this equation for d to get $d = \sqrt{(x_2 - x_1)^2 + (y_2 - y_1)^2}$.

> ### Distance Formula
> The **distance** d between the points (x_1, y_1) and (x_2, y_2) in the xy-plane is
> $$d = \sqrt{(x_2 - x_1)^2 + (y_2 - y_1)^2}.$$

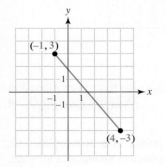

Figure 1.16

Figure 1.16 shows a line segment connecting the points $(-1, 3)$ and $(4, -3)$. Its length is
$$d = \sqrt{(4 - (-1))^2 + (-3 - 3)^2}$$
$$= \sqrt{61} \qquad \text{(exact length)}$$
$$\approx 7.81.$$

EXAMPLE 5   Finding the distance between two points

Find the exact distance between $(3, -4)$ and $(-2, 7)$. Then approximate this distance to the nearest hundredth.

SOLUTION In the distance formula, let (x_1, y_1) be $(3, -4)$ and (x_2, y_2) be $(-2, 7)$.

$$\sqrt{(x_2 - x_1)^2 + (y_2 - y_1)^2} = \sqrt{(-2 - 3)^2 + (7 - (-4))^2} \qquad \text{Distance formula}$$
$$= \sqrt{(-5)^2 + 11^2} \qquad \text{Subtract.}$$
$$= \sqrt{146} \qquad \text{Simplify.}$$
$$\approx 12.08 \qquad \text{Approximate.}$$

The exact distance is $\sqrt{146}$, and the approximate distance, rounded to the nearest hundredth, is 12.08. Note that we would obtain the *same* answer if we let (x_1, y_1) be $(-2, 7)$ and (x_2, y_2) be $(3, -4)$.

Now Try Exercise 15 ◄

In the next example the distance between two moving cars is found.

EXAMPLE 6   Finding the distance between two moving cars

Suppose that at noon car A is traveling south at 20 miles per hour and is located 80 miles north of car B. Car B is traveling east at 40 miles per hour.
(a) Let $(0, 0)$ be the initial coordinates of car B in the xy-plane, where units are in miles. Plot the location of each car at noon and at 1:30 P.M.
(b) Find the distance between the cars at 1:30 P.M.

SOLUTION
(a) If the initial coordinates of car B are $(0, 0)$, then the initial coordinates of car A are $(0, 80)$, because car A is 80 miles north of car B. After 1 hour and 30 minutes, or 1.5 hours, car A has traveled $1.5 \times 20 = 30$ miles south, and so it is located 50 miles north of the initial location of car B. Thus its coordinates are $(0, 50)$ at 1:30 P.M. Car B traveled $1.5 \times 40 = 60$ miles east, so its coordinates are $(60, 0)$ at 1:30 P.M. See Figure 1.17, where these points are plotted.
(b) To find the distance between the cars at 1:30 P.M. we must find the distance d between the points $(0, 50)$ and $(60, 0)$.

$$d = \sqrt{(60 - 0)^2 + (0 - 50)^2} \approx 78.1 \text{ miles}$$

Now Try Exercise 31 ◄

Figure 1.17

The Midpoint Formula

A common way to make estimations is to average data values. For example, in 1980 the average cost of tuition and fees at public colleges and universities was $800, whereas in 1982 it was $1000. One might estimate the cost of tuition and fees in 1981 to be $900. This type of averaging is referred to as finding the midpoint. If a line segment is drawn between two data points, then its *midpoint* is the unique point on the line segment that is equidistant from the endpoints.

On a real number line, the midpoint M of two data points x_1 and x_2 is calculated by averaging their coordinates, as shown in Figure 1.18. For example, the midpoint of -3 and 5 is $M = \dfrac{-3 + 5}{2} = 1$.

$$M = \frac{x_1 + x_2}{2}$$

$x_1 \qquad M \qquad x_2$

Figure 1.18

Figure 1.19

The midpoint formula in the *xy*-plane is similar to the formula for the real number line, except that both coordinates are averaged. Figure 1.19 shows midpoint M located on the line segment connecting the two data points (x_1, y_1) and (x_2, y_2). The *x*-coordinate of the midpoint M is located halfway between x_1 and x_2 and is $\frac{x_1 + x_2}{2}$. Similarly, the *y*-coordinate of M is the average of y_1 and y_2. For example, if we let (x_1, y_1) be $(-3, 1)$ and (x_2, y_2) be $(-1, 3)$ in Figure 1.19, then the midpoint is computed by

$$\left(\frac{-3 + -1}{2}, \frac{1 + 3}{2}\right) = (-2, 2).$$

Midpoint Formula in the *xy*-Plane

The **midpoint** of the line segment with endpoints (x_1, y_1) and (x_2, y_2) in the *xy*-plane is

$$\left(\frac{x_1 + x_2}{2}, \frac{y_1 + y_2}{2}\right).$$

EXAMPLE 7 Finding the midpoint

Find the midpoint of the line segment connecting the points $(6, -7)$ and $(-4, 3)$.

SOLUTION In the midpoint formula, let (x_1, y_1) be $(6, -7)$ and (x_2, y_2) be $(-4, 3)$. Then the midpoint M can be found as follows.

$$M = \left(\frac{x_1 + x_2}{2}, \frac{y_1 + y_2}{2}\right) \qquad \text{Midpoint formula}$$

$$= \left(\frac{6 + (-4)}{2}, \frac{-7 + 3}{2}\right) \qquad \text{Substitute.}$$

$$= (1, -2) \qquad \text{Simplify.} \quad \textbf{\textit{Now Try Exercise 37}} \blacktriangleleft$$

The population of the United States from 1800 to 1990 is shown by the blue curve in Figure 1.20. The red line segment connecting the points $(1800, 5)$ and $(1990, 249)$ does not accurately describe the population over the time period from 1800 to 1990. However, over a shorter time interval this type of line segment may be more accurate.

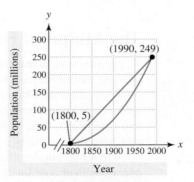

Figure 1.20 U.S. Population

EXAMPLE 8   Using the midpoint formula

In 1970 the population of the United States was 203 million, and by 1990 it had increased to 249 million.   (**Source:** Bureau of the Census.)
(a) Use the midpoint formula to estimate the population in 1980.
(b) Describe this approximation graphically.

SOLUTION
(a) The U.S. populations in 1970 and 1990 are given by the data points (1970, 203) and (1990, 249). The midpoint M of the line segment connecting these points is

$$M = \left(\frac{1970 + 1990}{2}, \frac{203 + 249}{2} \right) = (1980, 226).$$

The midpoint formula estimates a population of 226 million in 1980. (The actual population was 228 million.)

(b) Figure 1.21 shows a blue curve for the U.S. population and a red line segment with midpoint M connecting the data points (1970, 203) and (1990, 249). The line segment appears to be an accurate approximation over a relatively short period of 20 years.

Now Try Exercise 33 ◄

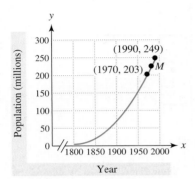

Figure 1.21   A Linear Approximation

Circles

A **circle** consists of the set of points in a plane that are equidistant from a fixed point. The distance is called the **radius** of the circle, and the fixed point is called the **center**. If we let the center of the circle be (h, k), the radius be r, and (x, y) be any point on the circle, then the distance between (x, y) and (h, k) must equal r. See Figure 1.22. By the distance formula we have

$$\sqrt{(x - h)^2 + (y - k)^2} = r.$$

Squaring each side gives

$$(x - h)^2 + (y - k)^2 = r^2.$$

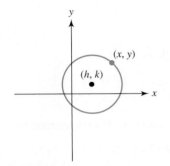

Figure 1.22

Standard Equation of a Circle

The circle with center (h, k) and radius r has equation

$$(x - h)^2 + (y - k)^2 = r^2.$$

NOTE If the center of a circle is $(0, 0)$, then the equation simplifies to $x^2 + y^2 = r^2$. For example, the equation of the circle with center $(0, 0)$ and radius 7 is $x^2 + y^2 = 49$.

EXAMPLE 9   Finding the center and radius of a circle

Find the center and radius of the circle with the given equation. Graph each circle.
(a) $x^2 + y^2 = 9$ (b) $(x - 1)^2 + (y + 2)^2 = 4$

SOLUTION
(a) Because the equation $x^2 + y^2 = 9$ can be written as $(x - 0)^2 + (y - 0)^2 = 3^2$, the center is $(0, 0)$ and the radius is 3. The graph of this circle is shown in Figure 1.23.
(b) For $(x - 1)^2 + (y + 2)^2 = 4$, the center is $(1, -2)$ and the radius is 2. The graph is shown in Figure 1.24.

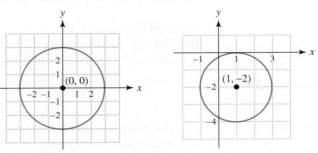

Figure 1.23   **Figure 1.24**

Now Try Exercises 59 and 63 ◀

EXAMPLE 10   Finding the equation of a circle

Find the equation of the circle that satisfies the conditions. Graph each circle.
(a) Radius 4, center $(-3, 5)$
(b) Center $(6, -3)$ with the point $(1, 2)$ on the circle

SOLUTION
(a) Let $r = 4$ and $(h, k) = (-3, 5)$. The equation of this circle is

$$(x - (-3))^2 + (y - 5)^2 = 4^2 \quad \text{or} \quad (x + 3)^2 + (y - 5)^2 = 16$$

A graph of the circle is shown in Figure 1.25.
(b) First we must find the distance between the points $(6, -3)$ and $(1, 2)$ to determine r.

$$r = \sqrt{(6 - 1)^2 + (-3 - 2)^2} = \sqrt{50} \approx 7.1$$

Since $r^2 = 50$, the equation of the circle is $(x - 6)^2 + (y + 3)^2 = 50$. Its graph is shown in Figure 1.26.

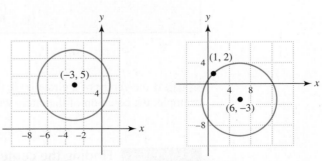

Figure 1.25   **Figure 1.26**

Now Try Exercises 71 and 75 ◀

EXAMPLE 11 Finding the equation of a circle

The endpoints of a diameter of a circle are $(-3, 4)$ and $(5, 6)$. Find the standard equation of the circle.

SOLUTION

Getting Started First find the center of the circle, which is the midpoint of a diameter. Then find the radius by calculating the distance between the center and one of the given endpoints of the diameter. ▶

Find the center C of the circle by applying the midpoint formula to the endpoints of the diameter $(-3, 4)$ and $(5, 6)$.

$$C = \left(\frac{-3 + 5}{2}, \frac{4 + 6}{2} \right) = (1, 5)$$

Use the distance formula to find the radius, which equals the distance from the center $(1, 5)$ to the endpoint $(5, 6)$.

$$r = \sqrt{(5 - 1)^2 + (6 - 5)^2} = \sqrt{17}$$

Thus r^2 equals 17, and the standard equation is $(x - 1)^2 + (y - 5)^2 = 17$.

Now Try Exercise 77 ◀

Graphing with a Calculator (Optional)

Graphing calculators can be used to create tables, scatterplots, line graphs, and other types of graphs. The **viewing rectangle**, or **window**, on a graphing calculator is similar to the view finder in a camera. A camera cannot take a picture of an entire scene; it must be centered on a portion of the available scenery and then it can capture different views of the same scene by zooming in and out. Graphing calculators have similar capabilities. The calculator screen can show only a finite, rectangular region of the xy-plane, which is infinite. The viewing rectangle must be specified by setting minimum and maximum values for both the x- and y-axes before a graph can be drawn.

We will use the following terminology to describe a viewing rectangle. **Xmin** is the minimum x-value and **Xmax** is the maximum x-value along the x-axis. Similarly, **Ymin** is the minimum y-value and **Ymax** is the maximum y-value along the y-axis. Most graphs show an x-scale and a y-scale using tick marks on the respective axes. The distance represented by consecutive tick marks on the x-axis is called **Xscl**, and the distance represented by consecutive tick marks on the y-axis is called **Yscl**. See Figure 1.27. This information can be written concisely as **[Xmin, Xmax, Xscl] by [Ymin, Ymax, Yscl]**. For example, $[-10, 10, 1]$ by $[-10, 10, 1]$ means that Xmin $= -10$, Xmax $= 10$, Xscl $= 1$, Ymin $= -10$, Ymax $= 10$, and Yscl $= 1$. This setting is referred to as the **standard viewing rectangle**.

Figure 1.27

EXAMPLE 12   Setting the viewing rectangle

Show the standard viewing rectangle and the viewing rectangle given by $[-30, 40, 10]$ by $[-400, 800, 100]$ on your calculator.

SOLUTION The required window settings and viewing rectangles are displayed in Figures 1.28–1.31. Notice that in Figure 1.29, there are 10 tick marks on the positive x-axis, since its length is 10 and the distance between consecutive tick marks is 1.

Calculator Help

To set a viewing rectangle or window, see Appendix A (page AP-3).

Figure 1.28

$[-10, 10, 1]$ by $[-10, 10, 1]$

Figure 1.29

```
WINDOW
 Xmin=-10
 Xmax=10
 Xscl=1
 Ymin=-10
 Ymax=10
 Yscl=1
 Xres=1
```

```
WINDOW
 Xmin=-30
 Xmax=40
 Xscl=10
 Ymin=-400
 Ymax=800
 Yscl=100
 Xres=1
```

Figure 1.30

$[-30, 40, 10]$ by $[-400, 800, 100]$

Figure 1.31

Now Try Exercise 81 ◄

EXAMPLE 13   Making a scatterplot with a graphing calculator

Plot the points $(-5, -5)$, $(-2, 3)$, $(1, -7)$, and $(4, 8)$ in the standard viewing rectangle.

SOLUTION The standard viewing rectangle is given by $[-10, 10, 1]$ by $[-10, 10, 1]$. The points $(-5, -5)$, $(-2, 3)$, $(1, -7)$, and $(4, 8)$ are plotted in Figure 1.32.

$[-10, 10, 1]$ by $[-10, 10, 1]$

Calculator Help

To make the scatterplot in Figure 1.32, see Appendix A (page AP-3).

Figure 1.32

Now Try Exercise 89 ◄

EXAMPLE 14 Creating a line graph with a graphing calculator

Table 1.5 lists the percentage of music sales accounted for by compact discs (CDs) from 1990 to 2005. Make a line graph of these sales in [1988, 2007, 2] by [0, 100, 20].

Table 1.5

Year	1990	1995	2000	2005
CDs (% share)	31	65	89	90

Source: Recording Industry Association of America.

SOLUTION Enter the data points (1990, 31), (1995, 65), (2000, 89), and (2005, 90). A line graph can be created by selecting this option on your graphing calculator. The graph is shown in Figure 1.33.

Calculator Help

To make a line graph, see Appendix A (page AP-3).

[1988, 2007, 2] by [0, 100, 20]

Figure 1.33

Now Try Exercise 93 ◄

1.2 Putting It All Together

The following table lists basic concepts in this section.

Concept	Explanation	Examples
Mean, or average	To find the mean, or average, of n numbers, divide their sum by n.	The mean of the four numbers $-3, 5, 6, 9$ is $$\frac{-3 + 5 + 6 + 9}{4} = 4.25.$$
Median	The median of a sorted list of numbers equals the value that is located in the middle of the list. Half the data are greater than or equal to the median, and half the data are less than or equal to the median.	The median of 2, 3, 6, 9, 11 is 6, the middle data item. The median of 2, 3, 6, 9 is the average of the two middle values: 3 and 6. Therefore the median is $\frac{3 + 6}{2} = 4.5$.
Relation, domain, and range	A relation is a set of ordered pairs (x, y). The set of x-values is called the domain, and the set of y-values is called the range.	The relation $S = \{(1, 3), (2, 5), (1, 6)\}$ has domain $D = \{1, 2\}$ and range $R = \{3, 5, 6\}$.

continued on next page

continued from previous page

Concept	Explanation	Examples
Distance formula	The distance between (x_1, y_1) and (x_2, y_2) is $$d = \sqrt{(x_2 - x_1)^2 + (y_2 - y_1)^2}.$$	The distance between $(2, -1)$ and $(-1, 3)$ is $$d = \sqrt{(-1 - 2)^2 + (3 - (-1))^2} = 5.$$
Midpoint formula	The midpoint of the line segment connecting (x_1, y_1) and (x_2, y_2) is $$M = \left(\frac{x_1 + x_2}{2}, \frac{y_1 + y_2}{2}\right).$$	The midpoint of the line segment connecting $(4, 3)$ and $(-2, 5)$ is $$M = \left(\frac{4 + (-2)}{2}, \frac{3 + 5}{2}\right) = (1, 4).$$
Standard equation of a circle	The circle with center (h, k) and radius r has the equation $$(x - h)^2 + (y - k)^2 = r^2.$$	The circle with center $(-3, 4)$ and radius 5 has the equation $$(x + 3)^2 + (y - 4)^2 = 25.$$

The next table summarizes some basic concepts related to one-variable and two-variable data.

Type of Data	Methods of Visualization	Comments
One-variable data	Number line, list, one-column or one-row table	The data items are the same type and can be described using x-values. Computations of the mean and median are performed on one-variable data.
Two-variable data	Two-column or two-row table, scatter-plot, line graph or other type of graph in the xy-plane	Two types of data are related and can be described by using ordered pairs (x, y).

1.2   Exercises

Data Involving One Variable

Exercises 1–4: For the table of data, complete the following.

   (a) Plot the numbers on a number line.
   (b) Find the maximum and minimum of the data.
   (c) Determine the mean of the data.

1.

3	-2	5	0	6	-1

2.

5	-3	4	-2	1	6

3.

-10	20	30	-20	0	10

4.

0.5	-1.5	2.0	4.5	-3.5	-1.0

Exercises 5–8: Sort the list of numbers from smallest to largest and display the result in a table.

   (a) Determine the maximum and minimum values.
   (b) Calculate the mean and median. Round each result to the nearest hundredth when appropriate.

5. $-10, 25, 15, -30, 55, 61, -30, 45, 5$

6. $-1.25, 4.75, -3.5, 1.5, 2.5, 4.75, 1.5$

7. $\sqrt{15}, 2^{2.3}, \sqrt[3]{69}, \pi^2, 2^\pi, 4.1$

8. $\frac{22}{7}, 3.14, \sqrt[3]{28}, \sqrt{9.4}, 4^{0.9}, 3^{1.2}$

Exercises 9 and 10: Geography   The set of numbers contains data about geographic features of the world.

 (a) *Plot the numbers on a number line.*
 (b) *Calculate the mean and median for the set of numbers. Interpret your results.*
 (c) *Try to identify the geographic feature associated with the largest number in the set.*

9. {31.7, 22.3, 12.3, 26.8, 24.9, 23.0} (Areas of largest freshwater lakes in thousands of square miles) (**Source:** U.S. National Oceanic and Atmospheric Administration.)

10. {19.3, 18.5, 29.0, 7.31, 16.1, 22.8, 20.3} (Highest elevations of the continents in thousands of feet) (**Source:** *National Geographic Atlas of the World.*)

11. *Designing a Data Set*   Find a set of three numbers with a mean of 20 and a median of 18. Is your answer unique?

12. *Designing a Data Set*   Find a set of five numbers with a mean of 10 and a median of 9. Is your answer unique?

Distance Formula

Exercises 13–28: Find the exact distance between the two points. Where appropriate, also give approximate results to the nearest hundredth.

13. $(2, -2), (5, 2)$

14. $(0, -3), (12, -8)$

15. $(7, -4), (9, 1)$

16. $(-1, -6), (-8, -5)$

17. $(3.6, 5.7), (-2.1, 8.7)$

18. $(-6.5, 2.7), (3.6, -2.9)$

19. $(-3, 2), (-3, 10)$

20. $(7, 9), (-1, 9)$

21. $\left(\frac{1}{2}, -\frac{1}{2}\right), \left(\frac{3}{4}, \frac{1}{2}\right)$

22. $\left(-\frac{1}{3}, \frac{2}{3}\right), \left(\frac{1}{3}, -\frac{4}{3}\right)$

23. $\left(\frac{2}{5}, \frac{3}{10}\right), \left(-\frac{1}{10}, \frac{4}{5}\right)$

24. $\left(-\frac{1}{2}, \frac{2}{3}\right), \left(\frac{1}{3}, -\frac{5}{2}\right)$

25. $(20, 30), (-30, -90)$

26. $(40, 6), (-20, 17)$

27. $(a, 0), (0, -b)$

28. $(x, y), (1, 2)$

29. *Geometry*   An **isosceles triangle** has at least two sides of equal length. Determine whether the triangle with vertices $(0, 0), (3, 4), (7, 1)$ is isosceles.

30. *Geometry*   An **equilateral triangle** has sides of equal length. Determine whether the triangle with vertices $(-1, -1), (2, 3), (-4, 3)$ is equilateral.

31. *Distance between Cars*   (Refer to Example 6.) At 9:00 A.M. car A is traveling north at 50 miles per hour and is located 50 miles south of car B. Car B is traveling west at 20 miles per hour.

 (a) Let $(0, 0)$ be the initial coordinates of car B in the xy-plane, where units are in miles. Plot the locations of each car at 9:00 A.M. and at 11:00 A.M.

 (b) Find the distance d between the cars at 11:00 A.M.

32. *Distance between Ships*   Two ships leave a harbor at the same time. The first ship heads north at 20 miles per hour, and the second ship heads west at 15 miles per hour. Write an expression that gives the distance d between the ships after t hours.

Midpoint Formula

Exercises 33–36: Use the midpoint formula for the following.

33. *U.S. Average Life Expectancy*   The average life expectancy for a female born in the United States was 77.4 years in 1980 and 79.5 years in 2000. Estimate the average life expectancy for a female born in 1990. (Actual life expectancy was 78.8.) (**Source:** Bureau of the Census.)

34. *State and Federal Inmates*   In 1990 there were 773,919 inmates in state and federal prisons, and in 2000 there were 1,391,892. Estimate the number of inmates in 1995. (Actual number was 1,125,874.) (**Source:** Department of Justice.)

35. *Olympic Times*   In the Olympic Games, the 200-meter dash is run in approximately 20 seconds. Estimate the time to run the 100-meter dash.

36. *Real Numbers*   Between any two real numbers a and b there is always another real number. How could such a number be found?

Exercises 37–46: Find the midpoint of the line segment connecting the points.

37. $(1, 2), (5, -3)$

38. $(-6, 7), (9, -4)$

39. $(-30, 50), (50, -30)$

40. $(28, -33), (52, 38)$

41. $(1.5, 2.9), (-5.7, -3.6)$ **42.** $(9.4, -4.5), (-7.7, 9.5)$

43. $(\sqrt{2}, \sqrt{5}), (\sqrt{2}, -\sqrt{5})$ **44.** $(\sqrt{7}, 3\sqrt{3}), (-\sqrt{7}, -\sqrt{3})$

45. $(a, b), (-a, 3b)$ **46.** $(-a, b), (3a, b)$

Data Involving Two Variables

Exercises 47–50: For the table of data, complete the following.

    (a) Express the data as a relation S.
    (b) Find the domain and range of S.

47.

x	−1	2	3	5	9
y	5	2	−1	−4	−5

48.

x	−2	0	2	4	6
y	−4	−2	−1	0	4

49.

x	1	4	5	4	1
y	5	5	6	6	5

50.

x	−1	0	3	−1	−2
y	$\frac{1}{2}$	1	$\frac{3}{4}$	3	$-\frac{5}{6}$

Exercises 51–56: Complete the following.

    (a) Find the domain and range of the relation.
    (b) Determine the maximum and minimum of the x-values and then of the y-values.
    (c) Label appropriate scales on the xy-axes.
    (d) Plot the relation.

51. $\{(0, 5), (-3, 4), (-2, -5), (7, -3), (0, 0)\}$

52. $\{(1, 1), (3, 0), (-5, -5), (8, -2), (0, 3)\}$

53. $\{(2, 2), (-3, 1), (-4, -1), (-1, 3), (0, -2)\}$

54. $\{(1, 1), (2, -3), (-1, -1), (-1, 2), (-1, 0)\}$

55. $\{(10, 50), (-35, 45), (0, -55), (75, 25), (-25, -25)\}$

56. $\{(-1.2, 1.5), (1.0, 0.5), (-0.3, 1.1), (-0.8, -1.3)\}$

Exercises 57 and 58: Plotting Real Data Use the table to make a scatterplot and line graph of the data.

57. Cell phone subscribers (millions)

Year	1990	1995	2000	2005
Subscribers	5	34	109	208

Source: CTIA—The Wireless Association.

58. Atmospheric CO_2 levels (parts per million)

Year	1958	1975	1990	2005
CO_2 amounts	315	335	355	380

Source: Mauna Loa Observatory.

Circles

Exercises 59–66: Find the center and radius of the circle.

59. $x^2 + y^2 = 25$ **60.** $x^2 + y^2 = 100$

61. $x^2 + y^2 = 7$ **62.** $x^2 + y^2 = 20$

63. $(x - 2)^2 + (y + 3)^2 = 9$

64. $(x + 1)^2 + (y - 1)^2 = 16$

65. $x^2 + (y + 1)^2 = 100$ **66.** $(x - 5)^2 + y^2 = 19$

Exercises 67–70: Find the standard equation of the circle.

67.

68.

69.

70.

Exercises 71–78: Find the standard equation of a circle that satisfies the conditions.

71. Radius 8, center $(3, -5)$

72. Radius 5, center $(-1, 4)$

73. Radius 7, center $(3, 0)$ **74.** Radius 1, center $(0, 0)$

75. Center $(0, 0)$ with the point $(-3, -1)$ on the circle

76. Center $(3, -5)$ with the point $(4, 2)$ on the circle

77. Endpoints of a diameter $(-5, -7)$ and $(1, 1)$

78. Endpoints of a diameter $(-3, -2)$ and $(1, -4)$

Graphing Calculators

Exercises 79–84: Predict the number of tick marks on the positive x-axis and the positive y-axis. Then show the viewing rectangle on your graphing calculator.

79. Standard viewing rectangle

80. $[-4.7, 4.7, 1]$ by $[-3.1, 3.1, 1]$

81. $[0, 100, 10]$ by $[-50, 50, 10]$

82. $[-30, 30, 5]$ by $[-20, 20, 5]$

83. $[1980, 1995, 1]$ by $[12000, 16000, 1000]$

84. $[1800, 2000, 20]$ by $[5, 20, 5]$

Exercises 85–88: Match the settings for a viewing rectangle with the correct figure (a–d).

85. $[-9, 9, 1]$ by $[-6, 6, 1]$

86. $[-6, 6, 1]$ by $[-9, 9, 1]$

87. $[-2, 2, 0.5]$ by $[-4.5, 4.5, 0.5]$

88. $[-4, 8, 1]$ by $[-600, 600, 100]$

a.

b.

c.

d.

Exercises 89–92: Make a scatterplot of the relation.

89. $\{(1, 3), (-2, 2), (-4, 1), (-2, -4), (0, 2)\}$

90. $\{(6, 8), (-4, -10), (-2, -6), (2, -5)\}$

91. $\{(10, -20), (-40, 50), (30, 60), (-50, -80), (70, 0)\}$

92. $\{(-1.2, 0.6), (1.0, -0.5), (0.4, 0.2), (-2.8, 1.4)\}$

Exercises 93–96: The table contains real data.

(a) Determine the maximum and minimum values for each variable in the table.

(b) Use your results from part (a) to find an appropriate viewing rectangle.

(c) Make a scatterplot of the data.

(d) Make a line graph of the data.

93. Digital subscriber lines (millions)

Year	2005	2006	2007	2008
DSL users	19.7	22.0	23.7	25.0

Source: Federal Communications Commission.

94. Cable modems (millions)

Year	2005	2006	2007	2008
Modems	15.1	17.3	19.5	21.7

Source: Federal Communications Commission.

95. Worldwide cigarette sales (trillions)

Year	1950	1960	1970	1980	1990	2000
Cigarettes	1.7	2.2	3.1	4.4	5.4	5.5

Source: Department of Agriculture.

96. Asian-American population y (in millions) during year x

x	1998	2000	2002	2004	2006
y	10.5	11.2	12.0	12.8	14.0

Source: U.S. Census Bureau.

Writing about Mathematics

97. Give an example of a relation that has meaning in the real world. Give an example of an ordered pair (x, y) that is in your relation. Does the ordered pair (y, x) also have meaning? Explain your answers.

98. Do the mean and median represent the same thing? Explain your answer and give an example.

CHECKING BASIC CONCEPTS FOR SECTIONS 1.1 AND 1.2

1. Approximate each expression to the nearest hundredth.
   (a) $\sqrt{4.2(23.1 + 0.5^3)}$ (b) $\dfrac{23 + 44}{85.1 - 32.9}$

2. Evaluate the expression by hand.
   (a) $5 - (-4)^2 \cdot 3$ (b) $5 \div 5\sqrt{2} + 2$

3. Write each number using scientific notation.
   (a) 348,500,000   (b) -1237.4

   (c) 0.00198

4. Find the exact distance between the points $(-3, 1)$ and $(3, -5)$. Then round this distance to the nearest hundredth.

5. Find the midpoint of the line segment connecting the points $(-2, 3)$ and $(4, 2)$.

6. Find the standard equation of a circle with center $(-4, 5)$ and radius 8.

7. The average depths in feet of the four oceans are 13,215, 12,881, 13,002, and 3953. Calculate the mean and median of these depths.

8. Make a scatterplot and a line graph with the four points $(-5, -4)$, $(-1, 2)$, $(2, -2)$, and $(3, 6)$. State the quadrant in which each point lies.

1.3 Functions and Their Representations

- **Learn function notation**
- **Represent a function four different ways**
- **Define a function formally**
- **Identify the domain and range of a function**
- **Use calculators to represent functions (optional)**
- **Identify functions**
- **Represent functions with diagrams and equations**

Introduction

Because there are more than 300 million people in the United States who consume many natural resources, *going green* has become an important social and environmental issue. Collectively, we affect the world around us. The average person uses 2200 paper napkins a year. A washing machine uses about 40 gallons of water for each load of clothes, and a car puts 19.4 pounds of carbon dioxide into the atmosphere every time it burns a gallon of gasoline. See Exercises 99 and 100.

The mathematical concept of a function includes formulas, graphs, tables, and verbal descriptions. All of these representations can be used to analyze the impact of human consumption on the environment and also to describe natural phenomena, such as lightning. This section introduces the important concept of a function, which is used throughout the course.

Basic Concepts

Although thunder is caused by lightning, we sometimes see a flash of lightning before we hear the thunder. This is because light travels at 186,000 *miles* per second, whereas sound travels at about 1050 *feet* per second. Since 1 mile equals 5280 feet, sound takes about 5 seconds to travel 1 mile. It follows that the farther away lightning is, the greater the time lapse between seeing the flash of lightning and hearing the thunder. Table 1.6 lists the *approximate* distance y in miles between a person and a bolt of lightning when there is a time lapse of x seconds between seeing the lightning and hearing the thunder. Note that the value of y can be found by dividing the corresponding value of x by 5.

Table 1.6 Distance from Lightning

x (seconds)	5	10	15	20	25
y (miles)	1	2	3	4	5

This table establishes a special type of relation between two sets of numbers, where each valid input x in seconds determines *exactly one* output y in miles. We say that Table 1.6 *represents* or *defines* a function f, where function f *computes* the distance between an observer and a lightning bolt.

To emphasize that y is a function of x, the notation $y = f(x)$ is used. The expression $f(x)$ does *not* indicate that f and x are multiplied. Rather, the notation $y = f(x)$, called *function notation*, is read "y equals f of x" and denotes that function f with input x produces output y. That is,

$$f(\text{Input}) = \text{Output},$$

where the input is in seconds and the output is in miles. For example, we write $f(5) = 1$ because if there is a **5**-second delay between a lightning bolt and its thunder, then the lightning bolt was about **1** mile away. The expression $f(5)$ represents the output from f when the input is 5. Similarly, $f(10) = 2$ and $f(15) = 3$. Note that a function calculates a set of ordered pairs (x, y), where $y = f(x)$. For example, the ordered pairs $(5, 1)$, $(10, 2)$, and $(15, 3)$ all belong to the relation computed by f. We can think of these ordered pairs as input-output pairs with the form **(input, output)**.

The distance y *depends* on the time x, and so y is called the *dependent variable* and x is called the *independent variable*. For example, if we pick the independent variable x to be 10 seconds, then the value of y equals $\frac{10}{5} = 2$ miles. The y-values depend on the x-values that we choose. The following diagram will help you visualize the computation performed by f.

Input x **in seconds** $\longrightarrow$ **Compute** $y = f(x)$. $\longrightarrow$ **Output** y **in miles**
(independent variable)   (Divide x by 5.)   (dependent variable)

This discussion about function notation is summarized as follows.

Function Notation

The notation $y = f(x)$ is called **function notation**. The **input** is x, the **output** is y, and the *name* of the function is f.

$$\begin{array}{c} \text{Name} \\ \downarrow \\ y = f(x) \\ \nearrow \quad \nwarrow \\ \text{Output} \quad \text{Input} \end{array}$$

The variable y is called the **dependent variable**, and the variable x is called the **independent variable**. The expression $f(20) = 4$ is read "f of 20 equals 4" and indicates that f outputs 4 when the input is 20. A function computes *exactly one* output for each valid input. The letters f, g, and h are often used to denote names of functions.

NOTE In addition to calculating the distance to a lightning bolt, functions can be used to compute many different types of quantities, such as sales tax or drug dosages. However, every function is a *relation* that calculates exactly *one* output for each valid input.

The set of valid or meaningful inputs x is called the **domain** of the function, and the set of corresponding outputs y is the **range**. For example, suppose that a function f computes the height after x seconds of a ball thrown into the air. Then the domain of f might consist of all times while the ball was in flight, and the range would include all heights attained by the ball.

The following operations can be carried out by functions because they result in *one* output for each valid input.

- Calculating the square of a number x
- Finding the sale price when an item with regular price x is discounted 25%
- Naming the biological mother of person x

Not all computations can be done by functions. Suppose that we were given an eye color as an input and asked to output the name of each person in the class having this eye color. Typically we would *not* be computing a function. If, for example, several people in the class had brown eyes, there would not be a unique output (name) for each input (eye color).

MAKING CONNECTIONS

The Expressions f and $f(x)$ The italic letter f represents the *name* of a function, whereas the expression $f(x)$ represents the function f evaluated for input x. That is, $f(x)$ typically represents a formula for function f that can be used to evaluate f for various values of x. For example, if f represents the name of the squaring function, then $f(x) = x^2$. Thus we speak of the graph of f or the domain of f, but we evaluate $f(3)$ to be $3^2 = 9$.

Representations of Functions

Functions can be represented by verbal descriptions, tables, symbols, and graphs.

Verbal Representation (Words)   If function f gives the distance between an observer and a bolt of lightning, then we can verbally describe f with the following sentence: "Divide x seconds by 5 to obtain y miles." We call this a **verbal representation** of f.

Sometimes when the computation performed by a function has meaning, we can describe this computation verbally. For example, another verbal representation of function f is "f calculates the number of miles from a lighting bolt when the delay between thunder and lightning is x seconds."

Numerical Representation (Table of Values)   Table 1.6 gave a numerical representation for the function f that calculates the distance between a lightning bolt and an observer. A **numerical representation** is a *table of values* that lists input-output pairs for a function. A different numerical representation for f is shown in Table 1.7.

Table 1.7   Distance from a Bolt of Lightning

x (seconds)	1	2	3	4	5	6	7
y (miles)	0.2	0.4	0.6	0.8	1.0	1.2	1.4

One difficulty with numerical representations is that it is often either inconvenient or impossible to list all possible inputs x. For this reason we sometimes refer to a table of this type as a **partial numerical representation** as opposed to a **complete numerical representation**, which would include all elements from the domain of a function. For example, many valid inputs do not appear in Table 1.7, such as $x = 11$ or $x = 0.75$.

Some tables do not represent a function. For example, Table 1.8 represents a relation but *not a function* because input **1** produces two outputs, **3** and **12**.

Table 1.8 A Relation That Is Not a Function

x	1	2	3	1
y	3	6	9	12

Symbolic Representation (Formula)   A formula gives a **symbolic representation** of a function. The computation performed by f is expressed by $f(x) = \frac{x}{5}$, where $y = f(x)$. We say that function f is *represented by*, *defined by*, or *given by* $f(x) = \frac{x}{5}$. It follows that $f(6) = \frac{6}{5} = 1.2$.

Similarly, if a function g computes the square of a number x, then g is represented by $g(x) = x^2$. A formula is an efficient, but less visual, way to define a function.

Graphical Representation (Graph)   Leonhard Euler (1707–1783) invented the function notation $f(x)$. He also was the first to allow a function to be represented by a graph, rather than only by a formula.

A **graphical representation**, or **graph**, visually pairs an x-input with a y-output. In a graph of a function, the ordered pairs (x, y) are plotted in the xy-plane. The ordered pairs

$$(1, 0.2), (2, 0.4), (3, 0.6), (4, 0.8), (5, 1.0), (6, 1.2), \text{ and } (7, 1.4)$$

from Table 1.7 are plotted in Figure 1.34. This scatterplot suggests a line for the graph of f, as shown in Figure 1.35.

Figure 1.34

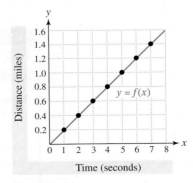

Figure 1.35

Note that since $f(5) = 1$, the point $(5, 1)$ lies on the graph of f. Similarly, since $(7, 1.4)$ lies on the graph of f, it follows that $f(7) = 1.4$.

MAKING CONNECTIONS

Functions, Points, and Graphs If $f(a) = b$, then the point (a, b) lies on the graph of f. Conversely, if the point (a, b) lies on the graph of f, then $f(a) = b$. Thus each point on the graph of f can be written in the form $(a, f(a))$.

EXAMPLE 1   Graphing the absolute value function by hand

Graph $f(x) = |x|$ by hand.

SOLUTION

Getting Started Unless you already know what the graph of a given function looks like, a good technique to use when graphing by hand is to first make a table of values. Then plot the points in the table and sketch a smooth curve (or line) between these points. ▶

Start by selecting convenient x-values and then substitute them into $f(x) = |x|$, as shown in Table 1.9. For example, when $x = -2$, then $f(-2) = |-2| = 2$, so the point $(-2, 2)$ is located on the graph of $y = f(x)$.

Next, plot the points $(-2, 2), (-1, 1), (0, 0), (1, 1)$, and $(2, 2)$ in the xy-plane, as shown in Figure 1.36. The points appear to be V-shaped, and the graph of f shown in Figure 1.37 results if all possible real number ordered pairs $(x, |x|)$ are plotted.

Table 1.9

x	$\lvert x \rvert$
-2	2
-1	1
0	0
1	1
2	2

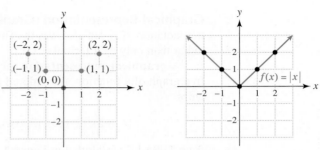

Figure 1.36   **Figure 1.37**

Now Try Exercise 13 ◀

MAKING CONNECTIONS

Four Representations of a Function

Symbolic Representation $f(x) = x^2 + 1$

Numerical Representation   *Graphical Representation*

x	y
-2	5
-1	2
0	1
1	2
2	5

Verbal Representation f squares input x and then adds 1 to produce output y.

Although there are *many* ways to represent a function, there is no *best* representation. Sometimes a formula works best—other times a graph or a table is preferable.

Formal Definition of a Function

Because the idea of a function is a fundamental concept in mathematics, it is important that we define a function precisely. This definition should allow for *all* representations of a function. The commonality among representations is the concept of an ordered pair. A relation is a set of ordered pairs, and a function is a special type of relation.

> ### Function
> A **function** is a relation in which *each* element in the domain corresponds to *exactly one* element in the range.

The set of ordered pairs for a function can be either finite or infinite. The function represented by $f = \{(1, 2), (3, 4), (5, 6)\}$ is a finite set of ordered pairs. In contrast, the function represented by $g(x) = x^2$ with all real numbers as its domain generates an infinite set of ordered pairs (x, x^2). Examples include $(1, 1)$, $(2, 4)$, and $(2.5, 6.25)$.

EXAMPLE 2 Computing median income as a function of educational attainment

The function f computes median 2004 individual annual earnings for males (in dollars) by educational attainment. This function is defined by $f(N) = 26,277$; $f(H) = 35,725$; $f(B) = 57,220$; $f(M) = 71,530$, where N denotes no diploma, H a high school diploma, B a bachelor's degree, and M a master's degree. (**Source:** *Digest of Education Statistics, 2005.*)
(a) Write f as a set of ordered pairs.
(b) Give the domain and range of f.
(c) Discuss the relationship between education and income.

SOLUTION
(a) $f = \{(N, 26277), (H, 35725), (B, 57220), (M, 71530)\}$.
(b) The domain D and range R of f are

$$D = \{N, H, B, M\} \quad \text{and} \quad R = \{26277, 35725, 57220, 71530\}.$$

(c) In general, the greater the educational attainment, the greater the annual earnings.
Now Try Exercise 97 ◄

CLASS DISCUSSION
Suppose a golf ball hit into the air reaches a maximum height of 144 feet and strikes the ground after 6 seconds. Let f be a function that computes the height y (in feet) after x seconds. Describe the domain and range of f.

MAKING CONNECTIONS

Relations and Functions Every function is a relation, whereas not every relation is a function. A function has exactly one output for each valid input.

Unless stated otherwise, the domain of a function f is the set of all *real* numbers for which its symbolic representation (formula) is defined. The domain can be thought of as the set of all valid inputs that make sense in the expression for $f(x)$. In this case the domain is often referred to as the **implied domain**. Other times the domain of a function must be restricted. For example, if an object falls for 5 seconds, then the distance d that it has fallen in feet after t seconds is given by $d(t) = 16t^2$ and the domain of d must be restricted to $0 \le t \le 5$.

NOTE When a function involves a square root, such as $f(x) = \sqrt{x}$, we assume that the output must be a real number. Thus the implied domain of f is the set of real numbers greater than or equal to 0, because the square root of a negative number is not a real number.

EXAMPLE 3   Evaluating a function and determining its domain

Let a function f be represented symbolically by $f(x) = \frac{x}{x^2 - 1}$.
(a) Evaluate $f(3)$ and $f(a + 1)$.   (b) Find the domain of f.

SOLUTION
(a) To evaluate $f(3)$, substitute 3 for x in the formula: $f(3) = \frac{3}{3^2 - 1} = \frac{3}{8}$.

To evaluate $f(a + 1)$, substitute $(a + 1)$ for x in the formula for $f(x)$.

Algebra Review
To square a binomial, see Chapter R (page R-18).

$$f(a + 1) = \frac{(a + 1)}{(a + 1)^2 - 1} \qquad \text{Let } x = (a + 1).$$

$$= \frac{a + 1}{a^2 + 2a + 1 - 1} \qquad \text{Square the binomial.}$$

$$= \frac{a + 1}{a^2 + 2a} \qquad \text{Simplify.}$$

(b) The expression for $f(x)$ is not defined when the denominator $x^2 - 1 = 0$. Because $(1)^2 - 1 = 0$ and $(-1)^2 - 1 = 0$, the domain of f is all real numbers except for -1 and 1. The domain can be expressed in *set-builder notation* as $\{x \mid x \neq -1, x \neq 1\}$.

Now Try Exercise 31 ◄

Set-Builder Notation

The expression $\{x \mid x \neq -1, x \neq 1\}$ is written in **set-builder notation** and represents the set of all real numbers x such that x does not equal -1 and x does not equal 1. Another example is $\{y \mid 1 < y < 5\}$, which represents the set of all real numbers y such that y is greater than 1 *and* less than 5.

Figure 1.38

EXAMPLE 4   Evaluating a function symbolically and graphically

A function g is given by $g(x) = x^2 - 2x$, and its graph is shown in Figure 1.38.
(a) Find the domain of g.
(b) Use $g(x)$ to evaluate $g(-1)$.
(c) Use the graph of g to evaluate $g(-1)$.

SOLUTION
(a) The domain for $g(x) = x^2 - 2x$ includes all real numbers because the formula is defined for all real number inputs x.
(b) To evaluate $g(-1)$, substitute -1 for x in $g(x) = x^2 - 2x$.

$$g(-1) = (-1)^2 - 2(-1) = 1 + 2 = 3$$

(c) Refer to Figure 1.39. Begin by finding $x = -1$ on the x-axis. Move upward until the graph of g is reached. Then move across to the y-axis. The y-value corresponding to an x-value of -1 is 3. Thus $g(-1) = 3$.

Now Try Exercise 37 ◄

Figure 1.39 $g(-1) = 3$

EXAMPLE 5 Finding the domain and range graphically

A graph of $f(x) = \sqrt{x - 2}$ is shown in Figure 1.40. Find the domain and range of f.

Figure 1.40

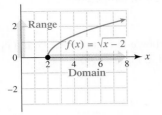

Figure 1.41

SOLUTION The arrow in the graph of f indicates that both the x-values and the y-values increase without reaching a maximum value. In Figure 1.41 the domain and range of f have been labeled by an arrow on each axis. Note that points appear on the graph for all x greater than or equal to 2. Thus the domain is $D = \{x \mid x \geq 2\}$. The minimum y-value on the graph of f is 0 and it occurs at the point (2, 0). There is no maximum y-value on the graph, so the range is $R = \{y \mid y \geq 0\}$.   *Now Try Exercise 45* ◀

Graphing Calculators and Functions (Optional)

Graphing calculators can be used to create graphs and tables of a function—usually more efficiently and reliably than pencil-and-paper techniques. However, a graphing calculator uses the same basic method that we might use to draw a graph. For example, one way to sketch a graph of $y = x^2$ is to first make a table of values, such as Table 1.10.

Table 1.10

x	-3	-2	-1	0	1	2	3
y	9	4	1	0	1	4	9

We can plot these points in the xy-plane, as shown in Figure 1.42. Next we might connect the points with a smooth curve, as shown in Figure 1.43. A graphing calculator typically plots numerous points and connects them to make a graph. In Figure 1.44, a graphing calculator has been used to graph $y = x^2$.

Figure 1.42   **Figure 1.43**   **Figure 1.44**

An Application   In the next example we use a graphing calculator to represent a function that computes the decrease in air temperature as altitude increases.

EXAMPLE 6   Representing a function

When the relative humidity is less than 100%, air cools at a rate of 3.6°F for every 1000-foot increase in altitude. Give verbal, symbolic, graphical, and numerical representations of a function f that computes this change in temperature for an increase in altitude of x thousand feet. Let the domain of f be $0 \leq x \leq 6$.   (**Source:** L. Battan, *Weather in Your Life.*)

SOLUTION
Verbal   Multiply the input x by -3.6 to obtain the change y in temperature.

Symbolic   Let $f(x) = -3.6x$.

Graphical   Since $f(x) = -3.6x$, enter $Y_1 = -3.6X$, as shown in Figure 1.45. Graph y_1 in a viewing rectangle such as $[0, 6, 1]$ by $[-25, 10, 5]$, used in Figure 1.46.

Calculator Help
To enter a formula and create a graph, see Appendix A (pages AP-4 and AP-5).

Figure 1.45

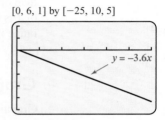

Figure 1.46

Numerical   It is impossible to list all inputs x, since $0 \leq x \leq 6$. However, Figures 1.47 and 1.48 show how to create a table for $Y_1 = -3.6X$ with $x = 0, 1, 2, 3, 4, 5, 6$. Other values for x in the domain of f are possible.

Calculator Help
To create a table similar to Figure 1.48, see Appendix A (page AP-6).

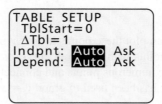

Figure 1.47

Figure 1.48

Now Try Exercise 101 ◄

NOTE   Each of the four representations presented in Example 6 represent the *same* function f.

An Application   Each year approximately 15 million people visit the doctor with foot, toe, and ankle problems. In the next example we use a graphing calculator to evaluate a function that computes the proper crutch length for some of these people.

EXAMPLE 7   Evaluating representations of a function

People who sustain leg injuries often require crutches. The formula $f(x) = 0.72x + 2$ computes the appropriate crutch length in inches for a person with a height of x inches. (**Source:** *Journal of the American Physical Therapy Association.*)

(a) Evaluate $f(65)$ symbolically. Interpret your result.

(b) Evaluate $f(65)$ graphically. Use the window $[60, 90, 10]$ by $[40, 80, 10]$.

(c) Make a table of f starting at $x = 60$ and incrementing by 1. Evaluate $f(65)$.

SOLUTION

(a) *Symbolically*   Since $f(x) = 0.72x + 2$, $f(65) = 0.72(65) + 2 = 48.8$. This result means that a person **65** inches tall needs about **49**-inch crutches.

(b) *Graphically*   Begin by graphing $Y_1 = 0.72X + 2$, as shown in Figure 1.49. Two ways to evaluate a function with a graphing calculator are to use either the "trace" or the "value" utility. The "value" utility is sometimes found under the CALCULATE menu. In Figures 1.50 and 1.51 this utility has been used to evaluate the graph at $x = 65$ to obtain 48.8.

Calculator Help

To make a graph and evaluate a function graphically, see Appendix A (page AP-5).

[60, 90, 10] by [40, 80, 10]

$y = 0.72x + 2$

Figure 1.49

```
CALCULATE
1:value
2:zero
3:minimum
4:maximum
5:intersect
6:dy/dx
7:∫f(x)dx
```

Figure 1.50

[60, 90, 10] by [40, 80, 10]

Y₁=0.72X+2

$f(65) = 48.8$

X=65   Y=48.8

Figure 1.51

(c) *Numerically*   To create a table on a graphing calculator, we retain the formula for y_1 and specify both the *starting value* for x (TblStart) and the *increment* between successive x-values (ΔTbl). Here we start the table at $x = 60$ and increment by 1. (There is no viewing rectangle to set.) Figures 1.52 and 1.53 illustrate these steps. When $x = 65$, $y = 48.8$, so $f(65) = 48.8$.

Calculator Help

To create the table in Figure 1.53, see Appendix A (page AP-6).

```
TABLE SETUP
 TblStart=60
 ΔTbl=1
Indpnt: Auto Ask
Depend: Auto Ask
```

Figure 1.52

X	Y₁
60	45.2
61	45.92
62	46.64
63	47.36
64	48.08
65	48.8 ← $f(65) = 48.8$
66	49.52

Y₁■0.72X+2

Figure 1.53

Now Try Exercise 57 ◀

NOTE   Regardless of the representation that we use to evaluate a function, the results should be equal. However, when we estimate from a graph, sometimes the result will be an approximation of the true value.

Identifying Functions

A function is a special type of relation where each valid input (x-value) produces exactly one output (y-value). We can use this concept to identify functions.

EXAMPLE 8 Determining if a set of ordered pairs is a function

Determine if each set of ordered pairs represents a function.
(a) $A = \{(-2, 3), (-1, 2), (0, -3), (-2, 4)\}$
(b) $B = \{(1, 4), (2, 5), (-3, -4), (-1, 7), (0, 4)\}$

SOLUTION
(a) Set A does not represent a function because input -2 results in two outputs: 3 and 4.
(b) Inputs 1 and 0 have the same output 4. However, set B represents a function because each input (x-value) results in one output (y-value). *Now Try Exercises 77 and 79* ◄

Figure 1.54

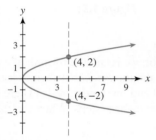

Figure 1.55

Vertical Line Test To conclude that a graph represents a function, we must be convinced that it is impossible for two distinct points with the same x-coordinate to lie on the graph. For example, the ordered pairs $(4, 2)$ and $(4, -2)$ are distinct points with the same x-coordinate. These two points could not lie on the graph of the same function because input 4 would result in two outputs: 2 and 2. When the points $(4, 2)$ and $(4, -2)$ are plotted, they lie on the same vertical line, as shown in Figure 1.54. A graph passing through these points intersects the line twice, as illustrated in Figure 1.55. Therefore the graph in Figure 1.55 does *not* represent a function.

To determine if a graph represents a function, simply visualize vertical lines in the xy-plane. If every vertical line intersects a graph at no more than one point, then it is a graph of a function. This procedure is called the **vertical line test** for a function.

Vertical Line Test

If every vertical line intersects a graph at no more than one point, then the graph represents a function.

NOTE If a vertical line intersects a graph more than once, then the graph does *not* represent a function.

EXAMPLE 9 Identifying a function graphically

Use the vertical line test to determine if the graph represents a function.

(a)

Figure 1.56

(b)

Figure 1.57

SOLUTION

(a) Note in Figure 1.56 that every vertical line that could be visualized would intersect the graph at most once. Therefore the graph represents a function.

(b) The graph in Figure 1.57 does not represent a function because it is possible for a vertical line to intersect the graph more than once.

Now Try Exercises 65 and 67 ◀

Functions Represented by Diagrams and Equations

Thus far we have discussed four representations of a function: verbal, numerical, symbolic, and graphical. Two other ways that we can represent, or define, a function are with diagrams and equations.

Diagrammatic Representation (Diagram) Functions are sometimes represented using **diagrammatic representations**, or **diagrams**. Figure 1.58 is a diagram of a function with domain $D = \{5, 10, 15, 20\}$ and range $R = \{1, 2, 3, 4\}$. An arrow is used to show that input x produces output y. For example, input 5 results in output 1, or $f(5) = 1$. Figure 1.59 shows a relation, but not a function, because input 2 results in two different outputs: 5 and 6.

Figure 1.58 Function

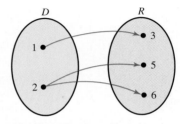

Figure 1.59 Not a Function

NOTE Functions are sometimes referred to as **mappings** between the domain and the range. If $f(5) = 1$, then we say that the range value 1 is the **image** of 5 and that the domain value 5 is the **preimage** of 1.

Functions Defined by Equations Equations can sometimes define functions. For example, the equation $x + y = 1$ defines a function f given by $f(x) = 1 - x$, where $y = f(x)$. Notice that for each input x, there is exactly one y output determined by $y = 1 - x$. In Figure 1.60 the graph passes the vertical line test. However, the graph of the equation $x^2 + y^2 = 4$ is a circle with center $(0, 0)$ and radius 2. Because a circle does not pass the vertical line test, this equation does not represent a function. See Figure 1.61.

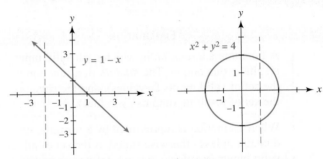

Figure 1.60 Function

Figure 1.61 Not a Function

EXAMPLE 10 Identifying a function

Determine if y is a function of x.
(a) $x = y^2$ **(b)** $y = x^2 - 2$

SOLUTION

(a) For y to be a function of x in the equation $x = y^2$, each valid x-value must result in one y-value. If we let $x = 4$, then y could be either -2 or 2 since

$$4 = (-2)^2 \quad \text{and} \quad 4 = (2)^2.$$

Therefore y is not a function of x. A graph of $x = y^2$ is shown in Figure 1.62. Note that this graph fails the vertical line test.

(b) In the equation $y = x^2 - 2$ each x-value determines exactly one y-value, and so y is a function of x. A graph of this equation is shown in Figure 1.63. This graph passes the vertical line test.

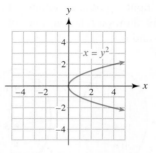

Figure 1.62   Figure 1.63

Now Try Exercises 83 and 89 ◀

1.3 Putting It All Together

The following table summarizes some important topics from this section.

Concept	Explanation	Examples
Function	A function is a *relation* in which each valid input results in one output. The *domain* of a function is the set of valid inputs (x-values), and the *range* is the set of resulting outputs (y-values).	$f = \{(1, 3), (2, 6), (3, 9), (4, 12)\}$ The domain is $D = \{1, 2, 3, 4\}$, and the range is $R = \{3, 6, 9, 12\}$.
Domain	When a function is represented by a formula, its domain, unless otherwise stated, is the set of all valid inputs (x-values) that are defined or make sense in the formula.	$f(x) = \dfrac{1}{\sqrt{x + 4}}$ Domain of f: $\{x \mid x > -4\}$

Concept	Explanation	Examples
Identifying graphs of functions	*Vertical Line Test*: If every vertical line intersects a graph at no more than one point, then the graph represents a function. (Otherwise the graph does not represent a function.)	 Not a function
Verbal representation of a function	Words describe precisely what is computed.	A verbal representation of $f(x) = x^2$ is "Square the input x to obtain the output."
Symbolic representation of a function	Mathematical formula	The squaring function is given by $f(x) = x^2$, and the square root function is given by $g(x) = \sqrt{x}$.
Numerical representation of a function	Table of values	A *partial* numerical representation of $f(x) = 3x$ is shown. | x	0	1	2	3 | |---|---|---|---|---|
$f(x)$	0	3	6	9	
Graphical representation of a function	Graph of ordered pairs (x, y) that satisfy $y = f(x)$	Each point on the graph satisfies $y = 2x$.

1.3 Exercises

Evaluating and Representing Functions

1. If $f(-2) = 3$, identify a point on the graph of f.

2. If $f(3) = -9.7$, identify a point on the graph of f.

3. If $(7, 8)$ lies on the graph of f, then $f(\underline{\quad}) = \underline{\quad}$.

4. If $(-3, 2)$ lies on the graph of f, then $f(\underline{\quad}) = \underline{\quad}$.

Exercises 5–20: Graph $y = f(x)$ by hand by first plotting points to determine the shape of the graph.

5. $f(x) = 3$

6. $f(x) = -2$

7. $f(x) = 2x$

8. $f(x) = x + 1$

9. $f(x) = 4 - x$

10. $f(x) = 3 + 2x$

11. $f(x) = \frac{1}{2}x - 2$

12. $f(x) = 2 - 2x$

13. $f(x) = |x - 1|$ **14.** $f(x) = |0.5x|$

15. $f(x) = |3x|$ **16.** $f(x) = |2x - 1|$

17. $f(x) = \frac{1}{2}x^2$ **18.** $f(x) = 2x^2$

19. $f(x) = x^2 - 2$ **20.** $f(x) = x^2 + 1$

Exercises 21–34: Complete the following.
  (a) Find f(x) for the indicated values of x, if possible.
  (b) Find the domain of f.

21. $f(x) = x^3$ for $x = -2, 5$

22. $f(x) = 2x - 1$ for $x = 8, -1$

23. $f(x) = \sqrt{x}$ for $x = -1, a + 1$

24. $f(x) = \sqrt{1 - x}$ for $x = -2, a + 2$

25. $f(x) = 6 - 3x$ for $x = -1, a + 1$

26. $f(x) = \dfrac{3x - 5}{x + 5}$ for $x = -1, a$

27. $f(x) = -7$ for $x = 6, a - 1$

28. $f(x) = x^2 - x + 1$ for $x = 1, -2$

29. $f(x) = \dfrac{1}{x^2}$ for $x = 4, -7$

30. $f(x) = \sqrt{x - 3}$ for $x = 4, a + 4$

31. $f(x) = \dfrac{1}{x^2 - 9}$ for $x = 4, a - 5$

32. $f(x) = \dfrac{1}{x^2 + 4}$ for $x = -2, a + 4$

33. $f(x) = \dfrac{1}{\sqrt{2 - x}}$ for $x = 1, a + 2$

34. $f(x) = \dfrac{1}{\sqrt{x - 1}}$ for $x = 0, a^2 - a + 1$

Exercises 35–40: (Refer to Example 4.) Use the graph to complete the following.
  (a) Find the domain of g.
  (b) Use the formula to evaluate g(−1) and g(2).
  (c) Use the graph of g to evaluate g(−1) and g(2).

35.

$g(x) = 2x - 1$

36.

$g(x) = -\frac{1}{2}x - 1$

37.
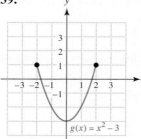
$g(x) = 2 - x^2$

38.
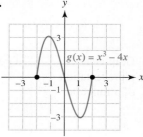
$g(x) = 3 - 2|x|$

39.
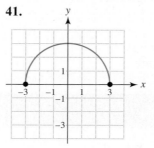
$g(x) = x^2 - 3$

40.
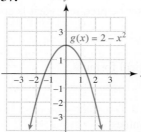
$g(x) = x^3 - 4x$

Exercises 41–46: Use the graph of the function f to estimate its domain and range. Evaluate f(0).

41.

42.

43.

44.

45.

46.

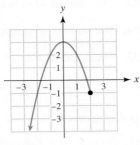

Exercises 47 and 48: Diagrams Complete the following.

  (a) *Evaluate* $f(2)$.
  (b) *Write f as a set of ordered pairs.*
  (c) *Find the domain and range of f.*

47.

48.

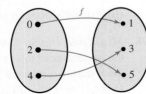

Exercises 49–52: Graph $y = f(x)$ *in the viewing rectangle* $[-4.7, 4.7, 1]$ *by* $[-3.1, 3.1, 1]$.

  (a) *Use the graph to evaluate* $f(2)$.
  (b) *Evaluate* $f(2)$ *symbolically.*
  (c) *Let* $x = -3, -2, -1, 0, 1, 2, 3$ *and make a table of values for* $f(x)$.

49. $f(x) = 0.25x^2$ **50.** $f(x) = 3 - 1.5x^2$

51. $f(x) = \sqrt{x + 2}$ **52.** $f(x) = |1.6x - 2|$

Exercises 53–60: Use $f(x)$ *to determine verbal, graphical, and numerical representations. For the numerical representation use a table with* $x = -2, -1, 0, 1, 2$. *Evaluate* $f(2)$.

53. $f(x) = x^2$ **54.** $f(x) = 2x - 5$

55. $f(x) = |2x + 1|$ **56.** $f(x) = 8$

57. $f(x) = 5 - x$ **58.** $f(x) = |x|$

59. $f(x) = \sqrt{x + 1}$ **60.** $f(x) = x^2 - 1$

Exercises 61 and 62: A function g is defined.
  (a) *Write g as a set of ordered pairs.*
  (b) *Give the domain and range of g.*

61. $g(-1) = 2, g(0) = 4, g(1) = -3, g(2) = 2$

62. $g(-4) = 5, g(0) = -5, g(4) = 5, g(8) = 0$

Exercises 63 and 64: Express a function f with the specified representation.

63. *Cost of Driving*   In 2008 the average cost of driving a new car was about 50 cents per mile. Give symbolic, graphical, and numerical representations of the cost in dollars of driving x miles. For the numerical representation use a table with $x = 1, 2, 3, 4, 5, 6$. (**Source:** Associated Press.)

64. *Counterfeit Money*   It is estimated that nine out of every one million bills are counterfeit. Give a numerical representation (table) of the predicted number of counterfeit bills in a sample of x million bills where $x = 0, 1, 2, \dots, 6$. (**Source:** Department of the Treasury.)

Identifying Functions

Exercises 65–70: Does the graph represent a function? If so, determine the function's domain and range.

65.

66.

67.

68.

69.

70.

Exercises 71–74: Determine if the following operation describes a function. Explain your answer.

71. Calculating the cube root of a number

72. Calculating your age

73. Listing the students who passed a given English exam

74. Finding the *x*-values in the domain of a relation

75. *Identification Numbers*   A relation takes a student's identification number at your college as input and outputs the student's name. Does this relation compute a function? Explain.

76. *Heights*   A relation takes a student's height rounded to the nearest inch as input and outputs the student's name with that height. Does this relation typically compute a function? Explain.

Exercises 77–82: Determine if S is a function.

77. $S = \{(1, 2), (2, 3), (4, 5), (1, 3)\}$

78. $S = \{(-3, 7), (-1, 7), (3, 9), (6, 7), (10, 0)\}$

79. $S = \{(a, 2), (b, 3), (c, 3), (d, 3), (e, 2)\}$

80. $S = \{(a, 2), (a, 3), (b, 5), (-b, 7)\}$

81. *S* is given by the table.

x	1	3	1
y	10.5	2	−0.5

82. *S* is given by the table.

x	1	2	3
y	1	1	1

Exercises 83–90: Determine if y is a function of x.

83. $x = y^4$

84. $y^2 = x + 1$

85. $\sqrt{x + 1} = y$

86. $x^2 = y - 7$

87. $x^2 + y^2 = 70$

88. $(x - 1)^2 + y^2 = 1$

89. $x + y = 2$

90. $y = |x|$

Exercises 91–96: Formulas   Write a symbolic representation (formula) for a function g that calculates the given quantity. Then evaluate g(10) and interpret the result.

91. The number of inches in *x* feet

92. The number of quarts in *x* gallons

93. The number of dollars in *x* quarters

94. The number of quarters in *x* dollars

95. The number of seconds in *x* days

96. The number of feet in *x* miles

Applications

97. *Income and Education*   (Refer to Example 2.) The function *I* computes median 2004 individual annual earnings for females (in dollars) by educational attainment. This function is defined by $I(N) = 19,162$, $I(H) = 26,029$, $I(B) = 41,681$, and $I(M) = 51,316$. (**Source:** *Digest of Education Statistics, 2005.*)
  (a) Write *I* as a set of ordered pairs.

  (b) Give the domain and range of *I*.

98. *Music and Digital Downloads*   Function *P* computes the percentage of total music sales that were digital downloads during a selected year. This function is defined by $P(2002) = 0.5\%$, $P(2003) = 1.3\%$, $P(2004) = 2.9\%$, $P(2005) = 5.7\%$, and $P(2006) = 9.4\%$. (**Source:** Recording Industry Association of America.)
  (a) Write *P* as a set of ordered pairs.

  (b) Give the domain and range of *P*.

99. *Going Green*   The average person uses 2200 paper napkins in one year. Write the formula for a function *N* that calculates the number of paper napkins that the average person uses in *x* years. Evaluate *N*(3) and interpret your result.

100. *Going Green*   The average top-loading washing machine uses about 40 gallons of water per load of clothes. Write the formula for a function *W* that calculates the number of gallons of water used while washing *x* loads of clothes. Evaluate *W*(30) and interpret your result.

101. *Air Temperature*   (Refer to Example 6.) When the relative humidity is 100%, air cools 5.8°F for every 1-mile increase in altitude. Give verbal, symbolic, graphical, and numerical representations of a function f that computes this change in temperature for an increase in altitude of x miles for $0 \leq x \leq 3$.   (**Source:** L. Battan.)

102. *Crutch Length*   (Refer to Example 7.) Determine the crutch length for someone 6 feet 3 inches tall. For each 1-inch increase in a person's height, by how much does the recommended crutch length increase?

103. *Distance to Lightning*   Find a formula for a function f that computes the distance between an observer and a lightning bolt when the speed of sound is 1150 feet per second. Evaluate $f(15)$ and interpret the result.

104. *Distance to Lightning*   Give a reasonable domain for the function f that you found in Exercise 103. Graph f over the domain that you selected. What is the range of your function? (Note that answers may vary.)

Writing about Mathematics

105. Explain how you could use a complete numerical representation (table) for a function to determine its domain and range.

106. Explain in your own words what a function is. How is a function different from a relation?

1.4 Types of Functions

- **Identify and use constant and linear functions**
- **Interpret slope as a rate of change**
- **Identify and use nonlinear functions**
- **Recognize linear and nonlinear data**

Introduction

Functions are used to describe, or *model*, everything from weather to new product "specs," global warming, and U.S. population. New functions are created each day in the dynamic field of mathematics. Finding new functions—whether to describe the wind speed in Hawaii or to calculate the memory requirements of an iPod—requires creativity. This section discusses three common types of functions: *constant, linear*, and *nonlinear*.

Constant Functions

The monthly average wind speeds in miles per hour at Hilo, Hawaii, from May through December are listed in Table 1.11.

Table 1.11

Month	May	June	July	Aug	Sept	Oct	Nov	Dec
Wind speed (mph)	7	7	7	7	7	7	7	7

Source: J. Williams, *The Weather Almanac.*

A Discrete Function   It is apparent that the monthly average wind speed is constant between May and December. These data can be described by a set f of ordered pairs (x, y), where x is the month and y is the wind speed. The months have been assigned the standard numbers.

$$f = \{(5, 7), (6, 7), (7, 7), (8, 7), (9, 7), (10, 7), (11, 7), (12, 7)\}$$

The function f is given by $f(x) = 7$, where $x = 5, 6, 7, \ldots, 12$. The output of f never changes. We say that f is a *constant function* and *models* the data in Table 1.11. The range of f is $R = \{7\}$, and the domain of f is $D = \{5, 6, 7, 8, 9, 10, 11, 12\}$. Since f is defined only at individual or discrete values of x, f is called a **discrete function**. The graph of a discrete function suggests a scatterplot, as shown in Figure 1.64.

Figure 1.64 A Discrete Constant Model

A Continuous Function   Sometimes it is more convenient to describe discrete data with a continuous graph. If the domain of $f(x) = 7$ is changed to $D = \{x \mid 5 \leq x \leq 12\}$, its graph becomes a continuous horizontal line (segment) without breaks. See Figure 1.65. The graph of a **continuous function** can be sketched without picking up the pencil. There are no breaks in the graph of a continuous function. The graph of a continuous constant function is a horizontal line (or in this case, a horizontal line segment).

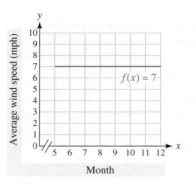

Figure 1.65   A Continuous Constant Model

If the domain includes $x = 6.5$, what does $f(6.5) = 7$ represent? The expression $f(6)$ computes the average wind speed in **June**, and $f(7)$ gives the average wind speed in **July**. We might interpret $f(6.5)$ to represent the average wind speed from **June 15 to July 15**. Other interpretations are possible.

A formal definition of a constant function is now given.

Constant Function

A function f represented by $f(x) = b$, where b is a constant (fixed number), is a **constant function**.

NOTE As simple as constant functions might appear, they occur frequently in applications. The following are examples of computations that can be described by constant functions. In both cases the independent variable (input) is time.

• A thermostat computes a constant function regardless of the weather outside by maintaining a set temperature. If the thermostat is set at 72°F, then $f(x) = 72$.
• A cruise control in a car computes a constant function by maintaining a fixed speed regardless of the type of road or terrain. If the cruise control is set at 55 miles per hour, then $f(x) = 55$.

Linear Functions

A car is initially located 30 miles north of the Texas border and is traveling north on Interstate 35 at 60 miles per hour. The distances between the automobile and the border are listed in Table 1.12 for various times.

Table 1.12

Elapsed time (hours)	0	1	2	3	4	5
Distance (miles)	30	90	150	210	270	330

Figure 1.66 A Discrete Linear Model

Figure 1.67 A Continuous Linear Model

The table shows that the distance increases by 60 miles every hour. These data can be given by a set f of ordered pairs (x, y), where x is the elapsed time and y is the distance.

$$f = \{(0, 30), (1, 90), (2, 150), (3, 210), (4, 270), (5, 330)\}$$

The set f is a function, but not a constant function. A scatterplot of f, shown in Figure 1.66, suggests a line that rises from left to right.

If the car travels for x hours, the distance traveled can be found by multiplying 60 times x and adding the initial distance of 30 miles. This computation can be expressed as $f(x) = 60x + 30$. For example, $f(1.5) = 60(1.5) + 30 = 120$ means that the car is **120** miles from the border after **1.5** hours. The formula is valid for nonnegative x. The graph of $f(x) = 60x + 30$ is a line (ray), shown in Figure 1.67. We call f a *linear function*.

Linear Function

A function f represented by $f(x) = ax + b$, where a and b are constants, is a **linear function**.

NOTE If $a = 0$, then $f(x) = b$, which defines a constant function. Thus every constant function is also a linear function.

Recognizing Linear Functions   In the example of the moving car, $f(x) = 60x + 30$, so $a = 60$ and $b = 30$. The value of a represents the speed of the car, and b is the initial distance of the car from the border. Other examples of linear functions include the following.

$$
\begin{array}{lll}
f(x) = 1.5x - 6 & a = 1.5, & b = -6 \\
g(t) = 8t & a = 8, & b = 0 \\
h(x) = 72 & a = 0, & b = 72 \\
k(t) = 1.9 - 3t & a = -3, & b = 1.9
\end{array}
$$

A distinguishing feature of a linear function f is that each time x increases by one unit, the value of $f(x)$ always changes by an amount equal to a. That is, a linear function has a **constant rate of change**. (The constant rate of change a is equal to the slope of the graph of f.) The following applications are modeled by linear functions. Try to determine the value of the constant a in each case.

- The wages earned by an individual working x hours at \$6.25 per hour
- The amount of tuition and fees due when registering for x credits if each credit costs \$75 and the fees are fixed at \$56

Slope as a Rate of Change

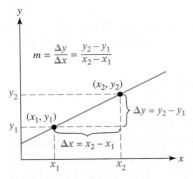

Figure 1.68

The graph of a (continuous) linear function is a line. Slope is a real number that measures the "tilt" of a line in the xy-plane. If the input x to a linear function increases by 1 unit, then the output y changes by a constant amount that is equal to the slope of its graph. In Figure 1.68 a line passes through the points (x_1, y_1) and (x_2, y_2). The *change in y* is $y_2 - y_1$, and the *change in x* is $x_2 - x_1$. The ratio of the change in y to the change in x is called the *slope*. We sometimes denote the change in y by Δy (delta y) and the change in x by Δx (delta x). That is, $\Delta y = y_2 - y_1$ and $\Delta x = x_2 - x_1$.

Slope

The **slope** m of the line passing through the points (x_1, y_1) and (x_2, y_2) is

$$m = \frac{\Delta y}{\Delta x} = \frac{y_2 - y_1}{x_2 - x_1}, \quad \text{where } x_1 \neq x_2.$$

If the slope of a line is positive, the line *rises* from left to right. If the slope is negative, the line *falls* from left to right. Slope 2 indicates that the line rises 2 units for every unit increase in x, and slope $-\frac{1}{2}$ indicates that the line *falls* $\frac{1}{2}$ unit for every unit increase in x. Slope 0 indicates that the line is horizontal. When $x_1 = x_2$, the line is vertical and the slope is undefined. Figures 1.69–1.72 illustrate these situations.

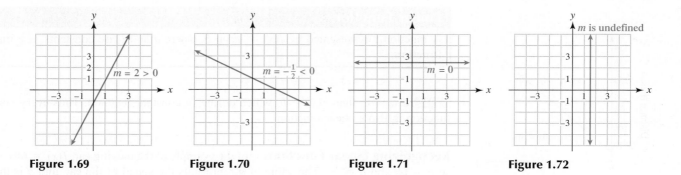

| Figure 1.69 | Figure 1.70 | Figure 1.71 | Figure 1.72 |

EXAMPLE 1 Calculating the slope of a line

Find the slope of the line passing through the points $(-2, 3)$ and $(1, -2)$. Plot these points together with the line. Interpret the slope.

SOLUTION The slope is

$$m = \frac{y_2 - y_1}{x_2 - x_1} = \frac{-2 - 3}{1 - (-2)} = -\frac{5}{3}.$$

Figure 1.73

A graph of the line passing through these two points is shown in Figure 1.73. The change in y is $\Delta y = -5$ and the change in x is $\Delta x = 3$, so $m = \frac{\Delta y}{\Delta x} = -\frac{5}{3}$ indicates that the line falls $\frac{5}{3}$ units for each unit increase in x, or equivalently, the line falls 5 units for each 3-unit increase in x.   ***Now Try Exercise 7*** ◄

Slope and the Value of a   The graph of $f(x) = ax + b$ is a line. Since $f(0) = b$ and $f(1) = a + b$, the graph of f passes through the points $(0, b)$ and $(1, a + b)$. The slope of this line is

$$m = \frac{y_2 - y_1}{x_2 - x_1} = \frac{a + b - b}{1 - 0} = \frac{a}{1} = a.$$

Interpreting Slope   In applications involving linear functions, slope sometimes is interpreted as a *(constant) rate of change*. In Example 7 of Section 1.3, the recommended crutch length for a person x inches tall was given by $f(x) = 0.72x + 2$. The slope of its graph is 0.72. One interpretation of this slope is that for each 1-inch increase in the height of a person, the crutch length should be increased by 0.72 inch.

Figure 1.74

EXAMPLE 2   Interpreting slope and iPod memory

Figure 1.74 shows the (approximate) number of songs that can be stored on x gigabytes of Classic iPod memory.   (**Source:** Apple Corporation.)
(a) Why is it reasonable for the graph to pass through the origin?
(b) Find the slope of the line segment.
(c) Interpret the slope as a rate of change.

SOLUTION
(a) Because 0 songs require no memory, the graph passes through the point (0, 0).
(b) The graph passes through the points (**0, 0**) and (**80, 20000**). The slope of the line is

$$m = \frac{20{,}000 - 0}{80 - 0} = 250.$$

(c) An iPod holds 250 songs per gigabyte.   *Now Try Exercise 31* ◄

MAKING CONNECTIONS

Units for Rates of Change The units for a rate of change can be found from a graph by placing the units from the vertical axis over the units from the horizontal axis. For example, in Figure 1.74 the units on the y-axis are *songs* and the units on the x-axis are *gigabytes*. Thus the units for the slope, or rate of change, are *songs per gigabyte*.

Nonlinear Functions

Table 1.13

Year	Population (millions)
1800	5
1820	10
1840	17
1860	31
1880	50
1900	76
1920	106
1940	132
1960	179
1980	226
2000	281
2020	336*

Source: Bureau of the Census.

*Projected

Table 1.13 shows the population of the United States at 20-year intervals from 1800 to 2020. Between 1800 and 1820 the population increased by 5 million, between 1820 and 1840 it increased by 7 million, and between 1840 and 1860 the increase was 14 million. Because the increases in population for the 20-year periods are not equal or nearly equal, we cannot describe (model) these data with a linear function. We need a *nonlinear function* instead.

The population data from Table 1.13 are shown in Figure 1.75. In Figure 1.76 the data together with a nonlinear function f have been plotted. The graph of f is curved rather than straight.

Figure 1.75 A Discrete Nonlinear Model

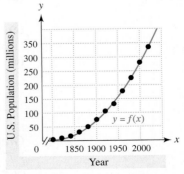

Figure 1.76 A Continuous Nonlinear Model

The time required to drive 100 miles depends on the average speed. Let $f(x)$ compute this time, given the average speed x as input. For example, $f(50) = 2$, because it would take 2 hours to travel 100 miles at an average speed of 50 miles per hour. Make a table of values for f. Is f linear or nonlinear?

Recognizing Nonlinear Functions   If a function is not linear, then it is called a **nonlinear function**. *The graph of a nonlinear function is not a (straight) line.* With a nonlinear function, it is possible for the input x to increase by 1 unit and the output y to change by different amounts. Nonlinear functions *cannot* be written in the form $f(x) = ax + b$.

Real-world phenomena often are modeled by using nonlinear functions. The following are two examples of quantities that can be described by nonlinear functions.

* The monthly average temperature in Chicago (Monthly average temperatures increase and decrease throughout the year.)
* The height of a child between the ages of 2 and 18 (A child grows faster at certain ages.)

EXAMPLE 3   Recognizing linear and nonlinear data

For each table decide whether the data are linear or nonlinear. If the data are linear, state the slope of the line that passes through the data points.

(a)

x	0	5	10	15	20
y	−4	−2	0	2	4

(b)

x	−3	0	3	6	9
y	5	7	10	14	19

(c)

x	0	1	2	3	4
y	11	11	11	11	11

(d)

x	0	1	3	6	10
y	3	6	9	12	15

SOLUTION

(a) The y-values increase 2 units for each 5-unit increase in x. Therefore the data are linear, but not constant. The slope of the line passing through the points is $m = \frac{2}{5}$.

(b) The y-values do not increase by a constant amount for each 3-unit increase in x. The data are nonlinear.

(c) These data are linear and constant. The slope of the line passing through these data points is 0.

(d) Although the y-values increase by 3 units between consecutive data points, the data are nonlinear because the corresponding increases in the x-values are not constant. For example, the slope of the line passing through $(0, 3)$ and $(1, 6)$ is 3, whereas the slope of the line passing through $(1, 6)$ and $(3, 9)$ is 1.5.   *Now Try Exercises 49 and 51* ◄

Graphs of Nonlinear Functions   There are many nonlinear functions. In Figures 1.77–1.80 graphs and formulas are given for four common nonlinear functions. Note that each graph is not a line. See Appendix B for more examples of functions.

Figure 1.77
Square Function

Figure 1.78
Square Root Function

Figure 1.79
Cube Function

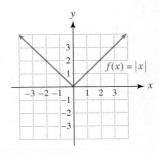

Figure 1.80
Absolute Value Function

The next example illustrates how to graph a nonlinear function by hand.

EXAMPLE 4   Graphing a nonlinear function

Graph $f(x) = \sqrt[3]{x}$.

SOLUTION

Getting Started Function f is the cube root function. For example, $\sqrt[3]{-8} = -2$, because $(-2)(-2)(-2) = (-2)^3 = -8$. Start by making a table of convenient values. See Table 1.14. Plot the points and then connect them with a smooth curve.

Table 1.14   Cube Root Function

x	-8	-1	0	1	8
$\sqrt[3]{x}$	-2	-1	0	1	2

The points $(-8, -2)$, $(-1, -1)$, $(0, 0)$, $(1, 1)$, and $(8, 2)$ and a smooth curve connecting them are shown in Figure 1.81. Notice that f is a nonlinear function because its graph is *not* a line.

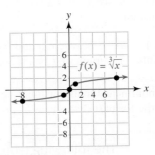

Figure 1.81 Cube Root Function

Now Try Exercise 45 ◀

Sketching a Nonlinear Model   The next example demonstrates how to model a physical situation with a graph of a nonlinear function.

EXAMPLE 5   Sketching a model

A hole is drilled in the bottom of a plastic tank that contains 100 gallons of water. Water begins to leak out, and after 10 minutes the tank is empty. Sketch a graph that shows the amount of water W in the tank after t minutes.

Figure 1.82

SOLUTION

Getting Started When the tank is full, water will rush out at a faster rate, and then the rate will be slower after a while. Because the rate at which water leaks out of the tank is not constant, a nonlinear graph is needed to show this situation. ▶

Let W be the number of gallons of water in the tank after t minutes. First plot the points $(0, 100)$ and $(10, 0)$, because 100 gallons of water are in the tank initially and the tank is empty after 10 minutes. Because the water level goes down faster at first, sketch a curve that "slopes" downward more at first and then starts to level off. See Figure 1.82.

Now Try Exercise 61 ◀

1.4 Putting It All Together

The following table summarizes important concepts from this section.

Concept	Formula	Examples
Slope of a line passing through (x_1, y_1) and (x_2, y_2)	$m = \dfrac{\Delta y}{\Delta x} = \dfrac{y_2 - y_1}{x_2 - x_1}$ $\Delta y = y_2 - y_1$ denotes the change in y. $\Delta x = x_2 - x_1$ denotes the change in x.	A line passing through $(-1, 3)$ and $(1, 7)$ has slope $m = \dfrac{7 - 3}{1 - (-1)} = \dfrac{4}{2} = 2$. This slope indicates that the line rises 2 units for each unit increase in x.
Constant function	$f(x) = b$, where b is a fixed number, or constant.	$f(x) = 12$, $g(x) = -2.5$, and $h(x) = 0$. Every constant function is also linear.
Linear function	$f(x) = ax + b$, where a and b are constants. The graph of f has slope a.	$f(x) = 3x - 1$, $g(x) = -5$, and $h(x) = \frac{1}{2} - \frac{3}{4}x$. Their graphs have slopes 3, 0, and $-\frac{3}{4}$.
Nonlinear function	A nonlinear function cannot be expressed in the form $f(x) = ax + b$.	$f(x) = \sqrt{x + 1}$, $g(x) = 4x^3$, and $h(x) = x^{1.01} + 2$

The following table summarizes important concepts related to constant, linear, and nonlinear functions.

Concept	Constant Function	Linear Function	Nonlinear Function
Slope of graph	Always zero	Always constant	No notion of one slope
Graph	Horizontal line	Nonvertical line	Not a line
Examples	$f(x) = 2$	$f(x) = 2x - 2$	$f(x) = x^2 - 3$

1.4   Exercises

Formulas for Linear Functions

Exercises 1–6: A linear function f can be written in the form
$f(x) = ax + b$. *Identify a and b for the given f(x).*

1. $f(x) = 5 - 2x$ **2.** $f(x) = 3 - 4x$

3. $f(x) = -8x$ **4.** $f(x) = 10x$

5. $f(x) = 7$ **6.** $f(x) = -6$

Slope

Exercises 7–22: If possible, find the slope of the line passing through each pair of points.

7. $(4, 6), (2, 5)$ **8.** $(-8, 5), (-3, -7)$

9. $(-1, 4), (5, -2)$ **10.** $(10, -4), (-15, 7)$

11. $(12, -8), (7, -8)$ **12.** $(8, -5), (8, 2)$

13. $(0.2, -0.1), (-0.3, 0.4)$ **14.** $(-0.3, 0.6), (-0.2, 1.1)$

15. $(-0.5, 9.2), (-0.3, 7.6)$ **16.** $(1.6, 12), (1.6, 5)$

17. $(1997, 5.6), (1994, 7.9)$ **18.** $(1824, 108), (1900, 380)$

19. $(-5, 6), (-5, 8)$ **20.** $(17, 7), (19, 7)$

21. $\left(\frac{1}{3}, -\frac{3}{5}\right), \left(-\frac{5}{6}, \frac{7}{10}\right)$ **22.** $\left(-\frac{13}{15}, -\frac{7}{8}\right), \left(\frac{1}{10}, \frac{3}{16}\right)$

Exercises 23–30: State the slope of the graph of f. Interpret this slope.

23. $f(x) = 2x + 7$ **24.** $f(x) = 6 - x$

25. $f(x) = -\frac{3}{4}x$ **26.** $f(x) = \frac{2}{3}x$

27. $f(x) = -5$ **28.** $f(x) = x + 5$

29. $f(x) = 9 - x$ **30.** $f(x) = 23$

Slope as a Rate of Change

31. *Price of Carpet*   The graph at the top of the next column shows the price of x square yards of carpeting.
  (a) Why is it reasonable for the graph to pass through the origin?

  (b) Find the slope of the graph.

  (c) Interpret the slope as a rate of change.

Carpet (square yards)

32. *Landscape Rock*   The figure shows the price of x tons of landscape rock.

Landscape rock (tons)

  (a) Why is it reasonable for the graph to pass through the origin?

  (b) Find the slope of the graph.

  (c) Interpret the slope as a rate of change.

33. *Going Green*   The function given by $P(x) = 19.4x$ calculates the pounds of CO_2 (carbon dioxide) released into the atmosphere by a car burning x gallons of gasoline.
  (a) Calculate $P(20)$ and interpret the result.

  (b) Find the slope of the graph of P. Interpret this slope as a rate of change.

34. *Velocity of a Train* The distance D in miles that a train is from a station after x hours is given by the formula $D(x) = 150 - 20x$.
(a) Calculate $D(5)$ and interpret the result.

(b) Find the slope of the graph of D. Interpret this slope as a rate of change.

35. *Velocity of a Car* A driver's distance D in miles from a rest stop after x hours is given by $D(x) = 75x$.
(a) How far is the driver from the rest stop after 2 hours?

(b) Find the slope of the graph of D. Interpret this slope as a rate of change.

36. *Age in the United States* The median age of the U.S. population for each year t between 1970 and 2010 can be approximated by the formula $A(t) = 0.243t - 450.8$. (**Source:** Bureau of the Census.)
(a) Compute the median ages in 1980 and 2000.

(b) What is the slope of the graph of A? Interpret the slope.

37. *Velocity* A car's distance from a service center along a straight highway can be described by a constant function of time. What can be said about the car's velocity?

38. *Velocity* A car's distance in miles from a rest stop along a straight stretch of an interstate highway after x hours can be modeled by $f(x) = ax$. Discuss what can be said about the car's velocity.

Linear and Nonlinear Functions

Exercises 39–48: Determine if f is a linear or nonlinear function. If f is a linear function, determine if f is a constant function. Support your answer by graphing f.

39. $f(x) = -2x + 5$ **40.** $f(x) = 3x - 2$

41. $f(x) = 1$ **42.** $f(x) = -2$

43. $f(x) = |x + 1|$ **44.** $f(x) = |2x - 1|$

45. $f(x) = x^2 - 1$ **46.** $f(x) = x^3$

47. $f(x) = 2\sqrt{x}$ **48.** $f(x) = \sqrt{x - 1}$

Linear and Nonlinear Data

Exercises 49–54: (Refer to Example 3.) Decide whether the data are linear or nonlinear. If the data are linear, state the slope m of the line passing through the data points.

49.

x	0	1	2	3	4
y	-1	3	7	11	15

50.

x	-4	-2	0	2	4
y	1	$-\frac{1}{2}$	-2	$-\frac{7}{2}$	-5

51.

x	-5	-3	1	3	5
y	-5	-2	1	4	7

52.

x	-4	-2	0	2	4
y	-1	-1	-1	-1	-1

53.

x	-4	0	1	2	5
y	5	3	$\frac{5}{2}$	2	$\frac{1}{2}$

54.

x	10	20	25	35	40
y	40	190	300	600	790

Exercises 55–58: Analyzing Real Data For the given data set complete the following.
(a) Make a line graph of the data. Let this graph represent a function f.
(b) Decide whether f is linear or nonlinear.

55. Toyota vehicles sold in the United States (millions)

Year	1998	2000	2002	2004
Vehicles	1.4	1.6	1.8	2.0

Source: Autodata.

56. Interest income after 1 year on an investment earning 7% per year

Investment	$500	$1000	$2000	$3500
Interest	$35	$70	$140	$245

57. Median incomes of full-time female workers

Year	1970	1980	1990	2000
Income	$5,440	$11,591	$20,591	$32,442

Source: Bureau of the Census, *Current Population Reports.*

58. Median incomes of full-time male workers

Year	1970	1980	1990	2000
Income	$9,184	$19,173	$28,979	$37,435

Source: Bureau of the Census, *Current Population Reports.*

59. *Average Wind Speed*   The table lists the average wind speed in miles per hour at Myrtle Beach, South Carolina. The months are assigned the standard numbers.

Month	1	2	3	4	5	6
Wind (mph)	7	8	8	8	7	7

Month	7	8	9	10	11	12
Wind (mph)	7	7	7	6	6	6

Source: J. Williams.

(a) Could these data be modeled exactly by a constant function?

(b) Determine a continuous, constant function f that models these data approximately.

(c) Graph f and the data.

60. *Marriage Rates*   The table lists the marriages per 1000 residents in Kentucky for selected years.

Year	2001	2002	2003	2004
Rate	9.0	9.0	9.1	8.9

Source: Bureau of the Census.

(a) Could these data be modeled exactly by a constant function?

(b) Determine a constant function f that models these data approximately.

(c) Graph f and the data.

Curve Sketching

Exercises 61–64: Sketch a graph that illustrates the motion of the person described. Let the x-axis represent time and the y-axis represent distance from home. Be sure to label each axis.

61. A person drives a car away from home for 2 hours at 50 miles per hour and then stops for 1 hour.

62. A person drives to a nearby park at 25 miles per hour for 1 hour, rests at the park for 2 hours, and then drives home at 50 miles per hour.

63. A person walks away from home at 4 miles per hour for 1 hour and then turns around and walks home at the same speed.

64. A person arrives at home after running at 8 miles per hour for 2 hours.

Writing about Mathematics

65. Describe two methods of determining if a data set can be modeled by a linear function. Create or find a data set that consists of four or more ordered pairs. Can your data set be modeled by a linear function?

66. Suppose you are given a graphical representation of a function f. Explain how you would determine whether f is constant, linear, or nonlinear. How would you determine the type if you were given a numerical or symbolic representation? Give examples.

EXTENDED AND DISCOVERY EXERCISE

1. *Geometry*   Suppose that the radius of a circle on a computer monitor is increasing at a constant rate of 1 inch per second.
(a) Does the circumference of the circle increase at a constant rate? If it does, find this rate.

(b) Does the area of the circle increase at a constant rate? Explain.

CHECKING BASIC CONCEPTS FOR SECTIONS 1.3 AND 1.4

1. Create symbolic, numerical, and graphical representations of a function f that computes the number of feet in x miles. For the numerical representation use a table and let $x = 1, 2, 3, 4, 5$.

2. Let $f(x) = \frac{2x}{x-4}$.
(a) Find $f(2)$ and $f(a + 4)$.

(b) Find the domain of f.

3. Graph $f(x) = x^2 - 2$. Identify the domain and range.

4. Find the slope of the line passing through the points $(-2, 4)$ and $(4, -5)$. If the graph of $f(x) = ax + b$ passes through these two points, what is the value of a?

5. Identify each function f as constant, linear, or nonlinear. Support your answer graphically.
(a) $f(x) = -1.4x + 5.1$

(b) $f(x) = 25$

(c) $f(x) = 2x^2 - 5$

1.5 Functions and Their Rates of Change

- **Identify where a function is increasing or decreasing**
- **Use interval notation**
- **Use and interpret average rate of change**
- **Calculate the difference quotient**

Introduction

Sales of rock music have not remained constant during the past two decades. In 1990, rock music accounted for 36% of all U.S. music sales. This percentage decreased to a low of 24% in 2001 and then increased to 34% in 2006.   (**Source:** Recording Industry Association of America.)

A linear function cannot be used to describe these data because the graph of a (non-constant) linear function either always rises or always falls. The concepts of increasing and decreasing are important to functions and their rates of change.

Increasing and Decreasing Functions

Figure 1.83 shows sales of rock music modeled by a continuous nonlinear function f. For example, $f(1990) = 36$ indicates that rock music accounted for 36% of all music sales in 1990. Rock music sales decreased from 1990 to 2001 and then increased from 2001 to 2006. Mathematically, we say that function f decreases for $1990 \leq x \leq 2001$ and increases for $2001 \leq x \leq 2006$.

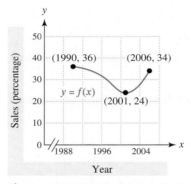

Figure 1.83 U.S. Rock Music Sales

> **NOTE** The inequality $a \leq x \leq b$ means that $x \geq a$ *and* $x \leq b$.

The concepts of increasing and decreasing are defined as follows.

Increasing and Decreasing Functions

Suppose that a function f is defined over an interval I on the number line. If x_1 and x_2 are in I,

(a) f **increases** on I if, whenever $x_1 < x_2$, $f(x_1) < f(x_2)$;

(b) f **decreases** on I if, whenever $x_1 < x_2$, $f(x_1) > f(x_2)$.

Figures 1.84 and 1.85 illustrate these concepts at the top of the next page.

The concepts of increasing and decreasing relate to whether the graph of a function rises or falls. If we could walk *from left to right* along the graph of an increasing function, it would be uphill. For a decreasing function, we would walk downhill. We speak of a function f increasing or decreasing *over an interval of its domain*. For example, in Figure 1.86 the function is decreasing (the graph falls) when $-2 \leq x \leq 0$ and increasing (the graph rises) when $0 \leq x \leq 2$.

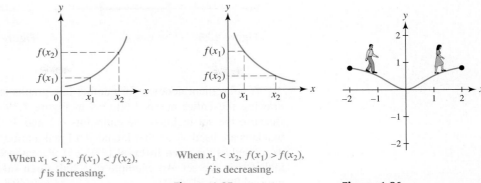

When $x_1 < x_2$, $f(x_1) < f(x_2)$,
f is increasing.

Figure 1.84

When $x_1 < x_2$, $f(x_1) > f(x_2)$,
f is decreasing.

Figure 1.85

Figure 1.86

EXAMPLE 1   Recognizing increasing and decreasing graphs

The graphs of three functions are shown in Figure 1.87. Determine intervals where each function is increasing or decreasing.

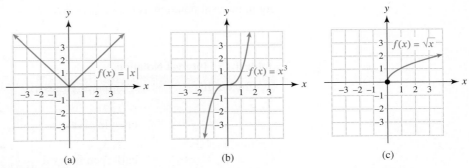

Figure 1.87

SOLUTION

(a) Moving from *left to right*, the graph of $f(x) = |x|$ is decreasing for $x \leq 0$ and increasing for $x \geq 0$.

(b) Moving from *left to right*, the graph of $f(x) = x^3$ is increasing for all real numbers x. Note that the y-values always increase as the x-values increase.

(c) The graph of $f(x) = \sqrt{x}$ is increasing for $x \geq 0$.   ***Now Try Exercises 17, 18, and 19*** ◄

NOTE When stating where a function is increasing and where it is decreasing, it is important to give x-intervals and not y-intervals.

Interval Notation

To describe intervals where functions are increasing or decreasing, a number line graph is sometimes used. The set $\{x \mid x > 2\}$, which includes all real numbers greater than 2, is graphed in Figure 1.88. Note that a parenthesis at $x = 2$ indicates that the endpoint *is not included*. The set $\{x \mid -1 \leq x \leq 4\}$ is shown in Figure 1.89 on the next page, and the set $\{x \mid -\frac{7}{2} < x < -\frac{1}{2}\}$ is shown in Figure 1.90 on the next page. Note that brackets, either [or], are used when endpoints *are included*.

Figure 1.88 $x > 2$

Figure 1.89 $-1 \le x \le 4$ **Figure 1.90** $-\frac{7}{2} < x < -\frac{1}{2}$

A convenient notation for number line graphs is called **interval notation**. Instead of drawing the entire number line, as in Figure 1.89, we can express the set as $[-1, 4]$. Because the set includes the endpoints -1 and 4, the interval is a **closed interval** and brackets are used. A set that included all real numbers satisfying $-\frac{7}{2} < x < -\frac{1}{2}$ would be expressed as the **open interval** $\left(-\frac{7}{2}, -\frac{1}{2}\right)$. Parentheses indicate that the endpoints are not included in the set. An example of a **half-open interval** is $[0, 4)$, which represents the interval $0 \le x < 4$.

Table 1.15 provides some examples of interval notation. The symbol ∞ refers to **infinity**; it does not represent a real number. The notation $(1, \infty)$ means $\{x \mid x > 1\}$, or simply $x > 1$. Since this interval has no maximum x-value, ∞ is used in the position of the right endpoint. A similar interpretation holds for the symbol $-\infty$, which represents **negative infinity**.

NOTE An inequality in the form $x < 1$ *or* $x > 3$ indicates the set of real numbers that are either less than 1 or greater than 3. The **union symbol** $\cup$ can be used to write this inequality in interval notation as $(-\infty, 1) \cup (3, \infty)$.

Table 1.15 Interval Notation

Inequality	Interval Notation	Graph
$-2 < x < 2$	$(-2, 2)$ open interval	
$-1 < x \le 3$	$(-1, 3]$ half-open interval	
$-3 \le x \le 2$	$[-3, 2]$ closed interval	
$x > -3$	$(-3, \infty)$ infinite interval	
$x \le 1$	$(-\infty, 1]$ infinite interval	
$x \le -2$ or $x > 1$	$(-\infty, -2] \cup (1, \infty)$ infinite intervals	
$-\infty < x < \infty$ (entire number line)	$(-\infty, \infty)$ infinite interval	

MAKING CONNECTIONS

Points and Intervals The expression $(2, 5)$ has two possible meanings. It may represent the ordered pair $(2, 5)$, which can be plotted as a point on the xy-plane, or it may represent the open interval $2 < x < 5$. To alleviate confusion, phrases like "the point $(2, 5)$" or "the interval $(2, 5)$" may be used.

Increasing, Decreasing, and Endpoints   There can be confusion as to whether to include the endpoints of an interval when stating where a function is increasing or decreasing. For example, is the graph of $f(x) = x^2$, shown in Figure 1.91, increasing on $[0, \infty)$ or just on $(0, \infty)$? The definition of increasing and decreasing allows us to include 0 as part of the interval I where f is increasing, because if we let $x_1 = 0$, then $f(0) < f(x_2)$ whenever $0 < x_2$. Thus $f(x) = x^2$ is increasing on $[0, \infty)$. Similarly, we can show that $f(x) = x^2$ is decreasing on $(-\infty, 0]$ by letting $x_2 = 0$. Do *not* confuse these concepts by saying that function f both increases and decreases at the point $(0, 0)$. The concepts of increasing and decreasing apply only to intervals of the real number line and not to individual points.

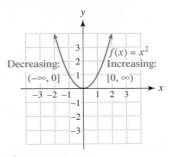

Figure 1.91

Note that it is *not* incorrect to say that $f(x) = x^2$ is increasing on $(0, \infty)$ because $f(x) = x^2$ also increases on $[1, \infty)$, $(10, \infty)$, and $[100, 200]$. However, we generally give the largest interval possible, $[0, \infty)$, when stating where a function is increasing or decreasing.   (**Reference:** J. Stewart, *Essential Calculus*, 2007, p. 8.)

NOTE Some definitions of increasing and decreasing functions require that the x-interval be an *open* interval. An open interval does *not* include the endpoints. For example, with these definitions, the graph of $f(x) = x^2$ is decreasing for $(-\infty, 0)$, not $(-\infty, 0]$, and increasing for $(0, \infty)$, not $[0, \infty)$.

> **EXAMPLE 2**   Determining where a function is increasing or decreasing

Use the graph of $f(x) = 4x - \frac{1}{3}x^3$ (shown in Figure 1.92) and interval notation to identify where f is increasing or decreasing.

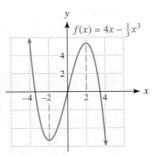

Figure 1.92

SOLUTION Moving from left to right on the graph of f, the y-values decrease until $x = -2$, increase until $x = 2$, and decrease thereafter. Thus $f(x) = 4x - \frac{1}{3}x^3$ is decreasing on $(-\infty, -2]$, increasing on $[-2, 2]$, and decreasing again on $[2, \infty)$. In interval notation f is decreasing on $(-\infty, -2] \cup [2, \infty)$.   *Now Try Exercise 21* ◀

Average Rate of Change

A nonlinear function can increase on one interval of its domain and decrease on another interval of its domain. The graphs of nonlinear functions are not lines, so there is no notion of a single slope. See Example 2. The slope of the graph of a linear function gives its rate of change. With a nonlinear function we speak of an *average* rate of change. Suppose that the points (x_1, y_1) and (x_2, y_2) lie on the graph of a nonlinear function f. See Figure 1.93. The slope of the line L passing through these two points represents the *average rate of change of f from x_1 to x_2*. The line L is referred to as a **secant line**. If different values for x_1 and x_2 are selected, then a different secant line and a different average rate of change usually result.

Figure 1.93   **Figure 1.94**

In applications the average rate of change measures how fast a quantity is changing over an interval of its domain, *on average*. For example, suppose the graph of the function f in Figure 1.94 represents the distance y in miles that a car has traveled on a straight highway (under construction) after x hours. The points $(0.5, 2)$ and $(1, 21)$ lie on this graph. Thus after 0.5 hour the car has traveled 2 miles and after 1 hour the car has traveled 21 miles. The slope of the line passing through these two points is

$$m = \frac{y_2 - y_1}{x_2 - x_1} = \frac{21 - 2}{1 - 0.5} = 38.$$

This means that during the half hour from 0.5 to 1 hour the average rate of change, or average velocity, was 38 miles per hour. These ideas lead to the following definition.

Average Rate of Change

Let (x_1, y_1) and (x_2, y_2) be distinct points on the graph of a function f. The **average rate of change of f from x_1 to x_2** is

$$\frac{y_2 - y_1}{x_2 - x_1}.$$

That is, the average rate of change from x_1 to x_2 equals the slope of the line passing through (x_1, y_1) and (x_2, y_2).

NOTE If $y = f(x)$, then average rate of change equals $\frac{f(x_2) - f(x_1)}{x_2 - x_1}$.

If f is a constant function, its average rate of change is 0. For a linear function defined by $f(x) = ax + b$, the average rate of change is a, the slope of its graph. The average rate of change for a nonlinear function varies.

Table 1.16

Year	Population
1800	5
1840	17
1900	76
1940	132

EXAMPLE 3 Calculating and interpreting average rates of change

Table 1.16 lists the U.S. population in millions for selected years.
(a) Calculate the average rates of change in the U.S. population from 1800 to 1840 and from 1900 to 1940. Interpret the results.
(b) Illustrate your results from part (a) graphically.

SOLUTION
(a) In **1800** the population was **5** million, and in **1840** it was **17** million. Therefore the average rate of change in the population from 1800 to 1840 was

$$\frac{17 - 5}{1840 - 1800} = 0.3.$$

In 1900 the population was 76 million, and in 1940 it was 132 million. Therefore the average rate of change in the population from 1900 to 1940 was

$$\frac{132 - 76}{1940 - 1900} = 1.4.$$

This means that from 1800 to 1840 the U.S. population increased, *on average*, by 0.3 million per year and from 1900 to 1940 the U.S. population increased, *on average*, by 1.4 million per year.
(b) These average rates of change can be illustrated graphically by sketching a line L_1 through the points (1800, 5) and (1840, 17) and another line L_2 through the points (1900, 76) and (1940, 132), as depicted in Figure 1.95. The slope of L_1 is 0.3 and the slope of L_2 is 1.4.   ***Now Try Exercises 51 and 59*** ◀

Figure 1.95

EXAMPLE 4 Modeling braking distance for a car

Highway engineers sometimes estimate the braking distance in feet for a car traveling at x miles per hour on wet, level pavement by using the formula $f(x) = \frac{1}{9}x^2$. (**Source:** L. Haefner, *Introduction to Transportation Systems.*)
(a) Evaluate $f(30)$ and $f(60)$. Interpret these results.
(b) Calculate the average rate of change of f from 30 to 60. Interpret this result.

SOLUTION
(a) $f(30) = \frac{1}{9}(30)^2 = 100$ and $f(60) = \frac{1}{9}(60)^2 = 400$. At 30 miles per hour the braking distance is 100 feet, and at 60 miles per hour the braking distance is 400 feet.
(b) The average rate of change of f from 30 to 60 is

$$\frac{400 - 100}{60 - 30} = 10.$$

Braking distance increases, on average, by 10 feet for each 1-mile-per-hour increase in speed between 30 and 60 miles per hour.   ***Now Try Exercise 61*** ◀

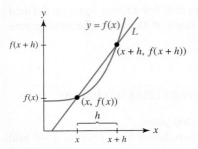

Figure 1.96

The Difference Quotient

The difference quotient often occurs in calculus. It uses function notation to calculate the average rate of change of a function f in general. Consider the graph of $y = f(x)$ shown in Figure 1.96, and let h be a real number. The points $(x, f(x))$ and $(x + h, f(x + h))$ denote the coordinates of two points on this graph. (Figure 1.96 shows $h > 0$.)

The line L passes through these two points and is a *secant line*. The slope m of L is

$$m = \frac{f(x + h) - f(x)}{(x + h) - x} = \frac{f(x + h) - f(x)}{h}.$$

This expression, written in function notation, is called the *difference quotient* and is equal to the average rate of change of f from x to $x + h$.

Difference Quotient

The **difference quotient of a function f** is an expression of the form

$$\frac{f(x + h) - f(x)}{h}, \quad \text{where } h \neq 0.$$

EXAMPLE 5   Calculating a difference quotient

Let $f(x) = x^2 - 2x$.
(a) Find $f(x + h)$.
(b) Find the difference quotient of f and simplify the result.

SOLUTION

Algebra Review
To square a binomial, see Chapter R (page R-18).

(a) To calculate $f(x + h)$, substitute $(x + h)$ for x in the expression $x^2 - 2x$.

$$f(x + h) = (x + h)^2 - 2(x + h) \qquad f(x) = x^2 - 2x$$
$$= x^2 + 2xh + h^2 - 2x - 2h \qquad \text{Square the binomial; apply the distributive property.}$$

(b) The difference quotient can be calculated as follows.

$$\frac{f(x + h) - f(x)}{h} = \frac{x^2 + 2xh + h^2 - 2x - 2h - (x^2 - 2x)}{h} \qquad \text{Substitute.}$$

$$= \frac{2xh + h^2 - 2h}{h} \qquad \text{Combine like terms.}$$

$$= \frac{h(2x + h - 2)}{h} \qquad \text{Factor out } h.$$

$$= 2x + h - 2 \qquad \text{Simplify.}$$

Now Try Exercise 75 ◀

EXAMPLE 6   Calculating a difference quotient

Let the distance d in feet that a racehorse runs in t seconds be $d(t) = 2t^2$ for $0 \leq t \leq 10$.
(a) Find $d(t + h)$.
(b) Find the difference quotient of d and simplify the result.

(c) Evaluate the difference quotient for $t = 7$ and $h = 0.1$. Interpret your results.

(d) Evaluate the difference quotient for $t = 4$ and $h = 1$. Then sketch a graph that illustrates the result.

SOLUTION

(a) To calculate $d(t + h)$, substitute $(t + h)$ for t in the expression $2t^2$.

$$
\begin{aligned}
d(t + h) &= 2(t + h)^2 && \text{Substitute } (t + h) \text{ for } t. \\
&= 2(t^2 + 2th + h^2) && \text{Square the binomial.} \\
&= 2t^2 + 4th + 2h^2 && \text{Distributive property}
\end{aligned}
$$

(b)

$$
\begin{aligned}
\frac{d(t + h) - d(t)}{h} &= \frac{2t^2 + 4th + 2h^2 - 2t^2}{h} && \text{Substitute for } d(t + h) \text{ and } d(t). \\[2mm]
&= \frac{4th + 2h^2}{h} && \text{Combine like terms.} \\[2mm]
&= \frac{h(4t + 2h)}{h} && \text{Factor out } h. \\[2mm]
&= 4t + 2h && \text{Simplify.}
\end{aligned}
$$

(c) If $t = 7$ and $h = 0.1$, then the difference quotient becomes

$$4t + 2h = 4(7) + 2(0.1) = 28.2.$$

The average rate of change, or average velocity, of the horse from 7 seconds to $7 + 0.1 = 7.1$ seconds is 28.2 feet per second.

(d) If $t = 4$ and $h = 1$, then $4t + 2h = 4(4) + 2(1) = 18$. Thus the slope of the line passing through $(4, 32)$ and $(5, 50)$ is $m = 18$. See Figure 1.97.

> **NOTE** Because $t = 4$ and $d = 2(4)^2 = 32$, the first point is $(4, 32)$. Because $t = 4$ and $h = 1$, it follows that $t + h = 4 + 1 = 5$ so $d = 2(5)^2 = 50$. The second point is $(5, 50)$.

Now Try Exercise 83 ◀

Figure 1.97

---

continued on next page

1.5 Putting It All Together

The following table summarizes the basic concepts in this section.

Concept	Explanation	Examples
Interval notation	An efficient notation for writing inequalities	$x \leq 6$ is equivalent to $(-\infty, 6]$. $x > 3$ is equivalent to $(3, \infty)$. $2 < x \leq 5$ is equivalent to $(2, 5]$.

continued from previous page

Concept	Explanation	Examples
Increasing and decreasing	f increases on an interval if, whenever $x_1 < x_2$, then $f(x_1) < f(x_2)$. f decreases on an interval if, whenever $x_1 < x_2$, then $f(x_1) > f(x_2)$.	 f is increasing on $(-\infty, -2] \cup [2, \infty)$. f is decreasing on $[-2, 2]$.
Average rate of change of f from x_1 to x_2	If (x_1, y_1) and (x_2, y_2) are distinct points on the graph of f, then the average rate of change from x_1 to x_2 equals $$\frac{y_2 - y_1}{x_2 - x_1}.$$	If $f(x) = 3x^2$, then the average rate of change from $x = 1$ to $x = 3$ is $$\frac{27 - 3}{3 - 1} = 12$$ because $f(3) = 27$ and $f(1) = 3$. This means that, on average, $f(x)$ increases by 12 units for each unit increase in x when $1 \le x \le 3$.
Difference quotient	Calculates average rate of change of f from x to $x + h$. $$\frac{f(x + h) - f(x)}{h}, h \ne 0$$	If $f(x) = 2x$, then the difference quotient equals $$\frac{2(x + h) - 2x}{h} = \frac{2h}{h} = 2.$$

1.5   Exercises

Interval Notation

Exercises 1–16: Express each of the following in interval notation.

1. $x \ge 5$ **2.** $x < 100$

3. $4 \le x < 19$ **4.** $-4 < x < -1$

5. $x \le -37$ **6.** $\{x \mid x \le -3\}$

7. $\{x \mid -1 \le x\}$ **8.** $17 > x \ge -3$

9. $\{x \mid x < 1 \text{ or } x \ge 3\}$ **10.** $\{x \mid x \le -2 \text{ or } x \ge 0\}$

11.

12.

13.

14.

15.

16.

Increasing and Decreasing Functions

Exercises 17–24: Use the graph of f to determine intervals where f is increasing and where f is decreasing.

17.

18.

19.

20.

21.

22.

23.

24.

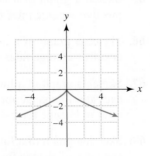

Exercises 25–44: Identify where f is increasing and where f is decreasing. (Hint: Consider the graph y = f(x).)

25. $f(x) = -3$

26. $f(x) = 5$

27. $f(x) = 2x - 1$

28. $f(x) = 4 - x$

29. $f(x) = x^2 - 2$

30. $f(x) = -\frac{1}{2}x^2$

31. $f(x) = 2x - x^2$

32. $f(x) = x^2 - 4x$

33. $f(x) = \sqrt{x - 1}$

34. $f(x) = -\sqrt{x + 1}$

35. $f(x) = |x + 3|$

36. $f(x) = |x - 1|$

37. $f(x) = x^3$

38. $f(x) = \sqrt[3]{x}$

39. $f(x) = \frac{1}{3}x^3 - 4x$

40. $f(x) = x^3 - 3x$

41. $f(x) = 2x^3 + 3x^2 - 12x$

42. $f(x) = -x^3 + 3x^2 + 9x - 10$

43. $f(x) = -\frac{1}{4}x^4 + \frac{1}{3}x^3 + x^2$

44. $f(x) = \frac{1}{4}x^4 - 2x^2$

*Exercises 45 and 46: The graph gives the tides at Clearwater Beach, Florida, x hours after midnight on a particular day, where $0 \le x \le 27$.   (**Source:** WWW Tide/Current Predictor.)*

45. When were water levels increasing?

46. When were water levels decreasing?

Average Rates of Change

Exercises 47–50: Use the graph and formula for f(x) to find the average rates of change of f from −3 to −1 and from 1 to 3.

47. $f(x) = 4$

48. $f(x) = 2x - 1$

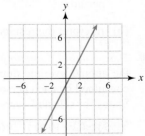

49. $f(x) = -0.3x^2 + 4$ **50.** $f(x) = 0.3x^2 - 4$

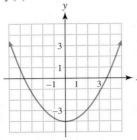

Exercises 51 and 52: (Refer to Example 3.) Use the given $f(x)$ to complete the following.

   (a) Calculate the average rate of change of f from $x = 1$ to $x = 2$.

   (b) Illustrate your result from part (a) graphically.

51. $f(x) = x^2$ **52.** $f(x) = 4 - x^2$

Exercises 53–58: Compute the average rate of change of f from x_1 to x_2. Round your answer to two decimal places when appropriate. Interpret your result graphically.

53. $f(x) = 7x - 2, x_1 = 1,$ and $x_2 = 4$

54. $f(x) = -8x + 5, x_1 = -2,$ and $x_2 = 0$

55. $f(x) = x^3 - 2x, x_1 = 2,$ and $x_2 = 4$

56. $f(x) = 0.5x^2 - 5, x_1 = -1,$ and $x_2 = 4$

57. $f(x) = \sqrt{2x - 1}, x_1 = 1,$ and $x_2 = 3$

58. $f(x) = \sqrt[3]{x + 1}, x_1 = 7,$ and $x_2 = 26$

59. *U.S. Cigarette Consumption*   The following table lists the number of cigarettes in billions consumed in the United States for selected years.

Year	1900	1940	1980	2006
Cigarettes	3	182	632	371

Source: Department of Health and Human Services.

   (a) Find the average rate of change during each time period.

   (b) Interpret the results.

60. *U.S. Alcohol Consumption*   The following table lists the annual *average* number of gallons of pure alcohol consumed by each person age 15 and older in the United States for selected years.

Year	1940	1960	1980	2000
Alcohol	1.56	2.07	2.76	2.18

Source: Department of Health and Human Services.

   (a) Find the average rate of change during each 20-year period.

   (b) Interpret the results.

61. *Falling Object*   The distance D in feet that an object has fallen after t seconds is given by $D(t) = 16t^2$.

   (a) Evaluate $D(2)$ and $D(4)$.

   (b) Calculate the average rate of change of D from 2 to 4. Interpret the result.

62. *Torricelli's Law*   A cylindrical tank contains 100 gallons of water. A plug is pulled from the bottom of the tank and the amount of water in gallons remaining in the tank after x minutes is given by

$$A(x) = 100\left(1 - \frac{x}{5}\right)^2.$$

   (a) Calculate the average rate of change of A from 1 to 1.5 and from 2 to 2.5. Interpret your results.

   (b) Are the two average rates of change the same or different? Explain why.

Curve Sketching

Exercises 63–68: Critical Thinking   Assume that each function is continuous. Do not use a graphing calculator.

63. On the same coordinate axes, sketch the graphs of a constant function f and a nonlinear function g that intersect exactly twice.

64. Sketch a graph of a linear function f that intersects a constant function g exactly once.

65. Sketch a graph of a nonlinear function f that has only positive average rates of change.

66. Sketch a graph of a nonlinear function f that has only negative average rates of change.

67. Sketch a graph of a function that has only positive average rates of change for $x \leq 0$ and only negative average rates of change for $x \geq 0$.

68. Sketch a graph of a function that has only positive average rates of change for $x \geq 1$ and only negative average rates of change for $x \leq 1$.

The Difference Quotient

Exercises 69–82: (Refer to Example 5.) Complete the following for the given $f(x)$.

   (a) Find $f(x + h)$.

   (b) Find the difference quotient of f and simplify.

69. $f(x) = 3$

70. $f(x) = -5$

71. $f(x) = -2x$

72. $f(x) = 10x$

73. $f(x) = 2x + 1$

74. $f(x) = -3x + 4$

75. $f(x) = 3x^2 + 1$

76. $f(x) = x^2 - 2$

77. $f(x) = -x^2 + 2x$

78. $f(x) = -4x^2 + 1$

79. $f(x) = 2x^2 - x + 1$

80. $f(x) = x^2 + 3x - 2$

81. $f(x) = x^3$

82. $f(x) = 1 - x^3$

83. *Speed of a Car* (Refer to Example 6.) Let the distance in feet that a car travels in t seconds be given by $d(t) = 8t^2$ for $0 \le t \le 6$.
(a) Find $d(t + h)$.

(b) Find the difference quotient for d and simplify.

(c) Evaluate the difference quotient when $t = 4$ and $h = 0.05$. Interpret your result.

84. *Draining a Pool* Let the number of gallons G of water in a pool after t hours be given by $G(t) = 4000 - 100t$ for $0 \le t \le 40$.
(a) Find $G(t + h)$.

(b) Find the difference quotient. Interpret your result.

Writing about Mathematics

85. What does the average rate of change represent for a linear function? What does it represent for a nonlinear function? Explain your answers.

86. What is the formula for the difference quotient? Given a formula for $f(x)$, explain how to find $f(x + h)$. Give an example.

87. Suppose that a function f has a positive average rate of change from 1 to 4. Is it correct to assume that function f only increases on the interval $[1, 4]$? Make a sketch to support your answer.

88. If $f(x) = ax + b$, what does the difference quotient for function f equal? Explain your reasoning.

EXTENDED AND DISCOVERY EXERCISE

1. *Velocity* (Refer to Example 6.) If the distance in feet run by a racehorse in t seconds is given by $d(t) = 2t^2$, then the difference quotient for d is $4t + 2h$. How could you estimate the velocity of the racehorse at exactly 7 seconds?

CHECKING BASIC CONCEPTS FOR SECTION 1.5

1. Write each expression in interval notation.
(a) $x \le 5$ (b) $1 \le x < 6$

2. Determine where $f(x) = x^2 - 2$ is increasing and where it is decreasing.

3. Find the average rate of change of $f(x) = x^2 - 3x$ from $x = -3$ to $x = -1$.

4. Find the difference quotient for $f(x) = 4x^2$.

---

1 Summary

CONCEPT	EXPLANATION AND EXAMPLES
SECTION 1.1 NUMBERS, DATA, AND PROBLEM SOLVING	
Sets of Numbers	*Natural numbers:* $N = \{1, 2, 3, 4, \ldots\}$
	Integers: $I = \{\ldots, -3, -2, -1, 0, 1, 2, 3, \ldots\}$
	Rational numbers: $\frac{p}{q}$, where p and q are integers with $q \ne 0$; includes repeating and terminating decimals
	Irrational numbers: Includes nonrepeating, nonterminating decimals
	Real numbers: Any number that can be expressed in decimal form; includes rational and irrational numbers

CONCEPT	EXPLANATION AND EXAMPLES

SECTION 1.1 NUMBERS, DATA, AND PROBLEM SOLVING (CONTINUED)

Order of Operations

Using the following order of operations, first perform all calculations within parentheses, square roots, and absolute value bars and above and below fraction bars. Then use the same order of operations to perform any remaining calculations.

1. Evaluate all exponents. Then do any negation *after* evaluating exponents.
2. Do all multiplication and division from *left to right*.
3. Do all addition and subtraction from *left to right*.

Example: $5 + 3 \cdot 2^3 = 5 + 3 \cdot 8 = 5 + 24 = 29$

Scientific Notation

A real number r is written as $c \times 10^n$, where $1 \le |c| < 10$.

Examples: $1234 = 1.234 \times 10^3$ $0.054 = 5.4 \times 10^{-2}$

SECTION 1.2 VISUALIZING AND GRAPHING DATA

Mean (Average) and Median

The mean represents the average of a set of numbers, and the median represents the middle of a sorted list.

Example: $4, 6, 9, 13, 15;$ Mean $= \dfrac{4 + 6 + 9 + 13 + 15}{5} = 9.4;$ Median $= 9$

Relation, Domain, and Range

A relation S is a set of ordered pairs. The domain D is the set of x-values, and the range R is the set of y-values.

Example: $S = \{(-1, 2), (4, -5), (5, 9)\};$ $D = \{-1, 4, 5\}, R = \{-5, 2, 9\}$

Cartesian (Rectangular) Coordinate System, or xy-Plane

The xy-plane has four quadrants and is used to graph ordered pairs.

Distance Formula

The distance d between the points (x_1, y_1) and (x_2, y_2) is
$$d = \sqrt{(x_2 - x_1)^2 + (y_2 - y_1)^2}.$$

Example: The distance between $(-3, 5)$ and $(2, -7)$ is
$$d = \sqrt{(2 - (-3))^2 + (-7 - 5)^2} = \sqrt{5^2 + (-12)^2} = 13.$$

CONCEPT	EXPLANATION AND EXAMPLES

SECTION 1.2 VISUALIZING AND GRAPHING DATA (CONTINUED)

Midpoint Formula

The midpoint M of the line segment with endpoints (x_1, y_1) and (x_2, y_2) is

$$M = \left(\frac{x_1 + x_2}{2}, \frac{y_1 + y_2}{2} \right).$$

Example: The midpoint of the line segment connecting $(1, 2)$ and $(-3, 5)$ is

$$M = \left(\frac{1 + (-3)}{2}, \frac{2 + 5}{2} \right) = \left(-1, \frac{7}{2} \right).$$

Standard Equation of a Circle

The circle with center (h, k) and radius r has the equation

$$(x - h)^2 + (y - k)^2 = r^2.$$

Example: A circle with center $(-2, 5)$ and radius 6 has the equation

$$(x + 2)^2 + (y - 5)^2 = 36.$$

Scatterplot and Line Graph

A scatterplot consists of a set of ordered pairs plotted in the xy-plane. When consecutive points are connected with line segments, a line graph results.

SECTION 1.3 FUNCTIONS AND THEIR REPRESENTATIONS

Function

A function computes exactly one output for each valid input. The set of valid inputs is called the domain D, and the set of outputs is called the range R.

Examples: $f(x) = \sqrt{1 - x}$

$D = \{x \mid x \le 1\}, R = \{y \mid y \ge 0\}$

$g = \{(-1, 0.5), (0, 4), (2, 4), (6, \pi)\}$

$D = \{-1, 0, 2, 6\}, R = \{0.5, \pi, 4\}$

Function Notation

Examples: $f(x) = x^2 - 4; f(3) = 3^2 - 4 = 5$

$f(a + 1) = (a + 1)^2 - 4 = a^2 + 2a - 3$

Representations of Functions

A function can be represented symbolically (formula), graphically (graph), numerically (table of values), and verbally (words). Other representations are possible.

Symbolic Representation $f(x) = x^2 - 1$

Numerical Representation   ***Graphical Representation***

x	y
-2	3
-1	0
0	-1
1	0
2	3

$f(x) = x^2 - 1$

Verbal Representation f computes the square of the input x and then subtracts 1.

CONCEPT	EXPLANATION AND EXAMPLES

SECTION 1.3 FUNCTIONS AND THEIR REPRESENTATIONS (CONTINUED)

Vertical Line Test	If any vertical line intersects a graph at most once, then the graph represents a function.

SECTION 1.4 TYPES OF FUNCTIONS

Slope

The slope m of the line passing through (x_1, y_1) and (x_2, y_2) is

$$m = \frac{\Delta y}{\Delta x} = \frac{y_2 - y_1}{x_2 - x_1}.$$

Example: The slope of the line passing through $(1, -1)$ and $(-2, 3)$ is

$$m = \frac{3 - (-1)}{-2 - 1} = -\frac{4}{3}.$$

Constant Function

Given by $f(x) = b$, where b is a constant; its graph is a horizontal line.

Linear Function

Given by $f(x) = ax + b$; its graph is a nonvertical line; the slope of its graph is equal to a, which is also equal to its constant rate of change.

Examples: The graph of $f(x) = -8x + 100$ has slope -8.
If $G(t) = -8t + 100$ calculates the number of gallons of water in a tank after t minutes, then water is *leaving* the tank at 8 gallons per minute.

Nonlinear Function

The graph of a nonlinear function is not a line and *cannot* be written as $f(x) = ax + b$.

Examples: $f(x) = x^2 - 4$; $g(x) = \sqrt[3]{x} - 2$; $h(t) = \dfrac{1}{t + 1}$

Linear and Nonlinear Data

If the slopes of the lines passing through consecutive data points are always equal (or nearly equal), then the data are linear. Otherwise the data are nonlinear.

Example: For each 2-unit increase in x, the y-values increase by 10 units. Consecutive slopes between points are $m = \frac{10}{2} = 5$ so the data are linear.

2 units 2 units 2 units

x	2	4	6	8
y	0	10	20	30

10 units 10 units 10 units

SECTION 1.5 FUNCTIONS AND THEIR RATES OF CHANGE

Interval Notation

A concise way to express intervals on the number line

Example: $x < 4$ is expressed as $(-\infty, 4)$.
$-3 \le x < 1$ is expressed as $[-3, 1)$.
$x \le 2$ or $x \ge 5$ is expressed as $(-\infty, 2] \cup [5, \infty)$.

CONCEPT	EXPLANATION AND EXAMPLES

SECTION 1.5 FUNCTIONS AND THEIR RATES OF CHANGE (CONTINUED)

Increasing/Decreasing

f increases on interval I if, whenever $x_1 < x_2$, $f(x_1) < f(x_2)$.
f decreases on interval I if, whenever $x_1 < x_2$, $f(x_1) > f(x_2)$.

Example: $f(x) = |x|$ increases on $[0, \infty)$ and decreases on $(-\infty, 0]$.

Average Rate of Change

If (x_1, y_1) and (x_2, y_2) are distinct points on the graph of f, then the average rate of change from x_1 to x_2 equals the slope of the line passing through these two points and equals

$$\frac{y_2 - y_1}{x_2 - x_1}.$$

Example: $f(x) = x^2$; because $f(2) = 4$ and $f(3) = 9$, the graph of f passes through the points $(2, 4)$ and $(3, 9)$, and the average rate of change from 2 to 3 is given by $\frac{9 - 4}{3 - 2} = 5$.

Difference Quotient

The difference quotient of a function f is an expression of the form

$$\frac{f(x + h) - f(x)}{h}, \quad \text{where } h \neq 0.$$

Example: Let $f(x) = x^2$. The difference quotient of f is

$$\frac{(x + h)^2 - x^2}{h} = \frac{x^2 + 2xh + h^2 - x^2}{h} = 2x + h.$$

1 Review Exercises

Exercises 1 and 2: Classify each number listed as one or more of the following: natural number, integer, rational number, or real number.

1. $-2, \frac{1}{2}, 0, 1.23, \sqrt{7}, \sqrt{16}$

2. $55, 1.5, \frac{104}{17}, 2^3, \sqrt{3}, -1000$

Exercises 3 and 4: Write each number in scientific notation.

3. 1,891,000   **4.** 0.0001001

Exercises 5 and 6: Write each number in standard form.

5. 1.52×10^4 **6.** -7.2×10^{-3}

7. Evaluate each expression with a calculator. Round answers to the nearest hundredth.

(a) $\sqrt[3]{1.2} + \pi^3$ (b) $\frac{3.2 + 5.7}{7.9 - 4.5}$

(c) $\sqrt{5^2 + 2.1}$ (d) $1.2(6.3)^2 + \frac{3.2}{\pi - 1}$

8. Evaluate each expression. Write your answer in scientific notation and in standard form.

(a) $(4 \times 10^3)(5 \times 10^{-5})$

(b) $\frac{3 \times 10^{-5}}{6 \times 10^{-2}}$

Exercises 9 and 10: Evaluate by hand.

9. $4 - 3^2 \cdot 5$

10. $3 \cdot 3^2 \div \dfrac{3 - 5}{6 + 2}$

Exercises 11 and 12: Sort the list of numbers from smallest to largest and display the result in a table.

    (a) Determine the maximum and minimum values.

    (b) Calculate the mean and median.

11. $-5, 8, 19, 24, -23$

12. $8.9, -1.2, -3.8, 0.8, 1.7, 1.7$

Exercises 13 and 14: Complete the following.

    (a) Express the data as a relation S.

    (b) Find the domain and range of S.

13.

x	−15	−10	0	5	20
y	−3	−1	1	3	5

14.

x	−0.6	−0.2	0.1	0.5	1.2
y	10	20	25	30	80

Exercises 15 and 16: Make a scatterplot of the relation. Determine if the relation is a function.

15. $\{(10, 13), (-12, 40), (-30, -23), (25, -22), (10, 20)\}$

16. $\{(1.5, 2.5), (0, 2.1), (-2.3, 3.1), (0.5, -0.8), (-1.1, 0)\}$

Exercises 17 and 18: Find the distance between the points.

17. $(-4, 5), (2, -3)$

18. $(1.2, -4), (0.2, 6)$

Exercises 19 and 20: Find the midpoint of the line segment with the given endpoints.

19. $(24, -16), (-20, 13)$

20. $\left(\frac{1}{2}, \frac{5}{4}\right), \left(\frac{1}{2}, -\frac{5}{2}\right)$

21. Determine if the triangle with vertices $(1, 2), (-3, 5)$, and $(0, 9)$ is isosceles. (*Hint:* An isosceles triangle has at least two sides with equal measure.)

22. Find the standard equation of a circle with center $(-5, 3)$ and radius 9.

23. A diameter of a circle has endpoints $(-2, 4)$ and $(6, 6)$. Find the standard equation of the circle.

24. Use the graph at the top of the next column to determine the domain and range of each function. Evaluate $f(-2)$.

(a)

(b)

Exercises 25–32: Graph $y = f(x)$ by first plotting points to determine the shape of the graph.

25. $f(x) = -2$

26. $f(x) = 3x$

27. $f(x) = -x + 1$

28. $f(x) = 2x - 3$

29. $f(x) = 4 - 2x^2$

30. $f(x) = \frac{1}{2}x^2 - 1$

31. $f(x) = |x + 3|$

32. $f(x) = \sqrt{3 - x}$

Exercises 33 and 34: Use the verbal representation to express the function f symbolically, graphically, and numerically. Let $y = f(x)$ with $0 \le x \le 100$. For the numerical representation, use a table with $x = 0, 25, 50, 75, 100$.

33. To convert x pounds to y ounces, multiply x by 16.

34. To find the area y of a square, multiply the length x of a side by itself.

Exercises 35–42: Complete the following for the function f.

    (a) Evaluate $f(x)$ at the indicated values of x.

    (b) Find the domain of f.

35. $f(x) = \sqrt[3]{x}$ for $x = -8, 1$

36. $f(x) = 3x + 2$ for $x = -2, 5$

37. $f(x) = 5$ for $x = -3, 1.5$

38. $f(x) = 4 - 5x$ for $x = -5, 6$

39. $f(x) = x^2 - 3$ for $x = -10, a + 2$

40. $f(x) = x^3 - 3x$ for $x = -10, a + 1$

41. $f(x) = \dfrac{1}{x^2 - 4}$ for $x = -3, a + 1$

42. $f(x) = \sqrt{x + 3}$ for $x = 1, a - 3$

43. Determine if y is a function of x in $x = y^2 + 5$.

44. Write $5 \le x < 10$ in interval notation.

Exercises 45 and 46: Determine if the graph represents a function.

45.   **46.**

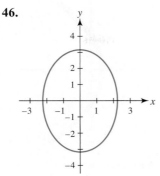

Exercises 47 and 48: Determine if S represents a function.

47. $S = \{(-3, 4), (-1, 2), (3, -5), (4, 2)\}$

48. $S = \{(-1, 3), (0, 2), (-1, 7), (3, -3)\}$

Exercises 49 and 50: State the slope of the graph of f.

49. $f(x) = 7$ **50.** $f(x) = \frac{1}{3}x - \frac{2}{3}$

Exercises 51–54: If possible, find the slope of the line passing through each pair of points.

51. $(-1, 7), (3, 4)$ **52.** $(1, -4), (2, 10)$

53. $(8, 4), (-2, 4)$ **54.** $\left(-\frac{1}{3}, \frac{2}{3}\right), \left(-\frac{1}{3}, -\frac{5}{6}\right)$

Exercises 55–58: Decide whether the function f is constant, linear, or nonlinear. Support your answer graphically.

55. $f(x) = 8 - 3x$ **56.** $f(x) = 2x^2 - 3x - 8$

57. $f(x) = |x + 2|$ **58.** $f(x) = 6$

59. Sketch a graph for a 2-hour period showing the distance between two cars meeting on a straight highway, each traveling 60 miles per hour. Assume that the cars are initially 120 miles apart.

60. Determine where the graph of $f(x) = |x - 3|$ is increasing and where it is decreasing.

61. Determine if the following data are modeled best by a constant, linear, or nonlinear function.

x	-2	0	2	4
y	50	42	34	26

62. Find the average rate of change of $f(x) = x^2 - x + 1$ from $x_1 = 1$ to $x_2 = 3$.

Exercises 63 and 64: Find the difference quotient for f(x).

63. $f(x) = 5x + 1$ **64.** $f(x) = 3x^2 - 2$

Applications

65. *Speed of Light*   The average distance between the planet Mars and the sun is approximately 228 million kilometers. Estimate the time required for sunlight, traveling at 300,000 kilometers per second, to reach Mars.   (**Source:** C. Ronan, *The Natural History of the Universe.*)

66. *Geometry*   Suppose that 0.25 cubic inch of paint is applied to a circular piece of plastic with a diameter of 20 inches. Estimate the thickness of the paint.

67. *Enclosing a Pool*   A rectangular swimming pool that is 25 feet by 50 feet has a 6-foot-wide sidewalk around it.
 (a) How much fencing would be needed to enclose the sidewalk?

 (b) Find the area of the sidewalk.

68. *Distance*   A driver's distance D in miles from a rest stop after t hours is given by $D(t) = 280 - 70t$.
 (a) How far is the driver from the rest stop after 2 hours?

 (b) Find the slope of the graph of D. Interpret this slope as a rate of change.

69. *Survival Rates*   The survival rates for song sparrows are shown in the table. The values listed are the numbers of song sparrows that attain a given age from 100 eggs. For example, 6 sparrows reach an age of 2 years from 100 eggs laid in the wild. (**Source:** S. Kress, *Bird Life.*)

Age	0	1	2	3	4
Number	100	10	6	3	2

 (a) Make a line graph of the data. Interpret the data.

 (b) Does this line graph represent a function?

 (c) Calculate and interpret the average rate of change for each 1-year period.

70. *Cost of Tuition*   The graph shows the cost of taking x credits at a university.

(a) Why is it reasonable for the graph to pass through the origin?

(b) Find the slope of the graph.

(c) Interpret the slope as a rate of change.

71. *Average Rate of Change*   Let $f(x) = 0.5x^2 + 50$ represent the outside temperature in degrees Fahrenheit at x P.M., where $1 \le x \le 5$.

(a) Graph f. Is f linear or nonlinear?

(b) Calculate the average rate of change of f from 1 P.M. to 4 P.M.

(c) Interpret this average rate of change verbally and graphically.

72. *Distance*   At noon car A is traveling north at 30 miles per hour and is located 20 miles north of car B. Car B is traveling west at 50 miles per hour. Approximate the distance between the cars at 12:45 P.M. to the nearest mile.

EXTENDED AND DISCOVERY EXERCISES

Because a parabolic curve becomes sharp gradually, as shown in the first figure, curves designed by engineers for highways and railroads frequently have parabolic, rather than circular, shapes. If railroad tracks changed abruptly from straight to circular, the momentum of the locomotive could cause a derailment. The second figure illustrates straight tracks connecting to a circular curve. (**Source:** F. Mannering and W. Kilareski, *Principles of Highway Engineering and Traffic Analysis*.)

In order to design a curve and estimate its cost, engineers determine the distance around the curve before it is built. In the third figure the distance along a parabolic curve from A to C is approximated by two line segments AB and BC. The distance formula can be used to calculate the length of each segment. The sum of these two lengths gives a crude estimate of the length of the curve.

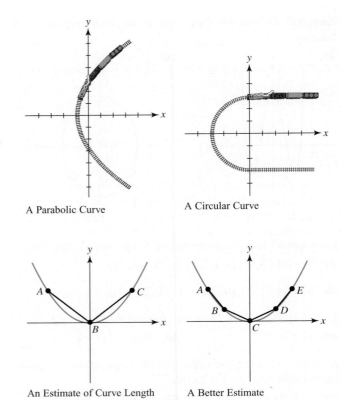

A Parabolic Curve   A Circular Curve

An Estimate of Curve Length   A Better Estimate

A better estimate can be made using four line segments, as shown in the fourth figure. As the number of segments increases, so does the accuracy of the approximation.

1. *Curve Length*   Suppose that a curve designed for railroad tracks is represented by the equation $y = 0.2x^2$, where the units are in kilometers. The points $(-3, 1.8)$, $(-1.5, 0.45)$, $(0, 0)$, $(1.5, 0.45)$, and $(3, 1.8)$ lie on the graph of $y = 0.2x^2$. Approximate the length of the curve from $x = -3$ to $x = 3$ by using line segments connecting these points.

Exercises 2–5: Curve Length   Use three line segments connecting the four points to estimate the length of the curve on the graph of f from $x = -1$ to $x = 2$. Graph f and a line graph of the four points in the indicated viewing rectangle.

2. $f(x) = x^2$; $(-1, 1)$, $(0, 0)$, $(1, 1)$, $(2, 4)$; $[-4.5, 4.5, 1]$ by $[-1, 5, 1]$

3. $f(x) = \sqrt[3]{x}$; $(-1, -1)$, $(0, 0)$, $(1, 1)$, $(2, \sqrt[3]{2})$; $[-3, 3, 1]$ by $[-2, 2, 1]$

4. $f(x) = 0.5x^3 + 2$; $(-1, 1.5)$, $(0, 2)$, $(1, 2.5)$, $(2, 6)$; $[-4.5, 4.5, 1]$ by $[0, 6, 1]$

5. $f(x) = 2 - 0.5x^2$; $(-1, 1.5)$, $(0, 2)$, $(1, 1.5)$, $(2, 0)$; $[-3, 3, 1]$ by $[-1, 3, 1]$

6. The distance along the curve of $y = x^2$ from $(0, 0)$ to $(3, 9)$ is about 9.747. Use this fact to estimate the distance along the curve of $y = 9 - x^2$ from $(0, 9)$ to $(3, 0)$.

7. Estimate the distance along the curve of $y = \sqrt{x}$ from $(1, 1)$ to $(4, 2)$. (The actual value is approximately 3.168.)

8. *Endangered Species*   The Florida scrub-jay is an endangered species that prefers to live in open landscape with short vegetation. NASA has attempted to create a habitat for these birds near Kennedy Space Center. The following table lists their population for selected years, where $x = 0$ corresponds to 1980, $x = 1$ to 1981, $x = 2$ to 1982, and so on.

x (1980 ↔ 0)	0	5	9
y (population)	3697	2512	2176
x (1980 ↔ 0)	11	15	19
y (population)	2100	1689	1127

Source: *Mathematics Explorations II*, NASA–AMATYC–NSF.

(a) Make a scatterplot of the data.

(b) Find a linear function f that models the data.

(c) Graph the data and f in the same viewing rectangle.

(d) Estimate the scrub-jay population in 1987 and in 2003.

9. *Global Warming*   If the global climate were to warm significantly as a result of the greenhouse effect or other climatic change, the Arctic ice cap would start to melt. It is estimated that this ice cap contains the equivalent of 680,000 cubic miles of water. Over 200 million people currently live on soil that is less than 3 feet above sea level. In the United States, several large cities have low average elevations, such as Boston (14 feet), New Orleans (4 feet), and San Diego (13 feet).   (**Sources:** Department of the Interior, Geological Survey.)

(a) Devise a plan to determine how much sea level would rise if the Arctic cap melted. (*Hint:* The radius of Earth is 3960 miles and 71% of its surface is covered by oceans.)

(b) Use your plan to estimate this rise in sea level.

(c) Discuss the implications of your calculation.

(d) Estimate how much sea level would rise if the 6,300,000 cubic miles of water in the Antarctic ice cap melted.

10. Prove that $\sqrt{2}$ is irrational by assuming that $\sqrt{2}$ is rational and arriving at a contradiction.

2 Linear Functions and Equations

For over two centuries people have been transferring carbon from below the surface of the earth into the atmosphere. Today, the burning of coal, oil, and natural gas releases 7 billion tons of carbon into the atmosphere each year. If the current rate of growth continues, the amount could double to 14 billion tons by 2056. This increase is modeled by a linear function *C* in the figure on the left below. The horizontal line $y = 7$ represents the 2006 level of emissions, and the horizontal line $y = 14$ represents a doubling of carbon emissions. Their points of intersection with the graph of *C* represent when these levels of emission could occur. The figure on the right illustrates what might happen if levels of carbon emission could be held at the 2006 rate of 7 billion tons per year for the next 50 years. In this case, emissions are expected to decline after 50 years. This graph, made up of line segments, is called a *piecewise-linear function*.

The human mind has never invented a labor-saving device greater than algebra.

—J. W. Gibbs

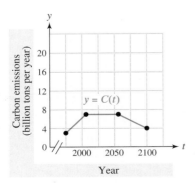

76

Whatever your point of view, mathematics plays an essential role in understanding the future of carbon emissions; without mathematical support, predictions lack credibility. To model carbon emissions, we need constant, linear, and piecewise-defined functions. All of these important concepts are discussed in this chapter.

Source: R. Socolow and S. Pacala, "A Plan to Keep Carbon in Check," *Scientific American,* September, 2006.

2.1 Linear Functions and Models

- **Understand how functions can be models**
- **Identify a graph or table of a linear function**
- **Model data with a linear function**
- **Evaluate and graph piecewise-defined functions**
- **Evaluate and graph the greatest integer function**
- **Use linear regression to model data (optional)**

Introduction

Throughout history, people have attempted to explain the world around them by creating models. A model is based on observations. It can be a diagram, a graph, an equation, a verbal expression, or some other form of communication. Models are used in diverse areas such as economics, physics, chemistry, astronomy, religion, and mathematics. Regardless of where it is used, a **model** is an *abstraction* with the following two characteristics:

1. A model is able to explain present phenomena. It should not contradict data and information already known to be correct.
2. A model is able to make predictions about data or results. It should use current information to forecast phenomena or create new information.

Mathematical models are used to forecast business trends, design the shapes of cars, estimate ecological trends, control highway traffic, describe epidemics, predict weather, and discover new information when human knowledge is inadequate.

Functions as Models

A function can sometimes be a model. Worldwide, people purchased and downloaded 420 million tracks of music in 2005, and this number increased to 795 million in 2007. (**Source:** International Federation of the Phonographic Industry.) If $t = 0$ corresponds to 2005, $t = 1$ to 2006, and so on, then the linear function D defined by $D(t) = 187.5t + 420$ accurately *models* these *known* data values because

$$D(0) = 187.5(0) + 420 = 420 \quad \text{and} \quad D(2) = 187.5(2) + 420 = 795.$$

We might use function D as a model to *predict* the downloads in 2006 by evaluating $D(1)$.

$$D(1) = 187.5(1) + 420 = 607.5$$

Thus D estimates that 607.5 million tracks of music were downloaded in 2006. Because 2006 is between 2005 and 2007, this calculated value is likely to be more accurate than if we used D to estimate the number of downloads many years into the future or the past. For example, to predict downloads in 2011 we might let $t = 6$ ($2005 + 6 = 2011$), and to predict downloads in 2001 we might let $t = -4$ ($2005 - 4 = 2001$).

$$D(6) = 187.5(6) + 420 = 1545$$
$$D(-4) = 187.5(-4) + 420 = -330$$

This model predicts that **1545** million tracks will be downloaded in 2011, which may or may not be correct. However, this model also estimates that -330 million tracks were downloaded in 2001, which is clearly incorrect.

Estimating values between data points, such as for 2006, is called **interpolation**, and estimating values "outside" of the given data points, such as for 2001 or 2011, is called **extrapolation**. Interpolation tends to be more accurate than extrapolation. Typically, for a model to be accurate, it must have limits on its domain. For example, it might be reasonable to limit the domain of D to $t = 0, 1, 2, 3,$ or 4.

Representations of Linear Functions

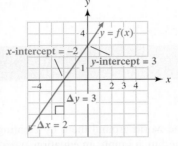

Figure 2.1 Linear Function

Any linear function can be written as $f(x) = ax + b$, where a equals the slope of the graph of f. Also, because $f(0) = a(0) + b = b$, the point $(0, b)$ lies on the graph of f and the value of b is the **y-intercept** of the graph of f.

Consider the graph of the linear function f shown in Figure 2.1. The graph is a line that intersects each axis once. From the graph we can see that when y increases by 3 units, x increases by 2 units. Thus the change in y is $\Delta y = 3$, the change in x is $\Delta x = 2$, and the slope is $\frac{\Delta y}{\Delta x} = \frac{3}{2}$. The graph f intersects the y-axis at the point $(0, 3)$, and so the y-intercept is 3.

Using this information, we can write a formula, or symbolic representation, of f as

$$f(x) = \frac{3}{2}x + 3.$$

A function can have at most one y-intercept because $f(0)$ can have at most one value.

The graph of f in Figure 2.1 intersects the x-axis at the point $(-2, 0)$. We say that the **x-intercept** on the graph of f is -2. When we evaluate $f(-2)$, we obtain

$$f(-2) = \frac{3}{2}(-2) + 3 = 0.$$

An x-intercept corresponds to an input that results in an output of 0. We also say that -2 is a *zero* of the function f, since $f(-2) = 0$. A **zero** of a function f corresponds to an x-intercept on the graph of f. If the slope of the graph of a linear function f is not 0, then the graph of f has exactly one x-intercept.

A table of values, or numerical representation, for $f(x) = \frac{3}{2}x + 3$ is shown in Table 2.1. Note that for every 1-unit increase in x, $f(x)$ increases by $\frac{3}{2}$, or 1.5, units.

Table 2.1 $f(x) = \frac{3}{2}x + 3$

x	-2	-1	0	1	2
$f(x)$	0	1.5	3	4.5	6

$$1.5 \quad 1.5 \quad 1.5 \quad 1.5$$

In general, for each 1-unit increase in x, a linear function g, given by $g(x) = ax + b$, changes by a units. When $a > 0$ the graph of g is increasing for all real numbers x, and when $a < 0$ the graph of g is decreasing for all real numbers x. Linear functions either *always increase* or *always decrease* on their domains, provided $a \neq 0$. If $a = 0$, then g is a constant function.

Figure 2.2

EXAMPLE 1 **Finding a formula from a graph**

Use the graph of a linear function f in Figure 2.2 to complete the following.
(a) Find the slope, y-intercept, and x-intercept.
(b) Write a formula for f.
(c) Find any zeros of f.

SOLUTION
(a) The line **falls 1** unit each time the x-values increase by **4** units. Therefore the slope is $-\frac{1}{4}$. The graph intersects the y-axis at the point $(0, -1)$ and intersects the x-axis at the point $(-4, 0)$. Therefore the y-intercept is -1, and the x-intercept is -4.
(b) Because the slope is $-\frac{1}{4}$ and the y-intercept is -1, it follows that

$$f(x) = -\frac{1}{4}x - 1.$$

(c) Zeros of f correspond to x-intercepts, so the only zero is -4.

Now Try Exercise 13 ◄

Modeling with Linear Functions

Linear functions can be used to model things that change at a constant rate. For example, the distance traveled by a car can be modeled by a linear function *if* the car is traveling at a constant speed.

Modeling with a Linear Function

To model a quantity that is changing at a constant rate with $f(x) = ax + b$, the following formula may be used.

$$f(x) = (\text{constant rate of change})\, x + (\text{initial amount})$$

The constant rate of change corresponds to the slope of the graph of f, and the initial amount corresponds to the y-intercept.

This method is illustrated in the next two examples.

EXAMPLE 2 **Writing formulas for functions**

Write the formula for a linear function that models each situation. Choose both an appropriate name and an appropriate variable for the function. State what the input variable represents and the domain of the function.
(a) In 2006 the average cost of attending a private college was $30,000, and it is projected to increase, on average, by $1750 per year until 2010. (**Source:** CNNMoney.com.)
(b) A car's speed is 50 miles per hour, and it begins to slow down at a constant rate of 10 miles per hour each second.

SOLUTION
(a) Getting Started To model cost with a linear function, we need to find two quantities: the initial amount and the rate of change. In this example the initial amount is $30,000 and the rate of change is $1750 per year. ►

Let C be the name of the function and x be the number of years after 2006. Then

$$C(x) = \textbf{(constant rate of change)}x + \textbf{(initial amount)}$$
$$= \mathbf{1750}x + \mathbf{30{,}000}$$

models the cost in dollars of attending a private college x years after 2006. Because this projection is valid only until 2010, or for 4 years past 2006, the domain D of function C is

$$D = \{x \mid x = 0, 1, 2, 3, \text{ or } 4\}.$$

Note that x represents a year, so it may be most appropriate to restrict the domain to integer values for x.

(b) Let S be the name of the function and t be the elapsed time in seconds that the car has been slowing down. Then

$$S(t) = \textbf{(constant rate of change)}t + \textbf{(initial speed)}$$
$$= \mathbf{-10}t + \mathbf{50}$$

models the speed of the car after an elapsed time of t seconds. Because the car's initial speed is **50** miles per hour and it **slows at 10** miles per hour per second, the car can slow down for at most 5 seconds before it comes to a stop. Thus the domain D of S is

$$D = \{t \mid 0 \le t \le 5\}.$$

Note that t represents time in seconds, so t does not need to be restricted to an integer.

Now Try Exercises 49 and 51 ◀

EXAMPLE 3   Finding a symbolic representation

A 100-gallon tank, initially full of water, is being drained at a rate of 5 gallons per minute.
(a) Write a formula for a linear function f that models the number of gallons of water in the tank after x minutes.
(b) How much water is in the tank after 4 minutes?
(c) Graph f. Identify the x- and y-intercepts and interpret each.
(d) Discuss the domain of f.

SOLUTION
(a) The amount of water in the tank is *decreasing* at 5 gallons per minute, so the constant rate of change is -5. The initial amount of water is 100 gallons.

$$f(x) = \textbf{(constant rate of change)}x + \textbf{(initial amount)}$$
$$= \mathbf{-5}x + \mathbf{100}$$

(b) After 4 minutes the tank contains $f(4) = -5(4) + 100 = 80$ gallons.
(c) Since $f(x) = -5x + 100$, the graph has y-intercept 100 and slope -5, as shown in Figure 2.3. The x-intercept is 20, which corresponds to the time in minutes that it takes to empty the tank. The y-intercept corresponds to the gallons of water initially in the tank.
(d) From the graph we see that the domain of f must be restricted to $0 \le x \le 20$. For example, 21 is not in the domain of f because $f(21) = -5(21) + 100 = -5$; the tank cannot hold -5 gallons. Similarly, -1 is not in the domain of f because $f(-1) = -5(-1) + 100 = 105$; the tank holds *at most* 100 gallons.

Now Try Exercise 55 ◀

Figure 2.3

If the slopes between consecutive pairs of data points are always the same, the data can be modeled exactly by a linear function. If the slopes between consecutive pairs of data points are nearly the same, then the data can be modeled approximately by a linear function. In the next example we model data approximately.

EXAMPLE 4 Modeling airliner CO_2 emissions

Airliners emit carbon dioxide into the atmosphere when they burn jet fuel. Table 2.2 shows the *average* number y of pounds of carbon dioxide (CO_2) emitted by an airliner for each passenger who flies a distance of x miles.

Table 2.2 Carbon Dioxide Emissions

x (miles)	240	360	680	800
y (pounds)	150	230	435	510

Source: E. Rogers and T. Kostigen, *The Green Book.*

(a) Calculate the slopes of the line segments that connect consecutive data points.
(b) Find a linear function f that models the data.
(c) Graph f and the data. What does the slope of the graph of f indicate?
(d) Calculate $f(1000)$ and interpret the result.

SOLUTION

(a) The slopes of the lines passing through the points (240, 150), (360, 230), (680, 435), and (800, 510) are as follows:

$$m_1 = \frac{230 - 150}{360 - 240} \approx 0.67, \quad m_2 = \frac{435 - 230}{680 - 360} \approx 0.64, \quad \text{and}$$

$$m_3 = \frac{510 - 435}{800 - 680} \approx 0.63.$$

(b) **Getting Started** A linear function can be written as $f(x) = ax + b$. We need to estimate values for a and b. One possibility for a is to find the average of m_1, m_2, and m_3. The value of b equals the y-intercept. ▶

The average of 0.67, 0.64, and 0.63 is 0.65, rounded to the nearest hundredth. Because traveling 0 miles produces 0 pounds of carbon dioxide, let the graph of f pass through (0, 0). Thus the y-intercept is 0 and $f(x) = 0.65x + 0$, where $a = 0.65$ and $b = 0$. Note that answers may vary slightly.

(c) A graph of the four data points and $f(x) = 0.65x$ is shown in Figure 2.4. The slope of 0.65 indicates that, on average, 0.65 pound of carbon dioxide is produced for each mile that a person travels in an airliner.

(d) $f(1000) = 0.65(1000) = 650$; thus **650** pounds of carbon dioxide are emitted into the atmosphere, on average, when a person flies **1000** miles.   ***Now Try Exercise 61*** ◀

Figure 2.4

Calculator Help

To make a scatterplot, see Appendix A (page AP-3). To plot data and graph an equation in the same viewing rectangle, see Appendix A (page AP-7).

MAKING CONNECTIONS

Slope and Approximately Linear Data Another way to obtain an initial value for a is to calculate the slope between the first and last data point in the table. The value for a can then be adjusted visually by graphing f and the data. In Example 4 this would have resulted in

$$a = \frac{510 - 150}{800 - 240} \approx 0.64,$$

which compares favorably with our decision to let $a = 0.65$.

Piecewise-Defined Functions

When a function f models data, there may not be one formula for $f(x)$ that works. In this case, the function is sometimes defined on pieces of its domain and is therefore called a **piecewise-defined function**. If each piece is linear, the function is a **piecewise-linear function**. An example of a piecewise-defined function is the *Fujita scale,* which classifies tornadoes by intensity. If a tornado has wind speeds between 40 and 72 miles per hour, it is an F1 tornado. Tornadoes with wind speeds greater than 72 miles per hour but not more than 112 miles per hour are F2 tornadoes. The Fujita scale is represented by the following function F, where the input x represents the maximum wind speed of a tornado and the output is the F-scale number from 1 to 5.

$$F(x) = \begin{cases} 1 & \text{if } 40 \leq x \leq 72 \\ 2 & \text{if } 72 < x \leq 112 \\ 3 & \text{if } 112 < x \leq 157 \\ 4 & \text{if } 157 < x \leq 206 \quad \leftarrow F(180) = 4 \\ 5 & \text{if } 206 < x \leq 260 \end{cases}$$

For example, if the maximum wind speed is 180 miles per hour, then $F(180) = 4$ because 180 is between 157 and 206; that is, $157 < 180 \leq 206$. Thus a tornado with a maximum wind speed of 180 miles per hour is an F4 tornado.

A graph of $y = F(x)$ is shown in Figure 2.5. It is composed of horizontal line segments. Because each piece is constant, F is sometimes called a **piecewise-constant function** or a **step function**. A solid dot occurs at the point (72, 1) and an open circle occurs at the point (72, 2), because technically a tornado with 72-mile-per-hour winds is an F1 tornado, not an F2 tornado. An open circle indicates that the point is *not* included in the graph of F.

You can draw the graph of a continuous function without picking up your pencil. Because there are breaks in the graph of F, function F is not continuous; rather, it is **discontinuous** at $x = 72, 112, 157,$ and 206.

Figure 2.5

EXAMPLE 5 Evaluating a graphical representation

Figure 2.6 depicts a graph of a piecewise-linear function f. It models the amount of water in thousands of gallons in a swimming pool after x hours have elapsed.
(a) Use the graph to evaluate $f(0)$, $f(25)$, and $f(40)$. Interpret the results.
(b) Discuss how the amount of water in the pool changed. Is f a continuous function on the interval [0, 50]?
(c) Interpret the slope of each line segment in the graph.
(d) Identify where f is increasing, decreasing, or constant.

SOLUTION
(a) From Figure 2.6, when $x = 0$, $y = 2$, so $f(0) = 2$. Initially, there are 2 thousand gallons of water in the pool. The points (25, 8) and (40, 4) lie on the graph of f. Therefore $f(25) = 8$ and $f(40) = 4$. After 25 hours there were 8 thousand gallons in the pool, and after 40 hours there were 4 thousand gallons.
(b) During the first 20 hours, water in the pool increased at a constant rate from 2 thousand to 8 thousand gallons. For the next 10 hours, the water level was constant. Between 30 and 50 hours, the pool was drained at a constant rate. Since there are no breaks in the graph, f is a continuous function on the interval [0, 50].
(c) Slope indicates the rate at which water is entering or leaving the pool. The first line segment connects the points (0, 2) and (20, 8). Its slope is $m = \frac{8-2}{20-0} = 0.3$. Since

Figure 2.6 Water in a Pool

the units are thousands of gallons and hours, water is entering the pool during this time at 300 gallons per hour. Between 20 and 30 hours the slope of the line segment is 0. Water is neither entering nor leaving the pool. The slope of the line segment connecting the points (30, 8) and (50, 0) is $m = \frac{0 - 8}{50 - 30} = -0.4$. On this interval water is *leaving* the pool at 400 gallons per hour.

(d) Function f is increasing for $0 \leq x \leq 20$, decreasing for $30 \leq x \leq 50$, and constant for $20 \leq x \leq 30$. These intervals correspond to when the water level in the pool is increasing, decreasing, or remaining constant. ***Now Try Exercise 67*** ◀

EXAMPLE 6 Evaluating and graphing a piecewise-defined function

Use $f(x)$ to complete the following.

$$f(x) = \begin{cases} x - 1 & \text{if } -4 \leq x < 2 \\ -2x & \text{if } 2 \leq x \leq 4 \end{cases}$$

(a) What is the domain of f?
(b) Evaluate $f(-3)$, $f(2)$, $f(4)$, and $f(5)$.
(c) Sketch a graph of f.
(d) Is f a continuous function on its domain?

SOLUTION

(a) Function f is defined for x-values satisfying either $-4 \leq x < 2$ or $2 \leq x \leq 4$. Thus the domain of f is $D = \{x \mid -4 \leq x \leq 4\}$, or $[-4, 4]$.

(b) For x-values satisfying $-4 \leq x < 2$, $f(x) = x - 1$ and so $f(-3) = -3 - 1 = -4$. Similarly, if $2 \leq x \leq 4$, then $f(x) = -2x$. Thus $f(2) = -2 \cdot 2 = -4$ and $f(4) = -2 \cdot 4 = -8$. The expression $f(5)$ is undefined because 5 is not in the domain of f.

Figure 2.7

(c) **Getting Started** Because each piece of $f(x)$ is linear, the graph of $y = f(x)$ consists of two line segments. Therefore we can find the endpoints of each line segment and then sketch the graph. ▶

The first piece is $y_1 = x - 1$ for $-4 \leq x < 2$. If $x = -4$, then $y_1 = -5$, and if $x = 2$, then $y_1 = 1$. Thus the endpoints for the first line segment are $(-4, -5)$ and $(2, 1)$. Plot a solid dot at $(-4, -5)$ because $(-4, -5)$ is included, and an open circle at $(2, 1)$, because $(2, 1)$ is *not included* in this piece of the function. Sketch a line segment between these points, as shown in Figure 2.7.

The second piece is $y_2 = -2x$ for $2 \leq x \leq 4$. If $x = 2$, then $y_2 = -4$, and if $x = 4$, then $y_2 = -8$. Thus the endpoints are $(2, -4)$ and $(4, -8)$. Plot a solid dot for each and connect these points with a line segment, as shown in Figure 2.8.

(d) The function f is not continuous because there is a break in its graph at $x = 2$.

Now Try Exercise 71 ◀

Figure 2.8

The Greatest Integer Function

A common piecewise-defined function used in mathematics is the greatest integer function, denoted $f(x) = [\![x]\!]$. The **greatest integer function** is defined as follows.

$[\![x]\!]$ is the greatest integer less than or equal to x.

Some examples of the evaluation of $[\![x]\!]$ include

$$[\![6.7]\!] = 6, \quad [\![3]\!] = 3, \quad [\![-2.3]\!] = -3, \quad [\![-10]\!] = -10, \quad \text{and} \quad [\![-\pi]\!] = -4.$$

The graph of $y = [\![x]\!]$ is shown in Figure 2.9 on the next page. The greatest integer function is both a piecewise-constant function and a step function.

Figure 2.9 The Greatest Integer Function

$$[x] = \begin{cases} \vdots \\ -2 & \text{if } -2 \le x < -1 \\ -1 & \text{if } -1 \le x < 0 \\ 0 & \text{if } 0 \le x < 1 \\ 1 & \text{if } 1 \le x < 2 \\ 2 & \text{if } 2 \le x < 3 \\ \vdots \end{cases}$$

Calculator Help

To access the greatest integer function or to set a calculator in dot mode, see Appendix A (pages AP-7 and AP-8).

[0, 10, 1] by [31, 35, 1]

Figure 2.10 Dot Mode

[0, 10, 1] by [31, 35, 1]

Figure 2.11 Connected Mode

In some applications, fractional parts are either not allowed or ignored. Framing lumber for houses is measured in 2-foot multiples, and mileage charges for rental cars may be calculated to the mile.

Suppose a car rental company charges \$31.50 per day plus \$0.25 for each mile driven, where fractions of a mile are ignored. The function given by $f(x) = 0.25[x] + 31.50$ calculates the cost of driving x miles in one day. For example, the cost of driving 100.4 miles is

$$f(100.4) = 0.25[100.4] + 31.50 = 0.25(100) + 31.50 = \$56.50.$$

On some calculators and computers, the greatest integer function is denoted int(X). A graph of $Y_1 = 0.25*\text{int}(X) + 31.5$ is shown in Figure 2.10.

MAKING CONNECTIONS

Connected and Dot Modes Graphing calculators often connect points to make a graph look continuous. However, if a graph has breaks in it, a graphing calculator may connect points where there should be breaks. In *dot mode*, points are plotted but not connected. Figure 2.11 is the same graph shown in Figure 2.10, except that it is plotted in *connected mode*. Note that connected mode generates an inaccurate graph of this step function.

Linear Regression (Optional)

We have used linear functions to model data involving the variables x and y. Unknown values for y were predicted at given values of x. Problems where one variable is used to predict the behavior of a second variable are called **regression** problems. If a linear function or line is used to approximate the data, then the technique is referred to as **linear regression**.

We have already solved problems by selecting a line that *visually* fits the data in a scatterplot. However, this technique has some disadvantages. First, it does not produce a unique line. Different people may arrive at different lines to fit the same data. Second, the line is not determined automatically by a calculator or computer. A person must view the data and adjust the line until it "fits." By contrast, a statistical method used to determine a unique linear function or line is based on **least squares**.

Correlation Coefficient Most graphing calculators have the capability to calculate the least-squares regression line automatically after the data points have been entered. When determining the least-squares line, calculators often compute a real number r, called the **correlation coefficient**, where $-1 \le r \le 1$. When r is positive and near 1, low x-values correspond to low y-values and high x-values correspond to high y-values. For example, there is a positive correlation between years of education x and income y. More years of education correlate with higher income. When r is near -1, the reverse is true. Low x-values correspond to high y-values and high x-values correspond to low y-values. An example is the relation between latitude and average yearly temperature. As latitude increases (moving toward either the north or the south pole), the average yearly temperature decreases. Therefore there will be a negative correlation between latitude and average yearly temperature. If $r \approx 0$, then there is little or no correlation between the data points. In this case, a linear function does not provide a suitable model. A summary of these concepts is shown in Table 2.3.

Table 2.3   Correlation Coefficient r ($-1 \le r \le 1$)

Value of r	Comments	Sample Scatterplot
$r = 1$	There is an exact linear fit. The line passes through all data points and has a positive slope.	
$r = -1$	There is an exact linear fit. The line passes through all data points and has a negative slope.	
$0 < r < 1$	There is a positive correlation. As the x-values increase, so do the y-values. The fit is not exact.	
$-1 < r < 0$	There is a negative correlation. As the x-values increase, the y-values decrease. The fit is not exact.	
$r = 0$	There is no correlation. The data has no tendency toward being linear. A regression line predicts poorly.	

MAKING CONNECTIONS

Correlation and Causation   When geese begin to fly north, summer is coming and the weather becomes warmer. Geese flying north correlates with warmer weather. However, geese flying north clearly does not *cause* warmer weather. It is important to remember that correlation does not always indicate the cause.

Calculator Help

To find a line of least-squares fit, see Appendix A (page AP-8).

In the next example we use a graphing calculator to find the line of least-squares fit that models three data points.

EXAMPLE 7   Determining a line of least-squares fit

Find the line of least-squares fit for the data points $(1, 1)$, $(2, 3)$, and $(3, 4)$. What is the correlation coefficient? Plot the data and graph the line.

SOLUTION   Begin by entering the three data points into the STAT EDIT menu. Refer to Figures 2.12–2.15. Select the LinReg(ax + b) option from the STAT CALC menu. From the home screen we can see that the line (linear function) of least-squares is given by the formula $y = \frac{3}{2}x - \frac{1}{3}$. The correlation coefficient is $r \approx 0.98$. Since $r \neq 1$, the line does not provide an *exact* model of the data.

[0, 5, 1] by [0, 5, 1]

Figure 2.12

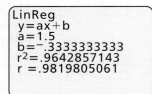

Figure 2.13

LinReg
y=ax+b
a=1.5
b=⁻.3333333333
r²=.9642857143
r =.9819805061

Figure 2.14

Figure 2.15

Now Try Exercise 85 ◀

An Application   In the next example we use regression to find a linear function that models numbers of airline passengers at some of the largest U.S. airports.

EXAMPLE 8   Predicting airline passengers

Gates A1 to A10 ↑

Table 2.4 lists the numbers in millions of airline passengers at some of the largest airports in the United States during 2002 and 2006.
(a) Graph the data by using the 2002 data for *x*-values and the corresponding 2006 data for *y*-values. Predict whether the correlation coefficient will be positive or negative.
(b) Use a calculator to find the linear function f based on least-squares regression that models the data. Graph $y = f(x)$ and the data in the same viewing rectangle.
(c) In 2002 Newark International Airport had 29.0 million passengers. Assuming that this airport followed a trend similar to that of the five airports listed in Table 2.4, use your linear function to estimate the number of passengers at Newark International in 2006. Compare this result to the actual value of 36.7 million passengers.

Table 2.4 Airline Passengers (millions)

Airport	2002	2006
Atlanta (Hartsfield)	76.9	84.4
Chicago (O'Hare)	66.5	77.0
Los Angeles (LAX)	56.2	61.0
Dallas/Fort Worth	52.8	60.2
Denver	35.7	47.3

Source: Airports Association Council International.

SOLUTION

(a) A scatterplot of the data is shown in Figure 2.16. Because increasing x-values correspond to increasing y-values, the correlation coefficient will be positive.

(b) Because $y = f(x)$, the formula for a linear function f that models the data is given by $f(x) = 0.9384x + 11.9098$, where coefficients have been rounded to four decimal places. See Figure 2.17. Graphs of f and the data are shown in Figure 2.18.

(c) We can use $f(x)$ to predict y when $x = 29.0$.

$$y = f(29.0) = 0.9384(29.0) + 11.9098 \approx 39.1 \text{ million}$$

This value is more than the actual value of 36.7 million.

Calculator Help

To copy the regression equation directly into Y_1, see Appendix A (page AP-13).

[30, 90, 10] by [30, 90, 10]

Figure 2.16

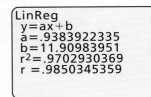

Figure 2.17

[30, 90, 10] by [30, 90, 10]

$y = 0.9384x + 11.9098$

Figure 2.18

Now Try Exercise 93 ◄

2.1 Putting It All Together

The following table summarizes important concepts.

Concept	Description
Models	A good model describes and explains current data. It should also make predictions and forecast phenomena.
Linear model	If a quantity experiences a constant rate of change, then it can be modeled by a linear function in the form $f(x) = ax + b$. $$f(x) = (\textbf{constant rate of change})x + (\textbf{initial value})$$

continued on next page

continued from previous page

Concept	Description
Graph of a linear function	The graph of a linear function is a line. If $f(x) = ax + b$, then the slope equals a and the y-intercept equals b. The following graph has slope $\frac{1}{2}$, y-intercept 1, and formula $f(x) = \frac{1}{2}x + 1$. The zero of f is the x-intercept, -2.
Piecewise-defined function	A function is piecewise-defined if it has different formulas on different intervals of its domain. Many times the domain is restricted. $$f(x) = \begin{cases} 2x - 3 & \text{if } -3 \le x < 1 \\ x + 5 & \text{if } 1 \le x \le 5 \end{cases}$$ When $x = 2$ then $f(x) = x + 5$, so $f(2) = 2 + 5 = 7$. The domain of f is $[-3, 5]$.
Correlation coefficient r	The values of r satisfy $-1 \le r \le 1$, where a line fits the data better if r is near -1 or 1. A value near 0 indicates a poor fit.
Least-squares regression line	The line of least-squares fit for the points $(1, 3)$, $(2, 5)$, and $(3, 6)$ is $y = \frac{3}{2}x + \frac{5}{3}$ and $r \approx 0.98$. Try verifying this with a calculator.

2.1 Exercises

Functions as Models

1. *U.S. Vehicle Production*   In 2000 there were 12.8 million vehicles produced in the United States, and in 2004 there were 12.0 million. The formula $V(t) = -0.2t + 12.8$ models these data exactly, where $t = 0$ corresponds to 2000, $t = 1$ to 2001, and so on.
   (a) Verify that $V(t)$ gives the exact values in millions for 2000 and 2004.

   (b) Use $V(t)$ to estimate the number of vehicles manufactured in 2002 and 2006. Do these estimates involve interpolation or extrapolation?

   (c) The actual value for 2002 was 12.3 million and for 2006 was 11.3 million. Discuss the accuracy of your results from part (b).

2. *U.S. Advertising Expenditures*   In 2002 $237 billion was spent on advertising in the United States, and in 2004 this amount was $264 billion. The formula $A(t) = 13.5t + 237$ models these data exactly, where $t = 0$ corresponds to 2002, $t = 1$ to 2003, and so on.
   (a) Verify that $A(t)$ gives the exact values in billions of dollars for 2002 and 2004.

   (b) Use $A(t)$ to estimate the advertising expenditures in 2000 and 2003. Do these estimates involve interpolation or extrapolation?

   (c) The actual value for 2000 was $244 billion and for 2003 was $245 billion. Discuss the accuracy of your results from part (b).

Exercises 3–6: A function f is given. Determine whether f models the data exactly or approximately.

3. $f(x) = 5x - 2$

x	1	2	3	4
y	3	8	13	18

4. $f(x) = 1 - 0.2x$

x	5	10	15	20
y	0	−1	−2	−4

5. $f(x) = 3.7 - 1.5x$

x	−6	0	1
y	12.7	3.7	2.1

6. $f(x) = 13.3x - 6.1$

x	1	2	5
y	7.2	20.5	60.4

Exercises 7–10: Find the formula for a linear function f that models the data in the table exactly.

7.

x	−2	0	4
f(x)	4	3	1

8.

x	−6	0	3
f(x)	−5	−1	1

9.

x	1	2	3
f(x)	7	9	11

10.

x	15	30	45
f(x)	40	30	20

Exercises 11 and 12: Write a symbolic representation (formula) for a function f that computes the following.

11. (a) The number of pounds in x ounces

(b) The number of dimes in x dollars

(c) The monthly electric bill in dollars if x kilowatt-hours are used at 6 cents per kilowatt-hour and there is a fee of \$6.50

(d) The cost of skiing x times with a \$500 season pass

12. (a) The distance traveled by a car moving at 50 miles per hour for x hours

(b) The total number of hours in day x

(c) The distance in miles between a runner and home after x hours if the runner starts 1 mile from home and jogs *away* from home at 6 miles per hour

(d) A car's speed in feet per second after x seconds if its tires are 2 feet in diameter and rotating 14 times per second

Graphs of Linear Functions

Exercises 13–18: The graph of a linear function f is shown.
(a) Identify the slope, y-intercept, and x-intercept.
(b) Write a formula for f.
(c) Estimate the zero of f.
(d) Is f increasing or decreasing on its domain?

13.

14.

15.

16.

17.

18.

Exercises 19–32: Graph the linear function by hand. Identify the slope and y-intercept.

19. $f(x) = 3x + 2$

20. $f(x) = -\frac{3}{2}x$

21. $f(x) = \frac{1}{2}x - 2$

22. $f(x) = 3 - x$

23. $g(x) = -2$

24. $g(x) = 20 - 10x$

25. $f(x) = 4 - \frac{1}{2}x$

26. $f(x) = 2x - 3$

27. $g(x) = \frac{1}{2}x$

28. $g(x) = 3$

29. $g(x) = 5 - 5x$

30. $g(x) = \frac{3}{4}x - 2$

31. $f(x) = 20x - 10$

32. $f(x) = -30x + 20$

Exercises 33–38: Write a formula for a linear function f whose graph satisfies the conditions.

33. Slope $-\frac{3}{4}$, y-intercept $\frac{1}{3}$

34. Slope -122, y-intercept 805

35. Slope 15, passing through the origin

36. Slope 1.68, passing through (0, 1.23)

37. Slope 0.5, passing through (1, 4.5)

38. Slope -2, passing through $(-1, 5)$

Exercises 39–44: Average Rate of Change Find the average rate of change of f from -2 to 2. What is the average rate of change of f from x_1 to x_2, where $x_1 \neq x_2$?

39. $f(x) = 10$

40. $f(x) = -5$

41. $f(x) = -\frac{1}{4}x$

42. $f(x) = \frac{5}{3}x$

43. $f(x) = 4 - 3x$

44. $f(x) = 5x + 1$

Modeling with Linear Functions

Exercises 45–48: Match the situation with the graph (a–d) that models it best, where x-values represent time.

45. Height of the Empire State Building from 1990 to 2000

46. Average cost of a new car from 1980 to 2000

47. Distance between a runner in a race and the finish line

48. Amount of money earned after x hours when working at an hourly rate of pay

a.

b.

c.

d.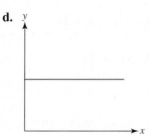

Exercises 49–54: Write a formula for a linear function that models the situation. Choose both an appropriate name and an appropriate variable for the function. State what the input variable represents and the domain of the function. Assume that the domain is an interval of the real numbers.

49. *U.S. Homes with Internet* In 2006 about 68% of U.S. homes had Internet access. This percentage was expected to increase, on average, by 1.5 percentage points per year for the next 4 years. (**Source:** 2007 Digital Future Report.)

50. *U.S. Cell Phones* In 2005 there were about 208 million U.S. cell phone subscribers. This number was expected to increase, on average, by 20 million per year for the next 3 years. (**Source:** CTIA Industry Survey.)

51. *Velocity of a Falling Object* A stone is dropped from a water tower and its velocity increases at a rate of 32 feet per second. The stone hits the ground with a velocity of 96 feet per second.

52. *Speed of a Car* A car is traveling at 30 miles per hour, and then it begins to slow down at a constant rate of 6 miles per hour every 4 seconds.

53. *Population Density* In 1900 the average number of people per square mile in the United States was 21.5, and it increased, on average, by 5.81 people every 10 years until 2000. (**Source:** Bureau of the Census.)

54. *Injury Rate* In 1992 the number of injury cases recorded in private industry per 100 full-time workers was 8.3, and it decreased, on average, by 0.32 injury every year until 2001. (**Source:** Bureau of Labor Statistics.)

55. *Draining a Water Tank* A 300-gallon tank is initially full of water and is being drained at a rate of 10 gallons per minute.
(**a**) Write a formula for a function W that gives the number of gallons of water in the tank after t minutes.

(**b**) How much water is in the tank after 7 minutes?

(**c**) Graph W and identify and interpret the intercepts.

(**d**) Find the domain of W.

56. *Filling a Tank* A 500-gallon tank initially contains 200 gallons of fuel oil. A pump is filling the tank at a rate of 6 gallons per minute.
(**a**) Write a formula for a linear function f that models the number of gallons of fuel oil in the tank after x minutes.

(b) Graph f. What is an appropriate domain for f?

(c) Identify the y-intercept and interpret it.

(d) Does the x-intercept of the graph of f have any physical meaning in this problem? Explain.

57. *HIV Infections*   In 2006 there were 40 million people worldwide who had been infected with HIV. At that time the infection rate was 4.3 million people per year. (**Source:** United Nations AIDS and World Health Organization.)

(a) Write a formula for a linear function f that models the total number of people in millions who were infected with HIV x years after 2006.

(b) Estimate the number of people who may have been infected by the year 2012.

58. *Birth Rate*   In 1990 the number of births per 1000 people in the United States was 16.7 and decreasing at 0.21 birth per 1000 people each year.   (**Source:** National Center for Health Statistics.)

(a) Write a formula for a linear function f that models the birth rate x years after 1990.

(b) Estimate the birth rate in 2003 and compare the estimate to the actual value of 14.

59. *Ice Deposits*   A roof has a 0.5-inch layer of ice on it from a previous storm. Another ice storm begins to deposit ice at a rate of 0.25 inch per hour.

(a) Find a formula for a linear function f that models the thickness of the ice on the roof x hours after the second ice storm started.

(b) How thick is the ice after 2.5 hours?

60. *Rainfall*   Suppose that during a storm rain is falling at a rate of 1 inch per hour. The water coming from a circular roof with a radius of 20 feet is running down a

downspout that can accommodate 400 gallons of water per hour. See the figure.

(a) Determine the number of cubic inches of water falling on the roof in 1 hour.

(b) One gallon equals about 231 cubic inches. Write a formula for a function g that computes the gallons of water landing on the roof in x hours.

(c) How many gallons of water land on the roof during a 2.5-hour rain storm?

(d) Will one downspout be sufficient to handle this type of rainfall? How many downspouts should there be?

Exercises 61 and 62: Modeling Fuel Consumption   The table shows the distance y in miles traveled by a vehicle using x gallons of gasoline.

(a) Calculate the slopes of the line segments that connect consecutive points.

(b) Find a linear function that models the data.

(c) Graph f and the data together. What does the slope indicate?

(d) Evaluate $f(30)$ and interpret the result.

61.

x (gallons)	5	10	15	20
y (miles)	84	169	255	338

62.

x (gallons)	5	10	15	20
y (miles)	194	392	580	781

Piecewise-Defined Functions

63. *Speed Limits*   The graph of $y = f(x)$ on the next page gives the speed limit y along a rural highway x miles from its starting point.

(a) What are the maximum and minimum speed limits along this stretch of highway?

(b) Estimate the miles of highway with a speed limit of 55 miles per hour.

(c) Evaluate $f(4)$, $f(12)$, and $f(18)$.

(d) At what x-values is the graph discontinuous? Interpret each discontinuity.

Distance (miles)

64. *ATM*   The graph of $y = f(x)$ depicts the amount of money y in dollars in an automatic teller machine (ATM) after x minutes.

(a) Determine the initial and final amounts of money in the ATM.

(b) Evaluate $f(10)$ and $f(50)$. Is f continuous?

(c) How many *withdrawals* occurred?

(d) When did the largest withdrawal occur? How much was it?

(e) How much was deposited into the machine?

Time (minutes)

65. *First-Class Mail*   In March 2008, the retail flat rate in dollars for first-class mail weighing up to 5 ounces could be computed by the piecewise-constant function P, where x is the number of ounces.

$$P(x) = \begin{cases} 0.80 & \text{if } 0 < x \leq 1 \\ 0.97 & \text{if } 1 < x \leq 2 \\ 1.14 & \text{if } 2 < x \leq 3 \\ 1.31 & \text{if } 3 < x \leq 4 \\ 1.48 & \text{if } 4 < x \leq 5 \end{cases}$$

(a) Evaluate $P(1.5)$ and $P(3)$. Interpret your results.

(b) Sketch a graph of P. What is the domain of P?

(c) Where is P discontinuous on its domain?

66. *Swimming Pool Levels*   The graph of $y = f(x)$ shows the amount of water y in thousands of gallons remaining in a swimming pool after x days.

(a) Estimate the initial and final amounts of water in the pool.

(b) When did the amount of water in the pool remain constant?

(c) Approximate $f(2)$ and $f(4)$.

(d) At what rate was water being drained from the pool when $1 \leq x \leq 3$?

Time (days)

Exercises 67 and 68: An individual is driving a car along a straight road. The graph shows the driver's distance from home after x hours.

(a) Use the graph to evaluate $f(1.5)$ and $f(4)$.
(b) Interpret the slope of each line segment.
(c) Describe the motion of the car.
(d) Identify where f is increasing, decreasing, or constant.

67.

Time (hours)

68.

Exercises 69–74: Complete the following for f(x).

(a) *Determine the domain of f.*
(b) *Evaluate f(−2), f(0), and f(3).*
(c) *Graph f.*
(d) *Is f continuous on its domain?*

69. $f(x) = \begin{cases} 2 & \text{if } -5 \le x \le -1 \\ x + 3 & \text{if } -1 < x \le 5 \end{cases}$

70. $f(x) = \begin{cases} 2x + 1 & \text{if } -3 \le x < 0 \\ x - 1 & \text{if } 0 \le x \le 3 \end{cases}$

71. $f(x) = \begin{cases} 3x & \text{if } -1 \le x < 1 \\ x + 1 & \text{if } 1 \le x \le 2 \end{cases}$

72. $f(x) = \begin{cases} -2 & \text{if } -6 \le x < -2 \\ 0 & \text{if } -2 \le x < 0 \\ 3x & \text{if } 0 \le x \le 4 \end{cases}$

73. $f(x) = \begin{cases} x & \text{if } -3 \le x \le -1 \\ 1 & \text{if } -1 < x < 1 \\ 2 - x & \text{if } 1 \le x \le 3 \end{cases}$

74. $f(x) = \begin{cases} 3 & \text{if } -4 \le x \le -1 \\ x - 2 & \text{if } -1 < x \le 2 \\ 0.5x & \text{if } 2 < x \le 4 \end{cases}$

Exercises 75 and 76: Graph f.

75. $f(x) = \begin{cases} -\frac{1}{2}x + 1 & \text{if } -4 \le x \le -2 \\ 1 - 2x & \text{if } -2 < x \le 1 \\ \frac{2}{3}x + \frac{4}{3} & \text{if } 1 < x \le 4 \end{cases}$

76. $f(x) = \begin{cases} \frac{3}{2} - \frac{1}{2}x & \text{if } -3 \le x < -1 \\ -2x & \text{if } -1 \le x \le 2 \\ \frac{1}{2}x - 5 & \text{if } 2 < x \le 3 \end{cases}$

77. Use $f(x)$ to complete the following:

$$f(x) = \begin{cases} 3x - 1 & \text{if } -5 \le x < 1 \\ 4 & \text{if } 1 \le x \le 3 \\ 6 - x & \text{if } 3 < x \le 5 \end{cases}$$

(a) Evaluate f at $x = -3, 1, 2,$ and 5.

(b) On what interval is f constant?

(c) Sketch a graph of f. Is f continuous on its domain?

78. Use $g(x)$ to complete the following.

$$g(x) = \begin{cases} -2x - 6 & \text{if } -8 \le x \le -2 \\ x & \text{if } -2 < x < 2 \\ 0.5x + 1 & \text{if } 2 \le x \le 8 \end{cases}$$

(a) Evaluate g at $x = -8, -2, 2,$ and 8.

(b) For what x-values is g increasing?

(c) Sketch a graph of g. Is g continuous on its domain?

Greatest Integer Function

Exercises 79–82: Complete the following.

(a) *Use dot mode to graph the function f in the standard viewing rectangle.*
(b) *Evaluate f(−3.1) and f(1.7).*

79. $f(x) = [\![2x - 1]\!]$ **80.** $f(x) = [\![x + 1]\!]$

81. $f(x) = 2[\![x]\!] + 1$ **82.** $f(x) = [\![-x]\!]$

83. *Lumber Costs* The lumber used to frame walls of houses is frequently sold in multiples of 2 feet. If the length of a board is not exactly a multiple of 2 feet, there is often no charge for the additional length. For example, if a board measures at least 8 feet but less than 10 feet, then the consumer is charged for only 8 feet.

(a) Suppose that the cost of lumber is \$0.80 for every 2 feet. Find a formula for a function f that computes the cost of a board x feet long for $6 \le x \le 18$.

(b) Graph f.

(c) Determine the costs of boards with lengths of 8.5 feet and 15.2 feet.

84. *Cost of Carpet* Each foot of carpet purchased from a 12-foot-wide roll costs $36. If a fraction of a foot is purchased, a customer does not pay for the extra amount. For example, if a customer wants 14 feet of carpet and the salesperson cuts off 14 feet 4 inches, the customer does not pay for the extra 4 inches.

(a) How much does 9 feet 8 inches of carpet from this roll cost?

(b) Using the greatest integer function, write a formula for the price P of x feet of carpet.

Linear Regression

Exercises 85 and 86: Find the line of least-squares fit for the given data points. What is the correlation coefficient? Plot the data and graph the line.

85. $(-2, 2), (1, 0), (3, -2)$ **86.** $(-1, -1), (1, 4), (2, 6)$

Exercises 87–90: Complete the following.

(a) *Conjecture whether the correlation coefficient r for the data will be positive, negative, or zero.*

(b) *Use a calculator to find the equation of the least-squares regression line and the value of r.*

(c) *Use the regression line to predict y when $x = 2.4$.*

87.

x	-1	0	1	2	3
y	-5.7	-2.6	1.1	3.9	7.3

88.

x	-4	-2	0	2	4
y	1.2	2.8	5.3	6.7	9.1

89.

x	1	3	5	7	10
y	5.8	-2.4	-10.7	-17.8	-29.3

90.

x	-4	-3	-1	3	5
y	37.2	33.7	27.5	16.4	9.8

91. *Distant Galaxies* In the late 1920s the famous observational astronomer Edwin P. Hubble (1889–1953) determined both the distance to several galaxies and the velocity at which they were receding from Earth. Four galaxies with their distances in light-years and velocities in miles per second are listed in the table at the top of the next column.

Galaxy	Distance	Velocity
Virgo	50	990
Ursa Minor	650	9,300
Corona Borealis	950	15,000
Bootes	1700	25,000

Source: A. Sharov and I. Novikov, *Edwin Hubble: The Discoverer of the Big Bang Universe.*

(a) Let x be distance and y be velocity. Plot the data points in $[-100, 1800, 100]$ by $[-1000, 28000, 1000]$.

(b) Find the least-squares regression line.

(c) If the galaxy Hydra is receding at a speed of 37,000 miles per second, estimate its distance.

92. *Cell Phones* One of the early problems with cell phones was the delay involved with placing a call when the system was busy. One study analyzed this delay. The table shows that as the number of calls increased by P percent, the average delay time D to put through a call also increased.

P (%)	0	20	40	60	80	100
D (minutes)	1	1.6	2.4	3.2	3.8	4.4

Source: A Mehrotra, *Cellular Radio: Analog and Digital Systems.*

(a) Let P correspond to x-values and D to y-values. Find the least-squares regression line that models these data. Plot the data and the regression line.

(b) Estimate the delay for a 50% increase in the number of calls.

93. *Passenger Travel* The table shows the number of miles (in trillions) traveled by passengers of all types for various years, where $x = 0$ corresponds to 1970, $x = 10$ to 1980, and so on.

Year (1970 ↔ 0)	0	10	20	30	35
Miles (trillions)	2.2	2.8	3.7	4.7	5.1

Source: Department of Transportation.

(a) Make a scatterplot of the data. Predict whether the correlation coefficient will be positive or negative.

(b) Use regression to find a formula $f(x) = ax + b$ so that f models the data.

(c) Graph f and the data. Interpret the slope.

(d) Predict the number of passenger miles in 2010.

94. *High School Enrollment*   The table lists the number of students (in millions) attending U.S. public school (grades 9–12) for selected years, where $x = 0$ corresponds to 2000, $x = 1$ to 2001, and so on.

x (year)	0	3	5	7
y (students)	13.5	14.3	14.8	15.1

Source: National Center for Education Statistics.

(a) Use regression to find a formula $f(x) = ax + b$ so that f models the data.

(b) Graph f and the data. Interpret the slope.

(c) Estimate enrollment in 2002 and compare the estimate to the actual value of 14.1 million.

Writing about Mathematics

95. How can you recognize a symbolic representation (formula) of a linear function? How can you recognize a graph or table of values of a linear function?

96. A student graphs $f(x) = x^2 - x$ in the viewing rectangle $[2, 2.1, 0.01]$ by $[1.9, 2.3, 0.1]$. Using the graph, the student decides that f is a linear function. How could you convince the student otherwise?

97. Explain how average rate of change relates to a linear function.

98. Find a real data set on the Internet that can be modeled by a linear function. Find the linear modeling function. Is your model exact or approximate? Explain.

99. Explain how you determine whether a linear function is increasing, decreasing, or constant. Give an example of each.

100. Explain what a piecewise-defined function is and why it is used. Sketch a graph of a continuous piecewise-linear function f that increases, decreases, and is constant. Let the domain of f be $-4 \le x \le 4$.

EXTENDED AND DISCOVERY EXERCISES

1. *Height and Shoe Size*   In this exercise you will determine if there is a relationship between height and shoe size.

(a) Have classmates write their sex, shoe size, and height in inches on a slip of paper. When you have enough information, complete the following table—one for adult males and one for adult females.

Height (inches)				
Shoe size				

(b) Make a scatterplot of each table, with height on the x-axis and shoe size on the y-axis. Is there any relationship between height and shoe size? Explain.

(c) Try to find a linear function that models each data set.

Exercises 2 and 3: Linear Approximation   Graph the function f in the standard viewing rectangle.

(a) Choose any curved portion of the graph of f and repeatedly zoom in. Describe how the graph appears. Repeat this process on different portions of the graph.

(b) Under what circumstances could a linear function be used to accurately model a nonlinear graph?

2. $f(x) = 4x - x^3$

3. $f(x) = x^4 - 5x^2$

2.2 Equations of Lines

- **Write the point-slope and slope-intercept forms**
- **Find the intercepts of a line**
- **Write equations for horizontal, vertical, parallel, and perpendicular lines**
- **Model data with lines and linear functions (optional)**
- **Use direct variation to solve problems**

Introduction

Apple Corporation sold approximately 4.4 million iPods in fiscal 2004 and 46.4 million iPods in fiscal 2006, making the iPod the fastest selling music player in history.   (**Source:** Apple Corporation.) Can we use this information to make estimates about future sales? Mathematics is often used to analyze data and to make predictions. One of the simplest ways to make estimates is to use linear functions and lines. This section discusses how to use data points to find equations of lines. See Example 4.

Forms for Equations of Lines

Point-Slope Form   Suppose that a nonvertical line with slope m passes through the point (x_1, y_1). If (x, y) is any point on this nonvertical line with $x \neq x_1$, then the change in y is $\Delta y = y - y_1$, the change in x is $\Delta x = x - x_1$, and the slope equals $m = \dfrac{\Delta y}{\Delta x} = \dfrac{y - y_1}{x - x_1}$, as illustrated in Figure 2.19.

With this slope formula, the equation of the line can be found.

$$m = \frac{y - y_1}{x - x_1} \qquad \text{Slope formula}$$

$$y - y_1 = m(x - x_1) \qquad \text{Cross multiply.}$$

$$y = m(x - x_1) + y_1 \qquad \text{Add } y_1 \text{ to each side.}$$

The equation $y - y_1 = m(x - x_1)$ is traditionally called the *point-slope form* of the equation of a line. Since we think of y as being a function of x, written $y = f(x)$, the equivalent form $y = m(x - x_1) + y_1$ will also be referred to as the point-slope form. The point-slope form is not unique, as any point on the line can be used for (x_1, y_1). However, these point-slope forms are *equivalent*—their graphs are identical.

Figure 2.19

> ### Point-Slope Form
>
> The line with slope m passing through the point (x_1, y_1) has an equation
>
> $$y = m(x - x_1) + y_1, \quad \text{or} \quad y - y_1 = m(x - x_1),$$
>
> the **point-slope form** of the equation of a line.

In the next example we find the equation of a line given two points.

EXAMPLE 1   Determining a point-slope form

Find an equation of the line passing through the points $(-2, -3)$ and $(1, 3)$. Plot the points and graph the line by hand.

SOLUTION Begin by finding the slope of the line.

$$m = \frac{3 - (-3)}{1 - (-2)} = \frac{6}{3} = 2$$

Figure 2.20

Algebra Review
To review the distributive property,
see Chapter R (page R-15).

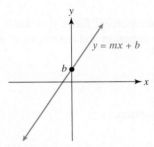

Figure 2.21

Substituting $(x_1, y_1) = (1, 3)$ and $m = 2$ into the point-slope form results in

$$y = 2(x - 1) + 3. \qquad y = m(x - x_1) + y_1$$

If we use the point $(-2, -3)$, the point-slope form is

$$y = 2(x + 2) - 3. \qquad \text{Note that } (x - (-2)) = (x + 2).$$

This line and the two points are shown in Figure 2.20.   ***Now Try Exercise 1*** ◀

Slope-Intercept Form   The two point-slope forms found in Example 1 are equivalent.

$y = 2(x - 1) + 3$	$y = 2(x + 2) - 3$	Point-slope form
$y = 2x - 2 + 3$	$y = 2x + 4 - 3$	Distributive property
$y = 2x + 1$	$y = 2x + 1$	Simplify.

Both point-slope forms simplify to the same equation.

The form $y = mx + b$ is called the *slope-intercept form*. Unlike the point-slope form, it is *unique*. The real number m represents the slope and the real number b represents the y-intercept, as illustrated in Figure 2.21.

> ## Slope-Intercept Form
> The line with slope m and y-intercept b is given by
> $$y = mx + b,$$
> the **slope-intercept form** of the equation of a line.

EXAMPLE 2   **Finding equations of lines**

Find the point-slope form for the line that satisfies the conditions. Then convert this equation into slope-intercept form.

(a) Slope $-\frac{1}{2}$, passing through the point $(-3, -7)$

(b) x-intercept -4, y-intercept 2

SOLUTION

(a) Let $m = -\frac{1}{2}$ and $(x_1, y_1) = (-3, -7)$ in the point-slope form.

$$y = m(x - x_1) + y_1 \qquad \text{Point-slope form}$$

$$y = -\frac{1}{2}(x + 3) - 7 \qquad \text{Substitute.}$$

The slope-intercept form can be found by simplifying.

$$y = -\frac{1}{2}(x + 3) - 7 \qquad \text{Point-slope form}$$

$$y = -\frac{1}{2}x - \frac{3}{2} - 7 \qquad \text{Distributive property}$$

$$y = -\frac{1}{2}x - \frac{17}{2} \qquad \text{Slope-intercept form}$$

(b) The line passes through the points $(-4, 0)$ and $(0, 2)$. Its slope is

$$m = \frac{2 - 0}{0 - (-4)} = \frac{1}{2}.$$

Thus a point-slope form for the line is $y = \frac{1}{2}(x + 4) + 0$, where the point $(-4, 0)$ is used for (x_1, y_1). The slope-intercept form is $y = \frac{1}{2}x + 2$.   *Now Try Exercises 5 and 9* ◄

The next example demonstrates how to find the slope-intercept form of a line without first finding the point-slope form.

EXAMPLE 3   Finding slope-intercept form

Find the slope-intercept form of the line passing through the points $(-2, 1)$ and $(2, 3)$.

SOLUTION
Getting Started We need to determine m and b in the slope-intercept form, $y = mx + b$. First find the slope m. Then substitute *either* point into the equation and determine b. ▶

$$m = \frac{3 - 1}{2 - (-2)} = \frac{2}{4} = \frac{1}{2}$$

Thus $y = \frac{1}{2}x + b$. To find b, we substitute $(2, 3)$ in this equation.

$$3 = \frac{1}{2}(2) + b \qquad \text{Let } x = 2 \text{ and } y = 3.$$

$$3 = 1 + b \qquad \text{Multiply.}$$

$$2 = b \qquad \text{Determine } b.$$

Thus $y = \frac{1}{2}x + 2$.   *Now Try Exercise 21* ◄

An Application   In the next example we model the data about iPods discussed in the introduction to this section.

EXAMPLE 4   Estimating iPod sales

Apple Corporation sold approximately 4.4 million iPods in fiscal 2004 and 46.4 million iPods in fiscal 2006.
(a) Find the point-slope form of the line passing through $(2004, 4.4)$ and $(2006, 46.4)$. Interpret the slope of the line as a rate of change.
(b) Sketch a graph of the data and the line connecting these points.
(c) Estimate sales in 2005 and compare the estimate to the true value of 23 million. Did your estimate involve interpolation or extrapolation?
(d) Estimate sales in 2003 and 2008. Discuss the accuracy of your answers.

SOLUTION
Getting Started First find the slope m of the line connecting the data points, and then substitute this value for m and either of the two data points in the point-slope form. We can use this equation to estimate sales by substituting the required year for x in the equation. ▶

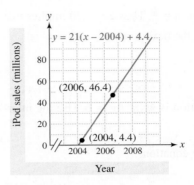

Figure 2.22 iPod Sales

(a) The slope of the line passing through (**2004**, **4.4**) and (**2006**, **46.4**) is

$$m = \frac{46.4 - 4.4}{2006 - 2004} = 21.$$

Thus sales of iPods increased, on average, by **21** million iPods per year from **2004** to **2006**. If we substitute **21** for m and (**2004**, **4.4**) for (x_1, y_1), the point-slope form is

$$y = 21(x - 2004) + 4.4.$$

(b) The requested line passing through the data points is shown in Figure 2.22.

(c) If $x = 2005$, then $y = 21(2005 - 2004) + 4.4 = 25.4$ million. This value is slightly high; its calculation involves interpolation because 2005 lies between 2004 and 2006.

(d) We can use the equation to estimate 2003 and 2008 sales as follows.

$$y = 21(2003 - 2004) + 4.4 = -16.6 \text{ million} \qquad \text{Let } x = 2003.$$
$$y = 21(2008 - 2004) + 4.4 = 88.4 \text{ million} \qquad \text{Let } x = 2008.$$

Both estimates involve extrapolation, because 2003 and 2008 are not between 2004 and 2006. The 2003 value is clearly incorrect because sales cannot be negative. The 2008 value appears to be more reasonable.   ***Now Try Exercise 81*** ◀

Finding Intercepts

The point-slope form and the slope-intercept form are not the only forms for the equation of a line. An equation of a line is in **standard form** when it is written as

$$ax + by = c,$$

where a, b, and c are constants. By using standard form, we can write the equation of any line, including vertical lines (which are discussed later in this section). Examples of equations of lines in standard form include

$$2x - 3y = -6, \qquad y = \frac{1}{4}, \qquad x = -3, \qquad \text{and} \qquad -3x + y = \frac{1}{2}.$$
$$ (a = 0) \qquad (b = 0)$$

Standard form is a convenient form for finding the x- and y-intercepts of a line. Once the intercepts have been found, we can graph the line. For example, to find the x-intercept for the line determined by $3x + 4y = 12$, we let $y = 0$ and solve for x to obtain

$$3x + 4(0) = 12, \quad \text{or} \quad x = 4.$$

The x-intercept is 4. To find the y-intercept, we let $x = 0$ and solve for y to obtain

$$3(0) + 4y = 12, \quad \text{or} \quad y = 3.$$

The y-intercept is 3. Thus the graph of $3x + 4y = 12$ passes through the points (4, 0) and (0, 3). Knowing these two points allows us to graph the line. This technique can be used to find intercepts on the graph of any equation, not just lines written in standard form.

Finding Intercepts

To find any x-intercepts, let $y = 0$ in the equation and solve for x.
To find any y-intercepts, let $x = 0$ in the equation and solve for y.

NOTE To solve $ax = b$, divide each side by a to obtain $x = \frac{b}{a}$. Thus $5x = 20$ implies that $x = \frac{20}{5} = 4$. Linear equations are solved in general in the next section.

EXAMPLE 5   Finding intercepts

Locate the x- and y-intercepts for the line whose equation is $4x + 3y = 6$. Use the intercepts to graph the equation.

SOLUTION To locate the x-intercept, let $y = 0$ in the equation.

$$4x + 3(0) = 6 \qquad \text{Let } y = 0.$$
$$x = 1.5 \qquad \text{Divide by 4.}$$

The x-intercept is **1.5**. Similarly, to find the y-intercept, substitute $x = 0$ into the equation.

$$4(0) + 3y = 6 \qquad \text{Let } x = 0.$$
$$y = 2 \qquad \text{Divide by 3.}$$

The y-intercept is **2**. Therefore the line passes through the points $(1.5, 0)$ and $(0, 2)$, as shown in Figure 2.23.   *Now Try Exercise 57* ◀

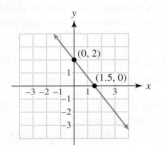

Figure 2.23

Horizontal, Vertical, Parallel, and Perpendicular Lines

Horizontal and Vertical Lines   The graph of a constant function f, defined by the formula $f(x) = b$, is a horizontal line having slope 0 and y-intercept b.

A vertical line cannot be represented by a function because distinct points on a vertical line have the same x-coordinate. In fact, this is the distinguishing feature of points on a vertical line—they all have the same x-coordinate. The vertical line shown in Figure 2.24 is $x = 3$. The equation of a vertical line with x-intercept k is given by $x = k$, as shown in Figure 2.25. Horizontal lines have slope 0, and vertical lines have an undefined slope.

Figure 2.24   **Figure 2.25**

Equations of Horizontal and Vertical Lines

An equation of the horizontal line with y-intercept b is $y = b$. An equation of the vertical line with x-intercept k is $x = k$.

EXAMPLE 6   Finding equations of horizontal and vertical lines

Find equations of vertical and horizontal lines passing through the point $(8, 5)$.

SOLUTION The x-coordinate of the point $(8, 5)$ is 8. The vertical line $x = 8$ passes through every point in the xy-plane with an x-coordinate of 8, including the point $(8, 5)$. Similarly, the horizontal line $y = 5$ passes through every point with a y-coordinate of 5, including $(8, 5)$.

Now Try Exercises 49 and 51 ◀

Parallel Lines   Slope is an important concept when determining whether two lines are parallel or perpendicular. Two nonvertical parallel lines have equal slopes.

Parallel Lines

Two lines with slopes m_1 and m_2, neither of which is vertical, are parallel if and only if their slopes are equal; that is, $m_1 = m_2$.

NOTE The phrase "if and only if" is used when two statements are mathematically equivalent. If two nonvertical lines are parallel, then it is true that $m_1 = m_2$. Conversely, if two nonvertical lines have equal slopes, then they are parallel. Either condition implies the other.

EXAMPLE 7   Finding parallel lines

Find the slope-intercept form of a line parallel to $y = -2x + 5$, passing through $(4, 3)$.

SOLUTION The line $y = -2x + 5$ has slope -2, so any parallel line also has slope $m = -2$. The line passing through $(4, 3)$ with slope -2 is determined as follows.

$$y = -2(x - 4) + 3 \qquad \text{Point-slope form}$$
$$y = -2x + 8 + 3 \qquad \text{Distributive property}$$
$$y = -2x + 11 \qquad \text{Slope-intercept form}$$

Now Try Exercise 35 ◀

Perpendicular Lines   Two lines with nonzero slopes are perpendicular if and only if the product of their slopes is equal to -1.

Perpendicular Lines

Two lines with nonzero slopes m_1 and m_2 are perpendicular if and only if their slopes have product -1; that is, $m_1 m_2 = -1$.

For perpendicular lines, m_1 and m_2 are *negative reciprocals*. That is, $m_1 = -\frac{1}{m_2}$ and $m_2 = -\frac{1}{m_1}$. Table 2.5 shows examples of values for m_1 and m_2 that result in perpendicular lines because $m_1 m_2 = -1$.

Table 2.5   Slopes of Perpendicular Lines

m_1	$\frac{1}{2}$	$\frac{6}{5}$	5	-1	$-\frac{2}{3}$
m_2	-2	$-\frac{5}{6}$	$-\frac{1}{5}$	1	$\frac{3}{2}$
$m_1 m_2$	-1	-1	-1	-1	-1

EXAMPLE 8 Finding perpendicular lines

Find the slope-intercept form of the line perpendicular to $y = -\frac{2}{3}x + 2$, passing through the point $(-2, 1)$. Graph the lines.

SOLUTION The line $y = -\frac{2}{3}x + 2$ has slope $-\frac{2}{3}$. The negative reciprocal of $m_1 = -\frac{2}{3}$ is $m_2 = \frac{3}{2}$. The slope-intercept form of a line having slope $\frac{3}{2}$ and passing through $(-2, 1)$ can be found as follows.

$$y = m(x - x_1) + y_1 \qquad \text{Point-slope form}$$

$$y = \frac{3}{2}(x + 2) + 1 \qquad \text{Let } m = \tfrac{3}{2}, x_1 = -2, \text{ and } y_1 = 1.$$

$$y = \frac{3}{2}x + 3 + 1 \qquad \text{Distributive property}$$

$$y = \frac{3}{2}x + 4 \qquad \text{Slope-intercept form}$$

Figure 2.26 shows graphs of these perpendicular lines. *Now Try Exercise 41* ◀

NOTE If a graphing calculator is used to graph these lines, a square viewing rectangle must be used for the lines to appear perpendicular.

Figure 2.26 Perpendicular Lines

EXAMPLE 9 Determining a rectangle

In Figure 2.27 a rectangle is outlined by four lines denoted y_1, y_2, y_3, and y_4. Find the equation of each line.

SOLUTION

Line y_1: This line passes through the points $(0, 0)$ and $(5, 3)$, so $m = \frac{3}{5}$ and the y-intercept is 0. Its equation is $y_1 = \frac{3}{5}x$.

Line y_2: This line passes through the point $(0, 0)$ and is perpendicular to y_1, so its slope is given by $m = -\frac{5}{3}$ and the y-intercept is 0. Its equation is $y_2 = -\frac{5}{3}x$.

Line y_3: This line passes through the point $(5, 3)$ and is parallel to y_2, so its slope is given by $m = -\frac{5}{3}$. In a point-slope form, its equation is $y_3 = -\frac{5}{3}(x - 5) + 3$, which is equivalent to $y_3 = -\frac{5}{3}x + \frac{34}{3}$.

Line y_4: This line passes through the point $(0, 5)$ and is parallel to y_1, so its slope is given by $m = \frac{3}{5}$. Its equation is $y_4 = \frac{3}{5}x + 5$. *Now Try Exercise 97* ◀

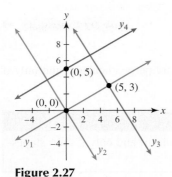

Figure 2.27

Calculator Help

To set a square viewing rectangle, see Appendix A (page AP-6).

CLASS DISCUSSION

Check the results from Example 9 by graphing the four equations in the same viewing rectangle. How does your graph compare with Figure 2.27? Why is it important to use a square viewing rectangle?

Modeling Data (Optional)

Point-slope form can sometimes be useful when modeling real data. In the next example we model the rise in the cost of tuition and fees at private colleges and universities.

EXAMPLE 10 Modeling data

Table 2.6 lists the average tuition and fees at private colleges for selected years.

Calculator Help

To make a scatterplot, see Appendix A (page AP-3). To plot data and graph an equation, see Appendix A (page AP-7).

Table 2.6   Tuition and Fees at Private Colleges

Year	1980	1985	1990	1995	2000	2005
Cost	$3617	$6121	$9340	$12,432	$16,233	$21,235

Source: The College Board.

(a) Make a scatterplot of the data.

(b) Find a linear function, given by $f(x) = m(x - x_1) + y_1$, that models the data. Interpret the slope m.

(c) Use f to estimate tuition and fees in 1998. Compare the estimate to the actual value of $14,709. Did your answer involve interpolation or extrapolation?

[1978, 2007, 5] by [0, 25000, 5000]

Figure 2.28

SOLUTION

(a) See Figure 2.28.

(b) The data table contains several points that could be used for (x_1, y_1). For example, we could choose the first data point, (1980, 3617), and then write

$$f(x) = m(x - 1980) + 3617.$$

To estimate a slope m we could choose two points that appear to lie on a line that models the data. For example, if we choose the first data point, (1980, 3617), and the fifth data point, (2000, 16233), then the slope m is

$$m = \frac{16{,}233 - 3617}{2000 - 1980} = 630.8.$$

[1978, 2007, 5] by [0, 25000, 5000]

$y = 630.8(x - 1980) + 3617$

Figure 2.29

This slope indicates that tuition and fees have risen, on average, $630.80 per year.

Figure 2.29 shows the graphs of $f(x) = 630.8(x - 1980) + 3617$ and the data. It is important to realize that *answers may vary* when modeling real data because if you choose different points, the resulting equation for $f(x)$ will be different. Also, you may choose to adjust the slope or use linear regression to obtain a better fit to the data.

(c) To estimate the tuition and fees in 1998, evaluate $f(1998)$.

$$f(1998) = 630.8(1998 - 1980) + 3617 = \$14{,}971.40$$

This value differs from the actual value by less than $300 and involves interpolation.

Now Try Exercise 85 ◀

MAKING CONNECTIONS

Modeling and Forms of Equations   In Example 10 we modeled college tuition and fees by using the formula

$$f(x) = 630.8(x - 1980) + 3617.$$

This point-slope form readily reveals that tuition and fees cost $3617 in 1980 and have risen, on average, $630.80 per year. In slope-intercept form, this formula becomes

$$f(x) = 630.8x - 1{,}245{,}367.$$

Although the slope is apparent in slope-intercept form, it is less obvious that the actual value of tuition in 1980 was $3617. Which form is more convenient often depends on the problem being solved.

Direct Variation

When a change in one quantity causes a proportional change in another quantity, the two quantities are said to *vary directly* or to *be directly proportional*. For example, if we work for $8 per hour, our pay is proportional to the number of hours that we work. Doubling the hours doubles the pay, tripling the hours triples the pay, and so on.

Direct Variation

Let x and y denote two quantities. Then y is **directly proportional** to x, or y **varies directly** with x, if there exists a nonzero number k such that

$$y = kx.$$

The number k is called the **constant of proportionality** or the **constant of variation**.

Figure 2.30 A Spring Being Stretched

If a person earns $57.75 working for 7 hours, the constant of proportionality k is the hourly pay rate. If y represents the pay in dollars and x the hours worked, then k is found by substituting values for x and y into the equation $y = kx$ and solving for k. That is,

$$57.75 = k(7), \quad \text{or} \quad k = \frac{57.75}{7} = 8.25,$$

so the hourly pay rate is $8.25 and, in general, $y = 8.25x$.

An Application   Hooke's law states that the distance that an elastic spring stretches beyond its natural length is *directly proportional* to the amount of weight hung on the spring, as illustrated in Figure 2.30. This law is valid whether the spring is stretched or compressed. The constant of proportionality is called the **spring constant**. Thus if a weight or force F is applied and the spring stretches a distance x beyond its natural length, then the equation $F = kx$ models this situation, where k is the spring constant.

EXAMPLE 11   Working with Hooke's Law

A 12-pound weight is hung on a spring, and it stretches 2 inches.
(a) Find the spring constant.
(b) Determine how far the spring will stretch when a 19-pound weight is hung on it.

SOLUTION
(a) Let $F = kx$, given that $F = 12$ pounds and $x = 2$ inches. Thus

$$12 = k(2), \quad \text{or} \quad k = 6,$$

and the spring constant equals **6**.
(b) Thus $F = 19$ and $F = 6x$ implies that $19 = 6x$, or $x = \frac{19}{6} \approx 3.17$ inches.

Now Try Exercise 113 ◀

The following four-step method can often be used to solve variation problems.

Solving a Variation Problem

When solving a variation problem, the following steps can be used.

STEP 1: Write the general equation for the type of variation problem that you are solving.

STEP 2: Substitute given values in this equation so the constant of variation k is the only unknown value in the equation. Solve for k.

STEP 3: Substitute the value of k in the general equation in Step 1.

STEP 4: Use this equation to find the requested quantity.

EXAMPLE 12 Solving a direct variation problem

Let T vary directly with x, and suppose that $T = 33$ when $x = 5$. Find T when $x = 31$.

SOLUTION

STEP 1: The equation for direct variation is $T = kx$.

STEP 2: Substitute 33 for T and 5 for x. Then solve for k.

$$T = kx \qquad \text{Direct variation equation}$$
$$33 = k(5) \qquad \text{Let } T = 33 \text{ and } x = 5.$$
$$\frac{33}{5} = k \qquad \text{Divide each side by 5.}$$

STEP 3: Thus $T = \frac{33}{5}x$, or $T = 6.6x$.

STEP 4: When $x = 31$, we have $T = 6.6(31) = 204.6$.   **Now Try Exercise 101** ◄

Suppose that for each point (x, y) in a data set the ratios $\frac{y}{x}$ are all equal to some constant k. That is, $\frac{y}{x} = k$ for each data point. Then $y = kx$, and so y varies directly with x and the constant of variation is k. In addition, the data points (x, y) all lie on the line $y = kx$, which has slope k and passes through the origin. These concepts are used in the next example.

EXAMPLE 13 Modeling memory requirements

Table 2.7 lists the megabytes (MB) x needed to record y seconds of music.

Table 2.7   Recording Digital Music

x (MB)	0.23	0.49	1.16	1.27
y (sec)	10.7	22.8	55.2	60.2

Source: Gateway 2000 System CD.

(a) Compute the ratios $\frac{y}{x}$ for the four data points. Does y vary directly with x? If it does, what is the constant of variation k?

(b) Estimate the seconds of music that can be stored on 5 megabytes.

(c) Graph the data in Table 2.7 and the line $y = kx$.

SOLUTION

(a) The four ratios $\frac{y}{x}$ from Table 2.7 are

$$\frac{10.7}{0.23} \approx 46.5, \quad \frac{22.8}{0.49} \approx 46.5, \quad \frac{55.2}{1.16} \approx 47.6, \quad \text{and} \quad \frac{60.2}{1.27} \approx 47.4.$$

[0, 1.5, 0.5] by [0, 70, 10]

$y = 47x$

Figure 2.31

Because the ratios are nearly equal and these are real data, it is reasonable to say that y is directly proportional to x. The constant of proportionality is about **47**, the average of the four ratios. This means that $y = 47x$ and we can store about **47** seconds of music per megabyte.

(b) Let $x = 5$ in the equation $y = 47x$, to obtain $y = 47(5) = 235$ seconds.

(c) Graphs of the data and the line $y = 47x$ are shown in Figure 2.31.

Now Try Exercise 115 ◀

2.2 Putting It All Together

The following table summarizes some important topics.

Concept	Comments	Examples
Point-slope form $$y = m(x - x_1) + y_1$$ or $$y - y_1 = m(x - x_1)$$	Used to find the equation of a line, given two points or one point and the slope	Given two points $(5, 1)$ and $(4, 3)$, first compute $m = \frac{3 - 1}{4 - 5} = -2$. An equation of this line is $$y = -2(x - 5) + 1.$$
Slope-intercept form $$y = mx + b$$	A unique equation for a line, determined by the slope m and the y-intercept b	An equation of the line with slope 5 and y-intercept -4 is $y = 5x - 4$.
Direct variation	The variable y is directly proportional to x or varies directly with x if $y = kx$ for some nonzero constant k. Constant k is the constant of proportionality or the constant of variation.	If the sales tax rate is 7%, the sales tax y on a purchase of x dollars is calculated by $y = 0.07x$, where $k = 0.07$. The sales tax on a purchase of $125 is $$y = 0.07(125) = \$8.75.$$

The following table summarizes the important concepts concerning special types of lines.

Concept	Equation(s)	Examples
Horizontal line	$y = b$, where b is a constant	A horizontal line with y-intercept 7 has the equation $y = 7$.
Vertical line	$x = k$, where k is a constant	A vertical line with x-intercept -8 has the equation $x = -8$.
Parallel lines	$y = m_1 x + b_1$ and $y = m_2 x + b_2$, where $m_1 = m_2$	The lines given by $y = -3x - 1$ and $y = -3x + 5$ are parallel because they both have slope -3.
Perpendicular lines	$y = m_1 x + b_1$ and $y = m_2 x + b_2$, where $m_1 m_2 = -1$	The lines $y = 2x - 5$ and $y = -\frac{1}{2}x + 2$ are perpendicular because $m_1 m_2 = 2\left(-\frac{1}{2}\right) = -1$.

2.2 Exercises

Equations of Lines

Exercises 1–4: Find the point-slope form of the line passing through the given points. Use the first point as (x_1, y_1). Plot the points and graph the line by hand.

1. $(1, 2), (3, -2)$ **2.** $(-2, 3), (1, 0)$

3. $(-3, -1), (1, 2)$ **4.** $(-1, 2), (-2, -3)$

Exercises 5–10: Find a point-slope form of the line satisfying the conditions. Use the first point given for (x_1, y_1). Then convert the equation to slope-intercept form.

5. Slope -2.4, passing through $(4, 5)$

6. Slope 1.7, passing through $(-8, 10)$

7. Passing through $(1, -2)$ and $(-9, 3)$

8. Passing through $(-6, 10)$ and $(5, -12)$

9. x-intercept 4, y-intercept -3

10. x-intercept -2, y-intercept 5

Exercises 11–14: Find the slope-intercept form for the line in the figure.

11.

12.

13.

14.

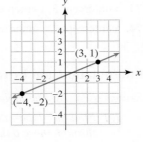

Exercises 15–20: Concepts Match the equation to its graph (a–f) shown in the next column.

15. $y = m(x - x_1) + y_1, m > 0$

16. $y = m(x - x_1) + y_1, m < 0$

17. $y = mx, m > 0$

18. $y = mx + b, m < 0$ and $b > 0$

19. $x = k, k > 0$ **20.** $y = b, b < 0$

a.

b.

c.

d.

e.

f.

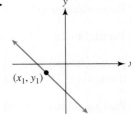

Exercises 21–48: Find the slope-intercept form for the line satisfying the conditions.

21. Passing through $(-1, -4)$ and $(1, 2)$

22. Passing through $(-1, 6)$ and $(2, -3)$

23. Passing through $(4, 5)$ and $(1, -3)$

24. Passing through $(8, -2)$ and $(-2, 3)$

25. y-intercept 5, slope -7.8

26. y-intercept -155, slope 5.6

27. y-intercept 45, x-intercept 90

28. x-intercept -6, y-intercept -8

29. Slope -3, passing through $(0, 5)$

30. Slope $\frac{1}{3}$, passing through $\left(\frac{1}{2}, -2\right)$

31. Passing through $(0, -6)$ and $(4, 0)$

32. Passing through $\left(\frac{3}{4}, -\frac{1}{4}\right)$ and $\left(\frac{5}{4}, \frac{7}{4}\right)$

33. Passing through $\left(\frac{1}{2}, \frac{3}{4}\right)$ and $\left(\frac{1}{5}, \frac{2}{3}\right)$

34. Passing through $\left(-\frac{7}{3}, \frac{5}{3}\right)$ and $\left(\frac{5}{6}, -\frac{7}{6}\right)$

35. Parallel to $y = 4x + 16$, passing through $(-4, -7)$

36. Parallel to the line $y = -\frac{3}{4}(x - 100) - 99$, passing through $(1, 3)$

37. Perpendicular to the line $y = -\frac{2}{3}(x - 1980) + 5$, passing through $(1980, 10)$

38. Perpendicular to $y = 6x - 10$, passing through $(15, -7)$

39. Parallel to $y = \frac{2}{3}x + 3$, passing through $(0, -2.1)$

40. Parallel to $y = -4x - \frac{1}{4}$, passing through $(2, -5)$

41. Perpendicular to $y = -2x$, passing through $(-2, 5)$

42. Perpendicular to $y = -\frac{6}{7}x + \frac{3}{7}$, passing through $(3, 8)$

43. Perpendicular to $x + y = 4$, passing through $(15, -5)$

44. Parallel to $2x - 3y = -6$, passing through $(4, -9)$

45. Passing through $(5, 7)$ and parallel to the line passing through $(1, 3)$ and $(-3, 1)$

46. Passing through $(1990, 4)$ and parallel to the line passing through $(1980, 3)$ and $(2000, 8)$

47. Passing through $(-2, 4)$ and perpendicular to the line passing through $\left(-5, \frac{1}{2}\right)$ and $\left(-3, \frac{2}{3}\right)$

48. Passing through $\left(\frac{3}{4}, \frac{1}{4}\right)$ and perpendicular to the line passing through $(-3, -5)$ and $(-4, 0)$

Exercises 49–56: Find an equation of the line satisfying the conditions.

49. Vertical, passing through $(-5, 6)$

50. Vertical, passing through $(1.95, 10.7)$

51. Horizontal, passing through $(-5, 6)$

52. Horizontal, passing through $(1.95, 10.7)$

53. Perpendicular to $y = 15$, passing through $(4, -9)$

54. Perpendicular to $x = 15$, passing through $(1.6, -9.5)$

55. Parallel to $x = 4.5$, passing through $(19, 5.5)$

56. Parallel to $y = -2.5$, passing through $(1985, 67)$

Finding Intercepts

Exercises 57–68: Determine the x- and y-intercepts on the graph of the equation. Graph the equation.

57. $4x - 5y = 20$

58. $-3x - 5y = 15$

59. $x - y = 7$

60. $15x - y = 30$

61. $6x - 7y = -42$

62. $5x + 2y = -20$

63. $y - 3x = 7$

64. $4x - 3y = 6$

65. $0.2x + 0.4y = 0.8$

66. $\frac{2}{3}y - x = 1$

67. $y = 8x - 5$

68. $y = -1.5x + 15$

*Exercises 69–72: The **intercept form of a line** is $\frac{x}{a} + \frac{y}{b} = 1$. Determine the x- and y-intercepts on the graph of the equation. Draw a conclusion about what the constants a and b represent in this form.*

69. $\dfrac{x}{5} + \dfrac{y}{7} = 1$

70. $\dfrac{x}{2} + \dfrac{y}{3} = 1$

71. $\dfrac{2x}{3} + \dfrac{4y}{5} = 1$

72. $\dfrac{5x}{6} - \dfrac{y}{2} = 1$

Exercises 73 and 74: (Refer to Exercises 69–72.) Write the intercept form for the line with the given intercepts.

73. x-intercept 5, y-intercept 9

74. x-intercept $\frac{2}{3}$, y-intercept $-\frac{5}{4}$

Interpolation and Extrapolation

Exercises 75–78: The table lists data that are exactly linear.

(a) *Find the slope-intercept form of the line that passes through these data points.*

(b) *Predict y when $x = -2.7$ and 6.3. Decide if these calculations involve interpolation or extrapolation.*

75.

x	−3	−2	−1	0	1
y	−7.7	−6.2	−4.7	−3.2	−1.7

76.

x	−2	−1	0	1	2
y	10.2	8.5	6.8	5.1	3.4

77.

x	5	23	32	55	61
y	94.7	56.9	38	−10.3	−22.9

78.

x	−11	−8	−7	−3	2
y	−16.1	−10.4	−8.5	−0.9	8.6

79. *Air Safety Inspectors*   The number of air safety inspectors for selected years is shown in the table.

Year	1998	1999	2000
Inspectors	3305	3185	3089

Source: Federal Aviation Administration.

(a) Find a linear function f that models these data. Is f exact or approximate?

(b) Use f to estimate the number of inspectors in 2005. Compare your answer to the actual value of 3450. Did your estimate involve interpolation or extrapolation?

(c) Explain the difficulty with trying to model these data with a linear function.

80. *Deaths on School Grounds*   Deaths on school grounds nationwide for school years ending in year x are shown in the table.

x (year)	1998	1999	2000
y (deaths)	43	26	9

Source: FBI.

(a) Find a linear function f that models these data. Is f exact or approximate?

(b) Use f to estimate the number of deaths on school grounds in 2003. Compare your answer to the actual value of 49. Did your estimate involve interpolation or extrapolation?

(c) Explain the difficulty with trying to model these data with a linear function.

Applications

81. *Projected Cost of College*   In 2003 the average annual cost of attending a private college or university, including tuition, fees, room, and board, was $25,000. This cost is projected to rise to $37,000 in 2010, as illustrated in the figure. (**Source:** Cerulli Associates.)

(a) Find a point-slope form of the line passing through the points (2003, 25000) and (2010, 37000). Interpret the slope.

(b) Use the equation to estimate the cost of attending a private college in 2007. Did your estimate involve interpolation or extrapolation?

(c) Find the slope-intercept form of this line.

82. *Distance*   A person is riding a bicycle along a straight highway. The graph shows the rider's distance y in miles from an interstate highway after x hours.

(a) How fast is the bicyclist traveling?

(b) Find the slope-intercept form of the line.

(c) How far was the bicyclist from the interstate highway initially?

(d) How far was the bicyclist from the interstate highway after 1 hour and 15 minutes?

83. *Music on the Internet* In 2002 sales of premium online music totaled $1.6 billion. In 2005 this revenue reached $3.6 billion. (**Source:** Jupiter Research.)

(a) Find a point-slope form of the line passing through (2002, 1.6) and (2005, 3.6). Interpret the slope.

(b) Use the equation to estimate projected sales in 2008. Did you use interpolation or extrapolation?

(c) Find the slope-intercept form of this line.

84. *Water in a Tank* The graph shows the amount of water in a 100-gallon tank after x minutes have elapsed.

(a) Is water entering or leaving the tank? How much water is in the tank after 3 minutes?

(b) Find the x- and y-intercepts. Interpret each intercept.

(c) Find the slope-intercept form of the equation of the line. Interpret the slope.

(d) Estimate the x-coordinate of the point $(x, 50)$ that lies on the line.

85. *E-mail Spam* The table lists the average number of worldwide spam messages *daily* in billions for selected years.

Year	1999	2000	2001	2002	2003	2004
Messages	1.0	2.3	4.0	5.6	7.3	8.8

Source: IDC.

(a) Make a scatterplot of the data.

(b) Find a linear function f that models these data. (Answers may vary.) Interpret the slope m.

(c) Use your function to estimate the average number of worldwide spam messages daily during 2007.

86. *Tuition and Fees* The table in the next column lists average tuition and fees at public 4-year colleges.

Year	1980	1985	1990
Tuition and fees	$804	$1318	$1908

Year	1995	2000	2005
Tuition and fees	$2811	$3487	$5491

Source: The College Board.

(a) Make a scatterplot of the data.

(b) Find a linear function that models the data. Interpret the slope m.

(c) Use this function to estimate tuition in 1992. Compare it to the actual value of $2334.

(d) Of the six data values, which one would you remove to make the data "more linear"? Explain.

87. *Toyota Vehicles Sold* The table lists the U.S. sales of Toyota vehicles in millions.

Year	1998	2000	2002	2004
Vehicles	1.4	1.6	1.8	2.0

Source: Autodata.

(a) Make a scatterplot of the data.

(b) Find $f(x) = m(x - x_1) + y_1$ so that $f(x)$ models these data. Interpret the slope m.

(c) Is $f(x)$ an exact or approximate model for the data listed in the table?

88. *Farm Pollution* In 1988 the number of farm pollution incidents reported in England and Wales was 4000. This number had increased at a rate of 280 per year since 1979. (**Source:** C. Mason, *Biology of Freshwater Pollution.*)

(a) Find an equation $y = m(x - x_1) + y_1$ that models these data, where y represents the number of pollution incidents during the year x.

(b) Estimate the number of incidents in 1975.

89. *Cost of Driving* The cost of driving a car includes both fixed costs and mileage costs. Assume that insurance and car payments cost $350 per month and gasoline, oil, and routine maintenance cost $0.29 per mile.

(a) Find a linear function f that gives the annual cost of driving this car x miles.

(b) What does the y-intercept on the graph of f represent?

90. *Average Wages* The average hourly wage (adjusted to 1982 dollars) was \$8.46 in 1970 and \$8.18 in 2005. (**Source:** Department of Commerce.)

(a) Find an equation of a line that passes through the points (1970, 8.46) and (2005, 8.18).

(b) Interpret the slope.

(c) Approximate the hourly wage in 2000. Compare the estimate to the actual value of \$8.04.

Exercises 91 and 92: Modeling Real Data The table contains data that can be modeled by a linear function f.

(a) *Make a scatterplot of the data. (Do not try to plot the undetermined point in the table.)*

(b) *Find a formula for f. Graph f together with the data.*

(c) *Interpret the slope m.*

(d) *Use f to approximate the undetermined value.*

91. Asian-American population in millions

Year	1996	1999	2002	2005	2008
Population	9.7	10.9	12.0	13.4	?

Source: Bureau of the Census.

92. Population of the western states in millions

Year	1950	1970	1990	2010
Population	20.2	34.8	52.8	?

Source: Bureau of the Census.

Perspectives and Viewing Rectangles

93. Graph $y = \frac{1}{1024}x + 1$ in [0, 3, 1] by [−2, 2, 1].

(a) Is the graph a horizontal line?

(b) Why does the calculator screen appear as it does?

94. Graph $y = 1000x + 1000$ in the standard window.

(a) Is the graph a vertical line?

(b) Explain why the calculator screen appears as it does.

95. *Square Viewing Rectangle* Graph the lines $y = 2x$ and $y = -\frac{1}{2}x$ in the standard viewing rectangle.

(a) Do the lines appear to be perpendicular?

(b) Graph the lines in the following viewing rectangles.

i. [−15, 15, 1] by [−10, 10, 1]
ii. [−10, 10, 1] by [−3, 3, 1]
iii. [−3, 3, 1] by [−2, 2, 1]

Do the lines appear to be perpendicular in any of these viewing rectangles?

(c) Determine the viewing rectangles where perpendicular lines will appear perpendicular. (Answers may vary.)

96. *Square Viewing Rectangle* Continuing with Exercise 95, make a conjecture about which viewing rectangles result in the graph of a circle with radius 5 and center at the origin appearing circular.

i. [−9, 9, 1] by [−6, 6, 1]
ii. [−5, 5, 1] by [−10, 10, 1]
iii. [−5, 5, 1] by [−5, 5, 1]
iv. [−18, 18, 1] by [−12, 12, 1]

Test your conjecture by graphing this circle in each viewing rectangle. (*Hint:* Graph $y_1 = \sqrt{25 - x^2}$ and $y_2 = -\sqrt{25 - x^2}$ to create the circle.)

Graphing a Rectangle

Exercises 97–100: (Refer to Example 9.) A rectangle is determined by the stated conditions. Find the slope-intercept form of the four lines that outline the rectangle.

97. Vertices (0, 0), (2, 2), and (1, 3)

98. Vertices (1, 1), (5, 1), and (5, 5)

99. Vertices (4, 0), (0, 4), (0, −4), and (−4, 0)

100. Vertices (1, 1) and (2, 3); the point (3.5, 1) lies on a side of the rectangle.

Direct Variation

Exercises 101–104: Let y be directly proportional to x. Complete the following.

101. Find y when $x = 5$, if $y = 7$ when $x = 14$.

102. Find y when $x = 2.5$, if $y = 13$ when $x = 10$.

103. Find y when $x = \frac{1}{2}$, if $y = \frac{3}{2}$ when $x = \frac{2}{3}$.

104. Find y when $x = 1.3$, if $y = 7.2$ when $x = 5.2$.

Exercises 105–108: Find the constant of proportionality k and the undetermined value in the table if y is directly proportional to x. Support your answer by graphing the equation $y = kx$ and the data points.

105.

x	3	5	6	8
y	7.5	12.5	15	?

106.

x	1.2	4.3	5.7	?
y	3.96	14.19	18.81	23.43

107. Sales tax y on a purchase of x dollars

x	$25	$55	?
y	$1.50	$3.30	$5.10

108. Cost y of buying x compact discs having the same price

x	3	4	5
y	$41.97	$55.96	?

109. *Cost of Tuition*   The cost of tuition is directly proportional to the number of credits taken. If 11 credits cost $720.50, find the cost of taking 16 credits. What is the constant of proportionality?

110. *Strength of a Beam*   The maximum load that a horizontal beam can carry is directly proportional to its width. If a beam 1.5 inches wide can support a load of 250 pounds, find the load that a beam of the same type can support if its width is 3.5 inches.

111. *Antarctic Ozone Layer*   Stratospheric ozone occurs in the atmosphere between altitudes of 12 and 18 miles. Ozone in the stratosphere is frequently measured in Dobson units, where 300 Dobson units corresponds to an ozone layer 3 millimeters thick. In 1991 the reported minimum in the Antarctic *ozone hole* was about 110 Dobson units. (**Source:** R. Huffman, *Atmospheric Ultraviolet Remote Sensing.*)

(a) The thickness y of the ozone layer is directly proportional to the number of Dobson units x. Find the constant of proportionality k.

(b) How thick was the ozone layer in 1991?

112. *Weight on Mars*   The weight of an object on Earth is directly proportional to the weight of an object on Mars. If a 25-pound object on Earth weighs 10 pounds on Mars, how much would a 195-pound astronaut weigh on Mars?

113. *Hooke's Law*   Suppose a 15-pound weight stretches a spring 8 inches, as shown in the figure.

8 in.

15 lb

(a) Find the spring constant.

(b) How far will a 25-pound weight stretch this spring?

114. *Hooke's Law*   If an 80-pound force compresses a spring 3 inches, how much force must be applied to compress the spring 7 inches?

115. *Force of Friction*   The table lists the force F needed to push a cargo box weighing x pounds on a wood floor.

x (lb)	150	180	210	320
F (lb)	26	31	36	54

(a) Compute the ratio $\frac{F}{x}$ for each data pair in the table. Interpret these ratios.

(b) Approximate a constant of proportionality k satisfying $F = kx$. (k is the *coefficient of friction.*)

(c) Graph the data and the equation together.

(d) Estimate the force needed to push a 275-pound cargo box on the floor.

116. *Electrical Resistance*   The electrical resistance of a wire varies directly with its length. If a 255-foot wire has a resistance of 1.2 ohms, find the resistance of 135 feet of the same type of wire. Interpret the constant of proportionality in this situation.

Writing about Mathematics

117. Compare the slope-intercept form with the point-slope form. Give examples of each.

118. Give an example of two quantities in real life that vary directly. Explain your answer. Use an equation to describe the relationship between the two quantities.

119. The graph of $f(x) = 3 - 2x$ passes through $(-1, 5)$ and $(3, -3)$. Evaluate $f(1)$ and find the midpoint of the two points. Compare and explain your results.

120. Explain how you would find the equation of a line passing through two points. Give an example.

EXTENDED AND DISCOVERY EXERCISES

Exercises 1 and 2: Estimating Populations   Biologists sometimes use direct variation to estimate the number of fish in small lakes. They start by tagging a small number of fish and then releasing them. They assume that over a period of time, the tagged fish distribute themselves evenly throughout the lake. Later, they collect a second sample. The total number of fish and the number of tagged fish in the second sample are counted. To determine the total population of fish in the lake, biologists assume that the proportion of tagged fish in the second sample is equal to the proportion of tagged fish in the entire lake. This technique can also be used to count other types of animals, such as birds, when they are not migrating.

1. Eighty-five fish are tagged and released into a pond. A later sample of 94 fish from the pond contains 13 tagged fish. Estimate the number of fish in the pond.

2. Sixty-three blackbirds are tagged and released. Later it is estimated that out of a sample of 32 blackbirds, only 8 are tagged. Estimate the population of blackbirds in the area.

CHECKING BASIC CONCEPTS FOR SECTIONS 2.1 AND 2.2

1. Graph $f(x) = 4 - 2x$ by hand. Identify the slope, the x-intercept, and the y-intercept.

2. The death rate from heart disease for people ages 15 through 24 is 2.7 per 100,000 people.
   (a) Write a function f that models the number of deaths in a population of x million people 15 to 24 years old.

   (b) There are about 39 million people in the United States who are 15 to 24 years old. Estimate the number of deaths from heart disease in this age group.

3. A driver of a car is initially 50 miles south of home, driving 60 miles per hour south. Write a function f that models the distance between the driver and home.

4. Find an equation of the line passing through the points $(-3, 4)$ and $(5, -2)$. Give equations of lines that are parallel and perpendicular to this line.

5. Find equations of horizontal and vertical lines that pass through the point $(-4, 7)$.

6. Write the slope-intercept form of the line.

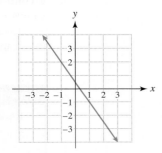

7. Find the x- and y-intercepts of the graph of the equation $-3x + 2y = -18$.

2.3 Linear Equations

- **Learn about equations and recognize a linear equation**
- **Solve linear equations symbolically**
- **Solve linear equations graphically and numerically**
- **Solve problems involving percentages**
- **Apply problem-solving strategies**

Introduction

In Example 4 of Section 2.2, we modeled iPod sales y in millions during year x with the equation of a line, or *linear function*, given by

$$f(x) = 21(x - 2004) + 4.4.$$

To predict the year when iPod sales might be **130** million, we could set the formula for $f(x)$ equal to **130** and solve the *linear equation*

$$130 = 21(x - 2004) + 4.4$$

for x. This section discusses linear equations and their solutions. See Example 4 in this section.

Equations

An **equation** is a statement that two mathematical expressions are equal. Equations always contain an equals sign. Some examples of equations include

$$x + 15 = 9x - 1, \quad x^2 - 2x + 1 = 2x, \quad z + 5 = 0,$$
$$xy + x^2 = y^3 + x, \quad \text{and} \quad 1 + 2 = 3.$$

The first three equations have one variable, the fourth equation has two variables, and the fifth equation contains only constants. For now, our discussion concentrates on equations with one variable.

To **solve** an equation means to find all values for the variable that make the equation a true statement. Such values are called **solutions**. The set of all solutions is the **solution set**. The solutions to the equation $x^2 - 1 = 0$ are 1 or -1, written as $x = \pm 1$. Either value for x **satisfies** the equation. The solution set is $\{-1, 1\}$. Two equations are **equivalent** if they have the same solution set. For example, the equations $x + 2 = 5$ and $x = 3$ are equivalent.

If an equation has no solutions, then its solution set is empty and the equation is called a **contradiction**. The equation $x + 2 = x$ has no solutions and is a contradiction. However, if every (meaningful) value for the variable is a solution, then the equation is an **identity**. The equation $x + x = 2x$ is an identity because every value for x makes the equation true. Any equation that is satisfied by some, but not all, values of the variable is a **conditional equation**. The equation $x^2 - 1 = 0$ is a conditional equation. Only the values -1 and 1 for x make this equation a true statement.

Like functions, equations can be either *linear* or *nonlinear*. A linear equation is one of the simplest types of equations.

Linear Equation in One Variable

A **linear equation** in one variable is an equation that can be written in the form

$$ax + b = 0,$$

where a and b are constants with $a \neq 0$.

If an equation is not linear, then we say that it is a **nonlinear equation**. The following are examples of linear equations. In each case, rules of algebra could be used to write the equation in the form $ax + b = 0$ with $a \neq 0$. A linear equation has *exactly one* solution: $-\frac{b}{a}$.

$$x - 12 = 0, \quad 2x - 4 = -x, \quad 2(1 - 4x) = 16, \quad x - 5 + 3(x - 1) = 0$$

Symbolic Solutions

Linear equations can be solved symbolically, and the solution is *always exact*. To solve a linear equation symbolically, we usually apply the *properties of equality* to the given equation and transform it into an equivalent equation that is simpler.

Properties of Equality

Addition Property of Equality

If a, b, and c are real numbers, then

$$a = b \quad \text{is equivalent to} \quad a + c = b + c.$$

Multiplication Property of Equality

If a, b, and c are real numbers with $c \neq 0$, then

$$a = b \quad \text{is equivalent to} \quad ac = bc.$$

Loosely speaking, the addition property states that "if equals are added to equals, the results are equal." For example, if $x + 5 = 15$, then we can add -5 to each side of the equation, or equivalently subtract 5 from each side, to obtain the following.

$$x + 5 = 15 \qquad \textit{Given equation}$$
$$x + 5 - 5 = 15 - 5 \qquad \textit{Subtract 5 from each side.}$$
$$x = 10 \qquad \textit{Simplify.}$$

Similarly, the multiplication property states that "if equals are multiplied by nonzero equals, the results are equal." For example, if $5x = 20$, then we can multiply each side by $\frac{1}{5}$, or equivalently divide each side by 5, to obtain the following.

$$5x = 20 \qquad \textit{Given equation}$$
$$\frac{5x}{5} = \frac{20}{5} \qquad \textit{Divide each side by 5.}$$
$$x = 4 \qquad \textit{Simplify.}$$

These two properties along with the distributive property are applied in the next two examples.

EXAMPLE 1   Solving a linear equation symbolically

Solve the equation $3(x - 4) = 2x - 1$. Check your answer.

SOLUTION

Getting Started First we apply the distributive property: $a(b - c) = ab - ac$. Thus

$$3(x - 4) = 3 \cdot x - 3 \cdot 4 = 3x - 12. \blacktriangleright$$

Solving the given equation results in the following.

$$3(x - 4) = 2x - 1 \qquad \textit{Given equation}$$
$$3x - 12 = 2x - 1 \qquad \textit{Distributive property}$$
$$3x - 2x - 12 + 12 = 2x - 2x - 1 + 12 \qquad \textit{Subtract 2x and add 12.}$$
$$3x - 2x = 12 - 1 \qquad \textit{Simplify.}$$
$$x = 11 \qquad \textit{Simplify.}$$

The solution is 11. We can check our answer as follows.

$$3(x - 4) = 2x - 1 \qquad \textit{Given equation}$$
$$3(11 - 4) \stackrel{?}{=} 2 \cdot 11 - 1 \qquad \textit{Let x = 11.}$$
$$21 = 21 \qquad \textit{The answer checks.}$$

Now Try Exercise 23 ◀

EXAMPLE 2   Solving a linear equation symbolically

Solve $3(2x - 5) = 10 - (x + 5)$. Check your answer.

SOLUTION

Getting Started In this problem subtraction must be distributed over the quantity $(x + 5)$. Thus

$$10 - (x + 5) = 10 - 1(x + 5) = 10 - x - 5. \blacktriangleright$$

Algebra Review
To review the distributive properties, see Chapter R (page R-15).

Solving the given equation results in the following.

$$3(2x - 5) = 10 - (x + 5)$$ *Given equation*

$$6x - 15 = 10 - x - 5$$ *Distributive property*

$$6x - 15 = 5 - x$$ *Simplify.*

$$7x - 15 = 5$$ *Add x to each side.*

$$7x = 20$$ *Add 15 to each side.*

$$x = \frac{20}{7}$$ *Divide each side by 7.*

The solution is $\frac{20}{7}$. To check this answer, let $x = \frac{20}{7}$ and simplify.

$$3(2x - 5) = 10 - (x + 5)$$ *Given equation*

$$3\left(2 \cdot \frac{20}{7} - 5\right) \stackrel{?}{=} 10 - \left(\frac{20}{7} + 5\right)$$ *Let x = $\frac{20}{7}$.*

$$\frac{15}{7} = \frac{15}{7}$$ *The answer checks.*

Now Try Exercise 27 ◀

Fractions and Decimals When fractions or decimals appear in an equation, we sometimes can make our work simpler by multiplying each side of the equation by the least common denominator (LCD) or a common denominator of all fractions in the equation. This method is illustrated in the next example.

EXAMPLE 3   Eliminating fractions and decimals

Solve each linear equation.

(a) $\dfrac{t - 2}{4} - \dfrac{1}{3}t = 5 - \dfrac{1}{12}(3 - t)$ **(b)** $0.03(z - 3) - 0.5(2z + 1) = 0.23$

Algebra Review
To review least common multiples and least common denominators, see Chapter R (page R-32).

SOLUTION

(a) To eliminate fractions, multiply each side (or term in the equation) by the LCD, 12.

$$\frac{t - 2}{4} - \frac{1}{3}t = 5 - \frac{1}{12}(3 - t)$$ *Given equation*

$$\frac{12(t - 2)}{4} - \frac{12}{3}t = 12(5) - \frac{12}{12}(3 - t)$$ *Multiply each side (term) by 12.*

$$3(t - 2) - 4t = 60 - (3 - t)$$ *Simplify.*

$$3t - 6 - 4t = 60 - 3 + t$$ *Distributive property*

$$-t - 6 = 57 + t$$ *Combine like terms on each side.*

$$-2t = 63$$ *Add −t and 6 to each side.*

$$t = -\frac{63}{2}$$ *Divide each side by −2.*

The solution is $-\frac{63}{2}$.

(b) To eliminate decimals, multiply each side (or term in the equation) by 100.

$$0.03(z - 3) - 0.5(2z + 1) = 0.23 \qquad \text{Given equation}$$

$$3(z - 3) - 50(2z + 1) = 23 \qquad \text{Multiply each side (term) by 100.}$$

$$3z - 9 - 100z - 50 = 23 \qquad \text{Distributive property}$$

$$-97z - 59 = 23 \qquad \text{Combine like terms.}$$

$$-97z = 82 \qquad \text{Add 59 to each side.}$$

$$z = -\frac{82}{97} \qquad \text{Divide each side by } -97.$$

The solution is $-\frac{82}{97}$.

Now Try Exercises 33 and 37 ◄

An Application   In the next example we solve the equation presented in the introduction to this section.

EXAMPLE 4   Estimating iPod sales

The linear function defined by $f(x) = 21(x - 2004) + 4.4$ estimates iPod sales (in millions of units) during fiscal year x. Use $f(x)$ to estimate when iPod sales could reach 130 million.

SOLUTION
We need to find x so that $f(x) = 130$.

$$21(x - 2004) + 4.4 = 130 \qquad \text{Equation to be solved}$$

$$21(x - 2004) = 125.6 \qquad \text{Subtract 4.4 from each side.}$$

$$21x - 42,084 = 125.6 \qquad \text{Distributive property}$$

$$21x = 42,209.6 \qquad \text{Add 42,084 to each side.}$$

$$x = \frac{42,209.6}{21} \qquad \text{Divide each side by 21.}$$

$$x \approx 2010 \qquad \text{Approximate.}$$

This model predicts that iPod sales could reach **130** million units in fiscal **2010**.

Now Try Exercise 91 ◄

Contradictions, Identities, and Conditional Equations   The next example illustrates how an equation can have no solutions (contradiction), one solution (conditional equation), or infinitely many solutions (identity).

EXAMPLE 5   Identifying contradictions, identities, and conditional equations

Identify each equation as a contradiction, identity, or conditional equation.
(a) $7 + 6x = 2(3x + 1)$ **(b)** $2x - 5 = 3 - (1 + 2x)$
(c) $2(5 - x) - 25 = 3(x - 5) - 5x$

SOLUTION
(a)

$$7 + 6x = 2(3x + 1) \qquad \text{Given equation}$$

$$7 + 6x = 6x + 2 \qquad \text{Distributive property}$$

$$7 = 2 \qquad \text{Subtract 6x from each side.}$$

The statement $7 = 2$ is false and there are *no* solutions. The equation is a contradiction.

(b)

$2x - 5 = 3 - (1 + 2x)$	*Given equation*
$2x - 5 = 3 - 1 - 2x$	*Distributive property*
$4x = 7$	*Add 2x and 5 to each side.*
$x = \dfrac{7}{4}$	*Divide each side by 4.*

There is one solution: $\frac{7}{4}$. This is a conditional equation.

(c)

$2(5 - x) - 25 = 3(x - 5) - 5x$	*Given equation*
$10 - 2x - 25 = 3x - 15 - 5x$	*Distributive property*
$-2x - 15 = -2x - 15$	*Simplify each side.*
$0 = 0$	*Add 2x and 15 to each side.*

The statement $0 = 0$ is true and the solution set includes *all real numbers*. The equation is an identity.   ***Now Try Exercises 39, 41, and 43*** ◀

Graphical and Numerical Solutions

Figure 2.32

Graphical Solutions   The equation $f(x) = g(x)$ results whenever the formulas for two functions f and g are set equal to each other. A solution to this equation corresponds to the x-coordinate of a point where the graphs of f and g intersect. This technique is called the *intersection-of-graphs method*. If the graphs of f and g are lines with different slopes, then their graphs intersect once. For example, if $f(x) = 2x + 1$ and $g(x) = -x + 4$, then the equation $f(x) = g(x)$ becomes $2x + 1 = -x + 4$. To apply the intersection-of-graphs method, we graph $y_1 = 2x + 1$ and $y_2 = -x + 4$, as shown in Figure 2.32.

Their graphs intersect at the point $(1, 3)$. Because the variable in the given equation $2x + 1 = -x + 4$ is x, the solution is 1, the x-coordinate of the point of intersection. When $x = 1$, the functions f and g both assume the value 3—that is, $f(1) = 2(1) + 1 = 3$ and $g(1) = -1 + 4 = 3$. The value 3 is the y-coordinate of the point of intersection $(1, 3)$.

The intersection-of-graphs method is summarized below.

Intersection-of-Graphs Method

The **intersection-of-graphs method** can be used to solve an equation graphically.

STEP 1:   Set y_1 equal to the left side of the equation, and set y_2 equal to the right side of the equation.

STEP 2:   Graph y_1 and y_2.

STEP 3:   Locate any points of intersection. The x-coordinates of these points correspond to solutions to the equation.

EXAMPLE 6   Solving an equation graphically and symbolically

Solve $2x - 1 = \frac{1}{2}x + 2$ graphically and symbolically.

SOLUTION

Graphical Solution   Graph $y_1 = 2x - 1$ and $y_2 = \frac{1}{2}x + 2$. Their graphs intersect at the point $(2, 3)$, as shown in Figure 2.33, so the solution is 2. Figure 2.34 shows these graphs as created by a graphing calculator.

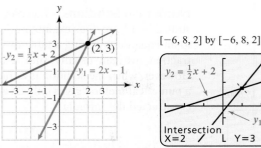

Figure 2.33 Intersection of Graphs

Figure 2.34

Symbolic Solution

$2x - 1 = \dfrac{1}{2}x + 2$		*Given equation*
$2x = \dfrac{1}{2}x + 3$		*Add 1 to each side.*
$\dfrac{3}{2}x = 3$		*Subtract $\frac{1}{2}x$ from each side.*
$\dfrac{2}{3} \cdot \dfrac{3}{2}x = \dfrac{2}{3} \cdot 3$		*Multiply each side by $\frac{2}{3}$.*
$x = 2$		*Multiply fractions.*

The solution is 2 and agrees with the graphical solution.   **Now Try Exercise 53** ◀

EXAMPLE 7   **Applying the intersection-of-graphs method**

During the 1990s, compact discs were a new technology that replaced cassette tapes. The percentage share of music sales (in dollars) held by compact discs from 1987 to 1998 could be modeled by $f(x) = 5.91x + 13.7$. During the same time period the percentage share of music sales held by cassette tapes could be modeled by $g(x) = -4.71x + 64.7$. In these formulas $x = 0$ corresponds to 1987, $x = 1$ to 1988, and so on. Use the intersection-of-graphs method to estimate the year when the percentage share of CDs equaled the percentage share of cassettes.   (**Source:** Recording Industry Association of America.)

SOLUTION   We must solve the linear equation $f(x) = g(x)$, or equivalently,

$$5.91x + 13.7 = -4.71x + 64.7.$$

Graph $Y_1 = 5.91X + 13.7$ and $Y_2 = -4.71X + 64.7$, as in Figure 2.35. In Figure 2.36 their graphs intersect near the point $(4.8, 42.1)$. Since $x = 0$ corresponds to 1987 and $1987 + 4.8 \approx 1992$, it follows that in 1992 sales of CDs and cassette tapes were approximately equal. Each had about 42.1% of the sales in 1992.

Calculator Help

To find the point of intersection in Figure 2.36, see Appendix A (page AP-8).

Figure 2.35

Figure 2.36

Now Try Exercise 93 ◀

Numerical Solutions Equations can be solved symbolically and graphically, as demonstrated in the previous examples. Sometimes it is also possible to find a numerical solution to an equation by using a table of values. When the solution is an integer or a convenient fraction, we can usually find it by using the table feature on a graphing calculator. However, when the solution is a fraction with a repeating decimal or an irrational number, a numerical method gives only an approximate solution. In Example 10 of Section 2.2, we constructed the formula

$$f(x) = 630.8(x - 1980) + 3617$$

to model tuition and fees at private colleges and universities during year x. To determine when tuition and fees were about \$15,600, we could solve the equation

$$630.8(x - 1980) + 3617 = 15{,}600$$

for x by making a table of values for $f(x)$, as shown in Figure 2.37. By scrolling through the x-values, we can see that tuition and fees at private schools were about \$15,600 in 1999. This equation could also be solved symbolically and graphically.

NOTE Regardless of whether we use a symbolic, graphical, or numerical method to solve an equation, we should find the same solution set. However, our answers may differ slightly because of rounding.

$f(1999) \approx 15{,}600$

X	Y1
1996	13710
1997	14341
1998	14971
1999	**15602**
2000	16233
2001	16864
2002	17495

X = 1999

Figure 2.37

Calculator Help

To make a table of values, see Appendix A (page AP-6).

EXAMPLE 8 Solving an equation numerically

Solve $\sqrt{3}(2x - \pi) + \frac{1}{3}x = 0$ numerically to the nearest tenth.

SOLUTION Enter $Y_1 = \sqrt{(3)}(2X - \pi) + X/3$ and make a table for y_1, incrementing by 1, as shown in Figure 2.38. This table shows that when $x = 1, y_1 < 0$, and when $x = 2, y_1 > 0$. Thus the solution is located in the interval $1 < x < 2$. To obtain a more accurate answer, make a table for y_1 starting at 1 and incrementing by 0.1, as shown in Figure 2.39. Now we see that the solution lies in the interval $1.4 < x < 1.5$. In Figure 2.40, which starts at 1.4 and increments by 0.01, we see that the solution lies in the interval $1.43 < x < 1.44$. To the nearest tenth, the solution is 1.4. *Now Try Exercise 71* ◀

X	Y1
−3	−16.83
−2	−13.04
−1	−9.239
0	−5.441
1	−1.644
2	**2.1535**
3	5.9509

X = 2

$y_1 < 0$
$y_1 > 0$

Figure 2.38

X	Y1
1	−1.644
1.1	−1.264
1.2	−.8845
1.3	−.5047
1.4	−.125
1.5	**.25475**
1.6	.6345

X = 1.5

$y_1 < 0$
$y_1 > 0$

Figure 2.39

X	Y1
1.4	−.125
1.41	−.087
1.42	−.049
1.43	−.0111
1.44	**.02691**
1.45	.06488
1.46	.10286

X = 1.44

$y_1 < 0$
$y_1 > 0$

Figure 2.40

MAKING CONNECTIONS

Symbolic, Graphical, and Numerical Solutions Linear equations can be solved symbolically, graphically, and numerically. Symbolic solutions to linear equations are *always exact*, whereas graphical and numerical solutions are *sometimes approximate*. The following example illustrates how to solve the equation $2x - 1 = 3$ with each method.

Symbolic Solution

$$2x - 1 = 3$$
$$2x = 4$$
$$x = 2$$

Check:

$$2(2) - 1 \overset{?}{=} 3$$
$$3 = 3$$

It checks.

Graphical Solution

The solution is **2**.

Numerical Solution

x	0	1	2	3
$2x - 1$	−1	1	3	5

Because $2x - 1$ equals 3 when $x = 2$, the solution to $2x - 1 = 3$ is **2**.

Percentages Applications involving percentages often result in linear equations because percentages can be computed by linear functions. A function for taking P percent of x is given by $f(x) = \frac{P}{100}x$, where $\frac{P}{100}$ is the decimal form for P percent. For example, to calculate 35% of x, let $f(x) = 0.35x$. Then 35% of $150 is $f(150) = 0.35(150) = 52.5$, or $52.50.

EXAMPLE 9 Solving an application involving percentages

A survey found that 76% of bicycle riders do not wear helmets. (**Source:** Opinion Research Corporation for Glaxo Wellcome, Inc.)
(a) Find a formula $f(x)$ for a function that computes the number of people who do not wear helmets among x bicycle riders.
(b) There are approximately 38.7 million riders of all ages who do not wear helmets. Find the total number of bicycle riders.

SOLUTION
(a) A function f that computes 76% of x is given by $f(x) = 0.76x$.
(b) We must find the x-value for which $f(x) = 38.7$ million, or solve the equation $0.76x = 38.7$. Solving gives $x = \frac{38.7}{0.76} \approx 50.9$ million bike riders.

Now Try Exercise 99 ◄

Solving for a Variable

The circumference C of a circle is given by $C = 2\pi r$, where r is the radius. This equation is *solved* for C. That is, given r, we can easily calculate C. For example, if $r = 4$, then $C = 2\pi(4)$, or $C = 8\pi$. However, if we are given C, then it is more work to calculate r. Solving the equation for r makes it simpler to calculate r.

$$C = 2\pi r \qquad \text{Given equation}$$

$$\frac{C}{2\pi} = r \qquad \text{Divide each side by } 2\pi.$$

The equation $r = \frac{C}{2\pi}$ is solved for r.

EXAMPLE 10 Solving for a variable

The area of a trapezoid with bases a and b and height h is given by $A = \frac{1}{2}h(a + b)$. Solve this equation for b.

SOLUTION
Getting Started If we multiply each side by 2 and divide each side by h, the right side of the equation becomes $a + b$. Subtracting a from each side isolates b. ▶

$$A = \frac{1}{2}h(a + b) \qquad \text{Given equation}$$

$$2A = h(a + b) \qquad \text{Multiply each side by 2.}$$

$$\frac{2A}{h} = a + b \qquad \text{Divide each side by } h.$$

$$\frac{2A}{h} - a = b \qquad \text{Subtract } a \text{ from each side.}$$

The equation $b = \frac{2A}{h} - a$ is solved for b.

Now Try Exercise 85 ◄

Problem-Solving Strategies

To become more proficient at solving problems, we need to establish a procedure to guide our thinking. The following steps may be helpful in solving application problems.

Solving Application Problems

STEP 1: Read the problem and make sure you understand it. Assign a variable to what you are being asked to find. If necessary, write other quantities in terms of this variable.

STEP 2: Write an equation that relates the quantities described in the problem. You may need to sketch a diagram and refer to known formulas.

STEP 3: Solve the equation and determine the solution.

STEP 4: Look back and check your solution. Does it seem reasonable?

These steps are applied in the next four examples.

EXAMPLE 11 Working together

A large pump can empty a tank of gasoline in 5 hours, and a smaller pump can empty the same tank in 9 hours. If both pumps are used to empty the tank, how long will it take?

SOLUTION

STEP 1: We are asked to find the time it takes for *both* pumps to empty the tank. Let this time be t.

$$t: \text{Time to empty the tank}$$

STEP 2: In 1 hour the large pump will empty $\frac{1}{5}$ of the tank and the smaller pump will empty $\frac{1}{9}$ of the tank. The fraction of the tank that they will empty together in 1 hour is given by $\frac{1}{5} + \frac{1}{9}$. In 2 hours the large pump will empty $\frac{2}{5}$ of the tank and the smaller pump will empty $\frac{2}{9}$ of the tank. The fraction of the tank that they will empty together in 2 hours is $\frac{2}{5} + \frac{2}{9}$. Similarly, in t hours the fraction of the tank that the two pumps can empty is $\frac{t}{5} + \frac{t}{9}$. Since the tank is empty when this fraction reaches 1, we must solve the following equation.

$$\frac{t}{5} + \frac{t}{9} = 1$$

STEP 3: Multiply by the LCD, 45, to eliminate fractions.

$$\frac{45t}{5} + \frac{45t}{9} = 45 \qquad \text{Multiply by LCD.}$$

$$9t + 5t = 45 \qquad \text{Simplify.}$$

$$14t = 45 \qquad \text{Add like terms.}$$

$$t = \frac{45}{14} \approx 3.21 \qquad \text{Divide by 14 and approximate.}$$

Working together, the two pumps can empty the tank in about 3.21 hours.

STEP 4: This sounds reasonable. Working together the two pumps should be able to empty the tank faster than the large pump working alone, but not twice as fast. Note that $\frac{3.21}{5} + \frac{3.21}{9} \approx 1$.

Now Try Exercise 101 ◄

EXAMPLE 12   Solving an application involving motion

In 1 hour an athlete traveled 10.1 miles by running first at 8 miles per hour and then at 11 miles per hour. How long did the athlete run at each speed?

SOLUTION

STEP 1: We are asked to find the time spent running at each speed. If we let x represent the time in hours spent running at 8 miles per hour, then $1 - x$ represents the time spent running at 11 miles per hour because the total running time was 1 hour.

x: Time spent running at 8 miles per hour

$1 - x$: Time spent running at 11 miles per hour

STEP 2: Distance d equals rate r times time t: that is, $d = rt$. In this example we have two rates (or speeds) and two times. The total distance must sum to 10.1 miles.

$$d = r_1 t_1 + r_2 t_2 \qquad \text{\textit{General equation}}$$
$$10.1 = 8x + 11(1 - x) \qquad \text{\textit{Substitute.}}$$

STEP 3: We can solve this equation symbolically.

$$10.1 = 8x + 11 - 11x \qquad \text{\textit{Distributive property}}$$
$$10.1 = 11 - 3x \qquad \text{\textit{Combine like terms.}}$$
$$3x = 0.9 \qquad \text{\textit{Add 3x; subtract 10.1.}}$$
$$x = 0.3 \qquad \text{\textit{Divide by 3.}}$$

The athlete runs 0.3 hour (18 minutes) at 8 miles per hour and 0.7 hour (42 minutes) at 11 miles per hour.

STEP 4: We can check this solution as follows.

$$8(0.3) + 11(0.7) = 10.1 \qquad \text{\textit{It checks.}}$$

This sounds reasonable. The runner's average speed was 10.1 miles per hour so the runner must have run longer at 11 miles per hour than at 8 miles per hour.

Now Try Exercise 103 ◄

Similar triangles are often used in applications involving geometry. Similar triangles are used to solve the next application.

EXAMPLE 13   Solving an application involving similar triangles

A person 6 feet tall stands 17 feet from the base of a streetlight, as illustrated in Figure 2.41. If the person's shadow is 8 feet, estimate the height of the streetlight.

SOLUTION

STEP 1: We are asked to find the height of the streetlight in Figure 2.41. Let x represent this height.

x: Height of the streetlight

6 ft

├─── 17 ft ───┤├─ 8 ft ─┤

Figure 2.41

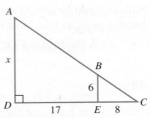

Figure 2.42

Geometry Review
To review similar triangles, see Chapter R (page R-5).

STEP 2: In Figure 2.42, triangle *ACD* is similar to triangle *BCE*. Thus ratios of corresponding sides are equal.

$$\frac{AD}{BE} = \frac{DC}{EC}$$

$$\frac{x}{6} = \frac{17 + 8}{8}$$

STEP 3: We can solve this equation symbolically.

$$\frac{x}{6} = \frac{25}{8} \qquad \textit{Simplify.}$$

$$x = \frac{6 \cdot 25}{8} \qquad \textit{Multiply by 6.}$$

$$x = 18.75 \qquad \textit{Simplify.}$$

The height of the streetlight is 18.75 feet.

STEP 4: One way to check this answer is to form a different proportion. Note that *x* is to 17 + 8 in triangle *ADC* as 6 is to 8 in triangle *BEC*. If $x = 18.75$, then $\frac{18.75}{25} = \frac{6}{8}$, which is true, and our answer checks.

Now Try Exercise 107 ◀

EXAMPLE 14 **Mixing acid in chemistry**

Pure water is being added to 153 milliliters of a 30% solution of hydrochloric acid. How much water should be added to dilute the solution to a 13% mixture?

SOLUTION

STEP 1: We are asked to find the amount of water that should be added to 153 milliliters of 30% acid to make a 13% solution. Let this amount of water be equal to *x*. See Figure 2.43.

$$x: \text{Amount of pure water to be added}$$

$$x + 153: \text{Final volume of 13\% solution}$$

Figure 2.43

STEP 2: Since only water is added, the total amount of acid in the solution after the water is added must equal the amount of acid before the water is added. The volume of pure acid after the water is added equals 13% of $x + 153$ milliliters, and the volume of pure acid before the water is added equals 30% of 153 milliliters. We must solve the following equation.

$$0.13(x + 153) = 0.30(153)$$

STEP 3: Begin by dividing each side by 0.13.

$$x + 153 = \frac{0.30(153)}{0.13} \qquad \textit{Divide by 0.13.}$$

$$x = \frac{0.30(153)}{0.13} - 153 \qquad \textit{Subtract 153.}$$

$$x \approx 200.08 \qquad \textit{Approximate.}$$

We should add about 200 milliliters of pure water.

STEP 4: Initially the solution contains $0.30(153) = 45.9$ milliliters of pure acid. If we add 200 milliliters of water to the 153 milliliters, the final solution is 353 milliliters, which includes 45.9 milliliters of pure acid. Its concentration is $\frac{45.9}{353} \approx 0.13$, or about 13%.

Now Try Exercise 111 ◄

2.3 Putting It All Together

A general four-step procedure for solving applications is found on page 122. The following table summarizes some of the important concepts in this section.

Concept	Explanation	Examples
Linear equation	A linear equation can be written as $ax + b = 0$, $a \neq 0$.	$4x + 5 = 0$ $3x - 1 = x + 2$
Addition property	$a = b$ is equivalent to $a + c = b + c$.	$x - 7 = 25$ $x - 7 + 7 = 25 + 7$ $x = 32$
Multiplication property	$a = b$ is equivalent to $ac = bc$, $c \neq 0$.	$\frac{1}{2}x = 4$ $2 \cdot \frac{1}{2}x = 4 \cdot 2$ $x = 8$
Distributive property	$a(b + c) = ab + ac$ $a(b - c) = ab - ac$	$2(5 + x) = 10 + 2x$ $-(2 - x) = -1(2 - x) = -2 + x$
Identity	An equation that is true for all (meaningful) values of the variable	$3(x - 2) = 3x - 6$ $2x + 3x = (2 + 3)x = 5x$
Contradiction	An equation that has no solutions	$x + 5 = x$ $2x - 2x = 5$
Conditional equation	An equation that is satisfied by some, but not all, of the values of the variable	$2x - 1 = 5$ Given equation $2x = 6$ Add 1. $x = 3$ Divide by 2.
Percentages	P percent of x equals $\frac{P}{100}x$, where $\frac{P}{100}$ is the decimal form for P percent.	35% of x is calculated by $f(x) = \frac{35}{100}x$, or $f(x) = 0.35x$.

continued on next page

continued from previous page

The following example illustrates how to solve $5x - 1 = 3$ symbolically, graphically, and numerically.

Symbolic Solution	*Graphical Solution*	*Numerical Solution*

Symbolic Solution

$$5x - 1 = 3$$
$$5x = 4$$
$$x = \frac{4}{5}$$

Check:

$$5\left(\frac{4}{5}\right) - 1 \overset{?}{=} 3$$
$$4 - 1 \overset{?}{=} 3$$
$$3 = 3$$

Graphical Solution

$[-9, 9, 1]$ by $[-6, 6, 1]$

The graphs of $y_1 = 5x - 1$ and $y_2 = 3$ intersect at $(0.8, 3)$, so the solution is **0.8**.

Numerical Solution

In the table, $y_1 = 3$ when $x = 0.8$.

2.3 Exercises

Concepts about Linear Equations

1. How many solutions are there to $ax + b = 0$ with $a \neq 0$?

2. How many times does the graph of $y = ax + b$ with $a \neq 0$ intersect the x-axis?

3. Apply the distributive property to $4 - (5 - 4x)$.

4. What property is used to solve $15x = 5$?

5. If $f(x) = ax + b$ with $a \neq 0$, how are the zero of f and the x-intercept of the graph of f related?

6. Distinguish between a contradiction and an identity.

Identifying Linear and Nonlinear Equations

Exercises 7–12: Determine whether the equation is linear or nonlinear by trying to write it in the form $ax + b = 0$.

7. $3x - 1.5 = 7$

8. $100 - 23x = 20x$

9. $2\sqrt{x} + 2 = 1$

10. $4x^3 - 7 = 0$

11. $7x - 5 = 3(x - 8)$

12. $2(x - 3) = 4 - 5x$

Solving Linear Equations Symbolically

Exercises 13–38: Solve the equation and check your answer.

13. $2x - 8 = 0$

14. $4x - 8 = 0$

15. $-5x + 3 = 23$

16. $-9x - 3 = 24$

17. $4(z - 8) = z$

18. $-3(2z - 1) = 2z$

19. $-5(3 - 4t) = 65$

20. $6(5 - 3t) = 66$

21. $k + 8 = 5k - 4$

22. $2k - 3 = k + 3$

23. $2(1 - 3x) + 1 = 3x$

24. $5(x - 2) = -2(1 - x)$

25. $-5(3 - 2x) - (1 - x) = 4(x - 3)$

26. $-3(5 - x) - (x - 2) = 7x - 2$

27. $-4(5x - 1) = 8 - (x + 2)$

28. $6(3 - 2x) = 1 - (2x - 1)$

29. $\frac{2}{7}n + \frac{1}{5} = \frac{4}{7}$

30. $\frac{6}{11} - \frac{2}{33}n = \frac{5}{11}n$

31. $\frac{1}{2}(d - 3) - \frac{2}{3}(2d - 5) = \frac{5}{12}$

32. $\frac{7}{3}(2d - 1) - \frac{2}{5}(4 - 3d) = \frac{1}{5}d$

33. $\frac{x - 5}{3} + \frac{3 - 2x}{2} = \frac{5}{4}$

34. $\frac{3x - 1}{5} - 2 = \frac{2 - x}{3}$

35. $0.1z - 0.05 = -0.07z$ 36. $1.1z - 2.5 = 0.3(z - 2)$

37. $0.15t + 0.85(100 - t) = 0.45(100)$

38. $0.35t + 0.65(10 - t) = 0.55(10)$

Exercises 39–48: Complete the following.

  (a) *Solve the equation symbolically.*
  (b) *Classify the equation as a contradiction, an identity, or a conditional equation.*

39. $5x - 1 = 5x + 4$

40. $7 - 9z = 2(3 - 4z) - z$

41. $3(x - 1) = 5$ 42. $22 = -2(2x + 1.4)$

43. $0.5(x - 2) + 5 = 0.5x + 4$

44. $\frac{1}{2}x - 2(x - 1) = -\frac{3}{2}x + 2$

45. $\frac{t + 1}{2} = \frac{3t - 2}{6}$ 46. $\frac{2x + 1}{3} = \frac{2x - 1}{3}$

47. $\frac{1 - 2x}{4} = \frac{3x - 1.5}{-6}$

48. $0.5(3x - 1) + 0.5x = 2x - 0.5$

Solving Linear Equations Graphically

Exercises 49 and 50: A linear equation is solved by using the intersection-of-graphs method. Find the solution by interpreting the graph. Assume that the solution is an integer.

49.

50.

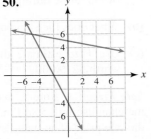

Exercises 51 and 52: Use the graph of $y = f(x)$ to solve each equation.

  (a) $f(x) = -1$ (b) $f(x) = 0$ (c) $f(x) = 2$

51.

52.

Exercises 53–58: Use the intersection-of-graphs method to solve the equation. Then solve symbolically.

53. $x + 4 = 1 - 2x$ 54. $2x = 3x - 1$

55. $-x + 4 = 3x$ 56. $1 - 2x = x + 4$

57. $2(x - 1) - 2 = x$ 58. $-(x + 1) - 2 = 2x$

Exercises 59–66: Solve the linear equation with the intersection-of-graphs method. Approximate the solution to the nearest thousandth whenever appropriate.

59. $5x - 1.5 = 5$ 60. $8 - 2x = 1.6$

61. $3x - 1.7 = 1 - x$ 62. $\sqrt{2}x = 4x - 6$

63. $3.1(x - 5) = \frac{1}{5}x - 5$ 64. $65 = 8(x - 6) - 5.5$

65. $\frac{6 - x}{7} = \frac{2x - 3}{3}$

66. $\pi(x - \sqrt{2}) = 1.07x - 6.1$

Solving Linear Equations Numerically

Exercises 67–74: Use tables to solve the equation numerically to the nearest tenth.

67. $2x - 7 = -1$ 68. $1 - 6x = 7$

69. $2x - 7.2 = 10$ 70. $5.8x - 8.7 = 0$

71. $\sqrt{2}(4x - 1) + \pi x = 0$

72. $\pi(0.3x - 2) + \sqrt{2}x = 0$

73. $0.5 - 0.1(\sqrt{2} - 3x) = 0$

74. $\sqrt{5} - \pi(\pi + 0.3x) = 0$

Solving Linear Equations by More Than One Method

Exercises 75–82: Solve the equation (to the nearest tenth)

   (a) *symbolically,*
   (b) *graphically, and*
   (c) *numerically.*

75. $5 - (x + 1) = 3$ **76.** $7 - (3 - 2x) = 1$

77. $\sqrt{3}(2 - \pi x) + x = 0$ **78.** $3(\pi - x) + \sqrt{2} = 0$

79. $x - 3 = 2x + 1$ **80.** $3(x - 1) = 2x - 1$

81. $6x - 8 = -7x + 18$

82. $5 - 8x = 3(x - 7) + 37$

Solving for a Variable

Exercises 83–90: Solve the equation for the specified variable.

83. $A = LW$ for W

84. $E = IR + 2$ for R

85. $P = 2L + 2W$ for L

86. $V = 2\pi rh + \pi r^2$ for h

87. $3x + 2y = 8$ for y

88. $5x - 4y = 20$ for y

89. $y = 3(x - 2) + x$ for x

90. $y = 4 - (8 - 2x)$ for x

Applications

91. *Income* The per capita (per person) income from 1980 to 2006 can be modeled by

$$f(x) = 1000(x - 1980) + 10{,}000,$$

where x is the year. Determine the year when the per capita income was $19,000. (**Source:** Bureau of the Census.)

92. *Tuition and Fees* In Example 10 of Section 2.2, we modeled tuition and fees in dollars during year x by

$$f(x) = 630.8(x - 1980) + 3617.$$

Use $f(x)$ to determine when tuition and fees reached $13,700.

93. *Vinyl and CD Sales* During the 1980s, sales of compact discs surpassed vinyl record sales. From 1985 to 1990, sales of compact discs in millions can be modeled by the formula $f(x) = 51.6(x - 1985) + 9.1$, whereas sales of vinyl LP records in millions can be modeled by $g(x) = -31.9(x - 1985) + 167.7$. Approximate the year x when sales of LP records and compact discs were equal by using the intersection-of-graphs method. (**Source:** Recording Industry Association of America.)

94. *Median Age* The median age A in the United States during year x, where $2000 \le x \le 2050$, is projected to be

$$A(x) = 0.07(x - 2000) + 35.3.$$

Use $A(x)$ to estimate when the median age may reach 37 years. (**Source:** Bureau of the Census.)

95. *Population Density* In 1980 the population density of the United States was 64 people per square mile, and in 2000 it was 80 people per square mile. Use a linear function to estimate when the U.S. population density reached 87 people per square mile.

96. *Value of a Home* In 1999 the value of a house was $180,000, and in 2009 it was $245,000.
  (a) Find a linear function V that approximates the value of the house during year x.

  (b) What does the slope of the graph of V represent?

  (c) Use V to estimate the year when the house was worth $219,000.

97. *Sale Price* A store is discounting all regularly priced merchandise by 25%. Find a function f that computes the sale price of an item having a regular price of x. If an item normally costs $56.24, what is its sale price?

98. *Sale Price* Continuing Exercise 97, use f to find the regular price of an item that costs $19.62 on sale.

99. *Skin Cancer* Approximately 4.5% of all cancer cases diagnosed in 2007 were skin cancer. (**Source:** American Cancer Society.)

    (a) If x cases of cancer were diagnosed, how many of these were skin cancer?

    (b) There were 65,000 cases of skin cancer diagnosed in 2007. Find the total number of cancer cases in 2007.

100. *Grades* In order to receive an A in a college course it is necessary to obtain an average of 90% correct on three 1-hour exams of 100 points each and on one final exam of 200 points. If a student scores 82, 88, and 91 on the 1-hour exams, what is the minimum score that the person can receive on the final exam and still earn an A?

101. *Working Together* Suppose that a lawn can be raked by one gardener in 3 hours and by a second gardener in 5 hours.

    (a) Mentally estimate how long it will take the two gardeners to rake the lawn working together.

    (b) Solve part (a) symbolically.

102. *Pumping Water* Suppose that a large pump can empty a swimming pool in 50 hours and a small pump can empty the pool in 80 hours. How long will it take to empty the pool if both pumps are used?

103. *Motion* A car went 372 miles in 6 hours, traveling part of the time at 55 miles per hour and part of the time at 70 miles per hour. How long did the car travel at each speed?

104. *Mixing Candy* Two types of candy sell for $2.50 per pound and $4.00 per pound. A store clerk is trying to make a 5-pound mixture worth $17.60. How much of each type of candy should be included in the mixture?

105. *Running* At 2:00 P.M. a runner heads north on a highway, jogging at 10 miles per hour. At 2:30 P.M. a driver heads north on the same highway to pick up the runner. If the car travels at 55 miles per hour, how long will it take the driver to catch the runner?

106. *Investments* A total of $5000 was invested in two accounts. One pays 5% annual interest, and the second pays 7% annual interest. If the first-year interest is $325, how much was invested in each account?

107. *Shadow Length* A person 66 inches tall is standing 15 feet from a streetlight. If the person casts a shadow 84 inches long, how tall is the streetlight?

108. *Height of a Tree* In the accompanying figure, a person 5 feet tall casts a shadow 4 feet long. A nearby tree casts a shadow 33 feet long. Find the height of the tree by solving a linear equation.

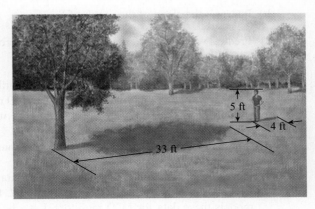

109. *Conical Water Tank* A water tank in the shape of an inverted cone has a height of 11 feet and a radius of 3.5 feet, as illustrated in the figure. If the volume of the cone is $V = \frac{1}{3}\pi r^2 h$, find the volume of the water in the tank when the water is 7 feet deep. (*Hint:* Consider using similar triangles.)

110. *Dimension of a Cone* (Refer to Exercise 109.) A conical water tank holds 100 cubic feet of water and has a diameter of 6 feet. Estimate its height to the nearest tenth of a foot.

111. *Chemistry* Determine how much pure water should be mixed with 5 liters of a 40% solution of sulfuric acid to make a 15% solution of sulfuric acid.

112. *Mixing Antifreeze* A radiator holds 5 gallons of fluid. If it is full with a 15% solution, how much fluid should be drained and replaced with a 65% antifreeze mixture to result in a 40% antifreeze mixture?

113. *Window Dimensions* A rectangular window has a length that is 18 inches more than its width. If its perimeter is 180 inches, find its dimensions.

114. *Online Holiday Shopping* In 2003 online holiday sales were $17 billion, and in 2006 they were $26 billion. (**Source:** Digital Lifestyles.)

(a) Find a linear function S that models these data.

(b) Interpret the slope of the graph of S.

(c) Predict when online holiday sales might reach $41 billion.

115. *Sales of CRT and LCD Screens* In 2002, 75 million CRT (cathode ray tube) monitors were sold and 29 million flat LCD (liquid crystal display) monitors were sold. In 2006 the numbers were 45 million for CRT monitors and 88 million for LCD monitors. (**Source:** International Data Corporation.)

(a) Find a linear function C that models these data for CRT monitors and another linear function L that models these data for LCD monitors. Let x be the year.

(b) Interpret the slopes of the graphs of C and of L.

▨ (c) Determine graphically the year when sales of these two types of monitors were equal.

(d) Solve part (c) symbolically.

▨ (e) Solve part (c) numerically.

116. *Geometry* A 174-foot-long fence is being placed around the perimeter of a rectangular swimming pool that has a 3-foot-wide sidewalk around it. The actual swimming pool without the sidewalk is twice as long as it is wide. Find the dimensions of the pool without the sidewalk.

117. *Temperature Scales* The Celsius and Fahrenheit scales are related by the equation $C = \frac{5}{9}(F - 32)$. These scales have the same temperature reading at a unique value where $F = C$. Find this temperature.

118. *Business* A company manufactures compact discs with recorded music. The master disc costs $2000 to produce and copies cost $0.45 each. If a company spent $2990 producing compact discs, how many copies did the company manufacture?

119. *Two-Cycle Engines* Two-cycle engines, used in snowmobiles, chain saws, and outboard motors, require a mixture of gasoline and oil. For certain engines the amount of oil in pints that should be added to x gallons

of gasoline is computed by $f(x) = 0.16x$. (**Source:** Johnson Outboard Motor Company.)

(a) Why is it reasonable to expect f to be linear?

(b) Evaluate $f(3)$ and interpret the answer.

(c) How much gasoline should be mixed with 2 pints of oil?

120. *Perimeter* Find the length of the longest side of the rectangle if its perimeter is 25 feet.

Modeling Data with Linear Functions

Exercises 121 and 122: The following data can be modeled by a linear function. Estimate the value of x when $y = 2.99$.

121.

x	2	4	6	8
y	0.51	1.23	1.95	2.67

122.

x	1	2	3	4
y	−1.66	2.06	5.78	9.50

Linear Regression

123. *Ring Size* The table lists ring size S for a finger with circumference x in centimeters.

x (cm)	4.65	5.40	5.66	6.41
S (size)	4	7	8	11

Source: Overstock.

(a) Find a linear function S that models the data.

(b) Find the circumference of a finger with a ring size of 6.

124. *Hat Size* The table lists hat size H for a head with circumference x in inches.

x (in.)	$21\frac{1}{8}$	$21\frac{7}{8}$	$22\frac{5}{8}$	25
S (size)	$6\frac{3}{4}$	7	$7\frac{1}{4}$	8

Source: Brentblack.

(a) Find a linear function S that models the data.

(b) Find the circumference of a head with a hat size of $7\frac{1}{2}$.

125. *Super Bowl Ads*   The table lists the cost in millions of dollars for a 30-second Super Bowl commercial for selected years.

Year	1990	1994	1998	2004	2008
Cost	0.8	1.2	1.6	2.3	2.7

Source: MSNBC.

(a) Find a linear function f that models the data.

(b) Estimate the cost in 1987 and compare the estimate to the actual value of $0.6 million. Did your estimate involve interpolation or extrapolation?

(c) Use f to predict the year when the cost could reach $3.2 million.

126. *Women in Politics*   The table lists the percentage P of women in state legislatures during year x.

x	1993	1997	2001	2005	2007
P	20.5	21.6	22.4	22.7	23.5

Source: National Women's Political Caucus.

(a) Find a linear function P that models the data.

(b) Estimate this percentage in 2003 and compare the estimate to the actual value of 22.4%. Did your estimate involve interpolation or extrapolation?

(c) Use P to predict the year when this percentage could reach 25%.

Writing about Mathematics

127. Describe a basic graphical method used to solve a linear equation. Give examples.

128. Describe verbally how to solve $ax + b = 0$. What assumptions have you made about the value of a?

EXTENDED AND DISCOVERY EXERCISES

1. *Geometry*   Suppose that two rectangles are similar and the sides of the first rectangle are twice as long as the corresponding sides of the second rectangle.

(a) Is the perimeter of the first rectangle twice the perimeter of the second rectangle? Explain.

(b) Is the area of the first rectangle twice the area of the second rectangle? Explain.

2. *Geometry*   Repeat the previous exercise for an equilateral triangle. Try to make a generalization. (*Hint:* The area of an equilateral triangle is $A = \frac{\sqrt{3}}{4}x^2$, where x is the length of a side.) What will happen to the circumference of a circle if the radius is doubled? What will happen to its area?

3. *Indoor Air Pollution*   Formaldehyde is an indoor air pollutant formerly found in plywood, foam insulation, and carpeting. When concentrations in the air reach 33 micrograms per cubic foot (μg/ft^3), eye irritation can occur. One square foot of new plywood could emit 140 μg per hour.   (**Source:** A. Hines, *Indoor Air Quality & Control.*)

(a) A room has 100 square feet of new plywood flooring. Find a linear function f that computes the amount of formaldehyde in micrograms that could be emitted in x hours.

(b) The room contains 800 cubic feet of air and has no ventilation. Determine how long it would take for concentrations to reach 33 μg/ft^3.

4. *Temperature and Volume*   The table shows the relationship between the temperature of a sample of helium and its volume.

Temperature (°C)	0	25	50	75	100
Volume (in^3)	30	32.7	35.4	38.1	40.8

(a) Make a scatterplot of the data.

(b) Write a formula for a function f that receives the temperature x as input and outputs the volume y of the helium.

(c) Find the volume when the temperature is 65°C.

(d) Find the temperature if the volume is 25 cubic inches. Did your answer involve interpolation or extrapolation? Do you believe that your answer is accurate?

2.4 Linear Inequalities

- **Understand basic terminology related to inequalities**
- **Solve linear inequalities symbolically**
- **Solve linear inequalities graphically and numerically**
- **Solve compound inequalities**

Introduction

If a person who weighs 143 pounds needs to purchase a life preserver for whitewater rafting, it is doubtful that there is one designed exactly for this weight. Life preservers are manufactured to support a range of body weights. A vest that is approved for weights between 120 and 160 pounds would be appropriate. Every airplane has a maximum weight allowance. It is important that this weight limit be accurately determined. However, most people feel more comfortable at takeoff if that maximum has not been reached, because any weight that is less than the maximum is also safe and allows a greater margin of error. Both of these situations involve the concept of inequality.

In mathematics much effort is expended in solving equations and determining equality. One reason is that equality is frequently a boundary between *greater than* and *less than*. The solution to an inequality often can be found by first locating where two expressions are equal. Since equality and inequality are closely related, many of the techniques used to solve equations also can be applied to inequalities.

Inequalities

Inequalities result whenever the equals sign in an equation is replaced with any one of the symbols $<, \leq, >,$ or $\geq$. Some examples of inequalities include

$$x + 15 < 9x - 1, \quad x^2 - 2x + 1 \geq 2x, \quad z + 5 > 0,$$
$$xy + x^2 \leq y^3 + x, \quad \text{and} \quad 2 + 3 > 1.$$

The first three inequalities have one variable, the fourth inequality contains two variables, and the fifth inequality has only constants. As with linear equations, our discussion focuses on inequalities with one variable.

To **solve** an inequality means to find all values for the variable that make the inequality a true statement. Such values are **solutions**, and the set of all solutions is the **solution set** to the inequality. Two inequalities are **equivalent** if they have the same solution set. It is common for an inequality to have infinitely many solutions. For instance, the inequality $x - 1 > 0$ has infinitely many solutions because any real number x satisfying $x > 1$ is a solution. The solution set is $\{x \mid x > 1\}$.

Like functions and equations, inequalities in one variable can be classified as *linear* or *nonlinear*.

Linear Inequality in One Variable

A **linear inequality** in one variable is an inequality that can be written in the form

$$ax + b > 0,$$

where $a \neq 0$. (The symbol $>$ may be replaced by $\geq, <,$ or $\leq$.)

Examples of linear inequalities include

$$3x - 4 < 0, \quad 7x + 5 \geq x, \quad x + 6 > 23, \quad \text{and} \quad 7x + 2 \leq -3x + 6.$$

Using techniques from algebra, we can transform these inequalities into one of the forms $ax + b > 0, ax + b \geq 0, ax + b < 0,$ or $ax + b \leq 0$. For example, by subtracting x

from each side of $7x + 5 \geq x$, we obtain the equivalent inequality $6x + 5 \geq 0$. If an inequality is not a linear inequality, it is called a **nonlinear inequality**.

MAKING CONNECTIONS

Linear Functions, Equations, and Inequalities   These concepts are closely related.

$$f(x) = ax + b \qquad \text{Linear function}$$
$$ax + b = 0, a \neq 0 \qquad \text{Linear equation}$$
$$ax + b > 0, a \neq 0 \qquad \text{Linear inequality}$$

NOTE   This relationship is used again in the next chapter to define quadratic functions, equations, and inequalities.

Properties of Inequalities

Let a, b, and c be real numbers.

1. $a < b$ and $a + c < b + c$ are equivalent.
   (The same number may be added to or subtracted from each side of an inequality.)
2. If $c > 0$, then $a < b$ and $ac < bc$ are equivalent.
   (Each side of an inequality may be multiplied or divided by the same positive number.)
3. If $c < 0$, then $a < b$ and $ac > bc$ are equivalent.
   (Each side of an inequality may be multiplied or divided by the same negative number provided the inequality symbol is reversed.)

Replacing $<$ with $\leq$ and $>$ with $\geq$ results in similar properties.

The following examples illustrate each property.

Property 1: To solve $x - 5 < 6$, add 5 to each side to obtain $x < 11$.

Property 2: To solve $5x < 10$, divide each side by 5 to obtain $x < 2$.

Property 3: To solve $-5x < 10$, divide each side by -5 to obtain $x > -2$. (Whenever you multiply or divide an inequality by a negative number, reverse the inequality symbol.)

Review of Interval Notation   In Section 1.5 interval notation was introduced as an efficient way to express intervals on the real number line. For example, the interval $3 \leq x \leq 5$ is written as $[3, 5]$, whereas the interval $3 < x < 5$ is written as $(3, 5)$. A bracket, [or], is used when an endpoint is included, and a parenthesis, (or), is used when an endpoint is not included. The interval $x \geq 2$ is written as $[2, \infty)$, where ∞ denotes infinity, and the interval $x < 2$ is written as $(-\infty, 2)$

In the next example we solve linear equalities and express the solution set in both set-builder and interval notation.

EXAMPLE 1   Solving linear inequalities symbolically

Solve each inequality. Write the solution set in set-builder and interval notation.

(a) $2x - 3 < \dfrac{x + 2}{-3}$ **(b)** $-3(4z - 4) \geq 4 - (z - 1)$

SOLUTION

(a) Use Property 3 by multiplying each side by -3 to clear fractions. Remember to reverse the inequality symbol when multiplying by a negative number.

$$2x - 3 < \frac{x + 2}{-3} \qquad \text{Given inequality}$$

$$-6x + 9 > x + 2 \qquad \text{Property 3: Multiply by } -3 \text{ and reverse the inequality symbol.}$$

$$9 > 7x + 2 \qquad \text{Property 1: Add } 6x.$$

$$7 > 7x \qquad \text{Property 1: Add } -2 \text{ (or subtract 2).}$$

$$1 > x \qquad \text{Property 2: Divide by 7.}$$

In set-builder notation the solution set is $\{x \mid x < 1\}$, and in interval notation it is written as $(-\infty, 1)$.

(b) Begin by applying the distributive property.

$$-3(4z - 4) \geq 4 - (z - 1) \qquad \text{Given inequality}$$

$$-12z + 12 \geq 4 - z + 1 \qquad \text{Distributive property}$$

$$-12z + 12 \geq -z + 5 \qquad \text{Simplify.}$$

$$-12z + z \geq 5 - 12 \qquad \text{Property 1: Add } z \text{ and } -12.$$

$$-11z \geq -7 \qquad \text{Simplify.}$$

$$z \leq \frac{7}{11} \qquad \text{Property 3: Divide by } -11 \text{ and reverse inequality symbol.}$$

In set-builder notation the solution set is $\left\{z \mid z \leq \frac{7}{11}\right\}$, and in interval notation it is written as $\left(-\infty, \frac{7}{11}\right]$. **Now Try Exercises 15 and 17** ◀

Graphical Solutions The intersection-of-graphs method can be extended to solve inequalities. Figure 2.44 shows the velocity of two cars in miles per hour after x minutes. V_A denotes the velocity of car A, and V_B denotes the velocity of car B. The domains of V_A and V_B are both assumed to be $0 \leq x \leq 6$.

At 3 minutes $V_A = V_B$, and both cars are traveling at 30 miles per hour. To the left of $x = 3$ the graph of V_A is below the graph of V_B, so car A is traveling slower than car B. Thus

$$V_A < V_B \quad \text{when} \quad 0 \leq x < 3.$$

To the right of $x = 3$ the graph of V_A is above the graph of V_B, so car A is traveling faster than car B. Thus

$$V_A > V_B \quad \text{when} \quad 3 < x \leq 6.$$

This technique is used in the next example.

Figure 2.44 Velocities of Two Cars

EXAMPLE 2 Solving a linear inequality graphically

Graph $y_1 = \frac{1}{2}x + 2$ and $y_2 = 2x - 1$ by hand. Use the graph to solve the linear inequality $\frac{1}{2}x + 2 > 2x - 1$.

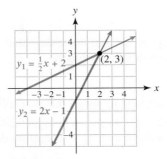

Figure 2.45

SOLUTION The graphs of $y_1 = \frac{1}{2}x + 2$ and $y_2 = 2x - 1$ are shown in Figure 2.45. The graphs intersect at the point $(2, 3)$. The graph of $y_1 = \frac{1}{2}x + 2$ is above the graph of $y_2 = 2x - 1$ to the left of the point of intersection, or when $x < 2$. Thus the solution set to the inequality $\frac{1}{2}x + 2 > 2x - 1$ is $\{x \mid x < 2\}$, or $(-\infty, 2)$.

Now Try Exercise 39 ◄

An Application In the next example we use graphical techniques to solve an application from meteorology.

EXAMPLE 3 Using the intersection-of-graphs method

When the air temperature reaches the dew point, fog may form. This phenomenon also causes clouds to form. See Figure 2.46. Both the air temperature and the dew point often decrease at a constant rate as the altitude above ground level increases. If the ground-level Fahrenheit temperature and dew point are T_0 and D_0, the air temperature can sometimes be approximated by $T(x) = T_0 - 19x$ and the dew point by $D(x) = D_0 - 5.8x$ at an altitude of x miles.
(a) If $T_0 = 75°F$ and $D_0 = 55°F$, determine the altitudes where clouds will not form.
(b) The slopes of the graphs for the functions T and D are called *lapse rates*. Interpret their meanings. Explain how these two slopes ensure a strong likelihood of clouds forming. **(Source:** A. Miller and R. Anthes, *Meteorology.*)

SOLUTION

(a) Since $T_0 = 75$ and $D_0 = 55$, let $T(x) = 75 - 19x$ and $D(x) = 55 - 5.8x$. Clouds will not form when the air temperature is greater than the dew point. Therefore we must solve the inequality $T(x) > D(x)$. Graph $T = 75 - 19x$ and $D = 55 - 5.8x$, as shown in Figure 2.47. The graphs intersect near $(1.52, 46.2)$. This means that the air temperature and dew point are both $46.2°F$ at about 1.52 miles above ground level. Clouds will not form below this altitude, or when the graph of T is above the graph of D. The solution set is $\{x \mid 0 \le x < 1.52\}$, where the endpoint 1.52 has been approximated.

Figure 2.46

[0, 5, 1] by [0, 80, 10]

Intersection
X=1.5151515_Y=46.212121⌟

Figure 2.47

(b) The slope of the graph of T is -19. This means that for each 1-mile increase in altitude, the air temperature *decreases* by $19°F$. Similarly, the slope of the graph of D is -5.8. The dew point *decreases* by $5.8°F$ for every 1-mile increase in altitude. As the altitude increases, the air temperature decreases at a faster rate than the dew point. As a result, the air temperature typically cools to the dew point at higher altitudes and clouds may form.

Now Try Exercise 85 ◄

CLASS DISCUSSION
How does the difference between the air temperature and the dew point at ground level affect the altitude at which clouds may form? Explain.

x-Intercept Method If a linear inequality can be written as $y_1 > 0$, where $>$ may be replaced by $\ge, \le,$ or $<$, then we can solve this inequality by using the *x*-intercept method. To apply this method for $y_1 > 0$, graph y_1 and find the *x*-intercept. The solution set includes *x*-values where the graph of y_1 is **above** the *x*-axis.

EXAMPLE 4   Applying the *x*-intercept method

Solve the inequality $1 - x > \frac{1}{2}x - 2$ by using the *x*-intercept method. Write the solution set in set-builder and interval notation. Then solve the inequality symbolically.

SOLUTION

Graphical Solution   Start by subtracting $\frac{1}{2}x - 2$ from each side to obtain the inequality $1 - x - \left(\frac{1}{2}x - 2\right) > 0$. Then graph $Y_1 = 1 - X - (X/2 - 2)$, as shown in Figure 2.48, where the *x*-intercept is 2. The graph of y_1 is above the *x*-axis when $x < 2$. Therefore the solution set to $y_1 > 0$ is $\{x \mid x < 2\}$, or $(-\infty, 2)$.

[−6, 6, 1] by [−4, 4, 1]

Figure 2.48

Calculator Help

To locate a zero or *x*-intercept on the graph of a function, see Appendix A (page AP-9).

Symbolic Solution

$$1 - x > \frac{1}{2}x - 2 \qquad \text{Given inequality}$$

$$-x > \frac{1}{2}x - 3 \qquad \text{Subtract 1 from each side.}$$

$$-\frac{3}{2}x > -3 \qquad \text{Subtract } \frac{1}{2}x \text{ from each side.}$$

$$x < 2 \qquad \text{Multiply by } -\frac{2}{3}; \text{ reverse inequality.}$$

The solution set is $\{x \mid x < 2\}$, or $(-\infty, 2)$.

Now Try Exercise 51 ◀

Visualizing Solutions   Example 4 suggests a general result about linear inequalities. The graph of the equation $y = ax + b$, $a \neq 0$, is a line that intersects the *x*-axis once and slopes either upward (if $a > 0$) or downward (if $a < 0$). If the *x*-intercept is k, then the solution set to $ax + b > 0$ satisfies either $x > k$ or $x < k$, as illustrated in Figures 2.49 and 2.50. (Similar remarks hold for linear inequalities that use the symbol $<$, $\leq$, or $\geq$.) Note that the solution to $ax + b = 0$ is k and the value of k is that *boundary* between *greater than* and *less than* in either situation.

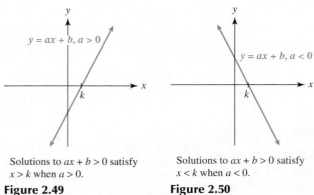

Solutions to $ax + b > 0$ satisfy $x > k$ when $a > 0$.

Figure 2.49

Solutions to $ax + b > 0$ satisfy $x < k$ when $a < 0$.

Figure 2.50

Numerical Solutions   Inequalities can sometimes be solved by using a table of values. The following example helps to explain the mathematical concept behind this method.

Suppose that it costs a company $5x + 200$ dollars to produce x pairs of headphones and the company receives $15x$ dollars for selling x pairs of headphones. Then the profit P from selling x pairs of headphones is $P = 15x - (5x + 200) = 10x - 200$. A value of $x = 20$ results in $P = 0$, and so $x = 20$ is called the **boundary number** because it represents the boundary between making money and losing money (the break-even point). To make money, the profit P must be positive, and the inequality

$$10x - 200 > 0$$

must be satisfied. The table of values for $y_1 = 10x - 200$ in Table 2.8 shows the boundary number $x = 20$ along with several **test values**. The test values of $x = 17, 18$, and 19 result in a loss. The test values of $x = 21, 22$, and 23 result in a profit. Therefore the solution set to $10x - 200 > 0$ is $\{x \mid x > 20\}$.

Table 2.8

Boundary number

x	17	18	19	20	21	22	23
$10x - 200$	-30	-20	-10	0	10	20	30

Less than 0   Greater than 0

EXAMPLE 5 Solving a linear inequality with test values

Solve $3(6 - x) + 5 - 2x < 0$ numerically.

SOLUTION We begin by making a table of $Y_1 = 3(6 - X) + 5 - 2X$, as shown in Figure 2.51. We can see that the boundary number for this inequality lies between $x = 4$ and $x = 5$. Changing the increment from 1 to 0.1 in Figure 2.52 shows that the boundary number for the inequality is $x = 4.6$. The test values of $x = 4.7, 4.8$, and 4.9 indicate that when $x > 4.6$, the inequality $y_1 < 0$ is true. The solution set is $\{x \mid x > 4.6\}$.

X	Y₁	
1	18	
2	13	Greater
3	8	than 0
4	**3**	
5	-2	Less
6	-7	than 0
7	-12	

X=4

X	Y₁	
4.3	1.5	
4.4	1	Greater
4.5	.5	than 0
4.6	**0**	
4.7	-.5	Less
4.8	-1	than 0
4.9	-1.5	

X=4.6

Figure 2.51   **Figure 2.52**   *Now Try Exercise 71* ◀

MAKING CONNECTIONS

Symbolic, Graphical, and Numerical Solutions Linear inequalities can be solved symbolically, graphically, or numerically. Each method is used in the following example to solve the inequality $3 - (x + 2) > 0$.

Symbolic Solution

$3 - (x + 2) > 0$

$-(x + 2) > -3$

$x + 2 < 3$

$x < 1$

Graphical Solution

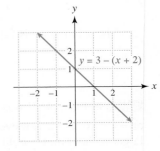

The graph of $y = 3 - (x + 2)$ is above the x-axis when $x < 1$.

Numerical Solution

x	$3 - (x + 2)$	
-2	3	Greater
-1	2	than 0
0	1	
1	0	Equal 0
2	-1	Less
3	-2	than 0

The values of $3 - (x + 2)$ are greater than 0 when $x < 1$.

Compound Inequalities

Sometimes a variable must satisfy two inequalities. For example, on a freeway there may be a minimum speed limit of 40 miles per hour and a maximum speed limit of 70 miles per hour. If x represents the speed of a vehicle, then x must satisfy the compound inequality

$$x \geq 40 \quad \text{and} \quad x \leq 70.$$

A **compound inequality** occurs when two inequalities are connected by the word *and* or *or*. When the word *and* connects two inequalities, the two inequalities can sometimes be written as a **three-part inequality**. For example, the previous compound inequality may be written as the three-part inequality

$$40 \leq x \leq 70.$$

Compound inequalities involving the word *or* are discussed in the next section.

EXAMPLE 6   Solving a three-part inequality symbolically

Solve the inequality. Write the solution set in set-builder and interval notation.

(a) $-4 \leq 5x + 1 < 21$ **(b)** $\dfrac{1}{2} < \dfrac{1 - 2t}{4} < 2$

SOLUTION
(a) Use properties of inequalities to simplify the three-part inequality.

$$-4 \leq 5x + 1 < 21 \qquad \text{Given inequality}$$
$$-5 \leq 5x < 20 \qquad \text{Add } -1 \text{ to each part.}$$
$$-1 \leq x < 4 \qquad \text{Divide each part by 5.}$$

The solution set is $\{x \mid -1 \leq x < 4\}$, or $[-1, 4)$.

(b) Begin by multiplying each part by 4 to clear fractions.

$$\frac{1}{2} < \frac{1 - 2t}{4} < 2 \qquad \text{Given inequality}$$
$$2 < 1 - 2t < 8 \qquad \text{Multiply each part by 4.}$$
$$1 < -2t < 7 \qquad \text{Add } -1 \text{ to each part.}$$
$$-\frac{1}{2} > t > -\frac{7}{2} \qquad \text{Divide by } -2; \text{ reverse inequalities.}$$
$$-\frac{7}{2} < t < -\frac{1}{2} \qquad \text{Rewrite the inequality.}$$

The solution set is $\left\{t \mid -\frac{7}{2} < t < -\frac{1}{2}\right\}$, or $\left(-\frac{7}{2}, -\frac{1}{2}\right)$.

Now Try Exercises 23 and 35 ◀

NOTE In Example 6, it is correct to write a three-part inequality as either $-\frac{1}{2} > t > -\frac{7}{2}$ or $-\frac{7}{2} < t < -\frac{1}{2}$. However, we usually write the smaller number on the left side and the larger number on the right side.

Three-part inequalities occur in many applications and can often be solved symbolically and graphically. This is demonstrated in the next example.

EXAMPLE 7 Modeling sunset times

In Boston, on the 82nd day (March 22) of 2008 the sun set at 7:00 P.M., and on the 136th day (May 15) the sun set at 8:00 P.M. Use a linear function S to estimate the days when the sun set between 7:15 P.M. and 7:45 P.M., inclusive. Do not consider any days of the year after May 15.   (**Source:** R. Thomas, *The Old Farmer's 2008 Almanac.*)

SOLUTION

Getting Started First, find a linear function S whose graph passes through the points $(82, 7)$ and $(136, 8)$. Then solve the compound inequality $7.25 \leq S(x) \leq 7.75$. Note that 7.25 hours past noon corresponds to 7:15 P.M. and 7.75 hours past noon corresponds to 7:45 P.M. ▶

Symbolic Solution   The slope of the line passing through $(82, 7)$ and $(136, 8)$ is given by $\frac{8 - 7}{136 - 82} = \frac{1}{54}$. The point-slope form of the line passing through $(82, 7)$ with slope $\frac{1}{54}$ is

$$S(x) = \frac{1}{54}(x - 82) + 7. \qquad \text{Point-slope form}$$

Now solve the following compound inequality.

$$7.25 \leq \frac{1}{54}(x - 82) + 7 \leq 7.75 \qquad \text{Given inequality}$$

$$0.25 \leq \frac{1}{54}(x - 82) \leq 0.75 \qquad \text{Subtract 7 from each part.}$$

$$13.5 \leq x - 82 \leq 40.5 \qquad \text{Multiply each part by 54.}$$

$$95.5 \leq x \leq 122.5 \qquad \text{Add 82 to each part.}$$

If we round 95.5 and 122.5 up to 96 and 123, then this model predicts that the sun set between 7:15 P.M. and 7:45 P.M. from the 96th day (April 5) to the 123rd day (May 2). (Note that the actual days were April 5 and May 1.)

Graphical Solution   Graph $y_1 = 7.25$, $y_2 = \frac{1}{54}(x - 82) + 7$, and $y_3 = 7.75$ and determine their points of intersection, **(95.5, 7.25)** and **(122.5, 7.75)**, as shown in Figures 2.53 and 2.54. The graph of y_2 is between the graphs of y_1 and y_3 for **$95.5 \leq x \leq 122.5$**. This agrees with the symbolic solution. A different graph showing this solution appears in Figure 2.55.

[80, 150, 10] by [6.5, 8.5, 1]

Figure 2.53

[80, 150, 10] by [6.5, 8.5, 1]

Figure 2.54

Figure 2.55

Now Try Exercise 93 ◀

EXAMPLE 8   Solving inequalities symbolically

Solve the linear inequalities symbolically. Express the solution set using interval notation.

(a) $-\dfrac{x}{2} + 1 \le 3$ **(b)** $-8 < \dfrac{3x-1}{2} \le 5$ **(c)** $5(x-6) < 2x - 2(1-x)$

SOLUTION

(a) Simplify the inequality as follows.

$$-\frac{x}{2} + 1 \le 3 \qquad \text{Given inequality}$$

$$-\frac{x}{2} \le 2 \qquad \text{Add } -1, \text{ or subtract } 1.$$

$$x \ge -4 \qquad \text{Multiply by } -2. \text{ Reverse the inequality.}$$

In interval notation the solution set is $[-4, \infty)$.

(b) The parts of this compound inequality can be solved simultaneously.

$$-8 < \frac{3x-1}{2} \le 5 \qquad \text{Given inequality}$$

$$-16 < 3x - 1 \le 10 \qquad \text{Multiply by 2.}$$

$$-15 < 3x \le 11 \qquad \text{Add 1.}$$

$$-5 < x \le \frac{11}{3} \qquad \text{Divide by 3.}$$

The solution set is $\left(-5, \frac{11}{3}\right]$.

(c) Start by applying the distributive property to each side of the inequality.

$$5(x - 6) < 2x - 2(1 - x) \qquad \text{Given inequality}$$

$$5x - 30 < 2x - 2 + 2x \qquad \text{Distributive property}$$

$$5x - 30 < 4x - 2 \qquad \text{Simplify.}$$

$$x - 30 < -2 \qquad \text{Subtract } 4x.$$

$$x < 28 \qquad \text{Add 30.}$$

The solution set is $(-\infty, 28)$.   **Now Try Exercises 13, 27, and 33** ◀

2.4 Putting It All Together

Any linear inequality can be written as $ax + b > 0$ with $a \ne 0$, where $>$ can be replaced by $\ge$, $<$, or $\le$. The following table includes methods for solving linear inequalities in the form $h(x) > 0$ or $f(x) > g(x)$. Inequalities involving $<$, $\le$, and $\ge$ are solved in a similar manner.

Concept	Explanation	Examples
Interval notation	An efficient notation for writing the solution set to inequalities To review interval notation, see Section 1.5.	$x < 2$ is equivalent to $(-\infty, 2)$. $x \ge 1$ is equivalent to $[1, \infty)$. $-1 \le x < 4$ is equivalent to $[-1, 4)$.

Concept	Explanation	Examples
Compound inequality	Two inequalities connected by the word *and* or *or*	$x \le 4$ or $x \ge 10$ $x \ge -3$ and $x < 4$ $x > 5$ and $x \le 20$ can be written as the three-part inequality $5 < x \le 20$.
Symbolic method	Use properties of inequalities to simplify $f(x) > g(x)$ to either $x > k$ or $x < k$ for some real number k.	$\frac{1}{2}x + 1 > 3 - \frac{3}{2}x$ Given inequality $2x + 1 > 3$ Add $\frac{3}{2}x$. $2x > 2$ Subtract 1. $x > 1$ Divide by 2.
Intersection-of-graphs method	To solve $f(x) > g(x)$, graph $y_1 = f(x)$ and $y_2 = g(x)$. Find the point of intersection. The solution set includes x-values where the graph of y_1 is above the graph of y_2.	$\frac{1}{2}x + 1 > 3 - \frac{3}{2}x$ Graph $y_1 = \frac{1}{2}x + 1$ and $y_2 = 3 - \frac{3}{2}x$. The solution set for $y_1 > y_2$ is $\{x \mid x > 1\}$.
The x-intercept method	Write the inequality as $h(x) > 0$. Graph $y_1 = h(x)$. Solutions occur where the graph is above the x-axis.	$\frac{1}{2}x + 1 > 3 - \frac{3}{2}x$ Graph $y_1 = \frac{1}{2}x + 1 - \left(3 - \frac{3}{2}x\right)$. The solution set for $y_1 > 0$ is $\{x \mid x > 1\}$.
Numerical method	Write the inequality as $h(x) > 0$. Create a table for $y_1 = h(x)$ and find the boundary number $x = k$ such that $h(k) = 0$. Use the test values in the table to determine if the solution set is $x > k$ or $x < k$.	$\frac{1}{2}x + 1 > 3 - \frac{3}{2}x$ Table $y_1 = \frac{1}{2}x + 1 - \left(3 - \frac{3}{2}x\right)$. The solution set for $y_1 > 0$ is $\{x \mid x > 1\}$. _see table below_

x	-1	0	1	2	3
y_1	-4	-2	0	2	4

Less than 0   Greater than 0

2.4   Exercises

Interval Notation

Exercises 1–8: Express the following in interval notation.

1. $x < 2$

2. $x > -3$

3. $x \geq -1$

4. $x \leq 7$

5. $\{x \mid 1 \leq x < 8\}$

6. $\{x \mid -2 < x \leq 4\}$

7. $\{x \mid x \leq 1\}$

8. $\{x \mid x > 5\}$

Solving Linear Inequalities Symbolically

Exercises 9–38: Solve the inequality symbolically. Express the solution set in set-builder or interval notation.

9. $2x + 6 \geq 10$

10. $-4x - 3 < 5$

11. $-2(x - 10) + 1 > 0$

12. $3(x + 5) \leq 0$

13. $\dfrac{t + 2}{3} \geq 5$

14. $\dfrac{2 - t}{6} < 0$

15. $4x - 1 < \dfrac{3 - x}{-3}$

16. $\dfrac{x + 5}{-10} > 2x + 3$

17. $-3(z - 4) \geq 2(1 - 2z)$

18. $-\frac{1}{4}(2z - 6) + z \geq 5$

19. $\dfrac{1 - x}{4} < \dfrac{2x - 2}{3}$

20. $\dfrac{3x}{4} < x - \dfrac{x + 2}{2}$

21. $2x - 3 > \frac{1}{2}(x + 1)$

22. $5 - (2 - 3x) \leq -5x$

23. $5 < 4t - 1 \leq 11$

24. $-1 \leq 2t \leq 4$

25. $3 \leq 4 - x \leq 20$

26. $-5 < 1 - 2x < 40$

27. $-7 \leq \dfrac{1 - 4x}{7} < 12$

28. $0 < \dfrac{7x - 5}{3} \leq 4$

29. $5 > 2(x + 4) - 5 > -5$

30. $\frac{8}{3} \geq \frac{4}{3} - (x + 3) \geq \frac{2}{3}$

31. $3 \leq \frac{1}{2}x + \frac{3}{4} \leq 6$

32. $-4 \leq 5 - \frac{4}{5}x < 6$

33. $5x - 2(x + 3) \geq 4 - 3x$

34. $3x - 1 < 2(x - 3) + 1$

35. $\dfrac{1}{2} \leq \dfrac{1 - 2t}{3} < \dfrac{2}{3}$

36. $-\dfrac{3}{4} < \dfrac{2 - t}{5} < \dfrac{3}{4}$

37. $\frac{1}{2}z + \frac{2}{3}(3 - z) - \frac{5}{4}z \geq \frac{3}{4}(z - 2) + z$

38. $\dfrac{2}{3}(1 - 2z) - \dfrac{3}{2}z + \dfrac{5}{6}z \geq \dfrac{2z - 1}{3} + 1$

Solving Linear Inequalities Graphically

Exercises 39–46: (Refer to Example 2.) Solve the inequality graphically. Use set-builder notation.

39. $x + 2 \geq 2x$

40. $2x - 1 \leq x$

41. $\frac{2}{3}x - 2 > -\frac{4}{3}x + 4$

42. $-2x \geq -\frac{5}{3}x + 1$

43. $-1 \leq 2x - 1 \leq 3$

44. $-2 < 1 - x < 2$

45. $-3 < x - 2 \leq 2$

46. $-1 \leq 1 - 2x < 5$

Exercises 47–50: Use the given graph of $y = ax + b$ to solve each equation and inequality. Write the solution set to each inequality in set-builder or interval notation.

    (a) $ax + b = 0$ (b) $ax + b < 0$ (c) $ax + b \geq 0$

47.

48.

49.

50.
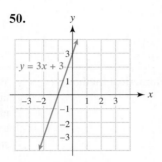

Exercises 51–54: x-Intercept Method   (Refer to Example 4.) Use the x-intercept method to solve the inequality. Write the solution set in set-builder or interval notation. Then solve the inequality symbolically.

51. $x - 3 \le \frac{1}{2}x - 2$ **52.** $x - 2 \le \frac{1}{3}x$

53. $2 - x < 3x - 2$ **54.** $\frac{1}{2}x + 1 > \frac{3}{2}x - 1$

Exercises 55–60: Solve the linear inequality graphically. Write the solution set in set-builder notation. Approximate endpoints to the nearest hundredth whenever appropriate.

55. $5x - 4 > 10$

56. $-3x + 6 \le 9$

57. $-2(x - 1990) + 55 \ge 60$

58. $\sqrt{2}x > 10.5 - 13.7x$

59. $\sqrt{5}(x - 1.2) - \sqrt{3}x < 5(x + 1.1)$

60. $1.238x + 0.998 \le 1.23(3.987 - 2.1x)$

Exercises 61–66; Solve the compound linear inequality graphically. Write the solution set in set-builder or interval notation, and approximate endpoints to the nearest tenth whenever appropriate.

61. $3 \le 5x - 17 < 15$ **62.** $-4 < \dfrac{55 - 3.1x}{4} < 17$

63. $1.5 \le 9.1 - 0.5x \le 6.8$ **64.** $0.2x < \dfrac{2x - 5}{3} < 8$

65. $x - 4 < 2x - 5 < 6$ **66.** $-3 \le 1 - x \le 2x$

67. The graphs of two linear functions f and g are shown.
  (a) Solve the equation $g(x) = f(x)$.

  (b) Solve the inequality $g(x) > f(x)$.

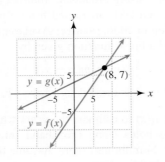

68. Use the figure to solve each equation or inequality.
  (a) $f(x) = g(x)$ **(b)** $g(x) = h(x)$

  (c) $f(x) < g(x) < h(x)$ **(d)** $g(x) > h(x)$

Solving Linear Inequalities Numerically

Exercises 69 and 70: Assume y_1 represents a linear function with the set of real numbers for its domain. Use the table to solve the inequalities. Use set-builder notation.

69. $y_1 > 0, y_1 \le 0$ **70.** $y_1 < 0, y_1 \ge 0$

X	Y1
0	220
1	165
2	110
3	55
4	0
5	-55
6	-110
X=0	

X	Y1
-5	-32
-4	-16
-3	0
-2	16
-1	32
0	48
1	64
X=-5	

Exercises 71–78: Solve each inequality numerically. Write the solution set in set-builder or interval notation, and approximate endpoints to the nearest tenth when appropriate.

71. $-4x - 6 > 0$ **72.** $1 - 2x \ge 9$

73. $1 \le 3x - 2 \le 10$ **74.** $-5 < 2x - 1 < 15$

75. $-\dfrac{3}{4} < \dfrac{2 - 5x}{3} \le \dfrac{3}{4}$ **76.** $\dfrac{3x - 1}{5} < 15$

77. $(\sqrt{11} - \pi)x - 5.5 \le 0$

78. $1.5(x - 0.7) + 1.5x < 1$

You Decide the Method

Exercises 79–82: Solve the inequality. Approximate the endpoints to the nearest thousandth when appropriate.

79. $2x - 8 > 5$ **80.** $5 < 4x - 2.5$

81. $\pi x - 5.12 \le \sqrt{2}x - 5.7(x - 1.1)$

82. $5.1x - \pi \ge \sqrt{3} - 1.7x$

Applications

83. *Distance between Cars* Cars A and B are both traveling in the same direction. Their distances in miles north of St. Louis after x hours are computed by the functions f_A and f_B, respectively. The graphs of f_A and f_B are shown in the figure for $0 \le x \le 10$.

(a) Which car is traveling faster? Explain.

(b) How many hours elapse before the two cars are the same distance from St. Louis? How far are they from St. Louis when this occurs?

(c) During what time interval is car B farther from St. Louis than car A?

Time (hours)

84. *Distance* Function f computes the distance y in miles between a car and the city of Omaha after x hours, where $0 \le x \le 6$. The graphs of f and the horizontal lines $y = 100$ and $y = 200$ are shown in the figure.

(a) Is the car moving toward or away from Omaha? Explain.

(b) Determine the times when the car is 100 miles and 200 miles from Omaha.

(c) Determine when the car is from 100 to 200 miles from Omaha.

(d) When is the car's distance from Omaha greater than 100 miles?

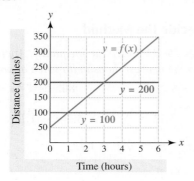

Time (hours)

85. *Clouds and Temperature* (Refer to Example 3.) Suppose the ground-level temperature is 65°F and the dew point is 50°F.

(a) Use the intersection-of-graphs method to estimate the altitudes where clouds will not form.

(b) Solve part (a) symbolically.

86. *Temperature and Altitude* Suppose the Fahrenheit temperature x miles above ground level is given by the formula $T(x) = 85 - 19x$.

(a) Use the intersection-of-graphs method to estimate the altitudes where the temperature is below freezing. Assume that the domain of T is $0 \le x \le 6$.

(b) What does the x-intercept on the graph of $y = T(x)$ represent?

(c) Solve part (a) symbolically.

87. *Prices of Homes* The median prices of a single-family home in the United States from 1990 to 2005 can be approximated by the formula $P(x) = 8667x + 90,000$, where $x = 0$ corresponds to 1990 and $x = 15$ to 2005. (**Source:** National Association of Realtors.)

(a) Interpret the slope of the graph of P.

(b) Estimate the years when the median price range was from \$142,000 to \$194,000.

88. *Population Density* The population density D of the United States in people per square mile during year x from 1900 to 2000 can be approximated by the formula $D(x) = 0.58x - 1080$. (**Source:** Bureau of the Census.)

(a) Interpret the slope of the graph of D.

(b) Estimate when the density varied between 50 and 75 people per square mile.

89. *Broadband Internet Connections* The number of households using broadband Internet connections, such as cable and DSL, increased from 6 million in 2000 to 30 million in 2004. (**Source:** eMarketer.)

(a) Find a linear function given by

$$B(x) = m(x - x_1) + y_1$$

that models these data, where x is the year.

(b) Use $B(x)$ to estimate the years when the number of households using broadband Internet connections was 24 million or more. Assume that the domain of B is 2000 to 2006.

90. *Online Betting* Consumer gambling losses from online betting were $4 billion in 2002 and $10 billion in 2005. (**Source:** Christiansen Capital Advisors.)

(a) Find a linear function given by

$$B(x) = m(x - x_1) + y_1$$

that models these data, where x is the year.

(b) Use $B(x)$ to estimate the years when consumer losses from online betting were more than $6 billion. Assume that the domain of B is 2002 to 2007.

91. *Consumer Spending* In 2005 consumers used credit and debit cards to pay for 40% of all purchases. This percentage is projected to be 55% in 2011. (**Source:** Bloomburg.)

(a) Find a linear function P that models the data.

(b) Estimate when this percentage was between 45% and 50%.

92. *VISA Cards* Annual transactions on VISA cards increased from $400 billion in 2002 to $635 billion in 2007. (**Source:** CardWeb.)

(a) Find a linear function V that models the data.

(b) Estimate when this number was between $450 billion and $540 billion.

93. *Modeling Sunrise Times* In Boston, on the 90th day (March 30) of 2008 the sun rose at 6:30 A.M., and on the 129th day (May 8) the sun rose at 5:30 A.M. Use a linear function to estimate the days when the sun rose between 5:45 A.M. and 6:00 A.M. Do not consider days after May 8. (**Source:** R. Thomas.)

94. *Modeling Sunrise Times* In Denver, on the 77th day (March 17) of 2008 the sun rose at 7:00 A.M., and on the 112th day (April 21) the sun rose at 6:00 A.M. Use a linear function to estimate the days when the sun rose between 6:10 A.M. and 6:40 A.M. Do not consider days after April 21. (**Source:** R. Thomas.)

95. *Error Tolerances* Suppose that an aluminum can is manufactured so that its radius r can vary from 1.99 inches to 2.01 inches. What range of values is possible for the circumference C of the can? Express your answer by using a three-part inequality.

96. *Error Tolerances* Suppose that a square picture frame has sides that vary between 9.9 inches and 10.1 inches. What range of values is possible for the perimeter P of the picture frame? Express your answer by using a three-part inequality.

97. *Modeling Data* The following data are exactly linear.

x	0	2	4	6
y	−1.5	4.5	10.5	16.5

(a) Find a linear function f that models the data.

(b) Solve the inequality $f(x) > 2.25$.

98. *Modeling Data* The following data are exactly linear.

x	1	2	3	4	5
y	0.4	3.5	6.6	9.7	12.8

(a) Find a linear function f that models the data.

(b) Solve the inequality $2 \leq f(x) \leq 8$.

Linear Regression

99. *Cell Phone Subscribers* The table lists the number N of cell phone subscribers worldwide in millions for selected years x.

x	2001	2002	2003	2004	2005
N	128	141	159	182	208

Source: CTIA—The Wireless Association.

(a) Find a linear function N that models the data.

(b) Estimate the years when this number was from 43 million to 83 million.

(c) Did your estimate involve interpolation or extrapolation?

100. *Home Ownership Rates* The table lists the percentage P of U.S. homes that are owned by their occupant rather than rented for selected years x.

x	1900	1950	1980	2006
P	47%	55%	64%	69%

Source: Bureau of the Census.

(a) Find a linear function P that models the data.

(b) Estimate the years when this percentage was from 58% to 60%.

(c) Did your estimate involve interpolation or extrapolation?

Writing about Mathematics

101. Suppose the solution to the equation $ax + b = 0$ with $a > 0$ is $x = k$. Discuss how the value of k can be used to help solve the linear inequalities $ax + b > 0$ and $ax + b < 0$. Illustrate this process graphically. How would the solution sets change if $a < 0$?

102. Describe how to numerically solve the linear inequality $ax + b \leq 0$. Give an example.

103. If you multiply each part of a three-part inequality by the same negative number, what must you make sure to do? Explain by using an example.

104. Explain how a linear function, a linear equation, and a linear inequality are related. Give an example.

EXTENDED AND DISCOVERY EXERCISES

1. *Arithmetic Mean*   The arithmetic mean of two numbers a and b is given by $\frac{a + b}{2}$. Use properties of inequalities to show that if $a < b$, then $a < \frac{a + b}{2} < b$.

2. *Geometric Mean*   The **geometric mean** of two numbers a and b is given by $\sqrt{ab}$. Use properties of inequalities to show that if $0 < a < b$, then $a < \sqrt{ab} < b$.

CHECKING BASIC CONCEPTS FOR SECTIONS 2.3 AND 2.4

1. Solve the linear equation $4(x - 2) = 2(5 - x) - 3$ by using each method. Compare your results.
   (a) Graphical   **(b)** Numerical   **(c)** Symbolic

2. Solve the inequality $2(x - 4) > 1 - x$. Express the solution set in set-builder notation.

3. Solve the compound inequality $-2 \leq 1 - 2x \leq 3$. Use set-builder or interval notation.

4. Use the graph to the right to solve each equation and inequality. Then solve each part symbolically. Use set-builder or interval notation when possible.
   (a) $-3(2 - x) - \frac{1}{2}x - \frac{3}{2} = 0$

(b) $-3(2 - x) - \frac{1}{2}x - \frac{3}{2} > 0$

(c) $-3(2 - x) - \frac{1}{2}x - \frac{3}{2} \leq 0$

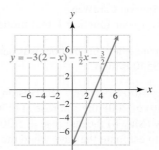

$y = -3(2 - x) - \frac{1}{2}x - \frac{3}{2}$

2.5 Absolute Value Equations and Inequalities

- **Evaluate and graph the absolute value function**
- **Solve absolute value equations**
- **Solve absolute value inequalities**

Introduction

A margin of error can be very important in many aspects of life, including being fired out of a cannon. The most dangerous part of the feat, first done by a human in 1875, is to land squarely on a net. For a human cannonball who wants to fly 180 feet in the air and then land in the center of a net with a 60-foot-long safe zone, there is a margin of error of ± 30 feet. That is, the horizontal distance D traveled by the human cannonball can vary between $180 - 30 = 150$ feet and $180 + 30 = 210$ feet.   (**Source:** Ontario Science Center.)

This margin of error can be expressed mathematically by using the *absolute value inequality*

$$|D - 180| \leq 30.$$

The absolute value is necessary because D can be either less than or greater than 180, but by not more than 30 feet.

The Absolute Value Function

Figure 2.56 The Absolute Value Function

The graph of $y = |x|$ is shown in Figure 2.56. It is V-shaped and cannot be represented by a single linear function. However, it can be represented by the lines $y = x$ (when $x \geq 0$) and $y = -x$ (when $x < 0$). This suggests that the absolute value function can be defined symbolically using a piecewise-linear function. The absolute value function is decreasing for $x \leq 0$ and increasing for $x \geq 0$.

$$|x| = \begin{cases} -x & \text{if } x < 0 \\ x & \text{if } x \geq 0 \end{cases}$$

There is another formula for $|x|$. Consider the following examples.

$$\sqrt{3^2} = \sqrt{9} = 3 \quad \text{and} \quad \sqrt{(-3)^2} = \sqrt{9} = 3.$$
$$\sqrt{7^2} = \sqrt{49} = 7 \quad \text{and} \quad \sqrt{(-7)^2} = \sqrt{49} = 7.$$

That is, regardless of whether a real number x is positive or negative, the expression $\sqrt{x^2}$ equals the *absolute value* of x. This statement is summarized by

$$\sqrt{x^2} = |x| \qquad \text{for all real numbers } x.$$

For example, $\sqrt{y^2} = |y|$, $\sqrt{(x-1)^2} = |x-1|$, and $\sqrt{(2x)^2} = |2x|$.

EXAMPLE 1   Analyzing the graph of $y = |ax + b|$

For each linear function f, graph $y = f(x)$ and $y = |f(x)|$ separately. Discuss how the absolute value affects the graph of f.
(a) $f(x) = x + 2$ **(b)** $f(x) = -2x + 4$

SOLUTION
(a) The graphs of $y_1 = x + 2$ and $y_2 = |x + 2|$ are shown in Figures 2.57 and 2.58, respectively. The graph of y_1 is a line with x-intercept -2. The graph of y_2 is V-shaped. The graphs are identical for $x > -2$. For $x < -2$, the graph of $y_1 = f(x)$ passes below the x-axis, and the graph of $y_2 = |f(x)|$ is the *reflection* of $y_1 = f(x)$ across the x-axis. The graph of $y_2 = |f(x)|$ does not dip below the x-axis because an absolute value is *never* negative.

Figure 2.57

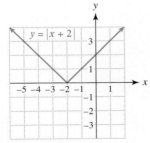

Figure 2.58

NOTE In general, the graph of $y = |f(x)|$ is a reflection of the graph of $y = f(x)$ across the x-axis whenever $f(x) < 0$. Otherwise (whenever $f(x) \geq 0$), their graphs are identical.

Calculator Help

To access the absolute value function, see Appendix A (page AP-10).

(b) The graphs of $y_1 = -2x + 4$ and $y_2 = |-2x + 4|$ are shown in Figures 2.59 and 2.60. Again, the graph of y_2 is V-shaped. The graph of $y_2 = |f(x)|$ is the reflection of f across the x-axis whenever the graph of $y_1 = f(x)$ is below the x-axis.

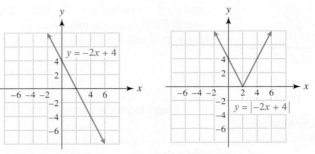

Figure 2.59   **Figure 2.60**

Now Try Exercises 13 and 17 ◀

Example 1 illustrates the fact that the graph of $y = |ax + b|$ with $a \neq 0$ is V-shaped and is never located below the x-axis. The vertex (or point) of the V-shaped graph corresponds to the x-intercept, which can be found by solving the linear equation $ax + b = 0$.

Absolute Value Equations

Figure 2.61

The equation $|x| = 5$ has two solutions: ± 5. This fact is shown visually in Figure 2.61, where the graph of $y = |x|$ intersects the horizontal line $y = 5$ at the points $(\pm 5, 5)$. In general, the solutions to $|x| = k$ with $k > 0$ are given by $x = \pm k$. Thus if $y = ax + b$, then $|ax + b| = k$ has two solutions given by $ax + b = \pm k$.

These concepts can be illustrated visually. The graph of $y = |ax + b|$ with $a \neq 0$ is V-shaped. It intersects the horizontal line $y = k$ twice whenever $k > 0$, as illustrated in Figure 2.62. Thus there are two solutions to the equation $|ax + b| = k$. This V-shaped graph intersects the line $y = 0$ once, as shown in Figure 2.63. As a result, equation $|ax + b| = 0$ has one solution, which corresponds to the x-intercept. When $k < 0$, the line $y = k$ lies *below* the x-axis and there are no points of intersection, as shown in Figure 2.64. Thus the equation $|ax + b| = k$ with $k < 0$ has no solutions.

Figure 2.62 Two Solutions   **Figure 2.63** One Solution   **Figure 2.64** No Solutions

---

Absolute Value Equations

Let k be a positive number. Then

$$|ax + b| = k \quad \text{is equivalent to} \quad ax + b = \pm k.$$

EXAMPLE 2 Solving absolute value equations

Solve each equation.

(a) $\left|\frac{3}{4}x - 6\right| = 15$ **(b)** $\left|1 - 2x\right| = -3$ **(c)** $\left|3x - 2\right| - 5 = -2$

SOLUTION

(a) The equation $\left|\frac{3}{4}x - 6\right| = 15$ is satisfied when $\frac{3}{4}x - 6 = \pm 15$.

$$\frac{3}{4}x - 6 = 15 \qquad \text{or} \qquad \frac{3}{4}x - 6 = -15 \qquad \textit{Equations to solve}$$
$$\frac{3}{4}x = 21 \qquad \text{or} \qquad \frac{3}{4}x = -9 \qquad \textit{Add 6 to each side.}$$
$$x = 28 \qquad \text{or} \qquad x = -12 \qquad \textit{Multiply by } \tfrac{4}{3}.$$

The solutions are -12 and 28.

(b) Because an absolute value is never negative, $\left|1 - 2x\right| \geq 0$ for all x and can never equal -3. There are no solutions. This is illustrated graphically in Figure 2.65.

(c) Because the right side of the equation is a negative number, it might appear at first glance that there were no solutions. However, if we add 5 to each side of the equation,

$$\left|3x - 2\right| - 5 = -2 \quad \text{becomes} \quad \left|3x - 2\right| = 3.$$

This equation is equivalent to $3x - 2 = \pm 3$ and has two solutions.

$$3x - 2 = 3 \qquad \text{or} \qquad 3x - 2 = -3 \qquad \textit{Equations to solve}$$
$$3x = 5 \qquad \text{or} \qquad 3x = -1 \qquad \textit{Add 2 to each side.}$$
$$x = \frac{5}{3} \qquad \text{or} \qquad x = -\frac{1}{3} \qquad \textit{Divide by 3.}$$

The solutions are $-\frac{1}{3}$ and $\frac{5}{3}$.

Now Try Exercises 21, 29, and 31 ◀

Figure 2.65 No Solutions

EXAMPLE 3 Solving an equation with technology

Solve the equation $\left|2x + 5\right| = 2$ graphically, numerically, and symbolically.

SOLUTION

Graphical Solution Graph $Y_1 = \text{abs}(2X + 5)$ and $Y_2 = 2$. The V-shaped graph of y_1 intersects the horizontal line at the points $(-3.5, 2)$ and $(-1.5, 2)$, as shown in Figures 2.66 and 2.67. The solutions are -3.5 and -1.5.

Numerical Solution Table $Y_1 = \text{abs}(2X + 5)$ and $Y_2 = 2$, as shown in Figure 2.68. The solutions to $y_1 = y_2$ are -3.5 and -1.5.

Calculator Help

To find a point of intersection, see Appendix A (page AP-8).

[−9, 9, 1] by [−6, 6, 1]

Figure 2.66

[−9, 9, 1] by [−6, 6, 1]

Figure 2.67

X	Y1	Y2
-4	3	2
-3.5	2	2 ←$y_1 = y_2$
-3	1	2
-2.5	0	2
-2	1	2
-1.5	2	2 ←$y_1 = y_2$
-1	3	2

Y1◨abs(2X+5)

Figure 2.68

Symbolic Solution The equation $|2x + 5| = 2$ is satisfied when $2x + 5 = \pm 2$.

$$2x + 5 = 2 \quad \text{or} \quad 2x + 5 = -2 \qquad \text{Equations to solve}$$
$$2x = -3 \quad \text{or} \quad 2x = -7 \qquad \text{Subtract 5 from each side.}$$
$$x = -\frac{3}{2} \quad \text{or} \quad x = -\frac{7}{2} \qquad \text{Divide by 2.}$$

Now Try Exercise 45 ◀

EXAMPLE 4 Describing speed limits with absolute values

The lawful speeds S on an interstate highway satisfy $|S - 55| \leq 15$. Find the maximum and minimum speed limits by solving the equation $|S - 55| = 15$.

SOLUTION The equation $|S - 55| = 15$ is equivalent to $S - 55 = \pm 15$.

$$S - 55 = 15 \quad \text{or} \quad S - 55 = -15 \qquad \text{Equations to solve}$$
$$S = 70 \quad \text{or} \quad S = 40 \qquad \text{Add 55 to each side.}$$

The maximum speed limit is 70 miles per hour and the minimum is 40 miles per hour.

Now Try Exercise 73 ◀

An Equation with Two Absolute Values Sometimes more than one absolute value sign occurs in an equation. For example, an equation might be in the form

$$|ax + b| = |cx + d|.$$

In this case there are two possibilities:

$$\text{either} \quad ax + b = cx + d \quad \text{or} \quad ax + b = -(cx + d).$$

This symbolic technique is demonstrated in the next example.

EXAMPLE 5 Solving an equation involving two absolute values

Solve the equation $|x - 2| = |1 - 2x|$.

SOLUTION We must solve both of the following equations.

$$x - 2 = 1 - 2x \quad \text{or} \quad x - 2 = -(1 - 2x)$$
$$3x = 3 \quad \text{or} \quad x - 2 = -1 + 2x$$
$$x = 1 \quad \text{or} \quad -1 = x$$

There are two solutions: -1 and 1.

Now Try Exercise 35 ◀

Absolute Value Inequalities

In Figure 2.69 the solutions to $|ax + b| = k$ are labeled s_1 and s_2. The V-shaped graph of $y = |ax + b|$ is below the horizontal line $y = k$ between s_1 and s_2, or when $s_1 < x < s_2$. The solution set for the inequality $|ax + b| < k$ is green on the x-axis. In Figure 2.70 the V-shaped graph is above the horizontal line $y = k$ left of s_1 or right of s_2, that is, when $x < s_1$ or $x > s_2$. The solution set for the inequality $|ax + b| > k$ is green on the x-axis.

Figure 2.69   **Figure 2.70**

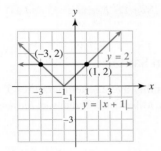

Figure 2.71

Note that in both figures equality (determined by s_1 and s_2) is the boundary between *greater than* and *less than*. For this reason, s_1 and s_2 are called *boundary numbers*.

For example, the graphs of $y = |x + 1|$ and $y = 2$ are shown in Figure 2.71. These graphs intersect at the points $(-3, 2)$ and $(1, 2)$. It follows that the two solutions to

$$|x + 1| = 2$$

are $s_1 = -3$ and $s_2 = 1$. The solutions to $|x + 1| < 2$ lie between $s_1 = -3$ and $s_2 = 1$, which can be written as $-3 < x < 1$. Furthermore, the solutions to $|x + 1| > 2$ lie "outside" $s_1 = -3$ and $s_2 = 1$. This can be written as $x < -3$ or $x > 1$.

These results are generalized as follows.

Absolute Value Inequalities

Let the solutions to $|ax + b| = k$ be s_1 and s_2, where $s_1 < s_2$ and $k > 0$.

1. $|ax + b| < k$ is equivalent to $s_1 < x < s_2$.
2. $|ax + b| > k$ is equivalent to $x < s_1$ or $x > s_2$.

Similar statements can be made for inequalities involving $\leq$ or $\geq$.

NOTE The union symbol $\cup$ may be used to write $x < s_1$ or $x > s_2$ in interval notation. For example, $x < -3$ or $x > 1$ is written as $(-\infty, -3) \cup (1, \infty)$ in interval notation. This indicates that the solution set includes all real numbers in either $(-\infty, -3)$ or $(1, \infty)$.

EXAMPLE 6   Solving inequalities involving absolute values symbolically

Solve each inequality symbolically. Write the solution set in interval notation.

(a) $|2x - 5| \leq 6$ **(b)** $|5 - x| > 3$

SOLUTION
(a) Begin by solving $|2x - 5| = 6$, or equivalently, $2x - 5 = \pm 6$.

$$
\begin{array}{ccc}
2x - 5 = 6 & \text{or} & 2x - 5 = -6 \\
2x = 11 & \text{or} & 2x = -1 \\
x = \dfrac{11}{2} & \text{or} & x = -\dfrac{1}{2}
\end{array}
$$

The solutions to $|2x - 5| = 6$ are $-\frac{1}{2}$ and $\frac{11}{2}$. The solution set for the inequality $|2x - 5| \leq 6$ includes all real numbers x satisfying $-\frac{1}{2} \leq x \leq \frac{11}{2}$. In interval notation this is written as $\left[-\frac{1}{2}, \frac{11}{2}\right]$.

(b) To solve $|5 - x| > 3$, begin by solving $|5 - x| = 3$, or equivalently, $5 - x = \pm 3$.

$$5 - x = 3 \quad \text{or} \quad 5 - x = -3$$
$$-x = -2 \quad \text{or} \quad -x = -8$$
$$x = 2 \quad \text{or} \quad x = 8$$

The solutions to $|5 - x| = 3$ are 2 and 8. The solution set for $|5 - x| > 3$ includes all real numbers x left of 2 or right of 8. Thus $|5 - x| > 3$ is equivalent to $x < 2$ or $x > 8$. In interval notation this is written as $(-\infty, 2) \cup (8, \infty)$.

Now Try Exercises 55 and 63 ◀

EXAMPLE 7 Analyzing the temperature range in Santa Fe

The inequality $|T - 49| \leq 20$ describes the range of monthly average temperatures T in degrees Fahrenheit for Santa Fe, New Mexico. (**Source:** A. Miller and J. Thompson, *Elements of Meteorology.*)

(a) Solve this inequality graphically and symbolically.
(b) The high and low monthly average temperatures satisfy the absolute value equation $|T - 49| = 20$. Use this fact to interpret the results from part (a).

SOLUTION

(a) *Graphical Solution* Graph $Y_1 = \text{abs}(X - 49)$ and $Y_2 = 20$, as in Figure 2.72. The V-shaped graph of y_1 intersects the horizontal line at the points $(29, 20)$ and $(69, 20)$. See Figures 2.73 and 2.74. The graph of y_1 is below the graph of y_2 between these two points. Thus the solution set consists of all temperatures T satisfying $29 \leq T \leq 69$.

[20, 80, 5] by [0, 35, 5]

Figure 2.72

[20, 80, 5] by [0, 35, 5]

Figure 2.73

[20, 80, 5] by [0, 35, 5]

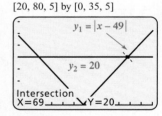

Figure 2.74

Symbolic Solution First solve the related equation $|T - 49| = 20$.

$$T - 49 = -20 \quad \text{or} \quad T - 49 = 20$$
$$T = 29 \quad \text{or} \quad T = 69$$

Thus by our previous discussion $|T - 49| \leq 20$ is equivalent to $29 \leq T \leq 69$.

(b) The solutions to $|T - 49| = 20$ are 29 and 69. Therefore the monthly average temperatures in Sante Fe vary between a low of 29°F (January) and a high of 69°F (July). The monthly averages are always within 20 degrees of 49°F. *Now Try Exercise 77* ◀

An Alternative Method There is a second symbolic method that can be used to solve absolute value inequalities. This method is often used in advanced mathematics courses, such as calculus. It is based on the following two properties.

Absolute Value Inequalities (Alternative Method)
Let k be a positive number. **1.** $\lvert ax + b \rvert < k$ is equivalent to $-k < ax + b < k$. **2.** $\lvert ax + b \rvert > k$ is equivalent to $ax + b < -k$ or $ax + b > k$. Similar statements can be made for inequalities involving $\leq$ or $\geq$.

EXAMPLE 8 Using an alternative method

Solve each absolute value inequality. Write your answer in interval notation.

(a) $\lvert 4 - 5x \rvert \leq 3$ **(b)** $\lvert -4x - 6 \rvert > 2$

SOLUTION

(a) $\lvert 4 - 5x \rvert \leq 3$ is equivalent to the following three-part inequality.

$$-3 \leq 4 - 5x \leq 3 \qquad \textit{Equivalent inequality}$$

$$-7 \leq -5x \leq -1 \qquad \textit{Subtract 4 from each part.}$$

$$\frac{7}{5} \geq x \geq \frac{1}{5} \qquad \textit{Divide each part by -5; reverse the inequality.}$$

CLASS DISCUSSION

In interval notation the solution is $\left[\frac{1}{5}, \frac{7}{5}\right]$.

Sketch the graphs of $y = ax + b$, $y = \lvert ax + b \rvert$, $y = -k$, and $y = k$ on one xy-plane. Now use these graphs to explain why the alternative method for solving absolute value inequalities is correct.

(b) $\lvert -4x - 6 \rvert > 2$ is equivalent to the following compound inequality.

$-4x - 6 < -2$	or	$-4x - 6 > 2$	*Equivalent compound inequality*
$-4x < 4$	or	$-4x > 8$	*Add 6 to each side.*
$x > -1$	or	$x < -2$	*Divide each by -4; reverse the inequality.*

In interval notation the solution set is $(-\infty, -2) \cup (-1, \infty)$.

Now Try Exercises 57 and 61 ◀

2.5 Putting It All Together

The following table summarizes some important concepts from this section.

Concept	Explanation	Examples
Absolute value function	$f(x) = \lvert x \rvert$: The output from the absolute value function is never negative. $f(x) = \lvert x \rvert$ is equivalent to $f(x) = \sqrt{x^2}$.	$f(-2) = \lvert -2 \rvert = 2$

continued on next page

continued from previous page

Concept	Explanation	Examples
Absolute value equations	**1.** If $k > 0$, then $\lvert ax + b \rvert = k$ has two solutions, given by $ax + b = \pm k$.	**1.** Solve $\lvert 3x - 5 \rvert = 4$. $\begin{array}{llll} 3x - 5 = -4 & \text{or} & 3x - 5 = 4 \\ 3x = 1 & \text{or} & 3x = 9 \\ x = \frac{1}{3} & \text{or} & x = 3 \end{array}$
	2. If $k = 0$, then $\lvert ax + b \rvert = k$ has one solution, given by $ax + b = 0$.	**2.** Solve $\lvert x - 1 \rvert = 0$. $x - 1 = 0$ implies $x = 1$.
	3. If $k < 0$, then $\lvert ax + b \rvert = k$ has no solutions.	**3.** $\lvert 4x - 9 \rvert = -2$ has no solutions.
Absolute value inequalities	To solve $\lvert ax + b \rvert < k$ or $\lvert ax + b \rvert > k$ with $k > 0$, first solve $\lvert ax + b \rvert = k$. Let these solutions be s_1 and s_2, where $s_1 < s_2$.	To solve $\lvert x - 5 \rvert < 4$ or $\lvert x - 5 \rvert > 4$, first solve $\lvert x - 5 \rvert = 4$ to obtain the solutions $s_1 = 1$ and $s_2 = 9$.
	1. $\lvert ax + b \rvert < k$ is equivalent to $s_1 < x < s_2$.	**1.** $\lvert x - 5 \rvert < 4$ is equivalent to $1 < x < 9$.
	2. $\lvert ax + b \rvert > k$ is equivalent to $x < s_1$ or $x > s_2$.	**2.** $\lvert x - 5 \rvert > 4$ is equivalent to $x < 1$ or $x > 9$.
Alternative method for solving absolute value inequalities	**1.** $\lvert ax + b \rvert < k$ with $k > 0$ is equivalent to $-k < ax + b < k.$	**1.** $\lvert x - 1 \rvert < 5$ is solved as follows. $-5 < x - 1 < 5$ $-4 < x < 6$
	2. $\lvert ax + b \rvert > k$ with $k > 0$ is equivalent to $ax + b < -k$ or $ax + b > k.$	**2.** $\lvert x - 1 \rvert > 5$ is solved as follows. $\begin{array}{llll} x - 1 < -5 & \text{or} & x - 1 > 5 \\ x < -4 & \text{or} & x > 6 \end{array}$

2.5 Exercises

Basic Concepts

Exercises 1–8: Let $a \neq 0$.

1. Solve $\lvert x \rvert = 3$.

2. Solve $\lvert x \rvert \leq 3$.

3. Solve $\lvert x \rvert > 3$.

4. Solve $\lvert ax + b \rvert \leq -2$.

5. Describe the graph of $y = \lvert ax + b \rvert$.

6. Solve $\lvert ax + b \rvert = 0$.

7. Rewrite $\sqrt{36a^2}$ by using an absolute value.

8. Rewrite $\sqrt{(ax + b)^2}$ by using an absolute value.

Absolute Value Graphs

Exercises 9–12: Graph by hand.

(a) *Find the x-intercept.*

(b) *Determine where the graph is increasing and where it is decreasing.*

9. $y = \lvert x + 1 \rvert$

10. $y = \lvert 1 - x \rvert$

11. $y = \lvert 2x - 3 \rvert$

12. $y = \lvert \frac{1}{2}x + 1 \rvert$

Exercises 13–18: (Refer to Example 1.) Do the following.

(a) *Graph $y = f(x)$.*

(b) *Use the graph of $y = f(x)$ to sketch a graph of the equation $y = \lvert f(x) \rvert$.*

(c) *Determine the x-intercept for the graph of the equation $y = \lvert f(x) \rvert$.*

13. $y = 2x$

14. $y = \frac{1}{2}x$

15. $y = 3x - 3$

16. $y = 2x - 4$

17. $y = 6 - 2x$

18. $y = 2 - 4x$

Absolute Value Equations and Inequalities

Exercises 19–40: Solve the absolute value equation.

19. $|-2x| = 4$ **20.** $|3x| = -6$

21. $|5x - 7| = 2$ **22.** $|-3x - 2| = 5$

23. $|3 - 4x| = 5$ **24.** $|2 - 3x| = 1$

25. $|-6x - 2| = 0$ **26.** $|6x - 9| = 0$

27. $|7 - 16x| = 0$ **28.** $|-x - 4| = 0$

29. $|17x - 6| = -3$ **30.** $|-8x - 11| = -7$

31. $|1.2x - 1.7| - 1 = 3$ **32.** $|3 - 3x| - 2 = 2$

33. $|4x - 5| + 3 = 2$ **34.** $|4.5 - 2x| + 1.1 = 9.7$

35. $|2x - 9| = |8 - 3x|$ **36.** $|x - 3| = |8 - x|$

37. $\left|\frac{3}{4}x - \frac{1}{4}\right| = \left|\frac{3}{4} - \frac{1}{4}x\right|$ **38.** $\left|\frac{1}{2}x + \frac{3}{2}\right| = \left|\frac{3}{2}x - \frac{7}{2}\right|$

39. $|15x - 5| = |35 - 5x|$

40. $|20x - 40| = |80x - 20|$

Exercises 41 and 42. The graphs of f and g are shown. Solve each equation and inequality.

41. **(a)** $f(x) = g(x)$

  (b) $f(x) < g(x)$

  (c) $f(x) > g(x)$

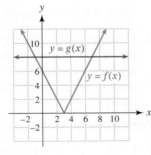

42. **(a)** $f(x) = g(x)$

  (b) $f(x) \leq g(x)$

  (c) $f(x) \geq g(x)$

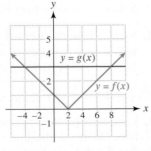

Exercises 43 and 44: Solve each equation or inequality.

43. **(a)** $|2x - 3| = 1$ **44.** **(a)** $|5 - x| = 2$

  (b) $|2x - 3| < 1$ **(b)** $|5 - x| \leq 2$

  (c) $|2x - 3| > 1$ **(c)** $|5 - x| \geq 2$

Exercises 45–48: Solve the equation

(a) *graphically,*
(b) *numerically, and*
(c) *symbolically.*
Then solve the related inequality.

45. $|2x - 5| = 10,\quad |2x - 5| < 10$

46. $|3x - 4| = 8,\quad |3x - 4| \leq 8$

47. $|5 - 3x| = 2,\quad |5 - 3x| > 2$

48. $|4x - 7| = 5,\quad |4x - 7| \geq 5$

Exercises 49–54: Solve the equation symbolically. Then solve the related inequality.

49. $|2.1x - 0.7| = 2.4,\quad |2.1x - 0.7| \geq 2.4$

50. $\left|\frac{1}{2}x - \frac{3}{4}\right| = \frac{7}{4},\quad \left|\frac{1}{2}x - \frac{3}{4}\right| \leq \frac{7}{4}$

51. $|3x| + 5 = 6,\quad |3x| + 5 > 6$

52. $|x| - 10 = 25,\quad |x| - 10 < 25$

53. $\left|\frac{2}{3}x - \frac{1}{2}\right| = -\frac{1}{4},\quad \left|\frac{2}{3}x - \frac{1}{2}\right| \leq -\frac{1}{4}$

54. $|5x - 0.3| = -4,\quad |5x - 0.3| > -4$

You Decide the Method

Exercises 55–66: Solve the inequality. Write the solution in interval notation.

55. $|3x - 1| < 8$ **56.** $|15 - x| < 7$

57. $|7 - 4x| \leq 11$ **58.** $|-3x + 1| \leq 5$

59. $|0.5x - 0.75| < 2$ **60.** $|2.1x - 5| \leq 8$

61. $|2x - 3| > 1$ **62.** $|5x - 7| > 2$

63. $|-3x + 8| \geq 3$ **64.** $|-7x - 3| \geq 5$

65. $|0.25x - 1| > 3$ **66.** $|-0.5x + 5| \geq 4$

Domain and Range

67. If $f(k) = -6$, what is the value of $|f(k)|$?

68. If $f(k) = 17$, what is the value of $|f(k)|$?

69. If the domain of $f(x)$ is given by $[-2, 4]$, what is the domain of $|f(x)|$?

70. If the domain of $f(x)$ is given by $(-\infty, 0]$, what is the domain of $|f(x)|$?

71. If the range of $f(x)$ is given by $(-\infty, 0]$, what is the range of $|f(x)|$?

72. If the range of $f(x)$ is given by $(-4, 5)$, what is the range of $|f(x)|$?

Applications

73. *Speed Limits* The lawful speeds S on an interstate highway satisfy $|S - 57.5| \le 17.5$. Find the maximum and minimum speed limits by solving the absolute value equation $|S - 57.5| = 17.5$.

74. *Human Cannonball* A human cannonball plans to travel 180 feet and land squarely on a net with a 70-foot-long safe zone.
  (a) What distances D can this performer travel and still land safely on the net?

  (b) Use an absolute value inequality to describe the restrictions on D.

75. *Temperature and Altitude* Air temperature decreases as altitude increases. If the ground temperature is 80°F, then the air temperature x miles high is $T = 80 - 19x$.
  (a) Determine the altitudes x where the air temperature T is between 0°F and 32°F, inclusive.

  (b) Use an absolute value inequality to describe these altitudes.

76. *Dew Point and Altitude* The dew point decreases as altitude increases. If the ground temperature is 80°F, then the dew point x miles high is $D = 80 - \frac{29}{5}x$.
  (a) Determine the altitudes x where the dew point D is between 50°F and 60°F, inclusive.

  (b) Use an absolute value inequality to describe these altitudes.

Exercises 77–82: Average Temperatures (Refer to Example 7.) The inequality describes the range of monthly average temperatures T in degrees Fahrenheit at a certain location.
  (a) Solve the inequality.
  (b) If the high and low monthly average temperatures satisfy equality, interpret the inequality.

77. $|T - 43| \le 24$, Marquette, Michigan

78. $|T - 62| \le 19$, Memphis, Tennessee

79. $|T - 50| \le 22$, Boston, Massachusetts

80. $|T - 10| \le 36$, Chesterfield, Canada

81. $|T - 61.5| \le 12.5$, Buenos Aires, Argentina

82. $|T - 43.5| \le 8.5$, Punta Arenas, Chile

83. *Error in Measurements* Products are often manufactured to be within a specified tolerance of a given size. For instance, if an aluminum can is supposed to have a diameter of 3 inches, either 2.99 inches or 3.01 inches might be acceptable. If the maximum error in the diameter d of a can is limited to 0.004 inch, then d must satisfy the absolute value inequality

$$|d - 3| \le 0.004.$$

Solve this inequality and interpret the results.

84. *Error in Measurements* (Refer to Exercise 83.) Suppose that a 12-inch ruler must have the correct length L to within 0.0002 inch.
  (a) Write an absolute value inequality for L that describes this requirement.

  (b) Solve this inequality and interpret the results.

85. *Relative Error* If a quantity is measured to be Q and its exact value is A, then the relative error in Q is

$$\left| \frac{Q - A}{A} \right|.$$

If the exact value is $A = 35$ and you want the relative error in Q to be less than or equal to 0.02 (or 2%), what values for Q are possible?

86. *Relative Error* (Refer to Exercise 85.) The exact perimeter P of a square is 50 feet. What measured lengths are possible for the side S of the square to have relative error in the perimeter that is less than or equal to 0.04 (or 4%)?

Writing about Mathematics

87. Explain how to solve $|ax + b| = k$ with $k > 0$ symbolically. Give an example.

88. Explain how you can use the solutions to $|ax + b| = k$ with $k > 0$ to solve the inequalities $|ax + b| < k$ and $|ax + b| > k$. Give an example.

EXTENDED AND DISCOVERY EXERCISES

1. Let δ be a positive number and let x and c be real numbers. Write an absolute value inequality that expresses that the distance between x and c on the number line is less than δ.

2. Let ϵ be a positive number, L be a real number, and f be a function. Write an absolute value inequality that expresses that the distance between $f(x)$ and L on the number line is less than ϵ.

CHECKING BASIC CONCEPTS FOR SECTION 2.5

1. Rewrite $\sqrt{4x^2}$ by using an absolute value.

2. Graph $y = |3x - 2|$ by hand.

3. (a) Solve the equation $|2x - 1| = 5$.

   (b) Use part (a) to solve the absolute value inequalities $|2x - 1| \leq 5$ and $|2x - 1| > 5$.

4. Solve each equation or inequality. For each inequality, write the solution set in interval notation.
   (a) $|2 - 5x| - 4 = -1$

   (b) $|3x - 5| \leq 4$

   (c) $|\frac{1}{2}x - 3| > 5$

5. Solve $|x + 1| = |2x|$.

2   Summary

CONCEPT	EXPLANATION AND EXAMPLES

SECTION 2.1 LINEAR FUNCTIONS AND MODELS

Linear Function

A linear function can be written as $f(x) = ax + b$. Its graph is a line.

Example: $f(x) = -2x + 5$; slope $= -2$, y-intercept $= 5$

Linear Model

If a quantity increases or decreases by a constant amount for each unit increase in x, then it can be modeled by a linear function given by

$$f(x) = (\text{constant rate of change})x + (\text{initial amount}).$$

Example: If water is pumped from a full 100-gallon tank at 7 gallons per minute, then $A(t) = 100 - 7t$ gives the gallons of water in the tank after t minutes.

Piecewise-Defined Function

A function defined by more than one formula on its domain

Examples: Step function, greatest integer function, absolute value function, and

$$f(x) = \begin{cases} 4 - x & \text{if } -4 \leq x < 1 \\ 3x & \text{if } 1 \leq x \leq 5 \end{cases}$$

It follows that $f(2) = 6$ because if $1 \leq x \leq 5$ then $f(x) = 3x$. Note that f is continuous on its domain of $[-4, 5]$.

Linear Regression

One way to determine a linear function or a line that models data is to use the method of least squares. This method determines a unique line that can be found with a calculator. The correlation coefficient r $(-1 \leq r \leq 1)$ measures how well a line fits the data.

Example: The line of least squares modeling the data $(1, 1)$, $(3, 4)$, and $(4, 6)$ is given by $y \approx 1.643x - 0.714$, with $r \approx 0.997$.

CONCEPT	EXPLANATION AND EXAMPLES

SECTION 2.2 EQUATIONS OF LINES

Point-Slope Form

If a line with slope m passes through (x_1, y_1), then
$$y = m(x - x_1) + y_1 \quad \text{or} \quad y - y_1 = m(x - x_1).$$

Example: $y = -\frac{3}{4}(x + 4) + 5$ has slope $-\frac{3}{4}$ and passes through $(-4, 5)$.

Slope-Intercept Form

If a line has slope m and y-intercept b, then $y = mx + b$.

Example: $y = 3x - 4$ has slope 3 and y-intercept -4.

Determining Intercepts

To find the x-intercept(s), let $y = 0$ in the equation and solve for x.
To find the y-intercept(s), let $x = 0$ in the equation and solve for y.

Examples: The x-intercept on the graph of $3x - 4y = 12$ is 4 because $3x - 4(0) = 12$ implies that $x = 4$.

The y-intercept on the graph of $3x - 4y = 12$ is -3 because $3(0) - 4y = 12$ implies that $y = -3$.

Horizontal and Vertical Lines

A horizontal line passing through the point (a, b) is given by $y = b$, and a vertical line passing through (a, b) is given by $x = a$.

Examples: The horizontal line $y = -3$ passes through $(6, -3)$.

The vertical line $x = 4$ passes through $(4, -2)$.

Parallel and Perpendicular Lines

Parallel lines have equal slopes satisfying $m_1 = m_2$, and perpendicular lines have slopes satisfying $m_1 m_2 = -1$, provided neither line is vertical.

Examples: The lines $y_1 = 3x - 1$ and $y_2 = 3x + 4$ are parallel.

The lines $y_1 = 3x - 1$ and $y_2 = -\frac{1}{3}x + 4$ are perpendicular.

Direct Variation

A quantity y is directly proportional to a quantity x, or y varies directly with x, if $y = kx$, where $k \neq 0$. If data vary directly, the ratios $\frac{y}{x}$ are equal to the constant of variation k.

Example: If a person works for \$8 per hour, then that person's pay P is directly proportional to, or varies directly with, the number of hours H that the person works by the equation $P = 8H$, where the constant of variation is $k = 8$.

SECTION 2.3 LINEAR EQUATIONS

Linear Equation

Can be written as $ax + b = 0$ with $a \neq 0$ and has one solution

Example: The solution to $2x - 4 = 0$ is 2 because $2(2) - 4 = 0$.

CONCEPT	EXPLANATION AND EXAMPLES

SECTION 2.3   LINEAR EQUATIONS (CONTINUED)

Properties of Equality

Addition: $a = b$ is equivalent to $a + c = b + c$.

Multiplication: $a = b$ is equivalent to $ac = bc$, provided $c \neq 0$.

Example: $\dfrac{1}{2}x - 4 = 3$ Given equation

$\dfrac{1}{2}x = 7$ Addition property; add 4.

$x = 14$ Multiplication property; multiply by 2.

Contradiction, Identity, and Conditional Equation

A contradiction has no solutions, an identity is true for all (meaningful) values of the variable, and a conditional equation is true for some, but not all, values of the variable.

Examples: $3(1 - 2x) = 3 - 6x$ Identity

$x + 5 = x$ Contradiction

$x - 1 = 4$ Conditional equation

Intersection-of-Graphs and x-Intercept Methods

Intersection-of-graphs method: Set y_1 equal to the left side of the equation and set y_2 equal to the right side. The x-coordinate of a point of intersection is a solution.

Example: The graphs of $y_1 = 2x$ and $y_2 = x + 1$ intersect at $(1, 2)$, so the solution to the linear equation $2x = x + 1$ is 1. See the figure below on the left.

x-intercept method: Move all terms to the left side of the equation. Set y_1 equal to the left side of the equation. The solutions are the x-intercepts.

Example: Write $2x = x + 1$ as $2x - (x + 1) = 0$. Graph $y_1 = 2x - (x + 1)$. The only x-intercept is 1, as shown in the figure on the right.

Point of Intersection (1, 2)

x-intercept: 1

Problem-Solving Strategies

STEP 1:   Read the problem and make sure you understand it. Assign a variable to what you are being asked to find. If necessary, write other quantities in terms of this variable.

STEP 2:   Write an equation that relates the quantities described in the problem. You may need to sketch a diagram and refer to known formulas.

STEP 3:   Solve the equation and determine the solution.

STEP 4:   Look back and check your solution. Does it seem reasonable?

CONCEPT	EXPLANATION AND EXAMPLES

SECTION 2.4 LINEAR INEQUALITIES

Linear Inequality

Can be written as $ax + b > 0$ with $a \neq 0$, where $>$ can be replaced by $<$, $\leq$, or $\geq$. If the solution to $ax + b = 0$ is k, then the solution to the linear inequality $ax + b > 0$ is either the interval $(-\infty, k)$ or the interval (k, ∞).

Example: $3x - 1 < 2$ is linear since it can be written as $3x - 3 < 0$. The solution set is $\{x \mid x < 1\}$, or $(-\infty, 1)$.

Properties of Inequality

Addition: $a < b$ is equivalent to $a + c < b + c$.
Multiplication: $a < b$ is equivalent to $ac < bc$ when $c > 0$.
$a < b$ is equivalent to $ac > bc$ when $c < 0$.

Example:

$-3x - 4 < 14$	*Given equation*
$-3x < 18$	*Addition property; add 4.*
$x > -6$	*Multiplication property; divide by -3. Reverse the inequality symbol.*

Compound Inequality

Example: $x \geq -2$ and $x \leq 4$, or equivalently, $-2 \leq x \leq 4$. This is called a three-part inequality.

SECTION 2.5 ABSOLUTE VALUE EQUATIONS AND INEQUALITIES

Absolute Value Function

An absolute value function is defined by $f(x) = |x|$. Its graph is V-shaped. An equivalent formula is $f(x) = \sqrt{x^2}$.

Examples: $f(-9) = |-9| = 9$; $\sqrt{(2x + 1)^2} = |2x + 1|$

Absolute Value Equations

$|ax + b| = k$ with $k > 0$ is equivalent to $ax + b = \pm k$.

Example: $|2x - 3| = 4$ is equivalent to $2x - 3 = 4$ or $2x - 3 = -4$.
The solutions are $\frac{7}{2}$ and $-\frac{1}{2}$.

Absolute Value Inequalities

Let the solutions to $|ax + b| = k$ be s_1 and s_2, where $s_1 < s_2$ and $k > 0$.

1. $|ax + b| < k$ is equivalent to $s_1 < x < s_2$.
2. $|ax + b| > k$ is equivalent to $x < s_1$ or $x > s_2$.

Similar statements can be made for inequalities involving $\leq$ or $\geq$.

Example: The solutions to $|2x + 1| = 5$ are given by $x = -3$ and $x = 2$.
The solutions to $|2x + 1| < 5$ are given by $-3 < x < 2$.
The solutions to $|2x + 1| > 5$ are given by $x < -3$ or $x > 2$.

2   Review Exercises

Exercises 1 and 2: The graph of a linear function f is shown.

  (a) *Identify the slope, y-intercept, and x-intercept.*
  (b) *Write a formula for f.*
  (c) *Find any zeros of f.*

1.

2.

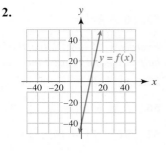

Exercises 3 and 4: Find f(x) = ax + b so that f models the data exactly.

3.

x	1	2	3	4
$f(x)$	2.5	0	−2.5	−5

4.

x	−3	6	15	24
$f(x)$	−1.65	−1.2	−0.75	−0.3

Exercises 5 and 6: Graph the linear function.

5. $f(x) = -\frac{2}{3}x$ **6.** $g(x) = 4 - 2x$

7. Write a formula for a linear function f whose graph has slope -2 and passes through $(-2, 3)$.

8. Find the average rate of change of $f(x) = -3x + 8$ from -2 to 3.

Exercises 9–14: Find the slope-intercept form of the equation of a line satisfying the conditions.

9. Slope 7, passing through $(-3, 9)$

10. Passing through $(2, -4)$ and $(7, -3)$

11. Passing through $(1, -1)$, parallel to $y = -3x + 1$

12. Passing through the point $(-2, 1)$, perpendicular to the line $y = 2(x + 5) - 22$

13. Parallel to the line segment connecting $(0, 3.1)$ and $(5.7, 0)$, passing through $(1, -7)$

14. Perpendicular to $y = -\frac{5}{7}x$, passing through $\left(\frac{6}{7}, 0\right)$

Exercises 15–20: Find an equation of the specified line.

15. Parallel to the y-axis, passing through $(6, -7)$

16. Parallel to the x-axis, passing through $(-3, 4)$

17. Horizontal, passing through $(1, 3)$

18. Vertical, passing through $(1.5, 1.9)$

19. Vertical with x-intercept 2.7

20. Horizontal with y-intercept -8

Exercises 21 and 22: Determine the x- and y-intercepts for the graph of the equation. Graph the equation.

21. $5x - 4y = 20$ **22.** $\dfrac{x}{3} - \dfrac{y}{2} = 1$

Exercises 23–28: Solve the linear equation either symbolically or graphically.

23. $5x - 22 = 10$ **24.** $5(4 - 2x) = 16$

25. $-2(3x - 7) + x = 2x - 1$

26. $5x - \frac{1}{2}(4 - 3x) = \frac{3}{2} - (2x + 3)$

27. $\pi x + 1 = 6$

28. $\dfrac{x - 4}{2} = x + \dfrac{1 - 2x}{3}$

Exercises 29 and 30: Use a table to solve each linear equation numerically to the nearest tenth.

29. $3.1x - 0.2 - 2(x - 1.7) = 0$

30. $\sqrt{7} - 3x - 2.1(1 + x) = 0$

Exercises 31–34: Complete the following.

  (a) *Solve the equation symbolically.*
  (b) *Classify the equation as a contradiction, an identity, or a conditional equation.*

31. $4(6 - x) = -4x + 24$

32. $\frac{1}{2}(4x - 3) + 2 = 3x - (1 + x)$

33. $5 - 2(4 - 3x) + x = 4(x - 3)$

34. $\dfrac{x - 3}{4} + \dfrac{3}{4}x - 5(2 - 7x) = 36x - \dfrac{43}{4}$

Exercises 35–38: Express the inequality in interval notation.

35. $x > -3$

36. $x \le 4$

37. $-2 \le x < \frac{3}{4}$

38. $x \le -2$ or $x > 3$

Exercises 39–44: Solve the linear inequality. Write the solution set in set-builder or interval notation.

39. $3x - 4 \le 2 + x$

40. $-2x + 6 \le -3x$

41. $\dfrac{2x - 5}{2} < \dfrac{5x + 1}{5}$

42. $-5(1 - x) > 3(x - 3) + \frac{1}{2}x$

43. $-2 \le 5 - 2x < 7$

44. $-1 < \dfrac{3x - 5}{-3} < 3$

Exercises 45 and 46: Solve the inequality graphically.

45. $2x > x - 1$

46. $-1 \le 1 + x \le 2$

47. The graphs of two linear functions f and g are shown in the figure. Solve each equation or inequality.

(a) $f(x) = g(x)$

(b) $f(x) < g(x)$

(c) $f(x) > g(x)$

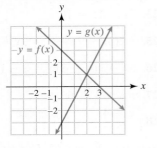

48. The graphs of three linear functions f, g, and h with domains $D = \{x \mid 0 \le x \le 7\}$ are shown in the figure. Solve each equation or inequality.

(a) $f(x) = g(x)$

(b) $g(x) = h(x)$

(c) $f(x) < g(x) < h(x)$

(d) $g(x) > h(x)$

49. Use $f(x)$ to complete the following.

$$f(x) = \begin{cases} 8 + 2x & \text{if } -3 \le x \le -1 \\ 5 - x & \text{if } -1 < x \le 2 \\ x + 1 & \text{if } 2 < x \le 5 \end{cases}$$

(a) Evaluate f at $x = -2, -1, 2,$ and 3.

(b) Sketch a graph of f. Is f continuous on its domain?

(c) Determine the x-value(s) where $f(x) = 3$.

50. If $f(x) = [\![2x - 1]\!]$, evaluate $f(-3.1)$ and $f(2.5)$.

Exercises 51–54: Solve the equation.

51. $|2x - 5| - 1 = 8$

52. $|3 - 7x| = 10$

53. $|6 - 4x| = -2$

54. $|9 + x| = |3 - 2x|$

Exercises 55–58: Solve the equation. Use the solutions to help solve the related inequality.

55. $|x| = 3,$ $\qquad |x| > 3$

56. $|-3x + 1| = 2,$ $\quad |-3x + 1| < 2$

57. $|3x - 7| = 10,$ $\quad |3x - 7| > 10$

58. $|4 - x| = 6,$ $\qquad |4 - x| \le 6$

Exercises 59–62: Solve the inequality.

59. $|3 - 2x| < 9$

60. $|-2x - 3| > 3$

61. $\left|\frac{1}{3}x - \frac{1}{6}\right| \ge 1$

62. $\left|\frac{1}{2}x\right| - 3 \le 5$

Applications

63. *Median Income*   The median U.S. family income between 1980 and 2005 can be modeled by the formula $f(x) = 1450(x - 1980) + 20,000$, where x is the year. (**Source:** Department of Commerce.)

 (a) Solve the equation $f(x) = 34,500$ graphically and interpret the solution.

(b) Solve part (a) symbolically.

64. *Course Grades*   In order to receive a B grade in a college course, it is necessary to have an overall average of 80% correct on two 1-hour exams of 75 points each and one final exam of 150 points. If a person scores 55 and 72 on the 1-hour exams, what is the minimum score that the person can receive on the final exam and still earn a B?

65. *Medicare Costs*   Estimates for Medicare costs in billions of dollars in year x can be modeled by the formula $f(x) = 18x - 35,750$, where $1995 \le x \le 2007$. Determine when Medicare costs were from \$268 to \$358 billion.   (**Source:** Office of Management and Budget.)

66. *Temperature Scales*   The table shows equivalent temperatures in degrees Celsius and degrees Fahrenheit.

°F	−40	32	59	95	212
°C	−40	0	15	35	100

(a) Plot the data with Fahrenheit temperature on the x-axis and Celsius temperature on the y-axis. What type of relation exists between the data?

(b) Find a function C that receives the Fahrenheit temperature x as input and outputs the corresponding Celsius temperature. Interpret the slope.

(c) If the temperature is 83°F, what is it in degrees Celsius?

67. *Distance from Home* The graph depicts the distance y that a person driving a car on a straight road is from home after x hours. Interpret the graph. What speeds did the car travel?

Time (hours)

68. *ACT Scores* The table lists the average composite ACT scores for selected years. Note that in 2007 the average score was 21.2. (**Source:** The American College Testing Program.)

Year	1989	1990	1991	1992	1993	1994
Score	20.6	20.6	20.6	20.6	20.7	20.8

(a) Make a line graph of the data.

(b) Let this line graph represent a piecewise-linear function f. Find a formula for f.

(c) What is the domain of f?

69. *Population Estimates* In 2004 the population of a city was 143,247, and in 2008 it was 167,933. Estimate the population in 2006.

70. *Distance* A driver of a car is initially 455 miles from home, traveling toward home on a straight freeway at 70 miles per hour.

(a) Write a formula for a linear function f that models the distance between the driver and home after x hours.

(b) Graph f. What is an appropriate domain?

(c) Identify the x- and y-intercepts. Interpret each.

71. *Working Together* Suppose that one worker can shovel snow from a storefront sidewalk in 50 minutes and another worker can shovel it in 30 minutes. How long will it take if they work together?

72. *Antifreeze* Initially, a tank contains 20 gallons of a 30% antifreeze solution. How many gallons of an 80% antifreeze solution should be added to the tank in order to increase the concentration of the antifreeze in the tank to 50%?

20 gal

73. *Running* An athlete traveled 13.5 miles in 1 hour and 48 minutes, jogging at 7 miles per hour and then at 8 miles per hour. How long did the runner jog at each speed?

74. *Least-Squares Fit* The table lists the actual annual cost y to drive a midsize car 15,000 miles per year for selected years x.

x	1960	1970	1980	1990	2000
y	$1394	$1763	$3176	$5136	$6880

Source: Runzheimer International.

(a) Predict whether the correlation coefficient is positive, negative, or zero.

(b) Find a least-squares regression line that models these data. What is the correlation coefficient?

(c) Estimate the cost of driving a midsize car in 1995.

(d) Estimate the year when the cost could reach $8000.

75. The table lists data that are exactly linear.

x	−3	−2	−1	1	2
y	6.6	5.4	4.2	1.8	0.6

(a) Determine the slope-intercept form of the line that passes through these data points.

(b) Predict y when x = −1.5 and 3.5. State whether these calculations involve interpolation or extrapolation.

(c) Predict x when y = 1.3.

76. *Geometry* A rectangle is twice as long as it is wide and has a perimeter of 78 inches. Find the width and length of this rectangle.

77. *Flow Rates* A water tank has an inlet pipe with a flow rate of 5 gallons per minute and an outlet pipe with a flow rate of 3 gallons per minute. A pipe can be either closed or completely open. The graph shows the number of gallons of water in the tank after x minutes have elapsed. Use the concept of slope to interpret each piece of this graph.

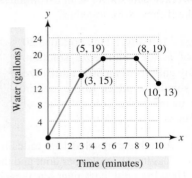

78. *Flow Rates* (Refer to Exercise 77.) Suppose the tank is modified so that it has a second inlet pipe, which flows at a rate of 2 gallons per minute. Interpret the graph by determining when each inlet and outlet pipe is open or closed.

79. *Air Temperature* For altitudes up to 4 kilometers, moist air will cool at a rate of about 6°C per kilometer. If the ground temperature is 25°C, at what altitudes would the air temperature be from 5°C to 15°C? (**Source:** A Miller and R. Anthes, *Meteorology.*)

80. *Water Pollution* At one time the Thames River in England supported an abundant community of fish. Pollution then destroyed all the fish in a 40-mile stretch near its mouth for a 45-year period beginning in 1915. Since then, improvement of sewage treatment facilities and other ecological steps have resulted in a dramatic increase in the number of different fish present. The number of species present from 1967 to 1978 can be modeled by $f(x) = 6.15x - 12{,}059$, where x is the year.

(a) Estimate the year when the number of species first exceeded 70.

(b) Estimate the years when the number of species was between 50 and 100.

81. *Relative Error* The actual length of a side of a building is 52.3 feet. How accurately must an apprentice carpenter measure this side to have the relative error in the measurement be less than 0.003 (0.3%)? (*Hint:* Use $\left|\frac{C-A}{A}\right|$, where C is the carpenter's measurement and A is the actual length.)

82. *Brown Trout* Due to acid rain, the percentage of lakes in Scandinavia that lost their population of brown trout increased dramatically between 1940 and 1975. Based on a sample of 2850 lakes, this percentage can be approximated by the following piecewise-linear function. (**Source:** C. Mason, *Biology of Freshwater Pollution.*)

$$f(x) = \begin{cases} \frac{11}{20}(x - 1940) + 7 & \text{if } 1940 \le x < 1960 \\ \frac{32}{15}(x - 1960) + 18 & \text{if } 1960 \le x \le 1975 \end{cases}$$

(a) Determine the percentage of lakes that lost brown trout by 1947 and by 1972.

(b) Sketch a graph of f.

(c) Is f a continuous function on its domain?

EXTENDED AND DISCOVERY EXERCISES

1. *Archeology* It is possible for archeologists to estimate the height of an adult based only on the length of the humerus, a bone located between the elbow and the shoulder. The approximate relationship between the height y of an individual and the length x of the humerus is shown in the table for both males and females. All measurements are in inches. Although individual values may vary, tables like this are the result of measuring bones from many skeletons.

x	8	9	10	11
y (females)	50.4	53.5	56.6	59.7
y (males)	53.0	56.0	59.0	62.0

x	12	13	14
y (females)	62.8	65.9	69.0
y (males)	65.0	68.0	71.0

(a) Find the estimated height of a female with a 12-inch humerus.

(b) Plot the ordered pairs (x, y) for both sexes. What type of relation exists between the data?

(c) For each 1-inch increase in the length of the humerus, what are the corresponding increases in the heights of females and of males?

(d) Determine linear functions f and g that model these data for females and males, respectively.

(e) Suppose a humerus from a person of unknown sex is estimated to be between 9.7 and 10.1 inches long. Use f and g to approximate the range for the height of a female and a male.

2. Continuing with Exercise 1, have members of the class measure their heights and the lengths of their humeri (plural of *humerus*) in inches.
   (a) Make a table of the results.

   (b) Find regression lines that fit the data points for males and females.

   (c) Compare your results with the table in Exercise 1.

3. *A Puzzle*   Three people leave for a city 15 miles away. The first person walks 4 miles per hour, and the other two people ride in a car that travels 28 miles per hour. After some time, the second person gets out of the car and walks 4 miles per hour to the city while the driver goes back and picks up the first person. The driver takes the first person to the city. If all three people arrive in the city at the same time, how far did each person walk?

4. *Comparing Ages*   The age of Earth is approximately 4.45 billion years. The earliest evidence of dinosaurs dates back 200 million years, whereas the earliest evidence of *Homo sapiens* dates back 300,000 years. If the age of Earth were condensed into one year, determine the approximate times when dinosaurs and *Homo sapiens* first appeared.

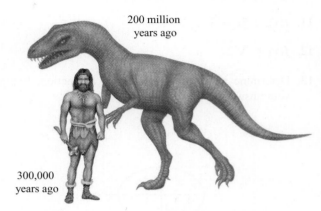

200 million years ago

300,000 years ago

5. *Limit Notation*   Let ϵ and δ be positive numbers; let x, c, and L be real numbers; and let f be a function. Consider the following: "If the distance between x and c is less than δ, then the distance between $f(x)$ and L is less than ϵ." Rewrite this sentence by using two absolute value inequalities.

1–2   Cumulative Review Exercises

1. Write 123,000 and 0.0051 in scientific notation.

2. Write 6.7×10^6 and 1.45×10^{-4} in standard form.

3. Evaluate $\dfrac{4 + \sqrt{2}}{4 - \sqrt{2}}$. Round your answer to the nearest hundredth.

4. The table represents a relation S.

x	−1	0	1	2	3
y	6	4	3	0	0

   (a) Does S represent a function?

   (b) Determine the domain and range of S.

5. Find the standard equation of a circle with center $(-2, 3)$ and radius 7.

6. Evaluate $-5^2 - 2 - \dfrac{10 - 2}{5 - 1}$ by hand.

7. Find the exact distance between $(-3, 5)$ and $(2, -3)$.

8. Find the midpoint of the line segment with endpoints $(5, -2)$ and $(-3, 1)$.

9. Find the domain and range of the function shown in the graph. Evaluate $f(-1)$.

(a)

(b)

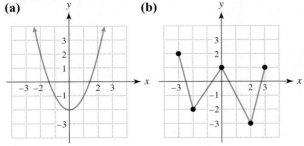

10. Graph f by hand.
   (a) $f(x) = 3 - 2x$ **(b)** $f(x) = |x + 1|$

   (c) $f(x) = x^2 - 3$ **(d)** $f(x) = \sqrt{x + 2}$

Exercises 11 and 12: Complete the following.
   (a) Evaluate $f(2)$ and $f(a - 1)$.
   (b) Determine the domain of f.

11. $f(x) = 5x - 3$

12. $f(x) = \sqrt{2x - 1}$

13. Determine if the graph represents a function. Explain your answer.

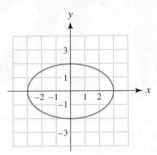

14. Write a formula for a function f that computes the cost of taking x credits if credits cost \$80 each and fees are fixed at \$89.

15. Find the average rate of change of $f(x) = x^2 - 2x + 1$ from $x = 1$ to $x = 2$.

16. Find the difference quotient for $f(x) = 2x^2 - x$.

Exercises 17 and 18: The graph of a linear function f is shown.
   (a) Identify the slope, y-intercept, and x-intercept.
   (b) Write a formula for f.
   (c) Find any zeros of f.

17.

18.
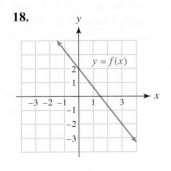

19. Write a formula for a linear function whose graph has slope -3 and passes through $\left(\frac{2}{3}, -\frac{2}{3}\right)$.

20. If $G(t) = 200 - 10t$ models the gallons of water in a tank after t minutes, interpret the numbers 200 and -10 in the formula for G.

Exercises 21–26: Write an equation of a line satisfying the given conditions. Use slope-intercept form whenever possible.

21. Passing through $(1, -5)$ and $\left(-3, \frac{1}{2}\right)$

22. Passing through the point $(-3, 2)$ and perpendicular to the line $y = \frac{2}{3}x - 7$

23. Parallel to the y-axis and passing through $(-1, 3)$

24. Slope 30, passing through $(2002, 50)$

25. Passing through $(-3, 5)$ and parallel to the line segment connecting $(2.4, 5.6)$ and $(3.9, 8.6)$

26. Perpendicular to the y-axis and passing through the origin

Exercises 27 and 28: Determine the x- and y-intercepts on the graph of the equation. Graph the equation.

27. $-2x + 3y = 6$ **28.** $x = 2y - 3$

Exercises 29–32: Solve the equation.

29. $4x - 5 = 1 - 2x$ **30.** $\dfrac{2x - 4}{2} = \dfrac{3x}{7} - 1$

31. $\frac{2}{3}(x - 2) - \frac{4}{5}x = \frac{4}{15} + x$

32. $-0.3(1 - x) - 0.1(2x - 3) = 0.4$

33. Solve $x + 1 = 2x - 2$ graphically and numerically.

34. Solve $2x - (5 - x) = \dfrac{1 - 4x}{2} + 5(x - 2)$. Is this equation either an identity or a contradiction?

Exercises 35–38: Express each inequality in interval notation.

35. $x < 5$ **36.** $-2 \leq x \leq 5$

37. $x < -2$ or $x > 2$ **38.** $x \geq -3$

Exercises 39 and 40: Solve the inequality. Write the solution set in set-builder or interval notation.

39. $-3(1 - 2x) + x \leq 4 - (x + 2)$

40. $\dfrac{1}{3} \leq \dfrac{2 - 3x}{2} < \dfrac{4}{3}$

41. The graphs of two linear functions f and g are shown. Solve each equation or inequality.

(a) $f(x) = g(x)$

(b) $f(x) > g(x)$

(c) $f(x) \leq g(x)$

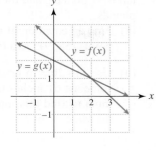

42. Graph f. Is f continuous on the interval $[-4, 4]$?

$$f(x) = \begin{cases} 2 - x & \text{if } -4 \leq x < -2 \\ \frac{1}{2}x + 5 & \text{if } -2 \leq x < 2 \\ 2x + 1 & \text{if } 2 \leq x \leq 4 \end{cases}$$

Exercises 43–46: Solve the equation.

43. $|d + 1| = 5$

44. $|3 - 2x| = 7$

45. $|2t| - 4 = 10$

46. $|11 - 2x| = |3x + 1|$

Exercises 47 and 48: Solve the inequality.

47. $|2t - 5| \leq 5$

48. $|5 - 5t| > 7$

Applications

49. *Volume of a Cylinder* The volume V of a cylinder is given by $V = \pi r^2 h$, where r is the radius of the cylinder and h is its height. If an aluminum can has a volume of 24 cubic inches and a radius of 1.5 inches, find its height to the nearest hundredth of an inch.

50. *Interpreting Slope* The figure shows the weight of a load of gravel in a dump truck. Interpret the slope of each line segment.

51. *Cost* A company's cost C in dollars for making x computers is $C(x) = 500x + 20,000$.

(a) Evaluate $C(1500)$. Interpret the result.

(b) Find the slope of the graph of C. Interpret the slope.

52. *Distance* At midnight car A is traveling north at 60 miles per hour and is located 40 miles south of car B. Car B is traveling west at 70 miles per hour. Approximate the distance between the cars at 1:15 A.M. to the nearest tenth of a mile.

53. *Average Rate of Change* On a warm summer day the Fahrenheit temperature x hours past noon is given by the formula $T(x) = 70 + \frac{3}{2}x^2$.

(a) Find the average rate of change of T from 2:00 P.M. to 4:00 P.M.

(b) Interpret this average rate of change.

54. *Distance from Home* A driver is initially 270 miles from home, traveling toward home on a straight interstate at 72 miles per hour.

(a) Write a formula for a function D that models the distance between the driver and home after x hours.

(b) What is an appropriate domain for D? Graph D.

(c) Identify the x- and y-intercepts. Interpret each.

55. *Working Together* Suppose one person can mow a large lawn in 5 hours with a riding mower and it takes another person 12 hours to mow the lawn with a push mower. How long will it take to mow the lawn if the two people work together?

56. *Running* An athlete traveled 15 miles in 1 hour and 45 minutes, running at 8 miles per hour and then 10 miles per hour. How long did the athlete run at each speed?

57. *Chicken Consumption* In 2001 Americans ate, on average, 56 pounds of chicken annually. This amount is expected to increase to 61 pounds in 2012. (**Source:** Department of Agriculture.)

(a) Determine a formula

$$f(x) = m(x - x_1) + y_1$$

that models these data. Let x be the year.

(b) Estimate the annual chicken consumption in 2007.

58. *Relative Error*   If the actual value of a quantity is A and its measured value is M, then the relative error in measurement M is $\left| \frac{M - A}{A} \right|$. If $A = 65$, determine the range of values for M to have a relative error that is less than or equal to 0.03 (3%).

59. The table lists per capita income.

Year	1970	1980	1990	2000
Income	$4095	$10,183	$19,572	$29,760

Source: Bureau of Economic Analysis.

(a) Find the least-squares regression line for the data.

(b) Estimate the per capita income in 1995. Did this calculation involve interpolation or extrapolation?

60. *Modeling Data*   According to government guidelines, the recommended minimum weight for a person 58 inches tall is 91 pounds, and for a person 64 inches tall it is 111 pounds.

(a) Find an equation of the line that passes through the points (58, 91) and (64, 111).

(b) Use this line to estimate the minimum weight for someone 61 inches tall. Then use the midpoint formula to find this minimum weight. Are the results the same? Why?

Quadratic Functions and Equations

3

The last basketball player in the NBA to shoot foul shots underhand was Rick Barry, who retired in 1980. On average, he was able to make about 9 out of 10 shots. Since then, every NBA player has used the overhand style of shooting foul shots—even though this style has often resulted in lower free-throw percentages.

According to Dr. Peter Brancazio, a physics professor emeritus from Brooklyn College and author of *Sports Science*, there are good reasons for shooting underhand. An underhand shot obtains a higher arc, and as the ball approaches the hoop, it has a better chance of going through the hoop than does a ball with a flatter arc. If a basketball is tossed at an angle of 32 degrees or less, it will likely hit the back of the rim and bounce out. The optimal angle is greater than 45 degrees and depends on the height at which the ball is released. Lower release points require steeper arcs and increase the chances of the ball passing through the hoop. (See the Extended and Discovery Exercise at the end of this chapter to model the arc of a basketball.)

Mathematics plays an important role in analyzing applied problems such as shooting foul shots. Whether NBA players choose to listen to Professor Brancazio is another question, but mathematics tells us that steeper arcs are necessary for accurate foul shooting.

Rick Barry

The work of mathematicians from even a millennium ago is still routinely used today, and the work of mathematicians from today will be part of the math of the future.

—Terry Tao, Fields Medal 2006

Source: Curtis Rist, "The Physics of Foul Shots," *Discover*, October 2000. (Photograph reprinted with permission.)

3.1 Quadratic Functions and Models

- Learn basic concepts about quadratic functions and their graphs
- Complete the square and apply the vertex formula
- Graph a quadratic function by hand
- Solve applications and model data
- Use quadratic regression to model data (optional)

Introduction

Sometimes when data lie on a line or nearly lie on a line, they can be modeled with a linear function ($f(x) = ax + b$) and are called *linear data*. Data that are not linear are called **nonlinear data** and must be modeled with a nonlinear function. There are many types of nonlinear data, and mathematicians can choose from a wide variety of nonlinear functions to create models. One of the simplest types of nonlinear functions is a *quadratic function,* which can be used to model the flight of a baseball, AIDS in America, or an athlete's heart rate. Figures 3.1–3.3 illustrate three sets of nonlinear data that can be modeled by a quadratic function. Although the complete graph of a quadratic function is either ∪-shaped or ∩-shaped, we often use only one side or a portion of the graph to model real-world data.

Figure 3.1

Figure 3.2

Figure 3.3

Basic Concepts

The formula for a quadratic function is different from that of a linear function because it contains an x^2-**term**. Examples of quadratic functions include

$$f(x) = 2x^2 - 4x - 1, \qquad g(x) = 4 - x^2, \qquad \text{and} \qquad h(x) = \frac{1}{3}x^2 + \frac{2}{3}x + 1.$$

The following box defines a *general form* for a quadratic function.

Quadratic Function

Let a, b, and c be constants with $a \neq 0$. A function represented by

$$f(x) = ax^2 + bx + c$$

is a **quadratic function.**

The domain of a quadratic function includes *all* real numbers. The leading coefficient of a quadratic function is a. For the previous three functions f, g, and h, the leading coefficients are $a = 2$ for $f(x) = 2x^2 - 4x - 1$, $a = -1$ for $g(x) = 4 - 1x^2$, and $a = \frac{1}{3}$ for $h(x) = \frac{1}{3}x^2 + \frac{2}{3}x + 1$.

The graph of a quadratic function is a **parabola**—a ∪-shaped graph that opens either upward or downward. Graphs of the three quadratic functions f, g, and h are shown in Figures 3.4–3.6, respectively. A parabola opens upward if a is positive and opens downward if a is negative. For example, since $a = -1$ for $g(x) = 4 - x^2$, the graph of g opens downward. The highest point on a parabola that opens downward or the lowest point on a parabola that opens upward is called the **vertex**. The vertical line passing through the vertex is called the **axis of symmetry**. The vertex and axis of symmetry are shown in each figure. Function f increases for $x \geq 1$ and decreases for $x \leq 1$, whereas function g increases for $x \leq 0$ and decreases for $x \geq 0$. Remember to move from left to right along the graph when determining where a function is increasing or decreasing. (See Section 1.5.)

Figure 3.4   **Figure 3.5**   **Figure 3.6**

The leading coefficient a of a quadratic function not only determines whether its graph opens upward or downward, but also controls the width of the parabola. Larger values of $|a|$ result in a narrower parabola, and smaller values of $|a|$ result in a wider parabola. This concept is illustrated in Figures 3.7 and 3.8.

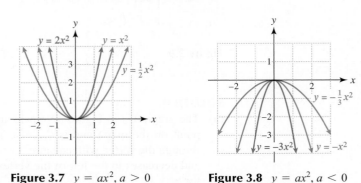

Figure 3.7 $y = ax^2, a > 0$ **Figure 3.8** $y = ax^2, a < 0$

CLASS DISCUSSION

What does the graph of a quadratic function resemble if its leading coefficient is nearly 0?

EXAMPLE 1   Identifying quadratic functions

Identify the function as linear, quadratic, or neither. If it is quadratic, identify the leading coefficient and evaluate the function at $x = 2$.

(a) $f(x) = 3 - 2^2x$ **(b)** $g(x) = 5 + x - 3x^2$ **(c)** $h(x) = \dfrac{3}{x^2 + 1}$

SOLUTION

Getting Started The formula for a quadratic function always has an x^2-term and may have an x-term and a constant. It does not have a variable raised to any other power or a variable in a denominator. ▶

(a) Because $f(x) = 3 - 2^2x$ can be written as $f(x) = -4x + 3$, f is linear.

(b) Because $g(x) = 5 + x - 3x^2$ can be written as $g(x) = -3x^2 + x + 5$, g is a quadratic function with $a = -3$, $b = 1$, and $c = 5$. The leading coefficient is $a = -3$ and

$$g(2) = 5 + 2 - 3(2)^2 = -5.$$

(c) $h(x) = \dfrac{3}{x^2 + 1}$ cannot be written as $h(x) = ax + b$ or as $h(x) = ax^2 + bx + c$, so h is neither a linear nor a quadratic function. *Now Try Exercises 1, 3, and 5* ◀

EXAMPLE 2 Analyzing graphs of quadratic functions

Use the graph of each quadratic function to determine the sign of the leading coefficient a, the vertex, and the equation of the axis of symmetry. Give the intervals where the function is increasing and where it is decreasing.

(a)

(b)

Figure 3.9   **Figure 3.10**

SOLUTION

(a) The graph of f in Figure 3.9 opens upward, so a is positive. The vertex is the lowest point on the graph and is $(1, -2)$. The axis of symmetry is a vertical line passing through the vertex with equation $x = 1$. Function f increases to the right of the vertex and decreases to the left of the vertex. Thus f is increasing for $x \geq 1$ and decreasing for $x \leq 1$.

(b) The graph of g in Figure 3.10 opens downward, so a is negative. The vertex is $(-2, 5)$, and the axis of symmetry is given by $x = -2$. Function g is increasing for $x \leq -2$ and decreasing for $x \geq -2$. *Now Try Exercises 7 and 9* ◀

Completing the Square and the Vertex Formula

When a quadratic function f is expressed as $f(x) = ax^2 + bx + c$, the coordinates of the vertex are not apparent. However, if f is written as $f(x) = a(x - h)^2 + k$, then the vertex is located at (h, k), as illustrated in Figure 3.11. For example, in Figure 3.12 the graph of $f(x) = -2(x - 1)^2 + 2$ opens downward with vertex $(1, 2)$.

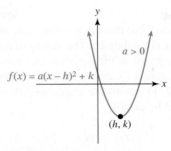

Figure 3.11 A Parabola with Vertex (h, k)

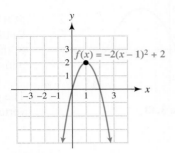

Figure 3.12 A Parabola with Vertex $(1, 2)$

To justify that the vertex is indeed (h, k), consider the following. If $a > 0$ in the form $f(x) = a(x - h)^2 + k$, then the term $a(x - h)^2$ is never negative and the minimum value of $f(x)$ is k. This value occurs when $x = h$ because

$$f(h) = a(h - h)^2 + k = 0 + k = k.$$

Thus the lowest point on the graph of $f(x) = a(x - h)^2 + k$ with $a > 0$ is (h, k), and because this graph is a parabola that opens upward, the vertex must be (h, k). A similar discussion can be used to justify that (h, k) is the vertex when $a < 0$.

The formula $f(x) = a(x - h)^2 + k$ is sometimes called the **standard form for a parabola with a vertical axis**. Because the vertex is apparent in this formula, we will call it simply the **vertex form**.

> ## Vertex Form
>
> The parabolic graph of $f(x) = a(x - h)^2 + k$ with $a \neq 0$ has vertex (h, k). Its graph opens upward when $a > 0$ and opens downward when $a < 0$.

EXAMPLE 3   Converting to $f(x) = ax^2 + bx + c$

Write $f(x) = 2(x - 1)^2 + 4$ in the form $f(x) = ax^2 + bx + c$.

Algebra Review
To review squaring a binomial, see Chapter R (page R-18).

SOLUTION   Begin by expanding the expression $(x - 1)^2$.

$$2(x - 1)^2 + 4 = 2(x^2 - 2x + 1) + 4 \qquad \text{Square binomial.}$$
$$= 2x^2 - 4x + 2 + 4 \qquad \text{Distributive property}$$
$$= 2x^2 - 4x + 6 \qquad \text{Add terms.}$$

Thus $f(x) = 2x^2 - 4x + 6$.

Now Try Exercise 15 ◄

The next example illustrates how to determine the vertex form given the graph of a parabola.

Figure 3.13

EXAMPLE 4 Writing the equation of a parabola

Find the vertex form for the graph shown in Figure 3.13.

SOLUTION

Getting Started We must determine a, h, and k in $f(x) = a(x - h)^2 + k$. The coordinates of the vertex correspond to the values of h and k. To find a, substitute the coordinates of a point on the graph in the equation and solve for a. ▶

From Figure 3.13, the vertex is $(2, 3)$. Thus $h = 2$ and $k = 3$ and $f(x) = a(x - 2)^2 + 3$. The point $(0, 1)$ lies on the graph, so $f(0) = 1$. (Any point on the graph of f other than the vertex could be used.)

$$f(x) = a(x - 2)^2 + 3 \qquad \text{Vertex form}$$
$$1 = a(0 - 2)^2 + 3 \qquad \text{Let } x = 0 \text{ and } f(0) = 1. \text{ Solve for } a.$$
$$1 = 4a + 3 \qquad \text{Simplify.}$$
$$-2 = 4a \qquad \text{Subtract 3 from each side.}$$
$$a = -\frac{1}{2} \qquad \text{Divide by 4 and rewrite equation.}$$

Thus $f(x) = -\frac{1}{2}(x - 2)^2 + 3$.

Now Try Exercise 25 ◀

Completing the Square We can convert the general form $f(x) = ax^2 + bx + c$ to vertex form by **completing the square.** If a quadratic expression can be written as $x^2 + kx + \left(\frac{k}{2}\right)^2$, then it is a perfect square trinomial and can be factored as

$$x^2 + kx + \left(\frac{k}{2}\right)^2 = \left(x + \frac{k}{2}\right)^2.$$

Note that the k used to complete the square is different from the k found in the vertex form. In general, a variable can have different meanings in different situations.

This technique of converting to vertex form by completing the square is illustrated in the next example.

EXAMPLE 5 Converting to vertex form

Write each formula in vertex form by completing the square.
(a) $f(x) = x^2 + 6x - 3$ **(b)** $f(x) = \frac{1}{3}x^2 - x + 2$

Algebra Review
To review perfect square trinomials, see Chapter R (page R-26).

SOLUTION
(a) Start by letting $y = f(x)$.

$$y = x^2 + 6x - 3 \qquad \text{Given formula}$$
$$y + 3 = x^2 + 6x \qquad \text{Add 3 to each side.}$$
$$y + 3 + 9 = x^2 + 6x + 9 \qquad \text{Let } k = 6; \text{ add } \left(\frac{k}{2}\right)^2 = \left(\frac{6}{2}\right)^2 = 9.$$
$$y + 12 = (x + 3)^2 \qquad \text{Factor perfect square trinomial.}$$
$$y = (x + 3)^2 - 12 \qquad \text{Subtract 12.}$$

The required form is $f(x) = (x + 3)^2 - 12$.

(b) Start by letting $y = f(x)$.

$$y = \frac{1}{3}x^2 - x + 2 \qquad \text{Given formula}$$

$$3y = x^2 - 3x + 6 \qquad \text{Make leading coefficient 1.}$$

$$3y - 6 = x^2 - 3x \qquad \text{Subtract 6 from each side.}$$

$$3y - 6 + \frac{9}{4} = x^2 - 3x + \frac{9}{4} \qquad \text{Let } k = -3; \text{ add } \left(\frac{k}{2}\right)^2 = \left(\frac{-3}{2}\right)^2 = \frac{9}{4}.$$

$$3y - \frac{15}{4} = \left(x - \frac{3}{2}\right)^2 \qquad \text{Factor perfect square trinomial.}$$

$$3y = \left(x - \frac{3}{2}\right)^2 + \frac{15}{4} \qquad \text{Add } \tfrac{15}{4} \text{ to each side.}$$

$$y = \frac{1}{3}\left(x - \frac{3}{2}\right)^2 + \frac{5}{4} \qquad \text{Multiply each side by } \tfrac{1}{3}.$$

The required form is $f(x) = \frac{1}{3}\left(x - \frac{3}{2}\right)^2 + \frac{5}{4}$.   ***Now Try Exercises 29 and 35*** ◀

Derivation of the Vertex Formula

The above procedure of completing the square can be done in general to derive a formula for determining the vertex of any parabola.

$$y = ax^2 + bx + c \qquad \text{General equation for a parabola}$$

$$\frac{y}{a} = x^2 + \frac{b}{a}x + \frac{c}{a} \qquad \text{Divide each side by } a \text{ to make leading coefficient 1.}$$

$$\frac{y}{a} - \frac{c}{a} = x^2 + \frac{b}{a}x \qquad \text{Subtract } \tfrac{c}{a} \text{ from each side.}$$

$$\frac{y}{a} - \frac{c}{a} + \frac{b^2}{4a^2} = x^2 + \frac{b}{a}x + \frac{b^2}{4a^2} \qquad \text{Add } \left(\frac{b/a}{2}\right)^2 = \frac{b^2}{4a^2}.$$

$$\frac{y}{a} + \frac{b^2 - 4ac}{4a^2} = \left(x + \frac{b}{2a}\right)^2 \qquad \text{Combine left terms; factor perfect square trinomial.}$$

$$\frac{y}{a} = \left(x + \frac{b}{2a}\right)^2 - \frac{b^2 - 4ac}{4a^2} \qquad \text{Isolate } y\text{-term on the left side.}$$

$$y = a\left(x + \frac{b}{2a}\right)^2 - \frac{b^2 - 4ac}{4a} \qquad \text{Multiply by } a.$$

$$y = a\left(x - \left(-\frac{b}{2a}\right)\right)^2 + \frac{4ac - b^2}{4a} \qquad \text{Write } y = a(x - h)^2 + k.$$

$$\underbrace{\phantom{x - \left(-\frac{b}{2a}\right)}}_{h} \qquad \underbrace{\phantom{\frac{4ac - b^2}{4a}}}_{k}$$

Because the coordinates of the vertex are (h, k), the x-coordinate is $-\frac{b}{2a}$. Note that it is *not* necessary to memorize the expression for k, because the y-coordinate can be found by evaluating $y = f(x)$ for $x = -\frac{b}{2a}$. This derivation of the *vertex formula* is now summarized.

Vertex Formula

The *vertex* of the graph of $f(x) = ax^2 + bx + c$ with $a \neq 0$ is the point $\left(-\frac{b}{2a}, f\left(-\frac{b}{2a}\right)\right)$.

NOTE If a parabola has two x-intercepts, then the x-coordinate of the vertex is equal to the midpoint of these two x-intercepts. For example, the x-intercepts in Figure 3.9 on page 172 are -1 and 3. Their midpoint is $\frac{-1 + 3}{2} = 1$, which is the x-coordinate of the vertex.

EXAMPLE 6   Using the vertex formula

Find the vertex of the graph of $f(x) = 1.5x^2 - 6x + 4$ symbolically. Find the intervals where f is increasing and where f is decreasing. Support your answer graphically and numerically.

SOLUTION

Symbolic   If $f(x) = 1.5x^2 - 6x + 4$, then the x-coordinate of the vertex is

$$x = -\frac{b}{2a} = -\frac{(-6)}{2(1.5)} = 2. \qquad \text{Let } a = 1.5 \text{ and } b = -6.$$

The y-coordinate of the vertex can be found by evaluating $f(2)$.

$$y = f(2) = 1.5(2)^2 - 6(2) + 4 = -2.$$

Thus the vertex is $(2, -2)$. The graph of f is $\cup$-shaped, so it is increasing for $x \geq 2$ and decreasing for $x \leq 2$. See Figure 3.14.

Graphical   This result can be supported by graphing $Y_1 = 1.5X^2 - 6X + 4$, as shown in Figure 3.14. The vertex is the lowest point on the graph.

Calculator Help

To find a minimum point on a graph, see Appendix A (page AP-10).

$[-4.7, 4.7, 1]$ by $[-3.1, 3.1, 1]$

Figure 3.14

Figure 3.15

Numerical   Numerical support is shown in Figure 3.15. The minimum y-value of -2 occurs when $x = 2$. Notice that if we move an equal distance left or right of the axis of symmetry, the y-values are equal. For example, when $x = 1$ or $x = 3$, $y_1 = -0.5$. Similarly, when $x = 0$ or $x = 4$, $y_1 = 4$.   *Now Try Exercise 45* ◄

EXAMPLE 7   Converting to $f(x) = a(x - h)^2 + k$

Use the vertex formula to write $f(x) = 3x^2 + 12x + 7$ in vertex form.

SOLUTION   Begin by finding the vertex. Let $a = 3$ and $b = 12$.

$$x = -\frac{b}{2a} = -\frac{12}{2(3)} = -2$$

Since $f(-2) = 3(-2)^2 + 12(-2) + 7 = -5$, the vertex is $(-2, -5)$. Because $a = 3$, $f(x)$ can be written as $f(x) = 3(x + 2)^2 - 5$.   *Now Try Exercise 37* ◄

Graphing Quadratic Functions   When sketching a parabola, it is important to determine the vertex, the axis of symmetry, and whether the parabola opens upward or downward. In the next example we sketch the graphs of two quadratic functions by hand.

EXAMPLE 8   Graphing quadratic functions by hand

Graph each quadratic function.
(a) $g(x) = 2(x - 1)^2 - 3$ **(b)** $h(x) = -\dfrac{1}{2}x^2 - x + 2$

SOLUTION

(a) The vertex is $(1, -3)$ and the axis of symmetry is $x = 1$. The parabola opens upward because $a = 2$ is positive. In Table 3.1 we list the vertex and a few other points located on either side of the vertex. Note the symmetry of the y-values on each side of the vertex. These points and a smooth $\cup$-shaped curve are plotted in Figure 3.16. When $x = 0$, $y = -1$, and so the y-intercept is -1.

Table 3.1

	x	y
	-1	5
	0	-1
Vertex →	1	-3
	2	-1
	3	5

Equal

Figure 3.16

(b) The given formula is not in vertex form, but we can find the vertex.

$$x = -\frac{b}{2a} = -\frac{-1}{2\left(-\frac{1}{2}\right)} = -1$$

The y-coordinate of the vertex is $h(-1) = -\frac{1}{2}(-1)^2 - (-1) + 2 = \frac{5}{2}$. Thus the vertex is $\left(-1, \frac{5}{2}\right)$, the axis of symmetry is $x = -1$, and the parabola opens downward because $a = -\frac{1}{2}$ is negative. In Table 3.2 we list the vertex and a few other points located on either side of the vertex. These points and a smooth $\cap$-shaped curve are plotted in Figure 3.17. When $x = 0$, $y = 2$, and so the y-intercept is 2.

Table 3.2

	x	y
	-4	-2
	-3	$\frac{1}{2}$
	-2	2
Vertex →	-1	$\frac{5}{2}$
	0	2
	1	$\frac{1}{2}$
	2	-2

Equal

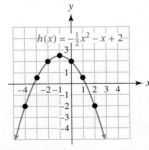

Figure 3.17

Now Try Exercises 59 and 71 ◀

Applications and Models

Sometimes when a quadratic function f is used in applications, the vertex provides important information. The reason is that the y-coordinate of the vertex is the minimum value of $f(x)$ when its graph opens upward and is the maximum value of $f(x)$ when its graph opens downward. This concept is applied in the next example.

Figure 3.18

EXAMPLE 9   Maximizing area

A rancher is fencing a rectangular area for cattle using the straight portion of a river as one side of the rectangle, as illustrated in Figure 3.18. If the rancher has 2400 feet of fence, find the dimensions of the rectangle that give the maximum area for the cattle.

SOLUTION

Getting Started Because the goal is to maximize the area, first write a formula for the area. If the formula is quadratic with $a < 0$ (parabola opening downward), then the maximum area will be the y-coordinate of the vertex. ▶

Let W be the width and L be the length of the rectangle. Because the 2400-foot fence does not go along the river, it follows that

$$W + L + W = 2400, \quad \text{or} \quad L = 2400 - 2W.$$

Area A of a rectangle equals length times width, so

$$
\begin{aligned}
A &= LW & &\text{Area of rectangle} \\
&= (2400 - 2W)W & &\text{Substitute for } L. \\
&= 2400W - 2W^2. & &\text{Distributive property}
\end{aligned}
$$

Thus the graph of $A = -2W^2 + 2400W$ is a parabola opening downward, and by the vertex formula, maximum area occurs when

$$W = -\frac{b}{2a} = -\frac{2400}{2(-2)} = 600 \text{ feet.}$$

The corresponding length is $L = 2400 - 2W = 2400 - 2(600) = 1200$ feet. The dimensions that maximize area are 600 feet by 1200 feet.   ***Now Try Exercise 91*** ◀

Another application of quadratic functions occurs in projectile motion, such as when a baseball is hit up in the air. If air resistance is ignored, then the formula

$$s(t) = -16t^2 + v_0 t + h_0$$

calculates the height s of the object above the ground in feet after t seconds. In this formula h_0 represents the initial height of the object in feet and v_0 represents its initial *vertical* velocity in feet per second. If the initial velocity is upward, then $v_0 > 0$, and if the initial velocity is downward, then $v_0 < 0$.

Figure 3.19

EXAMPLE 10   Modeling the flight of a baseball

A baseball is hit straight up with an initial velocity of $v_0 = 80$ feet per second (or about 55 miles per hour) and leaves the bat with an initial height of $h_0 = 3$ feet, as shown in Figure 3.19.

(a) Write a formula $s(t)$ that models the height of the baseball after t seconds.
(b) How high is the baseball after 2 seconds?
(c) Find the maximum height of the baseball. Support your answer graphically.

SOLUTION
(a) Because $v_0 = 80$ and the initial height is $h_0 = 3$,

$$s(t) = -16t^2 + v_0 t + h_0 = -16t^2 + 80t + 3.$$

(b) $s(2) = -16(2)^2 + 80(2) + 3 = 99$, so the baseball is 99 feet high after 2 seconds.
(c) Because $a = -16$, the graph of s is a parabola opening downward. The vertex is the highest point on the graph, with a t-coordinate of

$$t = -\frac{b}{2a} = -\frac{80}{2(-16)} = 2.5.$$

The corresponding y-coordinate of the vertex is

$$s(2.5) = -16(2.5)^2 + 80(2.5) + 3 = 103 \text{ feet}.$$

[0, 5, 1] by [−20, 120, 20]

Figure 3.20

Thus the vertex is $(2.5, 103)$ and the maximum height of the baseball is 103 feet after 2.5 seconds. Graphical support is shown in Figure 3.20, where $Y_1 = -16X^2 + 80X + 3$ and the vertex is located at $(2.5, 103)$.   *Now Try Exercise 87* ◄

Calculator Help

To find a maximum point on a graph, see Appendix A (page AP-10).

A well-conditioned athlete's heart rate can reach 200 beats per minute (bpm) during strenuous physical activity. Upon stopping an activity, a typical heart rate decreases rapidly at first and then gradually levels off, as shown in Table 3.3.

Table 3.3   Athlete's Heart Rate

Time (min)	0	2	4	6	8
Heart rate (bpm)	200	150	110	90	80

Adapted from: V. Thomas, *Science and Sport.*

Figure 3.21

The data are not linear because for each 2-minute interval the heart rate does not decrease by a fixed amount. In Figure 3.21 the data are modeled with a nonlinear function. Note that the graph of this function resembles the left half of a parabola that opens upward.

EXAMPLE 11 Modeling an athlete's heart rate

Find a quadratic function f expressed in vertex form that models the data in Table 3.3. Support your result by graphing f and the data together. What is the domain of your function?

SOLUTION To model the data we use the left half of a parabola. Since the minimum heart rate of 80 beats per minute occurs when $x = 8$, let $(8, 80)$ be the vertex and write

$$f(x) = a(x - 8)^2 + 80.$$

Next we must determine a value for the leading coefficient a. One possibility is to have the graph of f pass through the first data point $(0, 200)$, or equivalently, let $f(0) = 200$.

Table 3.3 **Athlete's Heart Rate (repeated)**

Time (min)	Heart rate (bpm)
0	200
2	150
4	110
6	90
8	80

Adapted from: V. Thomas, *Science and Sport.*

$$f(0) = 200 \qquad \text{Have the graph pass through } (0, 200).$$
$$a(0 - 8)^2 + 80 = 200 \qquad \text{Let } x = 0 \text{ in } f(x). \text{ Solve for } a.$$
$$a(0 - 8)^2 = 120 \qquad \text{Subtract 80.}$$
$$a = \frac{120}{64} \qquad \text{Divide by } (0 - 8)^2 = 64.$$
$$a = 1.875 \qquad \text{Write as a decimal.}$$

Thus $f(x) = 1.875(x - 8)^2 + 80$ can be used to model the athlete's heart rate. A graph of f and the data are shown in Figure 3.22, which is similar to Figure 3.21. Figure 3.23 shows a table of $Y_1 = 1.875(X - 8)^2 + 80$. Although the table in Figure 3.23 does not match Table 3.3 (repeated in the margin) exactly, it gives reasonable approximations. (Note that formulas for $f(x)$ may vary. For example, if we had selected the point (2, 150) rather than (0, 200), then $a \approx 1.94$. You may wish to verify this result.)

The domain D of $f(x) = 1.875(x - 8)^2 + 80$ needs to be restricted to $0 \le x \le 8$ because this interval corresponds to the domain of the data in Table 3.3.

Calculator Help

To create a table similar to Figure 3.23, see Appendix A (page AP-10).

$[-0.5, 8.5, 2]$ by $[0, 220, 20]$

Figure 3.22

Figure 3.23

Now Try Exercise 101 ◀

MAKING CONNECTIONS

General Form, Vertex Form, and Modeling   When modeling quadratic data by hand, it is often easier to use the vertex form, $f(x) = a(x - h)^2 + k$, rather than the general form, $f(x) = ax^2 + bx + c$. Because (h, k) corresponds to the vertex of a parabola, it may be appropriate to let (h, k) correspond to either the highest data point or the lowest data point in the scatterplot. Then a value for a can be found by substituting a data point into the formula for $f(x)$. In the next subsection least-squares regression provides a quadratic modeling function in general form.

Quadratic Regression (Optional)

In Chapter 2 we discussed how a regression line could be found by the method of least squares. This method can also be applied to quadratic data; the process is illustrated in the next example.

EXAMPLE 12   Finding a quadratic regression model

In one study the efficiency of photosynthesis in an Antarctic species of grass was investigated. Table 3.4 lists results for various temperatures. The temperature x is in degrees Celsius and the efficiency y is given as a percent. The purpose of the research was to determine the temperature at which photosynthesis is most efficient.   (**Source:** D. Brown and P. Rothery, *Models in Biology: Mathematics, Statistics and Computing.*)

Table 3.4   Efficiency of Photosynthesis

x (°C)	−1.5	0	2.5	5	7	10	12	15	17	20	22	25	27	30
y (%)	33	46	55	80	87	93	95	91	89	77	72	54	46	34

(a) Plot the data. Discuss reasons why a quadratic function might model the data.
(b) Find a least-squares function f given by $f(x) = ax^2 + bx + c$ that models the data.
(c) Graph f and the data in $[−5, 35, 5]$ by $[20, 110, 10]$. Discuss the fit.
(d) Determine the temperature at which f predicts that photosynthesis is most efficient. Compare the results with the findings of the experiment.

SOLUTION

(a) A plot of the data is shown in Figure 3.24. The y-values first increase and then decrease as the temperature x increases. The data suggest a parabolic shape opening downward.

(b) Enter the data into your calculator, and then select quadratic regression from the menu, as shown in Figure 3.25 and Figure 3.26. In Figure 3.27 the modeling function f is given (approximately) by $f(x) = −0.249x^2 + 6.77x + 46.37$.

$[−5, 35, 5]$ by $[20, 110, 10]$

Figure 3.24

Figure 3.25

EDIT **CALC** TESTS
1:1–Var Stats
2:2–Var Stats
3:Med–Med
4:LinReg(ax+b)
5:QuadReg
6:CubicReg
7↓QuartReg

Figure 3.26

QuadReg
y=ax²+bx+c
a=−.2487612655
b=6.76706594
c=46.37050717
R²=.956006327

Figure 3.27

Calculator Help

To find an equation of least-squares fit, see Appendix A (page AP-11).

(c) The graph of $Y_1 = −.249X^2 + 6.77X + 46.37$ with the data is shown in Figure 3.28. Although the model is not exact, the parabola describes the general trend in the data. The parabola opens downward because $a = −0.249$.

(d) The vertex is the highest point on the graph, located at approximately $(13.6, 92.4)$. See Figure 3.29. This model predicts that the highest efficiency is about 92.4%, and it occurs near 13.6°C. Although there are percentages in the table higher than 92.4%, 13.6°C is a reasonable estimate for the optimum temperature. The function f is attempting to model the general trend in the data and predict future results.

$[−5, 35, 5]$ by $[20, 110, 10]$

Figure 3.28

$[−5, 35, 5]$ by $[20, 110, 10]$

$y_1 = −0.249x^2 + 6.77x + 46.37$

Maximum
X=13.594374 . Y=92.386968

Figure 3.29

Now Try Exercise 109 ◀

3.1 Putting It All Together

$\mathbf{T}$he following table summarizes some important topics from this section.

Concept	Symbolic Representation	Comments and Examples
Quadratic function	$f(x) = ax^2 + bx + c$, where a, b, and c are constants with $a \neq 0$ (general form)	It models data that are not linear. Its graph is a parabola (∪-shaped) that opens either upward ($a > 0$) or downward ($a < 0$).
Parabola	The graph of $y = ax^2 + bx + c$, $a \neq 0$, is a parabola.	 Vertex: (h, k); axis of symmetry: $x = h$ Maximum (or minimum) y-value: k
Completing the square to find vertex form	To complete the square for $x^2 + kx$, add $\left(\frac{k}{2}\right)^2$ to make a perfect square trinomial.	$y = x^2 - 2x + 3$ $y - 3 = x^2 - 2x$ $k = -2$ $y - 3 + 1 = x^2 - 2x + 1$ Add $\left(\frac{-2}{2}\right)^2 = 1$. $y - 2 = (x - 1)^2$ $y = (x - 1)^2 + 2$
Vertex formula	The vertex for $f(x) = ax^2 + bx + c$ is the point $$\left(-\frac{b}{2a},\, f\left(-\frac{b}{2a}\right)\right).$$	If $f(x) = x^2 - 2x + 3$, then $$x = -\frac{-2}{2(1)} = 1.$$ y-value of vertex: $f(1) = 2$ Vertex: $(1, 2)$; axis of symmetry: $x = 1$
Vertex form (standard form for a parabola with vertical axis)	The vertex form for a quadratic function is $f(x) = a(x - h)^2 + k$, with vertex (h, k).	Let $f(x) = 2(x + 3)^2 - 5$. Parabola opens upward: $a > 0$ $a = 2$ Vertex: $(-3, -5)$ Axis of symmetry: $x = -3$ Minimum y-value on graph: -5

3.1   Exercises

Basics of Quadratic Functions

Exercises 1–6: Identify f as being linear, quadratic, or neither. If f is quadratic, identify the leading coefficient a and evaluate $f(-2)$.

1. $f(x) = 1 - 2x + 3x^2$ **2.** $f(x) = -5x + 11$

3. $f(x) = \dfrac{1}{x^2 - 1}$ **4.** $f(x) = (x^2 + 1)^2$

5. $f(x) = \frac{1}{2} - \frac{3}{10}x$ **6.** $f(x) = \frac{1}{5}x^2$

Exercises 7–10: Use the graph to find the following.

(a) *Sign of the leading coefficient*
(b) *Vertex*
(c) *Axis of symmetry*
(d) *Intervals where f is increasing and where f is decreasing*

7.

8.

9.

10.

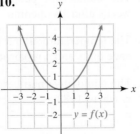

Exercises 11–14: The formulas for $f(x)$ and $g(x)$ are identical except for their leading coefficients a. Compare the graphs of f and g. You may want to support your answers by graphing f and g together.

11. $f(x) = x^2,$ $g(x) = 2x^2$

12. $f(x) = \frac{1}{2}x^2,$ $g(x) = -\frac{1}{2}x^2$

13. $f(x) = 2x^2 + 1,$ $g(x) = -\frac{1}{3}x^2 + 1$

14. $f(x) = x^2 + x,$ $g(x) = \frac{1}{4}x^2 + x$

Vertex Formula

Exercises 15–20: Identify the vertex and leading coefficient. Then write the expression as $f(x) = ax^2 + bx + c$.

15. $f(x) = -3(x - 1)^2 + 2$

16. $f(x) = 5(x + 2)^2 - 5$

17. $f(x) = 5 - 2(x - 4)^2$ **18.** $f(x) = \frac{1}{2}(x + 3)^2 - 5$

19. $f(x) = \frac{3}{4}(x + 5)^2 - \frac{7}{4}$ **20.** $f(x) = -5(x - 4)^2$

Exercises 21–28: Use the graph of the quadratic function f to write its formula as $f(x) = a(x - h)^2 + k$.

21.

22.

23.

24.

25.

26.

27.

28.

Exercises 29–40: Write the given expression in the form $f(x) = a(x - h)^2 + k$. *Identify the vertex.*

29. $f(x) = x^2 + 4x - 5$

30. $f(x) = x^2 + 10x + 7$

31. $f(x) = x^2 - 3x$

32. $f(x) = x^2 - 7x + 5$

33. $f(x) = 2x^2 - 5x + 3$

34. $f(x) = 3x^2 + 6x + 2$

35. $f(x) = \frac{1}{3}x^2 + x + 1$

36. $f(x) = -\frac{1}{2}x^2 - \frac{3}{2}x + 1$

37. $f(x) = 2x^2 - 8x - 1$

38. $f(x) = -\frac{1}{2}x^2 - x$

39. $f(x) = 2 - 9x - 3x^2$

40. $f(x) = 6 + 5x - 10x^2$

Exercises 41–52: Complete the following.

(a) *Use the vertex formula to find the vertex.*
(b) *Find the intervals where f is increasing and where f is decreasing.*

41. $f(x) = 6 - x^2$

42. $f(x) = 2x^2 - 2x + 1$

43. $f(x) = x^2 - 6x$

44. $f(x) = -2x^2 + 4x + 5$

45. $f(x) = 2x^2 - 4x + 1$

46. $f(x) = -3x^2 + x - 2$

47. $f(x) = \frac{1}{2}x^2 + 10$

48. $f(x) = \frac{9}{10}x^2 - 12$

49. $f(x) = -\frac{3}{4}x^2 + \frac{1}{2}x - 3$

50. $f(x) = -\frac{4}{5}x^2 - \frac{1}{5}x + 1$

51. $f(x) = 1.5 - 3x - 6x^2$

52. $f(x) = -4x^2 + 16x$

Graphing Quadratic Functions

Exercises 53–72: Sketch a graph of f.

53. $f(x) = x^2$

54. $f(x) = -2x^2$

55. $f(x) = -\frac{1}{2}x^2$

56. $f(x) = 4 - x^2$

57. $f(x) = x^2 - 3$

58. $f(x) = x^2 + 2$

59. $f(x) = (x - 2)^2 + 1$

60. $f(x) = (x + 1)^2 - 2$

61. $f(x) = -3(x + 1)^2 + 3$ **62.** $f(x) = -2(x - 1)^2 + 1$

63. $f(x) = x^2 - 2x - 2$ **64.** $f(x) = x^2 - 4x$

65. $f(x) = -x^2 + 4x - 2$ **66.** $f(x) = -x^2 + 2x + 1$

67. $f(x) = 2x^2 - 4x - 1$ **68.** $f(x) = 3x^2 + 6x$

69. $f(x) = -3x^2 - 6x + 1$ **70.** $f(x) = -2x^2 + 4x - 1$

71. $f(x) = -\frac{1}{2}x^2 + x + 1$ **72.** $f(x) = \frac{1}{2}x^2 - 2x + 2$

Exercises 73 and 74: ***Average Rate of Change***   *Find the average rate of change of f from 1 to 3.*

73. $f(x) = -3x^2 + 5x$ **74.** $f(x) = 4x^2 - 3x + 1$

Exercises 75 and 76: ***Difference Quotient***   *Find the difference quotient of f.*

75. $f(x) = 3x^2 - 2x$ **76.** $f(x) = 5 - 4x^2$

Exercises 77 and 78: Find the formula for a quadratic function that satisfies the given conditions.

77. Axis of symmetry $x = 3$, passing through the points $(3, 1)$ and $(1, 9)$

78. Vertex $(-3, 4)$, passing through $(-2, 1)$

Graphs and Models

Exercises 79–82: Match the situation with the graph of the quadratic function (a–d) that models it best.

79. The height y of a stone thrown from ground level after x seconds

80. The number of people attending a popular movie x weeks after its opening

81. The temperature after x hours in a house where the furnace quits and a repairperson fixes it

82. The cumulative number of reported AIDS cases in year x, where $1982 \leq x \leq 1994$

a.

b.

c.

d.

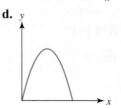

Applications and Models

83. *Maximizing Area*   A farmer has 1000 feet of fence to enclose a rectangular area. What dimensions for the rectangle result in the maximum area enclosed by the fence?

84. *Maximizing Area*   A homeowner has 80 feet of fence to enclose a rectangular garden. What dimensions for the garden give the maximum area?

85. *Maximizing Revenue*   Suppose the revenue R in thousands of dollars that a company receives from producing x thousand compact disc players is given by the formula $R(x) = x(40 - 2x)$.
    (a) Evaluate $R(2)$ and interpret the result.

    (b) How many CD players should the company produce to maximize its revenue?

    (c) What is the maximum revenue?

86. *Maximizing Revenue*   A large hotel is considering giving the following group discount on room rates: the regular price of $120 decreases by $2 for each room rented. For example, one room costs $118, two rooms cost $116 × 2 = $232, three rooms cost $114 × 3 = $342, and so on.
    (a) Write a formula for a function R that gives the revenue for renting x rooms.

    (b) Sketch a graph of R. What is a reasonable domain?

    (c) Determine the maximum revenue and the corresponding number of rooms rented.

87. *Hitting a Baseball*   A baseball is hit so that its height in feet after t seconds is $s(t) = -16t^2 + 44t + 4$.
    (a) How high is the baseball after 1 second?

    (b) Find the maximum height of the baseball. Support your answer graphically.

88. *Flight of a Baseball*   (Refer to Example 10.) A baseball is hit straight up with an initial velocity of $v_0 = 96$ feet per second (about 65 miles per hour) and leaves the bat with an initial height of $h_0 = 2.5$ feet.
    (a) Write a formula $s(t)$ that models the height after t seconds.

    (b) How high is the baseball after 4 seconds?

    (c) Find the maximum height of the baseball. Support your answer graphically.

89. *Throwing a Stone*   (Refer to Example 10.) A stone is thrown *downward* with a velocity of 66 feet per second (45 miles per hour) from a bridge that is 120 feet above a river, as illustrated in the figure.
    (a) Write a formula $s(t)$ that models the height of the stone after t seconds.

    (b) Does the stone hit the water within the first 2 seconds? Explain.

90. *Hitting a Golf Ball*   A golf ball is hit so that its height h in feet after t seconds is $h(t) = -16t^2 + 60t$.
    (a) What is the initial height of the golf ball?

    (b) How high is the golf ball after 1.5 seconds?

    (c) Find the maximum height of the golf ball.

91. *Maximizing Area*   (Refer to Example 9.) A farmer wants to fence a rectangular area by using the wall of a barn as one side of the rectangle and then enclosing the other three sides with 160 feet of fence. Find the dimensions of the rectangle that give the maximum area inside.

92. *Maximizing Area*   A rancher plans to fence a rectangular area for cattle using the straight portion of a river as one side of the rectangle. If the farmer has P feet of fence, find the dimensions of the rectangle that give the maximum area for the cattle.

Exercises 93–96: Maximizing Altitude   *If air resistance is ignored, the height h of a projectile above the ground after x seconds is given by* $h(x) = -\frac{1}{2}gx^2 + v_0x + h_0$, *where g is the acceleration due to gravity. This formula is also valid for other celestial bodies. Suppose a ball is thrown straight up at 88 feet per second from a height of 25 feet.*

   (a) For the given g, graphically estimate both the maximum height and the time when it occurs.
   (b) Solve part (a) symbolically.

93. $g = 32$ (Earth)   94. $g = 5.1$ (Moon)

95. $g = 13$ (Mars)   96. $g = 88$ (Jupiter)

Exercises 97 and 98: Find $f(x) = a(x - h)^2 + k$ so that f models the data exactly.

97.

x	−1	0	1	2	3
y	5	−1	−3	−1	5

98.

x	−2	−1	0	1	2
y	2	4	2	−4	−14

99. *Suspension Bridge*   The cables that support a suspension bridge, such as the Golden Gate Bridge, can be modeled by parabolas. Suppose that a 300-foot-long suspension bridge has at its ends towers that are 120 feet tall, as shown in the figure. If the cable comes within 20 feet of the road at the center of the bridge, find a function that models the height of the cable above the road a distance of *x* feet from the center of the bridge.

100. *Suspension Bridge*   Repeat Exercise 99 for a suspension bridge that has 100-foot towers, a length of 200 feet, and a cable that comes within 15 feet of the road at the center of the bridge.

101. *Heart Rate*   The table shows the heart rate of an athlete upon stopping a moderate activity.

Time (min)	0	1	2	3	4
Heart rate (bpm)	122	108	98	92	90

   (a) Model the data with $H(t) = a(t - h)^2 + k$. What is the domain of *H*?

   (b) Approximate the athlete's heart rate for $t = 1.5$ minutes.

102. *Heart Rate*   The heart rate of an athlete while weight training is recorded for 4 minutes. The table lists the heart rate after *x* minutes.

Time (min)	0	1	2	3	4
Heart rate (bpm)	84	111	120	110	85

(a) Explain why the data are not linear.

(b) Find a quadratic function *f* that models the data.

(c) What is the domain of your function?

Exercises 103 and 104: AIDS in America (1982–1994)   *In the early years of AIDS, the numbers of both AIDS cases and AIDS deaths could be modeled with quadratic functions. The tables list cumulative numbers for selected years.*

   (a) Find a quadratic function f that models the data.
   (b) Graph the data and f together.
   (c) Evaluate f(1991) and interpret the result.

103. Cumulative AIDS cases in thousands

Year	1982	1986	1990	1994
Cases	1.6	41.9	197	442

Source: Department of Health and Human Services.

104. Cumulative AIDS deaths in thousands

Year	1982	1986	1990	1994
Deaths	0.6	24.8	122	298

Source: Department of Health and Human Services.

Quadratic Regression

Exercises 105 and 106: Quadratic Models Use least-squares regression to find a quadratic function f that models the data given in the table. Estimate f(3.5) to the nearest hundredth.

105.

x	0	2	4	6
f(x)	−1	16	57	124

106.

x	10	20	30	40
f(x)	4.2	24.3	84.1	184

107. *Household Size*   The table lists the average number of people age 18 or older in a U.S. household.

Year	1940	1950	1960	1970
Household size	2.53	2.31	2.12	2.05

Year	1980	1990	2000	2006
Household size	1.97	1.93	1.93	1.92

Source: Bureau of the Census.

(a) Find a quadratic function that models the data. Support your result graphically.

(b) Estimate the average number of people age 18 or older in a household during 1975. Compare your estimate to the actual value of 2.01.

108. *Consumer Price Index* The table lists the consumer price index (CPI) for selected years.

Year	1950	1960	1970	1980	1990	2000
CPI	24.1	29.6	38.8	82.4	130.7	172.2

Source: Bureau of the Census.

(a) Find a quadratic function that models the data. Support your result graphically.

(b) Estimate the CPI in 1995 and compare your estimate to the actual value of 152.4.

109. *Head Start Enrollment* Head Start provides a wide range of services to children of low-income families. The table lists Head Start participation in thousands.

Year	1970	1980	1990	2006
Enrollment	447	376	541	909

Source: Department of Health and Human Services.
(a) Find a quadratic function that models the data.

(b) Estimate enrollment in 1985.

110. *U.S. Population* The table lists the U.S. population P in millions for select years x.

x	1840	1880	1920	1960	2000
P	17	50	106	179	281

Source: Bureau of the Census.
(a) Find a quadratic function that models the data.

(b) Estimate the population in 1980 and compare your estimate with the true value of 226 million.

Writing about Mathematics

111. How do the values of a, h, and k affect the graph of $f(x) = a(x - h)^2 + k$?

112. Explain why the vertex is important when you are trying to find either the maximum y-value or the minimum y-value on the graph of a quadratic function.

EXTENDED AND DISCOVERY EXERCISES

Exercises 1–4: If data are exactly linear, the slopes (or rates of change) between consecutive data points are equal. However,

this is not true of quadratic data. Suppose that, in a table containing quadratic data, consecutive x-values increase by the same amount. Then the slopes (or average rates of change) between consecutive data points will not be equal; rather, they will increase or decrease by a constant amount. For example, the quadratic data in the following table are determined by $y = x^2 + 2$. The slopes m (or average rates of change) between consecutive data points are 2, 6, 10, and 14. Note that consecutive values for m in this list always increase by the same constant 4, indicating that the data are quadratic.

x	0	2	4	6	8
y	2	6	18	38	66

$m = 2 \quad m = 6 \quad m = 10 \quad m = 14$

Use these concepts to determine if the data in the given table are linear, quadratic, or neither. If the data are linear or quadratic, find the next y-value in the table.

1.
x	1	2	3	4	5	6
y	−6	−9	−16	−27	−42	?

2.
x	5	10	15	20	25	30
y	9	8	7	6	5	?

3.
x	−2	0	2	4	6	8
y	−15	−7	1	18	43	?

4.
x	−4	−2	0	2	4	6
y	−9	−7	−1	9	23	?

Exercises 5–8: Difference Quotient Complete the following.
(a) Evaluate f(x) for each x-value in the table.

x	1	2	3	4	5
f(x)					

(b) Calculate the average rate of change of f between consecutive data points in the table.
(c) Find the difference quotient for f(x). Then let $h = 1$ in the difference quotient.
(d) Evaluate this difference quotient for x = 1, 2, 3, and 4. Compare these results to your results in part (b).

5. $f(x) = x^2 - 3$ **6.** $f(x) = 2x - x^2$

7. $f(x) = -2x^2 + 3x - 1$ **8.** $f(x) = 3x^2 + x + 2$

3.2 Quadratic Equations and Problem Solving

- Understand basic concepts about quadratic equations
- Use factoring, the square root property, completing the square, and the quadratic formula to solve quadratic equations
- Understand the discriminant
- Solve problems involving quadratic equations

Introduction

In Example 11 of Section 3.1 we modeled an athlete's heart rate x minutes after exercise stopped by using $f(x) = 1.875(x - 8)^2 + 80$. This vertex form can easily be changed to general form.

$$
\begin{aligned}
f(x) &= 1.875(x - 8)^2 + 80 && \text{Vertex form} \\
&= 1.875(x^2 - 16x + 64) + 80 && \text{Square the binomial.} \\
&= 1.875x^2 - 30x + 200 && \text{General form}
\end{aligned}
$$

To determine the length of time needed for the athlete's heart rate to slow from 200 beats per minute to **110** beats per minute, we can solve the *quadratic equation*

$$1.875x^2 - 30x + 200 = 110, \text{ or}$$

$$1.875x^2 - 30x + 90 = 0. \qquad \text{Subtract 110 from each side.}$$

(See Example 8.) A quadratic equation results when the formula for a quadratic function is set equal to a constant. Applications of quadratic functions often require solving quadratic equations.

Quadratic Equations

A quadratic equation can be defined as follows.

Quadratic Equation

A **quadratic equation** in one variable is an equation that can be written in the form

$$ax^2 + bx + c = 0,$$

where a, b, and c are constants with $a \neq 0$.

Examples of quadratic equations include

$$2x^2 - 3x - 4 = 0, \quad x^2 = 3, \quad -5x^2 + x = 0, \quad \text{and} \quad 3x + 1 = x^2.$$

Quadratic equations can have no real solutions, one real solution, or two real solutions. (Complex solutions are discussed in the next section.) Quadratic equations can be solved symbolically by a variety of methods: factoring, the square root property, completing the square, and the quadratic formula. They can also be solved graphically and numerically; however, the exact solution can *always* be obtained symbolically.

Factoring

Factoring is a common technique used to solve equations. It is based on the **zero-product property**, which states that if $ab = 0$, then $a = 0$ or $b = 0$ or both. It is important to remember that this property works only for 0. For example, if $ab = 1$, then this equation does *not* imply that either $a = 1$ or $b = 1$. For example, $a = \frac{1}{2}$ and $b = 2$ also satisfies $ab = 1$ and neither a nor b is 1.

EXAMPLE 1   Solving a quadratic equation with factoring

Solve each quadratic equation. Check your results.
(a) $2x^2 + 2x - 11 = 1$ **(b)** $12t^2 = t + 1$

SOLUTION
(a) Start by writing the equation in the form $ax^2 + bx + c = 0$.

$2x^2 + 2x - 11 = 1$	Given equation
$2x^2 + 2x - 12 = 0$	Subtract 1 from each side.
$x^2 + x - 6 = 0$	Divide each side by 2 (optional step).
$(x + 3)(x - 2) = 0$	Factor.
$x + 3 = 0$ or $x - 2 = 0$	Zero-product property
$x = -3$ or $x = 2$	Solve.

These solutions can be checked by substituting them in the given equation.

$$2(-3)^2 + 2(-3) - 11 \stackrel{?}{=} 1 \qquad 2(2)^2 + 2(2) - 11 \stackrel{?}{=} 1$$
$$1 = 1 \qquad\qquad\qquad 1 = 1$$

(b) Write the equation in the form $at^2 + bt + c = 0$.

$12t^2 = t + 1$	Given equation
$12t^2 - t - 1 = 0$	Subtract t and 1.
$(3t - 1)(4t + 1) = 0$	Factor.
$3t - 1 = 0$ or $4t + 1 = 0$	Zero-product property
$t = \dfrac{1}{3}$ or $t = -\dfrac{1}{4}$	Solve.

To check these solutions, substitute them into the given equation.

$$12\left(\frac{1}{3}\right)^2 = \frac{1}{3} + 1 \qquad 12\left(-\frac{1}{4}\right)^2 = -\frac{1}{4} + 1$$
$$\frac{4}{3} = \frac{4}{3} \qquad\qquad \frac{3}{4} = \frac{3}{4}$$

Now Try Exercises 1 and 7 ◄

We can also use factoring to find the *x*-intercepts of the graph of a quadratic function, because the *x*-intercepts of the graph of $y = ax^2 + bx + c$ correspond to the solutions to $ax^2 + bx + c = 0$.

EXAMPLE 2   Finding *x*-intercepts

Find the exact values for the *x*-intercepts shown in Figure 3.30.

SOLUTION From the graph it is difficult to determine the *exact* *x*-intercepts. However, they can be determined symbolically.

$24x^2 + 7x - 6 = 0$	Set expression equal to 0.
$(3x + 2)(8x - 3) = 0$	Factor.
$3x + 2 = 0$ or $8x - 3 = 0$	Zero-product property
$x = -\dfrac{2}{3}$ or $x = \dfrac{3}{8}$	Solve.

The *x*-intercepts are $-\frac{2}{3}$ and $\frac{3}{8}$.

Now Try Exercise 29 ◄

Algebra Review
To review factoring trinomials, see Chapter R (page R-23).

Figure 3.30

Figure 3.31

| EXAMPLE 3 | Solving a quadratic equation symbolically, graphically, and numerically |

On the shoreline of some recreational lakes in Minnesota, the Department of Natural Resources (DNR) restricts the size of buildings. Buildings built within 100 feet of the shoreline are sometimes limited to a maximum area of 120 square feet. Suppose a rectangular shed is being built that has an area of 120 square feet and is 7 feet longer than it is wide. See Figure 3.31. Determine its dimensions symbolically, graphically, and numerically.

SOLUTION

Symbolic Solution   Let x be the width of the shed. Then $x + 7$ represents the length. Since area equals width times length, we solve the following equation.

$$x(x + 7) = 120 \qquad \text{Area is equal to 120 square feet.}$$
$$x^2 + 7x = 120 \qquad \text{Distributive property}$$
$$x^2 + 7x - 120 = 0 \qquad \text{Subtract 120 from each side.}$$
$$(x + 15)(x - 8) = 0 \qquad \text{Factor.}$$
$$x + 15 = 0 \quad \text{or} \quad x - 8 = 0 \qquad \text{Zero-product property}$$
$$x = -15 \quad \text{or} \quad x = 8 \qquad \text{Solve.}$$

Since dimensions cannot be negative, the solution that has meaning is $x = 8$. The length is 7 feet longer than the width, so the dimensions of the shed are 8 feet by 15 feet.

Graphical Solution   The *intersection-of-graphs method* can be used to solve this equation by graphing $y_1 = x(x + 7)$ and $y_2 = 120$. Their graphs intersect at $(-15, 120)$ and $(8, 120)$, as shown in Figures 3.32 and 3.33. The solutions are -15 and 8.

Numerical Solution   Make a table for y_1 and y_2. The solution of 8 is shown in Figure 3.34, and the solution of -15 may be found by scrolling through the table. The equation $y_1 = y_2$ is satisfied when $x = -15$ or 8. The symbolic, graphical, and numerical solutions agree.

Calculator Help
To find a point of intersection, see Appendix A (page AP-8).

$[-20, 20, 5]$ by $[-20, 150, 10]$ $[-20, 20, 5]$ by $[-20, 150, 10]$

X	Y1	Y2
5	60	120
6	78	120
7	98	120
8	**120**	120 ← $y_1 = y_2$
9	144	120
10	170	120
11	198	120
X=8		

Figure 3.32   **Figure 3.33**   **Figure 3.34**

NOTE If the solutions are fractions or irrational numbers, it may be difficult or even impossible to determine them with a table of values.   ***Now Try Exercise 105*** ◀

The Square Root Property

Algebra Review
To review square roots, see Chapter R (page R-40).

Some types of quadratic equations can be written as $x^2 = k$, where k is a nonnegative number. The solutions to this equation are $\pm\sqrt{k}$. (Recall that the symbol $\pm$ represents *plus or minus*.) For example, $x^2 = 16$ has two solutions: ± 4. We refer to this as the **square root property**. See Extended and Discovery Exercise 7 for a justification of this property.

Square Root Property
Let k be a nonnegative number. Then the solutions to the equation
$$x^2 = k$$
are given by $x = \pm\sqrt{k}$.

EXAMPLE 4 Using the square root property

If a metal ball is dropped 100 feet from a water tower, its height h in feet above the ground after t seconds is given by $h(t) = 100 - 16t^2$. Determine how long it takes the ball to hit the ground.

SOLUTION The ball strikes the ground when the equation $100 - 16t^2 = 0$ is satisfied.

$$100 - 16t^2 = 0$$

$$100 = 16t^2 \qquad \text{Add } 16t^2 \text{ to each side.}$$

$$t^2 = \frac{100}{16} \qquad \text{Divide each side by 16. Rewrite.}$$

$$t = \pm\sqrt{\frac{100}{16}} \qquad \text{Square root property}$$

$$t = \pm\frac{10}{4} \qquad \text{Simplify.}$$

In this example only positive values for time are valid, so the ball strikes the ground after $\frac{10}{4}$, or 2.5, seconds. ***Now Try Exercise 101*** ◄

Functions can be defined by formulas, graphs, tables, and diagrams. Functions can also be defined by equations. In the next example we use the square root property to solve equations for y and then determine if y is a function of x, where $y = f(x)$.

EXAMPLE 5 Determining if equations represent functions

Solve each equation for y. Determine if y is a function of x.

(a) $x^2 + (y - 1)^2 = 4$ **(b)** $2y = \dfrac{x + y}{2}$

SOLUTION
(a) Start by subtracting x^2 from each side of the equation.

$$(y - 1)^2 = 4 - x^2 \qquad \text{Subtract } x^2 \text{ from each side.}$$

$$y - 1 = \pm\sqrt{4 - x^2} \qquad \text{Square root property}$$

$$y = 1 \pm \sqrt{4 - x^2} \qquad \text{Add 1 to each side.}$$

There are two formulas, $y = 1 + \sqrt{4 - x^2}$ and $y = 1 - \sqrt{4 - x^2}$, which indicates that y is *not* a function of x. That is, one x-input can produce two y-outputs.

NOTE The equation $x^2 + (y - 1)^2 = 4$ is in standard form for a circle that has center $(0, 1)$ and radius 2. A circle does not pass the vertical line test, so the equation does not represent a function. Use two functions to graph this circle by letting $y_1 = 1 + \sqrt{4 - x^2}$ (upper half) and $y_2 = 1 - \sqrt{4 - x^2}$ (lower half).

Algebra Review
To clear fractions, see Chapter R
(page R-35).

(b) Clear fractions by multiplying each side of $2y = \frac{x + y}{2}$ by 2.

$$4y = x + y \qquad \textit{Multiply each side by 2.}$$
$$3y = x \qquad \textit{Subtract y from each side.}$$
$$y = \frac{x}{3} \qquad \textit{Divide each side by 3.}$$

The equation $y = \frac{x}{3}$ defines a linear function, so y is a function of x.

Now Try Exercises 67 and 69 ◀

Completing the Square

Another technique that can be used to solve a quadratic equation is *completing the square*. If a quadratic equation is written in the form $x^2 + kx = d$, where k and d are constants, then the equation can be solved using

$$x^2 + kx + \left(\frac{k}{2}\right)^2 = \left(x + \frac{k}{2}\right)^2.$$

Algebra Review
To review factoring perfect square
trinomials, see Chapter R
(page R-26).

For example, $k = 6$ in $x^2 + 6x = 7$, so add $\left(\frac{k}{2}\right)^2 = \left(\frac{6}{2}\right)^2 = 9$ to each side.

$$x^2 + 6x = 7 \qquad \textit{Given equation}$$
$$x^2 + 6x + 9 = 7 + 9 \qquad \textit{Add 9 to each side.}$$
$$(x + 3)^2 = 16 \qquad \textit{Factor the perfect square.}$$
$$x + 3 = \pm 4 \qquad \textit{Square root property}$$
$$x = -3 \pm 4 \qquad \textit{Add −3 to each side.}$$
$$x = 1 \quad \text{or} \quad x = -7 \qquad \textit{Simplify.}$$

NOTE If the coefficient a of the x^2-term is not 1, we can divide each side of the equation by a so that it becomes 1. See Example 6(b).

Completing the square is useful when solving quadratic equations that do not factor easily.

EXAMPLE 6 Completing the square

Solve each equation
(a) $x^2 - 8x + 9 = 0$ **(b)** $2x^2 - 8x = 7$

SOLUTION
(a) Start by writing the equation in the form $x^2 + kx = d$.

$$x^2 - 8x + 9 = 0 \qquad \textit{Given equation}$$
$$x^2 - 8x = -9 \qquad \textit{Subtract 9 from each side.}$$
$$x^2 - 8x + 16 = -9 + 16 \qquad \textit{Add } \left(\frac{k}{2}\right)^2 = \left(\frac{-8}{2}\right)^2 = 16.$$
$$(x - 4)^2 = 7 \qquad \textit{Factor the perfect square.}$$
$$x - 4 = \pm\sqrt{7} \qquad \textit{Square root property}$$
$$x = 4 \pm \sqrt{7} \qquad \textit{Add 4 to each side.}$$

(b) Divide each side by 2 to obtain a 1 for the leading coefficient.

$$2x^2 - 8x = 7 \qquad \text{Given equation}$$

$$x^2 - 4x = \frac{7}{2} \qquad \text{Divide by 2.}$$

$$x^2 - 4x + 4 = \frac{7}{2} + 4 \qquad \text{Add } \left(\frac{-4}{2}\right)^2 = 4 \text{ to each side.}$$

$$(x - 2)^2 = \frac{15}{2} \qquad \text{Factor the perfect square.}$$

$$x - 2 = \pm\sqrt{\frac{15}{2}} \qquad \text{Square root property}$$

$$x = 2 \pm \sqrt{\frac{15}{2}} \qquad \text{Add 2 to each side.}$$

> **Now Try Exercises 49 and 51** ◀

Symbolic, Numerical, and Graphical Solutions   Quadratic equations can be solved symbolically, numerically, and graphically. The following example illustrates each technique for the equation $x(x - 2) = 3$.

Symbolic Solution	*Numerical Solution*	*Graphical Solution*

Symbolic Solution

$$x(x - 2) = 3$$
$$x^2 - 2x = 3$$
$$x^2 - 2x - 3 = 0$$
$$(x + 1)(x - 3) = 0$$

The solutions are -1 and 3.

Numerical Solution

x	$x(x - 2)$
-2	8
-1	3
0	0
1	-1
2	0
3	3
4	8

Let $y = x(x - 2)$. In the table $y = 3$ when $x = -1$ or $x = 3$.

Graphical Solution

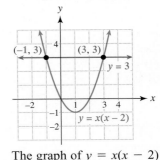

The graph of $y = x(x - 2)$ intersects the graph of $y = 3$ at $(-1, 3)$ and $(3, 3)$. The solutions are -1 and 3.

The Quadratic Formula

The quadratic formula can be used to find the solutions to *any* quadratic equation.

Quadratic Formula

The solutions to the quadratic equation $ax^2 + bx + c = 0$, where $a \neq 0$, are given by

$$x = \frac{-b \pm \sqrt{b^2 - 4ac}}{2a}.$$

NOTE When solving quadratic equations, a common strategy is to first write the equation in the form $ax^2 + bx + c = 0$ and to then try to factor the left side of the equation. If the factors are not easily found, the quadratic formula is used. Unlike graphical and numerical solutions, the quadratic formula always finds the exact solutions to a quadratic equation.

EXAMPLE 7   Using the quadratic formula

Solve the equation $3x^2 - 6x + 2 = 0$.

SOLUTION   Let $a = 3$, $b = -6$, and $c = 2$.

$$x = \frac{-b \pm \sqrt{b^2 - 4ac}}{2a} \qquad \text{Quadratic formula}$$

$$x = \frac{-(-6) \pm \sqrt{(-6)^2 - 4(3)(2)}}{2(3)} \qquad \text{Substitute for } a, b, \text{ and } c.$$

$$x = \frac{6 \pm \sqrt{12}}{6} \qquad \text{Simplify.}$$

$$x = 1 \pm \frac{1}{6}\sqrt{12} \qquad \text{Divide: } \frac{a \pm b}{c} = \frac{a}{c} \pm \frac{b}{c}.$$

Algebra Review
To review simplifying square roots, see Chapter R (page R-46).

NOTE Because $\sqrt{12} = \sqrt{4} \cdot \sqrt{3} = 2\sqrt{3}$, we can write $1 \pm \frac{1}{6}\sqrt{12}$ as $1 \pm \frac{1}{3}\sqrt{3}$.

Now Try Exercise 21 ◀

CLASS DISCUSSION

Use the results of Example 7 to evaluate each expression mentally.

$$3\left(1 + \tfrac{1}{6}\sqrt{12}\right)^2 - 6\left(1 + \tfrac{1}{6}\sqrt{12}\right) + 2, \quad 3\left(1 - \tfrac{1}{6}\sqrt{12}\right)^2 - 6\left(1 - \tfrac{1}{6}\sqrt{12}\right) + 2$$

EXAMPLE 8   Estimating an athlete's heart rate

Solve the quadratic equation $1.875x^2 - 30x + 90 = 0$ symbolically, graphically, and numerically, where $0 \le x \le 8$. (This equation was discussed in the introduction to this section.)

SOLUTION

Symbolic Solution   Let $a = 1.875$, $b = -30$, and $c = 90$ in the quadratic formula.

$$x = \frac{-b \pm \sqrt{b^2 - 4ac}}{2a} \qquad \text{Quadratic formula}$$

$$= \frac{-(-30) \pm \sqrt{(-30)^2 - 4(1.875)(90)}}{2(1.875)} \qquad a = 1.875, b = -30, \text{ and } c = 90$$

$$= \frac{30 \pm \sqrt{225}}{3.75} \qquad \text{Simplify.}$$

$$= 12 \text{ or } 4 \qquad \text{Simplify.}$$

The x-values are restricted to $0 \le x \le 8$, so the only valid solution is 4. Thus the athlete's heart rate reached 110 beats per minute 4 minutes after the athlete stopped exercising.

Graphical Solution   To use the *x-intercept method* to solve this quadratic equation, graph $y_1 = 1.875x^2 - 30x + 90$ and locate the *x*-intercepts, as shown in Figures 3.35 and 3.36. The *x*-intercepts are 4 and 12, in agreement with the symbolic solution.

Numerical Solution   Make a table of $y_1 = 1.875x^2 - 30x + 90$, as shown in Figure 3.37. The numerical solution agrees with the symbolic and graphical solutions because $y_1 = 0$ when $x = 4$ or $x = 12$.

Calculator Help

To find a zero, or *x*-intercept, see Appendix A (page AP-9).

[0, 16, 4] by [−125, 125, 25]

Figure 3.35

[0, 16, 4] by [−125, 125, 25]

Figure 3.36

Figure 3.37

Now Try Exercise 103 ◀

The Discriminant

One important difference between linear equations and nonlinear equations is that nonlinear equations, in general, can have *any* number of solutions. If the quadratic equation $ax^2 + bx + c = 0$ is solved graphically, the parabola $y = ax^2 + bx + c$ can intersect the *x*-axis zero, one, or two times, as illustrated in Figures 3.38–3.40, respectively. Each *x*-intercept is a *real* solution to the quadratic equation $ax^2 + bx + c = 0$.

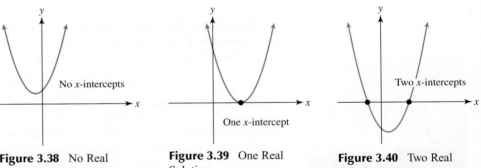

Figure 3.38   No Real Solutions

Figure 3.39   One Real Solution

Figure 3.40   Two Real Solutions

The quantity $b^2 - 4ac$ in the quadratic formula is called the **discriminant.** It provides information about the number of real solutions to a quadratic equation.

Quadratic Equations and the Discriminant

To determine the number of real solutions to $ax^2 + bx + c = 0$ with $a \neq 0$, evaluate the discriminant $b^2 - 4ac$.

1. If $b^2 - 4ac > 0$, there are two real solutions.
2. If $b^2 - 4ac = 0$, there is one real solution.
3. If $b^2 - 4ac < 0$, there are no real solutions.

NOTE When $b^2 - 4ac < 0$, the solutions to a quadratic equation may be expressed as two complex numbers. Complex numbers are discussed in the next section.

In Example 7 the discriminant is $b^2 - 4ac = (-6)^2 - 4(3)(2) = 12$. Because the discriminant is positive, there are two real solutions.

EXAMPLE 9 Using the discriminant

Use the discriminant to find the number of solutions to $9x^2 - 12.6x + 4.41 = 0$. Then solve the equation by using the quadratic formula. Support your result graphically.

SOLUTION

Symbolic Solution Let $a = 9$, $b = -12.6$, and $c = 4.41$. The discriminant is given by

$$b^2 - 4ac = (-12.6)^2 - 4(9)(4.41) = 0.$$

Since the discriminant is 0, there is one solution.

$$x = \frac{-b \pm \sqrt{b^2 - 4ac}}{2a} \qquad \text{Quadratic formula}$$

$$x = \frac{-(-12.6) \pm \sqrt{0}}{18} \qquad \text{Substitute.}$$

$$x = 0.7 \qquad \text{Simplify.}$$

The only solution is 0.7.

Graphical Solution A graph of $y = 9x^2 - 12.6x + 4.41$ is shown in Figure 3.41. The graph suggests that there is one x-intercept: 0.7. *Now Try Exercise 83* ◄

Figure 3.41

Factoring and the Discriminant If a, b, and c are *integers* and $b^2 - 4ac$ is a perfect square, then the trinomial $ax^2 + bx + c$ can be factored using only integer coefficients. For example, if $6x^2 + x - 2 = 0$, then

$$b^2 - 4ac = 1^2 - 4(6)(-2) = 49,$$

which *is* a perfect square ($49 = 7^2$). Thus we can factor $6x^2 + x - 2$ as $(2x - 1)(3x + 2)$ to solve $6x^2 + x - 2 = 0$. However, if $3x^2 - x - 1 = 0$, then $b^2 - 4ac = 13$, which is *not* a perfect square. This trinomial cannot be factored (by using traditional methods with integer coefficients) so either the quadratic formula or completing the square should be used to solve $3x^2 - x - 1 = 0$. See Extended and Discovery Exercises 1–4 at the end of this section.

An Application In Section 3.1 we learned that the height s of an object propelled into the air is modeled by $s(t) = -16t^2 + v_0 t + h_0$, where v_0 represents the object's initial (vertical) velocity in feet per second and h_0 represents the object's initial height in feet, as illustrated in Figure 3.42. In the next example we model the position of a projectile.

Figure 3.42

EXAMPLE 10 Modeling projectile motion

Table 3.5 shows the height of a projectile shot into the air.

Table 3.5 Height of a Projectile

t seconds	0	2	4	6	8
$s(t)$ (feet)	96	400	576	624	544

(a) Use $s(t) = -16t^2 + v_0 t + h_0$ to model the data.
(b) After how many seconds did the projectile strike the ground?

SOLUTION

Getting Started The value for h_0 can be determined by noting that $s(0) = 96$. The value for v_0 can be determined by using any other value in the table. We use $s(2) = 400$ to determine v_0. ▶

(a) Because $s(0) = 96$, $h_0 = 96$ and $s(t) = -16t^2 + v_0 t + 96$. Substituting $s(2) = 400$ gives the following result.

$$-16(2)^2 + v_0(2) + 96 = 400 \qquad s(2) = 400$$
$$2v_0 = 368 \qquad \text{Subtract 32 from each side.}$$
$$v_0 = 184 \qquad \text{Divide each side by 2.}$$

Thus $s(t) = -16t^2 + 184t + 96$ models the height of the projectile.

(b) The projectile strikes the ground when $s(t) = 0$.

$$-16t^2 + 184t + 96 = 0 \qquad s(t) = 0; \text{ equation to be solved}$$
$$2t^2 - 23t - 12 = 0 \qquad \text{Divide each term by } -8.$$
$$(2t + 1)(t - 12) = 0 \qquad \text{Factor the trinomial.}$$
$$t = -\frac{1}{2} \quad \text{or} \quad t = 12 \qquad \text{Solve the equation.}$$

Thus the projectile strikes the ground after 12 seconds. The solution of $-\frac{1}{2}$ has no meaning in this problem because it corresponds to a time before the projectile is shot into the air. ***Now Try Exercise 115*** ◀

Problem Solving

Many types of applications involve quadratic equations. To solve these problems, we use the steps for "Solving Application Problems" from Section 2.3 on page 122. In the next example the dimensions of a box are found by solving a quadratic equation.

> **EXAMPLE 11**   Solving a construction problem

A box is being constructed by cutting 2-inch squares from the corners of a rectangular piece of cardboard that is 6 inches longer than it is wide, as illustrated in Figure 3.43. If the box is to have a volume of 224 cubic inches, find the dimensions of the piece of cardboard.

Figure 3.43

SOLUTION

Geometry Review
To review formulas related to boxes, see Chapter R (page R-3).

STEP 1:   The rectangular piece of cardboard is 6 inches longer than it is wide. Let x be its width and $x + 6$ be its length.

$$x: \text{Width of the cardboard in inches}$$

$$x + 6: \text{Length of the cardboard in inches}$$

Figure 3.44

STEP 2: First make a drawing of the box with the appropriate labeling, as shown in Figure 3.44. The width of the *bottom* of the box is $x - 4$ inches, because two square corners with sides of 2 inches have been removed. Similarly, the length of the bottom of the box is $x + 2$ inches. Because the height times the width times the length must equal the volume, or 224 cubic inches, it follows that

$$2(x - 4)(x + 2) = 224, \quad \text{or} \quad (x - 4)(x + 2) = 112.$$

STEP 3: Write the quadratic equation in the form $ax^2 + bx + c = 0$ and factor.

$$\begin{aligned} x^2 - 2x - 8 &= 112 && \text{\textit{Equation to be solved}} \\ x^2 - 2x - 120 &= 0 && \text{\textit{Subtract 112.}} \\ (x - 12)(x + 10) &= 0 && \text{\textit{Factor.}} \\ x = 12 \quad \text{or} \quad x &= -10 && \text{\textit{Zero-product property}} \end{aligned}$$

Since the dimensions cannot be negative, the width of the cardboard is 12 inches and the length is 6 inches more, or 18 inches.

STEP 4: After the 2-inch-square corners are cut out, the dimensions of the bottom of the box are $12 - 4 = 8$ inches by $18 - 4 = 14$ inches. The volume of the box is then $2 \cdot 8 \cdot 14 = 224$ cubic inches, which checks. **_Now Try Exercise 107_** ◀

When DVDs are manufactured, a discount is sometimes given to a customer who makes a large order. Discounts affect the revenue that a company receives. We discuss this situation in the next example.

EXAMPLE 12 Determining revenue

A company charges $5 to burn one DVD, but it reduces this cost by $0.05 per DVD for each additional DVD ordered, up to a maximum of 60 DVDs. For example, the price for one DVD is $5, the price for two DVDs is 2($4.95) = $9.90, the price for 3 DVDs is 3($4.90) = $14.70, and so on. If the total price is $95, how many DVDs were ordered?

SOLUTION

[0, 100, 10] by [0, 150, 50]

Figure 3.45

[0, 100, 10] by [0, 150, 50]

Figure 3.46

STEP 1: We are asked to find the number of DVDs that results in an order costing $95. Let this number be x.

$$x: \text{Number of DVDs ordered}$$

STEP 2: Revenue equals the number of DVDs sold times the price of each DVD. If x DVDs are sold, then the price in dollars of each DVD is $5 - 0.05(x - 1)$. (Note that when $x = 1$ the price is $5 - 0.05(1 - 1) = \$5$.) The revenue R is given by

$$R(x) = x(5 - 0.05(x - 1)),$$

and we must solve the equation

$$x(5 - 0.05(x - 1)) = 95, \quad \text{or} \quad -0.05x^2 + 5.05x - 95 = 0.$$

STEP 3: Although this equation could be solved symbolically, we solve it graphically by letting $Y_1 = X(5 - 0.05(X - 1))$ and $Y_2 = 95$. In Figures 3.45 and 3.46, their graphs intersect at (25, 95) and (76, 95). The discount only applies to orders up to 60 DVDs, so the answer is 25 DVDs.

STEP 4: If 25 DVDs are ordered, then the cost of each DVD is $5 - 0.05(24) = \$3.80$ and the total revenue is $25(3.80) = \$95$. **_Now Try Exercise 113_** ◀

3.2 Putting It All Together

$\mathbf{T}$he following table summarizes important topics related to quadratic equations.

Concept	Explanation	Examples
Quadratic equation	$ax^2 + bx + c = 0$, where a, b, and c are constants with $a \neq 0$	A quadratic equation can have zero, one, or two real solutions. $x^2 = -5$ *No real solutions* $(x - 2)^2 = 0$ *One real solution* $x^2 - 4 = 0$ *Two real solutions*
Factoring	A symbolic technique for solving equations, based on the zero-product property: if $ab = 0$, then either $a = 0$ or $b = 0$.	$x^2 - 3x = -2$ $x^2 - 3x + 2 = 0$ $(x - 1)(x - 2) = 0$ $x - 1 = 0$ or $x - 2 = 0$ $x = 1$ or $x = 2$
Square root property	The solutions to $x^2 = k$ are $x = \pm\sqrt{k}$, where $k \geq 0$.	$x^2 = 9$ is equivalent to $x = \pm 3$. $x^2 = 11$ is equivalent to $x = \pm\sqrt{11}$.
Completing the square	To solve $x^2 + kx = d$ symbolically, add $\left(\frac{k}{2}\right)^2$ to each side to obtain a perfect square trinomial. Then apply the square root property.	$x^2 - 6x = 1$ $x^2 - 6x + 9 = 1 + 9$ $\left(\frac{-6}{2}\right)^2 = 9$ $(x - 3)^2 = 10$ $x - 3 = \pm\sqrt{10}$ $x = 3 \pm \sqrt{10}$
Quadratic formula	The solutions to $ax^2 + bx + c = 0$ are given by $$x = \frac{-b \pm \sqrt{b^2 - 4ac}}{2a}.$$ Always gives the *exact* solutions	To solve $2x^2 - x - 4 = 0$, let $a = 2$, $b = -1$, and $c = -4$. $$x = \frac{-(-1) \pm \sqrt{(-1)^2 - 4(2)(-4)}}{2(2)}$$ $$= \frac{1 \pm \sqrt{33}}{4} \approx 1.69 \text{ or } -1.19$$
Graphical solution	To use the *x-intercept method* to solve $ax^2 + bx + c = 0$, let y_1 equal the left side of the equation and graph y_1. The real solutions correspond to the *x*-intercepts. The *intersection-of-graphs* method can also be used when one side of the equation is *not* equal to 0. Let y_1 equal the left side of the equation and y_2 equal the right side of the equation. The real solutions correspond to the *x*-coordinates of any points of intersection.	To solve $x^2 + x - 2 = 0$, graph $y = x^2 + x - 2$. The *x*-intercepts are -2 and 1.

continued on next page

continued from previous page

Concept	Explanation	Examples
Numerical solution	To solve $ax^2 + bx + c = 0$, let y_1 equal the left side of the equation and create a table for y_1. The zeros of y_1 are the real solutions. May *not* be a good method when solutions are fractions or irrational numbers	To solve $2x^2 + x - 1 = 0$, make a table for $Y_1 = 2X^2 + X - 1$. The solutions are $-1, 0.5$.

3.2   Exercises

Quadratic Equations

Exercises 1–28: Solve the quadratic equation. Check your answers for Exercises 1–12.

1. $x^2 + x - 11 = 1$

2. $x^2 - 9x + 10 = -8$

3. $t^2 = 2t$

4. $t^2 - 7t = 0$

5. $3x^2 - 7x = 0$

6. $5x = 9x^2$

7. $2z^2 = 13z + 15$

8. $4z^2 = 7 - 27z$

9. $x(3x + 14) = 5$

10. $x(5x + 19) = 4$

11. $6x^2 + \frac{5}{2} = 8x$

12. $8x^2 + 63 = -46x$

13. $(t + 3)^2 = 5$

14. $(t - 2)^2 = 11$

15. $4x^2 - 13 = 0$

16. $9x^2 - 11 = 0$

17. $2(x - 1)^2 + 4 = 0$

18. $-3(x + 5)^2 - 6 = 0$

19. $\frac{1}{2}x^2 - 3x + \frac{1}{2} = 0$

20. $\frac{3}{4}x^2 + \frac{1}{2}x - \frac{1}{2} = 0$

21. $-3z^2 - 2z + 4 = 0$

22. $-4z^2 + z + 1 = 0$

23. $25k^2 + 1 = 10k$

24. $49k^2 + 4 = -28k$

25. $-0.3x^2 + 0.1x = -0.02$

26. $-0.1x^2 + 1 = 0.5x$

27. $2x(x + 2) = (x - 1)(x + 2)$

28. $(2x - 1)(x + 2) = (x + 3)(x + 1)$

Exercises 29–34: Find the exact values of any x-intercepts.

29.

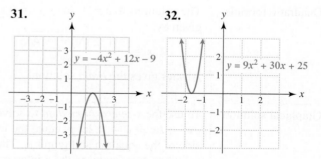

$y = 6x^2 + 13x - 5$

30.

$y = 6x^2 + 4x - 2$

31.

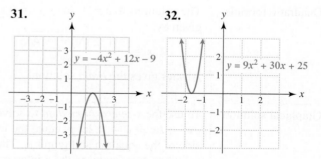

$y = -4x^2 + 12x - 9$

32.

$y = 9x^2 + 30x + 25$

33.

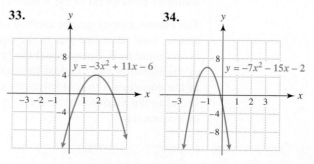

$y = -3x^2 + 11x - 6$

34.

$y = -7x^2 - 15x - 2$

Graphical and Numerical Solutions

Exercises 35–42: Solve each quadratic equation (a) graphically, (b) numerically, and (c) symbolically. Express graphical and numerical solutions to the nearest tenth when appropriate.

35. $x^2 + 2x = 0$

36. $x^2 - 4 = 0$

37. $x^2 - x - 6 = 0$

38. $2x^2 + 5x - 3 = 0$

39. $2x^2 = 6$

40. $x^2 - 225 = 0$

41. $4x(x - 3) = -9$

42. $-4x(x - 1) = 1$

Exercises 43–46: Solve the quadratic equation graphically.

43. $20x^2 + 11x = 3$

44. $-2x^2 + 4x = 1.595$

45. $2.5x^2 = 4.75x - 2.1$

46. $x(x + 24) = 6912$

Completing the Square

Exercises 47–60: Solve the equation by completing the square.

47. $x^2 + 4x - 6 = 0$

48. $x^2 - 10x = 1$

49. $x^2 + 5x = 4$

50. $x^2 + 6x - 5 = 0$

51. $3x^2 - 6x = 2$

52. $2x^2 - 3x + 1 = 0$

53. $x^2 - 8x = 10$

54. $x^2 - 2x = 2$

55. $\frac{1}{2}t^2 - \frac{3}{2}t = 1$

56. $\frac{1}{3}t^2 + \frac{1}{2}t = 2$

57. $-2z^2 + 3z + 1 = 0$

58. $-3z^2 - 5z + 3 = 0$

59. $-\frac{3}{2}z^2 - \frac{1}{4}z + 1 = 0$

60. $-\frac{1}{5}z^2 - \frac{1}{2}z + 2 = 0$

Finding Domains

Exercises 61–64: Find the domain of the function. Write your answer in set-builder notation.

61. $f(x) = \dfrac{1}{x^2 - 5}$

62. $f(x) = \dfrac{4x}{7 - x^2}$

63. $g(t) = \dfrac{5 - t}{t^2 - t - 2}$

64. $g(t) = \dfrac{t + 1}{2t^2 - 11t - 21}$

Solving for a Variable

Exercises 65–72: (Refer to Example 5.) Solve the equation for y. Determine if y is a function of x.

65. $4x^2 + 3y = \dfrac{y + 1}{3}$

66. $\dfrac{x^2 + y}{2} = y - 2$

67. $3y = \dfrac{2x - y}{3}$

68. $\dfrac{5 - y}{3} = \dfrac{x + 3y}{4}$

69. $x^2 + (y - 3)^2 = 9$

70. $(x + 2)^2 + (y + 1)^2 = 1$

71. $3x^2 + 4y^2 = 12$

72. $x - 25y^2 = 50$

Exercises 73–80: Solve for the specified variable.

73. $V = \frac{1}{3}\pi r^2 h$ for r

74. $V = \frac{1}{2}gt^2 + h$ for t

75. $K = \frac{1}{2}mv^2$ for v

76. $W = I^2R$ for I

77. $a^2 + b^2 = c^2$ for b

78. $S = 4\pi r^2 + x^2$ for r

79. $s = -16t^2 + 100t$ for t

80. $T^2 - kT - k^2 = 0$ for T

The Discriminant

Exercises 81–96: Complete the following.

   (a) Write the equation as $ax^2 + bx + c = 0$ with $a > 0$.

   (b) Calculate the discriminant $b^2 - 4ac$ and determine the number of real solutions.

   (c) Solve the equation.

81. $3x^2 = 12$

82. $8x^2 - 2 = 14$

83. $x^2 - 2x = -1$

84. $6x^2 = 4x$

85. $4x = x^2$

86. $16x^2 + 9 = 24x$

87. $x^2 + 1 = x$

88. $2x^2 + x = 2$

89. $2x^2 + 3x = 12 - 2x$

90. $3x^2 + 3 = 5x$

91. $9x(x - 4) = -36$

92. $\frac{1}{4}x^2 + 3x = x - 4$

93. $x\left(\frac{1}{2}x + 1\right) = -\frac{13}{2}$

94. $4x = 6 + x^2$

95. $3x^2 = 1 - x$

96. $x(5x - 3) = 1$

Exercises 97–100: The graph of $f(x) = ax^2 + bx + c$ is shown in the figure.

   (a) State whether $a > 0$ or $a < 0$.

   (b) Solve the equation $ax^2 + bx + c = 0$.

   (c) Is the discriminant positive, negative, or zero?

97.

98.

99.

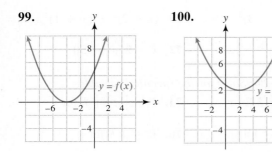

100.

Applications and Models

101. *Height of Baseball*   A baseball is dropped from a stadium seat that is 75 feet above the ground. Its height s in feet after t seconds is given by $s(t) = 75 - 16t^2$. Estimate to the nearest tenth of a second how long it takes for the baseball to strike the ground.

102. *Height of Baseball*   A baseball is thrown *downward* with an initial velocity of 30 feet per second from a stadium seat that is 80 feet above the ground. Estimate to the nearest tenth of a second how long it takes for the baseball to strike the ground.

103. *U.S. AIDS Deaths*   From 1984 to 1994 the equation

$$D(x) = 2375x^2 + 5134x + 5020$$

modeled the cumulative number of AIDS deaths x years after 1984. Estimate the year when there were 90,000 deaths.

104. *U.S. AIDS Cases*   From 1984 to 1994 the cumulative number of AIDS cases can be modeled by the equation

$$C(x) = 3034x^2 + 14{,}018x + 6400,$$

where x represents years after 1984. Estimate the year when 200,000 AIDS cases had been diagnosed.

105. *Screen Dimensions*   The width of a rectangular computer screen is 2.5 inches more than its height. If the area of the screen is 93.5 square inches, determine its dimensions symbolically, graphically, and numerically. Do your answers agree?

106. *Maximizing Area*   A rectangular pen for a pet is under construction using 100 feet of fence.
  (a) Find the dimensions that give an area of 576 square feet.
  (b) Find the dimensions that give maximum area.

107. *Construction*   (Refer to Example 11.) A box is being constructed by cutting 4-inch squares from the corners of a rectangular sheet of metal that is 10 inches longer than it is wide. If the box is to have a volume of 476 cubic inches, find the dimensions of the metal sheet.

108. *Construction*   A box is being constructed by cutting 2-inch squares from the corners of a *square* sheet of metal. If the box is to have a volume of 1058 cubic inches, find the dimensions of the metal sheet.

109. *Geometry*   A cylindrical aluminum can is being constructed to have a height h of 4 inches. If the can is to have a volume of 28 cubic inches, approximate its radius r. (*Hint:* $V = \pi r^2 h$.)

110. *Braking Distance*   Braking distance for cars on level pavement can be approximated by $D(x) = \frac{x^2}{30k}$. The input x is the car's velocity in miles per hour and the output $D(x)$ is the braking distance in feet. The positive constant k is a measure of the traction of the tires. Small values of k indicate a slippery road or worn tires.   (**Source:** L. Haefner, *Introduction to Transportation Systems.*)
  (a) Let $k = 0.3$. Evaluate $D(60)$ and interpret the result.

  (b) If $k = 0.25$, find the velocity x that corresponds to a braking distance of 300 feet.

111. *Window Dimensions*   A window comprises a square with sides of length x and a semicircle with diameter x, as shown in the figure. If the total area of the window is 463 square inches, estimate the value of x to the nearest hundredth of an inch.

112. *Picture Frame*   A frame for a picture is 2 inches wide. The picture inside the frame is 4 inches longer than it is wide. See the figure. If the area of the picture is 320 square inches, find the outside dimensions of the picture frame.

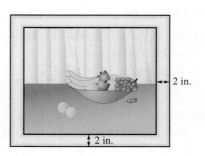

113. *Cost* (Refer to Example 12.) A company charges $20 to make one monogrammed shirt, but reduces this cost by $0.10 per shirt for each *additional* shirt ordered up to 100 shirts. If the cost of an order is $989, how many shirts were ordered?

114. *Ticket Prices* One airline ticket costs $250. For each additional airline ticket sold to a group, the price of every ticket is reduced by $2. For example, 2 tickets cost $2 \cdot 248 = \$496$ and 3 tickets cost $3 \cdot 246 = \$738$.
   (a) Write a quadratic function that gives the total cost of buying x tickets.

   (b) What is the cost of 5 tickets?

   (c) How many tickets were sold if the cost is $5200?

   (d) What number of tickets sold gives the greatest cost?

Modeling Quadratic Data

115. *Projectile Motion* The table shows the height of a projectile that is shot into the air.

t (seconds)	0	1	2	3	4
s (feet)	32	176	288	368	416

   (a) Use $s(t) = -16t^2 + v_0 t + h_0$ to model the data.

   (b) After how long did the projectile strike the ground?

116. *Falling Object* The table lists the velocity and distance traveled by a falling object for various elapsed times.

Time (sec)	0	1	2	3	4	5
Velocity (ft/sec)	0	32	64	96	128	160
Distance (ft)	0	16	64	144	256	400

   (a) Make a scatterplot of the ordered pairs determined by (time, velocity) and (time, distance) in the same viewing rectangle $[-1, 6, 1]$ by $[-10, 450, 50]$.

   (b) Find a function v that models the velocity.

   (c) The distance is modeled by $d(x) = ax^2$. Find a.

   (d) Find the time when the distance is 200 feet. Find the velocity at this time.

117. *Safe Runway Speed* The road (or taxiway) used by aircraft to exit a runway should not have sharp curves. The safe radius for any curve depends on the speed of the airplane. The table at the top of the next column lists the minimum radius R of the exit curves, where the taxiing speed of the airplane is x miles per hour.

x (mi/hr)	10	20	30	40	50	60
R (ft)	50	200	450	800	1250	1800

Source: Federal Aviation Administration.

   (a) If the taxiing speed x of the plane doubles, what happens to the minimum radius R of the curve?

   (b) The FAA used $R(x) = ax^2$ to compute the values in the table. Determine a.

   (c) If $R = 500$, find x. Interpret your results.

118. *Biology* Some types of worms have a remarkable capacity to live without moisture. The table shows the number of worms y surviving after x days in one study.

x (days)	0	20	40	80	120	160
y (worms)	50	48	45	36	20	3

Source: D. Brown and P. Rothery, *Models in Biology*.

   (a) Use regression to find a quadratic function f that models these data.

   (b) Graph f and the data in the same window.

   (c) Solve the quadratic equation $f(x) = 0$ graphically. Do both solutions have meaning? Explain.

119. *Wal-Mart Employees* The table lists numbers of Wal-Mart employees E in millions, x years after 1987.

x	0	5	10	15	20
E	0.20	0.38	0.68	1.4	2.2

Source: Wal-Mart.

   (a) Evaluate $E(15)$ and interpret the result.

   (b) Find a quadratic function f that models these data.

   (c) Graph the data and function f in the same xy-plane.

   (d) Use f to estimate the year when the number of employees may reach 3 million.

120. *Women in the Work Force* The number N of women in millions who were gainfully employed in the work force in selected years is shown in the table.

Year	1900	1910	1920	1930	1940	1950
N	5.3	7.4	8.6	10.8	12.8	18.4

Year	1960	1970	1980	1990	2000	2010
N	23.2	31.5	45.5	56.6	65.6	74.8

Source: Department of Labor.

(a) Use regression to find a quadratic function f that models the data. Support your result graphically.

(b) Predict the number of women in the labor force in 2020.

Writing about Mathematics

121. Discuss three symbolic methods for solving a quadratic equation. Make up a quadratic equation and use each method to find the solution set.

122. Explain how to solve a quadratic equation graphically.

EXTENDED AND DISCOVERY EXERCISES

Exercises 1–4: Discriminant and Factoring (Refer to page 196.) For each equation, calculate the discriminant. Use the discriminant to decide whether the equation can be solved by factoring. If it can, solve the equation by factoring. Otherwise, use the quadratic formula.

1. $8x^2 + 14x - 15 = 0$

2. $15x^2 - 17x - 4 = 0$

3. $5x^2 - 3x - 3 = 0$

4. $3x^2 - 2x - 4 = 0$

5. *Quadratic Formula* Prove the quadratic formula by completing the following.
   (a) Write $ax^2 + bx + c = 0$ as $x^2 + \frac{b}{a}x = -\frac{c}{a}$.
   (b) Complete the square to obtain $\left(x + \frac{b}{2a}\right)^2 = \frac{b^2 - 4ac}{4a^2}$.
   (c) Use the square root property and solve for x.

6. *Difference Quotient* If the difference quotient for the function $f(x) = ax^2 - bx + 1$ equals $2x + h - 4$, find values for a and b.

7. *Square Root Property* Use the fact that $\sqrt{x^2} = |x|$ to show that the solutions to $x^2 = k$ with $k > 0$ are given by $x = \pm\sqrt{k}$.

CHECKING BASIC CONCEPTS FOR SECTIONS 3.1 AND 3.2

1. Graph $f(x) = (x - 1)^2 - 4$. Identify the vertex, axis of symmetry, and x-intercepts.

2. A graph of $f(x) = ax^2 + bx + c$ is shown.
   (a) Is a positive, negative, or zero?
   (b) Find the vertex and axis of symmetry.
   (c) Solve $ax^2 + bx + c = 0$.
   (d) Is the discriminant positive, negative, or zero?

3. Use $f(x) = a(x - h)^2 + k$ to model the data exactly.

x	−3	−2	−1	0	1
$f(x)$	11	5	3	5	11

4. Find the vertex on the graph of $y = 3x^2 - 9x - 2$.

5. Write $f(x) = x^2 + 4x - 3$ as $f(x) = a(x - h)^2 + k$. What are the coordinates of the vertex? What is the minimum y-value of the graph of f?

6. Solve the quadratic equations.
   (a) $16x^2 = 81$
   (b) $2x^2 + 3x = 2$
   (c) $x^2 = x - 3$
   (d) $2x^2 = 3x + 4$

7. *Dimensions of a Rectangle* A rectangle is 4 inches longer than it is wide and has an area of 165 square inches. Find its dimensions.

8. *Height of a Baseball* The height s of a baseball in feet after t seconds is given by $s(t) = -16t^2 + 96t + 2$.
   (a) Find the height of the baseball after 1 second.
   (b) After how long is the baseball 142 feet high?
   (c) Find the maximum height of the baseball.
   (d) How long is the baseball in the air?

3.3 Complex Numbers

- Perform arithmetic operations on complex numbers
- Solve quadratic equations having complex solutions

Introduction

Throughout history, people have invented new numbers to solve equations and describe data. Often these new numbers were met with resistance and were regarded as being imaginary or unreal. The number 0 was not invented at the same time as the natural numbers. There was no Roman numeral for 0. No doubt there were skeptics who wondered why a number was needed to represent nothing. Negative numbers also met strong resistance. After all, how could one possibly have -6 apples? The same was true for complex numbers. However, complex numbers are no more imaginary than any other number created by mathematicians.

Basic Concepts

The graph of $y = x^2 + 1$ has no x-intercepts, as shown in Figure 3.47. Therefore the equation $x^2 + 1 = 0$ has no real solutions. If we try to solve $x^2 + 1 = 0$, we get $x^2 = -1$. Since $x^2 \geq 0$ for any real number x, there are no real number solutions. However, we can *invent* a solution.

$$x^2 = -1, \quad \text{or} \quad x = \pm\sqrt{-1}$$

We now define a new number called the **imaginary unit**, denoted i.

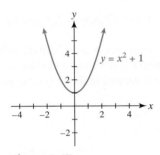

Figure 3.47

Properties of the Imaginary Unit i
$i = \sqrt{-1}, \qquad i^2 = -1$

Defining the number i allows us to say that the solutions to the equation $x^2 + 1 = 0$ are i and $-i$. Using the real numbers and the imaginary unit i, complex numbers can be defined. A **complex number** can be written in **standard form** as $a + bi$, where a and b are real numbers. The **real part** is a and the **imaginary part** is b. Every real number a is also a complex number because it can be written as $a + 0i$. A complex number $a + bi$ with $b \neq 0$ is an **imaginary number**. A complex number $a + bi$ with $a = 0$ and $b \neq 0$ is sometimes called a **pure imaginary number**. Examples of pure imaginary numbers include $3i$ and $-i$. Table 3.6 lists several complex numbers with their real and imaginary parts.

Table 3.6 Real and Imaginary Parts

$a + bi$	$-3 + 2i$	5	$-3i$	$-1 + 7i$	$-5 - 2i$	$4 + 6i$
a	-3	5	0	-1	-5	4
b	2	0	-3	7	-2	6

Using the imaginary unit i, square roots of negative numbers can be written as complex numbers.

The Expression $\sqrt{-a}$
If $a > 0$, then $\sqrt{-a} = i\sqrt{a}$.

EXAMPLE 1 Simplifying the expression $\sqrt{-a}$

Simplify each expression.

(a) $\sqrt{-16}$ **(b)** $\sqrt{-3}$ **(c)** $\dfrac{2 \pm \sqrt{-24}}{2}$

SOLUTION

(a) When $a > 0$, $\sqrt{-a} = i\sqrt{a}$. Thus $\sqrt{-16} = i\sqrt{16} = 4i$.

(b) $\sqrt{-3} = i\sqrt{3}$; we usually do not write $\sqrt{3}i$ because of the possibility of confusion about whether i is under the square root symbol.

(c) First note that $\sqrt{-24} = i\sqrt{24} = i\sqrt{4} \cdot \sqrt{6} = 2i\sqrt{6}$.

$$\frac{2 \pm \sqrt{-24}}{2} = \frac{2 \pm 2i\sqrt{6}}{2} \qquad \text{Simplify } \sqrt{-24}.$$

$$= \frac{2}{2} \pm \frac{2i\sqrt{6}}{2} \qquad \frac{a \pm b}{c} = \frac{a}{c} \pm \frac{b}{c}$$

$$= 1 \pm i\sqrt{6} \qquad \text{Simplify fractions.}$$

Now Try Exercises 1, 5, and 13 ◀

The property $\sqrt{a} \cdot \sqrt{b} = \sqrt{ab}$ is true only when *both* a and b are positive. When simplifying products containing square roots of negative numbers, it is important to first apply the property $\sqrt{-a} = i\sqrt{a}$, where $a > 0$. This technique is illustrated in the next example.

EXAMPLE 2 Simplifying complex expressions

Simplify each expression.

(a) $\sqrt{-3} \cdot \sqrt{-3}$ **(b)** $\sqrt{-2} \cdot \sqrt{-8}$

SOLUTION

(a) $\sqrt{-3} \cdot \sqrt{-3} = i\sqrt{3} \cdot i\sqrt{3} = i^2(\sqrt{3})^2 = -1(3) = -3$

NOTE $\sqrt{-3} \cdot \sqrt{-3} \neq \sqrt{(-3) \cdot (-3)} = \sqrt{9} = 3$

(b) $\sqrt{-2} \cdot \sqrt{-8} = i\sqrt{2} \cdot i\sqrt{8} = i^2\sqrt{16} = -1(4) = -4$

Now Try Exercises 15 and 17 ◀

Arithmetic Operations on Complex Numbers

Arithmetic operations are also defined for complex numbers.

Addition and Subtraction To add the complex numbers $(-2 + 3i)$ and $(4 - 6i)$, simply combine the real parts and the imaginary parts.

$$(-2 + 3i) + (4 - 6i) = -2 + 4 + 3i - 6i$$

$$= 2 - 3i$$

This same process works for subtraction.

$$(5 - 7i) - (8 + 3i) = 5 - 8 - 7i - 3i$$

$$= -3 - 10i$$

Multiplication   Two complex numbers can be multiplied. The property $i^2 = -1$ is applied when appropriate.

$$(-5 + i)(7 - 9i) = -5(7) + (-5)(-9i) + (i)(7) + (i)(-9i)$$
$$= -35 + 45i + 7i - 9i^2$$
$$= -35 + 52i - 9(-1)$$
$$= -26 + 52i$$

NOTE Express your results in the standard form $a + bi$.

Division   The **conjugate** of $a + bi$ is $a - bi$. To find the conjugate, change the sign of the imaginary part b. Table 3.7 lists examples of complex numbers and their conjugates.

Table 3.7   Complex Conjugates

$a + bi$	$2 + 5i$	$6 - 3i$	$-2 + 7i$	$-1 - i$	5	$-4i$
$a - bi$	$2 - 5i$	$6 + 3i$	$-2 - 7i$	$-1 + i$	5	$4i$

To simplify the quotient $\frac{3 + 2i}{5 - i}$, first multiply both the numerator and the denominator by the conjugate of the *denominator*.

$$\frac{3 + 2i}{5 - i} = \frac{(3 + 2i)(5 + i)}{(5 - i)(5 + i)}$$ Multiply by $\frac{conjugate}{conjugate}$.

$$= \frac{3(5) + (3)(i) + (2i)(5) + (2i)(i)}{(5)(5) + (5)(i) + (-i)(5) + (-i)(i)}$$ Expand.

$$= \frac{15 + 3i + 10i + 2i^2}{25 + 5i - 5i - i^2}$$ Simplify.

$$= \frac{15 + 13i + 2(-1)}{25 - (-1)}$$ $i^2 = -1$

$$= \frac{13 + 13i}{26}$$ Simplify.

$$= \frac{1}{2} + \frac{1}{2}i$$ $\frac{a + bi}{c} = \frac{a}{c} + \frac{b}{c}i$

The last step expresses the quotient as a complex number in standard form.

```
(-2+3i)+(4-6i)
                2-3i
(5-7i)-(8+3i)
              -3-10i
```

Figure 3.48

```
(-5+i)(7-9i)
            -26+52i
(3+2i)/(5-i)
              .5+.5i
Ans▶Frac
            1/2+1/2i
```

Figure 3.49

Evaluating Complex Arithmetic with Technology   Some calculators do complex arithmetic. The evaluation of the previous examples is shown in Figures 3.48 and 3.49.

EXAMPLE 3   Performing complex arithmetic

Write each expression in standard form. Support your results using a calculator.

(a) $(-3 + 4i) + (5 - i)$ **(b)** $(-7i) - (6 - 5i)$

(c) $(-3 + 2i)^2$ **(d)** $\dfrac{17}{4 + i}$

$$\begin{array}{l}(^-3+4\mathrm{i})+(5-\mathrm{i})\\\hspace{3.5cm}2+3\mathrm{i}\\(^-7\mathrm{i})-(6-5\mathrm{i})\\\hspace{3.5cm}^-6-2\mathrm{i}\end{array}$$

Figure 3.50

$$\begin{array}{l}(^-3+2\mathrm{i})2\\\hspace{2cm}5-12\mathrm{i}\\17/(4+\mathrm{i})\\\hspace{3cm}4-\mathrm{i}\end{array}$$

Figure 3.51

SOLUTION

(a) $(-3 + 4i) + (5 - i) = -3 + 5 + 4i - i = 2 + 3i$

(b) $(-7i) - (6 - 5i) = -6 - 7i + 5i = -6 - 2i$

(c) $(-3 + 2i)^2 = (-3 + 2i)(-3 + 2i)$

$$= 9 - 6i - 6i + 4i^2$$
$$= 9 - 12i + 4(-1)$$
$$= 5 - 12i$$

(d) $\dfrac{17}{4 + i} = \dfrac{17}{4 + i} \cdot \dfrac{4 - i}{4 - i}$ Multiply by $\dfrac{conjugate}{conjugate}$.

$$= \dfrac{68 - 17i}{16 - i^2}$$

$$= \dfrac{68 - 17i}{17} \qquad i^2 = -1$$

$$= 4 - i$$

Standard forms can be found using a calculator. See Figures 3.50 and 3.51.

Now Try Exercises 23, 25, 35, and 39 ◄

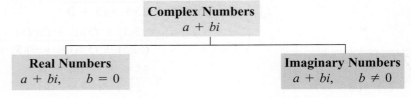

MAKING CONNECTIONS

Complex, Real, and Imaginary Numbers The following diagram illustrates the relationship among complex, real, and imaginary numbers, where a and b are real numbers. Note that complex numbers comprise two disjoint sets of numbers: the real numbers and the imaginary numbers.

Complex Numbers
$a + bi$

Real Numbers
$a + bi, \qquad b = 0$

Imaginary Numbers
$a + bi, \qquad b \neq 0$

NOTE *Every* real number is a complex number.

Quadratic Equations with Complex Solutions

We can use the quadratic formula to solve the quadratic equation $ax^2 + bx + c = 0$. If the discriminant, $b^2 - 4ac$, is negative, then there are no real solutions, and the graph of $y = ax^2 + bx + c$ does not intersect the x-axis. However, there are solutions that can be expressed as imaginary numbers. This is illustrated in the next example.

EXAMPLE 4   Solving quadratic equations with imaginary solutions

Solve the quadratic equation. Support your results graphically.
(a) $\frac{1}{2}x^2 + 17 = 5x$ (b) $x^2 + 3x + 5 = 0$ (c) $-2x^2 = 3$

SOLUTION

Getting Started Write each equation as $ax^2 + bx + c = 0$, and then apply the quadratic formula, which always "works." ►

Figure 3.52

Figure 3.53

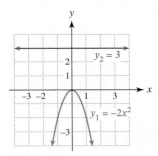

Figure 3.54

(a) Rewrite the equation as $\frac{1}{2}x^2 - 5x + 17 = 0$, and let $a = \frac{1}{2}$, $b = -5$, and $c = 17$.

$$x = \frac{-b \pm \sqrt{b^2 - 4ac}}{2a}$$

$$= \frac{5 \pm \sqrt{(-5)^2 - 4(0.5)(17)}}{2(0.5)}$$

$$= 5 \pm \sqrt{-9}$$

$$= 5 \pm 3i$$

In Figure 3.52 the graphs of $y_1 = \frac{1}{2}x^2 + 17$ and $y_2 = 5x$ do not intersect, which indicates that the equation $\frac{1}{2}x^2 + 17 = 5x$ has no real solutions. However, there are two complex solutions that are imaginary.

(b) Let $a = 1$, $b = 3$, and $c = 5$ and apply the quadratic formula.

$$x = \frac{-3 \pm \sqrt{3^2 - 4(1)(5)}}{2(1)}$$

$$= \frac{-3 \pm \sqrt{-11}}{2}$$

$$= \frac{-3 \pm i\sqrt{11}}{2}$$

$$= -\frac{3}{2} \pm \frac{i\sqrt{11}}{2}$$

In Figure 3.53 the graph of $y_1 = x^2 + 3x + 5$ does not intersect the x-axis, which indicates that the equation $x^2 + 3x + 5 = 0$ has no real solutions. However, there are two complex solutions that are imaginary.

(c) Rather than use the quadratic formula for this equation, we apply the square root property because the equation contains no x-term.

$$-2x^2 = 3 \qquad \textit{Given equation}$$

$$x^2 = -\frac{3}{2} \qquad \textit{Divide by } -2$$

$$x = \pm\sqrt{-\frac{3}{2}} \qquad \textit{Square root property}$$

$$x = \pm i\sqrt{\frac{3}{2}} \qquad \sqrt{-a} = i\sqrt{a}$$

In Figure 3.54 the graphs of $y_1 = -2x^2$ and $y_2 = 3$ do not intersect, which indicates that the equation $-2x^2 = 3$ has no real solutions. However, there are two complex solutions that are imaginary. **Now Try Exercises 59, 65, and 67** ◀

CLASS DISCUSSION

What is the result if the expression is evaluated? (See Example 4(b).)

$$\left(-\frac{3}{2} + \frac{i\sqrt{11}}{2}\right)^2 + 3\left(-\frac{3}{2} + \frac{i\sqrt{11}}{2}\right) + 5$$

3.3 Putting It All Together

Some of the important topics in this section are summarized in the following table.

Concept	Explanation	Comments and Examples
Imaginary unit	$i = \sqrt{-1},\ i^2 = -1$	The imaginary unit i allows us to define the complex numbers.
The expression $\sqrt{-a}$ with $a > 0$	$\sqrt{-a} = i\sqrt{a}$	$\sqrt{-4} = 2i$ $\sqrt{-5} = i\sqrt{5}$ $\sqrt{-32} = i\sqrt{32} = i\sqrt{16}\sqrt{2} = 4i\sqrt{2}$ $\sqrt{-5}\sqrt{-20} = i\sqrt{5}\,i\sqrt{20} = i^2\sqrt{100} = -10$
Complex number	$a + bi$, where a and b are real numbers	Every real number is a complex number. $5 - 4i, 5, 2 + i$, and $-9i$ are examples of complex numbers.
Standard form of a complex number	$a + bi$, where a and b are real numbers	Converting to standard form: $$\frac{3 \pm 4i}{2} = \frac{3}{2} + 2i \ \text{ or } \ \frac{3}{2} - 2i$$
Conjugates	The conjugate of $a + bi$ is $a - bi$.	$\begin{array}{cc} \textit{Number} & \textit{Conjugate} \\ 5 - 6i & 5 + 6i \\ -12i & 12i \\ -7 & -7 \\ 2 + 3i & 2 - 3i \end{array}$
Arithmetic operations on complex numbers	Complex numbers may be added, subtracted, multiplied, or divided.	$(2 + 3i) + (-3 - i) = -1 + 2i$ $(5 + i) - (3 - i) = 2 + 2i$ $(1 + i)(5 - i) = 5 - i + 5i - i^2 = 6 + 4i$ $\dfrac{3 + i}{1 - i} = \dfrac{(3 + i)(1 + i)}{(1 - i)(1 + i)}$ $\phantom{\dfrac{3+i}{1-i}} = 1 + 2i$
Complex solutions to equations	Complex numbers $a + bi$ with $b \neq 0$ can be solutions to equations that cannot be solved with only real numbers.	$x^2 + 5 = 0$ implies $x^2 = -5$; there are no real solutions, but $x = \pm i\sqrt{5}$ are two complex solutions. Note that the graph of $y = x^2 + 5$ has no x-intercepts. The quadratic formula can be used to find complex solutions.

3.3   Exercises

Complex Numbers

Exercises 1–20: Simplify by using the imaginary unit i.

1. $\sqrt{-4}$

2. $\sqrt{-16}$

3. $\sqrt{-100}$

4. $\sqrt{-49}$

5. $\sqrt{-23}$

6. $\sqrt{-11}$

7. $\sqrt{-12}$

8. $\sqrt{-32}$

9. $\sqrt{-54}$

10. $\sqrt{-28}$

11. $\dfrac{4 \pm \sqrt{-16}}{2}$

12. $\dfrac{-2 \pm \sqrt{-36}}{6}$

13. $\dfrac{-6 \pm \sqrt{-72}}{3}$

14. $\dfrac{2 \pm \sqrt{-8}}{4}$

15. $\sqrt{-5} \cdot \sqrt{-5}$

16. $\sqrt{-8} \cdot \sqrt{-8}$

17. $\sqrt{-18} \cdot \sqrt{-2}$

18. $\sqrt{-20} \cdot \sqrt{-5}$

19. $\sqrt{-3} \cdot \sqrt{-6}$

20. $\sqrt{-15} \cdot \sqrt{-5}$

Exercises 21–48: Write the expression in standard form.

21. $3i + 5i$

22. $-7i + 5i$

23. $(3 + i) + (-5 - 2i)$

24. $(-4 + 2i) + (7 + 35i)$

25. $2i - (-5 + 23i)$

26. $(12 - 7i) - (-1 + 9i)$

27. $3 - (4 - 6i)$

28. $(7 + i) - (-8 + 5i)$

29. $(2)(2 + 4i)$

30. $(-5)(-7 + 3i)$

31. $(1 + i)(2 - 3i)$

32. $(-2 + i)(1 - 2i)$

33. $(-3 + 2i)(-2 + i)$

34. $(2 - 3i)(1 + 4i)$

35. $(-2 + 3i)^2$

36. $(2 - 3i)^2$

37. $2i(1 - i)^2$

38. $-i(5 - 2i)^2$

39. $\dfrac{1}{1 + i}$

40. $\dfrac{1 - i}{2 + 3i}$

41. $\dfrac{4 + i}{5 - i}$

42. $\dfrac{10}{1 - 4i}$

43. $\dfrac{2i}{10 - 5i}$

44. $\dfrac{3 - 2i}{1 + 2i}$

45. $\dfrac{3}{-i}$

46. $\dfrac{4 - 2i}{i}$

47. $\dfrac{-2 + i}{(1 + i)^2}$

48. $\dfrac{3}{(2 - i)^2}$

Exercises 49–54: Evaluate the expression with a calculator.

49. $(23 - 5.6i) + (-41.5 + 93i)$

50. $(-8.05 - 4.67i) + (3.5 + 5.37i)$

51. $(17.1 - 6i) - (8.4 + 0.7i)$

52. $\left(\dfrac{3}{4} - \dfrac{1}{10}i\right) - \left(-\dfrac{1}{8} + \dfrac{4}{25}i\right)$

53. $(-12.6 - 5.7i)(5.1 - 9.3i)$

54. $(7.8 + 23i)(-1.04 + 2.09i)$

Exercises 55 and 56: Evaluate with a calculator. Round values to the nearest thousandth.

55. $\dfrac{17 - 135i}{18 + 142i}$

56. $\dfrac{141 + 52i}{102 - 31i}$

Quadratic Equations with Complex Solutions

Exercises 57–74: Solve. Write answers in standard form.

57. $x^2 + 5 = 0$

58. $4x^2 + 3 = 0$

59. $5x^2 + 1 = 3x^2$

60. $x(3x + 1) = -1$

61. $3x = 5x^2 + 1$

62. $4x^2 = x - 1$

63. $x(x - 4) = -5$

64. $2x^2 + x + 1 = 0$

65. $x^2 = 3x - 5$

66. $3x - x^2 = 5$

67. $x^2 + 2x + 4 = 0$

68. $x(x - 4) = -8$

69. $3x^2 - 4x = x^2 - 3$

70. $2x^2 + 3 = 1 - x$

71. $2x(x - 2) = x - 4$

72. $3x^2 + x = x(5 - x) - 2$

73. $3x(3 - x) - 8 = x(x - 2)$

74. $-x(7 - 2x) = -6 - (3 - x)$

Zeros of Quadratic Functions

Exercises 75–80: The graph of a function is given.

   (a) *Use the graph to predict the number of real zeros*
      and the number of imaginary zeros.
   (b) *Find these zeros using the quadratic formula.*

75.

$y = 2x^2 - x - 3$

76.

$y = -x^2 + 4.6x - 5.29$

77.

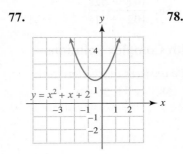

$y = x^2 + x + 2$

78.

$y = -2x^2 + 2x - 3$

79.

$y = -x^2 - 2$

80.

$y = x^2 - x + 1$

Writing about Mathematics

81. Could a quadratic function have one real zero and one imaginary zero? Explain.

82. Give an example of a quadratic function that has only real zeros and an example of one that has only imaginary zeros. How do their graphs compare? Explain how to determine from a graph whether a quadratic function has real zeros.

EXTENDED AND DISCOVERY EXERCISE

1. *Powers of i*   The properties of the imaginary unit are $i = \sqrt{-1}$ and $i^2 = -1$.

   (a) Begin simplifying the expressions $i, i^2, i^3, i^4, i^5, \ldots,$ until a simple pattern is discovered. For example, $i^3 = i \cdot i^2 = i \cdot (-1) = -i$.

   (b) Summarize your findings by describing how to simplify i^n for any natural number n.

3.4 Quadratic Inequalities

- **Understand basic concepts about quadratic inequalities**
- **Solve quadratic inequalities graphically**
- **Solve quadratic inequalities symbolically**

Introduction

Highway engineers often use quadratic functions to model safe stopping distances for cars. For example, $f(x) = \frac{1}{12}x^2 + \frac{11}{5}x$ is sometimes used to model the stopping distance for a car traveling at x miles per hour on dry, level pavement. If a driver can see only **200** feet ahead on a highway with a sharp curve, then safe driving speeds x satisfy the *quadratic inequality*

$$\frac{1}{12}x^2 + \frac{11}{5}x \leq 200,$$

or equivalently,

$$\frac{1}{12}x^2 + \frac{11}{5}x - 200 \leq 0.$$

(See Example 3.) This section discusses methods for solving quadratic inequalities.

Basic Concepts

A quadratic equation can be written as $ax^2 + bx + c = 0$ with $a \neq 0$. If the equals sign is replaced by $>$, $\geq$, $<$, or $\leq$, a **quadratic inequality** results. Examples of quadratic inequalities include

$$x^2 + 2x - 1 \geq 0, \quad 4x^2 < 1, \quad \text{and} \quad 2x^2 \leq 1 - 3x.$$

MAKING CONNECTIONS

Quadratic Function, Equation, and Inequality The three concepts are closely related.

$f(x) = ax^2 + bx + c, a \neq 0$	*Quadratic function*
$ax^2 + bx + c = 0, a \neq 0$	*Quadratic equation*
$ax^2 + bx + c > 0, a \neq 0$	*Quadratic inequality*

Because equality is (usually) the boundary between *greater than* and *less than*, a first step in solving a quadratic inequality is to determine the x-values where equality occurs. These x-values are the *boundary numbers*. We begin by discussing graphical solutions to quadratic inequalities.

Graphical and Numerical Solutions

The graph of a quadratic function is a parabola that opens either upward or downward. For example, Figure 3.55 shows the parabola $y = x^2 - x - 6$. Since $a = 1$, this parabola opens upward. It has x-intercepts -2 and 3, which satisfy the equation $x^2 - x - 6 = 0$. The parabola lies below the x-axis between the intercepts (or boundary numbers), so the solution set to the inequality $x^2 - x - 6 < 0$ is $\{x \mid -2 < x < 3\}$, or $(-2, 3)$ in interval notation. Similarly, the solutions to $x^2 - x - 6 > 0$ include x-values either left of $x = -2$ or right of $x = 3$, where the parabola is above the x-axis. This solution set is $\{x \mid x < -2 \text{ or } x > 3\}$, or $(-\infty, -2) \cup (3, \infty)$ in interval notation.

Figure 3.55

MAKING CONNECTIONS

Graphs and Inequalities   Suppose that $y = ax^2 + bx + c$. Then the solution set to $y > 0$ includes all x-values where the graph of y is *above* the x-axis. Similarly, the solution set to $y < 0$ includes all x-values where the graph of y is *below* the x-axis. This is because the x-axis corresponds to $y = 0$. This discussion is illustrated in Figures 3.56 and 3.57.

Figure 3.56

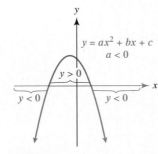

Figure 3.57

EXAMPLE 1 Solving quadratic inequalities graphically

Use the graphs of $y = ax^2 + bx + c$ in Figures 3.58–3.61 to solve the inequalities:

i. $ax^2 + bx + c \leq 0$ **ii.** $ax^2 + bx + c > 0$

(a)

Figure 3.58

(b)

Figure 3.59

(c)

Figure 3.60

(d)

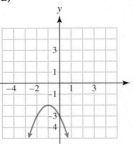

Figure 3.61

SOLUTION

(a) **i.** $ax^2 + bx + c \leq 0$ The x-intercepts, or boundary numbers, are -1 and 2. Between these x-values, the graph is below the x-axis and the y-values are negative. The solution set is $\{x \mid -1 \leq x \leq 2\}$.

ii. $ax^2 + bx + c > 0$ The graph is above the x-axis, and the y-values are positive, either left of $x = -1$ or right of $x = 2$. The solution set is $\{x \mid x < -1 \text{ or } x > 2\}$.

(b) **i.** $ax^2 + bx + c \leq 0$ The parabola opens downward with x-intercepts, or boundary numbers, of -4 and 2. The graph is below the x-axis, and the y-values are negative, either left of $x = -4$ or right of $x = 2$. The solution set is $\{x \mid x \leq -4 \text{ or } x \geq 2\}$.

ii. $ax^2 + bx + c > 0$ Between $x = -4$ and $x = 2$, the graph is above the x-axis and the y-values are positive. The solution set is $\{x \mid -4 < x < 2\}$.

(c) **i.** $ax^2 + bx + c \leq 0$ The graph never dips below the x-axis so the y-values are never negative. There is one x-intercept at $x = 3$, where $y = 0$. The solution set is $\{x \mid x = 3\}$.

ii. $ax^2 + bx + c > 0$ The graph is always above the x-axis except at $x = 3$. Thus every real number except $x = 3$ is a solution. The solution set is $\{x \mid x \neq 3\}$.

(d) **i.** $ax^2 + bx + c \leq 0$ The graph always lies below the x-axis so the y-values are negative and always satisfy the given inequality. The solution set is all real numbers, or $\{x \mid -\infty < x < \infty\}$.

ii. $ax^2 + bx + c > 0$ The graph is never above the x-axis so the solution set is empty.

Now Try Exercises 1, 3, and 5 ◄

EXAMPLE 2 Solving a quadratic inequality

Solve each equation or inequality.
(a) $2x^2 - 3x - 2 = 0$ **(b)** $2x^2 - 3x - 2 < 0$ **(c)** $2x^2 - 3x - 2 > 0$

SOLUTION

Getting Started By solving part (a) we can find the boundary numbers, which help determine the solution sets to parts (b) and (c). ►

(a) The equation $2x^2 - 3x - 2 = 0$ can be solved by factoring.

$$(2x + 1)(x - 2) = 0 \qquad \text{Factor trinomial.}$$

$$x = -\frac{1}{2} \quad \text{or} \quad x = 2 \qquad \text{Zero-product property}$$

The solutions are $-\frac{1}{2}$ and 2.

(b) The graph of $y = 2x^2 - 3x - 2$ is a parabola opening upward. Its x-intercepts are $-\frac{1}{2}$ and 2. See Figure 3.62. This parabola is below the x-axis ($y < 0$) for x-values between $-\frac{1}{2}$ and 2, so the solution set is $\left(-\frac{1}{2}, 2\right)$, or $\left\{x \mid -\frac{1}{2} < x < 2\right\}$.

(c) In Figure 3.63, the graph of $y = 2x^2 - 3x - 2$ is above the x-axis ($y > 0$) for x-values less than $-\frac{1}{2}$ or greater than 2, so the solution set is $\left(-\infty, -\frac{1}{2}\right) \cup \left(2, \infty\right)$, or $\left\{x \mid x < -\frac{1}{2} \text{ or } x > 2\right\}$.

The table of values in Figure 3.64 supports these graphical results. Note that $y_1 = 0$ for $x = -\frac{1}{2}$ and $x = 2$. For $-\frac{1}{2} < x < 2$, $y_1 < 0$, and for $x < -\frac{1}{2}$ or $x > 2$, $y_1 > 0$.

Figure 3.62

Figure 3.63

Figure 3.64

NOTE *It is not necessary to graph the parabola.* Instead you can simply *visualize* the parabola opening upward with x-intercepts $-\frac{1}{2}$ and 2.   *Now Try Exercise 11* ◄

CLASS DISCUSSION

Sketch a graph of $y = ax^2 + bx + c$ if the quadratic inequality $ax^2 + bx + c < 0$ satisfies the following conditions.

(a) $a > 0$, solution set: $\{x \mid -1 < x < 3\}$

(b) $a < 0$, solution set: $\{x \mid x \neq 1\}$

(c) $a < 0$, solution set: $\{x \mid x < -2 \text{ or } x > 2\}$

EXAMPLE 3 Determining safe speeds

In the introduction to this section the quadratic inequality

$$\frac{1}{12}x^2 + \frac{11}{5}x \le 200$$

was explained. Solve this inequality to determine safe speeds on a curve where a driver can see the road ahead for at most 200 feet. What might be a safe speed limit for this curve?

[−100, 100, 50] by [−300, 300, 100]

Figure 3.65

SOLUTION We can solve this inequality by graphing $Y_1 = X^2/12 + 11X/5$ and $Y_2 = 200$, as shown in Figure 3.65. Since we are interested in *positive* speeds, we need to locate only the point of intersection where x is positive. This is where $x \approx 37.5$. For positive x-values to the left of $x \approx 37.5$, $y_1 < 200$. Thus safe speeds are less than 37.5 miles per hour. A reasonable speed limit might be 35 miles per hour.

Now Try Exercise 61 ◄

Symbolic Solutions

Although it is usually easier to solve a quadratic inequality graphically by visualizing a parabola and its x-intercepts, we can also solve a quadratic inequality symbolically without the aid of a graph. The symbolic method, which involves a table or line graph, is often used to solve more complicated inequalities. (See Section 4.7.)

Solving Quadratic Inequalities

STEP 1: If necessary, write the inequality as $ax^2 + bx + c < 0$, where $<$ may be replaced by $>$, $\leq$, or $\geq$.

STEP 2: Solve the equation $ax^2 + bx + c = 0$. The solutions are called *boundary numbers*.

STEP 3: Use the boundary numbers to separate the number line into disjoint open intervals. Note that on each open interval, $y = ax^2 + bx + c$ is either always positive or always negative.

STEP 4: To solve the inequality, choose a convenient test value (an x-value) from each disjoint interval in Step 3. Evaluate $y = ax^2 + bx + c$ at each test point. If the result is positive, then $y > 0$ over that interval. If the result is negative, then $y < 0$ over that interval. You may want to use either a number line or a table of values to organize your work.

NOTE Do not pick a boundary number for a test value, because the result will be $y = 0$.

Algebra Review
To review factoring,
see Chapter R (page R-24).

For example, to solve $2x^2 - 5x - 12 < 0$ symbolically, replace $<$ with $=$ and solve the equation by factoring.

$$2x^2 - 5x - 12 = 0 \qquad \text{Quadratic equation}$$
$$(2x + 3)(x - 4) = 0 \qquad \text{Factor.}$$
$$2x + 3 = 0 \quad \text{or} \quad x - 4 = 0 \qquad \text{Zero-product property}$$
$$x = -\frac{3}{2} \quad \text{or} \quad x = 4 \qquad \text{Solve.}$$

The boundary numbers $-\frac{3}{2}$ and 4 separate the number line into three disjoint intervals:

$$\left(-\infty, -\tfrac{3}{2}\right), \quad \left(-\tfrac{3}{2}, 4\right), \quad \text{and } (4, \infty),$$

as illustrated in Figure 3.66.

Figure 3.66

The expression $2x^2 - 5x - 12$ is either always positive or always negative on a particular interval. To determine where $2x^2 - 5x - 12 < 0$, we can use the **test values** shown in Table 3.8. For example, since the test value -2 lies in the interval $\left(-\infty, -\frac{3}{2}\right)$ and $2x^2 - 5x - 12$ evaluated at $x = -2$ equals 6, which is greater than 0, it follows that the expression $2x^2 - 5x - 12$ is always positive for $\left(-\infty, -\frac{3}{2}\right)$. This interval has $+$ signs on the number line in Figure 3.67.

Figure 3.67

Table 3.8

Interval	Test Value x	$2x^2 - 5x - 12$	Positive or Negative?
$\left(-\infty, -\tfrac{3}{2}\right)$	-2	6	Positive
$\left(-\tfrac{3}{2}, 4\right)$	0	-12	Negative
$(4, \infty)$	6	30	Positive

From Table 3.8 we can see that the expression $2x^2 - 5x - 12$ is negative when $x = 0$, so it is always negative between the boundary numbers of $-\frac{3}{2}$ and 4. Negative signs are shown on the real number line in Figure 3.67 in the interval $\left(-\frac{3}{2}, 4\right)$. Finally, when $x = 6$, $2x^2 - 5x - 12 = 30$, which is greater than 0, and $+$ signs are placed along the x-axis in Figure 3.67 when $x > 4$. Therefore the solution set for $2x^2 - 5x - 12 < 0$ is $\left(-\frac{3}{2}, 4\right)$. Note that it is important to choose one test value less than $-\frac{3}{2}$, one test value between $-\frac{3}{2}$ and 4, and one test value greater than 4. You do *not* need to use both a table and a number line.

EXAMPLE 4   Solving a quadratic inequality

Solve $x^2 \geq 2 - x$ symbolically. Write the solution set in interval notation.

SOLUTION

STEP 1:   Rewrite the inequality as $x^2 + x - 2 \geq 0$.

STEP 2:   Solve the quadratic equation $x^2 + x - 2 = 0$.

$$(x + 2)(x - 1) = 0 \qquad \text{Factor.}$$
$$x = -2 \quad \text{or} \quad x = 1 \qquad \text{Zero-product property}$$

STEP 3:   These two boundary numbers separate the number line into three disjoint intervals:

$$(-\infty, -2), \quad (-2, 1), \quad \text{and} \quad (1, \infty).$$

STEP 4:   We choose the test values $x = -3$, $x = 0$, and $x = 2$. From Table 3.9 or Figure 3.68, the expression $x^2 + x - 2$ is positive when $x = -3$ and $x = 2$.

Figure 3.68

Table 3.9

Interval	Test Value x	$x^2 + x - 2$	Positive or Negative?
$(-\infty, -2)$	-3	4	Positive
$(-2, 1)$	0	-2	Negative
$(1, \infty)$	2	4	Positive

Therefore the expression $x^2 + x - 2$ is positive when $x < -2$ or $x > 1$. Thus the solution set is $(-\infty, -2] \cup [1, \infty)$. The boundary numbers, -2 and 1, are included because the inequality involves $\geq$ rather than $>$.   *Now Try Exercise 55* ◄

3.4 Putting It All Together

The following table summarizes concepts related to solving quadratic inequalities.

Concept	Description
Quadratic inequality	Can be written as $ax^2 + bx + c < 0$, where $<$ may be replaced by $>$, $\leq$, or $\geq$. **Example:** $-x^2 + x < -2$ is a quadratic inequality; it can be written as $-x^2 + x + 2 < 0$.

continued on next page

continued from previous page

Concept	Description
Graphical solution	Write the inequality as $ax^2 + bx + c < 0$, where $<$ may be $>$, $\leq$, or $\geq$. Graph $y = ax^2 + bx + c$, and use the x-intercepts, or boundary numbers, to determine x-values where the graph is below (above) the x-axis. In the figure, the inequality $-x^2 + x + 2 < 0$ is satisfied when either $x < -1$ or $x > 2$.
Symbolic solution	Write the inequality as $ax^2 + bx + c < 0$, where $<$ may be $>$, $\leq$, or $\geq$. Solve the equation $ax^2 + bx + c = 0$. To determine where $y = ax^2 + bx + c$ is positive or negative, use a table of test values or a number line. **Example:** Solve $-x^2 + x + 2 < 0$. Solving $-x^2 + x + 2 = 0$ results in $x = -1$ or $x = 2$. From the table the solution set is $\{x \mid x < -1 \text{ or } x > 2\}$, or $(-\infty, -1) \cup (2, \infty)$. <table><tr><th>Interval</th><th>Test Value x</th><th>$-x^2 + x + 2$</th><th>Positive or Negative?</th></tr><tr><td>$(-\infty, -1)$</td><td>-2</td><td>-4</td><td>Negative</td></tr><tr><td>$(-1, 2)$</td><td>0</td><td>2</td><td>Positive</td></tr><tr><td>$(2, \infty)$</td><td>3</td><td>-4</td><td>Negative</td></tr></table>

3.4 Exercises

Quadratic Inequalities

Exercises 1–6: The graph of $f(x) = ax^2 + bx + c$ is shown in the figure. Solve each inequality.

1. (a) $f(x) < 0$

   (b) $f(x) \geq 0$

2. (a) $f(x) > 0$

   (b) $f(x) < 0$

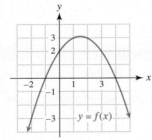

3. (a) $f(x) \leq 0$

   (b) $f(x) > 0$

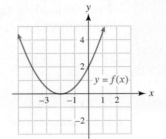

4. (a) $f(x) \geq 0$

   (b) $f(x) \leq 0$

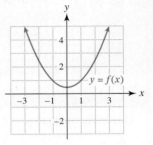

5. (a) $f(x) > 0$

   (b) $f(x) < 0$

6. (a) $f(x) \geq 0$

   (b) $f(x) < 0$

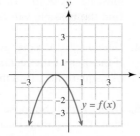

Exercises 7–10: Use the graph of $y = f(x)$ to solve each equation or inequality. Use set-builder or interval notation to write solution sets to the inequalities.

   (a) $f(x) = 0$ (b) $f(x) < 0$ (c) $f(x) > 0$

7.

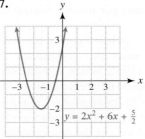

$y = 2x^2 + 6x + \frac{5}{2}$

8.

$y = 3x^2 + 4x - 4$

9.

$y = -5x^2 + 2x + 7$

10.

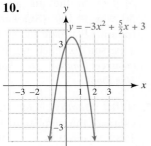

$y = -3x^2 + \frac{5}{2}x + 3$

Exercises 11–22: Solve each equation and inequality. Use set-builder or interval notation to write solution sets to the inequalities.

11. (a) $x^2 - x - 12 = 0$

    (b) $x^2 - x - 12 < 0$

    (c) $x^2 - x - 12 > 0$

12. (a) $x^2 - 8x + 12 = 0$

    (b) $x^2 - 8x + 12 < 0$

    (c) $x^2 - 8x + 12 > 0$

13. (a) $k^2 - 5 = 0$

    (b) $k^2 - 5 \leq 0$

    (c) $k^2 - 5 \geq 0$

14. (a) $n^2 - 17 = 0$

    (b) $n^2 - 17 \leq 0$

    (c) $n^2 - 17 \geq 0$

15. (a) $3x^2 + 8x = 0$

    (b) $3x^2 + 8x \leq 0$

    (c) $3x^2 + 8x \geq 0$

16. (a) $7x^2 - 4x = 0$

    (b) $7x^2 - 4x \leq 0$

    (c) $7x^2 - 4x \geq 0$

17. (a) $-4x^2 + 12x - 9 = 0$

    (b) $-4x^2 + 12x - 9 < 0$

    (c) $-4x^2 + 12x - 9 > 0$

18. (a) $x^2 + 2x + 1 = 0$

    (b) $x^2 + 2x + 1 < 0$

    (c) $x^2 + 2x + 1 > 0$

19. (a) $12z^2 - 23z + 10 = 0$

    (b) $12z^2 - 23z + 10 \leq 0$

    (c) $12z^2 - 23z + 10 \geq 0$

20. (a) $18z^2 + 9z - 20 = 0$

    (b) $18z^2 + 9z - 20 \leq 0$

    (c) $18z^2 + 9z - 20 \geq 0$

21. (a) $x^2 + 2x - 1 = 0$

    (b) $x^2 + 2x - 1 < 0$

    (c) $x^2 + 2x - 1 > 0$

22. (a) $x^2 + 4x - 3 = 0$

    (b) $x^2 + 4x - 3 < 0$

    (c) $x^2 + 4x - 3 > 0$

Exercises 23–26: The table contains test values for a quadratic function $f(x) = ax^2 + bx + c$. Solve each inequality.

   (a) $f(x) > 0$ (b) $f(x) \leq 0$

23.

x	-2	-1	0	1	2
$f(x)$	3	0	-1	0	3

24.

x	-6	-4	1	5	8
$f(x)$	-22	0	20	0	-36

25.

x	-6	-4	-2	0	2
$f(x)$	0	4	0	-12	-32

26.

x	-4	-2	0	2	4
$f(x)$	16	0	-8	-8	0

Exercises 27–52: Solve the inequality.

27. $2x^2 + 5x + 2 \le 0$

28. $x^2 - 3x - 4 < 0$

29. $x^2 + x > 6$

30. $-3x \ge 9 - 12x^2$

31. $x^2 \le 4$

32. $2x^2 > 16$

33. $x(x - 4) \ge -4$

34. $x^2 - 3x - 10 < 0$

35. $-x^2 + x + 6 \le 0$

36. $-x^2 - 2x + 8 > 0$

37. $6x^2 - x < 1$

38. $5x^2 \le 10 - 5x$

39. $(x + 4)(x - 10) \le 0$

40. $(x - 3.1)(x + 2.7) > 0$

41. $2x^2 + 4x + 3 < 0$

42. $2x^2 + x + 4 < 0$

43. $9x^2 + 4 > 12x$

44. $x^2 + 2x \ge 35$

45. $x^2 \ge x$

46. $x^2 \ge -3$

47. $x(x - 1) \ge 6$

48. $x^2 - 9 < 0$

49. $x^2 - 5 \le 0$

50. $0.5x^2 - 3.2x > -0.9$

51. $7x^2 + 515.2 \ge 179.8x$

52. $-10 < 3x - x^2$

Exercises 53–60: (Refer to Example 4.) Use a table to solve.

53. $x^2 - 9x + 14 \le 0$

54. $x^2 + 10x + 21 > 0$

55. $x^2 \ge 3x + 10$

56. $x^2 < 3x + 4$

57. $\frac{1}{8}x^2 + x + 2 \ge 0$

58. $x^2 - \frac{1}{2}x - 5 < 0$

59. $x^2 > 3 - 4x$

60. $2x^2 \le 1 - 4x$

Applications

61. *Stopping Distance* The stopping distance D in feet for a car traveling at x miles per hour on *wet* level pavement can be estimated by $D(x) = \frac{1}{9}x^2 + \frac{11}{3}x$. If a driver can see only 300 feet ahead on a curve, find a safe speed limit.

62. *Safe Driving Speeds* The stopping distance d in feet for a car traveling at x miles per hour is given by $d(x) = \frac{1}{12}x^2 + \frac{11}{9}x$. Determine the driving speeds that correspond to stopping distances between 300 and 500 feet, inclusive. Round speeds to the nearest mile per hour.

63. *Geometry* The volume of a cylinder is given by $V = \pi r^2 h$, where r is the radius and h is the height. If the height of a cylindrical can is 6 inches and the volume must be between 24π and 54π cubic inches, inclusive, find the possible values for the radius of the can.

64. *Geometry* A rectangle is 4 feet longer than it is wide. If the area of the rectangle must be less than or equal to 672 square feet, find the possible values for the width x.

65. *Heart Rate* Suppose that a person's heart rate, x minutes after vigorous exercise has stopped, can be modeled by $f(x) = \frac{4}{5}(x - 10)^2 + 80$. The output is in beats per minute, where the domain of f is $0 \le x \le 10$.
(a) Evaluate $f(0)$ and $f(2)$. Interpret the result.

(b) Estimate the times when the person's heart rate was between 100 and 120 beats per minute, inclusive.

66. *Carbon Monoxide Exposure* When a person breathes carbon monoxide (CO), it enters the bloodstream to form carboxyhemoglobin (COHb), which reduces the transport of oxygen to tissues. The formula given by $T(x) = 0.0079x^2 - 1.53x + 76$ approximates the number of hours T that it takes for a person's bloodstream to reach the 5% COHb level, where x is the concentration of CO in the air in parts per million (ppm) and $50 \le x \le 100$. (Smokers routinely have a 5% concentration.) Estimate the CO concentration x necessary for a person to reach the 5% COHb level in 4–5 hours. (*Source: Indoor Air Quality Environmental Information Handbook.*)

67. *AIDS Deaths* Let $f(x) = 2375x^2 + 5134x + 5020$ estimate the number of U.S. AIDS deaths x years after 1984, where $0 \le x \le 1994$. Estimate when the number of AIDS deaths was from 90,000 to 200,000.

68. *Air Density* As the altitude increases, air becomes thinner, or less dense. An approximation of the density of air at an altitude of x meters above sea level is given by

$$d(x) = (3.32 \times 10^{-9})x^2 - (1.14 \times 10^{-4})x + 1.22.$$

The output is the density of air in kilograms per cubic meter. The domain of d is $0 \le x \le 10,000$. (*Source: A. Miller and J. Thompson, Elements of Meteorology.*)
(a) Denver is sometimes referred to as the mile-high city. Compare the density of air at sea level and in Denver. (*Hint:* 1 ft $\approx$ 0.305 m.)

(b) Determine the altitudes where the density is greater than 1 kilogram per cubic meter.

69. *Modeling Water Flow* A cylindrical container measuring 16 centimeters high with a hole at the bottom was filled with water. As water leaked out, the height of the water inside the container was recorded. The results appear in the table.

16 cm

Time (sec)	0	15	30	45	60
Height (cm)	16	13.8	11.6	9.8	8.1

Time (sec)	90	105	120	150	180
Height (cm)	5.3	4.1	3.1	1.4	0.5

(a) Explain why the data cannot be modeled by a linear function.

(b) Use the table to estimate when the height of the water was from 5 to 10 centimeters.

(c) The height of the water can be modeled by

$$f(x) = 0.00036706x^2 - 0.1517x + 15.94,$$

where x represents the time. Graph f and the data in the same viewing rectangle.

(d) Solve part (b) using $f(x)$.

70. *Heart Rate* The table shows a person's heart rate after exercise has stopped.

Time (min)	0	2	4
Heart rate (bpm)	154	106	90

(a) Find values for the constants a, h, and k so that the formula $f(x) = a(x - h)^2 + k$ models the data, where x represents time and $0 \le x \le 4$.

(b) Evaluate $f(1)$ and interpret the result.

(c) Estimate the times when the heart rate was from 115 to 125 beats per minute.

Writing about Mathematics

71. Explain how a table of values can be used to help solve a quadratic inequality, provided that the boundary numbers are listed in the table.

72. Explain how to determine the solution set for the inequality $ax^2 + bx + c < 0$, where $a > 0$. How would the solution set change if $a < 0$?

CHECKING BASIC CONCEPTS FOR SECTIONS 3.3 AND 3.4

1. Simplify by using the imaginary unit i.
 (a) $\sqrt{-25}$ **(b)** $\sqrt{-3} \cdot \sqrt{-18}$
 (c) $\dfrac{7 \pm \sqrt{-98}}{14}$

2. Write each expression in standard form.
 (a) $-3i - (5 - 2i)$ **(b)** $(6 - 7i) + (-1 + i)$
 (c) $i(1 - i)(1 + i)$ **(d)** $\dfrac{1 + 2i}{4 - i}$

3. Use the graph of $y = f(x)$ to solve $f(x) \le 0$ and $f(x) > 0$. Write your answer in set-builder or interval notation.
 (a)   **(b)**

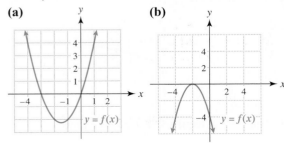

4. Solve each equation and inequality. Write the solution set for each inequality in set-builder or interval notation.
 (a) $2x^2 + 7x - 4 = 0$

 (b) $2x^2 + 7x - 4 < 0$

 (c) $2x^2 + 7x - 4 > 0$

5. Solve each inequality. Use set-builder or interval notation.
 (a) $x^2 - 5 \ge 0$

 (b) $4x^2 + 9 > 9x$

 (c) $2x(x - 1) \le 2$

6. *Safe Driving Speeds* The stopping distance d in feet for a car traveling x miles per hour on wet level pavement can be estimated by $d(x) = \frac{1}{9}x^2 + \frac{11}{3}x$. Determine the driving speeds that correspond to stopping distances between 80 and 180 feet, inclusive.

3.5 Transformations of Graphs

- **Graph functions using vertical and horizontal shifts**
- **Graph functions using stretching and shrinking**
- **Graph functions using reflections**
- **Combine transformations**
- **Model data with transformations (optional)**

Introduction

Graphs are often used to model different types of phenomena. For example, when a cold front moves across the United States, we might use a circular arc on a weather map to describe its shape. (See Exercise 7 in the Extended and Discovery Exercises on page 247.) If the front does not change its shape significantly, we could model the movement of the front on a television weather map by translating the circular arc. Before we can portray a cold front on a weather map, we need to discuss how to transform graphs of functions. (**Sources:** S. Hoggar, *Mathematics for Computer Graphics;* A. Watt, *3D Computer Graphics.*)

Vertical and Horizontal Shifts

Graphs of $f(x) = x^2$ and $g(x) = \sqrt{x}$ will be used to demonstrate shifts, or translations, in the xy-plane. Symbolic, numerical, and graphical representations of f and g are shown in Figures 3.69 and 3.70, respectively. Points listed in the table are plotted on the graph.

x	-2	-1	0	1	2
y	4	1	0	1	4

x	0	1	4
y	0	1	2

Figure 3.69

Figure 3.70

Vertical Shifts If 2 is added to the formula for each function, the original graphs are shifted upward 2 units. The graphs of $y = x^2 + 2$ and $y = \sqrt{x} + 2$ are shown in Figures 3.71 and 3.72. A table of points is included, together with the graph of the original function. Notice that the y-values in both the graphical and numerical representations increase by 2 units.

x	-2	-1	0	1	2
y	6	3	2	3	6

x	0	1	4
y	2	3	4

Figure 3.71 Upward 2 Units

Figure 3.72 Upward 2 Units

If 2 is subtracted from each of the original formulas, the graphs are shifted downward 2 units. Verify this by graphing $y = x^2 - 2$ and $y = \sqrt{x} - 2$. Translations of this type are called **vertical shifts**, or **vertical translations**. They do not alter the shape of the graph, only its position. The original and shifted graphs are congruent.

Horizontal Shifts   If the variable x is replaced by $(x - 2)$ in the formulas for f and g, a different type of shift results. Figures 3.73 and 3.74 show the graphs and tables of $y = (x - 2)^2$ and $y = \sqrt{(x - 2)}$, together with the graphs of the original functions.

x	0	1	2	3	4
y	4	1	0	1	4

x	2	3	6
y	0	1	2

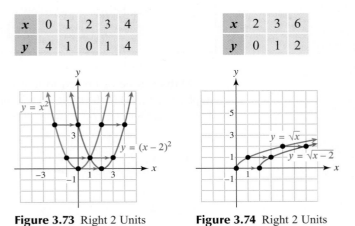

Figure 3.73 Right 2 Units   **Figure 3.74** Right 2 Units

Each new graph shows a shift of the original graph to the *right* by 2 units. Notice that a table for a graph shifted *right* 2 units can be obtained from the original table (such as the one on the preceding page) by *adding* 2 to each x-value.

If the variable x is replaced by $(x + 3)$ in each equation, the original graphs are translated to the *left* 3 units. The graphs of $y = (x + 3)^2$ and $y = \sqrt{(x + 3)}$ and their tables are shown in Figures 3.75 and 3.76. This type of translation is a **horizontal shift**, or **horizontal translation**. The table for a graph shifted *left* 3 units is obtained from the original table by *subtracting* 3 from each x-value.

x	-5	-4	-3	-2	-1
y	4	1	0	1	4

x	-3	-2	1
y	0	1	2

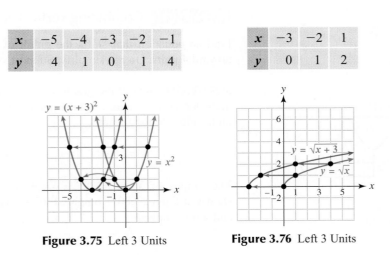

Figure 3.75 Left 3 Units   **Figure 3.76** Left 3 Units

These ideas are summarized in the following box.

Vertical and Horizontal Shifts

Let f be a function, and let c be a positive number.

To Graph	Shift the Graph of $y = f(x)$ by c Units
$y = f(x) + c$	upward
$y = f(x) - c$	downward
$y = f(x - c)$	right
$y = f(x + c)$	left

Shifts can be combined to translate a graph of $y = f(x)$. For example, to shift the graph of $y = f(x)$ to the right 2 units and downward 4 units, we graph $y = f(x - 2) - 4$. If $y = |x|$, then $y = f(x - 2) - 4 = |x - 2| - 4$, as shown in Figures 3.77–3.79.

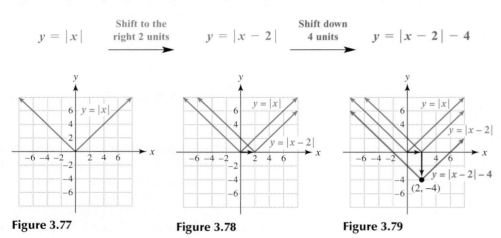

$$y = |x| \quad \xrightarrow[\text{right 2 units}]{\text{Shift to the}} \quad y = |x - 2| \quad \xrightarrow[\text{4 units}]{\text{Shift down}} \quad y = |x - 2| - 4$$

Figure 3.77 **Figure 3.78** **Figure 3.79**

EXAMPLE 1 Combining vertical and horizontal shifts

Find an equation that shifts the graph of $f(x) = \frac{1}{2}x^2 - 4x + 6$ to the left 8 units and upward 4 units. Do not simplify. Support your result by graphing.

SOLUTION To shift the graph of f to the left 8 units, replace x with $(x + 8)$ in the formula for $f(x)$; be sure to place parentheses around $x + 8$. To shift the graph upward 4 units, add 4 to the formula.

$$y = f(x + 8) + 4 = \frac{1}{2}(x + 8)^2 - 4(x + 8) + (6 + 4)$$

The graphs of $Y_1 = X^2/2 - 4X + 6$ and $Y_2 = (X + 8)^2/2 - 4(X + 8) + 10$ are shown in Figure 3.80. Compared to the vertex of y_1, the vertex of y_2 is 8 units to the left and 4 units upward. *Now Try Exercise 11* ◄

$[-9, 9, 2]$ by $[-6, 6, 2]$

Figure 3.80

EXAMPLE 2   Writing formulas

Write a formula for a function g whose graph is similar to that of $f(x) = 4x^2 - 2x + 1$ but is shifted right 1980 units and upward 50 units. Do not simplify the formula.

SOLUTION Replace x with $(x - 1980)$ in the formula for $f(x)$ and then add **50**.

$$g(x) = f(x - 1980) + 50$$
$$= 4(x - 1980)^2 - 2(x - 1980) + 1 + 50$$
$$= 4(x - 1980)^2 - 2(x - 1980) + 51 \qquad \textit{Now Try Exercise 19} \blacktriangleleft$$

In the next example we translate a circle that is centered at the origin.

EXAMPLE 3   Translating a circle

The equation of a circle having radius 3 and center $(0, 0)$ is $x^2 + y^2 = 9$. Write an equation that shifts this circle to the right 4 units and upward 2 units. What are the center and radius of this circle? (Note that a circle is not a function.)

SOLUTION
Getting Started The standard equation for a circle with center (h, k) and radius r is

$$(x - h)^2 + (y - k)^2 = r^2.$$

If we determine the new center and radius, we can apply the standard equation. ▶

If a circle with center $(0, 0)$ and radius 3 is translated to the right 4 units and upward 2 units, then the center of the new circle is $(0 + 4, 0 + 2)$, or $(4, 2)$; the radius remains the same. The standard equation for a circle with center $(4, 2)$ and radius 3 is

$$(x - 4)^2 + (y - 2)^2 = 9.$$

Figure 3.81 illustrates this translation. $\qquad$ *Now Try Exercise 23* ◀

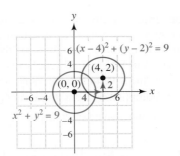

Figure 3.81 Translating a Circle

Example 3 illustrates that to translate a circle horizontally c units, replace x with $(x - c)$, and to translate a circle vertically c units, replace y with $(y - c)$.

Stretching and Shrinking

Vertical Stretching and Shrinking   The graph of a function can be transformed by vertical stretching or shrinking as described next.

> ### Vertical Stretching and Shrinking
>
> If the point (x, y) lies on the graph of $y = f(x)$, then the point (x, cy) lies on the graph of $y = cf(x)$. If $c > 1$, the graph of $y = cf(x)$ is a vertical stretching of the graph of $y = f(x)$, whereas if $0 < c < 1$, the graph of $y = cf(x)$ is a vertical shrinking of the graph of $y = f(x)$.

For example, if the point $(4, 2)$ is on the graph of $y = f(x)$, then the point $(4, 4)$ is on the graph of $y = 2f(x)$ and the point $(4, 1)$ is on the graph of $y = \frac{1}{2}f(x)$. The graph

Vertical Stretching and Shrinking

Vertical stretching "pulls" the graph away from the *x*-axis.

Vertical shrinking "pushes" the graph towards the *x*-axis. In both cases the *x*-intercepts do not change.

of $f(x) = \sqrt{x}$ in Figure 3.82 can be stretched or shrunk *vertically*. In Figure 3.83, the graph of $y = 2f(x)$, or $y = 2\sqrt{x}$, represents a vertical stretching of the graph of $y = \sqrt{x}$. In Figure 3.84, the graph of $y = \frac{1}{2}f(x)$, or $y = \frac{1}{2}\sqrt{x}$, represents a vertical shrinking of the graph of $y = \sqrt{x}$. Compared to the *y*-values in the table for $y = \sqrt{x}$, the *y*-values in the tables for $y = 2f(x)$ and $y = \frac{1}{2}f(x)$ have been multiplied by 2 and $\frac{1}{2}$, respectively. The *x*-values have not changed.

x	0	1	4
$f(x)$	0	1	2

x	0	1	4
$2f(x)$	0	2	4

x	0	1	4
$\frac{1}{2}f(x)$	0	$\frac{1}{2}$	1

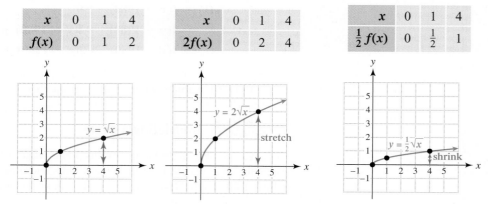

Figure 3.82 Given Function   **Figure 3.83** Vertical Stretching   **Figure 3.84** Vertical Shrinking

Horizontal Stretching and Shrinking   The line graph in Figure 3.85 can be stretched or shrunk *horizontally*. On one hand, if the line graph represents the graph of a function f, then the graph of $y = f\left(\frac{1}{2}x\right)$ in Figure 3.86 is a horizontal stretching of the graph of $y = f(x)$. On the other hand, the graph of $y = f(2x)$ in Figure 3.87 represents a horizontal shrinking of the graph of $y = f(x)$. Compared to the *x*-values in the table for $y = f(x)$, the *x*-values in the table for $y = f\left(\frac{1}{2}x\right)$ have been multiplied by 2 and the *x*-values in the table for $y = f(2x)$ have been multiplied by $\frac{1}{2}$. The *y*-values have not changed.

x	-2	-1	1	2
$f(x)$	3	-3	3	-3

x	-4	-2	2	4
$f\left(\frac{1}{2}x\right)$	3	-3	3	-3

x	-1	$-\frac{1}{2}$	$\frac{1}{2}$	1
$f(2x)$	3	-3	3	-3

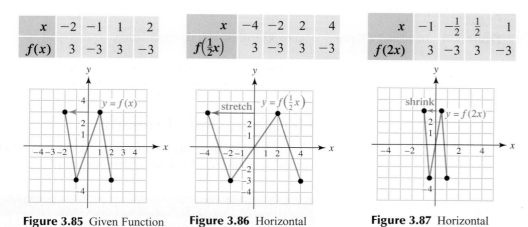

Figure 3.85 Given Function   **Figure 3.86** Horizontal Stretching   **Figure 3.87** Horizontal Shrinking

Imagine the graph of $y = f(x)$ as a flexible wire. Then a horizontal stretching would happen if the wire were pulled on each end, and a horizontal shrinking would happen if the wire were compressed. Note that horizontal stretching or shrinking does not change the height (maximum or minimum *y*-values) of the graph, nor does it change the *y*-intercept. Horizontal stretching and shrinking can be generalized for any function f.

MAKING CONNECTIONS

Horizontal Stretching and Shrinking

Horizontal stretching "pulls" the graph away from the *y*-axis.

Horizontal shrinking "pushes" the graph towards the *y*-axis. In both cases the *y*-intercepts do not change.

Horizontal Stretching and Shrinking

If the point (x, y) lies on the graph of $y = f(x)$, then the point $(\frac{x}{c}, y)$ lies on the graph of $y = f(cx)$. If $c > 1$, the graph of $y = f(cx)$ is a horizontal shrinking of the graph of $y = f(x)$, whereas if $0 < c < 1$, the graph of $y = f(cx)$ is a horizontal stretching of the graph of $y = f(x)$.

For example, if the point $(-2, 3)$ is on the graph of $y = f(x)$, then the point $(-1, 3)$ is on the graph of $y = f(2x)$ and the point $(-4, 3)$ is on the graph of $y = f\left(\frac{1}{2}x\right)$.

EXAMPLE 4   Stretching and shrinking of a graph

Use the graph and table of $y = f(x)$ in Figure 3.88 to sketch a graph of each equation.
(a) $y = 3f(x)$ **(b)** $y = f\left(\frac{1}{2}x\right)$

SOLUTION

(a) The graph of $y = 3f(x)$, shown in Figure 3.89, is a vertical stretching of the graph of $y = f(x)$, shown in Figure 3.88, and can be obtained by multiplying each *y*-coordinate on the graph of $y = f(x)$ by 3.

x	-1	0	2
$f(x)$	-2	1	-1

x	-1	0	2
$3f(x)$	-6	3	-3

x	-2	0	4
$f\left(\frac{1}{2}x\right)$	-2	1	-1

Figure 3.88

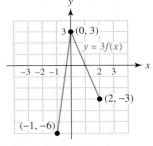

Figure 3.89 Vertical Stretching (*x*-intercepts unchanged)

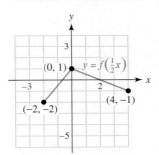

Figure 3.90 Horizontal Stretching (*y*-intercept unchanged)

(b) The graph of $y = f\left(\frac{1}{2}x\right)$, shown in Figure 3.90, is a horizontal stretching of the graph of $y = f(x)$, shown in Figure 3.88, and can be obtained by *dividing* each *x*-coordinate on the graph of $y = f(x)$ by $\frac{1}{2}$, which is equivalent to multiplying each *x*-coordinate by 2.

Now Try Exercises 33(a) and (b) ◀

Reflection of Graphs

Another type of translation is called a **reflection.** The reflection of the blue graph of $y = f(x)$ across the *x*-axis is shown in Figure 3.91 as a red curve.

If (x, y) is a point on the graph of f, then $(x, -y)$ lies on the graph of its reflection across the *x*-axis, as shown in Figure 3.91. Thus a reflection of $y = f(x)$ is given by the equation $-y = f(x)$, or equivalently, $y = -f(x)$. For example, a reflection of the graph of $f(x) = x^2 - x + 2$ is obtained by graphing $y = -f(x) = -(x^2 - x + 2)$.

Figure 3.91

If a point (x, y) lies on the graph of a function f, then the point $(-x, y)$ lies on the graph of its reflection across the y-axis, as shown in Figure 3.92. Thus a reflection of $y = f(x)$ across the y-axis is given by $y = f(-x)$. For example, a reflection of the graph of $f(x) = (x - 2)^2$ across the y-axis is shown in Figure 3.93 as a red curve. The reflection is given by the formula $f(-x) = (-x - 2)^2$.

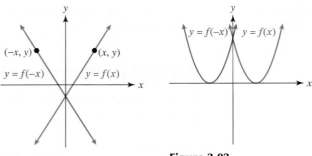

Figure 3.92   **Figure 3.93**

Calculator Help

To access the variable Y_1 as shown in Figure 3.94, see Appendix A (page AP-12).

These results are summarized in the following box.

Figure 3.94

[−10, 10, 1] by [−10, 10, 1]

Figure 3.95

Reflections of Graphs Across the x- and y-Axes

1. The graph of $y = -f(x)$ is a reflection of the graph of $y = f(x)$ across the x-axis.
2. The graph of $y = f(-x)$ is a reflection of the graph of $y = f(x)$ across the y-axis.

On a graphing calculator capable of using function notation, entering equations for reflections of a function f is easy. For example, if $f(x) = (x - 4)^2$, let $Y_1 = (X - 4)\wedge 2$, $Y_2 = -Y_1$, and $Y_3 = Y_1(-X)$. The graph of y_2 is the reflection of f across the x-axis, and y_3 is the reflection of y_1 across the y-axis. See Figures 3.94 and 3.95. However, it is *not* necessary to have this feature to graph reflections.

NOTE Compared to the y-values in a table of values for $y = f(x)$, the y-values in a table of values for $y = -f(x)$ are negated; the x-values do *not* change. Compared to the x-values in a table of values for $y = f(x)$, the x-values in a table of values for $y = f(-x)$ are negated; the y-values do *not* change.

EXAMPLE 5 Reflecting graphs of functions

For each representation of f, graph the reflection across the x-axis and across the y-axis.

(a) $f(x) = x^2 + 2x - 3$
(b) The graph of f is a line graph determined by Table 3.10.

Table 3.10

x	−2	−1	0	3
$f(x)$	1	−3	−1	2

SOLUTION

(a) The graph of $f(x) = x^2 + 2x - 3$ is a parabola with vertex $(-1, -4)$ and x-intercepts -3 and 1, as shown in Figure 3.96. To obtain its reflection across the x-axis, graph $y = -f(x)$, or $y = -(x^2 + 2x - 3)$, as shown in Figure 3.97. The vertex is now $(-1, 4)$; the x-intercepts have not changed. To obtain the reflection of f across the y-axis, let $y = f(-x)$, or $y = (-x)^2 + 2(-x) - 3$, and graph, as shown in Figure 3.98. The vertex is now $(1, -4)$, and the x-intercepts have changed to -1 and 3.

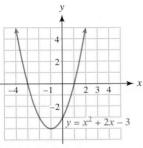

Figure 3.96 $y = f(x)$

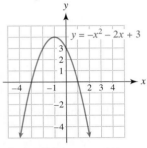

Figure 3.97 $y = -f(x)$

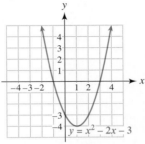

Figure 3.98 $y = f(-x)$

(b) The graph of $y = f(x)$ is a line graph, shown in Figure 3.99. To graph the reflection of f across the x-axis, start by making a table of values for $y = -f(x)$ by negating each y-value in the table for $f(x)$. Then plot these points and make a line graph, as in Figure 3.100.

x	-2	-1	0	3
$f(x)$	1	-3	-1	2

x	-2	-1	0	3
$-f(x)$	-1	3	1	-2

x	2	1	0	-3
$f(-x)$	1	-3	-1	2

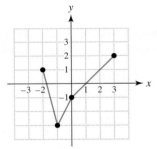

Figure 3.99 $y = f(x)$

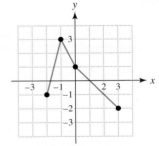

Figure 3.100 $y = -f(x)$

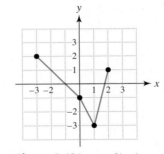

Figure 3.101 $y = f(-x)$

To graph the reflection of f across the y-axis, start by making a table of values for $y = f(-x)$ by negating each x-value in the table for $f(x)$. Then plot these points and make a line graph, as in Figure 3.101.

Now Try Exercises 69 and 73 ◀

Combining Transformations

Transformation of graphs can be combined to create new graphs. For example, the graph of $y = -2(x - 1)^2 + 3$ can be obtained by performing four transformations on the graph of $y = x^2$.

Transforming the graph of $y = x^2$ to the graph of $y = -2(x - 1)^2 + 3$:

1. Shift the graph of $y = x^2$ to the right 1 unit: $y = (x - 1)^2$.
2. Vertically stretch the graph of $y = (x - 1)^2$ by a factor of 2: $y = 2(x - 1)^2$.
3. Reflect the graph of $y = 2(x - 1)^2$ across the x-axis: $y = -2(x - 1)^2$.
4. Shift the graph of $y = -2(x - 1)^2$ upward 3 units: $y = -2(x - 1)^2 + 3$.

These steps are summarized as follows:

The resulting sequence of graphs is shown in Figures 3.102–3.105.

Figure 3.102 Shift Right

Figure 3.103 Vertical Stretch

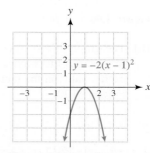

Figure 3.104 Reflect Across x-axis

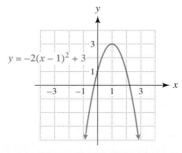

Figure 3.105 Shift Upward

NOTE *The order in which transformations are made can be important.* For example, changing the order of a stretch and shift can result in a different equation and graph.

1. Stretch by a factor of 2.

$y = 2|x| + 3$

2. Shift 3 units upward.

1. Shift 3 units upward.

$y = 2(|x| + 3)$, or $y = 2|x| + 6$

2. Stretch by a factor of 2.

Also be careful when performing reflections and shifts. On one hand, if we reflect the graph of $y = \sqrt{x}$ across the y-axis to obtain $y = \sqrt{-x}$ and then shift it right 2 units, we obtain $y = \sqrt{-(x - 2)}$. On the other hand, if we shift the graph of $y = \sqrt{x}$ right 2 units to obtain $y = \sqrt{x - 2}$ and then reflect it across the y-axis, we obtain $y = \sqrt{-x - 2}$. The final equations are different and so are their graphs. (Try sketching each graph.)

EXAMPLE 6 Combining transformations of graphs

Describe how the graph of each equation can be obtained by transforming the graph of $y = \sqrt{x}$. Then graph the equation.
(a) $y = -\frac{1}{2}\sqrt{x}$ **(b)** $y = \sqrt{-(x + 2)} - 1$

SOLUTION

(a) Vertically shrink the graph of $y = \sqrt{x}$ by a factor of $\frac{1}{2}$ and then reflect it across the x-axis. The graphs of $y = \sqrt{x}$ and $y = -\frac{1}{2}\sqrt{x}$ are shown in Figure 3.106.

(b) The following transformations can be used to obtain the graph of the equation $y = \sqrt{-(x + 2)} - 1$ from $y = \sqrt{x}$:

1. Reflect the graph of $y = \sqrt{x}$ across the y-axis: $y = \sqrt{-x}$.
2. Shift the graph of $y = \sqrt{-x}$ left 2 units: $y = \sqrt{-(x + 2)}$.
3. Shift the graph of $y = \sqrt{-(x + 2)}$ down 1 unit: $y = \sqrt{-(x + 2)} - 1$.

The graph of $y = \sqrt{x}$ and the resulting graph are shown in Figure 3.107.

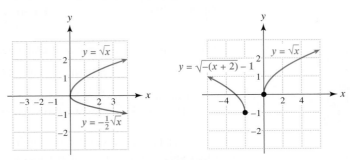

Figure 3.106　　　　　**Figure 3.107**

Now Try Exercises 41 and 55 ◀

Modeling with Transformations (Optional)

Transformations of the graph of $y = x^2$ can be used to model some types of nonlinear data. By shifting, stretching, and shrinking this graph, we can transform it into a *portion of a parabola* that has the desired shape and location. In the next example we demonstrate this technique by modeling numbers of Wal-Mart employees.

> **EXAMPLE 7** Modeling data with a quadratic function

Table 3.11 lists numbers of Wal-Mart employees in millions for selected years.
(a) Make a scatterplot of the data.
(b) Use transformations of graphs to determine $f(x) = a(x - h)^2 + k$ so that $f(x)$ models the data. Graph $y = f(x)$ together with the data.
(c) Use $f(x)$ to estimate the number of Wal-Mart employees in 2010.

Table 3.11
Wal-Mart Employees (in millions)

Year	Employees
1987	0.20
1992	0.37
1997	0.68
2002	1.4
2007	2.2

Source: Wal-Mart.

SOLUTION

(a) A scatterplot of the data is shown in Figure 3.108 on the next page. This plot suggests that the data could be modeled by the *right half* of a parabola that opens upward.

(b) Because the parabola opens upward, it follows that $a > 0$ and the vertex is the lowest point on the parabola. The minimum number of employees is 0.20 million in 1987. One possibility for the vertex (h, k) is $(1987, 0.20)$. Translate the graph of $y = x^2$ right 1987 units and upward 0.20 unit. Thus $f(x) = a(x - 1987)^2 + 0.20$.

To determine a, graph the data and $y = f(x)$ for different values of a. Figure 3.109 on the next page shows $y = f(x)$ with values of $a = 0.001$ and $a = 0.01$. From these graphs it can be seen that the desired value of a is between 0.001 and 0.01. With a little experimentation, a reasonable value for a near 0.005 can be found.

A scatterplot of the data and graph of $f(x) = 0.005(x - 1987)^2 + 0.2$ are shown in Figure 3.110. (Answers may vary.)

[1985, 2010, 5] by [0, 3, 0.5]   [1985, 2010, 5] by [0, 3, 0.5]   [1985, 2010, 5] by [0, 3, 0.5]

Figure 3.108   **Figure 3.109**   **Figure 3.110**

(c) To estimate the number of employees in 2010, evaluate $f(2010)$.

$$f(2010) = 0.005(2010 - 1987)^2 + 0.2 = 2.845$$

This model provides an estimate of about 2.8 million Wal-Mart employees in 2010. (The calculation involves extrapolation.)   ***Now Try Exercise 91*** ◄

Translations of graphs and figures play an important role in computer graphics. In older video games, the background often is translated to give the illusion that the player in the game is moving. A simple scene of a mountain and an airplane is shown in Figure 3.111. To make it appear to the player as though the airplane were flying to the right, the image of the mountain could be translated horizontally to the left, as shown in Figure 3.112. (**Source:** C. Pokorny and C. Gerald, *Computer Graphics.*)

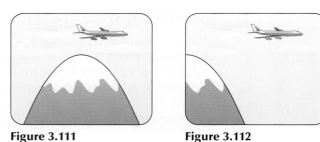

Figure 3.111   **Figure 3.112**

EXAMPLE 8   Using translations to model movement

Suppose that the mountain in Figure 3.111 can be described by the quadratic function represented by $f(x) = -0.4x^2 + 4$ and that the airplane is located at the point $(1, 5)$.
(a) Graph f in $[-4, 4, 1]$ by $[0, 6, 1]$, where the units are in kilometers. Plot a point (a scatterplot with one point) to mark the position of the airplane.
(b) Assume that the airplane is moving horizontally to the right at 0.4 kilometer per second. To give a video player the illusion that the airplane is moving, graph the image of the mountain and the position of the airplane after 5 seconds and then after 10 seconds.

SOLUTION
(a) The graph of $y = f(x) = -0.4x^2 + 4$ and the position of the airplane at $(1, 5)$ are shown in Figure 3.113. The "mountain" has been shaded to emphasize its position.

(b) Five seconds later, the airplane has moved $5(0.4) = 2$ kilometers right. In 10 seconds it has moved $10(0.4) = 4$ kilometers right. To graph these new positions, translate the graph of the mountain 2 and 4 kilometers (units) to the left. Replace x with $(x + 2)$ and graph

$$y = f(x + 2) = -0.4(x + 2)^2 + 4$$

together with the point $(1, 5)$. Next graph $y = f(x + 4) = -0.4(x + 4)^2 + 4$. The results are shown in Figures 3.114 and 3.115. The position of the airplane at $(1, 5)$ has not changed. However, it appears to have flown to the right.

Calculator Help

To shade below a parabola, see Appendix A (page AP-12).

$[-4, 4, 1]$ by $[0, 6, 1]$ $[-4, 4, 1]$ by $[0, 6, 1]$ $[-4, 4, 1]$ by $[0, 6, 1]$

       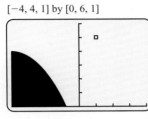

Figure 3.113   **Figure 3.114**   **Figure 3.115**

Now Try Exercise 99 ◄

CLASS DISCUSSION

Discuss how one might create the illusion of the airplane moving to the left and *gaining altitude* as it passes over the mountain.

3.5 Putting It All Together

In this section we discussed several transformations of graphs. The following table summarizes how these transformations affect the graph of $y = f(x)$.

Equation	Effect on Graph of $y = f(x)$
Let $c > 0$. $y = f(x) + c$ $y = f(x) - c$ $y = f(x + c)$ $y = f(x - c)$	Graph is shifted upward c units. Graph is shifted downward c units. Graph is shifted to the left c units. Graph is shifted to the right c units. **Examples:**

continued on next page

continued from previous page

Equation	Effect on Graph of $y = f(x)$
Let $c > 0$. $y = cf(x)$	If (x, y) lies on the graph of $y = f(x)$, then (x, cy) lies on the graph of $y = cf(x)$. The graph is vertically stretched if $c > 1$ and vertically shrunk if $0 < c < 1$. **Examples:**
Let $c > 0$. $y = f(cx)$	If (x, y) lies on the graph of $y = f(x)$, then $\left(\frac{x}{c}, y\right)$ lies on the graph of $y = f(cx)$. The graph is horizontally shrunk if $c > 1$ and horizontally stretched if $0 < c < 1$. **Examples:**
$y = -f(x)$ $y = f(-x)$	Graph is reflected across the x-axis. Graph is reflected across the y-axis. **Examples:**

3.5 Exercises

Vertical and Horizontal Translations

Exercises 1–8: Write the equation of the graph. (Note: The given graph is a translation of the graph of one of the following equations: $y = x^2$, $y = \sqrt{x}$, or $y = |x|$.)

1.

2.

3.

4.

5.

6.

7.

8.

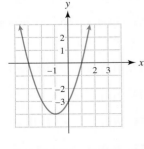

Exercises 9–14: Find an equation that shifts the graph of f by the desired amounts. Do not simplify. Graph f and the shifted graph in the same xy-plane.

9. $f(x) = x^2$; right 2 units, downward 3 units

10. $f(x) = 3x - 4$; left 3 units, upward 1 unit

11. $f(x) = x^2 - 4x + 1$; left 6 units, upward 4 units

12. $f(x) = x^2 - x - 2$; right 2 units, upward 3 units

13. $f(x) = \frac{1}{2}x^2 + 2x - 1$; left 3 units, downward 2 units

14. $f(x) = 5 - 3x - \frac{1}{2}x^2$; right 5 units, downward 8 units

Exercises 15–22: (Refer to Example 2.) Write a formula for a function g whose graph is similar to f(x) but satisfies the given conditions. Do not simplify the formula.

15. $f(x) = 3x^2 + 2x - 5$
   (a) Shifted left 3 units
   (b) Shifted downward 4 units

16. $f(x) = 2x^2 - 3x + 2$
   (a) Shifted right 8 units
   (b) Shifted upward 2 units

17. $f(x) = 2x^2 - 4x + 1$
   (a) Shifted right 2 units and upward 4 units
   (b) Shifted left 8 units and downward 5 units

18. $f(x) = 5x^2 - 3$
   (a) Shifted left 10 units and downward 6 units
   (b) Shifted right 1 unit and upward 10 units

19. $f(x) = 3x^2 - 3x + 2$
   (a) Shifted right 2000 units and upward 70 units
   (b) Shifted left 300 units and downward 30 units

20. $f(x) = |x| - 3$
   (a) Shifted right 4 units and downward 3 units
   (b) Shifted left 5 units and upward 2 units

21. $f(x) = \sqrt{x}$
   (a) Shifted right 4 units, reflected about the x-axis
   (b) Shifted left 2 units, reflected about the y-axis

22. $f(x) = \sqrt{x}$
   (a) Reflected about the x-axis, shifted left 2 units
   (b) Reflected about the y-axis, shifted right 3 units

Exercises 23–26: Translating Circles Write an equation that shifts the given circle in the specified manner. State the center and radius of the translated circle.

23. $x^2 + y^2 = 4$; right 3 units, downward 4 units

24. $x^2 + y^2 = 9$; right 2 units, downward 6 units

25. $x^2 + y^2 = 5$; left 5 units, upward 3 units

26. $x^2 + y^2 = 7$; left 3 units, downward 7 units

Transforming Graphical Representations

Exercises 27–34: Use the accompanying graph of $y = f(x)$ to sketch a graph of each equation.

27. **(a)** $y = f(x) + 2$ **28.** **(a)** $y = f(x + 1)$

  (b) $y = f(x - 2) - 1$ **(b)** $y = -f(x)$

  (c) $y = -f(x)$ **(c)** $y = 2f(x)$

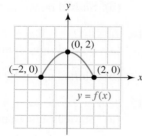

29. **(a)** $y = f(x + 3) - 2$ **30.** **(a)** $y = f(x - 1) - 2$

  (b) $y = f(-x)$ **(b)** $y = -f(x) + 1$

  (c) $y = \frac{1}{2}f(x)$ **(c)** $y = f\left(\frac{1}{2}x\right)$

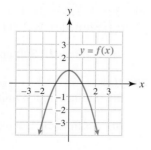

31. **(a)** $y = f(x + 1) + 1$ **32.** **(a)** $y = f(x) - 2$

  (b) $y = -f(x) - 1$ **(b)** $y = f(x - 1) + 2$

  (c) $y = \frac{1}{2}f(2x)$ **(c)** $y = 2f(-x)$

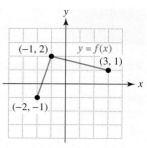

33. **(a)** $y = f(2x) + 1$ **34.** **(a)** $y = f(2x)$

  (b) $y = 2f\left(\frac{1}{2}x\right) + 1$ **(b)** $y = f\left(\frac{1}{2}x\right) - 1$

  (c) $y = \frac{1}{2}f(2 - x)$ **(c)** $y = 2f(1 - x) - 1$

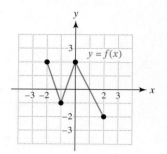

Graphing Transformations of Functions

Exercises 35–44. Use transformations to explain how the graph of f can be found by using the graph of $y = x^2$, $y = \sqrt{x}$, or $y = |x|$. You do not need to graph $y = f(x)$.

35. $f(x) = (x - 3)^2 + 1$ **36.** $f(x) = (x + 2)^2 - 3$

37. $f(x) = \frac{1}{4}(x + 1)^2$ **38.** $f(x) = 2(x - 4)^2$

39. $f(x) = -\sqrt{x + 5}$ **40.** $f(x) = -\sqrt{x} - 3$

41. $f(x) = 2\sqrt{-x}$ **42.** $f(x) = \sqrt{-\frac{1}{2}x}$

43. $f(x) = |-(x + 1)|$ **44.** $f(x) = |4 - x|$

Exercises 45–68: Use transformations to sketch a graph of f.

45. $f(x) = x^2 - 3$ **46.** $f(x) = -x^2$

47. $f(x) = (x - 5)^2 + 3$ **48.** $f(x) = (x + 4)^2$

49. $f(x) = -\sqrt{x}$ **50.** $f(x) = 2(x - 1)^2 + 1$

51. $f(x) = -x^2 + 4$ **52.** $f(x) = \sqrt{-x}$

53. $f(x) = |x| - 4$ **54.** $f(x) = \sqrt{x} + 1$

55. $f(x) = \sqrt{x - 3} + 2$ **56.** $f(x) = |x + 2| - 3$

57. $f(x) = |2x|$ **58.** $f(x) = \frac{1}{2}|x|$

59. $f(x) = 1 - \sqrt{x}$ **60.** $f(x) = 2\sqrt{(x-2)} - 1$

61. $f(x) = -\sqrt{1-x}$ **62.** $f(x) = \sqrt{-x} - 1$

63. $f(x) = \sqrt{-(x+1)}$ **64.** $f(x) = 2 + \sqrt{-(x-3)}$

65. $f(x) = (x-1)^3$ **66.** $f(x) = (x+2)^3$

67. $f(x) = -x^3$ **68.** $f(x) = (-x)^3 + 1$

Exercises 69–74: (Refer to Example 5.) For the given representation of a function f, graph the reflection across the x-axis and graph the reflection across the y-axis.

69. $f(x) = x^2 - 2x - 3$ **70.** $f(x) = 4 - 7x - 2x^2$

71. $f(x) = |x+1| - 1$ **72.** $f(x) = \frac{1}{2}|x-2| + 2$

73. Line graph determined by the table

x	-3	-1	1	2
$f(x)$	2	3	-1	-2

74. Line graph determined by the table

x	-4	-2	0	1
$f(x)$	-1	-4	2	2

Transforming Numerical Representations

Exercises 75–82: Two functions, f and g, are related by the given equation. Use the numerical representation of f to make a numerical representation of g.

75. $g(x) = f(x) + 7$

x	1	2	3	4	5	6
$f(x)$	5	1	6	2	7	9

76. $g(x) = f(x) - 10$

x	0	5	10	15	20
$f(x)$	-5	11	21	32	47

77. $g(x) = f(x-2)$

x	-4	-2	0	2	4
$f(x)$	5	2	-3	-5	-9

78. $g(x) = f(x+50)$

x	-100	-50	0	50	100
$f(x)$	25	80	120	150	100

79. $g(x) = f(x+1) - 2$

x	1	2	3	4	5	6
$f(x)$	2	4	3	7	8	10

80. $g(x) = f(x-3) + 5$

x	-3	0	3	6	9
$f(x)$	3	8	15	27	31

81. $g(x) = f(-x) + 1$

x	-2	-1	0	1	2
$f(x)$	11	8	5	2	-1

82. $g(x) = -f(x+2)$

x	-4	-2	0	2	4
$f(x)$	5	8	10	8	5

Exercises 83–90: The points $(-12, 6)$, $(0, 8)$, and $(8, -4)$ lie on the graph of $y = f(x)$. Determine three points that lie on the graph of $y = g(x)$.

83. $g(x) = f(x) + 2$

84. $g(x) = f(x) - 3$

85. $g(x) = f(x-2) + 1$

86. $g(x) = f(x+1) - 1$

87. $g(x) = -\frac{1}{2}f(x)$

88. $g(x) = -2f(x)$

89. $g(x) = f(-2x)$

90. $g(x) = f\left(-\frac{1}{2}x\right)$

Applications

Exercises 91–94: (Refer to Example 7.) Use transformations of graphs to model the table of data with the formula $f(x) = a(x-h)^2 + k$. (Answers may vary.)

91. Percentage of malpractice awards above $1 million

Year	1997	1998	1999	2000	2001
Percent	3.8	4.3	4.9	5.9	7.9

Source: Physician Insurers Association of America.

92. Number of titles released for DVD rentals

Year	1998	1999	2000	2001	2002
Titles	2049	4787	8723	14,321	21,260

Source: DVD Release Report.

93. Average sales price of a home in thousands of dollars

Year	1970	1980	1990	2000	2005
Price	30	80	150	210	300

Source: Bureau of the Census.

94. Average public college tuition and fees in dollars

Year	1998	2000	2002	2004	2006
Price	3247	3487	4081	5132	5836

Source: The College Board.

95. *U.S. Home Ownership*   The general trend in the percentage P of homes lived in by owners rather than renters between 1990 and 2006 is modeled by

$$P(x) = 0.00075x^2 + 0.17x + 44,$$

where $x = 0$ corresponds to 1990, $x = 1$ to 1991, and so on. Determine a function g that computes P, where x is the actual year. For example, $P(0) = 44$, so $g(1990) = 44$.

96. *U.S. AIDS Deaths*   The function D defined by

$$D(x) = 2375x^2 + 5134x + 5020$$

models AIDS deaths x years after 1984. Write a formula $g(x)$ that computes AIDS deaths during year x, where x is the actual year.

97. *Daylight Hours*   The figure shows a partial graph of the number of daylight hours at latitude 60°N. The variable x is measured in months, where $x = 0$ corresponds to December 21, the day with the least amount of daylight. For example, $x = 2$ represents February 21 and $x = -3$ represents September 21. The domain of this function is $-6 \leq x \leq 6$. (**Source:** J. Williams, *The Weather Almanac*.)

(a) Estimate daylight hours on February 21.

(b) Make a conjecture about the number of daylight hours on October 21, when $x = -2$.

(c) Sketch the left side of the graph for $-6 \leq x \leq 0$.

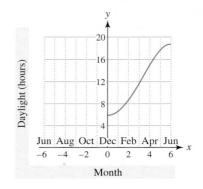

98. *Daylight Hours*   (Continuation of Exercise 97) Sketch a graph that shows the daylight hours at a latitude of 60°S, which is in the Southern Hemisphere rather than in the Northern Hemisphere.

Using Transformations to Model Motion

99. *Computer Graphics*   (Refer to Example 8.) Suppose that the airplane in Figure 3.111 is flying at 0.2 kilometer per second to the left, rather than to the right. If the position of the airplane is fixed at $(-1, 5)$, graph the image of the mountain and the position of the airplane after 15 seconds.

100. *Computer Graphics*   (Refer to Example 8.) Suppose that the airplane in Figure 3.111 is traveling to the right at 0.1 kilometer per second and gaining altitude at 0.05 kilometer per second. If the airplane's position is fixed at $(-1, 5)$, graph the image of the mountain and the position of the airplane after 20 seconds.

101. *Modeling a Weather Front*   Suppose a cold front is passing through the United States at noon, with a shape described by the function $y = \frac{1}{20}x^2$. Each unit represents 100 miles. Des Moines, Iowa, is located at $(0, 0)$, and the positive y-axis points north. See the figure.

(a) If the cold front moves south at 40 miles per hour and retains its present shape, graph its location at 4 P.M.

(b) Suppose that by midnight the vertex of the front, which is maintaining the same shape, has moved 250 miles south and 210 miles east of Des Moines. Columbus, Ohio, is located approximately 550 miles east and 80 miles south of Des Moines. Plot the locations of Des Moines and Columbus together with the new position of the cold front. Determine whether the cold front has reached Columbus by midnight.

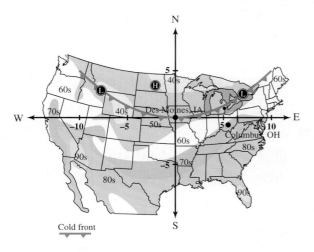

102. *Modeling Motion* The first figure below is a picture composed of lines and curves. In this exercise we will model only the semicircle that outlines the top of the silo. In order to make it appear that the person is walking to the right, the background must be translated horizontally to the left, as shown in the second figure.

The semicircle at the top of the silo in the first figure is described by $f(x) = \sqrt{9 - x^2} + 12$.

(a) Graph f in the window $[-12, 12, 1]$ by $[0, 16, 1]$.

(b) To give the illusion that the person is walking to the right at 2 units per second, graph the top of the silo after 1 second and after 4 seconds.

Writing about Mathematics

103. Explain how to graph the reflection of $y = f(x)$ across the x-axis. Give an example.

104. Let c be a positive number. Explain how to shift the graph of $y = f(x)$ upward, downward, left, or right c units. Give examples.

105. If the graph of $y = f(x)$ undergoes a vertical stretch or shrink to become the graph of $y = g(x)$, do these two graphs have the same x-intercepts? y-intercepts? Explain your answers.

106. If the graph of $y = f(x)$ undergoes a horizontal stretch or shrink to become the graph of $y = g(x)$, do these two graphs have the same x-intercepts? y-intercepts? Explain your answers.

EXTENDED AND DISCOVERY EXERCISES

Exercises 1–4: Commutative Property In the following exercises you will determine if transformations of a graph are commutative. To answer this question, start by determining if the two sequences of transformations are equivalent for all possible graphs.

1. Reflect across the x-axis, shift upward 1 unit.
Shift upward 1 unit, reflect across the x-axis.

2. Shift upward 2 units, shift left 3 units.
Shift left 3 units, shift upward 2 units.

3. Reflect across the x-axis, reflect across the y-axis.
Reflect across the y-axis, reflect across the x-axis.

4. Stretch in the vertical direction, shift downward 2 units.
Shift downward 2 units, stretch in the vertical direction.

CHECKING BASIC CONCEPTS FOR SECTION 3.5

1. Predict how the graph of each equation will appear compared to the graph of $f(x) = x^2$.
(a) $y = (x + 4)^2$ (b) $y = x^2 - 3$

(c) $y = (x - 5)^2 + 3$

2. Use the graph shown to sketch a graph of each equation.

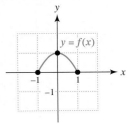

(a) $y = -2f(x)$ (b) $y = f\left(-\frac{1}{2}x\right)$
(c) $y = f(x - 1) + 1$

3. Write an equation that transforms the graph of $f(x) = x^2 - 4x + 1$ in the desired ways. Do not simplify.
(a) Right 3 units, downward 4 units

(b) Reflected about the x-axis

(c) Shifted left 6 units, reflected about the y-axis

(d) Reflected about the y-axis, shifted left 6 units

4. Use transformations to sketch a graph of the equation $y = \sqrt{x + 1} - 2$.

5. Use the table for $f(x)$ to make tables for $g(x)$ and $h(x)$.
(a) $g(x) = f(x - 2) + 3$ (b) $h(x) = -2f(x + 1)$

x	-4	-2	0	2	4
$f(x)$	1	3	6	8	9

3 Summary

CONCEPT	EXPLANATION AND EXAMPLES

SECTION 3.1 QUADRATIC FUNCTIONS AND MODELS

Quadratic Function

General form: $f(x) = ax^2 + bx + c, \quad a \neq 0$

Examples: $f(x) = x^2$ and $f(x) = -3x^2 + x + 5$

Parabola

The graph of a quadratic function is a parabola.

Vertex form: $f(x) = a(x - h)^2 + k$ (standard form for a parabola with a vertical axis)
Leading coefficient: a; vertex: (h, k); axis of symmetry: $x = h$

Example:

Completing the Square to Find the Vertex

The vertex of a parabola can be found by completing the square.

Example:
$$y = x^2 - 4x + 1$$
$$y - 1 = x^2 - 4x \qquad \text{Subtract 1.}$$
$$y - 1 + 4 = x^2 - 4x + 4 \qquad \text{Add } \left(\frac{-4}{2}\right)^2 = 4.$$
$$y + 3 = (x - 2)^2 \qquad \text{Perfect square trinomial}$$
$$y = (x - 2)^2 - 3 \qquad \text{Subtract 3.}$$

The vertex is $(2, -3)$.

Vertex Formula

x-coordinate: $x = -\dfrac{b}{2a}$; y-coordinate: $y = f\left(-\dfrac{b}{2a}\right)$

Example: $f(x) = 2x^2 + 4x - 4$

$$x = -\frac{4}{2(2)} = -1, \quad y = 2(-1)^2 + 4(-1) - 4 = -6$$

Vertex: $(-1, -6)$

SECTION 3.2 QUADRATIC EQUATIONS AND PROBLEM SOLVING

Quadratic Equation

Can be written as $ax^2 + bx + c = 0, \quad a \neq 0$
A quadratic equation can have zero, one, or two real solutions.

Examples: $x^2 + 1 = 0, \quad x^2 + 2x + 1 = 0, \quad$ and $\quad x(x - 1) = 20$

SECTION 3.2 QUADRATIC EQUATIONS AND PROBLEM SOLVING (CONTINUED)

Factoring

Write an equation in the form $ab = 0$ and apply the zero-product property.

Example:
$$x^2 - 3x = -2$$
$$x^2 - 3x + 2 = 0$$
$$(x - 1)(x - 2) = 0$$
$$x = 1 \quad \text{or} \quad x = 2$$

Square Root Property

If $x^2 = k$ and $k \geq 0$, then $x = \pm\sqrt{k}$.

Example: $x^2 = 16$ implies $x = \pm 4$.

Completing the Square

If $x^2 + kx = d$, then add $\left(\frac{k}{2}\right)^2$ to each side.

Example:
$$x^2 - 4x = 2 \qquad k = -4$$
$$x^2 - 4x + 4 = 2 + 4 \qquad \text{Add } \left(\frac{-4}{2}\right)^2 = 4.$$
$$(x - 2)^2 = 6 \qquad \text{Perfect square trinomial}$$
$$x - 2 = \pm\sqrt{6} \qquad \text{Square root property}$$
$$x = 2 \pm \sqrt{6} \qquad \text{Add 2.}$$

Quadratic Formula

$$x = \frac{-b \pm \sqrt{b^2 - 4ac}}{2a} \qquad \text{Always works}$$

Example: $2x^2 - 5x - 3 = 0$

$$x = \frac{-(-5) \pm \sqrt{(-5)^2 - 4(2)(-3)}}{2(2)} = \frac{5 \pm 7}{4} = 3, -\frac{1}{2}$$

Discriminant

The number of real solutions to $ax^2 + bx + c = 0$ with $a \neq 0$ can be found by evaluating the discriminant, $b^2 - 4ac$.

1. If $b^2 - 4ac > 0$, there are two real solutions.

2. If $b^2 - 4ac = 0$, there is one real solution.

3. If $b^2 - 4ac < 0$, there are no real solutions (two complex solutions).

SECTION 3.3 COMPLEX NUMBERS

Imaginary Unit

$$i = \sqrt{-1}, \qquad i^2 = -1$$

Examples: $\sqrt{-4} = 2i, \ \sqrt{-7} = i\sqrt{7}$

$$\sqrt{-3} \cdot \sqrt{-27} = i\sqrt{3} \cdot i\sqrt{27} = i^2\sqrt{81} = -9$$

CONCEPT	EXPLANATION AND EXAMPLES

SECTION 3.3 COMPLEX NUMBERS (CONTINUED)

Complex Conjugate

The conjugate of $a + bi$ is $a - bi$.

Examples:

Number	$5 - 2i$	$5i$	-7	$-1 + 4i$
Conjugate	$5 + 2i$	$-5i$	-7	$-1 - 4i$

Complex Number

$a + bi$, where a and b are real numbers (standard form)

Complex numbers include all real numbers. We can add, sutract, multiply, and divide complex numbers.

Examples: $(2 - 3i) + (1 + 5i) = (2 + 1) + (-3 + 5)i = 3 + 2i$ (Add)

$3i - (2 + i) = -2 + (3 - 1)i = -2 + 2i$ (Subtract)

$(3 - i)(1 + 2i) = 3(1) + 3(2i) - i(1) - i(2i) = 5 + 5i$ (Multiply)

$\dfrac{1 - i}{2 + i} = \dfrac{(1 - i)(2 - i)}{(2 + i)(2 - i)} = \dfrac{1 - 3i}{5} = \dfrac{1}{5} - \dfrac{3}{5}i$ (Divide)

Complex Solutions

The quadratic formula can be used to solve quadratic equations with complex solutions.

Example: The solutions to $x^2 - x + 2 = 0$ are

$$x = \frac{1 \pm \sqrt{(-1)^2 - 4(1)(2)}}{2(1)} = \frac{1}{2} \pm i\frac{\sqrt{7}}{2}.$$

SECTION 3.4 QUADRATIC INEQUALITIES

Quadratic Inequality

$ax^2 + bx + c < 0$ with $a \neq 0$, where $<$ may be replaced by $\leq$, $>$, or $\geq$.

Example: $3x^2 - x + 1 \leq 0$

Graphical Solution

Graph $y = ax^2 + bx + c$ and find the x-intercepts; then determine x-values where the inequality is satisfied.

Example: Solve $-x^2 - x + 2 > 0$.

The x-intercepts are -2 and 1.

Solution set is $\{x \mid -2 < x < 1\}$, or $(-2, 1)$.

CONCEPT	EXPLANATION AND EXAMPLES

SECTION 3.4 QUADRATIC INEQUALITIES (CONTINUED)

Symbolic Solution

Solve $ax^2 + bx + c = 0$ and use a table of values or number line to determine the x-intervals where the inequality is satisfied.

Example: Solve $x^2 - 4 \geq 0$.

$x^2 - 4 = 0$ implies $x = \pm 2$.

Solution set is $\{x \mid x \leq -2 \text{ or } x \geq 2\}$, or $(-\infty, -2] \cup [2, \infty)$.

Interval	Test Value x	$x^2 - 4$	Positive or Negative?
$(-\infty, -2)$	-3	5	Positive
$(-2, 2)$	0	-4	Negative
$(2, \infty)$	3	5	Positive

SECTION 3.5 TRANSFORMATIONS OF GRAPHS

Vertical Shifts with $c > 0$

$y = f(x) + c$ shifts the graph of $y = f(x)$ upward c units.
$y = f(x) - c$ shifts the graph of $y = f(x)$ downward c units.

Horizontal Shifts with $c > 0$

$y = f(x - c)$ shifts the graph of $y = f(x)$ to the right c units.
$y = f(x + c)$ shifts the graph of $y = f(x)$ to the left c units.

Vertical Stretching and Shrinking

$y = cf(x)$ vertically stretches the graph of $y = f(x)$ when $c > 1$ and shrinks the graph when $0 < c < 1$.

Horizontal Stretching and Shrinking

$y = f(cx)$ horizontally shrinks the graph of $y = f(x)$ when $c > 1$ and stretches the graph when $0 < c < 1$.

Reflections

$y = -f(x)$ is a reflection of $y = f(x)$ across the x-axis.
$y = f(-x)$ is a reflection of $y = f(x)$ across the y-axis.

3 | Review Exercises

Exercises 1 and 2: Use the graph to find the following.

(a) Sign of the leading coefficient
(b) Vertex
(c) Axis of symmetry
(d) Intervals where f is increasing and where f is decreasing

1.

2.

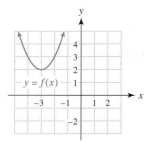

Exercises 3 and 4: Write $f(x)$ in the form $f(x) = ax^2 + bx + c$, and identify the leading coefficient.

3. $f(x) = -2(x - 5)^2 + 1$ **4.** $f(x) = \frac{1}{3}(x + 1)^2 - 2$

Exercises 5 and 6: Use the graph of the quadratic function f to write it as $f(x) = a(x - h)^2 + k$.

5.

6.

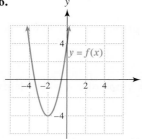

Exercises 7 and 8: Write $f(x)$ in the form $f(x) = a(x - h)^2 + k$, and identify the vertex.

7. $f(x) = x^2 + 6x - 1$ **8.** $f(x) = 2x^2 + 4x - 5$

Exercises 9 and 10: Use the vertex formula to determine the vertex on the graph of f.

9. $f(x) = -3x^2 + 2x - 4$ **10.** $f(x) = x^2 + 8x - 5$

Exercises 11–18: Sketch a graph of the function.

11. $f(x) = -3x^2 + 3$ **12.** $g(x) = 2(x - 1)^2 - 3$

13. $h(t) = t^2 - 4t - 1$ **14.** $f(x) = |x - 1| - 2$

15. $f(x) = -|x + 3|$ **16.** $g(t) = 3t - t^2$

17. $f(x) = -\sqrt{x + 1}$ **18.** $f(x) = \sqrt{2 - x}$

19. *Average Rate of Change*   Find the average rate of change of $f(x) = -6x^2 + 7x + 5$ from 2 to 4.

20. *Difference Quotient*   Find the difference quotient for $f(x) = x^2 - 2x$.

Exercises 21–32: Solve the quadratic equation.

21. $x^2 - x - 20 = 0$ **22.** $3x^2 + 4x = -1$

23. $x^2 = 4x$ **24.** $-5x^2 - 3x = 0$

25. $4z^2 - 7 = 0$ **26.** $25z^2 = 9$

27. $-2t^2 - 3t + 14 = 0$ **28.** $\frac{1}{2}t^2 + \frac{3}{4}t + \frac{1}{4} = 0$

29. $0.1x^2 - 0.3x = 1$ **30.** $x(6 - x) = -16$

31. $(k - 1)^2 = \frac{9}{4}$ **32.** $(k + 2)^2 = 7$

Exercises 33 and 34: The graph of $f(x) = ax^2 + bx + c$ is given.

(a) State whether $a > 0$ or $a < 0$.
(b) Estimate the real solutions to $ax^2 + bx + c = 0$.
(c) Is the discriminant positive, negative, or zero?

33.

34.

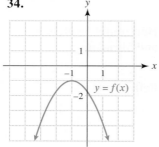

Exercises 35–40: Solve by completing the square.

35. $x^2 + 2x = 5$ **36.** $x^2 - 3x = 3$

37. $2z^2 - 6z - 1 = 0$ **38.** $-3z^2 - 2z + 2 = 0$

39. $\frac{1}{2}x^2 - 4x + 1 = 0$ **40.** $-\frac{1}{4}x^2 - \frac{1}{2}x + 1 = 0$

41. Solve the equation $2x^2 - 3y^2 = 6$ for y. Is y a function of x?

42. Solve $h = -\frac{1}{2}gt^2 + 100$ for t.

43. Use the imaginary unit i to simplify each expression.
  (a) $\sqrt{-16}$ (b) $\sqrt{-48}$ (c) $\sqrt{-5} \cdot \sqrt{-15}$

44. Write each expression in standard form.
  (a) $(2 - 3i) + (-3 + 3i)$

  (b) $(-5 + 3i) - (-3 - 5i)$

  (c) $(3 + 2i)(-4 - i)$ (d) $\frac{3 + 2i}{2 - i}$

Exercises 45–48: Use the graph and the given f(x) to complete the following.
  (a) *Find any x-intercepts.*
  (b) *Find the complex zeros of f.*

45.

$f(x) = -2x^2 - 4x + \frac{5}{2}$

46.

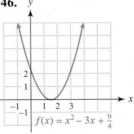

$f(x) = x^2 - 3x + \frac{9}{4}$

47.

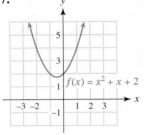

$f(x) = x^2 + x + 2$

48.

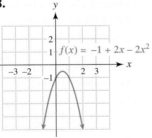

$f(x) = -1 + 2x - 2x^2$

Exercises 49 and 50: Find all complex solutions.

49. $4x^2 + 9 = 0$ **50.** $2x^2 + 3 = 2x$

51. Use the graph of $y = f(x)$ to solve the inequality.
  (a) $f(x) > 0$ (b) $f(x) \le 0$

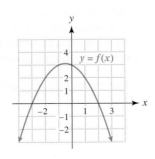

52. The table contains test values for a quadratic function. Solve each inequality.

x	-6	-5	0	3	4
$f(x)$	9	0	-15	0	9

  (a) $f(x) < 0$ (b) $f(x) \ge 0$

53. Solve the equation or inequality.
  (a) $x^2 - 3x + 2 = 0$ (b) $x^2 - 3x + 2 < 0$

  (c) $x^2 - 3x + 2 > 0$

54. Solve $2x^2 + 1.3x \le 0.4$ graphically.

Exercises 55–60: Solve the inequality. Use set-builder or interval notation to write a solution set to the inequality.

55. $x^2 - 3x + 2 \le 0$ **56.** $x^2 - 2x \ge 0$

57. $2x^2 + 3x + 1 < 0$ **58.** $9x^2 - 4 > 0$

59. $n(n - 2) \ge 15$ **60.** $n^2 + 4 \le 6n$

61. If $f(x) = 2x^2 - 3x + 1$, use transformations to graph $y = -f(x)$ and $y = f(-x)$.

62. Use the given graph of $y = f(x)$ to sketch a graph of each expression.
  (a) $y = f(x + 1) - 2$

  (b) $y = -2f(x)$

  (c) $y = f(2x)$

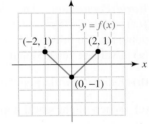

Exercises 63–68: Use transformations to sketch a graph of f.

63. $f(x) = x^2 - 4$ **64.** $f(x) = 3|x - 2| + 1$

65. $f(x) = \sqrt{x + 2}$ **66.** $f(x) = -4\sqrt{-x}$

67. $f(x) = -2(x - 2)^2 + 3$

68. $f(x) = -|x - 3|$

69. Use the table for $f(x)$ to write a table for $g(x)$, if $g(x) = -3f(x + 1) + 2$.

x	1	2	3	4
$f(x)$	-3	3	4	7

70. Sketch a graph of a quadratic function f that has axis of symmetry $x = 2$ and passes through the points $(2, 3)$ and $(4, -1)$. Find a formula for the function.

Applications

71. *Maximizing Area*   A homeowner has 44 feet of fence to enclose a rectangular garden. One side of the garden needs no fencing because it is along the wall of the house. What dimensions will maximize area?

72. *Maximizing Revenue*   The revenue R in dollars received from selling x radios is $R(x) = x(90 - x)$.
  (a) Evaluate $R(20)$ and interpret the result.

  (b) What number of radios sold will maximize revenue?

  (c) What is the maximum revenue?

  (d) What number of radios should be sold for revenue to be $2000 or more?

73. *Projectile*   A slingshot is used to propel a stone upward so that its height h in feet after t seconds is given by $h(t) = -16t^2 + 88t + 5$.
  (a) Evaluate $h(0)$ and interpret the result.

  (b) How high was the stone after 2 seconds?

  (c) Find the maximum height of the stone.

  (d) At what time(s) was the stone 117 feet high?

74. *World Population*   The function given by the formula $f(x) = 0.000478x^2 - 1.813x + 1720.1$ models world population in billions from 1950 to 2000 during year x.
  (a) Evaluate $f(1985)$ and interpret the result.

  (b) Estimate world population during the year 2000.

  (c) When might world population reach 7 billion?

75. *Construction*   A box is being constructed by cutting 3-inch squares from the corners of a rectangular sheet of metal that is 4 inches longer than it is wide. If the box is to have a volume of 135 cubic inches, find the dimensions of the metal sheet.

76. *Room Prices*   Room prices are regularly $100, but for each additional room rented by a group, the price is reduced by $3 for each room. For example, 1 room costs $100, 2 rooms cost 2 × $97 = $194, and so on.
  (a) Write a quadratic function C that gives the total cost of renting x rooms.

  (b) What is the total cost of renting 6 rooms?

  (c) How many rooms are rented if the cost is $730?

  (d) What number of rooms rented gives the greatest cost?

77. *Irrigation and Yield*   The table shows how irrigation of rice crops affects yield, where x represents the percent of total area that is irrigated and y is the rice yield in tons per hectare. (1 hectare ≈ 2.47 acres.)

x	0	20	40	60	80	100
y	1.6	1.8	2.2	3.0	4.5	6.1

Source: D. Grigg, *The World Food Problem*.

  (a) Use least-squares regression to find a quadratic function that models the data.

  (b) Solve the equation $f(x) = 3.7$. Interpret the results.

78. *Credit Card Debt*   The table lists outstanding balances on Visa and MasterCard credit cards in billions of dollars.

Year	1980	1984	1988	1992	1996
Debt	82	108	172	254	444

Source: Bankcard Holders of America.

  (a) Does a linear function model the data?

  (b) Find a quadratic function f that models these data. Solve the equation $f(x) = 212$. Interpret the results.

EXTENDED AND DISCOVERY EXERCISES

1. *Shooting a Foul Shot*   (Refer to the introduction to this chapter.) When a basketball player shoots a foul shot, the ball follows a parabolic arc. This arc depends on both the angle and velocity with which the basketball is released. If a person shoots the basketball overhand from a position 8 feet above the floor, then the path can sometimes be modeled by the parabola

$$y = \frac{-16x^2}{0.434v^2} + 1.15x + 8,$$

where v is the velocity of the ball in feet per second, as illustrated in the figure.   (**Source:** C. Rist, "The Physics of Foul Shots.")
  (a) If the basketball hoop is 10 feet high and located 15 feet away, what initial velocity v should the basketball have?

  (b) Check your answer from part (a) graphically. Plot the point $(0, 8)$ where the ball is released and the point $(15, 10)$ where the basketball hoop is. Does your graph pass through both points?

  (c) What is the maximum height of the basketball?

2. *Shooting a Foul Shot*   (Continuation of Exercise 1) If a person releases a basketball underhand from a position 3 feet above the floor, it often has a steeper arc than if it is released overhand and the path sometimes may be modeled by

$$y = \frac{-16x^2}{0.117v^2} + 2.75x + 3.$$

See the figure below. Complete parts (a), (b), and (c) from Exercise 1. Then compare the paths for an overhand shot and an underhand shot.

Exercises 3–6: Reflecting Functions   Computer graphics frequently use reflections. Reflections can speed up the generation of a picture or create a figure that appears perfectly symmetrical.   (**Source:** S. Hoggar, *Mathematics for Computer Graphics.*)

(a) *For the given f(x), constant k, and viewing rectangle, graph x = k, y = f(x), and y = f(2k − x).*

(b) *Generalize how the graph of y = f(2k − x) compares to the graph of y = f(x).*

3. $f(x) = \sqrt{x}, k = 2, [-1, 8, 1]$ by $[-4, 4, 1]$

4. $f(x) = x^2, k = -3, [-12, 6, 1]$ by $[-6, 6, 1]$

5. $f(x) = x^4 - 2x^2 + 1, k = -6, [-15, 3, 1]$ by $[-3, 9, 1]$

6. $f(x) = 4x - x^3, k = 5, [-6, 18, 1]$ by $[-8, 8, 1]$

7. *Modeling a Cold Front*   A weather map of the United States on April 22, 1996, is shown in the figure. There was a cold front roughly in the shape of a circular arc, with a radius of about 750 miles, passing north of Dallas and west of Detroit. The center of the arc was located near Pierre, South Dakota. If Pierre has the coordinates (0, 0) and the positive *y*-axis points north, then the equation of the front can be modeled by

$$f(x) = -\sqrt{750^2 - x^2},$$

where $0 \le x \le 750$.   (**Source:** AccuWeather, Inc.)

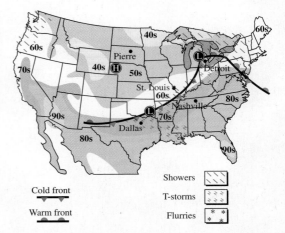

(a) St. Louis is located at (535, −400) and Nashville is at (730, −570), where units are in miles. Plot these points and graph *f* in the window [0, 1200, 100] by [−800, 0, 100]. Did the cold front reach these cities?

(b) During the next 12 hours, the center of the front moved approximately 110 miles south and 160 miles east. Assuming the cold front did not change shape, use transformations of graphs to determine an equation that models its new location.

(c) Use graphing to determine visually if the cold front reached both cities.

4 More Nonlinear Functions and Equations

The wonderful things you learn in schools are the work of many generations, produced by enthusiastic effort and infinite labor in every country.

—Albert Einstein

Mathematics can be both abstract and applied. Abstract mathematics is focused on axioms, theorems, and proofs. It can be derived independently of empirical evidence. Theorems that were proved centuries ago are still valid today. In this sense, abstract mathematics transcends time. Yet, even though mathematics can be developed in an abstract setting—separate from science and all measured data—it also has countless applications.

There is a common misconception that theoretical mathematics is unimportant, yet many of the ideas that eventually had great practical importance were first born in the abstract. For example, in 1854 George Boole published *Laws of Thought*, which outlined the basis for Boolean algebra. This was 85 years before the invention of the first digital computer. However, Boolean algebra became the basis on which modern computer hardware operates.

Much like Boolean algebra, the topic of complex numbers was at first theoretical. However, today complex numbers are used in the design of electrical circuits, ships, and airplanes. Basic quantum physics and certain features of the theory of relativity would not have been developed without complex numbers.

In this chapter we discuss some important topics in algebra that have had an impact on society. We are privileged to read in a few hours what took people centuries to discover. To ignore either the abstract beauty or the profound applicability of mathematics is like seeing a rose but never smelling one.

4.1 More Nonlinear Functions and Their Graphs

- **Learn terminology about polynomial functions**
- **Find extrema of a function**
- **Identify symmetry in a graph of a function**
- **Determine if a function is odd, even, or neither**

Introduction

Monthly average high temperatures at Daytona Beach are shown in Table 4.1.

Table 4.1   Monthly Average High Temperatures at Daytona Beach

Month	1	2	3	4	5	6	7	8	9	10	11	12
Temperature (°F)	69	70	75	80	85	88	90	89	87	81	76	70

Source: J. Williams, *The USA Weather Almanac.*

Figure 4.1 shows a scatterplot of the data. A linear function would not model these data because these data do not lie on a line. One possibility is to model the data with a quadratic function, as shown in Figure 4.2. However, a better fit can be obtained with the nonlinear function f whose graph is shown in Figure 4.3 and is given by

$$f(x) = 0.0145x^4 - 0.426x^3 + 3.53x^2 - 6.23x + 72,$$

where $x = 1$ corresponds to January, $x = 2$ to February, and so on. (Least-squares regression was used to determine $f(x)$.) Function f is a *polynomial function* with degree 4.

Figure 4.1 Temperature Data

Figure 4.2 Quadratic Model

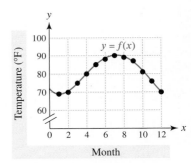

Figure 4.3 New Nonlinear Model

Polynomial Functions

The domain of a polynomial function is all real numbers, and its graph is continuous and smooth without breaks or sharp edges.

Polynomial Function

A **polynomial function f of degree n in the variable x** can be represented by

$$f(x) = a_n x^n + a_{n-1} x^{n-1} + \cdots + a_2 x^2 + a_1 x + a_0,$$

where each coefficient a_k is a real number, $a_n \neq 0$, and n is a nonnegative integer. The **leading coefficient** is a_n and the **degree** is n.

Examples of polynomial functions include the following.

Algebra Review
To review polynomials, see Chapter R
(page R-20).

Formula	Degree	Leading Coefficient
$f(x) = 10$	0	$a_0 = 10$
$g(x) = 2x - 3.7$	1	$a_1 = 2$
$h(x) = 1 - 1.4x + 3x^2$	2	$a_2 = 3$
$k(x) = -\dfrac{1}{2}x^6 + 4x^4 + x$	6	$a_6 = -\dfrac{1}{2}$

A polynomial function of degree 2 or higher is a *nonlinear* function. Functions f and g are linear, whereas functions h and k are nonlinear. As a result, polynomial functions are used to model both linear and nonlinear data.

NOTE Quadratic functions, which were discussed in Chapter 3, are examples of nonlinear functions. This chapter introduces *more* nonlinear functions.

Functions that contain radicals, ratios, or absolute values of variables are not polynomials. For example, $f(x) = 2\sqrt{x}$, $g(x) = \frac{1}{x-1}$, and $h(x) = |2x + 5|$ are *not* polynomials.

Identifying Extrema

In Figure 4.3 (page 249) the minimum monthly average temperature of 69°F occurs in January ($x = 1$) and the maximum monthly average temperature of 90°F occurs in July ($x = 7$). Minimum and maximum y-values on the graph of a function often represent important data. Graphs of polynomial functions often have "hills" or "valleys." For example, consider the two polynomial functions f and g given by

$$f(x) = \frac{1}{2}x^2 - 2x - 4 \quad \text{and} \quad g(x) = -\frac{1}{4}x^4 + \frac{2}{3}x^3 + \frac{5}{2}x^2 - 6x.$$

Their graphs are shown in Figures 4.4 and 4.5. (Some values have been rounded.)

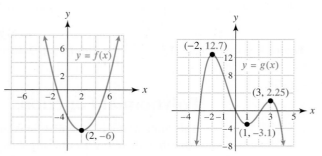

Figure 4.4   **Figure 4.5**

If one traces along the graph in Figure 4.4 from left to right, the y-values decrease until the vertex $(2, -6)$ is reached. To the right of the vertex, the y-values increase. The minimum y-value on the graph of f is -6. It is called the *absolute minimum* of f. Function f has no *absolute maximum*; there is no largest y-value on a parabola opening upward.

In Figure 4.5 the peak of the highest "hill" on the graph of g is $(-2, 12.7)$. Therefore the absolute maximum of g is 12.7. There is a smaller peak located at the point $(3, 2.25)$. In a small open interval near $x = 3$, the y-value of 2.25 is locally the largest. We say that g

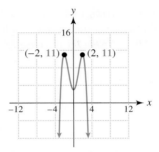

Figure 4.6

has a *local maximum* of **2.25**. Similarly, a "valley" occurs on the graph of *g*, where the lowest point is (1, −3.1). The value −**3.1** is not the smallest *y*-value on the entire graph of *g*. Therefore it is not an absolute minimum. Rather, −3.1 is a *local minimum*.

Maximum and minimum values that are either absolute or local are called **extrema** (plural of extremum). A function may have several local extrema, but at most one absolute maximum and one absolute minimum. However, it is possible for a function to assume an absolute extremum at two values of *x*. In Figure 4.6 the absolute maximum is **11**. It occurs at *x* = ±2. Note that 11 is also a local maximum, because near *x* = −2 and *x* = 2 it is the largest *y*-value.

Sometimes an absolute maximum (minimum) is called a *global maximum* (*minimum*). Similarly, sometimes a local maximum (minimum) is called a *relative maximum* (*minimum*).

Absolute and Local Extrema

Let *c* be in the domain of *f*.

f(*c*) is an **absolute (global) maximum** if *f*(*c*) ≥ *f*(*x*) *for all x in the domain of f*.
f(*c*) is an **absolute (global) minimum** if *f*(*c*) ≤ *f*(*x*) *for all x in the domain of f*.
f(*c*) is a **local (relative) maximum** if *f*(*c*) ≥ *f*(*x*) when *x* is *near c*.
f(*c*) is a **local (relative) minimum** if *f*(*c*) ≤ *f*(*x*) when *x* is *near c*.

NOTE The expression "near *c*" means that there is an open interval in the domain of *f* containing *c* where *f*(*c*) satisfies the inequality.

EXAMPLE 1 Identifying and interpreting extrema

Figure 4.7 shows the graph of a function *f* that models the volume of air in a person's lungs, measured in liters, after *x* seconds. (**Source:** V. Thomas, *Science and Sport.*)
(a) Find the absolute maximum and the absolute minimum of *f*. Interpret the results.
(b) Identify two local maxima (plural of maximum) and two local minima (plural of minimum) of *f*. Interpret the results.

SOLUTION

(a) The absolute maximum is 4 liters and occurs at C. The absolute minimum is 1 liter and occurs at D. At C a deep breath has been taken and the lungs are more inflated. After C, the person exhales until the lungs contain only 1 liter of air at D.

(b) One local maximum is 3 liters. It occurs at A and E and represents the amount of air in a person's lungs after inhaling normally. One local minimum is 2 liters. It occurs at B and F and represents the amount of air after exhaling normally. Another local maximum is 4 liters, which is also the absolute maximum. Similarly, 1 liter is a local minimum and also the absolute minimum.   *Now Try Exercise 91* ◀

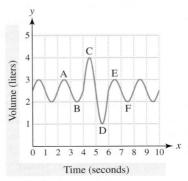

Figure 4.7 Volume of Air in a Person's Lungs

EXAMPLE 2 Identifying extrema

Use the graph of *f* in Figure 4.8 to estimate any local and absolute extrema.

SOLUTION

Local Extrema   The points (−2, 8) and (1, −19) on the graph of *f* correspond to the lowest point in a "valley." Thus there are local minima of **8** and −**19**. The point (−1, **13**) corresponds to the highest point on a "hill." Thus there is a local maximum of **13**.

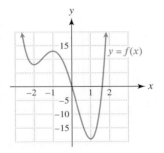

Figure 4.8

Absolute Extrema   Because the arrows point upward, there is no maximum y-value on the graph. Thus there is no absolute maximum. However, the minimum y-value on the graph of f occurs at the point $(1, -19)$.   The absolute minimum is -19.   **Now Try Exercise 15** ◄

Calculator Help

To find a minimum or maximum point on a graph, see Appendix A (page AP-10).

NOTE Extrema are y-values on the graph of a function, not x-values.

EXAMPLE 3   Modeling ocean temperatures

The monthly average ocean temperature in degrees Fahrenheit at Bermuda can be modeled by $f(x) = 0.0215x^4 - 0.648x^3 + 6.03x^2 - 17.1x + 76.4$, where $x = 1$ corresponds to January and $x = 12$ to December. The domain of f is $D = \{x \mid 1 \le x \le 12\}$.   (**Source:** J. Williams, *The Weather Almanac.*)

(a) Graph f in $[1, 12, 1]$ by $[50, 90, 10]$.
(b) Estimate the absolute extrema. Interpret the results.

SOLUTION

[1, 12, 1] by [50, 90, 10]

Minimum
X=2.0128359   Y=61.479564

Figure 4.9

(a) The graph of $Y_1 = .0215X^4 - .648X^3 + 6.03X^2 - 17.1X + 76.4$ is shown in Figure 4.9.
(b) Many graphing calculators have the capability to find maximum and minimum y-values. The points associated with absolute extrema are shown in Figures 4.9 and 4.10. An absolute minimum of about 61.5 corresponds to the point $(2.01, 61.5)$. This means that the monthly average ocean temperature is coldest during the month of February ($x \approx 2$) when it reaches a minimum of about 61.5°F.

    An absolute maximum of approximately 82 corresponds to $(7.61, 82.0)$. Rounding, we might say that the warmest average temperature occurs during August ($x \approx 8$) when it reaches a maximum of 82°F. (Or we might say that this maximum occurs in late July, since $x \approx 7.61$.)   **Now Try Exercise 93** ◄

[1, 12, 1] by [50, 90, 10]

Maximum
X=7.608694   Y=82.005128

Figure 4.10

Symmetry

Even Functions Symmetry is used frequently in art, mathematics, science, and computer graphics. Many objects are symmetric along a vertical line so that the left and right sides are mirror images. If an automobile is viewed from the front, the left side is typically a mirror image of the right side. Similarly, animals and people usually have an approximate left-right symmetry. Graphs of functions may also exhibit this type of symmetry, as shown in Figures 4.11–4.13.

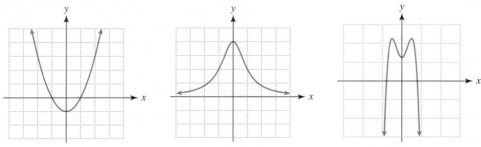

Figure 4.11   **Figure 4.12**   **Figure 4.13**

    If each graph were folded along the y-axis, the left and right halves would match. These graphs are **symmetric with respect to the y-axis**. A function whose graph satisfies this characteristic is called an *even function*.

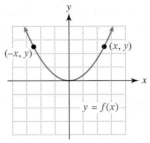

Figure 4.14 An Even Function

Figure 4.14 shows a graph of an even function f. Since the graph is symmetric with respect to the y-axis, the points (x, y) and $(-x, y)$ both lie on the graph of f. Thus $f(x) = y$ and $f(-x) = y$, and so $f(x) = f(-x)$ for an even function. This means that if we change the sign of the input, the output does not change. For example, if $g(x) = x^2$, then $g(2) = g(-2) = 4$. Since this is true for *every input*, g is an even function.

Even Function

A function f is an **even function** if $f(-x) = f(x)$ for every x in its domain. The graph of an even function is symmetric with respect to the y-axis.

Odd Functions   A second type of symmetry is shown in Figures 4.15–4.17. If we could spin or rotate the graph about the origin, the original graph would reappear after half a turn. These graphs are **symmetric with respect to the origin** and represent *odd functions*.

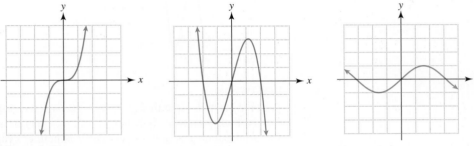

Figure 4.15   Figure 4.16   Figure 4.17

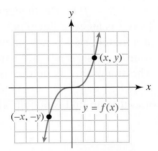

Figure 4.18 An Odd Function

In Figure 4.18 the point (x, y) lies on the graph of an odd function f. If this point spins half a turn, or $180°$, around the origin, its new location is $(-x, -y)$. Thus $f(x) = y$ and $f(-x) = -y$. It follows that $f(-x) = -y = -f(x)$ for any odd function f. Changing the sign of the input changes only the sign of the output. For example, if $g(x) = x^3$, then $g(3) = 27$ and $g(-3) = -27$. Since this is true for *every input*, g is an odd function.

Odd Function

A function f is an **odd function** if $f(-x) = -f(x)$ for every x in its domain. The graph of an odd function is symmetric with respect to the origin.

Identifying Odd and Even Functions   The terms *odd* and *even* have special meaning when they are applied to a polynomial function f. If $f(x)$ contains terms that have only odd powers of x, then f is an odd function. Similarly, if $f(x)$ contains terms that have only even powers of x (and possibly a constant term), then f is an even function. For example, $f(x) = x^6 - 4x^4 - 2x^2 + 5$ is an even function, whereas $g(x) = x^5 + 4x^3$ is an odd function. This can be shown *symbolically*.

$$f(-x) = (-x)^6 - 4(-x)^4 - 2(-x)^2 + 5 \qquad \text{Substitute } -x \text{ for } x.$$
$$= x^6 - 4x^4 - 2x^2 + 5 \qquad \text{Simplify.}$$
$$= f(x) \qquad f \text{ is an even function.}$$
$$g(-x) = (-x)^5 + 4(-x)^3 \qquad \text{Substitute } -x \text{ for } x.$$
$$= -x^5 - 4x^3 \qquad \text{Simplify.}$$
$$= -g(x) \qquad g \text{ is an odd function.}$$

NOTE It is important to remember that the graphs of many functions exhibit *no symmetry* with respect to either the *y*-axis or the origin. These functions are *neither* odd *nor* even.

EXAMPLE 4 Identifying odd and even functions

For each representation of a function f, identify whether f is odd, even, or neither.

(a)

x	-3	-2	-1	0	1	2	3
$f(x)$	10.5	2	-0.5	-2	-0.5	2	10.5

(b)

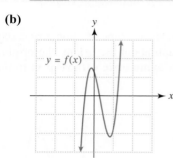

(c) $f(x) = x^3 - 5x$

(d) f is the cube root function.

SOLUTION

Getting Started If either $f(-x) = f(x)$ or its graph is symmetric with respect to the *y*-axis, then f is even. If either $f(-x) = -f(x)$ or its graph is symmetric with respect to the origin, then f is odd. Otherwise, f is neither even nor odd. ▶

(a) The function defined by the table has domain $D = \{-3, -2, -1, 0, 1, 2, 3\}$. Notice that $f(-3) = 10.5 = f(3)$, $f(-2) = 2 = f(2)$, and $f(-1) = -0.5 = f(1)$. The function f satisfies the statement $f(-x) = f(x)$ for every x in D. Thus f is an even function.

(b) If we fold the graph on the *y*-axis, the two halves do not match, so f is not an even function. Similarly, f is not an odd function since spinning its graph half a turn about the origin does not result in the same graph. The function f is neither odd nor even.

(c) Since f is a polynomial containing only odd powers of x, it is an odd function. This also can be shown symbolically as follows.

$$f(-x) = (-x)^3 - 5(-x) \qquad \text{Substitute } -x \text{ for } x.$$
$$= -x^3 + 5x \qquad \text{Simplify.}$$
$$= -(x^3 - 5x) \qquad \text{Distributive property}$$
$$= -f(x) \qquad f \text{ is an odd function.}$$

(d) Note that $\sqrt[3]{-8} = -2$ and that $\sqrt[3]{8} = 2$. In general, $\sqrt[3]{-x} = -\sqrt[3]{x}$, which indicates that $f(-x) = -f(x)$, where $f(x) = \sqrt[3]{x}$. Thus f is an odd function. This fact can also be seen by graphing $f(x) = \sqrt[3]{x}$, as shown in Figure 4.19. Spinning the graph of $f(x) = \sqrt[3]{x}$ a half a turn about the origin results in the same graph.

Now Try Exercises 47, 59, 63, and 71 ◀

Algebra Review

To review cube roots, see Chapter R (page R-40).

Figure 4.19

CLASS DISCUSSION

Discuss the possibility of the graph of a *function* being symmetric with respect to the *x*-axis.

4.1 Putting It All Together

The following table summarizes some important concepts related to the graphs of nonlinear functions.

Concept	Explanation	Graphical Example
Absolute, or global, maximum (minimum)	The maximum (minimum) y-value on the graph of $y = f(x)$ A graph of a function may or may not have an absolute maximum (minimum).	*(graph showing absolute maximum and absolute minimum)*
Local, or relative, maximum (minimum)	A maximum (minimum) y-value on the graph of $y = f(x)$ in an open interval of the domain of f A graph of a function may or may not have a local maximum (minimum). Note that it is possible for a y-value on the graph of f to be both an absolute maximum (minimum) *and* a local maximum (minimum).	*(graph showing local maximum and local minimum)*
Even function	$f(-x) = f(x)$ The graph is symmetric with respect to the y-axis. If the graph is folded on the y-axis, the left and right halves match. Changing the sign of the input does not change the output.	*(graph of even function $y = f(x)$)*
Odd function	$f(-x) = -f(x)$ The graph is symmetric with respect to the origin. If the graph is rotated about the origin, the graph reappears after half a turn. Changing the sign of the input changes only the sign of the output.	*(graph of odd function $y = f(x)$)*

4.1   Exercises

Note: Many of the answers in this section involve estimations. Your answers may vary slightly, particularly when you are reading a graph.

Polynomials

Exercises 1–10: Determine if the function is a polynomial function. If it is, state its degree and leading coefficient a.

1. $f(x) = 2x^3 - x + 5$ **2.** $f(x) = -x^4 + 1$

3. $f(x) = \sqrt{x}$ **4.** $f(x) = 2x^3 - \sqrt[3]{x}$

5. $f(x) = 1 - 4x - 5x^4$ **6.** $f(x) = 5 - 4x$

7. $g(t) = \dfrac{1}{t^2 + 3t - 1}$ **8.** $g(t) = \dfrac{1}{1 - t}$

9. $g(t) = 22$ **10.** $g(t) = |2t|$

Finding Extrema of Polynomials

Exercises 11–26: Use the graph of f to estimate the
   (a) local extrema   and   (b) absolute extrema.

11.

12.

13.

14.

15.

16.

17.

18.

19.

20.

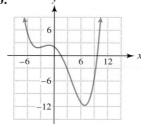

(Hint: *Exercises 21–26: Local extrema cannot occur at endpoints because they only occur on open intervals.*)

21.

22.

23.

24.

25.

26.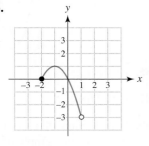

Exercises 27–38: Determine any
    (a) local extrema and (b) absolute extrema.
(Hint: *Consider the graph* $y = g(x)$.)

27. $g(x) = 1 - 3x$ **28.** $g(x) = \frac{1}{4}x$

29. $g(x) = x^2 + 1$ **30.** $g(x) = 1 - x^2$

31. $g(x) = -2(x + 3)^2 + 4$

32. $g(x) = \frac{1}{3}(x - 1)^2 - 2$

33. $g(x) = 2x^2 - 3x + 1$ **34.** $g(x) = -3x^2 + 4x - 1$

35. $g(x) = |x + 3|$ **36.** $g(x) = -|x| + 2$

37. $g(x) = \sqrt[3]{x}$ **38.** $g(x) = -x^3$

Exercises 39–46: Determine graphically any
    (a) local extrema and (b) absolute extrema.

39. $g(x) = 3x - x^3$ **40.** $g(x) = \dfrac{1}{1 + |x|}$

41. $f(x) = -3x^4 + 8x^3 + 6x^2 - 24x$

42. $f(x) = -x^4 + 4x^3 - 4x^2$

43. $f(x) = 0.5x^4 - 5x^2 + 4.5$

44. $f(x) = 0.01x^5 + 0.02x^4 - 0.35x^3 - 0.36x^2 + 1.8x$

45. $f(x) = \dfrac{8}{1 + x^2}$ **46.** $f(x) = \dfrac{6}{x^2 + 2x + 2}$

Symmetry

Exercises 47–50: Use the graph to determine if f is odd, even, or neither.

47.

48.

49.

50.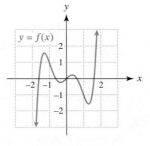

Exercises 51–70: Determine if f is odd, even, or neither.

51. $f(x) = 5x$ **52.** $f(x) = -3x$

53. $f(x) = x + 3$ **54.** $f(x) = 2x - 1$

55. $f(x) = x^2 - 10$ **56.** $f(x) = 8 - 2x^2$

57. $f(x) = x^4 - 6x^2 + 2$ **58.** $f(x) = -x^6 + 5x^2$

59. $f(x) = x^3 - 2x$ **60.** $f(x) = -x^5$

61. $f(x) = x^2 - x^3$ **62.** $f(x) = 3x^3 - 1$

63. $f(x) = \sqrt[3]{x^2}$ **64.** $f(x) = \sqrt{-x}$

65. $f(x) = \sqrt{1 - x^2}$ **66.** $f(x) = \sqrt{x^2}$

67. $f(x) = \dfrac{1}{1 + x^2}$ **68.** $f(x) = \dfrac{1}{x}$

69. $f(x) = |x + 2|$ **70.** $f(x) = \dfrac{1}{x + 1}$

71. The table is a complete representation of f. Decide if f is even, odd, or neither.

x	-100	-10	-1	0	1	10	100
$f(x)$	56	-23	5	0	-5	23	-56

72. The table is a complete representation of f. Decide if f is even, odd, or neither.

x	-5	-3	-1	1	2	3
$f(x)$	-4	-2	1	1	-2	-4

73. Complete the table if f is an even function.

x	-3	-2	-1	0	1	2	3
$f(x)$	21		-25			-12	

74. Complete the table if f is an odd function.

x	-5	-3	-2	0	2	3	5
$f(x)$	13		-5			-1	

75. If the points $(-5, -6)$ and $(-3, 4)$ lie on the graph of an odd function f, then what do $f(5)$ and $f(3)$ equal?

76. If the point $(1 - a, b + 1)$ lies on the graph of an even function f, then what does $f(a - 1)$ equal?

Concepts

77. Sketch a graph of an odd linear function.

78. Sketch a graph of an even linear function.

79. Does there exist a continuous odd function that is always increasing and whose graph passes through the points $(-3, -4)$ and $(2, 5)$? Explain.

80. Is there an even function whose domain is all real numbers and that is always decreasing? Explain.

81. Sketch a graph of a continuous function with an absolute minimum of -3 at $x = -2$ and a local minimum of -1 at $x = 2$.

82. Sketch a graph of a continuous function with no absolute extrema but with a local minimum of -2 at $x = -1$ and a local maximum of 2 at $x = 1$.

83. Sketch a graph of a continuous function that is increasing on $(-\infty, 2]$ and decreasing on $[2, \infty)$. Could this function be quadratic?

84. Sketch a graph of a continuous function with a local maximum of 2 at $x = -1$ and a local maximum of 0 at $x = 1$.

Translations of Graphs

Exercises 85–88: Use the graph of $f(x) = 4x - \frac{1}{3}x^3$ and translations of graphs to sketch the graph of the equation.

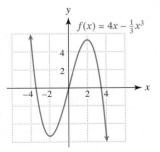

85. $y = f(x + 1)$

86. $y = f(x) - 2$

87. $y = 2f(x)$

88. $y = f\left(\frac{1}{2}x\right)$

89. If the graph of $y = f(x)$ is increasing on $[1, 4]$, then where is the graph of $y = f(x + 1) - 2$ increasing? Where is the graph of $y = -f(x - 2)$ decreasing?

90. If the graph of f is decreasing on $[0, \infty)$, then what can be said about the graph of $y = f(-x) + 1$? the graph of $y = -f(x) - 1$?

Applications

91. *Temperature in the Sun*   The graph shows the temperature readings of a thermometer in the sun (on a partly cloudy day) x hours past noon.

(a) Identify the absolute maximum and minimum. Interpret each.

(b) Identify any local maxima and minima. (Do not consider the endpoints.)

(c) For what x-values was the temperature increasing?

92. *Daytona Beach*   (Refer to the introduction to this section.) The graph at the top of the next page shows the monthly average high temperatures at Daytona Beach.

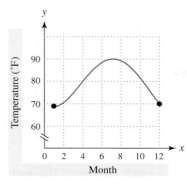

Month

(a) Identify the absolute maximum and minimum.

(b) Identify a local maximum.

(c) For what x-values was the temperature increasing?

93. *Heating Costs* In colder climates the cost for natural gas to heat homes can vary from one month to the next. The polynomial function given by

$$f(x) = -0.1213x^4 + 3.462x^3 - 29.22x^2 + 64.68x + 97.69$$

models the monthly cost in dollars of heating a typical home. The input x represents the month, where $x = 1$ corresponds to January and $x = 12$ to December. (**Source:** Minnegasco.)

(a) Where might the absolute extrema occur for $1 \le x \le 12$?

(b) Graph f in [1, 12, 1] by [0, 150, 10]. Find the absolute extrema and interpret the results.

94. *U.S. Birth Rate* From 1900 to 2005, the birth rate (births per 1000 people) x years after 1900 can be approximated by

$$f(x) = -0.0000285x^3 + 0.0057x^2 - 0.48x + 34.4.$$

(**Source:** National Center for Health Statistics.)

(a) Evaluate $f(65)$ and interpret your result.

(b) If the domain of f is $1900 \le x \le 2005$, identify the absolute extrema. Interpret each.

95. *Energy* The U.S. consumption of energy from 1950 to 1980 can be modeled by

$$f(x) = -0.00113x^3 + 0.0408x^2 - 0.0432x + 7.66,$$

where $x = 0$ corresponds to 1950 and $x = 30$ to 1980. Consumption is measured in quadrillion Btu. (**Source:** Department of Energy.)

(a) Evaluate $f(5)$ and interpret the result.

(b) Graph f in [0, 30, 5] by [6, 16, 1]. Describe the energy usage during this time period.

(c) Approximate the local maximum and interpret it.

96. *Natural Gas* The U.S. consumption of natural gas from 1965 to 1980 can be modeled by

$$f(x) = 0.0001234x^4 - 0.005689x^3 + 0.08792x^2 - 0.5145x + 1.514,$$

where $x = 6$ corresponds to 1966 and $x = 20$ to 1980. Consumption is measured in trillion cubic feet. (**Source:** Department of Energy.)

(a) Evaluate $f(10)$ and interpret the result.

(b) Graph f in [6, 20, 5] by [0.4, 0.8, 0.1]. Describe the energy usage during this time period.

(c) Determine the local extrema and interpret the results.

97. *Average Temperature* The graph approximates the monthly average temperatures in degrees Fahrenheit in Austin, Texas. In this graph x represents the month, where $x = 0$ corresponds to July.

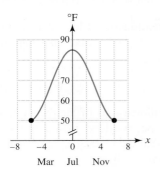

Mar   Jul   Nov

(a) Is this a graph of an odd or even function?

(b) June corresponds to $x = -1$ and August to $x = 1$. The average temperature in June is 83°F. What is the average temperature in August?

(c) March corresponds to $x = -4$ and November to $x = 4$. According to the graph, how do their average temperatures compare?

(d) Interpret what this type of symmetry implies about average temperatures in Austin.

98. *Height of a Projectile* When a projectile is shot into the air, it attains a maximum height and then falls back to the ground. Suppose that $x = 0$ corresponds to the time when the projectile's height is maximum. If air resistance is ignored, its height h above the ground at any time x may be modeled by $h(x) = -16x^2 + h_{max}$, where h_{max} is the projectile's maximum height above the ground. Height is measured in feet and time in seconds. Let $h_{max} = 400$ feet.

(a) Evaluate $h(-2)$ and $h(2)$. Interpret these results.

(b) Evaluate $h(-5)$ and $h(5)$. Interpret these results.

(c) Graph h for $-5 \le x \le 5$. Is h even or odd?

(d) How do $h(x)$ and $h(-x)$ compare when $-5 \le x \le 5$? What does this result indicate?

Writing about Mathematics

99. Explain the difference between a local and an absolute maximum. Are extrema x-values or y-values?

100. Describe ways to determine if a polynomial function is odd, even, or neither. Give examples.

101. If an odd function f has one local maximum of 5 at $x = 3$, then what else can be said about f? Explain.

102. If an even function f has an absolute minimum of -6 at $x = -2$, then what else can be said about f? Explain.

EXTENDED AND DISCOVERY EXERCISES

1. Find the dimensions of the rectangle of maximum area that can be inscribed in a semicircle with radius 3. Assume that the rectangle is positioned as shown in the accompanying figure.

2. *Minimizing Area* A piece of wire 20 inches long is cut into two pieces. One piece is bent into a square and the other is bent into an equilateral triangle, as illustrated.

(a) Write a formula that gives the area A of the two shapes in terms of x.

(b) Find the length of wire (to the nearest tenth of an inch) that should be used for the square if the combined area of the two shapes is to be minimized?

3. *Minimizing Time* A person is in a rowboat 3 miles from the closest point on a straight shoreline, as illustrated in the figure. The person would like to reach a cabin that is 8 miles down the shoreline. The person can row at 4 miles per hour and jog at 7 miles per hour.

(a) How long will it take to reach the cabin if the person rows straight toward shore at point A and then jogs to the cabin?

(b) How long will it take to reach the cabin if the person rows straight to the cabin and does no jogging?

(c) Find the minimum time to reach the cabin.

4.2 Polynomial Functions and Models

- **Understand the graphs of polynomial functions**
- **Evaluate and graph piecewise-defined functions**
- **Use polynomial regression to model data (optional)**

Introduction

The consumption of natural gas by the United States has varied over past decades. As shown in Table 4.2, energy consumption (in quadrillion Btu) increased, decreased, and then increased again. A scatterplot of the data is shown in Figure 4.20, and one possibility for a polynomial modeling function f is shown in Figure 4.21. Notice that f is neither linear nor quadratic. What degree of polynomial might we use to model these data? This question is answered in Example 4.

Table 4.2   Natural Gas Consumption

Year	1960	1970	1980	1990	2000
Consumption	12.4	21.8	20.4	19.3	24.0

Source: Department of Energy.

Figure 4.20 A Scatterplot

Figure 4.21 A Modeling Function

Graphs of Polynomial Functions

In Section 4.1 polynomial functions were defined. Their graphs are continuous and smooth; they have no breaks or sharp edges. The domain of a polynomial function is *all* real numbers. A polynomial function f of degree n can be expressed as

$$f(x) = a_n x^n + \cdots + a_2 x^2 + a_1 x + a_0,$$

where each coefficient a_k is a real number, $a_n \neq 0$, and n is a nonnegative integer. The *leading coefficient* is a_n, where n is the largest exponent of x.

The expression $a_n x^n + \cdots + a_2 x^2 + a_1 x + a_0$ is a **polynomial of degree n** and the equation $a_n x^n + \cdots + a_2 x^2 + a_1 x + a_0 = 0$ is a **polynomial equation of degree n**. Thus f is a *polynomial function*, $f(x)$ is a *polynomial,* and $f(x) = 0$ is a *polynomial equation.* For example, the function f, given by $f(x) = x^3 - 3x^2 + x - 5$, is a polynomial function, $x^3 - 3x^2 + x - 5$ is a polynomial, and $x^3 - 3x^2 + x - 5 = 0$ is a polynomial equation.

A **turning point** occurs whenever the graph of a polynomial function changes from increasing to decreasing or from decreasing to increasing. Turning points are associated with "hills" or "valleys" on a graph. The y-value at a turning point is either a local maximum or a local minimum of the function. In Figure 4.22 the graph has two turning points, $(-2, 8)$ and $(2, -8)$. A local maximum is 8 and a local minimum is -8.

We discuss the graphs of polynomial functions, starting with degree 0 and continuing to degree 5. Look for patterns in the graphs of these polynomial functions.

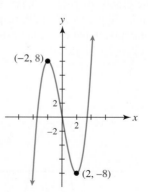

Figure 4.22

Constant Polynomial Functions   If $f(x) = a$ and $a \neq 0$, then f is both a constant function and a polynomial function of degree 0. (If $a = 0$, then f has an **undefined degree**.) Its graph is a horizontal line that does not coincide with the x-axis. Graphs of $f(x) = 4$ and $g(x) = -3$ are shown in Figures 4.23 and 4.24. A graph of a polynomial function of degree 0 has no x-intercepts or turning points.

Figure 4.23 Constant, $a > 0$ **Figure 4.24** Constant, $a < 0$

Linear Polynomial Functions   If $f(x) = ax + b$ and $a \neq 0$, then f is both a linear function and a polynomial function of degree 1. Its graph is a line that is neither horizontal nor vertical. The graphs of $f(x) = 2x - 3$ and $g(x) = -1.6x + 5$ are shown in Figures 4.25 and 4.26. A polynomial function of degree 1 has one x-intercept and no turning points.

The graph of $f(x) = ax + b$ with $a > 0$ is a line sloping upward from left to right. As one traces from left to right, the y-values become larger without a maximum. We say that the **end behavior** of the graph tends to $-\infty$ on the left and ∞ on the right. (Strictly speaking, the graph of a polynomial has infinite length and does not have an end.) More formally, we say that $f(x) \to -\infty$ as $x \to -\infty$ and $f(x) \to \infty$ as $x \to \infty$.

If $a < 0$, then the end behavior is switched. The line slopes downward from left to right. The y-values on the left side of the graph become large positive values without a maximum and the y-values on the right side become negative without a minimum. The end behavior tends to ∞ on the left and $-\infty$ on the right, or $f(x) \to \infty$ as $x \to -\infty$ and $f(x) \to -\infty$ as $x \to \infty$.

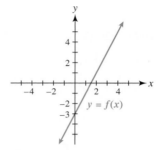

Figure 4.25 Linear, $a > 0$

Figure 4.26 Linear, $a < 0$

Quadratic Polynomial Functions   If $f(x) = ax^2 + bx + c$ and $a \neq 0$, then f is both a quadratic function and a polynomial function of degree 2. Its graph is a parabola that opens either upward ($a > 0$) or downward ($a < 0$). The graphs of $f(x) = 0.5x^2 + 2$, $g(x) = x^2 + 4x + 4$, and $h(x) = -x^2 + 3x + 4$ are shown in Figures 4.27–4.29, respectively. Quadratic functions can have zero, one, or two x-intercepts. A parabola has exactly one turning point, which is also the vertex.

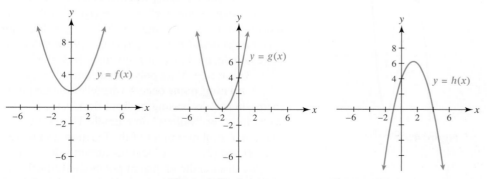

Figure 4.27 Quadratic, $a > 0$ **Figure 4.28** Quadratic, $a > 0$ **Figure 4.29** Quadratic, $a < 0$

If $a > 0$, as in Figure 4.27, then both sides of the graph go up. The end behavior tends to ∞ on both sides, or $f(x) \rightarrow \infty$ as $x \rightarrow \pm\infty$. If $a < 0$, as in Figure 4.29, then the end behavior is switched and tends to $-\infty$ on both sides, or $f(x) \rightarrow -\infty$ as $x \rightarrow \pm\infty$.

Cubic Polynomial Functions   If $f(x) = ax^3 + bx^2 + cx + d$ and $a \neq 0$, then f is both a **cubic function** and a polynomial function of degree 3. The graph of a cubic function can have zero or two turning points. The graph of $f(x) = -x^3 + 4x + 1$ in Figure 4.30 has two turning points, whereas the graph of $g(x) = \frac{1}{4}x^3$ in Figure 4.31 has no turning points.

Figure 4.30 Cubic, $a < 0$ **Figure 4.31** Cubic, $a > 0$ **Figure 4.32** Cubic, $a > 0$

If $a > 0$, the graph of a cubic function falls to the left and rises to the right, as in Figure 4.31. If $a < 0$, its graph rises to the left and falls to the right, as in Figure 4.30. The end behavior of a cubic function is similar to that of a linear function, tending to ∞ on one side and $-\infty$ on the other. Therefore its graph must cross the x-axis at least once. A cubic function can have up to three x-intercepts. The graph of $h(x) = x^3 - x^2 - x + 1$ in Figure 4.32 has two x-intercepts.

Quartic Polynomial Functions   If $f(x) = ax^4 + bx^3 + cx^2 + dx + e$ and $a \neq 0$, then f is both a **quartic function** and a polynomial function of degree 4. The graph of a quartic function can have up to four x-intercepts and three turning points; the graph of $f(x) = \frac{1}{2}x^4 + \frac{7}{2}x^3 + 7x^2 + 4x$ in Figure 4.33 is an example. The graph in Figure 4.34 of $g(x) = -x^4 + 2x^3$ has one turning point and two x-intercepts, and the graph in Figure 4.35 of $h(x) = -x^4 + x^3 + 3x^2 - x - 2$ has three turning points and three x-intercepts.

If $a > 0$, then both ends of the graph of a quartic function go up, as in Figure 4.33. If $a < 0$, then both ends of its graph go down, as in Figures 4.34 and 4.35. The end behaviors of quartic and quadratic functions are similar.

CLASS DISCUSSION

Can a quartic function have both an absolute maximum and an absolute minimum? Explain.

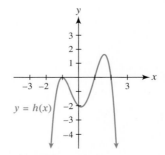

Figure 4.33 Quartic, $a > 0$ **Figure 4.34** Quartic, $a < 0$ **Figure 4.35** Quartic, $a < 0$

Quintic Polynomial Functions If $f(x) = ax^5 + bx^4 + cx^3 + dx^2 + ex + k$ and $a \neq 0$, then f is both a **quintic function** and a polynomial function of degree 5. The graph of a quintic function may have up to five x-intercepts and four turning points. An example is shown in Figure 4.36, given by $f(x) = x^5 - 5x^3 + 4x$. Other quintic functions are shown in Figures 4.37 and 4.38. They are defined by

$$g(x) = \frac{1}{5}x^5 - 3 \quad \text{and} \quad h(x) = -\frac{1}{2}x^5 - \frac{1}{2}x^4 + \frac{5}{2}x^3 + \frac{1}{2}x^2 - 4x + 2.$$

The function g has one x-intercept and no turning points. The graph of h appears to have two x-intercepts and two turning points. Notice that the end behavior of a quintic function is similar to that of linear and cubic functions.

Figure 4.36 Quintic, $a > 0$ **Figure 4.37** Quintic, $a > 0$ **Figure 4.38** Quintic, $a < 0$

The end behavior of polynomial functions is summarized in the following.

End Behavior of Polynomial Functions

Let f be a polynomial function with leading coefficient a and degree n.

1. $n \geq 2$ is even.
   $a > 0$ implies that the graph of f rises both to the left and to the right. That is, $f(x) \to \infty$ as $x \to \pm\infty$.
   $a < 0$ implies that the graph of f falls both to the left and to the right. That is, $f(x) \to -\infty$ as $x \to \pm\infty$.

2. $n \geq 1$ is odd.
   $a > 0$ implies that the graph of f falls to the left and rises to the right. That is, $f(x) \to -\infty$ as $x \to -\infty$ and $f(x) \to \infty$ as $x \to \infty$.
   $a < 0$ implies that the graph of f rises to the left and falls to the right. That is, $f(x) \to \infty$ as $x \to -\infty$ and $f(x) \to -\infty$ as $x \to \infty$.

The maximum numbers of x-intercepts and turning points on the graph of a polynomial function of degree n can be summarized as follows.

Degree, x-intercepts, and Turning Points

The graph of a polynomial function of degree n, with $n \geq 1$, has at most n x-intercepts and at most $n - 1$ turning points.

EXAMPLE 1 Analyzing the graph of a polynomial function

Figure 4.39 shows the graph of a polynomial function f.
(a) How many turning points and x-intercepts are there?
(b) Is the leading coefficient a positive or negative? Is the degree odd or even?
(c) Determine the minimum possible degree of f.

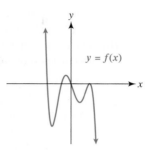

Figure 4.39

SOLUTION

(a) There are four turning points corresponding to the two "hills" and two "valleys." There appear to be four x-intercepts.
(b) The left side of the graph rises and the right side falls. Therefore $a < 0$ and the polynomial function has odd degree.
(c) The graph has four turning points. A polynomial of degree n can have at most $n - 1$ turning points. Therefore f must be at least degree 5. *Now Try Exercise 7* ◀

NOTE More examples of graphs of polynomials are found in the "Putting It All Together" for this section.

EXAMPLE 2 Analyzing the graph of a polynomial function

Graph $f(x) = x^3 - 2x^2 - 5x + 6$, and then complete the following.
(a) Identify the x-intercepts.
(b) Approximate the coordinates of any turning points to the nearest hundredth.
(c) Use the turning points to approximate any local extrema.

SOLUTION

(a) A graph of f, shown in Figure 4.40, appears to intersect the x-axis at the points $(-2, 0)$, $(1, 0)$, and $(3, 0)$. Therefore the x-intercepts are -2, 1, and 3.
(b) There are two turning points. From Figures 4.41 and 4.42 their coordinates are approximately $(-0.79, 8.21)$ and $(2.12, -4.06)$.
(c) There is a local maximum of about 8.21 and a local minimum of about -4.06.

Calculator Help

To find a minimum or a maximum point on a graph, see Appendix A (page AP-10).

$[-10, 10, 1]$ by $[-10, 10, 1]$

Figure 4.40

$[-10, 10, 1]$ by $[-10, 10, 1]$

Maximum
X=⁻.7862976 Y=8.2088207

Figure 4.41

$[-10, 10, 1]$ by $[-10, 10, 1]$

Minimum
X=2.1196339 Y=⁻4.060673

Figure 4.42

Now Try Exercise 25 ◀

Figure 4.43 Cubic with $a < 0$

EXAMPLE 3   Analyzing the end behavior of a graph

Let $f(x) = 2 + 3x - 3x^2 - 2x^3$.
(a) Give the degree and leading coefficient.
(b) State the end behavior of the graph of f.

SOLUTION
(a) Rewriting gives $f(x) = -2x^3 - 3x^2 + 3x + 2$. The term with highest degree is $-2x^3$, so the degree is 3 and the leading coefficient is -2.
(b) The degree of $f(x)$ is odd, and the leading coefficient is negative. Therefore the graph of f rises to the left and falls to the right. More formally,

$$f(x) \to \infty \text{ as } x \to -\infty \qquad \text{and} \qquad f(x) \to -\infty \text{ as } x \to \infty.$$

This conclusion is supported by Figure 4.43.   *Now Try Exercise 33* ◀

An Application   In the next example, we analyze the data presented in the introduction to this section.

EXAMPLE 4   Modeling natural gas consumption

Figure 4.20, which shows natural gas consumption from 1960 to 2000, is repeated in the margin.
(a) Could a linear or quadratic function model the data?
(b) What minimum degree polynomial might be appropriate to model the data?
(c) Should the leading coefficient a be positive or negative?

SOLUTION
(a) The data clearly do not lie on a line, so a linear function is not appropriate. Because natural gas consumption increases, decreases, and then increases, a quadratic function would not be a good choice either. The data are not ∪-shaped or ∩-shaped.
(b) Because the data increase, decrease, and then increase, a polynomial with at least two turning points would be appropriate. A cubic, or degree 3, polynomial is a possibility for a modeling function.
(c) The leading coefficient a should be positive because the data fall to the left and rise to the right. (If the data decreased, increased, and then decreased, a negative leading coefficient would be appropriate.)   *Now Try Exercises 39(a), (b), and (c)* ◀

Concavity (Optional)   Graphs of polynomial functions with degree 2 or greater are curves. **Concavity** is a mathematical description of how a curve bends. A line exhibits no, or zero, concavity because it is straight. A parabola that opens upward is said to be **concave upward** everywhere on its domain. See Figure 4.44. A parabola that opens downward is said to be **concave downward** everywhere on its domain. See Figure 4.45. A graph of a higher degree polynomial can be concave upward on one interval of its domain and concave downward on a different interval of its domain. See Figure 4.46, where the graph of $f(x) = 4x - x^3$ is concave upward on the interval $(-\infty, 0)$, shown in blue, and concave downward on $(0, \infty)$, shown in red. Concavity is usually defined for *open* intervals. Determining the exact x-value where a graph switches from concave upward to downward or vice versa can be difficult to do visually and often requires techniques learned in calculus. (See Extended and Discovery Exercises 2–7 for this section.)

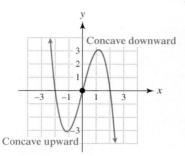

Figure 4.44   **Figure 4.45**   **Figure 4.46**

Piecewise-Defined Polynomial Functions

In Section 2.1 piecewise-defined functions were discussed. If each piece is a polynomial, then the function is a **piecewise-defined polynomial function** or **piecewise-polynomial function**. An example is given by $f(x)$.

$$f(x) = \begin{cases} x^3 & \text{if } x < 1 \\ x^2 - 1 & \text{if } x \geq 1 \end{cases}$$

One way to graph f is to first graph $y = x^3$ and $y = x^2 - 1$, as shown in Figures 4.47 and 4.48. Then the graph of f is found by using the portion of $y = x^3$ for $x < 1$ and the portion of $y = x^2 - 1$ for $x \geq 1$, as illustrated in Figure 4.49. At $x = 1$ there is a break in the graph, where the graph of f is discontinuous.

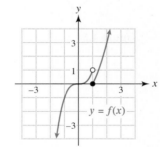

Figure 4.47   **Figure 4.48**   **Figure 4.49**

EXAMPLE 5   Evaluating a piecewise-defined polynomial function

Evaluate $f(x)$ at $x = -3, -2, 1,$ and 2.

$$f(x) = \begin{cases} x^2 - x & \text{if } -5 \leq x < -2 \\ -x^3 & \text{if } -2 \leq x < 2 \\ 4 - 4x & \text{if } 2 \leq x \leq 5 \end{cases}$$

SOLUTION To evaluate $f(-3)$ we use the formula $f(x) = x^2 - x$, because -3 is in the interval $-5 \leq x < -2$.

$$f(-3) = (-3)^2 - (-3) = 12$$

To evaluate $f(-2)$ we use $f(x) = -x^3$, because -2 is in the interval $-2 \le x < 2$.

$$f(-2) = -(-2)^3 = -(-8) = 8$$

Similarly, $f(1) = -1^3 = -1$ and $f(2) = 4 - 4(2) = -4$. **Now Try Exercise 73** ◀

EXAMPLE 6 Graphing a piecewise-defined function

Complete the following.
(a) Sketch a graph of f.
(b) Determine if f is continuous on its domain.
(c) Solve the equation $f(x) = 1$.

$$f(x) = \begin{cases} \frac{1}{2}x^2 - 2 & \text{if } -4 \le x \le 0 \\ 2x - 2 & \text{if } 0 < x < 2 \\ 2 & \text{if } 2 \le x \le 4 \end{cases}$$

SOLUTION

(a) For the first piece, graph the parabola determined by $y = \frac{1}{2}x^2 - 2$ on the interval $-4 \le x \le 0$. Place dots at the endpoints, which are $(-4, 6)$ and $(0, -2)$. See Figure 4.50. For the second piece, graph the line determined by $y = 2x - 2$. Place open circles at $(0, -2)$ and $(2, 2)$. Note that the left endpoint of the middle piece coincides with the right endpoint of the first piece. Finally, graph the horizontal line $y = 2$ from the points $(2, 2)$ to $(4, 2)$. Note that the left endpoint of the third piece coincides with the open circle on the right for the middle piece.

(b) The domain of f is $-4 \le x \le 4$. Because there are no breaks in the graph of f on its domain, the graph of f is continuous.

(c) The horizontal line $y = 1$ intersects the graph of $y = f(x)$ at two points, as shown in Figure 4.51. The x-coordinates of these two points of intersection can be found by solving the equations

$$\frac{1}{2}x^2 - 2 = 1 \quad \text{and} \quad 2x - 2 = 1.$$

The solutions are $-\sqrt{6} \approx -2.45$ and $\frac{3}{2}$. **Now Try Exercise 79** ◀

Figure 4.50

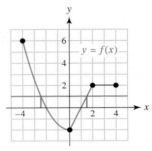

Figure 4.51

Polynomial Regression (Optional)

We now have the mathematical understanding to model the data presented in the introduction to this section. The polynomial modeling function f (shown in Figure 4.21 and repeated in the margin) falls to the left and rises to the right, so it has odd degree and the leading coefficient is positive. Since the graph of f has two turning points, it must be at least degree 3. A cubic polynomial $f(x)$ is a possible choice, where

$$f(x) = ax^3 + bx^2 + cx + d.$$

Using trial and error would be a difficult way to find values for a, b, c, and d. Instead, we can use least-squares regression, which was also discussed in Sections 2.1 and 3.1, for linear and quadratic functions. The next example illustrates *cubic regression*.

EXAMPLE 7 Determining a cubic modeling function

The data in Table 4.2 (repeated in the margin) list natural gas consumption.

Year	Consumption
1960	12.4
1970	21.8
1980	20.4
1990	19.3
2000	24.0

(a) Find a polynomial function of degree 3 that models the data.
(b) Graph f and the data together.
(c) Estimate natural gas consumption in 1974 and in 2007. Compare these estimates to the actual values of 21.2 and 23.0 quadrillion Btu, respectively.
(d) Did your estimates in part (c) involve interpolation or extrapolation? Is there a problem with using higher degree polynomials ($n \geq 3$) for extrapolation? Explain.

SOLUTION

(a) Enter the five data points (1960, 12.4), (1970, 21.8), (1980, 20.4), (1990, 19.3), and (2000, 24.0) into your calculator. Then select cubic regression, as shown in Figure 4.52. The equation for $f(x)$ is shown in Figure 4.53.
(b) A graph of f and a scatterplot of the data are shown in Figure 4.54.
(c) $f(1974) \approx 21.9$ and $f(2007) \approx 36.3$; the 1974 estimate is reasonably close to 21.2, whereas the 2007 estimate is not close to 23.0.
(d) The 1974 estimate uses interpolation, and the 2007 estimate uses extrapolation. Because the end behavior of a higher degree polynomial rapidly tends to either ∞ or $-\infty$, extrapolation-based estimates are usually inaccurate.

Calculator Help

To find an equation of least-squares fit, see Appendix A (page AP-11). To copy a regression equation into Y_1, see Appendix A (page AP-13.)

```
EDIT  CALC  TESTS
1:1-Var Stats
2:2-Var Stats
3:Med-Med
4:LinReg(ax+b)
5:QuadReg
6:CubicReg
7↓QuartReg
```

Figure 4.52

```
CubicReg
 y=ax³+bx²+cx+d
 a=.0013833333
 b=⁻8.2235
 c=16295.13667
 d=⁻10762915.92
 R²=.9941642351
```

Figure 4.53

[1955, 2005, 5] by [10, 25, 5]

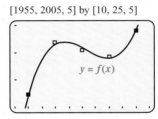

$y = f(x)$

Figure 4.54

Now Try Exercise 89 ◄

4.2 Putting It All Together	**H**igher degree polynomials generally have more complicated graphs. Each additional degree allows the graph to have possibly one more turning point and one more x-intercept. The graph of a polynomial function is continuous and smooth; it has no breaks or sharp edges. Its domain includes all real numbers. The end behavior of a polynomial always tends to either ∞ or $-\infty$. End behavior describes what happens to the y-values as $|x|$ becomes large.

A piecewise-defined function occurs when a function is defined by using two or more formulas for different intervals of the domain. If each formula is a polynomial, the function is called a piecewise-polynomial function.

The following summary shows important concepts regarding polynomial functions.

Function Type	Characteristics	Example Graphs
Constant (degree 0)	No x-intercepts and no turning points $f(x) = a,\ a \neq 0$	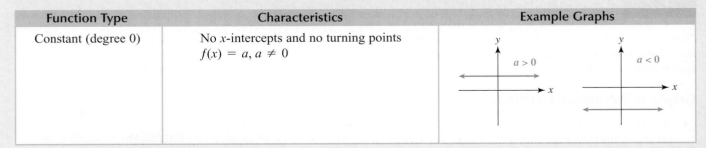

continued on next page

continued from previous page

Function Type	Characteristics	Example Graphs
Linear (degree 1)	One x-intercept and no turning points $f(x) = ax + b,\ a \neq 0$	
Quadratic (degree 2)	At most two x-intercepts and exactly one turning point $f(x) = ax^2 + bx + c,\ a \neq 0$	
Cubic (degree 3)	At most three x-intercepts and up to two turning points $f(x) = ax^3 + bx^2 + cx + d,\ a \neq 0$	
Quartic (degree 4)	At most four x-intercepts and up to three turning points $f(x) = ax^4 + bx^3 + cx^2 + dx + e,\ a \neq 0$	
Quintic (degree 5)	At most five x-intercepts and up to four turning points $f(x) = ax^5 + bx^4 + cx^3 + dx^2 + ex + k,$ $a \neq 0$	

4.2 Exercises

Note: Many of the answers in this section involve estimations. Your answers may vary slightly, particularly when you are reading a graph.

Graphs of Polynomial Functions

1. A runner is working out on a straight track. The graph at the top of the next page shows the runner's distance y in hundreds of feet from the starting line after t minutes.

(a) Estimate the turning points.

(b) Interpret each turning point.

Start

2. A stone is thrown into the air. Its height y in feet after t seconds is shown in the graph. Use the graph to complete the following.
   (a) Estimate the turning point.

   (b) Interpret this point.

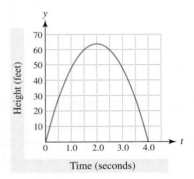

Exercises 3–12: Use the graph of the polynomial function f to complete the following. Let a be the leading coefficient of the polynomial $f(x)$.

   (a) Determine the number of turning points and esti-mate any x-intercepts.
   (b) State whether $a > 0$ or $a < 0$.
   (c) Determine the minimum degree of f.

3.

4.

5.

6.

7.

8.

9.

10.

11.

12.

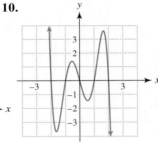

Exercises 13–18: Complete the following without a calculator.
   (a) Match the equation with its graph (a–f).
   (b) Identify the turning points.
   (c) Estimate the x-intercepts.
   (d) Estimate any local extrema.
   (e) Estimate any absolute extrema.

13. $f(x) = 1 - 2x + x^2$

14. $f(x) = 3x - x^3$

15. $f(x) = x^3 + 3x^2 - 9x$

16. $f(x) = x^4 - 8x^2$

17. $f(x) = 8x^2 - x^4$

18. $f(x) = x^5 + \frac{5}{2}x^4 - \frac{5}{3}x^3 - 5x^2$

a.

b.

c.

d.

e.

f.
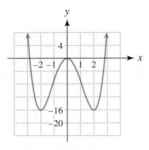

📱 *Exercises 19–26: Complete the following.*
   (a) Graph $y = f(x)$ in the standard viewing rectangle.
   (b) Approximate the coordinates of each turning point.
   (c) Estimate any local extrema.

19. $f(x) = \frac{1}{9}x^3 - 3x$

20. $f(x) = x^2 - 4x - 3$

21. $f(x) = 0.025x^4 - 0.45x^2 - 5$

22. $f(x) = -\frac{1}{8}x^4 + \frac{1}{3}x^3 + \frac{5}{4}x^2 - 3x + 3$

23. $f(x) = 1 - 2x + 3x^2$

24. $f(x) = 4x - \frac{1}{3}x^3$

25. $f(x) = \frac{1}{3}x^3 + \frac{1}{2}x^2 - 2x$

26. $f(x) = \frac{1}{4}x^4 + \frac{2}{3}x^3 - \frac{1}{2}x^2 - 2x + 1$

Exercises 27–38: Complete the following.
   (a) State the degree and leading coefficient of f.
   (b) State the end behavior of the graph of f.

27. $f(x) = -2x + 3$ **28.** $f(x) = \frac{2}{3}x - 2$

29. $f(x) = x^2 + 4x$ **30.** $f(x) = 5 - \frac{1}{2}x^2$

31. $f(x) = -2x^3$ **32.** $f(x) = 4x - \frac{1}{3}x^3$

33. $f(x) = x^2 - x^3 - 4$

34. $f(x) = x^4 - 4x^3 + 3x^2 - 3$

35. $f(x) = 0.1x^5 - 2x^2 - 3x + 4$

36. $f(x) = 3x^3 - 2 - x^4$

37. $f(x) = 4 + 2x - \frac{1}{2}x^2$

38. $f(x) = -0.2x^5 + 4x^2 - 3$

Modeling Data with Polynomials

Exercises 39–44: The data are modeled exactly by a linear, quadratic, cubic, or quartic function f with leading coefficient a. All zeros of f are real numbers located in the interval $[-3, 3]$.
   (a) Make a line graph of the data.
   (b) State the minimum degree of f.
   (c) Is $a > 0$ or is $a < 0$?
 📱 *(d) Find a formula for $f(x)$.*

39.

x	−3	−2	−1	0	1	2	3
$f(x)$	3	−8	−7	0	7	8	−3

40.

x	−3	−2	−1	0	1	2	3
$f(x)$	11	9	7	5	3	1	−1

41.

x	−3	−2	−1	0	1	2	3
$f(x)$	14	7	2	−1	−2	−1	2

42.

x	−3	−2	−1	0	1	2	3
$f(x)$	−13	−6	−1	2	3	2	−1

43.

x	−3	−2	−1	0	1	2	3
f(x)	−55	−5	1	−1	1	−5	−55

44.

x	−3	−2	−1	0	1	2	3
f(x)	−15	0	3	0	−3	0	15

Sketching Graphs of Polynomials

Exercises 45–56: If possible, sketch a graph of a polynomial that satisfies the conditions. Let a be the leading coefficient.

45. Degree 3 with three real zeros and $a > 0$

46. Degree 4 with four real zeros and $a < 0$

47. Linear with $a < 0$

48. Cubic with one real zero and $a > 0$

49. Degree 4 and an even function with four turning points

50. Degree 5 and symmetric with respect to the y-axis

51. Degree 3 and an odd function with no x-intercepts

52. Degree 6 and an odd function with five turning points

53. Degree 3 with turning points $(-1, 2)$ and $\left(1, \frac{2}{3}\right)$

54. Degree 4 with turning points $(-1, -1), (0, 0)$, and $(1, -1)$

55. Degree 2 with turning point $(-1, 2)$, passing through $(-3, 4)$ and $(1, 4)$

56. Degree 5 and an odd function with five x-intercepts and a negative leading coefficient.

Dominant Term of a Polynomial

Exercises 57 and 58: Graph the functions f, g, and h in the same viewing rectangle. What happens to their graphs as the size of the viewing rectangle increases? Explain why the term of highest degree in a polynomial is sometimes called the **dominant term**.

57. $f(x) = 2x^4, g(x) = 2x^4 - 5x^2 + 1$, and $h(x) = 2x^4 + 3x^2 - x - 2$
(a) $[-4, 4, 1]$ by $[-4, 4, 1]$

(b) $[-10, 10, 1]$ by $[-100, 100, 10]$

(c) $[-100, 100, 10]$ by $[-10^6, 10^6, 10^5]$

58. $f(x) = -x^3, g(x) = -x^3 + x^2 + 2$, and $h(x) = -x^3 - 2x^2 + x - 1$

(a) $[-4, 4, 1]$ by $[-4, 4, 1]$

(b) $[-10, 10, 1]$ by $[-100, 100, 10]$

(c) $[-100, 100, 10]$ by $[-10^5, 10^5, 10^4]$

Average Rates of Change

59. Compare the average rates of change from 0 to $\frac{1}{2}$ for $f(x) = x, g(x) = x^2$, and $h(x) = x^3$.

60. Compare the average rates of change from 1 to $\frac{3}{2}$ for $f(x) = x, g(x) = x^2$, and $h(x) = x^3$.

Exercises 61–64: Calculate the average rate of change of f on each interval. What happens to this average rate of change as the interval decreases in length?
    (a) [1.9, 2.1]   *(b)* [1.99, 2.01]   *(c)* [1.999, 2.001]

61. $f(x) = x^3$ **62.** $f(x) = 4x - \frac{1}{3}x^3$

63. $f(x) = \frac{1}{4}x^4 - \frac{1}{3}x^3$ **64.** $f(x) = 4x^2 - \frac{1}{2}x^4$

Exercises 65–68: Find the difference quotient of g.

65. $g(x) = 3x^3$ **66.** $g(x) = -2x^3$

67. $g(x) = 1 + x - x^3$ **68.** $g(x) = \frac{1}{2}x^3 - 2x$

Piecewise-Defined Functions

Exercises 69–76: Evaluate f(x) at the given values of x.

69. $x = -2$ and 1   **70.** $x = -1, 0$, and 3

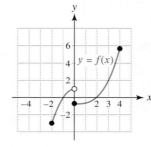

71. $x = -1, 1$, and 2   **72.** $x = -2, 0$, and 2

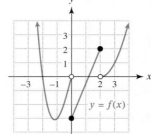

73. $x = -3$, 1, and 4

$$f(x) = \begin{cases} x^3 - 4x^2 & \text{if} & x \le -3 \\ 3x^2 & \text{if} -3 < x < 4 \\ x^3 - 54 & \text{if} & x \ge 4 \end{cases}$$

74. $x = -4$, 0, and 4

$$f(x) = \begin{cases} -4x & \text{if} & x \le -4 \\ x^3 + 2 & \text{if} -4 < x \le 2 \\ 4 - x^2 & \text{if} & x > 2 \end{cases}$$

75. $x = -2$, 1, and 2

$$f(x) = \begin{cases} x^2 + 2x + 6 & \text{if} -5 \le x < 0 \\ x + 6 & \text{if} \;\; 0 \le x < 2 \\ x^3 + 1 & \text{if} \;\; 2 \le x \le 5 \end{cases}$$

76. $x = 1975$, 1980, and 1998

$$f(x) = \begin{cases} 0.2(x - 1970)^3 + 60 & \text{if } 1970 \le x < 1980 \\ 190 - (x - 1980)^2 & \text{if } 1980 \le x < 1990 \\ 2(x - 1990) + 100 & \text{if } 1990 \le x \le 2000 \end{cases}$$

Exercises 77–82: Complete the following.
  (a) Sketch a graph of f.
  (b) Determine if f is continuous on its domain.
  (c) Solve f(x) = 0.

77. $f(x) = \begin{cases} 4 - x^2 & \text{if} -3 \le x \le 0 \\ x^2 - 4 & \text{if} \;\; 0 < x \le 3 \end{cases}$

78. $f(x) = \begin{cases} x^2 & \text{if} -2 \le x < 0 \\ x + 1 & \text{if} \;\; 0 \le x \le 2 \end{cases}$

79. $f(x) = \begin{cases} 2x & \text{if} -5 \le x < -1 \\ -2 & \text{if} -1 \le x < 0 \\ x^2 - 2 & \text{if} \;\; 0 \le x \le 2 \end{cases}$

80. $f(x) = \begin{cases} 0.5x^2 & \text{if} -4 \le x \le -2 \\ x & \text{if} -2 < x < 2 \\ x^2 - 4 & \text{if} \;\; 2 \le x \le 4 \end{cases}$

81. $f(x) = \begin{cases} x^3 + 3 & \text{if} -2 \le x \le 0 \\ x + 3 & \text{if} \;\; 0 < x < 1 \\ 4 + x - x^2 & \text{if} \;\; 1 \le x \le 3 \end{cases}$

82. $f(x) = \begin{cases} -2x & \text{if} -3 \le x < -1 \\ x^2 + 1 & \text{if} -1 \le x \le 2 \\ \frac{1}{2}x^3 + 1 & \text{if} \;\; 2 < x \le 3 \end{cases}$

Applications

83. *Electronics*   The **Heaviside function** H, used in the study of electrical circuits, is defined by

$$H(t) = \begin{cases} 0 & \text{if } t < 0 \\ 1 & \text{if } t \ge 0. \end{cases}$$

   (a) Evaluate $H(-2)$, $H(0)$, and $H(3.5)$.

   (b) Graph $y = H(t)$.

84. *A Strange Graph*   The following definition is discussed in advanced mathematics courses.

$$f(x) = \begin{cases} 0 & \text{if } x \text{ is a rational number} \\ 1 & \text{if } x \text{ is an irrational number} \end{cases}$$

   (a) Evaluate $f\left(-\frac{3}{4}\right)$, $f(-\sqrt{2})$, and $f(\pi)$.

   (b) Is f a function? Explain.

   (c) Discuss the difficulty with graphing $y = f(x)$.

85. *Modeling Temperature*   In the figure the monthly average temperature in degrees Fahrenheit from January to December in Minneapolis is modeled by a polynomial function f, where $x = 1$ corresponds to January and $x = 12$ to December.   (**Source:** A. Miller and J. Thompson, *Elements of Meteorology.*)

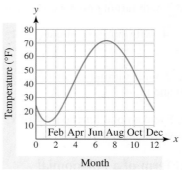

   (a) Estimate the turning points.

   (b) Interpret each turning point.

86. *Natural Gas Consumption*   Refer to Figure 4.21 and Table 4.2 in the introduction to this section.
   (a) Solve the equation $f(x) = 20$ graphically. Interpret the solution set.

   (b) Calculate the average rate of change in natural gas consumption from 1970 to 1980. Interpret the result.

87. *Endangered Species*   The total number y of endangered and threatened species is given in the table at the top of the next column for various years x.

x	1980	1985	1990	1995	2000	2005
y	786	941	1181	1599	1741	1827

Source: Fish and Wildlife Services.

(a) Use regression to find a polynomial function *f* that models the data.

(b) Use *f* to estimate the number of endangered and threatened species in 1998. Did your answer involve interpolation or extrapolation?

(c) Estimate the year when this number reached 1600.

88. *Marijuana Use*   The table lists the percentage *y* of high school seniors who had used marijuana within the previous month in the United States for various years *x*. In this table *x* = 0 corresponds to 1975 and *x* = 20 to 1995.

x (yr)	0	3	5	10	15	20
y (%)	27	37	33	25	14	21

Source: Health and Human Services Department.

(a) Use least-squares regression to find a polynomial function *f* that models the data.

(b) Use *f* to estimate marijuana use in 1997. Compare your answer to the actual value of 23%.

(c) Did your estimate in part (b) involve interpolation or extrapolation?

89. *Aging in America*   The table lists the number *N* (in thousands) of Americans over 100 years old for selected years *x*.

x	1960	1970	1980	1990	2000
N	3	5	15	37	75

(a) Use regression to find a polynomial of degree 3 that models the data.

(b) Graph *f* and the data.

(c) Estimate *N* in 1994 and in 2010. Compare these answers to the actual values of 50 thousand and 129 thousand, respectively.

(d) Did your estimates in part (c) involve interpolation or extrapolation?

90. *Modeling*   An object is lifted rapidly into the air at a constant speed and then dropped. Its height *h* in feet after *x* seconds is listed in the table.

x (sec)	0	1	2	3	4	5	6	7
h (ft)	0	36	72	108	144	128	80	0

(a) Make a line graph of the data. At what time does it appear that the object was dropped?

(b) Identify the time interval when the height could be modeled by a linear function. When could it be modeled by a nonlinear function?

(c) Determine values for the constants *m*, *a*, and *b* so that *f* models the data.

$$f(x) = \begin{cases} mx & \text{if } 0 \le x \le 4 \\ a(x-4)^2 + b & \text{if } 4 < x \le 7 \end{cases}$$

(d) Solve *f*(*x*) = 100 and interpret your answer.

91. *Modeling Water Flow*   A cylindrical container has a height of 16 centimeters. Water entered the container at a constant rate until it was completely filled. Then water was allowed to leak out through a small hole in the bottom. The height of the water in the container was recorded every half minute over a 5-minute period.

Time (min)	0	0.5	1.0	1.5	2.0	2.5
Height (cm)	0	4	8	12	16	11.6

Time (min)	3.0	3.5	4.0	4.5	5.0
Height (cm)	8.1	5.3	3.1	1.4	0.5

16 cm

(a) Plot the data.

(b) Find a piecewise-defined function that models the data. (*Hint:* Use regression.)

(c) Approximate the water level after 1.25 minutes and after 3.2 minutes.

(d) Estimate the time when water was flowing out of the tank and the water level was 5 centimeters.

92. *Modeling* A water tank is filled with a hose and then drained. The table shows the number of gallons y in the tank after t minutes.

t (min)	0	1	2	3	4	5	6	7
y (gal)	0	9	18	27	36	16	4	0

The following function f models the data in the table.

$$f(t) = \begin{cases} 9t & \text{if } 0 \leq t \leq 4 \\ 4t^2 - 56t + 196 & \text{if } 4 < t \leq 7 \end{cases}$$

Solve the equation $f(t) = 12$ and interpret the results.

Writing about Mathematics

Exercises 93–96: Discuss possible local or absolute extrema on the graph of f. Assume that $a > 0$.

93. $f(x) = ax + b$ **94.** $f(x) = ax^2 + bx + c$

95. $f(x) = ax^3 + bx^2 + cx + d$

96. $f(x) = a|x|$

EXTENDED AND DISCOVERY EXERCISES

1. *Torricelli's Law* A cylindrical tank contains 500 gallons of water. A plug is pulled from the bottom of the tank, and it takes 10 minutes to drain the tank. The amount A of water in gallons remaining in the tank after t minutes is approximated by

$$A(t) = 500\left(1 - \frac{t}{10}\right)^2.$$

(a) What is a reasonable domain for A?

(b) Evaluate $A(1)$ and interpret the result.

(c) What are the degree and leading coefficient of $A(t)$?

(d) Has half the water drained from the tank after 5 minutes? Does this agree with your intuition? Explain.

Exercises 2–7: Concavity Estimate the intervals where the graph of f is concave up and where the graph is concave downward. Use interval notation.

2.

3.

4.

5.

6.

7.
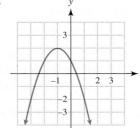

CHECKING BASIC CONCEPTS FOR SECTIONS 4.1 AND 4.2

1. Use the graph of f to complete the following.

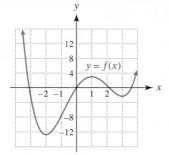

(a) Determine where f is increasing or decreasing.

(b) Identify any local extrema.

(c) Identify any absolute extrema.

(d) Approximate the x-intercepts and zeros of f. Then solve $f(x) = 0$. How are the x-intercepts, zeros, and solutions to $f(x) = 0$ related?

2. Use the graph to complete the following.

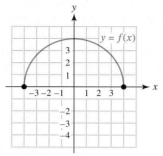

(a) Evaluate $f(-4)$, $f(0)$, and $f(4)$.

(b) What type of symmetry does the graph of f exhibit?

(c) Is f an odd function or an even function? Why?

(d) Find the domain and range of f.

3. If possible, sketch a graph of a cubic polynomial with a negative leading coefficient that satisfies each of the following conditions.

(a) Zero x-intercepts   (b) One x-intercept

(c) Two x-intercepts   (d) Four x-intercepts

4. Plot the data in the table.

x	-3.2	-2	0	2	3.2
y	-11	15	-10	15	-11

(a) What is the minimum degree of the polynomial function f that would be needed to model these data? Explain.

(b) Should function f be odd, even, or neither? Explain.

(c) Should the leading coefficient of f be positive or negative? Explain.

5. Use least-squares regression to find a polynomial that models the data in Exercise 4.

4.3 Division of Polynomials

- **Divide polynomials by monomials**
- **Divide polynomials by polynomials**
- **Apply the division algorithm**
- **Learn synthetic division**
- **Understand the remainder theorem**

Introduction

The area A of a rectangle with length L and width W is calculated by $A = LW$. If we are given the area and the width of a rectangle, we can find the length L by solving $A = LW$ for L to obtain $L = \frac{A}{W}$. For example, if the area is 48 square feet and the width is 6 feet, then the length of the rectangle equals $L = \frac{48}{6} = 8$ feet. Now consider the more general situation shown in Figure 4.55.

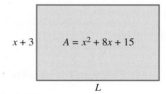

Figure 4.55

The area A is $x^2 + 8x + 15$ and the width W is $x + 3$, so we can find an expression for L in the same way by calculating

$$L = \frac{A}{W} = \frac{x^2 + 8x + 15}{x + 3}.$$

In this case, the calculation of L involves division of polynomials. (See Example 5.) This section discusses basic concepts related to division of polynomials.

Division by Monomials

Adding (or subtracting) fractions having like denominators is straightforward. For example, $\frac{3}{17} + \frac{7}{17} = \frac{3+7}{17}$, and so $\frac{3+7}{17} = \frac{3}{17} + \frac{7}{17}$. Note that when we reverse the process, the denominator of 17 is divided into each term in the numerator. (Terms are separated by addition or subtraction signs.) By reversing the process, we can sometimes simplify expressions.

$$\frac{3x^4 + 5x^3}{5x^3} = \frac{3x^4}{5x^3} + \frac{5x^3}{5x^3} = \frac{3}{5}x + 1 \qquad \text{Subtract exponents: } \frac{3x^4}{5x^3} = \frac{3}{5}x.$$

This process is used in the next example.

EXAMPLE 1   Dividing by a monomial

Divide $6x^3 - 3x^2 + 2$ by $2x^2$.

SOLUTION

Getting Started Remember to divide $2x^2$ into *every* term of $6x^3 - 3x^2 + 2$. ▶

Write the problem as $\frac{6x^3 - 3x^2 + 2}{2x^2}$. Then divide $2x^2$ into *every* term in the numerator.

$$\frac{6x^3 - 3x^2 + 2}{2x^2} = \frac{6x^3}{2x^2} - \frac{3x^2}{2x^2} + \frac{2}{2x^2}$$

$$= 3x - \frac{3}{2} + \frac{1}{x^2} \qquad \textbf{\textit{Now Try Exercise 3}} \blacktriangleleft$$

Algebra Review
To review simplification of rational expressions, see Chapter R (page R-30).

Division by Polynomials

Before dividing a polynomial by a binomial, we review division of natural numbers.

$$\text{quotient} \rightarrow 58$$
$$\text{divisor} \rightarrow 3\overline{)175} \leftarrow \text{dividend}$$
$$\underline{15}$$
$$25$$
$$\underline{24}$$
$$1 \leftarrow \text{remainder}$$

This result is checked as follows: $3 \cdot 58 + 1 = 175$. That is,

(Divisor) (Quotient) + (Remainder) = (Dividend).

The quotient and remainder can also be expressed as $58\frac{1}{3}$. Since 3 does not divide into 175 evenly, 3 is *not* a factor of 175. When the remainder is 0, the divisor is a *factor* of the dividend. Division of polynomials is similar to division of natural numbers.

EXAMPLE 2   Dividing polynomials

Divide $2x^3 - 3x^2 - 11x + 7$ by $x - 3$. Check the result.

SOLUTION Begin by dividing x into $2x^3$.

$$
\begin{array}{r}
2x^2 \\
x - 3\overline{)2x^3 - 3x^2 - 11x + 7} \\
\underline{2x^3 - 6x^2} \\
3x^2 - 11x
\end{array}
$$

$\frac{2x^3}{x} = 2x^2$

$2x^2(x - 3) = 2x^3 - 6x^2$
Subtract. Bring down $-11x$.

In the next step, divide x into $3x^2$.

$$
\begin{array}{r}
2x^2 + 3x \\
x - 3\overline{)2x^3 - 3x^2 - 11x + 7} \\
\underline{2x^3 - 6x^2} \\
3x^2 - 11x \\
\underline{3x^2 - 9x} \\
-2x + 7
\end{array}
$$

$\frac{3x^2}{x} = 3x$

$3x(x - 3) = 3x^2 - 9x$
Subtract. Bring down 7.

Now divide x into $-2x$.

$$
\begin{array}{r}
2x^2 + 3x - 2 \\
x - 3\overline{)2x^3 - 3x^2 - 11x + 7} \\
\underline{2x^3 - 6x^2} \\
3x^2 - 11x \\
\underline{3x^2 - 9x} \\
-2x + 7 \\
\underline{-2x + 6} \\
1
\end{array}
$$

$\frac{-2x}{x} = -2$

$-2(x - 3) = -2x + 6$
Subtract. Remainder is 1.

The quotient is $2x^2 + 3x - 2$ and the remainder is 1. Polynomial division is also checked by multiplying the divisor and quotient and then adding the remainder.

$$
\begin{aligned}
(x - 3)(2x^2 + 3x - 2) + 1 &= x(2x^2 + 3x - 2) - 3(2x^2 + 3x - 2) + 1 \\
&= 2x^3 + 3x^2 - 2x - 6x^2 - 9x + 6 + 1 \\
&= 2x^3 - 3x^2 - 11x + 7
\end{aligned}
$$

Now Try Exercise 9 ◄

Algebra Review
To review multiplication of polynomials, see Chapter R (page R-16).

A division problem, such as $\frac{175}{3}$, is typically given in the form $\frac{(\text{Dividend})}{(\text{Divisor})}$. If we divide each term in the equation

$$(\textbf{Dividend}) = (\textbf{Divisor})\,(\textbf{Quotient}) + (\textbf{Remainder})$$

by (Divisor), we obtain the equation

$$\frac{(\textbf{Dividend})}{(\textbf{Divisor})} = (\textbf{Quotient}) + \frac{(\textbf{Remainder})}{(\textbf{Divisor})}$$

For example, because **175** divided by **3** equals **58** remainder **1**, we can use this equation to justify writing $\frac{175}{3} = 58 + \frac{1}{3}$, or $58\frac{1}{3}$. We can use the results from Example 2 to write

$$\frac{2x^3 - 3x^2 - 11x + 7}{x - 3} = 2x^2 + 3x - 2 + \frac{1}{x - 3}.$$

This process is applied in the next example.

EXAMPLE 3 Dividing polynomials

Divide each expression. Check your answer.

(a) $\dfrac{6x^2 + 5x - 10}{2x + 3}$ **(b)** $(5x^3 - 4x^2 + 7x - 2) \div (x^2 + 1)$

SOLUTION

(a) Begin by dividing $2x$ into $6x^2$.

$$
\begin{array}{r}
3x \\
2x + 3 \overline{)\, 6x^2 + 5x - 10} \\
\underline{6x^2 + 9x } \\
-4x - 10
\end{array}
$$

$\dfrac{6x^2}{2x} = 3x$

$3x(2x + 3) = 6x^2 + 9x$

Subtract: $5x - 9x = -4x$.
Bring down the -10.

In the next step, divide $2x$ into $-4x$.

$$
\begin{array}{r}
3x - 2 \\
2x + 3 \overline{)\, 6x^2 + 5x - 10} \\
\underline{6x^2 + 9x } \\
-4x - 10 \\
\underline{-4x - 6} \\
-4
\end{array}
$$

$\dfrac{-4x}{2x} = -2$

$-2(2x + 3) = -4x - 6$

Subtract: $-10 - (-6) = -4$.

The quotient is $3x - 2$ with remainder -4. This result can also be written as follows.

$$ 3x - 2 + \dfrac{-4}{2x + 3} \qquad \text{(Quotient)} + \dfrac{\text{(Remainder)}}{\text{(Divisor)}} $$

To check this result use the equation

$$ \textbf{(Divisor)(Quotient)} + \textbf{(Remainder)} = \textbf{(Dividend)}. $$

This result can be checked as follows.

$$
\begin{aligned}
(2x + 3)(3x - 2) + (-4) &= 6x^2 + 5x - 6 - 4 \\
&= 6x^2 + 5x - 10 \qquad \text{The result checks.}
\end{aligned}
$$

(b) Begin by writing $x^2 + 1$ as $x^2 + 0x + 1$.

$$
\begin{array}{r}
5x - 4 \\
x^2 + 0x + 1 \overline{)\, 5x^3 - 4x^2 + 7x - 2} \\
\underline{5x^3 + 0x^2 + 5x } \\
-4x^2 + 2x - 2 \\
\underline{-4x^2 + 0x - 4} \\
2x + 2
\end{array}
$$

The quotient is $5x - 4$ with remainder of $2x + 2$. This result can also be written as

$$ 5x - 4 + \dfrac{2x + 2}{x^2 + 1}. $$

This result can be checked as follows.

$$
\begin{aligned}
(x^2 + 1)(5x - 4) + 2x + 2 &= 5x^3 - 4x^2 + 5x - 4 + 2x + 2 \\
&= 5x^3 - 4x^2 + 7x - 2 \qquad \text{The result checks.}
\end{aligned}
$$

Now Try Exercises 21 and 25 ◀

This process is summarized by the following *division algorithm for polynomials*.

Division Algorithm for Polynomials

Let $f(x)$ and $d(x)$ be two polynomials, with the degree of $d(x)$ greater than zero and less than the degree of $f(x)$. Then there exist unique polynomials $q(x)$ and $r(x)$ such that

$$f(x) \quad = \quad d(x) \quad \cdot \quad q(x) \quad + \quad r(x),$$

(Dividend) = (Divisor) · (Quotient) + (Remainder)

where either $r(x) = 0$ or the degree of $r(x)$ is less than the degree of $d(x)$. The polynomial $r(x)$ is called the remainder.

Synthetic Division

A shortcut called **synthetic division** can be used to divide $x - k$ into a polynomial. For example, to divide $x - 2$ into $3x^4 - 7x^3 - 4x + 5$, we perform the following steps. The equivalent steps involving long division are shown to the right.

$$
\begin{array}{r|rrrrr}
2 & 3 & -7 & 0 & -4 & 5 \\
  &   & 6 & -2 & -4 & -16 \\
\hline
  & 3 & -1 & -2 & -8 & -11
\end{array}
$$

$$
\require{enclose}
\begin{array}{r}
3x^3 - x^2 - 2x - 8 \\
x - 2 \enclose{longdiv}{3x^4 - 7x^3 + 0x^2 - 4x + 5} \\
\underline{3x^4 - 6x^3} \\
-1x^3 + 0x^2 \\
\underline{-1x^3 + 2x^2} \\
-2x^2 - 4x \\
\underline{-2x^2 + 4x} \\
-8x + 5 \\
\underline{-8x + 16} \\
-11
\end{array}
$$

Notice how the red and blue numbers in the expression for long division correspond to the third row in synthetic division. The remainder is -11, which is the last number in the third row. The degree of the quotient, $3x^3 - x^2 - 2x - 8$, is one less than the degree of $f(x)$. The quotient's coefficients are 3, -1, -2, and -8 and are found in the third row. The steps to divide a polynomial $f(x)$ by $x - k$ using synthetic division can be summarized as follows.

1. Write k to the left and the coefficients of $f(x)$ to the right in the top row. If any power of x does *not* appear in $f(x)$, include a 0 for that term. In this example, an x^2-term did not appear, so a 0 is included in the first row.
2. Copy the leading coefficient of $f(x)$ into the third row and multiply it by k. Write the result below the next coefficient of $f(x)$ in the second row. Add the numbers in the second column and place the result in the third row. Repeat the process. In this example, the leading coefficient is 3 and $k = 2$. Since $3 \cdot 2 = 6$, 6 is placed below -7. Then add to obtain $-7 + 6 = -1$. Multiply -1 by 2 and repeat.
3. The last number in the third row is the remainder. If the remainder is 0, then the binomial $x - k$ is a factor of $f(x)$. The other numbers in the third row are the coefficients of the quotient, with terms written in descending powers.

EXAMPLE 4   Performing synthetic division

Use synthetic division to divide $2x^3 + 4x^2 - x + 5$ by $x + 2$.

SOLUTION

Getting Started To find the value of k, write the divisor as $x - k$. Because $x + 2$ equals $x - (-2)$, the value of k is -2. ▶

Let $k = -2$ and perform synthetic division on the problem $\dfrac{2x^3 + 4x^2 - x + 5}{x + 2}$.

$$
\begin{array}{r|rrrr}
-2 & 2 & 4 & -1 & 5 \\
   &   & -4 & 0 & 2 \\
\hline
   & 2 & 0 & -1 & 7
\end{array}
$$

The remainder is **7** and the quotient is $2x^2 + 0x - 1 = 2x^2 - 1$. This result is expressed by the equation

$$\frac{2x^3 + 4x^2 - x + 5}{x + 2} = 2x^2 - 1 + \frac{7}{x + 2}.$$

Now Try Exercise 41 ◀

An Application from Geometry In the final example, we use division to solve the problem presented in the introduction to this section.

EXAMPLE 5 Finding the length of a rectangle

If the area of a rectangle is $x^2 + 8x + 15$ and its width is $x + 3$, use division to find its length.

SOLUTION We use synthetic division to divide $x^2 + 8x + 15$ by $x + 3$. However, long division could be used.

$$
\begin{array}{r|rrr}
-3 & 1 & 8 & 15 \\
   &   & -3 & -15 \\
\hline
   & 1 & 5 & 0
\end{array}
$$

The remainder is **0**, so $x + 3$ divides evenly into $x^2 + 8x + 15$, and the length is $x + 5$.

Now Try Exercise 51 ◀

Remainder Theorem If the divisor $d(x)$ is $x - k$, then the *division algorithm for polynomials* simplifies to

$$f(x) = (x - k)q(x) + r,$$

where r is a constant. If we let $x = k$ in this equation, then

$$f(k) = (k - k)q(k) + r = r.$$

Thus $f(k)$ is equal to the remainder obtained in synthetic division. In Example 4, when $f(x) = 2x^3 + 4x^2 - x + 5$ is divided by $x + 2$, the remainder is **7**. It follows that $f(-2) = 2(-2)^3 + 4(-2)^2 - (-2) + 5 = 7$. This result is summarized by the *remainder theorem*.

Remainder Theorem

If a polynomial $f(x)$ is divided by $x - k$, the remainder is $f(k)$.

4.3 Putting It All Together

The following table lists some important concepts related to a polynomial $f(x)$.

Concept	Explanation	Example
Division by a monomial	Be sure to divide the denominator into *every term* in the numerator.	$\dfrac{5a^3 - 10a^2}{5a^2} = \dfrac{5a^3}{5a^2} - \dfrac{10a^2}{5a^2} = a - 2$
Division by a polynomial	Division by a polynomial can be done in a manner similar to long division of natural numbers. See Examples 2 and 3.	When $6x^3 + 5x^2 - 8x + 4$ is divided by $2x - 1$, the quotient is $3x^2 + 4x - 2$ with remainder 2 and can be written as $$\dfrac{6x^3 + 5x^2 - 8x + 4}{2x - 1} = 3x^2 + 4x - 2 + \dfrac{2}{2x - 1}.$$
Division algorithm	(Dividend) = (Divisor)(Quotient) + (Remainder) This equation can be written as $$\dfrac{(\text{Dividend})}{(\text{Divisor})} = (\text{Quotient}) + \dfrac{(\text{Remainder})}{(\text{Divisor})}.$$	$\dfrac{x^3 - 1}{x + 1} = x^2 - x + 1 + \dfrac{-2}{x + 1}$ Dividend: $x^3 - 1$ Divisor: $x + 1$ Quotient: $x^2 - x + 1$ Remainder: -2
Synthetic division	An efficient method for dividing $x - k$ into a polynomial	Divide $2x^3 - 3x^2 + x + 2$ by $x + 1$. $$\begin{array}{r|rrrr} -1 & 2 & -3 & 1 & 2 \\ & & -2 & 5 & -6 \\ \hline & 2 & -5 & 6 & -4 \end{array}$$ The quotient is $2x^2 - 5x + 6$, and the remainder is -4.
Remainder theorem	If a polynomial $f(x)$ is divided by $x - k$, the remainder is $f(k)$.	If $f(x) = 3x^2 - 2x + 6$ is divided by $x - 2$, the remainder is $f(2) = 3(2)^2 - 2(2) + 6 = 14$.

4.3   Exercises

Division by Monomials
Exercises 1–8: Divide the expression.

1. $\dfrac{5x^4 - 15}{10x}$

2. $\dfrac{x^2 - 5x}{5x}$

3. $\dfrac{3x^4 - 2x^2 - 1}{3x^3}$

4. $\dfrac{5x^3 - 10x^2 + 5x}{15x^2}$

5. $\dfrac{x^3 - 4}{4x^3}$

6. $\dfrac{2x^4 - 3x^2 + 4x - 7}{-4x}$

7. $\dfrac{5x(3x^2 - 6x + 1)}{3x^2}$

8. $\dfrac{(1 - 5x^2)(x + 1) + x^2}{2x}$

Division by Polynomials

Exercises 9–14: Divide the first polynomial by the second. State the quotient and remainder.

9. $x^3 - 2x^2 - 5x + 6$ $x - 3$

10. $3x^3 - 10x^2 - 27x + 10$ $x + 2$

11. $2x^4 - 7x^3 - 5x^2 - 19x + 17$ $x + 1$

12. $x^4 - x^3 - 4x + 1$ $x - 2$

13. $3x^3 - 7x + 10$ $x - 1$

14. $x^4 - 16x^2 + 1$ $x + 4$

Exercises 15–22: Divide. Check your answer.

15. $\dfrac{x^4 - 3x^3 - x + 3}{x - 3}$

16. $\dfrac{x^3 - 2x^2 - x + 3}{x + 1}$

17. $\dfrac{4x^3 - x^2 - 5x + 6}{x - 1}$

18. $\dfrac{x^4 + 3x^3 - 4x + 1}{x + 2}$

19. $\dfrac{x^3 + 1}{x + 1}$

20. $\dfrac{x^5 + 3x^4 - x - 3}{x + 3}$

21. $\dfrac{6x^3 + 5x^2 - 8x + 4}{2x - 1}$

22. $\dfrac{12x^3 - 14x^2 + 7x - 7}{3x - 2}$

Exercises 23–30: Divide the expression.

23. $\dfrac{3x^4 - 7x^3 + 6x - 16}{3x - 7}$

24. $\dfrac{20x^4 + 6x^3 - 2x^2 + 15x - 2}{5x - 1}$

25. $\dfrac{5x^4 - 2x^2 + 6}{x^2 + 2}$

26. $\dfrac{x^3 - x^2 + 2x - 3}{x^2 + 3}$

27. $\dfrac{8x^3 + 10x^2 - 12x - 15}{2x^2 - 3}$

28. $\dfrac{3x^4 - 2x^2 - 5}{3x^2 - 5}$

29. $\dfrac{2x^4 - x^3 + 4x^2 + 8x + 7}{2x^2 + 3x + 2}$

30. $\dfrac{3x^4 + 2x^3 - x^2 + 4x - 3}{x^2 + x - 1}$

Division Algorithm

Exercises 31 and 32: Use the equation

$$(\text{Dividend}) = (\text{Divisor})(\text{Quotient}) + (\text{Remainder})$$

to complete the following.

31. $\dfrac{x^3 - 8x^2 + 15x - 6}{x - 2} = x^2 - 6x + 3$ implies

$(x - 2)(x^2 - 6x + 3) = \underline{\ ?\ }$.

32. $\dfrac{x^4 - 15}{x + 2} = x^3 - 2x^2 + 4x - 8 + \dfrac{1}{x + 2}$

implies $x^4 - 15 = (x + 2) \times \underline{\ ?\ } + \underline{\ ?\ }$.

Exercises 33–38: Use division to express the (Dividend) *as* (Divisor)(Quotient) + (Remainder).

33. $\dfrac{x^2 - 3x + 1}{x - 2}$

34. $\dfrac{2x^2 - x + 2}{x + 4}$

35. $\dfrac{2x^3 + x^2 - 2x}{2x + 1}$

36. $\dfrac{1 - x^2 + x^3}{x - 1}$

37. $\dfrac{x^3 - x^2 + x + 1}{x^2 + 1}$

38. $\dfrac{2x^3 + x^2 - x + 4}{x^2 + x}$

Synthetic Division

Exercises 39–46: Use synthetic division to divide the first polynomial by the second.

39. $x^3 + 2x^2 - 17x - 10$ $x + 5$

40. $x^3 - 2x + 1$ $x + 4$

41. $3x^3 - 11x^2 - 20x + 3$ $x - 5$

42. $x^4 - 3x^3 - 5x^2 + 2x - 16$ $x - 3$

43. $x^4 - 3x^3 - 4x^2 + 12x$ $x - 2$

44. $x^5 + \tfrac{1}{4}x^4 - x^3 - \tfrac{1}{4}x^2 + 3x - \tfrac{5}{4}$ $x + \tfrac{1}{4}$

45. $2x^5 - x^4 - x^3 + 4x + 3$ $x + \tfrac{1}{2}$

46. $x^4 - \tfrac{1}{2}x^3 + 3x^2 - \tfrac{5}{2}x + \tfrac{9}{2}$ $x - \tfrac{1}{2}$

Remainder Theorem

Exercises 47–50: Use the remainder theorem to find the remainder when $f(x)$ *is divided by the given* $x - k$.

47. $f(x) = 5x^2 - 3x + 1$ $x - 1$

48. $f(x) = -4x^2 + 6x - 7$ $x + 4$

49. $f(x) = 4x^3 - x^2 + 4x + 2 \qquad x + 2$

50. $f(x) = -x^4 + 4x^3 - x + 3 \qquad x - 3$

Applications

51. *Area of a Rectangle*   Use the figure to find the length L of the rectangle from its width and area A. Determine L when $x = 10$ feet.

$3x + 1$ | $A = 12x^2 + 13x + 3$

L

52. *Area of a Rectangle*   Use the figure to find the width W of the rectangle from its length and area A. Determine W when $x = 5$ inches.

W | $A = 3x^3 - 5x^2 + 3x - 5$

$x^2 + 1$

Writing about Mathematics

53. Compare division of integers to division of polynomials. Give examples.

54. When can you use synthetic division to divide two polynomials? Give one example where synthetic division can be used and one example where it cannot be used.

4.4 Real Zeros of Polynomial Functions

- **Understand the factor theorem**
- **Factor higher degree polynomials**
- **Analyze polynomials having multiple zeros**
- **Solve higher degree polynomial equations**
- **Understand the rational zeros test, Descartes' rule of signs, and the intermediate value property (optional)**

Introduction

Some species of birds, such as robins, have two nesting periods each summer. Because the survival rate for young birds is low, bird populations can fluctuate greatly during the summer months.   (**Source:** S. Kress, *Bird Life.*)

   The graph of $f(x) = x^3 - 61x^2 + 839x + 4221$ shown in Figure 4.56 models a population of birds in a small country, where $x = 1$ corresponds to June 1, $x = 2$ to June 2, $x = 3$ to June 3, and so on.

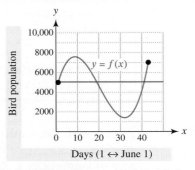

Figure 4.56 A Summer Bird Population

If we want to determine the dates when the population was 5000, we can solve

$$x^3 - 61x^2 + 839x + 4221 = 5000, \quad \text{or} \quad x^3 - 61x^2 + 839x - 779 = 0.$$

From the graph of f it appears that there were 5000 birds around June 1 ($x = 1$), June 20 ($x = 20$), and July 10 ($x = 40$). In Example 5 we find an exact symbolic solution.

Factoring Polynomials

The polynomial $f(x) = x^2 - 3x + 2$ can be factored as $f(x) = (x - 1)(x - 2)$. Note that $f(1) = 0$ and $(x - 1)$ is a factor of $f(x)$. Similarly, $f(2) = 0$ and $(x - 2)$ is a factor.

This discussion can be generalized. By the remainder theorem we know that

$$f(x) = (x - k)q(x) + r,$$

where r is the remainder. If $r = 0$, then $f(x) = (x - k)q(x)$ and $(x - k)$ is a factor of $f(x)$. Similarly, if $(x - k)$ is a factor of $f(x)$, then $r = 0$. That is,

$$f(x) = (x - k)q(x)$$

and $f(k) = (k - k)q(k) = 0 \cdot q(k) = 0$. This discussion justifies the *factor theorem*.

Factor Theorem

A polynomial $f(x)$ has a factor $x - k$ if and only if $f(k) = 0$.

NOTE If $f(-2) = 0$, then $(x - (-2))$, or $(x + 2)$, is a factor.

EXAMPLE 1 Applying the factor theorem

Use the graph in Figure 4.57 and the factor theorem to list the factors of $f(x)$.

SOLUTION Figure 4.57 shows that the zeros (or x-intercepts) of f are -2, 1, and 3. Since $f(-2) = 0$, the factor theorem states that $(x + 2)$ is a factor of $f(x)$. Similarly, $f(1) = 0$ implies that $(x - 1)$ is a factor, and $f(3) = 0$ implies that $(x - 3)$ is a factor. Thus the factors of $f(x)$ are $(x + 2)$, $(x - 1)$, and $(x - 3)$.   *Now Try Exercise 1* ◄

MAKING CONNECTIONS

x-Intercepts, Zeros, and Factors   Let $f(x)$ be a polynomial with degree 1 or more. The following statements are *equivalent*.

1. The graph of $y = f(x)$ has x-intercept k.
2. A real zero of $f(x)$ is k. That is, $f(k) = 0$.
3. A factor of $f(x)$ is $(x - k)$.

Figure 4.57

Figure 4.58

Zeros with Multiplicity   The polynomial $f(x) = x^2 + 4x + 4$ can be written as $f(x) = (x + 2)^2$. Since the factor $(x + 2)$ occurs twice in $f(x)$, the zero -2 is called a **zero of multiplicity 2**. The polynomial $g(x) = (x + 1)^3(x - 2)$ has zeros -1 and 2 with *multiplicities* 3 and 1, respectively. A graph of g is shown in Figure 4.58, where the x-intercepts coincide with the zeros of g. *Counting multiplicities,* a polynomial of degree n has at most n real zeros. For $g(x)$, the sum of the multiplicities is $3 + 1 = 4$, which equals its degree.

Complete Factored Form   The concepts discussed above, together with the factor theorem, can be used to find the *complete factored form* of a polynomial. The complete factored form of a polynomial is unique.

Complete Factored Form

Suppose a polynomial

$$f(x) = a_n x^n + \cdots + a_2 x^2 + a_1 x + a_0$$

has n real zeros $c_1, c_2, c_3, \ldots, c_n$, where a distinct zero is listed as many times as its multiplicity. Then $f(x)$ can be written in **complete factored form** as

$$f(x) = a_n(x - c_1)(x - c_2)(x - c_3)\cdots(x - c_n).$$

EXAMPLE 2 Finding a complete factorization

Write the complete factorization for each polynomial with the given zeros.

(a) $f(x) = 13x^2 - \frac{91}{2}x + 39;$ zeros: $\frac{3}{2}$ and 2

(b) $f(x) = 7x^3 - 21x^2 - 7x + 21;$ zeros: $-1, 1,$ and 3

SOLUTION

(a) The leading coefficient is **13** and the zeros are $\frac{3}{2}$ and **2**. By the factor theorem, $\left(x - \frac{3}{2}\right)$ and $(x - 2)$ are factors. The complete factorization is

$$f(x) = 13\left(x - \frac{3}{2}\right)(x - 2).$$

(b) The leading coefficient is 7 and the zeros are $-1, 1,$ and 3. The complete factorization is

$$f(x) = 7(x + 1)(x - 1)(x - 3).$$

Now Try Exercises 5 and 7 ◄

MAKING CONNECTIONS

Types of Factored Forms   If the leading coefficient of a polynomial $f(x)$ is 6 and the only zeros are $\frac{1}{3}$ and $\frac{1}{2}$, then the complete factored form is $f(x) = 6\left(x - \frac{1}{3}\right)\left(x - \frac{1}{2}\right)$ and it is *unique*. Sometimes we factor 6 as $3 \cdot 2$ and distribute the 3 over the first factor and the 2 over the second factor to obtain the slightly different factored form $f(x) = (3x - 1)(2x - 1)$.

EXAMPLE 3 Factoring a polynomial graphically

Use the graph of f in Figure 4.59 to factor $f(x) = 2x^3 - 4x^2 - 10x + 12$.

SOLUTION

Getting Started To factor $f(x)$ we need to determine the leading coefficient and the zeros of f. The zeros coincide with the x-intercepts of the graph of f. ►

The leading coefficient is **2**, and the zeros are $-2, 1,$ and 3. The complete factorization is

$$f(x) = 2(x + 2)(x - 1)(x - 3).$$

Now Try Exercise 13 ◄

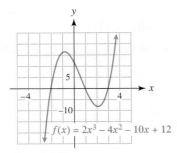

Figure 4.59

When factoring polynomials by hand, it is sometimes helpful to use the techniques of division discussed in Section 4.3.

EXAMPLE 4   Factoring a polynomial symbolically

The polynomial $f(x) = 2x^3 - 2x^2 - 34x - 30$ has a zero of -1. Express $f(x)$ in complete factored form.

SOLUTION   If -1 is a zero, then by the factor theorem $(x + 1)$ is a factor. To factor $f(x)$, divide $x + 1$ into $2x^3 - 2x^2 - 34x - 30$ by using synthetic division.

$$
\begin{array}{r|rrrr}
-1 & 2 & -2 & -34 & -30 \\
   &   & -2 & 4 & 30 \\
\hline
   & 2 & -4 & -30 & 0
\end{array}
$$

The remainder is 0, so $x + 1$ divides evenly into the dividend. By the division algorithm,
$$2x^3 - 2x^2 - 34x - 30 = (x + 1)(2x^2 - 4x - 30).$$

The quotient $2x^2 - 4x - 30$ can be factored further.
$$2x^2 - 4x - 30 = 2(x^2 - 2x - 15)$$
$$= 2(x + 3)(x - 5)$$

The complete factored form is $f(x) = 2(x + 1)(x + 3)(x - 5)$.   ***Now Try Exercise 31*** ◀

Algebra Review
To review factoring trinomials, see Chapter R (page R-23).

An Application   In the next example, we factor $g(x) = x^3 - 61x^2 + 839x - 779$, which was presented earlier, and then use the complete factorization to determine the days when the bird population was 5000.

EXAMPLE 5   Factoring a polynomial

Factor $g(x) = x^3 - 61x^2 + 839x - 779$. Use the zeros of $g(x)$ to determine when the bird population was 5000.

SOLUTION   From Figure 4.56, repeated in the margin, it appears that the bird population was 5000 when $x = 1$. If we substitute $x = 1$ in this polynomial, the result is 0.
$$g(1) = 1^3 - 61(1)^2 + 839(1) - 779 = 0$$
By the factor theorem, $(x - 1)$ is a factor of $g(x)$. We can use synthetic division to divide $x^3 - 61x^2 + 839x - 779$ by $x - 1$.

$$
\begin{array}{r|rrrr}
1 & 1 & -61 & 839 & -779 \\
  &   & 1 & -60 & 779 \\
\hline
  & 1 & -60 & 779 & 0
\end{array}
$$
    The remainder is 0.

By the division algorithm,
$$x^3 - 61x^2 + 839x - 779 = (x - 1)(x^2 - 60x + 779).$$
Since it is not obvious how to factor $x^2 - 60x + 779$, we can use the quadratic formula to find its zeros.

$$x = \frac{-b \pm \sqrt{b^2 - 4ac}}{2a}$$
$$= \frac{-(-60) \pm \sqrt{(-60)^2 - 4(1)(779)}}{2(1)}$$
$$= \frac{60 \pm 22}{2}$$
$$= 41 \text{ or } 19$$

Days (1 ↔ June 1)

The zeros of $g(x) = x^3 - 61x^2 + 839x - 779$ are 1, 19, and 41, and its leading coefficient is 1. The complete factorization is $g(x) = (x - 1)(x - 19)(x - 41)$. The bird population equals 5000 on June 1 ($x = 1$), June 19 ($x = 19$), and July 11 ($x = 41$).

Now Try Exercise 113 ◄

Graphs and Multiple Zeros

The polynomial $f(x) = 0.02(x + 3)^3(x - 3)^2$ has zeros -3 and 3 with multiplicities **3** and **2**, respectively. At the zero of even multiplicity the graph does not cross the x-axis, whereas the graph does cross the x-axis at the zero of odd multiplicity. See Figure 4.60.

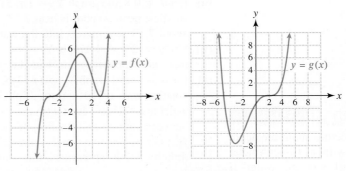

Figure 4.60   **Figure 4.61**

The graph of $g(x) = 0.03(x + 5)(x - 2)^3$ is shown in Figure 4.61. The zeros of g are -5 and 2 with multiplicities of 1 and 3, respectively. Since both zeros have odd multiplicity, the graph crosses the x-axis at -5 and 2. Notice that the zero 2 has a higher multiplicity and the graph levels off more near $x = 2$ than it does near $x = -5$. The higher the multiplicity of a zero, the more the graph of a polynomial levels off near the zero.

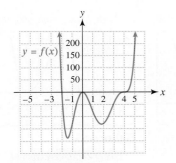

Figure 4.62

EXAMPLE 6   Finding multiplicities graphically

Figure 4.62 shows the graph of a sixth-degree polynomial $f(x)$ with leading coefficient 1. All zeros are integers. Write $f(x)$ in complete factored form.

SOLUTION   The x-intercepts or zeros of f are -2, 0, and 4. Since the graph crosses the x-axis at -2 and 4, these zeros have odd multiplicity. The graph of f levels off more at $x = 4$ than at $x = -2$, so 4 has a higher multiplicity than -2. At $x = 0$ the graph of f does not cross the x-axis. Thus 0 has even multiplicity. If -2 has multiplicity 1, 0 has multiplicity 2, and 4 has multiplicity 3, then the sum of the multiplicities is given by $1 + 2 + 3 = 6$, which equals the degree of $f(x)$. List the zeros as $-2, 0, 0, 4, 4$, and 4. The leading coefficient is 1, so the complete factorization of $f(x)$ is

$$f(x) = 1(x + 2)(x - 0)(x - 0)(x - 4)(x - 4)(x - 4), \quad \text{or}$$
$$f(x) = x^2(x + 2)(x - 4)^3.$$

Now Try Exercise 47 ◄

An Application: "Will It Float?"   Multiple zeros sometimes have physical significance. The next example shows how a multiple zero represents the boundary between an object's floating and sinking.

EXAMPLE 7 Interpreting a multiple zero

The polynomial $f(x) = \frac{\pi}{3}x^3 - 5\pi x^2 + \frac{500\pi d}{3}$ can be used to find the depth that a ball, 10 centimeters in diameter, sinks in water. The constant d is the density of the ball, where the density of water is 1. The smallest *positive* zero of $f(x)$ equals the depth that the sphere sinks. Approximate this depth for each material and interpret the results.

(a) A wood ball with $d = 0.8$
(b) A solid aluminum sphere with $d = 2.7$
(c) A water balloon with $d = 1$

SOLUTION

(a) Let $d = 0.8$ and graph $Y_1 = (\pi/3)X{\wedge}3 - 5\pi X{\wedge}2 + 500\pi(0.8)/3$. In Figure 4.63 the smallest positive zero is near 7.13. This means that the 10-centimeter wood ball sinks about 7.13 centimeters into the water.

(b) Let $d = 2.7$ and graph $Y_2 = (\pi/3)X{\wedge}3 - 5\pi X{\wedge}2 + 500\pi(2.7)/3$. In Figure 4.64 there is no positive zero. The aluminum sphere is more dense than water and sinks.

(c) Let $d = 1$ and graph $Y_3 = (\pi/3)X{\wedge}3 - 5\pi X{\wedge}2 + 500\pi/3$. In Figure 4.65 y_3 has one positive zero of 10 with multiplicity 2. The water balloon has the same density as water and "floats" even with the surface. The value of $d = 1$ represents the boundary between sinking and floating. If the ball floats, $f(x)$ has two positive zeros; if it sinks, $f(x)$ has no positive zeros. With the water balloon there is one positive zero with multiplicity 2 that represents a transition between floating and sinking.

CLASS DISCUSSION

Make a conjecture about the depth that a ball with a 10-centimeter diameter will sink in water if $d = 0.5$. Test your conjecture graphically.

$[-20, 20, 5]$ by $[-300, 500, 100]$ $[-20, 20, 5]$ by $[-500, 2000, 500]$ $[-20, 20, 5]$ by $[-300, 600, 100]$

Figure 4.63   **Figure 4.64**   **Figure 4.65**

Now Try Exercise 111 ◄

MAKING CONNECTIONS

Zeros and Multiplicity If a zero of a polynomial $f(x)$ has odd multiplicity, then its graph crosses the x-axis at the zero. If a zero has even multiplicity, then its graph intersects, but does not cross, the x-axis at the zero. The higher the multiplicity of a zero, the more the graph levels off at the zero. These concepts are illustrated in the following figures.

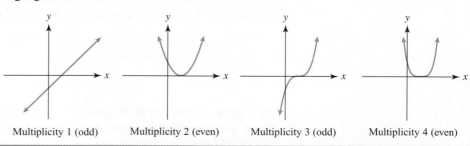

Multiplicity 1 (odd)   Multiplicity 2 (even)   Multiplicity 3 (odd)   Multiplicity 4 (even)

Rational Zeros (Optional)

If a polynomial has a rational zero, it can be found by using the *rational zero test*.

Rational Zero Test

Let $f(x) = a_n x^n + \cdots + a_2 x^2 + a_1 x + a_0$, where $a_n \neq 0$, represent a polynomial function f with *integer* coefficients. If $\frac{p}{q}$ is a rational number written in lowest terms and if $\frac{p}{q}$ is a zero of f, then p is a factor of the constant term a_0 and q is a factor of the leading coefficient a_n.

The following example illustrates how to find rational zeros by using this test.

EXAMPLE 8 Finding rational zeros of a polynomial

Find all rational zeros of $f(x) = 6x^3 - 5x^2 - 7x + 4$ and factor $f(x)$.

SOLUTION If $\frac{p}{q}$ is a rational zero in lowest terms, then p is a factor of the constant term 4 and q is a factor of the leading coefficient 6. The possible values for p and q are as follows.

$$p: \quad \pm 1, \quad \pm 2, \quad \pm 4$$
$$q: \quad \pm 1, \quad \pm 2, \quad \pm 3, \quad \pm 6$$

As a result, any rational zero of $f(x)$ in the form $\frac{p}{q}$ must occur in the list

$$\pm \frac{1}{6}, \quad \pm \frac{1}{3}, \quad \pm \frac{1}{2}, \quad \pm \frac{2}{3}, \quad \pm \frac{1}{1}, \quad \pm \frac{4}{3}, \quad \pm \frac{2}{1}, \quad \text{or} \quad \pm \frac{4}{1}.$$

Evaluate $f(x)$ at each value in the list. See Table 4.3.

Table 4.3

x	$f(x)$	x	$f(x)$	x	$f(x)$	x	$f(x)$
$\frac{1}{6}$	$\frac{49}{18}$	$\frac{1}{2}$	0	1	-2	2	18
$-\frac{1}{6}$	5	$-\frac{1}{2}$	$\frac{11}{2}$	-1	0	-2	-50
$\frac{1}{3}$	$\frac{4}{3}$	$\frac{2}{3}$	$-\frac{10}{9}$	$\frac{4}{3}$	0	4	280
$-\frac{1}{3}$	$\frac{50}{9}$	$-\frac{2}{3}$	$\frac{14}{3}$	$-\frac{4}{3}$	$-\frac{88}{9}$	-4	-432

From Table 4.3 there are three rational zeros: -1, $\frac{1}{2}$, and $\frac{4}{3}$. Since a third-degree polynomial has at most three zeros, the complete factored form of $f(x)$ is

$$f(x) = 6(x + 1)\left(x - \frac{1}{2}\right)\left(x - \frac{4}{3}\right),$$

which can also be written as $f(x) = (x + 1)(2x - 1)(3x - 4)$.   ***Now Try Exercise 57*** ◄

NOTE Although $f(x)$ in Example 8 had only rational zeros, it is important to realize that many polynomials have irrational zeros. Irrational zeros cannot be found using the rational zero test.

Descartes' Rule of Signs (Optional)

Descartes' rule of signs helps to determine the numbers of positive and negative real zeros of a polynomial function.

Descartes' Rule of Signs

Let $P(x)$ define a polynomial function with real coefficients and a nonzero constant term, with terms in descending powers of x.

(a) The number of positive real zeros either equals the number of variations in sign occurring in the coefficients of $P(x)$ or is less than the number of variations by a positive even integer.

(b) The number of negative real zeros either equals the number of variations in sign occurring in the coefficients of $P(-x)$ or is less than the number of variations by a positive even integer.

A *variation in sign* is a change from positive to negative or negative to positive in successive terms of the polynomial when written in descending powers of the variable. Missing terms (those with 0 coefficients) can be ignored.

EXAMPLE 9 Applying Descartes' rule of signs

Determine the possible numbers of positive real zeros and negative real zeros of $P(x) = x^4 - 6x^3 + 8x^2 + 2x - 1$.

SOLUTION We first consider the possible number of positive zeros by observing that $P(x)$ has three variations in sign.

$$+x^4 - 6x^3 + 8x^2 + 2x - 1$$

Thus by Descartes' rule of signs, $P(x)$ has either 3 or $3 - 2 = 1$ positive real zeros.

For negative zeros, consider the variations in sign for $P(-x)$.

$$P(-x) = (-x)^4 - 6(-x)^3 + 8(-x)^2 + 2(-x) - 1$$
$$= x^4 + 6x^3 + 8x^2 - 2x - 1$$

Since there is only one variation in sign, $P(x)$ has only one negative real zero.

Now Try Exercise 65 ◀

Polynomial Equations

In Section 3.2, factoring was used to solve quadratic equations. Factoring also can be used to solve polynomial equations with degree greater than 2.

EXAMPLE 10 Solving a cubic equation

Solve $x^3 + 3x^2 - 4x = 0$ symbolically. Support your answer graphically and numerically.

Figure 4.66

X	Y1
-4	0
-3	12
-2	12
-1	6
0	0
1	0
2	12

$Y_1 \equiv X^3+3X^2-4X$

Figure 4.67

SOLUTION
Symbolic Solution

$$x^3 + 3x^2 - 4x = 0 \quad \text{Given equation}$$
$$x(x^2 + 3x - 4) = 0 \quad \text{Factor out } x.$$
$$x(x + 4)(x - 1) = 0 \quad \text{Factor the quadratic expression.}$$
$$x = 0, \quad x + 4 = 0, \quad \text{or} \quad x - 1 = 0 \quad \text{Zero-product property}$$
$$x = 0, -4, \text{ or } 1 \quad \text{Solve.}$$

Graphical Solution   Graph $Y_1 = X^3 + 3X^2 - 4X$ as in Figure 4.66. The x-intercepts are -4, 0, and 1, which correspond to the solutions.

Numerical Solution   Table $Y_1 = X^3 + 3X^2 - 4X$ as in Figure 4.67. The zeros of Y_1 occur at $x = -4$, 0, and 1.   ***Now Try Exercise 71*** ◀

EXAMPLE 11   Solving a polynomial equation

Find all real solutions to each equation symbolically.
(a) $4x^4 - 5x^2 - 9 = 0$ **(b)** $2x^3 + 12 = 3x^2 + 8x$

SOLUTION
(a) The expression $4x^4 - 5x^2 - 9$ can be factored in a manner similar to the way quadratic expressions are factored.

$$4x^4 - 5x^2 - 9 = 0 \quad \text{Given equation}$$
$$(4x^2 - 9)(x^2 + 1) = 0 \quad \text{Factor.}$$
$$4x^2 - 9 = 0 \quad \text{or} \quad x^2 + 1 = 0 \quad \text{Zero-product property}$$
$$4x^2 = 9 \quad \text{or} \quad x^2 = -1 \quad \text{Add 9 or subtract 1.}$$
$$x^2 = \frac{9}{4} \quad \text{or} \quad x^2 = -1 \quad \text{Divide by 4.}$$
$$x = \pm\frac{3}{2} \quad \text{or} \quad x^2 = -1 \quad \text{Square root property}$$

The equation $x^2 = -1$ has no *real* solutions. The solutions are $-\frac{3}{2}$ and $\frac{3}{2}$.

(b) First transpose each term on the right side of the equation to the left side of the equation. Then use *grouping* to factor the polynomial.

Algebra Review
To review factoring by grouping, see Chapter R (page R-21).

$$2x^3 + 12 = 3x^2 + 8x \quad \text{Given equation}$$
$$2x^3 - 3x^2 - 8x + 12 = 0 \quad \text{Rewrite the equation.}$$
$$(2x^3 - 3x^2) + (-8x + 12) = 0 \quad \text{Associative property}$$
$$x^2(2x - 3) - 4(2x - 3) = 0 \quad \text{Factor.}$$
$$(x^2 - 4)(2x - 3) = 0 \quad \text{Factor out } 2x - 3.$$
$$x^2 - 4 = 0 \quad \text{or} \quad 2x - 3 = 0 \quad \text{Zero-product property}$$
$$x = \pm 2 \quad \text{or} \quad x = \frac{3}{2} \quad \text{Solve each equation.}$$

The solutions are -2, $\frac{3}{2}$, and 2.   ***Now Try Exercises 79 and 89*** ◀

Some types of polynomial equations cannot be solved easily by factoring. The next example illustrates how we can obtain an approximate solution graphically.

EXAMPLE 12   Finding a solution graphically

Solve the equation $\frac{1}{2}x^3 - 2x - 4 = 0$ graphically. Round to the nearest hundredth.

SOLUTION   A graph of $Y_1 = .5X^3 - 2X - 4$ is shown in Figure 4.68. Since there is only one x-intercept, the equation has one real solution: $x \approx 2.65$.

Calculator Help

To find a zero of a function, see Appendix A (page AP-9).

$[-9, 9, 1]$ by $[-6, 6, 1]$

$y = \frac{1}{2}x^3 - 2x - 4$

Zero
X=2.6494359   Y=0

Figure 4.68

Now Try Exercise 97 ◀

Intermediate Value Property (Optional)

In Example 12, we approximated a solution to $\frac{1}{2}x^3 - 2x - 4 = 0$ to be 2.65. How do we know for sure that there is indeed such a solution? The *intermediate value property* helps answer this question.

Intermediate Value Property

Let (x_1, y_1) and (x_2, y_2), with $y_1 \neq y_2$ and $x_1 < x_2$, be two points on the graph of a continuous function f. Then, on the interval $x_1 \leq x \leq x_2$, f assumes every value between y_1 and y_2 at least once.

From Example 12, let $f(x) = \frac{1}{2}x^3 - 2x - 4$. Because $f(0) = -4$ and $f(3) = 3.5$, the points $(0, -4)$ and $(3, 3.5)$ lie on the graph of f in Figure 4.68. By the intermediate value property, $f(x) = 0$ for at least one x-value, because 0 is between -4 and 3.5. Although we have not found the exact x-value, we know that a real zero of f does exist. Loosely speaking, the intermediate value property says that if $y_1 < 0$ and $y_2 > 0$, then we cannot draw a continuous graph of f without crossing the x-axis. The only way not to cross the x-axis would be to pick up the pencil, but this would create a discontinuous graph.

Applications   There are many examples of the intermediate value property. Physical motion is usually considered to be continuous. Suppose at one time a car is traveling at 20 miles per hour and at another time it is traveling at 40 miles per hour. It is logical to assume that the car traveled 30 miles per hour at least once between these times. In fact, by the intermediate value property, the car must have assumed all speeds between 20 and 40 miles per hour at least once. Similarly, if a jet airliner takes off and flies at an altitude of 30,000 feet, then by the intermediate value property we may conclude that the airliner assumed all altitudes between ground level and 30,000 feet at least once.

4.4 Putting It All Together

The following table summarizes important topics about real zeros of polynomial functions.

Concept	Explanation	Examples
Factor theorem	$(x - k)$ is a factor of $f(x)$ if and only if $f(k) = 0$.	$f(x) = x^2 + 3x - 4$ and $f(1) = 0$ implies that $(x - 1)$ is a factor of $f(x)$. That is, $f(x) = (x - 1)(x + 4)$.
x-intercepts, zeros, and factors	The following are *equivalent*: 1. The graph of f has x-intercept k. 2. A real zero of f is k. That is, $f(k) = 0$. 3. A factor of f is $(x - k)$.	Let $f(x) = x^2 - 2x - 3$. See the graph below. 1. The graph of f has x-intercepts -1 and 3. 2. $f(-1) = 0$ and $f(3) = 0$ 3. $f(x) = (x + 1)(x - 3)$ $f(x) = x^2 - 2x - 3$
Complete factored form	$f(x) = a_n(x - c_1) \cdots (x - c_n)$, where the c_k are zeros of f, with a distinct zero listed as many times as its multiplicity. This form is unique.	$f(x) = 3(x - 5)(x + 3)(x + 3)$ $= 3(x - 5)(x + 3)^2$ $a_n = 3, c_1 = 5, c_2 = -3, c_3 = -3$
Real zero with odd multiplicity	The graph of $y = f(x)$ crosses the x-axis at a real zero of odd multiplicity.	$f(x) = (x + 1)^3(x - 3)$ Both zeros, -1 and 3, have odd multiplicity.
Real zero with even multiplicity	The graph of $y = f(x)$ intersects but does not cross the x-axis at a real zero of even multiplicity.	$f(x) = (x + 1)^2(x - 3)^4$ Both zeros, -1 and 3, have even multiplicity.
Factoring a polynomial graphically (only real zeros)	Graph $y = f(x)$ and locate all the zeros or x-intercepts. If the leading coefficient is a and the zeros are c_1, c_2, and c_3, then $f(x) = a(x - c_1)(x - c_2)(x - c_3)$.	$f(x) = 2x^3 + 4x^2 - 2x - 4$ has zeros $-2, -1$, and 1 and leading coefficient 2. Thus $f(x) = 2(x + 2)(x + 1)(x - 1)$. $y = f(x)$

continued on next page

continued from previous page

Concept	Explanation	Examples
Solving polynomial equations	Polynomial equations can be solved symbolically, graphically, and numerically. Factoring is a useful symbolic technique.	Solve $x^3 - 4x^2 - 5x = 0$. $$x(x^2 - 4x - 5) = 0$$ $$x(x - 5)(x + 1) = 0$$ $$x = 0, 5, \text{ or } -1$$ See also Examples 10 and 11.

4.4 | Exercises

Factoring Polynomials

Exercises 1–4: Use the graph and the factor theorem to list the factors of $f(x)$.

1.

2.

3.

4.

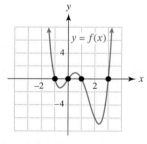

Exercises 5–12: Use the given zeros to write the complete factored form of $f(x)$.

5. $f(x) = 2x^2 - 25x + 77$; zeros: $\frac{11}{2}$ and 7

6. $f(x) = 6x^2 + 21x - 90$; zeros: -6 and $\frac{5}{2}$

7. $f(x) = x^3 - 2x^2 - 5x + 6$; zeros: -2, 1, and 3

8. $f(x) = x^3 + 6x^2 + 11x + 6$; zeros: -3, -2, and -1

9. $f(x) = -2x^3 + 3x^2 + 59x - 30$; zeros: $-5, \frac{1}{2}$, and 6

10. $f(x) = 3x^4 - 8x^3 - 67x^2 + 112x + 240$; zeros: -4, $-\frac{4}{3}$, 3, and 5

11. Let $f(x)$ be a quadratic polynomial with leading coefficient 7. Suppose that $f(-3) = 0$ and $f(2) = 0$. Write the complete factored form of $f(x)$.

12. Let $g(x)$ be a cubic polynomial with leading coefficient -4. Suppose that $g(-2) = 0$, $g(1) = 0$, and $g(4) = 0$. Write the complete factored form of $g(x)$.

Exercises 13 and 14: Use the graph to factor $f(x)$.

13. $f(x) = -2x^3 + 2x$

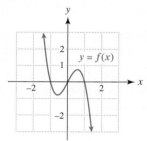

14. $f(x) = \frac{1}{4}x^4 - \frac{3}{2}x^3 + \frac{3}{4}x^2 + \frac{13}{2}x - 6$

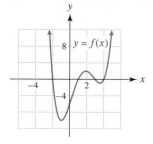

Exercises 15–18: The graph of a polynomial $f(x)$ with leading coefficient ± 1 and integer zeros is shown in the figure. Write its complete factored form.

15.

16.

17.

18.

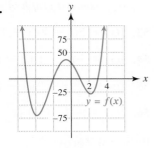

19. Let $f(x)$ be a cubic polynomial with zeros -1, 2, and 3. If the graph of f passes through the point $(0, 3)$, write the complete factored form of $f(x)$.

20. Let $g(x)$ be a quartic polynomial with zeros -2, -1, 1, and 2. If the graph of g passes through the point $(0, 8)$, write the complete factored form of $g(x)$.

Exercises 21–24: The graph of a polynomial $f(x)$ with integer zeros is shown in the figure. Write its complete factored form. Note that the leading coefficient of $f(x)$ is not ± 1.

21.

22.

23.

24.

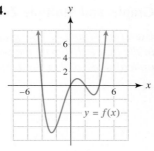

Exercises 25–30: Use graphing to factor $f(x)$.

25. $f(x) = 10x^2 + 17x - 6$

26. $f(x) = 2x^3 + 7x^2 + 2x - 3$

27. $f(x) = -3x^3 - 3x^2 + 18x$

28. $f(x) = \frac{1}{2}x^3 + \frac{5}{2}x^2 + x - 4$

29. $f(x) = x^4 + \frac{5}{2}x^3 - 3x^2 - \frac{9}{2}x$

30. $f(x) = 10x^4 + 7x^3 - 27x^2 + 2x + 8$

Exercises 31–36: (Refer to Example 4.) Write the complete factored form of the polynomial $f(x)$, given that k is a zero.

31. $f(x) = x^3 - 9x^2 + 23x - 15$ $k = 1$

32. $f(x) = 2x^3 + x^2 - 11x - 10$ $k = -2$

33. $f(x) = -4x^3 - x^2 + 51x - 36$ $k = -4$

34. $f(x) = 3x^3 - 11x^2 - 35x + 75$ $k = 5$

35. $f(x) = 2x^4 - x^3 - 13x^2 - 6x$ $k = -2$

36. $f(x) = 35x^4 + 48x^3 - 41x^2 + 6x$ $k = \frac{3}{7}$

Factor Theorem

Exercises 37–40: Use the factor theorem to decide if $x - k$ is a factor of $f(x)$ for the given k.

37. $f(x) = x^3 - 6x^2 + 11x - 6$ $k = 2$

38. $f(x) = x^3 + x^2 - 14x - 24$ $k = -3$

39. $f(x) = x^4 - 2x^3 - 13x^2 - 10x$ $k = 3$

40. $f(x) = 2x^4 - 11x^3 + 9x^2 + 14x$ $k = \frac{1}{2}$

Graphs and Multiple Zeros

Exercises 41 and 42: The graph of a polynomial $f(x)$ is shown in the figure. Estimate the zeros and state whether their multiplicities are odd or even. State the minimum degree of $f(x)$.

41.

42.

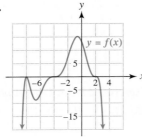

Exercises 43–46: Write a polynomial $f(x)$ in complete factored form that satisfies the conditions. Let the leading coefficient be 1.

43. Degree 3; zeros: -1 with multiplicity 2, and 6 with multiplicity 1

44. Degree 4; zeros: 5 and 7, both with multiplicity 2

45. Degree 4; zeros: 2 with multiplicity 3, and 6 with multiplicity 1

46. Degree 5; zeros: -2 with multiplicity 2, and 4 with multiplicity 3

Exercises 47–52: The graph of either a cubic, quartic, or quintic polynomial $f(x)$ with integer zeros is shown. Write the complete factored form of $f(x)$. (Hint: In Exercises 51 and 52 the leading coefficient is not ± 1.)

47.

48.

49.

50.

51.

52.

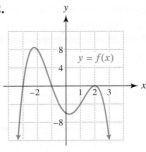

Exercises 53–56: Complete the following.

   (a) Find the x- and y-intercepts.
   (b) Determine the multiplicity of each zero of f.
   (c) Sketch a graph of $y = f(x)$ by hand.

53. $f(x) = 2(x + 2)(x + 1)^2$

54. $f(x) = -(x + 1)(x - 1)(x - 2)$

55. $f(x) = x^2(x + 2)(x - 2)$

56. $f(x) = -\frac{1}{2}(x + 2)^2(x - 1)^3$

Rational Zeros

Exercises 57–64: (Refer to Example 8.)

   (a) Use the rational zero test to find any rational zeros of the polynomial $f(x)$.
   (b) Write the complete factored form of $f(x)$.

57. $f(x) = 2x^3 + 3x^2 - 8x + 3$

58. $f(x) = x^3 - 7x + 6$

59. $f(x) = 2x^4 + x^3 - 8x^2 - x + 6$

60. $f(x) = 2x^4 + x^3 - 19x^2 - 9x + 9$

61. $f(x) = 3x^3 - 16x^2 + 17x - 4$

62. $f(x) = x^3 + 2x^2 - 3x - 6$

63. $f(x) = x^3 - x^2 - 7x + 7$

64. $f(x) = 2x^3 - 5x^2 - 4x + 10$

Descartes' Rule of Signs

Exercises 65–70: Use Descartes' rule of signs to determine the possible number of positive and negative real zeros for each function. Then, use a graph to determine the actual numbers of positive and negative real zeros.

65. $P(x) = 2x^3 - 4x^2 + 2x + 7$

66. $P(x) = x^3 + 2x^2 + x - 10$

67. $P(x) = 5x^4 + 3x^2 + 2x - 9$

68. $P(x) = 3x^4 + 2x^3 - 8x^2 - 10x - 1$

69. $P(x) = x^5 + 3x^4 - x^3 + 2x + 3$

70. $P(x) = 2x^5 - x^4 + x^3 - x^2 + x + 5$

Polynomial Equations

Exercises 71–76: Solve the equation

   (a) *symbolically,*
   (b) *graphically, and*
   (c) *numerically.*

71. $x^3 + x^2 - 6x = 0$ 72. $2x^2 - 8x + 6 = 0$

73. $x^4 - 1 = 0$ 74. $x^4 - 5x^2 + 4 = 0$

75. $-x^3 + 4x = 0$ 76. $6 - 4x - 2x^2 = 0$

Exercises 77–96: Solve the equation.

77. $x^3 - 25x = 0$ 78. $x^4 - x^3 - 6x^2 = 0$

79. $x^4 - x^2 = 2x^2 + 4$ 80. $x^4 + 5 = 6x^2$

81. $x^3 - 3x^2 - 18x = 0$ 82. $x^4 - x^2 = 0$

83. $2x^3 = 4x^2 - 2x$ 84. $x^3 = x$

85. $12x^3 = 17x^2 + 5x$ 86. $3x^3 + 3x = 10x^2$

87. $9x^4 + 4 = 13x^2$ 88. $4x^4 + 7x^2 - 2 = 0$

89. $4x^3 + 4x^2 - 3x - 3 = 0$

90. $9x^3 + 27x^2 - 2x - 6 = 0$

91. $2x^3 + 4 = x(x + 8)$ 92. $3x^3 + 18 = x(2x + 27)$

93. $8x^4 = 30x^2 - 27$ 94. $4x^4 - 21x^2 + 20 = 0$

95. $x^6 - 19x^3 = 216$ 96. $x^6 = 7x^3 + 8$

Exercises 97–102: (Refer to Example 12.) Solve the equation graphically. Round your answers to the nearest hundredth.

97. $x^3 - 1.1x^2 - 5.9x + 0.7 = 0$

98. $x^3 + x^2 - 18x + 13 = 0$

99. $-0.7x^3 - 2x^2 + 4x + 2.5 = 0$

100. $3x^3 - 46x^2 + 180x - 99 = 0$

101. $2x^4 - 1.5x^3 + 13 = 24x^2 + 10x$

102. $-x^4 + 2x^3 + 20x^2 = 22x + 41$

Intermediate Value Property

Exercises 103–106: Use the intermediate value property to show that $f(x) = 0$ for some x on the given interval.

103. $f(x) = x^2 - 5, 2 \le x \le 3$ (*Hint:* Evaluate $f(2)$ and $f(3)$ and then apply the intermediate value property.)

104. $f(x) = x^3 - x - 1, 1 \le x \le 2$

105. $f(x) = 2x^3 - 1, 0 \le x \le 1$

106. $f(x) = 4x^2 - x - 1, -1 \le x \le 0$

107. Let $f(x) = x^5 - x^2 + 4$. Evaluate $f(1)$ and $f(2)$. Is there a real number k such that $f(k) = 20$? Explain your answer.

108. Sketch a graph of a function f that passes through the points $(-2, 3)$ and $(1, -2)$ but never assumes a value of 0. What must be true about the graph of f?

Applications

109. *Winter Temperature*   The temperature T in degrees Fahrenheit on a cold night x hours past midnight can be approximated by $T(x) = x^3 - 6x^2 + 8x$, where $0 \le x \le 4$. Determine when the temperature was 0°F.

110. *Geometry*   A rectangular box has sides with lengths $x, x + 1$, and $x + 2$. If the volume of the box is 504 cubic inches, find the dimensions of the box.

111. *Floating Ball*   (Refer to Example 7.) If a ball has a 20-centimeter diameter, then

$$f(x) = \frac{\pi}{3}x^3 - 10\pi x^2 + \frac{4000\pi d}{3}$$

determines the depth that it sinks in water. Find the depth that this size ball sinks when $d = 0.6$.

112. *Floating Ball*   (Refer to Example 7.) Determine the depth that a pine ball with a 10-centimeter diameter sinks in water, if $d = 0.55$.

113. *Bird Populations*   (Refer to Example 5.) A bird population can be modeled by

$$f(x) = x^3 - 66x^2 + 1052x + 1652,$$

where $x = 1$ corresponds to June 1, $x = 2$ to June 2, and so on. Find the days when f estimates that there were 3500 birds.

114. *Insect Population*   An insect population P in thousands per acre x days past May 31 is approximated by $P(x) = 2x^3 - 18x^2 + 46x$, where $0 \le x \le 6$. Determine the dates when the insect population equaled 30 thousand per acre.

115. *Modeling Temperature*   Complete the following.
(a) Approximate the complete factored form of $f(x) = -0.184x^3 + 1.45x^2 + 10.7x - 27.9$.

(b) The cubic polynomial $f(x)$ models monthly average temperature at Trout Lake, Canada, in degrees Fahrenheit, where $x = 1$ corresponds to January and $x = 12$ represents December. Interpret the zeros of f.

116. *Average High Temperatures*   The monthly average high temperatures in degrees Fahrenheit at Daytona Beach can be modeled by

$$f(x) = 0.0151x^4 - 0.438x^3 + 3.60x^2 - 6.49x + 72.5,$$

where $x = 1$ corresponds to January and $x = 12$ represents December.
(a) Find the average high temperature during March and July.

(b) Graph f in [0.5, 12.5, 1] by [60, 100, 10]. Interpret the graph.

(c) Estimate graphically and numerically when the average high temperature is 80°F.

Polynomial Regression

117. *Water Pollution*   In one study, freshwater mussels were used to monitor copper discharge into a river from an electroplating works. Copper in high doses can be lethal to aquatic life. The table lists copper concentrations in mussels after 45 days at various distances downstream from the plant. The concentration C is measured in micrograms of copper per gram of mussel x kilometers downstream.

x	5	21	37	53	59
C	20	13	9	6	5

Source: R. Foster and J. Bates, "Use of mussels to monitor point source industrial discharges."

(a) Describe the relationship between x and C.

(b) Use regression to find a cubic polynomial function $f(x)$ that models the data.

(c) Graph C and the data.

(d) Concentrations above 10 are lethal to mussels. Locate this region in the river.

118. *Dog Years*   There is a saying that every year of a dog's life is equal to 7 years for a human. A more accurate approximation is given by the graph of f. Given a dog's age x, where $x \ge 1$, $f(x)$ models the equivalent age in human years. According to the Bureau of the Census, middle age for people begins at age 45.
(**Source:** J. Brearley and A. Nicholas, *This Is the Bichon Frise.*)

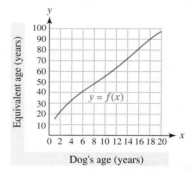

(a) Use the graph of f to estimate the equivalent age for dogs.

(b) Estimate $f(x)$ at $x = 2, 6, 10, 14,$ and 18.

(c) Use regression and the points you estimated to find a quartic polynomial function f that models the data points.

(d) Use $f(x)$ to solve part (a) either graphically or numerically.

Writing about Mathematics

119. Suppose that $f(x)$ is a quintic polynomial with distinct real zeros. Assuming you have access to technology, explain how to factor $f(x)$ approximately. Have you used the factor theorem? Explain.

120. Explain how to determine graphically whether a zero of a polynomial is a multiple zero. Sketch examples.

EXTENDED AND DISCOVERY EXERCISES

Exercises 1–6: Boundedness Theorem   *The boundedness theorem shows how the bottom row of a synthetic division is used to place upper and lower bounds on possible real zeros of a polynomial function.*

Boundedness Theorem

Let $P(x)$ define a polynomial function of degree $n \geq 1$ with real coefficients and with a positive leading coefficient. If $P(x)$ is divided synthetically by $x - c$ and

    (a) if $c > 0$ and all numbers in the bottom row of the synthetic division are nonnegative, then $P(x)$ has no zero greater than c;

    (b) if $c < 0$ and the numbers in the bottom row of the synthetic division alternate in sign (with 0 considered positive or negative, as needed), then $P(x)$ has no zero less than c.

Use the boundedness theorem to show that the real zeros of each polynomial function satisfy the given conditions.

1. $P(x) = x^4 - x^3 + 3x^2 - 8x + 8$;
no real zero greater than 2

2. $P(x) = 2x^5 - x^4 + 2x^3 - 2x^2 + 4x - 4$;
no real zero greater than 1

3. $P(x) = x^4 + x^3 - x^2 + 3$;
no real zero less than -2

4. $P(x) = x^5 + 2x^3 - 2x^2 + 5x + 5$;
no real zero less than -1

5. $P(x) = 3x^4 + 2x^3 - 4x^2 + x - 1$;
no real zero greater than 1

6. $P(x) = 3x^4 + 2x^3 - 4x^2 + x - 1$;
no real zero less than -2

CHECKING BASIC CONCEPTS FOR SECTIONS 4.3 AND 4.4

1. Simplify the expression $\dfrac{5x^4 - 10x^3 + 5x^2}{5x^2}$.

2. Divide the expression.

  (a) $\dfrac{x^3 - x^2 + 4x - 4}{x - 1}$

  (b) $\dfrac{2x^3 - 3x^2 + 4x + 4}{2x + 1}$

  (c) $\dfrac{x^4 - 3x^3 + 6x^2 - 13x + 9}{x^2 + 4}$

3. Use the graph of the cubic polynomial $f(x)$ in the next column to determine its complete factored form. State the multiplicity of each zero. Assume that all zeros are integers and that the leading coefficient is *not* ± 1.

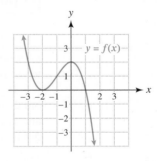

4. Solve $x^3 - 2x^2 - 15x = 0$.

5. Determine graphically the zeros of
$$f(x) = x^4 - x^3 - 18x^2 + 16x + 32.$$
Write $f(x)$ in complete factored form.

4.5 The Fundamental Theorem of Algebra

- Apply the fundamental theorem of algebra
- Factor polynomials having complex zeros
- Solve polynomial equations having complex solutions

Introduction

In Section 3.2 the quadratic formula was used to solve $ax^2 + bx + c = 0$. Are there similar formulas for higher degree polynomial equations? One of the most spectacular mathematical achievements during the sixteenth century was the discovery of formulas for solving cubic and quartic equations. This was accomplished by the Italian mathematicians Tartaglia, Cardano, Fior, del Ferro, and Ferrari between 1515 and 1545. These formulas are quite complicated and typically used only in computer software. Another spectacular result came from Carl Friedrich Gauss in 1797. He proved that all polynomials can be completely factored by using complex numbers. This result is called the *fundamental theorem of algebra*. Between 1750 and 1780 both Euler and Lagrange failed at finding symbolic solutions to the quintic equation $ax^5 + bx^4 + cx^3 + dx^2 + ex + k = 0$. Later, in about 1805, the Italian physician Ruffini proved that finding formulas for quintic or higher degree equations was impossible. His results make it necessary for us to rely on numerical and graphical methods.   (**Source:** H. Eves, *An Introduction to the History of Mathematics.*)

Fundamental Theorem of Algebra

One of the most brilliant mathematicians of all time, Carl Friedrich Gauss proved the fundamental theorem of algebra as part of his doctoral thesis at age 20. Although his theorem and proof were completed in 1797, they are still valid today.

Fundamental Theorem of Algebra

A polynomial $f(x)$ of degree n, with $n \geq 1$, has at least one complex zero.

The fundamental theorem of algebra guarantees that *every polynomial has a complete factorization,* if we are allowed to use complex numbers. (A complex number can be written as $a + bi$. If $b = 0$, then the complex number $a + bi$ is also a real number.)

If $f(x)$ is a polynomial of degree 1 or higher, then by the fundamental theorem of algebra there is a zero c_1 such that $f(c_1) = 0$. By the factor theorem, $(x - c_1)$ is a factor of $f(x)$ and $f(x) = (x - c_1) q_1(x)$ for some polynomial $q_1(x)$. If $q_1(x)$ has positive degree, then by the fundamental theorem of algebra there exists a zero c_2 of $q_1(x)$. By the factor theorem, $q_1(x)$ can be written as $q_1(x) = (x - c_2) q_2(x)$. Then

$$f(x) = (x - c_1)q_1(x) = (x - c_1)(x - c_2)q_2(x).$$

If $f(x)$ has degree n, this process can be continued until $f(x)$ is written in the complete factored form

$$f(x) = a_n(x - c_1)(x - c_2) \cdots (x - c_n),$$

where a_n is the leading coefficient and the c_k are complex zeros of $f(x)$. If each c_k is distinct, then $f(x)$ has n zeros. However, in general the c_k may not be distinct since multiple zeros are possible.

Number of Zeros Theorem

A polynomial of degree n has at most n distinct zeros.

EXAMPLE 1   Classifying zeros

All zeros for the given polynomials are distinct. Use Figures 4.69–4.71 to determine graphically the number of real zeros and the number of imaginary zeros.

(a) $f(x) = 3x^3 - 3x^2 - 3x - 5$ (b) $g(x) = 2x^2 + x + 1$
(c) $h(x) = -x^4 + 4x^2 + 4$

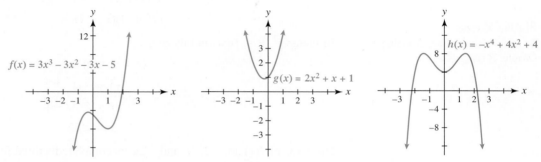

Figure 4.69   Figure 4.70   Figure 4.71

SOLUTION

Getting Started Each (distinct) real zero corresponds to an x-intercept. Imaginary zeros do *not* correspond to x-intercepts, but their number can be determined after the number of real zeros is known. ▶

(a) The graph of $f(x)$ in Figure 4.69 crosses the x-axis once so there is one real zero. Since f is degree 3 and all zeros are distinct, there are two imaginary zeros.
(b) The graph of $g(x)$ in Figure 4.70 never crosses the x-axis. Since g is degree 2, there are no real zeros and two imaginary zeros.
(c) The graph of $h(x)$ is shown in Figure 4.71. Since h is degree 4, there are two real zeros and the remaining two zeros are imaginary.   ***Now Try Exercises 1, 3, and 5*** ◀

EXAMPLE 2   Constructing a polynomial with prescribed zeros

Represent a polynomial $f(x)$ of degree 4 with leading coefficient 2 and zeros $-3, 5, i$, and $-i$ in (a) complete factored form and (b) expanded form.

SOLUTION

(a) Let $a_n = 2, c_1 = -3, c_2 = 5, c_3 = i$, and $c_4 = -i$. Then

$$f(x) = 2(x + 3)(x - 5)(x - i)(x + i).$$

(b) To expand this expression for $f(x)$, perform the following steps.

$$\begin{aligned}
2(x + 3)(x - 5)(x - i)(x + i) &= 2(x + 3)(x - 5)(x^2 + 1) \\
&= 2(x + 3)(x^3 - 5x^2 + x - 5) \\
&= 2(x^4 - 2x^3 - 14x^2 - 2x - 15) \\
&= 2x^4 - 4x^3 - 28x^2 - 4x - 30
\end{aligned}$$

Thus $f(x) = 2x^4 - 4x^3 - 28x^2 - 4x - 30$.   ***Now Try Exercise 13*** ◀

EXAMPLE 3   Factoring a cubic polynomial with imaginary zeros

Determine the complete factored form for $f(x) = x^3 + 2x^2 + 4x + 8$.

SOLUTION   We can use factoring by grouping to determine the complete factored form.

Algebra Review
To review factoring by grouping, see Chapter R (page R-21).

$$
\begin{aligned}
x^3 + 2x^2 + 4x + 8 &= (x^3 + 2x^2) + (4x + 8) && \text{Associative property}\\
&= x^2(x + 2) + 4(x + 2) && \text{Distributive property}\\
&= (x^2 + 4)(x + 2) && \text{Factor out } x + 2.
\end{aligned}
$$

To factor $x^2 + 4$, first find its zeros.

$$
\begin{aligned}
x^2 + 4 &= 0\\
x^2 &= -4\\
x &= \pm\sqrt{-4}\\
x &= \pm 2i
\end{aligned}
$$

The zeros of $f(x)$ are -2, $2i$, and $-2i$. Its complete factored form is

$$f(x) = (x + 2)(x - 2i)(x + 2i).$$ ***Now Try Exercise 29*** ◀

Conjugate Zeros   Notice that in Example 3 both $2i$ and $-2i$ were zeros of $f(x)$. The numbers $2i$ and $-2i$ are conjugates. This result can be generalized.

Conjugate Zeros Theorem

If a polynomial $f(x)$ has only real coefficients and if $a + bi$ is a zero of $f(x)$, then the conjugate $a - bi$ is also a zero of $f(x)$.

EXAMPLE 4   Constructing a polynomial with prescribed zeros

Find a cubic polynomial $f(x)$ with real coefficients, leading coefficient 2, and zeros 3 and $5i$. Express f in **(a)** complete factored form and **(b)** expanded form.

SOLUTION

(a) Since $f(x)$ has real coefficients, it must also have a third zero of $-5i$, the conjugate of $5i$. Let $c_1 = 3$, $c_2 = 5i$, $c_3 = -5i$, and $a_n = 2$. The complete factored form is

$$f(x) = 2(x - 3)(x - 5i)(x + 5i).$$

(b) To expand $f(x)$, perform the following steps.

$$
\begin{aligned}
2(x - 3)(x - 5i)(x + 5i) &= 2(x - 3)(x^2 + 25)\\
&= 2(x^3 - 3x^2 + 25x - 75)\\
&= 2x^3 - 6x^2 + 50x - 150
\end{aligned}
$$

Now Try Exercise 15 ◀

EXAMPLE 5   Finding imaginary zeros of a polynomial

Find the zeros of $f(x) = x^4 + x^3 + 2x^2 + x + 1$, given that one zero is $-i$.

SOLUTION   By the conjugate zeros theorem, it follows that i must also be a zero of $f(x)$. Therefore $(x - i)$ and $(x + i)$ are factors of $f(x)$. Because $(x - i)(x + i) = x^2 + 1$, we can use long division to find another quadratic factor of $f(x)$.

$$\begin{array}{r} x^2 + x + 1 \\ x^2 + 0x + 1 \overline{)x^4 + x^3 + 2x^2 + x + 1} \\ \underline{x^4 + 0x^3 + x^2} \\ x^3 + x^2 + x \\ \underline{x^3 + 0x^2 + x} \\ x^2 + 0x + 1 \\ \underline{x^2 + 0x + 1} \\ 0 \end{array}$$

The quotient is $x^2 + x + 1$ with remainder 0. By the division algorithm,

$$x^4 + x^3 + 2x^2 + x + 1 = (x^2 + 1)(x^2 + x + 1).$$

We can use the quadratic formula to find the zeros of $x^2 + x + 1$.

$$x = \frac{-b \pm \sqrt{b^2 - 4ac}}{2a}$$

$$= \frac{-1 \pm \sqrt{1^2 - 4(1)(1)}}{2(1)}$$

$$= -\frac{1}{2} \pm i\frac{\sqrt{3}}{2}$$

The four zeros of $f(x)$ are $\pm i$ and $-\frac{1}{2} \pm i\frac{\sqrt{3}}{2}$.

Now Try Exercise 21 ◀

Polynomial Equations with Complex Solutions

Every polynomial equation of degree n can be written in the form

$$a_n x^n + \cdots + a_2 x^2 + a_1 x + a_0 = 0.$$

If we let $f(x) = a_n x^n + \cdots + a_2 x^2 + a_1 x + a_0$ and write $f(x)$ in complete factored form as

$$f(x) = a_n(x - c_1)(x - c_2) \cdots (x - c_n),$$

then the solutions to the polynomial equation are the zeros $c_1, c_2, \ldots, c_n$ of $f(x)$. Solving cubic and quartic polynomial equations with this technique is illustrated in the next two examples.

EXAMPLE 6 Solving a polynomial equation

Solve $x^3 = 3x^2 - 7x + 21$.

SOLUTION Write the equation as $f(x) = 0$, where $f(x) = x^3 - 3x^2 + 7x - 21$. Although we could use factoring by grouping, as is done in Example 3, we use graphing instead to find one real zero of $f(x)$. Figure 4.72 shows that 3 is a zero of $f(x)$. By the factor theorem, $x - 3$ is a factor of $f(x)$. Using synthetic division, we divide $x - 3$ into $f(x)$.

$$\begin{array}{r} 3| \quad 1 \quad -3 \quad 7 \quad -21 \\ \quad \quad \quad 3 \quad 0 \quad 21 \\ \hline \quad 1 \quad 0 \quad 7 \quad 0 \end{array}$$

[−5, 5, 1] by [−30, 30, 10]

$y = x^3 - 3x^2 + 7x - 21$

Zero
X=3   Y=0

Figure 4.72

Thus $x^3 - 3x^2 + 7x - 21 = (x - 3)(x^2 + 7)$, and we can solve as follows.

$$x^3 - 3x^2 + 7x - 21 = 0 \qquad f(x) = 0$$
$$(x - 3)(x^2 + 7) = 0 \qquad \text{Factor.}$$
$$x - 3 = 0 \quad \text{or} \quad x^2 + 7 = 0 \qquad \text{Zero-product property}$$
$$x = 3 \quad \text{or} \quad x^2 = -7 \qquad \text{Solve.}$$
$$x = 3 \quad \text{or} \quad x = \pm i\sqrt{7} \qquad \text{Property of } i$$

The solutions are 3 and $\pm i\sqrt{7}$.

Now Try Exercise 33 ◀

EXAMPLE 7   Solving a polynomial equation

Solve $x^4 + x^2 = x^3$.

SOLUTION   Write the equation as $f(x) = 0$, where $f(x) = x^4 - x^3 + x^2$.

$$x^4 - x^3 + x^2 = 0 \qquad f(x) = 0$$
$$x^2(x^2 - x + 1) = 0 \qquad \text{Factor out } x^2.$$
$$x^2 = 0 \quad \text{or} \quad x^2 - x + 1 = 0 \qquad \text{Zero-product property}$$

The only solution to $x^2 = 0$ is 0. To solve $x^2 - x + 1 = 0$, use the quadratic formula, as in Example 5. The solutions are 0 and $\frac{1}{2} \pm i\frac{\sqrt{3}}{2}$.

The graphs of $Y_1 = X^4 + X^2$ and $Y_2 = X^3$ are shown in Figure 4.73. Notice that they appear to intersect only at the origin. This indicates that the only real solution is 0.

Now Try Exercise 37 ◀

[−3, 3, 1] by [−2, 2, 1]

Figure 4.73

4.5 Putting It All Together

Some of the important topics in this section are summarized in the following table.

Concept	Explanation	Comments and Examples
Number of zeros theorem	A polynomial of degree n has at most n distinct zeros. These zeros can be real or imaginary numbers.	The cubic polynomial, $$ax^3 + bx^2 + cx + d,$$ has *at most* 3 distinct zeros.
Fundamental theorem of algebra	A polynomial of degree n, with $n \geq 1$, has at least one complex zero.	This theorem guarantees that we can always factor a polynomial $f(x)$ into complete factored form: $$f(x) = a_n(x - c_1) \cdots (x - c_n),$$ where the c_k are complex numbers.
Conjugate zeros theorem	If a polynomial has *real* coefficients and $a + bi$ is a zero, then $a - bi$ is also a zero.	Since $\frac{1}{2} + \frac{1}{2}i$ is a zero of $2x^2 - 2x + 1$, it follows that $\frac{1}{2} - \frac{1}{2}i$ is also a zero.

4.5 Exercises

Zeros of Polynomials

Exercises 1–8: The graph and degree of a polynomial with real coefficients $f(x)$ are given. Determine the number of real zeros and the number of imaginary zeros. Assume that all zeros of $f(x)$ are distinct.

1. Degree 2

2. Degree 2

3. Degree 3

4. Degree 3

5. Degree 4

6. Degree 4

7. Degree 5

8. Degree 5

Exercises 9–18: Let a_n be the leading coefficient.

(a) *Find the complete factored form of a polynomial with real coefficients $f(x)$ that satisfies the conditions.*

(b) *Express $f(x)$ in expanded form.*

9. Degree 2; $a_n = 1$; zeros $6i$ and $-6i$

10. Degree 3; $a_n = 5$; zeros 2, i, and $-i$

11. Degree 3; $a_n = -1$; zeros -1, $2i$, and $-2i$

12. Degree 4; $a_n = 3$; zeros -2, 4, i, and $-i$

13. Degree 4; $a_n = 10$; zeros 1, -1, $3i$, and $-3i$

14. Degree 2; $a_n = -5$; zeros $1 + i$ and $1 - i$

15. Degree 4; $a_n = \frac{1}{2}$; zeros $-i$ and $2i$

16. Degree 3; $a_n = -\frac{3}{4}$; zeros $-3i$ and $\frac{2}{5}$

17. Degree 3; $a_n = -2$; zeros $1 - i$ and 3

18. Degree 4; $a_n = 7$; zeros $2i$ and $3i$

Exercises 19–22: (Refer to Example 5.) Find the zeros of $f(x)$, given that one zero is k.

19. $f(x) = 3x^3 - 5x^2 + 75x - 125$ $k = \frac{5}{3}$

20. $f(x) = x^4 + 2x^3 + 8x^2 + 8x + 16$ $k = 2i$

21. $f(x) = 2x^4 - x^3 + 19x^2 - 9x + 9$ $k = -3i$

22. $f(x) = 7x^3 + 5x^2 + 12x - 4$ $k = \frac{2}{7}$

Exercises 23–30: Complete the following.

(a) *Find all zeros of $f(x)$.*

(b) *Write the complete factored form of $f(x)$.*

23. $f(x) = x^2 + 25$ **24.** $f(x) = x^2 + 11$

25. $f(x) = 3x^3 + 3x$ **26.** $f(x) = 2x^3 + 10x$

27. $f(x) = x^4 + 5x^2 + 4$ **28.** $f(x) = x^4 + 4x^2$

29. $f(x) = x^3 + 2x^2 + 16x + 32$

30. $f(x) = x^4 + 2x^3 + x^2 + 8x - 12$

Exercises 31–42: Solve the polynomial equation.

31. $x^3 + x = 0$ **32.** $2x^3 - x + 1 = 0$

33. $x^3 = 2x^2 - 7x + 14$ **34.** $x^2 + x + 2 = x^3$

35. $x^4 + 5x^2 = 0$ **36.** $x^4 - 2x^3 + x^2 - 2x = 0$

37. $x^4 = x^3 - 4x^2$ **38.** $x^5 + 9x^3 = x^4 + 9x^2$

39. $x^4 + x^3 = 16 - 8x - 6x^2$

40. $x^4 + 2x^2 = x^3$ **41.** $3x^3 + 4x^2 + 6 = x$

42. $2x^3 + 5x^2 + x + 12 = 0$

Applications

Exercises 43–48: Electricity   Complex numbers are used in the study of electrical circuits. Impedance Z (or the opposition to the flow of electricity), voltage V, and current I can all be represented by complex numbers. They are related by the equation $Z = \frac{V}{I}$. Find the value of the missing variable.

43. $V = 50 + 98i$ $I = 8 + 5i$

44. $V = 30 + 60i$ $I = 8 + 6i$

45. $I = 1 + 2i$ $Z = 3 - 4i$

46. $I = \frac{1}{2} + \frac{1}{4}i$ $Z = 8 - 9i$

47. $Z = 22 - 5i$ $V = 27 + 17i$

48. $Z = 10 + 5i$ $V = 10 + 8i$

Writing about Mathematics

49. Could a cubic function with real coefficients have only imaginary zeros? Explain.

50. Give an example of a polynomial function that has only imaginary zeros and a polynomial function that has only real zeros. Explain how to determine graphically if a function has only imaginary zeros.

4.6 Rational Functions and Models

- **Identify a rational function and state its domain**
- **Identify asymptotes**
- **Interpret asymptotes**
- **Graph a rational function by using transformations**
- **Graph a rational function by hand (optional)**

Algebra Review
To review rational expressions, see Chapter R (page R-30).

Introduction

Rational functions are nonlinear functions that frequently occur in applications. For example, rational functions are used to design curves for railroad tracks, determine stopping distances on hills, and calculate the average number of people waiting in a line.

Rational Functions

A *rational number* can be expressed as a *ratio* $\frac{p}{q}$, where p and q are integers with $q \neq 0$. A rational function is defined similarly by using the concept of a polynomial.

Rational Function

A function f represented by $f(x) = \frac{p(x)}{q(x)}$, where $p(x)$ and $q(x)$ are polynomials and $q(x) \neq 0$, is a **rational function**.

The domain of a rational function includes all real numbers *except* the zeros of the denominator $q(x)$. The graph of a rational function is continuous except at *x*-values where $q(x) = 0$.

EXAMPLE 1 Identifying rational functions

Determine if the function is rational and state its domain.

(a) $f(x) = \dfrac{2x - 1}{x^2 + 1}$ **(b)** $g(x) = \dfrac{1}{\sqrt{x}}$ **(c)** $h(x) = \dfrac{x^3 - 2x^2 + 1}{x^2 - 3x + 2}$

SOLUTION

(a) Both the numerator, $2x - 1$, and the denominator, $x^2 + 1$, are polynomials, so f is a rational function. The domain of f includes all real numbers because $x^2 + 1 \neq 0$ for any real number x.

(b) $\sqrt{x}$ is not a polynomial, so g is not a rational function. The domain is $\{x \mid x > 0\}$.

(c) Both the numerator and the denominator are polynomials, so h is a rational function. Because

$$x^2 - 3x + 2 = (x - 1)(x - 2) = 0$$

when $x = 1$ or $x = 2$, the domain of h is $\{x \mid x \neq 1, x \neq 2\}$.

Now Try Exercises 1, 7, and 9 ◀

CLASS DISCUSSION

Is an integer a rational number? Is a polynomial function a rational function?

Vertical Asymptotes

A rational function given by $f(x) = \dfrac{p(x)}{q(x)}$ is undefined whenever $q(x) = 0$. If $q(k) = 0$ for some k, then a *vertical asymptote* of the graph of f *may* occur at $x = k$. Near a vertical asymptote, the y-values on the graph of f become very large (unbounded) in absolute value. Vertical asymptotes can have meaning in many types of applications.

An Application If cars leave a parking garage randomly and stop to pay the parking attendant on the way out, then the average length of the line depends on two factors: the average traffic rate at which cars are exiting the ramp and the average rate at which the parking attendant can wait on cars. For instance, if the average traffic rate is three cars per minute and the parking attendant can serve four cars per minute, then at times a line may form *if* cars arrive in a *random* manner. The **traffic intensity** x is the ratio of the average traffic rate to the average working rate of the attendant. In this example, $x = \frac{3}{4}$. (**Source:** F. Mannering and W. Kilareski, *Principles of Highway Engineering and Traffic Control.*)

EXAMPLE 2 Estimating the length of parking garage lines

If the traffic intensity is x, then the average number of cars waiting in line to exit a parking garage can be estimated by $N(x) = \dfrac{x^2}{2 - 2x}$, where $0 \leq x < 1$.

(a) Evaluate $N(0.5)$ and $N(0.9)$. Interpret the results.

(b) Use the graph of $y = N(x)$ in Figure 4.74 to explain what happens to the length of the line as the traffic intensity x increases to 1 from the left (denoted $x \to 1^-$.)

SOLUTION

(a) $N(0.5) = \dfrac{0.5^2}{2 - 2(0.5)} = 0.25$ and $N(0.9) = \dfrac{0.9^2}{2 - 2(0.9)} = 4.05$. This means that if the traffic intensity is 0.5, there is little waiting in line. As the traffic intensity increases to 0.9, the average line has more than four cars.

$N(x) = \dfrac{x^2}{2 - 2x}$

Average line length (cars)

Traffic intensity

Figure 4.74 Parking Garage Lines

$$N(x) = \frac{x^2}{2 - 2x}$$

Traffic intensity

(b) As the traffic intensity x approaches 1 from the left in Figure 4.74 (repeated in the margin), the graph of f increases rapidly without bound. Numerical support is given in Table 4.4. With a traffic intensity slightly less than 1, the attendant has difficulty keeping up. If cars occasionally arrive in groups, long lines will form. At $x = 1$ the denominator, $2 - 2x$, equals 0 and $N(x)$ is undefined.

Table 4.4 Traffic Intensity Approaching 1

x	0.94	0.95	0.96	0.97	0.98	0.99	1
$\dfrac{x^2}{2 - 2x}$	7.36	9.03	11.52	15.68	24.01	49.01	—

Now Try Exercise 93 ◀

Formal Definition   In Figure 4.74 (repeated in the margin), the red dashed vertical line $x = 1$ is a *vertical asymptote* of the graph of N. A graph of a different rational function f is shown in Figure 4.75. The $f(x)$-values *decrease without bound* as x approaches 2 from the left. This is denoted $f(x) \to -\infty$ as $x \to 2^-$. Similarly, the $f(x)$-values *increase without bound* as x approaches 2 from the right. This is expressed as $f(x) \to \infty$ as $x \to 2^+$. The red dashed line $x = 2$ is a vertical asymptote of the graph of f.

$y = f(x)$

Figure 4.75

NOTE A vertical asymptote is not part of the graph of a rational function; rather, it is an aid that is used to sketch and to better understand the graph of a rational function.

Vertical Asymptote

The line $x = k$ is a **vertical asymptote** of the graph of f if $f(x) \to \infty$ or $f(x) \to -\infty$ as x approaches k from either the left or the right.

If $x = k$ is a vertical asymptote of the graph of f, then k is not in the domain of f. Furthermore, the graph of a rational function f does *not* cross a vertical asymptote.

Horizontal Asymptotes

If the absolute value of x becomes large in the formula $f(x)$ for a rational function, then the graph of f *may* level off and begin to approximate a horizontal line. This horizontal line is called a *horizontal asymptote*. Horizontal asymptotes can have meaning in many types of applications.

$y = 25$

Figure 4.76 Size of a Small Fish

An Application   The graph of f in Figure 4.76 is an example of a von Bertalanffy growth curve. It models the length in millimeters of a small fish after x weeks. After several weeks the length of the fish begins to level off near 25 millimeters. Thus $y = 25$ is a horizontal asymptote of the graph of f. This is denoted by $f(x) \to 25$ as $x \to \infty$. **(Source:** D. Brown and P. Rothery, *Models in Biology*.)

In real-life applications, time does not actually approach infinity. For example, a fish does not live forever. However, the asymptote $y = 25$ does tend to model the length of the fish as it becomes older.

Formal Definition The graph of $f(x) = \frac{x^2}{x^2 + 1}$ is shown in Figure 4.77. The red dashed horizontal line is a horizontal asymptote of the graph of f and indicates that $f(x) \to 1$ as $x \to \infty$ or $x \to -\infty$. For example, $f(10) \approx 0.99$, $f(100) \approx 0.9999$, and $f(1000) \approx 0.999999$.

Figure 4.77

Horizontal Asymptote

The line $y = b$ is a **horizontal asymptote** of the graph of f if $f(x) \to b$ as x approaches either ∞ or $-\infty$.

NOTE Like a vertical asymptote, a horizontal asymptote is *not* part of the graph of a rational function; rather, it is an aid that is used to sketch and to better understand the graph of a rational function. However, unlike in the case of vertical asymptotes, it *is possible* for the graph of a rational function to *cross* a horizontal asymptote. See Examples 4(c) and 10.

Identifying Asymptotes

Asymptotes can be found visually from a graph and symbolically from a formula. The next two examples discuss these techniques.

EXAMPLE 3 Determining horizontal and vertical asymptotes visually

Use the graph of each rational function to determine any vertical or horizontal asymptotes.

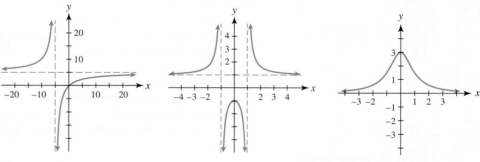

Figure 4.78   **Figure 4.79**   **Figure 4.80**

SOLUTION In Figure 4.78, $x = -5$ is a vertical asymptote and $y = 5$ is a horizontal asymptote. In Figure 4.79, $x = \pm 1$ are vertical asymptotes and $y = 1$ is a horizontal asymptote. In Figure 4.80, there are no vertical asymptotes. The x-axis ($y = 0$) is a horizontal asymptote.

Now Try Exercises 13, 15, and 17 ◄

The following technique can be used for rational functions to find vertical and horizontal asymptotes symbolically.

Finding Vertical and Horizontal Asymptotes

Let f be a rational function given by $f(x) = \dfrac{p(x)}{q(x)}$ written in *lowest* terms.

Vertical Asymptote

To find a vertical asymptote, set the denominator, $q(x)$, equal to 0 and solve. If k is a zero of $q(x)$, then $x = k$ is a vertical asymptote. *Caution:* If k is a zero of both $q(x)$ *and* $p(x)$, then $f(x)$ is *not* written in lowest terms, and $x - k$ is a common factor.

Horizontal Asymptote

(a) If the degree of the numerator is less than the degree of the denominator, then $y = 0$ (the x-axis) is a horizontal asymptote.

(b) If the degree of the numerator equals the degree of the denominator, then $y = \frac{a}{b}$ is a horizontal asymptote, where a is the leading coefficient of the numerator and b is the leading coefficient of the denominator.

(c) If the degree of the numerator is greater than the degree of the denominator, then there are no horizontal asymptotes.

EXAMPLE 4 Finding asymptotes

For each rational function, determine any horizontal or vertical asymptotes.

(a) $f(x) = \dfrac{6x - 1}{3x + 3}$ **(b)** $g(x) = \dfrac{x + 1}{x^2 - 4}$

(c) $h(x) = \dfrac{x^2 - 1}{x + 1}$

SOLUTION

(a) The degrees of the numerator and the denominator are both 1. Since the ratio of the leading coefficients is $\frac{6}{3} = 2$, the graph of f has a horizontal asymptote of $y = 2$. This is supported numerically in Figures 4.81 and 4.82, where the y-values approach 2 as the x-values increase or decrease. (That is, $y \to 2$ as $x \to \infty$ and $y \to 2$ as $x \to -\infty$.)

When $x = -1$, the denominator, $3x + 3$, equals 0 and the numerator, $6x - 1$, does not equal 0. Thus $x = -1$ is a vertical asymptote. A graph of f is shown in Figure 4.83. Note that it is *not* necessary to graph the function to determine the asymptotes.

(b) The degree of the numerator is one less than the degree of the denominator, so the x-axis, or $y = 0$, is a horizontal asymptote. When $x = \pm 2$, the denominator, $x^2 - 4$, equals 0 and the numerator, $x + 1$, does not equal 0. Thus $x = \pm 2$ are vertical asymptotes. See Figure 4.84. Note that the graph crosses the horizontal asymptote $y = 0$.

Figure 4.81

Figure 4.82

Algebra Review
To review simplifying rational expressions, see Chapter R (page R-30).

Figure 4.83 **Figure 4.84**

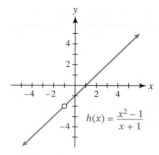

Figure 4.85

(c) The degree of the numerator is greater than the degree of the denominator, so there are no horizontal asymptotes. When $x = -1$, both numerator and denominator equal 0, so the expression is *not* in lowest terms. We can simplify $h(x)$ as follows.

$$h(x) = \frac{x^2 - 1}{x + 1} = \frac{(x + 1)(x - 1)}{x + 1} = x - 1, \qquad x \neq -1$$

The graph of $h(x)$ is the line $y = x - 1$ with the point $(-1, -2)$ missing. There are no vertical asymptotes. See Figure 4.85.   **Now Try Exercises 21, 23, and 31** ◀

Slant, or Oblique, Asymptotes   A third type of asymptote, which is neither vertical nor horizontal, occurs when the numerator of a rational function has degree *one more* than the degree of the denominator. For example, let $f(x) = \frac{x^2 + 2}{x - 1}$. If $x - 1$ is divided into $x^2 + 2$, the quotient is $x + 1$ with remainder 3. Thus

$$f(x) = x + 1 + \frac{3}{x - 1}$$

is an equivalent representation of f. For large values of $|x|$, the ratio $\frac{3}{x - 1}$ approaches 0 and the graph of f approaches $y = x + 1$. The line $y = x + 1$ is called a **slant asymptote**, or **oblique asymptote**, of the graph of f. A graph of f with vertical asymptote $x = 1$ and slant asymptote $y = x + 1$ is shown in Figure 4.86.

Figure 4.86

MAKING CONNECTIONS

Division Algorithm and Asymptotes   Suppose that the division algorithm is used to write a rational function f in the form

$$f(x) = (\text{Quotient}) + \frac{(\text{Remainder})}{(\text{Divisor})}.$$

1. If the quotient equals a constant k, then $y = k$ is a horizontal asymptote.
   Example: $f(x) = \frac{2x - 1}{x - 1} = 2 + \frac{1}{x - 1}$, so $y = 2$ is a horizontal asymptote.
2. If the quotient equals $ax + b$ with $a \neq 0$ (linear), then $y = ax + b$ is a slant asymptote.
   Example: $f(x) = \frac{x^2 + 2}{x - 1} = x + 1 + \frac{3}{x - 1}$, so $y = x + 1$ is a slant asymptote.

Graphs and Transformations of Rational Functions

Graphs of rational functions can vary greatly in complexity. We begin by graphing $y = \frac{1}{x}$ and then use transformations to graph other rational functions.

EXAMPLE 5   **Analyzing the graph of $y = \frac{1}{x}$**

Sketch a graph of $y = \frac{1}{x}$ and identify any asymptotes.

SOLUTION   Note that when $x = 0$, the denominator is 0 but the numerator is not. Thus $x = 0$ (the y-axis) is a vertical asymptote. Also, the degree of the numerator is less than the degree of the denominator, so $y = 0$ (the x-axis) is a horizontal asymptote. Table 4.5 on the next page lists points that lie on the graph of $y = \frac{1}{x}$. These points and the graph are shown in Figure 4.87 on the next page.

Table 4.5

x	$y = \frac{1}{x}$
-2	$-\frac{1}{2}$
-1	-1
$-\frac{1}{2}$	-2
0	—
$\frac{1}{2}$	2
1	1
2	$\frac{1}{2}$

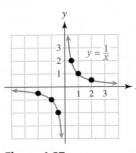

Figure 4.87

Now Try Exercise 41 ◀

$g(x) = \frac{1}{x-1} + 2$

Figure 4.88

Transformations Transformations of graphs can be used to graph some types of rational functions by hand. For example, we can graph $g(x) = \frac{1}{x-1} + 2$ by translating the graph of $f(x) = \frac{1}{x}$ right 1 unit and upward 2 units. That is, $g(x)$ can be written in terms of $f(x)$ by using the formula $g(x) = f(x-1) + 2$. Because the graph of f in Figure 4.87 has vertical asymptote $x = 0$ and horizontal asymptote $y = 0$, the graph of g in Figure 4.88 has vertical asymptote $x = 1$ and horizontal asymptote $y = $.

NOTE If we are given function g in the form $g(x) = \frac{2x-1}{x-1}$, then we can use the division algorithm to divide $x - 1$ into $2x - 1$ and obtain quotient 2 with remainder 1.

$$
\begin{array}{r}
2 \\
x-1 \overline{)2x-1} \\
2x-2 \\
\hline
1
\end{array}
$$

$\frac{2x}{x} = 2$

$2(x-1) = 2x - 2$

Subtract: $-1 - (-2) = 1$

Thus $g(x) = 2 + \frac{1}{x-1}$, and we can graph g as in Figure 4.88.

EXAMPLE 6 Using transformations to graph a rational function

Use the graph of $f(x) = \frac{1}{x^2}$ to sketch a graph of $g(x) = -\frac{1}{(x+2)^2}$. Include all asymptotes in your graph. Write $g(x)$ in terms of $f(x)$.

SOLUTION The graph of $y = \frac{1}{x^2}$ shown in Figure 4.89 has vertical asymptote $x = 0$ and horizontal asymptote $y = 0$. The graph of $g(x) = -\frac{1}{(x+2)^2}$ is a translation of the graph of $f(x) = \frac{1}{x^2}$ left 2 units and then a reflection across the x-axis. The vertical asymptote for $y = g(x)$ is $x = -2$ and the horizontal asymptote is $y = 0$, as shown in Figure 4.90. We can write $g(x)$ in terms of $f(x)$ as $g(x) = -f(x+2)$.

Figure 4.89

Figure 4.90

Now Try Exercise 51 ◀

Graphing with Technology   Although many graphing calculators have difficulty accurately showing some features of the graph of a rational function, they can be helpful when they are used in conjunction with symbolic techniques.

Calculator Help
To set dot mode, see Appendix A (page AP-8). To set a decimal window, see Appendix A (page AP-13).

NOTE Calculators often graph in connected or dot mode. (See Example 7.) If connected mode is used to graph a rational function, it may appear as though the calculator is graphing vertical asymptotes automatically. However, in most instances the calculator is connecting points inappropriately. Sometimes rational functions can be graphed in connected mode using a *decimal* or *friendly* viewing rectangle.

EXAMPLE 7   Analyzing a rational function with technology

Let $f(x) = \frac{2x^2 + 1}{x^2 - 4}$.

(a) Use a calculator to graph f. Find the domain of f.
(b) Identify any vertical or horizontal asymptotes.
(c) Sketch a graph of f that includes the asymptotes.

SOLUTION

$[-6, 6, 1]$ by $[-6, 6, 1]$

Figure 4.91 Dot Mode

(a) A calculator graph of f using dot mode is shown in Figure 4.91. The function is undefined when $x^2 - 4 = 0$, or when $x = \pm 2$. The domain of function f is given by $D = \{x \mid x \neq 2, x \neq -2\}$.
(b) When $x = \pm 2$, the denominator, $x^2 - 4$, equals 0 and the numerator, $2x^2 + 1$, does not equal 0. Therefore $x = \pm 2$ are vertical asymptotes. The degree of the numerator equals the degree of the denominator, and the ratio of the leading coefficients is $\frac{2}{1} = 2$. A horizontal asymptote of the graph of f is $y = 2$.
(c) A second graph of f and its asymptotes is shown in Figure 4.92.

Figure 4.92   *Now Try Exercise 57* ◄

Graphs with "Holes"   If $f(x) = \frac{p(x)}{q(x)}$, then it is possible that, for some number k, both $p(k) = 0$ and $q(k) = 0$. In this case, the graph of f may *not* have a vertical asymptote at $x = k$; rather, it may have a "hole" at $x = k$. See the next example and Figure 4.85.

EXAMPLE 8   Graphing a rational function having a "hole"

Graph $f(x) = \frac{2x^2 - 5x + 2}{x^2 - 3x + 2}$ by hand.

SOLUTION First factor the numerator and the denominator.

$$\frac{2x^2 - 5x + 2}{x^2 - 3x + 2} = \frac{(2x - 1)(x - 2)}{(x - 1)(x - 2)} = \frac{2x - 1}{x - 1}, \qquad x \neq 2$$

$y = \dfrac{1}{(x-1)} + 2$

$x \neq 2$

Figure 4.93

After factoring, it is apparent that both the numerator and the denominator equal 0 when $x = 2$. Therefore we simplify the rational expression to lowest terms and restrict the domain to $x \neq 2$.

From our previous work, long division can be used to show that $f(x) = \dfrac{2x - 1}{x - 1}$ is equivalent to $f(x) = 2 + \dfrac{1}{x - 1}$. Thus the graph of f is similar to Figure 4.88 except that the point $(2, 3)$ is missing and an open circle appears in its place. See Figure 4.93.

Now Try Exercise 69 ◄

Graphing Rational Functions by Hand (Optional)

To graph a rational function by hand, we sometimes need to solve a rational equation of the form $\dfrac{a}{b} = \dfrac{c}{d}$. One way to solve this equation is to **cross multiply** and obtain $ad = bc$, provided b and d are nonzero. Consider the following example.

$\dfrac{2x - 1}{3x + 2} = \dfrac{5}{4}$	Given equation
$4(2x - 1) = 5(3x + 2)$	Cross multiply.
$8x - 4 = 15x + 10$	Simplify.
$-7x = 14$	Subtract 15x; add 4.
$x = -2$	Divide by -7. The answer checks.

The following guidelines can be used to graph a rational function by hand.

Graphing a Rational Function

Let $f(x) = \dfrac{p(x)}{q(x)}$ define a rational function in *lowest* terms. To sketch its graph, follow these steps:

STEP 1: Find all vertical asymptotes.

STEP 2: Find all horizontal or oblique asymptotes.

STEP 3: Find the y-intercept, if possible, by evaluating $f(0)$.

STEP 4: Find the x-intercepts, if any, by solving $f(x) = 0$. (These will be the zeros of the numerator $p(x)$.)

STEP 5: Determine whether the graph will intersect its nonvertical asymptote $y = b$ by solving $f(x) = b$, where b is the y-value of the horizontal asymptote, or by solving $f(x) = mx + b$, where $y = mx + b$ is the equation of the oblique asymptote.

STEP 6: Plot selected points as necessary. Choose an x-value in each interval of the domain determined by the vertical asymptotes and x-intercepts.

STEP 7: Complete the sketch.

EXAMPLE 9   Graphing a rational function by hand

Graph $f(x) = \dfrac{2x + 1}{x - 3}$.

SOLUTION

Getting Started As you go through Steps 1 through 7, be sure to sketch all asymptotes first and then plot some key points. Finally, sketch the entire graph. ►

STEP 1: The vertical asymptote has equation $x = 3$.

STEP 2: The horizontal asymptote has equation $y = 2$.

STEP 3: $f(0) = -\frac{1}{3}$, so the y-intercept is $-\frac{1}{3}$.

STEP 4: Solve $f(x) = 0$ to find any x-intercepts.

$$\frac{2x + 1}{x - 3} = 0$$

$$2x + 1 = 0 \qquad \textit{If a fraction equals 0, its}$$

$$\qquad\qquad\qquad \textit{numerator must be 0.}$$

$$x = -\frac{1}{2}$$

The x-intercept is $-\frac{1}{2}$.

STEP 5: The graph does not intersect its horizontal asymptote, since $f(x) = 2$ has no solution. (Verify this.)

STEPS 6 AND 7: The points $(-4, 1)$, $\left(1, -\frac{3}{2}\right)$, and $\left(6, \frac{13}{3}\right)$ are on the graph and can be used to complete the sketch, shown in Figure 4.94.   *Now Try Exercise 87* ◀

Figure 4.94

EXAMPLE 10   Graphing a function that intersects its horizontal asymptote

Graph $f(x) = \dfrac{3x^2 - 3x - 6}{x^2 + 8x + 16}$.

SOLUTION

STEP 1: To find the vertical asymptote(s), solve $x^2 + 8x + 16 = 0$.

$$x^2 + 8x + 16 = 0$$
$$(x + 4)^2 = 0$$
$$x = -4$$

Since the numerator is not 0 when $x = -4$, the only vertical asymptote has equation $x = -4$.

STEP 2: Because the degrees of the numerator and denominator both equal 2, the ratio of the leading coefficients can be used to find the horizontal asymptote.

$$y = \frac{3}{1}, \qquad \begin{array}{l}\leftarrow \textit{Leading coefficient of numerator} \\ \leftarrow \textit{Leading coefficient of denominator}\end{array}$$

or $y = 3$.

STEP 3: The y-intercept is $f(0) = -\frac{3}{8}$.

STEP 4: To find the x-intercept(s), if any, solve $f(x) = 0$.

$$\frac{3x^2 - 3x - 6}{x^2 + 8x + 16} = 0$$

$$3x^2 - 3x - 6 = 0 \qquad \textit{Set the numerator equal to 0.}$$

$$x^2 - x - 2 = 0 \qquad \textit{Divide by 3.}$$

$$(x - 2)(x + 1) = 0 \qquad \textit{Factor.}$$

$$x = 2 \quad \text{or} \quad x = -1 \qquad \textit{Zero-product property}$$

The x-intercepts are -1 and 2.

STEP 5: Because the horizontal asymptote is $y = 3$, set $f(x) = 3$ and solve to locate the point where the graph intersects the horizontal asymptote.

Figure 4.95

$$\frac{3x^2 - 3x - 6}{x^2 + 8x + 16} = 3$$

$$3x^2 - 3x - 6 = 3x^2 + 24x + 48 \qquad \text{Let } 3 = \tfrac{3}{1}; \text{ cross multiply.}$$

$$-3x - 6 = 24x + 48 \qquad \text{Subtract } 3x^2.$$

$$-27x = 54 \qquad \text{Subtract } 24x; \text{ add } 6.$$

$$x = -2 \qquad \text{Divide by } -27.$$

The graph intersects its horizontal asymptote at $(-2, 3)$.

STEPS 6 AND 7: Some other points that lie on the graph are $(-10, 9)$, $\left(-8, 13\tfrac{1}{8}\right)$, and $\left(5, \tfrac{2}{3}\right)$. These can be used to complete the graph, shown in Figure 4.95.

Now Try Exercise 91 ◀

4.6 Putting It All Together

The following table summarizes some concepts about rational functions and equations. To determine vertical and horizontal asymptotes see the box on page 312.

Concept	Explanation	Examples
Rational function	$f(x) = \frac{p(x)}{q(x)}$, where $p(x)$ and $q(x)$ are polynomials with $q(x) \neq 0$.	$f(x) = \dfrac{x - 1}{x^2 + 2x + 1}$ $g(x) = 1 + \dfrac{1}{x}$ $\quad \left(\text{Note: } 1 + \dfrac{1}{x} = \dfrac{x + 1}{x}. \right)$
Vertical asymptote	If k is a zero of the denominator, but not of the numerator, then $x = k$ is a vertical asymptote.	The graph of $f(x) = \dfrac{2x + 1}{x - 2}$ has vertical asymptotes at $x = 2$ because 2 is a zero of $x - 2$, but not a zero of $2x + 1$.
Horizontal asymptote	A horizontal asymptote occurs when the degree of the numerator is less than or equal to the degree of the denominator.	$f(x) = \dfrac{1 - 4x^2}{3x^2 - x}$ Horizontal asymptote: $y = -\tfrac{4}{3}$ $g(x) = \dfrac{x}{4x^2 + 2x}$ Horizontal asymptote: $y = 0$
Graph of a rational function	The graph of a rational function is continuous, except at x-values where the denominator equals zero.	The graph of $f(x) = \dfrac{3x^2 + 1}{x^2 - 4}$ is discontinuous at $x = \pm2$. It has vertical asymptotes of $x = \pm2$ and a horizontal asymptote of $y = 3$.

Concept	Explanation	Examples
Basic rational equation	$\frac{a}{b} = \frac{c}{d}$ is equivalent to $ad = bc$, provided b and d are nonzero. Check your answer.	To solve the rational equation $$\frac{4}{2x - 1} = 8,$$ write 8 as $\frac{8}{1}$ and cross multiply. $$4 = 8(2x - 1)$$ $$12 = 16x$$ $$x = \frac{3}{4}$$

4.6   Exercises

Rational Functions

Exercises 1–12: Determine whether f is a rational function and state its domain.

1. $f(x) = \dfrac{x^3 - 5x + 1}{4x - 5}$ **2.** $f(x) = \dfrac{6}{x^2}$

3. $f(x) = x^2 - x - 2$ **4.** $f(x) = \dfrac{x^2 + 1}{\sqrt{x - 8}}$

5. $f(x) = \dfrac{|x - 1|}{x + 1}$ **6.** $f(x) = \dfrac{4}{x} + 1$

7. $f(x) = \dfrac{3x}{x^2 + 1}$ **8.** $f(x) = \dfrac{|x + 1|}{x + 1}$

9. $f(x) = \dfrac{3 - \sqrt{x}}{x^2 + x}$ **10.** $f(x) = \dfrac{x^3 - 3x + 1}{x^2 - 5}$

11. $f(x) = 4 - \dfrac{3}{x + 1}$ **12.** $f(x) = 5x^3 - 4x$

Asymptotes and Graphs

Exercises 13–18: Identify any horizontal or vertical asymptotes in the graph. State the domain of f.

13.

14.

15.

16.

17.

18.

Exercises 19 and 20: In the table, Y_1 is a rational function. Give a possible equation for a horizontal asymptote.

19.

X	Y₁
50	2.8654
100	2.9314
150	2.9539
200	2.9653
250	2.9722
300	2.9768
350	2.9801
X=50	

20.

X	Y₁
−10	4.8922
−20	4.9726
−30	4.9878
−40	4.9931
−50	4.9956
−60	4.9969
−70	4.9978
X=−10	

Exercises 21–32: Find any horizontal or vertical asymptotes.

21. $f(x) = \dfrac{4x + 1}{2x - 6}$ **22.** $f(x) = \dfrac{x + 6}{5 - 2x}$

23. $f(x) = \dfrac{3}{x^2 - 5}$

24. $f(x) = \dfrac{3x^2}{x^2 - 9}$

25. $f(x) = \dfrac{x^4 + 1}{x^2 + 3x - 10}$

26. $f(x) = \dfrac{4x^3 - 2}{x + 2}$

27. $f(x) = \dfrac{x^2 + 2x + 1}{2x^2 - 3x - 5}$

28. $f(x) = \dfrac{6x^2 - x - 2}{2x^2 + x - 6}$

29. $f(x) = \dfrac{3x(x + 2)}{(x + 2)(x - 1)}$

30. $f(x) = \dfrac{x}{x^3 - x}$

31. $f(x) = \dfrac{x^2 - 9}{x + 3}$

32. $f(x) = \dfrac{2x^2 - 3x + 1}{2x - 1}$

Exercises 33–36: Let a be a positive constant. Match f(x) with its graph (a–d) without using a calculator.

33. $f(x) = \dfrac{a}{x - 1}$

34. $f(x) = \dfrac{2x + a}{x - 1}$

35. $f(x) = \dfrac{x - a}{x + 2}$

36. $f(x) = \dfrac{-2x}{x^2 - a}$

a.

b.

c.

d.

Exercises 37–40: Write a formula f(x) for a rational function so that its graph has the specified asymptotes.

37. Vertical: $x = -3$; horizontal: $y = 1$

38. Vertical: $x = 4$; horizontal: $y = -3$

39. Vertical: $x = \pm 3$; horizontal: $y = 0$

40. Vertical: $x = -2$ and $x = 4$; horizontal: $y = 5$

Graphing Rational Functions

Exercises 41–44: Graph f and identify any asymptotes.

41. $f(x) = \dfrac{1}{x^2}$

42. $f(x) = -\dfrac{1}{x}$

43. $f(x) = -\dfrac{1}{2x}$

44. $f(x) = \dfrac{2}{x^2}$

Exercises 45–54: **Transformations** *Use transformations of the graph of either $f(x) = \frac{1}{x}$ or $h(x) = \frac{1}{x^2}$ to sketch a graph of $y = g(x)$ by hand. Show all asymptotes. Write g(x) in terms of either f(x) or h(x).*

45. $g(x) = \dfrac{1}{x - 3}$

46. $g(x) = \dfrac{1}{x + 2}$

47. $g(x) = \dfrac{1}{x} + 2$

48. $g(x) = 1 - \dfrac{2}{x}$

49. $g(x) = \dfrac{1}{x + 1} - 2$

50. $g(x) = \dfrac{1}{x - 2} + 1$

51. $g(x) = -\dfrac{2}{(x - 1)^2}$

52. $g(x) = \dfrac{1}{x^2} - 1$

53. $g(x) = \dfrac{1}{(x + 1)^2} - 2$

54. $g(x) = 1 - \dfrac{1}{(x - 2)^2}$

Exercises 55–62: Complete the following.

   (a) *Find the domain of f.*

  (b) *Graph f in an appropriate viewing rectangle.*

   (c) *Find any horizontal or vertical asymptotes.*

   (d) *Sketch a graph of f that includes any asymptotes.*

55. $f(x) = \dfrac{x + 3}{x - 2}$

56. $f(x) = \dfrac{6 - 2x}{x + 3}$

57. $f(x) = \dfrac{4x + 1}{x^2 - 4}$

58. $f(x) = \dfrac{0.5x^2 + 1}{x^2 - 9}$

59. $f(x) = \dfrac{4}{1 - 0.25x^2}$

60. $f(x) = \dfrac{x^2}{1 + 0.25x^2}$

61. $f(x) = \dfrac{x^2 - 4}{x - 2}$

62. $f(x) = \dfrac{4(x - 1)}{x^2 - x - 6}$

Exercises 63–72: Graph $y = f(x)$. You may want to use division, factoring, or transformations as an aid. Show all asymptotes and "holes."

63. $f(x) = \dfrac{x^2 - 2x + 1}{x - 1}$

64. $f(x) = \dfrac{4x^2 + 4x + 1}{2x + 1}$

65. $f(x) = \dfrac{x + 2}{x + 1}$ **66.** $f(x) = \dfrac{2x + 3}{x + 1}$

67. $f(x) = \dfrac{2x^2 - 3x - 2}{x^2 - 4x + 4}$ **68.** $f(x) = \dfrac{x^2 - x - 2}{x^2 - 2x - 3}$

69. $f(x) = \dfrac{2x^2 + 9x + 9}{2x^2 + 7x + 6}$ **70.** $f(x) = \dfrac{x^2 - 4}{x^2 - x - 6}$

71. $f(x) = \dfrac{-2x^2 + 11x - 14}{x^2 - 5x + 6}$

72. $f(x) = \dfrac{2x^2 - 3x - 14}{x^2 - 2x - 8}$

Slant Asymptotes

Exercises 73–80: Complete the following.

  (a) *Find any slant or vertical asymptotes.*
  (b) *Graph $y = f(x)$. Show all asymptotes.*

73. $f(x) = \dfrac{x^2 + 1}{x + 1}$ **74.** $f(x) = \dfrac{2x^2 - 5x - 2}{x - 2}$

75. $f(x) = \dfrac{0.5x^2 - 2x + 2}{x + 2}$ **76.** $f(x) = \dfrac{0.5x^2 - 5}{x - 3}$

77. $f(x) = \dfrac{x^2 + 2x + 1}{x - 1}$ **78.** $f(x) = \dfrac{2x^2 + 3x + 1}{x - 2}$

79. $f(x) = \dfrac{4x^2}{2x - 1}$ **80.** $f(x) = \dfrac{4x^2 + x - 2}{4x - 3}$

Basic Rational Equations

Exercises 81–86: Solve the equation.

81. $\dfrac{4}{x + 2} = -4$ **82.** $\dfrac{3}{2x + 1} = -1$

83. $\dfrac{x + 1}{x} = 2$ **84.** $\dfrac{2x}{x - 3} = -4$

85. $\dfrac{1 - x}{3x - 1} = -\dfrac{3}{5}$ **86.** $\dfrac{3 - 2x}{x + 2} = 12$

Graphing Rational Functions by Hand

Exercises 87–92: (Refer to Examples 9 and 10.) Graph f. Use the steps for graphing a rational function described in this section.

87. $f(x) = \dfrac{2x - 4}{x - 1}$ **88.** $f(x) = \dfrac{x + 3}{2x - 4}$

89. $f(x) = \dfrac{x^2 - 2x}{x^2 + 6x + 9}$ **90.** $f(x) = \dfrac{2x + 1}{x^2 + 6x + 8}$

91. $f(x) = \dfrac{x^2 + 2x + 1}{x^2 - x - 6}$ **92.** $f(x) = \dfrac{3x^2 + 3x - 6}{x^2 - x - 12}$

Applications

93. *Time Spent in Line*   If two parking attendants can wait on 8 vehicles per minute and vehicles are leaving the parking garage randomly at an average rate of x vehicles per minute, then the average time T in minutes spent waiting in line *and* paying the attendant is given by the formula $T(x) = -\dfrac{1}{x - 8}$, where $0 \le x < 8$. A graph of T is shown in the figure.

  (a) Evaluate $T(4)$ and $T(7.5)$. Interpret the results.

  (b) What happens to the wait as vehicles arrive at an average rate that approaches 8 cars per minute?

94. *Time Spent in Line*   (Refer to Exercise 93.) If the parking attendants can wait on 5 vehicles per minute, the average time T in minutes spent waiting in line *and* paying the attendant becomes $T(x) = -\dfrac{1}{x - 5}$.
  (a) What is a reasonable domain for T?

  (b) Graph $y = T(x)$. Include any vertical asymptotes.

  (c) Explain what happens to $T(x)$ as $x \to 5^-$.

95. *Length of Lines*   (Refer to Example 2.) Suppose that a parking attendant can wait on 40 cars per hour and that cars arrive randomly at a rate of x cars per hour. Then the average number of cars waiting in line can be estimated by

$$N(x) = \dfrac{x^2}{1600 - 40x}.$$

  (a) Evaluate $N(20)$ and $N(39)$.

  (b) Explain what happens to the length of the line as x approaches 40.

  (c) Find any vertical asymptotes of the graph of N.

96. *Construction Zone*   Suppose that a construction zone can allow 50 cars per hour to pass through and that cars arrive randomly at a rate of x cars per hour. Then the average number of cars waiting in line to get through the construction zone can be estimated by

$$N(x) = \frac{x^2}{2500 - 50x}.$$

(a) Evaluate $N(20)$, $N(40)$, and $N(49)$.

(b) Explain what happens to the length of the line as x approaches 50.

(c) Find any vertical asymptotes of the graph of N.

97. *Interpreting an Asymptote*   Suppose that an insect population in millions is modeled by $f(x) = \frac{10x + 1}{x + 1}$, where $x \geq 0$ is in months.

(a) Graph f in $[0, 14, 1]$ by $[0, 14, 1]$. Find the equation of the horizontal asymptote.

(b) Determine the initial insect population.

(c) What happens to the population over time?

(d) Interpret the horizontal asymptote.

98. *Interpreting an Asymptote*   Suppose that the population of a species of fish (in thousands) is modeled by $f(x) = \frac{x + 10}{0.5x^2 + 1}$, where $x \geq 0$ is in years.

(a) Graph f in $[0, 12, 1]$ by $[0, 12, 1]$. What is the horizontal asymptote?

(b) Determine the initial population.

(c) What happens to the population of this fish?

(d) Interpret the horizontal asymptote.

99. *Train Curves*   When curves are designed for trains, sometimes the outer rail is elevated or banked, so that a locomotive can safely negotiate the curve at a higher speed. See the figure. Suppose a circular curve is designed for 60 miles per hour. The formula $f(x) = \frac{2540}{x}$ computes the elevation y in inches of the outer track for a curve having a radius of x feet, where $y = f(x)$. (**Source:** L. Haefner, *Introduction to Transportation Systems*.)

(a) Evaluate $f(400)$ and interpret the result.

(b) Graph f in $[0, 600, 100]$ by $[0, 50, 5]$. How does the elevation change as the radius increases?

(c) Interpret the horizontal asymptote.

(d) Find the radius if the elevation is 12.7 inches.

100. *Probability*   A container holds x balls numbered 1 through x. Only one ball has the winning number.

(a) Find a function f that computes the probability, or likelihood, of *not* drawing the winning ball.

(b) What is the domain of f?

(c) What happens to the probability of *not* drawing the winning ball as the number of balls increases?

(d) Interpret the horizontal asymptote of the graph of f.

101. *Slippery Roads*   If a car is moving at 50 miles per hour on a level highway, then its braking distance depends on the road conditions. This distance in feet can be computed by $D(x) = \frac{2500}{30x}$, where x is the coefficient of friction between the tires and the road and $0 < x \leq 1$. A smaller value of x indicates that the road is more slippery.

(a) Identify and interpret the vertical asymptote.

(b) Estimate the coefficient of friction associated with a braking distance of 340 feet.

102. *Concentration of a Drug*   The concentration of a drug in a medical patient's bloodstream is given by the formula $f(t) = \frac{5}{t^2 + 1}$, where the input t is in hours, $t \geq 0$, and the output is in milligrams per liter.

(a) Does the concentration of the drug increase or decrease? Explain.

(b) The patient should not take a second dose until the concentration is below 1.5 milligrams per liter. How long should the patient wait before taking a second dose?

Writing about Mathematics

103. Let $f(x)$ be the formula for a rational function.

(a) Explain how to find any vertical or horizontal asymptotes of the graph of f.

(b) Discuss what a horizontal asymptote represents.

104. Discuss how to find the domain of a rational function symbolically and graphically.

Exercises 1–4: Rate of Change/Difference Quotient   Find the average rate of change of f from $x_1 = 1$ to $x_2 = 3$. Then find the difference quotient of f.

1. $f(x) = \dfrac{1}{x}$

2. $f(x) = \dfrac{1}{x^2}$

3. $f(x) = \dfrac{3}{2x}$

4. $f(x) = \dfrac{1}{5 - x}$

CHECKING BASIC CONCEPTS FOR SECTIONS 4.5 AND 4.6

1. Find a quadratic polynomial $f(x)$ with zeros $\pm 4i$ and leading coefficient 3. Write $f(x)$ in complete factored form and expanded form.

2. Sketch a graph of a quartic function (degree 4) with a negative leading coefficient, two real zeros, and two imaginary zeros.

3. Write $x^3 - x^2 + 4x - 4$ in complete factored form.

4. Solve each equation.
(a) $2x^3 + 45 = 5x^2 - 18x$

(b) $x^4 + 5x^2 = 36$

5. Let $f(x) = \dfrac{1}{x - 1} + 2$.
(a) Find the domain of f.

(b) Identify any vertical or horizontal asymptotes.

(c) Sketch a graph of f that includes all asymptotes.

6. Find any vertical or horizontal asymptotes for the graph of $f(x) = \dfrac{4x^2}{x^2 - 4}$. State the domain of f.

7. Sketch a graph of each rational function f. Include all asymptotes and any "holes" in your graph.
(a) $f(x) = \dfrac{3x - 1}{2x - 2}$ (b) $f(x) = \dfrac{1}{(x + 1)^2}$

(c) $f(x) = \dfrac{x + 2}{x^2 - 4}$ (d) $f(x) = \dfrac{x^2 + 1}{x^2 - 1}$

4.7 More Equations and Inequalities

- Solve rational equations
- Solve variation problems
- Solve polynomial inequalities
- Solve rational inequalities

Introduction

Waiting in line has become part of almost everyone's life. When people arrive randomly at a line, rational functions can be used to estimate the average number of people standing in line. For example, if an attendant at a ticket booth can wait on 30 customers per hour and if customers arrive at an average rate of x per hour, then the average number of customers waiting in line is computed by

$$f(x) = \frac{x^2}{900 - 30x},$$

where $0 \le x < 30$. Thus $f(28) \approx 13$ indicates that if customers arrive, on average, at 28 per hour, then the average number of people in line is 13. If a line length of **8** customers or fewer is acceptable, then we can use $f(x)$ to estimate customer arrival rates x that one attendant can accommodate by solving the *rational inequality*

$$\frac{x^2}{900 - 30x} \le 8.$$

(See Example 8.) Rational inequalities are discussed in this section along with other types of inequalities and equations.   (**Source:** N. Garber and L. Hoel, *Traffic and Highway Engineering*.)

Rational Equations

If $f(x)$ represents a rational function, then an equation that can be written in the form $f(x) = k$ for some constant k is a **rational equation**. Examples of rational equations include

$$\frac{x^2 - 1}{x^2 + x + 3} = 0, \quad \frac{3x}{x^3 + x} = \frac{3}{2}, \quad \text{and} \quad \frac{2}{x - 1} + \frac{1}{x} = -2.$$

An Application The *rational expression* given by $D(x) = \frac{2500}{30(0.3 + x)}$ calculates the stopping distance for a car traveling downhill at 50 miles per hour, where x is the **grade**, or slope, of the hill with $x < 0$. The solution to the *rational equation* $\frac{2500}{30(0.3 + x)} = 220$ gives the slope of a hill when the stopping distance equals **220** feet. (See Exercise 83.)

Rational equations can be solved symbolically, graphically, and numerically.

EXAMPLE 1 Solving a rational equation

Solve $\frac{4x}{x - 1} = 6$ symbolically, graphically, and numerically.

SOLUTION

Getting Started The equation $\frac{a}{b} = \frac{c}{d}$ with $b \neq 0$ and $d \neq 0$ is equivalent to $ad = bc$. In this example, you can think of 6 as the ratio $\frac{6}{1}$. This technique is sometimes called *cross multiplying*. ▶

Symbolic Solution

$$\frac{4x}{x - 1} = 6 \qquad \text{Given equation}$$

$$4x = 6(x - 1) \qquad \text{Cross multiply: } \frac{a}{b} = \frac{c}{d} \text{ implies } ad = bc.$$

$$4x = 6x - 6 \qquad \text{Distributive property}$$

$$-2x = -6 \qquad \text{Subtract 6x.}$$

$$x = 3 \qquad \text{Divide by } -2. \text{ (Check this answer.)}$$

Graphical Solution Graph $Y_1 = 4X/(X - 1)$ and $Y_2 = 6$. Their graphs intersect at $(3, 6)$, so the solution is 3. See Figure 4.96.

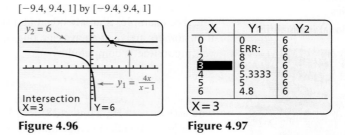

$[-9.4, 9.4, 1]$ by $[-9.4, 9.4, 1]$

Figure 4.96   **Figure 4.97**

Numerical Solution In Figure 4.97, $y_1 = y_2$ when $x = 3$.   *Now Try Exercise 1* ◀

Algebra Review
To review clearing fractions, see Chapter R (page R-35).

A common approach to solving rational equations symbolically is to multiply each side of the equation by a common denominator. This technique, which clears fractions from an equation, is used in Examples 2 and 3.

EXAMPLE 2 Solving a rational equation

Solve $\frac{6}{x^2} - \frac{5}{x} = 1$ symbolically.

SOLUTION The least common denominator for x^2 and x is x^2. Multiply each side of the equation by x^2.

$$\frac{6}{x^2} - \frac{5}{x} = 1 \qquad \textit{Given equation}$$

$$\frac{6}{x^2} \cdot x^2 - \frac{5}{x} \cdot x^2 = 1 \cdot x^2 \qquad \textit{Multiply each term by } x^2.$$

$$6 - 5x = x^2 \qquad \textit{Simplify.}$$

$$0 = x^2 + 5x - 6 \qquad \textit{Add 5x and subtract 6.}$$

$$0 = (x + 6)(x - 1) \qquad \textit{Factor.}$$

$$x + 6 = 0 \quad \text{or} \quad x - 1 = 0 \qquad \textit{Zero-product property}$$

$$x = -6 \quad \text{or} \quad x = 1 \qquad \textit{Solve.}$$

Algebra Review
To review finding a least common denominator, see Chapter R (page R-32).

Check to verify that each answer is correct. **Now Try Exercise 17** ◀

The next example illustrates the importance of checking possible solutions.

EXAMPLE 3 Solving a rational equation

Solve $\frac{1}{x + 3} + \frac{1}{x - 3} = \frac{6}{x^2 - 9}$ symbolically. Check the result.

SOLUTION The least common denominator is $(x + 3)(x - 3)$, or $x^2 - 9$.

$$\frac{1}{x + 3} + \frac{1}{x - 3} = \frac{6}{x^2 - 9} \qquad \textit{Given equation}$$

$$\frac{(x + 3)(x - 3)}{x + 3} + \frac{(x + 3)(x - 3)}{x - 3} = \frac{6(x + 3)(x - 3)}{x^2 - 9} \qquad \textit{Multiply by } (x + 3)(x - 3).$$

$$(x - 3) + (x + 3) = 6 \qquad \textit{Simplify.}$$

$$2x = 6 \qquad \textit{Combine terms.}$$

$$x = 3 \qquad \textit{Divide by 2.}$$

When 3 is substituted for x, two expressions in the *given* equation are undefined. There are no solutions. (The value 3 is called an **extraneous solution** because it does not satisfy the given equation.) **Now Try Exercise 21** ◀

An Application Rational equations are used in real-world applications such as the construction problem in the next example. Steps for solving application problems (see page 122) have been used to structure the solution.

EXAMPLE 4 Designing a box

A box with rectangular sides and a top is being designed to hold 324 cubic inches and to have a surface area of 342 square inches. If the length of the box is four times the height, find possible dimensions of the box.

SOLUTION

STEP 1: We are asked to find the dimensions of a box. If x is the height of the box and y is the width, then the length of the box is $4x$, or four times the height.

x: Height of the box y: Width of the box $4x$: Length of the box

STEP 2: To relate these variables to an equation, sketch a box as illustrated in Figure 4.98. The volume V of the box is height times width times length.

$$V = xy(4x) = 4x^2y$$

The surface area A of this box is determined by finding the area of the 6 rectangular sides: left and right sides, front and back, and top and bottom.

$$A = 2(4x \cdot x) + 2(x \cdot y) + 2(4x \cdot y)$$
$$= 8x^2 + 10xy$$

If we solve $V = 4x^2y$ for y and let $V = 324$, we obtain

$$y = \frac{V}{4x^2} = \frac{324}{4x^2} = \frac{81}{x^2}.$$

Substituting $y = \frac{81}{x^2}$ in the formula for A eliminates the y variable.

$$A = 8x^2 + 10xy \qquad \text{Area formula}$$
$$= 8x^2 + 10x \cdot \frac{81}{x^2} \qquad \text{Let } y = \frac{81}{x^2}.$$
$$= 8x^2 + \frac{810}{x} \qquad \text{Simplify.}$$

Since the surface area is $A = 342$ square inches, the height x can be determined by solving the rational equation

$$8x^2 + \frac{810}{x} = 342.$$

STEP 3: Figures 4.99 and 4.100 show the graphs of $Y_1 = 8X^2 + 810/X$ and $Y_2 = 342$. There are two *positive* solutions: $x = 3$ and $x = 4.5$.

[0, 6, 1] by [300, 400, 20]

$y_1 = 8x^2 + \dfrac{810}{x}$

$y_2 = 342$

Intersection
X=3.   ·Y=342

Figure 4.99

[0, 6, 1] by [300, 400, 20]

$y_1 = 8x^2 + \dfrac{810}{x}$

$y_2 = 342$

Intersection
X=4.5   ·Y=342

Figure 4.100

NOTE This equation can be written as $8x^3 - 342x + 810 = 0$, and the rational zeros test or factoring could be used to find the solutions, 3 and $\frac{9}{2}$.

If the height is $x = 3$ inches, then the length is $4 \cdot 3 = 12$ inches and the width is $y = \frac{81}{3^2} = 9$ inches. (Note that $y = \frac{81}{x^2}$.) If the height is 4.5 inches, then

Geometry Review
To review formulas related to box shapes, see Chapter R (page R-3).

Calculator Help
To find a point of intersection, see Appendix A (page AP-8).

Figure 4.98

the length is $4 \cdot 4.5 = 18$ inches and the width is $y = \frac{81}{4.5^2} = 4$ inches. Thus the dimensions of the box in inches can be either $3 \times 9 \times 12$ or $4.5 \times 4 \times 18$.

STEP 4:   We can check our results directly. If the dimensions are $3 \times 9 \times 12$, then

$$V = 3 \cdot 9 \cdot 12 = 324 \quad \text{and}$$

$$S = 2(3 \cdot 9) + 2(3 \cdot 12) + 2(9 \cdot 12) = 342.$$

If the dimensions are $4.5 \times 4 \times 18$, then

$$V = 4.5 \cdot 4 \cdot 18 = 324 \quad \text{and}$$

$$S = 2(4.5 \cdot 4) + 2(4.5 \cdot 18) + 2(4 \cdot 18) = 342.$$

In both cases our results check.   *Now Try Exercise 79* ◀

Variation

Direct Variation as the *n*th Power   In Section 2.2 direct variation was discussed. Sometimes a quantity y varies directly as a *power* of a variable. For example, the area A of a circle varies directly as the second power of the radius r. That is, $A = \pi r^2$.

Direct Variation as the *n*th Power

Let x and y denote two quantities and n be a positive number. Then y is **directly proportional to the *n*th power** of x, or y **varies directly as the *n*th power** of x, if there exists a nonzero number k such that

$$y = kx^n.$$

The number k is called the *constant of variation* or the *constant of proportionality*. In the formula $A = \pi r^2$, the constant of variation is π.

EXAMPLE 5   Modeling a pendulum

The time T required for a pendulum to swing back and forth once is called its *period*. The length L of a pendulum is directly proportional to the square of T. See Figure 4.101. A 2-foot pendulum has a 1.57-second period.
(a) Find the constant of proportionality k.
(b) Predict T for a pendulum having a length of 5 feet.

SOLUTION
(a) Because L is directly proportional to the square of T, we can write $L = kT^2$. If $L = 2$, then $T = 1.57$. Thus $k = \frac{L}{T^2} = \frac{2}{1.57^2} \approx 0.81$ and $L = 0.81T^2$.
(b) If $L = 5$, then $5 = 0.81T^2$. It follows that $T = \sqrt{5/0.81} \approx 2.48$ seconds.

Now Try Exercise 107 ◀

Figure 4.101

Inverse Variation as the *n*th Power   When two quantities vary inversely, an increase in one quantity results in a decrease in the second quantity. For example, it takes 4 hours to travel 100 miles at 25 miles per hour and 2 hours to travel 100 miles at 50 miles per hour. Greater speed results in less travel time. If s represents the average speed of a car and t is the time to travel 100 miles, then $s \cdot t = 100$, or $t = \frac{100}{s}$. Doubling the speed cuts the time in half; tripling the speed reduces the time by one-third. The quantities t and s are said to *vary inversely*. The constant of variation is 100.

Inverse Variation as the *n*th Power

Let x and y denote two quantities and n be a positive number. Then y is **inversely proportional to the *n*th power** of x, or y **varies inversely as the *n*th power** of x, if there exists a nonzero number k such that

$$y = \frac{k}{x^n}.$$

If $y = \frac{k}{x}$, then y is **inversely proportional** to x or y **varies inversely** as x.

NOTE To review steps for solving variation problems see the box on page 105.

Inverse variation occurs in measuring the intensity of light. If we increase our distance from a lightbulb, the intensity of the light decreases. Intensity I is inversely proportional to the second power of the distance d. The equation $I = \frac{k}{d^2}$ models this phenomenon.

EXAMPLE 6   Modeling the intensity of light

At a distance of 3 meters, a 100-watt bulb produces an intensity of 0.88 watt per square meter.   (**Source:** R. Weidner and R. Sells, *Elementary Classical Physics,* Volume 2.)
(a) Find the constant of proportionality k.
(b) Determine the intensity at a distance of 2 meters.

SOLUTION
(a) Substitute $d = 3$ and $I = 0.88$ in the equation $I = \frac{k}{d^2}$. Solve for k.

$$0.88 = \frac{k}{3^2}, \quad \text{or} \quad k = 7.92$$

(b) Let $I = \frac{7.92}{d^2}$ and $d = 2$. Then $I = \frac{7.92}{2^2} = 1.98$. The intensity at 2 meters is 1.98 watts per square meter.   ***Now Try Exercise 111*** ◄

Polynomial Inequalities

Graphical Solutions   In Section 3.4 a strategy for solving quadratic inequalities was presented. This strategy involves first finding boundary numbers (x-values) where equality holds. Once the boundary numbers are known, a graph or a table of test values can be used to determine the intervals where inequality holds. This strategy can be applied to other types of inequalities.

Algebra Review
To review interval notation, see Section 1.5.

For example, consider the inequality $p(x) > 0$, where $p(x) = -x^4 + 5x^2 - 4$. The graph of p is shown in Figure 4.102. The boundary numbers are $-2, -1, 1,$ and 2 and are shown in the graph and on the number line above it. The solution set to the inequality $p(x) > 0$ can be written in interval notation as $(-2, -1) \cup (1, 2)$ and corresponds to where the graph is above the x-axis. (The symbol $\cup$ means union and indicates that x can be in either interval.) The intervals $(-2, -1)$ and $(1, 2)$ are shaded green on the x-axis in Figure 4.103 and on the number line above it.

Similarly, the solution set to the inequality $p(x) < 0$ is $(-\infty, -2) \cup (-1, 1) \cup (2, \infty)$ and corresponds to where the graph of $y = p(x)$ is below the x-axis. These intervals are shaded red on the x-axis in Figure 4.104 and on the number line above it.

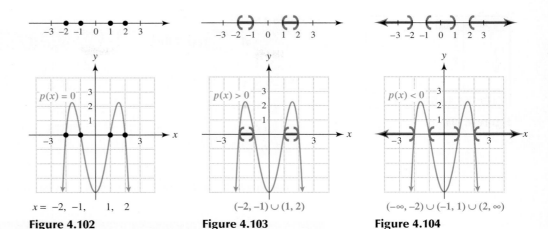

$x = -2, -1, \quad 1, \quad 2$

Figure 4.102

$(-2, -1) \cup (1, 2)$

Figure 4.103

$(-\infty, -2) \cup (-1, 1) \cup (2, \infty)$

Figure 4.104

MAKING CONNECTIONS

Visualization and Inequalities   A *precise* graph of p is *not* necessary to solve the polynomial inequality $p(x) > 0$ or $p(x) < 0$. Once the x-intercepts have been determined, we can use our knowledge about graphs of quartic polynomials (see Section 4.2) to visualize the M-shaped graph of p shown in Figure 4.102. We need to make only a rough sketch and then determine where the graph of p is above the x-axis and where it is below the x-axis.

Symbolic Solutions   Polynomial inequalities can also be solved symbolically. For example, to solve the inequality $-x^4 + 5x^2 - 4 > 0$, begin by finding the boundary numbers.

$$-x^4 + 5x^2 - 4 = 0 \qquad \text{Replace } > \text{ with } =.$$
$$x^4 - 5x^2 + 4 = 0 \qquad \text{Multiply by } -1.$$
$$(x^2 - 4)(x^2 - 1) = 0 \qquad \text{Factor.}$$
$$x^2 - 4 = 0 \quad \text{or} \quad x^2 - 1 = 0 \qquad \text{Zero-product property}$$
$$x = \pm 2 \quad \text{or} \quad x = \pm 1 \qquad \text{Square root property}$$

The boundary numbers -2, -1, 1, and 2 separate the number line into five intervals:

$$(-\infty, -2), (-2, -1), (-1, 1), (1, 2), \text{ and } (2, \infty),$$

Figure 4.105

as illustrated in Figure 4.105.

The polynomial $p(x) = -x^4 + 5x^2 - 4$ is either always positive or always negative on each of these intervals. To determine which is the case, we can choose one test value (x-value) from each interval and substitute this test value in $p(x)$. From Table 4.6 on the next page we see that the test value $x = -1.5$ results in

$$p(-1.5) = -(-1.5)^4 + 5(-1.5)^2 - 4 = 2.1875 > 0.$$

Figure 4.106 on the next page, which was created by a graphing calculator, is similar to Table 4.6. Since $x = -1.5$ is in the interval $(-2, -1)$, it follows that $p(x)$ is positive on this interval. The sign of $p(x)$ on the other intervals is determined similarly. Thus the solution set to $p(x) = -x^4 + 5x^2 - 4 > 0$ is $(-2, -1) \cup (1, 2)$, which is consistent with the graphical solution in Figure 4.103.

X	Y1
-3	-40
-1.5	2.1875
0	-4
1.5	2.1875
3	-40

Y1◼-X^4+5X²-4

Figure 4.106

Table 4.6

Interval	Test Value x	$-x^4 + 5x^2 - 4$	Positive or Negative?
$(-\infty, -2)$	-3	-40	Negative
$(-2, -1)$	-1.5	2.1875	Positive
$(-1, 1)$	0	-4	Negative
$(1, 2)$	1.5	2.1875	Positive
$(2, \infty)$	3	-40	Negative

These symbolic and graphical procedures can be summarized verbally as follows.

Solving Polynomial Inequalities

STEP 1: If necessary, write the inequality as $p(x) < 0$, where $p(x)$ is a polynomial and the inequality symbol $<$ may be replaced by $>$, $\leq$, or $\geq$.

STEP 2: Solve $p(x) = 0$. The solutions are called boundary numbers.

STEP 3: Use the boundary numbers to separate the number line into disjoint intervals. On each interval, $p(x)$ is either always positive or always negative.

STEP 4: To solve the inequality, either make a table of test values for $p(x)$ or use a graph of $y = p(x)$. For example, the solution set for $p(x) < 0$ corresponds to intervals where test values result in negative outputs or to intervals where the graph of $y = p(x)$ is below the x-axis.

EXAMPLE 7 Solving a polynomial inequality

Solve $x^3 \geq 2x^2 + 3x$ symbolically and graphically.

SOLUTION

Algebra Review

To review factoring, see Chapter R, (page R-20).

Figure 4.107

Symbolic Solution

STEP 1: Begin by writing the inequality as $x^3 - 2x^2 - 3x \geq 0$.

STEP 2: Replace the $\geq$ symbol with an equals sign and solve the resulting equation.

$$x^3 - 2x^2 - 3x = 0 \qquad \text{\small Replace } \geq \text{ with } =.$$
$$x(x^2 - 2x - 3) = 0 \qquad \text{\small Factor out } x.$$
$$x(x + 1)(x - 3) = 0 \qquad \text{\small Factor the trinomial.}$$
$$x = 0 \quad \text{or} \quad x = -1 \quad \text{or} \quad x = 3 \qquad \text{\small Zero-product property}$$

The boundary numbers are -1, 0, and 3.

STEP 3: The boundary numbers separate the number line into four disjoint intervals:

$$(-\infty, -1), (-1, 0), (0, 3), \text{ and } (3, \infty),$$

as illustrated in Figure 4.107.

STEP 4: In Table 4.7 the expression $x^3 - 2x^2 - 3x$ is evaluated at a test value from each interval. The solution set is $[-1, 0] \cup [3, \infty)$. In Figure 4.108 a graphing calculator has been used to evaluate the same test values. (*Note:* The boundary numbers are included in the solution set because the inequality involves $\geq$ rather than $>$.)

Figure 4.108

$[-5, 5, 1]$ by $[-7, 7, 1]$

Figure 4.109

Table 4.7

Interval	Test Value x	$x^3 - 2x^2 - 3x$	Positive or Negative?
$(-\infty, -1)$	-2	-10	Negative
$(-1, 0)$	-0.5	0.875	Positive
$(0, 3)$	1	-4	Negative
$(3, \infty)$	4	20	Positive

Graphical Solution   Graph $Y_1 = X^3 - 2X^2 - 3X$, as shown in Figure 4.109. The zeros or x-intercepts are located at -1, 0, and 3. The graph of y_1 is positive (or above the x-axis) for $-1 < x < 0$ or $3 < x < \infty$. If we include the boundary numbers, this result agrees with the symbolic solution.   ***Now Try Exercise 43*** ◀

MAKING CONNECTIONS

Functions, Equations, and Inequalities   The three concepts are related. For example,

$$f(x) = ax^3 + bx^2 + cx + d \qquad \textit{Cubic function}$$
$$ax^3 + bx^2 + cx + d = 0 \qquad \textit{Cubic equation}$$
$$ax^3 + bx^2 + cx + d < 0 \qquad \textit{Cubic inequality}$$

where $a \neq 0$. These concepts also apply to higher degree polynomials and to rational expressions.

Rational Inequalities

An Application   In the introduction we looked at how a rational inequality can be used to estimate the number of people standing in line at a ticket booth. In the next example, this inequality is solved graphically.

EXAMPLE 8   **Modeling customers in a line**

A ticket booth attendant can wait on 30 customers per hour. To keep the time waiting in line reasonable, the line length should not exceed 8 customers on average. Solve the inequality $\frac{x^2}{900 - 30x} \leq 8$ to determine the rates x at which customers can arrive before a second attendant is needed. Note that the x-values are limited to $0 \leq x < 30$.

$[0, 30, 5]$ by $[0, 10, 2]$

Figure 4.110

SOLUTION   Graph $Y_1 = X^2/(900 - 30X)$ and $Y_2 = 8$ for $0 \leq x \leq 30$, as shown in Figure 4.110. The only point of intersection on this interval is near $(26.97, 8)$. The graph of y_1 is below the graph of y_2 for x-values to the left of this point. We conclude that if the arrival rate is about 27 customers per hour or less, then the line length does not exceed 8 customers on average. If the arrival rate is more than 27 customers per hour, a second ticket booth attendant is needed.   ***Now Try Exercise 85*** ◀

Graphical and Symbolic Solutions   To solve rational inequalities, we can use the same basic techniques that we used to solve polynomial inequalities, with one important modification: boundary numbers also occur at x-values where the denominator of any rational expression in the inequality equals 0. The following steps can be used to solve a rational inequality.

Solving Rational Inequalities

STEP 1: If necessary, write the inequality in the form $\frac{p(x)}{q(x)} > 0$, where $p(x)$ and $q(x)$ are polynomials. Note that $>$ may be replaced by $<$, $\leq$, or $\geq$.

STEP 2: Solve $p(x) = 0$ and $q(x) = 0$. The solutions are boundary numbers.

STEP 3: Use the boundary numbers to separate the number line into disjoint intervals. On each interval, $\frac{p(x)}{q(x)}$ is either always positive or always negative.

STEP 4: Use a table of test values or a graph to solve the inequality in Step 1.

EXAMPLE 9 Solving a rational inequality

Solve $\frac{2-x}{2x} > 0$ symbolically. Support your answer graphically.

SOLUTION

Symbolic Solution The inequality is written in the form $\frac{p(x)}{q(x)} > 0$, so Step 1 is unnecessary.

STEP 2: Set the numerator and the denominator equal to 0 and solve.

Numerator	Denominator
$2 - x = 0$	$2x = 0$
$x = 2$	$x = 0$

STEP 3: The boundary numbers are 0 and 2, which separate the number line into three disjoint intervals: $(-\infty, 0)$, $(0, 2)$, and $(2, \infty)$.

STEP 4: Table 4.8 shows that the expression is positive between the two boundary numbers or when $0 < x < 2$. In interval notation the solution set is $(0, 2)$.

Table 4.8

Interval	Test Value x	$(2 - x)/(2x)$	Positive or Negative?
$(-\infty, 0)$	-1	-1.5	Negative
$(0, 2)$	1	0.5	Positive
$(2, \infty)$	4	-0.25	Negative

$[-4.7, 4.7, 1]$ by $[-3.1, 3.1, 1]$

Figure 4.111

Graphical Solution Graph $Y_1 = (2 - X)/(2X)$, as shown in Figure 4.111. The graph has a vertical asymptote at $x = 0$ and an x-intercept at $x = 2$. Between these boundary numbers the graph of y_1 is positive (or above the x-axis). The solution set is $(0, 2)$. This agrees with our symbolic solution.

Now Try Exercise 59 ◀

EXAMPLE 10 Solving a rational inequality symbolically

Solve $\frac{1}{x} \leq \frac{2}{x + 1}$.

SOLUTION

STEP 1: Begin by writing the inequality in the form $\frac{p(x)}{q(x)} \leq 0$.

$$\frac{1}{x} - \frac{2}{x + 1} \leq 0 \qquad \text{Subtract } \frac{2}{x + 1}.$$

Algebra Review
To review subtraction of rational expressions, see Chapter R (page R-34).

$$\frac{1}{x} \cdot \frac{(x+1)}{(x+1)} - \frac{2}{x+1} \cdot \frac{x}{x} \le 0 \qquad \text{Common denominator is } x(x+1).$$

$$\frac{x+1}{x(x+1)} - \frac{2x}{x(x+1)} \le 0 \qquad \text{Multiply.}$$

$$\frac{1-x}{x(x+1)} \le 0 \qquad \begin{array}{l}\text{Subtract numerators:}\\ x+1-2x=1-x.\end{array}$$

STEP 2: Find the zeros of the numerator and the denominator.

Numerator	**Denominator**
$1 - x = 0$	$x(x+1) = 0$
$x = 1$	$x = 0 \quad \text{or} \quad x = -1$

STEP 3: The boundary numbers are -1, 0, and 1, which separate the number line into four disjoint intervals: $(-\infty, -1)$, $(-1, 0)$, $(0, 1)$, and $(1, \infty)$.

STEP 4: Table 4.9 can be used to solve the inequality $\frac{1-x}{x(x+1)} \le 0$. The solution set is $(-1, 0) \cup [1, \infty)$. (*Note:* The boundary numbers -1 and 0 are not included in the solution set because the given inequality is undefined when $x = -1$ or $x = 0$.)

Table 4.9

Interval	Test Value x	$(1-x)/(x(x+1))$	Positive or Negative?
$(-\infty, -1)$	-2	1.5	Positive
$(-1, 0)$	-0.5	-6	Negative
$(0, 1)$	0.5	$0.\overline{6}$	Positive
$(1, \infty)$	2	$-0.1\overline{6}$	Negative

Now Try Exercise 75 ◀

Multiplying an Inequality by a Variable When solving a rational inequality, it is essential *not* to multiply or divide each side of the inequality by the LCD (least common denominator) if the LCD contains a *variable*. This technique often leads to an incorrect solution set.

For example, if each side of the rational inequality

$$\frac{1}{x} < 2$$

is multiplied by x to clear fractions, the inequality becomes

$$1 < 2x \text{ or } x > \frac{1}{2}.$$

However, this solution set is clearly incomplete because $x = -1$ is also a solution to the given inequality. In general, the variable x can be either negative or positive. If $x < 0$, the inequality symbol should be reversed, whereas if $x > 0$, the inequality symbol should not be reversed. Because we have no way of knowing ahead of time which is the case, this technique of multiplying by a variable should be *avoided*.

4.7 Putting It All Together

The following table outlines basic concepts for this section.

Concept	Description
Rational equation	Can be written as $f(x) = k$, where f is a rational function. To solve a rational equation, first clear fractions by multiplying each side by the LCD. *Check your answers.* **Example:** Solve $\frac{1}{x} + \frac{1}{x+2} = \frac{1}{x(x+2)}$. Multiply each side of the equation by the LCD: $x(x+2)$. $$\frac{x(x+2)}{x} + \frac{x(x+2)}{x+2} = \frac{x(x+2)}{x(x+2)}$$ $$(x+2) + x = 1$$ $$x = -\frac{1}{2} \qquad \text{It checks.}$$
Variation	*y varies directly as the nth power of x:* $y = kx^n$. **Example:** Let y vary directly as the cube (third power) of x. If the constant of variation is 5, then $y = 5x^3$. *y varies inversely as the nth power of x:* $y = \dfrac{k}{x^n}$. **Example:** Let y vary inversely as the square (second power) of x. If the constant of variation is 3, then $y = \frac{3}{x^2}$.
Polynomial inequality	Can be written as $p(x) < 0$, where $p(x)$ is a polynomial and $<$ may be replaced by $>, \leq,$ or $\geq$ **Examples:** $x^3 - x \leq 0,\ 2x^4 - 3x^2 \geq 5x + 1$
Solving a polynomial inequality	Follow the steps for solving a polynomial inequality presented on page 330. Either graphical or symbolic methods can be used. **Example:** A graph of $y = x^3 - 2x^2 - 5x + 6$ is shown. The boundary numbers are $-2, 1,$ and 3. The solution set to $x^3 - 2x^2 - 5x + 6 > 0$ is $(-2, 1) \cup (3, \infty)$.

Concept	Description
Rational inequality	Can be written as $\frac{p(x)}{q(x)} < 0$, where $p(x)$ and $q(x) \neq 0$ are polynomials and $<$ may be replaced by $>$, $\leq$, or $\geq$ **Examples:** $\frac{x - 3}{x + 2} \geq 0$, $2x - \frac{2}{x^2 - 1} > 5$
Solving a rational inequality	Follow the steps for solving a rational inequality presented on page 332. Either graphical or symbolic methods can be used. **Example:** A graph of $y = \frac{x - 1}{2x + 1}$ is shown. Boundary numbers occur where either the numerator or denominator equals zero: $x = 1$ or $x = -\frac{1}{2}$. The solution set to $\frac{x - 1}{2x + 1} < 0$ is $\left(-\frac{1}{2}, 1\right)$. $y = \frac{x - 1}{2x + 1}$

4.7   Exercises

Rational Equations

Exercises 1–6: Solve the rational equation

   (a) symbolically,
   (b) graphically, and
   (c) numerically.

1. $\dfrac{2x}{x + 2} = 6$ **2.** $\dfrac{3x}{2x - 1} = 3$

3. $2 - \dfrac{5}{x} + \dfrac{2}{x^2} = 0$ **4.** $\dfrac{1}{x^2} + \dfrac{1}{x} = 2$

5. $\dfrac{1}{x + 1} + \dfrac{1}{x - 1} = \dfrac{1}{x^2 - 1}$

6. $\dfrac{4}{x - 2} = \dfrac{3}{x - 1}$

Exercises 7–28: Find all real solutions. Check your results.

7. $\dfrac{x + 1}{x - 5} = 0$ **8.** $\dfrac{x - 2}{x + 3} = 1$

9. $\dfrac{6(1 - 2x)}{x - 5} = 4$ **10.** $\dfrac{2}{5(2x + 5)} + 3 = -1$

11. $\dfrac{1}{x + 2} + \dfrac{1}{x} = 1$ **12.** $\dfrac{2x}{x - 1} = 5 + \dfrac{2}{x - 1}$

13. $\dfrac{1}{x} - \dfrac{2}{x^2} = 5$ **14.** $\dfrac{1}{x^2 - 2} = \dfrac{1}{x}$

15. $\dfrac{x^3 - 4x}{x^2 + 1} = 0$

16. $\dfrac{1}{x + 2} + \dfrac{1}{x + 3} = \dfrac{2}{x^2 + 5x + 6}$

17. $\dfrac{35}{x^2} = \dfrac{4}{x} + 15$ **18.** $6 - \dfrac{35}{x} + \dfrac{36}{x^2} = 0$

19. $\dfrac{x + 5}{x + 2} = \dfrac{x - 4}{x - 10}$ **20.** $\dfrac{x - 1}{x + 1} = \dfrac{x + 3}{x - 4}$

21. $\dfrac{1}{x - 2} - \dfrac{2}{x - 3} = \dfrac{-1}{x^2 - 5x + 6}$

22. $\dfrac{1}{x - 1} + \dfrac{3}{x + 1} = \dfrac{4}{x^2 - 1}$

23. $\dfrac{2}{x - 1} + 1 = \dfrac{4}{x^2 - 1}$ **24.** $\dfrac{1}{x} + 2 = \dfrac{1}{x^2 + x}$

25. $\dfrac{1}{x + 2} = \dfrac{4}{4 - x^2} - 1$ **26.** $\dfrac{1}{x - 3} + 1 = \dfrac{6}{x^2 - 9}$

27. $\dfrac{1}{x - 1} + \dfrac{1}{x + 1} = \dfrac{2}{x^2 - 1}$

28. $\dfrac{1}{2x + 1} + \dfrac{1}{2x - 1} = \dfrac{2}{4x^2 - 1}$

Graphical Solutions to Inequalities

Exercises 29–34: Solve the equation and inequalities.

    (a) $f(x) = 0$
    (b) $f(x) > 0$
    (c) $f(x) < 0$

29.

30.

31.

32.

33.

34.
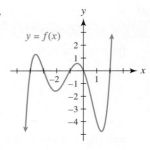

Exercises 35–40: Complete the following.

    (a) Identify where $f(x)$ *is undefined or* $f(x) = 0$.
    (b) Solve $f(x) > 0$.
    (c) Solve $f(x) < 0$.

35.

36.

37.

38.

39.

40.
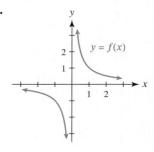

Polynomial Inequalities

Exercises 41–46: Solve the polynomial inequality

    (a) symbolically   and   (b) graphically.

41. $x^3 - x > 0$ **42.** $8x^3 < 27$

43. $x^3 + x^2 \geq 2x$ **44.** $2x^3 \leq 3x^2 + 5x$

45. $x^4 - 13x^2 + 36 < 0$ **46.** $4x^4 - 5x^2 - 9 \geq 0$

Exercises 47–52: Solve the polynomial inequality.

47. $7x^4 > 14x^2$ **48.** $3x^4 - 4x^2 < 7$

49. $(x - 1)(x - 2)(x + 2) \geq 0$

50. $-(x + 1)^2(x - 2) \geq 0$ **51.** $2x^4 + 2x^3 \leq 12x^2$

52. $x^3 + 6x^2 + 9x > 0$

Exercises 53–56: Solve the polynomial inequality graphically.

53. $x^3 - 7x^2 + 14x \leq 8$ **54.** $2x^3 + 3x^2 - 3x < 2$

55. $3x^4 - 7x^3 - 2x^2 + 8x > 0$

56. $x^4 - 5x^3 \leq 5x^2 + 45x + 36$

Rational Inequalities

Exercises 57–62: Solve the rational inequality
(a) symbolically and (b) graphically.

57. $\dfrac{1}{x} < 0$ **58.** $\dfrac{1}{x^2} > 0$

59. $\dfrac{4}{x + 3} \geq 0$ **60.** $\dfrac{x - 1}{x + 1} < 0$

61. $\dfrac{5}{x^2 - 4} < 0$ **62.** $\dfrac{x}{x^2 - 1} \geq 0$

Exercises 63–76: Solve the rational inequality.

63. $\dfrac{(x + 1)^2}{x - 2} \leq 0$ **64.** $\dfrac{2x}{(x - 2)^2} > 0$

65. $\dfrac{3 - 2x}{1 + x} < 0$ **66.** $\dfrac{x + 1}{4 - 2x} \geq 1$

67. $\dfrac{(x + 1)(x - 2)}{(x + 3)} < 0$ **68.** $\dfrac{x(x - 3)}{x + 2} \geq 0$

69. $\dfrac{2x - 5}{x^2 - 1} \geq 0$ **70.** $\dfrac{5 - x}{x^2 - x - 2} < 0$

71. $\dfrac{1}{x - 3} \leq \dfrac{5}{x - 3}$ **72.** $\dfrac{3}{2 - x} > \dfrac{x}{2 + x}$

73. $2 - \dfrac{5}{x} + \dfrac{2}{x^2} \geq 0$ **74.** $\dfrac{1}{x - 1} + \dfrac{1}{x + 1} > \dfrac{3}{4}$

75. $\dfrac{1}{x} \leq \dfrac{2}{x + 2}$ **76.** $\dfrac{1}{x + 1} < \dfrac{1}{x} + 1$

Applications

77. *Time Spent in Line*   Suppose the average number of vehicles arriving at the main gate of an amusement park is equal to 10 per minute, while the average number of vehicles being admitted through the gate per minute is equal to x. Then the average waiting time in minutes for each vehicle at the gate can be computed by $f(x) = \dfrac{x - 5}{x^2 - 10x}$, where $x > 10$.   (**Source:** F. Mannering.)

(a) Estimate the admittance rate x that results in an average wait of 15 seconds.

(b) If one attendant can serve 5 vehicles per minute, how many attendants are needed to keep the average wait to 15 seconds or less?

78. *Length of Lines*   (Refer to Example 2 in Section 4.6.) Determine the traffic intensity x when the average number of vehicles in line equals 3.

79. *Construction*   Find possible dimensions for a box with a volume of 196 cubic inches, a surface area of 280 square inches, and a length that is twice the width.

80. *Minimizing Surface Area*   An aluminum can is being designed to hold a volume of 100π cubic centimeters.

Geometry Review   To review formulas for cylinders, see Chapter R (page R-4).

(a) Find a formula for the volume V in terms of r and h.

(b) Write a formula for a function S that calculates the outside surface area of the can in terms of only r. Evaluate $S(2)$ and interpret the result.

(c) Find the dimensions that result in the least amount of aluminum being used in its construction.

81. *Minimizing Cost*   A cardboard box with no top and a square base is being constructed and must have a volume of 108 cubic inches. Let x be the length of a side of its base in inches.

(a) Write a formula $A(x)$ that calculates the outside surface area in square feet of the box.

(b) If cardboard costs $0.10 per square foot, write a formula $C(x)$ that gives the cost in dollars of the cardboard in the box.

(c) Find the dimensions of the box that would minimize the cost of the cardboard.

82. *Cost-Benefit*   A cost-benefit function C computes the cost in millions of dollars of implementing a city recycling project when x percent of the citizens participate, where $C(x) = \frac{1.2x}{100 - x}$.

(a) Graph C in $[0, 100, 10]$ by $[0, 10, 1]$. Interpret the graph as x approaches 100.

(b) If 75% participation is expected, determine the cost for the city.

(c) The city plans to spend $5 million on this recycling project. Estimate the percentage of participation that can be expected.

83. *Braking Distance*   The *grade x* of a hill is a measure of its steepness. For example, if a road rises 10 feet for every 100 feet of horizontal distance, then it has an uphill grade of $x = \frac{10}{100}$, or 10%. See the figure. Grades are typically kept quite small—usually less than 10%. The braking distance D for a car traveling at 50 miles per hour on a wet uphill grade is given by the formula $D(x) = \frac{2500}{30(0.3 + x)}$.   (**Source:** L. Haefner.)

(a) Evaluate $D(0.05)$ and interpret the result.

(b) Describe what happens to the braking distance as the hill becomes steeper. Does this agree with your driving experience?

(c) Estimate the grade associated with a braking distance of 220 feet.

84. *Braking Distance*   (Refer to Exercise 83.) If a car is traveling 50 miles per hour downhill, then the car's braking distance on a wet pavement is given by

$$D(x) = \frac{2500}{30(0.3 + x)},$$

where $x < 0$ for a downhill grade.

(a) Evaluate $D(-0.1)$ and interpret the result.

(b) What happens to the braking distance as the downhill grade becomes steeper? Does this agree with your driving experience?

(c) The graph of D has a vertical asymptote at $x = -0.3$. Give the physical significance of this asymptote.

(d) Estimate the grade associated with a braking distance of 350 feet.

85. *Waiting in Line*   (Refer to Example 8.) A parking garage attendant can wait on 40 cars per hour. If cars arrive randomly at a rate of x cars per hour, then the average line length is given by

$$f(x) = \frac{x^2}{1600 - 40x},$$

where the x-values are limited to $0 \le x < 40$.

(a) Solve the inequality $f(x) \le 8$.

(b) Interpret your answer from part (a).

86. *Time Spent in Line*   If a parking garage attendant can wait on 3 vehicles per minute and vehicles are leaving the ramp at x vehicles per minute, then the average wait in minutes for a car trying to exit is given by the formula $f(x) = \frac{1}{3 - x}$.

(a) Solve the three-part inequality $5 \le \frac{1}{3 - x} \le 10$.

(b) Interpret your result from part (a).

87. *Slippery Roads*   The coefficient of friction x measures the friction between the tires of a car and the road, where $0 < x \le 1$. A smaller value of x indicates that the road is more slippery. If a car is traveling at 60 miles per hour, then the braking distance D in feet is given by the formula $D(x) = \frac{120}{x}$.

(a) What happens to the braking distance as the coefficient of friction becomes smaller?

(b) Find values for the coefficient of friction x that correspond to a braking distance of 400 feet or more.

88. *Average Temperature*   The monthly average high temperature in degrees Fahrenheit at Daytona Beach, Florida, can be approximated by

$$f(x) = 0.0145x^4 - 0.426x^3 + 3.53x^2 - 6.22x + 72,$$

where $x = 1$ corresponds to January, $x = 2$ to February, and so on. Estimate graphically when the monthly average high temperature is 75°F or more.

89. *Geometry*   A cubical box is being manufactured to hold 213 cubic inches. If this measurement can vary between 212.8 cubic inches and 213.2 cubic inches inclusive, by how much can the length x of a side of the cube vary?

90. *Construction*   A cylindrical aluminum can is being manufactured so that its height h is 8 centimeters more than its radius r. Estimate values for the radius (to the nearest hundredth) that result in the can having a volume between 1000 and 1500 cubic centimeters inclusive.

Variation

Exercises 91–94: Find the constant of proportionality k.

91. $y = \frac{k}{x}$, and $y = 2$ when $x = 3$

92. $y = \frac{k}{x^2}$, and $y = \frac{1}{4}$ when $x = 8$

93. $y = kx^3$, and $y = 64$ when $x = 2$

94. $y = kx^{3/2}$, and $y = 96$ when $x = 16$

Exercises 95–98: Solve the variation problem.

95. Suppose T varies directly as the $\frac{3}{2}$ power of x. When $x = 4$, $T = 20$. Find T when $x = 16$.

96. Suppose y varies directly as the second power of x. When $x = 3$, $y = 10.8$. Find y when $x = 1.5$.

97. Let y be inversely proportional to x. When $x = 6$, $y = 5$. Find y when $x = 15$.

98. Let z be inversely proportional to the third power of t. When $t = 5$, $z = 0.08$. Find z when $t = 2$.

Exercises 99–102: Assume that the constant of proportionality is positive.

99. Let y be inversely proportional to x. If x doubles, what happens to y?

100. Let y vary inversely as the second power of x. If x doubles, what happens to y?

101. Suppose y varies directly as the third power of x. If x triples, what happens to y?

102. Suppose y is directly proportional to the second power of x. If x is halved, what happens to y?

Exercises 103 and 104: The data satisfy the equation $y = kx^n$, where n is a positive integer. Determine k and n.

103.

x	2	3	4	5
y	2	4.5	8	12.5

104.

x	3	5	7	9
y	32.4	150	411.6	874.8

Exercises 105 and 106: The data in the table satisfy the equation $y = \frac{k}{x^n}$, where n is a positive integer. Determine k and n.

105.

x	2	3	4	5
y	1.5	1	0.75	0.6

106.

x	2	4	6	8
y	9	2.25	1	0.5625

107. *Allometric Growth*   The weight y of a fiddler crab is directly proportional to the 1.25 power of the weight x of its claws. A crab with a body weight of 1.9 grams has claws weighing 1.1 grams. Estimate the weight of a fiddler crab with claws weighing 0.75 gram. (**Source:** D. Brown.)

108. *Gravity*   The weight of an object varies inversely as the second power of the distance from the *center* of Earth. The radius of Earth is approximately 4000 miles. If a person weighs 160 pounds on Earth's surface, what would this individual weigh 8000 miles above the surface of Earth?

109. *Hubble Telescope*   The brightness, or intensity, of starlight varies inversely as the square of its distance from Earth. The Hubble Telescope can see stars whose intensities are $\frac{1}{50}$ that of the faintest star now seen by ground-based telescopes. Determine how much farther the Hubble Telescope can see into space than ground-based telescopes. (**Source:** National Aeronautics and Space Administration.)

110. *Volume*   The volume V of a cylinder with a fixed height is directly proportional to the square of its radius r. If a cylinder with a radius of 10 inches has a volume of 200 cubic inches, what is the volume of a cylinder with the same height and a radius of 5 inches?

111. *Electrical Resistance*   The electrical resistance R of a wire varies inversely as the square of its diameter d. If a 25-foot wire with a diameter of 2 millimeters has a resistance of 0.5 ohm, find the resistance of a wire having the same length and a diameter of 3 millimeters.

112. *Strength of a Beam*   The strength of a rectangular wood beam varies directly as the square of the depth of its cross section. If a beam with a depth of 3.5 inches can support 1000 pounds, how much weight can the same type of beam hold if its depth is 12 inches?

Exercises 113 and 114: Violin String The frequency F of a vibrating string is directly proportional to the square root of the tension T on the string and inversely proportional to the length L of the string.

113. If both the tension and the length are doubled, what happens to F?

114. Give two ways to double the frequency F.

Writing about Mathematics

115. Describe the steps to graphically solve a polynomial inequality in the form $p(x) > 0$.

116. Describe the steps to symbolically solve a rational inequality in the form $f(x) > 0$.

4.8 Radical Equations and Power Functions

- **Learn properties of rational exponents**
- **Learn radical notation**
- **Solve equations involving radical expressions**
- **Understand properties and graphs of power functions**
- **Use power functions to model data**
- **Solve equations involving rational exponents**
- **Use power regression to model data (optional)**

Introduction

Johannes Kepler (1571–1630) was the first to recognize that the orbits of planets are elliptical, rather than circular. He also found that a power function models the relationship between a planet's distance from the sun and its period of revolution. Table 4.10 lists the average distance x from the sun and the time y in years for several planets to orbit the sun. The distance x has been normalized so that Earth is one unit away from the sun. For example, Jupiter is 5.2 times farther from the sun than Earth and requires 11.9 years to orbit the sun.

Table 4.10

Planet	Mercury	Venus	Earth	Mars	Jupiter	Saturn
x (distance)	0.387	0.723	1.00	1.52	5.20	9.54
y (period)	0.241	0.615	1.00	1.88	11.9	29.5

Source: C. Ronan, *The Natural History of the Universe.*

A scatterplot of the data in Table 4.10 is shown in Figure 4.112. To model these data, we might try a polynomial, such as $f(x) = x$ or $g(x) = x^2$. Figure 4.113 shows that $f(x) = x$ increases too slowly and $g(x) = x^2$ increases too fast. To model these data, a new type of function is required. That is, we need a function in the form $h(x) = x^b$, where $1 < b < 2$. Polynomials allow the exponent b to be only a nonnegative integer, whereas *power functions* allow b to be any real number. See Figure 4.114.

Figure 4.112 Orbital Data

Figure 4.113 Polynomial Model

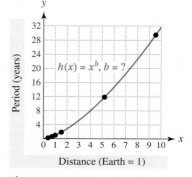

Figure 4.114 Power Model

In this section we consider power functions, and in Example 8 we use a power function to model Kepler's data. Because power functions often use rational numbers as exponents, we begin by reviewing properties of exponents and radical notation.

Rational Exponents and Radical Notation

The following properties can be used to simplify expressions with rational exponents.

Algebra Review
To review integer exponents, see Chapter R (page R-7). To review radical notation and rational exponents, see Chapter R (page R-41).

Properties of Rational Exponents

Let m and n be positive integers with $\frac{m}{n}$ in *lowest* terms and $n \geq 2$. Let r and p be rational numbers. Assume that b is a nonzero real number and that each expression is a real number.

Property	*Example*
1. $b^{m/n} = (b^m)^{1/n} = (b^{1/n})^m$	$4^{3/2} = (4^3)^{1/2} = (4^{1/2})^3 = 2^3 = 8$
2. $b^{m/n} = \sqrt[n]{b^m} = (\sqrt[n]{b})^m$	$8^{2/3} = \sqrt[3]{8^2} = (\sqrt[3]{8})^2 = 2^2 = 4$
3. $(b^r)^p = b^{rp}$	$(2^{3/2})^4 = 2^6 = 64$
4. $b^{-r} = \dfrac{1}{b^r}$	$4^{-1/2} = \dfrac{1}{4^{1/2}} = \dfrac{1}{2}$
5. $b^r b^p = b^{r+p}$	$3^{5/2} \cdot 3^{3/2} = 3^{(5/2)+(3/2)} = 3^4 = 81$
6. $\dfrac{b^r}{b^p} = b^{r-p}$	$\dfrac{5^{5/4}}{5^{3/4}} = 5^{(5/4)-(3/4)} = 5^{1/2}$

EXAMPLE 1 Applying properties of exponents

Simplify each expression by hand.

(a) $16^{3/4}$ **(b)** $\dfrac{4^{1/3}}{4^{5/6}}$ **(c)** $27^{-2/3} \cdot 27^{1/3}$ **(d)** $(5^{3/4})^{2/3}$ **(e)** $(-125)^{-4/3}$

SOLUTION
(a) $16^{3/4} = (\sqrt[4]{16})^3 = (2)^3 = 8$ $b^{m/n} = \sqrt[n]{b^m} = (\sqrt[n]{b})^m$

(b) $\dfrac{4^{1/3}}{4^{5/6}} = 4^{(1/3)-(5/6)} = 4^{-1/2} = \dfrac{1}{\sqrt{4}} = \dfrac{1}{2}$ $\dfrac{b^r}{b^p} = b^{r-p}$

(c) $27^{-2/3} \cdot 27^{1/3} = 27^{(-2/3)+(1/3)} = 27^{-1/3} = \dfrac{1}{\sqrt[3]{27}} = \dfrac{1}{3}$ $b^r b^p = b^{r+p}$

(d) $(5^{3/4})^{2/3} = 5^{(3/4)(2/3)} = 5^{1/2}$ or $\sqrt{5}$ $(b^r)^p = b^{rp}$

(e) $(-125)^{-4/3} = \dfrac{1}{(\sqrt[3]{-125})^4} = \dfrac{1}{(-5)^4} = \dfrac{1}{625}$ $b^{-r} = \frac{1}{b^r}$

Now Try Exercises 1, 7, 9, 11, and 13 ◀

EXAMPLE 2 Writing radicals with rational exponents

Use positive rational exponents to write each expression.

(a) $\sqrt{x}$ **(b)** $\sqrt[3]{x^2}$ **(c)** $(\sqrt[4]{z})^{-5}$ **(d)** $\sqrt{\sqrt[3]{y} \cdot \sqrt[4]{y}}$

SOLUTION

(a) $\sqrt{x} = x^{1/2}$ **(b)** $\sqrt[3]{x^2} = (x^2)^{1/3} = x^{2/3}$

(c) $(\sqrt[4]{z})^{-5} = (z^{1/4})^{-5} = z^{-5/4} = \dfrac{1}{z^{5/4}}$

(d) $\sqrt{\sqrt[3]{y} \cdot \sqrt[4]{y}} = \left(y^{1/3} \cdot y^{1/4}\right)^{1/2} = \left(y^{(1/3)+(1/4)}\right)^{1/2} = \left(y^{7/12}\right)^{1/2} = y^{7/24}$

Now Try Exercises 19, 21, 23, and 27 ◀

Equations Involving Radicals

In solving equations that contain square roots, it is common to square each side of an equation and then check the results. This is done in the next example.

EXAMPLE 3 Solving an equation containing a square root

Solve $x = \sqrt{15 - 2x}$. Check your answers.

SOLUTION Begin by squaring each side of the equation.

$$x = \sqrt{15 - 2x} \qquad \textit{Given equation}$$
$$x^2 = (\sqrt{15 - 2x})^2 \qquad \textit{Square each side.}$$
$$x^2 = 15 - 2x \qquad \textit{Simplify.}$$
$$x^2 + 2x - 15 = 0 \qquad \textit{Add 2x and subtract 15.}$$
$$(x + 5)(x - 3) = 0 \qquad \textit{Factor.}$$
$$x = -5 \quad \text{or} \quad x = 3 \qquad \textit{Solve.}$$

Check: Now substitute these values in the *given* equation $x = \sqrt{15 - 2x}$.

$$-5 \neq \sqrt{15 - 2(-5)} = 5, \qquad 3 = \sqrt{15 - 2(3)}$$

Thus 3 is the only solution. This result is supported graphically in Figure 4.115, where $Y_1 = X$ and $Y_2 = \sqrt{(15 - 2X)}$. Notice that *no* point of intersection occurs when $x = -5$.

Now Try Exercise 37 ◀

$[-9, 9, 1]$ by $[-6, 6, 1]$

$y_2 = \sqrt{15 - 2x}$

$y_1 = x$

Intersection
X=3   Y=3

Figure 4.115

The value -5 in Example 3 is called an **extraneous solution** because it does not satisfy the given equation. It is important to check results whenever *squaring* has been used to solve an equation.

NOTE If each side of an equation is raised to the same positive integer power, then any solutions to the given equation are *among* the solutions to the new equation. That is, the solutions to the equation $a = b$ are *among* the solutions to $a^n = b^n$. For this reason, we must check our answers.

In the next example, we cube each side of an equation that contains a cube root.

EXAMPLE 4 Solving an equation containing a cube root

Solve $\sqrt[3]{2x + 5} - 2 = 1$.

SOLUTION Start by adding 2 to each side. Then cube each side.

$$\sqrt[3]{2x + 5} = 3 \qquad \textit{Add 2 to each side.}$$
$$(\sqrt[3]{2x + 5})^3 = 3^3 \qquad \textit{Cube each side.}$$
$$2x + 5 = 27 \qquad \textit{Simplify.}$$

$$2x = 22 \qquad \text{Subtract 5 from each side.}$$
$$x = 11 \qquad \text{Divide by 2.}$$

The only solution is 11. The answer checks.   *Now Try Exercise 45* ◄

Squaring Twice   In the next example, we need to square twice to solve a radical equation.

> **EXAMPLE 5**   Squaring twice

Solve $\sqrt{2x + 3} - \sqrt{x + 1} = 1$.

SOLUTION

Getting Started When an equation contains two square root expressions, we frequently need to square twice. Start by isolating the more complicated radical and then square each side. ►

$$\sqrt{2x + 3} - \sqrt{x + 1} = 1 \qquad \text{Given equation}$$
$$\sqrt{2x + 3} = 1 + \sqrt{x + 1} \qquad \text{Isolate } \sqrt{2x + 3}.$$
$$\left(\sqrt{2x + 3}\right)^2 = \left(1 + \sqrt{x + 1}\right)^2 \qquad \text{Square each side.}$$
$$2x + 3 = 1 + 2\sqrt{x + 1} + x + 1 \qquad \text{Simplify.}$$
$$x + 1 = 2\sqrt{x + 1} \qquad \text{Isolate the remaining radical.}$$
$$(x + 1)^2 = \left(2\sqrt{x + 1}\right)^2 \qquad \text{Square each side again.}$$
$$x^2 + 2x + 1 = 4(x + 1) \qquad \text{Simplify.}$$
$$x^2 + 2x + 1 = 4x + 4 \qquad \text{Distributive property}$$
$$x^2 - 2x - 3 = 0 \qquad \text{Subtract 4x + 4.}$$
$$(x - 3)(x + 1) = 0 \qquad \text{Factor.}$$
$$x - 3 = 0 \quad \text{or} \quad x + 1 = 0 \qquad \text{Zero-product property}$$
$$x = 3 \quad \text{or} \qquad x = -1 \qquad \text{Solve each equation.}$$

Checking reveals that both -1 and 3 are solutions to the given equation.
                                                  Now Try Exercise 43 ◄

Power Functions and Models

Functions with rational exponents are often used to model physical characteristics of living organisms. For example, larger birds tend to have bigger wings. There is a relationship between a bird's weight and its wing size. (See Example 7.) This area of study in biology, in which the relative sizes of different characteristics of an organism are modeled, is called *allometry*.

Power functions often have rational exponents, and a special type of power function is a root function. These functions are defined as follows.

> **Power Function**
>
> A function f given by $f(x) = x^b$, where b is a constant, is a **power function**. If $b = \frac{1}{n}$ for some integer $n \geq 2$, then f is a **root function** given by $f(x) = x^{1/n}$, or equivalently, $f(x) = \sqrt[n]{x}$.

Examples of power functions include

$$f_1(x) = x^2, \qquad f_2(x) = x^{3/4}, \qquad f_3(x) = x^{0.4}, \qquad \text{and} \qquad f_4(x) = \sqrt[3]{x^2}.$$

Domains of Power Functions Suppose a positive rational number $\frac{p}{q}$ is written in lowest terms. Then the domain of $f(x) = x^{p/q}$ is all real numbers whenever q is odd and all *nonnegative* real numbers whenever q is even. (If b is a positive irrational number, the domain of $f(x) = x^b$ is all nonnegative real numbers.) For example, the domain of $f(x) = x^{1/3}$ $(f(x) = \sqrt[3]{x})$ is all real numbers, whereas the domain of $g(x) = x^{1/2}$ $(g(x) = \sqrt{x})$ is all nonnegative numbers.

Graphs of three common power functions are shown in Figures 4.116–4.118.

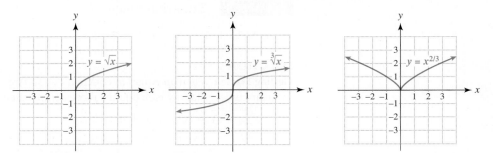

Figure 4.116 Square Root   **Figure 4.117** Cube Root   **Figure 4.118** Cube Root Square

EXAMPLE 6 Graphing power functions

Graph $f(x) = x^b$, where $b = 0.3$, 1, and 1.7, for $x \geq 0$. Discuss the effect that b has on the graph of f for $x \geq 1$.

SOLUTION The graphs of $y = x^{0.3}$, $y = x^1$, and $y = x^{1.7}$ are shown in Figure 4.119. For $x \geq 1$, larger values of b cause the graph of f to increase faster. Note that each graph passes through the point (1, 1). Why?   ***Now Try Exercises 53 and 54*** ◀

NOTE Graphs of power functions with negative exponents are investigated in the Extended and Discovery Exercises for this section.

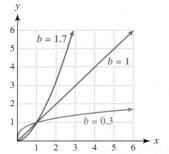

Figure 4.119 $f(x) = x^b, x \geq 0$

Modeling In the next two examples, we use power functions to determine weights of birds based on wing size and to describe planetary motion.

EXAMPLE 7 Modeling wing size of a bird

Heavier birds have larger wings with more surface area than do lighter birds. For some species of birds, this relationship can be modeled by $S(w) = 0.2w^{2/3}$, where w is the weight of the bird in kilograms, with $0.1 \leq w \leq 5$, and S is the surface area of the wings in square meters. (**Source:** C. Pennycuick, *Newton Rules Biology.*)
(a) Approximate $S(0.5)$ and interpret the result.
(b) What weight corresponds to a surface area of 0.25 square meter?

SOLUTION
(a) $S(0.5) = 0.2(0.5)^{2/3} \approx 0.126$. The wings of a bird that weighs 0.5 kilogram have a surface area of about 0.126 square meter.
(b) To answer this question we must solve the equation $0.2w^{2/3} = 0.25$.

$$0.2w^{2/3} = 0.25$$

$$w^{2/3} = \frac{0.25}{0.2} \qquad \text{Divide by 0.2.}$$

$$(w^{2/3})^3 = \left(\frac{0.25}{0.2}\right)^3 \qquad \text{Cube each side.}$$

$$w^2 = \left(\frac{0.25}{0.2}\right)^3 \qquad \text{Simplify.}$$

$$w = \pm\sqrt{\left(\frac{0.25}{0.2}\right)^3} \qquad \text{Square root property}$$

$$w \approx \pm 1.4 \qquad \text{Approximate.}$$

Since w must be positive, the wings of a 1.4-kilogram bird have a surface area of about 0.25 square meter. *Now Try Exercise 85* ◄

EXAMPLE 8 Modeling the period of planetary orbits

Use the data in Table 4.10 on page 340 to complete the following.
(a) Make a scatterplot of the data. Graphically estimate a value for b so that $f(x) = x^b$ models the data.
(b) Numerically check the accuracy of f.
(c) The average distances of Uranus, Neptune, and Pluto (no longer a planet) from the sun are 19.2, 30.1, and 39.5, respectively. Use f to estimate their periods of revolution. Compare these estimates to the actual values of 84.0, 164.8, and 248.5 years.

SOLUTION
(a) Graph the data and $y = x^b$ for different values of b. From the graphs of $y = x^{1.4}$, $y = x^{1.5}$, and $y = x^{1.6}$ in Figures 4.120–4.122, it can be seen that $b \approx 1.5$.

Calculator Help

To make a scatterplot, see Appendix A (page AP-3). To make a table like Figure 4.123, see Appendix A (page AP-10).

[0, 10, 1] by [0, 30, 10]   [0, 10, 1] by [0, 30, 10]   [0, 10, 1] by [0, 30, 10]

Figure 4.120

Figure 4.121

Figure 4.122

(b) Let $f(x) = x^{1.5}$ and table $Y_1 = X^{\wedge}1.5$. The values shown in Figure 4.123 model the data in Table 4.10 remarkably well.

X	Y1
.387	.24075
.723	.61476
1	1
1.52	1.874
5.2	11.858
9.54	29.466

X=

Figure 4.123

X	Y1
19.2	84.13
30.1	165.14
39.5	248.25

X=

Figure 4.124

(c) To approximate the number of years required for Uranus, Neptune, and Pluto to orbit the sun, evaluate $f(x) = x^{1.5}$ at $x = 19.2$, 30.1, and 39.5, as shown in Figure 4.124. These values are close to the actual values. *Now Try Exercise 89* ◄

Equations Involving Rational Exponents

Equations sometimes have rational exponents. The next example demonstrates a basic technique that can be used to solve some of these types of equations.

[−5, 5, 1] by [−40, 40, 10]

Intersection
X=2.9541769 Y=23

Figure 4.125

EXAMPLE 9 Solving an equation with rational exponents

Solve $2x^{5/2} - 7 = 23$. Round to the nearest hundredth and give graphical support.

SOLUTION

Symbolic Solution Start by adding 7 to each side.

$$2x^{5/2} = 30 \qquad \text{Add 7 to each side.}$$
$$x^{5/2} = 15 \qquad \text{Divide by 2.}$$
$$(x^{5/2})^2 = 15^2 \qquad \text{Square each side.}$$
$$x^5 = 225 \qquad \text{Properties of exponents}$$
$$x = 225^{1/5} \qquad \text{Take the fifth root.}$$
$$x \approx 2.95 \qquad \text{Approximate.}$$

Graphical Solution Graphical support is shown in Figure 4.125, where the graphs of $Y_1 = 2X^{\wedge}(5/2) - 7$ and $Y_2 = 23$ intersect near (2.95, 23).   **Now Try Exercise 69** ◀

MAKING CONNECTIONS

Solutions to $x^n = k$ (k a constant and $n \geq 2$ an integer)

n odd: The real solution to $x^n = k$ is $x = \sqrt[n]{k}$, or $x = k^{1/n}$, for all k. See Example 9.

n even: The real solutions to $x^n = k$ are $x = \pm\sqrt[n]{k}$, or $x = \pm k^{1/n}$, for $k \geq 0$. See Example 7.

Equations that have rational exponents are sometimes reducible to quadratic form.

EXAMPLE 10 Solving an equation having negative exponents

Solve $15n^{-2} - 19n^{-1} + 6 = 0$.

SOLUTION Two methods for solving this equation are presented.

Method 1:   Use the substitution $u = n^{-1} = \frac{1}{n}$ and $u^2 = n^{-2} = \frac{1}{n^2}$.

$$15n^{-2} - 19n^{-1} + 6 = 0 \qquad \text{Given equation}$$
$$15u^2 - 19u + 6 = 0 \qquad \text{Let } u = n^{-1} \text{ and } u^2 = n^{-2}.$$
$$(3u - 2)(5u - 3) = 0 \qquad \text{Factor.}$$
$$u = \frac{2}{3} \quad \text{or} \quad u = \frac{3}{5} \qquad \text{Zero-product property}$$

Because $u = \frac{1}{n}$, it follows that $n = \frac{1}{u}$. Thus $n = \frac{3}{2}$ or $n = \frac{5}{3}$.

Method 2:   Another way to solve this equation is to multiply each side by n^2 to clear fractions.

$$15n^{-2} - 19n^{-1} + 6 = 0 \qquad \text{Given equation}$$
$$n^2(15n^{-2} - 19n^{-1} + 6) = n^2(0) \qquad \text{Multiply each side by } n^2.$$
$$15n^2n^{-2} - 19n^2n^{-1} + 6n^2 = 0 \qquad \text{Distributive property}$$
$$15 - 19n + 6n^2 = 0 \qquad \text{Properties of exponents}$$

$$6n^2 - 19n + 15 = 0 \quad \text{Rewrite the equation.}$$
$$(2n - 3)(3n - 5) = 0 \quad \text{Factor.}$$
$$n = \frac{3}{2} \quad \text{or} \quad n = \frac{5}{3} \quad \text{Zero-product property}$$

Now Try Exercise 73 ◀

In the next example, we solve an equation with fractional exponents that can be written in quadratic form by using substitution.

EXAMPLE 11 Solving an equation having fractions for exponents

Solve $2x^{2/3} + 5x^{1/3} - 3 = 0$.

SOLUTION To solve this equation, use the substitution $u = x^{1/3}$.

$$2x^{2/3} + 5x^{1/3} - 3 = 0 \quad \text{Given equation}$$
$$2(x^{1/3})^2 + 5(x^{1/3}) - 3 = 0 \quad \text{Properties of exponents}$$
$$2u^2 + 5u - 3 = 0 \quad \text{Let } u = x^{1/3}.$$
$$(2u - 1)(u + 3) = 0 \quad \text{Factor.}$$
$$u = \frac{1}{2} \quad \text{or} \quad u = -3 \quad \text{Zero-product property}$$

Because $u = x^{1/3}$, it follows that $x = u^3$. Thus $x = \left(\frac{1}{2}\right)^3 = \frac{1}{8}$ or $x = (-3)^3 = -27$.

Now Try Exercise 77 ◀

Power Regression (Optional)

Rather than visually fit a curve to data, as was done in Example 8, we can use least-squares regression to fit the data. Least-squares regression was introduced in Section 2.1. In the next example, we apply this technique to data from biology.

EXAMPLE 12 Modeling the length of a bird's wing

Table 4.11 lists the weight W and the wingspan L for birds of a particular species.

Table 4.11 Weights and Wingspans

W (kilograms)	0.5	1.5	2.0	2.5	3.0
L (meters)	0.77	1.10	1.22	1.31	1.40

Source: C. Pennycuick.

(a) Use power regression to model the data with $L = aW^b$. Graph the data and the equation.
(b) Approximate the wingspan for a bird weighing 3.2 kilograms.

SOLUTION
(a) Let x be the weight W and y be the length L. See Figures 4.126–4.129 on the next page. Enter the data, and then select power regression (PwrReg), as shown in Figures 4.126 and 4.127. The results are shown in Figure 4.128. Let

$$y = 0.9674x^{0.3326} \quad \text{or} \quad L = 0.9674W^{0.3326}.$$

The data and equation are graphed in Figure 4.129.
(b) If a bird weighs **3.2** kilograms, this model predicts the wingspan to be

$$L = 0.9674(3.2)^{0.3326} \approx 1.42 \text{ meters.}$$

Calculator Help

To find an equation of least-squares fit, see Appendix A (page AP-11).

Figure 4.126

Figure 4.127

Figure 4.128

[0, 4, 1] by [0.5, 1.5, 0.5]

$y = 0.9674x^{0.3326}$

Figure 4.129

Now Try Exercise 95 ◄

4.8 Putting It All Together

The following table outlines important concepts in this section.

Concept	Explanation	Examples
Rational exponents	$x^{m/n} = (x^m)^{1/n}$ $= (x^{1/n})^m$	$9^{3/2} = (9^3)^{1/2} = (729)^{1/2} = 27$ $9^{3/2} = (9^{1/2})^3 = (3)^3 = 27$
Radical notation	$x^{1/2} = \sqrt{x}$ $x^{1/3} = \sqrt[3]{x}$ $x^{m/n} = \sqrt[n]{x^m}$ $= (\sqrt[n]{x})^m$	$25^{1/2} = \sqrt{25} = 5$ $27^{1/3} = \sqrt[3]{27} = 3$ $8^{2/3} = \sqrt[3]{8^2} = \sqrt[3]{64} = 4$ $4^{3/2} = \sqrt{4^3} = (\sqrt{4})^3 = (2)^3 = 8$
Solving radical equations	The solutions to $a = b$ are among the solutions to $a^n = b^n$ when n is a positive integer. Check your results.	Solve $\sqrt{2x + 3} = x$. $2x + 3 = x^2$ Square each side. $x^2 - 2x - 3 = 0$ Rewrite equation. $x = -1$ or $x = 3$ Factor and solve. Checking reveals that 3 is the only solution.
Power function	$f(x) = x^b$, where b is a constant	$f(x) = x^{5/4}$ $g(x) = x^{-3.14}$
Root function	$f(x) = x^{1/n}$, where $n \geq 2$ is an integer	$f(x) = x^{1/2}$ or $f(x) = \sqrt{x}$ $g(x) = x^{1/5}$ or $g(x) = \sqrt[5]{x}$

Several types of functions are listed in the following summary, which may be used as a reference for future work. Unless specified otherwise, each tick mark represents 1 unit.

Type of Function	Examples	Graphs
Linear function $f(x) = ax + b$	$f(x) = 0.5x - 1$ $g(x) = -3x + 2$ $h(x) = 2$	
Polynomial function $f(x) = a_nx^n + \cdots + a_2x^2 + a_1x + a_0$	$f(x) = x^2 - 1$ $g(x) = x^3 - 4x - 1$ $h(x) = -x^4 + 4x^2 - 2$	
Rational function $f(x) = \frac{p(x)}{q(x)}$, where $p(x)$ and $q(x)$ are polynomials with $q(x) \neq 0$	$f(x) = \dfrac{1}{x}$ $g(x) = \dfrac{2x - 1}{x + 2}$ $h(x) = \dfrac{1}{x^2 - 1}$	
Root function $f(x) = x^{1/n}$, where $n \geq 2$ is an integer	$f(x) = x^{1/2} = \sqrt{x}$ $g(x) = x^{1/3} = \sqrt[3]{x}$ $h(x) = x^{1/4} = \sqrt[4]{x}$	
Power function $f(x) = x^b$, where b is a constant	$f(x) = x^{2/3}$ $g(x) = x^{1.41}$ $h(x) = x^3$	

4.8 Exercises

Properties of Exponents

Exercises 1–18: Evaluate the expression by hand.

1. $8^{2/3}$

2. $-16^{3/2}$

3. $16^{-3/4}$

4. $25^{-3/2}$

5. $-81^{0.5}$

6. $32^{1/5}$

7. $(9^{3/4})^2$

8. $(4^{-1/2})^{-4}$

9. $\dfrac{8^{5/6}}{8^{1/2}}$

10. $\dfrac{4^{-1/2}}{4^{3/2}}$

11. $27^{5/6} \cdot 27^{-1/6}$

12. $16^{2/3} \cdot 16^{-1/6}$

13. $(-27)^{-5/3}$

14. $(-32)^{-3/5}$

15. $(0.5^{-2})^2$

16. $\left(2^{-2}\right)^{-3/2}$

17. $\left(\frac{2}{3}\right)^{-2}$

18. $(8^{-1/3} + 27^{-1/3})^2$

Exercises 19–28: Use positive exponents to rewrite.

19. $\sqrt{2x}$

20. $\sqrt{x+1}$

21. $\sqrt[3]{z^5}$

22. $\sqrt[5]{x^2}$

23. $(\sqrt[4]{y})^{-3}$

24. $(\sqrt[3]{y^2})^{-5}$

25. $\sqrt{x} \cdot \sqrt[3]{x}$

26. $(\sqrt[5]{z})^{-3}$

27. $\sqrt{y} \cdot \sqrt[4]{y}$

28. $\dfrac{\sqrt[3]{x}}{\sqrt{x}}$

Exercises 29–32: Use radical notation to rewrite.

29. $a^{-3/4}b^{1/2}$

30. $a^{-2/3}b^{3/5}$

31. $(a^{1/2} + b^{1/2})^{1/2}$

32. $(a^{3/4} - b^{3/2})^{1/3}$

Equations Involving Radicals

Exercises 33–50: Solve the equation. Check your answers.

33. $\sqrt{x+2} = x - 4$

34. $\sqrt{2x+1} = 13$

35. $\sqrt{3x+7} = 3x + 5$

36. $\sqrt{1-x} = x + 5$

37. $\sqrt{5x-6} = x$

38. $x - 5 = \sqrt{5x-1}$

39. $\sqrt{x+5} + 1 = x$

40. $\sqrt{4-3x} = x + 8$

41. $\sqrt{x+1} + 3 = \sqrt{3x+4}$

42. $\sqrt{x} = \sqrt{x-5} + 1$

43. $\sqrt{2x} - \sqrt{x+1} = 1$

44. $\sqrt{2x-4} + 2 = \sqrt{3x+4}$

45. $\sqrt[3]{z+1} = -3$

46. $\sqrt[3]{z+5} = 4$

47. $\sqrt[3]{x+1} = \sqrt[3]{2x-1}$

48. $\sqrt[3]{2x^2+1} = \sqrt[3]{1-x}$

49. $\sqrt[4]{x-2} + 4 = 20$

50. $\sqrt[4]{2x+3} = \sqrt{x+1}$

Power Functions

Exercises 51 and 52: Evaluate $f(x)$ at the given x. Approximate each result to the nearest hundredth.

51. $f(x) = x^{3/2} - x^{1/2}, \quad x = 50$

52. $f(x) = x^{5/4} - x^{-3/4}, \quad x = 7$

Exercises 53 and 54: Match $f(x)$ with its graph. Assume that a and b are constants with $0 < a < 1 < b$.

53. $f(x) = x^a$

54. $f(x) = x^b$

a.

b.
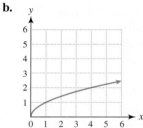

Exercises 55–60: Use translations to graph f.

55. $f(x) = \sqrt{x} + 1$

56. $f(x) = \sqrt[3]{x} - 1$

57. $f(x) = x^{2/3} - 1$

58. $f(x) = \sqrt{x} - 1$

59. $f(x) = \sqrt{x+2} - 1$

60. $f(x) = (x-1)^{2/3}$

Equations Involving Rational Exponents

Exercises 61–82: Solve the equation. Check your answers.

61. $x^3 = 8$

62. $x^4 = \frac{1}{81}$

63. $x^{1/4} = 3$

64. $x^{1/3} = \frac{1}{5}$

65. $x^{2/5} = 4$

66. $x^{2/3} = 16$

67. $2(x^{1/5} - 2) = 0$

68. $x^{1/2} + x^{1/2} = 8$

69. $4x^{3/2} + 5 = 21$

70. $2x^{1/3} - 5 = 1$

71. $n^{-2} + 3n^{-1} + 2 = 0$

72. $2n^{-2} - n^{-1} = 3$

73. $5n^{-2} + 13n^{-1} = 28$ 74. $3n^{-2} - 19n^{-1} + 20 = 0$

75. $x^{2/3} - x^{1/3} - 6 = 0$ 76. $x^{2/3} + 9x^{1/3} + 14 = 0$

77. $6x^{2/3} - 11x^{1/3} + 4 = 0$

78. $10x^{2/3} + 29x^{1/3} + 10 = 0$

79. $x^{3/4} - x^{1/2} - x^{1/4} + 1 = 0$

80. $x^{3/4} - 2x^{1/2} - 4x^{1/4} + 8 = 0$

81. $x^{-2/3} - 2x^{-1/3} - 3 = 0$

82. $6x^{-2/3} - 13x^{-1/3} - 5 = 0$

Exercises 83 and 84: Average Rate of Change Let the distance from home in miles of a person after t hours on a straight path be given by $s(t)$. Approximate the average rate of change of s from $t_1 = \frac{1}{2}$ to $t_2 = \frac{9}{2}$ to the nearest tenth and interpret the result.

83. $s(t) = \sqrt{96t}$

84. $s(t) = 3t^{3/4}$

Applications and Models

85. *Modeling Wing Size* Suppose that the surface area S of a bird's wings in square feet can be modeled by $S(w) = 1.27w^{2/3}$, where w is the weight of the bird in pounds, with $1 \le w \le 10$. Estimate the weight of a bird with wings having a surface area of 3 square feet.

86. *Modeling Wingspan* The wingspan L in feet of a bird weighing W pounds is given by $L = 2.43W^{0.3326}$. Estimate the wingspan of a bird that weighs 5.2 pounds.

87. *Modeling Planetary Orbits* The formula $f(x) = x^{1.5}$ calculates the number of years it takes for a planet to orbit the sun if its average distance from the sun is x times that of Earth. If there were a planet located 15 times as far from the sun as Earth, how many years would it take for the planet to orbit the sun?

88. *Modeling Planetary Orbits* (Refer to Exercise 87.) If there were a planet that took 200 years to orbit the sun, what would be its average distance x from the sun compared to that of Earth?

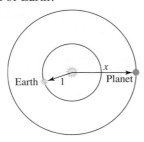

89. *Trout and Pollution* Rainbow trout are sensitive to zinc ions in the water. High concentrations are lethal. The average survival times x in minutes for trout in various concentrations of zinc ions y in milligrams per liter (mg/L) are listed in the table.

x (min)	0.5	1	2	3
y (mg/L)	4500	1960	850	525

Source: C. Mason, *Biology of Freshwater Pollution.*

(a) These data can be modeled by $f(x) = ax^b$, where a and b are constants. Determine an appropriate value for a. (*Hint:* Let $f(1) = 1960$.)

(b) Estimate b.

(c) Evaluate $f(4)$ and interpret the result.

90. *Lunar Orbits for Jupiter* Use the data in the table to complete the following.

Moons of Jupiter	Distance (10^3 km)	Period (days)
Metis	128	0.29
Almathea	181	0.50
Thebe	222	0.67
Europa	671	3.55
Ganymede	1070	7.16
Callisto	1883	16.69

(a) Make a scatterplot of the data. Estimate a value for b so that $f(x) = 0.0002x^b$ models the data.

(b) Numerically check the accuracy of f.

(c) The moon Io is 422 thousand kilometers from Jupiter. Estimate its period and compare the estimate to the actual value of 1.77 days. Did your estimate involve interpolation or extrapolation?

91. *Fiddler Crab Size*   Allometric relations often can be modeled by $f(x) = ax^b$, where a and b are constants. One study showed that for a male fiddler crab weighing over 0.75 gram, the weight of its claws can be estimated by $f(x) = 0.445x^{1.25}$. The input x is the weight of the crab in grams, and the output $f(x)$ is the weight of the claws in grams.   (**Sources:** J. Huxley, *Problems of Relative Growth*; D. Brown and P. Rothery, *Models in Biology: Mathematics, Statistics and Computing*.)

    (**a**) Predict the weight of the claws of a 2-gram crab.

    (**b**) Approximate graphically the weight of a crab that has 0.5-gram claws.

    (**c**) Solve part (b) symbolically.

92. *Weight and Height*   Allometry can be applied to height-weight relationships for humans. The average weight in pounds for men and women can sometimes be estimated by $f(x) = ax^{1.7}$, where x is a person's height in inches and a is a constant determined by the sex of the individual.

    (**a**) If the average weight of a 68-inch-tall man is 152 pounds, approximate a. Use f to estimate the average weight of a 66-inch-tall man.

    (**b**) If the average weight of a 68-inch-tall woman is 137 pounds, approximate a. Use f to estimate the average weight of a 70-inch-tall woman.

Exercises 93 and 94: Power Regression   *The table contains data that can be modeled by a function of the form $f(x) = ax^b$. Use regression to find the constants a and b to the nearest hundredth. Graph f and the data.*

93.

x	2	4	6	8
$f(x)$	3.7	4.2	4.6	4.9

94.

x	3	6	9	12
$f(x)$	23.8	58.5	99.2	144

95. *Wal-Mart Employees*   The table lists numbers N of Wal-Mart employees (in millions) x years *after 1980*.

x	7	12	17	22	27
N	0.20	0.37	0.68	1.4	2.2

Source: Wal-Mart.

    (**a**) Find a power function f that models the data in the table.

    (**b**) Use f to predict the number of employees in the year 2012. Did your answer involve interpolation or extrapolation?

    (**c**) When did the number of employees first reach 1 million?

96. *DVD Rentals*   The table lists numbers of titles T released for DVD rentals x years *after 1995*.

x	3	4	5	6	7
T	2049	4787	8723	14,321	21,260

Source: DVD Release Report.

    (**a**) Find a power function f that models the data in the table.

    (**b**) Use f to estimate the number of titles released in 2006. Did your answer involve interpolation or extrapolation?

    (**c**) When did the number of releases first surpass 45,000?

97. *Pulse Rate and Weight*   According to one model, the rate at which an animal's heart beats varies with its weight. Smaller animals tend to have faster pulses, whereas larger animals tend to have slower pulses. The table lists average pulse rates in beats per minute (bpm) for animals with various weights in pounds (lb). Use regression (or some other method) to find values for a and b so that $f(x) = ax^b$ models these data.

Weight (lb)	40	150	400	1000	2000
Pulse (bpm)	140	72	44	28	20

Source: C. Pennycuick.

98. *Pulse Rate and Weight*   (Continuation of Exercise 97) Use the results in the previous exercise to calculate the pulse rates for a 60-pound dog and a 2-ton whale.

Writing about Mathematics

99. Can a function be both a polynomial function and a power function? Explain.

100. Explain the basic steps needed to solve equations that contain square roots of variables.

EXTENDED AND DISCOVERY EXERCISES

1. *Odd Root Functions*   Graph $y = \sqrt[n]{x}$ for $n = 3, 5$, and 7. State some generalizations about a graph of an odd root function.

2. *Even Root Functions*   Graph $y = \sqrt[n]{x}$ for $n = 2, 4$, and 6. State some generalizations about a graph of an even root function.

3. *Power Functions*   Graph $y = x^b$ for $b = -1, -3$, and -5. State some generalizations about a graph of a power function, where b is a negative odd integer.

4. *Power Functions*   Graph $y = x^b$ for $b = -2, -4$, and -6. State some generalizations about a graph of a power function, where b is a negative even integer.

Exercises 5 and 6: Difference Quotient   Find the difference quotient of f.

5. $f(x) = \sqrt{x}$ 6. $f(x) = \frac{1}{x}$

Exercises 7–10: Negative Rational Exponents   Write the expression as one ratio without any negative exponents.

7. $\dfrac{x^{-2/3} + x^{1/3}}{x}$

8. $\dfrac{x^{1/4} - x^{-3/4}}{x}$

9. $\dfrac{\frac{2}{3}(x + 1)x^{-1/3} - x^{2/3}}{(x + 1)^2}$

10. $\dfrac{(x^2 + 1)^{1/2} - \frac{1}{2}x(x^2 + 1)^{-1/2}(2x)}{x^2 + 1}$

CHECKING BASIC CONCEPTS FOR SECTIONS 4.7 AND 4.8

(a) $\dfrac{3x - 1}{1 - x} = 1$ (b) $3 + \dfrac{8}{x} = \dfrac{35}{x^2}$

(c) $\dfrac{1}{x - 1} - \dfrac{1}{3(x + 2)} = \dfrac{1}{x^2 + x - 2}$

2. Solve $2x^3 + x^2 - 6x < 0$.

3. Solve $\dfrac{x^2 - 1}{x + 2} \geq 0$.

4. Let y vary inversely as the cube of x. If $x = \frac{1}{5}$, then $y = 150$. Find y if $x = \frac{1}{2}$.

5. Simplify each expression without a calculator.
   (a) $-4^{3/2}$ (b) $(8^{-2})^{1/3}$ (c) $\sqrt[3]{27^2}$

6. Solve the equation $4x^{3/2} - 3 = 29$.

7. Solve the equation $\sqrt{5x - 4} = x - 2$.

8. Solve each equation.
   (a) $n^{-2} + 6n^{-1} = 16$

   (b) $2x^{2/3} + 5x^{1/3} - 12 = 0$

9. Find a and b so that $f(x) = ax^b$ models the data.

x	1	2	3	4
$f(x)$	2	2.83	3.46	4

4   Summary

CONCEPT	EXPLANATION AND EXAMPLES

SECTION 4.1 MORE NONLINEAR FUNCTIONS AND THEIR GRAPHS

Polynomial Function

Can be represented by $f(x) = a_n x^n + a_{n-1} x^{n-1} + \cdots + a_2 x^2 + a_1 x + a_0$
The leading coefficient is $a_n \neq 0$ and the degree is n.

Example: $f(x) = -4x^3 - 2x^2 + 6x + \frac{1}{2}; \; a_n = -4; \; n = 3$

Absolute and Local Extrema

The accompanying graph has the following extrema.

Absolute maximum: none

Absolute minimum: -14.8

Local maximum: 1

Local minima: $-4.3, -14.8$

Symmetry

Even function: $f(-x) = f(x)$; the graph is symmetric with respect to the y-axis.

Odd function: $f(-x) = -f(x)$; the graph is symmetric with respect to the origin.

Examples: $f(x) = x^4 - 3x^2$ $f(x) = x - x^3$

SECTION 4.2 POLYNOMIAL FUNCTIONS AND MODELS

Graphs of Polynomial Functions

Their graphs are continuous with no breaks, and their domains include all real numbers. The graph of a polynomial function of degree $n \geq 1$ has at most n x-intercepts and at most $n - 1$ turning points. For a discussion of the end behavior for graphs of polynomial functions, see page 264.

Examples: See "Putting It All Together" for Section 4.2 on pages 269–270.

Piecewise-Polynomial Functions

Example:

$$f(x) = \begin{cases} x^2 - 2 & \text{if } x < 0 \\ 1 - 2x & \text{if } x \geq 0 \end{cases}$$

$f(-2) = (-2)^2 - 2 = 2$

$f(0) = 1 - 2(0) = 1$

$f(2) = 1 - 2(2) = -3$

f is discontinuous at $x = 0$.

CONCEPT	EXPLANATION AND EXAMPLES

SECTION 4.3 DIVISION OF POLYNOMIALS

Division Algorithm

Let $f(x)$ and $d(x)$ be two polynomials with the degree of $d(x)$ greater than zero and less than the degree of $f(x)$. Then there exist unique polynomials $q(x)$ and $r(x)$ such that

$$f(x) = d(x) \cdot q(x) + r(x),$$

where either $r(x) = 0$ or the degree of $r(x)$ is less than the degree of $d(x)$. That is,

(Dividend) = (Divisor) · (Quotient) + (Remainder).

Example: $\dfrac{x^2 - 4x + 5}{x - 1} = x - 3 + \dfrac{2}{x - 1}$. That is,

$x^2 - 4x + 5 = (x - 1)(x - 3) + 2.$

Remainder Theorem

If a polynomial $f(x)$ is divided by $x - k$, the remainder is $f(k)$.

Example: If $x^2 - 4x + 5$ is divided by $x - 1$, the remainder is

$$f(1) = 1^2 - 4(1) + 5 = 2.$$

SECTION 4.4 REAL ZEROS OF POLYNOMIAL FUNCTIONS

Factor Theorem

A polynomial $f(x)$ has a factor $x - k$ if and only if $f(k) = 0$.

Example: $f(x) = x^2 - 3x + 2$;
$f(1) = 0$ implies that $(x - 1)$ is a factor of $x^2 - 3x + 2$.

Complete Factored Form

$f(x) = a_n(x - c_1)(x - c_2) \cdots (x - c_n)$, where a_n is the leading coefficient of the polynomial $f(x)$ and the c_k are its zeros. This form is unique.

Example: $f(x) = -2x^3 + 8x$ has zeros of -2, 0, and 2.
$f(x) = -2(x + 2)(x - 0)(x - 2)$

Multiple Real Zeros

Odd multiplicity: Graph crosses the x-axis.

Even multiplicity: Graph touches but does not cross the x-axis.

Example: Let $f(x) = -2(x + 1)^3(x - 4)^2$; $f(x)$
has a zero of -1 with odd multiplicity
3 and a zero of 4 with even multiplicity 2.

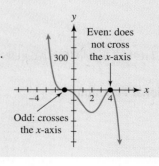

CONCEPT	EXPLANATION AND EXAMPLES

SECTION 4.4 REAL ZEROS OF POLYNOMIAL FUNCTIONS (CONTINUED)

Polynomial Equations

Polynomial equations can be solved symbolically, graphically, and numerically. A common symbolic technique is factoring.

Example: Solve $x^3 - 4x = 0$ symbolically and graphically.

Symbolic Solution

$$x(x^2 - 4) = 0$$
$$x(x - 2)(x + 2) = 0$$
$$x = 0, \quad x = 2, \quad \text{or} \quad x = -2$$

Graphical Solution

The x-intercepts are -2, 0, and 2.

SECTION 4.5 THE FUNDAMENTAL THEOREM OF ALGEBRA

Fundamental Theorem of Algebra

A polynomial $f(x)$ of degree $n \geq 1$ has at least one complex zero.
Explanation: With complex numbers, any polynomial can be written in complete factored form.

Examples: $x^2 + 1 = (x + i)(x - i)$
$3x^2 - 3x - 6 = 3(x + 1)(x - 2)$

Number of Zeros Theorem

A polynomial of degree n has at most n distinct zeros.

Example: A cubic polynomial has at most 3 distinct zeros.

Polynomial Equations with Complex Solutions

Polynomial equations can have both real and imaginary solutions.

Example: Solve $x^4 - 1 = 0$.
$$(x^2 - 1)(x^2 + 1) = 0$$
$$(x - 1)(x + 1)(x^2 + 1) = 0$$
$$x = 1, \quad x = -1, \quad \text{or} \quad x^2 = -1$$
$$x = \pm 1, \pm i$$

$y = x^4 - 1$
The x-intercepts are -1 and 1. Imaginary solutions cannot be found from a graph.

SECTION 4.6 RATIONAL FUNCTIONS AND MODELS

Rational Functions

$f(x) = \dfrac{p(x)}{q(x)}$, where $p(x)$ and $q(x) \neq 0$ are polynomials

Example: $f(x) = \dfrac{2x - 3}{x - 1}$

Horizontal asymptote: $y = 2$
Vertical asymptote: $x = 1$

To find vertical and horizontal asymptotes, see page 312.
To graph rational functions by hand, see page 316.

CONCEPT	EXPLANATION AND EXAMPLES

SECTION 4.7 MORE EQUATIONS AND INEQUALITIES

Solving Rational Equations

Multiply each side by the LCD. Check your results.

Example: $\dfrac{-24}{x-3} - 4 = x + 3$

$$-24 - 4(x-3) = (x+3)(x-3) \qquad \text{Multiply by } x-3.$$
$$-24 - 4x + 12 = x^2 - 9 \qquad \text{Multiply.}$$
$$0 = x^2 + 4x + 3 \qquad \text{Combine terms.}$$
$$0 = (x+3)(x+1) \qquad \text{Factor.}$$
$$x = -3 \quad \text{or} \quad x = -1 \qquad \text{Both solutions check.}$$

Direct Variation

Let x and y denote two quantities and n be a positive number. Then y is *directly proportional to the nth power of x*, or y *varies directly as the nth power of x*, if there exists a nonzero number k such that $y = kx^n$.

Example: Because $V = \frac{4}{3}\pi r^3$, the volume of a sphere varies directly as the third power of the radius. The constant of variation is $\frac{4}{3}\pi$.

Inverse Variation

Let x and y denote two quantities and n be a positive number. Then y is *inversely proportional to the nth power of x*, or y *varies inversely as the nth power of x*, if there exists a nonzero number k such that $y = \frac{k}{x^n}$.

Example: Because $I = \frac{k}{d^2}$, the intensity of a light source varies inversely as the square of the distance from the light source.

Polynomial Inequality

Write the inequality as $p(x) > 0$, where $>$ may be replaced by $\geq$, $<$, or $\leq$. Replace the inequality sign with an equals sign, and solve this equation. The solutions are called boundary numbers. Then use a graph or table to find the solution set to the given inequality.

Example: $4x - x^3 > 0$; Boundary numbers: $-2, 0, 2$
Solution set: $(-\infty, -2) \cup (0, 2)$

Rational Inequality

As with polynomial inequalities, find the boundary numbers, including x-values where any expressions are undefined.

Example: $\dfrac{(x+2)(x-3)}{x} \geq 0$

Boundary numbers: $-2, 0, 3$
Solution set: $[-2, 0) \cup [3, \infty)$

CONCEPT	EXPLANATION AND EXAMPLES

SECTION 4.8 RADICAL EQUATIONS AND POWER FUNCTIONS

Rational Exponents

$$x^{m/n} = \sqrt[n]{x^m} = (\sqrt[n]{x})^m$$

Example: $25^{3/2} = \sqrt{25^3} = (\sqrt{25})^3 = 5^3 = 125$

Solving Radical Equations

When an equation contains a square root, isolate the square root and then square each side. *Be sure to check your results.*

Example:	$x + \sqrt{3x - 3} = 1$	Given equation
	$\sqrt{3x - 3} = 1 - x$	Subtract x.
	$3x - 3 = (1 - x)^2$	Square each side.
	$3x - 3 = 1 - 2x + x^2$	Square binomial.
	$0 = x^2 - 5x + 4$	Combine terms.
	$0 = (x - 1)(x - 4)$	Factor.
	$x = 1 \quad \text{or} \quad x = 4$	Solve.

Check: 1 is a solution, but 4 is not.

$$1 + \sqrt{3(1) - 3} = 1 \qquad 4 + \sqrt{3(4) - 3} \neq 1$$

Power Function

$f(x) = x^b$, where b is a constant

Example: $f(x) = x^{4/3}, \qquad g(x) = x^{1.72}$

---

4   Review Exercises

1. State the degree and leading coefficient of the polynomial $f(x) = 4 + x - 2x^2 - 7x^3$.

Exercises 2 and 3: Use the graph of f to estimate any
  (a) local extrema   and   (b) absolute extrema.

2.

3.

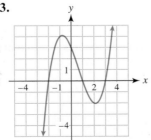

4. Graph $f(x) = -0.25x^4 + 0.67x^3 + 9.5x^2 - 20x - 50$.

(a) Approximate any local extrema.

(b) Approximate any absolute extrema.

(c) Determine where f is increasing or decreasing.

5. Graph $f(x) = x^4 + 2x^3 - 9x^2 - 2x + 20$. Count the local maxima, local minima, and x-intercepts.

6. Graph f. Estimate any turning points.
$$f(x) = 0.03x^5 - 0.21x^4 + 0.21x^3 + 0.57x^2$$
$$- 0.48x - 0.6$$

Exercises 7–10: Determine if f is even, odd, or neither.

7. $f(x) = 2x^6 - 5x^4 - x^2$ **8.** $f(x) = -5x^3 - 18$

9. $f(x) = 7x^5 + 3x^3 - x$ **10.** $f(x) = \dfrac{1}{1 + x^2}$

Exercises 11 and 12: The table is a complete representation of f. Decide if f is even, odd, or neither.

11.

x	-4	-2	0	2	4
f(x)	13	7	0	-7	-13

12.

x	-5	-3	-1	1	3	5
f(x)	-6	2	7	7	2	-6

Exercises 13 and 14: Sketch a graph of a polynomial function that satisfies the given conditions.

13. Cubic polynomial, two x-intercepts, and a positive leading coefficient

14. Degree 4 with a positive leading coefficient, three turning points, and one x-intercept

Exercises 15 and 16: Use the graph of the polynomial function f to complete the following.

(a) Determine the number of turning points and estimate any x-intercepts.
(b) State whether $a > 0$ or $a < 0$.
(c) Determine the minimum degree of f.

15.

16.

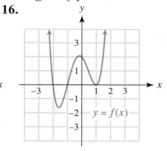

Exercises 17 and 18: State the end behavior of f.

17. $f(x) = -2x^3 + 4x - 2$

18. $f(x) = 1 - 2x - x^4$

19. Find the average rate of change of $f(x) = x^3 + 1$ from $x = -2$ to $x = -1$.

20. Find the difference quotient for $g(x) = 4x^3$.

21. Let $f(x)$ be given by

$$f(x) = \begin{cases} 2x & \text{if } 0 \le x < 2 \\ 8 - x^2 & \text{if } 2 \le x \le 4. \end{cases}$$

(a) Sketch a graph of f. Is f continuous on its domain?

(b) Evaluate $f(1)$ and $f(3)$.

(c) Solve the equation $f(x) = 2$.

22. Determine the type of symmetry that the graph of $g(x) = x^5 - 4x^3$ exhibits.

Exercises 23–26: Divide the expression.

23. $\dfrac{14x^3 - 21x^2 - 7x}{7x}$

24. $\dfrac{2x^3 - x^2 - 4x + 1}{x + 2}$

25. $\dfrac{4x^3 - 7x + 4}{2x + 3}$

26. $\dfrac{3x^3 - 5x^2 + 13x - 18}{x^2 + 4}$

27. The polynomial given by $f(x) = \frac{1}{2}x^3 - 3x^2 + \frac{11}{2}x - 3$ has zeros 1, 2, and 3. Write its complete factored form.

28. Find the complete factored form of the polynomial $f(x) = x^4 + 2x^3 - 13x^2 - 14x + 24$ given that 1 and 3 are zeros.

29. Find the complete factored form of the polynomial $f(x) = 2x^3 + 3x^2 - 18x + 8$ graphically.

30. Write a complete factored form of a quintic (degree 5) polynomial $f(x)$ that has zeros -2 and 2 with multiplicities 2 and 3, respectively.

31. Use the graph of $y = f(x)$ to write its complete factored form. (Do not assume that the leading coefficient is ± 1.)

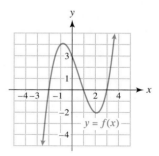

32. What is the maximum number of times that a horizontal line can intersect the graph of each type of polynomial?
(a) linear (degree 1)   (b) quadratic   (c) cubic

Exercises 33 and 34: Use the rational zero test to determine any rational zeros of f(x).

33. $f(x) = 2x^3 + x^2 - 13x + 6$

34. $f(x) = x^3 + x^2 - 11x - 11$

Exercises 35–38: Solve the equation.

35. $9x = 3x^3$

36. $x^3 - x^2 - 6x = 0$

37. $x^4 - 3x^2 + 2 = 0$

38. $2x^3 + x^2 = 6x + 3$

Exercises 39 and 40: Solve the equation graphically. Round your answers to the nearest hundredth.

39. $x^3 - 3x + 1 = 0$

40. $x^4 - 2x = 2$

Exercises 41–44: Write the expression in standard form.

41. $(2 - 2i) + (-3 + 2i)$ **42.** $(-5 + 4i) - (-3 - 4i)$

43. $(3 + 2i)^2$

44. $\dfrac{3 + i}{1 + i}$

Exercises 45 and 46: Find all real and imaginary solutions.

45. $x^3 + x = 0$

46. $x^4 + 3x^2 + 2 = 0$

47. Determine graphically the number of real zeros and the number of imaginary zeros of the polynomial function $f(x) = x^3 - 3x^2 + 3x - 9$.

48. Write a polynomial $f(x)$ in complete factored form that has degree 3, leading coefficient 4, and zeros 1, $3i$, and $-3i$. Then write $f(x)$ in expanded form.

49. Find the complete factorization of $f(x) = 2x^2 + 4$.

50. Use the graph of $f(x) = 2x^4 - x^2 - 1$ to predict the number of real zeros and the number of imaginary zeros of f. Find these zeros symbolically.

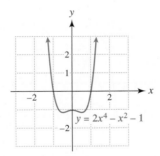

51. If a zero of f is i, find the complete factored form of $f(x) = x^4 + x^3 + 2x^2 + x + 1$.

52. State the domain of $f(x) = \frac{3x - 2}{5x + 4}$. Identify any horizontal or vertical asymptotes in the graph of f.

53. Find any horizontal or vertical asymptotes in the graph of

$$f(x) = \frac{2x^2 + x - 15}{3x^2 + 8x - 3}.$$

54. Let $f(x) = \dfrac{2x^2}{x^2 - 4}$.

  (a) Find the domain of f.

  (b) Identify any horizontal or vertical asymptotes.

  (c) Graph f with a graphing calculator.

  (d) Sketch a graph of f that includes all asymptotes.

Exercises 55–58: Graph $y = g(x)$ by hand.

55. $g(x) = \dfrac{1}{x + 1} - 2$ **56.** $g(x) = \dfrac{x}{x - 1}$

57. $g(x) = \dfrac{x^2 - 1}{x^2 + 2x + 1}$ **58.** $g(x) = \dfrac{2x - 3}{2x^2 + x - 6}$

59. Sketch a graph of a function f with vertical asymptote $x = -2$ and horizontal asymptote $y = 2$.

60. Solve the equation $\frac{3x}{x - 2} = 2$ symbolically, graphically, and numerically.

Exercises 61–66: Solve the equation. Check your results.

61. $\dfrac{5x + 1}{x + 3} = 3$ **62.** $\dfrac{1}{x} - \dfrac{1}{x^2} + 2 = 0$

63. $\dfrac{1}{x + 2} + \dfrac{1}{x - 2} = \dfrac{4}{x^2 - 4}$

64. $x - \dfrac{1}{x} = 4$

65. $\dfrac{x + 5}{x - 2} = \dfrac{x - 1}{x + 1}$ **66.** $\dfrac{2x - 5}{3x + 1} = 5$

Exercises 67 and 68: Use the graph of f to solve each inequality.

   (a) $f(x) > 0$ (b) $f(x) < 0$

67.   **68.**

Exercises 69–72: Solve the inequality.

69. $x^3 + x^2 - 6x > 0$ **70.** $x^4 + 4 < 5x^2$

71. $\dfrac{2x - 1}{x + 2} > 0$

72. $\dfrac{1}{x} + \dfrac{1}{x + 2} \le \dfrac{4}{3}$

Exercises 73–76: Evaluate the radical expression by hand.

73. $(36^{3/4})^2$

74. $(9^{-3/2})^{-2}$

75. $(2^{-3/2} \cdot 2^{1/2})^{-3}$

76. $\left(\dfrac{4}{9}\right)^{-3/2}$

Exercises 77–80: Write the radical expression using positive exponents.

77. $\sqrt[3]{x^4}$

78. $\left(\sqrt[4]{z}\right)^{-1/2}$

79. $\sqrt[3]{y} \cdot \sqrt{y}$

80. $\sqrt{x} \cdot \sqrt[3]{x^2} \cdot \sqrt[4]{x^3}$

Exercises 81 and 82: Give the domain of the power function. Approximate $f(3)$ to the nearest hundredth.

81. $f(x) = x^{5/2}$

82. $f(x) = x^{-2/3}$

Exercises 83–96: Solve the equation. Check your results.

83. $x^5 = 1024$

84. $x^{1/3} = 4$

85. $\sqrt{x - 2} = x - 4$

86. $x^{3/2} = 27$

87. $2x^{1/4} + 3 = 6$

88. $\sqrt{x - 2} = 14 - x$

89. $\sqrt[3]{2x - 3} + 1 = 4$

90. $x^{1/3} + 3x^{1/3} = -2$

91. $2n^{-2} - 5n^{-1} = 3$

92. $m^{-3} + 2m^{-2} + m^{-1} = 0$

93. $k^{2/3} - 4k^{1/3} - 5 = 0$

94. $x^{3/4} - 16x^{1/4} = 0$

95. $\sqrt{x + 1} + 1 = \sqrt{2x}$

96. $\sqrt{x - 2} = 5 - \sqrt{x + 3}$

Applications

97. *Allometry*   One of the earliest studies in allometry was performed by Bryan Robinson during the eighteenth century. He found that the pulse rate of an animal could be approximated by $f(x) = 1607x^{-0.75}$. The input x is the length of the animal in inches, and the output $f(x)$ is the approximate number of heartbeats per minute.   (**Source:** H. Lancaster, *Quantitative Methods in Biology and Medical Sciences.*)

(a) Use f to estimate the pulse rates of a 2-foot dog and a 5.5-foot person.

(b) What length corresponds to a pulse rate of 400 beats per minute?

98. *Time Spent in Line*   Suppose a parking garage attendant can wait on 4 vehicles per minute and vehicles are leaving the ramp randomly at an average rate of x vehicles

per minute. Then the average time T in minutes spent waiting in line and paying the attendant is given by

$$T(x) = \dfrac{1}{4 - x},$$

where $0 \le x < 4$.   (**Source:** N. Garber and L. Hoel, *Traffic and Highway Engineering.*)

(a) Evaluate $T(2)$ and interpret the result.

(b) Graph T for $0 \le x < 4$.

(c) What happens to the waiting time as x increases from 0 to (nearly) 4?

(d) Find x if the waiting time is 5 minutes.

99. *Modeling Ocean Temperatures*   The formula

$$T(m) = -0.064m^3 + 0.56m^2 + 2.9m + 61$$

approximates the ocean temperature in degrees Fahrenheit at Naples, Florida. In this formula m is the month, with $m = 1$ corresponding to January.

(a) What is the average ocean temperature in May?

(b) Estimate the absolute maximum of T on the closed interval $[1, 12]$ and interpret the result.

100. *Minimizing Surface Area*   Find possible dimensions that minimize the surface area of a box with no top that has a volume of 96 cubic inches and a length that is three times the width.

101. *Falling Object*   If an object is dropped from a height h, then the time t required for the object to strike the ground is directly proportional to the square root of h. If it requires 1 second for an object to fall 16 feet, how long does it take for an object to fall 256 feet?

102. *Animals and Trotting Speeds*   Taller animals tend to take longer, but fewer, steps per second than shorter animals. The relationship between the shoulder height h in meters of an animal and an animal's stepping frequency F in steps per second, while *trotting*, is shown in the table.

h	0.5	1.0	1.5	2.0	2.5
F	2.6	1.8	1.5	1.3	1.2

Source: C. Pennycuick, *Newton Rules Biology.*

(a) Find values for constants a and b so that the formula $f(x) = ax^b$ models the data.

(b) Estimate the stepping frequency for an elephant with a 3-meter shoulder height.

EXTENDED AND DISCOVERY EXERCISES

Exercises 1 and 2: Velocity *Suppose that a person is riding a bicycle on a straight road and that $f(t)$ computes the total distance in feet that the rider has traveled after t seconds. To calculate the person's average velocity between time t_1 and time t_2, we can evaluate the difference quotient*

$$\frac{f(t_2) - f(t_1)}{t_2 - t_1}.$$

(a) *For the given $f(t)$ and the indicated values of t_1 and t_2, calculate the average velocity of the bike rider. Make a table to organize your work.*

(b) *Make a conjecture about the velocity of the bike rider precisely at time t_1.*

1. $f(t) = t^2, t_1 = 10$
    (i) $t_2 = 11$
    (ii) $t_2 = 10.1$
    (iii) $t_2 = 10.01$
    (iv) $t_2 = 10.001$

2. $f(t) = \sqrt{t}, t_1 = 4$
    (i) $t_2 = 5$
    (ii) $t_2 = 4.1$
    (iii) $t_2 = 4.01$
    (iv) $t_2 = 4.001$

Exercises 3–7: Instantaneous Velocity *In Exercises 1 and 2 the average velocity of a bike rider was calculated over smaller and smaller time intervals. If we calculate the average velocity as the time interval approaches 0, we say that we are calculating the **instantaneous velocity**. To find the instantaneous velocity, we can calculate the average velocity between times $t_1 = a$ and $t_2 = a + h$. The difference quotient becomes*

$$\frac{f(t_2) - f(t_1)}{t_2 - t_1} = \frac{f(a + h) - f(a)}{h}.$$

For example, if $f(t) = t^2$, then

$$\frac{f(a + h) - f(a)}{h} = \frac{(a + h)^2 - a^2}{h}$$
$$= 2a + h.$$

If we let h approach 0 ($h \to 0$), then the instantaneous velocity becomes $2a + 0 = 2a$ at time $t_1 = a$. For example, if $a = 10$, then $v = 2(10) = 20$ feet per second. Does this result agree with Exercise 1?

(a) *For the given $f(t)$, calculate the difference quotient $\frac{f(a + h) - f(a)}{h}$. Be sure to simplify completely.*

(b) *Let h approach 0 and determine a formula for the instantaneous velocity at time $t_1 = a$.*

(c) *Calculate the instantaneous velocity at times $a = 5$, 10, and 15.*

3. $f(t) = 5t$

4. $f(t) = t^2 + 2t$

5. $f(t) = \dfrac{200}{t}$

6. $f(t) = t^3$

7. $f(t) = \sqrt{t}$ (*Hint:* To simplify the ratio, multiply the numerator and denominator by $\sqrt{a + h} + \sqrt{a}$. This procedure is called **rationalizing the numerator**.)

Algebra Review To review rationalizing the denominator, see Chapter R (page R-50).

Exercises 8–11: Average Rates of Change *These exercises investigate the relationship between polynomial functions and their average rates of change. For example, the average rate of change of $f(x) = x^2$ from x to $x + 0.001$ for any x can be calculated and graphed as shown in the figures. The graph of f is a parabola, and the graph of its average rate of change is a line. Try to discover what this relationship is by completing the following.*

$[-10, 10, 1]$ by $[-10, 10, 1]$

(a) *Graph each function and its average rate of change from x to $x + 0.001$.*

(b) *Compare the graphs. How are turning points on the graph of a function related to its average rate of change?*

(c) *Generalize your results. Test your generalization.*

8. *Linear Functions*
    $f_1(x) = 3x + 1$ $f_2(x) = -2x + 6$
    $f_3(x) = 1.5x - 5$ $f_4(x) = -4x - 2.5$

9. *Quadratic Functions*
    $f_1(x) = 2x^2 - 3x + 1$ $f_2(x) = -0.5x^2 + 2x + 2$
    $f_3(x) = x^2 + x - 2$ $f_4(x) = -1.5x^2 - 4x + 6$

10. *Cubic Functions*
    $f_1(x) = 0.5x^3 - x^2 - 2x + 1$
    $f_2(x) = -x^3 + x^2 + 3x - 5$
    $f_3(x) = 2x^3 - 5x^2 + x - 3$
    $f_4(x) = -x^3 + 3x - 4$

11. *Quartic Functions*
    $f_1(x) = 0.05x^4 + 0.2x^3 - x^2 - 2.4x$
    $f_2(x) = -0.1x^4 + 0.1x^3 + 1.3x^2 - 0.1x - 1.2$
    $f_3(x) = 0.1x^4 + 0.4x^3 - 0.2x^2 - 2.4x - 2.4$

1-4   Cumulative Review Exercises

1. Find the percent change if a quantity Q changes from 45 to 54.

2. Write 0.065 in scientific notation and 7.88×10^5 in standard notation.

3. Let $S = \{(-3, 4), (-1, -2), (0, 4), (1, -2), (-1, 5)\}$.
   (a) Find the domain and range of S.

   (b) Is S a function?

4. Find the exact distance between $(-1, 4)$ and $(3, -9)$.

5. Use the graph to express the domain and range of f. Then evaluate $f(0)$.

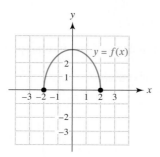

6. Graph $y = g(x)$ by hand.
   (a) $g(x) = 2 - 3x$ (b) $g(x) = |2x - 1|$

   (c) $g(x) = \frac{1}{2}(x - 2)^2 + 2$ (d) $g(x) = x^3 - 1$

   (e) $g(x) = \sqrt{-x}$ (f) $g(x) = \sqrt[3]{x}$

   (g) $g(x) = \dfrac{1}{x - 4} + 2$ (h) $g(x) = x^2 - x$

Exercises 7 and 8: Complete the following.
   (a) Determine the domain of f.
   (b) Evaluate f(2).

7. $f(x) = \sqrt{x^2 - 4}$ 8. $\dfrac{2x - 3}{3x^2 + 11x - 4}$

9. The monthly cost of driving a car is $200 for maintenance plus $0.25 a mile. Write a formula for a function C that calculates the monthly cost of driving a car x miles. Evaluate $C(2000)$ and interpret the result.

10. The graph of a linear function f is shown.

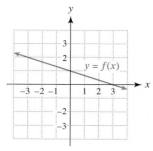

   (a) Identify the slope, y-intercept, and x-intercept.

   (b) Write a formula for $f(x)$.

   (c) Evaluate $f(-3)$ symbolically and graphically.

   (d) Find any zeros of f.

11. Find the average rate of change of $f(x) = x^3 - x$ from $x = -3$ to $x = -2$.

12. Find the difference quotient for $f(x) = x^2 + 6x$.

13. Write the slope-intercept form for a line that passes through $(-2, 5)$ and $(3, -4)$.

14. Write the slope-intercept form for a line that passes through $(-1, 4)$ and is perpendicular to the line $3x - 4y = 12$.

15. Write an equation of a line that is parallel to the x-axis and passes through $(4, -5)$.

16. Determine the x- and y-intercepts on the graph of $5x - 4y = 10$. Graph the equation.

17. If $C(x) = 15x + 2000$ calculates the cost in dollars of producing x radios, interpret the numbers 15 and 2000 in the formula for $C(x)$.

18. Solve $-2.4x - 2.1 = \sqrt{3}x + 1.7$ both graphically and numerically. Round your answer to the nearest tenth.

Exercises 19–30: Solve the equation.

19. $-3(2 - 3x) - (-x - 1) = 1$

20. $\dfrac{5 - 3x}{6} = \dfrac{x - (3 - 4x)}{2}$

21. $|3x - 4| + 1 = 5$ **22.** $x^3 + 5 = 5x^2 + x$

23. $7x^2 + 9x = 10$ **24.** $2x^2 + x + 2 = 0$

25. $2x^3 + 4x^2 = 6x$ **26.** $x^4 + 9 = 10x^2$

27. $3x^{2/3} + 5x^{1/3} - 2 = 0$ **28.** $\sqrt{5 + 2x} + 4 = x + 5$

29. $\dfrac{2x - 3}{5 - x} = \dfrac{4x - 3}{1 - 2x}$ **30.** $\sqrt[3]{x - 4} - 1 = 3$

31. Solve $\frac{1}{2}x - (4 - x) + 1 = \frac{3}{2}x - 5$. Is this equation either an identity or a contradiction?

32. Graph f. Is f continuous on its domain? Evaluate $f(1)$.

$$f(x) = \begin{cases} x^2 - 1 & \text{if } -3 \le x \le -1 \\ x + 1 & \text{if } -1 < x < 1 \\ 1 - x^2 & \text{if } 1 \le x \le 3 \end{cases}$$

Exercises 33–38: Solve the inequality.

33. $-\frac{1}{3}x - (1 + x) > \frac{2}{3}x$ **34.** $-4 \le 4x - 6 < \frac{5}{2}$

35. $|5x - 7| \ge 3$ **36.** $5x^2 + 13x - 6 < 0$

37. $x^3 - 9x \le 0$ **38.** $\dfrac{4x - 3}{x + 2} > 0$

39. The graph of a nonlinear function f is shown. Solve each equation or inequality.
   (a) $f(x) = 0$ **(b)** $f(x) > 0$ **(c)** $f(x) \le 0$

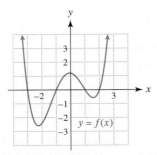

40. Write the quadratic polynomial $f(x) = 2x^2 - 4x + 1$ in the form $f(x) = a(x - h)^2 + k$.

41. Find the vertex on the graph of $f(x) = -\frac{1}{2}x^2 + 3x - 2$.

42. Solve $x^2 - 3x = 1$ by completing the square.

43. Use the given graph of $y = f(x)$ at the top of the next column to sketch a graph of each equation.

(a) $y = f(x + 2) - 1$ **(b)** $y = -2f(x)$

(c) $y = f(-x) + 1$ **(d)** $y = f\left(\frac{1}{2}x\right)$

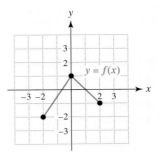

44. Use transformations of graphs to sketch a graph of $y = 2\sqrt{x} + 1$.

45. Use the graph of f to estimate each of the following.
   (a) Where f is increasing or decreasing

   (b) The zeros of f

   (c) The coordinates of any turning points

   (d) Any local extrema

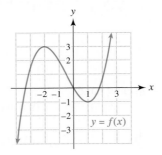

46. Are $f(x) = x^4 - 5x^3 - 7$ and $g(x) = \sqrt{9 - x^2}$ even, odd, or neither?

47. Sketch a graph of a quartic (degree 4) function with a negative leading coefficient, three x-intercepts, and three turning points.

48. State the end behavior of $f(x) = 4 + 3x - x^3$.

49. Divide each expression.
   (a) $\dfrac{4a^3 - 8a^2 + 12}{4a^2}$ **(b)** $\dfrac{2x^3 - 4x + 1}{x - 1}$

   (c) $\dfrac{x^4 + 2x^3 - x^2 + 5x - 2}{x^2 + 2}$

50. A cubic function f has zeros $-1, 0,$ and 2 and leading coefficient -5. Write the complete factored form.

51. A degree 6 function f has zeros $-3, 1,$ and 4 with multiplicities 1, 2, and 3, respectively. If the leading coefficient is 4, write the complete factored form of $f(x)$.

52. A quintic (degree 5) function f with real coefficients has leading coefficient $\frac{1}{2}$ and zeros $-2, i,$ and $-2i$. Write $f(x)$ in complete factored form and expanded form.

53. Write the complete factored form of the polynomial given by $f(x) = 2x^3 - x^2 - 6x + 3$.

54. Use the graph to write the complete factored form of the cubic polynomial $f(x)$.

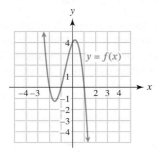

55. Write $\frac{3 + 4i}{1 - i}$ in standard form.

56. Find all solutions, real or imaginary, to $x^4 - 25 = 0$.

57. State the domain of $f(x) = \frac{2x - 5}{x^2 - 3x - 4}$. Find any vertical or horizontal asymptotes.

58. Write $\sqrt[3]{x^5}$ using rational exponents. Evaluate the expression for $x = 8$.

Applications

59. *Water in a Pool*   The graph shows the amount of water in a swimming pool x hours past noon. Find the slope of each line segment and interpret each slope.

Time (hours)

60. *Distance*   At noon, one runner is heading south at 8 miles per hour and is located 2 miles north of a second runner, who is heading west at 7 miles per hour. Approximate the distance between the runners to the nearest tenth of a mile at 12:30 P.M.

61. *Average Rate of Change*   The total distance D in feet traveled by a car after t seconds is given by $D(t) = 4t^2$ for $0 \leq t \leq 7$. Find the average rate of change of D from 0 to 2 and 2 to 4. Interpret these average rates of change.

62. *Working Together*   Suppose one person can paint a room in 10 hours and another person can paint the same room in 8 hours. How long will it take to paint the room if they work together?

63. *Mixing Acid*   Two liters of a 35% sulfuric acid solution need to be diluted to a 20% solution. How many liters of a 12% sulfuric acid solution should be mixed with the 2-liter solution?

64. *U.S. Population*   In 1990 the population of the United States was about 250 million, and in 2000 it was about 280 million.   (**Source:** Bureau of the Census.)
    (a) Find a linear function P that models the data during year t.

    (b) Estimate the year when the population of the United States reached 300 million.

65. *Maximizing Revenue*   The revenue R in dollars from selling x thousand toy figures is given by the formula $R(x) = x(800 - x)$. How many toy figures should be sold to maximize revenue?

66. *Construction*   A box is being constructed by cutting 2-inch squares from the corners of a rectangular sheet of metal that is 6 inches longer than it is wide. If the box is to have a volume of 270 cubic inches, find the dimensions of the metal sheet.

67. *Group Rates*   Round-trip airline tickets to Hawaii are regularly $800, but for each additional ticket purchased the price is reduced by $5. For example, 1 ticket costs $800, 2 tickets cost $2(795) = 1590, and 3 tickets cost $3(790) = 2370.
    (a) Write a quadratic function C that gives the total cost of purchasing t tickets.

    (b) Solve $C(t) = 17,000$ and interpret the result.

    (c) Find the absolute maximum for C and interpret your result. Assume that t must be an integer.

68. *Modeling Data* Find a quadratic function in the form $f(x) = a(x - h)^2 + k$ that models the data in the table. Graph $y = f(x)$ and the data if a graphing calculator is available.

x	4	6	8	10
y	6	15	37	80

69. *Health Care Cost* Average health care cost per employee has risen dramatically since 1998. The table lists the annual percent change from 1998 to 2002.

Year	1998	1999	2000	2001	2002
Increase (%)	6.1	7.3	8.1	11.2	14.7

Source: Mercer U.S. Health Care Survey, 2002.

(a) Find a polynomial function that models these data. (Answers may vary.)

(b) Estimate the annual percent change in 2004.

70. *Americans and Weight* The percentage of Americans who are 30 or more pounds above a healthy weight has increased during the past 40 years. The table lists this percentage for various years.

Year	1961	1978	1991	2000
Percentage	13	15	23	31

Source: National Health and Nutrition Examination Survey.

(a) Find a polynomial function that models these data. (Answers may vary.)

(b) Estimate the year when this percentage may reach 40%. Compare your prediction with the experts' prediction of 2009.

71. *Minimizing Surface Area* A cylindrical can is being constructed to have a volume of 10π cubic inches. Find the dimensions of the can that result in the least amount of aluminum being used in its construction.

Exponential and Logarithmic Functions

5

In 1900 the population of the world was approximately 1.6 billion. At that time the Swedish scientist Svante Arrhenius first predicted a greenhouse effect resulting from emissions of carbon dioxide by the industrialized countries. His classic calculation, which made use of logarithms, predicted that a doubling of the carbon dioxide concentration in the atmosphere would raise the average global temperature by 7°F to 11°F. Since then, the world population has increased dramatically to approximately 6.7 billion. With this increase in population, there has been a corresponding increase in emissions of greenhouse gases. As a result, there is more discussion today than ever about the impact and causes of global warming.

When quantities such as population and pollution increase rapidly, nonlinear functions and equations are used to model their growth. In this chapter, exponential and logarithmic functions are introduced. They occur in a wide variety of applications. Examples include acid rain, the decline of the bluefin tuna, air pollution, global warming, salinity of the ocean, demand for organ transplants, diversity of bird species, the relationship between caloric intake and land ownership in developing countries, hurricanes, earthquakes, and increases in skin cancer due to a decrease in the ozone layer. Understanding these issues and discovering solutions will require creativity, innovation, and mathematics. This chapter presents some of the mathematical tools necessary for modeling these trends.

The important thing is not to stop questioning.
—Albert Einstein

Source: **M. Kraljic,** *The Greenhouse Effect.*

5.1 Combining Functions

- **Perform arithmetic operations on functions**
- **Review function notation**
- **Perform composition of functions**

Introduction

Addition, subtraction, multiplication, and division can be performed on numbers and variables. These arithmetic operations can also be used to combine functions. For example, to model the stopping distance of a car, we compute two quantities. The first quantity is the *reaction distance,* which is the distance that a car travels between the time when a driver first recognizes a hazard and the time when the brakes are applied. The second quantity is *braking distance,* which is the distance that a car travels after the brakes have been applied. *Stopping distance* is equal to the sum of the reaction distance and the braking distance. One way to determine stopping distance is to find one function that models the reaction distance and a second function that calculates the braking distance. The stopping distance is then found by adding the two functions.

Arithmetic Operations on Functions

Highway engineers frequently assume that drivers have a reaction time of 2.5 seconds or less. During this time, a car continues to travel at a constant speed, until the driver is able to move his or her foot from the accelerator to the brake pedal. If a car is traveling at x miles per hour, then $r(x) = \left(\frac{11}{3}\right)x$ computes the distance in feet that the car travels in 2.5 seconds. For example, a driver traveling at **55** miles per hour might have a reaction distance of $r(55) = \left(\frac{11}{3}\right)(55) \approx$ **201.7** feet.

The braking distance in feet for a car traveling on wet level pavement at x miles per hour can be approximated by $b(x) = x^2/9$. For instance, a car traveling at 55 miles per hour would need $b(55) = 55^2/9 \approx$ **336.1** feet to stop after the brakes have been applied. The estimated stopping distance s at **55** miles per hour is the sum of these distances: **201.7** + **336.1** = **537.8** feet. See Figure 5.1. (**Source:** L. Haefner, *Introduction to Transportation Systems.*)

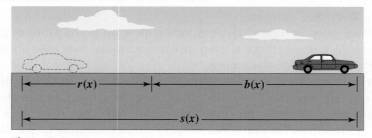

Figure 5.1 $s(x) = r(x) + b(x)$

The concept of finding the sum of two functions can be represented symbolically, graphically, and numerically, as illustrated in the next example.

EXAMPLE 1 Representing stopping distance

A driver with a reaction time of 2.5 seconds traveling at 60 miles per hour attempts to stop on wet level pavement in order to avoid an accident.

(a) Write a symbolic representation for a function s using r and b that computes the stopping distance for a car traveling at x miles per hour. Evaluate $s(60)$.

(b) Graph r, b, and s. Interpret the graph.

(c) Illustrate the relationship among r, b, and s numerically.

SOLUTION

(a) *Symbolic Representation*   Let $s(x)$ be the sum of the two formulas for r and b.

$$s(x) = r(x) + b(x) = \frac{11}{3}x + \frac{1}{9}x^2$$

The stopping distance for a car traveling at 60 miles per hour is

$$s(60) = \frac{11}{3}(60) + \frac{1}{9}(60)^2 = 620 \text{ feet.}$$

(b) *Graphical Representation*   Graph $y_1 = \frac{11}{3}x$, $y_2 = \frac{1}{9}x^2$, and $y_3 = \frac{11}{3}x + \frac{1}{9}x^2$ on the same axes, as shown in Figure 5.2. For any x-value in the graph, the sum of y_1 and y_2 equals y_3. Figure 5.2 illustrates that

$$s(60) = r(60) + b(60) = 220 + 400 = 620.$$

(c) *Numerical Representation*   Table 5.1 shows numerical representations of $r(x)$, $b(x)$, and $s(x) = r(x) + b(x)$. Notice that values for $s(x)$ can be found by adding $r(x)$ and $b(x)$. For example, when $x = 60$, $r(60) = 220$ and $b(60) = 400$. It follows that $s(60) = 220 + 400 = 620$.

Figure 5.2

Table 5.1

x	0	12	24	36	48	60	← Speed
$r(x)$	0	44	88	132	176	220	← Reaction distance
$b(x)$	0	16	64	144	256	400	← Braking distance
$s(x)$	0	60	152	276	432	620	← Stopping distance: $s(x) = r(x) + b(x)$

Now Try Exercise 99 ◀

We now formally define arithmetic operations on functions.

Operations on Functions

If $f(x)$ and $g(x)$ both exist, the sum, difference, product, and quotient of two functions f and g are defined by

$$(f + g)(x) = f(x) + g(x),$$
$$(f - g)(x) = f(x) - g(x),$$
$$(fg)(x) = f(x) \cdot g(x), \quad \text{and}$$
$$\left(\frac{f}{g}\right)(x) = \frac{f(x)}{g(x)}, \quad \text{where } g(x) \neq 0.$$

Operations on Functions and Domains   The domains of the sum, difference, and product of f and g include x-values that are in *both* the domain of f and the domain of g. The domain of the quotient f/g includes all x-values in both the domain of f and the domain of g, where $g(x) \neq 0$.

EXAMPLE 2 Performing arithmetic operations on functions symbolically

Let $f(x) = 2 + \sqrt{x - 1}$ and $g(x) = x^2 - 4$.
(a) Find the domains of $f(x)$ and $g(x)$. Then find the domains of $(f + g)(x)$, $(f - g)(x)$, $(fg)(x)$, and $(f/g)(x)$.
(b) If possible, evaluate $(f + g)(5)$, $(f - g)(1)$, $(fg)(0)$, and $(f/g)(3)$.
(c) Write expressions for $(f + g)(x)$, $(f - g)(x)$, $(fg)(x)$, and $(f/g)(x)$.

SOLUTION
(a) Whenever $x \geq 1$, $f(x) = 2 + \sqrt{x - 1}$ is defined. Therefore the domain of f is $\{x \mid x \geq 1\}$. The domain of $g(x) = x^2 - 4$ is all real numbers. The domains of $f + g$, $f - g$, and fg include all x-values in *both* the domain of f and the domain of g. Thus their domains are $\{x \mid x \geq 1\}$.

To determine the domain of f/g, we must also exclude x-values for which $g(x) = x^2 - 4 = 0$. This occurs when $x = \pm 2$. Thus the domain of f/g is $\{x \mid x \geq 1, x \neq 2\}$. (Note that $x \neq -2$ is satisfied if $x \geq 1$.)

(b) The expressions can be evaluated as follows.

$$(f + g)(5) = f(5) + g(5) = (2 + \sqrt{5 - 1}) + (5^2 - 4) = 4 + 21 = 25$$
$$(f - g)(1) = f(1) - g(1) = (2 + \sqrt{1 - 1}) - (1^2 - 4) = 2 - (-3) = 5$$

$(fg)(0)$ is undefined, since 0 is not in the domain of $f(x)$. $(f/g)(3)$ equals

$$\frac{f(3)}{g(3)} = \frac{2 + \sqrt{3 - 1}}{3^2 - 4} = \frac{2 + \sqrt{2}}{5}.$$

(c) The sum, difference, product, and quotient of f and g are calculated as follows.

$$(f + g)(x) = f(x) + g(x) = (2 + \sqrt{x - 1}) + (x^2 - 4) = \sqrt{x - 1} + x^2 - 2$$
$$(f - g)(x) = f(x) - g(x) = (2 + \sqrt{x - 1}) - (x^2 - 4) = \sqrt{x - 1} - x^2 + 6$$
$$(fg)(x) = f(x) \cdot g(x) = (2 + \sqrt{x - 1})(x^2 - 4)$$
$$\left(\frac{f}{g}\right)(x) = \frac{f(x)}{g(x)} = \frac{2 + \sqrt{x - 1}}{x^2 - 4}$$

Now Try Exercises 9 and 15 ◄

Graphical, Numerical, and Symbolic Evaluation In the next example, we evaluate operations on functions in three ways.

EXAMPLE 3 Evaluating combinations of functions

If possible, use each representation of f and g to evaluate $(f + g)(4)$, $(f - g)(-2)$, $(fg)(1)$, and $(f/g)(0)$.

(a)

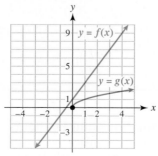

(b)

x	−2	0	1	4
$f(x)$	−3	1	3	9

x	−2	0	1	4
$g(x)$	—	0	1	2

(c) $f(x) = 2x + 1$, $g(x) = \sqrt{x}$

SOLUTION

(a) *Graphical Evaluation*   From the graph, $f(4) = 9$ and $g(4) = 2$. Thus

$$(f + g)(4) = f(4) + g(4) = 9 + 2 = 11.$$

Although $f(-2) = -3$, $g(-2)$ is undefined because -2 is not in the domain of g. Thus $(f - g)(-2)$ is undefined. The domains of f and g include 1, and it follows that $(fg)(1) = f(1)g(1) = 3(1) = 3$. The graph of g intersects the origin, so $g(0) = 0$. Thus $(f/g)(0) = \frac{f(0)}{g(0)}$ is undefined.

(b) *Numerical Evaluation*   From the tables, $f(4) = 9$ and $g(4) = 2$. As in part (a),

$$(f + g)(4) = f(4) + g(4) = 9 + 2 = 11.$$

A dash in the table indicates that $g(-2)$ is undefined, so $(f - g)(-2)$ is also undefined. The calculations of $(fg)(1)$ and $(f/g)(0)$ are done in a similar manner.

(c) *Symbolic Evaluation*   Use the formulas $f(x) = 2x + 1$ and $g(x) = \sqrt{x}$.

$$(f + g)(4) = f(4) + g(4) = (2 \cdot 4 + 1) + \sqrt{4} = 9 + 2 = 11$$
$$(f - g)(-2) = f(-2) - g(-2) = (2 \cdot (-2) + 1) - \sqrt{-2} \text{ is undefined.}$$
$$(fg)(1) = f(1)g(1) = (2 \cdot 1 + 1)\sqrt{1} = 3(1) = 3$$
$$\left(\frac{f}{g}\right)(0) = \frac{f(0)}{g(0)} \text{ is undefined, since } g(0) = 0.$$

Now Try Exercises 7, 31, and 35 ◄

An Application   The next example is an application from business involving the difference between two functions.

EXAMPLE 4   Finding the difference of two functions

Once the sound track for a DVD has been recorded, the cost of producing the master disc can be significant. After this disc has been made, additional discs can be produced inexpensively. A typical cost for producing the master disc is $2000, while additional discs cost approximately $2 each.   (**Source:** Windcrest Productions.)

(a) Assuming no other expenses, find a function C that outputs the cost of producing the master disc plus x additional DVDs. Find the cost of making the master disc and 1500 additional DVDs.

(b) Suppose that each DVD is sold for $12. Find a function R that computes the revenue from selling x DVDs. Find the revenue from selling 1500 DVDs.

(c) Assuming that the master disc is not sold, determine a function P that outputs the profit from selling x DVDs. How much profit is there from selling 1500 DVDs?

SOLUTION

(a) The cost of producing the master disc for $2000 plus x additional discs at $2 each is given by $C(x) = 2x + 2000$. The $2000 cost is sometimes called a *fixed cost*. The cost of manufacturing the master disc and 1500 additional DVDs is

$$C(1500) = 2(1500) + 2000 = \$5000.$$

(b) The revenue from x discs at $12 each is computed by $R(x) = 12x$. The revenue from selling 1500 DVDs is $R(1500) = 12(1500) = \$18,000$.

Algebra Review
To review arithmetic operations on polynomials, see Chapter R (page R-13).

(c) Profit P is equal to revenue minus cost. This can be written using function notation.

$$P(x) = R(x) - C(x)$$
$$= 12x - (2x + 2000)$$
$$= 12x - 2x - 2000$$
$$= 10x - 2000$$

If $x = 1500$, $P(1500) = 10(1500) - 2000 = \$13,000$.   *Now Try Exercise 95* ◀

Review of Function Notation

In the next example, we review how to evaluate function notation before we discuss composition of functions.

> **EXAMPLE 5**   Evaluating function notation
>
> Let $g(x) = 3x^2 - 6x + 2$. Evaluate each expression.
> **(a)** $g(2)$ **(b)** $g(k)$ **(c)** $g(x^2)$ **(d)** $g(x + 2)$
>
> **SOLUTION**
> **(a)** $g(2) = 3(2)^2 - 6(2) + 2 = 12 - 12 + 2 = 2$
> **(b)** $g(k) = 3k^2 - 6k + 2$
> **(c)** $g(x^2) = 3(x^2)^2 - 6(x^2) + 2 = 3x^4 - 6x^2 + 2$
> **(d)** $g(x + 2) = 3(x + 2)^2 - 6(x + 2) + 2$
> $$= 3(x^2 + 4x + 4) - 6(x + 2) + 2$$
> $$= 3x^2 + 12x + 12 - 6x - 12 + 2$$
> $$= 3x^2 + 6x + 2$$ *Now Try Exercise 45* ◀

Algebra Review
To review squaring a binomial, see Chapter R (page R-18).

Composition of Functions

Many tasks in life are performed in sequence. For example, to go to a movie we might get into a car, drive to the movie theater, and get out of the car. A similar situation occurs with functions. For example, to convert miles to inches we might first convert miles to feet and then convert feet to inches. Since there are 5280 feet in a mile, $f(x) = 5280x$ converts x miles to an equivalent number of feet. Then $g(x) = 12x$ changes feet to inches. To convert x miles to inches, we combine the functions f and g in sequence. Figure 5.3 illustrates how to convert 5 miles to inches. First, $f(5) = 5280(5) = 26,400$. Then the output of 26,400 feet from f is used as input for g. The number of inches in 26,400 feet is $g(26,400) = 12(26,400) = 316,800$. This computation is called the *composition* of g and f.

The composition of g and f shown in Figure 5.3 can be expressed symbolically. The symbol $\circ$ is used to denote composition of two functions.

$$(g \circ f)(5) = g(f(5)) \qquad \text{First compute } f(5).$$
$$= g(5280 \cdot 5) \qquad f(x) = 5280x$$
$$= g(26,400) \qquad \text{Simplify.}$$
$$= 12(26,400) \qquad g(x) = 12x$$
$$= 316,800 \qquad \text{Simplify.}$$

5   26,400   316,800

Miles   Feet   Inches

Figure 5.3

A distance of 5 miles is equivalent to 316,800 inches. The concept of composition of two functions is now defined formally.

Composition of Functions

If f and g are functions, then the **composite function** $g \circ f$, or **composition** of g and f is defined by

$$(g \circ f)(x) = g(f(x)).$$

We read $g(f(x))$ as "g of f of x."

The domain of $g \circ f$ is all x in the domain of f such that $f(x)$ is in the domain of g.

Symbolic Evaluation of Composite Functions   The next three examples discuss how to evaluate composite functions and find their domains symbolically.

EXAMPLE 6   **Finding a symbolic representation of a composite function**

Find a formula for the composite function $g \circ f$ that converts x miles into inches.

SOLUTION   Let $f(x) = 5280x$ and $g(x) = 12x$.

$$
\begin{aligned}
(g \circ f)(x) &= g(f(x)) && \text{Definition of composition} \\
&= g(5280x) && f(x) = 5280x \text{ is the input for } g. \\
&= 12(5280x) && g \text{ multiplies the input by 12.} \\
&= 63{,}360x && \text{Simplify.}
\end{aligned}
$$

Thus $(g \circ f)(x) = 63{,}360x$ converts x miles to inches.   **Now Try Exercise 97** ◀

NOTE   Converting units is only one application of composition of functions. Another application involves examining how a decrease in the ozone layer causes an increase in ultraviolet sunlight, which in turn causes increases in the number of skin cancer cases. See Example 10.

MAKING CONNECTIONS

Product and Composition of Two Functions   Computing a product of two functions is fundamentally different from computing a composition of two functions. With the product $(fg)(x) = f(x) \cdot g(x)$, both f and g receive the *same* input x. Then their outputs are multiplied. However, with the composition $(f \circ g)(x) = f(g(x))$, the output $g(x)$ provides the input for f. These concepts are illustrated in Figure 5.4.

Figure 5.4 Comparison of Multiplication and Composition

EXAMPLE 7 Evaluating a composite function symbolically

Let $f(x) = x^2 + 3x + 2$ and $g(x) = \frac{1}{x}$.
(a) Evaluate $(f \circ g)(2)$ and $(g \circ f)(2)$. How do they compare?
(b) Find the composite functions defined by $(f \circ g)(x)$ and $(g \circ f)(x)$. Are they equivalent expressions?
(c) Find the domains of $(f \circ g)(x)$ and $(g \circ f)(x)$.

SOLUTION
(a) $(f \circ g)(2) = f(g(2)) = f\left(\frac{1}{2}\right) = \left(\frac{1}{2}\right)^2 + 3\left(\frac{1}{2}\right) + 2 = \frac{15}{4} = 3.75$
$(g \circ f)(2) = g(f(2)) = g(2^2 + 3 \cdot 2 + 2) = g(12) = \frac{1}{12} \approx 0.0833$
The results are not equal.

(b) $(f \circ g)(x) = f(g(x)) = f\left(\frac{1}{x}\right) = \left(\frac{1}{x}\right)^2 + 3\left(\frac{1}{x}\right) + 2 = \frac{1}{x^2} + \frac{3}{x} + 2$

$(g \circ f)(x) = g(f(x)) = g(x^2 + 3x + 2) = \frac{1}{x^2 + 3x + 2}$

The expressions for $(f \circ g)(x)$ and $(g \circ f)(x)$ are not equivalent.

(c) The domain of f is all real numbers, and the domain of g is $\{x \mid x \neq 0\}$. The domain of $(f \circ g)(x) = f(g(x))$ consists of all x in the domain of g such that $g(x)$ is in the domain of f. Thus the domain of $(f \circ g)(x) = \frac{1}{x^2} + \frac{3}{x} + 2$ is $\{x \mid x \neq 0\}$.

The domain of $(g \circ f)(x) = g(f(x))$ consists of all x in the domain of f such that $f(x)$ is in the domain of g. Since $x^2 + 3x + 2 = 0$ when $x = -1$ or $x = -2$, the domain of $(g \circ f)(x) = \frac{1}{x^2 + 3x + 2}$ is $\{x \mid x \neq -1, x \neq -2\}$.

Now Try Exercises 53 and 57 ◀

MAKING CONNECTIONS

Composition and Domains To find the domain of a composition of two functions, it is sometimes helpful not to immediately simplify the resulting expression. For example, if $f(x) = x^2$ and $g(x) = \sqrt{x - 1}$, then

$$(f \circ g)(x) = \left(\sqrt{x - 1}\right)^2.$$

From this unsimplified expression, we can see that the domain (input) of $f \circ g$ must be restricted to $x \geq 1$ for the output to be a real number. As a result,

$$(f \circ g)(x) = x - 1, \quad x \geq 1.$$

EXAMPLE 8 Finding symbolic representations for composite functions

Find $(f \circ g)(x)$ and $(g \circ f)(x)$.
(a) $f(x) = x + 2,$ $\qquad g(x) = x^3 - 2x^2 - 1$
(b) $f(x) = \sqrt{2x},$ $\qquad g(x) = \dfrac{1}{x + 1}$
(c) $f(x) = 2x - 3,$ $\qquad g(x) = x^2 + 5$

SOLUTION
Getting Started When simplifying a composition of functions, the first step is often to write $(f \circ g)(x) = f(g(x))$ or $(g \circ f)(x) = g(f(x))$. ▶

(a) Begin by writing $(f \circ g)(x) = f(g(x)) = f(x^3 - 2x^2 - 1)$. Function f adds 2 to the input. That is, $f(\textbf{input}) = (\textbf{input}) + 2$ because $f(x) = x + 2$. Thus

$$f(x^3 - 2x^2 - 1) = (x^3 - 2x^2 - 1) + 2 = x^3 - 2x^2 + 1.$$

To find $(g \circ f)(x)$, begin by writing $(g \circ f)(x) = g(f(x)) = g(x + 2)$. Function g does the following to its input: $g(\textbf{input}) = (\textbf{input})^3 - 2(\textbf{input})^2 - 1$. Thus

$$g(x + 2) = (x + 2)^3 - 2(x + 2)^2 - 1.$$

(b) $(f \circ g)(x) = f(g(x)) = f\left(\dfrac{1}{x + 1}\right) = \sqrt{2 \cdot \dfrac{1}{x + 1}} = \sqrt{\dfrac{2}{x + 1}}$

$(g \circ f)(x) = g(f(x)) = g(\sqrt{2x}) = \dfrac{1}{\sqrt{2x} + 1}$

(c) $(f \circ g)(x) = f(g(x)) = f(x^2 + 5) = 2(x^2 + 5) - 3 = 2x^2 + 7$

$(g \circ f)(x) = g(f(x)) = g(2x - 3) = (2x - 3)^2 + 5 = 4x^2 - 12x + 14$

Now Try Exercises 59, 63, and 65 ◀

Graphical Evaluation of Composite Functions   The next example shows how to evaluate a composition of functions graphically.

EXAMPLE 9   Evaluating a composite function graphically

Use the graphs of f and g shown in Figure 5.5 to evaluate each expression.
(a) $(f \circ g)(2)$ **(b)** $(g \circ f)(-3)$ **(c)** $(f \circ f)(-3)$

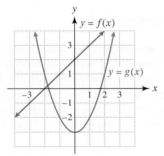

Figure 5.5

SOLUTION
(a) Because $(f \circ g)(2) = f(g(2))$, first evaluate $g(2)$. From Figure 5.6, $g(2) = 1$ and

$$(f \circ g)(2) = f(g(2)) = f(1).$$

To complete the evaluation of $(f \circ g)(2)$, use Figure 5.7 to determine that $f(1) = 3$. Thus $(f \circ g)(2) = 3$.
(b) Because $(g \circ f)(-3) = g(f(-3))$, first evaluate the expression $f(-3)$. Figure 5.8 shows that $f(-3) = -1$, so

$$(g \circ f)(-3) = g(f(-3)) = g(-1).$$

From Figure 5.9, $g(-1) = -2$. Thus $(g \circ f)(-3) = -2$.
(c) Similarly, $(f \circ f)(-3) = f(f(-3)) = f(-1) = 1$.

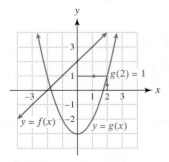

Figure 5.6 $g(2) = 1$

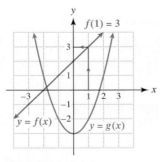

Figure 5.7 $f(1) = 3$

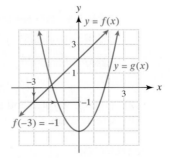

Figure 5.8 $f(-3) = -1$

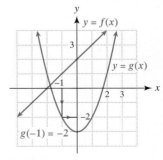

Figure 5.9 $g(-1) = -2$

Now Try Exercise 75 ◀

Numerical Evaluation of Composite Functions   Tables 5.2 and 5.3 are numerical representations of functions f and g.

Table 5.2

x	1	2	3	4
$f(x)$	2	3	4	1

Table 5.3

x	1	2	3	4
$g(x)$	4	3	2	1

We can use these tables to evaluate $g \circ f$ and $f \circ g$. For example, $(g \circ f)(3)$ can be evaluated as follows.

$$(g \circ f)(3) = g(f(3)) \qquad \text{Definition of composition}$$
$$= g(4) \qquad f(3) = 4 \text{ in Table 5.2}$$
$$= 1 \qquad g(4) = 1 \text{ in Table 5.3}$$

Also, we can find that $(f \circ g)(3) = f(g(3)) = f(2) = 3$ by using Table 5.3 and then Table 5.2.

NOTE   Composition of functions is *not* commutative. That is, $(g \circ f)(x) \neq (f \circ g)(x)$ in general. For example, from above, $(g \circ f)(3) \neq (f \circ g)(3)$.

An Application   The next example illustrates how composition of functions occurs in the analysis of the ozone layer, ultraviolet (UV) radiation, and cases of skin cancer.

EXAMPLE 10   Evaluating a composite function numerically

Depletion of the ozone layer has caused an increase in the amount of UV radiation reaching the surface of the earth. An increase in UV radiation is associated with skin cancer. In Table 5.4 the function f computes the approximate percent *increase* in UV radiation resulting from an x percent *decrease* in the thickness of the ozone layer. The function g shown in Table 5.5 computes the expected percent increase in cases of skin cancer resulting from an x percent increase in UV radiation.   (**Source:** R. Turner, D. Pearce, and I. Bateman, *Environmental Economics*.)

Table 5.4   Percent Increase in UV Radiation

x	0	1	2	3	4	5	6	← Percent decrease in the ozone layer
$f(x)$	0	1.5	3.0	4.5	6.0	7.5	9.0	← Percent increase in UV radiation

Table 5.5   Percent Increase in Skin Cancer

x	0	1.5	3.0	4.5	6.0	7.5	9.0	← Percent increase in UV radiation
$g(x)$	0	5.25	10.5	15.75	21.0	26.25	31.5	← Percent increase in cases of skin cancer

(a) Find $(g \circ f)(2)$ and interpret this calculation.
(b) Create a table for $g \circ f$. Describe what $(g \circ f)(x)$ computes.

SOLUTION
(a) $(g \circ f)(2) = g(f(2)) = g(3.0) = 10.5$. This means that a 2% decrease in the thickness of the ozone layer results in a 3% increase in UV radiation, which could cause a 10.5% increase in skin cancer.

(b) The values for $(g \circ f)(x)$ can be found in a similar manner. See Table 5.6.

Table 5.6

x	0	1	2	3	4	5	6	Percent decrease in the ozone layer
$(g \circ f)(x)$	0	5.25	10.5	15.75	21.0	26.25	31.5	Percent increase in cases of skin cancer

The composition $(g \circ f)(x)$ computes the percent increase in cases of skin cancer resulting from an x percent decrease in the ozone layer. **Now Try Exercise 101** ◀

Writing Compositions When you are solving problems, it is sometimes helpful to recognize a function as the composition of two simpler functions. For example, $h(x) = \sqrt[3]{x^2}$ can be thought of as the composition of the cube root function, $g(x) = \sqrt[3]{x}$, and the squaring function, $f(x) = x^2$. Then h can be written as $h(x) = g(f(x)) = \sqrt[3]{x^2}$. This concept is demonstrated in the next example. Note that answers may vary.

EXAMPLE 11 Writing a function as a composition of two functions

Find functions f and g so that $h(x) = (g \circ f)(x)$.

(a) $h(x) = (x + 3)^2$ **(b)** $h(x) = \sqrt{2x - 7}$ **(c)** $h(x) = \dfrac{1}{x^2 + 2x}$

SOLUTION

(a) Let $f(x) = x + 3$ and $g(x) = x^2$. Then

$$(g \circ f)(x) = g(f(x)) = g(x + 3) = (x + 3)^2.$$

(b) Let $f(x) = 2x - 7$ and $g(x) = \sqrt{x}$. Then

$$(g \circ f)(x) = g(f(x)) = g(2x - 7) = \sqrt{2x - 7}.$$

(c) Let $f(x) = x^2 + 2x$ and $g(x) = \frac{1}{x}$. Then

$$(g \circ f)(x) = g(f(x)) = g(x^2 + 2x) = \dfrac{1}{x^2 + 2x}.$$

Now Try Exercises 81, 85, and 91 ◀

5.1 Putting It All Together

Addition, subtraction, multiplication, and division can be used to combine functions. Composition of functions is another way to combine functions, and it is fundamentally different from arithmetic operations on functions. In the composition $(g \circ f)(x)$, the *output* $f(x)$ is used as *input* for g.

The following table summarizes some concepts involved with combining functions.

continued on next page

continued from previous page

Concept	Notation	Examples
Sum of two functions	$(f + g)(x) = f(x) + g(x)$	$f(x) = x^2, g(x) = 2x + 1$ $(f + g)(3) = f(3) + g(3) = 9 + 7 = 16$ $(f + g)(x) = f(x) + g(x) = x^2 + 2x + 1$
Difference of two functions	$(f - g)(x) = f(x) - g(x)$	$f(x) = 3x, g(x) = 2x + 1$ $(f - g)(1) = f(1) - g(1) = 3 - 3 = 0$ $(f - g)(x) = f(x) - g(x) = 3x - (2x + 1)$ $= x - 1$
Product of two functions	$(fg)(x) = f(x) \cdot g(x)$	$f(x) = x^3, g(x) = 1 - 3x$ $(fg)(-2) = f(-2) \cdot g(-2) = (-8)(7) = -56$ $(fg)(x) = f(x) \cdot g(x) = x^3(1 - 3x) = x^3 - 3x^4$
Quotient of two functions	$\left(\dfrac{f}{g}\right)(x) = \dfrac{f(x)}{g(x)}, g(x) \neq 0$	$f(x) = x^2 - 1, g(x) = x + 2$ $\left(\dfrac{f}{g}\right)(2) = \dfrac{f(2)}{g(2)} = \dfrac{3}{4}$ $\left(\dfrac{f}{g}\right)(x) = \dfrac{f(x)}{g(x)} = \dfrac{x^2 - 1}{x + 2}, x \neq -2$
Composition of two functions	$(g \circ f)(x) = g(f(x))$	$f(x) = x^3, g(x) = x^2 - 2x + 1$ $(g \circ f)(2) = g(f(2)) = g(8)$ $= 64 - 16 + 1 = 49$ $(g \circ f)(x) = g(f(x)) = g(x^3)$ $= (x^3)^2 - 2(x^3) + 1$ $= x^6 - 2x^3 + 1$

5.1   Exercises

Concepts

1. If $f(3) = 2$ and $g(3) = 5$, $(f + g)(3) = $ _____.

2. If $f(3) = 2$ and $g(2) = 5$, $(g \circ f)(3) = $ _____.

3. If $f(x) = x^2$ and $g(x) = 4x$, $(fg)(x) = $ _____.

4. If $f(x) = x^2$ and $g(x) = 4x$, $(f \circ g)(x) = $ _____.

5. *Cost of Carpet*   If $f(x)$ calculates the number of square feet in x square yards and $g(x)$ calculates the cost in dollars of x square feet of carpet, what does $(g \circ f)$ calculate?

6. *Time Conversion*   If $f(x)$ calculates the number of days in x hours and $g(x)$ calculates the number of years in x days, what does $(g \circ f)$ calculate?

Arithmetic Operations on Functions

Exercises 7–10: Use $f(x)$ and $g(x)$ to evaluate each expression symbolically.

7. $f(x) = 2x - 3, g(x) = 1 - x^2$
 (a) $(f + g)(3)$ **(b)** $(f - g)(-1)$

 (c) $(fg)(0)$ **(d)** $(f/g)(2)$

8. $f(x) = 4x - x^3, g(x) = x + 3$
   (a) $(g + g)(-2)$ **(b)** $(f - g)(0)$

   (c) $(gf)(1)$ **(d)** $(g/f)(-3)$

9. $f(x) = 2x + 1, g(x) = \dfrac{1}{x}$
   (a) $(f + g)(2)$ **(b)** $(f - g)\left(\frac{1}{2}\right)$

   (c) $(fg)(4)$ **(d)** $(f/g)(0)$

10. $f(x) = \sqrt[3]{x^2}, g(x) = |x - 3|$
   (a) $(f + g)(-8)$ **(b)** $(f - g)(-1)$

   (c) $(fg)(0)$ **(d)** $(f/g)(27)$

Exercises 11–30: Use f(x) and g(x) to find a formula for each expression. Identify its domain.
   (a) $(f + g)(x)$ *(b)* $(f - g)(x)$
   (c) $(fg)(x)$ *(d)* $(f/g)(x)$

11. $f(x) = 2x,$ $g(x) = x^2$

12. $f(x) = 1 - 4x,$ $g(x) = 3x + 1$

13. $f(x) = x^2 - 1,$ $g(x) = x^2 + 1$

14. $f(x) = 4x^3 - 8x^2,$ $g(x) = 4x^2$

15. $f(x) = x - \sqrt{x - 1},$ $g(x) = x + \sqrt{x - 1}$

16. $f(x) = 3 + \sqrt{2x + 9},$ $g(x) = 3 - \sqrt{2x + 9}$

17. $f(x) = \sqrt{x - 1},$ $g(x) = \sqrt{x + 1}$

18. $f(x) = \sqrt{1 - x},$ $g(x) = x^3$

19. $f(x) = \dfrac{1}{x + 1},$ $g(x) = \dfrac{3}{x + 1}$

20. $f(x) = x^{1/2},$ $g(x) = 3$

21. $f(x) = \dfrac{1}{2x - 4},$ $g(x) = \dfrac{x}{2x - 4}$

22. $f(x) = \dfrac{1}{x},$ $g(x) = x^3$

23. $f(x) = x^2 - 1,$ $g(x) = |x + 1|$

24. $f(x) = |2x - 1|,$ $g(x) = |2x + 1|$

25. $f(x) = \dfrac{x^2 - 3x + 2}{x + 1},$ $g(x) = \dfrac{x^2 - 1}{x - 2}$

26. $f(x) = \dfrac{4x - 2}{x + 2},$ $g(x) = \dfrac{2x - 1}{3x + 6}$

27. $f(x) = \dfrac{2}{x^2 - 1},$ $g(x) = \dfrac{x + 1}{x^2 - 2x + 1}$

28. $f(x) = \dfrac{1}{x + 2},$ $g(x) = x^2 + x - 2$

29. $f(x) = x^{5/2} - x^{3/2},$ $g(x) = x^{1/2}$

30. $f(x) = x^{2/3} - 2x^{1/3} + 1,$ $g(x) = x^{1/3} - 1$

Exercises 31–34: Use the graph to evaluate each expression.

31.

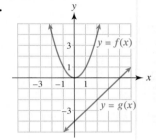

   (a) $(f + g)(2)$

   (b) $(f - g)(1)$

   (c) $(fg)(0)$

   (d) $(f/g)(1)$

32.

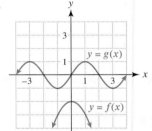

   (a) $(f + g)(1)$

   (b) $(f - g)(0)$

   (c) $(fg)(-1)$

   (d) $(f/g)(1)$

33.

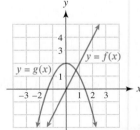

   (a) $(f + g)(0)$

   (b) $(f - g)(-1)$

   (c) $(fg)(1)$

   (d) $(f/g)(2)$

34.

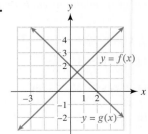

   (a) $(f + g)(-1)$

   (b) $(f - g)(-2)$

   (c) $(fg)(0)$

   (d) $(f/g)(2)$

Exercises 35 and 36: Use the tables to evaluate each expression, if possible.

    (a) $(f + g)(-1)$ (b) $(g - f)(0)$

    (c) $(gf)(2)$ (d) $(f/g)(2)$

35.

x	-1	0	2
$f(x)$	-3	5	1

x	-1	0	2
$g(x)$	-2	3	0

36.

x	-1	0	2
$f(x)$	4	1	3

x	-1	0	2
$g(x)$	2	0	1

Exercises 37 and 38: Use the table to evaluate each expression, if possible.

    (a) $(f + g)(2)$ (b) $(f - g)(4)$

    (c) $(fg)(-2)$ (d) $(f/g)(0)$

37.

x	-2	0	2	4
$f(x)$	0	5	7	10
$g(x)$	6	0	-2	5

38.

x	-2	0	2	4
$f(x)$	-4	8	5	0
$g(x)$	2	-1	4	0

39. Use the table in Exercise 37 to complete the following table.

x	-2	0	2	4
$(f + g)(x)$				
$(f - g)(x)$				
$(fg)(x)$				
$(f/g)(x)$				

40. Use the table in Exercise 38 to complete the table in Exercise 39.

Review of Function Notation

Exercises 41–52: For the given g(x), evaluate each of the following.

    (a) $g(-3)$ (b) $g(b)$ (c) $g(x^3)$ (d) $g(2x - 3)$

41. $g(x) = 2x + 1$ **42.** $g(x) = 5 - \frac{1}{2}x$

43. $g(x) = 2(x + 3)^2 - 4$ **44.** $g(x) = -(x - 1)^2$

45. $g(x) = \frac{1}{2}x^2 + 3x - 1$ **46.** $g(x) = 2x^2 - x - 9$

47. $g(x) = \sqrt{x + 4}$ **48.** $g(x) = \sqrt{2 - x}$

49. $g(x) = |3x - 1| + 4$ **50.** $g(x) = 2|1 - x| - 7$

51. $g(x) = \dfrac{4x}{x + 3}$ **52.** $g(x) = \dfrac{x + 3}{2}$

Composition of Functions

Exercises 53–56: Use the given f(x) and g(x) to evaluate each expression.

53. $f(x) = \sqrt{x + 5}$, $g(x) = x^2$

    (a) $(f \circ g)(2)$ (b) $(g \circ f)(-1)$

54. $f(x) = |x^2 - 4|$, $g(x) = 2x^2 + x + 1$

    (a) $(f \circ g)(1)$ (b) $(g \circ f)(-3)$

55. $f(x) = 5x - 2$, $g(x) = |x|$

    (a) $(f \circ g)(-4)$ (b) $(g \circ f)(5)$

56. $f(x) = \dfrac{1}{x - 4}$, $g(x) = 5$

    (a) $(f \circ g)(3)$ (b) $(g \circ f)(8)$

Exercises 57–72: Use the given f(x) and g(x) to find each of the following. Identify its domain.

    (a) $(f \circ g)(x)$ (b) $(g \circ f)(x)$ (c) $(f \circ f)(x)$

57. $f(x) = x^3$, $g(x) = x^2 + 3x - 1$

58. $f(x) = 2 - x$, $g(x) = \dfrac{1}{x^2}$

59. $f(x) = x + 2$, $g(x) = x^4 + x^2 - 3x - 4$

60. $f(x) = x^2$, $g(x) = \sqrt{1 - x}$

61. $f(x) = 2 - 3x$, $g(x) = x^3$

62. $f(x) = \sqrt{x}$, $g(x) = 1 - x^2$

63. $f(x) = \dfrac{1}{x + 1}$, $g(x) = 5x$

64. $f(x) = \dfrac{1}{3x}$, $g(x) = \dfrac{2}{x - 1}$

65. $f(x) = x + 4$, $g(x) = \sqrt{4 - x^2}$

66. $f(x) = 2x + 1$, $g(x) = 4x^3 - 5x^2$

67. $f(x) = \sqrt{x - 1}$, $g(x) = 3x$

68. $f(x) = \dfrac{x - 3}{2}$, $g(x) = 2x + 3$

69. $f(x) = 1 - 5x,$ $\qquad$ $g(x) = \dfrac{1 - x}{5}$

70. $f(x) = \sqrt[3]{x - 1},$ $\qquad$ $g(x) = x^3 + 1$

71. $f(x) = \dfrac{1}{kx}, k > 0,$ $\qquad$ $g(x) = \dfrac{1}{kx}, k > 0$

72. $f(x) = ax^2, a > 0,$ $\qquad$ $g(x) = \sqrt{ax}, a > 0$

Exercises 73–76: Use the graph to evaluate each expression.

73.

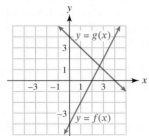

(a) $(f \circ g)(4)$

(b) $(g \circ f)(3)$

(c) $(f \circ f)(2)$

74.

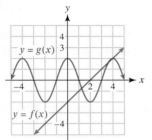

(a) $(f \circ g)(2)$

(b) $(g \circ g)(0)$

(c) $(g \circ f)(4)$

75.

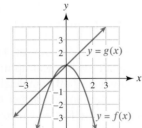

(a) $(f \circ g)(1)$

(b) $(g \circ f)(-2)$

(c) $(g \circ g)(-2)$

76.

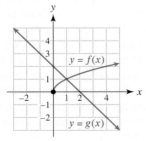

(a) $(f \circ g)(-2)$

(b) $(g \circ f)(1)$

(c) $(f \circ f)(0)$

Exercises 77 and 78: Numerical representations for the functions f and g are given. Evaluate the expression, if possible.

$\quad$ *(a)* $(g \circ f)(1)$ $\qquad$ *(b)* $(f \circ g)(4)$ $\qquad$ *(c)* $(f \circ f)(3)$

77.

x	1	2	3	4
$f(x)$	4	3	1	2

x	1	2	3	4
$g(x)$	2	3	4	5

78.

x	1	3	4	6
$f(x)$	2	6	5	7

x	2	3	5	7
$g(x)$	4	2	6	0

79. Use the tables for $f(x)$ and $g(x)$ in Exercise 77 to complete the composition shown in the diagram.

80. Use the tables for $f(x)$ and $g(x)$ in Exercise 78 to complete the composition shown in the diagram.

Exercises 81–94: (Refer to Example 11.) Find functions f and g so that $h(x) = (g \circ f)(x)$. Answers may vary.

81. $h(x) = \sqrt{x - 2}$ $\qquad$ **82.** $h(x) = (x + 2)^4$

83. $h(x) = \dfrac{1}{x + 2}$ $\qquad$ **84.** $h(x) = 5(x + 2)^2 - 4$

85. $h(x) = 4(2x + 1)^3$ $\qquad$ **86.** $h(x) = \sqrt[3]{x^2 + 1}$

87. $h(x) = (x^3 - 1)^2$ $\qquad$ **88.** $h(x) = 4(x - 5)^{-2}$

89. $h(x) = -4|x + 2| - 3$ $\quad$ **90.** $h(x) = 5\sqrt{x - 1}$

91. $h(x) = \dfrac{1}{(x - 1)^2}$ $\qquad$ **92.** $h(x) = \dfrac{2}{x^2 - x + 1}$

93. $h(x) = x^{3/4} - x^{1/4}$ $\qquad$ **94.** $h(x) = x^{2/3} - 5x^{1/3} + 4$

Applications

95. *Profit* (Refer to Example 4.) Determine a profit function P that results if the DVDs are sold for \$15 each. Find the profit from selling 3000 DVDs.

96. *Revenue, Cost, and Profit* Suppose that it costs $150,000 to produce a master disc for a music video and $1.50 to produce each copy.

(a) Write a cost function C that outputs the cost of producing the master disc and x copies.

(b) If the music videos are sold for $6.50 each, find a function R that outputs the revenue received from selling x music videos. What is the revenue from selling 8000 videos?

(c) Assuming that the master disc is not sold, find a function P that outputs the profit from selling x music videos. What is the profit from selling 40,000 videos?

(d) How many videos must be sold to break even? That is, how many videos must be sold for the revenue to equal the cost?

97. *Converting Units* There are 36 inches in a yard and 2.54 centimeters in an inch.

(a) Write a function I that converts x yards to inches.

(b) Write a function C that converts x inches to centimeters.

(c) Express a function F that converts x yards to centimeters as a composition of two functions.

(d) Write a formula for F.

98. *Converting Units* There are 4 quarts in 1 gallon, 4 cups in 1 quart, and 16 tablespoons in 1 cup.

(a) Write a function Q that converts x gallons to quarts.

(b) Write a function C that converts x quarts to cups.

(c) Write a function T that converts x cups to tablespoons.

(d) Express a function F that converts x gallons to tablespoons as a composition of *three* functions.

(e) Write a formula for F.

99. *Stopping Distance* (Refer to Example 1.) A car is traveling at 60 miles per hour. The driver has a reaction time of 1.25 seconds.

(a) Determine a function r that computes the reaction distance for this driver.

(b) Find a formula $s(x)$ that computes the stopping distance for this driver traveling at x miles per hour.

(c) Evaluate $s(60)$ and interpret the result.

100. *Stopping Distance* (Refer to Example 1.) If a driver with a reaction time of 1.5 seconds attempts to stop while traveling at x miles per hour on dry level pavement, the reaction distance is $r(x) = \frac{11}{5}x$ and the braking distance is $b(x) = \frac{1}{11}x^2$, where both distances are in feet.

(a) Write a formula for a function s in terms of $r(x)$ and $b(x)$ that gives the stopping distance when driving at x miles per hour. Evaluate $s(55)$.

(b) Graph r, b, and s on the same axes. Explain how the graph of s can be found using the graphs of r and b.

(c) Make tables for $r(x)$ and $b(x)$ at $x = 11, 22, 33, 44,$ and 55. Then use these tables to construct a table for $s(x)$.

101. *Skin Cancer* (Refer to Example 10 and Tables 5.4 and 5.5.) If possible, calculate the composition and interpret the result.

(a) $(g \circ f)(1)$ (b) $(f \circ g)(21)$

102. *Skin Cancer* In Example 10, f and g are both linear.

(a) Find symbolic representations for f and g.

(b) Determine $(g \circ f)(x)$.

(c) Evaluate $(g \circ f)(3.5)$ and interpret the result.

103. *Urban Heat Island* Cities are made up of large amounts of concrete and asphalt, which heat up in the daytime from sunlight but do not cool off completely at night. As a result, urban areas tend to be warmer than the surrounding rural areas. This effect is called the *urban heat island*. In the first figure at the top of the next page, f computes the average increase in nighttime summer temperatures in degrees Celsius at Sky Harbor Airport in Phoenix from 1948 to 1990. In this graph 1948 is the base year with a zero temperature increase. The rise in urban temperature increased peak demand for electricity. In the second figure at the top of the next page, g computes the percent increase in electrical demand for an average nighttime temperature increase of x degrees Celsius. (**Source:** W. Cotton and R. Pielke, *Human Impacts on Weather and Climate.*)

Year

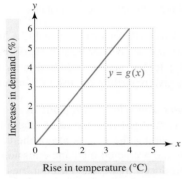

Rise in temperature (°C)

(a) Evaluate $(g \circ f)(1975)$ graphically.

(b) Interpret $(g \circ f)(x)$.

104. *Urban Heat Island*   (Refer to Exercise 103.) If possible, calculate the composition and interpret the result.
(a) $(g \circ f)(1980)$ **(b)** $(f \circ g)(3)$

105. *Urban Heat Island*   (Refer to Exercise 103.) The functions f and g are given by $f(x) = 0.11(x - 1948)$ and $g(x) = 1.5x$.
(a) Evaluate $(g \circ f)(1960)$.

(b) Find $(g \circ f)(x)$.

(c) What type of functions are f, g, and $g \circ f$?

106. *Swimming Pools*   In the figures, f computes the cubic feet of water in a pool after x days, and g converts cubic feet to gallons.

Time (days)

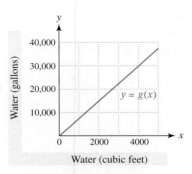

Water (cubic feet)

(a) Find the gallons of water in the pool after 2 days.

(b) Interpret $(g \circ f)(x)$.

107. *Temperature*   The function f computes the temperature on a summer day after x hours, and g converts Fahrenheit temperature to Celsius temperature. See the figures.

Time (hours)

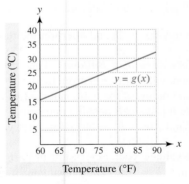

Temperature (°F)

(a) Evaluate $(g \circ f)(2)$.

(b) Interpret $(g \circ f)(x)$.

108. *Circular Wave*   A marble is dropped into a lake, resulting in a circular wave whose radius increases at a rate of 6 inches per second. Write a formula for C that gives the circumference of the circular wave in inches after t seconds.

109. *Circular Wave*   (Refer to Exercise 108.) Write a function *A* that gives the area contained inside the circular wave in square inches after *t* seconds.

110. *Geometry*   The surface area of a cone (excluding the bottom) is given by $S = \pi r \sqrt{r^2 + h^2}$, where *r* is its radius and *h* is its height, as shown in the figure. If the height is twice the radius, write a formula for *S* in terms of *r*.

111. *Equilateral Triangle* The area of an equilateral triangle with sides of length *s* is given by

$$A(s) = \frac{\sqrt{3}}{4}s^2.$$

(a) Find $A(4s)$ and interpret the result.

(b) Find $A(s + 2)$ and interpret the result.

112. *Surface Area of a Balloon*   The surface area *A* of a balloon with radius *r* is given by $A(r) = 4\pi r^2$. Suppose that the radius of the balloon increases from *r* to *r* + *h*, where *h* is a small positive number.
(a) Find $A(r + h) - A(r)$. Interpret your answer.

(b) Evaluate your expression in part (a) when *r* = 3 and *h* = 0.1. Then evaluate it for *r* = 6 and *h* = 0.1.

(c) If the radius of the balloon increases by 0.1, does the surface area always increase by a fixed amount or does the amount depend on the value of *r*?

113. *Acid Rain*   An air pollutant responsible for acid rain is sulfur dioxide (SO_2). In the table, emissions of SO_2 from burning coal during year *x* are computed by *C*(*x*), and emissions of SO_2 from burning oil are computed by *O*(*x*). Amounts are in millions of tons.

x	1860	1900	1940	1970	2000
C(*x*)	2.4	12.6	24.2	32.4	55.0
O(*x*)	0.0	0.2	2.3	17.6	23.0

Source: B. Freedman.

(a) Evaluate $(C + O)(1970)$.

(b) Interpret $(C + O)(x)$.

(c) Make a table for $(C + O)(x)$.

114. *Acid Rain*   (Refer to Exercise 113.) Make a table for a function *h* defined by $h(x) = (C/O)(x)$. Round values for *h*(*x*) to the nearest hundredth. What information does *h* give regarding the use of coal and oil?

115. *Methane Emissions*   Methane is a greenhouse gas that lets sunlight into the atmosphere but blocks heat from escaping the earth's atmosphere. In the table, *f* models the predicted methane emissions in millions of tons produced by *developed* countries. The function *g* models the same emissions for *developing* countries.

x	1990	2000	2010	2020	2030
f(*x*)	27	28	29	30	31
g(*x*)	5	7.5	10	12.5	15

Source: A. Nilsson, *Greenhouse Earth*.

(a) Make a table for a function *h* that models the total predicted methane emissions for developed *and* developing countries.

(b) Write an equation that relates *f*(*x*), *g*(*x*), and *h*(*x*).

116. *Methane Emissions*   (Refer to Exercise 115.) The figure shows graphs of the functions *f* and *g* that model methane emissions. Use these graphs to sketch a graph of the function *h*.

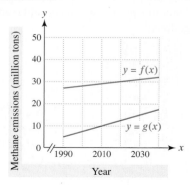

117. *Methane Emissions*   (Refer to Exercises 115 and 116.) Formulas for f and g are $f(x) = 0.1x - 172$ and $g(x) = 0.25x - 492.5$, where x is the year. Find a symbolic representation for h.

118. *AIDS in the U.S. (1982–1994)*   During year x, where $1982 \leq x \leq 1994$, the cumulative number of AIDS cases is modeled by $f(x) = 3200(x - 1982)^2 + 1586$ and the cumulative number of AIDS deaths is modeled by $g(x) = 1900(x - 1982)^2 + 619$. Graph $h(x) = \frac{g(x)}{f(x)}$ in [1982, 1994, 2] by [0, 1, 0.1]. Interpret the graph. (**Source:** Department of Health and Human Services.)

119. *Energy of a Falling Object*   A ball with mass m is dropped from an initial height of h_0 and lands with a final velocity of v_f. The kinetic energy of the ball is $K(v) = \frac{1}{2}mv^2$, where v is its velocity, and the potential energy of the ball is $P(h) = mgh$, where h is its height and g is a constant.
  (a) Show that $P(h_0) = K(v_f)$. (*Hint:* $v_f = \sqrt{2gh_0}$.)

  (b) Interpret your result from part (a).

120. *Sphere*   The volume V of a sphere with radius r is given by $V = \frac{4}{3}\pi r^3$, and the surface area S is given by $S = 4\pi r^2$. Show that $V = \frac{4}{3}\pi \left(\frac{S}{4\pi}\right)^{3/2}$.

Applying Concepts

121. Show that the sum of two linear functions is a linear function.

122. Show that if f and g are odd functions, then the composition $g \circ f$ is also an odd function.

123. Let $f(x) = k$ and $g(x) = ax + b$, where k, a, and b are constants.
  (a) Find $(f \circ g)(x)$. What type of function is $f \circ g$?

  (b) Find $(g \circ f)(x)$. What type of function is $g \circ f$?

124. Show that if $f(x) = ax + b$ and $g(x) = cx + d$, then $(g \circ f)(x)$ also represents a linear function. Find the slope of the graph of $(g \circ f)(x)$.

Writing about Mathematics

125. Describe differences between $(fg)(x)$ and $(f \circ g)(x)$. Give examples.

126. Describe differences between $(f \circ g)(x)$ and $(g \circ f)(x)$. Give examples.

5.2   Inverse Functions and Their Representations

- **Calculate inverse operations**
- **Identify one-to-one functions**
- **Find inverse functions symbolically**
- **Use other representations to find inverse functions**

Introduction

Many actions are reversible. A closed door can be opened—an open door can be closed. One hundred dollars can be withdrawn from and deposited into a savings account. These actions undo or cancel each other. But not all actions are reversible. Explosions and weather are two examples. In mathematics the concept of reversing a calculation and arriving at the original value is associated with an *inverse*.

Actions and their inverses occur in everyday life. Suppose a person opens a car door, gets in, and starts the engine. What are the inverse actions? They are to turn off the engine, get out, and close the car door. Notice that we must reverse the order as well as applying the inverse operation at each step.

Inverse Operations and Inverse Functions

Inverse Operations   In mathematics there are basic operations that can be considered inverse operations. For example, if we begin with 10 and add 5, the result is 15. To undo this operation, subtract 5 from 15 to obtain 10. Addition and subtraction are inverse operations. The same is true for multiplication and division. If we multiply a number by 2 and then divide by 2, the original number is obtained. Division and multiplication are inverse operations.

> **EXAMPLE 1** Finding inverse actions and operations

For each of the following, state the inverse actions or operations.
(a) Put on a coat and go outside.
(b) Subtract 7 from x and divide the result by 2.

SOLUTION
(a) To find the inverse actions, reverse the order and apply the inverse action at each step. The inverse actions would be to come inside and take off the coat.
(b) We must reverse the order and apply the inverse operation at each step. The inverse operations would be to multiply x by 2 and add 7. The original operations could be expressed as $\frac{x-7}{2}$, and the inverse operations could be written as $2x + 7$.

Now Try Exercises 3 and 7 ◀

Inverse Functions For an introduction to the basic concept of an inverse function, refer to Table 5.7, which can be used to convert gallons to pints. It is apparent that there are 8 pints in a gallon, and $f(x) = 8x$ converts x gallons to an equivalent number of pints. However, if we want to reverse the computation and convert pints to gallons, a different function g is required. This conversion is calculated by $g(x) = \frac{x}{8}$, where x is the number of pints. Function g performs the inverse operation of f. We say that f and g are *inverse functions* and write this as $g(x) = f^{-1}(x)$. See Table 5.8. We read f^{-1} as "f inverse."

Table 5.7 Gallons to Pints

x	1	2	3	4
$f(x)$	8	16	24	32

Table 5.8 Pints to Gallons

x	8	16	24	32
$f^{-1}(x)$	1	2	3	4

Multiplication and division are inverse operations. Addition and subtraction are also inverse operations. If $f(x) = x + 5$, then the *inverse function* of f is given by $f^{-1}(x) = x - 5$. For example, $f(5) = 10$, and $f^{-1}(10) = 5$. If input x produces output y with function f, input y produces output x with function f^{-1}. This can be seen numerically in Tables 5.9 and 5.10.

Table 5.9 Add 5

x	0	5	10	15
$f(x)$	5	10	15	20

Table 5.10 Subtract 5

x	5	10	15	20
$f^{-1}(x)$	0	5	10	15

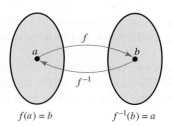

$f(a) = b$ $f^{-1}(b) = a$

Figure 5.10 Inverse Functions

From Tables 5.9 and 5.10, note that $f(0) = 5$ and $f^{-1}(5) = 0$. In general, if $f(a) = b$, then $f^{-1}(b) = a$. That is, if f outputs b with input a, then f^{-1} must output a with input b. *Inputs and outputs* (domains and ranges) *are interchanged for inverse functions*. This statement is illustrated in Figure 5.10.

When $f(x) = x + 5$ and $f^{-1}(x) = x - 5$ are applied in sequence, the output of f is used as input for f^{-1}. This is composition of functions.

$$
\begin{aligned}
(f^{-1} \circ f)(x) &= f^{-1}(f(x)) && \text{Definition of composition} \\
&= f^{-1}(x + 5) && f(x) = x + 5 \\
&= (x + 5) - 5 && f^{-1} \text{ subtracts 5 from its input.} \\
&= x && \text{Simplify.}
\end{aligned}
$$

The Notation f^{-1} and Negative Exponents If a represents a real number, then $a^{-1} = \frac{1}{a}$. For example, $4^{-1} = \frac{1}{4}$. On the other hand, if f represents a function, note that $f^{-1}(x) \neq \frac{1}{f(x)}$. Instead, $f^{-1}(x)$ represents the inverse function of f. For instance, if $f(x) = 5x$, then $f^{-1}(x) = \frac{x}{5} \neq \frac{1}{5x}$.

The composition $f^{-1} \circ f$ with input x produces output x. The same action occurs when computing the composition $f \circ f^{-1}$.

$$
\begin{aligned}
(f \circ f^{-1})(x) &= f(f^{-1}(x)) && \text{Definition of composition}\\
&= f(x - 5) && f^{-1}(x) = x - 5\\
&= (x - 5) + 5 && f \text{ adds 5 to its input.}\\
&= x && \text{Simplify.}
\end{aligned}
$$

Another pair of inverse functions is given by $f(x) = x^3$ and $f^{-1}(x) = \sqrt[3]{x}$. Before a formal definition of inverse functions is given, we must discuss one-to-one functions.

One-to-One Functions

Does every function have an inverse function? The next example answers this question.

EXAMPLE 2 Determining if a function has an inverse function

Table 5.11 represents a function C that computes the percentage of the time that the sky is cloudy in Augusta, Georgia, where x corresponds to the standard numbers for the months. Determine if C has an inverse function.

Table 5.11   Cloudy Skies in Augusta

x (month)	1	2	3	4	5	6	7	8	9	10	11	12
C(x) (%)	43	40	39	29	28	26	27	25	30	26	31	39

Source: J. Williams, *The Weather Almanac*.

SOLUTION For each input, C computes exactly one output. For example, $C(3) = 39$ means that during March the sky is cloudy 39% of the time. If C has an inverse function, the inverse must receive 39 as input and produce exactly one output. Both March and December have cloudy skies 39% of the time. Given an input of 39, it is impossible for an inverse *function* to output both 3 and 12. Therefore C does *not* have an inverse function.

Now Try Exercise 19 ◄

If *different inputs* of a function f produce the *same output*, then an inverse function of f does *not* exist. However, if different inputs always produce different outputs, f is a *one-to-one function*. *Every one-to-one function has an inverse function.* For example, $f(x) = x^2$ is not one-to-one because $f(-2) = 4$ and $f(2) = 4$. Therefore $f(x) = x^2$ does not have an inverse function because an inverse *function* cannot receive input **4** and produce both **−2** and **2** as outputs. However, $g(x) = 5x$ is one-to-one because different inputs always result in different outputs. Therefore g has an inverse function: $g^{-1}(x) = \frac{x}{5}$.

One-to-One Function

A function f is a **one-to-one function** if, for elements c and d in the domain of f,

$$c \neq d \quad \text{implies} \quad f(c) \neq f(d).$$

That is, different inputs always result in different outputs.

A function f is one-to-one if equal outputs always have the same input. This statement can be written as $f(c) = f(d)$ **implies** $c = d$, which is an equivalent definition.

EXAMPLE 3 Determining if a function is one-to-one graphically

Use each graph to determine if f is one-to-one and if f has an inverse function.

(a)

(b)

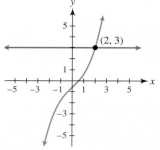

Figure 5.11

Figure 5.12

SOLUTION

(a) To decide if the graph represents a one-to-one function, determine if different inputs (x-values) always correspond to different outputs (y-values). In Figure 5.11 the horizontal line $y = 2$ intersects the graph of f at $(-1, 2)$, $(1, 2)$, and $(3, 2)$. This means that $f(-1) = f(1) = f(3) = 2$. Three distinct inputs, -1, 1, and 3, produce the same output, 2. Therefore f is not one-to-one and does not have an inverse function.

(b) Any horizontal line will intersect the graph of f at most once. For example, the line $y = 3$ intersects the graph of f at one point, $(2, 3)$. See Figure 5.12. Only input 2 results in output 3. This is true in general. Therefore f is one-to-one and has an inverse function.

Now Try Exercises 13 and 15 ◀

NOTE To show that f is not one-to-one, it is not necessary to find the actual points of intersection—we have to show only that a horizontal line can intersect the graph of f more than once.

The technique of visualizing horizontal lines to determine if a graph represents a one-to-one function is called the *horizontal line test*.

Use the horizontal line test to explain why a nonconstant linear function has an inverse function, whereas a quadratic function does not.

Horizontal Line Test

If every horizontal line intersects the graph of a function f at most once, then f is a one-to-one function.

Increasing, Decreasing, and One-to-One Functions If a continuous function f is increasing on its domain, then every horizontal line will intersect the graph of f at most once. By the horizontal line test, f is a one-to-one function. For example, the function f shown in Example 3(b) is always increasing on its domain and so it is one-to-one. Similarly, if a continuous function g is only decreasing on its domain, then g is a one-to-one function.

Symbolic Representations of Inverse Functions

If a function f is one-to-one, then an inverse function f^{-1} exists. Therefore $f(a) = b$ implies $f^{-1}(b) = a$ for every a in the domain of f. That is,

$$(f^{-1} \circ f)(a) = f^{-1}(f(a)) = f^{-1}(b) = a.$$

Similarly, $f^{-1}(b) = a$ implies $f(a) = b$ for every b in the domain of f^{-1} and so

$$(f \circ f^{-1})(b) = f(f^{-1}(b)) = f(a) = b.$$

These two properties can be used to define an inverse function.

Inverse Function

Let f be a one-to-one function. Then f^{-1} is the **inverse function** of f if

$$(f^{-1} \circ f)(x) = f^{-1}(f(x)) = x \quad \text{for every } x \text{ in the domain of } f \text{ and}$$
$$(f \circ f^{-1})(x) = f(f^{-1}(x)) = x \quad \text{for every } x \text{ in the domain of } f^{-1}.$$

In the next two examples, we find an inverse function and verify that it is correct.

EXAMPLE 4 Finding and verifying an inverse function

Let f be the one-to-one function given by $f(x) = x^3 - 2$.
(a) Find a formula for $f^{-1}(x)$. **(b)** Identify the domain and range of f^{-1}.
(c) Verify that your result from part (a) is correct.

SOLUTION
(a) Since $f(x) = x^3 - 2$, function f cubes the input x and then subtracts 2. To reverse this calculation, the inverse function must add 2 to the input x and then take the cube root. That is, $f^{-1}(x) = \sqrt[3]{x + 2}$. An important symbolic technique for finding $f^{-1}(x)$ is to solve the equation $y = f(x)$ for x.

$$y = x^3 - 2 \qquad \text{\textit{y = f(x); now solve for x.}}$$
$$y + 2 = x^3 \qquad \text{\textit{Add 2.}}$$
$$\sqrt[3]{y + 2} = x \qquad \text{\textit{Take the cube root.}}$$

Interchange x and y to obtain $y = \sqrt[3]{x + 2}$. This gives us the formula for $f^{-1}(x)$.
(b) Both the domain and the range of the cube root function include all real numbers. The graph of $f^{-1}(x) = \sqrt[3]{x + 2}$ is the graph of the cube root function shifted left 2 units. Therefore the domain and range of f^{-1} also include all real numbers.
(c) To verify that $f^{-1}(x) = \sqrt[3]{x + 2}$ is indeed the inverse of $f(x) = x^3 - 2$, we must show that $f^{-1}(f(x)) = x$ and that $f(f^{-1}(x)) = x$.

Algebra Review
To review rational exponents and radical notation, see Chapter R (page R-41).

$$f^{-1}(f(x)) = f^{-1}(x^3 - 2) \qquad \text{\textit{f(x) = x}}^3 \text{\textit{ − 2}}$$
$$= \sqrt[3]{(x^3 - 2) + 2} \qquad \text{\textit{f}}^{-1}\text{\textit{(x) = }}\sqrt[3]{x + 2}$$
$$= \sqrt[3]{x^3} \qquad \text{\textit{Combine terms.}}$$
$$= x \qquad \text{\textit{Simplify.}}$$
$$f(f^{-1}(x)) = f(\sqrt[3]{x + 2}) \qquad \text{\textit{f}}^{-1}\text{\textit{(x) = }}\sqrt[3]{x + 2}$$
$$= (\sqrt[3]{x + 2})^3 - 2 \qquad \text{\textit{f(x) = x}}^3 \text{\textit{ − 2}}$$
$$= (x + 2) - 2 \qquad \text{\textit{Cube the expression.}}$$
$$= x \qquad \text{\textit{Combine terms.}}$$

These calculations verify that our result is correct. *Now Try Exercise 73* ◄

The symbolic technique used in Example 4(a) is now summarized verbally.

> ### Finding a Symbolic Representation for f^{-1}
>
> To find a formula for f^{-1}, perform the following steps.
>
> **STEP 1:** Verify that f is a one-to-one function.
>
> **STEP 2:** Solve the equation $y = f(x)$ for x, obtaining the equation $x = f^{-1}(y)$.
>
> **STEP 3:** Interchange x and y to obtain $y = f^{-1}(x)$.
>
> To verify $f^{-1}(x)$, show that $(f^{-1} \circ f)(x) = x$ and $(f \circ f^{-1})(x) = x$.

NOTE One reason for interchanging the variables in Step 3 is to make it easier to graph $y = f^{-1}(x)$ in the xy-plane.

$f(x)$

x

EXAMPLE 5   Finding and verifying an inverse function

Suppose that a person x inches tall sprains an ankle. An approximation for the length of the crutch (in inches) that this person might need is calculated by $f(x) = \frac{18}{25}x + 2$.
(a) Explain why f is a one-to-one function.
(b) Find a formula for $f^{-1}(x)$.
(c) Verify that this is the inverse function of f.
(d) Interpret the meaning of $f^{-1}(56)$.

SOLUTION
(a) Since f is a linear function, its graph is a line with a nonzero slope of $\frac{18}{25}$. Every horizontal line intersects it at most once. By the horizontal line test, f is one-to-one.
(b) To find $f^{-1}(x)$, solve the equation $y = f(x)$ for x.

$$y = \frac{18}{25}x + 2 \qquad y = f(x)$$

$$y - 2 = \frac{18}{25}x \qquad \text{Subtract 2.}$$

$$\frac{25}{18}(y - 2) = x \qquad \text{Multiply by } \tfrac{25}{18}, \text{ the reciprocal of } \tfrac{18}{25}.$$

Now interchange x and y to obtain $y = \frac{25}{18}(x - 2)$. The formula for the inverse is, therefore,

$$f^{-1}(x) = \frac{25}{18}(x - 2).$$

(c) We must show that $(f^{-1} \circ f)(x) = x$ and $(f \circ f^{-1})(x) = x$.

$$(f^{-1} \circ f)(x) = f^{-1}(f(x)) \qquad \text{Definition of composition}$$

$$= f^{-1}\left(\frac{18}{25}x + 2\right) \qquad f(x) = \tfrac{18}{25}x + 2$$

$$= \frac{25}{18}\left(\left(\frac{18}{25}x + 2\right) - 2\right) \qquad f^{-1}(x) = \tfrac{25}{18}(x - 2)$$

$$= \frac{25}{18}\left(\frac{18}{25}x\right) \qquad \text{Simplify.}$$

$$= x \qquad \text{It checks.}$$

Similarly,

$$(f \circ f^{-1})(x) = f(f^{-1}(x)) \qquad \text{Definition of composition}$$

$$= f\left(\frac{25}{18}(x - 2)\right) \qquad f^{-1}(x) = \frac{25}{18}(x - 2)$$

$$= \frac{18}{25}\left(\frac{25}{18}(x - 2)\right) + 2 \qquad f(x) = \frac{18}{25}x + 2$$

$$= (x - 2) + 2 \qquad \text{Simplify.}$$

$$= x \qquad \text{It checks.}$$

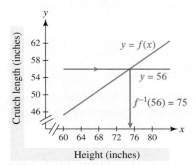

Figure 5.13

(d) The expression $f(x)$ calculates the proper crutch length for a person x inches tall, and $f^{-1}(x)$ computes the height of a person requiring a crutch x inches long. Therefore $f^{-1}(56) = \frac{25}{18}(56 - 2) = 75$ means that a 56-inch crutch would be appropriate for a person 75 inches tall. See Figure 5.13.   *Now Try Exercise 123* ◀

EXAMPLE 6   Restricting the domain of a function

Let $f(x) = (x - 1)^2$.
(a) Does f have an inverse function? Explain.
(b) Restrict the domain of f so that f^{-1} exists.
(c) Find $f^{-1}(x)$ for the restricted domain.

SOLUTION
(a) The graph of $f(x) = (x - 1)^2$, shown in Figure 5.14, does not pass the horizontal line test. Therefore f is not one-to-one and does not have an inverse function.

Figure 5.14 Not One-to-One

Figure 5.15 Restricting the Domain

(b) If we restrict the domain of f to $D = \{x \mid x \geq 1\}$, then f becomes a one-to-one function. To illustrate this, the graph of $y = (x - 1)^2$ for $x \geq 1$ is shown in Figure 5.15. This graph passes the horizontal line test and f^{-1} exists on the restricted domain.
(c) Assume that $x \geq 1$ and solve the equation $y = f(x)$ for x.

$$y = (x - 1)^2 \qquad y = f(x)$$

$$\sqrt{y} = x - 1 \qquad \text{Take the positive square root.}$$
$$\qquad\qquad\qquad \text{Note: } x \geq 1 \text{ implies that } x - 1 \geq 0.$$

$$\sqrt{y} + 1 = x \qquad \text{Add 1.}$$

Thus $f^{-1}(x) = \sqrt{x} + 1$.   *Now Try Exercise 63* ◀

Figure 5.16

NOTE The graph of $f(x) = x^5 + 3x^3 - x^2 + 5x - 3$ is shown in Figure 5.16. By the horizontal line test, f is one-to-one and f^{-1} exists. However, it would be difficult to obtain a formula for f because we would need to solve $y = x^5 + 3x^3 - x^2 + 5x - 3$ for x. It is important to realize that many functions have inverses that are either difficult or impossible to find symbolically.

Other Representations of Inverse Functions

Given a symbolic representation of a one-to-one function f, we can often find its inverse f^{-1} by using the techniques discussed in the previous subsection. Numerical and graphical representations of a one-to-one function can also be used to find its inverse.

Numerical Representations In Table 5.12, f has domain $D = \{1940, 1970, 2000\}$ and it computes the percentage of the U.S. population with 4 or more years of college in year x.

Table 5.12

x	1940	1970	2000
$f(x)$	5	11	27

Table 5.13

x	5	11	27
$f^{-1}(x)$	1940	1970	2000

Function f is one-to-one because different inputs always produce different outputs. Therefore f^{-1} exists. Since $f(1940) = 5$, it follows that $f^{-1}(5) = 1940$. Similarly, $f^{-1}(11) = 1970$ and $f^{-1}(27) = 2000$. Table 5.13 shows a numerical representation of f^{-1}.

The domain of f is $\{\mathbf{1940}, \mathbf{1970}, \mathbf{2000}\}$ and the range of f is $\{\mathbf{5}, \mathbf{11}, \mathbf{27}\}$. The domain of f^{-1} is $\{\mathbf{5}, \mathbf{11}, \mathbf{27}\}$ and the range of f^{-1} is $\{\mathbf{1940}, \mathbf{1970}, \mathbf{2000}\}$. The functions f and f^{-1} interchange domains and ranges. This result is true in general for inverse functions.

The diagrams in Figures 5.17 and 5.18 demonstrate this property. To obtain f^{-1} from f, we simply reverse the arrows for f. This reversal causes the domains and ranges to be interchanged.

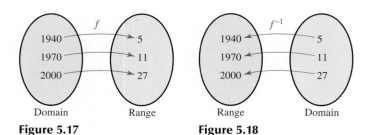

Figure 5.17   **Figure 5.18**

Figure 5.19 shows a function f that is not one-to-one. In Figure 5.20 the arrows defining f have been reversed. This is a relation. However, this relation does not represent the inverse *function* because input 4 produces two outputs, 1 and 2.

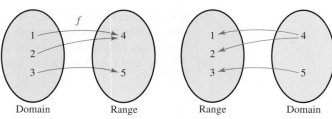

Figure 5.19 Function   **Figure 5.20** Not a Function

The relationship between domains and ranges is summarized in the following box.

Domains and Ranges of Inverse Functions

The domain of f equals the range of f^{-1}.
The range of f equals the domain of f^{-1}.

Graphical Representations   If the point $(2, 5)$ lies on the graph of f, then $f(2) = 5$ and $f^{-1}(5) = 2$. Therefore the point $(5, 2)$ must lie on the graph of f^{-1}. In general, if the point (a, b) lies on the graph of f, then the point (b, a) lies on the graph of f^{-1}. Refer to Figure 5.21. If a line segment is drawn between the points (a, b) and (b, a), the line $y = x$ is a perpendicular bisector of this line segment. Figure 5.22 shows pairs of points in the form (a, b) and (b, a). Figure 5.23 shows continuous graphs of f and f^{-1} passing through these points. The graph of f^{-1} is a *reflection* of the graph of f across the line $y = x$.

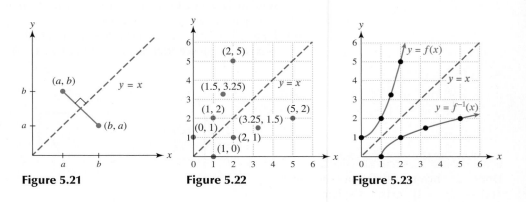

Figure 5.21   **Figure 5.22**   **Figure 5.23**

Graphs of Functions and Their Inverses

The graph of f^{-1} is a reflection of the graph of f across the line $y = x$.

EXAMPLE 7   **Representing an inverse function graphically**

Let $f(x) = x^3 + 2$. Graph f. Then sketch a graph of f^{-1}.

SOLUTION   Figure 5.24 shows a graph of f. To sketch a graph of f^{-1}, reflect the graph of f across the line $y = x$. The graph of f^{-1} appears as though it were the "reflection" of the graph of f in a mirror located along $y = x$. See Figure 5.25.

Calculator Help

To graph an inverse function, see Appendix A (page AP-13).

$[-5, 5, 1]$ by $[-5, 5, 1]$

$y = x^3 + 2$

Figure 5.24   **Figure 5.25**   *Now Try Exercise 113* ◀

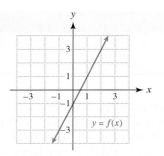

Figure 5.26

EXAMPLE 8   Evaluating f and f^{-1} graphically

Use the graph of f in Figure 5.26 to evaluate each expression.
(a) $f(2)$ **(b)** $f^{-1}(3)$ **(c)** $f^{-1}(-3)$

SOLUTION

Getting Started To evaluate $f(a)$ graphically, find a on the x-axis. Move upward or downward to the graph of f and determine the corresponding y-value. To evaluate $f^{-1}(b)$ graphically, find b on the y-axis. Move left or right to the graph of f and determine the corresponding x-value. ▶

(a) To evaluate $f(2)$, find 2 on the x-axis, move upward to the graph of f, and then move across to the y-axis to obtain $f(2) = 3$, as shown in Figure 5.27.

(b) Start by finding 3 on the y-axis, move right to the graph of f, and then move downward to the x-axis to obtain $f^{-1}(3) = 2$, as shown in Figure 5.28. Notice that $f(2) = 3$ from part (a) and $f^{-1}(3) = 2$ here.

(c) Find -3 on the y-axis, move left to the graph of f, and then move upward to the x-axis. We can see from Figure 5.29 that $f^{-1}(-3) = -1$.

CLASS DISCUSSION

Does an inverse function for $f(x) = |2x - 1|$ exist? Explain. What difficulties would you encounter if you tried to evaluate $f^{-1}(3)$ graphically?

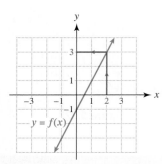

Figure 5.27 $f(2) = 3$

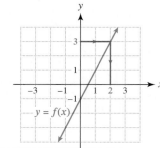

Figure 5.28 $f^{-1}(3) = 2$

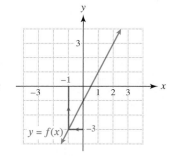

Figure 5.29 $f^{-1}(-3) = -1$

Now Try Exercise 101 ◀

5.2 Putting It All Together

The following table summarizes some important concepts about inverse functions.

Concept	Comments	Examples
One-to-one function	f is one-to-one if different inputs always result in different outputs. That is, $a \neq b$ implies $f(a) \neq f(b)$.	$f(x) = x^2 - 4x$ is not one-to-one because $f(0) = 0$ and $f(4) = 0$. With this function, *different* inputs can result in the *same* output.

Concept	Comments	Examples
Horizontal line test	If every horizontal line intersects the graph of f at most once, then f is one-to-one.	 Not one-to-one
Inverse function	If a function f is one-to-one, it has an inverse function f^{-1} that satisfies both $$(f^{-1} \circ f)(x) = f^{-1}(f(x)) = x$$ and $$(f \circ f^{-1})(x) = f(f^{-1}(x)) = x.$$	$f(x) = 3x - 1$ is one-to-one and has inverse function $f^{-1}(x) = \frac{x+1}{3}$. $$f^{-1}(f(x)) = f^{-1}(3x - 1)$$ $$= \frac{(3x - 1) + 1}{3}$$ $$= x$$ Similarly, $$f(f^{-1}(x)) = x.$$
Domains and ranges of inverse functions	The domain of f equals the range of f^{-1}. The range of f equals the domain of f^{-1}.	Let $f(x) = (x + 2)^2$ with restricted domain $x \geq -2$ and range $y \geq 0$. It follows that $f^{-1}(x) = \sqrt{x} - 2$ with domain $x \geq 0$ and range $y \geq -2$.

Both functions and their inverses have verbal, symbolic, numerical, and graphical representations. The following table summarizes these representations.

Concept	Examples
Verbal representations	Given a verbal representation of f, apply the inverse operations in reverse order to find f^{-1}. *Example:* The function f multiplies 2 times x and then adds 25. The function f^{-1} subtracts 25 from x and divides the result by 2.
Symbolic representations	If possible, solve the equation $y = f(x)$ for x. For example, let $f(x) = 3x - 5$. $$y = 3x - 5 \quad \text{is equivalent to} \quad \frac{y + 5}{3} = x.$$ Interchange x and y to obtain $f^{-1}(x) = \frac{x + 5}{3}$.
Numerical representations	<table><tr><td>x</td><td>1</td><td>2</td><td>3</td></tr><tr><td>$f(x)$</td><td>0</td><td>5</td><td>7</td></tr></table> <table><tr><td>x</td><td>0</td><td>5</td><td>7</td></tr><tr><td>$f^{-1}(x)$</td><td>1</td><td>2</td><td>3</td></tr></table> (Interchange inputs and outputs.)

continued on next page

continued from previous page

Concept	Examples
Graphical representations	The graph of f^{-1} can be obtained by reflecting the graph of f across the line $y = x$.

5.2 Exercises

Inverse Operations

Exercises 1–4: State the inverse action or actions.

1. Opening a window

2. Climbing up a ladder

3. Walking into a classroom, sitting down, and opening a book

4. Opening the door and turning on the lights

Exercises 5–12: Describe verbally the inverse of the statement. Then express both the given statement and its inverse symbolically.

5. Add 2 to x.

6. Multiply x by 5.

7. Subtract 2 from x and multiply the result by 3.

8. Divide x by 20 and then add 10.

9. Take the cube root of x and add 1.

10. Multiply x by -2 and add 3.

11. Take the reciprocal of a nonzero number x.

12. Take the square root of a positive number x.

One-to-One Functions

Exercises 13–18: Use the graph of $y = f(x)$ to determine if f is one-to-one.

13.

14.

15.

16.

17.

18.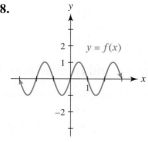

Exercises 19–22: The table is a complete representation of f. Use the table to determine if f is one-to-one and has an inverse.

19.

x	1	2	3	4
$f(x)$	4	3	3	5

20.

x	-2	0	2	4
$f(x)$	4	2	0	-2

21.

x	0	2	4	6	8
$f(x)$	-1	0	4	1	-3

22.

x	-2	-1	0	1	2
$f(x)$	4	1	0	1	4

Exercises 23–36: Determine if f is one-to-one. You may want to graph $y = f(x)$ and apply the horizontal line test.

23. $f(x) = 2x - 7$

24. $f(x) = x^2 - 1$

25. $f(x) = -2x^2 + x$

26. $f(x) = 4 - \frac{3}{4}x$

27. $f(x) = x^4$

28. $f(x) = |2x - 5|$

29. $f(x) = |x - 1|$

30. $f(x) = x^3$

31. $f(x) = \frac{1}{1 + x^2}$

32. $f(x) = \frac{1}{x}$

33. $f(x) = 3x - x^3$

34. $f(x) = x^{2/3}$

35. $f(x) = x^{1/2}$

36. $f(x) = x^3 - 4x$

Exercises 37–40: Modeling   Decide if the situation could be modeled by a one-to-one function.

37. The distance between the ground and a person who is riding a Ferris wheel after x seconds

38. The cumulative numbers of AIDS cases from 1980 to present

39. The population of the United States from 1980 to 2000

40. The height y of a stone thrown upward after x seconds

Symbolic Representations of Inverse Functions

Exercises 41–62: Find a symbolic representation for $f^{-1}(x)$.

41. $f(x) = \sqrt[3]{x}$

42. $f(x) = 2x$

43. $f(x) = -2x + 10$

44. $f(x) = x^3 + 2$

45. $f(x) = 3x - 1$

46. $f(x) = \frac{x - 1}{2}$

47. $f(x) = 2x^3 - 5$

48. $f(x) = 1 - \frac{1}{2}x^3$

49. $f(x) = x^2 - 1, x \geq 0$

50. $f(x) = (x + 2)^2, x \leq -2$

51. $f(x) = \frac{1}{2x}$

52. $f(x) = \frac{2}{\sqrt{x}}$

53. $f(x) = \frac{1}{2}(4 - 5x) + 1$

54. $f(x) = 6 - \frac{3}{4}(2x - 4)$

55. $f(x) = \frac{x}{x + 2}$

56. $f(x) = \frac{3x}{x - 1}$

57. $f(x) = \frac{2x + 1}{x - 1}$

58. $f(x) = \frac{1 - x}{3x + 1}$

59. $f(x) = \frac{1}{x} - 3$

60. $f(x) = \frac{1}{x + 5} + 2$

61. $f(x) = \frac{1}{x^3 - 1}$

62. $f(x) = \frac{2}{2 - x^3}$

Exercises 63–70: Restrict the domain of f(x) so that f is one-to-one. Then find $f^{-1}(x)$. Answers may vary.

63. $f(x) = 4 - x^2$

64. $f(x) = 2(x + 3)^2$

65. $f(x) = (x - 2)^2 + 4$

66. $f(x) = x^4 - 1$

67. $f(x) = x^{2/3} + 1$

68. $f(x) = 2(x + 3)^{2/3}$

69. $f(x) = \sqrt{9 - 2x^2}$

70. $f(x) = \sqrt{25 - x^2}$

Exercises 71–84: Find a formula for $f^{-1}(x)$. Identify the domain and range of f^{-1}. Verify that f and f^{-1} are inverses.

71. $f(x) = 5x - 15$

72. $f(x) = (x + 3)^2, x \geq -3$

73. $f(x) = \sqrt[3]{x} - 5$

74. $f(x) = 6 - 7x$

75. $f(x) = \dfrac{x - 5}{4}$

76. $f(x) = \dfrac{x + 2}{9}$

77. $f(x) = \sqrt{x - 5}, x \geq 5$

78. $f(x) = \sqrt{5 - 2x}, x \leq \frac{5}{2}$

79. $f(x) = \dfrac{1}{x + 3}$

80. $f(x) = \dfrac{2}{x - 1}$

81. $f(x) = 2x^3$

82. $f(x) = 1 - 4x^3$

83. $f(x) = x^2, x \geq 0$

84. $f(x) = \sqrt[3]{1 - x}$

Numerical Representations of Inverse Functions

Exercises 85–88: Use the table for $f(x)$ to find a table for $f^{-1}(x)$. Identify the domains and ranges of f and f^{-1}.

85.

x	1	2	3
f(x)	5	7	9

86.

x	1	10	100
f(x)	0	1	2

87.

x	0	2	4
f(x)	0	4	16

88.

x	0	1	2
f(x)	1	2	4

Exercises 89 and 90: Use $f(x)$ to complete the table.

89. $f(x) = 4x$

x	0	2	4	6
f⁻¹(x)				

90. $f(x) = x^3$

x	−8	−1	8	27
f⁻¹(x)				

Exercises 91–98: Use the tables to evaluate the following.

x	0	1	2	3	4
f(x)	1	3	5	4	2

x	−1	1	2	3	4
g(x)	0	2	1	4	5

91. $f^{-1}(3)$

92. $f^{-1}(5)$

93. $g^{-1}(4)$

94. $g^{-1}(0)$

95. $(f \circ g^{-1})(1)$

96. $(g^{-1} \circ g^{-1})(2)$

97. $(g \circ f^{-1})(5)$

98. $(f^{-1} \circ g)(4)$

Graphs and Inverse Functions

99. *Interpreting an Inverse*   The graph of f computes the balance in a savings account after x years. Estimate each expression. Interpret what $f^{-1}(x)$ computes.

(a) $f(1)$ (b) $f^{-1}(110)$ (c) $f^{-1}(160)$

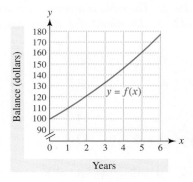

100. *Interpreting an Inverse*   The graph of f computes the Celsius temperature of a pan of water after x minutes. Estimate each expression. Interpret what the expression $f^{-1}(x)$ computes.

(a) $f(4)$ (b) $f^{-1}(90)$ (c) $f^{-1}(80)$

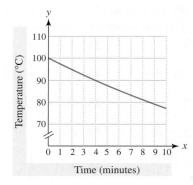

Exercises 101–104: Use the graph to evaluate the expression.

101.

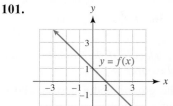

(a) $f(-1)$

(b) $f^{-1}(-2)$

(c) $f^{-1}(0)$

(d) $(f^{-1} \circ f)(3)$

102.

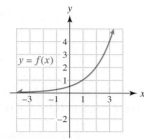

(a) $f(1)$

(b) $f^{-1}(1)$

(c) $f^{-1}(4)$

(d) $(f \circ f^{-1})(2.5)$

103.

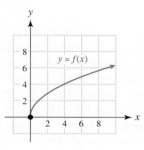

(a) $f(4)$

(b) $f^{-1}(0)$

(c) $f^{-1}(6)$

(d) $(f \circ f^{-1})(4)$

104.

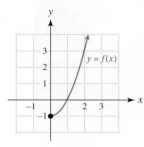

(a) $f(1)$

(b) $f^{-1}(-1)$

(c) $f^{-1}(3)$

(d) $(f \circ f^{-1})(1)$

Exercises 105–110: Use the graph of $y = f(x)$ to sketch a graph of $y = f^{-1}(x)$.

105.

106.

107.

108.

109.

110.

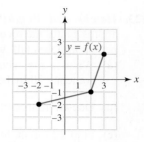

Exercises 111–116: Graph $y = f(x)$ and $y = x$. Then graph $y = f^{-1}(x)$.

111. $f(x) = 2x - 1$

112. $f(x) = -\frac{1}{2}x + 1$

113. $f(x) = x^3 - 1$

114. $f(x) = \sqrt[3]{x} - 1$

115. $f(x) = (x + 1)^2, x \geq -1$

116. $f(x) = \sqrt{x} + 1$

Exercises 117–120: Graph $y = f(x)$, $y = f^{-1}(x)$, and $y = x$ in a square viewing rectangle such as $[-4.7, 4.7, 1]$ by $[-3.1, 3.1, 1]$.

117. $f(x) = 3x - 1$

118. $f(x) = \dfrac{3 - x}{2}$

119. $f(x) = \frac{1}{3}x^3 - 1$

120. $f(x) = \sqrt[3]{x} - 1$

Calculator Help

To learn about a square viewing rectangle, see Appendix A (page AP-6).

Applications

121. *Volume* The volume V of a sphere with radius r is given by $V = \frac{4}{3}\pi r^3$.

(a) Does V represent a one-to-one function?

(b) What does the inverse of V compute?

(c) Find a formula for the inverse.

(d) Normally we interchange x and y to find the inverse function. Does it make sense to interchange V and r in part (c) of this exercise? Explain.

122. *Temperature* The formula $F = \frac{9}{5}C + 32$ converts a Celsius temperature to Fahrenheit temperature.

(a) Find a formula for the inverse.

(b) Normally we interchange x and y to find the inverse function. Does it makes sense to interchange F and C in part (a) of this exercise? Explain.

(c) What Celsius temperature is equivalent to 68°F?

123. *Height and Weight* The formula $W = \frac{25}{7}h - \frac{800}{7}$ approximates the recommended minimum weight for a person h inches tall, where $62 \le h \le 76$.

   (a) What is the recommended minimum weight for someone 70 inches tall?

   (b) Does W represent a one-to-one function?

   (c) Find a formula for the inverse.

   (d) Evaluate the inverse for 150 pounds and interpret the result.

   (e) What does the inverse compute?

124. *Planetary Orbits* The formula $T(x) = x^{3/2}$ calculates the time in years that it takes a planet to orbit the sun if the planet is x times farther from the sun than Earth is.
   (a) Find the inverse of T.

   (b) What does the inverse of T calculate?

125. *Converting Units* The tables represent a function F that converts yards to feet and a function Y that converts miles to yards. Evaluate each expression and interpret the results.

x (yd)	1760	3520	5280	7040	8800
$F(x)$ (ft)	5280	10,560	15,840	21,120	26,400

x (mi)	1	2	3	4	5
$Y(x)$ (yd)	1760	3520	5280	7040	8800

   (a) $(F \circ Y)(2)$ (b) $F^{-1}(26{,}400)$

   (c) $(Y^{-1} \circ F^{-1})(21{,}120)$

126. *Converting Units* (Refer to Exercise 125.)
   (a) Find formulas for $F(x)$, $Y(x)$, and $(F \circ Y)(x)$.

   (b) Find a formula for $(Y^{-1} \circ F^{-1})(x)$. What does this function compute?

127. *Converting Units* The tables represent a function C that converts tablespoons to cups and a function Q that converts cups to quarts. Evaluate each expression and interpret the results.

x (tbsp)	32	64	96	128
$C(x)$ (c)	2	4	6	8

x (c)	2	4	6	8
$Q(x)$ (qt)	0.5	1	1.5	2

   (a) $(Q \circ C)(96)$

   (b) $Q^{-1}(2)$

   (c) $(C^{-1} \circ Q^{-1})(1.5)$

128. *Converting Units* (Refer to Exercise 127.)
   (a) Find formulas for $C(x)$, $Q(x)$, and $(Q \circ C)(x)$.

   (b) Find a formula for $(C^{-1} \circ Q^{-1})(x)$. What does this function compute?

129. *Air Pollution* Tiny particles suspended in the air are necessary for clouds to form. Experts believe that air pollutants may cause an increase in cloud cover. From 1930 to 1980 the percentage of cloud cover over the world's oceans was monitored. The formula $f(x) = 0.06(x - 1930) + 62.5$ for $1930 \le x \le 1980$, approximates this percentage. (**Source:** W. Cotton and R. Pielke, *Human Impacts on Weather and Climate*.)

   (a) Evaluate the expressions $f(1930)$ and $f(1980)$. How did the amount of cloud cover over the oceans change during this 50-year period?

   (b) What does $f^{-1}(x)$ compute?

   (c) Use part (a) to evaluate $f^{-1}(62.5)$ and $f^{-1}(65.5)$.

   (d) Find $f^{-1}(x)$.

130. *Rise in Sea Level* Because of the greenhouse effect, the global sea level could rise due to partial melting of the polar ice caps. The table represents a function R that models this expected rise in sea level in centimeters for the year t. (This model assumes no changes in current trends.)

t (yr)	1990	2000	2030	2070	2100
$R(t)$ (cm)	0	1	18	44	66

Source: A. Nilsson, *Greenhouse Earth*.

(a) Is R a one-to-one function? Explain.

(b) Use $R(t)$ to find a table for $R^{-1}(t)$. Interpret R^{-1}.

Writing about Mathematics

131. Explain how to find verbal, numerical, graphical, and symbolic representations of an inverse function. Give examples.

132. Can a one-to-one function have more than one x-intercept or more than one y-intercept? Explain.

133. If the graphs of $y = f(x)$ and $y = f^{-1}(x)$ intersect at a point (a, b), what can be said about this point? Explain.

134. If $f(x) = ax^2 + bx + c$ with $a \neq 0$, does $f^{-1}(x)$ exist? Explain.

EXTENDED AND DISCOVERY EXERCISES

1. *Interpreting an Inverse*   Let $f(x)$ compute the height in feet of a rocket after x seconds of upward flight.
(a) Explain what $f^{-1}(x)$ computes.

(b) Interpret the solution to the equation $f(x) = 5000$.

(c) Explain how to solve the equation in part (b) using $f^{-1}(x)$.

2. If the graph of f lies entirely in quadrants I and II, in which quadrant(s) does the graph of f^{-1} lie?

CHECKING BASIC CONCEPTS FOR SECTIONS 5.1 AND 5.2

1. Use the table to evaluate each expression, if possible.

x	-2	-1	0	1	2
$f(x)$	0	1	-2	-1	2
$g(x)$	1	-2	-1	2	0

(a) $(f + g)(1)$ (b) $(f - g)(-1)$

(c) $(fg)(0)$ (d) $(f/g)(2)$

(e) $(f \circ g)(2)$ (f) $(g \circ f)(-2)$

2. Use the graph to evaluate each expression, if possible.

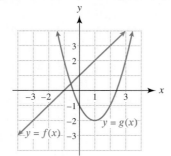

(a) $(f + g)(1)$ (b) $(g - f)(0)$

(c) $(fg)(2)$ (d) $(g/f)(-1)$

(e) $(f \circ g)(2)$ (f) $(g \circ f)(1)$

3. Let $f(x) = x^2 + 3x - 2$ and $g(x) = 3x - 1$. Find each expression.
(a) $(f + g)(x)$ (b) $(f/g)(x)$ (c) $(f \circ g)(x)$

4. If $f(x) = 5 - 2x$, find $f^{-1}(x)$.

5. Use the graph in Exercise 2 to answer the following.
(a) Is f one-to-one?   Does f^{-1} exist?   If so, find it.

(b) Is g one-to-one?   Does g^{-1} exist?   If so, find it.

6. Graph $f(x) = \sqrt[3]{x}$ and $y = x$. Then graph $y = f^{-1}(x)$.

7. Use the table in Exercise 1 to evaluate the following. (Assume that f^{-1} exists.)
(a) $f^{-1}(-2)$ (b) $(f^{-1} \circ g)(1)$

8. Use the graph in Exercise 2 to evaluate the following.
(a) $f^{-1}(2)$ (b) $(f^{-1} \circ g)(0)$

5.3 Exponential Functions and Models

- **Distinguish between linear and exponential growth**
- **Recognize exponential growth and decay**
- **Calculate compound interest**
- **Use the natural exponential function in applications**
- **Model data with exponential functions**

Introduction

Though modern exponential notation was not developed until 1637 by the great French mathematician René Descartes, some of the earliest applications involving exponential functions occurred in the calculation of interest. The custom of charging interest dates back to at least 2000 B.C. in ancient Babylon, where interest rates ran as high as 33%. Today, exponential functions are used not only to calculate interest, but also to model a wide variety of phenomena in business, biology, medicine, engineering, and education. This section discusses exponential functions and their representations.   (**Source:** *Historical Topics for the Mathematics Classroom, Thirty-first Yearbook*, NCTM.)

Linear and Exponential Growth

A linear function g can be written as $g(x) = mx + b$, where m represents the rate of change in $g(x)$ for each unit increase in x. For example, Table 5.14 shows a numerical representation of a linear function g. Each time x increases by 1 unit, $g(x)$ increases by 2 units. We can write a formula $g(x) = 2x + 3$ because the rate of change is 2 and $g(0) = 3$.

Table 5.14   A Linear Function

x	0	1	2	3	4	5
$y = g(x)$	3	5	7	9	11	13

An exponential function is fundamentally different from a linear function. Rather than *adding* a fixed amount to the previous y-value for each unit increase in x, an exponential function *multiplies* the previous y-value by a fixed amount for each unit increase in x. Table 5.15 shows an exponential function f. Note that in Table 5.15 consecutive y-values are found by multiplying the previous y-value by 2.

Table 5.15   An Exponential Function

x	0	1	2	3	4	5
$y = f(x)$	3	6	12	24	48	96

Compare the following patterns for calculating $g(x)$ and $f(x)$.

Linear Growth

$g(0) = 3$

$g(1) = \underset{g(0)}{\underbrace{3}} + 2 = 3 + 2 \cdot 1 = 5$

$g(2) = \underset{g(1)}{\underbrace{3 + 2}} + 2 = 3 + 2 \cdot 2 = 7$

$g(3) = \underset{g(2)}{\underbrace{3 + 2 + 2}} + 2 = 3 + 2 \cdot 3 = 9$

$g(4) = \underset{g(3)}{\underbrace{3 + 2 + 2 + 2}} + 2 = 3 + 2 \cdot 4 = 11$

$g(5) = \underset{g(4)}{\underbrace{3 + 2 + 2 + 2 + 2}} + 2 = 3 + 2 \cdot 5 = 13$

Exponential Growth

$f(0) = 3$

$f(1) = \underset{f(0)}{\underbrace{3}} \cdot 2 = 3 \cdot 2^1 = 6$

$f(2) = \underset{f(1)}{\underbrace{3 \cdot 2}} \cdot 2 = 3 \cdot 2^2 = 12$

$f(3) = \underset{f(2)}{\underbrace{3 \cdot 2 \cdot 2}} \cdot 2 = 3 \cdot 2^3 = 24$

$f(4) = \underset{f(3)}{\underbrace{3 \cdot 2 \cdot 2 \cdot 2}} \cdot 2 = 3 \cdot 2^4 = 48$

$f(5) = \underset{f(4)}{\underbrace{3 \cdot 2 \cdot 2 \cdot 2 \cdot 2}} \cdot 2 = 3 \cdot 2^5 = 96$

Notice that if x is a positive integer then

$$g(x) = 3 + \underbrace{(2 + 2 + \cdots + 2)}_{x \text{ terms}} \quad \text{and} \quad f(x) = 3 \cdot \underbrace{(2 \cdot 2 \cdot \cdots \cdot 2)}_{x \text{ factors}}.$$

Using these patterns, we can write formulas for $g(x)$ and $f(x)$ as follows.

$$g(x) = 3 + 2x \quad \text{and} \quad f(x) = 3 \cdot 2^x$$

This discussion gives motivation for the following definition.

Exponential Function

A function f represented by

$$f(x) = Ca^x, \quad a > 0, \quad a \neq 1, \quad \text{and} \quad C > 0,$$

is an **exponential function with base a and coefficient C.**

NOTE Some definitions for an exponential function require that $C = 1$. In this case, an exponential function is defined as $f(x) = a^x$.

Examples of exponential functions include

$$f(x) = 3^x, \quad g(x) = 5(1.7)^x, \quad \text{and} \quad h(x) = 4\left(\frac{1}{2}\right)^x.$$

Their bases are 3, 1.7, and $\frac{1}{2}$, respectively. When evaluating exponential functions, remember that any *nonzero* number raised to the 0 power equals 1. Thus $f(0) = 3^0 = 1$, $g(0) = 5(1.7)^0 = 5$, and $h(0) = 4\left(\frac{1}{2}\right)^0 = 4$.

NOTE If $f(x) = Ca^x$, then $f(0) = Ca^0 = C(1) = C$. That is, C equals the value of the function at $x = 0$. If x represents time, then C often equals the *initial value* of the quantity being modeled by f.

For large values of x, an exponential function with $a > 1$ grows faster than any linear function. Figures 5.30 and 5.31 show graphs of $f(x) = 2^x$ and $g(x) = 2x$, respectively. The points shown in the corresponding tables have also been plotted. Notice that the graph of the exponential function f increases more rapidly than the graph of the linear function g for large values of x.

x	-1	0	1	2	3	4
2^x	$\frac{1}{2}$	1	2	4	8	16

x	-1	0	1	2	3	4
$2x$	-2	0	2	4	6	8

Figure 5.30 Exponential Function

Figure 5.31 Linear Function

EXAMPLE 1 Recognizing linear and exponential data

For each table, find either a linear or an exponential function that models the data.

(a)

x	0	1	2	3
y	-3	-1.5	0	1.5

(b)

x	0	1	2	3	4
y	16	4	1	$\frac{1}{4}$	$\frac{1}{16}$

(c)

x	0	1	2	3
y	3	4.5	6.75	10.125

(d)

x	0	1	2	3	4
y	16	12	8	4	0

SOLUTION

(a) For each unit increase in x, the y-values increase by 1.5, so the data are linear. Because $y = -3$ when $x = 0$, it follows that the data can be modeled by $f(x) = 1.5x - 3$.

(b) For each unit increase in x, the y-values are multiplied by $\frac{1}{4}$. This is an exponential function given by $f(x) = Ca^x$ with $C = f(0) = 16$ and $a = \frac{1}{4}$, so $f(x) = 16\left(\frac{1}{4}\right)^x$.

(c) Since the data do not change by a fixed amount *for each unit increase in x*, the data are not linear. To determine if the data are exponential, calculate ratios of consecutive y-values.

$$\frac{4.5}{3} = 1.5, \quad \frac{6.75}{4.5} = 1.5, \quad \frac{10.125}{6.75} = 1.5$$

For each unit increase in x, the next y-value in the table can be found by multiplying the previous y-value by 1.5, so let $a = 1.5$. Since $y = 3$ when $x = 0$, let $C = 3$. Thus $f(x) = 3(1.5)^x$.

(d) For each unit increase in x, the next y-value is found by adding -4 to the previous y-value, so $f(x) = -4x + 16$. **Now Try Exercises 15 and 17** ◀

Determining an Exponential Function Two points can be used to determine a line. In a similar way, two points can be used to determine C and a in the formula $f(x) = Ca^x$. This technique is demonstrated in the next example.

EXAMPLE 2 Finding exponential functions

Find values for C and a so that $f(x) = Ca^x$ satisfies the conditions.
(a) $f(0) = 4$ and $f(1) = 6$ (b) $f(-1) = 8$ and $f(2) = 1$

SOLUTION

(a) $f(0) = 4$, so $C = 4$. Because $\frac{f(1)}{f(0)} = \frac{6}{4} = 1.5$, it follows that for each unit increase in x, the output is multiplied by 1.5. Thus $a = 1.5$ and $f(x) = 4(1.5)^x$.

(b) Note that $\frac{f(2)}{f(-1)} = \frac{Ca^2}{Ca^{-1}} = a^3$, and so $a^3 = \frac{1}{8}$, or $a = \sqrt[3]{\frac{1}{8}} = \frac{1}{2}$. Thus $f(x) = C\left(\frac{1}{2}\right)^x$. We can determine C as follows. Because $f(2) = 1$ and $f(2) = C\left(\frac{1}{2}\right)^2 = \frac{1}{4}C$, it follows that $\frac{1}{4}C = 1$, or $C = 4$. Thus $f(x) = 4\left(\frac{1}{2}\right)^x$. **Now Try Exercises 29 and 31** ◀

Exponential Growth and Decay

If an exponential function is written as $f(x) = Ca^x$ with $a > 1$, then $f(x)$ experiences **exponential growth**, as illustrated in Figure 5.32. If $a > 1$, then the y-values increase by a factor of a for each unit increase in x. We say that the **growth factor** is a. For example, if $f(x) = 5(4)^x$, then the growth factor is 4. If $0 < a < 1$, then the y-values decrease by a

Figure 5.32 Exponential Growth

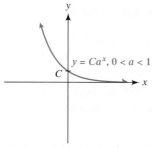

Figure 5.33 Exponential Decay

factor of a for each unit increase in x. In this case, $f(x) = Ca^x$ experiences **exponential decay** with **decay factor** a. See Figure 5.33. Exponential decay was modeled in Example 1(b), where the decay factor was $a = \frac{1}{4}$.

Properties of Exponential Functions

An *exponential function* f, defined by $f(x) = Ca^x$ with $a > 0$, $a \neq 1$, and $C > 0$, has the following properties.

1. The domain of f is $(-\infty, \infty)$ and the range of f is $(0, \infty)$.
2. The graph of f is continuous with no breaks. The x-axis is a horizontal asymptote. There are no x-intercepts and the y-intercept is C.
3. If $a > 1$, f is increasing on its domain; if $0 < a < 1$, f is decreasing on its domain.
4. f is one-to-one and therefore has an inverse. (See Section 5.4.)

Graphs of Exponential Functions To investigate further the effect that a has on the graph of an exponential function, we can let $C = 1$ in $f(x) = Ca^x$ and graph $y = 1^x$, $y = 1.3^x$, $y = 1.7^x$, and $y = 2.5^x$, as shown in Figure 5.34. (Note that $f(x) = 1^x$ does *not* represent an exponential function. However, the graph of $y = 1^x$ represents the boundary between exponential growth and decay.) As a increases, the graph of $y = a^x$ increases at a faster rate. On the other hand, the graphs of the equations $y = 0.7^x$, $y = 0.5^x$, and $y = 0.15^x$ decrease faster as a decreases. See Figure 5.35. The graph of $y = a^x$ is **increasing** when $a > 1$ and **decreasing** when $0 < a < 1$. Note that the graph of $y = a^x$, $a > 0$, always passes through the point $(0, 1)$.

Algebra Review
To review properties of exponents, see Chapter R (page R-7) and Section 4.8.

Figure 5.34

Figure 5.35

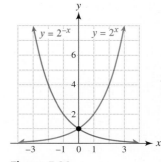

Figure 5.36

In Section 3.5 we saw that the graph of $y = f(-x)$ is a reflection of the graph of $y = f(x)$ across the y-axis. As a result, the graph of $y = a^{-x}$ is a reflection of $y = a^x$ across the y-axis. For example, the graphs of $y = 2^x$ and $y = 2^{-x}$ are reflections across the y-axis. See Figure 5.36. Note that, by properties of exponents, $2^{-x} = \left(\frac{1}{2}\right)^x$.

MAKING CONNECTIONS

Exponential Functions and Polynomial Functions An exponential function has a *variable* for an exponent, whereas a polynomial function has *constants* for exponents. For example, $f(x) = 3^x$ represents an exponential function, and $g(x) = x^3$ represents a polynomial function.

EXAMPLE 3   Comparing exponential and polynomial functions

Compare $f(x) = 3^x$ and $g(x) = x^3$ graphically and numerically for $x \geq 0$.

SOLUTION

Graphical Comparison   The graphs $Y_1 = 3\char`^X$ and $Y_2 = X\char`^3$ are shown in Figure 5.37. For $x \geq 6$, the graph of the exponential function y_1 increases significantly faster than the graph of the polynomial function y_2.

Numerical Comparison   Tables for y_1 and y_2 are shown in Figure 5.38. The values for y_1 increase faster than the values for y_2.

[0, 12, 1] by [0, 10000, 1000]

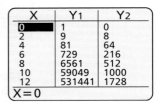

X	Y₁	Y₂
0	1	0
2	9	8
4	81	64
6	729	216
8	6561	512
10	59049	1000
12	531441	1728

X=0

Figure 5.37   **Figure 5.38**   *Now Try Exercise 23* ◀

The results of Example 3 are true in general; that is, for large enough inputs, exponential functions with $a > 1$ eventually become greater than *any* polynomial function.

Compound Interest

Suppose $1000 is deposited in an account paying 10% annual interest. At the end of 1 year, the account will contain $1000 plus 10% of $1000, or $100, in interest. Let P represent the **principal**, or initial amount deposited. Then the amount A_1 in the account after 1 year can be computed as follows.

$$A_1 = P + (0.10)P \qquad \text{Principal plus interest}$$
$$= P(1 + 0.10) \qquad \text{Factor.}$$
$$= 1000(1.10) \qquad P = \$1000$$
$$= 1100 \qquad \text{Simplify.}$$

The sum of the principal and interest is $1100.

During the second year, the account earns interest on $1100. The amount A_2 after the second year will be equal to A_1 plus 10% of A_1.

$$A_2 = A_1 + (0.10)A_1$$
$$= A_1(1 + 0.10) \qquad \text{Factor.}$$
$$= P(1 + 0.10)(1 + 0.10) \qquad A_1 = P(1 + 0.10)$$
$$= P(1 + 0.10)^2 \qquad \text{Simplify.}$$
$$= 1000(1.10)^2 \qquad P = \$1000$$
$$= 1210 \qquad \text{Simplify.}$$

After 2 years, there will be $1210.

We would like to determine a general formula for A, the amount in the account after t years. To do this, compute A_3 and observe a pattern.

$$
\begin{aligned}
A_3 &= A_2 + (0.10)A_2 \\
&= A_2(1 + 0.10) && \text{Factor.} \\
&= P(1 + 0.10)^2(1 + 0.10) && A_2 = P(1 + 0.10)^2 \\
&= P(1 + 0.10)^3 && \text{Simplify.} \\
&= 1000(1.10)^3 && P = \$1000 \\
&= 1331 && \text{Simplify.}
\end{aligned}
$$

The amount is \$1331. We can see that, in general after t years, $A = P(1 + 0.10)^t$. This type of interest is said to be *compounded annually,* because it is paid once a year.

If the interest rate is r, *expressed in decimal form,* then $A = P(1 + r)^t$. Notice that in the expression $P(1 + r)^t$ the variable t occurs as an exponent.

EXAMPLE 4 Calculating an account balance

If the principal is \$2000 and the interest rate is 8% compounded annually, calculate the account balance after 4 years.

SOLUTION The principal is $P = 2000$, the interest rate is $r = 0.08$, and the number of years is $t = 4$.

$$
A = P(1 + r)^t = 2000(1 + 0.08)^4 \approx 2720.98
$$

After 4 years, the account contains \$2720.98. *Now Try Exercise 63* ◀

In most savings accounts, interest is paid more often than once a year. In this case a smaller amount of interest is paid more frequently. For example, suppose \$1000 is deposited in an account paying 10% annual interest, compounded quarterly. After 3 months, the interest would amount to one-fourth of 10%, or 2.5%, of \$1000. The account balance would be $1000(1 + 0.025) = \$1025$. After the next 3-month period, interest would be paid on the \$1025. In a manner similar to that used for annual compounding, the balance would be $\$1000(1 + 0.025)^2 \approx \$1050.63$ after 6 months, $\$1000(1 + 0.025)^3 \approx \$1076.89$ after 9 months, and $\$1000(1 + 0.025)^4 \approx \$1103.81$ after a year. With annual compounding the amount is \$1100. The difference of \$3.81 is due to compounding quarterly. Although this amount is small after 1 year, compounding more frequently can have a dramatic effect over a long time.

Compound Interest

If a principal of P dollars is deposited in an account paying an annual rate of interest r (expressed in decimal form) compounded (paid) n times per year, then after t years the account will contain A dollars, where

$$
A = P\left(1 + \frac{r}{n}\right)^{nt}.
$$

EXAMPLE 5 Comparing compound interest

Suppose $1000 is deposited by a 20-year-old worker in an Individual Retirement Account (IRA) that pays an annual interest rate of 12%. Describe the effect on the balance after 45 years at age 65 if interest were compounded quarterly rather than annually.

SOLUTION

Compounded Annually Let $P = 1000$, $r = 0.12$, $n = 1$, and $t = 65 - 20 = 45$.

$$A = P\left(1 + \frac{r}{n}\right)^{nt} = 1000(1 + 0.12)^{45} \approx \$163{,}987.60$$

Compounded Quarterly Let $P = 1000$, $r = 0.12$, $n = 4$, and $t = 45$.

$$A = 1000\left(1 + \frac{0.12}{4}\right)^{4(45)} = 1000(1 + 0.03)^{180} \approx \$204{,}503.36$$

Quarterly compounding results in an increase of $40,515.76!   *Now Try Exercise 71* ◀

CLASS DISCUSSION

Make a conjecture about the effect on the IRA balance after 45 years if the interest rate in Example 5 were 6% instead of 12%. Test your conjecture.

The Natural Exponential Function

In Example 5, compounding interest quarterly rather than annually made a significant difference in the balance after 45 years. What would happen if interest were compounded daily or even hourly? Would there be a limit to the amount of interest that could be earned? To answer these questions, suppose $1 was deposited in an account at the very high interest rate of 100%. Table 5.16 shows the amount of money after 1 year, compounding n times during the year. The first column represents the time interval between compound interest payments. In Table 5.16, the formula $A = P\left(1 + \frac{r}{n}\right)^{nt}$ simplifies to $A = \left(1 + \frac{1}{n}\right)^{n}$, since $P = t = r = 1$. Notice that A levels off near $2.72. The graph $y = \left(1 + \frac{1}{x}\right)^{x}$ is shown in Figure 5.39. The graph approaches the horizontal asymptote $y = e$. This means that the y-values never exceed the value of $e \approx 2.7183$.

Table 5.16

Time	n	A
Year	1	2.000000
Month	12	2.613035
Day	365	2.714567
Hour	8760	2.718127
Minute	525,600	2.718279
Second	31,536,000	2.718282

Figure 5.39 $y = \left(1 + \frac{1}{x}\right)^{x} \to e$

Compounding that is done more frequently, by letting n become large without bound, is called **continuous compounding**. The exponential expression $\left(1 + \frac{1}{n}\right)^{n}$ reaches a limit of approximately 2.718281828 as $n \to \infty$. This value is so important in mathematics that it has been given its own symbol, e, sometimes called **Euler's number**. The number e has many of the same characteristics as π. Its decimal expansion never terminates or repeats in a pattern. It is an irrational number.

Value of e

To eleven decimal places, $e \approx 2.71828182846$.

Continuous compounding can be applied to population growth. Compounding annually would mean that all births and deaths occurred on December 31. Similarly, compounding quarterly would mean that births and deaths occurred at the end of March, June, September, and December. In large populations, births and deaths occur *continuously* throughout the year. Compounding continuously is a *natural* way to model large populations.

Calculator Help

When evaluating e^x, be sure to use the built-in key for e^x, rather than using an approximation for e, such as 2.72.

The Natural Exponential Function

The function f, represented by $f(x) = e^x$, is the **natural exponential function**.

EXAMPLE 6 Evaluating the natural exponential function

Approximate to four decimal places $f(x) = e^x$ when $x = 1, 0.5,$ and -2.56.

SOLUTION In Figure 5.40 these values are approximated as follows: $f(1) = e^1 \approx 2.7183$, $f(0.5) = e^{0.5} \approx 1.6487$, and $f(-2.56) = e^{-2.56} \approx 0.0773$.

```
e^(1)
          2.718281828
e^(.5)
          1.648721271
e^(-2.56)
           .0773047404
```

Figure 5.40

Now Try Exercise 41 ◄

When interest is compounded continuously, the following formula can be used to compute the amount of money in an account.

Continuously Compounded Interest

If a principal of P dollars is deposited in an account paying an annual rate of interest r (expressed in decimal form), compounded continuously, then after t years the account will contain A dollars, where

$$A = Pe^{rt}.$$

EXAMPLE 7 Calculating continuously compounded interest

The principal in an IRA is $1000 and the interest rate is 12%, compounded continuously.
(a) How much money will there be after 45 years?
(b) Determine graphically and numerically the 10-year period when the account balance increases the most.

SOLUTION

(a) Let $P = 1000$, $r = 0.12$, and $t = 45$. Then $A = 1000e^{(0.12)45} \approx \$221,406.42$. This is more than the \$204,503.36 that resulted from compounding quarterly in Example 5.

(b) Graph $Y_1 = 1000e\wedge(.12X)$, as shown in Figure 5.41. From the graph we can see that the account balance increases the most during the last 10 years. Numerical support for this conclusion is shown in Figure 5.42.

[0, 45, 5] by [0, 250000, 100000]

$y_1 = 1000e^{0.12x}$

Figure 5.41

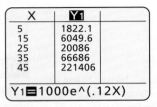

X	Y1
5	1822.1
15	6049.6
25	20086
35	66686
45	221406

Y1■1000e^(.12X)

Figure 5.42

Now Try Exercise 75 ◀

Calculator Help

To make a table like Figure 5.42, see Appendix A (page AP-10).

Natural Exponential Growth and Decay   The natural exponential function is often used to model growth of a quantity. If A_0 is the initial amount of a quantity A at time $t = 0$ and if k is a *positive* constant, then exponential growth of A can be modeled by

$$A(t) = A_0 e^{kt}. \quad \text{Growth } (k > 0)$$

Figure 5.43 illustrates this type of growth graphically. Similarly,

$$A(t) = A_0 e^{-kt} \quad \text{Decay } (k > 0)$$

can be used to model exponential decay provided $k > 0$. Figure 5.44 illustrates this type of decay graphically. A larger value of k causes the graph of A to increase or decrease more rapidly. That is, a larger value of k causes the rate of change in A to be greater in absolute value.

$A(t) = A_0 e^{kt}, k > 0$

Figure 5.43 Exponential Growth

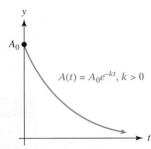

$A(t) = A_0 e^{-kt}, k > 0$

Figure 5.44 Exponential Decay

Sometimes variables such as N or Q are used instead of the variable A in these two formulas. In the next example, we model the number of bacteria N in a sample that undergoes exponential growth.

EXAMPLE 8   Modeling the growth of *E. coli* bacteria

E. coli (*Escherichia coli*) is a type of bacteria that inhabits the intestines of animals. These bacteria are capable of rapid growth and can be dangerous to humans—especially children.

In one study, *E. coli* bacteria were found to be capable of doubling in number about every 49.5 minutes. Their number N after t minutes could be modeled by $N(t) = N_0 e^{0.014t}$. Suppose that $N_0 = 500,000$ is the initial number of bacteria per milliliter.   (**Source:** G. S. Stent, *Molecular Biology of Bacterial Viruses.*)

(a) Make a conjecture about the number of bacteria per milliliter after 99 minutes. Verify your conjecture.

(b) Determine graphically the elapsed time when there were 25 million bacteria per milliliter.

[0, 400, 100] by [0, 3 × 10^7, 1 × 10^7]

Figure 5.45

SOLUTION

(a) Since the bacteria double every 49.5 minutes, there would be 1,000,000 per milliliter after 49.5 minutes and 2,000,000 after 99 minutes. This is verified by evaluating

$$N(99) = 500,000 e^{0.014(99)} \approx 2,000,000.$$

(b) *Graphical Solution*   Solve $N(t) = 25,000,000$ by graphing $Y_1 = 500000e^{\wedge}(0.014X)$ and $Y_2 = 25000000$. Their graphs intersect near (279.4, 25,000,000), as shown in Figure 5.45. Thus in a 1-milliliter sample, half a million *E. coli* bacteria could increase to 25 million in approximately 279 minutes, or 4 hours and 39 minutes.

Now Try Exercise 83 ◀

NOTE You will learn how to solve this equation symbolically in Section 5.6. See Exercise 75 in that section.

Exponential Models

Exponential functions have many applications. The next example analyzes the increase in atmospheric carbon dioxide (CO_2). Carbon dioxide is a greenhouse gas.

EXAMPLE 9   Modeling atmospheric CO_2 concentrations

Predicted concentrations of atmospheric carbon dioxide (CO_2) in parts per million (ppm) are shown in Table 5.17. (These concentrations assume that current trends continue.) The CO_2 levels in the year 2000 were greater than they had been at any time in the previous 160,000 years.

Table 5.17   Concentrations of Atmospheric CO_2

Year	2000	2050	2100	2150	2200
CO_2 (ppm)	364	467	600	769	987

Source: R. Turner, *Environmental Economics.*

(a) Let $x = 0$ correspond to 2000 and $x = 200$ to 2200. Find values for C and a so that $f(x) = Ca^x$ models these data.

(b) Estimate CO_2 concentrations for the year 2025.

SOLUTION

(a) **Getting Started** When modeling data by hand with $f(x) = Ca^x$, one strategy is to let C equal $f(0)$. Then substitute a data point into this formula and find a. ▶

The concentration is 364 when $x = 0$, so $C = 364$. This gives $f(x) = 364a^x$. One possibility for determining a is to use the last data point and require that the graph of f pass through the point $(200, 987)$. It then follows that $f(200) = 987$.

$$364a^{200} = 987 \qquad \text{\textit{f(200) = 987}}$$

$$a^{200} = \frac{987}{364} \qquad \text{\textit{Divide by 364.}}$$

$$(a^{200})^{1/200} = \left(\frac{987}{364}\right)^{1/200} \qquad \text{\textit{Take the } $\frac{1}{200}$\textit{th power.}}$$

$$a = \left(\frac{987}{364}\right)^{1/200} \qquad \text{\textit{Properties of exponents}}$$

$$a \approx 1.005 \qquad \text{\textit{Approximate.}}$$

Thus $f(x) = 364(1.005)^x$. Answers for $f(x)$ may vary slightly.

(b) Since 2025 corresponds to $x = 25$, evaluate $f(25)$.

$$f(25) = 364(1.005)^{25} \approx 412$$

Concentration of CO_2 could reach 412 ppm by 2025.   *Now Try Exercise 85* ◀

EXAMPLE 10   Modeling traffic flow

Cars arrive randomly at an intersection with an average rate of 30 cars per hour. Highway engineers estimate the likelihood, or probability, that at least one car will enter the intersection within a period of x minutes with $f(x) = 1 - e^{-0.5x}$.   (**Source:** F. Mannering and W. Kilareski, *Principles of Highway Engineering and Traffic Analysis*.)

(a) Evaluate $f(2)$ and interpret the answer.

(b) Graph f for $0 \leq x \leq 60$. What is the likelihood that at least one car will enter the intersection during a 60-minute period?

[0, 60, 10] by [0, 1.2, 0.2]

$y = 1 - e^{-0.5x}$

Figure 5.46

SOLUTION

(a) $f(2) = 1 - e^{-0.5(2)} = 1 - e^{-1} \approx 0.63$. There is a 63% chance that at least one car will enter the intersection during a 2-minute period.

(b) Graph $Y_1 = 1 - e^{\wedge}(-0.5X)$, as shown in Figure 5.46. As time progresses, the probability increases and begins to approach 1. It is almost certain that at least one car will enter the intersection during a 60-minute period. (Note that a horizontal asymptote is $y = 1$.)   *Now Try Exercise 87* ◀

Radioactive Decay   Radioactivity is an application of exponential decay. When an element such as uranium undergoes radioactive decay, atoms change from one element to another. The time it takes for half of the atoms to decay into a different element is called the **half-life**, and this time varies for different elements. Radioactive carbon-14, which is found in all living things, has a half-life of about 5700 years and can be used to date fossils.

If the initial amount of carbon-14 equals C grams, then $f(x) = Ca^x$ models the amount of carbon-14 present after x years, where a needs to be determined for carbon-14. After 5700 years, there will be $\frac{1}{2}C$ grams, so $f(5700) = \frac{1}{2}C$ implies that

$$Ca^{5700} = \frac{1}{2}C.$$

To solve this equation for a, begin by dividing each side by C.

$$a^{5700} = \frac{1}{2} \qquad \text{Divide each side by } C.$$

$$(a^{5700})^{1/5700} = \left(\frac{1}{2}\right)^{1/5700} \qquad \text{Raise to the } \tfrac{1}{5700}\text{th power.}$$

$$a = \left(\frac{1}{2}\right)^{1/5700} \qquad \text{Properties of exponents}$$

Using properties of exponents, we write $f(x) = Ca^x$ as

$$f(x) = C\left(\left(\frac{1}{2}\right)^{1/5700}\right)^x = C\left(\frac{1}{2}\right)^{x/5700}.$$

This discussion is summarized in the following box.

Modeling Radioactive Decay

If a radioactive sample containing C units has a half-life of k years, then the amount A remaining after x years is given by

$$A(x) = C\left(\frac{1}{2}\right)^{x/k}.$$

The half-life of radium-226 is about **1600** years. After **9600** years, a **2**-gram sample decays to

$$A(9600) = 2\left(\frac{1}{2}\right)^{9600/1600} = 0.03125 \text{ gram.}$$

EXAMPLE 11   Finding the age of a fossil

A fossil contains 5% of the amount of carbon-14 that the organism contained when it was alive. Graphically estimate its age. The situation is illustrated in Figure 5.47.

SOLUTION   The initial amount of carbon-14 is 100% (or 1), the final amount is 5% (or 0.05), and the half-life is 5700 years, so $A = 0.05$, $C = 1$, and $k = 5700$ in $A(x) = C\left(\frac{1}{2}\right)^{x/k}$. To determine the age of the fossil, solve

$$0.05 = 1\left(\frac{1}{2}\right)^{x/5700}$$

for x. Graph $Y_1 = 0.05$ and $Y_2 = 0.5^{\wedge}(X/5700)$, as shown in Figure 5.48. Their graphs intersect near $(24635, 0.05)$, so the fossil is about 24,635 years old.

Figure 5.47

$[0, 50000, 10000]$ by $[0, 0.1, 0.01]$

$y_2 = 0.5^{x/5700}$

$y_1 = 0.05$

Intersection
X=24634.99   Y=.05

Figure 5.48

Now Try Exercise 89 ◀

5.3 Putting It All Together

The following table summarizes some important concepts about exponential functions and types of growth and decay.

Concept	Comments	Examples
Exponential function $$f(x) = Ca^x,$$ where $a > 0$, $a \neq 1$, and $C > 0$	Exponential growth occurs when $a > 1$, and exponential decay occurs when $0 < a < 1$. C often represents the initial amount present because $f(0) = C$.	$f(x) = 5(0.8)^x$ *Decay* $g(x) = 3^x$ *Growth* *Decay: $0 < a < 1$* *Growth: $a > 1$*
Linear growth $$y = mx + b$$	If data increase by a fixed amount m for each unit increase in x, they can be modeled by a linear function.	The following data can be modeled by $f(x) = 3x + 2$ because the data increase 3 units for each unit increase in x and because $y = 2$ when $x = 0$. x: 0, 1, 2, 3 y: 2, 5, 8, 11 $+3$ $+3$ $+3$
Exponential growth $$y = Ca^x$$	If data increase by a constant factor a for each unit increase in x, they can be modeled by an exponential function.	The following data can be modeled by $f(x) = 2(3)^x$ because the data increase by a factor of 3 for each unit increase in x and because $y = 2$ when $x = 0$. x: 0, 1, 2, 3 y: 2, 6, 18, 54 $\times 3$ $\times 3$ $\times 3$
Interest compounded n times per year $$A = P\left(1 + \frac{r}{n}\right)^{nt}$$	P is the principal, r is the interest rate (expressed in decimal form), n is the number of times interest is paid each year, t is the number of years, and A is the amount after t years.	$\$500$ at 8%, compounded monthly, for 3 years yields $$500\left(1 + \frac{0.08}{12}\right)^{12(3)} \approx \$635.12.$$
The number e	The number e is an irrational number important in mathematics, much like π.	$e \approx 2.718282$

Concept	Comments	Examples
Natural exponential function	This function is an exponential function with base e and $C = 1$.	$f(x) = e^x$
Interest compounded continuously $A = Pe^{rt}$	P is the principal, r is the interest rate (expressed in decimal form), t is the number of years, and A is the amount after t years.	\$500 at 8% compounded continuously for 3 years yields $500e^{0.08(3)} \approx \$635.62$.
Radioactive decay	If a radioactive sample has a half-life of k years and contains C units, then the amount A remaining after x years is given by $$A(x) = C\left(\frac{1}{2}\right)^{x/k}.$$	A 5-gram sample of radioactive material with a half-life of 300 years is modeled by $$A(x) = 5\left(\frac{1}{2}\right)^{x/300}.$$

5.3 Exercises

Exponents

Exercises 1–14: Simplify the expression without a calculator.

1. 2^{-3}

2. $(-3)^{-2}$

3. $3(4)^{1/2}$

4. $5\left(\frac{1}{2}\right)^{-3}$

5. $-2(27)^{2/3}$

6. $-4(8)^{-2/3}$

7. $4^{1/6}4^{1/3}$

8. $\frac{9^{5/6}}{9^{1/3}}$

9. $e^x e^x$

10. $e^{3x}e^{1+x}$

11. 3^0

12. $5\left(\frac{3}{4}\right)^0$

13. $(5^{101})^{1/101}$

14. $(8^{27})^{1/27}$

Linear and Exponential Growth

Exercises 15–20: (Refer to Example 1.) Find either a linear or an exponential function that models the data in the table.

15.

x	0	1	2	3	4
y	2	0.8	−0.4	−1.6	−2.8

16.

x	0	1	2	3	4
y	2	8	32	128	512

17.

x	−3	−2	−1	0	1
y	64	32	16	8	4

18.

x	−2	−1	0	1	2
y	3	5.5	8	10.5	13

19.

x	−4	−2	0	2	4
y	0.3125	1.25	5	20	80

20.

x	−15	−5	5	15	25
y	22	24	26	28	30

21. *Job Offer*   A company offers a college graduate \$40,000 per year with a guaranteed 8% raise each year. Is this an example of linear or exponential growth? Find a function f that computes the salary during the nth year.

22. *Job Offer*   A new employee is offered \$35,000 per year with a guaranteed \$5000 raise each year. Is this an example of linear or exponential growth? Find a function f that computes the salary during the nth year.

23. *Comparing Growth*   Which function becomes larger for $0 \le x \le 10$: $f(x) = 2^x$ or $g(x) = x^2$?

24. *Comparing Growth*   Which function becomes larger for $0 \le x \le 10$: $f(x) = 4 + 3x$ or $g(x) = 4(3)^x$?

25. *Comparing Growth*   Which function becomes larger for $0 \le x \le 10$: $f(x) = 2x + 1$ or $g(x) = 2^{-x}$?

26. *Salaries*   If you were offered 1¢ for the first week of work, 3¢ for the second week, 5¢ for the third week, 7¢ for the fourth week, and so on for a year, would you accept the offer? Would you accept an offer that pays 1¢ for the first week of work, 2¢ for the second week, 4¢ for the third week, 8¢ for the fourth week, and so on for a year? Explain your answers.

Exponential Functions

Exercises 27–34: Find C and a so that $f(x) = Ca^x$ satisfies the given conditions.

27. $f(0) = 5$ and for each unit increase in x, the output is multiplied by 1.5.

28. $f(1) = 3$ and for each unit increase in x, the output is multiplied by $\frac{3}{4}$.

29. $f(0) = 10$ and $f(1) = 20$

30. $f(0) = 7$ and $f(-1) = 1$

31. $f(1) = 9$ and $f(2) = 27$

32. $f(-1) = \frac{1}{4}$ and $f(1) = 4$

33. $f(-2) = \frac{9}{2}$ and $f(2) = \frac{1}{18}$

34. $f(-2) = \frac{3}{4}$ and $f(2) = 12$

Exercises 35–38: Find C and a so that $f(x) = Ca^x$ models the situation described. State what the variable x represents in your formula. (Answers may vary.)

35. There are initially 5000 bacteria, and this sample doubles in size every hour.

36. Fifteen hundred dollars is deposited in an account that triples in value every decade.

37. In 2000 a house was worth $200,000, and its value *decreases* by 5% each year thereafter.

38. A fish population is initially 6000 and decreases by half each year.

39. *Tire Pressure*   The pressure in a tire with a leak is 30 pounds per square inch initially and can be modeled by $f(x) = 30(0.9)^x$ after x minutes. What is the tire's pressure after 9.5 minutes?

40. *Population*   The population of California was about 38 million in 2007 and increasing by 1.6% each year. Estimate the population of California in 2012.

Exercises 41–44: Approximate f(x) to four decimal places.

41. $f(x) = 4e^{-1.2x}$, $x = -2.4$

42. $f(x) = -2.1e^{-0.71x}$, $x = 1.9$

43. $f(x) = \frac{1}{2}(e^x - e^{-x})$, $x = -0.7$

44. $f(x) = 4(e^{-0.3x} - e^{-0.6x})$, $x = 1.6$

Graphs of Exponential Functions

Exercises 45–52: Sketch a graph of $y = f(x)$.

45. $f(x) = 2^x$ **46.** $f(x) = 4^x$

47. $f(x) = 3^{-x}$ **48.** $f(x) = 3(2^{-x})$

49. $f(x) = 2\left(\frac{1}{3}\right)^x$ **50.** $f(x) = 2(3^x)$

51. $f(x) = \left(\frac{1}{2}\right)^x$ **52.** $f(x) = \left(\frac{1}{4}\right)^x$

Exercises 53–56: Use the graph of $y = Ca^x$ to determine values for C and a.

53.

54.

55.

56.

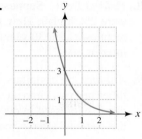

57. Let $f(x) = 7\left(\frac{1}{8}\right)^x$.

(a) What are the domain and range of f?

(b) Is f either increasing or decreasing on its domain?

(c) Find any asymptotes on the graph of f.

(d) Find any x- or y-intercepts on the graph of f.

(e) Is f a one-to-one function? Does f have an inverse?

58. Repeat Exercise 57 with $f(x) = e^x$.

59. Match the symbolic representation of f with its graphical representation (a–d). Do *not* use a calculator.

(i) $f(x) = e^x$ (ii) $f(x) = 3^{-x}$

(iii) $f(x) = 1.5^x$ (iv) $f(x) = 0.99^x$

a.

b.

c.

d.

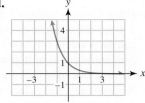

60. *Modeling Phenomena* Match the situation with the graph (a–d) at the top of the next column that models it best.

(i) Balance in an account after x years earning 10% interest compounded continuously

(ii) Balance in an account after x years earning 5% interest compounded annually

(iii) Air pressure in a car tire with a large hole in it after x minutes

(iv) Air pressure in a car tire with a pinhole in it after x minutes

a.

b.

c.

d.

Exercises 61 and 62: The graph of $y = f(x)$ is shown in the figure. Sketch a graph of each equation using translations of graphs and reflections. Do not use a graphing calculator.

61. $f(x) = 2^x$

(a) $y = 2^x - 2$

(b) $y = 2^{x-1}$

(c) $y = 2^{-x}$

(d) $y = -2^x$

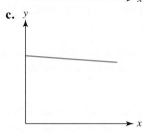

62. $f(x) = e^{-0.5x}$

(a) $y = -e^{-0.5x}$

(b) $y = e^{-0.5x} - 3$

(c) $y = e^{-0.5(x-2)}$

(d) $y = e^{0.5x}$

Compound Interest

Exercises 63–70: Use the compound interest formula to determine the final value of each amount.

63. $600 at 7% compounded annually for 5 years

64. $2300 at 11% compounded semiannually for 10 years

65. $950 at 3% compounded daily for 20 years

66. $3300 at 8% compounded quarterly for 2 years

67. $2000 at 10% compounded continuously for 8 years

68. $100 at 19% compounded continuously for 50 years

69. $1600 at 10.4% compounded monthly for 2.5 years

70. $2000 at 8.7% compounded annually for 5 years

71. *Investments*   Compare investing $2000 at 10% compounded monthly for 20 years with investing $2000 at 13% compounded monthly for 20 years.

72. *Lake Property*   In some states, lake shore property is increasing in value by 15% per year. Determine the value of a $90,000 lake lot after 5 years.

73. *College Tuition*   If college tuition is currently $8000 per year, inflating at 6% per year, what will be the cost of tuition in 10 years?

74. *Doubling Time*   How long does it take for an investment to double its value if the interest is 12% compounded annually? 6% compounded annually?

Applications

75. *Saving for Retirement*   Suppose $1500 is deposited into an IRA with an interest rate of 6%, compounded continuously.
(a) How much money will there be after 30 years?

(b) Determine graphically or numerically the 10-year period when the account balance increases the most.

76. *Population Growth*   The population of Phoenix, Arizona, was 1.3 million in 2000 and growing continuously at a 3% rate.
(a) Assuming this trend continues, estimate the population of Phoenix in 2010.

(b) Determine graphically or numerically when this population might reach 2 million.

77. *Federal Debt*   In fiscal year 2008 the federal budget deficit was about $340 billion. At the same time, 30-year treasury bonds were paying 4.54% interest. Suppose the American taxpayer loaned $340 billion to the federal government at 4.54% compounded annually. If the federal government waited 30 years to pay the entire amount back, including the interest, how much would this be?   (**Source:** Department of the Treasury.)

78. *Federal Debt*   Suppose that interest rates in Exercise 77 were 2% higher. How much would the federal government owe after 30 years? Is the national debt sensitive to interest rates?

79. *Annuity*   If x dollars is deposited every 2 weeks (26 times per year) into an account paying an annual interest rate r, expressed in decimal form, then the amount A in the account after t years can be approximated by the formula
$$A = x\left(\frac{(1 + r/26)^{26t} - 1}{(r/26)}\right).$$
If $50 is deposited every 2 weeks into an account paying 8% interest, approximate the amount after 10 years.

80. *Annuity*   (Refer to Exercise 79.) Suppose a retirement account pays 10% annual interest. Determine how much a 20-year-old worker should deposit in this account every 2 weeks in order to have $1 million at age 65.

81. *Continuous Compounding*   Over 5 years, the total value of a mutual fund account decreases continuously by 15%. Find a formula $A(x)$ that calculates the amount of money in the account after x years.

82. *Continuous Compounding*   A sum of money P in an account receives continuous interest and triples in 15 years. Find a formula $A(x)$ that calculates the amount of money in the account after x years.

83. *E. Coli Growth*   (Refer to Example 8.)
(a) Approximate the number of *E. coli* after 3 hours.

(b) Estimate graphically the elapsed time when there are 10 million bacteria per milliliter.

84. *Drug Concentrations*   Sometimes after a patient takes a drug, the amount of medication A in the bloodstream can be modeled by $A = A_0e^{-rt}$, where A_0 is the initial concentration in milligrams per liter, r is the hourly percentage decrease (in decimal form) of the drug in the bloodstream, and t is the elapsed time in hours. Suppose that a drug's concentration is initially 2 milligrams per liter and that $r = 0.2$.
(a) Find the drug concentration after 3.5 hours.

(b) When did the drug concentration reach 1.5 milligrams per liter?

85. *Greenhouse Gases*   (Refer to Example 9.) Chlorofluorocarbons (CFCs) are gases created by people that increase the greenhouse effect and damage the ozone layer. The following table lists future concentrations of CFC-12 in parts per billion (ppb) if current trends continue.

Year	2000	2005	2010	2015	2020
CFC-12 (ppb)	0.72	0.88	1.07	1.31	1.60

Source: R. Turner, Environmental Economics.

(a) Find values for C and a so that $f(x) = Ca^x$ models these data, where x is years after 2000.

(b) Estimate the CFC-12 concentration in 2013.

86. *Bacteria Growth*   The table lists the concentration in a sample of E. coli bacteria B (in billions per liter) after x hours.

x	0	3	5	8
B	0.5	6.2	33.3	414

Source: G. S. Stent, *Molecular Biology of Bacterial Viruses.*

(a) Find values for C and a so that $f(x) = Ca^x$ models these data.

(b) Estimate the bacteria concentration after 6.2 hours.

87. *Modeling Traffic Flow*   Cars arrive randomly at an intersection with an average rate of 50 cars per hour. The likelihood, or probability, that at least one car will enter the intersection within a period of x minutes can be estimated by $P(x) = 1 - e^{-5x/6}$.

(a) Find the likelihood that at least one car enters the intersection during a 3-minute period.

(b) Graphically determine the value of x that gives a 50–50 chance of at least one car entering the intersection during an interval of x minutes.

88. *Tree Density*   Ecologists studied the spacing between individual trees in a forest in British Columbia. The probability, or likelihood, that there is at least one tree located in a circle with a radius of x feet can be estimated by $P(x) = 1 - e^{-0.1144x}$. For example, $P(7) \approx 0.55$ means that if a person picks a point at random in the forest, there is a 55% chance that at least one tree will be located within 7 feet. See the figure.   (**Source:** E. Pielou, *Populations and Community Ecology.*)

(a) Evaluate $P(2)$ and $P(20)$, and interpret the results.

(b) Graph P. Explain verbally why it is logical for P to be an increasing function. Does the graph have a horizontal asymptote?

(c) Solve $P(x) = 0.5$ and interpret the result.

89. *Radioactive Carbon-14*   (Refer to Example 11.) A fossil contains 10% of the carbon-14 that the organism contained when it was alive. Graphically estimate its age.

90. *Radioactive Carbon-14*   A fossil contains 20% of the carbon-14 that the organism contained when it was alive. Estimate its age.

91. *Radioactive Radium-226*   The half-life of radium-226 is about 1600 years. After 3000 years, what percentage P of a sample of radium remains?

92. *Radioactive Strontium-90*   Radioactive strontium-90 has a half-life of about 28 years and sometimes contaminates the soil. After 50 years, what percentage of a sample of radioactive strontium would remain?

93. *Swimming Pool Maintenance*   Chlorine is frequently used to disinfect swimming pools. The chlorine concentration should remain between 1.5 and 2.5 parts per million. On warm sunny days with many swimmers agitating the water, 30% of the chlorine can dissipate into the air or combine with other chemicals.   (**Source:** D. Thomas, *Swimming Pool Operators Handbook.*)

(a) Find C and a so that $f(x) = Ca^x$ models the amount of chlorine in the pool after x days. Assume that the initial amount is 2.5 parts per million.

(b) What is the chlorine concentration after 2 days if no chlorine is added?

(c) Estimate graphically or numerically the number of days before chlorine should be added.

94. *Radioactive Cesium-137* Radioactive cesium-137 was emitted in large amounts in the Chernobyl nuclear power station accident in Russia. The amount of a 100-milligram sample of cesium remaining after x years can be described by $A(x) = 100e^{-0.02295x}$. (**Source:** C. Mason, *Biology of Freshwater Pollution.*)

(a) How much remains after 50 years? Is the half-life of cesium more or less than 50 years?

(b) Estimate graphically the half-life of cesium-137.

95. *Thickness of Runways* Heavier aircraft require runways with thicker pavement for landings and takeoffs. A pavement 6 inches thick can accommodate an aircraft weighing 80,000 pounds, whereas a 12-inch-thick pavement is necessary for a 350,000-pound plane. The relation between pavement thickness t in inches and gross weight W in thousands of pounds can be modeled by $W = Ca^t$. (**Source:** FAA.)

(a) Find values for C and a.

(b) How heavy an airplane can a 9-inch-thick runway accommodate?

(c) What is the minimum thickness for a 242,000-pound plane?

96. *Filters* Impurities in water are frequently removed using filters. Suppose that a 1-inch filter allows 10% of the impurities to pass through it. The other 90% is trapped in the filter.

(a) Find a formula in the form $f(x) = 100a^x$ that calculates the percentage of impurities passing through x inches of this type of filter.

(b) Use $f(x)$ to estimate the percentage of impurities passing through 2.3 inches of the filter.

97. *Trains* The faster a locomotive travels, the more horsepower is needed. The formula $H(x) = 0.157(1.033)^x$ calculates this horsepower for a level track. The input x is in miles per hour and the output $H(x)$ is the horsepower required per ton of cargo. (**Source:** L. Haefner, *Introduction to Transportation Systems.*)

(a) Evaluate $H(30)$ and interpret the result.

(b) Determine the horsepower needed to move a 5000-ton train 30 miles per hour.

(c) Some types of locomotives are rated for 1350 horsepower. How many locomotives of this type would be needed in part (b)?

98. *Survival of Reindeer* For all types of animals, the percentage that survive into the next year decreases. In one study, the survival rate of a sample of reindeer was modeled by $S(t) = 100(0.999993)^{t^5}$. The function S outputs the percentage of reindeer that survive t years. (**Source:** D. Brown.)

(a) Evaluate $S(4)$ and $S(15)$. Interpret the results.

(b) Graph S in [0, 15, 5] by [0, 110, 10]. Interpret the graph. Does the graph have a horizontal asymptote?

Writing about Mathematics

99. Explain how linear and exponential functions differ. Give examples.

100. Discuss the domain and range of an exponential function f. Is f one-to-one? Explain.

EXTENDED AND DISCOVERY EXERCISES

Exercises 1–4: Present Value In the compound interest formula $A = P(1 + r/n)^{nt}$, we can think of P as the present value of an investment and A as the future value of an investment after t years. For example, if you were saving for college and needed a future value of A dollars, then P would represent the amount needed in an account today to reach your goal in t years at an interest rate of r, compounded n times per year. If we solve the equation for P, it results in

$$P = A(1 + r/n)^{-nt}.$$

1. Verify that the two formulas are equivalent by transforming the first equation into the second.

2. What should the present value of a savings account be to cover $30,000 of college expenses in 12.5 years, if the account pays 7.5% interest compounded quarterly?

3. Suppose you want to have $15,000 to buy a car in 3 years. What should the present value of a savings account be to reach this goal, if the account pays 5% compounded monthly?

4. A parent expects college costs to reach $40,000 in 6 years. To cover the $40,000 in future expenses, how much should the parent deposit in an account that pays 6% interest compounded *continuously*?

Exercises 5–8: Average Rate of Change of e^x Complete the following. Round your answers to two decimal places.

(a) Find the average rate of change of $f(x) = e^x$ from x to $x + 0.001$ for the given x.

(b) Approximate $f(x) = e^x$ for the given x.

(c) Compare your answers in parts (a) and (b).

5. $x = 0$ **6.** $x = -2$

7. $x = -0.5$ **8.** $x = 1.5$

9. *Average Rate of Change*   What is the pattern in the results from Exercises 5–8? You may want to test your conjecture by trying different values of x.

10. *Average Rate of Change*   (Refer to Exercises 5–8.) For any real number k, what is a good approximation for the average rate of change of $f(x) = e^x$ on a small interval near $x = k$? Explain how your answer relates to the graph of $f(x) = e^x$.

5.4 Logarithmic Functions and Models

- **Evaluate the common logarithmic function**
- **Evaluate logarithms with other bases**
- **Solve basic exponential and logarithmic equations**
- **Solve general exponential and logarithmic equations**
- **Convert between exponential and logarithmic forms**

Introduction

Bacteria growth is often exponential. For example, Table 5.18 and Figure 5.49 show the concentration B (in thousands per milliliter) in a bacteria sample. The symbolic representation is $B(x) = 10^x$, where x represents time in elapsed days. To determine how long it takes for the concentration to reach 500 thousand per millimeter, we can solve the equation $10^x = 500$. The graphical solution in Figure 5.50 is about 2.7, but finding this value symbolically requires *logarithms*.

Table 5.18

x	0	1	2	3
$B(x)$	1	10	100	1000

Figure 5.49

Figure 5.50 $B^{-1}(500) \approx 2.7$

The Common Logarithmic Function

Notice that $B(x) = 10^x$ in Figure 5.49 is increasing and one-to-one. Its graph passes the horizontal line test. Therefore B has an inverse function. Because $B(1) = 10$, it follows that $B^{-1}(10) = 1$. Similarly, $B(2) = 100$ implies that $B^{-1}(100) = 2$, and in general, $B(k) = 10^k$ implies that $B^{-1}(10^k) = k$. Table 5.19 lists values for $B^{-1}(x)$. Notice that, unlike B, B^{-1} increases *very slowly* for large values of x.

The solution to $10^x = 500$ represents the number of days needed for the bacteria concentration to reach 500 thousand per milliliter. To solve this equation, we must find an exponent k such that $10^k = 500$. Because $B^{-1}(10^k) = k$, this is equivalent to evaluating $B^{-1}(500)$. From Table 5.19, k is between 2 and 3 because $100 \leq 500 \leq 1000$. (This fact is also apparent from Figure 5.50.)

The inverse of $y = 10^x$ is defined to be the *common (base-10) logarithm,* denoted $\log x$ or $\log_{10} x$. Thus $10^2 = 100$ implies $\log 100 = 2$, $10^3 = 1000$ implies $\log 1000 = 3$, and in general, $10^a = b$ implies $\log b = a$. Thus $B^{-1}(500) = \log 500$.

Table 5.19

x	1	10	100	1000
$B^{-1}(x)$	0	1	2	3

The common exponential function and the common logarithmic function, represented in Tables 5.20 and 5.21, are inverse functions. (In our example, $B^{-1}(x) = \log x$.)

Table 5.20 Common Exponential Function

x	-4	-3	-2	-1	0	0.5	1	2	π
10^x	10^{-4}	10^{-3}	10^{-2}	10^{-1}	10^0	$10^{0.5}$	10^1	10^2	10^π

Table 5.21 Common Logarithmic Function

x	10^{-4}	10^{-3}	10^{-2}	10^{-1}	10^0	$10^{0.5}$	10^1	10^2	10^π
$\log x$	-4	-3	-2	-1	0	0.5	1	2	π

If a positive number x can be written as 10^k, then $\log x = k$. *A logarithm is an exponent k.* This concept can be illustrated visually.

Common Logarithm

The Graph of the Common Logarithmic Function The points $(10^{-1}, -1)$, $(10^0, 0)$, $(10^{0.5}, 0.5)$, and $(10^1, 1)$ are located on the graph of $y = \log x$. They are plotted in Figure 5.51. Any *positive* real number x can be expressed as $x = 10^k$ for some real number k. In this case $\log x = \log 10^k = k$. The graph of $y = \log x$ is continuous, as shown in Figure 5.51. The y-axis is a vertical asymptote. The common logarithmic function is one-to-one and always increasing. Its domain is all positive real numbers, and its range is all real numbers.

We have shown *verbal*, *numerical*, *graphical*, and *symbolic* representations of the common logarithmic function. A formal definition of the common logarithm is now given.

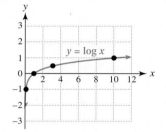

Figure 5.51 The Common Logarithmic Function

Common Logarithm

The **common logarithm of a positive number** x, denoted $\log x$, is defined by

$$\log x = k \quad \text{if and only if} \quad x = 10^k,$$

where k is a real number. The function given by

$$f(x) = \log x$$

is called the **common logarithmic function**.

NOTE The common logarithmic function outputs an *exponent k,* which may be positive, negative, or zero. However, a valid input must be positive. Thus its range is $(-\infty, \infty)$ and its domain is $(0, \infty)$.

EXAMPLE 1 Evaluating common logarithms

Simplify each logarithm by hand.

(a) $\log 1$ **(b)** $\log \dfrac{1}{1000}$ **(c)** $\log \sqrt{10}$ **(d)** $\log(-2)$

Algebra Review
To review properties of exponents, see Section 4.8 and Chapter R (page R-7).

SOLUTION

(a) Since $1 = 10^0$, $\log 1 = \log 10^0 = 0$.

(b) $\log \dfrac{1}{1000} = \log 10^{-3} = -3$

(c) $\log \sqrt{10} = \log (10^{1/2}) = \dfrac{1}{2}$

(d) The domain of $f(x) = \log x$ is $(0, \infty)$, so $\log (-2)$ is undefined because the input is negative.

Now Try Exercise 3 ◀

```
log(12)
            1.079181246
10^1.079181246
                      12
```

Figure 5.52

Square Roots and Logarithms   Much like the square root function, the common logarithmic function does not have an easy-to-evaluate formula. For example, $\sqrt{4} = 2$ and $\sqrt{100} = 10$ can be calculated mentally, but for $\sqrt{2}$ we usually rely on a calculator. Similarly, $\log 100 = 2$ can be found mentally since $100 = 10^2$, whereas $\log 12$ can be approximated using a calculator. See Figure 5.52. To check that $\log 12 \approx 1.079181246$, evaluate $10^{1.079181246} \approx 12$. Another similarity between the square root function and the common logarithmic function is that their domains do not include negative numbers. If outputs are restricted to real numbers, both $\sqrt{-3}$ and $\log (-3)$ are undefined expressions.

> **MAKING CONNECTIONS**
>
> **Logarithms and Exponents**   *A logarithm is an exponent.* For example, $\log 1000$ represents the exponent k such that $10^k = 1000$. Thus $\log 1000 = 3$ because $10^3 = 1000$.

Figure 5.53

Graphs and Inverse Properties   Because $f(x) = 10^x$ and $f^{-1}(x) = \log x$ represent inverse functions, it follows that their graphs are reflections across the line $y = x$, as shown in Figure 5.53. For inverse functions, $f^{-1}(f(x)) = x$ and $f(f^{-1}(x)) = x$, so the following inverse properties hold for all valid inputs x.

> ### Inverse Properties of the Common Logarithm
>
> The following inverse properties hold for the common logarithm.
>
> $$\log 10^x = x \quad \text{for any real number } x \text{ and}$$
> $$10^{\log x} = x \quad \text{for any positive number } x$$

A graphing calculator has been used to illustrate these properties in Figures 5.54 and 5.55. (If you have a calculator available, try some other examples.)

```
log(10^5)
                    5
log(10^1.6)
                  1.6
log(10^(-2.5))
                 -2.5
```

Figure 5.54

```
10^log(2)
                    2
10^log(3.7)
                  3.7
10^log(0.12)
                  .12
```

Figure 5.55

An Application The human ear is extremely sensitive; it can detect pressures on the eardrum that range from 10^{-16} w/cm² (watts per square centimeter) to 10^{-4} w/cm². Sound with an intensity of 10^{-4} w/cm² is painful to the human eardrum. Because of this wide range of intensities, scientists use logarithms to measure sound in decibels.

EXAMPLE 2 Applying common logarithms to sound

Sound levels in decibels (db) can be computed by $D(x) = 10 \log (10^{16}x)$, where x is the intensity of the sound in watts per square centimeter. (**Source:** R. Weidner and R. Sells, *Elementary Classical Physics*.)
(a) At what decibel level does the threshold for pain occur?
(b) Make a table of D for the intensities $x = 10^{-13}, 10^{-12}, 10^{-11}, \ldots, 10^{-4}$.
(c) If x increases by a factor of 10, what is the corresponding increase in decibels?

SOLUTION
(a) The threshold for pain occurs when $x = 10^{-4}$.

$$\begin{aligned} D(10^{-4}) &= 10 \log (10^{16} \cdot 10^{-4}) && \textit{Substitute x = 10}^{-4}. \\ &= 10 \log (10^{12}) && \textit{Add exponents when multiplying like bases.} \\ &= 10(12) && \textit{Evaluate the logarithm.} \\ &= 120 && \textit{Multiply.} \end{aligned}$$

CLASS DISCUSSION
If the sound level increases by 60 db, by what factor does the intensity increase?

The human eardrum begins to hurt at 120 decibels.
(b) Other values for x in Table 5.22 are computed in a similar manner.

Table 5.22

x (w/cm²)	10^{-13}	10^{-12}	10^{-11}	10^{-10}	10^{-9}	10^{-8}	10^{-7}	10^{-6}	10^{-5}	10^{-4}
$D(x)$ (db)	30	40	50	60	70	80	90	100	110	120

(c) From Table 5.22, a tenfold increase in x results in an increase of 10 decibels.

Now Try Exercise 117 ◀

Logarithms with Other Bases

It is possible to develop base-a logarithms with any positive base $a \neq 1$. For example, in computer science base-2 logarithms are frequently used. A numerical representation of the base-2 logarithmic function, denoted $f(x) = \log_2 x$, is shown in Table 5.23. If x can be expressed in the form $x = 2^k$ for some k, then $\log_2 x = k$. Thus $\log_2 x$ is an exponent. A graph of $y = \log_2 x$ is shown in Figure 5.56.

Figure 5.56

Table 5.23 Base-2 Logarithm

x	$2^{-3.1}$	2^{-2}	$2^{-0.5}$	2^0	$2^{0.5}$	2^2	$2^{3.1}$
$\log_2 x$	-3.1	-2	-0.5	0	0.5	2	3.1

In a similar manner, a numerical representation for the base-e logarithm is presented in Table 5.24. The base-e logarithm is referred to as the **natural logarithm** and denoted either $\log_e x$ or $\ln x$. Natural logarithms are used in mathematics, science, economics, and technology. A graph of $y = \ln x$ is shown in Figure 5.57.

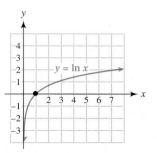

Figure 5.57

Table 5.24 Natural Logarithm

x	$e^{-3.1}$	e^{-2}	$e^{-0.5}$	e^0	$e^{0.5}$	e^2	$e^{3.1}$
$\ln x$	-3.1	-2	-0.5	0	0.5	2	3.1

A base-a logarithm is now defined. Its domain is $(0, \infty)$ and its range is $(-\infty, \infty)$.

Logarithm

The **logarithm with base a of a positive number x**, denoted by $\log_a x$, is defined by

$$\log_a x = k \quad \text{if and only if} \quad x = a^k,$$

where $a > 0$, $a \neq 1$, and k is a real number. The function, given by

$$f(x) = \log_a x,$$

is called the **logarithmic function with base a**.

EXAMPLE 3   Finding the domain of a logarithmic function

State the domain of f.
(a) $f(x) = \log_2 (x - 4)$ **(b)** $f(x) = \ln (10^x)$

SOLUTION
(a) *The input for a logarithmic function must be positive.* Any element of the domain of f must satisfy $x - 4 > 0$, or equivalently, $x > 4$. Thus $D = (4, \infty)$.
(b) The expression 10^x is positive for all real numbers x. (See Figure 5.53, where the red graph of $y = 10^x$ is above the x-axis for all values of x.) Thus $D = (-\infty, \infty)$, or all real numbers.   ***Now Try Exercises 13 and 17*** ◄

CLASS DISCUSSION
Make a numerical representation for a base-4 logarithm. Evaluate $\log_4 16$.

Remember that *a logarithm is an exponent*. The expression $\log_a x$ is the exponent k such that $a^k = x$. Logarithms with base a also satisfy inverse properties.

Inverse Properties

The following inverse properties hold for logarithms with base a.

$$\log_a a^x = x \quad \text{for any real number } x \text{ and}$$
$$a^{\log_a x} = x \quad \text{for any positive number } x$$

EXAMPLE 4   Applying inverse properties

Use inverse properties to evaluate each expression.

(a) $\log_6 6^{-1.3}$ **(b)** $5^{\log_5 (x+8)}$ **(c)** $\log_{1/2} \left(\frac{1}{2}\right)^{45}$

SOLUTION
(a) $\log_a a^x = x$, so $\log_6 6^{-1.3} = -1.3$.
(b) $a^{\log_a x} = x$, so $5^{\log_5 (x+8)} = x + 8$, provided $x > -8$.
(c) $\log_a a^x = x$, so $\log_{1/2} \left(\frac{1}{2}\right)^{45} = 45$. Note that the base of a logarithmic function can be a positive fraction less than 1.   ***Now Try Exercises 21, 23, and 25*** ◄

EXAMPLE 5   Evaluating logarithms

Evaluate each logarithm.

(a) $\log_2 8$ **(b)** $\log_5 \dfrac{1}{25}$ **(c)** $\log_7 49$ **(d)** $\ln e^{-7}$

SOLUTION

(a) To determine $\log_2 8$, express 8 as 2^k for some k. Since $8 = 2^3$, $\log_2 8 = \log_2 2^3 = 3$.

(b) $\log_5 \dfrac{1}{25} = \log_5 \dfrac{1}{5^2} = \log_5 5^{-2} = -2$

(c) $\log_7 49 = \log_7 7^2 = 2$

(d) $\ln e^{-7} = \log_e e^{-7} = -7$ *Now Try Exercises 33, 37, 39, and 49* ◀

Figure 5.58

Graphs and Inverse Functions   The inverse function of $f(x) = a^x$ is $f^{-1}(x) = \log_a x$. Therefore the graph of $y = \log_a x$ is a reflection of the graph of $y = a^x$ across the line $y = x$. Figure 5.58 shows graphs of $f(x) = 2^x$ and $f^{-1}(x) = \log_2 x$. The graph of f^{-1} is a reflection of f in the line $y = x$. The rapid growth in f is called *exponential growth*. On the other hand, the graph of f^{-1} begins to level off. This slower rate of growth is called **logarithmic growth**. Since the range of f is all positive numbers, the domain of f^{-1} is all positive numbers. Notice that when $0 < x < 1$, $f^{-1}(x) = \log_2 x$ outputs negative numbers, and when $x > 1$, $f^{-1}(x)$ outputs positive numbers.

The graphs of $g(x) = e^x$ and $g^{-1}(x) = \ln x$ are shown in Figure 5.59. Because $e \approx 2.7183 > 2$, the graph of g increases faster than the graph of $f(x) = 2^x$, and the graph of g^{-1} levels off faster than the graph of $f^{-1}(x) = \log_2 x$.

Figure 5.59

MAKING CONNECTIONS

Exponential and Logarithmic Functions   The inverse of an exponential function is a logarithmic function, and the inverse of a logarithmic function is an exponential function. For example,

if $f(x) = 10^x$ then $f^{-1}(x) = \log x$,

if $g(x) = \ln x$ then $g^{-1}(x) = e^x$,   and

if $h(x) = 2^x$ then $h^{-1}(x) = \log_2 x$.

Basic Equations

Base-10 Exponential Equations   To solve the equation $10 + x = 100$, we subtract 10 from each side because addition and subtraction are inverse operations.

$$10 + x - 10 = 100 - 10$$

$$x = 90$$

To solve the equation $10x = 100$, we divide each side by 10 because multiplication and division are inverse operations.

$$\frac{10x}{10} = \frac{100}{10}$$

$$x = 10$$

Now suppose that we want to solve the exponential equation

$$10^x = 100.$$

What is new about this type of equation is that the variable x is an *exponent*. The inverse operation of 10^x is log x. Rather than subtracting 10 from each side or dividing each side by 10, we take the base-10 logarithm of each side. Doing so results in

$$\log 10^x = \log 100.$$

Because $\log 10^x = x$ for all real numbers x, the equation becomes

$$x = \log 100, \text{ or equivalently, } x = 2.$$

These concepts are applied in the next example. Note that if $m = n$ (m and n positive), then $\log m = \log n$.

EXAMPLE 6 Solving equations of the form $10^x = k$

Solve each equation, if possible.
(a) $10^x = 0.001$ **(b)** $10^x = 55$ **(c)** $10^x = -1$

SOLUTION
(a) Take the common logarithm of each side of the equation $10^x = 0.001$. Then

$$\log 10^x = \log 0.001, \quad \text{or} \quad x = \log 10^{-3} = -3.$$

(b) In a similar manner, $10^x = 55$ is equivalent to $x = \log 55 \approx 1.7404$.
(c) The equation $10^x = -1$ has no real solutions, because -1 is not in the range of 10^x. Figure 5.60 shows that the graphs of $y_1 = 10^x$ and $y_2 = -1$ do not intersect. Note that $\log(-1)$ is undefined.

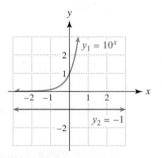

Figure 5.60

Now Try Exercise 53 ◀

EXAMPLE 7 Solving a base-10 exponential equation

Solve $4(10^{3x}) = 244$.

SOLUTION Begin by dividing each side by 4.

$$4(10^{3x}) = 244 \qquad \text{Given equation}$$
$$10^{3x} = 61 \qquad \text{Divide by 4.}$$
$$\log 10^{3x} = \log 61 \qquad \text{Take the common logarithm.}$$
$$3x = \log 61 \qquad \text{Inverse property: } \log 10^k = k$$
$$x = \frac{\log 61}{3} \qquad \text{Divide by 3.}$$
$$x \approx 0.595 \qquad \text{Approximate.}$$

Now Try Exercise 65 ◀

Converting $10^{3x} = 61$ to the equivalent equation $3x = \log 61$ is sometimes referred to as changing **exponential form** to **logarithmic form**. In Example 7, the common logarithm is taken of each side of the equation. We usually include this step to emphasize the fact that *inverse properties* are being used to solve these equations.

Common Logarithmic Equations A logarithmic equation contains logarithms. To solve logarithmic equations we *exponentiate* each side of the equation and then apply the inverse property $10^{\log x} = x$. Note that if $m = n$, then $10^m = 10^n$.

$$\log x = 2.5 \qquad \textit{Given logarithmic equation}$$
$$10^{\log x} = 10^{2.5} \qquad \textit{Exponentiate each side; base 10.}$$
$$x = 10^{2.5} \qquad \textit{Inverse property: } 10^{\log k} = k$$
$$x \approx 316.23 \qquad \textit{Approximate.}$$

Converting the equation $\log x = k$ to the equivalent equation $x = 10^k$ is called changing *logarithmic form* to *exponential form*.

EXAMPLE 8 Solving equations of the form $\log x = k$

Solve each equation.
(a) $\log x = 3$ **(b)** $\log x = -2$ **(c)** $\log x = 2.7$

SOLUTION
(a) $\log x = 3 \qquad \textit{Given equation}$
$$10^{\log x} = 10^3 \qquad \textit{Exponentiate each side; base 10.}$$
$$x = 10^3 \qquad \textit{Inverse property: } 10^{\log k} = k$$
$$x = 1000 \qquad \textit{Simplify.}$$

(b) Similarly, $\log x = -2$ is equivalent to $x = 10^{-2} = 0.01$.
(c) $\log x = 2.7$ is equivalent to $x = 10^{2.7} \approx 501.19$. *Now Try Exercise 77* ◄

EXAMPLE 9 Solving a common logarithmic equation

Solve $5 \log 2x = 16$.

SOLUTION Begin by dividing each side by 5.

$$5 \log 2x = 16 \qquad \textit{Given equation}$$
$$\log 2x = 3.2 \qquad \textit{Divide each side by 5.}$$
$$10^{\log 2x} = 10^{3.2} \qquad \textit{Exponentiate each side; base 10.}$$
$$2x = 10^{3.2} \qquad \textit{Inverse property: } 10^{\log k} = k$$
$$x = \frac{10^{3.2}}{2} \qquad \textit{Divide each side by 2.}$$
$$x \approx 792.45 \qquad \textit{Approximate.} \qquad \textit{Now Try Exercise 87} ◄$$

Modeling with Logarithms Some types of data grow slowly and can be modeled by $f(x) = a + b \log x$. For example, a larger area of land tends to have a wider variety of birds. However, if the land area doubles, the number of species of birds does not double; the land area has to more than double before the number of species doubles.

EXAMPLE 10 Modeling data with logarithms

Table 5.25 lists the number of species of tropical birds on islands of different sizes near New Guinea.
(a) Find values for a and b so that $f(x) = a + b \log x$ models the data.
(b) Determine symbolically and graphically the size of an island that might have 50 different species of birds.

Table 5.25

Area (mi²)	0.1	1	10	100	1000
Species of birds	26	39	52	65	78

Source: B. Freedman, *Environmental Ecology*.

SOLUTION

(a) Getting Started When modeling data by hand with $f(x) = a + b \log x$, one strategy is to let a equal $f(1)$. (Note that $\log 1 = 0$.) Then substitute a data point into this formula and find b. This technique is illustrated below. ▶

Begin by substituting $x = 1$ into the formula to determine a.

$$a + b \log 1 = 39 \qquad f(1) = 39$$
$$a + b(0) = 39 \qquad \log 1 = \log 10^0 = 0$$
$$a = 39 \qquad \text{Simplify.}$$

Figure 5.61

Thus $f(x) = 39 + b \log x$. Although we can use any data point in Table 5.25 to find b, we let $x = 10$.

$$39 + b \log 10 = 52 \qquad f(10) = 52$$
$$b \log 10 = 13 \qquad \text{Subtract 39.}$$
$$b = 13 \qquad \log 10 = 1$$

The data are modeled by $f(x) = 39 + 13 \log x$, a result verified in Figure 5.61.

(b) *Symbolic Solution* We must solve the equation $f(x) = 50$.

$$39 + 13 \log x = 50 \qquad f(x) = 50$$
$$13 \log x = 11 \qquad \text{Subtract 39.}$$
$$\log x = \frac{11}{13} \qquad \text{Divide by 13.}$$
$$10^{\log x} = 10^{11/13} \qquad \text{Exponentiate each side; base 10.}$$
$$x = 10^{11/13} \qquad \text{Inverse property: } 10^{\log k} = k$$
$$x \approx 7 \qquad \text{Approximate.}$$

To have 50 species of birds, an island should be about 7 square miles.

[0, 15, 5] by [0, 70, 10]

Figure 5.62

Graphical Solution The graphs of $Y_1 = 39 + 13 \log (X)$ and $Y_2 = 50$ in Figure 5.62 intersect near the point (7.02, 50). This result agrees with our symbolic solution.

Now Try Exercise 119 ◀

General Equations

Base-a Exponential Equations   The exponential function $f(x) = a^x$ is one-to-one and therefore has an inverse function $f^{-1}(x) = \log_a x$. Because $f^{-1}(f(x)) = x$ for all real numbers x, or equivalently $\log_a a^x = x$, it follows that to solve an exponential equation we can take the base-a logarithm of each side. Note that if $m = n$ (m and n positive), then $\log_a m = \log_a n$.

EXAMPLE 11   Solving exponential equations

Solve each equation.

(a) $3^x = \dfrac{1}{27}$ **(b)** $e^x = 5$ **(c)** $3(2^x) - 7 = 20$

SOLUTION

(a) $\log_3 3^x = \log_3 \dfrac{1}{27}$ Take the base-3 logarithm of each side.

$\log_3 3^x = \log_3 3^{-3}$ Properties of exponents

$x = -3$ Inverse property: $\log_a a^k = k$

(b) Take the natural logarithm of each side. Then

$$\ln e^x = \ln 5 \quad \text{is equivalent to} \quad x = \ln 5 \approx 1.609.$$

Many calculators are able to compute natural logarithms. The evaluation of $\ln 5$ is shown in Figure 5.63. Notice that $e^{1.609437912} \approx 5$.

(c) $3(2^x) - 7 = 20$ Given equation

$3(2^x) = 27$ Add 7 to each side.

$2^x = 9$ Divide each side by 3.

$\log_2 2^x = \log_2 9$ Take the base-2 logarithm of each side.

$x = \log_2 9$ Inverse property: $\log_a a^k = k$

Now Try Exercises 55 and 67 ◄

```
ln(5)
            1.609437912
e^(1.609437912)
            4.999999998
```

Figure 5.63

Introduction to the Change of Base Formula   Figure 5.64 gives a graphical solution of about 3.17 for Example 11(c). The exact solution of $\log_2 9$ can be approximated by using the following **change of base formula**. (Its derivation is given in Section 5.5.)

$$\log_a x = \frac{\log_b x}{\log_b a}$$

Because calculators typically have only $\log x$ and $\ln x$ keys, this formula is often written as

$$\log_a x = \frac{\log x}{\log a} \quad \text{or} \quad \log_a x = \frac{\ln x}{\ln a}.$$

For example, $\log_2 9 = \dfrac{\log 9}{\log 2} \approx 3.17$, which agrees with the graphical solution.

$[-5, 5, 1]$ by $[-10, 30, 10]$

```
y₂ = 20

y₁ = 3(2^x) − 7

Intersection
X=3.169925   Y=20
```

Figure 5.64

Base-a Logarithmic Equations   To solve the equation $\log_a x = k$, we exponentiate each side of the equation by using base a. That is, we use the inverse property: $a^{\log_a x} = x$ for all positive x. This is illustrated in the next example. Note that if $m = n$, then $a^m = a^n$.

EXAMPLE 12 Solving logarithmic equations

Solve each equation.

(a) $\log_2 x = 5$ **(b)** $\log_5 x = -2$ **(c)** $\ln x = 4.3$ **(d)** $3 \log_2 5x = 9$

SOLUTION

(a) $\log_2 x = 5$ Given equation

$\quad\quad 2^{\log_2 x} = 2^5$ Exponentiate each side; base 2.

$\quad\quad\quad x = 2^5$ Inverse property: $a^{\log_a k} = k$

$\quad\quad\quad x = 32$ Simplify.

(b) Similarly, $\log_5 x = -2$ is equivalent to $x = 5^{-2} = \frac{1}{25}$.

(c) $\ln x = 4.3$ is equivalent to $x = e^{4.3} \approx 73.7$.

(d) $3 \log_2 5x = 9$ Given equation

$\quad\quad \log_2 5x = 3$ Divide each side by 3.

$\quad\quad 2^{\log_2 5x} = 2^3$ Exponentiate each side; base 2.

$\quad\quad\quad\quad 5x = 8$ Inverse property: $a^{\log_a k} = k$

$\quad\quad\quad\quad x = \frac{8}{5}$ Divide each side by 5.   *Now Try Exercises 79 and 83* ◄

Exponential and Logarithmic Forms   We can solve $3^x = 7$ by taking the base-3 logarithm of each side.

$$3^x = 7 \quad\quad \text{Given equation}$$

$$\log_3 3^x = \log_3 7 \quad\quad \text{Take the base-3 logarithm of each side.}$$

$$x = \log_3 7 \quad\quad \text{Inverse property: } \log_a a^k = k$$

As they become more proficient, many people skip the second step. Converting $3^x = 7$ to $x = \log_3 7$ is called *changing exponential form to logarithmic form*. In a similar manner, we can solve $\log_3 x = 7$ by exponentiating each side base-3.

$$\log_3 x = 7 \quad\quad \text{Given equation}$$

$$3^{\log_3 x} = 3^7 \quad\quad \text{Exponentiate each side; base-3.}$$

$$x = 3^7 \quad\quad \text{Inverse property: } a^{\log_a k} = k$$

Converting $\log_3 x = 7$ to $x = 3^7$ is called *changing logarithmic form to exponential form*. Again, as they become more proficient, many people skip the second step by applying the definition of a logarithm for $a > 0, x > 0$, and $a \neq 1$:

$$\log_a x = k \quad \text{is equivalent to} \quad x = a^k.$$

EXAMPLE 13 Converting exponential and logarithmic forms

Complete each of the following.

(a) Change $5^{3x} = 8$ to logarithmic form.

(b) Change $\log_4 (4 - x) = 3$ to exponential form.

SOLUTION

(a) $5^{3x} = 8$ is equivalent to $3x = \log_5 8$.

(b) $\log_4 (4 - x) = 3$ is equivalent to $4 - x = 4^3$.   *Now Try Exercises 113 and 115* ◄

5.4 Putting It All Together

The following table summarizes some important concepts about base-a logarithms. Common and natural logarithms satisfy the same properties.

Concept	Explanation	Examples
Base-a logarithm $a > 0, a \neq 1$	The base-a logarithm of a positive number x is $$\log_a x = k \quad \text{if and only if} \quad x = a^k.$$ That is, a logarithm is an exponent k.	$\log 100 = \log 10^2 = 2$ $\log_2 8 = \log_2 2^3 = 3$ $\log_3 \sqrt[3]{3} = \log_3 3^{1/3} = \frac{1}{3}$ $\ln 5 \approx 1.609$ (using a calculator)
Graph of $y = \log_a x$	The graph of the base-a logarithmic function *always* passes through the point $(1, 0)$ because $\log_a 1 = \log_a a^0 = 0$. The y-axis is a vertical asymptote. The graph passes the horizontal line test, so $f(x) = \log_a x$ is one-to-one and has an inverse given by $f^{-1}(x) = a^x$.	
Inverse properties	a^x and $\log_a x$ represent inverse operations. That is, $$\log_a a^x = x$$ and $$a^{\log_a x} = x \quad \text{for } x > 0.$$	$\log 10^{-3} = -3$ $\log_4 4^6 = 6$ $10^{\log x} = x, x > 0$ $4^{\log_4 2x} = 2x, x > 0$ $e^{\ln 5} = 5$
Inverse functions	The inverse function of $f(x) = a^x$ is $$f^{-1}(x) = \log_a x.$$	$f(x) = 10^x \quad \leftrightarrow \quad f^{-1}(x) = \log x$ $g(x) = e^x \quad \leftrightarrow \quad g^{-1}(x) = \ln x$ $h(x) = \log_2 x \quad \leftrightarrow \quad h^{-1}(x) = 2^x$
Exponential equations	To solve $a^x = k$, take the base-a logarithm of each side.	$10^x = 15 \qquad\quad e^x = 20$ $\log 10^x = \log 15 \quad \ln e^x = \ln 20$ $x = \log 15 \qquad\quad x = \ln 20$
Logarithmic equations	To solve $\log_a x = k$, exponentiate each side; use base a.	$\log x = 3 \qquad\quad \ln x = 5$ $10^{\log x} = 10^3 \qquad e^{\ln x} = e^5$ $x = 1000 \qquad\quad x = e^5$
Exponential and logarithmic forms	$\log_a x = k$ is equivalent to $x = a^k$.	$\log_5 25 = 2$ is equivalent to $25 = 5^2$. $4^3 = 64$ is equivalent to $\log_4 64 = 3$.

5.4 Exercises

Common Logarithms

Exercises 1 and 2: Complete the table.

1.

x	10^0	10^4	10^{-8}	$10^{1.26}$
$\log x$		4		

2.

x	10^{-2}	$10^{-\pi}$	10^5	$10^{7.89}$
$\log x$			5	

Exercises 3–8: Evaluate each expression by hand, if possible.

3. (a) $\log(-3)$ (b) $\log \frac{1}{100}$

   (c) $\log \sqrt{0.1}$ (d) $\log 5^0$

4. (a) $\log 10{,}000$ (b) $\log(-\pi)$

   (c) $\log \sqrt{0.001}$ (d) $\log 8^0$

5. (a) $\log 10$ (b) $\log 10{,}000$

   (c) $20 \log 0.1$ (d) $\log 10 + \log 0.001$

6. (a) $\log 100$ (b) $\log 1{,}000{,}000$

   (c) $5 \log 0.01$ (d) $\log 0.1 - \log 1000$

7. (a) $2 \log 0.1 + 4$ (b) $\log 10^{1/2}$

   (c) $3 \log 100 - \log 1000$ (d) $\log(-10)$

8. (a) $\log(-4)$ (b) $\log 1$

   (c) $\log 0$ (d) $-6 \log 100$

Exercises 9 and 10: Determine mentally an integer n so that the logarithm is between n and n + 1. Check your result with a calculator.

9. (a) $\log 79$ (b) $\log 500$

   (c) $\log 5$ (d) $\log 0.5$

10. (a) $\log 63$ (b) $\log 5000$

    (c) $\log 9$ (d) $\log 0.04$

Exercises 11 and 12: Find the exact value of each expression.

11. (a) $\log \sqrt{1000}$ (b) $\log \sqrt[3]{10}$

    (c) $\log \sqrt[5]{0.1}$ (d) $\log \sqrt{0.01}$

12. (a) $\log \sqrt{100{,}000}$ (b) $\log \sqrt[3]{100}$

    (c) $2 \log \sqrt{0.1}$ (d) $10 \log \sqrt[3]{10}$

Domains of Logarithmic Functions

Exercises 13–20: Find the domain of f and write it in set-builder or interval notation.

13. $f(x) = \log(x + 3)$ 14. $f(x) = \ln(2x - 4)$

15. $f(x) = \log_2(x^2 - 1)$ 16. $f(x) = \log_4(4 - x^2)$

17. $f(x) = \log_3(4^x)$ 18. $f(x) = \log_5(5^x - 25)$

19. $f(x) = \ln(\sqrt{3 - x} - 1)$

20. $f(x) = \log(4 - \sqrt{2 - x})$

General Logarithms

Exercises 21–50: Simplify the expression.

21. $\log_8 8^{-5.7}$ 22. $\log_4 4^{-1.23}$

23. $7^{\log_7 2x}$ 24. $6^{\log_6(x+1)}$

25. $\log_{1/3}\left(\frac{1}{3}\right)^{64}$ 26. $\log_{0.4}\left(\frac{2}{5}\right)^{-3}$

27. $\ln e^{-4}$ 28. $2^{\log_2 k}$

29. $\log_5 5^{\pi}$ 30. $\log_6 6^9$

31. $3^{\log_3(x-1)}$ 32. $8^{\log_8(\pi+1)}$

33. $\log_2 64$ 34. $\log_2 \frac{1}{4}$

35. $\log_4 2$ 36. $\log_3 9$

37. $\ln e^{-3}$ 38. $\ln e$

39. $\log_8 64$ 40. $\ln \sqrt[3]{e}$

41. $\log_{1/2}\left(\frac{1}{4}\right)$ 42. $\log_{1/3}\left(\frac{1}{27}\right)$

43. $\log_{1/6} 36$ 44. $\log_{1/4} 64$

45. $\log_a \frac{1}{a}$ 46. $\log_a(a^2 \cdot a^3)$

47. $\log_5 5^0$ 48. $\ln \sqrt{e}$

49. $\log_2 \frac{1}{16}$ 50. $\log_8 8^k$

Exercises 51 and 52: Complete the table by hand.

51. $f(x) = 2 \log_2(x - 5)$ **52.** $f(x) = 2 \log_3(2x)$

x	6	7	21
$f(x)$			

x	$\frac{1}{18}$	$\frac{3}{2}$	$\frac{9}{2}$
$f(x)$			

Solving Exponential Equations

Exercises 53–76: Solve each equation. Use the change of base formula to approximate exact answers to the nearest hundredth when appropriate.

53. (a) $10^x = 0.01$ (b) $10^x = 7$ (c) $10^x = -4$

54. (a) $10^x = 1000$ (b) $10^x = 5$ (c) $10^x = -2$

55. (a) $4^x = \frac{1}{16}$ (b) $e^x = 2$ (c) $5^x = 125$

56. (a) $2^x = 9$ (b) $10^x = \frac{1}{1000}$ (c) $e^x = 8$

57. (a) $9^x = 1$ (b) $10^x = \sqrt{10}$ (c) $4^x = \sqrt[3]{4}$

58. (a) $2^x = \sqrt{8}$ (b) $7^x = 1$ (c) $e^x = \sqrt[3]{e}$

59. $e^{-x} = 3$

60. $e^{-x} = \frac{1}{2}$

61. $10^x - 5 = 95$

62. $2 \cdot 10^x = 66$

63. $10^{3x} = 100$

64. $4 \cdot 10^{2x} + 1 = 21$

65. $5(10^{4x}) = 65$

66. $3(10^{x-2}) = 72$

67. $4(3^x) - 3 = 13$

68. $5(7^x) + 3 = 83$

69. $e^x + 1 = 24$

70. $1 - 2e^x = -5$

71. $2^x + 1 = 15$

72. $3 \cdot 5^x = 125$

73. $5e^x + 2 = 20$

74. $6 - 2e^{3x} = -10$

75. $8 - 3(2)^{0.5x} = -40$

76. $2(3)^{-2x} + 5 = 167$

Solving Logarithmic Equations

Exercises 77–98: Solve each equation. Approximate answers to four decimal places when appropriate.

77. (a) $\log x = 2$ (b) $\log x = -3$ (c) $\log x = 1.2$

78. (a) $\log x = 1$ (b) $\log x = -4$ (c) $\log x = 0.3$

79. (a) $\log_2 x = 6$ (b) $\log_3 x = -2$ (c) $\ln x = 2$

80. (a) $\log_4 x = 2$ (b) $\log_8 x = -1$ (c) $\ln x = -2$

81. $\log_2 x = 1.2$

82. $\log_4 x = 3.7$

83. $5 \log_7 2x = 10$

84. $2 \log_4 x = 3.4$

85. $2 \log x = 6$

86. $\log 4x = 2$

87. $2 \log 5x = 4$

88. $6 - \log x = 3$

89. $4 \ln x = 3$

90. $\ln 5x = 8$

91. $5 \ln x - 1 = 6$

92. $2 \ln 3x = 8$

93. $4 \log_2 x = 16$

94. $\log_3 5x = 10$

95. $5 \ln(2x) + 6 = 12$

96. $16 - 4 \ln 3x = 2$

97. $9 - 3 \log_4 2x = 3$

98. $7 \log_6(4x) + 5 = -2$

Exercises 99 and 100: Find values for a and b so that f(x) models the data exactly.

99. $f(x) = a + b \log x$ **100.** $f(x) = a + b \log_2 x$

x	1	10	100
y	5	7	9

x	1	2	4
y	3.1	6	8.9

Graphs of Logarithmic Functions

Exercises 101–104: Use the graph of f to sketch a graph of f^{-1}. Give a symbolic representation of f^{-1}.

101.

102.

103.

104.

Exercises 105–108: Graph f and state its domain.

105. $f(x) = \log(x + 1)$ **106.** $f(x) = \log(x - 3)$

107. $f(x) = \ln(-x)$ 📱 **108.** $f(x) = \ln(x^2 + 1)$

Exercises 109 and 110: Graph y = f(x). Is f increasing or decreasing on its domain?

109. $f(x) = \log_{1/2} x$ **110.** $f(x) = \log_{1/3} x$

Exercises 111 and 112: Complete the following.

    (a) Graph $y = f(x)$, $y = f^{-1}(x)$, and $y = x$.
    (b) Determine the intervals where f and f^{-1} are increasing or decreasing.

111. $f(x) = \log_3 x$ **112.** $f(x) = \log_{1/2} x$

Exponential and Logarithmic Forms

Exercises 113 and 114: Change each equation to its equivalent logarithmic form.

113. (a) $7^{4x} = 4$ (b) $e^x = 7$ (c) $c^x = b$

114. (a) $5^{2x} = 9$ (b) $b^x = a$ (c) $d^{2x} = b$

Exercises 115 and 116: Change each equation to its equivalent exponential form.

115. (a) $\log_8 x = 3$ (b) $\log_9(2 + x) = 5$ (c) $\log_k b = c$

116. (a) $\log x = 4$ (b) $\ln 8x = 7$ (c) $\log_a x = b$

Applications

117. *Decibels* (Refer to Example 2.) Use the formula $D(x) = 10 \log(10^{16}x)$ to determine the decibels when the intensity of a sound is $x = 10^{-11/2}$ watt per square centimeter.

118. *Decibels* (Refer to Example 2.) If the decibels of a sound increased by 15, by what factor did the intensity of the sound increase?

119. *Diversity of Birds* (Refer to Example 10.) The table lists the number of species of birds on islands of various sizes. Find values for a and b so that $f(x) = a + b \log x$ models these data. Estimate the size of an island that might have 16 species of birds.

Area (km²)	0.1	1	10	100	1000
Species of birds	3	7	11	15	19

120. *Diversity of Insects* The table lists the number of types of insects found in wooded regions with various acreages. Find values for a and b so that $f(x) = a + b \log x$ models these data. Then use f to estimate an acreage that might have 1200 types of insects.

Area (acres)	10	100	1000	10,000	100,000
Insect Types	500	800	1100	1400	1700

121. *Growth of Bacteria* The table lists the number of bacteria y in millions after an elapsed time of x days.

x	0	1	2	3	4
y	3	6	12	24	48

    (a) Find values for C and a so that $f(x) = Ca^x$ models the data.

    (b) Estimate when there were 16 million bacteria.

122. *Growth of an Investment* The growth of an investment is shown in the table.

x (years)	0	5	10	15	20
y (dollars)	100	300	900	2700	8100

    (a) Find values for C and a so that $f(x) = Ca^x$ models the data.

    (b) Estimate when the account contained $2000.

123. *Runway Length* There is a relation between an airplane's weight x and the runway length L required for takeoff. For some airplanes the minimum runway length L in thousands of feet may be modeled by $L(x) = 3 \log x$, where x is measured in thousands of pounds. (**Source:** L. Haefner, *Introduction to Transportation Systems.*)

    (a) Graph L for $0 < x \le 50$. Interpret the graph.

    (b) If the weight of an airplane increases tenfold from 10,000 to 100,000 pounds, does the length of the required runway also increase by a factor of 10? Explain.

    (c) Generalize your answer from part (b).

124. *Runway Length* (Refer to Exercise 123.) Estimate the maximum weight of a plane that can take off from a runway that is 5 thousand feet long.

125. *Acid Rain* Air pollutants frequently cause acid rain. A measure of the acidity is pH, which ranges between 1 and 14. Pure water is neutral and has a pH of 7. Acidic solutions have a pH less than 7, whereas alkaline solutions have a pH greater than 7. A pH value can be computed by pH $= -\log x$, where x represents the hydrogen ion concentration in moles per liter. In rural areas of Europe, rainwater typically has $x = 10^{-4.7}$. (**Source:** G. Howells, *Acid Rain and Acid Water*.)
   (a) Find its pH.

   (b) Seawater has a pH of 8.2. How many times greater is the hydrogen ion concentration in rainwater from rural Europe than in seawater?

126. *Acid Rain* (Refer to Exercise 125.) Find the hydrogen ion concentration for the following pH levels of acid rain. (**Source:** G. Howells.)
   (a) 4.92 (pH of rain at Amsterdam Islands)

   (b) 3.9 (pH of some rain in the eastern United States)

127. *Earthquakes* The Richter scale is used to measure the intensity of earthquakes, where intensity corresponds to the amount of energy released by an earthquake. If an earthquake has an intensity of x, then its *magnitude*, as computed by the Richter scale, is given by the formula $R(x) = \log \frac{x}{I_0}$, where I_0 is the intensity of a small measurable earthquake.
   (a) On July 26, 1963, an earthquake in Yugoslavia had a magnitude of 6.0 on the Richter scale, and on August 19, 1977, an earthquake in Indonesia measured 8.0. Find the intensity x for each of these earthquakes if $I_0 = 1$.

   (b) How many times more intense was the Indonesian earthquake than the Yugoslavian earthquake?

128. *Earthquakes* (Refer to Exercise 127.) If the intensity x of an earthquake increases by a *factor* of 10^3, by how much does the Richter number R increase? Generalize your results.

129. *Hurricanes* Hurricanes are some of the largest storms on earth. They are very low pressure areas with diameters of over 500 miles. The barometric air pressure in inches of mercury at a distance of x miles from the eye of a severe hurricane is modeled by the formula $f(x) = 0.48 \ln (x + 1) + 27$. (**Source:** A. Miller and R. Anthes, *Meteorology*.)
   (a) Evaluate $f(0)$ and $f(100)$. Interpret the results.

   (b) Graph f in [0, 250, 50] by [25, 30, 1]. Describe how air pressure changes as one moves away from the eye of the hurricane.

   (c) At what distance from the eye of the hurricane is the air pressure 28 inches of mercury?

130. *Predicting Wind Speed* Wind speed typically varies in the first 20 meters above the ground. Close to the ground, wind speed is often less than it is at 20 meters above the ground. For this reason, the National Weather Service usually measures wind speeds at heights between 5 and 10 meters. For a particular day, let the formula $f(x) = 1.2 \ln x + 2.3$ compute the wind speed in meters per second at a height x meters above the ground for $x \geq 1$. (**Source:** A. Miller.)

   (a) Find the wind speed at a height of 5 meters.

   (b) Graph f in the window [0, 20, 5] by [0, 7, 1]. Interpret the graph.

   (c) Estimate the height where the wind speed is 5 meters per second.

131. *Cooling an Object* A pot of boiling water with a temperature of 100°C is set in a room with a temperature of 20°C. The temperature T of the water after x hours is given by $T(x) = 20 + 80e^{-x}$.
   (a) Estimate the temperature of the water after 1 hour.

   (b) How long did it take the water to cool to 60°C?

132. *Warming an Object* A can of soda with a temperature of 5°C is set in a room with a temperature of 20°C. The temperature T of the soda after x minutes is given by $T(x) = 20 - 15(10)^{-0.05x}$.
  (a) Estimate the temperature of the soda after 5 minutes.

  (b) After how many minutes was the temperature of the soda 15°C?

133. *Traffic Flow* (Refer to Example 10, Section 5.3.) Cars arrive randomly at an intersection with an average rate of 20 cars per hour. The likelihood, or probability, that no car enters the intersection within a period of x minutes can be estimated by $f(x) = e^{-x/3}$.
  (a) What is the probability that no car enters the intersection during a 5-minute period?

  (b) Determine the value of x that gives a 30% chance that no car enters the intersection during an interval of x minutes.

134. *Population Growth* The population of Tennessee in millions is given by $P(x) = 4.88e^{0.0133x}$, where $x = 0$ corresponds to 1990. (**Source:** Bureau of the Census.)
  (a) Determine symbolically the year when the population of Tennessee was 5.4 million.

  📅 **(b)** Solve part (a) graphically.

Writing about Mathematics

135. Describe the relationship among exponential functions and logarithmic functions. Explain why logarithms are needed to solve exponential equations.

136. Give verbal, numerical, graphical, and symbolic representations of a base-5 logarithmic function.

EXTENDED AND DISCOVERY EXERCISES

1. *Average Rate of Change of* $\ln x$ Find the average rate of change of $f(x) = \ln x$ from x to $x + 0.001$ for each value of x. Round your answers to two decimal places.
  (a) $x = 1$ **(b)** $x = 2$

  (c) $x = 3$ **(d)** $x = 4$

2. *Average Rate of Change of* $\ln x$ (Refer to Exercise 1.) Compare each average rate of change of $\ln x$ to x. What is the pattern? Make a generalization.

3. *Modeling Algae Growth* When sewage was accidentally dumped into Lake Tahoe, the concentration of the algae *Selenastrum* increased from 1000 cells per milliliter to approximately 1,000,000 cells per milliliter within 6 days. Let f model the algae concentration after x days have elapsed. (**Source:** A. Payne, "Responses of the three test algae of the algal assay procedure: bottle test.")
  (a) Compute the average rate of change in f from 0 to 6 days.

  (b) The *specific growth rate r* is defined by the formula $r = \dfrac{\log N_2 - \log N_1}{x_2 - x_1}$, where N_1 is the algae concentration at time x_1 and N_2 is the algae concentration at time x_2. Compute r in this example.

  (c) Discuss why environmental scientists might use the specific growth rate r, rather than the average rate of change, to describe algae growth.

4. *Greenhouse Effect* According to one model, the future increases in average global temperatures (due to carbon dioxide levels exceeding 280 parts per million) can be estimated using $T = 6.5 \ln (C/280)$, where C is the concentration of atmospheric carbon dioxide in parts per million (ppm) and T is in degrees Fahrenheit. Let future amounts of carbon dioxide be modeled by the formula $C(x) = 364(1.005)^x$, where $x = 0$ corresponds to the year 2000, $x = 100$ to 2100, and so on. (**Source:** W. Clime, *The Economics of Global Warming.*)
  (a) Use composition of functions to write T as a function of x. Evaluate T when $x = 100$ and interpret the result.

  📅 **(b)** Graph $C(x)$ in [0, 200, 50] by [0, 1000, 100] and $T(x)$ in [0, 200, 50] by [0, 10, 1]. Describe each graph.

  (c) How does an exponential growth in carbon dioxide concentrations affect the increase in global temperature?

CHECKING BASIC CONCEPTS FOR SECTIONS 5.3 AND 5.4

1. If the principal is $1200 and the interest rate is 9.5% compounded monthly, calculate the account balance after 4 years. Determine the balance if the interest is compounded continuously.

2. Find values for C and a so that $f(x) = Ca^x$ models the data in the table.

x	0	1	2	3
y	4	2	1	0.5

3. Explain verbally what $\log_2 15$ represents. Is it equal to an integer? (Do not use a calculator.)

4. Evaluate each of the following logarithms by hand.
   (a) $\log_6 36$ (b) $\log\sqrt{10} + \log 0.01$ (c) $\ln\dfrac{1}{e^2}$

5. Solve each equation.
   (a) $e^x = 5$ (b) $10^x = 25$ (c) $\log x = 1.5$

6. Solve each equation.
   (a) $2e^x + 1 = 25$ (b) $\log 2x = 2.3$

   (c) $\log x^2 = 1$

7. *Population*   In July 1994, the population of New York state in millions was modeled by $f(x) = 18.2e^{0.001x}$, and the population of Florida in millions was modeled by $g(x) = 14e^{0.0168x}$. In both formulas x is the year, where $x = 0$ corresponds to July 1994.   (**Source:** Bureau of the Census.)

(a) Find the population of each state in July 1994.

(b) Assuming these trends continue, estimate the year when the population of Florida will equal the population of New York. What will their populations be at this time?

8. *Growth in Salary*   Suppose that a person's salary is initially \$30,000 and is modeled by $f(x)$, where x represents the number of years of experience. Use $f(x)$ to approximate the years of experience when the salary first exceeds \$60,000.
   (a) $f(x) = 30{,}000(1.1)^x$

   (b) $f(x) = 30{,}000 \log(10 + x)$

Would most people prefer that their salaries increase exponentially or logarithmically?

5.5 Properties of Logarithms

- **Apply basic properties of logarithms**
- **Expand and combine logarithmic expressions**
- **Use the change of base formula**

Introduction

The discovery of logarithms by John Napier (1550–1617) played an important role in the history of science. Logarithms were instrumental in allowing Johannes Kepler (1571–1630) to calculate the positions of the planet Mars, which led to his discovery of the laws of planetary motion. Kepler's laws were used by Isaac Newton (1643–1727) to discover the universal laws of gravitation. Although calculators and computers have made tables of logarithms obsolete, applications involving logarithms still play an important role in modern-day computation. One reason is that logarithms possess several important properties. For example, the loudness of a sound can be measured in decibels by the formula $f(x) = 10 \log(10^{16}x)$, where x is the intensity of the sound in watts per square centimeter. In Example 4, we use properties of logarithms to simplify this formula.

Basic Properties of Logarithms

Logarithms possess several important properties. One property of logarithms states that the sum of the logarithms of two numbers equals the logarithm of their product. For example, we see in Figure 5.65 that

$$\log 5 + \log 2 = \log 10 \qquad 5 \cdot 2 = 10$$

and in Figure 5.66 that

$$\log 4 + \log 25 = \log 100. \qquad 4 \cdot 25 = 100$$

These calculations illustrate a basic property of logarithms: $\log_a m + \log_a n = \log_a(mn)$.

$$\boxed{\begin{array}{ll} \log(5)+\log(2) & \\ & 1 \\ \log(10) & \\ & 1 \end{array}}$$

Figure 5.65

$$\boxed{\begin{array}{ll} \log(4)+\log(25) & \\ & 2 \\ \log(100) & \\ & 2 \end{array}}$$

Figure 5.66

Four properties of logarithms are as follows.

Properties of Logarithms

For positive numbers m, n, and $a \neq 1$ and any real number r:

1. $\log_a 1 = 0$ and $\log_a a = 1$
2. $\log_a m + \log_a n = \log_a (mn)$
3. $\log_a m - \log_a n = \log_a \left(\dfrac{m}{n} \right)$
4. $\log_a (m^r) = r \log_a m$

Algebra Review

To review properties of exponents, see Section 4.8 and Chapter R (page R-7).

The properties of logarithms are a direct result of the properties of exponents and the inverse property $\log_a a^k = k$, as shown below.

Property 1:   This property is a direct result of the inverse property: $\log_a a^x = x$.

$$\log_a 1 = \log_a a^0 = 0 \quad \text{and} \quad \log_a a = \log_a a^1 = 1$$

Examples: $\log 1 = 0$ and $\ln e = 1$

Property 2:   If m and n are positive numbers, then we can write $m = a^c$ and $n = a^d$ for some real numbers c and d.

$$\log_a m + \log_a n = \log_a a^c + \log_a a^d = c + d$$
$$\log_a (mn) = \log_a (a^c a^d) = \log_a (a^{c+d}) = c + d$$

Thus $\log_a m + \log_a n = \log_a (mn)$.

Example: Let $m = 100$ and $n = 1000$.

$$\log m + \log n = \log 100 + \log 1000 = \log 10^2 + \log 10^3 = 2 + 3 = 5$$
$$\log (mn) = \log (100 \cdot 1000) = \log 100{,}000 = \log 10^5 = 5$$

Property 3:   Let $m = a^c$ and $n = a^d$ for some real numbers c and d.

$$\log_a m - \log_a n = \log_a a^c - \log_a a^d = c - d$$
$$\log_a \left(\frac{m}{n} \right) = \log_a \left(\frac{a^c}{a^d} \right) = \log_a (a^{c-d}) = c - d$$

Thus $\log_a m - \log_a n = \log_a \left(\frac{m}{n} \right)$.

Example: Let $m = 100$ and $n = 1000$.

$$\log m - \log n = \log 100 - \log 1000 = \log 10^2 - \log 10^3 = 2 - 3 = -1$$
$$\log \left(\frac{m}{n} \right) = \log \left(\frac{100}{1000} \right) = \log \left(\frac{1}{10} \right) = \log (10^{-1}) = -1$$

Property 4:   Let $m = a^c$ and r be any real number.

$$\log_a m^r = \log_a (a^c)^r = \log_a (a^{cr}) = cr$$
$$r \log_a m = r \log_a a^c = rc$$

Thus $\log_a (m^r) = r \log_a m$.

Example: Let $m = 100$ and $r = 3$.

$$\log m^r = \log 100^3 = \log 1{,}000{,}000 = \log 10^6 = 6$$
$$r \log m = 3 \log 100 = 3 \log 10^2 = 3 \cdot 2 = 6$$

Caution: $\log_a (m + n) \neq \log_a m + \log_a n$; $\log_a (m - n) \neq \log_a m - \log_a n$

EXAMPLE 1   Recognizing properties of logarithms

Use a calculator to evaluate each pair of expressions. Then state which property of logarithms this calculation illustrates.

(a) $\ln 5 + \ln 4, \ln 20$ **(b)** $\log 10 - \log 5, \log 2$ **(c)** $\log 5^2, 2 \log 5$

SOLUTION

(a) From Figure 5.67, we see that the two expressions are equal. These calculations illustrate Property 2 because $\ln 5 + \ln 4 = \ln (5 \cdot 4) = \ln 20$.

(b) The two expressions are equal in Figure 5.68, and these calculations illustrate Property 3 because $\log 10 - \log 5 = \log \frac{10}{5} = \log 2$.

(c) The two expressions are equal in Figure 5.69, and these calculations illustrate Property 4 because $\log 5^2 = 2 \log 5$.

```
ln(5)+ln(4)
           2.995732274
ln(20)
           2.995732274
```

```
log(10)−log(5)
           .3010299957
log(2)
           .3010299957
```

```
log(5²)
           1.397940009
2log(5)
           1.397940009
```

Figure 5.67   **Figure 5.68**   **Figure 5.69**

Now Try Exercises 1, 3, and 5 ◀

Expanding and Combining Logarithmic Expressions

The next two examples demonstrate how properties of logarithms can be used to expand logarithmic expressions.

EXAMPLE 2   Expanding logarithmic expressions

Use properties of logarithms to expand each expression. Write your answers without exponents.

(a) $\log xy$ **(b)** $\ln \dfrac{5}{z}$ **(c)** $\log_4 \dfrac{\sqrt[3]{x}}{\sqrt{k}}$

SOLUTION

(a) By Property 2, $\log xy = \log x + \log y$.

(b) By Property 3, $\ln \frac{5}{z} = \ln 5 - \ln z$.

(c) Begin by using Property 3.

$$\log_4 \frac{\sqrt[3]{x}}{\sqrt{k}} = \log_4 \sqrt[3]{x} - \log_4 \sqrt{k} \qquad \text{Property 3}$$

$$= \log_4 x^{1/3} - \log_4 k^{1/2} \qquad \text{Properties of exponents}$$

$$= \tfrac{1}{3} \log_4 x - \tfrac{1}{2} \log_4 k \qquad \text{Property 4}$$

Now Try Exercises 7, 11, and 21 ◀

EXAMPLE 3 Applying properties of logarithms

Expand each expression. Write your answers without exponents.

(a) $\log_2 2x^4$ (b) $\ln \frac{7x^3}{k^2}$ (c) $\log \frac{\sqrt{x+1}}{(x-2)^3}$

SOLUTION

(a) $\log_2 2x^4 = \log_2 2 + \log_2 x^4 \qquad$ Property 2

$\qquad\quad = 1 + 4 \log_2 x \qquad\qquad$ Properties 1 and 4

(b) $\ln \frac{7x^3}{k^2} = \ln 7x^3 - \ln k^2 \qquad$ Property 3

$\qquad\quad = \ln 7 + \ln x^3 - \ln k^2 \qquad$ Property 2

$\qquad\quad = \ln 7 + 3 \ln x - 2 \ln k \qquad$ Property 4

(c) $\log \frac{\sqrt{x+1}}{(x-2)^3} = \log \sqrt{x+1} - \log (x-2)^3 \qquad$ Property 3

$\qquad\quad = \log (x+1)^{1/2} - \log (x-2)^3 \qquad$ Properties of exponents

$\qquad\quad = \tfrac{1}{2} \log (x+1) - 3 \log (x-2) \qquad$ Property 4

Now Try Exercises 9, 15, and 27 ◀

An Application   Sometimes properties of logarithms are used in applications to simplify a formula. This is illustrated in the next example.

EXAMPLE 4 Analyzing sound with decibels

Sound levels in decibels (db) can be computed by $D(x) = 10 \log (10^{16}x)$.

(a) Use properties of logarithms to simplify the formula for D.

(b) Ordinary conversation has an intensity of $x = 10^{-10}$ w/cm^2. Find the decibel level.

SOLUTION

(a) To simplify the formula, use Property 2.

$$D(x) = 10 \log (10^{16}x)$$

$$= 10(\log 10^{16} + \log x) \qquad \text{Property 2}$$

$$= 10(16 + \log x) \qquad\qquad \text{Evaluate the first logarithm.}$$

$$= 160 + 10 \log x \qquad\qquad \text{Distributive property}$$

(b) $D(10^{-10}) = 160 + 10 \log (10^{-10}) = 160 + 10(-10) = 160 - 100 = 60$

Ordinary conversation occurs at about 60 decibels.   *Now Try Exercise 83* ◀

The next two examples demonstrate how properties of logarithms can be used to combine logarithmic expressions.

EXAMPLE 5 Applying properties of logarithms

Write each expression as the logarithm of a single expression.

(a) $\ln 2e + \ln \dfrac{1}{e}$ **(b)** $\log_2 27 + \log_2 x^3$ **(c)** $\log x^3 - \log x^2$

SOLUTION

(a) By Property 2, $\ln 2e + \ln \frac{1}{e} = \ln \left(2e \cdot \frac{1}{e}\right) = \ln 2$.

(b) By Property 2, $\log_2 27 + \log_2 x^3 = \log_2 (27x^3)$.

(c) By Property 3, $\log x^3 - \log x^2 = \log \frac{x^3}{x^2} = \log x$.

Now Try Exercises 41, 43, and 49 ◄

EXAMPLE 6 Combining terms in logarithmic expressions

Write each expression as the logarithm of a single expression.

(a) $\log 5 + \log 15 - \log 3$ **(b)** $2 \ln x - \frac{1}{2} \ln y - 3 \ln z$

(c) $5 \log_3 x + \log_3 2x - \log_3 y$

SOLUTION

(a) $\log 5 + \log 15 - \log 3 = \log (5 \cdot 15) - \log 3$ Property 2

$$= \log \left(\frac{5 \cdot 15}{3}\right)$$ Property 3

$$= \log 25$$ Simplify.

Algebra Review
To review rational exponents and radical notation, see Chapter R (page R-41).

(b) $2 \ln x - \dfrac{1}{2} \ln y - 3 \ln z = \ln x^2 - \ln y^{1/2} - \ln z^3$ Property 4

$$= \ln \left(\frac{x^2}{y^{1/2}}\right) - \ln z^3$$ Property 3

$$= \ln \frac{x^2}{y^{1/2}z^3}$$ Property 3

$$= \ln \frac{x^2}{z^3 \sqrt{y}}$$ Properties of exponents

(c) $5 \log_3 x + \log_3 2x - \log_3 y = \log_3 x^5 + \log_3 2x - \log_3 y$ Property 4

$$= \log_3 (x^5 \cdot 2x) - \log_3 y$$ Property 2

$$= \log_3 \frac{2x^6}{y}$$ Property 3

Now Try Exercises 39, 47, and 51 ◄

Change of Base Formula

Occasionally it is necessary to evaluate a logarithm with a base other than 10 or e. This computation can be accomplished by using a change of base formula.

Change of Base Formula

Let x, $a \neq 1$, and $b \neq 1$ be positive real numbers. Then

$$\log_a x = \frac{\log_b x}{\log_b a}.$$

The change of base formula can be derived as follows.

$$y = \log_a x$$
$$a^y = a^{\log_a x} \qquad \text{Exponentiate each side; base } a.$$
$$a^y = x \qquad \text{Inverse property}$$
$$\log_b a^y = \log_b x \qquad \text{Take base-}b \text{ logarithm of each side.}$$
$$y \log_b a = \log_b x \qquad \text{Property 4}$$
$$y = \frac{\log_b x}{\log_b a} \qquad \text{Divide by } \log_b a.$$
$$\log_a x = \frac{\log_b x}{\log_b a} \qquad \text{Substitute } \log_a x \text{ for } y. \text{ (First equation)}$$

To calculate $\log_2 5$, evaluate $\frac{\log 5}{\log 2} \approx 2.322$. The change of base formula was used with $x = 5$, $a = 2$, and $b = 10$. We could also have evaluated $\frac{\ln 5}{\ln 2} \approx 2.322$.

EXAMPLE 7 Applying the change of base formula

Use a calculator to approximate each expression to the nearest thousandth.
(a) $\log_4 20$ **(b)** $\log_2 125 + \log_7 39$

SOLUTION
(a) Using the change of base formula, we have $\log_4 20 = \frac{\log 20}{\log 4} \approx 2.161$. We could also evaluate $\frac{\ln 20}{\ln 4}$ to obtain the same result, as shown in Figure 5.70.
(b) $\log_2 125 + \log_7 39 = \frac{\log 125}{\log 2} + \frac{\log 39}{\log 7} \approx 8.848$. See Figure 5.71.

```
log(20)/log(4)
          2.160964047
ln(20)/ln(4)
          2.160964047
```

```
log(125)/log(2)+
log(39)/log(7)
          8.848482542
```

Figure 5.70   **Figure 5.71**   *Now Try Exercises 67 and 71* ◀

The change of base formula can be used to graph base-a logarithmic functions.

EXAMPLE 8 Using the change of base formula

Solve the equation $\log_2 (x^3 + x - 1) = 5$ graphically.

SOLUTION Graph $Y_1 = \log (X^3 + X - 1)/\log (2)$ and $Y_2 = 5$. See Figure 5.72. Their graphs intersect near the point $(3.104, 5)$. The solution is given by $x \approx 3.104$.

Now Try Exercise 77 ◀

$[-10, 10, 1]$ by $[-10, 10, 1]$

Figure 5.72

5.5 Putting It All Together

The following table summarizes some properties of logarithms.

Concept	Explanation	Examples
Properties of logarithms	1. $\log_a 1 = 0$ and $\log_a a = 1$ 2. $\log_a m + \log_a n = \log_a (mn)$ 3. $\log_a m - \log_a n = \log_a\left(\frac{m}{n}\right)$ 4. $\log_a (m^r) = r \log_a m$	1. $\ln 1 = 0$ and $\log_2 2 = 1$ 2. $\log 3 + \log 6 = \log (3 \cdot 6) = \log 18$ 3. $\log_3 8 - \log_3 2 = \log_3 \frac{8}{2} = \log_3 4$ 4. $\log 6^7 = 7 \log 6$
Change of base formula	Let x, $a \neq 1$, and $b \neq 1$ be positive real numbers. Then $$\log_a x = \frac{\log_b x}{\log_b a}.$$	$\log_3 6 = \dfrac{\log 6}{\log 3} = \dfrac{\ln 6}{\ln 3} \approx 1.631$
Graphing logarithmic functions	Use the change of base formula to graph $y = \log_a x$ whenever $a \neq 10$ and $a \neq e$.	To graph $y = \log_2 x$, let $Y_1 = \log (X)/\log (2)$ or $Y_1 = \ln (X)/\ln (2)$.

5.5 Exercises

Note: When applying properties of logarithms, assume that all variables are positive.

Properties of Logarithms

Exercises 1–6: (Refer to Example 1.) Use a calculator to approximate each pair of expressions. Then state which property of logarithms this calculation illustrates.

1. $\log 4 + \log 7$, $\log 28$ **2.** $\ln 12 + \ln 5$, $\ln 60$

3. $\ln 72 - \ln 8$, $\ln 9$ **4.** $3 \log 4$, $\log 4^3$

5. $10 \log 2$, $\log 1024$

6. $\log_2 100 - \log_2 20$, $\log_2 5$

Exercises 7–32: (Refer to Examples 2 and 3.) Expand the expression. If possible, write your answer without exponents.

7. $\log_2 ab$ **8.** $\ln 3x$

9. $\ln 7a^4$ **10.** $\log \dfrac{a^3}{3}$

11. $\log \dfrac{6}{z}$ **12.** $\ln \dfrac{xy}{z}$

13. $\log \dfrac{x^2}{3}$ **14.** $\log 3x^6$

15. $\ln \dfrac{2x^7}{3k}$

16. $\ln \dfrac{kx^3}{5}$

17. $\log_2 4k^2x^3$

18. $\log \dfrac{5kx^2}{11}$

19. $\log_5 \dfrac{25x^3}{y^4}$

20. $\log_2 \dfrac{32}{xy^2}$

21. $\ln \dfrac{x^4}{y^2\sqrt{z^3}}$

22. $\ln \dfrac{x\sqrt[3]{y^2}}{z^6}$

23. $\log_4 (0.25(x + 2)^3)$

24. $\log (0.001(a - b)^{-3})$

25. $\log_5 \dfrac{x^3}{(x - 4)^4}$

26. $\log_8 \dfrac{(3x - 2)^2}{x^2 + 1}$

27. $\log_2 \dfrac{\sqrt{x}}{z^2}$

28. $\log \sqrt{\dfrac{xy^2}{z}}$

29. $\ln \sqrt[3]{\dfrac{2x + 6}{(x + 1)^5}}$

30. $\log \dfrac{\sqrt{x^2 + 4}}{\sqrt[3]{x - 1}}$

31. $\log_2 \dfrac{\sqrt[3]{x^2 - 1}}{\sqrt{1 + x^2}}$

32. $\log_8 \sqrt[3]{\dfrac{x + y^2}{2z + 1}}$

Exercises 33–56: (Refer to Examples 5 and 6.) Write the expression as a logarithm of a single expression.

33. $\log 2 + \log 3$

34. $\log \sqrt{2} + \log \sqrt[3]{2}$

35. $\ln \sqrt{5} - \ln 25$

36. $\ln 33 - \ln 11$

37. $\log 20 + \log \frac{1}{10}$

38. $\log_2 24 + \log_2 \frac{1}{48}$

39. $\log 4 + \log 3 - \log 2$

40. $\log_3 5 - \log_3 10 - \log_3 \frac{1}{2}$

41. $\log_7 5 + \log_7 k^2$

42. $\log_6 45 + \log_6 b^3$

43. $\ln x^6 - \ln x^3$

44. $\log 10x^5 - \log 5x$

45. $\log \sqrt{x} + \log x^2 - \log x$

46. $\log \sqrt[4]{x} + \log x^4 - \log x^2$

47. $3 \ln x - \frac{3}{2} \ln y + 4 \ln z$

48. $\frac{2}{3} \ln y - 4 \ln x - \frac{1}{2} \ln z$

49. $\ln \dfrac{1}{e^2} + \ln 2e$

50. $\ln 4e^3 - \ln 2e^2$

51. $2 \ln x - 4 \ln y + \frac{1}{2} \ln z$

52. $\frac{1}{3} \log_5 (x + 1) + \frac{1}{3} \log_5 (x - 1)$

53. $\log 4 - \log x + 7 \log \sqrt{x}$

54. $\ln 3e - \ln \dfrac{1}{4e}$

55. $2 \log (x^2 - 1) + 4 \log (x - 2) - \frac{1}{2} \log y$

56. $\log_3 x + \log_3 \sqrt{x + 3} - \frac{1}{3} \log_3 (x - 4)$

Exercises 57–62: Complete the following.

(a) Use a table of $f(x)$ and $g(x)$ to determine whether $f(x) = g(x)$.

(b) If possible, use properties of logarithms to show that $f(x) = g(x)$.

57. $f(x) = \log 3x + \log 2x$, $g(x) = \log 6x^2$

58. $f(x) = \ln 3x - \ln 2x$, $g(x) = \ln x$

59. $f(x) = \ln 2x^2 - \ln x$, $g(x) = \ln 2x$

60. $f(x) = \log x^2 + \log x^3$, $g(x) = 5 \log x$

61. $f(x) = \ln x^4 - \ln x^2$, $g(x) = 2 \ln x$

62. $f(x) = (\ln x)^2$, $g(x) = 2 \ln x$

Exercises 63–66: Sketch a graph of f.

63. $f(x) = \log_2 x$

64. $f(x) = \log_2 x^2$

65. $f(x) = \log_3 |x|$

66. $f(x) = \log_4 2x$

Change of Base Formula

Exercises 67–76: Use the change of base formula to approximate the logarithm to the nearest thousandth.

67. $\log_2 25$

68. $\log_3 67$

69. $\log_5 130$

70. $\log_6 0.77$

71. $\log_2 5 + \log_2 7$

72. $\log_9 85 + \log_7 17$

73. $\sqrt{\log_4 46}$

74. $2 \log_5 15 + \sqrt[3]{\log_3 67}$

75. $\dfrac{\log_2 12}{\log_2 3}$

76. $\dfrac{\log_7 125}{\log_7 25}$

Exercises 77–80: Solve the equation graphically. Express any solutions to the nearest thousandth.

77. $\log_2 (x^3 + x^2 + 1) = 7$

78. $\log_3 (1 + x^2 + 2x^4) = 4$

79. $\log_2 (x^2 + 1) = 5 - \log_3 (x^4 + 1)$

80. $\ln (x^2 + 2) = \log_2 (10 - x^2)$

Applications

81. *Runway Length* (Refer to Exercise 123, Section 5.4.) Use a natural logarithm (instead of a common logarithm) to write the formula $L(x) = 3 \log x$. Evaluate $L(50)$ for each formula. Do your answers agree?

82. *Allometry* The equation $y = bx^a$ is used in applications involving biology and allometry. Another form of this equation is $\log y = \log b + a \log x$. Use properties of logarithms to obtain this second equation from the first. (**Source:** H. Lancaster, *Quantitative Methods in Biological and Medical Sciences.*)

83. *Decibels* (Refer to Example 4.)   If the intensity x of a sound increases by a factor of 10, by how much does the decibel level increase?

84. *Decibels* (Refer to Example 4.) Use a natural logarithm to write the formula $f(x) = 160 + 10 \log x$. Evaluate $f(5 \times 10^{-8})$ for each formula. Do your answers agree?

85. *Light Absorption*   When sunlight passes through lake water, its initial intensity I_0 decreases to a weaker intensity I at a depth of x feet according to the formula

$$\ln I - \ln I_0 = -kx,$$

where k is a positive constant. Solve this equation for I.

x ft

86. *Dissolving Salt*   If C grams of salt are added to a sample of water, the amount A of undissolved salt is modeled by $A = Ca^x$, where x is time. Solve the equation for x.

87. *Population Growth*   The population P (in millions) of California x years after 2000 can be modeled by $P = 34e^{0.013x}$.
   (a) Use properties of logarithms to solve this equation for x.
   (b) Use your equation to find x when $P = 38$. Interpret your answer.

88. *Population Growth*   The population P (in millions) of Georgia x years after 2000 can be modeled by $P = 8e^{0.023x}$.
   (a) Use properties of logarithms to solve this equation for x.
   (b) Use your equation to find x when $P = 10$. Interpret your answer.

89. Solve $A = Pe^{rt}$ for t.

90. Solve $P = P_0 e^{r(t-t_0)} + 5$ for t.

91. Write the sum

$$\log 1 + 2 \log 2 + 3 \log 3 + 4 \log 4 + 5 \log 5$$

as a logarithm of a single expression.

92. Show that

$$\log_2 (x + \sqrt{x^2 - 4}) + \log_2 (x - \sqrt{x^2 - 4}) = 2$$

is an identity. What is the domain of the expression on the left side of the equation?

Writing about Mathematics

93. A student insists that $\log (x + y)$ and $\log x + \log y$ are equal. How could you convince the student otherwise?

94. A student insists that $\log \left(\frac{x}{y} \right)$ and $\frac{\log x}{\log y}$ are equal. How could you convince the student otherwise?

5.6 Exponential and Logarithmic Equations

- **Solve exponential equations**
- **Solve logarithmic equations**

Introduction

The population of the world has grown rapidly during the past century. As a result, heavy demands have been made on the world's resources. Exponential functions and equations are often used to model this rapid growth, whereas logarithms are used to model slower growth.

Exponential Equations

The population P of the world was 3 billion in 1960, was 6.7 billion in 2008, and can be modeled by $P(x) = 3(1.017)^{x-1960}$, where x is the year. We can use P to estimate the year when world population reached **5** billion by solving the *exponential equation*

$$3(1.017)^{x-1960} = 5.$$

An equation in which one or more variables occur in the exponent of an expression is called an **exponential equation**. In the next example, we use Property 4 of logarithms, given by $\log_a (m^r) = r \log_a m$, to solve this equation.

> **EXAMPLE 1** Modeling world population
>
> World population in billions during year x can be modeled by $P(x) = 3(1.017)^{x-1960}$, shown in Figure 5.73. Solve the equation $3(1.017)^{x-1960} = 5$ symbolically to estimate the year when world population reached 5 billion.
>
> **SOLUTION** First divide each side by 3, and then take the common logarithm of each side. (The natural logarithm could also be used.)
>
> $$3(1.017)^{x-1960} = 5 \qquad \text{Given equation}$$
>
> $$(1.017)^{x-1960} = \frac{5}{3} \qquad \text{Divide by 3.}$$
>
> $$\log (1.017)^{x-1960} = \log \frac{5}{3} \qquad \text{Take the common logarithm.}$$
>
> $$(x - 1960) \log (1.017) = \log \frac{5}{3} \qquad \text{Property 4: } \log (m^r) = r \log m$$
>
> $$x - 1960 = \frac{\log (5/3)}{\log (1.017)} \qquad \text{Divide by log (1.017).}$$
>
> $$x = 1960 + \frac{\log (5/3)}{\log (1.017)} \qquad \text{Add 1960.}$$
>
> $$x \approx 1990.3 \qquad \text{Approximate.}$$
>
> This model predicts that world population reached 5 billion during 1990.
>
> *Now Try Exercise 71* ◀

Figure 5.73

(Graph: World population (billions) vs. Year, showing $P(x) = 3(1.017)^{x-1960}$ with years 1960, 1970, 1980, 1990, 2000, 2010 on the horizontal axis and 0 through 7 on the vertical axis.)

CLASS DISCUSSION

What is the growth factor for $P(x) = 3(1.017)^{x-1960}$? By what percentage has world population grown, on average, each year from 1960 to 2008?

Exponential equations occur in a variety of applications and can be solved symbolically, graphically, and numerically. In the next example, these techniques are used to solve an exponential equation that describes the minimum thickness of the pavement for an airport runway.

EXAMPLE 2   Calculating the thickness of a runway

Heavier aircraft require runways with thicker pavement for landings and takeoffs. The relation between the thickness of the pavement t in inches and gross weight W in thousands of pounds can be approximated by

$$W(t) = 18.29e^{0.246t}.$$

(a) Determine the required thickness of the runway for a 130,000-pound plane.
(b) Solve part (a) graphically and numerically.

Calculator Help

To find a point of intersection, see Appendix A (page AP-8).

SOLUTION
(a) *Symbolic Solution*   Because the unit for W is thousands of pounds, we solve the equation $W(t) = 130$.

$$18.29e^{0.246t} = 130 \qquad W(t) = 130$$

$$e^{0.246t} = \frac{130}{18.29} \qquad \text{Divide by 18.29.}$$

$$\ln e^{0.246t} = \ln \frac{130}{18.29} \qquad \text{Take the natural logarithm of each side.}$$

$$0.246t = \ln \frac{130}{18.29} \qquad \text{Inverse property: } \ln e^k = k.$$

$$t = \frac{\ln (130/18.29)}{0.246} \qquad \text{Divide by 0.246.}$$

$$t \approx 7.97 \qquad \text{Approximate.}$$

[0, 10, 1] by [0, 200, 50]

$y_2 = 130$

$y_1 = 18.29e^{0.246t}$

Intersection
X=7.9722764 Y=130

Figure 5.74

The runway should be about 8 inches thick.

(b) *Graphical Solution*   Let $Y_1 = 18.29e^{\wedge}(.246X)$ and $Y_2 = 130$. In Figure 5.74, their graphs intersect near (7.97, 130).

Numerical Solution   Numerical support of this result is shown in Figure 5.75, where $y_1 \approx y_2$ when $x = 8$.

Now Try Exercise 73 ◀

X	Y1	Y2
5	62.574	130
6	80.026	130
7	102.35	130
8	130.89	130 ← $y_1 \approx y_2$
9	167.39	130
10	214.08	130
11	273.79	130

X=8

Figure 5.75

To solve an exponential equation with a base other than 10 or e, Property 4 of logarithms can be used. This technique was demonstrated in Example 1 and is also used in the next example.

EXAMPLE 3   Modeling the decline of bluefin tuna

Bluefin tuna are large fish that can weigh 1500 pounds and swim at speeds of 55 miles per hour. Because they are used for sushi, a prime fish was worth over $50,000 in 2008. As a result, the western Atlantic bluefin tuna have been exploited, and their numbers have declined exponentially. Their numbers in thousands from 1974 to 1991 can be modeled by the formula $f(x) = 230(0.881)^x$, where x is years after 1974. (In more recent years, environmental controls have helped slow this exponential decline.) (**Source:** B. Freedman, *Environmental Ecology*.)

(a) Estimate the number of bluefin tuna in 1974 and 1991.
(b) Determine symbolically the year when they numbered 50 thousand.

SOLUTION

(a) To determine their numbers in 1974 and 1991, evaluate $f(0)$ and $f(17)$.

$$f(0) = 230(0.881)^0 = 230(1) = 230$$
$$f(17) = 230(0.881)^{17} \approx 26.7$$

Bluefin tuna decreased from 230 thousand in 1974 to fewer than 27 thousand in 1991.

(b) Solve the equation $f(x) = 50$.

$$230(0.881)^x = 50 \qquad \text{\small $f(x) = 50$}$$

$$0.881^x = \frac{5}{23} \qquad \text{\small Divide by 230; simplify.}$$

$$\ln 0.881^x = \ln \frac{5}{23} \qquad \text{\small Take the natural logarithm of each side. (The common logarithm could also be used.)}$$

$$x \ln 0.881 = \ln \frac{5}{23} \qquad \text{\small Property 4: } \ln m^r = r \ln m$$

$$x = \frac{\ln (5/23)}{\ln 0.881} \qquad \text{\small Divide by } \ln 0.881.$$

$$x \approx 12.04 \qquad \text{\small Approximate.}$$

They numbered about 50 thousand in $1974 + 12.04 \approx 1986$. **Now Try Exercise 77** ◀

In real applications, a modeling function is seldom provided with the data. However, it is not uncommon to be given the general form of a function that describes a data set.

EXAMPLE 4 Finding a modeling function

The gap between organs available for transplant and people who need them has widened. In 1999 about 60,000 people were waiting for organ transplants, and in 2007 this number had increased to 100,000. This trend can be modeled by $T(x) = Ca^{x-1999}$, where C and a are constants. (**Source:** Scientific Registry of Transplant Recipients.)

(a) Approximate C and a if the unit for T is thousands of people.

(b) Use $T(x)$ to estimate the number of people who were waiting for organ transplants in 2004. Compare the estimate to the actual value of 86,000.

SOLUTION

(a) In 1999 there were 60,000 people waiting, so $T(1999) = 60$. This equation can be used to find C.

$$T(1999) = Ca^{1999-1999} = Ca^0 = C(1) = 60$$

Thus $C = 60$ and $T(x) = 60a^{x-1999}$. In 2007 the number waiting was 100,000, so we can solve the equation $T(2007) = 100$ to find the value of a.

$$60a^{2007-1999} = 100 \qquad \text{\small $T(2007) = 100$}$$

$$a^8 = \frac{5}{3} \qquad \text{\small Divide by 60; simplify.}$$

$$a = \sqrt[8]{\frac{5}{3}} \qquad \text{\small Take the eighth root of each side.}$$

$$a \approx 1.066 \qquad \text{\small Approximate.}$$

Thus T is given by $T(x) = 60(1.066)^{x-1999}$.

(b) For 2004 this model estimates that

$$T(2004) = 60(1.066)^{2004-1999} \approx 82.6$$

thousand people were waiting for an organ transplant, which is about 3400 less than the actual value. *Now Try Exercise 79* ◀

Exponential equations can occur in many forms. Although some types of exponential equations cannot be solved symbolically, Example 5 shows four equations that can.

EXAMPLE 5 Solving exponential equations symbolically

Solve each equation.

(a) $10^{x+2} = 10^{3x}$ **(b)** $5(1.2)^x + 1 = 26$ **(c)** $\left(\dfrac{1}{4}\right)^{x-1} = \dfrac{1}{9}$ **(d)** $5^{x-3} = e^{2x}$

SOLUTION
(a) Start by taking the common logarithm of each side.

$10^{x+2} = 10^{3x}$	Given equation
$\log 10^{x+2} = \log 10^{3x}$	Take the common logarithm.
$x + 2 = 3x$	Inverse property: $\log 10^k = k$
$2 = 2x$	Subtract x.
$x = 1$	Divide by 2; rewrite.

> **MAKING CONNECTIONS**
>
> **Equations and One-to-One Functions** If a function f is one-to-one, then $f(b) = f(a)$ implies that $a = b$. Because $f(x) = 10^x$ is a one-to-one function, it follows that $10^{x+2} = 10^{3x}$ implies that $x + 2 = 3x$. This step is equivalent to taking the common logarithm of each side of the given equation.

(b) **Getting Started** When solving an exponential equation, first isolate the exponential term on one side of the equation, if possible. Then take the logarithm of each side. ▶

$5(1.2)^x = 25$	Subtract 1.
$(1.2)^x = 5$	Divide by 5.
$\log (1.2)^x = \log 5$	Take the common logarithm.
$x \log (1.2) = \log 5$	Property 4: $\log m^r = r \log m$
$x = \dfrac{\log 5}{\log (1.2)}$	Divide by log (1.2).
$x \approx 8.827$	Approximate.

(c) Begin by taking the common logarithm of each side.

$\left(\dfrac{1}{4}\right)^{x-1} = \dfrac{1}{9}$	Given equation
$\log \left(\dfrac{1}{4}\right)^{x-1} = \log \dfrac{1}{9}$	Take the common logarithm.
$(x - 1) \log \left(\dfrac{1}{4}\right) = \log \dfrac{1}{9}$	Property 4: $\log m^r = r \log m$
$x - 1 = \dfrac{\log (1/9)}{\log (1/4)}$	Divide by log $\frac{1}{4}$.
$x = 1 + \dfrac{\log (1/9)}{\log (1/4)} \approx 2.585$	Add 1 and approximate.

This equation could also be solved by taking the natural logarithm of each side.

(d) Begin by taking the natural (or common) logarithm of each side.

$$5^{x-3} = e^{2x}$$ Given equation

$$\ln 5^{x-3} = \ln e^{2x}$$ Take the natural logarithm.

$$(x - 3) \ln 5 = 2x$$ Property 4; inverse property

$$x \ln 5 - 3 \ln 5 = 2x$$ Distributive property

$$x \ln 5 - 2x = 3 \ln 5$$ Subtract 2x; add 3 ln 5.

$$x(\ln 5 - 2) = 3 \ln 5$$ Factor out x.

$$x = \frac{3 \ln 5}{\ln 5 - 2}$$ Divide by ln 5 − 2.

$$x \approx -12.36$$ Approximate.

Now Try Exercises 13, 17, 23, and 25 ◀

If a hot object is put in a room with temperature T_0, then according to **Newton's law of cooling**, the temperature of the object after time t is modeled by

$$T(t) = T_0 + Da^t,$$

where $0 < a < 1$ and D is the initial temperature *difference* between the object and the room. This phenomenon is discussed in the next example.

EXAMPLE 6   Modeling coffee cooling

A pot of coffee with a temperature of 100°C is set down in a room with a temperature of 20°C. The coffee cools to 60°C after 1 hour.
(a) Find values for T_0, D, and a so that $T(t) = T_0 + Da^t$ models the data.
(b) Find the temperature of the coffee after half an hour.
(c) How long did it take for the coffee to reach 50°C? Support your result graphically.

SOLUTION
(a) The room has temperature $T_0 = 20$°C, and the initial temperature difference between the coffee and the room is $D = 100 - 20 = 80$°C. It follows that $T(t) = 20 + 80a^t$. We can use the fact that the temperature of the coffee after **1** hour was **60°C** to determine the value of a.

$$T(1) = 60$$

$$20 + 80a^1 = 60$$ Let t = 1 in T(t) = 20 + 80a^t.

$$80a = 40$$ Subtract 20.

$$a = \frac{1}{2}$$ Divide by 80.

Thus $T(t) = 20 + 80\left(\frac{1}{2}\right)^t$.
(b) After half an hour, the temperature is

$$T\left(\frac{1}{2}\right) = 20 + 80\left(\frac{1}{2}\right)^{1/2} \approx 76.6°C.$$

(c) *Symbolic Solution*   To determine when the coffee reached 50°C, solve $T(t) = 50$.

$$20 + 80\left(\frac{1}{2}\right)^t = 50 \qquad \textcolor{gray}{T(t) = 50}$$

$$80\left(\frac{1}{2}\right)^t = 30 \qquad \textcolor{gray}{\text{Subtract 20.}}$$

$$\left(\frac{1}{2}\right)^t = \frac{3}{8} \qquad \textcolor{gray}{\text{Divide by 80; simplify.}}$$

$$\log\left(\frac{1}{2}\right)^t = \log\frac{3}{8} \qquad \textcolor{gray}{\text{Take the common logarithm.}}$$

$$t\,\log\left(\frac{1}{2}\right) = \log\frac{3}{8} \qquad \textcolor{gray}{\text{Property 4}}$$

$$t = \frac{\log(3/8)}{\log(1/2)} \qquad \textcolor{gray}{\text{Divide by } \log\tfrac{1}{2}.}$$

$$t \approx 1.415 \qquad \textcolor{gray}{\text{Approximate.}}$$

The temperature reaches 50°C after about 1.415 hours, or 1 hour and 25 minutes.

Graphical Solution   The graphs of $Y_1 = 20 + 80(1/2)^\wedge X$ and $Y_2 = 50$ intersect near (1.415, 50), as shown in Figure 5.76. This result agrees with the symbolic solution.

Now Try Exercise 83 ◀

[0, 3, 1] by [0, 100, 10]

Figure 5.76

NOTE   Newton's law of cooling can also model the temperature of a cold object that is brought into a warm room. In this case, the temperature difference D is negative.

Some exponential equations *cannot* be solved symbolically but can be solved graphically. This is demonstrated in the next example.

EXAMPLE 7   **Solving an exponential equation graphically**

Solve $e^{-x} + 2x = 3$ graphically. Approximate all solutions to the nearest hundredth.

SOLUTION   The graphs of $Y_1 = e^\wedge(-X) + 2X$ and $Y_2 = 3$ intersect near the points $(-1.92, 3)$ and $(1.37, 3)$, as shown in Figures 5.77 and 5.78. Thus the solutions are approximately -1.92 and 1.37.

Figure 5.77   **Figure 5.78**   *Now Try Exercise 65* ◀

Logarithmic Equations

Logarithmic equations contain logarithms. Like exponential equations, logarithmic equations also occur in applications. To solve a logarithmic equation, we use the inverse property $a^{\log_a x} = x$. This technique is illustrated in the next example.

EXAMPLE 8   Solving a logarithmic equation

Solve $3 \log_3 x = 12$.

SOLUTION   Begin by dividing each side by 3.

$$\log_3 x = 4 \qquad \text{Divide by 3.}$$
$$3^{\log_3 x} = 3^4 \qquad \text{Exponentiate each side; base 3.}$$
$$x = 81 \qquad \text{Inverse property: } a^{\log_a k} = k \qquad \boxed{\textit{Now Try Exercise 49}} \blacktriangleleft$$

EXAMPLE 9   Solving a logarithmic equation symbolically

In developing countries, there is a relationship between the amount of land a person owns and the average daily calories consumed. This relationship is modeled by the formula $C(x) = 280 \ln (x + 1) + 1925$, where x is the amount of land owned in acres and $0 \le x \le 4$.   (**Source:** D. Grigg, *The World Food Problem.*)
(a) Find the average caloric intake for a person who owns no land.
(b) A graph of C is shown in Figure 5.79. Interpret the graph.
(c) Determine symbolically the number of acres owned by someone whose average intake is 2000 calories per day.

Figure 5.79

SOLUTION
(a) Since $C(0) = 280 \ln (0 + 1) + 1925 = 1925$, a person without land consumes an average of 1925 calories per day.
(b) As the amount of land x increases, the caloric intake y also increases. However, the rate of increase slows. This would be expected because there is a limit to the number of calories an average person would eat, regardless of his or her economic status.
(c) Solve the equation $C(x) = 2000$.

$$280 \ln (x + 1) + 1925 = 2000 \qquad C(x) = 2000$$
$$280 \ln (x + 1) = 75 \qquad \text{Subtract 1925.}$$
$$\ln (x + 1) = \frac{75}{280} \qquad \text{Divide by 280.}$$
$$e^{\ln (x+1)} = e^{75/280} \qquad \text{Exponentiate each side; base } e.$$
$$x + 1 = e^{75/280} \qquad \text{Inverse property: } e^{\ln k} = k$$
$$x = e^{75/280} - 1 \qquad \text{Subtract 1.}$$
$$x \approx 0.307 \qquad \text{Approximate.}$$

A person who owns about 0.3 acre has an average intake of 2000 calories per day.
$$\boxed{\textit{Now Try Exercise 87}} \blacktriangleleft$$

Like exponential equations, logarithmic equations can occur in many forms. The next example illustrates three equations that can be solved symbolically.

EXAMPLE 10   Solving logarithmic equations symbolically

Solve each equation.
(a) $\log (2x + 1) = 2$ (b) $\log_2 4x = 2 - \log_2 x$
(c) $2 \ln (x + 1) = \ln (1 - 2x)$

SOLUTION

(a) To solve the equation, exponentiate each side of the equation using base 10.

$$\log (2x + 1) = 2 \qquad \textit{Given equation}$$
$$10^{\log (2x+1)} = 10^2 \qquad \textit{Exponentiate each side; base 10.}$$
$$2x + 1 = 100 \qquad \textit{Inverse property: } 10^{\log k} = k$$
$$x = 49.5 \qquad \textit{Solve for x.}$$

(b) To solve this equation, apply properties of logarithms.

$$\log_2 4x = 2 - \log_2 x \qquad \textit{Given equation}$$
$$\log_2 4x + \log_2 x = 2 \qquad \textit{Add } \log_2 x.$$
$$\log_2 4x^2 = 2 \qquad \textit{Property 2: } \log_a m + \log_a n = \log_a (mn)$$
$$2^{\log_2 4x^2} = 2^2 \qquad \textit{Exponentiate each side; base 2.}$$
$$4x^2 = 4 \qquad \textit{Inverse property: } a^{\log_a k} = k$$
$$x = \pm 1 \qquad \textit{Solve for x.}$$

However, -1 is not a solution since $\log_2 x$ is undefined for negative values of x. Thus the only solution is 1.

(c) Start by applying Property 4 to the given equation.

$$2 \ln (x + 1) = \ln (1 - 2x) \qquad \textit{Given equation}$$
$$\ln (x + 1)^2 = \ln (1 - 2x) \qquad \textit{Property 4}$$
$$e^{\ln (x+1)^2} = e^{\ln (1-2x)} \qquad \textit{Exponentiate; base e.}$$
$$(x + 1)^2 = 1 - 2x \qquad \textit{Inverse property: } a^{\log_a k} = k$$
$$x^2 + 2x + 1 = 1 - 2x \qquad \textit{Expand the binomial.}$$
$$x^2 + 4x = 0 \qquad \textit{Combine terms.}$$
$$x(x + 4) = 0 \qquad \textit{Factor.}$$
$$x = 0 \quad \text{or} \quad x = -4 \qquad \textit{Zero-product property}$$

Substituting $x = 0$ and $x = -4$ in the given equation shows that 0 is a solution but -4 is not a solution. **_Now Try Exercises 53, 57, and 61_** ◀

EXAMPLE 11 Modeling the life span of a robin

In one study, the life spans of 129 robins were monitored over a 4-year period. The equation $y = \dfrac{2 - \log (100 - x)}{0.42}$ can be used to calculate the number of years y required for x percent of the robin population to die. For example, to find the time when 40% of the robins had died, substitute $x = 40$ into the equation. The result is $y \approx 0.53$, or about half a year. (**Source:** D. Lack, *The Life of a Robin.*)

(a) Graph y in [0, 100, 10] by [0, 5, 1]. Interpret the graph.

(b) Estimate graphically the percentage of the robins that had died after 2 years.

SOLUTION

(a) The graph of $Y_1 = (2 - \log (100 - X))/.42$ is shown in Figure 5.80. At first, the graph increases slowly between 0% and 60%. Since years are listed on the y-axis and percentages on the x-axis, this slow increase in f indicates that not many years pass

[0, 100, 10] by [0, 5, 1]

$$y_1 = \frac{2 - \log (100 - x)}{0.42}$$

Figure 5.80

$[0, 100, 10]$ by $[0, 5, 1]$

$y_1 = \dfrac{2 - \log(100 - x)}{0.42}$

$y_2 = 2$

Intersection
X=85.545602 Y=2

Figure 5.81

before a large percentage of the robins die. When $x \geq 60$, the graph begins to increase dramatically. This indicates that a small percentage have a comparatively long life span.

(b) To determine the percentage that had died after 2 years, graph y_1 and $y_2 = 2$. Their graphs intersect near $(85.5, 2)$, as shown in Figure 5.81. After 2 years, approximately 85.5% of the robins had died—a surprisingly high percentage. ***Now Try Exercise 89*** ◀

MAKING CONNECTIONS

Solving Exponential and Logarithmic Equations   At some point in the process of solving an exponential equation, we often take a logarithm of each side of the equation. Similarly, when solving a logarithmic equation, we often exponentiate each side of the equation.

5.6 Putting It All Together

The following table summarizes techniques that can be used to solve some types of exponential and logarithmic equations symbolically.

Concept	Explanation	Example
Exponential equations	Typical form: $Ca^x = k$ Solve for a^x. Then take a base-a logarithm of each side. Use the inverse property: $$\log_a a^x = x.$$	$4e^x = 24$ $e^x = 6$ $\ln e^x = \ln 6$ $x = \ln 6 \approx 1.79$
Logarithmic equations	*Equation 1:* $C \log_a x = k$ Solve for $\log_a x$. Then exponentiate each side with base a. Use the inverse property: $$a^{\log_a x} = x.$$ *Equation 2:* $\log_a bx \pm \log_a cx = k$ When more than one logarithm with the same base occurs, use properties of logarithms to combine logarithms. Be sure to check any solutions.	**1.** $4 \log x = 10$ $\log x = 2.5$ $10^{\log x} = 10^{2.5}$ $x = 10^{2.5} \approx 316$ **2.** $\log x + \log 4x = 2$ $\log 4x^2 = 2$ $4x^2 = 10^2$ $x^2 = 25$ $x = \pm 5$ The only solution is 5.

5.6 Exercises

Solving Exponential Equations

Exercises 1–4: The graphical and symbolic representations of f and g are shown.

    (a) Use the graph to solve $f(x) = g(x)$.
    (b) Solve $f(x) = g(x)$ symbolically.

1.

2.

3.

4.

Exercises 5–34: Solve the exponential equation.

5. $4e^x = 5$

6. $2e^{-x} = 8$

7. $2(10)^x + 5 = 45$

8. $100 - 5(10)^x = 7$

9. $2.5e^{-1.2x} = 1$

10. $9.5e^{0.005x} = 19$

11. $1.2(0.9)^x = 0.6$

12. $0.05(1.15)^x = 5$

13. $4(1.1)^{x-1} = 16$

14. $3(2)^{x-2} = 99$

15. $5(1.2)^{3x-2} + 94 = 100$

16. $1.4(2)^{x+3} = 2.8$

17. $5^{3x} = 5^{1-2x}$

18. $7^{x^2} = 7^{4x-3}$

19. $10^{(x^2)} = 10^{3x-2}$

20. $e^{2x} = e^{5x-3}$

21. $\left(\frac{1}{5}\right)^x = -5$

22. $2^x = -4$

23. $\left(\frac{2}{5}\right)^{x-2} = \frac{1}{3}$

24. $\left(\frac{3}{2}\right)^{x+1} = \frac{7}{3}$

25. $4^{x-1} = 3^{2x}$

26. $3^{1-2x} = e^{0.5x}$

27. $e^{x-3} = 2^{3x}$

28. $6^{x+1} = 4^{2x-1}$

29. $3(1.4)^x - 4 = 60$

30. $2(1.05)^x + 3 = 10$

31. $5(1.015)^{x-1980} = 8$

32. $30 - 3(0.75)^{x-1} = 29$

33. $4\left(\frac{3}{4}\right)^{x+1} = \frac{1}{81}\left(\frac{3}{2}\right)^{5+x}$

34. $5\left(\frac{2}{5}\right)^{x+1} = \frac{1}{125}\left(\frac{4}{5}\right)^{x-3}$

Solving Logarithmic Equations

Exercises 35–64: Solve the logarithmic equation.

35. $3 \log x = 2$

36. $5 \ln x = 10$

37. $\ln 2x = 5$

38. $\ln 4x = 1.5$

39. $\log 2x^2 = 2$

40. $\log (2 - x) = 0.5$

41. $\log_2 (3x - 2) = 4$

42. $\log_3 (1 - x) = 1$

43. $\log_5 (8 - 3x) = 3$

44. $\log_6 (2x + 4) = 2$

45. $160 + 10 \log x = 50$

46. $160 + 10 \log x = 120$

47. $\ln x + \ln x^2 = 3$

48. $\log x^5 = 4 + 3 \log x$

49. $2 \log_2 x = 4.2$

50. $3 \log_2 3x = 1$

51. $\log x + \log 2x = 2$

52. $\ln 2x + \ln 3x = \ln 6$

53. $\log (2 - 3x) = 3$

54. $\log (x^2 + 1) = 2$

55. $\ln x + \ln (3x - 1) = \ln 10$

56. $\log x + \log (2x + 5) = \log 7$

57. $2 \ln x = \ln (2x + 1)$

58. $\log (x^2 + 3) = 2 \log (x + 1)$

59. $\log (x + 1) + \log (x - 1) = \log 3$

60. $\ln (x^2 - 4) - \ln (x + 2) = \ln (3 - x)$

61. $\log_2 2x = 4 - \log_2 (x + 2)$

62. $\log_3 x + \log_3 (x + 2) = \log_3 24$

63. $\log_5 (x + 1) + \log_5 (x - 1) = \log_5 15$

64. $\log_7 4x - \log_7 (x + 3) = \log_7 x$

Solving Equations Graphically

Exercises 65–70: The following equations cannot be solved symbolically. Solve these equations graphically and round your answers to the nearest hundredth.

65. $2x + e^x = 2$

66. $xe^x - 1 = 0$

67. $x^2 - x \ln x = 2$

68. $x \ln |x| = -2$

69. $xe^{-x} + \ln x = 1$

70. $2^{x-2} = \log x^4$

Applications

71. *Population Growth* World population P in billions during year x is modeled by $P(x) = 3(1.017)^{x-1960}$, where $1960 \le x \le 2010$. Estimate the year when world population reached 4 billion.

72. *Population of Arizona* The population P of Arizona has been increasing at an annual rate of 3.5%. In 1990 the population of Arizona was 3.7 million.
(a) Write a formula for $P(x)$, where x is the year and P is in millions.
(b) Estimate the population of Arizona in 2005.

73. *Light Absorption* When light passes through water, its intensity I decreases according to the formula $I(x) = I_0 e^{-kx}$, where I_0 is the initial intensity of the light and x is the depth in feet. If $I_0 = 1000$ lumens per square meter and $k = 0.12$, determine symbolically and graphically the depth at which the intensity is 25% of I_0.

74. *Light Absorption* (Refer to Exercise 73.) Let $I(x) = 500e^{-0.2x}$ and determine symbolically and graphically the depth x at which the intensity I is 1% of $I_0 = 500$.

75. *Modeling Bacteria* (Refer to Section 5.3, Example 8.) The number N of *E. coli* bacteria in millions per milliliter after t minutes can be modeled by $N(t) = 0.5e^{0.014t}$.

Determine symbolically the elapsed time required for the concentration of bacteria to reach 25 million per milliliter.

76. *Credit Cards* From 1987 to 1996 the number of Visa cards and MasterCards was up 80% to 376 million. The formula $f(x) = 36.2e^{0.14x}$ models the amount of credit card spending from Thanksgiving to Christmas in billions of dollars, x years after 1987. (**Source:** National Credit Counseling Services.)
(a) Determine symbolically the year when this amount reached $55 billion.

(b) Solve part (a) graphically or numerically.

77. *Bluefin Tuna* The number of Atlantic bluefin tuna in thousands x years after 1974 can be modeled by $f(x) = 230(0.881)^x$. Estimate the year when the number of bluefin tuna reached 95 thousand.

78. *U.S. Liver Transplants* The number of individuals waiting for liver transplants can be modeled by $f(x) = 2340(1.124)^{x-1988}$, where x is the year. (**Source:** United Network for Organ Sharing.)
(a) Estimate this number in 2005.

(b) Determine symbolically when this number might reach 30,000.

79. *Population Growth* In 2000 the population of India reached 1 billion, and in 2025 it is projected to be 1.4 billion. (**Source:** Bureau of the Census.)
(a) Find values for C and a so that $P(x) = Ca^{x-2000}$ models the population of India in year x.

(b) Estimate India's population in 2010.

(c) Use P to determine the year when India's population might reach 1.5 billion.

80. *Population of Pakistan* In 2007 the population of Pakistan was 164 million, and it is expected to be 250 million in 2025. (**Source:** United Nations.)
(a) Approximate C and a so that $P(x) = Ca^{x-2007}$ models these data, where P is in millions and x is the year.

(b) Estimate the population of Pakistan in 2015, and compare your estimate to the predicted value of 204 million.

(c) Estimate when this population could reach 212 million.

81. *Organ Transplants* (Refer to Example 4.) Determine when the number of individuals waiting for organ transplants is expected to reach 140,000.

82. *Midair Near Collisions* The table shows the number of airliner near collisions y in year x.

x	1989	1991	1993	1995
y	131	78	44	34

Source: FAA.

(a) Approximate constants C and a so that the formula $f(x) = Ca^{(x-1989)}$ models the data.

📷 (b) Support your answer by graphing f and the data.

83. *Newton's Law of Cooling* A pan of boiling water with a temperature of 212°F is set in a bin of ice with a temperature of 32°F. The pan cools to 70°F in 30 minutes.

(a) Find T_0, D, and a so that $T(t) = T_0 + Da^t$ models the data, where t is in hours.

(b) Find the temperature of the pan after 10 minutes.

(c) How long did it take the pan to reach 40°F? Support your result graphically.

84. *Warming an Object* A pan of cold water with a temperature of 35°F is brought into a room with a temperature of 75°F. After 1 hour, the temperature of the pan of water is 45°F.

(a) Find T_0, D, and a so that $T(t) = T_0 + Da^t$ models the data, where t is in hours.

(b) Find the temperature of the water after 3 hours.

(c) How long would it take the water to reach 60°F?

85. *Warming a Soda Can* Suppose that a can of soda, initially at 5°C, warms to 18°C after 2 hours in a room that has a temperature of 20°C.

(a) Find the temperature of the soda can after 1.5 hours.

(b) How long did it take for the soda to warm to 15°C?

86. *Cooling a Soda Can* A soda can at 80°F is put into a cooler containing ice at 32°F. The temperature of the soda after t minutes is given by $T(t) = 32 + 48(0.9)^t$.

(a) Evaluate $T(30)$ and interpret your results.

(b) How long did it take for the soda to cool to 50°F?

87. *Caloric Intake* (Refer to Example 9.) The formula

$$C(x) = 280 \ln (x + 1) + 1925$$

models the number of calories consumed daily by a person owning x acres of land in a developing country. Estimate the number of acres owned for which average intake is 2300 calories per day.

88. *Salinity* The salinity of the oceans changes with latitude and with depth. In the tropics, the salinity increases on the surface of the ocean due to rapid evaporation. In the higher latitudes, there is less evaporation and rainfall causes the salinity to be less on the surface than at lower depths. The function given by

$$S(x) = 31.5 + 1.1 \log (x + 1)$$

models salinity to depths of 1000 meters at a latitude of 57.5°N. The input x is the depth in meters and the output $S(x)$ is in grams of salt per kilogram of seawater. (**Source:** D. Hartman, *Global Physical Climatology.*)

(a) Evaluate $S(500)$. Interpret your result.

📷 (b) Graph S. Discuss any trends.

(c) Find the depth where the salinity equals 33.

89. *Life Span* Solve Example 11, part (b) symbolically.

90. *Life Span of Sparrows* (Refer to Example 11.) The life span of a sample of sparrows was studied. The equation $y = \dfrac{2 - \log (100 - x)}{0.37}$ calculates the number of years y required for x percent of the sparrows to die, where $0 \le x \le 95$.

(a) Find y when $x = 40$. Interpret your answer.

(b) Find x when $y = 1.5$. Interpret your answer.

91. *Bacteria Growth* The concentration of bacteria in a sample can be modeled by $B(t) = B_0 e^{kt}$, where t is in hours and B is the concentration in billions of bacteria per liter.

(a) If the concentration increases by 15% in 6 hours, find k.

(b) If $B_0 = 1.2$, find B after 8.2 hours.

(c) By what percentage does the concentration increase each hour?

92. *Voltage* The voltage in a circuit can be modeled by $V(t) = V_0 e^{kt}$, where t is in milliseconds.

(a) If the voltage decreases by 85% in 5 milliseconds, find k.

(b) If $V_0 = 4.5$ volts, find V after 2.3 milliseconds.

(c) By what percentage does the voltage decrease each millisecond?

Exercises 93 and 94: For the given annual interest rate r, estimate the time for P dollars to double.

93. $P = \$1000$, $r = 8.5\%$ compounded quarterly

94. $P = \$750$, $r = 12\%$ compounded continuously

*Exercises 95 and 96: **Continuous Compounding** Suppose that P dollars is deposited in a savings account paying 9% interest compounded continuously. After t years, the account will contain $A(t) = Pe^{0.09t}$ dollars.*

  (a) *Solve $A(t) = b$ for the given values of P and b.*
  (b) *Interpret your results.*

95. $P = 500$ and $b = 750$

96. $P = 1000$ and $b = 2000$

97. *Radioactive Carbon-14*   The percentage P of radioactive carbon-14 remaining in a fossil after t years is given by $P = 100\left(\frac{1}{2}\right)^{t/5700}$. Suppose a fossil contains 35% of the carbon-14 that the organism contained when it was alive. Estimate the age of the fossil.

98. *Radioactive Radium-226*   The amount A of radium in milligrams remaining in a sample after t years is given by $A(t) = 0.02\left(\frac{1}{2}\right)^{t/1600}$. How many years will it take for the radium to decay to 0.004 milligram?

99. *Traffic Flow*   (Refer to Section 5.3, Example 10.) The probability that a car will enter an intersection within a period of x minutes is given by $P(x) = 1 - e^{-0.5x}$. Determine symbolically the elapsed time x when there is a 50–50 chance that a car has entered the intersection. (*Hint:* Solve $P(x) = 0.5$.)

100. *Modeling Traffic Flow*   Cars arrive randomly at an intersection with an average traffic volume of 1 car per minute. The likelihood, or probability, that at least one car enters the intersection during a period of x minutes can be estimated by $f(x) = 1 - e^{-x}$.
  (a) What is the probability that at least one car enters the intersection during a 5-minute period?

  (b) Determine the value of x that gives a 40% chance that at least one car enters the intersection during an interval of x minutes.

101. *Modeling Bacteria Growth*   Suppose that the concentration of a bacteria sample is 100,000 bacteria per milliliter. If the concentration doubles every 2 hours, how long will it take for the concentration to reach 350,000 bacteria per milliliter?

102. *Modeling Bacteria Growth*   Suppose that the concentration of a bacteria sample is 50,000 bacteria per milliliter. If the concentration triples in 4 days, how long will it take for the concentration to reach 85,000 bacteria per milliliter?

103. *Continuous Compounding*   Suppose that $2000 is deposited in an account and the balance increases to $2300 after 4 years. How long will it take for the account to grow to $3200? Assume continuous compounding.

104. *Modeling Radioactive Decay*   Suppose that a 0.05-gram sample of a radioactive substance decays to 0.04 gram in 20 days. How long will it take for the sample to decay to 0.025 gram?

105. *Drug Concentrations*   The concentration of a drug in a patient's bloodstream after t hours is modeled by the formula $C(t) = 11(0.72)^t$, where C is measured in milligrams per liter.
  (a) What is the initial concentration of the drug?

  (b) How long does it take for the concentration to decrease to 50% of its initial level?

106. *Reducing Carbon Emissions*   When fossil fuels are burned, carbon is released into the atmosphere. Governments could reduce carbon emissions by placing a tax on fossil fuels. The **cost-benefit** equation

$$\ln(1 - P) = -0.0034 - 0.0053x$$

estimates the relationship between a tax of x dollars per ton of carbon and the percent P reduction in emissions of carbon, where P is in decimal form. Determine P when $x = 60$. Interpret the result.   (**Source:** W. Clime, *The Economics of Global Warming.*)

107. *Investments*   The formula $A = P\left(1 + \frac{r}{n}\right)^{nt}$ can be used to calculate the future value of an investment. Solve the equation for t.

108. *Decibels*   The formula $D = 160 + 10\log x$ can be used to calculate loudness of a sound. Solve the equation for x.

Writing about Mathematics

109. Explain how to solve the equation $Ca^x = k$ symbolically for x. Demonstrate your method.

110. Explain how to solve the equation $b\log_a x = k$ symbolically for x. Demonstrate your method.

EXTENDED AND DISCOVERY EXERCISE

1. *Exponential Functions* Show that any exponential function in the form $f(x) = Ca^x$ can be written as $f(x) = Ce^{kx}$. That is, write k in terms of a. Use your method to write $g(x) = 2^x$ in the form e^{kx} for some k.

CHECKING BASIC CONCEPTS FOR SECTIONS 5.5 AND 5.6

1. Use properties of logarithms to expand $\log \frac{x^2 y^3}{\sqrt[3]{z}}$. Write your answer without exponents.

2. Write the expression $\frac{1}{2} \ln x - 3 \ln y + \ln z$ as a logarithm of a single expression.

3. Solve each equation.
   (a) $5(1.4)^x - 4 = 25$ (b) $4^{2-x} = 4^{2x+1}$

4. Solve each equation.
   (a) $5 \log_2 2x = 25$

   (b) $\ln (x + 1) + \ln (x - 1) = \ln 3$

5. The temperature T of a cooling object in degrees Fahrenheit after x minutes is given by

$$T = 80 + 120(0.9)^x.$$

   (a) What happens to T after a long time?

   (b) After how long is the object's temperature 100°F?

5.7 Constructing Nonlinear Models

- **Find an exponential model**
- **Find a logarithmic model**
- **Find a logistic model**
- **Select a model**

Introduction

If data change at a constant rate, then they can be modeled with a linear function. However, real-life data often change at a nonconstant rate. For example, a tree grows slowly when it is small and then gradually grows faster as it becomes larger. Finally, when the tree is mature, its height begins to level off. This type of growth is nonlinear.

　　Three types of nonlinear data are shown in Figures 5.82–5.84, where t represents time. In Figure 5.82 the data increase rapidly, and an exponential function might be an appropriate modeling function. In Figure 5.83 the data are growing, but at a slower rate than in Figure 5.82. These data could be modeled by a logarithmic function. Finally, in Figure 5.84 the data increase slowly, then increase faster, and finally level off. These data might represent the height of a tree over a 50-year period. To model these data, we need a new type of function called a *logistic function*.

Figure 5.82 Exponential Data

Figure 5.83 Logarithmic Data

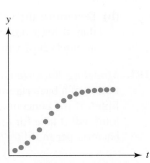

Figure 5.84 Logistic Data

Exponential Model

Both world population and bacteria growth can sometimes be modeled by an exponential function that *increases*. Exponential functions can also be used to model data that *decrease*. In the next example, an exponential function is used to model atmospheric pressure.

EXAMPLE 1 Modeling atmospheric pressure

As altitude increases, air pressure decreases. The atmospheric pressure P in millibars (mb) at a given altitude x in meters is listed in Table 5.26.

Table 5.26 Altitude and Air Pressure

x (m)	0	5000	10,000	15,000	20,000	25,000	30,000
P (mb)	1013	541	265	121	55	26	12

Source: A. Miller and J. Thompson, *Elements of Meteorology.*

(a) Make a scatterplot of the data. What type of function might model the data?
(b) Use regression to find an exponential function given by $f(x) = ab^x$. Graph the data and f in the same viewing rectangle.
(c) Use f to estimate the air pressure at an altitude of 23,000 feet.

Calculator Help

To find an equation of least-squares fit, see Appendix A (page AP-11). To copy a regression equation directly into Y_1, see Appendix A (page AP-13).

SOLUTION

(a) The data are shown in Figure 5.85. A *decreasing* exponential function might model the data.

NOTE It is possible for a different function, such as a portion of a polynomial graph, to model the data. Answers may vary.

(b) Figures 5.86 and 5.87 show that values of $a \approx 1104.9$ and $b \approx 0.99985$ are obtained from exponential regression, where $f(x) = ab^x$. Figure 5.88 illustrates that f models the data quite accurately.
(c) $f(23{,}000) = 1104.9(0.99985)^{23,000} \approx 35.1$ millibars

$[-5000, 35000, 5000]$ by
$[-100, 1200, 100]$

Figure 5.85

Figure 5.86

Figure 5.87

$[-5000, 35000, 5000]$ by
$[-100, 1200, 100]$

Figure 5.88

Now Try Exercise 15 ◀

Logarithmic Model

An investor buying a certificate of deposit (CD) usually gets a higher interest rate if the money is deposited for a longer period of time. However, the interest rate for a 2-year CD is not usually twice the rate for a 1-year CD. Instead the rate of interest gradually increases with a longer-term CD. In the next example, we model interest rates with a logarithmic function.

EXAMPLE 2 Modeling interest rates

Table 5.27 lists the interest rates for certificates of deposit. Use the data to complete the following.

Table 5.27 Yield on Certificates of Deposit

Time	6 mo	1 yr	2.5 yr	5 yr
Yield	4.75%	5.03%	5.25%	5.54%

Source: *USA Today.*

(a) Make a scatterplot of the data. What type of function might model these data?
(b) Use least-squares regression to find a formula $f(x) = a + b \ln x$ that models the data.
(c) Graph f and the data in the same viewing rectangle.

SOLUTION
(a) Enter the data points (0.5, 4.75), (1, 5.03), (2.5, 5.25), and (5, 5.54) into your calculator. A scatterplot of the data is shown in Figure 5.89. The data increase but gradually level off. A logarithmic modeling function may be appropriate.
(b) In Figures 5.90 and 5.91 least-squares regression has been used to find a logarithmic function f given (approximately) by $f(x) = 5 + 0.33 \ln x$.
(c) A graph of f and the data are shown in Figure 5.92.

[0, 6, 1] by [4, 6, 1]

Figure 5.89

Figure 5.90

Figure 5.91

[0, 6, 1] by [4, 6, 1]

Figure 5.92

Now Try Exercise 13 ◀

Logistic Model

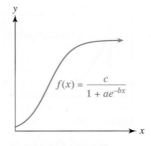

Figure 5.93 Sigmoidal Curve

In real life, populations of bacteria, insects, and animals do not continue to grow indefinitely. Initially, population growth may be slow. Then, as their numbers increase, so does the rate of growth. After a region has become heavily populated or saturated, the population usually levels off because of limited resources.

This type of growth may be modeled by a **logistic function** represented by $f(x) = \frac{c}{1 + ae^{-bx}}$, where a, b, and c are positive constants. A typical graph of a logistic function f is shown in Figure 5.93. The graph of f is referred to as a **sigmoidal curve**. The next example demonstrates how a logistic function can be used to describe the growth of a yeast culture.

EXAMPLE 3 Modeling logistic growth

One of the earliest studies about population growth was done using yeast plants in 1913. A small amount of yeast was placed in a container with a fixed amount of nourishment. The units of yeast were recorded every 2 hours. The data are listed in Table 5.28.

Table 5.28

Time	0	2	4	6	8	10	12	14	16	18
Yeast	9.6	29.0	71.1	174.6	350.7	513.3	594.8	640.8	655.9	661.8

Source: T. Carlson, *Biochem.;* D. Brown, *Models in Biology.*

(a) Make a scatterplot of the data in Table 5.28. Describe the growth.
(b) Use least-squares regression to find a logistic function f that models the data.
(c) Graph f and the data in the same viewing rectangle.
(d) Approximate graphically the time when the amount of yeast was 200 units.

[−2, 20, 1] by [−100, 800, 100]

Figure 5.94

SOLUTION

(a) A scatterplot of the data is shown in Figure 5.94. The yeast increase slowly at first. Then they grow more rapidly, until the amount of yeast gradually levels off. The limited amount of nourishment causes this leveling off.
(b) In Figure 5.95 and 5.96 we see least-squares regression being used to find a logistic function f given by

$$f(x) = \frac{661.8}{1 + 74.46e^{-0.552x}}.$$

(c) In Figure 5.97 the data and f are graphed in the same viewing rectangle. The fit for the *real* data is remarkably good.

[−2, 20, 1] by [−100, 800, 100]   [−2, 20, 1] by [−100, 800, 100]

Figure 5.95

Figure 5.96

Figure 5.97

Figure 5.98

(d) The graphs of $Y_1 = f(x)$ and $Y_2 = 200$ intersect near (6.29, 200), as shown in Figure 5.98. The amount of yeast reached 200 units after about 6.29 hours.

Now Try Exercise 17 ◄

CLASS DISCUSSION

In Example 3, suppose that after 18 hours the experiment had been extended and more nourishment had been provided for the yeast plants. Sketch a possible graph of the amount of yeast.

MAKING CONNECTIONS

Logistic Functions and Horizontal Asymptotes   If a logistic function is given by $f(x) = \frac{c}{1 + ae^{-bx}}$, where a, b, and c are positive constants, then the graph of f has a horizontal asymptote of $y = c$. (Try to explain why this is true.) In Example 3, the value of c was 661.8. This means that the amount of yeast leveled off at about 661.8 units.

Selecting a Model

In real-data applications, a modeling function is seldom given. Many times we must choose the type of modeling function and then find it using least-squares regression. Thus far in this section, we have used exponential, logarithmic, and logistic functions to model data. In the next two examples, we select a modeling function.

EXAMPLE 4 Modeling highway design

To allow enough distance for cars to pass on two-lane highways, engineers calculate minimum sight distances between curves and hills. See the figure. Table 5.29 shows the minimum sight distance y in feet for a car traveling at x miles per hour.

Table 5.29

x (mph)	20	30	40	50	60	65	70
y (ft)	810	1090	1480	1840	2140	2310	2490

Source: L. Haefner, *Introduction to Transportation Systems.*

(a) Find a modeling function for the data.
(b) Graph the data and your modeling function.
(c) Estimate the minimum sight distance for a car traveling at 43 miles per hour.

SOLUTION

Getting Started One strategy is to plot the data and then decide if the data are linear or nonlinear. If the data are approximately linear, use linear regression to find the modeling function. If the data are nonlinear, think about how the data increase or decrease. You may want to try several types of modeling functions, such as quadratic, cubic, power, exponential, or logarithmic, before making a final decision. ▶

(a) A scatterplot is shown in Figure 5.99. The data appear to be (nearly) linear, so linear regression has been used to obtain $f(x) = 33.93x + 113.4$. See Figure 5.100.
(b) The data and f are graphed in Figure 5.101. Function f gives a good fit.

[0, 80, 10] by [0, 3000, 1000]

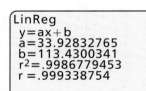

[0, 80, 10] by [0, 3000, 1000]

Figure 5.99   **Figure 5.100**   **Figure 5.101**

(c) $f(43) = 33.93(43) + 113.4 \approx 1572$ feet

Now Try Exercise 19 ◀

You may have seen asbestos being removed from buildings. Before 1960 people were generally unaware of its health hazards. As a result, insulation workers who worked with asbestos experienced higher rates of lung cancer. The following example models data from this era.

EXAMPLE 5 Modeling asbestos and cancer

Table 5.30 lists the number N of lung cancer cases occurring within a group of asbestos insulation workers with a cumulative total of 100,000 years of work experience, with their first date of employment x years ago.

Table 5.30

x (years)	10	15	20	25	30
N (cases)	6.9	25.4	63.6	130	233

Source: A. Walker, *Observation and Inference.*

(a) Find a modeling function for the data.
(b) Graph the data and your modeling function.
(c) Estimate the number of lung cancer cases for $x = 23$ years.

SOLUTION

Getting Started The data are nonlinear and increasing, so there are a number of functions you can try. Three possibilities are quadratic, power, and exponential. ▶

(a) A scatterplot is shown in Figure 5.102. To model the data we have used a power function given by $f(x) = 0.004334x^{3.2}$. See Figure 5.103. (Answers may vary.)
(b) The data and f are graphed in Figure 5.104. Function f gives a good fit.

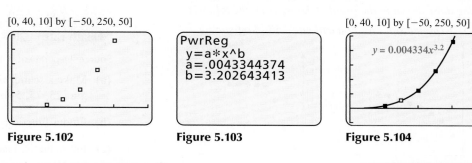

[0, 40, 10] by [−50, 250, 50]

PwrReg
y=a*x^b
a=.0043344374
b=3.202643413

[0, 40, 10] by [−50, 250, 50]

$y = 0.004334x^{3.2}$

Figure 5.102   **Figure 5.103**   **Figure 5.104**

(c) $f(23) = 0.004334(23)^{3.2} \approx 99$ cases

Now Try Exercise 21 ◀

5.7 Putting It All Together

The following table summarizes the basics of exponential, logarithmic, and logistic models. Least-squares regression can be used to determine the constants a, b, c, and C.

Concept	Explanation	Examples
Exponential model	$f(x) = Ca^x$, $f(x) = ab^x$, or $A(t) = A_0 e^{kt}$	Exponential functions can be used to model data that increase or decrease rapidly over time.
Logarithmic model	$f(x) = a + b \log x$ or $f(x) = a + b \ln x$	Logarithmic functions can be used to model data that increase gradually over time.
Logistic model	$f(x) = \dfrac{c}{1 + ae^{-bx}}$	Logistic functions can be used to model data that at first increase slowly, then increase rapidly, and finally level off.

5.7　Exercises

Note: Because different functions can be used to model the same data, your answers may vary from the given answers. You can check the validity of your answer by graphing the data and your modeling function in the same viewing rectangle.

Selecting a Model

Exercises 1–4: Select an appropriate type of modeling function for the data shown in the graph. Choose from the following.

　i. Exponential　ii. Logarithmic　iii. Logistic

Exercises 5–10: Make a scatterplot of the data. Then find an exponential, logarithmic, or logistic function f that best models the data.

5.

x	1	2	3	4
y	2.04	3.47	5.90	10.02

6.

x	1	2	3	4	5
y	1.98	2.35	2.55	2.69	2.80

7.

x	1	2	3	4	5
y	1.1	3.1	4.3	5.2	5.8

8.

x	1	2	3	4	5	6
y	1	2	4	7	9	10

9.

x	0	1	2	3	4	5
y	0.3	1.3	4.0	7.5	9.3	9.8

10.

x	1	2	3	4	5
y	2.0	1.6	1.3	1.0	0.82

Applications

11. *Heart Disease Death Rates*　The table contains heart disease death rates per 100,000 people for selected ages.

Age	30	40	50	60	70
Death rate	30.5	108.2	315	776	2010

Source: Department of Health and Human Services.

(a) Make a scatterplot of the data in the viewing rectangle [25, 75, 5] by [−100, 2100, 200].

(b) Find a function f that models the data.

(c) Estimate the heart disease death rate for people who are 80 years old.

12. *Female Mechanics*　The number of females working in automotive repair is increasing. The table shows the number of female ASE-certified technicians for selected years.

Year	1988	1989	1990	1991
Total	556	614	654	737

Year	1992	1993	1994	1995
Total	849	1086	1329	1592

Source: National Institute for Automotive Service Excellence.

(a) What type of function might model these data?

(b) Use least-squares regression to find an exponential function given by $f(x) = ab^x$ that models the data. Let $x = 0$ correspond to 1988.

(c) Use f to estimate the number of certified female technicians in 2005. Round the result to the nearest hundred.

13. *Telecommuting*　Some workers use technology such as e-mail, computers, and multiple phone lines to work at home, rather than in the office. However, because of the need for teamwork and collaboration in the workplace, fewer employees are telecommuting than expected. The table lists telecommuters T in millions during year x.

x	1997	1998	1999	2000	2001
T	9.2	9.6	10.0	10.4	10.6

x	2002	2003	2004	2005	2006
T	11.0	11.1	11.2	11.3	11.4

Source: *USA Today.*

Find a function f that models the data, where $x = 1$ corresponds to 1997, $x = 2$ to 1998, and so on.

📠 **14.** *Hurricanes*   The table shows the air pressure y in inches of mercury x miles from the eye of a hurricane.

x	2	4	8	15	30	100
y	27.3	27.7	28.04	28.3	28.7	29.3

Source: A. Miller and R. Anthes, *Meteorology.*

(a) Make a scatterplot of the data.

(b) Find a function f that models the data.

(c) Estimate the air pressure at 50 miles.

📠 **15.** *Atmospheric Density*   The table lists the atmospheric density y in kilograms per cubic meter (kg/m^3) at an altitude of x meters.

x (m)	0	5000	10,000	15,000
y (kg/m³)	1.2250	0.7364	0.4140	0.1948

x (m)	20,000	25,000	30,000
y (kg/m³)	0.0889	0.0401	0.0184

Source: A. Miller.

(a) Find a function f that models the data.

(b) Predict the density at 7000 meters. (The actual value is 0.59 kg/m³.)

16. *Modeling Data*   Use the table to complete the following.

x	1	2	5	10
y	2.5	2.1	1.6	1.2

(a) Find a function f that models the data.

(b) Solve the equation $f(x) = 1.8$.

📠 **17.** *Insect Population*   The table at the top of the next column shows the density y of a species of insect measured in thousands per acre after x days.

x	2	4	6	8	10	12	14
y	0.38	1.24	2.86	4.22	4.78	4.94	4.98

(a) Find a function f that models the data.

(b) Use f to estimate the insect density after a long time.

18. *Heart Disease*   As age increases, so does the likelihood of coronary heart disease (CHD). The percentage P of people x years old with signs of CHD is shown in the table.

x	15	25	35	45	55	65	75
P (%)	2	7	19	43	68	82	87

Source: D. Hosmer and S. Lemeshow, *Applied Logistics Regression.*

(a) Evaluate $P(25)$ and interpret the answer.

(b) Find a function that models the data.

(c) Graph P and the data.

(d) At what age does a person have a 50% chance of having signs of CHD?

19. *U.S. Radio Stations*   The numbers N of radio stations on the air for selected years x are listed in the table.

x	1960	1970	1980	1990	2000
N(x)	4133	6760	8566	10,788	12,808

Source: M. Street Corporation.

(a) Find a function that models the data.

(b) Graph N and the data.

(c) When did the number of radio stations on the air reach 9500?

20. *College Tuition and Fees*   The table lists average public tuition and fees T in constant (2006) dollars during year x.

x	1980	1985	1990	1995	2000
T(x)	1707	2227	2620	3346	3662

Source: The College Board.

(a) Find a function that models the data.

(b) Graph T and the data.

(c) When did tuition and fees first reach $3000?

21. *Wing Size* Heavier birds tend to have larger wings than smaller birds. For one species of bird, the table lists the area A of the bird's wing in square inches if the bird weighs w pounds.

w (lb)	2	6	10	14	18
$A(w)$ (in^2)	160	330	465	580	685

Source: C. Pennycuick, *Newton Rules Biology.*

(a) Find a function that models the data.

(b) Graph A and the data.

(c) What weight corresponds to a wing area of 500 square inches?

22. *Wing Span* Heavier birds tend to have a longer wing span than smaller birds. For one species of bird, the table lists the length L of the bird's wing span in feet if the bird weighs w pounds.

w (lb)	0.22	0.88	1.76	2.42
$L(w)$ (ft)	1.38	2.19	2.76	3.07

Source: C. Pennycuick, *Newton Rules Biology.*

(a) Find a function that models the data.

(b) Graph L and the data.

(c) What weight corresponds to a wing span of 2 feet?

23. *Tree Growth* (Refer to the introduction to this section.) The height H of a tree in feet after x years is listed in the table.

x (yr)	1	5	10	20	30	40
$H(x)$ (ft)	1.3	3	8	32	47	50

(a) Evaluate $H(5)$ and interpret the answer.

(b) Find a function that models the data.

(c) Graph H and the data.

(d) What is the age of the tree when its height is 25 feet?

(e) Did your answer involve interpolation or extrapolation?

24. *Bird Populations* Near New Guinea there is a relationship between the number of bird species found on an island and the size of the island. The table lists the number of species of birds y found on an island with an area of x square kilometers.

x (km^2)	0.1	1	10	100	1000
y (species)	10	15	20	25	30

Source: B. Freedman, *Environmental Ecology.*

(a) Find a function f that models the data.

(b) Predict the number of bird species on an island of 5000 square kilometers.

(c) Did your answer involve interpolation or extrapolation?

25. *Fertilizer Usage* Between 1950 and 1980 the use of chemical fertilizers increased. The table lists worldwide average usage y in kilograms per hectare of cropland, x years after 1950. (*Note:* 1 hectare $\approx$ 2.47 acres.)

x	0	13	22	29
y	12.4	27.9	54.3	77.1

Source: D. Grigg.

(a) Graph the data. Are the data linear?

(b) Find a function f that models the data.

(c) Predict fertilizer usage in 1989. The actual value was 98.7 kilograms per hectare. What does this indicate about usage of fertilizer during the 1980s?

26. *Social Security* If major reform occurs in the Social Security system, individuals may be able to invest some of their contributions into individual accounts. These accounts would be managed by financial firms, which often charge fees. The table lists the amount in billions of dollars that may be collected if fees are 0.93% of the assets each year.

Year	2005	2010	2015	2020
Fees ($ billions)	20	41	80	136

Source: Social Security Advisory Council.

(a) Use exponential regression to find a and b so that $f(x) = ab^x$ models the data x years after 2000.

(b) Graph f and the data.

(c) Estimate the fees in 2013.

(d) Did your answer involve interpolation or extrapolation?

Writing about Mathematics

27. How can you distinguish data that illustrate exponential growth from data that illustrate logarithmic growth?

28. Give an example of data that could be modeled by a logistic function and explain why.

EXTENDED AND DISCOVERY EXERCISE

1. For medical reasons, dyes may be injected into the bloodstream to determine the health of internal organs. In one study involving animals, the dye BSP was injected to assess the blood flow in the liver. The results are listed in the table, where x represents the elapsed time in minutes and y is the concentration of the dye in the bloodstream in milligrams per milliliter.

x	1	2	3	4	5	7
y	0.102	0.077	0.057	0.045	0.036	0.023

x	9	13	16	19	22
y	0.015	0.008	0.005	0.004	0.003

Source: F. Harrison, "The measurement of liver blood flow in conscious calves."

(a) Find a function that models the data.

(b) Estimate the elapsed time when the concentration of the dye reaches 30% of its initial concentration of 0.133 mg/ml.

(c) Let $g(x) = 0.133(0.878(0.73^x) + 0.122(0.92^x))$. This formula was used by the researchers to model the data. Compare the accuracy of your formula to that of $g(x)$.

CHECKING BASIC CONCEPTS FOR SECTION 5.7

Exercises 1–4: Find a function that models the data. Choose from exponential, logarithmic, or logistic functions.

1.

x	2	3	4	5	6	7
y	0.72	0.86	1.04	1.24	1.49	1.79

2.

x	2	3	4	5	6	7
y	0.08	1.30	2.16	2.83	3.38	3.84

3.

x	2	3	4	5	6	7
y	0.25	0.86	2.19	3.57	4.23	4.43

4. *World Population* The table lists the actual or projected world population y (in billions) for selected years x.

x	1950	1960	2000	2050	2075
y	2.5	3.0	6.1	8.9	9.2

Source: U.N. Dept. of Economic and Social Affairs.

---

5 Summary

CONCEPT	EXPLANATION AND EXAMPLES

SECTION 5.1 COMBINING FUNCTIONS

Arithmetic Operations on Functions

Addition: $(f + g)(x) = f(x) + g(x)$
Subtraction: $(f - g)(x) = f(x) - g(x)$
Multiplication: $(fg)(x) = f(x) \cdot g(x)$
Division: $(f/g)(x) = \dfrac{f(x)}{g(x)}, g(x) \neq 0$

Examples: Let $f(x) = x^2 - 5, g(x) = x^2 - 4$.
$(f + g)(x) = (x^2 - 5) + (x^2 - 4) = 2x^2 - 9$
$(f - g)(x) = (x^2 - 5) - (x^2 - 4) = -1$
$(fg)(x) = (x^2 - 5)(x^2 - 4) = x^4 - 9x^2 + 20$
$(f/g)(x) = \dfrac{x^2 - 5}{x^2 - 4}, x \neq 2, x \neq -2$

CONCEPT	EXPLANATION AND EXAMPLES

SECTION 5.1 COMBINING FUNCTIONS (CONTINUED)

Composition of Functions

Composition: $(g \circ f)(x) = g(f(x))$
$(f \circ g)(x) = f(g(x))$

Examples: Let $f(x) = 3x + 2$, $g(x) = 2x^2 - 4x + 1$.
$$g(f(x)) = g(3x + 2)$$
$$= 2(3x + 2)^2 - 4(3x + 2) + 1$$
$$f(g(x)) = f(2x^2 - 4x + 1)$$
$$= 3(2x^2 - 4x + 1) + 2$$
$$= 6x^2 - 12x + 5$$

SECTION 5.2 INVERSE FUNCTIONS AND THEIR REPRESENTATIONS

Inverse Function

The inverse function of f is f^{-1} if
$f^{-1}(f(x)) = x$ for every x in the domain of f and
$f(f^{-1}(x)) = x$ for every x in the domain of f^{-1}.
Note: If $f(a) = b$, then $f^{-1}(b) = a$.

Example: Find the inverse function of $f(x) = 4x - 5$.

$y = 4x - 5$ is equivalent to $\frac{y + 5}{4} = x$. (Solve for x.)
Thus $f^{-1}(x) = \frac{x + 5}{4}$.

One-to-One Function

If different inputs always result in different outputs, then f is one-to-one. That is, $a \neq b$ implies $f(a) \neq f(b)$.
Note: If f is one-to-one, then f has an inverse denoted f^{-1}.

Example: $f(x) = x^2 + 1$ is not one-to-one because $f(2) = f(-2) = 5$.

Horizontal Line Test

If every horizontal line intersects the graph of a function f at most once, then f is a one-to-one function.

SECTION 5.3 EXPONENTIAL FUNCTIONS AND MODELS

Exponential Function

$f(x) = Ca^x$, $a > 0$, $a \neq 1$, and $C > 0$
Exponential growth: $a > 1$; exponential decay: $0 < a < 1$

Examples: $f(x) = 3(2)^x$ (growth); $f(x) = 1.2(0.5)^x$ (decay)

Exponential Data

For each unit increase in x, the y-values increase (or decrease) by a constant factor a.

Example: The data in the table are modeled by $y = 5(2)^x$.

x	0	1	2	3
y	5	10	20	40

CONCEPT	EXPLANATION AND EXAMPLES

SECTION 5.3 EXPONENTIAL FUNCTIONS AND MODELS (CONTINUED)

Natural Exponential Function

$f(x) = e^x$, where $e \approx 2.718282$

Compound Interest

$A = P\left(1 + \frac{r}{n}\right)^{nt}$, where P is the principal, r is the interest expressed as a decimal, n is the number of times interest is paid each year, and t is the number of years

Example: $A = 2000\left(1 + \dfrac{0.10}{12}\right)^{12(4)} \approx \$2978.71$

calculates the future value of \$2000 invested at 10% compounded monthly for 4 years.

Continuously Compounded Interest

$A = Pe^{rt}$, where P is the principal, r is the interest expressed as a decimal, and t is the number of years

Example: $A = 2000e^{0.10(4)} \approx \$2983.65$

calculates the future value of \$2000 invested at 10% compounded continuously for 4 years.

SECTION 5.4 LOGARITHMIC FUNCTIONS AND MODELS

Common Logarithm

$\log x = k$ if and only if $x = 10^k$

Natural Logarithm

$\ln x = k$ if and only if $x = e^k$

General Logarithm

$\log_a x = k$ if and only if $x = a^k$

Examples: $\log 100 = 2$ because $100 = 10^2$.

$\ln \sqrt{e} = \dfrac{1}{2}$ because $\sqrt{e} = e^{1/2}$.

$\log_2 \frac{1}{8} = -3$ because $\frac{1}{8} = 2^{-3}$

Inverse Properties

$\log 10^k = k$, $\quad 10^{\log k} = k$, $\quad k > 0$
$\ln e^k = k$, $\quad e^{\ln k} = k$, $\quad k > 0$
$\log_a a^k = k$, $\quad a^{\log_a k} = k$, $\quad k > 0$

Examples: $10^{\log 100} = 100$; $e^{\ln 23} = 23$; $\log_4 64 = \log_4 4^3 = 3$

Inverse Functions

The inverse function of $f(x) = a^x$ is $f^{-1}(x) = \log_a x$.

Examples: If $f(x) = 10^x$, then $f^{-1}(x) = \log x$.
If $f(x) = \ln x$, then $f^{-1}(x) = e^x$.
If $f(x) = \log_5 x$, then $f^{-1}(x) = 5^x$.

Exponential and Logarithm Forms

$\log_a x = k$ is equivalent to $x = a^k$.

Examples: $\log_2 16 = 4$ is equivalent to $16 = 2^4$.
$81 = 3^4$ is equivalent to $\log_3 81 = 4$.

CONCEPT	EXPLANATION AND EXAMPLES

SECTION 5.5 PROPERTIES OF LOGARITHMS

Properties of Logarithms

1. $\log_a 1 = 0$ and $\log_a a = 1$
2. $\log_a m + \log_a n = \log_a (mn)$
3. $\log_a m - \log_a n = \log_a \left(\dfrac{m}{n}\right)$
4. $\log_a (m^r) = r \log_a m$

Examples:
1. $\log_4 1 = 0$ and $\log_4 4 = 1$
2. $\log 2 + \log 5 = \log (2 \cdot 5) = \log 10 = 1$
3. $\log 500 - \log 5 = \log (500/5) = \log 100 = 2$
4. $\log_2 2^3 = 3 \log_2 2 = 3(1) = 3$

Change of Base Formula

$$\log_a x = \frac{\log_b x}{\log_b a}$$

Example: $\log_3 23 = \dfrac{\log 23}{\log 3} \approx 2.854$

SECTION 5.6 EXPONENTIAL AND LOGARITHMIC EQUATIONS

Solving Exponential Equations

To solve an exponential equation we typically take the logarithm of each side.

Example:

$4e^x = 48$	*Given equation*
$e^x = 12$	*Divide by 4.*
$\ln e^x = \ln 12$	*Take the natural logarithm.*
$x = \ln 12$	*Inverse property*
$x \approx 2.485$	*Approximate.*

Solving Logarithmic Equations

To solve a logarithmic equation we typically need to exponentiate each side.

Example:

$5 \log_3 x = 10$	*Given equation*
$\log_3 x = 2$	*Divide by 5.*
$3^{\log_3 x} = 3^2$	*Exponentiate; base 3.*
$x = 9$	*Inverse property*

SECTION 5.7 CONSTRUCTING NONLINEAR MODELS

Exponential Model

$f(x) = Ca^x$, $f(x) = ab^x$, or $A(t) = A_0 e^{kt}$
Models data that increase or decrease rapidly

Logarithmic Model

$f(x) = a + b \ln x$ or $f(x) = a + b \log x$
Models data that increase slowly

Logistic Model

$f(x) = \dfrac{c}{1 + ae^{-bx}}$, a, b, and c are positive constants

Models data that increase slowly at first, then increase more rapidly, and finally level off near the value of c. Its graph is a sigmoidal curve. See Figure 5.93.

5   Review Exercises

1. Use the tables to evaluate, if possible.

x	−1	0	1	3
f(x)	3	5	7	9

x	−1	0	1	3
g(x)	−2	0	1	9

(a) $(f + g)(1)$

(b) $(f - g)(3)$

(c) $(fg)(-1)$

(d) $(f/g)(0)$

2. Use the graph to evaluate each expression.
   (a) $(f - g)(2)$
   (b) $(fg)(0)$

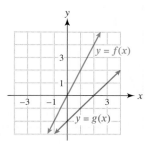

3. Let $f(x) = x^2$ and $g(x) = 1 - x$ and evaluate.
   (a) $(f + g)(3)$
   (b) $(f - g)(-2)$

   (c) $(fg)(1)$
   (d) $(f/g)(3)$

4. Use $f(x) = x^2 + 3x$ and $g(x) = x^2 - 1$ to find each expression. Identify its domain.
   (a) $(f + g)(x)$
   (b) $(f - g)(x)$

   (c) $(fg)(x)$
   (d) $(f/g)(x)$

5. Numerical representations for f and g are given by the tables. Evaluate each expression.

x	−2	0	2	4
f(x)	1	4	3	2

x	1	2	3	4
g(x)	2	4	−2	0

(a) $(g \circ f)(-2)$ (b) $(f \circ g)(3)$ (c) $f^{-1}(3)$

6. Use the graph at the top of the next column to evaluate each expression.
   (a) $(f \circ g)(2)$ (b) $(g \circ f)(0)$ (c) $f^{-1}(1)$

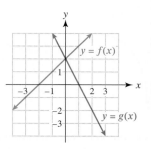

7. Let $f(x) = \sqrt{x}$ and $g(x) = x^2 + x$ and evaluate.
   (a) $(f \circ g)(2)$
   (b) $(g \circ f)(9)$

8. Use $f(x) = x^2 + 1$ and $g(x) = x^3 - x^2 + 2x + 1$ to find each expression.
   (a) $(f \circ g)(x)$
   (b) $(g \circ f)(x)$

Exercises 9–12: Find $(f \circ g)(x)$ and identify its domain.

9. $f(x) = x^3 - x^2 + 3x - 2$ $g(x) = x^{-1}$

10. $f(x) = \sqrt{x + 3}$ $g(x) = 1 - x^2$

11. $f(x) = \sqrt[3]{2x - 1}$ $g(x) = \frac{1}{2}x^3 + \frac{1}{2}$

12. $f(x) = \dfrac{2}{x - 5}$ $g(x) = \dfrac{1}{x + 1}$

Exercises 13 and 14: Find f and g so that $h(x) = (g \circ f)(x)$.

13. $h(x) = \sqrt{x^2 + 3}$ 14. $h(x) = \dfrac{1}{(2x + 1)^2}$

Exercises 15 and 16: Describe the inverse operations of the given statement. Then express both the statement and its inverse symbolically.

15. Divide x by 10 and add 6.

16. Subtract 5 from x and take the cube root.

Exercises 17 and 18: Determine if f is one-to-one.

17. $f(x) = 3x - 1$ 18. $f(x) = 3x^2 - 2x + 1$

Exercises 19 and 20: Determine if f is one-to-one.

19.

20.

21. The table is a complete representation of f. Use the table of f to determine a table for f^{-1}. Identify the domains and ranges of f and f^{-1}.

x	−1	0	4	6
$f(x)$	6	4	3	1

22. Use the graph of f to sketch a graph of f^{-1}.

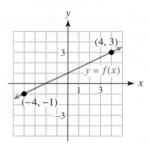

Exercises 23 and 24: Find $f^{-1}(x)$.

23. $f(x) = 3x - 5$

24. $f(x) = \frac{3x}{x + 7}$

25. Verify that $f(x) = 2x - 1$ and $f^{-1}(x) = \frac{x + 1}{2}$ are inverses.

26. Restrict the domain of $f(x) = 2(x - 4)^2 + 3$ so that f is one-to-one. Then find $f^{-1}(x)$.

Exercises 27 and 28: Use the tables to evaluate the given expression.

x	0	1	2	3
$f(x)$	4	3	2	1

x	0	1	2	3
$g(x)$	0	2	3	4

27. $(f \circ g^{-1})(4)$

28. $(g^{-1} \circ f^{-1})(1)$

29. Find $f^{-1}(x)$ if $f(x) = \sqrt{x + 1}, x \geq -1$. Identify the domain and range of f and of f^{-1}.

30. Simplify $e^x e^{-2x}$.

Exercises 31 and 32: Find C and a so that $f(x) = Ca^x$ satisfies the given conditions.

31. $f(0) = 3$ and $f(3) = 24$

32. $f(-1) = 8$ and $f(1) = 2$

Exercises 33–36: Sketch a graph of $y = f(x)$. Identify the domain of f.

33. $f(x) = 4(2)^{-x}$

34. $f(x) = 3^{x-1}$

35. $f(x) = \log_4 x$

36. $f(x) = \log(x + 1)$

Exercises 37 and 38: Use the graph of $f(x) = Ca^x$ to determine values for C and a.

37.

38.

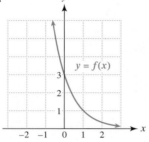

39. Determine the final value of $1200 invested at 9% compounded semiannually for 3 years.

40. Determine the final value of $500 invested at 6.5% compounded continuously for 8 years.

41. Solve $e^x = 19$ symbolically.

42. Solve $2^x - x^2 = x$ graphically. Round each solution to the nearest thousandth.

Exercises 43–46: Evaluate the expression without a calculator.

43. $\log 1000$

44. $\log 0.001$

45. $10 \log 0.01 + \log \frac{1}{10}$

46. $\log 100 + \log \sqrt[3]{10}$

Exercises 47–50: Evaluate the logarithm without a calculator.

47. $\log_3 9$

48. $\log_5 \frac{1}{25}$

49. $\ln e$

50. $\log_2 32$

Exercises 51 and 52: Approximate to the nearest thousandth.

51. $\log_3 18$

52. $\log_2 173$

Exercises 53–60: Solve the equation.

53. $10^x = 125$

54. $1.5^x = 55$

55. $e^{0.1x} = 5.2$

56. $4e^{2x} - 5 = 3$

57. $5^{-x} = 10$

58. $3(10)^{-x} = 6$

59. $50 - 3(0.78)^{x-10} = 21$ **60.** $5(1.3)^x + 4 = 104$

Exercises 61 and 62: Find either a linear or an exponential function that models the data in the table.

61.

x	0	1	2
y	1.5	3	6

62.

x	0	1	2
y	3	4.5	6

Exercises 63–66: Solve the equation.

63. $\log x = 1.5$

64. $\log_3 x = 4$

65. $\ln x = 3.4$

66. $4 - \ln(5 - x) = \frac{5}{2}$

Exercises 67 and 68: Use properties of logarithms to write the expression as a logarithm of a single expression.

67. $\log 6 + \log 5x$

68. $\log \sqrt{3} - \log \sqrt[3]{3}$

69. Expand $\ln \dfrac{y}{x^2}$.

70. Expand $\log \dfrac{4x^3}{k}$.

Exercises 71–76: Solve the logarithmic equation.

71. $8 \log x = 2$

72. $\ln 2x = 2$

73. $2 \log 3x + 5 = 15$

74. $5 \log_2 x = 25$

75. $2 \log_5(x + 2) = \log_5(x + 8)$

76. $\ln(5 - x) - \ln(5 + x) = -\ln 9$

77. Suppose that b is the y-intercept on the graph of a one-to-one function f. What is the x-intercept on the graph of f^{-1}? Explain your reasoning.

78. Let $f(x) = ax + b$ with $a \neq 0$.
   (a) Show that f^{-1} is also linear by finding $f^{-1}(x)$.

   (b) How is the slope of the graph of f related to the slope of the graph of f^{-1}?

Applications

79. *Bacteria Growth*   There are initially 4000 bacteria per milliliter in a sample, and after 1 hour their concentration increases to 6000 bacteria per milliliter. Assume exponential growth.
   (a) How many bacteria are there after 2.5 hours?

   (b) After how long are there 8500 bacteria per milliliter?

80. *Newton's Law of Cooling*   A pan of boiling water with a temperature of 100°C is set in a room with a temperature of 20°C. The water cools to 50°C in 40 minutes.
   (a) Find values for T_0, D, and a so that the formula $T(t) = T_0 + Da^t$ models the data, where t is in hours.

   (b) Find the temperature of the water after 90 minutes.

   (c) How long does it take the water to reach 30°C?

81. *Combining Functions*   The total number of gallons of water passing through a pipe after x seconds is computed by $f(x) = 10x$. Another pipe delivers $g(x) = 5x$ gallons after x seconds. Find a function h that gives the volume of water passing through both pipes in x seconds.

82. *Test Scores*   Let scores on a standardized test be modeled by $f(x) = 36e^{-(x-20)^2/49}$. The function f computes the number in thousands of people that received score x. Solve the equation $f(x) = 30$. Interpret your result.

83. *Modeling Growth*   The function given by
$$W(x) = 175.6(1 - 0.66e^{-0.24x})^3$$
models the weight in milligrams of a small fish called the *Lebistes reticulatus* after x weeks, where $0 \leq x \leq 14$. Solve the equation $W(x) = 50$. Interpret the result. (**Source:** D. Brown and P. Rothery, *Models in Biology.*)

84. *Radioactive Decay*   After 23 days, a 10-milligram sample of a radioactive material decays to 5 milligrams. After how many days will there be 1 milligram of the material?

85. *Tire Pressure*   A car tire has a small leak, and the tire pressure in pounds per square inch after t minutes is given by $P(t) = 32e^{-0.2t}$. After how many minutes is the pressure 15 pounds per square inch?

86. *Converting Units*   The figures show graphs of a function f that converts fluid ounces to pints and a function g that converts pints to quarts. Evaluate each expression. Interpret the results.

(a) $(g \circ f)(32)$

(b) $f^{-1}(1)$

(c) $(f^{-1} \circ g^{-1})(1)$

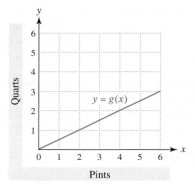

87. *Modeling Epidemics*   In 1666 the village of Eyam, located in England, experienced an outbreak of the Great Plague. Out of 261 people in the community, only 83 people survived. The tables show a function f that computes the number of people who were infected after x days.

x	0	15	30	45
$f(x)$	7	21	57	111

x	60	75	90	125
$f(x)$	136	158	164	178

Source: G. Raggett, "Modeling the Eyam plague."

Find a function f that models the given data. (Answers may vary.)

88. *Greenhouse Gases*   Methane is a greenhouse gas that is produced when fossil fuels are burned. In 1600 methane had an atmospheric concentration of 700 parts per billion (ppb), whereas in 2000 its concentration was about 1700 ppb.   (**Source:** D. Wuebbles and J. Edmonds, *Primer on Greenhouse Gases.*)

(a) Find values for C and a so that $f(x) = Ca^x$ models the data, where x is the year.

(b) Solve $f(x) = 1000$ and interpret the answer.

89. *Exponential Regression*   The data in the table can be modeled by $f(x) = ab^x$. Use regression to estimate the constants a and b. Graph f and the data.

x	1	2	3	4
y	2.59	1.92	1.42	1.05

90. *Logarithmic Regression*   The data in the table can be modeled by $f(x) = a + b \ln x$. Use regression to estimate the constants a and b. Graph f and the data.

x	2	3	4	5
y	2.93	3.42	3.76	4.03

EXTENDED AND DISCOVERY EXERCISES

1. *Modeling Data with Power Functions*   There is a procedure to determine whether data can be modeled by $y = ax^b$, where a and b are constants. Start by taking the natural logarithm of each side of this equation.

$$\ln y = \ln (ax^b)$$
$$\ln y = \ln a + \ln x^b \qquad \text{ln } (mn) = \ln m + \ln n$$
$$\ln y = \ln a + b \ln x \qquad \text{ln } (m^r) = r \ln m$$

If we let $z = \ln y$, $d = \ln a$, and $w = \ln x$, then the equation $\ln y = \ln a + b \ln x$ becomes $z = d + bw$. Thus the data points $(w, z) = (\ln x, \ln y)$ lie on the line having a slope of b and y-intercept $d = \ln a$. The following steps provide a procedure for finding the constants a and b.

Modeling Data with the Equation $y = ax^b$

If a data set (x, y) can be modeled by the (power) equation $y = ax^b$, then the following procedure can be applied to determine the constants a and b.

STEP 1: Let $w = \ln x$ and $z = \ln y$ for each data point. Graph the points (w, z). If these data are not linear, then do *not* use this procedure.

STEP 2: Find an equation of a line in the form $z = bw + d$ that models the data points (w, z). (Linear regression may be used.)

STEP 3: The slope of the line equals the constant b. The value of a is given by $a = e^d$.

Apply this procedure to the table of data for the orbital distances and periods of the moons of Jupiter. Let the distance be x and the period be y.

Moons of Jupiter	Distance (10^3 km)	Period (days)
Metis	128	0.29
Almathea	181	0.50
Thebe	222	0.67
Io	422	1.77
Europa	671	3.55
Ganymede	1070	7.16
Callisto	1883	16.69

2. *Global Warming*   Greenhouse gases such as carbon dioxide trap heat from the sun. Presently, the net incoming solar radiation reaching Earth's surface is approximately 240 watts per square meter (w/m²). Any portion of this amount that is due to greenhouse gases is called *radiative forcing*. The table lists the estimated increase in radiative forcing R over the levels in 1750.

x (year)	1800	1850	1900	1950	2000
$R(x)$ (w/m²)	0.2	0.4	0.6	1.2	2.4

Source: A. Nilsson, *Greenhouse Earth.*

(a) Estimate constants C and k so that $R(x) = Ce^{kx}$ models the data. Let $x = 0$ correspond to 1800.

(b) Estimate the year when the additional radiative forcing could reach 3 w/m².

3. *Global Warming*   (Refer to Exercise 2.) The relationship between radiative forcing R and the increase in average global temperature T in degrees Fahrenheit can be modeled by $T(R) = 1.03R$. For example, $T(2) = 1.03(2) = 2.06$ means that if Earth's atmosphere traps an additional 2 watts per square meter, then the average global temperature may increase by 2.06°F *if* all other factors remained constant.   (**Source:** W. Clime, *The Economics of Global Warming.*)

(a) Use $R(x)$ from Exercise 2 to express $(T \circ R)(x)$ symbolically, where x is the year and $x = 0$ corresponds to the year 1800.

(b) Evaluate $(T \circ R)(100)$ and interpret its meaning.

Exercises 4 and 5: Try to decide if the expression is precisely an integer. (Hint: Use computer software capable of calculating a large number of decimal places.)

4. $\left(\dfrac{1}{\pi}\ln(640{,}320^3 + 744)\right)^2$

   (**Source:** I. J. Good, "What is the most amazing approximate integer in the universe?")

5. $e^{\pi\sqrt{163}}$ (**Source:** W. Cheney and D. Kincaid, *Numerical Mathematics and Computing.*)

6 Trigonometric Functions

Industrial robots comprise a multi-billion-dollar industry in the United States. Robots paint our automobiles, visit the *Titanic* deep in the ocean, recognize voices, and even perform surgery that is more precise and less invasive than traditional techniques. In the future, robots will affect our quality of life by being able to assist the elderly, handle hazardous material, and help repair over 2 million miles of underground piping in our deteriorating infrastructure. Spectacular advances in microelectronic mechanical systems (MEMS) may make it possible for tiny robots to repair human tissue at the cellular level.

The trigonometry that has played a key role in the design of robots goes back to the 19th century. Some of the most important results in robotics have come from mathematicians and their interactions with engineers. Trigonometry is also used in applications involving monthly average temperatures, tidal currents, the Global Positioning System, electricity, highways, orbits of satellites, phases of the moon, and even music. In this chapter, we discuss the six basic trigonometric functions and many of their applications.

Source: National Science Foundation, *The Interplay between Mathematics and Robotics: Summary of a Workshop.* Arlington, Virginia: 2000.

> It is not enough to have a good mind; the main thing is to use it well.
>
> —René Descartes

6.1 Angles and Their Measure

- **Learn basic concepts about angles**
- **Apply degree measure to problems**
- **Apply radian measure to problems**
- **Calculate arc length**
- **Calculate the area of a sector**

Introduction

In 2007 at the French Open, Venus Williams hit the fastest serve recorded in a women's tour main-draw match, reaching 128 miles per hour. Much of the speed gained in a serve comes from flexing the wrist and rotating the shoulder joint. These rotations create *angular speed*, which transfers *linear speed* to the tennis ball. Trigonometry can be used to understand how fast it is possible to hit a tennis ball. (**Source:** J. Cooper and R. Glassow, *Kinesiology*.)

Angles

An **angle** is formed by rotating a ray about its endpoint. The starting position of the ray is called the **initial side**, and the final position of the ray is the **terminal side**. If the rotation of the ray is counterclockwise, the angle has *positive measure*; if the rotation is clockwise, the angle has *negative measure*. For simplicity we will refer to an angle as being **positive** or **negative**. The endpoint of the ray is called the **vertex** of the angle. See Figures 6.1 and 6.2, where the Greek letter θ (theta) has been used to denote an angle.

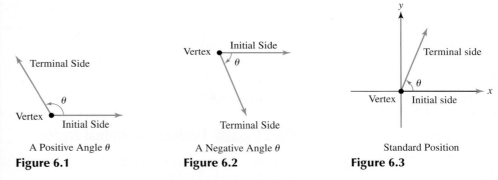

A Positive Angle θ	A Negative Angle θ	Standard Position
Figure 6.1	**Figure 6.2**	**Figure 6.3**

If the vertex is positioned so that it corresponds to the origin in the *xy*-plane and the initial side coincides with the positive *x*-axis, then the angle is in **standard position**, as shown in Figure 6.3.

There are two common systems for measuring the size of an angle: *degree measure* and *radian measure*.

Degree Measure

In **degree measure**, one complete rotation of a ray about its endpoint contains 360 degrees. One degree, denoted 1°, represents $\frac{1}{360}$ of a complete rotation. Figure 6.4 shows some examples of angles with their degree measure.

Figure 6.4 Degree Measure

A **right angle** has measure 90°, and a **straight angle** has measure 180°. The measure of an **acute angle** is greater than 0° but less than 90°, whereas the measure of an **obtuse angle** is greater than 90° but less than 180°. Examples are shown in Figure 6.5.

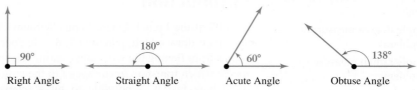

Figure 6.5 Types of Angles

We will use the Greek letters α (alpha), β (beta), γ (gamma), and θ (theta) to denote angles. For simplicity, we sometimes refer to an angle θ having measure 45° as a 45° angle, or an angle of 45°. This may be expressed as $\theta = 45°$. Two angles with the same initial and terminal sides are **coterminal angles**. Examples of coterminal angles are shown in Figure 6.6.

Figure 6.6 Coterminal Angles

EXAMPLE 1 Finding coterminal angles

Find three angles coterminal with $\theta = 45°$, where θ is in standard position. Sketch these angles in standard position.

SOLUTION We can find coterminal angles by either adding multiples of 360° to θ or subtracting multiples of 360° from θ.

i. $45° + 360° = 405°$ **ii.** $45° + 2(360°) = 765°$ **iii.** $45° - 360° = -315°$

The angles 405°, 765°, and −315° are all coterminal with a 45° angle. These three angles and θ are sketched in Figure 6.7.

Figure 6.7

Now Try Exercise 13 ◄

Two positive angles are **complementary angles** if their sum is 90° and **supplementary angles** if their sum is 180°. For example, $\alpha = 35°$ and $\beta = 55°$ are complementary angles, whereas $\alpha = 60°$ and $\beta = 120°$ are supplementary angles. See Figures 6.8 and 6.9.

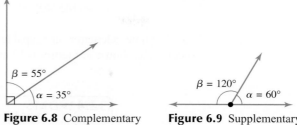

Figure 6.8 Complementary Angles   **Figure 6.9** Supplementary Angles

Fractions of a degree may be measured using **minutes** and **seconds**. One minute, written $1'$, equals $\frac{1}{60}$ of a degree, and one second, written $1''$, equals $\frac{1}{60}$ of a minute or $\frac{1}{3600}$ of a degree. The measurement $25°45'30''$ represents 25 degrees, 45 minutes, 30 seconds. Expressed in decimal degrees, this measurement is

$$25°45'30'' = 25° + \left(\frac{45}{60}\right)° + \left(\frac{30}{3600}\right)° = 25.758\overline{3}°.$$

We can convert decimal degrees to degrees, minutes, and seconds as illustrated in the next example.

EXAMPLE 2   Converting to degrees, minutes, and seconds

Convert $34.41°$ to degrees, minutes, and seconds.

SOLUTION

$$
\begin{aligned}
34.41° &= 34° + 0.41° && \text{\textit{Rewrite the expression.}}\\
&= 34° + 0.41(60') && \text{\textit{1° = 60'}}\\
&= 34° + 24.6' && \text{\textit{Multiply.}}\\
&= 34° + 24' + 0.6' && \text{\textit{Rewrite the expression.}}\\
&= 34° + 24' + 0.6(60'') && \text{\textit{1' = 60''}}\\
&= 34° + 24' + 36'' && \text{\textit{Multiply.}}\\
&= 34°24'36'' && \text{\textit{Rewrite.}}
\end{aligned}
$$

Now Try Exercise 27 ◀

In the next example, we find complementary and supplementary angles using degree measure.

EXAMPLE 3   Finding complementary and supplementary angles

Find angles that are complementary and supplementary to $\alpha = 34°19'42''$.

SOLUTION   If angle β is complementary to α, then $\beta = 90° - \alpha$.

$$
\begin{aligned}
\beta &= 90° - 34°19'42''\\
&= 89°59'60'' - 34°19'42'' && \text{\textit{90° = 89°59'60''}}\\
&= 55°40'18''
\end{aligned}
$$

An angle supplementary to $\alpha = 34°19'42''$ is given by $\gamma = 180° - \alpha$.

$$\gamma = 180° - 34°19'42''$$
$$= 179°59'60'' - 34°19'42'' \qquad 180° = 179°59'60''$$
$$= 145°40'18''$$

Now Try Exercise 31 ◀

NOTE Some calculators are capable of performing arithmetic using degrees, minutes, and seconds, as shown in Figures 6.10 and 6.11. (See the ANGLE menu.)

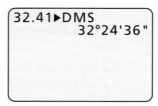

Figure 6.10 Degree Mode   **Figure 6.11** Degree Mode

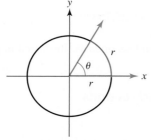

Figure 6.12 One Radian

Radian Measure

A second system of angle measure is based on *radians*. Radians are a common unit of measurement in many technical fields, including calculus. Using radian measure often makes formulas simpler and easier to use. Two examples are the formulas for arc length and area of a sector, which are introduced later in this section.

Angle θ in Figure 6.12 has a measure of one radian. The vertex of θ is located at the center of the circle, and its initial and terminal sides intercept an arc whose length is equal to the radius of the circle.

Radian Measure

An angle that has its vertex at the center of a circle and intercepts an arc on the circle equal in length to the radius of the circle has a measure of **one radian**.

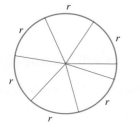

Figure 6.13 One Revolution Contains 2π Radians

The circumference of a circle is $C = 2\pi r$. If we mark off distances of r along the circumference of a circle, it will appear as in Figure 6.13, where $2\pi \approx 6.28$ distances of r are shown. Therefore one rotation contains $2\pi \approx 6.28$ radians, and so $360°$ is equivalent to 2π radians. Radian measure can be compared to degree measure using proportions. Since $180°$ is equivalent to π radians, it follows that

$$\frac{\text{radian measure}}{\text{degree measure}} = \frac{\pi}{180°}.$$

Solving for radian measure results in

$$\text{radian measure} = \text{degree measure} \times \frac{\pi}{180°},$$

and solving for degree measure results in

$$\text{degree measure} = \text{radian measure} \times \frac{180°}{\pi}.$$

The following statements summarize the preceding discussion.

Converting Between Degrees and Radians

To convert *degrees to radians*, multiply a degree measure by $\frac{\pi}{180°}$.

To convert *radians to degrees*, multiply a radian measure by $\frac{180°}{\pi}$.

NOTE One radian is equivalent to approximately 57.3°.

EXAMPLE 4   Converting degrees to radians

Convert each degree measure to radian measure.
(a) 90°   **(b)** 225°

SOLUTION

(a) To convert degrees to radians, multiply by $\frac{\pi}{180°}$.

$$90° \times \frac{\pi}{180°} = \frac{\pi}{2} \text{ radians}$$

Thus 90° is equivalent to $\frac{\pi}{2}$ radians.

(b) $225° \times \frac{\pi}{180°} = \frac{5\pi}{4}$ radians   *Now Try Exercise 39* ◄

Table 6.1 shows some equivalent measures in degrees and radians.

Table 6.1

Degrees	0°	30°	45°	60°	90°	180°	360°
Radians	0	$\frac{\pi}{6}$	$\frac{\pi}{4}$	$\frac{\pi}{3}$	$\frac{\pi}{2}$	π	2π

EXAMPLE 5   Converting radians to degrees

Convert each radian measure to degree measure.

(a) $\frac{4\pi}{3}$ **(b)** $\frac{5\pi}{6}$

SOLUTION

(a) To convert radians to degrees, multiply by $\frac{180°}{\pi}$.

$$\frac{4\pi}{3} \times \frac{180°}{\pi} = 240°$$

Thus $\frac{4\pi}{3}$ radians is equivalent to 240°.

(b) $\frac{5\pi}{6} \times \frac{180°}{\pi} = 150°$ *Now Try Exercise 43* ◄

NOTE Some calculators can convert degrees to radians and radians to degrees, as shown in Figures 6.14 and 6.15, respectively. When converting from degrees to radians, many calculators give only decimal approximations rather than exact values. For example, $\frac{\pi}{2}$ may be expressed as 1.570796327.

```
90°
           1.570796327
225°
           3.926990817
```

Figure 6.14 Radian Mode

```
(4π/3)ʳ
              240
(5π/6)ʳ
              150
```

Figure 6.15 Degree Mode

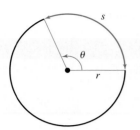

Figure 6.16 Arc Length s

Arc Length

From geometry we know that the arc length s on a circle is proportional to the measure of the central angle θ. See Figure 6.16. A central angle of 2π radians corresponds to an arc length that equals the circumference $C = 2\pi r$. Using proportions yields

$$\frac{s}{\theta} = \frac{2\pi r}{2\pi},$$

which simplifies to $s = r\theta$.

Arc Length

The **arc length** s intercepted on a circle of radius r by a central angle of θ *radians* is given by

$$s = r\theta.$$

NOTE Angle θ *must* be in radian measure to use the arc length formula $s = r\theta$.

EXAMPLE 6   Finding arc length

A circle has a radius of 25 inches. Find the length of an arc intercepted by a central angle of 45°.

SOLUTION First convert 45° to radian measure.

$$45° \times \frac{\pi}{180°} = \frac{\pi}{4}$$

Figure 6.17

The arc length s is given by

$$s = r\theta = 25\left(\frac{\pi}{4}\right) = 6.25\pi \text{ inches.}$$

The arc length s shown in Figure 6.17 is $6.25\pi \approx 19.6$ inches.   *Now Try Exercise 47* ◄

EXAMPLE 7   Finding distance between cities

Albuquerque, New Mexico, and Glasgow, Montana, have the same longitude of 106°37′ W. The latitude of Albuquerque is 35°03′ N, and the latitude of Glasgow is 48°13′ N. If the radius of Earth is approximately 3955 miles, estimate the distance between Albuquerque and Glasgow. See Figure 6.18.   (**Source:** J. Williams, *The Weather Almanac.*)

SOLUTION This distance can be estimated using the arc length formula. Start by converting $\theta = 48°13' - 35°03' = 13°10'$ to radian measure.

$$\left[13° + \left(\frac{10}{60}\right)°\right] \times \frac{\pi}{180°} \approx 0.2298 \text{ radian}$$

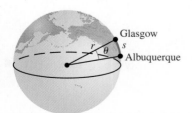

Figure 6.18 Distance between Two Cities (not to scale)

The distance between Albuquerque and Glasgow is approximated by

$$s = r\theta \approx 3955(0.2298) \approx 909 \text{ miles.} \quad \textit{Now Try Exercise 81} ◄$$

Figure 6.19 Flexing the Wrist in Tennis

Rotation and Angular Speed The human joint that can be flexed the fastest is the wrist, which can rotate through 90°, or $\frac{\pi}{2}$ radians, in 0.045 second while holding a tennis racket. See Figure 6.19. **Angular speed** ω (omega) measures the speed of rotation and is defined by

$$\omega = \frac{\theta}{t}, \qquad \text{Angular speed}$$

where θ is the angle of rotation and t is time. The angular speed of a human wrist holding a tennis racket is

$$\frac{\pi/2}{0.045} \approx 34.9 \text{ rad/sec},$$

or about 35 radians per second. The **linear speed** v at which the tip of the racket travels as a result of flexing the wrist is given by

$$v = r\omega, \qquad \text{Linear speed}$$

where r is the radius (distance) from the tip of the racket to the wrist joint and ω is in radians per unit of time. If $r = 2$ feet, then the speed at the tip of the racket is

$$v = r\omega \approx (2)(35) = 70 \text{ ft/sec},$$

or about 48 miles per hour. In a tennis serve the arm rotates at the shoulder, so the final speed of the racket is considerably faster.   (**Source:** J. Cooper and R. Glassow, *Kinesiology.*)

CLASS DISCUSSION

A human shoulder can rotate at about 25 radians per second. Estimate how much this shoulder rotation increases the speed of a racket if the length of the (straight) arm and racket together is 4 feet.

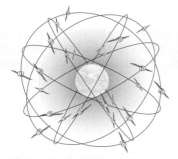

Figure 6.20 GPS Satellite Coverage

EXAMPLE 8 Finding the speed of a GPS satellite

Each of the 24 satellites used in the Global Positioning System is located 16,526 miles from the center of Earth and has a nearly circular orbit with a period of 12 hours. See Figure 6.20.   (**Source:** Y. Zhao, *Vehicle Location and Navigation Systems.*)
(a) Find the angular speed of a satellite.
(b) Estimate the linear speed of a satellite.

SOLUTION
(a) A GPS satellite circles Earth once every 12 hours. Its angular speed is

$$\omega = \frac{2\pi}{t} = \frac{2\pi}{12} = \frac{\pi}{6} \approx 0.5236 \text{ rad/hr}.$$

(b) Its linear speed is $v = r\omega = (16{,}526)(0.5236) \approx 8653$ miles per hour.

Now Try Exercise 79 ◄

Area of a Sector

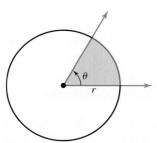

Figure 6.21 Sector of a Circle

The **sector of a circle** is the portion of the interior of a circle intercepted by a central angle. The shaded region in Figure 6.21 shows a sector of a circle with radius r and central angle θ. The area of a sector is proportional to the measure of the central angle. If the central angle is 2π radians, then the area of the sector is the entire interior of the circle, which has an area of πr^2. Using proportions yields

$$\frac{\text{area of a sector}}{\theta} = \frac{\pi r^2}{2\pi}.$$

Solving the equation for the area of the sector results in

$$\text{area of a sector} = \frac{1}{2} r^2 \theta.$$

Area of a Sector

The **area of a sector** A of a circle of radius r and central angle θ in *radians* is given by

$$A = \frac{1}{2} r^2 \theta.$$

NOTE Angle θ *must* be in radian measure to use the area formula $A = \frac{1}{2} r^2 \theta$.

EXAMPLE 9 Finding the area of a sector

A circle has a radius of 6 inches. Find the area of the sector if its central angle is $60°$.

SOLUTION Since $60°$ is equivalent to $\frac{\pi}{3}$ radians, the area of the sector is given by

$$A = \frac{1}{2} r^2 \theta = \frac{1}{2} (6)^2 \left(\frac{\pi}{3} \right) = 6\pi \text{ square inches.}$$

This region of $6\pi \approx 18.8$ square inches is illustrated in Figure 6.22.

Now Try Exercise 63 ◄

Figure 6.22

An Application Consider the robotic arm shown in Figure 6.23. The *work space* of the robotic arm is the shaded region and corresponds to the places that the hand can reach either by rotating or by changing the length of the arm. (**Source:** W. Stadler, *Analytical Robotics and Mechatronics.*)

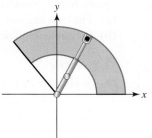

Figure 6.23 Work Space of a Robotic Arm

EXAMPLE 10 Finding the area of the work space for a robotic arm

Suppose that a robotic arm similar to the one in Figure 6.23 can rotate between $\theta = 10°$ and $\theta = 130°$. If the length of the robotic arm can vary between 5 inches and 20 inches, find the area of its work space.

Figure 6.24

SOLUTION The work space can be thought of as a large sector having radius $r_1 = 20$ inches with a small sector of radius $r_2 = 5$ inches removed. The arm can rotate through $130° - 10° = 120°$, or $\frac{2\pi}{3}$ radians. See Figure 6.24. The area A of the work space is computed as follows.

$$A = \frac{1}{2}r_1^2\theta - \frac{1}{2}r_2^2\theta$$

$$= \frac{1}{2}\theta\left(r_1^2 - r_2^2\right)$$

$$= \frac{1}{2}\left(\frac{2\pi}{3}\right)(20^2 - 5^2)$$

$$= 125\pi$$

The work space is $125\pi \approx 392.7$ square inches.

Now Try Exercise 71 ◄

6.1 Putting It All Together

Some concepts involving angles are summarized in the following table.

Concept	Explanation or Formula	Examples
Degree measure	One complete rotation contains 360°.	A right angle contains 90°. A straight angle contains 180°. An acute angle α satisfies $0° < \alpha < 90°$. An obtuse angle β satisfies $90° < \beta < 180°$.
Radian measure	One complete rotation contains 2π radians.	2π radians is equivalent to 360°. π radians is equivalent to 180°. $\frac{\pi}{2}$ radians is equivalent to 90°. $\frac{\pi}{3}$ radians is equivalent to 60°. $\frac{\pi}{4}$ radian is equivalent to 45°. $\frac{\pi}{6}$ radian is equivalent to 30°.
Arc length	$s = r\theta$, where θ is in radians	If $r = 12$ feet and $\theta = 90°$, then $s = (12)\frac{\pi}{2} = 6\pi \approx 18.8$ feet.
Area of a sector	$A = \frac{1}{2}r^2\theta$, where θ is in radians	If $r = 6$ inches and $\theta = 45°$, then $A = \frac{1}{2}(6)^2\left(\frac{\pi}{4}\right) = 4.5\pi \approx 14.1$ square inches.
Angular speed	$\omega = \frac{\theta}{t}$, where θ is the angle of rotation and t is time	If $\theta = 5$ radians and $t = 0.1$ second, then $\omega = \frac{5}{0.1} = 50$ radians per second.
Linear speed of a rotating object	$v = r\omega$, where r is the radius and ω is the angular speed in *radians* per unit of time	If $r = 3$ feet and $\omega = 5$ radians per second, then $v = (3)(5) = 15$ feet per second.

6.1 Exercises

Angles

Exercises 1 and 2: Sketch the following angles in standard position.

1. (a) 45°   (b) −150°

   (c) $\frac{\pi}{3}$ (d) $-\frac{3\pi}{4}$

2. (a) −90°   (b) 225°

   (c) $-\frac{2\pi}{3}$ (d) $\frac{\pi}{6}$

Exercises 3–10: Sketch an angle θ in standard position that satisfies the conditions. Assume α is in standard position.

3. Acute   **4.** Obtuse

5. A positive straight angle **6.** Complementary to 60°

7. Positive and the terminal side lies in quadrant III

8. Negative and the terminal side lies in quadrant IV

9. Negative and coterminal with $\alpha = 90°$

10. Positive and coterminal with $\alpha = -135°$

11. What fraction of a complete revolution is each of the following angles?
   (a) 90°   (b) 30°   (c) $\frac{\pi}{3}$ (d) $\frac{\pi}{4}$

12. What angle is its own complement? What angle is its own supplement?

Exercises 13–20: Find a positive angle and a negative angle that are coterminal with the given angle.

13. 150°   **14.** 65°

15. −72°   **16.** −330°

17. $\frac{\pi}{2}$ **18.** $\frac{5\pi}{6}$

19. $-\frac{\pi}{5}$ **20.** $-\frac{2\pi}{3}$

Exercises 21–24: Express the angle in decimal degrees.

21. 125°15′   **22.** 15°30′

23. 108°45′36″   **24.** 256°06′12″

Exercises 25–28: Convert the given angle to degrees, minutes, and seconds.

25. 125.3°   **26.** 15.25°

27. 51.36°   **28.** 22.46°

Exercises 29–34: Find the complementary angle α and the supplementary angle β to θ.

29. $\theta = 55.9°$ **30.** $\theta = 71.5°$

31. $\theta = 85°23'45''$ **32.** $\theta = 5°45'30''$

33. $\theta = 23°40'35''$ **34.** $\theta = 67°25'10''$

Radian Measure

Exercises 35–38: Use the figure to determine the radian measure of angle θ. Then approximate the degree measure of θ to the nearest tenth of a degree.

35.

36.

37.

38.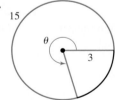

Exercises 39–42: Convert each angle from degree measure to radian measure. Round to the nearest hundredth of a radian when appropriate.

39. (a) 45°   (b) 135°   (c) −120°   (d) −210°

40. (a) 105°   (b) 245°   (c) −255°   (d) −80°

41. (a) 37°   (b) 123.4°   (c) −92°25′ (d) 230°17′

42. (a) 56°   (b) 88.7°   (c) 122°15′ (d) −7°48′

Exercises 43–46: Convert each angle from radian measure to degree measure. Round to the nearest hundredth of a degree when appropriate.

43. (a) $\frac{\pi}{6}$ (b) $\frac{\pi}{15}$ (c) $-\frac{5\pi}{3}$ (d) $-\frac{7\pi}{6}$

44. (a) $-\frac{\pi}{12}$ (b) $-\frac{5\pi}{2}$ (c) $\frac{17\pi}{15}$ (d) $\frac{5\pi}{6}$

45. (a) $\frac{\pi}{4}$ (b) $\frac{\pi}{7}$ (c) 3.1 (d) $-\frac{5}{2}$

46. (a) $\frac{7\pi}{2}$ (b) $\frac{2\pi}{5}$ (c) −4.1 (d) $-\frac{2}{3}$

Arc Length

Exercises 47–52: Use the formula $s = r\theta$ to determine the missing value in the figure.

47.

48.

49.

50.

51.

52.

Exercises 53–58: Find the length of the arc intercepted by a central angle θ in a circle of radius r.

53. $r = 3$ m, $\theta = \frac{\pi}{12}$ **54.** $r = 7.3$ mm, $\theta = \frac{7\pi}{4}$

55. $r = 12$ ft, $\theta = 15°$ **56.** $r = 5$ cm, $\theta = 240°$

57. $r = 2$ mi, $\theta = 1°45'$ **58.** $r = 3$ mi, $\theta = 4°15'09''$

Exercises 59–62: A minute hand on a clock is 4 inches long. Determine how far the tip of the minute hand travels between the given times. Find the linear speed of the tip.

59. 10:15 A.M., 10:30 A.M.   **60.** 1:00 P.M., 1:40 P.M.

61. 3:00 P.M., 4:15 P.M.   **62.** 11:00 A.M., 1:25 P.M.

Area of a Sector

Exercises 63–66: Find the area of the shaded sector.

63.

64.

65.

66.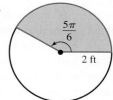

Exercises 67–70: Find the area of the sector of a circle having radius r and central angle θ.

67. $r = 13.1$ cm, $\theta = \frac{\pi}{15}$ **68.** $r = 7.3$ m, $\theta = \frac{5\pi}{4}$

69. $r = 1.5$ ft, $\theta = 30°$ **70.** $r = 5.5$ in., $\theta = 225°$

*Exercises 71–74: **Robotics**   (Refer to Example 10.) Find the area of the work space for a robotic arm that can rotate between angles θ_1 and θ_2 and can change its length from r_1 to r_2. See the figure.*

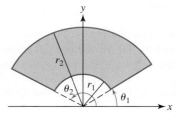

71. $\theta_1 = -45°$, $\theta_2 = 90°$, $r_1 = 6$ in., $r_2 = 26$ in.

72. $\theta_1 = -60°$, $\theta_2 = 60°$, $r_1 = 0.5$ ft, $r_2 = 2.5$ ft

73. $\theta_1 = 15°$, $\theta_2 = 195°$, $r_1 = 21$ cm, $r_2 = 95$ cm

74. $\theta_1 = 43°$, $\theta_2 = 178°$, $r_1 = 0.4$ m, $r_2 = 1.8$ m

Applications

75. *Bicycle Tire*   A bicycle has a tire 26 inches in diameter that is rotating at 15 radians per second. Approximate the speed of the bicycle in feet per second and in miles per hour.

76. *Skateboard Wheel*   The wheels on a skateboard have a diameter of 2.25 inches. If a skateboarder is traveling downhill at 15 miles per hour, determine the angular velocity of the wheels in radians per second.

77. *Location of the North Star*   Presently the North Star, Polaris, is located near the true North Pole. However, because Earth is inclined 23.5°, the moon's gravitational pull on Earth is uneven. As a result, Earth precesses like a spinning top and the direction of the celestial North Pole traces out a circular path once every 26,000 years, as shown in the figure. For example, in the year 14,000 the star Vega and not Polaris will be located at the celestial North Pole. Calculate the angle in seconds that the celestial North Pole moves each year, as viewed from the center *C* of this circular path.   (**Source:** M. Zeilik et al., *Introductory Astronomy and Astrophysics.*)

78. *Ferris Wheel*   A large Ferris wheel has a diameter of 140 feet. It completes 1 revolution every 420 seconds.
(a) Find the angular velocity in radians per second.

(b) What is the linear speed of a person who is riding this Ferris wheel?

79. *Fan Speed*   The blades of a fan have a 30-inch diameter and rotate at 500 revolutions per minute.
(a) Find the angular velocity of a fan blade.

(b) Estimate the linear speed at the tip of a fan blade.

80. *Earth's Rotation*   Earth rotates 1 complete revolution every 24 hours and has an equatorial radius of about 3963 miles.
(a) Find the angular velocity of a person standing at the equator in radians per hour.

(b) Estimate the linear speed in miles per hour at the equator due to Earth's rotation.

81. *Distance between Cities*   (Refer to Example 7.) Daytona Beach, Florida, and Akron, Ohio, have nearly the same longitude of 81° W. The latitude of Daytona Beach is 29°11′, and the latitude of Akron is 40°55′. Approximate the distance between these two cities if the average radius of Earth is 3955 miles.   (**Source:** J. Williams.)

82. *Nautical Miles*   Nautical miles are used by ships and airplanes. They are different from statute miles, which equal 5280 feet. A nautical mile is defined to be the arc length along the equator intercepted by a central angle *AOB* of 1 minute, as illustrated in the figure. If the equatorial radius of Earth is 3963 statute miles, use the arc length formula to approximate the number of statute miles in 1 nautical mile. Round your answer to two decimal places.

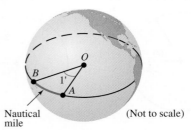

83. *Tennis Serve*   (Refer to the discussion before Example 8.) Suppose that a tennis player rotates her wrist 85° in 0.05 second. If the tennis racket is 28 inches in length, estimate the speed of the tip of the racket in feet per second.

84. *Club Speed in Golf*   The shoulder joint can rotate at about 25 radians per second. Assuming that a golfer's arm is straight and the distance from the shoulder to the club head is 5 feet, estimate the linear speed of the club head from shoulder rotation.   (**Source:** J. Cooper and R. Glassow, *Kinesiology.*)

85. *Pulleys*   Approximate how many inches the weight in the figure will rise if *r* = 11 inches and the pulley is rotated through an angle of 75.3°.

86. *Pulleys*   Use the figure in Exercise 85 to estimate the angle *θ* through which the pulley should be rotated to raise the weight 5 inches.

87. *Bicycle Chain Drive*   The figure shows the chain drive of a bicycle. The radius of the sprocket wheel that the pedals are attached to is 3.75 inches, and the radius of the other sprocket wheel is 1.5 inches.
   (a) Determine the number of revolutions that the bicycle tire rotates when the pedals are rotated one revolution.

   (b) If the bicycle has a tire with a 26-inch diameter, determine how fast the bicycle travels in feet per second when the pedals turn through two revolutions per second.

88. *Wind Speed*   One of the most common ways to measure wind speed is with a *three-cup anemometer*, as shown in the figure. The cups catch the wind and cause the vertical shaft to rotate. At lower wind speeds the cups move at approximately the same speed as the wind. If the cups are rotating 5 times per second with a radius of 6 inches, estimate the wind speed in miles per hour. (**Source:** J. Navarra, *Atmosphere, Weather and Climate.*)

89. *Velocity of Planets*   The average distance D in millions of miles from the sun and the orbital period P in years are given for various planets. Assuming that the orbits are circular, approximate the average orbital velocity in miles per hour for each planet. Discuss the effect that average distance from the sun has on orbital velocity. (**Source:** C. Ronan, *The Natural History of the Universe.*)
   (a) Venus: $D = 67.2$, $P = 0.615$

   (b) Earth: $D = 92.9$, $P = 1$

   (c) Jupiter: $D = 483.6$, $P = 11.86$

   (d) Neptune: $D = 2794$, $P = 164.8$

90. *Speed of a Propeller*   When a 90-horsepower outboard motor is at full throttle, its propeller makes 5000 revolutions per minute. Find the angular velocity of the propeller in radians per second. What is the linear speed in inches per second of a point at the tip of the propeller if its diameter is 10 inches?

91. *Surveying*   The *subtense bar method* is a technique used in surveying to measure distances. A subtense bar is shown in the figure connecting points P and Q. If the distance d from the surveyor to the bar is large, then there is little difference between the length of the subtense bar, which is usually 2 meters, and that of the arc connecting P and Q. Similarly, there is little difference between d and the radius r of the arc intercepted by the subtense bar. If θ is measured to be $0.835°$, approximate d using the arc length formula.   (**Source:** I. Mueller and K. Ramsayer, *Introduction to Surveying.*)

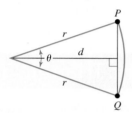

92. *Diameter of the Moon*   (Refer to Exercise 91.) The distance to the moon is approximately 238,900 miles. Use the arc length formula to estimate the diameter d of the moon if angle θ in the accompanying figure is measured to be $0.517°$.

(Not to scale)

93. *Solar Power Plant*   A 150-megawatt solar power plant requires approximately 475,000 square meters of land to collect the required amount of energy from sunlight. (**Source:** C. Winter, *Solar Power Plants.*)
   (a) If this land area is circular, approximate its radius.

   (b) If this land area is a sector of a circle with $\theta = 70°$, approximate its radius.

94. *Measuring the Circumference of Earth*   The first accurate estimate of the distance around Earth was made by the Greek astronomer Eratosthenes (276–195 B.C.), who noted that the noontime position of the sun at the summer solstice differed by $7°12'$ from the city of Syene to the city of Alexander. See the figure. The distance between these two cities is 496 miles. Use the arc length formula to estimate the radius of Earth. Then find the circumference of Earth.   (**Source:** M. Zeilik.)

Writing about Mathematics

95. Give definitions for 1 degree and 1 radian. Compare these two units of angle measure. Which unit of measure do you prefer? Explain why.

96. Suppose a central angle θ of a circle remains fixed. Describe what happens to the arc length intercepted by θ and the area of the corresponding sector as the radius r doubles and triples.

EXTENDED AND DISCOVERY EXERCISES

1. *Arc Length Formula*   Modify the arc length formula $s = r\theta$ so that angle θ can be given in degrees rather than radians. Which of the two formulas is simpler?

2. *Area of a Sector Formula*   Modify the area formula $A = \frac{1}{2}r^2\theta$ so that angle θ can be given in degrees rather than radians. Which of the two formulas is simpler?

6.2 Right Triangle Trigonometry

- **Learn basic concepts about trigonometric functions**
- **Apply right triangle trigonometry**
- **Understand complementary angles and cofunctions**

Introduction

A right triangle is a basic geometric shape that occurs in many applications such as astronomy, surveying, construction, highway design, GPS, and aerial photography. Trigonometric functions are used to *solve* triangles. Solving a triangle involves finding the measure of each side and angle in the triangle. The sine function is one of the earliest trigonometric functions; it dates back to the ancient Greeks, who invented a similar function called the *chord function*. However, it was not until 1550 that the sine function was formally defined in terms of right triangles by Georg Rhaeticus. Like other functions that we have encountered previously, the six trigonometric functions can be used to model data and a variety of physical phenomena.   (**Source:** NCTM, *Historical Topics for the Mathematics Classroom, Thirty-first Yearbook.*)

Basic Concepts of Trigonometric Functions

The **standard labeling** used to designate vertices, angles, and sides of a triangle ABC is shown in Figure 6.25. The vertices are denoted A, B, and C, the angles α, β, and γ, and the lengths of the sides opposite these angles a, b, and c. If triangle ABC is a right triangle, then we let $\gamma = 90°$.

Many important ideas in trigonometry depend on the properties of similar triangles. **Similar triangles** have congruent corresponding angles, but similar triangles are not necessarily the same size. Two similar triangles are shown in Figure 6.26. Corresponding sides of similar triangles are proportional.

The right triangles ABC and $AB'C'$ shown in Figure 6.27 are similar triangles. By the properties of similar triangles, the following ratios are equal.

$$\frac{BC}{AB} = \frac{B'C'}{AB'}$$

Figure 6.25 Standard Labeling

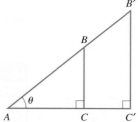

Figure 6.26 Similar Triangles   **Figure 6.27** Similar Triangles

Figure 6.28

That is, the ratio of the side opposite angle θ to the hypotenuse is constant for a given angle θ and does not depend on the size of the right triangle. If the measure of θ changes, then the ratio of the side opposite θ to the hypotenuse also changes. This concept can be used to define a new function called the *sine function*. That is, if θ is an acute angle in a right triangle, as shown in Figure 6.28, then we define the sine of θ as

$$\sin \theta = \frac{\text{side opposite}}{\text{hypotenuse}},$$

where $\sin \theta$ denotes the sine function with input θ.

CLASS DISCUSSION

Is it possible that $\sin \theta > 1$ for some acute angle θ? Explain your reasoning.

EXAMPLE 1   Evaluating the sine function

Find $\sin 30°$. Support your answer by using a calculator.

SOLUTION Since the sine function depends only on the measure of θ, we can choose any size right triangle to evaluate $\sin \theta$. For convenience, let the length of the hypotenuse equal 2, as shown in Figure 6.29. From geometry we know that the length of the shortest leg in a 30°–60° right triangle is half the hypotenuse. Thus the side opposite equals 1 and

$$\sin 30° = \frac{\text{side opposite}}{\text{hypotenuse}} = \frac{1}{2}.$$

This result is supported in Figure 6.30, where the sine function has been evaluated at 30° using a calculator set in degree mode.   *Now Try Exercise 5* ◀

Figure 6.29

Figure 6.30 Degree Mode

Using Figure 6.27 and 6.28, we can define other trigonometric functions. Since

$$\frac{AC}{AB} = \frac{AC'}{AB'},$$

the ratio of the side adjacent to θ to the hypotenuse is constant for a fixed angle θ and does not depend on the size of the right triangle. We define the *cosine function* to be

$$\cos \theta = \frac{\text{side adjacent}}{\text{hypotenuse}}.$$

The six trigonometric functions of angle θ are called **sine, cosine, tangent, cosecant, secant,** and **cotangent**. The customary abbreviation for each trigonometric function is shown below.

Right Triangle-Based Definitions of Trigonometric Functions

Let θ be an acute angle in a right triangle. Then the six trigonometric functions of θ may be evaluated as follows.

$$\sin \theta = \frac{\text{side opposite}}{\text{hypotenuse}} \qquad \cos \theta = \frac{\text{side adjacent}}{\text{hypotenuse}} \qquad \tan \theta = \frac{\text{side opposite}}{\text{side adjacent}}$$

$$\csc \theta = \frac{\text{hypotenuse}}{\text{side opposite}} \qquad \sec \theta = \frac{\text{hypotenuse}}{\text{side adjacent}} \qquad \cot \theta = \frac{\text{side adjacent}}{\text{side opposite}}$$

The next example illustrates how to evaluate the trigonometric functions.

EXAMPLE 2   Evaluating trigonometric functions

Consider the right triangle shown in Figure 6.31. Find the six trigonometric functions of θ.

SOLUTION In triangle ABC, the side opposite angle θ is $a = 8$ and the hypotenuse is $c = 17$. To find the adjacent side b, we apply the Pythagorean theorem.

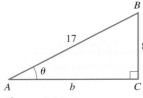

Figure 6.31

$$\begin{array}{ll} c^2 = a^2 + b^2 & \text{Pythagorean theorem} \\ b^2 = c^2 - a^2 & \text{Solve for } b^2. \\ b^2 = 17^2 - 8^2 & \text{Let } c = 17 \text{ and } a = 8. \\ b^2 = 225 & \text{Simplify.} \\ b = 15 & \text{Solve for } b, \text{ where } b > 0. \end{array}$$

Thus the six trigonometric functions of θ are as follows.

$$\sin \theta = \frac{\text{side opposite}}{\text{hypotenuse}} = \frac{8}{17} \qquad \csc \theta = \frac{\text{hypotenuse}}{\text{side opposite}} = \frac{17}{8}$$

$$\cos \theta = \frac{\text{side adjacent}}{\text{hypotenuse}} = \frac{15}{17} \qquad \sec \theta = \frac{\text{hypotenuse}}{\text{side adjacent}} = \frac{17}{15}$$

$$\tan \theta = \frac{\text{side opposite}}{\text{side adjacent}} = \frac{8}{15} \qquad \cot \theta = \frac{\text{side adjacent}}{\text{side opposite}} = \frac{15}{8}$$

Now Try Exercise 17 ◀

Figure 6.32 Angle of Elevation

If an object is located above the horizontal, then the angle between the horizontal and the line of sight XY is called the **angle of elevation**. See Figure 6.32. If an object is located below the horizontal, then the angle between the horizontal and the line of sight XY is called the **angle of depression**. See Figure 6.33.

An Application   Trigonometry allows people to determine distances and heights without measuring them directly. For example, the altitude of the cloud base is particularly important at airports. (The cloud base is where the lowest layer of clouds begins to form.) Although it is not practical to measure the altitude of the cloud base directly, trigonometry can indirectly determine this height at nighttime. In Figure 6.34 a bright spotlight is directed vertically upward. It creates a bright spot on the cloud base. From a known horizontal distance d from the spotlight, the angle of elevation θ is measured. The side adjacent to θ is d,

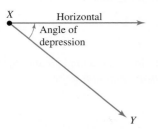

Figure 6.33 Angle of Depression

and the side opposite θ is h, where h represents the height of the cloud base.   (**Source:** F. Cole, *Introduction to Meteorology*.) It follows that

$$\tan \theta = \frac{\text{side opposite}}{\text{side adjacent}} = \frac{h}{d}.$$

Figure 6.34

EXAMPLE 3   Determining the cloud base

Suppose that in Figure 6.34 $\theta = 55°$ and $d = 1150$ feet. Estimate the height of the cloud base. (Neglect the height of the telescope and spotlight in Figure 6.34.)

SOLUTION   Solve the equation $\tan \theta = \frac{h}{d}$ for h and then substitute values for θ and d.

$$\tan \theta = \frac{h}{d}$$

$$h = d \tan \theta \qquad \text{Cross multiply.}$$

$$= 1150 \tan 55° \qquad \text{Substitute for } d \text{ and } \theta.$$

$$\approx 1150 \,(1.4281) \qquad \text{Approximate } \tan 55°.$$

$$\approx 1642 \text{ feet} \qquad \text{Multiply.}$$

Thus the cloud base is about 1642 feet high. A calculator was used to approximate $\tan 55°$, as shown in Figure 6.35.   *Now Try Exercise 57* ◀

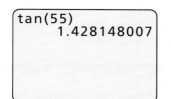

Figure 6.35 Degree Mode

Solving a Triangle   We can use trigonometric functions to find unknown sides of a right triangle. This process, sometimes referred to as *solving a triangle*, is demonstrated in the next example.

EXAMPLE 4   Solving a triangle

Find the lengths of the unknown sides a and c for the right triangle shown in Figure 6.36. Round each value to the nearest hundredth.

SOLUTION   We are given angle $\theta = 40°$ and side $b = 35$, which is adjacent to angle θ. Side a is opposite angle θ. Because the tangent function involves the opposite and adjacent sides, we use it to find side a.

$$\tan 40° = \frac{a}{35} \qquad \tan \theta = \frac{\text{side opposite}}{\text{side adjacent}}$$

$$35 \tan 40° = a \qquad \text{Multiply by 35.}$$

$$a \approx 29.37 \qquad \text{Rewrite; approximate (if desired).}$$

Figure 6.36

To find the length of hypotenuse c, we could use the Pythagorean theorem. However, we use $\cos \theta$ instead, which involves the adjacent side and the hypotenuse.

$$\cos 40° = \frac{35}{c} \qquad \cos \theta = \frac{\text{side adjacent}}{\text{hypotenuse}}$$

$$c \cos 40° = 35 \qquad \text{Multiply by } c.$$

$$c = \frac{35}{\cos 40°} \qquad \text{Divide by } \cos 40°.$$

$$c \approx 45.69 \qquad \text{Approximate (if desired).}$$

Now Try Exercise 37 ◀

Finding Exact Values   In most applications, calculators are used to approximate values of the trigonometric functions. However, with the aid of geometry we can determine exact values for the trigonometric functions of some special angles such as 30°, 45°, and 60°.

EXAMPLE 5   Evaluating trigonometric functions

Evaluate the six trigonometric functions of $\theta = 45°$.

SOLUTION   Begin by drawing a right triangle with a 45° angle, as shown in Figure 6.37. The lengths of the legs in this triangle are equal. Since the size of the right triangle does *not* affect the values of the trigonometric functions, let the lengths of both legs equal 1. Using the Pythagorean theorem, we can find the length of the hypotenuse as follows.

$$c^2 = a^2 + b^2 \qquad \text{Pythagorean theorem}$$

$$c^2 = 1^2 + 1^2 \qquad a = b = 1$$

$$c^2 = 2 \qquad \text{Simplify.}$$

$$c = \sqrt{2} \qquad \text{Solve for } c, \text{ where } c > 0.$$

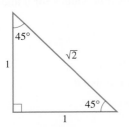

Figure 6.37 A 45°–45° Right Triangle

The hypotenuse has length $\sqrt{2}$. Evaluating the six trigonometric functions gives the following.

$$\sin 45° = \frac{\text{side opposite}}{\text{hypotenuse}} = \frac{1}{\sqrt{2}} \qquad \csc 45° = \frac{\text{hypotenuse}}{\text{side opposite}} = \frac{\sqrt{2}}{1} = \sqrt{2}$$

$$\cos 45° = \frac{\text{side adjacent}}{\text{hypotenuse}} = \frac{1}{\sqrt{2}} \qquad \sec 45° = \frac{\text{hypotenuse}}{\text{side adjacent}} = \frac{\sqrt{2}}{1} = \sqrt{2}$$

$$\tan 45° = \frac{\text{side opposite}}{\text{side adjacent}} = \frac{1}{1} = 1 \qquad \cot 45° = \frac{\text{side adjacent}}{\text{side opposite}} = \frac{1}{1} = 1$$

Now Try Exercise 21 ◀

In Example 5, $\sin 45° = \frac{1}{\sqrt{2}}$ and $\csc 45° = \frac{\sqrt{2}}{1}$. Since

$$\sin \theta = \frac{\text{side opposite}}{\text{hypotenuse}} \quad \text{and} \quad \csc \theta = \frac{\text{hypotenuse}}{\text{side opposite}},$$

it follows that $\csc \theta = \frac{1}{\sin \theta}$ in general. That is, the values of $\csc \theta$ and $\sin \theta$ are *reciprocals*. In a similar manner,

$$\sec \theta = \frac{1}{\cos \theta} \quad \text{and} \quad \cot \theta = \frac{1}{\tan \theta}.$$

```
1/sin(45)
        1.414213562
1/cos(45)
        1.414213562
1/tan(45)
                   1
```

Figure 6.38 Degree Mode

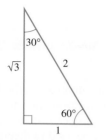

Figure 6.39 A 30°–60° Right Triangle

Calculators   Most calculators have keys to evaluate the sine, cosine, and tangent functions, but do not have keys to evaluate the cosecant, secant, and cotangent functions. These three functions may be evaluated by using the following *reciprocal identities*.

$$\csc\theta = \frac{1}{\sin\theta}, \qquad \sec\theta = \frac{1}{\cos\theta}, \qquad \cot\theta = \frac{1}{\tan\theta}$$

Results from Example 5 are supported in Figure 6.38, where these reciprocal identities have been applied to evaluate csc 45°, sec 45°, and cot 45°. Note that $\sqrt{2} \approx 1.414213562$.

NOTE Do not use the $\sin^{-1}$, $\cos^{-1}$, and $\tan^{-1}$ keys on your calculator to evaluate reciprocals—they represent inverse functions, which will be discussed in Section 6.6. You may want to use the x^{-1} key to evaluate reciprocals instead.

Using the right triangles in Figures 6.37 and 6.39, we can evaluate the six trigonometric functions at 30°, 45°, and 60° without the aid of a calculator. Table 6.2 lists the values of the six trigonometric functions for selected angles.

Table 6.2

θ	$\sin\theta$	$\cos\theta$	$\tan\theta$	$\csc\theta$	$\sec\theta$	$\cot\theta$
30°	$\frac{1}{2}$	$\frac{\sqrt{3}}{2}$	$\frac{1}{\sqrt{3}}$	2	$\frac{2}{\sqrt{3}}$	$\sqrt{3}$
45°	$\frac{1}{\sqrt{2}}$	$\frac{1}{\sqrt{2}}$	1	$\sqrt{2}$	$\sqrt{2}$	1
60°	$\frac{\sqrt{3}}{2}$	$\frac{1}{2}$	$\sqrt{3}$	$\frac{2}{\sqrt{3}}$	2	$\frac{1}{\sqrt{3}}$

NOTE If we rationalize the denominators in Table 6.2, then

$$\frac{1}{\sqrt{2}} = \frac{1}{\sqrt{2}} \cdot \frac{\sqrt{2}}{\sqrt{2}} = \frac{\sqrt{2}}{2}.$$

Similarly, $\frac{1}{\sqrt{3}} = \frac{\sqrt{3}}{3}$ and $\frac{2}{\sqrt{3}} = \frac{2\sqrt{3}}{3}$.

Applications of Right Triangle Trigonometry

For centuries astronomers wanted to know how far it was to the stars. Not until 1838 did the astronomer Friedrich Bessel determine the distance to a star called 61 Cygni. He used a *parallax method* that relied on the measurement of very small angles. See Figure 6.40. As Earth revolves around the sun, the observed parallax of 61 Cygni is $\theta \approx 0.0000811°$. Because stars are so distant, parallax angles are very small. (**Sources:** H. Freebury, *A History of Mathematics;* M. Zeilik et al., *Introductory Astronomy and Astrophysics.*)

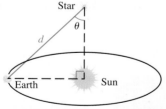

Figure 6.40 Parallax of a Star (not to scale)

EXAMPLE 6   Calculating the distance to a star

One of the nearest stars is Alpha Centauri, which has a parallax of $\theta \approx 0.000212°$. (**Source:** M. Zeilik et al.)
(a) Find the distance to Alpha Centauri if the Earth–Sun distance is 93,000,000 miles.
(b) A light-year, defined to be the distance that light travels in 1 year, equals about 5.9 trillion miles. Find the distance to Alpha Centauri in light-years.

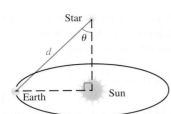

Star

θ

d

Earth

Sun

SOLUTION

(a) Let d represent the distance between Earth and Alpha Centauri. From Figure 6.40 (repeated in the margin) it can be seen that

$$\sin \theta = \frac{93{,}000{,}000}{d}, \quad \text{or} \quad d = \frac{93{,}000{,}000}{\sin \theta}.$$

Substituting for θ gives the following result.

$$d = \frac{93{,}000{,}000}{\sin 0.000212°} \approx 2.51 \times 10^{13} \text{ miles}$$

(b) This distance equals $\dfrac{2.51 \times 10^{13}}{5.9 \times 10^{12}} \approx 4.3$ light-years. *Now Try Exercise 73* ◀

An Application from Surveying Water is often an obstacle to surveyors in the field when measuring distances between two points. For example, to measure the distance between points P and Q in Figure 6.41, a baseline PR, perpendicular to PQ, is determined. Angle PRQ is then measured. Then right triangle trigonometry can be used to determine the length of PQ. (**Source:** P. Kissam, *Surveying Practice.*)

Figure 6.41

EXAMPLE 7 Finding distance

Suppose in Figure 6.41 the length of PR is 94.75 feet and angle PRQ has measure 41.6°. Estimate the distance between points P and Q.

SOLUTION Let angle PRQ be θ. Since $\tan \theta = \frac{PQ}{PR}$, it follows that

$$PQ = PR \tan \theta = 94.75 \tan 41.6° \approx 84.12 \text{ feet.} \quad \textit{Now Try Exercise 67} ◀$$

Highway Design Next we derive a formula that is used in the design of highways.

EXAMPLE 8 Deriving a formula for the design of highway curves

One common type of highway curve is a *simple horizontal curve*. It consists of two straight segments of highway connected by a circular arc with radius r, as shown in Figure 6.42. The distance d is called the *external distance*. (**Source:** F. Mannering and W. Kilareski, *Principles of Highway Engineering and Traffic Analysis.*)

(a) Derive a formula for d that involves r and θ.

(b) Find d for a curve with a 750-foot radius and $\theta = 36°$.

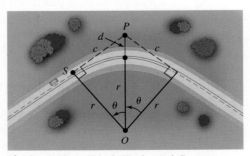

Figure 6.42 A Simple Horizontal Curve

SOLUTION

(a)
$$\cos\theta = \frac{r}{r+d} \qquad \text{Use triangle } OSP.$$

$$(r+d)\cos\theta = r \qquad \text{Multiply by } r+d.$$

$$r+d = \frac{r}{\cos\theta} \qquad \text{Divide by } \cos\theta.$$

$$d = \frac{r}{\cos\theta} - r \qquad \text{Subtract } r.$$

$$d = r\left(\frac{1}{\cos\theta} - 1\right) \qquad \text{Factor out } r.$$

(b) $d = 750\left(\dfrac{1}{\cos 36°} - 1\right) \approx 177$ feet

Now Try Exercise 81 ◀

Complementary Angles and Cofunctions

In Figure 6.43, α and β are complementary angles since their measures sum to 90°. The six trigonometric functions for α and β can be expressed as follows.

Figure 6.43

$$\sin\alpha = \frac{a}{c} = \cos\beta \qquad \cos\alpha = \frac{b}{c} = \sin\beta$$

$$\tan\alpha = \frac{a}{b} = \cot\beta \qquad \cot\alpha = \frac{b}{a} = \tan\beta$$

$$\sec\alpha = \frac{c}{b} = \csc\beta \qquad \csc\alpha = \frac{c}{a} = \sec\beta$$

Notice that the value of a trigonometric function for α equals the value of the trigonometric cofunction for β. For example, $\sin\alpha = \cos\beta$ and $\sin\beta = \cos\alpha$. This is how cofunctions were named. In 1620 Edmund Gunter combined the words "complement" and "sine" to obtain *co*sine. Similarly, the cosecant and cotangent functions are the "complementary functions" of the secant and tangent functions, respectively, and their names were shortened to *co*secant and *co*tangent.

Cofunction Formulas

$$\sin\theta = \cos(90° - \theta) \qquad \cos\theta = \sin(90° - \theta)$$

$$\tan\theta = \cot(90° - \theta) \qquad \cot\theta = \tan(90° - \theta)$$

$$\sec\theta = \csc(90° - \theta) \qquad \csc\theta = \sec(90° - \theta)$$

EXAMPLE 9 Evaluating trigonometric functions using complementary angles

Write an equivalent expression using a cofunction. Then evaluate the expression using a calculator.
(a) $\cot 23°$ **(b)** $\sec 70°$ **(c)** $\cos 12°$

SOLUTION

(a) The complementary angle of $23°$ is $90° - 23° = 67°$. Thus

$$\cot 23° = \tan 67° \approx 2.3559.$$

(b) $\sec 70° = \csc(90° - 70°) = \csc 20° = \frac{1}{\sin 20°} \approx 2.9238$

(c) $\cos 12° = \sin(90° - 12°) = \sin 78° \approx 0.9781$ *Now Try Exercise 53* ◀

6.2 Putting It All Together

The following table summarizes some properties of right triangle trigonometry.

Concept	Formulas and Figures	
Trigonometric functions	Let θ be an acute angle in a right triangle ABC. $$\sin\theta = \frac{\text{side opposite}}{\text{hypotenuse}} \qquad \csc\theta = \frac{\text{hypotenuse}}{\text{side opposite}}$$ $$\cos\theta = \frac{\text{side adjacent}}{\text{hypotenuse}} \qquad \sec\theta = \frac{\text{hypotenuse}}{\text{side adjacent}}$$ $$\tan\theta = \frac{\text{side opposite}}{\text{side adjacent}} \qquad \cot\theta = \frac{\text{side adjacent}}{\text{side opposite}}$$	
Cofunction formulas	Let α and β be complementary angles. $$\sin\alpha = \cos(90° - \alpha) = \cos\beta$$ $$\tan\alpha = \cot(90° - \alpha) = \cot\beta$$ $$\sec\alpha = \csc(90° - \alpha) = \csc\beta$$	

6.2 Exercises

Basic Concepts

Exercises 1–4: Sketch a right triangle with the following properties. Label the measure of each angle and side.

1. Acute angles of $30°$ and $60°$ and a hypotenuse with length 2

2. Acute angle of $45°$ and a leg with length 1

3. Isosceles and a hypotenuse with length 4

4. Acute angle of $60°$ and the shorter leg with length 3

Evaluating Trigonometric Functions

Exercises 5–10: (Refer to Example 1.) Use a 30°–60° right triangle to find the exact value of the trigonometric expression.

5. sin 60°

6. tan 30°

7. cos 30°

8. cot 30°

9. sec 60°

10. csc 60°

Exercises 11–16: Use a 45°–45° right triangle to find the exact value of the trigonometric expression.

11. tan 45°

12. sec 45°

13. cot 45°

14. csc 45°

15. sin 45°

16. cos 45°

Exercises 17–20: Find the six trigonometric functions of θ.

17.

18.

19.

20.

Exercises 21–30: Find the six trigonometric functions of the given angle. Approximate to three decimal places when appropriate.

21. 60°

22. 45°

23. 25°

24. 30°

25. 5°35′

26. 85°35′33″

27. 13°45′30″

28. 45°44′

29. 1.05°

30. 0.161°

Exercises 31–36: Let θ be an acute angle. Find the unknown trigonometric value, using the given information.

31. sec θ if cos θ = $\frac{1}{3}$

32. cot θ if tan θ = 5

33. csc θ if sin θ = $\frac{12}{13}$

34. sin θ if csc θ = $\frac{5}{4}$

35. tan θ if cot θ = $\frac{7}{24}$

36. cos θ if sec θ = $\frac{7}{5}$

Solving Triangles

Exercises 37–44: Find the lengths of the unknown sides in the right triangle. Round values to the nearest hundredth.

37.

38.

39.

40.

41.

42.

43.

44.

Exercises 45–48: Find the exact length of each side labeled with a variable in the figure.

45.

46.

47.

48.

Exercises 49–52: Complete the following for right triangle ABC having the standard labeling shown in Figure 6.25. Approximate the answer to the nearest hundredth.

49. Find a if $b = 12$ and $\alpha = 60°$.

50. Find b if $c = 23$ and $\beta = 45°$.

51. Find c if $a = 100$ and $\beta = 53°43'$.

52. Find a if $b = 64$ and $\alpha = 78°15'$.

Cofunctions

Exercises 53–56: (Refer to Example 9.) Write an equivalent expression using a cofunction. Approximate the expression to four decimal places using a calculator.

53. (a) sin 70°   **(b)** cos 40°

54. (a) cot 23°   **(b)** tan 48°

55. (a) csc 49°   **(b)** sec 63°

56. (a) cot 87°   **(b)** sec 72°

Applications

57. *Height of the Cloud Base*   (Refer to Example 3 and Figure 6.34.) From a distance of 1500 feet from the spotlight, the angle of elevation θ equals 37°30′. Find the height of the cloud base.   (**Source:** F. Cole.)

58. *Height of a Tree*   One hundred feet from the trunk of a tree on level ground, the angle of elevation of the top of the tree is 35°. Find the height of the tree to the nearest foot.

59. *Length of a Shadow*   The angle of elevation of the sun is 34°. Find the length of a shadow cast by a person who is 5 feet 3 inches tall. Round your answer to the nearest tenth of a foot.

60. *Height of a Tower*   The shadow of a vertical tower is 40.6 meters long when the angle of elevation of the sun is 34.6°. Find the height of the tower.

61. *Aerial Photography*   An aerial photograph is taken directly above a building. The length of the building's shadow is 48 feet when the angle of elevation of the sun is 35.3°. Estimate the height of the building.

62. *Angle of Depression of a Floodlight*   A company safety committee recommended that a floodlight be mounted in a parking lot as shown in the figure, so as to illuminate the employee exit. Find h to the nearest tenth of a foot.

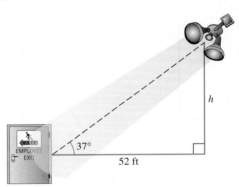

63. *Angle of Depression*   An airplane is flying near a football stadium at 12,000 feet above level ground. The angle of depression from the airplane to the stadium is 13°. How far horizontally must the airplane fly to be directly over the football stadium? Round your answer to the nearest thousand feet.

64. *Angle of Depression*   An airplane is flying 10,500 feet above the level ground. The angle of depression from the plane to the base of a tree is 13°50′. How far horizontally must the plane fly to be directly over the tree?

65. *Height of a Building*   From a window 30 feet above the street, the angle of elevation to the top of the building across the street is 50° and the angle of depression to the base of this building is 20°. See the figure at the top of the next page. Find the height of the building across the street.

66. *Height of a Building* The angle of elevation from the top of a small building to the top of a nearby taller building is 46°40′, while the angle of depression to the bottom is 14°10′. If the smaller building is 28.0 meters high, find the height *x* of the taller building.

67. *Surveying* (Refer to Example 7 and Figure 6.41.) Find the distance from *P* to *Q* if *PR* is 85.62 feet and angle *PQR* is 23.76°.

68. *Weather Tower* A 410-foot weather tower used to measure wind speed has a guy wire attached to it 175 feet above the ground. The angle between the wire and the vertical tower is 57°, as shown in the figure. Approximate the length of the guy wire. (**Source:** Brookhaven National Laboratory.)

69. *Height of a Mountain* From a point *A* the angle of elevation of Mount Kilimanjaro in Africa is 13.7°, and from a point *B*, directly behind *A*, the angle of elevation is 10.4°. See the figure. If the distance between *A* and *B* is

5 miles, approximate the height of Mount Kilimanjaro to the nearest hundred feet.

70. *Height of Mt. Whitney* The angle of elevation from Lone Pine to the top of Mt. Whitney is 10°50′. Exactly 7 kilometers from Lone Pine along a straight, level road toward Mt. Whitney, the angle of elevation is 22°40′. Find the height of the top of Mt. Whitney above the level of the road to the nearest hundredth of a kilometer.

71. *Height of a Pyramid* The angle of elevation from a point on the ground to the top of a pyramid is 35°30′. The angle of elevation from a point 135 feet farther back to the top of the pyramid is 21°10′. Find the height of the pyramid to the nearest foot.

72. *Height of an Antenna* An antenna is on top of the center of a house. The angle of elevation from a point on the ground 28.0 meters from the center of the house to the top of the antenna is 27°10′, and the angle of elevation to the bottom of the antenna is 18°10′. Find the height of the antenna.

73. *Distance to Nearby Stars* (Refer to Example 6 and Figure 6.40.) The table lists the parallax θ in degrees for some nearby stars. Approximate the distance from Earth to each star in miles and in light-years.

Star	θ (degrees)
Barnard's Star	1.52×10^{-4}
Sirius	1.05×10^{-4}
61 Cygni	8.11×10^{-5}
Procyon	7.97×10^{-5}

Source: M. Zeilik et al.

74. *Parallax and Distance* (Refer to Exercise 73.) When the parallax θ is equal in measure to 1 second, a star is said to have a distance from Earth of 1 parsec. If the distance between Earth and the sun is 93,000,000 miles, approximate the number of miles in 1 parsec. How many light-years is this? (**Source:** M. Zeilik.)

75. *Observing Mercury* The planet Mercury is closer to the sun than Earth. For this reason it can only be observed low in the horizon around sunset or sunrise. See the figure, where angle θ is called the *elongation*. Because Mercury's orbit is not circular, the elongation varies between 18° and 28°. Approximate the minimum and maximum distances between Mercury and the sun. (**Source:** M. Zeilik.)

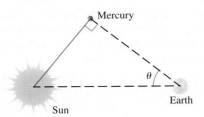

76. *Observing Venus* (Refer to Exercise 75.) The orbit of Venus is nearly circular with an elongation of 48°. Estimate the distance between Venus and the sun.

77. *Orbital Height of a GPS Satellite* The figure illustrates a satellite in the Global Positioning System (GPS) orbiting over the equator, where $r = 3963$ miles and $\theta = 76.1°$. Use $d = r\left(\frac{1}{\cos\theta} - 1\right)$ from Example 8 to determine the altitude of the GPS satellite above Earth's surface to the nearest mile. (**Source:** Y. Zhao, *Vehicle Location and Navigation Systems.*)

78. *Heights of Lunar Mountains* The lunar mountain peak Huygens has a height of 21,000 feet. The shadow of Huygens on a photograph was 2.8 mm, and the nearby mountain Bradley had a shadow of 1.8 mm on the same photograph. Use similar triangles to calculate

the height of Bradley to the nearest hundred feet. (**Source:** T. Webb, *Celestial Objects for Common Telescopes.*)

79. *Highway Curve Design* Highway curves are sometimes banked so that the outside of the curve is slightly elevated or inclined above the inside of the curve, as shown in the figure. This inclination is called the *superelevation*. Both the curve's radius and the superelevation must be correct for a given speed limit. The relationship among a car's velocity v in feet per second, the safe radius r of the curve in feet, and the superelevation θ in degrees is given by $r = \frac{v^2}{4.5 + 32.2\tan\theta}$. (**Source:** F. Mannering and W. Kilareski.)

(a) A curve has a speed limit of 66 feet per second (45 mi/hr) and a superelevation of $\theta = 3°$. Approximate the safe radius r.

(b) Find r if $\theta = 5°$ and $v = 66$.

(c) Make a conjecture about how increasing θ affects the safe radius r. Verify your conjecture by tabling r, starting at $\theta = 0$ and incrementing by 1. Let $v = 66$.

80. *Highway Design* (Refer to Exercise 79.) A highway curve has a radius of $r = 1150$ feet and a superelevation of $\theta = 2.1°$. What should be the speed limit (in miles per hour) for this curve?

81. *Highway Design* (Refer to Example 8 and Figure 6.42.) Find the external distance d for a highway curve with $r = 625$ feet and $\theta = 54°$.

82. *Highway Design*   (Refer to Example 8.) A simple horizontal curve is shown in the figure. The points P and S mark the beginning and end of the curve. Let Q be the point of intersection where the two straight sections of highway leading into the curve would meet if extended. The radius of the curve is r, and the angle θ denotes how many degrees the curve turns. If $r = 765$ feet and $\theta = 83°$, find the distance between P and Q.   (**Source:** F. Mannering and W. Kilareski.)

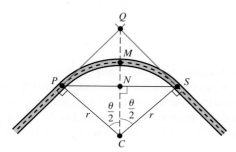

83. *Area of an Equilateral Triangle*   Find the area of the equilateral triangle shown in the figure in terms of s.

84. *Area of a Hexagon*   Write the area of the hexagon shown in the figure in terms of x. Assume that the six triangles that comprise the hexagon are equilateral and congruent.

Writing about Mathematics

85. Most calculators have built-in keys to compute the sine, cosine, and tangent functions, but not the secant, cosecant, and cotangent functions. Is it possible to evaluate all of the trigonometric functions with this type of calculator? Explain and include examples.

86. The sine function is defined in terms of right triangles as $\sin \theta = \frac{\text{side opposite}}{\text{hypotenuse}}$. Suppose that a fixed angle θ occurs in two right triangles. If the hypotenuse in the first triangle has twice the length of the hypotenuse in the second triangle, what can be said about the sides opposite in each triangle? How does the value $\sin \theta$ compare in each triangle? Explain.

CHECKING BASIC CONCEPTS FOR SECTIONS 6.1 AND 6.2

1. Find the radian measure of each angle.
   (a) $45°$
   (b) $75°$

2. Find the degree measure of each angle.
   (a) $\frac{\pi}{6}$
   (b) $\frac{5\pi}{4}$

3. Find the arc length intercepted by a central angle of $\theta = 30°$ in a circle with radius $r = 12$ inches. Calculate the area of the sector determined by r and θ.

4. Evaluate the six trigonometric functions of $\theta = 60°$ by hand. Support your results by using a calculator.

5. Use the right triangle in the figure to find the six trigonometric functions of θ.

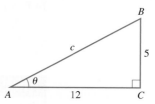

6. If $\alpha = 63°$ and $a = 9$ in right triangle ABC, approximate the length of the hypotenuse to the nearest tenth.

6.3 The Sine and Cosine Functions and Their Graphs

- **Define the sine and cosine functions for any angle**
- **Define the sine and cosine functions for any real number by using the unit circle and the wrapping function**
- **Represent the sine and cosine functions**
- **Use the sine and cosine functions in applications**
- **Model with the sine function (optional)**

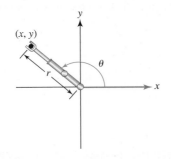

Introduction

The sine and cosine functions are used not only in applications involving right triangles, but also to model phenomena involving rotation and periodic motion. Extending the domains of the trigonometric functions from acute angles to angles of any measure will allow us to model a wide variety of phenomena such as biorhythms, weather, tides, electricity, robotic arms, and the design of highways.

Definitions

Robotics is a rapidly growing field that requires extensive mathematics. One basic problem in designing a robotic arm is determining the location of the robot's hand. Suppose we have a robotic arm that rotates at the shoulder and is controlled by changing the angle θ and the length of the arm r, as illustrated in Figure 6.44. We would like to find a relation between the xy-coordinates of the hand and the values for r and θ. (**Source:** W. Stadler, *Analytical Robotics and Mechatronics.*)

Notice that for a fixed angle θ, if the length of the arm is changed from r_1 to r_2, triangle ABC in Figure 6.45 and triangle DEF in Figure 6.46 are similar triangles. Thus the following ratios are equal and depend only on the measure of θ.

$$\frac{x_1}{r_1} = \frac{x_2}{r_2} \quad \text{and} \quad \frac{y_1}{r_1} = \frac{y_2}{r_2}$$

Figure 6.44

Figure 6.45

Figure 6.46

Figure 6.47

In Figure 6.47 the Pythagorean theorem gives $r^2 = x^2 + y^2$. Since $r > 0$, it follows that $r = \sqrt{x^2 + y^2}$. The ratios $\frac{y}{r}$ and $\frac{x}{r}$ can be used to define the *sine* and *cosine* functions of any angle θ.

The Sine and Cosine Functions of Any Angle θ

Let angle θ be in standard position with the point (x, y) lying on the angle's terminal side. If $r = \sqrt{x^2 + y^2}$, then

$$\sin \theta = \frac{y}{r} \quad \text{and} \quad \cos \theta = \frac{x}{r} \quad (r \neq 0).$$

Figure 6.48

Although the terminal side of θ was shown in the second quadrant, these definitions are valid for any angle θ.

MAKING CONNECTIONS

Right Triangle Trigonometry If $0° < \theta < 90°$, then x corresponds to the length of the adjacent side, y corresponds to the length of the opposite side, and r corresponds to the length of the hypotenuse. See Figure 6.48. These new definitions for sine and cosine are consistent with the definitions when θ is an acute angle, presented in Section 6.2.

EXAMPLE 1 Evaluating sine and cosine for coterminal angles

Suppose a robotic hand is located at the point $(15, -8)$, where all units are in inches.
(a) Find the length of the arm.
(b) Let α satisfy $0° \le \alpha < 360°$ and represent the angle between the positive x-axis and the robotic arm. Find $\sin \alpha$ and $\cos \alpha$.
(c) Let β satisfy $-360° \le \beta < 0°$ and represent the angle between the positive x-axis and the robotic arm. Find $\sin \beta$ and $\cos \beta$. How do the values for $\sin \beta$ and $\cos \beta$ compare with the values for $\sin \alpha$ and $\cos \alpha$?

SOLUTION
(a) Since $r = \sqrt{15^2 + (-8)^2} = 17$, the length of the arm is 17 inches. See Figure 6.49.
(b) Let $x = 15$, $y = -8$, and $r = 17$. Then

$$\sin \alpha = \frac{y}{r} = -\frac{8}{17} \quad \text{and} \quad \cos \alpha = \frac{15}{17}.$$

(c) In Figure 6.50, β satisfies $-360° \le \beta < 0°$. Since the values of x, y, and r do *not* change, the trigonometric values for β are the *same* as those for α in part (b).

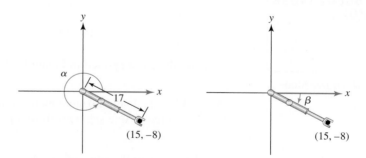

Figure 6.49   **Figure 6.50**

Now Try Exercise 1 ◀

The results of Example 1 can be generalized. If α and β are coterminal angles, then

$$\sin \alpha = \sin \beta \quad \text{and} \quad \cos \alpha = \cos \beta.$$

The angles θ and $\theta + 360°$ are coterminal for any θ. Their terminal sides pass through the same point (x, y), as shown in Figure 6.51. Therefore for all θ, $\sin \theta = \sin (\theta + 360°)$ and $\cos \theta = \cos (\theta + 360°)$. In general, if n is any integer, then

$$\sin \theta = \sin (\theta + n \cdot 360°) \quad \text{and} \quad \cos \theta = \cos (\theta + n \cdot 360°).$$

Figure 6.51

sin (90)
                                    1
sin (90+360)
                                    1
sin (90−2∗360)
                                    1

Figure 6.52 Degree Mode

As a result, we say that the sine and cosine functions are **periodic** with **period** $360°$ (or 2π radians). For example,

$$\sin 90° = \sin (90° + 360°) = \sin (90° - 2 \cdot 360°).$$

See Figure 6.52.

If an angle is a multiple of $30°$ or $45°$, exact evaluation of the sine function or cosine function is possible by hand. However, in most applications calculators are used to obtain numerical approximations. The next example illustrates an angle for which the sine and cosine functions can be evaluated exactly by hand.

EXAMPLE 2 Finding values of $\sin \theta$ and $\cos \theta$

Find $\sin 120°$ and $\cos 120°$. Support your answer using a calculator.

SOLUTION When evaluating the sine or cosine function by hand, we can select any positive value for r and not change the resulting values for $\sin \theta$ and $\cos \theta$. For convenience, we let $r = 2$, as shown in Figure 6.53. The length of the shortest leg in a $30°$–$60°$ right triangle is half the hypotenuse. Since the terminal side of θ is in the second quadrant, $x < 0$ and so $x = -1$. Next we find y.

Figure 6.53

$$x^2 + y^2 = r^2 \qquad \text{Pythagorean theorem}$$
$$y^2 = r^2 - x^2 \qquad \text{Subtract } x^2.$$
$$y^2 = 2^2 - (-1)^2 \qquad \text{Let } r = 2 \text{ and } x = -1.$$
$$y = \pm\sqrt{3} \qquad \text{Square root property}$$

Since the terminal side of θ is in the second quadrant, $y > 0$ and so $y = \sqrt{3}$. Thus

$$\sin 120° = \frac{y}{r} = \frac{\sqrt{3}}{2} \approx 0.8660 \qquad \text{and}$$

$$\cos 120° = \frac{x}{r} = -\frac{1}{2} = -0.5.$$

These results are supported in Figure 6.54.   *Now Try Exercise 13* ◀

sin (120)
            .8660254038
cos (120)
                       −.5

Figure 6.54 Degree Mode

NOTE When evaluating trigonometric functions by hand as in Example 2, *always* draw the perpendicular line segment from a point on the terminal side to the x-axis, *not* the y-axis.

The values of the sine and cosine functions for the **quadrantal angles** $0°$, $90°$, $180°$, and $270°$ are shown in Table 6.3. Try to verify these values.

Table 6.3

	0°	90°	180°	270°
sin θ	0	1	0	−1
cos θ	1	0	−1	0

NOTE If the terminal side of an angle θ in standard position lies on the y-axis, $\sin \theta = \pm 1$ and $\cos \theta = 0$. If the terminal side of an angle θ in standard position lies on the x-axis, $\sin \theta = 0$ and $\cos \theta = \pm 1$.

Figure 6.55 Radian Mode

Evaluation in Radian Mode   Trigonometric functions can also be evaluated using radian mode. For example, since 90° is equivalent to $\frac{\pi}{2}$ radians, it follows that

$$\sin \frac{\pi}{2} = 1 \quad \text{and} \quad \cos \frac{\pi}{2} = 0.$$

In Figure 6.55 a calculator has been used to perform these evaluations in *radian mode*.

EXAMPLE 3   Finding values of sin t and cos t

Find the exact values of $\sin \left(-\frac{3\pi}{4}\right)$ and $\cos \left(-\frac{3\pi}{4}\right)$.

SOLUTION

Getting Started   Because there is no degree symbol, it is assumed that $-\frac{3\pi}{4}$ is measured in radians. Its terminal side lies in quadrant III, as shown in Figure 6.56. ▶

From a point on the terminal side, draw a perpendicular line segment to the x-axis, forming a 45°–45° right triangle. Label the resulting triangle conveniently so that the length of each leg equals 1. (Each side is labeled -1, because the terminal side lies in quadrant III.) By the Pythagorean theorem, the hypotenuse is $\sqrt{2}$. Thus

$$\sin \left(-\frac{3\pi}{4}\right) = \frac{y}{r} = -\frac{1}{\sqrt{2}} \quad \text{and} \quad \cos \left(-\frac{3\pi}{4}\right) = \frac{x}{r} = -\frac{1}{\sqrt{2}}.$$

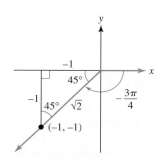

Figure 6.56

NOTE If we rationalize the denominator, each result becomes $-\frac{\sqrt{2}}{2}$.

Now Try Exercise 27 ◀

An Application from Highway Design   *Grade,* or slope, is a measure of steepness and indicates whether a highway is uphill or downhill. A 5% grade indicates that a road is increasing 5 vertical feet for each 100-foot increase in horizontal distance. *Grade resistance R* is the gravitational force acting on a vehicle and is given by

$$R = W \sin \theta,$$

where W is the weight of the vehicle and θ is the angle associated with the grade. See Figure 6.57. For an uphill grade $\theta > 0$, and for a downhill grade $\theta < 0$.   (**Source:** F. Mannering and W. Kilareski, *Principles of Highway Engineering and Traffic Analysis.*)

Figure 6.57

Figure 6.58 A Grade of −6%

EXAMPLE 4   Calculating the grade resistance

A downhill highway grade is modeled by the line $y = -0.06x$ in the fourth quadrant.
(a) Find the grade of the road.
(b) Determine the grade resistance for a 3000-pound car. Interpret the result.

SOLUTION
(a) The slope of the line is −0.06, so when x *increases* by 100 feet, y *decreases* by 6 feet. See Figure 6.58. Thus this road has a grade of −6%.
(b) First we must find $\sin \theta$. From Figure 6.58 we see that the point $(100, -6)$ lies on the terminal side of θ. Since

$$r = \sqrt{100^2 + (-6)^2} = \sqrt{10{,}036},$$

it follows that

$$\sin \theta = \frac{y}{r} = \frac{-6}{\sqrt{10{,}036}}.$$

The grade resistance is

$$R = W \sin \theta = 3000\left(\frac{-6}{\sqrt{10{,}036}}\right) \approx -179.7 \text{ lb}.$$

On this stretch of highway, gravity would pull a 3000-pound vehicle *downhill* with a force of about 180 pounds. Note that a downhill grade results in a negative grade resistance.   *Now Try Exercise 87* ◀

The Unit Circle

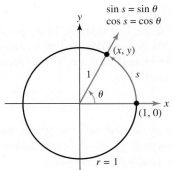

$\sin s = \sin \theta$
$\cos s = \cos \theta$

Figure 6.59 Unit Circle

Trigonometric functions were first associated with angles. As applications became more diverse, real numbers were included in the domains of the six trigonometric functions. Real numbers were needed to represent quantities such as time and distance. To extend the domains of the sine and cosine functions to include all real numbers, we will introduce the unit circle.

The **unit circle** has radius 1 and equation $x^2 + y^2 = 1$. Let (x, y) be a point on the unit circle, and let s be the arc length along the unit circle from the point $(1, 0)$ to the point (x, y) determined by a counterclockwise rotation. See Figure 6.59. Since $r = 1$, the arc length formula $s = r\theta$ reduces to $s = \theta$. That is, the real number s representing arc length is numerically equal to the radian measure of θ. Thus if s is a real number and θ is an angle measured in radians, as shown in Figure 6.59, we define

$$\sin s = \sin \theta \quad \text{and} \quad \cos s = \cos \theta.$$

If $s < 0$, then the red arc in Figure 6.59 is wrapped around the unit circle in a *clockwise* direction and θ has negative measure. This discussion suggests how trigonometric functions of a real number are evaluated.

Trigonometric Functions of Real Numbers

The value of a trigonometric function for the real number s is equal to its value for s radians.

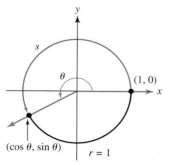

Figure 6.60 Unit Circle

NOTE To evaluate a trigonometric function of a real number s with a calculator, use *radian* mode.

If (x, y) is a point on the unit circle lying on the terminal side of an angle θ in standard position, then

$$\sin \theta = \frac{y}{r} = y \qquad \text{and} \qquad \cos \theta = \frac{x}{r} = x$$

because $r = 1$. Points (x, y) on the unit circle can be expressed as $(\cos \theta, \sin \theta)$, as illustrated in Figure 6.60. Because trigonometric functions can be defined using the unit circle, they are also referred to as **circular functions**.

The unit circle is described by $x^2 + y^2 = 1$. Since $x = \cos \theta$ and $y = \sin \theta$, we have

$$(\cos \theta)^2 + (\sin \theta)^2 = 1, \qquad \text{or equivalently,}$$

$$\sin^2 \theta + \cos^2 \theta = 1.$$

This equation is an example of a trigonometric identity. An identity is true for all meaningful values of the variable. In Figures 6.61 and 6.62 this identity is evaluated for different values of θ. In every case the result is 1, regardless of whether θ is measured in degrees or radians.

```
(sin(80))2+(cos(
80))2
                1
(sin(‾12))2+(cos
(‾12))2
                1
```

```
(sin(π/7))2+(cos
(π/7))2
                1
(sin(1.5))2+(cos
(1.5))2
                1
```

Figure 6.61 Degree Mode   **Figure 6.62** Radian Mode

The Wrapping Function (Optional)

Trigonometric functions of real numbers can be defined more formally by using the concept of the wrapping function. To define the **wrapping function** W on the unit circle, let the input s be any positive real number. Next cut a string of length s and place one end of the string at the **initial point** $(1, 0)$ on the unit circle. See Figure 6.63. Then *wrap* the string, shown as a red arc in Figure 6.64, counterclockwise along the circumference of the unit circle. If the other end of the string reaches or stops at the **terminal point** with coordinates (x, y) on the unit circle, then the wrapping function outputs (x, y). That is, given any real number input s, the wrapping function outputs the coordinates of a point on the unit

Figure 6.63 Unit Circle

Figure 6.64 Unit Circle

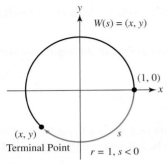

Figure 6.65

circle. If s is positive, the string is wrapped counterclockwise, and if s is negative, the string has length $|s|$ and is wrapped clockwise. See Figure 6.65. Note that the initial point is *always* $(1, 0)$ but the terminal point varies and depends on s.

NOTE The wrapping function is different from other functions that we have encountered because its output is a *point* on the unit circle.

EXAMPLE 5 Evaluating the wrapping function

For each real number s, evaluate $W(s)$.

(a) 2π **(b)** π **(c)** $-\dfrac{\pi}{2}$ **(d)** $-\dfrac{\pi}{4}$

SOLUTION

(a) Because the radius of the unit circle is 1, the circumference C of the unit circle is $C = 2\pi r = 2\pi$. If a string of length $s = 2\pi$ is wrapped counterclockwise around the unit circle, it will make one complete revolution and the terminal point will coincide with the initial point $(1, 0)$. Thus $W(2\pi) = (1, 0)$. See Figure 6.66.

(b) Half the circumference of the unit circle is π. The string wraps halfway around the unit circle, and the terminal point is $(-1, 0)$. Thus $W(\pi) = (-1, 0)$. See Figure 6.67.

(c) Because s is negative, the string wraps clockwise around the unit circle, as shown in Figure 6.68. A length of $\frac{\pi}{2}$ represents a fourth of the circumference of the unit circle, so the terminal point is $(0, -1)$ and $W\left(-\frac{\pi}{2}\right) = (0, -1)$.

(d) A distance of $\frac{\pi}{4}$ represents an eighth of the circumference of the unit circle. Thus the terminal point lies on the line $y = -x$, as shown in Figure 6.69. The $45°$–$45°$ right triangle formed has a hypotenuse with length 1 and legs both with length a.

$$a^2 + a^2 = 1 \qquad \text{Pythagorean theorem: } a = b$$

$$2a^2 = 1 \qquad \text{Add like terms.}$$

$$a^2 = \frac{1}{2} \qquad \text{Divide by 2.}$$

$$a = \pm\frac{1}{\sqrt{2}} \qquad \text{Square root property}$$

Because the terminal point is located in quadrant IV, the x-coordinate is $\dfrac{1}{\sqrt{2}}$ and the y-coordinate is $-\dfrac{1}{\sqrt{2}}$. Thus $W\left(-\frac{\pi}{4}\right) = \left(\dfrac{1}{\sqrt{2}}, -\dfrac{1}{\sqrt{2}}\right)$.

Figure 6.66

Figure 6.67

Figure 6.68

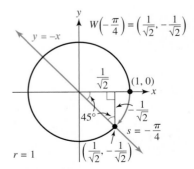

Figure 6.69

Now Try Exercises 51, 53, 55, and 59 ◀

NOTE The concept of string is used only to help you visualize the wrapping function. Instead you could visualize moving a distance s around the unit circle.

Evaluating the Sine and Cosine with the Wrapping Function The wrapping function on the unit circle can be used to define the sine and cosine functions for any real number s by letting

$$\sin s = y \quad \text{and} \quad \cos s = x,$$

where $W(s) = (x, y)$. The next example illustrates how to apply this definition.

EXAMPLE 6 Evaluating the sine and cosine functions

Find $\sin s$ and $\cos s$ for each real number s.

(a) 2π **(b)** π **(c)** $-\dfrac{\pi}{2}$ **(d)** $-\dfrac{\pi}{4}$

SOLUTION

Getting Started First note that the values for s are the same as those used in Example 5. Thus if $W(s) = (x, y)$, we need only apply the fact that $\sin s = y$ and $\cos s = x$. ▶

(a) $W(2\pi) = (1, 0)$ by Example 5(a), so $\sin 2\pi = 0$ and $\cos 2\pi = 1$.
(b) $W(\pi) = (-1, 0)$ by Example 5(b), so $\sin \pi = 0$ and $\cos \pi = -1$.
(c) $W\left(-\frac{\pi}{2}\right) = (0, -1)$ by Example 5(c), so $\sin\left(-\frac{\pi}{2}\right) = -1$ and $\cos\left(-\frac{\pi}{2}\right) = 0$.
(d) $W\left(-\frac{\pi}{4}\right) = \left(\dfrac{1}{\sqrt{2}}, -\dfrac{1}{\sqrt{2}}\right)$ by Example 5(d), so $\sin\left(-\frac{\pi}{4}\right) = -\dfrac{1}{\sqrt{2}}$ and

$\cos\left(-\frac{\pi}{4}\right) = \dfrac{1}{\sqrt{2}}$. *Now Try Exercises 67, 69, 71, and 75* ◀

NOTE The results obtained by using the wrapping function to evaluate trigonometric functions are equivalent to those arrived at by the other methods presented earlier. The approach of the wrapping function on a unit circle has the advantage that it is not necessary to use the concept of an angle and so the method applies to any real number s.

Representations of the Sine and Cosine Functions

Like other functions that we have studied, the sine and cosine functions also have symbolic, numerical, and graphical representations. Both functions are nonlinear.

The Sine Function A *symbolic representation* of the sine function is $f(t) = \sin t$. The domain of the sine function is all real numbers. There is no simple formula that can be used to evaluate the sine function. Instead we generally rely on other methods for its evaluation.

A *numerical representation* of $f(t) = \sin t$ is shown in Table 6.4. (These values can be found by hand as in Example 3 or with a calculator.) Since outputs from the sine function correspond to a y-coordinate on the unit circle, the range of the sine function is $-1 \leq y \leq 1$.

Table 6.4

t	0	$\frac{\pi}{4}$	$\frac{\pi}{2}$	$\frac{3\pi}{4}$	π	$\frac{5\pi}{4}$	$\frac{3\pi}{2}$	$\frac{7\pi}{4}$	2π
$\sin t$	0	$\dfrac{1}{\sqrt{2}}$	1	$\dfrac{1}{\sqrt{2}}$	0	$-\dfrac{1}{\sqrt{2}}$	-1	$-\dfrac{1}{\sqrt{2}}$	0

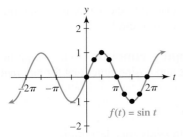

Figure 6.70 The Sine Function

A graphical representation of $f(t) = \sin t$ is shown in Figure 6.70, where the points in Table 6.4 are also plotted. The graph increases from 0 to 1 for $0 \leq t \leq \frac{\pi}{2}$, decreases from 1 to -1 for $\frac{\pi}{2} \leq t \leq \frac{3\pi}{2}$, and then increases from -1 to 0 for $\frac{3\pi}{2} \leq t \leq 2\pi$. The graph repeats itself both to the left and to the right every 2π units on the t-axis, since the sine function has period 2π, or $360°$. That is, $\sin t = \sin(t + 2\pi n)$, where n is any integer.

MAKING CONNECTIONS

The Unit Circle and the Sine Graph The graph of the sine function may be easier to understand if we use a unit circle. If a point P on the unit circle starts at $(1, 0)$ and rotates counterclockwise, then the y-coordinate of P corresponds to $\sin \theta$. Refer to Figure 6.71. As a result, $\sin \theta$ first increases from 0 to 1, then decreases from 1 to -1, and finally increases from -1 to 0, where P completes one rotation and returns to the point $(1, 0)$. If P continues to rotate, it passes through the same points a second time, and so the sine function has period 2π, or $360°$.

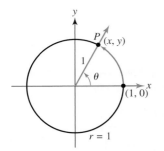

Figure 6.71

EXAMPLE 7   Evaluating the sine function

Evaluate $f(t) = \sin t$ at $t = \frac{3\pi}{2}$ by hand.

SOLUTION An angle θ of $\frac{3\pi}{2}$ radians in standard position has a terminal side that intersects the unit circle at the point $(0, -1)$. See Figure 6.72. Therefore $\sin\left(\frac{3\pi}{2}\right) = -1$.

Now Try Exercise 41 ◀

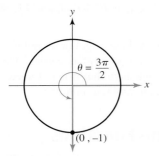

Figure 6.72

The Cosine Function   The cosine function, *represented symbolically* by $f(t) = \cos t$, has numerical and graphical representations similar to those of the sine function. Its domain is all real numbers. Like the sine function, the cosine function does not have a simple formula.

A *numerical representation* of $f(t) = \cos t$ is shown in Table 6.5. Since an output from the cosine function corresponds to an x-coordinate on the unit circle, the range of the cosine function is from -1 to 1. Like the sine function, the cosine function has period 2π, or $360°$.

Table 6.5

t	0	$\frac{\pi}{4}$	$\frac{\pi}{2}$	$\frac{3\pi}{4}$	π	$\frac{5\pi}{4}$	$\frac{3\pi}{2}$	$\frac{7\pi}{4}$	2π
$\cos t$	1	$\frac{1}{\sqrt{2}}$	0	$-\frac{1}{\sqrt{2}}$	-1	$-\frac{1}{\sqrt{2}}$	0	$\frac{1}{\sqrt{2}}$	1

A graphical representation of $f(t) = \cos t$ is shown in Figure 6.73, where the points in Table 6.5 are also plotted. Notice that the overall shape of the graph of the cosine function is similar to that of the graph of the sine function. However, the cosine function decreases from 1 to -1 for $0 \leq t \leq \pi$ and increases from -1 to 1 for $\pi \leq t \leq 2\pi$.

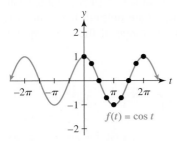

Figure 6.73 The Cosine Function

CLASS DISCUSSION

If a point P on the unit circle in Figure 6.71 starts at $(1, 0)$ and rotates counterclockwise, then the x-coordinate of P corresponds to $\cos \theta$. Explain how this relates to the graph of the cosine function.

EXAMPLE 8 Evaluating the cosine function

Evaluate $f(t) = \cos t$ at $t = \frac{5\pi}{6}$ by hand.

SOLUTION An angle of $\frac{5\pi}{6}$ radians in standard position has a terminal side that intersects the unit circle in the second quadrant. See Figure 6.74. To find the point of intersection (x, y), notice that the hypotenuse of the 30°–60° right triangle has length 1 and the shorter leg has length $y = \frac{1}{2}$. We can determine x symbolically.

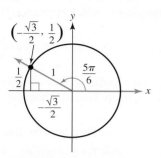

$$x^2 + y^2 = 1 \qquad \textit{Equation of the unit circle}$$

$$x^2 + \left(\frac{1}{2}\right)^2 = 1 \qquad y = \frac{1}{2}$$

$$x^2 = \frac{3}{4} \qquad \textit{Solve for } x^2.$$

$$x = \pm\frac{\sqrt{3}}{2} \qquad \textit{Square root property}$$

Figure 6.74

The point (x, y) is located in the second quadrant, so choose $x = -\frac{\sqrt{3}}{2}$. Because $\cos t = x$, it follows that $\cos \frac{5\pi}{6} = -\frac{\sqrt{3}}{2} \approx -0.8660$. **Now Try Exercise 43** ◄

Applications of the Sine and Cosine Functions

Periodic graphs that are similar in shape to the graphs of the sine and cosine functions are **sinusoidal**. Of all the periodic graphs, sinusoidal graphs are the most important in applications because they occur in nearly every aspect of physical science.

Because the moon orbits Earth, we observe different phases of the moon during the period of a month. In Figure 6.75, angle θ is called the *phase angle*. The *phase F* of the moon is computed by

$$F(\theta) = \frac{1}{2}(1 - \cos \theta)$$

and gives the fraction of the moon's face that is illuminated by the sun. (**Source:** P. Duffet-Smith, *Practical Astronomy with Your Calculator.*)

Figure 6.75 Phase Angle θ (not to scale)

EXAMPLE 9 Modeling the phases of the moon

Let $F(\theta) = \frac{1}{2}(1 - \cos \theta)$.
(a) A graph of F is shown in Figure 6.76 on the next page. Discuss how the graph relates to the phases of the moon.
(b) Evaluate $F(0)$, $F\left(\frac{\pi}{2}\right)$, $F(\pi)$, and $F\left(\frac{3\pi}{2}\right)$. Interpret each result.

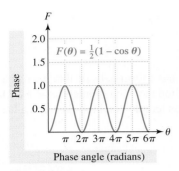

Figure 6.76

SOLUTION

(a) The phases of the moon are periodic and have a period of 2π. Each peak represents a full moon, and each valley represents a new moon.

(b) $F(0) = \frac{1}{2}(1 - \cos 0) = \frac{1}{2}(1 - 1) = 0$. When $\theta = 0$, the moon is located between Earth and the sun. Since $F = 0$, the face of the moon is not visible, which corresponds to a *new moon*.

$F\left(\frac{\pi}{2}\right) = \frac{1}{2}\left(1 - \cos \frac{\pi}{2}\right) = \frac{1}{2}(1 - 0) = \frac{1}{2}$. When $\theta = \frac{\pi}{2}$, $F = \frac{1}{2}$. Thus half the face of the moon is visible. This phase is called the *first quarter.*

$F(\pi) = \frac{1}{2}(1 - \cos \pi) = \frac{1}{2}(1 - (-1)) = 1$. When $\theta = \pi$, Earth is between the moon and the sun. Since $F = 1$, the face of the moon is completely visible, which corresponds to a *full moon.*

$F\left(\frac{3\pi}{2}\right) = \frac{1}{2}\left(1 - \cos \frac{3\pi}{2}\right) = \frac{1}{2}(1 - 0) = \frac{1}{2}$. When $\theta = \frac{3\pi}{2}$, $F = \frac{1}{2}$. Thus half the face of the moon is visible. This phase is called the *last quarter.*   ***Now Try Exercise 89*** ◀

An Application from Electricity   Common household current is called *alternating current* (AC) because the voltage and amperage change direction 120 times per second. Sinusoidal curves are often used to model the voltage in a common household electrical outlet, as demonstrated in the next example.

EXAMPLE 10   **Analyzing household current**

The voltage V in a household outlet can be modeled by $V(t) = 160 \sin(120\pi t)$, where t represents time in seconds.

(a) A graph of $y = V(t)$ is shown in Figure 6.77. Describe the voltage.

(b) Evaluate $V\left(\frac{1}{240}\right)$ and interpret the result.

(c) One way to estimate the "average" voltage in a circuit is to use the **root mean square** voltage, which equals the maximum voltage divided by $\sqrt{2}$. Find this "average" voltage.

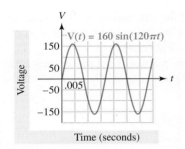

Figure 6.77

SOLUTION

(a) The graph of V in Figure 6.77 is sinusoidal and varies between -160 volts and 160 volts, corresponding to the fact that the current is changing direction in a household outlet. When the graph is above the t-axis the current is one direction, and when it is below the t-axis the current is in the opposite direction.

(b) $V\left(\frac{1}{240}\right) = 160 \sin \frac{\pi}{2} = 160(1) = 160$. After $\frac{1}{240} \approx 0.004$ second, the voltage is 160 volts. Figure 6.77 supports this result.

(c) The maximum voltage is 160 volts. The "average" voltage is $\left(\frac{160}{\sqrt{2}}\right) \approx 113$ volts.

Common household electricity is often rated at 110–120 volts.   ***Now Try Exercise 91*** ◀

NOTE In Example 10, the input to the function V was time, not an angle. This example illustrates why it is necessary to extend the domain of trigonometric functions to include real numbers.

Modeling with the Sine Function (Optional)

The study of *biological clocks* is a fascinating field. Many living organisms undergo regular biological rhythms. A simple example is a flower that opens during daylight and closes at nighttime. One amazing result is that flowers often continue to open and close even when they are placed in continual darkness.   (**Source:** F. Brown et al., *The Biological Clock.*)

EXAMPLE 11   Modeling biological rhythms

Some types of water plants are *luminescent*—they radiate a type of "cold" light that is similar to the light emitted from a firefly. A sample of a luminescent plant (*Gonyaulax polyedra*) was put into continual dim light. The luminescence was measured at 6-hour intervals, and the data in Table 6.6 summarize the results. Noon corresponds to $t = 0$ and the *y*-units of luminescence are arbitrary.

[−6, 54, 6] by [0, 8, 1]

Figure 6.78

Table 6.6

t (hour)	0	6	12	18	24	30	36	42	48
y (luminescence)	1	4	7	4	1	4	7	4	1

Source: E. Bünning, *The Physiological Clock.*

(a) Make a scatterplot of the data in [−6, 54, 6] by [0, 8, 1]. Interpret the data.
(b) Graph the data and $f(t) = 3 \sin (0.27t - 1.7) + 4$. How well does f model the data?
(c) Estimate the luminescence when $t = 33$.

[−6, 54, 6] by [0, 8, 1]

Figure 6.79

SOLUTION

(a) A scatterplot of the data is shown in Figure 6.78. Luminescence appears to be periodic, increasing and decreasing at regular intervals even though the lighting was always dim. The plant was most luminescent during times that correspond to midnight: $t = 12$ and 36. It was least luminescent at times corresponding to noon: $t = 0, 24$, and 48.

(b) $Y_1 = 3 \sin (.27X - 1.7) + 4$ and the data are graphed in Figure 6.79. The graph of f models the periodic data quite well.

(c) Evaluate $f(t)$ at $t = 33$ using radian mode.

$$f(33) = 3 \sin (0.27(33) - 1.7) + 4 \approx 6.4$$

The luminescence was about 6.4 after 33 hours, or at 9 P.M.   ***Now Try Exercise 97*** ◄

Calculator Help

To make a scatterplot and graph an equation, see Appendix A (pages AP-3 and AP-7).

6.3 Putting It All Together

In this section we extended the domains of the sine and cosine functions to include any angle θ and any real number t. Both of these functions are nonlinear functions, and their ranges include values satisfying $-1 \leq y \leq 1$. Some concepts about the sine and cosine functions are summarized in the following table.

Concept	Formulas and Figures
Trigonometric functions of angles	 $\sin \theta = \dfrac{y}{r}$ and $\cos \theta = \dfrac{x}{r}$, where $r = \sqrt{x^2 + y^2}$

continued on next page

continued from previous page

Concept	Formulas and Figures
The unit circle	$x^2 + y^2 = 1$ $(r = 1)$ $$\sin t = y \text{ and } \cos t = x$$ The period of the sine and cosine functions is 2π, or $360°$. Graphs of the sine and cosine functions are shown in Figures 6.70 and 6.73. Their graphs repeat every 2π units along the x-axis and have period 2π. For example, $\sin t = \sin(t + 2\pi n)$ for all real numbers t and any integer n.

6.3 Exercises

Basic Concepts

Exercises 1–4: The xy-coordinates of the hand for a robotic arm are shown in the figure.

(a) *Find the length of the arm.*

(b) *Find the sine and cosine functions for the angle θ.*

1.

2.

3.

4.

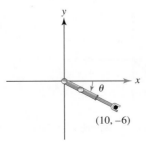

Exercises 5–8: Find sin θ and cos θ.

5.

6.

7.

8.

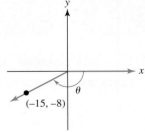

Exercises 9–12: The terminal side of an angle θ in standard position lies on the line in the given quadrant. Find the sine and cosine functions of θ. (Hint: Find a point (x, y) lying on the terminal side of θ.)

9. $y = 2x,$ Quadrant I

10. $y = -\frac{1}{2}x,$ Quadrant II

11. $y = -3x,$ Quadrant IV

12. $y = \frac{3}{4}x,$ Quadrant III

Evaluating Sine and Cosine

Exercises 13–28: (Refer to Examples 2 and 3.) Find the sine and cosine functions by hand for the given angle. Then support your answer using a calculator.

13. 45°   **14.** 150°

15. −30°   **16.** −180°

17. 225°   **18.** 510°

19. −420°   **20.** −225°

21. $\frac{\pi}{3}$ **22.** $\frac{5\pi}{4}$

23. $-\frac{\pi}{2}$ **24.** -2π

25. $\frac{7\pi}{6}$ **26.** $\frac{4\pi}{3}$

27. $-\frac{9\pi}{4}$ **28.** $-\frac{19\pi}{6}$

Exercises 29–36: Approximate the sine and cosine of each angle to four decimal places.

29. 93.2°   **30.** −43°

31. 123°50′   **32.** 12°40′45″

33. −4   **34.** 1.56

35. $\frac{11\pi}{7}$ **36.** $-\frac{7\pi}{5}$

The Unit Circle

Exercises 37–40: The figure shows angle θ in standard position with its terminal side intersecting the unit circle. Evaluate sin θ and cos θ.

37.

38.

39.

40.

Exercises 41–50: (Refer to Examples 7 and 8.) Use a unit circle to evaluate sin t and cos t by hand.

41. $t = \frac{\pi}{2}$ **42.** $t = \pi$

43. $t = \frac{7\pi}{6}$ **44.** $t = -\frac{\pi}{4}$

45. $t = -\frac{3\pi}{4}$ **46.** $t = \frac{5\pi}{3}$

47. $t = \frac{5\pi}{2}$ **48.** $t = \frac{11\pi}{6}$

49. $t = -\frac{\pi}{3}$ **50.** $t = -\frac{3\pi}{2}$

The Wrapping Function

Exercises 51–66: (Refer to Example 5.) For each real number s, evaluate W(s).

51. 3π **52.** 4π

53. -2π **54.** -5π

55. $\frac{3\pi}{2}$ **56.** $\frac{7\pi}{2}$

57. $-\frac{5\pi}{2}$ **58.** $-\frac{9\pi}{2}$

59. $\frac{5\pi}{4}$ **60.** $\frac{7\pi}{4}$

61. $-\frac{5\pi}{4}$ **62.** $-\frac{3\pi}{4}$

63. $\frac{11\pi}{6}$ **64.** $\frac{7\pi}{3}$

65. $-\frac{7\pi}{3}$ **66.** $-\frac{5\pi}{6}$

Exercises 67–82: (Refer to Exercises 51–66.) Use the wrapping function to evaluate sin s *and* cos s *for each real number s.*

67. 3π **68.** 4π

69. -2π **70.** -5π

71. $\frac{3\pi}{2}$ **72.** $\frac{7\pi}{2}$

73. $-\frac{5\pi}{2}$ **74.** $-\frac{9\pi}{2}$

75. $\frac{5\pi}{4}$ **76.** $\frac{7\pi}{4}$

77. $-\frac{5\pi}{4}$ **78.** $-\frac{3\pi}{4}$

79. $\frac{11\pi}{6}$ **80.** $\frac{7\pi}{3}$

81. $-\frac{7\pi}{3}$ **82.** $-\frac{5\pi}{6}$

Graphs of Trigonometric Functions

Exercises 83–86: Graph the function f in the viewing rectangle $[-2\pi, 2\pi, \pi/2]$ by $[-4, 4, 1]$. Identify the range of f and then evaluate $f\left(\frac{3\pi}{2}\right)$.

83. (a) $f(t) = 3 \sin t$ **(b)** $f(t) = \sin(3t)$

84. (a) $f(t) = 2 \cos t$ **(b)** $f(t) = \cos(2t)$

85. (a) $f(t) = 2 \cos(t) + 1$ **(b)** $f(t) = \cos(2t) - 1$

86. (a) $f(t) = 3 \sin(t) + 1$ **(b)** $f(t) = \sin(3t) - 1$

Applications

87. *Highway Grade*   (Refer to Example 4.) Suppose an uphill grade of a highway can be modeled by the line $y = 0.03x$ in the first quadrant.
(a) Find the grade of the hill.

(b) Determine the grade resistance for a gravel truck weighing 25,000 pounds.

88. *Highway Grade*   (Refer to Example 4.) A downhill graph is modeled by the line $y = -0.04x$ in the fourth quadrant.
(a) Find the grade of the road.

(b) Determine the grade resistance for a 6000-pound pickup truck. Interpret the result.

89. *Phases of the Moon*   (Refer to Example 9.) Find all phase angles θ that correspond to the first quarter phase of the moon. Assume that θ can be any angle measured in radians.

90. *Phases of the Moon*   (Refer to Example 9.) Find all phase angles θ that correspond to a full moon and all phase angles that correspond to a new moon. Assume that θ can be any angle measured in radians.

91. *Voltage*   (Refer to Example 10.) Electric ranges and ovens often use a higher voltage than that found in normal household outlets. This voltage can be modeled by $V(t) = 310 \sin(120\pi t)$, where t represents time in seconds.
(a) A graph of $y = V(t)$ is shown in the given figure. Describe the voltage in the circuit.

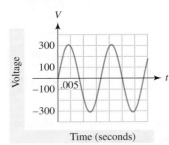

(b) Evaluate $V\left(\frac{1}{120}\right)$ and interpret the result.

(c) Approximate the root mean square voltage.

92. *Amperage*   (Refer to Example 10.) The amperage I in an electrical circuit after t seconds is modeled by $I(t) = 10 \cos(120\pi t)$.
(a) Find the range of I.

(b) Evaluate $I\left(\frac{1}{60}\right)$ and interpret the result.

93. *Braking Distance and Grade*   The braking distance D in feet required for a typical automobile on dry pavement to change its velocity from V_1 to V_2 can be estimated using the equation

$$D = \frac{1.05\left(V_1^2 - V_2^2\right)}{27 + 64.4 \sin \theta}.$$

In this equation, θ represents the angle of the grade of the highway and velocity is measured in feet per second. See Figure 6.57.   (**Source:** F. Mannering.)

(a) Compute the number of feet required to slow a car from 88 feet per second (60 mi/hr) to 44 feet per second (30 mi/hr) while it is traveling uphill with $\theta = 3°$.

(b) Repeat part (a) with $\theta = -3°$.

(c) How is braking distance affected by θ? Does this result agree with your driving experience?

94. *Braking Distance*   (Refer to Exercise 93.) An automobile is traveling at 88 feet per second (60 mi/hr) on a highway with $\theta = -4°$. The driver sees a stalled truck in the road and applies the brakes 250 feet from the vehicle. Assuming that a collision cannot be avoided, how fast is the car traveling when it collides with the truck?

95. *Highway Design*   When an automobile travels along a circular curve with radius r, trees and buildings situated on the inside of the curve can obstruct the driver's vision. See the figure. To ensure a safe stopping distance S, the *minimum* distance d that should be cleared on the inside of a highway curve is given by the equation $d = r\left(1 - \cos \frac{\beta}{2}\right)$, where β in radians is determined by $\beta = \frac{S}{r}$.   (**Source:** F. Mannering.)

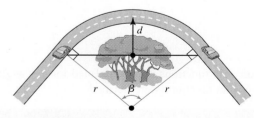

(a) At 45 miles per hour, $S = 390$ feet. If $r = 600$ feet, approximate d.

(b) At 60 miles per hour, $S = 620$ feet. Approximate d for the same curve.

(c) Discuss how the speed limit affects the amount of land that should be cleared on the inside of the curve.

96. *Dead Reckoning*   Suppose an airplane flies the path shown in the figure, where distance is measured in hundreds of miles. Approximate the final coordinates of the airplane if its initial coordinates are (0, 0).

97. *Modeling Biological Rhythms*   (Refer to Example 11.) The table shows the time y in the afternoon when a typical flying squirrel becomes active during month x.

x (month)	1	2	3	4
y (P.M.)	4:30	5:15	5:45	6:30

x (month)	5	6	7	8
y (P.M.)	7:00	7:30	7:45	7:30

x (month)	9	10	11	12
y (P.M.)	6:45	6:15	5:00	4:15

Source: J. Harker, *The Physiology of Diurnal Rhythms.*

(a) Make a scatterplot of the data. (*Hint:* Represent 4:30 P.M. by 4.5.)

(b) The formula $f(x) = 1.9 \sin (0.42x - 1.2) + 5.7$ models the time of sunset, where x represents the month. Graph f and the data together. Interpret the graph.

98. *Modeling Biological Rhythms*   (Refer to Example 11.) The table lists the luminescence y of a plant after t hours, where $t = 0$ corresponds to midnight.

t	0	6	12	18	24	30	36	42	48
y	10	6	2	6	10	6	2	6	10

(a) Graph the data and $L(t) = 4 \cos (0.27t) + 6$. How well does the function model the data?

(b) Estimate the luminescence when $t = 15$.

(c) Use the midpoint formula to complete part (b). Are your answers the same? Why?

Writing about Mathematics

99. Discuss whether the sine and cosine functions are linear or nonlinear functions. Use graphical and numerical representations to justify your reasoning.

100. Describe two ways to define the sine and cosine functions. Give examples.

1. Graph $f(t) = \sin t$ and $g(t) = \cos t$ in the same viewing rectangle. Then translate the graph of f to the left $\frac{\pi}{2}$ units by graphing $y = f\left(t + \frac{\pi}{2}\right)$. How does this translated graph compare to the graph of g?

2. (Continuation of Exercise 1) Translate the graph of $g(t) = \cos t$ to the right $\frac{3\pi}{2}$ units by graphing the function given by $y = g\left(t - \frac{3\pi}{2}\right)$. How does this translated graph compare to the graph of $f(t) = \sin t$?

6.4 Other Trigonometric Functions and Their Graphs

- **Learn definitions and basic identities**
- **Define the other trigonometric functions for any angle and any real number**
- **Represent other trigonometric functions**
- **Solve applications**

Introduction

Unlike the sine function, the tangent and cotangent functions did not originate with astronomy. Rather, they were developed as part of surveying land, finding heights of objects, and determining time. The cotangent function was computed as early as 1500 B.C. in Egypt, where sundials were used to determine time. Depending on the position of the sun in the sky, a vertical stick casts shadows of different lengths. See Figure 6.80. If θ represents the angle of elevation of the sun and the stick is 1 unit long, then the length of the shadow equals $\cot \theta$. This is a simple device for evaluating the cotangent function. (**Source:** NCTM, *Historical Topics for the Mathematics Classroom, Thirty-first Yearbook.*)

Figure 6.80 Modeling Shadow Length

Definitions

Using Figure 6.81, the other trigonometric functions may be defined for any angle θ in a manner similar to the way sine and cosine functions were defined.

Figure 6.81

Trigonometric Functions of Any Angle θ

Let (x, y) be a point other than the origin on the terminal side of an angle θ in standard position. If $r = \sqrt{x^2 + y^2}$, then the six trigonometric functions are as follows.

$$\sin \theta = \frac{y}{r} \qquad \csc \theta = \frac{r}{y} \ (y \neq 0)$$

$$\cos \theta = \frac{x}{r} \qquad \sec \theta = \frac{r}{x} \ (x \neq 0)$$

$$\tan \theta = \frac{y}{x} \ (x \neq 0) \qquad \cot \theta = \frac{x}{y} \ (y \neq 0)$$

The domains of both the sine and the cosine functions include all angles. However, the cosecant and cotangent functions are undefined when $y = 0$, which corresponds to angles whose terminal sides lie on the *x-axis*. Examples include $0°$, $\pm 180°$, and $\pm 360°$. The domains of the cosecant and cotangent functions are as follows.

$$D = \{\theta \mid \theta \neq 180° \cdot n\}, \quad n \text{ an integer}$$

Similarly, the tangent and secant functions are undefined whenever $x = 0$. This corresponds to angles whose terminal sides lie on the *y-axis*. Examples include $\pm 90°$, $\pm 270°$, and $\pm 450°$. The domains of the tangent and secant functions are as follows.

$$D = \{\theta \mid \theta \neq 90° + 180° \cdot n\}, \quad n \text{ an integer}$$

In some situations we can evaluate the six trigonometric functions without a calculator, as illustrated in the next example.

EXAMPLE 1   Finding values of trigonometric functions

The point $(5, -12)$ is located on the terminal side of an angle θ in standard position. Find the six trigonometric functions of θ.

SOLUTION   There are many coterminal angles that have the point $(5, -12)$ located on their terminal sides. However, the trigonometric values for coterminal angles are equal. In Figure 6.82 one possibility for θ is shown. Begin by calculating r.

$$r = \sqrt{x^2 + y^2} = \sqrt{5^2 + (-12)^2} = 13$$

Since $x = 5$, $y = -12$, and $r = 13$, the values of the six trigonometric functions are as follows.

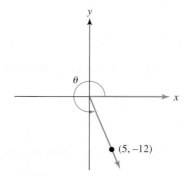

Figure 6.82

$$\sin \theta = \frac{y}{r} = -\frac{12}{13} \qquad \csc \theta = \frac{r}{y} = -\frac{13}{12}$$

$$\cos \theta = \frac{x}{r} = \frac{5}{13} \qquad \sec \theta = \frac{r}{x} = \frac{13}{5}$$

$$\tan \theta = \frac{y}{x} = -\frac{12}{5} \qquad \cot \theta = \frac{x}{y} = -\frac{5}{12}$$ *Now Try Exercise 1* ◀

NOTE   If an angle is a multiple of $30°$ or $45°$, then we can determine its six trigonometric functions by hand.

EXAMPLE 2   Evaluating the trigonometric functions

Find the six trigonometric functions for each angle by hand.
(a) $330°$ **(b)** $-\frac{5\pi}{4}$

SOLUTION
Getting Started   First sketch the given angle in standard position and draw a perpendicular line segment from the terminal side to the *x-axis*. Choose a convenient value for x, y, or r. Apply properties of right triangles to find the lengths of the remaining sides and a point (x, y) on the terminal side of the angle. The values of x, y, and r can be used to determine the value of each trigonometric function. ▶

Figure 6.83

Figure 6.84 A 30°–60°
Right Triangle

Figure 6.85

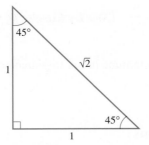

Figure 6.86 A 45°–45°
Right Triangle

(a) An angle of 330° is sketched in Figure 6.83, with a perpendicular line segment drawn to the x-axis. A 30°–60° right triangle is formed. See also Figure 6.84. If we choose r to have length 2, then the short leg has length 1, which is half the hypotenuse. By the Pythagorean theorem, the longer leg has length $\sqrt{3}$. From Figure 6.83 we can see that $x = \sqrt{3}$, $y = -1$, and $r = 2$.

$$\sin 330° = \frac{y}{r} = -\frac{1}{2} \qquad \csc 330° = \frac{r}{y} = -\frac{2}{1} = -2$$

$$\cos 330° = \frac{x}{r} = \frac{\sqrt{3}}{2} \qquad \sec 330° = \frac{r}{x} = \frac{2}{\sqrt{3}}$$

$$\tan 330° = \frac{y}{x} = -\frac{1}{\sqrt{3}} \qquad \cot 330° = \frac{x}{y} = \frac{\sqrt{3}}{-1} = -\sqrt{3}$$

(b) An angle of $-\frac{5\pi}{4}$ (radians) is sketched in Figure 6.85, with a perpendicular line segment drawn to the x-axis. A 45°–45° right triangle is formed. See also Figure 6.86. If we choose each leg to have length 1, then by the Pythagorean theorem the hypotenuse has length $\sqrt{2}$. From Figure 6.85 we can see that $x = -1$, $y = 1$, and $r = \sqrt{2}$.

$$\sin\left(-\frac{5\pi}{4}\right) = \frac{y}{r} = \frac{1}{\sqrt{2}} \qquad \csc\left(-\frac{5\pi}{4}\right) = \frac{r}{y} = \frac{\sqrt{2}}{1} = \sqrt{2}$$

$$\cos\left(-\frac{5\pi}{4}\right) = \frac{x}{r} = -\frac{1}{\sqrt{2}} \qquad \sec\left(-\frac{5\pi}{4}\right) = \frac{r}{x} = -\frac{\sqrt{2}}{1} = -\sqrt{2}$$

$$\tan\left(-\frac{5\pi}{4}\right) = \frac{y}{x} = -\frac{1}{1} = -1 \qquad \cot\left(-\frac{5\pi}{4}\right) = \frac{x}{y} = -\frac{1}{1} = -1$$

> **Now Try Exercises 7 and 17** ◀

Some Basic Trigonometric Identities

Since $\sin\theta = \frac{y}{r}$ and $\csc\theta = \frac{r}{y}$, it follows that $\sin\theta$ and $\csc\theta$ are reciprocals. This may be expressed using either of the *reciprocal identities*,

$$\sin\theta = \frac{1}{\csc\theta} \qquad \text{or} \qquad \csc\theta = \frac{1}{\sin\theta}.$$

In Example 1 it was shown that $\sin\theta = -\frac{12}{13}$. It follows that

$$\csc\theta = \frac{1}{-12/13} = -\frac{13}{12}.$$

Identities are equations that are true for all meaningful values of a variable. Reciprocal identities also hold for $\tan\theta$ and $\cot\theta$ as well as $\cos\theta$ and $\sec\theta$.

Reciprocal Identities		
$\sin\theta = \dfrac{1}{\csc\theta}$	$\cos\theta = \dfrac{1}{\sec\theta}$	$\tan\theta = \dfrac{1}{\cot\theta}$
$\csc\theta = \dfrac{1}{\sin\theta}$	$\sec\theta = \dfrac{1}{\cos\theta}$	$\cot\theta = \dfrac{1}{\tan\theta}$

The expressions $\cot \theta$ and $\tan \theta$ can be written in terms of $\sin \theta$ and $\cos \theta$. For example,

$$\cot \theta = \frac{x}{y} = \frac{x/r}{y/r} = \frac{\cos \theta}{\sin \theta}.$$

A *quotient identity* can be obtained for $\tan \theta$ similarly.

Quotient Identities

$$\tan \theta = \frac{\sin \theta}{\cos \theta} \qquad \cot \theta = \frac{\cos \theta}{\sin \theta}$$

EXAMPLE 3 Using identities

If $\sin \theta = \frac{3}{5}$ and $\cos \theta = -\frac{4}{5}$, find the other four trigonometric functions of θ.

SOLUTION To find the other four trigonometric functions, apply the quotient and reciprocal identities.

$$\tan \theta = \frac{\sin \theta}{\cos \theta} = \frac{3/5}{-4/5} = -\frac{3}{4}$$

$$\cot \theta = \frac{\cos \theta}{\sin \theta} = \frac{-4/5}{3/5} = -\frac{4}{3}$$

$$\sec \theta = \frac{1}{\cos \theta} = \frac{1}{-4/5} = -\frac{5}{4}$$

$$\csc \theta = \frac{1}{\sin \theta} = \frac{1}{3/5} = \frac{5}{3} \qquad \textit{Now Try Exercise 29} \blacktriangleleft$$

In Section 6.3 we showed that $\sin^2 \theta + \cos^2 \theta = 1$ for all real numbers (or angles) θ. In the next example, we apply this identity to find the values of the trigonometric functions.

EXAMPLE 4 Applying the identity $\sin^2 \theta + \cos^2 \theta = 1$

If $\sin \theta = \frac{4}{5}$ and $\cos \theta < 0$, find the values of the other five trigonometric functions.

SOLUTION The identity $\sin^2 \theta + \cos^2 \theta = 1$ can be used to determine $\cos \theta$.

$$\sin^2 \theta + \cos^2 \theta = 1 \qquad \text{Identity}$$

$$\cos^2 \theta = 1 - \sin^2 \theta \qquad \text{Subtract } \sin^2 \theta.$$

$$\cos \theta = \pm \sqrt{1 - \sin^2 \theta} \qquad \text{Square root property}$$

$$\cos \theta = \pm \sqrt{1 - \left(\frac{4}{5}\right)^2} \qquad \text{Let } \sin \theta = \frac{4}{5}.$$

$$\cos \theta = \pm \sqrt{\frac{9}{25}} \qquad \text{Simplify.}$$

$$\cos \theta = -\frac{3}{5} \qquad \cos \theta < 0$$

Now we can use $\sin\theta = \frac{4}{5}$ and $\cos\theta = -\frac{3}{5}$ to determine the other four trigonometric functions.

$$\tan\theta = \frac{\sin\theta}{\cos\theta} = \frac{4/5}{-3/5} = -\frac{4}{3}$$

$$\cot\theta = \frac{1}{\tan\theta} = \frac{1}{-4/3} = -\frac{3}{4}$$

$$\sec\theta = \frac{1}{\cos\theta} = \frac{1}{-3/5} = -\frac{5}{3}$$

$$\csc\theta = \frac{1}{\sin\theta} = \frac{1}{4/5} = \frac{5}{4} \qquad \text{\textbf{\textit{Now Try Exercise 35}}} \blacktriangleleft$$

Unit Circle Evaluation   Suppose the terminal side of an angle θ in standard position intersects the unit circle at the point (x, y). See Figure 6.87. In this case $r = 1$ and the definitions of the six trigonometric functions become as follows.

$$\sin\theta = y \qquad \cos\theta = x \qquad \tan\theta = \frac{y}{x}$$

$$\csc\theta = \frac{1}{y} \qquad \sec\theta = \frac{1}{x} \qquad \cot\theta = \frac{x}{y}$$

As with the sine and cosine functions, evaluating the other trigonometric functions for a real number t is equivalent to evaluating these functions for t radians.

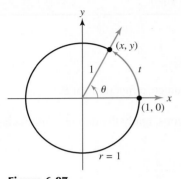

Figure 6.87

The Wrapping Function (Optional)

The six trigonometric functions can be evaluated for real numbers by using the wrapping function presented in Section 6.3. For any real number s, if $W(s) = (x, y)$, then the six trigonometric functions can be defined as follows.

$$\sin s = y \qquad \cos s = x \qquad \tan s = \frac{y}{x}$$

$$\csc s = \frac{1}{y} \qquad \sec s = \frac{1}{x} \qquad \cot s = \frac{x}{y}$$

The next example illustrates how the wrapping function can be used to evaluate the six trigonometric functions.

EXAMPLE 5   Applying the wrapping function

For $s = -5\pi$, evaluate $W(s)$. If possible, find the six trigonometric functions of s.

SOLUTION   The circumference of the unit circle is 2π. A string of length 5π wraps (clockwise) around the unit circle two and a half times. Because the initial point is always $(1, 0)$, the terminal point is $(-1, 0)$, as shown in Figure 6.88. Thus $W(-5\pi) = (-1, 0)$. The six trigonometric functions can be evaluated by using $x = -1$ and $y = 0$. Note that the cosecant and cotangent are undefined because $y = 0$.

$$\sin(-5\pi) = y = 0 \qquad\qquad \csc(-5\pi) = \frac{1}{y} \text{ (undefined)}$$

$$\cos(-5\pi) = x = -1 \qquad\qquad \sec(-5\pi) = \frac{1}{x} = \frac{1}{-1} = -1$$

$$\tan(-5\pi) = \frac{y}{x} = \frac{0}{-1} = 0 \qquad\qquad \cot(-5\pi) = \frac{x}{y} \text{ (undefined)}$$

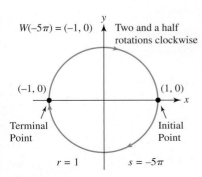

Figure 6.88

$$\text{\textbf{\textit{Now Try Exercise 39}}} \blacktriangleleft$$

Figure 6.89

Representations of Other Trigonometric Functions

Like sin t and cos t, the other four trigonometric functions cannot be evaluated with simple formulas. However, these functions do have numerical and graphical representations. We begin by discussing the tangent function.

The Tangent Function   Figure 6.89 shows angle θ in *standard position* with its terminal side passing through the point (x, y). The slope of this terminal side is

$$m = \frac{y - 0}{x - 0} = \frac{y}{x} = \tan \theta.$$

It can be shown that the slope of the terminal side of θ equals tan θ, regardless of the quadrant containing θ. Using this fact, we will analyze the tangent function by letting θ vary from $-\frac{\pi}{2}$ to $\frac{\pi}{2}$, or equivalently, $-90°$ to $90°$.

When $\theta = -\frac{\pi}{2}$, the slope of the terminal side of θ is undefined, and so tan $\left(-\frac{\pi}{2}\right)$ is undefined. Refer to Figure 6.90. For values of θ slightly greater than $-\frac{\pi}{2}$, the slope of the terminal side of θ is both negative and large in absolute value. At $\theta = -\frac{\pi}{4}$, the slope of the terminal side is $m = -1$ and tan $\left(-\frac{\pi}{4}\right) = -1$. The slope is negative and increasing for $-\frac{\pi}{2} < \theta < 0$. When $\theta = 0$, $m = 0$ and tan $0 = 0$.

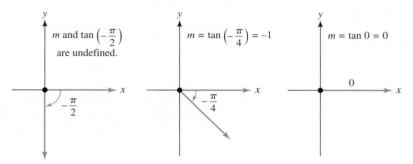

Figure 6.90

As θ increases from 0 to $\frac{\pi}{2}$, the slope m is increasing and positive. Refer to Figures 6.91 and 6.92. At $\theta = \frac{\pi}{4}$, the slope is $m = 1$ and tan $\left(\frac{\pi}{4}\right) = 1$. As θ increases toward $\frac{\pi}{2}$, the slope of the terminal side increases without bound and becomes undefined when $\theta = \frac{\pi}{2}$. Thus tan $\left(\frac{\pi}{2}\right)$ is undefined.

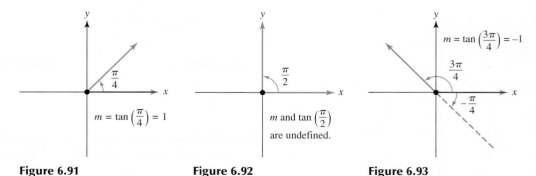

Figure 6.91   **Figure 6.92**   **Figure 6.93**

As θ becomes greater than $\frac{\pi}{2}$, the values of tan θ repeat. For example, when $\theta = \frac{3\pi}{4}$, the slope of the terminal side of θ is equal to the slope of the terminal side of $\theta = -\frac{\pi}{4}$. See Figure 6.93. As a result, the tangent function has period π, or $180°$.

Figure 6.94

Figure 6.95 The Tangent Function

A *numerical representation* of $f(t) = \tan t$ for selected (real number) values of t satisfying $-\frac{\pi}{2} \leq t \leq \frac{\pi}{2}$ is shown in Table 6.7. (A dash indicates that $\tan t$ is undefined.)

Table 6.7

t	$-\frac{\pi}{2}$	$-\frac{\pi}{3}$	$-\frac{\pi}{4}$	$-\frac{\pi}{6}$	0	$\frac{\pi}{6}$	$\frac{\pi}{4}$	$\frac{\pi}{3}$	$\frac{\pi}{2}$
$\tan t$	—	$-\sqrt{3}$	-1	$-\frac{1}{\sqrt{3}}$	0	$\frac{1}{\sqrt{3}}$	1	$\sqrt{3}$	—

Graphs of $y = \tan t$ and the data in Table 6.7 appear in Figure 6.94, where t is measured in radians. Vertical asymptotes occur when $\tan t$ is undefined at $t = \pm\frac{\pi}{2}$. Since the tangent function has period π (or 180°), we can extend the graph of $y = \tan t$ as shown in Figure 6.95. Notice that vertical asymptotes occur at $t = \pm\frac{\pi}{2}, \pm\frac{3\pi}{2}, \pm\frac{5\pi}{2}, \ldots$, where $\tan t$ is undefined. The domain of the tangent function is $D = \left\{ t \mid t \neq \frac{\pi}{2} + \pi n \right\}$.

EXAMPLE 6 Evaluating the tangent function

Evaluate $f(t) = \tan t$ at $t = -\frac{\pi}{4}$ by hand.

SOLUTION If angle $-\frac{\pi}{4}$ is in standard position, the terminal side intersects the unit circle at the point $\left(\frac{1}{\sqrt{2}}, -\frac{1}{\sqrt{2}} \right)$. See Figure 6.96. Therefore

$$\tan\left(-\frac{\pi}{4}\right) = \frac{y}{x} = \frac{-1/\sqrt{2}}{1/\sqrt{2}} = -1.$$

Note that the 45°–45° right triangle in Figure 6.97 can be used as an aid in determining the length of the sides of the triangle in Figure 6.96. Also note that the slope of the terminal side of the angle is -1 and equals $\tan\left(-\frac{\pi}{4}\right)$.

Figure 6.96

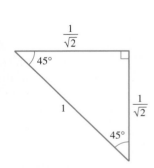

Figure 6.97

Now Try Exercise 51 ◄

The Cosecant Function The cosecant function can also be evaluated by hand. For example, if $t = \frac{7\pi}{6}$, then the terminal side of $\theta = \frac{7\pi}{6}$ intersects the unit circle at the point $\left(-\frac{\sqrt{3}}{2}, -\frac{1}{2} \right)$. See Figure 6.98. It follows that

$$\csc\frac{7\pi}{6} = \frac{1}{y} = \frac{1}{-1/2} = -2.$$

Figure 6.98

Figure 6.99

Note that the 30°–60° right triangle in Figure 6.99 can be used as an aid in determining the length of the sides of the triangle in Figure 6.98.

A *numerical representation* of $f(t) = \csc t$ is shown in Table 6.8 for selected values of t satisfying $-\pi \le t \le \pi$. Note that $\csc t = \frac{1}{\sin t}$. (A dash indicates that $\csc t$ is undefined.)

Table 6.8

t	$-\pi$	$-\frac{3\pi}{4}$	$-\frac{\pi}{2}$	$-\frac{\pi}{4}$	0	$\frac{\pi}{4}$	$\frac{\pi}{2}$	$\frac{3\pi}{4}$	π
csc t	—	$-\sqrt{2}$	-1	$-\sqrt{2}$	—	$\sqrt{2}$	1	$\sqrt{2}$	—

A graph of $\csc t$ is given in Figure 6.100. The domain of the cosecant function is $D = \{t \mid t \ne \pi n\}$, where n is an integer and vertical asymptotes occur whenever $t = \pi n$. The period of the cosecant function equals 2π.

Figure 6.100 The Cosecant Function

Figure 6.101 The Sine and Cosecant Functions

Since $\csc t = \frac{1}{\sin t}$, we can sketch the graph of $y = \csc t$ by using the graph of $y = \sin t$ as an aid. See Figure 6.101. When $\sin t = 0$, $\csc t$ is undefined and a vertical asymptote occurs on the graph of $\csc t$. When $\sin t = \pm 1$, $\csc t = \pm 1$. Whenever $\sin t$ increases, $\csc t$ decreases, and whenever $\sin t$ decreases, $\csc t$ increases. Because $|\sin t| \le 1$ for all t, it follows that $|\csc t| \ge 1$ for all $t \ne \pi n$.

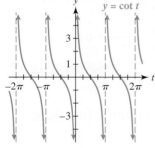

Figure 6.102 The Cotangent Function

The Cotangent Function A graph of $y = \cot t$ is shown in Figure 6.102. Its period is π, and its domain is $D = \{t \mid t \ne \pi n\}$, where n is an integer. Vertical asymptotes occur at $t = \pi n$.

Trigonometric functions may be represented graphically in either radian or degree mode. If a friendly window is selected using degree mode, a trigonometric function may be accurately graphed in connected mode. Graphs of the cosecant and cotangent functions are shown in Figures 6.103 and 6.104.

[−352.5°, 352.5°, 90°] by [−4, 4, 1]   [−352.5°, 352.5°, 90°] by [−4, 4, 1]

Figure 6.103 The Cosecant Function

Figure 6.104 The Cotangent Function

The Secant Function A graph of $y = \sec t$ is shown in Figure 6.105. Its period is 2π, and its domain is $D = \{t \mid t \neq \frac{\pi}{2} + \pi n\}$, where n is an integer. Vertical asymptotes occur at $t = \frac{\pi}{2} + \pi n$. The graph of the cosine function may be used as an aid in graphing the secant function, since the zeros of $y = \cos t$ correspond to the vertical asymptotes on the graph of $y = \sec t$. See Figure 6.106.

Figure 6.105 The Secant Function

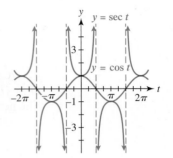

Figure 6.106 The Cosine and Secant Functions

The next two examples illustrate how to evaluate the six trigonometric functions first by hand and then with a calculator.

EXAMPLE 7 Evaluating the trigonometric functions

If $\theta = \frac{3\pi}{2}$, find the six trigonometric functions of θ.

SOLUTION If θ is in standard position, then its terminal side intersects the unit circle at the point $(0, -1)$. See Figure 6.107. Thus $x = 0$ and $y = -1$. The values of the six trigonometric functions are as follows.

Figure 6.107

$$\sin \frac{3\pi}{2} = y = -1 \qquad\qquad \csc \frac{3\pi}{2} = \frac{1}{y} = \frac{1}{-1} = -1$$

$$\cos \frac{3\pi}{2} = x = 0 \qquad\qquad \sec \frac{3\pi}{2} = \frac{1}{x} \text{ is undefined, since } x = 0$$

$$\tan \frac{3\pi}{2} = \frac{y}{x} \text{ is undefined, since } x = 0 \qquad \cot \frac{3\pi}{2} = \frac{x}{y} = \frac{0}{-1} = 0$$

Now Try Exercise 57 ◄

EXAMPLE 8 Using a calculator to evaluate trigonometric functions

Find values of the six trigonometric functions of θ.

(a) $\theta = 102.6°$ **(b)** $\theta = 2.56$

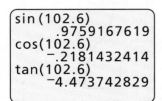

Figure 6.108 Degree Mode

SOLUTION

(a) Most calculators do not have special keys for secant, cosecant, and cotangent. To evaluate these functions, we use the reciprocal identities. For example, to find $\sec(102.6°)$, evaluate $1/\cos(102.6°)$. The values of the trigonometric functions are shown in Figures 6.108 and 6.109 (on the next page).

1/sin(102.6)
          1.024677553
1/cos(102.6)
          ⁻4.584143857
1/tan(102.6)
          ⁻.2235264829

Figure 6.109 Degree Mode

(b) Since there is no degree symbol, use radian mode. With the aid of a calculator, the values of the six trigonometric functions are approximated as follows.

$$\sin 2.56 \approx 0.5494 \qquad \cos 2.56 \approx -0.8356 \qquad \tan 2.56 \approx -0.6574$$

$$\csc 2.56 \approx 1.8203 \qquad \sec 2.56 \approx -1.1968 \qquad \cot 2.56 \approx -1.5210$$

Now Try Exercises 63 and 67 ◀

Applications of Trigonometric Functions

Highway Design A *sag curve* occurs when a highway goes downhill and then uphill. Improperly designed sag curves can be dangerous at night because a vehicle's headlights point downward and may not illuminate the uphill portion of the highway, as illustrated in Figure 6.110. The minimum safe length L for a typical sag curve with a 40-mile-per-hour speed limit is computed by the equation

$$L = \frac{2700}{h + 3 \tan \alpha}.$$

The variable h represents the height of the headlights above the road surface, and α represents a small angle associated with the vertical alignment of the headlight, shown in Figure 6.110.   (**Source:** F. Mannering and W. Kilareski, *Principles of Highway Engineering and Traffic Analysis.*)

Figure 6.110 Designing a Safe Sag Curve (not to scale)

EXAMPLE 9 Designing a sag curve

Calculate L for a car with headlights 2.5 feet above the ground and $\alpha = 1°$. Repeat the calculations for a truck with headlights 4 feet above the ground and $\alpha = 2°$.

SOLUTION For the car, let $h = 2.5$ and $\alpha = 1°$.

$$L = \frac{2700}{2.5 + 3 \tan 1°} \approx 1058 \text{ feet}$$

For the truck, let $h = 4$ and $\alpha = 2°$.

$$L = \frac{2700}{4 + 3 \tan 2°} \approx 658 \text{ feet}$$

Since both cars and trucks use the same highways, engineers typically use the larger of the two distances to design a safe sag curve.

Now Try Exercise 91 ◀

Bending of Starlight   When a stick is put partially into a glass of water, it appears to bend at the surface of the water. A similar phenomenon occurs when starlight enters Earth's atmosphere. To an observer on the ground, a star's apparent position α is different from its true position. This phenomenon is referred to as *atmospheric refraction*. See Figure 6.111 below. The amount that starlight is bent is given by angle θ measured in seconds, where $\theta = 57.3 \tan \alpha$ with $0° \leq \alpha \leq 45°$.   (**Source:** W. Schlosser, *Challenges of Astronomy.*)

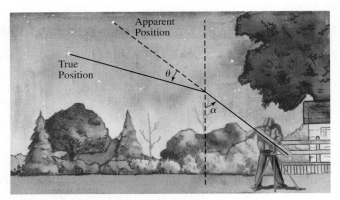

Apparent Position

True Position

θ

α

Figure 6.111 Atmospheric Refraction

EXAMPLE 10   Calculating refraction of starlight

Referring to Figure 6.111 and the formula $\theta = 57.3 \tan \alpha$, calculate θ when $\alpha = 32°35'21''$.

SOLUTION $\theta = 57.3 \tan (32°35'21'') \approx 36.6''$. The star is actually 36.6″ lower in the sky than it appears.   *Now Try Exercise 89* ◀

Time of Day and Sun Tanning   The shortest path through Earth's atmosphere for the sun's rays occurs when the sun is directly overhead. As the sun moves lower in the horizon, sunlight travels through more atmosphere. See Figure 6.112, where $d = h \csc \theta$. Explain why. If the angle of elevation of the sun is θ, then the path length d of sunlight through the atmosphere increases by a factor of $\csc \theta$ from its noontime value of h. (Because of the curvature of Earth, this model is not accurate when $\theta < 20°$ and the sun is positioned near the horizon. See Exercise 95.) When sunlight passes through more atmosphere, less ultraviolet light reaches Earth's surface. This is one reason why some experts recommend sun tanning either earlier or later in the day.   (**Source:** C. Winter, *Solar Power Plants.*)

Atmosphere

h

d

θ

(Not to scale)

Figure 6.112

EXAMPLE 11   Measuring the intensity of the sun

Assuming that the sun is directly overhead at noon, calculate the percent increase between the atmospheric distance that sunlight passes through at noon and at 10:00 A.M.

SOLUTION From Figure 6.112, $\theta = 90°$ at noon. The sun moves 360° in the sky in 24 hours, or 15° per hour. Therefore 2 hours earlier $\theta = 60°$. Since $\csc 90° = 1$ and $\csc 60° = 1/\sin 60° \approx 1.15$, this means that sunlight travels through about 15% more atmosphere at 10:00 A.M. than at noon.   *Now Try Exercise 93* ◀

6.4 Putting It All Together

There are six trigonometric functions. The following table summarizes the domain, range, and period of each trigonometric function. In this table n is an integer and t is a real number.

Function	Domain	Range	Period
$\sin t$	$-\infty < t < \infty$	$-1 \leq \sin t \leq 1$	2π
$\cos t$	$-\infty < t < \infty$	$-1 \leq \cos t \leq 1$	2π
$\tan t$	$t \neq \frac{\pi}{2} + \pi n$	$-\infty < \tan t < \infty$	π
$\cot t$	$t \neq \pi n$	$-\infty < \cot t < \infty$	π
$\sec t$	$t \neq \frac{\pi}{2} + \pi n$	$\left| \sec t \right| \geq 1$	2π
$\csc t$	$t \neq \pi n$	$\left| \csc t \right| \geq 1$	2π

Graphs of Trigonometric Functions

Graphs of the six trigonometric functions, including any asymptotes, are as follows.

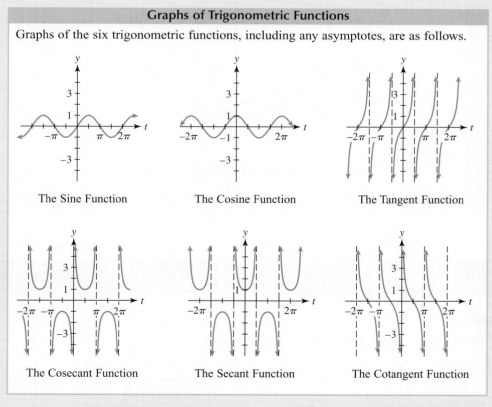

The Sine Function

The Cosine Function

The Tangent Function

The Cosecant Function

The Secant Function

The Cotangent Function

continued on next page

continued from previous page

Evaluating Trigonometric Functions

The following figure can be used to evaluate the trigonometric functions for certain angles. For example, if $\theta = 120°$ (or $\frac{2\pi}{3}$ radians) is in standard position, then its terminal side intersects the unit circle at the point $\left(-\frac{1}{2}, \frac{\sqrt{3}}{2}\right)$. It follows that $\cos 120° = -\frac{1}{2}$ and $\sin 120° = \frac{\sqrt{3}}{2}$. Other trigonometric values can be found using these values. Note that $\frac{\sqrt{2}}{2} = \frac{1}{\sqrt{2}}$.

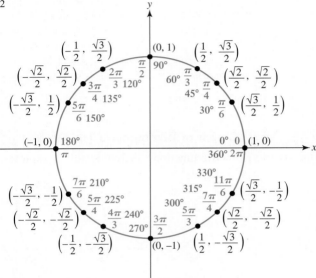

The Unit Circle and Special Angles

6.4 Exercises

Basic Concepts

Exercises 1–4: Find the six trigonometric functions of θ.

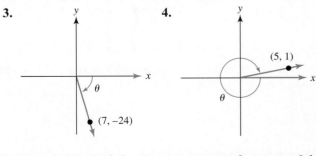

1.

2.

3.

4.

Exercises 5–24: Find the six trigonometric functions of the given angle by hand.

5. 90°

6. 135°

7. $-45°$ **8.** $-180°$

9. π **10.** $\frac{3\pi}{4}$

11. $-\frac{\pi}{3}$ **12.** $-\frac{5\pi}{6}$

13. $-\frac{\pi}{2}$ **14.** $\frac{3\pi}{2}$

15. $360°$ **16.** $60°$

17. $\frac{\pi}{6}$ **18.** $-\frac{2\pi}{3}$

19. $\frac{4\pi}{3}$ **20.** $\frac{2\pi}{3}$

21. $-225°$ **22.** $-315°$

23. $-\frac{13\pi}{6}$ **24.** $-\frac{7\pi}{6}$

Exercises 25–28: The terminal side of an angle θ lies on the line in the given quadrant. Find the six trigonometric functions of θ. How does the slope of the line compare to tan θ?

25. $y = -4x$, Quadrant II   **26.** $y = \frac{1}{2}x$, Quadrant I

27. $y = 6x$, Quadrant III   **28.** $y = -\frac{2}{3}x$, Quadrant IV

Identities

Exercises 29–38: Determine the values of the trigonometric functions of θ by using the given information.

29. $\sin\theta = \frac{3}{5}$ and $\cos\theta = \frac{4}{5}$

30. $\sin\theta = -\frac{7}{25}$ and $\cos\theta = \frac{24}{25}$

31. $\csc\theta = -\frac{17}{15}$ and $\sec\theta = -\frac{17}{8}$

32. $\csc\theta = 2$ and $\sec\theta = -\frac{2}{\sqrt{3}}$

33. $\tan\theta = \frac{5}{12}$ and $\cos\theta = \frac{12}{13}$ *(Hint: $\sin\theta = \tan\theta\cos\theta$.)*

34. $\sin\theta = \frac{3}{5}$ and $\cot\theta = -\frac{4}{3}$

35. $\sin\theta = -\frac{3}{5}$ and $\cos\theta > 0$

36. $\sin\theta = -\frac{12}{13}$ and $\cos\theta < 0$

37. $\cos\theta = -\frac{4}{5}$ and $\sin\theta < 0$

38. $\cos\theta = \frac{7}{25}$ and $\sin\theta > 0$

Wrapping Function

Exercises 39–46: Use the wrapping function to determine the six trigonometric functions of the given number.

39. -7π **40.** 4π

41. $\frac{7\pi}{2}$ **42.** $-\frac{3\pi}{2}$

43. $-\frac{3\pi}{4}$ **44.** $\frac{7\pi}{4}$

45. $\frac{7\pi}{6}$ **46.** $-\frac{11\pi}{6}$

Unit Circle Evaluation

Exercises 47–50: The figure shows angle θ in standard position with its terminal side intersecting the unit circle. Evaluate the six trigonometric functions of θ.

47.

48.

49.

50.

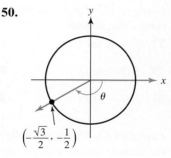

Exercises 51–56: Evaluate the trigonometric function by using the unit circle.

51. $\tan\left(-\frac{5\pi}{4}\right)$ **52.** $\tan\left(-\frac{7\pi}{4}\right)$

53. $\csc\frac{\pi}{6}$ **54.** $\csc\frac{7\pi}{6}$

55. $\sec\frac{\pi}{6}$ **56.** $\cot\frac{7\pi}{4}$

Exercises 57–62: Evaluate the six trigonometric functions of the given number by using the unit circle.

57. $\frac{9\pi}{2}$ **58.** $-\frac{9\pi}{4}$

59. $-\pi$ **60.** 7π

61. $\frac{4\pi}{3}$ **62.** $\frac{5\pi}{6}$

Approximations and Graphs

Exercises 63–70: Approximate to four decimal places.

63. (a) sin 93.2°   **(b)** csc 93.2°

64. (a) cos (−43°)   **(b)** sec (−43°)

65. (a) tan 234°33′   **(b)** cot 234°33′

66. (a) sec 123°44′25″   **(b)** cos 123°44′25″

67. (a) cot (−4)   **(b)** tan (−4)

68. (a) csc 1.56   **(b)** sin 1.56

69. (a) $\cos\left(\frac{11\pi}{7}\right)$ **(b)** $\sec\left(\frac{11\pi}{7}\right)$

70. (a) $\tan\left(\frac{7\pi}{5}\right)$ **(b)** $\cot\left(\frac{7\pi}{5}\right)$

Exercises 71–76: Graph $y = f(t)$. Discuss the symmetry of the graph.

71. $f(t) = \sin t$ **72.** $f(t) = \cos t$

73. $f(t) = \tan t$ **74.** $f(t) = \cot t$

75. $f(t) = \sec t$ **76.** $f(t) = \csc t$

Exercises 77–82: Use the graph of f to identify its domain, range, and period.

77. $f(t) = \sin t$ **78.** $f(t) = \cos t$

79. $f(t) = \tan t$ **80.** $f(t) = \cot t$

81. $f(t) = \sec t$ **82.** $f(t) = \csc t$

Exercises 83–86: Average Rate of Change   Approximate, to the nearest thousandth, the average rate of change of f from 0 to $\frac{\pi}{4}$.

83. $f(x) = \sin x$ **84.** $f(x) = \cos x$

85. $f(x) = \tan \frac{1}{2}x$ **86.** $f(x) = \sin 2x$

Applications

87. *Shadow Length*   The introduction showed how shadows can be used to compute the cotangent function.
  (a) Graph $f(t) = \cot t$ for $0 \le t \le \pi$.

  (b) Let $t = 0$ correspond to sunrise, $t = \frac{\pi}{2}$ to noon, and $t = \pi$ to sunset. Explain how the graph of f models the length of a shadow cast by a vertical stick with length 1.

88. *Shadow Length*   (Refer to Figure 6.80 on page 522.) Calculate the shadow length of a 2-foot stick when the angle of elevation of the sun is $\theta = 27°31′$.

89. *Refraction*   (Refer to Example 10.) Use the formula $\theta = 57.3 \tan \alpha$ to calculate the refraction θ in seconds when $\alpha = 17°23′43″$. Use an identity to rewrite the formula in terms of sin α and cos α.

90. *Refraction*   (Refer to Example 10.) Use the formula $\theta = 57.3 \tan \alpha$ to calculate the refraction θ in seconds when $\alpha = 5°15′50″$. Use an identity to rewrite the formula in terms of cot α.

91. *Highway Design*   (Refer to Example 9.) Calculate the minimum length L of a typical sag curve on a highway with a 40-mile-per-hour speed limit for a car with headlights 2 feet above the ground and alignment set at $\alpha = 1.5°$.

92. *Highway Design*   Repeat the Exercise 91 for a truck with headlights 3.5 feet above the ground and alignment set at $\alpha = 2.5°$.

93. *Intensity of the Sun*   (Refer to Example 11.) If the sun is directly overhead at noon, calculate the percent increase between the atmospheric distance that sunlight must pass through at noon and at 3:00 P.M.

94. *Intensity of the Sun*   (Refer to Example 11.) If the sun is directly overhead at 1:00 P.M., calculate the percent increase between the atmospheric distance that sunlight must pass through at 1:00 P.M. and at 2:30 P.M.

95. *Intensity of the Sun*   (Refer to Example 11.) The formula $y_1 = \csc \theta = \frac{1}{\sin \theta}$, presented in Example 11 to calculate the path length of sunlight through the atmosphere relative to the path length at noon, is not accurate when the sun's elevation is less than 20°. A more accurate formula for small values of θ is given by

$$y_2 = \frac{1}{\sin \theta + 0.5(6° + \theta)^{-1.64}},$$

where θ is measured in degrees. Make a table of y_1 and y_2 starting at $\theta = 2°$ and incrementing by 1°. How do the values of y_1 and y_2 compare as θ increases? (**Source:** C. Winter.)

96. *GPS Satellite Communication*   Artificial satellites that orbit Earth often use VHF signals to communicate with the ground. Because VHF signals travel in straight lines, a satellite orbiting Earth can communicate with a fixed location on the ground only during certain times. The height h in miles of an orbit with communication time T is given by

$$h = 3955 (\sec (\pi T/P) - 1),$$

where P is the period for the satellite to orbit Earth. Suppose a GPS satellite orbit has a period of $P = 12$ hours and can communicate with a person at the North Pole for $T = 5.08$ hours during each orbit. Approximate the height h of its orbit. (**Sources:** W. Schlosser, *Challenges of Astronomy*; Y. Zhao.)

Exercises 97 and 98: Projectile Flight   If a projectile is fired with an initial velocity of v feet per second at an angle θ with the horizontal, it will follow a parabolic path described by $y = \frac{-16x^2}{v^2(\cos \theta)^2} + x \tan \theta$. *See the figure.*

97. If $v = 750$ and $\theta = 30°$, graph the path of the projectile.
 (**a**) Approximate the coordinates when the maximum height occurs.

 (**b**) Assuming the ground is flat, find the total horizontal distance traveled by the projectile graphically.

98. If $v = 500$ and $\theta = 45°$, graph the path of the projectile.
 (**a**) Determine the maximum height graphically.

 (**b**) Assume that the projectile is fired from the base of a hill that rises with a constant slope of $\frac{1}{4}$. Approximate graphically the total horizontal distance traveled by the projectile.

Writing about Mathematics

99. Discuss whether any of the six trigonometric functions are linear functions. Justify your reasoning.

100. If α and β are coterminal angles, what can be said about the six trigonometric functions of α and β? Explain your answer using an example.

EXTENDED AND DISCOVERY EXERCISE

1. The figure shows the unit circle and an acute angle θ in standard position. Explain why the trigonometric functions $\sin \theta$, $\cos \theta$, $\tan \theta$, and $\sec \theta$ are equal to the lengths of the line segments shown.

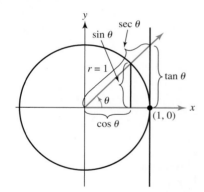

CHECKING BASIC CONCEPTS FOR SECTIONS 6.3 AND 6.4

1. Find the six trigonometric functions of θ if θ is in standard position and its terminal side passes through the point $(-7, 6)$.

2. Evaluate $\sin 45°$ by hand. Check your results with a calculator.

3. Sketch a graph of the sine, cosine, and tangent functions.

4. Evaluate the six trigonometric functions by hand at the given real number. Support your results using a calculator.
 (**a**) $-\pi$ (**b**) $\frac{3\pi}{4}$ (**c**) $\frac{7\pi}{6}$

6.5 Graphing Trigonometric Functions

- **Learn basic transformations of trigonometric graphs**
- **Graph trigonometric functions by hand**
- **Understand simple harmonic motion**
- **Model real data with trigonometric functions (optional)**

Introduction

Trigonometric graphs can be used to model periodic data. For example, monthly average temperatures and precipitation are usually periodic. They can vary dramatically between January and December, but tend to be periodic from one year to the next. Tides are also periodic. By performing basic transformations of graphs we can model a variety of phenomena. See Example 8.

Transformations of Trigonometric Graphs

We begin by discussing how the constants a and b affect the graphs of $y = a \sin bx$ and $y = a \cos bx$. The constant a controls the *amplitude* of a sinusoidal wave. If $y = 3 \sin x$, then the amplitude is $|a| = 3$. The graph of $f(x) = \sin x$ oscillates between -1 and 1, whereas the graph of $g(x) = 3 \sin x$ oscillates between -3 and 3. See Figure 6.113.

When the constant a is negative, the graphs are reflected across the x-axis. For example, Figure 6.114 shows the graphs of $f(x) = 2 \cos x$ and $g(x) = -2 \cos x$. The graph of g is a reflection of the graph of f across the x-axis.

Figure 6.113   **Figure 6.114**

Figure 6.115

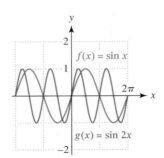

Figure 6.116

The constant b controls the number of oscillations in each interval of length 2π. For example, if $b = 2$, then there are two complete oscillations in every interval of length 2π. As a result, the graph repeats every π units and the period of both $y = \sin 2x$ and $y = \cos 2x$ is π. The *period* P of a sinusoidal graph can be computed by using the formula $P = \frac{2\pi}{b}$, where $b > 0$.

Figure 6.115 shows graphs of $f(x) = \sin x$ and $g(x) = \sin 2x$. Note that the graph of g oscillates "twice as fast" as the graph of f. On the other hand, if $b = \frac{1}{2}$, the graph of $g(x) = \sin \frac{1}{2}x$ oscillates "half as fast" as the graph of $f(x) = \sin x$ and has period 4π. That is, the graph of g completes only half an oscillation every interval of length 2π, or equivalently, the graph of g completes one oscillation every interval of length 4π. See Figure 6.116.

This discussion can also be understood in terms of stretching and shrinking graphs. When $b > 1$, the graph of $y = \sin bx$ is horizontally *shrunk* compared to the graph of $y = \sin x$. When $0 < b < 1$, the graph of $y = \sin bx$ is horizontally *stretched* compared to the graph of $y = \sin x$.

Predict how the graph of the given equation will compare with the graph of $f(x) = \cos x$. Check your prediction by graphing both f and the equation.

i. $y = 2 \cos x$ **ii.** $y = -\frac{1}{2} \cos x$ **iii.** $y = -3 \cos x$

EXAMPLE 1   Sketching graphs of trigonometric functions

Sketch a graph of each equation. Identify the period and amplitude.

(a) $y = 3 \cos \frac{1}{2}x$ **(b)** $y = -2 \sin 3x$

SOLUTION

(a) The graph of $y = 3 \cos \frac{1}{2}x$ is similar to that of $y = \cos x$, except that its period P is

$$P = \frac{2\pi}{b} = \frac{2\pi}{1/2} = 4\pi$$

rather than 2π and its amplitude is 3 rather than 1. A graph of $y = 3 \cos \frac{1}{2}x$ is shown in Figure 6.117, and a graph $y = \cos x$ is shown in Figure 6.118 for comparison.

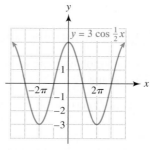

Figure 6.117   **Figure 6.118**

(b) First consider the graph of $y = \sin 3x$, shown in Figure 6.119, which has period

$$P = \frac{2\pi}{b} = \frac{2\pi}{3}.$$

It oscillates three times every interval of length 2π, whereas the graph of $y = \sin x$, shown in Figure 6.120, oscillates once every interval of length 2π. The graph of $y = -2 \sin 3x$ is similar to the graph of $y = \sin 3x$, except that its amplitude is 2 rather than 1 and its graph is reflected across the x-axis. Figure 6.121 shows the required graph of $y = -2 \sin 3x$.

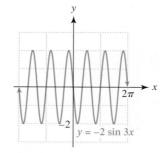

Figure 6.119   **Figure 6.120**   **Figure 6.121**

Now Try Exercises 1 and 3 ◄

Vertical and Horizontal Shifts   Trigonometric graphs can be shifted vertically or horizontally. For example, to shift the graph of $f(x) = \sin x$ **upward 1** unit, graph $g(x) = \sin(x) + 1$, as shown in Figure 6.122. Similarly, to shift the graph of $f(x) = \sin x$ right $\frac{\pi}{2}$ units, graph the equation $g(x) = \sin\left(x - \frac{\pi}{2}\right)$, as shown in Figure 6.123. A horizontal shift of a trigonometric graph is called a *phase shift*. The phase shift for $g(x) = \sin\left(x - \frac{\pi}{2}\right)$ is $\frac{\pi}{2}$ because the graph of g is shifted to the *right* $\frac{\pi}{2}$ units.

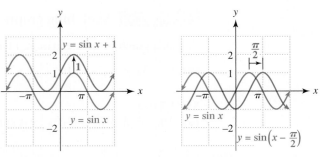

Figure 6.122   **Figure 6.123**

These concepts are applied in the next example.

EXAMPLE 2   Transforming sinusoidal graphs

Explain how the graph of f can be obtained from the graph of either the sine or the cosine function. Then graph f.

(a) $f(x) = 3\cos(2x) + 1$ **(b)** $f(x) = -2\sin\left(x - \frac{\pi}{2}\right)$

SOLUTION

(a) The graph of f can be obtained from the graph of the cosine function by performing the following steps: Shorten the period to π, increase the amplitude to 3, and shift the graph upward 1 unit. These three steps are shown in Figures 6.124–6.126, where the equations $y_1 = \cos 2x$, $y_2 = 3\cos 2x$, and $y_3 = 3\cos(2x) + 1$ are graphed, respectively.

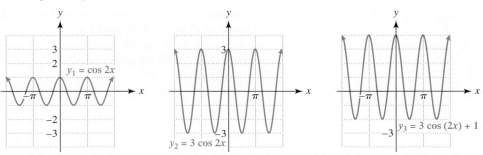

Figure 6.124 Shorten Period   **Figure 6.125** Increase Amplitude   **Figure 6.126** Shift Upward

(b) The graph of f can be obtained from the graph of the sine function by shifting the graph right $\frac{\pi}{2}$ units and increasing the amplitude to 2. The negative sign will cause the graph to be reflected across the x-axis. These steps are shown in Figures 6.127–6.129, where $y_1 = \sin\left(x - \frac{\pi}{2}\right)$, $y_2 = 2\sin\left(x - \frac{\pi}{2}\right)$, and $y_3 = -2\sin\left(x - \frac{\pi}{2}\right)$ are graphed, respectively.

Figure 6.127 Shift Right

Figure 6.128 Increase Amplitude

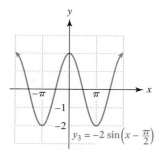

Figure 6.129 Reflect Graph

Now Try Exercises 7 and 9 ◀

The discussion above is summarized in the following box.

Figure 6.130 Transforming the Sine Graph

Amplitude, Period, Phase Shift, and Vertical Shift

The **amplitude**, **period**, **phase shift**, and **vertical shift** for the graphs of

$$y = a \sin (b(x - c)) + d \quad \text{and} \quad y = a \cos (b(x - c)) + d$$

with $b > 0$ may be determined as follows.

$$\text{Amplitude} = |a|, \quad \text{Period} = \frac{2\pi}{b}, \quad \text{Phase shift} = c, \quad \text{Vertical shift} = d$$

The vertical shift is $|d|$ units upward when $d > 0$ and $|d|$ units downward when $d < 0$. The graph of $y = a \sin (b(x - c)) + d$ in Figure 6.130 illustrates the role of each constant, assuming that all constants are positive.

EXAMPLE 3   Identifying amplitude, period, phase shift, and vertical shift

For the graph of each equation, identify the amplitude, period, phase shift, and vertical shift.

(a) $y = 7 \sin \left(4\left(x + \frac{\pi}{3}\right)\right) + 2$ **(b)** $y = -3 \cos (5x - 3) - 6$

SOLUTION

(a) The amplitude is 7, and the period is

$$P = \frac{2\pi}{b} = \frac{2\pi}{4} = \frac{\pi}{2}.$$

The phase shift is $-\frac{\pi}{3}$, and the vertical shift is 2.

(b) Start by applying the distributive property to rewrite the equation as

$$y = -3 \cos \left(5\left(x - \frac{3}{5}\right)\right) - 6.$$

The amplitude is 3, and the period is $P = \frac{2\pi}{b} = \frac{2\pi}{5}$. The phase shift is $\frac{3}{5}$, and the vertical shift is -6.

Now Try Exercises 13 and 17 ◀

Graphing Trigonometric Functions by Hand

Transformations can be used to graph trigonometric functions by hand, as was shown in the previous subsection. A second technique for graphing sinusoidal functions by hand is to first locate five **key points**. These key points are labeled for the graph of $y = \sin x$ in Figure 6.131. Notice that they are equally spaced along the interval $[0, 2\pi]$ on the x-axis.

$(0, 0)$	$\left(\frac{\pi}{2}, 1\right)$	$(\pi, 0)$	$\left(\frac{3\pi}{2}, -1\right)$	$(2\pi, 0)$
x-intercept	Maximum	x-intercept	Minimum	x-intercept

Five key points on the graph of $y = \cos x$ are shown in Figure 6.132.

$(0, 1)$	$\left(\frac{\pi}{2}, 0\right)$	$(\pi, -1)$	$\left(\frac{3\pi}{2}, 0\right)$	$(2\pi, 1)$
Maximum	x-intercept	Minimum	x-intercept	Maximum

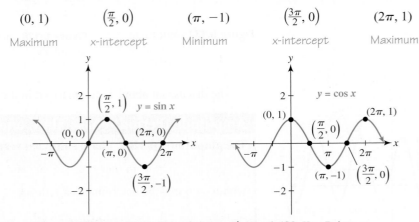

Figure 6.131 Key Points on the Sine Graph

Figure 6.132 Key Points on the Cosine Graph

One way to sketch a sinusoidal graph by hand is to find the *transformed* key points. You can use these points and the fact that the graph is periodic to make a sketch. This process is illustrated for the sine function in the next example.

EXAMPLE 4 Sketching a sinusoidal graph by hand

Find the amplitude, period, and phase shift of the graph of $f(x) = 3 \sin 2x$. Sketch a graph of f by hand on the interval $[-2\pi, 2\pi]$.

SOLUTION If $f(x) = 3 \sin 2x$, then the amplitude is 3 and the period is π. There is no horizontal or vertical shifting. The graph of f passes through the point $(0, 0)$ and completes one oscillation in π units. Thus the graph passes through the point $(\pi, 0)$. An amplitude of 3 will change the maximum and minimum y-values to 3 and -3, respectively. The transformed key points are equally spaced on the graph of f.

$(0, 0)$	$\left(\frac{\pi}{4}, 3\right)$	$\left(\frac{\pi}{2}, 0\right)$	$\left(\frac{3\pi}{4}, -3\right)$	$(\pi, 0)$
x-intercept	Maximum	x-intercept	Minimum	x-intercept

These points and the graph of f are plotted in Figure 6.133. **Now Try Exercise 33** ◄

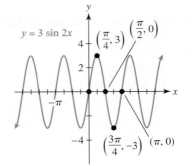

Figure 6.133

EXAMPLE 5 Sketching a graph by hand

Graph $y = 2 \cos (4x + \pi)$ by hand.

SOLUTION First write the equation in the form $y = a \cos (b(x - c))$ by factoring out 4.

$$y = 2 \cos \left(4\left(x + \frac{\pi}{4}\right)\right)$$

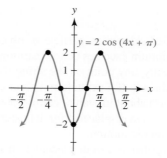

Figure 6.134

How would you modify the graph in Figure 6.134 to obtain the graph of

$$y = 2 \cos(4x + \pi) - 1?$$

What are the key points for this graph?

The amplitude is 2, and the period is $\frac{2\pi}{4} = \frac{\pi}{2}$. The graph is translated $\frac{\pi}{4}$ unit to the left compared to the graph of $y = 2 \cos 4x$. Because the graph is translated $\frac{\pi}{4}$ unit to the left, start the x-values in the key points at $0 - \frac{\pi}{4} = -\frac{\pi}{4}$. The first period ends at $-\frac{\pi}{4} + \frac{\pi}{2} = \frac{\pi}{4}$. The key points for this cosine graph are as follows.

$$\left(-\frac{\pi}{4}, 2\right) \qquad \left(-\frac{\pi}{8}, 0\right) \qquad (0, -2) \qquad \left(\frac{\pi}{8}, 0\right) \qquad \left(\frac{\pi}{4}, 2\right)$$

Maximum   x-intercept   Minimum   x-intercept   Maximum

Plot these key points to complete one period of the graph. The graph can be extended to show two periods, ranging from $-\frac{\pi}{2}$ to $\frac{\pi}{2}$, as shown in Figure 6.134.

Now Try Exercise 45 ◀

Graphs of Other Trigonometric Functions   The graphs of the other four trigonometric functions can also be transformed. The period of the tangent function and the cotangent function is π. As a result, the graphs of $y = \tan(b(x - c))$ and $y = \cot(b(x - c))$ have a period of $P = \frac{\pi}{b}$ and a phase shift of c. Since the ranges of the tangent and cotangent functions include all real numbers, their graphs do not have an amplitude. The next example illustrates tangent and cotangent graphs. Graphs of the secant and cosecant functions are done in Exercises 65–70.

EXAMPLE 6   Graphing other trigonometric functions

Find the period and phase shift for the graph of f. Graph f on the interval $[-2\pi, 2\pi]$. Identify where asymptotes occur in the graph.

(a) $f(x) = \tan \frac{1}{2}x$ **(b)** $f(x) = \cot\left(x + \frac{\pi}{2}\right) + 1$

SOLUTION

(a) If $f(x) = \tan \frac{1}{2}x$, then $b = \frac{1}{2}$ and $c = 0$. The period is $P = \frac{\pi}{b} = 2\pi$, and there is no phase shift. Graph $y = \tan \frac{1}{2}x$, as shown in Figure 6.135. On the interval $[-2\pi, 2\pi]$, vertical asymptotes occur at $x = \pm\pi$. The graph of f is horizontally stretched compared to the graph of the tangent function.

(b) If $f(x) = \cot\left(x + \frac{\pi}{2}\right) + 1$, then $b = 1$ and $c = -\frac{\pi}{2}$. The period is π, the phase shift is $-\frac{\pi}{2}$, and the graph is shifted upward 1 unit. Vertical asymptotes occur at $x = \pm\frac{\pi}{2}$ and $x = \pm\frac{3\pi}{2}$. The graph of $y = \cot\left(x + \frac{\pi}{2}\right) + 1$ is shown in Figure 6.136.

Figure 6.135

Figure 6.136

Now Try Exercises 57 and 63 ◀

Simple Harmonic Motion

Harmonic motion occurs frequently in nature and physical science. A particle or object undergoing small oscillations about a point of stable equilibrium executes simple harmonic motion. For example, if a string on a guitar is plucked gently, any point on the string will vibrate back and forth about its natural position of equilibrium. If a pendulum on a clock is pulled to one side, the pendulum will swing back and forth about its stable, vertical position. A small weight on a spring will bounce up and down when displaced from its natural length. All of these situations are examples of simple harmonic motion.

When an object undergoes simple harmonic motion, its distance or displacement from its stable or natural position can be modeled by either

$$s(t) = a \sin bt \quad \text{or} \quad s(t) = a \cos bt,$$

where $s(t)$ represents the displacement being experienced at time t. To better understand this, consider the spring and weight shown in Figure 6.137. If $s = 0$ corresponds to the spring's natural length, then $s = -2$ inches indicates that the spring is stretched 2 units beyond its natural length and $s = 2$ indicates that the spring is compressed 2 units. When a spring is stretched and let go, it will oscillate up and down. If the displacement s is plotted after t seconds, a sinusoidal graph results, as shown in Figure 6.138.

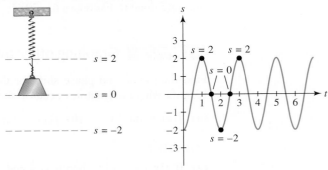

Figure 6.137
Displacement of a Spring

Figure 6.138 Modeling
Oscillations of a Spring

The amplitude of each oscillation equals $|a|$, and the period P is given by $\frac{2\pi}{b}$. The number of oscillations per unit time is the **frequency**. The frequency F equals the reciprocal of the period P. That is, $F = \frac{1}{P}$. Since $P = \frac{2\pi}{b}$, it follows that $F = \frac{b}{2\pi}$, or equivalently, $b = 2\pi F$. Substituting for b in the formulas for $s(t)$ results in

$$s(t) = a \sin(2\pi Ft) \quad \text{or} \quad s(t) = a \cos(2\pi Ft).$$

EXAMPLE 7 Modeling simple harmonic motion

Suppose that the weight and spring in Figure 6.137 have a period of 0.4 second. Initially the weight is lifted 3 inches above its natural length and then let go.
(a) Find an equation $s(t) = a \cos bt$ that models the displacement s of the weight.
(b) Estimate s after 0.92 second. Determine if the weight is moving upward or downward at this time.

SOLUTION
(a) Let $s(t) = a \cos(2\pi Ft)$. The spring is initially compressed 3 inches, so the amplitude is 3. Let $a = 3$. The period is $P = 0.4$, so the frequency is

$$F = \frac{1}{0.4} = 2.5.$$

This indicates that the weight and spring oscillate up and down 2.5 times per second. It follows that $b = 2\pi F = 5\pi$ and $s(t) = 3\cos(5\pi t)$. Note that the initial position is $s(0) = 3$ inches.

(b) After 0.92 second, the displacement is

$$s(0.92) = 3\cos(5\pi(0.92)) \approx -0.927. \qquad \text{Use radian mode.}$$

The weight is about 0.93 inch *below* its natural position after 0.92 second.

The weight initially moves downward and oscillates with a period of 0.4 second. During the time intervals $(0, 0.2)$, $(0.4, 0.6)$, and $(0.8, 1.0)$ the weight is moving downward, and during the intervals $(0.2, 0.4)$, $(0.6, 0.8)$, and $(1.0, 1.2)$ the weight is moving upward. Thus the weight is moving *downward* when $t = 0.92$. Try verifying this graphically if you have a calculator.   *Now Try Exercise 91* ◀

Models Involving Trigonometric Functions (Optional)

Modeling Temperature   The monthly average temperatures for Prince George, Canada, are shown in Table 6.9, where the months have been assigned the standard numbers.

Table 6.9

Month	1	2	3	4	5	6	7	8	9	10	11	12
Temperature (°F)	15	19	28	41	50	55	59	57	50	41	28	19

Source: A. Miller and J. Thompson, *Elements of Meteorology.*

Since the data are periodic, they are plotted over a 2-year period in Figure 6.139. For example, both $x = 1$ and $x = 13$ correspond to January. Notice that the scatterplot suggests that a sinusoidal graph might model these data.

Data of the type shown in Figure 6.139 sometimes can be modeled by

$$f(x) = a\sin(b(x - c)) + d,$$

where a, b, c, and d are constants. To estimate values for these constants, we will perform transformations on the graph of $y = \sin x$.

The addition of a constant d causes a vertical shift to the graph of f. The monthly average temperatures at Prince George vary from 15°F to 59°F. The average of these two temperatures is 37°F. If we let $d = 37$ and graph $y = \sin(x) + 37$, the resulting graph is as shown in Figure 6.140. Compared to the graph of the sine function, this graph is shifted upward 37 units. (Be sure to use radian mode.)

From Figure 6.140 it is apparent that the amplitude of the oscillations in the graph of $y = \sin(x) + 37$ is not large enough to model the data. The monthly average temperatures have a range of $59°F - 15°F = 44°F$. If we let $a = \frac{44}{2} = 22$, the peaks and valleys on the graph of $y = 22\sin(x) + 37$ will model the data better. See Figure 6.141.

In Figure 6.141 the oscillations are too frequent, which indicates that the period is too small. Since the temperature cycles every 12 months, the period is $P = \frac{2\pi}{b} = 12$. Thus

$$b = \frac{2\pi}{12} = \frac{\pi}{6} \approx 0.524.$$

The graph of $y = 22\sin\left(\frac{\pi}{6}x\right) + 37$ is shown in Figure 6.142.

We can obtain a reasonable fit by shifting the graph horizontally to the right. The maximum of $y = 22\sin\left(\frac{\pi}{6}x\right) + 37$ is at $x = 3$, which corresponds to March. We would like

[0, 25, 2] by [−5, 70, 10]

Figure 6.139

[0, 25, 2] by [−5, 70, 10]

$y = \sin(x) + 37$

Figure 6.140

[0, 25, 2] by [−5, 70, 10]

$y = 22\sin(x) + 37$

Figure 6.141

[0, 25, 2] by [−5, 70, 10]

$y = 22\sin\left(\frac{\pi}{6}x\right) + 37$

Figure 6.142

Concept	Explanation	Examples
Sinusoidal Model $f(x) = a\sin(b(x - c)) + d$ or $f(x) = a\cos(b(x - c)) + d$ with $b > 0$	Amplitude $= \lvert a\rvert$ Period $= \dfrac{2\pi}{b}$ Phase shift $= c$ Vertical shift $= d$, upward if $d > 0$ and downward if $d < 0$ If $b = 2\pi F$, then F represents the frequency.	Let $f(x) = 3\sin(2(x - \pi)) - 1$. Amplitude $= 3$ Period $= \frac{2\pi}{2} = \pi$ Phase shift $= \pi$ Vertical shift downward 1 unit The frequency is $F = \frac{b}{2\pi} = \frac{1}{\pi}$, which is approximately 0.32 oscillation per unit of time.
Simple Harmonic Motion $s(t) = a\sin bt$ or $s(t) = a\cos bt$	An object that oscillates about a stable equilibrium point undergoes simple harmonic motion.	A pendulum on a clock A weight on a spring

Is the value of $c = 4$ the only phase shift that can be used to model the temperature data? Explain your reasoning.

this maximum to occur in July ($x = 7$), so we translate the graph 4 units to the right by replacing x with $x - 4$ to obtain

$$y = 22 \sin\left(\frac{\pi}{6}(x - 4)\right) + 37.$$

This equation is graphed in Figure 6.143. The phase shift is $c = 4$.

[0, 25, 2] by [−5, 70, 10]

Figure 6.143

6.5 Exercises

Graphs of Trigonometric Functions

Exercises 1–6: Sketch a graph of the equation on the interval $[-4\pi, 4\pi]$. Identify the period and amplitude.

1. $y = 3 \sin \frac{1}{2}x$ **2.** $y = 2 \cos \frac{1}{3}x$

3. $y = -2 \cos 3x$ **4.** $y = -3 \sin 2x$

5. $y = \frac{1}{2} \sin \pi x$ **6.** $y = -\frac{3}{2} \cos \frac{\pi}{2}x$

Exercises 7–12: (Refer to Example 2.) Explain how the graph of f can be obtained from the graph of either the sine or the cosine function. Then sketch a graph of f on the interval $[-2\pi, 2\pi]$.

7. $y = 3 \sin(2x) - 1$ **8.** $y = -2 \cos(3x) + 1$

9. $y = -2 \cos\left(x + \frac{\pi}{2}\right)$ **10.** $y = -\frac{1}{2} \sin(x + \pi)$

11. $y = \frac{1}{2} \cos(\pi x - 1)$ **12.** $y = \sin\left(\frac{2}{3}x\right) + 2$

Exercises 13–18: For the graph of the equation, identify the amplitude, period, phase shift, and vertical shift. Do not graph the equation.

13. $y = 3 \sin\left(4\left(x - \frac{\pi}{4}\right)\right) - 4$

14. $y = -5 \sin\left(\frac{1}{2}(x - \pi)\right) + 7$

15. $y = -4 \cos\left(\frac{\pi}{2}(x - 1)\right) + 6$

16. $y = \frac{4}{5} \cos\left(\pi x + \frac{\pi}{3}\right) - \frac{2}{3}$

17. $y = \frac{2}{3} \sin(6x + 3\pi) - \frac{5}{2}$

18. $y = 20 \cos\left(\frac{2}{3}x + \pi\right) + 2$

Exercises 19 and 20: A graph of the equation $y = a \sin bx$ is shown, where b is a positive constant. Estimate the values for a and b.

19.

20.

Exercises 21 and 22: The graph of a trigonometric function f represented by $f(x) = a \sin(b(x - c))$ is shown, where a, b, and c are nonnegative. State the amplitude, period, and phase shift.

21.

22.

Exercises 23–28: Match the function f with its graph (a–f). Do not use a calculator.

23. $f(t) = 2 \sin\left(\frac{1}{2}t\right)$ **24.** $f(t) = -\sin(2t)$

25. $f(t) = 3 \cos(\pi t)$ **26.** $f(t) = 2 \sin\left(t - \frac{\pi}{4}\right)$

27. $f(t) = \cos\left(t + \frac{\pi}{2}\right)$ **28.** $f(t) = -3 \cos t$

a.

b.

c.

d.

e.

f.

Exercises 29 and 30: Determine an equation $y = a \sin(b(x - c))$ for the graph shown in the exercise. Assume that a, b, and c are nonnegative.

29. Exercise 21   **30.** Exercise 22

Exercises 31–48: Find the amplitude, period, and phase shift of f. Then graph f by hand on the interval $[-2\pi, 2\pi]$.

31. $f(t) = 2 \sin t$

32. $f(x) = -3 \sin x$

33. $f(x) = \sin \frac{1}{2}x$

34. $f(t) = \cos 2t$

35. $f(x) = 1 + \sin x$

36. $f(x) = -2 + 2 \cos x$

37. $f(t) = \cos(\pi t) + 2$

38. $f(t) = 2 \sin\left(t - \frac{\pi}{2}\right)$

39. $f(t) = -\sin(2(t + \pi))$ **40.** $f(t) = -3 \cos \frac{1}{2}t$

41. $f(x) = -\cos\left(x - \frac{\pi}{2}\right)$ **42.** $f(x) = -2 \cos(x + \pi)$

43. $f(x) = -\frac{1}{2} \sin 2x$ **44.** $f(x) = 3 \sin 4x$

45. $f(x) = 2 \cos\left(2x + \frac{\pi}{2}\right) - 1$

46. $f(x) = -\frac{1}{2} \sin(3x + \pi) + 2$

47. $f(t) = \cos\left(2\left(t - \frac{\pi}{2}\right)\right)$ **48.** $f(t) = 3 \sin\left(\frac{1}{2}(t - \pi)\right)$

Exercises 49–56: Graph f in $[-2\pi, 2\pi, \pi/2]$ by $[-4, 4, 1]$. State the amplitude, period, and phase shift.

49. $f(t) = 2 \sin(2t)$ **50.** $f(t) = -3 \sin(t - \pi)$

51. $f(t) = \frac{1}{2} \cos\left(3\left(t + \frac{\pi}{3}\right)\right)$

52. $f(t) = 1.5 \cos\left(\frac{1}{2}\left(t + \frac{\pi}{2}\right)\right)$

53. $f(t) = -2.5 \cos\left(2t + \frac{\pi}{2}\right)$

54. $f(t) = -2.5 \sin\left(\pi t + \frac{\pi}{2}\right) - 1$

55. $f(x) = -2 \cos\left(2\pi x + \frac{\pi}{4}\right) + 1$

56. $f(x) = \frac{3}{4} \sin(\pi x + \pi) - 2$

Exercises 57–64: (Refer to Example 6.) Find the period and phase shift for the graph of f. Graph f on the interval $[-2\pi, 2\pi]$. Identify where asymptotes occur in the graph.

57. $f(t) = \tan 2t$ **58.** $f(t) = \tan \frac{1}{2}t$

59. $f(t) = \tan\left(t - \frac{\pi}{2}\right)$ **60.** $f(t) = \cot\left(\frac{1}{3}\left(t - \frac{\pi}{2}\right)\right)$

61. $f(t) = -\cot 2t$ **62.** $f(t) = -\cot\left(t + \frac{\pi}{2}\right)$

63. $f(x) = \cot\left(2\left(x - \frac{\pi}{4}\right)\right) - 1$

64. $f(x) = \tan\left(\frac{1}{2}x - \frac{\pi}{2}\right)$

Exercises 65–70: Use the directions for Exercises 57–64 to graph f.

65. $f(t) = \sec \frac{1}{2}t$ **66.** $f(t) = \sec\left(2\left(t - \frac{\pi}{2}\right)\right)$

67. $f(t) = \csc(t - \pi)$ **68.** $f(t) = -\csc 2t$

69. $f(x) = \sec\left(\frac{1}{3}\left(x - \frac{\pi}{6}\right)\right)$ **70.** $f(x) = \csc(\pi(x - 1))$

Applications

71. *Average Temperatures*   The graph models the monthly average temperature y in degrees Fahrenheit for a city in Canada, where x is the month.

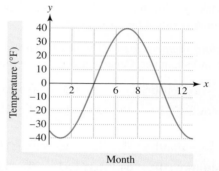

(a) Find the maximum and minimum monthly average temperatures.

(b) Find the amplitude and period. Interpret the results.

(c) Explain what the x-intercepts represent.

72. *Average Temperatures*   The graph in Exercise 71 is given by $y = 40 \cos\left(\frac{\pi}{6}(x - 7)\right)$. Modify this equation to model the following situations.

(a) The maximum monthly average temperature is 50°F and the minimum is −50°F.

(b) The maximum monthly average temperature is 60°F and the minimum is −20°F.

(c) The maximum monthly average temperature occurs in August and the minimum occurs in February.

73. *Average Temperatures*   The monthly average temperatures in degrees Fahrenheit at Mould Bay, Canada, may be modeled by $f(x) = 34 \sin\left(\frac{\pi}{6}(x - 4.3)\right)$, where x is the month and $x = 1$ corresponds to January.   (**Source:** A. Miller and J. Thompson.)

(a) Find the amplitude, period, and phase shift.

(b) Approximate the average temperature during May and December.

(c) Estimate the *yearly* average temperature at Mould Bay.

74. *Average Temperatures* The monthly average temperatures in degrees Fahrenheit at Austin, Texas, are given by $f(x) = 17.5 \sin\left(\frac{\pi}{6}(x - 4)\right) + 67.5$, where x is the month and $x = 1$ corresponds to January. (**Source:** A. Miller and J. Thompson.)

(a) Find the amplitude, period, phase shift, and vertical shift.

(b) Determine the maximum and minimum monthly average temperature and the months when they occur.

(c) Make a conjecture as to how the *yearly* average temperature might be related to $f(x)$.

75. *Modeling Temperatures* The monthly average temperatures in Vancouver, Canada, are shown in the table.

Month	1	2	3	4	5	6
Temperature (°F)	36	39	43	48	55	59

Month	7	8	9	10	11	12
Temperature (°F)	64	63	57	50	43	39

Source: A. Miller and J. Thompson.

(a) Plot the average monthly temperature over a 24-month period by letting $x = 1$ and $x = 13$ correspond to January.

(b) Find the constants a, b, c, and d so that the function $f(x) = a \sin(b(x - c)) + d$ models the data.

(c) Graph f together with the data.

76. *Modeling Temperatures* The monthly average temperatures in Chicago, Illinois, are shown in the table.

Month	1	2	3	4	5	6
Temperature (°F)	25	28	36	48	61	72

Month	7	8	9	10	11	12
Temperature (°F)	74	75	66	55	39	28

Source: A. Miller and J. Thompson.

(a) Plot the monthly average temperature over a 24-month period by letting $x = 1$ and $x = 13$ correspond to January.

(b) Find the constants a, b, c, and d so that the function $f(x) = a \sin(b(x - c)) + d$ models the data.

(c) Graph f and the data together.

77. *Modeling Temperatures* The monthly average high temperatures in Augusta, Georgia, are shown in the table.

Month	1	2	3	4	5	6
Temperature (°F)	58	60	68	77	82	90

Month	7	8	9	10	11	12
Temperature (°F)	92	91	83	77	68	60

Source: J. Williams.

(a) Use $f(x) = a \cos(b(x - c)) + d$ to model these data.

(b) Are different values for c possible? Explain.

78. *Modeling Temperatures* The maximum monthly average temperature in Anchorage, Alaska, is 57°F and the minimum is 12°F.

(a) Using only these two temperatures, determine $f(x) = a \cos(b(x - c)) + d$ so that $f(x)$ models the monthly average temperatures in Anchorage.

(b) Graph f and the actual data in the table over a 2-year period.

Month	1	2	3	4	5	6
Temperature (°F)	12	18	23	36	46	55

Month	7	8	9	10	11	12
Temperature (°F)	57	55	48	36	23	16

Source: A. Miller and J. Thompson.

79. *Daylight Hours* The graph models the daylight hours at 60°N latitude, where $x = 1$ corresponds to January 1, $x = 2$ to February 1, and so on.

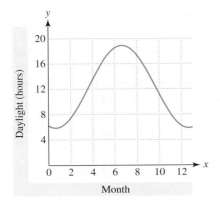

(a) Estimate the maximum number of daylight hours. When does this occur?

(b) Estimate the minimum number of daylight hours. When does this occur?

(c) Interpret the amplitude and period.

80. *Daylight Hours*   The graph in Exercise 79 is given by $y = 6.5 \sin\left(\frac{\pi}{6}(x - 3.65)\right) + 12.4$. Modify this equation to model the following situations.
(a) At 50°N latitude, the maximum daylight is about 16.3 hours and the minimum is about 8.3 hours.

(b) The daylight hours at 60°S latitude

(c) The daylight hours at the equator

81. *Average Precipitation*   The graph models the monthly average precipitation in inches at Mount Adams, Washington, over a 3-year period, where x is the month.

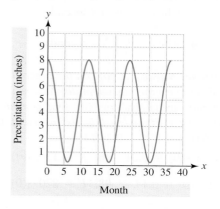

Month

(a) Find the maximum and minimum monthly average precipitation.

(b) Find the amplitude and interpret the result.

(c) Use $f(x) = a \cos(b(x - c)) + d$ to model the graph.

82. *Daylight Hours*   San Antonio, Texas, has a latitude of 29.5°N. The table lists the number of daylight hours on the first day of each month at San Antonio.

Month	1	2	3	4	5	6
Daylight (hr)	10.2	10.7	11.5	12.5	13.3	13.9

Month	7	8	9	10	11	12
Daylight (hr)	14.1	13.6	12.7	11.9	11.0	10.4

Source: J. Williams, *The Weather Almanac.*

(a) Plot the data over a 2-year period.

(b) Use $f(x) = a \cos(b(x - c)) + d$ to model these data.

(c) Estimate the daylight hours on February 15.

83. *Average Precipitation*   Suppose that the monthly average precipitation at a particular location varies sinusoidally between a maximum of 6 inches in January and a minimum of 2 inches in July. Let $t = 1$ correspond to January and $t = 12$ correspond to December.
(a) Use $f(t) = a \cos(b(t - c)) + d$ to model these conditions.

(b) Evaluate $f(1)$ and $f(7)$.

84. *Modeling Tidal Currents*   Tides cause ocean currents to flow into and out of harbors and canals. The table shows the speed of the ocean current at Cape Cod Canal in bogo-knots (bk) x hours after midnight on August 26, 1998. (Note that to change bogo-knots to knots, take the square root of the absolute value of the number of bogo-knots.)

Time (hr)	3.7	6.75	9.8
Current (bk)	−18	0	18

Time (hr)	13.0	16.1	22.2
Current (bk)	0	−18	18

Source: *WWW Tide and Current Predictor.*

(a) Use $f(x) = a \cos(b(x - c)) + d$ to model the data.

(b) Graph f and the data in [0, 24, 4] by [−20, 20, 5]. Interpret the graph.

85. *Modeling Ocean Temperature*   The following table lists the monthly average ocean temperatures in degrees Fahrenheit at Veracruz, Mexico.

Month	1	2	3	4	5	6
Temperature (°F)	72	73	74	78	81	83

Month	7	8	9	10	11	12
Temperature (°F)	84	85	84	82	78	74

Source: J. Williams.

(a) Make a scatterplot of the data over a 2-year period.

(b) Use $f(x) = a \sin(b(x - c)) + d$ to model the data.

86. *Ocean Temperatures* The water temperature in degrees Fahrenheit at St. Petersburg, Florida, can be modeled by $f(x) = 12.4 \sin\left(\frac{\pi}{6}(x - 4.2)\right) + 75$. Modify this formula to model the following situations.

(a) The monthly average water temperatures vary between 60°F and 90°F.

(b) The monthly average water temperatures vary between 50°F and 70°F.

87. *Interpreting a Model* Graph $y = 20 + 15 \sin\frac{\pi t}{12}$ for $0 \le t \le 12$. Let y represent the outdoor temperature in degrees Celsius at time t in hours, where $t = 0$ corresponds to 9 A.M. Interpret the graph.

88. *Carbon Dioxide Levels in Hawaii* At Mauna Loa, Hawaii, atmospheric carbon dioxide levels in parts per million (ppm) have been measured regularly since 1958. The equation

$$L(x) = 0.022x^2 + 0.55x + 316 + 3.5 \sin(2\pi x)$$

may be used to model these levels, where x is the year and $x = 0$ corresponds to 1960.

(a) Graph L in [20, 35, 5] by [320, 370, 10] and interpret the graph.

(b) The function L is represented by the sum of a quadratic function and a sine function. How does each function affect the shape of the graph? Discuss reasons for each function. (**Source:** A. Nilsson, *Greenhouse Earth.*)

89. *Music and the Sine Function* A *pure tone* can be modeled by a sine wave. Pure tones typically sound dull and uninteresting. An example of a pure tone is the sound heard when a tuning fork is lightly struck. The pure tone of the first A-note above middle C can be modeled by $f(t) = \sin(880\pi t)$. (**Source:** J. Pierce, *The Science of Musical Sound.*)

(a) In [0, 1/100, 1/880] by [−1.5, 1.5, 0.5], graph the function f.

(b) Find the period P of this tone.

(c) Frequency gives the number of vibrations or cycles per second in a sinusoidal graph. The human ear can hear frequencies from 16.4 to 16,000 cycles per second. Frequency F may be determined using the equation $F = \frac{1}{P}$, where P is the period. Find the frequency of this A-note.

90. *Music* (Continuation of Exercise 89.) Middle C has a frequency of 261.6 cycles per second and can be modeled by $g(t) = \sin(523.2\pi t)$. (**Source:** J. Pierce.)

(a) Estimate the period of middle C.

(b) Graph f from Exercise 89 and g in the window [0, 1/100, 1/880] by [−1.5, 1.5, 0.5]. Compare their graphs.

Simple Harmonic Motion

Exercises 91–94: Springs (*Refer to Example 7.*) *Suppose that a weight on a spring has an initial position of s(0) and a period of P.*

(a) *Find a function s given by s(t) = a cos(2πFt) that models the displacement of the weight.*

(b) *Evaluate s(1). Is the weight moving upward, downward, or neither when t = 1?*

91. $s(0) = 2$ inches, $P = 0.5$ second

92. $s(0) = 5$ inches, $P = 1.5$ seconds

93. $s(0) = -3$ inches, $P = 0.8$ second

94. $s(0) = -4$ inches, $P = 1.2$ seconds

Exercises 95–98: Music *A note on the piano has the given frequency F. Suppose the maximum displacement at the center of the piano wire is given by s(0). Find constants a and b so that s(t) = a cos bt models this displacement. Graph s(t) in [0, 0.05, 0.01] by [−0.3, 0.3, 0.1].*

95. $F = 27.5$, $s(0) = 0.21$ **96.** $F = 110$, $s(0) = 0.11$

97. $F = 55$, $s(0) = 0.14$ **98.** $F = 220$, $s(0) = 0.06$

Sine Regression

Exercises 99–102: Modeling Data *Use regression to find a formula $f(x) = a \sin(bx + c) + d$ that models the real data given in the exercise. Graph the data and f together.*

99. Exercise 75 **100.** Exercise 76

101. Exercise 77 **102.** Exercise 78

Writing about Mathematics

103. Discuss how the constants a, b, c, and d affect the graph of $y = a \sin(b(x - c)) + d$. Give an example.

104. Discuss some types of real data that could be modeled by $y = a \cos(b(x - c)) + d$. Give an example.

6.6 Inverse Trigonometric Functions

- **Define and use the inverse sine function**
- **Define and use the inverse cosine function**
- **Define and use the inverse tangent function**
- **Solve triangles and equations**

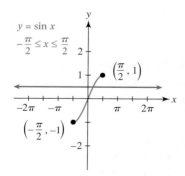

Introduction

In construction it is sometimes necessary to determine angles. For example, the pitch, or slope, of a roof frequently is expressed as the ratio $\frac{k}{12}$, where k represents a k-foot rise for every 12 feet of run in horizontal distance. See Figure 6.148. A typical roof pitch for homes is $\frac{6}{12}$. To correctly cut the rafters, a carpenter needs to know the measure of angle θ. This problem can be solved easily using inverse trigonometric functions. See Exercise 78.

Figure 6.148

The Inverse Sine Function

A numerical representation of the sine function is shown in Table 6.10.

Table 6.10

x	0	$\frac{\pi}{6}$	$\frac{\pi}{4}$	$\frac{\pi}{3}$	$\frac{\pi}{2}$	$\frac{2\pi}{3}$	$\frac{3\pi}{4}$	$\frac{5\pi}{6}$	π
$f(x)$	0	$\frac{1}{2}$	$\frac{1}{\sqrt{2}}$	$\frac{\sqrt{3}}{2}$	1	$\frac{\sqrt{3}}{2}$	$\frac{1}{\sqrt{2}}$	$\frac{1}{2}$	0

Figure 6.149 The Sine Function

Notice that different inputs do not always result in different outputs. Therefore the sine function is not one-to-one and so an inverse function does not exist. For example, $f(0) = 0$ and $f(\pi) = 0$, so $f^{-1}(0)$ cannot be defined as a single output, which is necessary for an inverse *function* to exist. By the horizontal line test we also can see that the sine function is not one-to-one. See Figure 6.149, where it is possible for a horizontal line to intersect the sine graph infinitely many times.

If we restrict the domain of $f(x) = \sin x$ to $-\frac{\pi}{2} \le x \le \frac{\pi}{2}$, as shown in Figure 6.150, the graph of $y = \sin x$ is one-to-one, since a horizontal line intersects it at most once. On this restricted domain, the sine function has a unique inverse called the *inverse sine function*.

Figure 6.150 Restricting the Domain

Inverse Sine Function

The **inverse sine function**, denoted $\sin^{-1} x$ or arcsin x, is defined as follows: For $-1 \le x \le 1$ and y in the interval $\left[-\frac{\pi}{2}, \frac{\pi}{2}\right]$,

$$y = \sin^{-1} x \quad \text{or} \quad y = \arcsin x \qquad \text{means} \qquad x = \sin y.$$

NOTE When evaluating the inverse sine function, it may be helpful to think of $\sin^{-1} x$ as an *angle* θ, where $\sin \theta = x$ and θ satisfies $-\frac{\pi}{2} \le \theta \le \frac{\pi}{2}$.

The next example illustrates how to evaluate the inverse sine function.

EXAMPLE 1 Evaluating the inverse sine function

Evaluate each of the following by hand and then support your results with a calculator.

(a) $\sin^{-1} 1$ **(b)** $\arcsin\left(-\frac{1}{2}\right)$

SOLUTION

(a) The expression $\sin^{-1} 1$ represents the angle (or real number) θ whose sine equals 1 and that satisfies $-\frac{\pi}{2} \leq \theta \leq \frac{\pi}{2}$. Thus $\theta = \sin^{-1} 1 = \frac{\pi}{2} \approx 1.57$. In degrees, $\sin^{-1} 1 = 90°$.

(b) The expression $\arcsin\left(-\frac{1}{2}\right)$ represents the angle (or real number) θ whose sine equals $-\frac{1}{2}$ and that satisfies $-\frac{\pi}{2} \leq \theta \leq \frac{\pi}{2}$. Thus $\theta = \arcsin\left(-\frac{1}{2}\right) = -\frac{\pi}{6} \approx -0.52$. In degrees, $\sin^{-1}\left(-\frac{1}{2}\right) = -30°$. Figures 6.151 and 6.152 support these results in both radian mode and degree mode.

Now Try Exercise 13 ◀

```
sin⁻¹(1)
          1.570796327
sin⁻¹(-1/2)
          -.5235987756
```

Figure 6.151 Radian Mode

```
sin⁻¹(1)
                    90
sin⁻¹(-1/2)
                   -30
```

Figure 6.152 Degree Mode

Functions and their inverses interchange domains and ranges. The range of the sine function is $-1 \leq y \leq 1$. Therefore the domain of the inverse sine function is $-1 \leq x \leq 1$. Since the domain of the sine function has been restricted to $-\frac{\pi}{2} \leq x \leq \frac{\pi}{2}$, the range of the inverse sine function is $-\frac{\pi}{2} \leq y \leq \frac{\pi}{2}$. That is, $\sin^{-1} x$ outputs angles (numbers) only in the interval $\left[-\frac{\pi}{2}, \frac{\pi}{2}\right]$. The following are properties of the inverse sine function.

$$\sin^{-1}(\sin x) = x \quad \text{for} \quad -\frac{\pi}{2} \leq x \leq \frac{\pi}{2}$$
$$\sin(\sin^{-1} x) = x \quad \text{for} \quad -1 \leq x \leq 1$$

NOTE $\sin^{-1} 2$ is undefined because $-1 \leq \sin x \leq 1$ for all angles (or numbers) x.

MAKING CONNECTIONS

Notation and Inverse Functions In Section 5.2, we learned that $f^{-1}(x) \neq \frac{1}{f(x)}$. The same is true for the inverse sine function: $\sin^{-1} x \neq \frac{1}{\sin x}$. For example, $\sin^{-1} 1 = \frac{\pi}{2} \approx 1.57$ and $\frac{1}{\sin 1} \approx 1.19$. Note that $(\sin x)^{-1} = \frac{1}{\sin x}$.

EXAMPLE 2 Finding representations of the inverse sine function

Represent the inverse sine function verbally, numerically, graphically, and symbolically.

SOLUTION

Verbal Representation To compute $\sin^{-1} x$ for $-1 \leq x \leq 1$, determine the angle (or real number) θ such that $\sin \theta = x$ and $-\frac{\pi}{2} \leq \theta \leq \frac{\pi}{2}$.

Numerical Representation Table 6.11 shows a numerical representation of $\sin x$ on the interval $\left[-\frac{\pi}{2}, \frac{\pi}{2}\right]$. It follows that a numerical representation of $\sin^{-1} x$ is as shown in Table 6.12. Notice that if $\sin a = b$, then $\sin^{-1} b = a$, provided $-\frac{\pi}{2} \leq a \leq \frac{\pi}{2}$.

Table 6.11

x	$-\frac{\pi}{2}$	$-\frac{\pi}{3}$	$-\frac{\pi}{4}$	$-\frac{\pi}{6}$	0	$\frac{\pi}{6}$	$\frac{\pi}{4}$	$\frac{\pi}{3}$	$\frac{\pi}{2}$
sin x	-1	$-\frac{\sqrt{3}}{2}$	$-\frac{1}{\sqrt{2}}$	$-\frac{1}{2}$	0	$\frac{1}{2}$	$\frac{1}{\sqrt{2}}$	$\frac{\sqrt{3}}{2}$	1

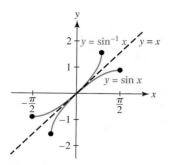

Figure 6.153 The Inverse Sine Function

Table 6.12

x	-1	$-\dfrac{\sqrt{3}}{2}$	$-\dfrac{1}{\sqrt{2}}$	$-\dfrac{1}{2}$	0	$\dfrac{1}{2}$	$\dfrac{1}{\sqrt{2}}$	$\dfrac{\sqrt{3}}{2}$	1
$\sin^{-1}x$	$-\dfrac{\pi}{2}$	$-\dfrac{\pi}{3}$	$-\dfrac{\pi}{4}$	$-\dfrac{\pi}{6}$	0	$\dfrac{\pi}{6}$	$\dfrac{\pi}{4}$	$\dfrac{\pi}{3}$	$\dfrac{\pi}{2}$

Graphical Representation   The graph of $y = \sin^{-1}x$ can be found by reflecting the graph of $y = \sin x$ for $-\frac{\pi}{2} \leq x \leq \frac{\pi}{2}$ across the line $y = x$, as shown in Figure 6.153.

Symbolic Representation   A symbolic representation of the inverse sine function can be written as either $f(x) = \sin^{-1}x$ or $f(x) = \arcsin x$. There is no simple formula that can be used to evaluate $\sin^{-1}x$.

Now Try Exercise 31 ◀

An Application   In track and field, when an athlete throws the shot, the distance that the shot travels depends on the angle θ that the initial direction of the shot makes with the horizontal. Angle θ in Figure 6.154 is called the *projection angle*. The optimal projection angle θ, which results in maximum distance for the shot, may be calculated by

$$\theta = \sin^{-1}\sqrt{\dfrac{v^2}{2v^2 + 64.4h}},$$

where v is the initial speed in feet per second of the shot and h is the height in feet of the shot when it is released.   (**Source:** J. Cooper and R. Glassow, *Kinesiology.*)

CLASS DISCUSSION

If a cannonball is shot from ground level, what is the optimal projection angle? Does this angle depend on v?

Figure 6.154 Projection Angle θ

EXAMPLE 3   Finding the optimal projection angle for a shot-putter

Suppose that an athlete releases a shot 8 feet above the ground with velocity v. Give a numerical representation of the optimum projection angle θ. Interpret the results.

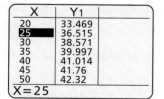

Figure 6.155 Degree Mode

SOLUTION Make a table for $Y_1 = \sin^{-1}(\sqrt{(X^2/(2X^2 + 64.4*8))})$, as shown in Figure 6.155. We can see that the faster a person throws the shot, the greater the optimal projection angle θ becomes. For example, if a shot is thrown at 25 feet per second, then $\theta \approx 36.5°$, whereas if the shot is thrown at 50 feet per second, then $\theta \approx 42.3°$.

Now Try Exercise 79 ◀

The Inverse Cosine Function

By the horizontal line test, the cosine function is not one-to-one, as illustrated in Figure 6.156 on the next page. If we restrict the domain of $f(x) = \cos x$ to $0 \leq x \leq \pi$, then the resulting function is one-to-one and has an inverse function. See Figure 6.157 on the next page. This inverse function is called the *inverse cosine function*.

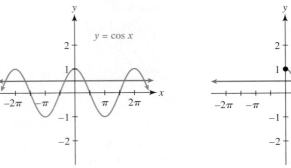

Figure 6.156 The Cosine Function   **Figure 6.157** Restricting the Domain

Inverse Cosine Function

The **inverse cosine function**, denoted $\cos^{-1} x$ or arccos x, is defined as follows: For $-1 \le x \le 1$ and y in the interval $[0, \pi]$,

$$y = \cos^{-1} x \quad \text{or} \quad y = \arccos x \qquad \text{means} \qquad x = \cos y.$$

EXAMPLE 4   Evaluating the inverse cosine function

Evaluate each of the following.
(a) $\cos^{-1} 1$ **(b)** $\arccos(-0.75)$

SOLUTION
(a) The expression $\cos^{-1} 1$ represents the angle (or real number) θ whose cosine equals 1 and that satisfies $0 \le \theta \le \pi$. Thus $\theta = \cos^{-1} 1 = 0$, or $0°$.
(b) A calculator is often necessary to evaluate inverse trigonometric functions approximately. In Figure 6.158, $\cos^{-1}(-0.75) \approx 2.42$ radians, or about $138.6°$.

> *Now Try Exercises 15 and 19(c)* ◀

The following are properties of the inverse cosine function.

$$\cos^{-1}(\cos x) = x \quad \text{for} \quad 0 \le x \le \pi$$
$$\cos(\cos^{-1} x) = x \quad \text{for} \quad -1 \le x \le 1$$

```
cos⁻¹(‾.75)
          2.418858406
Ans*180/π
          138.5903779
```

Figure 6.158 Radian Mode

EXAMPLE 5   Finding representations of the inverse cosine function

Represent the inverse cosine function verbally, numerically, graphically, and symbolically.

SOLUTION
Verbal Representation   To compute $\cos^{-1} x$ for $-1 \le x \le 1$, determine the angle θ (or real number) such that $\cos \theta = x$ and $0 \le \theta \le \pi$.

Numerical Representation   A numerical representation of $\cos x$ on the interval $[0, \pi]$ is shown in Table 6.13.

Table 6.13

x	0	$\frac{\pi}{6}$	$\frac{\pi}{4}$	$\frac{\pi}{3}$	$\frac{\pi}{2}$	$\frac{2\pi}{3}$	$\frac{3\pi}{4}$	$\frac{5\pi}{6}$	π
$\cos x$	1	$\frac{\sqrt{3}}{2}$	$\frac{1}{\sqrt{2}}$	$\frac{1}{2}$	0	$-\frac{1}{2}$	$-\frac{1}{\sqrt{2}}$	$-\frac{\sqrt{3}}{2}$	-1

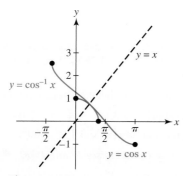

Figure 6.159 The Inverse Cosine Function

A numerical representation of $\cos^{-1} x$ is shown in Table 6.14. Notice that if $\cos a = b$, then $\cos^{-1} b = a$, provided that $0 \le a \le \pi$.

Table 6.14

x	-1	$-\dfrac{\sqrt{3}}{2}$	$-\dfrac{1}{\sqrt{2}}$	$-\dfrac{1}{2}$	0	$\dfrac{1}{2}$	$\dfrac{1}{\sqrt{2}}$	$\dfrac{\sqrt{3}}{2}$	1
$\cos^{-1} x$	π	$\dfrac{5\pi}{6}$	$\dfrac{3\pi}{4}$	$\dfrac{2\pi}{3}$	$\dfrac{\pi}{2}$	$\dfrac{\pi}{3}$	$\dfrac{\pi}{4}$	$\dfrac{\pi}{6}$	0

Graphical Representation The graph of $y = \cos^{-1} x$ can be found by reflecting the graph of $y = \cos x$ for $0 \le x \le \pi$ across the line $y = x$, as shown in Figure 6.159.

Symbolic Representation A symbolic representation of the inverse cosine function can be written as either $f(x) = \cos^{-1} x$ or $f(x) = \arccos x$. There is no simple formula that can be used to evaluate $\cos^{-1} x$.

Now Try Exercise 33 ◄

CLASS DISCUSSION

Explain the results in Figure 6.160.

$$\begin{array}{l} \cos^{-1}(\cos(90)) \\ \qquad\qquad\qquad 90 \\ \cos^{-1}(\cos(-270)) \\ \qquad\qquad\qquad 90 \end{array}$$

Figure 6.160 Degree Mode

The Inverse Tangent Function

By the horizontal line test, the tangent function is not one-to-one, as shown in Figure 6.161. If we restrict the domain of $f(x) = \tan x$ to $-\frac{\pi}{2} < x < \frac{\pi}{2}$, then the resulting function is one-to-one and has an inverse function. See Figure 6.162. This inverse function is called the *inverse tangent function*.

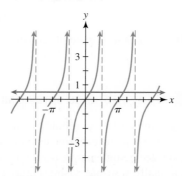

Figure 6.161 The Tangent Function

Figure 6.162 Restricting the Domain

Inverse Tangent Function

The **inverse tangent function**, denoted $\tan^{-1} x$ or $\arctan x$, is defined as follows: For y in the interval $\left(-\frac{\pi}{2}, \frac{\pi}{2}\right)$,

$$y = \tan^{-1} x \quad \text{or} \quad y = \arctan x \qquad \text{means} \qquad x = \tan y.$$

The following are properties of the inverse tangent function.

$$\tan^{-1}(\tan x) = x \quad \text{for } -\tfrac{\pi}{2} < x < \tfrac{\pi}{2}$$
$$\tan(\tan^{-1} x) = x \quad \text{for all real numbers } x$$

EXAMPLE 6 Evaluating the inverse tangent function

Evaluate each of the following. Support your answer using a calculator.

(a) $\tan^{-1} 1$ **(b)** $\arctan\left(-\sqrt{3}\right)$

```
tan⁻¹(1)
        .7853981634
tan⁻¹(-√(3))
       ⁻1.047197551
```

Figure 6.163 Radian Mode

SOLUTION

(a) The expression $\tan^{-1} 1$ represents the angle (or real number) θ whose tangent equals 1 and that satisfies $-\frac{\pi}{2} < \theta < \frac{\pi}{2}$. Thus $\theta = \tan^{-1}(1) = \frac{\pi}{4} \approx 0.7854$. See Figure 6.163. In degrees, $\tan^{-1}(1) = 45°$.

(b) The expression $\arctan\left(-\sqrt{3}\right)$ represents the angle (or real number) θ whose tangent is $-\sqrt{3}$ and that satisfies $-\frac{\pi}{2} < \theta < \frac{\pi}{2}$. Thus $\theta = \arctan\left(-\sqrt{3}\right) = -\frac{\pi}{3} \approx -1.047$. Numerical support is shown in Figure 6.163. In degrees, $\arctan\left(-\sqrt{3}\right) = -60°$.

Now Try Exercises 17 and 19(b) ◄

EXAMPLE 7 Evaluating inverse trigonometric functions

Let θ be an acute angle. Write $\cos \theta$ in terms of x, if $\theta = \sin^{-1} x$.

SOLUTION

Getting Started First sketch a right triangle so that one of its angles is θ. Because $\theta = \sin^{-1} x$, it follows that $\sin \theta = x$. Label the sides of the triangle so that $\sin \theta = \frac{x}{1}$. Use the triangle to find $\cos \theta$. ►

Figure 6.164 shows angle θ with $\sin \theta = \frac{x}{1}$. By the Pythagorean theorem, the other leg has length $\sqrt{1 - x^2}$. Because $\cos \theta$ equals the side adjacent divided by the hypotenuse, we have $\cos \theta = \sqrt{1 - x^2}$.

Now Try Exercise 41 ◄

Figure 6.164

An Application The next example illustrates the use of the inverse tangent function in robotics.

EXAMPLE 8 Using robots to spray paint

In industry it is common to use robots to spray paint. The robotic arm in Figure 6.165 is being used to paint a flat surface. Because the spray gun must move at a constant speed v, parallel to the surface being painted, the angle of the arm θ_1 and the angle of the spray gun θ_2 must be continually adjusted. Using Figure 6.165, it can be shown that

$$\theta_1 = \arctan \frac{h}{vt} \quad \text{and} \quad \theta_2 = 90° - \theta_1,$$

where $t > 0$ is time in seconds. (Try to verify this.) Let $v = 3$ inches per second and $h = 24$ inches. Determine the degree measure of θ_1 and θ_2 after 10 seconds. (**Source:** W. Stadler, *Analytical Robotics and Mechatronics.*)

CLASS DISCUSSION

Give verbal, numerical, and graphical representation of $y = \tan^{-1} x$.

Figure 6.165

Figure 6.166

SOLUTION Substitute $h = 24$ and $v = 3$. When $t = 10$, $\theta_1 = \arctan \frac{24}{3(10)} \approx 38.7°$ and $\theta_2 = 90° - \theta_1 \approx 51.3°$.

> *Now Try Exercise 83* ◄

Solving Triangles and Equations

Figure 6.166 shows *standard labeling* used to denote the vertices, sides, and angles of a right triangle. The next example illustrates the process of finding the measures of the angles and sides in a triangle, called *solving a triangle*.

EXAMPLE 9 Solving a right triangle

Solve triangle ABC in Figure 6.167 if $a = 5$ and $c = 13$. Round values to the nearest tenth.

SOLUTION We are given $a = 5$, $c = 13$, and $\gamma = 90°$. We must find b, α, and β. We begin by finding b using the Pythagorean theorem.

$$a^2 + b^2 = c^2$$
$$b^2 = c^2 - a^2$$
$$b^2 = 13^2 - 5^2$$
$$b^2 = 144$$

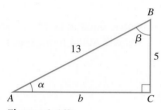

Figure 6.167

Thus $b = 12$. We can find angle α as follows.

$$\sin \alpha = \frac{5}{13}$$

$$\alpha = \sin^{-1} \frac{5}{13}$$

$$\alpha \approx 22.6°$$

Since β is complementary to α, $\beta \approx 90° - 22.6° = 67.4°$.

> *Now Try Exercise 37* ◄

NOTE There is more than one way to solve the triangle in Example 9. We could have let $\alpha = \tan^{-1} \frac{5}{12} \approx 22.6°$ and $\beta = \cos^{-1} \frac{5}{13} \approx 67.4°$. There are other possibilities.

EXAMPLE 10 Finding angles in a triangle

Approximate the degree measure of the angles α and β shown in Figure 6.168.

SOLUTION From Figure 6.168 we see that $\tan \alpha = \frac{9}{17}$. Thus $\alpha = \tan^{-1} \frac{9}{17} \approx 27.9°$. Since α and β are complementary angles, $\beta \approx 90° - 27.9° = 62.1°$.

> *Now Try Exercise 35* ◄

Figure 6.168

An Application   Grade resistance is the force F that causes a car to roll down a hill. It can be calculated by $F = W \sin \theta$, where θ represents the angle of the grade and W represents the weight of the vehicle. (Refer to Figure 6.57 in Section 6.3 on page 509.)

EXAMPLE 11 Calculating highway grade

Find the angle θ for which a 3000-pound car has grade resistance of 500 pounds.

SOLUTION Solve the equation $F = W \sin \theta$ for θ.

$$F = W \sin \theta \qquad \text{Given equation}$$

$$500 = 3000 \sin \theta \qquad \text{Let } W = 3000 \text{ and } F = 500.$$

$$\sin \theta = \frac{1}{6} \qquad \text{Solve for } \sin \theta.$$

$$\theta = \sin^{-1} \frac{1}{6} \qquad \text{Property of inverse sine}$$

$$\theta \approx 9.6° \qquad \text{Approximate.}$$

Thus if a road is inclined at approximately 9.6°, a 3000-pound car would experience a force of 500 pounds pulling downhill. *Now Try Exercise 73* ◄

Solving an Equation There are a wide variety of trigonometric equations. In the next example, we solve one basic type of equation. In Chapter 7, we will solve more types of trigonometric equations.

EXAMPLE 12 Solving a trigonometric equation

Solve $9 \cos^2 \theta = 4$, where θ is an acute angle.

SOLUTION Begin by dividing each side of the equation by 9.

$$9 \cos^2 \theta = 4 \qquad \text{Given equation}$$

$$\cos^2 \theta = \frac{4}{9} \qquad \text{Divide by 9.}$$

$$\cos \theta = \pm \frac{2}{3} \qquad \text{Square root property}$$

Because θ is an acute angle, $\cos \theta$ must be positive. Thus $\cos \theta = \frac{2}{3}$, and the solution to the equation is $\theta = \cos^{-1} \frac{2}{3} \approx 48.2°$. *Now Try Exercise 61* ◄

6.6 Putting It All Together

The six trigonometric functions are not one-to-one functions. However, by *restricting* their domains to appropriate intervals, inverse trigonometric functions can be defined. In this section we defined inverse functions for the sine, cosine, and tangent functions. Inverse functions can be used to solve triangles and equations.

The following table summarizes some properties of inverse trigonometric functions.

Function	Explanation	Examples and Graphs
Inverse sine	*Description:* $f(x) = \sin^{-1} x$ or $f(x) = \arcsin x$ computes the angle or number in $\left[-\frac{\pi}{2}, \frac{\pi}{2}\right]$ whose sine equals x, where $-1 \le x \le 1$. *Domain:* $\{x \mid -1 \le x \le 1\}$ *Range:* $\left\{y \mid -\frac{\pi}{2} \le y \le \frac{\pi}{2}\right\}$ *Inverse Properties:* $\quad \sin^{-1}(\sin x) = x \quad$ for $\quad -\frac{\pi}{2} \le x \le \frac{\pi}{2}$ $\quad \sin(\sin^{-1} x) = x \quad$ for $\quad -1 \le x \le 1$	$\sin^{-1} 1 = 90°$, or $\frac{\pi}{2}$ $\arcsin\left(-\frac{1}{2}\right) = -30°$, or $-\frac{\pi}{6}$ $\sin^{-1} 0 = 0°$, or 0 $\sin^{-1} 4$ is undefined. $\sin^{-1} 0.3 \approx 17.5°$, or 0.305
Inverse cosine	*Description:* $f(x) = \cos^{-1} x$ or $f(x) = \arccos x$ computes the angle or number in $[0, \pi]$ whose cosine equals x, where $-1 \le x \le 1$. *Domain:* $\{x \mid -1 \le x \le 1\}$ *Range:* $\{y \mid 0 \le y \le \pi\}$ *Inverse Properties:* $\quad \cos^{-1}(\cos x) = x \quad$ for $\quad 0 \le x \le \pi$ $\quad \cos(\cos^{-1} x) = x \quad$ for $-1 \le x \le 1$	$\cos^{-1} 1 = 0°$, or 0 $\arccos\left(-\frac{1}{2}\right) = 120°$, or $\frac{2\pi}{3}$ $\cos^{-1} 0 = 90°$, or $\frac{\pi}{2}$ $\cos^{-1}(-5)$ is undefined. $\cos^{-1} 0.8 \approx 36.9°$, or 0.644
Inverse tangent	*Description:* $f(x) = \tan^{-1} x$ or $f(x) = \arctan x$ computes the angle or number in $\left(-\frac{\pi}{2}, \frac{\pi}{2}\right)$ whose tangent equals x, where x is any real number. *Domain:* $\{x \mid -\infty < x < \infty\}$ (all real numbers) *Range:* $\left\{y \mid -\frac{\pi}{2} < y < \frac{\pi}{2}\right\}$ *Inverse Properties:* $\quad \tan^{-1}(\tan x) = x \quad$ for $-\frac{\pi}{2} < x < \frac{\pi}{2}$ $\quad \tan(\tan^{-1} x) = x \quad$ for all real numbers x	$\tan^{-1} 1 = 45°$, or $\frac{\pi}{4}$ $\tan^{-1} 0 = 0°$, or 0 $\tan^{-1} 8 \approx 82.9°$, or 1.446 $\arctan(-1) = -45°$, or $-\frac{\pi}{4}$

6.6   Exercises

Review of Inverses

1. For a function f to have an inverse, f must be _____.

2. A function is one-to-one if different inputs always result in _____ outputs.

3. If $f(\pi) = -1$, then $f^{-1}(-1) = $ _____.

4. If $f(c) = d$, then $f^{-1}(d) = $ _____.

5. If $f^{-1}(0) = 1$, then $f(1) = $ _____.

6. If $f^{-1}(b) = a$, then $f(a) = $ _____.

Inverse Trigonometric Functions

7. Since $\sin \frac{\pi}{2} = 1$ and $\frac{\pi}{2}$ is in the interval $\left[-\frac{\pi}{2}, \frac{\pi}{2}\right]$, $\sin^{-1} 1 = $ _____.

8. Since $\cos \frac{\pi}{3} = \frac{1}{2}$ and $\frac{\pi}{3}$ is in the interval $[0, \pi]$, $\cos^{-1} \frac{1}{2} = $ _____.

9. Since $\tan\left(-\frac{\pi}{4}\right) = -1$ and $-\frac{\pi}{4}$ is in the interval $\left(-\frac{\pi}{2}, \frac{\pi}{2}\right)$, $\tan^{-1}(-1) = $ _____.

10. Since $\sin\left(-\frac{\pi}{6}\right) = -\frac{1}{2}$ and $-\frac{\pi}{6}$ is in the interval $\left[-\frac{\pi}{2}, \frac{\pi}{2}\right]$, $\sin^{-1}\left(-\frac{1}{2}\right) = $ _____.

11. Since $\cos\left(\frac{2\pi}{3}\right) = -\frac{1}{2}$ and $\frac{2\pi}{3}$ is in the interval $[0, \pi]$, $\cos^{-1}\left(-\frac{1}{2}\right) = $ _____.

12. Since $\tan\left(\frac{\pi}{3}\right) = \sqrt{3}$ and $\frac{\pi}{3}$ is in the interval $\left(-\frac{\pi}{2}, \frac{\pi}{2}\right)$, $\tan^{-1} \sqrt{3} = $ _____.

Exercises 13–18: Evaluate each of the following, if possible. Give results in both radians and degrees.

13. (a) $\sin^{-1} 1$ (b) $\arcsin 4$

   (c) $\arcsin\left(-\frac{\sqrt{3}}{2}\right)$

14. (a) $\arcsin \frac{1}{2}$ (b) $\sin^{-1}(-2)$

   (c) $\sin^{-1}(-1)$

15. (a) $\cos^{-1} 0$ (b) $\arccos(-1)$

   (c) $\cos^{-1} \frac{1}{2}$

16. (a) $\arccos \frac{\sqrt{3}}{2}$ (b) $\cos^{-1}\left(-\frac{1}{2}\right)$

   (c) $\arccos 1$

17. (a) $\tan^{-1} 1$ (b) $\arctan(-1)$

   (c) $\tan^{-1} \sqrt{3}$

18. (a) $\arctan(-\sqrt{3})$ (b) $\tan^{-1} 0$

   (c) $\tan^{-1}\left(-\frac{1}{\sqrt{3}}\right)$

Exercises 19 and 20: Approximate the following to a hundredth of a radian and a tenth of a degree.

19. (a) $\sin^{-1} 1.5$ (b) $\tan^{-1} 10$

   (c) $\arccos(-0.25)$

20. (a) $\cos^{-1}\left(-\frac{1}{3}\right)$ (b) $\arcsin(-0.54)$

   (c) $\arctan(-2.5)$

Exercises 21 and 22: Evaluate each expression using the figure to obtain either α or β.

21. (a) $\tan^{-1} \frac{4}{3}$

   (b) $\sin^{-1} \frac{3}{5}$

   (c) $\arccos \frac{3}{5}$

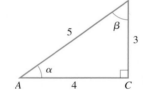

22. (a) $\arcsin \frac{12}{13}$

   (b) $\cos^{-1} \frac{5}{13}$

   (c) $\tan^{-1} \frac{5}{12}$

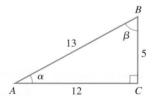

Exercises 23–28: Evaluate each expression.

23. $\sin(\sin^{-1} 1)$ 24. $\sin^{-1}\left(\sin \frac{\pi}{4}\right)$

25. $\cos^{-1}\left(\cos \frac{5\pi}{4}\right)$ 26. $\cos(\cos^{-1}(-3))$

27. $\tan(\tan^{-1}(-3))$ 28. $\tan^{-1}\left(\tan \frac{\pi}{5}\right)$

Exercises 29 and 30: Evaluate the expression in degree mode. Make a generalization about the result and then test your conjecture.

29. (a) $\sin^{-1}\frac{3}{5} + \cos^{-1}\frac{3}{5}$ (b) $\sin^{-1}\frac{1}{3} + \cos^{-1}\frac{1}{3}$

(c) $\sin^{-1}\frac{2}{7} + \cos^{-1}\frac{2}{7}$

30. (a) $\tan^{-1}\frac{3}{4} + \tan^{-1}\frac{4}{3}$ (b) $\tan^{-1}\frac{5}{12} + \tan^{-1}\frac{12}{5}$

(c) $\tan^{-1}\frac{1}{4} + \tan^{-1}4$

Exercises 31–34: Represent the given f(x) verbally, numerically, and graphically.

31. $f(x) = \sin^{-1}2x$

32. $f(x) = \sin^{-1}\frac{1}{2}x$

33. $f(x) = \cos^{-1}\frac{1}{2}x$

34. $f(x) = \cos^{-1}2x$

Solving Triangles

Exercises 35–40: (Refer to Example 9.) Solve the right triangle.

35.

36.

37.

38.

39.

40.

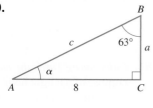

Exercises 41–44: (Refer to Example 7.) Evaluate the indicated trigonometric function of θ, where θ is an acute angle determined by an inverse trigonometric function. (Hint: Make a sketch of a right triangle containing angle θ.)

41. $\tan\theta$, if $\theta = \sin^{-1}x$

42. $\sin\theta$, if $\theta = \tan^{-1}\dfrac{x}{\sqrt{1 - x^2}}$

43. $\cos\theta$, if $\theta = \sin^{-1}\dfrac{x}{\sqrt{1 + x^2}}$

44. $\tan\theta$, if $\theta = \cos^{-1}\dfrac{1}{x}$

Exercises 45–50: (Refer to Example 7.) Write the expression as an algebraic expression of u if $0 < u < 1$.

45. $\sin(\cos^{-1}u)$ **46.** $\cos(\sin^{-1}u)$

47. $\tan(\cos^{-1}u)$ **48.** $\sin(\tan^{-1}u)$

49. $\cot\left(\tan^{-1}\dfrac{1}{u}\right)$ **50.** $\sec(\sin^{-1}u)$

Solving Trigonometric Equations

Exercises 51–56: Solve the trigonometric equation for θ, where $0° \le \theta \le 90°$.

51. $\sin\theta = 1$ **52.** $\cos\theta = \frac{1}{2}$

53. $\tan\theta = 1$ **54.** $\sin\theta = \dfrac{\sqrt{3}}{2}$

55. $\cos\theta = 0$ **56.** $\tan\theta = \dfrac{1}{\sqrt{3}}$

Exercises 57–62: Solve the equation for θ, where θ is an acute angle. Approximate θ to the nearest tenth of a degree.

57. $2\cos\theta = \frac{1}{4}$ **58.** $3\sin\theta = \frac{4}{5}$

59. $\tan\theta - 1 = 5$ **60.** $4\cos\theta + 1 = 6$

61. $\sin^2\theta = 0.87$ **62.** $\tan^3\theta - 2 = 1.65$

Exercises 63–70: Solve the equation for t, where t is a real number in the given interval. Approximate t to three decimal places.

63. $\tan t = -\frac{1}{5}, \left(-\frac{\pi}{2}, \frac{\pi}{2}\right)$ **64.** $\sin t = -\frac{1}{3}, \left[-\frac{\pi}{2}, \frac{\pi}{2}\right]$

65. $\cos t = 0.452, [0, \pi]$ **66.** $\tan t = 5.67, \left(-\frac{\pi}{2}, \frac{\pi}{2}\right)$

67. $2\sin t = -0.557, \left[-\frac{\pi}{2}, \frac{\pi}{2}\right]$

68. $3\cos t + 1 = 0.333, [0, \pi]$

69. $\cos^2 t = \frac{1}{25}, [0, \pi]$ **70.** $\sin^2 t = \frac{1}{16}, \left[-\frac{\pi}{2}, \frac{\pi}{2}\right]$

Applications

71. *Angle of Elevation* Find the angle of elevation θ of the top of a 50-foot tree at a distance of 85 feet. See the figure.

72. *Angle of Elevation* A 28-foot building casts a 40-foot shadow on level ground. Find the angle of elevation θ of the sun to the nearest tenth of a degree. See the figure.

28 ft

40 ft

73. *Grade Resistance* (Refer to Example 11.) Approximate θ to the nearest tenth of a degree for the given grade resistance F and vehicle weight W. Use the equation $F = W \sin \theta$.

(a) $F = 400$ lb, $W = 5000$ lb

(b) $F = 130$ lb, $W = 3500$ lb

(c) $F = -200$ lb, $W = 4000$ lb

74. *Robotics* Approximate the angle θ if the robotic hand is located at the following points, where $-90° < \theta < 90°$. See the figure.

(a) $(5, 11)$ **(b)** $(1, -3)$

y

x

θ

75. *Designing Steps* Steps are being attached to a deck as shown in the figure at the top of the next column. The bottom of the steps should land 10 feet from the deck. If the deck is 4 feet above the ground, estimate angle θ between the ground and the side board of the step.

4 ft

θ

10 ft

76. *Designing Steps* (Refer to Exercise 75.) If the length of the side boards for the steps is 4 feet and the deck is 2 feet above the ground, find angle θ.

77. *Phases of the Moon* (Refer to Example 9 in Section 6.3.) Find the phase angle θ for the given phase F. Assume that $0° \le \theta \le 180°$ and use $F(\theta) = \frac{1}{2}(1 - \cos \theta)$.

(a) $F = \frac{1}{4}$ **(b)** $F = \frac{3}{4}$

78. *Roof Pitch* The pitch, or slope, of a roof may be expressed in the form $k/12$, where k represents a k-foot rise for every 12 feet of run in horizontal distance. Determine angle θ in Figure 6.148 for each pitch.

(a) $\frac{3}{12}$ **(b)** $\frac{4}{12}$ **(c)** $\frac{6}{12}$ **(d)** $\frac{12}{12}$

79. *Shot Put* (Refer to Example 3.) Suppose that a shot is released 7 feet above the ground with a velocity of 43 feet per second. Find the optimal projection angle.

80. *Shot-Putting on the Moon* Repeat Exercise 79 with $v = 50$ feet per second if the shot is thrown on the moon and the optimal projection angle is given by

$$\theta = \sin^{-1} \sqrt{\frac{v^2}{2v^2 + 10.2h}}.$$

81. *Calculating Daylight Hours* The ability to calculate the number of daylight hours H at any location is important for estimating potential solar energy production. The value of H on the longest day can be calculated using the formula

$$\cos (0.1309 H) = -0.4336 \tan L,$$

where L is the latitude. Using *radian* mode, calculate the greatest number of daylight hours H during the year for the various cities and their latitudes. (**Source:** C. Winter, *Solar Power Plants.*)

(a) Akron, Ohio; $L = 40°55'$

(b) Corpus Christi, Texas; $L = 27°46'$

(c) Richmond, Virginia; $L = 37°30'$

82. *Shortest Day*   (Refer to Exercise 81.) The value of H on the shortest day can be calculated using the formula

$$\cos(0.1309\, H) = 0.4336 \tan L.$$

Find the least number of daylight hours at the following locations.   (**Source:** C. Winter.)

(a) Anchorage, Alaska; $L = 61°10'$

(b) Atlantic City, New Jersey; $L = 39°27'$

(c) Honolulu, Hawaii; $L = 21°20'$

83. *Robotics*   (Refer to Example 8.) Let $v = 5$ inches per second and $h = 18$. Determine the degree measure of θ_1 and θ_2 after 5 seconds.

84. *Snell's Law*   When a ray of light enters water, it is bent. This change in direction can be calculated using Snell's law. See the figure. The angles θ_1 and θ_2 are related by the equation $n_1 \sin \theta_1 = n_2 \sin \theta_2$, where n_1 and n_2 are constants called *indexes of refraction*. For air $n_1 = 1$, and for water $n_2 = 1.33$. If a ray of light enters the water with $\theta_1 = 40°$, estimate θ_2. (**Source:** R. Weidner and R. Sells, *Elementary Classical Physics*, Vol. 2.)

85. *Landscaping Formula*   A shrub is planted in a 100-foot-wide space between buildings measuring 75 feet and 150 feet tall. The location of the shrub determines how much sun it receives each day. Show that if θ is the angle in the figure and x is the distance of the shrub from the taller building, then the value of θ (in radians) is given by

$$\theta = \pi - \arctan\left(\frac{75}{100 - x}\right) - \arctan\left(\frac{150}{x}\right).$$

150 ft

75 ft

x

100 ft

86. *Communications Satellite Coverage*   The figure shows a stationary communications satellite positioned 20,000 miles above the equator. What percent of the equator can be seen from the satellite? The diameter of Earth is 7927 miles at the equator.

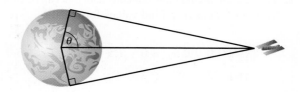

Writing about Mathematics

87. Explain verbally what each expression computes. Give examples.
   (a) $\sin^{-1} x$ **(b)** $\cos^{-1} x$ **(c)** $\tan^{-1} x$

88. Explain why

$$\sin^{-1}\left(\sin \tfrac{\pi}{2}\right) = \tfrac{\pi}{2} \quad \text{but} \quad \sin^{-1}\left(\sin \tfrac{5\pi}{2}\right) \neq \tfrac{5\pi}{2}.$$

Give a similar example using $\cos x$ and $\cos^{-1} x$.

1. *Movie Screen*   A 10-foot-high movie screen is mounted on a vertical wall so that the bottom of the screen is 6 feet above a horizontal floor. A person sits on a level floor x feet from the screen. If eye level is 3 feet above the floor, then angle θ shown in the figure can be expressed as

$$\theta = \tan^{-1}\left(\frac{10x}{x^2 + 39}\right).$$

Graph θ in $[0, 50, 10]$ by $[0, 50, 10]$ using degree mode. Determine where a person should sit to maximize θ.

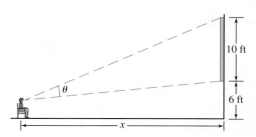

10 ft

6 ft

θ

x

1. Graph $f(t) = 3 \sin\left(2\left(t - \frac{\pi}{4}\right)\right)$ for $-\pi \le t \le \pi$. State the amplitude, period, and phase shift.

2. The accompanying table contains data that can be modeled by $f(t) = a \cos(bt)$. Find values for a and b.

t	0	0.5	1.0	1.5	2.0	2.5	3.0
$f(t)$	2	0	-2	0	2	0	-2

3. Evaluate the following, expressing your answer in degrees.
   (a) $\sin^{-1} 0$ **(b)** $\cos^{-1}(-1)$

   (c) $\tan^{-1}(-1)$ **(d)** $\sin^{-1}\frac{1}{2}$

   (e) $\tan^{-1}\sqrt{3}$ **(f)** $\cos^{-1}\frac{1}{2}$

4. Solve a right triangle if $a = 30$ and $b = 40$.

5. Use a calculator to solve each equation, where t is in the indicated interval.
   (a) $\sin t = 0.55,\ \left[-\frac{\pi}{2}, \frac{\pi}{2}\right]$

   (b) $\cos t = -0.35,\ [0, \pi]$

   (c) $\tan t = -2.9,\ \left(-\frac{\pi}{2}, \frac{\pi}{2}\right)$

6   Summary

CONCEPT	EXPLANATION AND EXAMPLES

SECTION 6.1   ANGLES AND THEIR MEASURE

Angle Measure

Degree measure: $360° =$ one revolution

$1° = 60'$ (minutes)

$1' = 60''$ (seconds)

Radian measure: 2π radians $=$ one revolution

radian measure $\times \frac{180°}{\pi} =$ degree measure

degree measure $\times \frac{\pi}{180°} =$ radian measure

Arc Length

$s = r\theta$, where θ is in *radians*

Example: The arc length intercepted by a central angle of 120° with a radius of 5 inches is

$$s = 5\left(\frac{2\pi}{3}\right) = \frac{10\pi}{3} \approx 10.47 \text{ inches.}$$

Area of Sector

$A = \frac{1}{2}r^2\theta$, where θ is in *radians*

Example: The area of the sector determined by a central angle of 45° with a radius of 10 inches is

$$A = \frac{1}{2}\left(10^2\right)\left(\frac{\pi}{4}\right) = \frac{25\pi}{2} \approx 39.27 \text{ square inches.}$$

CONCEPT	EXPLANATION AND EXAMPLES

SECTION 6.2 RIGHT TRIANGLE TRIGONOMETRY

Trigonometric Functions (Right Triangles)

$$\sin \theta = \frac{\text{side opposite}}{\text{hypotenuse}} \qquad \csc \theta = \frac{\text{hypotenuse}}{\text{side opposite}}$$

$$\cos \theta = \frac{\text{side adjacent}}{\text{hypotenuse}} \qquad \sec \theta = \frac{\text{hypotenuse}}{\text{side adjacent}}$$

$$\tan \theta = \frac{\text{side opposite}}{\text{side adjacent}} \qquad \cot \theta = \frac{\text{side adjacent}}{\text{side opposite}}$$

Example: The six trigonometric functions of θ in the figure are as follows.

$$\sin \theta = \frac{11}{61} \qquad \csc \theta = \frac{61}{11}$$

$$\cos \theta = \frac{60}{61} \qquad \sec \theta = \frac{61}{60}$$

$$\tan \theta = \frac{11}{60} \qquad \cot \theta = \frac{60}{11}$$

SECTION 6.3 THE SINE AND COSINE FUNCTIONS AND THEIR GRAPHS

Sine and Cosine

If angle θ is in standard position and its terminal side passes through the point (x, y), then

$$\sin \theta = \frac{y}{r} \quad \text{and} \quad \cos \theta = \frac{x}{r},$$

where $r = \sqrt{x^2 + y^2}$. In the figure to the right, $x = 3, y = -4,$ and $r = \sqrt{3^2 + (-4)^2} = 5$.

$$\sin \theta = -\frac{4}{5} \quad \text{and} \quad \cos \theta = \frac{3}{5}$$

Unit Circle

If the terminal side of an angle t intersects the unit circle at the point (x, y), then $\sin t = y$ and $\cos t = x$. The domains of the sine and cosine functions include all real numbers, and their ranges include all real numbers y such that $-1 \leq y \leq 1$.

In the figure, $x = -\frac{1}{2}$ and $y = \frac{\sqrt{3}}{2}$. Thus

$$\sin t = \frac{\sqrt{3}}{2} \quad \text{and} \quad \cos t = -\frac{1}{2}.$$

SECTION 6.4 OTHER TRIGONOMETRIC FUNCTIONS AND THEIR GRAPHS

Trigonometric Functions

The domains, ranges, periods, and graphs of the six trigonometric functions are discussed in Putting It All Together in Section 6.4.

If angle θ is in standard position and its terminal side passes through the point (x, y), then

$$\tan \theta = \frac{y}{x}, \quad \cot \theta = \frac{x}{y}, \quad \sec \theta = \frac{r}{x}, \quad \text{and} \quad \csc \theta = \frac{r}{y}, \quad \text{where } r = \sqrt{x^2 + y^2}.$$

CONCEPT	EXPLANATION AND EXAMPLES

SECTION 6.4 OTHER TRIGONOMETRIC FUNCTIONS AND THEIR GRAPHS (CONTINUED)

In the figure, $x = -2$, $y = 3$, and

$$r = \sqrt{(-2)^2 + 3^2} = \sqrt{13}.$$

$$\tan \theta = -\frac{3}{2} \qquad \cot \theta = -\frac{2}{3}$$

$$\sec \theta = -\frac{\sqrt{13}}{2} \qquad \csc \theta = \frac{\sqrt{13}}{3}$$

Unit Circle

If the terminal side of an angle t intersects the unit circle at the point (x, y), then

$$\tan t = \frac{y}{x}, \quad \cot t = \frac{x}{y}, \quad \sec t = \frac{1}{x}, \quad \text{and} \quad \csc t = \frac{1}{y}.$$

In the figure, $x = 0$ and $y = 1$. Thus

$\tan t$ is undefined; $\cot t = 0$;

$\sec t$ is undefined; $\csc t = 1$.

SECTION 6.5 GRAPHING TRIGONOMETRIC FUNCTIONS

Modeling with Sine and Cosine Functions

$$f(x) = a \sin (b(x - c)) + d \quad \text{or} \quad f(x) = a \cos (b(x - c)) + d$$

Amplitude $= |a|$ Period $= \dfrac{2\pi}{b}, b > 0$

Phase shift $= c$ Vertical shift $= d$

If $b = 2\pi F$, then F represents the frequency of the sinusoidal graph.

Simple Harmonic Motion

Can be modeled by $s(t) = a \sin bt$ or $s(t) = a \cos bt$

Example: If a spring is initially compressed 3 inches and oscillates 4 times per second, then $a = 3$, $b = 2\pi F = 8\pi$, and its motion can be modeled by $s(t) = 3 \cos (8\pi t)$.

SECTION 6.6 INVERSE TRIGONOMETRIC FUNCTIONS

Inverse Sine Function

$\theta = \sin^{-1} x$ implies that $\sin \theta = x$ and either $-\frac{\pi}{2} \le \theta \le \frac{\pi}{2}$ or $-90° \le \theta \le 90°$.

$\sin^{-1} x$ is also denoted arcsin x.

Examples: $\sin^{-1} 1 = \frac{\pi}{2}$, $\sin^{-1} \left(-\frac{1}{2}\right) = -30°$, arcsin $0 = 0°$

Inverse Cosine Function

$\theta = \cos^{-1} x$ implies that $\cos \theta = x$ and either $0 \le \theta \le \pi$ or $0° \le \theta \le 180°$.

$\cos^{-1} x$ is also denoted arccos x.

Examples: $\cos^{-1} 1 = 0$, $\cos^{-1} (-1) = 180°$, arccos $\frac{1}{2} = 60°$

CONCEPT	EXPLANATION AND EXAMPLES

SECTION 6.6 INVERSE TRIGONOMETRIC FUNCTIONS (CONTINUED)

Inverse Tangent Function

$\theta = \tan^{-1} x$ implies that $\tan \theta = x$ and either $-\frac{\pi}{2} < \theta < \frac{\pi}{2}$ or $-90° < \theta < 90°$.

$\tan^{-1} x$ is also denoted arctan x.

Examples: $\tan^{-1} 1 = \frac{\pi}{4}$, $\tan^{-1}(-1) = -45°$,   arctan $\sqrt{3} = 60°$

Solving Triangles

Inverse trigonometric functions can be used to solve equations and find angles in triangles.

Example: In triangle ABC, inverse trigonometric functions can be used to find α and β.

$$\alpha = \sin^{-1}\frac{3}{5} \approx 36.9°$$

$$\alpha = \tan^{-1}\frac{3}{4} \approx 36.9°$$

$$\beta = \cos^{-1}\frac{3}{5} \approx 53.1°$$

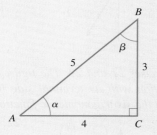

---

6   Review Exercises

1. Sketch the following angles in standard position.
   (a) $60°$ (b) $-\frac{5\pi}{6}$

2. Find the complementary angle and the supplementary angle to $\theta = 61°40'$.

3. Convert each angle from radians to degrees.
   (a) $\frac{\pi}{3}$ (b) $\frac{\pi}{36}$ (c) $-\frac{5\pi}{6}$ (d) $-\frac{7\pi}{4}$

4. Convert each angle from degrees to radians.
   (a) $30°$ (b) $165°$ (c) $-90°$ (d) $-105°$

5. Find the length of the arc intercepted by a central angle $\theta = 60°$ and a radius $r = 6$ feet.

6. Find the area of the sector of a circle having a radius $r = 5$ inches and a central angle $\theta = 150°$.

Exercises 7–12: Find the exact value of each trigonometric expression by hand.

7. $\sin 30°$

8. $\tan 45°$

9. $\cot 60°$

10. $\cos 60°$

11. $\sec \frac{\pi}{4}$

12. $\csc \frac{\pi}{6}$

13. Find the six trigonometric functions of θ.

14. Solve triangle ABC.

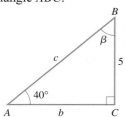

15. Find $\csc \theta$ if $\sin \theta = \frac{1}{3}$.

16. Find $\cot \theta$ if $\sin \theta = \frac{5}{13}$ and $\cos \theta = -\frac{12}{13}$.

Exercises 17 and 18: Approximate the six trigonometric functions of θ to three decimal places.

17. $\theta = 25°$

18. $\theta = -\frac{6\pi}{7}$

Exercises 19 and 20: Find the six trigonometric functions of angle θ.

19.

20.

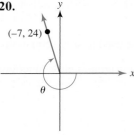

Exercises 21 and 22: The figure shows angle θ in standard position with its terminal side intersecting the unit circle. Find the six trigonometric functions of θ.

21.

22.

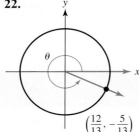

Exercises 23–26: Evaluate the function f at the given value of t.

23. $f(t) = \sin t,\ t = -\frac{\pi}{2}$ **24.** $f(t) = \cos t,\ t = \pi$

25. $f(t) = \tan t,\ t = -3\pi$ **26.** $f(t) = \csc t,\ t = \frac{\pi}{2}$

27. Find the other trigonometric functions of θ if $\sin \theta = -\frac{4}{5}$ and $\cos \theta = \frac{3}{5}$.

28. Convert $65°45'36''$ to decimal degrees.

Exercises 29–32: Graph f for $-2\pi \le t \le 2\pi$. State the amplitude, period, and phase shift.

29. $f(t) = 3 \cos 2t$ **30.** $f(t) = -2 \sin (2t + \pi)$

31. $f(t) = -3 \sin\left(\frac{1}{2}(t - \pi)\right) + 1$

32. $f(t) = \frac{1}{2} \cos\left(\frac{\pi}{2}(t - 1)\right) - 2$

Exercises 33 and 34: A graph of $y = a \cos bx$ is shown, where b is a positive constant. Estimate the values for a and b.

33.

34.

Exercises 35 and 36: Graph f for $-2\pi \le t \le 2\pi$. State the period and phase shift.

35. $f(t) = \cot 2t$ **36.** $f(t) = \sec 2t$

Exercises 37 and 38: If possible, evaluate each of the following in both radians and degrees.

37. (a) $\sin^{-1}(-1)$ (b) $\arccos \frac{1}{2}$ (c) $\tan^{-1} 1$

38. (a) $\arcsin 3$ (b) $\cos^{-1} 0$ (c) $\arctan\left(-\sqrt{3}\right)$

39. Approximate the following to a hundredth of a radian and a tenth of a degree.
    (a) $\sin^{-1}(-0.6)$ (b) $\tan^{-1} 5$ (c) $\arccos 0.12$

40. Evaluate each expression.
    (a) $\sin(\sin^{-1} 0.5)$ (b) $\tan^{-1}(\tan 45°)$

Exercises 41 and 42: Solve the right triangle ABC.

41.

42.

Exercises 43 and 44: Solve the equation for θ, where $0° \le \theta \le 90°$.

43. $\tan \theta = \dfrac{1}{\sqrt{3}}$ **44.** $\cos \theta = 1$

Exercises 45 and 46: Solve the equation for θ, where θ is an acute angle. Approximate θ to the nearest tenth of a degree.

45. $\cos \theta = \frac{1}{5}$ **46.** $3 \sin \theta = \frac{15}{13}$

Exercises 47 and 48: Solve the equation for t, where t is a real number located in the indicated interval. Approximate t to four decimal places.

47. $\tan t = -\frac{3}{4}, \left(-\frac{\pi}{2}, \frac{\pi}{2}\right)$ **48.** $\sin t = -\frac{3}{5}, \left[-\frac{\pi}{2}, \frac{\pi}{2}\right]$

Applications

49. *Ferris Wheel* A Ferris wheel has a diameter of 50 feet and completes 1 revolution every 50 seconds.
(a) Find the angular velocity of the Ferris wheel in radians per second.

(b) What is the linear speed in feet per second of a person who is riding this Ferris wheel?

50. *Fan Speed* The blades of a fan have a 25-inch diameter and rotate at 400 revolutions per minute.
(a) Find the angular velocity of a fan blade.

(b) Estimate the linear speed in inches per second at the tip of a fan blade.

51. *Height of a Tree* Eighty feet from the trunk of a tree on level ground, the angle of elevation of the top of the tree is 48°. Estimate the height of the tree to the nearest foot.

52. *Angle of Elevation* Find the angle of elevation of the top of a 35-foot building at a horizontal distance of 52 feet.

53. *Grade Resistance* Approximate θ to the nearest tenth of a degree for the given grade resistance F and vehicle weight W, where $F = W \sin \theta$.
(a) $F = 350$ lb, $W = 6000$ lb

(b) $F = 160$ lb, $W = 4500$ lb

54. *Highway Grade* (Refer to Exercise 53.) Suppose an uphill grade of a highway can be modeled by the line $y = 0.05x$.
(a) Find the grade of the hill.

(b) Determine the grade resistance for a gravel truck that weighs 30,000 pounds.

55. *Distance between Cities* Cheyenne, Wyoming, and Colorado Springs, Colorado, have nearly the same longitude of 104°45′ W. The latitude of Cheyenne is 41°09′, and the latitude of Colorado Springs is 38°49′. Approximate the distance between these two cities if the average radius of Earth is 3955 miles. (**Source:** J. Williams, *The Weather Almanac.*)

56. *Safe Distance for a Tree* From a distance of 45 feet from the base of a tree, the angle of elevation to the top of the tree is 57°, as shown in the figure. A building is located

52 feet from the base of the tree. Determine if the tree could fall in a storm and damage the building.

57. *Modeling Temperatures* The monthly average low temperatures in Green Bay, Wisconsin, are shown in the table.

Month	1	2	3	4	5	6
Temperature (°F)	6	10	22	35	45	52

Month	7	8	9	10	11	12
Temperature (°F)	58	56	48	38	26	11

Source: A. Miller and J. Thompson, *Elements of Meteorology.*

(a) Plot the monthly average temperature over a 24-month period by letting $x = 1$ and $x = 13$ correspond to January.

(b) Use $f(x) = a \cos (b(x - c)) + d$ to model the data.

(c) Graph f together with the data.

58. *Phases of the Moon* If the phase angle of the moon is θ, then the phase F of the moon is given by

$$F = \frac{1}{2}(1 - \cos \theta).$$

Solve this equation for θ.

EXTENDED AND DISCOVERY EXERCISES

1. *Surveying* The first fundamental problem of surveying is to determine the coordinates of a point Q given the coordinates of a point P, the distance between P and Q, and the bearing θ from P to Q. See the figure. (**Source:** I. Mueller and K. Ramsayer, *Introduction to Surveying.*)

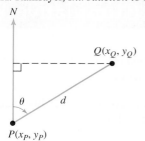

(a) Find a formula for the coordinates (x_Q, y_Q) of the point Q given θ, the coordinates (x_P, y_P) of P, and the distance d between P and Q.

(b) Use your formula to find the coordinates (x_Q, y_Q) if $(x_P, y_P) = (152, 186)$, $\theta = 23.2°$, and $d = 208$ feet.

2. *Highway Grade* (Refer to Example 4, Section 6.3.) Complete the table for the trigonometric values of θ to five decimal places.

θ	$\sin \theta$	$\tan \theta$
0°		
1°		
2°		
3°		
4°		

(a) How do $\sin \theta$ and $\tan \theta$ compare for small values of θ?

(b) Highway grades are usually small. Give an approximation to the grade resistance given by $F = W \sin \theta$ that uses the tangent function instead of the sine function.

(c) A stretch of highway has a 4-foot vertical rise for every 100 feet of horizontal run. Use your approximation to estimate the grade resistance for a 3000-pound car.

(d) Compare your result to the exact answer using $F = W \sin \theta$.
(**Source:** F. Mannering and W. Kilareski, *Principles of Highway Engineering and Traffic Analysis.*)

3. *Average Temperature* The maximum monthly average temperature in Buenos Aires, Argentina, is 74°F and the minimum monthly average temperature is 49°F.

(a) Using these two temperatures, find values for a, b, c, and d so that

$$f(x) = a \cos (b(x - c)) + d$$

models the monthly average temperature.

(b) On the same coordinate axes, graph f for a 2-year period together with the actual data values found in the table. Are your results as good as you expected? Explain.

(c) Buenos Aires is located in the Southern Hemisphere. Discuss the effect that this has on the graph of f compared to a city in the Northern Hemisphere.

Month	1	2	3	4	5	6
Temperature (°F)	74	73	69	61	55	50

Month	7	8	9	10	11	12
Temperature (°F)	49	51	55	60	66	71

Source: A. Miller and J. Thompson, *Elements of Meteorology.*

4. Suppose two cities, one in the Northern Hemisphere and one in the Southern Hemisphere, both have monthly average temperatures that can be modeled by

$$f(x) = a \cos (b(x - c)) + d,$$

where x is the month. Suppose also that for both cities the warmest monthly average temperature is 80°F and the coldest is 40°F.

(a) Find $f(x)$ for each city. (Assume that the warmest month for the northern city is July and the coldest month is January, and the opposite is true for the southern city.)

(b) Explain how $f(x)$ for one city could be used to obtain $f(x)$ for the other city.

1–6 Cumulative Review Exercises

1. Write 125,000 in scientific notation and 4.67×10^{-3} in standard notation.

2. Find the midpoint of the line segment connecting $(-3, 2)$ and $(-1, 6)$.

3. Graph $y = g(x)$ by hand.
   (a) $g(x) = 2x - 3$
   (b) $g(x) = |x + 2|$
   (c) $g(x) = 2^x$
   (d) $g(x) = \sqrt{x - 2}$
   (e) $g(x) = \dfrac{1}{x + 1}$
   (f) $g(x) = x^3$

Exercises 4 and 5: Complete the following.
   (a) Determine the domain of f.
   (b) Evaluate f(−1) and f(2a).

4. $f(x) = \sqrt{4 - x}$ **5.** $f(x) = \dfrac{x - 2}{4x^2 - 16}$

6. The graph of a linear function f is shown.
   (a) Identify the slope, y-intercept, and x-intercept.

   (b) Write a formula for f(x).

   (c) Evaluate f(−2) symbolically and graphically.

   (d) Find any zeros of f.

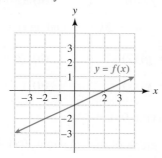

7. Find the average rate of change of $f(x) = 10^x$ from $x = 1$ to $x = 2$.

8. Find the difference quotient for $f(x) = 3x^2$.

9. Write the slope-intercept form of a line that passes through $(2, -3)$ and is parallel to the line $2x + 3y = 6$.

10. Determine the x- and y-intercepts on the graph of the equation $-2x + 5y = 20$.

11. *Solve each equation.*
   (a) $|4 - 5x| = 8$ **(b)** $2e^x - 1 = 27$

   (c) $x^3 - 3x^2 + 2x = 0$ **(d)** $x^4 + 8 = 6x^2$

   (e) $x^2 - x - 2 = 0$ **(f)** $\sqrt{2x - 1} = x - 2$

   (g) $\dfrac{x}{x - 2} = \dfrac{2x - 1}{x + 1}$ **(h)** $\log_2 (x + 1) = 16$

12. Graph f. Is f continuous on its domain? Evaluate f(1).
$$f(x) = \begin{cases} 1 - x & \text{if } -4 \le x \le -1 \\ -2x & \text{if } -1 < x < 2 \\ \frac{1}{2}x^2 & \text{if } 2 \le x \le 4 \end{cases}$$

13. Solve the inequality. Write the solution set in interval notation.
   (a) $-3(2 - x) < 4 - (2x + 1)$

   (b) $-3 \le 4 - 3x < 6$

   (c) $|4x - 3| \ge 9$ **(d)** $x^2 - 5x + 4 \le 0$

   (e) $t^3 - t > 0$ **(f)** $\dfrac{1}{t + 2} - 3 \ge 0$

14. Write $f(x) = -2x^2 + 6x - 1$ in the vertex form given by $f(x) = a(x - h)^2 + k$.

15. Find the vertex on the graph of $f(x) = 3x^2 - 4x + 1$.

16. A graph of a quadratic function $f(x) = ax^2 + bx + c$ is shown.
   (a) State whether $a > 0$ or $a < 0$.

   (b) Estimate the real solutions to $ax^2 + bx + c = 0$.

   (c) Determine if the discriminant, $b^2 - 4ac$, is positive, negative, or zero.

   (d) Write the complete factored form of $f(x)$.

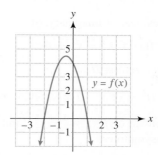

17. Solve $2x^2 + 4x = 1$ by completing the square.

18. Use the given graph of $y = f(x)$ to sketch a graph of each expression.
   (a) $y = f(x - 1) + 2$ **(b)** $y = \frac{1}{2}f(x)$

   (c) $y = -f(-x)$ **(d)** $y = f(2x)$

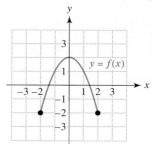

19. Use the graph of f to estimate each of the following.
   (a) Where f is increasing or decreasing

   (b) The zeros of f

   (c) The coordinates of any turning points

   (d) Any local extrema

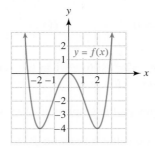

20. Predict the end behavior of $f(x) = -x^4 + 2x + 5$.

21. Divide each expression.

   (a) $\dfrac{5a^4 - 2a^2 + 4}{2a^2}$ **(b)** $\dfrac{x^3 - 3x^2 + x + 1}{x^2 + 1}$

22. Write the complete factored form for the polynomial given by $f(x) = 2x^3 + x^2 - 8x - 4$.

23. Find all zeros, real or imaginary, of

$$f(x) = x^3 - x^2 + 4x - 4,$$

given that one zero is $2i$.

24. Write $(2 + i)^2 + (2 - 3i)$ in standard form.

25. State the domain of $f(x) = \frac{2x + 5}{3x - 7}$. Find any vertical or horizontal asymptotes.

26. Write $x^{2/3}$ in radical notation. Evaluate the expression for $x = 27$.

27. Use the tables to evaluate each expression, if possible.

x	0	1	2	3	4
$f(x)$	4	3	2	1	0

x	0	1	2	3	4
$g(x)$	0	4	3	2	1

   (a) $(f + g)(2)$ **(b)** $(g/f)(4)$

   (c) $(f \circ g)(3)$ **(d)** $(f^{-1} \circ g)(1)$

28. Use the graphs of f and g to complete the following.

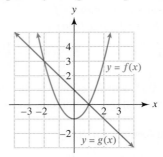

   (a) $(f - g)(-1)$ **(b)** $(fg)(2)$

   (c) $(g \circ f)(0)$ **(d)** $(g^{-1} \circ f)(2)$

29. Let $f(x) = x^2 + 3x - 2$ and $g(x) = x - 2$. Find the following.

   (a) $(f + g)(2)$ **(b)** $(g \circ f)(1)$

   (c) $(f - g)(x)$ **(d)** $(f \circ g)(x)$

30. Find $f^{-1}(x)$ if $f(x) = 2\sqrt[3]{x} + 1$.

31. Sketch a graph of the equations $f(x) = 3x - 2, y = x$, and $y = f^{-1}(x)$.

32. Find either a linear or an exponential function f that models the data in the table.

x	0	1	2	3	4
$f(x)$	9	6	4	$\frac{8}{3}$	$\frac{16}{9}$

33. Use the graph of $y = Ca^x$ to determine values for C and a.

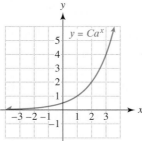

34. Five hundred dollars is deposited in an account that pays 5% annual interest compounded monthly. Find the amount in the account after 10 years.

35. Simplify each logarithm by hand.
   (a) $\log 100$ **(b)** $\log_2 16$

   (c) $\ln \dfrac{1}{e^2}$ **(d)** $\log_4 \sqrt[3]{4}$

   (e) $\log 4 + \log 25$ **(f)** $\log_6 24 - \log_6 4$

36. Write $3 \log x - 4 \log y + \frac{1}{2} \log z$ as a logarithm of a single expression.

37. Approximate $\log_3 125$ to three decimal places.

38. Solve each equation.
   (a) $3(2)^{-2x} + 4 = 100$ **(b)** $2 \log_3 3x = 4$

   (c) $\log 4x + \log \left(x + \frac{7}{4}\right) = \log 2$

39. Find the complementary angle α and the supplementary angle β to $10°34'$.

40. Convert $152.24°$ to degrees, minutes, and seconds.

41. Convert $150°$ to radians.

42. Convert $\frac{5\pi}{4}$ radians to degrees.

43. Find the length of the arc intercepted by a central angle of $15°$ in a circle with a radius of 3 feet.

44. Find the six trigonometric functions of θ.

45. Find the length of a in the figure.

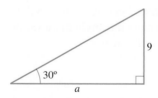

46. Approximate $\sec (1.24)$ to the nearest hundredth.

47. Find the six trigonometric functions of θ.

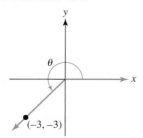

48. Find the exact values of the six trigonometric functions of $\theta = \frac{7\pi}{6}$.

49. Find the exact values of the six trigonometric functions of θ if $\cos \theta = \frac{5}{13}$ and $\sin \theta < 0$.

50. State the amplitude, period, phase shift, and vertical shift for the graph of $f(x) = 5 \sin \left(\frac{\pi}{2} (x - 1)\right) - 2$.

51. Sketch a graph of f on the interval $[-2\pi, 2\pi]$.
   (a) $f(x) = 2 \sin 3x$ **(b)** $f(x) = \sec x$

   (c) $f(x) = -2 \cos \left(2x + \frac{\pi}{2}\right) + 1$

52. Evaluate $\sin^{-1} \left(-\frac{1}{2}\right)$.

53. Solve the right triangle.

54. Solve $2 \cos \theta = -1$ if $0 \le \theta \le 180°$.

Applications

55. *Average Rate of Change*   The total distance D in feet that an object falls after t seconds is given by the formula $D(t) = 16t^2$ for $0 \le t \le 3$.
   (a) Find the average rate of change of D from 0 to 1 and 1 to 2.

   (b) Interpret these average rates of change.

   (c) Find the difference quotient of D.

56. *Working Together*   Suppose one person can mow a large lawn in 4 hours and another person can mow the same lawn in 6 hours. How long will it take to mow the lawn if they work together?

57. *Inverse Variation*   The force of gravity F varies inversely as the square of the distance d from the *center* of Earth. If a person weighs 150 pounds on the surface of Earth ($d = 4000$ miles), how much would this person weigh at a distance of 10,000 miles from the center of Earth?

58. *Linear Programming*   Find the minimum value of $C = 3x + y$ subject to the following constraints.

$$x + y \ge 1$$
$$2x + 3y \le 6$$
$$x \ge 0, y \ge 0$$

59. *College Tuition*   In 1980 the annual cost of tuition and fees at public colleges and universities was about $800, and in 2000 it was about $3500.   (**Source:** The College Board.)
  (a) Determine a formula, $C(t) = m(t - t_1) + y_1$, that models these data. Let C be in dollars and t be the year.

  (b) Estimate the year when tuition and fees reached $3000, and compare the estimate to the actual year, 1996.

60. *Volume of a Balloon*   The radius r in inches of a spherical balloon after t seconds is given by $r = \sqrt{t}$.
  (a) Is the radius increasing or decreasing?

  (b) Write a formula for a function V that calculates the volume of the sphere after t seconds.

  (c) Evaluate $V(4)$ and interpret the result.

61. *Inverse Function*   The function given by the formula $f(x) = \frac{5}{9}(x - 32)$ converts degrees Fahrenheit to degrees Celsius.
  (a) Find $f^{-1}(x)$.

  (b) What does f^{-1} compute?

62. *Bacteria Growth*   There are initially 200,000 bacteria per milliliter in a sample. The number of bacteria grows exponentially and reaches 300,000 per milliliter after 3 hours.
  (a) Use the formula $N(t) = N_0 e^{kt}$ to model the concentration of bacteria after t hours.

  (b) Evaluate $N(5)$ and interpret the result.

  (c) After how long did the concentration reach 500,000 per milliliter?

63. *Bicycle*   A bicycle has a tire that is 24 inches in diameter and is rotating at 20 radians per second. Approximate the speed of the bicycle in feet per second.

64. *Length of a Shadow*   If the angle of elevation of the sun is 43°, find the length of the shadow cast by a person who is 6 feet tall. Round your answer to the nearest tenth of a foot.

65. *Modeling Temperature*   The monthly average high temperatures in New Orleans are shown in the table. Model these data by using a function of the form

$$f(x) = a \sin(b(x - c)) + d.$$

Month	1	2	3	4	5	6
Temperature (°F)	61	65	70	78	85	89

Month	7	8	9	10	11	12
Temperature (°F)	90	89	88	80	70	63

Source: J. Williams, *The Weather Almanac.*

66. *Angle of Elevation*   A 52-foot building cast a 63-foot shadow. Estimate the angle of elevation of the sun to the nearest tenth of a degree.

Trigonometric Identities and Equations

Music is both art and science. During the Greek and Roman eras, music played an important role in philosophy and science. Although Pythagoras is usually associated with the Pythagorean theorem, in 500 B.C. he also discovered the mathematical relationships between lengths of strings and musical intervals. This discovery of the mathematical ratios that governed pitch and motion was the beginning of the science of musical sound. In the Middle Ages, music was studied with arithmetic, geometry, and astronomy as part of the liberal arts curriculum. In 1862, the psychologist and scientist Hermann von Helmholtz published his classic work that opened a new direction for music using mathematics and technology. Then in 1957, Max Mathews created complex musical sounds with a computer. Mathematics has had an essential role in the development and reproduction of music.

The communications industry also uses mathematics. For example, each number on a touch-tone phone has a unique sound that is a combination of two different tones. These unique tones are easily distinguished when transmitted over phone lines. Trigonometric functions play an important role in clear, reliable communication systems.

Many applications of mathematics began as theoretical mathematics. This is why both pure and applied mathematics are important to society. This chapter introduces the concept of verifying an identity. Identities are important because they allow us to write trigonometric expressions in simpler and more convenient forms.

Music is the pleasure the human mind experiences from counting without being aware that it is counting.

—Gottfried Leibniz

Source: J. Pierce, *The Science of Musical Sound.*

7.1 Fundamental Identities

- **Learn and apply the reciprocal and quotient identities**
- **Learn and apply the Pythagorean identities**
- **Learn and apply the negative-angle identities**

Introduction

Trigonometric expressions can often be written in more than one way. For example, in Section 6.4 we discussed that $\cot \theta$ is equivalent to $\frac{\cos \theta}{\sin \theta}$. The equation

$$\cot \theta = \frac{\cos \theta}{\sin \theta}$$

is a *trigonometric identity*. This identity is true for every value of θ, provided $\sin \theta \neq 0$. Trigonometric identities are used to help solve equations and model physical phenomena in music, science, and electronics. They are also used in calculus. We begin our discussion with the reciprocal and quotient identities.

Reciprocal and Quotient Identities

In Section 6.4, the following definitions were presented for the trigonometric functions of any angle θ. See Figure 7.1.

Figure 7.1

Trigonometric Functions of Any Angle θ

Let (x, y) be a point other than the origin on the terminal side of an angle θ in standard position. If $r = \sqrt{x^2 + y^2}$, then the six trigonometric functions are as follows.

$$\sin \theta = \frac{y}{r} \qquad \csc \theta = \frac{r}{y} \ (y \neq 0)$$

$$\cos \theta = \frac{x}{r} \qquad \sec \theta = \frac{r}{x} \ (x \neq 0)$$

$$\tan \theta = \frac{y}{x} \ (x \neq 0) \qquad \cot \theta = \frac{x}{y} \ (y \neq 0)$$

NOTE These definitions are valid whether angle θ is measured in radians or degrees.

These definitions allow us to write several identities. For example, since

$$\cos \theta = \frac{x}{r} \qquad \text{and} \qquad \sec \theta = \frac{r}{x},$$

it follows that

$$\sec \theta = \frac{1}{x/r} = \frac{1}{\cos \theta} \qquad \text{and} \qquad \cos \theta = \frac{1}{r/x} = \frac{1}{\sec \theta}.$$

These identities are examples of *reciprocal identities*.

Reciprocal Identities

$$\sin \theta = \frac{1}{\csc \theta} \qquad \cos \theta = \frac{1}{\sec \theta} \qquad \tan \theta = \frac{1}{\cot \theta}$$

$$\csc \theta = \frac{1}{\sin \theta} \qquad \sec \theta = \frac{1}{\cos \theta} \qquad \cot \theta = \frac{1}{\tan \theta}$$

For an example of a reciprocal identity, let $\csc \theta = \frac{5}{3}$. Because $\sin \theta = \frac{1}{\csc \theta}$, it follows that $\sin \theta = \frac{3}{5}$. In the next example, we use two reciprocal identities to simplify a trigonometric expression.

EXAMPLE 1 Applying reciprocal identities

Use reciprocal identities to rewrite

$$\frac{1}{\csc^2 \theta} + \frac{1}{\sec^2 \theta}$$

in terms of $\sin \theta$ and $\cos \theta$ and then simplify if possible.

SOLUTION

Getting Started First note that $\csc^2 \theta = (\csc \theta)^2$ and $\sec^2 \theta = (\sec \theta)^2$. The reciprocal identities $\csc \theta = \frac{1}{\sin \theta}$ and $\sec \theta = \frac{1}{\cos \theta}$ can be used to rewrite the given identity as follows. ▶

$$
\begin{aligned}
\frac{1}{\csc^2 \theta} + \frac{1}{\sec^2 \theta} &= \left(\frac{1}{\csc \theta}\right)^2 + \left(\frac{1}{\sec \theta}\right)^2 && \text{Rewrite: } \frac{1}{x^2} = \left(\frac{1}{x}\right)^2. \\
&= (\sin \theta)^2 + (\cos \theta)^2 && \text{Reciprocal identities} \\
&= \sin^2 \theta + \cos^2 \theta && \text{Rewrite.} \\
&= 1 && \text{Identity}
\end{aligned}
$$

(The identity $\sin^2 \theta + \cos^2 \theta = 1$ was introduced in Section 6.3 on page 511. We discuss this identity later in this section.)   ***Now Try Exercise 11*** ◀

NOTE When writing the square of $\sin \theta$, it is common to write $\sin^2 \theta$ instead of $(\sin \theta)^2$. This is also true for the other trigonometric functions.

Because $\sin \theta = \frac{y}{r}$ and $\cos \theta = \frac{x}{r}$, it is possible to write the other four trigonometric functions in terms of $\sin \theta$ and $\cos \theta$. For example,

$$\tan \theta = \frac{y}{x} = \frac{y/r}{x/r} = \frac{\sin \theta}{\cos \theta}.$$

This identity is an example of a *quotient identity*.

Quotient Identities

$$\tan \theta = \frac{\sin \theta}{\cos \theta} \qquad \cot \theta = \frac{\cos \theta}{\sin \theta}$$

NOTE The reciprocal and quotient identities are valid not only for angles but also for any real number t. For example, $\tan t = \frac{\sin t}{\cos t}$ for all real numbers t such that $\cos t \neq 0$.

Evaluating Trigonometric Functions   If $\sin \theta$ and $\cos \theta$ are known, the reciprocal and quotient identities can be used to find the other four trigonometric functions of θ, as illustrated in the next example.

EXAMPLE 2 Using reciprocal and quotient identities

If $\sin \theta = \frac{7}{25}$ and $\cos \theta = -\frac{24}{25}$, find the other four trigonometric functions of θ.

SOLUTION We can use identities to find $\tan \theta$, $\cot \theta$, $\sec \theta$, and $\csc \theta$.

$$\tan \theta = \frac{\sin \theta}{\cos \theta} = \frac{7/25}{-24/25} = -\frac{7}{24}$$

$$\cot \theta = \frac{\cos \theta}{\sin \theta} = \frac{-24/25}{7/25} = -\frac{24}{7}$$

$$\sec \theta = \frac{1}{\cos \theta} = \frac{1}{-24/25} = -\frac{25}{24}$$

$$\csc \theta = \frac{1}{\sin \theta} = \frac{1}{7/25} = \frac{25}{7}$$

Now Try Exercise 13 ◀

EXAMPLE 3 Using identities to find trigonometric values

If $\tan \theta = -\frac{8}{15}$ and $\cos \theta = -\frac{15}{17}$, find the other four trigonometric functions of θ.

SOLUTION Using the reciprocal identities, we can find $\cot \theta$ and $\sec \theta$.

$$\cot \theta = \frac{1}{\tan \theta} = \frac{1}{-8/15} = -\frac{15}{8}$$

$$\sec \theta = \frac{1}{\cos \theta} = \frac{1}{-15/17} = -\frac{17}{15}$$

To find $\sin \theta$, consider the following.

$$\tan \theta \cos \theta = \frac{\sin \theta}{\cos \theta} \cdot \cos \theta = \sin \theta$$

Thus

$$\sin \theta = \tan \theta \cos \theta = \left(-\frac{8}{15}\right)\left(-\frac{15}{17}\right) = \frac{8}{17}$$

and, using a reciprocal identity, we find

$$\csc \theta = \frac{1}{\sin \theta} = \frac{1}{8/17} = \frac{17}{8}.$$

Now Try Exercise 15 ◀

Pythagorean Identities

In Section 6.3, the unit circle (shown in Figure 7.2) was used to define the sine and cosine functions.

$$\sin \theta = y \quad \text{and} \quad \cos \theta = x$$

An equation for the unit circle is given by $x^2 + y^2 = 1$. By substitution, it follows that

$$(\cos \theta)^2 + (\sin \theta)^2 = 1,$$

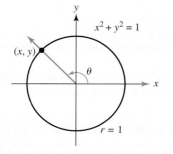

Figure 7.2 Unit Circle

or equivalently, $\cos^2 \theta + \sin^2 \theta = 1$. This identity can be supported graphically and numerically by letting

$$Y_1 = (\cos(X))^2 + (\sin(X))^2,$$

as shown in Figures 7.3 and 7.4. The graph of Y_1 is the horizontal line $y = 1$. Either radian or degree mode may be used. Notice that regardless of the value of x, $Y_1 = 1$.

[−352.5, 352.5, 90] by [−2, 2, 1]

Figure 7.3 Degree Mode   **Figure 7.4** Degree Mode   **Figure 7.5** Degree Mode

NOTE On some older graphing calculators, $\cos^2 x$ must be entered as $(\cos(X))^2$, rather than $\cos(X)^2$, since $\cos(X)^2$ may be interpreted as $\cos(x^2)$. See Figure 7.5.

Using the right triangle shown in Figure 7.6, we also can verify the Pythagorean identity $\cos^2 \theta + \sin^2 \theta = 1$ geometrically when θ is acute.

Figure 7.6

$$x^2 + y^2 = r^2 \qquad \text{Pythagorean theorem}$$

$$\frac{x^2}{r^2} + \frac{y^2}{r^2} = \frac{r^2}{r^2} \qquad \text{Divide each side by } r^2.$$

$$\left(\frac{x}{r}\right)^2 + \left(\frac{y}{r}\right)^2 = 1 \qquad \text{Properties of exponents}$$

$$\cos^2 \theta + \sin^2 \theta = 1 \qquad \cos\theta = \tfrac{x}{r}, \sin\theta = \tfrac{y}{r}$$

For this reason, $\sin^2 \theta + \cos^2 \theta = 1$ is an example of a Pythagorean identity. Two other Pythagorean identities can be derived as follows.

$$\sin^2 \theta + \cos^2 \theta = 1$$

$$\frac{\sin^2 \theta}{\cos^2 \theta} + \frac{\cos^2 \theta}{\cos^2 \theta} = \frac{1}{\cos^2 \theta} \qquad \text{Divide by } \cos^2\theta.$$

$$\tan^2 \theta + 1 = \sec^2 \theta \qquad \tan\theta = \tfrac{\sin\theta}{\cos\theta}, \sec\theta = \tfrac{1}{\cos\theta}$$

In a similar manner, a third Pythagorean identity can be found.

$$\sin^2 \theta + \cos^2 \theta = 1$$

$$\frac{\sin^2 \theta}{\sin^2 \theta} + \frac{\cos^2 \theta}{\sin^2 \theta} = \frac{1}{\sin^2 \theta} \qquad \text{Divide by } \sin^2\theta.$$

$$1 + \cot^2 \theta = \csc^2 \theta \qquad \cot\theta = \tfrac{\cos\theta}{\sin\theta}, \csc\theta = \tfrac{1}{\sin\theta}$$

This discussion is summarized in the following box.

CLASS DISCUSSION

Use the triangle in Figure 7.6 to justify the Pythagorean identity

$$1 + \tan^2 \theta = \sec^2 \theta.$$

Pythagorean Identities

$$\sin^2 \theta + \cos^2 \theta = 1 \qquad 1 + \tan^2 \theta = \sec^2 \theta \qquad 1 + \cot^2 \theta = \csc^2 \theta$$

In the next example, we use a reciprocal, quotient, and Pythagorean identity to simplify a trigonometric expression.

EXAMPLE 4 Applying identities to an expression

Simplify the expression $1 + \sin^2 \theta \sec^2 \theta$.

SOLUTION Begin by applying a reciprocal identity.

$$
\begin{aligned}
1 + \sin^2 \theta \sec^2 \theta &= 1 + \sin^2 \theta \cdot \frac{1}{\cos^2 \theta} && \text{Reciprocal identity: } \sec \theta = \frac{1}{\cos \theta} \\
&= 1 + \frac{\sin^2 \theta}{\cos^2 \theta} && \text{Multiply.} \\
&= 1 + \left(\frac{\sin \theta}{\cos \theta} \right)^2 && \text{Properties of exponents} \\
&= 1 + \tan^2 \theta && \text{Quotient identity: } \tan \theta = \frac{\sin \theta}{\cos \theta} \\
&= \sec^2 \theta && \text{Pythagorean identity: } 1 + \tan^2 \theta = \sec^2 \theta
\end{aligned}
$$

Now Try Exercise 33 ◀

An Application The next example illustrates the use of identities in electronics.

EXAMPLE 5 Applying a Pythagorean identity to radios

Tuners in radios select a radio station by adjusting the frequency. These tuners may contain an inductor L and a capacitor C, as illustrated in Figure 7.7. The energy stored in the inductor at time t is given by $L(t) = k \sin^2 (2\pi F t)$, and the energy stored in the capacitor is given by $C(t) = k \cos^2(2\pi F t)$, where F is the frequency of the radio station and k is a constant. The total energy E in the circuit is given by $E(t) = L(t) + C(t)$. Show that E is a constant function. (**Source:** R. Weidner and R. Sells, *Elementary Classical Physics*, Vol. 2.)

Figure 7.7 An Inductor L and a Capacitor C

SOLUTION

$$
\begin{aligned}
E(t) &= L(t) + C(t) && \text{Given equation} \\
&= k \sin^2 (2\pi F t) + k \cos^2 (2\pi F t) && \text{Substitute.} \\
&= k (\sin^2 (2\pi F t) + \cos^2 (2\pi F t)) && \text{Factor.} \\
&= k (1) && \sin^2 \theta + \cos^2 \theta = 1 \, (\theta = 2\pi F t) \\
&= k && k \text{ is constant.}
\end{aligned}
$$

Now Try Exercise 85 ◀

Figure 7.8 Angle θ in Quadrant II

Quadrants and Signs of Trigonometric Functions If an angle θ is in standard position and its terminal side lies in quadrant II, as shown in Figure 7.8, then θ is *contained in quadrant II*, or θ is a *second quadrant angle*. Similar statements can be made for angles whose terminal sides lie in other quadrants.

Let the point (x, y) lie on the terminal side of angle θ with $r = \sqrt{x^2 + y^2} > 0$. Then $\sin \theta = \frac{y}{r}$ is positive when $y > 0$ and negative when $y < 0$. As a result, $\sin \theta$ is positive for first and second quadrant angles and negative for third and fourth quadrant angles.

Similarly, a point (x, y) has a positive x-coordinate in either quadrant I or IV and a negative x-coordinate in either quadrant II or III. Thus $\cos \theta = \frac{x}{r}$ is positive in quadrants I and IV and negative in quadrants II and III. The signs of the six trigonometric functions for angles in each quadrant are listed in Figure 7.9. Notice that the six trigonometric functions are all positive for angles in the first quadrant.

QUADRANT II

$\sin \theta > 0$
$\cos \theta < 0$
$\tan \theta < 0$
$\csc \theta > 0$
$\sec \theta < 0$
$\cot \theta < 0$

QUADRANT I

$\sin \theta > 0$
$\cos \theta > 0$
$\tan \theta > 0$
$\csc \theta > 0$
$\sec \theta > 0$
$\cot \theta > 0$

QUADRANT III

$\sin \theta < 0$
$\cos \theta < 0$
$\tan \theta > 0$
$\csc \theta < 0$
$\sec \theta < 0$
$\cot \theta > 0$

QUADRANT IV

$\sin \theta < 0$
$\cos \theta > 0$
$\tan \theta < 0$
$\csc \theta < 0$
$\sec \theta > 0$
$\cot \theta < 0$

Figure 7.9

EXAMPLE 6 Finding the quadrant containing an angle

If $\sin \theta > 0$ and $\cos \theta < 0$, find the quadrant that contains θ. Support your results graphically and numerically.

SOLUTION If $\sin \theta > 0$ and $\cos \theta < 0$, then any point (x, y) on the terminal side of θ must satisfy $y > 0$ and $x < 0$. Thus θ is contained in quadrant II.

Graphical Support The graphs of $Y_1 = \sin (X)$ and $Y_2 = \cos (X)$ are shown in Figure 7.10. Notice that when $\theta = 135°$ (a second quadrant angle), the graph of $\sin \theta$ is above the x-axis and the graph of $\cos \theta$ is below the x-axis.

Numerical Support Numerical support is given in Figure 7.11, where angles in quadrant II have positive sine values listed under Y_1 and negative cosine values listed under Y_2.

[0, 352.5, 90] by [−2, 2, 1]

$y_1 = \sin x$

$y_2 = \cos x$

X=135   Y=.70710678

Figure 7.10 Degree Mode

X	Y₁	Y₂
90	1	0
105	.96593	⁻.2588
120	.86603	⁻.5
135	.70711	⁻.7071
150	.5	⁻.866
165	.25882	⁻.9659
180	0	⁻1

X=90

Figure 7.11 Degree Mode

Now Try Exercise 47 ◀

Evaluating Trigonometric Functions If a trigonometric function of θ is known and the quadrant containing θ is also known, then we can find the other five trigonometric functions of θ, as illustrated in the next example.

EXAMPLE 7 Using identities to find trigonometric values

If $\sin\theta = -\frac{3}{5}$ and θ is a third quadrant angle, find the values of the other trigonometric functions.

SOLUTION

Getting Started If we can find $\cos\theta$, then we can determine the other four trigonometric functions by using reciprocal and quotient identities. To determine $\cos\theta$, we can use the identity $\cos^2\theta + \sin^2\theta = 1$ and the fact that θ is a third quadrant angle. ▶

$$\sin^2\theta + \cos^2\theta = 1 \qquad \text{Pythagorean identity}$$

$$\cos^2\theta = 1 - \sin^2\theta \qquad \text{Subtract } \sin^2\theta.$$

$$\cos\theta = \pm\sqrt{1 - \sin^2\theta} \qquad \text{Square root property}$$

$$\cos\theta = \pm\sqrt{1 - (-3/5)^2} \qquad \text{Substitute for } \sin\theta.$$

$$\cos\theta = \pm\frac{4}{5} \qquad \text{Simplify.}$$

In quadrant III, $x < 0$, and so $\cos\theta = -\frac{4}{5}$. The other four trigonometric functions of θ can be found by letting $\sin\theta = -\frac{3}{5}$ and $\cos\theta = -\frac{4}{5}$ and applying identities.

$$\tan\theta = \frac{\sin\theta}{\cos\theta} = \frac{-3/5}{-4/5} = \frac{3}{4}$$

$$\cot\theta = \frac{\cos\theta}{\sin\theta} = \frac{-4/5}{-3/5} = \frac{4}{3}$$

$$\sec\theta = \frac{1}{\cos\theta} = \frac{1}{-4/5} = -\frac{5}{4}$$

$$\csc\theta = \frac{1}{\sin\theta} = \frac{1}{-3/5} = -\frac{5}{3} \qquad \textit{Now Try Exercise 61} \;◀$$

EXAMPLE 8 Using identities to find trigonometric values

If $\tan\theta = -\frac{7}{3}$ and $\sin\theta > 0$, find the values of the other trigonometric functions.

SOLUTION Since $\tan\theta = -\frac{7}{3}$, it follows that $\cot\theta = -\frac{3}{7}$. Next we find $\sec\theta$.

$$\sec^2\theta = 1 + \tan^2\theta \qquad \text{Pythagorean identity}$$

$$\sec\theta = \pm\sqrt{1 + \tan^2\theta} \qquad \text{Square root property}$$

$$= \pm\sqrt{1 + (-7/3)^2} \qquad \text{Substitute for } \tan\theta.$$

$$= \pm\frac{\sqrt{58}}{3} \qquad \text{Simplify.}$$

Since $\tan\theta < 0$ and $\sin\theta > 0$, θ is a second quadrant angle. It follows that

$$\sec\theta = -\frac{\sqrt{58}}{3} \quad \text{and} \quad \cos\theta = -\frac{3}{\sqrt{58}}.$$

Since $\sin \theta = \tan \theta \cos \theta$,

$$\sin \theta = \left(-\frac{7}{3}\right)\left(-\frac{3}{\sqrt{58}}\right) = \frac{7}{\sqrt{58}}.$$

Using a reciprocal identity gives

$$\csc \theta = \frac{1}{\sin \theta} = \frac{\sqrt{58}}{7}.$$

Now Try Exercise 55 ◀

We can also express any trigonometric function in terms of any other trigonometric function. This fact is demonstrated in the next example.

EXAMPLE 9 **Expressing one function in terms of another**

Write $\cos x$ and $\cot x$ in terms of $\sin x$, if $\sec x < 0$.

SOLUTION We can write $\cos x$ in terms of $\sin x$ by applying a Pythagorean identity. Note that $\sec x < 0$ implies that $\cos x < 0$ because $\cos x = \frac{1}{\sec x}$ by a reciprocal identity.

$$\sin^2 x + \cos^2 x = 1 \qquad \text{Pythagorean identity}$$
$$\cos^2 x = 1 - \sin^2 x \qquad \text{Subtract } \sin^2 x.$$
$$\cos x = \pm\sqrt{1 - \sin^2 x} \qquad \text{Square root property}$$

Because $\cos x < 0$, we let $\cos x = -\sqrt{1 - \sin^2 x}$. Now we can write $\cot x$ in terms of $\sin x$ by using a quotient identity.

$$\cot x = \frac{\cos x}{\sin x} \qquad \text{Quotient identity}$$

$$= -\frac{\sqrt{1 - \sin^2 x}}{\sin x} \qquad \text{Substitute for } \cos x.$$

Now Try Exercise 65 ◀

EXAMPLE 10 **Using fundamental identities to find trigonometric expressions**

If $\sin \theta = x$ and θ is a fourth quadrant angle, find an expression for $\sec \theta$ in terms of x. Approximate $\sec \theta$ if $\sin \theta = -0.7813$.

SOLUTION We begin by writing $\sec \theta$ in terms of $\sin \theta$.

$$\sin^2 \theta + \cos^2 \theta = 1 \qquad \text{Pythagorean identity}$$
$$\cos^2 \theta = 1 - \sin^2 \theta \qquad \text{Subtract } \sin^2 \theta.$$
$$\cos \theta = \pm\sqrt{1 - \sin^2 \theta} \qquad \text{Square root property}$$
$$\sec \theta = \pm\frac{1}{\sqrt{1 - \sin^2 \theta}} \qquad \sec \theta = \frac{1}{\cos \theta}$$
$$\sec \theta = \pm\frac{1}{\sqrt{1 - x^2}} \qquad \sin \theta = x$$

Angle θ is in quadrant IV. Thus $\sec \theta > 0$ and

$$\sec \theta = \frac{1}{\sqrt{1 - x^2}}.$$

Since $\sin \theta = x$, let $x = -0.7813$. Then

$$\sec \theta = \frac{1}{\sqrt{1 - (-0.7813)^2}} \approx 1.602. \qquad \text{\textit{Now Try Exercise 71}} \blacktriangleleft$$

Negative-Angle Identities

In Section 4.1, we discussed odd and even functions. The graphs of odd functions are symmetric with respect to the origin, and the graphs of even functions are symmetric with respect to the y-axis. Symbolically, we say that an odd function f satisfies $f(-x) = -f(x)$ for all x in the domain of f. Similarly, an even function f satisfies $f(-x) = f(x)$ for all x in its domain. The graphs of all six trigonometric functions have symmetry, as shown in Figures 7.12–7.17.

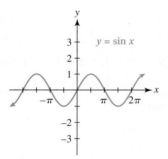

Figure 7.12 The Sine Function (An Odd Function)

Figure 7.13 The Cosecant Function (An Odd Function)

Figure 7.14 The Cosine Function (An Even Function)

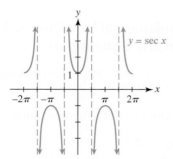

Figure 7.15 The Secant Function (An Even Function)

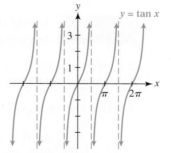

Figure 7.16 The Tangent Function (An Odd Function)

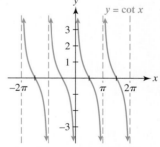

Figure 7.17 The Cotangent Function (An Odd Function)

The cosine and secant functions are even functions, and the other four trigonometric functions are odd functions. For even functions the sign of the input does not affect the output, whereas for odd functions changing the sign of the input only changes the sign of the output. These results can be expressed using the negative-angle identities.

Negative-Angle Identities

$\sin(-\theta) = -\sin\theta$	$\cos(-\theta) = \cos\theta$	$\tan(-\theta) = -\tan\theta$
$\csc(-\theta) = -\csc\theta$	$\sec(-\theta) = \sec\theta$	$\cot(-\theta) = -\cot\theta$

Negative-angle identities are often used to simplify expressions. For example, the expression $\cos(-x) + \sin(-x)$ can be simplified to $\cos x - \sin x$.

An Application   In the final example, we see how the symmetry of a trigonometric function can be used to model temperature.

EXAMPLE 11   Modeling temperature

The monthly average high temperatures in degrees Fahrenheit at Chattanooga, Tennessee, can be modeled by

$$f(x) = 21\cos\left(\frac{\pi x}{6}\right) + 70,$$

where x is the month with $x = -6$ corresponding to January, $x = 0$ to July, and $x = 5$ to December.   (**Source:** J. Williams, *The Weather Almanac.*)
(a) Graph f in $[-6, 6, 1]$ by $[40, 100, 10]$. Interpret any symmetry in the graph.
(b) Make a table of f and discuss whether f is an even or odd function.
(c) Express any symmetry symbolically.

$[-6, 6, 1]$ by $[40, 100, 10]$

$y = 21\cos\left(\frac{\pi x}{6}\right) + 70$

Figure 7.18 Radian Mode

X	Y1
‾3	70
‾2	80.5
‾1	88.187
0	91
1	88.187
2	80.5
3	70

X = 0

Figure 7.19 Radian Mode

SOLUTION
(a) Graph $Y_1 = 21\cos(\pi X/6) + 70$, as shown in Figure 7.18 in radian mode. The graph is symmetric with respect to the y-axis. This type of symmetry implies that the monthly average high temperatures x months before July and x months after July are equal. For example, April ($x = -3$), which is 3 months before July, and October ($x = 3$), which is 3 months after July, both have the same average high temperature of 70°.
(b) A table of Y_1 is shown in Figure 7.19. Notice that f is an even function since the sign of the input does not affect the output.
(c) Symbolically, this symmetry can be expressed as $f(-x) = f(x)$.

Now Try Exercise 87 ◄

7.1 Putting It All Together

This section discussed the reciprocal identities, the quotient identities, the Pythagorean identities, and the negative-angle identities. Collectively, they are called fundamental identities. If $\sin\theta$ and $\cos\theta$ are known, then the reciprocal and quotient identities can be used to find the remaining four trigonometric functions of θ. The Pythagorean identities can be used to find trigonometric values when only one trigonometric value and the quadrant are known. The negative-angle identities can be used to determine if a trigonometric function is an odd function or an even function. Cosine and secant are even functions, and sine, cosecant, tangent, and cotangent are odd functions.

continued on next page

continued from previous page

Identity	General Form		
Reciprocal	$\sin \theta = \dfrac{1}{\csc \theta}$	$\cos \theta = \dfrac{1}{\sec \theta}$	$\tan \theta = \dfrac{1}{\cot \theta}$
	$\csc \theta = \dfrac{1}{\sin \theta}$	$\sec \theta = \dfrac{1}{\cos \theta}$	$\cot \theta = \dfrac{1}{\tan \theta}$
Quotient	$\tan \theta = \dfrac{\sin \theta}{\cos \theta}$	$\cot \theta = \dfrac{\cos \theta}{\sin \theta}$	
Pythagorean	$\sin^2 \theta + \cos^2 \theta = 1$	$1 + \tan^2 \theta = \sec^2 \theta$	$1 + \cot^2 \theta = \csc^2 \theta$
Negative-angle	$\sin(-\theta) = -\sin \theta$	$\cos(-\theta) = \cos \theta$	$\tan(-\theta) = -\tan \theta$
	$\csc(-\theta) = -\csc \theta$	$\sec(-\theta) = \sec \theta$	$\cot(-\theta) = -\cot \theta$

7.1 Exercises

Reciprocal Identities

Exercises 1–6: Use a reciprocal identity to find the indicated trigonometric function of θ.

1. $\cot \theta$ if $\tan \theta = \frac{1}{2}$

2. $\csc \theta$ if $\sin \theta = -\frac{5}{6}$

3. $\sec \theta$ if $\cos \theta = \frac{2}{7}$

4. $\tan \theta$ if $\cot \theta = -\frac{3}{7}$

5. $\cos \theta$ if $\sec \theta = -4$

6. $\sin \theta$ if $\csc \theta = 7$

Exercises 7–12: Use reciprocal identities to rewrite the expression in terms of $\sin \theta$ and $\cos \theta$.

7. $\dfrac{\tan \theta}{\cot \theta}$

8. $\dfrac{\sec \theta}{\csc \theta}$

9. $\dfrac{\cot^2 \theta}{\csc^2 \theta}$

10. $\dfrac{\tan^2 \theta}{\sec^2 \theta}$

11. $\dfrac{1}{\csc \theta} + \dfrac{1}{\sec \theta}$

12. $\csc \theta \sec \theta \tan \theta$

Fundamental Identities

Exercises 13–20: Find the other trigonometric functions of θ.

13. $\sin \theta = \frac{3}{5}$ and $\cos \theta = -\frac{4}{5}$

14. $\tan \theta = -\frac{12}{5}$ and $\cos \theta = \frac{5}{13}$

15. $\cot \theta = \frac{7}{24}$ and $\sin \theta = -\frac{24}{25}$

16. $\sin \theta = \frac{12}{13}$ and $\cos \theta = \frac{5}{13}$

17. $\sin \theta = -\frac{60}{61}$ and $\cos \theta = -\frac{11}{61}$

18. $\sin \theta = \frac{2}{3}$ and $\cos \theta = -\frac{\sqrt{5}}{3}$

19. $\csc \theta = \sqrt{2}$ and $\sec \theta = -\sqrt{2}$

20. $\csc \theta = -\frac{13}{12}$ and $\sec \theta = -\frac{13}{5}$

Exercises 21–42: Simplify each expression.

21. $\sec \theta \cos \theta$

22. $\tan \theta \cot \theta$

23. $\sin \theta \csc \theta$

24. $\tan \theta \cos \theta$

25. $(\sin^2 \theta + \cos^2 \theta)^3$

26. $(1 + \tan^2 \theta) \cos^2 \theta$

27. $1 - \sin^2 \theta$

28. $1 - \cos^2(-\theta)$

29. $\sec^2 \theta - 1$

30. $1 + \dfrac{\cos^2 \theta}{\sin^2 \theta}$

31. $\dfrac{\sin(-\theta)}{\cos(-\theta)}$

32. $\sin\theta\,(\csc\theta + \sec\theta)$

33. $\dfrac{\sin^2\theta + \cos^2\theta}{\cos\theta}$

34. $\dfrac{1 + \tan^2\theta}{\sec^2\theta}$

35. $\dfrac{\sec^2(-\theta)}{\csc^2\theta}$

36. $\dfrac{1 - \cos^2\theta}{\sin^2\theta + \cos^2\theta}$

37. $\dfrac{\cot x}{\csc x}$

38. $\dfrac{\tan x}{\sec x}$

39. $(\sin^2 x)(1 + \cot^2 x)$

40. $\dfrac{\cos x}{\sin x \cot x}$

41. $\sec(-x) + \csc(-x)$

42. $-\cos(-x)\sec(-x)$

Exercises 43–46: Determine if the equation represents an identity. If you have a graphing calculator, support your answer by making a table of the left and right sides of the equation.

43. $\sec\theta\cot\theta = \csc\theta$

44. $(\sin\theta + \cos\theta)^2 = 1$

45. $\cot^2\theta - \csc^2\theta = 1$

46. $1 + \tan^2\theta = \dfrac{1}{\cos^2\theta}$

Exercises 47–52: (Refer to Example 6.) Determine the quadrant that contains θ. If you have a graphing calculator, support your result either graphically or numerically.

47. $\sin\theta < 0$ and $\cos\theta > 0$

48. $\tan\theta > 0$ and $\cos\theta < 0$

49. $\sec\theta < 0$ and $\sin\theta < 0$

50. $\csc\theta > 0$ and $\tan\theta > 0$

51. $\cot\theta < 0$ and $\sin\theta > 0$

52. $\cos\theta > 0$ and $\cot\theta < 0$

Exercises 53–64: Find the other trigonometric functions of θ.

53. $\cos\theta = \frac{1}{2}$ and $\sin\theta < 0$

54. $\csc\theta = \sqrt{3}$ and $\cos\theta < 0$

55. $\tan\theta = -\frac{11}{60}$ and $\csc\theta < 0$

56. $\sec\theta = -\frac{5}{4}$ and $\sin\theta > 0$

57. $\sin\theta = \frac{7}{25}$ and $\cos\theta > 0$

58. $\cot\theta = \frac{12}{5}$ and $\sec\theta < 0$

59. $\sin\theta = -\frac{1}{3}$ and $\sec\theta < 0$

60. $\tan\theta = -\frac{1}{2}$ and $\sin\theta > 0$

61. $\sec\theta = \frac{37}{12}$ and θ in quadrant IV

62. $\tan\theta = \frac{3}{4}$ and θ in quadrant I

63. $\csc\theta = \frac{7}{3}$ and θ in quadrant II

64. $\cos\theta = -\frac{3}{5}$ and θ in quadrant III

Writing Trigonometric Expressions

65. Write $\sin x$ and $\tan x$ in terms of $\cos x$, if $\csc x > 0$.

66. Write $\cos x$ and $\sec x$ in terms of $\sin x$, if $\cos x < 0$.

67. Write $\sin x$ and $\sec x$ in terms of $\tan x$, if $\cos x < 0$.

68. Write $\cos x$ and $\csc x$ in terms of $\cot x$, if $\sin x > 0$.

69. Write $\cot x$ and $\cos x$ in terms of $\csc x$, if $\cot x < 0$.

70. Write $\tan x$ and $\sin x$ in terms of $\sec x$, if $\tan x > 0$.

Exercises 71–74: (Refer to Example 10.) Write the trigonometric function in terms of x. Then evaluate this trigonometric function if $x = 0.5126$.

71. $\cos\theta$ if $\sin\theta = x$ and θ is acute

72. $\sec\theta$ if $\cos\theta = x$

73. $\sin\theta$ if $\cot\theta = x$ and θ is in quadrant III

74. $\tan\theta$ if $\cos\theta = x$ and θ is in quadrant IV

75. Write $\tan\theta$ in terms of x if $\sin\theta = x$ and θ is acute.

76. Write $\sin\theta$ in terms of x if $\sec\theta = x$ and θ is acute.

Negative-Angle Identities

Exercises 77–82: Use a negative-angle identity to write an equivalent trigonometric expression involving a positive angle.

77. $\sin(-13°)$

78. $\cos\left(-\frac{\pi}{7}\right)$

79. $\tan\left(-\frac{\pi}{11}\right)$

80. $\cot(-75°)$

81. $\sec\left(-\frac{2\pi}{5}\right)$

82. $\csc(-160°)$

Applications

83. *Distance to the Stars* In Example 6 of Section 6.2, the distance d to a star was found by using

$$d = \frac{93,000,000}{\sin \theta},$$

where θ is the parallax of the star. Use a reciprocal identity to rewrite this formula.

84. *Height of a Building* If the angle of elevation of the sun is θ, then a building 40 feet high will cast a shadow x feet long, where $x = \frac{40}{\tan \theta}$. Use a reciprocal identity to rewrite this formula.

85. *Radio Tuners* (Refer to Example 5.) Let the energy stored in the inductor L be given by the formula $L(t) = 3 \cos^2 (6,000,000t)$ and the energy in the capacitor C be given by $C(t) = 3 \sin^2 (6,000,000t)$, where t is time in seconds. The total energy E in the circuit is given by $E(t) = L(t) + C(t)$.

(a) Graph L, C, and E in the window $[0, 10^{-6}, 10^{-7}]$ by $[-1, 4, 1]$. Interpret the graph.

(b) Make a table of L, C, and E, starting at $t = 0$ and incrementing by 10^{-7}. Interpret your results.

(c) Use a fundamental identity to derive a simplified expression for $E(t)$.

86. *Intensity of a Lamp* According to Lambert's law, the intensity of light from a single source on a flat surface at point P is given by $I = k \cos^2 \theta$, where k is a constant. See the figure. (**Source:** C. Winter, *Solar Power Plants.*)

(a) Let $k = 1$ and use degree mode to graph I in the window $[-90, 90, 45]$ by $[-1, 2, 1]$. For what value of θ is I maximum?

(b) Write I in terms of the sine function.

Exercises 87 and 88: Temperature (Refer to Example 11.) Suppose that the monthly high temperature at a location is modeled by $f(x)$, where x is the month with $x = -6$ corresponding to January, $x = 0$ to July, and $x = 5$ to December.

(a) Graph f in $[-6, 6, 1]$ by $[0, 100, 10]$. Interpret any symmetry in the graph.

(b) Make a table of f and discuss whether f is an even or odd function.

(c) Express any symmetry symbolically.

87. $f(x) = 40 \cos\left(\frac{\pi x}{6}\right) + 50$ **88.** $f(x) = -15 \cos\left(\frac{\pi x}{6}\right) + 60$

89. *Oscillating Spring* The distance or displacement y of a weight attached to an oscillating spring from its natural position is modeled by $y = 4 \cos(2\pi t)$, where t is in seconds. See the figure. Potential energy is the energy of position and is given by $P = ky^2$, where k is a constant. The weight has the greatest potential energy when the spring is stretched or compressed the most. (**Source:** R. Weidner and R. Sells, *Elementary Classical Physics*, Vol. 1.)

(a) Write an expression for P that involves the cosine function.

(b) Let $k = 2$ and graph P in the viewing rectangle $[0, 2, 0.5]$ by $[-1, 40, 8]$. For $0 \le t \le 2$, at what times is P maximum and at what times is P minimum? Interpret your result.

(c) Use a fundamental identity to write P in terms of the sine function.

90. *Energy in an Oscillating Spring* (Refer to Exercise 89.) Two types of mechanical energy are kinetic energy and potential energy. Kinetic energy is the energy of motion, and potential energy is the energy of position. A stretched spring has potential energy, which is converted to kinetic energy when it is released. If the potential energy of a weight attached to a spring is $P(t) = k \cos^2 (4\pi t)$, where k is a constant and t is in seconds, then its kinetic energy is given by $K(t) = k \sin^2 (4\pi t)$. The total mechanical energy E is $E(t) = P(t) + K(t)$.

(a) If $k = 2$, graph P, K, and E in [0, 0.5, 0.25] by [−1, 3, 1]. Interpret the graph.

(b) Make a table of K, P, and E, starting at $t = 0$ and incrementing by 0.05. Interpret the results.

(c) Use a fundamental identity to derive a simplified expression for $E(t)$.

Writing about Mathematics

91. Explain in your own words what a trigonometric identity is. Give two examples.

92. Answer each of the following.
   (a) Give two characteristics of an even function. Which of the trigonometric functions are even?

   (b) Give two characteristics of an odd function. Which of the trigonometric functions are odd?

93. A student writes "$\cos^2 + \sin^2 = 1$." Comment on the correctness of this expression.

94. Since $\sec^2 \theta = 1 + \tan^2 \theta$ is an identity, does it follow that $\sec \theta = 1 + \tan \theta$? Explain your reasoning.

7.2 Verifying Identities

- **Simplify trigonometric expressions**
- **Learn how to verify identities**

Introduction

In Example 5 of the previous section, we used the identity $\sin^2 \theta + \cos^2 \theta = 1$ to show that the energy stored in a tuner in a radio is constant. Trigonometric identities are used in both applications and calculus. Before we can use an identity, we must verify that it is correct. Although the equality of two trigonometric expressions can be *supported* graphically and numerically, symbolic *verification* is necessary to be certain that an equation is indeed an identity.

Simplifying Trigonometric Expressions

Many of the algebraic skills that you have already learned can be used to simplify trigonometric expressions. For example, suppose we wanted to multiply the following expression.

$$(1 - \cos \theta)(1 + \cos \theta)$$

In algebra we learned that

$$(1 - x)(1 + x) = 1 + x - x - x^2$$
$$= 1 - x^2.$$

If we substitute $\cos \theta$ for x, then

$$(1 - \cos \theta)(1 + \cos \theta) = 1 + \cos \theta - \cos \theta - \cos^2 \theta$$
$$= 1 - \cos^2 \theta.$$

In algebra we do not simplify $1 - x^2$ further. However, since $\sin^2 \theta + \cos^2 \theta = 1$, it follows that $\sin^2 \theta = 1 - \cos^2 \theta$. As a result,

$$(1 - \cos \theta)(1 + \cos \theta) = \sin^2 \theta.$$

MAKING CONNECTIONS

Algebraic and Trigonometric Expressions Many of the techniques used to rewrite algebraic expressions can also be used to rewrite trigonometric expressions. Here are some examples.

1. $\tan^2 \theta - 4$
$$= (\tan \theta - 2)(\tan \theta + 2) \qquad \text{is similar to} \qquad x^2 - 4 = (x - 2)(x + 2).$$

2. $\cos \theta (\sin \theta + \cos \theta)$
$$= \cos \theta \sin \theta + \cos^2 \theta \qquad \text{is similar to} \qquad x(y + x) = xy + x^2.$$

3. $\dfrac{\sin \theta}{\cos \theta} + \dfrac{1}{\cos \theta} = \dfrac{\sin \theta + 1}{\cos \theta} \qquad \text{is similar to} \qquad \dfrac{y}{x} + \dfrac{1}{x} = \dfrac{y + 1}{x}.$

The next example illustrates the addition of two trigonometric expressions.

EXAMPLE 1 Adding two trigonometric expressions

Write $\tan t + \cot t$ as a product of two trigonometric functions.

SOLUTION

Getting Started We begin by writing $\tan t$ and $\cot t$ as ratios involving $\sin t$ and $\cos t$.

$$\tan t + \cot t = \frac{\sin t}{\cos t} + \frac{\cos t}{\sin t}$$

Algebra Review
To review addition of rational expressions, see Chapter R (page R-34).

In algebra we combine $\frac{y}{x} + \frac{x}{y}$ by using the common denominator xy as follows.

$$\frac{y}{x} + \frac{x}{y} = \frac{y}{x} \cdot \frac{y}{y} + \frac{x}{y} \cdot \frac{x}{x} \qquad \text{Multiply each ratio by 1.}$$

$$= \frac{y^2}{xy} + \frac{x^2}{xy} \qquad \text{Simplify.}$$

$$= \frac{y^2 + x^2}{xy} \qquad \text{Add.}$$

Now substitute $\cos t$ for x and $\sin t$ for y. ▶

$$\tan t + \cot t = \frac{\sin t}{\cos t} + \frac{\cos t}{\sin t} \qquad \text{Quotient identities}$$

$$= \frac{\sin t}{\cos t} \cdot \frac{\sin t}{\sin t} + \frac{\cos t}{\sin t} \cdot \frac{\cos t}{\cos t} \qquad \text{Multiply each ratio by 1.}$$

$$= \frac{\sin^2 t}{\cos t \sin t} + \frac{\cos^2 t}{\cos t \sin t} \qquad \text{Simplify.}$$

$$= \frac{\sin^2 t + \cos^2 t}{\cos t \sin t} \qquad \text{Add.}$$

$$= \frac{1}{\cos t \sin t} \qquad \sin^2 t + \cos^2 t = 1$$

$$= \sec t \csc t \qquad \sec t = \frac{1}{\cos t}; \csc t = \frac{1}{\sin t}$$

Thus $\tan t + \cot t$ is equivalent to $\sec t \csc t$. **Now Try Exercise 11** ◀

NOTE To simplify a trigonometric expression, it is *not* necessary to first write the equation using x and y, as was done in Example 1. However, sometimes you may find this technique helpful.

EXAMPLE 2 Factoring a trigonometric expression

Factor each expression.
(a) $\sec^2 \theta - 1$ **(b)** $2 \sin^2 t + \sin t - 1$

SOLUTION
(a) In algebra we factor $x^2 - 1$ as $(x - 1)(x + 1)$. This technique can be applied to the given expression.

$$\sec^2 \theta - 1 = (\sec \theta - 1)(\sec \theta + 1)$$

Algebra Review
To review factoring polynomials, see Chapter R (page R-20).

(b) Since $2y^2 + y - 1$ can be factored as $(2y - 1)(y + 1)$, it follows that

$$2 \sin^2 t + \sin t - 1 = (2 \sin t - 1)(\sin t + 1).$$

Now Try Exercises 25 and 27 ◄

An Application from Electricity   Trigonometric expressions are used in applications involving electricity, as illustrated in the next example.

EXAMPLE 3 Analyzing electromagnets

Electromagnets are used in a variety of situations, such as lifting scrap metal, ringing door bells, and opening door locks in apartments. Let the wattage W consumed by an electromagnet at t seconds be

$$W(t) = 100 \sin^2 (120\pi t)$$

and the voltage V in the circuit be

$$V(t) = 160 \cos (120\pi t).$$

(**Source:** A. Howatson, *Electrical Circuits and Systems.*)
(a) Express $W(t)$ in terms of the cosine function. When V is maximum or minimum, what is the value of W? Explain.
(b) Support your answer in part (a) by graphing W and V in the viewing rectangle $[0, 1/30, 1/60]$ by $[-180, 180, 20]$.

SOLUTION
(a) Let $\theta = 120\pi t$. Since $\sin^2 \theta = 1 - \cos^2 \theta$, W can be expressed as

$$W(t) = 100(1 - \cos^2 (120\pi t)).$$

Since $V(t) = 160 \cos (120\pi t)$, the voltage is maximum (160) or minimum (-160) whenever $\cos (120 \pi t) = \pm 1$. When $\cos (120 \pi t) = \pm 1$, it follows that the value of W is given by $W(t) = 100(1 - (\pm 1)^2) = 0$.

(b) Graph $Y_1 = 100(\sin (120\pi X))^2$ and $Y_2 = 160 \cos (120\pi X)$, as shown in Figure 7.20. The wattage y_1 is 0 whenever the voltage y_2 is ± 160.   *Now Try Exercise 77* ◄

$[0, 1/30, 1/60]$ by $[-180, 180, 20]$

Figure 7.20 Radian Mode

Verification of Identities

When verifying that an equation is an identity, we usually begin with one side of the equation and write a sequence of equivalent expressions until it is transformed into the other side. This technique is illustrated in the following examples.

EXAMPLE 4 Verifying an identity

Verify that $(\cos \theta + \sin \theta)^2 = 1 + 2 \sin \theta \cos \theta$.

SOLUTION
Getting Started In algebra the expression $(x + y)^2$ can be expanded as follows.

$$(x + y)^2 = (x + y)(x + y)$$
$$= x^2 + xy + yx + y^2$$
$$= x^2 + 2xy + y^2$$

We can perform similar steps with the left side of the trigonometric equation. ▶

$$(\cos \theta + \sin \theta)^2 = (\cos \theta + \sin \theta)(\cos \theta + \sin \theta)$$
$$= \cos^2 \theta + \cos \theta \sin \theta + \sin \theta \cos \theta + \sin^2 \theta$$
$$= \cos^2 \theta + 2 \sin \theta \cos \theta + \sin^2 \theta$$
$$= 1 + 2 \sin \theta \cos \theta$$

The last step is true since $\sin^2 \theta + \cos^2 \theta = 1$. *Now Try Exercise 33* ◀

EXAMPLE 5 Verifying identities

Verify each identity.

(a) $\dfrac{\sin^2 \theta}{1 + \cos \theta} = 1 - \cos \theta$ (b) $\dfrac{\csc t}{\cot t} - \dfrac{\cot t}{\csc t} = \tan t \sin t$

SOLUTION
(a) Begin by applying a Pythagorean identity.

$$\frac{\sin^2 \theta}{1 + \cos \theta} = \frac{1 - \cos^2 \theta}{1 + \cos \theta} \qquad \sin^2 \theta + \cos^2 \theta = 1$$

$$= \frac{(1 - \cos \theta)(1 + \cos \theta)}{1 + \cos \theta} \qquad \text{Factor.}$$

$$= 1 - \cos \theta \qquad \text{Simplify.}$$

(b) Begin by finding a common denominator for the expressions on the left side of the equation.

$$\frac{\csc t}{\cot t} - \frac{\cot t}{\csc t} = \frac{\csc t}{\cot t} \cdot \frac{\csc t}{\csc t} - \frac{\cot t}{\csc t} \cdot \frac{\cot t}{\cot t} \qquad \text{Find a common denominator.}$$

$$= \frac{\csc^2 t}{\cot t \csc t} - \frac{\cot^2 t}{\csc t \cot t} \qquad \text{Multiply.}$$

$$= \frac{\csc^2 t - \cot^2 t}{\cot t \csc t} \qquad \text{Subtract the expressions.}$$

$$= \frac{1 + \cot^2 t - \cot^2 t}{\cot t \csc t} \qquad \csc^2 t = 1 + \cot^2 t$$

$$= \frac{1}{\cot t \csc t} \qquad \text{Simplify.}$$

$$= \frac{1}{\cot t} \cdot \frac{1}{\csc t} \qquad \text{Rewrite the expression.}$$

$$= \tan t \sin t \qquad \text{Reciprocal identities}$$

Now Try Exercises 49 and 59 ◀

EXAMPLE 6 Verifying an identity

Verify that $\frac{\sin t}{1 - \cos t} = \frac{1 + \cos t}{\sin t}$ is an identity.

SOLUTION

Getting Started Earlier, we saw that $(1 - \cos \theta)(1 + \cos \theta) = \sin^2 \theta$. This result can be used to verify the given identity by multiplying the numerator and denominator by 1 written as $\frac{1 + \cos t}{1 + \cos t}$. This technique is demonstrated below. ▶

$$\frac{\sin t}{1 - \cos t} = \frac{\sin t}{1 - \cos t} \cdot \frac{1 + \cos t}{1 + \cos t} \qquad \text{Multiply the ratio by 1.}$$

$$= \frac{\sin t (1 + \cos t)}{1 - \cos^2 t} \qquad \text{Simplify.}$$

$$= \frac{\sin t (1 + \cos t)}{\sin^2 t} \qquad \sin^2 t = 1 - \cos^2 t$$

$$= \frac{1 + \cos t}{\sin t} \qquad \text{Simplify.} \quad \textbf{Now Try Exercise 61} ◀$$

EXAMPLE 7 Verifying an identity

Verify that $\frac{\sec x + \tan x}{\sec x - \tan x} = \frac{1 + \sin x}{1 - \sin x}$.

SOLUTION

Getting Started Because $\sec x = \frac{1}{\cos x}$ and $\tan x = \frac{\sin x}{\cos x}$, we start to simplify the expression by multiplying the numerator and denominator by $\cos x$. ▶

$$\frac{\sec x + \tan x}{\sec x - \tan x} = \frac{(\sec x + \tan x) \cos x}{(\sec x - \tan x) \cos x} \qquad \text{Multiply the ratio by 1.}$$

$$= \frac{\sec x \cos x + \tan x \cos x}{\sec x \cos x - \tan x \cos x} \qquad \text{Distributive property}$$

$$= \frac{\frac{1}{\cos x} \cdot \cos x + \frac{\sin x}{\cos x} \cdot \cos x}{\frac{1}{\cos x} \cdot \cos x - \frac{\sin x}{\cos x} \cdot \cos x} \qquad \text{Reciprocal and quotient identities}$$

$$= \frac{1 + \sin x}{1 - \sin x} \qquad \text{Simplify.} \quad \textbf{Now Try Exercise 57} ◀$$

In the final example, we show how to give graphical and numerical support.

EXAMPLE 8 Verifying an identity

Verify that $\frac{\tan \theta}{\sec \theta} = \sin \theta$ symbolically. Give graphical and numerical support.

SOLUTION

Symbolic Verification **Getting Started** We will start with the more complicated expression $\frac{\tan \theta}{\sec \theta}$ and simplify it to $\sin \theta$. We begin by writing $\tan \theta$ and $\sec \theta$ in terms of $\sin \theta$ and $\cos \theta$. ▶

$$\frac{\tan \theta}{\sec \theta} = \frac{\sin \theta / \cos \theta}{1 / \cos \theta} \qquad \text{Quotient and reciprocal identities}$$

$$= \frac{\sin \theta}{\cos \theta} \cdot \frac{\cos \theta}{1} \qquad \text{Invert and multiply.}$$

$$= \sin \theta \qquad \text{Simplify.}$$

These steps verify symbolically that $\frac{\tan \theta}{\sec \theta} = \sin \theta$ is an identity.

Graphical Support   Graph $Y_1 = \tan(X)/(1/\cos(X))$ and $Y_2 = \sin(X)$, as shown below in Figures 7.21 and 7.22. Their graphs appear to be identical. Note that because $y_1 = \frac{\tan \theta}{\sec \theta}$ and $\sec \theta = \frac{1}{\cos \theta}$, it follows that $y_1 = \frac{\tan \theta}{1/\cos \theta}$.

Numerical Support   See Figure 7.23. Note that when $\theta = \frac{\pi}{2} \approx 1.5708$, the ratio $\frac{\tan \theta}{\sec \theta}$ is undefined, whereas $\sin \theta = 1$. However, the equation $\frac{\tan \theta}{\sec \theta} = \sin \theta$ is nonetheless an identity because $y_1 = y_2$ whenever both expressions are defined. (The step size for this table is $\frac{\pi}{6}$.)

$[-2\pi, 2\pi, \pi/2]$ by $[-2, 2, 1]$ $[-2\pi, 2\pi, \pi/2]$ by $[-2, 2, 1]$

X	Y₁	Y₂
0	0	0
.5236	.5	.5
1.0472	.86603	.86603
1.5708	ERR:	1
2.0944	.86603	.86603
2.618	.5	.5
3.1416	0	0

X=0

Figure 7.21 Radian Mode   **Figure 7.22** Radian Mode   **Figure 7.23** Radian Mode

Now Try Exercise 69 ◀

7.2 Putting It All Together

Becoming proficient at verifying identities requires practice. Many of the skills learned in algebra can be used to help verify identities. The following table lists some suggestions that may be helpful.

Suggestions for Verifying Identities

1. Become familiar with the fundamental identities found in Section 7.1.
2. Use your knowledge of how to simplify algebraic expressions to guide you, particularly when factoring or combining ratios.
3. When verifying an identity, start by simplifying the more complicated side of the equation. Otherwise, choose a side of the equation that you can transform into a different expression.
4. If you are simplifying the left side of the equation, work toward making the left side appear more like the right. For example, if the left side contains an addition sign but the right side does not, add the terms on the left side.
5. If you are uncertain how to proceed, one strategy is to write each trigonometric function in terms of sine and cosine and then simplify. Another strategy is to apply fundamental identities, if possible.
6. If a ratio contains $1 + \sin \theta$, it is sometimes helpful to multiply the numerator and denominator by $1 - \sin \theta$. Then

$$(1 + \sin \theta)(1 - \sin \theta) = 1 - \sin^2 \theta = \cos^2 \theta.$$

Similar statements can be made for $1 - \sin \theta$, $1 + \cos \theta$, and $1 - \cos \theta$. See Example 6.

7.2   Exercises

Simplifying Expressions

Exercises 1–4: Multiply the algebraic expression. Then multiply the corresponding trigonometric expression. If possible, simplify the resulting trigonometric expression.

1. (a) $(1 + x)(1 - x)$ **(b)** $(1 + \sin \theta)(1 - \sin \theta)$

2. (a) $(x - 1)(x + 1)$ **(b)** $(\csc \theta - 1)(\csc \theta + 1)$

3. (a) $x(x - 1)$ **(b)** $\sec \theta (\sec \theta - 1)$

4. (a) $(x + 1)(2x - 1)$ **(b)** $(\tan \theta + 1)(2 \tan \theta - 1)$

Exercises 5–8: Factor the algebraic expression. Then factor the corresponding trigonometric expression.

5. (a) $x^2 + 2x + 1$ **(b)** $\cos^2 \theta + 2 \cos \theta + 1$

6. (a) $2x^2 - 3x + 1$ **(b)** $2 \sin^2 t - 3 \sin t + 1$

7. (a) $x^2 - 2x$ **(b)** $\sec^2 t - 2 \sec t$

8. (a) $3x - 9x^2$ **(b)** $3 \tan \theta - 9 \tan^2 \theta$

Exercises 9–16: Simplify the algebraic expression. Then simplify the corresponding trigonometric expression completely.

9. (a) $\dfrac{1}{1 - x} + \dfrac{1}{1 + x}$ **(b)** $\dfrac{1}{1 - \cos \theta} + \dfrac{1}{1 + \cos \theta}$

10. (a) $x + \dfrac{1}{x}$ **(b)** $\tan t + \dfrac{1}{\tan t}$

11. (a) $\dfrac{x}{y} + \dfrac{y}{x}$ **(b)** $\dfrac{\cos t}{\sin t} + \dfrac{\sin t}{\cos t}$

12. (a) $\dfrac{1}{y} - \dfrac{x^2}{y}$ **(b)** $\dfrac{1}{\sin \theta} - \dfrac{\cos^2 \theta}{\sin \theta}$

13. (a) $\dfrac{1}{1/y^2} + \dfrac{1}{1/x^2}$ **(b)** $\dfrac{1}{\csc^2 t} + \dfrac{1}{\sec^2 t}$

14. (a) $\left(\dfrac{1}{x} + x\right)^2$ **(b)** $(\cot \theta + \tan \theta)^2$

15. (a) $\dfrac{x/y}{1/y}$ **(b)** $\dfrac{\cot \theta}{\csc \theta}$

16. (a) $\dfrac{1 - x^2}{1 + x}$ **(b)** $\dfrac{1 - \cos^2 \theta}{1 + \cos \theta}$

Exercises 17–24: Perform the indicated operations and simplify.

17. $\cos \theta \tan \theta$

18. $\sin^2 \theta \csc \theta$

19. $\tan \theta (\cos \theta - \csc \theta)$

20. $(\sin \theta - \cos \theta)^2$

21. $(1 + \tan t)^2$

22. $(\sin t - 1)(\sin t + 1)$

23. $\dfrac{\csc^2 \theta - 1}{\csc^2 \theta}$

24. $\sin^2 t (1 + \cot^2 t)$

Exercises 25–30: Factor the trigonometric expression and simplify, if possible.

25. $1 - \tan^2 \theta$ **26.** $\sin^2 t - \cos^2 t$

27. $\sec^2 t - \sec t - 6$ **28.** $\cos \theta \sin^2 \theta + \cos^3 \theta$

29. $\tan^4 \theta + 3 \tan^2 \theta + 2$ **30.** $\sin^4 t - \cos^4 t$

Verifying Identities

Exercises 31–68: Verify the identity.

31. $\csc^2 \theta - \cot^2 \theta = 1$ **32.** $\dfrac{\tan^2 \theta + 1}{\sec \theta} = \sec \theta$

33. $(1 - \sin t)^2 = 1 - 2 \sin t + \sin^2 t$

34. $\dfrac{\sin^2 t}{\cos t} = \sec t - \cos t$ **35.** $\dfrac{\sin t + \cos t}{\sin t} = 1 + \cot t$

36. $\sec^4 \theta - \sec^2 \theta = \tan^4 \theta + \tan^2 \theta$

37. $\sec^2 \theta - 1 = \tan^2 \theta$ **38.** $\dfrac{\csc^2 \theta}{\cot \theta} = \csc \theta \sec \theta$

39. $\dfrac{\tan^2 t}{\sec t} = \sec t - \cos t$ **40.** $\dfrac{\sec^2 \theta - 1}{\sec^2 \theta} = \sin^2 \theta$

41. $\cot x + 1 = \csc x (\cos x + \sin x)$

42. $\dfrac{1 + \sin x}{\cos x} = \dfrac{\cos x}{1 - \sin x}$ **43.** $\dfrac{\sec t}{1 + \sec t} = \dfrac{1}{\cos t + 1}$

44. $\sec^2 t + \csc^2 t = \sec^2 t \csc^2 t$

45. $(\sec t - 1)(\sec t + 1) = \tan^2 t$

46. $\csc^4 \theta - \cot^4 \theta = \csc^2 \theta + \cot^2 \theta$

47. $\dfrac{1 - \sin^2 \theta}{\cos \theta} = \cos \theta$ **48.** $\dfrac{\tan^2 t - 1}{1 + \tan^2 t} = 1 - 2 \cos^2 t$

49. $\dfrac{\sec t}{\tan t} - \dfrac{\tan t}{\sec t} = \cos t \cot t$

50. $\dfrac{\sin^4 t - \cos^4 t}{\sin^2 t - \cos^2 t} = 1$ **51.** $\dfrac{\cot^2 t}{\csc t + 1} = \csc t - 1$

52. $\sec \theta - \cos \theta = \tan \theta \sin \theta$

53. $\dfrac{\cot t}{\cot t + 1} = \dfrac{1}{1 + \tan t}$

54. $\cos^4 t - \sin^4 t = 2 \cos^2 t - 1$

55. $\dfrac{1}{1 - \sin t} + \dfrac{1}{1 + \sin t} = 2 \sec^2 t$

56. $\cot \theta + \tan \theta = \csc \theta \sec \theta$

57. $\dfrac{\csc t + \cot t}{\csc t - \cot t} = (\csc t + \cot t)^2$

58. $\dfrac{\csc t}{1 + \csc t} - \dfrac{\csc t}{1 - \csc t} = 2 \sec^2 t$

59. $\dfrac{\cos^2 t}{1 - \sin t} = 1 + \sin t$

60. $\csc t + \dfrac{\sec t}{\tan t} = \dfrac{2}{\sin t}$ **61.** $\dfrac{1}{1 + \sin \theta} = \dfrac{1 - \sin \theta}{\cos^2 \theta}$

62. $\dfrac{2 \sin^2 t + 3 \sin t - 2}{\sin t + 2} = 2 \sin t - 1$

63. $\sqrt{1 - \sin^2 \theta} = \cos \theta$, where θ is acute

64. $\sqrt{\sec^2 \theta - 1} = \tan \theta$, where θ is acute

65. $\dfrac{1 + 2 \sin x + \sin^2 x}{\cos^2 x} = \dfrac{1 + \sin x}{1 - \sin x}$

66. $\dfrac{\tan t - \cot t}{\sin t \cos t} = \sec^2 t - \csc^2 t$

67. $(1 - \cos^2 x)(1 + \cos^2 x) = 2 \sin^2 x - \sin^4 x$

68. $\sin^4 x - \cos^4 x = 2 \sin^2 x - 1$

Exercises 69–76: Verify the identity. If you have a graphing calculator, give graphical or numerical support.

69. $\cot \theta \sin \theta = \cos \theta$ **70.** $\tan \theta \cos \theta = \sin \theta$

71. $(1 - \cos^2 \theta)(1 + \tan^2 \theta) = \tan^2 \theta$

72. $\cos^2 \theta (1 + \cot^2 \theta) = \cot^2 \theta$

73. $\cos t (\tan t - \sec t) = \sin t - 1$

74. $\dfrac{\cos \theta}{1 - \sin \theta} = \sec \theta + \tan \theta$

75. $\dfrac{\tan (-\theta)}{\sin (-\theta)} = \sec \theta$

76. $\tan^2 t - \sin^2 t = \tan^2 t \sin^2 t$

Applications

77. *Electromagnets*   (Refer to Example 3.) Let the wattage consumed by an electromagnet be given by the formula $W(t) = 5 \cos^2 (120\pi t)$ and the voltage be given by the formula $V(t) = 25 \sin (120\pi t)$, where t is in seconds.
 (a) Express $W(t)$ in terms of the sine function. When V is maximum or minimum, what is the value of W?

 (b) Support your answer in part (a) by graphing W and V in [0, 1/15, 1/60] by [−30, 30, 10].

78. *Oscillating Spring*   (Refer to Exercise 90, Section 7.1.) The potential energy P of a weight on a spring is given by $P(t) = 5 \cos^2 (4\pi t)$, and its kinetic energy K is given by $K(t) = 5 \sin^2 (4\pi t)$, where t is in seconds.
 (a) Express $P(t)$ in terms of the sine function. When K is maximum or minimum, what is the value of P?

 (b) Support your answer in part (a) by graphing P and K in [0, 0.5, 0.25] by [−1, 5, 1]. Interpret the graph.

Writing about Mathematics

79. Create a trigonometric identity of your own. Verify the identity symbolically and then give graphical and numerical support.

80. Explain how to show that an equation is not an identity. Give an example.

CHECKING BASIC CONCEPTS FOR SECTIONS 7.1 AND 7.2

1. Determine the quadrant containing θ if $\cot \theta > 0$ and $\sin \theta < 0$.

2. Determine the other trigonometric functions of θ using the given information.
 (a) $\sin \theta = \dfrac{5}{13}$ and $\cos \theta = -\dfrac{12}{13}$

 (b) $\sec \theta = \dfrac{5}{4}$ and $\sin \theta < 0$

 (c) $\tan \theta = -\dfrac{1}{2}$ and $\cos \theta = \dfrac{2}{\sqrt{5}}$

3. Simplify each expression.
 (a) $(1 - \sin \theta)(1 + \sin \theta)$

 (b) $\tan^2 t \csc^2 t - 1$

4. Factor the trigonometric expression.
 (a) $\tan^2 t - 1$ **(b)** $3 \sin^2 t + \sin t - 2$

5. Verify each identity.
 (a) $(1 - \sin^2 \theta)(1 + \cot^2 \theta) = \cot^2 \theta$

 (b) $\dfrac{\cot^2 t}{\csc t} = \csc t - \sin t$

7.3 Trigonometric Equations

- **Find and use reference angles**
- **Solve trigonometric equations and applications**
- **Solve inverse trigonometric equations**

Introduction

In previous chapters we saw how applications that involve functions result in the need to solve equations. In a similar manner, applications that involve trigonometric functions result in the need to solve trigonometric equations. For example, the number of daylight hours near 30°N latitude can be modeled by

$$f(x) = 1.95 \cos\left(\frac{\pi}{6}(x - 6.6)\right) + 12.15,$$

where $x = 1$ corresponds to January 1, $x = 2$ to February 1, and so on. (See Section 6.5, Exercise 82.) To estimate when there are 11 hours of daylight, we can solve the *trigonometric equation* $f(x) = 11$, or

$$1.95 \cos\left(\frac{\pi}{6}(x - 6.6)\right) + 12.15 = 11.$$

Like other types of equations, trigonometric equations can be solved graphically, numerically, and symbolically. We begin by discussing reference angles, which will be used to solve trigonometric equations.

Reference Angles

A **reference angle** for an angle θ, written θ_R, is the acute angle made by the terminal side of θ and the x-axis. It is assumed that θ is in standard position and its terminal side does not lie on either the x- or the y-axis. Examples of reference angles in the four quadrants are shown in Figures 7.24–7.27.

Figure 7.24 Quadrant I

Figure 7.25 Quadrant II

Figure 7.26 Quadrant III

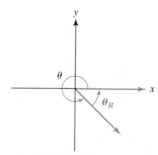

Figure 7.27 Quadrant IV

> **EXAMPLE 1** Finding reference angles

Find the reference angle for θ.

(a) $\theta = 43°$ **(b)** $\theta = \dfrac{2\pi}{3}$ **(c)** $\theta = -55°$ **(d)** $\theta = -\dfrac{3\pi}{4}$

SOLUTION

(a) Since θ is in quadrant I, θ and θ_R are equal. Thus $\theta_R = 43°$.

(b) The terminal side of $\theta = \frac{2\pi}{3}$ lies in quadrant II, as in Figure 7.25. In this case, the acute angle between the terminal side of θ and the x-axis is given by

$$\theta_R = \pi - \theta = \pi - \frac{2\pi}{3} = \frac{\pi}{3}.$$

(c) The terminal side of $\theta = -55°$ lies in quadrant IV, and the acute angle between it and the x-axis is $\theta_R = 55°$. See Figure 7.28.

(d) The terminal side of $\theta = -\frac{3\pi}{4}$ (or $-135°$) lies in quadrant III, and the acute angle between it and the x-axis is $\theta_R = \frac{\pi}{4}$. See Figure 7.29.

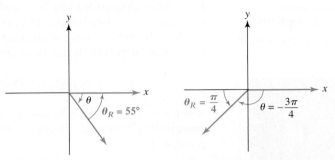

Figure 7.28   **Figure 7.29**

Now Try Exercises 3, 5, 7, and 9 ◀

The reference angle is important because it can help determine trigonometric values.

Reference Angles and Trigonometric Functions

Let θ be an angle in standard position with reference angle θ_R. Then

$$|\sin \theta| = \sin \theta_R \qquad |\cos \theta| = \cos \theta_R \qquad |\tan \theta| = \tan \theta_R$$

$$|\csc \theta| = \csc \theta_R \qquad |\sec \theta| = \sec \theta_R \qquad |\cot \theta| = \cot \theta_R.$$

The signs of the trigonometric functions of θ are determined by the quadrant that contains θ. (See Figure 7.9 on page 583.)

Figure 7.30

From Example 1(d), $\theta = -\frac{3\pi}{4}$ has reference angle $\theta_R = \frac{\pi}{4}$. Therefore $\sin\left(-\frac{3\pi}{4}\right) = \pm\sin\frac{\pi}{4}$, $\cos\left(-\frac{3\pi}{4}\right) = \pm\cos\frac{\pi}{4}$, and so on. Because $-\frac{3\pi}{4}$ is a quadrant III angle, $x < 0$ and $y < 0$. It follows that both the sine and cosine functions are negative in this quadrant. Thus $\sin\left(-\frac{3\pi}{4}\right) = -\sin\frac{\pi}{4}$ and $\cos\left(-\frac{3\pi}{4}\right) = -\cos\frac{\pi}{4}$. Reference angles are used to solve trigonometric equations.

Basic Trigonometric Equations

To solve various types of trigonometric equations, it is important to be able to solve basic types of equations such as $\cos \theta = -\frac{\sqrt{3}}{2}$, $\sin \theta = -\frac{1}{2}$, and $\tan \theta = 1$. For example, to solve $\cos \theta = -\frac{\sqrt{3}}{2}$ for θ in $[0°, 360°)$, begin by calculating

$$\theta_R = \cos^{-1}\frac{\sqrt{3}}{2} = 30°.$$

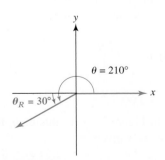

Figure 7.31

Because $\cos \theta < 0$, angle θ must be a second or third quadrant angle. Make a sketch of the two possible angles θ in $[0°, 360°)$ with reference angle $\theta_R = 30°$. From Figures 7.30 and 7.31 the solutions are given by $\theta = 150°$ and $\theta = 210°$.

EXAMPLE 2 Solving trigonometric equations using reference angles

Solve the following equations.

(a) $\sin \theta = -\frac{1}{2}$ for θ in $[0°, 360°)$ **(b)** $\tan \theta = 1$ for θ in $[0, 2\pi)$

SOLUTION

(a) We start by solving the equation $\sin \theta_R = \frac{1}{2}$. The solution to this equation is $\theta_R = \sin^{-1} \frac{1}{2} = 30°$. The sine function is negative in quadrants III and IV. (Note that on the unit circle $\sin \theta = y$ and y is negative in quadrants III and IV.) Therefore the solution to $\sin \theta = -\frac{1}{2}$ is an angle θ, located in either quadrant III or quadrant IV, that has a reference angle of 30°. There are two such angles: 210° and 330°. See Figures 7.32 and 7.33. Thus 210° and 330° are solutions.

(b) The solution to $\tan \theta_R = 1$ is $\theta_R = \tan^{-1} 1 = \frac{\pi}{4}$. The tangent function is positive in quadrants I and III. See Figure 7.9. Thus $\frac{\pi}{4}$ and $\frac{5\pi}{4}$ are solutions. See Figures 7.34 and 7.35.

Figure 7.32

Figure 7.33

Figure 7.34

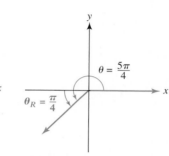

Figure 7.35

Now Try Exercises 11 and 13 ◄

Solving Trigonometric Equations

In the previous section we verified trigonometric *identities*. Identities are equations that are true for all meaningful values of the variable. In this section we discuss trigonometric equations that are *conditional*. Conditional equations are satisfied by some but not all values of the variable. For example,

$$\cos \theta = 1$$

is a conditional trigonometric equation since the cosine function equals 1 only for certain values of θ, such as $\theta = 0°$ or $\theta = 360°$.

EXAMPLE 3 Solving a trigonometric equation

Solve $2 \sin \theta - 1 = 1$ on the interval $[0°, 360°)$ symbolically and graphically.

SOLUTION

Symbolic Solution **Getting Started** To solve the linear equation $2x - 1 = 1$, we first add 1 to each side and then divide each side by 2 to obtain $x = 1$. This technique can be applied to the equation $2 \sin \theta - 1 = 1$ to obtain $\sin \theta = 1$. ►

[0, 360, 90] by [−4, 4, 1]

Figure 7.36 Degree Mode

Begin by solving the given equation for sin θ.

$$2 \sin \theta - 1 = 1 \qquad \textit{Given equation}$$
$$2 \sin \theta = 2 \qquad \textit{Add 1.}$$
$$\sin \theta = 1 \qquad \textit{Divide by 2.}$$

The only solution to $\sin \theta = 1$ on the interval $[0°, 360°)$ is $\theta = \sin^{-1} 1 = 90°$.

Graphical Solution Graph $Y_1 = 2 \sin(X) - 1$ and $Y_2 = 1$, as in Figure 7.36. Their graphs intersect at $x = 90°$.   ***Now Try Exercise 29*** ◀

An Application from Astronomy In the next example, we find all the phase angles associated with a particular phase F of the moon. The fraction of the moon that appears illuminated is called the *phase F*. The *phase angle θ* is shown in Figure 7.37. (**Source:** M. Zeilik et al., *Introductory Astronomy and Astrophysics.*)

Sun

Earth

θ

Moon

Figure 7.37 Phase Angle θ

EXAMPLE 4 **Finding phase angles for the moon**

The phase F associated with a phase angle θ is given by

$$F = \frac{1}{2}(1 - \cos \theta).$$

Find all phase angles θ in degrees when $F = 0.75$. (Note that $F = 0.5$ corresponds to a first quarter moon or last quarter moon and $F = 1$ corresponds to a full moon.)

SOLUTION Let $F = 0.75$ and solve the given equation.

$$0.75 = \frac{1}{2}(1 - \cos \theta) \qquad \textit{Let F = 0.75.}$$
$$1.5 = 1 - \cos \theta \qquad \textit{Multiply by 2.}$$
$$\cos \theta = -0.5 \qquad \textit{Solve for cos } \theta.$$

Start by solving the equation $\cos \theta_R = 0.5$. The solution is $\theta_R = \cos^{-1} 0.5 = 60°$. The cosine function is negative in quadrants II and III. Angles in these quadrants with a 60° reference angle are 120° and 240°, where $0° \leq \theta \leq 360°$. Verify this fact. Since the cosine function has period 360°, all solutions can be written in the form

$$\theta = 120° + 360° \cdot n \qquad \text{or} \qquad \theta = 240° + 360° \cdot n,$$

where n is an integer. For example, 120°, 120° ± 360°, and 120° ± 720° are solutions, as well as 240°, 240° ± 360°, and 240° ± 720°.   ***Now Try Exercise 115*** ◀

Finding All Solutions Many of the techniques used to solve polynomial equations can be applied to trigonometric equations, as illustrated in the next example.

> **EXAMPLE 5** Solving trigonometric equations
>
> Find all solutions to each equation. Express your results in radians.
> **(a)** $2 \cot t + 1 = -1$ **(b)** $2 \sin^2 t - 5 \sin t + 2 = 0$
>
> **SOLUTION**
> **(a)** In algebra the equation $2x + 1 = -1$ implies $x = -1$. In a similar manner,
>
> $$2 \cot t + 1 = -1 \quad \text{implies} \quad \cot t = -1.$$
>
> If $\cot t = -1$, then $\tan t = \frac{1}{\cot t} = -1$ and t has a reference angle of $\tan^{-1} 1 = \frac{\pi}{4}$. The cotangent is negative in quadrants II and IV, so the solutions to $\cot t = -1$ in $[0, 2\pi)$ are $t = \frac{3\pi}{4}$ and $t = \frac{7\pi}{4}$. Note that these angles both have a reference angle of $\frac{\pi}{4}$. Since cotangent has a period of π, all solutions can be expressed in the form
>
> $$t = \frac{3\pi}{4} + \pi n \qquad \text{or} \qquad t = \frac{7\pi}{4} + \pi n,$$
>
> where n is an integer. These solutions are equivalent to just $t = \frac{3\pi}{4} + \pi n$ because the difference between $\frac{7\pi}{4}$ and $\frac{3\pi}{4}$ is π.
>
> **(b) Getting Started** In algebra the equation $2x^2 - 5x + 2 = 0$ can be solved by factoring.
>
> $$2x^2 - 5x + 2 = (2x - 1)(x - 2) = 0$$
>
> The solutions are $\frac{1}{2}$ and 2. We can factor a trigonometric expression in a manner similar to that above. ▶
>
> $$2 \sin^2 t - 5 \sin t + 2 = (2 \sin t - 1)(\sin t - 2) = 0$$
>
> We must solve the equations $\sin t = \frac{1}{2}$ and $\sin t = 2$. If $\sin t = \frac{1}{2}$, the reference angle is $\sin^{-1} \frac{1}{2} = \frac{\pi}{6}$. The sine function is positive in quadrants I and II, so the solutions in $[0, 2\pi)$ are $t = \frac{\pi}{6}$ and $t = \frac{5\pi}{6}$. Since $-1 \le \sin t \le 1$ for all t, the equation $\sin t = 2$ has no solutions. The sine function has period 2π, and all solutions to the given equation can be expressed as
>
> $$t = \frac{\pi}{6} + 2\pi n \qquad \text{or} \qquad t = \frac{5\pi}{6} + 2\pi n.$$
>
> **Now Try Exercises 53 and 57** ◀

Algebra Review
To review factoring of trinomials, see Chapter R (page R-23).

MAKING CONNECTIONS

Polynomial and Trigonometric Equations A polynomial equation of degree n has at most n solutions. However, a trigonometric equation typically has an infinite number of solutions, as demonstrated in Examples 4 and 5.

Approximating Solutions Sometimes equations have solutions that are not multiples of common angles such as $\frac{\pi}{6}$ or $\frac{\pi}{4}$. In the next example, we find and then approximate solutions to a trigonometric equation.

EXAMPLE 6 Approximating solutions

Solve $1.7 \csc t + 2.3 = 0$ to the nearest thousandth for t in $[0, 2\pi)$.

SOLUTION Begin by solving the equation for $\csc t$.

$$1.7 \csc t + 2.3 = 0 \qquad \textit{Given equation}$$

$$1.7 \csc t = -2.3 \qquad \textit{Subtract 2.3 from each side.}$$

$$\csc t = -\frac{2.3}{1.7} \qquad \textit{Divide each side by 1.7.}$$

Because $\csc t = \frac{1}{\sin t}$, it follows that $\sin t = -\frac{1.7}{2.3}$, or $\sin t = -\frac{17}{23}$. The reference number is $t_R = \sin^{-1} \frac{17}{23} \approx 0.832$. The sine function is negative in quadrants III and IV. For t in $[0, 2\pi)$, there are two solutions given by

$$\pi + \sin^{-1} \frac{17}{23} \approx 3.973 \qquad \text{and} \qquad 2\pi - \sin^{-1} \frac{17}{23} \approx 5.451.$$

See Figures 7.38 and 7.39. (Note that 0.832 radian is approximately equivalent to 47.7°.)

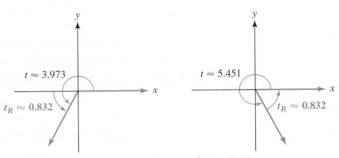

Figure 7.38　　　　　　　**Figure 7.39**

Now Try Exercise 91 ◀

An Application　Solar power companies are interested in the number of daylight hours during different times of the year and at different latitudes. In the next example, we solve the trigonometric equation presented in the introduction to determine when there are 11 hours of daylight at 30°N latitude.

EXAMPLE 7 Analyzing daylight hours

The number of daylight hours at 30°N latitude can be modeled by

$$f(x) = 1.95 \cos\left(\frac{\pi}{6}(x - 6.6)\right) + 12.15,$$

where $x = 1$ corresponds to January 1, $x = 2$ to February 1, and so on. Estimate graphically and numerically when there are 11 hours of daylight.

SOLUTION

Graphical Solution Graph $Y_1 = 1.95 \cos(\pi/6(X - 6.6)) + 12.15$ and $Y_2 = 11$ in radian mode. Their graphs intersect near $x = 2.4$ and $x = 10.8$. See Figures 7.40 and 7.41. These values correspond to about February 11 and October 25. (Note that four tenths of February is $0.4 \times 28 \approx 11$ days and eight tenths of October is $0.8 \times 31 \approx 25$ days.)

Numerical Solution Make a table of Y_1, starting at $x = 2$ and incrementing by 0.1. The table in Figure 7.42 shows that $Y_1 \approx 11$ when $x = 2.4$. Scrolling down the table would also show that $Y_1 \approx 11$ when $x = 10.8$.

[0, 13, 1] by [8, 16, 1]   [0, 13, 1] by [8, 16, 1]

X	Y1	Y2
2	10.701	11
2.1	10.771	11
2.2	10.845	11
2.3	10.923	11
2.4	11.004	11 ←
2.5	11.088	11
2.6	11.175	11
X=2.4		

Figure 7.40 Radian Mode   **Figure 7.41** Radian Mode   **Figure 7.42** Radian Mode

Now Try Exercise 119 ◀

More Trigonometric Equations In the next example, we solve an equation for all real numbers t, where the argument of the trigonometric function is $4t$.

EXAMPLE 8 Solving a trigonometric equation

Solve $-0.6 \sin 4t = 0.3$, where t is any real number.

SOLUTION First we let $\theta = 4t$. Then the given equation becomes

$$-0.6 \sin \theta = 0.3.$$

Next we solve this modified equation for all real numbers θ.

$$-0.6 \sin \theta = 0.3 \qquad \text{Let } \theta = 4t.$$

$$\sin \theta = -\frac{1}{2} \qquad \text{Divide by } -0.6.$$

From Example 2(a), the solutions to $\sin \theta = -\frac{1}{2}$ on $[0°, 360°)$ are $210°$ and $330°$. In radian measure these solutions are $\theta = \frac{7\pi}{6}$ and $\frac{11\pi}{6}$. Thus all real-number solutions to the equation $-0.6 \sin \theta = 0.3$ are

$$\theta = \frac{7\pi}{6} + 2\pi n \qquad \text{or} \qquad \theta = \frac{11\pi}{6} + 2\pi n,$$

where n is an integer. Because $\theta = 4t$, we can determine t by substituting $4t$ for θ.

$$4t = \frac{7\pi}{6} + 2\pi n \qquad \text{or} \qquad 4t = \frac{11\pi}{6} + 2\pi n$$

Finally we divide each equation by 4 to obtain

$$t = \frac{7\pi}{24} + \frac{\pi n}{2} \qquad \text{or} \qquad t = \frac{11\pi}{24} + \frac{\pi n}{2}. \qquad \textit{Now Try Exercise 73} ◀$$

Some trigonometric equations contain more than one type of trigonometric function. In these situations it is sometimes helpful to use trigonometric identities to rewrite the equation in terms of one trigonometric function. This is illustrated in the next example.

EXAMPLE 9 Solving a trigonometric equation

Solve $2 \tan \theta = \sec^2 \theta$ symbolically on the interval $[0, 2\pi)$.

SOLUTION This equation contains two different trigonometric functions. We begin by applying the identity $1 + \tan^2 \theta = \sec^2 \theta$ to rewrite the equation only in terms of $\tan \theta$.

$2 \tan \theta = \sec^2 \theta$	Given equation
$2 \tan \theta = 1 + \tan^2 \theta$	$\sec^2 \theta = 1 + \tan^2 \theta$
$\tan^2 \theta - 2 \tan \theta + 1 = 0$	Rewrite the equation.
$(\tan \theta - 1)(\tan \theta - 1) = 0$	Factor.
$\tan \theta = 1$	Solve for $\tan \theta$.

The solutions are $\frac{\pi}{4}$ and $\frac{5\pi}{4}$. See Example 2(b).   **Now Try Exercise 43** ◀

In the next example, we solve a trigonometric equation by squaring each side. When squaring each side of an equation, it is important to *check the answers*.

EXAMPLE 10 Solving a trigonometric equation by squaring

Solve $\sec t = 1 + \tan t$ for $0 \le t < 2\pi$.

SOLUTION Begin by squaring each side of the equation.

Algebra Review
To review squaring a binomial, see Chapter R (page R-18).

$\sec t = 1 + \tan t$	Given equation
$\sec^2 t = (1 + \tan t)^2$	Square each side.
$\sec^2 t = 1 + 2 \tan t + \tan^2 t$	Square the expression.
$\sec^2 t = \sec^2 t + 2 \tan t$	$1 + \tan^2 t = \sec^2 t$
$0 = 2 \tan t$	Subtract $\sec^2 t$ from each side.
$\tan t = 0$	Divide by 2 and rewrite.
$t = 0$ or $t = \pi$	Solve for t when $0 \le t < 2\pi$.

Since we squared each side of the equation, we check $t = 0$ and $t = \pi$ in the given equation.

$$\sec 0 \overset{?}{=} 1 + \tan 0 \qquad \sec \pi \overset{?}{=} 1 + \tan \pi$$
$$1 = 1 + 0 \qquad\qquad -1 \ne 1 + 0$$

The only solution is 0. The value of π is an *extraneous solution*.

Now Try Exercise 71 ◀

More Applications

Highway Curves Highway curves are sometimes banked so that the outside of the curve is slightly elevated or inclined, as shown in Figure 7.43. This inclination is called the *superelevation*. The relationship among a car's velocity v in feet per second, the safe radius r of the curve in feet, and the superelevation θ in degrees is given by

Figure 7.43

$$r = \frac{v^2}{4.5 + 32.2 \tan \theta}.$$

(**Source:** F. Mannering and W. Kilareski, *Principles of Highway Engineering and Traffic Analysis.*)

EXAMPLE 11 Determining superelevation for a highway curve

A highway curve with a radius of 700 feet and a speed limit of 88 feet per second (60 mi/hr) is being designed. Find the appropriate superelevation for the curve.

SOLUTION Let $r = 700$ and $v = 88$ and then solve the equation for θ.

$$700 = \frac{88^2}{4.5 + 32.2 \tan \theta} \qquad \text{Let } r = 700 \text{ and } v = 88.$$

$$4.5 + 32.2 \tan \theta = \frac{88^2}{700} \qquad \text{Properties of ratios}$$

$$32.2 \tan \theta = \frac{88^2}{700} - 4.5 \qquad \text{Subtract 4.5.}$$

$$\tan \theta = \frac{88^2/700 - 4.5}{32.2} \qquad \text{Divide by 32.2.}$$

$$\tan \theta \approx 0.2038 \qquad \text{Approximate.}$$

$$\theta \approx \tan^{-1} 0.2038 \approx 11.5° \qquad \text{Apply the inverse tangent.}$$

The superelevation should be about 11.5°.   ***Now Try Exercise 117*** ◀

Locating the Position of a Planet One step in the process used by astronomers to calculate the position of a planet as it orbits the sun involves finding the solution of Kepler's equation. Kepler's equation is a trigonometric equation that *cannot* be solved symbolically. It must be solved either graphically or numerically. In real applications, it is not uncommon to encounter equations that cannot be solved symbolically.

EXAMPLE 12 Solving the equation of Kepler

One example of Kepler's equation is $\theta = 0.087 + 0.093 \sin \theta$. Solve this equation graphically.   (**Source:** J. Meeus, *Astronomical Algorithms*.)

SOLUTION The given equation is equivalent to $\theta - 0.087 - 0.093 \sin \theta = 0$. To solve this equation graphically, let $Y_1 = X - 0.087 - 0.093 \sin X$ and graph. (Be sure to use radian mode.) The x-intercept is near 0.096, as shown in Figure 7.44.

Now Try Exercise 125 ◀

[0, 0.2, 0.05] by [−0.2, 0.2, 0.1]

$y = x - 0.087 - 0.093 \sin x$

Zero
X=.09590555 Y=0

Figure 7.44 Radian Mode

Solving Inverse Trigonometric Equations

Some types of equations contain inverse trigonometric functions. To solve these equations we often make use of the following inverse properties.

$$\sin (\sin^{-1} x) = x \qquad \text{for} \quad -1 \le x \le 1$$

$$\cos (\cos^{-1} x) = x \qquad \text{for} \quad -1 \le x \le 1$$

$$\tan (\tan^{-1} x) = x \qquad \text{for} \quad -\infty < x < \infty$$

These types of equations are solved in the next example.

EXAMPLE 13 Solving inverse trigonometric equations

Solve each equation symbolically.

(a) $\cos^{-1} x = \pi$ (b) $\dfrac{\pi}{2} - \tan^{-1} x = \dfrac{\pi}{4}$ (c) $\sin^{-1} 2x = \dfrac{\pi}{3}$

SOLUTION

(a) Begin by taking the cosine of each side of the given equation $\cos^{-1} x = \pi$.

$$\cos (\cos^{-1} x) = \cos \pi \qquad \text{Take the cosine of each side.}$$

$$x = -1 \qquad \text{Simplify.}$$

(b) Begin solving the given equation $\dfrac{\pi}{2} - \tan^{-1} x = \dfrac{\pi}{4}$ for $\tan^{-1} x$.

$$\dfrac{\pi}{4} = \tan^{-1} x \qquad \text{Subtract } \tfrac{\pi}{4}; \text{ add } \tan^{-1} x.$$

$$\tan \dfrac{\pi}{4} = \tan (\tan^{-1} x) \qquad \text{Take the tangent of each side.}$$

$$1 = x \qquad \text{Simplify.}$$

(c) Begin by taking the sine of each side of the given equation $\sin^{-1} 2x = \dfrac{\pi}{3}$.

$$\sin (\sin^{-1} 2x) = \sin \dfrac{\pi}{3} \qquad \text{Take the sine of each side.}$$

$$2x = \dfrac{\sqrt{3}}{2} \qquad \text{Simplify.}$$

$$x = \dfrac{\sqrt{3}}{4} \qquad \text{Divide by 2.}$$

Now Try Exercises 105, 111, and 113 ◀

7.3 Putting It All Together

$\mathbf{A}$lgebraic skills such as factoring and solving equations can also be used to solve trigonometric equations. Unlike polynomial equations, which always have a finite number of solutions, trigonometric equations frequently have infinitely many solutions.

The following table gives examples of the concepts presented in this section.

Concept	Comments	Examples
Reference angle	If an angle θ is in standard position, then its reference angle is the acute angle made by the terminal side of θ and the x-axis.	If $\theta = \dfrac{11\pi}{6}$, then $\theta_R = \dfrac{\pi}{6}$.

Concept	Comments	Examples
Trigonometric equation	First, use techniques from algebra to isolate any trigonometric functions. Then solve these simpler equations for the given variable.	$\sqrt{3}\tan\theta = -1$ $\tan\theta = -\dfrac{1}{\sqrt{3}}$ The reference angle is $\theta_R = \tan^{-1}\dfrac{1}{\sqrt{3}} = 30°$. Since $\tan\theta$ is negative in quadrants II and IV, the solutions in $[0°, 360°)$ are $150°$ and $330°$.
Inverse trigonometric equations	Equations that involve inverse trigonometric functions can sometimes be solved by using the following properties. $\sin(\sin^{-1}x) = x, \; -1 \le x \le 1$ $\cos(\cos^{-1}x) = x, \; -1 \le x \le 1$ $\tan(\tan^{-1}x) = x, \; -\infty < x < \infty$	$\tan^{-1}x = \dfrac{\pi}{3}$ $\tan(\tan^{-1}x) = \tan\dfrac{\pi}{3}$ $x = \sqrt{3}$

7.3 Exercises

Reference Angles

Exercises 1–10: Find the reference angle for θ.

1. $\theta = 120°$ **2.** $\theta = 230°$

3. $\theta = 85°$ **4.** $\theta = -130°$

5. $\theta = -65°$ **6.** $\theta = 340°$

7. $\theta = \dfrac{5\pi}{6}$ **8.** $\theta = \dfrac{7\pi}{4}$

9. $\theta = -\dfrac{2\pi}{3}$ **10.** $\theta = -\dfrac{5\pi}{4}$

Basic Trigonometric Equations

Exercises 11–18: Solve for θ in $[0°, 360°)$ and in $[0, 2\pi)$.

11. (a) $\sin\theta = 1$ (b) $\sin\theta = -1$

12. (a) $\cos\theta = \dfrac{1}{2}$ (b) $\cos\theta = -\dfrac{1}{2}$

13. (a) $\tan\theta = \sqrt{3}$ (b) $\tan\theta = -\sqrt{3}$

14. (a) $\cot\theta = 1$ (b) $\cot\theta = -1$

15. (a) $\sec\theta = 2$ (b) $\sec\theta = -2$

16. (a) $\csc\theta = \sqrt{2}$ (b) $\csc\theta = -\sqrt{2}$

17. (a) $\sin\theta = 3$ (b) $\sin\theta = -3$

18. (a) $\cos\theta = \dfrac{\sqrt{3}}{2}$ (b) $\cos\theta = -\dfrac{\sqrt{3}}{2}$

Exercises 19–24: Find all solutions. Express your answer in radians.

19. (a) $\sin t = \dfrac{1}{2}$ (b) $\sin t = -\dfrac{1}{2}$

20. (a) $\cos t = 1$ (b) $\cos t = -1$

21. (a) $\tan t = 1$ (b) $\tan t = -1$

22. (a) $\csc t = \dfrac{1}{4}$ (b) $\csc t = -\dfrac{1}{4}$

23. (a) $\sec t = 2$ (b) $\sec t = -2$

24. (a) $\cot t = \sqrt{3}$ (b) $\cot t = -\sqrt{3}$

Solving Trigonometric Equations

Exercises 25–28: Use the graph to estimate any solutions to the given equation for $0 \le t < 2\pi$. Then solve the equation symbolically.

25. $\sin t = \cos t$

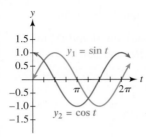

26. $\csc t = \sec t$

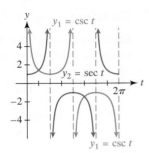

27. $3 \cot t = 2 \sin t$

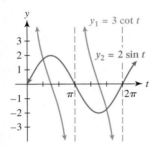

28. $2 \cos^2 t = 1 - \cos t$

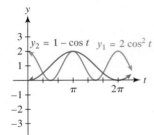

Exercises 29–36: Solve the algebraic equation for x. Then solve the trigonometric equation for $0° \le \theta < 360°$.

29. (a) $2x - 1 = 0$ **(b)** $2 \sin \theta - 1 = 0$

30. (a) $x - 1 = 0$ **(b)** $\cot \theta - 1 = 0$

31. (a) $x^2 = x$ **(b)** $\sin^2 \theta = \sin \theta$

32. (a) $x^2 - x = 0$ **(b)** $\cos^2 \theta - \cos \theta = 0$

33. (a) $x^2 + 1 = 2$ **(b)** $\tan^2 \theta + 1 = 2$

34. (a) $(x - 1)(x + 1) = 0$ **(b)** $(\sin \theta - 1)(\sin \theta + 1) = 0$

35. (a) $x^2 + x = 2$ **(b)** $\cos^2 \theta + \cos \theta = 2$

36. (a) $2x^2 + 3x = -1$ **(b)** $2 \sin^2 \theta + 3 \sin \theta = -1$

Exercises 37–52: Solve the equation for t in $[0, 2\pi)$.

37. $\tan^2 t - 3 = 0$ **38.** $2 \sin t = \sqrt{3}$

39. $3 \cos t + 4 = 0$ **40.** $\cos^2 t + \cos t - 6 = 0$

41. $\sin t \cos t = \cos t$ **42.** $\cos^2 t - \sin^2 t = 0$

43. $\csc^2 t = 2 \cot t$ **44.** $2 \sin^2 t - 3 \cos t = 3$

45. $\sin^2 t = \frac{1}{4}$ **46.** $\cos^2 t = -\frac{1}{2}$

47. $\sin t \cos t = 0$ **48.** $\tan t - \cot t = 0$

49. $2 \sec t = \tan^2 t + 1$ **50.** $\cos^2 t = 3 \sin^2 t$

51. $\tan t + \sec t = 1$ **52.** $\cos t - \sin t = 1$

Exercises 53–72: Find all solutions to the equation. Express your results in radians.

53. $2 \tan t - 1 = 1$ **54.** $\sqrt{3} \cot t - 1 = 0$

55. $2 \sin t + 2 = 3$ **56.** $2 \cos t - 1 = 0$

57. $2 \sin^2 t - 3 \sin t = -1$ **58.** $2 \cos^2 t + 3 \cos t + 1 = 0$

59. $\sec^2 t + 3 \sec t + 2 = 0$ **60.** $\csc^2 t - 3 = -1$

61. $\tan^2 t - 1 = 0$ **62.** $2 \cos t = -1$

63. $\sin^2 t + \sin t - 20 = 0$ **64.** $3 \cos t - 5 = 0$

65. $\cos t \sin t = \sin t$ **66.** $2 \cos^2 t - 1 = 0$

67. $\sec^2 t = 2 \tan t$ **68.** $\cos^2 t - 2 \sin t - 1 = 0$

69. $\sin^2 t \cos^2 t = 0$ **70.** $2 \cot^2 t \sin t - \cot^2 t = 0$

71. $\sin t + \cos t = 1$ **72.** $\sin t - \cos t = 1$

Exercises 73–90: (Refer to Example 8.) Solve the equation, where t is any real number. Approximate t to three decimal places when appropriate.

73. $\sin 3t = \frac{1}{2}$ **74.** $\cos 2t = -\frac{1}{2}$

75. $\cos 4t = -\frac{\sqrt{3}}{2}$ **76.** $\sin 4t = \frac{1}{\sqrt{2}}$

77. $\tan 5t = 1$ **78.** $\cot 3t = -\sqrt{3}$

79. $2 \sin 4t = -1$ **80.** $5 \cos 6t = 2.5$

81. $-\sec 4t = \sqrt{2}$ **82.** $\sqrt{3} \csc 3t = -2$

83. $2 \sin 8t - 3 = -1$ **84.** $3 \cos 8t - 4 = -4$

85. $\cot 4t + 5 = 6$ **86.** $\sqrt{3} \tan 2t + 2 = 3$

87. $5 \cos 3t = 1$ **88.** $7 \cos 5t = -2$

89. $\sin 2t = \frac{1}{3}$ **90.** $\frac{1}{7} \sin 7t = \frac{1}{20}$

Approximate Solutions

Exercises 91–96: Solve the equation to the nearest thousandth for t in $[0, 2\pi)$.

91. $2.1 \sec t - 4.5 = 0$ **92.** $2 \csc t + 2 = 8.3$

93. $5.8 \sin t - 3.7 = 0.2$ **94.** $6 \cos t + 2 = 0$

95. $5 \tan^2 t - 3 = 0$ **96.** $7 \cot^2 t + 1.2 = 6$

Exercises 97–102: The following equations cannot be solved symbolically. Approximate to two decimal places any solutions on $[0, 2\pi)$ graphically or numerically.

97. $\tan x = x$ **98.** $x - \cos x = 0$

99. $\sin x = (x - 1)^2$ **100.** $\sin^2 x - \ln x = 0$

101. $2x \cos(x + 1) = \sin(\cos x)$

102. $e^{-0.1x} \cos x = x \sin x$

Numerical Solutions

Exercises 103 and 104: Use the table to find the solutions to the given equation on $[0°, 360°)$. Then write all solutions to the equation.

103. $\tan \theta - \dfrac{1}{\sqrt{3}} = 0$ **104.** $\tan \theta - \sin \theta = 0$

X	Y1
30	0
90	ERR:
150	‾1.155
210	0
270	ERR:
330	‾1.155
390	0

Y1 ≣ tan(X)−1/√(3)

X	Y1
0	0
60	.86603
120	‾2.598
180	0
240	2.5981
300	‾.866
360	0

Y1 ≣ tan(X)−sin(X)

Solving Inverse Trigonometric Equations

Exercises 105–114: Solve the equation.

105. $\sin^{-1} x = \dfrac{\pi}{2}$ **106.** $\sin^{-1} 2x = -\dfrac{\pi}{4}$

107. $2 \cos^{-1} x = \dfrac{5\pi}{3}$ **108.** $\cos^{-1} x = 0$

109. $\pi + \tan^{-1} x = \dfrac{3\pi}{4}$ **110.** $\tan^{-1} x = -\dfrac{\pi}{3}$

111. $\tan^{-1}(3x + 1) = \dfrac{\pi}{4}$ **112.** $\dfrac{\pi}{4} + \sin^{-1}(x + 1) = \dfrac{\pi}{2}$

113. $\cos^{-1} x + 3 \cos^{-1} x = \pi$

114. $\dfrac{\pi}{6} + \sin^{-1} 4x = \dfrac{\pi}{3}$

Applications

115. *First Quarter* (Refer to Example 4.) Let $F = 0.25$ and solve the equation $F = \frac{1}{2}(1 - \cos \theta)$ for all θ to determine the phase angles for when 25% of the moon is illuminated.

116. *Full Moon* (Refer to Example 4.) Let $F = 1$ and solve the equation $F = \frac{1}{2}(1 - \cos \theta)$ for all θ to determine the phase angles for a full moon.

117. *Designing Highway Curves* (Refer to Example 11.) A highway curve with a radius of $r = 200$ feet and a speed limit of $v = 44$ feet per second (30 mi/hr) is being designed. Use the formula

$$r = \frac{v^2}{4.5 + 32.2 \tan \theta}$$

to find an appropriate superelevation θ for the given curve.

118. *Designing Highway Curves* (Refer to Example 11.) A highway curve with a radius of $r = 800$ feet and a speed limit of $v = 66$ feet per second (45 mi/hr) is being designed. Find the appropriate superelevation for the curve.

119. *Daylight Hours* (Refer to Example 7.) The number of daylight hours y at 60°N latitude can be modeled by

$$y = 6.5 \sin\left(\frac{\pi}{6}(x - 3.65)\right) + 12.4,$$

where $x = 1$ corresponds to January 1, $x = 2$ to February 1, and so on. Estimate graphically or numerically when there are 9 hours of daylight. (**Source:** J. Williams.)

120. *Average Temperatures* The monthly average high temperature y in degrees Fahrenheit at Phoenix, Arizona, can be modeled by

$$y = 20.3 \sin(0.53x - 2.18) + 83.8,$$

where $x = 1$ corresponds to January, $x = 2$ to February, and so on. Estimate graphically or numerically when the average monthly high temperature is 93°F. (**Source:** J. Williams.)

121. *Daylight Hours* Solve Exercise 119 symbolically.

122. *Monthly Average Temperatures* Solve Exercise 120 symbolically.

123. *Maximum Monthly Sunshine*   The maximum number of hours of sunshine each month is listed in the table for 50°N latitude.

Month	1	2	3	4	5	6
Hours of Sunshine	261	279	363	407	471	482
Month	7	8	9	10	11	12
Hours of Sunshine	486	442	374	329	267	246

Source: C. Winter, *Solar Power Plants.*

(a) Find a function f that models the data.

(b) Estimate graphically any solutions to the inequality $f(x) \geq 350$ on the interval $[1, 12]$. Interpret the result.

124. *Music and Pure Tones*   A pure tone can be described by a sinusoidal graph. The graph of the function $P = 0.004 \sin(100\pi t)$ shown in the figure represents the pressure of a pure tone on an eardrum in pounds per square foot at time t in seconds.   (**Source:** J. Roederer, *Introduction to the Physics and Psychophysics of Music.*)

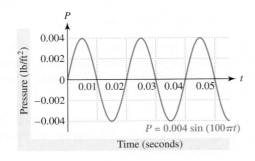

$P = 0.004 \sin(100\pi t)$

Time (seconds)

(a) Estimate all solutions to the equation $P = 0.004$ on the interval $[0, 0.05]$.

(b) Interpret these solutions.

Exercises 125 and 126: Equation of Kepler (Refer to Example 12.) Solve Kepler's equation to within two decimal places graphically or numerically.

125. $\theta = 0.26 + 0.017 \sin \theta$

126. $\theta = 0.18 + 0.249 \sin \theta$

127. *Sums of Pure Tones*   If two loudspeakers located at different positions produce the same pure tone, the human ear may hear one sound that is equal to the sum of the individual tones. Since the sources are at different locations, their sinusoidal waves will have different phase angles.   (**Source:** N. Fletcher and T. Rossing, *The Physics of Musical Instruments.*)

(a) Let two musical tones be given by
$$P_1 = 0.003 \sin(880\pi t - 0.7) \quad \text{and}$$
$$P_2 = 0.002 \sin(880\pi t + 0.6).$$
Graph P_1, P_2, and their sum $P = P_1 + P_2$ separately in the viewing rectangle $[0, 0.01, 0.005]$ by $[-0.005, 0.005, 0.001]$.

(b) Determine the maximum pressure for P.

(c) Is the maximum pressure for P equal to the sum of the maximums of P_1 and P_2? Explain.

128. *Sound*   (Refer to Exercise 127.) Suppose that two loudspeakers located at different positions produce pure tones given by
$$P_1 = A_1 \sin(2\pi Ft + \alpha) \quad \text{and}$$
$$P_2 = A_2 \sin(2\pi Ft + \beta),$$
where F is their common frequency. Then the resulting tone heard by a listener may be written as $P = A \sin(2\pi Ft + \theta)$, where
$$A = \sqrt{(A_1 \cos \alpha + A_2 \cos \beta)^2 + (A_1 \sin \alpha + A_2 \sin \beta)^2}$$
and
$$\theta = \arctan\left[\frac{A_1 \sin \alpha + A_2 \sin \beta}{A_1 \cos \alpha + A_2 \cos \beta}\right].$$

(**Source:** N. Fletcher.)

(a) Find A and θ for P if $F = 440$, $A_1 = 0.003$, $\alpha = -0.7$, $A_2 = 0.002$, and $\beta = 0.6$.

(b) Graph $P = A \sin(2\pi Ft + \theta)$ and $y = P_1 + P_2$ in $[0, 0.01, 0.005]$ by $[-0.005, 0.005, 0.001]$. Do the graphs appear to be identical?

Writing about Mathematics

129. Explain the difference between a conditional equation and an identity. Give one example of each.

130. Explain why knowledge of algebra is important when solving trigonometric equations. Give one example of how knowledge of algebra can be applied to solving a trigonometric equation.

7.4   Sum and Difference Identities

Introduction

Music is made up of vibrations that create pressure on our eardrums. Musical tones can sometimes be modeled with sinusoidal graphs. When more than one tone is played, the resulting pressure is equal to the sum of the individual pressures. Sum and difference identities are sometimes helpful in the analysis of music. This section introduces several trigonometric identities and some of their applications.

Sum and Difference Identities for Cosine

The graph of $y = \cos\left(t - \frac{\pi}{2}\right)$ is translated to the right $\frac{\pi}{2}$ units compared to the graph of $y = \cos t$, as shown in Figure 7.45. If the graph of $y = \cos t$ is translated right $\frac{\pi}{2}$ units, it coincides with the graph of $y = \sin t$, which is shown in Figure 7.46. This discussion suggests that the equation

$$\cos\left(t - \frac{\pi}{2}\right) = \sin t$$

is true for all real numbers t.

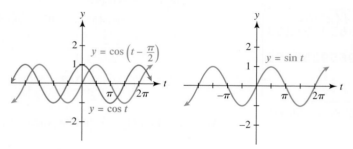

Figure 7.45   **Figure 7.46**

It is important to note that

$$\cos\left(t - \frac{\pi}{2}\right) \neq \cos t - \cos\frac{\pi}{2} = \cos t - 0 = \cos t.$$

To verify this result symbolically, a new identity is needed. Suppose that α and β represent any two angles or real numbers. Then the following identity can be used to calculate the cosine of their difference. (Its proof is given at the end of this section.)

$$\cos(\alpha - \beta) = \cos\alpha\cos\beta + \sin\alpha\sin\beta$$

The next example demonstrates how to apply this identity.

EXAMPLE 1   Using the cosine difference identity

Verify the identity

$$\cos\left(t - \frac{\pi}{2}\right) = \sin t.$$

SOLUTION

Getting Started Start by letting $\alpha = t$ and $\beta = \frac{\pi}{2}$ in the cosine difference identity

$$\cos (\alpha - \beta) = \cos \alpha \cos \beta + \sin \alpha \sin \beta. \blacktriangleright$$

This substitution gives the following result.

$$\cos \left(t - \frac{\pi}{2} \right) = \cos t \cos \frac{\pi}{2} + \sin t \sin \frac{\pi}{2}$$

$$= \cos t (0) + \sin t (1)$$

$$= \sin t$$

Now Try Exercise 29 ◄

In the next example, we use a difference identity to find the exact value of cos 15°.

EXAMPLE 2 Applying the cosine difference identity

Find the exact value of cos 15°. Use a calculator to support your result.

SOLUTION Since $45° - 30° = 15°$ and the exact trigonometric values for 45° and 30° are known, we proceed as follows.

$$\cos 15° = \cos (45° - 30°)$$

$$= \cos 45° \cos 30° + \sin 45° \sin 30° \qquad \textit{Difference identity for cosine}$$

$$= \frac{\sqrt{2}}{2} \cdot \frac{\sqrt{3}}{2} + \frac{\sqrt{2}}{2} \cdot \frac{1}{2} \qquad \textit{Evaluate each function.}$$

$$= \frac{\sqrt{6} + \sqrt{2}}{4} \qquad \textit{Simplify the exact value.}$$

We see in Figure 7.47 that the value of cos 15° agrees with the symbolic result.

Now Try Exercise 5 ◄

```
(√(6)+√(2))/4
          .9659258263
cos(15)
          .9659258263
```

Figure 7.47 Degree Mode

With the aid of the difference identity for cosine, we can derive a sum identity.

$$\cos (\alpha + \beta) = \cos (\alpha - (-\beta))$$

$$= \cos \alpha \cos (-\beta) + \sin \alpha \sin (-\beta) \qquad \textit{Difference identity for cosine}$$

$$= \cos \alpha \cos \beta - \sin \alpha \sin \beta \qquad \textit{cos} (-\beta) = \cos \beta;$$
$$\qquad\qquad\qquad\qquad\qquad\qquad\qquad\qquad \textit{sin} (-\beta) = -\sin \beta$$

Sum and difference identities for cosine are as follows.

Cosine of Sum or Difference

$$\cos (\alpha + \beta) = \cos \alpha \cos \beta - \sin \alpha \sin \beta$$
$$\cos (\alpha - \beta) = \cos \alpha \cos \beta + \sin \alpha \sin \beta$$

NOTE $\cos (\alpha + \beta) \neq \cos \alpha + \cos \beta$ and $\cos (\alpha - \beta) \neq \cos \alpha - \cos \beta$

Cofunction Identities Section 6.2 introduced the cofunction identities for an acute angle θ. These identities are true for any real number t. We verify one of these identities.

EXAMPLE 3 Verifying a cofunction identity

Verify that $\cos\left(\frac{\pi}{2} - t\right) = \sin t$.

SOLUTION Let $\alpha = \frac{\pi}{2}$ and $\beta = t$ in the cosine difference identity.

$$\cos\left(\frac{\pi}{2} - t\right) = \cos\frac{\pi}{2}\cos t + \sin\frac{\pi}{2}\sin t$$

$$= (0)\cos t + (1)\sin t$$

$$= \sin t$$

Now Try Exercise 19 ◀

The following cofunction identities are valid for *any* real number t.

Cofunction Identities for Any Real Number t

$$\cos\left(\frac{\pi}{2} - t\right) = \sin t \qquad \sin\left(\frac{\pi}{2} - t\right) = \cos t$$

$$\cot\left(\frac{\pi}{2} - t\right) = \tan t \qquad \tan\left(\frac{\pi}{2} - t\right) = \cot t$$

$$\csc\left(\frac{\pi}{2} - t\right) = \sec t \qquad \sec\left(\frac{\pi}{2} - t\right) = \csc t$$

Sum and Difference Identities for Sine

There are also sum and difference identities for sine.

$$\sin(\alpha + \beta) = \cos\left(\frac{\pi}{2} - (\alpha + \beta)\right) \qquad \text{Cofunction identity}$$

$$= \cos\left(\left(\frac{\pi}{2} - \alpha\right) - \beta\right) \qquad \text{Associative property}$$

$$= \cos\left(\frac{\pi}{2} - \alpha\right)\cos\beta + \sin\left(\frac{\pi}{2} - \alpha\right)\sin\beta \qquad \text{Difference identity for cosine}$$

$$= \sin\alpha\cos\beta + \cos\alpha\sin\beta \qquad \text{Cofunction identities}$$

The difference identity for sine can be derived in a similar manner.

The following box gives the sum and difference identities for sine.

Sine of Sum or Difference

$$\sin(\alpha + \beta) = \sin\alpha\cos\beta + \cos\alpha\sin\beta$$

$$\sin(\alpha - \beta) = \sin\alpha\cos\beta - \cos\alpha\sin\beta$$

NOTE $\sin(\alpha + \beta) \neq \sin\alpha + \sin\beta$ and $\sin(\alpha - \beta) \neq \sin\alpha - \sin\beta$

Figure 7.48

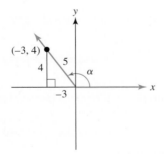

Figure 7.49

EXAMPLE 4 Analyzing an identity graphically, verbally, and symbolically

Give graphical and verbal support for $\sin(\theta + \pi) = -\sin\theta$. Then verify the identity symbolically.

SOLUTION

Graphical Support   Graph $y = \sin(\theta + \pi)$ and $y = -\sin\theta$, as shown in Figures 7.48 and 7.49. Their graphs appear to be identical.

Verbal Support   The graph of $y = \sin(\theta + \pi)$ is similar to the graph of $y = \sin\theta$ except that it is translated left π units. The graph of $y = -\sin\theta$ is similar to the graph of $y = \sin\theta$ except that it is reflected across the x-axis. Translating the sine graph left π units or reflecting the sine graph across the x-axis results in the same graph. Therefore we suspect that $\sin(\theta + \pi) = -\sin\theta$ is an identity.

Symbolic Verification   Let $\alpha = \theta$ and $\beta = \pi$ in the sum identity for sine.

$$\sin(\theta + \pi) = \sin\theta\cos\pi + \cos\theta\sin\pi$$

$$= \sin\theta\,(-1) + \cos\theta\,(0)$$

$$= -\sin\theta \qquad \textit{\textbf{Now Try Exercise 11}} \blacktriangleleft$$

EXAMPLE 5 Applying sum identities for sine and cosine

Let $\sin\alpha = \frac{4}{5}$ and $\cos\beta = \frac{3}{5}$. If α is in quadrant II and β is in quadrant IV, find each of the following.

(a) $\sin(\alpha + \beta)$ **(b)** $\cos(\alpha + \beta)$ **(c)** $\tan(\alpha + \beta)$
(d) The quadrant containing $\alpha + \beta$

SOLUTION

(a) First sketch possible angles for α and for β, as shown in Figures 7.50 and 7.51. We can see that $\cos\alpha = -\frac{3}{5}$ and $\sin\beta = -\frac{4}{5}$. Now apply the sum identities for sine.

$$\sin(\alpha + \beta) = \sin\alpha\cos\beta + \cos\alpha\sin\beta$$

$$= \left(\frac{4}{5}\right)\left(\frac{3}{5}\right) + \left(-\frac{3}{5}\right)\left(-\frac{4}{5}\right)$$

$$= \frac{24}{25}$$

Figure 7.50

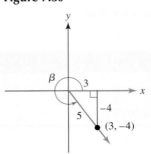

Figure 7.51

(b) To find $\cos(\alpha + \beta)$, apply the sum identity for cosine.

$$\cos(\alpha + \beta) = \cos\alpha\cos\beta - \sin\alpha\sin\beta$$

$$= \left(-\frac{3}{5}\right)\left(\frac{3}{5}\right) - \left(\frac{4}{5}\right)\left(-\frac{4}{5}\right)$$

$$= \frac{7}{25}$$

(c) $\tan(\alpha + \beta) = \dfrac{\sin(\alpha + \beta)}{\cos(\alpha + \beta)} = \dfrac{24/25}{7/25} = \dfrac{24}{7}$

(d) Since both $\sin(\alpha + \beta)$ and $\cos(\alpha + \beta)$ are positive, $\alpha + \beta$ is in quadrant I.

\textbf{Now Try Exercise 21} \blacktriangleleft

EXAMPLE 6 Verifying an identity

Verify the identity $\dfrac{\sin(\alpha - \beta)}{\sin\alpha\sin\beta} = \cot\beta - \cot\alpha$.

SOLUTION Begin by expanding the expression $\sin(\alpha - \beta)$.

$$\dfrac{\sin(\alpha - \beta)}{\sin\alpha\sin\beta} = \dfrac{\sin\alpha\cos\beta - \cos\alpha\sin\beta}{\sin\alpha\sin\beta} \qquad \text{Difference identity}$$

$$= \dfrac{\sin\alpha\cos\beta}{\sin\alpha\sin\beta} - \dfrac{\cos\alpha\sin\beta}{\sin\alpha\sin\beta} \qquad \dfrac{a-b}{c} = \dfrac{a}{c} - \dfrac{b}{c}$$

$$= \dfrac{\cos\beta}{\sin\beta} - \dfrac{\cos\alpha}{\sin\alpha} \qquad \text{Simplify each ratio.}$$

$$= \cot\beta - \cot\alpha \qquad \text{Quotient identity}$$

Now Try Exercise 45 ◀

Applications

Back Stress Because human joints both bend and rotate, trigonometry frequently is applied to human physiology. The next example shows how to calculate the force exerted by a person's back muscles and gives a rather amazing result.

EXAMPLE 7 Analyzing stress on a person's back

If a person with weight W bends at the waist with a straight back, then the force F exerted by the lower back muscles may be approximated using $F = 2.89W\sin\left(\theta + \frac{\pi}{2}\right)$, where θ is the angle between a person's torso and the horizontal. See Figure 7.52. (**Source:** H. Metcalf, *Topics in Classical Biophysics.*)

(a) Let $W = 155$ pounds. The graph of $F = 2.89W\sin\left(\theta + \frac{\pi}{2}\right)$ is shown in Figure 7.53. Interpret the graph.

Figure 7.52

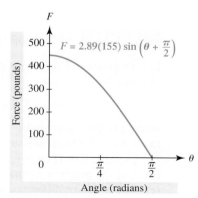

$F = 2.89(155)\sin\left(\theta + \frac{\pi}{2}\right)$

Force (pounds)

Angle (radians)

Figure 7.53

(b) Show that $F = 2.89W\cos\theta$.
(c) For what value of θ does F equal 400 pounds?

SOLUTION

(a) When $\theta = 0$, the person's back is parallel to the ground and force F exerted by the back muscles has a maximum of about 450 pounds. This is nearly *three times* the person's weight! As the person straightens up, θ increases, while F decreases. When $\theta = \frac{\pi}{2}$, the person is standing straight up and $F = 0$.

(b) To show $F = 2.89W \cos \theta$, apply a sum identity for sine.

$$F = 2.89W \sin\left(\theta + \frac{\pi}{2}\right)$$

$$= 2.89W\left(\sin \theta \cos \frac{\pi}{2} + \cos \theta \sin \frac{\pi}{2}\right)$$

$$= 2.89W\left(\sin \theta\,(0) + \cos \theta\,(1)\right)$$

$$= 2.89W \cos \theta$$

(c) Let $F = 400$, $W = 155$, and solve for θ.

$$400 = 2.89\,(155) \cos \theta \qquad \textit{F = 400}$$

$$\cos \theta = \frac{400}{2.89\,(155)} \qquad \textit{Solve for cos }\theta.$$

$$\theta = \cos^{-1}\left(\frac{400}{2.89\,(155)}\right) \qquad \textit{Solve for }\theta.$$

$$\theta \approx 0.467, \text{ or } 26.8° \qquad \textit{Approximate }\theta.$$

Now Try Exercise 53 ◀

Music and Mathematics Music is composed of tones with various frequencies. Pressure exerted on the eardrum by a pure tone may be modeled by either $P(t) = a \cos bt$ or $P(t) = a \sin bt$, where a and b are constants and t represents time. When two tuning forks produce the same pure tone, the human ear hears only one sound that is equal to the sum of the individual tones. Trigonometry can be used to model this situation. (**Source:** N. Fletcher and T. Rossing, *The Physics of Musical Instruments.*)

EXAMPLE 8 Modeling musical tones

Let the pressure P in grams per square meter exerted on the eardrum by two sources be modeled by

$$P_1(t) = 5 \cos(440\pi t) \qquad \text{and} \qquad P_2(t) = 3 \sin(440\pi t),$$

where t is time in seconds.

(a) Graph the total pressure, $P = P_1 + P_2$, on the eardrum in the viewing rectangle [0, 0.01, 0.001] by [−8, 8, 1].

(b) Use the graph to estimate values for a and k such that $P = a \sin(440\pi t + k)$.

(c) Use a sum or difference identity for sine to verify that $P \approx P_1 + P_2$.

SOLUTION

(a) Let $Y_1 = 5 \cos(440\pi X)$, $Y_2 = 3 \sin(440\pi X)$, and $Y_3 = Y_1 + Y_2$. The graph of $y_3 = P_1 + P_2$ is shown in Figure 7.54 on the next page.

[0, 0.01, 0.001] by [−8, 8, 1]

Figure 7.54 Radian Mode

Calculator Help

To find a maximum point on a graph, see Appendix A (page AP-10).

(b) A maximum *y*-value occurs near (0.00039116, 5.831). (Note that the *x*- and *y*-values shown in Figure 7.54 may vary slightly.) Thus let *a* = 5.831. Since sin θ is maximum when $\theta = \frac{\pi}{2}$, we let *t* = 0.00039116 and solve the following equation for *k*.

$$440\pi(0.00039116) + k = \frac{\pi}{2} \qquad\qquad 440\pi t + k = \frac{\pi}{2}$$

$$k = \frac{\pi}{2} - 440\pi(0.00039116)$$

$$\approx 1.0301$$

Thus let $P \approx 5.831 \sin(440\pi t + 1.0301)$. (Other values for *k* are possible.)

(c) Apply the sum identity for sine.

$$P \approx 5.831 \sin(440\pi t + 1.0301)$$
$$= 5.831(\sin(440\pi t)\cos(1.0301) + \cos(440\pi t)\sin(1.0301))$$
$$\approx 5.831(\sin(440\pi t)(0.5147) + \cos(440\pi t)(0.8574))$$
$$\approx 3.00 \sin(440\pi t) + 5.00\cos(440\pi t))$$
$$= P_1 + P_2$$

Now Try Exercise 55 ◀

Sum and Difference Identities for Tangent

Sum and difference identities can also be found for the tangent function.

Tangent of Sum or Difference

$$\tan(\alpha + \beta) = \frac{\tan\alpha + \tan\beta}{1 - \tan\alpha\tan\beta}$$

$$\tan(\alpha - \beta) = \frac{\tan\alpha - \tan\beta}{1 + \tan\alpha\tan\beta}$$

NOTE $\tan(\alpha + \beta) \neq \tan\alpha + \tan\beta$ and $\tan(\alpha - \beta) \neq \tan\alpha - \tan\beta$

These identities are a result of the sum and difference identities for sine and cosine. For example, the difference identity for tangent can be verified as follows.

$$\tan(\alpha - \beta) = \frac{\sin(\alpha - \beta)}{\cos(\alpha - \beta)} \qquad\qquad \text{Use } \tan\theta = \tfrac{\sin\theta}{\cos\theta} \text{ with } \theta = \alpha - \beta.$$

$$= \frac{\sin\alpha\cos\beta - \cos\alpha\sin\beta}{\cos\alpha\cos\beta + \sin\alpha\sin\beta} \qquad\qquad \text{Apply difference identities.}$$

$$= \frac{\dfrac{\sin\alpha\cos\beta}{\cos\alpha\cos\beta} - \dfrac{\cos\alpha\sin\beta}{\cos\alpha\cos\beta}}{\dfrac{\cos\alpha\cos\beta}{\cos\alpha\cos\beta} + \dfrac{\sin\alpha\sin\beta}{\cos\alpha\cos\beta}} \qquad\qquad \text{Divide each term by } \cos\alpha\cos\beta.$$

$$= \frac{\tan\alpha - \tan\beta}{1 + \tan\alpha\tan\beta} \qquad\qquad \text{Simplify.}$$

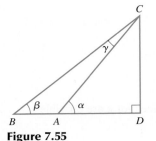

Figure 7.55

EXAMPLE 9 Using the tangent difference identity

Use Figure 7.55 to find $\tan \gamma$ if $\tan \alpha = \frac{4}{3}$ and $\tan \beta = \frac{3}{4}$.

SOLUTION From Figure 7.55, α and $\beta + \gamma$ are both supplements of angle BAC. Thus $\alpha = \beta + \gamma$ or $\gamma = \alpha - \beta$.

$$
\begin{aligned}
\tan \gamma &= \tan(\alpha - \beta) && \gamma = \alpha - \beta \\
&= \frac{\tan \alpha - \tan \beta}{1 + \tan \alpha \tan \beta} && \text{Tangent difference identity} \\
&= \frac{\frac{4}{3} - \frac{3}{4}}{1 + \frac{4}{3} \cdot \frac{3}{4}} && \tan \alpha = \frac{4}{3}; \tan \beta = \frac{3}{4} \\
&= \frac{7}{24} && \text{Simplify.}
\end{aligned}
$$

Now Try Exercise 47 ◀

Derivation of an Identity

We conclude this section by deriving the difference identity for cosine. Begin by considering the angles α and β in standard position and the unit circle, as shown in Figure 7.56. The terminal side of α intersects the unit circle at $(\cos \alpha, \sin \alpha)$, and the terminal side of β intersects the unit circle at $(\cos \beta, \sin \beta)$. The angle formed between the terminal sides of α and β equals $\alpha - \beta$. Now consider angle $\alpha - \beta$ in standard position. Its terminal side intersects the unit circle at the point $(\cos(\alpha - \beta), \sin(\alpha - \beta))$. This is shown in Figure 7.57.

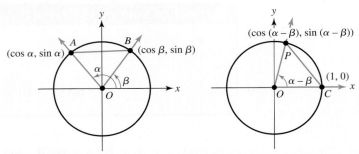

Figure 7.56 Unit Circle   **Figure 7.57** Unit Circle

Since triangles ABO and PCO are congruent, the distance from A to B in Figure 7.56 equals the distance from P to C in Figure 7.57.

$$
\sqrt{(\cos \alpha - \cos \beta)^2 + (\sin \alpha - \sin \beta)^2} = \sqrt{(\cos(\alpha - \beta) - 1)^2 + (\sin(\alpha - \beta) - 0)^2}
$$

Squaring each side and clearing parentheses produces

$$
\cos^2 \alpha - 2 \cos \alpha \cos \beta + \cos^2 \beta + \sin^2 \alpha - 2 \sin \alpha \sin \beta + \sin^2 \beta =
$$
$$
\cos^2(\alpha - \beta) - 2 \cos(\alpha - \beta) + 1 + \sin^2(\alpha - \beta).
$$

Since $\sin^2 \theta + \cos^2 \theta = 1$ for any θ, the above equation simplifies to

$$
2 - 2 \cos \alpha \cos \beta - 2 \sin \alpha \sin \beta = 2 - 2 \cos(\alpha - \beta).
$$

Solving this equation for $\cos(\alpha - \beta)$ gives the cosine difference identity

$$
\cos(\alpha - \beta) = \cos \alpha \cos \beta + \sin \alpha \sin \beta.
$$

7.4 Putting It All Together

Sum and difference identities occur in a variety of applications such as music, physiology, and electronics. The sum and difference identities are also used in calculus. The following table lists the important identities in this section.

Identities	General Form
Cosine sum and difference	$\cos(\alpha + \beta) = \cos\alpha \cos\beta - \sin\alpha \sin\beta$ $\cos(\alpha - \beta) = \cos\alpha \cos\beta + \sin\alpha \sin\beta$
Sine sum and difference	$\sin(\alpha + \beta) = \sin\alpha \cos\beta + \cos\alpha \sin\beta$ $\sin(\alpha - \beta) = \sin\alpha \cos\beta - \cos\alpha \sin\beta$
Tangent sum and difference	$\tan(\alpha + \beta) = \dfrac{\tan\alpha + \tan\beta}{1 - \tan\alpha \tan\beta}$ $\tan(\alpha - \beta) = \dfrac{\tan\alpha - \tan\beta}{1 + \tan\alpha \tan\beta}$
Cofunction	$\cos\left(\frac{\pi}{2} - t\right) = \sin t \qquad \sin\left(\frac{\pi}{2} - t\right) = \cos t$ $\cot\left(\frac{\pi}{2} - t\right) = \tan t \qquad \tan\left(\frac{\pi}{2} - t\right) = \cot t$ $\csc\left(\frac{\pi}{2} - t\right) = \sec t \qquad \sec\left(\frac{\pi}{2} - t\right) = \csc t$

7.4 Exercises

Sum and Difference Identities

Exercises 1–10: Find the exact value for each expression. Use a calculator to support your result numerically.

1. $\sin 15°$ **2.** $\sin 105°$

3. $\tan 15°$ **4.** $\sin 75°$

5. $\cos 75°$ **6.** $\cos 105°$

7. $\sin \frac{\pi}{12}$ $\left(Hint: \frac{\pi}{3} - \frac{\pi}{4} = \frac{\pi}{12}\right)$

8. $\cos \frac{5\pi}{12}$

9. $\sin \frac{5\pi}{12}$ **10.** $\tan \frac{\pi}{12}$

Exercises 11–16: Complete the following for the identity.

(a) Give graphical and verbal support for the identity.
(b) Verify the identity symbolically.

11. $\sin\left(t + \frac{\pi}{2}\right) = \cos t$ **12.** $\sin\left(t + \frac{3\pi}{2}\right) = -\cos t$

13. $\cos(t + \pi) = -\cos t$ **14.** $\cos\left(t + \frac{3\pi}{2}\right) = \sin t$

15. $\sec\left(t - \frac{\pi}{2}\right) = \csc t$ **16.** $\tan\left(t + \frac{\pi}{2}\right) = -\cot t$

Exercises 17–20: Use a difference identity to verify the cofunction identity.

17. $\sin\left(\frac{\pi}{2} - t\right) = \cos t$ **18.** $\tan\left(\frac{\pi}{2} - t\right) = \cot t$

19. $\sec\left(\frac{\pi}{2} - t\right) = \csc t$ **20.** $\csc\left(\frac{\pi}{2} - t\right) = \sec t$

Exercises 21–28: (Refer to Example 5.) Find the following.

(a) $\sin(\alpha + \beta)$ (b) $\cos(\alpha + \beta)$
(c) $\tan(\alpha + \beta)$ (d) *The quadrant containing* $\alpha + \beta$

21. $\sin\alpha = \frac{3}{5}$ and $\sin\beta = \frac{5}{13}$, α and β in quadrant I

22. $\cos\alpha = -\frac{12}{13}$ and $\cos\beta = -\frac{5}{13}$, α and β in quadrant II

23. $\sin\alpha = -\frac{8}{17}$ and $\cos\beta = \frac{11}{61}$, α in quadrant III and β in quadrant I

24. $\cos \alpha = -\frac{24}{25}$ and $\sin \beta = \frac{4}{5}$, α in quadrant II and β in quadrant I

25. $\cos \alpha = -\frac{3}{5}$ and $\cos \beta = \frac{12}{13}$, α in quadrant III and β in quadrant IV

26. $\tan \alpha = \frac{3}{4}$ and $\cos \beta = -\frac{4}{5}$, α in quadrant I and β in quadrant III

27. $\tan \alpha = -\frac{5}{12}$ and $\sec \beta = -\frac{61}{11}$, α and β in quadrant II

28. $\cot \alpha = \frac{3}{4}$ and $\csc \beta = \frac{25}{24}$, α and β in quadrant I

Exercises 29–46: Verify the identity.

29. $\cos\left(t - \frac{\pi}{4}\right) = \frac{\sqrt{2}}{2}(\cos t + \sin t)$

30. $\sin\left(t + \frac{\pi}{4}\right) = \frac{\sqrt{2}}{2}(\cos t + \sin t)$

31. $\tan\left(t + \frac{\pi}{4}\right) = \dfrac{1 + \tan t}{1 - \tan t}$

32. $\tan(45° - \theta) = \dfrac{1 - \tan \theta}{1 + \tan \theta}$

33. $\dfrac{\cos(x - y)}{\cos(x + y)} = \dfrac{1 + \tan x \tan y}{1 - \tan x \tan y}$

34. $\dfrac{\sin(x - y)}{\sin(x + y)} = \dfrac{\tan x - \tan y}{\tan x + \tan y}$

35. $\dfrac{\cos(\alpha - \beta)}{\cos \alpha \sin \beta} = \tan \alpha + \cot \beta$

36. $\cos(\theta + \theta) = 1 - 2\sin^2 \theta$

37. $\sin 2t = 2\sin t \cos t$

38. $\cos 2t = \cos^2 t - \sin^2 t$

39. $\sin(\alpha + \beta) + \sin(\alpha - \beta) = 2\sin \alpha \cos \beta$

40. $\cos(\alpha + \beta) + \cos(\alpha - \beta) = 2\cos \alpha \cos \beta$

41. $\tan(\pi - \theta) = -\tan \theta$

42. $\tan(\theta + \pi) = \tan \theta$

43. $\tan(x - y) - \tan(y - x) = \dfrac{2(\tan x - \tan y)}{1 + \tan x \tan y}$

44. $\dfrac{\sin(x - y)}{\sin y} + \dfrac{\cos(x - y)}{\cos y} = \dfrac{\sin x}{\sin y \cos y}$

45. $\dfrac{\sin(x + y)}{\cos x \cos y} = \tan x + \tan y$

46. $\dfrac{\tan(x + y) - \tan y}{1 + \tan(x + y)\tan y} = \tan x$

Exercises 47 and 48: Solve Example 9 by using the given information.

47. $\tan \alpha = \frac{6}{7}$ and $\tan \beta = \frac{5}{7}$

48. $\cot \alpha = \frac{8}{13}$ and $\cot \beta = \frac{11}{13}$

Lines and Slopes

Exercises 49–52: Suppose two lines, l_1 and l_2, intersect the x-axis making angles α and β, as shown in the figure. Then the slopes of l_1 and l_2 satisfy $m_1 = \tan \alpha$ and $m_2 = \tan \beta$, respectively. If l_1 and l_2 intersect with angle θ as shown, then it follows that $\beta = \alpha + \theta$, or equivalently, $\theta = \beta - \alpha$.

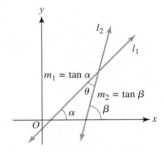

49. Use a difference identity for tangent to show that
$$\tan \theta = \frac{m_2 - m_1}{1 + m_1 m_2}.$$

50. Is the formula in Exercise 49 valid if β is an obtuse angle? Explain.

51. Find θ for the two intersecting lines given by $y = 2x - 3$ and $y = \frac{3}{5}x + 1$.

52. Find θ for the two intersecting lines given by $y = \frac{1}{2}x + 1$ and $y = 3 - x$.

Applications

53. *Back Stress*   (Refer to Example 7.) Let $F = 2.89W\cos \theta$.
   (a) Suppose a 200-pound person bends at the waist so that $\theta = \frac{\pi}{4}$. Estimate the force F exerted by the person's back muscles.

   (b) Approximate the value of θ that results in the back muscles exerting a force F of 400 pounds.

54. *Sound Waves*   Sound is a result of waves applying pressure to a person's eardrum. For a particular sound wave radiating outward, the trigonometric function $P(r) = \frac{a}{r}\cos(\pi r - 1000t)$ can be used to express the pressure at a radius of r feet from the source after t seconds. In this formula, a is the maximum sound pressure at the source, measured in pounds per square foot. (**Source:** L. Beranek, *Noise and Vibration Control.*)

(a) Let $a = 0.4$, $t = 1$, and graph the sound pressure for $0 \le r \le 20$. What happens to the pressure P as the radius r increases?

(b) Use a difference identity to simplify the expression for $P(r)$ when r is an even integer.

55. *Modeling Musical Tones* (Refer to Example 8.) Let the pressure exerted by two sound waves in grams per square meter be given by $P_1(t) = 4 \cos(220\pi t)$ and $P_2(t) = 3 \sin(220\pi t)$, where t is in seconds.
(a) Graph the total pressure $P = P_1 + P_2$ in the window $[0, 0.02, 0.001]$ by $[-6, 6, 1]$.

(b) Use the graph to estimate values for a and k such that $P = a \sin(220\pi t + k)$.

(c) Use a sum or difference identity for sine to verify that
$$a \sin(220\pi t + k) \approx 4 \cos(220\pi t) + 3 \sin(220\pi t).$$

56. *Electricity* When voltages $V_1 = 50 \sin(120\pi t)$ and $V_2 = 120 \cos(120\pi t)$ are applied to the same circuit, the resulting voltage V is equal to their sum. (**Source:** D. Bell, *Fundamentals of Electric Circuits.*)
(a) Graph $V = V_1 + V_2$ in the viewing rectangle $[0, 0.05, 0.01]$ by $[-160, 160, 40]$.

(b) Use the graph to estimate values for a and k so that $V = a \sin(120\pi t + k)$.

(c) Use a sum or difference identity for sine to verify part (b).

Writing about Mathematics

57. Are $\sin(45° + 30°)$ and $\sin 45° + \sin 30°$ equivalent expressions? Explain your answer.

58. Are the expressions $\cos(\alpha - \beta)$ and $\cos\alpha - \cos\beta$ equal? Give an example to justify your answer.

EXTENDED AND DISCOVERY EXERCISES

Modeling Musical Beats Musicians sometimes tune instruments by playing the same tone on two instruments and listening for a phenomenon known as *beats*. Beats occur when two musical tones vary slightly in frequency. When the two instruments are in tune, the beats will disappear. The human ear hears beats because the sound pressure slowly rises and falls as a result of this slight variation in the frequency. The pressure P on an eardrum can be modeled by $P = a \sin(2\pi F t)$, where F is the frequency of the tone, t is time in seconds, and P is in pounds per square foot. This phenomenon can be modeled with a graphing calculator. (**Source:** J. Pierce, *The Science of Musical Sound.*)

1. Consider two tones with similar frequencies of 440 and 443 cycles per second and pressures
$$P_1 = 0.006 \sin(880\pi t) \text{ and } P_2 = 0.004 \sin(886\pi t),$$
respectively.
(a) Graph $P = P_1 + P_2$ in $[0.15, 1.15, 0.05]$ by $[-0.01, 0.01, 0.001]$, where P is the total pressure exerted by the tones on an eardrum. How many beats are there in this 1-second interval?

(b) Repeat part (a) with frequencies of 220 and 224.

(c) Determine a way to find the number of beats per second if the frequencies of the tones are F_1 and F_2.

Exercises 2 and 3: Music and Beats (Refer to Exercise 1.) Given two musical tones P_1 and P_2, graph their sum in $[0.2, 1.2, 0.05]$ by $[-0.01, 0.01, 0.001]$. Count the number of beats in 1 second.

2. $P_1 = 0.007 \sin(450\pi t)$, $P_2 = 0.005 \sin(454\pi t)$

3. $P_1 = 0.004 \cos(830\pi t)$, $P_2 = 0.005 \sin(836\pi t)$

CHECKING BASIC CONCEPTS FOR SECTIONS 7.3 AND 7.4

1. Find the reference angle of each angle.
(a) $225°$ **(b)** $\frac{5\pi}{6}$

2. Solve each equation for θ in $[0°, 360°)$.
(a) $\cos\theta = \frac{1}{2}$ **(b)** $\sin\theta = -\frac{\sqrt{3}}{2}$

3. Find all solutions where t is a real number.
(a) $\sin t = -\cos t$ **(b)** $2\sin^2 t = 1 - \cos t$

4. Use a sum or difference identity to find $\cos\frac{\pi}{12}$.

5. Verify the identity $\sin(t - \pi) = -\sin t$ symbolically. If you have a graphing calculator, give graphical or numerical support.

7.5 Multiple-Angle Identities

- **Learn and use the double-angle identities**
- **Learn and use power-reducing identities**
- **Learn and use the half-angle formulas**
- **Solve equations**
- **Learn and use product-to-sum and sum-to-product identities**

Introduction

In 1831, Michael Faraday discovered that when a wire is passed near a magnet, a small electric current is produced in the wire. This property is used to generate electric current for homes, schools, and businesses throughout the world. By rotating thousands of wires near large electromagnets, massive amounts of electricity can be produced. In 1 year, utilities in the United States generate enough electricity to power a 100-watt light bulb for over 3 billion years!

Voltage, amperage, and wattage are quantities that can be modeled by sinusoidal graphs and functions. To model electricity and other phenomena, trigonometric functions and identities are used. This section introduces several important multiple-angle identities. **(Sources:** R. Weidner and R. Sells, *Elementary Classical Physics*, Vol. 2; J. Wright, *The Universal Almanac.)*

Double-Angle Identities

The sum identities can be used to derive double-angle identities for sine, cosine, and tangent. In each identity we let $\alpha = \theta$ and $\beta = \theta$.

Sine Double-Angle Identity

$$\sin 2\theta = \sin(\theta + \theta)$$
$$= \sin\theta\cos\theta + \cos\theta\sin\theta \qquad \sin(\alpha + \beta) = \sin\alpha\cos\beta + \cos\alpha\sin\beta$$
$$= 2\sin\theta\cos\theta$$

Cosine Double-Angle Identities

$$\cos 2\theta = \cos(\theta + \theta)$$
$$= \cos\theta\cos\theta - \sin\theta\sin\theta \qquad \cos(\alpha + \beta) = \cos\alpha\cos\beta - \sin\alpha\sin\beta$$
$$= \cos^2\theta - \sin^2\theta$$

Applying the Pythagorean identity $\sin^2\theta + \cos^2\theta = 1$, we can write the expression $\cos^2\theta - \sin^2\theta$ as

$$\cos^2\theta - \sin^2\theta = \cos^2\theta - (1 - \cos^2\theta)$$
$$= 2\cos^2\theta - 1$$

or as

$$\cos^2\theta - \sin^2\theta = (1 - \sin^2\theta) - \sin^2\theta$$
$$= 1 - 2\sin^2\theta.$$

Tangent Double-Angle Identity

$$\tan 2\theta = \tan(\theta + \theta)$$
$$= \frac{\tan\theta + \tan\theta}{1 - \tan\theta\tan\theta} \qquad \tan(\alpha + \beta) = \frac{\tan\alpha + \tan\beta}{1 - \tan\alpha\tan\beta}$$
$$= \frac{2\tan\theta}{1 - \tan^2\theta}$$

A summary of these double-angle identities is given below.

Double-Angle Identities

$$\sin 2\theta = 2 \sin \theta \cos \theta$$
$$\cos 2\theta = \cos^2 \theta - \sin^2 \theta = 2 \cos^2 \theta - 1 = 1 - 2 \sin^2 \theta$$
$$\tan 2\theta = \frac{2 \tan \theta}{1 - \tan^2 \theta}$$

The next example illustrates that $\sin 2\theta \neq 2 \sin \theta$.

EXAMPLE 1 **Using double-angle identities**

Verify graphically and symbolically that $\sin 2\theta$ and $2 \sin \theta$ are *not* equivalent expressions.

SOLUTION

Graphical Verification Graph $y = \sin 2\theta$ and $y = 2 \sin \theta$, as shown in Figures 7.58 and 7.59. Notice that the graphs are different.

Symbolic Verification The expressions $\sin 2\theta$ and $2 \sin \theta$ are not equivalent; rather,

$$\sin 2\theta = 2 \sin \theta \cos \theta \neq 2 \sin \theta.$$

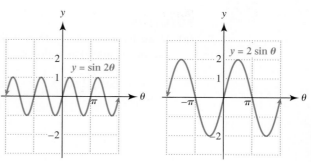

Figure 7.58   **Figure 7.59**

Now Try Exercise 15 ◄

EXAMPLE 2 **Using double-angle identities**

Given $\cos \theta = -\frac{12}{13}$ and $\sin \theta > 0$, find $\sin 2\theta$, $\cos 2\theta$, and $\tan 2\theta$. Use a calculator to support your result.

SOLUTION

Symbolic Solution Since $\cos \theta < 0$ and $\sin \theta > 0$, θ is contained in quadrant II. One possibility for θ is shown in Figure 7.60. We see that $\sin \theta = \frac{5}{13}$. Using double-angle identities, we obtain the following results.

$$\sin 2\theta = 2 \sin \theta \cos \theta = 2 \cdot \frac{5}{13} \cdot \left(-\frac{12}{13} \right) = -\frac{120}{169}$$

$$\cos 2\theta = \cos^2 \theta - \sin^2 \theta = \left(-\frac{12}{13} \right)^2 - \left(\frac{5}{13} \right)^2 = \frac{119}{169}$$

$$\tan 2\theta = \frac{\sin 2\theta}{\cos 2\theta} = \frac{-120/169}{119/169} = -\frac{120}{119}$$

Figure 7.60

Calculator Support Since θ is contained in quadrant II, we can support these results by letting $\theta = \arccos\left(-\frac{12}{13}\right) \approx 157.38°$ and performing the calculations shown in Figure 7.61 and in Figure 7.62.

```
sin(2cos⁻¹(⁻12/13
))▶Frac
              ⁻120/169
cos(2cos⁻¹(⁻12/13
))▶Frac
              119/169
```

Figure 7.61

```
tan(2cos⁻¹(⁻12/13
))▶Frac
              ⁻120/119
```

Figure 7.62

Now Try Exercise 19 ◀

EXAMPLE 3 Evaluating expressions with double-angle identities

Use a double-angle identity to evaluate each expression.

(a) $\cos\left(2\sin^{-1}\frac{1}{3}\right)$ **(b)** $\sin\left(2\cos^{-1}\left(-\frac{3}{5}\right)\right)$ **(c)** $\sin\left(2\tan^{-1}x\right), x > 0$

SOLUTION

(a) If we let $\theta = \sin^{-1}\frac{1}{3}$, then it follows that $\sin\theta = \frac{1}{3}$. The expression $\cos\left(2\sin^{-1}\frac{1}{3}\right)$ can be evaluated as follows.

$$\cos\left(2\sin^{-1}\frac{1}{3}\right) = \cos(2\theta) \qquad \theta = \sin^{-1}\frac{1}{3}$$

$$= 1 - 2\sin^2\theta \qquad \text{Double-angle identity}$$

$$= 1 - 2\left(\frac{1}{3}\right)^2 \qquad \sin\theta = \frac{1}{3}$$

$$= \frac{7}{9} \qquad \text{Simplify.}$$

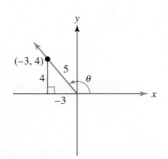

Figure 7.63

(b) Begin by letting $\theta = \cos^{-1}\left(-\frac{3}{5}\right)$ and sketching angle θ in standard position, as shown in Figure 7.63. Notice that $\cos\theta = -\frac{3}{5}$ and that θ is a second quadrant angle. From Figure 7.63 it follows that $\sin\theta = \frac{4}{5}$. The expression $\sin\left(2\cos^{-1}\left(-\frac{3}{5}\right)\right)$ can be evaluated as follows.

$$\sin\left(2\cos^{-1}\left(-\frac{3}{5}\right)\right) = \sin(2\theta) \qquad \theta = \cos^{-1}\left(-\frac{3}{5}\right)$$

$$= 2\sin\theta\cos\theta \qquad \text{Double-angle identity}$$

$$= 2\left(\frac{4}{5}\right)\left(-\frac{3}{5}\right) \qquad \sin\theta = \frac{4}{5}, \cos\theta = -\frac{3}{5}$$

$$= -\frac{24}{25} \qquad \text{Simplify.}$$

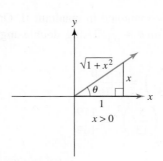

Figure 7.64

(c) Begin by letting $\theta = \tan^{-1}x$ and sketching angle θ in standard position, as shown in Figure 7.64. Since $x > 0$, we have drawn a first quadrant angle θ whose tangent function equals $\frac{x}{1}$. That is, $\tan\theta = x$. From Figure 7.64 it can be concluded that

$\sin \theta = \dfrac{x}{\sqrt{1 + x^2}}$ and $\cos \theta = \dfrac{1}{\sqrt{1 + x^2}}$. The trigonometric expression $\sin (2 \tan^{-1} x)$ can be evaluated as follows.

$$\sin (2 \tan^{-1} x) = \sin (2\theta) \qquad\qquad \theta = \tan^{-1} x$$

$$= 2 \sin \theta \cos \theta \qquad\qquad \textit{Double-angle identity}$$

$$= 2 \left(\dfrac{x}{\sqrt{1 + x^2}} \right) \left(\dfrac{1}{\sqrt{1 + x^2}} \right) \qquad \textit{Substitute.}$$

$$= \dfrac{2x}{1 + x^2} \qquad\qquad \textit{Simplify.}$$

Now Try Exercises 25, 27, and 31 ◀

In the next example, we verify an identity by using a double-angle identity.

EXAMPLE 4 Verifying an identity

Verify the identity $\dfrac{\sec^2 \theta}{1 - \tan^2 \theta} = \sec 2\theta$.

SOLUTION Begin by applying a reciprocal identity.

$$\dfrac{\sec^2 \theta}{1 - \tan^2 \theta} = \dfrac{1}{\cos^2 \theta \, (1 - \tan^2 \theta)} \qquad \textit{Reciprocal identity}$$

$$= \dfrac{1}{\cos^2 \theta \left(1 - \dfrac{\sin^2 \theta}{\cos^2 \theta} \right)} \qquad \textit{Quotient identity}$$

$$= \dfrac{1}{\cos^2 \theta - \sin^2 \theta} \qquad \textit{Distributive property}$$

$$= \dfrac{1}{\cos 2\theta} \qquad \textit{Double-angle identity}$$

$$= \sec 2\theta \qquad \textit{Reciprocal identity}$$

Now Try Exercise 67 ◀

EXAMPLE 5 Deriving a triple-angle identity

Write $\cos 3\theta$ in terms of $\cos \theta$.

SOLUTION

$$\cos 3\theta = \cos (2\theta + \theta) \qquad\qquad 3\theta = 2\theta + \theta$$

$$= \cos 2\theta \cos \theta - \sin 2\theta \sin \theta \qquad \textit{Sum identity for cosine}$$

$$= (2 \cos^2 \theta - 1) \cos \theta - (2 \sin \theta \cos \theta) \sin \theta \qquad \textit{Double-angle identities}$$

$$= 2 \cos^3 \theta - \cos \theta - 2 \sin^2 \theta \cos \theta \qquad \textit{Multiply.}$$

$$= 2 \cos^3 \theta - \cos \theta - 2 (1 - \cos^2 \theta) \cos \theta \qquad \textit{Apply } \sin^2 \theta + \cos^2 \theta = 1.$$

$$= 2 \cos^3 \theta - \cos \theta - 2 \cos \theta + 2 \cos^3 \theta \qquad \textit{Distributive property}$$

$$= 4 \cos^3 \theta - 3 \cos \theta \qquad \textit{Combine like terms.}$$

Now Try Exercise 71 ◀

Power-Reducing Identities

Power-reducing identities for sine, cosine, and tangent can be derived using the double-angle identities.

Sine Power-Reducing Identity   We can solve $\cos 2\theta = 1 - 2\sin^2\theta$ for $\sin^2\theta$ to obtain

$$\sin^2\theta = \frac{1 - \cos 2\theta}{2}.$$

Cosine Power-Reducing Identity   Solving $\cos 2\theta = 2\cos^2\theta - 1$ for $\cos^2\theta$ gives

$$\cos^2\theta = \frac{1 + \cos 2\theta}{2}.$$

Tangent Power-Reducing Identity   We can use the power-reducing identities

$$\sin^2\theta = \frac{1 - \cos 2\theta}{2} \quad \text{and} \quad \cos^2\theta = \frac{1 + \cos 2\theta}{2}$$

to derive

$$\tan^2\theta = \frac{\sin^2\theta}{\cos^2\theta} = \frac{1 - \cos 2\theta}{1 + \cos 2\theta}.$$

A summary of these identities is given in the box below.

Power-Reducing Identities

$$\sin^2\theta = \frac{1 - \cos 2\theta}{2} \qquad \cos^2\theta = \frac{1 + \cos 2\theta}{2} \qquad \tan^2\theta = \frac{1 - \cos 2\theta}{1 + \cos 2\theta}$$

In the next example, we write known identities in terms of inputs other than θ.

EXAMPLE 6   Writing identities

Complete each statement.

(a) $\sin^2\theta = \frac{1 - \cos 2\theta}{2}$, so $\sin^2 4t = $ _____.

(b) $\tan^2\theta = \frac{1 - \cos 2\theta}{1 + \cos 2\theta}$, so $\tan^2\left(\frac{1}{2}x\right) = $ _____.

(c) $\sin 2\theta = 2\sin\theta\cos\theta$, so $\sin 8t = $ _____.

SOLUTION

(a) Getting Started   Substitute $4t$ for θ on each side of the given equation. ▶

$$\sin^2 4t = \frac{1 - \cos(2 \cdot 4t)}{2}$$

$$= \frac{1 - \cos 8t}{2}$$

(b) Substitute $\frac{1}{2}x$ for θ on each side of the equation.

$$\tan^2\left(\frac{1}{2}x\right) = \frac{1 - \cos\left(2 \cdot \frac{1}{2}x\right)}{1 + \cos\left(2 \cdot \frac{1}{2}x\right)}$$

$$= \frac{1 - \cos x}{1 + \cos x}$$

(c) Let $2\theta = 8t$, so $\theta = 4t$. Thus $\sin 2\theta = 2 \sin \theta \cos \theta$ implies

$$\sin 8t = 2 \sin 4t \cos 4t.$$ *Now Try Exercises 1, 3, and 5* ◀

EXAMPLE 7 **Using a power-reducing identity**

Find the exact value of $\sin^2 22.5°$. Use a calculator to support your results.

SOLUTION Let $\theta = 22.5°$ and $2\theta = 45°$ and apply the sine power-reducing identity.

$$\sin^2 22.5° = \frac{1 - \cos 45°}{2} \qquad \sin^2 \theta = \frac{1 - \cos 2\theta}{2}$$

$$= \frac{1 - \sqrt{2}/2}{2} \qquad \cos 45° = \frac{\sqrt{2}}{2}$$

$$= \frac{2 - \sqrt{2}}{4} \qquad \text{Multiply by } \frac{2}{2}.$$

```
(2-√(2))/4
          .1464466094
(sin(22.5))^2
          .1464466094
```

Figure 7.65 Degree Mode

Support for this result is shown in Figure 7.65.   *Now Try Exercise 45* ◀

An Application   Next we apply a power-reducing identity to electrical circuits.

EXAMPLE 8 **Using a power-reducing identity to analyze wattage**

Amperage I is a measure of the amount of electricity passing through a wire, and voltage V is a measure of the force "pushing" the electricity. The wattage W consumed by an electrical device can be calculated using the equation $W = VI$. **(Source:** G. Wilcox and C. Hesselberth, *Electricity for Engineering Technology.*)

(a) Voltage in a household circuit is given by $V = 160 \sin (120\pi t)$, where t is in seconds. Suppose that the amperage flowing through a toaster is given by $I = 12 \sin (120\pi t)$. Graph the wattage W consumed by the toaster in [0, 0.04, 0.01] by [−200, 3000, 200].

(b) Write the wattage as $W = a \cos (k\pi t) + d$, where a, k, and d are constants.

(c) Compare the periods of the voltage, amperage, and wattage.

(d) The wattage of this toaster equals half the maximum of W. Find the wattage.

SOLUTION

(a) Since $W = VI$, graph the equation $Y_3 = Y_1 * Y_2$, where $Y_1 = 160 \sin (120\pi X)$ and $Y_2 = 12 \sin (120\pi X)$, as shown in Figure 7.66, where *radian* mode is used.

(b)

$$W = VI$$

$$= 160 \sin (120\pi t) \, 12 \sin (120\pi t) \qquad \text{Substitute for } V \text{ and } I.$$

$$= 1920 \sin^2 (120\pi t) \qquad \text{Multiply.}$$

$$= 1920 \cdot \frac{1 - \cos (240\pi t)}{2} \qquad \text{Power-reducing identity}$$

$$= 960 - 960 \cos (240\pi t) \qquad \text{Simplify.}$$

[0, 0.04, 0.01] by [−200, 3000, 200]

$y_3 = y_1 * y_2$

Figure 7.66 Wattage W

Thus let $a = -960$, $k = 240$, and $d = 960$. Then the wattage can be written as

$$W = -960 \cos (240\pi t) + 960.$$

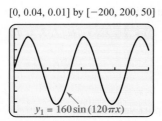

[0, 0.04, 0.01] by [−200, 200, 50]

$y_1 = 160 \sin (120\pi x)$

Figure 7.67 Voltage V

(c) The period for both V and I is $\frac{2\pi}{120\pi} = \frac{1}{60}$ second, and the period for W is $\frac{2\pi}{240\pi} = \frac{1}{120}$ second. This result is supported in Figure 7.67, where the graph of V requires twice as much time as W to complete one oscillation.

(d) The maximum wattage is 1920 watts whenever $\cos (240\pi t) = -1$. Half this amount is 960 watts, which is the wattage rating for the toaster. ***Now Try Exercise 119*** ◄

Half-Angle Formulas

We can obtain half-angle formulas by using the power-reducing identities.

$$\sin^2 x = \frac{1 - \cos 2x}{2} \qquad \textit{Power-reducing identity}$$

$$\sin x = \pm\sqrt{\frac{1 - \cos 2x}{2}} \qquad \textit{Square root property}$$

$$\sin\frac{\theta}{2} = \pm\sqrt{\frac{1 - \cos \theta}{2}} \qquad \textit{Let } x = \frac{\theta}{2} \textit{ and } 2x = \theta.$$

The following box gives some half-angle formulas. (The second and third half-angle formulas for tangent are verified in Exercises 83 and 84.)

Half-Angle Formulas

$$\sin\frac{\theta}{2} = \pm\sqrt{\frac{1 - \cos \theta}{2}} \qquad \cos\frac{\theta}{2} = \pm\sqrt{\frac{1 + \cos \theta}{2}}$$

$$\tan\frac{\theta}{2} = \pm\sqrt{\frac{1 - \cos \theta}{1 + \cos \theta}} \qquad \tan\frac{\theta}{2} = \frac{1 - \cos \theta}{\sin \theta} \qquad \tan\frac{\theta}{2} = \frac{\sin \theta}{1 + \cos \theta}$$

To decide whether a positive or negative sign should be used in a half-angle formula, we must determine the quadrant containing $\frac{\theta}{2}$. This is illustrated in the next example.

EXAMPLE 9 Using a half-angle formula to find an exact value

Find the exact value of $\sin (-15°)$.

SOLUTION We can use the fact that $\cos (-30°) = \frac{\sqrt{3}}{2}$ to find $\sin (-15°)$. We choose the negative sign in the half-angle formula since the sine function is negative in quadrant IV.

$$\sin (-15°) = \sin \left(\frac{-30°}{2}\right) \qquad \textit{Let } \frac{\theta}{2} = -15° \textit{ and } \theta = -30°.$$

$$= -\sqrt{\frac{1 - \cos (-30°)}{2}} \qquad \textit{Half-angle formula}$$

$$= -\sqrt{\frac{1 - \sqrt{3}/2}{2}} \qquad \cos (-30°) = \frac{\sqrt{3}}{2}$$

$$= -\sqrt{\frac{2 - \sqrt{3}}{4}} \qquad \textit{Multiply numerator and denominator by 2.}$$

$$= -\frac{\sqrt{2 - \sqrt{3}}}{2} \qquad \sqrt{4} = 2 \qquad \qquad \textit{Now Try Exercise 49} ◄$$

EXAMPLE 10   Using half-angle formulas to find exact values

If $\cos\theta = -\frac{3}{5}$ and $90° \le \theta \le 180°$, find $\sin\frac{\theta}{2}$, $\cos\frac{\theta}{2}$, and $\tan\frac{\theta}{2}$.

SOLUTION   Since $90° \le \theta \le 180°$, it follows that $45° \le \frac{\theta}{2} \le 90°$. In quadrant I the values for $\sin\frac{\theta}{2}$, $\cos\frac{\theta}{2}$, and $\tan\frac{\theta}{2}$ are all positive.

$$\sin\frac{\theta}{2} = \sqrt{\frac{1 - \cos\theta}{2}} = \sqrt{\frac{1 + 3/5}{2}} = \frac{2}{\sqrt{5}}, \text{ or } \frac{2\sqrt{5}}{5}$$

$$\cos\frac{\theta}{2} = \sqrt{\frac{1 + \cos\theta}{2}} = \sqrt{\frac{1 - 3/5}{2}} = \frac{1}{\sqrt{5}}, \text{ or } \frac{\sqrt{5}}{5}$$

$$\tan\frac{\theta}{2} = \sqrt{\frac{1 - \cos\theta}{1 + \cos\theta}} = \sqrt{\frac{1 + 3/5}{1 - 3/5}} = 2 \qquad \textit{Now Try Exercise 59} \blacktriangleleft$$

Solving Equations

When trigonometric functions are used in modeling to make predictions, there is often a need to solve trigonometric equations.

EXAMPLE 11   Solving a trigonometric equation

Solve the trigonometric equation $\cos\theta - \sin 2\theta = 0$ symbolically, graphically, and numerically for $0° \le \theta < 360°$.

SOLUTION

Symbolic Solution   We begin by applying the identity $\sin 2\theta = 2\sin\theta\cos\theta$.

$\cos\theta - \sin 2\theta = 0$	*Given equation*
$\cos\theta - 2\sin\theta\cos\theta = 0$	*Double-angle identity*
$\cos\theta(1 - 2\sin\theta) = 0$	*Factor out $\cos\theta$.*
$\cos\theta = 0 \qquad$ or $\qquad 1 - 2\sin\theta = 0$	*Zero-product property*
$\cos\theta = 0 \qquad$ or $\qquad \sin\theta = \dfrac{1}{2}$	*Solve for $\sin\theta$.*
$\theta = 90°, 270° \qquad$ or $\qquad \theta = 30°, 150°$	*Solve for θ.*

On the interval $[0°, 360°)$, the solutions are $30°$, $90°$, $150°$, and $270°$.

Graphical Solution   Graphical support is given in Figure 7.68, where the graph of $Y_1 = \cos(X) - \sin(2X)$ is shown. Note that the four x-intercepts $30°$, $90°$, $150°$, and $270°$ correspond to the four symbolic solutions.

Numerical Solution   Numerical support is shown in Figure 7.69.

Now Try Exercise 95 ◀

[0, 352.5, 30] by [−2, 2, 1]

$y_1 = \cos x - \sin 2x$

X=30   Y=0

Figure 7.68 Degree Mode

X	Y1
30	0
90	0
150	0
210	⁻1.732
270	0
330	1.7321
390	0

X=30

Figure 7.69 Degree Mode

EXAMPLE 12   Solving a trigonometric equation

Solve the equation $4\cos\frac{x}{2} - 2 = 0$ for $0° \le x < 360°$.

SOLUTION Begin by solving $4 \cos \frac{x}{2} - 2 = 0$ for $\cos \frac{x}{2}$.

$$4 \cos \frac{x}{2} - 2 = 0 \qquad \textit{Given equation}$$

$$\cos \frac{x}{2} = \frac{1}{2} \qquad \textit{Solve for } \cos \frac{x}{2}.$$

$$\frac{x}{2} = 60° \quad \text{or} \quad \frac{x}{2} = 300° \qquad \textit{Solve for } \frac{x}{2}.$$

$$x = 120° \quad \text{or} \quad x = 600° \qquad \textit{Multiply by 2.}$$

The only solution on $[0°, 360°)$ is $120°$. **Now Try Exercise 99** ◀

EXAMPLE 13 Solving trigonometric equations

Find all solutions, expressed in radians.
(a) $\cos 2t + 2 \cos^2 t = 0$ **(b)** $2 \sin 2t = \sqrt{3}$

SOLUTION
(a) Begin by applying the double-angle identity $\cos 2\theta = 2 \cos^2 \theta - 1$.

$$\cos 2t + 2 \cos^2 t = 0 \qquad \textit{Given equation}$$

$$2 \cos^2 t - 1 + 2 \cos^2 t = 0 \qquad \textit{Double-angle identity}$$

$$4 \cos^2 t - 1 = 0 \qquad \textit{Combine terms.}$$

$$\cos^2 t = \frac{1}{4} \qquad \textit{Solve for } \cos^2 t.$$

$$\cos t = \pm \frac{1}{2} \qquad \textit{Square root property}$$

On the interval $[0, 2\pi)$, there are four angles whose cosines equal either $\frac{1}{2}$ or $-\frac{1}{2}$. They are $\frac{\pi}{3}, \frac{2\pi}{3}, \frac{4\pi}{3}$, and $\frac{5\pi}{3}$. Since $\cos t$ has period 2π, all solutions can be written as

$$\frac{\pi}{3} + 2\pi n, \quad \frac{2\pi}{3} + 2\pi n, \quad \frac{4\pi}{3} + 2\pi n, \quad \text{or} \quad \frac{5\pi}{3} + 2\pi n,$$

where n is an integer. Note that since $\frac{4\pi}{3} - \frac{\pi}{3} = \pi$, the solutions $\frac{\pi}{3} + 2\pi n$ and $\frac{4\pi}{3} + 2\pi n$ can be combined and written as $\frac{\pi}{3} + \pi n$. Similarly, the solutions $\frac{2\pi}{3} + 2\pi n$ and $\frac{5\pi}{3} + 2\pi n$ can be written as $\frac{2\pi}{3} + \pi n$. As a result, all solutions can also be written as

$$\frac{\pi}{3} + \pi n \quad \text{or} \quad \frac{2\pi}{3} + \pi n.$$

(b) Begin by dividing each side by 2.

$$2 \sin 2t = \sqrt{3} \qquad \textit{Given equation}$$

$$\sin 2t = \frac{\sqrt{3}}{2} \qquad \textit{Divide by 2.}$$

Let $\theta = 2t$ and find all values of θ on the interval $[0, 2\pi)$, where $\sin \theta = \frac{\sqrt{3}}{2}$. Because the reference angle for θ is $\theta_R = \sin^{-1} \frac{\sqrt{3}}{2} = \frac{\pi}{3}$ and the sine function is positive in quadrants I and II, it follows that $\theta = \frac{\pi}{3}$ or $\theta = \frac{2\pi}{3}$. Thus all possible solutions in terms of θ can be written as

$$\theta = \frac{\pi}{3} + 2\pi n \qquad \text{or} \qquad \theta = \frac{2\pi}{3} + 2\pi n.$$

Since $\theta = 2t$, we can write the solutions to the given equation in terms of t.

$$2t = \frac{\pi}{3} + 2\pi n \qquad \text{or} \qquad 2t = \frac{2\pi}{3} + 2\pi n$$

$$t = \frac{\pi}{6} + \pi n \qquad \text{or} \qquad t = \frac{\pi}{3} + \pi n \qquad \textit{Divide by 2.}$$

Now Try Exercises 101 and 107 ◀

Product-to-Sum and Sum-to-Product Identities

The sum and difference identities for sine and cosine can be used to derive several identities that make it possible to rewrite a product as a sum. For example, adding the identities for $\cos(\alpha + \beta)$ and $\cos(\alpha - \beta)$ results in the following.

$$\cos(\alpha + \beta) = \cos\alpha\cos\beta - \sin\alpha\sin\beta$$
$$\cos(\alpha - \beta) = \cos\alpha\cos\beta + \sin\alpha\sin\beta$$

$$\overline{\cos(\alpha + \beta) + \cos(\alpha - \beta) = 2\cos\alpha\cos\beta}$$

Rewriting gives $\cos\alpha\cos\beta = \frac{1}{2}(\cos(\alpha + \beta) + \cos(\alpha - \beta))$. Four product-to-sum identities are given below. The other three identities are derived in a similar manner.

Product-to-Sum Identities

$$\cos\alpha\cos\beta = \frac{1}{2}(\cos(\alpha + \beta) + \cos(\alpha - \beta))$$

$$\sin\alpha\sin\beta = \frac{1}{2}(\cos(\alpha - \beta) - \cos(\alpha + \beta))$$

$$\sin\alpha\cos\beta = \frac{1}{2}(\sin(\alpha + \beta) + \sin(\alpha - \beta))$$

$$\cos\alpha\sin\beta = \frac{1}{2}(\sin(\alpha + \beta) - \sin(\alpha - \beta))$$

EXAMPLE 14 Using a product-to-sum identity

Write the product $\cos 5\theta \cos 3\theta$ as a sum.

SOLUTION We begin by applying the first product-to-sum identity with the substitution $\alpha = 5\theta$ and $\beta = 3\theta$.

$$\cos 5\theta \cos 3\theta = \frac{1}{2}(\cos(5\theta + 3\theta) + \cos(5\theta - 3\theta))$$

$$= \frac{1}{2}(\cos 8\theta + \cos 2\theta)$$

Now Try Exercise 85 ◀

By rewriting the product-to-sum identities, we can derive four sum-to-product identities. (Sum-to-product identities have applications in areas such as musical sounds and touch-tone phones.) If we let $a = \alpha + \beta$ and $b = \alpha - \beta$, and we multiply by 2, then the identity

$$\cos \alpha \cos \beta = \frac{1}{2}(\cos(\alpha + \beta) + \cos(\alpha - \beta))$$

becomes

$$2 \cos \frac{a+b}{2} \cos \frac{a-b}{2} = \cos a + \cos b.$$

Note that

$$\frac{a+b}{2} = \frac{\alpha + \beta + \alpha - \beta}{2} = \alpha \quad \text{and}$$

$$\frac{a-b}{2} = \frac{(\alpha + \beta) - (\alpha - \beta)}{2} = \beta.$$

The other three sum-to-product identities can be derived in a similar manner.

Sum-to-Product Identities

$$\cos a + \cos b = 2 \cos \frac{a+b}{2} \cos \frac{a-b}{2}$$

$$\cos a - \cos b = -2 \sin \frac{a+b}{2} \sin \frac{a-b}{2}$$

$$\sin a + \sin b = 2 \sin \frac{a+b}{2} \cos \frac{a-b}{2}$$

$$\sin a - \sin b = 2 \cos \frac{a+b}{2} \sin \frac{a-b}{2}$$

For example, we can write the sum $\cos 70° + \cos 40°$ as follows.

$$\cos 70° + \cos 40° = 2 \cos \frac{70° + 40°}{2} \cos \frac{70° - 40°}{2} = 2 \cos 55° \cos 15°$$

An Application from Telephone Technology   Touch-tone phones are often used to make business transactions. One reason this can be done is that each number has a unique pair of frequencies. For example, when 1 is pressed, two frequencies of 697 hertz and 1209 hertz are simultaneously transmitted through the phone line. A *hertz* (Hz) is equal to one cycle per second. When 2 is pressed, the pair of frequencies transmitted is 697 hertz and 1336 hertz. As a result, 2 has a different tone from 1. Table 7.1 shows the frequency pairs for the numbers 0 through 9.

Table 7.1   Frequencies Used in Touch-Tone Phones

Number	0	1	2	3	4	5	6	7	8	9
Frequency 1 (Hz)	941	697	697	697	770	770	770	852	852	852
Frequency 2 (Hz)	1336	1209	1336	1477	1209	1336	1477	1209	1336	1477

A tone with frequencies F_1 and F_2 can be modeled by

$$a_1 \cos (2\pi F_1 t) + a_2 \cos (2\pi F_2 t).$$

If both tones have the same intensity, then we can let $a_1 = a_2 = 1$.

EXAMPLE 15 Analyzing touch-tone phones

(a) Write an expression that models the sound of a 5 on a touch-tone phone.
(b) Rewrite the expression in part (a) as a product of trigonometric expressions.

SOLUTION

(a) From Table 7.1 we can see that for number 5, $F_1 = 770$ and $F_2 = 1336$. Thus

$$y = \cos (1540\pi t) + \cos (2672\pi t).$$

(b) Let $a = 1540\pi t$ and $b = 2672\pi t$ in the first sum-to-product identity.

$$\cos 1540\pi t + \cos 2672\pi t$$

$$= 2 \cos \frac{1540\pi t + 2672\pi t}{2} \cos \frac{1540\pi t - 2672\pi t}{2}$$

$$= 2 \cos (2106\pi t) \cos (-566\pi t)$$

$$= 2 \cos (2106\pi t) \cos (566\pi t) \qquad \cos (-\theta) = \cos \theta$$

Now Try Exercise 125 ◄

7.5 Putting It All Together

This section introduced several multiple-angle identities. If the trigonometric functions of θ are known, these identities can be used to find the trigonometric functions of $\frac{\theta}{2}$ and 2θ. Some multiple-angle identities or formulas are summarized in the following table.

Identity or Formula	General Form
Double-angle	$\sin 2\theta = 2 \sin \theta \cos \theta$ $\cos 2\theta = \cos^2 \theta - \sin^2 \theta = 2 \cos^2 \theta - 1 = 1 - 2 \sin^2 \theta$ $\tan 2\theta = \dfrac{2 \tan \theta}{1 - \tan^2 \theta}$
Power-reducing	$\sin^2 \theta = \dfrac{1 - \cos 2\theta}{2} \qquad \cos^2 \theta = \dfrac{1 + \cos 2\theta}{2} \qquad \tan^2 \theta = \dfrac{1 - \cos 2\theta}{1 + \cos 2\theta}$
Half-angle	$\sin \dfrac{\theta}{2} = \pm\sqrt{\dfrac{1 - \cos \theta}{2}} \qquad \cos \dfrac{\theta}{2} = \pm\sqrt{\dfrac{1 + \cos \theta}{2}}$ $\tan \dfrac{\theta}{2} = \pm\sqrt{\dfrac{1 - \cos \theta}{1 + \cos \theta}} \qquad \tan \dfrac{\theta}{2} = \dfrac{1 - \cos \theta}{\sin \theta} \qquad \tan \dfrac{\theta}{2} = \dfrac{\sin \theta}{1 + \cos \theta}$

7.5   Exercises

Writing Identities

Exercises 1–8: Complete each statement.

1. $\sin^2 \theta = \frac{1 - \cos 2\theta}{2}$, so $\sin^2 10t =$ _____ .

2. $\cos^2 \theta = \frac{1 + \cos 2\theta}{2}$, so $\cos^2 8x =$ _____ .

3. $\tan^2 \theta = \frac{1 - \cos 2\theta}{1 + \cos 2\theta}$, so $\tan^2 5t =$ _____ .

4. $\tan 2\theta = \frac{2 \tan \theta}{1 - \tan^2 \theta}$, so $\tan t =$ _____ .

5. $\sin 2\theta = 2 \sin \theta \cos \theta$, so $\sin 20x =$ _____ .

6. $\cos 2\theta = \cos^2 \theta - \sin^2 \theta$, so $\cos 16t =$ _____ .

7. $\tan \frac{\theta}{2} = \frac{1 - \cos \theta}{\sin \theta}$, so $\tan 5x =$ _____ .

8. $\tan \frac{\theta}{2} = \frac{\sin \theta}{1 + \cos \theta}$, so $\tan 4t =$ _____ .

Double-Angle Identities

Exercises 9–14: If possible, evaluate expressions (a) and (b) and compare their values.

9. **(a)** $\sin 30° + \sin 30°$ **(b)** $\sin 60°$

10. **(a)** $\sin 45° + \sin 45°$ **(b)** $\sin 90°$

11. **(a)** $\cos 60° + \cos 60°$ **(b)** $\cos 120°$

12. **(a)** $\cos 90° + \cos 90°$ **(b)** $\cos 180°$

13. **(a)** $\tan 45° + \tan 45°$ **(b)** $\tan 90°$

14. **(a)** $\tan 30° + \tan 30°$ **(b)** $\tan 60°$

Exercises 15 and 16: (Refer to Example 1.) Verify graphically and symbolically that the two expressions are not equivalent.

15. $\tan 2\theta,\ 2 \tan \theta$ 16. $\cos 3\theta,\ 3 \cos \theta$

Exercises 17–24: (Refer to Example 2.) Do the following.
   (a) Find $\sin 2\theta$, $\cos 2\theta$, and $\tan 2\theta$.
   (b) Use a calculator to support your results.

17. $\cos \theta = \frac{4}{5}$ and $\sin \theta = \frac{3}{5}$

18. $\sin \theta = \frac{12}{13}$ and $\cos \theta = \frac{5}{13}$

19. $\sin \theta = -\frac{24}{25}$ and $\cos \theta > 0$

20. $\cos \theta = -\frac{7}{25}$ and $\tan \theta > 0$

21. $\sin \theta = -\frac{11}{61}$ and $\sec \theta > 0$

22. $\csc \theta = -2$ and $\sec \theta > 0$

23. $\tan \theta = \frac{7}{24}$ and $\cos \theta < 0$

24. $\cot \theta = \frac{5}{12}$ and $\sin \theta > 0$

Exercises 25–34: (Refer to Example 3.) Use an identity to evaluate the expression.

25. $\sin \left(2 \cos^{-1} 1\right)$ 26. $\cos \left(2 \sin^{-1} \frac{1}{2}\right)$

27. $\cos \left(2 \sin^{-1} \frac{7}{25}\right)$ 28. $\sin \left(2 \tan^{-1} \frac{3}{4}\right)$

29. $\cos \left(3 \sin^{-1} \frac{5}{13}\right)$ *(Hint: Use the results from Example 5.)*

30. $\tan \left(2 \tan^{-1} \frac{1}{2}\right)$

31. $\sin \left(2 \tan^{-1} x\right), x < 0$

32. $\cos \left(2 \sin^{-1} x\right), x > 0$

33. $\cos \left(\sin^{-1} \frac{3}{5} - \sin^{-1} \frac{4}{5}\right)$ *(Hint: Let $\alpha = \sin^{-1} \frac{3}{5}$, $\beta = \sin^{-1} \frac{4}{5}$, and apply a difference identity.)*

34. $\sin \left(\tan^{-1} \frac{12}{5} + \cos^{-1} \frac{12}{13}\right)$

Exercises 35–40: Rewrite using a double-angle identity.

35. $2 \cos \theta \sin \theta$ 36. $2 \sin 2\theta \cos 2\theta$

37. $\sin \theta \cos \theta$

38. $(\sin \theta - \cos \theta)(\sin \theta + \cos \theta)$

39. $2 \cos^2 2\theta - 1$ 40. $1 - 2 \sin^2 3\theta$

Exercises 41–44: Write the expression as one term.

41. $\sin^2 3\theta + \cos^2 3\theta$ 42. $1 + \tan^2 2\theta$

43. $\csc^2 5x - 1$ 44. $\sin^2 8x + \cos^2 8x$

Power-Reducing Identities and Half-Angle Formulas

Exercises 45–48: Find the exact value of the expression. Use a calculator to support your result.

45. $\cos^2 22.5°$ 46. $\sin^2 15°$

47. $\tan^2 75°$ 48. $\csc^2 105°$

Exercises 49–52: Use a half-angle formula to find the exact value of the expression. Use a calculator to support your result.

49. (a) $\cos 15°$ **(b)** $\tan(-15°)$

50. (a) $\sin 67.5°$ **(b)** $\cos(-67.5°)$

51. (a) $\tan \frac{\pi}{8}$ **(b)** $\sin\left(-\frac{\pi}{8}\right)$

52. (a) $\cos \frac{5\pi}{12}$ **(b)** $\cos\left(-\frac{5\pi}{12}\right)$

Exercises 53–58: Use a half-angle formula to simplify the expression. Use a calculator to support your result.

53. $\sqrt{\dfrac{1 - \cos 60°}{2}}$ **54.** $\sqrt{\dfrac{1 + \cos 60°}{2}}$

55. $\sqrt{\dfrac{1 + \cos 50°}{2}}$ **56.** $\sqrt{\dfrac{1 - \cos 50°}{2}}$

57. $\sqrt{\dfrac{1 - \cos 40°}{1 + \cos 40°}}$ **58.** $\sqrt{\dfrac{1 + \cos 26°}{1 - \cos 26°}}$

Exercises 59–64: (Refer to Example 10.) Find $\sin \frac{\theta}{2}$, $\cos \frac{\theta}{2}$, and $\tan \frac{\theta}{2}$.

59. $\cos \theta = \frac{4}{5}$ and $0° < \theta < 90°$

60. $\cos \theta = \frac{1}{3}$ and $0° < \theta < 90°$

61. $\tan \theta = -\frac{5}{12}$ and $-90° < \theta < 0°$

62. $\sec \theta = -2$ and $90° < \theta < 180°$

63. $\csc \theta = \frac{25}{24}$ and $90° < \theta < 180°$

64. $\sin \theta = \frac{4}{5}$ and $0° < \theta < 90°$

Verifying Identities

Exercises 65–74: Verify the identity. Give graphical or numerical support.

65. $4 \sin 2x = 8 \sin x \cos x$

66. $\cos 4\theta = 1 - 2 \sin^2 2\theta$

67. $\dfrac{2 - \sec^2 x}{\sec^2 x} = \cos 2x$

68. $(\sin x + \cos x)^2 = \sin 2x + 1$

69. $\sec 2x = \dfrac{1}{1 - 2 \sin^2 x}$ **70.** $2 \csc 2t = \csc t \sec t$

71. $\sin 3\theta = 3 \sin \theta - 4 \sin^3 \theta$

72. $\dfrac{2 \tan x}{1 + \tan^2 x} = \sin 2x$

73. $\sin 4\theta = 4 \sin \theta \cos \theta \cos 2\theta$

74. $\cos 4t = 8 \cos^4 t - 8 \cos^2 t + 1$

Exercises 75–84: Verify the identity.

75. $\dfrac{\sin 2\theta}{\sin \theta} = 2 \cos \theta$ **76.** $2 \sin^2 4\theta = 1 - \cos 8\theta$

77. $2 \cos^2 \frac{\theta}{2} = 1 + \cos \theta$

78. $\dfrac{\sin^2 2\theta}{1 + \cos 2\theta} = 2 \sin^2 \theta$ **79.** $\cos^4 \theta - \sin^4 \theta = \cos 2\theta$

80. $\dfrac{1 - \tan^2 x}{1 + \tan^2 x} = \cos 2x$ **81.** $\csc 2t = \dfrac{\csc t}{2 \cos t}$

82. $\tan \theta + \cot \theta = \dfrac{2}{\sin 2\theta}$

83. $\tan \dfrac{x}{2} = \dfrac{\sin x}{1 + \cos x}$ $\left(\text{Hint: Let } \tan \dfrac{x}{2} = \dfrac{\sin \frac{x}{2}}{\cos \frac{x}{2}}. \right)$

84. $\tan \dfrac{x}{2} = \dfrac{1 - \cos x}{\sin x}$ *(Hint: Use the identity in Exercise 83.)*

Product-to-Sum and Sum-to-Product Identities

Exercises 85–88: Write each expression as a sum or difference of trigonometric functions.

85. (a) $\cos 50° \sin 20°$ **(b)** $\cos 2x \cos x$

86. (a) $2 \sin 74° \sin 24°$ **(b)** $8 \sin 18x \cos 13x$

87. (a) $\sin 7\theta \cos 3\theta$ **(b)** $\sin 8x \sin 4x$

88. (a) $2 \cos 5x \cos 7x$ **(b)** $4 \cos 9\theta \sin 2\theta$

Exercises 89–92: Write each expression as a product of trigonometric functions.

89. (a) $\sin 40° + \sin 30°$ **(b)** $\cos 45° + \cos 35°$

90. (a) $\cos 104° - \cos 24°$ **(b)** $\sin 32° - \sin 64°$

91. (a) $\cos 6\theta + \cos 4\theta$ **(b)** $\sin 7x + \sin 4x$

92. (a) $\sin 3x - \sin 5x$ **(b)** $\cos 3\theta - \cos \theta$

Solving Equations

Exercises 93–100: Find the solutions to the equation in the interval $[0°, 360°)$

(a) symbolically,   (b) graphically,   and (c) numerically.

93. $\cos 2\theta = 1$ **94.** $\sin 2\theta = \frac{1}{2}$

95. $\sin 2\theta + \cos \theta = 0$ **96.** $\cos \frac{\theta}{2} = -\frac{\sqrt{3}}{2}$

97. $\sin \frac{\theta}{2} = 1$ **98.** $\cos 2\theta + \cos \theta = 0$

99. $\sqrt{2} \sin \frac{\theta}{2} - 1 = 0$ **100.** $2 \cos \frac{\theta}{2} + 1 = 0$

Exercises 101–116: Find all solutions, expressed in radians.

101. $2 \cos 2t = \sqrt{3}$ **102.** $2 \sin 2t = -1$

103. $\sin 2t + \sin t = 0$ **104.** $\sin t - \cos 2t = 0$

105. $2 \sin \frac{t}{2} - 1 = 0$ **106.** $\sin 2t = 2 \cos^2 t$

107. $\cos 2t = \sin t$ **108.** $\cos 2t - \cos t = 0$

109. $\tan 2t = 1$ **110.** $\cot 2t = \sqrt{3}$

111. $2 \cos \frac{t}{2} = 1$ **112.** $\tan \frac{t}{2} = 1$

113. $\cos 2t = 2 \sin t \cos t$

114. $2 \cos^2 2t = 1 - \cos 2t$

115. $2 \sin^2 2t + \sin 2t - 1 = 0$

116. $2 \cos \frac{t}{2} + 1 = 0$

Exercises 117 and 118: Approximate to the nearest thousandth all solutions on $[0, 2\pi)$.

117. $\sin t + \sin 2t = \cos t$

118. $\sin 3t + \sin 2t = 2 \cos t$

Applications

119. *Electricity* (Refer to Example 8.) Suppose that the voltage in a 220-volt electrical circuit is modeled by $V(t) = 310 \sin(120\pi t)$ and that the amperage flowing through a heater is $I = 7 \sin(120\pi t)$. (**Source:** G. Wilcox and C. Hesselberth.)
 (a) Graph the wattage $W = VI$ consumed by the heater in $[0, 0.04, 0.01]$ by $[-500, 2500, 500]$.

 (b) Find values for the constants a, k, and d so that $W = a \cos(k\pi t) + d$.

120. *Electricity* If a toaster is plugged into a common household outlet, the wattage W used varies according to the equation $W = \frac{V^2}{R}$, where V is the voltage and R is a constant that measures the resistance of the toaster in ohms. (**Source:** D. Bell, *Fundamentals of Electric Circuits.*)
 (a) Graph W if $R = 15$ and $V = 163 \sin(120\pi t)$ in $[0, 0.05, 0.01]$ by $[-500, 2000, 500]$.

 (b) Approximate the maximum wattage consumed by the toaster.

 (c) Use a power-reducing identity to express the wattage as $W = a \cos(240\pi t) + d$, where a and d are constants.

Exercises 121 and 122: Electricity Let the voltage in an electrical outlet be given by $V(t) = 320 \sin(120\pi t)$ at time t in seconds. Find the times when V equals the following values.

121. 160 volts

122. $160\sqrt{3}$ volts

123. *Highway Curves* When an automobile travels along a circular curve, objects like trees and buildings situated on the inside of the curve can obstruct the driver's vision. If the cars in the figure are a safe stopping distance apart, then the distance d that should be cleared on the inside of the curve is $d = r\left(1 - \cos \frac{\beta}{2}\right)$, where r is the radius of the curve and β is the central angle between the cars. (**Source:** F. Mannering and W. Kilareski, *Principles of Highway Engineering and Traffic Analysis.*)

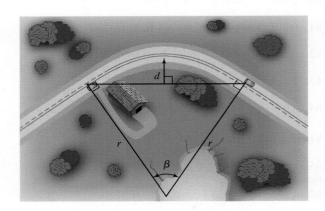

 (a) Find d if $\beta = 80°$ and $r = 600$ feet.

 (b) Use the figure to justify this formula.

 (c) Is the given formula equivalent to the formula $d = r\left(1 - \frac{1}{2} \cos \beta\right)$? Explain.

124. *Highway Curves* The figure on the next page represents a circular curve with radius r and central angle θ. The tangent length T is an important distance used by surveyors. (**Source:** F. Mannering.)

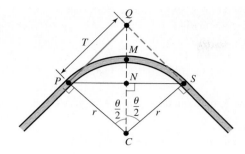

(a) Show that $T = r \tan \frac{\theta}{2}$.

(b) Find T for a curve with a 1500-foot radius and $\theta = 80°$.

125. *Touch-Tone Phones*   (Refer to Example 15.) For the numbers 3 and 4 on a touch-tone phone, complete the following.

(a) Write formulas for each number that are the *sum* of two cosine functions and that model the tone generated by each number.

(b) Write the formulas for each tone from part (a) as a *product* of two trigonometric functions.

126. *Musical Tones and Beats*   If two musical tones with nearly the same frequency are played simultaneously, a phenomenon called *beats* occurs. Let $P_1 = 0.04 \cos(110\pi t)$ and $P_2 = 0.04 \cos(116\pi t)$ represent these tones.

(a) Graph $P = P_1 + P_2$ in the viewing rectangle $[0.2, 1.2, 0.2]$ by $[-0.08, 0.08, 0.01]$. How many beats are there?

(b) Use a sum-to-product identity to write $P_1 + P_2$ as a product of trigonometric expressions.

Writing about Mathematics

127. Suppose a student believes that an equation is an identity but cannot verify it symbolically. Discuss techniques that the student could use to support this belief.

128. Does the equation $\sin^2 3\theta + \cos^2 2\theta = 1$ represent an identity? Explain your reasoning.

CHECKING BASIC CONCEPTS FOR SECTION 7.5

1. Find values for $\sin 2\theta$ and $\cos 2\theta$ if $\cos \theta = -\frac{7}{25}$ and $\sin \theta = \frac{24}{25}$.

2. Use a half-angle formula to find the exact value of $\sin 22.5°$.

3. Find $\sin \frac{\theta}{2}$ and $\cos \frac{\theta}{2}$ if $\sin \theta = \frac{4}{5}$ and θ is acute.

4. Verify that $\dfrac{\cos 2\theta}{\cos^2 \theta} = 2 - \sec^2 \theta$.

5. Solve $\sin 2\theta = 2 \cos \theta$ for $0° \leq \theta < 360°$.

---

7   Summary

CONCEPT	EXPLANATION AND EXAMPLES

SECTION 7.1 FUNDAMENTAL IDENTITIES

Reciprocal Identities

$$\sin \theta = \frac{1}{\csc \theta} \qquad \cos \theta = \frac{1}{\sec \theta} \qquad \tan \theta = \frac{1}{\cot \theta}$$

$$\csc \theta = \frac{1}{\sin \theta} \qquad \sec \theta = \frac{1}{\cos \theta} \qquad \cot \theta = \frac{1}{\tan \theta}$$

Example: If $\sin \theta = \frac{3}{5}$, then $\csc \theta = \frac{5}{3}$.

CONCEPT	EXPLANATION AND EXAMPLES

SECTION 7.1 FUNDAMENTAL IDENTITIES (CONTINUED)

Quotient Identities

$$\tan \theta = \frac{\sin \theta}{\cos \theta} \qquad \cot \theta = \frac{\cos \theta}{\sin \theta}$$

Example: If $\sin \theta = -\frac{3}{5}$ and $\cos \theta = \frac{4}{5}$, then $\tan \theta = \frac{-3/5}{4/5} = -\frac{3}{4}$.

Pythagorean Identities

$$\sin^2 \theta + \cos^2 \theta = 1 \qquad 1 + \tan^2 \theta = \sec^2 \theta \qquad 1 + \cot^2 \theta = \csc^2 \theta$$

Examples: $\sin^2 30° + \cos^2 30° = 1$

$$1 + \tan^2 \frac{\pi}{4} = \sec^2 \frac{\pi}{4}$$

Negative-Angle Identities

$$\sin(-\theta) = -\sin \theta \qquad \cos(-\theta) = \cos \theta \qquad \tan(-\theta) = -\tan \theta$$

$$\csc(-\theta) = -\csc \theta \qquad \sec(-\theta) = \sec \theta \qquad \cot(-\theta) = -\cot \theta$$

Note: Cosine and secant are even functions, having graphs that are symmetric with respect to the *y*-axis. Sine, cosecant, tangent, and cotangent are odd functions, having graphs that are symmetric with respect to the origin. See page 586.

Examples: $\cos(-45°) = \cos 45°, \quad \sin(-60°) = -\sin 60°$

SECTION 7.2 VERIFYING IDENTITIES

Verifying Identities

To verify an identity, simplify one side of the equation until it equals the other side of the equation. Make use of the fundamental identities from Section 7.1.

Example: Verify that $\dfrac{1 - \sin^2 \theta}{\cos \theta} = \cos \theta$.

$$\frac{1 - \sin^2 \theta}{\cos \theta} = \frac{\cos^2 \theta}{\cos \theta} \qquad \sin^2 \theta + \cos^2 \theta = 1$$

$$= \cos \theta \qquad \text{Simplify.}$$

SECTION 7.3 TRIGONOMETRIC EQUATIONS

Reference Angle

The reference angle for an angle θ in standard position is the acute angle between the terminal side of θ and the *x*-axis.

Example: The reference angle for $\theta = 120°$ is $\theta_R = 60°$.

Trigonometry Equations

Unlike polynomial equations, which have a finite number of solutions, trigonometric equations can have infinitely many solutions.

Example: Find all solutions to $2 \sin \theta - 1 = 0$.

$$2 \sin \theta - 1 = 0$$

$$2 \sin \theta = 1$$

$$\sin \theta = \frac{1}{2}$$

CONCEPT	EXPLANATION AND EXAMPLES

SECTION 7.3 TRIGONOMETRIC EQUATIONS (CONTINUED)

Since $\theta_R = \sin^{-1}\frac{1}{2} = 30°$ and the sine function is positive in quadrants I and II, the solutions to $\sin\theta = \frac{1}{2}$ for $0° \le \theta < 360°$ are 30° and 150°. Since $\sin\theta$ has period 360°, all solutions to the equation can be written as

$\theta = 30° + 360° \cdot n$ or $\theta = 150° + 360° \cdot n$,   where n is an integer.

SECTION 7.4 SUM AND DIFFERENCE IDENTITIES

Cosine Sum and Difference

$\cos(\alpha + \beta) = \cos\alpha\cos\beta - \sin\alpha\sin\beta$

$\cos(\alpha - \beta) = \cos\alpha\cos\beta + \sin\alpha\sin\beta$

Example: $\cos(60° - 45°) = \cos 60°\cos 45° + \sin 60°\sin 45°$

Sine Sum and Difference

$\sin(\alpha + \beta) = \sin\alpha\cos\beta + \cos\alpha\sin\beta$

$\sin(\alpha - \beta) = \sin\alpha\cos\beta - \cos\alpha\sin\beta$

Example: $\sin(60° - 45°) = \sin 60°\cos 45° - \cos 60°\sin 45°$

Tangent Sum and Difference

$\tan(\alpha + \beta) = \dfrac{\tan\alpha + \tan\beta}{1 - \tan\alpha\tan\beta}$, $\tan(\alpha - \beta) = \dfrac{\tan\alpha - \tan\beta}{1 + \tan\alpha\tan\beta}$

Example: $\tan(60° - 45°) = \dfrac{\tan 60° - \tan 45°}{1 + \tan 60°\tan 45°}$

Cofunction Identities

$\cos\left(\dfrac{\pi}{2} - t\right) = \sin t$ $\sin\left(\dfrac{\pi}{2} - t\right) = \cos t$

$\cot\left(\dfrac{\pi}{2} - t\right) = \tan t$ $\tan\left(\dfrac{\pi}{2} - t\right) = \cot t$

$\csc\left(\dfrac{\pi}{2} - t\right) = \sec t$ $\sec\left(\dfrac{\pi}{2} - t\right) = \csc t$

Example: $\tan\dfrac{\pi}{3} = \tan\left(\dfrac{\pi}{2} - \dfrac{\pi}{6}\right) = \cot\dfrac{\pi}{6}$

SECTION 7.5 MULTIPLE-ANGLE IDENTITIES

Double-Angle Identities

$\sin 2\theta = 2\sin\theta\cos\theta$

$\cos 2\theta = \cos^2\theta - \sin^2\theta = 2\cos^2\theta - 1 = 1 - 2\sin^2\theta$

$\tan 2\theta = \dfrac{2\tan\theta}{1 - \tan^2\theta}$

Example: $\sin 120° = \sin(2 \cdot 60°) = 2\sin 60°\cos 60°$

CONCEPT	EXPLANATION AND EXAMPLES

SECTION 7.5 MULTIPLE-ANGLE IDENTITIES (CONTINUED)

Power-Reducing Identities $\sin^2\theta = \dfrac{1 - \cos 2\theta}{2}$ $\cos^2\theta = \dfrac{1 + \cos 2\theta}{2}$ $\tan^2\theta = \dfrac{1 - \cos 2\theta}{1 + \cos 2\theta}$

Example: $\sin^2 30° = \dfrac{1 - \cos 60°}{2}$

Half-Angle Formulas

$$\sin\frac{\theta}{2} = \pm\sqrt{\frac{1 - \cos\theta}{2}} \qquad \cos\frac{\theta}{2} = \pm\sqrt{\frac{1 + \cos\theta}{2}}$$

$$\tan\frac{\theta}{2} = \pm\sqrt{\frac{1 - \cos\theta}{1 + \cos\theta}} \qquad \tan\frac{\theta}{2} = \frac{1 - \cos\theta}{\sin\theta} \qquad \tan\frac{\theta}{2} = \frac{\sin\theta}{1 + \cos\theta}$$

Example: $\cos 15° = \cos\dfrac{30°}{2} = \sqrt{\dfrac{1 + \cos 30°}{2}}$

Product-to-Sum Identities

$$\cos\alpha\cos\beta = \tfrac{1}{2}(\cos(\alpha + \beta) + \cos(\alpha - \beta))$$

$$\sin\alpha\sin\beta = \tfrac{1}{2}(\cos(\alpha - \beta) - \cos(\alpha + \beta))$$

$$\sin\alpha\cos\beta = \tfrac{1}{2}(\sin(\alpha + \beta) + \sin(\alpha - \beta))$$

$$\cos\alpha\sin\beta = \tfrac{1}{2}(\sin(\alpha + \beta) - \sin(\alpha - \beta))$$

Example: $\cos 45°\cos 60° = \tfrac{1}{2}(\cos(45° + 60°) + \cos(45° - 60°))$

Sum-to-Product Identities

$$\cos a + \cos b = 2\cos\frac{a + b}{2}\cos\frac{a - b}{2}$$

$$\cos a - \cos b = -2\sin\frac{a + b}{2}\sin\frac{a - b}{2}$$

$$\sin a + \sin b = 2\sin\frac{a + b}{2}\cos\frac{a - b}{2}$$

$$\sin a - \sin b = 2\cos\frac{a + b}{2}\sin\frac{a - b}{2}$$

Example: $\cos 60° + \cos 30° = 2\cos\dfrac{60° + 30°}{2}\cos\dfrac{60° - 30°}{2}$

---

7   Review Exercises

Exercises 1 and 2: Determine the quadrant that contains θ.

1. $\sec\theta < 0$ and $\sin\theta > 0$

2. $\cot\theta > 0$ and $\cos\theta < 0$

Exercises 3–6: Use the given information to find the other trigonometric functions of θ.

3. $\sin\theta = \dfrac{3}{5}$ and $\cos\theta = -\dfrac{4}{5}$

4. $\sec\theta = -\dfrac{13}{12}$ and $\csc\theta = -\dfrac{13}{5}$

5. $\tan\theta = -\dfrac{7}{24}$ and $\cos\theta = \dfrac{24}{25}$

6. $\cot\theta = -\dfrac{1}{2}$ and $\sin\theta > 0$

Exercises 7–10: Use a negative-angle identity to write an equivalent trigonometric expression involving a positive angle.

7. $\sin(-13°)$

8. $\cos(-106°)$

9. $\sec\left(-\frac{3\pi}{7}\right)$

10. $\tan\left(-\frac{5\pi}{11}\right)$

Exercises 11–16: Simplify each expression.

11. $\sec\theta\cot\theta\sin\theta$

12. $\sin\theta\csc\theta$

13. $(\sec^2 t - 1)(\csc^2 t - 1)$

14. $\dfrac{\sec\theta}{\csc\theta} + \dfrac{\sin\theta}{\cos\theta}$

15. $\dfrac{\csc\theta\sin\theta}{\sec\theta}$

16. $\dfrac{\cos^2\theta}{1 - \sin\theta}$

Exercises 17–20: Use a calculator to approximate the other trigonometric functions of θ to four decimal places.

17. $\tan\theta = 1.2367$ and θ is acute

18. $\sin\theta = -0.3434$ and θ is in quadrant IV

19. $\cos\theta = -0.4544$ and θ is in quadrant II

20. $\tan\theta = -0.8595$ and θ is in quadrant IV

Exercises 21–24: Factor the trigonometric expression.

21. $\sin^2\theta + 2\sin\theta + 1$

22. $2\cos^2 t - 3\cos t + 1$

23. $\tan^2\theta - 9$

24. $2\sec^2\theta - 3\sec\theta - 5$

Exercises 25–38: Verify the identity.

25. $(\sec\theta - 1)(\sec\theta + 1) = \tan^2\theta$

26. $(\cos\theta + \sin\theta)^2 + (\cos\theta - \sin\theta)^2 = 2$

27. $(1 + \tan t)^2 = \sec^2 t + 2\tan t$

28. $(1 - \cos^2 t)(1 + \tan^2 t) = \tan^2 t$

29. $\sin(x - \pi) = -\sin x$ **30.** $\cos(\pi + x) = -\cos x$

31. $\sin 8x = 2\sin 4x\cos 4x$

32. $\cos^4 x - \sin^4 x = \cos 2x$

33. $\sec 2x = \dfrac{1}{2\cos^2 x - 1}$ **34.** $\dfrac{1 + \tan^2 x}{\sin^2 x + \cos^2 x} = \sec^2 x$

35. $\cos^4 x\sin^3 x = (\cos^4 x - \cos^6 x)\sin x$

36. $\sin^4 x = \frac{3}{8} - \frac{1}{2}\cos 2x + \frac{1}{8}\cos 4x$

37. $\sec^4\theta - \tan^4\theta = 1 + 2\tan^2\theta$

38. $\dfrac{1 + \cos\theta}{\sin\theta} + \dfrac{\sin\theta}{1 + \cos\theta} = 2\csc\theta$

Exercises 39–42: Find the reference angle for θ.

39. $\theta = 240°$

40. $\theta = 320°$

41. $\theta = \dfrac{9\pi}{7}$

42. $\theta = -\dfrac{7\pi}{6}$

Exercises 43 and 44: Use the graph to estimate any solutions to the trigonometric equation for $[0, 2\pi)$. Then solve the equation symbolically.

43. $\cos^2\theta - 2\cos\theta = 0$ **44.** $2\sin t\cos t - \cos t = 0$

Exercises 45 and 46: Solve the given trigonometric equation for $0° \le \theta < 360°$.

45. (a) $\tan\theta = \sqrt{3}$ **(b)** $\cot\theta = -\sqrt{3}$

46. (a) $\sin\theta = 1$ **(b)** $\cos\theta = -1$

Exercises 47–50: Solve the equation for $0° \le \theta < 360°$.

47. $2\cos\theta - 1 = 0$ **48.** $\cot^2\theta = 1$

49. $2\sin^2\theta + \sin\theta - 3 = 0$

50. $\sin^2\theta + 2\cos\theta = 1$

Exercises 51 and 52: Solve the equation for t on the interval $[0, 2\pi)$.

51. $\tan^2 t - 2\tan t + 1 = 0$

52. $2\sin t = \sqrt{3}$

Exercises 53–56: Find all solutions to the equation. Express your results in both degrees and radians.

53. $3\tan^2 t - 1 = 0$ **54.** $2\sin^2 t - \sin t - 1 = 0$

55. $\sin 2t + 3\cos t = 0$ **56.** $\cos 2t = 1$

Exercises 57 and 58: Use a half-angle formula to find the exact value for the expression. Use a calculator to support your result.

57. $\cos 105°$ **58.** $\sin\dfrac{\pi}{12}$

Exercises 59 and 60: Estimate any solutions in the interval $[-2\pi, 2\pi]$ graphically to two decimal places.

59. $\tan x = x + 1$ **60.** $\sin(\cos x) = \tan x$

Exercises 61 and 62: Find the following.
   (a) $\sin(\alpha + \beta)$ (b) $\cos(\alpha + \beta)$
   (c) $\tan(\alpha + \beta)$ (d) The quadrant containing $\alpha + \beta$

61. $\cos\alpha = \frac{3}{5}$ and $\cos\beta = \frac{12}{13}$, α and β are in quadrant I.

62. $\sin\alpha = -\frac{12}{13}$ and $\tan\beta = -\frac{4}{3}$, α and β are in quadrant IV.

Exercises 63 and 64: Complete the following.
   (a) Find $\sin 2\theta$, $\cos 2\theta$, and $\tan 2\theta$.
   (b) Use a calculator to support your results.

63. $\sin\theta = \frac{4}{5}$ and $\tan\theta = -\frac{4}{3}$

64. $\sin\theta = -\frac{12}{13}$ and $\cos\theta = -\frac{5}{13}$

Exercises 65 and 66: Complete the following.
   (a) Find $\sin\frac{\theta}{2}$, $\cos\frac{\theta}{2}$, and $\tan\frac{\theta}{2}$.
   (b) Use a calculator to support your results.

65. $\cos\theta = \frac{1}{4}$ and $0° < \theta < 90°$

66. $\tan\theta = -\frac{8}{15}$ and $-90° < \theta < 0°$

Exercises 67 and 68: Evaluate the expression.
67. $\cos\left(2\tan^{-1}\frac{11}{60}\right)$ **68.** $\sin\left(2\sin^{-1}x\right)$, $x > 0$

Applications

69. *Daylight Hours*   The number of daylight hours y at 20° S latitude can be modeled by

$$y = 1.2\cos\left(\frac{\pi}{6}(x - 0.7)\right) + 12.1,$$

where $x = 1$ corresponds to January 1, $x = 2$ to February 1, and so on. Estimate graphically or numerically when the number of daylight hours equals 11.5 hours.   (**Source:** J. Williams.)

70. *Music and Pure Tones*   A pure tone is modeled by the graph of $P(t) = 0.006\cos(50\pi t)$, where P represents the pressure on an eardrum in pounds per square foot at time t in seconds.   (**Source:** J. Roederer, *Introduction to the Physics and Psychophysics of Music.*)
   (a) Graph P in $[0, 0.1, 0.01]$ by $[-0.008, 0.008, 0.001]$.

   (b) Estimate all solutions to the equation $P = 0$ on the interval $0 \leq t \leq 0.1$.

71. *Modeling Musical Tones*   Let the pressure exerted on the eardrum by two sound waves in pounds per square foot be given by
$$P_1(t) = 0.006\cos(100\pi t) \quad \text{and}$$
$$P_2(t) = 0.008\sin(100\pi t).$$

   (a) Graph the pressure on the eardrum $P = P_1 + P_2$ in the window $[0, 0.06, 0.01]$ by $[-0.012, 0.012, 0.002]$.

   (b) Use the graph to find values for a and k such that $P = a\sin(100\pi t + k)$.

   (c) Use a sum or difference identity for sine to verify your result in part (b).

72. *Electricity*   Suppose the voltage in an electrical circuit is given by $V(t) = 17\sin(120\pi t)$ and the amperage flowing through a heater is given by $I(t) = 2\sin(120\pi t)$, where t is in seconds.   (**Source:** G. Wilcox and C. Hesselberth, *Electricity for Engineering Technology.*)
   (a) Graph the wattage $W = VI$ consumed by the heater in $[0, 0.04, 0.01]$ by $[-40, 40, 10]$.

   (b) Use a power-reducing identity to express the wattage in the form $W = a\cos(k\pi t) + d$, where a, k, and d are constants.

73. *Electromagnets*   Let the wattage consumed by an electromagnet be given by $W(t) = 7\cos^2(240\pi t)$ and the voltage be given by $V(t) = 50\sin(240\pi t)$, where t is in seconds. Express $W(t)$ in terms of the sine function. When V is maximum or minimum, what is the value of W? Explain.

74. *Average Temperatures*   Let the monthly average high temperature y in degrees Fahrenheit at a location be given by $y = 15\sin\left(\frac{\pi}{6}(x - 4)\right) + 60$, where x is the month and $x = 1$ corresponds to January. Estimate when the monthly average high temperature is 60° F.

75. *Back Stress*   The force exerted by the back muscles of a 100-pound person can be estimated by $F = 289\cos\theta$. (See Figure 7.52 on page 617.) Find θ in degrees and radians when $F = 250$ pounds.

76. *Electricity*   Let $V(t) = 80\sin(120\pi t)$ denote the voltage in an electrical outlet at time t in seconds, where t is a real number. Find all times when the voltage is 40 volts.

EXTENDED AND DISCOVERY EXERCISES

1. *Piano Strings*   A piano string vibrates at more than one frequency when it is struck. It produces a complex wave that can be modeled by a sum of several pure tones. If a piano key with a frequency of f_1 is played, then the corresponding string will not only vibrate at f_1 but also vibrate at the higher frequencies of $2f_1$, $3f_1$, $4f_1$, and so on. The *fundamental frequency* of the string is f_1, and the higher frequencies are called the *upper harmonics*. The human ear will hear the sum of these frequencies as one complex tone.   (**Source:** J. Roederer, *Introduction to the Physics and Psychophysics of Music.*)

(a) If the A note above middle C is played, its fundamental frequency is $f_1 = 440$ hertz. (One hertz equals one cycle per second.) The piano string also vibrates at frequencies of $f_2 = 2(440) = 880$, $f_3 = 3(440) = 1320$, $f_4 = 4(440) = 1760$, and so on. The pressure for each frequency in pounds per square foot is modeled by

$$P_1 = 0.002 \sin (2\pi(440)t),$$

$$P_2 = \frac{0.002}{2} \sin (2\pi(880)t),$$

$$P_3 = \frac{0.002}{3} \sin (2\pi(1320)t),$$

$$P_4 = \frac{0.002}{4} \sin (2\pi(1760)t), \quad \text{and}$$

$$P_5 = \frac{0.002}{5} \sin (2\pi(2200)t),$$

where t is in seconds. Graph each of the following expressions for P in the viewing rectangle $[0, 0.01, 0.002]$ by $[-0.005, 0.005, 0.001]$.

  i. $P = P_1$
  ii. $P = P_1 + P_2$
  iii. $P = P_1 + P_2 + P_3$
  iv. $P = P_1 + P_2 + P_3 + P_4$
  v. $P = P_1 + P_2 + P_3 + P_4 + P_5$

(b) The final graph of P models what the human ear hears. Describe this graph.

(c) Estimate the maximum pressure of
$$P = P_1 + P_2 + P_3 + P_4 + P_5.$$

(d) A pure tone with a frequency of 440 hertz is modeled by $P = P_1$, whereas a piano generates the graph of $P = P_1 + P_2 + P_3 + P_4 + P_5$. Compare and contrast these two graphs.

2. *Low Tones and Small Speakers*   Small speakers found in older radios and telephones often cannot vibrate at frequencies less than 200 hertz. Thirty-five keys on a piano have frequencies less than 200 hertz. Nonetheless

these notes can still be heard on these speakers. When a piano string creates a tone of 110 hertz, it also creates tones at 220, 330, 440, 550, and 660 hertz. A small speaker cannot reproduce the 110 hertz vibration, but it can reproduce the higher frequencies, which are called the upper harmonics. The low tones can still be heard because the speaker produces *difference tones* of the upper harmonics. The difference between consecutive frequencies is 110 hertz, and this difference tone will be heard on a small speaker even though the speaker cannot vibrate at 110 hertz. This phenomenon can be visualized with a graphing calculator.   (**Source:** A. Benade.)

(a) Graph the upper harmonics represented by the pressure wave
$$P = \tfrac{1}{2} \sin (2\pi(220)t) + \tfrac{1}{3} \sin (2\pi(330)t)$$
$$+ \tfrac{1}{4} \sin (2\pi(440)t)$$
in $[0, 0.03, 0.01]$ by $[-1.2, 1.2, 0.5]$.

(b) Estimate the t-values on the interval $0 \le t \le 0.03$ where P is maximum.

(c) Approximate the frequency of these maximum values. What does a person hear in addition to the frequencies of 220, 330, and 440 hertz?

(d) Discuss the advantage of having large speakers instead of smaller ones. (*Hint:* Try graphing the pressure produced by a speaker that can vibrate both at 110 hertz and at the upper harmonics.)

3. *Piano Strings*   When a string is set into vibration by striking it, the amplitude A of the vibrations decreases over time, while the frequency of the vibration remains constant. This phenomenon is called *exponential decay* and can be modeled by $A = A_0 e^{-kt} \sin (2\pi Ft)$, where F is the frequency, t is time in seconds, and k and A_0 are positive constants.   (**Source:** J. Roederer.)

(a) Graph A when $A_0 = 0.1$, $F = 15$, and $k = 1.2$ in the window $[0, 1, 0.1]$ by $[-0.15, 0.15, 0.05]$.

(b) Now graph the equations $y_1 = -0.1e^{-1.2t}$ and $y_2 = 0.1e^{-1.2t}$ with A. Describe how the graphs of y_1 and y_2 relate to the graph of A.

(c) The *decay half-time* is the time it takes for the maximum amplitude of A_0 to decrease to $\tfrac{1}{2}A_0$. Estimate this time graphically. (The decay half-time for a typical piano string is about 0.4 second.)

8 Further Topics in Trigonometry

There is no branch of mathematics, however abstract, which may not some day be applied to the real world.

—Nikolai Lobachevsky

World technology took a giant step forward when the Large Hadron Collider (LHC) was put into operation in September 2008. The LHC, a 17-mile-long nuclear particle accelerator located 100 meters beneath the border between France and Switzerland, will propel particles near light speed. The LHC is expected to unlock secrets of the universe by penetrating deeper into the heart of matter and further back into time. It may even create tiny black holes and determine the existence of extra dimensions. These discoveries will affect our lives by providing new medical treatments for disease and new technologies.

Mathematics has played an essential role not only in the design of the LHC, but also in predicting possible outcomes for experiments. As the complexity of machines and experiments increases, mathematics plays a greater and greater role in analyzing data and predicting outcomes. Many unusual scientific concepts defy intuition, and mathematics is necessary for their understanding.

Angles, triangles, vectors, and complex numbers help make new technologies, such as the LHC, the Global Positioning System, aerial photography, and navigation, possible. In this chapter, we use these mathematical concepts to solve a variety of applications.

Source: CNN.com

8.1 Law of Sines

- Learn about oblique triangles
- Derive the law of sines
- Solve triangles
- Solve the ambiguous case

Introduction

In Chapter 6, we solved right triangles. Solving a triangle involves finding the length of each side and the measure of each angle in the triangle. In many areas of study, triangles without right angles often occur. For example, the height of the Gateway to the West Arch in St. Louis can be found from the ground by using triangles without right angles. (See Exercise 55.) To solve these *oblique triangles*, we will use the law of sines and the law of cosines. In this section, the law of sines is derived and used to solve several problems.

Oblique Triangles

If a triangle is not a right triangle, then it is called an **oblique triangle**. There are four different situations, or cases, that can occur when attempting to solve an oblique triangle.

1. All three sides are given. This situation determines a unique triangle and is referred to as SSS. See Figure 8.1. Note that the length of any one side must be less than the sum of the lengths of the other two sides.
2. Two sides and the angle included are given. This situation determines a unique triangle and is referred to as SAS. See Figure 8.2.
3. One side and two angles are given. This situation determines a unique triangle and is referred to as AAS or ASA. See Figure 8.3. Note that whenever two angles of a triangle are known, the third angle can be found by using the fact that the sum of the measures of the angles equals 180°.
4. Two sides and an angle opposite one of the sides are given. This situation does *not always* determine a unique triangle and is referred to as SSA. There may be zero, one, or two triangles that can satisfy SSA. As a result, we call SSA the **ambiguous case**. See Figures 8.4–8.6.

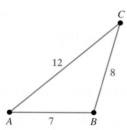

Figure 8.1 Case 1: SSS

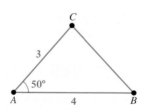

Figure 8.2 Case 2: SAS

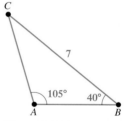

Figure 8.3 Case 3: AAS or ASA

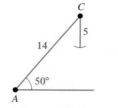

Figure 8.4 Case 4: SSA (No Triangles)

Figure 8.5 Case 4: SSA (One Triangle)

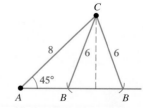

Figure 8.6 Case 4: SSA (Two Triangles)

Cases 1 and 2 are solved in Section 8.2 using the *law of cosines*, and Cases 3 and 4 are solved in this section using the *law of sines*.

Solving Triangles with the Law of Sines

We will label triangles as shown in Figure 8.7 and refer to this labeling as the **standard labeling**. For example, angle α is located at vertex A and side a is opposite angle α. Note that angle γ need *not* be a right angle.

The law of sines can be derived using the oblique triangle shown in Figure 8.8.

$$\sin \alpha = \frac{h}{c}, \quad \text{or} \quad h = c \sin \alpha$$

$$\sin \gamma = \frac{h}{a}, \quad \text{or} \quad h = a \sin \gamma$$

Since $h = c \sin \alpha$ and $h = a \sin \gamma$, it follows that

$$c \sin \alpha = a \sin \gamma,$$

or, if we divide each side by ac,

$$\frac{\sin \alpha}{a} = \frac{\sin \gamma}{c}.$$

In a similar manner it can be shown that

$$\frac{\sin \alpha}{a} = \frac{\sin \beta}{b}.$$

This discussion supports the following result.

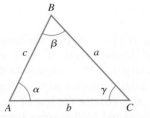

B

Figure 8.7 Standard Labeling

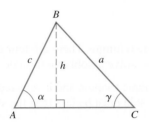

Figure 8.8

Law of Sines

Any triangle with standard labeling satisfies

$$\frac{\sin \alpha}{a} = \frac{\sin \beta}{b} = \frac{\sin \gamma}{c}, \quad \text{or equivalently,} \quad \frac{a}{\sin \alpha} = \frac{b}{\sin \beta} = \frac{c}{\sin \gamma}.$$

An Application from Aerial Photography The next example illustrates how the law of sines is used in aerial photography.

EXAMPLE 1 Using aerial photography to find distances (ASA)

Trigonometry is used extensively in aerial photography. Figure 8.9 depicts a situation in which a camera lens has an angular coverage of 75°. As a picture is taken over level ground, the airplane's distance is 4800 feet from a house located on the edge of the photograph and the angle of elevation of the airplane from the house is 48°. Find the ground distance a shown in the photograph.   (**Source:** F. Moffitt, *Photogrammetry.*)

SOLUTION In this example we are given two angles and the side included (ASA), so let $\alpha = 75°$, $\beta = 48°$, and $c = 4800$, as shown in Figure 8.10 at the top of the next page. To find the third angle γ we can use the fact that the angles sum to 180°.

$$\gamma = 180° - \alpha - \beta$$
$$= 180° - 75° - 48°$$
$$= 57°$$

Figure 8.9

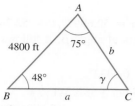

Figure 8.10

Side a corresponds to the ground distance shown in the photograph and can be found by using the law of sines.

$$\frac{a}{\sin \alpha} = \frac{c}{\sin \gamma} \qquad \text{Law of sines}$$

$$\frac{a}{\sin 75°} = \frac{4800}{\sin 57°} \qquad \text{Substitute.}$$

$$a = \frac{4800 \sin 75°}{\sin 57°} \qquad \text{Solve for } a.$$

$$a \approx 5528 \qquad \text{Approximate } a.$$

The photograph will show about 5528 feet of ground distance from one edge of the photograph to the other.   *Now Try Exercises 11 and 37* ◀

EXAMPLE 2   Solving a triangle (AAS)

If $\beta = 85°$, $\gamma = 40°$, and $b = 26$, solve triangle ABC.

SOLUTION Sketch triangle ABC as shown in Figure 8.11.

$$\alpha = 180° - \beta - \gamma = 180° - 85° - 40° = 55°$$

Side a can be found by using the law of sines.

Figure 8.11

$$\frac{a}{\sin \alpha} = \frac{b}{\sin \beta} \qquad \text{Law of sines}$$

$$\frac{a}{\sin 55°} = \frac{26}{\sin 85°} \qquad \text{Substitute.}$$

$$a = \frac{26 \sin 55°}{\sin 85°} \qquad \text{Solve for } a.$$

$$a \approx 21.4 \qquad \text{Approximate } a.$$

Side c can be found in a similar manner.

$$\frac{c}{\sin \gamma} = \frac{b}{\sin \beta} \qquad \text{Law of sines}$$

$$\frac{c}{\sin 40°} = \frac{26}{\sin 85°} \qquad \text{Substitute.}$$

$$c = \frac{26 \sin 40°}{\sin 85°} \qquad \text{Solve for } c.$$

$$c \approx 16.8 \qquad \text{Approximate } c. \qquad \textit{Now Try Exercise 19} ◀$$

EXAMPLE 3   Estimating the distance to the moon (ASA)

Since the moon is a relatively close celestial object, its distance can be measured directly using trigonometry. To find this distance, two photographs of the moon were taken at precisely the same time from two locations. On April 29, 1976, at 11:35 A.M., the lunar angles of elevation during a partial solar eclipse at Bochum in upper Germany and at Donaueschingen in lower Germany were measured as $\alpha = 52.6997°$ and $\theta = 52.7430°$, respectively. See Figure 8.12 on the next page. If the two cities are 398.02 kilometers apart, approximate the distance to the moon. Disregard the curvature of Earth in this calculation. (**Source:** W. Schlosser, *Challenges of Astronomy.*)

Moon
(Not to scale)

Bochum
α
398.02 km
θ
Donaueschingen

Figure 8.12

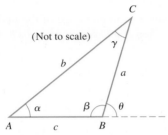

C
(Not to scale)
γ
b
a
α β θ
A c B

Figure 8.13

SOLUTION

Getting Started Consider triangle ABC in Figure 8.13, where $\alpha = 52.6997°$, $\theta = 52.7430°$, and $c = 398.02$. Because $\alpha + \gamma$ and θ are both supplements of angle β, it follows that $\alpha + \gamma = \theta$ and so

$$\gamma = \theta - \alpha = 52.7430° - 52.6997° = 0.0433°.$$

The distance to the moon can be approximated by finding either a or b. Why? ▶

We find a by applying the law of sines.

$$\frac{a}{\sin \alpha} = \frac{c}{\sin \gamma} \qquad \text{Law of sines}$$

$$\frac{a}{\sin 52.6997°} = \frac{398.02}{\sin 0.0433°} \qquad \text{Substitute.}$$

$$a = \frac{398.02 \sin 52.6997°}{\sin 0.0433°} \qquad \text{Solve for } a.$$

$$a \approx 419{,}000 \text{ km} \qquad \text{Approximate } a.$$

The distance to the moon on that day was about 419,000 kilometers.

Now Try Exercise 39 ◀

Bearings Bearings are used in both surveying and aerial navigation to determine directions. If a single angle is used for a **bearing**, then it is understood that the bearing is measured *clockwise* from due north. Some examples of bearings are shown in Figure 8.14.

N 40°

N 130°

N 325°

Figure 8.14 Examples of Bearings

In the next example, we locate a fire by using bearings. Locating fires precisely is particularly important for firefighters in rugged areas with limited access. (**Source:** I. Mueller and K. Ramsayer, *Introduction to Surveying*.)

EXAMPLE 4 Determining the location of a forest fire

A fire is spotted from two ranger stations that are 4 miles apart, as illustrated in Figure 8.15. From station A the bearing of the fire is 35°, and from station B the bearing of the fire is 335°. Find the distance between the fire and each ranger station if station A lies directly west of station B. Approximate the results to the nearest hundredth of a mile.

N 35° A 4 miles 335° B C N

Figure 8.15

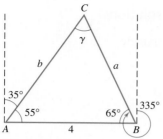

Figure 8.16

SOLUTION See triangle ABC in Figure 8.16, where $\alpha = 55°$, $\beta = 335° - 270° = 65°$, and $c = 4$. Thus

$$\gamma = 180° - 55° - 65° = 60°.$$

Using the law of sines, we can find a.

$$\frac{a}{\sin \alpha} = \frac{c}{\sin \gamma} \qquad \text{Law of sines}$$

$$\frac{a}{\sin 55°} = \frac{4}{\sin 60°} \qquad \text{Substitute.}$$

$$a = \frac{4 \sin 55°}{\sin 60°} \qquad \text{Solve for } a.$$

$$a \approx 3.78 \text{ mi} \qquad \text{Approximate } a.$$

In a similar manner, we can find b.

$$\frac{b}{\sin \beta} = \frac{c}{\sin \gamma} \qquad \text{Law of sines}$$

$$\frac{b}{\sin 65°} = \frac{4}{\sin 60°} \qquad \text{Substitute.}$$

$$b = \frac{4 \sin 65°}{\sin 60°} \qquad \text{Solve for } b.$$

$$b \approx 4.19 \text{ mi} \qquad \text{Approximate } b.$$

The fire is 4.19 miles from station A and 3.78 miles from station B.

Now Try Exercise 41 ◀

The Ambiguous Case (SSA)

If we are given two sides and an angle opposite one of the sides (SSA), there may be zero, one, or two triangles that satisfy these conditions. Refer to Figures 8.4–8.6. For this reason SSA is called the *ambiguous case*. The following box summarizes the steps needed to recognize and solve each situation of the ambiguous case.

Solving the Ambiguous Case (SSA)

Given two sides and an angle opposite one side, the following steps can be used to determine whether there are zero, one, or two triangles that satisfy the conditions. For simplicity, assume that a, b, and α are given.

STEP 1: Use the law of sines to find $\sin \beta$, where β is the unknown angle opposite one of the given sides.

No Solutions: If $\sin \beta > 1$, there are no possible triangles.

One Solution: If $\sin \beta = 1$, then $\beta = 90°$ and $\gamma = 90° - \alpha$. Use the law of sines or the Pythagorean theorem to find the unknown side c. There is one right triangle.

STEP 2: If $\sin \beta = k$, where $0 < k < 1$, calculate $\sin^{-1} k = \beta_1$ and $\beta_2 = 180° - \beta_1$. Note that β_1 and β_2 are the two possible solutions to $\sin \beta = k$.

One Solution: If $\alpha + \beta_2 \geq 180°$, there is one possible triangle determined by β_1. Let $\gamma = 180° - \alpha - \beta_1$. Use the law of sines to find c.

Two Solutions: If $\alpha + \beta_2 < 180°$, there are two possible triangles. Let $\gamma_1 = 180° - \alpha - \beta_1$ and $\gamma_2 = 180° - \alpha - \beta_2$. Use the law of sines to determine c_1 and c_2.

EXAMPLE 5 Solving the ambiguous case (no solutions)

Let $\alpha = 62°$, $a = 6$, and $b = 10$. If possible, solve the triangle.

SOLUTION

STEP 1: We begin by using the law of sines to find $\sin \beta$.

$$\frac{\sin \beta}{b} = \frac{\sin \alpha}{a} \qquad \text{Law of sines}$$

$$\frac{\sin \beta}{10} = \frac{\sin 62°}{6} \qquad \text{Substitute.}$$

$$\sin \beta = \frac{10 \sin 62°}{6} \qquad \text{Solve for } \sin \beta.$$

$$\sin \beta \approx 1.47 > 1 \qquad \text{Approximate } \sin \beta.$$

Figure 8.17

Since the sine function is never greater than 1, there are no solutions for β. No such triangle exists. See Figure 8.17. *Now Try Exercise 29* ◀

An Application from Construction Trusses are used in construction to support roofs, radio towers, bridges, and aircraft frames. A truss is composed of straight segments joined together at their ends. Many times the smaller shapes within a truss are triangles. See Figure 8.18. If trusses are designed properly, they can be relatively light and very strong. In the next example, we determine that a particular truss design results in a unique truss. (**Source:** W. Riley, *Statics and Mechanics of Materials.*)

Figure 8.18 A Bridge Truss

EXAMPLE 6 Solving the ambiguous case (one solution)

Suppose that an engineer has designed the truss in Figure 8.18 and specified that in triangle ABC, $BC = 17$ feet, $AC = 22$ feet, and angle $ABC = 32°$. Determine the length of AB to the nearest tenth of a foot. Is this value for AB unique?

SOLUTION

STEP 1: Using standard labeling, let $a = 17$, $b = 22$, and $\beta = 32°$. Start by finding $\sin \alpha$.

$$\frac{\sin \alpha}{a} = \frac{\sin \beta}{b} \qquad \text{Law of sines}$$

$$\frac{\sin \alpha}{17} = \frac{\sin 32°}{22} \qquad \text{Substitute.}$$

$$\sin \alpha = \frac{17 \sin 32°}{22} \qquad \text{Solve for } \sin \alpha.$$

$$\sin \alpha \approx 0.4095 \qquad \text{Approximate } \sin \alpha.$$

STEP 2: There are two values possible for angle α if $\sin \alpha \approx 0.4095$. Angle α lies in quadrant I or II with reference angle $\alpha_R \approx \sin^{-1}(0.4095) \approx 24.2°$. Thus

$$\alpha_1 \approx 24.2° \quad \text{and} \quad \alpha_2 \approx 180° - 24.2° = 155.8°.$$

However, if $\alpha_2 \approx 155.8°$, then $\alpha_2 + \beta = 155.8° + 32° \geq 180°$, which is not possible in a triangle. Therefore $\alpha_1 \approx 24.2°$ is the only possibility and

$$\gamma \approx 180° - 24.2° - 32° = 123.8°.$$

Suppose we are given a, b, and α and calculate $\sin \beta = 1$. Discuss the number of solutions for triangle ABC.

The law of sines allows us to find AB, or side c.

$$\frac{c}{\sin \gamma} = \frac{b}{\sin \beta} \qquad \text{Law of sines}$$

$$\frac{c}{\sin 123.8°} \approx \frac{22}{\sin 32°} \qquad \text{Substitute.}$$

$$c \approx \frac{22 \sin 123.8°}{\sin 32°} \qquad \text{Solve for } c.$$

$$c \approx 34.5 \text{ ft} \qquad \text{Approximate } c.$$

Thus AB is about 34.5 feet long, and this value is unique.

Now Try Exercise 25 ◀

EXAMPLE 7 **Solving the ambiguous case (two solutions)**

Let $\beta = 55°$, $a = 8.5$, and $b = 7.3$. Solve the triangle. Round to the nearest tenth.

SOLUTION

STEP 1: Begin by finding $\sin \alpha$.

$$\frac{\sin \alpha}{a} = \frac{\sin \beta}{b} \qquad \text{Law of sines}$$

$$\frac{\sin \alpha}{8.5} = \frac{\sin 55°}{7.3} \qquad \text{Substitute.}$$

$$\sin \alpha = \frac{8.5 \sin 55°}{7.3} \qquad \text{Solve for } \sin \alpha.$$

$$\sin \alpha \approx 0.9538 \qquad \text{Approximate } \sin \alpha.$$

Suppose you are given a, b, and α. If $a > b$, what can be said about the number of solutions? Explain your reasoning.

STEP 2: Two angles that satisfy $\sin \alpha \approx 0.9538$ in quadrants I or II are

$$\alpha_1 \approx \sin^{-1}(0.9538) \approx 72.5° \qquad \text{and} \qquad \alpha_2 \approx 180° - 72.5° = 107.5°.$$

Both of these values for α are valid, since they do not result in the sum of the angles exceeding 180°. There are *two* solutions.

Solution 1 Let $\alpha_1 \approx 72.5°$. Then $\gamma_1 \approx 180° - 72.5° - 55° = 52.5°$. Side c_1 can then be found.

$$\frac{c_1}{\sin \gamma_1} = \frac{b}{\sin \beta} \qquad \text{Law of sines}$$

$$c_1 \approx \frac{7.3 \sin 52.5°}{\sin 55°} \qquad \text{Substitute and solve for } c_1.$$

$$c_1 \approx 7.1 \qquad \text{Approximate } c_1.$$

A sketch of triangle ABC is shown in Figure 8.19.

Figure 8.19 Solution 1

Solution 2 Let $\alpha_2 \approx 107.5°$. Then $\gamma_2 \approx 180° - 107.5° - 55° = 17.5°$. Then side c_2 can be found.

$$\frac{c_2}{\sin \gamma_2} = \frac{b}{\sin \beta} \qquad \text{Law of sines}$$

$$c_2 \approx \frac{7.3 \sin 17.5°}{\sin 55°} \qquad \text{Substitute and solve for } c_2.$$

$$c_2 \approx 2.7 \qquad \text{Approximate } c_2.$$

A sketch of triangle ABC is shown in Figure 8.20.

Figure 8.20 Solution 2

Now Try Exercise 21 ◀

8.1 Putting It All Together

The law of sines can be expressed as either

$$\frac{\sin \alpha}{a} = \frac{\sin \beta}{b} = \frac{\sin \gamma}{c} \quad \text{or} \quad \frac{a}{\sin \alpha} = \frac{b}{\sin \beta} = \frac{c}{\sin \gamma}.$$

When using the law of sines, a good strategy is to select an equation with the unknown variable in the numerator. We can use the law of sines to solve triangles when we are given ASA, AAS, or SSA. The case where we are given SSA is called the ambiguous case, since there may be zero, one, or two triangles that satisfy the conditions. Several of these situations for the ambiguous case are shown in the accompanying table.

The Ambiguous Case	Examples
No solutions for triangle *ABC*	$\sin \beta > 1$ (α acute) $\sin \beta > 1$ (α obtuse)
One solution for triangle *ABC*	$\sin \beta = 1$ (α acute) $a > b$ (α acute) $a > b$ (α obtuse)
Two solutions for triangle *ABC*	$\sin \beta < 1$ and $a < b$ (α acute)

8.1   Exercises

Recognizing the Ambiguous Case (SSA)

Exercises 1–8: Let triangle ABC have standard labeling. Given the following angles and sides, decide if solving the triangle results in the ambiguous case.

1. α, β, and a

2. α, γ, and c

3. a, b, and c

4. α, a, and b

5. β, b, and c

6. α, b, and c

7. γ, a, and c

8. β, b, and α

Solving Triangles

Exercises 9–18: Solve the triangle, if possible. If the triangle represents the ambiguous case, solve for all possible triangles. Approximate values to the nearest tenth.

9.

10.

11.

12.

13.

14.

15.

16.

17.

18.

Exercises 19–36: Solve the triangle, if possible. Approximate values to the nearest tenth when appropriate.

19. $\alpha = 32°$, $\beta = 55°$, $b = 12$

20. $\beta = 20°$, $\gamma = 67°$, $c = 9$

21. $\alpha = 20°$, $b = 9$, $a = 7$

22. $\alpha = 20°$, $b = 7$, $a = 9$

23. $b = 10$, $\beta = 30°$, $c = 20$

24. $a = 13.5$, $\alpha = 46°$, $c = 27.8$

25. $\gamma = 102°$, $c = 51.6$, $a = 42.1$

26. $\beta = 43°$, $b = 22.1$, $c = 30.7$

27. $\alpha = 55.2°$, $\gamma = 114.8°$, $b = 19.5$

28. $c = 225$, $\alpha = 103.2°$, $\beta = 62.5°$

29. $b = 6.2$, $c = 7.4$, $\beta = 73°$

30. $\alpha = 45°$, $a = 5$, $b = 5\sqrt{2}$

31. $\alpha = 35°15'$, $a = 5$, $b = 12$

32. $\gamma = 71°35'$, $c = 6$, $b = 9$

33. $\beta = 46°45'$, $a = 6$, $b = 5$

34. $\alpha = 54°12'$, $c = 12$, $a = 10$

35. $\alpha = 56°30'$, $\beta = 23°45'$, $c = 100$

36. $\beta = 56°48'$, $\gamma = 10°12'$, $a = 55$

Applications

37. *Aerial Photography* (Refer to Example 1.) The plane shown in the figure is taking an aerial photograph with a camera lens that has an angular coverage of 70°. The ground below is inclined at 7°. If the angle of elevation of the plane at *B* is 52° and distance *BC* is 3500 feet, estimate the ground distance *AB* (to the nearest foot) that will appear in the picture. (**Source:** F. Moffit.)

38. *Aerial Photography* As a picture is taken over level ground, the airplane's distance from a building located at the edge of the photograph is 7500 feet and the angle of depression to the building is 56°. See the figure. Find the ground distance *b* shown in the photograph to the nearest hundred feet.

39. *Distance to the Moon* (Refer to Example 3 and Figure 8.12.) Suppose that the lunar angle at Bochum in upper Germany had been measured as $\alpha = 52.6901°$ instead of 52.6997°. Determine the effect that this would have on the estimation of the distance to the moon. Interpret the result.

40. *Distance to the Moon* (Refer to Example 3 and Figure 8.12.) Suppose that the distance between two locations is 452.45 kilometers and that their angles of elevation to the moon are $\alpha = 47.8981°$ and $\theta = 47.9443°$. Estimate the distance to the moon (to the nearest thousand kilometers).

41. *Locating a Ship* The figure shows the bearings of a ship on Lake Ontario from two observation points located on a straight shoreline. The bearing from the first observation point is 54.3°, and the bearing from the second observation point is 325.2°. If the distance between these points is 15 miles, how far is it from the ship to shore? Assume that the first observation point is directly west of the second observation point.

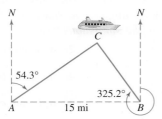

42. *Distance* An ore ship is traveling west toward Duluth on Lake Superior at 18 miles per hour. The bearing of Split Rock Lighthouse is 285°. After 1 hour, the bearing of the lighthouse is 340°. Find the distance between the ship and the lighthouse when the second bearing was determined.

43. *Truss Construction* For the truss shown in the figure, *AB* is 24.2 feet, angle *ABD* is 118°, and angle *BDF* is 28°. Find the length of *BD*. (**Source:** W. Riley.)

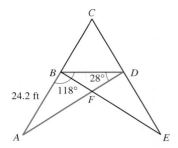

44. *Truss Construction* Use the results of Example 6 to solve triangle *ACD* in Figure 8.18 if angle *BAD* is 90°.

45. *Distance Across a River* To find the distance *AB* across a river, a distance *BC* = 354 meters is measured off on one side of the river. See the figure. It is found that angle *ABC* = 112°10′ and that angle *BCA* = 15°20′. Find the distance *AB*.

46. *Distance Across a Canyon*   To find the distance *RS* across a canyon, a distance *TR* = 582 yards is measured off on one side of the canyon. See the figure. It is found that angle *TRS* = 102°20′ and that angle *RTS* = 32°50′. Find the distance *RS*.

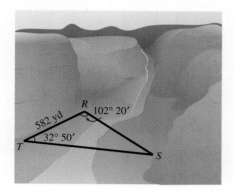

47. *Height of a Helicopter*   A helicopter is sighted at the same time by two ground observers who are 3 miles apart on the same side of the helicopter. See the figure. They report angles of elevation of 20.5° and 27.8°. How high is the helicopter?

48. *Height of a Hot-Air Balloon*   Two observation points *A* and *B* are 1500 feet apart. From these points the angles of elevation of a hot-air balloon are 43° and 47°, as illustrated in the figure. Find the height of the balloon to the nearest foot.

49. *Height of a Balloon*   A balloonist is directly above a straight road 1.5 miles long that joins two towns. She

finds that the town closer to her is at an angle of depression of 35° and the farther town is at an angle of depression of 31°. How high above the ground is the balloon?

50. *Highway Construction*   In a *reverse curve*, or S-curve, two circular curves are used to connect two straight portions of highway that are offset, as illustrated in the figure. Angles α and β will not be equal if the two straight portions of highway have different directions. Typically the same radius *r* is used for both portions of the reverse curve. If *r* = 480 feet, α = 38°, β = 15°, and θ = 75°, find the distance between *A* and *B* to the nearest foot.   (**Source:** P. Kissam.)

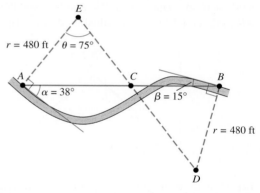

51. *Surveying*   To find the distance between two points *A* and *B* on opposite sides of a small pond, a surveyor determines that *AC* is 97.3 feet, angle *ACB* is 55.1°, and angle *CAB* is 75.7°, as illustrated in the figure. Find the distance between *A* and *B*.

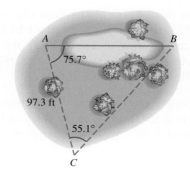

52. *Trigonometric Leveling* In surveying it is often necessary to determine the height of an inaccessible point P, as illustrated in the figure. Points A, B, and C lie on level ground. (**Source:** P. Kissam, *Surveying Practice.*)

(a) If angle ABP is $50°$, angle PAB is $53.3°$, and AB is 102 feet, find PB.

(b) If angle PBC is $47°$ and C is directly below P, find PC.

53. *Locating a Ship* From two observation points A and B, a sinking ship is spotted at point C. Angle CAB and angle ABC are measured as $28°$ and $60°$, and the distance AB is about 4.12 miles, as illustrated in the figure.

(a) How far is the ship from point A, to the nearest hundredth of a mile?

(b) If the coordinates of A are $(0, 0)$ and the coordinates of B are $(4, 1)$, find the bearing of the ship from point A to the nearest degree.

54. *Airplane Navigation* An airplane takes off with a bearing of $55°$ and flies 480 miles, after which it changes its course to a bearing of $285°$. Finally the airplane flies back to its starting point with a bearing of $180°$. Find the total mileage flown by the airplane.

55. *Height of the Gateway Arch* The tallest monument in the world is the Gateway to the West Arch in St. Louis. From point A the top of the arch has an angle of eleva-

tion of $64.91°$, and from point B the angle of elevation is $60.81°$. See the figure. If distance AB is 57 feet, find the height of this monument to the nearest foot. (**Source:** *The Guinness Book of Records.*)

56. *Height of a Tower* A vertical tower supporting a cable for chairlifts to transport skiers up a mountain is located on a ski slope inclined at $28°$, as illustrated in the figure. If the length of the tower's shadow is 21 feet along the mountain side when the angle of the sun is $57°$ with respect to the ski slope, calculate the height of the tower.

Writing about Mathematics

57. In your own words, describe two situations where the law of sines can be applied. Give an example of each situation.

58. Suppose that you are given α, a, and b for triangle ABC. If α is obtuse, what is the maximum number of triangles that could satisfy these conditions? What is the maximum number if α is acute? Explain your reasoning and give examples.

8.2 Law of Cosines

- Derive the law of cosines
- Solve triangles
- Find areas of triangles

Introduction

Surveying has been used for centuries in construction and in the determination of boundaries. Today the Global Positioning System (GPS) is being used to determine distances on Earth. The signal from a GPS satellite contains the information necessary for hand-held receivers to calculate both the position of a GPS satellite and its distance from the receiver. This information can be used to accurately calculate distances and angles between points on the ground. The law of cosines, which is a generalization of the Pythagorean theorem, is used in GPS calculations. In this section, the law of cosines is introduced and used to solve several applications.   (**Source:** J. Van Sickle, *GPS for Land Surveyors.*)

Derivation of the Law of Cosines

The law of cosines can be used to solve a triangle given either all three sides (SSS) or two sides and the angle included (SAS). In both cases a unique triangle is formed, as illustrated in Figures 8.21 and 8.22.

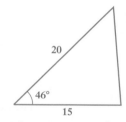

Figure 8.21 Given **SSS**   **Figure 8.22** Given **SAS**

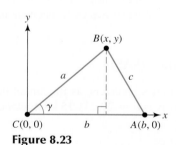

Figure 8.23

Consider triangle *ABC* shown in Figure 8.23, with point *B* having coordinates (x, y). Using definitions of sine and cosine, we find

$$\cos \gamma = \frac{x}{a} \quad \text{and} \quad \sin \gamma = \frac{y}{a},$$

or equivalently,

$$x = a \cos \gamma \quad \text{and} \quad y = a \sin \gamma.$$

As a result, the coordinates of *B* are $(a \cos \gamma, a \sin \gamma)$. Since the coordinates of point *A* are $(b, 0)$, the distance *c* between points *A* and *B* can be found.

$$c = \sqrt{(a \cos \gamma - b)^2 + (a \sin \gamma - 0)^2} \qquad \text{Distance formula}$$
$$c^2 = (a \cos \gamma - b)^2 + (a \sin \gamma - 0)^2 \qquad \text{Square each side.}$$
$$c^2 = a^2 \cos^2 \gamma - 2ab \cos \gamma + b^2 + a^2 \sin^2 \gamma \qquad \text{Expand each expression.}$$
$$c^2 = a^2 (\cos^2 \gamma + \sin^2 \gamma) - 2ab \cos \gamma + b^2 \qquad \text{Distributive property}$$
$$c^2 = a^2 + b^2 - 2ab \cos \gamma \qquad \cos^2 \gamma + \sin^2 \gamma = 1$$

Algebra Review
To review squaring a binomial, see Chapter R (page R-18).

This result is valid for any triangle *ABC* and is known as the *law of cosines*. Since the vertices in Figure 8.23 could be rearranged, three possible equations are associated with the law of cosines.

Figure 8.24

> ### Law of Cosines
>
> Any triangle with standard labeling satisfies
> $$a^2 = b^2 + c^2 - 2bc \cos \alpha$$
> $$b^2 = a^2 + c^2 - 2ac \cos \beta$$
> $$c^2 = a^2 + b^2 - 2ab \cos \gamma.$$

Solving Triangles

A common problem in surveying is to find the distance between two points A and B situated on opposite sides of a building, as illustrated in Figure 8.24. This distance can be found by applying the law of cosines.   (**Source:** P. Kissam, *Surveying Practice.*)

> **EXAMPLE 1**   **Finding the distance between two points (SAS)**

The surveyor in Figure 8.24 determines that CA is 75 feet, CB is 58 feet, and angle ACB is 83°. Find distance AB to the nearest foot.

SOLUTION Note that $CA = b$, $CB = a$, and angle $ACB = \gamma$. So let $b = 75$, $a = 58$, and $\gamma = 83°$. To find c, apply the law of cosines.

$$\begin{aligned} c^2 &= a^2 + b^2 - 2ab \cos \gamma & \text{Law of cosines} \\ &= (58)^2 + (75)^2 - 2(58)(75) \cos 83° & \text{Substitute.} \\ &\approx 7929 & \text{Approximate.} \\ c &\approx 89.04 & \text{Take the square root.} \end{aligned}$$

The points A and B are about 89 feet apart.   *Now Try Exercises 13 and 53* ◄

Global Positioning System   In the next example, the distance between two GPS receivers is found. Finding the distance between two points might be important to surveyors or to search parties looking for lost hikers. The distance between two GPS receivers is sometimes called the *baseline*.

> **EXAMPLE 2**   **Using GPS to find a baseline distance (SAS)**

A search party and an injured hiker both have hand-held GPS receivers, as illustrated in Figure 8.25. The distance from the satellite to the search party is $b = 20,231.15$ kilometers,

Figure 8.25

and the distance from the satellite to the hiker is $c = 20{,}231.57$ kilometers. If it is determined that $\alpha = 0.01456°$, estimate the baseline a between the search party and the hiker to the nearest hundredth of a kilometer.

SOLUTION We can use the law of cosines to find a.

$$a^2 = b^2 + c^2 - 2bc \cos \alpha$$
$$= (20{,}231.15)^2 + (20{,}231.57)^2 - 2(20{,}231.15)(20{,}231.57) \cos 0.01456°$$
$$\approx 26.61$$
$$a \approx 5.16$$

The distance between the search party and the hiker is about 5.16 kilometers.

Now Try Exercises 19 and 55 ◀

An Application from Construction Trusses are frequently used to support roofs on buildings, as illustrated in Figure 8.26. The simplest type of roof truss is a triangle, as shown in Figure 8.27. One basic task when constructing a roof truss is to cut the ends of the rafters so that the roof has the correct slope. (**Source:** W. Riley, *Statics and Mechanics of Materials*.)

Figure 8.26

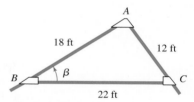

Figure 8.27 Roof Truss

EXAMPLE 3 Designing a roof truss (SSS)

Find β for the truss shown in Figure 8.27.

SOLUTION Begin by letting $a = 22$, $b = 12$, and $c = 18$, and then use the law of cosines to find β to the nearest degree.

$$b^2 = a^2 + c^2 - 2ac \cos \beta \qquad \text{Law of cosines}$$
$$2ac \cos \beta = a^2 + c^2 - b^2 \qquad \text{Transpose terms.}$$
$$\cos \beta = \frac{a^2 + c^2 - b^2}{2ac} \qquad \text{Divide by } 2ac.$$
$$\cos \beta = \frac{22^2 + 18^2 - 12^2}{2(22)(18)} \qquad \text{Substitute.}$$
$$\cos \beta \approx 0.8384 \qquad \text{Approximate.}$$

Thus $\beta \approx \cos^{-1}(0.8384) \approx 33°$.

Now Try Exercise 63 ◀

EXAMPLE 4 Solving a triangle (SSS)

Find α, β, and γ in triangle ABC to the nearest tenth of a degree if $a = 5$, $b = 6$, and $c = 9$.

SOLUTION

Getting Started The largest angle, γ, is opposite the largest side, c. We start by finding γ. ▶

$$c^2 = a^2 + b^2 - 2ab \cos \gamma \qquad \text{Law of cosines}$$

$$2ab \cos \gamma = a^2 + b^2 - c^2 \qquad \text{Transpose terms.}$$

$$\cos \gamma = \frac{a^2 + b^2 - c^2}{2ab} \qquad \text{Divide by 2ab.}$$

$$\cos \gamma = \frac{5^2 + 6^2 - 9^2}{2(5)(6)} \qquad \text{Let } a = 5, b = 6, \text{ and } c = 9.$$

$$\cos \gamma = -\frac{1}{3} \qquad \text{Simplify.}$$

Thus $\gamma \approx \cos^{-1}\left(-\frac{1}{3}\right) \approx 109.5°$. The law of cosines could be used again to find either α or β. However, we use the law of sines to find α.

$$\frac{\sin \alpha}{a} = \frac{\sin \gamma}{c} \qquad \text{Law of sines}$$

$$\sin \alpha = \frac{a \sin \gamma}{c} \qquad \text{Multiply by a.}$$

$$\sin \alpha \approx \frac{5 \sin 109.5°}{9} \qquad \text{Let } a = 5, c = 9, \text{ and } \gamma \approx 109.5°.$$

$$\sin \alpha \approx 0.5237 \qquad \text{Simplify.}$$

Because γ is the largest angle, it follows that α must be an acute angle and that $\alpha \approx \sin^{-1}(0.5237) \approx 31.6°$. To find β we use the fact that the measures of the angles sum to 180° in a triangle.

$$\beta \approx 180° - 109.5° - 31.6° = 38.9° \qquad \textbf{\textit{Now Try Exercise 21}} \blacktriangleleft$$

NOTE In solving the case SSS, first find the largest angle, which is opposite the longest side. There can be at most one obtuse angle in a triangle, and the law of cosines finds an obtuse angle, if it exists, because the cosine of an obtuse angle is negative, as was the case in Example 4.

Area Formulas

One task that is frequently performed by surveyors is to find the acreage of a lot using a technique called triangulation. *Triangulation* divides a parcel of land into triangles. The area of the lot equals the sum of the areas of the triangles. We begin our discussion by developing some area formulas for triangles.

The area K of any triangle is given by $K = \frac{1}{2}bh$, where b is its base and h is its height. Using trigonometry, we can find a formula for the area of the triangle shown in Figure 8.28.

Figure 8.28

$$\sin \alpha = \frac{h}{c}, \qquad \text{or} \qquad h = c \sin \alpha$$

Thus the area equals

$$K = \frac{1}{2}bh = \frac{1}{2}bc \sin \alpha.$$

Since the labels for the vertices in triangle ABC could be rearranged, three area formulas can be written as follows. Notice that these formulas can be applied when we are given SAS. (We use K for area rather than A so as not to cause confusion with vertex A.)

Area of a Triangle

For any triangle with standard labeling, the area K is given by

$$K = \frac{1}{2}ab \sin \gamma, \qquad K = \frac{1}{2}ac \sin \beta, \qquad \text{or} \qquad K = \frac{1}{2}bc \sin \alpha.$$

EXAMPLE 5   Finding the area of a triangle (SAS)

Find the area of triangle ABC in Figure 8.29 to the nearest square foot.

SOLUTION We are given $\beta = 55°$, $a = 34$ feet, and $c = 42$ feet. Thus the area K is given by the following.

$$K = \frac{1}{2}ac \sin \beta = \frac{1}{2}(34)(42) \sin 55° \approx 585 \text{ square feet}$$

Now Try Exercise 37 ◀

Figure 8.29

Heron's Formula   The next formula can be used to find the area of a triangle when the lengths of three sides are known. It is named after the Greek mathematician Heron.

Heron's Formula

If a triangle has sides with lengths a, b, and c, then its area K is given by

$$K = \sqrt{s(s - a)(s - b)(s - c)},$$

where $s = \frac{1}{2}(a + b + c)$ and s is called the **semiperimeter**.

EXAMPLE 6   Finding the area of a triangle (SSS)

Approximate the area of triangle ABC with sides $a = 4$, $b = 5$, and $c = 7$.

SOLUTION

Getting Started Begin by calculating $s = \frac{1}{2}(4 + 5 + 7) = 8$. ▶

Then the area is

$$K = \sqrt{8(8 - 4)(8 - 5)(8 - 7)} = \sqrt{96} \approx 9.8$$

Now Try Exercise 39 ◀

An Application from Surveying   One method for finding the area of a lot is called the *distance method*. This method can be used to find the area of an irregular lot, as illustrated in the next example. The distance method does not measure angles—it measures only distances between points. Triangulation and Heron's formula can be used to find the area of the lot.   (**Source:** I. Mueller and K. Ramsayer, *Introduction to Surveying.*)

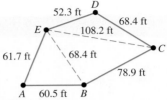

Figure 8.30

EXAMPLE 7 Applying the distance method to find the area of a lot

Find the area of the parcel of land determined by $ABCDE$ in Figure 8.30.

SOLUTION For triangle ABE, $s = \frac{1}{2}(60.5 + 68.4 + 61.7) = 95.3$, and its area is

$$K_1 = \sqrt{95.3(95.3 - 60.5)(95.3 - 68.4)(95.3 - 61.7)} \approx 1731.$$

For triangle BCE, $s = \frac{1}{2}(78.9 + 108.2 + 68.4) = 127.75$, and its area is

$$K_2 = \sqrt{127.75(127.75 - 78.9)(127.75 - 108.2)(127.75 - 68.4)} \approx 2691.$$

For triangle CDE, $s = \frac{1}{2}(68.4 + 52.3 + 108.2) = 114.45$, and its area is

$$K_3 = \sqrt{114.45(114.45 - 68.4)(114.45 - 52.3)(114.45 - 108.2)} \approx 1431.$$

The area of the lot is

$$K_1 + K_2 + K_3 \approx 1731 + 2691 + 1431 = 5853 \text{ square feet.}$$

Now Try Exercise 77 ◀

8.2 Putting It All Together

The law of cosines can be used to solve triangles when either three sides (SSS) or two sides and the angle included (SAS) are given. Heron's formula can be used to find the area of a triangle when the lengths of all three sides are known. Important formulas found in this section are given in the table, where it is assumed that triangle ABC has standard labeling.

Concept	Explanation	Examples
Law of cosines	$a^2 = b^2 + c^2 - 2bc \cos \alpha$ $b^2 = a^2 + c^2 - 2ac \cos \beta$ $c^2 = a^2 + b^2 - 2ab \cos \gamma$ Can be used to solve a triangle given either SSS or SAS	If $b = 3$, $c = 4$, and $\alpha = 60°$, then $$a^2 = 3^2 + 4^2 - 2(3)(4) \cos 60° = 13$$ and $a = \sqrt{13}$.
Area formulas	$K = \frac{1}{2}ab \sin \gamma$ $K = \frac{1}{2}ac \sin \beta$ $K = \frac{1}{2}bc \sin \alpha$ Can be used to find the area of a triangle given SAS	If $a = 2$ feet, $b = 3$ feet, and $\gamma = 30°$, then the area of the triangle is $$K = \frac{1}{2}(2)(3) \sin 30° = 1.5 \text{ ft}^2.$$

Concept	Explanation	Examples
Heron's formula	$K = \sqrt{s(s-a)(s-b)(s-c)}$, where $s = \frac{1}{2}(a+b+c)$ Can be used to find the area of a triangle given SSS	If $a = 3$ feet, $b = 5$ feet, and $c = 4$ feet, then $$s = \tfrac{1}{2}(3+5+4) = 6,$$ and the area of the triangle is $$K = \sqrt{6(6-3)(6-5)(6-4)} = 6 \text{ ft}^2.$$

8.2   Exercises

Determining a Method to Solve a Triangle

Exercises 1–8: Assume triangle ABC has standard labeling and complete the following.

(a) Determine if AAS, ASA, SSA, SAS, or SSS is given.
(b) Decide if the law of sines or the law of cosines should be used first to solve the triangle.

1. $a, b,$ and γ **2.** $\alpha, \gamma,$ and c

3. $a, b,$ and α **4.** $a, b,$ and c

5. $\alpha, \beta,$ and c **6.** $a, c,$ and α

7. $\beta, a,$ and γ **8.** $b, c,$ and α

Solving Triangles

Exercises 9 and 10: Find the length of the remaining side of each triangle. Do not use a calculator.

9.

10.

Exercises 11 and 12: Find the value of θ in each triangle. Do not use a calculator.

11.

12.

Exercises 13–18: Solve the triangle. Approximate values to the nearest tenth.

13.

14.

15.

16.

17.

18.
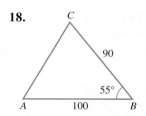

Exercises 19–30: Solve the triangle. Round values to the nearest tenth when appropriate.

19. $a = 45, \gamma = 35°, b = 24$

20. $c = 7.9, \beta = 52°, a = 9.6$

21. $a = 2.4, b = 1.7, c = 1.4$

22. $a = 43, b = 41, c = 34$

23. $\alpha = 10°30', b = 24.1, c = 15.8$

24. $a = 12.8, b = 15.8, \gamma = 36°$

25. $a = 10.6, b = 25.8, c = 20.6$

26. $a = 104, b = 121, c = 111$

27. $\beta = 122°10', a = 20, c = 15$

28. $b = 9.1, \alpha = 43°30', c = 12.5$

29. $a = 5.3, b = 6.7, c = 7.1$

30. $a = 4.2, b = 5.1, c = 3.7$

Does This Triangle Exist?

Exercises 31–36: Decide if a triangle exists that satisfies the conditions. Justify your answer.

31. $a = 10, b = 12, c = 25$

32. $a = 10, \beta = 51°, c = 5$

33. $\alpha = 89°, b = 63, \gamma = 112°$

34. $a = 2, b = 10, \alpha = 50°$

35. $\gamma = 54°, b = 63, \alpha = 63°$

36. $a = 5, b = 6, c = 8$

Area of Triangles

Exercises 37–40: Approximate the area of the triangle to the nearest tenth.

37.

38.

39.

40.

Exercises 41–52: Approximate the area of the triangle to the nearest tenth.

41. $a = 10, b = 12, \gamma = 58°$

42. $\alpha = 40°, b = 5.8, c = 8.8$

43. $\beta = 78°, a = 5.5, c = 6.8$

44. $\alpha = 23°, \gamma = 47°, b = 53$

45. $\beta = 31°, \alpha = 54°, a = 2.6$

46. $a = 7, b = 8, c = 9$

47. $a = 5.5, b = 6.7, c = 9.2$

48. $a = 104, b = 98, c = 112$

49. $a = 11, b = 13, c = 20$

50. $a = 13, b = 14, c = 15$

51. $a = 21, \alpha = 42°, c = 16$

52. $b = 35, c = 38, \gamma = 50°48'$

Applications

53. *Obstructed View*   (Refer to Example 1.) In the figure, a surveyor is attempting to find the distance between points A and B. A grove of trees is obstructing the view, so the surveyor determines that AC is 143 feet, BC is 123 feet, and angle ACB is $78°35'$. Find the distance between A and B to the nearest foot.

54. *Distance across a Lake*   Points A and B are on opposite sides of a lake. From a third point, C, the angle between the lines of sight to A and B is $46.3°$. If AC is 350 meters and BC is 286 meters, find AB.

55. *Ship Navigation*   Two ships set sail with bearings of 52° and 121°, traveling at 20 miles per hour and 14 miles per hour, respectively. Find the approximate distance between the ships after 1.5 hours.

56. *Distance between Two Boats*   Two boats leave a dock together. Each travels in a straight line. The angle between their courses measures 54°10′. One boat travels 36.2 kilometers per hour and the other 45.6 kilometers per hour. How far apart will they be after 3 hours?

57. *Diagonals of a Parallelogram*   One side of a parallelogram is 3.5 feet and another side is 5.2 feet. The angle between these two sides is 56°. Find the lengths of the diagonals to the nearest tenth of a foot.

58. *Diagonals of a Parallelogram*   The sides of a parallelogram are 4.0 centimeters and 6.0 centimeters. One angle is 58° while another is 122°. Find the lengths of the diagonals.

59. *Air Navigation*   An airplane flies in the triangular course shown in the figure. To the nearest degree, find the bearings of the plane while traveling from A to B and from B to C.

60. *Flight Distance*   Airports A and B are 450 kilometers apart, on an east–west line. A pilot flies in a northeast direction from A to airport C. From C, the pilot flies 359 kilometers on a bearing of 128°40′ to B. How far is C from A?

61. *Area of a Lot*   A surveyor measures the sides of a triangular lot to be $a = 145.2$, $b = 136.8$, and $c = 95.3$, where measurements are in feet.
   (a) Approximate angles α, β, and γ.

   (b) What is the area of the lot to the nearest square foot?

62. *Angle in a Parallelogram*   One side of a parallelogram is 6.4 yards and another side is 5.3 yards. The shorter diagonal is 3.5 yards. Find the angle opposite the shorter diagonal to the nearest tenth of a degree.

63. *Truss Construction*   (Refer to Example 3.) Find angle θ for the triangular truss shown in the figure.

64. *Robotics*   The figure illustrates the MIT Scheinman robotic arm. Suppose the length of the upper arm is 20 centimeters and the combined length of the forearm and hand is 30 centimeters. If the arm is positioned so that $\theta = 126°$, find the distance between the hand at point A and the shoulder joint at point B.   (**Source:** G. Beni and S. Hackwood, *Recent Advances in Robotics.*)

65. *Distance between Airports*   Airports A and B are 515 miles apart, and airport A is directly west of airport B. Airport C is located in a northeasterly direction from airport A and is 357 miles from airport B. See the figure. If the bearing from airport C to airport B is 125°, find the distance between airports A and C to the nearest mile.

66. *Navigation* A ship is sailing east. At one point, the bearing of a submerged rock is 38°45′. After the ship sails 20.4 miles, the bearing of the rock is 291°15′. Find the distance between the ship and the rock (to the nearest tenth of a mile) when the second bearing is taken.

67. *Distance between a Ship and a Submarine* From an airplane flying over the ocean, the angle of depression to a submarine lying just under the surface is 24°10′. At the same moment, the angle of depression from the airplane to a battleship is 17°30′. See the figure. The distance from the airplane to the battleship is 5120 feet. Find the distance between the battleship and the submarine. (Assume the airplane, submarine, and battleship are in a vertical plane.)

68. *Distance between a Ship and a Rock* A ship is sailing east. At one point, the bearing of a submerged rock is 45°20′. After the ship has sailed 15.2 miles, the bearing of the rock has become 308°40′. Find the distance of the ship from the rock at the latter point.

69. *Distance between Two Ships* Two ships leave a harbor together, traveling on courses that have an angle of 135°40′ between them. If they each travel 402 miles, how far apart are they?

70. *Highway Curve* The most common highway curve consists of a circular arc connecting two sections of straight road, as illustrated in the figure. Find the straight-line distance between *PC* (point of curve) and *PT* (point of tangency). (**Source:** P. Kissam.)

71. *Painting* A painter needs to cover a triangular region with sides of 25 feet and 15 feet. The angle between these two sides is 128°. Find the area of the region to the nearest tenth of a square foot.

72. *Painting* A painter needs to cover a triangular region with sides of 30 feet, 40 feet, and 38 feet. If each can of paint covers 125 square feet, how many cans of paint are needed?

73. *Area of Regular Polygons* If a regular polygon has *n* sides of equal length *L*, then its area *A* is computed by

$$A = \frac{nL^2}{4} \cot \frac{\pi}{n}.$$

(**Source:** M. Mortenson, *Computer Graphics.*)
(a) Using this formula, find the area of an equilateral triangle with sides of 6 inches.

(b) Using Heron's formula, find the area of this triangle. Compare answers.

74. *Area of Regular Polygons* (Refer to Exercise 73.) The measure of an interior angle in a regular polygon with *n* sides is given by $180°\left(1 - \frac{2}{n}\right)$. For example, a square is a regular polygon with *n* = 4 and each interior angle equals $180°\left(1 - \frac{2}{4}\right) = 90°$.
(a) Find the area of a regular pentagon with sides of length 8 inches, using the formula given in Exercise 73. See the figure.

(b) Using triangulation and Heron's formula, find the area of this regular pentagon.

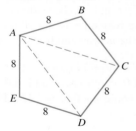

75. *Area of a Lot* Find the area of the lot in the figure to the nearest square foot.

76. *Area of a Lot* Find the area of the quadrangular lot shown in the figure to the nearest square foot.

77. *Area of a Lot* Apply the distance method discussed in Example 7 to find the area of the lot in the figure to the nearest square foot.

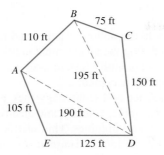

78. *Area of the Bermuda Triangle* Find the area of the Bermuda Triangle (to the nearest thousand square miles) if the sides of the triangle have approximate lengths of 850 miles, 925 miles, and 1300 miles.

Writing about Mathematics

79. Describe two different situations where the law of cosines can be applied. Give an example of each situation.

80. Describe two methods to find the area of a triangle. What information do you need to apply each method? Give an example of each situation.

EXTENDED AND DISCOVERY EXERCISES

1. *Distance between a Satellite and a Tracking Station* A satellite traveling in a circular orbit 1600 kilometers above Earth is due to pass directly over a tracking station at noon. See the figure. Assume that the satellite takes 2 hours to make an orbit and that the radius of Earth is 6400 kilometers. Find the distance between the satellite and the tracking station at 12:03 P.M. (**Source:** *Space Mathematics* by B. Kastner, Ph.D. Copyright © 1972 by the National Aeronautics and Space Administration. Courtesy of NASA.)

2. *Computer Graphics* When a triangle is shaded on a computer screen, three pixels are used to determine the vertices of the triangle. Many computer screens are 1024 pixels by 800 pixels, and each pixel represents 0.015 inch by 0.015 inch, as illustrated in the figure, where pixel (7, 3) is shown. If three pixels $A(100, 300)$, $B(500, 200)$, and $C(320, 600)$ represent vertices of a triangle, approximate the area of triangle ABC on the computer screen to the nearest tenth of a square inch. (**Source:** J. Foley, *Introduction to Computer Graphics.*)

CHECKING BASIC CONCEPTS FOR SECTIONS 8.1 AND 8.2

1. Solve triangle ABC if $\alpha = 44°$, $\gamma = 62°$, and $a = 12$.

2. Solve triangle ABC if $\alpha = 32°$, $a = 6$, and $b = 8$. How many solutions are there?

3. Use the law of cosines to solve the triangles.

(a)

(b)

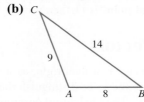

4. Find the area of triangle ABC to the nearest hundredth.
(a) $a = 4.5, b = 5.2, \gamma = 55°$

(b) $a = 6, b = 7, c = 9$

8.3 Vectors

- **Learn basic concepts about vectors**
- **Learn representations of vectors**
- **Find the magnitude, direction angle, and components of a vector**
- **Perform operations on vectors**
- **Learn and apply the dot product**
- **Use vectors to calculate work**

Introduction

The beginnings of vectors go back centuries to the notion of a directed line segment, but the formal development of vectors occurred during the nineteenth and twentieth centuries, after the invention of complex numbers. It was not until Einstein used vectors in his theory of relativity that their importance became readily accepted.

Vectors are a profound invention. In science and technology they provide a simple model for visualizing difficult concepts such as force, velocity, and electric fields. Vectors are essential to creating today's amazing computer graphics. This section discusses some of the important properties and applications of vectors. (**Sources:** *Historical Topics for the Mathematical Classroom, Thirty-first Yearbook, NCTM; M. Mortenson, Computer Graphics.*)

Basic Concepts

Many quantities in mathematics can be described using real numbers, or **scalars**. Examples include a person's weight, the cost of a CD player, and the gas mileage of a car. Other quantities must be represented using vector quantities. A **vector quantity** involves both *magnitude* and *direction*. Magnitude can be interpreted as size or length. For example, if a car is traveling north at 50 miles per hour, then the direction *north* coupled with a *speed* of 50 miles per hour represents a vector quantity called *velocity*. In science a distinction is made between speed and velocity—speed is the magnitude of velocity.

A vector quantity can be represented by a directed line segment called a **vector**. A vector **v** representing the velocity of a car traveling 50 miles per hour north is shown in Figure 8.31; the vector **u** represents a velocity of 25 miles per hour east. Notice that the length of **u** is half the length of **v**. Vectors do *not* have position—rather, they have magnitude and direction. A vector can be translated, provided its direction and magnitude (length) do not change. Two vectors are **equal** if they have the same magnitude and direction. In Figure 8.32 each directed line segment represents the same vector **v**.

Figure 8.31

Figure 8.32

Figure 8.33
Vector $\mathbf{v} = \overrightarrow{PQ}$

A vector is usually represented symbolically by a letter printed in boldface type, such as **a**, **b**, **v**, or **F**. A second way to denote a vector is to use two points. If the **initial point** of a vector **v** is P and its **terminal point** is Q, then $\mathbf{v} = \overrightarrow{PQ}$, as illustrated in Figure 8.33.

Representations of Vectors

If we place the initial point of vector **v** at the origin, as in Figure 8.34, then its terminal point (a_1, a_2) can be used to determine **v**. To distinguish the *point* (a_1, a_2) from the *vector* **v**, we use the notation $\mathbf{v} = \langle a_1, a_2 \rangle$. The **horizontal component** of **v** is a_1 and the **vertical component** of **v** is a_2. See Figure 8.35.

Figure 8.34
Vector $\mathbf{v} = \langle a_1, a_2 \rangle$

Figure 8.35

A vector with its initial point at the origin in the rectangular coordinate system is called a **position vector**. Figure 8.34 shows the position vector **v**. The *positive* angle θ between the x-axis and the position vector is called the **direction angle** for the vector. In Figure 8.35, θ is the direction angle for vector **v**. If $\mathbf{v} = \langle a_1, a_2 \rangle$, then its direction angle θ satisfies

$$\tan \theta = \frac{a_2}{a_1}, \quad \text{where } a_1 \neq 0.$$

The length of a vector equals its magnitude. If $\mathbf{v} = \langle a_1, a_2 \rangle$, then the *magnitude* of **v** is denoted $\|\mathbf{v}\|$. Applying the Pythagorean theorem to Figure 8.35 gives $\|\mathbf{v}\| = \sqrt{a_1{}^2 + a_2{}^2}$.

Magnitude of a Vector

If $\mathbf{v} = \langle a_1, a_2 \rangle$, then the **magnitude** (or length) of **v** is given by

$$\|\mathbf{v}\| = \sqrt{a_1{}^2 + a_2{}^2}.$$

If $\|\mathbf{v}\| = 1$, then **v** is a **unit vector**.

EXAMPLE 1 Finding magnitude and direction angle

Find the magnitude and direction angle of $\mathbf{u} = \langle 3, -2 \rangle$.

SOLUTION The magnitude of vector **u** is

$$\|\mathbf{u}\| = \sqrt{3^2 + (-2)^2} = \sqrt{13}.$$

To find the direction angle θ, start with $\tan \theta = \frac{-2}{3} = -\frac{2}{3}$. Vector **u** has a positive horizontal component and a negative vertical component, placing the position vector in quadrant IV. A calculator reveals that $\tan^{-1}\left(-\frac{2}{3}\right) \approx -33.7°$. Adding 360° yields the direction angle $\theta \approx 326.3°$. See Figure 8.36. *Now Try Exercise 27* ◀

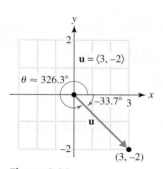

Figure 8.36

If the magnitude and the direction angle of a vector **v** are known, then horizontal and vertical components of **v** can be found by using the following equations. (Explain why.)

Horizontal and Vertical Components

The horizontal and vertical components for a vector $\mathbf{v} = \langle a_1, a_2 \rangle$ having direction angle θ are given by

$$a_1 = \|\mathbf{v}\| \cos \theta \quad \text{and} \quad a_2 = \|\mathbf{v}\| \sin \theta.$$

That is, $\mathbf{v} = \langle \|\mathbf{v}\| \cos \theta, \|\mathbf{v}\| \sin \theta \rangle$.

EXAMPLE 2 Finding horizontal and vertical components

Vector **w** in Figure 8.37 has magnitude 25.0 and direction angle 41.7°. Find the horizontal and vertical components. Round to the nearest tenth.

SOLUTION Use the two formulas in the box, with $\|\mathbf{w}\| = 25.0$ and $\theta = 41.7°$.

$$a_1 = 25.0 \cos 41.7° \qquad a_2 = 25.0 \sin 41.7°$$
$$a_1 \approx 18.7 \qquad a_2 \approx 16.6$$

Therefore $\mathbf{w} \approx \langle 18.7, 16.6 \rangle$. The horizontal component is 18.7, and the vertical component is 16.6 (rounded to the nearest tenth). *Now Try Exercise 35* ◀

Figure 8.37

Figure 8.38

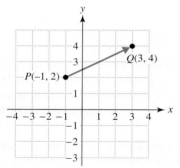

Figure 8.39

EXAMPLE 3 Writing vectors in the form $\langle a_1, a_2 \rangle$

Write each vector in Figure 8.38 in the form $\langle a_1, a_2 \rangle$.

SOLUTION

$$\mathbf{u} = \langle 5 \cos 60°, 5 \sin 60° \rangle = \langle 5 \cdot \tfrac{1}{2}, 5 \cdot \tfrac{\sqrt{3}}{2} \rangle = \langle \tfrac{5}{2}, \tfrac{5\sqrt{3}}{2} \rangle$$
$$\mathbf{v} = \langle 2 \cos 180°, 2 \sin 180° \rangle = \langle 2(-1), 2(0) \rangle = \langle -2, 0 \rangle$$
$$\mathbf{w} = \langle 6 \cos 280°, 6 \sin 280° \rangle \approx \langle 1.04, -5.91 \rangle$$

Now Try Exercises 39 and 43 ◀

Using Two Points to Determine a Vector If a vector has initial point P with coordinates (a_1, b_1) and terminal point Q with coordinates (a_2, b_2), then vector $\overrightarrow{PQ}$ is given by $\overrightarrow{PQ} = \langle a_2 - a_1, b_2 - b_1 \rangle$. See the next example. Note that $\overrightarrow{QP} = \langle a_1 - a_2, b_1 - b_2 \rangle$.

EXAMPLE 4 Finding a vector graphically and symbolically

Let P have coordinates $(-1, 2)$ and Q have coordinates $(3, 4)$. Find vector $\overrightarrow{PQ}$ graphically and symbolically. Calculate the magnitude of $\overrightarrow{PQ}$.

SOLUTION To graph $\overrightarrow{PQ}$, plot the points P and Q. Then sketch a directed line segment from P to Q, as shown in Figure 8.39. We can see that the horizontal component is 4 and the vertical component is 2. A symbolic representation of $\overrightarrow{PQ}$ is given by

$$\overrightarrow{PQ} = \langle 3 - (-1), 4 - 2 \rangle = \langle 4, 2 \rangle.$$

The magnitude, or length, of $\overrightarrow{PQ}$ is

$$\|\overrightarrow{PQ}\| = \sqrt{4^2 + 2^2} = \sqrt{20} \approx 4.47.$$ ***Now Try Exercise 49*** ◀

Operations on Vectors

Vector Addition Suppose that a swimmer heads directly across a river at 3 miles per hour. If the current is 4 miles per hour, then the person will be carried a distance downstream before reaching the other side, as illustrated in Figure 8.40. We can use vectors to visually find the direction and speed that the swimmer will travel across the river.

Figure 8.40 A Swimmer in a Current

Figure 8.41 Vector Addition

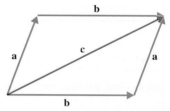

Figure 8.42 The Parallelogram Rule: **c** = **a** + **b**

Let vector **a** represent the speed and direction of the swimmer with no current, vector **b** represent the direction and speed of the current, and vector **c** represent the final direction and speed of the swimmer. We can find the length and direction of **c** by applying the **parallelogram rule**, as shown in Figure 8.41. The speed and direction of the swimmer are represented by the diagonal **c** of the parallelogram (rectangle), which is determined by **a** and **b**. Vector **c** is called the **sum** or **resultant** of vectors **a** and **b**.

Symbolically, we can represent the velocity of the swimmer with no current by $\mathbf{a} = \langle 0, 3 \rangle$, the velocity of the current by $\mathbf{b} = \langle 4, 0 \rangle$, and the velocity of the swimmer in the current by $\mathbf{c} = \langle 4, 3 \rangle$. Vector **c** is the sum of vectors **a** and **b** and can be found as follows.

$$\mathbf{a} + \mathbf{b} = \langle 0, 3 \rangle + \langle 4, 0 \rangle = \langle 0 + 4, 3 + 0 \rangle = \langle 4, 3 \rangle = \mathbf{c}$$

Since $\|\mathbf{c}\| = \sqrt{4^2 + 3^2} = 5$, the swimmer moves 5 miles per hour in the direction of **c**.

Figure 8.42 illustrates graphically how to find $\mathbf{c} = \mathbf{a} + \mathbf{b}$ in general by using the parallelogram rule. The following box defines vector addition symbolically.

Vector Addition

If $\mathbf{a} = \langle a_1, a_2 \rangle$ and $\mathbf{b} = \langle b_1, b_2 \rangle$, then the **sum** of **a** and **b** is given by

$$\mathbf{a} + \mathbf{b} = \langle a_1, a_2 \rangle + \langle b_1, b_2 \rangle = \langle a_1 + b_1, a_2 + b_2 \rangle.$$

Suppose that vector **a** represents a force of 80 pounds pulling on a water-ski towrope and **b** represents a force of 60 pounds pulling on a second towrope. See Figure 8.43. The resultant force $\mathbf{c} = \mathbf{a} + \mathbf{b}$ is given by the diagonal of the parallelogram shown in Figure 8.44. Vector **c** represents the net force exerted by the two water skiers.

Figure 8.43

Figure 8.44

EXAMPLE 5 Applying the parallelogram rule

Find the magnitude of the resultant force on the ski boat in the preceding discussion if the angle between the towropes is 25°.

SOLUTION The magnitude of the force equals the length of the diagonal AC in Figure 8.44. Angle BAD equals 25°. Since angle ABC is $180° - 25° = 155°$, we find AC by applying the law of cosines.

$$AC^2 = 60^2 + 80^2 - 2(60)(80) \cos 155° \approx 18{,}701$$
$$AC \approx 137 \text{ pounds}$$

Now Try Exercise 53 ◄

Vector Notation Using i and j   A second type of vector notation involves the vectors $\mathbf{i} = \langle 1, 0 \rangle$ and $\mathbf{j} = \langle 0, 1 \rangle$. A vector $\mathbf{a} = \langle a_1, a_2 \rangle$ can be expressed as

$$\mathbf{a} = a_1\mathbf{i} + a_2\mathbf{j}.$$

For example, $\langle 3, -4 \rangle$ and $3\mathbf{i} - 4\mathbf{j}$ represent the same vector.

MAKING CONNECTIONS

Imaginary unit i and unit vector i   Section 3.3 discussed the *imaginary unit i*, where $i = \sqrt{-1}$ and $i^2 = -1$. This section introduces the *vector* $\mathbf{i} = \langle 1, 0 \rangle$. Each represents a different mathematical concept.

EXAMPLE 6   Finding resultant forces

Forces $\mathbf{F}_1 = 5\mathbf{i} + 12\mathbf{j}$ and $\mathbf{F}_2 = 4\mathbf{i} - 3\mathbf{j}$ act at the same point. Find the resultant force $\mathbf{F}$ and its magnitude.

SOLUTION   The position vectors for $\mathbf{F}_1$ and $\mathbf{F}_2$ and the resultant force $\mathbf{F}$ are shown in Figure 8.45. The resultant force $\mathbf{F}$ equals $\mathbf{F}_1 + \mathbf{F}_2$.

$$\begin{aligned}
\mathbf{F}_1 + \mathbf{F}_2 &= (5\mathbf{i} + 12\mathbf{j}) + (4\mathbf{i} - 3\mathbf{j}) \\
&= (5 + 4)\mathbf{i} + (12 - 3)\mathbf{j} \\
&= 9\mathbf{i} + 9\mathbf{j} \\
\|\mathbf{F}\| &= \sqrt{9^2 + 9^2} = \sqrt{162} \approx 12.7
\end{aligned}$$

Now Try Exercise 57 ◀

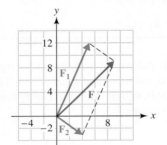

Figure 8.45

Vector Subtraction   Subtraction can be defined both graphically and symbolically. The difference $\mathbf{a} - \mathbf{b}$ is shown graphically in Figure 8.46. Notice that by the parallelogram rule $\mathbf{b} + (\mathbf{a} - \mathbf{b}) = \mathbf{a}$.

Vector subtraction can be defined symbolically as follows.

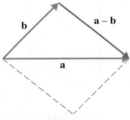

Figure 8.46

Vector Subtraction

If $\mathbf{a} = \langle a_1, a_2 \rangle$ and $\mathbf{b} = \langle b_1, b_2 \rangle$, then the **difference** of $\mathbf{a}$ and $\mathbf{b}$ is given by

$$\mathbf{a} - \mathbf{b} = \langle a_1, a_2 \rangle - \langle b_1, b_2 \rangle = \langle a_1 - b_1, a_2 - b_2 \rangle.$$

EXAMPLE 7   Adding and subtracting vectors

Let $\mathbf{a} = \langle -3, 4 \rangle$ and $\mathbf{b} = \langle 5, -6 \rangle$. Find $\mathbf{a} + \mathbf{b}$ and $\mathbf{a} - \mathbf{b}$.

SOLUTION   To add two vectors, we add corresponding components.

$$\mathbf{a} + \mathbf{b} = \langle -3, 4 \rangle + \langle 5, -6 \rangle = \langle -3 + 5, 4 + (-6) \rangle = \langle 2, -2 \rangle$$

To subtract two vectors, we subtract corresponding components.

$$\mathbf{a} - \mathbf{b} = \langle -3, 4 \rangle - \langle 5, -6 \rangle = \langle -3 - 5, 4 - (-6) \rangle = \langle -8, 10 \rangle$$

Now Try Exercise 63 ◀

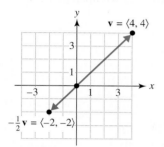

Figure 8.47
North Winds

Figure 8.48

Scalar Multiplication Scalar multiplication occurs when a vector **v** is multiplied by a real number, or *scalar*, *k* to form *k***v**. Vectors **v** and *k***v** are parallel if $k \neq 0$. Vector *k***v** points in the *same* direction as **v** if $k > 0$, and *k***v** points in the *opposite* direction of **v** if $k < 0$. The magnitude of *k***v** is $|k|$ times the magnitude of **v**. For example, suppose that a 10-mile-per-hour wind is blowing from the north. Since the wind is blowing toward the south, the wind may be represented by $\mathbf{v} = \langle 0, -10 \rangle$. If the wind doubles in speed but does not change direction, then the *scalar product* 2**v** models this situation.

$$2\mathbf{v} = 2\langle 0, -10 \rangle = \langle 2 \cdot 0, 2 \cdot (-10) \rangle = \langle 0, -20 \rangle$$

The wind is now blowing at 20 miles per hour toward the south. See Figure 8.47.

Next, suppose that a wind from the southwest is modeled by $\mathbf{v} = \langle 4, 4 \rangle$. Then

$$-\frac{1}{2}\mathbf{v} = -\frac{1}{2}\langle 4, 4 \rangle = \left\langle -\frac{1}{2} \cdot 4, -\frac{1}{2} \cdot 4 \right\rangle = \langle -2, -2 \rangle$$

represents a wind from the northeast in the opposite direction of **v** with half the speed. See Figure 8.48. This discussion motivates the following definition of scalar multiplication.

Scalar Multiplication

If $\mathbf{v} = \langle v_1, v_2 \rangle$ and *k* is a real number, then the **scalar product** *k***v** is given by

$$k\mathbf{v} = k\langle v_1, v_2 \rangle = \langle kv_1, kv_2 \rangle.$$

Sums, differences, and scalar products can be calculated graphically and symbolically, as illustrated in the next example.

EXAMPLE 8 **Performing operations on vectors**

Find each of the following expressions graphically and symbolically if $\mathbf{a} = \langle -3, 4 \rangle$ and $\mathbf{b} = \langle -1, -2 \rangle$.

(a) $\|\mathbf{a}\|$ **(b)** $-2\mathbf{b}$ **(c)** $\mathbf{a} + 2\mathbf{b}$

SOLUTION

(a) Graph $\mathbf{a} = \langle -3, 4 \rangle$, as shown in Figure 8.49. The length of **a** appears to be about 5. This can be verified symbolically.

$$\|\mathbf{a}\| = \sqrt{(-3)^2 + (4)^2} = 5$$

(b) Graph $\mathbf{b} = \langle -1, -2 \rangle$. The scalar product $-2\mathbf{b}$ points in the opposite direction of **b** with twice the length. See Figure 8.50, where $-2\mathbf{b} = \langle 2, 4 \rangle$. Symbolically this is given by

$$-2\mathbf{b} = -2\langle -1, -2 \rangle = \langle -2 \cdot (-1), -2 \cdot (-2) \rangle = \langle 2, 4 \rangle.$$

Figure 8.49

Figure 8.50

Figure 8.51

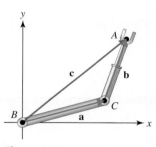

Figure 8.52

(c) Graph $\mathbf{a} = \langle -3, 4 \rangle$ and $2\mathbf{b} = \langle -2, -4 \rangle$. See Figure 8.51. By the parallelogram rule, the diagonal represents $\mathbf{a} + 2\mathbf{b} = \langle -5, 0 \rangle$. This can be verified symbolically.

$$\mathbf{a} + 2\mathbf{b} = \langle -3, 4 \rangle + 2\langle -1, -2 \rangle = \langle -3, 4 \rangle + \langle -2, -4 \rangle = \langle -5, 0 \rangle$$

Now Try Exercise 79 ◀

On some graphing calculators the list feature can be used to perform operations on vectors. In Figure 8.52 a calculator has been used to evaluate the expressions in Example 8. On other calculators vectors can be represented by ordered pairs with parentheses.

EXAMPLE 9 Performing operations on vectors

Find $-3\mathbf{a} + 5\mathbf{b}$ if $\mathbf{a} = 2\mathbf{i} + 3\mathbf{j}$ and $\mathbf{b} = 6\mathbf{i} - 7\mathbf{j}$.

SOLUTION

$$\begin{aligned} -3\mathbf{a} + 5\mathbf{b} &= -3(2\mathbf{i} + 3\mathbf{j}) + 5(6\mathbf{i} - 7\mathbf{j}) \\ &= (-6\mathbf{i} - 9\mathbf{j}) + (30\mathbf{i} - 35\mathbf{j}) \\ &= (-6 + 30)\mathbf{i} + (-9 - 35)\mathbf{j} \\ &= 24\mathbf{i} - 44\mathbf{j} \end{aligned}$$

Now Try Exercise 81 ◀

An Application from Robotics Robotic arms are sometimes modeled using vectors. Consider the *planar two-arm manipulator* in Figure 8.53. If $\overrightarrow{BC} = \mathbf{a}$ and $\overrightarrow{CA} = \mathbf{b}$, then the position of the hand is given by $\overrightarrow{BA} = \mathbf{c}$. Since $\mathbf{c} = \mathbf{a} + \mathbf{b}$, we can easily locate the position of the hand if $\mathbf{a}$ and $\mathbf{b}$ are known. (**Source:** J. Craig, *Introduction to Robotics.*)

EXAMPLE 10 Using vectors to locate a robotic hand

Let $\mathbf{a} = \langle 3.1, 1.5 \rangle$ and $\mathbf{b} = \langle 1.4, 2.4 \rangle$ in Figure 8.53.
(a) Find the position of the robotic hand.
(b) Suppose the upper arm represented by $\mathbf{a}$ doubles its length and the forearm represented by $\mathbf{b}$ reduces its length by half. Find the new position of the hand.

Figure 8.53

SOLUTION
(a) To find the position of the hand, evaluate $\mathbf{a} + \mathbf{b}$.

$$\mathbf{a} + \mathbf{b} = \langle 3.1, 1.5 \rangle + \langle 1.4, 2.4 \rangle = \langle 4.5, 3.9 \rangle$$

The hand is located at the point $(4.5, 3.9)$.
(b) The new position is represented by

$$2\mathbf{a} + \frac{1}{2}\mathbf{b} = 2\langle 3.1, 1.5 \rangle + \frac{1}{2}\langle 1.4, 2.4 \rangle$$

$$= \langle 6.2, 3.0 \rangle + \langle 0.7, 1.2 \rangle$$

$$= \langle 6.9, 4.2 \rangle.$$

The new coordinates of the robotic hand are $(6.9, 4.2)$. *Now Try Exercise 117* ◀

An Application from Navigation   Vectors are frequently used to describe air and fluid flow. The next example uses vectors to describe the motion of an airplane.

EXAMPLE 11   Using vectors in navigation

An airplane is flying with an airspeed of 300 miles per hour and a bearing of 40° in a 30-mile-per-hour west wind.

(a) Find vectors $\mathbf{v}$ and $\mathbf{u}$ that model the velocity of the airplane and the velocity of the wind, respectively.

(b) Use vectors to determine the groundspeed of the plane.

(c) Find the final bearing of the plane in the wind.

SOLUTION

(a) Consider Figure 8.54, which shows vectors $\mathbf{v}$ and $\mathbf{u}$ graphically. Since $\mathbf{u}$ models a west wind, it points to the right with length 30 and can be represented symbolically by $\mathbf{u} = \langle 30, 0 \rangle$. Let a_1 be the horizontal component and a_2 be the vertical component of $\mathbf{v}$. Since $\|\mathbf{v}\| = 300$, it follows that $a_1 = 300 \cos 50°$ and $a_2 = 300 \sin 50°$. Thus

$$\mathbf{v} = \langle 300 \cos 50°, 300 \sin 50° \rangle \approx \langle 192.8, 229.8 \rangle.$$

(b) The true course of the plane is given by $\mathbf{c} = \mathbf{v} + \mathbf{u}$.

$$\mathbf{c} = \mathbf{v} + \mathbf{u}$$
$$= \langle 300 \cos 50°, 300 \sin 50° \rangle + \langle 30, 0 \rangle$$
$$= \langle 300 \cos 50° + 30, 300 \sin 50° \rangle$$

The groundspeed of the plane equals $\|\mathbf{c}\|$.

$$\|\mathbf{c}\| = \sqrt{(300 \cos 50° + 30)^2 + (300 \sin 50°)^2} \approx 320.1$$

The groundspeed of the airplane is approximately 320 miles per hour.

(c) Since $\mathbf{c} = \langle 300 \cos 50° + 30, 300 \sin 50° \rangle$, the direction angle θ is determined by the vector $\mathbf{c}$ and the positive x-axis (East). Thus $\tan \theta = \frac{300 \sin 50°}{300 \cos 50° + 30} \approx 1.0313$ and $\theta \approx \tan^{-1} 1.0313 \approx 45.9°$. The final bearing of the plane in the wind equals $90° - 45.9° = 44.1°$.   ***Now Try Exercise 109*** ◄

Figure 8.54

The Dot Product

Thus far we have discussed addition, subtraction, and scalar multiplication of vectors. Another operation on vectors, called the *dot product*, is important because it can be used to find angles between vectors. The dot product has applications in computer graphics, solar energy, and physics. We begin by defining the dot product.

Dot Product

Let $\mathbf{a} = \langle a_1, a_2 \rangle$ and $\mathbf{b} = \langle b_1, b_2 \rangle$. The **dot product** of $\mathbf{a}$ and $\mathbf{b}$, denoted $\mathbf{a} \cdot \mathbf{b}$, is a *real number* given by

$$\mathbf{a} \cdot \mathbf{b} = a_1 b_1 + a_2 b_2.$$

In the next example, we calculate dot products. Notice that the dot product of two vectors is a real number, rather than a vector.

EXAMPLE 12 Calculating dot products

Calculate $\mathbf{a} \cdot \mathbf{b}$.
(a) $\mathbf{a} = \langle 4, -3 \rangle, \mathbf{b} = \langle -1, 2 \rangle$ **(b)** $\mathbf{a} = 2\mathbf{i} + 5\mathbf{j}, \mathbf{b} = -3\mathbf{i} + 2\mathbf{j}$

SOLUTION
(a) $\mathbf{a} \cdot \mathbf{b} = \langle 4, -3 \rangle \cdot \langle -1, 2 \rangle = (4)(-1) + (-3)(2) = -10$
(b) $\mathbf{a} \cdot \mathbf{b} = (2\mathbf{i} + 5\mathbf{j}) \cdot (-3\mathbf{i} + 2\mathbf{j}) = (2)(-3) + (5)(2) = 4$

Now Try Exercises 87(a) and 91(a) ◀

Figure 8.55

In Figure 8.55 the *angle between vectors **a** and **b** is* θ, where $0° \leq \theta \leq 180°$. If $\theta = 90°$ the vectors are **perpendicular**, and if $\theta = 0°$ or $180°$ the vectors are **parallel**. If $\theta = 0°$ the vectors point in the *same* direction, and if $\theta = 180°$ they point in *opposite* directions.

It is shown in Exercise 5 of the Extended Exercises at the end of the chapter that for any two nonzero vectors **a** and **b**,

$$\mathbf{a} \cdot \mathbf{b} = \|\mathbf{a}\| \|\mathbf{b}\| \cos \theta.$$

This result can be used to find the angle θ between **a** and **b**.

Angle Between Two Vectors

If **a** and **b** are nonzero vectors, then the **angle θ between a and b** is given by

$$\theta = \cos^{-1} \frac{\mathbf{a} \cdot \mathbf{b}}{\|\mathbf{a}\| \|\mathbf{b}\|}.$$

Vectors **a** and **b** are perpendicular if and only if $\mathbf{a} \cdot \mathbf{b} = 0$.

EXAMPLE 13 Finding the angle between two vectors

Sketch the vectors **a** and **b**. Then find the angle θ between **a** and **b**.
(a) $\mathbf{a} = 2\mathbf{i} - 3\mathbf{j}, \mathbf{b} = 3\mathbf{i} + 2\mathbf{j}$ **(b)** $\mathbf{a} = \langle -4, 3 \rangle, \mathbf{b} = \langle 1, -2 \rangle$

SOLUTION
(a) Vectors **a** and **b** appear to be perpendicular in Figure 8.56. Since

$$\mathbf{a} \cdot \mathbf{b} = (2)(3) + (-3)(2) = 0,$$

the vectors are perpendicular and $\theta = 90°$.

Figure 8.56

(b) A sketch of the vectors is shown in Figure 8.57. They are neither perpendicular nor parallel. Since

$$\|\mathbf{a}\| = \sqrt{(-4)^2 + (3)^2} = 5 \quad \text{and} \quad \|\mathbf{b}\| = \sqrt{(1)^2 + (-2)^2} = \sqrt{5},$$

it follows that

$$\theta = \cos^{-1} \frac{(-4)(1) + (3)(-2)}{5\sqrt{5}} \approx 153.4°.$$

Now Try Exercise 87(b) and (c) ◀

Work

In science a force does work only when an object moves. For example, a person pushing against a brick wall does no work, whereas a person lifting a 20-pound weight does work.

Figure 8.57

Work equals force times distance, *provided* the force is in the same direction as the movement of the object. If a 150-pound person climbs up a 20-foot rope, then the work W done is

$$W = 150 \times 20 = 3000 \text{ foot-pounds.}$$

A **foot-pound** equals the work required to lift 1 pound a vertical distance of 1 foot. If the force is not in the same direction as the movement, then we must use trigonometry to determine work.

Consider a person pulling a wagon, as shown in Figure 8.58, where **F** represents the force on the handle and **D** represents the distance and direction that the wagon is pulled. The force **F** can be expressed as the sum of a horizontal vector in the direction of **D** and a vertical vector perpendicular to **D**, as illustrated in Figure 8.59. Using the right triangle in Figure 8.60, we see that the horizontal component of **F** is given by $\|\mathbf{F}\| \cos \theta$ and the vertical component of **F** is given by $\|\mathbf{F}\| \sin \theta$.

Figure 8.58

Figure 8.59

Figure 8.60

CLASS DISCUSSION

Suppose a force vector **F** is perpendicular to **D**. How much work is done? Interpret your answer.

The work done pulling the wagon is equal to the horizontal component $\|\mathbf{F}\| \cos \theta$ times the distance, which is given by $\|\mathbf{D}\|$. That is,

$$W = \|\mathbf{F}\| \|\mathbf{D}\| \cos \theta = \mathbf{F} \cdot \mathbf{D}.$$

Work equals the dot product of the force vector **F** and the displacement vector **D**.

Work

If a constant force **F** is applied to an object that moves along a vector **D**, then the work W done is

$$W = \mathbf{F} \cdot \mathbf{D}.$$

EXAMPLE 14   Calculating work

Find the work done when a force $\mathbf{F} = \langle 3, -2 \rangle$ moves an object from point $P = (-2, 1)$ to point $Q = (3, -1)$, where force is measured in pounds and distance in feet.

SOLUTION   First we must find the displacement vector $\mathbf{D} = \overrightarrow{PQ}$, where

$$\overrightarrow{PQ} = \langle 3 - (-2), -1 - 1 \rangle = \langle 5, -2 \rangle.$$

The work W done can be calculated as follows.

$$W = \mathbf{F} \cdot \mathbf{D} = \langle 3, -2 \rangle \cdot \langle 5, -2 \rangle = 15 + 4 = 19 \text{ foot-pounds}$$

Now Try Exercise 103 ◀

Figure 8.61

EXAMPLE 15 Calculating work

A 150-pound person walks 500 feet up a hiking trail that is inclined at 20°. Use vectors to compute the work done by the person, as illustrated in Figure 8.61.

SOLUTION Vector **D** is 500 feet long and is given by $\mathbf{D} = \langle 500 \cos 20°, 500 \sin 20° \rangle$. Since gravity pulls downward, the force exerted by the person against gravity is $\mathbf{F} = \langle 0, 150 \rangle$. The work done is

$$W = \mathbf{F} \cdot \mathbf{D} = (0)(500 \cos 20°) + (150)(500 \sin 20°) \approx 25{,}650 \text{ foot-pounds.}$$

Now Try Exercise 121 ◀

8.3 Putting It All Together

Vectors are an important invention of the nineteenth and twentieth centuries that has enabled science and technology to model a wide variety of phenomena. The following table summarizes some basic concepts regarding vectors.

Concept	Explanation	Examples
Vectors	Vector quantities denote both magnitude and direction. A vector **a** can be expressed as either $\mathbf{a} = \langle a_1, a_2 \rangle$ or $\mathbf{a} = a_1\mathbf{i} + a_2\mathbf{j}$. Its magnitude is given by $\|\mathbf{a}\| = \sqrt{a_1^2 + a_2^2}$. The numbers a_1 and a_2 are called the horizontal and vertical components of **a**, respectively. A vector with its initial point at the origin is called a position vector. The positive angle between the x-axis and the position vector is called the direction angle θ. If vector **a** has direction angle θ, then $$\mathbf{a} = \langle \|\mathbf{a}\| \cos\theta, \|\mathbf{a}\| \sin\theta \rangle.$$	$\mathbf{a} = \langle 1, 2 \rangle$ and $\mathbf{a} = \mathbf{i} + 2\mathbf{j}$ represent the same vector. The magnitude of **a** is given by $$\|\mathbf{a}\| = \sqrt{1^2 + 2^2} = \sqrt{5}.$$ The horizontal component is 1, and the vertical component is 2.
Operations on vectors	Let $\mathbf{a} = \langle a_1, a_2 \rangle$, $\mathbf{b} = \langle b_1, b_2 \rangle$, and k be a real number. $$\mathbf{a} + \mathbf{b} = \langle a_1, a_2 \rangle + \langle b_1, b_2 \rangle$$ $$= \langle a_1 + b_1, a_2 + b_2 \rangle \quad \text{Sum}$$ $$\mathbf{a} - \mathbf{b} = \langle a_1, a_2 \rangle - \langle b_1, b_2 \rangle$$ $$= \langle a_1 - b_1, a_2 - b_2 \rangle \quad \text{Difference}$$ $$k\mathbf{a} = k\langle a_1, a_2 \rangle = \langle ka_1, ka_2 \rangle \quad \text{Scalar product}$$	Let $\mathbf{a} = \langle 4, 1 \rangle$, $\mathbf{b} = \langle 3, 2 \rangle$. $$\mathbf{a} + \mathbf{b} = \langle 4, 1 \rangle + \langle 3, 2 \rangle$$ $$= \langle 7, 3 \rangle$$ $$\mathbf{a} - \mathbf{b} = \langle 4, 1 \rangle - \langle 3, 2 \rangle$$ $$= \langle 1, -1 \rangle$$ $$3\mathbf{a} = 3\langle 4, 1 \rangle = \langle 12, 3 \rangle$$
Dot product	If $\mathbf{a} = \langle a_1, a_2 \rangle$ and $\mathbf{b} = \langle b_1, b_2 \rangle$, then $\mathbf{a} \cdot \mathbf{b} = a_1 b_1 + a_2 b_2$.	Let $\mathbf{a} = \langle 2, -2 \rangle$, $\mathbf{b} = \langle 3, 1 \rangle$. $\mathbf{a} \cdot \mathbf{b} = (2)(3) + (-2)(1) = 4$

Concept	Explanation	Examples
Angle θ between two vectors	If $\mathbf{a} = \langle a_1, a_2 \rangle$ and $\mathbf{b} = \langle b_1, b_2 \rangle$, then $$\theta = \cos^{-1} \frac{\mathbf{a} \cdot \mathbf{b}}{\|\mathbf{a}\| \|\mathbf{b}\|}.$$ Vectors $\mathbf{a}$ and $\mathbf{b}$ are perpendicular ($\theta = 90°$) if and only if $\mathbf{a} \cdot \mathbf{b} = 0$.	If $\mathbf{a} = \langle 1, 0 \rangle$ and $\mathbf{b} = \langle 3, 4 \rangle$, then $$\theta = \cos^{-1} \frac{\langle 1, 0 \rangle \cdot \langle 3, 4 \rangle}{\|\langle 1, 0 \rangle\| \|\langle 3, 4 \rangle\|}$$ $$= \cos^{-1} \frac{3}{(1)(5)} \approx 53.1°.$$ If $\mathbf{a} = \langle 4, -3 \rangle$ and $\mathbf{b} = \langle 3, 4 \rangle$, then $\mathbf{a} \cdot \mathbf{b} = (4)(3) + (-3)(4) = 0$. Thus $\mathbf{a}$ and $\mathbf{b}$ are perpendicular.
Work	If a constant force $\mathbf{F}$ is applied to an object that moves along a vector $\mathbf{D}$, then the work done is $W = \mathbf{F} \cdot \mathbf{D}$.	If $\mathbf{F} = 3\mathbf{i} - 4\mathbf{j}$ and $\mathbf{D} = 10\mathbf{i} - 20\mathbf{j}$, then $$W = \mathbf{F} \cdot \mathbf{D}$$ $$= (3)(10) + (-4)(-20)$$ $$= 110 \text{ foot-pounds},$$ where units are in feet and pounds.

8.3 Exercises

Representing Vectors and Their Magnitudes

Exercises 1–4: Use the graphical representation of **v** *to complete the following.*

    (a) Estimate integer values for components a_1 and a_2 so that $\mathbf{v} = \langle a_1, a_2 \rangle$.

    (b) Calculate $\|\mathbf{v}\|$.

1.

2.

3.

4.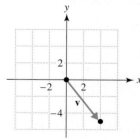

Exercises 5–10: Complete the following.

    (a) Sketch a vector **v** *that models the situation.*

    (b) Express **v** *as* $\langle a_1, a_2 \rangle$.

    (c) Find $2\mathbf{v}$ *and* $-\frac{1}{2}\mathbf{v}$. *Interpret each result.*

5. A 20-mile-per-hour north wind

6. A 10-mile-per-hour west wind

7. A 5-mile-per-hour northwest wind

8. A 7-mile-per-hour southeast wind

9. A 30-pound force upward

10. A 15-pound force pulling at a 45° angle in standard position

Exercises 11–18: Complete the following for vector **v.**
    (a) *Find the horizontal and vertical components.*
    (b) *Calculate* $\|\mathbf{v}\|$ *and decide if* **v** *is a unit vector.*
    (c) *Graph* **v** *and interpret* $\|\mathbf{v}\|$.

11. $\mathbf{v} = \langle 1, 1 \rangle$ **12.** $\mathbf{v} = \langle -1, 0 \rangle$

13. $\mathbf{v} = \langle 3, -4 \rangle$ **14.** $\mathbf{v} = \langle -2, -2 \rangle$

15. $\mathbf{v} = \mathbf{i}$ **16.** $\mathbf{v} = -3\mathbf{j}$

17. $\mathbf{v} = 5\mathbf{i} + 12\mathbf{j}$ **18.** $\mathbf{v} = -\frac{3}{5}\mathbf{i} - \frac{4}{5}\mathbf{j}$

Direction Angles and Components

Exercises 19–30: Find the magnitude and the direction angle θ for the given vector. Let $0° \le \theta < 360°$ *and round θ to the nearest tenth when appropriate.*

19. $\langle 3, 0 \rangle$ **20.** $\langle 0, -2 \rangle$

21. $\langle -1, 1 \rangle$ **22.** $\langle 2, 2 \rangle$

23. $\langle \sqrt{3}, -1 \rangle$ **24.** $\langle -1, \sqrt{3} \rangle$

25. $\langle -5, -12 \rangle$ **26.** $\langle 7, -24 \rangle$

27. $\langle 13, -84 \rangle$ **28.** $\langle -11, -60 \rangle$

29. $\langle -20, 21 \rangle$ **30.** $\langle 16, -63 \rangle$

Exercises 31–38: Find the horizontal and vertical components for **v,** *given magnitude* $\|\mathbf{v}\|$ *and direction angle θ. Round values to the nearest tenth.*

31. $\|\mathbf{v}\| = 4, \theta = 180°$ **32.** $\|\mathbf{v}\| = 1, \theta = 270°$

33. $\|\mathbf{v}\| = \sqrt{2}, \theta = 135°$ **34.** $\|\mathbf{v}\| = \sqrt{2}, \theta = 225°$

35. $\|\mathbf{v}\| = 23, \theta = 54°$ **36.** $\|\mathbf{v}\| = 71, \theta = 163°$

37. $\|\mathbf{v}\| = 34, \theta = 312°$ **38.** $\|\mathbf{v}\| = 25, \theta = 73°$

Exercises 39–42: Write the vector in the form $\langle a_1, a_2 \rangle$.

39.   **40.**

41.   **42.**

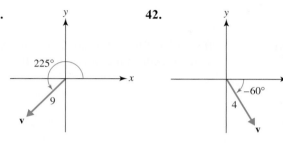

Exercises 43–46: Write the vector in the form $\langle a_1, a_2 \rangle$. *Round values to the nearest hundredth.*

43.   **44.**

45.   **46.**

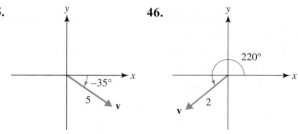

Exercises 47–52: A vector **v** *has initial point P and terminal point Q.*
    (a) *Graph* $\overrightarrow{PQ}$.
    (b) *Write* $\overrightarrow{PQ}$ *as* $\mathbf{v} = \langle a_1, a_2 \rangle$.
    (c) *Find the magnitude of* $\overrightarrow{PQ}$.

47. $P = (0, 0),$ $Q = (-1, 2)$

48. $P = (0, 0),$ $\qquad Q = (4, -6)$

49. $P = (1, 2),$ $\qquad Q = (3, 6)$

50. $P = (-1, -2),$ $\qquad Q = (4, 4)$

51. $P = (-2, 4),$ $\qquad Q = (3, -2)$

52. $P = (0, -4),$ $\qquad Q = (1, 3)$

Resultant Forces

Exercises 53–56: Use the parallelogram rule to find the magnitude of the resultant force for the two forces shown in the figure (to the nearest tenth of a pound).

53.

54.

55.

56.

Exercises 57–62: Forces $\mathbf{F}_1$ and $\mathbf{F}_2$ act at the same point. Find the resultant force $\mathbf{F}$ and its magnitude.

57. $\mathbf{F}_1 = 3\mathbf{i} - 4\mathbf{j}, \mathbf{F}_2 = -8\mathbf{i} + 16\mathbf{j}$

58. $\mathbf{F}_1 = -7\mathbf{i} + 3\mathbf{j}, \mathbf{F}_2 = -\mathbf{i} + 12\mathbf{j}$

59. $\mathbf{F}_1 = \langle 8, 9 \rangle, \mathbf{F}_2 = \langle 1, -31 \rangle$

60. $\mathbf{F}_1 = \langle 48, 0 \rangle, \mathbf{F}_2 = \langle 0, -55 \rangle$

61. $\mathbf{F}_1 = 0.5\mathbf{i} + 0.7\mathbf{j}, \mathbf{F}_2 = -1.5\mathbf{i} - 5.7\mathbf{j}$

62. $\mathbf{F}_1 = \frac{1}{4}\mathbf{i} + \frac{1}{2}\mathbf{j}, \mathbf{F}_2 = -\frac{1}{2}\mathbf{i} + \frac{3}{4}\mathbf{j}$

Operations on Vectors

Exercises 63–70: Evaluate each of the following.

$\quad$ *(a)* $\mathbf{a} + \mathbf{b}$ $\qquad$ *(b)* $\mathbf{a} - \mathbf{b}$

63. $\mathbf{a} = \langle 0, 2 \rangle,$ $\qquad \mathbf{b} = \langle 3, 0 \rangle$

64. $\mathbf{a} = \langle 1, 1 \rangle,$ $\qquad \mathbf{b} = \langle -2, 3 \rangle$

65. $\mathbf{a} = 2\mathbf{i} + \mathbf{j},$ $\qquad \mathbf{b} = \mathbf{i} - 2\mathbf{j}$

66. $\mathbf{a} = \mathbf{i} + 2\mathbf{j},$ $\qquad \mathbf{b} = -2\mathbf{i} + 3\mathbf{j}$

67. $\mathbf{a} = \left\langle -\sqrt{2}, \frac{1}{2} \right\rangle,$ $\qquad \mathbf{b} = \left\langle \sqrt{2}, -\frac{3}{4} \right\rangle$

68. $\mathbf{a} = \left\langle \frac{4}{5}, -\frac{5}{6} \right\rangle,$ $\qquad \mathbf{b} = \left\langle \frac{3}{10}, \frac{2}{3} \right\rangle$

69. $\mathbf{a} = \left(\cos \frac{\pi}{4}\right)\mathbf{i} + \left(\sin \frac{\pi}{4}\right)\mathbf{j},$ $\quad \mathbf{b} = \left(\cos \frac{\pi}{2}\right)\mathbf{i} + \left(\sin \frac{\pi}{2}\right)\mathbf{j}$

70. $\mathbf{a} = \left(\cos \frac{3\pi}{2}\right)\mathbf{i} + \left(\sin \frac{3\pi}{2}\right)\mathbf{j},$ $\mathbf{b} = (\cos \pi)\mathbf{i} + (\sin \pi)\mathbf{j}$

Exercises 71–76: Use the figure to evaluate each of the following.

$\quad$ *(a)* $\mathbf{a} + \mathbf{b}$ $\qquad$ *(b)* $\mathbf{a} - \mathbf{b}$ $\qquad$ *(c)* $-\mathbf{a}$

71.

72.

73.

74.

75.

76.

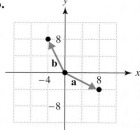

Exercises 77–80: Evaluate each of the following graphically and symbolically.

$\quad$ *(a)* $\|\mathbf{a}\|$ $\qquad$ *(b)* $2\mathbf{a}$ $\qquad$ *(c)* $2\mathbf{a} + 3\mathbf{b}$

77. $\mathbf{a} = 2\mathbf{i},$ $\qquad\qquad \mathbf{b} = \mathbf{i} + \mathbf{j}$

78. $\mathbf{a} = -\mathbf{i} + \mathbf{j},$ $\qquad\quad \mathbf{b} = \mathbf{i} - \mathbf{j}$

79. $a = \langle -1, 2 \rangle$, $b = \langle 3, 0 \rangle$

80. $a = \langle -2, -1 \rangle$, $b = \langle -3, 2 \rangle$

Exercises 81–86: Given vectors a *and* b, *evaluate each of the following.*

    (a) $-a + 4b$ (b) $2a - 3b$

81. $a = i - 2j$, $b = -5i + 2j$

82. $a = -6i + j$, $b = 7i - 3j$

83. $a = \langle 1, -4 \rangle$, $b = \langle -3, 5 \rangle$

84. $a = \langle -5, 13 \rangle$, $b = \langle 0, -1 \rangle$

85. $a = \langle -7, -2 \rangle$, $b = \langle 9, -1 \rangle$

86. $a = \langle 5, 1 \rangle$, $b = \langle -2, -4 \rangle$

Dot Product and Work

Exercises 87–94: Complete the following for vectors a *and* b.

    (a) *Find* $a \cdot b$.
    (b) *Approximate the angle* θ *between* a *and* b *to the nearest tenth of a degree.*
    (c) *State if vectors* a *and* b *are perpendicular, parallel, or neither. If* a *and* b *are parallel, state whether they point in the same direction or in opposite directions.*

87. $a = \langle 1, -2 \rangle$, $b = \langle 3, 1 \rangle$

88. $a = \langle 4, -5 \rangle$, $b = \langle 2, -2 \rangle$

89. $a = \langle 6, 8 \rangle$, $b = \langle -4, 3 \rangle$

90. $a = \langle 1, -2 \rangle$, $b = \langle -2, 4 \rangle$

91. $a = 5i + 6j$, $b = 10i + 12j$

92. $a = -2i + 6j$, $b = 3i + j$

93. $a = i + 3j$, $b = 0.5i - 1.5j$

94. $a = -12i + 16j$, $b = -3i + 4j$

Exercises 95–98: Find the work done in each situation.

95. Lifting a 30-pound weight 5 feet into the air

96. Lifting a 15-pound bucket 8 feet into the air

97. Pushing a stalled car on level ground with a force of 100 pounds for 1000 feet

98. A 150-pound person running up 5 flights of steps with 10 feet between floors

Exercises 99–102: Find the work done when a constant force F *is applied to an object that moves along the vector* D, *where units are in pounds and feet. Find the magnitude of* F.

99. $F = \langle 10, 20 \rangle$, $D = \langle 15, 22 \rangle$

100. $F = \langle 64, 36 \rangle$, $D = \langle 22, -33 \rangle$

101. $F = 5i - 3j$, $D = 3i - 4j$

102. $F = 7i - 24j$, $D = -2i - 5j$

Exercises 103–106: Calculate the work done when the force $F = 5i + 3j$ *moves an object from P to Q.*

103. $P = (-2, 3)$, $Q = (1, 6)$

104. $P = (-2, -1)$, $Q = (1, 3)$

105. $P = (2, -3)$, $Q = (4, -5)$

106. $P = (1, 1)$, $Q = (-1, 6)$

Applications

107. *Swimming in a Current* A swimmer heads directly north across a river at 3 miles per hour, in a current that is 2 miles per hour, as illustrated in the figure. Find a vector that models the resulting direction and speed of the swimmer. With what speed is the swimmer moving in the river?

2 mi/hr   3 mi/hr

108. *Wind and Vectors* A wind can be described by $v = 6i + 8j$, where vector j points north and represents a south wind of 1 mile per hour.
(a) What is the speed of the wind?

(b) Find $3v$. Interpret the result.

(c) Interpret the wind if it switches to $u = -8i + 8j$.

109. *Air Navigation* (Refer to Example 11.) An airplane heads west at 400 miles per hour in a 50-mile-per-hour northwest wind. Find a vector that models the resulting direction and speed of the airplane. Find the ground-speed and bearing of the airplane in the wind.

110. *Air Navigation* A plane with an airspeed of 240 miles per hour is headed on a bearing of 110°. A north wind is blowing (from the north) at 18 miles per hour. Find the groundspeed and the final bearing of the plane in the wind.

111. *Course and Groundspeed* A plane flies 450 miles per hour on a bearing of 160°. A 20-mile-per-hour wind is blowing from the south. Find the groundspeed and the final bearing of the plane in the wind.

112. *Course and Groundspeed* A plane flies on a bearing of 230° at 350 miles per hour. A wind is blowing from the west at 30 miles per hour. Find the groundspeed and the final bearing of the plane in the wind.

113. *Airspeed and Groundspeed* A pilot wants to fly on a course of 75°. By flying due east, the pilot finds that a 40-mile-per-hour wind, blowing from the south, puts the plane on course. Find the airspeed and the groundspeed.

114. *Force and Water-Ski Towropes* (Refer to Example 5.) Forces of 65 pounds and 110 pounds are exerted by two water-ski towropes. If the angle between the towropes is 19°, find the magnitude of the resultant force.

115. *Measuring Rainfall* Suppose that vector **R** models the amount of rainfall in inches and the direction it falls, and vector **A** models the area in square inches and orientation of the opening of a rain gauge, as illustrated in the figure. The total volume V of water collected in the rain gauge is given by $V = |\mathbf{R} \cdot \mathbf{A}|$. This formula calculates the volume of water collected even if the wind is blowing the rain in a slanted direction or the rain gauge is not exactly vertical. Let $\mathbf{R} = \mathbf{i} - 2\mathbf{j}$ and $\mathbf{A} = 0.5\mathbf{i} + \mathbf{j}$.
(a) Find $\|\mathbf{R}\|$ and $\|\mathbf{A}\|$. Interpret your results.

(b) Calculate V and interpret this result.

(c) For the rain gauge to collect the maximum amount of water, what must be true about vectors **R** and **A**?

116. *Solar Panels* Suppose that the sun's intensity (in watts per square centimeter) and direction are given by vector **I**, and a solar panel's area (in square centimeters) and orientation are given by vector **A**, as illustrated in the figure. Then the total number of watts W that are collected by the solar panel is given by $W = |\mathbf{I} \cdot \mathbf{A}|$. Let $\mathbf{I} = 0.01\mathbf{i} - 0.02\mathbf{j}$ and $\mathbf{A} = 400\mathbf{i} + 300\mathbf{j}$.

(a) Find $\|\mathbf{I}\|$ and $\|\mathbf{A}\|$. Interpret your results.

(b) Calculate W and interpret this result.

(c) For the solar panel to absorb maximum wattage, what must be true about vectors **I** and **A**?

117. *Robotics* (Refer to Example 10.) Consider the planar two-arm manipulator shown in the figure. Let the upper arm be modeled by $\mathbf{a} = \langle 3, 2 \rangle$ and the forearm be modeled by $\mathbf{b} = \langle -2, 2 \rangle$, where units are in feet. (**Source:** J. Craig.)

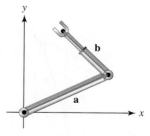

(a) Find a vector **c** that represents the position of the hand.

(b) How far is the hand from the origin?

(c) Find the position of the hand if the length of the upper arm triples and the length of the forearm is reduced by half.

118. *Robotics* A planar three-arm manipulator is shown in the figure on the next page, with joint angles measured relative to a positive horizontal axis. (**Source:** R. Murray, *A Mathematical Introduction to Robotic Manipulation*.)

(a) Find vectors **a**, **b**, and **c** that represent each part of the robotic arm.

(b) Find a vector **d** that represents the position of the hand. How far is the hand from the origin?

119. *Translations in Computer Graphics* Vectors are used in computer graphics to compute translations of points. For example, suppose we would like to translate the point $(-1, 2)$ by $\mathbf{v} = \langle 2, 1 \rangle$, as illustrated in the figure, where the *point* $(-1, 2)$ has been represented by the *vector* $\mathbf{a} = \langle -1, 2 \rangle$. The new location of $(-1, 2)$ is modeled by

$$\mathbf{b} = \mathbf{a} + \mathbf{v} = \langle -1, 2 \rangle + \langle 2, 1 \rangle = \langle 1, 3 \rangle.$$

Thus $(-1, 2)$ has been translated by $\mathbf{v}$ to $(1, 3)$. (**Source:** J. Foley, *Introduction to Computer Graphics.*)

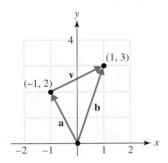

(a) Find the new coordinates of $(-2, 4)$ if it is translated by $\mathbf{v} = \langle 4, -2 \rangle$.

(b) Represent this situation graphically.

120. *Translations in Computer Graphics* (Refer to Exercise 119.) Let triangle *ABC* have vertices $(1, 1)$, $(3, 0)$, and $(4, 3)$.

(a) Find the new vertices if the triangle *ABC* is translated by $\mathbf{v} = \langle -2, 1 \rangle$.

(b) Describe the change in triangle *ABC* if it were translated by $-2\mathbf{v}$.

121. *Work* (Refer to Example 15.) A 145-pound person walks 1.5 miles up a hiking trail inclined at 15°. Use a dot product to calculate the work *W* done (in foot-pounds).

122. *Work* A wagon is pulled 500 feet using a force of 10 pounds applied to the handle, which makes a 40° angle with the horizontal. See Figure 8.58. Use a dot product to calculate the work *W* done.

123. *Air Navigation* A pilot would like to fly to a city that is 200 miles away and has a bearing of 135°. The wind is blowing from the north at 30 miles per hour, and the trip is to take 1 hour. Find the direction and speed that the pilot should adopt to accomplish this.

124. *Work* Calculate the work required to push an 1800-pound car up a 7° incline for 0.1 mile.

Writing about Mathematics

125. State the basic properties of a vector. Does a vector have position? Explain. Give two examples of how to write a vector.

126. State one application of vectors. Give a specific example of a vector for this application and explain how the vector models that application.

EXTENDED AND DISCOVERY EXERCISES

1. *Computer Graphics* Vectors frequently are used to determine the color of pixels on computer screens. For example, suppose that we would like to color the right side of the screen blue and the left side yellow, where the boundary is vector $\overrightarrow{PQ}$, as illustrated in the figure below.

First find $\overrightarrow{PR}$ perpendicular to $\overrightarrow{PQ}$. Then to determine if a pixel at point *S* should be blue or yellow, consider the angle θ between vectors $\overrightarrow{PS}$ and $\overrightarrow{PR}$. If θ is acute, then *S* must be on the same side of $\overrightarrow{PQ}$ as *R* and is colored blue, as illustrated in the left figure below. If angle θ is obtuse, then S is on the opposite side of $\overrightarrow{PQ}$ from R and is colored yellow, as shown in the right figure below.

 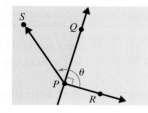

We can determine whether θ is acute or obtuse by calculating the dot product $\overrightarrow{PS} \cdot \overrightarrow{PR}$. If the dot product is positive, then

$$\theta = \cos^{-1} \left(\frac{\overrightarrow{PS} \cdot \overrightarrow{PR}}{\|\overrightarrow{PS}\| \, \|\overrightarrow{PR}\|} \right)$$

is acute and S should be blue. If the dot product is negative, then θ is obtuse and S should be yellow. (**Source:** J. Foley, *Introduction to Computer Graphics.*)

Let P be $(-1, 2)$, S be $(-2, -5)$, $\overrightarrow{PQ} = \mathbf{i} + 5\mathbf{j}$, and $\overrightarrow{PR} = 5\mathbf{i} - \mathbf{j}$. Determine the appropriate color at point S. See the figure below.

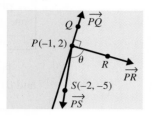

2. *Dot Products in Computer Graphics* A computer screen is gray to the left of vector $\overrightarrow{PQ} = 2\mathbf{i} - 3\mathbf{j}$ and blue to the right, where vector $\overrightarrow{PR} = 3\mathbf{i} + 2\mathbf{j}$ is perpendicular to $\overrightarrow{PQ}$. Let point P be $(2, 1)$. Determine the color of a pixel located at S.
   (**a**) $S = (3, -2)$ (**b**) $S = (2, 2)$

3. *Dot Products in Computer Graphics* A computer screen is to be blue above vector $\overrightarrow{PQ} = 5\mathbf{i} - \mathbf{j}$ and white below, where point P is $(2, 1)$. Determine the color of a pixel located at S.
   (**a**) $S = (100, -10)$ (**b**) $S = (-500, 50)$

8.4 Parametric Equations

- **Learn basic concepts about parametric equations**
- **Graph parametric equations**
- **Use parametric equations to solve applications**

Introduction

We have used functions to model curves in the *xy*-plane. Sometimes a curve cannot be modeled by a function. For example, a circle cannot be described by a single function because a circle fails the vertical line test. Parametric equations represent a different approach to describing curves in the *xy*-plane. They are used in industry to draw complicated curves and surfaces, such as the hood of an automobile. Parametric equations are also used in computer graphics, engineering, and physics. (**Source:** F. Hill, *Computer Graphics.*)

Basic Concepts

In previous chapters we graphed curves in the *xy*-plane determined by $y = f(x)$. Some curves cannot be represented by $y = f(x)$, but they can be represented by parametric equations. Some sample graphs of parametric equations are shown in Figures 8.62–8.64.

[−6, 6, 1] by [−4, 4, 1]   [−6, 6, 1] by [−4, 4, 1]   [−6, 6, 1] by [−4, 4, 1]

Figure 8.62   **Figure 8.63**   **Figure 8.64**

We now define parametric equations of a plane curve.

Parametric Equations of a Plane Curve

A **plane curve** is a set of points (x, y) such that $x = f(t)$ and $y = g(t)$, where f and g are continuous functions on an interval $a \leq t \leq b$. The equations $x = f(t)$ and $y = g(t)$ are **parametric equations** with **parameter** t.

Parametric equations can be represented symbolically, numerically, and graphically and described verbally. This is illustrated in the next example.

EXAMPLE 1   Representing parametric equations

Let $x = t + 3$ and $y = t^2$ for $-3 \leq t \leq 3$.
(a) Make a table of values for x and y with $t = -3, -2, -1, \ldots, 3$.
(b) Plot the points in the table and graph the curve. Add arrows to show how the curve is traced out.
(c) Describe the curve.

SOLUTION
(a) *Numerical Representation*   A numerical representation of the parametric equations is shown in Table 8.1. For example, if $t = 2$, then $x = 2 + 3 = 5$ and $y = 2^2 = 4$.

Table 8.1

t	-3	-2	-1	0	1	2	3
x	0	1	2	3	4	5	6
y	9	4	1	0	1	4	9

(b) *Graphical Representation*   Each ordered pair (x, y) in Table 8.1 is plotted in Figure 8.65, and then the points are connected to obtain the curve. The curve starts at $(0, 9)$ and ends at $(6, 9)$. The curve is traced out from left to right as t increases from -3 to 3.

(c) *Verbal Representation*   The curve in Figure 8.65 appears to be the lower portion of a parabola with vertex $(3, 0)$. See Example 2.   *Now Try Exercise 1* ◀

Graphing Calculators   Graphing calculators are capable of using parametric equations to make tables and graphs. In addition to setting values for the viewing rectangle, we must specify the interval for t. A window setting, table, and graph for the parametric equations in Example 1 are shown in Figures 8.66–8.68. The variable Tstep represents the increment in the parameter t and has a value of 0.1 in this case.

Figure (left margin):

x = t + 3
y = t²
−3 ≤ t ≤ 3

Figure 8.65

Calculator Help
To create the graph shown in Figure 8.68 with parametric equations, see Appendix A (page AP-20).

```
WINDOW
 Tmin=-3
 Tmax=3
 Tstep=.1
 Xmin=-2
 Xmax=10
 Xscl=1
↓Ymin=-2
```

Figure 8.66

T	X₁T	Y₁T
-3	0	9
-2	1	4
-1	2	1
0	3	0
1	4	1
2	5	4
3	6	9

T=-3

Figure 8.67

$[-2, 10, 1]$ by $[-2, 10, 1]$

x = t + 3
y = t²
−3 ≤ t ≤ 3

Figure 8.68

Converting and Graphing Equations

We can verify symbolically that the curve in Figure 8.65 is indeed a portion of a parabola, as demonstrated in the next example.

EXAMPLE 2   Finding an equivalent rectangular equation

Find an equivalent rectangular equation for $x = t + 3$ and $y = t^2$, where $-3 \leq t \leq 3$. Note that these parametric equations were discussed in Example 1.

SOLUTION   Begin by solving $x = t + 3$ for t to obtain $t = x - 3$. Substituting for t in $y = t^2$ results in

$$y = (x - 3)^2,$$

which represents a parabola with vertex $(3, 0)$. When $t = -3$ then $x = 0$, and when $t = 3$ then $x = 6$. Thus the domain is restricted to $0 \leq x \leq 6$.   *Now Try Exercise 15* ◀

EXAMPLE 3   Finding an equivalent rectangular equation

Find an equivalent rectangular equation for each pair of parametric equations. Use the rectangular equation to help graph the parametric equation. Add arrows to show how the parametric curve is traced out.
(a) $x = 4t$, $y = t - 3$; $-\infty < t < \infty$
(b) $x = \sqrt{4 - t^2}$, $y = t$; $-2 \leq t \leq 2$

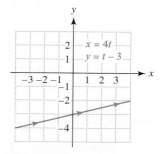

Figure 8.69

SOLUTION
(a) Start by solving $x = 4t$ for t to obtain $t = \frac{1}{4}x$. Substitute for t in the given parametric equation for y.

$$y = t - 3 \qquad \text{Given parametric equation}$$

$$y = \frac{1}{4}x - 3 \qquad \text{Let } t = \frac{1}{4}x.$$

Because $y = \frac{1}{4}x - 3$, these parametric equations trace out a line with slope $\frac{1}{4}$ and y-intercept -3. As t increases, x also increases, so this line is traced out from left to right, as illustrated in Figure 8.69. Note that t can be any real number.
(b) Because $y = t$, it follows that a rectangular equation is $x = \sqrt{4 - y^2}$. To determine the graph of this equation, square each side.

$$x = \sqrt{4 - y^2} \qquad \text{Rectangular equation}$$

$$x^2 = 4 - y^2 \qquad \text{Square each side.}$$

$$x^2 + y^2 = 4 \qquad \text{Add } y^2 \text{ to each side.}$$

The equation $x^2 + y^2 = 4$ is a circle with center $(0, 0)$ and radius 2. Because $\sqrt{4 - y^2}$ is never negative, it follows that $x \geq 0$. Thus the parametric equation traces out only the right half of this circle. See Figure 8.70. Because $y = t$, this semicircle is traced from bottom to top as y increases from -2 to 2.

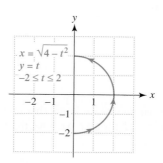

Figure 8.70

Now Try Exercises 9 and 13 ◀

Parametric equations can model a circle, as shown in the next example.

EXAMPLE 4   Graphing a circle with parametric equations

Graph $x = 2 \cos t$ and $y = 2 \sin t$ for $0 \leq t \leq 2\pi$. Find an equivalent equation by using rectangular coordinates.

Figure 8.71

$[-3, 3, 1]$ by $[-2, 2, 1]$

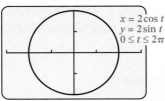

$x = 2\cos t$
$y = 2\sin t$
$0 \le t \le 2\pi$

Figure 8.72

SOLUTION Let $X_1 = 2 \cos (T)$ and $Y_1 = 2 \sin (T)$, and graph these parametric equations for $0 \le t \le 2\pi$, as shown in Figures 8.71 and 8.72. The circle is traced out once. Be sure to have the mode of the calculator set for parametric equations. A square viewing rectangle is necessary for the curve to appear circular rather than elliptical. (Note that Tstep $= 0.1$.)

To verify that this is a circle, consider the following.

$$x^2 + y^2 = (2 \cos t)^2 + (2 \sin t)^2 \qquad x = 2\cos t, y = 2 \sin t$$
$$= 4 \cos^2 t + 4 \sin^2 t \qquad \textit{Properties of exponents}$$
$$= 4(\cos^2 t + \sin^2 t) \qquad \textit{Distributive property}$$
$$= 4 \qquad \cos^2 t + \sin^2 t = 1$$

The parametric equations are equivalent to $x^2 + y^2 = 4$, which is a circle with center $(0, 0)$ and radius 2. **Now Try Exercise 19** ◀

In the next two examples, an equation written in terms of x and y is converted to parametric equations.

EXAMPLE 5   Converting to parametric equations

Convert $x = y^2 - 4y + 4$ to parametric equations.

SOLUTION There is more than one way to convert this equation to parametric equations. One simple way is to let $y = t$ and then write the parametric equations as

$$x = t^2 - 4t + 4, \quad y = t,$$

where t is any real number. To write a different pair of parametric equations, note that

$$x = y^2 - 4y + 4 = (y - 2)^2.$$

Let $t = y - 2$, or $y = t + 2$, and then another pair of parametric equations is

$$x = t^2, \quad y = t + 2. \qquad \textbf{\textit{Now Try Exercise 53}} ◀$$

EXAMPLE 6   Converting to parametric equations

Given the equation $x^2 + y^2 = 1$, complete the following.
(a) Find parametric equations for this equation.
(b) What portion of the graph appears for $0 \le t \le \pi$?

SOLUTION
(a) The graph of $x^2 + y^2 = 1$ is the unit circle. From trigonometry we know that on the unit circle $x = \cos t$ and $y = \sin t$. Since $\cos^2 t + \sin^2 t = 1$ for all t, we have the following result.

$$x^2 + y^2 = \cos^2 t + \sin^2 t = 1$$

Thus parametric equations for the unit circle are

$$x = \cos t, \quad y = \sin t; \quad 0 \le t \le 2\pi.$$

(b) When t increases from 0 to π, the upper half of the circle is graphed, moving from the point $(1, 0)$ to the point $(-1, 0)$. See Figure 8.73. **Now Try Exercise 49** ◀

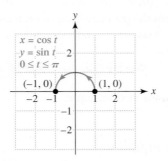

$x = \cos t$
$y = \sin t$
$0 \le t \le \pi$

Figure 8.73

Applications of Parametric Equations

Parametric equations are used to simulate motion. If a ball or shot is thrown with a velocity of v feet per second at an angle θ with the horizontal, its flight can be modeled by the parametric equations

$$x = (v \cos \theta)t \quad \text{and} \quad y = (v \sin \theta)t - 16t^2 + h,$$

where t is in seconds and h is the ball's initial height above the ground. The term $-16t^2$ occurs because gravity is pulling downward. See Figure 8.74. (These equations ignore air resistance.)

Figure 8.74

EXAMPLE 7 Simulating motion with parametric equations

Three golf balls are hit simultaneously into the air at 132 feet per second (90 miles per hour), making angles of 30°, 50°, and 70° with the horizontal.
(a) Assuming the ground is level, determine graphically which ball travels the farthest. Estimate this distance.
(b) Which ball reaches the greatest height? Estimate this height.

SOLUTION
(a) The three sets of parametric equations determined by the three golf balls are as follows, since $h = 0$.

$$X_1 = 132 \cos (30)T, \quad Y_1 = 132 \sin (30)T - 16T^{\wedge}2$$
$$X_2 = 132 \cos (50)T, \quad Y_2 = 132 \sin (50)T - 16T^{\wedge}2$$
$$X_3 = 132 \cos (70)T, \quad Y_3 = 132 \sin (70)T - 16T^{\wedge}2$$

The graphs of the three sets of parametric equations are shown in Figures 8.75 and 8.76, where $0 \le t \le 9$. A graphing calculator in *simultaneous mode* has been used so that we can view all three balls in flight at the same time. From the second graph we can see that the ball hit at 50° travels the farthest distance. Using the trace feature, we estimate this distance to be about 540 feet. (Use degree mode.)
(b) The ball hit at 70° reaches the greatest height of about 240 feet.

Now Try Exercise 71 ◀

[0, 600, 50] by [0, 400, 50]

Figure 8.75

[0, 600, 50] by [0, 400, 50]

Figure 8.76

CLASS DISCUSSION

If a golf ball is hit at 88 feet per second (60 mi/hr), use trial and error to find the angle θ that results in a maximum distance for the ball.

EXAMPLE 8 Modeling the flight of a baseball

A baseball is hit from a height of 4 feet at a 30° angle above the horizontal. Its initial velocity is 128 feet per second.
(a) Write parametric equations that model the flight of the baseball.
(b) Determine the horizontal distance that the ball travels in the air, assuming that the ground is level.
(c) What is the maximum height of the baseball?
(d) Would the ball clear a 4-foot-high fence that is 400 feet from the batter?

SOLUTION
(a) Let $v = 128$, $\theta = 30°$, and $h = 4$. Then the parametric equations become

$$x = (128 \cos 30°)t \quad \text{and} \quad y = (128 \sin 30°)t - 16t^2 + 4.$$

Since $\cos 30° = \dfrac{\sqrt{3}}{2}$ and $\sin 30° = \dfrac{1}{2}$, these equations can be rewritten as

$$x = (64\sqrt{3})t \quad \text{and} \quad y = 64t - 16t^2 + 4.$$

(b) To find how far the ball travels, we first determine the length of time that the ball is in flight. The ball hits the ground when $y = 0$.

$64t - 16t^2 + 4 = 0$	Substitute for y.
$16t^2 - 64t - 4 = 0$	Rewrite quadratic equation.
$t = \dfrac{64 \pm \sqrt{(-64)^2 - 4(16)(-4)}}{2(16)}$	Quadratic formula
$t \approx 4.0616 \quad \text{or} \quad t \approx -0.0616$	Approximate.

After 4.0616 seconds, the ball traveled *horizontally* $x = 64\sqrt{3}(4.0616) \approx 450.2$ feet.
(c) The graph of $y = 64t - 16t^2 + 4$ is a parabola that opens downward. Using the vertex formula, we find that the maximum height of the ball occurs after

$$t = -\frac{b}{2a} = -\frac{64}{2(-16)} = 2 \text{ seconds.}$$

The maximum height is $y = 64(2) - 16(2)^2 + 4 = 68$ feet.
(d) To find how long it takes the ball to reach the fence, solve the equation $x = 400$ for t.

$$(64\sqrt{3})t = 400, \quad \text{or} \quad t = \frac{400}{64\sqrt{3}} \approx 3.61 \text{ seconds}$$

After 3.61 seconds, the ball has traveled horizontally 400 feet and is

$$y = 64(3.61) - 16(3.61)^2 + 4 \approx 27 \text{ feet}$$

high. The baseball easily clears the 4-foot fence.   *Now Try Exercise 75* ◀

An Application from Computer Graphics   Parametric equations are used frequently in computer graphics to design a variety of figures and letters. Computer fonts are sometimes designed using parametric equations. In the next example, we use parametric equations to design a "smiley" face consisting of a head, two eyes, and a mouth.   (**Source:** F. Hill.)

EXAMPLE 9 Creating drawings with parametric equations

Graph a "smiley" face using parametric equations. Answers may vary.

SOLUTION

Head We can use a circle centered at the origin for the head. If the radius is 2, then let $x = 2 \cos t$ and $y = 2 \sin t$ for $0 \le t \le 2\pi$. These equations are graphed in Figure 8.77.

Eyes For the eyes we can use two small circles. The eye in the first quadrant can be modeled by $x = 1 + 0.3 \cos t$ and $y = 1 + 0.3 \sin t$ for $0 \le t \le 2\pi$. This represents a circle centered at $(1, 1)$ with radius 0.3. The eye in the second quadrant can be modeled by $x = -1 + 0.3 \cos t$ and $y = 1 + 0.3 \sin t$ for $0 \le t \le 2\pi$, which is a circle centered at $(-1, 1)$ with radius 0.3. These equations are graphed in Figure 8.78.

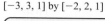

[−3, 3, 1] by [−2, 2, 1]   [−3, 3, 1] by [−2, 2, 1]   [−3, 3, 1] by [−2, 2, 1]

$x = 2 \cos t$
$y = 2 \sin t$
$0 \le t \le 2\pi$

Figure 8.77   **Figure 8.78**   **Figure 8.79**

CLASS DISCUSSION

Modify the face in Example 9 so that it is frowning. Try to find a way to make the right eye shut rather than open.

Mouth For the smile we can use the lower half of a circle. Using trial and error, we might arrive at $x = 0.5 \cos \frac{1}{2} t$ and $y = -0.5 - 0.5 \sin \frac{1}{2} t$. This is a semicircle centered at $(0, -0.5)$ with radius 0.5. Since we are letting $0 \le t \le 2\pi$, the term $\frac{1}{2} t$ ensures that only half a circle (a semicircle) is drawn. The minus sign before $0.5 \sin \frac{1}{2} t$ in the y-equation causes the lower half of the semicircle to be drawn rather than the upper half. The final result is shown in Figure 8.79. The pupils have been added by plotting the points $(1, 1)$ and $(-1, 1)$, and the coordinate axes have been turned off.   *Now Try Exercise 69* ◀

8.4 Putting It All Together

Parametric equations can be used to model a wide variety of curves in the xy-plane that cannot be represented by a single function. Some important concepts about parametric equations are summarized in the following table.

Concept	Explanation	Examples
Plane curve and parametric equations	A plane curve is a set of points (x, y) such that $x = f(t)$ and $y = g(t)$, where f and g are continuous on an interval $a \le t \le b$. The equations $x = f(t)$ and $y = g(t)$ are parametric equations with parameter t.	If $x = 2t$ and $y = t^2$ for $-1 \le t \le 2$, then the resulting graph is a portion of a parabola.

continued on next page

continued from previous page

Concept	Explanation	Examples
Writing parametric equations in terms of x and y	Solve one of the parametric equations for t and substitute into the second equation.	If $x = t^3$ and $y = t^2 - 2$, solve $x = t^3$ for t to obtain $t = x^{1/3}$. Substituting gives $y = x^{2/3} - 2$.
Converting to parametric equations	If possible, solve the equation for one of the variables. Let the other variable equal t. Now write the first variable in terms of t. Answers may vary.	If $4x = y^2$, solve the equation for x to obtain $x = \frac{1}{4}y^2$. Let $y = t$. Then $$x = \frac{1}{4}t^2 \quad \text{and} \quad y = t.$$ Another possibility is $$x = t^2 \quad \text{and} \quad y = 2t.$$

8.4 Exercises

Graphs of Parametric Equations

Exercises 1–8: Use the parametric equations to complete the following.
  (a) *Make a table of values for $t = 0, 1, 2, 3$.*
  (b) *Plot the points from the table and graph the curve for $0 \le t \le 3$. Add arrows to show how the curve is traced out.*
  (c) *Describe the curve.*

1. $x = t - 1$, $y = 2t$

2. $x = t + 1$, $y = t - 2$

3. $x = t + 2$, $y = (t - 2)^2$

4. $x = \frac{1}{3}t^2$, $y = t - 1$

5. $x = \sqrt{9 - t^2}$, $y = t$

6. $x = t^2$, $y = 2t + 1$

7. $x = t$, $y = \sqrt{9 - t^2}$

8. $x = 3t$, $y = t^2 + 2$

Exercises 9–24: Find a rectangular equation for each curve and describe the curve. Support your result by graphing the parametric equations.

9. $x = 3t$, $y = t - 1$; $-\infty < t < \infty$

10. $x = t + 3$, $y = 2t$; $-\infty < t < \infty$

11. $x = 3t^2$, $y = t + 1$; $-\infty < t < \infty$

12. $x = t^2 - 2t + 1$, $y = t - 1$; $-\infty < t < \infty$

13. $x = \sqrt{1 - t^2}$, $y = t$; $-1 \le t \le 1$

14. $x = \frac{1}{2}t$, $y = \sqrt{4 - t^2}$; $-2 \le t \le 2$

15. $x = t$, $y = \frac{1}{2}t^2$; $-2 \le t \le 2$

16. $x = \sqrt[3]{t}$, $y = t$; $-2 \le t \le 2$

17. $x = t - 2$, $y = t^2 + 1$; $-1 \le t \le 2$

18. $x = 2t$, $y = t^2 + 1$; $-1 \le t \le 2$

19. $x = 3 \sin t$, $y = 3 \cos t$; $-\pi \le t \le \pi$

20. $x = 4 \cos t$, $y = 4 \sin t$; $0 \le t \le 2\pi$

21. $x = 2 \sin t$, $y = -2 \cos t$; $0 \le t \le 2\pi$

22. $x = \cos 2t$, $y = \sin 2t$; $0 \le t \le \pi$

23. $x = 3 \cos 2t$, $y = 3 \sin 2t$; $0 \le t \le \pi$

24. $x = 2 \cos^2 t$, $y = 2 \sin^2 t$; $0 \le t \le \frac{\pi}{2}$

Exercises 25–42: Graph the parametric equations.

25. $x = \frac{1}{3}t,$ $y = \frac{2}{3}t + 1;$ $-\infty < t < \infty$

26. $x = t + 3,$ $y = 2t - 1;$ $-\infty < t < \infty$

27. $x = t^2,$ $y = 2t;$ $-\infty < t < \infty$

28. $x = \frac{1}{2}(t + 2)^2,$ $y = t + 2;$ $-\infty < t < \infty$

29. $x = \cos t,$ $y = \sin t;$ $0 \le t \le \pi$

30. $x = 2 \sin t,$ $y = 2 \cos t;$ $-\pi \le t \le 0$

31. $x = t^3,$ $y = t^2;$ $-2 \le t \le 2$

32. $x = e^t,$ $y = t - 1;$ $-2 \le t \le 2$

33. $x = t^2,$ $y = \ln t;$ $0 < t \le 2$

34. $x = t^3 - t,$ $y = e^t;$ $-1.5 \le t \le 1.5$

35. $x = t - \sin t,$ $y = 1 - \cos t;$ $0 \le t \le 6\pi$

36. $x = t^3 + 3t,$ $y = 2 \cos t;$ $-1 \le t \le 1$

37. $x = 2 + \cos t,$ $y = \sin t - 1;$ $0 \le t \le 2\pi$

38. $x = -2 + \cos t,$ $y = \sin t + 1;$ $0 \le t \le 2\pi$

39. $x = \cos^3 t,$ $y = \sin^3 t;$ $0 \le t \le 2\pi$

40. $x = \cos^5 t,$ $y = \sin^5 t;$ $0 \le t \le 2\pi$

41. $x = |3 \sin t|,$ $y = |3 \cos t|;$ $0 \le t \le \pi$

42. $x = 3 \sin 2t,$ $y = 3 \cos t;$ $0 \le t \le 2\pi$

Exercises 43–54: Convert the given equation to parametric equations. Answers may vary.

43. $2x + y = 4$

44. $5x - 4y = 20$

45. $y = 4 - x^2$

46. $x = y^2 - 2$

47. $x = y^2 + y - 3$

48. $5x = y^3 + 1$

49. $x^2 + y^2 = 4$

50. $x^2 + y^2 = 9$

51. $\ln y = 0.1x^2$

52. $e^x = |1 - y|$

53. $x = y^2 - 2y + 1$

54. $x = 4y^2 + 4y + 1$

 Exercises 55–60: Graph each pair of parametric equations for $0 \le t \le 2\pi$. Describe any differences in the two graphs.

55. (a) $x = 3 \cos t,$ $y = 3 \sin t$
    (b) $x = 3 \cos 2t,$ $y = 3 \sin 2t$

56. (a) $x = 2 \cos t,$ $y = 2 \sin t$
    (b) $x = 2 \cos t,$ $y = -2 \sin t$

57. (a) $x = 3 \cos t,$ $y = 3 \sin t$
    (b) $x = 3 \sin t,$ $y = 3 \cos t$

58. (a) $x = t,$ $y = t^2$
    (b) $x = t^2,$ $y = t$

59. (a) $x = -1 + \cos t,$ $y = 2 + \sin t$
    (b) $x = 1 + \cos t,$ $y = 2 + \sin t$

60. (a) $x = 2 \cos \frac{1}{2}t,$ $y = 2 \sin \frac{1}{2}t$
    (b) $x = 2 \cos t,$ $y = 2 \sin t$

Designing Shapes and Figures

 Exercises 61–64: Graph the following set of parametric equations for $0 \le t \le 2\pi$ in the viewing rectangle $[0, 6, 1]$ by $[0, 4, 1]$. Identify the letter of the alphabet that is being graphed.

61. $x_1 = 1,$ $y_1 = 1 + t/\pi$
    $x_2 = 1 + t/(3\pi),$ $y_2 = 2$
    $x_3 = 1 + t/(2\pi),$ $y_3 = 3$

62. $x_1 = 1,$ $y_1 = 1 + t/\pi$
    $x_2 = 1 + t/(3\pi),$ $y_2 = 2$
    $x_3 = 1 + t/(2\pi),$ $y_3 = 3$
    $x_4 = 1 + t/(2\pi),$ $y_4 = 1$

63. $x_1 = 1,$ $y_1 = 1 + t/\pi$
    $x_2 = 1 + 1.3 \sin (0.5t),$ $y_2 = 2 + \cos (0.5t)$

64. $x_1 = 2 + 0.8 \cos (0.85t),$ $y_1 = 2 + \sin (0.85t)$
    $x_2 = 1.2 + t/(1.3\pi),$ $y_2 = 2$

 Exercises 65–68: Designing Letters   Find a set of parametric equations that results in a letter similar to the one shown in the figure. Use the viewing rectangle given by $[-4.7, 4.7, 1]$ by $[-3.1, 3.1, 1]$ and turn off the coordinate axes. Answers may vary.

65.

66.

67.

68.

69. *Designing a Face*   (Refer to Example 9.) Use parametric equations to create your own "smiley" face. This face should have a head, a mouth, and eyes.

70. *Designing a Face*   Add a nose to the face that you designed in Exercise 69.

Applications

71. *Flight of a Golf Ball*   (Refer to Example 7.) Two golf balls are hit into the air at 66 feet per second (45 mi/hr), making angles of 35° and 50° with the horizontal. If the ground is level, estimate the horizontal distance traveled by each golf ball.

72. *Flight of a Golf Ball*   Solve Exercise 71 if, instead of the ground being level, the ground is inclined with a slope of $m = 0.1$.

73. *Flight of a Golf Ball*   If a golf ball is hit at 88 feet per second (60 mi/hr), making an angle of 45° with the horizontal, will it go over a fence 10 feet high that is 200 feet away on level ground?

74. *Simulating Gravity on the Moon*   (Refer to Example 7.) If an object is thrown on the moon, the parametric equations of flight are

$$x = (v \cos \theta)t \quad \text{and} \quad y = (v \sin \theta)t - 2.66t^2 + h.$$

Estimate the horizontal distance that a golf ball hit 88 feet per second (60 mi/hr) at an angle of 45° with the horizontal travels on the moon if the moon's surface is level.

75. *Flight of a Baseball*   (Refer to Example 8.) A baseball is hit with an angle of elevation of 45°, from the top of a ridge that is 50 feet above an area of level ground. The initial velocity of the ball is 88 feet per second, or 60 miles per hour. Find the horizontal distance traveled by the ball in the air.

76. *Flight of a Baseball*   A baseball is hit from a height of 3 feet at a 60° angle above the horizontal. Its initial velocity is 64 feet per second.
   (a) Write parametric equations that model the flight of the baseball.

   (b) Determine the horizontal distance traveled by the ball in the air. Assume that the ground is level.

   (c) What is the maximum height of the baseball? At that time, how far has the ball traveled horizontally?

   (d) Would the ball clear a 5-foot high fence that is 100 feet from the batter?

Exercises 77–80: Lissajous Figures   Lissajous figures occur in electronics and may be used to find the frequency of an unknown voltage. (See Extended and Discovery Exercises 1–4 below.) Graph the Lissajous figure for $0 \le t \le 6.5$ in the viewing rectangle $[-6, 6, 1]$ by $[-4, 4, 1]$.

77. $x = 2 \cos t, \qquad y = 3 \sin 2t$

78. $x = 3 \cos 2t, \qquad y = 3 \sin 3t$

79. $x = 3 \sin 4t, \qquad y = 3 \cos 3t$

80. $x = 4 \sin 4t, \qquad y = 3 \sin 5t$

Writing about Mathematics

81. Describe the basic form of parametric equations. Give an example. Explain how graphs of parametric equations can differ from graphs of functions.

82. Suppose that a function is defined by $y = f(x)$, where the domain of f is $a \le x \le b$. Explain how we could represent f with parametric equations. Apply your method to $f(x) = x^2 + 1$, where $-2 \le x \le 2$.

EXTENDED AND DISCOVERY EXERCISES

*Exercises 1–4: Electronic Technology   Parametric equations have applications in electricity. If two sinusoidal voltages, denoted by $x = V_1(t)$ and $y = V_2(t)$, are applied to an oscilloscope, a stationary pattern called a Lissajous figure may appear, as shown in the figure below. If the frequency F_1 of V_1 is known and the frequency F_2 of V_2 is unknown, then a Lissajous figure may be used to find F_2. The ratio $\frac{F_1}{F_2}$ is equal to the ratio of the corresponding number of tangents to the enclosing rectangle. The number of tangents along a vertical side of the rectangle corresponds to F_1, and the number of tangents along a horizontal side corresponds to F_2. In this figure $\frac{F_1}{F_2} = \frac{3}{2}$. Therefore $F_2 = \frac{2}{3}F_1$. Determine F_2 given the Lissajous figure and the frequency F_1 in cycles per second.   (**Source:** R. Smith and R. Dorf, Circuits, Devices, and Systems.)*

1. $F_1 = 150$

2. $F_1 = 60$

3. $F_1 = 400$

4. $F_1 = 1200$

CHECKING BASIC CONCEPTS FOR SECTIONS 8.3 AND 8.4

1. Let the point P be $(-1, 3)$ and the point Q be $(3, 7)$. Find the following.
    (a) $\mathbf{v} = \overrightarrow{PQ}$ **(b)** $\|\mathbf{v}\|$ **(c)** $\overrightarrow{PQ} + \overrightarrow{QP}$

2. Let $\mathbf{v} = 2\mathbf{i} - \mathbf{j}$ and $\mathbf{u} = -3\mathbf{i} + 2\mathbf{j}$. Find the following graphically and symbolically.
    (a) $2\mathbf{v} + \mathbf{u}$ **(b)** $2\mathbf{v}$ **(c)** $\mathbf{v} - 3\mathbf{u}$

3. Let $\mathbf{a} = \langle 3, -2 \rangle$ and $\mathbf{b} = \langle -1, 3 \rangle$. Find the following.
    (a) $\mathbf{a} \cdot \mathbf{b}$

    (b) The angle θ between $\mathbf{a}$ and $\mathbf{b}$ rounded to the nearest tenth of a degree

4. Graph the parametric equations given by $x = t + 1$ and $y = (t - 1)^2$ for $-1 \le t \le 5$. Write these parametric equations in terms of x and y.

8.5 Polar Equations

- **Learn the polar coordinate system**
- **Graph polar equations**
- **Graph polar equations with graphing calculators (optional)**
- **Solve polar equations**

Introduction

Many times a change in a frame of reference can have a profound effect on the solution of a problem. Thus far we have graphed functions only in the xy-plane. Many interesting curves, such as a spiral, cannot be represented by a function since these curves fail the vertical line test. Creating a new coordinate system makes some types of equations simpler. For example, the equation $x^2 + y^2 = 1$ describes the unit circle in the rectangular coordinate system. Every point lying on the unit circle is 1 unit from the origin. If we specify a new variable r that represents the radius of the circle, then $r = 1$ also describes the unit circle. A change of variable has resulted in a simpler equation.

     This section introduces a new coordinate system called the *polar coordinate system*, which originated with Jakob Bernoulli (1654–1705). This new system allows us to express many curves with relatively simple equations. (**Source:** H. Eves, *An Introduction to the History of Mathematics.*)

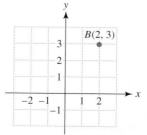

Figure 8.80

Figure 8.81

The Polar Coordinate System

In Figure 8.80 point A is located at $x = -2$. One number completely determines the location of a point on a one-dimensional number line. In two dimensions, two numbers are necessary to locate a point. For example, two numbers are needed to locate point B at $(2, 3)$ in Figure 8.81. In the xy-plane we are accustomed to identifying points using (x, y), where x and y are real numbers. However, using the xy-plane is not the only way to locate a point in a plane.

     The **polar coordinate system** uses r and θ instead of x and y to locate a point P, as shown in Figure 8.82 on the next page. The distance OP between P and the origin is represented by $|r|$. Angle θ can be expressed in either degrees or radians and is assumed to be in standard position. If $r > 0$, point P lies on the terminal side of θ, and if $r < 0$, point P lies on the ray pointing in the *opposite* direction of the terminal side of θ, a distance of $|r|$ from

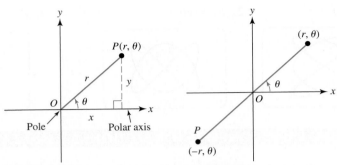

Figure 8.82 $r > 0$ **Figure 8.83**

the origin. See Figure 8.83. In the polar coordinate system, the origin is called the **pole** and the positive x-axis corresponds to the **polar axis**.

Using Figure 8.82 and trigonometry, we can establish the following relationships between rectangular and polar coordinates.

Rectangular and Polar Coordinates

If a point has rectangular coordinates (x, y) and polar coordinates (r, θ), then these coordinates are related as follows.

$$x = r \cos \theta, \qquad y = r \sin \theta$$

$$r^2 = x^2 + y^2, \qquad \tan \theta = \frac{y}{x} (x \neq 0)$$

EXAMPLE 1 **Plotting points in polar coordinates**

Plot the points (r, θ) on a polar grid.

(a) $(2, 45°)$ **(b)** $(-3, 150°)$ **(c)** $\left(3.5, -\frac{\pi}{3} \right)$

SOLUTION

(a) Let $r = 2$ and $\theta = 45°$. Plot a point 2 units from the pole on the terminal side of $\theta = 45°$, as shown in Figure 8.84. Note that θ is in standard position.

(b) Since $r = -3 < 0$ and $\theta = 150°$, begin by locating the terminal side of θ in the second quadrant. Next plot a point 3 units from the pole in the opposite direction of the terminal side of θ, as shown in Figure 8.85.

(c) Since $r = 3.5$ and $\theta = -\frac{\pi}{3}$ (radians), the point is in the fourth quadrant. See Figure 8.86.

Figure 8.84   **Figure 8.85**   **Figure 8.86**

Now Try Exercises 1 and 3 ◀

MAKING CONNECTIONS

Vectors and Polar Coordinates If a vector $\mathbf{v} = \langle a, b \rangle$ has magnitude $\|\mathbf{v}\|$ and direction angle θ, then

$$a = \|\mathbf{v}\| \cos \theta \qquad \text{and} \qquad b = \|\mathbf{v}\| \sin \theta,$$

where $\|\mathbf{v}\| = \sqrt{a^2 + b^2}$ and $\tan \theta = \frac{b}{a}$, $a \neq 0$. Similarly, if a point has rectangular coordinates (a, b) and polar coordinates (r, θ) with $r > 0$, then

$$a = r \cos \theta \qquad \text{and} \qquad b = r \sin \theta,$$

where $r = \sqrt{a^2 + b^2}$ and $\tan \theta = \frac{b}{a}$, $a \neq 0$.

CLASS DISCUSSION

What can be said about a point (r, θ) if $r = 0$?

NOTE Unlike xy-coordinates, polar coordinates are *not* unique. For example, the $r\theta$-coordinates of $(2, 0°)$, $(2, 360°)$, $(2, -360°)$, and $(-2, 180°)$ all represent the same point.

In the next example, we convert polar coordinates to rectangular coordinates.

EXAMPLE 2   **Converting to rectangular coordinates**

Convert each point from polar coordinates to rectangular coordinates.

(a) $(5, 180°)$ **(b)** $\left(3, -\dfrac{\pi}{3}\right)$

SOLUTION

(a) To convert $(5, 180°)$ to rectangular coordinates, use the equations $x = r \cos \theta$ and $y = r \sin \theta$ with $r = 5$ and $\theta = 180°$.

$$x = 5 \cos (180°) = -5$$
$$y = 5 \sin (180°) = 0$$

The corresponding rectangular coordinates are $(-5, 0)$.

(b) To convert $\left(3, -\dfrac{\pi}{3}\right)$ to rectangular coordinates, let $r = 3$ and $\theta = -\dfrac{\pi}{3}$, where θ is in radians.

$$x = 3 \cos\left(-\frac{\pi}{3}\right) = \frac{3}{2}, \quad y = 3 \sin\left(-\frac{\pi}{3}\right) = -\frac{3\sqrt{3}}{2} \approx -2.6$$

The rectangular coordinates are $\left(\dfrac{3}{2}, -\dfrac{3\sqrt{3}}{2}\right)$.   *Now Try Exercises 11 and 15* ◀

In the next example, points expressed in rectangular coordinates are converted to polar coordinates. However, this conversion is *not* unique because, unlike rectangular coordinates, polar coordinates are not unique.

EXAMPLE 3   **Expressing points in polar coordinates**

Given the point $(1, \sqrt{3})$ in rectangular coordinates, find polar coordinates (r, θ) that satisfy each condition.

(a) $r > 0$, $0° \leq \theta < 360°$
(b) $r > 0$, $-360° \leq \theta < 0°$
(c) $r < 0$, $0° \leq \theta < 360°$

SOLUTION

(a) $r > 0$, $0° \leq \theta < 360°$ Because $r^2 = x^2 + y^2$, let $r = \sqrt{1 + 3} = 2$. The point $(1, \sqrt{3})$ is located in the first quadrant of the xy-plane. Therefore $\tan \theta = \frac{y}{x} = \frac{\sqrt{3}}{1}$ and $\theta = \tan^{-1} \sqrt{3} = 60°$. Let $(r, \theta) = (2, 60°)$. See Figure 8.87.

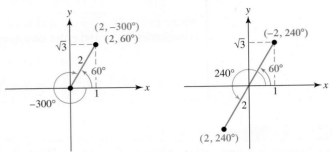

Figure 8.87   Figure 8.88

(b) $r > 0$, $-360° \leq \theta < 0°$ Rather than let $\theta = 60°$, we can use $\theta = -300°$, so $(r, \theta) = (2, -300°)$. See Figure 8.87. Notice that $60°$ and $-300°$ are coterminal angles.

(c) $r > 0$, $0° \leq \theta < 360°$ We can let $r = -2$, but then we need to let angle $\theta = 60° + 180° = 240°$. Thus $(r, \theta) = (-2, 240°)$, as illustrated in Figure 8.88.

Now Try Exercise 23 ◀

Graphs of Polar Equations

When we graph $y = f(x)$ in the xy-coordinate system, we are graphing a function. A vertical line can intersect the graph of a function at most once. As a result, many shapes such as circles, hearts, and leaves cannot be represented by a function. Like parametric equations, polar equations can be valuable for representing curves.

EXAMPLE 4 Representing polar equations

Make a table of values and graph each curve. Then describe the curve.

(a) $\theta = \dfrac{\pi}{4}$ (b) $r = 3$ (c) $r = 2 + 2 \cos \theta$

SOLUTION

(a) *Numerical Representation* Since $\theta = \frac{\pi}{4}$, every point lying on this graph is of the form $\left(r, \frac{\pi}{4}\right)$, where r can be any real number. Five of these points are listed in Table 8.2.

Table 8.2

θ	$\frac{\pi}{4}$	$\frac{\pi}{4}$	$\frac{\pi}{4}$	$\frac{\pi}{4}$	$\frac{\pi}{4}$
r	-2	-1	0	1	2

Figure 8.89 A Line

Graphical Representation A graph of $\theta = \frac{\pi}{4}$ and the points in Table 8.2 are shown in Figure 8.89.

Verbal Representation The graph is a line with slope 1 passing through the pole.

(b) *Numerical Representation* A table of values is shown in Table 8.3.

Table 8.3

θ	0°	60°	120°	180°	240°	300°
r	3	3	3	3	3	3

Figure 8.90 A Circle

Figure 8.91 A Cardioid

Graphical Representation A graph, with these points, is shown in Figure 8.90.

Verbal Representation The polar equation represents a circle with radius 3.

(c) **Numerical Representation** A table of values for $r = 2 + 2\cos\theta$ is shown in Table 8.4. For example, when $\theta = 60°$, then

$$r = 2 + 2\cos 60° = 2 + 2(0.5) = 3.$$

Table 8.4

θ	0°	60°	120°	180°	240°	300°	360°
r	4	3	1	0	1	3	4

Graphical Representation To help graph the equation we can plot the points in Table 8.4. See Figure 8.91. Notice that since the cosine function has period 360°, the graph repeats after 360°.

Verbal Representation The polar equation represents a heart-shaped graph called a **cardioid**.
Now Try Exercises 31, 33, and 37 ◄

EXAMPLE 5 Graphing in polar coordinates

Graph each polar equation.
(a) $r = 2\sin\theta$ (b) $r = 2 + \cos\theta$

SOLUTION
(a) The polar equation $r = 2\sin\theta$ can be converted to rectangular coordinates by first multiplying each side of the equation by r.

$r = 2\sin\theta$	*Given polar equation*
$r^2 = 2r\sin\theta$	*Multiply by r.*
$x^2 + y^2 = 2y$	*Convert to xy-coordinates.*
$x^2 + y^2 - 2y = 0$	*Subtract 2y.*
$x^2 + y^2 - 2y + 1 = 1$	*Complete the square by adding 1.*
$x^2 + (y - 1)^2 = 1$	*Perfect square trinomial*

Figure 8.92

This final equation represents a circle with center $(0, 1)$ and radius 1. The graph of $r = 2\sin\theta$ is shown in Figure 8.92.

(b) To graph this polar equation, start by making a table of values as shown in Table 8.5, where values are rounded to the nearest tenth. (Degree measure could also be used.)

Table 8.5

θ	0	$\frac{\pi}{4}$	$\frac{\pi}{2}$	$\frac{3\pi}{4}$	π	$\frac{5\pi}{4}$	$\frac{3\pi}{2}$	$\frac{7\pi}{4}$
$r = 2 + \cos\theta$	3	2.7	2	1.3	1	1.3	2	2.7

The points in Table 8.5 and the equation $r = 2 + \cos\theta$ are graphed in Figure 8.93. The graph is called a **limaçon** *without an inner loop.*

Figure 8.93

Now Try Exercises 47 and 51 ◄

Rose Curves   Polar equations in the form $r = a \sin n\theta$ or $r = a \cos n\theta$ result in graphs of rose curves. It can be shown that when n is odd there are n leaves and when n is even there are $2n$ leaves. In the next example, we graph this type of equation.

EXAMPLE 6   Graphing a four-leaved rose

Graph $r = 3 \cos 2\theta$.

SOLUTION

Getting Started   To graph this polar equation by hand, start by making a table of values like the one shown in Table 8.6. Degree measure has been used, and values for r have been rounded to the nearest tenth. Notice that the values repeat themselves starting at 180°.

Table 8.6

θ	0°	15°	30°	45°	60°	75°	90°
$r = 3 \cos 2\theta$	3	2.6	1.5	0	−1.5	−2.6	−3

θ	105°	120°	135°	150°	165°	180°
$r = 3 \cos 2\theta$	−2.6	−1.5	0	1.5	2.6	3

Figure 8.94

Plotting these points in order gives the graph, called a **four-leaved rose**. Note in Figure 8.94 that the graph is developed with a continuous curve, beginning with the upper half of the right horizontal leaf and ending with the lower half of that leaf. As the graph is traced, the curve passes through the pole four times. Each leaf has length 3.

Now Try Exercise 55 ◀

EXAMPLE 7   Writing polar equations in rectangular form

Write the polar equation in terms of x and y. Describe its graph.

(a) $r = 3 \csc \theta$ **(b)** $r = \dfrac{2}{4 \cos \theta - 3 \sin \theta}$

SOLUTION
(a) Begin by applying a reciprocal identity.

$$r = 3 \csc \theta \qquad \text{Given equation}$$

$$r = \frac{3}{\sin \theta} \qquad \text{Reciprocal identity}$$

$$r \sin \theta = 3 \qquad \text{Multiply by } \sin \theta.$$

$$y = 3 \qquad y = r \sin \theta$$

Its graph is a horizontal line.
(b) Start by cross-multiplying.

$$r = \frac{2}{4 \cos \theta - 3 \sin \theta} \qquad \text{Given equation}$$

$$4r \cos \theta - 3r \sin \theta = 2 \qquad \text{Cross-multiply.}$$

$$4x - 3y = 2 \qquad \text{Substitute.}$$

$$y = \frac{4}{3}x - \frac{2}{3} \qquad \text{Solve for } y.$$

Its graph is a line with slope $\frac{4}{3}$ and y-intercept $-\frac{2}{3}$.   *Now Try Exercises 69 and 71* ◀

Figure 8.95 A Logarithmic Spiral

Logarithmic Spiral In 1638, René Descartes described a *logarithmic spiral* using the complicated equation $y = x \tan (\ln (x^2 + y^2))$. This curve, shown in Figure 8.95, cannot be represented by a function. With polar coordinates, this rectangular equation reduces to the much simpler equation $r = e^{\theta/2}$. Johann Bernoulli was so entranced by this remarkable curve that he ordered it carved on his tombstone. (**Source:** H. Resnikoff and R. Wells, *Mathematics in Civilization.*)

Graphing Calculators and Polar Equations (Optional)

Technology can be used to make tables and graphs in polar coordinates. The table and graph in Example 4(c) are shown in Figures 8.96 and 8.97.

Calculator Help

To create the graph shown in Figure 8.97 with a polar equation, see Appendix A (page AP-21).

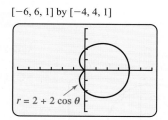

[-6, 6, 1] by [-4, 4, 1]

Figure 8.96 Degree Mode   **Figure 8.97** A Cardioid

As is the case with rectangular and parametric equations, a viewing rectangle must be selected before graphing a polar equation. First choose whether the graph should be plotted in radian or degree mode. Next determine an interval for θ. Many times the interval $0° \leq \theta \leq 360°$ is sufficient to have the entire graph generated. Then select a square viewing rectangle, if possible.

EXAMPLE 8 **Representing polar equations**

Graph each curve. Then describe the curve.
(a) $r = 3 \cos 2\theta$ **(b)** $r = 1 - 2 \sin \theta$

SOLUTION

(a) This equation was graphed by hand in Example 6. In Figure 8.98 a graphing calculator has been used to create a graph of $r = 3 \cos 2\theta$. Turn on the polar grid and try tracing the graph to see how each leaf of the rose is generated by the polar equation. Notice, for example, the location of the point $(-1.5, 60°)$ in Figure 8.98.

[-6, 6, 1] by [-4, 4, 1]

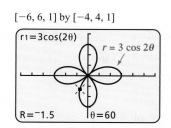

[-6, 6, 1] by [-4, 4, 1]

Figure 8.98 A Four-Leaved Rose   **Figure 8.99** Degree Mode   **Figure 8.100** A Limaçon

(b) A graph of $r = 1 - 2 \sin \theta$ is shown in Figure 8.99. The graph is called a **limaçon** *with an inner loop*. A graph with better resolution is shown in Figure 8.100. Notice that the inner loop occurs when $30° \leq \theta \leq 150°$ and $r \leq 0$.

Now Try Exercises 79 and 83 ◀

Table 8.7 Distances and Eccentricities

Planet	a	e
Jupiter	5.20	0.048
Saturn	9.54	0.056
Uranus	19.2	0.047
Neptune	30.1	0.009
Pluto	39.4	0.249

Polar Equations of Conics The polar equation

$$r = \frac{a(1 - e^2)}{1 + e \cos \theta}$$

can be used to model the orbits of planets and comets, where a is the average distance of the celestial body from the sun in astronomical units and e is a constant called the *eccentricity*. (Smaller values of e indicate that an orbit is more circular, whereas larger values indicate a more elliptical orbit. Note that values for e vary between 0 and 1. One astronomical unit equals 93 million miles.) The sun is located at the pole. Table 8.7 lists a and e for the outer planets and Pluto. (**Source:** H. Karttunen et al., *Fundamental Astronomy.*)

EXAMPLE 9 Determining orbits

Use graphing to determine if Pluto is always farther from the sun than Neptune is.

SOLUTION The orbital equations, given below, are graphed in Figure 8.101.

$$\text{Neptune: } r_1 = \frac{30.1(1 - 0.009^2)}{1 + 0.009 \cos \theta}$$

$$\text{Pluto: } r_2 = \frac{39.4(1 - 0.249^2)}{1 + 0.249 \cos \theta}$$

[−60, 60, 10] by [−40, 40, 10]

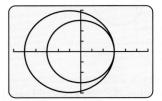

Figure 8.101

[27, 33, 1] by [−2, 2, 1]

Figure 8.102

The graph shows that their orbits pass near each other. By zooming in we can determine that the orbit of Pluto actually passes inside the orbit of Neptune. See Figure 8.102. Therefore there are times when Neptune—not Pluto—is farther from the sun. However, Pluto's average distance from the sun is considerably greater than Neptune's average distance. Neptune was farther from the sun than Pluto was for a 20-year period that ended in 1999. ***Now Try Exercise 95*** ◀

Solving Polar Equations

We can solve polar equations symbolically, graphically, and numerically.

EXAMPLE 10 Solving a polar equation

Find values for θ where the circle $r = 3$ intersects the cardioid $r = 2 + 2 \cos \theta$. Let $0° \le \theta \le 360°$.

SOLUTION
Symbolic Solution Begin by setting the two equations equal.

$$3 = 2 + 2 \cos \theta \qquad \text{Set equations equal.}$$

$$\cos \theta = \frac{1}{2} \qquad \text{Solve for } \cos \theta.$$

$$\theta = 60° \text{ or } 300° \qquad \text{Solve for } \theta.$$

(Be sure to check symbolic solutions.)

Graphical Solution Using the intersection-of-graphs method, let $r_1 = 3$ and let $r_2 = 2 + 2 \cos (\theta)$. Their graphs intersect when $\theta = 60°$ and $300°$, as shown in Figures 8.103 and 8.104.

Figure 8.103 Degree Mode

Figure 8.104 Degree Mode

θ	r1	r2
0	3	4
60	3	3 ←
120	3	1
180	3	0
240	3	1
300	3	3 ←
360	3	4

r2▄2+2cos(θ)

Figure 8.105 Degree Mode

Numerical Solution   Numerical support is shown in Figure 8.105, where $r_1 = r_2 = 3$ when $\theta = 60°$ and $300°$.

Now Try Exercise 91 ◀

8.5 Putting It All Together

Polar equations and graphs can be used to describe curves that are not easily represented by equations involving x and y. Polar coordinates are not unique for a given point, whereas rectangular coordinates are unique for a given point. Polar equations can be solved graphically, numerically, and symbolically. Some concepts regarding polar coordinates and graphs of polar equations are summarized in the following table.

Concept	Examples and Explanations
Polar coordinates	A point is determined by r and θ. To convert between rectangular and polar coordinates, use $x = r \cos \theta$, $y = r \sin \theta$, $\tan \theta = \frac{y}{x}$, and $r^2 = x^2 + y^2$. *(figure showing point (r, θ) with $y = r \sin \theta$, $x = r \cos \theta$)*
Graphs in polar coordinates	**Circle** $r = a$ **Cardioid** $r = a \pm a \sin \theta$ or $r = a \pm a \cos \theta$

continued on next page

continued from previous page

Concept	Examples and Explanations
Graphs in polar coordinates	**Rose Curve** $r = a \sin n\theta$ or $r = a \cos n\theta$ The number of leaves equals n when n is odd and is twice n when n is even. **Limaçon** ($b \neq a$) $r = a \pm b \sin \theta$ or $r = a \pm b \cos \theta$ $a > b$; no inner loop $a < b$; one inner loop

8.5   Exercises

Polar Coordinates

Exercises 1–4: Plot the points (r, θ).

1. (a) $(2, 0°)$ **(b)** $(3, 120°)$ **(c)** $(-1, 135°)$

2. (a) $(-2, 60°)$ **(b)** $(1, 120°)$ **(c)** $(2, 270°)$

3. (a) $\left(2, \frac{\pi}{3}\right)$ **(b)** $\left(-3, -\frac{\pi}{6}\right)$ **(c)** $\left(0, \frac{3\pi}{4}\right)$

4. (a) $(4, \pi)$ **(b)** $\left(1, \frac{\pi}{4}\right)$ **(c)** $\left(-3, -\frac{3\pi}{2}\right)$

Exercises 5–10: Determine if the pair of polar coordinates represents the same point.

5. $(2, 180°), (2, -180°)$ **6.** $(1, 90°), (-1, -90°)$

7. $(3, 45°), (3, -45°)$ **8.** $(-2, 135°), (2, -135°)$

9. $(0, 40°), (0, 50°)$ **10.** $(-3, 30°), (3, 210°)$

Exercises 11–20: Change the polar coordinates (r, θ) to rectangular coordinates (x, y).

11. $(3, 45°)$ **12.** $(-4, 225°)$

13. $(10, 90°)$ **14.** $\left(-1, \frac{\pi}{3}\right)$

15. $(5, 2\pi)$ **16.** $\left(-3, -\frac{\pi}{2}\right)$

17. $(-3, 60°)$ **18.** $(4, \pi)$

19. $\left(-2, -\frac{3\pi}{2}\right)$ **20.** $\left(10, \frac{2\pi}{3}\right)$

Exercises 21–26: For the point given in rectangular coordinates, find equivalent polar coordinates (r, θ) that satisfy the conditions.

(a) $r > 0$, $0° \leq \theta < 360°$
(b) $r < 0$, $-180° < \theta \leq 180°$

21. $(0, 3)$ **22.** $(-3, 0)$

23. $(-1, -\sqrt{3})$ **24.** $(\sqrt{3}, -1)$

25. $(3, -3)$ **26.** $(2, 2)$

Exercises 27–30: For the point given in rectangular coordinates, find equivalent polar coordinates (r, θ) that satisfy $r > 0$ and $0 \le \theta < 2\pi$. Approximate θ to the nearest hundredth of a radian.

27. $(7, 24)$ **28.** $(3, -4)$

29. $(-5, 12)$ **30.** $(11, -60)$

Graphs of Polar Equations

Exercises 31 and 32: Graph the equation.

31. $\theta = 60°$ **32.** $\theta = -135°$

Exercises 33–40: Complete the following.
   (a) Make a table of values with $\theta = 0°, 90°, 180°, 270°$.
   (b) Plot the points from the table and graph the curve for $0° \le \theta \le 360°$.

33. $r = 2$ **34.** $r = 1$

35. $r = 3 \sin \theta$ **36.** $r = 2 - 2 \sin \theta$

37. $r = 2 + 2 \sin \theta$ **38.** $r = 3 - 2 \cos \theta$

39. $r = 2 - \cos \theta$ **40.** $r = 2 + \sin \theta$

Exercises 41–58: Graph the polar equation by hand.

41. $\theta = 60°$ **42.** $\theta = -\frac{\pi}{4}$

43. $r = 3$ **44.** $r = \cos \theta$

45. $r \sin \theta = 3$ **46.** $r \cos \theta = -2$

47. $r = 2 \cos \theta$ **48.** $r = -2 \sin \theta$

49. $r = \sin 2\theta$ **50.** $r = \cos 2\theta$

51. $r = 3 + \cos \theta$ **52.** $r = 3 - \sin \theta$

53. $r = 1 - 2 \sin \theta$ **54.** $r = 1 + 2 \cos \theta$

55. $r = 3 \sin 2\theta$ **56.** $r = 2 \cos 2\theta$

57. $r = 2 \cos 3\theta$ **58.** $r = 4 \sin 3\theta$

Exercises 59–66: Write the equation in polar form.

59. $y = 3$ **60.** $x = -5$

61. $y = x$ **62.** $y = -\sqrt{3}$

63. $x^2 + y^2 = 9$ **64.** $x^2 + y^2 = 36$

65. $x^2 + y^2 = 2x$ **66.** $x^2 + y^2 = -4y$

Exercises 67–74: (Refer to Example 7.) Write the polar equation in terms of x and y.

67. $r = 3$ **68.** $r = 5$

69. $r = 2 \sec \theta$ **70.** $r = 2 \csc \theta$

71. $r = \dfrac{3}{2 \cos \theta + 4 \sin \theta}$ **72.** $r = \dfrac{2}{5 \cos \theta - \sin \theta}$

73. $r = \cos \theta$ **74.** $r = 2 \sin \theta$

Exercises 75–88: Graph the curve.

75. $r = 3 + 3 \cos \theta$ (cardioid)

76. $r = 2 - 2 \sin \theta$ (cardioid)

77. $r = 3 - 2 \sin \theta$ (limaçon)

78. $r = 4 + \cos \theta$ (limaçon)

79. $r = 2 - 4 \cos \theta$ (limaçon with a loop)

80. $r = 1 + 2 \sin \theta$ (limaçon with a loop)

81. $r = 4 \sin \theta$ (circle)

82. $r = 2 \cos 3\theta$ (three-leaved rose)

83. $r = 2 \cos 5\theta$ (five-leaved rose)

84. $r = 3 \sin 4\theta$ (eight-leaved rose)

85. $r = \frac{\theta}{2}$ (spiral)

86. $r = e^{\theta/4}$ (logarithmic spiral)

87. $r^2 = 2 \sin 2\theta$ (lemniscate)

88. $r^2 = 4 \cos 2\theta$ (lemniscate)

Solving Equations in Polar Coordinates

Exercises 89 and 90: Find values for θ that satisfy the equation $r_1 = r_2$, where $0° \le \theta \le 360°$. Check any solutions.

89. $r_1 = 3, r_2 = 2 + 2 \sin \theta$

90. $r_1 = 1, r_2 = 2 \cos \theta$

Exercises 91–94: Solve the polar equation $r_1 = r_2$, where $0° \le \theta \le 360°$,

(a) *symbolically,* (b) *graphically, and* (c) *numerically.*

91. $r_1 = 3, r_2 = 2 - 2 \sin \theta$

92. $r_1 = 3, r_2 = 2 - \sin \theta$ **93.** $r_1 = 1, r_2 = 2 \sin \theta$

94. $r_1 = 2 - \sin \theta, r_2 = 2 + \cos \theta$

Applications

*Exercises 95 and 96: **Planetary Orbits** (Refer to Example 9 and Table 8.7.) Graph the planetary orbits in a square viewing rectangle using polar coordinates.*

95. Saturn and Uranus **96.** Jupiter and Neptune

*Exercises 97 and 98: **Planetary Orbits** (Refer to Example 9.)*

Planet	a	e
Mercury	0.39	0.206
Venus	0.78	0.007
Earth	1.00	0.017
Mars	1.52	0.093

Source: H. Karttunen.

97. Graph the orbits of the four inner planets in the same square viewing rectangle.

98. NASA is planning future missions to Mars. Estimate graphically the closest distance possible between Earth and Mars.

*Exercises 99 and 100: **Broadcasting Patterns** Many times radio stations do not broadcast in all directions with the same intensity. To avoid interference with an existing station to the north, a new station may be licensed to broadcast only east and west. To create an east-west signal, two radio towers are sometimes used, as illustrated in the figure. Locations where the radio signal is received correspond to the interior*

of the curve defined by $r^2 = 40,000 \cos 2\theta$, where the polar axis (or positive x-axis) points east. (**Source:** R. Weidner and R. Sells, *Elementary Classical Physics, Vol. 2.*)

99. Graph $r^2 = 40,000 \cos 2\theta$ for $0° \le \theta \le 360°$, where units are in miles. Assuming the radio towers are located near the pole, use the graph to describe the regions where the signal can be received and where the signal cannot be received.

100. (Refer to Exercise 99.) Suppose a radio signal pattern is given by $r^2 = 22,500 \sin 2\theta$. Graph this pattern and interpret the results.

Writing about Mathematics

101. Explain why (r, θ) and $(-r, \theta + 180°)$ represent the same points in polar coordinates. Give two examples.

102. Give an example of a curve other than a circle that is more convenient to express in polar coordinates than in rectangular coordinates. Give an example of a curve that is more convenient to express using rectangular coordinates. Explain your reasoning.

EXTENDED AND DISCOVERY EXERCISES

1. *Logarithmic Spiral* Figure 8.95 shows a logarithmic spiral that can be described in rectangular coordinates by $y = x \tan (\ln (x^2 + y^2))$ and can be described in polar coordinates by the simpler equation $r = e^{\theta/2}$. Show that the first equation reduces to the second equation by assuming that $-\frac{\pi}{2} < \theta < \frac{\pi}{2}$. (*Hint:* Let $\tan \theta = \frac{y}{x}$ and $r^2 = x^2 + y^2$.)

2. *Polar Graphs* Consider the graphs of $r = a \cos n\theta$ and $r = a \sin n\theta$, where n is a positive integer and θ is given by $0° \le \theta < 360°$.

(a) How many times are these graphs traced over when n is even?

(b) How many times are these graphs traced over when n is odd?

8.6 Trigonometric Form and Roots of Complex Numbers

- Learn trigonometric form
- Find products and quotients of complex numbers
- Apply De Moivre's theorem
- Find roots of complex numbers

Introduction

One of the earliest encounters with the square root of a negative number was in A.D. 50, when Heron of Alexandria derived the expression $\sqrt{81 - 144}$. Square roots of negative numbers resulted in the invention of complex numbers. As late as the sixteenth and seventeenth centuries, mathematicians felt uneasy about negative numbers and square roots of negative numbers. The famous mathematician René Descartes rejected complex numbers and coined the term "imaginary" numbers.

In the historical development of our present-day number system, the introduction of new numbers was often met with resistance. Today, complex numbers are readily accepted and play an important role in the design of airplanes, ships, fractals, and electrical circuits. See Exercises 67 and 68. (**Sources:** M. Kline, *Mathematics: The Loss of Certainty; Historical Topics for the Mathematics Classroom, Thirty-first Yearbook,* NCTM.)

Trigonometric Form

Complex numbers were introduced in Section 3.3. Any complex number can be expressed in *standard form* as $a + bi$, where a and b are real numbers. The *real part* is a and the *imaginary part* is b.

Real numbers can be plotted on a number line. Since complex numbers are determined by both the real part a and the imaginary part b, we use the **complex plane** to plot complex numbers. The horizontal axis is the **real axis** and the vertical axis is the **imaginary axis**. For example, $2 + 3i$ can be plotted in the complex plane as the point $(2, 3)$, and $3 - 4i$ can be plotted as the point $(3, -4)$. See Figure 8.106.

A second form for complex numbers is called *trigonometric form*, which uses the variables r and θ to locate a complex number. In Figure 8.107 we see that

Figure 8.106
The Complex Plane

$$\cos \theta = \frac{a}{r} \qquad \text{and} \qquad \sin \theta = \frac{b}{r}.$$

Solving for a and b gives

$$a = r \cos \theta \qquad \text{and} \qquad b = r \sin \theta.$$

As a result, we can write the complex number $a + bi$ as follows.

$$a + bi = r \cos \theta + (r \sin \theta)i$$
$$= r(\cos \theta + i \sin \theta)$$

By the Pythagorean theorem, $r = \sqrt{a^2 + b^2}$. It also follows that

Figure 8.107
Trigonometric Form

$$\tan \theta = \frac{b}{a} \quad (a \neq 0).$$

Trigonometric Form of a Complex Number

The expression

$$r(\cos \theta + i \sin \theta)$$

is called a **trigonometric form** of the complex number $a + bi$, where $a = r \cos \theta$ and $b = r \sin \theta$. The number $r = \sqrt{a^2 + b^2}$ is the **modulus** of $a + bi$, and θ is the **argument** of $a + bi$.

NOTE The expression $\cos\theta + i\sin\theta$ can be written as cis θ. The expression $|a + bi|$ is sometimes used to denote the *modulus* of the complex number $a + bi$.

EXAMPLE 1 Converting standard form to trigonometric form

Find the trigonometric form for each complex number, where $0° \le \theta \le 360°$.
(a) $1 + i$ **(b)** $-1 - i\sqrt{3}$

SOLUTION

Figure 8.108

(a) Plot $1 + i$ in the complex plane, as shown in Figure 8.108. The modulus r is

$$r = \sqrt{1^2 + 1^2} = \sqrt{2}.$$

We can see that $\tan\theta = \frac{b}{a} = \frac{1}{1}$. Therefore $\theta = \tan^{-1}1 = 45°$. The trigonometric form is

$$\sqrt{2}\,(\cos 45° + i\sin 45°).$$

(b) Plot $-1 - i\sqrt{3}$, as shown in Figure 8.109. The modulus r is

$$r = \sqrt{(-1)^2 + (-\sqrt{3})^2} = 2.$$

The argument θ is in quadrant III and satisfies $\tan\theta = \frac{-\sqrt{3}}{-1} = \sqrt{3}$. The reference angle for θ is $\theta_R = \tan^{-1}(\sqrt{3}) = 60°$. Thus $\theta = 240°$ and the trigonometric form is $2(\cos 240° + i\sin 240°)$.

Now Try Exercises 13 and 19 ◀

Figure 8.109

MAKING CONNECTIONS

Vectors, Polar Coordinates, and Trigonometric Form The following are equivalent concepts.

1. Finding the magnitude $\|\mathbf{v}\|$ and direction angle θ for $\mathbf{v} = \langle a, b \rangle$
2. Finding polar coordinates (r, θ), where $r > 0$, for the point (a, b)
3. Finding the trigonometric form $r(\cos\theta + i\sin\theta)$ for $a + bi$

For example:

1. Given $\mathbf{v} = \langle 3, 4 \rangle$, then $\|\mathbf{v}\| = \sqrt{3^2 + 4^2} = 5$ and $\theta = \tan^{-1}\frac{4}{3} \approx 53.1°$. It follows that $\mathbf{v} \approx \langle 5\cos 53.1°, 5\sin 53.1° \rangle$.
2. Given $P = (3, 4)$, then $r = \sqrt{3^2 + 4^2} = 5$ and $\theta = \tan^{-1}\frac{4}{3} \approx 53.1°$. It follows that $P \approx (5, 53.1°)$ in polar coordinates.
3. Given $3 + 4i$, then $r = \sqrt{3^2 + 4^2} = 5$ and $\theta = \tan^{-1}\frac{4}{3} \approx 53.1°$. It follows that $3 + 4i \approx 5(\cos 53.1° + i\sin 53.1°)$ in trigonometric form.

See Extended and Discovery Exercises 1–4 at the end of this section.

Graphing Calculators Some calculators have the capability to convert the complex number $a + bi$ to trigonometric or **polar form**. This is illustrated in Figures 8.110 and 8.111, where the first computation gives the modulus and the second gives the argument. See Example 1. Notice that angles of 240° and $-120°$ are coterminal angles. The value of θ is *not* unique in a trigonometric form. If θ_1 and θ_2 are *coterminal angles*, then

$$r(\cos\theta_1 + i\sin\theta_1) \quad \text{and} \quad r(\cos\theta_2 + i\sin\theta_2)$$

represent equivalent trigonometric forms.

```
R▶Pr(1,1)
            1.414213562
√(2)
            1.414213562
R▶Pθ(1,1)
                      45
```

Figure 8.110 Degree Mode

```
R▶Pr(⁻1, ⁻√(3))
                       2
R▶Pθ(⁻1, ⁻√(3))
                   ⁻120
```

Figure 8.111 Degree Mode

EXAMPLE 2 Converting trigonometric form to standard form

Write the complex number as $a + bi$, where a and b are real numbers.

(a) $4\left(\cos \frac{\pi}{2} + i \sin \frac{\pi}{2}\right)$ (b) $\sqrt{3}(\cos 150° + i \sin 150°)$

SOLUTION

(a) $4\left(\cos \frac{\pi}{2} + i \sin \frac{\pi}{2}\right) = 4(0 + i(1)) = 4i$

(b) $\sqrt{3}(\cos 150° + i \sin 150°) = \sqrt{3}\left(-\frac{\sqrt{3}}{2} + \frac{1}{2}i\right) = -\frac{3}{2} + \frac{\sqrt{3}}{2}i$

Now Try Exercises 27 and 31 ◄

Products and Quotients of Complex Numbers

If two complex numbers, z_1 and z_2, are expressed in trigonometric form, it is straightforward to find either their product or their quotient. Let

$$z_1 = r_1(\cos \theta_1 + i \sin \theta_1) \quad \text{and} \quad z_2 = r_2(\cos \theta_2 + i \sin \theta_2).$$

Then

$$z_1 z_2 = r_1(\cos \theta_1 + i \sin \theta_1) \cdot r_2(\cos \theta_2 + i \sin \theta_2)$$
$$= r_1 r_2(\cos \theta_1 \cos \theta_2 + i \cos \theta_1 \sin \theta_2 + i \sin \theta_1 \cos \theta_2 + i^2 \sin \theta_1 \sin \theta_2)$$
$$= r_1 r_2((\cos \theta_1 \cos \theta_2 - \sin \theta_1 \sin \theta_2) + i(\cos \theta_1 \sin \theta_2 + \sin \theta_1 \cos \theta_2)).$$

Using the sum identities for cosine and sine, we can simplify the last expression to

$$z_1 z_2 = r_1 r_2(\cos (\theta_1 + \theta_2) + i \sin (\theta_1 + \theta_2)).$$

Using similar reasoning, it can be shown that

$$\frac{z_1}{z_2} = \frac{r_1}{r_2}(\cos (\theta_1 - \theta_2) + i \sin (\theta_1 - \theta_2)).$$

These results are summarized in the following box.

Products and Quotients of Complex Numbers

Let $z_1 = r_1(\cos \theta_1 + i \sin \theta_1)$ and $z_2 = r_2(\cos \theta_2 + i \sin \theta_2)$. Then

$$z_1 z_2 = r_1 r_2(\cos (\theta_1 + \theta_2) + i \sin (\theta_1 + \theta_2))$$

$$\frac{z_1}{z_2} = \frac{r_1}{r_2}(\cos (\theta_1 - \theta_2) + i \sin (\theta_1 - \theta_2)), \quad r_2 \neq 0.$$

EXAMPLE 3   Finding products and quotients

Find the product and quotient of

$$z_1 = 4(\cos 45° + i \sin 45°) \quad \text{and} \quad z_2 = 2(\cos 135° + i \sin 135°).$$

Express the answer in standard form.

SOLUTION

$$\begin{aligned}
z_1 z_2 &= 4(\cos 45° + i \sin 45°) \cdot 2(\cos 135° + i \sin 135°) \\
&= (4 \cdot 2)(\cos (45° + 135°) + i \sin (45° + 135°)) \\
&= 8(\cos 180° + i \sin 180°) \\
&= 8(-1 + 0) \\
&= -8
\end{aligned}$$

$$\begin{aligned}
\frac{z_1}{z_2} &= \frac{4(\cos 45° + i \sin 45°)}{2(\cos 135° + i \sin 135°)} \\
&= \frac{4}{2}(\cos (45° - 135°) + i \sin (45° - 135°)) \\
&= 2(\cos (-90°) + i \sin (-90°)) \\
&= 2(0 + -1i) \\
&= -2i
\end{aligned}$$

Now Try Exercise 35 ◀

De Moivre's Theorem

If a complex number z is expressed in trigonometric form, then z^n for any positive integer n can be computed easily. Let $z = r(\cos \theta + i \sin \theta)$ and consider the following.

$$\begin{aligned}
z^2 &= r(\cos \theta + i \sin \theta) \cdot r(\cos \theta + i \sin \theta) \\
&= r^2(\cos (\theta + \theta) + i \sin (\theta + \theta)) \\
&= r^2(\cos 2\theta + i \sin 2\theta) \\
z^3 &= zz^2 \\
&= r(\cos \theta + i \sin \theta) \cdot r^2(\cos 2\theta + i \sin 2\theta) \\
&= r^3(\cos 3\theta + i \sin 3\theta)
\end{aligned}$$

In general it can be shown that

$$z^n = r^n(\cos n\theta + i \sin n\theta).$$

This result is summarized in the following theorem, which is due to Abraham De Moivre (1667–1754), a French Huguenot who was a close friend of Isaac Newton. This theorem has become a keystone of analytic trigonometry.   (**Source:** H. Eves, *An Introduction to the History of Mathematics.*)

De Moivre's Theorem

Let $z = r(\cos \theta + i \sin \theta)$ and n be a positive integer. Then

$$z^n = r^n(\cos n\theta + i \sin n\theta).$$

EXAMPLE 4   Finding a power of a complex number

Use De Moivre's theorem to evaluate $(1 + i)^8$ and express the result in standard form.

SOLUTION From Example 1(a), the trigonometric form of $z = 1 + i$ is

$$z = \sqrt{2}\,(\cos 45° + i \sin 45°).$$

By De Moivre's theorem,

$$z^8 = (\sqrt{2})^8 (\cos (8 \cdot 45°) + i \sin (8 \cdot 45°))$$
$$= 16(\cos 360° + i \sin 360°)$$
$$= 16(1 + 0i)$$
$$= 16.$$

Now Try Exercise 47 ◄

Fractals and Complex Numbers   During the past 20 years, computer graphics and complex numbers have made it possible to produce many beautiful fractals. In 1977, Benoit B. Mandelbrot first used the term *fractal*. Largely because of his efforts, fractal geometry has become a new field of study. A fractal is an enchanting geometric figure with an endless self-similarity property, repeating itself infinitely with ever decreasing dimensions. If you look at smaller and smaller portions of the figure, you will continue to see the whole—much like when you look into two parallel mirrors that are facing each other. Not only do fractals have aesthetic appeal; they also have applications in science. An example of a fractal is the *Mandelbrot set* shown in Figure 8.112.   (**Source:** B. Mandelbrot, *The Fractal Geometry of Nature*.)

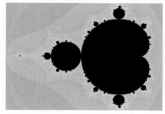

Figure 8.112

EXAMPLE 5   Analyzing the Mandelbrot set

The fractal called the Mandelbrot set is shown in Figure 8.112. To determine if a complex number $z = a + bi$ is in the Mandelbrot set, we can perform the following sequence of calculations. Let

$$z_0 = z$$
$$z_1 = z_0{}^2 + z_0$$
$$z_2 = z_1{}^2 + z_0$$
$$z_3 = z_2{}^2 + z_0$$

and so on. If the modulus of any z_k ever exceeds 2, then z is not in the Mandelbrot set; otherwise, z is in the Mandelbrot set. Determine if the complex number belongs to the Mandelbrot set.   (**Source:** F. Hill.)
(a) $z = 1 + i$ **(b)** $z = 0.5i$

SOLUTION
(a) Let $z_0 = 1 + i$. Then

$$z_1 = (1 + i)^2 + (1 + i) = 1 + 3i.$$

Since the modulus of z_1 is

$$|1 + 3i| = \sqrt{1^2 + 3^2} = \sqrt{10} > 2,$$

the complex number $1 + i$ is not in the Mandelbrot set.

```
(.5i)^2+(.5i)
          -.25+.5i
Ans^2+(.5i)
          -.1875+.25i
-.02734375+.406...
```

Figure 8.113

(b) Let $z_0 = 0.5i$. Then

$$z_1 = (0.5i)^2 + 0.5i = -0.25 + 0.5i$$

$$z_2 = (-0.25 + 0.5i)^2 + 0.5i = -0.1875 + 0.25i$$

$$z_3 = (-0.1875 + 0.25i)^2 + 0.5i \approx -0.0273 + 0.406i.$$

The modulus of each consecutive z_k never exceeds 2. Thus $0.5i$ is in the Mandelbrot set. You may find it helpful to use a calculator to perform these calculations. See Figure 8.113.

Now Try Exercises 63 and 65 ◀

Roots of Complex Numbers

A number w is the **nth root** of a number z if $w^n = z$. De Moivre's theorem can be used to find roots of complex numbers. To see this, let the trigonometric forms of w and z be

$$w = s(\cos \alpha + i \sin \alpha) \quad \text{and} \quad z = r(\cos \theta + i \sin \theta).$$

Then, by De Moivre's theorem, $w^n = z$ implies that

$$s^n(\cos n\alpha + i \sin n\alpha) = r(\cos \theta + i \sin \theta).$$

Thus $s^n = r$, or $s = \sqrt[n]{r}$. Furthermore the following two equations must be satisfied.

$$\cos n\alpha = \cos \theta \quad \text{and} \quad \sin n\alpha = \sin \theta$$

Since the cosine and sine functions have period 360°, $n\alpha = \theta + 360° \cdot k$ for some integer k, or

$$\alpha = \frac{\theta + 360° \cdot k}{n}.$$

Substituting these results in the trigonometric form for w gives

$$w_k = \sqrt[n]{r} \left(\cos \frac{\theta + 360° \cdot k}{n} + i \sin \frac{\theta + 360° \cdot k}{n} \right).$$

We obtain a unique value of w_k for $k = 0, 1, 2, \ldots, n - 1$. This discussion is summarized in the following box.

Roots of a Complex Number

Let $z = r(\cos \theta + i \sin \theta)$ be a nonzero complex number and n be any positive integer. Then z has exactly n distinct nth roots given by

$$w_k = \sqrt[n]{r} \left(\cos \frac{\theta + 360° \cdot k}{n} + i \sin \frac{\theta + 360° \cdot k}{n} \right),$$

where $k = 0, 1, 2, \ldots, n - 1$. If radian measure is used, then let

$$w_k = \sqrt[n]{r} \left(\cos \frac{\theta + 2\pi k}{n} + i \sin \frac{\theta + 2\pi k}{n} \right).$$

EXAMPLE 6 **Finding cube roots of a complex number**

Find the three cube roots of $8i$. Check your results with a calculator.

SOLUTION

Getting Started First write the complex number $8i$ in trigonometric form.

$$8i = 8(\cos 90° + i \sin 90°)$$

The three cube roots of $8i$ can be found by letting $n = 3, r = 8, \theta = 90°$, and $k = 0, 1, 2.$ ▶

$$w_0 = \sqrt[3]{8}\left(\cos\frac{90° + 360° \cdot \mathbf{0}}{3} + i\sin\frac{90° + 360° \cdot \mathbf{0}}{3}\right)$$

$$= 2(\cos 30° + i \sin 30°)$$

$$= 2\left(\frac{\sqrt{3}}{2} + \frac{1}{2}i\right)$$

$$= \sqrt{3} + i$$

$$w_1 = \sqrt[3]{8}\left(\cos\frac{90° + 360° \cdot \mathbf{1}}{3} + i\sin\frac{90° + 360° \cdot \mathbf{1}}{3}\right)$$

$$= 2(\cos 150° + i \sin 150°)$$

$$= 2\left(-\frac{\sqrt{3}}{2} + \frac{1}{2}i\right)$$

$$= -\sqrt{3} + i$$

$$w_2 = \sqrt[3]{8}\left(\cos\frac{90° + 360° \cdot \mathbf{2}}{3} + i\sin\frac{90° + 360° \cdot \mathbf{2}}{3}\right)$$

$$= 2(\cos 270° + i \sin 270°)$$

$$= 2(0 - i)$$

$$= -2i$$

The three cube roots of $8i$ are $\sqrt{3} + i$, $-\sqrt{3} + i$, and $-2i$. These can be checked using a calculator, as shown in Figure 8.114.

Now Try Exercise 59 ◀

```
(√(3)+i)^3
                8i
(⁻√(3)+i)^3
                8i
(⁻2i)^3
                8i
```

Figure 8.114

CLASS DISCUSSION

One cube root of the complex number $z = -1$ is $w = -1$. Find the other two cube roots of z graphically.

Graphical Interpretation If the three cube roots of $8i$ are plotted in the complex plane, they lie on a circle of radius 2, equally spaced 120° apart, as shown in Figure 8.115. In general, the nth roots of a complex number $z = r(\cos \theta + i \sin \theta)$ will lie equally spaced on a circle of radius $\sqrt[n]{r}$.

Figure 8.115 The Cube Roots of $8i$

EXAMPLE 7 Finding square roots of a complex number

Find the two square roots of $1 + i\sqrt{3}$.

SOLUTION First write the complex number $1 + i\sqrt{3}$ in trigonometric form.

$$1 + i\sqrt{3} = 2\left(\cos\frac{\pi}{3} + i\sin\frac{\pi}{3}\right)$$

The two square roots of $1 + i\sqrt{3}$ can be found by letting $n = 2, r = 2, \theta = \frac{\pi}{3}$, and $k = 0, 1$.

$$w_0 = \sqrt{2}\left(\cos\frac{\frac{\pi}{3} + 2\pi \cdot 0}{2} + i\sin\frac{\frac{\pi}{3} + 2\pi \cdot 0}{2}\right) \qquad w_1 = \sqrt{2}\left(\cos\frac{\frac{\pi}{3} + 2\pi \cdot 1}{2} + i\sin\frac{\frac{\pi}{3} + 2\pi \cdot 1}{2}\right)$$

$$= \sqrt{2}\left(\cos\frac{\pi}{6} + i\sin\frac{\pi}{6}\right) \qquad\qquad = \sqrt{2}\left(\cos\frac{7\pi}{6} + i\sin\frac{7\pi}{6}\right)$$

$$= \sqrt{2}\left(\frac{\sqrt{3}}{2} + \frac{1}{2}i\right) \qquad\qquad = \sqrt{2}\left(-\frac{\sqrt{3}}{2} - \frac{1}{2}i\right)$$

$$= \frac{\sqrt{6}}{2} + \frac{\sqrt{2}}{2}i \qquad\qquad = -\frac{\sqrt{6}}{2} - \frac{\sqrt{2}}{2}i$$

Thus the two square roots of $1 + i\sqrt{3}$ are $\frac{\sqrt{6}}{2} + \frac{\sqrt{2}}{2}i$ and $-\frac{\sqrt{6}}{2} - \frac{\sqrt{2}}{2}i$.

Now Try Exercise 55 ◀

8.6 Putting It All Together

The following table summarizes some of the important topics in this section.

Concept	Explanation	Examples
Trigonometric form	If $z = a + bi$, then its trigonometric form is $z = r(\cos\theta + i\sin\theta)$, where $r = \sqrt{a^2 + b^2}$ and $\tan\theta = \frac{b}{a}(a \neq 0)$. The modulus is r and the argument is θ.	If $z = 1 + i\sqrt{3}$, then $$r = \sqrt{1^2 + (\sqrt{3})^2} = 2$$ and $\theta = \tan^{-1}\frac{\sqrt{3}}{1} = 60°$. Thus $z = 2(\cos 60° + i\sin 60°)$.
Products and quotients	Let $z_1 = r_1(\cos\theta_1 + i\sin\theta_1)$ and $z_2 = r_2(\cos\theta_2 + i\sin\theta_2)$. Then $$z_1 z_2 = r_1 r_2(\cos(\theta_1 + \theta_2) + i\sin(\theta_1 + \theta_2))$$ and $$\frac{z_1}{z_2} = \frac{r_1}{r_2}(\cos(\theta_1 - \theta_2) + i\sin(\theta_1 - \theta_2)), r_2 \neq 0.$$	If $z_1 = 3(\cos 66° + i\sin 66°)$ and $z_2 = 2(\cos 22° + i\sin 22°)$, then $$z_1 z_2 = 6(\cos 88° + i\sin 88°)$$ and $$\frac{z_1}{z_2} = \frac{3}{2}(\cos 44° + i\sin 44°).$$

Concept	Explanation	Examples
De Moivre's theorem	Let $z = r(\cos \theta + i \sin \theta)$. Then $$z^n = r^n(\cos n\theta + i \sin n\theta).$$	If $z = 2(\cos 7° + i \sin 7°)$, then $$z^3 = 2^3(\cos 21° + i \sin 21°).$$
Roots of complex numbers	Let $z = r(\cos \theta + i \sin \theta)$ and n be any positive integer. Then the nth roots of z are given by $$w_k = \sqrt[n]{r}\left(\cos \frac{\theta + 360° \cdot k}{n} + i \sin \frac{\theta + 360° \cdot k}{n}\right),$$ where $k = 0, 1, 2, \ldots, n - 1$.	The three cube roots of $8 = 8(\cos 0° + i \sin 0°)$ are as follows: $w_0 = \sqrt[3]{8}\left(\cos \frac{0° + 360° \cdot 0}{3} + i \sin \frac{0° + 360° \cdot 0}{3}\right)$ $\quad = 2$ $w_1 = \sqrt[3]{8}\left(\cos \frac{0° + 360° \cdot 1}{3} + i \sin \frac{0° + 360° \cdot 1}{3}\right)$ $\quad = 2(\cos 120° + i \sin 120°) = -1 + i\sqrt{3}$ $w_2 = \sqrt[3]{8}\left(\cos \frac{0° + 360° \cdot 2}{3} + i \sin \frac{0° + 360° \cdot 2}{3}\right)$ $\quad = 2(\cos 240° + i \sin 240°) = -1 - i\sqrt{3}$

8.6   Exercises

The Complex Plane

Exercises 1–4: Plot the numbers in the complex plane.

1. (a) $3 + 2i$ (b) $-1 + i$ (c) $3i$

2. (a) $-2i$ (b) $2 + 2i$ (c) $2 - 2i$

3. (a) -3 (b) $4 - 2i$ (c) $-1 - 3i$

4. (a) $-1 - i$ (b) $4 + 3i$ (c) 4

Trigonometric Form

Exercises 5–12: Find the modulus of the number.

5. $1 + i$ 6. $3 - 4i$

7. $12 - 5i$ 8. $-24 + 7i$

9. -6 10. $15i$

11. $2 - 3i$ 12. $11 - 60i$

Exercises 13–22: Write the number in trigonometric form. Let $0° \le \theta < 360°$.

13. $-1 + i$ 14. $1 - i$

15. 5 16. -3

17. $4i$ 18. $-i$

19. $-1 + i\sqrt{3}$ 20. $-\sqrt{2} - i\sqrt{2}$

21. $\sqrt{3} + i$ 22. $-\frac{\sqrt{3}}{2} + \frac{1}{2}i$

Exercises 23–26: Write the number in trigonometric form. Let $0 \le \theta < 2\pi$.

23. -2 24. $4i$

25. $-2 + 2i$ 26. $1 + i\sqrt{3}$

Exercises 27–34: Write the number in standard form.

27. $5(\cos 180° + i \sin 180°)$

28. $3(\cos 90° + i \sin 90°)$ 29. $2(\cos 45° + i \sin 45°)$

30. $\cos 150° + i \sin 150°$ 31. $2\left(\cos \frac{\pi}{6} + i \sin \frac{\pi}{6}\right)$

32. $4\left(\cos \frac{3\pi}{2} + i \sin \frac{3\pi}{2}\right)$ **33.** $3(\cos 2\pi + i \sin 2\pi)$

34. $5\left(\cos \frac{3\pi}{4} + i \sin \frac{3\pi}{4}\right)$

Exercises 35–40: Find $z_1 z_2$ and $\frac{z_1}{z_2}$. Express your answer in standard form.

35. $z_1 = 9(\cos 45° + i \sin 45°)$,
$z_2 = 3(\cos 15° + i \sin 15°)$

36. $z_1 = 5(\cos 90° + i \sin 90°)$,
$z_2 = 2(\cos 30° + i \sin 30°)$

37. $z_1 = 6\left(\cos \frac{3\pi}{4} + i \sin \frac{3\pi}{4}\right)$,
$z_2 = \cos \frac{\pi}{4} + i \sin \frac{\pi}{4}$

38. $z_1 = 4(\cos 300° + i \sin 300°)$,
$z_2 = 2(\cos 60° + i \sin 60°)$

39. $z_1 = \cos 15° + i \sin 15°$,
$z_2 = \cos\left(-\frac{\pi}{4}\right) + i \sin\left(-\frac{\pi}{4}\right)$

40. $z_1 = 11\left(\cos \frac{2\pi}{3} + i \sin \frac{2\pi}{3}\right)$,
$z_2 = 22(\cos 30° + i \sin 30°)$

Powers of Complex Numbers

Exercises 41–46: Use De Moivre's theorem to evaluate the expression. Write the result in standard form.

41. $(2(\cos 30° + i \sin 30°))^3$

42. $(3(\cos 45° + i \sin 45°))^4$

43. $(\cos 10° + i \sin 10°)^{36}$

44. $(\cos 1° + i \sin 1°)^{90}$

45. $(5(\cos 60° + i \sin 60°))^2$

46. $(2(\cos 90° + i \sin 90°))^5$

Exercises 47–50: Use De Moivre's theorem to evaluate the expression. Write the result in standard form and check it using a calculator.

47. $(1 + i)^3$ **48.** $(3i)^4$

49. $(\sqrt{3} + i)^5$ **50.** $(2 - 2i)^6$

Roots of Complex Numbers

Exercises 51–62: Find the following roots and express them in standard form. Check your results with a calculator.

51. The square roots of $4(\cos 120° + i \sin 120°)$

52. The cube roots of $27(\cos 180° + i \sin 180°)$

53. The cube roots of $\cos 180° + i \sin 180°$

54. The fourth roots of $16(\cos 240° + i \sin 240°)$

55. The square roots of i

56. The cube roots of 1

57. The cube roots of -8

58. The square roots of $-4i$

59. The cube roots of $64i$

60. The fourth roots of -1

61. The fourth roots of 81

62. The square roots of $-1 + i\sqrt{3}$

Fractals

Exercises 63–66: Mandelbrot Set   (Refer to Example 5.) Determine if the complex number belongs to the Mandelbrot set.

63. $-0.4i$ **64.** $0.5 + i$

65. $1 + i$ **66.** $-0.2 + 0.2i$

Applications

67. *Electrical Circuits   Impedance* is a measure of the opposition to the flow of current in an electrical circuit. It consists of two parts called the *resistance* and the *reactance*. Light bulbs add resistance to an electrical circuit, and reactance occurs when electricity passes through coils of wire like those found in electric motors. Impedance Z in ohms (Ω) may be expressed as a complex number, where the real part represents the resistance and the imaginary part represents the reactance. For example, if the resistive part is 3 ohms and the reactive part is 4 ohms, then the impedance could be described by the complex number $Z = 3 + 4i$. The modulus of Z gives the total impedance in ohms. In a series circuit like the one shown in the figure at the top of the next page, the total impedance is the sum of the individual impedances.   (**Source:** R. Smith and R. Dorf, *Circuits, Devices and Systems.*)

(a) The circuit contains two light bulbs and two electric motors. If it is assumed that the light bulbs represent resistance and the motors represent reactance, express impedance as $Z = a + bi$.

(b) Find total impedance in ohms by calculating the modulus of Z.

68. *Electrical Circuits* (Continuation of Exercise 67.) In the parallel electrical circuit shown in the figure, impedance Z is given by

$$Z = \frac{1}{\frac{1}{Z_1} + \frac{1}{Z_2}},$$

where Z_1 and Z_2 represent the impedances for the two branches of the circuit. (**Source:** G. Wilcox and C. Hesselberth, *Electricity for Engineering Technology.*)

(a) Find Z.

(b) Find total impedance to the nearest tenth of an ohm by calculating the modulus of Z.

Writing about Mathematics

69. Explain how to find a trigonometric form of a complex number $a + bi$. Give an example. Is trigonometric form unique for a given complex number z? Explain.

70. Suppose that one fourth root w of a complex number z is known. Explain how to find the other fourth roots of z graphically.

EXTENDED AND DISCOVERY EXERCISES

Exercises 1–4: Making Connections Complete the following. Choose angles in $[0°, 360°)$ and round to the nearest tenth.

(a) Write the vector in terms of its magnitude and direction angle.
(b) Write the point (a, b) in polar coordinates.
(c) Write the complex number $a + bi$ in trigonometric form.

1. $\langle \sqrt{3}, 1 \rangle, (\sqrt{3}, 1), \sqrt{3} + i$

2. $\langle 5, 12 \rangle, (5, 12), 5 + 12i$

3. $\langle 4, -3 \rangle, (4, -3), 4 - 3i$

4. $\langle -7, 24 \rangle, (-7, 24), -7 + 24i$

CHECKING BASIC CONCEPTS FOR SECTIONS 8.5 AND 8.6

1. Plot the following points (r, θ) on a polar grid.
   (a) $(2, 30°)$ **(b)** $(3, -60°)$ **(c)** $(-4, 120°)$

2. Graph the equation.
   (a) $r = 2$ **(b)** $\theta = -\frac{\pi}{4}$
   (c) $r = 3 + 3\cos\theta$ **(d)** $r = 3\cos 2\theta$

3. Plot the numbers in the complex plane.
   (a) $-3 + 2i$ **(b)** $-4 - 3i$

4. Find the trigonometric form of $1 + i\sqrt{3}$.

5. Find the three cube roots of i.

8   Summary

CONCEPT	EXPLANATION AND EXAMPLES

SECTION 8.1   LAW OF SINES

Law of Sines

$$\frac{\sin \alpha}{a} = \frac{\sin \beta}{b} = \frac{\sin \gamma}{c}, \quad \text{or} \quad \frac{a}{\sin \alpha} = \frac{b}{\sin \beta} = \frac{c}{\sin \gamma}$$

The law of sines can be used to solve triangles given ASA, AAS, or SSA. SSA is called the ambiguous case and can have zero, one, or two solutions.

Example: Given $\beta = 32°$, $\gamma = 46°$, and $c = 10$, find b.
We are given AAS, as illustrated in the figure.

$$\frac{b}{\sin \beta} = \frac{c}{\sin \gamma} \quad \text{implies that} \quad b = \frac{10 \sin 32°}{\sin 46°} \approx 7.37.$$

SECTION 8.2   LAW OF COSINES

Law of Cosines

$$a^2 = b^2 + c^2 - 2bc \cos \alpha$$
$$b^2 = a^2 + c^2 - 2ac \cos \beta$$
$$c^2 = a^2 + b^2 - 2ab \cos \gamma$$

The law of cosines can be used to solve triangles given SAS or SSS. Each situation results in a unique solution.

Example: Given $a = 5$, $b = 6$, and $c = 7$, find α.
We are given SSS, as illustrated in the figure.
$a^2 = b^2 + c^2 - 2bc \cos \alpha$ implies that

$$\cos \alpha = \frac{b^2 + c^2 - a^2}{2bc} = \frac{6^2 + 7^2 - 5^2}{2(6)(7)} \approx 0.714.$$

Thus $\alpha \approx \cos^{-1}(0.714) \approx 44.4°$.

SECTION 8.3   VECTORS

Vectors

A vector is a directed line segment that has both magnitude (length) and direction. Three different representations for a vector are

$$\mathbf{v} = \langle a_1, a_2 \rangle, \quad \mathbf{v} = a_1 \mathbf{i} + a_2 \mathbf{j}, \quad \text{and} \quad \mathbf{v} = \overrightarrow{PQ}.$$

Horizontal component $= a_1$; vertical component $= a_2$

Magnitude: $\|\mathbf{v}\| = \sqrt{a_1^2 + a_2^2}$

Direction angle: The positive angle θ between the x-axis and the position vector $\mathbf{v}$, where

$$\mathbf{v} = \langle \|\mathbf{v}\| \cos \theta, \|\mathbf{v}\| \sin \theta \rangle$$

CONCEPT	EXPLANATION AND EXAMPLES

SECTION 8.3 VECTORS (CONTINUED)

Calculations Involving Vectors

Let $\mathbf{a} = \langle a_1, a_2 \rangle$ and $\mathbf{b} = \langle b_1, b_2 \rangle$.

Sum: $\mathbf{a} + \mathbf{b} = \langle a_1 + b_1, a_2 + b_2 \rangle$

Difference: $\mathbf{a} - \mathbf{b} = \langle a_1 - b_1, a_2 - b_2 \rangle$

Scalar Product: $k\mathbf{a} = \langle ka_1, ka_2 \rangle$

Dot Product: $\mathbf{a} \cdot \mathbf{b} = a_1 b_1 + a_2 b_2$

Angle θ between $\mathbf{a}$ and $\mathbf{b}$: $\theta = \cos^{-1}\left(\dfrac{\mathbf{a} \cdot \mathbf{b}}{\|\mathbf{a}\| \, \|\mathbf{b}\|} \right)$

Work

If a constant force $\mathbf{F}$ is applied to an object that moves along a vector $\mathbf{D}$, then the work done is $W = \mathbf{F} \cdot \mathbf{D}$.

SECTION 8.4 PARAMETRIC EQUATIONS

Plane Curve and Parametric Equations

Curve defined by the parametric equations $x = f(t)$ and $y = g(t)$, where f and g are continuous and t is the parameter

Example: $x = \cos t, y = \sin t; 0 \le t \le 2\pi$

Since $x^2 + y^2 = \cos^2 t + \sin^2 t = 1$, this curve describes the unit circle. See the figure.

SECTION 8.5 POLAR EQUATIONS

Polar Coordinates and Equations

Points are identified using r and θ instead of x and y. Polar equations are plotted in the polar plane, where the pole corresponds to the origin and the polar axis corresponds to the positive x-axis. Polar coordinates are not unique.

Example: $r = \cos 5\theta$ (rose curve with 5 leaves)

SECTION 8.6 TRIGONOMETRIC FORM AND ROOTS OF COMPLEX NUMBERS

Trigonometric Form and Complex Numbers

The expression $r(\cos \theta + i \sin \theta)$ is the trigonometric form of $a + bi$, where $a = r \cos \theta$ and $b = r \sin \theta$.

Modulus: $|z| = r = \sqrt{a^2 + b^2}$; argument: θ

Example: $z = 2(\cos 30° + i \sin 30°) = \sqrt{3} + i$

Modulus: $r = 2$; argument: $\theta = 30°$

CONCEPT	EXPLANATION AND EXAMPLES

SECTION 8.6 TRIGONOMETRIC FORM AND ROOTS OF COMPLEX NUMBERS (CONTINUED)

Operations on Complex Numbers and Trigonometric Form

Let $z_1 = r_1(\cos\theta_1 + i\sin\theta_1)$ and $z_2 = r_2(\cos\theta_2 + i\sin\theta_2)$.

$$z_1 z_2 = r_1 r_2(\cos(\theta_1 + \theta_2) + i\sin(\theta_1 + \theta_2))$$

$$\frac{z_1}{z_2} = \frac{r_1}{r_2}(\cos(\theta_1 - \theta_2) + i\sin(\theta_1 - \theta_2))$$

$$z_1^n = r_1^n(\cos(n\theta_1) + i\sin(n\theta_1)) \qquad \text{De Moivre's theorem}$$

Example: Let $z_1 = 6(\cos 120° + i\sin 120°)$ and $z_2 = 2(\cos 80° + i\sin 80°)$.

$$z_1 z_2 = 12(\cos 200° + i\sin 200°)$$

$$\frac{z_1}{z_2} = 3(\cos 40° + i\sin 40°)$$

$$z_1^4 = 6^4(\cos(4 \cdot 120°) + i\sin(4 \cdot 120°))$$

Roots of Complex Numbers

If $z = r(\cos\theta + i\sin\theta)$, then the nth roots of z are given by

$$w_k = \sqrt[n]{r}\left(\cos\frac{\theta + 360° \cdot k}{n} + i\sin\frac{\theta + 360° \cdot k}{n}\right),$$

where $k = 0, 1, 2, \ldots, n-1$.

Example: Let $z = 4i = 4(\cos 90° + i\sin 90°)$.
The square roots of z are as follows.

$$w_k = \sqrt{4}\left(\cos\frac{90° + 360° \cdot k}{2} + i\sin\frac{90° + 360° \cdot k}{2}\right) \quad \text{for } k = 0 \text{ and } 1$$

Simplifying gives

$$w_0 = 2(\cos 45° + i\sin 45°) = \sqrt{2} + i\sqrt{2} \quad \text{and}$$
$$w_1 = 2(\cos 225° + i\sin 225°) = -\sqrt{2} - i\sqrt{2}.$$

8   Review Exercises

Exercises 1–4: Solve the triangle. Approximate values to the nearest tenth.

1.

2.

3.

4.

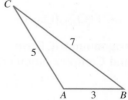

Exercises 5–10: Solve the triangle. Approximate values to the nearest tenth.

5. $\alpha = 19°, \beta = 46°, b = 13$

6. $\alpha = 30°, b = 10, a = 8$

7. $\gamma = 20°, b = 8, c = 11$

8. $\alpha = 70°, b = 17, a = 5$

9. $b = 23, \gamma = 35°, a = 18$

10. $a = 65, b = 45, c = 32$

Exercises 11–14: Approximate the area of each triangle to the nearest tenth.

11. $a = 12.3, b = 13.7, \gamma = 39°$

12. $\alpha = 40°, \beta = 55°, c = 67$

13. $a = 34, b = 67, c = 53$

14. $a = 2.1, b = 1.7, c = 2.2$

Exercises 15 and 16: Complete the following for vector **v***.*

  (a) Give the horizontal and vertical components.
  (b) Find $\|\mathbf{v}\|$*.*
  (c) Graph **v** *and interpret* $\|\mathbf{v}\|$*.*

15. $\mathbf{v} = \langle 3, 4 \rangle$ **16.** $\mathbf{v} = -5\mathbf{i} + 12\mathbf{j}$

Exercises 17 and 18: A vector **v** *has initial point P and terminal point Q.*

  (a) Graph $\overrightarrow{PQ}$*.*
  (b) Write $\overrightarrow{PQ}$ *as* $\mathbf{v} = a_1\mathbf{i} + a_2\mathbf{j}$*.*
  (c) Find $\|\overrightarrow{PQ}\|$*.*

17. $P = (0, 0), Q = (-2, -4)$

18. $P = (3, 2), Q = (-3, -1)$

Exercises 19–22: Find each of the following.

  (a) 2a   (b) a − 3b   (c) a · b
  (d) The angle θ *between* **a** *and* **b** *rounded to a tenth of a degree*

19. $\mathbf{a} = \langle 3, -2 \rangle, \mathbf{b} = \langle 1, 1 \rangle$

20. $\mathbf{a} = \langle 3, 2 \rangle, \mathbf{b} = \langle -2, -3 \rangle$

21. $\mathbf{a} = 2\mathbf{i} + 2\mathbf{j}, \mathbf{b} = \mathbf{i} + \mathbf{j}$

22. $\mathbf{a} = \mathbf{i} - 2\mathbf{j}, \mathbf{b} = 2\mathbf{i} + \mathbf{j}$

23. *Resultant Force*   Use the parallelogram rule to find the magnitude of the resultant force of the two forces shown in the figure.

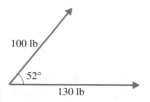

24. *Work*   Find the work done when a constant force $\mathbf{F} = 300\mathbf{i} + 400\mathbf{j}$ is applied to an object that moves along the vector $\mathbf{D} = 10\mathbf{i} - 2\mathbf{j}$, where units are in pounds and feet. Approximate the magnitude of $\mathbf{F}$ and interpret the result.

Exercises 25–27: Graph the parametric equations.

25. $x = t + 2, \qquad y = t^2 - 3; \qquad -2 \le t \le 2$

26. $x = t^3 - 4, \qquad y = t - 1; \qquad 0 \le t \le 2$

27. $x = 2 \cos t, \qquad y = -2 \sin t; \qquad 0 \le t \le 2\pi$

28. Change the polar coordinates (r, θ) to rectangular coordinates (x, y).
  (a) $(2, 135°)$ **(b)** $(-1, 60°)$

Exercises 29–34: Graph each polar equation.

29. $r = 1 + \cos \theta$ **30.** $r = \sin \theta$

31. $r = 3 \sin 3\theta$ **32.** $r = 2 - \cos \theta$

33. $r = 3 + 3 \sin \theta$ **34.** $r = 1 - 2 \sin \theta$

35. Plot the numbers in the complex plane.
  (a) $4 - i$ **(b)** $-2 + 2i$

  (c) $-2i$ **(d)** -4

36. Write the complex number in trigonometric form. Let θ satisfy $0° \le \theta < 360°$.
  (a) $-2 + 2i$ **(b)** $\sqrt{3} + i$

  (c) $5i$ **(d)** -6

37. Find $z_1 z_2$ and $\dfrac{z_1}{z_2}$ in standard form, if

$$z_1 = 4(\cos 150° + i \sin 150°) \text{ and}$$
$$z_2 = 2(\cos 30° + i \sin 30°).$$

38. Use De Moivre's theorem to evaluate z^4 if the trigonometric form of z is $z = 2(\cos 45° + i \sin 45°)$. Write the result in standard form.

Exercises 39 and 40: Find the following roots. Check your results with a calculator.

39. The square roots of $4(\cos 60° + i \sin 60°)$

40. The cube roots of $27i$

Applications

41. *Airplane Navigation* An airplane takes off with a bearing of 130° and flies 350 miles. Then it changes its course to a bearing of 60° and flies for 500 miles. Determine how far the plane is from its takeoff point.

42. *Obstructed View* To find the distance between two points A and B on opposite sides of a small building, a surveyor measures AC as 63.15 feet, angle ACB as 43.56°, and CB as 103.53 feet. Find the distance between A and B to the nearest tenth of a foot.

43. *Height of an Airplane* Two observation points A and B are 950 feet apart. From these points the angles of elevation of an airplane are 52° and 57°, as illustrated in the figure. Find the height of the airplane to the nearest foot.

44. *Area of a Lot* A surveyor measures two sides and the included angle of a triangular lot as $a = 93.6$ feet, $b = 110.6$ feet, and $\gamma = 51.8°$. Find the area of the lot.

45. *Area of a Lot* Find the area of the quadrangular lot shown in the figure to the nearest square foot.

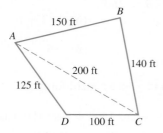

46. *Interpreting a Vector* A boat is heading west at 20 miles per hour in a current that is flowing south at 6 miles per hour. Find a vector $\mathbf{v}$ that models the direction and speed of the boat. What does $\|\mathbf{v}\|$ represent?

47. *Robotics* Consider the planar two-arm manipulator shown in the figure, where units are in centimeters. Let the upper arm be modeled by the vector $\mathbf{a} = 40\mathbf{i} - 20\mathbf{j}$ and the forearm be modeled by the vector $\mathbf{b} = 20\mathbf{i} + 30\mathbf{j}$. (**Source:** J. Craig, *Introduction to Robotics.*)

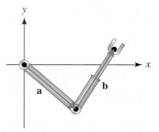

(a) Find a vector $\mathbf{c}$ that gives the position of the hand.

(b) How far is the hand from the origin?

(c) Find the position of the hand if the length of the forearm doubles.

48. *Work* A 200-pound person walks 0.75 mile up a hiking trail inclined at 15°. Use a dot product to compute the work done in foot-pounds.

49. *Flight of a Golf Ball* A golf ball is hit at 50 feet per second, making an angle of 45° with the horizontal as it leaves the club. If the ground is level, estimate the horizontal distance traveled by the golf ball in the air.

50. *Aerial Photography* A camera lens has an angular coverage of 86°. Suppose an aerial photograph is taken vertically with no tilt at an altitude of 3500 feet over ground with an increasing slope of 5°, as shown in the figure. Calculate the ground distance CB that would appear in the resulting photograph (to the nearest foot). (**Source:** F. Moffitt.)

EXTENDED AND DISCOVERY EXERCISES

1. *Velocity of a Star* The velocity vector $\mathbf{v}$ of a star relative to the sun can be expressed as the resultant vector of two perpendicular vectors—the radial velocity $\mathbf{v}_r$ and the tangential velocity $\mathbf{v}_t$, where $\mathbf{v} = \mathbf{v}_r + \mathbf{v}_t$, as illustrated in the figure on the next page. If a star is located near the sun and its velocity is large, then its motion across the sky will also be large. Barnard's Star is relatively close to the sun with a distance of 35 trillion miles. It moves across the sky through an angle of 10.34″ per year, which is the largest of any known star. Its radial velocity is $v_r = 67$ miles per second toward the sun. (**Sources:** A. Acker and C. Jaschek, *Astronomical Methods and Calculations;* M. Zeilik, *Introductory Astronomy and Astrophysics.*)

(a) Approximate $\|\mathbf{v}_t\|$ for Barnard's Star in miles per second. (*Hint:* Use $s = r\theta$.)

(b) Compute $\|\mathbf{v}\|$.

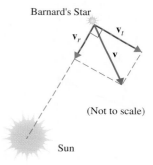

Barnard's Star

(Not to scale)

Sun

2. *Fractals*   The fractal called the *Julia set* is shown in the figure. To determine if a complex number $z = a + bi$ belongs to this set, repeatedly compute the sequence of values

$$z_1 = z^2 - 1, \quad z_2 = z_1^2 - 1, \quad z_3 = z_2^2 - 1,$$

and so on. If the modulus of any of the resulting complex numbers exceeds 2, then the complex number z is not in the Julia set. Otherwise z is in this set. Determine if the complex numbers belong to the Julia set. (**Source:** R. Crownover, *Introduction to Fractals and Chaos. Reprinted with permission.*)

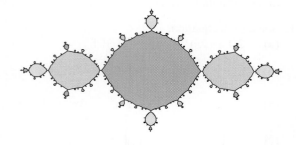

(a) $z = 0 + 0i$ (b) $z = 1 + i$

(c) $z = -0.2i$

3. *Aerial Photography*   Aerial photography from satellites and planes has become important to many applications such as map-making, national security, and surveying. If a photograph is taken from a plane with the camera tilted at an angle θ, then trigonometry can be used to find the ground coordinates of the object, as illustrated in the figure. If an object's photographic coordinates in inches are (x, y), then its ground coordinates (X, Y) in feet can be computed using the formulas

$$X = \frac{ax}{f \sec \theta - y \sin \theta},$$

$$Y = \frac{ay \cos \theta}{f \sec \theta - y \sin \theta},$$

where f is the focal length of the camera in inches and a is the altitude of the airplane in feet. Suppose the photographic coordinates of a house and nearby forest fire are $(x_H, y_H) = (0.9, 3.5)$ and $(x_F, y_F) = (2.1, -2.4)$, respectively. (**Source:** F. Moffitt, *Photogrammetry.*)

(a) Find the distance between the house and the fire on the photograph to the nearest hundredth of an inch.

(b) If the photograph was taken at 7400 feet by a camera with a focal length of 6 inches and a tilt of $\theta = 4.1°$, find the ground distance in feet between the house and the fire.

4. *Shadows in Computer Graphics*   Vectors are used frequently in computer graphics to simulate realistic shadows. For example, suppose an airplane is taking off from a runway, as illustrated in the figure on the next page. Let the length and direction of the airplane at takeoff be given by vector **L**. If the sunlight is assumed to be perpendicular to the runway, then the length of the airplane's shadow cast on the runway equals $\|\mathbf{L}\| \cos \theta$. From previous work, we know that if vector **R** points in the direction of the runway, then

$$\mathbf{L} \cdot \mathbf{R} = \|\mathbf{L}\| \|\mathbf{R}\| \cos \theta.$$

Solving for $\|\mathbf{L}\| \cos \theta$ results in

$$\|\mathbf{L}\| \cos \theta = \frac{\mathbf{L} \cdot \mathbf{R}}{\|\mathbf{R}\|}.$$

The expression $\frac{\mathbf{L} \cdot \mathbf{R}}{\|\mathbf{R}\|}$ represents the **component of L in the direction of R**. Find the length of the shadow on the runway for each **L** and **R**. Assume units are in feet. (**Source:** C. Pokorny and C. Gerald, *Computer Graphics.*)

(a) $\mathbf{L} = 40\mathbf{i} + 10\mathbf{j},$ $\mathbf{R} = \mathbf{i}$

(b) $\mathbf{L} = 35\mathbf{i} + 5\mathbf{j},$ $\mathbf{R} = 10\mathbf{i} + \mathbf{j}$

(c) $\mathbf{L} = 100\mathbf{i} + 8\mathbf{j},$ $\mathbf{R} = 30\mathbf{i} + 2\mathbf{j}$

5. *The Dot Product*   In the figure,

$$\mathbf{a} = \langle a_1, a_2 \rangle, \qquad \mathbf{b} = \langle b_1, b_2 \rangle, \qquad \text{and}$$
$$\mathbf{a} - \mathbf{b} = \langle a_1 - b_1, a_2 - b_2 \rangle.$$

Apply the law of cosines to the triangle and derive the equation $\mathbf{a} \cdot \mathbf{b} = \|\mathbf{a}\| \|\mathbf{b}\| \cos \theta.$

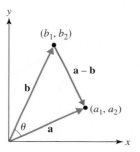

1–8   Cumulative Review Exercises

1. Write 91,200 in scientific notation and 6.734×10^{-3} in standard notation.

2. Evaluate $\sqrt{\sin^2 (1.2) + 5}.$ Round your answer to three decimal places.

3. Find the exact distance between $(3, -2)$ and $(7, -9)$.

4. Graph $y = g(x)$ by hand.

(a) $g(x) = \frac{1}{2}x - 1$ (b) $g(x) = |x + 2|$

(c) $g(x) = \frac{1}{2}(x + 1)^2 - 2$

(d) $g(x) = \sqrt{x + 1}$ (e) $g(x) = \dfrac{1}{x + 1}$

(f) $g(x) = -3 \cos \left(2 \left(x - \frac{\pi}{2} \right) \right)$

(g) $g(x) = \tan \frac{1}{2}x$ (h) $g(x) = \sec x$

5. Find the domain of $f(x) = \sqrt{4 - x}$ and evaluate $f(-5)$.

6. Find the average rate of change of $f(x) = 3x^2 - 2x$ from $x = 1$ to $x = 3$.

7. Find the difference quotient for $f(x) = 4x^2$.

8. Write the slope-intercept form for a line that passes through $(-1, 4)$ and $(1, -3)$.

9. Determine the x- and y-intercepts on the graph of $4x + 3y = -12$. Graph the equation.

10. If $G(t) = 300 - 10t$ calculates the gallons of water in a tank after t seconds, interpret the numbers 300 and -10 in the formula for $G(t)$.

11. Solve each equation.

(a) $-2(1 - 2x) = -(5 - x) + 1$

(b) $|2x - 5| = 6$ (c) $6x^2 + 22x = 8$

(d) $x^3 = x$ (e) $x^4 - 2x^2 - 3 = 0$

(f) $\dfrac{1}{2x} - \dfrac{2}{x^2} = -7$ (g) $3x^{2/3} = 12$

(h) $2e^{3x} - 1 = 50$ (i) $\ln x - \ln (x - 1) = \ln 2$

(j) $\sin t = \frac{1}{2}, 0 \le t < 2\pi$

(k) $\tan 2t = -\sqrt{3}$ **(l)** $2\cos^2 t + \cos t = 1$

12. Evaluate $f(-1)$ and graph $y = f(x)$. Is f continuous on its domain?

$$f(x) = \begin{cases} x + 2 & \text{if } -3 \le x \le -1 \\ -2x - 1 & \text{if } -1 < x < 1 \\ x^2 - 4 & \text{if } 1 \le x \le 3 \end{cases}$$

13. Solve each inequality. Use interval notation.
   (a) $4(x - 3) > 1 - x$ **(b)** $|2x - 1| \le 3$

   (c) $x^2 - 2x - 3 > 0$ **(d)** $x^3 - 4x > 0$

   (e) $\dfrac{x}{x - 1} \le 0$ **(f)** $-4 \le 4 - 3x \le 12$

14. The graph of a cubic polynomial function f is shown. Solve each equation or inequality. Write the solution set for each inequality in interval notation.
   (a) $f(x) = 0$ **(b)** $f(x) > 0$ **(c)** $f(x) \le 0$

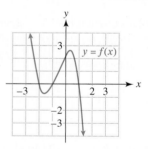

15. Write $f(x) = \frac{1}{2}x^2 + 2x - 1$ in the vertex form given by $f(x) = a(x - h)^2 + k$.

16. Find the vertex on the graph of $f(x) = 2x^2 - 4x + 1$.

17. Let $f(x) = x^2 + 2x - 3$. If the graph of $y = g(x)$ is similar to the graph of $y = f(x)$ except that it is shifted 2 units left and 3 units down, write a formula for $g(x)$. Simplify your formula.

18. Use the graph of f at the top of the next column to estimate each of the following.
   (a) Where f is increasing or decreasing

   (b) The zeros of f

   (c) The coordinates of any turning points

   (d) Any local extrema

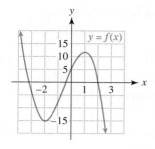

19. Determine if the function defined by the formula $f(x) = x^3 - 2x + 1$ is even, odd, or neither.

20. Predict the end behavior of $f(x) = 4 + 3x + x^2 - 4x^4$.

21. Divide each expression.
   (a) $\dfrac{3x^3 - x + 2}{x + 2}$ **(b)** $\dfrac{2x^3 - 3x^2 + x - 1}{2x - 1}$

22. A cubic function f has zeros $-2, 3$, and 5 and leading coefficient 4. Write the complete factored form of $f(x)$.

23. A quartic (degree 4) function f with real coefficients has leading coefficient 6 and zeros $1, -1$, and i. Write $f(x)$ in complete factored form and expanded form.

24. Write $(2 + 3i)(4 - 2i)$ in standard form.

25. State the domain of $f(x) = \frac{3x + 4}{2 - 3x}$. Find any vertical or horizontal asymptotes on the graph of $y = f(x)$.

26. Let $f(x) = \frac{1}{x^2 - 3}$ and $g(x) = 2x + 1$. Find each of the following.
   (a) $(f + g)(2)$ **(b)** $(g \circ f)(2)$

   (c) $(g/f)(x)$ **(d)** $(f \circ g)(x)$

27. Find $f^{-1}(x)$ if $f(x) = 3x - 2$.

28. Find an exponential function given by $f(x) = Ca^x$ that models the data in the table.

x	0	1	2	3
$f(x)$	2	6	18	54

29. There are initially 5000 bacteria, and this number doubles in size every 1.5 hours. Find C and a so that $f(t) = Ca^t$ models the number of bacteria after t hours.

30. One thousand dollars is deposited in an account that pays 7% annual interest compounded quarterly. Find the amount in the account after 8 years.

31. Simplify each logarithm by hand.

(a) $\log_3 \frac{1}{27}$ (b) $\ln \frac{1}{e^3}$

(c) $\log \sqrt[3]{10}$ (d) $\log_4 32 - \log_4 \frac{1}{2}$

32. Expand the expression $\ln \sqrt[3]{\frac{x^3 y}{z^2}}$.

33. Write $2 \log_2 x + 3 \log_2 y + \frac{1}{2} \log_2 z$ as a logarithm of a single expression.

34. Convert $75.25°$ to degrees and minutes.

35. Convert $225°$ to radians.

36. Convert $\frac{11\pi}{6}$ radians to degrees.

37. Find the six trigonometric functions of θ.

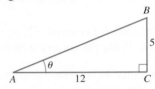

38. Approximate $\cos \frac{\pi}{7}$ to the nearest hundredth.

39. Find the values of the six trigonometric functions of an angle θ in standard position with its terminal side passing through the point $(11, -60)$.

40. Find the exact values of the six trigonometric functions of $\theta = \frac{2\pi}{3}$.

41. Find the values of the six trigonometric functions of θ if $\sin \theta = -\frac{7}{25}$ and $\sec \theta < 0$.

42. Evaluate $\tan^{-1} \sqrt{3}$.

43. Solve the right triangle shown in Exercise 37.

44. Simplify $(1 - \cos t)(1 + \cos t)$.

45. Factor $\cot^2 \theta - 2 \cot \theta + 1$.

46. Verify each identity.

(a) $(1 - \cos^2 \theta)(1 + \tan^2 \theta) = \tan^2 \theta$

(b) $\dfrac{\sin(\alpha + \beta)}{\cos \alpha \cos \beta} = \tan \alpha + \tan \beta$

(c) $\dfrac{\csc^2 \theta}{1 - \cot^2 \theta} = -\sec 2\theta$

47. Find $\cos(\alpha - \beta)$ if $\sin \alpha = \frac{5}{13}$, $\cos \beta = -\frac{4}{5}$, and α and β are in quadrant II.

48. Solve $3 \tan \frac{t}{2} = \sin 2t$ graphically to the nearest hundredth on the interval $[0, 2\pi)$.

49. Solve triangle ABC. Approximate to the nearest tenth.

(a) $\alpha = 31°, \gamma = 53°, b = 15$

(b) $\alpha = 31°, a = 6, b = 5$

(c) $\beta = 56°, a = 6, c = 8$

(d) $a = 6, b = 7, c = 8$

50. Find the area of a triangle with sides of length 7, 10, and 15 feet.

51. Let $\mathbf{a} = \langle -5, 12 \rangle$ and $\mathbf{b} = \langle 7, -24 \rangle$. Find the following.

(a) $\|\mathbf{b}\|$ (b) $2\mathbf{a} - 3\mathbf{b}$ (c) $\mathbf{a} \cdot \mathbf{b}$

(d) The angle between $\mathbf{a}$ and $\mathbf{b}$

52. Graph the parametric equations $x = \frac{1}{2}t$, $y = (t - 1)^2$ for any real number t.

53. Graph the polar equation $r = 3 - 2 \sin \theta$ for θ satisfying $0° \le \theta < 360°$.

54. Find the three cube roots of $27i$.

Applications

55. *Average Rate of Change*   The total distance D in feet traveled by a race car after t seconds is given by $D(t) = 12t^2$ for $0 \le t \le 7$.

(a) Find the average rate of change of D from 1 to 3 and 4 to 6.

(b) Interpret these average rates of change.

56. *Mixing Candy*   A 5-pound mixture of two types of candy costs \$21.25. The two types of candy cost \$3.50 per pound and \$4.50 per pound. How much of each type of candy is in the mixture?

57. *Construction*   A box is being constructed by cutting 3-inch squares from the corners of a rectangular sheet of metal that is 4 inches longer than it is wide. If the box is to have a volume of 351 cubic inches, find the dimensions of the metal sheet.

58. *Modeling Data*   Find a quadratic function in the form $f(x) = a(x - h)^2 + k$ that models the data in the table *exactly*.

x	−1	0	1	2	3	4
y	−26	−11	−2	1	−2	−11

59. *Designing a Box*   A box with rectangular sides and a top is being designed to hold 288 cubic inches and to have a surface area of 288 square inches. If the width is half the length, find possible dimensions for the box.

60. *Inverse Variation*   The force of gravity F varies inversely as the square of the distance r from the *center* of the moon. If a rock weighs 50 pounds on the surface of the moon ($r = 1750$ kilometers), how much would this rock weigh at a distance of 7000 kilometers from the center of the moon?

61. *College Tuition*   In 1980 the cost of tuition and fees at private colleges and universities was about $3600, and in 2000 it was about $16,200.   (**Source:** The College Board.)
   (**a**) Determine a formula $C(t) = m(t - t_1) + y_1$ that models these data. Let C be in dollars and t be the year.

   (**b**) Estimate the year when tuition and fees reached $11,000, and compare your estimate to the actual year of 1993.

62. *Bicycle*   A bicycle tire is 2 feet in diameter and rotating at 25 radians per second. Approximate the speed of the bicycle in feet per second.

63. *Length of a Shadow*   The angle of elevation of the sun is 63°. Find the length of the shadow cast by a person who is 5 feet tall. Round to the nearest tenth of a foot.

64. *Modeling Temperature*   The monthly average high temperatures for a location are shown in the table. Model these data using $f(x) = a \sin(b(x - c)) + d$.

Month	1	2	3	4	5	6
Temperature (°F)	25	28	37	50	63	72

Month	7	8	9	10	11	12
Temperature (°F)	75	72	62	50	38	28

65. *Angle of Elevation*   An 85-foot tree casts a 57-foot shadow. Estimate the angle of elevation of the sun to the nearest tenth of a degree.

66. *Distance*   An ore ship is traveling east at 20 miles per hour. The bearing of a submerged rock is 75°. After 2 hours, the bearing of the rock is 305°. Find the distance between the ship and the rock when the second bearing is determined.

67. *Surveyor*   A surveyor measures two sides of a triangular lot to be $a = 242$ feet and $b = 165$ feet. The angle between these sides is $\gamma = 72°$.
   (**a**) Find the length of the third side c.

   (**b**) Estimate the area of the lot.

68. *Work*   A wagon is pulled 1000 feet using a force of 15 pounds applied to the handle, which makes a 35° angle with the horizontal. Use a dot product to calculate the work done.

69. *Flight of a Golf Ball*   A golf ball is hit into the air at 96 feet per second, making an angle of 60° with the horizontal. Use parametric equations to estimate the horizontal distance traveled by the golf ball before it strikes the ground.

9 Systems of Equations and Inequalities

Go deep enough into anything and you will find mathematics.

—Dean Schlicter

A recent breakthrough in visualizing our world is the invention of digital images. One of the first digital pictures was created by James Blinn of NASA in 1981. Digital images are represented by numbers rather than film, and so they can be transmitted for millions of miles without loss of clarity. The first consumer-oriented digital cameras were sold in 1994. Starting in 2009, TV stations are required by the Federal Communications Commission (FCC) to transmit digital signals.

A digital image is essentially a long string of numbers that can be placed in an array, or a matrix. In order to compress the large files involved with digital pictures and movies, computer scientists use mathematics to develop clever compression formulas, which make it possible to place songs and movies on CDs and DVDs.

Mathematics plays a central role in new technology. In this chapter mathematics is used to solve systems of equations, represent digital pictures, compute movement in computer graphics, and even represent color on computer monitors. Throughout history, many important discoveries and inventions have been based on the creative insights of a few people. These insights have had a profound impact on our society.

Source: A. Tucker, *Fundamentals of Computing I; USA Today; NASA/JPL.*

9.1   Functions and Systems of Equations in Two Variables

* **Evaluate functions of two variables**
* **Understand basic concepts about systems of equations**
* **Recognize types of linear systems**
* **Apply the method of substitution**
* **Apply the elimination method**
* **Apply graphical and numerical methods**
* **Solve problems involving joint variation**

Introduction

Many quantities in everyday life depend on more than one variable. Calculating the area of a rectangular room requires both its length and its width. The heat index is a function of temperature and humidity, and windchill is determined using temperature and wind speed. Information about grades and credit hours is necessary to compute a grade point average.

In earlier chapters we discussed functions of one input or one variable. Quantities determined by more than one variable often are computed by a function of more than one variable. The mathematical concepts concerning functions of one input also apply to functions of more than one input. One unifying concept about every function is that it produces *at most one output* each time it is evaluated.

Functions of Two Variables

The arithmetic operations of addition, subtraction, multiplication, and division are computed by functions of two inputs. In order to perform addition, two numbers must be provided. The addition of x and y results in one output, z. The addition function f can be represented symbolically by $f(x, y) = x + y$, where $z = f(x, y)$. For example, the addition of 3 and 4 can be written as $z = f(3, 4) = 3 + 4 = 7$. In this case, $f(x, y)$ is a **function of two inputs** or a **function of two variables**. The **independent variables** are x and y, and z is the **dependent variable**. The output z depends on the inputs x and y. In a similar manner, a division function can be defined by $g(x, y) = \frac{x}{y}$, where $z = g(x, y)$.

EXAMPLE 1   Evaluating functions of more than one input

For each function, evaluate the expression and interpret the result.
(a) $f(3, -4)$, where $f(x, y) = xy$ represents the multiplication function
(b) $M(120, 5)$, where $M(m, g) = \frac{m}{g}$ computes the gas mileage when traveling m miles on g gallons of gasoline
(c) $V(0.5, 2)$, where $V(r, h) = \pi r^2 h$ calculates the volume of a cylindrical barrel with radius r feet and height h feet. (See Figure 9.1.)

SOLUTION
(a) $f(3, -4) = (3)(-4) = -12$. The product of 3 and -4 is -12.
(b) $M(120, 5) = \frac{120}{5} = 24$. If a car travels 120 miles on 5 gallons of gasoline, its gas mileage is 24 miles per gallon.
(c) $V(0.5, 2) = \pi(0.5)^2(2) = 0.5\pi \approx 1.57$. If a barrel has a radius of 0.5 foot and a height of 2 feet, it holds about 1.57 cubic feet of liquid.   *Now Try Exercises 1, 3, and 5* ◀

Figure 9.1

Systems of Equations in Two Variables

A **linear equation in two variables** can be written in the form

$$ax + by = k,$$

where a, b, and k are constants and a and b are not equal to 0. Examples of linear equations in two variables include

$$2x - 3y = 4, \qquad -x - 5y = 0, \qquad \text{and} \qquad 5x - y = 10.$$

The average of two numbers x and y can be found with the formula $\frac{x+y}{2}$. To find two numbers whose average is 10, we solve the linear equation in two variables

$$\frac{x+y}{2} = 10.$$

(Note that $\frac{x+y}{2} = 10$ is a linear equation because it can be written as $\frac{1}{2}x + \frac{1}{2}y = 10$.) The numbers 5 and 15 average to 10, as do the numbers 0 and 20. In fact, any pair of numbers whose sum is 20 will satisfy this equation. There are *infinitely many solutions* to this linear equation. For a unique solution, another restriction, or constraint, must be placed on the variables x and y.

Many situations involving two variables result in the need to determine values for x and y that satisfy two equations. For example, suppose that we would like to find a pair of numbers whose average is 10 and whose difference is 2. The function $f(x, y) = \frac{x+y}{2}$ calculates the average of two numbers, and $g(x, y) = x - y$ computes their difference. The solution could be found by solving two linear equations.

$$f(x, y) = \frac{x+y}{2} = 10$$

$$g(x, y) = x - y = 2$$

This pair of equations is called a **system of linear equations** because we are solving more than one linear equation at once. A **solution** to a system of equations consists of an x-value *and* a y-value that satisfy *both* equations simultaneously. The set of all solutions is called the **solution set**. Using trial and error, we see that $x = 11$ and $y = 9$ satisfy both equations. This is the only solution and can be expressed as the *ordered pair* (**11, 9**).

Systems of equations that have at least one nonlinear equation are called **nonlinear systems of equations**. The following are two examples of nonlinear systems of equations:

$$\begin{array}{cc} x^2 - 5y = 8 \\ 2x + 3y = 3 \end{array} \quad \text{and} \quad \begin{array}{cc} x^2 + y^2 = 5 \\ \sqrt{x} - y = -2. \end{array}$$

Types of Linear Systems in Two Variables

Any system of linear equations in two variables can be written in the form

$$a_1x + b_1y = c_1$$
$$a_2x + b_2y = c_2,$$

where a_1, b_1, c_1, a_2, b_2, and c_2 are constants. The graph of this system consists of *two* lines in the xy-plane. The lines can intersect at a point, coincide, or be parallel, as illustrated in Figures 9.2–9.4. **Coincident lines** are identical lines and indicate that the two equations are equivalent and have the same graph.

Figure 9.2 Intersecting Lines   **Figure 9.3** Coincident Lines   **Figure 9.4** Parallel Lines

CLASS DISCUSSION

Explain why a system of linear equations in two variables cannot have two or three solutions.

A **consistent** system of linear equations has either one solution or infinitely many solutions. If the system has one solution, then the equations are **independent**, and they can be represented by lines that intersect at one point, as shown in Figure 9.2. If the system has infinitely many solutions, then the equations are **dependent**, and they can be represented by coincident lines, as illustrated in Figure 9.3. An **inconsistent** system has no solutions and can be represented by parallel lines, as shown in Figure 9.4.

This discussion is summarized in the following box.

Graphs of a System of Two Linear Equations in Two Variables

1. **One Solution:** The graphs of the two equations are distinct lines that intersect at one point. The system is *consistent*. There is one solution, which is given by the coordinates of the point of intersection. In this case the equations are *independent*.
2. **Infinitely Many Solutions:** The graphs of the two equations are the same line. The system is *consistent*. There are infinitely many solutions, and the equations are *dependent*.
3. **No Solutions:** The graphs of the two equations are distinct parallel lines. The system is *inconsistent*. There are no solutions.

EXAMPLE 2 Recognizing types of linear systems

Graph each system of equations and find any solutions. Identify the system as consistent or inconsistent. If the system is consistent, state whether the equations are dependent or independent.

(a) $\begin{aligned} x - y &= 2 \\ -x + y &= 1 \end{aligned}$ (b) $\begin{aligned} 4x - y &= 2 \\ x - 2y &= -3 \end{aligned}$ (c) $\begin{aligned} 2x - y &= 1 \\ -4x + 2y &= -2 \end{aligned}$

SOLUTION

Getting Started Start by solving each equation for y. Use the resulting slope-intercept form to graph each line. Determine any points of intersection. ▶

(a) Graph $y = x - 2$ and $y = x + 1$, as shown in Figure 9.5. Their graphs do not intersect, so there are no solutions. The system is inconsistent.
(b) Graph $y = 4x - 2$ and $y = \frac{1}{2}x + \frac{3}{2}$, as shown in Figure 9.6. Their graphs intersect at $(1, 2)$. There is one solution, so the equations are consistent and independent. Because graphical solutions can be approximate, we check this solution by substituting 1 for x and 2 for y in the given system.

$$4(1) - 2 = 2 \qquad \text{True}$$
$$(1) - 2(2) = -3 \qquad \text{True}$$

Because $(1, 2)$ satisfies *both* equations, it is the solution.

Figure 9.5　　　　　**Figure 9.6**

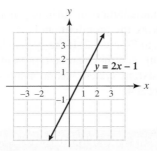

Figure 9.7

(c) Solving the equations $2x - y = 1$ and $-4x + 2y = -2$ for y results in the same equation: $y = 2x - 1$. Therefore their graphs coincide, as shown in Figure 9.7. (Note that the second equation results when the first equation is multiplied by -2, so the equations are equivalent.) Any point on this line is a solution to both equations. Thus the system has infinitely many solutions of the form $\{(x, y) \,|\, 2x - y = 1\}$ and is consistent. The equations are dependent.   *Now Try Exercises 35, 37, and 41* ◀

The Method of Substitution

The **method of substitution** is often used to solve systems of equations symbolically. It is summarized by the following steps.

The Method of Substitution

To use the method of substitution to solve a system of two equations in two variables, perform the following steps.

STEP 1: Choose a variable in one of the two equations. Solve the equation for that variable.

STEP 2: Substitute the result from Step 1 into the other equation and solve for the remaining variable.

STEP 3: Use the value of the variable from Step 2 to determine the value of the other variable. To do this, you may want to use the equation you found in Step 1.

To check your answer, substitute the value of each variable into the *given* equations. These values should satisfy *both* equations.

An Application   In the next example, we use the method of substitution to solve an example involving real data.

EXAMPLE 3   Applying the method of substitution

In fiscal 2007, Apple Corporation sold a combined total of 53.1 million iPods and iPhones. There were 50.3 million more iPods sold than iPhones.   (**Source:** Apple Corporation.)
(a) Write a system of equations whose solution gives the individual sales of iPods and of iPhones.
(b) Solve the system of equations. Interpret the results.
(c) Is your system consistent or inconsistent? If it is consistent, state whether the equations are dependent or independent.

SOLUTION

Getting Started When setting up a system of equations, it is important to identify what each variable represents. Then express the situation with equations. Finally, apply the method of substitution. ▶

(a) Let x be the number of iPods sold in millions and y be the number of iPhones sold in millions. The combined total is 53.1 million, so let $x + y = 53.1$. Because iPod sales exceeded iPhone sales by 50.3 million, let $x - y = 50.3$. Thus the system of equations is as follows.

$$x + y = 53.1 \qquad \text{Total sales are 53.1 million.}$$
$$x - y = 50.3 \qquad \text{iPod sales exceeded iPhone sales by 50.3 million.}$$

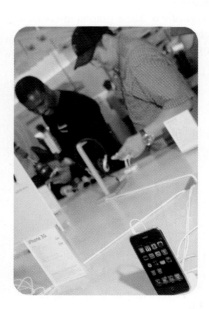

(b) STEP 1:   With this system, we can solve for either variable in either equation. If we solve for x in the second equation, we obtain the equation $x = y + 50.3$.

STEP 2:   Substitute $(y + 50.3)$ for x in the first equation, $x + y = 53.1$, and solve.

$$(y + 50.3) + y = 53.1 \qquad \text{Substitute } (y + 50.3) \text{ for } x.$$

$$2y = 2.8 \qquad \text{Subtract 50.3; combine terms.}$$

$$y = 1.4 \qquad \text{Divide by 2.}$$

STEP 3:   To find x, substitute 1.4 for y in the equation $x = y + 50.3$ from Step 1 to obtain $x = 1.4 + 50.3 = 51.7$. The solution is $(51.7, 1.4)$.

Thus in 2007, there were 51.7 million iPods and 1.4 million iPhones sold.

(c) There is one solution, so the system is consistent and the equations are independent.

Now Try Exercise 113 ◄

In the next example, we solve a system and check our result.

EXAMPLE 4   Using the method of substitution

Solve the system symbolically. Check your answer.

$$5x - 2y = -16$$
$$x + 4y = \quad -1$$

SOLUTION

STEP 1:   Begin by solving one of the equations for one of the variables. One possibility is to solve the second equation for x.

$$x + 4y = -1 \qquad \text{Second equation}$$

$$x = -4y - 1 \qquad \text{Subtract 4y from each side.}$$

STEP 2:   Next, substitute $(-4y - 1)$ for x in the first equation and solve the resulting equation for y.

$$5x - 2y = -16 \qquad \text{First equation}$$

$$5(-4y - 1) - 2y = -16 \qquad \text{Let } x = -4y - 1.$$

$$-20y - 5 - 2y = -16 \qquad \text{Distributive property}$$

$$-5 - 22y = -16 \qquad \text{Combine like terms.}$$

$$-22y = -11 \qquad \text{Add 5 to each side.}$$

$$y = \frac{1}{2} \qquad \text{Divide each side by } -22.$$

STEP 3:   Now find the value of x by using the equation $x = -4y - 1$ from Step 1. Since $y = \frac{1}{2}$, it follows that $x = -4\left(\frac{1}{2}\right) - 1 = -3$. The solution can be written as an ordered pair: $\left(-3, \frac{1}{2}\right)$.

Check: Substitute $x = -3$ and $y = \frac{1}{2}$ in both given equations.

$$5(-3) - 2\left(\frac{1}{2}\right) \stackrel{?}{=} -16 \qquad \text{True}$$

$$-3 + 4\left(\frac{1}{2}\right) \stackrel{?}{=} -1 \qquad \text{True}$$

Both equations are satisfied, so the solution is $\left(-3, \frac{1}{2}\right)$.   *Now Try Exercise 43* ◄

Nonlinear Systems of Equations The method of substitution can also be used to solve nonlinear systems of equations. In the next example, we solve a nonlinear system of equations having two solutions. In general, a nonlinear system of equations can have *any number of solutions*.

EXAMPLE 5 Solving a nonlinear system of equations

Solve the system symbolically.

$$6x + 2y = 10$$
$$2x^2 - 3y = 11$$

SOLUTION

STEP 1: Begin by solving one of the equations for one of the variables. One possibility is to solve the first equation for y.

$$6x + 2y = 10 \qquad \text{First equation}$$
$$2y = 10 - 6x \qquad \text{Subtract } 6x \text{ from each side.}$$
$$y = 5 - 3x \qquad \text{Divide each side by 2.}$$

STEP 2: Next, substitute $(5 - 3x)$ for y in the second equation and solve the resulting quadratic equation for x.

$$2x^2 - 3y = 11 \qquad \text{Second equation}$$
$$2x^2 - 3(5 - 3x) = 11 \qquad \text{Let } y = 5 - 3x.$$
$$2x^2 - 15 + 9x = 11 \qquad \text{Distributive property}$$
$$2x^2 + 9x - 26 = 0 \qquad \text{Subtract 11 and simplify.}$$
$$(2x + 13)(x - 2) = 0 \qquad \text{Factor.}$$
$$x = -\frac{13}{2} \quad \text{or} \quad x = 2 \qquad \text{Zero-product property}$$

STEP 3: Now find the corresponding y-values for each x-value. From Step 1 we know that $y = 5 - 3x$, so it follows that $y = 5 - 3\left(-\frac{13}{2}\right) = \frac{49}{2}$ or $y = 5 - 3(2) = -1$. Thus the solutions are $\left(-\frac{13}{2}, \frac{49}{2}\right)$ and $(2, -1)$. **Now Try Exercise 55** ◀

EXAMPLE 6 Solving a nonlinear system of equations

Use the method of substitution to determine the points where the line $y = 2x$ intersects the circle $x^2 + y^2 = 5$. Sketch a graph that illustrates the solutions.

SOLUTION Substitute $(2x)$ for y in the equation $x^2 + y^2 = 5$.

$$x^2 + y^2 = 5 \qquad \text{Second equation}$$
$$x^2 + (2x)^2 = 5 \qquad y = 2x$$
$$x^2 + 4x^2 = 5 \qquad \text{Square the expression.}$$
$$5x^2 = 5 \qquad \text{Add like terms.}$$
$$x^2 = 1 \qquad \text{Divide by 5.}$$
$$x = \pm 1 \qquad \text{Square root property}$$

Algebra Review
To review factoring, see Chapter R (page R-23).

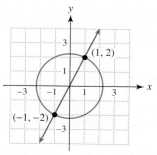

Figure 9.8

Since $y = 2x$ we see that when $x = 1, y = 2$ and when $x = -1, y = -2$. The graphs of $x^2 + y^2 = 5$ and $y = 2x$ intersect at the points $(1, 2)$ and $(-1, -2)$. This nonlinear system has two solutions, which are shown in Figure 9.8.

Now Try Exercise 99 ◀

EXAMPLE 7 Identifying a system with zero or infinitely many solutions

If possible, solve each system of equations.

(a) $x^2 + y = 1$ (b) $2x - 4y = 5$
    $x^2 - y = -2$ $-x + 2y = -\frac{5}{2}$

SOLUTION

(a) **STEP 1:** Solve the second equation for y, which gives $y = x^2 + 2$.

STEP 2: Substitute $(x^2 + 2)$ for y in the first equation and then solve for x, if possible.

$$x^2 + y = 1 \qquad \text{First equation}$$
$$x^2 + (x^2 + 2) = 1 \qquad \text{Let } y = x^2 + 2.$$
$$2x^2 + 2 = 1 \qquad \text{Combine like terms.}$$
$$2x^2 = -1 \qquad \text{Subtract 2.}$$

Because $2x^2 \geq 0$, it follows that there are no real solutions and the system is inconsistent.

STEP 3: This step is not necessary because there are no real solutions for x. Figure 9.9 shows that the graphs, which are parabolas, do not intersect.

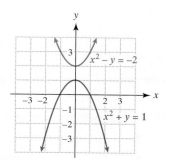

Figure 9.9

(b) **STEP 1:** First solve the second equation for x to obtain $x = 2y + \frac{5}{2}$.

STEP 2: Substitute $\left(2y + \frac{5}{2}\right)$ for x in the first equation and then solve for y.

$$2x - 4y = 5 \qquad \text{First equation}$$
$$2\left(2y + \frac{5}{2}\right) - 4y = 5 \qquad \text{Let } x = 2y + \frac{5}{2}.$$
$$4y + 5 - 4y = 5 \qquad \text{Distributive property}$$
$$5 = 5 \qquad \text{Combine like terms.}$$

The equation $5 = 5$ is an identity and indicates that there are infinitely many solutions. The system is consistent and the equations are dependent. Note that we can multiply each side of the second equation by -2 to obtain the first equation.

$$-2(-x + 2y) = -2\left(-\frac{5}{2}\right) \qquad \text{Multiply second equation by } -2.$$
$$2x - 4y = 5 \qquad \text{Distributive property}$$

STEP 3: In Figure 9.10 the graphs of the equations are identical lines. The solution set is $\{(x, y) \mid 2x - 4y = 5\}$ and includes all points on this line; $\left(0, -\frac{5}{4}\right)$ and $\left(2, -\frac{1}{4}\right)$ are examples of solutions to this system.

Now Try Exercises 49 and 61 ◀

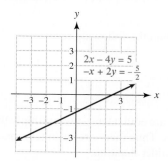

Figure 9.10

The Elimination Method

The **elimination method** is another way to solve systems of equations symbolically. This method is based on the property that *equals added to equals are equal*. That is, if

$$a = b \quad \text{and} \quad c = d, \quad \text{then}$$
$$a + c = b + d.$$

The goal of this method is to obtain an equation from which one of the two variables has been eliminated. This task is sometimes accomplished by adding two equations. The elimination method is demonstrated in the next example for three types of systems.

EXAMPLE 8 Using elimination to solve a system

Use elimination to solve each system of equations, if possible. Identify the system as consistent or inconsistent. If the system is consistent, state whether the equations are dependent or independent. Support your results graphically.

(a) $2x - y = -4$
$3x + y = -1$

(b) $4x - y = 10$
$-4x + y = -10$

(c) $x - y = 6$
$x - y = 3$

SOLUTION

(a) *Symbolic Solution* We can eliminate the y-variable by adding the equations.

$$2x - y = -4$$
$$\underline{3x + y = -1}$$
$$5x = -5 \quad \text{or} \quad x = -1 \qquad \textit{Add equations.}$$

Now the y-variable can be determined by substituting $x = -1$ in either equation.

$$2x - y = -4 \qquad \textit{First equation}$$
$$2(-1) - y = -4 \qquad \textit{Let x = -1.}$$
$$-y = -2 \qquad \textit{Add 2.}$$
$$y = 2 \qquad \textit{Multiply by -1.}$$

The solution is $(-1, 2)$. There is a unique solution so the system is consistent and the equations are independent.

Graphical Solution For a graphical solution, start by solving each equation for y.

$$2x - y = -4 \quad \text{is equivalent to} \quad y = 2x + 4.$$
$$3x + y = -1 \quad \text{is equivalent to} \quad y = -3x - 1.$$

The graphs of $y = 2x + 4$ and $y = -3x - 1$ intersect at the point $(-1, 2)$, as shown in Figure 9.11.

(b) *Symbolic Solution* If we add the equations, we obtain the following result.

$$4x - y = 10$$
$$\underline{-4x + y = -10}$$
$$0 = 0 \qquad \textit{Add equations.}$$

The equation $0 = 0$ is an identity. The two equations are equivalent: if we multiply the first equation by -1, we obtain the second equation. Thus there are infinitely many solutions, and we can write the solution set in set-builder notation.

$$\{(x, y) \,|\, 4x - y = 10\}$$

Some examples of solutions are $(3, 2)$, $(4, 6)$, and $(1, -6)$. The system is consistent and the equations are dependent.

Graphical Solution For a graphical solution, start by solving each equation for y. Both equations are equivalent to $y = 4x - 10$. Their graphs are identical and coincide. The graph of $y = 4x - 10$ is shown in Figure 9.12. Any point on the line represents a solution to the system. For example, $(3, 2)$ is a solution.

Figure 9.11

Figure 9.12

(c) *Symbolic Solution*   If we subtract the second equation from the first, we obtain the following result. (Note that subtracting the second equation from the first is equivalent to multiplying the second equation by -1 and then adding it to the first.)

Figure 9.13

$$x - y = 6$$
$$\underline{x - y = 3}$$
$$0 = 3 \qquad \text{Subtract.}$$

The equation $0 = 3$ is a contradiction. Therefore there are no solutions, and the system is inconsistent.

Graphical Solution   For a graphical solution, start by solving each equation for y to obtain $y = x - 6$ and $y = x - 3$. The graphs of $y = x - 6$ and $y = x - 3$, shown in Figure 9.13, are parallel lines that never intersect, so there are no solutions.

Now Try Exercises 71, 75, and 77 ◀

Sometimes multiplication is performed before elimination is used, as illustrated in the next example.

EXAMPLE 9 **Multiplying before using elimination**

Solve each system of equations by using elimination.
(a) $2x - 3y = 18$ **(b)** $5x + 10y = 10$
     $5x + 2y = 7$ $x + 2y = 2$

SOLUTION
(a) If we multiply the first equation by 2 and the second equation by 3, then the y-coefficients become -6 and 6. Addition eliminates the y-variable.

$$4x - 6y = 36 \qquad \text{Multiply by 2.}$$
$$\underline{15x + 6y = 21} \qquad \text{Multiply by 3.}$$
$$19x = 57, \quad \text{or} \quad x = 3 \qquad \text{Add.}$$

Substituting $x = 3$ in $2x - 3y = 18$ results in

$$2(3) - 3y = 18, \quad \text{or} \quad y = -4.$$

The solution is $(3, -4)$.
(b) If the second equation is multiplied by -5, addition eliminates both variables.

$$5x + 10y = 10$$
$$\underline{-5x - 10y = -10} \qquad \text{Multiply by } -5.$$
$$0 = 0 \qquad \text{Add.}$$

The statement $0 = 0$ is an identity. The equations are dependent and there are infinitely many solutions. The solution set is $\{(x, y) \mid x + 2y = 2\}$.

Now Try Exercises 81 and 83 ◀

Elimination and Nonlinear Systems   Elimination can also be used to solve some nonlinear systems of equations, as illustrated in the next example.

EXAMPLE 10 Using elimination to solve a nonlinear system

Solve the system of equations.

$$x^2 + y^2 = 4$$
$$2x^2 - y = 7$$

SOLUTION If we multiply each side of the first equation by 2, multiply each side of the second equation by -1, and then add, the x variable is eliminated.

$2x^2 + 2y^2 = 8$	Multiply by 2.
$-2x^2 + y = -7$	Multiply by -1.
$2y^2 + y = 1$	Add.

Next we solve $2y^2 + y - 1 = 0$ for y.

$$(2y - 1)(y + 1) = 0 \qquad \text{Factor.}$$

$$y = \frac{1}{2} \quad \text{or} \quad y = -1 \qquad \text{Solve.}$$

Solving $x^2 + y^2 = 4$ for x results in $x = \pm\sqrt{4 - y^2}$. If $y = \frac{1}{2}$, then $x = \pm\sqrt{\frac{15}{4}}$, which can be written as $\pm\frac{\sqrt{15}}{2}$. If $y = -1$, then $x = \pm\sqrt{3}$. Thus there are four solutions: $\left(\pm\frac{\sqrt{15}}{2}, \frac{1}{2}\right)$ and $(\pm\sqrt{3}, -1)$.

A graph of the system of equations is shown in Figure 9.14. The four points of intersection correspond to the four solutions. In Figure 9.15 the four points of intersection are labeled.

Figure 9.14   **Figure 9.15**   *Now Try Exercise 93* ◄

Graphical and Numerical Methods

An Application   The next example illustrates how a system with two variables can be solved symbolically, graphically, and numerically.

EXAMPLE 11 Modeling roof trusses

Linear systems occur in the design of roof trusses for homes and buildings. See Figure 9.16. One of the simplest types of roof trusses is an equilateral triangle. If a 200-pound force is applied to the peak of a truss, as shown in Figure 9.17, then the weights W_1 and W_2 exerted on each rafter of the truss are determined by the following system of linear equations. (**Source:** R. Hibbeler, *Structural Analysis*.)

$$W_1 - W_2 = 0$$
$$\frac{\sqrt{3}}{2}(W_1 + W_2) = 200$$

Estimate the solution symbolically, graphically, and numerically.

Figure 9.16   **Figure 9.17**

SOLUTION

Symbolic Solution   The system of equations can be written as follows.

$$W_1 - W_2 = 0 \qquad \text{First equation}$$
$$\frac{\sqrt{3}}{2}W_1 + \frac{\sqrt{3}}{2}W_2 = 200 \qquad \text{Distributive property}$$

We can apply elimination by multiplying the first equation by $\frac{\sqrt{3}}{2}$ and then adding.

$$\frac{\sqrt{3}}{2}W_1 - \frac{\sqrt{3}}{2}W_2 = 0 \qquad \text{Multiply by } \frac{\sqrt{3}}{2}.$$
$$\frac{\sqrt{3}}{2}W_1 + \frac{\sqrt{3}}{2}W_2 = 200$$

$$\overline{\sqrt{3}\,W_1 = 200} \qquad \text{Add equations.}$$

Dividing by $\sqrt{3}$ gives $W_1 = \frac{200}{\sqrt{3}} \approx 115.47$ pounds. From the first equation, it follows that $W_1 = W_2$, and so $W_2 \approx 115.47$ pounds.

Graphical Solution   Begin by solving each equation for the variable W_2.

$$W_2 = W_1$$
$$W_2 = \frac{400}{\sqrt{3}} - W_1$$

Graph the equations $Y_1 = X$ and $Y_2 = (400/\sqrt{3}) - X$. Their graphs intersect near the point $(115.47, 115.47)$, as shown in Figure 9.18. This means that each rafter supports a weight of approximately 115 pounds.

Numerical Solution   In Figure 9.19, $y_1 \approx y_2$ for $x = 115$.   **Now Try Exercise 115** ◀

[0, 200, 50] by [0, 200, 50]

$y_2 = \frac{400}{\sqrt{3}} - x$

$y_1 = x$

Intersection
X=115.47005 Y=115.47005

Figure 9.18

X	Y1	Y2
112	112	118.94
113	113	117.94
114	114	116.94
115	115	115.94
116	116	114.94
117	117	113.94
118	118	112.94

X=115

Figure 9.19

EXAMPLE 12   Determining the dimensions of a cylinder

The volume V of a cylindrical container with a radius r and height h is computed by $V(r, h) = \pi r^2 h$. See Figure 9.20. The lateral surface area S of the container, *excluding* the circular top and bottom, is computed by $S(r, h) = 2\pi rh$.

$V(r, h) = \pi r^2 h$ $S(r, h) = 2\pi rh$

Figure 9.20

Geometry Review
To review formulas related to cylinders, see Chapter R (page R-4).

(a) Write a system of equations whose solution is the dimensions for a cylinder with a volume of 38 cubic inches and a lateral surface area of 63 square inches.

(b) Solve the system of equations graphically and symbolically.

SOLUTION

(a) The equations $V(r, h) = 38$ and $S(r, h) = 63$ must be satisfied. This results in the following system of nonlinear equations.

$$\pi r^2 h = 38$$
$$2\pi rh = 63$$

(b) *Graphical Solution*   To find the solution graphically, we can solve each equation for h and then apply the intersection-of-graphs method.

$$h = \frac{38}{\pi r^2}$$

$$h = \frac{63}{2\pi r}$$

[0, 4, 1] by [0, 20, 5]

Figure 9.21

Let r correspond to x and h to y. Graph $Y_1 = 38/(\pi X^2)$ and $Y_2 = 63/(2\pi X)$. Their graphs intersect near the point (1.206, 8.312), as shown in Figure 9.21. Therefore a cylinder with a radius of $r \approx 1.206$ inches and height of $h \approx 8.312$ inches has a volume of 38 cubic inches and lateral surface area of 63 square inches.

Symbolic Solution   Because $h = \frac{38}{\pi r^2}$ and $h = \frac{63}{2\pi r}$, we can determine r by solving the following equation.

$$\frac{38}{\pi r^2} = \frac{63}{2\pi r} \qquad \text{Equation to be solved}$$

$$2\pi r^2 \left(\frac{38}{\pi r^2} \right) = 2\pi r^2 \left(\frac{63}{2\pi r} \right) \qquad \text{Multiply by the LCD, } 2\pi r^2.$$

$$76 = 63r \qquad \text{Simplify.}$$

$$\frac{76}{63} = r \qquad \text{Divide by 63.}$$

Because $r = \frac{76}{63} \approx 1.206$, $h = \frac{63}{2\pi r} = \frac{63}{2\pi(76/63)} \approx 8.312$, the symbolic result verifies our graphical result.

Now Try Exercise 117 ◄

An Example That Requires a Graphical Solution   Sometimes it is either difficult or impossible to solve a nonlinear system of equations symbolically. However, it might be possible to solve such a system graphically.

EXAMPLE 13   Solving a nonlinear system of equations graphically

Solve the system graphically to the nearest thousandth.

$$2x^3 - y = 2$$
$$\ln x^2 - 3y = -1$$

SOLUTION   Begin by solving both equations for y. The first equation becomes $y = 2x^3 - 2$. Solving the second equation for y gives the following results.

$$\ln x^2 - 3y = -1 \qquad \textit{Second equation}$$
$$\ln x^2 + 1 = 3y \qquad \textit{Add 3y and 1 to each side.}$$
$$\frac{\ln x^2 + 1}{3} = y \qquad \textit{Divide each side by 3.}$$

The graphs of $Y_1 = 2X^3 - 2$ and $Y_2 = (\ln (X^2) + 1)/3$ in Figure 9.22 intersect at one point. To the nearest thousandth, the solution is $(1.058, 0.371)$.

$[-6, 6, 1]$ by $[-4, 4, 1]$

Figure 9.22

Now Try Exercise 107 ◀

Joint Variation

A quantity may depend on more than one variable. For example, the volume V of a cylinder is given by $V = \pi r^2 h$. We say that V *varies jointly* as h and the square of r. The *constant of variation* is π.

Joint Variation

Let m and n be real numbers. Then z **varies jointly** as the mth power of x and the nth power of y if a nonzero real number k exists such that

$$z = kx^m y^n.$$

EXAMPLE 14   Modeling the amount of wood in a tree

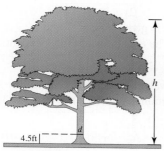

Figure 9.23

To estimate the volume of timber in a given area of forest, formulas have been developed to find the amount of wood contained in a tree with height h in feet and diameter d in inches. See Figure 9.23. One study concluded that the volume V of wood in a tree varies jointly as the 1.12 power of h and the 1.98 power of d. (The diameter is measured 4.5 feet above the ground.)   (**Source:** B. Ryan, B. Joiner, and T. Ryan, *Minitab Handbook.*)

(a) Write an equation that relates V, h, and d.

(b) A tree with a 13.8-inch diameter and a 64-foot height has a volume of 25.14 cubic feet. Estimate the constant of variation k.

(c) Estimate the volume of wood in a tree with $d = 11$ inches and $h = 47$ feet.

SOLUTION

(a) $V = kh^{1.12}d^{1.98}$, where k is the constant of variation

(b) Substitute $d = 13.8$, $h = 64$, and $V = 25.14$ into the equation and solve for k.

$$25.14 = k(64)^{1.12}(13.8)^{1.98}$$

$$k = \frac{25.14}{(64)^{1.12}(13.8)^{1.98}} \approx 0.00132$$

Thus let $V = \mathbf{0.00132}\, h^{1.12}d^{1.98}$.

(c) $V = 0.00132(47)^{1.12}(11)^{1.98} \approx 11.4$ cubic feet   *Now Try Exercise 145* ◀

9.1 Putting It All Together

The following table summarizes some mathematical concepts involved with functions and equations in two variables.

Concept	Comments	Example
Function of two inputs or variables	$z = f(x, y)$, where x and y are inputs and z is the output	$f(x, y) = x^2 + 5y$ $f(2, 3) = 2^2 + 5(3) = 19$
System of two linear equations	The equations can be written as $ax + by = k$. A solution is an ordered pair (x, y) that satisfies both equations.	$2x - 3y = 6$ $5x + 4y = -8$ Solution: $(0, -2)$
Nonlinear system of two equations	A system of equations that is not linear is nonlinear. A solution is an ordered pair (x, y) that satisfies both equations. A nonlinear system of equations can have any number of solutions.	$5x^2 - 4xy = -3$ $\dfrac{5}{x} - 2y = 1$ Solutions: $(1, 2)$, $\left(-\frac{7}{5}, -\frac{16}{7}\right)$
Consistent system of linear equations in two variables	A consistent linear system has either one or infinitely many solutions. Its graph is either distinct, intersecting lines or identical lines.	$\begin{array}{r} x + y = 10 \\ x - y = 4 \\ \hline 2x\quad\ = 14 \end{array}$ Solution is given by $x = 7$ and $y = 3$. The equations are independent.
System of dependent linear equations in two variables	A system of dependent linear equations has infinitely many solutions. The graph consists of two identical lines.	$2x + 2y = 2$ and $x + y = 1$ are equivalent (dependent) equations. The solution set is $\{(x, y) \mid x + y = 1\}$.
Inconsistent system of linear equations in two variables	An inconsistent linear system has no solutions. The graph is two parallel lines.	$\begin{array}{rl} x + y = & 1 \\ x + y = & 2 \qquad \text{Subtract.} \\ \hline 0 = & -1 \qquad \text{Always false} \end{array}$

Concept	Comments	Example
Method of substitution	Solve one equation for a variable. Then substitute the result in the second equation and solve.	$x - y = 1$ $x + y = 5$ If $x - y = 1$, then $x = 1 + y$. Substitute in the second equation: $(1 + y) + y = 5$. This results in $y = 2$ and $x = 3$. The solution is $(3, 2)$.
Elimination method	By performing arithmetic operations on a system, a variable is eliminated.	$\begin{aligned} 2x + y &= 5 \\ x - y &= 1 \quad \text{Add.} \\ \hline 3x &= 6 \end{aligned}$ so $x = 2$ and $y = 1$.
Graphical method for two equations	Solve both equations for the same variable. Then apply the intersection-of-graphs method.	If $x + y = 3$, then $y = 3 - x$. If $4x - y = 2$, then $y = 4x - 2$. Graph and locate the point of intersection at $(1, 2)$.

9.1 Exercises

Functions of More Than One Input

Exercises 1 and 2: Evaluate the function for the indicated inputs and interpret the result.

1. $A(5, 8)$, where $A(b, h) = \frac{1}{2}bh$ (A computes the area of a triangle with base b and height h.)

2. $A(20, 35)$, where $A(w, l) = wl$ (A computes the area of a rectangle with width w and length l.)

Exercises 3–8: Evaluate the expression for the given $f(x, y)$.

3. $f(2, -3)$ if $f(x, y) = x^2 + y^2$

4. $f(-1, 3)$ if $f(x, y) = 2x^2 - y^2$

5. $f(-2, 3)$ if $f(x, y) = 3x - 4y$

6. $f(5, -2)$ if $f(x, y) = 6y - \frac{1}{2}x$

7. $f\left(\frac{1}{2}, -\frac{7}{4}\right)$ if $f(x, y) = \frac{2x}{y + 3}$

8. $f(0.2, 0.5)$ if $f(x, y) = \frac{5x}{2y + 1}$

Exercises 9–12: Write a symbolic representation for $f(x, y)$ if the function f computes the following quantity.

9. The sum of y and twice x

10. The product of x^2 and y^2

11. The product of x and y divided by $1 + x$

12. The square root of the sum of x and y

Exercises 13–18: Solve the equation for x and then solve it for y.

13. $3x - 4y = 7$

14. $-x - 5y = 4$

15. $x - y^2 = 5$

16. $2x^2 + y = 4$

17. $\dfrac{2x - y}{3y} = 1$

18. $\dfrac{x + y}{x - y} = 2$

Systems of Equations

Exercises 19–22: Determine which ordered pairs are solutions to the given system of equations. State whether the system is linear or nonlinear.

19. $(2, 1), (-2, 1), (1, 0)$
$2x + y = 5$
$x + y = 3$

20. $(3, 2), (3, -4), (5, 0)$
$x - y = 5$
$2x + y = 10$

21. $(4, -3), (0, 5), (4, 3)$
$x^2 + y^2 = 25$
$2x + 3y = -1$

22. $(4, 8), (8, 4), (-4, -8)$
$xy = 32$
$x + y = 12$

Exercises 23–26: The figure shows the graph of a system of two linear equations. Use the graph to estimate the solution to this system of equations. Then solve the system symbolically.

23.

24.

25.

26.
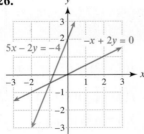

Exercises 27–30: Write a system of linear equations with two variables whose solution satisfies the problem. State what each variable represents. Then solve the system.

27. *Screen Dimensions*   The screen of a rectangular television set is 2 inches wider than it is high. If the perimeter of the screen is 38 inches, find its dimensions.

28. *Numbers*   The sum of two numbers is 300 and their difference is 8. Find the two numbers.

29. *Tickets*   Admission prices to a movie are $4 for children and $7 for adults. If 75 tickets were sold for $456, how many of each type of ticket were sold?

30. *Coins*   A sample of 16 dimes and quarters has a value of $2.65. How many of each type of coin are there?

Consistent and Inconsistent Linear Systems

Exercises 31–34: The figure represents a system of linear equations. Classify the system as consistent or inconsistent. Solve the system graphically and symbolically, if possible.

31.

32.

33.

34.
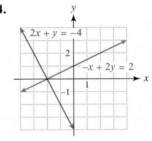

Exercises 35–42: Graph each system of equations and find any solutions. Check your answers. Identify the system as consistent or inconsistent. If the system is consistent, state whether the equations are dependent or independent.

35. $2x + y = 3$
$-2x - y = 4$

36. $x - 4y = 4$
$2x - 8y = 4$

37. $3x - y = 7$
$-2x + y = -5$

38. $-x + 2y = 3$
$3x - y = 1$

39. $x - 2y = -6$
$-2x + y = 6$

40. $2x - 3y = 1$
$x + y = -2$

41. $2x - y = -4$
$-4x + 2y = 8$

42. $3x - y = -2$
$-3x + y = 2$

The Method of Substitution

Exercises 43–54: If possible, solve the system of linear equations and check your answer.

43. $x + 2y = 0$
$3x + 7y = 1$

44. $-2x - y = -2$
$3x + 4y = -7$

45. $2x - 9y = -17$
$8x + 5y = 14$

46. $3x + 6y = 0$
$4x - 2y = -5$

47. $\frac{1}{2}x - y = -5$
$x + \frac{1}{2}y = 10$

48. $-x - \frac{1}{3}y = -4$
$\frac{1}{3}x + 2y = 7$

49. $3x - 2y = 5$
$-6x + 4y = -10$

50. $\frac{1}{2}x - \frac{3}{4}y = \frac{1}{2}$
$\frac{1}{5}x - \frac{3}{10}y = \frac{1}{5}$

51. $2x - 7y = 8$
$-3x + \frac{21}{2}y = 5$

52. $0.6x - 0.2y = 2$
$-1.2x + 0.4y = 3$

53. $0.2x - 0.1y = 0.5$
$0.4x + 0.3y = 2.5$

54. $100x + 200y = 300$
$200x + 100y = 0$

Exercises 55–68: If possible, solve the nonlinear system of equations.

55. $x^2 - y = 0$
$2x + y = 0$

56. $x^2 - y = 3$
$x + y = 3$

57. $xy = 8$
$x + y = 6$

58. $2x - y = 0$
$2xy = 4$

59. $x^2 + y^2 = 20$
$y = 2x$

60. $x^2 + y^2 = 9$
$x + y = 3$

61. $\sqrt{x} - 2y = 0$
$x - y = -2$

62. $x^2 + y^2 = 4$
$2x^2 + y = -3$

63. $2x^2 - y = 5$
$-4x^2 + 2y = -10$

64. $-6\sqrt{x} + 2y = -3$
$2\sqrt{x} - \frac{2}{3}y = 1$

65. $x^2 - y = 4$
$x^2 + y = 4$

66. $x^2 + x = y$
$2x^2 - y = 2$

67. $x^3 - x = 3y$
$x - y = 0$

68. $x^4 + y = 4$
$3x^2 - y = 0$

*Exercises 69 and 70: **Area and Perimeter**   The area of a rectangle with length l and width w is computed by $A(l, w) = lw$, and its perimeter is calculated by $P(l, w) = 2l + 2w$. Assume that $l > w$ and use the method of substitution to solve the system of equations for l and w.*

69. $A(l, w) = 35$
$P(l, w) = 24$

70. $A(l, w) = 300$
$P(l, w) = 70$

The Elimination Method

Exercises 71–80: Use elimination to solve the system of equations, if possible. Identify the system as consistent or inconsistent. If the system is consistent, state whether the equations are dependent or independent. Support your results graphically or numerically.

71. $x + y = 20$
$x - y = 8$

72. $2x + y = 15$
$x - y = 0$

73. $x + 3y = 10$
$x - 2y = -5$

74. $4x + 2y = 10$
$-2x - y = 10$

75. $x + y = 500$
$-x - y = -500$

76. $2x + 3y = 5$
$5x - 2y = 3$

77. $2x + 4y = 7$
$-x - 2y = 5$

78. $4x - 3y = 5$
$3x + 4y = 2$

79. $2x + 3y = 2$
$x - 2y = -5$

80. $x - 3y = 1$
$2x - 6y = 2$

Exercises 81–92: Solve the system, if possible.

81. $\frac{1}{2}x - y = 5$
$x - \frac{1}{2}y = 4$

82. $\frac{1}{2}x - \frac{1}{3}y = 1$
$\frac{1}{3}x - \frac{1}{2}y = 1$

83. $7x - 3y = -17$
$-21x + 9y = 51$

84. $-\frac{1}{3}x + \frac{1}{6}y = -1$
$2x - y = 6$

85. $\frac{2}{3}x + \frac{4}{3}y = \frac{1}{3}$
$-2x - 4y = 5$

86. $5x - 2y = 7$
$10x - 4y = 6$

87. $0.2x + 0.3y = 8$
$-0.4x + 0.2y = 0$

88. $2x - 3y = 1$
$3x - 2y = 2$

89. $2x + 3y = 7$
$-3x + 2y = -4$

90. $5x + 4y = -3$
$3x - 6y = -6$

91. $7x - 5y = -15$
$-2x + 3y = -2$

92. $-5x + 3y = -36$
$4x - 5y = 34$

Exercises 93–98: Use elimination to solve the nonlinear system of equations.

93. $x^2 + y = 12$
$x^2 - y = 6$

94. $x^2 + 2y = 15$
$2x^2 - y = 10$

95. $x^2 + y^2 = 25$
$x^2 + 7y = 37$

96. $x^2 + y^2 = 36$
$x^2 - 6y = 36$

97. $x^2 + y^2 = 4$
$2x^2 + y^2 = 8$

98. $x^2 + y^2 = 4$
$x^2 - y^2 = 4$

Using More Than One Method

Exercises 99–102: Solve the nonlinear system of equations
(a) symbolically and (b) graphically.

99. $x^2 + y^2 = 16$
$x - y = 0$

100. $x^2 - y = 1$
$3x + y = -1$

101. $xy = 12$
$x - y = 4$

102. $x^2 + y^2 = 2$
$x^2 - y = 0$

Exercises 103–106: Solve the system of linear equations
(a) graphically, (b) numerically, and
(c) symbolically.

103. $2x + y = 1$
$x - 2y = 3$

104. $3x + 2y = -2$
$2x - y = -6$

105. $-2x + y = 0$
$7x - 2y = 3$

106. $x - 4y = 15$
$3x - 2y = 15$

Finding Approximate Solutions

Exercises 107–112: Approximate, to the nearest thousandth, any solutions to the nonlinear system of equations graphically.

107. $x^3 - 3x + y = 1$
$x^2 + 2y = 3$

108. $x^2 + y = 5$
$x + y^2 = 6$

109. $2x^3 - x^2 = 5y$
$2^{-x} - y = 0$

110. $x^4 - 3x^3 = y$
$\log x^2 - y = 0$

111. $e^{2x} + y = 4$
$\ln x - 2y = 0$

112. $3x^2 + y = 3$
$(0.3)^x + 4y = 1$

Applications

113. *Population* In 2000, the combined population of Minneapolis/St. Paul, Minnesota, was 670,000. The population of Minneapolis was 96,000 greater than the population of St. Paul. (**Source:** Census Bureau.)
  (a) Write a system of equations whose solution gives the population of each city in thousands.

  (b) Solve the system of equations.

  (c) Is your system consistent or inconsistent? If it is consistent, state whether the equations are dependent or independent.

114. *U.S. Energy Consumption* In 2006, the commercial sector used 3.74 quadrillion (10^{15}) Btu of energy from natural gas and petroleum. It used 3.02 quadrillion Btu more natural gas than petroleum. (**Source:** Department of Energy.)

  (a) Write a system of equations whose solution gives the consumption of natural gas and petroleum (in quadrillion Btu).

  (b) Solve the system of equations.

  (c) Is your system consistent or inconsistent? If it is consistent, state whether the equations are dependent or independent.

115. *Roof Truss* (Refer to Example 11.) The weights W_1 and W_2 exerted on each rafter for the roof truss shown in the figure are determined by the system of linear equations. Solve the system.

$$W_1 + \sqrt{2}W_2 = 300$$
$$\sqrt{3}W_1 - \sqrt{2}W_2 = 0$$

116. *Time on the Internet* From 2001 to 2005 the average number of hours that a user spent on the Internet each week increased by 110%. This percent increase amounted to 11 hours. Find the average number of hours that a user spent on the Internet each week in 2001 and 2005. (**Source:** UCLA Center for Communications Policy.)

117. *Geometry* (Refer to Example 12.) Find the radius and height of a cylindrical container with a volume of 50 cubic inches and a lateral surface area of 65 square inches.

118. *Geometry* (Refer to Example 12.) Determine if it is possible to construct a cylindrical container, *including* the top and bottom, with a volume of 38 cubic inches and a surface area of 38 square inches.

119. *Dimensions of a Box* A box has an *open* top, rectangular sides, and a square base. Its volume is 576 cubic inches, and its outside surface area is 336 square inches. Find the dimensions of the box.

120. *Dimensions of a Box*   A box has rectangular sides, and its rectangular top and base are twice as long as they are wide. Its volume is 588 cubic inches, and its outside surface area is 448 square inches. Find its dimensions.

121. *Robberies*   The total number of robberies in 2000 and 2001 was 831,000. From 2000 to 2001 the number of robberies declined by 15,000.   (**Source:** FBI.)
 (a) Write a system of equations whose solution represents the number of robberies committed in each of these years.

 (b) Solve the system symbolically.

 (c) Solve the system graphically.

122. *Vaccinations*   During the first four months of 2004, a total of 6767 people were vaccinated for smallpox in Florida and Texas. There were 273 more people vaccinated in Florida than in Texas.   (**Source:** CDC.)
 (a) Write a system of equations whose solution represents the number of vaccinations given in each state.

 (b) Solve the system symbolically.

 (c) Solve the system graphically.

123. *Student Loans*   A student takes out two loans totaling $3000 to help pay for college expenses. One loan is at 8% interest, and the other is at 10%. Interest for both loans is compounded annually.
 (a) If the first-year interest is $264, write a system of equations whose solution is the amount of each loan.

 (b) Find the amount of each loan.

124. *Student Loans*   (Refer to Exercise 123.) Suppose that both loans have an interest rate of 10% and the total first-year interest is $300. If possible, determine the amount of each loan. Interpret your results.

125. *Student Loans*   (Refer to Exercises 123 and 124.) Suppose that both loans are at 10% and the total annual interest is $264. If possible, determine the amount of each loan. Interpret your results.

126. *Investments*   A student invests $5000 at two annual interest rates, 5% and 7%. After 1 year the student receives a total of $325 in interest. How much did the student invest at each interest rate?

127. *Maximizing Area*   Suppose a rectangular pen for a pet is to be made using 40 feet of fence. Let *l* represent its length and *w* its width, with $l \geq w$.

 (a) Find *l* and *w* if the area is 91 square feet.

 (b) Write a formula for the area *A* in terms of *w*.

 (c) What is the maximum area possible for the pen? Interpret this result.

128. *River Current*   A tugboat can pull a barge 60 miles upstream in 15 hours. The same tugboat and barge can make the return trip downstream in 6 hours. Determine the speed of the current in the river.

129. *Air Speed*   A jet airliner travels 1680 miles in 3 hours with a tail wind. The return trip, into the wind, takes 3.5 hours. Find both the speed of the jet with no wind and the wind speed. (*Hint:* First find the ground speed of the airplane in each direction.)

130. *Plane Speed*   A plane flies 1500 miles against the wind in 3 hours and 45 minutes. The return trip with the wind takes 3 hours. Assume that the wind speed stays constant. Find the speed of the wind and the speed of the airplane with no wind.

131. *The Toll of War*   American battlefield deaths in World Wars I and II totaled about 345,000. There were about 5.5 times as many deaths in World War II as World War I. Find the number of American battlefield deaths in each war. Round your answers to the nearest whole number. (**Source:** Defense Department.)

132. *Cigarettes Sold Overseas*   The total number of cigarettes sold overseas in 1990 and 2000 was 10.9 trillion cigarettes. The number sold in 2000 was 100 billion more than the number sold in 1990. How many cigarettes were sold each year?   (**Source:** Department of Agriculture.)

133. *Height and Weight*   The relationship between a professional basketball player's height *h* in inches and weight *w* in pounds was modeled using two samples of players. The resulting modeling equations for the two samples were $w = 7.46h - 374$ and $w = 7.93h - 405$. Assume that $65 \leq h \leq 85$.
 (a) Use each equation to predict the weight of a professional basketball player who is 6′11″.

 (b) Determine graphically the height where the two models give the same weight.

 (c) For each model, what change in weight is associated with a 1-inch increase in height?

134. *Heart Rate* In one study a group of athletes were exercised to exhaustion. Let x and y represent an athlete's heart rate 5 seconds and 10 seconds after stopping exercise, respectively. It was found that the maximum heart rate H for these athletes satisfied the following two equations.

$$H = \quad 0.491x + 0.468y + 11.2$$
$$H = -0.981x + 1.872y + 26.4$$

If an athlete had a maximum heart rate of $H = 180$, determine x and y graphically. Interpret your answer. (**Source:** V. Thomas, *Science and Sport.*)

135. *Surface Area and the Human Body* The surface area of the skin covering the human body is a function of more than one variable. A taller person tends to have a larger surface area, as does a heavier person. Both height and weight influence the surface area of a person's body. A formula to determine the surface area of a person's body in square meters is given by $S(w, h) = 0.007184w^{0.425}h^{0.725}$, where w is weight in kilograms and h is height in centimeters. Use S to estimate the surface area of a person who is 65 inches (165.1 centimeters) tall and weighs 154 pounds (70 kilograms). (**Source:** H. Lancaster, *Quantitative Methods in Biological and Medical Sciences.*)

Exercises 136–138: Skin and the Human Body (Refer to Exercise 135.) Estimate, to the nearest tenth, the surface area of a person with weight w and height h.

136. $w = 86$ kilograms, $h = 185$ centimeters

137. $w = 132$ pounds, $h = 62$ inches

138. $w = 220$ pounds, $h = 75$ inches

Joint Variation

Exercises 139 and 140: Approximate the constant of variation to the nearest hundredth.

139. The variable z varies jointly as the second power of x and the third power of y. When $x = 2$ and $y = 2.5, z = 31.9$.

140. The variable z varies jointly as the 1.5 power of x and the 2.1 power of y. When $x = 4$ and $y = 3.5, z = 397$.

141. The variable z varies jointly as the square root of x and the cube root of y. If $z = 10.8$ when $x = 4$ and $y = 8$, find z when $x = 16$ and $y = 27$.

142. The variable z varies jointly as the third powers of x and y. If $z = 2160$ when $x = 3$ and $y = 4$, find z when $x = 2$ and $y = 5$.

143. *Wind Power* The electrical power generated by a windmill varies jointly as the square of the diameter of the area swept out by the blades and the cube of the wind velocity. If a windmill with an 8-foot diameter and a 10-mile-per-hour wind generates 2405 watts, how much power would be generated if the blades swept out an area 6 feet in diameter and the wind was 20 miles per hour?

144. *Strength of a Beam* The strength of a rectangular beam varies jointly as its width and the square of its thickness. If a beam 5.5 inches wide and 2.5 inches thick supports 600 pounds, how much can a similar beam that is 4 inches wide and 1.5 inches thick support?

145. *Volume of Wood* (Refer to Example 14.) One cord of wood contains 128 cubic feet. Estimate the number of cords in a tree that is 105 feet tall and has a diameter of 38 inches.

146. *Carpeting* The cost of carpet for a rectangular room varies jointly as its width and length. If a room 10 feet wide and 12 feet long costs $1560 to carpet, find the cost to carpet a room that is 11 feet by 23 feet. Interpret the constant of variation.

147. *Surface Area* Use the results of Exercise 135 to find a formula for $S(w, h)$ that calculates the surface area of a person if w is given in *pounds* and h is given in *inches*.

148. *Surface Area* Use the results of Exercise 147 to solve Exercises 137 and 138.

Writing about Mathematics

149. Give an example of a quantity occurring in everyday life that can be computed by a function of more than one input. Identify the inputs and the output.

150. Give an example of a system of linear equations with two variables. Explain how to solve the system graphically and symbolically.

9.2 Systems of Inequalities in Two Variables

- Solve inequalities in two variables graphically
- Solve systems of inequalities in two variables
- Learn basic properties of linear programming in two variables

Introduction

There is no single ideal weight for a person. Instead there is a range of recommended weights, which often depend on the person's height. Figure 9.24 shows one relationship between recommended weights w and heights h. To describe the shaded region in this graph, we need a *system of linear inequalities*. (See Exercises 37–40.) This section discusses inequalities in two variables and their graphs. (**Source:** Department of Agriculture.)

Figure 9.24

Systems of Linear and Nonlinear Inequalities

A linear inequality in two variables can be written as

$$ax + by \leq c,$$

where a, b, and c are constants with a and b not equal to zero. (The symbol $\leq$ can be replaced by $\geq$, $<$, or $>$.) If an ordered pair (x, y) makes the inequality a true statement, then (x, y) is a solution. The set of all solutions is called the *solution set*. The graph of an inequality includes all points (x, y) in the solution set.

The graph of a linear inequality is a **half-plane**, which may include the boundary. For example, to graph the linear inequality $3x - 2y \leq 6$, solve the inequality for y. This results in $y \geq \frac{3}{2}x - 3$. Then graph the line $y = \frac{3}{2}x - 3$, which is the boundary. The solution set to the inequality includes the line and the half-plane above the line, as shown in Figure 9.25.

A second approach is to write the equation $3x - 2y = 6$ as $y = \frac{3}{2}x - 3$ and graph. To determine which half-plane to shade, select a **test point**. For example, the point $(0, 0)$ satisfies the inequality $3x - 2y \leq 6$, so shade the region containing $(0, 0)$.

Figure 9.25

$3x - 2y \leq 6$

$(0, 0)$

EXAMPLE 1 Graphing inequalities

Graph the solution set to each inequality.
(a) $2x - 3y \leq -6$ **(b)** $x^2 + y^2 < 9$

SOLUTION

Getting Started When graphing an inequality, start by replacing the inequality sign with an equals sign. Graph the resulting equation by using a solid line (or curve) if equality is part

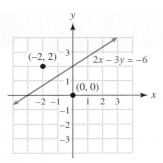

Figure 9.26

of the given inequality or a dashed line (or curve) if it is not. Then use a test point to determine which region of the graph to shade. ▶

(a) For $2x - 3y \le -6$ start by graphing the line $2x - 3y = -6$, as shown in Figure 9.26. Note that this line is solid because equality is included. We can determine which side of the line to shade by using test points. For example, we find that the test point $(-2, 2)$ lies above the line and the test point $(0, 0)$ lies below the line. See Table 9.1.

Table 9.1

Test Point	$2x - 3y \le -6$	True or False?
$(-2, 2)$	$2(-2) - 3(2) \overset{?}{\le} -6$	True
$(0, 0)$	$2(0) - 3(0) \overset{?}{\le} -6$	False

The test point $(-2, 2)$ satisfies the given inequality, so shade the region above the line that contains the point $(-2, 2)$. See Figure 9.27.

Figure 9.27

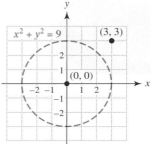

Figure 9.28

(b) For $x^2 + y^2 < 9$ start by graphing the circle $x^2 + y^2 = 9$, as shown in Figure 9.28. Note that this circle is dashed because equality is *not* included. The test point $(3, 3)$ lies outside the circle and the test point $(0, 0)$ lies inside the circle. See Table 9.2.

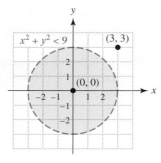

Figure 9.29

Table 9.2

Test Point	$x^2 + y^2 < 9$	True or False?
$(3, 3)$	$3^2 + 3^2 \overset{?}{<} 9$	False
$(0, 0)$	$0^2 + 0^2 \overset{?}{<} 9$	True

The test point $(0, 0)$ satisfies the given inequality, so shade the region inside the circle. Note that the actual circle is not part of the solution set. See Figure 9.29.

Now Try Exercises 7 and 9 ◀

In Section 9.1, we saw that systems of equations could be linear or nonlinear. In an analogous manner, systems of inequalities can be linear or nonlinear. The next example illustrates a system of each type. Both are solved graphically.

EXAMPLE 2 Solving systems of inequalities graphically

Solve each system of inequalities by shading the solution set. Identify one solution.

(a) $\quad y > x^2$ (b) $\quad x + 3y \le 9$
$\quad\quad x + y < 4$ $2x - y \le -1$

SOLUTION

(a) This is a nonlinear system. Graph the parabola $y = x^2$ and the line $y = 4 - x$. Since $y > x^2$ and $y < 4 - x$, the region satisfying the system lies above the parabola and below the line. It does not include the boundaries, which are shown using a dashed line and curve. See Figure 9.30.

Any point in the shaded region represents a solution. For example, $(0, 2)$ lies in the shaded region and is a solution, since $x = 0$ and $y = 2$ satisfy *both* inequalities.

(b) Begin by solving each linear inequality for y.

$$y \le -\frac{1}{3}x + 3$$

$$y \ge 2x + 1$$

Graph $y = -\frac{1}{3}x + 3$ and $y = 2x + 1$. The region satisfying the system is below the first (red) line and above the second (blue) line. Because equality is included, the boundaries, which are shown as solid lines in Figure 9.31, are part of the region. The test point $(-3, 0)$ is a solution, since it satisfies both inequalities.

Figure 9.30

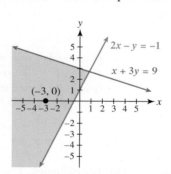

Figure 9.31

Now Try Exercises 17 and 19 ◄

Graphing Calculators Graphing calculators can be used to shade regions in the xy-plane. See Figure 9.32. The solution set shown in Figure 9.30 is also shown in Figure 9.33, where a graphing calculator has been used. However, the boundary is not dashed.

Figures 9.34 and 9.35 show a different method of shading a solution set; we have shaded the area below $Y_1 = (-1/3)X + 3$ and above $Y_2 = 2X + 1$. The solution set is the region shaded with both vertical *and* horizontal lines and corresponds to the shaded region in Figure 9.31.

Calculator Help

To shade a graph, see Appendix A (pages AP-12 and AP-14).

$[-5, 5, 1]$ by $[-2, 8, 1]$

$[-6, 6, 1]$ by $[-6, 6, 1]$

Figure 9.32

Figure 9.33

Figure 9.34

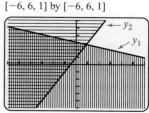

Figure 9.35

An Application of Inequalities The next example discusses how a system of inequalities (based on average temperature and precipitation) can be used to determine where forests, grasslands, and deserts will occur.

EXAMPLE 3 Modeling plant growth

Two factors that have a critical effect on plant growth are temperature and precipitation. If a region has too little precipitation, it will be a desert. Forests tend to exist in regions where temperatures are relatively low and there is sufficient rainfall. At other levels of precipitation and temperature, grasslands may prevail. Figure 9.36 illustrates the relationship among forests, grasslands, and deserts, as suggested by annual average temperature T in degrees Fahrenheit and precipitation P in inches. (**Source:** A. Miller and J. Thompson, *Elements of Meteorology.*)

(a) Determine a system of linear inequalities that describes where grasslands occur.
(b) Bismarck, North Dakota, has an annual average temperature of 40°F and precipitation of 15 inches. According to the graph, what type of plant growth would you expect near Bismarck? Do these values satisfy the system of inequalities from part (a)?

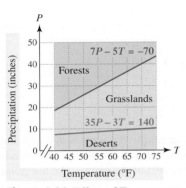

Figure 9.36 Effect of Temperature and Precipitation on Plant Growth

SOLUTION

(a) Grasslands occur for ordered pairs (T, P) lying between the two lines in Figure 9.36. The boundary between deserts and grasslands is determined by $35P - 3T = 140$. Solving for P (the variable on the vertical axis) results in

$$P = \frac{3}{35}T + \frac{140}{35}.$$

Grasslands grow where values of P are above the line. This region is described by $P > \frac{3}{35}T + \frac{140}{35}$, or equivalently, $35P - 3T > 140$. In a similar manner, the region below the boundary between grasslands and forests is represented by the inequality $7P - 5T < -70$. Thus grasslands satisfy the following system of inequalities.

$$35P - 3T > 140$$
$$7P - 5T < -70$$

(b) For Bismarck, $T = 40$ and $P = 15$. From Figure 9.36, it appears that the point $(40, 15)$ lies between the two lines, so the graph predicts that grasslands will exist around Bismarck. Substituting these values for T and P into the system of inequalities results in the following true statements.

$$35(15) - 3(40) = 405 > 140 \quad \text{True}$$
$$7(15) - 5(40) = -95 < -70 \quad \text{True}$$

The temperature and precipitation values for Bismarck satisfy the system of inequalities for grasslands. *Now Try Exercises 37 and 39* ◀

Linear Programming

An important application of mathematics in business and social sciences is linear programming. **Linear programming** is a procedure used to optimize quantities such as cost and profit. It was developed during World War II as a method of efficiently allocating supplies. Linear programming applications frequently contain thousands of variables and are solved by computers. However, here we focus on problems involving two variables.

A linear programming problem consists of a linear **objective function** and a system of linear inequalities called **constraints**. The solution set for the system of linear inequalities is called the set of **feasible solutions**. The objective function describes a quantity that is to be optimized. For example, linear programming is often used to maximize profit or minimize cost. The following example illustrates these concepts.

EXAMPLE 4 Finding maximum profit

Suppose a small company manufacturers two products—radios and CD players. Each radio results in a profit of $15, and each CD player provides a profit of $35. Due to demand, the company must produce at least 5 and not more than 25 radios per day. The number of radios cannot exceed the number of CD players, and the number of CD players cannot exceed 30. How many of each should the company manufacture to obtain maximum profit?

SOLUTION Let x be the number of radios produced daily and y be the number of CD players produced daily. Since the profit from x radios is $15x$ dollars and the profit from y CD players is $35y$ dollars, the total daily profit P is given by

$$P = 15x + 35y.$$

The company produces from 5 to 25 radios per day, so the inequalities

$$x \geq 5 \quad \text{and} \quad x \leq 25$$

must be satisfied. The requirements that the number of radios cannot exceed the number of CD players and the number of CD players cannot exceed 30 indicate that

$$x \leq y \quad \text{and} \quad y \leq 30.$$

Since the numbers of radios and CD players cannot be negative, we have

$$x \geq 0 \quad \text{and} \quad y \geq 0.$$

Listing all the constraints on production gives

$$x \geq 5, \quad x \leq 25, \quad y \leq 30, \quad x \leq y, \quad x \geq 0, \quad \text{and} \quad y \geq 0.$$

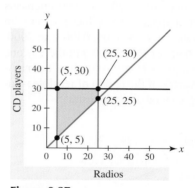

Figure 9.37

Graphing these constraints results in the shaded region shown in Figure 9.37. This shaded region is the set of feasible solutions. The vertices (or corners) of this region are (5, 5), (25, 25), (25, 30), and (5, 30).

It can be shown that maximum profit occurs at a vertex of the region of feasible solutions. Thus we evaluate P at each vertex, as shown in Table 9.3.

Table 9.3

Vertex	$P = 15x + 35y$	
(5, 5)	$15(5) + 35(5) = 250$	
(25, 25)	$15(25) + 35(25) = 1250$	
(25, 30)	$15(25) + 35(30) = 1425$	← **Maximum profit**
(5, 30)	$15(5) + 35(30) = 1125$	

The maximum value of P is 1425 at vertex (25, 30). Thus the maximum profit is $1425 and it occurs when 25 radios and 30 CD players are manufactured.

Now Try Exercise 57 ◀

The following theorem holds for linear programming problems.

Fundamental Theorem of Linear Programming

If the optimal value for a linear programming problem exists, then it occurs at a vertex of the region of feasible solutions.

Figure 9.38

Justification of the Fundamental Theorem   To better understand the fundamental theorem of linear programming, consider the following example. Suppose that we want to maximize $P = 30x + 70y$ subject to the following four constraints:

$$x \geq 10, \qquad x \leq 50, \qquad y \geq x, \qquad \text{and} \qquad y \leq 60.$$

The corresponding region of feasible solutions is shown in Figure 9.38.

  Each value of P determines a unique line. For example, if $P = 7000$, then the equation for P becomes $30x + 70y = 7000$. The resulting line, shown in Figure 9.39, does not intersect the region of feasible solutions. Thus there are no values for x and y that lie in this region and result in a profit of 7000. Figure 9.39 also shows the lines that result from letting $P = 0, 1000,$ and 3000. If $P = 1000$, then the line intersects the region of feasible solutions only at the vertex $(10, 10)$. This means that if $x = 10$ and $y = 10$, then $P = 30(10) + 70(10) = 1000$. If $P = 3000$, then the line $30x + 70y = 3000$ intersects the region of feasibility infinitely many times. However, it appears that values greater than 3000 are possible for P. In Figure 9.40 lines are drawn for $P = 5700, 6300,$ and 7000. Notice that there are no points of intersection for $P = 6300$ or $P = 7000$, but there is one vertex in the region of feasible solutions at $(50, 60)$ that gives $P = 5700$. Thus the maximum value of P is 5700 and this maximum occurs at a vertex of the region of feasible solutions. The fundamental theorem of linear programming generalizes this result.

CLASS DISCUSSION

What is the minimum value for P subject to the given constraints?

Figure 9.39

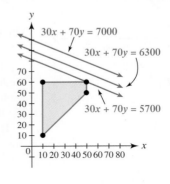

Figure 9.40

EXAMPLE 5   Finding the minimum of an objective function

Find the minimum value of $C = 2x + 3y$ subject to the following constraints.

$$x + y \geq 4$$
$$2x + y \leq 8$$
$$x \geq 0, \quad y \geq 0$$

SOLUTION Sketch the region determined by the constraints and find all vertices, as shown in Figure 9.41.

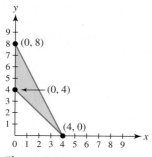

Figure 9.41

Evaluate the objective function C at each vertex, as shown in Table 9.4.

Table 9.4

Vertex	$C = 2x + 3y$	
(4, 0)	$2(4) + 3(0) = 8$	← **Minimum value**
(0, 8)	$2(0) + 3(8) = 24$	
(0, 4)	$2(0) + 3(4) = 12$	

The minimum value for C is 8 and it occurs at vertex (4, 0), or when $x = 4$ and $y = 0$.

Now Try Exercise 53 ◀

The following procedure describes how to solve a linear programming problem.

Solving a Linear Programming Problem

STEP 1: Read the problem carefully. Consider making a table to display the given information.

STEP 2: Use the table to write the objective function and all the constraints.

STEP 3: Sketch a graph of the region of feasible solutions. Identify all vertices, or corner points.

STEP 4: Evaluate the objective function at each vertex. A maximum (or a minimum) occurs at a vertex. If the region is unbounded, a maximum (or minimum) may not exist.

EXAMPLE 6 Minimizing cost

A breeder is buying two brands of food, A and B, for her animals. Each serving is a mixture of the two foods and should contain at least 40 grams of protein and at least 30 grams of fat. Brand A costs 90 cents per unit, and Brand B costs 60 cents per unit. Each unit of Brand A contains 20 grams of protein and 10 grams of fat, whereas each unit of Brand B contains 10 grams of protein and 10 grams of fat. Determine how much of each brand should be bought to obtain a minimum cost per serving.

SOLUTION

STEP 1: After reading the problem carefully, begin by listing the information, as illustrated in Table 9.5. (Your table may be different.)

Table 9.5

Brand	Units	Protein/Unit	Total Protein	Fat/Unit	Total Fat	Cost
A	x	20	$20x$	10	$10x$	$90x$
B	y	10	$10y$	10	$10x$	$60y$
		Minimum Total Protein →	40		30	←Minimum Total Fat

STEP 2: If x units of Brand A are purchased at 90¢ per unit and y units of Brand B are purchased at 60¢ per unit, then the cost C is given by $C = 90x + 60y$. Each serving requires at least 40 grams of protein. If x units of Brand A are bought (each containing 20 grams of protein), y units of Brand B are bought (each containing 10 grams

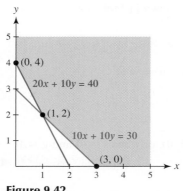

Figure 9.42

of protein), and each serving requires at least 40 grams of protein, then we can write $20x + 10y \geq 40$. Similarly, since each serving requires at least 30 grams of fat, we can write $10x + 10y \geq 30$. The linear programming problem can be written as follows.

$$\begin{array}{lll} \text{Minimize:} & C = 90x + 60y & \textit{Cost} \\ \text{Subject to:} & 20x + 10y \geq 40 & \textit{Protein} \\ & 10x + 10y \geq 30 & \textit{Fat} \\ & x \geq 0, \quad y \geq 0 & \end{array}$$

STEP 3: The region that contains the feasible solutions is unbounded and shown in Figure 9.42. The vertices for this region are $(0, 4)$, $(1, 2)$, and $(3, 0)$.

STEP 4: Evaluate the objective function C at each vertex, as shown in Table 9.6.

Table 9.6

Vertex	$C = 90x + 60y$	
$(0, 4)$	240	
$(1, 2)$	210	← **Minimum cost**
$(3, 0)$	270	

The minimum cost occurs when 1 unit of Brand A and 2 units of Brand B are mixed, at a cost of $2.10 per serving. ***Now Try Exercise 59*** ◄

9.2 Putting It All Together

The following table summarizes some important mathematical concepts from this section.

Concept	Comments	Example
Linear inequality in two variables	$ax + by \leq c$ ($\leq$ may be replaced by $<$, $>$, or $\geq$) The solution set is typically a shaded region in the xy-plane.	$2x - 3y \leq 12$ $-2x + y \leq 4$

Concept	Comments	Example
Linear programming	In a linear programming problem, the maximum or minimum of an objective function is found, subject to constraints. If a solution exists, it occurs at a vertex of the region of feasible solutions.	Maximize the objective function $$P = 2x + 3y,$$ subject to the following constraints. $$2x + y \leq 6$$ $$x + 2y \leq 6$$ $$x \geq 0, y \geq 0$$ The maximum of $P = 10$ occurs at vertex $(2, 2)$.

9.2 Exercises

Inequalities

Exercises 1–12: Graph the solution set to the inequality.

1. $x \geq y$ **2.** $y > -3$

3. $x < 1$ **4.** $y > 2x$

5. $x + y \leq 2$ **6.** $x + y > -3$

7. $2x + y > 4$ **8.** $2x + 3y \leq 6$

9. $x^2 + y^2 > 4$ **10.** $x^2 + y^2 \leq 1$

11. $x^2 + y \leq 2$ **12.** $2x^2 - y < 1$

Exercises 13–16: Match the system of inequalities with the appropriate graph (a–d). Use the graph to identify one solution.

13. $x + y \geq 2$
    $x - y \leq 1$

14. $2x - y > 0$
    $x - 2y \leq 1$

15. $\frac{1}{2}x^3 - y > 0$
    $2x - y \leq 1$

16. $x^2 + y \leq 4$
    $x^2 - y \leq 2$

a.

b.

c.

d.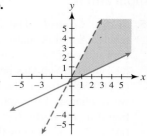

Exercises 17–22: Graph the solution set to the system of inequalities. Use the graph to identify one solution.

17. $y \geq x^2$
$x + y \leq 6$

18. $y \leq \sqrt{x}$
$y \geq 1$

19. $x + 2y > -2$
$x + 2y < 5$

20. $x - y \leq 3$
$x + y \leq 3$

21. $x^2 + y^2 \leq 16$
$x + y < 2$

22. $x^2 + y \leq 4$
$x^2 - y \leq 3$

Exercises 23–34: Graph the solution set to the system of inequalities.

23. $x + 2y \leq 4$
$2x - y \geq 6$

24. $3x - y \leq 3$
$x + 2y \leq 2$

25. $3x + 2y < 6$
$x + 3y \leq 6$

26. $4x + 3y \geq 12$
$2x + 6y \geq 4$

27. $x - 2y \geq 0$
$x - 3y \leq 3$

28. $2x - 4y \geq 4$
$x + y \leq 0$

29. $x^2 + y^2 \leq 4$
$y \geq 1$

30. $x^2 - y \leq 0$
$x^2 + y^2 \leq 6$

31. $2x^2 + y \leq 0$
$x^2 - y \leq 3$

32. $x^2 + 2y \leq 4$
$x^2 - y \leq 0$

33. $x^2 + y^2 \leq 4$
$x^2 + 2y \leq 2$

34. $2x + 3y \leq 6$
$\frac{1}{2}x^2 - y \leq 2$

Applications

35. *Traffic Control* The figure shows two intersections, labeled A and B, that involve one-way streets. The numbers and variables represent the average traffic flow rates measured in vehicles per hour. For example, an average of 500 vehicles per hour enter intersection A from the west, whereas 150 vehicles per hour enter this intersection from the north. A stoplight will control the unknown traffic flow denoted by the variables x and y. Use the fact that the number of vehicles entering an intersection must equal the number leaving to determine x and y.

36. *Traffic Control* (Refer to Exercise 35.) Suppose that the number of vehicles entering intersection A from the west varies between 400 and 600. If all other traffic flows remain the same as in the figure, what effect does this have on the ranges of the values for x and y?

Exercises 37–40: Weight and Height (Refer to the introduction to this section.) The following graph shows a weight and height chart. The weight w is listed in pounds and the height h in inches. The shaded area is a recommended region. (**Source:** Department of Agriculture.)

37. What does this chart indicate about an individual who weighs 125 pounds and is 70 inches tall?

38. Use the graph to estimate the recommended weight range for a person 74 inches tall.

39. Use the graph to find a system of linear inequalities that describes the recommended region.

40. Explain why inequalities are more appropriate than equalities for describing recommended weight and height combinations.

Linear Programming

Exercises 41–44: Shade the region of feasible solutions for the following constraints.

41. $x + y \leq 4$
$x + y \geq 1$
$x \geq 0, y \geq 0$

42. $x + 2y \leq 8$
$2x + y \geq 2$
$x \geq 0, y \geq 0$

43. $3x + 2y \leq 12$
$2x + 3y \leq 12$
$x \geq 0, y \geq 0$

44. $x + y \leq 4$
$x + 4y \geq 4$
$x \geq 0, y \geq 0$

Exercises 45 and 46: The graph shows a region of feasible solutions for P. Find the maximum and minimum values of P.

45. $P = 3x + 5y$

46. $P = 6x + y$

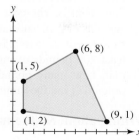

Exercises 47–50: The graph shows a region of feasible solutions for C. Find the maximum and minimum values of C.

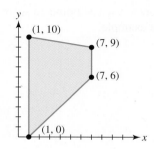

47. $C = 3x + 5y$

48. $C = 5x + 5y$

49. $C = 10y$

50. $C = 3x - y$

Exercises 51 and 52: Write a system of linear inequalities that describes the shaded region.

51.

52.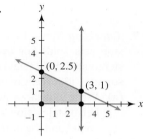

53. Find the minimum value of $C = 4x + 2y$ subject to the following constraints.

$$x + y \geq 3$$
$$2x + 3y \leq 12$$
$$x \geq 0, y \geq 0$$

54. Find the maximum value of $P = 3x + 5y$ subject to the following constraints.

$$3x + y \leq 8$$
$$x + 3y \leq 8$$
$$x \geq 0, y \geq 0$$

Exercises 55 and 56: If possible, maximize and minimize z subject to the given constraints.

55. $z = 7x + 6y$

$$x + y \leq 8$$
$$x + y \geq 4$$
$$x \geq 0, y \geq 0$$

56. $z = 8x + 3y$

$$4x + y \geq 12$$
$$x + 2y \geq 6$$
$$x \geq 0, y \geq 0$$

57. *Maximizing Profit*   Rework Example 4 if the profit from each radio is $20 and the profit from each CD player is $15.

58. *Maximizing Revenue*   A refinery produces both gasoline and fuel oil, and sells gasoline for $4.00 per gallon and fuel oil for $3.60 per gallon. The refinery can produce at most 600,000 gallons a day, but must produce at least 2 gallons of fuel oil for every gallon of gasoline. At least 150,000 gallons of fuel oil must be made each day for the coming winter. Determine how much of each type of fuel should be produced to maximize revenue.

59. *Minimizing Cost*   (Refer to Example 6.) A breeder is mixing Brand A and Brand B. Each serving should contain at least 60 grams of protein and 30 grams of fat. Brand A costs 80 cents per unit, and Brand B costs 50 cents per unit. Each unit of Brand A contains 15 grams of protein and 10 grams of fat, whereas each unit of Brand B contains 20 grams of protein and 5 grams of fat. Determine how much of each food should be bought to achieve a minimum cost per serving.

60. *Pet Food Cost*   A pet owner is buying two brands of food, X and Y, for his animals. Each serving of the mixture of the two foods should contain at least 60 grams of protein and 40 grams of fat. Brand X costs 75 cents per unit, and Brand Y costs 50 cents per unit. Each unit of Brand X contains 20 grams of protein and 10 grams of fat, whereas each unit of Brand Y contains 10 grams of protein and 10 grams of fat. How much of each brand should be bought to obtain a minimum cost per serving?

61. *Raising Animals*   A breeder can raise no more than 50 hamsters and mice and no more than 20 hamsters. If she sells the hamsters for $15 each and the mice for $10 each, find the maximum revenue produced.

62. *Maximizing Storage*   A manager wants to buy filing cabinets. Cabinet X costs $100, requires 6 square feet of floor space, and holds 8 cubic feet. Cabinet Y costs $200, requires 8 square feet of floor space, and holds 12 cubic feet. No more than $1400 can be spent, and the office has room for no more than 72 square feet of cabinets. The manager wants the maximum storage capacity within the limits imposed by funds and space. How many of each type of cabinet should be bought?

63. *Maximizing Profit*   A business manufactures two parts, X and Y. Machines A and B are needed to make each part. To make part X, machine A is needed for 4 hours and machine B is needed for 2 hours. To make part Y, machine A is needed for 1 hour and machine B is needed for 3 hours. Machine A is available for 40 hours each week and machine B is available for 30 hours. The profit from part X is $500 and the profit from part Y is $600. How many parts of each type should be made to maximize weekly profit?

64. *Minimizing Cost*   Two substances, X and Y, are found in pet food. Each substance contains the ingredients A and B. Substance X is 20% ingredient A and 50% ingredient B. Substance Y is 50% ingredient A and 30% ingredient B. The cost of substance X is $2 per pound, and the cost of substance Y is $3 per pound. The pet store needs at least 251 pounds of ingredient A and at least 200 pounds of ingredient B. If cost is to be minimal, how many pounds of each substance should be ordered? Find the minimum cost.

Writing about Mathematics

65. Give the general form of a system of linear inequalities in two variables. Discuss what distinguishes a system of linear inequalities from a nonlinear system of inequalities.

66. Discuss how to use test points to solve a linear inequality. Give an example.

CHECKING BASIC CONCEPTS FOR SECTIONS 9.1 AND 9.2

1. Evaluate $d(13, 18)$ if
$$d(x, y) = \sqrt{(x - 1)^2 + (y - 2)^2}.$$

2. Solve the nonlinear system of equations using the method of substitution.
$$2x^2 - y = 0$$
$$3x + 2y = 7$$

3. Solve $z = x^2 + y^2$ for y.

4. Solve the system of linear equations by using elimination.
$$3x - 2y = 4$$
$$-x + 6y = 8$$

5. Graph the solution set to $3x - 2y \leq 6$.

6. Graph the solution set to the system of inequalities. Use the graph to identify one solution.
$$x^2 - y < 3$$
$$x - y \geq 1$$

7. *Television Sets*   In 1999 there were 32 million television sets sold. For every 10 television sets sold with stereo sound, 19 television sets were sold without stereo sound.   (**Source:** Consumer Electronics.)
   (a) Write a linear system whose solution gives the number of television sets sold (in millions) with stereo sound and the number of television sets sold (in millions) without stereo sound.

   (b) Solve the system and interpret the result.

9.3 Systems of Linear Equations in Three Variables

- Learn basic concepts about systems in three variables
- Solve systems using elimination and substitution
- Identify systems with no solutions
- Solve systems with infinitely many solutions

Introduction

In Section 9.1 we discussed how to solve systems of linear equations in two variables. Systems of linear equations can have any number of variables. For example, in designing modern aircraft, engineers routinely solve systems of linear equations containing 100,000 variables. Computers are necessary to solve these systems efficiently. However, in this section we discuss solving systems of linear equations containing three variables by hand.

Basic Concepts

When writing systems of linear equations in three variables it is common, but not necessary, to use the variables x, y, and z. For example,

$$2x - 3y + 4z = 4$$
$$-y + 2z = 0$$
$$x + 5y - 6z = 7$$

represents a system of linear equations in three variables. The solution to this system is given by $x = 3$, $y = 2$, and $z = 1$ because each equation is satisfied when these values are substituted for the variables in the system of linear equations.

$$2(3) - 3(2) + 4(1) \overset{?}{=} 4 \qquad \text{True}$$
$$-(2) + 2(1) \overset{?}{=} 0 \qquad \text{True}$$
$$(3) + 5(2) - 6(1) \overset{?}{=} 7 \qquad \text{True}$$

The solution to this system can be written as the **ordered triple** $(3, 2, 1)$. This system of linear equations has exactly one solution. In general, systems of linear equations can have zero, one, or infinitely many solutions.

EXAMPLE 1 Checking for solutions

Determine whether $(-1, -3, 2)$ or $(1, -10, -13)$ is a solution to the system of equations.

$$x - 4y + 2z = 15$$
$$4x - y + z = 1$$
$$6x - 2y - 3z = -6$$

SOLUTION First substitute $x = -1$, $y = -3$, and $z = 2$ in the system of linear equations, and then substitute $x = 1$, $y = -10$, and $z = -13$.

$(-1) - 4(-3) + 2(2) \overset{?}{=} 15$ True $\qquad (1) - 4(-10) + 2(-13) \overset{?}{=} 15$ True

$4(-1) - (-3) + (2) \overset{?}{=} 1$ True $\qquad 4(1) - (-10) + (-13) \overset{?}{=} 1$ True

$6(-1) - 2(-3) - 3(2) \overset{?}{=} -6$ True $\qquad 6(1) - 2(-10) - 3(-13) \overset{?}{=} -6$ False

The ordered triple $(-1, -3, 2)$ satisfies all three equations, so it is a solution to the system of equations. The ordered triple $(1, -10, -13)$ is not a solution to the system of equations because it satisfies only two of the three equations.

Now Try Exercise 5 ◀

Solving with Elimination and Substitution

We can solve systems of linear equations in three variables by hand. The following procedure uses substitution and elimination and assumes that the variables are x, y, and z.

Solving a System of Linear Equations in Three Variables

STEP 1: Eliminate one variable, such as x, from two of the equations.

STEP 2: Apply the techniques discussed in Section 9.1 to solve the two resulting equations in two variables from Step 1. If x is eliminated, then solve these equations to find y and z.

If there are no solutions for y and z, then the given system also has no solutions. If there are infinitely many solutions for y and z, then write y in terms of z and proceed to Step 3.

STEP 3: Substitute the values for y and z in one of the given equations to find x. The solution is (x, y, z). Check your answer as in Example 1.

> **EXAMPLE 2** Solving a linear system in three variables

Solve the following system.

$$\begin{aligned} x - y + 2z &= 6 \\ 2x + y - 2z &= -3 \\ -x - 2y + 3z &= 7 \end{aligned}$$

SOLUTION

STEP 1: We begin by eliminating the variable x from the second and third equations. To eliminate x from the second equation, we multiply the first equation by -2 and then add it to the second equation. To eliminate x from the third equation, we add the first and third equations.

$$\begin{array}{ll} -2x + 2y - 4z = -12 & \text{First equation times } -2 \\ \underline{2x + y - 2z = -3} & \text{Second equation} \\ 3y - 6z = -15 & \text{Add.} \end{array}$$

$$\begin{array}{ll} x - y + 2z = 6 & \text{First equation} \\ \underline{-x - 2y + 3z = 7} & \text{Third equation} \\ -3y + 5z = 13 & \text{Add.} \end{array}$$

STEP 2: Take the two resulting equations from Step 1 and eliminate either variable. Here we add the two equations to eliminate y.

$$\begin{array}{ll} 3y - 6z = -15 & \\ \underline{-3y + 5z = 13} & \\ -z = -2 & \text{Add the equations.} \\ z = 2 & \text{Multiply by } -1. \end{array}$$

Now we can use substitution to find the value of y. We let $z = 2$ in either equation used in Step 2 to find y.

$$\begin{array}{ll} 3y - 6z = -15 & \text{Equation from Step 2} \\ 3y - 6(2) = -15 & \text{Substitute } z = 2. \\ 3y - 12 = -15 & \text{Multiply.} \\ 3y = -3 & \text{Add 12.} \\ y = -1 & \text{Divide by 3.} \end{array}$$

STEP 3:   Finally, we substitute $y = -1$ and $z = 2$ in any of the *given* equations to find x.

$$x - y + 2z = 6 \qquad \text{First given equation}$$
$$x - (-1) + 2(2) = 6 \qquad \text{Let } y = -1 \text{ and } z = 2.$$
$$x + 1 + 4 = 6 \qquad \text{Simplify.}$$
$$x = 1 \qquad \text{Subtract 5.}$$

The solution is $(1, -1, 2)$. Check this solution.   *Now Try Exercise 9* ◀

In the next example, we determine numbers of tickets sold for a play.

EXAMPLE 3   **Finding numbers of tickets sold**

One thousand tickets were sold for a play, generating \$3800 in revenue. The prices of the tickets were \$3 for children, \$4 for students, and \$5 for adults. There were 100 fewer student tickets sold than adult tickets. Find the number of each type of ticket sold.

SOLUTION   Let x be the number of tickets sold to children, y be the number of tickets sold to students, and z be the number of tickets sold to adults. The total number of tickets sold was 1000, so

$$x + y + z = 1000.$$

Each child's ticket cost \$3, so the revenue generated from selling x tickets would be $3x$. Similarly, the revenue generated from students would be $4y$, and the revenue from adults would be $5z$. Total ticket sales were \$3800, so

$$3x + 4y + 5z = 3800.$$

The equation $z - y = 100$, or $y - z = -100$, must also be satisfied, as 100 fewer tickets were sold to students than adults.

To find the price of a ticket, we need to solve the following system of linear equations.

$$x + y + z = 1000$$
$$3x + 4y + 5z = 3800$$
$$y - z = -100$$

STEP 1:   We begin by eliminating the variable x from the first equation. To do so, we multiply the first equation by 3 and subtract the second equation.

$$3x + 3y + 3z = 3000 \qquad \text{First given equation times 3}$$
$$\underline{3x + 4y + 5z = 3800} \qquad \text{Second equation}$$
$$-y - 2z = -800 \qquad \text{Subtract.}$$

STEP 2:   We then use the equation that resulted from Step 1 and the third given equation to eliminate y.

$$-y - 2z = -800 \qquad \text{Equation from Step 1}$$
$$\underline{y - z = -100} \qquad \text{Third given equation}$$
$$-3z = -900 \qquad \text{Add the equations.}$$

Thus $z = 300$. To find y, we can substitute $z = 300$ in the third equation.

$$
\begin{aligned}
y - z &= -100 && \text{\textit{Third given equation}} \\
y - 300 &= -100 && \text{\textit{Let z = 300.}} \\
y &= 200 && \text{\textit{Add 300.}}
\end{aligned}
$$

STEP 3: Finally, we substitute $y = 200$ and $z = 300$ in the first equation.

$$
\begin{aligned}
x + y + z &= 1000 && \text{\textit{First given equation}} \\
x + 200 + 300 &= 1000 && \text{\textit{Let y = 200 and z = 300.}} \\
x &= 500 && \text{\textit{Subtract 500.}}
\end{aligned}
$$

Thus **500** tickets were sold to children, **200** to students, and **300** to adults. Check this answer.   ***Now Try Exercise 31*** ◄

Systems with No Solutions

Regardless of the number of variables, a system of linear equations can have zero, one, or infinitely many solutions. In the next example, a system of linear equations has no solutions.

> **EXAMPLE 4** Identifying a system with no solutions

Three students buy lunch in the cafeteria. One student buys 2 hamburgers, 1 order of fries, and 1 soda for $9. Another student buys 1 hamburger, 2 orders of fries, and 1 soda for $8. The third student buys 3 hamburgers, 3 orders of fries, and 2 sodas for $18. If possible, find the cost of each item. Interpret the results.

SOLUTION Let x be the cost of a hamburger, y be the cost of an order of fries, and z be the cost of a soda. Then the purchases of the three students can be expressed as a system of linear equations.

$$
\begin{aligned}
2x + y + z &= 9 \\
x + 2y + z &= 8 \\
3x + 3y + 2z &= 18
\end{aligned}
$$

STEP 1: We can eliminate z in the first equation by subtracting the second equation from the first equation. We can eliminate z in the third equation by subtracting twice the second equation from the third equation.

$$
\begin{array}{ll}
\begin{aligned}
2x + y + z &= 9 \\
\underline{x + 2y + z} &= \underline{8} \\
x - y \phantom{{}+ z} &= 1
\end{aligned}
&
\begin{aligned}
3x + 3y + 2z &= 18 \\
\underline{2x + 4y + 2z} &= \underline{16} \\
x - y \phantom{{}+ 2z} &= 2
\end{aligned}
\end{array}
$$

First equation   *Third equation*
Second equation   *Twice second equation*
Subtract.   *Subtract.*

STEP 2: The equations $x - y = 1$ and $x - y = 2$ are *inconsistent* because the difference between two numbers cannot be both 1 and 2. Step 3 is not necessary—the system of equations has no solutions.

> **NOTE** In this problem the third student bought the same amount of food as the first and second students bought together. Therefore the third student should have paid $9 + $8 = $17 rather than $18. *Inconsistent pricing* led to an *inconsistent system* of linear equations.   ***Now Try Exercise 33*** ◄

Systems with Infinitely Many Solutions

Some systems of linear equations have infinitely many solutions. In this case, we say that the system of linear equations is consistent, but the equations are dependent. A system of dependent equations is solved in the next example.

EXAMPLE 5   Solving a system with infinitely many solutions

Solve the following system of linear equations.

$$\begin{aligned} x + y - z &= -2 \\ x + 2y - 2z &= -3 \\ y - z &= -1 \end{aligned}$$

SOLUTION

STEP 1:   Because x does not appear in the third equation, begin by eliminating x from the first equation. To do this, subtract the second equation from the first equation.

$$\begin{array}{ll} x + y - z = -2 & \text{First equation} \\ \underline{x + 2y - 2z = -3} & \text{Second equation} \\ -y + z = 1 & \text{Subtract.} \end{array}$$

STEP 2:   Adding the resulting equation from Step 1 and the third given equation gives the equation $0 = 0$, which indicates that there are infinitely many solutions.

$$\begin{array}{ll} -y + z = 1 \\ \underline{y - z = -1} \\ 0 = 0 & \text{Add.} \end{array}$$

The variable y can be written in terms of z as $y = z - 1$.

STEP 3:   To find x, substitute the results from Step 2 in the first given equation.

$$\begin{array}{ll} x + y - z = -2 & \text{First given equation} \\ x + (z - 1) - z = -2 & \text{Let } y = z - 1. \\ x = -1 & \text{Solve for } x. \end{array}$$

Solutions to the given system are of the form $(-1, z - 1, z)$, where z is any real number. For example, if $z = 2$, then $(-1, 1, 2)$ is one possible solution.   ***Now Try Exercise 15*** ◄

CLASS DISCUSSION

Three students buy lunch in the cafeteria. One student buys 1 hamburger, 1 order of fries, and 1 soda for $5. Another student buys 2 hamburgers, 2 orders of fries, and 2 sodas for $10. The third student buys 3 hamburgers, 3 orders of fries, and 3 sodas for $15. Can you find the cost of each item? Interpret your answer.

9.3 Putting It All Together

In this section we discussed how to solve a system of three linear equations in three variables by hand. Systems of linear equations can have no solutions, one solution, or infinitely many solutions. The following table summarizes some of the important concepts presented in this section.

Concept	Explanation
System of linear equations in three variables	The following is a system of three linear equations in three variables. $$\begin{aligned} x - 2y + z &= 0 \\ -x + y + z &= 4 \\ -y + 4z &= 10 \end{aligned}$$
Solution to a linear system in three variables	The solution to a linear system in three variables is an ordered triple, expressed as (x, y, z). The solution to the preceding system is $(1, 2, 3)$ because substituting $x = 1$, $y = 2$, and $z = 3$ in each equation results in a true statement. $$\begin{aligned} (1) - 2(2) + (3) &= 0 \quad \text{True} \\ -(1) + (2) + (3) &= 4 \quad \text{True} \\ -(2) + 4(3) &= 10 \quad \text{True} \end{aligned}$$
Solving a linear system with substitution and elimination	Refer to Example 2. **STEP 1:** Eliminate one variable, such as x, from two of the equations. **STEP 2:** Apply the techniques discussed in Section 9.1 to solve the two resulting equations in two variables from Step 1. If x is eliminated, then solve these equations to find y and z. **STEP 3:** Substitute the values for y and z in one of the given equations to find x. The solution is (x, y, z). Check your answer.

9.3 Exercises

1. Can a system of linear equations have exactly three solutions?

2. Does the ordered triple $(1, 2, 3)$ satisfy the equation $3x + 2y + z = 10$?

3. To solve a system of linear equations in two variables, how many equations do you usually need?

4. To solve a system of linear equations in three variables, how many equations do you usually need?

Exercises 5–8: Determine whether each ordered triple is a solution to the system of linear equations.

5. $(0, 2, -2), (-1, 3, -2)$
$$\begin{aligned} x + y - z &= 4 \\ -x + y + z &= 2 \\ x + y + z &= 0 \end{aligned}$$

6. $(5, 2, 2), (2, -1, 1)$
$$\begin{aligned} 2x - 3y + 3z &= 10 \\ x - 2y - 3z &= 1 \\ 4x - y + z &= 10 \end{aligned}$$

7. $\left(-\frac{5}{11}, \frac{20}{11}, -2\right), (1, 2, -1)$
$$\begin{aligned} x + 3y - 2z &= 9 \\ -3x + 2y + 4z &= -3 \\ -2x + 5y + 2z &= 6 \end{aligned}$$

8. $(1, 2, 3), (11, 16, -3)$
$$\begin{aligned} 4x - 2y + 2z &= 6 \\ 2x - 4y - 6z &= -24 \\ -3x + 3y + 2z &= 9 \end{aligned}$$

Exercises 9–30: If possible, solve the system.

9. $x + y + z = 6$
   $-x + 2y + z = 6$
   $y + z = 5$

10. $x - y + z = -2$
    $x - 2y + z = 0$
    $y - z = 1$

11. $x + 2y + 3z = 4$
    $2x + y + 3z = 5$
    $x - y + z = 2$

12. $x - y + z = 2$
    $3x - 2y + z = -1$
    $x + y = -3$

13. $3x + y + z = 0$
    $4x + 2y + z = 1$
    $2x - 2y - z = 2$

14. $-x - 5y + 2z = 2$
    $x + y + 2z = 2$
    $3x + y - 4z = -10$

15. $x + 3y + z = 6$
    $3x + y - z = 6$
    $x - y - z = 0$

16. $2x - y + 2z = 6$
    $-x + y + z = 0$
    $-x - 3z = -6$

17. $x - 4y + 2z = -2$
    $x + 2y - 2z = -3$
    $x - y = 4$

18. $2x + y + 3z = 4$
    $-3x - y - 4z = 5$
    $x + y + 2z = 0$

19. $4a - b + 2c = 0$
    $2a + b - c = -11$
    $2a - 2b + c = 3$

20. $a - 4b + 3c = 2$
    $-a - 2b + 5c = 9$
    $a + 2b + c = 6$

21. $a + b + c = 0$
    $a - b - c = 3$
    $a + 3b + 3c = 5$

22. $a - 2b + c = -1$
    $a + 5b = -3$
    $2a + 3b + c = -2$

23. $3x + 2y + z = -1$
    $3x + 4y - z = 1$
    $x + 2y + z = 0$

24. $x - 2y + z = 1$
    $x + y + 2z = 2$
    $2x + 3y + z = 6$

25. $-x + 3y + z = 3$
    $2x + 7y + 4z = 13$
    $4x + y + 2z = 7$

26. $x + 2y + z = 0$
    $3x + 2y - z = 4$
    $-x + 2y + 3z = -4$

27. $-x + 2z = -9$
    $y + 4z = -13$
    $3x + y = 13$

28. $x + y + z = -1$
    $2x + z = -6$
    $2y + 3z = 0$

29. $\frac{1}{2}x - y + \frac{1}{2}z = -4$
    $x + 2y - 3z = 20$
    $-\frac{1}{2}x + 3y + 2z = 0$

30. $\frac{3}{4}x + y + \frac{1}{2}z = -3$
    $x + y - z = -8$
    $\frac{1}{4}x - 2y + z = -4$

Applications

31. *Tickets Sold* Five hundred tickets were sold for a play, generating $3560. The prices of the tickets were $5 for children, $7 for students, and $10 for adults. There were 180 more student tickets sold than adult tickets. Find the number of each type of ticket sold.

32. *Tickets Sold* One thousand tickets were sold for a baseball game. There were one hundred more adult tickets sold than student tickets, and there were four times as many tickets sold to students as to children. How many of each type of ticket were sold?

33. *Buying Lunch* Three students buy lunch in the cafeteria. One student buys 2 hamburgers, 2 orders of fries, and 1 soda for $9. Another student buys 1 hamburger, 1 order of fries, and 1 soda for $5. The third student buys 1 hamburger and 1 order of fries for $5. If possible, find the cost of each item. Interpret the results.

34. *Cost of CDs* The table shows the total cost of purchasing various combinations of differently priced CDs. The types of CDs are labeled A, B, and C.

A	B	C	Total Cost
2	1	1	$48
3	2	1	$71
1	1	2	$53

(a) Let a be the cost of a CD of type A, b be the cost of a CD of type B, and c be the cost of a CD of type C. Write a system of three linear equations whose solution gives the cost of each type of CD.

(b) Solve the system of equations and check your answer.

35. *Geometry* The largest angle in a triangle is 25° more than the smallest angle. The sum of the measures of the two smaller angles is 30° more than the measure of the largest angle.

(a) Let x, y, and z be the measures of the three angles from largest to smallest. Write a system of three linear equations whose solution gives the measure of each angle.

(b) Solve the system of equations and check your answer.

36. *Geometry* The perimeter of a triangle is 105 inches. The longest side is 22 inches longer than the shortest side. The sum of the lengths of the two shorter sides is 15 inches more than the length of the longest side. Find the lengths of the sides of the triangle.

37. *Investment Mixture* A sum of $20,000 is invested in three mutual funds. In one year the first fund grew by 5%, the second by 7%, and the third by 10%. Total earnings for the year were $1650. The amount invested in the third fund was 4 times the amount invested in the first fund. Find the amount invested in each fund.

38. *Home Prices* Prices of homes can depend on several factors such as size and age. The table shows the selling prices for three homes. In this table, price P is given in thousands of dollars, age A in years, and home size S in thousands of square feet. These data may be modeled by $P = a + bA + cS$.

Price (P)	Age (A)	Size (S)
190	20	2
320	5	3
50	40	1

   (a) Write a system of linear equations whose solution gives a, b, and c.

   (b) Solve this system of linear equations.

   (c) Predict the price of a home that is 10 years old and has 2500 square feet.

39. *Mixture Problem* One type of lawn fertilizer consists of a mixture of nitrogen, N; phosphorus, P; and potassium, K. An 80-pound sample contains 8 more pounds of nitrogen and phosphorus than of potassium. There is 9 times as much potassium as phosphorus.

   (a) Write a system of three equations whose solution gives the amount of nitrogen, phosphorus, and potassium in this sample.

   (b) Solve the system of equations.

40. *Business Production* A business has three machines that manufacture containers. Together they can make 100 containers per day, whereas the two fastest machines can make 80 containers per day. The fastest machine makes 34 more containers per day than the slowest machine.

   (a) Let x, y, and z be the numbers of containers that the machines make from fastest to slowest. Write a system of three equations whose solution gives the number of containers each machine can make.

   (b) Solve the system of equations.

Writing about Mathematics

41. When using elimination and substitution, explain how to recognize a system of linear equations that has no solutions.

42. When using elimination and substitution, explain how to recognize a system of linear equations that has infinitely many solutions.

9.4 Solutions to Linear Systems Using Matrices

- **Represent systems of linear equations with matrices**
- **Learn row-echelon form**
- **Perform Gaussian elimination**
- **Learn reduced row-echelon form**
- **Perform Gauss-Jordan elimination**
- **Solve systems of linear equations with technology (optional)**

Introduction

In 2004 total iPod sales were 3 million. By 2006 this number had increased to 55 million, and by 2008 this cumulative total had reached 150 million. Because three distinct points (that are not collinear) determine the graph of a quadratic function, we can model these data by finding a unique parabola that passes through the points (2004, 3), (2006, 55), and (2008, 150), as illustrated in Figure 9.43.

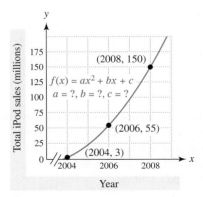

Figure 9.43

One way to accomplish this task is to set up a linear system of equations in three variables and solve it by using a matrix. This section discusses matrices and how they can be used to solve systems of linear equations. (See Example 11 and Exercise 83.)

Representing Systems of Linear Equations with Matrices

Arrays of numbers occur frequently in many different situations. Spreadsheets often make use of arrays, where data are displayed in a tabular format. A **matrix** is a rectangular array of elements. The following are examples of matrices whose elements are real numbers.

$$\begin{bmatrix} 4 & -7 \\ -2 & 9 \end{bmatrix} \quad \begin{bmatrix} -1 & -5 & 3 \\ 1.2 & 0 & -1.3 \\ 4.1 & 5 & 7 \end{bmatrix} \quad \begin{bmatrix} -3 & -6 & 9 & 5 \\ \sqrt{2} & -8 & -8 & 0 \\ 3 & 0 & 19 & -7 \\ -11 & -3 & 7 & 8 \end{bmatrix} \quad \begin{bmatrix} 5 & -2 \\ -2 & \pi \\ 1 & -1 \end{bmatrix} \quad \begin{bmatrix} 1 & -0.5 & 9 \\ 5 & 0.4 & -3 \end{bmatrix}$$

$$2 \times 2 \qquad\qquad 3 \times 3 \qquad\qquad\qquad 4 \times 4 \qquad\qquad\quad 3 \times 2 \qquad\qquad 2 \times 3$$

The dimension of a matrix is given much like the dimensions of a rectangular room. We might say a room is m feet long and n feet wide. The **dimension** of a matrix is $m \times n$ (m by n) if it has m rows and n columns. For example, the last matrix has a dimension of 2×3 because it has 2 rows and 3 columns. If the numbers of rows and columns are equal, the matrix is a **square matrix**. The first three matrices are square matrices.

Matrices are frequently used to represent systems of linear equations. A linear system with three equations and three variables can be written as follows.

$$a_1 x + b_1 y + c_1 z = d_1$$
$$a_2 x + b_2 y + c_2 z = d_2$$
$$a_3 x + b_3 y + c_3 z = d_3$$

CLASS DISCUSSION

Give a general form of a system of linear equations with four equations and four variables. Write its augmented matrix.

The a_k, b_k, c_k, and d_k are constants and x, y, and z are variables. The coefficients of the variables can be represented using a 3×3 matrix. This matrix is called the **coefficient matrix** of the linear system.

$$\begin{bmatrix} a_1 & b_1 & c_1 \\ a_2 & b_2 & c_2 \\ a_3 & b_3 & c_3 \end{bmatrix}$$

If we enlarge the matrix to include the constants d_k, the system may be written as

$$\left[\begin{array}{ccc|c} a_1 & b_1 & c_1 & d_1 \\ a_2 & b_2 & c_2 & d_2 \\ a_3 & b_3 & c_3 & d_3 \end{array} \right].$$

Because this matrix is an enlargement of the coefficient matrix, it is commonly called an **augmented matrix**. The vertical line between the third and fourth columns corresponds to where the equals sign occurs in each equation.

EXAMPLE 1   Representing a linear system with an augmented matrix

Express each linear system with an augmented matrix. State the dimension of the matrix.

(a) $\begin{aligned} 3x - 4y &= 6 \\ -5x + y &= -5 \end{aligned}$

(b) $\begin{aligned} 2x - 5y + 6z &= -3 \\ 3x + 7y - 3z &= 8 \\ x + 7y &= 5 \end{aligned}$

SOLUTION

(a) This system has two equations with two variables. It can be represented by an augmented matrix having dimension 2×3.

$$\begin{bmatrix} 3 & -4 & | & 6 \\ -5 & 1 & | & -5 \end{bmatrix} \qquad \begin{array}{r} 3x - 4y = 6 \\ -5x + y = -5 \end{array}$$

(b) This system has three equations with three variables. Note that variable z does not appear in the third equation. A value of 0 is inserted for its coefficient.

$$\begin{bmatrix} 2 & -5 & 6 & | & -3 \\ 3 & 7 & -3 & | & 8 \\ 1 & 7 & 0 & | & 5 \end{bmatrix} \qquad \begin{array}{r} 2x - 5y + 6z = -3 \\ 3x + 7y - 3z = 8 \\ x + 7y \phantom{{}- 3z} = 5 \end{array}$$

This matrix has dimension 3×4.   ***Now Try Exercises 3 and 5*** ◀

EXAMPLE 2 Converting an augmented matrix into a linear system

Write the linear system represented by the augmented matrix. Let the variables be x, y, and z.

(a) $$\begin{bmatrix} 1 & 0 & 2 & | & -3 \\ 2 & 2 & 10 & | & 3 \\ -1 & 2 & 3 & | & 5 \end{bmatrix}$$ (b) $$\begin{bmatrix} 1 & 2 & 3 & | & -4 \\ 0 & 1 & -6 & | & 7 \\ 0 & 0 & 1 & | & 8 \end{bmatrix}$$

SOLUTION

(a) **Getting Started** The first column corresponds to x, the second to y, and the third to z. When a 0 appears, the variable for that column does not appear in the equation. The vertical line gives the location of the equals sign. The last column represents the constant terms. ▶

$$\begin{array}{r} x \phantom{{}+ 2y} + 2z = -3 \\ 2x + 2y + 10z = 3 \\ -x + 2y + 3z = 5 \end{array}$$

(b) The augmented matrix represents the following linear system.

$$\begin{array}{r} x + 2y + 3z = -4 \\ y - 6z = 7 \\ z = 8 \end{array}$$

     Now Try Exercises 7 and 9 ◀

Row-Echelon Form

To solve linear systems with augmented matrices, a convenient form is row-echelon form. The following matrices are in row-echelon form.

$$\begin{bmatrix} 1 & 3 & 0 & -1 \\ 0 & 1 & -6 & 1 \\ 0 & 0 & 1 & -2 \end{bmatrix} \quad \begin{bmatrix} 1 & 2 & 0 \\ 0 & 1 & 4 \end{bmatrix} \quad \begin{bmatrix} 1 & 3 & -1 & 5 \\ 0 & 1 & -1 & 3 \\ 0 & 0 & 1 & 0 \end{bmatrix} \quad \begin{bmatrix} 1 & 3 & -1 & 5 \\ 0 & 0 & 1 & 3 \\ 0 & 0 & 0 & 0 \end{bmatrix} \quad \begin{bmatrix} 1 & 3 & 5 \\ 0 & 0 & 1 \end{bmatrix}$$

The elements of the **main diagonal** are blue in each matrix. Scanning down the main diagonal of a matrix in row-echelon form, we see that this diagonal first contains only 1's, and then possibly 0's. The first nonzero element in each nonzero row is 1. This 1 is called a **leading 1**. If two rows contain a leading 1, then the row with the left-most leading 1 is listed first in the matrix. Rows containing only 0's occur last in the matrix. All elements below the main diagonal are 0.

The next example demonstrates a technique called **backward substitution**. It can be used to solve linear systems represented by an augmented matrix in row-echelon form.

EXAMPLE 3 Solving a linear system with backward substitution

Solve the system of linear equations represented by the augmented matrix.

(a) $\begin{bmatrix} 1 & 1 & 3 & | & 12 \\ 0 & 1 & -2 & | & -4 \\ 0 & 0 & 1 & | & 3 \end{bmatrix}$ (b) $\begin{bmatrix} 1 & -1 & 5 & | & 5 \\ 0 & 1 & 3 & | & 3 \\ 0 & 0 & 0 & | & 0 \end{bmatrix}$

SOLUTION
(a) The matrix represents the linear system

$$x + y + 3z = 12$$
$$y - 2z = -4$$
$$z = 3.$$

Since $z = 3$, substitute this value in the second equation to find y.

$$y - 2(3) = -4, \quad \text{or} \quad y = 2$$

Then $y = 2$ and $z = 3$ can be substituted in the first equation to determine x.

$$x + 2 + 3(3) = 12, \quad \text{or} \quad x = 1$$

The solution is given by $x = 1$, $y = 2$, and $z = 3$ and can be expressed as the *ordered triple* (**1, 2, 3**).

(b) The matrix represents the linear system

$$x - y + 5z = 5$$
$$y + 3z = 3$$
$$0 = 0.$$

The last equation, $0 = 0$, is an identity. Its presence usually indicates infinitely many solutions. Use the second equation to write y in terms of z.

$$y = 3 - 3z$$

Next, substitute $(3 - 3z)$ for y in the first equation and write x in terms of z.

$$x - (3 - 3z) + 5z = 5$$
$$x - 3 + 3z + 5z = 5 \qquad \text{Distributive property}$$
$$x = 8 - 8z. \qquad \text{Solve for x}$$

All solutions can be written as the ordered triple $(8 - 8z, 3 - 3z, z)$, where z is any real number. There are infinitely many solutions. Sometimes we say that all solutions can be written in terms of the **parameter** z, where z is any real number. For example, if we let $z = 1$, then $y = 3 - 3(1) = 0$ and $x = 8 - 8(1) = 0$. Thus one solution to the system is (0, 0, 1).   ***Now Try Exercises 17 and 19*** ◄

Gaussian Elimination

The methods of elimination and substitution from Section 9.1 can be applied to linear systems that involve more than three variables. As the number of variables increases, these methods can become cumbersome and inefficient. However, properly combining these two methods creates a state-of-the-art numerical method capable of solving systems of linear equations that contain thousands of variables. This numerical method is called *Gaussian elimination with backward substitution.* Even though this method dates back to Carl Friedrich Gauss (1777–1855), it continues to be one of the most efficient methods for solving systems of linear equations.

If an augmented matrix is not in row-echelon form, it can be transformed into row-echelon form using *Gaussian elimination.* This method uses three basic matrix row transformations.

Matrix Row Transformations

For any augmented matrix representing a system of linear equations, the following row transformations result in an equivalent system of linear equations.

1. Any two rows may be interchanged.
2. The elements of any row may be multiplied by a nonzero constant.
3. Any row may be changed by adding to (or subtracting from) its elements a multiple of the corresponding elements of another row.

When we transform a matrix into row-echelon form, we also are transforming a system of linear equations. The next two examples illustrate how Gaussian elimination with backward substitution is performed.

EXAMPLE 4 Transforming a matrix into row-echelon form

Use Gaussian elimination with backward substitution to solve the linear system of equations.

$$x + y + z = 1$$
$$-x + y + z = 5$$
$$y + 2z = 5$$

SOLUTION

Getting Started The goal is to apply matrix row transformations that transform the given matrix into row-echelon form. Then we can perform backward substitution to determine the solution. ▶

The linear system is written to the right to illustrate how each row transformation affects the corresponding system of linear equations. Note that it is *not* necessary to write the system of equations to the right of the augmented matrix.

Augmented Matrix	*Linear System*
$\begin{bmatrix} 1 & 1 & 1 & \vert & 1 \\ -1 & 1 & 1 & \vert & 5 \\ 0 & 1 & 2 & \vert & 5 \end{bmatrix}$	$\begin{aligned} x + y + z &= 1 \\ -x + y + z &= 5 \\ y + 2z &= 5 \end{aligned}$

The first step is to add the first equation to the second equation to obtain a 0 where the coefficient of x in the second row is highlighted. This row operation is denoted $R_2 + R_1$, and the result becomes the new row 2. It is important to write down each row operation so that we can check our work easily. The row that is changing is written first.

$$R_2 + R_1 \rightarrow \begin{bmatrix} 1 & 1 & 1 & | & 1 \\ 0 & 2 & 2 & | & 6 \\ 0 & 1 & 2 & | & 5 \end{bmatrix} \qquad \begin{array}{rcl} x + y + z &=& 1 \\ 2y + 2z &=& 6 \\ y + 2z &=& 5 \end{array}$$

To have the matrix in row-echelon form, we need the highlighted 2 in the second row to be a 1. Multiply each element in row 2 by $\frac{1}{2}$ and denote the operation $\frac{1}{2}R_2$.

$$\frac{1}{2}R_2 \rightarrow \begin{bmatrix} 1 & 1 & 1 & | & 1 \\ 0 & 1 & 1 & | & 3 \\ 0 & 1 & 2 & | & 5 \end{bmatrix} \qquad \begin{array}{rcl} x + y + z &=& 1 \\ y + z &=& 3 \\ y + 2z &=& 5 \end{array}$$

For the matrix to be in row-echelon form, we need a 0 where the 1 is highlighted in row 3. Subtract row 2 from row 3 and denote the operation $R_3 - R_2$.

$$R_3 - R_2 \rightarrow \begin{bmatrix} 1 & 1 & 1 & | & 1 \\ 0 & 1 & 1 & | & 3 \\ 0 & 0 & 1 & | & 2 \end{bmatrix} \qquad \begin{array}{rcl} x + y + z &=& 1 \\ y + z &=& 3 \\ z &=& 2 \end{array}$$

Because we have a 1 in the highlighted box, the matrix is now in row-echelon form, and we see that $z = 2$. Backward substitution may be applied now to find the solution. Substituting $z = 2$ in the second equation gives

$$y + 2 = 3, \quad \text{or} \quad y = 1.$$

Finally, let $y = 1$ and $z = 2$ in the first equation to determine x.

$$x + 1 + 2 = 1, \quad \text{or} \quad x = -2$$

The solution of the system is given by $x = -2$, $y = 1$, and $z = 2$, or $(-2, 1, 2)$.

Now Try Exercise 29 ◀

EXAMPLE 5 **Transforming a matrix into row-echelon form**

Use Gaussian elimination with backward substitution to solve the linear system of equations.

$$\begin{array}{rcl} 2x + 4y + 4z &=& 4 \\ x + 3y + z &=& 4 \\ -x + 3y + 2z &=& -1 \end{array}$$

SOLUTION The initial linear system and augmented matrix are written first.

Augmented Matrix   *Linear System*

$$\begin{bmatrix} 2 & 4 & 4 & | & 4 \\ 1 & 3 & 1 & | & 4 \\ -1 & 3 & 2 & | & -1 \end{bmatrix} \qquad \begin{array}{rcl} 2x + 4y + 4z &=& 4 \\ x + 3y + z &=& 4 \\ -x + 3y + 2z &=& -1 \end{array}$$

First we obtain a 1 where the coefficient of x in the first row is highlighted. This can be accomplished by either multiplying the first equation by $\frac{1}{2}$ or interchanging rows 1 and 2. We multiply row 1 by $\frac{1}{2}$. This operation is denoted $\frac{1}{2}R_1$, to indicate that row 1 in the previous augmented matrix is multiplied by $\frac{1}{2}$.

$$\frac{1}{2}R_1 \rightarrow \begin{bmatrix} 1 & 2 & 2 & | & 2 \\ 1 & 3 & 1 & | & 4 \\ -1 & 3 & 2 & | & -1 \end{bmatrix} \qquad \begin{array}{rcl} x + 2y + 2z &=& 2 \\ x + 3y + z &=& 4 \\ -x + 3y + 2z &=& -1 \end{array}$$

The next step is to eliminate the x-variable in rows 2 and 3 by obtaining zeros in the highlighted positions. To do this, subtract row 1 from row 2, and add row 1 to row 3.

$$\begin{aligned}
& \\
R_2 - R_1 \rightarrow \\
R_3 + R_1 \rightarrow
\end{aligned}
\begin{bmatrix}
1 & 2 & 2 & 2 \\
0 & 1 & -1 & 2 \\
0 & 5 & 4 & 1
\end{bmatrix}
\qquad
\begin{aligned}
x + 2y + 2z &= 2 \\
y - z &= 2 \\
5y + 4z &= 1
\end{aligned}$$

Since we have a 1 for the coefficient of y in the second row, the next step is to eliminate the y-variable in row 3, and obtain a zero where the y-coefficient of 5 is highlighted. Multiply row 2 by 5, and subtract the result from row 3.

$$\begin{aligned}
& \\
& \\
R_3 - 5R_2 \rightarrow
\end{aligned}
\begin{bmatrix}
1 & 2 & 2 & 2 \\
0 & 1 & -1 & 2 \\
0 & 0 & 9 & -9
\end{bmatrix}
\qquad
\begin{aligned}
x + 2y + 2z &= 2 \\
y - z &= 2 \\
9z &= -9
\end{aligned}$$

Finally, make the coefficient of z in the third row equal 1 by multiplying row 3 by $\frac{1}{9}$.

$$\begin{aligned}
& \\
& \\
\tfrac{1}{9}R_3 \rightarrow
\end{aligned}
\begin{bmatrix}
1 & 2 & 2 & 2 \\
0 & 1 & -1 & 2 \\
0 & 0 & 1 & -1
\end{bmatrix}
\qquad
\begin{aligned}
x + 2y + 2z &= 2 \\
y - z &= 2 \\
z &= -1
\end{aligned}$$

The final matrix is in row-echelon form. Backward substitution may be applied to find the solution. Substituting $z = -1$ in the second equation gives

$$y - (-1) = 2, \quad \text{or} \quad y = 1.$$

Next, substitute $y = 1$ and $z = -1$ in the first equation to determine x.

$$x + 2(1) + 2(-1) = 2, \quad \text{or} \quad x = 2$$

The solution to the system is $(2, 1, -1)$.

Now Try Exercise 33 ◀

A Geometric Interpretation In geometry, the graph of a linear equation in three variables is a plane in three-dimensional space. Considering the possible intersections of the planes representing three equations in three unknowns, the solution set of such a system may be either a single ordered triple (x, y, z), an infinite set of ordered triples (dependent equations), or the empty set (an inconsistent system). See Figure 9.44.

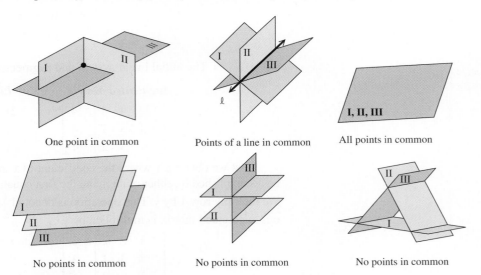

One point in common   Points of a line in common   All points in common

No points in common   No points in common   No points in common

Figure 9.44

Reduced Row-Echelon Form Sometimes it is convenient to express a matrix in reduced row-echelon form. A matrix in row-echelon form is in **reduced row-echelon form** if every element above and below a leading 1 in a column is 0. The following matrices are examples of reduced row-echelon form.

$$\begin{bmatrix} 1 & 0 \\ 0 & 1 \end{bmatrix} \quad \begin{bmatrix} 1 & 0 \\ 0 & 0 \end{bmatrix} \quad \begin{bmatrix} 1 & 0 & 0 \\ 0 & 1 & 0 \\ 0 & 0 & 1 \end{bmatrix} \quad \begin{bmatrix} 1 & 0 & 3 \\ 0 & 1 & -2 \end{bmatrix} \quad \begin{bmatrix} 1 & 0 & 0 & 3 \\ 0 & 1 & 0 & 1 \\ 0 & 0 & 1 & -1 \end{bmatrix} \quad \begin{bmatrix} 1 & 0 & 4 & 8 \\ 0 & 1 & -1 & 2 \\ 0 & 0 & 0 & 0 \end{bmatrix}$$

If an augmented matrix is in reduced row-echelon form, solving the system of linear equations is often straightforward.

EXAMPLE 6 Determining a solution from a matrix in reduced row-echelon form

Each matrix represents a system of linear equations. Find the solution.

(a) $\left[\begin{array}{cc|c} 1 & 0 & 6 \\ 0 & 1 & -5 \end{array}\right]$

(b) $\left[\begin{array}{ccc|c} 1 & 0 & 0 & 3 \\ 0 & 1 & 0 & -1 \\ 0 & 0 & 1 & 2 \end{array}\right]$

(c) $\left[\begin{array}{ccc|c} 1 & 0 & 0 & 4 \\ 0 & 1 & 0 & 3 \\ 0 & 0 & 0 & 2 \end{array}\right]$

(d) $\left[\begin{array}{ccc|c} 1 & 0 & -2 & -3 \\ 0 & 1 & 2 & 1 \\ 0 & 0 & 0 & 0 \end{array}\right]$

SOLUTION

(a) The system involves two equations and two unknowns. The solution is $(6, -5)$.

(b) The top row represents $1x + 0y + 0z = 3$, or $x = 3$. Using similar reasoning for the second and third rows yields $y = -1$ and $z = 2$. The solution is $(3, -1, 2)$.

(c) The last row represents $0x + 0y + 0z = 2$, which has no solutions because $0 \neq 2$. Therefore there are no solutions to the system of equations.

(d) The last row simplifies to $0 = 0$, which is an identity and is always true. The second row gives $y + 2z = 1$, or $y = -2z + 1$. The first row represents $x - 2z = -3$, or $x = 2z - 3$. Thus this system of linear equations has infinitely many solutions. All solutions can be written as an ordered triple in the form $(2z - 3, -2z + 1, z)$, where z can be any real number. ***Now Try Exercises 51, 53, 55, and 57*** ◀

Gauss-Jordan Elimination Matrix row transformations can be used to transform an augmented matrix into reduced row-echelon form. This approach requires more effort than transforming a matrix into row-echelon form, but often eliminates the need for backward substitution. The technique is sometimes called **Gauss-Jordan elimination**.

EXAMPLE 7 Transforming a matrix into reduced row-echelon form

Use Gauss-Jordan elimination to solve the linear system.

$$\begin{aligned} 2x + \quad y + 2z &= 10 \\ x \qquad\quad + 2z &= 5 \\ x - 2y + 2z &= 1 \end{aligned}$$

SOLUTION The linear system has been written to the right for illustrative purposes.

$$
\begin{array}{cc}
\textbf{\textit{Augmented Matrix}} & \textbf{\textit{Linear System}}
\end{array}
$$

$$
\left[\begin{array}{ccc|c}
2 & 1 & 2 & 10 \\
1 & 0 & 2 & 5 \\
1 & -2 & 2 & 1
\end{array}\right]
\qquad
\begin{array}{rcr}
2x + y + 2z &=& 10 \\
x + 2z &=& 5 \\
x - 2y + 2z &=& 1
\end{array}
$$

Start by obtaining a leading 1 in the first row by interchanging rows 1 and 2.

$$
\begin{array}{c}
R_2 \rightarrow \\
R_1 \rightarrow \\

\end{array}
\left[\begin{array}{ccc|c}
1 & 0 & 2 & 5 \\
2 & 1 & 2 & 10 \\
1 & -2 & 2 & 1
\end{array}\right]
\qquad
\begin{array}{rcr}
x + 2z &=& 5 \\
2x + y + 2z &=& 10 \\
x - 2y + 2z &=& 1
\end{array}
$$

Next subtract 2 times row 1 from row 2. Then subtract row 1 from row 3. This eliminates the x-variable from the second and third equations.

$$
\begin{array}{c}
 \\
R_2 - 2R_1 \rightarrow \\
R_3 - R_1 \rightarrow
\end{array}
\left[\begin{array}{ccc|c}
1 & 0 & 2 & 5 \\
0 & 1 & -2 & 0 \\
0 & -2 & 0 & -4
\end{array}\right]
\qquad
\begin{array}{rcr}
x + 2z &=& 5 \\
y - 2z &=& 0 \\
-2y &=& -4
\end{array}
$$

To eliminate the y-variable in row 3, add 2 times row 2 to row 3.

$$
\begin{array}{c}
 \\
 \\
R_3 + 2R_2 \rightarrow
\end{array}
\left[\begin{array}{ccc|c}
1 & 0 & 2 & 5 \\
0 & 1 & -2 & 0 \\
0 & 0 & -4 & -4
\end{array}\right]
\qquad
\begin{array}{rcr}
x + 2z &=& 5 \\
y - 2z &=& 0 \\
-4z &=& -4
\end{array}
$$

To obtain a leading 1 in row 3, multiply by $-\frac{1}{4}$.

$$
\begin{array}{c}
 \\
 \\
-\dfrac{1}{4}R_3 \rightarrow
\end{array}
\left[\begin{array}{ccc|c}
1 & 0 & 2 & 5 \\
0 & 1 & -2 & 0 \\
0 & 0 & 1 & 1
\end{array}\right]
\qquad
\begin{array}{rcr}
x + 2z &=& 5 \\
y - 2z &=& 0 \\
z &=& 1
\end{array}
$$

The matrix is in row-echelon form. It can be transformed into reduced row-echelon form by subtracting 2 times row 3 from row 1, and adding 2 times row 3 to row 2.

$$
\begin{array}{c}
R_1 - 2R_3 \rightarrow \\
R_2 + 2R_3 \rightarrow \\

\end{array}
\left[\begin{array}{ccc|c}
1 & 0 & 0 & 3 \\
0 & 1 & 0 & 2 \\
0 & 0 & 1 & 1
\end{array}\right]
\qquad
\begin{array}{rcr}
x &=& 3 \\
y &=& 2 \\
z &=& 1
\end{array}
$$

This matrix is in reduced row-echelon form. The solution is $(3, 2, 1)$.

Now Try Exercise 61 ◀

EXAMPLE 8   Transforming a system that has no solutions

Solve the system of linear equations, if possible.

$$
\begin{array}{rcr}
x - 2y + 3z &=& 2 \\
2x + 3y + 2z &=& 7 \\
4x - y + 8z &=& 8
\end{array}
$$

SOLUTION Because it is not necessary to write the linear system next to the matrix, we write the matrices in a horizontal format in this example.

$$\begin{bmatrix} 1 & -2 & 3 & | & 2 \\ 2 & 3 & 2 & | & 7 \\ 4 & -1 & 8 & | & 8 \end{bmatrix} \begin{matrix} \\ R_2 - 2R_1 \rightarrow \\ R_3 - 4R_1 \rightarrow \end{matrix} \begin{bmatrix} 1 & -2 & 3 & | & 2 \\ 0 & 7 & -4 & | & 3 \\ 0 & 7 & -4 & | & 0 \end{bmatrix} \begin{matrix} \\ \\ R_3 - R_2 \rightarrow \end{matrix} \begin{bmatrix} 1 & -2 & 3 & | & 2 \\ 0 & 7 & -4 & | & 3 \\ 0 & 0 & 0 & | & -3 \end{bmatrix}$$

The last row of the last matrix represents $0x + 0y + 0z = -3$, which has no solutions because $0 \neq -3$. There are no solutions.   *Now Try Exercise 65* ◄

Solving Systems of Linear Equations with Technology (Optional)

Gaussian elimination is a numerical method. If the arithmetic at each step is done exactly, then it may be thought of as an exact symbolic procedure. However, in real applications the augmented matrix is usually quite large and its elements are not all integers. As a result, calculators and computers often are used to solve systems of equations. Their solutions usually are approximate. The next three examples use a graphing calculator to solve systems of linear equations.

EXAMPLE 9 Solving a system of equations using technology

Use a graphing calculator to solve the system of linear equations in Example 7.

SOLUTION To solve this system, enter the augmented matrix

$$A = \begin{bmatrix} 2 & 1 & 2 & | & 10 \\ 1 & 0 & 2 & | & 5 \\ 1 & -2 & 2 & | & 1 \end{bmatrix},$$

as shown in Figures 9.45–9.47.

Calculator Help

To enter the elements of a matrix, see Appendix A (page AP-14).

Figure 9.45

Figure 9.46

Figure 9.47

A graphing calculator can transform matrix A into reduced row-echelon form, as illustrated in Figures 9.48 and 9.49. Notice that the reduced row-echelon form obtained from the graphing calculator agrees with our results from Example 7. The solution is $(3, 2, 1)$.

Calculator Help

To transform a matrix into reduced row-echelon form, see Appendix A (page AP-15).

Figure 9.48   **Figure 9.49**   *Now Try Exercise 67* ◄

EXAMPLE 10 Transforming a matrix into reduced row-echelon form

For three food shelters operated by a charitable organization, three different quantities are computed: monthly food costs F in dollars, number of people served per month N, and monthly charitable receipts R in dollars. The data are shown in Table 9.7.

Table 9.7

Food Costs (F)	Number Served (N)	Charitable Receipts (R)
3000	2400	8000
4000	2600	10,000
8000	5900	14,000

Source: D. Sanders, *Statistics: A First Course.*

(a) Model these data by using $F = aN + bR + c$, where a, b, and c are constants.
(b) Predict the food costs for a shelter that serves 4000 people and receives charitable receipts of $12,000. Round your answer to the nearest hundred dollars.

SOLUTION
(a) Since $F = aN + bR + c$, the constants a, b, and c satisfy the following equations.

$$3000 = a(2400) + b(8000) + c$$
$$4000 = a(2600) + b(10{,}000) + c$$
$$8000 = a(5900) + b(14{,}000) + c$$

This system can be rewritten as

$$2400a + \quad 8000b + c = 3000$$
$$2600a + 10{,}000b + c = 4000$$
$$5900a + 14{,}000b + c = 8000.$$

The resulting augmented matrix is

$$A = \begin{bmatrix} 2400 & 8000 & 1 & | & 3000 \\ 2600 & 10{,}000 & 1 & | & 4000 \\ 5900 & 14{,}000 & 1 & | & 8000 \end{bmatrix}.$$

Figure 9.50 shows the matrix A. The fourth column of A may be viewed by using the arrow keys. In Figure 9.51, A has been transformed into reduced row-echelon form. From Figure 9.51, we see that $a \approx 0.6897$, $b \approx 0.4310$, and $c \approx -2103$. Thus let $F = 0.6897N + 0.431R - 2103$.

Figure 9.50

Figure 9.51

(b) To predict the food costs for a shelter that serves 4000 people and receives charitable receipts of $12,000, let $N = 4000$ and $R = 12,000$ and evaluate F.

$$F = 0.6897(4000) + 0.431(12,000) - 2103 = 5827.8.$$

This model predicts monthly food costs of about $5800.   *Now Try Exercise 73* ◀

Determining a Quadratic Function   The introduction to this section discussed how three points can be used to determine a quadratic function whose graph passes through these points. The next example illustrates this method.

EXAMPLE 11   Determining a quadratic function

More than half of private-sector employees cannot carry vacation days into a new year. The average number y of paid days off for full-time workers at medium to large companies after x years of employment is listed in Table 9.8.

Table 9.8

x (years)	1	15	30
y (days)	9.4	18.8	21.9

Source: Bureau of Labor Statistics.

(a) Determine the coefficients for $f(x) = ax^2 + bx + c$ so that f models these data.
(b) Graph f with the data in $[-4, 32, 5]$ by $[8, 23, 2]$.
(c) Estimate the number of paid days off after 3 years of employment. Compare it to the actual value of 11.2 days.

Calculator Help

To plot data and to graph an equation, see Appendix A (page AP-7).

```
rref([A])
...1 0 0  -.016026...
...0 1 0  .9278489...
...0 0 1  8.488177...
```

Figure 9.52

$[-4, 32, 5]$ by $[8, 23, 2]$

Figure 9.53

SOLUTION
(a) For f to model the data, the equations $f(1) = 9.4$, $f(15) = 18.8$, and $f(30) = 21.9$ must be satisfied.

$$f(1) = a(1)^2 + b(1) + c = 9.4$$
$$f(15) = a(15)^2 + b(15) + c = 18.8$$
$$f(30) = a(30)^2 + b(30) + c = 21.9$$

The associated augmented matrix is

$$\begin{bmatrix} 1^2 & 1 & 1 & 9.4 \\ 15^2 & 15 & 1 & 18.8 \\ 30^2 & 30 & 1 & 21.9 \end{bmatrix}.$$

Figure 9.52 shows a portion of the matrix represented in reduced row-echelon form. The solution is $a \approx -0.016026$, $b \approx 0.92785$, and $c \approx 8.4882$.
(b) Graph $Y_1 = -0.016026X^2 + 0.92785X + 8.4882$ along with the points $(1, 9.4)$, $(15, 18.8)$, and $(30, 21.9)$. The graph of f passes through the points, as expected. See Figure 9.53.
(c) To estimate the number of paid days off after 3 years of employment, evaluate $f(3)$.

$$f(3) = -0.016026(3)^2 + 0.92785(3) + 8.4882 \approx 11.1$$

This is quite close to the actual value of 11.2 days.   *Now Try Exercise 83* ◀

9.4 Putting It All Together

One of the most common methods for solving a system of linear equations is Gaussian elimination with backward substitution. Through a sequence of matrix row operations, an augmented matrix is transformed into row-echelon form or reduced row-echelon form. Backward substitution is frequently used to find the solution when a matrix is in row-echelon form.

The following table summarizes some mathematical concepts involved with solving systems of linear equations in three variables.

Augmented Matrix

A linear system can be represented by an augmented matrix.

$$\left[\begin{array}{ccc|c} 2 & 0 & -3 & 2 \\ -1 & 2 & -2 & -5 \\ 1 & -2 & -1 & 7 \end{array}\right] \qquad \begin{array}{rcrcrcr} 2x & & & - & 3z & = & 2 \\ -x & + & 2y & - & 2z & = & -5 \\ x & - & 2y & - & z & = & 7 \end{array}$$

Row-Echelon Form

The following matrices are in row-echelon form. They represent three possible situations: no solutions, one solution, and infinitely many solutions.

$$\left[\begin{array}{ccc|c} 1 & -2 & 1 & 0 \\ 0 & 1 & 2 & 3 \\ 0 & 0 & 0 & 1 \end{array}\right] \qquad \left[\begin{array}{ccc|c} 1 & -2 & 1 & 0 \\ 0 & 1 & 2 & 3 \\ 0 & 0 & 1 & 1 \end{array}\right] \qquad \left[\begin{array}{ccc|c} 1 & -2 & 1 & 0 \\ 0 & 1 & 2 & 3 \\ 0 & 0 & 0 & 0 \end{array}\right]$$

No solutions   One solution   Infinitely many solutions
                   $(1, 1, 1)$ $(6 - 5z, 3 - 2z, z)$

Backward Substitution

Backward substitution can be used to solve a system of linear equations represented by an augmented matrix in row-echelon form.

$$\left[\begin{array}{ccc|c} 1 & -2 & 1 & 3 \\ 0 & 1 & -2 & -3 \\ 0 & 0 & 1 & 2 \end{array}\right]$$

From the last row, $z = 2$.
Substitute $z = 2$ in the second row: $y - 2(2) = -3$, or $y = 1$.
Let $z = 2$ and $y = 1$ in the first row: $x - 2(1) + 2 = 3$, or $x = 3$.
The solution is $(3, 1, 2)$.

Reduced Row-Echelon Form

The Gauss-Jordan elimination method can be used to transform an augmented matrix into reduced row-echelon form, which eliminates the need for backward substitution. The following matrices are in reduced row-echelon form. The solution is given below the matrix.

$$\left[\begin{array}{ccc|c} 1 & 0 & 0 & 4 \\ 0 & 1 & 0 & 5 \\ 0 & 0 & 0 & 2 \end{array}\right] \qquad \left[\begin{array}{ccc|c} 1 & 0 & 0 & 1 \\ 0 & 1 & 0 & 2 \\ 0 & 0 & 1 & 3 \end{array}\right] \qquad \left[\begin{array}{ccc|c} 1 & 0 & 3 & 2 \\ 0 & 1 & 2 & 3 \\ 0 & 0 & 0 & 0 \end{array}\right]$$

No solutions   One solution   Infinitely many solutions
                   $(1, 2, 3)$ $(2 - 3z, 3 - 2z, z)$

9.4 Exercises

Dimensions of Matrices and Augmented Matrices

Exercises 1 and 2: State the dimension of each matrix.

1. (a) $\begin{bmatrix} 1 \\ 2 \\ 3 \end{bmatrix}$ **(b)** $\begin{bmatrix} a & b & c \\ d & e & b \end{bmatrix}$ **(c)** $\begin{bmatrix} 3 & 0 \\ 1 & -4 \end{bmatrix}$

2. (a) $\begin{bmatrix} -1 & 1 \end{bmatrix}$ **(b)** $\begin{bmatrix} 1 & -1 \\ 7 & 5 \\ -4 & 0 \end{bmatrix}$

    (c) $\begin{bmatrix} 1 & 3 & 8 & -3 \\ 1 & -1 & 1 & -2 \\ 4 & 5 & 0 & -1 \end{bmatrix}$

Exercises 3–6: Represent the linear system by an augmented matrix, and state the dimension of the matrix.

3. $\begin{array}{r} 5x - 2y = 3 \\ -x + 3y = -1 \end{array}$ **4.** $\begin{array}{r} 3x + y = 4 \\ -x + 4y = 5 \end{array}$

5. $\begin{array}{r} -3x + 2y + z = -4 \\ 5x \quad\quad - z = 9 \\ x - 3y - 6z = -9 \end{array}$ **6.** $\begin{array}{r} x + 2y - z = 2 \\ -2x + y - 2z = -3 \\ 7x + y - z = 7 \end{array}$

Exercises 7–10: Write the system of linear equations that the augmented matrix represents.

7. $\left[\begin{array}{rr|r} 3 & 2 & 4 \\ 0 & 1 & 5 \end{array}\right]$ **8.** $\left[\begin{array}{rr|r} -2 & 1 & 5 \\ 7 & 9 & 2 \end{array}\right]$

9. $\left[\begin{array}{rrr|r} 3 & 1 & 4 & 0 \\ 0 & 5 & 8 & -1 \\ 0 & 0 & -7 & 1 \end{array}\right]$ **10.** $\left[\begin{array}{rrr|r} 1 & -1 & 3 & 2 \\ -2 & 1 & 1 & -2 \\ -1 & 0 & -2 & 1 \end{array}\right]$

Row-Echelon Form

Exercises 11 and 12: Is the matrix in row-echelon form?

11. (a) $\left[\begin{array}{rr|r} 1 & 3 & 2 \\ 0 & 1 & -1 \end{array}\right]$ **(b)** $\left[\begin{array}{rrr|r} 1 & 4 & -1 & 0 \\ 0 & -1 & 1 & 3 \\ 0 & 2 & 1 & 7 \end{array}\right]$

    (c) $\left[\begin{array}{rrr|r} 1 & 6 & -8 & 5 \\ 0 & 1 & 7 & 9 \\ 0 & 0 & 1 & 11 \end{array}\right]$

12. (a) $\left[\begin{array}{rr|r} 1 & 3 & 2 \\ 0 & -1 & -1 \end{array}\right]$ **(b)** $\left[\begin{array}{rrr|r} 1 & 3 & -1 & 8 \\ 0 & 1 & 5 & 3 \\ 0 & 0 & 0 & 0 \end{array}\right]$

    (c) $\left[\begin{array}{rrr|r} 0 & 0 & 1 & 1 \\ 0 & 1 & 7 & 9 \\ 1 & 2 & -1 & 11 \end{array}\right]$

Exercises 13–22: The augmented matrix is in row-echelon form and represents a linear system. Solve the system by using backward substitution, if possible. Write the solution as either an ordered pair or an ordered triple.

13. $\left[\begin{array}{rr|r} 1 & 2 & 3 \\ 0 & 1 & -1 \end{array}\right]$ **14.** $\left[\begin{array}{rr|r} 1 & -5 & 6 \\ 0 & 0 & 1 \end{array}\right]$

15. $\left[\begin{array}{rr|r} 1 & -1 & 2 \\ 0 & 1 & 0 \end{array}\right]$ **16.** $\left[\begin{array}{rr|r} 1 & 4 & -2 \\ 0 & 1 & 3 \end{array}\right]$

17. $\left[\begin{array}{rrr|r} 1 & 1 & -1 & 4 \\ 0 & 1 & -1 & 2 \\ 0 & 0 & 1 & 1 \end{array}\right]$ **18.** $\left[\begin{array}{rrr|r} 1 & -2 & -1 & 0 \\ 0 & 1 & -3 & 1 \\ 0 & 0 & 1 & 2 \end{array}\right]$

19. $\left[\begin{array}{rrr|r} 1 & 2 & -1 & 5 \\ 0 & 1 & -2 & 1 \\ 0 & 0 & 0 & 0 \end{array}\right]$ **20.** $\left[\begin{array}{rrr|r} 1 & -1 & 2 & 8 \\ 0 & 1 & -4 & 2 \\ 0 & 0 & 0 & 0 \end{array}\right]$

21. $\left[\begin{array}{rrr|r} 1 & 2 & 1 & -3 \\ 0 & 1 & -3 & \frac{1}{2} \\ 0 & 0 & 0 & 4 \end{array}\right]$ **22.** $\left[\begin{array}{rrr|r} 1 & 0 & -4 & \frac{3}{4} \\ 0 & 1 & 2 & 1 \\ 0 & 0 & 0 & -3 \end{array}\right]$

Solving Systems with Gaussian Elimination

Exercises 23–26: Perform each row operation on the given matrix by completing the matrix at the right.

23. $\left[\begin{array}{rrr|r} 2 & -4 & 6 & 10 \\ -3 & 5 & 3 & 2 \\ 4 & 8 & 4 & -8 \end{array}\right] \begin{array}{l} \frac{1}{2}R_1 \to \\ \\ \frac{1}{4}R_3 \to \end{array} \left[\begin{array}{rrr|r} 1 & & & \\ -3 & 5 & 3 & 2 \\ & & 1 & \end{array}\right]$

24. $\left[\begin{array}{rrr|r} 1 & -2 & 1 & 3 \\ 1 & 4 & 0 & -1 \\ 2 & 0 & 1 & 5 \end{array}\right] \begin{array}{l} \\ R_2 - R_1 \to \\ R_3 - 2R_1 \to \end{array} \left[\begin{array}{rrr|r} 1 & -2 & 1 & 3 \\ & 6 & & \\ & & & -1 \end{array}\right]$

25. $\begin{bmatrix} 1 & -1 & 1 & | & 2 \\ -1 & 2 & -2 & | & 0 \\ 1 & 7 & 0 & | & 5 \end{bmatrix} \begin{matrix} \\ R_2 + R_1 \rightarrow \\ R_3 - R_1 \rightarrow \end{matrix} \begin{bmatrix} 1 & -1 & 1 & | & 2 \\ & & & | & \\ & & & | & \end{bmatrix}$

26. $\begin{bmatrix} 1 & -2 & 3 & | & 6 \\ 2 & 1 & 4 & | & 5 \\ -3 & 5 & 3 & | & 2 \end{bmatrix} \begin{matrix} \\ R_2 - 2R_1 \rightarrow \\ R_3 + 3R_1 \rightarrow \end{matrix} \begin{bmatrix} 1 & -2 & 3 & | & 6 \\ & & & | & \\ & & & | & \end{bmatrix}$

Exercises 27–38: Use Gaussian elimination with backward substitution to solve the system of linear equations. Write the solution as an ordered pair or an ordered triple whenever possible.

27. $\begin{aligned} x + 2y &= 3 \\ -x - y &= 7 \end{aligned}$

28. $\begin{aligned} 2x + 4y &= 10 \\ x - 2y &= -3 \end{aligned}$

29. $\begin{aligned} x + 2y + z &= 3 \\ x + y - z &= 3 \\ -x - 2y + z &= -5 \end{aligned}$

30. $\begin{aligned} x + y + z &= 6 \\ 2x + 3y - z &= 3 \\ x + y + 2z &= 10 \end{aligned}$

31. $\begin{aligned} x + 2y - z &= -1 \\ 2x - y + z &= 0 \\ -x - y + 2z &= 7 \end{aligned}$

32. $\begin{aligned} x + 3y - 2z &= -4 \\ 2x + 6y + z &= -3 \\ x + y - 4z &= -2 \end{aligned}$

33. $\begin{aligned} 3x + y + 3z &= 14 \\ x + y + z &= 6 \\ -2x - 2y + 3z &= -7 \end{aligned}$

34. $\begin{aligned} x + 3y - 2z &= 3 \\ -x - 2y + z &= -2 \\ 2x - 7y + z &= 1 \end{aligned}$

35. $\begin{aligned} 2x + 5y + z &= 8 \\ x + 2y - z &= 2 \\ 3x + 7y &= 5 \end{aligned}$

36. $\begin{aligned} x + y + z &= 3 \\ x + y + 2z &= 4 \\ 2x + 2y + 3z &= 7 \end{aligned}$

37. $\begin{aligned} -x + 2y + 4z &= 10 \\ 3x - 2y - 2z &= -12 \\ x + 2y + 6z &= 8 \end{aligned}$

38. $\begin{aligned} 4x - 2y + 4z &= 8 \\ 3x - 7y + 6z &= 4 \\ -x - 5y + 2z &= 7 \end{aligned}$

Exercises 39–50: Solve the system, if possible.

39. $\begin{aligned} x - y + z &= 1 \\ x + 2y - z &= 2 \\ y - z &= 0 \end{aligned}$

40. $\begin{aligned} x - y - 2z &= -11 \\ x - 2y - z &= -11 \\ -x + y + 3z &= 14 \end{aligned}$

41. $\begin{aligned} 2x - 4y + 2z &= 11 \\ x + 3y - 2z &= -9 \\ 4x - 2y + z &= 7 \end{aligned}$

42. $\begin{aligned} x - 4y + z &= 9 \\ 3y - 2z &= -7 \\ -x + z &= 0 \end{aligned}$

43. $\begin{aligned} 3x - 2y + 2z &= -18 \\ -x + 2y - 4z &= 16 \\ 4x - 3y - 2z &= -21 \end{aligned}$

44. $\begin{aligned} 2x - y - z &= 0 \\ x - y - z &= -2 \\ 3x - 2y - 2z &= -2 \end{aligned}$

45. $\begin{aligned} x - 4y + 3z &= 26 \\ -x + 3y - 2z &= -19 \\ -y + z &= 10 \end{aligned}$

46. $\begin{aligned} 4x - y - z &= 0 \\ 4x - 2y &= 0 \\ 2x + z &= 1 \end{aligned}$

47. $\begin{aligned} 5x + 4z &= 7 \\ 2x - 4y &= 6 \\ 3y + 3z &= 3 \end{aligned}$

48. $\begin{aligned} y + 2z &= -5 \\ 3x - 2z &= -6 \\ -x - 4y &= 11 \end{aligned}$

49. $\begin{aligned} 5x - 2y + z &= 5 \\ x + y - 2z &= -2 \\ 4x - 3y + 3z &= 7 \end{aligned}$

50. $\begin{aligned} 2x - 4y - z &= 2 \\ x + y - 3z &= 10 \\ -x - 7y + 8z &= 2 \end{aligned}$

Exercises 51–58: (Refer to Example 6.) The augmented matrix is in reduced row-echelon form and represents a system of linear equations. If possible, solve the system.

51. $\begin{bmatrix} 1 & 0 & | & 12 \\ 0 & 1 & | & 3 \end{bmatrix}$

52. $\begin{bmatrix} 1 & -1 & | & 1 \\ 0 & 0 & | & 0 \end{bmatrix}$

53. $\begin{bmatrix} 1 & 0 & 0 & | & -2 \\ 0 & 1 & 0 & | & 4 \\ 0 & 0 & 1 & | & \frac{1}{2} \end{bmatrix}$

54. $\begin{bmatrix} 1 & 0 & 0 & | & 7 \\ 0 & 1 & 0 & | & -9 \\ 0 & 0 & 1 & | & 3 \end{bmatrix}$

55. $\begin{bmatrix} 1 & 0 & 2 & | & 4 \\ 0 & 1 & -1 & | & -3 \\ 0 & 0 & 0 & | & 0 \end{bmatrix}$

56. $\begin{bmatrix} 1 & 0 & 1 & | & -2 \\ 0 & 1 & 3 & | & 5 \\ 0 & 0 & 0 & | & 0 \end{bmatrix}$

57. $\begin{bmatrix} 1 & 0 & 0 & | & \frac{3}{4} \\ 0 & 1 & 0 & | & -1 \\ 0 & 0 & 0 & | & \frac{2}{3} \end{bmatrix}$

58. $\begin{bmatrix} 1 & 0 & 0 & | & 10 \\ 0 & 1 & 0 & | & 21 \\ 0 & 0 & 0 & | & -2 \end{bmatrix}$

*Exercises 59–66: **Reduced Row-Echelon Form** Use Gauss-Jordan elimination to solve the system of equations.*

59. $\begin{aligned} x - y &= 1 \\ x + y &= 5 \end{aligned}$

60. $\begin{aligned} 2x + 3y &= 1 \\ x - 2y &= -3 \end{aligned}$

61. $\begin{aligned} x + 2y + z &= 3 \\ y - z &= -2 \\ -x - 2y + 2z &= 6 \end{aligned}$

62. $\begin{aligned} x + z &= 2 \\ x - y - z &= 0 \\ -2x + y &= -2 \end{aligned}$

63. $\begin{aligned} x - y + 2z &= 7 \\ 2x + y - 4z &= -27 \\ -x + y - z &= 0 \end{aligned}$

64. $\begin{aligned} 2x - 4y - 6z &= 2 \\ x - 3y + z &= 12 \\ 2x + y + 3z &= 5 \end{aligned}$

65. $\begin{aligned} 2x + y - z &= 2 \\ x - 2y + z &= 0 \\ x + 3y - 2z &= 4 \end{aligned}$

66. $\begin{aligned} -2x - y + z &= 3 \\ x + y - 3z &= 1 \\ x - 2y - 4z &= 2 \end{aligned}$

📵 *Exercises 67–72: Technology Use technology to find the solution. Approximate values to the nearest thousandth.*

67. $\begin{aligned} 5x - 7y + 9z &= 40 \\ -7x + 3y - 7z &= 20 \\ 5x - 8y - 5z &= 15 \end{aligned}$

68. $\begin{aligned} 12x - 4y - 7z &= 8 \\ -8x - 6y + 9z &= 7 \\ 34x + 6y - 2z &= 5 \end{aligned}$

69. $\begin{aligned} 2.1x + 0.5y + 1.7z &= 4.9 \\ -2x + 1.5y - 1.7z &= 3.1 \\ 5.8x - 4.6y + 0.8z &= 9.3 \end{aligned}$

70.
$$53x + 95y + 12z = 108$$
$$81x - 57y - 24z = -92$$
$$-9x + 11y - 78z = 21$$

71.
$$0.1x + 0.3y + 1.7z = 0.6$$
$$0.6x + 0.1y - 3.1z = 6.2$$
$$2.4y + 0.9z = 3.5$$

72.
$$103x - 886y + 431z = 1200$$
$$-55x + 981y = 1108$$
$$-327x + 421y + 337z = 99$$

Applications

73. *Food Shelters* (Refer to Example 10.) For three food shelters, monthly food costs F in dollars, number of people served per month N, and monthly charitable receipts R in dollars are as shown in the table.

Food Costs (F)	Number Served (N)	Charitable Receipts (R)
1300	1800	5000
5300	3200	12,000
6500	4500	13,000

(a) Model these data using $F = aN + bR + c$, where a, b, and c are constants.

(b) Predict the food costs for a shelter that serves 3500 people and receives charitable receipts of $12,500. Round your answer to the nearest hundred dollars.

74. *Estimating the Weight of a Bear* The following table shows the weight W, neck size N, and chest size C for a representative sample of black bears.

W (pounds)	N (inches)	C (inches)
100	17	27
272	25	36
381	30	43

Source: M. Triola, *Elementary Statistics.*

(a) Find values for a, b, and c so that the equation $W = a + bN + cC$ models these data.

(b) Estimate the weight of a bear with a 20-inch neck and a 31-inch chest size.

(c) Explain why it is reasonable for the coefficients b and c to be positive.

75. *Pumping Water* Three pumps are being used to empty a small swimming pool. The first pump is twice as fast as the second pump. The first two pumps can empty the pool in 8 hours, while all three pumps can empty it in 6 hours. How long would it take each pump to empty the pool individually? (*Hint:* Let x represent the fraction of the pool that the first pump can empty in 1 hour. Let y and z represent this fraction for the second and third pumps, respectively.)

76. *Pumping Water* Suppose in Exercise 75 that the first pump is three times as fast as the third pump, the first and second pumps can empty the pool in 6 hours, and all three pumps can empty the pool in 8 hours.

(a) Are these data realistic? Explain your reasoning.

(b) Make a conjecture about a mathematical solution to these data.

(c) Test your conjecture by solving the problem.

77. *Electricity* In the study of electrical circuits, the application of Kirchoff's rules frequently results in systems of linear equations. To determine the current (in amperes) in each branch of the circuit shown in the figure, solve the system of linear equations. Round values to the nearest hundredth.

$$I_1 = I_2 + I_3$$
$$15 + 4I_3 = 14I_2$$
$$10 + 4I_3 = 5I_1$$

78. *Electricity* (Refer to Exercise 77.) Find the current (in amperes) in each branch of the circuit shown in the figure by solving the system of linear equations. Round values to the nearest hundredth.

$$I_1 = I_2 + I_3$$
$$20 = 4I_1 + 7I_3$$
$$10 + 7I_3 = 6I_2$$

79. *Investment* A sum of $5000 is invested in three mutual funds that pay 8%, 11%, and 14% annual interest rates. The amount of money invested in the fund paying 14% equals the total amount of money invested in the other two funds, and the total annual interest from all three funds is $595.

(a) Write a system of equations whose solution gives the amount invested in each mutual fund. Be sure to state what each variable represents.

(b) Solve the system of equations.

80. *Investment*   A sum of $10,000 is invested in three accounts that pay 6%, 8%, and 10% interest. Twice as much money is invested in the account paying 10% as in the account paying 6%, and the total annual interest from all three accounts is $842.

(a) Write a system of equations whose solution gives the amount invested in each account. Be sure to state what each variable represents.

(b) Solve the system of equations.

Exercises 81 and 82: Traffic Flow   *The figure shows three one-way streets with intersections A, B, and C. Numbers indicate the average traffic flow in vehicles per minute. The variables x, y, and z denote unknown traffic flows that need to be determined for timing of stoplights.*

(a) *If the number of vehicles per minute entering an intersection must equal the number exiting an intersection, verify that the accompanying system of linear equations describes the traffic flow.*

(b) *Rewrite the system and solve.*

(c) *Interpret your solution.*

81. A: $x + 5 = y + 7$
     B: $z + 6 = x + 3$
     C: $y + 3 = z + 4$

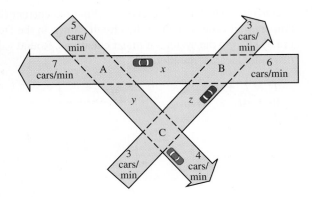

82. A: $x + 7 = y + 4$
     B: $4 + 5 = x + z$
     C: $y + 8 = 9 + 4$

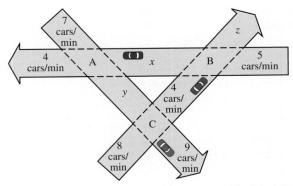

Exercises 83–86: Each set of data can be modeled by $f(x) = ax^2 + bx + c$.

(a) *Write a linear system whose solution represents values of a, b, and c.*

(b) *Use technology to find the solution.*

(c) *Graph f and the data in the same viewing rectangle.*

(d) *Make your own prediction using f.*

83. *Estimating iPod Sales*   (Refer to the introduction to this section.) The table lists total iPod sales y in millions x years after 2004.

x	0	2	4
y	3	55	150

Source: Apple Corporation.

84. *Head Start Enrollment*   The table lists annual enrollment in thousands for the Head Start program x years after 1980.

x	0	10	26
y	376	541	909

Source: Dept. of Health and Human Services.

85. *Chronic Health Care*   A large percentage of the U.S. population will require chronic health care in the coming decades. The average age of caregivers is 50–64, while the typical person needing chronic care is 85 or older. The ratio y of potential caregivers to those needing chronic health care will shrink in the coming years x, as shown in the table.

x	1990	2010	2030
y	11	10	6

Source: Robert Wood Johnson Foundation, *Chronic Care in America: A 21st Century Challenge.*

86. *Carbon Dioxide Levels*   Carbon dioxide (CO_2) is a greenhouse gas. The table at the top of the next page lists its concentration y in parts per million (ppm) measured at Mauna Loa, Hawaii, for three selected years x.

x	1958	1973	2003
y	315	325	376

Source: A. Nilsson, *Greenhouse Earth*.

Writing about Mathematics

87. A linear equation in three variables can be represented by a flat plane. Describe geometrically situations that can occur when a system of three linear equations has either no solution or an infinite number of solutions.

88. Give an example of an augmented matrix in row-echelon form that represents a system of linear equations that has no solution. Explain your reasoning.

EXTENDED AND DISCOVERY EXERCISES

Exercises 1 and 2: Solve the system of four equations with four variables.

1.
$$w + x + 2y - z = 4$$
$$2w + x + 2y + z = 5$$
$$-w + 3x + y - 2z = -2$$
$$3w + 2x + y + 3z = 3$$

2.
$$2w - 5x + 3y - 2z = -13$$
$$3w + 2x + 4y - 9z = -28$$
$$4w + 3x - 2y - 4z = -13$$
$$5w - 4x - 3y + 3z = 0$$

CHECKING BASIC CONCEPTS FOR SECTIONS 9.3 AND 9.4

1. If possible, solve the system of linear equations.

(a)
$$x - 2y + z = -2$$
$$x + y + 2z = 3$$
$$2x - y - z = 5$$

(b)
$$x - 2y + z = -2$$
$$x + y + 2z = 3$$
$$2x - y + 3z = 1$$

(c)
$$x - 2y + z = -2$$
$$x + y + 2z = 3$$
$$2x - y + 3z = 5$$

2. *Tickets Sold* Two thousand tickets were sold for a play, generating $19,700. The prices of the tickets were $5 for children, $10 for students, and $12 for adults. There were 100 more adult tickets sold than student tickets. Find the number of each type of ticket sold.

3. Solve the system of linear equations using Gaussian elimination and backward substitution.
$$x + z = 2$$
$$x + y - z = 1$$
$$-x - 2y - z = 0$$

4. Use technology to solve the system of linear equations in Exercise 3.

9.5 Properties and Applications of Matrices

- **Learn matrix notation**
- **Find sums, differences, and scalar multiples of matrices**
- **Find matrix products**
- **Use technology (optional)**

Introduction

Matrices occur in many fields of study. New technologies, such as digital imaging and computer graphics, frequently utilize matrices to achieve their goals. Many of these technologies require substantial interdisciplinary skills, including mathematics. In this section we discuss properties of matrices and some of their applications.

Matrix Notation

The following notation is used to denote elements in a matrix A.

$$\begin{bmatrix} a_{11} & a_{12} \\ a_{21} & a_{22} \end{bmatrix} \quad \begin{bmatrix} a_{11} & a_{12} & a_{13} \\ a_{21} & a_{22} & a_{23} \\ a_{31} & a_{32} & a_{33} \end{bmatrix} \quad \begin{bmatrix} a_{11} & a_{12} & a_{13} & a_{14} \\ a_{21} & a_{22} & a_{23} & a_{24} \\ a_{31} & a_{32} & a_{33} & a_{34} \\ a_{41} & a_{42} & a_{43} & a_{44} \end{bmatrix} \quad \begin{bmatrix} a_{11} & a_{12} \\ a_{21} & a_{22} \\ a_{31} & a_{32} \end{bmatrix} \quad \begin{bmatrix} a_{11} & a_{12} & a_{13} \\ a_{21} & a_{22} & a_{23} \end{bmatrix}$$

A general element is denoted by a_{ij}. This refers to the element in the ith row, jth column. For example, a_{23} would be the element of A located in the second row, third column.

Two m by n matrices A and B are **equal** if corresponding elements are equal. If A and B have different dimensions, they cannot be equal. For example,

$$\begin{bmatrix} 3 & -3 & 7 \\ 2 & 6 & -2 \\ 4 & 2 & 5 \end{bmatrix} = \begin{bmatrix} 3 & -3 & 7 \\ 2 & 6 & -2 \\ 4 & 2 & 5 \end{bmatrix}$$

because all corresponding elements are equal. However,

$$\begin{bmatrix} 1 & 4 \\ -3 & 2 \\ 4 & -7 \end{bmatrix} \neq \begin{bmatrix} 1 & 4 \\ -3 & 2 \\ 5 & -7 \end{bmatrix} \qquad a_{31} \neq b_{31}$$

because $4 \neq 5$ in row 3 and column 1, and

$$\begin{bmatrix} 1 & 2 & 3 \\ 4 & 5 & 6 \end{bmatrix} \neq \begin{bmatrix} 1 & 2 \\ 4 & 5 \end{bmatrix}$$
$$2 \times 3 \qquad\qquad 2 \times 2$$

because the matrices have different dimensions.

EXAMPLE 1 Determining matrix elements

Let a_{ij} denote a general element in A and b_{ij} a general element in B, where

$$A = \begin{bmatrix} 3 & -3 & 7 \\ 1 & 6 & -2 \\ 4 & 2 & 5 \end{bmatrix} \quad \text{and} \quad B = \begin{bmatrix} 3 & x & 7 \\ 1 & 6 & -2 \\ 4 & 5 & 2 \end{bmatrix}.$$

(a) Identify a_{12}, b_{32}, and a_{13}.
(b) Compute $a_{31}b_{13} + a_{32}b_{23} + a_{33}b_{33}$.
(c) Is there a value for x that will make the statement $A = B$ true?

SOLUTION
(a) The element a_{12} is located in the first row, second column of A. Thus, $a_{12} = -3$. In a similar manner, we find that $b_{32} = 5$ and $a_{13} = 7$.
(b) $a_{31}b_{13} + a_{32}b_{23} + a_{33}b_{33} = (4)(7) + (2)(-2) + (5)(2) = 34$
(c) No, since $a_{32} = 2 \neq 5 = b_{32}$ and $a_{33} = 5 \neq 2 = b_{33}$. Even if we let $x = -3$, other corresponding elements in A and B are not equal. *Now Try Exercise 9* ◀

Sums, Differences, and Scalar Multiples of Matrices

An Application The FCC has mandated that all television stations transmit digital signals starting in the year 2009. Digital photography is a technology in which matrices play an important role. Figure 9.54 shows a black-and-white photograph. In Figure 9.55 a grid has been laid over this photograph. Each square in the grid represents a pixel in a digitized photograph.

Figure 9.56 is a matrix with dimension 20×20 in which the shades of gray in the photo are stored according to the gray levels shown in Figure 9.57. In this scale, 0 corresponds to white and 11 represents black. As the scale increases from 0 to 11, the shade of gray darkens. Notice how the dark pole in the picture is represented by the fifth and sixth

Figure 9.54 Photograph to Be Digitized

Figure 9.55 Placing a Grid

Figure 9.56 Digitized Photograph

Source (Figures 9.54–9.57): *Visualization* by Friedhoff and Benzon, © 1991 by W. H. Freeman and Company. Used with permission.

Figure 9.57 Gray Levels

columns in the matrix. These columns contain the numbers 10 and 11, which indicates a dark portion in the photograph. Because it is a numerical, or digital, representation of the photograph, the matrix can be stored and transmitted over the Internet. In real applications, digital photographs often use a grid with 2592 rows and 3872 columns. The gray scale typically contains 256 levels or more instead of the 12 in our example.

Matrix Addition and Subtraction   To simplify the concept of a digital image, we reduce the grid to 3 × 3 and have four gray levels, where 0 represents white, 1 light gray, 2 dark gray, and 3 black. Suppose that we would like to digitize the letter T shown in Figure 9.58, using the gray levels shown in Figure 9.59. Since the T is dark gray and the background is white, Figure 9.58 can be represented by

Figure 9.58

$$A = \begin{bmatrix} 2 & 2 & 2 \\ 0 & 2 & 0 \\ 0 & 2 & 0 \end{bmatrix}.$$

Suppose that we want to make the entire picture darker. If we changed every element in A to 3, the entire picture would be black. A more acceptable solution would be to darken each pixel by one gray level. This corresponds to adding 1 to each element in the matrix A and can be accomplished efficiently using matrix notation.

Figure 9.59 Gray Levels

$$\begin{bmatrix} 2 & 2 & 2 \\ 0 & 2 & 0 \\ 0 & 2 & 0 \end{bmatrix} + \begin{bmatrix} 1 & 1 & 1 \\ 1 & 1 & 1 \\ 1 & 1 & 1 \end{bmatrix} = \begin{bmatrix} 2+1 & 2+1 & 2+1 \\ 0+1 & 2+1 & 0+1 \\ 0+1 & 2+1 & 0+1 \end{bmatrix} = \begin{bmatrix} 3 & 3 & 3 \\ 1 & 3 & 1 \\ 1 & 3 & 1 \end{bmatrix}$$

To add two matrices of equal dimension, add corresponding elements. The result is shown in picture form in Figure 9.60. Notice that the background is now light gray and the T is black. The entire picture is darker.

To lighten the picture in Figure 9.60, subtract 1 from each element. *To subtract two matrices of equal dimension, subtract corresponding elements.*

Figure 9.60

$$\begin{bmatrix} 3 & 3 & 3 \\ 1 & 3 & 1 \\ 1 & 3 & 1 \end{bmatrix} - \begin{bmatrix} 1 & 1 & 1 \\ 1 & 1 & 1 \\ 1 & 1 & 1 \end{bmatrix} = \begin{bmatrix} 3-1 & 3-1 & 3-1 \\ 1-1 & 3-1 & 1-1 \\ 1-1 & 3-1 & 1-1 \end{bmatrix} = \begin{bmatrix} 2 & 2 & 2 \\ 0 & 2 & 0 \\ 0 & 2 & 0 \end{bmatrix}$$

EXAMPLE 2 Adding and subtracting matrices

If $A = \begin{bmatrix} 7 & 8 & -1 \\ 0 & -1 & 6 \end{bmatrix}$ and $B = \begin{bmatrix} 5 & -2 & 10 \\ -3 & 2 & 4 \end{bmatrix}$, find the following.

(a) $A + B$ **(b)** $B + A$ **(c)** $A - B$

SOLUTION

(a) $A + B = \begin{bmatrix} 7 & 8 & -1 \\ 0 & -1 & 6 \end{bmatrix} + \begin{bmatrix} 5 & -2 & 10 \\ -3 & 2 & 4 \end{bmatrix}$

$= \begin{bmatrix} 7 + 5 & 8 + (-2) & -1 + 10 \\ 0 + (-3) & -1 + 2 & 6 + 4 \end{bmatrix}$

$= \begin{bmatrix} 12 & 6 & 9 \\ -3 & 1 & 10 \end{bmatrix}$

(b) $B + A = \begin{bmatrix} 5 & -2 & 10 \\ -3 & 2 & 4 \end{bmatrix} + \begin{bmatrix} 7 & 8 & -1 \\ 0 & -1 & 6 \end{bmatrix}$

$= \begin{bmatrix} 5 + 7 & -2 + 8 & 10 + (-1) \\ -3 + 0 & 2 + (-1) & 4 + 6 \end{bmatrix}$

$= \begin{bmatrix} 12 & 6 & 9 \\ -3 & 1 & 10 \end{bmatrix}$

Notice that $A + B = B + A$. The commutative property for matrix addition holds in general, provided that A and B have the same dimension.

(c) $A - B = \begin{bmatrix} 7 & 8 & -1 \\ 0 & -1 & 6 \end{bmatrix} - \begin{bmatrix} 5 & -2 & 10 \\ -3 & 2 & 4 \end{bmatrix}$

$= \begin{bmatrix} 7 - 5 & 8 - (-2) & -1 - 10 \\ 0 - (-3) & -1 - 2 & 6 - 4 \end{bmatrix}$

$= \begin{bmatrix} 2 & 10 & -11 \\ 3 & -3 & 2 \end{bmatrix}$

Now Try Exercise 13 ◀

An Application Increasing the contrast in a picture causes a light area to become lighter and a dark area to become darker. As a result, there are fewer pixels with intermediate gray levels. Changing contrast is different from making the entire picture lighter or darker.

EXAMPLE 3 Applying addition of matrices to digital photography

Increase the contrast of the $+$ sign in Figure 9.61 by changing light gray to white and dark gray to black. Use matrices to represent this computation.

SOLUTION Figure 9.61 can be represented by the matrix A.

$$A = \begin{bmatrix} 1 & 2 & 1 \\ 2 & 2 & 2 \\ 1 & 2 & 1 \end{bmatrix}$$

Figure 9.61

To change the contrast, we reduce each 1 in matrix A to 0 and increase each 2 to 3. The addition of matrix B can accomplish this task.

$$A + B = \begin{bmatrix} 1 & 2 & 1 \\ 2 & 2 & 2 \\ 1 & 2 & 1 \end{bmatrix} + \begin{bmatrix} -1 & 1 & -1 \\ 1 & 1 & 1 \\ -1 & 1 & -1 \end{bmatrix} = \begin{bmatrix} 0 & 3 & 0 \\ 3 & 3 & 3 \\ 0 & 3 & 0 \end{bmatrix}$$

The picture corresponding to $A + B$ is shown in Figure 9.62.

Figure 9.62   *Now Try Exercises 27 and 29* ◄

Multiplication of a Matrix by a Scalar   The matrix

$$B = \begin{bmatrix} 1 & 1 & 1 \\ 1 & 1 & 1 \\ 1 & 1 & 1 \end{bmatrix}$$

can be used to darken a digital picture. Suppose that a photograph is represented by a matrix A with gray levels 0 through 11. Every time matrix B is added to A, the picture becomes slightly darker. For example, if

$$A = \begin{bmatrix} 0 & 5 & 0 \\ 5 & 5 & 5 \\ 0 & 5 & 0 \end{bmatrix}$$

then the addition of $A + B + B$ would darken the picture by two gray levels and could be computed by

$$A + B + B = \begin{bmatrix} 0 & 5 & 0 \\ 5 & 5 & 5 \\ 0 & 5 & 0 \end{bmatrix} + \begin{bmatrix} 1 & 1 & 1 \\ 1 & 1 & 1 \\ 1 & 1 & 1 \end{bmatrix} + \begin{bmatrix} 1 & 1 & 1 \\ 1 & 1 & 1 \\ 1 & 1 & 1 \end{bmatrix} = \begin{bmatrix} 2 & 7 & 2 \\ 7 & 7 & 7 \\ 2 & 7 & 2 \end{bmatrix}.$$

A simpler way to write the expression $A + B + B$ is $A + 2B$. Multiplying B by 2 to obtain $2B$ is called **scalar multiplication**.

$$2B = 2\begin{bmatrix} 1 & 1 & 1 \\ 1 & 1 & 1 \\ 1 & 1 & 1 \end{bmatrix} = \begin{bmatrix} 2(1) & 2(1) & 2(1) \\ 2(1) & 2(1) & 2(1) \\ 2(1) & 2(1) & 2(1) \end{bmatrix} = \begin{bmatrix} 2 & 2 & 2 \\ 2 & 2 & 2 \\ 2 & 2 & 2 \end{bmatrix}$$

Every element of B is multiplied by the real number 2.

Sometimes a matrix B is denoted $B = [b_{ij}]$, where b_{ij} represents the element in the ith row, jth column. In this way, we could write $2B$ as $2[b_{ij}] = [2b_{ij}]$. This indicates that to calculate $2B$, multiply each b_{ij} by 2. In a similar manner, a matrix A is sometimes denoted by $[a_{ij}]$.

Some operations on matrices are now summarized.

Operations on Matrices

Matrix Addition

The sum of two $m \times n$ matrices A and B is the $m \times n$ matrix $A + B$, in which each element is the sum of the corresponding elements of A and B. This is written as $A + B = [a_{ij}] + [b_{ij}] = [a_{ij} + b_{ij}]$. If A and B have different dimensions, then $A + B$ is undefined.

Matrix Subtraction

The difference of two $m \times n$ matrices A and B is the $m \times n$ matrix $A - B$, in which each element is the difference of the corresponding elements of A and B. This is written as $A - B = [a_{ij}] - [b_{ij}] = [a_{ij} - b_{ij}]$. If A and B have different dimensions, then $A - B$ is undefined.

Multiplication of a Matrix by a Scalar

The product of a scalar (real number) k and an $m \times n$ matrix A is the $m \times n$ matrix kA, in which each element is k times the corresponding element of A. This is written as $kA = k[a_{ij}] = [ka_{ij}]$.

EXAMPLE 4 Performing scalar multiplication

If $A = \begin{bmatrix} 2 & 7 & 11 \\ -1 & 3 & -5 \\ 0 & 9 & -12 \end{bmatrix}$, find $-4A$.

SOLUTION

$$-4A = -4\begin{bmatrix} 2 & 7 & 11 \\ -1 & 3 & -5 \\ 0 & 9 & -12 \end{bmatrix} = \begin{bmatrix} -4(2) & -4(7) & -4(11) \\ -4(-1) & -4(3) & -4(-5) \\ -4(0) & -4(9) & -4(-12) \end{bmatrix} = \begin{bmatrix} -8 & -28 & -44 \\ 4 & -12 & 20 \\ 0 & -36 & 48 \end{bmatrix}$$

Now Try Exercise 15(b). ◀

EXAMPLE 5 Performing operations on matrices

If possible, perform the indicated operations using

$$A = \begin{bmatrix} 4 & -2 \\ 3 & 5 \end{bmatrix}, B = \begin{bmatrix} 0 & 1 \\ -2 & 3 \end{bmatrix}, C = \begin{bmatrix} 1 & -1 \\ 0 & 7 \\ -4 & 2 \end{bmatrix}, \text{ and } D = \begin{bmatrix} -1 & -3 \\ 9 & -7 \\ 1 & 8 \end{bmatrix}.$$

(a) $A + 3B$ **(b)** $A - C$ **(c)** $-2C - 3D$

SOLUTION

(a) $A + 3B = \begin{bmatrix} 4 & -2 \\ 3 & 5 \end{bmatrix} + 3\begin{bmatrix} 0 & 1 \\ -2 & 3 \end{bmatrix}$

$\qquad = \begin{bmatrix} 4 & -2 \\ 3 & 5 \end{bmatrix} + \begin{bmatrix} 0 & 3 \\ -6 & 9 \end{bmatrix} = \begin{bmatrix} 4 & 1 \\ -3 & 14 \end{bmatrix}$

(b) $A - C$ is undefined because the dimension of A is 2×2 and unequal to the dimension of C, which is 3×2.

(c) $-2C - 3D = -2\begin{bmatrix} 1 & -1 \\ 0 & 7 \\ -4 & 2 \end{bmatrix} - 3\begin{bmatrix} -1 & -3 \\ 9 & -7 \\ 1 & 8 \end{bmatrix}$

$= \begin{bmatrix} -2 & 2 \\ 0 & -14 \\ 8 & -4 \end{bmatrix} - \begin{bmatrix} -3 & -9 \\ 27 & -21 \\ 3 & 24 \end{bmatrix} = \begin{bmatrix} 1 & 11 \\ -27 & 7 \\ 5 & -28 \end{bmatrix}$

Now Try Exercises 17 and 19 ◄

Matrix Products

Addition, subtraction, and multiplication can be performed on numbers, variables, and functions. The same operations apply to matrices. Matrix multiplication is different from scalar multiplication.

An Application   Suppose two students are taking day classes at one college and night classes at another, in order to graduate on time. Tables 9.9 and 9.10 list the number of credits taken by the students and the cost per credit at each college.

The cost of tuition is computed by multiplying the number of credits times the cost of each credit. Student 1 is taking 10 credits at $60 each and 7 credits at $80 each. The total tuition for Student 1 is $10(\$60) + 7(\$80) = \$1160$. In a similar manner, the tuition for Student 2 is given by $11(\$60) + 4(\$80) = \$980$. The information in these tables can be represented by matrices. Let A represent Table 9.9 and B represent Table 9.10.

$$A = \begin{bmatrix} 10 & 7 \\ 11 & 4 \end{bmatrix} \quad \text{and} \quad B = \begin{bmatrix} 60 \\ 80 \end{bmatrix}$$

The matrix product AB calculates total tuition for each student.

$$AB = \begin{bmatrix} 10 & 7 \\ 11 & 4 \end{bmatrix}\begin{bmatrix} 60 \\ 80 \end{bmatrix} = \begin{bmatrix} 10(60) + 7(80) \\ 11(60) + 4(80) \end{bmatrix} = \begin{bmatrix} 1160 \\ 980 \end{bmatrix}$$

Generalizing from this example provides the following definition of matrix multiplication.

Table 9.9   Credits

	College A	College B
Student 1	10	7
Student 2	11	4

Table 9.10

	Cost per Credit
College A	$60
College B	$80

Dimension of AB

Figure 9.63

Matrix Multiplication

The **product** of an $m \times n$ matrix A and an $n \times k$ matrix B is the $m \times k$ matrix AB, which is computed as follows. To find the element of AB in the ith row and jth column, multiply each element in the ith row of A by the corresponding element in the jth column of B. The sum of these products will give the element in row i, column j of AB.

NOTE   In order to compute the product of two matrices, the number of columns in the first matrix must equal the number of rows in the second matrix, as illustrated in Figure 9.63.

EXAMPLE 6   Multiplying matrices

If possible, compute each product using

$$A = \begin{bmatrix} 1 & -1 \\ 0 & 3 \\ 4 & -2 \end{bmatrix}, B = \begin{bmatrix} -1 \\ -2 \end{bmatrix}, C = \begin{bmatrix} 1 & 2 & 3 \\ 4 & 5 & 6 \end{bmatrix}, \text{ and } D = \begin{bmatrix} 1 & -1 & 2 \\ 0 & 3 & -2 \\ -3 & 4 & 5 \end{bmatrix}.$$

(a) AB **(b)** CA **(c)** DC **(d)** CD

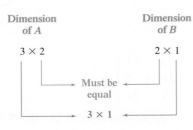

Dimension of A

Dimension of B

3×2 2×1

Must be equal

3×1

Dimension of AB

Figure 9.64

SOLUTION

(a) The dimension of A is 3×2 and the dimension of B is 2×1. The dimension of AB is 3×1 and can be found, as shown in Figure 9.64. The product AB is found as follows.

$$AB = \begin{bmatrix} 1 & -1 \\ 0 & 3 \\ 4 & -2 \end{bmatrix} \begin{bmatrix} -1 \\ -2 \end{bmatrix} = \begin{bmatrix} (1)(-1) + (-1)(-2) \\ (0)(-1) + (3)(-2) \\ (4)(-1) + (-2)(-2) \end{bmatrix} = \begin{bmatrix} 1 \\ -6 \\ 0 \end{bmatrix}$$

(b) The dimension of C is 2×3 and the dimension of A is 3×2. Thus CA is 2×2.

$$CA = \begin{bmatrix} 1 & 2 & 3 \\ 4 & 5 & 6 \end{bmatrix} \begin{bmatrix} 1 & -1 \\ 0 & 3 \\ 4 & -2 \end{bmatrix}$$

$$= \begin{bmatrix} 1(1) + 2(0) + 3(4) & 1(-1) + 2(3) + 3(-2) \\ 4(1) + 5(0) + 6(4) & 4(-1) + 5(3) + 6(-2) \end{bmatrix}$$

$$= \begin{bmatrix} 13 & -1 \\ 28 & -1 \end{bmatrix}$$

(c) The dimension of D is 3×3 and the dimension of C is 2×3. Therefore DC is undefined. Note that D has **3** columns and C has only **2** rows.

(d) The dimension of C is 2×3 and the dimension of D is 3×3. Thus CD is 2×3.

$$CD = \begin{bmatrix} 1 & 2 & 3 \\ 4 & 5 & 6 \end{bmatrix} \begin{bmatrix} 1 & -1 & 2 \\ 0 & 3 & -2 \\ -3 & 4 & 5 \end{bmatrix}$$

$$= \begin{bmatrix} 1(1) + 2(0) + 3(-3) & 1(-1) + 2(3) + 3(4) & 1(2) + 2(-2) + 3(5) \\ 4(1) + 5(0) + 6(-3) & 4(-1) + 5(3) + 6(4) & 4(2) + 5(-2) + 6(5) \end{bmatrix}$$

$$= \begin{bmatrix} -8 & 17 & 13 \\ -14 & 35 & 28 \end{bmatrix}$$

Now Try Exercises 31, 35, 37, and 47 ◀

MAKING CONNECTIONS

The Commutative Property and Matrix Multiplication Example 6 shows that $CD \neq DC$. Unlike multiplication of numbers, variables, and functions, matrix multiplication is *not* commutative. Instead, matrix multiplication is similar to function composition, where for a general pair of functions $f \circ g \neq g \circ f$.

Square matrices have the same number of rows as columns and have dimension $n \times n$ for some natural number n. When we multiply two square matrices, both having dimension $n \times n$, the resulting matrix also has dimension $n \times n$.

EXAMPLE 7 Multiplying square matrices

If $A = \begin{bmatrix} 1 & 0 & 7 \\ 3 & 2 & -1 \\ -5 & -2 & 5 \end{bmatrix}$ and $B = \begin{bmatrix} 4 & -6 & 7 \\ 8 & 9 & 10 \\ 0 & 1 & -3 \end{bmatrix}$, find AB.

SOLUTION

$$AB = \begin{bmatrix} 1 & 0 & 7 \\ 3 & 2 & -1 \\ -5 & -2 & 5 \end{bmatrix} \begin{bmatrix} 4 & -6 & 7 \\ 8 & 9 & 10 \\ 0 & 1 & -3 \end{bmatrix}$$

$$= \begin{bmatrix} 1(4) + 0(8) + 7(0) & 1(-6) + 0(9) + 7(1) & 1(7) + 0(10) + 7(-3) \\ 3(4) + 2(8) - 1(0) & 3(-6) + 2(9) - 1(1) & 3(7) + 2(10) - 1(-3) \\ -5(4) - 2(8) + 5(0) & -5(-6) - 2(9) + 5(1) & -5(7) - 2(10) + 5(-3) \end{bmatrix}$$

$$= \begin{bmatrix} 4 & 1 & -14 \\ 28 & -1 & 44 \\ -36 & 17 & -70 \end{bmatrix}$$

Now Try Exercise 41 ◀

Real numbers satisfy the commutative, associative, and distributive properties for various arithmetic operations. Matrices also satisfy some of these properties, provided that their dimensions are valid so the resulting expressions are defined. The following box summarizes these basic properties.

Properties of Matrices

Let A, B, and C be matrices. Assume that each matrix operation is defined.

1. $A + B = B + A$ — Commutative property for matrix addition (No commutative property for matrix multiplication)

2. $(A + B) + C = A + (B + C)$ — Associative property for matrix addition

3. $(AB)C = A(BC)$ — Associative property for matrix multiplication

4. $A(B + C) = AB + AC$ — Distributive property

Technology and Matrices (Optional)

Computing arithmetic operations on large matrices by hand can be a difficult task. Many graphing calculators have the capability to perform addition, subtraction, multiplication, and scalar multiplication with matrices, as the next two examples demonstrate.

EXAMPLE 8 Multiplying matrices with technology

Use a graphing calculator to find the product AB from Example 7.

SOLUTION First enter the matrices A and B into your calculator, as illustrated in Figures 9.65 and 9.66. Then find their product on the home screen, as shown in Figure 9.67. Notice that the answer agrees with our results from Example 7.

Calculator Help

To enter the elements of a matrix, see Appendix A (page AP-14). To multiply two matrices, see Appendix A (page AP-15).

```
MATRIX[A]   3 ×3
[ 1      0      7    ]
[ 3      2     ¯1    ]
[ ¯5    ¯2      5    ]
```

Figure 9.65

```
MATRIX[B]   3 ×3
[ 4     ¯6      7    ]
[ 8      9     10    ]
[ 0      1     ¯3    ]
```

Figure 9.66

```
[A]∗[B]
[ [4      1     ¯14 ]
  [28    ¯1      44 ]
  [¯36   17     ¯70]]
```

Figure 9.67

Now Try Exercise 49 ◀

EXAMPLE 9   Using technology to evaluate a matrix expression

Evaluate the expression $2A + 3B^3$, where

$$A = \begin{bmatrix} 3 & -1 & 2 \\ -1 & 6 & -1 \\ 2 & -1 & 9 \end{bmatrix} \quad \text{and} \quad B = \begin{bmatrix} 1 & -2 & 5 \\ 3 & 1 & -1 \\ 5 & 2 & 1 \end{bmatrix}.$$

```
2[A]+3[B]^3
[[300   -32    322]
 [133    63     73 ]
 [358    28    348]]
```

Figure 9.68

SOLUTION In the expression $2A + 3B^3$, B^3 is equal to BBB. Enter each matrix into a calculator and evaluate the expression. Figure 9.68 shows the result of this computation.

Now Try Exercise 51 ◀

9.5 Putting It All Together

Addition, subtraction, and multiplication can be performed on numbers, variables, and functions. In this section we looked at how these operations also apply to matrices. The following table provides examples of these operations.

Matrix Addition

$$\begin{bmatrix} 1 & 2 & 3 \\ 5 & 6 & 7 \end{bmatrix} + \begin{bmatrix} -1 & 0 & 8 \\ 9 & -2 & 10 \end{bmatrix} = \begin{bmatrix} 1 + (-1) & 2 + 0 & 3 + 8 \\ 5 + 9 & 6 + (-2) & 7 + 10 \end{bmatrix} = \begin{bmatrix} 0 & 2 & 11 \\ 14 & 4 & 17 \end{bmatrix}$$

The matrices must have the same dimension for their sum to be defined.

Matrix Subtraction

$$\begin{bmatrix} 1 & -4 \\ -3 & 4 \\ 2 & 7 \end{bmatrix} - \begin{bmatrix} 5 & 1 \\ 3 & 6 \\ 8 & -9 \end{bmatrix} = \begin{bmatrix} 1 - 5 & -4 - 1 \\ -3 - 3 & 4 - 6 \\ 2 - 8 & 7 - (-9) \end{bmatrix} = \begin{bmatrix} -4 & -5 \\ -6 & -2 \\ -6 & 16 \end{bmatrix}$$

The matrices must have the same dimension for their difference to be defined.

Scalar Multiplication

$$3 \begin{bmatrix} 3 & -2 \\ 0 & 1 \end{bmatrix} = \begin{bmatrix} 3(3) & 3(-2) \\ 3(0) & 3(1) \end{bmatrix} = \begin{bmatrix} 9 & -6 \\ 0 & 3 \end{bmatrix}$$

Matrix Multiplication

$$\begin{bmatrix} 0 & 1 \\ 2 & -3 \end{bmatrix} \begin{bmatrix} 3 & -5 \\ 4 & 6 \end{bmatrix} = \begin{bmatrix} 0(3) + 1(4) & 0(-5) + 1(6) \\ 2(3) + (-3)(4) & 2(-5) + (-3)(6) \end{bmatrix} = \begin{bmatrix} 4 & 6 \\ -6 & -28 \end{bmatrix}$$

For a matrix product to be defined, the number of columns in the first matrix must equal the number of rows in the second matrix. Matrix multiplication is not commutative. That is, $AB \neq BA$ in general.

9.5   Exercises

Elements of Matrices

Exercises 1–4: Determine each of the following for the given matrix A, if possible.

(a) $a_{12}, a_{21},$ and a_{32} (b) $a_{11}a_{22} + 3a_{23}$

1. $\begin{bmatrix} 1 & 2 & 3 \\ 4 & 5 & 6 \end{bmatrix}$

2. $\begin{bmatrix} 1 & 2 & 3 & 4 \\ 5 & 6 & 7 & 8 \\ 9 & 10 & 11 & 12 \end{bmatrix}$

3. $\begin{bmatrix} 1 & -1 & 4 \\ 3 & -2 & 5 \\ 7 & 0 & -6 \end{bmatrix}$

4. $\begin{bmatrix} 1 & -2 \\ -4 & 5 \end{bmatrix}$

Exercises 5–8: If possible, find values for x and y so that the matrices A and B are equal.

5. $A = \begin{bmatrix} x & 2 \\ -2 & 1 \end{bmatrix}$, $B = \begin{bmatrix} 1 & 2 \\ -2 & y \end{bmatrix}$

6. $A = \begin{bmatrix} 1 & x+y & 3 \\ 4 & -1 & 6 \\ 3 & 7 & -2 \end{bmatrix}$, $B = \begin{bmatrix} 1 & 2 & 3 \\ 4 & -1 & 6 \\ 3 & y & -2 \end{bmatrix}$

7. $A = \begin{bmatrix} x & 3 \\ 6 & -2 \end{bmatrix}$, $B = \begin{bmatrix} 1 & y & 0 \\ 6 & -2 & 0 \\ 0 & 0 & 0 \end{bmatrix}$

8. $A = \begin{bmatrix} 4 & -2 \\ 3 & -4 \\ x & y \end{bmatrix}$, $B = \begin{bmatrix} 4 & -2 & -2 \\ 3 & -4 & -4 \\ 7 & 8 & 8 \end{bmatrix}$

Exercises 9 and 10: Let a_{ij} and b_{ij} be general elements for the given matrices A and B.

(a) Identify $a_{12}, b_{32},$ and b_{22}.
(b) Compute $a_{11}b_{11} + a_{12}b_{21} + a_{13}b_{31}$.
(c) If possible, find a value for x that makes A = B.

9. $A = \begin{bmatrix} 1 & 3 & -4 \\ 3 & 0 & 7 \\ x & 1 & -1 \end{bmatrix}$, $B = \begin{bmatrix} 1 & x & -4 \\ 3 & 0 & 7 \\ 3 & 1 & -1 \end{bmatrix}$

10. $A = \begin{bmatrix} 0 & -1 & 6 \\ 2 & x & -1 \\ 9 & -2 & 1 \end{bmatrix}$, $B = \begin{bmatrix} 0 & -1 & x \\ 2 & 6 & -1 \\ 7 & -2 & 1 \end{bmatrix}$

Addition, Subtraction, and Scalar Multiplication

Exercises 11–14: For the given matrices A and B find each of the following.

(a) $A + B$ (b) $B + A$ (c) $A - B$

11. $A = \begin{bmatrix} 4 & -1 \\ -1 & 4 \end{bmatrix}$, $B = \begin{bmatrix} -1 & 4 \\ 4 & -1 \end{bmatrix}$

12. $A = \begin{bmatrix} 2 & -4 \\ -1 & \frac{1}{2} \\ 3 & -2 \end{bmatrix}$, $B = \begin{bmatrix} 5 & 0 \\ 3 & \frac{1}{2} \\ -1 & 1 \end{bmatrix}$

13. $A = \begin{bmatrix} 3 & 4 & -1 \\ 0 & -3 & 2 \\ -2 & 5 & 10 \end{bmatrix}$, $B = \begin{bmatrix} 11 & 5 & -2 \\ 4 & -7 & 12 \\ 6 & 6 & 6 \end{bmatrix}$

14. $A = \begin{bmatrix} 1 & 6 & 1 & -2 \\ 0 & 1 & 3 & 5 \\ 0 & 0 & 1 & -2 \end{bmatrix}$, $B = \begin{bmatrix} 1 & 0 & 0 & 9 \\ 3 & 1 & 0 & 3 \\ -1 & 4 & 1 & -2 \end{bmatrix}$

Exercises 15–20: If possible, find each of the following.

(a) $A + B$ (b) $3A$ (c) $2A - 3B$

15. $A = \begin{bmatrix} 2 & -6 \\ 3 & 1 \end{bmatrix}$, $B = \begin{bmatrix} -1 & 0 \\ -2 & 3 \end{bmatrix}$

16. $A = \begin{bmatrix} 1 & -2 & 5 \\ 3 & -4 & -1 \end{bmatrix}$, $B = \begin{bmatrix} 0 & -1 & -5 \\ -3 & 1 & 2 \end{bmatrix}$

17. $A = \begin{bmatrix} 1 & -1 & 0 \\ 1 & 5 & 9 \\ -4 & 8 & -5 \end{bmatrix}$, $B = \begin{bmatrix} 2 & 8 & -1 \\ 6 & -1 & 3 \end{bmatrix}$

18. $A = \begin{bmatrix} 6 & 2 & 9 \\ 3 & -2 & 0 \\ -1 & 4 & 8 \end{bmatrix}$, $B = \begin{bmatrix} 1 & 0 & -1 \\ 3 & 0 & 7 \\ 0 & -2 & -5 \end{bmatrix}$

19. $A = \begin{bmatrix} -2 & -1 \\ -5 & 1 \\ 2 & -3 \end{bmatrix}$, $B = \begin{bmatrix} 2 & -1 \\ 3 & 1 \\ 7 & -5 \end{bmatrix}$

20. $A = \begin{bmatrix} 0 & 1 \\ 3 & 2 \\ 4 & -9 \end{bmatrix}$, $B = \begin{bmatrix} 5 & 2 & -7 \\ 8 & -2 & 0 \end{bmatrix}$

Exercises 21–26: Evaluate the matrix expression.

21. $2\begin{bmatrix} 2 & -1 \\ 5 & 1 \\ 0 & 3 \end{bmatrix} + \begin{bmatrix} 5 & 0 \\ 7 & -3 \\ 1 & 1 \end{bmatrix} - \begin{bmatrix} 9 & -4 \\ 4 & 4 \\ 1 & 6 \end{bmatrix}$

22. $-3\begin{bmatrix} 3 & 8 \\ -1 & -9 \end{bmatrix} + 5\begin{bmatrix} 4 & -8 \\ 1 & 6 \end{bmatrix}$

23. $\begin{bmatrix} 4 & 6 \\ 3 & -7 \end{bmatrix} - 2\begin{bmatrix} 1 & 0 \\ -4 & 1 \end{bmatrix}$

24. $\begin{bmatrix} 5 & -1 & 6 \\ -2 & 10 & 12 \\ 5 & 2 & 9 \end{bmatrix} - \begin{bmatrix} -1 & 2 & 2 \\ 2 & -1 & 2 \\ 2 & 2 & -1 \end{bmatrix}$

25. $2\begin{bmatrix} 2 & -1 & -1 \\ -1 & 2 & -1 \\ -1 & -1 & 2 \end{bmatrix} + 3\begin{bmatrix} 1 & 2 & 3 \\ 2 & 1 & 3 \\ 2 & 3 & 1 \end{bmatrix}$

26. $3\begin{bmatrix} 1 & 0 & 3 & -1 \\ 0 & 1 & 2 & -1 \\ 1 & 0 & -3 & 1 \end{bmatrix} - 4\begin{bmatrix} -1 & 0 & 0 & 4 \\ 0 & -1 & 3 & 2 \\ 2 & 0 & 1 & -1 \end{bmatrix}$

Matrices and Digital Photography

*Exercises 27–30: **Digital Photography** (Refer to the discussion of digital photography in this section.) Consider the following simplified digital photograph, which has a 3 × 3 grid with four gray levels numbered from 0 to 3. It shows the number 1 in dark gray on a light gray background. Let A be the 3 × 3 matrix that represents this figure digitally.*

27. Find the matrix A.

28. Find a matrix B such that adding B to A will cause the entire picture to become one gray level darker. Evaluate the expression $A + B$.

29. (Refer to Example 3.) Find a matrix B such that adding B to A will enhance the contrast of A by one gray level. Evaluate $A + B$.

30. Find a matrix B such that subtracting B from A will cause the entire picture to become lighter by one gray level. Evaluate the expression $A - B$.

Matrix Multiplication

Exercises 31–48: If possible, find AB and BA.

31. $A = \begin{bmatrix} 1 & -1 \\ 2 & 0 \end{bmatrix}$, $B = \begin{bmatrix} -2 & 3 \\ 1 & 2 \end{bmatrix}$

32. $A = \begin{bmatrix} -3 & 5 \\ 2 & 7 \end{bmatrix}$, $B = \begin{bmatrix} -1 & 2 \\ 0 & 7 \end{bmatrix}$

33. $A = \begin{bmatrix} 5 & -7 & 2 \\ 0 & 1 & 5 \end{bmatrix}$, $B = \begin{bmatrix} 9 & 8 & 7 \\ 1 & -1 & -2 \end{bmatrix}$

34. $A = \begin{bmatrix} 2 & 1 & -1 \\ 0 & 2 & 1 \\ 3 & 2 & -1 \end{bmatrix}$, $B = \begin{bmatrix} 1 & 0 \\ 2 & -1 \\ 3 & 1 \end{bmatrix}$

35. $A = \begin{bmatrix} 3 & -1 \\ 1 & 0 \\ -2 & -4 \end{bmatrix}$, $B = \begin{bmatrix} -2 & 5 & -3 \\ 9 & -7 & 0 \end{bmatrix}$

36. $A = \begin{bmatrix} -1 & 0 & -2 \\ 4 & -2 & 1 \end{bmatrix}$, $B = \begin{bmatrix} 2 & -2 \\ 5 & -1 \\ 0 & 1 \end{bmatrix}$

37. $A = \begin{bmatrix} 1 & -1 & 0 \\ 2 & -1 & 5 \\ 6 & 1 & -4 \end{bmatrix}$, $B = \begin{bmatrix} -1 & 3 & -1 \\ 7 & -7 & 1 \end{bmatrix}$

38. $A = \begin{bmatrix} 2 & -1 & -5 \\ 4 & -1 & 6 \\ -2 & 0 & 9 \end{bmatrix}$, $B = \begin{bmatrix} 1 & 2 \\ -1 & -1 \\ 2 & 0 \end{bmatrix}$

39. $A = \begin{bmatrix} 2 & -3 \\ 5 & 3 \end{bmatrix}$, $B = \begin{bmatrix} -3 \\ 4 \\ 1 \end{bmatrix}$

40. $A = \begin{bmatrix} 3 & -1 \\ 2 & -2 \\ 0 & 4 \end{bmatrix}$, $B = \begin{bmatrix} 1 & -4 & 0 \\ -1 & 3 & 2 \end{bmatrix}$

41. $A = \begin{bmatrix} 2 & -1 & 3 \\ 0 & 1 & 0 \\ 2 & -2 & 3 \end{bmatrix}$, $B = \begin{bmatrix} 1 & 5 & -1 \\ 0 & 1 & 3 \\ -1 & 2 & 1 \end{bmatrix}$

42. $A = \begin{bmatrix} 1 & -2 & 5 \\ 1 & 0 & -2 \\ 1 & 3 & 2 \end{bmatrix}$, $B = \begin{bmatrix} -1 & 4 & 2 \\ -3 & 0 & 1 \\ 5 & 1 & 0 \end{bmatrix}$

43. $A = \begin{bmatrix} 2 & -1 \\ 3 & 1 \end{bmatrix}$, $B = \begin{bmatrix} 1 \\ 3 \end{bmatrix}$

44. $A = [5 \quad -3]$, $\qquad$ $B = \begin{bmatrix} 1 \\ 3 \end{bmatrix}$

45. $A = \begin{bmatrix} -3 & 1 \\ 2 & -4 \end{bmatrix}$, $\qquad$ $B = \begin{bmatrix} 1 & 0 & -2 \\ -4 & 8 & 1 \end{bmatrix}$

46. $A = \begin{bmatrix} 6 & 1 & 0 \\ -2 & 5 & 1 \\ 4 & -7 & 10 \end{bmatrix}$, $\qquad$ $B = \begin{bmatrix} 10 \\ 20 \\ 30 \end{bmatrix}$

47. $A = \begin{bmatrix} 1 & 0 & -2 \\ 3 & -4 & 1 \\ 2 & 0 & 5 \end{bmatrix}$, $\qquad$ $B = \begin{bmatrix} 1 \\ -1 \\ 3 \end{bmatrix}$

48. $A = \begin{bmatrix} 1 & -1 & 3 & -2 \\ 1 & 0 & 3 & 4 \\ 2 & -2 & 0 & 8 \end{bmatrix}$, $\qquad$ $B = \begin{bmatrix} 1 & -1 \\ 0 & 5 \\ 2 & 3 \\ -5 & 4 \end{bmatrix}$

Technology and Matrices

Exercises 49–52: Use the given A and B to evaluate each expression.

$$A = \begin{bmatrix} 3 & -2 & 4 \\ 5 & 2 & -3 \\ 7 & 5 & 4 \end{bmatrix}, \quad B = \begin{bmatrix} 1 & 1 & -5 \\ -1 & 0 & -7 \\ -6 & 4 & 3 \end{bmatrix}$$

49. AB $\qquad$ **50.** BA

51. $3A^2 + 2B$ $\qquad$ **52.** $B^2 - 3A$

Exercises 53–56: Properties of Matrices Use a graphing calculator to evaluate the expression with the given matrices A, B, and C. Compare your answers for parts (a) and (b). Then interpret the results.

$$A = \begin{bmatrix} 2 & -1 & 3 \\ 1 & 3 & -5 \\ 0 & -2 & 1 \end{bmatrix}, B = \begin{bmatrix} 6 & 2 & 7 \\ 3 & -4 & -5 \\ 7 & 1 & 0 \end{bmatrix},$$

$$C = \begin{bmatrix} 1 & 4 & -3 \\ 8 & 1 & -1 \\ 4 & 6 & -2 \end{bmatrix}$$

53. (a) $A(B + C)$ $\qquad$ **(b)** $AB + AC$

54. (a) $(A - B)C$ $\qquad$ **(b)** $AC - BC$

55. (a) $(A - B)^2$ $\qquad$ **(b)** $A^2 - AB - BA + B^2$

56. (a) $(AB)C$ $\qquad$ **(b)** $A(BC)$

Matrices and Negatives of Digital Images

57. *Negative Image* The negative image of a picture interchanges black and white. The number 1 is represented by the matrix A. Determine a matrix B such that $B - A$ represents the negative image of the picture represented by A. Evaluate $B - A$.

$$A = \begin{bmatrix} 0 & 3 & 0 \\ 0 & 3 & 0 \\ 0 & 3 & 0 \end{bmatrix}$$

58. *Negative Image* (Refer to Exercise 57.) Matrix A represents a digital photograph. Find a matrix B that represents the negative image of this picture.

$$A = \begin{bmatrix} 0 & 3 & 0 \\ 1 & 3 & 1 \\ 2 & 3 & 2 \end{bmatrix}$$

Exercises 59 and 60: Digital Photography The digital image represents the letter F using 20 pixels in a 5 × 4 grid. Assume that there are four gray levels from 0 to 3.

59. Find a matrix A that represents this digital image of the letter F.

60. (Continuation of Exercise 59)
  (a) Find a matrix B such that $B - A$ represents the negative image of the picture represented by A.

  (b) Find a matrix C such that $A + C$ represents a decrease in the contrast of A by one gray level.

Exercises 61–64: Digitizing Letters Complete the following.
  (a) Design a matrix A with dimension 4 × 4 that represents a digital image of the given letter. Assume that there are four gray levels from 0 to 3.
  (b) Find a matrix B such that B − A represents the negative image of the picture represented by matrix A from part (a).

61. Z $\qquad\qquad\qquad$ **62.** N

63. L $\qquad\qquad\qquad$ **64.** O

Applications

Exercises 65–68: Tuition Costs   (Refer to the discussion before Example 6.)

(a) *Find a matrix A and a column matrix B that describe the following tables.*

(b) *Find the matrix product AB, and interpret the result.*

65.

	College A	College B
Student 1	12	4
Student 2	8	7

	Cost per Credit
College A	$55
College B	$70

66.

	College A	College B
Student 1	15	2
Student 2	12	4

	Cost per Credit
College A	$90
College B	$75

67.

	College A	College B
Student 1	10	5
Student 2	9	8
Student 3	11	3

	Cost per Credit
College A	$60
College B	$70

68.

	College A	College B	College C
Student 1	6	0	3
Student 2	11	3	0
Student 3	0	12	3

	Cost per Credit
College A	$50
College B	$65
College C	$60

69. *Auto Parts*   A store owner makes two separate orders for three types of auto parts: I, II, and III. The numbers of parts ordered are represented by the matrix A.

$$A = \begin{matrix} & \text{I} & \text{II} & \text{III} \\ & \begin{bmatrix} 3 & 4 & 8 \\ 5 & 6 & 2 \end{bmatrix} & \begin{matrix} \text{Order 1} \\ \text{Order 2} \end{matrix} \end{matrix}$$

For example, Order 1 called for 4 parts of type II. The cost in dollars of each part can be represented by the matrix B.

$$B = \begin{matrix} & \text{Cost} \\ \begin{bmatrix} 10 \\ 20 \\ 30 \end{bmatrix} & \begin{matrix} \text{Part I} \\ \text{Part II} \\ \text{Part III} \end{matrix} \end{matrix}$$

Find AB and interpret the result.

70. *Car Sales*   Two car dealers buy four different makes of cars: I, II, III, and IV. The number of each make of automobile bought by each dealer is represented by the matrix A.

$$A = \begin{matrix} & \text{I} & \text{II} & \text{III} & \text{IV} \\ & \begin{bmatrix} 1 & 3 & 8 & 4 \\ 3 & 5 & 7 & 0 \end{bmatrix} & \begin{matrix} \text{Dealer 1} \\ \text{Dealer 2} \end{matrix} \end{matrix}$$

For example, Dealer 2 bought 7 cars of type III. The cost in thousands of dollars of each type of car can be represented by the matrix B.

$$B = \begin{matrix} & \text{Cost} \\ \begin{bmatrix} 15 \\ 21 \\ 28 \\ 38 \end{bmatrix} & \begin{matrix} \text{Make I} \\ \text{Make II} \\ \text{Make III} \\ \text{Make IV} \end{matrix} \end{matrix}$$

Find AB and interpret the result.

Writing about Mathematics

71. Discuss whether matrix multiplication is more like multiplication of functions or composition of functions. Explain your reasoning.

72. Describe one application of matrices.

EXTENDED AND DISCOVERY EXERCISES

Exercises 1–4: Representing Colors   Colors for computer monitors are often described using ordered triples. One model, called the RGB system, uses red, green, and blue to generate all colors. The figure describes the relationships of these colors in this system. Red is (1, 0, 0), green is (0, 1, 0), and blue is (0, 0, 1). *Since equal amounts of red and green combine to form yellow, yellow is represented by* (1, 1, 0). *Similarly, magenta (a deep reddish purple) is a mixture of blue and red and is represented by* (1, 0, 1). *Cyan is* (0, 1, 1), *since it is a mixture of blue and green.*

Another color model uses cyan, magenta, and yellow. Referred to as the CMY model, it is used in the four-color printing process for textbooks like this one. In this system, cyan is (1, 0, 0), *magenta is* (0, 1, 0), *and yellow is* (0, 0, 1). *In the CMY model, red is created by mixing magenta and yellow. Thus, red is* (0, 1, 1) *in this system. To convert ordered triples in the RGB model to ordered triples in the CMY model, we can use the following matrix equation. In both of these systems, color intensities vary between 0 and 1.* (**Sources:** I. Kerlow, *The Art of 3-D Computer Animation and Imaging;* R. Wolff.)

$$\begin{bmatrix} C \\ M \\ Y \end{bmatrix} = \begin{bmatrix} 1 \\ 1 \\ 1 \end{bmatrix} - \begin{bmatrix} R \\ G \\ B \end{bmatrix}$$

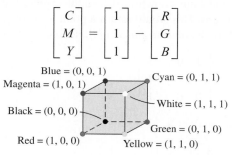

Blue = (0, 0, 1)
Magenta = (1, 0, 1)
Cyan = (0, 1, 1)
Black = (0, 0, 0)
White = (1, 1, 1)
Red = (1, 0, 0)
Green = (0, 1, 0)
Yellow = (1, 1, 0)

1. In the RGB model, aquamarine is (0.631, 1, 0.933). Use the matrix equation to determine the mixture of cyan, magenta, and yellow that makes aquamarine in the CMY model.

2. In the RGB model, rust is (0.552, 0.168, 0.066). Use the matrix equation to determine the mixture of cyan, magenta, and yellow that makes rust in the CMY model.

3. Use the given matrix equation to find a matrix equation that changes colors represented by ordered triples in the CMY model into ordered triples in the RGB model.

4. In the CMY model, (0.012, 0, 0.597) is a cream color. Use the matrix equation from Exercise 3 to determine the mixture of red, green, and blue that makes a cream color in the RGB model.

9.6 Inverses of Matrices

- **Understand matrix inverses and the identity matrices**
- **Find inverses symbolically**
- **Represent linear systems with matrix equations**
- **Solve linear systems with matrix inverses**

Introduction

Section 5.2 discussed how the inverse function f^{-1} will undo or cancel the computation performed by the function f. Like functions, some matrices have inverses. The inverse of a matrix A will undo or cancel the computation performed by A. For example, in computer graphics, if a matrix A rotated a figure on the screen 90° clockwise, then the inverse matrix would rotate the figure 90° counterclockwise. Similarly, if a matrix B translated a figure 3 units right, then B^{-1} would restore the figure to its original position by translating it 3 units left. This section discusses matrix inverses and some of their applications.

Understanding Matrix Inverses

An Application   In computer graphics, the matrix

$$A = \begin{bmatrix} 1 & 0 & h \\ 0 & 1 & k \\ 0 & 0 & 1 \end{bmatrix}$$

is used to translate a point (x, y) horizontally h units and vertically k units. The translation is to the right if $h > 0$ and to the left if $h < 0$. Similarly, the translation is upward if $k > 0$ and downward if $k < 0$. A point (x, y) is represented by the 3×1 **column matrix**

$$X = \begin{bmatrix} x \\ y \\ 1 \end{bmatrix}.$$

The third element in X is always equal to 1. For example, the point $(-1, 2)$ could be translated 3 units right and 4 units downward by computing the matrix product

$$AX = \begin{bmatrix} 1 & 0 & 3 \\ 0 & 1 & -4 \\ 0 & 0 & 1 \end{bmatrix} \begin{bmatrix} -1 \\ 2 \\ 1 \end{bmatrix} = \begin{bmatrix} 2 \\ -2 \\ 1 \end{bmatrix} = Y.$$

Its new location is $(2, -2)$. In the matrix $A, h = 3$ and $k = -4$. See Figure 9.69. (**Source:** C. Pokorny and C. Gerald, *Computer Graphics.*)

If A translates a point 3 units right and 4 units downward, then the inverse matrix translates a point 3 units left and 4 units upward. This would return a point to its original position after being translated by A. Therefore the *inverse matrix of A*, denoted A^{-1}, is given by

$$A^{-1} = \begin{bmatrix} 1 & 0 & -3 \\ 0 & 1 & 4 \\ 0 & 0 & 1 \end{bmatrix}.$$

In $A^{-1}, h = -3$ and $k = 4$. The matrix product $A^{-1}Y$ results in

$$A^{-1}Y = \begin{bmatrix} 1 & 0 & -3 \\ 0 & 1 & 4 \\ 0 & 0 & 1 \end{bmatrix} \begin{bmatrix} 2 \\ -2 \\ 1 \end{bmatrix} = \begin{bmatrix} -1 \\ 2 \\ 1 \end{bmatrix} = X.$$

The matrix A^{-1} translates $(2, -2)$ to its original coordinates of $(-1, 2)$. The two translations acting on the point $(-1, 2)$ can be represented by the following computation.

$$A^{-1}AX = \begin{bmatrix} 1 & 0 & -3 \\ 0 & 1 & 4 \\ 0 & 0 & 1 \end{bmatrix} \begin{bmatrix} 1 & 0 & 3 \\ 0 & 1 & -4 \\ 0 & 0 & 1 \end{bmatrix} \begin{bmatrix} -1 \\ 2 \\ 1 \end{bmatrix} = \begin{bmatrix} 1 & 0 & 0 \\ 0 & 1 & 0 \\ 0 & 0 & 1 \end{bmatrix} \begin{bmatrix} -1 \\ 2 \\ 1 \end{bmatrix} = \begin{bmatrix} -1 \\ 2 \\ 1 \end{bmatrix} = X$$

That is, the action of A followed by A^{-1} on the point $(-1, 2)$ results in $(-1, 2)$. In a similar manner, if we reverse the order of A^{-1} and A to compute $AA^{-1}X$, the result is again X.

$$AA^{-1}X = \begin{bmatrix} 1 & 0 & 3 \\ 0 & 1 & -4 \\ 0 & 0 & 1 \end{bmatrix} \begin{bmatrix} 1 & 0 & -3 \\ 0 & 1 & 4 \\ 0 & 0 & 1 \end{bmatrix} \begin{bmatrix} -1 \\ 2 \\ 1 \end{bmatrix} = \begin{bmatrix} 1 & 0 & 0 \\ 0 & 1 & 0 \\ 0 & 0 & 1 \end{bmatrix} \begin{bmatrix} -1 \\ 2 \\ 1 \end{bmatrix} = \begin{bmatrix} -1 \\ 2 \\ 1 \end{bmatrix} = X$$

Notice that both matrix products $A^{-1}A$ and AA^{-1} resulted in a matrix with 1's on its main diagonal and 0's elsewhere.

The Identity Matrix

An $n \times n$ matrix with 1's on its main diagonal and 0's elsewhere is called the $n \times n$ *identity matrix*. This matrix is important because its product with any $n \times n$ matrix A always equals A.

Figure 9.69 Translation of a Point

The $n \times n$ Identity Matrix

The $n \times n$ **identity matrix,** denoted I_n, has only 1's on its main diagonal and 0's elsewhere.

Some examples of identity matrices are

$$I_2 = \begin{bmatrix} 1 & 0 \\ 0 & 1 \end{bmatrix}, \qquad I_3 = \begin{bmatrix} 1 & 0 & 0 \\ 0 & 1 & 0 \\ 0 & 0 & 1 \end{bmatrix}, \qquad \text{and} \qquad I_4 = \begin{bmatrix} 1 & 0 & 0 & 0 \\ 0 & 1 & 0 & 0 \\ 0 & 0 & 1 & 0 \\ 0 & 0 & 0 & 1 \end{bmatrix}.$$

If A is any $n \times n$ matrix, then $I_n A = A$ and $AI_n = A$. For instance, if

$$A = \begin{bmatrix} 2 & 3 \\ 4 & 5 \end{bmatrix},$$

then

$$I_2 A = \begin{bmatrix} 1 & 0 \\ 0 & 1 \end{bmatrix} \begin{bmatrix} 2 & 3 \\ 4 & 5 \end{bmatrix} = \begin{bmatrix} 2 & 3 \\ 4 & 5 \end{bmatrix} = A \qquad \text{and}$$

$$AI_2 = \begin{bmatrix} 2 & 3 \\ 4 & 5 \end{bmatrix} \begin{bmatrix} 1 & 0 \\ 0 & 1 \end{bmatrix} = \begin{bmatrix} 2 & 3 \\ 4 & 5 \end{bmatrix} = A.$$

Matrix Inverses

Next we formally define the inverse of an $n \times n$ matrix A, whenever it exists.

Inverse of a Square Matrix

Let A be an $n \times n$ matrix. If there exists an $n \times n$ matrix, denoted A^{-1}, that satisfies

$$A^{-1}A = I_n \qquad \text{and} \qquad AA^{-1} = I_n,$$

then A^{-1} is the **inverse** of A.

If A^{-1} exists, then A is **invertible** or **nonsingular**. On the other hand, if a matrix A is not invertible, then it is **singular**. Not every matrix has an inverse. For example, the **zero matrix** with dimension 3×3 is given by

$$O_3 = \begin{bmatrix} 0 & 0 & 0 \\ 0 & 0 & 0 \\ 0 & 0 & 0 \end{bmatrix}.$$

The matrix O_3 does not have an inverse. The product of O_3 with any 3×3 matrix B would again be O_3, rather than the identity matrix I_3.

EXAMPLE 1 Verifying an inverse

Determine if B is the inverse of A, where

$$A = \begin{bmatrix} 5 & 3 \\ -3 & -2 \end{bmatrix} \qquad \text{and} \qquad B = \begin{bmatrix} 2 & 3 \\ -3 & -5 \end{bmatrix}.$$

SOLUTION For B to be the inverse of A, it must satisfy $AB = I_2$ and $BA = I_2$.

$$AB = \begin{bmatrix} 5 & 3 \\ -3 & -2 \end{bmatrix}\begin{bmatrix} 2 & 3 \\ -3 & -5 \end{bmatrix} = \begin{bmatrix} 1 & 0 \\ 0 & 1 \end{bmatrix} = I_2$$

$$BA = \begin{bmatrix} 2 & 3 \\ -3 & -5 \end{bmatrix}\begin{bmatrix} 5 & 3 \\ -3 & -2 \end{bmatrix} = \begin{bmatrix} 1 & 0 \\ 0 & 1 \end{bmatrix} = I_2$$

Thus B is the inverse of A. That is, $B = A^{-1}$. **Now Try Exercise 1** ◄

An Application The next example discusses the significance of an inverse matrix in computer graphics.

EXAMPLE 2 Interpreting an inverse matrix

The matrix A can be used to rotate a point $90°$ clockwise about the origin, where

$$A = \begin{bmatrix} 0 & 1 & 0 \\ -1 & 0 & 0 \\ 0 & 0 & 1 \end{bmatrix} \quad \text{and} \quad A^{-1} = \begin{bmatrix} 0 & -1 & 0 \\ 1 & 0 & 0 \\ 0 & 0 & 1 \end{bmatrix}.$$

(a) Use A to rotate the point $(-2, 0)$ clockwise $90°$ about the origin.
(b) Make a conjecture about the effect of A^{-1} on the resulting point.
(c) Test this conjecture.

SOLUTION
(a) First, let the point $(-2, 0)$ be represented by the column matrix

$$X = \begin{bmatrix} -2 \\ 0 \\ 1 \end{bmatrix}.$$

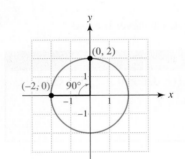

Then compute

$$AX = \begin{bmatrix} 0 & 1 & 0 \\ -1 & 0 & 0 \\ 0 & 0 & 1 \end{bmatrix}\begin{bmatrix} -2 \\ 0 \\ 1 \end{bmatrix} = \begin{bmatrix} 0 \\ 2 \\ 1 \end{bmatrix} = Y.$$

Figure 9.70 Rotating a Point about the Origin

If the point $(-2, 0)$ is rotated $90°$ clockwise about the origin, its new location is $(0, 2)$. See Figure 9.70.

(b) Since A^{-1} represents the inverse operation of A, A^{-1} will rotate the point located at $(0, 2)$ counterclockwise $90°$, back to $(-2, 0)$.

(c) This conjecture is correct, since

$$A^{-1}Y = \begin{bmatrix} 0 & -1 & 0 \\ 1 & 0 & 0 \\ 0 & 0 & 1 \end{bmatrix}\begin{bmatrix} 0 \\ 2 \\ 1 \end{bmatrix} = \begin{bmatrix} -2 \\ 0 \\ 1 \end{bmatrix} = X.$$

Now Try Exercise 67 ◄

Finding Inverses Symbolically

The inverse matrix of an $n \times n$ matrix A can be found symbolically by first forming the augmented matrix $[A\,|\,I_n]$ and then performing matrix row operations, until the left side of the augmented matrix becomes the identity matrix. The resulting augmented matrix can be written as $[I_n\,|\,A^{-1}]$, where the right side of the matrix is A^{-1}.

In Example 3, we find A^{-1} for ourselves.

EXAMPLE 3 Finding an inverse symbolically

Find A^{-1} if

$$A = \begin{bmatrix} 0 & 1 & 0 \\ -1 & 0 & 0 \\ 0 & 0 & 1 \end{bmatrix}.$$

SOLUTION

Getting Started Begin by forming the following 3×6 augmented matrix with the 3×3 identity matrix on the right half.

$$\begin{bmatrix} 0 & 1 & 0 & | & 1 & 0 & 0 \\ -1 & 0 & 0 & | & 0 & 1 & 0 \\ 0 & 0 & 1 & | & 0 & 0 & 1 \end{bmatrix}$$

Next use row transformations to obtain the 3×3 identity on the left side. ▶

To obtain a 1 in the first row and first column, we negate the elements in row 2 and then interchange row 1 and row 2. The same row transformations are also applied to the right side of the augmented matrix.

$$\begin{bmatrix} 0 & 1 & 0 & | & 1 & 0 & 0 \\ -1 & 0 & 0 & | & 0 & 1 & 0 \\ 0 & 0 & 1 & | & 0 & 0 & 1 \end{bmatrix} \begin{matrix} -R_2 \to \\ R_1 \to \\ R_3 \to \end{matrix} \begin{bmatrix} 1 & 0 & 0 & | & 0 & -1 & 0 \\ 0 & 1 & 0 & | & 1 & 0 & 0 \\ 0 & 0 & 1 & | & 0 & 0 & 1 \end{bmatrix}$$

Because the left side of the augmented matrix is now the 3×3 identity, we stop. The right side of the augmented matrix is A^{-1}. Thus

$$A^{-1} = \begin{bmatrix} 0 & -1 & 0 \\ 1 & 0 & 0 \\ 0 & 0 & 1 \end{bmatrix},$$

and our result agrees with the information in Example 2.   ***Now Try Exercise 21*** ◀

Many times finding inverses requires several steps of row transformations. In the next two examples, we find the inverse of a 2×2 matrix and a 3×3 matrix.

EXAMPLE 4 Finding the inverse of a 2×2 matrix symbolically

Find A^{-1} if

$$A = \begin{bmatrix} 1 & 4 \\ 2 & 9 \end{bmatrix}.$$

SOLUTION Begin by forming a 2×4 augmented matrix. Perform matrix row operations to obtain the identity matrix on the left side, and perform the same operation on the right side of this matrix.

$$\begin{bmatrix} 1 & 4 & | & 1 & 0 \\ 2 & 9 & | & 0 & 1 \end{bmatrix} \begin{matrix} \\ R_2 - 2R_1 \to \end{matrix} \begin{bmatrix} 1 & 4 & | & 1 & 0 \\ 0 & 1 & | & -2 & 1 \end{bmatrix} \begin{matrix} R_1 - 4R_2 \to \\ \end{matrix} \begin{bmatrix} 1 & 0 & | & 9 & -4 \\ 0 & 1 & | & -2 & 1 \end{bmatrix}$$

Since the 2×2 identity matrix appears on the left side, it follows that the right side equals A^{-1}. That is,

$$A^{-1} = \begin{bmatrix} 9 & -4 \\ -2 & 1 \end{bmatrix}.$$

Furthermore, it can be verified that $A^{-1}A = I_2 = AA^{-1}$.　　　*Now Try Exercise 15* ◀

EXAMPLE 5 Finding the inverse of a 3 × 3 matrix symbolically

Find A^{-1} if

$$A = \begin{bmatrix} 1 & 0 & 1 \\ 2 & 1 & 3 \\ -1 & 1 & 1 \end{bmatrix}.$$

SOLUTION Begin by forming the following 3×6 augmented matrix. Perform matrix row operations to obtain the identity matrix on the left side, and perform the same operation on the right side of this matrix.

$$\begin{bmatrix} 1 & 0 & 1 & | & 1 & 0 & 0 \\ 2 & 1 & 3 & | & 0 & 1 & 0 \\ -1 & 1 & 1 & | & 0 & 0 & 1 \end{bmatrix} \begin{matrix} \\ R_2 - 2R_1 \rightarrow \\ R_3 + R_1 \rightarrow \end{matrix} \begin{bmatrix} 1 & 0 & 1 & | & 1 & 0 & 0 \\ 0 & 1 & 1 & | & -2 & 1 & 0 \\ 0 & 1 & 2 & | & 1 & 0 & 1 \end{bmatrix}$$

$$\begin{matrix} \\ \\ R_3 - R_2 \rightarrow \end{matrix} \begin{bmatrix} 1 & 0 & 1 & | & 1 & 0 & 0 \\ 0 & 1 & 1 & | & -2 & 1 & 0 \\ 0 & 0 & 1 & | & 3 & -1 & 1 \end{bmatrix} \begin{matrix} R_1 - R_3 \rightarrow \\ R_2 - R_3 \rightarrow \\ \end{matrix} \begin{bmatrix} 1 & 0 & 0 & | & -2 & 1 & -1 \\ 0 & 1 & 0 & | & -5 & 2 & -1 \\ 0 & 0 & 1 & | & 3 & -1 & 1 \end{bmatrix}$$

The right side is equal to A^{-1}. That is,

$$A^{-1} = \begin{bmatrix} -2 & 1 & -1 \\ -5 & 2 & -1 \\ 3 & -1 & 1 \end{bmatrix}.$$

It can be verified that $A^{-1}A = I_3 = AA^{-1}$.　　　*Now Try Exercise 25* ◀

NOTE If it is not possible to obtain the identity matrix on the left side of the augmented matrix by using matrix row operations, then A^{-1} does *not* exist.

Representing Linear Systems with Matrix Equations

In Section 9.4 linear systems were solved using Gaussian elimination with backward substitution. This method used an augmented matrix to represent a system of linear equations. A system of linear equations can also be represented by a matrix equation.

$$3x - 2y + 4z = 5$$
$$2x + y + 3z = 9$$
$$-x + 5y - 2z = 5$$

Let A, X, and B be matrices defined as

| *Coefficient Matrix* | *Variable Matrix* | | *Constant Matrix* |

$$A = \begin{bmatrix} 3 & -2 & 4 \\ 2 & 1 & 3 \\ -1 & 5 & -2 \end{bmatrix}, \quad X = \begin{bmatrix} x \\ y \\ z \end{bmatrix}, \quad \text{and} \quad B = \begin{bmatrix} 5 \\ 9 \\ 5 \end{bmatrix}.$$

The matrix product AX is given by

$$AX = \begin{bmatrix} 3 & -2 & 4 \\ 2 & 1 & 3 \\ -1 & 5 & -2 \end{bmatrix} \begin{bmatrix} x \\ y \\ z \end{bmatrix} = \begin{bmatrix} 3x + (-2)y + 4z \\ 2x + 1y + 3z \\ (-1)x + 5y + (-2)z \end{bmatrix} = \begin{bmatrix} 3x - 2y + 4z \\ 2x + y + 3z \\ -x + 5y - 2z \end{bmatrix}.$$

Thus the matrix equation $AX = B$ simplifies to

$$\begin{bmatrix} 3x - 2y + 4z \\ 2x + y + 3z \\ -x + 5y - 2z \end{bmatrix} = \begin{bmatrix} 5 \\ 9 \\ 5 \end{bmatrix}.$$

This matrix equation $AX = B$ is equivalent to the original system of linear equations. Any system of *linear* equations can be represented by a matrix equation.

EXAMPLE 6 Representing linear systems with matrix equations

Represent each system of linear equations in the form $AX = B$.

(a) $\begin{aligned} 3x - 4y &= 7 \\ -x + 6y &= -3 \end{aligned}$ **(b)** $\begin{aligned} x - 5y &= 2 \\ -3x + 2y + z &= -7 \\ 4x + 5y + 6z &= 10 \end{aligned}$

SOLUTION

(a) This linear system comprises two equations and two variables. The equivalent matrix equation is

$$AX = \begin{bmatrix} 3 & -4 \\ -1 & 6 \end{bmatrix} \begin{bmatrix} x \\ y \end{bmatrix} = \begin{bmatrix} 7 \\ -3 \end{bmatrix} = B.$$

(b) The equivalent matrix equation is

$$AX = \begin{bmatrix} 1 & -5 & 0 \\ -3 & 2 & 1 \\ 4 & 5 & 6 \end{bmatrix} \begin{bmatrix} x \\ y \\ z \end{bmatrix} = \begin{bmatrix} 2 \\ -7 \\ 10 \end{bmatrix} = B.$$

Now Try Exercises 39 and 43 ◀

Solving Linear Systems with Inverses

The matrix equation $AX = B$ can be solved by using A^{-1}, if it exists.

$$\begin{aligned} AX &= B && \text{Linear system} \\ A^{-1}AX &= A^{-1}B && \text{Multiply each side by } A^{-1}. \\ I_nX &= A^{-1}B && A^{-1}A = I_n \\ X &= A^{-1}B && I_nX = X \text{ for any } n \times 1 \text{ matrix } X \end{aligned}$$

To solve a linear system, multiply each side of the matrix equation $AX = B$ by A^{-1}, if it exists. The solution to the system is unique and can be written as $X = A^{-1}B$.

NOTE Since matrix multiplication is not commutative, it is essential to multiply each side of the equation on the *left* by A^{-1}. That is, $X = A^{-1}B \neq BA^{-1}$ in general.

EXAMPLE 7   Solving a linear system using the inverse of a 2×2 matrix

Write the linear system as a matrix equation in the form $AX = B$. Find A^{-1} and solve for X.

$$x + 4y = 3$$
$$2x + 9y = 5$$

SOLUTION   The linear system can be written as

$$AX = \begin{bmatrix} 1 & 4 \\ 2 & 9 \end{bmatrix} \begin{bmatrix} x \\ y \end{bmatrix} = \begin{bmatrix} 3 \\ 5 \end{bmatrix} = B.$$

The matrix A^{-1} was found in Example 4. Thus we can solve for X as follows.

$$X = A^{-1}B = \begin{bmatrix} 9 & -4 \\ -2 & 1 \end{bmatrix} \begin{bmatrix} 3 \\ 5 \end{bmatrix} = \begin{bmatrix} 7 \\ -1 \end{bmatrix}$$

The solution to the system is $(7, -1)$. Check this.   ***Now Try Exercise 47*** ◀

Technology and Inverse Matrices   In the next two examples, we use technology to solve the system of linear equations. Technology is especially helpful when finding A^{-1}.

EXAMPLE 8   Solving a linear system using the inverse of a 3×3 matrix

Write the linear system as the matrix equation $AX = B$. Find A^{-1} and solve for X.

$$x + 3y - z = 6$$
$$-2y + z = -2$$
$$-x + y - 3z = 4$$

SOLUTION   The linear system can be written as

$$AX = \begin{bmatrix} 1 & 3 & -1 \\ 0 & -2 & 1 \\ -1 & 1 & -3 \end{bmatrix} \begin{bmatrix} x \\ y \\ z \end{bmatrix} = \begin{bmatrix} 6 \\ -2 \\ 4 \end{bmatrix} = B.$$

The matrix A^{-1} can be found by hand or with a graphing calculator, as shown in Figure 9.71. The solution to the system is given by $x = 4.5$, $y = -0.5$, and $z = -3$. See Figure 9.72.

Calculator Help

To find the inverse of a matrix, see Appendix A (page AP-16). To solve a linear system with a matrix inverse, see Appendix A (page AP-17).

```
[A]⁻¹
[ [1.25    2    .25 ]
  [-.25   -1   -.25]
  [-.5    -1    .5 ]]
```

Figure 9.71

```
[A]⁻¹*[B]
              [ [4.5]
                [-.5]
                [-3 ]]
```

Figure 9.72   ***Now Try Exercise 59*** ◀

EXAMPLE 9   Modeling blood pressure

In one study of adult males, the effect of both age A in years and weight W in pounds on systolic blood pressure P was found to be modeled by $P(A, W) = a + bA + cW$, where a, b, and c are constants. Table 9.11 lists three individuals with representative blood pressures.   (**Source:** C. H. Brase and C. P. Brase, *Understandable Statistics*.)

Table 9.11

P	A	W
113	39	142
138	53	181
152	65	191

```
[A]⁻¹*[B]
 [[32.7804878  ]
  [.9024390244 ]
  [.3170731707 ]]
```

Figure 9.73

(a) Use Table 9.11 to approximate values for the constants a, b, and c.

(b) Estimate a typical systolic blood pressure for an individual who is 55 years old and weighs 175 pounds.

SOLUTION

(a) Determine the constants a, b, and c in $P(A, W) = a + bA + cW$ by solving the following three equations.

$$P(39, 142) = a + b(39) + c(142) = 113$$
$$P(53, 181) = a + b(53) + c(181) = 138$$
$$P(65, 191) = a + b(65) + c(191) = 152$$

These three equations can be rewritten as follows.

$$a + 39b + 142c = 113$$
$$a + 53b + 181c = 138$$
$$a + 65b + 191c = 152$$

This system can be represented by the matrix equation $AX = B$.

$$AX = \begin{bmatrix} 1 & 39 & 142 \\ 1 & 53 & 181 \\ 1 & 65 & 191 \end{bmatrix} \begin{bmatrix} a \\ b \\ c \end{bmatrix} = \begin{bmatrix} 113 \\ 138 \\ 152 \end{bmatrix} = B$$

The solution, $X = A^{-1}B$, is shown in Figure 9.73. The values for the constants are $a \approx 32.78$, $b \approx 0.9024$, and $c \approx 0.3171$. Thus it follows that P is given by the equation $P(A, W) = 32.78 + 0.9024A + 0.3171W$.

(b) Evaluate $P(55, 175) = 32.78 + 0.9024(55) + 0.3171(175) \approx 137.9$. This model predicts that a typical (male) individual 55 years old, weighing 175 pounds, has a systolic blood pressure of approximately 138. Clearly, this could vary greatly among individuals.   *Now Try Exercise 73* ◀

9.6 Putting It All Together

The following table summarizes some of the mathematical concepts presented in this section.

Concept	Comments	Examples
Identity matrix	The $n \times n$ identity matrix I_n has only 1's on the main diagonal and 0's elsewhere. When it is multiplied by any $n \times n$ matrix A, the result is A.	$\begin{bmatrix} 1 & 0 \\ 0 & 1 \end{bmatrix}\begin{bmatrix} 2 & 3 \\ 4 & 5 \end{bmatrix} = \begin{bmatrix} 2 & 3 \\ 4 & 5 \end{bmatrix}$ and $\begin{bmatrix} 2 & 3 \\ 4 & 5 \end{bmatrix}\begin{bmatrix} 1 & 0 \\ 0 & 1 \end{bmatrix} = \begin{bmatrix} 2 & 3 \\ 4 & 5 \end{bmatrix}$, where $I_2 = \begin{bmatrix} 1 & 0 \\ 0 & 1 \end{bmatrix}$

continued on next page

continued from previous page

Concept	Comments	Examples
Matrix inverse	If an $n \times n$ matrix A has an inverse, it is unique, is denoted A^{-1}, and satisfies the equations $AA^{-1} = I_n$ and $A^{-1}A = I_n$. Matrix inverses can be found by using technology. They can also be found with pencil and paper by performing matrix row operations on the augmented matrix $[A \mid I_n]$ until it is transformed to $[I_n \mid A^{-1}]$.	If $A = \begin{bmatrix} 2 & 3 \\ 3 & 5 \end{bmatrix}$, then $A^{-1} = \begin{bmatrix} 5 & -3 \\ -3 & 2 \end{bmatrix}$ because $$AA^{-1} = \begin{bmatrix} 2 & 3 \\ 3 & 5 \end{bmatrix}\begin{bmatrix} 5 & -3 \\ -3 & 2 \end{bmatrix} = \begin{bmatrix} 1 & 0 \\ 0 & 1 \end{bmatrix}$$ $$A^{-1}A = \begin{bmatrix} 5 & -3 \\ -3 & 2 \end{bmatrix}\begin{bmatrix} 2 & 3 \\ 3 & 5 \end{bmatrix} = \begin{bmatrix} 1 & 0 \\ 0 & 1 \end{bmatrix}.$$
Systems of linear equations	Systems of linear equations can be written by using the matrix equation $AX = B$. If A is invertible, then there will be a unique solution given by $X = A^{-1}B$. If A is not invertible, then there could be either no solution or infinitely many solutions. In the latter case, Gaussian elimination should be applied.	The linear system $\begin{aligned} 2x - y &= 3 \\ x + 2y &= 4 \end{aligned}$ can be written as $AX = B$, where $$A = \begin{bmatrix} 2 & -1 \\ 1 & 2 \end{bmatrix}, X = \begin{bmatrix} x \\ y \end{bmatrix}, \text{ and } B = \begin{bmatrix} 3 \\ 4 \end{bmatrix}.$$ The solution to the system is given by $$X = A^{-1}B = \begin{bmatrix} 0.4 & 0.2 \\ -0.2 & 0.4 \end{bmatrix}\begin{bmatrix} 3 \\ 4 \end{bmatrix} = \begin{bmatrix} 2 \\ 1 \end{bmatrix}.$$ The solution is $(2, 1)$.

9.6   Exercises

Inverse and Identity Matrices

Exercises 1–6: Determine if B is the inverse matrix of A by calculating AB and BA.

1. $A = \begin{bmatrix} 4 & 3 \\ 5 & 4 \end{bmatrix}$, $B = \begin{bmatrix} 4 & -3 \\ -5 & 4 \end{bmatrix}$

2. $A = \begin{bmatrix} -1 & 2 \\ -3 & 8 \end{bmatrix}$, $B = \begin{bmatrix} -4 & 1 \\ -2 & 0.5 \end{bmatrix}$

3. $A = \begin{bmatrix} 1 & -1 & 2 \\ 0 & 1 & -1 \\ 1 & 0 & 2 \end{bmatrix}$, $B = \begin{bmatrix} 2 & 2 & -1 \\ -1 & 0 & 1 \\ -1 & -1 & 1 \end{bmatrix}$

4. $A = \begin{bmatrix} 2 & 1 & 1 \\ -1 & 0 & -1 \\ 0 & 2 & -1 \end{bmatrix}$, $B = \begin{bmatrix} 2 & 3 & -1 \\ -1 & -2 & 1 \\ -2 & -4 & 1 \end{bmatrix}$

5. $A = \begin{bmatrix} 2 & 1 & -1 \\ 3 & 0 & 2 \\ -1 & 0 & 1 \end{bmatrix}$, $B = \begin{bmatrix} 0 & 1 & -2 \\ 1 & -3 & 7 \\ 0 & -1 & 3 \end{bmatrix}$

6. $A = \begin{bmatrix} 1 & -1 & 1 \\ 0 & 1 & 0 \\ 1 & 1 & 2 \end{bmatrix}$, $B = \begin{bmatrix} 2 & 3 & -1 \\ 0 & 1 & 0 \\ -1 & -2 & 1 \end{bmatrix}$

Exercises 7–10: Find the value of the constant k in A^{-1}.

7. $A = \begin{bmatrix} 1 & 1 \\ 1 & 2 \end{bmatrix}$, $A^{-1} = \begin{bmatrix} 2 & -1 \\ -1 & k \end{bmatrix}$

8. $A = \begin{bmatrix} -2 & 2 \\ 1 & -2 \end{bmatrix}$, $A^{-1} = \begin{bmatrix} -1 & k \\ -0.5 & -1 \end{bmatrix}$

9. $A = \begin{bmatrix} 1 & 3 \\ -1 & -5 \end{bmatrix}$, $A^{-1} = \begin{bmatrix} k & 1.5 \\ -0.5 & -0.5 \end{bmatrix}$

10. $A = \begin{bmatrix} -2 & 5 \\ -3 & 4 \end{bmatrix}$, $A^{-1} = \begin{bmatrix} \frac{4}{7} & -\frac{5}{7} \\ k & -\frac{2}{7} \end{bmatrix}$

Exercises 11–14: Predict the results of $I_n A$ and $A I_n$. Then verify your prediction.

11. $I_2 = \begin{bmatrix} 1 & 0 \\ 0 & 1 \end{bmatrix}$, $A = \begin{bmatrix} 1 & -2 \\ 4 & 3 \end{bmatrix}$

12. $I_3 = \begin{bmatrix} 1 & 0 & 0 \\ 0 & 1 & 0 \\ 0 & 0 & 1 \end{bmatrix}$, $A = \begin{bmatrix} 1 & -4 & 3 \\ 1 & 9 & 5 \\ 3 & -5 & 0 \end{bmatrix}$

13. $I_3 = \begin{bmatrix} 1 & 0 & 0 \\ 0 & 1 & 0 \\ 0 & 0 & 1 \end{bmatrix}$, $A = \begin{bmatrix} 0 & 0 & 0 \\ 0 & 0 & 0 \\ 0 & 0 & 0 \end{bmatrix}$

14. $I_4 = \begin{bmatrix} 1 & 0 & 0 & 0 \\ 0 & 1 & 0 & 0 \\ 0 & 0 & 1 & 0 \\ 0 & 0 & 0 & 1 \end{bmatrix}$, $A = \begin{bmatrix} 5 & -2 & 6 & -3 \\ 0 & 1 & 4 & -1 \\ -5 & 7 & 9 & 8 \\ 0 & 0 & 3 & 1 \end{bmatrix}$

Calculating Inverses

Exercises 15–28: (Refer to Examples 3–5.) Let A be the given matrix. Find A^{-1} without a calculator.

15. $\begin{bmatrix} 1 & 2 \\ 1 & 3 \end{bmatrix}$

16. $\begin{bmatrix} 1 & 0 \\ 1 & -1 \end{bmatrix}$

17. $\begin{bmatrix} -1 & 2 \\ 3 & -5 \end{bmatrix}$

18. $\begin{bmatrix} 1 & 3 \\ 2 & 5 \end{bmatrix}$

19. $\begin{bmatrix} 8 & 5 \\ 2 & 1 \end{bmatrix}$

20. $\begin{bmatrix} -2 & 4 \\ -5 & 9 \end{bmatrix}$

21. $\begin{bmatrix} 0 & 0 & 1 \\ 1 & 0 & 0 \\ 0 & 1 & 0 \end{bmatrix}$

22. $\begin{bmatrix} 1 & 0 & 0 \\ 1 & 1 & 0 \\ 0 & 1 & 1 \end{bmatrix}$

23. $\begin{bmatrix} 1 & 0 & 1 \\ 2 & 1 & 3 \\ -1 & 1 & 1 \end{bmatrix}$

24. $\begin{bmatrix} -2 & 1 & 0 \\ 1 & 0 & 1 \\ -1 & 1 & 0 \end{bmatrix}$

25. $\begin{bmatrix} 1 & 2 & -1 \\ 2 & 5 & 0 \\ -1 & -1 & 2 \end{bmatrix}$

26. $\begin{bmatrix} 2 & -2 & 1 \\ 1 & 3 & 2 \\ 4 & -2 & 4 \end{bmatrix}$

27. $\begin{bmatrix} -2 & 1 & -3 \\ 0 & 1 & 2 \\ 1 & -2 & 1 \end{bmatrix}$

28. $\begin{bmatrix} 1 & -1 & 1 \\ -1 & 2 & 1 \\ 0 & 2 & 1 \end{bmatrix}$

Exercises 29–38: Let A be the given matrix. Find A^{-1}.

29. $\begin{bmatrix} 0.5 & -1.5 \\ 0.2 & -0.5 \end{bmatrix}$

30. $\begin{bmatrix} -0.5 & 0.5 \\ 3 & 2 \end{bmatrix}$

31. $\begin{bmatrix} 1 & 2 & 0 \\ -1 & 4 & -1 \\ 2 & -1 & 0 \end{bmatrix}$

32. $\begin{bmatrix} -2 & 0 & 1 \\ 5 & -4 & 1 \\ 1 & -2 & 0 \end{bmatrix}$

33. $\begin{bmatrix} 2 & -2 & 1 \\ 0 & 5 & 8 \\ 0 & 0 & -1 \end{bmatrix}$

34. $\begin{bmatrix} 2 & 0 & 2 \\ 1 & 5 & 0 \\ -1 & 0 & 2 \end{bmatrix}$

35. $\begin{bmatrix} 3 & -1 & -1 \\ -1 & 3 & -1 \\ -1 & -1 & 3 \end{bmatrix}$

36. $\begin{bmatrix} 2 & -3 & 1 \\ 5 & -6 & 3 \\ 3 & 2 & 0 \end{bmatrix}$

37. $\begin{bmatrix} 1 & -1 & 0 & 0 \\ -1 & 5 & -1 & 0 \\ 0 & -1 & 5 & -1 \\ 0 & 0 & -1 & 1 \end{bmatrix}$

38. $\begin{bmatrix} 3 & 1 & 0 & 0 \\ 1 & 3 & 1 & 0 \\ 0 & 1 & 3 & 1 \\ 0 & 0 & 1 & 3 \end{bmatrix}$

Matrices and Linear Systems

Exercises 39–46: Represent the system of linear equations in the form $AX = B$.

39. $\begin{aligned} 2x - 3y &= 7 \\ -3x - 4y &= 9 \end{aligned}$

40. $\begin{aligned} -x + 3y &= 10 \\ 2x - 6y &= -1 \end{aligned}$

41. $\begin{aligned} \tfrac{1}{2}x - \tfrac{3}{2}y &= \tfrac{1}{4} \\ -x + 2y &= 5 \end{aligned}$

42. $\begin{aligned} -1.1x + 3.2y &= -2.7 \\ 5.6x - 3.8y &= -3.0 \end{aligned}$

43. $\begin{aligned} x - 2y + z &= 5 \\ 3y - z &= 6 \\ 5x - 4y - 7z &= 0 \end{aligned}$

44. $\begin{aligned} 4x - 3y + 2z &= 8 \\ -x + 4y + 3z &= 2 \\ -2x - 5z &= 2 \end{aligned}$

45. $\begin{aligned} 4x - y + 3z &= -2 \\ x + 2y + 5z &= 11 \\ 2x - 3y &= -1 \end{aligned}$

46. $\begin{aligned} x - 2y + z &= 12 \\ 4y + 3z &= 13 \\ -2x + 7y &= -2 \end{aligned}$

Solving Linear Systems

Exercises 47–54: Complete the following.

(a) Write the system in the form $AX = B$.

(b) Solve the system by finding A^{-1} and then using the equation $X = A^{-1}B$. (Hint: Some of your answers from Exercises 15–28 may be helpful.)

47. $\begin{aligned} x + 2y &= 3 \\ x + 3y &= 6 \end{aligned}$

48. $\begin{aligned} 2x + y &= 4 \\ -x + 2y &= -1 \end{aligned}$

49. $\begin{aligned} -x + 2y &= 5 \\ 3x - 5y &= -2 \end{aligned}$

50. $\begin{aligned} x + 3y &= -3 \\ 2x + 5y &= -2 \end{aligned}$

51. $\begin{aligned} x + \qquad z &= -7 \\ 2x + y + 3z &= -13 \\ -x + y + z &= -4 \end{aligned}$

52. $\begin{aligned} -2x + y \qquad &= -5 \\ x \qquad + z &= -5 \\ -x + y \qquad &= -4 \end{aligned}$

53. $\begin{aligned} x + 2y - z &= 2 \\ 2x + 5y \qquad &= -1 \\ -x - y + 2z &= 0 \end{aligned}$

54. $\begin{aligned} 2x - 2y + z &= 1 \\ x + 3y + 2z &= 3 \\ 4x - 2y + 4z &= 4 \end{aligned}$

Exercises 55–62: Complete the following for the given system of linear equations.

   (a) *Write the system in the form $AX = B$.*

   (b) *Solve the linear system by computing $X = A^{-1}B$ with a calculator. Approximate the solution to the nearest hundredth when appropriate.*

55. $\begin{aligned} 1.5x + 3.7y &= 0.32 \\ -0.4x - 2.1y &= 0.36 \end{aligned}$

56. $\begin{aligned} 31x + 18y &= 64.1 \\ 5x - 23y &= -59.6 \end{aligned}$

57. $\begin{aligned} 0.08x - 0.7y &= -0.504 \\ 1.1x - 0.05y &= 0.73 \end{aligned}$

58. $\begin{aligned} -231x + 178y &= -439 \\ 525x - 329y &= 2282 \end{aligned}$

59. $\begin{aligned} 3.1x + 1.9y - z &= 1.99 \\ 6.3x \qquad - 9.9z &= -3.78 \\ -x + 1.5y + 7z &= 5.3 \end{aligned}$

60. $\begin{aligned} 17x - 22y - 19z &= -25.2 \\ 3x + 13y - 9z &= 105.9 \\ x - 2y + 6.1z &= -23.55 \end{aligned}$

61. $\begin{aligned} 3x - y + z &= 4.9 \\ 5.8x - 2.1y \qquad &= -3.8 \\ -x \qquad + 2.9z &= 3.8 \end{aligned}$

62. $\begin{aligned} 1.2x - 0.3y - 0.7z &= -0.5 \\ -0.4x + 1.3y + 0.4z &= 0.9 \\ 1.7x + 0.6y + 1.1z &= 1.3 \end{aligned}$

Interpreting Inverses

Exercises 63 and 64: Translations (Refer to the discussion in this section about translating a point.) The matrix product AX performs a translation on the point (x, y), where

$$A = \begin{bmatrix} 1 & 0 & h \\ 0 & 1 & k \\ 0 & 0 & 1 \end{bmatrix} \text{ and } X = \begin{bmatrix} x \\ y \\ 1 \end{bmatrix}.$$

   (a) *Predict the new location of the point (x, y) when it is translated by A. Compute $Y = AX$ to verify your prediction.*

   (b) *Make a conjecture as to what $A^{-1}Y$ represents. Find A^{-1} and calculate $A^{-1}Y$ to test your conjecture.*

   (c) *What will AA^{-1} and $A^{-1}A$ equal?*

63. $A = \begin{bmatrix} 1 & 0 & 2 \\ 0 & 1 & 3 \\ 0 & 0 & 1 \end{bmatrix}$, $(x, y) = (0, 1)$, and $X = \begin{bmatrix} 0 \\ 1 \\ 1 \end{bmatrix}$

64. $A = \begin{bmatrix} 1 & 0 & -4 \\ 0 & 1 & 5 \\ 0 & 0 & 1 \end{bmatrix}$, $(x, y) = (4, 2)$, and $X = \begin{bmatrix} 4 \\ 2 \\ 1 \end{bmatrix}$

Exercises 65 and 66: Translations (Refer to the discussion in this section about translating a point.) Find a 3×3 matrix A that performs the following translation of a point (x, y) represented by X. Find A^{-1} and describe what it computes.

65. 3 units to the left and 5 units downward

66. 6 units to the right and 1 unit upward

67. *Rotation* (Refer to Example 2.) The matrix B rotates the point (x, y) clockwise about the origin $45°$, where

$$B = \begin{bmatrix} \dfrac{1}{\sqrt{2}} & \dfrac{1}{\sqrt{2}} & 0 \\ -\dfrac{1}{\sqrt{2}} & \dfrac{1}{\sqrt{2}} & 0 \\ 0 & 0 & 1 \end{bmatrix} \text{ and } B^{-1} = \begin{bmatrix} \dfrac{1}{\sqrt{2}} & -\dfrac{1}{\sqrt{2}} & 0 \\ \dfrac{1}{\sqrt{2}} & \dfrac{1}{\sqrt{2}} & 0 \\ 0 & 0 & 1 \end{bmatrix}.$$

   (a) Let X represent the point $(-\sqrt{2}, -\sqrt{2})$. Compute $Y = BX$.

   (b) Find $B^{-1}Y$. Interpret what B^{-1} computes.

68. *Rotation* (Refer to Exercise 67.) Predict the result of the computations $BB^{-1}X$ and $B^{-1}BX$ for any point (x, y) represented by X. Explain this result geometrically.

69. *Translations* The matrix A translates a point to the right 4 units and downward 2 units, and the matrix B translates a point to the left 3 units and upward 3 units, where

$$A = \begin{bmatrix} 1 & 0 & 4 \\ 0 & 1 & -2 \\ 0 & 0 & 1 \end{bmatrix} \text{ and } B = \begin{bmatrix} 1 & 0 & -3 \\ 0 & 1 & 3 \\ 0 & 0 & 1 \end{bmatrix}.$$

   (a) Let X represent the point $(1, 1)$. Predict the result of $Y = ABX$. Check your prediction.

   (b) Find AB mentally, and then compute AB.

   (c) Would you expect $AB = BA$? Verify your answer.

   (d) Find $(AB)^{-1}$ mentally. Explain your reasoning.

70. *Rotation*   (Refer to Exercises 63 and 67 for A and B.)

   (a) Let X represent the point $(0, \sqrt{2})$. If this point is rotated about the origin $45°$ clockwise and then translated 2 units to the right and 3 units upward, determine its new coordinates geometrically.

   (b) Compute $Y = ABX$, and explain the result.

   (c) Is ABX equal to BAX? Interpret your answer.

   (d) Find a matrix that translates Y back to X.

Applications

71. *Cost of CDs*   A music store marks its compact discs A, B, or C to indicate one of three selling prices. The last column in the table shows the total cost of a purchase. Use this information to determine the cost of one CD of each type by setting up a matrix equation and solving it with an inverse.

A	B	C	Total
2	3	4	$120.91
1	4	0	$62.95
2	1	3	$79.94

72. *Traffic Flow*   (Refer to Exercises 81 and 82 in Section 9.4.) The figure shows four one-way streets with intersections A, B, C, and D. Numbers indicate the average traffic flow in vehicles per minute. The variables x_1, x_2, x_3, and x_4 denote unknown traffic flows.

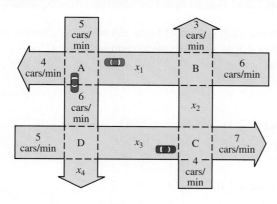

   (a) The number of vehicles per minute entering an intersection equals the number exiting an intersection. Verify that the given system of linear equations describes the traffic flow.

     A: $x_1 + 5 = 4 + 6$ B: $x_2 + 6 = x_1 + 3$

     C: $x_3 + 4 = x_2 + 7$ D: $6 + 5 = x_3 + x_4$

   (b) Write the system as $AX = B$ and solve using A^{-1}.

   (c) Interpret your results.

73. *Home Prices*   The selling price of a home depends on several factors. Real-estate companies sometimes study how the selling price of a home is related to its size and condition. The table contains representative data on sales of three homes. Price P is measured in thousands of dollars, home size S is in square feet, and condition C is rated on a scale from 1 to 10, where 10 represents excellent condition. The variables were found to be related by the equation $P = a + bS + cC$.

P	S	C
122	1500	8
130	2000	5
158	2200	10

   (a) Use the table to write a system of linear equations whose solution gives a, b, and c. Solve this system of linear equations.

   (b) Estimate the selling price of a home with 1800 square feet and a condition of 7.

74. *Tire Sales*   A study investigated the relationship among annual tire sales T in thousands, automobile registrations A in millions, and personal disposable income I in millions of dollars. Representative data for three different years are shown in the table. The data were modeled by $T = aA + bI + c$, where a, b, and c are constants. (**Source:** J. Jarrett, *Business Forecasting Methods.*)

T	A	I
10,170	113	308
15,305	133	622
21,289	155	1937

   (a) Use the data to write a system of linear equations whose solution gives a, b, and c.

   (b) Solve this linear system. Write a formula for T, as $T = aA + bI + c$.

   (c) If $A = 118$ and $I = 311$, predict T. (The actual value for T was 11,314.)

75. *Leontief Economic Model* Suppose that a closed economic region has three industries: service, electrical power, and tourism. The service industry uses 20% of its own production, 40% of the electrical power, and 80% of the tourism. The power company use 40% of the service industry, 20% of the electrical power, and 10% of the tourism. The tourism industry uses 40% of the service industry, 40% of the electrical power, and 10% of the tourism.

(a) Let S, E, and T be the numbers of units produced by the service, electrical, and tourism industries, respectively. The following system of linear equations can be used to determine the relative number of units each industry needs to produce. (This model assumes that all production is consumed by the region.)

$$0.2S + 0.4E + 0.8T = S$$
$$0.4S + 0.2E + 0.1T = E$$
$$0.4S + 0.4E + 0.1T = T$$

Solve the system and write the solution in terms of T.

(b) If tourism produces 60 units, how many units should the service and electrical industries produce?

76. *Plate Glass Sales* Plate glass sales G can be affected by the number of new building contracts B issued and the number of automobiles A produced, since plate glass is used in buildings and cars. To forecast sales, a plate glass company in California collected data for three consecutive years, shown in the table. All units are in millions. The data were modeled by $G = aA + bB + c$, where a, b, and c are constants. (**Source:** S. Makridakis and S. Wheelwright, *Forecasting Methods for Management.*)

G	A	B
603	5.54	37.1
657	6.93	41.3
779	7.64	45.6

(a) Write a system of linear equations whose solution gives a, b, and c.

(b) Solve this linear system. Write a formula for G.

(c) For the following year, it was estimated that $A = 7.75$ and $B = 47.4$. Predict G. (The actual value for G was 878.)

Writing about Mathematics

77. Discuss how to solve the matrix equation $AX = B$ if A^{-1} exists.

78. Give an example of a 2×2 matrix A with only nonzero elements that does not have an inverse. Explain what happens if one attempts to find A^{-1} symbolically.

CHECKING BASIC CONCEPTS FOR SECTIONS 9.5 AND 9.6

1. Perform the operations on the given matrices A and B.

$$A = \begin{bmatrix} 1 & 0 & 1 \\ -1 & 1 & 2 \\ 1 & 3 & 0 \end{bmatrix}, \quad B = \begin{bmatrix} -1 & 1 & 2 \\ 0 & 4 & 1 \\ 1 & -2 & 0 \end{bmatrix}$$

(a) $A + B$ **(b)** $2A - B$ **(c)** AB

2. Find the inverse of the matrix A by hand.

$$A = \begin{bmatrix} 0 & 0 & 1 \\ 1 & 1 & 0 \\ 1 & 0 & 1 \end{bmatrix}$$

3. Write each system of linear equations as a matrix equation $AX = B$. Solve the system utilizing A^{-1}.

(a) $\quad x - 2y = 13$
$\quad\ 2x + 3y = 5$

(b) $\quad x - y + z = 2$
$\quad -x + y + z = 4$
$\quad\quad\quad y - z = -1$

(c) $\quad 3.1x - 5.3y = -2.682$
$\quad -0.1x + 1.8y = 0.787$

4. Find A^{-1} if $A = \begin{bmatrix} 2 & -3 & 5 \\ 4 & -3 & 2 \\ 1 & 5 & -4 \end{bmatrix}$.

9.7 Determinants

- **Define and calculate determinants**
- **Apply Cramer's rule**
- **Use determinants to find areas of regions**

Introduction

Determinants are used in mathematics for theoretical purposes. However, they also are used to test if a matrix is invertible and to find the area of certain geometric figures, such as triangles. A *determinant* is a real number associated with a square matrix. We begin our discussion by defining a determinant for a 2×2 matrix.

Definition and Calculation of Determinants

Finding the determinant of a matrix with dimension 2×2 is a straightforward arithmetic calculation.

Determinant of a 2×2 Matrix

The **determinant** of

$$A = \begin{bmatrix} a & b \\ c & d \end{bmatrix}$$

is a real number defined by

$$\det A = ad - cb.$$

Later we define determinants for any $n \times n$ matrix. The following theorem can be used to determine if a matrix is invertible.

Invertible Matrix

A square matrix A is invertible if and only if $\det A \neq 0$.

EXAMPLE 1 Determining if a 2×2 matrix is invertible

Determine if A^{-1} exists by computing the determinant of the matrix A.

(a) $A = \begin{bmatrix} 3 & -4 \\ -5 & 9 \end{bmatrix}$ **(b)** $A = \begin{bmatrix} 52 & -32 \\ 65 & -40 \end{bmatrix}$

SOLUTION

(a) The determinant of the 2×2 matrix A is calculated as follows.

$$\det A = \det \begin{bmatrix} 3 & -4 \\ -5 & 9 \end{bmatrix} = (3)(9) - (-5)(-4) = 7$$

Since $\det A = 7 \neq 0$, the matrix A is invertible and A^{-1} exists.

(b) Similarly,

$$\det A = \det \begin{bmatrix} 52 & -32 \\ 65 & -40 \end{bmatrix} = (52)(-40) - (65)(-32) = 0.$$

Since $\det A = 0$, A^{-1} does not exist. Try finding A^{-1}. What happens?

Now Try Exercises 1 and 3 ◀

We can use determinants of 2×2 matrices to find determinants of larger square matrices. In order to do this, we first define the concepts of a *minor* and a *cofactor*.

Minors and Cofactors

The **minor**, denoted by M_{ij}, for element a_{ij} in the square matrix A is the real number computed by performing the following steps.

STEP 1: Delete the ith row and jth column from the matrix A.

STEP 2: Compute the determinant of the resulting matrix, which is equal to M_{ij}.

The **cofactor**, denoted A_{ij}, for a_{ij} is defined by $A_{ij} = (-1)^{i+j}M_{ij}$.

EXAMPLE 2 Calculating minors and cofactors

Find the following minors and cofactors for the matrix A.

$$A = \begin{bmatrix} 2 & -3 & 1 \\ -2 & 1 & 0 \\ 0 & -1 & 4 \end{bmatrix}$$

(a) M_{11} and M_{21} **(b)** A_{11} and A_{21}

SOLUTION

(a) To obtain the minor M_{11}, begin by crossing out the first row and first column of A.

$$A = \begin{bmatrix} 2 & -3 & 1 \\ -2 & 1 & 0 \\ 0 & -1 & 4 \end{bmatrix}$$

The remaining elements form the 2×2 matrix

$$B = \begin{bmatrix} 1 & 0 \\ -1 & 4 \end{bmatrix}.$$

The minor M_{11} is equal to $\det B = (1)(4) - (-1)(0) = 4$.

M_{21} is found by crossing out the second row and first column of A.

$$A = \begin{bmatrix} 2 & -3 & 1 \\ -2 & 1 & 0 \\ 0 & -1 & 4 \end{bmatrix}$$

The resulting matrix is

$$B = \begin{bmatrix} -3 & 1 \\ -1 & 4 \end{bmatrix}.$$

Thus $M_{21} = \det B = (-3)(4) - (-1)(1) = -11$.

(b) Since $A_{ij} = (-1)^{i+j} M_{ij}$, A_{11} and A_{21} can be computed as follows.

$$A_{11} = (-1)^{1+1}M_{11} = (-1)^2(4) = 4$$

$$A_{21} = (-1)^{2+1}M_{21} = (-1)^3(-11) = 11 \quad \textbf{\textit{Now Try Exercise 5}} \blacktriangleleft$$

Using the concept of a cofactor, we can calculate the determinant of *any* square matrix.

Calculating the Determinant of a Square Matrix Using the Method of Cofactors

Multiply each element in any row or column of the matrix by its cofactor. The sum of the products is equal to the determinant.

To compute the determinant of a 3×3 matrix A, begin by selecting a row or column.

$$A = \begin{bmatrix} a_{11} & a_{12} & a_{13} \\ a_{21} & a_{22} & a_{23} \\ a_{31} & a_{32} & a_{33} \end{bmatrix}$$

For example, if the second row of A is selected, the elements are a_{21}, a_{22}, and a_{23}. Then

$$\det A = a_{21}A_{21} + a_{22}A_{22} + a_{23}A_{23}.$$

On the other hand, utilizing the elements of a_{11}, a_{21}, and a_{31} in the first column gives

$$\det A = a_{11}A_{11} + a_{21}A_{21} + a_{31}A_{31}.$$

Regardless of the row or column selected, the value of $\det A$ is the same. The calculation is easier if some elements in the selected row or column equal 0.

EXAMPLE 3 **Evaluating the determinant of a 3 × 3 matrix**

Find $\det A$ if

$$A = \begin{bmatrix} 2 & -3 & 1 \\ -2 & 1 & 0 \\ 0 & -1 & 4 \end{bmatrix}.$$

SOLUTION To find the determinant of A, we can select any row or column. If we begin *expanding* about the first column of A, then

$$\det A = a_{11}A_{11} + a_{21}A_{21} + a_{31}A_{31}.$$

In the first column, $a_{11} = 2$, $a_{21} = -2$, $a_{31} = 0$. In Example 2, the cofactors A_{11} and A_{21} were computed as 4 and 11, respectively. Since A_{31} is multiplied by $a_{31} = 0$, its value is unimportant. Thus

$$\begin{aligned} \det &= a_{11}A_{11} + a_{21}A_{21} + a_{31}A_{31} \\ &= 2(4) + (-2)(11) + (0)A_{31} \qquad \text{\small First column of A} \\ &= -14. \end{aligned}$$

We could also have expanded about the second row.

$$\begin{aligned} \det A &= a_{21}A_{21} + a_{22}A_{22} + a_{23}A_{23} \\ &= (-2)A_{21} + (1)A_{22} + (0)A_{23} \qquad \text{\small Second row of A} \end{aligned}$$

CLASS DISCUSSION

If a row or column in matrix A contains only zeros, what is $\det A$?

To complete this computation we need to determine only A_{22}, since A_{21} is known to be 11 and A_{23} is multiplied by 0. To compute A_{22}, delete the second row and column of A to obtain M_{22}.

$$M_{22} = \det \begin{bmatrix} 2 & 1 \\ 0 & 4 \end{bmatrix} = 8 \qquad \text{and} \qquad A_{22} = (-1)^{2+2}(8) = 8$$

Thus $\det A = (-2)(11) + (1)(8) + (0)A_{23} = -14$. The same value for $\det A$ is obtained in both calculations.

Now Try Exercise 17 ◀

Instead of calculating $(-1)^{i+j}$ for each cofactor, we can use the following **sign matrix** to find determinants of 3×3 matrices. The checkerboard pattern can be expanded to include larger square matrices.

$$\begin{bmatrix} + & - & + \\ - & + & - \\ + & - & + \end{bmatrix} \quad \text{Sign matrix}$$

For example, if

$$A = \begin{bmatrix} 2 & 3 & 7 \\ -3 & -2 & -1 \\ 4 & 0 & 2 \end{bmatrix},$$

we can compute det A by expanding about the second column to take advantage of the 0. The second column contains $-$, $+$, and $-$ signs. Therefore

$$\det A = -(3) \det \begin{bmatrix} -3 & -1 \\ 4 & 2 \end{bmatrix} + (-2) \det \begin{bmatrix} 2 & 7 \\ 4 & 2 \end{bmatrix} - (0) \det \begin{bmatrix} 2 & 7 \\ -3 & -1 \end{bmatrix}$$
$$= -3(-2) + (-2)(-24) - (0)(19)$$
$$= 54.$$

Graphing calculators can evaluate determinants, as shown in the next example.

EXAMPLE 4   **Using technology to find a determinant**

Find the determinant of A.

(a) $A = \begin{bmatrix} 2 & -3 & 1 \\ -2 & 1 & 0 \\ 0 & -1 & 4 \end{bmatrix}$ **(b)** $A = \begin{bmatrix} 2 & -3 & 1 & 5 \\ 7 & 1 & -8 & 0 \\ 5 & 4 & 9 & 7 \\ -2 & 3 & 3 & 0 \end{bmatrix}$

SOLUTION
(a) The determinant of this matrix was calculated in Example 3 by hand. To use technology, enter the matrix and evaluate its determinant, as shown in Figure 9.74. The result is det $A = -14$, which agrees with our earlier calculation.

Calculator Help
To calculate a determinant, see
Appendix A (page AP-17).

Figure 9.74   **Figure 9.75**

(b) The determinant of a 4×4 matrix can be computed using cofactors. However, it is considerably easier to use technology. From Figure 9.75 we see that det $A = 966$.

Now Try Exercises 23 and 24 ◀

Cramer's Rule

Determinants can be used to solve *linear* systems by employing a method called **Cramer's rule**. Cramer's rule for linear systems in three variables is discussed in the Extended and Discovery Exercises at the end of this section.

Cramer's Rule for Linear Systems in Two Variables

The solution to the linear system

$$a_1 x + b_1 y = c_1$$
$$a_2 x + b_2 y = c_2$$

is given by $x = \frac{E}{D}$ and $y = \frac{F}{D}$, where

$$E = \det \begin{bmatrix} c_1 & b_1 \\ c_2 & b_2 \end{bmatrix}, \quad F = \det \begin{bmatrix} a_1 & c_1 \\ a_2 & c_2 \end{bmatrix}, \quad \text{and} \quad D = \det \begin{bmatrix} a_1 & b_1 \\ a_2 & b_2 \end{bmatrix} \neq 0.$$

EXAMPLE 5 Using Cramer's rule to solve a linear system in two variables

Use Cramer's rule to solve the linear system

$$4x + y = 146$$
$$9x + y = 66.$$

SOLUTION In this system $a_1 = 4$, $b_1 = 1$, $c_1 = 146$, $a_2 = 9$, $b_2 = 1$, and $c_2 = 66$. By Cramer's rule, the solution can be found as follows.

$$E = \det \begin{bmatrix} c_1 & b_1 \\ c_2 & b_2 \end{bmatrix} = \det \begin{bmatrix} 146 & 1 \\ 66 & 1 \end{bmatrix} = (146)(1) - (66)(1) = \mathbf{80}$$

$$F = \det \begin{bmatrix} a_1 & c_1 \\ a_2 & c_2 \end{bmatrix} = \det \begin{bmatrix} 4 & 146 \\ 9 & 66 \end{bmatrix} = (4)(66) - (9)(146) = \mathbf{-1050}$$

$$D = \det \begin{bmatrix} a_1 & b_1 \\ a_2 & b_2 \end{bmatrix} = \det \begin{bmatrix} 4 & 1 \\ 9 & 1 \end{bmatrix} = (4)(1) - (9)(1) = \mathbf{-5}$$

The solution is

$$x = \frac{E}{D} = \frac{80}{-5} = -16 \quad \text{and} \quad y = \frac{F}{D} = \frac{-1050}{-5} = 210.$$

Now Try Exercise 25 ◀

Area of Regions

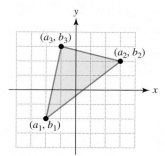

Figure 9.76

Determinants may be used to find the area of a triangle. If a triangle has vertices (a_1, b_1), (a_2, b_2), and (a_3, b_3), as shown in Figure 9.76, then its area is equal to the *absolute value* of D, where

$$D = \frac{1}{2} \det \begin{bmatrix} a_1 & a_2 & a_3 \\ b_1 & b_2 & b_3 \\ 1 & 1 & 1 \end{bmatrix}.$$

If the vertices are entered into the columns of D in a *counterclockwise* direction, then D will be positive. (**Source:** W. Taylor, *The Geometry of Computer Graphics*.)

EXAMPLE 6   Computing the area of a parallelogram

Use determinants to calculate the area of the parallelogram in Figure 9.77.

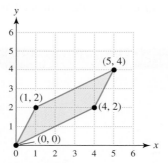

Figure 9.77

SOLUTION   To find the area of the parallelogram, we view the parallelogram as comprising two triangles. One triangle has vertices at $(0, 0)$, $(4, 2)$, and $(1, 2)$, and the other triangle has vertices at $(4, 2)$, $(5, 4)$, and $(1, 2)$. The area of the parallelogram is equal to the sum of the areas of the two triangles. Since these triangles are congruent, we can calculate the area of one triangle and double it. The area of one triangle is equal to D.

$$D = \frac{1}{2} \det \begin{bmatrix} 0 & 4 & 1 \\ 0 & 2 & 2 \\ 1 & 1 & 1 \end{bmatrix} = \frac{1}{2}(6) = 3$$

Since the vertices were entered in a counterclockwise direction, D is positive. The area of one triangle is equal to 3. Therefore the area of the parallelogram is twice this value, or 6 square units.

Now Try Exercise 35 ◄

CLASS DISCUSSION

Suppose we are given three distinct vertices and $D = 0$. What must be true about the three points?

9.7 Putting It All Together

The determinant of a square matrix A is a real number, denoted $\det A$. If $\det A \neq 0$, then the matrix A is invertible. Cramer's rule is a method for solving systems of linear equations that involves determinants. The following table summarizes the calculation of 2×2 and 3×3 determinants by hand.

Determinants of 2 × 2 Matrices

The determinant of a 2×2 matrix A is given by

$$\det A = \det \begin{bmatrix} a & b \\ c & d \end{bmatrix} = ad - cb.$$

Example: $\det \begin{bmatrix} 6 & -2 \\ 3 & 7 \end{bmatrix} = (6)(7) - (3)(-2) = 48$

Determinants of 3 × 3 Matrices

Finding the determinant of a 3×3 matrix A can be reduced to calculating the determinants of three 2×2 matrices. This calculation can be performed using cofactors.

$$\det A = \det \begin{bmatrix} a_1 & b_1 & c_1 \\ a_2 & b_2 & c_2 \\ a_3 & b_3 & c_3 \end{bmatrix}$$

$$= a_1 \det \begin{bmatrix} b_2 & c_2 \\ b_3 & c_3 \end{bmatrix} - a_2 \det \begin{bmatrix} b_1 & c_1 \\ b_3 & c_3 \end{bmatrix} + a_3 \det \begin{bmatrix} b_1 & c_1 \\ b_2 & c_2 \end{bmatrix}$$

$$= a_1(b_2 c_3 - b_3 c_2) - a_2(b_1 c_3 - b_3 c_1) + a_3(b_1 c_2 - b_2 c_1)$$

| Determinants of 3 × 3 Matrices   (continued) |

Example: $\det \begin{bmatrix} 1 & -2 & 3 \\ 4 & 5 & -1 \\ -3 & 7 & 8 \end{bmatrix} = (1) \det \begin{bmatrix} 5 & -1 \\ 7 & 8 \end{bmatrix} - (4) \det \begin{bmatrix} -2 & 3 \\ 7 & 8 \end{bmatrix} + (-3) \det \begin{bmatrix} -2 & 3 \\ 5 & -1 \end{bmatrix}$

$$= (1)(47) - 4(-37) - 3(-13) = 234$$

9.7 Exercises

Calculating Determinants

Exercises 1–4: Determine if the matrix A is invertible by calculating det A.

1. $A = \begin{bmatrix} 4 & 3 \\ 5 & 4 \end{bmatrix}$ **2.** $A = \begin{bmatrix} 1 & -3 \\ 2 & 6 \end{bmatrix}$

3. $A = \begin{bmatrix} -4 & 6 \\ -8 & 12 \end{bmatrix}$ **4.** $A = \begin{bmatrix} 10 & -20 \\ -5 & 10 \end{bmatrix}$

Exercises 5–8: Find the specified minor and cofactor for A.

5. M_{12} and A_{12} if $A = \begin{bmatrix} 1 & -1 & 3 \\ 2 & 3 & -2 \\ 0 & 1 & 5 \end{bmatrix}$

6. M_{23} and A_{23} if $A = \begin{bmatrix} 1 & 2 & -1 \\ 4 & 6 & -3 \\ 2 & 3 & 9 \end{bmatrix}$

7. M_{22} and A_{22} if $A = \begin{bmatrix} 7 & -8 & 1 \\ 3 & -5 & 2 \\ 1 & 0 & -2 \end{bmatrix}$

8. M_{31} and A_{31} if $A = \begin{bmatrix} 0 & 0 & -1 \\ 6 & -7 & 1 \\ 8 & -9 & -1 \end{bmatrix}$

Exercises 9–12: Let A be the given matrix. Find det A by expanding about the first column. State whether A^{-1} exists.

9. $\begin{bmatrix} 1 & 4 & -7 \\ 0 & 2 & -3 \\ 0 & -1 & 3 \end{bmatrix}$ **10.** $\begin{bmatrix} 0 & 2 & 8 \\ -1 & 3 & 5 \\ 0 & 4 & 1 \end{bmatrix}$

11. $\begin{bmatrix} 5 & 1 & 6 \\ 0 & -2 & 0 \\ 0 & 4 & 0 \end{bmatrix}$ **12.** $\begin{bmatrix} 3 & 2 & 3 \\ 2 & 2 & 2 \\ 1 & 3 & 1 \end{bmatrix}$

Exercises 13–20: Let A be the given matrix. Find det A by using the method of cofactors.

13. $\begin{bmatrix} 2 & 0 & 0 \\ 0 & 3 & 0 \\ 0 & 0 & 5 \end{bmatrix}$ **14.** $\begin{bmatrix} 0 & 0 & 2 \\ 0 & 3 & 0 \\ 5 & 0 & 0 \end{bmatrix}$

15. $\begin{bmatrix} 0 & 0 & 0 \\ -8 & 3 & -9 \\ 15 & 5 & 9 \end{bmatrix}$ **16.** $\begin{bmatrix} 1 & 1 & 5 \\ -3 & -3 & 0 \\ 7 & 0 & 0 \end{bmatrix}$

17. $\begin{bmatrix} 3 & -1 & 2 \\ 0 & 5 & 7 \\ 1 & 0 & -1 \end{bmatrix}$ **18.** $\begin{bmatrix} 3 & 0 & -1 \\ 2 & 3 & -4 \\ 6 & -5 & 1 \end{bmatrix}$

19. $\begin{bmatrix} 1 & -5 & 2 \\ -7 & 1 & 3 \\ 0 & 4 & -2 \end{bmatrix}$ **20.** $\begin{bmatrix} 1 & -1 & 2 \\ -2 & 0 & 1 \\ 1 & 1 & -1 \end{bmatrix}$

Exercises 21–24: Let A be the given matrix. Use technology to calculate det A.

21. $\begin{bmatrix} 11 & -32 \\ 1.2 & 55 \end{bmatrix}$ **22.** $\begin{bmatrix} 17 & -4 & 3 \\ 11 & 5 & -15 \\ 7 & -9 & 23 \end{bmatrix}$

23. $\begin{bmatrix} 2.3 & 5.1 & 2.8 \\ 1.2 & 4.5 & 8.8 \\ -0.4 & -0.8 & -1.2 \end{bmatrix}$ **24.** $\begin{bmatrix} 1 & -1 & 3 & 7 \\ 9 & 2 & -7 & -4 \\ 5 & -7 & 1 & -9 \\ 7 & 1 & 3 & 6 \end{bmatrix}$

Cramer's Rule

Exercises 25–32: Use Cramer's rule to solve the system of linear equations.

25. $-x + 2y = 5$
$3x + 3y = 1$

26. $2x + y = -3$
$-4x - 6y = -7$

27. $-2x + 3y = 8$
$4x - 5y = 3$

28. $5x - 3y = 4$
$-3x - 7y = 5$

29. $7x + 4y = 23$
$11x - 5y = 70$

30. $-7x + 5y = 8.2$
$6x + 4y = -0.4$

31. $1.7x - 2.5y = -0.91$
$-0.4x + 0.9y = 0.423$

32. $-2.7x + 1.5y = -1.53$
$1.8x - 5.5y = -1.68$

Calculating Area

Exercises 33–36: Use a determinant to find the area of the shaded region.

33.

34.

35.

36.

Applying a Concept

Exercises 37–40: Use the concept of the area of a triangle to determine if the three points are collinear.

37. $(1, 3), (-3, 11), (2, 1)$

38. $(3, 6), (-1, -6), (5, 11)$

39. $(-2, -5), (4, 4), (2, 3)$

40. $(4, -5), (-2, 10), (6, -10)$

Equations of Lines

Exercises 41–44: If a line passes through the points (x_1, y_1) and (x_2, y_2), then an equation of this line can be found by calculating the determinant.

$$\det \begin{bmatrix} x & y & 1 \\ x_1 & y_1 & 1 \\ x_2 & y_2 & 1 \end{bmatrix} = 0$$

Find the standard form $ax + by = c$ of the line passing through the given points.

41. $(2, 1)$ and $(-1, 4)$

42. $(-1, 3)$ and $(4, 2)$

43. $(6, -7)$ and $(4, -3)$

44. $(5, 1)$ and $(2, -2)$

Writing about Mathematics

45. Choose two matrices A and B with dimension 2×2. Calculate det A, det B, and det (AB). Repeat this process until you are able to discover how these three determinants are related. Summarize your results.

46. Calculate det A and det A^{-1} for different matrices. Compare the determinants. Try to generalize your results.

EXTENDED AND DISCOVERY EXERCISES

Exercises 1–6: Cramer's Rule Cramer's rule can be applied to systems of three linear equations in three variables. For the system of equations

$$a_1x + b_1y + c_1z = d_1$$
$$a_2x + b_2y + c_2z = d_2$$
$$a_3x + b_3y + c_3z = d_3,$$

the solution can be written as follows.

$$D = \det \begin{bmatrix} a_1 & b_1 & c_1 \\ a_2 & b_2 & c_2 \\ a_3 & b_3 & c_3 \end{bmatrix}, \quad E = \det \begin{bmatrix} d_1 & b_1 & c_1 \\ d_2 & b_2 & c_2 \\ d_3 & b_3 & c_3 \end{bmatrix}$$

$$F = \det \begin{bmatrix} a_1 & d_1 & c_1 \\ a_2 & d_2 & c_2 \\ a_3 & d_3 & c_3 \end{bmatrix}, \quad G = \det \begin{bmatrix} a_1 & b_1 & d_1 \\ a_2 & b_2 & d_2 \\ a_3 & b_3 & d_3 \end{bmatrix}$$

If $D \neq 0$, a unique solution exists and is given by

$$x = \frac{E}{D}, \quad y = \frac{F}{D}, \quad z = \frac{G}{D}.$$

Use Cramer's rule to solve the system of equations.

1. $x + y + z = 6$
$2x + y + 2z = 9$
$y + 3z = 9$

2. $y + z = 1$
$2x - y - z = -1$
$x + y - z = 3$

3. $x + z = 2$
$x + y = 0$
$y + 2z = 1$

4. $x + y + 2z = 1$
$-x - 2y - 3z = -2$
$y - 3z = 5$

5. $x + 2z = 7$
$-x + y + z = 5$
$2x - y + 2z = 6$

6. $x + 2y + 3z = -1$
$2x - 3y - z = 12$
$x + 4y - 2z = -12$

Exercises 7–10: Equations of Circles Given three distinct points on a circle $(x_1, y_1), (x_2, y_2),$ and $(x_3, y_3),$ we can find the equation of the circle by using the following determinant.

$$\det \begin{bmatrix} x^2 + y^2 & x & y & 1 \\ x_1^2 + y_1^2 & x_1 & y_1 & 1 \\ x_2^2 + y_2^2 & x_2 & y_2 & 1 \\ x_3^2 + y_3^2 & x_3 & y_3 & 1 \end{bmatrix} = 0$$

Find the equation of the circle through the given points.

7. $(0, 2) (2, 0),$ and $(-2, 0)$

8. $(0, 0), (4, 0),$ and $(2, -2)$

9. $(0, 1), (1, -1),$ and $(2, 2)$

10. $(1, 0), (-1, 2),$ and $(3, 2)$

CHECKING BASIC CONCEPTS FOR SECTION 9.7

1. Find the determinant of the matrix A by using the method of cofactors. Is A invertible?

$$A = \begin{bmatrix} 1 & -1 & 2 \\ 2 & 3 & 1 \\ 0 & -2 & 5 \end{bmatrix}$$

2. Use Cramer's rule to solve the system of equations.
$$3x - 4y = 7$$
$$-4x + 3y = 5$$

---

9 Summary

CONCEPT	EXPLANATION AND EXAMPLES

SECTION 9.1 FUNCTIONS AND SYSTEMS OF EQUATIONS IN TWO VARIABLES

Functions of Two Variables

$z = f(x, y),$ where x and y are inputs to f

Example: $f(x, y) = 2x - 3y$
$f(4, -1) = 2(4) - 3(-1) = 11$

System of Linear Equations in Two Variables

General form: $a_1 x + b_1 y = c_1$
$a_2 x + b_2 y = c_2$

A linear system can have zero, one, or infinitely many solutions. A solution can be written as an ordered pair. A linear system may be solved symbolically, graphically, or numerically.

Example: $x - y = 2$
$2x + y = 7$ Solution: (3, 1)

CONCEPT	EXPLANATION AND EXAMPLES

SECTION 9.1 FUNCTIONS AND SYSTEMS OF EQUATIONS IN TWO VARIABLES (CONTINUED)

Types of Linear Systems with Two Variables

Consistent system: Has either one solution (independent equations) or infinitely many solutions (dependent equations)

Inconsistent system: Has no solutions

One Solution   Infinitely Many   No Solutions
                          Solutions

Method of Substitution for Two Equations

Can be used to solve systems of linear or nonlinear equations

Example: $x - y = -3$
$x + 4y = 17$

Solve the first equation for x to obtain $x = y - 3$. Substitute this result in the second equation and solve for y.

$$(y - 3) + 4y = 17 \quad \text{implies that} \quad y = 4.$$

Then $x = 4 - 3 = 1$ and the solution is $(1, 4)$.

Method of Elimination

Can be used to solve systems of linear or nonlinear equations

Example: $2x - 3y = 4$
$\underline{\ x + 3y = 11\ }$
$3x \qquad = 15, \quad \text{or} \quad x = 5 \qquad \text{Add.}$

Substituting $x = 5$ in the first equation gives $y = 2$.
The solution is $(5, 2)$.

Joint Variation

Let m and n be real numbers. Then z *varies jointly* as the mth power of x and the nth power of y if a nonzero real number k exists such that $z = kx^my^n$.

Example: The area of a triangle varies jointly as the base b and the height h because $A = \frac{1}{2}bh$. Note that $k = \frac{1}{2}$, $m = 1$, and $n = 1$ in this example.

CONCEPT	EXPLANATION AND EXAMPLES

SECTION 9.2 SYSTEMS OF INEQUALITIES IN TWO VARIABLES

System of Inequalities in Two Variables

The solution set is often a shaded region in the xy-plane.

Example: $x + y \leq 4$
$y \geq 0$
$x \geq 0$

Linear Programming

Method for maximizing (or minimizing) an objective function subject to a set of constraints

Example: Maximize $P = 2x + 4y$, subject to

$x + y \leq 4, x \geq 0, y \geq 0.$

The maximum of $P = 16$ occurs at the vertex $(0, 4)$, in the region of feasible solutions. See the figure above.

SECTION 9.3 SYSTEMS OF LINEAR EQUATIONS IN THREE VARIABLES

Solution to a System of Linear Equations in Three Variables

An ordered triple (x, y, z) that satisfies *every* equation

Example: $x - 2y + 3z = 6$
$-x + 3y + 4z = 17$
$3x + 4y - 5z = -4$

The solution is $(1, 2, 3)$ because $x = 1$, $y = 2$, and $z = 3$ satisfies all three equations. (Check this fact.)

Elimination and Substitution

Systems of linear equations in three variables can be solved by using elimination and substitution. The following three steps outline this process.

STEP 1: Eliminate one variable, such as x, from two of the equations.

STEP 2: Apply the techniques discussed in Section 9.1 to solve the two equations in two variables resulting from Step 1. If x is eliminated, then solve these equations to find y and z.

If there are no solutions for y and z, then the given system has no solutions. If there are infinitely many solutions for y and z, then write y in terms of z and go to Step 3.

STEP 3: Substitute the values for y and z in one of the given equations to find x. The solution is (x, y, z).

CONCEPT	EXPLANATION AND EXAMPLES

SECTION 9.4 SOLUTIONS TO LINEAR SYSTEMS USING MATRICES

Matrices and Systems of Linear Equations

An augmented matrix can be used to represent a system of linear equations.

Augmented Matrix

Example:

$$\begin{aligned} x - 2y + z &= 0 \\ -x + 4y - z &= 4 \\ 2x + y - 3z &= -5 \end{aligned} \qquad \left[\begin{array}{ccc|c} 1 & -2 & 1 & 0 \\ -1 & 4 & -1 & 4 \\ 2 & 1 & -3 & -5 \end{array}\right]$$

Row-Echelon Form

Examples:

$$\begin{bmatrix} 1 & 2 & -1 \\ 0 & 1 & 2 \end{bmatrix} \qquad \left[\begin{array}{ccc|c} 1 & 3 & -2 & 7 \\ 0 & 1 & 4 & 5 \\ 0 & 0 & 1 & -3 \end{array}\right]$$

Gaussian Elimination with Backward Substitution

Gaussian elimination can be used to transform a system of linear equations into row-echelon form. Then backward substitution can be used to solve the resulting system of linear equations. Graphing calculators can also be used to solve systems of equations.

SECTION 9.5 PROPERTIES AND APPLICATIONS OF MATRICES

Operations on Matrices

Matrices can be added, subtracted, and multiplied, but there is *no* division of matrices.

Addition

$$\begin{bmatrix} 2 & 4 \\ 5 & 6 \end{bmatrix} + \begin{bmatrix} -2 & 1 \\ 7 & 3 \end{bmatrix} = \begin{bmatrix} 0 & 5 \\ 12 & 9 \end{bmatrix}$$

Subtraction

$$\begin{bmatrix} -3 & 0 \\ 4 & -4 \end{bmatrix} - \begin{bmatrix} 1 & 2 \\ 6 & -7 \end{bmatrix} = \begin{bmatrix} -4 & -2 \\ -2 & 3 \end{bmatrix}$$

Scalar Multiplication

$$3\begin{bmatrix} 5 & 1 & 6 & -1 \\ 0 & -2 & 3 & 2 \end{bmatrix} = \begin{bmatrix} 15 & 3 & 18 & -3 \\ 0 & -6 & 9 & 6 \end{bmatrix}$$

Multiplication

$$\begin{bmatrix} 2 & -1 \\ 0 & 3 \\ -7 & 1 \end{bmatrix} \cdot \begin{bmatrix} 1 & -1 & 0 \\ 3 & -5 & -4 \end{bmatrix} = \begin{bmatrix} -1 & 3 & 4 \\ 9 & -15 & -12 \\ -4 & 2 & -4 \end{bmatrix}$$

SECTION 9.6 INVERSES OF MATRICES

Matrix Inverses

The inverse of an $n \times n$ matrix A, denoted A^{-1}, satisfies $A^{-1}A = I_n$ and $AA^{-1} = I_n$, where I_n is the $n \times n$ identity matrix. The inverse of a matrix can be found by hand or with technology.

Example: $A = \begin{bmatrix} 5 & 2 \\ 2 & 1 \end{bmatrix}$ and $A^{-1} = \begin{bmatrix} 1 & -2 \\ -2 & 5 \end{bmatrix}$

$$\begin{bmatrix} 5 & 2 \\ 2 & 1 \end{bmatrix}\begin{bmatrix} 1 & -2 \\ -2 & 5 \end{bmatrix} = \begin{bmatrix} 1 & 0 \\ 0 & 1 \end{bmatrix} = I_2$$

$$\begin{bmatrix} 1 & -2 \\ -2 & 5 \end{bmatrix}\begin{bmatrix} 5 & 2 \\ 2 & 1 \end{bmatrix} = \begin{bmatrix} 1 & 0 \\ 0 & 1 \end{bmatrix} = I_2$$

CONCEPT	EXPLANATION AND EXAMPLES

SECTION 9.6 INVERSES OF MATRICES (CONTINUED)

Matrix Equations

A system of linear equations can be written as the matrix equation $AX = B$.

Example:
$$2x - 2y = 3$$
$$-3x + 4y = 2$$

$$AX = \begin{bmatrix} 2 & -2 \\ -3 & 4 \end{bmatrix} \begin{bmatrix} x \\ y \end{bmatrix} = \begin{bmatrix} 3 \\ 2 \end{bmatrix} = B$$

The solution can be found as follows.

$$X = A^{-1}B = \begin{bmatrix} 2 & 1 \\ 1.5 & 1 \end{bmatrix} \begin{bmatrix} 3 \\ 2 \end{bmatrix} = \begin{bmatrix} 8 \\ 6.5 \end{bmatrix}$$

SECTION 9.7 DETERMINANTS

Determinant of a 2 × 2 Matrix

$$\det A = \det \begin{bmatrix} a & b \\ c & d \end{bmatrix} = ad - cb$$

Example: $\det \begin{bmatrix} 1 & 4 \\ 3 & 5 \end{bmatrix} = (1)(5) - (3)(4) = -7$

Determinant of a 3 × 3 Matrix

Finding a 3 × 3 determinant can be reduced to calculating the determinants of three 2 × 2 matrices by using cofactors. If $\det A \neq 0$, then A^{-1} exists.

Example: $\det \begin{bmatrix} 3 & 1 & -1 \\ 2 & 2 & 0 \\ 0 & 1 & -3 \end{bmatrix}$

$$= 3 \begin{bmatrix} 2 & 0 \\ 1 & -3 \end{bmatrix} - 2 \begin{bmatrix} 1 & -1 \\ 1 & -3 \end{bmatrix} + 0 \begin{bmatrix} 1 & -1 \\ 2 & 0 \end{bmatrix}$$

$$= 3(-6) - 2(-2) + 0(2)$$

$$= -14$$

Cramer's Rule

Cramer's rule makes use of determinants to solve systems of linear equations. However, Gaussian elimination with backward substitution is usually more efficient.

9   Review Exercises

Exercises 1 and 2: Evaluate the function for the inputs.

1. $A(3, 6)$, where $A(b, h) = \frac{1}{2}bh$

2. $V(2, 5)$, where $V(r, h) = \pi r^2 h$

Exercises 3 and 4: Solve the system of equations
    (a) *graphically and*   (b) *symbolically.*

3. $3x + y = 1$
    $2x - 3y = 8$

4. $x^2 - y = 1$
    $x + y = 1$

Exercises 5–8: Use the elimination method to solve each system of linear equations, if possible. Identify the system as consistent or inconsistent.

5. $2x + y = 7$
$x - 2y = -4$

6. $3x + 3y = 15$
$-x - y = -4$

7. $6x - 15y = 12$
$-4x + 10y = -8$

8. $3x - 4y = -10$
$4x + 3y = -30$

Exercises 9 and 10: Use elimination to solve the nonlinear system of equations.

9. $x^2 - 3y = 3$
$x^2 + 2y^2 = 5$

10. $2x - 3y = 1$
$2x^2 + y = 1$

Exercises 11 and 12: Graph the solution set to the inequality.

11. $y \geq -1$

12. $2x - y < 4$

Exercises 13 and 14: Graph the solution set to the system of inequalities. Use the graph to find one solution.

13. $x^2 + y^2 < 9$
$x + y > 3$

14. $x + 3y \geq 3$
$x + y \leq 4$

Exercises 15–18: If possible, solve the system.

15. $x - y + z = -2$
$x + 2y - z = 2$
$2y + 3z = 7$

16. $x - 3y + 2z = -10$
$2x - y + 3z = -9$
$-x - y + z = -1$

17. $-x + 2y + 2z = 9$
$x + y - 3z = 6$
$3y - z = 8$

18. $2x - y - 3z = -9$
$x - 8z = -23$
$-3x + 2y - 2z = -5$

Exercises 19–22: The augmented matrix represents a system of linear equations. Solve the system.

19. $\begin{bmatrix} 1 & 5 & | & 6 \\ 0 & 1 & | & 3 \end{bmatrix}$

20. $\begin{bmatrix} 1 & 2 & -2 & | & 8 \\ 0 & 1 & 1 & | & 5 \\ 0 & 0 & 0 & | & 0 \end{bmatrix}$

21. $\begin{bmatrix} 1 & 0 & 0 & | & -2 \\ 0 & 1 & 0 & | & 3 \\ 0 & 0 & 1 & | & 0 \end{bmatrix}$

22. $\begin{bmatrix} 1 & 0 & 0 & | & -5 \\ 0 & 1 & -4 & | & 1 \\ 0 & 0 & 0 & | & 5 \end{bmatrix}$

Exercises 23 and 24: Use Gaussian elimination with backward substitution to solve the system of linear equations.

23. $2x - y + 2z = 10$
$x - 2y + z = 8$
$3x - y + 2z = 11$

24. $x - 2y + z = 1$
$2x - 5y + 3z = 4$
$2x - 3y + z = 0$

Exercises 25 and 26: Let a_{ij} denote the general term for the matrix A. Find each of the following.

    (a) $a_{12} + a_{22}$ (b) $a_{11} - 2a_{23}$

25. $A = \begin{bmatrix} -2 & 3 & -1 \\ 5 & 2 & 4 \end{bmatrix}$

26. $\begin{bmatrix} -1 & 2 & 5 \\ 1 & -3 & 7 \\ 0 & 7 & -2 \end{bmatrix}$

Exercises 27 and 28: Evaluate the following.

    (a) $A + 2B$ (b) $A - B$ (c) $-4A$

27. $A = \begin{bmatrix} 1 & -3 \\ 2 & -1 \end{bmatrix}$, $B = \begin{bmatrix} 3 & 2 \\ -5 & 1 \end{bmatrix}$

28. $A = \begin{bmatrix} 4 & 0 & 1 \\ -2 & 8 & 9 \end{bmatrix}$, $B = \begin{bmatrix} -5 & 3 & 2 \\ -4 & 0 & 7 \end{bmatrix}$

Exercises 29–32: If possible, find AB and BA.

29. $A = \begin{bmatrix} 2 & 0 \\ -5 & 3 \end{bmatrix}$, $B = \begin{bmatrix} -1 & -2 \\ 4 & 7 \end{bmatrix}$

30. $A = \begin{bmatrix} 1 & -2 \\ 2 & 3 \end{bmatrix}$, $B = \begin{bmatrix} 1 & 0 & 2 \\ -1 & 3 & 4 \end{bmatrix}$

31. $A = \begin{bmatrix} 2 & -1 & 3 \\ 2 & 4 & 0 \end{bmatrix}$, $B = \begin{bmatrix} 1 & 0 \\ -1 & 2 \\ 0 & 3 \end{bmatrix}$

32. $A = \begin{bmatrix} 1 & -1 & 2 \\ 0 & 3 & 4 \\ 1 & 0 & 2 \end{bmatrix}$, $B = \begin{bmatrix} -1 & 0 & 0 \\ 2 & 0 & -1 \\ 1 & 4 & 2 \end{bmatrix}$

Exercises 33 and 34: Determine if B is the inverse matrix of A by evaluating AB and BA.

33. $A = \begin{bmatrix} 8 & 5 \\ 6 & 4 \end{bmatrix}$, $B = \begin{bmatrix} 2 & -2.5 \\ -3 & 4 \end{bmatrix}$

34. $A = \begin{bmatrix} -1 & 1 & 2 \\ 1 & 0 & -1 \\ 0 & 1 & 2 \end{bmatrix}$, $B = \begin{bmatrix} -1 & 0 & 1 \\ 2 & 2 & -1 \\ -1 & -1 & -1 \end{bmatrix}$

Exercises 35–37: Complete the following.

    (a) Write the system in the form $AX = B$.
    (b) Solve the linear system by computing $X = A^{-1}B$.

35. $x - 3y = 4$
$2x - y = 3$

36. $x - 2y + z = 0$
$2x + y + 2z = 10$
$y + z = 3$

37. $\begin{aligned} 12x + 7y - 3z &= 14.6 \\ 8x - 11y + 13z &= -60.4 \\ -23x \quad\quad + 9z &= -14.6 \end{aligned}$

38. If possible, graphically approximate the solution of each system of equations to the nearest thousandth. Identify each system as consistent or inconsistent. If the system is consistent, determine if the equations are dependent or independent.

(a) $\begin{aligned} 3.1x + 4.2y &= 6.4 \\ 1.7x - 9.1y &= 1.6 \end{aligned}$

(b) $\begin{aligned} 6.3x - 5.1y &= 9.3 \\ 4.2x - 3.4y &= 6.2 \end{aligned}$

(c) $\begin{aligned} 0.32x - 0.64y &= 0.96 \\ -0.08x + 0.16y &= -0.72 \end{aligned}$

Exercises 39 and 40: Let A be the given matrix. Find A^{-1}.

39. $\begin{bmatrix} 1 & -2 \\ -1 & 1 \end{bmatrix}$

40. $\begin{bmatrix} 1 & 0 & 1 \\ 1 & 1 & 1 \\ 0 & 1 & -1 \end{bmatrix}$

Exercises 41 and 42: Let A be the given matrix. Find det A by using the method of cofactors.

41. $\begin{bmatrix} 2 & 1 & 3 \\ 0 & 3 & 4 \\ 1 & 0 & 5 \end{bmatrix}$

42. $\begin{bmatrix} 3 & 0 & 2 \\ 1 & 3 & 5 \\ -5 & 2 & 0 \end{bmatrix}$

Exercises 43 and 44: Let A be the given matrix. Use technology to find det A. State whether A is invertible.

43. $\begin{bmatrix} 13 & 22 \\ 55 & -57 \end{bmatrix}$

44. $\begin{bmatrix} 6 & -7 & -1 \\ -7 & 3 & -4 \\ 23 & 54 & 77 \end{bmatrix}$

Applications

45. *Area and Perimeter* Let l represent the length of a rectangle and w its width, where $l \geq w$. Then its area can be computed by $A(l, w) = lw$ and its perimeter by $P(l, w) = 2l + 2w$. Solve the system of equations determined by $A(l, w) = 77$ and $P(l, w) = 36$.

46. *Cylinder* Approximate the radius r and height h of a cylindrical container with a volume V of 30 cubic inches and a lateral (side) surface area S of 45 square inches.

47. *Student Loans* A student takes out two loans totaling $2000 to help pay for college expenses. One loan is at 7% interest, and the other is at 9%. Interest for both loans is compounded annually.

(a) If the combined total interest for the first year is $156, find the amount of each loan symbolically.

(b) Determine the amount of each loan graphically or numerically.

48. *Dimensions of a Screen* The screen of a rectangular television set is 3 inches wider than it is high. If the perimeter of the screen is 42 inches, find its dimensions by writing a system of linear equations and solving.

49. *CD Prices* A music store marks its compact discs A or B to indicate one of two selling prices. Each row in the table represents a purchase. Determine the cost of each type of CD by using a matrix inverse.

A	B	Total
1	2	$37.47
2	3	$61.95

50. *Digital Photography* Design a 3×3 matrix A that represents a digital photograph of the letter T in black on a white background. Find a matrix B such that adding B to A darkens only the white background by one gray level.

51. *Area* Use a determinant to find the area of the triangle whose vertices are $(0, 0)$, $(5, 2)$, and $(2, 5)$.

52. *Voter Turnout* The table shows the percent y of voter turnout in the United States for the presidential election in year x, where $x = 0$ corresponds to 1900. Find a quadratic function defined by $f(x) = ax^2 + bx + c$ that models these data. Graph f together with the data.

x	24	60	96
y	48.9	62.8	48.8

Source: Committee for the Study of the American Electorate.

53. *Joint Variation* Suppose P varies jointly as the square of x and the cube of y. If $P = 432$ when $x = 2$ and $y = 3$, find P when $x = 3$ and $y = 5$.

54. *Linear Programming* Find the maximum value of $P = 3x + 4y$ subject to the following constraints.

$$x + 3y \leq 12$$
$$3x + y \leq 12$$
$$x \geq 0, y \geq 0$$

EXTENDED AND DISCOVERY EXERCISES

1. To form the **transpose** of a matrix A, denoted A^T, let the first row of A be the first column of A^T, the second row of A be the second column of A^T, and so on, for each row of A. The following are examples of A and A^T. If A has dimension $m \times n$, then A^T has dimension $n \times m$.

$$A = \begin{bmatrix} 3 & -3 & 7 \\ 1 & 6 & -2 \\ 4 & 2 & 5 \end{bmatrix}, \quad A^T = \begin{bmatrix} 3 & 1 & 4 \\ -3 & 6 & 2 \\ 7 & -2 & 5 \end{bmatrix}$$

$$A = \begin{bmatrix} 1 & 2 \\ 3 & 4 \\ 5 & 6 \end{bmatrix}, \quad A^T = \begin{bmatrix} 1 & 3 & 5 \\ 2 & 4 & 6 \end{bmatrix}$$

Find the transpose of each matrix A.

(a) $A = \begin{bmatrix} 3 & -3 \\ 2 & 6 \\ 4 & 2 \end{bmatrix}$
(b) $A = \begin{bmatrix} 0 & 1 & -2 \\ 2 & 5 & 4 \\ -4 & 3 & 9 \end{bmatrix}$

(c) $A = \begin{bmatrix} 5 & 7 \\ 1 & -7 \\ 6 & 3 \\ -9 & 2 \end{bmatrix}$

Least-Squares Fit and Matrices

Exercises 2–4: Least-Square Models *The table shows the average cost of tuition and fees y in dollars at 4-year public colleges. In this table $x = 0$ represents 1980 and $x = 20$ corresponds to 2000.*

x	0	5	10	15	20
y	804	1318	1908	2860	3487

Source: The College Board.

These data can be modeled by using linear regression. Ideally, we would like $f(x) = ax + b$ to satisfy the following five equations.

$$f(0) = a(0) + b = 804$$
$$f(5) = a(5) + b = 1318$$
$$f(10) = a(10) + b = 1908$$
$$f(15) = a(15) + b = 2860$$
$$f(20) = a(20) + b = 3487$$

Since the data points are not collinear, it is impossible for the graph of a line to pass through all five points. These five equations can be written as

$$AX = \begin{bmatrix} 0 & 1 \\ 5 & 1 \\ 10 & 1 \\ 15 & 1 \\ 20 & 1 \end{bmatrix} \begin{bmatrix} a \\ b \end{bmatrix} = \begin{bmatrix} 804 \\ 1318 \\ 1908 \\ 2860 \\ 3487 \end{bmatrix} = B.$$

*The least-squares solution is found by solving the **normal equations***

$$A^T A X = A^T B$$

for X. The solution is $X = (A^T A)^{-1} A^T B$. Using technology, we find $a = 138.16$ and $b = 693.8$. Thus f is given by the formula $f(x) = 138.16x + 693.8$. The function f and the data can be graphed. See the figure.

$[-5, 25, 5]$ by $[0, 4000, 1000]$

Solve the normal equations to model the data with the line determined by $f(x) = ax + b$. Plot the data and f in the same viewing rectangle.

2. *Tuition and Fees* The table shows average cost of tuition and fees y in dollars at private 4-year colleges. In this table $x = 0$ corresponds to 1980 and $x = 20$ to 2000.

x	0	5	10	15	20
y	3617	6121	9340	12,216	16,233

Source: The College Board.

3. *Satellite TV* The table lists the number of satellite television subscribers y in millions. In this table $x = 0$ corresponds to 1995 and $x = 5$ to the year 2000.

x	0	1	2	3	4	5
y	2.2	4.5	7.9	10.5	13	15

Source: USA Today.

4. *Federal Deficit* The table lists the annual federal deficit y in billions of dollars from 1992 to 1996, where $x = 2$ corresponds to 1992 and $x = 6$ to 1996.

x	2	3	4	5	6
y	290	255	203	164	107

Source: Department of the Treasury.

Cryptography

5. *Codes*   Businesses and government agencies frequently send classified messages in code. One cryptographic technique that involves matrices is the **polygraphic system**. In this system each letter in the alphabet is associated with a number between 1 and 26. The following table gives a common example.

A	B	C	D	E	F	G	H	I	J
1	2	3	4	5	6	7	8	9	10

K	L	M	N	O	P	Q	R	S	T
11	12	13	14	15	16	17	18	19	20

U	V	W	X	Y	Z
21	22	23	24	25	26

Source: A. Sinkov, *Elementary Cryptanalysis: A Mathematical Approach.*

For example, the word MATH is coded as 13 1 20 8. Enter these numbers in a 2×2 matrix B.

$$B = \begin{bmatrix} 13 & 20 \\ 1 & 8 \end{bmatrix}$$

To code these letters, a 2×2 matrix, such as

$$A = \begin{bmatrix} 2 & 1 \\ -5 & -2 \end{bmatrix},$$

is multiplied times B to form the product AB.

$$AB = \begin{bmatrix} 2 & 1 \\ -5 & -2 \end{bmatrix} \begin{bmatrix} 13 & 20 \\ 1 & 8 \end{bmatrix} = \begin{bmatrix} 27 & 48 \\ -67 & -116 \end{bmatrix}$$

Since the resulting elements of AB are less than 1 or greater than 26, they may be scaled between 1 and 26 by adding or subtracting multiples of 26.

$$27 - 1(26) = 1 \qquad 48 - 1(26) = 22$$
$$-67 + 3(26) = 11 \qquad -116 + 5(26) = 14$$

Thus the word MATH is coded as **1 11 22 14**, or AKVN. The advantage of this coding technique is that a particular letter is not always coded the same each time. Use the matrix A to code the following words.

(a) HELP   **(b)** LETTER

6. *Decoding a Message*   (Refer to Exercise 5.) To decode a message we reverse the process by using A^{-1}. Decode each of the following.

(a) UBNL   **(b)** QNABMV

10 Conic Sections

Enthusiasm is the most important thing in life.

—Tennessee Williams

Throughout history, people have been fascinated by the universe around them and compelled to try to understand it. Conic sections have played an important role in gaining this understanding. Although conic sections were described and named by the Greek astronomer Apollonius in 200 B.C., it was not until much later that they were used to model motion in the universe. In the sixteenth century Tycho Brahe, the greatest observational astronomer of the age, recorded precise data on planetary movement in the sky. In 1619, using Brahe's data, Johannes Kepler determined that planets move in elliptical orbits around the sun. In 1686 Newton used Kepler's work to show that elliptical orbits are the result of his famous theory of gravitation. We now know that all celestial objects—including planets, comets, asteroids, and satellites—travel in paths described by conic sections. Today scientists search the sky for information about the universe with enormous radio telescopes in the shape of parabolic dishes.

The search to understand the universe has been directed toward both the infinite and the infinitesimal. In 1911 Ernest Rutherford determined the basic structure of the atom. Small atomic particles are capable of traveling in trajectories described by conic sections.

Parabolas, ellipses, and hyperbolas have had a profound influence on our understanding of ourselves and the cosmos around us. In this chapter we consider these age-old curves.

Source: *Historical Topics for the Mathematics Classroom, Thirty-first Yearbook,* NCTM.

10.1 Parabolas

- **Find equations of parabolas**
- **Graph parabolas**
- **Learn the reflective property of parabolas**
- **Translate parabolas**

Introduction

Conic sections are named after the different ways in which a plane can intersect a cone. See Figure 10.1. Three basic conic sections are parabolas, ellipses, and hyperbolas. A circle is also an example of a conic section.

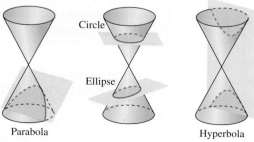

Figure 10.1

Figures 10.2–10.4 show examples of conic sections in the xy-plane.

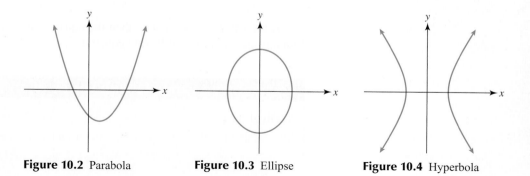

Figure 10.2 Parabola   **Figure 10.3** Ellipse   **Figure 10.4** Hyperbola

In this section we discuss parabolas.

Equations and Graphs of Parabolas

In Chapter 3 we saw that a parabola with vertex $(0, 0)$ can be represented symbolically by the equation $y = ax^2$. With this representation, a parabola can open either upward when $a > 0$ or downward when $a < 0$. The following definition of a parabola allows it to open in *any* direction.

Parabola

A **parabola** is the set of points in a plane that are equidistant from a fixed point and a fixed line. The fixed point is called the **focus** and the fixed line is called the **directrix** of the parabola.

Figure 10.5 Vertical Axis

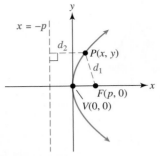

Figure 10.6 Horizontal Axis

Figures 10.5 and 10.6 show two parabolas with vertex V and focus F. The first parabola has a vertical axis (of symmetry) and directrix $y = -p$, and the second parabola has a horizontal axis (of symmetry) and directrix $x = -p$. For any point P located at (x, y) on the parabola, the distance d_1 from F to P is equal to the perpendicular distance d_2 from P to the directrix. (By the vertical line test, the parabola in Figure 10.6 cannot be represented by a function.)

MAKING CONNECTIONS

Functions and Points In Figures 10.5 and 10.6, the point P is labeled $P(x, y)$. This notation resembles the notation for a function involving two inputs, since the point P is determined by x and y.

We can derive an equation of the parabola shown in Figure 10.5. Since $d_1 = d_2$, the distance formula can be used to express the variables x, y, and p in an equation.

$$d_1 = d_2$$
$$\sqrt{(x - 0)^2 + (y - p)^2} = \sqrt{(x - x)^2 + (y - (-p))^2} \qquad \text{Distance formula}$$
$$x^2 + (y - p)^2 = 0^2 + (y + p)^2 \qquad \text{Square each side.}$$
$$x^2 + y^2 - 2py + p^2 = y^2 + 2py + p^2 \qquad \text{Expand binomials.}$$
$$x^2 - 2py = 2py \qquad \text{Subtract } y^2 \text{ and } p^2.$$
$$x^2 = 4py \qquad \text{Add 2py.}$$

If the value of p is known, then the equation of a parabola with vertex $(0, 0)$ can be found using one of the following equations.

Equation of a Parabola with Vertex (0, 0)

Vertical Axis

The parabola with a focus at $(0, p)$ and directrix $y = -p$ has equation

$$x^2 = 4py.$$

The parabola opens upward if $p > 0$ and downward if $p < 0$.

Horizontal Axis

The parabola with a focus at $(p, 0)$ and directrix $x = -p$ has equation

$$y^2 = 4px.$$

The parabola opens to the right if $p > 0$ and to the left if $p < 0$.

EXAMPLE 1 Sketching graphs of parabolas

Sketch a graph of each parabola. Label the vertex, focus, and directrix.
(a) $x^2 = 8y$ **(b)** $y^2 = -2x$

SOLUTION
(a) Getting Started The equation $x^2 = 8y$ is in the form $x^2 = 4py$, where $8 = 4p$. Therefore the parabola has a vertical axis with $p = 2$. The vertex is $(0, 0)$. ▶

Since $p > 0$, the parabola opens **upward**. The focus is located at $(0, p)$, or $(0, 2)$, and the directrix is $y = -p$, or $y = -2$. The graph of the parabola is shown in Figure 10.7.

Figure 10.7   **Figure 10.8**

(b) The equation $y^2 = -2x$ has the form $y^2 = 4px$, where $-2 = 4p$. Therefore the parabola has a horizontal axis with $p = -0.5$. Since $p < 0$, the parabola opens to the **left**. The focus is located at $(-0.5, 0)$, and the directrix is $x = 0.5$. See Figure 10.8.

Now Try Exercises 17 and 21 ◀

In the next example, we find the equation of a parabola, given its focus and directrix.

Figure 10.9

EXAMPLE 2 **Finding the equation of a parabola**

Find the equation of the parabola with focus $(-1.5, 0)$ and directrix $x = 1.5$, as shown in Figure 10.9. Sketch a graph of the parabola.

SOLUTION A parabola always *opens toward the focus* and *away from the directrix*. From Figure 10.9 we see that the parabola should open to the left. It follows that $p < 0$ in the equation $y^2 = 4px$. The distance between the focus at $(-1.5, 0)$ and the vertex at $(0, 0)$ is 1.5, and so $p = -1.5 < 0$. (Note that the vertex of the parabola is $(0, 0)$ because the vertex always lies *midway* between the focus and the directrix.) The equation of the parabola is $y^2 = 4(-1.5)x$, or $y^2 = -6x$, and a graph of the parabola is shown in Figure 10.10.

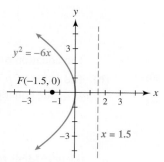

Figure 10.10   *Now Try Exercise 29* ◀

Reflective Property of Parabolas

When a parabola is rotated about its axis, it sweeps out a shape called a **paraboloid**, as shown in Figure 10.11. Paraboloids have a special reflective property. When incoming, parallel rays of light from the sun or distant stars strike the surface of a paraboloid, each ray is reflected toward the focus. See Figure 10.12. If the rays are sunlight, intense heat is produced, which can be used to generate solar heat. Radio signals from distant space also concentrate at the focus. Scientists can analyze these signals by placing a receiver at the focus. This property of a paraboloid can also be used in reverse. If a light source is placed at the focus, then the light is reflected straight ahead, as shown in Figure 10.13. Searchlights, flashlights, and car headlights make use of this property.

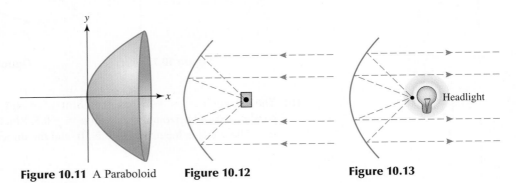

Figure 10.11 A Paraboloid   **Figure 10.12**   **Figure 10.13**

An Application   The next example illustrates a use of this reflective property.

EXAMPLE 3   Locating the receiver for a radio telescope

The giant Arecibo telescope in Puerto Rico has a parabolic dish with a diameter of 300 meters and a depth of 50 meters. See Figure 10.14.   (**Source:** National Astronomy and Ionosphere Center, Arecibo Observatory.)

(a) Find an equation in the form $y = ax^2$ that describes a cross section of this dish.

(b) If the receiver is located at the focus, how far should it be from the vertex?

Figure 10.14

SOLUTION

(a) Locate a parabola that passes through $(-150, 50)$ and $(150, 50)$, as in Figure 10.15. Substitute either point into $y = ax^2$.

$$y = ax^2$$
$$50 = a(150)^2$$
$$a = \frac{50}{150^2} = \frac{1}{450}$$

The equation of the parabola is $y = \frac{1}{450}x^2$, where $-150 \leq x \leq 150$.

(b) The value of p represents the distance from the vertex to the focus. To determine p, write the equation in the form $x^2 = 4py$. Then

$$y = \frac{1}{450}x^2 \quad \text{is equivalent to} \quad x^2 = 450y.$$

It follows that $4p = 450$, or $p = 112.5$. Therefore the receiver should be located about 112.5 meters from the vertex.

Figure 10.15

Now Try Exercise 89 ◀

Figure 10.16 Vertex (h, k); Vertical Axis

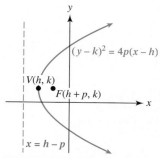

Figure 10.17 Vertex (h, k); Horizontal Axis

Algebra Review
To review translations, or shifts, see Section 3.5.

Translations of Parabolas

If the equation of a parabola is given by either $x^2 = 4py$ or $y^2 = 4px$, then its vertex is $(0, 0)$. We can use translations of graphs to find the equation of a parabola with vertex (h, k). This translation can be obtained by replacing x with $(x - h)$ and y with $(y - k)$.

$$(x - h)^2 = 4p(y - k) \qquad \text{Vertex } (h, k); \text{ vertical axis}$$
$$(y - k)^2 = 4p(x - h) \qquad \text{Vertex } (h, k); \text{ horizontal axis}$$

These two parabolas with $p > 0$ are shown in Figures 10.16 and 10.17, respectively. These results are summarized as follows.

Equation of a Parabola with Vertex (h, k)

$(x - h)^2 = 4p(y - k)$ — Vertical axis; vertex: (h, k)
$p > 0$: opens upward; $p < 0$: opens downward
Focus: $(h, k + p)$; directrix: $y = k - p$

$(y - k)^2 = 4p(x - h)$ — Horizontal axis; vertex: (h, k)
$p > 0$: opens to the right; $p < 0$: opens to the left
Focus: $(h + p, k)$; directrix: $x = h - p$

EXAMPLE 4 Graphing a parabola with vertex (h, k)

Graph the parabola given by $x = -\frac{1}{8}(y + 3)^2 + 2$. Label the vertex, focus, and directrix.

SOLUTION Rewrite the equation in the form $(y - k)^2 = 4p(x - h)$.

$$x = -\frac{1}{8}(y + 3)^2 + 2 \qquad \text{Given equation}$$
$$x - 2 = -\frac{1}{8}(y + 3)^2 \qquad \text{Subtract 2.}$$
$$-8(x - 2) = (y + 3)^2 \qquad \text{Multiply by } -8.$$
$$(y + 3)^2 = -8(x - 2) \qquad \text{Rewrite equation.}$$

The vertex is $(2, -3)$; $4p = -8$, or $p = -2$; and the parabola opens to the left. The focus is 2 units left of the vertex, and the directrix is 2 units right of the vertex. Thus the focus is $(0, -3)$, and the directrix is $x = 4$. See Figure 10.18.

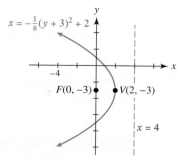

Figure 10.18

Now Try Exercise 55 ◄

EXAMPLE 5 Finding the equation of a parabola with vertex (h, k)

Find the equation of the parabola with focus $(3, -4)$ and directrix $y = 2$. Sketch a graph of the parabola. Label the focus, directrix, and vertex.

SOLUTION The focus and directrix are shown in Figure 10.19. The parabola opens downward ($p < 0$), and its equation has the form $(x - h)^2 = 4p(y - k)$. The vertex is located midway between the focus and the directrix, so its coordinates are $(3, -1)$. The distance between the focus $(3, -4)$ and the vertex $(3, -1)$ is 3, so $p = -3 < 0$. The equation of the parabola is

$$(x - 3)^2 = -12(y + 1),$$

and its graph is shown in Figure 10.20.

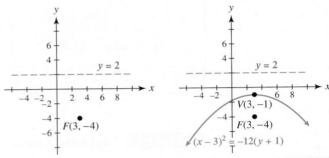

Figure 10.19   **Figure 10.20**   *Now Try Exercise 61* ◀

EXAMPLE 6 Finding the equation of a parabola

Write $2x = y^2 + 4y + 12$ in the form $(y - k)^2 = a(x - h)$.

SOLUTION Write the given equation in the required form by completing the square.

$$2x = y^2 + 4y + 12 \qquad \text{Given equation}$$
$$2x - 12 = y^2 + 4y \qquad \text{Subtract 12 from each side.}$$

To complete the square on the right side of the equation, add $\left(\frac{4}{2}\right)^2 = 4$ to each side.

$$2x - 12 + 4 = y^2 + 4y + 4 \qquad \text{Add 4 to each side.}$$
$$2x - 8 = (y + 2)^2 \qquad \text{Perfect square trinomial}$$
$$2(x - 4) = (y + 2)^2 \qquad \text{Distributive property}$$

The given equation is equivalent to $(y + 2)^2 = 2(x - 4)$.   *Now Try Exercise 67* ◀

CLASS DISCUSSION

Sketch a graph of the parabola in Example 6. Identify the vertex, focus, and directrix.

Using Technology Graphing calculators can graph parabolas with horizontal axes, as illustrated in the next example.

EXAMPLE 7 Graphing a parabola with technology

Graph the equation $(y - 1)^2 = -0.5(x - 2)$ with a graphing calculator.

[−4.7, 4.7, 1] by [−3.1, 3.1, 1]

Figure 10.21

SOLUTION Begin by solving the equation for y.

$$(y - 1)^2 = -0.5(x - 2)$$ Given equation

$$y - 1 = \pm\sqrt{-0.5(x - 2)}$$ Square root property

$$y = 1 \pm \sqrt{-0.5(x - 2)}$$ Add 1.

Let $Y_1 = 1 + \sqrt{(-0.5(X - 2))}$ and $Y_2 = 1 - \sqrt{(-0.5(X - 2))}$. The graph of y_1 creates the upper portion of the parabola, and the graph of y_2 creates the lower portion of the parabola, as shown in Figure 10.21.

Now Try Exercise 77 ◀

10.1 Putting It All Together

The following table summarizes some important concepts about parabolas.

Concept	Equation	Example
Parabola with vertex $(0, 0)$ and vertical axis	$x^2 = 4py$ $p > 0$: opens upward $p < 0$: opens downward Focus: $(0, p)$ Directrix: $y = -p$	$x^2 = -2y$ has $4p = -2$, or $p = -\frac{1}{2}$. The parabola opens downward with vertex $(0, 0)$, focus $\left(0, -\frac{1}{2}\right)$, and directrix $y = \frac{1}{2}$.
Parabola with vertex $(0, 0)$ and horizontal axis	$y^2 = 4px$ $p > 0$: opens to the right $p < 0$: opens to the left Focus: $(p, 0)$ Directrix: $x = -p$	$y^2 = 4x$ has $4p = 4$, or $p = 1$. The parabola opens to the right with vertex $(0, 0)$, focus $(1, 0)$, and directrix $x = -1$.

continued on next page

continued from previous page

Concept	Equation	Example
Parabola with vertex (h, k) and vertical axis	$(x - h)^2 = 4p(y - k)$ $p > 0$: opens upward $p < 0$: opens downward Focus: $(h, k + p)$ Directrix: $y = k - p$	$(x - 1)^2 = 8(y - 3)$ has $p = 2$. The parabola opens upward with vertex $(1, 3)$, focus $(1, 5)$, and directrix $y = 1$.
Parabola with vertex (h, k) and horizontal axis	$(y - k)^2 = 4p(x - h)$ $p > 0$: opens to the right $p < 0$: opens to the left Focus: $(h + p, k)$ Directrix: $x = h - p$	$(y + 1)^2 = -2(x + 2)$ has $p = -\frac{1}{2}$. The parabola opens to the left with vertex $(-2, -1)$, focus $\left(-\frac{5}{2}, -1\right)$, and directrix $x = -\frac{3}{2}$.

10.1 Exercises

Parabolas with Vertex (0, 0)

Exercises 1–10: Sketch a graph of the parabola.

1. $x^2 = y$

2. $x^2 = -y$

3. $y^2 = -x$

4. $y^2 = x$

5. $4x^2 = -2y$

6. $y^2 = -3x$

7. $y^2 = -4x$

8. $x^2 = 4y$

9. $y^2 = -\frac{1}{2}x$

10. $8x = y^2$

Exercises 11–16: Match the equation with its graph (a–f).

11. $x^2 = 2y$

12. $x^2 = -2y$

13. $y^2 = -8x$

14. $y^2 = 4x$

15. $x = \frac{1}{2}y^2$

a.
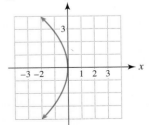

16. $y = -2x^2$

b.

c.

d.

e.

f.

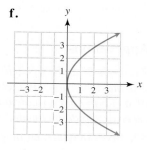

Exercises 17–26: *Graph the parabola. Label the vertex, focus, and directrix.*

17. $16y = x^2$

18. $y = -2x^2$

19. $x = \frac{1}{8}y^2$

20. $-y^2 = 6x$

21. $-4x = y^2$

22. $\frac{1}{2}y^2 = 3x$

23. $x^2 = -8y$

24. $x^2 = -4y$

25. $2y^2 = -8x$

26. $-3x = \frac{1}{4}y^2$

Exercises 27–30: *Sketch a parabola with focus and directrix as shown in the figure. Find an equation of the parabola.*

27.

28.

29.

30.

Exercises 31–40: *Find an equation of the parabola with vertex* (0, 0) *that satisfies the given conditions.*

31. Focus $\left(0, \frac{3}{4}\right)$

32. Directrix $y = 2$

33. Directrix $x = 2$

34. Focus $(-1, 0)$

35. Focus $(1, 0)$

36. Focus $\left(0, -\frac{1}{2}\right)$

37. Directrix $x = \frac{1}{4}$

38. Directrix $y = -1$

39. Horizontal axis, passing through $(1, -2)$

40. Vertical axis, passing through $(-2, 3)$

Exercises 41–44: *Find an equation of a parabola that satisfies the given conditions.*

41. Focus $(0, -3)$ and directrix $y = 3$

42. Focus $(0, 2)$ and directrix $y = -2$

43. Focus $(-1, 0)$ and directrix $x = 1$

44. Focus $(3, 0)$ and directrix $x = -3$

Parabolas with Vertex (h, k)

Exercises 45–48: *Sketch a graph of the parabola.*

45. $(x - 1)^2 = (y - 2)$ **46.** $(x - 2)^2 = -(y + 1)$

47. $(y - 1)^2 = -(x + 1)$ **48.** $(y + 2)^2 = 2x$

Exercises 49–52: *Match the equation with its graph (a–d).*

49. $(x - 1)^2 = 4(y - 1)$ **50.** $(x + 1)^2 = -4(y - 2)$

51. $(y - 2)^2 = -8x$ **52.** $(y + 1)^2 = 8(x + 3)$

a.

b.

c.

d.

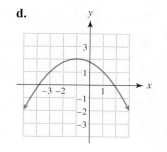

Exercises 53–58: Graph the parabola. Label the vertex, focus, and directrix.

53. $(x - 2)^2 = 8(y + 2)$ **54.** $\frac{1}{16}(x + 4)^2 = -(y - 4)$

55. $x = -\frac{1}{4}(y + 3)^2 + 2$ **56.** $x = 2(y - 2)^2 - 1$

57. $y = -\frac{1}{4}(x + 2)^2$ **58.** $-2(y + 1) = (x + 3)^2$

Exercises 59–62: Find an equation of a parabola that satisfies the given conditions. Sketch a graph of the parabola. Label the focus, directrix, and vertex.

59. Focus $(0, 2)$ and vertex $(0, 1)$

60. Focus $(-1, 2)$ and vertex $(3, 2)$

61. Focus $(0, 0)$ and directrix $x = -2$

62. Focus $(2, 1)$ and directrix $x = -1$

Exercises 63–66: Find an equation of a parabola that satisfies the given conditions.

63. Focus $(-1, 3)$ and directrix $y = 7$

64. Focus $(1, 2)$ and directrix $y = 4$

65. Horizontal axis, vertex $(-2, 3)$, passing through $(-4, 0)$

66. Horizontal axis, vertex $(-1, 2)$, passing through $(2, 3)$

Exercises 67–74: Write the given equation either in the form $(y - k)^2 = a(x - h)$ or in the form $(x - h)^2 = a(y - k)$.

67. $-2x = y^2 + 6x + 10$ **68.** $y^2 + 8x - 8 = 4x$

69. $x = 2y^2 + 4y - 1$ **70.** $x = 3y^2 - 6y - 2$

71. $x^2 - 3x + 4 = 2y$ **72.** $-3y = -x^2 + 4x - 6$

73. $4y^2 + 4y - 5 = 5x$ **74.** $-2y^2 + 5y + 1 = -x$

Graphing Parabolas with Technology

 Exercises 75–80: Graph the parabola.

75. $(y + 0.75)^2 = -3x$ **76.** $(y - 3)^2 = \frac{1}{7}x$

77. $(y - 0.5)^2 = 3.1(x + 1.3)$

78. $1.4(y - 1.5)^2 = 0.5(x + 2.1)$

79. $x = 2.3(y + 1)^2$ **80.** $(y - 2.5)^2 = 4.1(x + 1)$

Solving Nonlinear Systems

Exercises 81–86: Solve each system.

81. $x^2 = 2y$
$x^2 = y + 1$

82. $x^2 = -3y$
$-x^2 = 2y - 2$

83. $\frac{1}{3}y^2 = -3x$
$y^2 = x + 1$

84. $-2y^2 = x - 5$
$y^2 = 2x$

85. $(y - 1)^2 = x + 1$
$(y + 2)^2 = -x + 4$

86. $(y + 1)^2 = -x$
$-(y - 1)^2 = x + 4$

Applications

Exercises 87 and 88: Satellite Dishes   (Refer to Example 3.) Use the dimensions of a television satellite dish in the shape of a paraboloid to calculate how far from the vertex the receiver should be located.

87. Six-foot diameter, nine inches deep

88. Nine-inch radius, two inches deep

89. *Radio Telescope*   (Refer to Example 3.) The Parkes radio telescope, shown in the figure, has the shape of a parabolic dish with a diameter of 210 feet and a depth of 32 feet.   (**Source:** J. Mar, *Structure Technology for Large Radio and Radar Telescope Systems.*)

(a) Determine an equation of the form $y = ax^2$ with $a > 0$ describing a cross section of the dish.

(b) The receiver is placed at the focus. How far from the vertex is the receiver located?

90. *Radio Telescope*   (Refer to Example 3.) A radio telescope is being designed in the shape of a parabolic dish with a diameter of 180 feet and a depth of 25 feet.
(a) Determine an equation of the form $x = ay^2$ with $a > 0$ describing a cross section of the dish.

(b) The receiver is placed at the focus. How far from the vertex should the receiver be located?

91. *Comets*   A comet sometimes travels along a parabolic path as it passes the sun. In this case the sun is located at the focus of the parabola and the comet passes the sun once, rather than orbiting the sun. Suppose the path of a comet is given by $y^2 = 100x$, where units are in millions of miles.
(a) Find the coordinates of the sun.

(b) Find the minimum distance between the sun and the comet.

92. *Headlight*   A headlight is being constructed in the shape of a paraboloid with a depth of 4 inches and a diameter of 5 inches, as illustrated in the figure. Find the distance d that the bulb should be from the vertex in order to have the beam of light shine straight ahead.

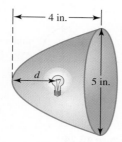

93. *Solar Heater*   A solar heater is being designed to heat a pipe that will contain water, as illustrated in the figure. A cross section of the heater is described by the equation

$x^2 = ky$, where k is a constant and all units are in feet. If the pipe is to be placed 18 inches from the vertex of this cross section, find the value of k.

$x^2 = ky$

94. *Solar Heater*   If the figure in Exercise 93 is rotated 90° clockwise, then the front edge of the reflector can be described by the equation $y^2 = kx$. If the pipe is placed 2 feet from the vertex of this cross section, find the value of k.

Writing about Mathematics

95. Explain how the distance between the focus and the vertex of a parabola affects the shape of the parabola.

96. Explain how to determine the direction that a parabola opens, given the focus and the directrix.

10.2 Ellipses

- **Find equations of ellipses**
- **Graph ellipses**
- **Learn the reflective property of ellipses**
- **Translate ellipses**
- **Solve nonlinear systems of equations and inequalities**

Introduction

When planets travel around the sun, they usually do not travel in circular orbits. Instead they travel in elliptical orbits, which look more like ovals than circles. This discovery by Johannes Kepler made it possible for astronomers to determine the precise positions of all types of celestial objects, such as asteroids, comets, and moons, and made it easier to predict both solar and lunar eclipses. Ellipses are also used in construction and medicine. In this section we examine the basic properties of ellipses.   (**Source:** *Historical Topics for the Mathematics Classroom, Thirty-first Yearbook,* NCTM.)

Equations and Graphs of Ellipses

One method for sketching an ellipse is to tie a string to two nails driven into a flat board. If a pencil is placed inside the loop formed by the string, the curve it traces (shown in Figure 10.22 on the next page) is an ellipse. The sum of the distances d_1 and d_2 between the pencil and each of the nails is always fixed by the string. The locations of the nails correspond to the *foci* of the ellipse. If the two nails coincide, the ellipse becomes a circle. As the nails spread farther apart, the ellipse becomes more elongated, or *eccentric*.

Figure 10.22

This method of sketching an ellipse suggests the following definition.

> ### Ellipse
> An **ellipse** is the set of points in a plane, the sum of whose distances from two fixed points is constant. Each fixed point is called a **focus** (plural **foci**) of the ellipse.

In Figures 10.23 and 10.24 the **major axis** and **minor axis** are labeled for each ellipse. The major axis is the longer line segment connecting V_1 and V_2, and the minor axis is the shorter line segment connecting U_1 and U_2. Figure 10.23 shows an ellipse with a **horizontal major axis**, and Figure 10.24 illustrates an ellipse with a **vertical major axis**. The **vertices** V_1 and V_2 of each ellipse are located at the endpoints of the major axis.

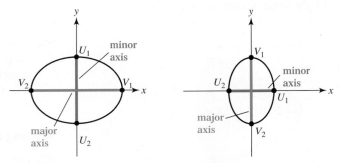

Figure 10.23   **Figure 10.24**

Since a vertical line can intersect the graph of an ellipse more than once, an ellipse cannot be described by a function. However, some ellipses can be represented by the following equations.

> ### Standard Equations for Ellipses Centered at (0, 0)
> The ellipse with center at the origin, *horizontal* major axis, and equation
> $$\frac{x^2}{a^2} + \frac{y^2}{b^2} = 1 \qquad (a > b > 0)$$
> has vertices $(\pm a, 0)$, endpoints of the minor axis $(0, \pm b)$, and foci $(\pm c, 0)$, where $c^2 = a^2 - b^2$ and $c \geq 0$.
>
> The ellipse with center at the origin, *vertical* major axis, and equation
> $$\frac{x^2}{b^2} + \frac{y^2}{a^2} = 1 \qquad (a > b > 0)$$
> has vertices $(0, \pm a)$, endpoints of the minor axis $(\pm b, 0)$, and foci $(0, \pm c)$, where $c^2 = a^2 - b^2$ and $c \geq 0$.

NOTE If $a = b$, then the ellipse becomes a circle with radius $r = a$ and center $(0, 0)$.

Figures 10.25 and 10.26 show two ellipses. The first has a horizontal major axis and the second has a vertical major axis. The coordinates of the vertices V_1 and V_2, foci F_1 and F_2, and endpoints of the minor axis U_1 and U_2 are labeled. In each figure $a > b > 0$.

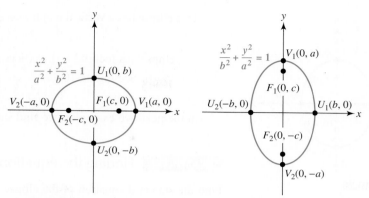

Figure 10.25   **Figure 10.26**

MAKING CONNECTIONS

Intercept Form of a Line and Standard Form for an Ellipse If the equation of a line can be written as

$$\frac{x}{a} + \frac{y}{b} = 1, \qquad \text{Intercept form}$$

then the x-intercept is a and the y-intercept is b. If the equation of an ellipse centered at (0, 0) can be written as

$$\frac{x^2}{a^2} + \frac{y^2}{b^2} = 1, \qquad \text{Standard form}$$

then the x-intercepts are $\pm a$ and the y-intercepts are $\pm b$. See Extended and Discovery Exercises 1–6 at the end of this section.

EXAMPLE 1   Sketching graphs of ellipses

Sketch a graph of each ellipse. Label the vertices, foci, and endpoints of the minor axis.

(a) $\dfrac{x^2}{9} + \dfrac{y^2}{4} = 1$ **(b)** $25x^2 + 16y^2 = 400$

SOLUTION

(a) The equation $\frac{x^2}{9} + \frac{y^2}{4} = 1$ describes an ellipse with $a = 3$ and $b = 2$. The ellipse has a horizontal major axis with vertices $(\pm 3, 0)$. The endpoints of the minor axis are $(0, \pm 2)$. To locate the foci, find c.

$$c^2 = a^2 - b^2 = 9 - 4 = 5, \quad \text{or} \quad c = \sqrt{5} \approx 2.24.$$

The foci are located on the major axis, with coordinates $\left(\pm\sqrt{5}, 0\right)$. See Figure 10.27.

(b) The given equation can be written in standard form by dividing each side by 400.

$$25x^2 + 16y^2 = 400 \qquad \text{Given equation}$$

$$\frac{25x^2}{400} + \frac{16y^2}{400} = \frac{400}{400} \qquad \text{Divide by 400.}$$

$$\frac{x^2}{16} + \frac{y^2}{25} = 1 \qquad \text{Simplify each fraction.}$$

Figure 10.27

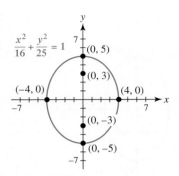

$$\frac{x^2}{16} + \frac{y^2}{25} = 1$$

Figure 10.28

This ellipse has a vertical major axis with $a = 5$ and $b = 4$. The value of c is given by

$$c^2 = 5^2 - 4^2 = 9, \quad \text{or} \quad c = 3.$$

The ellipse has foci $(0, \pm 3)$, vertices $(0, \pm 5)$, and endpoints of the minor axis located at $(\pm 4, 0)$. See Figure 10.28. **Now Try Exercises 1 and 5** ◄

In the next two examples, we find standard equations for ellipses.

EXAMPLE 2 Finding the equation of an ellipse

Find the standard equation of the ellipse shown in Figure 10.29. Identify the coordinates of the vertices and the foci.

SOLUTION The ellipse is centered at $(0, 0)$ and has a horizontal major axis. Its standard equation has the form

$$\frac{x^2}{a^2} + \frac{y^2}{b^2} = 1.$$

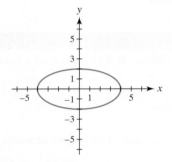

Figure 10.29

The endpoints of the major axis are $(\pm 4, 0)$, and the endpoints of the minor axis are $(0, \pm 2)$. It follows that $a = 4$ and $b = 2$, and the standard equation is

$$\frac{x^2}{16} + \frac{y^2}{4} = 1.$$

The foci lie on the horizontal major axis and can be determined as follows.

$$c^2 = a^2 - b^2 = 16 - 4 = 12$$

Thus $c = \sqrt{12} \approx 3.46$, and the coordinates of the foci are $\left(\pm \sqrt{12}, 0\right)$. A graph of the ellipse with the vertices and foci plotted is shown in Figure 10.30.

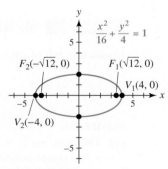

Figure 10.30 **Now Try Exercise 13** ◄

EXAMPLE 3 Finding the equation of an ellipse

Find the standard equation of the ellipse with foci $(0, \pm 1)$ and vertices $(0, \pm 3)$.

SOLUTION Since the foci and vertices lie on the y-axis, the ellipse has a vertical major axis. Its standard equation has the form

$$\frac{x^2}{b^2} + \frac{y^2}{a^2} = 1.$$

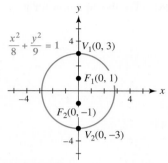

$\frac{x^2}{8} + \frac{y^2}{9} = 1$

$V_1(0, 3)$

$F_1(0, 1)$

$F_2(0, -1)$

$V_2(0, -3)$

Figure 10.31

Because the foci are $(0, \pm 1)$ and the vertices are $(0, \pm 3)$, it follows that $c = 1$ and $a = 3$. The value of b^2 can be found by using the equation $c^2 = a^2 - b^2$.

$$b^2 = a^2 - c^2 = 9 - 1 = 8$$

Thus the equation of the ellipse is $\frac{x^2}{8} + \frac{y^2}{9} = 1$. Its graph is shown in Figure 10.31.

Now Try Exercise 21 ◀

Eccentricity and Applications   The planets travel around the sun in elliptical orbits. Although their orbits are nearly circular, many have a slight eccentricity to them. The **eccentricity** e of an ellipse is defined by

$$e = \frac{\sqrt{a^2 - b^2}}{a} = \frac{c}{a}. \qquad a > b > 0$$

Since the foci of an ellipse lie *inside* the ellipse, $0 < c < a$ and $0 < \frac{c}{a} < 1$. Therefore the eccentricity e of an ellipse satisfies $0 < e < 1$. If $a = b$, then $e = 0$ and the ellipse becomes a circle. See Figure 10.32. As e increases, the foci spread apart and the ellipse becomes more elongated. See Figures 10.33 and 10.34.

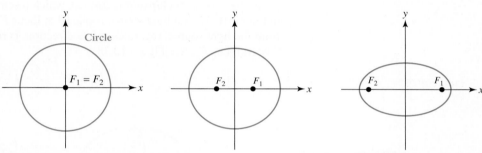

Figure 10.32 $e = 0$ **Figure 10.33** $e = 0.4$ **Figure 10.34** $e = 0.8$

MAKING CONNECTIONS

The Number e and the Variable e Do not confuse the *variable e*, which is used to denote the eccentricity of an ellipse, with the irrational *number $e \approx 2.72$*, which is the base of the natural exponential function $f(x) = e^x$ and of the natural logarithmic function $g(x) = \ln x$.

Astronomers have measured values of a and e for the eight major planets and Pluto. With this information and the fact that the sun is located at one focus of the ellipse, the equations of their orbits can be found.

EXAMPLE 4   Finding the orbital equation for Pluto

Pluto has $a = 39.44$ and $e = 0.249$. (For Earth, $a = 1$.) Graph the orbit of Pluto and the position of the sun in $[-60, 60, 10]$ by $[-40, 40, 10]$. (**Source:** M. Zeilik, *Introductory Astronomy and Astrophysics*.)

Calculator Help

To learn how to access the variable Y_1, see Appendix A (page AP-12).

SOLUTION   Let the orbit of Pluto be given by $\frac{x^2}{a^2} + \frac{y^2}{b^2} = 1$. Then

$$e = \frac{c}{a} = 0.249 \qquad \text{implies} \qquad c = 0.249a = 0.249(39.44) \approx 9.821.$$

To find b, solve the equation $c^2 = a^2 - b^2$ for b.

$$b = \sqrt{a^2 - c^2}$$
$$= \sqrt{39.44^2 - 9.821^2} \approx 38.20$$

Pluto's orbit is modeled by $\frac{x^2}{39.44^2} + \frac{y^2}{38.20^2} = 1$. Since $c \approx 9.821$, the foci are $(\pm 9.821, 0)$. The sun could be located at either focus. We locate the sun at $(9.821, 0)$.

Figure 10.35

$[-60, 60, 10]$ by $[-40, 40, 10]$

$y_1 = 38.2\sqrt{1 - x^2/39.44^2}$

$y_2 = -38.2\sqrt{1 - x^2/39.44^2}$

Figure 10.36

To graph this ellipse on a graphing calculator, we must solve the equation for y.

$$\frac{x^2}{39.44^2} + \frac{y^2}{38.20^2} = 1$$

$$\frac{y^2}{38.20^2} = 1 - \frac{x^2}{39.44^2}$$

$$\frac{y}{38.20} = \pm\sqrt{1 - \frac{x^2}{39.44^2}}$$

$$y = \pm 38.20\sqrt{1 - \frac{x^2}{39.44^2}}$$

See Figures 10.35 and 10.36.

Now Try Exercise 89 ◀

Reflective Property of Ellipses

Like parabolas, ellipses also have an important reflective property. If an ellipse is rotated about the x-axis, an **ellipsoid** is formed, which resembles the shell of an egg, as illustrated in Figure 10.37. If a light source is placed at focus F_1, then every beam of light emanating from the light source, regardless of its direction, is reflected at the surface of the ellipsoid toward focus F_2. See Figure 10.38.

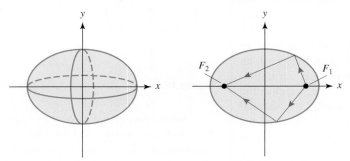

Figure 10.37 Ellipsoid

Figure 10.38

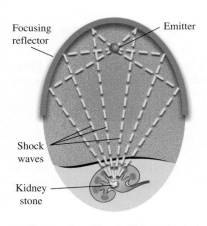

The top of an ellipse is illustrated in this depiction of how a lithotripter crushes a kidney stone.

Figure 10.39

An Application A *lithotripter* is a machine designed to break up kidney stones without surgery, by producing powerful shock waves. See Figure 10.39. To focus these shock waves accurately, a lithotripter uses the reflective property of ellipses. A patient is carefully positioned so that the kidney stone is at one focus, while the source of the shock waves is located at the other focus. When shock waves are emitted, they are reflected directly toward the kidney stone. As a result, it absorbs all of the energy from the shock wave and is broken up without harming the patient. This nonsurgical procedure reduces both risk and recovery time. In Exercises 91–93 this reflective property of ellipses is applied.

Translations of Ellipses

If the equation of an ellipse is given by either $\frac{x^2}{a^2} + \frac{y^2}{b^2} = 1$ or $\frac{x^2}{b^2} + \frac{y^2}{a^2} = 1$, then the center of the ellipse is $(0, 0)$. We can use translations of graphs to find the equation of an ellipse centered at (h, k) by replacing x with $(x - h)$ and y with $(y - k)$.

Standard Equations for Ellipses Centered at (h, k)

An ellipse with center (h, k) and either a horizontal or a vertical major axis satisfies one of the following equations, where $a > b > 0$ and $c^2 = a^2 - b^2$ with $c \geq 0$.

$$\frac{(x - h)^2}{a^2} + \frac{(y - k)^2}{b^2} = 1$$ Major axis: horizontal; foci: $(h \pm c, k)$
Vertices: $(h \pm a, k)$

$$\frac{(x - h)^2}{b^2} + \frac{(y - k)^2}{a^2} = 1$$ Major axis: vertical; foci: $(h, k \pm c)$
Vertices: $(h, k \pm a)$

EXAMPLE 5   Translating an ellipse

Translate the ellipse with equation $\frac{x^2}{9} + \frac{y^2}{4} = 1$ so that it is centered at $(-1, 2)$. Find the new equation and sketch its graph.

SOLUTION To translate the center from $(0, 0)$ to $(-1, 2)$, replace x with $(x + 1)$ and y with $(y - 2)$. The new equation is

$$\frac{(x + 1)^2}{9} + \frac{(y - 2)^2}{4} = 1.$$

The given ellipse is shown in Figure 10.40, and the translated ellipse is shown in Figure 10.41.

Figure 10.40

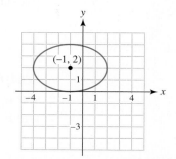

Figure 10.41   *Now Try Exercise 29* ◀

EXAMPLE 6   Graphing an ellipse with center (h, k)

Graph the ellipse given by $\frac{(x + 2)^2}{16} + \frac{(y - 2)^2}{25} = 1$. Label the vertices and the foci.

SOLUTION The ellipse has a vertical major axis, and its center is $(-2, 2)$. Since $a^2 = 25$ and $b^2 = 16$, it follows that $c^2 = a^2 - b^2 = 25 - 16 = 9$. Thus $a = 5, b = 4$, and $c = 3$. The vertices are located **5** units above and below the center of the ellipse, and the foci are located **3** units above and below the center of the ellipse. That is, the vertices are $(-2, 2 \pm 5)$, or $(-2, 7)$ and $(-2, -3)$, and the foci are $(-2, 2 \pm 3)$, or $(-2, 5)$ and $(-2, -1)$. A graph of the ellipse is shown in Figure 10.42.   *Now Try Exercise 43* ◀

Figure 10.42

EXAMPLE 7   Finding the standard equation of an ellipse

Write $4x^2 - 16x + 9y^2 + 54y + 61 = 0$ in the standard form for an ellipse centered at (h, k). Identify the center and the vertices.

SOLUTION We can write the given equation in standard form by completing the square.

$4x^2 - 16x + 9y^2 + 54y + 61 = 0$	*Given equation*
$4(x^2 - 4x + \underline{\quad}) + 9(y^2 + 6y + \underline{\quad}) = -61$	*Distributive property*
$4(x^2 - 4x + \underline{4}) + 9(y^2 + 6y + \underline{9}) = -61 + 16 + 81$	*Complete the square.*
$4(x - 2)^2 + 9(y + 3)^2 = 36$	*Perfect square trinomials*
$\dfrac{(x - 2)^2}{9} + \dfrac{(y + 3)^2}{4} = 1$	*Divide each side by 36.*

The center is $(2, -3)$. Because the major axis is horizontal and $a = 3$, the vertices of the ellipse are $(2 \pm 3, -3)$, or $(5, -3)$ and $(-1, -3)$.   *Now Try Exercise 53* ◄

More Nonlinear Systems of Equations

In Sections 9.1 and 9.2 we discussed systems of equations and inequalities. In this and the next subsections we revisit these topics.

EXAMPLE 8   Solving a nonlinear system of equations

Use substitution to solve the following system of equations. Give graphical support.

$$9x^2 + 4y^2 = 36$$
$$12x^2 + y^2 = 12$$

SOLUTION

STEP 1: Begin by solving the second equation for y^2 to obtain $y^2 = 12 - 12x^2$.

STEP 2: Next, substitute $(12 - 12x^2)$ for y^2 in the first equation and solve for x.

$9x^2 + 4y^2 = 36$	*First equation*
$9x^2 + 4(12 - 12x^2) = 36$	*Let $y^2 = 12 - 12x^2$.*
$9x^2 + 48 - 48x^2 = 36$	*Distributive property*
$-39x^2 = -12$	*Subtract 48; simplify.*
$x^2 = \dfrac{4}{13}$	*Divide by -39; simplify.*
$x = \pm\sqrt{\dfrac{4}{13}}$	*Square root property*

STEP 3: To determine the y-values, substitute $x^2 = \frac{4}{13}$ in $y^2 = 12 - 12x^2$.

$$y^2 = 12 - 12\left(\frac{4}{13}\right) = \frac{108}{13}, \quad \text{or} \quad y = \pm\sqrt{\frac{108}{13}}$$

There are four solutions: $\left(\pm\sqrt{\frac{4}{13}}, \pm\sqrt{\frac{108}{13}}\right)$.

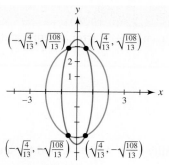

$\left(-\sqrt{\frac{4}{13}}, \sqrt{\frac{108}{13}}\right)$ $\left(\sqrt{\frac{4}{13}}, \sqrt{\frac{108}{13}}\right)$

$\left(-\sqrt{\frac{4}{13}}, -\sqrt{\frac{108}{13}}\right)$ $\left(\sqrt{\frac{4}{13}}, -\sqrt{\frac{108}{13}}\right)$

Figure 10.43

To graph the system of equations by hand, put each equation in standard form by dividing the first given equation by 36 and the second given equation by 12.

$$\frac{x^2}{4} + \frac{y^2}{9} = 1 \quad \text{and} \quad x^2 + \frac{y^2}{12} = 1$$

The graphs of these ellipses and the four solutions are shown in Figure 10.43.

Now Try Exercise 67 ◀

NOTE The system of equations in Example 8 could also be solved by elimination. To do this, multiply the second equation by -4 and add.

More Nonlinear Systems of Inequalities

In the next example, we solve a nonlinear system of inequalities whose graph involves a parabola and an ellipse.

EXAMPLE 9 Solving a nonlinear inequality

Shade the region in the *xy*-plane that satisfies the system of inequalities.

$$36x^2 + 25y^2 \leq 900$$
$$x + (y + 2)^2 \leq 4$$

SOLUTION Before sketching a graph, rewrite these two inequalities as follows.

First Inequality: $\dfrac{36}{900}x^2 + \dfrac{25}{900}y^2 \leq \dfrac{900}{900}$ Divide each term by 900.

$\dfrac{x^2}{25} + \dfrac{y^2}{36} \leq 1$ Simplify to standard form.

Second Inequality: $(y + 2)^2 \leq -x + 4$ Subtract x.

$(y + 2)^2 \leq -(x - 4)$ Distributive property

The first inequality represents the region inside an ellipse, as shown in Figure 10.44. The second inequality represents the region left of a parabola that opens to the left with vertex $(4, -2)$, as shown in Figure 10.45. The solution set for the system, which satisfies both inequalities, is shaded in Figure 10.46. (To verify this, try the test point $(-2, -2)$.)

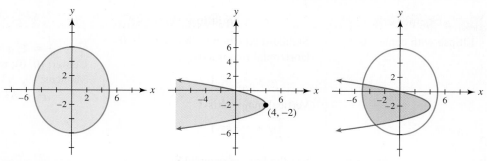

Figure 10.44   **Figure 10.45**   **Figure 10.46**

Now Try Exercise 83 ◀

Area Inside an Ellipse   The following formula can be used to calculate the area inside an ellipse.

Explain why the area formula for an ellipse is a generalization of the area formula for a circle.

Area Inside an Ellipse

The area A of the region contained inside an ellipse is given by $A = \pi ab$, where a and b are from the standard equation of the ellipse.

This formula is applied in the next example.

EXAMPLE 10   **Finding the area inside an ellipse**

Shade the region in the xy-plane that satisfies the inequality $x^2 + 4y^2 \leq 4$. Find the area of this region if units are in inches.

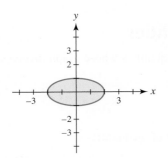

Figure 10.47

SOLUTION   Begin by dividing each term in the given inequality by 4.

$$x^2 + 4y^2 \leq 4 \qquad \text{Given inequality}$$

$$\frac{x^2}{4} + \frac{4y^2}{4} \leq \frac{4}{4} \qquad \text{Divide by 4.}$$

$$\frac{x^2}{4} + \frac{y^2}{1} \leq 1 \qquad \text{Simplify the fractions.}$$

The boundary of the region is the ellipse $\frac{x^2}{4} + \frac{y^2}{1} = 1$. The region *inside* the ellipse satisfies the inequality. To verify this fact, note that the test point $(0, 0)$, which is located inside the ellipse, satisfies the inequality. The solution set is shaded in Figure 10.47. The area of this region is

$$A = \pi ab = \pi(2)(1) = 2\pi \approx 6.28 \text{ square inches.}$$

Now Try Exercise 85 ◀

10.2 Putting It All Together

The following table summarizes some important concepts about ellipses.

Concept	Equation	Example
Ellipse with center $(0, 0)$	Standard equation with $a > b > 0$ Horizontal major axis: $$\frac{x^2}{a^2} + \frac{y^2}{b^2} = 1$$ Vertical major axis: $$\frac{x^2}{b^2} + \frac{y^2}{a^2} = 1$$ See the box on page 844.	$\frac{x^2}{4} + \frac{y^2}{9} = 1; a = 3, b = 2$ Center: $(0, 0)$; major axis: vertical Vertices; $(0, \pm3)$; foci: $(0, \pm\sqrt{5})$ $(c^2 = a^2 - b^2 = 9 - 4 = 5, \text{ so } c = \sqrt{5}.)$

Concept	Equation	Example
Ellipse with center (h, k)	Standard equation with $a > b > 0$ Horizontal major axis: $$\frac{(x - h)^2}{a^2} + \frac{(y - k)^2}{b^2} = 1$$ Vertical major axis: $$\frac{(x - h)^2}{b^2} + \frac{(y - k)^2}{a^2} = 1$$ See the box on page 849.	$\frac{(x - 1)^2}{4} + \frac{(y + 1)^2}{9} = 1$; $a = 3, b = 2$ Center: $(1, -1)$; major axis: vertical Vertices: $(1, -1 \pm 3)$; foci: $\left(1, -1 \pm \sqrt{5}\right)$ $\left(c^2 = a^2 - b^2 = 9 - 4 = 5, \text{ so } c = \sqrt{5}.\right)$ 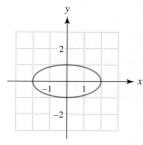
Area inside an ellipse	$A = \pi ab$	The area inside the ellipse given by $\frac{x^2}{49} + \frac{y^2}{9} = 1$ is $$A = \pi(7)(3) = 21\pi \text{ square units.}$$

10.2 Exercises

Ellipses with Center (0, 0)

Exercises 1–8: Graph the ellipse. Label the foci and the endpoints of each axis.

1. $\dfrac{x^2}{4} + \dfrac{y^2}{9} = 1$ **2.** $\dfrac{x^2}{9} + \dfrac{y^2}{4} = 1$

3. $\dfrac{x^2}{36} + \dfrac{y^2}{16} = 1$ **4.** $x^2 + \dfrac{y^2}{4} = 1$

5. $x^2 + 4y^2 = 400$ **6.** $9x^2 + 5y^2 = 45$

7. $25x^2 + 9y^2 = 225$ **8.** $5x^2 + 4y^2 = 20$

Exercises 9–12: Match the equation with its graph (a–d).

9. $\dfrac{x^2}{16} + \dfrac{y^2}{36} = 1$ **10.** $\dfrac{x^2}{4} + y^2 = 1$

11. $\dfrac{x^2}{16} + \dfrac{y^2}{4} = 1$ **12.** $\dfrac{x^2}{9} + \dfrac{y^2}{9} = 1$

a.

b.

c.

d.
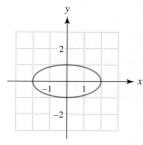

Exercises 13–16: Find the standard equation of the ellipse shown in the figure. Identify the coordinates of the vertices, endpoints of the minor axis, and the foci.

13.

14.

15.

16.

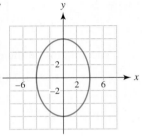

Exercises 17–20: The foci F_1 and F_2, vertices V_1 and V_2, and endpoints U_1 and U_2 of the minor axis of an ellipse are labeled in the figure. Graph the ellipse and find its standard equation. (The coordinates of V_1, V_2, F_1, and F_2 are integers.)

17.

18.

19.

20.

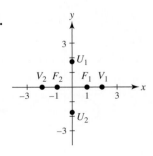

Exercises 21–28: Find an equation of the ellipse, centered at the origin, satisfying the conditions.

21. Foci $(0, \pm 2)$, vertices $(0, \pm 4)$

22. Foci $(0, \pm 3)$, vertices $(0, \pm 5)$

23. Foci $(\pm 5, 0)$, vertices $(\pm 6, 0)$

24. Foci $(\pm 4, 0)$, vertices $(\pm 6, 0)$

25. Horizontal major axis of length 8, minor axis of length 6

26. Vertical major axis of length 12, minor axis of length 8

27. Eccentricity $\frac{2}{3}$, horizontal major axis of length 6

28. Eccentricity $\frac{3}{4}$, vertices $(0, \pm 8)$

Ellipses with Center (h, k)

Exercises 29–32: Translate the ellipse with the given equation so that it is centered at the given point. Find the new equation and sketch its graph.

29. $\dfrac{x^2}{4} + \dfrac{y^2}{3} = 1; (2, -1)$ **30.** $\dfrac{x^2}{9} + \dfrac{y^2}{2} = 1; (-3, 7)$

31. $\dfrac{x^2}{2} + \dfrac{y^2}{9} = 1; (-3, -4)$ **32.** $\dfrac{x^2}{15} + \dfrac{y^2}{16} = 1; (5, -6)$

Exercises 33–38: Sketch a graph of the ellipse.

33. $\dfrac{(x-2)^2}{4} + \dfrac{(y-1)^2}{9} = 1$

34. $\dfrac{(x+1)^2}{16} + \dfrac{(y+3)^2}{9} = 1$

35. $\dfrac{(x+1)^2}{16} + \dfrac{(y+2)^2}{25} = 1$

36. $\dfrac{(x-4)^2}{9} + \dfrac{y^2}{4} = 1$ **37.** $\dfrac{(x+2)^2}{4} + y^2 = 1$

38. $x^2 + \dfrac{(y-3)^2}{4} = 1$

Exercises 39–42: Match the equation with its graph (a–d).

39. $\dfrac{(x-2)^2}{16} + \dfrac{(y+4)^2}{36} = 1$

40. $\dfrac{(x+1)^2}{4} + \dfrac{y^2}{9} = 1$

41. $\dfrac{(x+1)^2}{9} + \dfrac{(y-1)^2}{4} = 1$

42. $\dfrac{x^2}{25} + \dfrac{(y+1)^2}{10} = 1$

a.

b.

c.

d.

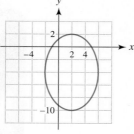

Exercises 43–46: Sketch a graph of the ellipse. Identify the foci and vertices.

43. $\dfrac{(x-1)^2}{9} + \dfrac{(y-1)^2}{25} = 1$

44. $\dfrac{(x+2)^2}{25} + \dfrac{(y+1)^2}{16} = 1$

45. $\dfrac{(x+4)^2}{16} + \dfrac{(y-2)^2}{9} = 1$

46. $\dfrac{x^2}{4} + \dfrac{(y-1)^2}{9} = 1$

Exercises 47–50: Find an equation of an ellipse that satisfies the given conditions.

47. Center $(2, 1)$, focus $(2, 3)$, and vertex $(2, 4)$

48. Center $(-3, -2)$, focus $(-1, -2)$, and vertex $(1, -2)$

49. Vertices $(\pm 3, 2)$ and foci $(\pm 2, 2)$

50. Vertices $(-1, \pm 3)$ and foci $(-1, \pm 1)$

Exercises 51 and 52: Find an (approximate) equation of the ellipse shown in the figure.

51.

52.

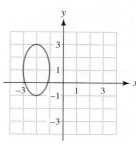

Exercises 53–60: (Refer to Example 7.) Write the equation in standard form for an ellipse centered at (h, k). Identify the center and the vertices.

53. $9x^2 + 18x + 4y^2 - 8y - 23 = 0$

54. $9x^2 - 36x + 16y^2 - 64y - 44 = 0$

55. $4x^2 + 8x + y^2 + 2y + 1 = 0$

56. $x^2 - 6x + 9y^2 = 0$

57. $4x^2 + 16x + 5y^2 - 10y + 1 = 0$

58. $2x^2 + 4x + 3y^2 - 18y + 23 = 0$

59. $16x^2 - 16x + 4y^2 + 12y = 51$

60. $16x^2 + 48x + 4y^2 - 20y + 57 = 0$

Graphing Ellipses with Technology

Exercises 61–64: Graph the ellipse.

61. $\dfrac{x^2}{15} + \dfrac{y^2}{10} = 1$

62. $\dfrac{(x-1.2)^2}{7.1} + \dfrac{y^2}{3.5} = 1$

63. $4.1x^2 + 6.3y^2 = 25$

64. $\frac{1}{2}x^2 + \frac{1}{3}y^2 = \frac{1}{6}$

Solving Equations and Inequalities

Exercises 65–70: Solve the system of equations. Give graphical support by making a sketch.

65. $\dfrac{x^2}{4} + \dfrac{y^2}{9} = 1$

$x + y = 3$

66. $\dfrac{x^2}{16} + \dfrac{y^2}{25} = 1$

$-2x + y = 5$

67. $4x^2 + 16y^2 = 64$
$x^2 + y^2 = 9$

68. $4x^2 + y^2 = 4$
$x^2 + y^2 = 2$

69. $x^2 + y^2 = 9$
$2x^2 + 3y^2 = 18$

70. $x^2 + y^2 = 4$
$(x - 1)^2 + y^2 = 4$

Exercises 71–76: Solve the system of equations.

71. $\dfrac{x^2}{2} + \dfrac{y^2}{4} = 1$
$-x^2 + 2y = 4$

72. $x^2 + \frac{1}{9}y^2 = 1$
$x + y = 3$

73. $\dfrac{x^2}{2} + \dfrac{y^2}{4} = 1$
$\dfrac{x^2}{4} + \dfrac{y^2}{2} = 1$

74. $\dfrac{x^2}{5} + \dfrac{y^2}{10} = 1$
$\dfrac{x^2}{10} + \dfrac{y^2}{5} = 1$

75. $(x - 2)^2 + y^2 = 9$
$x^2 + y^2 = 9$

76. $(x - 2) - y^2 = 0$
$\dfrac{x^2}{4} + \dfrac{y^2}{9} = 1$

Exercises 77–84: Shade the solutions set to the system.

77. $(x - 1)^2 + (y + 1)^2 < 4$
$(x + 1)^2 + y^2 > 1$

78. $\dfrac{x^2}{16} + \dfrac{y^2}{25} < 1$
$\dfrac{x^2}{4} + \dfrac{y^2}{9} > 1$

79. $\dfrac{x^2}{4} + \dfrac{y^2}{9} \le 1$
$x + y \ge 2$

80. $\dfrac{x^2}{16} + \dfrac{y^2}{25} \le 1$
$-x + y \le 4$

81. $x^2 + y^2 \le 4$
$x^2 + (y - 2)^2 \le 4$

82. $x^2 + (y + 1)^2 \le 9$
$(x + 1)^2 + y^2 \le 9$

83. $x^2 + y^2 \le 4$
$(x + 1)^2 - y \le 0$

84. $4x^2 + 9y^2 \le 36$
$x - (y - 2)^2 \ge 0$

Exercises 85–88: (Refer to Example 10.) Shade the region in the xy-plane that satisfies the given inequality. Find the area of this region if units are in feet.

85. $4x^2 + 9y^2 \le 36$

86. $9x^2 + y^2 \le 9$

87. $\dfrac{(x - 1)^2}{25} + \dfrac{(y + 2)^2}{16} \le 1$

88. $\dfrac{(x + 3)^2}{4} + \dfrac{(y - 2)^2}{8} \le 1$

Applications

Exercises 89 and 90: Orbits of Planets   (Refer to Example 4.) Find an equation of the orbit for the planet. Graph its orbit and the location of the sun at a focus on the positive x-axis.

89. Mercury: $e = 0.206$, $a = 0.387$

90. Mars: $e = 0.093$, $a = 1.524$

91. *Lithotripter*   (Refer to the discussion in this section.) The source of a shock wave is placed at one focus of an ellipsoid with a major axis of 8 inches and a minor axis of 5 inches. Estimate, to the nearest thousandth of an inch, how far a kidney stone should be positioned from the source.

92. *Shape of a Lithotripter*   A patient's kidney stone is placed 12 units away from the source of the shock waves of a lithotripter. The lithotripter is based on an ellipse with a minor axis that measures 16 units. Find the equation of an ellipse that would satisfy this situation.

93. *Whispering Gallery*   A large room constructed in the shape of the upper half of an ellipsoid has a unique property. Two people, standing at the two foci, can hear each other whispering, even though there may be considerable distance between them. Any sound emanating from one focus is reflected directly toward the other focus. See the figure. If the foci are 100 feet apart and the maximum height of the ceiling is 40 feet, estimate the area of the floor of the room.

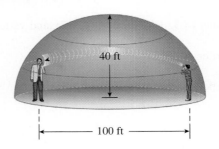

40 ft

100 ft

94. *Halley's Comet*   Halley's comet travels in an elliptical orbit with $a = 17.95$ and $b = 4.44$ and passes by Earth roughly every 76 years. Note that each unit represents one astronomical unit, or 93 million miles. The comet most recently passed by Earth in February 1986. (**Source:** M. Zeilik, *Introductory Astronomy and Astrophysics*.)

(a) Write an equation for this orbit, centered at $(0, 0)$ with major axis on the *x*-axis.

(b) If the sun lies (at the focus) on the positive *x*-axis, approximate its coordinates.

(c) Determine the maximum and minimum distances between Halley's comet and the sun.

95. *The Roman Colosseum*   The perimeter of the Roman Colosseum is an ellipse with major axis 620 feet and minor axis 513 feet. Find the distance between the foci of this ellipse.

96. *Orbit of Earth*   (Refer to Example 4.) Earth has a nearly circular orbit with $e \approx 0.0167$ and $a = 93$ million miles. Approximate the minimum and maximum distances between Earth and the sun.   (**Source:** M. Zeilik, *Introductory Astronomy and Astrophysics*.)

97. *Arch Bridge*   An elliptical arch under a bridge is constructed so that it is 60 feet wide and has a maximum height of 25 feet, as illustrated in the figure. Find the height of the arch 15 feet from the center of the arch.

98. *Perimeter of an Ellipse*   The perimeter P of an ellipse can be approximated by

$$P \approx 2\pi \sqrt{\frac{a^2 + b^2}{2}}.$$

(a) Approximate the distance in miles that Mercury travels in one orbit of the sun if $a = 36.0$, $b = 35.2$, and the units are in millions of miles.

(b) If a planet has a circular orbit, does this formula give the *exact* perimeter? Explain.

99. *Satellite Orbit*   The orbit of *Explorer VII* and the outline of Earth's surface are shown in the figure. This orbit can be described by the equation $\frac{x^2}{a^2} + \frac{y^2}{b^2} = 1$, where $a = 4464$ and $b = 4462$. The surface of Earth can be described by $(x - 164)^2 + y^2 = 3960^2$. Find the maximum and minimum heights of the satellite above Earth's surface if all units are in miles.   (**Sources:** W. Loh; W. Thomson.)

100. *Orbital Velocity*   The maximum and minimum velocities in kilometers per second of a celestial body moving in an elliptical orbit can be calculated by

$$v_{\max} = \frac{2\pi a}{P} \sqrt{\frac{1 + e}{1 - e}} \quad \text{and} \quad v_{\min} = \frac{2\pi a}{P} \sqrt{\frac{1 - e}{1 + e}}.$$

In these equations, a is half the length of the major axis of the orbit in kilometers, P is the orbital period in seconds, and e is the eccentricity of the orbit.   (**Source:** M. Zeilik.)

(a) Calculate $v_{\max}$ and $v_{\min}$ for Pluto if $a = 5.913 \times 10^9$ kilometers, the period is $P = 2.86 \times 10^{12}$ seconds, and the eccentricity is $e = 0.249$.

(b) If a planet has a circular orbit, what can be said about its orbital velocity?

Writing about Mathematics

101. Explain how the distance between the foci of an ellipse affects the shape of the ellipse.

102. Given the standard equation of an ellipse, explain how to determine the length of the major axis. How can you determine whether the major axis is vertical or horizontal?

EXTENDED AND DISCOVERY EXERCISES

Exercises 1–4: (Refer to Making Connections on page 845.)
Write the equation of the line satisfying the conditions in
intercept form. Then find the x- and y-intercepts of the line.

1. Passing through $(-2, 6)$ and $(4, -3)$

2. Passing through $(-6, -4)$ and $(3, 8)$

3. Slope -2, passing through $(3, -1)$

4. Slope 4, passing through $(-2, 1)$

Exercises 5 and 6: (Refer to Making Connections on page 845.)
Find the x- and y-intercepts of the ellipse.

5. $\dfrac{x^2}{25} + \dfrac{y^2}{9} = 1$

6. Vertices $(0, \pm 13)$ and foci $(0, \pm 12)$

CHECKING BASIC CONCEPTS FOR SECTIONS 10.1 AND 10.2

1. Graph the parabola defined by $x = \frac{1}{2}y^2$. Include the focus and directrix.

2. Find an equation of the parabola with focus $(-1, 0)$ and directrix $y = 3$.

3. Graph the ellipse defined by $\dfrac{x^2}{36} + \dfrac{y^2}{100} = 1$. Include the foci and label the major and minor axes.

4. Find an equation of the ellipse centered at $(3, -2)$ with a vertical major axis of length 6 and minor axis of length 4. What are the coordinates of the foci?

5. A parabolic reflector for a searchlight has a diameter of 4 feet and a depth of 1 foot. How far from the vertex should the filament of the light bulb be located?

6. Solve the nonlinear system of equations.
$$\begin{aligned} x^2 + y^2 &= 10 \\ 2x^2 + 3y^2 &= 29 \end{aligned}$$

7. Write $x^2 - 4x + 4y^2 + 8y - 8 = 0$ in the standard form for an ellipse centered at (h, k). Identify the center and the vertices.

10.3 Hyperbolas

- **Find equations of hyperbolas**
- **Graph hyperbolas**
- **Learn the reflective property of hyperbolas**
- **Translate hyperbolas**

Introduction

Hyperbolas have several interesting properties. For example, if a comet passes by the sun with a high velocity, then the sun's gravity may not be strong enough to cause the comet to go into orbit; instead, the comet will pass by the sun just once and follow a trajectory that can be described by a hyperbola. Hyperbolas also have a reflective property, which is used in telescopes. In this section we look at some basic properties of hyperbolas.

Equations and Graphs of Hyperbolas

A third type of conic section is a hyperbola.

Hyperbola

A **hyperbola** is the set of points in a plane, the difference of whose distances from two fixed points is constant. Each fixed point is called a **focus** of the hyperbola.

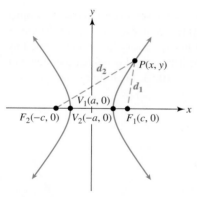

Figure 10.48

Figure 10.48 shows a point $P(x, y)$ on a hyperbola, with distance d_1 from focus $F_1(c, 0)$ and distance d_2 from focus $F_2(-c, 0)$. Regardless of the location of the point $P(x, y)$, $|d_2 - d_1| = 2a$. The **transverse axis** is the line segment connecting the **vertices** $V_1(a, 0)$ and $V_2(-a, 0)$, and its length equals $2a$.

Two hyperbolas are shown in Figures 10.49 and 10.50. The vertices, $(\pm a, 0)$ or $(0, \pm a)$, and foci, $(\pm c, 0)$ or $(0, \pm c)$, are labeled. The two parts of the hyperbola in Figure 10.49 are the **left branch** and the **right branch**, whereas in Figure 10.50 the hyperbola has an **upper branch** and a **lower branch**. A line segment connecting the points $(0, \pm b)$ in Figure 10.49 and $(\pm b, 0)$ in Figure 10.50 is the **conjugate axis**. The dashed lines $y = \pm \frac{b}{a} x$ and $y = \pm \frac{a}{b} x$ are **asymptotes** for the respective hyperbolas. They can be used as an aid in graphing. The dashed rectangle is called the **fundamental rectangle**.

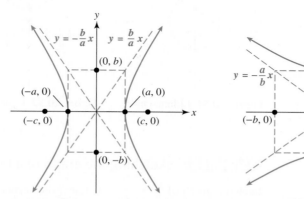

Figure 10.49 Horizontal Transverse Axis   **Figure 10.50** Vertical Transverse Axis

By the vertical line test, a hyperbola cannot be represented by a function, but many can be described by the following equations. The constants a, b, and c are *positive*.

Standard Equations for Hyperbolas Centered at (0, 0)

The hyperbola with center at the origin, *horizontal* transverse axis, and equation

$$\frac{x^2}{a^2} - \frac{y^2}{b^2} = 1$$

has asymptotes $y = \pm \frac{b}{a} x$, vertices $(\pm a, 0)$, and foci $(\pm c, 0)$, where $c^2 = a^2 + b^2$. The hyperbola with center at the origin, *vertical* transverse axis, and equation

$$\frac{y^2}{a^2} - \frac{x^2}{b^2} = 1$$

has asymptotes $y = \pm \frac{a}{b} x$, vertices $(0, \pm a)$, and foci $(0, \pm c)$, where $c^2 = a^2 + b^2$.

NOTE A hyperbola consists of two solid (blue) curves, or branches. The asymptotes, foci, transverse axis, conjugate axis, and fundamental (dashed) rectangle are not part of the hyperbola, but are aids for sketching its graph.

An Application of Conic Sections   One interpretation of an asymptote relates to trajectories of comets as they approach the sun. Comets travel in parabolic, elliptic, or hyperbolic trajectories. If the speed of a comet is too slow, the gravitational pull of the sun will capture the comet in an elliptical orbit. See Figure 10.51. If the speed of the comet is too fast, the comet will pass by the sun once in a hyperbolic trajectory; farther from the sun, gravity becomes weaker and the comet will eventually return to a straight-line trajectory that is determined by the *asymptote* of the hyperbola. See Figure 10.52. Finally, if the speed is neither too slow nor too fast, the comet will travel in a parabolic path. See Figure 10.53. In all three cases, the sun is located at a focus of the conic section.

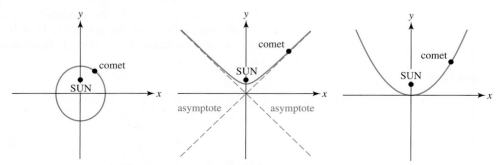

Figure 10.51 Elliptic Orbit   **Figure 10.52** Hyperbolic Path   **Figure 10.53** Parabolic Path

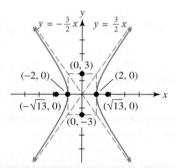

Figure 10.54

EXAMPLE 1   Sketching the graph of a hyperbola

Sketch a graph of $\frac{x^2}{4} - \frac{y^2}{9} = 1$. Label the vertices, foci, and asymptotes.

SOLUTION The equation is in standard form with $a = 2$ and $b = 3$. It has a horizontal transverse axis with vertices $(\pm 2, 0)$. The endpoints of the conjugate axis are $(0, \pm 3)$. To locate the foci, find c.

$$c^2 = a^2 + b^2 = 4 + 9 = 13, \quad \text{or} \quad c = \sqrt{13} \approx 3.61.$$

The foci are $\left(\pm\sqrt{13}, 0\right)$. The asymptotes are $y = \pm\frac{b}{a}x$, or $y = \pm\frac{3}{2}x$. See Figure 10.54.

Now Try Exercise 1 ◀

EXAMPLE 2   Finding the equation of a hyperbola

Find the equation of the hyperbola centered at the origin with a vertical transverse axis of length 6 and focus $(0, 5)$. Also find the equations of its asymptotes.

SOLUTION Since the hyperbola is centered at the origin with a vertical axis, its equation is $\frac{y^2}{a^2} - \frac{x^2}{b^2} = 1$. The transverse axis has length $6 = 2a$, so $a = 3$. Since one focus is located at $(0, 5)$, $c = 5$. We can find b by using the following equation.

$$b^2 = c^2 - a^2$$
$$b = \sqrt{c^2 - a^2}$$
$$b = \sqrt{5^2 - 3^2} = 4$$

The standard equation of this hyperbola is $\frac{y^2}{9} - \frac{x^2}{16} = 1$. Its asymptotes are $y = \pm\frac{a}{b}x$, or $y = \pm\frac{3}{4}x$.

Now Try Exercise 19 ◀

EXAMPLE 3   Finding the equation of a hyperbola

Find the standard equation of the hyperbola shown in Figure 10.55. Identify the vertices, foci, and asymptotes.

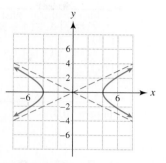

Figure 10.55

SOLUTION

Getting Started Because the hyperbola is centered at (0, 0) with a horizontal transverse axis, its equation has the form

$$\frac{x^2}{a^2} - \frac{y^2}{b^2} = 1.$$

Sketch the fundamental rectangle first. Half of its length equals a, and half of its width equals b. ▶

In Figure 10.56 the fundamental rectangle is determined by the four points $(\pm 4, 0)$ and $(0, \pm 2)$, and its diagonals correspond to the asymptotes. It follows that $a = 4$ and $b = 2$. Thus the standard equation of the hyperbola is

$$\frac{x^2}{16} - \frac{y^2}{4} = 1.$$

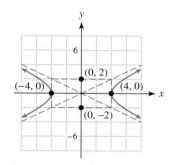

Figure 10.56

The vertices are $(\pm 4, 0)$, and the asymptotes are $y = \pm \frac{1}{2}x$. To find the foci, find c.

$$c^2 = a^2 + b^2 = 4^2 + 2^2 = 20, \quad \text{or} \quad c = \sqrt{20}.$$

The foci are $\left(\pm \sqrt{20}, 0 \right)$.

Now Try Exercise 37 ◀

Using Technology   In the next example, a graphing calculator is used to graph a hyperbola.

EXAMPLE 4   Graphing a hyperbola with technology

Use a graphing calculator to graph $\frac{y^2}{4.2} - \frac{x^2}{8.4} = 1$.

SOLUTION   Begin by solving the given equation for y.

$$\frac{y^2}{4.2} = 1 + \frac{x^2}{8.4} \qquad \text{Add } \tfrac{x^2}{8.4}.$$

$$y^2 = 4.2\left(1 + \frac{x^2}{8.4} \right) \qquad \text{Multiply by 4.2.}$$

$$y = \pm\sqrt{4.2\left(1 + \frac{x^2}{8.4} \right)} \qquad \text{Square root property}$$

Graph $Y_1 = \sqrt{(4.2(1 + X^2/8.4))}$ and $Y_2 = -\sqrt{(4.2(1 + X^2/8.4))}$. See Figures 10.57 and 10.58.

Figure 10.57

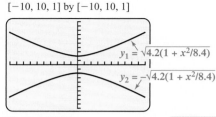

$[-10, 10, 1]$ by $[-10, 10, 1]$

$y_1 = \sqrt{4.2(1 + x^2/8.4)}$

$y_2 = -\sqrt{4.2(1 + x^2/8.4)}$

Figure 10.58

Now Try Exercise 55 ◀

Reflective Property of Hyperbolas

Hyperbolas have an important reflective property. If a hyperbola is rotated about the x-axis, a **hyperboloid** is formed, as illustrated in Figure 10.59. Any beam of light that is directed toward focus F_1 will be reflected by the hyperboloid toward focus F_2. See Figure 10.60.

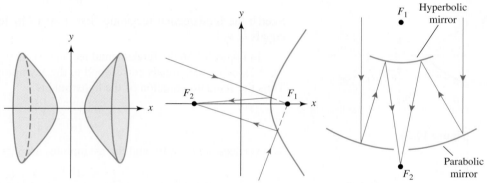

Figure 10.59 Hyperboloid   **Figure 10.60**   **Figure 10.61**

An Application   Telescopes sometimes make use of both parabolic and hyperbolic mirrors, as shown in Figure 10.61. When parallel rays of light from distant stars strike the large parabolic (primary) mirror, they are reflected toward its focus, F_1. A smaller hyperbolic (secondary) mirror is placed so that its focus is also located at F_1. Light rays striking the hyperbolic mirror are reflected toward its other focus, F_2, through a small hole in the parabolic mirror, and into an eye piece. See Exercise 68.

Translations of Hyperbolas

If the equation of a hyperbola is either $\dfrac{x^2}{a^2} - \dfrac{y^2}{b^2} = 1$ or $\dfrac{y^2}{a^2} - \dfrac{x^2}{b^2} = 1$ then the center of the hyperbola is $(0, 0)$. We can use translations of graphs to find the equation of a hyperbola centered at (h, k) by replacing x with $(x - h)$ and y with $(y - k)$. The constants a, b, and c are positive.

Standard Equations for Hyperbolas Centered at (h, k)

A hyperbola with center (h, k) and either a horizontal or a vertical transverse axis satisfies one of the following equations, where $c^2 = a^2 + b^2$.

$$\frac{(x - h)^2}{a^2} - \frac{(y - k)^2}{b^2} = 1$$

Transverse axis: horizontal
Vertices: $(h \pm a, k)$; foci: $(h \pm c, k)$
Asymptotes: $y = \pm \frac{b}{a}(x - h) + k$

$$\frac{(y - k)^2}{a^2} - \frac{(x - h)^2}{b^2} = 1$$

Transverse axis: vertical
Vertices: $(h, k \pm a)$; foci: $(h, k \pm c)$
Asymptotes: $y = \pm \frac{a}{b}(x - h) + k$

EXAMPLE 5   Graphing a hyperbola with center (h, k)

Graph the hyperbola whose equation is $\frac{(y + 2)^2}{9} - \frac{(x - 2)^2}{16} = 1$. Label the vertices, foci, and asymptotes.

SOLUTION The hyperbola has a vertical transverse axis, and its center is $(2, -2)$. Since $a^2 = 9$ and $b^2 = 16$, it follows that $c^2 = a^2 + b^2 = 9 + 16 = 25$. Thus $a = 3$, $b = 4$, and $c = 5$. The vertices are located 3 units above and below the center of the hyperbola, and the foci are located 5 units above and below the center of the hyperbola. That is, the vertices are $(2, -2 \pm 3)$, or $(2, 1)$ and $(2, -5)$, and the foci are $(2, -2 \pm 5)$, or $(2, 3)$ and $(2, -7)$. The asymptotes are given by

$$y = \pm \frac{a}{b}(x - h) + k, \quad \text{or} \quad y = \pm \frac{3}{4}(x - 2) - 2.$$

A graph of the hyperbola is shown in Figure 10.62.   ***Now Try Exercise 41*** ◀

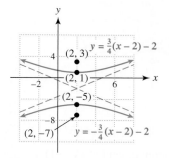

Figure 10.62

EXAMPLE 6   Finding the standard equation of a hyperbola

Write $9x^2 - 18x - 4y^2 - 16y = 43$ in the standard form for a hyperbola centered at (h, k). Identify the center and the vertices.

SOLUTION We can write the given equation in standard form by completing the square.

$$9x^2 - 18x - 4y^2 - 16y = 43 \qquad \text{Given equation}$$
$$9(x^2 - 2x + \underline{\quad}) - 4(y^2 + 4y + \underline{\quad}) = 43 \qquad \text{Distributive property}$$
$$9(x^2 - 2x + \underline{1}) - 4(y^2 + 4y + \underline{4}) = 43 + 9 - 16 \qquad \text{Complete the square.}$$
$$9(x - 1)^2 - 4(y + 2)^2 = 36 \qquad \text{Perfect square trinomials}$$
$$\frac{(x - 1)^2}{4} - \frac{(y + 2)^2}{9} = 1 \qquad \text{Divide each side by 36.}$$

The center is $(1, -2)$. Because $a = 2$ and the transverse axis is horizontal, the vertices of the hyperbola are $(h \pm a, k) = (1 \pm 2, -2)$.   ***Now Try Exercise 47*** ◀

10.3 Putting It All Together

The following table summarizes some important concepts about hyperbolas.

Concept	Equation	Example
Hyperbola with center $(0, 0)$	Standard equation Transverse axis: horizontal $$\frac{x^2}{a^2} - \frac{y^2}{b^2} = 1$$ Transverse axis: vertical $$\frac{y^2}{a^2} - \frac{x^2}{b^2} = 1$$ See the box on page 859.	$\frac{y^2}{4} - \frac{x^2}{9} = 1$; $a = 2, b = 3$ Transverse axis: vertical Vertices: $(0, \pm 2)$; foci: $\left(0, \pm \sqrt{13}\right)$ $\left(c^2 = a^2 + b^2 = 4 + 9 = 13, \text{ so } c = \sqrt{13}.\right)$ Asymptotes: $y = \pm\frac{2}{3}x$
Hyperbola with center (h, k)	Standard equation Transverse axis: horizontal $$\frac{(x - h)^2}{a^2} - \frac{(y - k)^2}{b^2} = 1$$ Transverse axis: vertical $$\frac{(y - k)^2}{a^2} - \frac{(x - h)^2}{b^2} = 1$$ See the box on page 863.	$\frac{(x - 1)^2}{4} - \frac{(y + 1)^2}{9} = 1$; $a = 2, b = 3$ Transverse axis: horizontal; center: $(1, -1)$ Vertices: $(1 \pm 2, -1)$; foci: $\left(1 \pm \sqrt{13}, -1\right)$ $\left(c^2 = a^2 + b^2 = 4 + 9 = 13, \text{ so } c = \sqrt{13}.\right)$ Asymptotes: $y = \pm\frac{3}{2}(x - 1) - 1$

10.3 Exercises

Hyperbolas with Center (0, 0)

Exercises 1–8: Sketch a graph of the hyperbola, including the asymptotes. Give the coordinates of the foci and vertices.

1. $\dfrac{x^2}{9} - \dfrac{y^2}{49} = 1$

2. $\dfrac{x^2}{16} - \dfrac{y^2}{4} = 1$

3. $\dfrac{y^2}{36} - \dfrac{x^2}{16} = 1$

4. $\dfrac{y^2}{4} - \dfrac{x^2}{4} = 1$

5. $x^2 - y^2 = 9$

6. $49y^2 - 25x^2 = 1225$

7. $9y^2 - 16x^2 = 144$

8. $4x^2 - 4y^2 = 100$

Exercises 9–12: Match the equation with its graph (a–d).

9. $\dfrac{x^2}{4} - \dfrac{y^2}{6} = 1$

10. $\dfrac{x^2}{9} - y^2 = 1$

11. $\dfrac{y^2}{9} - \dfrac{x^2}{16} = 1$

12. $\dfrac{y^2}{4} - \dfrac{x^2}{4} = 1$

a.

b.

c.

d.
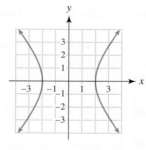

Exercises 13–16: Sketch a graph of a hyperbola, centered at the origin, with the foci, vertices, and asymptotes shown in the figure. Find an equation of the hyperbola. (The coordinates of the foci and vertices are integers.)

13.

14.

15.

16.

Exercises 17–26: Determine an equation of the hyperbola, centered at the origin, satisfying the conditions. Give the equations of its asymptotes.

17. Foci $(0, \pm 13)$, vertices $(0, \pm 12)$

18. Foci $(\pm 13, 0)$, vertices $(\pm 5, 0)$

19. Vertical transverse axis of length 4, foci $(0, \pm 5)$

20. Horizontal transverse axis of length 12, foci $(\pm 10, 0)$

21. Vertices $(\pm 3, 0)$, asymptotes $y = \pm \dfrac{2}{3} x$

22. Vertices $(0, \pm 4)$, asymptotes $y = \pm \dfrac{1}{2} x$

23. Endpoints of conjugate axis $(0, \pm 3)$, vertices $(\pm 4, 0)$

24. Endpoints of conjugate axis $(\pm 4, 0)$, vertices $(0, \pm 2)$

25. Vertices $\left(\pm \sqrt{10}, 0 \right)$, passing through $(10, 9)$

26. Vertices $\left(0, \pm \sqrt{5} \right)$, passing through $(4, 5)$

Hyperbolas with Center (*h, k*)

Exercises 27–32: Sketch a graph of the hyperbola. Identify the vertices, foci, and asymptotes.

27. $\dfrac{(x-1)^2}{16} - \dfrac{(y-2)^2}{4} = 1$

28. $\dfrac{(y+1)^2}{16} - \dfrac{(x+3)^2}{9} = 1$

29. $\dfrac{(y-2)^2}{36} - \dfrac{(x+2)^2}{4} = 1$

30. $\dfrac{(x+1)^2}{4} - \dfrac{(y-1)^2}{4} = 1$

31. $\dfrac{x^2}{4} - (y-1)^2 = 1$

32. $(y+1)^2 - \dfrac{(x-3)^2}{4} = 1$

Exercises 33–36: Match the equation with its graph (a–d).

33. $\dfrac{(x-2)^2}{4} - \dfrac{(y+4)^2}{4} = 1$ **34.** $\dfrac{(x+1)^2}{4} - \dfrac{y^2}{9} = 1$

35. $\dfrac{(y+1)^2}{16} - \dfrac{(x-2)^2}{16} = 1$ **36.** $\dfrac{y^2}{25} - \dfrac{(x+1)^2}{9} = 1$

a.

b.

c.

d.

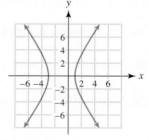

Exercises 37 and 38: Find an (approximate) equation of the hyperbola shown in the graph at the top of the next column. Identify the vertices, foci, and asymptotes.

37.

38.

Exercises 39–42: Sketch a graph of the hyperbola, including the asymptotes. Give the coordinates of the vertices and foci.

39. $\dfrac{(x-1)^2}{4} - \dfrac{(y-1)^2}{4} = 1$

40. $\dfrac{(x+2)^2}{4} - \dfrac{(y+1)^2}{16} = 1$

41. $\dfrac{(y+1)^2}{16} - \dfrac{(x-1)^2}{9} = 1$

42. $y^2 - \dfrac{(x-2)^2}{4} = 1$

Exercises 43–46: Find the standard equation of a hyperbola with center (h, k) that satisfies the given conditions.

43. Center $(2, -2)$, focus $(4, -2)$, and vertex $(3, -2)$

44. Center $(-1, 1)$, focus $(-1, 4)$, and vertex $(-1, 3)$

45. Vertices $(-1, \pm 1)$ and foci $(-1, \pm 3)$

46. Vertices $(2 \pm 1, 1)$ and foci $(2 \pm 3, 1)$

Exercises 47–54: Write the standard form for a hyperbola centered at (h, k). Identify the center and the vertices.

47. $x^2 - 2x - y^2 + 2y = 4$

48. $y^2 + 4y - x^2 + 2x = 6$

49. $3y^2 + 24y - 2x^2 + 12x + 24 = 0$

50. $4x^2 + 16x - 9y^2 + 18y = 29$

51. $x^2 - 6x - 2y^2 + 7 = 0$

52. $y^2 + 8y - 3x^2 + 13 = 0$

53. $4y^2 + 32y - 5x^2 - 10x + 39 = 0$

54. $5x^2 + 10x - 7y^2 + 28y = 58$

Graphing Hyperbolas with Technology

Exercises 55–58: Graph the hyperbola.

55. $\dfrac{(y-1)^2}{11} - \dfrac{x^2}{5.9} = 1$ **56.** $\dfrac{x^2}{5.3} - \dfrac{y^2}{6.7} = 1$

57. $3y^2 - 4x^2 = 15$ **58.** $2.1x^2 - 6y^2 = 12$

Solving Equations

Exercises 59–66: Solve the system of equations.

59. $x^2 - y^2 = 4$ **60.** $x^2 - 4y^2 = 16$
   $x^2 + y^2 = 9$ $x^2 + 4y^2 = 16$

61. $\dfrac{x^2}{4} - \dfrac{y^2}{9} = 1$ **62.** $x^2 - y^2 = 4$
   $x + y = 2$ $x + y = 2$

63. $8x^2 - 6y^2 = 24$ **64.** $3y^2 - 4x^2 = 12$
   $5x^2 + 3y^2 = 24$ $y^2 + 2x^2 = 34$

65. $\dfrac{y^2}{3} - \dfrac{x^2}{4} = 1$ **66.** $x^2 - 4y^2 = 16$
   $3x - y = 0$ $y^2 - 4x^2 = 4$

Applications

67. *Satellite Orbits*   The trajectory of a satellite near Earth can trace a hyperbola, parabola, or ellipse. If the satellite follows either a hyperbolic or a parabolic path, it escapes Earth's gravitational influence after a single pass. The path that a satellite travels near Earth depends on both its velocity V in meters per second and its distance D in meters from the center of Earth. Its path is hyperbolic if $V > \dfrac{k}{\sqrt{D}}$, parabolic if $V = \dfrac{k}{\sqrt{D}}$, and elliptic if $V < \dfrac{k}{\sqrt{D}}$, where $k = 2.82 \times 10^7$ is a constant.   (**Sources:** W. Loh, *Dynamics and Thermodynamics of Planetary Entry;* W. Thomson, *Introduction to Space Dynamics.*)

   (a) When *Explorer IV* was at a maximum distance of 42.5×10^6 meters from Earth's center, it had a velocity of 2090 meters/second. Determine the shape of its trajectory.

   (b) If an orbiting satellite is to escape Earth's gravity so that it can travel to another planet, its velocity must be increased so that its trajectory changes from elliptic to hyperbolic. What range of velocities would allow *Explorer IV* to leave Earth's influence when it is at a maximum distance?

   (c) Explain why it is easier to change a satellite's trajectory from an ellipse to a hyperbola when D is maximum rather than minimum.

68. *Telescopes*   (Refer to Figure 10.61.) Suppose that the coordinates of F_1 are $(0, 5.2)$ and the coordinates of F_2 are $(0, -5.2)$. If the coordinates of the vertex of the hyperbolic mirror are $(0, 4.1)$, find the standard equation of a hyperbola whose upper branch coincides with the hyperbolic mirror.

Writing about Mathematics

69. Explain how the center and the asymptotes of a hyperbola are related to the fundamental rectangle.

70. Given the standard equation of a hyperbola, explain how to determine the length of the transverse axis. How can you determine whether the transverse axis is vertical or horizontal?

EXTENDED AND DISCOVERY EXERCISES

1. *Structure of an Atom*   In 1911, Ernest Rutherford discovered the basic structure of the atom by "shooting" positively charged alpha particles with a speed of 10^7 meter per second at a piece of gold foil 6×10^{-7} meter thick. Only a small percentage of the alpha particles struck a gold nucleus head on and were deflected directly back toward their source. The rest of the particles often followed a *hyperbolic* trajectory because they were repelled by positively charged gold nuclei. The figure shows an alpha particle A initially approaching a gold nucleus N and being deflected at an angle $\theta = 90°$. N is located at a focus of the hyperbola, and the trajectory of A passes through a vertex of the hyperbola. (**Source:** H. Semat and J. Albright, *Introduction to Atomic and Nuclear Physics.*)

   (a) Determine the equation of the trajectory of the alpha particle if $d = 5 \times 10^{-14}$ meter.

   (b) What was the minimum distance between the centers of the alpha particle and the gold nucleus?

2. *Sound Detection* Microphones are placed at points $(-c, 0)$ and $(c, 0)$. An explosion occurs at point $P(x, y)$, which has a positive x-coordinate. See the figure.

    The sound is detected at the closer microphone t seconds before being detected at the farther microphone. Assume that sound travels at a speed of 330 meters per second, and show that P must be on the hyperbola

$$\frac{x^2}{330^2 t^2} - \frac{y}{4c^2 - 330^2 t^2} = \frac{1}{4}.$$

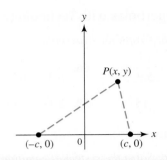

CHECKING BASIC CONCEPTS FOR SECTION 10.3

1. Use the graph to find the standard form of the hyperbola.

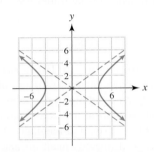

2. Graph the hyperbola defined by $\frac{x^2}{9} - \frac{y^2}{16} = 1$. Include the foci and asymptotes.

3. Find an equation of the hyperbola centered at $(1, 3)$ with a horizontal transverse axis of length 6 and a conjugate axis of length 4. Identify the foci.

4. Write $9y^2 - 54y - 16x^2 - 32x = 79$ in the standard form for a hyperbola centered at (h, k). Identify the center and the vertices.

10   Summary

CONCEPT	EXPLANATION AND EXAMPLES

SECTION 10.1   PARABOLAS

Conic Sections	Basic types: parabola, ellipse, circle, and hyperbola.
Parabolas with Vertex (0, 0)	***Standard Forms*** $x^2 = 4py$ (vertical axis) or $y^2 = 4px$ (horizontal axis) ***Meaning of p*** Both the vertex-focus distance and the vertex-directrix distance are p. The sign of p determines if the parabola opens upward or downward—or left or right. The focus is either $(0, p)$ or $(p, 0)$, and the directrix is either $y = -p$ or $x = -p$. In the figures below $p > 0$ but $p < 0$ is also possible.

Vertical Axis: $x^2 = 4py$

Horizontal Axis: $y^2 = 4px$

CONCEPT	EXPLANATION AND EXAMPLES

SECTION 10.1 PARABOLAS (CONTINUED)

Parabolas with Vertex (h, k)

$(x - h)^2 = 4p(y - k)$ — Vertical axis; focus: $(h, k + p)$
Directrix: $y = k - p$

$(y - k)^2 = 4p(x - h)$ — Horizontal axis; focus: $(h + p, k)$
Directrix: $x = h - p$

SECTION 10.2 ELLIPSES

Ellipses with Center (0, 0)

Standard Forms with $a > b > 0$

$\dfrac{x^2}{a^2} + \dfrac{y^2}{b^2} = 1$ (horizontal major axis)   or $\dfrac{x^2}{b^2} + \dfrac{y^2}{a^2} = 1$ (vertical major axis)

Meaning of a, b, and c

The distance from the center to a vertex is a, the distance from the center to an endpoint of the minor axis is b, and the distance from the center to a focus is c. The ratio $\frac{c}{a}$ equals the eccentricity e. The values of a, b, and c are related by $c^2 = a^2 - b^2$. The foci are either $(\pm c, 0)$ or $(0, \pm c)$, and the vertices are either $(\pm a, 0)$ or $(0, \pm a)$. If $a = b$, then the ellipse becomes a circle.

Horizontal Major Axis

Vertical Major Axis

Ellipses with Center (h, k)

$a > b > 0$; $c^2 = a^2 - b^2$

$\dfrac{(x - h)^2}{a^2} + \dfrac{(y - k)^2}{b^2} = 1$ — Major axis: horizontal; foci: $(h \pm c, k)$
Vertices: $(h \pm a, k)$

$\dfrac{(x - h)^2}{b^2} + \dfrac{(y - k)^2}{a^2} = 1$ — Major axis: vertical; foci: $(h, k \pm c)$
Vertices: $(h, k \pm a)$

Area Inside an Ellipse

The area A of the region contained inside an ellipse is given by $A = \pi ab$, where a and b are from the standard equation of the ellipse.

SECTION 10.3 HYPERBOLAS

Hyperbolas with Center (0, 0)

Standard Forms with both a and b positive

$\dfrac{x^2}{a^2} - \dfrac{y^2}{b^2} = 1$ (horizontal transverse axis)   or $\dfrac{y^2}{a^2} - \dfrac{x^2}{b^2} = 1$ (vertical transverse axis)

CONCEPT	EXPLANATION AND EXAMPLES

SECTION 10.3 HYPERBOLAS (CONTINUED)

Meaning of a, b, and c

The distance from the center to a vertex is a, and the distance from the center to a focus is c. The asymptotes are $y = \pm\frac{b}{a}x$ if the transverse axis is horizontal and $y = \pm\frac{a}{b}x$ if the transverse axis is vertical. The values of a, b, and c are related by $c^2 = a^2 + b^2$. The foci are either $(\pm c, 0)$ or $(0, \pm c)$, and the vertices are either $(\pm a, 0)$ or $(0, \pm a)$.

Horizontal Transverse Axis

Vertical Transverse Axis

| **Hyperbolas with Center (h, k)** | $a > 0, b > 0; c^2 = a^2 + b^2$ |

$$\frac{(x - h)^2}{a^2} - \frac{(y - k)^2}{b^2} = 1$$

Transverse axis: horizontal
Foci: $(h \pm c, k)$; vertices: $(h \pm a, k)$
Asymptotes: $y = \pm\frac{b}{a}(x - h) + k$

$$\frac{(y - k)^2}{a^2} - \frac{(x - h)^2}{b^2} = 1$$

Transverse axis: vertical
Foci: $(h, k \pm c)$; vertices: $(h, k \pm a)$
Asymptotes: $y = \pm\frac{a}{b}(x - h) + k$

10   Review Exercises

Exercises 1–6: Sketch a graph of the equation.

1. $-x^2 = y$

2. $y^2 = 2x$

3. $\dfrac{x^2}{25} + \dfrac{y^2}{49} = 1$

4. $\dfrac{y^2}{4} + \dfrac{x^2}{2} = 1$

5. $\dfrac{y^2}{4} - \dfrac{x^2}{9} = 1$

6. $x^2 - y^2 = 4$

Exercises 7–12: Match the equation with its graph (a–f).

7. $x^2 = 2y$

8. $y^2 = -3x$

9. $x^2 + y^2 = 4$

10. $\dfrac{x^2}{36} + \dfrac{y^2}{49} = 1$

11. $\dfrac{x^2}{4} - \dfrac{y^2}{9} = 1$

12. $\dfrac{y^2}{36} - \dfrac{x^2}{25} = 1$

a.

b.

c.

d.

e.

f.

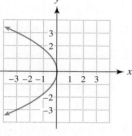

Exercises 13–18: Determine an equation of the conic section that satisfies the given conditions.

13. Parabola with focus $(2, 0)$ and vertex $(0, 0)$

14. Parabola with vertex $(5, 2)$ and focus $(5, 0)$

15. Ellipse with foci $(\pm 4, 0)$ and vertices $(\pm 5, 0)$

16. Ellipse centered at the origin with vertical major axis of length 14 and minor axis of length 8

17. Hyperbola with foci $(0, \pm 10)$ and endpoints of the conjugate axis $(\pm 6, 0)$

18. Hyperbola with vertices $(-2 \pm 3, 3)$ and foci given by $(-2 \pm 4, 3)$

Exercises 19–25: Sketch a graph of the conic section. Give the coordinates of any foci.

19. $-4y = x^2$

20. $y^2 = 8x$

21. $\dfrac{x^2}{25} + \dfrac{y^2}{4} = 1$

22. $49x^2 + 36y^2 = 1764$

23. $\dfrac{x^2}{16} - \dfrac{y^2}{9} = 1$

24. $\dfrac{y^2}{4} - x^2 = 1$

25. $(x - 3)^2 + (y + 1)^2 = 9$

Exercises 26–28: Sketch a graph of the conic section. Identify the coordinates of its center when appropriate.

26. $\dfrac{(y - 2)^2}{4} + \dfrac{(x + 1)^2}{16} = 1$

27. $\dfrac{(x - 1)^2}{4} - \dfrac{(y + 1)^2}{4} = 1$

28. $(x + 2) = 4(y - 1)^2$

29. Sketch a graph of $(y - 4)^2 = -8(x - 8)$. Include the focus and the directrix.

Exercises 30–32: Graph the equation.

30. $y^2 = \frac{3}{4}x$

31. $7.1x^2 + 8.2y^2 = 60$

32. $\dfrac{(y - 1.4)^2}{7} - \dfrac{(x + 2.3)^2}{11} = 1$

Exercises 33 and 34: Write the equation in the form given by $(y - k)^2 = a(x - h)$.

33. $-2x = y^2 + 8x + 14$

34. $2y^2 - 12y + 16 = x$

Exercises 35–38: Write the equation in standard form for an ellipse or a hyperbola centered at (h, k). Identify the center and the vertices.

35. $4x^2 + 8x + 25y^2 - 250y = -529$

36. $5x^2 + 20x + 2y^2 - 8y = -18$

37. $x^2 + 4x - 4y^2 + 24y = 36$

38. $4y^2 + 8y - 3x^2 + 6x = 11$

Exercises 39 and 40: Solve the system of equations.

39. $\dfrac{x^2}{4} + \dfrac{y^2}{4} = 1$
    $\dfrac{x^2}{8} + \dfrac{y^2}{2} = 1$

40. $x^2 - y^2 = 1$
    $x + y = 2$

Exercises 41 and 42: Shade the solution set to the system.

41. $\dfrac{x^2}{9} + \dfrac{y^2}{4} \le 1$
    $x + y \le 3$

42. $y^2 - x^2 \le 9$
    $y - x \le 0$

Applications

43. *Comets* A comet travels along an elliptical orbit around the sun. Its path can be described by the equation

$$\frac{x^2}{70^2} + \frac{y^2}{500^2} = 1,$$

where units are in millions of miles.

(a) What are the comet's minimum and maximum distance from the sun?

(b) Estimate the distance that the comet travels in one orbit around the sun. (Refer to Exercise 98 from Section 10.2.)

44. *Searchlight*   A searchlight is constructed in the shape of a paraboloid with a depth of 7 inches and a diameter of 20 inches, as illustrated in the figure. Determine the distance d that the bulb should be from the vertex in order to have the beam of light shine straight ahead.

45. *Arch Bridge*   An elliptical arch under a bridge is constructed so that it is 80 feet wide and has a maximum height of 30 feet, as illustrated in the figure. Find the height of the arch 10 feet from the center of the arch.

EXTENDED AND DISCOVERY EXERCISES

Exercises 1–5: Neptune and Pluto   *Both Neptune and Pluto travel around the sun in elliptical orbits. For Neptune's orbit, $a = 30.10$, and for Pluto's orbit, $a = 39.44$, where the variable a represents their average distances from the sun in astronomical units. (One astronomical unit equals 93 million miles.) The value of the variable a also corresponds to half the length of the major axis. Pluto has a highly eccentric orbit with $e = 0.249$, and Neptune has a nearly circular orbit with $e = 0.009$.*

1. Calculate the value of c for Neptune and Pluto.

2. Position the sun at the *origin* of the xy-plane. Find the coordinates of the center of Neptune's orbit and the coordinates of the center of Pluto's orbit. Assume that the centers lie on the positive x-axis.

3. Find equations for Neptune's orbit and for Pluto's orbit.

4. Graph both orbits in the same xy-plane.

5. Is Pluto always farther from the sun than Neptune? Explain.

Further Topics in Algebra

Mathematics permeates the fabric of modern society. Its presence is subtle at times and frequently missed by the average person, but its influence is nonetheless profound. Mathematics is the language of technology—mathematics allows society to quantify its experiences.

While technology and visualization provide new ways for us to investigate mathematical problems, they are *not* replacements for mathematical understanding. The human mind is capable of mathematical insight and decision making, but does not usually perform long arithmetic calculations proficiently. On the other hand, computers and calculators are incapable of mathematical insight, but are excellent at performing arithmetic and other routine computation. Together technology and the human mind can create amazing inventions.

In previous chapters we saw hundreds of examples of the use of mathematics to describe physical phenomena and events. Mathematics is diverse in its ability to adapt to new situations and solve complex problems. If any subject area is studied in enough detail, mathematics usually appears.

This chapter introduces further topics in mathematics. It represents only a small fraction of the topics found in mathematics. Although it may be difficult to predict exactly what the future will bring, one thing is certain—mathematics will continue to play a very important role.

Source: R. Wolff and L. Yaeger, *Visualization of Natural Phenomena.*

The art of asking the right questions in mathematics is more important than the art of solving them.
—Georg Cantor

11.1 Sequences

- **Understand basic concepts about sequences**
- **Learn how to represent sequences**
- **Identify and use arithmetic sequences**
- **Identify and use geometric sequences**

Introduction

Sequences are a fundamental concept in mathematics and have many applications. A sequence is a function that computes an ordered list. For example, the average person in the United States uses about 100 gallons of water each day. The function $f(n) = 100n$ generates the terms of the *sequence*

$$100, 200, 300, 400, 500, 600, 700, \ldots$$

when $n = 1, 2, 3, 4, 5, 6, 7, \ldots$. This *ordered list* represents the gallons of water used by the average person after n days.

Basic Concepts

A second example of a sequence involves investing money. If $100 is deposited into a savings account paying 5% interest compounded annually, then the function defined by $g(n) = 100(1.05)^n$ calculates the account balance after n years, which is given by

$$g(1), g(2), g(3), g(4), g(5), g(6), g(7), \ldots.$$

These terms can be approximated as

$$105, 110.25, 115.76, 121.55, 127.63, 134.01, 140.71, \ldots.$$

We now define a sequence formally.

Sequence

An **infinite sequence** is a function that has the set of natural numbers as its domain. A **finite sequence** is a function with domain $D = \{1, 2, 3, \ldots, n\}$ for some fixed natural number n.

Since sequences are functions, many of the concepts discussed in previous chapters apply to sequences. Instead of letting y represent the output, it is common to write $a_n = f(n)$, where n is a natural number in the domain of the sequence. The **terms** of a sequence are

$$a_1, a_2, a_3, \ldots, a_n, \ldots.$$

The first term is $a_1 = f(1)$, the second term is $a_2 = f(2)$, and so on. The **nth term**, or **general term**, of a sequence is $a_n = f(n)$.

EXAMPLE 1   Finding terms of sequences

Write the first four terms a_1, a_2, a_3, and a_4 of each sequence, where $a_n = f(n)$.

(a) $f(n) = 2n - 5$ **(b)** $f(n) = 4(2)^{n-1}$ **(c)** $f(n) = (-1)^n \left(\dfrac{n}{n+1} \right)$

SOLUTION

(a) Evaluate the following.

$$a_1 = f(1) = 2(1) - 5 = -3$$
$$a_2 = f(2) = 2(2) - 5 = -1$$

In a similar manner, $a_3 = f(3) = 1$ and $a_4 = f(4) = 3$.

(b) Since $f(n) = 4(2)^{n-1}$,

$$a_1 = f(1) = 4(2)^{1-1} = 4.$$

Similarly, $a_2 = 8$, $a_3 = 16$, and $a_4 = 32$.

(c) Let $f(n) = (-1)^n \left(\frac{n}{n+1}\right)$, and substitute $n = 1, 2, 3,$ and 4.

$$a_1 = f(1) = (-1)^1 \left(\frac{1}{1+1}\right) = -\frac{1}{2}$$

$$a_2 = f(2) = (-1)^2 \left(\frac{2}{2+1}\right) = \frac{2}{3}$$

$$a_3 = f(3) = (-1)^3 \left(\frac{3}{3+1}\right) = -\frac{3}{4}$$

$$a_4 = f(4) = (-1)^4 \left(\frac{4}{4+1}\right) = \frac{4}{5}$$

Note that the factor $(-1)^n$ causes the terms of the sequence to alternate signs.

Now Try Exercises 1, 3, and 9 ◀

Graphing Calculators and Sequences   Graphing calculators may be used to calculate the terms of a sequence. Figures 11.1–11.3 show the first four terms of each sequence in Example 1. The graphing calculator is in sequence mode.

```
seq(2n-5,n,1,4)
        {-3 -1 1 3}
```

Figure 11.1

```
seq(4(2)^(n-1),n
,1,4)
        {4 8 16 32}
```

Figure 11.2

```
seq((-1)^n*(n/(n
+1)),n,1,4)▶Frac
{-1/2 2/3 -3/4 ...
```

Figure 11.3

Recursive Sequences   Some sequences are not defined using a general term. Instead they are defined *recursively*. With a **recursive sequence**, we must find terms a_1 through a_{n-1} before we can find a_n.

EXAMPLE 2   Finding the terms of a recursive sequence

Find the first four terms of the recursive sequence that is defined by

$$a_n = 2a_{n-1} + 1; \qquad a_1 = 3.$$

SOLUTION   The sequence is defined recursively, so we must find the terms in order.

$$a_1 = 3$$

$$a_2 = 2a_1 + 1 = 2(3) + 1 = 7$$

$$a_3 = 2a_2 + 1 = 2(7) + 1 = 15$$

$$a_4 = 2a_3 + 1 = 2(15) + 1 = \mathbf{31}$$

The first four terms are 3, 7, 15, and 31.

Now Try Exercise 15(a) ◀

An Application A population model for an insect with a life span of 1 year can be described using a recursive sequence. Suppose each adult female insect produces r female offspring that survive to reproduce the following year. Let $f(n)$ calculate the insect population during year n. Then the number of female insects is given recursively by

$$f(n) = rf(n - 1) \quad \text{for } n > 1.$$

The number of female insects in the year n is equal to r times the number of female insects in the previous year $n - 1$. This symbolic representation of a function is fundamentally different from any other equation we have encountered thus far. The reason is that the function f is defined symbolically in terms of *itself*. To evaluate $f(n)$, we evaluate $f(n - 1)$. To evaluate $f(n - 1)$, we evaluate $f(n - 2)$, and so on. If we know the number of adult female insects during the first year, then we can determine the sequence. That is, if $f(1)$ is given, we can determine $f(n)$ by first computing

$$f(1), f(2), f(3), \ldots, f(n - 1).$$

The next example illustrates this recursively defined sequence. (**Source:** D. Brown and P. Rothery, *Models in Biology: Mathematics, Statistics and Computing.*)

EXAMPLE 3 Modeling an insect population

Suppose that the initial density of adult female insects is 1000 per acre and $r = 1.1$. Then the density of female insects during the year n is described by

$$f(1) = 1000$$
$$f(n) = 1.1f(n - 1), \quad n > 1.$$

(a) Rewrite this symbolic representation in terms of a_n.
(b) Find a_4 and interpret the result. Is the density of female insects increasing or decreasing?
(c) A general term for this sequence is given by $f(n) = 1000(1.1)^{n-1}$. Use this representation to find a_4.

SOLUTION
(a) Since $a_n = f(n)$ for all n, $a_{n-1} = f(n - 1)$. The sequence can be expressed as

$$a_1 = 1000$$
$$a_n = 1.1a_{n-1}, \quad n > 1.$$

(b) In order to calculate the fourth term, a_4, we must first determine a_1, a_2, and a_3.

$$a_1 = 1000$$
$$a_2 = 1.1a_1 = 1.1(1000) = 1100$$
$$a_3 = 1.1a_2 = 1.1(1100) = 1210$$
$$a_4 = 1.1a_3 = 1.1(1210) = 1331$$

CLASS DISCUSSION

How does the value of r in Example 3 affect the population density in future years?

The fourth term is $a_4 = 1331$. The female population density is increasing and reaches 1331 per acre during the fourth year.
(c) Since $a_4 = f(4)$,

$$a_4 = 1000(1.1)^{4-1} = 1331.$$

It is less work to find a_n using a formula for a general term rather than a recursive formula—particularly if n is large. *Now Try Exercise 89* ◀

Representations of Sequences

Sequences are functions. Therefore they have graphical, numerical, and symbolic representations.

EXAMPLE 4 Representing a sequence numerically and graphically

Let a recursive sequence be defined as follows.

$$a_1 = 3$$
$$a_n = 2a_{n-1} - 2, \quad n > 1$$

(a) Give a numerical representation (list each term in a table) for $n = 1, 2, 3, 4, 5$.
(b) Graph the first five terms of this sequence.

SOLUTION

(a) *Numerical Representation* Start by calculating the first five terms of the sequence.

$$a_1 = 3$$
$$a_2 = 2a_1 - 2 = 2(3) - 2 = 4$$
$$a_3 = 2a_2 - 2 = 2(4) - 2 = 6$$
$$a_4 = 2a_3 - 2 = 2(6) - 2 = 10$$
$$a_5 = 2a_4 - 2 = 2(10) - 2 = 18$$

The first five terms are 3, 4, 6, 10, and 18. A numerical representation of the sequence is shown in Table 11.1.

Table 11.1

n	1	2	3	4	5
a_n	3	4	6	10	18

(b) *Graphical Representation* To represent these terms graphically, plot the points $(1, 3)$, $(2, 4)$, $(3, 6)$, $(4, 10)$, and $(5, 18)$, as shown in Figure 11.4. Because the domain of a sequence contains only natural numbers, the graph of a sequence is a scatterplot.

Now Try Exercise 19 ◄

A graphing calculator set in sequence mode may be used to calculate the terms of the recursive sequence in Example 4. See Figures 11.5 and 11.6.

Figure 11.4

Calculator Help
To make a table or graph of a sequence, see Appendix A (page AP-18).

Figure 11.5

[0, 6, 1] by [0, 20, 4]

Figure 11.6 Dot Mode

An Application The next example illustrates numerical and graphical representations for a sequence involving population growth.

EXAMPLE 5 Representing a sequence numerically and graphically

Frequently the population of a particular insect does not continue to grow indefinitely, as it does in Example 3. Instead, the population grows rapidly at first and then levels off because of competition for limited resources. In one study, the behavior of the winter moth was modeled with a sequence similar to the following, where a_n represents the population density in thousands per acre at the beginning of year n. (**Source:** G. Varley and G. Gradwell, "Population models for the winter moth.")

$$a_1 = 1$$
$$a_n = 2.85a_{n-1} - 0.19a_{n-1}^2, \qquad n \geq 2$$

(a) Give a numerical representation for $n = 1, 2, 3, \ldots, 10$. Describe what happens to the population density of the winter moth.

(b) Use the numerical representation to graph the sequence.

n	u(n)
1	1
2	2.66
3	6.2366
4	10.384
5	9.1069
6	10.197
7	9.3056

$n = 1$

Figure 11.7

SOLUTION

(a) Evaluate $a_1, a_2, a_3, \ldots, a_{10}$ recursively. Since $a_1 = 1$,

$$a_2 = 2.85a_1 - 0.19a_1^2 = 2.85(1) - 0.19(1)^2 = 2.66 \qquad \text{and}$$
$$a_3 = 2.85a_2 - 0.19a_2^2 = 2.85(2.66) - 0.19(2.66)^2 \approx 6.24.$$

Approximate values for other terms are shown in Table 11.2. Figure 11.7 shows the sequence computed using a calculator. The sequence is denoted $u(n)$ rather than a_n.

Table 11.2

n	1	2	3	4	5	6	7	8	9	10
a_n	1	2.66	6.24	10.4	9.11	10.2	9.31	10.1	9.43	9.98

(b) The graph of a sequence is a set of discrete points. Plot the points

$$(1, 1), (2, 2.66), (3, 6.24), \ldots, (10, 9.98),$$

as shown in Figure 11.8. The insect population increases rapidly at first and then oscillates about the line $y = 9.7$. (See the Class Discussion in the margin.) The oscillations become smaller as n increases, indicating that the population density may stabilize near 9.7 thousand per acre. Some calculators can plot sequences, as shown in Figure 11.9. In this figure, the first 20 terms have been plotted in dot mode.

Figure 11.8 Insect Population

[0, 21, 1] by [0, 14, 1]

Figure 11.9   *Now Try Exercise 91* ◀

Arithmetic Sequences

Suppose that a person's starting salary is $20,000 per year. Thereafter this person receives a $1000 raise each year. The salary *after* n years of experience is represented by

$$f(n) = 1000n + 20,000,$$

where f is a linear function. After 10 years of experience, the annual salary would be

$$f(10) = 1000(10) + 20,000 = \$30,000.$$

If a sequence can be defined by a linear function, it is an *arithmetic sequence*.

Algebra Review
To review linear functions, see
Section 2.1.

Infinite Arithmetic Sequence

An **infinite arithmetic sequence** is a linear function whose domain is the set of natural numbers.

An arithmetic sequence can be defined recursively by $a_n = a_{n-1} + d$, where d is a constant. Since $d = a_n - a_{n-1}$ for each valid n, d is called the **common difference**. If $d = 0$, then the sequence is a **constant sequence**. A **finite arithmetic sequence** is similar to an infinite arithmetic sequence except that its domain is $D = \{1, 2, 3, \ldots, n\}$, where n is a fixed natural number.

EXAMPLE 6 Determining arithmetic sequences

Determine if f is an arithmetic sequence for each situation.
(a) $f(n) = n^2 + 3n$
(b) f as graphed in Figure 11.10
(c) The function f given in Table 11.3

Table 11.3

n	1	2	3	4	5	6	7
$f(n)$	-1.5	0	1.5	3	4.5	6	7.5

Figure 11.10

SOLUTION

Getting Started The formula for an arithmetic sequence is given by $f(n) = dn + c$, its graph lies on a line with slope d, and consecutive terms in its numerical representation always change by a common difference d. ▶

(a) This sequence is not arithmetic because $f(n) = n^2 + 3n$ is nonlinear.
(b) The sequence in Figure 11.10 represents an arithmetic sequence because the points lie on a line. A linear function could generate these points. Notice that the slope between points is always 2, which equals the common difference d.
(c) The successive terms, $-1.5, 0, 1.5, 3, 4.5, 6, 7.5$, increase by precisely 1.5. Therefore the common difference is $d = 1.5$. Since $a_n = a_{n-1} + 1.5$ for each valid n, the sequence is arithmetic.   ***Now Try Exercises 61, 65, and 67*** ◀

Since an arithmetic sequence is a linear function, it can always be represented by $f(n) = dn + c$, where d is the common difference and c is a constant.

EXAMPLE 7 Finding symbolic representations for arithmetic sequences

Find a general term $a_n = f(n)$ for each arithmetic sequence.

(a) $a_1 = 3$ and $d = -2$ **(b)** $a_3 = 4$ and $a_9 = 17$

SOLUTION
(a) Let $f(n) = dn + c$. Since $d = -2$, $f(n) = -2n + c$.

$$a_1 = f(1) = -2(1) + c = 3, \quad \text{or} \quad c = 5$$

Thus $a_n = -2n + 5$.

(b) Since $a_3 = 4$ and $a_9 = 17$, we find a linear function $f(n) = dn + c$ that satisfies the equations $f(3) = 4$ and $f(9) = 17$. The common difference is equal to the slope between the points $(3, 4)$ and $(9, 17)$.

$$d = \frac{17 - 4}{9 - 3} = \frac{13}{6}$$

It follows that $f(n) = \frac{13}{6}n + c$.

$$a_3 = f(3) = \frac{13}{6}(3) + c = 4, \quad \text{or} \quad c = -\frac{5}{2}$$

Thus $a_n = \frac{13}{6}n - \frac{5}{2}$.

Now Try Exercises 41 and 45 ◀

Finding the nth Term If a_1 is the first term of an arithmetic sequence and d is the common difference, then consecutive terms of the sequence are given by

$$a_2 = a_1 + d$$
$$a_3 = a_2 + d = a_1 + d + d = a_1 + 2d$$
$$a_4 = a_3 + d = a_1 + 2d + d = a_1 + 3d$$
$$a_5 = a_4 + d = a_1 + 3d + d = a_1 + 4d$$

and, in general, $a_n = a_1 + (n - 1)d$. This result is summarized in the following box.

---

nth Term of an Arithmetic Sequence

In an arithmetic sequence with first term a_1 and common difference d, the nth term, a_n, is given by

$$a_n = a_1 + (n - 1)d.$$

EXAMPLE 8 **Finding a formula for an arithmetic sequence**

Find a symbolic representation (formula) for the arithmetic sequence given by

$$9, 8.5, 8, 7.5, 7, 6.5, 6, \ldots.$$

SOLUTION The first term is 9. Successive terms can be found by subtracting 0.5 from (or adding -0.5 to) the previous term. Therefore $a_1 = 9$ and $d = -0.5$, and it follows that

$$a_n = a_1 + (n - 1)d \qquad \text{General formula}$$
$$= 9 + (n - 1)(-0.5) \qquad \text{Substitute.}$$
$$= -0.5n + 9.5. \qquad \text{Simplify.} \quad \textbf{Now Try Exercise 31(c)} ◀$$

Geometric Sequences

Suppose that a person with a starting salary of $20,000 per year receives a 5% raise each year. If $a_n = f(n)$ computes this salary at the *beginning* of the nth year, then

$$f(1) = 20,000$$
$$f(2) = 20,000(1.05) = 21,000$$
$$f(3) = 21,000(1.05) = 22,050$$
$$f(4) = 22,050(1.05) = 23,152.50$$

are salaries for the first 4 years. Each salary results from multiplying the previous salary by 1.05. A general term in the sequence can be written as

$$f(n) = 20{,}000(1.05)^{n-1}.$$

During the 10th year, the annual salary is

$$f(10) = 20{,}000(1.05)^{10-1} \approx \$31{,}027.$$

This type of sequence is a *geometric sequence* given by $f(n) = cr^{n-1}$, where c and r are constants. Geometric sequences are capable of either rapid growth or rapid decay. The first four terms from some geometric sequences are shown in Table 11.4. The corresponding values of c and r have been included.

The terms of a geometric sequence can be found by multiplying the previous term by r. We now define a geometric sequence formally.

Table 11.4 Terms of Geometric Sequences

c	r	a_1, a_2, a_3, a_4
1	2	1, 2, 4, 8
1	$\frac{1}{2}$	$1, \frac{1}{2}, \frac{1}{4}, \frac{1}{8}$
2	-4	$2, -8, 32, -128$
3	$\frac{1}{10}$	3, 0.3, 0.03, 0.003

Infinite Geometric Sequence

An **infinite geometric sequence** is a function defined by $f(n) = cr^{n-1}$, where c and r are nonzero constants. The domain of f is the set of natural numbers.

A geometric sequence can be defined recursively by $a_n = ra_{n-1}$, where $a_n = f(n)$ and the first term is $a_1 = c$. Since $r = \frac{a_n}{a_{n-1}}$ for each valid n, r is called the **common ratio**.

The next example illustrates how to recognize symbolic, graphical, and numerical representations of geometric sequences.

EXAMPLE 9   Determining geometric sequences

Decide which of the following represents a geometric sequence.
(a) $a_n = 4(0.5)^n$
(b) The sequence a_n represented graphically in Figure 11.11
(c) The numerical representation of a_n in Table 11.5

Table 11.5

n	1	2	3	4	5	6
a_n	1	-3	9	-27	81	-243

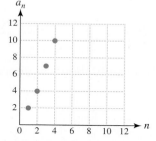

Figure 11.11

MAKING CONNECTIONS

Exponential Functions and Geometric Sequences If the domain of an exponential function $f(x) = Ca^x$ is restricted to the natural numbers, then f represents a geometric sequence. For example, $f(x) = 3(2)^x$ generates the geometric sequence $6, 12, 24, 48, 96, \ldots$ when $x = 1, 2, 3, 4, 5, \ldots$.

SOLUTION

Getting Started The formula for a geometric sequence is given by $a_n = cr^{n-1}$, and consecutive terms on its graph change by a common ratio r, as do consecutive terms in its numerical representation. ▶

(a) The formula for a_n can be written as

$$a_n = 4(0.5)^n = 4(0.5)(0.5)^{n-1} = 2(0.5)^{n-1}.$$

Thus a_n represents a geometric sequence with $c = 2$ and $r = 0.5$.

(b) The points on the graph are $(1, 2)$, $(2, 4)$, $(3, 7)$, and $(4, 10)$. Thus $a_1 = 2, a_2 = 4, a_3 = 7$, and $a_4 = 10$. Taking ratios of successive terms results in

$$\frac{a_2}{a_1} = 2, \qquad \frac{a_3}{a_2} = \frac{7}{4}, \qquad \text{and} \qquad \frac{a_4}{a_3} = \frac{10}{7}.$$

Since these ratios are not equal, there is no *common* ratio. The sequence is not geometric.

(c) The terms in Table 11.5 are $a_1 = 1$, $a_2 = -3$, $a_3 = 9$, $a_4 = -27$, $a_5 = 81$, and $a_6 = -243$. Note that these terms result from multiplying the previous term by -3. This sequence can be written either as

$$a_n = -3a_{n-1} \quad \text{with} \quad a_1 = 1$$

or as

$$a_n = (-3)^{n-1}.$$

Therefore the sequence is geometric. *Now Try Exercises 69, 73, and 75* ◀

EXAMPLE 10 Finding symbolic representations for geometric sequences

Find a general term a_n for each geometric sequence.
(a) $a_1 = 5$ and $r = 1.12$ **(b)** $a_2 = 8$ and $a_5 = 512$

SOLUTION
(a) Since the first term is $a_1 = c = 5$ and the common ratio is 1.12, $a_n = 5(1.12)^{n-1}$.
(b) We need to find $a_n = cr^{n-1}$ so that $a_2 = 8$ and $a_5 = 512$. Start by determining the common ratio r. Since

$$\frac{a_5}{a_2} = \frac{cr^{5-1}}{cr^{2-1}} = \frac{r^4}{r^1} = r^3 \quad \text{and} \quad \frac{a_5}{a_2} = \frac{512}{8} = 64,$$

it follows that $r^3 = 64$, or $r = 4$. So $a_n = c(4)^{n-1}$. Now

$$a_2 = c(4)^{2-1} = 8, \quad \text{or} \quad c = 2.$$

Thus $a_n = 2(4)^{n-1}$. *Now Try Exercises 51 and 55* ◀

11.1 Putting It All Together

The following table summarizes some fundamental concepts about sequences.

Concept	Explanation	Example
Infinite sequence	A function f whose domain is the set of natural numbers; denoted $a_n = f(n)$; the terms are $a_1, a_2, a_3, \ldots$.	$f(n) = n^2 - 2n$, where $a_n = f(n)$ The first three terms are $a_1 = 1^2 - 2(1) = -1$ $a_2 = 2^2 - 2(2) = 0$ $a_3 = 3^2 - 2(3) = 3$. Graphs of sequences are scatterplots.
Recursive sequence	Defined in terms of previous terms; a_1 through a_{n-1} must be calculated before a_n can be found.	$a_n = 2a_{n-1}$, $a_1 = 1$ $a_1 = 1$, $a_2 = 2$, $a_3 = 4$, and $a_4 = 8$ A new term is found by multiplying the previous term by 2.

Concept	Explanation	Example
Arithmetic sequence	A *linear* function whose domain is the natural numbers; $a_n = dn + c$ or $a_n = a_{n-1} + d$, with common difference d. General term is $a_n = a_1 + (n - 1)d$.	$f(n) = 2n - 1$, where $a_n = f(n)$ $a_1 = 1$, $a_2 = 3$, $a_3 = 5$, and $a_4 = 7$ Consecutive terms increase by the common difference $d = 2$. The points on the graph of this sequence lie on a line with slope 2.
Geometric sequence	$f(n) = cr^{n-1}$, where c is a nonzero constant and r is the nonzero common ratio; may also be written as $a_n = ra_{n-1}$	$f(n) = 2(3)^{n-1}$, where $a_n = f(n)$ $a_1 = 2$, $a_2 = 6$, $a_3 = 18$, and $a_4 = 54$ Consecutive terms are found by multiplying the previous term by the common ratio $r = 3$.

11.1 Exercises

Finding Terms of Sequences

Exercises 1–12: Find the first four terms of the sequence.

1. $a_n = 2n + 1$

2. $a_n = 3(n - 1) + 5$

3. $a_n = 4(-2)^{n-1}$

4. $a_n = 2(3)^n$

5. $a_n = \dfrac{n}{n^2 + 1}$

6. $a_n = 5 - \dfrac{1}{n^2}$

7. $a_n = (-1)^n \left(\dfrac{1}{2}\right)^n$

8. $a_n = (-1)^n \left(\dfrac{1}{n}\right)$

9. $a_n = (-1)^{n-1} \left(\dfrac{2^n}{1 + 2^n}\right)$

10. $a_n = (-1)^{n-1} \left(\dfrac{1}{3^n}\right)$

11. $a_n = 2^n + n^2$

12. $a_n = \dfrac{1}{n} + \dfrac{1}{3n}$

Exercises 13 and 14: Use the graphical representation to list the terms of the sequence.

13.

14.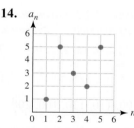

Exercises 15–28: Complete the following for the recursively defined sequence.

    (a) Find the first four terms.

    (b) Graph these terms.

15. $a_n = 2a_{n-1}; a_1 = 1$

16. $a_n = a_{n-1} + 5; a_1 = -4$

17. $a_n = a_{n-1} + 3; a_1 = -3$

18. $a_n = 2a_{n-1} + 1; a_1 = 1$

19. $a_n = 3a_{n-1} - 1; a_1 = 2$

20. $a_n = \frac{1}{2}a_{n-1}; a_1 = 16$

21. $a_n = a_{n-1} - a_{n-2}; a_1 = 2, a_2 = 5$

22. $a_n = 2a_{n-1} + a_{n-2}; a_1 = 0, a_2 = 1$

23. $a_n = a_{n-1}^2; a_1 = 2$

24. $a_n = \frac{1}{2}a_{n-1}^3 + 1; a_1 = 0$

25. $a_n = a_{n-1} + n; a_1 = 1$

26. $a_n = 3a_{n-1}^2; a_1 = 2$

27. $a_n = a_{n-1}a_{n-2}; a_1 = 2, a_2 = 3$

28. $a_n = 2a_{n-1}^2 + a_{n-2}; a_1 = 2, a_2 = 1$

Representations of Sequences

Exercises 29–34: The first five terms of an arithmetic sequence are given. Find

    (a) numerical,   (b) graphical,   and   (c) symbolic

representations of the sequence. Include at least eight terms of the sequence for the graphical and numerical representations.

29. $1, 3, 5, 7, 9$ **30.** $4, 1, -2, -5, -8$

31. $7.5, 6, 4.5, 3, 1.5$ **32.** $5.1, 5.5, 5.9, 6.3, 6.7$

33. $\frac{1}{2}, 2, \frac{7}{2}, 5, \frac{13}{2}$ **34.** $2, 4, 6, 8, 10$

Exercises 35–40: The first five terms of a geometric sequence are given. Find

    (a) numerical,   (b) graphical,   and   (c) symbolic

representations of the sequence. Include at least eight terms of the sequence for the graphical and numerical representations.

35. $8, 4, 2, 1, \frac{1}{2}$ **36.** $32, -8, 2, -\frac{1}{2}, \frac{1}{8}$

37. $\frac{3}{4}, \frac{3}{2}, 3, 6, 12$ **38.** $\frac{1}{27}, \frac{1}{9}, \frac{1}{3}, 1, 3$

39. $-\frac{1}{4}, -\frac{1}{2}, -1, -2, -4$ **40.** $9, 6, 4, \frac{8}{3}, \frac{16}{9}$

Exercises 41–50: Find a general term a_n for the arithmetic sequence.

41. $a_1 = 5, d = -2$ **42.** $a_1 = -3, d = 5$

43. $a_3 = 1, d = 3$ **44.** $a_4 = 12, d = -10$

45. $a_2 = 5, a_6 = 13$ **46.** $a_3 = 22, a_{17} = -20$

47. $a_1 = 8, a_4 = 17$ **48.** $a_1 = -2, a_5 = 8$

49. $a_5 = -4, a_8 = -2.5$ **50.** $a_3 = 10, a_7 = -4$

Exercises 51–60: Find a general term a_n for the geometric sequence.

51. $a_1 = 2, r = \frac{1}{2}$ **52.** $a_1 = 0.8, r = -3$

53. $a_3 = \frac{1}{32}, r = -\frac{1}{4}$ **54.** $a_4 = 3, r = 3$

55. $a_3 = 2, a_6 = \frac{1}{4}$ **56.** $a_2 = 6, a_4 = 24, r > 0$

57. $a_1 = -5, a_3 = -125, r < 0$

58. $a_1 = 10, a_2 = 2$

59. $a_2 = -1, a_7 = -32$ **60.** $a_2 = \frac{9}{4}, a_4 = \frac{81}{4}, r < 0$

Identifying Types of Sequences

Exercises 61–68: Determine if f is an arithmetic sequence.

61. $f(n) = 4 - 3n^3$ **62.** $f(n) = 2(n - 1)$

63. $f(n) = 4n - (3 - n)$ **64.** $f(n) = n^2 - n + 2$

65.   **66.**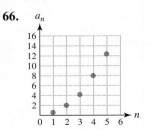

67.

n	1	2	3	4	5
$f(n)$	3	1	-1	-3	-5

68.

n	1	2	3	4	5
$f(n)$	1	4	9	16	25

Exercises 69–76: Determine if f is a geometric sequence.

69. $f(n) = 4(2)^{n-1}$ **70.** $f(n) = -3(0.25)^n$

71. $f(n) = -3(n)^2$ **72.** $f(n) = 2(n - 1)^n$

73.   **74.**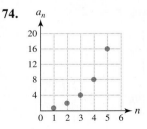

75.

n	1	2	3	4	5
$f(n)$	$\frac{1}{2}$	$\frac{3}{4}$	1	$\frac{5}{4}$	$\frac{5}{2}$

76.

n	1	2	3	4	5
$f(n)$	9	3	1	$\frac{1}{3}$	$\frac{1}{9}$

Exercises 77–82: Given the terms of a finite sequence, classify it as arithmetic, geometric, or neither.

77. $-5, 2, 9, 16, 23, 30$ **78.** $5, 2, -2, -6, -11$

79. $2, 8, 32, 128, 512$ **80.** $5.75, 5.5, 5.25, 5, 4.75, 4.5$

81. $100, 110, 130, 160, 200$

82. $0.7, 0.21, 0.063, 0.0189, 0.00567$

Exercises 83–86: Use the graph to determine if the sequence is arithmetic or geometric. If the sequence is arithmetic, state the sign of the common difference d and estimate its value. If the sequence is geometric, give the sign of the common ratio r and state if $|r| < 1$ or $|r| > 1$.

83.

84.

85.

86.

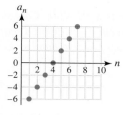

Modeling Insect and Bacteria Populations

Exercises 87 and 88: Insect Population The annual population density of a species of insect after n years is modeled by a sequence. Use the graph to discuss trends in the insect population.

87.

88.

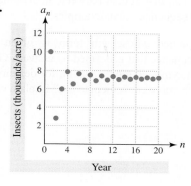

89. *Insect Population* (Refer to Example 3.) Suppose that the density of female insects during the first year is 500 per acre with $r = 0.8$.

  (a) Write a recursive sequence that describes these data, where a_n denotes the female insect density during year n.

  (b) Find the six terms $a_1, a_2, a_3, \ldots, a_6$. Interpret the results.

  (c) Find a formula for a_n.

90. *Bacteria Growth* Some strains of bacteria can double their size and divide every 40 minutes. (**Source:** F. Hoppensteadt and C. Peskin, *Mathematics in Medicine and the Life Sciences.*)

  (a) Write a recursive sequence that describes this growth where each value of n represents a 40-minute interval. Let $a_1 = 300$ represent the initial number of bacteria per milliliter. Find the first five terms.

  (b) Determine the number of bacteria per milliliter after 10 hours have elapsed.

  (c) Is this sequence arithmetic or geometric? Explain.

91. *Insect Population* (Refer to Example 5.) Suppose an insect population density at the beginning of year n can be modeled by the recursively defined sequence

$$a_1 = 8$$
$$a_n = 2.9a_{n-1} - 0.2a_{n-1}^2, \qquad n > 1.$$

  (a) Find the population for $n = 1, 2, 3$.

  (b) Graph the given sequence for $n = 1, 2, 3, \ldots, 20$. Interpret the graph.

92. *Bacteria Growth* (Refer to Exercise 90.) If bacteria are cultured in a medium with limited nutrients, competition ensues and growth slows. According to *Verhulst's model,* the number of bacteria at 40-minute intervals is given by

$$a_n = \left(\frac{2}{1 + a_{n-1}/K}\right)a_{n-1},$$

where K is a constant.

  (a) Let $a_1 = 200$ and $K = 10,000$. Graph the sequence for $n = 1, 2, 3, \ldots, 20$.

  (b) Describe the growth of these bacteria.

  (c) Trace the graph of the sequence. Make a conjecture as to why K is called the **saturation constant**. Test your conjecture by changing the value of K.

Applications

93. *Fibonacci Sequence* The *Fibonacci sequence* dates back to 1202. It is one of the most famous sequences in mathematics and can be defined recursively by

$$a_1 = 1, a_2 = 1$$
$$a_n = a_{n-1} + a_{n-2} \quad \text{for } n > 2.$$

(a) Find the first 12 terms of this sequence.

(b) Compute $\frac{a_n}{a_{n-1}}$ when $n = 2, 3, 4, \ldots, 12$. What happens to this ratio?

(c) Show that for $n = 2$, 3, and 4 the terms of the Fibonacci sequence satisfy the equation

$$a_{n-1} \cdot a_{n+1} - a_n^2 = (-1)^n.$$

94. *Bouncing Ball* If a tennis ball is dropped, it bounces, or rebounds, to 80% of its initial height.

(a) Write the first five terms of a sequence that gives the maximum height attained by the tennis ball on each rebound when it is dropped from an initial height of 5 feet. Let $a_1 = 5$. What type of sequence is this?

(b) Give a graphical representation of these terms.

(c) Find a general term a_n.

95. *Salary Increases* Suppose an employee's initial salary is $30,000.

(a) If this person receives a $2000 raise for each year of experience, determine a sequence that gives the salary at the beginning of the nth year. What type of sequence is this?

(b) Suppose another employee has the same starting salary and receives a 5% raise after each year. Find a sequence that computes the salary at the beginning of the nth year. What type of sequence is this?

(c) Which salary is higher at the beginning of the 10th year and the 20th year?

(d) Graph both sequences in the same viewing rectangle. Compare the two salaries.

96. *Area* A sequence of smaller squares is formed by connecting the midpoints of the sides of a larger square, as

shown in the figure. If the area of the largest square is one square unit, give the first five terms of a sequence that describes the area of each successive square. What type of sequence is this? Write an expression for the area of the nth square.

Exercises 97–100: Computing Square Roots The following recursively defined sequence can be used to compute $\sqrt{k}$ for any positive number k.

$$a_1 = k; \quad a_n = \frac{1}{2}\left(a_{n-1} + \frac{k}{a_{n-1}}\right)$$

This sequence was known to Sumerian mathematicians 4000 years ago, but it is still used today. Use this sequence to approximate the given square root by finding a_6. Compare your result with the actual value. (**Source:** P. Heinz-Otto, *Chaos and Fractals.*)

97. $\sqrt{2}$ **98.** $\sqrt{11}$

99. $\sqrt{21}$ **100.** $\sqrt{41}$

101. Suppose that a_n and b_n represent arithmetic sequences. Show that their sum, $c_n = a_n + b_n$, is also an arithmetic sequence.

102. Explain why the sequence log 2, log 4, log 8, log 16, … is an arithmetic sequence.

Writing about Mathematics

103. Explain how we can distinguish between an arithmetic and a geometric sequence. Give examples.

104. Compare a sequence whose nth term is given by $a_n = f(n)$ to a sequence that is defined recursively. Give examples. Which symbolic representation for defining a sequence is usually more convenient to use? Explain why.

11.2 Series

- Understand basic concepts about series
- Identify and find the sum of arithmetic series
- Identify and find the sum of geometric series
- Learn and use summation notation

Introduction

Although the terms *sequence* and *series* are sometimes used interchangeably in everyday English, they represent different mathematical concepts. In mathematics, a sequence is an *ordered list* (a function whose domain is the set of natural numbers), whereas a series is a summation of the terms in a sequence. Series have played a central role in the development of modern mathematics. Today series are often used to approximate functions that are too complicated to have simple formulas. Series are also instrumental in calculating approximations of numbers like π and e.

Basic Concepts

Suppose a person has a starting salary of $30,000 per year and receives a $2000 raise each year. Then

$$30,000, \ 32,000, \ 34,000, \ 36,000, \ 38,000$$

are terms of the finite sequence that describe this person's salaries over a 5-year period. The total earned is given by the finite *series*

$$30,000 + 32,000 + 34,000 + 36,000 + 38,000,$$

whose sum is $170,000. Any sequence can be used to define a series. For example, the *infinite sequence*

$$1, \frac{1}{3}, \frac{1}{9}, \frac{1}{27}, \frac{1}{81}, \frac{1}{243}, \ldots$$

defines the terms of the *infinite series*

$$1 + \frac{1}{3} + \frac{1}{9} + \frac{1}{27} + \frac{1}{81} + \frac{1}{243} + \cdots.$$

We now define the concept of a series, where $a_1, a_2, a_3, \ldots, a_n, \ldots$ represent terms of a sequence.

Series

A **finite series** is an expression of the form

$$a_1 + a_2 + a_3 + \cdots + a_n,$$

and an **infinite series** is an expression of the form

$$a_1 + a_2 + a_3 + \cdots + a_n + \cdots.$$

Partial Sums   An infinite series contains infinitely many terms. We must define what is meant by the sum of an *infinite* series. Let the following be a **sequence of partial sums**.

$$S_1 = a_1$$
$$S_2 = a_1 + a_2$$
$$S_3 = a_1 + a_2 + a_3$$
$$\vdots$$
$$S_n = a_1 + a_2 + a_3 + \cdots + a_n$$

If S_n approaches a real number S as $n \to \infty$, then the sum of the infinite series is S. For example, let $S_1 = 0.3$, $S_2 = 0.3 + 0.03$, $S_3 = 0.3 + 0.03 + 0.003$, and so on. Then, as $n \to \infty$, $S_n \to \frac{1}{3}$. We say that the infinite series

$$0.3 + 0.03 + 0.003 + 0.0003 + \cdots = 0.3333\ldots = 0.\overline{3}$$

has sum $\frac{1}{3}$. Some infinite series do not have a sum S. For example, the series given by $1 + 2 + 3 + 4 + 5 + \cdots$ would have an **unbounded**, or "infinite," sum.

EXAMPLE 1 Finding partial sums

For each a_n, calculate S_4.
(a) $a_n = 2n + 1$ **(b)** $a_n = n^2$

SOLUTION
(a) Because $S_4 = a_1 + a_2 + a_3 + a_4$, start by calculating the first four terms of the sequence $a_n = 2n + 1$.

$$a_1 = 2(1) + 1 = 3; \qquad a_2 = 2(2) + 1 = 5;$$
$$a_3 = 2(3) + 1 = 7; \qquad a_4 = 2(4) + 1 = 9$$

Thus $S_4 = \underbrace{3 + 5 + 7 + 9}_{4 \text{ terms}} = 24.$

(b) $a_1 = 1^2 = 1; \qquad a_2 = 2^2 = 4; \qquad a_3 = 3^2 = 9; \qquad a_4 = 4^2 = 16$
Thus $S_4 = 1 + 4 + 9 + 16 = 30.$ ***Now Try Exercises 7 and 11*** ◀

Calculating π   The techniques used to calculate π throughout history make a fascinating story. Since π is an irrational number, it cannot be represented exactly by a fraction. Its decimal expansion neither repeats nor has a discernible pattern. The ability to compute π was essential to the development of a society, because π appears in formulas used in construction, surveying, and geometry. In early historical records, π was given the value of 3. Later the Egyptians used a value of $\frac{256}{81} \approx 3.1605$. It was not until the discovery of series that exceedingly accurate decimal approximations of π were possible. In 2002, after 400 hours of supercomputer time, π was computed to 1.24 trillion digits. Why would anyone want to compute π to so many decimal places? One practical reason is to test electrical circuits in new computers. If a computer has a small defect in its hardware, there is a good chance that an error will appear after trillions of arithmetic calculations are performed during the computation of π.   (**Sources:** P. Beckmann, *A History of Pi;* P. Heinz-Otto, *Chaos and Fractals.*)

EXAMPLE 2 Computing π with a series

The infinite series given by

$$\frac{\pi^4}{90} = \frac{1}{1^4} + \frac{1}{2^4} + \frac{1}{3^4} + \frac{1}{4^4} + \frac{1}{5^4} + \cdots + \frac{1}{n^4} + \cdots$$

can be used to estimate π.
(a) Approximate π by finding the sum of the first four terms.
(b) Use technology to approximate π by summing the first 50 terms. Compare the result to the actual value of π.

SOLUTION

(a) Summing the first four terms results in the following.

$$\frac{\pi^4}{90} \approx \frac{1}{1^4} + \frac{1}{2^4} + \frac{1}{3^4} + \frac{1}{4^4} \approx 1.078751929$$

This approximation can be solved for π by multiplying by 90 and then taking the fourth root. Thus

$$\pi \approx \sqrt[4]{90(1.078751929)} \approx 3.139.$$

Figure 11.12

(b) Some calculators are capable of summing the terms of a sequence, as shown in Figure 11.12. (Summing the terms of a sequence is equivalent to finding the sum of a series.) The first 50 terms of the series provide an approximation of $\pi \approx 3.141590776$. This computation matches the actual value of π for the first five decimal places.

Now Try Exercise 107 ◀

Arithmetic Series

Calculator Help

To find the sum of the terms of a series, see Appendix A (page AP-19).

Summing the terms of an arithmetic sequence results in an **arithmetic series.** For example, the sequence defined by $a_n = 2n - 1$ for $n = 1, 2, 3, \ldots, 7$ is the arithmetic sequence

$$1, 3, 5, 7, 9, 11, 13.$$

The corresponding arithmetic series is

$$1 + 3 + 5 + 7 + 9 + 11 + 13.$$

The following formula can be used to sum the first n terms of an arithmetic sequence. (For a proof, see Exercise 3 in the Extended and Discovery Exercises at the end of this chapter.)

Sum of the First n Terms of an Arithmetic Sequence

The **sum of the first n terms of an arithmetic sequence**, denoted S_n, is found by averaging the first and nth terms and then multiplying by n. That is,

$$S_n = a_1 + a_2 + a_3 + \cdots + a_n = n\left(\frac{a_1 + a_n}{2}\right).$$

Since $a_n = a_1 + (n - 1)d$, S_n can also be written as follows.

$$S_n = n\left(\frac{a_1 + a_n}{2}\right)$$

$$= \frac{n}{2}(a_1 + a_1 + (n - 1)d)$$

$$= \frac{n}{2}(2a_1 + (n - 1)d)$$

EXAMPLE 3 Finding the sum of a finite arithmetic series

Use a formula to find the sum of the arithmetic series

$$2 + 4 + 6 + 8 + \cdots + 100.$$

SOLUTION The series $2 + 4 + 6 + 8 + \cdots + 100$ has $n = 50$ terms with $a_1 = 2$ and $a_{50} = 100$. We can use the formula

$$S_n = n\left(\frac{a_1 + a_n}{2}\right)$$

to find its sum.

$$S_{50} = 50\left(\frac{2 + 100}{2}\right) = 2550$$

We can also use the formula

$$S_n = \frac{n}{2}(2a_1 + (n - 1)d)$$

with common difference $d = 2$ to find this sum.

$$S_{50} = \frac{50}{2}(2(2) + (50 - 1)2) = 2550$$

The two answers agree, as expected.   *Now Try Exercise 15* ◀

EXAMPLE 4   Finding the sum of a finite arithmetic series

A person has a starting annual salary of \$30,000 and receives a \$1500 raise each year.
(a) Calculate the total amount earned over 10 years.
(b) Verify this value using a calculator.

SOLUTION
(a) The arithmetic sequence describing the salary during year n is computed by

$$a_n = 30,000 + 1500(n - 1).$$

The first and tenth year's salaries are

$$a_1 = 30,000 + 1500(1 - 1) = 30,000$$
$$a_{10} = 30,000 + 1500(10 - 1) = 43,500.$$

Thus the total amount earned during this 10-year period is

$$S_{10} = 10\left(\frac{30,000 + 43,500}{2}\right) = \$367,500.$$

This sum can also be found using $S_n = \frac{n}{2}(2a_1 + (n - 1)d)$.

$$S_{10} = \frac{10}{2}(2 \cdot 30,000 + (10 - 1)1500) = \$367,500$$

(b) To verify this result with a calculator, compute the sum

$$a_1 + a_2 + a_3 + \cdots + a_{10},$$

where $a_n = 30,000 + 1500(n - 1)$. This calculation is shown in Figure 11.13. The result of 367,500 agrees with part (a).   *Now Try Exercise 95* ◀

```
sum(seq(30000+15
00(n−1),n,1,10))
              367500
```

Figure 11.13

EXAMPLE 5   Finding a term of an arithmetic series

The sum of an arithmetic series with 15 terms is 285. If $a_{15} = 40$, find a_1.

SOLUTION To find a_1, we apply the sum formula

$$S_n = n\left(\frac{a_1 + a_n}{2}\right) = a_1 + a_2 + a_3 + \cdots + a_n$$

with $n = 15$ and $a_{15} = 40$.

$$15\left(\frac{a_1 + 40}{2}\right) = 285$$

$$15(a_1 + 40) = 570 \qquad \text{Multiply by 2.}$$

$$a_1 + 40 = 38 \qquad \text{Divide by 15.}$$

$$a_1 = -2 \qquad \text{Subtract 40.} \qquad \boxed{\textit{Now Try Exercise 23}} \blacktriangleleft$$

CLASS DISCUSSION

Explain why a formula for the sum of an infinite arithmetic series is not given.

Geometric Series

What will happen if we attempt to find the sum of the terms of an infinite geometric sequence? The mathematical concept of infinity dates back to at least the paradoxes of Zeno (450 B.C.). The following problem illustrates a difficulty that occurred in attempting to comprehend infinity.

Suppose that a person walked 1 mile on the first day, $\frac{1}{2}$ mile the second day, $\frac{1}{4}$ mile the third day, and so on. How far down the road would this person travel? This distance is described by the infinite series

$$1 + \frac{1}{2} + \frac{1}{4} + \frac{1}{8} + \frac{1}{16} + \frac{1}{32} + \frac{1}{64} + \cdots.$$

Does the sum of an infinite number of positive values always become infinitely large? Problems like this took centuries for mathematicians to solve. (**Source:** H. Eves, *An Introduction to the History of Mathematics.*)

Finite Geometric Series In a manner analogous to the way an arithmetic series was defined, a **geometric series** is defined as the sum of the terms of a geometric sequence. In order to calculate sums of infinite geometric series, we begin by finding sums of finite geometric series.

Any finite geometric sequence can be written as

$$a_1, \ a_1 r, \ a_1 r^2, \ a_1 r^3, \ \ldots, \ a_1 r^{n-1}.$$

The summation of these n terms is a finite geometric series. Its sum S_n is expressed by

$$S_n = a_1 + a_1 r + a_1 r^2 + a_1 r^3 + \cdots + a_1 r^{n-1}.$$

To find the value of S_n, multiply this equation by r.

$$r S_n = a_1 r + a_1 r^2 + a_1 r^3 + \cdots + a_1 r^{n-1} + a_1 r^n$$

Subtracting this equation from the previous equation results in

$$S_n - r S_n = a_1 - a_1 r^n$$

$$S_n(1 - r) = a_1(1 - r^n)$$

$$S_n = a_1\left(\frac{1 - r^n}{1 - r}\right), \qquad \text{provided } r \neq 1.$$

This formula can be used to find the sum of the first n terms of a geometric sequence.

Sum of the First n Terms of a Geometric Sequence

If a geometric sequence has first term a_1 and common ratio r, then the sum of the first n terms is given by

$$S_n = a_1 \left(\frac{1 - r^n}{1 - r} \right), \qquad r \neq 1.$$

EXAMPLE 6 Finding the sum of finite geometric series

Approximate the sum for the given values of n.

(a) $1 + \frac{1}{2} + \frac{1}{4} + \cdots + \left(\frac{1}{2}\right)^{n-1}$; $n = 5, 10,$ and 20

(b) $3 - 6 + 12 - 24 + 48 - \cdots + 3(-2)^{n-1}$; $n = 3, 8,$ and 13

SOLUTION

(a) This geometric series has $a_1 = 1$ and $r = \frac{1}{2} = 0.5$.

$$S_5 = 1 \left(\frac{1 - 0.5^5}{1 - 0.5} \right) = 1.9375$$

$$S_{10} = 1 \left(\frac{1 - 0.5^{10}}{1 - 0.5} \right) \approx 1.998047$$

$$S_{20} = 1 \left(\frac{1 - 0.5^{20}}{1 - 0.5} \right) \approx 1.999998$$

(b) This geometric series has $a_1 = 3$ and $r = -2$.

$$S_3 = 3 \left(\frac{1 - (-2)^3}{1 - (-2)} \right) = 9$$

$$S_8 = 3 \left(\frac{1 - (-2)^8}{1 - (-2)} \right) = -255$$

$$S_{13} = 3 \left(\frac{1 - (-2)^{13}}{1 - (-2)} \right) = 8193$$

Now Try Exercises 41 and 43 ◀

Annuities A sequence of deposits made at equal time intervals is called an **annuity**. Suppose A_0 dollars is deposited at the end of each year into an account that pays an annual interest rate i compounded annually. At the end of the first year, the account contains A_0 dollars. At the end of the second year, A_0 dollars would be deposited again. In addition, the first deposit of A_0 dollars would have received interest during the second year. Therefore the value of the annuity after 2 years is

$$A_0 + A_0(1 + i).$$

After 3 years the balance is

$$A_0 + A_0(1 + i) + A_0(1 + i)^2,$$

and after n years this amount is given by

$$A_0 + A_0(1 + i) + A_0(1 + i)^2 + \cdots + A_0(1 + i)^{n-1}.$$

This is a geometric series with first term $a_1 = A_0$ and common ratio $r = (1 + i)$. The sum of the first n terms is given by

$$S_n = A_0\left(\frac{1 - (1 + i)^n}{1 - (1 + i)}\right) = A_0\left(\frac{(1 + i)^n - 1}{i}\right).$$

EXAMPLE 7 Finding the future value of an annuity

Suppose that a 20-year-old worker deposits $1000 into an account at the end of each year until age 65. If the interest rate is 12%, find the future value of the annuity.

SOLUTION Let $A_0 = 1000$, $i = 0.12$, and $n = 45$. The future value of the annuity is given by

$$S_n = A_0\left(\frac{(1 + i)^n - 1}{i}\right)$$

$$= 1000\left(\frac{(1 + 0.12)^{45} - 1}{0.12}\right)$$

$$\approx \$1,358,230.$$

Now Try Exercise 97 ◀

Infinite Geometric Series In Example 6 the value of r affects the sum of a finite geometric series. If $|r| > 1$, as in part (b), then r^n becomes large in absolute value for increasing n. As a result, the sum of the series also becomes large in absolute value. On the other hand, in part (a) the common ratio satisfies $|r| < 1$. In this case the values of r^n become closer to 0 as n increases. For large values of n and $|r| < 1$,

$$S_n = a_1\left(\frac{1 - r^n}{1 - r}\right) \approx a_1\left(\frac{1 - 0}{1 - r}\right) = \frac{a_1}{1 - r}.$$

This result is summarized in the following box.

Sum of an Infinite Geometric Sequence

The sum of the infinite geometric sequence with first term a_1 and common ratio r is given by

$$S = \frac{a_1}{1 - r},$$

provided $|r| < 1$. If $|r| \geq 1$, then this sum does not exist.

Infinite series can be used to describe repeating decimals. For example, the fraction $\frac{5}{9}$ can be written as the repeating decimal $0.555555\ldots$. This decimal can be expressed as an infinite series.

$$\frac{5}{9} = 0.5 + 0.05 + 0.005 + 0.0005 + \cdots + 0.5(0.1)^{n-1} + \cdots$$

In this series $a_1 = 0.5$ and $r = 0.1$. Since $|r| < 1$, the sum exists and is given by

$$S = \frac{0.5}{1 - 0.1} = \frac{5}{9},$$

as expected.

We are now able to answer the question concerning how far a person will walk if he or she travels 1 mile on the first day, $\frac{1}{2}$ mile the second day, $\frac{1}{4}$ mile the third day, and so on.

EXAMPLE 8 Finding the sum of an infinite geometric series

Find the sum of the infinite geometric series

$$1 + \frac{1}{2} + \frac{1}{4} + \frac{1}{8} + \cdots.$$

SOLUTION In this series, the first term is $a_1 = 1$ and the common ratio is $\frac{1}{2} = 0.5$. Its sum is

$$S = \frac{a_1}{1 - r} = \frac{1}{1 - 0.5} = 2.$$

If it were possible to walk in the prescribed manner, the total distance traveled after each day would always be less than 2 miles.

Now Try Exercise 45 ◄

MAKING CONNECTIONS

Sequences and Series A sequence is an *ordered* list, whereas a series is a sum of the terms of a sequence. For example,

$$1, 3, 5, 7, 9, 11, 13, 15$$

is a sequence, and

$$1 + 3 + 5 + 7 + 9 + 11 + 13 + 15$$

is a series.

Summation Notation

Summation notation is used to write series efficiently. The symbol Σ, the uppercase Greek letter *sigma*, indicates a sum.

Summation Notation

$$\sum_{k=1}^{n} a_k = a_1 + a_2 + a_3 + \cdots + a_n$$

The letter k is called the **index of summation**. The numbers 1 and n represent the subscripts of the first and last terms in the series. They are called the **lower limit** and **upper limit** of the summation, respectively.

EXAMPLE 9 Using summation notation

Evaluate each series.

(a) $\displaystyle\sum_{k=1}^{5} k^2$ **(b)** $\displaystyle\sum_{k=1}^{4} 5$ **(c)** $\displaystyle\sum_{k=3}^{6} (2k - 5)$

SOLUTION

(a) $\displaystyle\sum_{k=1}^{5} k^2 = 1^2 + 2^2 + 3^2 + 4^2 + 5^2 = 55$

(b) $\displaystyle\sum_{k=1}^{4} 5 = 5 + 5 + 5 + 5 = 20$

(c) $\displaystyle\sum_{k=3}^{6} (2k - 5) = \underset{k\,=\,3}{(2(3) - 5)} + \underset{k\,=\,4}{(2(4) - 5)} + \underset{k\,=\,5}{(2(5) - 5)} + \underset{k\,=\,6}{(2(6) - 5)}$

$$= 1 + 3 + 5 + 7 = 16$$

Now Try Exercises 61, 63, and 67 ◄

EXAMPLE 10 Writing a series in summation notation

Write the series using summation notation. Let the lower limit equal 1.

(a) $\dfrac{1}{2^3} + \dfrac{1}{3^3} + \dfrac{1}{4^3} + \dfrac{1}{5^3} + \dfrac{1}{6^3} + \dfrac{1}{7^3} + \dfrac{1}{8^3}$

(b) $\dfrac{1}{2} + \dfrac{2}{3} + \dfrac{3}{4} + \dfrac{4}{5} + \dfrac{5}{6} + \dfrac{6}{7} + \dfrac{7}{8}$

SOLUTION

(a) The terms of the series can be written as $\dfrac{1}{(k + 1)^3}$ for $k = 1, 2, 3, \ldots, 7$. Thus

$$\frac{1}{2^3} + \frac{1}{3^3} + \frac{1}{4^3} + \frac{1}{5^3} + \frac{1}{6^3} + \frac{1}{7^3} + \frac{1}{8^3} = \sum_{k=1}^{7} \frac{1}{(k + 1)^3}.$$

MAKING CONNECTIONS

Notation The expressions

$$\sum_{k=1}^{n} a_k \text{ and } \Sigma_{k=1}^{n}\, a_k$$

are equivalent.

(b) The terms of the series can be written as $\dfrac{k}{k + 1}$ for $k = 1, 2, 3, \ldots, 7$. Thus

$$\frac{1}{2} + \frac{2}{3} + \frac{3}{4} + \frac{4}{5} + \frac{5}{6} + \frac{6}{7} + \frac{7}{8} = \sum_{k=1}^{7} \frac{k}{k + 1}.$$

Now Try Exercises 69 and 71 ◄

EXAMPLE 11 Shifting the index of a series

Rewrite each summation so that the index starts with $n = 1$.

(a) $\displaystyle\sum_{k=4}^{7} k^2$ (b) $\displaystyle\sum_{k=8}^{30} (2k - 3)$

SOLUTION

(a) **Getting Started** Because $\Sigma_{k=4}^{7} k^2 = 4^2 + 5^2 + 6^2 + 7^2$, the summation has four terms. Instead of letting $k = 4, 5, 6, 7$, we must rewrite the sum so that $n = 1, 2, 3, 4$. It follows that $n + 3 = k$, so substitute $n + 3$ for k in the expression. ▶

$$\sum_{k=4}^{7} k^2 = \sum_{n=1}^{4} (n + 3)^2$$

$$= (1 + 3)^2 + (2 + 3)^2 + (3 + 3)^2 + (4 + 3)^2$$

$$= 4^2 + 5^2 + 6^2 + 7^2$$

(b) Instead of letting $k = 8, 9, 10, \ldots, 30$, we must rewrite the given sum so that $n = 1, 2, 3, \ldots, 23$. Thus $n + 7 = k$.

$$\sum_{k=8}^{30} (2k - 3) = \sum_{n=1}^{23} (2(n + 7) - 3)$$

$$= \sum_{n=1}^{23} (2n + 11) \qquad \textit{Now Try Exercises 75 and 77} \blacktriangleleft$$

The following box lists properties for summation notation.

Properties for Summation Notation

Let $a_1, a_2, a_3, \ldots, a_n$ and $b_1, b_2, b_3, \ldots, b_n$ be sequences, and c be a constant.

1. $\displaystyle\sum_{k=1}^{n} ca_k = c \sum_{k=1}^{n} a_k$ **4.** $\displaystyle\sum_{k=1}^{n} c = nc$

2. $\displaystyle\sum_{k=1}^{n} (a_k + b_k) = \sum_{k=1}^{n} a_k + \sum_{k=1}^{n} b_k$ **5.** $\displaystyle\sum_{k=1}^{n} k = \frac{n(n + 1)}{2}$

3. $\displaystyle\sum_{k=1}^{n} (a_k - b_k) = \sum_{k=1}^{n} a_k - \sum_{k=1}^{n} b_k$ **6.** $\displaystyle\sum_{k=1}^{n} k^2 = \frac{n(n + 1)(2n + 1)}{6}$

These properties can be used to find sums, as illustrated in the next example.

EXAMPLE 12 Applying summation notation

Use properties for summation notation to find each sum.

(a) $\displaystyle\sum_{k=1}^{40} 5$ **(b)** $\displaystyle\sum_{k=1}^{22} 2k$ **(c)** $\displaystyle\sum_{k=1}^{14} (2k^2 - 3)$

SOLUTION

(a) $\displaystyle\sum_{k=1}^{40} 5 = 40(5) = 200$ *Property 4 with $n = 40$ and $c = 5$*

(b) $\displaystyle\sum_{k=1}^{22} 2k = 2 \sum_{k=1}^{22} k$ *Property 1 with $c = 2$ and $a_k = k$*

$\qquad = 2 \cdot \dfrac{22(22 + 1)}{2}$ *Property 5 with $n = 22$*

$\qquad = 506$

(c) $\displaystyle\sum_{k=1}^{14} (2k^2 - 3) = \sum_{k=1}^{14} 2k^2 - \sum_{k=1}^{14} 3$ *Property 3 with $a_k = 2k^2$ and $b_k = 3$*

$\qquad = 2 \sum_{k=1}^{14} k^2 - \sum_{k=1}^{14} 3$ *Property 1 with $c = 2$ and $a_k = k^2$*

$\qquad = 2 \cdot \dfrac{14(14 + 1)(2 \cdot 14 + 1)}{6} - 14(3)$ *Properties 6 and 4*

$\qquad = 1988$

$\textit{Now Try Exercises 81, 83, and 89} \blacktriangleleft$

11.2 Putting It All Together

The following table summarizes concepts related to series.

Concept	Explanation	Example
Series	A series is the summation of the terms of a sequence. A finite series always has a sum, but an infinite series may not have a sum.	$2 + 4 + 6 + 8 + \cdots + 20$ $S_4 = a_1 + a_2 + a_3 + a_4$ (partial sum) $= 2 + 4 + 6 + 8$ $= 20$
Arithmetic series	An arithmetic series is the summation of the terms of an arithmetic sequence. The sum of the first n terms is given by $$S_n = n\left(\frac{a_1 + a_n}{2}\right)$$ or $$S_n = \frac{n}{2}(2a_1 + (n-1)d).$$	$1 + 4 + 7 + 10 + 13 + 16 + 19$ (7 terms) $S_7 = 7\left(\frac{a_1 + a_7}{2}\right) = 7\left(\frac{1 + 19}{2}\right) = 70$ or $S_7 = \frac{7}{2}(2a_1 + 6d) = \frac{7}{2}(2(1) + 6(3)) = 70$
Geometric series	A geometric series is the summation of the terms of a geometric sequence. The sum of the first n terms is given by $$S_n = a_1\left(\frac{1 - r^n}{1 - r}\right).$$ If $|r| < 1$, then an infinite geometric series has the sum $$S = \frac{a_1}{1 - r}.$$	$3 + 6 + 12 + 24 + 48 + 96$ (6 terms) $S_6 = 3\left(\frac{1 - 2^6}{1 - 2}\right) = 189$ The infinite geometric series $1 + \frac{1}{4} + \frac{1}{16} + \frac{1}{64} + \cdots$ has a sum $S = \frac{1}{1 - \frac{1}{4}} = \frac{4}{3}.$
Summation notation	The series $a_1 + a_2 + \cdots + a_n$ can be written as $$\sum_{k=1}^{n} a_k.$$	$1^2 + 2^2 + 3^2 + 4^2 + \cdots + 10^2$ can be written as $$\sum_{k=1}^{10} k^2.$$

11.2   Exercises

Concepts

1. Write a sequence whose terms represent the first six positive even integers.

2. Write a sequence whose terms represent the first seven positive odd integers.

3. Write a series that represents the sum of the first six positive even integers. Find its sum.

4. Write a series that represents the sum of the first seven positive odd integers. Find its sum.

Exercises 5 and 6: Let A_n represent the number of U.S. AIDS deaths reported n years after 2000.

5. Write a series whose sum gives the cumulative number of AIDS deaths from 2005 to 2009.

6. Explain what S_6 represents.

Finding Sums of Series

Exercises 7–14: For the given a_n, calculate S_5.

7. $a_n = 3n$

8. $a_n = n + 4$

9. $a_n = 2n - 1$

10. $a_n = 4n + 1$

11. $a_n = n^2 + 1$

12. $a_n = 2n^2$

13. $a_n = \dfrac{n}{n + 1}$

14. $a_n = \dfrac{1}{2n}$

Exercises 15–22: Use a formula to find the sum of the arithmetic series.

15. $3 + 5 + 7 + 9 + 11 + 13 + 15 + 17$

16. $7.5 + 6 + 4.5 + 3 + 1.5 + 0 + (-1.5)$

17. $1 + 2 + 3 + 4 + \cdots + 50$

18. $1 + 3 + 5 + 7 + \cdots + 97$

19. $-7 + (-4) + (-1) + 2 + 5 + \cdots + 98 + 101$

20. $89 + 84 + 79 + 74 + \cdots + 9 + 4$

21. The first 40 terms of the series defined by $a_n = 5n$

22. The first 50 terms of the series defined by $a_n = 1 - 3n$

23. The sum of an arithmetic series with 15 terms is 255. If $a_1 = 3$, find a_{15}.

24. The sum of an arithmetic series with 20 terms is 610. If $a_{20} = 59$, find a_1.

Exercises 25–34: Use a formula to find the sum of the first 20 terms for the arithmetic sequence.

25. $a_1 = 4, d = 2$

26. $a_1 = -3, d = \frac{2}{3}$

27. $a_1 = 10, d = -\frac{1}{2}$

28. $a_1 = 0, d = -4$

29. $a_1 = 4, a_{20} = 190.2$

30. $a_1 = -4, a_{20} = 15$

31. $a_1 = -2, a_{11} = 50$

32. $a_1 = 6, a_5 = -30$

33. $a_2 = 6, a_{12} = 31$

34. $a_8 = 4, a_{10} = 14$

Exercises 35–40: Use a formula to find the sum of the finite geometric series.

35. $1 + 2 + 4 + 8 + 16 + 32 + 64 + 128$

36. $2 + \frac{1}{2} + \frac{1}{8} + \frac{1}{32} + \frac{1}{128} + \frac{1}{512}$

37. $0.5 + 1.5 + 4.5 + 13.5 + 40.5 + 121.5 + 364.5$

38. $0.6 + 0.3 + 0.15 + 0.075 + 0.0375$

39. The first 20 terms of the series defined by $a_n = 3(2)^{n-1}$

40. The first 15 terms of the series defined by $a_n = 2\left(\frac{1}{3}\right)^n$

Exercises 41–44: Use a formula to approximate the sum for $n = 4, 7$, and 10.

41. $1 - \frac{1}{2} + \frac{1}{4} - \frac{1}{8} + \cdots + \left(-\frac{1}{2}\right)^{n-1}$

42. $3 - 1 + \frac{1}{3} - \frac{1}{9} + \cdots + 3\left(-\frac{1}{3}\right)^{n-1}$

43. $\frac{1}{3} + \frac{2}{3} + \frac{4}{3} + \frac{8}{3} + \cdots + \frac{1}{3}(2)^{n-1}$

44. $4 + \frac{8}{3} + \frac{16}{9} + \frac{32}{27} + \cdots + 4\left(\frac{2}{3}\right)^{n-1}$

Exercises 45–50: Find the sum of the infinite geometric series.

45. $1 + \frac{1}{3} + \frac{1}{9} + \frac{1}{27} + \frac{1}{81} + \cdots$

46. $5 + \frac{5}{2} + \frac{5}{4} + \frac{5}{8} + \frac{5}{16} + \cdots$

47. $6 - 4 + \frac{8}{3} - \frac{16}{9} + \frac{32}{27} - \frac{64}{81} + \cdots$

48. $-2 + \frac{1}{2} - \frac{1}{8} + \frac{1}{32} - \frac{1}{128} + \cdots$

49. $1 - \frac{1}{10} + \frac{1}{100} - \frac{1}{1000} + \cdots + \left(-\frac{1}{10}\right)^{n-1} + \cdots$

50. $25 - 5 + 1 - \frac{1}{5} + \cdots + 25\left(-\frac{1}{5}\right)^{n-1} + \cdots$

Decimal Numbers and Geometric Series

Exercises 51–56: Write each rational number in the form of an infinite geometric series.

51. $\frac{2}{3}$ **52.** $\frac{1}{9}$

53. $\frac{9}{11}$ **54.** $\frac{14}{33}$

55. $\frac{1}{7}$ **56.** $\frac{23}{99}$

Exercises 57–60: Write the sum of each geometric series as a rational number.

57. $0.8 + 0.08 + 0.008 + 0.0008 + \cdots$

58. $0.9 + 0.09 + 0.009 + 0.0009 + \cdots$

59. $0.45 + 0.0045 + 0.000045 + \cdots$

60. $0.36 + 0.0036 + 0.000036 + \cdots$

Summation Notation

Exercises 61–68: Write out the terms of the series and then evaluate it.

61. $\sum_{k=1}^{4}(k + 1)$ **62.** $\sum_{k=1}^{6}(3k - 1)$

63. $\sum_{k=1}^{8}4$ **64.** $\sum_{k=2}^{6}(5 - 2k)$

65. $\sum_{k=1}^{7}k^3$ **66.** $\sum_{k=1}^{4}5(2)^{k-1}$

67. $\sum_{k=4}^{5}(k^2 - k)$ **68.** $\sum_{k=1}^{5}\log k$

Exercises 69–74: Write the series with summation notation. Let the lower limit equal 1.

69. $1^4 + 2^4 + 3^4 + 4^4 + 5^4 + 6^4$

70. $1 + \frac{1}{5} + \frac{1}{25} + \frac{1}{125} + \frac{1}{625}$

71. $1 + \frac{4}{3} + \frac{6}{4} + \frac{8}{5} + \frac{10}{6} + \frac{12}{7} + \frac{14}{8}$

72. $2 + \frac{5}{8} + \frac{10}{27} + \frac{17}{64} + \frac{26}{125} + \frac{37}{216}$

73. $1 + \frac{1}{2^2} + \frac{1}{3^2} + \frac{1}{4^2} + \frac{1}{5^2} + \cdots$

74. $1 + \frac{1}{10} + \frac{1}{100} + \frac{1}{1000} + \frac{1}{10,000} + \cdots$

Exercises 75–80: (Refer to Example 11.) Rewrite each summation so that the index starts with $n = 1$.

75. $\sum_{k=6}^{9}k^3$ **76.** $\sum_{k=5}^{10}(k^2 - 2)$

77. $\sum_{k=9}^{32}(3k - 2)$ **78.** $\sum_{k=8}^{21}(4k + 1)$

79. $\sum_{k=16}^{52}(k^2 - 3k)$ **80.** $\sum_{k=25}^{59}(k^2 + 4k)$

Exercises 81–92: (Refer to Example 12.) Use properties for summation notation to find the sum.

81. $\sum_{k=1}^{60}9$ **82.** $\sum_{k=1}^{43}-4$

83. $\sum_{k=1}^{15}5k$ **84.** $\sum_{k=1}^{22}-2k$

85. $\sum_{k=1}^{31}(3k - 3)$ **86.** $\sum_{k=1}^{17}(1 - 4k)$

87. $\sum_{k=1}^{25}k^2$ **88.** $\sum_{k=1}^{12}3k^2$

89. $\sum_{k=1}^{16}(k^2 - k)$ **90.** $\sum_{k=1}^{18}(k^2 - 4k + 3)$

91. $\sum_{k=5}^{24}k$ **92.** $\sum_{k=7}^{19}(k^2 + 1)$

93. Verify the formula $\sum_{k=1}^{n} k = \frac{n(n + 1)}{2}$ by using the formula for the sum of the first n terms of a finite arithmetic sequence.

94. Use Exercise 93 to find the sum of the series $\sum_{k=1}^{200} k$.

Applications

Exercises 95 and 96: Salaries A person has the given starting salary S and receives a raise R each year thereafter.

    (a) Use a formula to calculate the total amount earned over 15 years.

    (b) Use a calculator to verify this value.

95. $S = \$42,000, R = \$1800$

96. $S = \$35,000, R = \$2500$

Exercises 97–100: Annuities (Refer to Example 7.) Determine the future value of each annuity.

97. $A_0 = \$2000$, $i = 0.08, n = 20$

98. $A_0 = \$500$, $i = 0.15, n = 10$

99. $A_0 = \$10,000$, $i = 0.11, n = 5$

100. $A_0 = \$3000$, $i = 0.19, n = 45$

101. *Stacking Logs* Logs are stacked in layers, with one fewer log in each layer. See the figure. If the top layer has 7 logs and the bottom layer has 15 logs, what is the total number of logs in the pile? Use a formula to find the sum.

102. *Stacking Logs* (Refer to Exercise 101.) Suppose a stack of logs has 13 logs in the top layer and a total of 7 layers. How many logs are in the stack?

103. *Filter* Suppose that one filter removes half of the impurities in a water supply.
 (a) Find a series that represents the amount of impurities removed by a sequence of n filters. Express your answer in summation notation.

 (b) How many filters would be necessary to remove all of the impurities?

104. *Walking* Suppose that a person walks 1 mile on the first day, $\frac{1}{3}$ mile on the second day, $\frac{1}{9}$ mile on the third day, and so on. Assuming that a person could walk each distance precisely, estimate how far the person would have traveled after a very long time.

105. *Area* (Refer to Exercise 96, Section 11.1.) Use a geometric series to find the sum of the areas of the squares if they continue indefinitely.

106. *Perimeter* (Refer to Exercise 96, Section 11.1.) Use a geometric series to find the sum of the perimeters of the squares if they continue indefinitely.

Exercises 107 and 108: The Natural Exponential Function The following series can be used to estimate the value of e^a for any real number a:

$$e^a \approx 1 + a + \frac{a^2}{2!} + \frac{a^3}{3!} + \cdots + \frac{a^n}{n!},$$

where $n! = 1 \cdot 2 \cdot 3 \cdot 4 \cdot \cdots \cdot n$. Use the first eight terms of this series to approximate the given expression. Compare this estimate with the actual value.

107. e **108.** e^{-1}

Computing Partial Sums

Exercises 109–112: Use a_k and n to find $S_n = \sum_{k=1}^{n} a_k$. (Refer to Example 6.) Then evaluate the infinite geometric series $S = \sum_{k=1}^{\infty} a_k$. Compare S to the values for S_n.

109. $a_k = \left(\frac{1}{3}\right)^{k-1}$; $n = 2, 4, 8, 16$

110. $a_k = 3\left(\frac{1}{2}\right)^{k-1}$; $n = 5, 10, 15, 20$

111. $a_k = 4\left(-\frac{1}{10}\right)^{k-1}$; $n = 1, 2, 3, 4, 5, 6$

112. $a_k = 2(-0.02)^{k-1}$; $n = 1, 2, 3, 4, 5, 6$

Writing about Mathematics

113. Discuss the difference between a sequence and a series. Give examples.

114. Under what circumstances can we find the sum of a geometric series? Give examples.

115. Explain how to write the series

$$\log 1 + \log 2 + \log 3 + \cdots + \log n$$

as one term.

116. Explain how to write the series

$$\log 2 - \log 4 + \log 6 - \log 8 + \cdots + (-1)^{n-1}\log 2n$$

as one term. Assume n is even.

1. Give graphical and numerical representations of the sequence defined by $a_n = -2n + 3$, where $a_n = f(n)$. Include the first six terms.

2. Determine if the sequence is arithmetic or geometric. If it is arithmetic, state the common difference; if it is geometric, give the common ratio.
   (a) $2, -4, 8, -16, 32, -64, 128, \ldots$

   (b) $-3, 0, 3, 6, 9, 12, \ldots$

   (c) $4, 2, 1, \frac{1}{2}, \frac{1}{4}, \frac{1}{8}, \frac{1}{16}, \ldots$

3. Determine if the series is arithmetic or geometric. Use a formula to find its sum.
   (a) $1 + 5 + 9 + 13 + \cdots + 37$

   (b) $3 + 1 + \frac{1}{3} + \frac{1}{9} + \frac{1}{27} + \frac{1}{81}$

   (c) $2 + \frac{1}{2} + \frac{1}{8} + \frac{1}{32} + \cdots$

   (d) $0.9 + 0.09 + 0.009 + 0.0009 + \cdots$

4. Write each series in Exercise 3 in summation notation.

5. Use properties of summation notation to find the sum.

   (a) $\displaystyle\sum_{k=1}^{15}(k + 2)$ (b) $\displaystyle\sum_{k=1}^{21} 2k^2$

6. *Bouncing Ball* A ball is dropped from a height of 6 feet. On each bounce the ball returns to $\frac{2}{3}$ of its previous height. How far does the ball travel (up and down) before it comes to rest?

11.3 Counting

- **Apply the fundamental counting principle**
- **Calculate and apply permutations**
- **Calculate and apply combinations**

Introduction

The notion of *counting* in mathematics includes much more than simply counting from 1 to 100. It also includes determining the number of ways that an event can occur. For example, how many ways are there to answer a true-false quiz with ten questions? The answer involves counting the different ways that a student could answer such a quiz. Counting is an important concept that is used to calculate probabilities. Probability is discussed in Section 11.6.

Fundamental Counting Principle

Suppose that a quiz has only two questions. The first is a multiple-choice question with four choices, A, B, C, or D, and the second is a true-false (T-F) question. The **tree diagram** in Figure 11.14 can be used to count the ways that this quiz can be answered.

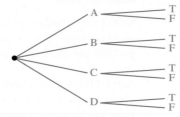

Figure 11.14 Different Ways to Answer a Quiz

A tree diagram is a systematic way of listing every possibility. From Figure 11.14 we can see that there are eight ways to answer the test. They are

AT, AF, BT, BF, CT, CF, DT, and DF.

For instance, CF indicates a quiz with answers of C on the first question and F on the second question.

A tree diagram is not always practical, because it can quickly become very large. For this reason mathematicians have developed more efficient ways of counting. Since the multiple-choice question has four possible answers, after which the true-false question has two possible answers, there are $4 \cdot 2 = 8$ possible ways of answering the test. This is an application of the *fundamental counting principle*, which applies to independent events. Two events are **independent** if neither event influences the outcome of the other.

Fundamental Counting Principle

Let $E_1, E_2, E_3, \ldots, E_n$ be a sequence of n independent events. If event E_k can occur m_k ways for $k = 1, 2, 3, \ldots, n$, then there are

$$m_1 \cdot m_2 \cdot m_3 \cdot \cdots \cdot m_n$$

ways for all n events to occur.

EXAMPLE 1 Counting ways to answer an exam

An exam contains four true-false questions and six multiple-choice questions. Each multiple-choice question has five possible answers. Count the number of ways that the exam can be answered.

SOLUTION

Getting Started Answering these ten questions can be thought of as a sequence of ten independent events. There are two ways to answer each of the first four questions and five ways of answering each of the next six questions. ▶

The number of ways to answer the exam is

$$\underbrace{2 \cdot 2 \cdot 2 \cdot 2}_{\text{4 factors}} \cdot \underbrace{5 \cdot 5 \cdot 5 \cdot 5 \cdot 5 \cdot 5}_{\text{6 factors}} = 2^4 5^6 = 250{,}000.$$

Now Try Exercise 3 ◀

In the next example, we count the number of different license plates possible with a given format.

EXAMPLE 2 Counting license plates

Sometimes a license plate is limited to 3 uppercase letters (A through Z) followed by 3 digits (0 through 9). For example, ABB 112 would be a valid license plate. Would this format provide enough license plates for a state with 8 million vehicles?

SOLUTION Since there are 26 letters of the alphabet, it follows that there are 26 ways to choose each of the first three letters of the license plate. Similarly, there are 10 digits, so there are 10 ways to choose each of the three digits in the license plate. By the fundamental counting principle, there are

$$26 \cdot 26 \cdot 26 \cdot 10 \cdot 10 \cdot 10 = 17{,}576{,}000$$

unique license plates that could be issued. This format for license plates could accommodate more than 8 million vehicles. *Now Try Exercise 5* ◀

An Application   At one time, toll-free numbers always began with 800. Because there were not enough toll-free 800 numbers to meet demand, phone companies started to use toll-free numbers that began with 888 and 877. In the next example, we count the number of valid 800 numbers.   (**Source:** Database Services Management.)

EXAMPLE 3   Counting toll-free 800 telephone numbers

Count the total number of 800 numbers if the local portion of a telephone number (the last seven digits) does not start with a 0 or 1.

SOLUTION   A toll-free 800 number assumes the following form.

Local Number

We can think of choosing the remaining digits for the local number as seven independent events. Since the local number cannot begin with a 0 or 1, there are eight possibilities (2 to 9) for the first digit. The remaining six digits can be any number from 0 to 9, so there are ten possibilities for each of these digits. The total is given by

$$\underset{\text{First digit}}{\underline{8}} \cdot \underset{\text{Last 6 digits}}{\underline{10 \cdot 10 \cdot 10 \cdot 10 \cdot 10 \cdot 10}} = 8 \cdot 10^6 = 8{,}000{,}000.$$

Now Try Exercise 47 ◀

Permutations

A **permutation** is an *ordering* or *arrangement*. For example, suppose that three students are scheduled to give speeches in a class. The different arrangements in which these speeches can be ordered are called permutations. Initially, any one of the three students could give the first speech. After the first speech, there are two students remaining for the second speech. For the third speech there is only one possibility. By the fundamental counting principle, the total number of permutations is equal to

$$3 \cdot 2 \cdot 1 = 6.$$

If the students are denoted as A, B, and C, then these six permutations are ABC, ACB, BAC, BCA, CAB, and CBA. In a similar manner, if there were ten students scheduled to give speeches, the number of permutations would increase to

$$10 \cdot 9 \cdot 8 \cdot 7 \cdot 6 \cdot 5 \cdot 4 \cdot 3 \cdot 2 \cdot 1 = 3{,}628{,}800.$$

A more efficient way of writing the previous two products is to use *factorial notation*. The number $n!$ (read "n-factorial") is defined as follows.

---

n-Factorial

For any natural number n,

$$n! = n(n - 1)(n - 2) \cdots (3)(2)(1)$$

and

$$0! = 1.$$

Factorials grow rapidly and can be computed on most calculators.

EXAMPLE 4 Calculating factorials

Compute $n!$ for $n = 0, 1, 2, 3, 4,$ and 5 by hand. Use a calculator to find $8!$, $13!$, and $25!$.

SOLUTION The values for $n!$ can be calculated as

$$0! = 1, \quad 1! = 1, \quad 2! = 2 \cdot 1 = 2, \quad 3! = 3 \cdot 2 \cdot 1 = 6,$$
$$4! = 4 \cdot 3 \cdot 2 \cdot 1 = 24, \quad \text{and} \quad 5! = 5 \cdot 4 \cdot 3 \cdot 2 \cdot 1 = 120.$$

Figure 11.15 shows the values of $8!$, $13!$, and $25!$. The value for $25!$ is an approximation. Notice how rapidly $n!$ increases.   ***Now Try Exercise 27*** ◀

```
8!
                        40320
13!
             6227020800
25!
     1.551121004E25
```

Figure 11.15

A Famous Unsolved Problem One of the most famous unanswered questions in computing today is called the *traveling salesperson problem*. It is a relatively simple problem to state, but if someone could design a procedure to solve this problem *efficiently*, he or she would not only become famous but also provide a valuable way for businesses to save millions of dollars on scheduling problems, such as assigning bus routes and truck deliveries.

Calculator Help

To calculate $n!$, see Appendix A (page AP-19).

One instance of the traveling salesperson problem can be stated as follows. A salesperson must begin and end at home and travel to three cities. Assuming that the salesperson can travel between any pair of cities, what route would minimize the salesperson's mileage? In Figure 11.16, the four cities are labeled A, B, C, and D. Let the salesperson live in city A. There are six routes that could be tried. They are listed in Table 11.6 with the appropriate mileage for each.

Table 11.6

Route	Mileage
A B C D A	$200 + 480 + 500 + 350 = 1530$
A B D C A	$200 + 100 + 500 + 300 = 1100$
A C B D A	$300 + 480 + 100 + 350 = 1230$
A C D B A	$300 + 500 + 100 + 200 = 1100$
A D B C A	$350 + 100 + 480 + 300 = 1230$
A D C B A	$350 + 500 + 480 + 200 = 1530$

Figure 11.16 Traveling Salesperson Problem

The shortest route of 1100 miles occurs when the salesperson either (1) starts at A and travels through B, D, and C and back to A or (2) reverses this route. Currently, this method of listing all possible routes to find the minimum distance is the only known way to consistently find the optimal solution for any general map containing n cities. In fact, people have not been able to determine whether a *significantly* faster method even exists. (**Source:** J. Smith, *Design and Analysis of Algorithms*.)

Counting the number of routes involves the fundamental counting principle. At the first step, the salesperson can travel to any one of three cities. Once this city has been selected, there are two possible cities to choose, and then one. Finally the salesperson returns home. The total number of routes is given by

$$3! = 3 \cdot 2 \cdot 1 = 6.$$

If the salesperson must travel to 30 cities, then there are

$$30! \approx 2.7 \times 10^{32}$$

routes to check, far too many to check even with the largest supercomputers.

Next suppose that a salesperson must visit three of eight possible cities. At first there are eight cities to choose. After the first city has been visited, there are seven cities to select. Since the salesperson travels to only three of the eight cities, there are

$$8 \cdot 7 \cdot 6 = 336$$

possible routes. This number of permutations is denoted $P(8, 3)$. It represents the number of arrangements that can be made using three elements taken from a sample of eight.

Calculator Help

To calculate $P(n, r)$, see Appendix A (page AP-20).

Permutations of *n* Elements Taken *r* at a Time

If $P(n, r)$ denotes the number of permutations of n elements taken r at a time, with $r \le n$, then

$$P(n, r) = \frac{n!}{(n - r)!} = \underbrace{n(n - 1)(n - 2) \cdots (n - r + 1)}_{r \text{ factors}}.$$

EXAMPLE 5 Calculating $P(n, r)$

Calculate each of the following by hand. Then support your answers by using a calculator.
(a) $P(7, 3)$ **(b)** $P(100, 2)$

SOLUTION
(a) $P(7, 3) = \underbrace{7 \cdot 6 \cdot 5}_{3 \text{ factors}} = 210$

(b) $P(100, 2) = 100 \cdot 99 = 9900$. We also can compute this number as follows.

$$P(100, 2) = \frac{100!}{(100 - 2)!} = \frac{100 \cdot 99 \cdot 98!}{98!} = 100 \cdot 99 = 9900$$

In this case, it is helpful to cancel 98! before performing the arithmetic. Both of these computations are performed by a calculator in Figure 11.17.

Now Try Exercises 31 and 37 ◀

```
7 nPr 3
                   210
100 nPr 2
                  9900
```

Figure 11.17

EXAMPLE 6 Calculating permutations

For a class of 30 students, how many arrangements are there in which 4 students each give a speech?

SOLUTION The number of permutations of 30 elements taken 4 at a time is given by

$$P(30, 4) = \underbrace{30 \cdot 29 \cdot 28 \cdot 27}_{4 \text{ factors}} = 657,720.$$

Thus there are 657,720 ways to arrange the four speeches.   *Now Try Exercise 43* ◀

In the next example, we determine the number of ways that 4 people can have different birthdays. For example, if the birthdays of 4 people are February 29, May 2, June 30, and July 11, then these dates would be one way that 4 people could have different birthdays.

EXAMPLE 7 Counting birthdays

Count the possible ways that 4 people can have different birthdays.

SOLUTION Counting February 29, there are 366 possible birthdays. The first person could have any of the 366 possible birthdays. The second person could have any of 365 birthdays, because the first person's birthday cannot be duplicated. Similarly, the third person could have any of 364 birthdays, and the fourth person could have any of 363 birthdays. The total number of ways that 4 people could have different birthdays equals

$$P(366, 4) = \underbrace{366 \cdot 365 \cdot 364 \cdot 363}_{4 \text{ factors}} \approx 1.77 \times 10^{10}.$$

Now Try Exercise 53 ◀

CLASS DISCUSSION

Count the number of arrangements of 52 cards in a standard deck. Is it likely that there are arrangements that no one has ever shuffled at any time in the history of the world? Explain.

Combinations

Unlike a permutation, a **combination** is not an ordering or arrangement, but rather a subset of a set of elements. Order is unimportant in finding combinations. For example, suppose we want to select a tennis team of two players from four people. The order in which the selection is made does not affect the final team of two players. From a set of four people, we select a subset of two players. The number of possible subsets, or combinations, is denoted either $C(4, 2)$ or $\binom{4}{2}$.

To calculate $C(4, 2)$, we first consider $P(4, 2)$. If we denote the four players by the letters A, B, C, and D, there are $P(4, 2) = 4 \cdot 3 = 12$ permutations given by

AB, BA, AC, CA, AD, DA, BC, CB, BD, DB, CD, DC.

However, the team that comprises person A and person B is equivalent to the team with person B and person A. The sets {AB} and {BA} are equal. The valid combinations are the following two-element subsets of {A, B, C, D}.

{AB}   {AC}   {AD}   {BC}   {BD}   {CD}

That is, $C(4, 2) = \frac{P(4, 2)}{2!} = 6$. The relationship between $P(n, r)$ and $C(n, r)$ is given below.

Calculator Help

To calculate $C(n, r)$, see Appendix A (page AP-20).

Combinations of *n* Elements Taken *r* at a Time

If $C(n, r)$ denotes the number of combinations of n elements taken r at a time, with $r \leq n$, then

$$C(n, r) = \frac{P(n, r)}{r!} = \frac{n!}{(n - r)! \, r!}.$$

EXAMPLE 8 Calculating $C(n, r)$

Calculate each of the following. Support your answers by using a calculator.

(a) $C(7, 3)$ **(b)** $\binom{50}{47}$

SOLUTION

(a) $C(7, 3) = \dfrac{7!}{(7 - 3)!3!} = \dfrac{7!}{4!3!} = \dfrac{7 \cdot 6 \cdot 5 \cdot 4!}{4!3!} = \dfrac{7 \cdot 6 \cdot 5}{3!} = \dfrac{210}{6} = 35$

Figure 11.18

(b) The notation $\left(\frac{50}{47}\right)$ is equivalent to $C(50, 47)$.

$$\binom{50}{47} = \frac{50!}{(50 - 47)!\, 47!} = \frac{50!}{3!\, 47!} = \frac{50 \cdot 49 \cdot 48 \cdot 47!}{3!\, 47!}$$

$$= \frac{50 \cdot 49 \cdot 48}{3!} = \frac{117,600}{6} = 19,600$$

These computations are performed by a calculator in Figure 11.18.

Now Try Exercises 59 and 65 ◀

EXAMPLE 9 Counting combinations

A college student has 5 courses left in her major and plans to take 2 of them this semester. Assuming that this student has the prerequisites for all 5 courses, determine how many ways these 2 courses can be selected.

SOLUTION The order in which the courses are selected is unimportant. From a set of 5 courses, the student selects a subset of 2 courses. The number of subsets is

$$C(5, 2) = \frac{5!}{(5 - 2)!\, 2!} = \frac{5!}{3!\, 2!} = 10.$$

There are 10 ways to select 2 courses from a set of 5. *Now Try Exercise 67* ◀

EXAMPLE 10 Calculating the number of ways to play the lottery

To win the jackpot in a lottery, a person must select 5 different numbers from 1 to 55 and then pick the powerball, which has a number from 1 to 42. Count the ways to play the game. (**Source:** Minnesota State Lottery.)

SOLUTION From 55 numbers a player picks 5 numbers. There are $C(55, 5)$ ways of doing this. There are 42 ways to choose the powerball. By the fundamental counting principle, the number of ways to play the game equals $C(55, 5) \cdot 42 = 146,107,962$.

Now Try Exercise 71 ◀

MAKING CONNECTIONS

Permutations and Combinations A *permutation* is an arrangement (or list) of objects in which the ordering of the objects is important. For example, the expression $P(20, 9)$ would give the number of batting orders possible for 9 players selected from a team of 20 players. A *combination* is a subset of a set of objects in which the ordering of the objects is *not* important. For example, the expression $C(20, 9)$ would give the number of committees possible when 9 people are selected from a group of 20 people. The order in which committee members are selected does not affect the makeup of the committee.

EXAMPLE 11 Counting committees

How many committees of 6 people can be selected from 6 women and 3 men if a committee must have at least 2 men?

SOLUTION

Getting Started Because there are 3 men and each committee must have at least 2 men, a committee can include either 2 or 3 men. The order of selection is not important. Therefore we need to consider combinations of committee members rather than permutations. ▶

Two Men: This committee would consist of 4 women and 2 men. We can select 2 men from a group of 3 in $C(3, 2) = 3$ ways. Four women can be selected from a group of 6 in $C(6, 4) = 15$ ways. By the fundamental counting principle, a total of

$$C(3, 2) \cdot C(6, 4) = 3 \cdot 15 = \mathbf{45}$$

committees have 2 men.

Three Men: This committee would have 3 women and 3 men. We can select 3 men from a group of 3 in $C(3, 3) = 1$ way. We can select 3 women from a group of 6 in $C(6, 3) = 20$ ways. By the fundamental counting principle, a total of

$$C(3, 3) \cdot C(6, 3) = 1 \cdot 20 = \mathbf{20}$$

committees have 3 men.

The total number of possible committees would be $\mathbf{45 + 20 = 65}$.

Now Try Exercise 69 ◀

11.3 Putting It All Together

The fundamental counting principle can be used to determine the number of ways a sequence of independent events can occur. Permutations are arrangements or listings, whereas combinations are subsets of a set of events.

The following table summarizes some concepts related to counting in mathematics.

Notation	Explanation	Examples
n-factorial: $n!$	$n!$ represents the product $$n(n-1)\cdots(3)(2)(1).$$	$6! = 6 \cdot 5 \cdot 4 \cdot 3 \cdot 2 \cdot 1 = 720$ $0! = 1$
$P(n, r) = \dfrac{n!}{(n-r)!}$	$P(n, r)$ represents the number of permutations of n elements taken r at a time.	The number of two-letter strings that can be formed using the four letters A, B, C, and D exactly once is given by $$P(4, 2) = \frac{4!}{(4-2)!} = \frac{4!}{2!} = 12,$$ or $P(4, 2) = 4 \cdot 3 = 12.$
$C(n, r) = \dfrac{n!}{(n-r)!\,r!}$	$C(n, r)$ represents the number of combinations of n elements taken r at a time.	The number of committees of 3 people that can be formed from 5 people is $$C(5, 3) = \frac{5!}{(5-3)!\,3!} = \frac{5!}{2!3!} = 10.$$

11.3 Exercises

Counting

Exercises 1–4: Exam Questions   Count the number of ways that the questions on an exam could be answered.

1. Ten true-false questions

2. Ten multiple-choice questions with five choices each

3. Five true-false questions and ten multiple-choice questions with four choices each

4. One question involving matching ten items in one column with ten items in another column, using a one-to-one correspondence

Exercises 5–8: License Plates   Count the number of possible license plates with the given constraints.

5. Three digits followed by three letters

6. Two letters followed by four digits

7. Three letters followed by three digits or letters

8. Two letters followed by either three or four digits

Exercises 9–12: Counting Strings   Count the number of five-letter strings that can be formed with the given letters, assuming a letter can be used more than once.

9. A, B, C   10. W, X, Y, Z

11. D, E, F, G, H   12. A, C

Exercises 13–16: Counting Strings   Count the number of strings that can be formed with the given letters, assuming each letter is used exactly once.

13. A, B   14. A, B, C

15. W, X, Y, Z   16. V, W, X, Y, Z

17. *Combination Lock*   A briefcase has two locks. The combination to each lock consists of a three-digit number, in which digits may be repeated. See the figure. How many combinations are possible? (*Hint:* The word *combination* is a misnomer. Lock combinations are permutations in which the arrangement of the numbers is important.)

18. *Combination Lock*   A typical combination for a padlock consists of three numbers from 0 to 39. Count the combinations possible with this type of lock if a number may be repeated.

19. *Garage Door Openers*   The code for some garage door openers consists of 12 electrical switches that can be set to either 0 or 1 by the owner. With this type of opener, how many codes are possible?   (**Source:** Promax.)

20. *Lottery*   To win the jackpot in a lottery game, a person must pick three numbers from 0 to 9 in the correct order. If a number can be repeated, how many ways are there to play the game?

21. *Radio Stations*   Call letters for a radio station usually begin with either a K or a W, followed by three letters. In 2005, there were 13,517 radio stations on the air. Is there any shortage of call letters for new radio stations? (**Source:** M. Street Corporation.)

22. *Access Codes*   An ATM access code often consists of a four-digit number. How many codes are possible without giving two accounts the same access code?

23. *Computer Package*   A computer store offers a package in which buyers choose 1 of 2 monitors, 1 of 3 printers, and 1 of 4 types of software. How many different packages can be purchased?

24. *Dice*   A red die and a blue die are thrown. How many ways are there for both dice to show an even number?

25. *Telephone Numbers*   How many different 7-digit telephone numbers are possible if the first digit cannot be a 0 or a 1?

26. *Dinner Choices*   A menu offers 5 different salads, 10 different entrées, and 4 different desserts. How many ways are there to order a salad, an entrée, and a dessert?

Exercises 27–30: Evaluate the expression.

27. 6!   28. 0!

29. 10!   30. 7!

Permutations

Exercises 31–40: Evaluate the expression.

31. $P(5, 3)$ 32. $P(10, 2)$

33. $P(8, 1)$ 34. $P(6, 6)$

35. $P(7, 3)$ **36.** $P(12, 3)$

37. $P(25, 2)$ **38.** $P(20, 1)$

39. $P(10, 4)$ **40.** $P(34, 2)$

41. *Standing in Line*   How many ways can 4 people stand in a line?

42. *Books on a Shelf*   How many arrangements are there of 6 different books on a shelf?

43. *Giving a Speech*   In how many arrangements can 3 students from a class of 15 each give a speech?

44. *Introductions*   How many ways could 5 basketball players be introduced at a game?

45. *Traveling Salesperson*   (Refer to the discussion after Example 4.) A salesperson must travel to 3 of 7 cities. Direct travel is possible between every pair of cities. How many arrangements are there in which the salesperson could visit these 3 cities? Assume that traveling a route in reverse order constitutes a different arrangement.

46. *Traveling Salesperson*   A salesperson must start at city A, stop in each city once, and then return to city A. Use the figure to find the route with the least mileage.

47. *Phone Numbers*   How many 7-digit phone numbers are there if the first three numbers must be 387, 388, or 389?

48. *Keys*   How many distinguishable ways can 4 keys be put on a key ring?

49. *Sitting at a Round Table*   How many ways can 7 people sit at a round table? (For a way to be different, at least one person must be sitting next to someone different.)

50. *Batting Orders*   A softball team has 10 players. How many batting orders are possible?

51. *Baseball Positions*   In how many ways can 9 players be assigned to the 9 positions on a baseball team, assuming that any player can play any position?

52. *Musical Chairs Seating*   In a game of musical chairs, 7 children will sit in 6 chairs arranged in a circle. One child will be left out. How many (different) ways can the children sit in the chairs? (For a way to be different, at least one child must be sitting next to someone different.)

53. *Birthdays*   In how many ways can 5 people have different birthdays?

54. *Course Schedule*   A scheduling committee has 1 room in which to offer 5 mathematics courses. In how many ways can the committee arrange the 5 courses over the day?

55. *Telephone Numbers*   How many 10-digit telephone numbers are there if the first digit and the fourth digit cannot be a 0 or a 1?

56. *Car Designs*   There are 10 basic colors available for a new car, along with 5 basic styles of trim. In how many ways can a person pick the color and trim?

Combinations

Exercises 57–66: Evaluate the expression.

57. $C(3, 1)$ **58.** $C(4, 3)$

59. $C(6, 3)$ **60.** $C(7, 5)$

61. $C(5, 0)$ **62.** $C(10, 2)$

63. $\binom{8}{2}$ **64.** $\binom{9}{4}$

65. $\binom{20}{18}$ **66.** $\binom{100}{2}$

67. *Lottery*   To win the jackpot in a lottery, one must select 5 different numbers from 1 to 39. How many ways are there to play this game?

68. *Selecting a Committee*   How many ways can a committee of 5 be selected from 8 people?

69. *Selecting a Committee*   How many committees of 4 people can be selected from 5 women and 3 men if a committee must have 2 people of each sex on it?

70. *Essay Questions*   On a test with 6 essay questions, students are asked to answer 4 questions. How many ways can the essay questions be selected?

71. *Test Questions*   A test consists of two parts. In the first part a student must choose 3 of 5 essay questions, and in the second part a student must choose 4 of 5 essay questions. How many ways can the essay questions be selected?

72. *Cards*   How many ways are there to draw a 5-card hand from a 52-card deck?

73. *Selecting Marbles*   How many ways are there to draw 3 red marbles and 2 blue marbles from a jar that contains 10 red marbles and 12 blue marbles?

74. *Book Arrangements*   A professor has 3 copies of an algebra book and 4 copies of a calculus text. How many distinguishable ways can the books be placed on a shelf?

75. *Peach Samples*   How many samples of 3 peaches can be drawn from a crate of 24 peaches? (Assume that the peaches are distinguishable.)

76. *Flower Samples*   A bouquet of flowers contains 3 red roses, 4 yellow roses, and 5 white roses. In how many ways can a person choose 1 flower of each type? (Assume that the flowers are distinguishable.)

77. *Permutations*   Show that $P(n, n - 1) = P(n, n)$. Give an example that supports your result.

78. *Combinations*   Show that $\binom{n}{r} = \binom{n}{n - r}$. Give an example that supports your result.

Writing about Mathematics

79. Explain the difference between a permutation and a combination. Give examples.

80. Explain what counting is, as presented in this section.

11.4 The Binomial Theorem

- Derive the binomial theorem
- Use the binomial theorem
- Apply Pascal's triangle

Introduction

In this section we discuss how to expand expressions in the form $(a + b)^n$, where n is a natural number. These expressions occur in statistics, finite mathematics, computer science, and calculus.

Derivation of the Binomial Theorem

Combinations play a central role in the development of the binomial theorem. The binomial theorem can be used to expand expressions of the form $(a + b)^n$. Before stating the binomial theorem, we begin by counting the number of strings of a given length that can be formed with only the variables a and b.

EXAMPLE 1   Calculating distinguishable strings

Count the number of distinguishable strings that can be formed with the given number of a's and b's. List these strings.
(a) Two a's, one b **(b)** Two a's, three b's   **(c)** Four a's, no b's

SOLUTION
(a) Using two a's and one b, we can form strings of length three. Once the b has been positioned, the string is determined. For example, if the b is placed in the middle position,

,

then the string must be *aba*. From a set of three slots, we choose one slot in which to place the b. This is computed by $C(3, 1) = 3$. The strings are *aab*, *aba*, and *baa*.

(b) With two *a*'s and three *b*'s we can form strings of length five. Once the locations of the three *b*'s have been selected, the string is determined. For instance, if the *b*'s are placed in the first, third, and fifth positions,

$$b \quad \boxed{} \quad b \quad \boxed{} \quad b,$$

then the string becomes *babab*. From a set of five slots, we select three in which to place the *b*'s. This is computed by $C(5, 3) = 10$. The ten strings are

bbbaa, bbaba, bbaab, babba, babab, baabb, abbba, abbab, ababb, and *aabbb*.

(c) There is only one string of length four that contains no *b*'s. This is *aaaa* and is computed by

$$C(4, 0) = \frac{4!}{(4 - 0)!0!} = \frac{24}{24(1)} = 1.$$

Now Try Exercises 9, 11, and 13 ◀

Next we expand $(a + b)^n$ for a few values of *n*, without simplifying.

$$(a + b)^1 = a + b$$
$$(a + b)^2 = (a + b)(a + b)$$
$$= aa + ab + ba + bb$$
$$(a + b)^3 = (a + b)(a + b)^2$$
$$= (a + b)(aa + ab + ba + bb)$$
$$= aaa + aab + aba + abb + baa + bab + bba + bbb$$

Notice that $(a + b)^1$ is the sum of all possible strings of length one that can be formed using *a* and *b*. The only possibilities are *a* and *b*. The expression $(a + b)^2$ is the sum of all possible strings of length two using *a* and *b*. The strings are *aa, ab, ba,* and *bb*. Similarly, $(a + b)^3$ is the sum of all possible strings of length three using *a* and *b*. This pattern continues for higher powers of $(a + b)$.

Strings with equal numbers of *a*'s and equal numbers of *b*'s can be combined into one term. For example, in $(a + b)^2$ the terms *ab* and *ba* can be combined as 2*ab*. Notice that there are $C(2, 1) = 2$ distinguishable strings of length two containing one *b*. Similarly, in the expansion of $(a + b)^3$, the terms containing one *a* and two *b*'s can be combined as

$$abb + bab + bba = 3ab^2.$$

There are $C(3, 2) = 3$ strings of length three that contain two *b*'s.

We can use these concepts to expand $(a + b)^4$. The expression $(a + b)^4$ consists of the sum of all strings of length four using only the letters *a* and *b*: $C(4, 0) = 1$ string containing no *b*'s, $C(4, 1) = 4$ strings containing one *b*, and so on, up to $C(4, 4) = 1$ string containing four *b*'s. Thus

$$(a + b)^4 = \binom{4}{0}a^4b^0 + \binom{4}{1}a^3b^1 + \binom{4}{2}a^2b^2 + \binom{4}{3}a^1b^3 + \binom{4}{4}a^0b^4$$

$$= a^4 + 4a^3b + 6a^2b^2 + 4ab^3 + b^4.$$

These results are summarized by the binomial theorem.

Binomial Theorem

For any positive integer n and numbers a and b,

$$(a + b)^n = \binom{n}{0}a^n + \binom{n}{1}a^{n-1}b^1 + \cdots + \binom{n}{n-1}a^1b^{n-1} + \binom{n}{n}b^n.$$

We can use the binomial theorem to expand $(a + b)^n$. For example,

$$(a + b)^3 = \binom{3}{0}a^3 + \binom{3}{1}a^2b^1 + \binom{3}{2}a^1b^2 + \binom{3}{3}b^3$$

$$= 1a^3 + 3a^2b + 3ab^2 + 1b^3$$

$$= a^3 + 3a^2b + 3ab^2 + b^3.$$

Since $\binom{n}{r} = C(n, r)$, we can use the combination formula $C(n, r) = \frac{n!}{(n-r)!\,r!}$ to evaluate the binomial coefficients.

EXAMPLE 2 Applying the binomial theorem

Use the binomial theorem to expand the expression $(2a + 1)^5$.

SOLUTION Using the binomial theorem, we arrive at the following result.

$$(2a + 1)^5 = \binom{5}{0}(2a)^5 + \binom{5}{1}(2a)^41^1 + \binom{5}{2}(2a)^31^2 + \binom{5}{3}(2a)^21^3 + \binom{5}{4}(2a)^11^4 + \binom{5}{5}1^5$$

$$= \frac{5!}{5!0!}(32a^5) + \frac{5!}{4!1!}(16a^4) + \frac{5!}{3!2!}(8a^3) + \frac{5!}{2!3!}(4a^2) + \frac{5!}{1!4!}(2a) + \frac{5!}{0!5!}$$

$$= 32a^5 + 80a^4 + 80a^3 + 40a^2 + 10a + 1$$

Now Try Exercise 21 ◀

Pascal's Triangle

Expanding $(a + b)^n$ for increasing values of n gives the following results.

$$(a + b)^0 = \qquad\qquad\qquad 1$$
$$(a + b)^1 = \qquad\qquad\qquad 1a + 1b$$
$$(a + b)^2 = \qquad\qquad\qquad 1a^2 + 2ab + 1b^2$$
$$(a + b)^3 = \qquad\qquad 1a^3 + 3a^2b + 3ab^2 + 1b^3$$
$$(a + b)^4 = \qquad\quad 1a^4 + 4a^3b + 6a^2b^2 + 4ab^3 + 1b^4$$
$$(a + b)^5 = \quad 1a^5 + 5a^4b + 10a^3b^2 + 10a^2b^3 + 5ab^4 + 1b^5$$

Notice that $(a + b)^1$ has two terms starting with a and ending with b, $(a + b)^2$ has three terms starting with a^2 and ending with b^2, and in general $(a + b)^n$ has $n + 1$ terms starting with a^n and ending with b^n. The exponent on a decreases by 1 each successive term, and the exponent on b increases by 1 each successive term.

The triangle formed by the highlighted numbers is called **Pascal's triangle**. It can be used to efficiently compute the binomial coefficients, $C(n, r)$. The triangle consists of 1's along the sides. Each element inside the triangle is the sum of the two numbers above it. Pascal's triangle is usually written without variables, as in Figure 11.19. It can be extended to include as many rows as needed.

Figure 11.19 Pascal's Triangle

$$\begin{array}{ccccccccccc}
& & & & & 1 & & & & & \\
& & & & 1 & & 1 & & & & \\
& & & 1 & & 2 & & 1 & & & \\
& & 1 & & 3 & & 3 & & 1 & & \\
& 1 & & 4 & & 6 & & 4 & & 1 & \\
1 & & 5 & & 10 & & 10 & & 5 & & 1
\end{array}$$

We can use this triangle to expand powers of binomials in the form $(a + b)^n$, where n is a natural number. For example, the expression $(m + n)^4$ consists of five terms written as follows.

$$(m + n)^4 = \_\ m^4 + \_\ m^3n^1 + \_\ m^2n^2 + \_\ m^1n^3 + \_\ n^4$$

Because there are five terms, the coefficients can be found in the fifth row of Pascal's triangle, which from Figure 11.19 (repeated in the margin) is

$$1\ \ 4\ \ 6\ \ 4\ \ 1.$$

Thus

$$(m + n)^4 = \underline{1}\ m^4 + \underline{4}\ m^3n^1 + \underline{6}\ m^2n^2 + \underline{4}\ m^1n^3 + \underline{1}\ n^4$$
$$= m^4 + 4m^3n + 6m^2n^2 + 4mn^3 + n^4.$$

EXAMPLE 3   **Expanding expressions with Pascal's triangle**

Expand each of the following.
(a) $(2x + 1)^5$ **(b)** $(3x - y)^3$

SOLUTION
(a) To expand $(2x + 1)^5$, let $a = 2x$ and $b = 1$ in the binomial theorem. We can use the sixth row of Pascal's triangle to obtain the coefficients 1, 5, 10, 10, 5, and 1. Compare this solution with the solution for Example 2.

$$(2x + 1)^5 = 1(2x)^5 + 5(2x)^4(1)^1 + 10(2x)^3(1)^2 + 10(2x)^2(1)^3 + 5(2x)^1(1)^4 + 1(1)^5$$
$$= 32x^5 + 80x^4 + 80x^3 + 40x^2 + 10x + 1$$

(b) Let $a = 3x$ and $b = -y$ in the binomial theorem. Use the coefficients 1, 3, 3, and 1 from the fourth row of Pascal's triangle.

$$(3x - y)^3 = 1(3x)^3 + 3(3x)^2(-y)^1 + 3(3x)^1(-y)^2 + 1(-y)^3$$
$$= 27x^3 - 27x^2y + 9xy^2 - y^3 \quad \textit{Now Try Exercises 33 and 35} \blacktriangleleft$$

Finding the kth Term   The binomial theorem gives *all* of the terms of $(a + b)^n$. However, we can find any individual term by noting that the $(r + 1)$st term in the binomial expansion for $(a + b)^n$ is given by the formula $\binom{n}{r}a^{n-r}b^r$ for $0 \le r \le n$. The next example shows how to use this formula to find the $(r + 1)$st term of $(a + b)^n$.

EXAMPLE 4   **Finding the kth term in a binomial expansion**

Find the third term of $(x - y)^5$.

SOLUTION   In this example the $(r + 1)$st term is the *third* term in the expansion of $(x - y)^5$. That is, $r + 1 = 3$, or $r = 2$. Also, the exponent in the expression is $n = 5$. To get this binomial into the form $(a + b)^n$, we note that the first term in the binomial is $a = x$ and that the second term in the binomial is $b = -y$. Substituting the values for r, n, a, and b in the formula for the $(r + 1)$st term yields

$$\binom{5}{2}(x)^{5-2}(-y)^2 = 10x^3y^2.$$

The third term in the binomial expansion of $(x - y)^5$ is $10x^3y^2$.   ***Now Try Exercise 45*** ◀

11.4 Putting It All Together

The following table summarizes topics related to the binomial theorem.

Concept	Explanation	Example
Binomial coefficient	$\binom{n}{r} = C(n, r) = \dfrac{n!}{(n-r)!\, r!}$	$\binom{5}{3} = \dfrac{5!}{(5-3)!\, 3!} = \dfrac{120}{2 \cdot 6} = 10$ $\binom{4}{0} = \dfrac{4!}{(4-0)!\, 0!} = \dfrac{24}{24 \cdot 1} = 1$
Binomial theorem	$(a + b)^n =$ $\binom{n}{0}a^n + \binom{n}{1}a^{n-1}b + \cdots + \binom{n}{n}b^n$ for any positive integer n and real numbers a and b.	$(a + b)^3 = \binom{3}{0}a^3 + \binom{3}{1}a^2b + \binom{3}{2}ab^2 + \binom{3}{3}b^3$ $= a^3 + 3a^2b + 3ab^2 + b^3$ The binomial coefficients can also be found using the fourth row of Pascal's triangle, which is shown below.
Pascal's triangle	A triangle of numbers that can be used to find the binomial coefficients needed to expand an expression of the form $(a + b)^n$. To expand $(a + b)^n$, use row $n + 1$ of Pascal's triangle.	$$\begin{array}{ccccccccccc} & & & & & 1 & & & & & \\ & & & & 1 & & 1 & & & & \\ & & & 1 & & 2 & & 1 & & & \\ & & 1 & & 3 & & 3 & & 1 & & \\ & 1 & & 4 & & 6 & & 4 & & 1 & \\ 1 & & 5 & & 10 & & 10 & & 5 & & 1 \end{array}$$
Finding the $(r + 1)$st term of $(a + b)^n$	The $(r + 1)$st term of $(a + b)^n$ is given by $$\binom{n}{r}a^{n-r}b^r$$ for $0 \le r \le n$.	To find the fifth term of $(x + y)^6$, let $r + 1 = 5$, or $r = 4$, and $n = 6$. $$\binom{n}{r}a^{n-r}b^r = \binom{6}{4}x^{6-4}y^4$$ $$= 15x^2y^4$$

11.4 Exercises

Binomial Coefficients

Exercises 1–8: Evaluate the expression.

1. $\binom{5}{4}$

2. $\binom{6}{2}$

3. $\binom{4}{0}$

4. $\binom{4}{2}$

5. $\binom{6}{5}$

6. $\binom{6}{3}$

7. $\binom{3}{3}$

8. $\binom{5}{2}$

Binomial Theorem

Exercises 9–16: (Refer to Example 1.) Calculate the number of distinguishable strings that can be formed with the given number of a's and b's.

9. Three a's, two b's

10. Five a's, three b's

11. Four a's, four b's

12. One a, five b's

13. Five a's, no b's

14. No a's, three b's

15. Four a's, one b

16. Four a's, two b's

Exercises 17–30: Use the binomial theorem to expand each expression.

17. $(x + y)^2$

18. $(x + y)^4$

19. $(m + 2)^3$

20. $(m + 2n)^5$

21. $(2x - 3)^3$

22. $(x + y^2)^3$

23. $(p - q)^6$

24. $(p^2 - 3)^4$

25. $(2m + 3n)^3$

26. $(3a - 2b)^5$

27. $(1 - x^2)^4$

28. $(2 + 3x^2)^3$

29. $(2p^3 - 3)^3$

30. $(2r + 3t)^4$

Pascal's Triangle

Exercises 31–44: Use Pascal's triangle to help expand the expression.

31. $(x + y)^2$

32. $(m + n)^3$

33. $(3x + 1)^4$

34. $(2x - 1)^4$

35. $(2 - x)^5$

36. $(2a + 3b)^3$

37. $(x^2 + 2)^4$

38. $(5 - x^2)^3$

39. $(4x - 3y)^4$

40. $(3 - 2x)^5$

41. $(m + n)^6$

42. $(2m - n)^4$

43. $(2x^3 - y^2)^3$

44. $(3x^2 + y^3)^4$

Finding the *k*th Term

Exercises 45–52: Find the specified term.

45. The fourth term of $(a + b)^9$

46. The second term of $(m - n)^9$

47. The fifth term of $(x + y)^8$

48. The third term of $(a + b)^7$

49. The fourth term of $(2x + y)^5$

50. The eighth term of $(2a - b)^9$

51. The sixth term of $(3x - 2y)^6$

52. The seventh term of $(2a + b)^9$

Writing about Mathematics

53. Explain how to find the numbers in Pascal's triangle.

54. Compare the expansion of $(a + b)^n$ with the expansion of $(a - b)^n$. Give an example.

CHECKING BASIC CONCEPTS FOR SECTIONS 11.3 AND 11.4

1. Count the ways to answer a quiz that consists of 8 true-false questions.

2. Count the number of 5-card poker hands that can be dealt using a standard deck of 52 cards.

3. How many distinct license plates could be made using a letter followed by 5 digits or letters?

4. Expand each expression.
   (a) $(2x + 1)^4$ (b) $(4 - 3x)^3$

11.5 Mathematical Induction

- **Learn basic concepts about mathematical induction**
- **Use mathematical induction to prove statements**
- **Apply the generalized principle of mathematical induction**

Introduction

The brilliant mathematician Carl Friedrich Gauss (1777–1855) proved the fundamental theorem of algebra at age 20. When he was a young child, he amazed his teacher by showing that

$$1 + 2 + 3 + 4 + \cdots + 100 = \frac{100(101)}{2}.$$

With *mathematical induction* we will be able to show, more generally, that

$$1 + 2 + 3 + 4 + \cdots + n = \frac{n(n + 1)}{2}.$$

Mathematical induction is a powerful method of proof. It is used not only in mathematics; it is also used in computer science, to prove that programs and basic concepts are correct.

Mathematical Induction

Many results in mathematics are claimed true for every positive integer. Any of these results could be checked for $n = 1$, $n = 2$, $n = 3$, and so on, but since the set of positive integers is infinite, it would be impossible to check every possible case. For example, let S_n represent the statement that the sum of the first n positive integers is $\frac{n(n + 1)}{2}$; that is,

$$S_n: 1 + 2 + 3 + \cdots + n = \frac{n(n + 1)}{2}.$$

The truth of this statement can be checked quickly for the first few values of n.

If $n = 1$, S_1 is $\quad\quad\quad\quad\quad 1 = \frac{1(1 + 1)}{2}$, a true statement, since $1 = 1$.

If $n = 2$, S_2 is $\quad\quad\quad\quad 1 + 2 = \frac{2(2 + 1)}{2}$, a true statement, since $3 = 3$.

If $n = 3$, S_3 is $\quad\quad 1 + 2 + 3 = \frac{3(3 + 1)}{2}$, a true statement, since $6 = 6$.

If $n = 4$, S_4 is $1 + 2 + 3 + 4 = \frac{4(4 + 1)}{2}$, a true statement, since $10 = 10$.

Since the statement is true for $n = 1, 2, 3, 4$, and so on, can we conclude that the statement is true for all positive integers by checking a finite number of examples? The answer is *no*. To prove that such a statement is true for every positive integer, we use the following principle.

Principle of Mathematical Induction

Let S_n be a statement concerning the positive integer n. Suppose that

1. S_1 is true;

2. for any positive integer k, if S_k is true, then S_{k+1} is also true.

Then S_n is true for every positive integer n.

A proof by mathematical induction can be explained as follows. By assumption (1), the statement is true when $n = 1$. By assumption (2), the fact that the statement is true for $n = 1$ implies that it is true for $n = 1 + 1 = 2$. Using (2) again, the statement is thus true for $2 + 1 = 3$, for $3 + 1 = 4$, for $4 + 1 = 5$, and so on. Continuing in this way shows that the statement must be true for *every* positive integer.

The situation is similar to that of an infinite number of dominoes lined up. If the first domino is pushed over, it pushes the next, which pushes the next, and so on, indefinitely.

Another example of the principle of mathematical induction might be an infinite ladder. Suppose the rungs are spaced so that, whenever you are on a rung, you know you can move to the next rung. Then *if* you can get to the first rung, you can go as high up the ladder as you wish.

Two separate steps are required for a proof by mathematical induction.

Proof by Mathematical Induction

STEP 1: Prove that the statement is true for $n = 1$.

STEP 2: Show that for any positive integer k, if S_k is true, then S_{k+1} is also true.

Proving Statements

Mathematical induction is used in the next example to prove the statement S_n discussed earlier.

EXAMPLE 1 **Proving an equality statement**

Let S_n represent the statement

$$1 + 2 + 3 + \cdots + n = \frac{n(n + 1)}{2}.$$

Prove that S_n is true for every positive integer n.

SOLUTION The proof by mathematical induction is as follows.

STEP 1: Show that the statement is true when $n = 1$. If $n = 1$, S_1 becomes

$$1 = \frac{1(1 + 1)}{2},$$

which is true.

STEP 2: Show that if S_k is true, then S_{k+1} is also true, where S_k is the statement

$$1 + 2 + 3 + \cdots + k = \frac{k(k + 1)}{2}$$

and S_{k+1} is the statement

$$1 + 2 + 3 + \cdots + k + (k + 1) = \frac{(k + 1)[(k + 1) + 1]}{2}.$$

Start with S_k and assume it is a true statement.

$$1 + 2 + 3 + \cdots + k = \frac{k(k + 1)}{2} \qquad S_k \text{ is true.}$$

Add $k + 1$ to each side of equation S_k and simplify to obtain S_{k+1}.

$$1 + 2 + 3 + \cdots + k + (k + 1) = \frac{k(k + 1)}{2} + (k + 1)$$

$$= (k + 1)\left(\frac{k}{2} + 1\right) \qquad \text{Factor out } k + 1.$$

$$= (k + 1)\left(\frac{k + 2}{2}\right)$$

$$= \frac{(k + 1)[(k + 1) + 1]}{2}$$

This final result is the statement S_{k+1}. Therefore, if S_k is true, then S_{k+1} is also true. The two steps required for a proof by mathematical induction have been completed, so the statement S_n is true for every positive integer n.   ***Now Try Exercise 1*** ◄

EXAMPLE 2   **Proving an equality statement**

Let S_n represent the statement

$$2^1 + 2^2 + 2^3 + 2^4 + \cdots + 2^n = 2^{n+1} - 2.$$

Prove that S_n is true for every positive integer n.

SOLUTION

STEP 1:   Show that the statement S_1 is true, where S_1 is

$$2^1 = 2^{1+1} - 2.$$

Since $2 = 4 - 2$, S_1 is a true statement.

STEP 2:   Show that if S_k is true, then S_{k+1} is also true, where S_k is

$$2^1 + 2^2 + 2^3 + \cdots + 2^k = 2^{k+1} - 2$$

and S_{k+1} is

$$2^1 + 2^2 + 2^3 + \cdots + 2^k + 2^{k+1} = 2^{(k+1)+1} - 2.$$

Start with the given equation S_k and add 2^{k+1} to each side of this equation. Then algebraically change the right side to look like the right side of S_{k+1}.

$$2^1 + 2^2 + 2^3 + \cdots + 2^k + 2^{k+1} = 2^{k+1} - 2 + 2^{k+1}$$

$$= 2 \cdot 2^{k+1} - 2$$

$$= 2^{(k+1)+1} - 2$$

The final result is the statement S_{k+1}. Therefore, if S_k is true, then S_{k+1} is also true. The two steps required for a proof by mathematical induction have been completed, so the statement S_n is true for every positive integer n.   ***Now Try Exercise 5*** ◄

EXAMPLE 3   **Proving an inequality statement**

Prove that if x satisfies $0 < x < 1$, then for every positive integer n,

$$0 < x^n < 1.$$

SOLUTION

STEP 1: Here S_1 is the statement

$$\text{if } 0 < x < 1, \text{ then } 0 < x^1 < 1,$$

which is true.

STEP 2: S_k is the statement

$$\text{if } 0 < x < 1, \text{ then } 0 < x^k < 1.$$

To show that S_k implies that S_{k+1} is true, multiply all three parts of $0 < x^k < 1$ by x to get

$$x \cdot 0 < x \cdot x^k < x \cdot 1.$$

(Here the fact that $0 < x$ is used. Why?) Simplify to obtain

$$0 < x^{k+1} < x.$$

Because $x < 1$,

$$0 < x^{k+1} < 1,$$

which implies that S_{k+1} is true. Therefore, if S_k is true, then S_{k+1} is true. Since both steps for a proof by mathematical induction have been completed, the given statement is true for every positive integer n.   ***Now Try Exercise 23*** ◀

Generalized Principle of Mathematical Induction

Some statements S_n are not true for the first few values of n, but are true for all values of n that are greater than or equal to some fixed integer j. The following slightly generalized form of the principle of mathematical induction addresses these cases.

Generalized Principle of Mathematical Induction

Let S_n be a statement concerning the positive integer n. Let j be a fixed positive integer. Suppose that

1. S_j is true;
2. for any positive integer k, $k \geq j$, S_k implies S_{k+1}.

Then S_n is true for all positive integers n, where $n \geq j$.

EXAMPLE 4 Using the generalized principle

Let S_n represent the statement $2^n > 2n + 1$. Show that S_n is true for all values of n such that $n \geq 3$.

SOLUTION (Check that S_n is false for $n = 1$ and $n = 2$.)

STEP 1: Show that S_n is true for $n = 3$. If $n = 3$, S_3 is

$$2^3 > 2 \cdot 3 + 1, \qquad \text{or}$$
$$8 > 7.$$

Thus S_3 is true.

STEP 2: Now show that S_k implies S_{k+1} for $k \geq 3$, where

$$S_k \text{ is } 2^k > 2k + 1 \qquad \text{and}$$
Assume S_k is true.

$$S_{k+1} \text{ is } 2^{k+1} > 2(k + 1) + 1.$$
Show S_{k+1} is true.

Multiply each side of $2^k > 2k + 1$ by 2, obtaining

$$2 \cdot 2^k > 2(2k + 1), \qquad \text{or}$$
$$2^{k+1} > 4k + 2.$$

Rewrite $4k + 2$ as $2(k + 1) + 2k$, to get

$$2^{k+1} > 2(k + 1) + 2k.$$

Since k is a positive integer greater than 3,

$$2k > 1.$$

It follows that

$$2^{k+1} > 2(k + 1) + 2k > 2(k + 1) + 1, \qquad \text{or}$$
$$2^{k+1} > 2(k + 1) + 1,$$

as required. Thus S_k implies S_{k+1}, and this, together with the fact that S_3 is true, shows that S_n is true for every positive integer n greater than or equal to 3.

Now Try Exercise 27 ◄

11.5 Putting It All Together

Some important concepts about mathematical induction are summarized in the following table.

Concept	Explanation
Principle of mathematical induction	Let S_n be a statement concerning the positive integer n. Suppose that **1.** S_1 is true; **2.** for any positive integer k, if S_k is true, then S_{k+1} is also true. Then S_n is true for every positive integer n.
Proof by mathematical induction	**STEP 1:** Prove that the statement is true for $n = 1$. **STEP 2:** Show that for any positive integer k, if S_k is true, then S_{k+1} is also true.
Generalized principle of mathematical induction	Let S_n be a statement concerning the positive integer n. Let j be a fixed positive integer. Suppose that **1.** S_j is true; **2.** for any positive integer k, $k \geq j$, S_k implies S_{k+1}. Then S_n is true for all positive integers n, where $n \geq j$.

11.5   Exercises

Mathematical Induction

Exercises 1–14: Use mathematical induction to prove the statement. Assume that n is a positive integer.

1. $3 + 6 + 9 + \cdots + 3n = \dfrac{3n(n + 1)}{2}$

2. $1 + 3 + 5 + \cdots + (2n - 1) = n^2$

3. $5 + 10 + 15 + \cdots + 5n = \dfrac{5n(n + 1)}{2}$

4. $4 + 7 + 10 + \cdots + (3n + 1) = \dfrac{n(3n + 5)}{2}$

5. $3 + 3^2 + 3^3 + \cdots + 3^n = \dfrac{3(3^n - 1)}{2}$

6. $1^2 + 2^2 + 3^2 + \cdots + n^2 = \dfrac{n(n + 1)(2n + 1)}{6}$

7. $1^3 + 2^3 + 3^3 + \cdots + n^3 = \dfrac{n^2(n + 1)^2}{4}$

8. $5 \cdot 6 + 5 \cdot 6^2 + 5 \cdot 6^3 + \cdots + 5 \cdot 6^n = 6(6^n - 1)$

9. $\dfrac{1}{1 \cdot 2} + \dfrac{1}{2 \cdot 3} + \dfrac{1}{3 \cdot 4} + \cdots + \dfrac{1}{n(n + 1)} = \dfrac{n}{n + 1}$

10. $7 \cdot 8 + 7 \cdot 8^2 + 7 \cdot 8^3 + \cdots + 7 \cdot 8^n = 8(8^n - 1)$

11. $\dfrac{4}{5} + \dfrac{4}{5^2} + \dfrac{4}{5^3} + \cdots + \dfrac{4}{5^n} = 1 - \dfrac{1}{5^n}$

12. $\dfrac{1}{2} + \dfrac{1}{2^2} + \dfrac{1}{2^3} + \cdots + \dfrac{1}{2^n} = 1 - \dfrac{1}{2^n}$

13. $\dfrac{1}{1 \cdot 4} + \dfrac{1}{4 \cdot 7} + \cdots + \dfrac{1}{(3n - 2)(3n + 1)} = \dfrac{n}{3n + 1}$

14. $x^{2n} + x^{2n-1}y + \cdots + xy^{2n-1} + y^{2n} = \dfrac{x^{2n+1} - y^{2n+1}}{x - y}$

Exercises 15–18: Find all positive integers n for which the given statement is not true.

15. $3^n > 6n$ **16.** $3^n > 2n + 1$

17. $2^n > n^2$ **18.** $n! > 2n$

Exercises 19–28: Prove the statement by mathematical induction.

19. $(a^m)^n = a^{mn}$ (Assume a and m are constants.)

20. $(ab)^n = a^n b^n$ (Assume a and b are constants.)

21. $2^n > 2n$ if $n \geq 3$

22. $3^n > 2n + 1$, if $n \geq 2$

23. If $a > 1$, then $a^n > 1$.

24. If $a > 1$, then $a^n > a^{n-1}$.

25. If $0 < a < 1$, then $a^n < a^{n-1}$.

26. $2^n > n^2$ for $n > 4$

27. If $n \geq 4$, then $n! > 2^n$, where

$$n! = n(n - 1)(n - 2) \cdots (3)(2)(1).$$

28. $4^n > n^4$ for $n \geq 5$

Applications

29. *Number of Handshakes*   Suppose that each of the n ($n \geq 2$) people in a room shakes hands with everyone else, but not with himself. Show that the number of handshakes is $\dfrac{n^2 - n}{2}$.

30. *Sides of a Polygon*   The series of sketches starts with an equilateral triangle having sides of length 1. In the following steps, equilateral triangles are constructed on each side of the preceding figure. The length of the sides of each new triangle is $\frac{1}{3}$ the length of the sides of the preceding triangles. Develop a formula for the number of sides of the nth figure. Use mathematical induction to prove your answer.

31. *Perimeter* Find the perimeter of the *n*th figure in Exercise 30.

32. *Area* Show that the area of the *n*th figure in Exercise 30 is given by

$$\sqrt{3}\left[\frac{2}{5} - \frac{3}{20}\left(\frac{4}{9}\right)^{n-1}\right].$$

33. *Tower of Hanoi* A pile of *n* rings, each ring smaller than the one below it, is on a peg. Two other pegs are attached to the same board as this peg. In a game called the *Tower of Hanoi* puzzle, all the rings must be moved to a different peg, with only one ring moved at a time and with no ring ever placed on top of a smaller ring. Find the least number of moves required. Prove your result with mathematical induction.

Writing about Mathematics

34. Explain the principle of mathematical induction.

35. Explain how the generalized principle of mathematical induction differs from the principle of mathematical induction.

36. When using mathematical induction, why is it important to prove that the statement holds for $n = 1$?

11.6 Probability

- **Learn the basic concepts about probability**
- **Calculate the probability of compound events**
- **Calculate the probability of independent and dependent events**

Introduction

Questions of chance have no doubt engaged the minds of people since antiquity. However, the mathematical treatment of probability did not begin until the fifteenth century. The birth of probability theory as a mathematical discipline occurred in the seventeenth century with the work of Blaise Pascal and Pierre Fermat. Today probability pervades society. It is used not only to determine outcomes in gambling, but also to predict weather, genetic outcomes, and the risk involved with various types of substances and behaviors.

Risk is the chance, or probability, that a harmful event will occur. The following activities increase the annual risk of death by one chance in a million: flying 1000 miles in a jet, traveling 300 miles in a car, riding 10 miles on a bicycle, smoking 1.4 cigarettes, living 2 days in New York City, having one chest X-ray, or living 2 months with a cigarette smoker. In Section 11.3 we saw that the likelihood of winning the jackpot in a lottery was one chance in 146,107,962.

Ideally, we would live in a risk-free world. However, almost every action or substance exposes people to some risk. Recognizing relative risks is important for a long and healthy life. Probability provides us with a measure of the likelihood that an event will occur. Knowledge about probability allows individuals to make informed decisions about their lives. (**Sources:** NCTM, *Historical Topics for the Mathematics Classroom, Thirty-first Yearbook;* J. Rodricks, *Calculated Risk.*)

Definition of Probability

In the study of probability, experiments often are performed. An experiment might involve tossing a coin or measuring the cholesterol level of a patient who has had a heart attack. A result from an experiment is called an **outcome**. The set of all possible outcomes is the **sample space**. Any subset of a sample space is an **event**.

For example, if an experiment involves rolling a (fair) die, then the sample space consists of $S = \{1, 2, 3, 4, 5, 6\}$. The event $E = \{1, 6\}$ contains the outcomes of 1 or 6 showing on the die. If $n(E)$ and $n(S)$ denote the number of outcomes in E and S, then $n(E) = 2$ and $n(S) = 6$. The probability of rolling a 1 or a 6 is given by $P(E) = \frac{2}{6}$. That is, the likelihood of event E occurring is two chances in six. These concepts are summarized in the following box.

Probability of an Event

If the outcomes of a finite sample space S are equally likely and if E is an event in S, then the **probability of E** is given by

$$P(E) = \frac{n(E)}{n(S)},$$

where $n(E)$ and $n(S)$ represent the number of outcomes in E and S, respectively.

Since $n(E) \leq n(S)$, the probability of an event E satisfies $0 \leq P(E) \leq 1$. If $P(E) = 1$, then event E is *certain* to occur. If $P(E) = 0$, then event E is *impossible*.

EXAMPLE 1 Drawing a card

One card is drawn at random from a standard deck of 52 cards. Find the probability that the card is an ace.

SOLUTION The sample space S consists of 52 outcomes that correspond to drawing any one of 52 cards. Each outcome is equally likely. Let E represent the event of drawing an ace. There are four aces in the deck, so event E contains four outcomes. Therefore $n(S) = 52$ and $n(E) = 4$. The probability of drawing an ace is given by

$$P(E) = \frac{n(E)}{n(S)} = \frac{4}{52} = \frac{1}{13}.$$ ***Now Try Exercise 15*** ◀

EXAMPLE 2 Estimating probability of organ transplants

Table 11.7

Heart	2608
Kidney	76,203
Liver	16,287
Lung	2103

Source: OPTN.

In 2008, there were 98,941 patients waiting for an organ transplant. Table 11.7 lists the numbers of patients waiting for the *most common types* of transplants. None of these people need two or more transplants. Approximate the probability that a transplant patient chosen at random will need

(a) a kidney or a heart   **(b)** neither a kidney nor a heart

SOLUTION
(a) Let each patient represent an outcome in a sample space S. The event E of a transplant patient needing either a kidney or a heart contains $76,203 + 2608 = 78,811$ outcomes. The desired probability is

$$P(E) = \frac{n(E)}{n(S)} = \frac{78,811}{98,941} \approx 0.80.$$

In 2008, about 80% of transplant patients needed either a kidney or a heart.

(b) Let F be the event of a patient waiting for an organ other than a kidney or a heart. Then

$$n(F) = n(S) - n(E) = 98,941 - 78,811 = 20,130.$$

The probability of F is

$$P(F) = \frac{n(F)}{n(S)} = \frac{20,130}{98,941} \approx 0.20, \quad \text{or} \quad 20\%. \quad \textbf{\textit{Now Try Exercise 37}} ◀$$

Notice that $P(E) + P(F) = 1$ in Example 2. The events E and F are **complements** because $E \cap F = \varnothing$ and $E \cup F = S$, where $\varnothing$ denotes the *empty set*. That is, a transplant patient is either waiting for a kidney or a heart (event E) or not waiting for a kidney or a heart (event F). The complement of E may be denoted by E'.

Venn Diagrams   Probability concepts can be illustrated using **Venn diagrams**. In Figure 11.20 the sample space S of an experiment is the union of the disjoint sets E and its complement E'. That is, $E \cup E' = S$ and $E \cap E' = \varnothing$.

If $P(E)$ is known, then $P(E')$ can be calculated as follows.

$$P(E') = \frac{n(E')}{n(S)} = \frac{n(S) - n(E)}{n(S)} = 1 - \frac{n(E)}{n(S)} = 1 - P(E)$$

In Example 2(b), the probability of $F = E'$ could also have been calculated by using

$$P(F) = 1 - P(E) \approx 1 - 0.80 = 0.20.$$

Figure 11.20 $E \cup E' = S$

Probability of a Complement

Let E be an event and E' be its complement. If the probability of E is $P(E)$, then the probability of its complement is given by

$$P(E') = 1 - P(E).$$

EXAMPLE 3   Finding probabilities of human eye color

In 1865, Gregor Mendel performed important research in genetics. His work led to a better understanding of dominant and recessive genetic traits. One example is human eye color, which is determined by a pair of genes called a *genotype*. According to Mendel's research, brown eye color B is dominant over blue eye color b. If a person has the genotype BB, Bb, or bB, he or she will have brown eyes. The genotype bb will result in blue eyes. A person receives one gene (B or b) from each parent. Table 11.8 shows how these two genes can be paired.   (**Source:** H. Lancaster, *Quantitative Methods in Biology and Medical Sciences.*)

(a) Assuming that each genotype is equally likely, find the probability of blue eyes.

(b) What is the probability that a person has brown eyes?

Table 11.8

	B	b
B	BB	Bb
b	bB	bb

SOLUTION

(a) The sample space S consists of four equally likely outcomes denoted BB, Bb, bB, and bb. The event E of blue eye color (bb) occurs once. Therefore

$$P(E) = \frac{n(E)}{n(S)} = \frac{1}{4} = 0.25.$$

(b) In Table 11.8 brown eyes are the complement of blue eyes. The probability of brown eyes is

$$P(E') = 1 - P(E) = 1 - 0.25 = 0.75.$$

This probability also could be computed as

$$P(E') = \frac{n(E')}{n(S)} = \frac{3}{4} = 0.75,$$

since there are three genotypes that result in brown eye color.   ***Now Try Exercise 41*** ◀

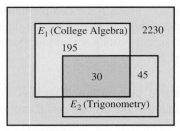

Figure 11.21

Compound Events

Frequently the probability of more than one event is needed. For example, suppose a college with a total of 2500 students has 225 students enrolled in college algebra, 75 in trigonometry, and 30 in both. Let E_1 denote the event that a student is enrolled in college algebra and E_2 the event that a student is enrolled in trigonometry. Then the Venn diagram in Figure 11.21 visually describes the situation.

In this Venn diagram, it is important that the 30 students taking both courses not be counted twice. Set E_1 has a total of $195 + 30 = 225$ students, and set E_2 contains $45 + 30 = 75$ students.

EXAMPLE 4 Calculating the probability of a union

In the preceding scenario, suppose a student is selected at random. What is the probability that this student is enrolled in college algebra, trigonometry, or both?

SOLUTION We would like to find the probability $P(E_1$ or $E_2)$, which is usually denoted $P(E_1 \cup E_2)$. Since $n(E_1 \cup E_2) = 195 + 30 + 45 = 270$ and $n(S) = 2500$,

$$P(E_1 \cup E_2) = \frac{n(E_1 \cup E_2)}{n(S)} = \frac{270}{2500} = 0.108.$$

There is a 10.8% chance that a student selected at random will be taking college algebra, trigonometry, or both. *Now Try Exercise 35* ◀

In the previous example, it would have been incorrect to simply add the probability of a student taking algebra and the probability of a student taking trigonometry. Their sum would be

$$P(E_1) + P(E_2) = \frac{n(E_1)}{n(S)} + \frac{n(E_2)}{n(S)}$$

$$= \frac{225}{2500} + \frac{75}{2500}$$

$$= \frac{300}{2500}$$

$$= 0.12, \quad \text{or} \quad 12\%.$$

This sum is greater than $P(E_1 \cup E_2)$ because the 30 students taking both courses are counted *twice* in this calculation. In order to find the correct probability for $P(E_1 \cup E_2)$, we must subtract the probability of the intersection $E_1 \cap E_2$.

$$P(E_1 \cup E_2) = P(E_1) + P(E_2) - P(E_1 \cap E_2)$$

$$= \frac{n(E_1)}{n(S)} + \frac{n(E_2)}{n(S)} - \frac{n(E_1 \cap E_2)}{n(S)}$$

$$= \frac{225}{2500} + \frac{75}{2500} - \frac{30}{2500}$$

$$= \frac{270}{2500}$$

$$= 0.108, \quad \text{or} \quad 10.8\%$$

This is the same result obtained in Example 4 and suggests the following property.

Probability of the Union of Two Events

For any two events E_1 and E_2,

$$P(E_1 \cup E_2) = P(E_1) + P(E_2) - P(E_1 \cap E_2).$$

EXAMPLE 5   Rolling dice

Suppose two dice are rolled. Find the probability that the dice show either a sum of eight or a pair.

SOLUTION   In Table 11.9 the roll of the dice is represented by an ordered pair. For example, the ordered pair (3, 6) represents the first die showing 3 and the second die 6.

Table 11.9   Rolling Two Dice

(1, 1)	(1, 2)	(1, 3)	(1, 4)	(1, 5)	(1, 6)
(2, 1)	**(2, 2)**	(2, 3)	(2, 4)	(2, 5)	**(2, 6)**
(3, 1)	(3, 2)	**(3, 3)**	(3, 4)	**(3, 5)**	(3, 6)
(4, 1)	(4, 2)	(4, 3)	**(4, 4)**	(4, 5)	(4, 6)
(5, 1)	(5, 2)	**(5, 3)**	(5, 4)	**(5, 5)**	(5, 6)
(6, 1)	**(6, 2)**	(6, 3)	(6, 4)	(6, 5)	**(6, 6)**

Because each die can show six different outcomes, there are a total of $6 \cdot 6 = 36$ outcomes in the sample space S. Let E_1 denote the event of rolling a sum of eight and E_2 the event of rolling a pair. Then

$$E_1 = \{(6, 2), (5, 3), (4, 4), (3, 5), (2, 6)\} \quad \text{and}$$
$$E_2 = \{(1, 1), (2, 2), (3, 3), (4, 4), (5, 5), (6, 6)\}.$$

The intersection of E_1 and E_2 is

$$E_1 \cap E_2 = \{(4, 4)\}.$$

Since $n(S) = 36$, $n(E_1) = 5$, $n(E_2) = 6$, and $n(E_1 \cap E_2) = 1$, the following can be computed.

$$P(E_1 \cup E_2) = P(E_1) + P(E_2) - P(E_1 \cap E_2)$$
$$= \frac{n(E_1)}{n(S)} + \frac{n(E_2)}{n(S)} - \frac{n(E_1 \cap E_2)}{n(S)}$$
$$= \frac{5}{36} + \frac{6}{36} - \frac{1}{36}$$
$$= \frac{10}{36}, \quad \text{or} \quad \frac{5}{18}$$

This result can be verified by counting the number of boldfaced outcomes in Table 11.9. Of the 36 possible outcomes, 10 satisfy the conditions, so the probability is $\frac{10}{36}$, or $\frac{5}{18}$.

Now Try Exercise 43 ◄

Mutually Exclusive Events   If $E_1 \cap E_2 = \varnothing$, then the events E_1 and E_2 are **mutually exclusive**. Mutually exclusive events have no outcomes in common, so $P(E_1 \cap E_2) = 0$. In this case, $P(E_1 \cup E_2) = P(E_1) + P(E_2)$.

EXAMPLE 6   Drawing cards

Find the probability of drawing either an ace or a king from a standard deck of 52 cards.

SOLUTION The event E_1 of drawing an ace and the event E_2 of drawing a king are mutually exclusive. No card can be both an ace and a king. Therefore $P(E_1 \cap E_2) = 0$. The probability of drawing an ace is $P(E_1) = \frac{4}{52}$, since there are 4 aces in 52 cards. Similarly, the probability of drawing a king is $P(E_2) = \frac{4}{52}$.

$$P(E_1 \cup E_2) = P(E_1) + P(E_2)$$
$$= \frac{4}{52} + \frac{4}{52}$$
$$= \frac{8}{52}, \quad \text{or} \quad \frac{2}{13}$$

Thus the probability of drawing either an ace or a king is $\frac{2}{13}$. **Now Try Exercise 45** ◀

The next example uses concepts from both counting and probability.

EXAMPLE 7 Drawing a poker hand

A standard deck of cards contains 52 cards, consisting of 13 different cards from each of four suits: hearts, diamonds, spades, and clubs. A poker hand consists of 5 cards drawn from a standard deck of cards, and a flush occurs when the 5 cards are all from the same suit. Find an expression for the probability of drawing 5 cards of the same suit in one try. Assume that the cards are not replaced.

Calculator Help

To calculate $C(n, r)$, see Appendix A (page AP-20).

```
4(13 nCr 5)/(52
nCr 5)
      .0019807923
```

Figure 11.22

SOLUTION Let E be the event of drawing 5 cards of the same suit. To determine $n(E)$, start by calculating the number of ways to draw a flush in a particular suit, such as hearts. From a set of 13 hearts, 5 hearts need to be drawn. There are $\binom{13}{5}$ ways to draw this hand. Because there are 4 suits, there are $4\binom{13}{5}$ ways to draw 5 cards of the same suit. Thus $n(E) = 4\binom{13}{5}$. The sample space S consists of all 5-card poker hands that can be drawn from a deck of 52 cards. There are $\binom{52}{5}$ different poker hands. Thus $n(S) = \binom{52}{5}$. The probability of a flush can now be calculated as follows.

$$P(E) = \frac{n(E)}{n(S)} = \frac{4\binom{13}{5}}{\binom{52}{5}}$$

In Figure 11.22, we see that $P(E) \approx 0.00198$. Thus there is about a 0.2% chance of drawing 5 cards of the same suit (in one try) from a standard deck of 52 cards.

Now Try Exercise 31 ◀

Independent and Dependent Events

Two events are **independent** if they do not influence each other. Otherwise they are **dependent**. An example of independent events would be one coin being tossed twice. The result of the first toss does not affect the second toss.

Probability of Independent Events

If E_1 and E_2 are independent events, then

$$P(E_1 \cap E_2) = P(E_1) \cdot P(E_2).$$

EXAMPLE 8 Tossing a coin

Suppose a coin is tossed twice. Determine the probability that the result is two heads.

SOLUTION Let E_1 be the event of a head on the first toss, and let E_2 be the event of a head on the second toss. Then $P(E_1) = P(E_2) = \frac{1}{2}$. The two events are independent. The probability of two heads occurring is

$$P(E_1 \cap E_2) = \frac{1}{2} \cdot \frac{1}{2} = \frac{1}{4}.$$

This probability of $\frac{1}{4}$ also can be found using a tree diagram, as shown in Figure 11.23. There are four equally likely outcomes. Tosses resulting in two heads occur once.

Now Try Exercise 21 ◄

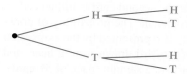

Figure 11.23

EXAMPLE 9 Rolling dice

What is the probability of rolling a sum of 12 with two dice?

SOLUTION The roll of one die does not influence the roll of the other. They are independent events. To obtain a sum of 12, both dice must show a 6. Let E_1 be the event of rolling a 6 with the first die and E_2 the event of rolling a 6 with the second die. Then $P(E_1) = P(E_2) = \frac{1}{6}$. The probability of rolling a 12 is

$$P(E_1 \cap E_2) = P(E_1) \cdot P(E_2) = \frac{1}{6} \cdot \frac{1}{6} = \frac{1}{36}.$$

This can be verified by using Table 11.9. There is only 1 outcome out of 36 that results in a sum of 12.

Now Try Exercise 25 ◄

CLASS DISCUSSION

When a pair of dice are rolled, what sum is most likely to appear?

If events E_1 and E_2 influence each other, they are *dependent*. The probability of dependent events is given as follows.

Probability of Dependent Events

If E_1 and E_2 are dependent events, then

$$P(E_1 \cap E_2) = P(E_1) \cdot P(E_2, \text{ given that } E_1 \text{ occurred}).$$

NOTE $P(E_2, \text{ given that } E_1 \text{ occurred})$ is called the **conditional probability of E_2, given E_1**.

EXAMPLE 10 Drawing cards

Find the probability of drawing two hearts from a standard deck of 52 cards when the first card is
(a) replaced before drawing the second card   **(b)** not replaced

SOLUTION
(a) Let E_1 denote the event of the first card being a heart and E_2 the event of the second card being a heart. If the first card is replaced before the second card is drawn, the two

events are independent. Since there are 13 hearts in a standard deck of 52 cards, the probability of two hearts being drawn is

$$P(E_1 \cap E_2) = P(E_1) \cdot P(E_2) = \frac{13}{52} \cdot \frac{13}{52} = \frac{1}{16}.$$

(b) If the first card is not replaced, then the outcome for the second card is influenced by the first card. Therefore the events E_1 and E_2 are dependent. The probability of drawing a second heart, given that the first card is a heart, is represented by the expression $P(E_2$, given E_1 has occurred) $= \frac{12}{51}$. That is, if the first card drawn were a heart and removed from the deck, then there would be 12 hearts in a sample space of 51 cards.

$$P(E_1 \cap E_2) = P(E_1) \cdot P(E_2, \text{given that } E_1 \text{ occurred}) = \frac{13}{52} \cdot \frac{12}{51} = \frac{1}{17}$$

Thus the probability of drawing two hearts is slightly less if the first card is not replaced.

Now Try Exercise 55 ◀

EXAMPLE 11   Calculating the probability of dependent events

Table 11.10 shows the number of students (by gender) registered for either a Spanish class or a French class. No student is taking both languages.

Table 11.10

	Spanish	French	Totals
Females	20	25	45
Males	40	25	65
Totals	60	50	110

If one student is selected at random, calculate each of the following.
(a) The probability that the student is female
(b) The probability that the student is taking Spanish, given that the student is female
(c) The probability that the student is female and taking Spanish

SOLUTION
(a) Let F represent the event that the student is female. Since 45 of the 110 students are female, $P(F) = \frac{45}{110} = \frac{9}{22}$.
(b) Let S be the event that the student is taking Spanish. The probability that the student is taking Spanish, given that the student is female, is $P(S$, given $F) = \frac{20}{45} = \frac{4}{9}$, because 20 of the 45 female students are taking Spanish.
(c) The probability that the student is female and taking Spanish is calculated by

$$P(F \cap S) = P(F) \cdot P(S, \text{given } F) = \frac{45}{110} \cdot \frac{20}{45} = \frac{2}{11}.$$

Table 11.10 shows that 20 of the 110 students are female and taking Spanish. Thus the required probability is $\frac{20}{110} = \frac{2}{11}$, which is in agreement with our previous calculation.

Now Try Exercise 65 ◀

EXAMPLE 12   Analyzing a polygraph test

Suppose there is a 6% chance that a polygraph test will incorrectly say a person is lying when he or she is actually telling the truth. If a person tells the truth 95% of the time, what percentage of the time will the polygraph test incorrectly indicate a lie for this person?

SOLUTION Let E_1 be the event that the person is telling the truth and E_2 be the event that the polygraph test is incorrect. Then

$$P(E_1 \cap E_2) = P(E_1) \cdot P(E_2, \text{ given the person is telling the truth})$$
$$= (0.95)(0.06)$$
$$= 0.057, \quad \text{or} \quad 5.7\%.$$

Now Try Exercise 61 ◄

11.6 Putting It All Together

The following table summarizes some concepts about probability. In this table, S denotes a finite sample space of equally likely outcomes, and E, E_1, and E_2 denote events in a sample space S.

Concept	Explanation	Example
Probability P	A number P satisfying $0 \leq P \leq 1$	$P(E) = 1$ indicates that an event E is certain to occur, $P(E) = 0$ indicates that an event E is impossible, and $P(E) = 0.3$ indicates that an event E has a 30% chance of occurring.
Probability of an event	The probability of event E is $$P(E) = \frac{n(E)}{n(S)},$$ where $n(E)$ and $n(S)$ denote the number of equally likely outcomes in E and S, respectively.	The probability of rolling two dice and their sum being 3 is $\frac{2}{36}$. This is because there are $6 \cdot 6 = 36$ ways to roll two dice and only 2 ways to roll a sum of 3. (They are 1 and 2 or 2 and 1.)
Probability of a complement	If E and E' are complementary events, then $E \cap E' = \varnothing$, $E \cup E' = S$, and $P(E') = 1 - P(E)$.	The probability of rolling a 6 with one die is $\frac{1}{6}$. Therefore the probability of *not* rolling a 6 is $1 - \frac{1}{6} = \frac{5}{6}$.
Probability of the union of two events	The probability of E_1 or E_2 (or both) occurring is $P(E_1 \cup E_2) = P(E_1) + P(E_2) - P(E_1 \cap E_2)$.	If $P(E_1) = 0.5$, $P(E_2) = 0.2$, and $P(E_1 \cap E_2) = 0.1$, then $P(E_1 \cup E_2) = 0.5 + 0.2 - 0.1 = 0.6$.
Probability of independent events	Two events E_1 and E_2 are independent if they do not influence one another. $$P(E_1 \cap E_2) = P(E_1) \cdot P(E_2)$$	If E_1 represents a head on the first toss of a coin and E_2 represents a tail on the second toss, then E_1 and E_2 are independent events and $$P(E_1 \cap E_2) = \frac{1}{2} \cdot \frac{1}{2} = \frac{1}{4}.$$
Probability of dependent events	Two events E_1 and E_2 are dependent if they are not independent. $P(E_1 \cap E_2) =$ $P(E_1) \cdot P(E_2, \text{ given that } E_1 \text{ occurred})$	Drawing two hearts without replacement from a standard deck of cards is an example of dependent events. See Example 10.

11.6 Exercises

Probability of an Event

Exercises 1–8: Does the number represent a probability?

1. $\frac{11}{13}$

2. 0.995

3. 2.5

4. 1

5. 0

6. 110%

7. −0.375

8. $\frac{9}{8}$

Exercises 9–18: Find the probability of each event.

9. Tossing a head with a fair coin

10. Tossing a tail with a fair coin

11. Rolling a 2 with a fair die

12. Rolling a 5 or 6 with a fair die

13. Guessing the correct answer to a true-false question

14. Guessing the correct answer to a multiple-choice question with five choices

15. Randomly drawing a king from a standard deck of 52 cards

16. Randomly drawing a club from a standard deck of 52 cards

17. Randomly guessing a four-digit ATM access code

18. Randomly picking the winning team at a basketball game

19. The table shows people's favorite pizza toppings.

Pepperoni	43%
Sausage	19%
Mushrooms	14%
Vegetables	13%

Source: *USA Today*.

(a) If a person is selected at random, what is the probability that pepperoni is not his or her favorite topping?

(b) Find the probability that a person's favorite topping is either mushrooms or sausage.

20. (Refer to Example 2.) Find the probability that a transplant patient in 2008 was waiting for the following.

(a) A lung

(b) A heart or a liver

Probability of Compound Events

Exercises 21–30: Find the probability of the compound event.

21. Tossing a coin twice with the outcomes of two tails

22. Tossing a coin three times with the outcomes of three heads

23. Rolling a die three times and obtaining a 5 or 6 on each roll

24. Rolling a sum of 7 with two dice

25. Rolling a sum of 2 with two dice

26. Rolling a sum other than 7 with two dice

27. Rolling a die four times without obtaining a 6

28. Rolling a die four times and obtaining at least one 6

29. Drawing four consecutive aces from a standard deck of 52 cards without replacement

30. Drawing a pair (two cards with the same value) from a standard deck of 52 cards without replacement

31. *Poker Hands* (Refer to Example 7.) Calculate the probability of drawing 3 hearts and 2 diamonds in a 5-card poker hand. Assume that drawn cards are not replaced and that the 5 cards are drawn only once.

32. *Poker Hands* (Refer to Example 7.) Calculate the probability of drawing 3 kings and 2 queens in a 5-card poker hand. Assume that drawn cards are not replaced and that the 5 cards are drawn only once.

33. *Quality Control* A quality-control experiment involves selecting one string of decorative lights from a box of 20. If the string is defective, the entire box of 20 is rejected. Suppose the box contains four defective strings of lights. What is the probability of rejecting the box?

34. *Quality Control* (Refer to Exercise 33.) Suppose three strings of lights are tested. If any of the strings are defective, the entire box of 20 is rejected. What is the

probability of rejecting a box if there are four defective strings of lights in the box? (*Hint:* Start by finding the probability that the box is not rejected.)

35. *Entrance Exams* A group of students is preparing for college entrance exams. It is estimated that 50% need help with mathematics, 45% with English, and 25% with both.

(a) Draw a Venn diagram representing these data.

(b) Use this diagram to find the probability that a student needs help with mathematics, English, or both.

(c) Solve part (b) symbolically by applying a probability formula.

36. *College Classes* In a college of 5500 students, 950 students are enrolled in English classes, 1220 in business classes, and 350 in both. If a student is chosen at random, find the probability that he or she is enrolled in an English class, a business class, or both.

37. *New Books Published* In 2002, a total of 119,923 new books and editions were published. The table lists the numbers of books published in specific areas. If a new book or edition is selected at random, find the probability that its subject area is one of those specified.

Art	4481
Business	4539
History	6818
Music	1614
Religion	6659
Science	7032

Source: R. R. Bowker Co.

(a) Art or music   (b) Neither science nor religion

38. *Death Rates* In 2004, the U.S. death rate per 100,000 people was 817. What is the probability that a person selected at random died during 2004? (**Source:** Department of Health and Human Services.)

39. *Death Rates* In 2004, the death rate per 100,000 people between the ages of 20 and 24 was 94. What is the probability that a person selected at random from this age group died during 2004? (**Source:** Department of Health and Human Services.)

40. *Tossing a Coin* Find the probability of tossing a coin n times and obtaining n heads. What happens to this probability as n increases? Does this agree with your intuition? Explain.

41. *U.S. AIDS* By 2005, a total of 984,155 cases of AIDS had been diagnosed. The table lists AIDS cases diagnosed in certain cities. Estimate the probability that a person diagnosed with AIDS satisfied the following conditions.

New York	158,502
Los Angeles	49,913
San Francisco	30,277
Miami	29,092

Source: Department of Health and Human Services.

(a) Resided in New York

(b) Did not reside in New York

(c) Resided in Los Angeles or Miami

42. *Rolling Dice* Find the probability of rolling a die five times and obtaining a 6 on the first two rolls, a 5 on the third roll, and a 1, 2, 3, or 4 on the last two rolls.

43. *Rolling Dice* (Refer to Example 5.) Two dice are rolled. Find the probability that the dice show a sum of either 5 or 6.

44. *Rolling Dice* Two dice are rolled. Find the probability that the dice show a sum other than 7 or 11.

45. *Drawing Cards* (Refer to Example 6.) Find the probability of drawing a 2, 3, or 4 from a standard deck of 52 playing cards.

46. *Drawing Cards* Find the probability of drawing two cards, neither of which is an ace or a queen.

47. *Unfair Die* Suppose a die is not fair, but instead the probability P of each number n is as listed in the table. Find the probability of each event.

n	1	2	3	4	5	6
P	0.1	0.1	0.1	0.2	0.2	0.3

(a) Rolling a number that is 4 or higher

(b) Rolling a 6 twice on consecutive rolls

48. Unfair Coin Suppose a coin is not fair, but instead the probability of obtaining a head (H) is $\frac{3}{4}$ and a tail (T) is $\frac{1}{4}$. What is the probability of each event?

(a) HT   (b) HH

(c) HHT   (d) THT

49. Dice Suppose there are two dice, one red and one blue, having the probabilities shown in the table in Exercise 47. If both dice are rolled, find the probability of the given sum.

(a) 12   (b) 11

50. Garage Door Code The code for some garage door openers consists of 12 electrical switches that can be set to either 0 or 1 by the owner. Each setting represents a different code. What is the probability of guessing someone's code at random? (**Source:** Promax.)

51. Lottery To win a lottery, a person must pick 3 numbers from 0 to 9 in the correct order. If a number may be repeated, what is the probability of winning this game with one play?

52. Lottery To win the jackpot in a lottery, a person must pick 5 numbers from 1 to 55 and then pick the power-ball, which has a number from 1 to 42. If the numbers are picked at random, what is the probability of winning this game with one play?

53. Marbles A jar contains 22 red marbles, 18 blue marbles, and 10 green marbles. If a marble is drawn from the jar at random, find the probability that the color is the following.

(a) Red   (b) Not red   (c) Blue or green

54. Marbles A jar contains 55 red marbles and 45 blue marbles. If 2 marbles are drawn from the jar at random without replacement, find the probability that the marbles satisfy the following.

(a) Both are blue.   (b) Neither is blue.

(c) The first marble is red and the second marble is blue.

Conditional Probability and Dependent Events

55. Drawing a Card Find the probability of drawing a queen from a standard deck of cards given that one card, a queen, has already been drawn and not replaced.

56. Drawing a Card Find the probability of drawing a king from a standard deck of cards given that two cards, both kings, have already been drawn and not replaced.

57. Drawing a Card A card is drawn from a standard deck of 52 cards. Given that the card is a face card, what is the probability that the card is a king? (*Hint:* A face card is a jack, queen, or king.)

58. Drawing a Card Three cards are drawn from a deck without replacement. Find the probability that the three cards are an ace, king, and queen in that order.

59. Drawing Marbles A jar initially contains 10 red marbles and 23 blue marbles. What is the probability of drawing a blue marble, given that 2 red marbles and 4 blue marbles have already been drawn?

60. Volleyball Serve The probability that the first serve of a volleyball is out of bounds is 0.3, and the probability that the second serve of a volleyball is in bounds, given that the first serve was out of bounds, is 0.8. Find the probability that the first serve is out of bounds and the second serve is in bounds.

61. Cloudy and Windy The probability of a day being cloudy is 30%, and the probability of it being cloudy and windy is 12%. Given that the day is cloudy, what is the probability that it will be windy?

62. Rainy and Windy The probability of a day being rainy is 80%, and the probability of it being windy and rainy is 72%. Given that the day is rainy, what is the probability that it will be windy?

63. Rolling Dice Two dice are rolled. If the first die shows a 2, find the probability that the sum of the dice is 7 or more.

64. Rolling Dice Three dice are rolled. If the first die shows a 4, find the probability that the sum of the three dice is less than 12.

65. Defective Parts The table shows numbers of automobile parts that are either defective or not defective.

	Type A	Type B	Totals
Defective	7	11	18
Not defective	123	94	217
Totals	130	105	235

If one part is selected at random, calculate each of the following.

(a) The probability that the part is defective

(b) The probability that the part is type A, given that it is defective

(c) The probability that the part is type A and defective

66. *Health* The table shows numbers of patients with two different diseases by gender. (Assume that a person cannot have both diseases.)

	Disease A	Disease B	Totals
Females	145	851	996
Males	256	355	611
Totals	401	1206	1607

If one patient is selected at random, find the probability that the patient is female and has disease B.

67. *Prime Numbers* Suppose a number from 1 to 15 is selected at random. Find the probability of each event.
(a) The number is odd.

(b) The number is even.

(c) The number is prime. (*Hint:* A natural number greater than 1 that has only itself and 1 as factors is called a **prime number**.)

(d) The number is prime and odd.

(e) The number is prime and even.

68. *Students and Classes* (Refer to Example 11.) If one student is selected at random, use Table 11.10 to calculate each of the following.
(a) The probability that the student is male

(b) The probability that the student is taking French, given that the student is male

(c) The probability that the student is male and taking French

Writing about Mathematics

69. What values are possible for a probability? Interpret different probabilities and give examples.

70. Discuss the difference between dependent and independent events. How are their probabilities calculated?

CHECKING BASIC CONCEPTS FOR SECTIONS 11.5 AND 11.6

1. Use mathematical induction to prove that

$$4 + 8 + 12 + 16 + \cdots + 4n = 2n(n + 1).$$

2. Use mathematical induction to prove that $n^2 \leq 2^n$ for $n \geq 4$.

3. Find the probability of tossing a coin four times and obtaining a head every time.

4. Find the probability of rolling a sum of 11 with two dice.

5. Find the probability of drawing four aces and a queen from a standard deck of 52 cards.

6. In 2003, there were 2.7 million high school graduates, of which 1.3 million were male. If a 2003 high school graduate is selected at random, estimate the probability that this graduate is female. (**Source:** The American College Testing Program.)

---

11   Summary

CONCEPT	EXPLANATION AND EXAMPLES

SECTION 11.1 SEQUENCES

Sequences

An infinite sequence is a function whose domain is the natural numbers. Its graph is a scatterplot.

Example: $a_n = \frac{1}{2}n^2 - 2$; the first 4 terms are as follows.

$$a_1 = \frac{1}{2}(1)^2 - 2 = -\frac{3}{2}, \qquad a_2 = \frac{1}{2}(2)^2 - 2 = 0$$

$$a_3 = \frac{1}{2}(3)^2 - 2 = \frac{5}{2}, \qquad a_4 = \frac{1}{2}(4)^2 - 2 = 6$$

CONCEPT	EXPLANATION AND EXAMPLES

SECTION 11.1 SEQUENCES (CONTINUED)

Arithmetic Sequence

Recursive Definition:
$a_n = a_{n-1} + d$, where d is the common difference

Function Definition:
$f(n) = dn + c$, or equivalently, $f(n) = a_1 + d(n-1)$, where $a_n = f(n)$ and d is the common difference

Example: $a_n = a_{n-1} + 3$, $a_1 = 4$ and $f(n) = 3n + 1$ describe the same sequence. The common difference is $d = 3$. The terms of the sequence are

$$4, 7, 10, 13, 16, 19, 22, \ldots.$$

Geometric Sequence

Recursive Definition:
$a_n = ra_{n-1}$, where r is the common ratio

Function Definition:
$f(n) = cr^{n-1}$, where $c = a_1$ and r is the common ratio

Example: $a_n = -2a_{n-1}$, $a_1 = 3$ and $f(n) = 3(-2)^{n-1}$ describe the same sequence. The common ratio is $r = -2$. The terms of the sequence are

$$3, -6, 12, -24, 48, -96, 192, \ldots.$$

SECTION 11.2 SERIES

Series

A series is the summation of the terms of a sequence.

Examples: $1 + \dfrac{1}{2} + \dfrac{1}{4} + \dfrac{1}{8} + \cdots + \dfrac{1}{2^{n-1}} + \cdots$ Infinite series

$$\sum_{k=1}^{5} 2k = 2 + 4 + 6 + 8 + 10 \qquad \text{Finite series}$$

Arithmetic Series

Finite Arithmetic Series:

$$\sum_{k=1}^{n} a_k = a_1 + a_2 + a_3 + \cdots + a_n,$$

where $a_k = dk + c$ for some constants c and d and d is the common difference. Summing the terms of an arithmetic sequence results in an arithmetic series.

Sum of the First n Terms:

$$S_n = n\left(\frac{a_1 + a_n}{2}\right) \qquad \text{or} \qquad S_n = \frac{n}{2}(2a_1 + (n-1)d)$$

Example: The series $4 + 7 + 10 + 13 + 16 + 19 + 22$ is defined by $a_k = 3k + 1$. Its sum is $S_7 = 7\left(\frac{4 + 22}{2}\right) = 91$.

CONCEPT	EXPLANATION AND EXAMPLES

SECTION 11.2 SERIES (CONTINUED)

Geometric Series

Infinite Geometric Series:

$$\sum_{k=1}^{\infty} a_k = a_1 + a_2 + a_3 + \cdots + a_n + \cdots,$$

where $a_k = a_1 r^{k-1}$ for some nonzero constants a_1 and r
Summing the terms of a geometric sequence results in a geometric series.

Sum of First n Terms:

$S_n = a_1\left(\frac{1 - r^n}{1 - r}\right)$, where a_1 is the first term and r is the common ratio

Example: The series $3 + 6 + 12 + 24 + 48 + 96$ has $a_1 = 3$ and $r = 2$.

Its sum is $S_6 = 3\left(\frac{1 - 2^6}{1 - 2}\right) = 189$.

Sum of Infinite Geometric Series:

$S = \frac{a_1}{1 - r}$, if $|r| < 1$. S does not exist if $|r| \geq 1$.

Example: $4 + 1 + \frac{1}{4} + \frac{1}{16} + \frac{1}{64} + \cdots$ has sum $S = \dfrac{4}{1 - \frac{1}{4}} = \dfrac{16}{3}$.

SECTION 11.3 COUNTING

Fundamental Counting Principle

Let E_1 and E_2 be independent events. If event E_1 can occur m_1 ways and if event E_2 can occur m_2 ways, then there are $m_1 \cdot m_2$ ways for both events to occur.

Example: If two multiple-choice questions have 5 choices each, then there are $5 \cdot 5 = 25$ ways to answer the two questions.

Factorial Notation

$n! = n(n - 1)(n - 2) \cdots (3)(2)(1)$

Examples: $0! = 1$; $1! = 1$; $5! = 5 \cdot 4 \cdot 3 \cdot 2 \cdot 1 = 120$

Permutations

$P(n, r) = \frac{n!}{(n - r)!}$ represents the number of permutations, or arrangements, of n elements taken r at a time. Order is important in calculating a permutation.

Example: $P(5, 3) = \dfrac{5!}{(5 - 3)!} = \dfrac{120}{2} = 60$

Combinations

$C(n, r) = \binom{n}{r} = \frac{n!}{(n - r)! \, r!}$ represents the number of combinations of n elements taken r at a time. Order is unimportant in calculating a combination.

Example: $C(5, 3) = \dbinom{5}{3} = \dfrac{5!}{(5 - 3)!3!} = \dfrac{120}{2 \cdot 6} = 10$

CONCEPT	EXPLANATION AND EXAMPLES

SECTION 11.4 THE BINOMIAL THEOREM

Binomial Theorem

$$(a + b)^n = \binom{n}{0}a^n + \binom{n}{1}a^{n-1}b + \cdots + \binom{n}{n-1}ab^{n-1} + \binom{n}{n}b^n$$

Example: $(x + y)^4 = 1x^4 + 4x^3y + 6x^2y^2 + 4xy^3 + 1y^4$

The coefficients can also be found in the fifth row of Pascal's triangle.

Pascal's Triangle

Pascal's triangle can be used to calculate the binomial coefficients when expanding the expression $(a + b)^n$.

SECTION 11.5 MATHEMATICAL INDUCTION

Principle of Mathematical Induction

Let S_n be a statement concerning the positive integer n. Suppose that

1. S_1 is true;
2. for any positive integer k, if S_k is true, then S_{k+1} is also true.

Then S_n is true for every positive integer n.

SECTION 11.6 PROBABILITY

Probability

$P(E) = \frac{n(E)}{n(S)}$, where $n(E)$ is the number of outcomes in event E and $n(S)$ is the number of equally likely outcomes in the finite sample space S. Note that $0 \leq P(E) \leq 1$.

Compound Events

Probability of either E_1 or E_2 (or both) occurring:

$$P(E_1 \cup E_2) = P(E_1) + P(E_2) - P(E_1 \cap E_2)$$

If E_1 and E_2 are *mutually exclusive*, then $E_1 \cap E_2 = \varnothing$ and

$$P(E_1 \cup E_2) = P(E_1) + P(E_2).$$

Probability of *both* E_1 and E_2 occurring:
If E_1 and E_2 are *independent*, then

$$P(E_1 \cap E_2) = P(E_1) \cdot P(E_2).$$

If E_1 and E_2 are *dependent*, then

$$P(E_1 \cap E_2) = P(E_1) \cdot P(E_2, \text{ given that } E_1 \text{ occurred}).$$

11   Review Exercises

Exercises 1–4: Find the first four terms of the sequence.

1. $a_n = -3n + 2$

2. $a_n = n^2 + n$

3. $a_n = 2a_{n-1} + 1; a_1 = 0$

4. $a_n = a_{n-1} + 2a_{n-2}; a_1 = 1, a_2 = 4$

Exercises 5 and 6: Use the graphical representation to identify the terms of the finite sequence.

5.

6.
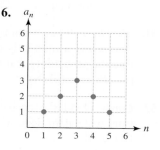

Exercises 7 and 8: The first five terms of an infinite arithmetic or geometric sequence are given. Find

(a) numerical,   (b) graphical,   and   (c) symbolic

representations of the sequence. Include at least the first eight terms of the sequence in your table and graph.

7. $3, 1, -1, -3, -5$

8. $1.5, -3, 6, -12, 24$

9. Find a general term a_n for the arithmetic sequence with $a_3 = -3$ and $d = 4$.

10. Find a general term a_n for the geometric sequence with $a_1 = 2.5$ and $a_6 = -80$.

Exercises 11–14: Determine if the sequence is arithmetic, geometric, or neither.

11. $f(n) = 5 - 2n$

12. $f(n) = 3n^2$

13. $f(n) = 3(2)^n$

14. $f(n) = n + \left(\frac{1}{2}\right)^{n-1}$

Exercises 15 and 16: For the given a_n, calculate S_5.

15. $a_n = 4n + 1$

16. $a_n = 3(4)^{n-1}$

Exercises 17–20: Use a formula to find the sum of the series.

17. $-2 + 1 + 4 + 7 + 10 + 13 + 16 + 19 + 22$

18. $2 + 4 + 6 + 8 + \cdots + 98 + 100$

19. $1 + 3 + 9 + 27 + 81 + 243 + 729 + 2187$

20. $64 + 16 + 4 + 1 + \frac{1}{4} + \frac{1}{16}$

Exercises 21 and 22: Find the sum of the infinite geometric series.

21. $4 - \frac{4}{3} + \frac{4}{9} - \frac{4}{27} + \frac{4}{81} - \frac{4}{243} + \cdots$

22. $0.2 + 0.02 + 0.002 + 0.0002 + 0.00002 + \cdots$

Exercises 23 and 24: Write out the terms of the series.

23. $\sum_{k=1}^{5}(5k + 1)$ **24.** $\sum_{k=1}^{4}(2 - k^2)$

Exercises 25 and 26: Write the series using summation notation.

25. $1^3 + 2^3 + 3^3 + 4^3 + 5^3 + 6^3$

26. $1 + \frac{1}{10} + \frac{1}{100} + \frac{1}{1000} + \frac{1}{10,000}$

Exercises 27 and 28: Use a formula to find the sum of the first thirty terms of the arithmetic sequence.

27. $a_1 = 5, d = -3$ **28.** $a_1 = -2, a_{10} = 16$

29. Write $\frac{2}{11}$ as an infinite geometric series.

30. Write the infinite series

$$0.23 + 0.0023 + 0.000023 + \cdots$$

as a rational number.

Exercises 31 and 32: Evaluate the expression.

31. $P(6, 3)$ **32.** $C(7, 4)$

Exercises 33 and 34: Use mathematical induction to prove that the statement is true for every positive integer.

33. $1 + 3 + 5 + 7 + \cdots + (2n - 1) = n^2$

34. $2 + 2^2 + 2^3 + \cdots + 2^n = 2(2^n - 1)$

Exercises 35 and 36: Find the probability of each event.

35. A 1, 2, or 3 appears when a die is rolled

36. Tossing three heads in a row using a fair coin

Applications

37. *Standing in Line*   In how many arrangements can 5 people stand in a line?

38. *Giving a Speech* How many arrangements are possible in which 4 students out of a class of 15 each give a speech?

39. *Exam Questions* Count the ways that an exam, consisting of 20 multiple-choice questions with 4 choices each, could be answered.

40. *License Plates* Count the different license plates having 4 numeric digits followed by 2 letters.

41. *Combination Lock* A combination lock consists of 4 numbers from 0 to 49. If a number may be repeated, find the number of possible combinations.

42. *Dice* A red die and a blue die are rolled. How many ways are there for the sum to equal 4?

43. *Height of a Ball* When a Ping-Pong ball is dropped, it rebounds to 90% of its initial height.
   (a) Write the first five terms of a sequence that gives the maximum height attained by the ball on each rebound when it is dropped from an initial height of 4 feet. Let $a_1 = 4$. What type of sequence describes these maximum heights?

   (b) Give a graphical representation of this sequence for the first five terms.

   (c) Find a formula for a_n.

44. *Falling Object* If air resistance is ignored, an object falls 16, 48, 80, and 112 feet during each successive 1-second interval.
   (a) What type of sequence describes these distances?

   (b) Determine how far an object falls during the sixth second.

   (c) Find a formula for the *n*th term of this sequence.

45. *Committees* In how many ways can a committee of 3 be selected from 6 people?

46. *Committees* Find the number of committees with 3 women and 3 men that can be selected from a group of 7 women and 5 men.

47. *Test Questions* On a test with 10 essay questions, students are asked to answer 6 questions. How many ways can the essay questions be selected?

48. *Binomial Theorem* Use the binomial theorem to expand the expression $(2x - y)^4$.

49. *Quality Control* A quality-control experiment involves selecting 2 batteries from a pack of 16. If either battery is defective, the entire pack of 16 is rejected. Suppose a pack contains 2 defective batteries. What is the probability that this pack will not be rejected?

50. *Venn Diagram* Of a group of 82 students, 19 are enrolled in music, 22 in art, and 10 in both.
   (a) Draw a Venn diagram representing these data.

   (b) Use this diagram to determine the probability that a student selected at random is enrolled in music, art, or both.

   (c) Solve part (b) symbolically by applying a probability formula.

51. *Marbles* A jar contains 13 red, 27 blue, and 20 green marbles. If a marble is drawn from the jar at random, find the probability that the marble is the following.
   (a) Blue   **(b)** Not blue   **(c)** Red

52. *Cards* Find the probability of drawing 2 diamonds from a standard deck of 52 cards without replacing the first card.

53. *Insect Populations* The monthly density of an insect population, measured in thousands per acre, is described by the recursive sequence

$$a_1 = 100; \quad a_n = \frac{2a_{n-1}}{1 + (a_{n-1}/4000)}, \quad n > 1.$$

Use your calculator to graph the sequence in the window [0, 16, 1] by [0, 5000, 1000]. Include the first 15 terms and discuss any trends illustrated by the graph.

EXTENDED AND DISCOVERY EXERCISES

Exercises 1 and 2: Antibiotic Resistance *Because of the frequent use of antibiotics in society, many strains of bacteria are becoming resistant. Some types of haploid bacteria contain genetic material called plasmids. Plasmids are capable of making a strain of bacterium resistant to antibiotic drugs. Genetic engineers want to predict the resistance of various bacteria after many generations.*

1. Suppose a strain of bacterium contains two plasmids R_1 and R_2. Plasmid R_1 is resistant to the antibiotic ampicillin, whereas plasmid R_2 is resistant to the antibiotic tetracycline. When bacteria reproduce through cell division, the type of plasmids passed on to each new cell is random. For example, a daughter cell could have two plasmids of type R_1 and no plasmid of type R_2, one of each type, or no plasmid of type R_1 and two plasmids of type R_2. The probability $P_{k,j}$ that a mother cell with k plasmids of type R_1 produces a daughter cell with j plasmids of type R_1 can be calculated by the formula

$$P_{k,j} = \frac{\binom{2k}{j}\binom{4-2k}{2-j}}{\binom{4}{2}}.$$

(**Source:** F. Hoppensteadt and C. Peskin, *Mathematics in Medicine and the Life Sciences.*)

(a) Compute $P_{k,j}$ for $0 \le k$, $j \le 2$. Assume that $\binom{0}{0} = 1$ and $\binom{k}{j} = 0$ whenever $k < j$. Record your results in the matrix

$$P = \begin{bmatrix} P_{00} & P_{01} & P_{02} \\ P_{10} & P_{11} & P_{12} \\ P_{20} & P_{21} & P_{22} \end{bmatrix}.$$

(b) Which elements in P are the greatest? Interpret the result.

2. (Continuation of Exercise 1.) The genetic makeup of future generations of the haploid bacterium can be modeled using matrices. Let $A = [a_1, a_2, a_3]$ be a 1×3 matrix containing three probabilities. The value of a_1 is the probability that it has two R_1 plasmids and no R_2

plasmid; a_2 is the probability that it has one R_1 plasmid and one R_2 plasmid; a_3 is the probability that it has no R_1 plasmid and two R_2 plasmids. If an entire generation of the bacterium has one plasmid of each type, then $A_1 = [0, 1, 0]$. In this case the bacterium is resistant to both antibiotics. The probabilities A_n for plasmids R_1 and R_2 in the nth generation of the bacterium can be calculated with the matrix recurrence equation $A_n = A_{n-1}P$, where $n > 1$ and P is the 3×3 matrix determined in Exercise 1. The resulting phenomenon was not well understood until quite recently. It is now used in the genetic engineering of plasmids.

(a) If an entire strain of the bacterium is resistant to both the antibiotics ampicillin and tetracycline, make a conjecture as to the drug resistance of future generations of this bacterium.

(b) Test your conjecture by repeatedly computing the matrix product $A_n = A_{n-1}P$. Let $A_1 = [0, 1, 0]$ and $n = 2, 3, \ldots, 12$. Interpret the result. (It may surprise you.)

3. The sum of the first n terms of an arithmetic series is given by $S_n = n\left(\frac{a_1 + a_n}{2}\right)$. Justify this formula using a geometric discussion. (*Hint:* Start by graphing an arithmetic sequence where n is an odd number.)

1-11   Cumulative Review Exercises

1. Write 34,500 in scientific notation and 1.52×10^{-4} in standard form.

2. Evaluate $\dfrac{5 - \sqrt[3]{4}}{\pi^2 - (\sqrt{3} + 1)}$. Round your answer to the nearest hundredth.

3. Find the exact distance between $(-4, 2)$ and $(1, -2)$.

4. Graph f.
   (a) $f(x) = 4x - 2$
   (b) $f(x) = |2x - 1|$
   (c) $f(x) = x^2 + 2x$
   (d) $f(x) = \sqrt{x - 1}$
   (e) $f(x) = \dfrac{1}{x + 2}$
   (f) $f(x) = x^3 - 2$
   (g) $f(x) = \log_2 x$
   (h) $f(x) = 3\left(\frac{1}{2}\right)^x$

Exercises 5 and 6: Complete the following.
   (a) *Evaluate $f(-3)$ and $f(a + 1)$.*
   (b) *Determine the domain of f.*

5. $f(x) = \sqrt{1 - x}$
6. $f(x) = \dfrac{1}{x^2 - 4}$

7. Find the average rate of change of $f(x) = x^3 - 4$ from $x = -2$ to $x = -1$.

8. Find the difference quotient for $f(x) = x^2 - 3x$.

Exercises 9 and 10: Write an equation of a line satisfying the conditions. Use slope-intercept form whenever possible.

9. Passing through $(2, -4)$ and $(-3, 2)$

10. Passing through the point $(-1, 3)$ and perpendicular to the line $y = -\frac{3}{4}x + 1$

11. The graph of a linear function f is shown.
  (a) Identify the slope, y-intercept, and x-intercept.

  (b) Write a formula for f.

  (c) Find any zeros of f.

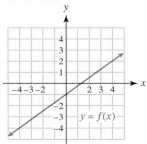

12. Determine the x- and y-intercepts on the graph of $-3x + 4y = 12$. Graph the equation.

13. Solve each equation.
  (a) $4(x - 2) + 1 = 3 - \frac{1}{2}(2x + 3)$

  (b) $6x^2 = 13x + 5$ **(c)** $x^2 - x - 3 = 0$

  (d) $x^3 + x^2 = 4x + 4$ **(e)** $x^4 - 4x^2 + 3 = 0$

  (f) $\dfrac{1}{x - 3} = \dfrac{4}{x + 5}$ **(g)** $3e^{2x} - 5 = 23$

  (h) $2\log(x + 1) - 1 = 2$

  (i) $\sqrt{x + 3} + 4 = x + 1$

  (j) $|3x - 1| = 5$

14. Solve each inequality. Write your answer in set-builder or interval notation.
  (a) $3x - 5 < x + 1$ **(b)** $x^2 - 4x - 5 \le 0$

  (c) $(x + 1)(x - 2)(x - 3) > 0$

  (d) $\dfrac{2}{x - 1} < 0$ **(e)** $|3x - 5| \le 4$

  (f) $|4 - x| > 0$ **(g)** $\dfrac{3}{4} \le \dfrac{1 - 2x}{3} < \dfrac{5}{2}$

15. Graph f. Is f continuous on its domain?
$$f(x) = \begin{cases} 2x + 3 & \text{if } -3 \le x < -1 \\ x^2 & \text{if } -1 \le x < 1 \\ 2 - x & \text{if } 1 \le x \le 3 \end{cases}$$

16. Solve $-2.3x + 3.4 = \sqrt{2x^2 - 1}$ graphically. Round your answers to the nearest tenth.

17. The graph of a nonlinear function f is shown. Solve each equation or inequality. Write the solution set to each inequality in set-builder or interval notation.
  (a) $f(x) = 0$ **(b)** $f(x) > 0$ **(c)** $f(x) \le 0$

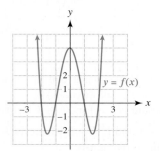

18. Write $f(x) = 3x^2 + 24x + 43$ in the vertex form given by $f(x) = a(x - h)^2 + k$.

19. Find the vertex on the graph of $f(x) = -3x^2 + 9x + 1$.

20. Use the given graph of $y = f(x)$ to sketch a graph of each equation.
  (a) $y = f(x - 1) + 2$ **(b)** $y = \frac{1}{2}f(x)$

  (c) $y = -f(x)$ **(d)** $y = f(2x)$

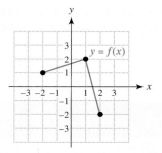

21. Use the graph of f to estimate each of the following.
  (a) Intervals where f is increasing or decreasing

  (b) The zeros of f

  (c) The coordinates of any turning points

  (d) Any local extrema

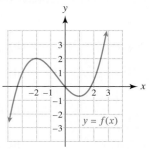

22. Determine if the function defined by $f(x) = x^3 - 5x$ is even, odd, or neither.

23. Divide each expression.

(a) $\dfrac{6x^4 - 2x^2 + 1}{2x^2}$ (b) $\dfrac{2x^4 - 3x^3 - x + 2}{x + 1}$

24. A degree 4 function f has zeros $-2, -1, 1,$ and 2 and leading coefficient 6. Write the complete factored form of $f(x)$.

25. A degree 3 function f with real coefficients has leading coefficient 3 and zeros -1 and $3i$. Write $f(x)$ in complete factored form and expanded form.

26. Write $(2 - i)(2 + 3i)$ in standard form.

27. Find all solutions, real or imaginary, to the quadratic equation $x^2 + 2x + 5 = 0$.

28. State the domain of $f(x) = \frac{2x - 5}{x + 5}$. Find any vertical or horizontal asymptotes for the graph of f.

29. Write $\sqrt[5]{(x + 1)^3}$ using rational exponents. Evaluate the expression for $x = 31$.

30. Use the tables to evaluate each expression, if possible.

x	0	1	2	3
$f(x)$	1	2	4	5

x	0	1	2	3
$g(x)$	4	3	2	1

(a) $(f - g)(1)$ (b) $(f/g)(2)$

(c) $(g \circ f)(3)$ (d) $(g \circ f^{-1})(5)$

31. Use the graphs of f and g to complete the following.

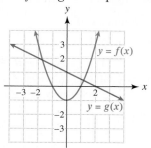

(a) $(f + g)(2)$ (b) $(fg)(0)$

(c) $(g \circ f)(1)$ (d) $(g^{-1} \circ f)(-2)$

32. Let $f(x) = \frac{1}{x + 2}$ and $g(x) = x^2 + x - 4$. Find each of the following.

(a) $(f - g)(0)$ (b) $(g \circ f)(-1)$

(c) $(fg)(x)$ (d) $(g \circ f)(x)$

33. Find $f^{-1}(x)$ if $f(x) = \frac{x}{x + 1}$.

34. Graph $f(x) = x^3 - 1, y = x$, and $y = f^{-1}(x)$ on the same axes.

35. Find either a linear or an exponential function f that models the data in the table.

x	0	1	2	3
$f(x)$	2	6	18	54

36. There are initially 1000 bacteria in a sample, and the number doubles every 2 hours.

(a) Find C and a so that $f(x) = Ca^x$ models the number of bacteria after x hours.

(b) Estimate the number of bacteria after 5.2 hours.

(c) When are there 9000 bacteria in the sample?

37. Five hundred dollars is deposited in an account that pays 6% annual interest compounded quarterly. Find the amount in the account after 15 years.

38. Simplify each logarithm by hand.

(a) $\log_2 \frac{1}{16}$ (b) $\log \sqrt{10}$

(c) $\ln e^4$ (d) $\log_4 2 + \log_4 32$

39. Find the domain and range of each function f.

(a) $f(x) = x^2 - 2x + 1$

(b) $f(x) = 10^x$ (c) $f(x) = \ln x$

(d) $f(x) = \dfrac{1}{x}$

40. Expand the expression $\log \sqrt{\frac{x + 1}{yz}}$.

41. Write $2 \log x + 3 \log y - \frac{1}{3} \log z$ as a logarithm of a single expression.

42. Approximate $\log_4 52$ to three decimal places.

43. Graph $y = f(x)$.

(a) $f(x) = \csc x$

(b) $f(x) = -2 \sin \frac{1}{2} x$

(c) $f(x) = 3 \cos \left(2\left(x - \frac{\pi}{2}\right)\right)$

(d) $f(x) = \tan (\pi x)$

44. Find the domain of $f(x) = \sec 2x$.

45. Solve each equation for all real numbers t.

(a) $\cos t = -\dfrac{\sqrt{3}}{2}$

(b) $\tan^2 t - 1 = 0$

(c) $\sin^2 t + \dfrac{1}{2} \sin t = 0$

(d) $\cos 2t = -1$

46. Find the complementary angle α and the supplementary angle β to $54°35'12''$.

47. Convert $5.54°$ to degrees, minutes, and seconds.

48. Convert $135°$ to radians.

49. Convert $\dfrac{5\pi}{4}$ radians to degrees.

50. Approximate $\cot \dfrac{2\pi}{9}$ to the nearest hundredth.

51. Find the values of the six trigonometric functions of an angle θ in standard position having its terminal side pass through the point $(-7, 24)$.

52. Find the exact values of the six trigonometric functions of $\theta = \dfrac{5\pi}{6}$.

53. Find the values of the six trigonometric functions of θ if $\cos \theta = -\dfrac{11}{61}$ and $\cos \theta < 0$.

54. Evaluate $\cos^{-1}\left(\dfrac{1}{2}\right)$.

55. Solve the right triangle shown in the figure.

56. Simplify $(\sec t - 1)(\sec t + 1)$.

57. Verify the identity.
$1 - \sin^2 \theta + \cot^2 \theta - \sin^2 \theta \cot^2 \theta = \cot^2 \theta$

58. Solve $2 \sin 2x = \ln x$ graphically for $0 \le x < 2\pi$.

59. Solve triangle ABC. Approximate values to the nearest tenth.

(a) $\beta = 42°, \gamma = 31°, a = 22$

(b) $\gamma = 50°, c = 7, b = 8$

(c) $\alpha = 44°, b = 7, c = 8$

(d) $a = 10, b = 11, c = 12$

60. Find the area of a triangle with $\alpha = 30°, b = 15$ feet, and $c = 20$ feet.

61. Let $\mathbf{a} = \langle 3, -4 \rangle$ and $\mathbf{b} = \langle -5, 12 \rangle$. Find the following.

(a) $\|\mathbf{a}\|$

(b) $4\mathbf{b} - 2\mathbf{a}$

(c) $\mathbf{a} \cdot \mathbf{b}$

(d) The angle between $\mathbf{a}$ and $\mathbf{b}$

62. Graph the parametric equations $x = 2 \cos t, y = 2 \sin t$ for $0 \le t \le 2\pi$.

63. Graph the polar equation given by $r = 1 + \sin \theta$ for $0 \le \theta < 2\pi$.

64. Find the two square roots of $-16i$.

65. Solve the system of equations.

(a) $\begin{aligned} 2x + 3y &= 4 \\ 2x - 5y &= -12 \end{aligned}$

(b) $\begin{aligned} -2x + \tfrac{1}{2}y &= 1 \\ 4x - y &= -2 \end{aligned}$

(c) $\begin{aligned} x^2 + y^2 &= 16 \\ 2x^2 - y^2 &= 11 \end{aligned}$

(d) $\begin{aligned} x + y - 2z &= -6 \\ 2x - y - 3z &= -18 \\ 3y - z &= 6 \end{aligned}$

66. The variable z varies inversely as the square of x. If $z = 8$ when $x = 50$, find z when $x = 36$.

67. Graph the solution set to each inequality or system of inequalities.

(a) $-2x + 3y < 6$

(b) $\begin{aligned} x + y &\le 4 \\ x - 2y &> 6 \end{aligned}$

68. Find $2A + B$ and AB if

$$A = \begin{bmatrix} -1 & 0 & 2 \\ 1 & -3 & 1 \\ 0 & -3 & 4 \end{bmatrix} \quad \text{and} \quad B = \begin{bmatrix} 1 & 5 & 1 \\ -2 & 2 & 1 \\ 0 & 1 & -2 \end{bmatrix}.$$

69. Find A^{-1} if $A = \begin{bmatrix} 1 & -2 \\ -3 & 4 \end{bmatrix}$.

70. Solve by using A^{-1}.

$$\begin{aligned} x - y - 2z &= 5 \\ -x + 2y + 3z &= -7 \\ 2y + z &= -2 \end{aligned}$$

71. Calculate the determinant of each matrix.

(a) $\begin{bmatrix} -1 & 4 \\ 2 & 3 \end{bmatrix}$ **(b)** $\begin{bmatrix} 2 & 3 & -1 \\ 3 & -1 & 5 \\ 0 & 0 & -2 \end{bmatrix}$

72. Sketch a graph of each equation. Label any foci.

(a) $y^2 = 2x$ **(b)** $\dfrac{x^2}{9} + \dfrac{y^2}{25} = 1$

(c) $9x^2 - 18x + 4y^2 + 16y + 25 = 36$

(d) $\dfrac{(x + 1)^2}{16} - \dfrac{(y - 2)^2}{9} = 1$

73. Find an equation of a parabola with vertex $(0, 0)$ and focus $\left(\frac{3}{4}, 0\right)$.

74. Find an equation of an ellipse with vertices $(\pm 3, 1)$ and foci $(\pm 2, 1)$.

75. Find an equation of a hyperbola with foci $(0, \pm 13)$ and vertices $(0, \pm 5)$.

76. Find the first four terms of each sequence.
(a) $a_n = (-1)^{n-1}(3)^n$

(b) $a_n = a_{n-1}a_{n-2}; a_1 = 2, a_2 = 3$

77. Find a general term for the arithmetic sequence given that $a_1 = 4$ and $a_3 = 12$.

78. Find a general term for the geometric sequence given that $a_2 = 6$ and $r = \frac{1}{2}$.

79. Use a formula to find the sum of each sequence.
(a) $2 + 5 + 8 + 11 + \cdots + 74$

(b) $0.2 + 0.02 + 0.002 + 0.0002 + \cdots$

80. Find the sum $\sum_{k=1}^{7} (k^2 + k)$.

81. Count the number of license plates that can be formed by three letters followed by four digits.

82. Evaluate $P(4, 2)$ and $\binom{6}{3}$.

83. Expand $(2x - 1)^4$.

84. Use mathematical induction to show that
$$5 + 7 + 9 + 11 + \cdots + (2n + 3) = n(n + 4).$$

85. Find the probability of drawing a heart or an ace from a standard deck of 52 cards.

86. Find the probability of rolling a sum of 7 with two dice.

87. A number from 1 to 20 is drawn at random. Find the probability that the number is prime.

88. The probability of a day being cloudy is 40%, and the probability of it being cloudy *and* windy is 15%. Given that the day is cloudy, what is the probability that it will be windy?

Applications

89. *Distance*   At noon, car A is traveling north at 50 miles per hour and is located 30 miles north of car B. Car B is traveling east at 50 miles per hour. Approximate the distance between the cars at 1:45 P.M. to the nearest tenth of a mile.

90. *Distance from Home*   A driver is initially 240 miles from home, traveling toward home on a straight interstate at 60 miles per hour.
(a) Write a formula for a linear function D that models the distance between the driver and home after x hours.

(b) What is an appropriate domain for D?

(c) Graph D.

(d) Identify the x- and y-intercepts. Interpret each.

91. *Running*   An athlete traveled 10.5 miles in 1.3 hours, jogging at 7 miles per hour and 9 miles per hour. How long did the athlete run at each speed?

92. *Average Rate of Change*   The total distance D in feet traveled by a racehorse after t seconds is given by $D(t) = 3t^2$ for $0 \le t \le 5$.
(a) Find the average rate of change of D from 0 to 1 and 3 to 4.

(b) Interpret these average rates of change.

93. *Working Together*   Suppose one person can mow a lawn in 5 hours and another person can mow the same lawn in 4 hours. How long will it take to mow the lawn if they work together?

94. *Maximum Height*   A stone is shot upward with a slingshot. Its height s in feet after t seconds is given by $s(t) = -16t^2 + 96t + 4$. Find its maximum height.

95. *Airline Tickets* Tickets for a charter flight are regularly $400, but for each additional ticket bought the cost of each ticket is reduced by $5.

(a) Write a quadratic function C that gives the total cost of buying x tickets.

(b) Solve $C(x) = 7000$ and interpret the result.

(c) Find the absolute maximum for C and interpret your result. Assume that x is an integer.

96. *Dimensions of a Rectangle* A rectangle has a perimeter of 48 inches and an area of 143 square inches. Find its dimensions.

97. *Satellite Dish* A parabolic satellite dish has a 3-foot diameter and is 6 inches deep. How far from the vertex should the receiver be so that it is located at the focus?

98. *Bouncing Ball* A Ping-Pong ball is dropped from 4 feet. Each time the ball bounces, it rebounds to 75% of its previous height. Use an infinite geometric series to estimate the *total* distance that the ball travels before coming to rest on the floor.

99. *Marbles* A jar contains 15 red, 28 blue, and 34 green marbles. If a marble is drawn at random, find the probability that the marble is not blue.

100. *Bicycle* A wheel on a trailer is 21 inches in diameter and rotating at 45 radians per second. Find the speed of the trailer in feet per second.

101. *Length of a Shadow* The angle of elevation of the sun is 42°. Find the length of the shadow cast by a building that is 105 feet tall. Round your answer to the nearest tenth of a foot.

102. *Daylight Hours* The average number of daylight hours for each month at a location are shown in the table. Model these data by using a function of the form

$$f(t) = a \cos(b(t - c)) + d.$$

Month	1	2	3	4	5	6
Daylight Hours	9.4	10.5	12	13.5	14.6	15

Month	7	8	9	10	11	12
Daylight Hours	14.6	13.5	12	10.5	9.4	9

103. *Angle of Elevation* A 6-foot person casts a 7-foot shadow. Estimate the angle of elevation of the sun to the nearest tenth of a degree.

104. *Surveyor* A surveyor measures two sides of a lot in the shape of a parallelogram to be $a = 153$ feet and $c = 167$ feet. The angle between these sides is $\beta = 54°$.

(a) Find the length of the shortest diagonal to the nearest tenth of a foot.

(b) Estimate the area of the lot.

Reference: Basic Concepts from Algebra and Geometry

R

Throughout the text there are algebra and geometry review notes that direct students to "see Chapter R." This reference chapter contains seven sections, which provide a review of important topics from algebra and geometry. Students can refer to these sections for more explanation or extra practice. Instructors can use these sections to emphasize a variety of mathematical skills.

R.1 Formulas from Geometry

- Use formulas for shapes in a plane
- Find sides of right triangles by applying the Pythagorean theorem
- Apply formulas to three-dimensional objects
- Use similar triangles to solve problems

Geometric Shapes in a Plane

This subsection discusses formulas related to rectangles, triangles, and circles.

Rectangles   The distance around the boundary of a geometric shape in a plane is called its **perimeter**. The perimeter of a rectangle equals the sum of the lengths of its four sides. For example, the perimeter of the rectangle shown in Figure R.1 is $5 + 4 + 5 + 4 = 18$ feet. The perimeter P of a rectangle with length L and width W is $P = 2L + 2W$.

Figure R.1 Rectangle

The area A of a rectangle equals the product of its length and width: $A = LW$. So the rectangle in Figure R.1 has an area of $5 \cdot 4 = 20$ square feet.

Many times the perimeter or area of a rectangle is written in terms of variables, as demonstrated in the next example.

EXAMPLE 1   Finding the perimeter and area of a rectangle

The length of a rectangle is three times greater than its width. If the width is x inches, write expressions that give the perimeter and area.

Figure R.2

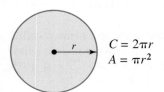

Figure R.3 Triangle

SOLUTION The width of the rectangle is x inches, so its length is $3x$ inches. A sketch is shown in Figure R.2. The perimeter is

$$
\begin{aligned}
P &= 2L + 2W \\
&= 2(3x) + 2(x) \\
&= 8x \text{ inches.}
\end{aligned}
$$

The area is $A = LW = 3x \cdot x = 3x^2$ square inches.

Now Try Exercise 7 ◀

Triangles If the base of a triangle is b and its height is h, as illustrated in Figure R.3, then the area A of the triangle is given by

$$
A = \frac{1}{2}bh.
$$

EXAMPLE 2 Finding the area of a triangle

Calculate the area of the triangle.

SOLUTION The triangle has a base of 8 feet and a height of 5 feet. Therefore its area is

$$
A = \frac{1}{2}bh = \frac{1}{2} \cdot 8 \cdot 5 = 20 \text{ square feet.}
$$

Now Try Exercise 11 ◀

$C = 2\pi r$
$A = \pi r^2$

Figure R.4 Circle

Circles The perimeter of a circle is called its **circumference** C and is given by $C = 2\pi r$, where r is the radius of the circle. The area A of a circle is $A = \pi r^2$. See Figure R.4.

EXAMPLE 3 Finding the circumference and area of a circle

A circle has a radius of 12.5 inches. Approximate its circumference and area.

SOLUTION

Circumference: $C = 2\pi r = 2\pi(12.5) = 25\pi \approx 78.5$ inches

Area: $A = \pi r^2 = \pi(12.5)^2 = 156.25\pi \approx 490.9$ square inches

Now Try Exercise 21 ◀

The Pythagorean Theorem

One of the most famous theorems in mathematics is the Pythagorean theorem. It states that a triangle with legs a and b and hypotenuse c is a right triangle if and only if

$$
a^2 + b^2 = c^2,
$$

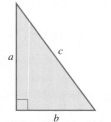

Figure R.5 $a^2 + b^2 = c^2$

as illustrated in Figure R.5.

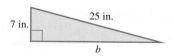

Figure R.6

EXAMPLE 4   Finding the perimeter of a right triangle

Find the perimeter of the triangle shown in Figure R.6.

SOLUTION Given one leg and the hypotenuse of a right triangle, we can use the Pythagorean theorem to find the other leg. Let $a = 7$, $c = 25$, and find b.

$$\begin{aligned}
a^2 + b^2 &= c^2 &&\text{Pythagorean theorem} \\
b^2 &= c^2 - a^2 &&\text{Subtract } a^2. \\
b^2 &= 25^2 - 7^2 &&\text{Let } a = 7 \text{ and } c = 25. \\
b^2 &= 576 &&\text{Simplify.} \\
b &= 24 &&\text{Solve for } b > 0.
\end{aligned}$$

The perimeter of the triangle is

$$a + b + c = 7 + 24 + 25 = 56 \text{ inches.}$$ **Now Try Exercise 29** ◄

Three-Dimensional Objects

Objects that occupy space have both volume and surface area. This subsection discusses rectangular boxes, spheres, cylinders, and cones.

Rectangular Boxes   The volume V of a rectangular box with length L, width W, and height H equals $V = LWH$. See Figure R.7. The surface area S of the box equals the sum of the areas of the six sides: $S = 2LW + 2WH + 2LH$.

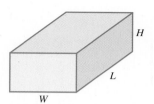

Figure R.7   Rectangular Box

EXAMPLE 5   Finding the volume and surface area of a box

The box in Figure R.8 has dimensions x by $2x$ by y. Find its volume and surface area.

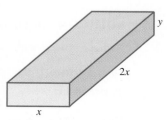

Figure R.8

SOLUTION

Volume: $LWH = 2x \cdot x \cdot y = 2x^2y$ cubic units

Surface Area: Base and top: $2x \cdot x + 2x \cdot x = 4x^2$
Front and back: $xy + xy = 2xy$
Left and right sides: $2xy + 2xy = 4xy$
Total surface area: $4x^2 + 2xy + 4xy = 4x^2 + 6xy$ square units

Now Try Exercise 41 ◄

Spheres   The volume V of a sphere with radius r is $V = \frac{4}{3}\pi r^3$, and its surface area S is $S = 4\pi r^2$. See Figure R.9.

Figure R.9   Sphere

EXAMPLE 6   Finding the volume and surface area of a sphere

Estimate, to the nearest tenth, the volume and surface area of a sphere with a radius of 5.1 feet.

SOLUTION

Volume: $V = \frac{4}{3}\pi r^3 = \frac{4}{3}\pi(5.1)^3 \approx 555.6$ cubic feet

Surface Area: $S = 4\pi r^2 = 4\pi(5.1)^2 \approx 326.9$ square feet   *Now Try Exercise 47* ◀

Cylinders   A soup can is usually made in the shape of a cylinder. The volume of a cylinder with radius r and height h is $V = \pi r^2 h$. See Figure R.10. To find the total surface area of a cylinder, we add the area of the top and bottom to the area of the side. Figure R.11 illustrates a can cut open to determine its surface area. The top and bottom are circular with areas of πr^2 each, and the side has a surface area of $2\pi rh$. The total surface area is $S = 2\pi r^2 + 2\pi rh$. The side surface area is called the **lateral surface area**.

Figure R.10 Cylinder   **Figure R.11** A Cylinder Cut Open

EXAMPLE 7   Finding the volume and surface area of a cylinder

A cylinder has radius $r = 3$ inches and height $h = 2.5$ feet. Find its volume and total surface area to the nearest tenth.

SOLUTION   Begin by changing 2.5 feet to 30 inches so that all units are in inches.

Volume: $V = \pi r^2 h = \pi(3)^2(30) = 270\pi \approx 848.2$ cubic inches

Total Surface Area: $S = 2\pi r^2 + 2\pi rh = 2\pi(3)^2 + 2\pi(3)(30) = 198\pi \approx 622.0$
square inches
                                            Now Try Exercise 51 ◀

Cones   The volume of a cone with radius r and height h is $V = \frac{1}{3}\pi r^2 h$, as shown in Figure R.12. (Compare this formula with the formula for the volume of a cylinder.) Excluding the bottom of the cone, the side (or lateral) surface area is $S = \pi r\sqrt{r^2 + h^2}$. The bottom of the cone is circular and has a surface area of πr^2.

Figure R.12 Cone

EXAMPLE 8   Finding the volume and surface area of a cone

Approximate, to the nearest tenth, the volume and surface area (side only) of a cone with a radius of 1.45 inches and a height of 5.12 inches.

SOLUTION

Volume: $V = \frac{1}{3}\pi r^2 h = \frac{1}{3}\pi(1.45)^2(5.12) \approx 11.3$ cubic inches

Surface Area (side only): $S = \pi r\sqrt{r^2 + h^2} = \pi(1.45)\sqrt{(1.45)^2 + (5.12)^2} \approx 24.2$
square inches
                                            Now Try Exercise 55 ◀

Similar Triangles

The corresponding angles of **similar triangles** have equal measure, but similar triangles are not necessarily the same size. Two similar triangles are shown in Figure R.13. Notice that both triangles have angles of 30°, 60°, and 90°. Corresponding sides are not equal in length; however, corresponding ratios are equal. For example, in triangle *ABC* the ratio of the shortest leg to the hypotenuse equals $\frac{2}{4} = \frac{1}{2}$, and in triangle *DEF* this ratio is $\frac{3}{6} = \frac{1}{2}$.

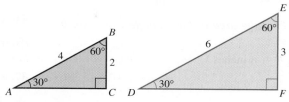

Figure R.13   Similar Triangles

EXAMPLE 9   Using similar triangles

Find the length of *BC* in Figure R.14.

SOLUTION   Notice that triangle *ABC* and triangle *ADE* are both right triangles. These triangles share an angle at vertex *A*. Therefore triangles *ABC* and *ADE* have two corresponding angles that are congruent. Because the sum of the angles in a triangle equals 180°, all three corresponding angles in these two triangles are congruent. Thus triangles *ABC* and *ADE* are similar.

Since corresponding ratios are equal, we can find *BC* as follows.

$$\frac{BC}{AC} = \frac{DE}{AE}$$

$$\frac{BC}{5} = \frac{6}{7}$$

Solving this equation for *BC* gives $BC = \frac{30}{7} \approx 4.3$.   ***Now Try Exercise 61*** ◄

NOTE   For a summary of these formulas, see the back endpapers of this text.

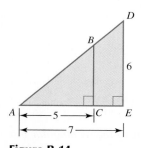

Figure R.14

R.1   Exercises

Rectangles

Exercises 1–6: Find the area and perimeter of the rectangle with length L and width W.

   1. $L = 15$ feet, $W = 7$ feet

   2. $L = 16$ inches, $W = 10$ inches

   3. $L = 100$ meters, $W = 35$ meters

   4. $L = 80$ yards, $W = 13$ yards

   5. $L = 3x$, $W = y$ **6.** $L = a + 5$, $W = a$

Exercises 7–10: Find the area and perimeter of the rectangle in terms of the width W.

   7. The width W is half the length.

   8. Triple the width W minus 3 equals the length.

   9. The length equals the width W plus 5.

   10. The length is 2 less than twice the width W.

Triangles

Exercises 11 and 12: Find the area of the triangle shown in the figure.

11.

12.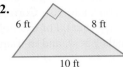

Exercises 13–20: Find the area of the triangle with base b and height h.

13. $b = 5$ inches, $h = 8$ inches

14. $b = 24$ inches, $h = 9$ feet

15. $b = 10.1$ meters, $h = 730$ meters

16. $b = 52$ yards, $h = 102$ feet

17. $b = 2x$, $h = 6x$ **18.** $b = x$, $h = x + 4$

19. $b = z$, $h = 5z$ **20.** $b = y + 1$, $h = 2y$

Circles

Exercises 21–26: Find the circumference and area of the circle. Approximate each value to the nearest tenth when appropriate.

21. $r = 4$ meters   **22.** $r = 1.5$ feet

23. $r = 19$ inches   **24.** $r = 22$ miles

25. $r = 2x$ **26.** $r = 5z$

Pythagorean Theorem

Exercises 27–32: Use the Pythagorean theorem to find the missing side of the right triangle with legs a and b and hypotenuse c. Then calculate the perimeter. Approximate values to the nearest tenth when appropriate.

27. $a = 60$ feet, $b = 11$ feet

28. $a = 21$ feet, $b = 11$ yards

29. $a = 5$ centimeters, $c = 13$ centimeters

30. $a = 6$ meters, $c = 15$ meters

31. $b = 7$ millimeters, $c = 10$ millimeters

32. $b = 1.2$ miles, $c = 2$ miles

Exercises 33–36: Find the area of the right triangle that satisfies the conditions. Approximate values to the nearest tenth when appropriate.

33. Legs with lengths 3 feet and 6 feet

34. Hypotenuse 10 inches and leg 6 inches

35. Hypotenuse 15 inches and leg 11 inches

36. Shorter leg 40 centimeters and hypotenuse twice the shorter leg

Rectangular Boxes

Exercises 37–44: Find the volume and surface area of a rectangular box with length L, width W, and height H.

37. $L = 4$ feet, $W = 3$ feet, $H = 2$ feet

38. $L = 6$ meters, $W = 4$ meters, $H = 1.5$ meters

39. $L = 4.5$ inches, $W = 4$ inches, $H = 1$ foot

40. $L = 9.1$ yards, $W = 8$ yards, $H = 6$ feet

41. $L = 3x$, $W = 2x$, $H = x$

42. $L = 6z$, $W = 5z$, $H = 7z$

43. $L = x$, $W = 2y$, $H = 3z$

44. $L = 8x$, $W = y$, $H = z$

Exercises 45 and 46: Find the volume of the rectangular box in terms of the width W.

45. The length is twice the width W, and the height is half the width.

46. The width W is three times the height and one-third of the length.

Spheres

Exercises 47–50: Find the volume and surface area of the sphere satisfying the given condition, where r is the radius and d is the diameter. Approximate values to the nearest tenth.

47. $r = 3$ feet   **48.** $r = 4.1$ inches

49. $d = 6.4$ meters   **50.** $d = 16$ feet

Cylinders

Exercises 51–54: Find the volume, the surface area of the side, and the total surface area of the cylinder that satisfies the given conditions, where r is the radius and h is the height. Approximate values to the nearest tenth.

51. $r = 0.5$ foot, $h = 2$ feet

52. $r = \frac{1}{3}$ yard, $h = 2$ feet

53. $r = 12$ millimeters, and h is twice r

54. r is one-fourth of h, and $h = 2.1$ feet

Cones

Exercises 55–60: Approximate, to the nearest tenth, the volume and surface area (side only) of the cone satisfying the given conditions, where r is the radius and h is the height.

55. $r = 5$ centimeters, $h = 6$ centimeters

56. $r = 8$ inches, $h = 30$ inches

57. $r = 24$ inches, $h = 3$ feet

58. $r = 100$ centimeters, $h = 1.3$ meters

59. Three times r equals h, and $r = 2.4$ feet

60. Twice h equals r, and $h = 3$ centimeters

Similar Triangles

Exercises 61–64: Use the fact that triangles ABC and DEF are similar to find the value of x.

61.

62.

63.

64.

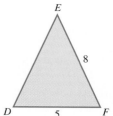

R.2 Integer Exponents

- Use bases and exponents
- Use zero and negative exponents
- Apply the product, quotient, and power rules

Bases and Positive Exponents

The expression 8^2 is an exponential expression with base **8** and exponent **2**. Table R.1 contains examples of other exponential expressions.

Table R.1

Expression	Base	Exponent
$2 \cdot 2 \cdot 2 = 2^3$	2	3
$6 \cdot 6 \cdot 6 \cdot 6 = 6^4$	6	4
$7 = 7^1$	7	1
$0.5 \cdot 0.5 = 0.5^2$	0.5	2
$x \cdot x \cdot x = x^3$	x	3

We read 0.5^2 as "0.5 squared," 2^3 as "2 cubed," and 6^4 as "6 to the fourth power."

EXAMPLE 1   Writing numbers in exponential notation

Use the given base to write each number as an exponential expression. Check your results with a calculator.

(a) 10,000   (base 10)   **(b)** 27   (base 3)   **(c)** 32   (base 2)

SOLUTION

(a) $10{,}000 = 10 \cdot 10 \cdot 10 \cdot 10 = 10^4$ **(b)** $27 = 3 \cdot 3 \cdot 3 = 3^3$

(c) $32 = 2 \cdot 2 \cdot 2 \cdot 2 \cdot 2 = 2^5$

These values are supported in Figure R.15, where exponential expressions are evaluated with a graphing calculator, using the $\boxed{\wedge}$ key.   ***Now Try Exercises 11, 13, and 15*** ◀

```
10^4
                10000
3^3
                   27
2^5
                   32
```

Figure R.15

Zero and Negative Exponents

Exponents can be defined for any integer. The following box lists some properties for integer exponents.

Integer Exponents

Let a and b be nonzero real numbers and m and n be positive integers. Then

1. $a^n = a \cdot a \cdot a \cdot \cdots \cdot a$ (n factors of a)

2. $a^0 = 1$ (*Note:* 0^0 is undefined.)

3. $a^{-n} = \dfrac{1}{a^n}$ and $\dfrac{1}{a^{-n}} = a^n$

4. $\dfrac{a^{-n}}{b^{-m}} = \dfrac{b^m}{a^n}$

5. $\left(\dfrac{a}{b}\right)^{-n} = \left(\dfrac{b}{a}\right)^{n}$

EXAMPLE 2   Evaluating expressions

Evaluate each expression.

(a) 3^{-4} **(b)** $\dfrac{1}{2^{-3}}$ **(c)** $\left(\dfrac{5}{7}\right)^{-2}$ **(d)** $\dfrac{1}{(xy)^{-1}}$ **(e)** $\dfrac{2^{-2}}{3t^{-3}}$

SOLUTION

(a) $3^{-4} = \dfrac{1}{3^4} = \dfrac{1}{3 \cdot 3 \cdot 3 \cdot 3} = \dfrac{1}{81}$ **(b)** $\dfrac{1}{2^{-3}} = 2^3 = 2 \cdot 2 \cdot 2 = 8$

(c) $\left(\dfrac{5}{7}\right)^{-2} = \left(\dfrac{7}{5}\right)^{2} = \dfrac{7}{5} \cdot \dfrac{7}{5} = \dfrac{49}{25}$ **(d)** $\dfrac{1}{(xy)^{-1}} = (xy)^1 = xy$

⎵ Base is xy.

(e) Note that only t, not $3t$, is raised to the power of -3.

$$\dfrac{2^{-2}}{3t^{-3}} = \dfrac{t^3}{3(2^2)} = \dfrac{t^3}{3 \cdot 4} = \dfrac{t^3}{12}$$

⎵ Base is t.   ***Now Try Exercises 27, 29, 31, 65, and 67*** ◀

Product, Quotient, and Power Rules

We can calculate products and quotients of exponential expressions *provided their bases are the same*. For example,

$$3^2 \cdot 3^3 = (3 \cdot 3) \cdot (3 \cdot 3 \cdot 3) = 3^5.$$

This expression has a total of $2 + 3 = 5$ factors of 3, so the result is 3^5. To multiply exponential expressions with *like bases*, add exponents.

The Product Rule

For any nonzero number a and integers m and n,

$$a^m \cdot a^n = a^{m+n}.$$

The product rule holds for negative exponents. For example,

$$10^5 \cdot 10^{-2} = 10^{5+(-2)} = 10^3.$$

EXAMPLE 3   Using the product rule

Multiply and simplify.

(a) $7^3 \cdot 7^{-4}$ **(b)** $x^3 x^{-2} x^4$ **(c)** $(3y^2)(2y^{-4})$

SOLUTION

(a) $7^3 \cdot 7^{-4} = 7^{3+(-4)} = 7^{-1} = \dfrac{1}{7}$ **(b)** $x^3 x^{-2} x^4 = x^{3+(-2)+4} = x^5$

(c) $(3y^2)(2y^{-4}) = 3 \cdot 2 \cdot y^2 \cdot y^{-4} = 6y^{2+(-4)} = 6y^{-2} = \dfrac{6}{y^2}$

Note that 6 is not raised to the power of -2 in the expression $6y^{-2}$.

Now Try Exercises 35, 37, and 39 ◀

Consider division of exponential expressions using the following example.

$$\frac{6^5}{6^3} = \frac{6 \cdot 6 \cdot \cancel{6} \cdot \cancel{6} \cdot \cancel{6}}{\cancel{6} \cdot \cancel{6} \cdot \cancel{6}} = 6 \cdot 6 = 6^2$$

After simplifying, there are two 6s left in the numerator. The result is $6^{5-3} = 6^2 = 36$. To divide exponential expressions with *like bases*, subtract exponents.

The Quotient Rule

For any nonzero number a and integers m and n,

$$\frac{a^m}{a^n} = a^{m-n}.$$

The quotient rule holds true for negative exponents. For example,

$$\frac{2^{-6}}{2^{-4}} = 2^{-6-(-4)} = 2^{-2} = \frac{1}{2^2} = \frac{1}{4}.$$

Figure R.16

This result is supported by Figure R.16.

EXAMPLE 4   Using the quotient rule

Simplify the expression. Use positive exponents.

(a) $\dfrac{10^4}{10^6}$ **(b)** $\dfrac{x^5}{x^2}$ **(c)** $\dfrac{15x^2y^3}{5x^4y}$

SOLUTION

(a) $\dfrac{10^4}{10^6} = 10^{4-6} = 10^{-2} = \dfrac{1}{10^2} = \dfrac{1}{100}$ **(b)** $\dfrac{x^5}{x^2} = x^{5-2} = x^3$

(c) $\dfrac{15x^2y^3}{5x^4y} = \dfrac{15}{5} \cdot \dfrac{x^2}{x^4} \cdot \dfrac{y^3}{y^1} = 3 \cdot x^{(2-4)}y^{(3-1)} = 3x^{-2}y^2 = \dfrac{3y^2}{x^2}$

Now Try Exercises 45, 47, and 51 ◀

How should we evaluate $(4^3)^2$? To answer this question, consider

$$(4^3)^2 = 4^3 \cdot 4^3 = 4^{3+3} = 4^6.$$

Similarly,

$$(x^4)^3 = x^4 \cdot x^4 \cdot x^4 = x^{4+4+4} = x^{12}.$$

These results suggest that to raise a power to a power, we must multiply the exponents.

Raising Powers to Powers

For any nonzero real number a and integers m and n,

$$(a^m)^n = a^{mn}.$$

EXAMPLE 5   Raising powers to powers

Simplify each expression. Use positive exponents.
(a) $(5^2)^3$ **(b)** $(2^4)^{-2}$ **(c)** $(b^{-7})^5$

SOLUTION

(a) $(5^2)^3 = 5^{2 \cdot 3} = 5^6 = 15{,}625$ **(b)** $(2^4)^{-2} = 2^{4(-2)} = 2^{-8} = \dfrac{1}{2^8} = \dfrac{1}{256}$

(c) $(b^{-7})^5 = b^{-7 \cdot 5} = b^{-35} = \dfrac{1}{b^{35}}$ *Now Try Exercises 55, 56, and 57* ◀

How can we simplify the expression $(2x)^3$? Consider the following.

$$(2x)^3 = 2x \cdot 2x \cdot 2x = (2 \cdot 2 \cdot 2) \cdot (x \cdot x \cdot x) = 2^3x^3$$

This result suggests that to cube a product, we can cube each factor.

Raising Products to Powers

For any nonzero real numbers a and b and integer n,

$$(ab)^n = a^n b^n.$$

EXAMPLE 6   Raising products to powers

Simplify each expression. Use positive exponents.
(a) $(6y)^2$ **(b)** $(x^2y)^{-2}$ **(c)** $(2xy^3)^4$

SOLUTION

(a) $(6y)^2 = 6^2y^2 = 36y^2$ **(b)** $(x^2y)^{-2} = (x^2)^{-2}y^{-2} = x^{-4}y^{-2} = \dfrac{1}{x^4y^2}$

(c) $(2xy^3)^4 = 2^4x^4(y^3)^4 = 16x^4y^{12}$ *Now Try Exercises 59, 60, and 73* ◀

To simplify a power of a quotient, use the following rule.

Raising Quotients to Powers

For nonzero numbers a and b and any integer n,

$$\left(\frac{a}{b}\right)^n = \frac{a^n}{b^n}.$$

EXAMPLE 7   Raising quotients to powers

Simplify each expression. Use positive exponents.

(a) $\left(\dfrac{3}{x}\right)^3$ **(b)** $\left(\dfrac{1}{2^3}\right)^{-2}$ **(c)** $\left(\dfrac{3x^{-3}}{y^2}\right)^4$

SOLUTION

(a) $\left(\dfrac{3}{x}\right)^3 = \dfrac{3^3}{x^3} = \dfrac{27}{x^3}$ **(b)** $\left(\dfrac{1}{2^3}\right)^{-2} = \dfrac{1^{-2}}{(2^3)^{-2}} = \dfrac{1}{2^{-6}} = 2^6 = 64$

(c) $\left(\dfrac{3x^{-3}}{y^2}\right)^4 = \dfrac{3^4(x^{-3})^4}{(y^2)^4} = \dfrac{81x^{-12}}{y^8} = \dfrac{81}{x^{12}y^8}$ *Now Try Exercises 61, 63, and 81* ◀

In the next example, we use several properties of exponents to simplify expressions.

EXAMPLE 8   Simplifying expressions

Write each expression using positive exponents. Simplify the result completely.

(a) $\left(\dfrac{x^2y^{-3}}{3z^{-4}}\right)^{-2}$ **(b)** $\dfrac{(rt^3)^{-3}}{(r^2t^3)^{-2}}$

SOLUTION

(a) $\left(\dfrac{x^2y^{-3}}{3z^{-4}}\right)^{-2} = \left(\dfrac{3z^{-4}}{x^2y^{-3}}\right)^2$ **(b)** $\dfrac{(rt^3)^{-3}}{(r^2t^3)^{-2}} = \dfrac{(r^2t^3)^2}{(rt^3)^3}$

$= \left(\dfrac{3y^3}{x^2z^4}\right)^2$ $= \dfrac{r^4t^6}{r^3t^9}$

$= \dfrac{9y^6}{x^4z^8}$ $= \dfrac{r}{t^3}$ *Now Try Exercises 77 and 79* ◀

R.2   Exercises

Concepts

1. Are the expressions 2^3 and 3^2 equal? Explain your answer.

2. Are the expressions -4^2 and $(-4)^2$ equal? Explain your answer.

3. $7^{-n} =$ _____

4. $6^m \cdot 6^n =$ _____

5. $\dfrac{5^m}{5^n} =$ _____

6. $(3x)^k =$ _____

7. $(2^m)^k =$ _____

8. $\left(\dfrac{x}{y}\right)^m =$ _____

9. $5 \times 10^3 =$ _____

10. $5 \times 10^{-3} =$ _____

Properties of Exponents

Exercises 11–16: (Refer to Example 1.) Write the number as an exponential expression, using the base shown. Check your result with a calculator.

11. 8 (base 2)

12. 1000 (base 10)

13. 256 (base 4)

14. $\frac{1}{64}$ (base 4)

15. 1 (base 3)

16. $\frac{1}{49}$ (base 7)

Exercises 17–34: Evaluate the expression by hand. Check your result with a calculator.

17. 5^3

18. 5^{-3}

19. -2^4

20. $(-2)^4$

21. 5^0

22. $\left(-\frac{2}{3}\right)^{-3}$

23. $\left(\frac{2}{3}\right)^3$

24. $\frac{1}{4^{-2}}$

25. $\left(-\frac{1}{2}\right)^4$

26. $\left(-\frac{3}{4}\right)^3$

27. 4^{-3}

28. 10^{-4}

29. $\frac{1}{2^{-4}}$

30. $\frac{1}{3^{-2}}$

31. $\left(\frac{3}{4}\right)^{-3}$

32. $\left(\frac{1}{2}\right)^0$

33. $\frac{3^{-2}}{2^{-3}}$

34. $\frac{10^{-4}}{4^{-3}}$

Exercises 35–44: Use the product rule to simplify.

35. $6^3 \cdot 6^{-4}$

36. $10^2 \cdot 10^5 \cdot 10^{-3}$

37. $2x^2 \cdot 3x^{-3} \cdot x^4$

38. $3y^4 \cdot 6y^{-4} \cdot y$

39. $10^0 \cdot 10^6 \cdot 10^2$

40. $y^3 \cdot y^{-5} \cdot y^4$

41. $5^{-2} \cdot 5^3 \cdot 2^{-4} \cdot 2^3$

42. $2^{-3} \cdot 3^4 \cdot 3^{-2} \cdot 2^5$

43. $(2a^3)(b^2)(a^{-4})(4b^{-5})$

44. $(3x^{-4})(2x^2)(5y^4)(y^{-3})$

Exercises 45–54: Use the quotient rule to simplify the expression. Use positive exponents to write your answer.

45. $\frac{5^4}{5^2}$

46. $\frac{6^2}{6^{-7}}$

47. $\frac{a^{-3}}{a^2 \cdot a}$

48. $\frac{y^0 \cdot y \cdot y^5}{y^{-2} \cdot y^{-3}}$

49. $\frac{24x^3}{6x}$

50. $\frac{10x^5}{5x^{-3}}$

51. $\frac{12a^2b^3}{18a^4b^2}$

52. $\frac{-6x^7y^3}{3x^2y^{-5}}$

53. $\frac{21x^{-3}y^4}{7x^4y^{-2}}$

54. $\frac{32x^3y}{-24x^5y^{-3}}$

Exercises 55–64: Use the power rules to simplify the expression. Use positive exponents to write your answer.

55. $(5^{-1})^3$

56. $(-4^2)^3$

57. $(y^4)^{-2}$

58. $(x^2)^4$

59. $(4y^2)^3$

60. $(-2xy^3)^{-4}$

61. $\left(\dfrac{4}{x}\right)^3$

62. $\left(\dfrac{-3}{x^3}\right)^2$

63. $\left(\dfrac{2x}{z^4}\right)^{-5}$

64. $\left(\dfrac{2xy}{3z^5}\right)^{-1}$

Exercises 65–90: Use rules of exponents to simplify the expression. Use positive exponents to write your answer.

65. $\dfrac{2}{(ab)^{-1}}$

66. $\dfrac{5a^2}{(xy)^{-1}}$

67. $\dfrac{2^{-3}}{2t^{-2}}$

68. $\dfrac{t^{-3}}{2t^{-1}}$

69. $\dfrac{6a^2b^{-3}}{4ab^{-2}}$

70. $\dfrac{20a^{-2}b}{4a^{-2}b^{-1}}$

71. $\dfrac{5r^2st^{-3}}{25rs^{-2}t^2}$

72. $\dfrac{36r^{-1}(st)^2}{9(rs)^2t^{-1}}$

73. $(3x^2y^{-3})^{-2}$

74. $(-2x^{-3}y^{-2})^3$

75. $\dfrac{(d^3)^{-2}}{(d^{-2})^3}$

76. $\dfrac{(b^2)^{-1}}{(b^{-4})^3}$

77. $\left(\dfrac{3t^2}{2t^{-1}}\right)^3$

78. $\left(\dfrac{-2t}{4t^{-2}}\right)^{-1}$

79. $\dfrac{(-m^2n^{-1})^{-2}}{(mn)^{-1}}$

80. $\dfrac{(-mn^4)^{-1}}{(m^2n)^{-3}}$

81. $\left(\dfrac{2a^3}{6b}\right)^4$

82. $\left(\dfrac{-3a^2}{9b^3}\right)^4$

83. $\dfrac{8x^{-3}y^{-2}}{4x^{-2}y^{-4}}$

84. $\dfrac{6x^{-1}y^{-1}}{9x^{-2}y^3}$

85. $\dfrac{(r^2t^2)^{-2}}{(r^3t)^{-1}}$

86. $\dfrac{(2rt)^2}{(rt^4)^{-2}}$

87. $\dfrac{4x^{-2}y^3}{(2x^{-1}y)^2}$

88. $\dfrac{(ab)^3}{a^4b^{-4}}$

89. $\left(\dfrac{15r^2t}{3r^{-3}t^4}\right)^3$

90. $\left(\dfrac{4(xy)^2}{(2xy^{-2})^3}\right)^{-2}$

R.3 Polynomial Expressions

- **Perform addition and subtraction on monomials**
- **Perform addition and subtraction on polynomials**
- **Apply the distributive property**
- **Perform multiplication on polynomials**
- **Find the product of a sum and difference**
- **Square a binomial**

Addition and Subtraction of Monomials

A **term** is a number, a variable, or a *product* of numbers and variables raised to powers. Examples of terms include

$$-15, \quad y, \quad x^4, \quad 3x^3z, \quad x^{-1/2}y^{-2}, \quad \text{and} \quad 6x^{-1}y^3.$$

If the variables in a term have only *nonnegative integer* exponents, the term is called a *monomial*. Examples of monomials include

$$-4, \quad 5y, \quad x^2, \quad 5x^2z^6, \quad -xy^7, \quad \text{and} \quad 6xy^3.$$

If two terms contain the same variables raised to the same powers, we call them like terms. We can add or subtract *like* terms, but not *unlike* terms. For example, the terms

$$x^3 + y^3$$

cannot be combined because they are *unlike terms*. However,

$$x^3 + 3x^3 = (1 + 3)x^3 = 4x^3$$

because x^3 and $3x^3$ are *like terms*. To add or subtract monomials, we simply combine like terms, as illustrated in the next example.

> **EXAMPLE 1**   Adding and subtracting monomials
>
> Simplify each of the following expressions by combining like terms.
> **(a)** $8x^2 - 4x^2 + x^3$ **(b)** $9x - 6xy^2 + 2xy^2 + 4x$
>
> **SOLUTION**
> **(a)** The terms $8x^2$ and $-4x^2$ are like terms, so they may be combined.
>
> $$8x^2 - 4x^2 + x^3 = (8 - 4)x^2 + x^3 \qquad \text{Combine like terms.}$$
> $$= 4x^2 + x^3 \qquad \text{Subtract.}$$
>
> However, $4x^2$ and x^3 are unlike terms and cannot be combined.

(b) The terms $9x$ and $4x$ may be combined, as may $-6xy^2$ and $2xy^2$.

$$9x - 6xy^2 + 2xy^2 + 4x = 9x + 4x - 6xy^2 + 2xy^2 \qquad \textit{Commutative property}$$
$$= (9 + 4)x + (-6 + 2)xy^2 \qquad \textit{Combine like terms.}$$
$$= 13x - 4xy^2 \qquad \textit{Add.}$$

Now Try Exercises 5 and 11 ◀

Addition and Subtraction of Polynomials

A polynomial is either a monomial or a sum (or difference) of monomials. Examples of polynomials include

$$5x^4z^2, \quad 9x^4 - 5, \quad 4x^2 + 5xy - y^2, \quad \text{and} \quad 4 - y^2 + 5y^4 + y^5.$$

$$\text{1 term} \quad\quad \text{2 terms} \quad\quad\quad \text{3 terms} \quad\quad\quad\quad\quad \text{4 terms}$$

Polynomials containing one variable are called polynomials of one variable. The second and fourth polynomials shown above are examples of polynomials of one variable. The leading coefficient of a polynomial of one variable is the coefficient of the monomial with highest degree. The degree of a polynomial with one variable equals the exponent of the monomial with the highest power. Table R.2 shows several polynomials of one variable along with their degrees and leading coefficients. A polynomial of degree 1 is a *linear* polynomial, a polynomial of degree 2 is a *quadratic* polynomial, and a polynomial of degree 3 is a *cubic* polynomial.

Table R.2

Polynomial	Degree	Leading Coefficient	Type
-98	0	-98	Constant
$2x - 7$	1	2	Linear
$-5z + 9z^2 + 7$	2	9	Quadratic
$-2x^3 + 4x^2 + x - 1$	3	-2	Cubic
$7 - x + 4x^2 + x^5$	5	1	Fifth Degree

To add two polynomials, we combine like terms.

EXAMPLE 2 Adding polynomials

Simplify.
(a) $(2x^2 - 3x + 7) + (3x^2 + 4x - 2)$ **(b)** $(z^3 + 4z + 8) + (4z^2 - z + 6)$

SOLUTION
(a) $(2x^2 - 3x + 7) + (3x^2 + 4x - 2) = 2x^2 + 3x^2 - 3x + 4x + 7 - 2$
$$= (2 + 3)x^2 + (-3 + 4)x + (7 - 2)$$
$$= 5x^2 + x + 5$$

(b) $(z^3 + 4z + 8) + (4z^2 - z + 6) = z^3 + 4z^2 + 4z - z + 8 + 6$
$$= z^3 + 4z^2 + (4 - 1)z + (8 + 6)$$
$$= z^3 + 4z^2 + 3z + 14$$

Now Try Exercises 21 and 27 ◀

To subtract integers, we add the first integer and the **opposite**, or **additive inverse**, of the second integer. For example, to evaluate $3 - 5$, we perform the following operations.

$$3 - 5 = 3 + (-5) \qquad \text{Add the opposite.}$$
$$= -2 \qquad \text{Simplify.}$$

Similarly, to subtract two polynomials, we add the first polynomial and the opposite, or additive inverse, of the second polynomial. To find the opposite of a polynomial, we simply negate each term. Table R.3 shows three polynomials and their opposites.

Table R.3

Polynomial	Opposite
$9 - x$	$-9 + x$
$5x^2 + 4x - 1$	$-5x^2 - 4x + 1$
$-x^4 + 5x^3 - x^2 + 5x - 1$	$x^4 - 5x^3 + x^2 - 5x + 1$

EXAMPLE 3   Subtracting polynomials

Simplify.
(a) $(y^5 + 3y^3) - (-y^4 + 2y^3)$ **(b)** $(5x^3 + 9x^2 - 6) - (5x^3 - 4x^2 - 7)$

SOLUTION
(a) The opposite of $(-y^4 + 2y^3)$ is $(y^4 - 2y^3)$.

$$(y^5 + 3y^3) - (-y^4 + 2y^3) = (y^5 + 3y^3) + (y^4 - 2y^3)$$
$$= y^5 + y^4 + (3 - 2)y^3$$
$$= y^5 + y^4 + y^3$$

(b) The opposite of $(5x^3 - 4x^2 - 7)$ is $(-5x^3 + 4x^2 + 7)$.

$$(5x^3 + 9x^2 - 6) - (5x^3 - 4x^2 - 7) = (5x^3 + 9x^2 - 6) + (-5x^3 + 4x^2 + 7)$$
$$= (5 - 5)x^3 + (9 + 4)x^2 + (-6 + 7)$$
$$= 0x^3 + 13x^2 + 1$$
$$= 13x^2 + 1 \qquad \textit{Now Try Exercises 39 and 41} \blacktriangleleft$$

Distributive Properties

Distributive properties are used frequently in the multiplication of polynomials. For all real numbers a, b, and c,

$$a(b + c) = ab + ac \qquad \text{and}$$
$$a(b - c) = ab - ac.$$

In the next example, we use these distributive properties to multiply expressions.

EXAMPLE 4   Using distributive properties

Multiply.
(a) $4(5 + x)$ **(b)** $-3(x - 4y)$ **(c)** $(2x - 5)(6)$

SOLUTION

(a) $4(5 + x) = 4 \cdot 5 + 4 \cdot x = 20 + 4x$

(b) $-3(x - 4y) = -3 \cdot x - (-3) \cdot (4y) = -3x + 12y$

(c) $(2x - 5)(6) = 2x \cdot 6 - 5 \cdot 6 = 12x - 30$ *Now Try Exercises 47, 51, and 53* ◀

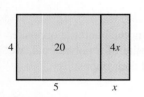

Figure R.17 Area: $20 + 4x$

You can visualize the solution in part (a) of Example 4 by using areas of rectangles. If a rectangle has width 4 and length $5 + x$, its area is $20 + 4x$, as shown in Figure R.17.

Multiplying Polynomials

A polynomial with two terms is a **binomial**, and one with three terms is a **trinomial**. Examples are shown in Table R.4.

Table R.4

Monomials	$2x^2$	$-3x^4$	9
Binomials	$3x - 1$	$2x^3 - x$	$x^2 + 5$
Trinomials	$x^2 - 3x + 5$	$5x^2 - 2x + 10$	$2x^3 - x^2 - 2$

In the next example, we multiply two binomials.

EXAMPLE 5   Multiplying binomials

Multiply $(x + 1)(x + 3)$.

SOLUTION   To multiply $(x + 1)(x + 3)$, we apply the distributive property.

$$(x + 1)(x + 3) = (x + 1)(x) + (x + 1)(3)$$
$$= x \cdot x + 1 \cdot x + x \cdot 3 + 1 \cdot 3$$
$$= x^2 + x + 3x + 3$$
$$= x^2 + 4x + 3 \qquad \textit{Now Try Exercise 55} ◀$$

To multiply the bionomials $(x + 1)$ and $(x + 3)$, we multiplied every term in $x + 1$ by every term in $x + 3$. That is,

$$(x + 1)(x + 3) = x^2 + 3x + x + 3$$
$$= x^2 + 4x + 3.$$

NOTE This process of multiplying binomials is called *FOIL*. You may use the name to remind yourself to multiply the first terms (F), outside terms (O), inside terms (I), and last terms (L).

Multiply the *First terms* to obtain x^2. $(x + 1)(x + 3)$

Multiply the *Outside terms* to obtain $3x$. $(x + 1)(x + 3)$

Multiply the *Inside terms* to obtain x. $(x + 1)(x + 3)$

Multiply the *Last terms* to obtain 3. $(x + 1)(x + 3)$

The following box summarizes how to multiply two polynomials in general.

> ### Multiplying Polynomials
>
> The product of two polynomials may be found by multiplying every term in the first polynomial by every term in the second polynomial and then combining like terms.

EXAMPLE 6   Multiplying binomials

Multiply each binomial.

(a) $(2x - 1)(x + 2)$ **(b)** $(1 - 3x)(2 - 4x)$ **(c)** $(x^2 + 1)(5x - 3)$

SOLUTION

(a) $(2x - 1)(x + 2) = 2x \cdot x + 2x \cdot 2 - 1 \cdot x - 1 \cdot 2$

$$= 2x^2 + 4x - x - 2$$

$$= 2x^2 + 3x - 2$$

(b) $(1 - 3x)(2 - 4x) = 1 \cdot 2 - 1 \cdot 4x - 3x \cdot 2 + 3x \cdot 4x$

$$= 2 - 4x - 6x + 12x^2$$

$$= 2 - 10x + 12x^2$$

(c) $(x^2 + 1)(5x - 3) = x^2 \cdot 5x - x^2 \cdot 3 + 1 \cdot 5x - 1 \cdot 3$

$$= 5x^3 - 3x^2 + 5x - 3 \qquad \textit{Now Try Exercises 57, 59, and 77} \blacktriangleleft$$

EXAMPLE 7   Multiplying polynomials

Multiply each expression.

(a) $3x(x^2 + 5x - 4)$ **(b)** $-x^2(x^4 - 2x + 5)$ **(c)** $(x + 2)(x^2 + 4x - 3)$

SOLUTION

(a) $3x(x^2 + 5x - 4) = 3x \cdot x^2 + 3x \cdot 5x - 3x \cdot 4$

$$= 3x^3 + 15x^2 - 12x$$

(b) $-x^2(x^4 - 2x + 5) = -x^2 \cdot x^4 + x^2 \cdot 2x - x^2 \cdot 5$

$$= -x^6 + 2x^3 - 5x^2$$

(c) $(x + 2)(x^2 + 4x - 3) = x \cdot x^2 + x \cdot 4x - x \cdot 3 + 2 \cdot x^2 + 2 \cdot 4x - 2 \cdot 3$

$$= x^3 + 4x^2 - 3x + 2x^2 + 8x - 6$$

$$= x^3 + 6x^2 + 5x - 6 \qquad \textit{Now Try Exercises 67, 69, and 73} \blacktriangleleft$$

Some Special Products

The following special product often occurs in mathematics.

$$(a - b)(a + b) = a \cdot a + a \cdot b - b \cdot a - b \cdot b$$

$$= a^2 + ab - ba - b^2$$

$$= a^2 - b^2$$

That is, the product of a sum of two numbers and their difference equals the difference of their squares.

> **EXAMPLE 8**   Finding the product of a sum and difference

Multiply.
(a) $(x - 3)(x + 3)$ **(b)** $(5 + 4x^2)(5 - 4x^2)$

SOLUTION

(a) If we let $a = x$ and $b = 3$, we can apply $(a - b)(a + b) = a^2 - b^2$. Thus

$$(x - 3)(x + 3) = (x)^2 - (3)^2$$
$$= x^2 - 9.$$

(b) Similarly,

$$(5 + 4x^2)(5 - 4x^2) = (5)^2 - (4x^2)^2$$
$$= 25 - 16x^4.$$

Now Try Exercises 79 and 95 ◄

Two other special products involve *squaring a binomial:*

$$(a + b)^2 = (a + b)(a + b)$$
$$= a^2 + ab + ba + b^2$$
$$= a^2 + 2ab + b^2$$

and

$$(a - b)^2 = (a - b)(a - b)$$
$$= a^2 - ab - ba + b^2$$
$$= a^2 - 2ab + b^2.$$

Note that to obtain the middle term, we multiply the two terms in the binomial and double the result.

> **EXAMPLE 9**   Squaring a binomial

Multiply.
(a) $(x + 5)^2$ **(b)** $(3 - 2x)^2$

SOLUTION

(a) If we let $a = x$ and $b = 5$, we can apply $(a + b)^2 = a^2 + 2ab + b^2$. Thus

$$(x + 5)^2 = (x)^2 + 2(x)(5) + (5)^2$$
$$= x^2 + 10x + 25.$$

(b) Applying the formula $(a - b)^2 = a^2 - 2ab + b^2$, we find

$$(3 - 2x)^2 = (3)^2 - 2(3)(2x) + (2x)^2$$
$$= 9 - 12x + 4x^2.$$

Now Try Exercises 85 and 91 ◄

R.3 Exercises

Monomials and Polynomials

Exercises 1–12: Combine like terms whenever possible.

1. $3x^3 + 5x^3$

2. $-9z + 6z$

3. $5y^7 - 8y^7$

4. $9x - 7x$

5. $5x^2 + 8x + x^2$

6. $5x + 2x + 10x$

7. $9x^2 - x + 4x - 6x^2$

8. $-y^2 - \frac{1}{2}y^2$

9. $x^2 + 9x - 2 + 4x^2 + 4x$

10. $6y + 4y^2 - 6y + y^2$

11. $7y + 9x^2y - 5y + x^2y$

12. $5ab - b^2 + 7ab + 6b^2$

Exercises 13–18: Identify the degree and leading coefficient of the polynomial.

13. $5x^2 - 4x + \frac{3}{4}$

14. $-9y^4 + y^2 + 5$

15. $5 - x + 3x^2 - \frac{2}{5}x^3$

16. $7x + 4x^4 - \frac{4}{3}x^3$

17. $8x^4 + 3x^3 - 4x + x^5$

18. $5x^2 - x^3 + 7x^4 + 10$

Exercises 19–28: Add the polynomials.

19. $(5x + 6) + (-2x + 6)$

20. $(5y^2 + y^3) + (12y^2 - 5y^3)$

21. $(2x^2 - x + 7) + (-2x^2 + 4x - 9)$

22. $(x^3 - 5x^2 + 6) + (5x^2 + 3x + 1)$

23. $(4x) + (1 - 4.5x)$

24. $(y^5 + y) + \left(5 - y + \frac{1}{3}y^2\right)$

25. $(x^4 - 3x^2 - 4) + \left(-8x^4 + x^2 - \frac{1}{2}\right)$

26. $(3z + z^4 + 2) + (-3z^4 - 5 + z^2)$

27. $(2z^3 + 5z - 6) + (z^2 - 3z + 2)$

28. $(z^4 - 6z^2 + 3) + (5z^3 + 3z^2 - 3)$

Exercises 29–34: Find the opposite of the polynomial.

29. $7x^3$

30. $-3z^8$

31. $19z^5 - 5z^2 + 3z$

32. $-x^2 - x + 6$

33. $z^4 - z^2 - 9$

34. $1 - 8x + 6x^2 - \frac{1}{6}x^3$

Exercises 35–42: Subtract the polynomials.

35. $(5x - 3) - (2x + 4)$

36. $(10x + 5) - (-6x - 4)$

37. $(x^2 - 3x + 1) - (-5x^2 + 2x - 4)$

38. $(-x^2 + x - 5) - (x^2 - x + 5)$

39. $(4x^4 + 2x^2 - 9) - (x^4 - 2x^2 - 5)$

40. $(8x^3 + 5x^2 - 3x + 1) - (-5x^3 + 6x - 11)$

41. $(x^4 - 1) - (4x^4 + 3x + 7)$

42. $(5x^4 - 6x^3 + x^2 + 5) - (x^3 + 11x^2 + 9x - 3)$

Exercises 43–54: Apply the distributive property.

43. $5x(x - 5)$

44. $3x^2(-2x + 2)$

45. $-5(3x + 1)$

46. $-(-3x + 1)$

47. $5(y + 2)$

48. $4(x - 7)$

49. $-2(5x + 9)$

50. $-3x(5 + x)$

51. $(y - 3)6y$

52. $(2x - 5)8x^3$

53. $-4(5x - y)$

54. $-6(3y - 2x)$

Exercises 55–66: Multiply the binomials.

55. $(y + 5)(y - 7)$

56. $(3x + 1)(2x + 1)$

57. $(3 - 2x)(3 + x)$

58. $(7x - 3)(4 - 7x)$

59. $(-2x + 3)(x - 2)$

60. $(z - 2)(4z + 3)$

61. $\left(x - \frac{1}{2}\right)\left(x + \frac{1}{4}\right)$

62. $\left(z - \frac{1}{3}\right)\left(z - \frac{1}{6}\right)$

63. $(x^2 + 1)(2x^2 - 1)$

64. $(x^2 - 2)(x^2 + 4)$

65. $(x + y)(x - 2y)$

66. $(x^2 + y^2)(x - y)$

Exercises 67–78: Multiply the polynomials.

67. $3x(2x^2 - x - 1)$

68. $-2x(3 - 2x + 5x^2)$

69. $-x(2x^4 - x^2 + 10)$

70. $-2x^2(5x^3 + x^2 - 2)$

71. $(2x^2 - 4x + 1)(3x^2)$

72. $(x - y + 5)(xy)$

73. $(x + 1)(x^2 + 2x - 3)$

74. $(2x - 1)(3x^2 - x + 6)$

75. $(2 - 3x)(5 - 2x)(x^2 - 1)$

76. $(3 + z)(6 - 4z)(4 + 2z^2)$

77. $(x^2 + 2)(3x - 2)$

78. $(4 + x)(2x^2 - 3)$

Exercises 79–96: Multiply the expressions.

79. $(x - 7)(x + 7)$ **80.** $(x + 9)(x - 9)$

81. $(3x + 4)(3x - 4)$ **82.** $(9x - 4)(9x + 4)$

83. $(2x - 3y)(2x + 3y)$ **84.** $(x + 2y)(x - 2y)$

85. $(x + 4)^2$ **86.** $(z + 9)^2$

87. $(2x + 1)^2$ **88.** $(3x + 5)^2$

89. $(x - 1)^2$ **90.** $(x - 7)^2$

91. $(2 - 3x)^2$ **92.** $(5 - 6x)^2$

93. $3x(x + 1)(x - 1)$ **94.** $-4x(3x - 5)^2$

95. $(2 - 5x^2)(2 + 5x^2)$ **96.** $(6y - x^2)(6y + x^2)$

R.4 Factoring Polynomials

- **Use common factors**
- **Factor by grouping**
- **Factor $x^2 + bx + c$**
- **Factor trinomials by grouping**
- **Factor trinomials with FOIL**
- **Factor the difference of two squares**
- **Factor perfect square trinomials**
- **Factor the sum and difference of two cubes**

Common Factors

When factoring a polynomial, we first look for factors that are common to each term in an expression. By applying a distributive property, we can write a polynomial as two factors. For example, each term in $2x^2 + 4x$ contains a factor of $2x$.

$$2x^2 = 2x \cdot x$$
$$4x = 2x \cdot 2$$

Thus the polynomial $2x^2 + 4x$ can be factored as follows.

$$2x^2 + 4x = 2x(x + 2)$$

EXAMPLE 1   Finding common factors

Factor.
(a) $6z^3 - 2z^2 + 4z$ **(b)** $4x^3y^2 + x^2y^3$

SOLUTION
(a) Each of the terms $6z^3$, $2z^2$, and $4z$ contains a common factor of $2z$. That is,

$$6z^3 = 2z \cdot 3z^2, \quad 2z^2 = 2z \cdot z, \quad \text{and} \quad 4z = 2z \cdot 2.$$

Thus $6z^3 - 2z^2 + 4z = 2z(3z^2 - z + 2)$.
(b) Both $4x^3y^2$ and x^2y^3 contain a common factor of x^2y^2. That is,

$$4x^3y^2 = x^2y^2 \cdot 4x \quad \text{and} \quad x^2y^3 = x^2y^2 \cdot y.$$

Thus $4x^3y^2 + x^2y^3 = x^2y^2(4x + y)$.   *Now Try Exercises 5 and 15* ◄

Many times we factor out the *greatest common factor*. For example, the polynomial $15x^4 - 5x^2$ has a common factor of $5x$. We could write this polynomial as

$$15x^4 - 5x^2 = 5x(3x^3 - x).$$

However, we can also factor out $5x^2$ to obtain

$$15x^4 - 5x^2 = 5x^2(3x^2 - 1).$$

Because $5x^2$ is the common factor with the highest degree and largest coefficient, we say that $5x^2$ is the **greatest common factor** (GCF) of $15x^4 - 5x^2$.

EXAMPLE 2 Factoring greatest common factors

Factor.
(a) $6m^3n^2 - 3mn^2 + 9m$ **(b)** $-9x^4 + 6x^3 - 3x^2$

SOLUTION
(a) The GCF of $6m^3n^2$, $3mn^2$, and $9m$ is $3m$.

$$6m^3n^2 = 3m \cdot 2m^2n^2, \quad 3mn^2 = 3m \cdot n^2, \quad \text{and} \quad 9m = 3m \cdot 3$$

Thus $6m^3n^2 - 3mn^2 + 9m = 3m(2m^2n^2 - n^2 + 3)$.
(b) Rather than factoring out $3x^2$, we can factor out $-3x^2$ and make the leading coefficient of the remaining expression positive.

$$-9x^4 = -3x^2 \cdot 3x^2, \quad 6x^3 = -3x^2 \cdot -2x, \quad \text{and} \quad -3x^2 = -3x^2 \cdot 1$$

Thus $-9x^4 + 6x^3 - 3x^2 = -3x^2(3x^2 - 2x + 1)$. *Now Try Exercises 11 and 17* ◀

Factoring by Grouping

Factoring by grouping is a technique that makes use of the associative and distributive properties. The next example illustrates the first step in this factoring technique.

Consider the polynomial

$$3t^3 + 6t^2 + 2t + 4.$$

We can factor this polynomial by first grouping it into two binomials.

$(3t^3 + 6t^2) + (2t + 4)$ Associative property

$3t^2(t + 2) + 2(t + 2)$ Factor out common factors.

$(3t^2 + 2)(t + 2).$ Factor out $(t + 2)$.

The following steps summarize factoring four terms by grouping.

Factoring by Grouping

STEP 1: Use parentheses to group the terms into binomials with common factors. Begin by writing the expression with a plus sign between the binomials.

STEP 2: Factor out the common factor in each binomial.

STEP 3: Factor out the common binomial. If there is no common binomial, try a different grouping or a different method of factoring.

EXAMPLE 3   Factoring by grouping

Factor each polynomial.
(a) $12x^3 - 9x^2 - 8x + 6$ (b) $2x - 2y + ax - ay$

SOLUTION

(a) $12x^3 - 9x^2 - 8x + 6 = (12x^3 - 9x^2) + (-8x + 6)$ *Write with a plus sign between binomials.*

$= 3x^2(4x - 3) - 2(4x - 3)$ *Factor out $3x^2$ and -2.*

$= (3x^2 - 2)(4x - 3)$ *Factor out $4x - 3$.*

(b) $2x - 2y + ax - ay \quad = (2x - 2y) + (ax - ay)$ *Group terms.*

$= 2(x - y) + a(x - y)$ *Factor out 2 and a.*

$= (2 + a)(x - y)$ *Factor out $x - y$.*

Now Try Exercises 21 and 31 ◄

Factoring $x^2 + bx + c$

The product $(x + 3)(x + 4)$ can be found as follows.

$$(x + 3)(x + 4) = x^2 + 4x + 3x + 12$$
$$= x^2 + \quad 7x \quad + 12$$

The middle term $7x$ is found by calculating the sum $4x + 3x$, and the last term is found by calculating the product $3 \cdot 4 = 12$.

When we factor polynomials, we are *reversing* the process of multiplication. To factor $x^2 + 7x + 12$, we must find m and n that satisfy

$$x^2 + 7x + 12 = (x + m)(x + n).$$

Because

$$(x + m)(x + n) = x^2 + (m + n)x + mn,$$

it follows that $mn = 12$ and $m + n = 7$. To determine m and n, we list factors of 12 and their sum, as shown in Table R.5.

Because $3 \cdot 4 = 12$ and $3 + 4 = 7$, we can write the factored form as

$$x^2 + 7x + 12 = (x + 3)(x + 4).$$

This result can always be checked by multiplying the two binomials.

$$(x + 3)(x + 4) = x^2 + 7x + 12$$

Table R.5   Factor Pairs for 12

Factors	1, 12	2, 6	3, 4
Sum	13	8	7

$3x$
$+4x$
$7x$ ←— The middle term checks.

Factoring $x^2 + bx + c$

To factor the trinomial $x^2 + bx + c$, find integers m and n that satisfy

$$m \cdot n = c \quad \text{and} \quad m + n = b.$$

Then $x^2 + bx + c = (x + m)(x + n)$.

EXAMPLE 4   Factoring the form $x^2 + bx + c$

Factor each trinomial.
(a) $x^2 + 10x + 16$ (b) $x^2 + 7x - 30$

Table R.6   Factor Pairs for 16

Factors	1, 16	2, 8	4, 4
Sum	17	10	8

SOLUTION

(a) We need to find a factor pair for 16 whose sum is 10. From Table R.6 the required factor pair is $m = 2$ and $n = 8$. Thus

$$x^2 + 10x + 16 = (x + 2)(x + 8).$$

(b) Factors of -30 whose sum equals 7 are -3 and 10. Thus

$$x^2 + 7x - 30 = (x - 3)(x + 10).$$

Now Try Exercises 33 and 37 ◀

EXAMPLE 5   Removing common factors first

Factor completely.
(a) $3x^2 + 15x + 18$ (b) $5x^3 + 5x^2 - 60x$

SOLUTION

(a) If we first factor out the common factor of 3, the resulting trinomial is easier to factor.

$$3x^2 + 15x + 18 = 3(x^2 + 5x + 6)$$

Now we find m and n such that $mn = 6$ and $m + n = 5$. These numbers are 2 and 3.

$$3x^2 + 15x + 18 = 3(x^2 + 5x + 6)$$
$$= 3(x + 2)(x + 3)$$

(b) First, we factor out the common factor of $5x$. Then we factor the resulting trinomial.

$$5x^3 + 5x^2 - 60x = 5x(x^2 + x - 12)$$
$$= 5x(x - 3)(x + 4)$$

Now Try Exercises 51 and 53 ◀

Factoring Trinomials by Grouping

In this subsection we use grouping to factor trinomials in the form $ax^2 + bx + c$ with $a \neq 1$. For example, one way to factor $3x^2 + 14x + 8$ is to find two numbers m and n such that $mn = 3 \cdot 8 = 24$ and $m + n = 14$. Because $2 \cdot 12 = 24$ and $2 + 12 = 14$, $m = 2$ and $n = 12$. Using grouping, we can now factor this trinomial.

$$3x^2 + 14x + 8 = 3x^2 + 2x + 12x + 8 \qquad \text{Write 14x as 2x + 12x.}$$
$$= (3x^2 + 2x) + (12x + 8) \qquad \text{Associative property}$$
$$= x(3x + 2) + 4(3x + 2) \qquad \text{Factor out x and 4.}$$
$$= (x + 4)(3x + 2) \qquad \text{Distributive property}$$

Writing the polynomial as $3x^2 + 12x + 2x + 8$ would also work.

Factoring $ax^2 + bx + c$ by Grouping

To factor $ax^2 + bx + c$, perform the following steps. (Assume that a, b, and c have no factor in common.)

1. Find numbers m and n such that $mn = ac$ and $m + n = b$. This step may require trial and error.
2. Write the trinomial as $ax^2 + mx + nx + c$.
3. Use grouping to factor this expression as two binomials.

EXAMPLE 6 Factoring $ax^2 + bx + c$ by grouping

Factor each trinomial.
(a) $12y^2 + 5y - 3$ **(b)** $6r^2 - 19r + 10$

SOLUTION

(a) In this trinomial $a = 12$, $b = 5$, and $c = -3$. Because $mn = ac$ and $m + n = b$, the numbers m and n satisfy $mn = -36$ and $m + n = 5$. Thus $m = 9$ and $n = -4$.

$$
\begin{aligned}
12y^2 + 5y - 3 &= 12y^2 + 9y - 4y - 3 && \text{Write } 5y \text{ as } 9y - 4y.\\
&= (12y^2 + 9y) + (-4y - 3) && \text{Associative property}\\
&= 3y(4y + 3) - 1(4y + 3) && \text{Factor out } 3y \text{ and } -1.\\
&= (3y - 1)(4y + 3) && \text{Distributive property}
\end{aligned}
$$

(b) In this trinomial $a = 6$, $b = -19$, and $c = 10$. Because $mn = ac$ and $m + n = b$, the numbers m and n satisfy $mn = 60$ and $m + n = -19$. Thus $m = -4$ and $n = -15$.

$$
\begin{aligned}
6r^2 - 19r + 10 &= 6r^2 - 4r - 15r + 10 && \text{Write } -19r \text{ as } -4r - 15r.\\
&= (6r^2 - 4r) + (-15r + 10) && \text{Associative property}\\
&= 2r(3r - 2) - 5(3r - 2) && \text{Factor out } 2r \text{ and } -5.\\
&= (2r - 5)(3r - 2) && \text{Distributive property}
\end{aligned}
$$

Now Try Exercises 41 and 43. ◄

Factoring Trinomials with FOIL

An alternative to factoring trinomials by grouping is to use FOIL in *reverse*. For example, the factors of $3x^2 + 7x + 2$ are two binomials.

$$3x^2 + 7x + 2 \stackrel{?}{=} (\underline{} + \underline{})(\underline{} + \underline{})$$

The expressions to be placed in the four blanks are yet to be found. By the FOIL method, we know that the product of the first terms is $3x^2$. Because $3x^2 = 3x \cdot x$, we can write

$$3x^2 + 7x + 2 \stackrel{?}{=} (\underline{\;3x\;} + \underline{})(\underline{\;x\;} + \underline{}).$$

The product of the last terms in each binomial must equal 2. Because $2 = 1 \cdot 2$, we can put the 1 and 2 in the blanks, but we must be sure to place them correctly so that the product of the *outside terms* plus the product of the *inside terms* equals $7x$.

$$(3x + 1)(x + 2) = 3x^2 + 7x + 2$$

1x
+6x
7x   ← Middle term checks.

If we had interchanged the 1 and 2, we would have obtained an incorrect result.

$$(3x + 2)(x + 1) = 3x^2 + 5x + 2$$

2x
+3x
5x   ← Middle term is *not* $7x$.

In the next example, we factor expressions of the form $ax^2 + bx + c$, where $a \neq 1$. In this situation, we may need to *guess and check* or use *trial and error* a few times to find the correct factors.

EXAMPLE 7 Factoring the form $ax^2 + bx + c$

Factor each trinomial.
(a) $6x^2 - x - 2$ **(b)** $4x^3 - 14x^2 + 6x$

SOLUTION

(a) The factors of $6x^2$ are either $2x$ and $3x$ or $6x$ and x. The factors of -2 are either -1 and 2 or 1 and -2. To obtain a middle term of $-x$, we use the following factors.

$$(3x - 2) \ (2x + 1) = 6x^2 - x - 2$$

$-4x$
$+3x$
$-x$ ⟵ It checks.

To find the correct factorization, we may need to guess and check a few times.

(b) Each term contains a common factor of $2x$, so we do the following step first.

$$4x^3 - 14x^2 + 6x = 2x(2x^2 - 7x + 3)$$

Next we factor $2x^2 - 7x + 3$. The factors of $2x^2$ are $2x$ and x. Because the middle term is negative, we use -1 and -3 for factors of 3.

$$4x^3 - 14x^2 + 6x = 2x(2x^2 - 7x + 3)$$
$$= 2x(2x - 1)(x - 3)$$

Now Try Exercises 55 and 57 ◀

Difference of Two Squares

When we factor polynomials, we are *reversing* the process of multiplying polynomials. In Section R.3 we discussed the equation

$$(a - b)(a + b) = a^2 - b^2.$$

We can use this equation to factor a difference of two squares.

Difference of Two Squares
For any real numbers a and b, $$a^2 - b^2 = (a - b)(a + b).$$

NOTE The sum of two squares *cannot* be factored (using real numbers). For example, $x^2 + y^2$ cannot be factored, whereas $x^2 - y^2$ can be factored. It is important to remember that $x^2 + y^2 \neq (x + y)^2$.

EXAMPLE 8 Factoring the difference of two squares

Factor each polynomial, if possible.
(a) $9x^2 - 64$ **(b)** $4x^2 + 9y^2$ **(c)** $4a^3 - 4a$

SOLUTION

(a) Note that $9x^2 = (3x)^2$ and $64 = 8^2$.

$$9x^2 - 64 = (3x)^2 - (8)^2$$
$$= (3x - 8)(3x + 8)$$

(b) Because $4x^2 + 9y^2$ is the *sum* of two squares, it *cannot* be factored.

(c) Start by factoring out the common factor of $4a$.

$$4a^3 - 4a = 4a(a^2 - 1)$$
$$= 4a(a - 1)(a + 1)$$

Now Try Exercises 61, 65, and 69 ◄

EXAMPLE 9 Applying the difference of two squares

Factor each expression.
(a) $x^4 - y^4$ **(b)** $6r^2 - 24t^4$

SOLUTION
(a) Use $a^2 - b^2 = (a - b)(a + b)$, with $a = x^2$ and $b = y^2$.

$$x^4 - y^4 = (x^2)^2 - (y^2)^2 \qquad \text{Write as squares.}$$
$$= (x^2 - y^2)(x^2 + y^2) \qquad \text{Difference of squares}$$
$$= (x - y)(x + y)(x^2 + y^2) \qquad \text{Difference of squares}$$

(b) Start by factoring out the common factor of 6.

$$6r^2 - 24t^4 = 6(r^2 - 4t^4) \qquad \text{Factor out 6.}$$
$$= 6\left(r^2 - (2t^2)^2\right) \qquad \text{Write as squares.}$$
$$= 6(r - 2t^2)(r + 2t^2) \qquad \text{Difference of squares}$$

Now Try Exercises 66 and 67 ◄

Perfect Square Trinomials

In Section R.3 we expanded $(a + b)^2$ and $(a - b)^2$ as follows.

$$(a + b)^2 = a^2 + 2ab + b^2 \quad \text{and} \quad (a - b)^2 = a^2 - 2ab + b^2$$

The expressions $a^2 + 2ab + b^2$ and $a^2 - 2ab + b^2$ are called **perfect square trinomials**. We can use the following formulas to factor them.

Perfect Square Trinomials

For any real numbers a and b,

$$a^2 + 2ab + b^2 = (a + b)^2 \quad \text{and}$$
$$a^2 - 2ab + b^2 = (a - b)^2.$$

EXAMPLE 10 Factoring perfect square trinomials

Factor each expression.
(a) $x^2 + 6x + 9$ **(b)** $81x^2 - 72x + 16$

SOLUTION
(a) Let $a^2 = x^2$ and $b^2 = 3^2$. In a perfect square trinomial, the middle term is $2ab$.

$$2ab = 2(x)(3) = 6x,$$

which equals the given middle term. Thus $a^2 + 2ab + b^2 = (a + b)^2$ implies

$$x^2 + 6x + 9 = (x + 3)^2.$$

(b) Let $a^2 = (9x)^2$ and $b^2 = 4^2$. In a perfect square trinomial, the middle term is $2ab$.

$$2ab = 2(9x)(4) = 72x,$$

which equals the given middle term. Thus $a^2 - 2ab + b^2 = (a - b)^2$ implies

$$81x^2 - 72x + 16 = (9x - 4)^2.$$

Now Try Exercises 77 and 81 ◀

EXAMPLE 11 Factoring perfect square trinomials

Factor $25a^3 + 10a^2b + ab^2$.

SOLUTION Start by factoring out the common factor of a. Then factor the resulting perfect square trinomial.

$$25a^3 + 10a^2b + ab^2 = a(25a^2 + 10ab + b^2)$$
$$= a(5a + b)^2$$

Now Try Exercise 89 ◀

Sum and Difference of Two Cubes

The sum or difference of two cubes may be factored. This fact is justified by the following two equations.

$$(a + b)(a^2 - ab + b^2) = a^3 + b^3 \quad \text{and}$$
$$(a - b)(a^2 + ab + b^2) = a^3 - b^3$$

These equations can be verified by multiplying the left side to obtain the right side. For example,

$$(a + b)(a^2 - ab + b^2) = a \cdot a^2 - a \cdot ab + a \cdot b^2 + b \cdot a^2 - b \cdot ab + b \cdot b^2$$
$$= a^3 - a^2b + ab^2 + a^2b - ab^2 + b^3$$
$$= a^3 + b^3.$$

Sum and Difference of Two Cubes

For any real numbers a and b,

$$a^3 + b^3 = (a + b)(a^2 - ab + b^2) \quad \text{and}$$
$$a^3 - b^3 = (a - b)(a^2 + ab + b^2).$$

EXAMPLE 12 Factoring the sum and difference of two cubes

Factor each polynomial.
(a) $x^3 + 8$ **(b)** $27x^3 - 64y^3$ **(c)** $27p^9 - 8q^6$

SOLUTION

(a) Because $x^3 = (x)^3$ and $8 = 2^3$, we let $a = x$, $b = 2$ and factor. Substituting in

$$a^3 + b^3 = (a + b)(a^2 - ab + b^2)$$

gives

$$x^3 + 2^3 = (x + 2)(x^2 - x \cdot 2 + 2^2)$$
$$= (x + 2)(x^2 - 2x + 4).$$

Note that the quadratic expression does not factor further.

(b) Here, $27x^3 = (3x)^3$ and $64y^3 = (4y)^3$, so

$$27x^3 - 64y^3 = (3x)^3 - (4y)^3.$$

Substituting $a = 3x$ and $b = 4y$ in

$$a^3 - b^3 = (a - b)(a^2 + ab + b^2)$$

gives

$$(3x)^3 - (4y)^3 = (3x - 4y)((3x)^2 + 3x \cdot 4y + (4y)^2)$$
$$= (3x - 4y)(9x^2 + 12xy + 16y^2).$$

(c) Let $a^3 = (3p^3)^3$ and $b^3 = (2q^2)^3$. Then $a^3 - b^3 = (a - b)(a^2 + ab + b^2)$ implies

$$27p^9 - 8q^6 = (3p^3 - 2q^2)(9p^6 + 6p^3q^2 + 4q^4).$$

Now Try Exercises 93, 95, and 99 ◀

R.4 Exercises

Greatest Common Factor

Exercises 1–18: Factor out the greatest common factor.

1. $10x - 15$

2. $32 - 16x$

3. $2x^3 - 5x$

4. $3y - 9y^2$

5. $8x^3 - 4x^2 + 16x$

6. $-5x^3 + x^2 - 4x$

7. $5x^4 - 15x^3 + 15x^2$

8. $28y + 14y^3 - 7y^5$

9. $15x^3 + 10x^2 - 30x$

10. $14a^4 - 21a^2 + 35a$

11. $6r^5 - 8r^4 + 12r^3$

12. $15r^6 + 20r^4 - 10r^3$

13. $8x^2y^2 - 24x^2y^3$

14. $36xy - 24x^3y^3$

15. $18mn^2 - 12m^2n^3$

16. $24m^2n^3 + 12m^3n^2$

17. $-4a^2 - 2ab + 6ab^2$

18. $-5a^2 + 10a^2b^2 - 15ab$

Factoring by Grouping

Exercises 19–32: Use grouping to factor the polynomial.

19. $x^3 + 3x^2 + 2x + 6$

20. $4x^3 + 3x^2 + 8x + 6$

21. $6x^3 - 4x^2 + 9x - 6$

22. $x^3 - 3x^2 - 5x + 15$

23. $z^3 - 5z^2 + z - 5$

24. $y^3 - 7y^2 + 8y - 56$

25. $y^4 + 2y^3 - 5y^2 - 10y$ **26.** $4z^4 + 4z^3 + z^2 + z$

27. $2x^3 - 3x^2 + 2x - 3$

28. $8x^3 - 2x^2 + 12x - 3$

29. $2x^4 - x^3 + 4x - 2$

30. $2x^4 - 5x^3 + 10x - 25$

31. $ab - 3a + 2b - 6$

32. $2ax - 6bx - ay + 3by$

Factoring Trinomials

Exercises 33–58: Factor the expression completely.

33. $x^2 + 7x + 10$

34. $x^2 + 3x - 10$

35. $x^2 + 8x + 12$

36. $x^2 - 8x + 12$

37. $z^2 + z - 42$

38. $z^2 - 9z + 20$

39. $z^2 + 11z + 24$

40. $z^2 + 15z + 54$

41. $24x^2 + 14x - 3$

42. $25x^2 - 5x - 6$

43. $6x^2 - x - 2$

44. $10x^2 + 3x - 1$

45. $1 + x - 2x^2$

46. $3 - 5x - 2x^2$

47. $20 + 7x - 6x^2$ **48.** $4 + 13x - 12x^2$

49. $5x^3 + x^2 - 6x$ **50.** $2x^3 + 8x^2 - 24x$

51. $3x^3 + 12x^2 + 9x$ **52.** $12x^3 - 8x^2 - 20x$

53. $2x^2 - 14x + 20$ **54.** $7x^2 + 35x + 42$

55. $60t^4 + 230t^3 - 40t^2$ **56.** $24r^4 + 8r^3 - 80r^2$

57. $4m^3 + 10m^2 - 6m$ **58.** $30m^4 + 3m^3 - 9m^2$

Difference of Two Squares

Exercises 59–76: Factor the expression completely, if possible.

59. $x^2 - 25$ **60.** $z^2 - 169$

61. $4x^2 - 25$ **62.** $36 - y^2$

63. $36x^2 - 100$ **64.** $9x^2 - 4y^2$

65. $64z^2 - 25z^4$ **66.** $100x^3 - x$

67. $16x^4 - y^4$ **68.** $x^4 - 9y^2$

69. $a^2 + 4b^2$ **70.** $9r^4 + 25t^4$

71. $4 - r^2t^2$ **72.** $25 - x^4y^2$

73. $(x - 1)^2 - 16$ **74.** $(y + 2)^2 - 1$

75. $4 - (z + 3)^2$ **76.** $64 - (t - 3)^2$

Perfect Square Trinomials

Exercises 77–90: Factor the expression.

77. $x^2 + 2x + 1$ **78.** $x^2 - 6x + 9$

79. $4x^2 + 20x + 25$ **80.** $x^2 + 10x + 25$

81. $x^2 - 12x + 36$ **82.** $16z^4 - 24z^3 + 9z^2$

83. $9z^3 - 6z^2 + z$ **84.** $49y^2 + 42y + 9$

85. $9y^3 + 30y^2 + 25y$ **86.** $25y^3 - 20y^2 + 4y$

87. $4x^2 - 12xy + 9y^2$ **88.** $25a^2 + 60ab + 36b^2$

89. $9a^3b - 12a^2b + 4ab$ **90.** $16a^3 + 8a^2b + ab^2$

Sum and Difference of Two Cubes

Exercises 91–102: Factor the expression.

91. $x^3 - 1$ **92.** $x^3 + 1$

93. $y^3 + z^3$ **94.** $y^3 - z^3$

95. $8x^3 - 27$ **96.** $8 - z^3$

97. $x^4 + 125x$ **98.** $3x^4 - 81x$

99. $8r^6 - t^3$ **100.** $125r^6 + 64t^3$

101. $10m^9 - 270n^6$ **102.** $5t^6 + 40r^3$

General Factoring

Exercises 103–158: Factor the expression completely.

103. $16x^2 - 25$ **104.** $25x^2 - 30x + 9$

105. $x^3 - 64$ **106.** $1 + 8y^3$

107. $x^2 + 16x + 64$ **108.** $12x^2 + x - 6$

109. $5x^2 - 38x - 16$ **110.** $125x^3 - 1$

111. $x^4 + 8x$ **112.** $2x^3 - 12x^2 + 18x$

113. $64x^3 + 8y^3$ **114.** $54 - 16x^3$

115. $3x^2 - 5x - 8$ **116.** $15x^2 - 11x + 2$

117. $7a^3 + 20a^2 - 3a$ **118.** $b^3 - b^2 - 2b$

119. $2x^3 - x^2 + 6x - 3$ **120.** $3x^3 - 5x^2 + 3x - 5$

121. $2x^4 - 5x^3 - 25x^2$ **122.** $10x^3 + 28x^2 - 6x$

123. $2x^4 + 5x^2 + 3$ **124.** $2x^4 + 2x^2 - 4$

125. $x^3 + 3x^2 + x + 3$ **126.** $x^3 + 5x^2 + 4x + 20$

127. $5x^3 - 5x^2 + 10x - 10$

128. $5x^4 - 20x^3 + 10x - 40$

129. $ax + bx - ay - by$ **130.** $ax - bx - ay + by$

131. $18x^2 + 12x + 2$ **132.** $-3x^2 + 30x - 75$

133. $-4x^3 + 24x^2 - 36x$ **134.** $18x^3 - 60x^2 + 50x$

135. $27x^3 - 8$ **136.** $27x^3 + 8$

137. $-x^4 - 8x$ **138.** $x^5 - 27x^2$

139. $x^4 - 2x^3 - x + 2$ **140.** $x^4 + 3x^3 + x + 3$

141. $r^4 - 16$ **142.** $r^4 - 81$

143. $25x^2 - 4a^2$ **144.** $9y^2 - 16z^2$

145. $2x^4 - 2y^4$ **146.** $a^4 - b^4$

147. $9x^3 + 6x^2 - 3x$ **148.** $8x^3 + 28x^2 - 16x$

149. $(z - 2)^2 - 9$ **150.** $(y + 2)^2 - 4$

151. $3x^5 - 27x^3 + 3x^2 - 27$

152. $2x^5 - 8x^3 - 16x^2 + 64$

153. $(x + 2)^2(x + 4)^4 + (x + 2)^3(x + 4)^3$

154. $(x - 3)(2x + 1)^3 + (x - 3)^2(2x + 1)^2$

155. $(6x + 1)(8x - 3)^4 - (6x + 1)^2(8x - 3)^3$

156. $(2x + 3)^4(x + 1)^4 - (2x + 3)^3(x + 1)^5$

157. $4x^2(5x - 1)^5 + 2x(5x - 1)^6$

158. $x^4(7x + 3)^3 + x^5(7x + 3)^2$

R.5 Rational Expressions

- **Simplify rational expressions**
- **Multiply and divide fractions**
- **Perform multiplication and division on rational expressions**
- **Find least common multiples and denominators**
- **Add and subtract fractions**
- **Perform addition and subtraction on rational expressions**
- **Clear fractions from equations**
- **Simplify complex fractions**

Simplifying Rational Expressions

When simplifying fractions, we sometimes use the **basic principle of fractions**, which states that

$$\frac{a \cdot c}{b \cdot c} = \frac{a}{b}.$$

This principle holds because $\frac{c}{c} = 1$ and $\frac{a}{b} \cdot 1 = \frac{a}{b}$. It can be used to simplify a fraction.

$$\frac{6}{44} = \frac{3 \cdot 2}{22 \cdot 2} = \frac{3}{22}$$

This same principle can also be used to simplify rational expressions.

Simplifying Rational Expressions

The following principle can be used to simplify rational expressions, where A, B, and C are polynomials.

$$\frac{A \cdot C}{B \cdot C} = \frac{A}{B}, \quad B \text{ and } C \text{ are nonzero.}$$

EXAMPLE 1   Simplifying rational expressions

Simplify each expression.

(a) $\dfrac{9x}{3x^2}$ (b) $\dfrac{2z^2 - 3z - 9}{z^2 + 2z - 15}$ (c) $\dfrac{a^2 - b^2}{a + b}$

SOLUTION

(a) First factor out the greatest common factor, $3x$, in the numerator and denominator.

$$\frac{9x}{3x^2} = \frac{3 \cdot 3x}{x \cdot 3x} = \frac{3}{x}$$

(b) Start by factoring the numerator and denominator.

$$\frac{2z^2 - 3z - 9}{z^2 + 2z - 15} = \frac{(2z + 3)(z - 3)}{(z + 5)(z - 3)} = \frac{2z + 3}{z + 5}$$

(c) Start by factoring the numerator as the difference of squares.

$$\frac{a^2 - b^2}{a + b} = \frac{(a - b)(a + b)}{a + b} = a - b$$

Now Try Exercises 1, 5, and 11 ◄

Review of Multiplication and Division of Fractions

Recall that to multiply two fractions we use the property

$$\frac{a}{b} \cdot \frac{c}{d} = \frac{ac}{bd}.$$

For example, $\frac{2}{5} \cdot \frac{3}{7} = \frac{2 \cdot 3}{5 \cdot 7} = \frac{6}{35}$.

EXAMPLE 2 Multiplying fractions

Multiply and simplify the product.

(a) $\frac{4}{9} \cdot \frac{3}{8}$ (b) $\frac{2}{3} \cdot \frac{3}{4} \cdot \frac{5}{6}$

SOLUTION

(a) $\frac{4}{9} \cdot \frac{3}{8} = \frac{4 \cdot 3}{9 \cdot 8} = \frac{12}{72} = \frac{1 \cdot 12}{6 \cdot 12} = \frac{1}{6}$ (b) $\frac{2}{3} \cdot \frac{3}{4} \cdot \frac{5}{6} = \frac{6 \cdot 5}{12 \cdot 6} = \frac{5}{12}$

Now Try Exercises 15 and 17 ◄

Recall that to divide two fractions we "invert and multiply." That is, we change a division problem to a multiplication problem, using the property

$$\frac{a}{b} \div \frac{c}{d} = \frac{a}{b} \cdot \frac{d}{c}.$$

For example, $\frac{3}{4} \div \frac{5}{4} = \frac{3}{4} \cdot \frac{4}{5} = \frac{3 \cdot 4}{5 \cdot 4} = \frac{3}{5}$.

Multiplication and Division of Rational Expressions

Multiplying and dividing rational expressions is similar to multiplying and dividing fractions.

Products and Quotients of Rational Expressions

To multiply two rational expressions, multiply numerators and multiply denominators.

$$\frac{A}{B} \cdot \frac{C}{D} = \frac{AC}{BD}, \qquad B \text{ and } D \text{ are nonzero.}$$

To divide two rational expressions, multiply by the reciprocal of the divisor.

$$\frac{A}{B} \div \frac{C}{D} = \frac{A}{B} \cdot \frac{D}{C}, \qquad B, C, \text{ and } D \text{ are nonzero.}$$

EXAMPLE 3 Multiplying rational expressions

Multiply.

(a) $\frac{1}{x} \cdot \frac{2x}{x + 1}$ (b) $\frac{x - 1}{x} \cdot \frac{x - 1}{x + 2}$

SOLUTION

(a) $\dfrac{1}{x} \cdot \dfrac{2x}{x+1} = \dfrac{1 \cdot 2x}{x(x+1)} = \dfrac{2}{x+1}$ **(b)** $\dfrac{x-1}{x} \cdot \dfrac{x-1}{x+2} = \dfrac{(x-1)(x-1)}{x(x+2)}$

Now Try Exercises 33 and 37 ◀

EXAMPLE 4 Dividing two rational expressions

Divide and simplify.

(a) $\dfrac{2}{x} \div \dfrac{2x-1}{4x}$ **(b)** $\dfrac{x^2-1}{x^2+x-6} \div \dfrac{x-1}{x+3}$

SOLUTION

(a) $\dfrac{2}{x} \div \dfrac{2x-1}{4x} = \dfrac{2}{x} \cdot \dfrac{4x}{2x-1}$ "Invert and multiply."

$= \dfrac{8x}{x(2x-1)}$ Multiply.

$= \dfrac{8}{2x-1}$ Simplify.

(b) $\dfrac{x^2-1}{x^2+x-6} \div \dfrac{x-1}{x+3} = \dfrac{x^2-1}{x^2+x-6} \cdot \dfrac{x+3}{x-1}$ "Invert and multiply."

$= \dfrac{(x+1)(x-1)}{(x-2)(x+3)} \cdot \dfrac{x+3}{x-1}$ Factor.

$= \dfrac{(x+1)(x-1)(x+3)}{(x-2)(x-1)(x+3)}$ Commutative property

$= \dfrac{x+1}{x-2}$ Simplify.

Now Try Exercises 45 and 53 ◀

Least Common Multiples and Denominators

To add or subtract fractions and rational expressions, we need to find a common denominator. The **least common denominator** (LCD) is equivalent to the **least common multiple** (LCM) of the denominators. The following procedure can be used to find the least common multiple.

Finding the Least Common Multiple

The least common multiple (LCM) of two polynomials can be found as follows.

STEP 1: Factor each polynomial completely.

STEP 2: List each factor the greatest number of times that it occurs in either factorization.

STEP 3: Find the product of this list of factors. The result is the LCM.

The next example illustrates how to use this procedure.

EXAMPLE 5 Finding least common multiples

Find the least common multiple for each pair of expressions.

(a) $4x, 5x^3$ **(b)** $x^2 + 4x + 4, x^2 + 3x + 2$

SOLUTION

(a) **STEP 1:**   Factor each polynomial completely.

$$4x = 2 \cdot 2 \cdot x \quad \text{and} \quad 5x^3 = 5 \cdot x \cdot x \cdot x$$

STEP 2:   The factor 2 occurs twice, the factor 5 occurs once, and the factor x occurs at most three times. The list then is **2, 2, 5, x, x,** and x.

STEP 3:   The LCM is the product $2 \cdot 2 \cdot 5 \cdot x \cdot x \cdot x$, or $20x^3$.

(b) **STEP 1:**   Factor each polynomial as follows.

$$x^2 + 4x + 4 = (x + 2)(x + 2) \quad \text{and} \quad x^2 + 3x + 2 = (x + 1)(x + 2)$$

STEP 2:   The factor $(x + 1)$ occurs once, and $(x + 2)$ occurs at most twice.

STEP 3:   The LCM is the product $(x + 1)(x + 2)^2$, which is left in factored form.

> *Now Try Exercises 61 and 65* ◀

EXAMPLE 6   **Finding a least common denominator**

Find the LCD for the expressions $\dfrac{1}{x^2 + 4x + 4}$ and $\dfrac{5}{x^2 + 3x + 2}$.

SOLUTION   From Example 5(b), the LCM for $x^2 + 4x + 4$ and $x^2 + 3x + 2$ is

$$(x + 1)(x + 2)^2.$$

Therefore the LCD is *also* $(x + 1)(x + 2)^2$.

> *Now Try Exercise 69* ◀

Review of Addition and Subtraction of Fractions

Recall that to add two fractions we use the property $\frac{a}{c} + \frac{b}{c} = \frac{a + b}{c}$. This property requires *like* denominators. For example, $\frac{1}{5} + \frac{3}{5} = \frac{1 + 3}{5} = \frac{4}{5}$. When the denominators are not alike, we must find a common denominator. Before adding two fractions, such as $\frac{2}{3}$ and $\frac{1}{4}$, we write them with 12 as their common denominator. That is, we multiply each fraction by 1 written in an appropriate form. For example, to write $\frac{2}{3}$ with a denominator of 12, we multiply $\frac{2}{3}$ by 1, written as $\frac{4}{4}$.

$$\frac{2}{3} = \frac{2}{3} \cdot \frac{4}{4} = \frac{8}{12} \quad \text{and} \quad \frac{1}{4} = \frac{1}{4} \cdot \frac{3}{3} = \frac{3}{12}$$

Once the fractions have a common denominator, we can add them, as in

$$\frac{2}{3} + \frac{1}{4} = \frac{8}{12} + \frac{3}{12} = \frac{11}{12}.$$

The *least common denominator* (LCD) for $\frac{2}{3}$ and $\frac{1}{4}$ is equal to the *least common multiple* (LCM) of 3 and 4. Thus the least common denominator is 12.

EXAMPLE 7   **Adding fractions**

Simplify $\dfrac{3}{5} + \dfrac{2}{7}$.

SOLUTION   The LCD is 35.

$$\frac{3}{5} + \frac{2}{7} = \frac{3}{5} \cdot \frac{7}{7} + \frac{2}{7} \cdot \frac{5}{5} = \frac{21}{35} + \frac{10}{35} = \frac{31}{35}$$

> *Now Try Exercise 27* ◀

Recall that subtraction is similar to addition. To subtract two fractions with *like* denominators, we use the property $\frac{a}{c} - \frac{b}{c} = \frac{a-b}{c}$. For example, $\frac{3}{11} - \frac{7}{11} = \frac{3-7}{11} = -\frac{4}{11}$.

EXAMPLE 8   Subtracting fractions

Simplify $\frac{3}{8} - \frac{5}{6}$.

SOLUTION The LCD is 24.

$$\frac{3}{8} - \frac{5}{6} = \frac{3}{8} \cdot \frac{3}{3} - \frac{5}{6} \cdot \frac{4}{4} = \frac{9}{24} - \frac{20}{24} = -\frac{11}{24}$$

Now Try Exercise 29 ◀

Addition and Subtraction of Rational Expressions

Addition and subtraction of rational expressions with like denominators are performed in the following manner.

Sums and Differences of Rational Expressions

To add (or subtract) two rational expressions with like denominators, add (or subtract) their numerators. The denominator does not change.

$$\frac{A}{C} + \frac{B}{C} = \frac{A + B}{C}$$

$$\frac{A}{C} - \frac{B}{C} = \frac{A - B}{C}, \quad C \neq 0$$

NOTE If the denominators are not alike, begin by writing each rational expression, using a common denominator. The LCD equals the LCM of the denominators.

EXAMPLE 9   Adding rational expressions

Add and simplify.

(a) $\dfrac{x}{x + 2} + \dfrac{3x + 1}{x + 2}$ **(b)** $\dfrac{1}{x - 1} + \dfrac{2x}{x + 1}$

SOLUTION
(a) The denominators are alike, so we add the numerators and keep the same denominator.

$$\frac{x}{x + 2} + \frac{3x + 1}{x + 2} = \frac{x + 3x + 1}{x + 2} \qquad \text{Add numerators.}$$

$$= \frac{4x + 1}{x + 2} \qquad \text{Combine like terms.}$$

(b) The LCM for $x - 1$ and $x + 1$ is their product, $(x - 1)(x + 1)$.

$$\frac{1}{x - 1} + \frac{2x}{x + 1} = \frac{1}{x - 1} \cdot \frac{x + 1}{x + 1} + \frac{2x}{x + 1} \cdot \frac{x - 1}{x - 1} \qquad \textit{Change to a common denominator.}$$

$$= \frac{x + 1}{(x - 1)(x + 1)} + \frac{2x(x - 1)}{(x + 1)(x - 1)} \qquad \textit{Multiply.}$$

$$= \frac{x + 1 + 2x^2 - 2x}{(x - 1)(x + 1)} \qquad \textit{Add numerators; distributive property}$$

$$= \frac{2x^2 - x + 1}{(x - 1)(x + 1)} \qquad \textit{Combine like terms.}$$

Now Try Exercises 73 and 81 ◀

Subtraction of rational expressions is similar.

EXAMPLE 10   Subtracting rational expressions

Subtract and simplify the expression $\dfrac{x - 1}{x} - \dfrac{5}{x + 5}$.

SOLUTION   The LCD is $x(x + 5)$.

$$\frac{x - 1}{x} - \frac{5}{x + 5} = \frac{x - 1}{x} \cdot \frac{x + 5}{x + 5} - \frac{5}{x + 5} \cdot \frac{x}{x} \qquad \textit{Change to a common denominator.}$$

$$= \frac{(x - 1)(x + 5)}{x(x + 5)} - \frac{5x}{x(x + 5)} \qquad \textit{Multiply.}$$

$$= \frac{(x - 1)(x + 5) - 5x}{x(x + 5)} \qquad \textit{Subtract numerators.}$$

$$= \frac{x^2 + 4x - 5 - 5x}{x(x + 5)} \qquad \textit{Multiply binomials.}$$

$$= \frac{x^2 - x - 5}{x(x + 5)} \qquad \textit{Combine like terms.}$$

Now Try Exercise 79 ◀

Clearing Fractions

To solve rational equations, it is sometimes advantageous to multiply each side by the LCD to clear fractions. For example, the LCD for the equation $\frac{1}{x + 2} + \frac{1}{x - 2} = 0$ is $(x + 2)(x - 2)$. Multiplying each side by the LCD results in the following.

$$(x + 2)(x - 2)\left(\frac{1}{x + 2} + \frac{1}{x - 2}\right) = 0 \qquad \textit{Multiply by LCD.}$$

$$\frac{(x + 2)(x - 2)}{x + 2} + \frac{(x + 2)(x - 2)}{x - 2} = 0 \qquad \textit{Distributive property}$$

$$(x - 2) + (x + 2) = 0 \qquad \textit{Simplify.}$$

$$x = 0 \qquad \textit{Combine like terms and solve.}$$

This technique is applied in the next example.

EXAMPLE 11   Clearing fractions

Clear fractions from the equation and solve.

$$\frac{3}{x} + \frac{x}{x^2 - 1} - \frac{4}{x + 1} = 0$$

SOLUTION   The LCD is $x(x^2 - 1) = x(x - 1)(x + 1)$.

$$x(x^2 - 1)\left(\frac{3}{x} + \frac{x}{x^2 - 1} - \frac{4}{x + 1}\right) = x(x^2 - 1) \cdot 0$$

$$\frac{3x(x^2 - 1)}{x} + \frac{x(x)(x^2 - 1)}{x^2 - 1} - \frac{4x(x^2 - 1)}{x + 1} = 0$$

$$\frac{3x(x^2 - 1)}{x} + \frac{x^2(x^2 - 1)}{x^2 - 1} - \frac{4x(x - 1)(x + 1)}{x + 1} = 0$$

$$3(x^2 - 1) + x^2 - 4x(x - 1) = 0$$

$$3x^2 - 3 + x^2 - 4x^2 + 4x = 0$$

$$4x - 3 = 0$$

$$x = \frac{3}{4}$$

The solution is $\frac{3}{4}$. Check this answer.   *Now Try Exercise 111* ◀

Complex Fractions

A complex fraction is a rational expression that contains fractions in its numerator, denominator, or both. One strategy for simplifying a complex fraction is to multiply the numerator and denominator by the LCD of the fractions in the numerator and denominator. For example, the LCD for the complex fraction

$$\frac{1 - \dfrac{1}{x}}{1 + \dfrac{1}{2x}}$$

is $2x$. To simplify, multiply the complex fraction by 1, expressed in the form $\frac{2x}{2x}$.

$$\frac{\left(1 - \dfrac{1}{x}\right) \cdot 2x}{\left(1 + \dfrac{1}{2x}\right) \cdot 2x} = \frac{2x - \dfrac{2x}{x}}{2x + \dfrac{2x}{2x}} \qquad \text{Distributive property}$$

$$= \frac{2x - 2}{2x + 1} \qquad \text{Simplify.}$$

In the next example, we simplify a complex fraction.

EXAMPLE 12   Simplifying a complex fraction

Simplify the complex fraction.

$$\frac{\dfrac{3}{x - 1} - \dfrac{2}{x}}{\dfrac{1}{x - 1} + \dfrac{3}{x}}$$

SOLUTION The LCD is the product, $x(x - 1)$. Multiply the expression by $\frac{x(x-1)}{x(x-1)}$.

$$\frac{\left(\dfrac{3}{x-1} - \dfrac{2}{x}\right)}{\left(\dfrac{1}{x-1} + \dfrac{3}{x}\right)} \cdot \frac{x(x-1)}{x(x-1)} = \frac{\dfrac{3x(x-1)}{x-1} - \dfrac{2x(x-1)}{x}}{\dfrac{x(x-1)}{x-1} + \dfrac{3x(x-1)}{x}}$$ Distributive property

$$= \frac{3x - 2(x-1)}{x + 3(x-1)}$$ Simplify.

$$= \frac{3x - 2x + 2}{x + 3x - 3}$$ Distributive property

$$= \frac{x + 2}{4x - 3}$$ Combine like terms.

Now Try Exercise 115 ◄

R.5   Exercises

Simplifying Rational Expressions

Exercises 1–14: Simplify the expression.

1. $\dfrac{10x^3}{5x^2}$

2. $\dfrac{24t^3}{6t^2}$

3. $\dfrac{(x-5)(x+5)}{x-5}$

4. $-\dfrac{5-a}{a-5}$

5. $\dfrac{x^2 - 16}{x - 4}$

6. $\dfrac{(x+5)(x-4)}{(x+7)(x+5)}$

7. $\dfrac{x+3}{2x^2 + 5x - 3}$

8. $\dfrac{2x^2 - 9x + 4}{6x^2 + 7x - 5}$

9. $-\dfrac{z+2}{4z+8}$

10. $\dfrac{x^2 - 25}{x^2 + 10x + 25}$

11. $\dfrac{x^2 + 2x}{x^2 + 3x + 2}$

12. $\dfrac{x^2 - 3x - 10}{x^2 - 6x + 5}$

13. $\dfrac{a^3 + b^3}{a + b}$

14. $\dfrac{a^3 - b^3}{a - b}$

Review of Fractions

Exercises 15–32: Simplify.

15. $\dfrac{5}{8} \cdot \dfrac{4}{15}$

16. $\dfrac{7}{2} \cdot \dfrac{4}{21}$

17. $\dfrac{5}{6} \cdot \dfrac{3}{10} \cdot \dfrac{8}{3}$

18. $\dfrac{9}{5} \cdot \dfrac{10}{3} \cdot \dfrac{1}{27}$

19. $\dfrac{4}{7} \div \dfrac{8}{7}$

20. $\dfrac{5}{12} \div \dfrac{10}{9}$

21. $\dfrac{1}{2} \div \dfrac{3}{4} \div \dfrac{5}{6}$

22. $\dfrac{3}{4} \div \dfrac{7}{8} \div \dfrac{5}{14}$

23. $\dfrac{3}{8} + \dfrac{5}{8}$

24. $\dfrac{5}{9} + \dfrac{2}{9}$

25. $\dfrac{3}{7} - \dfrac{4}{7}$

26. $\dfrac{8}{11} - \dfrac{9}{11}$

27. $\dfrac{2}{3} + \dfrac{5}{11}$

28. $\dfrac{9}{13} + \dfrac{3}{2}$

29. $\dfrac{4}{5} - \dfrac{1}{10}$

30. $\dfrac{3}{4} - \dfrac{7}{12}$

31. $\dfrac{1}{3} + \dfrac{3}{4} - \dfrac{3}{7}$

32. $\dfrac{6}{11} - \dfrac{1}{2} + \dfrac{3}{8}$

Multiplication and Division of Rational Expressions

Exercises 33–58: Simplify the expression.

33. $\dfrac{1}{x^2} \cdot \dfrac{3x}{2}$

34. $\dfrac{6a}{5} \cdot \dfrac{5}{12a^2}$

35. $\dfrac{5x}{3} \div \dfrac{10x}{6}$

36. $\dfrac{2x^2 + x}{3x + 9} \div \dfrac{x}{x + 3}$

37. $\dfrac{x+1}{2x-5} \cdot \dfrac{x}{x+1}$

38. $\dfrac{4x+8}{2x} \cdot \dfrac{x^2}{x+2}$

39. $\dfrac{(x-5)(x+3)}{3x-1} \cdot \dfrac{x(3x-1)}{(x-5)}$

40. $\dfrac{b^2+1}{b^2-1} \cdot \dfrac{b-1}{b+1}$

41. $\dfrac{x^2-2x-35}{2x^3-3x^2} \cdot \dfrac{x^3-x^2}{2x-14}$

42. $\dfrac{2x+4}{x+1} \cdot \dfrac{x^2+3x+2}{4x+2}$

43. $\dfrac{6b}{b+2} \div \dfrac{3b^4}{2b+4}$

44. $\dfrac{5x^5}{x-2} \div \dfrac{10x^3}{5x-10}$

45. $\dfrac{3a+1}{a^7} \div \dfrac{a+1}{3a^8}$

46. $\dfrac{x^2-16}{x+3} \div \dfrac{x+4}{x^2-9}$

47. $\dfrac{x+5}{x^3-x} \div \dfrac{x^2-25}{x^3}$

48. $\dfrac{x^2+x-12}{2x^2-9x-5} \div \dfrac{x^2+7x+12}{2x^2-7x-4}$

49. $\dfrac{x-2}{x^3-x} \div \dfrac{x^2-2x}{x^2-1}$

50. $\dfrac{x^2+3x+2}{2x^2+7x+3} \div \dfrac{x^2-4}{2x^2-x-1}$

51. $\dfrac{x^2-3x+2}{x^2+5x+6} \div \dfrac{x^2+x-2}{x^2+2x-3}$

52. $\dfrac{2x^2+x-1}{6x^2+x-2} \div \dfrac{2x^2+5x+3}{6x^2+13x+6}$

53. $\dfrac{x^2-4}{x^2+x-2} \div \dfrac{x-2}{x-1}$

54. $\dfrac{x^2+2x+1}{x-2} \div \dfrac{x+1}{2x-4}$

55. $\dfrac{3y}{x^2} \div \dfrac{y^2}{x} \div \dfrac{y}{5x}$

56. $\dfrac{x+1}{y-2} \div \dfrac{2x+2}{y-2} \div \dfrac{x}{y}$

57. $\dfrac{x-3}{x-1} \div \dfrac{x^2}{x-1} \div \dfrac{x-3}{x}$

58. $\dfrac{2x}{x-2} \div \dfrac{x+2}{x} \div \dfrac{7x}{x^2-4}$

Least Common Multiples

Exercises 59–66: Find the least common multiple.

59. 12, 18

60. 9, 15

61. $5a^3$, $10a$

62. $6a^2$, $9a^5$

63. z^2-4z, $(z-4)^2$

64. z^2-1, z^2+2z+1

65. x^2-6x+9, x^2-5x+6

66. x^2-4, x^2-4x+4

Common Denominators

Exercises 67–72: Find the LCD for the rational expressions.

67. $\dfrac{1}{x+1}, \dfrac{1}{7}$

68. $\dfrac{1}{2x-1}, \dfrac{1}{x+1}$

69. $\dfrac{1}{x+4}, \dfrac{1}{x^2-16}$

70. $\dfrac{4}{2x^2}, \dfrac{1}{2x+2}$

71. $\dfrac{3}{2}, \dfrac{x}{2x+1}, \dfrac{x}{2x-4}$

72. $\dfrac{1}{x}, \dfrac{1}{x^2-4x}, \dfrac{1}{2x}$

Addition and Subtraction of Rational Expressions

Exercises 73–102: Simplify.

73. $\dfrac{4}{x+1} + \dfrac{3}{x+1}$

74. $\dfrac{2}{x^2} + \dfrac{5}{x^2}$

75. $\dfrac{2}{x^2-1} - \dfrac{x+1}{x^2-1}$

76. $\dfrac{2x}{x^2+x} - \dfrac{2x}{x+1}$

77. $\dfrac{x}{x+4} - \dfrac{x+1}{x(x+4)}$

78. $\dfrac{4x}{x+2} + \dfrac{x-5}{x-2}$

79. $\dfrac{2}{x^2} - \dfrac{4x-1}{x}$

80. $\dfrac{2x}{x-5} - \dfrac{x}{x+5}$

81. $\dfrac{x+3}{x-5} + \dfrac{5}{x-3}$

82. $\dfrac{x}{2x-1} + \dfrac{1-x}{3x}$

83. $\dfrac{3}{x-5} - \dfrac{1}{x-3} - \dfrac{2x}{x-5}$

84. $\dfrac{2x+1}{x-1} - \dfrac{3}{x+1} + \dfrac{x}{x-1}$

85. $\dfrac{x}{x^2-9} + \dfrac{5x}{x-3}$

86. $\dfrac{a^2+1}{a^2-1} + \dfrac{a}{1-a^2}$

87. $\dfrac{b}{2b-4} - \dfrac{b-1}{b-2}$ **88.** $\dfrac{y^2}{2-y} - \dfrac{y}{y^2-4}$

89. $\dfrac{2x}{x-5} + \dfrac{2x-1}{3x^2-16x+5}$

90. $\dfrac{x+3}{2x-1} + \dfrac{3}{10x^2-5x}$

91. $\dfrac{x}{(x-1)^2} - \dfrac{1}{(x-1)(x+3)}$

92. $\dfrac{3}{x^2-x-6} - \dfrac{2}{x^2+5x+6}$

93. $\dfrac{x}{x^2-5x+4} + \dfrac{2}{x^2-2x-8}$

94. $\dfrac{3}{x^2-2x+1} + \dfrac{1}{x^2-3x+2}$

95. $\dfrac{x}{x^2-4} - \dfrac{1}{x^2+4x+4}$

96. $\dfrac{3x}{x^2+2x-3} + \dfrac{1}{x^2-2x+1}$

97. $\dfrac{3x}{x-y} - \dfrac{3y}{x^2-2xy+y^2}$

98. $\dfrac{4c}{ab} + \dfrac{3b}{ac} - \dfrac{2a}{bc}$ **99.** $x + \dfrac{1}{x-1} - \dfrac{1}{x+1}$

100. $5 - \dfrac{6}{n^2-36} + \dfrac{3}{n-6}$

101. $\dfrac{6}{t-1} + \dfrac{2}{t-2} + \dfrac{1}{t}$

102. $\dfrac{3}{x-5} - \dfrac{1}{x-3} - \dfrac{2x}{x-5}$

Clearing Fractions

Exercises 103–112: (Refer to Example 11.) Clear fractions and solve. Check your answers.

103. $\dfrac{1}{x} + \dfrac{3}{x^2} = 0$ **104.** $\dfrac{1}{x-2} + \dfrac{3}{x+1} = 0$

105. $\dfrac{1}{x} + \dfrac{3x}{2x-1} = 0$ **106.** $\dfrac{x}{2x-5} + \dfrac{4}{x} = 0$

107. $\dfrac{2x}{9-x^2} + \dfrac{1}{3-x} = 0$

108. $\dfrac{1}{1-x^2} + \dfrac{1}{1+x} = 0$

109. $\dfrac{1}{2x} + \dfrac{1}{2x^2} - \dfrac{1}{x^3} = 0$

110. $\dfrac{1}{x^2-16} + \dfrac{4}{x+4} - \dfrac{5}{x-4} = 0$

111. $\dfrac{1}{x} - \dfrac{2}{x+5} + \dfrac{1}{x-5} = 0$

112. $\dfrac{1}{x-2} + \dfrac{1}{x-3} - \dfrac{2}{x} = 0$

Complex Fractions

Exercises 113–124: Simplify the expression.

113. $\dfrac{1+\dfrac{1}{x}}{1-\dfrac{1}{x}}$ **114.** $\dfrac{\dfrac{1}{2}-x}{\dfrac{1}{x}-2}$

115. $\dfrac{\dfrac{1}{x-5}}{\dfrac{4}{x}-\dfrac{1}{x-5}}$ **116.** $\dfrac{1+\dfrac{1}{x-3}}{\dfrac{1}{x-3}-1}$

117. $\dfrac{\dfrac{1}{x}+\dfrac{2-x}{x^2}}{\dfrac{3}{x^2}-\dfrac{1}{x}}$ **118.** $\dfrac{\dfrac{1}{x-1}+\dfrac{2}{x}}{2-\dfrac{1}{x}}$

119. $\dfrac{\dfrac{1}{x+3}+\dfrac{2}{x-3}}{2-\dfrac{1}{x-3}}$ **120.** $\dfrac{\dfrac{1}{x}+\dfrac{2}{x}}{\dfrac{1}{x-1}+\dfrac{x}{2}}$

121. $\dfrac{\dfrac{4}{x-5}}{\dfrac{1}{x+5}+\dfrac{1}{x}}$ **122.** $\dfrac{\dfrac{2}{x-4}}{1-\dfrac{1}{x+4}}$

123. $\dfrac{\dfrac{1}{2a}-\dfrac{1}{2b}}{\dfrac{1}{a^2}-\dfrac{1}{b^2}}$ **124.** $\dfrac{\dfrac{1}{2x^2}-\dfrac{1}{2y^2}}{\dfrac{1}{3y^2}+\dfrac{1}{3x^2}}$

R.6 Radical Notation and Rational Exponents

- Use radical notation
- Apply rational exponents
- Use properties of rational exponents

Radical Notation

Square Root Recall the definition of the square root of a number a.

Square Root

The number b is a *square root* of a if $b^2 = a$.

Every positive number a has two square roots, one positive and one negative. For example, the square roots of 100 are 10 and -10. Recall that the *positive* square root is called the *principal square root* and is denoted $\sqrt{a}$. The *negative square root* is denoted $-\sqrt{a}$. To identify both square roots, we write $\pm\sqrt{a}$. The symbol $\pm$ is read "plus or minus." The symbol $\sqrt{}$ is called the **radical sign**. The expression under the radical sign is called the **radicand**, and an expression containing a radical sign is called a **radical expression**. Examples of radical expressions include

$$\sqrt{6}, \quad 5 + \sqrt{x + 1}, \quad \text{and} \quad \sqrt{\frac{3x}{2x - 1}}.$$

In the next example, we show how to find the principal square root of an expression.

EXAMPLE 1 Finding principal square roots

Find the principal square root of each expression.

(a) 25 **(b)** 17 **(c)** $\dfrac{4}{9}$ **(d)** $c^2, c > 0$

SOLUTION
(a) Because $5 \cdot 5 = 25$, the principal, or positive, square root of 25 is $\sqrt{25} = 5$.
(b) The principal square root of 17 is $\sqrt{17}$. This value is not an integer, but we can approximate it. Figure R.18 shows that $\sqrt{17} \approx 4.12$, rounded to the nearest hundredth. Note that calculators do not give exact answers when approximating many radical expressions; they give decimal approximations.
(c) Because $\frac{2}{3} \cdot \frac{2}{3} = \frac{4}{9}$, the principal square root of $\frac{4}{9}$ is $\sqrt{\frac{4}{9}} = \frac{2}{3}$.
(d) The principal square root of c^2 is $\sqrt{c^2} = c$, as it is given that c is positive.

Now Try Exercises 7, 9, 11, and 13 ◀

```
√(17)
         4.123105626
```

Figure R.18

Cube Root Another common radical expression is the cube root of a number a, denoted $\sqrt[3]{a}$.

Cube Root

The number b is a *cube root* of a if $b^3 = a$.

Although the square root of a negative number is not a real number, the cube root of a negative number is a negative real number. *Every real number has one real cube root.*

EXAMPLE 2   Finding cube roots

Find the cube root of each expression.

(a) 8 (b) -27 (c) 16 (d) $\dfrac{1}{64}$ (e) d^6

SOLUTION

(a) $\sqrt[3]{8} = 2$ because $2^3 = 2 \cdot 2 \cdot 2 = 8$.

(b) $\sqrt[3]{-27} = -3$ because $(-3)^3 = (-3)(-3)(-3) = -27$.

(c) $\sqrt[3]{16}$ is not an integer. Figure R.19 shows that $\sqrt[3]{16} \approx 2.52$.

(d) $\sqrt[3]{\frac{1}{64}} = \frac{1}{4}$ because $\left(\frac{1}{4}\right)^3 = \frac{1}{4} \cdot \frac{1}{4} \cdot \frac{1}{4} = \frac{1}{64}$.

(e) $\sqrt[3]{d^6} = d^2$ because $(d^2)^3 = d^2 \cdot d^2 \cdot d^2 = d^{2+2+2} = d^6$.

Now Try Exercises 15, 17, 19, 21, and 31 ◀

```
3√(16)
           2.5198421
```

Figure R.19

nth Root   We can generalize square roots and cube roots to include the *n*th root of a number *a*. The number *b* is an **nth root** of *a* if $b^n = a$, where *n* is a positive integer, and the principal *n*th root is denoted $\sqrt[n]{a}$. The number *n* is called the **index**. For the square root, the index is 2, although we usually write $\sqrt{a}$ rather than $\sqrt[2]{a}$. When *n* is odd, we are finding an **odd root**, and when *n* is even, we are finding an **even root**. The square root $\sqrt{a}$ is an example of an even root, and the cube root $\sqrt[3]{a}$ is an example of an odd root.

NOTE An odd root of a negative number is a negative number, but the even root of a negative number is *not* a real number. For example, $\sqrt[3]{-8} = -2$, whereas $\sqrt[4]{-81}$ is *not* a real number.

EXAMPLE 3   Finding *n*th roots

Find each root.

(a) $\sqrt[4]{16}$ (b) $\sqrt[5]{-32}$

SOLUTION

(a) $\sqrt[4]{16} = 2$ because $2^4 = 2 \cdot 2 \cdot 2 \cdot 2 = 16$. Note that when *n* is even the principal *n*th root is positive.

(b) $\sqrt[5]{-32} = -2$ because $(-2)^5 = (-2)(-2)(-2)(-2)(-2) = -32$.

Now Try Exercises 37 and 38 ◀

Rational Exponents

When *m* and *n* are integers, the product rule states that $a^m a^n = a^{m+n}$. This rule can be extended to include exponents that are fractions. For example,

$$4^{1/2} \cdot 4^{1/2} = 4^{1/2+1/2} = 4^1 = 4.$$

That is, if we multiply $4^{1/2}$ by itself, the result is 4. Because we also know that $\sqrt{4} \cdot \sqrt{4} = 4$, this discussion suggests that $4^{1/2} = \sqrt{4}$ and motivates the following definition.

The Expression $a^{1/n}$

If *n* is an integer greater than 1, then

$$a^{1/n} = \sqrt[n]{a}.$$

If $a < 0$ and *n* is an even positive integer, then $a^{1/n}$ is not a real number.

The next two examples show how to interpret rational exponents.

EXAMPLE 4 Interpreting rational exponents

Write each expression in radical notation. Then evaluate the expression (to the nearest hundredth when appropriate).

(a) $36^{1/2}$ **(b)** $23^{1/5}$ **(c)** $(5x)^{1/2}$

SOLUTION

(a) The exponent $\frac{1}{2}$ indicates a square root. Thus $36^{1/2} = \sqrt{36}$, which evaluates to 6.

(b) The exponent $\frac{1}{5}$ indicates a fifth root. Thus $23^{1/5} = \sqrt[5]{23}$, which is not an integer. Figure R.20 shows this expression approximated in both exponential and radical notation. In either case $23^{1/5} \approx 1.87$.

(c) The exponent $\frac{1}{2}$ indicates a square root, so $(5x)^{1/2} = \sqrt{5x}$.

Now Try Exercises 41, 43, and 59 ◀

```
23^(1/5)
          1.872171231
5×√(23)
          1.872171231
```

Figure R.20

Suppose that we want to define the expression $8^{2/3}$. On the one hand, using properties of exponents we have

$$8^{1/3} \cdot 8^{1/3} = 8^{1/3+1/3} = 8^{2/3}.$$

On the other hand, we have

$$8^{1/3} \cdot 8^{1/3} = \sqrt[3]{8} \cdot \sqrt[3]{8} = 2 \cdot 2 = 4.$$

Thus $8^{2/3} = 4$, and that value is obtained whether we interpret $8^{2/3}$ as either

$$8^{2/3} = (8^2)^{1/3} = \sqrt[3]{8^2} = \sqrt[3]{64} = 4$$

or

$$8^{2/3} = \left(8^{1/3}\right)^2 = \left(\sqrt[3]{8}\right)^2 = 2^2 = 4.$$

This result illustrates that $8^{2/3} = \sqrt[3]{8^2} = \left(\sqrt[3]{8}\right)^2 = 4$ and suggests the following definition.

The Expression $a^{m/n}$

If m and n are positive integers with $\frac{m}{n}$ in lowest terms, then

$$a^{m/n} = \sqrt[n]{a^m} = \left(\sqrt[n]{a}\right)^m.$$

If $a < 0$ and n is an even integer, then $a^{m/n}$ is not a real number.

EXAMPLE 5 Interpreting rational exponents

Write each expression in radical notation. Then evaluate the expression when the result is an integer.

(a) $(-27)^{2/3}$ **(b)** $12^{3/5}$

SOLUTION

(a) The exponent $\frac{2}{3}$ indicates either that we take the cube root of -27 and then square it or that we square -27 and then take the cube root. In either case the result will be the same. Thus

$$(-27)^{2/3} = \left(\sqrt[3]{-27}\right)^2 = (-3)^2 = 9$$

or

$$(-27)^{2/3} = \sqrt[3]{(-27)^2} = \sqrt[3]{729} = 9.$$

(b) The exponent $\frac{3}{5}$ indicates either that we take the fifth root of 12 and then cube it or that we cube 12 and then take the fifth root. Thus

$$12^{3/5} = \left(\sqrt[5]{12}\right)^3 \quad \text{or} \quad 12^{3/5} = \sqrt[5]{12^3}.$$

This result is not an integer.   *Now Try Exercises 47 and 61* ◀

From properties of exponents we know that $a^{-n} = \frac{1}{a^n}$, where n is a positive integer. We now define this property for negative rational exponents.

The Expression $a^{-m/n}$

If m and n are positive integers with $\frac{m}{n}$ in lowest terms, then

$$a^{-m/n} = \frac{1}{a^{m/n}}, \quad a \neq 0.$$

EXAMPLE 6   Interpreting negative rational exponents

Write each expression in radical notation and then evaluate.
(a) $(64)^{-1/3}$ **(b)** $(81)^{-3/4}$

SOLUTION

(a) $(64)^{-1/3} = \dfrac{1}{64^{1/3}} = \dfrac{1}{\sqrt[3]{64}} = \dfrac{1}{4}$

(b) $(81)^{-3/4} = \dfrac{1}{81^{3/4}} = \dfrac{1}{\left(\sqrt[4]{81}\right)^3} = \dfrac{1}{3^3} = \dfrac{1}{27}$ *Now Try Exercises 51 and 53* ◀

Properties of Rational Exponents

Any rational number can be written as a ratio of two integers. That is, if p is a rational number, then $p = \frac{m}{n}$, where m and n are integers. Properties for integer exponents also apply to rational exponents—with one exception. If n is even in the expression $a^{m/n}$ and $\frac{m}{n}$ is written in lowest terms, then a must be nonnegative (not negative) for the result to be a real number.

Properties of Exponents

Let p and q be rational numbers written in lowest terms. For all real numbers a and b for which the expressions are real numbers, the following properties hold.

1. $a^p \cdot a^q = a^{p+q}$ Product rule for exponents

2. $a^{-p} = \dfrac{1}{a^p}, \quad \dfrac{1}{a^{-p}} = a^p$ Negative exponents

3. $\left(\dfrac{a}{b}\right)^{-p} = \left(\dfrac{b}{a}\right)^p$ Negative exponents for quotients

4. $\dfrac{a^p}{a^q} = a^{p-q}$ Quotient rule for exponents

5. $(a^p)^q = a^{pq}$ Power rule for exponents
6. $(ab)^p = a^p b^p$ Power rule for products

7. $\left(\dfrac{a}{b}\right)^p = \dfrac{a^p}{b^p}$ Power rule for quotients

---

In the next example, we apply these properties.

EXAMPLE 7 Applying properties of exponents

Write each expression using rational exponents and simplify. Write the answer with a positive exponent. Assume that all variables are positive numbers.

(a) $\sqrt{x} \cdot \sqrt[3]{x}$ (b) $\sqrt[3]{27x^2}$ (c) $\left(\dfrac{x^2}{81}\right)^{-1/2}$

SOLUTION

(a) $\sqrt{x} \cdot \sqrt[3]{x} = x^{1/2} \cdot x^{1/3}$ Use rational exponents.
$= x^{1/2+1/3}$ Product rule for exponents
$= x^{5/6}$ Simplify.

(b) $\sqrt[3]{27x^2} = (27x^2)^{1/3}$ Use rational exponents.
$= 27^{1/3}(x^2)^{1/3}$ Power rule for products
$= 3x^{2/3}$ Power rule for exponents

(c) $\left(\dfrac{x^2}{81}\right)^{-1/2} = \left(\dfrac{81}{x^2}\right)^{1/2}$ Negative exponents for quotients

$= \dfrac{(81)^{1/2}}{(x^2)^{1/2}}$ Power rule for quotients

$= \dfrac{9}{x}$ Power rule for exponents; simplify.

Now Try Exercises 81, 89, and 99 ◀

R.6 Exercises

Square Roots and Cube Roots

Exercises 1–6: Find the square roots of the number. Approximate your answers to the nearest hundredth whenever appropriate.

1. 25
2. 49
3. $\frac{16}{25}$
4. $\frac{64}{81}$
5. 11
6. 17

Exercises 7–14: Find the principal square root of the number. Approximate your answer to the nearest hundredth whenever appropriate.

7. 144
8. 100
9. 23
10. 45
11. $\frac{4}{49}$
12. $\frac{16}{121}$
13. $b^2, b < 0$
14. $(xy)^2, xy > 0$

Exercises 15–22: Find the cube root of the number.

15. 27
16. 64
17. -8
18. -125
19. $\frac{1}{27}$
20. $-\frac{1}{64}$
21. b^9
22. $8x^6$

Radical Notation

Exercises 23–40: If possible, simplify the expression by hand. If you cannot, approximate the answer to the nearest hundredth. Variables represent any real number.

23. $\sqrt{9}$
24. $\sqrt{121}$
25. $-\sqrt{5}$
26. $\sqrt{11}$
27. $\sqrt[3]{27}$
28. $\sqrt[3]{64}$
29. $\sqrt[3]{-64}$
30. $-\sqrt[3]{-1}$
31. $\sqrt[3]{5}$
32. $\sqrt[3]{-13}$
33. $-\sqrt[3]{x^9}$
34. $\sqrt[3]{(x+1)^6}$
35. $\sqrt[3]{(2x)^6}$
36. $\sqrt[3]{9x^3}$
37. $\sqrt[4]{81}$
38. $\sqrt[5]{-1}$
39. $\sqrt[5]{-7}$
40. $\sqrt[4]{6}$

Rational Exponents

Exercises 41–46: Write the expression in radical notation.

41. $6^{1/2}$
42. $7^{1/3}$

43. $(xy)^{1/2}$

44. $x^{2/3}y^{1/5}$

45. $y^{-1/5}$

46. $\left(\dfrac{x}{y}\right)^{-2/7}$

Exercises 47–54: Write the expression in radical notation. Then evaluate the expression when the result is an integer.

47. $27^{2/3}$

48. $8^{4/3}$

49. $(-1)^{4/3}$

50. $81^{3/4}$

51. $8^{-1/3}$

52. $16^{-3/4}$

53. $13^{-3/5}$

54. $23^{-1/2}$

Exercises 55–76: Evaluate the expression by hand. Approximate the answer to the nearest hundredth when appropriate.

55. $16^{1/2}$

56. $8^{1/3}$

57. $256^{1/4}$

58. $4^{3/2}$

59. $32^{1/5}$

60. $(-32)^{1/5}$

61. $(-8)^{4/3}$

62. $(-1)^{3/5}$

63. $2^{1/2} \cdot 2^{2/3}$

64. $5^{3/5} \cdot 5^{1/10}$

65. $\left(\dfrac{4}{9}\right)^{1/2}$

66. $\left(\dfrac{27}{64}\right)^{1/3}$

67. $\dfrac{4^{2/3}}{4^{1/2}}$

68. $\dfrac{6^{1/5} \cdot 6^{3/5}}{6^{2/5}}$

69. $4^{-1/2}$

70. $9^{-3/2}$

71. $(-8)^{-1/3}$

72. $49^{-1/2}$

73. $\left(\dfrac{1}{16}\right)^{-1/4}$

74. $\left(\dfrac{16}{25}\right)^{-3/2}$

75. $(2^{1/2})^3$

76. $(5^{6/5})^{-1/2}$

Exercises 77–106: Simplify the expression and write it with rational exponents. Assume that all variables are positive.

77. $(x^2)^{3/2}$

78. $(y^4)^{1/2}$

79. $(x^2y^8)^{1/2}$

80. $(y^{10}z^4)^{1/4}$

81. $\sqrt[3]{x^3y^6}$

82. $\sqrt{16x^4}$

83. $\sqrt{\dfrac{y^4}{x^2}}$

84. $\sqrt[3]{\dfrac{x^{12}}{z^6}}$

85. $\sqrt{y^3} \cdot \sqrt[3]{y^2}$

86. $\left(\dfrac{x^6}{81}\right)^{1/4}$

87. $\left(\dfrac{x^6}{27}\right)^{2/3}$

88. $\left(\dfrac{1}{x^8}\right)^{-1/4}$

89. $\left(\dfrac{x^2}{y^6}\right)^{-1/2}$

90. $\dfrac{\sqrt{x}}{\sqrt[3]{27x^6}}$

91. $\sqrt{\sqrt{y}}$

92. $\sqrt{\sqrt[3]{(3x)^2}}$

93. $(a^{-1/2})^{4/3}$

94. $(x^{-3/2})^{2/3}$

95. $(a^3b^6)^{1/3}$

96. $(64x^3y^{18})^{1/6}$

97. $\dfrac{(k^{1/2})^{-3}}{(k^2)^{1/4}}$

98. $\dfrac{(b^{3/4})^4}{(b^{4/5})^{-5}}$

99. $\sqrt{b} \cdot \sqrt[4]{b}$

100. $\sqrt[3]{t} \cdot \sqrt[5]{t}$

101. $\sqrt{z} \cdot \sqrt[3]{z^2} \cdot \sqrt[4]{z^3}$

102. $\sqrt{b} \cdot \sqrt[3]{b} \cdot \sqrt[5]{b}$

103. $p^{1/2}\left(p^{3/2} + p^{1/2}\right)$

104. $d^{3/4}(d^{1/4} - d^{-1/4})$

105. $\sqrt[3]{x}\left(\sqrt{x} - \sqrt[3]{x^2}\right)$

106. $\dfrac{1}{2}\sqrt{x}\left(\sqrt{x} + \sqrt[4]{x^2}\right)$

R.7 Radical Expressions

- Apply the product rule
- Simplify radical expressions
- Apply the quotient rule
- Perform addition, subtraction, and multiplication on radical expressions
- Rationalize the denominator

Product Rule for Radical Expressions

The product of two (like) roots is equal to the root of their product.

Product Rule for Radical Expressions

Let a and b be real numbers, where $\sqrt[n]{a}$ and $\sqrt[n]{b}$ are both defined. Then

$$\sqrt[n]{a} \cdot \sqrt[n]{b} = \sqrt[n]{a \cdot b}.$$

> **NOTE** The product rule works only when the radicals have the *same* index.

We apply the product rule in the next two examples.

EXAMPLE 1   Multiplying radical expressions

Multiply each pair of radical expressions.

(a) $\sqrt{5} \cdot \sqrt{20}$ **(b)** $\sqrt[3]{-3} \cdot \sqrt[3]{9}$

SOLUTION

(a) $\sqrt{5} \cdot \sqrt{20} = \sqrt{5 \cdot 20} = \sqrt{100} = 10$

(b) $\sqrt[3]{-3} \cdot \sqrt[3]{9} = \sqrt[3]{-3 \cdot 9} = \sqrt[3]{-27} = -3$

Now Try Exercises 3 and 5 ◄

EXAMPLE 2   Multiplying radical expressions containing variables

Multiply each pair of radical expressions. Assume that all variables are positive.

(a) $\sqrt{x} \cdot \sqrt{x^3}$ **(b)** $\sqrt[3]{2a} \cdot \sqrt[3]{5a}$ **(c)** $\sqrt[5]{\dfrac{2x}{y}} \cdot \sqrt[5]{\dfrac{16y}{x}}$

SOLUTION

(a) $\sqrt{x} \cdot \sqrt{x^3} = \sqrt{x \cdot x^3} = \sqrt{x^4} = x^2$

(b) $\sqrt[3]{2a} \cdot \sqrt[3]{5a} = \sqrt[3]{2a \cdot 5a} = \sqrt[3]{10a^2}$

(c) $\sqrt[5]{\dfrac{2x}{y}} \cdot \sqrt[5]{\dfrac{16y}{x}} = \sqrt[5]{\dfrac{2x}{y} \cdot \dfrac{16y}{x}}$ Product rule

$= \sqrt[5]{\dfrac{32xy}{xy}}$ Multiply fractions.

$= \sqrt[5]{32}$ Simplify.

$= 2$ $2^5 = 32$

Now Try Exercises 27, 33, and 35 ◄

Simplifying Radicals   An integer a is a **perfect nth power** if there exists an integer b such that $b^n = a$. Thus 36 is a **perfect square** because $6^2 = 36$, 8 is a **perfect cube** because $2^3 = 8$, and 81 is a *perfect fourth power* because $3^4 = 81$.

The product rule for radicals can be used to simplify radical expressions. For example, because the largest perfect square factor of 50 is 25, the expression $\sqrt{50}$ can be simplified as

$$\sqrt{50} = \sqrt{25 \cdot 2} = \sqrt{25} \cdot \sqrt{2} = 5\sqrt{2}.$$

This procedure is generalized as follows.

Simplifying Radicals (nth Roots)

STEP 1:   Determine the largest perfect nth power factor of the radicand.

STEP 2:   Use the product rule to factor out and simplify this perfect nth power.

EXAMPLE 3   Simplifying radical expressions

Simplify each expression.

(a) $\sqrt{300}$ **(b)** $\sqrt[3]{16}$ **(c)** $\sqrt[4]{512}$

SOLUTION

(a) First note that $300 = 100 \cdot 3$ and that 100 is the largest perfect square factor of 300.

$$\sqrt{300} = \sqrt{100} \cdot \sqrt{3} = 10\sqrt{3}$$

(b) The largest perfect cube factor of 16 is 8. Thus $\sqrt[3]{16} = \sqrt[3]{8} \cdot \sqrt[3]{2} = 2\sqrt[3]{2}$.

(c) $\sqrt[4]{512} = \sqrt[4]{256} \cdot \sqrt[4]{2} = 4\sqrt[4]{2}$ because $4^4 = 256$.

Now Try Exercises 37, 39, and 41 ◀

NOTE To simplify a cube root of a negative number, we usually factor out the negative of the largest perfect cube factor. For example, because $-16 = -8 \cdot 2$, it follows that $\sqrt[3]{-16} = \sqrt[3]{-8} \cdot \sqrt[3]{2} = -2\sqrt[3]{2}$. This procedure can be used with any odd root of a negative number.

EXAMPLE 4 Simplifying radical expressions

Simplify each expression. Assume that all variables are positive.

(a) $\sqrt{25x^4}$ **(b)** $\sqrt{32n^3}$ **(c)** $\sqrt[3]{-16x^3y^5}$ **(d)** $\sqrt[3]{2a} \cdot \sqrt[3]{4a^2b}$

SOLUTION

(a) $\sqrt{25x^4} = 5x^2$ $\qquad\qquad\qquad$ $(5x^2)^2 = 25x^4$

(b) $\sqrt{32n^3} = \sqrt{(16n^2)2n}$ $\qquad\quad$ $16n^2$ is the largest perfect square factor.

$\qquad\quad = \sqrt{16n^2} \cdot \sqrt{2n}$ $\qquad\quad$ Product rule

$\qquad\quad = 4n\sqrt{2n}$ $\qquad\qquad\quad$ $(4n)^2 = 16n^2$

(c) $\sqrt[3]{-16x^3y^5} = \sqrt[3]{(-8x^3y^3)2y^2}$ $\quad$ $8x^3y^3$ is the largest perfect cube factor.

$\qquad\quad = \sqrt[3]{-8x^3y^3} \cdot \sqrt[3]{2y^2}$ $\quad$ Product rule

$\qquad\quad = -2xy\sqrt[3]{2y^2}$ $\qquad\quad$ $(-2xy)^3 = -8x^3y^3$

(d) $\sqrt[3]{2a} \cdot \sqrt[3]{4a^2b} = \sqrt[3]{(2a)(4a^2b)}$ $\quad$ Product rule

$\qquad\quad = \sqrt[3]{(8a^3)b}$ $\qquad\quad$ $8a^3$ is the largest perfect cube factor.

$\qquad\quad = \sqrt[3]{8a^3} \cdot \sqrt[3]{b}$ $\qquad\quad$ Product rule

$\qquad\quad = 2a\sqrt[3]{b}$ $\qquad\qquad\quad$ $(2a)^3 = 8a^3$

Now Try Exercises 45, 47, 49, and 51 ◀

The product rule for radical expressions cannot be used if the radicals do not have the same indexes. In this case we use rational exponents, as illustrated in the next example.

EXAMPLE 5 Multiplying radicals with different indexes

Simplify each expression. Write your answer in radical notation.

(a) $\sqrt{2} \cdot \sqrt[3]{4}$ **(b)** $\sqrt[3]{x} \cdot \sqrt[4]{x}$

SOLUTION

(a) First note that $\sqrt[3]{4} = \sqrt[3]{2^2} = 2^{2/3}$. Thus

$$\sqrt{2} \cdot \sqrt[3]{4} = 2^{1/2} \cdot 2^{2/3} = 2^{1/2+2/3} = 2^{7/6}.$$

In radical notation, $2^{7/6} = \sqrt[6]{2^7} = \sqrt[6]{2^6 \cdot 2^1} = \sqrt[6]{2^6} \cdot \sqrt[6]{2} = 2\sqrt[6]{2}$.

(b) $\sqrt[3]{x} \cdot \sqrt[4]{x} = x^{1/3} \cdot x^{1/4} = x^{7/12} = \sqrt[12]{x^7}$ $\qquad$ *Now Try Exercises 57 and 59* ◀

Quotient Rule for Radical Expressions

The root of a quotient is equal to the quotient of the roots.

> **Quotient Rule for Radical Expressions**
>
> Let a and b be real numbers, where $\sqrt[n]{a}$ and $\sqrt[n]{b}$ are both defined and $b \neq 0$. Then
> $$\sqrt[n]{\frac{a}{b}} = \frac{\sqrt[n]{a}}{\sqrt[n]{b}}.$$

EXAMPLE 6 Simplifying quotients

Simplify each radical expression. Assume that all variables are positive.

(a) $\sqrt[3]{\dfrac{5}{8}}$ **(b)** $\sqrt{\dfrac{16}{y^2}}$

SOLUTION

(a) $\sqrt[3]{\dfrac{5}{8}} = \dfrac{\sqrt[3]{5}}{\sqrt[3]{8}} = \dfrac{\sqrt[3]{5}}{2}$

(b) $\sqrt{\dfrac{16}{y^2}} = \dfrac{\sqrt{16}}{\sqrt{y^2}} = \dfrac{4}{y}$ because $y > 0$.

Now Try Exercises 7 and 21 ◀

EXAMPLE 7 Simplifying radical expressions

Simplify each radical expression. Assume that all variables are positive.

(a) $\dfrac{\sqrt{40}}{\sqrt{10}}$ **(b)** $\sqrt[4]{\dfrac{16x^3}{y^4}}$ **(c)** $\sqrt{\dfrac{5a^2}{8}} \cdot \sqrt{\dfrac{5a^3}{2}}$

SOLUTION

(a) $\dfrac{\sqrt{40}}{\sqrt{10}} = \sqrt{\dfrac{40}{10}} = \sqrt{4} = 2$

(b) $\sqrt[4]{\dfrac{16x^3}{y^4}} = \dfrac{\sqrt[4]{16x^3}}{\sqrt[4]{y^4}} = \dfrac{\sqrt[4]{16} \cdot \sqrt[4]{x^3}}{\sqrt[4]{y^4}} = \dfrac{2\sqrt[4]{x^3}}{y}$

(c) To simplify this expression, we use both the product and quotient rules.

$$\sqrt{\frac{5a^2}{8}} \cdot \sqrt{\frac{5a^3}{2}} = \sqrt{\frac{25a^5}{16}} \qquad \text{Product rule}$$

$$= \frac{\sqrt{25a^5}}{\sqrt{16}} \qquad \text{Quotient rule}$$

$$= \frac{\sqrt{25a^4} \cdot \sqrt{a}}{\sqrt{16}} \qquad \text{Factor out largest perfect square.}$$

$$= \frac{5a^2\sqrt{a}}{4} \qquad (5a^2)^2 = 25a^4$$

Now Try Exercises 13, 22, and 35 ◀

Addition and Subtraction

We can add $2x^2$ and $5x^2$ to obtain $7x^2$ because they are *like* terms. That is,

$$2x^2 + 5x^2 = (2 + 5)x^2 = 7x^2.$$

We can add and subtract **like radicals**, which have the same index and the same radicand. For example, we can add $3\sqrt{2}$ and $5\sqrt{2}$ because they are like radicals.

$$3\sqrt{2} + 5\sqrt{2} = (3 + 5)\sqrt{2} = 8\sqrt{2}$$

Sometimes two radical expressions that are not alike can be added by changing them to like radicals. For example, $\sqrt{20}$ and $\sqrt{5}$ are unlike radicals. However,

$$\sqrt{20} = \sqrt{4 \cdot 5} = \sqrt{4} \cdot \sqrt{5} = 2\sqrt{5},$$

so

$$\sqrt{20} + \sqrt{5} = 2\sqrt{5} + \sqrt{5} = 3\sqrt{5}.$$

We cannot combine $x + x^2$ because they are unlike terms. Similarly, we cannot combine $\sqrt{2} + \sqrt{5}$ because they are unlike radicals.

EXAMPLE 8   Adding radical expressions

Add the expressions and simplify.

(a) $10\sqrt{11} + 4\sqrt{11}$ **(b)** $5\sqrt[3]{6} + \sqrt[3]{6}$ **(c)** $\sqrt{12} + 7\sqrt{3}$

SOLUTION

(a) $10\sqrt{11} + 4\sqrt{11} = (10 + 4)\sqrt{11} = 14\sqrt{11}$

(b) $5\sqrt[3]{6} + \sqrt[3]{6} = (5 + 1)\sqrt[3]{6} = 6\sqrt[3]{6}$

(c) $\begin{aligned} \sqrt{12} + 7\sqrt{3} &= \sqrt{4 \cdot 3} + 7\sqrt{3} \\ &= \sqrt{4} \cdot \sqrt{3} + 7\sqrt{3} \\ &= 2\sqrt{3} + 7\sqrt{3} \\ &= 9\sqrt{3} \end{aligned}$

Now Try Exercises 63, 68, and 69 ◄

EXAMPLE 9   Adding radical expressions

Add the expressions and simplify. Assume that all variables are positive.

(a) $-2\sqrt{4x} + \sqrt{x}$ **(b)** $3\sqrt{3k} + 5\sqrt{12k} + 9\sqrt{48k}$

SOLUTION

(a) Note that $\sqrt{4x} = \sqrt{4} \cdot \sqrt{x} = 2\sqrt{x}$.

$$-2\sqrt{4x} + \sqrt{x} = -2(2\sqrt{x}) + \sqrt{x} = -4\sqrt{x} + \sqrt{x} = -3\sqrt{x}$$

(b) Note that $\sqrt{12k} = \sqrt{4} \cdot \sqrt{3k} = 2\sqrt{3k}$ and that $\sqrt{48k} = \sqrt{16} \cdot \sqrt{3k} = 4\sqrt{3k}$.

$$\begin{aligned} 3\sqrt{3k} + 5\sqrt{12k} + 9\sqrt{48k} &= 3\sqrt{3k} + 5(2\sqrt{3k}) + 9(4\sqrt{3k}) \\ &= (3 + 10 + 36)\sqrt{3k} \\ &= 49\sqrt{3k} \end{aligned}$$

Now Try Exercises 77 and 83 ◄

Subtraction of radical expressions is similar to addition, as illustrated in the next example.

EXAMPLE 10   Subtracting radical expressions

Subtract and simplify. Assume that all variables are positive.

(a) $3\sqrt[3]{xy^2} - 2\sqrt[3]{xy^2}$ **(b)** $\sqrt{16x^3} - \sqrt{x^3}$

SOLUTION

(a) $3\sqrt[3]{xy^2} - 2\sqrt[3]{xy^2} = (3-2)\sqrt[3]{xy^2} = \sqrt[3]{xy^2}$

(b) $\sqrt{16x^3} - \sqrt{x^3} = \sqrt{16} \cdot \sqrt{x^3} - \sqrt{x^3}$

$$= 4\sqrt{x^3} - \sqrt{x^3}$$

$$= 3\sqrt{x^3}$$

$$= 3x\sqrt{x}$$ *Now Try Exercises 75 and 81* ◀

Multiplication

Some types of radical expressions can be multiplied like binomials. The next example demonstrates this technique.

EXAMPLE 11   Multiplying radical expressions

Multiply and simplify.

(a) $(\sqrt{b} - 4)(\sqrt{b} + 5)$ **(b)** $(4 + \sqrt{3})(4 - \sqrt{3})$

SOLUTION

(a) This expression can be multiplied and then simplified.

$$(\sqrt{b} - 4)(\sqrt{b} + 5) = \sqrt{b} \cdot \sqrt{b} + 5\sqrt{b} - 4\sqrt{b} - 4 \cdot 5$$

$$= b + \sqrt{b} - 20$$

Compare this product to $(b - 4)(b + 5) = b^2 + b - 20$.

(b) This expression is in the form $(a + b)(a - b)$, which equals $a^2 - b^2$.

$$(4 + \sqrt{3})(4 - \sqrt{3}) = (4)^2 - (\sqrt{3})^2$$

$$= 16 - 3$$

$$= 13$$ *Now Try Exercises 89 and 95* ◀

Rationalizing the Denominator

Quotients containing radical expressions can appear to be different but actually be equal. For example, $\dfrac{1}{\sqrt{3}}$ and $\dfrac{\sqrt{3}}{3}$ represent the same real number even though they do not look equal. To show this fact, we multiply the first quotient by 1 in the form $\dfrac{\sqrt{3}}{\sqrt{3}}$.

$$\frac{1}{\sqrt{3}} \cdot \frac{\sqrt{3}}{\sqrt{3}} = \frac{1 \cdot \sqrt{3}}{\sqrt{3} \cdot \sqrt{3}} = \frac{\sqrt{3}}{3}$$

NOTE $\sqrt{b} \cdot \sqrt{b} = \sqrt{b^2} = b$ for any *positive* number b.

One way to standardize radical expressions is to remove any radical expressions from the denominator. This process is called **rationalizing the denominator**. The next example demonstrates how to rationalize the denominator of two quotients.

EXAMPLE 12   Rationalizing the denominator

Rationalize each denominator. Assume that all variables are positive.

(a) $\dfrac{3}{5\sqrt{3}}$ **(b)** $\sqrt{\dfrac{x}{24}}$

SOLUTION

(a) We multiply this expression by **1** in the form $\dfrac{\sqrt{3}}{\sqrt{3}}$.

$$\frac{3}{5\sqrt{3}}\cdot\frac{\sqrt{3}}{\sqrt{3}}=\frac{3\sqrt{3}}{5\sqrt{9}}=\frac{3\sqrt{3}}{5\cdot 3}=\frac{\sqrt{3}}{5}$$

(b) Because $\sqrt{24}=\sqrt{4}\cdot\sqrt{6}=2\sqrt{6}$, we start by simplifying the expression.

$$\sqrt{\frac{x}{24}}=\frac{\sqrt{x}}{\sqrt{24}}=\frac{\sqrt{x}}{2\sqrt{6}}$$

To rationalize the denominator, we multiply this expression by **1** in the form $\dfrac{\sqrt{6}}{\sqrt{6}}$.

$$\frac{\sqrt{x}}{2\sqrt{6}}=\frac{\sqrt{x}}{2\sqrt{6}}\cdot\frac{\sqrt{6}}{\sqrt{6}}=\frac{\sqrt{6x}}{12}$$

Now Try Exercises 99 and 101 ◄

If the denominator consists of two terms, at least one of which contains a radical expression, then the **conjugate** of the denominator is found by changing a + sign to a − sign or vice versa. For example, the conjugate of $\sqrt{2}+\sqrt{3}$ is $\sqrt{2}-\sqrt{3}$, and the conjugate of $\sqrt{3}-1$ is $\sqrt{3}+1$. In the next example, we multiply the numerator and denominator by the conjugate of the *denominator* to rationalize the denominator of fractions that contain radicals.

EXAMPLE 13   Rationalizing the denominator

Rationalize the denominator. Assume that all variables are positive.

(a) $\dfrac{3+\sqrt{5}}{2-\sqrt{5}}$ **(b)** $\dfrac{\sqrt{x}}{\sqrt{x}-2}$

SOLUTION

(a) The conjugate of the denominator is $2+\sqrt{5}$.

$$\frac{3+\sqrt{5}}{2-\sqrt{5}}=\frac{(3+\sqrt{5})}{(2-\sqrt{5})}\cdot\frac{(2+\sqrt{5})}{(2+\sqrt{5})} \qquad \text{Multiply by } \frac{conjugate}{conjugate}.$$

$$=\frac{6+3\sqrt{5}+2\sqrt{5}+(\sqrt{5})^2}{(2)^2-(\sqrt{5})^2} \qquad \text{Multiply.}$$

$$=\frac{11+5\sqrt{5}}{4-5} \qquad \text{Combine terms.}$$

$$=-11-5\sqrt{5} \qquad \text{Simplify.}$$

(b) The conjugate of the denominator is $\sqrt{x} + 2$.

$$\frac{\sqrt{x}}{\sqrt{x} - 2} = \frac{\sqrt{x}}{(\sqrt{x} - 2)} \cdot \frac{(\sqrt{x} + 2)}{(\sqrt{x} + 2)} \quad \text{Multiply by } \frac{conjugate}{conjugate}.$$

$$= \frac{x + 2\sqrt{x}}{x - 4} \quad \text{Multiply.}$$

Now Try Exercises 103 and 109 ◀

R.7   Exercises

Multiplying and Dividing

Exercises 1–36: Simplify the expression. Assume that all variables are positive.

1. $\sqrt{3} \cdot \sqrt{3}$

2. $\sqrt{2} \cdot \sqrt{18}$

3. $\sqrt{2} \cdot \sqrt{50}$

4. $\sqrt[3]{-2} \cdot \sqrt[3]{-4}$

5. $\sqrt[3]{4} \cdot \sqrt[3]{16}$

6. $\sqrt[3]{x} \cdot \sqrt[3]{x^2}$

7. $\sqrt{\dfrac{9}{25}}$

8. $\sqrt[3]{\dfrac{x}{8}}$

9. $\sqrt{\dfrac{1}{2}} \cdot \sqrt{\dfrac{1}{8}}$

10. $\sqrt{\dfrac{5}{3}} \cdot \sqrt{\dfrac{1}{3}}$

11. $\sqrt{\dfrac{x}{2}} \cdot \sqrt{\dfrac{x}{8}}$

12. $\sqrt{\dfrac{4}{y}} \cdot \sqrt{\dfrac{y}{5}}$

13. $\dfrac{\sqrt{45}}{\sqrt{5}}$

14. $\dfrac{\sqrt{7}}{\sqrt{28}}$

15. $\sqrt[4]{9} \cdot \sqrt[4]{9}$

16. $\sqrt[5]{16} \cdot \sqrt[5]{-2}$

17. $\dfrac{\sqrt[5]{64}}{\sqrt[5]{-2}}$

18. $\dfrac{\sqrt[4]{324}}{\sqrt[4]{4}}$

19. $\dfrac{\sqrt{a^2b}}{\sqrt{b}}$

20. $\dfrac{\sqrt{4xy^2}}{\sqrt{x}}$

21. $\sqrt[3]{\dfrac{x^3}{8}}$

22. $\sqrt{\dfrac{36}{z^4}}$

23. $\sqrt{4x^4}$

24. $\sqrt[3]{-8y^3}$

25. $\sqrt[4]{16x^4y}$

26. $\sqrt[3]{8xy^3}$

27. $\sqrt{3x} \cdot \sqrt{12x}$

28. $\sqrt{6x^5} \cdot \sqrt{6x}$

29. $\sqrt[3]{8x^6y^3z^9}$

30. $\sqrt{16x^4y^6}$

31. $\sqrt[4]{\dfrac{3}{4}} \cdot \sqrt[4]{\dfrac{27}{4}}$

32. $\sqrt[5]{\dfrac{4}{-9}} \cdot \sqrt[5]{\dfrac{8}{-27}}$

33. $\sqrt[4]{25z} \cdot \sqrt[4]{25z}$

34. $\sqrt[5]{3z^2} \cdot \sqrt[5]{7z}$

35. $\sqrt[5]{\dfrac{7a}{b^2}} \cdot \sqrt[5]{\dfrac{b^2}{7a^6}}$

36. $\sqrt[3]{\dfrac{8m}{n}} \cdot \sqrt[3]{\dfrac{n^4}{m^2}}$

Exercises 37–54: Simplify the radical expression by factoring out the largest perfect nth power. Assume that all variables are positive.

37. $\sqrt{200}$

38. $\sqrt{72}$

39. $\sqrt[3]{81}$

40. $\sqrt[3]{256}$

41. $\sqrt[4]{64}$

42. $\sqrt[5]{27 \cdot 81}$

43. $\sqrt[5]{-64}$

44. $\sqrt[3]{-81}$

45. $\sqrt{8n^3}$

46. $\sqrt{32a^2}$

47. $\sqrt{12a^2b^5}$

48. $\sqrt{20a^3b^2}$

49. $\sqrt[3]{-125x^4y^5}$

50. $\sqrt[3]{-81a^5b^2}$

51. $\sqrt[3]{5t} \cdot \sqrt[3]{125t}$

52. $\sqrt[4]{4bc^3} \cdot \sqrt[4]{64ab^3c^2}$

53. $\sqrt[4]{\dfrac{9t^5}{r^8}} \cdot \sqrt[4]{\dfrac{9r}{5t}}$

54. $\sqrt[5]{\dfrac{4t^6}{r}} \cdot \sqrt[5]{\dfrac{8t}{r^6}}$

Exercises 55–62: Simplify the expression. Assume that all variables are positive and write your answer in radical notation.

55. $\sqrt{3} \cdot \sqrt[3]{3}$

56. $\sqrt{5} \cdot \sqrt[3]{5}$

57. $\sqrt[4]{8} \cdot \sqrt[3]{4}$

58. $\sqrt[5]{16} \cdot \sqrt{2}$

59. $\sqrt[4]{x^3} \cdot \sqrt[3]{x}$

60. $\sqrt[4]{x^3} \cdot \sqrt{x}$

61. $\sqrt[4]{rt} \cdot \sqrt[3]{r^2t}$

62. $\sqrt[3]{a^3b^2} \cdot \sqrt{a^2b}$

Exercises 63–88: Simplify the expression. Assume that all variables are positive.

63. $2\sqrt{3} + 7\sqrt{3}$

64. $8\sqrt{7} + 2\sqrt{7}$

65. $\sqrt{x} + \sqrt{x} - \sqrt{y}$

66. $\sqrt{xy^2} - \sqrt{x}$

67. $2\sqrt[3]{6} - 7\sqrt[3]{6}$

68. $18\sqrt[3]{3} + 3\sqrt[3]{3}$

69. $3\sqrt{28} + 3\sqrt{7}$

70. $9\sqrt{18} - 2\sqrt{8}$

71. $\sqrt{44} - 4\sqrt{11}$

72. $\sqrt[4]{5} + 2\sqrt[4]{5}$

73. $2\sqrt[3]{16} + \sqrt[3]{2} - \sqrt{2}$

74. $5\sqrt[3]{x} - 3\sqrt[3]{x}$

75. $\sqrt[3]{xy} - 2\sqrt[3]{xy}$

76. $3\sqrt{x^3} - \sqrt{x}$

77. $\sqrt{4x + 8} + \sqrt{x + 2}$

78. $\sqrt{2a + 1} + \sqrt{8a + 4}$

79. $\dfrac{15\sqrt{8}}{4} - \dfrac{2\sqrt{2}}{5}$

80. $\dfrac{23\sqrt{11}}{2} - \dfrac{\sqrt{44}}{8}$

81. $20\sqrt[3]{b^4} - 4\sqrt[3]{b}$

82. $2\sqrt[4]{64} - \sqrt[4]{324} + \sqrt[4]{4}$

83. $2\sqrt{3z} + 3\sqrt{12z} + 3\sqrt{48z}$

84. $\sqrt{64x^3} - \sqrt{x} + 3\sqrt{x}$

85. $\sqrt[4]{81a^5b^5} - \sqrt[4]{ab}$

86. $\sqrt[4]{xy^5} + \sqrt[4]{x^5y}$

87. $5\sqrt[3]{\dfrac{n^4}{125}} - 2\sqrt[3]{n}$

88. $\sqrt[3]{\dfrac{8x}{27}} - \dfrac{2\sqrt[3]{x}}{3}$

Exercises 89–96: Multiply and simplify.

89. $(3 + \sqrt{7})(3 - \sqrt{7})$

90. $(5 - \sqrt{5})(5 + \sqrt{5})$

91. $(\sqrt{x} + 8)(\sqrt{x} - 8)$

92. $(\sqrt{ab} - 3)(\sqrt{ab} + 3)$

93. $(\sqrt{ab} - \sqrt{c})(\sqrt{ab} + \sqrt{c})$

94. $(\sqrt{2x} + \sqrt{3y})(\sqrt{2x} - \sqrt{3y})$

95. $(\sqrt{x} - 7)(\sqrt{x} + 8)$

96. $(\sqrt{ab} - 1)(\sqrt{ab} - 2)$

Exercises 97–112: Rationalize the denominator.

97. $\dfrac{4}{\sqrt{3}}$

98. $\dfrac{8}{\sqrt{2}}$

99. $\dfrac{5}{3\sqrt{5}}$

100. $\dfrac{6}{11\sqrt{3}}$

101. $\sqrt{\dfrac{b}{12}}$

102. $\sqrt{\dfrac{5b}{72}}$

103. $\dfrac{1}{3 - \sqrt{2}}$

104. $\dfrac{1}{\sqrt{3} - 2}$

105. $\dfrac{\sqrt{2}}{\sqrt{5} + 2}$

106. $\dfrac{\sqrt{3} - 1}{\sqrt{3} + 1}$

107. $\dfrac{1}{\sqrt{7} - \sqrt{6}}$

108. $\dfrac{1}{\sqrt{8} - \sqrt{7}}$

109. $\dfrac{\sqrt{z}}{\sqrt{z} - 3}$

110. $\dfrac{2\sqrt{z}}{2 - \sqrt{z}}$

111. $\dfrac{\sqrt{a} + \sqrt{b}}{\sqrt{a} - \sqrt{b}}$

112. $\dfrac{1}{\sqrt{a + 1} + \sqrt{a}}$

Appendix A: Using the Graphing Calculator

Overview of the Appendix

The intent of this appendix is to provide instruction in the TI-83, TI-83 Plus, and TI-84 Plus graphing calculators that may be used in conjunction with this textbook. It includes specific keystrokes needed to work several examples from the text. Students are also advised to consult the *Graphing Calculator Guidebook* provided by the manufacturer.

The following is a listing of the topics covered in this appendix.

Figure A.1

Figure A.2

Figure A.3 Normal Mode

Displaying Numbers in Scientific Notation

To display numbers in scientific notation when the calculator is in normal mode, set the graphing calculator in scientific mode (SCI) by using the following keystrokes. See Figure A.1.

(MODE) (▷) (ENTER) (2nd) (MODE [QUIT])

Figure A.2 shows the numbers 5432 and 0.00001234 displayed in scientific notation.

> **Summary: Setting Scientific Mode**
>
> If your calculator is in normal mode, it can be set in scientific mode by pressing
>
> (MODE) (▷) (ENTER) (2nd) (MODE [QUIT]).
>
> These keystrokes return the graphing calculator to the home screen.

Entering Numbers in Scientific Notation

Numbers can be entered in scientific notation. For example, to enter 4.2×10^{-3} in scientific notation, use the following keystrokes. (Be sure to use the negation key ($-$) rather than the subtraction key.)

(4) (.) (2) (2nd) (,[EE]) ((−)) (3)

This number can also be entered using the following keystrokes. See Figure A.3.

(4) (.) (2) (×) (1) (0) (^) (() ((−)) (3) ())

> **Summary: Entering Numbers in Scientific Notation**
>
> One way to enter a number in scientific notation is to use the keystrokes
>
> (2nd) (,)
>
> to access an exponent (EE) of 10.

Entering Mathematical Expressions

Several expressions are evaluated in Example 6, Section 1.1. To evaluate $\sqrt[3]{131}$, use the following keystrokes from the home screen.

(MATH) (4) (1) (3) (1) ()) (ENTER)

To calculate $\pi^3 + 1.2^2$, use the following keystrokes. (Do *not* use 3.14 for π.)

(2nd) (^[π]) (^) (3) (+) (1) (.) (2) (x^2) (ENTER)

To calculate $|\sqrt{3} - 6|$, use the following keystrokes.

(MATH) (▷) (1) (2nd) (x^2[√]) (3) ()) (−) (6) ()) (ENTER)

ZOOM MEMORY
1:ZBox
2:Zoom In
3:Zoom Out
4:ZDecimal
5:ZSquare
6:ZStandard
7↓ZTrig

Figure A.4

WINDOW
 Xmin = ‾10
 Xmax = 10
 Xscl = 1
 Ymin = ‾10
 Ymax = 10
 Yscl = 1
 Xres = 1

Figure A.5

WINDOW
 Xmin = ‾30
 Xmax = 40
 Xscl = 10
 Ymin = ‾400
 Ymax = 800
 Yscl = 100
 Xres = 1

Figure A.6

L1	L2	L3	1
1	4	------	
2	5		
3	6		

L1 = {1,2,3}

Figure A.7

L1	L2	L3	1
‾5	‾5	------	
‾2	3		
1	‾7		
4	8	------	

L1(5) =

Figure A.8

STAT PLOTS
1:Plot1...Off
  L1   L2
2:Plot2...Off
  L1   L2
3:Plot3...Off
  L1   L2
4↓PlotsOff

Figure A.9

Summary: Entering Common Mathematical Expressions

To calculate a cube root, use the keystrokes (MATH) (4).

To access the number π, use the keystrokes (2nd) (^[π]).

To access the absolute value, use the keystrokes (MATH) (▷) (1).

To access the square root, use the keystrokes (2nd) (x²[√]).

Setting the Viewing Rectangle

In Example 12, Section 1.2, there are at least two ways to set the standard viewing rectangle to $[-10, 10, 1]$ by $[-10, 10, 1]$. The first method involves pressing (ZOOM) followed by (6). (See Figure A.4.) The second method is to press (WINDOW) and enter the following keystrokes. (See Figure A.5.)

(Be sure to use the negation key $(-)$ rather than the subtraction key.) The viewing rectangle $[-30, 40, 10]$ by $[-400, 800, 100]$ can be set in a similar manner, as shown in Figure A.6. To see the viewing rectangle, press (GRAPH).

Summary: Setting the Viewing Rectangle

To set the standard viewing rectangle, press (ZOOM) (6). To set any viewing rectangle, press (WINDOW) and enter the necessary values. To see the viewing rectangle, press (GRAPH).

Note: You do not need to change "Xres".

Making a Scatterplot or a Line Graph

In Example 13, Section 1.2, we are asked to make a scatterplot with $(-5, -5)$, $(-2, 3)$, $(1, -7)$, and $(4, 8)$. Begin this task by following these steps.

1. Press (STAT) followed by (1).
2. If list L1 is not empty, use the arrow keys to place the cursor on L1, as shown in Figure A.7. Then press (CLEAR) followed by (ENTER). This deletes all elements in the list. Similarly, if L2 is not empty, clear the list.
3. Input each x-value into list L1, followed by (ENTER). Input each y-value into list L2, followed by (ENTER). See Figure A.8.

It is essential that both lists have the same number of values—otherwise an error message will appear when a scatterplot is attempted. Before these four points can be plotted, STAT-PLOT must be turned on. It is accessed by pressing

(2nd) (Y = [STAT PLOT]),

as shown in Figure A.9.

   There are three possible STATPLOTS, numbered 1, 2, and 3. Any one of the three can be selected. The first plot is selected by pressing (1). Next, place the cursor over "On"

Figure A.10

[−10, 10, 1] by [−10, 10, 1]

Figure A.11

and press (ENTER) to turn Plot1 on. There are six types of plots that can be selected. The first type is a *scatterplot* and the second type is a *line graph*, so place the cursor over the first type of plot and press (ENTER) to select a scatterplot. (To make the line graph in Example 14, Section 1.2, be sure to select the line graph.) The *x*-values are stored in list L1, so select L1 for "Xlist" by pressing (2nd)(1). Similarly, press (2nd)(2) for "Ylist," since the *y*-values are stored in list L2. Finally, there are three styles of marks that can be used to show data points in the graph. We usually use the first, because it is largest and shows up the best. Make the screen appear as in Figure A.10. Before plotting the four data points, be sure to set an appropriate viewing rectangle. Then press (GRAPH). The data points will appear as in Figure A.11.

NOTE 1 A fast way to set the viewing rectangle for any scatterplot is to select the ZOOM-STAT feature by pressing (ZOOM)(9). This feature automatically scales the viewing rectangle so that all data points are shown.

NOTE 2 If an equation has been entered into the (Y =) menu and selected, it will be graphed with the data. Throughout this textbook, this feature is used frequently in modeling data.

NOTE 3 When the error message "ERR: DIM MISMATCH" appears, it usually means that there are not equal numbers of *x*-values and *y*-values in lists L1 and L2.

Summary: Making a Scatterplot or a Line Graph

The following are basic steps necessary to make either a scatterplot or a line graph.

STEP 1: Use (STAT)(1) to access lists L1 and L2.

STEP 2: If list L1 is not empty, place the cursor on L1 and press (CLEAR)(ENTER). Repeat for list L2 if it is not empty.

STEP 3: Enter the *x*-values into list L1 and the *y*-values into list L2.

STEP 4: Use (2nd)(Y = [STAT PLOT]) to set the appropriate parameters for the scatterplot or line graph.

STEP 5: Either set an appropriate viewing rectangle or press (ZOOM)(9). This feature automatically sets the viewing rectangle and plots the data.

Note: (ZOOM)(9) *cannot* be used to set a viewing rectangle for the graph of a function.

Deleting and Inserting a List   A list, such as L2, can be deleted. Press (STAT)(1) and then place the cursor on L2 and press (DEL). If you want to insert a deleted list, press (STAT)(1) and then place the cursor where you want to insert the list. For example, to insert L2, place the cursor on L3. Press (2nd)(DEL [INS])(2nd)(2[L2]).

Figure A.12

Entering a Formula for a Function

To enter the formula for a function f, press (Y =). For example, use the following key-strokes after "Y₁ = " to enter $f(x) = 2x^2 - 3x + 7$. See Figure A.12.

(Y =)(CLEAR)(2)(X, T, θ, n)(^)(2)(−)(3)(X, T, θ, n)(+)(7)

Figure A.13

Figure A.14

$[-10, 10, 1]$ by $[-10, 10, 1]$

Figure A.15

Note that there is a built-in key for entering the variable X. If "$Y_1 =$ " does not appear after you press (Y =), press (MODE) and make sure the calculator is set in function mode, denoted "Func". See Figure A.13.

Summary: Entering a Formula for a Function

To enter the formula for a function, press (Y =). To delete an existing formula, press (CLEAR). Then enter the symbolic representation for the function.

Graphing a Function

To graph a function such as $f(x) = x^2 - 4$, start by pressing (Y =) and then enter $Y_1 = X^2 - 4$. If there is an equation already entered, remove it by pressing (CLEAR). The equals sign in "$Y_1 =$ " should be in reverse video (a dark rectangle surrounding a white equals sign), which indicates that the equation will be graphed. If the equals sign is not in reverse video, place the cursor over it and press (ENTER). Set an appropriate viewing rectangle and then press (GRAPH). The graph of f will appear in the specified viewing rectangle. See Figures A.14 and A.15.

NOTE If the error message "ERR: DIM MISMATCH" appears when you try to graph a function, check to see if one of the STATPLOTS is turned on. If it is, turn it off and then try graphing the function.

Summary: Graphing a Function

Use the (Y =) menu to enter the formula for the function and the (WINDOW) menu to set an appropriate viewing rectangle. Then press (GRAPH).

ZoomFit   The ZoomFit feature can be used to find an appropriate window when graphing a function. ZoomFit leaves the current Xmin and Xmax settings unchanged and adjusts the current Ymin and Ymax values so that they are equal to the smallest and largest y-values on the graph of the function between Xmin and Xmax. To use ZoomFit, press (ZOOM) (0).

Evaluating a Function Graphically

In Example 7, Section 1.3, we are asked to evaluate the function $f(x) = 0.72x + 2$ graphically at $x = 65$. Begin by entering the formula $Y_1 = .72X + 2$ into the (Y =) menu. Then graph f in the appropriate viewing rectangle, as shown in Figure 1.49. To evaluate $f(65)$, use the following keystrokes, which access the CALCULATE menu.

(2nd) (TRACE [CALC]) (1) (6) (5) (ENTER)

See Figures 1.50 and 1.51.

> ### Summary: Evaluating a Function Graphically
>
> To evaluate a function graphically, begin by graphing the function. Then use the following keystrokes to use the value routine in the CALCULATE menu.
>
>
>
> Then enter the x-value where the function should be evaluated and press (ENTER). The x-value must be between Xmin and Xmax.

Making a Table

Figure A.16

In Example 7, Section 1.3, a table of values is requested. Start by pressing (Y =) and then entering the formula $Y_1 = .72X + 2$, as shown in Figure A.16. To set the table parameters, use the following keystrokes. (See Figure A.17.)

$$\boxed{2nd}\ \boxed{\text{WINDOW [TBLSET]}}\ \boxed{6}\ \boxed{0}\ \boxed{\text{ENTER}}\ \boxed{1}$$

These keystrokes specify a table that starts at $x = 60$ and increments the x-values by 1. Therefore the values of Y_1 at $x = 60, 61, 62, \ldots$ appear in the table. To create this table, press the following keys.

$$\boxed{2nd}\ \boxed{\text{GRAPH [TABLE]}}$$

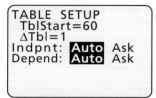

Figure A.17

One can scroll through x- and y-values by using the arrow keys. See Figure A.18. Note that there is no first or last x-value in the table.

> ### Summary: Making a Table of a Function
>
> Enter the formula for the function using (Y =). Then press
>
> $$\boxed{2nd}\ \boxed{\text{WINDOW [TBLSET]}}$$
>
> to set the starting x-value and the increment between x-values appearing in the table. Create the table by pressing
>
> $$\boxed{2nd}\ \boxed{\text{GRAPH [TABLE]}}.$$

X	Y1
60	45.2
61	45.92
62	46.64
63	47.36
64	48.08
65	48.8
66	49.52

$Y_1 = .72X + 2$

Figure A.18

Squaring a Viewing Rectangle

In a square viewing rectangle, the graph of $y = x$ is a line that makes a 45° angle with the positive x-axis, a circle appears circular, and all sides of a square have the same length. An approximately square viewing rectangle can be set if the distance along the x-axis is 1.5 times the distance along the y-axis. Examples of viewing rectangles that are (approximately) square include

$$[-6, 6, 1]\ \text{by}\ [-4, 4, 1] \quad \text{and} \quad [-9, 9, 1]\ \text{by}\ [-6, 6, 1].$$

Square viewing rectangles can be set automatically by pressing either

$$\boxed{\text{ZOOM}}\ \boxed{4} \quad \text{or} \quad \boxed{\text{ZOOM}}\ \boxed{5}.$$

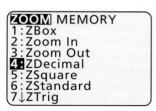

Figure A.19

ZOOM 4 provides a decimal window, which is discussed later. See Figure A.19.

Summary: Squaring a Viewing Rectangle

Either (ZOOM) (4) or (ZOOM) (5) may be used to produce a square viewing rectangle. An (approximately) square viewing rectangle has the form

$$[-1.5k, 1.5k, 1] \text{ by } [-k, k, 1],$$

where k is a positive number.

Plotting Data and an Equation

Figure A.20

In Example 4, Section 2.1, we are asked to plot data and graph a modeling function in the same xy-plane. (You may want to refer to the subsection on making a scatterplot and line graph in this appendix.) Start by entering the x-values into list L1 and the y-values into list L2, as shown in Figure A.20. Then press (Y =) and enter the formula $Y_1 = .65X$ for $f(x)$. Make sure that STATPLOT is on, and set an appropriate viewing rectangle. See Figures A.21 and A.22, and note that Figure A.21 shows "Plot1" in reverse video, which indicates that the scatterplot is on. Now press (GRAPH) to have both the scatterplot and the graph of Y_1 appear in the same viewing rectangle, as shown in Figure A.23.

Figure A.21   **Figure A.22**

[0, 900, 100] by [0, 800, 100]

Figure A.23

Summary: Plotting Data and an Equation

STEP 1:   Enter the x-values into list L1 and the y-values into list L2 using the STAT EDIT menu. Turn on Plot1 so that the scatterplot appears.

STEP 2:   Use the (Y =) menu to enter the equation to be graphed.

STEP 3:   Use (WINDOW) or (ZOOM) to set an appropriate viewing rectangle.

STEP 4:   Press (GRAPH) to graph both the scatterplot and the equation in the same viewing rectangle.

Accessing the Greatest Integer Function

To access the greatest integer function, enter the following keystrokes from the home screen.

(MATH) (▷) (5)

See Figure A.24.

Figure A.24

Summary: Accessing the Greatest Integer Function

STEP 1:   Press (MATH).

STEP 2:   Position the cursor over "NUM".

STEP 3:   Press (5) to select the greatest integer function, which is denoted "int(".

Figure A.25

Setting Connected and Dot Mode

In Figure 2.10 of Section 2.1 a form of the greatest integer function is graphed in dot mode, and in Figure 2.11 it is graphed in connected mode. To set your graphing calculator in dot mode, press (MODE), position the cursor over "Dot", and press (ENTER). See Figure A.25. Graphs will now appear in dot mode rather than connected mode.

Summary: Setting Connected or Dot Mode
STEP 1:   Press (MODE).
STEP 2:   Position the cursor over "Connected" or "Dot". Press (ENTER).

Finding the Line of Least-Squares Fit

Figure A.26

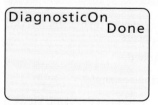

Figure A.27

In Example 7, Section 2.1, the line of least-squares fit for the points (1, 1), (2, 3), and (3, 4) is found. Begin by entering the points in the same way as for a scatterplot. See Figure 2.12, where the x-values are in list L1 and the y-values are in list L2.

After the data have been entered, perform the following keystrokes from the home screen.

(CLEAR) (STAT) (▷) (4)

(See Figure 2.13.) This causes "LinReg(ax+b)" to appear on the home screen, as shown in Figure A.26. The graphing calculator assumes that the x-values are in list L1 and the y-values are in list L2. Now press (ENTER). The result is shown in Figure 2.14.

If the correlation coefficient r does not appear, enter the keystrokes

(2nd) (0 [CATALOG])

DiagnosticOn
              Done

Figure A.28

and scroll down until you find "DiagnosticsOn". Press (ENTER) twice. See Figures A.27 and A.28. The graphs of the data and the least-squares regression line are shown in Figure 2.15.

Summary: Linear Least-Squares Fit
STEP 1:   Enter the data using (STAT) (1), as is done for a scatterplot. Input the x-values into list L1 and the y-values into list L2.
STEP 2:   Press (STAT) (▷) (4) from the home screen to access the least-squares regression line. Press (ENTER) to start the computation. See page AP-13 to learn how to copy a regression equation into Y_1.

Locating a Point of Intersection

Figure A.29

In Example 7, Section 2.3, we are asked to find the point of intersection for two lines. To find the point of intersection for the graphs of

$$f(x) = 5.91x + 13.7 \quad \text{and} \quad g(x) = -4.71x + 64.7,$$

start by entering Y_1 and Y_2, as shown in Figure A.29. Set the viewing rectangle to [0, 12, 2] by [0, 100, 10], and graph both equations in the same viewing rectangle, as shown in Figure 2.35. Then press the following keys to find the intersection point.

(2nd) (TRACE [CALC]) (5)

Figure A.30

[0, 12, 2] by [0, 100, 10]

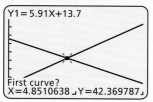

Figure A.31

[0, 12, 2] by [0, 100, 10]

Figure A.32

Figure A.33

See Figure A.30, where the intersect utility is being selected. The calculator prompts for the first curve, as shown in Figure A.31. Use the arrow keys to locate the cursor near the point of intersection and press (ENTER). Repeat these steps for the second curve, as shown in Figure A.32. Finally the calculator prompts for a guess. For each of the three prompts, place the free-moving cursor near the point of intersection and press (ENTER). The approximate coordinates of the point of intersection are shown in Figure 2.36.

Summary: Finding a Point of Intersection

STEP 1:   Graph the two functions in an appropriate viewing rectangle.

STEP 2:   Press (2nd) (TRACE [CALC]) (5).

STEP 3:   Use the arrow keys to select an approximate location for the point of intersection. Press (ENTER) to make the three selections for "First curve?", "Second curve?", and "Guess?". (If the cursor is near the point of intersection, you usually do not need to move the cursor for each selection. Just press (ENTER) three times.)

Locating a Zero of a Function

In Example 4, Section 2.4, we are asked to locate an x-intercept, or *zero*, of the function f given by $f(x) = 1 - x - \left(\frac{1}{2}x - 2\right)$. Start by entering $Y_1 = 1 - X - (X/2 - 2)$ into the (Y =) menu. Set the viewing rectangle to $[-6, 6, 1]$ by $[-4, 4, 1]$ and graph Y_1. Afterwards, press the following keys to invoke the zero finder. (See Figure A.33.)

(2nd) (TRACE [CALC]) (2)

The calculator prompts for a left bound. Use the arrow keys to set the cursor to the left of the x-intercept and press (ENTER). The calculator then prompts for a right bound. Set the cursor to the right of the x-intercept and press (ENTER). Finally the calculator prompts for a guess. Set the cursor roughly at the x-intercept and press (ENTER). See Figures A.34–A.36. The calculator then approximates the x-intercept, or zero, automatically, as shown in Figure 2.48.

[−6, 6, 1] by [−4, 4, 1]

Figure A.34

[−6, 6, 1] by [−4, 4, 1]

Figure A.35

[−6, 6, 1] by [−4, 4, 1]

Figure A.36

Summary: Locating a Zero of a Function

STEP 1:   Graph the function in an appropriate viewing rectangle.

STEP 2:   Press (2nd) (TRACE [CALC]) (2).

STEP 3:   Select the left and right bounds, followed by a guess. Press (ENTER) after each selection. The calculator then approximates the zero.

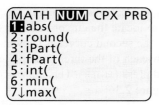

Figure A.37

Accessing the Absolute Value

In Example 1, Section 2.5, the absolute value is used to graph $f(x) = |x + 2|$. To graph f, begin by entering $Y_1 = \text{abs}(X + 2)$. The absolute value (abs) is accessed by pressing

$$\boxed{\text{MATH}}\ \boxed{\triangleright}\ \boxed{1}.$$

See Figure A.37.

Summary: Accessing the Absolute Value
STEP 1: Press $\boxed{\text{MATH}}$.
STEP 2: Position the cursor over "NUM".
STEP 3: Press $\boxed{1}$ to select the absolute value.

Figure A.38

Finding Extrema (Minima and Maxima)

In Example 6, Section 3.1, we are asked to find a minimum point (or vertex) on the graph of $f(x) = 1.5x^2 - 6x + 4$. Start by entering $Y_1 = 1.5X^2 - 6X + 4$ into the $\boxed{Y=}$ menu. Set the viewing rectangle to $[-4.7, 4.7, 1]$ by $[-3.1, 3.1, 1]$ by entering $\boxed{\text{ZOOM}}\ \boxed{4}$. Then perform the following keystrokes to find the minimum y-value.

$$\boxed{\text{2nd}}\ \boxed{\text{TRACE [CALC]}}\ \boxed{3}$$

See Figure A.38.

$[-4.7, 4.7, 1]$ by $[-3.1, 3.1, 1]$

Figure A.39

The calculator prompts for a left bound. Use the arrow keys to position the cursor to the left of the vertex and press $\boxed{\text{ENTER}}$. Similarly, position the cursor to the right of the vertex for the right bound and press $\boxed{\text{ENTER}}$. Finally the calculator asks for a guess between the left and right bounds. Place the cursor near the minimum point and press $\boxed{\text{ENTER}}$. See Figures A.39–A.41. The minimum point (or vertex) is shown in Figure 3.14.

To find a maximum of the function f on an interval, use a similar approach, except enter

$$\boxed{\text{2nd}}\ \boxed{\text{TRACE [CALC]}}\ \boxed{4}.$$

$[-4.7, 4.7, 1]$ by $[-3.1, 3.1, 1]$

The calculator prompts for left and right bounds, followed by a guess. Press $\boxed{\text{ENTER}}$ after the cursor has been located appropriately for each prompt. An example of a maximum point is displayed in Figure 3.20.

Figure A.40

Summary: Finding Extrema (Maxima and Minima)
STEP 1: Graph the function in an appropriate viewing rectangle.
STEP 2: Press $\boxed{\text{2nd}}\ \boxed{\text{TRACE [CALC]}}\ \boxed{3}$ to find a minimum point or $\boxed{\text{2nd}}\ \boxed{\text{TRACE [CALC]}}\ \boxed{4}$ to find a maximum point.
STEP 3: Use the arrow keys to locate the left and right x-bounds, followed by a guess. Press $\boxed{\text{ENTER}}$ to select each position of the cursor.

$[-4.7, 4.7, 1]$ by $[-3.1, 3.1, 1]$

Figure A.41

Using the Ask Table Feature

In Example 11, Section 3.1, a table with x-values of 0, 2, 4, 6, and 8 is created. Start by entering $Y_1 = 1.875(X - 8)^2 + 80$. To obtain the table shown in Figure 3.23, use the Ask feature rather than the Auto feature for the independent variable (Indpnt:). Press $\boxed{\text{2nd}}$

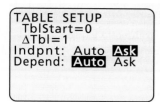

Figure A.42

X	Y1
0	200
2	147.5
4	110

Y1◼1.875(X−8)²+...

Figure A.43

EDIT **CALC** TESTS
1:1−Var Stats
2:2−Var Stats
3:Med−Med
4:LinReg(ax+b)
5:QuadReg
6◼CubicReg
7↓QuartReg

Figure A.44

(GRAPH [TABLE]). Whenever an x-value is entered, the corresponding y-value is calculated automatically. See Figures A.42 and A.43.

Summary: Using the Ask Feature for a Table

STEP 1: Enter the formula for $f(x)$ into Y₁ by using the (Y=) menu.

STEP 2: Press (2nd)(WINDOW [TBLSET]) to access "TABLE SETUP" and then select "Ask" for the independent variable (Indpnt:). "TblStart" and "ΔTbl" do not need to be set.

STEP 3: Enter x-values of your choice. The corresponding y-values will be calculated automatically.

Finding a Nonlinear Function of Least-Squares Fit

In Example 12, Section 3.1, a quadratic function of least-squares fit is found in a manner similar to the way a linear function of least-squares fit is found. To solve Example 12, start by pressing (STAT)(1) and then enter the data points from Table 3.4, as shown in Figure 3.25. Input the x-values into list L1 and the y-values into list L2. To find the equation for a quadratic polynomial of least-squares fit, perform the following keystrokes from the home screen.

(CLEAR)(STAT)(▷)(5)

This causes "Quadreg" to appear on the home screen. The calculator assumes that the x-values are in list L1 and the y-values are in list L2, unless otherwise designated. Press (ENTER) to obtain the quadratic regression equation, as shown in Figure 3.27. Graphs of the data and the regression equation are shown in Figure 3.28.

Other types of regression equations, such as cubic, quartic, power, and exponential, can be selected from the STAT CALC menu. See Figure A.44.

Summary: Nonlinear Least-Squares Fit

STEP 1: Enter the data using (STAT)(1). Input the x-values into list L1 and the y-values into list L2, as is done for a scatterplot.

STEP 2: From the home screen, press (STAT)(▷) and select a type of least-squares modeling function from the menu. Press (ENTER) to initiate the computation.

Evaluating Complex Arithmetic

Complex arithmetic can be performed in much the same way as other arithmetic expressions are evaluated. The imaginary unit i is obtained by entering

(2nd)(. [i])

from the home screen. For example, to add the numbers $(-2 + 3i) + (4 - 6i)$, perform the following keystrokes on the home screen.

The result is shown in the first two lines of Figure 3.48 in Section 3.3. Other complex arithmetic operations are done similarly.

Summary: Evaluating Complex Arithmetic

Enter a complex expression in the same way as any other arithmetic expression. To obtain the complex number i, use $\boxed{\text{2nd}}\,\boxed{.\,[i]}$.

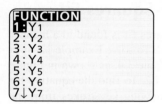

Figure A.45

Accessing the Variable Y_1

In Figure 3.94, Section 3.5, the expressions $-Y_1$ and $Y_1(-X)$ in the $\boxed{Y=}$ menu are used to graph reflections. The Y_1 variable can be found by pressing the following keys. (See Figures A.45 and A.46.)

$$\boxed{\text{VARS}}\ \boxed{\triangleright}\ \boxed{1}\ \boxed{1}$$

Figure A.46

Summary: Accessing the Variable Y_1

STEP 1:   Press $\boxed{\text{VARS}}$.

STEP 2:   Position the cursor over "Y-VARS".

STEP 3:   Press $\boxed{1}$ twice.

These keystrokes will make Y_1 appear on the screen.

```
DRAW  POINTS STO
1:ClrDraw
2:Line(
3:Horizontal
4:Vertical
5:Tangent(
6:DrawF
7:Shade(
```

Figure A.47

Shading between Two Graphs

In Example 8, Section 3.5, the region below the graph of $f(x) = -0.4x^2 + 4$ is shaded to make it look like a mountain, as illustrated in Figure 3.113. One way to shade below the graph of f is to begin by entering $Y_1 = -.4X^{\wedge}2 + 4$ after pressing $\boxed{Y=}$. Then use the following keystrokes from the home screen.

The expression Shade(-5, Y_1) should appear on your home screen. See Figures A.47 and A.48. The shading utility, accessed from the DRAW menu, requires a lower function and then an upper function, separated by a comma. When $\boxed{\text{ENTER}}$ is pressed, the graphing calculator shades between the graph of the lower function and the graph of the upper function. For the lower function we have arbitrarily selected $y = -5$ because its graph lies below the graph of f and does not appear in the viewing rectangle in Figure 3.113. Instead of entering the variable Y_1, we could enter the formula $-.4X^{\wedge}2 + 4$ for the upper function.

```
Shade(-5, Y1)
```

Figure A.48

Summary: Shading a Graph

STEP 1:   Press $\boxed{\text{2nd}}\,\boxed{\text{PRGM [DRAW]}}\,\boxed{7}$ from the home screen.

STEP 2:   Enter a formula or a variable such as Y_1 for the lower function, followed by a comma.

STEP 3:   Enter a formula or a variable such as Y_2 for the upper function, followed by a right parenthesis.

STEP 4:   Set an appropriate viewing rectangle.

STEP 5:   Press $\boxed{\text{ENTER}}$. The region between the two graphs will be shaded.

Figure A.49

Figure A.50

Figure A.51

Copying a Regression Equation into $Y_1 =$

In Example 7, Section 4.2, we are asked to use cubic regression to model real data. The resulting formula for the cubic function, shown in Figure 4.53, is quite complicated and tedious to enter into $Y_1 =$ by hand. A graphing calculator has the capability to copy this equation into Y_1 automatically. To do this, clear the equation for $Y_1 =$. Then enter Y_1 after "CubicReg", as shown in Figure A.49. When (ENTER) is pressed, the regression equation will be calculated and then copied into $Y_1 =$, as shown in Figure A.50. The following keystrokes may be used from the home screen. (Be sure to enter the data into lists L1 and L2.)

(STAT) (▷) (6) (VARS) (▷) (1) (1) (ENTER)

> **Summary: Copying a Regression Equation into $Y_1 =$**
>
> **STEP 1:** Clear Y_1 in the (Y =) menu if an equation is present. Return to the home screen.
>
> **STEP 2:** Select a type of regression from the STAT CALC menu.
>
> **STEP 3:** Press (VARS) (▷) (1) (1) (ENTER).

Setting a Decimal Window

In Example 1, Section 4.7, a decimal (or friendly) window is used to trace the graph of f. With a decimal window, the cursor stops on convenient x-values. In the decimal window $[-9.4, 9.4, 1]$ by $[-6.2, 6.2, 1]$, the cursor stops on x-values that are multiples of 0.2. If we reduce the viewing rectangle to $[-4.7, 4.7, 1]$ by $[-3.1, 3.1, 1]$, the cursor stops on x-values that are multiples of 0.1. To set this smaller window automatically, press (ZOOM) (4). See Figure A.51. Decimal windows are useful when graphing rational functions with asymptotes in connected mode.

> **Summary: Setting a Decimal Window**
>
> Press (ZOOM) (4) to set the viewing rectangle $[-4.7, 4.7, 1]$ by $[-3.1, 3.1, 1]$. A convenient larger decimal window is $[-9.4, 9.4, 1]$ by $[-6.2, 6.2, 1]$.

Graphing an Inverse Function

In Example 7, Section 5.2, the inverse function of $f(x) = x^3 + 2$ is graphed. A graphing calculator can graph the inverse of a function without a formula for $f^{-1}(x)$. Begin by entering $Y_1 = X^3 + 2$ into the (Y =) menu. Then return to the home screen by pressing

(2nd) (MODE [QUIT]).

The DrawInv utility may be accessed by pressing

(2nd) (PRGM [DRAW]) (8),

followed by

(VARS) (▷) (1) (1)

to obtain the variable Y_1. See Figure A.52. Pressing (ENTER) causes both Y_1 and its inverse to be graphed, as shown in Figure A.53.

Figure A.52

$[-6, 6, 1]$ by $[-4, 4, 1]$

Figure A.53

> ### Summary: Graphing an Inverse Function
>
> **STEP 1:** Enter the formula for $f(x)$ into Y_1 using the $(Y=)$ menu.
>
> **STEP 2:** Set an appropriate viewing rectangle by pressing $(WINDOW)$.
>
> **STEP 3:** Return to the home screen by pressing $(2nd)(MODE [QUIT])$.
>
> **STEP 4:** Press $(2nd)(PRGM [DRAW])(8)(VARS)(\triangleright)(1)(1)(ENTER)$ to create the graphs of f and f^{-1}.

Shading a System of Inequalities

Figure A.54

In Example 2(b), Section 9.2, we are asked to shade the solution set for the system of linear inequalities $x + 3y \leq 9, 2x - y \leq -1$. Begin by solving each system for y to obtain $y \leq (9 - x)/3$ and $y \geq 2x + 1$. Then let $Y_1 = (9 - X)/3$ and $Y_2 = 2X + 1$, as shown in Figure 9.34. Position the cursor to left of Y_1 and press $(ENTER)$ three times. The triangle that appears indicates that the calculator will shade the region below the graph of Y_1. Next locate the cursor to the left of Y_2 and press $(ENTER)$ twice. This triangle indicates that the calculator will shade the region above the graph of Y_2. After setting the viewing rectangle to $[-6, 6, 1]$ by $[-6, 6, 1]$, press $(GRAPH)$. The result is shown in Figure 9.35. The solution set could also be shaded using Shade(Y_2, Y_1) from the home screen. See Figures A.54 and A.55.

$[-6, 6, 1]$ by $[-6, 6, 1]$

Figure A.55

> ### Summary: Shading a System of Equations
>
> **STEP 1:** Solve each inequality for y.
>
> **STEP 2:** Enter the formulas as Y_1 and Y_2 in the $(Y=)$ menu.
>
> **STEP 3:** Locate the cursor to the left of Y_1 and press $(ENTER)$ two or three times, to shade either above or below the graph of Y_1. Repeat for Y_2.
>
> **STEP 4:** Set an appropriate viewing rectangle.
>
> **STEP 5:** Press $(GRAPH)$.
>
> *Note:* The Shade utility under the DRAW menu can also be used to shade the region *between* two graphs.

Entering the Elements of a Matrix

In Example 9, Section 9.4, the augmented matrix A is given by

$$A = \begin{bmatrix} 2 & 1 & 2 & 10 \\ 1 & 0 & 2 & 5 \\ 1 & -2 & 2 & 1 \end{bmatrix}.$$

On the TI-83 Plus and TI-84 Plus, use the following keystrokes to define a matrix A with dimension 3×4. (On the TI-83 graphing calculator, the matrix menu is found by pressing $(MATRIX)$.)

See Figures 9.45 and 9.46. Then input the 12 elements of the matrix A, row by row, as shown in Figure 9.46. Finish each entry by pressing (ENTER). After these elements have been entered, press

(2nd) (MODE [QUIT])

to return to the home screen. To display the matrix A, press

(2nd) (x^{-1} [MATRIX]) (1) (ENTER).

See Figure 9.47.

Summary: Entering the Elements of a Matrix A

STEP 1: Begin by accessing the matrix A by pressing (2nd) (x^{-1} [MATRIX]) (▷) (▷) (1).

STEP 2: Enter the dimension of A by pressing (m) (ENTER) (n) (ENTER), where the dimension of the matrix is $m \times n$.

STEP 3: Input each element of the matrix, row by row. Finish each entry by pressing (ENTER). Use (2nd) (MODE [QUIT]) to return to the home screen.

Note: On the TI-83, replace the keystrokes (2nd) (x^{-1} [MATRIX]) with (MATRIX).

Reduced Row-Echelon Form

In Example 9, Section 9.4, the reduced row-echelon form of a matrix is found. To find this reduced row-echelon form on the TI-83 Plus and TI-84 Plus, use the following keystrokes from the home screen.

(2nd) (x^{-1} [MATRIX]) (▷) (ALPHA) (APPS [B]) (2nd) (x^{-1} [MATRIX]) (1) ( )) (ENTER)

The resulting matrix is shown in Figure 9.49. On the TI-83 graphing calculator, use the following keystrokes to find the reduced row-echelon form.

(MATRIX) (▷) (ALPHA) (MATRIX [B]) (MATRIX) (1) ( )) (ENTER)

Summary: Finding the Reduced Row-Echelon Form of a Matrix

STEP 1: To make rref([A]) appear on the home screen, use the following keystrokes.

(2nd) (x^{-1} [MATRIX]) (▷) (ALPHA) (APPS [B]) (2nd) (x^{-1} [MATRIX]) (1) ( )) (ENTER)

STEP 2: Press (ENTER) to calculate the reduced row-echelon form. Use arrow keys to access elements that do not appear on the screen.

Note: On the TI-83, replace the keystrokes (2nd) (x^{-1} [MATRIX]) with (MATRIX) and (APPS [B]) with (MATRIX [B]).

Performing Arithmetic Operations on Matrices

In Example 8, Section 9.5, the matrices A and B are multiplied. Begin by entering the elements for the matrices A and B. The following keystrokes can be used to define a matrix A with dimension 3×3.

(2nd) (x^{-1} [MATRIX]) (▷) (▷) (1) (3) (ENTER) (3) (ENTER)

Next input the 9 elements in the matrix A, row by row. Finish each entry by pressing (ENTER). See Figure 9.65. Repeat this process to define a matrix B with dimension 3×3.

(2nd) (x^{-1} [MATRIX]) (▷) (▷) (2) (3) (ENTER) (3) (ENTER)

Enter the 9 elements in B. See Figure 9.66. After the elements of A and B have been entered, press

(2nd) (MODE [QUIT])

to return to the home screen. To multiply the expression AB, use the following keystrokes from the home screen.

(2nd) (x^{-1} [MATRIX]) (1) (×) (2nd) (x^{-1} [MATRIX]) (2) (ENTER)

The result is shown in Figure 9.67.

Summary: Performing Arithmetic Operations on Matrices

STEP 1: Enter the elements of each matrix, beginning with the keystrokes

(2nd) (x^{-1} [MATRIX]) (▷) (▷) (k) (m) (ENTER) (n) (ENTER),

where k is the menu number of the matrix and the dimension of the matrix is $m \times n$.

STEP 2: Return to the home screen by pressing (2nd) (MODE [QUIT]).

STEP 3: Enter the matrix expression, followed by (ENTER). Use the keystrokes

(2nd) (x^{-1} [MATRIX]) (k)

to access the matrix with menu number k.

Note: On the TI-83, replace the keystrokes (2nd) (x^{-1} [MATRIX]) with (MATRIX).

MATRIX[A]   3 ×3
[1   3   ⁻1   ]
[0   ⁻2   1   ]
[⁻1   1   ⁻3   ]

Figure A.56

Finding the Inverse of a Matrix

In Example 8, Section 9.6, the inverse of A, denoted A^{-1}, is displayed in Figure 9.71. To calculate A^{-1}, start by entering the elements of the matrix A, as shown in Figure A.56. To compute A^{-1}, perform the following keystrokes from the home screen.

(2nd) (x^{-1} [MATRIX]) (1) (x^{-1}) (ENTER)

The results are shown in Figure 9.71.

Summary: Finding the Inverse of a Square Matrix

STEP 1: Enter the elements of the square matrix A.

STEP 2: Return to the home screen by pressing

(2nd) (MODE [QUIT]).

STEP 3: Perform the following keystrokes from the home screen to display A^{-1}.

(2nd) (x^{-1} [MATRIX]) (1) (x^{-1}) (ENTER)

Note: On the TI-83, replace the keystrokes (2nd) (x^{-1} [MATRIX]) with (MATRIX).

Solving a Linear System with a Matrix Inverse

In Example 8, Section 9.6, the solution to a system of equations is found. The matrix equation $AX = B$ has the solution $X = A^{-1}B$, provided A^{-1} exists, and is given by

$$AX = \begin{bmatrix} 1 & 3 & -1 \\ 0 & -2 & 1 \\ -1 & 1 & -3 \end{bmatrix} \begin{bmatrix} x \\ y \\ z \end{bmatrix} = \begin{bmatrix} 6 \\ -2 \\ 4 \end{bmatrix} = B.$$

To solve this equation, start by entering the elements of the matrices A and B. To compute the solution $A^{-1}B$, perform the following keystrokes from the home screen.

(2nd) (x^{-1} [MATRIX]) (1) (x^{-1}) (×) (2nd) (x^{-1} [MATRIX]) (2) (ENTER)

The results are shown in Figure 9.72.

Summary: Solving a Linear System with a Matrix Inverse

STEP 1:   Write the system of equations as $AX = B$.

STEP 2:   Enter the elements of the matrices A and B.

STEP 3:   Return to the home screen by pressing

(2nd) (MODE [QUIT]).

STEP 4:   Perform the following keystrokes.

(2nd) (x^{-1} [MATRIX]) (1) (x^{-1}) (×) (2nd) (x^{-1} [MATRIX]) (2) (ENTER)

Note: On the TI-83, replace the keystrokes with (2nd) (x^{-1} [MATRIX]) with (MATRIX).

```
MATRIX[A]   3 ×3
[ 2      -3     1     ]
[ -2      1     0     ]
[ 0      -1     4     ]
```

Figure A.57

Evaluating a Determinant

In Example 4(a), Section 9.7, a graphing calculator is used to evaluate a determinant of a matrix. Start by entering the 9 elements of the 3×3 matrix A, as shown in Figure A.57. To compute det A, perform the following keystrokes from the home screen.

(2nd) (x^{-1} [MATRIX]) (▷) (1) (2nd) (x^{-1} [MATRIX]) (1) ( )) (ENTER)

The results are shown in Figure 9.74.

Summary: Evaluating a Determinant of a Square Matrix

STEP 1:   Enter the elements of the matrix A.

STEP 2:   Return to the home screen by pressing

(2nd) (MODE [QUIT]).

STEP 3:   Perform the following keystrokes.

(2nd) (x^{-1} [MATRIX]) (▷) (1) (2nd) (x^{-1} [MATRIX]) (1) ( )) (ENTER)

Note: On the TI-83, replace the keystrokes (2nd) (x^{-1} [MATRIX]) with (MATRIX).

Figure A.58

Figure A.59

Creating a Sequence

A graphing calculator can be used to calculate the terms of the sequence given by $f(n) = 2n - 5$ for $n = 1, 2, 3, 4$. See Example 1(a), Section 11.1. Start by setting the mode of the calculator to sequence ("Seq") using the following keystrokes. (See Figure A.58.)

Then enter the following from the home screen.

(2nd) (STAT [LIST]) (▷) (5)

On the home screen, "seq(" will appear, as shown in Figure A.59. This sequence utility requires that four things be entered—all separated by commas. They are the formula, the variable, the subscript of the first term, and the subscript of the last term. Use the following keystrokes to obtain the first four terms (a_1, a_2, a_3, a_4) of the sequence $a_n = 2n - 5$, as shown in Figure 11.1.

Summary: Creating a Sequence

STEP 1: To create a sequence, use the keystrokes

(2nd) (STAT [LIST]) (▷) (5) .

STEP 2: Enter the formula, the variable, the subscript of the first term, and the subscript of the last term—all separated by commas. For example, if you want the first 10 terms ($a_1, a_2, a_3, \ldots, a_{10}$) of $a_n = n^2$, enter seq ($n^2, n, 1, 10$). Be sure to set your calculator in sequence mode.

STEP 3: Press (ENTER) to get the terms of the sequence to appear.

Entering, Tabling, and Graphing a Sequence

In Example 5, Section 11.1, a table and a graph of a sequence are created with a graphing calculator. The calculator should be set to sequence mode by entering the following keystrokes.

(MODE) (▽) (▽) (▽) (▷) (▷) (▷) (ENTER)

Figure A.60

To enter the formula for a sequence, press (Y =). See Figure A.60. Let nMin $= 1$, since the initial value of n is equal to 1. To enter $a_n = 2.85a_{n-1} - .19a_{n-1}^2$, use the following keystrokes, after clearing out any old formula. (Notice that the graphing calculator uses u instead of a to denote a term of the sequence.)

Since $a_1 = 1$, let $u(n$Min$) = \{1\}$. This can be done as follows.

(CLEAR) (2nd) (( ) (1) (2nd) ( ))

See Figure A.60.

```
┌─────────────────────────────┐
│ TABLE SETUP                 │
│   TblStart=1                │
│   ΔTbl=1                    │
│ Indpnt: ▮Auto▮  Ask         │
│ Depend: ▮Auto▮  Ask         │
│                             │
│                             │
└─────────────────────────────┘
```

Figure A.61

```
┌─────────────────────────────┐
│ WINDOW                      │
│   nMin=1                    │
│   nMax=20                   │
│   PlotStart=1               │
│   PlotStep=1                │
│   Xmin=0                    │
│   Xmax=21                   │
│ ↓Xscl=1                     │
└─────────────────────────────┘
```

Figure A.62

To create a table for this sequence, starting with a_1 and incrementing n by 1, perform the following keystrokes.

(2nd) (WINDOW [TBLSET]) (1) (ENTER) (1) (2nd) (GRAPH [TABLE])

See Figure A.61 and Figure 11.7.

To graph the first 20 terms of this sequence, start by selecting (WINDOW). Since we want the first 20 terms plotted, let nMin $= 1$, nMax $= 20$, PlotStart $= 1$, and PlotStep $= 1$. The window can be set as [0, 21, 1] by [0, 14, 1]. See Figure A.62. To graph the sequence, press (GRAPH). The resulting graph uses dot mode and is shown in Figure 11.9.

Summary: Entering, Tabling, and Graphing a Sequence

STEP 1: Set the mode to "Seq" by using the (MODE) menu.

STEP 2: Enter the formula for the sequence by pressing (Y =).

STEP 3: To create a table of a sequence, set the start and increment values with

(2nd) (WINDOW [TBLSET])

and then press

(2nd) (GRAPH [TABLE]).

STEP 4: To graph a sequence, set the viewing rectangle by using (WINDOW) and then press (GRAPH). Be sure to use dot mode.

Summing a Series

In Example 2, Section 11.2, the sum of the series $\sum_{n=1}^{50}\left(\frac{1}{n^4}\right)$ is found by using a graphing calculator. Use the following keystrokes from the home screen.

The results are shown in the first three lines of Figure 11.12.

Summary: Summing a Series

STEP 1: Use (2nd) (STAT [LIST]) (▷) (▷) (5) to access the sum utility.

STEP 2: Use (2nd) (STAT [LIST]) (▷) (5) to access the sequence utility. (To use the sequence utility, see "Creating a Sequence" in this appendix.)

Calculating Factorial Notation

In Example 4, Section 11.3, factorial notation is evaluated with a graphing calculator. The factorial utility is found under the MATH PRB menus. To calculate 8!, use the following keystrokes from the home screen.

The results are shown in the first two lines of Figure 11.15.

Summary: Calculating Factorial Notation

To calculate n factorial, use the following keystrokes.

The value of n should be entered as a number, not a variable.

Calculating Permutations and Combinations

In Example 5(a), Section 11.3, the permutation $P(7, 3)$ is evaluated. To perform this calculation, use the following keystrokes from the home screen.

The results are shown in the first two lines of Figure 11.17.

In Example 8(a), Section 11.3, the combination $C(7, 3)$ can be calculated by using the following keystrokes.

$$\boxed{7}\ \boxed{\text{MATH}}\ \boxed{\triangleright}\ \boxed{\triangleright}\ \boxed{\triangleright}\ \boxed{3}\ \boxed{3}\ \boxed{\text{ENTER}}$$

The results are shown in the first two lines of Figure 11.18.

Summary: Calculating Permutations and Combinations

STEP 1:   To calculate $P(n, r)$, use $\boxed{\text{MATH}}$ and select "PRB" followed by $\boxed{2}$.

STEP 2:   To calculate $C(n, r)$, use $\boxed{\text{MATH}}$ and select "PRB" followed by $\boxed{3}$.

Graphing Parametric Equations

In Figure 8.68, Section 8.4, the parametric equations $x = t + 3$, $y = t^2$ for $-3 \le t \le 3$ are graphed. To set your graphing calculator in parametric mode, press $\boxed{\text{MODE}}$, position the cursor over "Par", and press $\boxed{\text{ENTER}}$. See Figure A.63. Next Press $\boxed{Y =}$ and enter the equations for x and y, as shown in Figure A.64.

Figure A.63

Figure A.64

To set a viewing rectangle, press $\boxed{\text{WINDOW}}$. In addition to setting Xmin, Xmax, Xscl, Ymin, Ymax, and Yscl, you must set values for Tmin, Tmax, and Tstep. Tmin refers to the minimum value of t in the graph, and Tmax refers to the maximum value of t. It is given that $-3 \le t \le 3$, so it follows that Tmin $= -3$ and Tmax $= 3$. However, when an interval for t is not given, it may take a little experimentation to determine an appropriate interval for t. Tstep represents the increment between consecutive t-values on the graph. If Tstep is too large, the graph appears more like a line graph than a smooth curve. If Tstep is too small, the graphing calculator will take a long time to create the graph. Many times a reasonable value is Tstep = 0.1. See Figure 8.66. A parametric graph can be created by pressing ⟨GRAPH⟩.

Tables for parametric equations can be created. Press

⟨2nd⟩ ⟨WINDOW [TBLSET]⟩

and proceed in the usual manner. Note that the variables TblStart and ΔTbl refer to t and not x. See Figure 8.67.

Summary: Graphing Parametric Equations

1. Press ⟨MODE⟩, move the cursor to "Par", and press ⟨ENTER⟩.
2. Press ⟨Y =⟩ and enter the equations for x and y.
3. Press ⟨WINDOW⟩ and set the viewing rectangle. Be sure to set Tmin, Tmax, and Tstep. When in doubt, let Tstep = 0.1.
4. To make the graph appear, press ⟨GRAPH⟩.

Graphing in Polar Coordinates

In Figure 8.97, Section 8.5, the polar equation $r = 2 + 2 \cos \theta$ for $0° \le \theta \le 360°$ is graphed. To set your graphing calculator in polar coordinate mode, press ⟨MODE⟩, position the cursor over "Pol", and press ⟨ENTER⟩. See Figure A.65. Polar equations can be graphed in either degree or radian mode. To set your calculator in degree mode, position the cursor over "Degree" and press ⟨ENTER⟩. See Figure A.66. Next press ⟨Y =⟩ and enter the equation for "r_1 =", as shown in Figure A.67. Note that the polar equation must be solved for the variable r.

Figure A.65

Figure A.66

```
 Plot1  Plot2  Plot3
\r1⬛2+2cos(θ)
\r2=
\r3=
\r4=
\r5=
\r6=
```

Figure A.67

```
WINDOW
 θmin = 0
 θmax = 360
 θstep = 7.5
 Xmin = ‾6
 Xmax = 6
 Xscl = 1
↓Ymin = ‾4
```

Figure A.68 Degree Mode

To set a viewing rectangle, press ⟨WINDOW⟩. In addition to setting Xmin, Xmax, Xscl, Ymin, Ymax, and Yscl, you must set values for θmin, θmax, and θstep. The variable θmin refers to the minimum value of θ, and θmax refers to the maximum value of θ. Since $\cos \theta$ is periodic with 360°, the entire graph will appear if we let $0° \le \theta \le 360°$. Let θmin = 0 and θmax = 360. The variable θstep represents the increment between consecutive θ-values on the polar graph. If θstep is too large, the graph appears more like a line graph than a smooth curve. If θstep is too small, the graphing calculator will take a long time to create the graph. In degree mode a reasonable value for θstep is 7.5°, and in radian mode a reasonable value for θstep is 0.1 radian. See Figure A.68. A polar graph can be created by pressing ⟨GRAPH⟩.

Tables for polar coordinates can be created. Press

$$\boxed{\text{2nd}}\ \boxed{\text{WINDOW [TBLSET]}}$$

and proceed in the usual manner. Note that the variables TblStart and ΔTbl refer to θ and not x. See Figure 8.96.

Summary: Graphing in Polar Coordinates

1. Press $\boxed{\text{MODE}}$, move the cursor to "Pol", and press $\boxed{\text{ENTER}}$. Set the calculator to either degree or radian mode.
2. Press $\boxed{\text{Y} =}$ and enter the polar equation.
3. Press $\boxed{\text{WINDOW}}$ and set the viewing rectangle. Be sure to set θmin, θmax, and θstep. When in doubt, let θstep $= 7.5$ in degree mode and θstep $= 0.1$ in radian mode.
4. To make the graph appear, press $\boxed{\text{GRAPH}}$.

Appendix B: A Library of Functions

Basic Functions

The following are symbolic, numerical, and graphical representations of several functions used in algebra. Their domains D and ranges R are given in interval notation.

Identity Function: $f(x) = x$

x	-2	-1	0	1	2
$y = x$	-2	-1	0	1	2

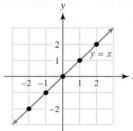

$D = (-\infty, \infty)$
$R = (-\infty, \infty)$

Absolute Value Function: $f(x) = |x|$

x	-2	-1	0	1	2
$y = |x|$	2	1	0	1	2

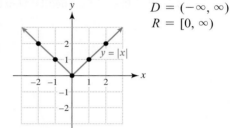

$D = (-\infty, \infty)$
$R = [0, \infty)$

Square Function: $f(x) = x^2$

x	-2	-1	0	1	2
$y = x^2$	4	1	0	1	4

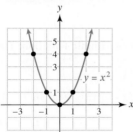

$D = (-\infty, \infty)$
$R = [0, \infty)$

Cube Function: $f(x) = x^3$

x	-2	-1	0	1	2
$y = x^3$	-8	-1	0	1	8

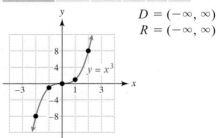

$D = (-\infty, \infty)$
$R = (-\infty, \infty)$

Square Root Function: $f(x) = \sqrt{x}$

x	0	1	4	9
$y = \sqrt{x}$	0	1	2	3

$D = [0, \infty)$
$R = [0, \infty)$

Cube Root Function: $f(x) = \sqrt[3]{x}$

x	-8	-1	0	1	8
$y = \sqrt[3]{x}$	-2	-1	0	1	2

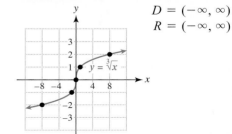

$D = (-\infty, \infty)$
$R = (-\infty, \infty)$

Greatest Integer Function: $f(x) = [x]$

x	-2.5	-1.5	0	1.5	2.5
$y = [x]$	-3	-2	0	1	2

$D = (-\infty, \infty)$

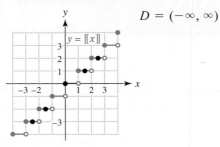

Reciprocal Function: $f(x) = \frac{1}{x}$

x	-2	-1	0	1	2
$y = \frac{1}{x}$	$-\frac{1}{2}$	-1	—	1	$\frac{1}{2}$

$D = (-\infty, 0) \cup (0, \infty)$
$R = (-\infty, 0) \cup (0, \infty)$

Base-2 Exponential Function: $f(x) = 2^x$

x	-2	-1	0	1	2
$y = 2^x$	$\frac{1}{4}$	$\frac{1}{2}$	1	2	4

$D = (-\infty, \infty)$
$R = (0, \infty)$

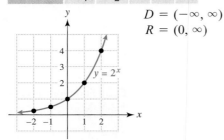

Natural Exponential Function: $f(x) = e^x$

x	-2	-1	0	1	2
$y = e^x$	e^{-2}	e^{-1}	1	e^1	e^2

$D = (-\infty, \infty)$
$R = (0, \infty)$

Common Logarithmic Function: $f(x) = \log x$

x	0.1	1	4	7	10
$y = \log x$	-1	0	$\log 4$	$\log 7$	1

$D = (0, \infty)$
$R = (-\infty, \infty)$

Natural Logarithmic Function: $f(x) = \ln x$

x	$\frac{1}{2}$	1	2	e	e^2
$y = \ln x$	$\ln \frac{1}{2}$	0	$\ln 2$	1	2

$D = (0, \infty)$
$R = (-\infty, \infty)$

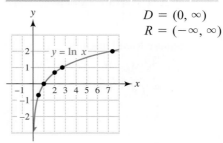

Sine Function: $f(x) = \sin x$

x	0	$\frac{\pi}{2}$	π	$\frac{3\pi}{2}$	2π
$y = \sin x$	0	1	0	-1	0

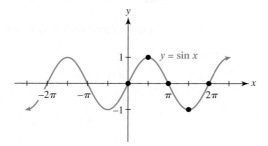

Cosine Function: $f(x) = \cos x$

x	0	$\frac{\pi}{2}$	π	$\frac{3\pi}{2}$	2π
$y = \cos x$	1	0	-1	0	1

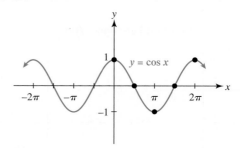

Tangent Function: $f(x) = \tan x$

x	$-\frac{\pi}{3}$	$-\frac{\pi}{4}$	0	$\frac{\pi}{4}$	$\frac{\pi}{3}$
$y = \tan x$	$-\sqrt{3}$	-1	0	1	$\sqrt{3}$

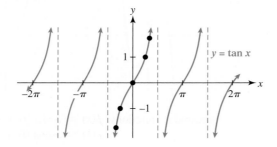

Cotangent Function: $f(x) = \cot x$

x	$\frac{\pi}{6}$	$\frac{\pi}{4}$	$\frac{\pi}{2}$	$\frac{3\pi}{4}$	$\frac{5\pi}{6}$
$y = \cot x$	$\sqrt{3}$	1	0	-1	$-\sqrt{3}$

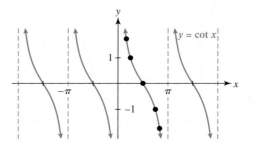

Cosecant Function: $f(x) = \csc x$

x	$\frac{\pi}{6}$	$\frac{\pi}{4}$	$\frac{\pi}{2}$	$\frac{3\pi}{4}$	$\frac{5\pi}{6}$
$y = \csc x$	2	$\sqrt{2}$	1	$\sqrt{2}$	2

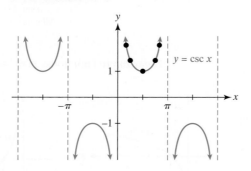

Secant Function: $f(x) = \sec x$

x	$-\frac{\pi}{3}$	$-\frac{\pi}{4}$	0	$\frac{\pi}{4}$	$\frac{\pi}{3}$
$y = \sec x$	2	$\sqrt{2}$	1	$\sqrt{2}$	2

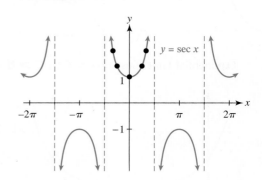

Families of Functions

This subsection shows the formulas and graphs of some families of functions, such as linear, quadratic, and exponential functions. Notice that the appearance of the graphs of these functions depends on the value of k, m, or a.

Constant Functions: $f(x) = k$

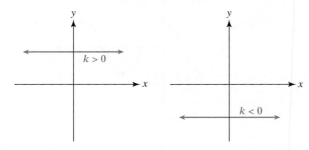

Linear Functions: $f(x) = mx + b$

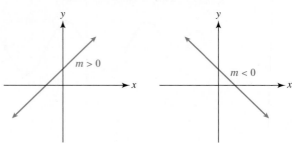

Quadratic Functions: $f(x) = ax^2 + bx + c$

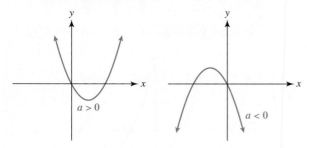

Cubic Functions: $f(x) = ax^3 + bx^2 + cx + d$

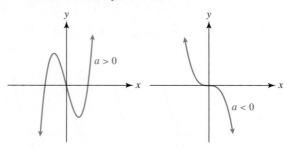

Power Functions: $f(x) = x^a$, $x > 0$

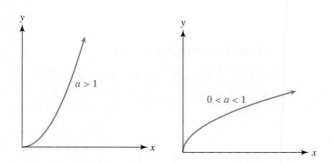

Sinusoidal Functions: $f(x) = a \sin(b(x - c)) + d$ or
$$f(x) = a \cos(b(x - c)) + d$$

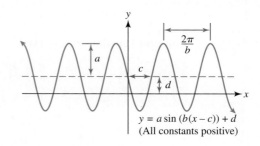

$y = a \sin(b(x - c)) + d$
(All constants positive)

Exponential Functions: $f(x) = Ca^x$, $C > 0$

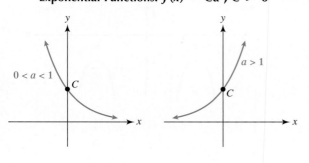

Logarithmic Functions: $f(x) = \log_a x$

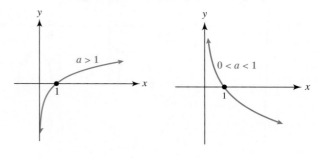

Appendix C: Partial Fractions

Decomposition of Rational Expressions

The sums of rational expressions are found by combining two or more rational expressions into one rational expression. Here, the reverse process is considered: Given one rational expression, express it as the sum of two or more rational expressions. A special type of sum of rational expressions is called the **partial fraction decomposition**; each term in the sum is a **partial fraction**. The technique of finding partial fraction decompositions is useful in calculus and can be accomplished by using the following steps.

Partial Fraction Decomposition of $\frac{f(x)}{g(x)}$

STEP 1: If $\frac{f(x)}{g(x)}$ is not a proper fraction (a fraction with the numerator of lower degree than the denominator), divide $f(x)$ by $g(x)$. For example,

$$\frac{x^4 - 3x^3 + x^2 + 5x}{x^2 + 3} = x^2 - 3x - 2 + \frac{14x + 6}{x^2 + 3}.$$

Then apply the following steps to the remainder, which is a proper fraction.

STEP 2: Factor $g(x)$ completely into factors of the form $(ax + b)^m$ or $(cx^2 + dx + e)^n$, where $cx^2 + dx + e$ is *irreducible* and m and n are positive integers.

STEP 3: **(a)** For each distinct linear factor $(ax + b)$, include in the decomposition the term

$$\frac{A}{ax + b}.$$

(b) For each repeated linear factor $(ax + b)^m$, include in the decomposition the terms

$$\frac{A_1}{ax + b} + \frac{A_2}{(ax + b)^2} + \cdots + \frac{A_m}{(ax + b)^m}.$$

STEP 4: **(a)** For each distinct quadratic factor $(cx^2 + dx + e)$, include in the decomposition the term

$$\frac{Bx + C}{cx^2 + dx + e}.$$

(b) For each repeated quadratic factor $(cx^2 + dx + e)^n$, include in the decomposition the terms

$$\frac{B_1x + C_1}{cx^2 + dx + e} + \frac{B_2x + C_2}{(cx^2 + dx + e)^2} + \cdots + \frac{B_nx + C_n}{(cx^2 + dx + e)^n}.$$

STEP 5: Use algebraic techniques to solve for the constants in the numerators.

To find the constants in Step 5, the goal is to get a system of equations with as many equations as there are unknowns in the numerators. One method for finding these equations is to substitute values for x on each side of the rational equation formed in Steps 3 and 4.

Distinct Linear Factors

EXAMPLE 1 Finding a partial fraction decomposition

Find the partial fraction decomposition of

$$\frac{2x^4 - 8x^2 + 5x - 2}{x^3 - 4x}.$$

SOLUTION The given fraction is not a proper fraction; the numerator has higher degree than the denominator. Perform the following division.

$$
\begin{array}{r}
2x \\
x^3 - 4x\overline{)2x^4 - 8x^2 + 5x - 2} \\
\underline{2x^4 - 8x^2} \\
5x - 2
\end{array}
$$

Algebra Review
To review clearing fractions from rational equations, see chapter R (page R-35).

The quotient is $\frac{2x^4 - 8x^2 + 5x - 2}{x^3 - 4x} = 2x + \frac{5x - 2}{x^3 - 4x}$. Now work with the remainder fraction. Factor the denominator as $x^3 - 4x = x(x + 2)(x - 2)$. Since the factors are distinct linear factors, use Step 3(a) to write the decomposition as

$$\frac{5x - 2}{x^3 - 4x} = \frac{A}{x} + \frac{B}{x + 2} + \frac{C}{x - 2},\qquad \text{Equation 1}$$

where A, B, and C are constants that need to be found. Multiply each side of equation 1 by $x(x + 2)(x - 2)$ to get

$$5x - 2 = A(x + 2)(x - 2) + Bx(x - 2) + Cx(x + 2).\qquad \text{Equation 2}$$

Equation 1 is an identity, since both sides represent the same rational expression. Thus equation 2 is also an identity. Equation 1 holds for all values of x except 0, -2, and 2. However, equation 2 holds for all values of x. In particular, substituting 0 for x in equation 2 gives $-2 = -4A$, so $A = \frac{1}{2}$. Similarly, choosing $x = -2$ gives $-12 = 8B$, so $B = -\frac{3}{2}$. Finally, choosing $x = 2$ gives $8 = 8C$, so $C = 1$. The remainder rational expression can be written as the sum of partial fractions

$$\frac{5x - 2}{x^3 - 4x} = \frac{1}{2x} + \frac{-3}{2(x + 2)} + \frac{1}{x - 2},$$

and the given rational expression can be written as

$$\frac{2x^4 - 8x^2 + 5x - 2}{x^3 - 4x} = 2x + \frac{1}{2x} + \frac{-3}{2(x + 2)} + \frac{1}{x - 2}.$$

Check the work by combining the terms on the right.   ***Now Try Exercise 11*** ◀

Repeated Linear Factors

EXAMPLE 2 Finding a partial fraction decomposition

Find the partial fraction decomposition of

$$\frac{2x}{(x - 1)^3}.$$

SOLUTION This is a proper fraction. The denominator is already factored with repeated linear factors. We write the decomposition as shown, using Step 3(b).

$$\frac{2x}{(x-1)^3} = \frac{A}{x-1} + \frac{B}{(x-1)^2} + \frac{C}{(x-1)^3}$$

We clear denominators by multiplying each side of this equation by $(x-1)^3$.

$$2x = A(x-1)^2 + B(x-1) + C$$

Substituting 1 for x leads to $C = 2$, so

$$2x = A(x-1)^2 + B(x-1) + 2. \qquad \text{Equation 1}$$

We found C and we still need to find values for A and B. *Any* number can be substituted for x. For example, when we choose $x = -1$ (because it is easy to substitute), equation 1 becomes

$$-2 = 4A - 2B + 2$$
$$-4 = 4A - 2B$$
$$-2 = 2A - B. \qquad \text{Equation 2}$$

Substituting 0 for x in equation 1 gives

$$0 = A - B + 2$$
$$2 = -A + B. \qquad \text{Equation 3}$$

Now we solve the system of equations 2 and 3 to get $A = 0$ and $B = 2$. The partial fraction decomposition is

$$\frac{2x}{(x-1)^3} = \frac{2}{(x-1)^2} + \frac{2}{(x-1)^3}.$$

We needed three substitutions because there were three constants to evaluate: A, B, and C. To check this result, we could combine the terms on the right.   **Now Try Exercise 13** ◀

Distinct Linear and Quadratic Factors

EXAMPLE 3   Finding a partial fraction decomposition

Find the partial fraction decomposition of

$$\frac{x^2 + 3x - 1}{(x+1)(x^2+2)}.$$

SOLUTION This denominator has distinct linear and quadratic factors, neither of which is repeated. Since $x^2 + 2$ cannot be factored, it is *irreducible*. The partial fraction decomposition is

$$\frac{x^2 + 3x - 1}{(x+1)(x^2+2)} = \frac{A}{x+1} + \frac{Bx + C}{x^2+2}.$$

Multiply each side by $(x+1)(x^2+2)$ to get

$$x^2 + 3x - 1 = A(x^2+2) + (Bx+C)(x+1). \qquad \text{Equation 1}$$

First substitute -1 for x to get

$$(-1)^2 + 3(-1) - 1 = A((-1)^2 + 2) + 0$$
$$-3 = 3A$$
$$A = -1.$$

Replace A with -1 in equation 1 and substitute any value for x. If $x = 0$, then

$$0^2 + 3(0) - 1 = -1(0^2 + 2) + (B \cdot 0 + C)(0 + 1)$$
$$-1 = -2 + C$$
$$C = 1.$$

Now, letting $A = -1$ and $C = 1$, substitute again in equation 1, using another number for x. For $x = 1$,

$$3 = -3 + (B + 1)(2)$$
$$6 = 2B + 2$$
$$B = 2.$$

With $A = -1$, $B = 2$, and $C = 1$, the partial fraction decomposition is

$$\frac{x^2 + 3x - 1}{(x + 1)(x^2 + 2)} = \frac{-1}{x + 1} + \frac{2x + 1}{x^2 + 2}.$$

This work can be checked by combining terms on the right.   **Now Try Exercise 21** ◀

For fractions with denominators that have quadratic factors, another method is often more convenient. The system of equations is formed by equating coefficients of like terms on each side of the partial fraction decomposition. For instance, in Example 3, after each side was multiplied by the common denominator, equation 1 was

$$x^2 + 3x - 1 = A(x^2 + 2) + (Bx + C)(x + 1). \qquad \text{Equation 1}$$

Multiplying on the right and collecting like terms, we have

$$x^2 + 3x - 1 = Ax^2 + 2A + Bx^2 + Bx + Cx + C$$
$$x^2 + 3x - 1 = (A + B)x^2 + (B + C)x + (C + 2A).$$

Now equating the coefficients of like powers of x gives three equations:

$$1 = A + B$$
$$3 = B + C$$
$$-1 = C + 2A.$$

Solving this system of equations for A, B, and C would give the partial fraction decomposition. The next example uses a combination of the two methods.

Repeated Quadratic Factors

EXAMPLE 4   Finding a partial fraction decomposition

Find the partial fraction decomposition of

$$\frac{2x}{(x^2 + 1)^2(x - 1)}.$$

SOLUTION This expression has both a linear factor and a repeated quadratic factor. By Steps 3(a) and 4(b),

$$\frac{2x}{(x^2 + 1)^2(x - 1)} = \frac{Ax + B}{x^2 + 1} + \frac{Cx + D}{(x^2 + 1)^2} + \frac{E}{x - 1}.$$

Multiplying each side by $(x^2 + 1)^2(x - 1)$ leads to

$$2x = (Ax + B)(x^2 + 1)(x - 1) + (Cx + D)(x - 1) + E(x^2 + 1)^2. \qquad \text{Equation 1}$$

If $x = 1$, equation 1 reduces to $2 = 4E$, or $E = \frac{1}{2}$. Substituting $\frac{1}{2}$ for E in equation 1 and combining terms on the right gives

$$2x = \left(A + \frac{1}{2}\right)x^4 + (-A + B)x^3 + (A - B + C + 1)x^2$$

$$+ (-A + B + D - C)x + \left(-B - D + \frac{1}{2}\right). \qquad \text{Equation 2}$$

To get additional equations involving the unknowns, equate the coefficients of like powers of x on each side of equation 2. Setting corresponding coefficients of x^4 equal gives $0 = A + \frac{1}{2}$, or $A = -\frac{1}{2}$. From the corresponding coefficients of x^3, $0 = -A + B$, which means that since $A = -\frac{1}{2}$, $B = -\frac{1}{2}$. From the coefficients of x^2, $0 = A - B + C + 1$. Since $A = -\frac{1}{2}$ and $B = -\frac{1}{2}$, it follows that $C = -1$. Finally, from the coefficients of x, $2 = -A + B + D - C$. Substituting for A, B, and C gives $D = 1$. With

$$A = -\frac{1}{2}, \quad B = -\frac{1}{2}, \quad C = -1, \quad D = 1, \quad \text{and} \quad E = \frac{1}{2},$$

the given fraction has the partial fraction decomposition

$$\frac{2x}{(x^2 + 1)^2(x - 1)} = \frac{-\frac{1}{2}x - \frac{1}{2}}{x^2 + 1} + \frac{-x + 1}{(x^2 + 1)^2} + \frac{\frac{1}{2}}{x - 1},$$

or

$$\frac{2x}{(x^2 + 1)^2(x - 1)} = \frac{-(x + 1)}{2(x^2 + 1)} + \frac{-x + 1}{(x^2 + 1)^2} + \frac{1}{2(x - 1)}.$$

Now Try Exercise 25 ◄

In summary, to solve for the constants in the numerators of a partial fraction decomposition, use either of the following methods or a combination of the two.

Techniques for Decomposition into Partial Fractions

Method 1 for Linear Factors

STEP 1: Multiply each side of the rational expression by the common denominator.

STEP 2: Substitute the zero of each factor in the resulting equation. For repeated linear factors, substitute as many other numbers as necessary to find all the constants in the numerators. The number of substitutions required will equal the number of constants.

Method 2 for Quadratic Factors

STEP 1: Multiply each side of the rational expression by the common denominator.

STEP 2: Collect terms on the right side of the resulting equation.

STEP 3: Equate the coefficients of like terms to get a system of equations.

STEP 4: Solve the system to find the constants in the numerators.

C   Exercises

Exercises 1–30: Find the partial fraction decomposition for the rational expression.

1. $\dfrac{5}{3x(2x + 1)}$

2. $\dfrac{3x - 1}{x(x + 1)}$

3. $\dfrac{4x + 2}{(x + 2)(2x - 1)}$

4. $\dfrac{x + 2}{(x + 1)(x - 1)}$

5. $\dfrac{x}{x^2 + 4x - 5}$

6. $\dfrac{5x - 3}{(x + 1)(x - 3)}$

7. $\dfrac{2x}{(x + 1)(x + 2)^2}$

8. $\dfrac{2}{x^2(x + 3)}$

9. $\dfrac{4}{x(1 - x)}$

10. $\dfrac{4x^2 - 4x^3}{x^2(1 - x)}$

11. $\dfrac{4x^2 - x - 15}{x(x + 1)(x - 1)}$

12. $\dfrac{2x + 1}{(x + 2)^3}$

13. $\dfrac{x^2}{x^2 + 2x + 1}$

14. $\dfrac{3}{x^2 + 4x + 3}$

15. $\dfrac{2x^5 + 3x^4 - 3x^3 - 2x^2 + x}{2x^2 + 5x + 2}$

16. $\dfrac{6x^5 + 7x^4 - x^2 + 2x}{3x^2 + 2x - 1}$

17. $\dfrac{x^3 + 4}{9x^3 - 4x}$

18. $\dfrac{x^3 + 2}{x^3 - 3x^2 + 2x}$

19. $\dfrac{-3}{x^2(x^2 + 5)}$

20. $\dfrac{2x + 1}{(x + 1)(x^2 + 2)}$

21. $\dfrac{3x - 2}{(x + 4)(3x^2 + 1)}$

22. $\dfrac{3}{x(x + 1)(x^2 + 1)}$

23. $\dfrac{1}{x(2x + 1)(3x^2 + 4)}$

24. $\dfrac{x^4 + 1}{x(x^2 + 1)^2}$

25. $\dfrac{3x - 1}{x(2x^2 + 1)^2}$

26. $\dfrac{3x^4 + x^3 + 5x^2 - x + 4}{(x - 1)(x^2 + 1)^2}$

27. $\dfrac{-x^4 - 8x^2 + 3x - 10}{(x + 2)(x^2 + 4)^2}$

28. $\dfrac{x^2}{x^4 - 1}$

29. $\dfrac{5x^5 + 10x^4 - 15x^3 + 4x^2 + 13x - 9}{x^3 + 2x^2 - 3x}$

30. $\dfrac{3x^6 + 3x^4 + 3x}{x^4 + x^2}$

Appendix D: Rotation of Axes

Derivation of Rotation Equations

If we begin with an *xy*-coordinate system having origin *O* and rotate the axes about *O* through an angle θ, the new coordinate system is called a **rotation** of the *xy*-system. Trigonometric identities can be used to obtain equations for converting the coordinates of a point from the *xy*-system to the rotated *x′y′*-system. Let *P* be any point other than the origin, with coordinates (x, y) in the *xy*-system and (x', y') in the *x′y′*-system. See Figure D.1. Let $OP = r$, and let α represent the angle made by *OP* and the *x′*-axis. As shown in Figure D.1,

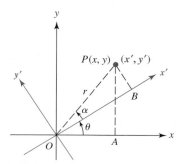

Figure D.1

$$\cos(\theta + \alpha) = \frac{OA}{r} = \frac{x}{r}, \quad \sin(\theta + \alpha) = \frac{AP}{r} = \frac{y}{r},$$

$$\cos\alpha = \frac{OB}{r} = \frac{x'}{r}, \quad \sin\alpha = \frac{BP}{r} = \frac{y'}{r}.$$

These four statements can be rewritten as

$$x = r\cos(\theta + \alpha), \quad y = r\sin(\theta + \alpha), \quad x' = r\cos\alpha, \quad y' = r\sin\alpha.$$

Using the trigonometric identity for the cosine of the sum of two angles gives

$$x = r\cos(\theta + \alpha)$$
$$= r(\cos\theta\cos\alpha - \sin\theta\sin\alpha)$$
$$= (r\cos\alpha)\cos\theta - (r\sin\alpha)\sin\theta$$
$$= x'\cos\theta - y'\sin\theta.$$

Using the identity for the sine of the sum of two angles in the same way gives $y = x'\sin\theta + y'\cos\theta$. This proves the following result.

Rotation Equations

If the rectangular coordinate axes are rotated about the origin through an angle θ and if the coordinates of a point *P* are (x, y) and (x', y') with respect to the *xy*-system and the *x′y′*-system, respectively, then the **rotation equations** are

$$x = x'\cos\theta - y'\sin\theta \quad \text{and} \quad y = x'\sin\theta + y'\cos\theta.$$

Applying a Rotation Equation

EXAMPLE 1   Finding an equation after a rotation

The equation of a curve is $x^2 + y^2 + 2xy + 2\sqrt{2}x - 2\sqrt{2}y = 0$. Find the resulting equation if the axes are rotated 45°. Graph the equation.

SOLUTION If $\theta = 45°$, then $\sin \theta = \frac{\sqrt{2}}{2}$ and $\cos \theta = \frac{\sqrt{2}}{2}$, and the rotation equations become

$$x = \frac{\sqrt{2}}{2}x' - \frac{\sqrt{2}}{2}y' \quad \text{and} \quad y = \frac{\sqrt{2}}{2}x' + \frac{\sqrt{2}}{2}y'.$$

Substituting these values into the given equation yields

$$x^2 + y^2 + 2xy + 2\sqrt{2}x - 2\sqrt{2}y = 0$$

$$\left(\frac{\sqrt{2}}{2}x' - \frac{\sqrt{2}}{2}y'\right)^2 + \left(\frac{\sqrt{2}}{2}x' + \frac{\sqrt{2}}{2}y'\right)^2$$

$$+ 2\left(\frac{\sqrt{2}}{2}x' - \frac{\sqrt{2}}{2}y'\right)\left(\frac{\sqrt{2}}{2}x' + \frac{\sqrt{2}}{2}y'\right)$$

$$+ 2\sqrt{2}\left(\frac{\sqrt{2}}{2}x' - \frac{\sqrt{2}}{2}y'\right) - 2\sqrt{2}\left(\frac{\sqrt{2}}{2}x' + \frac{\sqrt{2}}{2}y'\right) = 0.$$

Expanding these terms yields

$$\frac{1}{2}x'^2 - x'y' + \frac{1}{2}y'^2 + \frac{1}{2}x'^2 + x'y' + \frac{1}{2}y'^2 + x'^2 - y'^2$$

$$+ 2x' - 2y' - 2x' - 2y' = 0.$$

Collecting terms gives

$$2x'^2 - 4y' = 0$$

$$x'^2 - 2y' = 0 \qquad \text{Divide by 2.}$$

or, finally,

$$x'^2 = 2y',$$

the equation of a parabola. The graph is shown in Figure D.2.   ***Now Try Exercise 13*** ◀

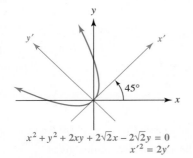

$$x^2 + y^2 + 2xy + 2\sqrt{2}x - 2\sqrt{2}y = 0$$
$$x'^2 = 2y'$$

Figure D.2

We have graphed equations written in the form $Ax^2 + Cy^2 + Dx + Ey + F = 0$. As we saw in the preceding example, the rotation of axes eliminated the xy-term. Thus, to graph by hand an equation that has an xy-term, it is necessary to find an appropriate **angle of rotation** to eliminate the xy-term. The necessary angle of rotation can be determined by using the following result. The proof is quite lengthy and is not presented here.

Angle of Rotation

The xy-term is removed from the general equation

$$Ax^2 + Bxy + Cy^2 + Dx + Ey + F = 0$$

by a rotation of the axes through an angle θ, $0° < \theta < 90°$, where

$$\cot 2\theta = \frac{A - C}{B}.$$

This result can be used to find the appropriate angle of rotation, θ. To find the rotation equations, first find $\sin \theta$ and $\cos \theta$. The following example illustrates a way to obtain $\sin \theta$ and $\cos \theta$ from $\cot 2\theta$ without first identifying the angle θ.

EXAMPLE 2 Rotating and graphing

Rotate the axes and graph $52x^2 - 72xy + 73y^2 = 200$.

Figure D.3

SOLUTION Here $A = 52$, $B = -72$, and $C = 73$. By substitution,

$$\cot 2\theta = \frac{52 - 73}{-72} = \frac{-21}{-72} = \frac{7}{24}.$$

To find $\sin \theta$ and $\cos \theta$, use the trigonometric identities

$$\sin \theta = \sqrt{\frac{1 - \cos 2\theta}{2}} \quad \text{and} \quad \cos \theta = \sqrt{\frac{1 + \cos 2\theta}{2}}.$$

Sketch a right triangle and label it as in Figure D.3, to see that $\cos 2\theta = \frac{7}{25}$. (Recall that in the two quadrants with which we are concerned, $0° \leq 2\theta \leq 180°$, cosine and cotangent have the same sign.) Then

$$\sin \theta = \sqrt{\frac{1 - \frac{7}{25}}{2}} = \sqrt{\frac{9}{25}} = \frac{3}{5} \quad \text{and} \quad \cos \theta = \sqrt{\frac{1 + \frac{7}{25}}{2}} = \sqrt{\frac{16}{25}} = \frac{4}{5}.$$

Use these values for $\sin \theta$ and $\cos \theta$ to obtain

$$x = \frac{4}{5}x' - \frac{3}{5}y' \quad \text{and} \quad y = \frac{3}{5}x' + \frac{4}{5}y'.$$

Substituting these expressions for x and y into the original equation yields

$$52\left(\frac{4}{5}x' - \frac{3}{5}y'\right)^2 - 72\left(\frac{4}{5}x' - \frac{3}{5}y'\right)\left(\frac{3}{5}x' + \frac{4}{5}y'\right) + 73\left(\frac{3}{5}x' + \frac{4}{5}y'\right)^2 = 200.$$

This becomes

$$52\left(\frac{16}{25}x'^2 - \frac{24}{25}x'y' + \frac{9}{25}y'^2\right) - 72\left(\frac{12}{25}x'^2 + \frac{7}{25}x'y' - \frac{12}{25}y'^2\right)$$

$$+ 73\left(\frac{9}{25}x'^2 + \frac{24}{25}x'y' + \frac{16}{25}y'^2\right) = 200.$$

Combining terms gives

$$25x'^2 + 100y'^2 = 200.$$

Divide each side by 200 to get

$$\frac{x'^2}{8} + \frac{y'^2}{2} = 1,$$

an equation of an ellipse having x'-intercepts $\pm\sqrt{8}$ and y'-intercepts $\pm\sqrt{2}$. The graph is shown in Figure D.4. To find θ, use the fact that

$$\frac{\sin \theta}{\cos \theta} = \frac{\frac{3}{5}}{\frac{4}{5}} = \frac{3}{4} = \tan \theta,$$

from which $\theta = \tan^{-1}\frac{3}{4} \approx 37°$.

Now Try Exercise 17 ◀

$\sin \theta = \frac{3}{5}$
$\cos \theta = \frac{4}{5}$
$\tan \theta = \frac{3}{4}$

$52x^2 - 72xy + 73y^2 = 200$
$\frac{x'^2}{8} + \frac{y'^2}{2} = 1$

Figure D.4

Summary of Conics with an xy-Term

The following summary enables us to use the general equation to decide on the type of graph to expect.

Equations of Conics with xy-Term

If the general second-degree equation

$$Ax^2 + Bxy + Cy^2 + Dx + Ey + F = 0$$

has a graph, it will be one of the following:

(a) a circle or an ellipse (or a point) if $B^2 - 4AC < 0$;
(b) a parabola (or one line or two parallel lines) if $B^2 - 4AC = 0$;
(c) a hyperbola (or two intersecting lines) if $B^2 - 4AC > 0$;
(d) a straight line if $A = B = C = 0$ and $D \neq 0$ or $E \neq 0$.

D   Exercises

Exercises 1–6: Use the summary in this section to predict the graph of the second-degree equation.

1. $4x^2 + 3y^2 + 2xy - 5x = 8$

2. $x^2 + 2xy - 3y^2 + 2y = 12$

3. $2x^2 + 3xy - 4y^2 = 0$

4. $x^2 - 2xy + y^2 + 4x - 8y = 0$

5. $4x^2 + 4xy + y^2 + 15 = 0$

6. $-x^2 + 2xy - y^2 + 16 = 0$

Exercises 7–12: Find the angle of rotation θ that will remove the xy-term in the equation.

7. $2x^2 + \sqrt{3}xy + y^2 + x = 5$

8. $4\sqrt{3}x^2 + xy + 3\sqrt{3}y^2 = 10$

9. $3x^2 + \sqrt{3}xy + 4y^2 + 2x - 3y = 12$

10. $4x^2 + 2xy + 2y^2 + x - 7 = 0$

11. $x^2 - 4xy + 5y^2 = 18$

12. $3\sqrt{3}x^2 - 2xy + \sqrt{3}y^2 = 25$

Exercises 13–16: Use the given angle of rotation to remove the xy-term and graph the equation.

13. $x^2 - xy + y^2 = 6; \theta = 45°$

14. $2x^2 - xy + 2y^2 = 25; \theta = 45°$

15. $8x^2 - 4xy + 5y^2 = 36; \sin\theta = \dfrac{2}{\sqrt{5}}$

16. $5y^2 + 12xy = 10; \sin\theta = \dfrac{3}{\sqrt{13}}$

Exercises 17–24: Remove the xy-term from the equation by performing a suitable rotation. Graph the equation.

17. $3x^2 - 2xy + 3y^2 = 8$ **18.** $x^2 + xy + y^2 = 3$

19. $x^2 - 4xy + y^2 = -5$

20. $x^2 + 2xy + y^2 + 4\sqrt{2}x - 4\sqrt{2}y = 0$

21. $7x^2 + 6\sqrt{3}xy + 13y^2 = 64$

22. $7x^2 + 2\sqrt{3}xy + 5y^2 = 24$

23. $3x^2 - 2\sqrt{3}xy + y^2 - 2x - 2\sqrt{3}y = 0$

24. $2x^2 + 2\sqrt{3}xy + 4y^2 = 5$

Exercises 25–30: In the equation, remove the xy-term by rotation. Then translate the axes and sketch the graph.

25. $x^2 + 3xy + y^2 - 5\sqrt{2}y = 15$

26. $x^2 - \sqrt{3}xy + 2\sqrt{3}x - 3y - 3 = 0$

27. $4x^2 + 4xy + y^2 - 24x + 38y - 19 = 0$

28. $12x^2 + 24xy + 19y^2 - 12x - 40y + 31 = 0$

29. $16x^2 + 24xy + 9y^2 - 130x + 90y = 0$

30. $9x^2 - 6xy + y^2 - 12\sqrt{10}x - 36\sqrt{10}y = 0$

Bibliography

Acker, A., and C. Jaschek. *Astronomical Methods and Calculations.* New York: John Wiley and Sons, 1986.

Battan, L. *Weather in Your Life.* San Francisco: W. H. Freeman, 1983.

Beckmann, P. *A History of Pi.* New York: Barnes and Noble, 1993.

Benade, A. *Fundamentals of Musical Acoustics.* New York: Oxford University Press, 1976.

Beni, G., and S. Hackwood. *Recent Advances in Robotics.* New York: John Wiley and Sons, 1985.

Beranek, L. *Noise and Vibration Control.* Washington, D.C.: Institute of Noise Control Engineering, 1988.

Brase, C. H., and C. P. Brase. *Understandable Statistics.* 5th ed. Lexington, Mass.: D. C. Heath, 1995.

Brearley, J., and A. Nicholas. *This Is the Bichon Frise.* Hong Kong: TFH Publication, 1973.

Brown, D., and P. Rothery. *Models in Biology: Mathematics, Statistics and Computing.* West Sussex, England: John Wiley and Sons, 1993.

Brown, F., J. Hastings, and J. Palmer. *The Biological Clock.* New York: Academic Press, 1970.

Carlson, T. "Über Geschwindigkeit und Grösse der Hefevermehrung in Würze." *Biochem. A.* 57:313–334.

Cheney, W., and D. Kincaid. *Numerical Mathematics and Computing.* 3rd ed. Pacific Grove, Calif.: Brooks/Cole Publishing Company, 1994.

Clime, W. *The Economics of Global Warming.* Washington, D.C.: Institute for International Economics, 1992.

Cole, F. *Introduction to Meteorology.* New York: Wiley, 1980.

Cooper, J., and R. Glassow. *Kinesiology.* 2nd ed. St. Louis: The C. V. Mosby Company, 1968.

Cotton, W., and R. Pielke. *Human Impacts on Weather and Climate.* Geophysical Science Series, vol. 2. Fort Collins, Colo.: *ASTeR Press, 1992.

Craig, J. *Introduction to Robotics: Mechanics and Control.* Reading, Mass.: Addison-Wesley Publishing Company, 1989.

Crownover, R. *Introduction to Fractals and Chaos.* Boston: Jones and Bartlett, 1995.

Duffet-Smith, P. *Practical Astronomy with Your Calculator.* New York: Cambridge University Press, 1988.

Eves, H. *An Introduction to the History of Mathematics.* 5th ed. Philadelphia: Saunders College Publishing, 1983.

Fletcher, N., and T. Rossing. *The Physics of Musical Instruments.* New York: Springer-Verlag, 1991.

Foley, J., A. van Dam, S. Feiner, J. Hughes, and R. Phillips. *Introduction to Computer Graphics.* Reading, Mass.: Addison-Wesley Publishing Company, 1994.

Foster, R., and J. Bates. "Use of mussels to monitor point source industrial discharges." *Environ. Sci. Technol.* 12:958–962.

Frank, A. "Plug-in Hybrid Vehicles for a Sustainable Future." *American Scientist*, March–April 2007.

Freebury, H. *A History of Mathematics.* New York: Macmillan, 1961.

Freedman, B. *Environmental Ecology: The Ecological Effects of Pollution, Disturbance, and Other Stresses.* 2nd ed. San Diego: Academic Press, 1995.

Friedhoff, M., and W. Benzon. *The Second Computer Revolution: Visualization.* New York: W. H. Freeman, 1991.

Garber, N., and L. Hoel. *Traffic and Highway Engineering.* Boston, Mass.: PWS Publishing Co., 1997.

Good, I. J. "What is the most amazing approximate integer in the universe?" *Pi Mu Epsilon Journal* 5 (1972): 314–315.

Grigg, D. *The World Food Problem.* Oxford: Blackwell Publishers, 1993.

Haber-Schaim, U., J. Cross, G. Abegg, J. Dodge, and J. Walter. *Introductory Physical Science.* Englewood Cliffs, N.J.: Prentice Hall, 1972.

Haefner, L. *Introduction to Transportation Systems.* New York: Holt, Rinehart and Winston, 1986.

Harker, J. *The Physiology of Diurnal Rhythms.* New York: Cambridge University Press, 1964.

Harrison, F., F. Hills, J. Paterson, and R. Saunders. "The measurement of liver blood flow in conscious calves." *Quarterly Journal of Experimental Physiology* 71:235–247.

Hartman, D. *Global Physical Climatology.* San Diego: Academic Press, 1994.

Heinz-Otto, P., H. Jürgens, and D. Saupe. *Chaos and Fractals: New Frontiers in Science.* New York: Springer-Verlag, 1993.

Hibbeler, R. *Structural Analysis.* Englewood Cliffs, N.J.: Prentice-Hall, 1995.

Hill, F. *Computer Graphics.* New York: Macmillan Publishing Company, 1990.

Hines, A., T. Ghosh, S. Loyalka, and R. Warder, Jr. *Indoor Air Quality and Control.* Englewood Cliffs, N.J.: Prentice-Hall, 1993.

Hoggar, S. *Mathematics for Computer Graphics.* New York: Cambridge University Press, 1993.

Hoppensteadt, F., and C. Peskin. *Mathematics in Medicine and the Life Sciences.* New York: Springer-Verlag, 1992.

Hosmer, D., and S. Lemeshow. *Applied Logistic Regression.* New York: John Wiley and Sons, 1989.

Howells, G. *Acid Rain and Acid Waters.* 2nd ed. New York: Ellis Horwood, 1995.

Huffman, R. *Atmospheric Ultraviolet Remote Sensing.* San Diego: Academic Press, 1992.

Huxley, J. *Problems of Relative Growth.* London: Methuen and Co., 1932.

Jarrett, J. *Business Forecasting Methods.* Oxford: Basil Blackwell, 1991.

Karttunen, H., P. Kroger, H. Oja, M. Poutanen, and K. Donner, eds. *Fundamental Astronomy.* 2nd ed. New York: Springer-Verlag, 1994.

Kerlow, I. *The Art of 3-D Computer Animation and Imaging.* New York: Van Nostrand Riehold, 1996.

Kissam, P. *Surveying Practice.* 3rd ed. New York: McGraw-Hill, 1978.

Kraljic, M. *The Greenhouse Effect.* New York: The H. W. Wilson Company, 1992.

Kress, S. *Bird Life—A Guide to the Behavior and Biology of Birds.* Racine, Wisc.: Western Publishing Company, 1991.

Lack, D. *The Life of a Robin.* London: Collins, 1965.

Lancaster, H. *Quantitative Methods in Biological and Medical Sciences: A Historical Essay.* New York: Springer-Verlag, 1994.

Loh, W. *Dynamics and Thermodynamics of Planetary Entry.* Englewood Cliffs, N.J.: Prentice-Hall, 1963.

Makridakis, S., and S. Wheelwright. *Forecasting Methods for Management.* New York: John Wiley and Sons, 1989.

Mandelbrot, B. *The Fractal Geometry of Nature.* New York: W. H. Freeman Company, 1982.

Mannering, F., and W. Kilareski. *Principles of Highway Engineering and Traffic Analysis.* New York: John Wiley and Sons, 1990.

Mar, J., and H. Liebowitz. *Structure Technology for Large Radio and Radar Telescope Systems.* Cambridge, Mass.: The MIT Press, 1969.

Mason, C. *Biology of Freshwater Pollution.* New York: Longman and Scientific and Technical, John Wiley and Sons, 1991.

Meeus, J. *Astronomical Algorithms.* Richmond, Va.: Willman-Bell, 1991.

Mehrotra, A. *Cellular Radio: Analog and Digital Systems.* Boston: Artech House, 1994.

Metcalf, H. *Topics in Classical Biophysics.* Englewood Cliffs, N.J.: Prentice-Hall, 1980.

Miller, A., and J. Thompson. *Elements of Meteorology.* 2nd ed. Columbus, Ohio: Charles E. Merrill Publishing Company, 1975.

Miller, A., and R. Anthes. *Meteorology.* 5th ed. Columbus, Ohio: Charles E. Merrill Publishing Company, 1985.

Moffitt, F. *Photogrammetry.* Scranton, Pa.: International Textbook Company, 1967.

Motz, L., and J. Weaver. *The Story of Mathematics.* New York: Plenum Press, 1993.

Mueller, I., and K. Ramsayer. *Introduction to Surveying.* New York: Frederick Ungar Publishing Company, 1979.

Murray, R., S. Sastry, and L. Zexiang. *A Mathematical Introduction to Robotic Manipulation.* Boca Raton, FL: CRC Press, Inc., 1994.

National Council of Teachers of Mathematics. *Historical Topics for the Mathematics Classroom, Thirty-first Yearbook*, 1969.

Navarra, J. *Atmosphere, Weather and Climate.* Philadelphia: W. B. Saunders, 1979.

Nilsson, A. *Greenhouse Earth.* New York: John Wiley and Sons, 1992.

Paetsch, M. *Mobile Communications in the U.S. and Europe: Regulation, Technology, and Markets.* Norwood, Mass.: Artech House, 1993.

Payne, A. "Responses of the three test Algae of the algal assay procedure: bottle test." *Water Res.* 9: 437–445.

Pennycuick, C. *Newton Rules Biology.* New York: Oxford University Press, 1992.

Pielou, E. *Population and Community Ecology: Principles and Methods.* New York: Gordon and Breach Science Publishers, 1974.

Pierce, J. *The Science of Musical Sound.* New York: W. H. Freeman, 1992.

Pokorny, C., and C. Gerald. *Computer Graphics: The Principles behind the Art and Science.* Irvine, Calif.: Franklin, Beedle, and Associates, 1989.

Raggett, G. "Modeling the Eyam plague." *The Institute of Mathematics and Its Applications* 18: 221–226.

Resnikoff, H., and R. Wells, Jr. *Mathematics in Civilization.* New York: Dover Publications, Inc., 1984.

Riley, W., L. Sturges, and D. Morris. *Statics and Mechanics of Materials: An Integrated Approach.* New York: John Wiley and Sons, Inc., 1995.

Rist, Curtis. "The Physics of Foul Shots." *Discover*, October 2000.

Robert Wood Johnson Foundation. *Chronic Care in America: A 21st Century Challenge*, 1996.

Rodricks, J. *Calculated Risk.* New York: Cambridge University Press, 1992.

Roederer, J. *Introduction to the Physics and Psychophysics of Music.* New York: Springer-Verlag, 1973.

Rogers, E., and T. Kostigen. *The Green Book.* New York: Random House, 2007.

Ronan, C. *The Natural History of the Universe.* New York: MacMillan Publishing Company, 1991.

Ryan, B., B. Joiner, and T. Ryan. *Minitab Handbook.* Boston: Duxbury Press, 1985.

Sanders, D. *Statistics: A First Course.* 5th ed. New York: McGraw-Hill, 1995.

Schlosser, W. *Challenges of Astronomy.* New York: Springer-Verlag, 1991.

Semat, H., and J. Albright. *Introduction to Atomic and Nuclear Physics.* Austin, Tex.: Holt, Rinehart and Winston, 1972.

Sharov, A., and I. Novikov. *Edwin Hubble: The Discoverer of the Big Bang Universe.* New York: Cambridge University Press, 1993.

Sinkov, A. *Elementary Cryptanalysis: A Mathematical Approach.* New York: Random House, 1968.

Smith, R., and R. Dorf. *Circuits, Devices and Systems.* 5th ed. New York: John Wiley and Sons, Inc., 1992.

Socolow, R., and S. Pacala. "A Plan to Keep Carbon in Check." *Scientific American,* September 2006.

Stadler, W. *Analytical Robotics and Mechatronics.* New York: McGraw-Hill, Inc., 1995.

Stent, G. S. *Molecular Biology of Bacterial Viruses.* San Francisco: W. H. Freeman, 1963.

Stewart, J. *Essentials of Calculus.* Pacific Grove, Calif.: Books/Cole Publishing, 2007.

Taylor, W. *The Geometry of Computer Graphics.* Pacific Grove, Calif.: Wadsworth and Brooks/Cole, 1992.

Teutsch, S., and R. Churchill. *Principles and Practice of Public Health Surveillance.* New York: Oxford University Press, 1994.

Thomas, D. *Swimming Pool Operators Handbook.* Washington, D.C.: National Swimming Pool Foundation, 1972.

Thomas, R. *The Old Farmer's 2008 Almanac.* Dublin, N.H.: The Old Farmer's Almanac, 2007.

Thomas, V. *Science and Sport.* London: Faber and Faber, 1970.

Thomson, W. *Introduction to Space Dynamics.* New York: John Wiley and Sons, 1961.

Triola, M. *Elementary Statistics.* Pearson Education, 2004.

Tucker, A., A. Bernat, W. Bradley, R. Cupper, and G. Scragg. *Fundamentals of Computing 1. Logic: Problem Solving, Programs, and Computers.* New York: McGraw-Hill, 1995.

Turner, R. K., D. Pierce, and I. Bateman. *Environmental Economics: An Elementary Approach.* Baltimore: The Johns Hopkins University Press, 1993.

Van Sickle, J. *GPS for Land Surveyors.* Chelsea, Mich.: Ann Arbor Press, 1996.

Varley, G., and G. Gradwell. "Population models for the winter moth." *Symposium of the Royal Entomological Society of London* 4:132–142.

Walker, A. *Observation and Inference: An Introduction to the Methods of Epidemiology.* Newton Lower Falls, Mass.: Epidemiology Resources, 1991.

Wang, Z. "Self-Powered Nanotech." *Scientific American,* January 2008.

Watt, A. *3D Computer Graphics.* Reading, Mass.: Addison-Wesley Publishing Company, 1993.

Webb, T. *Celestial Objects for Common Telescopes.* New York: Dover Publications Inc., 1962.

Weidner, R., and R. Sells. *Elementary Classical Physics,* vol. 2. Boston: Allyn and Bacon, 1965.

Wilcox, G., and C. Hesselberth. *Electricity for Engineering Technology.* Boston: Allyn and Bacon, 1970.

Williams, J. *The Weather Almanac 1995.* New York: Vintage Books, 1994.

Winter, C. *Solar Power Plants.* New York: Springer-Verlag, 1991.

Wolff, R., and L. Yaeger. *Visualization of Natural Phenomena.* New York: Springer-Verlag, 1993.

Wuebbles, D., and J. Edmonds. *Primer on Greenhouse Gases.* Chelsea, Mich.: Lewis Publishers, 1991.

Zeilik, M., S. Gregory, and D. Smith. *Introductory Astronomy and Astrophysics.* 3rd ed. Philadelphia: Saunders College Publishers, 1992.

Zhao, Y. *Vehicle Location and Navigation Systems.* Boston, Mass.: Artech House, 1997.

Answers to Selected Exercises

CHAPTER 1: Introduction to Functions and Graphs

SECTION 1.1 (pp. 9–12)

1. Rational number, real number   **3.** Rational number, real number   **5.** Real number
7. Natural number: $\sqrt{9}$; integers: $-3, \sqrt{9}$; rational numbers: $-3, \frac{2}{9}, \sqrt{9}, 1.\overline{3}$; irrational numbers: $\pi, -\sqrt{2}$
9. Natural numbers: none; integer: $-\sqrt{4}$; rational numbers: $\frac{1}{3}, 5.1 \times 10^{-6}, -2.33, 0.\overline{7}, -\sqrt{4}$; irrational number: $\sqrt{13}$
11. Rational numbers   **13.** Rational numbers
15. Integers   **17.** 51   **19.** -84 **21.** 0
23. 5   **25.** 8   **27.** -32 **29.** 1.848×10^5
31. 4.361×10^{-2} **33.** 2.45×10^3
35. 5.6×10^{-1} **37.** -8.7×10^{-3}
39. 2.068×10^2 **41.** 0.000001   **43.** 200,000,000
45. 156.7   **47.** 500,000   **49.** 4500   **51.** 67,000
53. 8×10^8; 800,000,000   **55.** 3.5×10^{-1}; 0.35
57. 2.1×10^{-3}; 0.0021   **59.** 5×10^{-3}; 0.005
61. 4.24×10^{19} **63.** 8.72×10^4 **65.** 7.67×10^{11}
67. 5.769   **69.** 0.058   **71.** 0.419   **73.** -1.235
75. 15.819   **77.** 62.5%   **79.** -39.3%
81. \$106 per credit; no, it is \$99.64.
83. 1.4×10^{-1} watt
85. About 53,794 miles per hour
87. (a) Approximately \$1820   **(b)** About \$19,715
89. 2.9×10^{-4} cm
91. (a) 18,466,667 ft   **(b)** Yes
93. (a) $7.436\pi \approx 23.4$ in^3 **(b)** Yes
95. 0.25 foot, or 3 inches

SECTION 1.2 (pp. 24–27)

1. (a)
(b) Max: 6; min: -2 **(c)** Mean: $\frac{11}{6} = 1.8\overline{3}$
3. (a)
(b) Max: 30; min: -20 **(c)** Mean: 5
5.

-30	-30	-10	5	15	25	45	55	61

(a) Max: 61; min: -30
(b) Mean: $\frac{136}{9} \approx 15.11$; median: 15

7. $\sqrt{15} \approx 3.87$, $2^{2.3} \approx 4.92$, $\sqrt[3]{69} \approx 4.102$, $\pi^2 \approx 9.87$, $2^\pi \approx 8.82$, 4.1

$\sqrt{15}$	4.1	$\sqrt[3]{69}$	$2^{2.3}$	2^π	π^2

(a) Max: π^2; min: $\sqrt{15}$ **(b)** Mean: 5.95; median: 4.51
9. (a)
(b) Mean: 23.5; median: 23.95
The average area of the six largest lakes is 23,500 square miles. Half of these lakes have areas below 23,950 square miles and half have areas above.   **(c)** Lake Superior
11. 16, 18, 26; no   **13.** 5   **15.** $\sqrt{29} \approx 5.39$
17. $\sqrt{41.49} \approx 6.44$ **19.** 8
21. $\frac{\sqrt{17}}{4} \approx 1.03$ **23.** $\frac{\sqrt{2}}{2} \approx 0.71$ **25.** 130
27. $\sqrt{a^2 + b^2}$ **29.** Yes
31. (a)

(b) $d = \sqrt{4100} \approx 64.0$ miles
33. 78.45 years   **35.** 10 seconds   **37.** $(3, -0.5)$
39. $(10, 10)$ **41.** $(-2.1, -0.35)$ **43.** $(\sqrt{2}, 0)$
45. $(0, 2b)$
47. (a) $S = \{(-1, 5), (2, 2), (3, -1), (5, -4), (9, -5)\}$
(b) $D = \{-1, 2, 3, 5, 9\}$, $R = \{-5, -4, -1, 2, 5\}$
49. (a) $S = \{(1, 5), (4, 5), (5, 6), (4, 6), (1, 5)\}$
(b) $D = \{1, 4, 5\}$, $R = \{5, 6\}$
51. (a) $D = \{-3, -2, 0, 7\}$, $R = \{-5, -3, 0, 4, 5\}$
(b) x-min: -3; x-max: 7; y-min: -5; y-max: 5
(c) & (d)

53. (a) $D = \{-4, -3, -1, 0, 2\}$, $R = \{-2, -1, 1, 2, 3\}$
(b) x-min: -4; x-max: 2; y-min: -2; y-max: 3
(c) & (d)

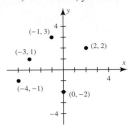

55. (a) $D = \{-35, -25, 0, 10, 75\}$,
$R = \{-55, -25, 25, 45, 50\}$
(b) x-min: -35; x-max: 75; y-min: -55; y-max: 50
(c) & (d)

57. Scatterplot   Line graph

59. Center: $(0, 0)$; radius: 5   **61.** Center: $(0, 0)$; radius: $\sqrt{7}$
63. Center: $(2, -3)$; radius: 3
65. Center: $(0, -1)$; radius: 10
67. $(x - 1)^2 + (y + 2)^2 = 1$
69. $(x + 2)^2 + (y - 1)^2 = 4$
71. $(x - 3)^2 + (y + 5)^2 = 64$
73. $(x - 3)^2 + y^2 = 49$ **75.** $x^2 + y^2 = 10$
77. $(x + 2)^2 + (y + 3)^2 = 25$
79. x-axis: 10; y-axis: 10   **81.** x-axis: 10; y-axis: 5

83. x-axis: 16; y-axis: 5

85. b   **87.** a
89. $[-5, 5, 1]$ by $[-5, 5, 1]$ **91.** $[-100, 100, 10]$ by $[-100, 100, 10]$

93. (a) x-min: 2005; x-max: 2008; y-min: 19.7; y-max: 25.0
(b) $[2003, 2010, 1]$ by $[15, 30, 5]$ (answers may vary)
(c)   (d)

95. (a) x-min: 1950; x-max: 2000; y-min: 1.7; y-max: 5.5
(b) $[1940, 2010, 10]$ by $[0, 7, 1]$ (answers may vary)
(c)   (d)

CHECKING BASIC CONCEPTS FOR SECTIONS 1.1 AND 1.2 (p. 28)

1. (a) 9.88   (b) 1.28
3. (a) 3.485×10^8 (b) -1.2374×10^3
(c) 1.98×10^{-3}
5. $\left(1, \frac{5}{2}\right)$
7. Mean $= 10,762.75$; median $= 12,941.5$

SECTION 1.3 (pp. 41–45)

1. $(-2, 3)$ **3.** $f(7) = 8$
5.   **7.**

9.   **11.**

13.   **15.**

17.   **19.**

21. (a) $f(-2) = -8, f(5) = 125$
(b) All real numbers
23. (a) $f(-1)$ is undefined, $f(a + 1) = \sqrt{a + 1}$
(b) Nonnegative real numbers
25. (a) $f(-1) = 9, f(a + 1) = 3 - 3a$
(b) All real numbers
27. (a) $f(6) = -7, f(a - 1) = -7$
(b) All real numbers
29. (a) $f(4) = \frac{1}{16}, f(-7) = \frac{1}{49}$ **(b)** $x \neq 0$
31. (a) $f(4) = \frac{1}{7}, f(a - 5) = \frac{1}{a^2 - 10a + 16}$
(b) $x \neq 3, x \neq -3$
33. (a) $f(1) = 1, f(a + 2) = \frac{1}{\sqrt{-a}}$ **(b)** $x < 2$
35. (a) All real numbers   **(b)** $g(-1) = -3; g(2) = 3$
(c) $g(-1) = -3; g(2) = 3$
37. (a) All real numbers   **(b)** $g(-1) = 1; g(2) = -2$
(c) $g(-1) = 1; g(2) = -2$
39. (a) $D = \{x \mid -2 \leq x \leq 2\}$
(b) $g(-1) = -2; g(2) = 1$
(c) $g(-1) = -2; g(2) = 1$
41. $D = \{x \mid -3 \leq x \leq 3\}, R = \{y \mid 0 \leq y \leq 3\};$
$f(0) = 3$
43. $D = $ all real numbers, $R = \{y \mid y \leq 2\}; f(0) = 2$
45. $D = \{x \mid x \geq -1\}, R = \{y \mid y \leq 2\}; f(0) = 0$
47. (a) $f(2) = 7$ **(b)** $f = \{(1, 7), (2, 7), (3, 8)\}$
(c) $D = \{1, 2, 3\}, R = \{7, 8\}$
49. (a) $[-4.7, 4.7, 1]$ by $[-3.1, 3.1, 1]$ **(b)** $f(2) = 1$
(c)

51. (a) $[-4.7, 4.7, 1]$ by $[-3.1, 3.1, 1]$ **(b)** $f(2) = 2$
    (c)

53. Verbal: Square the input x.
Graphical: Numerical:
$[-10, 10, 1]$ by $[-10, 10, 1]$

x	-2	-1	0	1	2
y	4	1	0	1	4

$f(2) = 4$
55. Verbal: Multiply the input by 2, add 1, and then take the absolute value.
Graphical: Numerical:
$[-6, 6, 1]$ by $[-4, 4, 1]$

x	-2	-1	0	1	2
y	3	1	1	3	5

$f(2) = 5$
57. Verbal: Subtract x from 5.
Graphical: Numerical:
$[-10, 10, 1]$ by $[-10, 10, 1]$

x	-2	-1	0	1	2
y	7	6	5	4	3

$f(2) = 3$
59. Verbal: Add 1 to the input and then take the square root of the result.
Graphical: Numerical:
$[-6, 6, 1]$ by $[-4, 4, 1]$

x	-2	-1	0	1	2
y	—	0	1	$\sqrt{2}$	$\sqrt{3}$

$f(2) = \sqrt{3}$
61. (a) $g = \{(-1, 2), (0, 4), (1, -3), (2, 2)\}$
(b) $D = \{-1, 0, 1, 2\}, R = \{-3, 2, 4\}$
63. Symbolic: $f(x) = 0.50x$
Graphical: Numerical:

Miles	1	2	3	4	5	6
Cost	0.50	1.00	1.50	2.00	2.50	3.00

65. Yes. Domain and range include all real numbers.
67. No
69. Yes. $D: \{x \mid -4 \le x \le 4\}$; $R: \{y \mid 0 \le y \le 4\}$
71. Yes. Each real number input x has exactly one cube root.
73. No. Usually, several students pass a given exam.
75. Yes, because IDs are unique.
77. No   **79.** Yes   **81.** No   **83.** No   **85.** Yes
87. No   **89.** Yes
91. $g(x) = 12x$; $g(10) = 120$; there are 120 inches in 10 feet.
93. $g(x) = \frac{x}{4}$; $g(10) = 2.5$; there are 2.5 dollars in 10 quarters.
95. $g(x) = 86,400x$; $g(10) = 864,000$; there are 864,000 seconds in 10 days.
97. (a) $I = \{(N, 19162), (H, 26029), (B, 41681), (M, 51316)\}$ **(b)** $D = \{N, H, B, M\}$,
$R = \{19162, 26029, 41681, 51316\}$
99. $N(x) = 2200x$; $N(3) = 6600$; in 3 years the average person uses 6600 napkins.
101. Verbal: Multiply the input x by -5.8 to obtain the change in temperature.
Symbolic: $f(x) = -5.8x$.
Graphical: Numerical:

$[0, 3, 1]$ by $[-20, 20, 5]$

103. $f(x) = \frac{115}{528}x$, $f(15) \approx 3.3$; with a 15-second delay the lightning bolt was about 3.3 miles away.

SECTION 1.4 (pp. 53–55)

1. $a = -2$, $b = 5$ **3.** $a = -8$, $b = 0$
5. $a = 0$, $b = 7$ **7.** $\frac{1}{2}$ **9.** -1 **11.** 0 **13.** -1
15. -8 **17.** $-\frac{23}{30} \approx -0.7667$ **19.** Undefined
21. $-\frac{39}{35} \approx -1.1143$
23. Slope $= 2$; the graph rises 2 units for every unit increase in x.
25. Slope $= -\frac{3}{4}$; the graph falls $\frac{3}{4}$ unit for every unit increase in x, or equivalently, the line falls 3 units for every 4-unit increase in x.
27. Slope $= 0$; the graph neither falls nor rises for every unit increase in x.
29. Slope $= -1$; the graph falls 1 unit for every unit increase in x.
31. (a) Zero square yards of carpet would cost $0.
(b) Slope $= 20$ **(c)** The carpet costs $20 per square yard.

33. (a) $P(20) = 388$; burning 20 gallons of gas produces 388 pounds of CO_2.   **(b)** 19.4; 19.4 pounds of CO_2 are produced for every gallon of gas that is burned.
35. (a) 150 miles   **(b)** Slope $= 75$; the car is traveling *away* from the rest stop at 75 miles per hour.
37. The velocity of the car is zero.
39. Linear, but not constant   **41.** Linear and constant
43. Nonlinear   **45.** Nonlinear   **47.** Nonlinear
49. Linear, $m = 4$ **51.** Nonlinear
53. Linear, $m = -\frac{1}{2}$
55. (a)   **57. (a)**

(b) Linear   **(b)** Nonlinear
59. (a) No   **(b)** $f(x) = 7$
(c) $[0, 15, 3]$ by $[0, 10, 1]$

61.   **63.**

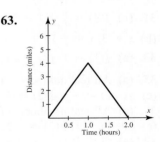

1.4 EXTENDED AND DISCOVERY EXERCISE (p. 55)

1. (a) Yes; 2π inches per second   **(b)** No; because the area function depends on the radius squared, the area function is not a linear function and does not increase at a constant rate.

CHECKING BASIC CONCEPTS FOR SECTIONS 1.3 AND 1.4 (p. 55)

1. Symbolic: $f(x) = 5280x$
Numerical:

x	1	2	3	4	5
$f(x)$	5280	10,560	15,840	21,120	26,400

Graphical:

3.

D = all real numbers, $R = \{y \,|\, y \geq -2\}$
5. (a) Linear **(b)** Constant (and linear)
(c) Nonlinear

SECTION 1.5 (pp. 64–67)

1. $[5, \infty)$ **3.** $[4, 19)$ **5.** $(-\infty, -37]$
7. $[-1, \infty)$ **9.** $(-\infty, 1) \cup [3, \infty)$
11. $(-3, 5]$ **13.** $(-\infty, -2)$
15. $(-\infty, -2) \cup [1, \infty)$
17. Incr: never; decr: $(-\infty, \infty)$, $(\{x \,|\, -\infty < x < \infty\})$
19. Incr: $[2, \infty)$, $(\{x \,|\, x \geq 2\})$; decr: $(-\infty, 2]$, $(\{x \,|\, x \leq 2\})$
21. Incr: $(-\infty, -2] \cup [1, \infty)$, $(\{x \,|\, x \leq -2 \text{ or } x \geq 1\})$; decr: $[-2, 1]$, $(\{x \,|\, -2 \leq x \leq 1\})$
23. Incr: $[-8, 0] \cup [8, \infty)$, $(\{x \,|\, -8 \leq x \leq 0 \text{ or } x \geq 8\})$; decr: $(-\infty, -8] \cup [0, 8]$, $(\{x \,|\, x \leq -8 \text{ or } 0 \leq x \leq 8\})$
25. Neither
27. Incr: $(-\infty, \infty)$, $(\{x \,|\, -\infty < x < \infty\})$; decr: never
29. Incr: $[0, \infty)$, $(\{x \,|\, x \geq 0\})$; decr: $(-\infty, 0]$, $(\{x \,|\, x \leq 0\})$
31. Incr: $(-\infty, 1]$, $(\{x \,|\, x \leq 1\})$; decr: $[1, \infty)$, $(\{x \,|\, x \geq 1\})$
33. Incr: $[1, \infty)$, $(\{x \,|\, x \geq 1\})$; decr: never
35. Incr: $[-3, \infty)$, $(\{x \,|\, x \geq -3\})$; decr: $(-\infty, -3]$, $(\{x \,|\, x \leq -3\})$
37. Incr: $(-\infty, \infty)$, $(\{x \,|\, -\infty < x < \infty\})$; decr: never
39. Incr: $(-\infty, -2] \cup [2, \infty)$, $(\{x \,|\, x \leq -2 \text{ or } x \geq 2\})$; decr: $[-2, 2]$, $(\{x \,|\, -2 \leq x \leq 2\})$
41. Incr: $(-\infty, -2] \cup [1, \infty)$, $(\{x \,|\, x \leq -2 \text{ or } x \geq 1\})$; decr: $[-2, 1]$, $(\{x \,|\, -2 \leq x \leq 1\})$
43. Incr: $(-\infty, -1] \cup [0, 2]$, $(\{x \,|\, x \leq -1 \text{ or } 0 \leq x \leq 2\})$; decr: $[-1, 0] \cup [2, \infty)$, $(\{x \,|\, -1 \leq x \leq 0 \text{ or } x \geq 2\})$
45. $[0, 2.4] \cup [8.7, 14.7] \cup [21, 27]$, $(\{x \,|\, 0 \leq x \leq 2.4 \text{ or } 8.7 \leq x \leq 14.7 \text{ or } 21 \leq x \leq 27\})$
47. From -3 to -1: 0; from 1 to 3: 0
49. From -3 to -1: 1.2; from 1 to 3: -1.2

51. (a) 3 **(b)**

53. 7; the slope of the line passing through $(1, f(1))$ and $(4, f(4))$ is 7.
55. 26; the slope of the line passing through $(2, f(2))$ and $(4, f(4))$ is 26.
57. 0.62; the slope of the line passing through $(1, f(1))$ and $(3, f(3))$ is approximately 0.62.
59. (a) From 1900 to 1940: 4.475; from 1940 to 1980: 11.25; from 1980 to 2006: about -10.04.
(b) From 1900 to 1940 cigarette consumption increased by 4.475 billion cigarettes per year, on average. The other rates may be interpreted similarly.
61. (a) $D(2) = 64$; $D(4) = 256$ **(b)** 96; the object's average speed from 2 to 4 seconds is 96 feet per second.

For Exercises 63–67, answers may vary.
63.

65.

67.

69. (a) 3 **(b)** 0 **71. (a)** $-2x - 2h$ **(b)** -2
73. (a) $2x + 2h + 1$ **(b)** 2
75. (a) $3x^2 + 6xh + 3h^2 + 1$ **(b)** $6x + 3h$
77. (a) $-x^2 - 2xh - h^2 + 2x + 2h$ **(b)** $-2x - h + 2$
79. (a) $2x^2 + 4xh + 2h^2 - x - h + 1$
(b) $4x + 2h - 1$
81. (a) $x^3 + 3x^2h + 3h^2x + h^3$ **(b)** $3x^2 + 3hx + h^2$
83. (a) $8t^2 + 16th + 8h^2$ **(b)** $16t + 8h$
(c) 64.4; the average speed of the car from 4 to 4.05 seconds is 64.4 feet per second.

1.5 EXTENDED AND DISCOVERY EXERCISE (p. 67)

1. Let $t = 7$ and $h = 0$.

CHECKING BASIC CONCEPTS FOR SECTION 1.5 (p. 67)

1. (a) $(-\infty, 5]$ **(b)** $[1, 6)$ **3.** -7

CHAPTER 1 REVIEW EXERCISES (pp. 71–74)

1. Natural number: $\sqrt{16}$; integer: $-2, 0, \sqrt{16}$;
rational number: $-2, \frac{1}{2}, 0, 1.23, \sqrt{16}$; real number:
$-2, \frac{1}{2}, 0, 1.23, \sqrt{7}, \sqrt{16}$
3. 1.891×10^6 **5.** 15,200   **7. (a)** 32.07
(b) 2.62   **(c)** 5.21   **(d)** 49.12   **9.** -41
11.

-23	-5	8	19	24

(a) Max: 24; min: -23
(b) Mean: 4.6; median: 8
13. (a) $S = \{(-15, -3), (-10, -1), (0, 1), (5, 3), (20, 5)\}$
(b) $D = \{-15, -10, 0, 5, 20\}, R = \{-3, -1, 1, 3, 5\}$
15. Not a function
$[-50, 50, 10]$ by $[-50, 50, 10]$

17. 10
19. $\left(2, -\frac{3}{2}\right)$ **21.** Yes   **23.** $(x - 2)^2 + (y - 5)^2 = 17$
25.

27.

29.

31.

33. $f(x) = 16x$
$[0, 100, 10]$ by $[0, 1800, 300]$

x	0	25	50	75	100
$f(x)$	0	400	800	1200	1600

35. (a) $f(-8) = -2, f(1) = 1$ **(b)** All real numbers
37. (a) $f(-3) = 5, f(1.5) = 5$ **(b)** All real numbers
39. (a) $f(-10) = 97, f(a + 2) = a^2 + 4a + 1$
(b) All real numbers

41. (a) $f(-3) = \frac{1}{5}, f(a + 1) = \dfrac{1}{a^2 + 2a - 3}$
(b) $D = \{x \mid x \neq 2, x \neq -2\}$ **43.** No   **45.** Yes
47. Yes   **49.** 0   **51.** $-\frac{3}{4}$ **53.** 0   **55.** Linear
57. Nonlinear
59.

61. Linear, but not constant   **63.** 5   **65.** 760 seconds
67. (a) 198 feet   **(b)** 1044 sq ft
69. (a) The data decrease rapidly, indicating a very high
mortality rate during the first year.
$[-1, 5, 1]$ by $[0, 110, 10]$

(b) Yes   **(c)** From 0 to 1: -90; from 1 to 2: -4; from 2
to 3: -3; from 3 to 4: -1. During the first year the population decreased, on average, by 90 birds. The other average
rates of change can be interpreted similarly.
71. (a) $[1, 5, 1]$ by $[40, 70, 5]$

Nonlinear
(b) 2.5   **(c)** The average rate of change in outside temperature from 1 P.M. to 4 P.M. was 2.5°F per hour. The slope of the
line segment from $(1, 50.5)$ to $(4, 58)$ is 2.5.
$[1, 5, 1]$ by $[50, 63, 5]$

CHAPTER 1 EXTENDED AND DISCOVERY EXERCISES (pp. 74–75)

1. About 7.17 km
3. About 3.862   **5.** About 4.039
$[-3, 3, 1]$ by $[-2, 2, 1]$ $[-3, 3, 1]$ by $[-1, 3, 1]$

7. About 3.16; estimates will be less than the true value.
9. (a) Determine the number of square miles of Earth's surface that are covered by the oceans. Then divide the total volume of the water from the ice cap by the surface area of the oceans to get the rise in sea level.   **(b)** About 25.7 feet
(c) Since the average elevations of Boston, New Orleans, and San Diego are all less than 25 feet, these cities would be under water without some type of dike system.
(d) About 238 feet

CHAPTER 2: Linear Functions and Equations

SECTION 2.1 (pp. 88–95)

1. (a) $V(0) = 12.8$; $V(4) = 12.0$ **(b)** 12.4 million in 2002 (interpolation); 11.6 million in 2006 (extrapolation)
(c) Interpolation was more accurate.
3. Exactly   **5.** Approximately
7. $f(x) = -\frac{1}{2}x + 3$ **9.** $f(x) = 2x + 5$
11. (a) $f(x) = \frac{x}{16}$ **(b)** $f(x) = 10x$
(c) $f(x) = 0.06x + 6.50$ **(d)** $f(x) = 500$
13. (a) Slope: 2; y-int: -1; x-int: 0.5
(b) $f(x) = 2x - 1$ **(c)** 0.5   **(d)** Increasing
15. (a) Slope: $-\frac{1}{3}$; y-int: 2; x-int: 6
(b) $f(x) = -\frac{1}{3}x + 2$ **(c)** 6   **(d)** Decreasing
17. (a) Slope: 20; y-int: -50; x-int: 2.5
(b) $f(x) = 20x - 50$ **(c)** 2.5   **(d)** Increasing
19.

Slope: 3; y-int: 2

21.

Slope: $\frac{1}{2}$; y-int: -2

23.

Slope: 0; y-int: -2

25.

Slope: $-\frac{1}{2}$; y-int: 4

27.
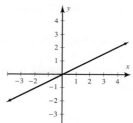
Slope: $\frac{1}{2}$; y-int: 0

29.

Slope: -5; y-int: 5

31.

Slope: 20; y-int: -10

33. $f(x) = -\frac{3}{4}x + \frac{1}{3}$ **35.** $f(x) = 15x$
37. $f(x) = 0.5x + 4$ **39.** 0; 0   **41.** $-\frac{1}{4}$; $-\frac{1}{4}$
43. -3; -3 **45.** d   **47.** c
49. $I(t) = 1.5t + 68$; t represents years after 2006;
$D = \{t \mid 0 \le t \le 4\}$ **51.** $V(t) = 32t$; t represents time in seconds; $D = \{t \mid 0 \le t \le 3\}$ **53.** $P(t) = 21.5 + 0.581t$; t represents years after 1900; $D = \{t \mid 0 \le t \le 100\}$
55. (a) $W(t) = -10t + 300$ **(b)** 230 gal
(c) x-int: 30, the tank is empty after 30 minutes; y-int: 300, the tank contains 300 gallons of water initially.
(d) $D = \{t \mid 0 \le t \le 30\}$

57. (a) $f(x) = 4.3x + 40$ **(b)** 65,800,000
59. (a) $f(x) = 0.25x + 0.5$ **(b)** 1.125 inches
61. (a) 17; 17.2; 16.6   **(b)** $f(x) = 17x$
(c) The vehicle is getting 17 miles per gallon.

(d) $f(30) = 510$; the vehicle travels 510 miles on 30 gallons of gas.

63. (a) Max: 55 mi/hr; min: 30 mi/hr   **(b)** 12 miles
(c) $f(4) = 40, f(12) = 30, f(18) = 55$
(d) $x = 4, 6, 8, 12,$ and 16. The speed limit changes at each discontinuity.
65. (a) $P(1.5) = 0.97, P(3) = 1.14$; it costs \$0.97 to mail 1.5 ounces and \$1.14 to mail 3 ounces.
(b) $D = \{x \,|\, 0 < x \le 5\}$

(c) $x = 1, 2, 3, 4$
67. (a) $f(1.5) = 30; f(4) = 10$
(b) $m_1 = 20$ indicates that the car is moving away from home at 20 mi/hr; $m_2 = -30$ indicates that the car is moving toward home at 30 mi/hr; $m_3 = 0$ indicates that the car is not moving; $m_4 = -10$ indicates that the car is moving toward home at 10 mi/hr.   **(c)** The driver starts at home and drives away from home at 20 mi/hr for 2 hours until the car is 40 miles from home. The driver then travels toward home at 30 mi/hr for 1 hour until the car is 10 miles from home. Then the car does not move for 1 hour. Finally, the driver returns home in 1 hour at 10 mi/hr.
(d) Incr: $0 \le x \le 2$; decr: $2 \le x \le 3$ or $4 \le x \le 5$; const: $3 \le x \le 4$
69. (a) $-5 \le x \le 5$
(b) $f(-2) = 2, f(0) = 3, f(3) = 6$

(c)   **(d)** f is continuous.

71. (a) $-1 \le x \le 2$
(b) $f(-2) =$ undefined, $f(0) = 0, f(3) =$ undefined
(c)   **(d)** f is not continuous.

73. (a) $-3 \le x \le 3$
(b) $f(-2) = -2, f(0) = 1, f(3) = -1$
(c)   **(d)** f is not continuous.

75.

77. (a) $f(-3) = -10, f(1) = 4, f(2) = 4, f(5) = 1$
(b) $[1, 3]$ **(c)** f is not continuous.

79. (a) $[-10, 10, 1]$ by $[-10, 10, 1]$

(b) $f(-3.1) = -8, f(1.7) = 2$
81. (a) $[-10, 10, 1]$ by $[-10, 10, 1]$

(b) $f(-3.1) = -7, f(1.7) = 3$
83. (a) $f(x) = 0.8 \lceil x/2 \rceil$ for $6 \le x \le 18$
(b) $[6, 18, 1]$ by $[0, 8, 1]$

(c) $f(8.5) = \$3.20, f(15.2) = \$5.60$

85. $y \approx -0.789x + 0.526$; $r \approx -0.993$
[-3, 4, 1] by [-3, 3, 1]

87. (a) Positive   **(b)** $y = ax + b$, where $a = 3.25$ and $b = -2.45$; $r \approx 0.9994$ **(c)** $y = 5.35$
89. (a) Negative   **(b)** $y = ax + b$, where $a \approx -3.8857$ and $b \approx 9.3254$; $r \approx -0.9996$ **(c)** $y \approx -0.00028$
(because of rounding, answers may vary slightly)
91. (a) [-100, 1800, 100] by [-1000, 28,000, 1000]

(b) $y = ax + b$, where $a \approx 14.680$ and $b \approx 277.82$
(c) 2500 light-years
93. (a) Positive
[-5, 40, 5] by [0, 6, 1]

(b) $f(x) \approx 0.0854x + 2.078$
(c) The number of passenger miles increased by about 0.0854 trillion per year, on average.
[-5, 40, 5] by [0, 6, 1]

(d) About 5.5 trillion

2.1 EXTENDED AND DISCOVERY EXERCISES (p. 95)

1. Answers will vary.
3. (a) [1.580, 1.584, 0.001] by [-6.252, -6.248, 0.001]

(b) A linear function would be a good approximation over a small interval.

SECTION 2.2 (pp. 107–112)

1. $y = -2(x - 1) + 2$ **3.** $y = \frac{3}{4}(x + 3) - 1$

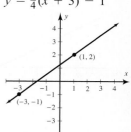

5. $y = -2.4(x - 4) + 5$; $y = -2.4x + 14.6$
7. $y = -\frac{1}{2}(x - 1) - 2$; $y = -\frac{1}{2}x - \frac{3}{2}$
9. $y = \frac{3}{4}(x - 4) + 0$; $y = \frac{3}{4}x - 3$
11. $y = \frac{2}{3}x - 1$ **13.** $y = -\frac{3}{5}x + \frac{3}{5}$ **15.** c
17. b   **19.** e   **21.** $y = 3x - 1$ **23.** $y = \frac{8}{3}x - \frac{17}{3}$
25. $y = -7.8x + 5$ **27.** $y = -\frac{1}{2}x + 45$
29. $y = -3x + 5$ **31.** $y = \frac{3}{2}x - 6$
33. $y = \frac{5}{18}x + \frac{11}{18}$ **35.** $y = 4x + 9$
37. $y = \frac{3}{2}x - 2960$ **39.** $y = \frac{2}{3}x - 2.1$
41. $y = \frac{1}{2}x + 6$ **43.** $y = x - 20$
45. $y = \frac{1}{2}x + \frac{9}{2}$ **47.** $y = -12x - 20$
49. $x = -5$ **51.** $y = 6$ **53.** $x = 4$ **55.** $x = 19$
57. x-int: 5; y-int: -4   **59.** x-int: 7; y-int: -7

61. x-int: -7; y-int: 6   **63.** x-int: $-\frac{7}{3}$; y-int: 7

65. x-int: 4; y-int: 2

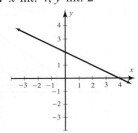

67. x-int: $\frac{5}{8}$; y-int: -5

69. x-int: 5; y-int: 7; a and b represent the x- and y-intercepts, respectively.
71. x-int: $\frac{3}{2}$; y-int: $\frac{5}{4}$; a and b represent the x- and y-intercepts, respectively.   **73.** $\frac{x}{5} + \frac{y}{9} = 1$
75. (a) $y = 1.5x - 3.2$ **(b)** When $x = -2.7$, $y = -7.25$ (interpolation); when $x = 6.3$, $y = 6.25$ (extrapolation)
77. (a) $y = -2.1x + 105.2$ **(b)** When $x = -2.7$, $y = 110.87$ (extrapolation); when $x = 6.3$, $y = 91.97$ (interpolation)
79. (a) $f(x) = -108(x - 1998) + 3305$, or $f(x) = -108x + 219,089$ (answers may vary); approximate   **(b)** 2549; the estimated value is too low; extrapolation   **(c)** The numbers were decreasing but increased after 9/11.
81. (a) $y = \frac{12,000}{7}(x - 2003) + 25,000$, or $y = \frac{12,000}{7}(x - 2010) + 37,000$; the cost of attending a private college or university is increasing by $\frac{12,000}{7} \approx \$1714$ per year, on average.   **(b)** About \$31,857; interpolation
(c) $y = 1714x - 3,408,714$ (approximate)
83. (a) $y = \frac{2}{3}(x - 2002) + 1.6$; sales increased by \$0.67 billion per year, on average.   **(b)** \$5.6 billion; extrapolation   **(c)** $y = \frac{2}{3}x - \frac{19,996}{15}$
85. (a) [1998, 2005, 1] by [0, 10, 1]

(b) Using the first and last points gives $f(x) = 1.56(x - 1999) + 1$; the average number of worldwide spam messages daily increased by 1.56 billion per year, on average (answers may vary).
(c) About 13.5 billion
87. (a) [1997, 2005, 1] by [0, 2.2, 0.2]

(b) $f(x) = 0.1(x - 1998) + 1.4$; U.S. sales of Toyota vehicles increased by 0.1 million per year.   **(c)** Exact

89. (a) $f(x) = 0.29x + 4200$ **(b)** It represents the annual fixed costs.
91. (a) [1995, 2003, 1] by [9, 13, 1]

(b) $f(x) \approx 0.4167x - 802.04$ (answers may vary)
(c) The Asian-American population increased by about 0.42 million each year, on average.
(d) $f(2008) \approx 14.62$ million
93. [0, 3, 1] by [−2, 2, 1]

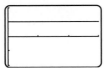

(a) No, the slope is not zero.   **(b)** The resolution of most calculator screens is not high enough to show the slight increase in the y-values.
95. (a) They do not appear to be perpendicular in the standard viewing rectangle (answers may vary for different calculators).
(b) In [−15, 15, 1] by [−10, 10, 1] and [−3, 3, 1] by [−2, 2, 1] they appear to be perpendicular.

[−10, 10, 1] by [−10, 10, 1]   [−15, 15, 1] by [−10, 10, 1]

[−10, 10, 1] by [−3, 3, 1]   [−3, 3, 1] by [−2, 2, 1]

(c) The lines appear perpendicular when the distance shown along the x-axis is approximately 1.5 times the distance along the y-axis.
97. $y_1 = x, y_2 = -x, y_3 = x + 2, y_4 = -x + 4$
99. $y_1 = x + 4, y_2 = x - 4, y_3 = -x + 4, y_4 = -x - 4$
101. 2.5   **103.** $\frac{9}{8}$ **105.** $k = 2.5, y = 20$ when $x = 8$
107. $k = 0.06, x = \$85$ when $y = \$5.10$
109. \$1048, $k = 65.5$ **111. (a)** $k = 0.01$ **(b)** 1.1 mm
113. (a) $k = \frac{15}{8}$ **(b)** $13\frac{1}{3}$ inches
115. (a) For (150, 26), $\frac{F}{x} \approx 0.173$; for (180, 31), $\frac{F}{x} \approx 0.172$; for (210, 36), $\frac{F}{x} \approx 0.171$; for (320, 54), $\frac{F}{x} \approx 0.169$; the ratios give the force needed to push a 1-pound box.   **(b)** $k \approx 0.17$ (answers may vary)

(c) [125, 350, 25] by [0, 75, 5]

(d) 46.75 pounds

2.2 EXTENDED AND DISCOVERY EXERCISES (p. 112–113)

1. About 615 fish

CHECKING BASIC CONCEPTS FOR SECTIONS 2.1 AND 2.2 (p. 113)

1. Slope: -2; y-int: 4; x-int: 2

3. $f(t) = 60t + 50$, where t is in hours
5. Horizontal: $y = 7$; vertical: $x = -4$
7. x-int: 6; y-int: -9

SECTION 2.3 (pp. 126–131)

1. One   **3.** $4x - 1$ **5.** They are equal.   **7.** Linear
9. Nonlinear   **11.** Linear   **13.** 4   **15.** -4
17. $\frac{32}{3}$ **19.** 4   **21.** 3   **23.** $\frac{1}{3}$ **25.** $\frac{4}{7}$ **27.** $-\frac{2}{19}$
29. $\frac{13}{10}$ **31.** $\frac{17}{10}$ **33.** $-\frac{17}{8}$ **35.** $\frac{5}{17}$ **37.** $\frac{400}{7}$
39. (a) No solutions   **(b)** Contradiction
41. (a) $\frac{8}{3}$ **(b)** Conditional
43. (a) All real numbers   **(b)** Identity
45. (a) No solutions   **(b)** Contradiction
47. (a) All real numbers   **(b)** Identity
49. 3   **51. (a)** 4   **(b)** 2   **(c)** -2
53. -1 **55.** 1   **57.** 4   **59.** 1.3   **61.** 0.675
63. 3.621   **65.** 2.294   **67.** 3   **69.** 8.6
71. 0.2   **73.** -1.2 **75.** 1   **77.** 0.8   **79.** -4
81. 2   **83.** $W = \frac{A}{L}$ **85.** $L = \frac{1}{2}P - W$
87. $y = 4 - \frac{3}{2}x$ **89.** $x = \frac{1}{4}y + \frac{3}{2}$ **91.** 1989
93. About 1987   **95.** About 2009
97. $f(x) = 0.75x$; $42.18
99. (a) $0.045x$ **(b)** About 1,444,000
101. (a) About 2 hours (answers may vary)
(b) $\frac{15}{8} = 1.875$ hours
103. 3.2 hours at 55 mi/hr and 2.8 hours at 70 mi/hr
105. $\frac{1}{9}$ hour, or $6\frac{2}{3}$ minutes   **107.** About 17.29 feet

109. About 36.4 cubic feet   **111.** About 8.33 liters
113. 36 inches by 54 inches
115. (a) $C(x) = -7.5x + 15,090$;
$L(x) = 14.75x - 29,500.5$ **(b)** Sales of CRT monitors decreased by 7.5 million monitors per year, on average. Sales of LCD monitors increased by 14.75 million monitors per year, on average.   **(c)** 2004   **(d)** 2004   **(e)** 2004
117. $-40°F$ is equivalent to $-40°C$
119. (a) f is linear because the amount of oil is mixed at a constant rate.   **(b)** 0.48 pint; 0.48 pint of oil should be added to 3 gallons of gasoline to get the correct mixture.
(c) 12.5 gallons   **121.** $\frac{80}{9} \approx 8.89$
123. (a) $S(x) \approx 3.974x - 14.479$ (answers may vary)
(b) About 5.15 cm
125. (a) $f(x) \approx 0.10677x - 211.69$ (answers may vary)
(b) $0.5 million; extrapolation   **(c)** About 2013

2.3 EXTENDED AND DISCOVERY EXERCISES (p. 131)

1. (a) Yes; since multiplication distributes over addition, doubling the lengths gives double the sum of the lengths.
(b) No; for example, in the case of a square (a type of rectangle), the square of twice a side is four times the square of the side.
3. (a) $f(x) = 14,000x$ **(b)** About 1.9 hours

SECTION 2.4 (pp. 142–146)

1. $(-\infty, 2)$ **3.** $[-1, \infty)$ **5.** $[1, 8)$ **7.** $(-\infty, 1]$
9. $\{x \mid x \geq 2\}$, or $[2, \infty)$
11. $\{x \mid x < 10.5\}$, or $(-\infty, 10.5)$
13. $\{x \mid x \geq 13\}$, or $[13, \infty)$ **15.** $\{x \mid x < 0\}$, or $(-\infty, 0)$
17. $\{x \mid x \geq -10\}$, or $[-10, \infty)$
19. $\{x \mid x > 1\}$, or $(1, \infty)$
21. $\{x \mid x > \frac{7}{3}\}$, or $\left(\frac{7}{3}, \infty\right)$
23. $\{x \mid \frac{3}{2} < x \leq 3\}$, or $\left(\frac{3}{2}, 3\right]$
25. $\{x \mid -16 \leq x \leq 1\}$, or $[-16, 1]$
27. $\{x \mid -20.75 < x \leq 12.5\}$, or $(-20.75, 12.5]$
29. $\{x \mid -4 < x < 1\}$, or $(-4, 1)$
31. $\{x \mid \frac{9}{2} \leq x \leq \frac{21}{2}\}$, or $\left[\frac{9}{2}, \frac{21}{2}\right]$
33. $\{x \mid x \geq \frac{5}{3}\}$, or $\left[\frac{5}{3}, \infty\right)$
35. $\{x \mid -\frac{1}{2} < x \leq -\frac{1}{4}\}$, or $\left(-\frac{1}{2}, -\frac{1}{4}\right]$
37. $\{x \mid x \leq \frac{21}{19}\}$, or $\left(-\infty, \frac{21}{19}\right)$
39. $\{x \mid x \leq 2\}$ **41.** $\{x \mid x > 3\}$
43. $\{x \mid 0 \leq x \leq 2\}$ **45.** $\{x \mid -1 < x \leq 4\}$
47. (a) 2   **(b)** $\{x \mid x < 2\}$, or $(-\infty, 2)$
(c) $\{x \mid x \geq 2\}$, or $[2, \infty)$
49. (a) -2 **(b)** $\{x \mid x > -2\}$, or $(-2, \infty)$
(c) $\{x \mid x \leq -2\}$, or $(-\infty, -2]$
51. $\{x \mid x \leq 2\}$, or $(-\infty, 2]$
53. $\{x \mid x > 1\}$, or $(1, \infty)$

55. $\{x \mid x > 2.8\}$ **57.** $\{x \mid x \le 1987.5\}$

59. $\{x \mid x > -1.82\}$

61. $\{x \mid 4 \le x < 6.4\}$, or $[4, 6.4)$

63. $\{x \mid 4.6 \le x \le 15.2\}$, or $[4.6, 15.2]$

65. $\{x \mid 1 < x < 5.5\}$, or $(1, 5.5)$

67. (a) 8 **(b)** $\{x \mid x < 8\}$

69. $\{x \mid x < 4\}$; $\{x \mid x \ge 4\}$

71. $\left\{x \mid x < -\frac{3}{2}\right\}$, or $\left(-\infty, -\frac{3}{2}\right)$

73. $\{x \mid 1 \le x \le 4\}$, or $[1, 4]$

75. $\left\{x \mid -\frac{1}{20} \le x < \frac{17}{20}\right\}$, or $\left[-\frac{1}{20}, \frac{17}{20}\right)$

77. $\{x \mid x \le 31.4\}$, or $(-\infty, 31.4]$

79. $\left\{x \mid x > \frac{13}{2}\right\}$, or $\left(\frac{13}{2}, \infty\right)$

81. $\{x \mid x \le 1.534\}$, or $(-\infty, 1.534]$

83. (a) Car A is traveling faster since its graph has the greater slope. **(b)** 2.5 hours; 225 miles **(c)** $0 \le x < 2.5$

85. (a) $0 \le x < 1.14$ (approximately)

(b) $0 \le x < \frac{15}{13.2} \approx 1.14$

87. (a) The median price of a single-family home increased, on average, by \$8667 per year.

(b) From 1996 to 2002 (approximately)

89. (a) $B(x) = 6(x - 2000) + 6$ or $B(x) = 6(x - 2004) + 30$ **(b)** From 2003 to 2006

91. (a) $P(x) = 2.5x - 4972.5$ **(b)** From 2007 to 2009

93. About day 110 (April 19) to day 119 (April 28)

95. $3.98\pi \le C \le 4.02\pi$

97. (a) $f(x) = 3x - 1.5$ **(b)** $x > 1.25$

99. (a) $N(x) = 20.1x - 40,096.7$ **(b)** From 1997 to 1999 **(c)** Extrapolation

2.4 EXTENDED AND DISCOVERY EXERCISES (p. 146)

1. $a < b \Rightarrow 2a < a + b < 2b \Rightarrow a < \frac{a + b}{2} < b$

CHECKING BASIC CONCEPTS FOR SECTIONS 2.3 AND 2.4 (p. 146)

1. (a) 2.5 **(b)** 2.5 **(c)** 2.5

3. $\left\{x \mid -1 \le x \le \frac{3}{2}\right\}$, or $\left[-1, \frac{3}{2}\right]$

SECTION 2.5 (pp. 154–156)

1. $-3, 3$ **3.** $x < -3$ or $x > 3$, or $(-\infty, -3) \cup (3, \infty)$

5. It is V-shaped with the vertex on the x-axis. **7.** $|6a|$

9.

(a) -1 **(b)** Incr: $x \ge -1$, or $[-1, \infty)$; decr: $x \le -1$, or $(-\infty, -1]$

11.

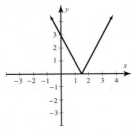

(a) $\frac{3}{2}$ **(b)** Incr: $x \ge \frac{3}{2}$, or $\left[\frac{3}{2}, \infty\right)$; decr: $x \le \frac{3}{2}$, or $\left(-\infty, \frac{3}{2}\right]$

13. (a)

(b)

(c) 0

15. (a)

(b)

(c) 1

17 (a)

(b)

(c) 3

19. $-2, 2$ **21.** $1, \frac{9}{5}$ **23.** $-\frac{1}{2}, 2$ **25.** $-\frac{1}{3}$ **27.** $\frac{7}{16}$

29. No solutions **31.** $-\frac{23}{12}, \frac{19}{4}$ **33.** No solutions

35. $-1, \frac{17}{5}$ **37.** $-1, 1$ **39.** $-3, 2$

41. (a) $-1, 7$ **(b)** $-1 < x < 7$, or $(-1, 7)$

(c) $x < -1$ or $x > 7$, or $(-\infty, -1) \cup (7, \infty)$

43. (a) $1, 2$ **(b)** $1 < x < 2$, or $(1, 2)$

(c) $x < 1, x > 2$, or $(-\infty, 1) \cup (2, \infty)$

45. $-\frac{5}{2}, \frac{15}{2}$; $-\frac{5}{2} < x < \frac{15}{2}$, or $\left(-\frac{5}{2}, \frac{15}{2}\right)$

47. $\frac{7}{3}, 1$; $x < 1$ or $x > \frac{7}{3}$, or $\left(-\infty, 1\right) \cup \left(\frac{7}{3}, \infty\right)$

49. $-\frac{17}{21}, \frac{31}{21}$; $x \le -\frac{17}{21}$ or $x \ge \frac{31}{21}$, or
$\left(-\infty, -\frac{17}{21}\right] \cup \left[\frac{31}{21}, \infty\right)$

51. $-\frac{1}{3}, \frac{1}{3}$; $x < -\frac{1}{3}$ or $x > \frac{1}{3}$, or $\left(-\infty, -\frac{1}{3}\right) \cup \left(\frac{1}{3}, \infty\right)$

53. There are no solutions for the equation or inequality.

55. $\left(-\frac{7}{3}, 3\right)$, or $-\frac{7}{3} < x < 3$

57. $\left[-1, \frac{9}{2}\right]$, or $-1 \le x \le \frac{9}{2}$

59. $\left(-\frac{5}{2}, \frac{11}{2}\right)$, or $-\frac{5}{2} < x < \frac{11}{2}$

61. $(-\infty, 1) \cup (2, \infty)$, or $x < 1$ or $x > 2$

63. $\left(-\infty, \frac{5}{3}\right] \cup \left[\frac{11}{3}, \infty\right)$, or $x \le \frac{5}{3}$ or $x \ge \frac{11}{3}$

65. $(-\infty, -8) \cup (16, \infty)$, or $x < -8$ or $x > 16$

67. 6   **69.** $[-2, 4]$ **71.** $[0, \infty)$

73. Max: 75 mi/hr; min: 40 mi/hr

75. (a) $\frac{48}{19} \le x \le \frac{80}{19}$ (b) $\left|x - \frac{64}{19}\right| \le \frac{16}{19}$

77. (a) $19 \le T \le 67$ (b) The monthly average temperatures in Marquette vary between a low of 19°F and a high of 67°F. The monthly averages are always within 24 degrees of 43°F.

79. (a) $28 \le T \le 72$ (b) The monthly average temperatures in Boston vary between a low of 28°F and a high of 72°F. The monthly averages are always within 22 degrees of 50°F.

81. (a) $49 \le T \le 74$ (b) The monthly average temperatures in Buenos Aires vary between a low of 49°F (possibly in July) and a high of 74°F (possibly in January). The monthly averages are always within 12.5 degrees of 61.5°F.

83. $2.996 \le d \le 3.004$; diameters between 2.996 and 3.004 inches are acceptable.

85. $34.3 \le Q \le 35.7$

2.5 EXTENDED AND DISCOVERY EXERCISES (p. 156)

1. $|x - c| < \delta$

CHECKING BASIC CONCEPTS FOR SECTION 2.5 (p. 157)

1. $|2x|$ **3.** (a) $-2, 3$ (b) $[-2, 3]$, or $-2 \le x \le 3$; $(-\infty, -2) \cup (3, \infty)$, or $x < -2$ or $x > 3$ **5.** $-\frac{1}{3}, 1$

CHAPTER 2: REVIEW EXERCISES (pp. 161–164)

1. (a) Slope: -2; y-int: 6; x-int: 3   (b) $f(x) = -2x + 6$
(c) 3   **3.** $f(x) = -2.5x + 5$
5.

7. $f(x) = -2x - 1$ **9.** $y = 7x + 30$

11. $y = -3x + 2$ **13.** $y = -\frac{31}{57}x - \frac{368}{57}$

15. $x = 6$ **17.** $y = 3$ **19.** $x = 2.7$

21. x-int: 4; y-int: -5

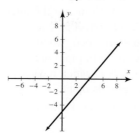

23. 6.4   **25.** $\frac{15}{7} \approx 2.143$ **27.** $\frac{5}{\pi} \approx 1.592$ **29.** -2.9

31. (a) All real numbers   (b) Identity

33. (a) -3 (b) Conditional

35. $(-3, \infty)$ **37.** $\left[-2, \frac{3}{4}\right)$

39. $\{x \mid x \le 3\}$, or $(-\infty, 3]$

41. $\{x \mid -\infty < x < \infty\}$, or $(-\infty, \infty)$

43. $\left\{x \mid -1 < x \le \frac{7}{2}\right\}$, or $\left(-1, \frac{7}{2}\right]$

45. $\{x \mid x > -1\}$, or $(-1, \infty)$

47. (a) 2   (b) $x > 2$ (c) $x < 2$

49. (a) $f(-2) = 4, f(-1) = 6, f(2) = 3, f(3) = 4$

(b) f is continuous.   (c) $x = -\frac{5}{2}$ or 2

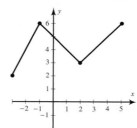

51. $-2, 7$ **53.** No solutions

55. ± 3; $x < -3$ or $x > 3$, or $(-\infty, -3) \cup (3, \infty)$

57. $\frac{17}{3}, -1$; $x < -1$ or $x > \frac{17}{3}$, or $(-\infty, -1) \cup \left(\frac{17}{3}, \infty\right)$

59. $-3 < x < 6$, or $(-3, 6)$

61. $x \le -\frac{5}{2}$ or $x \ge \frac{7}{2}$, or $\left(-\infty, -\frac{5}{2}\right] \cup \left[\frac{7}{2}, \infty\right)$

63. (a) 1990; in 1990 the median U.S. family income was about \$34,500.   (b) 1990

65. From 2001 to 2006

67. Initially the car is at home. After traveling 30 mi/hr for 1 hour, the car is 30 miles away from home. During the second hour the car travels 20 mi/hr until it is 50 miles away. During the third hour the car travels toward home at 30 mi/hr until it is 20 miles away. During the fourth hour the car travels away from home at 40 mi/hr until it is 60 miles away from home. During the last hour the car travels 60 miles at 60 mi/hr until it arrives home.

69. 155,590   **71.** 18.75 minutes

73. 0.9 hour at 7 mi/hr and 0.9 hour at 8 mi/hr
75. (a) $y = -1.2x + 3$
(b) $y = 4.8$ when $x = -1.5$, interpolation; $y = -1.2$ when $x = 3.5$, extrapolation   (c) $\frac{17}{12}$
77. When $0 \le x \le 3$, the slope is 5, which means the inlet pipe is open and the outlet pipe is closed; when $3 < x \le 5$, the slope is 2, which means both pipes are open; when $5 < x \le 8$, the slope is 0, which means both pipes are closed; when $8 < x \le 10$, the slope is -3, which means the inlet pipe is closed and the outlet pipe is open.
79. The distance above the ground is between $1\frac{2}{3}$ kilometers and $3\frac{1}{3}$ kilometers.
81. Between 52.1431 feet and 52.4569 feet

CHAPTER 2: EXTENDED AND DISCOVERY EXERCISES (pp. 164–165)

1. (a) 62.8 inches
(b) $[7, 15, 1]$ by $[45, 75, 5]$ $[7, 15, 1]$ by $[45, 75, 5]$

    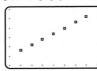

Both sets of data are linear.
(c) Female: 3.1 inches; male: 3.0 inches
(d) $f(x) = 3.1(x - 8) + 50.4$;
$g(x) = 3.0(x - 8) + 53$
(e) $55.67 \le$ female height ≤ 56.91;
$58.1 \le$ male height ≤ 59.3
3. 3 miles   **5.** If $|x - c| < \delta$, then $|f(x) - L| < \epsilon$.

CHAPTERS 1–2: CUMULATIVE REVIEW EXERCISES (pp. 165–168)

1. 1.23×10^5; 5.1×10^{-3} **3.** 2.09
5. $(x + 2)^2 + (y - 3)^2 = 49$ **7.** $\sqrt{89}$
9. (a) $D = \{x \mid -\infty < x < \infty\}$, $R = \{y \mid y \ge -2\}$;
$f(-1) = -1$
(b) $D = \{x \mid -3 \le x \le 3\}$, $R = \{y \mid -3 \le y \le 2\}$;
$f(-1) = -\frac{1}{2}$
11. (a) $f(2) = 7$; $f(a - 1) = 5a - 8$
(b) $D = \{x \mid -\infty < x < \infty\}$
13. No. The graph does not pass the vertical line test.
15. 1   **17.** (a) $\frac{2}{3}$; -2; 3   (b) $f(x) = \frac{2}{3}x - 2$ (c) 3
19. $f(x) = -3x + \frac{4}{3}$ **21.** $y = -\frac{11}{8}x - \frac{29}{8}$
23. $x = -1$ **25.** $y = 2x + 11$

27. x-int: -3; y-int: 2

29. 1   **31.** $-\frac{24}{17}$ **33.** 3   **35.** $(-\infty, 5)$
37. $(-\infty, -2) \cup (2, \infty)$ **39.** $\{x \mid x \le \frac{5}{8}\}$, or $\left(-\infty, \frac{5}{8}\right]$
41. (a) 2   (b) $x < 2$ (c) $x \ge 2$
43. $-6, 4$ **45.** $-7, 7$
47. $\{x \mid 0 \le x \le 5\}$, or $[0, 5]$ **49.** 3.40 inches
51. (a) 770,000; it costs \$770,000 to manufacture 1500 computers.   (b) 500; each additional computer costs \$500 to manufacture.
53. (a) 9°F per hour   (b) On average, the temperature increased by 9°F per hour over this 2-hour period.
55. $\frac{60}{17} \approx 3.53$ hours
57. (a) $f(x) = \frac{5}{11}(x - 2001) + 56$
or $f(x) = \frac{5}{11}(x - 2012) + 61$ (b) About 58.7 pounds
59. (a) $y = 863.84x - 1{,}698{,}819.9$
(b) \$24,541; interpolation

CHAPTER 3: Quadratic Functions and Equations

SECTION 3.1 (pp. 183–187)

1. Quadratic; leading coefficient: 3; $f(-2) = 17$
3. Neither linear nor quadratic   **5.** Linear
7. (a) $a > 0$ (b) $(1, 0)$ (c) $x = 1$
(d) Incr: $x \ge 1$, or $[1, \infty)$; decr: $x \le 1$, or $(-\infty, 1]$
9. (a) $a < 0$ (b) $(-3, -2)$ (c) $x = -3$
(d) Incr: $x \le -3$, or $(-\infty, -3]$; decr: $x \ge -3$, or $[-3, \infty)$
11. The graph of g is narrower than the graph of f.
13. The graph of g is wider than the graph of f and opens downward rather than upward.
15. Vertex: $(1, 2)$; leading coefficient: -3;
$f(x) = -3x^2 + 6x - 1$
17. Vertex: $(4, 5)$; leading coefficient: -2;
$f(x) = -2x^2 + 16x - 27$
19. Vertex: $\left(-5, -\frac{7}{4}\right)$; leading coefficient: $\frac{3}{4}$;
$f(x) = \frac{3}{4}x^2 + \frac{15}{2}x + 17$
21. $f(x) = (x - 2)^2 - 2$ **23.** $f(x) = \frac{1}{2}(x - 2)^2 - 3$
25. $f(x) = -2(x + 1)^2 + 3$
27. $f(x) = -3(x - 2)^2 + 6$
29. $f(x) = (x + 2)^2 - 9$; vertex: $(-2, -9)$

31. $f(x) = \left(x - \frac{3}{2}\right)^2 - \frac{9}{4}$; vertex: $\left(\frac{3}{2}, -\frac{9}{4}\right)$

33. $f(x) = 2\left(x - \frac{5}{4}\right)^2 - \frac{1}{8}$; vertex: $\left(\frac{5}{4}, -\frac{1}{8}\right)$

35. $f(x) = \frac{1}{3}\left(x + \frac{3}{2}\right)^2 + \frac{1}{4}$; vertex: $\left(-\frac{3}{2}, \frac{1}{4}\right)$

37. $f(x) = 2(x - 2)^2 - 9$; vertex: $(2, -9)$

39. $f(x) = -3(x + 1.5)^2 + 8.75$; vertex: $(-1.5, 8.75)$

41. (a) $(0, 6)$ **(b)** Incr: $x \le 0$, or $(-\infty, 0]$; decr: $x \ge 0$, or $[0, \infty)$

43. (a) $(3, -9)$ **(b)** Incr: $x \ge 3$, or $[3, \infty)$; decr: $x \le 3$, or $(-\infty, 3]$

45. (a) $(1, -1)$ **(b)** Incr: $x \ge 1$, or $[1, \infty)$; decr: $x \le 1$, or $(-\infty, 1]$

47. (a) $(0, 10)$ **(b)** Incr: $x \ge 0$, or $[0, \infty)$; decr: $x \le 0$, or $(-\infty, 0]$

49. (a) $\left(\frac{1}{3}, -\frac{35}{12}\right)$ **(b)** Incr: $x \le \frac{1}{3}$, or $\left(-\infty, \frac{1}{3}\right]$; decr: $x \ge \frac{1}{3}$, or $\left[\frac{1}{3}, \infty\right)$

51. (a) $\left(-\frac{1}{4}, \frac{15}{8}\right)$ **(b)** Incr: $x \le -\frac{1}{4}$, or $\left(-\infty, -\frac{1}{4}\right]$; decr: $x \ge -\frac{1}{4}$, or $\left[-\frac{1}{4}, \infty\right)$

53. **55.**

57. **59.**

61. **63.**

65. **67.**

69. **71.**

73. -7 **75.** $6x + 3h - 2$ **77.** $f(x) = 2(x - 3)^2 + 1$

79. d **81.** a **83.** 250 ft by 250 ft

85. (a) $R(2) = 72$; the company receives $72,000 for producing 2000 CD players. **(b)** 10,000 **(c)** $200,000

87. (a) 32 ft **(b)** 34.25 ft

89. (a) $s(t) = -16t^2 - 66t + 120$

(b) Yes, because $s(2) < 0$.

91. 40 ft by 80 ft **93.** 146 ft after 2.75 sec

95. 323 ft after 6.77 sec **97.** $f(x) = 2(x - 1)^2 - 3$

99. $f(x) = \frac{1}{225} x^2 + 20$, or $f(x) \approx 0.0044x^2 + 20$

101. (a) $H(t) = 2(t - 4)^2 + 90$; $D = \{t \mid 0 \le t \le 4\}$

(b) $H(1.5) = 102.5$ beats per minute

103. (a) $f(x) = 3.06(x - 1982)^2 + 1.6$ (answers may vary)

(b) $[1980, 1996, 2]$ by $[-50, 500, 100]$

(c) About 250; by 1991 a total of about 250 thousand AIDS cases had been reported.

105. $f(x) = 3.125x^2 + 2.05x - 0.9$; $f(3.5) \approx 44.56$

107. (a) $f(x) \approx 0.00019838x^2 - 0.79153x + 791.46$

(b) $f(1975) \approx 1.99$, which is close to the actual value.

109. (a) $f(x) \approx 0.59462x^2 - 2350.82x + 2,323,895$

(b) About 454 thousand

3.1 EXTENDED AND DISCOVERY EXERCISES (p. 187)

1. Quadratic; $y = -61$ when $x = 6$

3. Neither

5. (a)

x	1	2	3	4	5
$f(x)$	-2	1	6	13	22

(b) 3; 5; 7; 9 **(c)** $2x + h$; $2x + 1$

(d) 3; 5; 7; 9; the results are the same.

7. (a)

x	1	2	3	4	5
$f(x)$	0	-3	-10	-21	-36

(b) $-3; -7; -11; -15$

(c) $-4x + 3 - 2h$; $-4x + 1$

(d) $-3; -7; -11; -15$; the results are the same.

SECTION 3.2 (pp. 200–204)

1. $-4, 3$ **3.** $0, 2$ **5.** $0, \frac{7}{3}$ **7.** $-1, \frac{15}{2}$ **9.** $-5, \frac{1}{3}$

11. $\frac{1}{2}, \frac{5}{6}$ **13.** $-3 \pm \sqrt{5}$ **15.** $\pm \frac{\sqrt{13}}{2}$

17. No real solutions   **19.** $3 \pm 2\sqrt{2}$

21. $\frac{-1 \pm \sqrt{13}}{3}$ **23.** $\frac{1}{5}$ **25.** $\frac{5 \pm \sqrt{85}}{30}$ **27.** $-2, -1$

29. $-\frac{5}{2}, \frac{1}{3}$ **31.** $\frac{3}{2}$ **33.** $\frac{2}{3}, 3$ **35.** $-2, 0$ **37.** $-2, 3$

39. $\pm\sqrt{3} \approx \pm 1.7$ **41.** 1.5 **43.** $-0.75, 0.2$

45. $0.7, 1.2$ **47.** $-2 \pm \sqrt{10}$

49. $-\frac{5}{2} \pm \frac{1}{2}\sqrt{41}$ **51.** $1 \pm \frac{\sqrt{15}}{3}$ **53.** $4 \pm \sqrt{26}$

55. $\frac{3 \pm \sqrt{17}}{2}$ **57.** $\frac{3 \pm \sqrt{17}}{4}$ **59.** $\frac{-1 \pm \sqrt{97}}{12}$

61. $\{x \mid x \neq \sqrt{5}, x \neq -\sqrt{5}\}$

63. $\{t \mid t \neq -1, t \neq 2\}$ **65.** $y = \frac{-12x^2 + 1}{8}$; yes

67. $y = \frac{x}{5}$; yes   **69.** $y = 3 \pm \sqrt{9 - x^2}$; no

71. $y = \pm\frac{\sqrt{12 - 3x^2}}{2}$; no   **73.** $r = \pm\sqrt{\frac{3V}{\pi h}}$

75. $v = \pm\sqrt{\frac{2K}{m}}$ **77.** $b = \pm\sqrt{c^2 - a^2}$

79. $t = \frac{25 \pm \sqrt{625 - 4s}}{8}$

81. (a) $3x^2 - 12 = 0$ (b) $b^2 - 4ac = 144 > 0$. There are two real solutions.   (c) ± 2

83. (a) $x^2 - 2x + 1 = 0$ (b) $b^2 - 4ac = 0$. There is one real solution.   (c) 1

85. (a) $x^2 - 4x = 0$ (b) $b^2 - 4ac = 16 > 0$. There are two real solutions.   (c) $0, 4$

87. (a) $x^2 - x + 1 = 0$ (b) $b^2 - 4ac = -3 < 0$. There are no real solutions.

89. (a) $2x^2 + 5x - 12 = 0$ (b) $b^2 - 4ac = 121 > 0$. There are two real solutions.   (c) $-4, \frac{3}{2}$

91. (a) $9x^2 - 36x + 36 = 0$ (b) $b^2 - 4ac = 0$. There is one real solution.   (c) 2

93. (a) $\frac{1}{2}x^2 + x + \frac{13}{2} = 0$ (b) $b^2 - 4ac = -12 < 0$. There are no real solutions.

95. (a) $3x^2 + x - 1 = 0$ (b) $b^2 - 4ac = 13 > 0$. There are two real solutions. (c) $\frac{1 \pm \sqrt{13}}{6}$

97. (a) $a > 0$ (b) $-6, 2$ (c) Positive

99. (a) $a > 0$ (b) -4 (c) Zero

101. 2.2 seconds   **103.** 1989   **105.** 8.5 in. by 11 in.; yes

107. 15 in. by 25 in.   **109.** About 1.49 in.

111. About 18.23 in.

113. 86 shirts   **115.** (a) $s(t) = -16t^2 + 160t + 32$ (b) About 10.2 sec

117. (a) R quadruples.   (b) $a = 0.5$ (c) $x = \sqrt{1000} \approx 31.6$. The safe speed for a curve with radius 500 ft is about 31 mi/hr or slower.

119. (a) $E(15) = 1.4$; in 2002 there were 1.4 million Wal-Mart employees. (b) $f(x) \approx 0.00474x^2 + 0.00554x + 0.205$ (answers may vary)

(c) [0, 25, 5] by [0, 2.6, 0.2]

(d) About 2011 (answers may vary)

3.2 EXTENDED AND DISCOVERY EXERCISES (p. 204)

1. $676 = 26^2$; yes; $-\frac{5}{2}, \frac{3}{4}$ **3.** 69; no; $\frac{3 \pm \sqrt{69}}{10}$

7. $x^2 = k \Rightarrow \sqrt{x^2} = \sqrt{k} \Rightarrow |x| \Rightarrow \sqrt{k} \Rightarrow x = \pm\sqrt{k}$

CHECKING BASIC CONCEPTS FOR SECTIONS 3.1 AND 3.2 (p. 204)

1. Vertex; $(1, -4)$; axis of symmetry: $x = 1$; x-intercepts: $-1, 3$

3. $f(x) = 2(x + 1)^2 + 3$

5. $f(x) = (x + 2)^2 - 7; (-2, -7); -7$

7. 11 in. by 15 in.

SECTION 3.3 (pp. 211–212)

1. $2i$ **3.** $10i$ **5.** $i\sqrt{23}$ **7.** $2i\sqrt{3}$ **9.** $3i\sqrt{6}$

11. $2 \pm 2i$ **13.** $-2 \pm 2i\sqrt{2}$ **15.** -5 **17.** -6

19. $-3\sqrt{2}$ **21.** $8i$ **23.** $-2 - i$ **25.** $5 - 21i$

27. $-1 + 6i$ **29.** $4 + 8i$ **31.** $5 - i$ **33.** $4 - 7i$

35. $-5 - 12i$ **37.** 4 **39.** $\frac{1}{2} - \frac{1}{2}i$ **41.** $\frac{19}{26} + \frac{9}{26}i$

43. $-\frac{2}{25} + \frac{4}{25}i$ **45.** $3i$ **47.** $\frac{1}{2} + i$

49. $-18.5 + 87.4i$ **51.** $8.7 - 6.7i$

53. $-117.27 + 88.11i$ **55.** $-0.921 - 0.236i$

57. $\pm i\sqrt{5}$ **59.** $\pm i\sqrt{\frac{1}{2}}$ **61.** $\frac{3}{10} \pm \frac{i\sqrt{11}}{10}$ **63.** $2 \pm i$

65. $\frac{3}{2} \pm \frac{i\sqrt{11}}{2}$ **67.** $-1 \pm i\sqrt{3}$ **69.** $1 \pm \frac{i\sqrt{2}}{2}$

71. $\frac{5}{4} \pm \frac{i\sqrt{7}}{4}$ **73.** $\frac{11}{8} \pm \frac{i\sqrt{7}}{8}$

75. (a) Two real zeros   (b) $-1, \frac{3}{2}$

77. (a) Two imaginary zeros   (b) $-\frac{1}{2} \pm \frac{i\sqrt{7}}{2}$

79. (a) Two imaginary zeros   (b) $\pm i\sqrt{2}$

3.3 EXTENDED AND DISCOVERY EXERCISE (p. 212)

1. (a) $i^1 = i, i^2 = -1, i^3 = -i, i^4 = 1, i^5 = i, i^6 = -1, i^7 = -i, i^8 = 1$, and so on.   (b) Divide n by 4. If the remainder is r, then $i^n = i^r$, where $i^0 = 1, i^1 = i, i^2 = -1$, and $i^3 = -i$.

SECTION 3.4 (pp. 218–221)

1. (a) $-3 < x < 2$ **(b)** $x \le -3$ or $x \ge 2$
3. (a) $x = -2$ **(b)** $x \ne -2$
5. (a) No solutions   **(b)** All real numbers
7. (a) $-\frac{5}{2}, -\frac{1}{2}$ **(b)** $\left(-\frac{5}{2}, -\frac{1}{2}\right)$, or $\left\{x \mid -\frac{5}{2} < x < -\frac{1}{2}\right\}$
(c) $\left(-\infty, -\frac{5}{2}\right) \cup \left(-\frac{1}{2}, \infty\right)$, or
$\left\{x \mid x < -\frac{5}{2} \text{ or } x > -\frac{1}{2}\right\}$
9. (a) $-1, \frac{7}{5}$ **(b)** $(-\infty, -1) \cup \left(\frac{7}{5}, \infty\right)$, or
$\left\{x \mid x < -1 \text{ or } x > \frac{7}{5}\right\}$ **(c)** $\left(-1, \frac{7}{5}\right)$, or
$\left\{x \mid -1 < x < \frac{7}{5}\right\}$
11. (a) $-3, 4$ **(b)** $(-3, 4)$, or $\{x \mid -3 < x < 4\}$
(c) $(-\infty, -3) \cup (4, \infty)$, or $\{x \mid x < -3 \text{ or } x > 4\}$
13. (a) $\pm\sqrt{5}$ **(b)** $\left[-\sqrt{5}, \sqrt{5}\right]$, or
$\left\{x \mid -\sqrt{5} \le x \le \sqrt{5}\right\}$
(c) $\left(-\infty, -\sqrt{5}\right] \cup \left[\sqrt{5}, \infty\right)$, or
$\left\{x \mid x \le -\sqrt{5} \text{ or } x \ge \sqrt{5}\right\}$
15. (a) $-\frac{8}{3}, 0$ **(b)** $\left[-\frac{8}{3}, 0\right]$, or $\left\{x \mid -\frac{8}{3} \le x \le 0\right\}$
(c) $\left(-\infty, -\frac{8}{3}\right] \cup [0, \infty)$, or $\left\{x \mid x \le -\frac{8}{3} \text{ or } x \ge 0\right\}$
17. (a) $\frac{3}{2}$ **(b)** $\left(-\infty, \frac{3}{2}\right) \cup \left(\frac{3}{2}, \infty\right)$, or
$\left\{x \mid x < \frac{3}{2} \text{ or } x > \frac{3}{2}\right\}$ **(c)** No solutions
19. (a) $\frac{2}{3}, \frac{5}{4}$ **(b)** $\left[\frac{2}{3}, \frac{5}{4}\right]$, or $\left\{x \mid \frac{2}{3} \le x \le \frac{5}{4}\right\}$
(c) $\left(-\infty, \frac{2}{3}\right] \cup \left[\frac{5}{4}, \infty\right)$, or $\left\{x \mid x \le \frac{2}{3} \text{ or } x \ge \frac{5}{4}\right\}$
21. (a) $-1 \pm \sqrt{2}$ **(b)** $\left(-1 - \sqrt{2}, -1 + \sqrt{2}\right)$, or
$\left\{x \mid -1 - \sqrt{2} < x < -1 + \sqrt{2}\right\}$
(c) $\left(-\infty, -1 - \sqrt{2}\right) \cup \left(-1 + \sqrt{2}, \infty\right)$, or
$\left\{x \mid x < -1 - \sqrt{2} \text{ or } x > -1 + \sqrt{2}\right\}$
23. (a) $x < -1$ or $x > 1$ **(b)** $-1 \le x \le 1$
25. (a) $-6 < x < -2$ **(b)** $x \le -6$ or $x \ge -2$
27. $-2 \le x \le -0.5$ **29.** $x < -3$ or $x > 2$
31. $-2 \le x \le 2$ **33.** All real numbers
35. $x \le -2$ or $x \ge 3$ **37.** $-\frac{1}{3} < x < \frac{1}{2}$
39. $-4 \le x \le 10$ **41.** No solutions
43. All real numbers except $\frac{2}{3}$ **45.** $x \le 0$ or $x \ge 1$
47. $x \le -2$ or $x \ge 3$ **49.** $-\sqrt{5} \le x \le \sqrt{5}$
51. $x \le \frac{23}{7}$ or $x \ge 22.4$ **53.** $2 \le x \le 7$
55. $x \le -2$ or $x \ge 5$ **57.** All real numbers
59. $x < -2 - \sqrt{7}$ or $x > -2 + \sqrt{7}$
61. About 35 mi/hr, but not more than 38 mi/hr
63. $2 \le r \le 3$ (inches)
65. (a) $f(0) = 160$, $f(2) = 131.2$; initially the heart rate is
160 bpm, and after 2 minutes it is about 131 bpm.
(b) About $2.9 \le x \le 5$ (minutes)
67. From 1989 to 1992
69. (a) The height does not change by the same amount in
each 15-second interval.

(b) From 43 to 95 seconds (approximately)
(c) $[-25, 200, 25]$ by $[-5, 20, 5]$

(d) From 43.8 to 93 seconds (approximately)

**CHECKING BASIC CONCEPTS FOR
SECTIONS 3.3 AND 3.4 (p. 221)**

1. (a) $5i$ **(b)** $-3\sqrt{6}$ **(c)** $\frac{1}{2} \pm \frac{i\sqrt{2}}{2}$
3. (a) $[-3, 0]$, or $\{x \mid -3 \le x \le 0\}$; $(-\infty, -3) \cup (0, \infty)$,
or $\{x \mid x < -3 \text{ or } x > 0\}$
(b) $(-\infty, \infty)$, or $\{x \mid -\infty < x < \infty\}$; no solutions
5. (a) $\left(-\infty, -\sqrt{5}\right] \cup \left[\sqrt{5}, \infty\right)$, or
$\left\{x \mid x \le -\sqrt{5} \text{ or } x \ge \sqrt{5}\right\}$
(b) $(-\infty, \infty)$, or $\{x \mid -\infty < x < \infty\}$
(c) $\left[\frac{1 - \sqrt{5}}{2}, \frac{1 + \sqrt{5}}{2}\right]$, or
$\left\{x \mid \frac{1 - \sqrt{5}}{2} \le x \le \frac{1 + \sqrt{5}}{2}\right\}$

SECTION 3.5 (pp. 235–239)

1. $y = (x + 2)^2$ **3.** $y = \sqrt{x + 3}$
5. $y = |x + 2| - 1$
7. $y = \sqrt{x + 2} - 3$ **9.** $y = (x - 2)^2 - 3$
$[-10, 10, 1]$ by $[-10, 10, 1]$

11. $y = (x + 6)^2 - 4(x + 6) + 5$
$[-10, 10, 1]$ by $[-10, 10, 1]$

13. $y = \frac{1}{2}(x + 3)^2 + 2(x + 3) - 3$
$[-10, 10, 1]$ by $[-10, 10, 1]$

15. (a) $g(x) = 3(x + 3)^2 + 2(x + 3) - 5$
(b) $g(x) = 3x^2 + 2x - 9$
17. (a) $g(x) = 2(x - 2)^2 - 4(x - 2) + 5$
(b) $g(x) = 2(x + 8)^2 - 4(x + 8) - 4$

19. (a) $g(x) = 3(x - 2000)^2 - 3(x - 2000) + 72$
(b) $g(x) = 3(x + 300)^2 - 3(x + 300) - 28$
21. (a) $g(x) = -\sqrt{x} - 4$ **(b)** $g(x) = \sqrt{-x} + 2$
23. $(x - 3)^2 + (y + 4)^2 = 4$; center: $(3, -4)$; $r = 2$
25. $(x + 5)^2 + (y - 3)^2 = 5$; center: $(-5, 3)$; $r = \sqrt{5}$
27. (a)

(b)

(c)

29. (a)

(b)

(c)

31. (a)

(b)

(c)

33. (a)

(b)

(c)

35. Shift the graph of $y = x^2$ right 3 units and upward 1 unit.
37. Shift the graph of $y = x^2$ left 1 unit and vertically shrink it with factor $\frac{1}{4}$.
39. Reflect the graph of $y = \sqrt{x}$ across the x-axis and shift it left 5 units.
41. Reflect the graph of $y = \sqrt{x}$ across the y-axis and vertically stretch it with factor 2.
43. Reflect the graph of $y = |x|$ across the y-axis and shift it left 1 unit.

45.

47.

49.

51.

53.

55.

57.

59.

61.

63.

65.

67.

69. x-axis:

y-axis:

71. x-axis:

y-axis:

73. x-axis:

y-axis:

75.

x	1	2	3	4	5	6
$g(x)$	12	8	13	9	14	16

77.

x	-2	0	2	4	6
$g(x)$	5	2	-3	-5	-9

79.

x	0	1	2	3	4	5
$g(x)$	0	2	1	5	6	8

81.

x	-2	-1	0	1	2
$g(x)$	0	3	6	9	12

83. $(-12, 8)$, $(0, 10)$, and $(8, -2)$

85. $(-10, 7)$, $(2, 9)$, and $(10, -3)$

87. $(-12, -3)$, $(0, -4)$, and $(8, 2)$

89. $(6, 6)$, $(0, 8)$, and $(-4, -4)$

91. $f(x) = \frac{1}{4}(x - 1997)^2 + 3.8$

93. $f(x) = 0.22(x - 1970)^2 + 30$

95. $g(x) = 0.00075(x - 1990)^2 + 0.17(x - 1990) + 44$

97. (a) When $x = 2$, the y-value is about 9. There are about 9 hours of daylight on February 21 at 60°N latitude.
(b) Since February 21 is 2 months after the shortest day and October 21 is 2 months before, they both have approximately the same number of daylight hours. A reasonable conjecture would be 9 hours.
(c) There are approximately the same number of daylight hours x months before and after December 21. The left side of the graph should be a reflection about the y-axis, as shown in the figure.

99. $y = -0.4(x - 3)^2 + 4$ (mountain)
$[-4, 4, 1]$ by $[0, 6, 1]$

101. (a) $y = \frac{1}{20}x^2 - 1.6$
$[-15, 15, 1]$ by $[-10, 10, 1]$

(b) $y = \frac{1}{20}(x - 2.1)^2 - 2.5$.
The front has reached Columbus by midnight.
$[-15, 15, 1]$ by $[-10, 10, 1]$

3.5 EXTENDED AND DISCOVERY EXERCISES (p. 239)

1. No, they are not commutative.
3. Yes, they are commutative.

CHECKING BASIC CONCEPTS FOR SECTION 3.5 (p. 239)

1. (a) Shifted 4 units to the left
(b) Shifted 3 units down
(c) Shifted 5 units to the right and 3 units up
3. (a) $y = (x - 3)^2 - 4(x - 3) - 3$
(b) $y = -x^2 + 4x - 1$
(c) $y = (-x + 6)^2 - 4(-x + 6) + 1$
(d) $y = (-(x + 6))^2 - 4(-(x + 6)) + 1$
5. (a)

x	-2	0	2	4	6
$g(x)$	4	6	9	11	12

(b)

x	-5	-3	-1	1	3
$h(x)$	-2	-6	-12	-16	-18

CHAPTER 3 REVIEW EXERCISES (pp. 244–247)

1. (a) $a < 0$ **(b)** $(2, 4)$ **(c)** $x = 2$
(d) Incr: $x \le 2$, or $(-\infty, 2]$; Decr: $x \ge 2$, or $[2, \infty)$
3. $f(x) = -2x^2 + 20x - 49$; leading coefficient: -2
5. $f(x) = -(x + 1)^2 + 2$
7. $f(x) = (x + 3)^2 - 10$; vertex: $(-3, -10)$ **9.** $\left(\frac{1}{3}, -\frac{11}{3}\right)$

11.   **13.**

15.   **17.**

19. -29 **21.** $-4, 5$ **23.** $0, 4$

25. $\pm\frac{\sqrt{7}}{2}$ **27.** $-\frac{7}{2}, 2$ **29.** $-2, 5$ **31.** $-\frac{1}{2}, \frac{5}{2}$

33. (a) $a > 0$ **(b)** $-3, 2$ **(c)** Positive

35. $-1 \pm \sqrt{6}$ **37.** $\frac{3 \pm \sqrt{11}}{2}$ **39.** $4 \pm \sqrt{14}$

41. $y = \pm\sqrt{\frac{2x^2 - 6}{3}}$; no

43. (a) $4i$ **(b)** $4i\sqrt{3}$ **(c)** $-5\sqrt{3}$
45. (a) $-\frac{5}{2}, \frac{1}{2}$ **(b)** $-\frac{5}{2}, \frac{1}{2}$
47. (a) None   **(b)** $-\frac{1}{2} \pm \frac{i\sqrt{7}}{2}$ **49.** $\pm\frac{3}{2}i$

51. (a) $-3 < x < 2$, or $(-3, 2)$
(b) $x \le -3$ or $x \ge 2$, or $(-\infty, -3] \cup [2, \infty)$
53. (a) $1, 2$ **(b)** $1 < x < 2$, or $(1, 2)$
(c) $x < 1$ or $x > 2$, or $(-\infty, 1) \cup (2, \infty)$
55. $\{x \mid 1 \le x \le 2\}$, or $[1, 2]$
57. $\{x \mid -1 < x < -\frac{1}{2}\}$, or $\left(-1, -\frac{1}{2}\right)$
59. $\{x \mid x \le -3 \text{ or } x \ge 5\}$, or $(-\infty, -3] \cup [5, \infty)$
61. $y = -f(x)$ $y = f(-x)$

63.   **65.**

67.   **69.**

x	0	1	2	3
$g(x)$	11	-7	-10	-19

71. 11 ft by 22 ft
73. (a) $h(0) = 5$; the stone was 5 ft above the ground when it was released.   **(b)** 117 ft   **(c)** 126 ft
(d) After 2 seconds and 3.5 seconds
75. 11 in. by 15 in.
77. (a) $f(x) \approx 0.00051x^2 - 0.00604x + 1.6$
(b) About 70.4; when 70.4% of the area is irrigated, the yield will be 3.7 tons per hectare.

CHAPTER 3 EXTENDED AND DISCOVERY EXERCISES (pp. 246–247)

1. (a) 23.32 ft/sec
(b) $[-1, 16, 1]$ by $[-1, 16, 1]$ Yes   **(c)** 12.88 ft

3. (a) $[-1, 8, 1]$ by $[-4, 4, 1]$

(b) The graph of $y = f(2k - x) = f(4 - x)$ is a reflection of $y = f(x)$ across the line $x = 2$.

5. (a) $[-15, 3, 1]$ by $[-3, 9, 1]$

(b) The graph of $f(2k - x) = f(-12 - x)$ is a reflection of $y = f(x)$ across the line $x = -6$.

7. (a) $[0, 1200, 100]$ by $[-800, 0, 100]$

The front reached St. Louis, but not Nashville.

(b) $g(x) = -\sqrt{750^2 - (x - 160)^2} - 110$

(c) $[0, 1200, 100]$ by $[-800, 0, 100]$

The cold front reached both cities in less than 12 hours.

CHAPTER 4: More Nonlinear Functions and Equations

SECTION 4.1 (pp. 256–260)

1. Yes; degree: 3; a: 2 **3.** No
5. Yes; degree: 4; a: -5
7. No **9.** Yes; degree: 0; a: 22
11. (a) Local maximum: approximately 5.5; local minimum: approximately -5.5 **(b)** No absolute extrema
13. (a) Local maxima: approximately 17 and 27; local minima: approximately -10 and 24 **(b)** No absolute extrema
15. (a) Local maxima: approximately 0.5 and 2.8; local minimum: approximately 0 **(b)** Absolute maximum: 2.8; no absolute minimum
17. (a) Local maximum: 0; local minimum: approximately -1000 **(b)** No absolute maximum; absolute minimum: -1000

19. (a) Local maximum: 1; local minimum: -1
(b) Absolute maximum: 1; absolute minimum: -1
21. (a) No local maxima; local minimum: approximately -3.2 **(b)** Absolute maximum: 3; absolute minimum: approximately -3.2
23. (a) Local maxima: approximately 0.5 and 2; local minima: approximately -2 and -0.5 **(b)** Absolute maximum: 2; absolute minimum: -2
25. (a) No local maxima; local minimum: -2
(b) No absolute maximum; absolute minimum: -2
27. (a) No local extrema **(b)** No absolute extrema
29. (a) Local minimum: 1; no local maxima
(b) Absolute minimum: 1; no absolute maximum
31. (a) Local maximum: 4; no local minima
(b) Absolute maximum: 4; no absolute minimum
33. (a) Local minimum: $-\frac{1}{8}$; no local maxima
(b) Absolute minimum: $-\frac{1}{8}$; no absolute maximum
35. (a) Local minimum: 0; no local maxima
(b) Absolute minimum: 0; no absolute maximum
37. (a) No local extrema **(b)** No absolute extrema
39. (a) Local minimum: -2; local maximum: 2
(b) No absolute extrema
41. (a) Local maxima: 19, -8; local minimum: -13
(b) Absolute maximum: 19; no absolute minimum
43. (a) Local minimum: -8; local maximum: 4.5
(b) Absolute minimum: -8; no absolute maximum
45. (a) Local maximum: 8; no local minima
(b) Absolute maximum: 8; no absolute minimum
47. Neither **49.** Even
51. Odd **53.** Neither **55.** Even **57.** Even
59. Odd **61.** Neither **63.** Even
65. Even **67.** Even **69.** Neither **71.** Odd
73. Note that $f(0)$ can be any number.

x	-3	-2	-1	0	1	2	3
$f(x)$	21	-12	-25	1	-25	-12	21

75. $f(5) = 6$; $f(3) = -4$
77. Answers may vary.

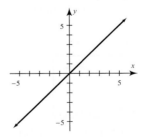

79. No. If $(2, 5)$ is on the graph of an odd function f, then so is $(-2, -5)$. Since f would pass through $(-3, -4)$ and then $(-2, -5)$, it could not always be increasing.

81. Answers may vary.

83. Answers may vary; yes, but it does not have to be quadratic.

85.

87.

89. On [0, 3]; on [3, 6]

91. (a) Absolute maximum: 84°F; absolute minimum: 63°F; the high temperature was 84°F and the low was 63°F.   **(b)** Local maxima: approximately 78°F and 84°F; local minima: approximately 63°F and 72°F
(c) $1.6 \le x \le 2.9$; $3.8 \le x \le 5$ (approximate)
93. (a) Possible absolute maximum in January and absolute minimum in July   **(b)** Absolute maximum: $140; absolute minimum: $15; the maximum cost, $140, occurs in January and the minimum cost, $15, occurs in July.

[1, 12, 1] by [0, 150, 10]   [1, 12, 1] by [0, 150, 10]

95. (a) $f(5) \approx 8.32$; in 1955, U.S. consumption of energy was about 8.32 quadrillion Btu.
(b) The energy consumption increased, reached a maximum value, and then decreased.

[0, 30, 5] by [6, 16, 1]   [0, 30, 5] by [6, 16, 1]

(c) Local maximum: approximately 14.5; in 1973 or 1974, energy use peaked at 14.5 quadrillion Btu.
97. (a) Even   **(b)** 83°F   **(c)** They are equal.
(d) Monthly average temperatures are symmetric about July. July has the highest average and January the lowest.

The pairs June-August, May-September, April-October, March-November, and February-December have approximately the same average temperatures.

4.1 EXTENDED AND DISCOVERY EXERCISES (p. 260)

1. The maximum area occurs when the figure is a rectangle with length about 4.24 and height about 2.12.
3. (a) About 1 hour 54 minutes   **(b)** About 2 hours 8 minutes   **(c)** About 1 hour 46 minutes

SECTION 4.2 (pp. 270–276)

1. (a) The turning points are approximately (1.6, 3.6), (3, 1.2), (4.4, 3.6).   **(b)** After 1.6 minutes, the runner is 360 feet from the starting line. The runner turns and jogs toward the starting line. After 3 minutes, the runner is 120 feet from the starting line. The runner turns and jogs away from the starting line. After 4.4 minutes, the runner is again 360 feet from the starting line. The runner turns and jogs back to the starting line.
3. (a) 0; 0.5   **(b)** $a > 0$ **(c)** 1
5. (a) 3; $-6, -1$, and 6   **(b)** $a < 0$ **(c)** 4
7. (a) 4; $-3, -1, 0, 1$, and 2   **(b)** $a > 0$ **(c)** 5
9. (a) 2; -3 **(b)** $a > 0$ **(c)** 3
11. (a) 1; -1 and 2   **(b)** $a > 0$ **(c)** 2
13. (a) d   **(b)** (1, 0)   **(c)** 1   **(d)** No local maxima; local minimum: 0   **(e)** No absolute maxima; absolute minimum: 0
15. (a) b   **(b)** $(-3, 27), (1, -5)$ **(c)** $x \approx -4.9$, $x = 0, x \approx 1.9$ **(d)** Local maximum: 27; local minimum: -5 **(e)** No absolute maximum; no absolute minimum
17. (a) a   **(b)** $(-2, 16), (0, 0), (2, 16)$
(c) $x \approx -2.8, x = 0, \approx 2.8$ **(d)** Local maximum: 16; local minimum: 0   **(e)** Absolute maximum: 16; no absolute minimum
19. (a) $[-10, 10, 1]$ by $[-10, 10, 1]$

(b) $(-3, 6), (3, -6)$ **(c)** Local minimum: -6; local maximum: 6
21. (a) $[-10, 10, 1]$ by $[-10, 10, 1]$

(b) There are three turning points located at $(-3, -7.025), (0, -5)$, and $(3, -7.025)$.
(c) Local minimum: -7.025; local maximum: -5

23. (a) [−10, 10, 1] by [−10, 10, 1]

(b) $\left(\frac{1}{3}, \frac{2}{3}\right) \approx (0.333, 0.667)$

(c) Local minimum: $\frac{2}{3} \approx 0.667$; no local maximum

25. (a) [−10, 10, 1] by [−10, 10, 1]

(b) $\left(-2, \frac{10}{3}\right) \approx (-2, 3.333), \left(1, -\frac{7}{6}\right) \approx (1, -1.167)$

(c) Local minimum: $-\frac{7}{6} \approx -1.167$; local maximum: $\frac{10}{3} \approx 3.333$

27. (a) Degree: 1; leading coefficient: −2

(b) Up on left end, down on right end; $f(x) \rightarrow \infty$ as $x \rightarrow -\infty$, $f(x) \rightarrow -\infty$ as $x \rightarrow \infty$

29. (a) Degree: 2; leading coefficient: 1

(b) Up on both ends; $f(x) \rightarrow \infty$ as $x \rightarrow -\infty$, $f(x) \rightarrow \infty$ as $x \rightarrow \infty$

31. (a) Degree: 3; leading coefficient: −2

(b) Up on left end, down on right end; $f(x) \rightarrow \infty$ as $x \rightarrow -\infty$, $f(x) \rightarrow -\infty$ as $x \rightarrow \infty$

33. (a) Degree: 3; leading coefficient: −1

(b) Up on left end, down on right end; $f(x) \rightarrow \infty$ as $x \rightarrow -\infty$, $f(x) \rightarrow -\infty$ as $x \rightarrow \infty$

35. (a) Degree: 5; leading coefficient: 0.1

(b) Down on left end, up on right end; $f(x) \rightarrow -\infty$ as $x \rightarrow -\infty$, $f(x) \rightarrow \infty$ as $x \rightarrow \infty$

37. (a) Degree: 2; leading coefficient: $-\frac{1}{2}$

(b) Down on both ends; $f(x) \rightarrow -\infty$ as $x \rightarrow -\infty$, $f(x) \rightarrow -\infty$ as $x \rightarrow \infty$

39. (a)

41. (a)

(b) Degree 3

(c) $a < 0$

(d) $f(x) = -x^3 + 8x$

(b) Degree 2

(c) $a > 0$

(d) $f(x) = x^2 - 2x - 1$

43. (a)

(b) Degree 4   **(c)** $a < 0$ **(d)** $f(x) = -x^4 + 3x^2 - 1$

45. Answers may vary   **47.** Answers may vary.

49. Not possible   **51.** Not possible

53. Answers may vary.   **55.**

57. As the viewing rectangle increases in size, the graphs begin to look alike. Each formula contains the term $2x^4$, which determines the end behavior of the graph for large values of $|x|$.

(a) [−4, 4, 1] by [−4, 4, 1]   **(b)** [−10, 10, 1] by [−100, 100, 10]

(c) [−100, 100, 10] by [-10^6, 10^6, 10^5]

59. For f: 1; for g: 0.5; for h: 0.25. On the interval $[0, 0.5]$, the higher the degree of the function, the smaller the average rate of change.

61. (a) 12.01   **(b)** 12.0001   **(c)** 12.000001

The average rate of change is approaching 12.

63. (a) $4.01\overline{6}$ **(b)** $4.0001\overline{6}$ **(c)** $4.000001\overline{6}$
The average rate of change is approaching 4.
65. $9x^2 + 9xh + 3h^2$
67. $-3x^2 - 3xh - h^2 + 1$
69. $f(-2) \approx 5, f(1) \approx 0$
71. $f(-1) \approx -1, f(1) \approx 1, f(2) \approx -2$
73. $f(-3) = -63, f(1) = 3, f(4) = 10$
75. $f(-2) = 6, f(1) = 7, f(2) = 9$
77. (a)   **79. (a)**

(b) f is not continuous.   **(b)** f is continuous.
(c) ± 2 **(c)** $\sqrt{2}$
81. (a)   **83. (a)** $H(-2) = 0;$
                    $H(0) = 1; H(3.5) = 1$
                    (b)

(b) f is continuous.

(c) $-\sqrt[3]{3}, \dfrac{\sqrt{17} + 1}{2}$

85. (a) Approximately $(1, 13)$ and $(7, 72)$
(b) The low monthly average temperature of $13°F$ occurs in January. The high monthly average temperature of $72°F$ occurs in July.
87. (a) $f(x) \approx -0.15311111x^3 + 914.695238x^2 - 1,821,416.18x + 1,208,942,069.3$ (answers may vary)
(b) $f(1998) \approx 1704$; interpolation   **(c)** About 1996
89. (a) $f(x) \approx \frac{2}{3000}x^3 - 3.9x^2 + 7604.7333x - 4,942,725$
(b) $[1950, 2010, 10]$ by $[0, 80, 20]$

(c) $f(1994) \approx 50$; $f(2010) \approx 133$; the 1994 value is correct and the 2010 value is 4 thousand too high.
(d) 1994, interpolation; 2010, extrapolation

91. (a)

(b) $f(x) = \begin{cases} 8x & \text{if } 0 \le x \le 2 \\ 1.32x^2 - 14.4x + 39.39 & \text{if } 2 < x \le 5 \end{cases}$

(c) $f(1.25) = 10$ cm, $f(3.2) \approx 7$ cm   (answers may vary)
(d) About 3.55 minutes (answers may vary)

4.2 EXTENDED AND DISCOVERY EXERCISES (p. 276)

1. (a) $D = [0, 10]$, or $\{x \,|\, 0 \le x \le 10\}$
(b) $A(1) = 405$; after 1 min, the tank contains 405 gal of water.   **(c)** Degree: 2; leading coefficient: 5   **(d)** No, more than half; yes, because the water will drain faster at first.
3. Conc. down: $(-\infty, \infty)$
5. Conc. up: $(1, \infty)$; conc. down: $(-\infty, 1)$
7. Conc. up: $(-2, 2)$; conc. down: $(-\infty, -2) \cup (2, \infty)$

CHECKING BASIC CONCEPTS FOR SECTIONS 4.1 AND 4.2 (pp. 276–277)

1. (a) Incr: $[-2, 1] \cup [3, \infty)$, or $\{x \,|\, -2 \le x \le 1 \text{ or } x \ge 3\}$; decr: $(-\infty, -2] \cup [1, 3]$, or $\{x \,|\, x \le -2 \text{ or } 1 \le x \le 3\}$
(b) Local maximum: approximately 3; local minima: approximately -13 and -2
(c) No absolute maximum; absolute minimum: approximately -13 **(d)** Approximately $-3.1, 0, 2.2,$ and 3.6; they are the same values.
3. (a) Not possible
(b) Answers may vary.   **(c)** Answers may vary.

(d) Not possible
5. $f(x) \approx -1.01725x^4 + 10.319x^2 - 10$

SECTION 4.3 (pp. 283–285)

1. $\frac{x^3}{2} - \frac{3}{2x}$ **3.** $x - \frac{2}{3x} - \frac{1}{3x^3}$ **5.** $-\frac{1}{x^3} + \frac{1}{4}$

7. $5x - 10 + \frac{5}{3x}$

9. Quotient: $x^2 + x - 2$; remainder: 0

11. Quotient: $2x^3 - 9x^2 + 4x - 23$; remainder: 40

13. Quotient: $3x^2 + 3x - 4$; remainder: 6

15. $x^3 - 1$

17. $4x^2 + 3x - 2 + \frac{4}{x - 1}$ **19.** $x^2 - x + 1$

21. $3x^2 + 4x - 2 + \frac{2}{2x - 1}$ **23.** $x^3 + 2 + \frac{-2}{3x - 7}$

25. $5x^2 - 12 + \frac{30}{x^2 + 2}$ **27.** $4x + 5$

29. $x^2 - 2x + 4 + \frac{-1}{2x^2 + 3x + 2}$

31. $x^3 - 8x^2 + 15x - 6$

33. $(x - 2)(x - 1) - 1$

35. $(2x + 1)(x^2 - 1) + 1$

37. $(x^2 + 1)(x - 1) + 2$

39. $x^2 - 3x - 2$

41. $3x^2 + 4x + \frac{3}{x - 5}$ **43.** $x^3 - x^2 - 6x$

45. $2x^4 - 2x^3 + 4 + \frac{1}{x + 0.5}$

47. 3 **49.** -42 **51.** $L = 4x + 3$; 43 ft

SECTION 4.4 (pp. 296–301)

1. $(x + 2), (x + 1), (x - 1)$

3. $(x + 2), (x + 1), (x - 1), (x - 2)$

5. $f(x) = 2\left(x - \frac{11}{2}\right)(x - 7)$

7. $f(x) = (x + 2)(x - 1)(x - 3)$

9. $f(x) = -2(x + 5)\left(x - \frac{1}{2}\right)(x - 6)$

11. $f(x) = 7(x + 3)(x - 2)$

13. $f(x) = -2x(x + 1)(x - 1)$

15. $f(x) = (x + 4)(x - 2)(x - 8)$

17. $f(x) = -1(x + 8)(x + 4)(x + 2)(x - 4)$

19. $f(x) = \frac{1}{2}(x + 1)(x - 2)(x - 3)$

21. $f(x) = \frac{1}{2}(x + 1)(x - 1)(x - 2)$

23. $f(x) = -2(x + 2)(x + 1)(x - 1)(x - 2)$

25. $f(x) = 10(x + 2)\left(x - \frac{3}{10}\right)$

27. $f(x) = -3x(x - 2)(x + 3)$

29. $f(x) = x(x + 1)(x + 3)\left(x - \frac{3}{2}\right)$

31. $f(x) = (x - 1)(x - 3)(x - 5)$

33. $f(x) = -4(x + 4)\left(x - \frac{3}{4}\right)(x - 3)$

35. $f(x) = 2x(x + 2)\left(x + \frac{1}{2}\right)(x - 3)$

37. Yes **39.** No

41. -2 (odd), 4 (even); minimum degree: 5

43. $f(x) = (x + 1)^2(x - 6)$

45. $f(x) = (x - 2)^3(x - 6)$

47. $f(x) = (x + 2)^2(x - 4)$

49. $f(x) = -1(x + 3)^2(x - 3)^2$

51. $f(x) = 2(x + 1)^2(x - 1)^3$

53. **(a)** x-int: $-2, -1$; y-int: 4

(b) -2 has multiplicity 1; -1 has multiplicity 2

(c)

55. **(a)** x-int: $-2, 0, 2$; y-int: 0

(b) 0 has multiplicity 2; -2 and 2 each have multiplicity 1

(c)

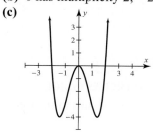

57. **(a)** $-3, \frac{1}{2}, 1$ **(b)** $f(x) = 2(x + 3)\left(x - \frac{1}{2}\right)(x - 1)$

59. **(a)** $-2, -1, 1, \frac{3}{2}$

(b) $f(x) = 2(x + 2)(x + 1)(x - 1)\left(x - \frac{2}{3}\right)$

61. **(a)** $\frac{1}{3}, 1, 4$ **(b)** $f(x) = 3\left(x - \frac{1}{3}\right)(x - 1)(x - 4)$

63. **(a)** 1 **(b)** $f(x) = \left(x + \sqrt{7}\right)(x - 1)\left(x - \sqrt{7}\right)$

65. Possible: 0 or 2 positive, 1 negative; actual: 0 positive, 1 negative

67. Possible: 1 positive, 1 negative; actual: 1 positive, 1 negative

69. Possible: 0 or 2 positive, 1 or 3 negative; actual: 0 positive, 1 negative

71. $-3, 0, 2$ **73.** $-1, 1$ **75.** $-2, 0, 2$

77. $-5, 0, 5$ **79.** ± 2 **81.** $-3, 0, 6$

83. 0, 1 **85.** $-\frac{1}{4}, 0, \frac{5}{3}$

87. $\pm \frac{2}{3}, \pm 1$ **89.** $-1, \pm \frac{\sqrt{3}}{2}$ **91.** $\pm 2, \frac{1}{2}$

93. $\pm \frac{3}{2}, \pm \frac{\sqrt{6}}{2}$ **95.** $-2, 3$ **97.** $-2.01, 0.12, 2.99$

99. $-4.05, -0.52, 1.71$

101. $-2.69, -1.10, 0.55, 3.98$

103. Because $f(2) = -1 < 0$ and $f(3) = 4 > 0$, the intermediate value property states that there exists an x-value between 2 and 3 where $f(x) = 0$.

105. Because $f(0) = -1 < 0$ and $f(1) = 1 > 0$, the intermediate value property states that there exists an x-value between 0 and 1 where $f(x) = 0$.

107. 4, 32; yes, by the intermediate value property

109. 12 A.M., 2 A.M., 4 A.M.

111. Approximately 11.34 cm

113. June 2, June 22, and July 12

115. (a) $f(x) \approx -0.184(x + 6.01)(x - 2.15)(x - 11.7)$
(b) The zero of -6.01 has no significance. The zeros of $2.15 \approx 2$ and $11.7 \approx 12$ indicate that during February and December the average temperature is $0°F$.

117. (a) As x increases, C decreases.
(b) $C(x) \approx -0.000068x^3 + 0.0099x^2 - 0.653x + 23$
(c) $[0, 70, 10]$ by $[0, 22, 5]$

(d) $0 \le x < 32.1$ (approximately)

4.4 EXTENDED AND DISCOVERY EXERCISES (p. 301)

1. Dividing $P(x)$ by $x - 2$ synthetically results in the following bottom row: 1   1   5   2   12.
Since $2 > 0$ and the bottom row values are all nonnegative, there is no real zero greater than 2.

3. Dividing $P(x)$ by $x + 2$ synthetically results in the following bottom row: 1 -1 1 -2 7.
Since $-2 < 0$ and the bottom row values alternate in sign, there is no real zero less than -2.

5. Divding $P(x)$ by $x - 1$ synthetically results in the following bottom row: 3   5   1   2   1.
Since $1 > 0$ and the bottom row values are all nonnegative, there is no real zero greater than 1.

CHECKING BASIC CONCEPTS FOR SECTIONS 4.3 AND 4.4 (p. 301)

1. $x^2 - 2x + 1$

3. $f(x) = -\frac{1}{2}(x + 2)^2(x - 1)$;
-2 has multiplicity 2; 1 has multiplicity 1

5. The zeros are $-4, -1, 2,$ and 4;
$f(x) = (x + 4)(x + 1)(x - 2)(x - 4)$

SECTION 4.5 (pp. 307–308)

1. Two imaginary zeros

3. One real zero; two imaginary zeros

5. Two real zeros; two imaginary zeros

7. Three real zeros; two imaginary zeros

9. (a) $f(x) = (x - 6i)(x + 6i)$ **(b)** $f(x) = x^2 + 36$

11. (a) $f(x) = -1(x + 1)(x - 2i)(x + 2i)$
(b) $f(x) = -x^3 - x^2 - 4x - 4$

13. (a) $f(x) = 10(x - 1)(x + 1)(x - 3i)(x + 3i)$
(b) $f(x) = 10x^4 + 80x^2 - 90$

15. (a) $f(x) = \frac{1}{2}(x + i)(x - i)(x + 2i)(x - 2i)$
(b) $f(x) = \frac{1}{2}x^4 + \frac{5}{2}x^2 + 2$

17. (a) $f(x) = -2(x - (1 - i))(x - (1 + i))(x - 3)$
(b) $f(x) = -2x^3 + 10x^2 - 16x + 12$

19. $\frac{5}{3}, \pm 5i$ **21.** $\pm 3i, \frac{1}{4} \pm \frac{i\sqrt{7}}{4}$

23. (a) $\pm 5i$ **(b)** $f(x) = (x - 5i)(x + 5i)$

25. (a) $0, \pm i$ **(b)** $f(x) = 3(x - 0)(x - i)(x + i)$, or
$f(x) = 3x(x - i)(x + i)$

27. (a) $\pm i, \pm 2i$
(b) $f(x) = (x - i)(x + i)(x - 2i)(x + 2i)$

29. (a) $-2, \pm 4i$ **(b)** $f(x) = (x + 2)(x + 4i)(x - 4i)$

31. $0, \pm i$ **33.** $2, \pm i\sqrt{7}$ **35.** $0, \pm i\sqrt{5}$

37. $0, \frac{1}{2} \pm \frac{i\sqrt{15}}{2}$ **39.** $-2, 1, \pm i\sqrt{8}$

41. $-2, \frac{1}{3} \pm \frac{i\sqrt{8}}{3}$ **43.** $Z = 10 + 6i$

45. $V = 11 + 2i$ **47.** $I = 1 + i$

SECTION 4.6 (pp. 319–322)

1. Yes; $D = \left\{x \mid x \ne \frac{5}{4}\right\}$

3. Yes; $D =$ all real numbers

5. No; $D = \{x \mid x \ne -1\}$

7. Yes; $D =$ all real numbers

9. No; $D = \{x \mid x \ne -1, x \ne 0\}$

11. Yes; $D = \{x \mid x \ne -1\}$

13. Horizontal: $y = 4$; vertical: $x = 2$; $D = \{x \mid x \ne 2\}$

15. Horizontal: $y = -4$; vertical: $x = \pm 2$;
$D = \{x \mid x \ne 2, x \ne -2\}$

17. Horizontal: $y = 0$; vertical: none; $D =$ all real numbers

19. $y = 3$ **21.** Horizontal: $y = 2$; vertical: $x = 3$

23. Horizontal: $y = 0$; vertical: $x = \pm\sqrt{5}$

25. Horizontal: none; vertical: $x = -5$ or 2

27. Horizontal: $y = \frac{1}{2}$; vertical: $x = \frac{5}{2}$

29. Horizontal: $y = 3$; vertical: $x = 1$

31. Horizontal: none; vertical: none, since $f(x) = x - 3$
for $x \ne -3$ **33.** b   **35.** d

37. $f(x) = \frac{x + 1}{x + 3}$ (answers may vary)

39. $f(x) = \frac{1}{x^2 - 9}$ (answers may vary)

41. Horizontal: $y = 0$;
vertical: $x = 0$

43. Horizontal: $y = 0$
vertical: $x = 0$

45. $g(x) = f(x - 3)$

47. $g(x) = f(x) + 2$

59. (a) $D = \{x \mid x \neq 2, x \neq -2\}$ **(b)** $[-9.4, 9.4, 1]$ by
(c) Horizontal: $y = 0$; $[-9.3, 9.3, 1]$
vertical: $x = \pm 2$
(d)

49. $g(x) = f(x + 1) - 2$

51. $g(x) = -2h(x - 1)$

61. (a) $D = \{x \mid x \neq 2\}$ **(b)** $[-4.7, 4.7, 1]$ by
(c) Horizontal: none; vertical: $[-6.2, 6.2, 1]$
none, since $f(x) = x + 2$ for $x \neq 2$
(d)

53. $g(x) = h(x + 1) - 2$

63.

65.

55. (a) $D = \{x \mid x \neq 2\}$ **(b)** $[-9.4, 9.4, 1]$ by
(c) Horizontal: $y = 1$; $[-6.2, 6.2, 1]$
vertical: $x = 2$
(d)

67.

69.

57. (a) $D = \{x \mid x \neq 2, x \neq -2\}$ **(b)** $[-9.4, 9.4, 1]$ by
(c) Horizontal: $y = 0$; $[-6.2, 6.2, 1]$
vertical: $x = \pm 2$
(d)

71.

73. (a) Slant: $y = x - 1$; vertical: $x = -1$

(b)

75. (a) Slant: $y = \frac{1}{2}x - 3$; vertical: $x = -2$

(b)

(b) When $x = 0$, there are 1 million insects.   **(c)** It starts to level off at 10 million.   **(d)** The horizontal asymptote $y = 10$ represents the limiting population after a long time.

99. (a) $f(400) = \frac{2540}{400} = 6.35$ inches. A curve designed for 60 miles per hour with a radius of 400 ft should have the outer rail elevated 6.35 in.   **(b)** As the radius x of the curve increases, the elevation of the outer rail decreases.

[0, 600, 100] by [0, 50, 5]

(c) The horizontal asymptote is $y = 0$. As the radius of the curve increases without bound ($x \to \infty$), the tracks become straight and no elevation or banking ($y \to 0$) is necessary.   **(d)** 200 ft

101. (a) The vertical asymptote is $x = 0$. As the coefficient of friction x decreases to 0, the braking distances become larger without a maximum.   **(b)** $\frac{25}{102}$

77. (a) Slant: $y = x + 3$; vertical: $x = 1$

(b)

79. (a) Slant: $y = 2x + 1$; vertical: $x = \frac{1}{2}$

(b)

81. -3 **83.** 1   **85.** $-\frac{1}{2}$
87.

89.

91.

4.6 EXTENDED AND DISCOVERY EXERCISES (p. 323)

1. $-\frac{1}{3}$; $-\frac{1}{x(x + h)}$ **3.** $-\frac{1}{2}$; $-\frac{3}{2x(x + h)}$

CHECKING BASIC CONCEPTS FOR SECTIONS 4.5 AND 4.6 (p. 323)

1. $f(x) = 3(x - 4i)(x + 4i) = 3x^2 + 48$
3. $f(x) = (x - 1)(x - 2i)(x + 2i)$
5. (a) $D = \{x \mid x \neq 1\}$ **(b)** Vertical asymptote: $x = 1$; horizontal asymptote: $y = 2$

(c)

93. (a) $T(4) = 0.25$; when vehicles leave the ramp at an average rate of 4 vehicles per minute, the wait is 0.25 minute or 15 seconds. $T(7.5) = 2$; when vehicles leave the ramp at an average rate of 7.5 vehicles per minute, the wait is 2 minutes.   **(b)** The wait increases dramatically.
95. (a) $N(20) = 0.5$, $N(39) \approx 38$
(b) It increases dramatically.   **(c)** $x = 40$
97. (a) $y = 10$

[0, 14, 1] by [0, 14, 1]

7. (a)

(b)

(c)

(d)

SECTION 4.7 (pp. 335–340)

1. -3 **3.** $\frac{1}{2}, 2$ **5.** $\frac{1}{2}$ **7.** -1 **9.** $\frac{13}{8}$

11. $\pm\sqrt{2}$ **13.** No real solutions   **15.** $0, \pm 2$

17. $-\frac{5}{3}, \frac{7}{5}$ **19.** -14

21. No real solutions (extraneous: 2)

23. -3 (extraneous: 1)

25. 1 (extraneous: -2)

27. No real solutions (extraneous: 1)

29. **(a)** $-4, -2,$ or 2
(b) $(-4, -2) \cup (2, \infty)$, or $\{x \mid -4 < x < -2 \text{ or } x > 2\}$
(c) $(-\infty, -4) \cup (-2, 2)$, or $\{x \mid x < -4 \text{ or } -2 < x < 2\}$

31. **(a)** $-4, -2, 0,$ or 2
(b) $(-4, -2) \cup (0, 2)$, or $\{x \mid -4 < x < -2 \text{ or } 0 < x < 2\}$
(c) $(-\infty, -4) \cup (-2, 0) \cup (2, \infty)$, or
$\{x \mid x < -4 \text{ or } -2 < x < 0 \text{ or } x > 2\}$

33. **(a)** $-2, 1,$ or 2   **(b)** $(-\infty, -2) \cup (-2, 1)$, or
$\{x \mid x < -2 \text{ or } -2 < x < 1\}$
(c) $(1, 2) \cup (2, \infty)$, or $\{x \mid 1 < x < 2 \text{ or } x > 2\}$

35. **(a)** 0   **(b)** $(-\infty, 0) \cup (0, \infty)$, or $\{x \mid x < 0 \text{ or } x > 0\}$
(c) No solutions

37. **(a)** 0 or 1
(b) $(-\infty, 0) \cup (1, \infty)$, or $\{x \mid x < 0 \text{ or } x > 1\}$
(c) $(0, 1)$, or $\{x \mid 0 < x < 1\}$

39. **(a)** $-2, 0,$ or 2
(b) $(-\infty, -2) \cup (2, \infty)$, or $\{x \mid x < -2 \text{ or } x > 2\}$
(c) $(-2, 0) \cup (0, 2)$, or $\{x \mid -2 < x < 0 \text{ or } 0 < x < 2\}$

41. $(-1, 0) \cup (1, \infty)$, or $\{x \mid -1 < x < 0 \text{ or } x > 1\}$

43. $[-2, 0] \cup [1, \infty)$, or $\{x \mid -2 \le x \le 0 \text{ or } x \ge 1\}$

45. $(-3, -2) \cup (2, 3)$, or $\{x \mid -3 < x < -2 \text{ or } 2 < x < 3\}$

47. $\left(-\infty, -\sqrt{2}\right) \cup \left(\sqrt{2}, \infty\right)$, or $\left\{x \mid x < -\sqrt{2} \text{ or } x > \sqrt{2}\right\}$

49. $[-2, 1] \cup [2, \infty)$, or $\{x \mid -2 \le x \le 1 \text{ or } x \ge 2\}$

51. $[-3, 2]$, or $\{x \mid -3 \le x \le 2\}$

53. $(-\infty, 1] \cup [2, 4]$, or $\{x \mid x \le 1 \text{ or } 2 \le x \le 4\}$

55. $(-\infty, -1) \cup \left(0, \frac{4}{3}\right) \cup (2, \infty)$, or $\left\{x \mid x < -1 \text{ or } \right.$
$\left. 0 < x < \frac{4}{3} \text{ or } x > 2\right\}$

57. $(-\infty, 0)$, or $\{x \mid x < 0\}$

59. $(-3, \infty)$, or $\{x \mid x > -3\}$

61. $(-2, 2)$, or $\{x \mid -2 < x < 2\}$

63. $(-\infty, 2)$, or $\{x \mid x < 2\}$

65. $(-\infty, -1) \cup \left(\frac{3}{2}, \infty\right)$, or $\left\{x \mid x < -1 \text{ or } x > \frac{3}{2}\right\}$

67. $(-\infty, -3) \cup (-1, 2)$, or $\{x \mid x < -3 \text{ or } -1 < x < 2\}$

69. $(-1, 1) \cup \left[\frac{5}{2}, \infty\right)$, or $\left\{x \mid -1 < x < 1 \text{ or } x \ge \frac{5}{2}\right\}$

71. $(3, \infty)$, or $\{x \mid x > 3\}$

73. $(-\infty, 0) \cup \left(0, \frac{1}{2}\right] \cup [2, \infty)$, or $\left\{x \mid x < 0 \text{ or } \right.$
$\left. 0 < x \le \frac{1}{2} \text{ or } x \ge 2\right\}$

75. $(-2, 0) \cup [2, \infty)$, or $\{x \mid -2 < x < 0 \text{ or } x \ge 2\}$

77. **(a)** About 12.4 cars per minute   **(b)** 3

79. Two possible solutions: width $= 7$ in., length $= 14$ in.,
height $= 2$ in.; width ≈ 2.266 in., length ≈ 4.532 in.,
height ≈ 19.086 in.

81. **(a)** $A(x) = \frac{x^2}{144} + \frac{3}{x}$ **(b)** $C(x) = 0.1\left(\frac{x^2}{144} + \frac{3}{x}\right)$
(c) 6 in. $\times$ 6 in. $\times$ 3 in.

83. **(a)** $D(0.05) \approx 238$; the braking distance for a car traveling at 50 miles per hour on a 5% uphill grade is about
238 ft.   **(b)** As the uphill grade x increases, the braking
distance decreases, which agrees with driving experience.
(c) $x = \frac{13}{165} \approx 0.079$, or 7.9%

85. **(a)** $x \le 36$ (approximately)   **(b)** The average line
length is less than or equal to 8 cars when the average
arrival rate is about 36 cars per hour or less.

87. **(a)** The braking distance increases.   **(b)** $0 < x \le 0.3$

89. $\sqrt[3]{212.8} \le x \le \sqrt[3]{213.2}$, or (approximately)
$5.97022 \le x \le 5.97396$ inches

91. $k = 6$ **93.** $k = 8$ **95.** $T = 160$ **97.** $y = 2$

99. Becomes half as much   **101.** Becomes 27 times as much

103. $k = 0.5, n = 2$ **105.** $k = 3, n = 1$

107. 1.18 grams   **109.** $\sqrt{50} \approx 7$ times as far

111. $\frac{2}{9}$ ohm   **113.** F decreases by a factor of $\frac{\sqrt{2}}{2}$.

SECTION 4.8 (pp. 350–352)

1. 4   **3.** $\frac{1}{8}$ **5.** -9 **7.** 27   **9.** 2   **11.** 9   **13.** $-\frac{1}{243}$

15. 16   **17.** $\frac{9}{4}$ **19.** $(2x)^{1/2}$ **21.** $z^{5/3}$ **23.** $\frac{1}{y^{3/4}}$

25. $x^{5/6}$ **27.** $y^{3/4}$ **29.** $\frac{\sqrt{b}}{\sqrt[4]{a^3}}$ **31.** $\sqrt{\sqrt{a} + \sqrt{b}}$

33. 7   **35.** -1 **37.** 2, 3   **39.** 4   **41.** 15   **43.** 8

45. -28 **47.** 2   **49.** 65,538

51. $50^{3/2} - 50^{1/2} \approx 346.48$ **53.** b

55. **57.**

59.

61. 2   **63.** 81   **65.** ± 32 **67.** 32   **69.** $\sqrt[3]{16}$

71. $-1, -\frac{1}{2}$ **73.** $-\frac{1}{4}, \frac{5}{7}$ **75.** $-8, 27$ **77.** $\frac{1}{8}, \frac{64}{27}$

79. 1   **81.** $-1, \frac{1}{27}$

83. About 3.5; average speed is about 3.5 mi/hr

85. $w \approx 3.63$ lb   **87.** About 58.1 yr

89. (a) $a = 1960$ **(b)** $b \approx -1.2$
(c) $f(4) = 1960(4)^{-1.2} \approx 371$. If the zinc ion concentration reaches 371 milligrams per liter, a rainbow trout will survive, on average, 4 minutes.
91. (a) $f(2) \approx 1.06$ grams
(b) & (c) Approximately 1.1 grams
93. $a \approx 3.20, b \approx 0.20$
   $[1, 9, 1]$ by $[0, 6, 1]$

95. (a) $f(x) = 0.005192x^{1.7902}$ (answers may vary)
(b) About 2.6 million; extrapolation   **(c)** About 1999
97. $a \approx 874.54, b \approx -0.49789$

4.8 EXTENDED AND DISCOVERY EXERCISES (p. 353)

1. The graph of an odd root function (not shown) is always increasing; the function is negative for $x < 0$, positive for $x > 0$, and zero at $x = 0$. It is an odd function.
3. The graph of a power function in which the exponent is a negative odd integer (not shown) has the y-axis as a vertical asymptote and the x-axis as a horizontal asymptote; the function is undefined at $x = 0$; the function is decreasing on $(-\infty, 0)$ and on $(0, \infty)$. The function is symmetric with respect to the origin. It is an odd function.
5. $\dfrac{1}{\sqrt{x + h} + \sqrt{x}}$ **7.** $\dfrac{x + 1}{x^{5/3}}$ **9.** $\dfrac{2 - x}{3x^{1/3}(x + 1)^2}$

CHECKING BASIC CONCEPTS FOR SECTIONS 4.7 AND 4.8 (p. 353)

1. (a) $\frac{1}{2}$ **(b)** $-5, \frac{7}{3}$
(c) No real solutions (extraneous: -2)
3. $(-2, -1] \cup [1, \infty)$, or $\{x \mid -2 < x < -1 \text{ or } x > 1\}$
5. (a) -8 **(b)** $\frac{1}{4}$ **(c)** 9   **7.** 8 (extraneous: 1)
9. $a = 2, b = \frac{1}{2}$

CHAPTER 4 REVIEW EXERCISES (pp. 358–361)

1. Degree: 3; leading coefficient: -7
3. (a) Local minimum: -2; local maximum: 4
(b) No absolute minimum; no absolute maximum
5. One local maximum, two local minima, two x-intercepts
$[-10, 10, 1]$ by $[-100, 100, 10]$

7. Even   **9.** Odd   **11.** Odd
13. Answers may vary.

15. (a) $2; -2, 0, 1$ **(b)** $a < 0$ **(c)** 3
17. Up on left end, down on right end;
$f(x) \to \infty$ as $x \to -\infty$; $f(x) \to -\infty$ as $x \to \infty$
19. 7
21. (a)

f is continuous.
(b) $f(1) = 2; f(3) = -1$ **(c)** $1, \sqrt{6}$
23. $2x^2 - 3x - 1$
25. $2x^2 - 3x + 1 + \dfrac{1}{2x + 3}$
27. $f(x) = \frac{1}{2}(x - 1)(x - 2)(x - 3)$
29. $f(x) = 2(x + 4)\left(x - \frac{1}{2}\right)(x - 2)$
31. $f(x) = \frac{1}{2}(x + 2)(x - 1)(x - 3)$
33. $-3, \frac{1}{2}, 2$ **35.** $0, \pm\sqrt{3}$ **37.** $\pm 1, \pm\sqrt{2}$
39. $-1.88, 0.35, 1.53$ **41.** -1 **43.** $5 + 12i$
45. $0, \pm i$ **47.** One real zero, two imaginary zeros
49. $f(x) = 2(x - i\sqrt{2})(x + i\sqrt{2})$
51. $f(x) = (x + i)(x - i)\left(x - \left(-\frac{1}{2} + \frac{i\sqrt{3}}{2}\right)\right) \times$
$\left(x - \left(-\frac{1}{2} - \frac{i\sqrt{3}}{2}\right)\right)$
53. Horizontal: $y = \frac{2}{3}$; vertical: $x = \frac{1}{3}$
55.

57.

59. Answers may vary.

61. 4   **63.** No real solutions (extraneous: 2)   **65.** $-\frac{1}{3}$

67. (a) $(-\infty, -4) \cup (-2, 3)$, or
$\{x \mid x < -4 \text{ or } -2 < x < 3\}$

(b) $(-4, -2) \cup (3, \infty)$, or $\{x \mid -4 < x < -2 \text{ or } x > 3\}$

69. $(-3, 0) \cup (2, \infty)$, or $\{x \mid -3 < x < 0 \text{ or } x > 2\}$

71. $(-\infty, -2) \cup \left(\frac{1}{2}, \infty\right)$, or $\left\{x \mid x < -2 \text{ or } x > \frac{1}{2}\right\}$

73. 216   **75.** 8   **77.** $x^{4/3}$ **79.** $y^{1/2}$

81. $D = \{x \mid x \geq 0\}; f(3) \approx 15.59$

83. 4   **85.** 6   **87.** $\frac{81}{16}$ **89.** 15

91. $-2, \frac{1}{3}$ **93.** $-1, 125$ **95.** 8

97. (a) Dog: 148; person: 69   **(b)** 6.4 in.

99. (a) 81.5°F   **(b)** 87.3; the ocean reaches a maximum temperature of about 87.3°F in late July.

101. 4 sec

CHAPTER 4 EXTENDED AND DISCOVERY EXERCISES (p. 362)

1. (a)

$f(t) = t^2$	$t_1 = 10$ $t_2 = 11$	$t_1 = 10$ $t_2 = 10.1$	$t_1 = 10$ $t_2 = 10.01$	$t_1 = 10$ $t_2 = 10.001$
average velocity (ft/sec)	21	20.1	20.01	20.001

(b) The velocity of the bike rider is 20 ft/sec at 10 sec.

3. (a) 5   **(b)** 5   **(c)** 5, 5, 5

5. (a) $\dfrac{-200}{a(a + h)}$ **(b)** $\dfrac{-200}{a^2}$ **(c)** $-8, -2, -\frac{8}{9}$

7. (a) $\dfrac{1}{\sqrt{a + h} + \sqrt{a}}$ **(b)** $\dfrac{1}{2\sqrt{a}}$

(c) $\dfrac{1}{2\sqrt{5}} \approx 0.224, \dfrac{1}{2\sqrt{10}} \approx 0.158, \dfrac{1}{2\sqrt{15}} \approx 0.129$

9. (a)

f_1 in $[-10, 10, 1]$ by $[-10, 10, 1]$ f_2 in $[-10, 10, 1]$ by $[-10, 10, 1]$

f_3 in $[-10, 10, 1]$ by $[-10, 10, 1]$ f_4 in $[-10, 10, 1]$ by $[-10, 10, 1]$

(b) The graph of each quadratic function has one turning point, whereas the graph of its average rate of change has no turning points. The x-intercept on the graph of the average rate of change corresponds to the x-coordinate of the vertex of the quadratic function.   **(c)** For any quadratic function, the graph of its average rate of change is a linear function. If the leading coefficient of the quadratic function is negative, the slope of the linear function is negative. If the leading coefficient of the quadratic function is positive, the slope of the linear function is positive.

11. (a)

f_1 in $[-10, 10, 1]$ by $[-10, 10, 1]$ f_2 in $[-10, 10, 1]$ by $[-10, 10, 1]$

f_3 in $[-10, 10, 1]$ by $[-10, 10, 1]$

(b) The graph of a quartic function has one or three turning points; the graph of its average rate of change generally has two turning points. The x-coordinate of a turning point of the function corresponds to an x-intercept on the graph of the average rate of change.   **(c)** For any quartic function, the graph of its average rate of change is a cubic function. The leading coefficients of a quartic function and its average rate of change have the same sign.

CHAPTERS 1–4 CUMULATIVE REVIEW EXERCISES (pp. 363–366)

1. 20%

3. (a) $D = \{-3, -1, 0, 1\}; R = \{-2, 4, 5\}$ **(b)** No

5. $D = \{x \mid -2 \leq x \leq 2\}; R = \{x \mid 0 \leq x \leq 3\}; f(0) = 3$

7. (a) $D = \{x \mid x \leq -2 \text{ or } x \geq 2\}$ **(b)** $f(2) = 0$

9. $C(x) = 0.25x + 200; C(2000) = 700$; the cost of driving 2000 miles in one month is $700.

11. 18   **13.** $y = -\frac{9}{5}x + \frac{7}{5}$ **15.** $y = -5$

17. Each radio costs $15 to manufacture. The fixed cost is $2000.   **19.** $\frac{3}{5}$ **21.** $0, \frac{8}{3}$ **23.** $-2, \frac{5}{7}$ **25.** $-3, 0, 1$

27. $\frac{1}{27}; -8$ **29.** $\frac{4}{5}$ **31.** No solutions; contradiction

33. $\left(-\infty, -\frac{1}{2}\right)$, or $\left\{x \mid x < -\frac{1}{2}\right\}$

35. $\left(-\infty, \frac{4}{5}\right] \cup [2, \infty)$, or $\left\{x \mid x \leq \frac{4}{5} \text{ or } x \geq 2\right\}$

37. $(-\infty, -3] \cup [0, 3]$, or $\{x \mid x \leq -3 \text{ or } 0 \leq x \leq 3\}$

39. (a) $-3, -1, 1, 2$ **(b)** $(-\infty, -3) \cup (-1, 1) \cup (2, \infty)$, or $\{x \mid x < -3 \text{ or } -1 < x < 1 \text{ or } x > 2\}$

(c) $[-3, -1] \cup [1, 2]$, or $\{x \mid -3 \leq x \leq -1 \text{ or } 1 \leq x \leq 2\}$ **41.** $\left(3, \frac{5}{2}\right)$

43. (a)

(b)

(c)

(d)

45. (a) Incr: $(-\infty, -2] \cup [1, \infty]$, or $\{x \,|\, x \le -2 \text{ or } x \ge 1\}$; decr: $[-2, 1]$, or $\{x \,|\, -2 \le x \le 1\}$
(b) Approximately -3.3, 0, and 1.8
(c) $(-2, 3)$ and $(1, -1)$ **(d)** Local maximum: 3; local minimum: -1
47. Answers may vary.

49. (a) $a - 2 + \frac{3}{a^2}$ **(b)** $2x^2 + 2x - 2 + \frac{-1}{x-1}$
(c) $x^2 + 2x - 3 + \frac{x+4}{x^2+2}$
51. $f(x) = 4(x+3)(x-1)^2(x-4)^3$
53. $f(x) = 2\left(x - \sqrt{3}\right)\left(x + \sqrt{3}\right)\left(x - \frac{1}{2}\right)$ **55.** $-\frac{1}{2} + \frac{7}{2}i$
57. $D = \{x \,|\, x \ne -1, x \ne 4\}$; vertical: $x = -1, x = 4$; horizontal: $y = 0$
59. $m_1 = 20$ thousand: the pool is being filled at a rate of 20 thousand gallons per hour. $m_2 = 10$ thousand: the pool is being filled at a rate of 10 thousand gallons per hour. $m_3 = 0$: the pool is neither being filled nor being emptied. $m_4 = -15$ thousand: the pool is being emptied at a rate of 15 thousand gallons per hour.
61. 8; 24. Between 0 and 2 sec, the car travels at an average rate of 8 ft per sec. Between 2 and 4 sec, the car travels at an average rate of 24 ft per sec.
63. 3.75 liters   **65.** 400 thousand toy figures
67. (a) $C(t) = t(805 - 5t) = 805t - 5t^2$ **(b)** 25, 136; the cost is \$17,000 when either 25 or 136 tickets are purchased. **(c)** \$32,400 when 80 or 81 tickets are sold

69. (a) $f(x) = 0.5(x - 1998)^2 + 6.1$
(b) 24.1% (answers may vary)
71. $r \approx 1.7$ in., $h \approx 3.5$ in.

CHAPTER 5: Exponential and Logarithmic Functions

SECTION 5.1 (pp. 378–385)

1. 7   **3.** $4x^3$ **5.** The cost of x square yards of carpet
7. (a) -5 **(b)** -5 **(c)** -3 **(d)** $-\frac{1}{3}$
9. (a) $\frac{11}{2}$ **(b)** 0   **(c)** $\frac{9}{4}$ **(d)** Undefined
11. (a) $(f + g)(x) = 2x + x^2$; all real numbers
(b) $(f - g)(x) = 2x - x^2$; all real numbers
(c) $(fg)(x) = 2x^3$; all real numbers
(d) $(f/g)(x) = \frac{2}{x}$; $D = \{x \,|\, x \ne 0\}$
13. (a) $(f + g)(x) = 2x^2$; all real numbers
(b) $(f - g)(x) = -2$; all real numbers
(c) $(fg)(x) = x^4 - 1$; all real numbers
(d) $(f/g)(x) = \frac{x^2 - 1}{x^2 + 1}$; all real numbers
15. (a) $(f + g)(x) = 2x$; $D = \{x \,|\, x \ge 1\}$
(b) $(f - g)(x) = -2\sqrt{x - 1}$; $D = \{x \,|\, x \ge 1\}$
(c) $(fg)(x) = x^2 - x + 1$; $D = \{x \,|\, x \ge 1\}$
(d) $(f/g)(x) = \frac{x - \sqrt{x - 1}}{x + \sqrt{x - 1}}$; $D = \{x \,|\, x \ge 1\}$
17. (a) $(f + g)(x) = 2\sqrt{x}$; $D = \{x \,|\, x \ge 0\}$
(b) $(f - g)(x) = -2$; $D = \{x \,|\, x \ge 0\}$
(c) $(fg)(x) = x - 1$; $D = \{x \,|\, x \ge 0\}$
(d) $(f/g)(x) = \frac{\sqrt{x} - 1}{\sqrt{x} + 1}$; $D = \{x \,|\, x \ge 0\}$
19. (a) $(f + g)(x) = \frac{4}{x + 1}$; $D = \{x \,|\, x \ne -1\}$
(b) $(f - g)(x) = -\frac{2}{x + 1}$; $D = \{x \,|\, x \ne -1\}$
(c) $(fg)(x) = \frac{3}{(x + 1)^2}$; $D = \{x \,|\, x \ne -1\}$
(d) $(f/g)(x) = \frac{1}{3}$; $D = \{x \,|\, x \ne -1\}$
21. (a) $(f + g)(x) = \frac{x + 1}{2x - 4}$; $D = \{x \,|\, x \ne 2\}$
(b) $(f - g)(x) = \frac{1 - x}{2x - 4}$; $D = \{x \,|\, x \ne 2\}$
(c) $(fg)(x) = \frac{x}{(2x - 4)^2}$; $D = \{x \,|\, x \ne 2\}$
(d) $(f/g)(x) = \frac{1}{x}$; $D = \{x \,|\, x \ne 0, x \ne 2\}$
23. (a) $(f + g)(x) = x^2 - 1 + |x + 1|$; all real numbers
(b) $(f - g)(x) = x^2 - 1 - |x + 1|$; all real numbers
(c) $(fg)(x) = (x^2 - 1)|x + 1|$; all real numbers
(d) $(f/g)(x) = \frac{x^2 - 1}{|x + 1|}$; $D = \{x \,|\, x \ne -1\}$

25. (a) $(f + g)(x) = \frac{(x - 1)(2x^2 - 2x + 5)}{(x + 1)(x - 2)}$;

$D = \{x | x \neq -1, x \neq 2\}$

(b) $(f - g)(x) = \frac{-3(x - 1)(2x - 1)}{(x + 1)(x - 2)}$;

$D = \{x | x \neq -1, x \neq 2\}$

(c) $(fg)(x) = (x - 1)^2; D = \{x | x \neq -1, x \neq 2\}$

(d) $(f/g)(x) = \frac{(x - 2)^2}{(x + 1)^2}$;

$D = \{x | x \neq 1, x \neq -1, \text{ and } x \neq 2\}$

27. (a) $(f + g)(x) = \frac{x^2 + 4x - 1}{(x - 1)^2(x + 1)}$;

$D = \{x | x \neq 1, x \neq -1\}$

(b) $(f - g)(x) = \frac{-x^2 - 3}{(x - 1)^2(x + 1)}; D = \{x | x \neq 1, x \neq -1\}$

(c) $(fg)(x) = \frac{2}{(x - 1)^3}; D = \{x | x \neq 1, x \neq -1\}$

(d) $(f/g)(x) = \frac{2(x - 1)}{(x + 1)^2}; D = \{x | x \neq 1, x \neq -1\}$

29. (a) $(f + g)(x) = x^{1/2}(x^2 - x + 1)$;

$D = \{x | x \geq 0\}$

(b) $(f - g)(x) = x^{1/2}(x^2 - x - 1); D = \{x | x \geq 0\}$

(c) $(fg)(x) = x^2(x - 1); D = \{x | x \geq 0\}$

(d) $(f/g)(x) = x(x - 1); D = \{x | x > 0\}$

31. (a) 2   (b) 4   (c) 0   (d) $-\frac{1}{3}$

33. (a) 2   (b) -3 (c) 2   (d) -2

35. (a) -5 (b) -2 (c) 0   (d) Undefined

37. (a) 5   (b) 5   (c) 0   (d) Undefined

39.

x	-2	0	2	4
$(f + g)(x)$	6	5	5	15
$(f - g)(x)$	-6	5	9	5
$(fg)(x)$	0	0	-14	50
$(f/g)(x)$	0	—	-3.5	2

41. (a) $g(-3) = -5$ (b) $g(b) = 2b + 1$

(c) $g(x^3) = 2x^3 + 1$ (d) $g(2x - 3) = 4x - 5$

43. (a) $g(-3) = -4$ (b) $g(b) = 2(b + 3)^2 - 4$

(c) $g(x^3) = 2(x^3 + 3)^2 - 4$

(d) $g(2x - 3) = 8x^2 - 4$

45. (a) $g(-3) = -\frac{11}{2}$ (b) $g(b) = \frac{1}{2}b^2 + 3b - 1$

(c) $g(x^3) = \frac{1}{2}x^6 + 3x^3 - 1$ (d) $g(2x - 3) = 2x^2 - \frac{11}{2}$

47. (a) $g(-3) = 1$ (b) $g(b) = \sqrt{b + 4}$

(c) $g(x^3) = \sqrt{x^3 + 4}$ (d) $g(2x - 3) = \sqrt{2x + 1}$

49. (a) $g(-3) = 14$ (b) $g(b) = |3b - 1| + 4$

(c) $g(x^3) = |3x^3 - 1| + 4$

(d) $g(2x - 3) = |6x - 10| + 4$

51. (a) $g(-3)$ is undefined   (b) $g(b) = \frac{4b}{b + 3}$

(c) $g(x^3) = \frac{4x^3}{x^3 + 3}$ (d) $g(2x - 3) = \frac{2(2x - 3)}{x}$

53. (a) 3   (b) 4   **55.** (a) 18   (b) 23

57. (a) $(f \circ g)(x) = (x^2 + 3x - 1)^3$; all real numbers

(b) $(g \circ f)(x) = x^6 + 3x^3 - 1$; all real numbers

(c) $(f \circ f)(x) = x^9$; all real numbers

59. (a) $(f \circ g)(x) = x^4 + x^2 - 3x - 2$; all real numbers

(b) $(g \circ f)(x) = (x + 2)^4 + (x + 2)^2 - 3(x + 2) - 4$; all real numbers   (c) $(f \circ f)(x) = x + 4$; all real numbers

61. (a) $(f \circ g)(x) = 2 - 3x^3$; all real numbers

(b) $(g \circ f)(x) = (2 - 3x)^3$; all real numbers

(c) $(f \circ f)(x) = 9x - 4$; all real numbers

63. (a) $(f \circ g)(x) = \frac{1}{5x + 1}; D = \{x | x \neq -\frac{1}{5}\}$

(b) $(g \circ f)(x) = \frac{5}{x + 1}; D = \{x | x \neq -1\}$

(c) $(f \circ f)(x) = \frac{x + 1}{x + 2}; D = \{x | x \neq -1, x \neq -2\}$

65. (a) $(f \circ g)(x) = \sqrt{4 - x^2} + 4$;

$D = \{x | -2 \leq x \leq 2\}$

(b) $(g \circ f)(x) = \sqrt{4 - (x + 4)^2}; D = \{x | -6 \leq x \leq -2\}$

(c) $(f \circ f)(x) = x + 8$; all real numbers

67. (a) $(f \circ g)(x) = \sqrt{3x - 1}; D = \{x | x \geq \frac{1}{3}\}$

(b) $(g \circ f)(x) = 3\sqrt{x - 1}; D = \{x | x \geq 1\}$

(c) $(f \circ f)(x) = \sqrt{\sqrt{x - 1} - 1}; D = \{x | x \geq 2\}$

69. (a) $(f \circ g)(x) = x$; all real numbers

(b) $(g \circ f)(x) = x$; all real numbers

(c) $(f \circ f)(x) = 25x - 4$; all real numbers

71. (a) $(f \circ g)(x) = x; D = \{x | x \neq 0\}$

(b) $(g \circ f)(x) = x; D = \{x | x \neq 0\}$

(c) $(f \circ f)(x) = x; D = \{x | x \neq 0\}$

73. (a) -4 (b) 2   (c) -4

75. (a) -3 (b) -2 (c) 0

77. (a) 5   (b) Undefined   (c) 4   **79.** 4; 2

Answers may vary for Exercises 81–93.

81. $f(x) = x - 2, g(x) = \sqrt{x}$

83. $f(x) = x + 2, g(x) = \frac{1}{x}$

85. $f(x) = 2x + 1, g(x) = 4x^3$

87. $f(x) = x^3 - 1, g(x) = x^2$

89. $f(x) = x + 2, g(x) = -4|x| - 3$

91. $f(x) = x - 1, g(x) = \frac{1}{x^2}$

93. $f(x) = x^{1/4}, g(x) = x^3 - x$

95. $P(x) = 13x - 2000; P(3000) = \$37,000$

97. (a) $I(x) = 36x$ (b) $C(x) = 2.54x$

(c) $F(x) = (C \circ I)(x)$ (d) $F(x) = 91.44x$

99. (a) $r(x) = \frac{11}{6}x$ (b) $s(x) = \frac{11}{6}x + \frac{1}{9}x^2$

(c) $s(60) = 510$; it takes 510 feet to stop when traveling 60 mi/hr.

101. (a) $(g \circ f)(1) = 5.25$; a 1% decrease in the ozone layer could result in a 5.25% increase in skin cancer.

(b) Not possible using the given tables

103. (a) 4.5%   (b) $(g \circ f)(x)$ computes the percent increase in peak electrical demand during year x.

105. (a) $(g \circ f)(1960) = 1.98$

(b) $(g \circ f)(x) = 0.165(x - 1948)$

(c) $f, g,$ and $g \circ f$ are all linear functions.

107. (a) $(g \circ f)(2) \approx 25°C$ **(b)** $(g \circ f)(x)$ computes the Celsius temperature after x hours.

109. $A = 36\pi t^2$

111. (a) $A(4s) = 16\frac{\sqrt{3}}{4}s^2 = 16A(s)$; if the length of a side is quadrupled, the area increases by a factor of 16.

(b) $A(s + 2) = \frac{\sqrt{3}}{4}(s^2 + 4s + 4) = A(s) + \sqrt{3}(s + 1)$; if the length of a side increases by 2, the area increases by $\sqrt{3}(s + 1)$.

113. (a) 50   **(b)** $(C + O)(x)$ computes the total SO_2 emissions from coal and oil during year x.

(c)

x	1860	1900	1940	1970	2000
$(C + O)(x)$	2.4	12.8	26.5	50.0	78.0

115. (a)

x	1990	2000	2010	2020	2030
$h(x)$	32	35.5	39	42.5	46

(b) $h(x) = f(x) + g(x)$

117. $h(x) = 0.35x - 664.5$

119. (a) $P(h_0) = mgh_0 = \frac{1}{2}m \cdot 2gh_0 = \frac{1}{2}m(\sqrt{2gh_0})^2 = \frac{1}{2}m(v_f)^2 = K(v_f)$

(b) The potential energy of the ball before it is dropped is equal to the kinetic energy of the ball when it lands.

121. Let $f(x) = ax + b$ and $g(x) = cx + d$. Then $f(x) + g(x) = (ax + b) + (cx + d) = (a + c)x + (b + d)$, which is linear.

123. (a) $(f \circ g)(x) = k$; a constant function
(b) $(g \circ f)(x) = ak + b$; a constant function

SECTION 5.2 (pp. 396–401)

1. Closing a window
3. Closing a book, standing up, and walking out of the classroom   **5.** Subtract 2 from x; $x + 2$ and $x - 2$
7. Divide x by 3 and then add 2; $3(x - 2)$ and $\frac{x}{3} + 2$
9. Subtract 1 from x and cube the result; $\sqrt[3]{x} + 1$ and $(x - 1)^3$ **11.** Take the reciprocal of x; $\frac{1}{x}$ and $\frac{1}{x}$
13. One-to-one   **15.** Not one-to-one
17. Not one-to-one
19. Not one-to-one; does not have an inverse
21. One-to-one; does have an inverse
23. One-to-one   **25.** Not one-to-one
27. Not one-to-one   **29.** Not one-to-one
31. Not one-to-one   **33.** Not one-to-one
35. One-to-one   **37.** No   **39.** Yes
41. $f^{-1}(x) = x^3$ **43.** $f^{-1}(x) = -\frac{1}{2}x + 5$
45. $f^{-1}(x) = \frac{x + 1}{3}$ **47.** $f^{-1}(x) = \sqrt[3]{\frac{x + 5}{2}}$
49. $f^{-1}(x) = \sqrt{x + 1}$ **51.** $f^{-1}(x) = \frac{1}{2x}$
53. $f^{-1}(x) = -\frac{2(x - 3)}{5}$ **55.** $f^{-1}(x) = -\frac{2x}{x - 1}$
57. $f^{-1}(x) = \frac{x + 1}{x - 2}$ **59.** $f^{-1}(x) = \frac{1}{x + 3}$
61. $f^{-1}(x) = \sqrt[3]{\frac{1 + x}{x}}$

63. If the domain of f is restricted to $x \geq 0$, then $f^{-1}(x) = \sqrt{4 - x}$.
65. If the domain of f is restricted to $x \geq 2$, then $f^{-1}(x) = 2 + \sqrt{x - 4}$.
67. If the domain of f is restricted to $x \geq 0$, then $f^{-1}(x) = (x - 1)^{3/2}$.
69. If the domain of f is restricted to $0 \leq x \leq \frac{3}{\sqrt{2}}$ then $f^{-1}(x) = \sqrt{\frac{9 - x^2}{2}}$.
71. $f^{-1}(x) = \frac{x + 15}{5}$; D and R are all real numbers.
73. $f^{-1}(x) = x^3 + 5$; D and R are all real numbers.
75. $f^{-1}(x) = 4x + 5$; D and R are all real numbers.
77. $f^{-1}(x) = x^2 + 5$; $D = \{x \mid x \geq 0\}$ and $R = \{y \mid y \geq 5\}$
79. $f^{-1}(x) = \frac{1}{x} - 3$; $D = \{x \mid x \neq 0\}$ and $R = \{y \mid y \neq -3\}$
81. $f^{-1}(x) = \sqrt[3]{\frac{x}{2}}$; D and R are all real numbers.
83. $f^{-1}(x) = \sqrt{x}$; D and R include all nonnegative real numbers.

85.

x	5	7	9
$f^{-1}(x)$	1	2	3

For f: $D = \{1, 2, 3\}$, $R = \{5, 7, 9\}$;
for f^{-1}: $D = \{5, 7, 9\}$, $R = \{1, 2, 3\}$

87.

x	0	4	16
$f^{-1}(x)$	0	2	4

For f: $D = \{0, 2, 4\}$, $R = \{0, 4, 16\}$;
for f^{-1}: $D = \{0, 4, 16\}$, $R = \{0, 2, 4\}$

89.

x	0	2	4	6
$f^{-1}(x)$	0	$\frac{1}{2}$	1	$\frac{3}{2}$

91. 1   **93.** 3   **95.** 5   **97.** 1
99. (a) $f(1) \approx \$110$ **(b)** $f^{-1}(110) \approx 1$ year
(c) $f^{-1}(160) \approx 5$ years; $f^{-1}(x)$ computes the years necessary for the account to accumulate x dollars.
101. (a) 2   **(b)** 3   **(c)** 1   **(d)** 3
103. (a) 4   **(b)** 0   **(c)** 9   **(d)** 4

105.   **107.**

109.   **111.**

113.

115.

117. $Y_1 = 3X - 1$, $Y_2 = (X + 1)/3$, $Y_3 = X$

$[-4.7, 4.7, 1]$ by $[-3.1, 3.1, 1]$

119. $Y_1 = X^3/3 - 1$, $Y_2 = \sqrt[3]{(3X + 3)}$, $Y_3 = X$

$[-4.7, 4.7, 1]$ by $[-3.1, 3.1, 1]$

121. (a) Yes   **(b)** The radius r of a sphere with volume V
(c) $r = \sqrt[3]{\dfrac{3V}{4\pi}}$ **(d)** No. If V and r were interchanged, then
r would represent the volume and V would represent the
radius.   **123. (a)** 135.7 pounds   **(b)** Yes
(c) $h = \dfrac{7}{25}\left(W + \dfrac{800}{7}\right) = \dfrac{7}{25}W + 32$ **(d)** 74; the maximum recommended height for a person weighing 150 lb is
74 in.   **(e)** The inverse formula computes the maximum
recommended height for a person of a given weight.
125. (a) $(F \circ Y)(2) = 10{,}560$ represents the number of feet
in 2 miles.   **(b)** $F^{-1}(26{,}400) = 8800$ represents the number
of yards in 26,400 ft.   **(c)** $(Y^{-1} \circ F^{-1})(21{,}120) = 4$ represents the number of miles in 21,120 ft.
127. (a) $(Q \circ C)(96) = 1.5$ represents the number of
quarts in 96 tbsp.   **(b)** $Q^{-1}(2) = 8$ represents the number
of cups in 2 qt.   **(c)** $(C^{-1} \circ Q^{-1})(1.5) = 96$ represents
the number of tablespoons in 1.5 qt.
129. (a) $f(1930) = 62.5$, $f(1980) = 65.5$; increased by
3%   **(b)** $f^{-1}(x)$ computes the year when the cloud cover
was x percent.   **(c)** $f^{-1}(62.5) = 1930$, $f^{-1}(65.5) = 1980$
(d) $f^{-1}(x) = \dfrac{50}{3}(x - 62.5) + 1930$

5.2 EXTENDED AND DISCOVERY EXERCISES (p. 401)

1. (a) $f^{-1}(x)$ computes the elapsed time in seconds when
the rocket was x ft above the ground.   **(b)** The solution to
the equation $f(x) = 5000$ is the elapsed time in seconds
when the rocket reached 5000 ft above the ground.
(c) Evaluate $f^{-1}(5000)$.

**CHECKING BASIC CONCEPTS FOR SECTIONS 5.1
AND 5.2 (p. 401)**

1. (a) $(f + g)(1) = 1$ **(b)** $(f - g)(-1) = 3$
(c) $(fg)(0) = 2$ **(d)** $(f/g)(2)$ is undefined.
(e) $(f \circ g)(2) = -2$ **(f)** $(g \circ f)(-2) = -1$
3. (a) $(f + g)(x) = x^2 + 6x - 3$
(b) $(f/g)(x) = \dfrac{x^2 + 3x - 2}{3x - 1}, x \neq \dfrac{1}{3}$
(c) $(f \circ g)(x) = 9x^2 + 3x - 4$
5. (a) Yes; yes; $f^{-1}(x) = x - 1$ **(b)** No; no
7. (a) 0   **(b)** 2

SECTION 5.3 (pp. 415–420)

1. $\dfrac{1}{8}$ **3.** 6   **5.** -18 **7.** 2   **9.** e^{2x}
11. 1   **13.** 5   **15.** Linear; $f(x) = -1.2x + 2$
17. Exponential; $f(x) = 8\left(\dfrac{1}{2}\right)^x$
19. Exponential; $f(x) = 5(2^x)$
21. Exponential; $f(n) = 40{,}000(1.08)^{n-1}$
23. $f(x) = 2^x$ **25.** $f(x) = 2x + 1$
27. $C = 5, a = 1.5$ **29.** $C = 10, a = 2$
31. $C = 3, a = 3$ **33.** $C = \dfrac{1}{2}, a = \dfrac{1}{3}$
35. $C = 5000, a = 2$; x represents time in hours.
37. $C = 200{,}000, a = 0.95$; x represents the number of
years after 2000.
39. About 11 pounds per square inch
41. 71.2571   **43.** -0.7586
45.

47.

49.

51.

53. $C = 1, a = \dfrac{1}{2}$ **55.** $C = \dfrac{1}{2}, a = 4$
57. (a) $D: (-\infty, \infty)$ or $\{x \mid -\infty < x < \infty\}$; $R: (0, \infty)$ or
$\{x \mid x > 0\}$ **(b)** Decreasing   **(c)** $y = 0$
(d) y-intercept: 7; no x-intercept   **(e)** Yes; yes

59. (i) b   **(ii)** d   **(iii)** a   **(iv)** c
61. (a)   **(b)**

(c)   **(d)**

63. $841.53   **65.** $1730.97   **67.** $4451.08
69. $2072.76
71. 10%: $14,656.15; 13%: $26,553.58; a 13% rate results in $11,897.43 more interest than a 10% rate.
73. $14,326.78
75. (a) $9074.47   **(b)** The last 10 years
77. About $1288 billion, or $1.288 trillion
79. $19,870.65
81. $A(x) = Pe^{-0.0325x}$, where P is the initial value
83. (a) About 6,214,000 bacteria per milliliter
(b) There will be 10 million *E. coli* bacteria per milliliter after about 214 min, or about 3.6 hr.
85. (a) $C \approx 0.72, a \approx 1.041$ (answers may vary slightly)
(b) 1.21 ppb (answers may vary slightly)
87. (a) About 92%   **(b)** About 0.83 min
89. About 18,935 yr
91. About 27.3%
93. (a) $C = 2.5, a = 0.7$ **(b)** 1.225 ppm
(c) About 1.4 days   **95. (a)** $C = 18.29, a = 1.279$
(b) 168,000 pounds   **(c)** About 10.5 in.
97. (a) $H(30) \approx 0.42$; about 0.42 horsepower is required for each ton pulled at 30 mi/hr.
(b) About 2100 horsepower   **(c)** 2

5.3 EXTENDED AND DISCOVERY EXERCISES (pp. 420–421)

1. The formulas are equivalent.   **3.** $12,914.64
5. (a) About 1.0005   **(b)** 1   **(c)** They are very similar.
7. (a) About 0.6068   **(b)** About 0.6065
(c) They are very similar.
9. The average rate of change near x and the value of the function at x are approximately equal.

SECTION 5.4 (pp. 433–437)

1.

x	10^0	10^4	10^{-8}	$10^{1.26}$
$\log x$	0	4	-8	1.26

3. (a) Undefined   **(b)** -2 **(c)** $-\frac{1}{2}$ **(d)** 0
5. (a) 1   **(b)** 4   **(c)** -20 **(d)** -2
7. (a) 2   **(b)** $\frac{1}{2}$ **(c)** 3   **(d)** Undefined
9. (a) $n = 1$, $\log 79 \approx 1.898$ **(b)** $n = 2$, $\log 500 \approx 2.699$
(c) $n = 0$, $\log 5 \approx 0.699$ **(d)** $n = -1$, $\log 0.5 \approx -0.301$
11. (a) $\frac{3}{2}$ **(b)** $\frac{1}{3}$ **(c)** $-\frac{1}{5}$ **(d)** -1
13. $\{x \mid x > -3\}$, or $(-3, \infty)$
15. $\{x \mid x < -1 \text{ or } x > 1\}$, or $(-\infty, -1) \cup (1, \infty)$
17. $\{x \mid -\infty < x < \infty\}$, or $(-\infty, \infty)$
19. $\{x \mid x < 2\}$, or $(-\infty, 2)$
21. -5.7 **23.** $2x, x > 0$ **25.** 64
27. -4 **29.** π **31.** $x - 1$ for $x > 1$
33. 6   **35.** $\frac{1}{2}$ **37.** -3 **39.** 2   **41.** 2   **43.** -2
45. -1 **47.** 0   **49.** -4
51.

x	6	7	21
$f(x)$	0	2	8

53. (a) -2 **(b)** $\log 7 \approx 0.85$ **(c)** No solutions
55. (a) -2 **(b)** $\ln 2 \approx 0.69$ **(c)** 3
57. (a) 0   **(b)** $\frac{1}{2}$ **(c)** $\frac{1}{3}$
59. $-\ln 3 \approx -1.10$ **61.** 2   **63.** $\frac{2}{3}$
65. $\frac{\log 13}{4} \approx 0.28$ **67.** $\log_3 4 \approx 1.26$ **69.** $\ln 23 \approx 3.14$
71. $\log_2 14 \approx 3.81$ **73.** $\ln \left(\frac{18}{5}\right) \approx 1.28$
75. 8   **77. (a)** $10^2 = 100$ **(b)** $10^{-3} = 0.001$
(c) $10^{1.2} \approx 15.8489$
79. (a) $2^6 = 64$ **(b)** $3^{-2} = \frac{1}{9}$ **(c)** $e^2 \approx 7.3891$
81. $2^{1.2} \approx 2.2974$ **83.** $\frac{49}{2}$ **85.** 1000   **87.** 20
89. $e^{3/4} \approx 2.1170$ **91.** $e^{7/5} \approx 4.0552$ **93.** 16
95. $\frac{1}{2} e^{6/5} \approx 1.6601$ **97.** 8   **99.** $a = 5, b = 2$
101. $f^{-1}(x) = \ln x$ **103.** $f^{-1}(x) = 2^x$

105. $D = \{x \mid x > -1\}$ **107.** $D = \{x \mid x < 0\}$
$[-6, 6, 1]$ by $[-4, 4, 1]$ $[-6, 6, 1]$ by $[-4, 4, 1]$

109. Decreasing

111. (a)

(b) f incr: $(0, \infty)$; f^{-1}
incr: $(-\infty, \infty)$

113. (a) $4x = \log_7 4$ **(b)** $x = \ln 7$ **(c)** $x = \log_c b$
115. (a) $x = 8^3$ **(b)** $2 + x = 9^5$ **(c)** $b = k^c$
117. 105 decibels   **119.** $a = 7, b = 4$; about 178 km²
121. (a) $C = 3, a = 2$ **(b)** After about 2.4 days
123. (a) Since L is increasing, heavier planes generally require longer runways.
$[0, 50, 10]$ by $[0, 6, 1]$

(b) No. It increases by 3000 ft.   **(c)** If the weight increases tenfold, the runway length increases by 3000 ft.
125. (a) 4.7   **(b)** $10^{3.5} \approx 3162$
127. (a) Yugoslavia: $x = 1,000,000$; Indonesia: $x = 100,000,000$ **(b)** 100
129. (a) $f(0) = 27$, $f(100) \approx 29.2$. At the eye, the barometric air pressure is 27 in.; 100 mi away, it is 29.2 in.
(b) The air pressure increases rapidly at first and then starts to level off.
$[0, 250, 50]$ by $[25, 30, 1]$

(c) About 7 miles
131. (a) About 49.4°C   **(b)** About 0.69 hr, or 41.4 min
133. (a) About 0.189, or an 18.9% chance
(b) $x = -3 \ln (0.3) \approx 3.6$ min

5.4 EXTENDED AND DISCOVERY EXERCISES (p. 437)
1. (a) 1.00   **(b)** 0.50   **(c)** 0.33   **(d)** 0.25
3. (a) 166,500   **(b)** $r = 0.5$ **(c)** The specific growth rate is a smaller, more convenient number to use.

CHECKING BASIC CONCEPTS FOR SECTIONS 5.3 AND 5.4 (pp. 437–438)
1. \$1752.12; \$1754.74
3. $\log_2 15$ represents the exponent k such that $2^k = 15$. No.

5. (a) $\ln 5 \approx 1.609$ **(b)** $\log 25 \approx 1.398$
(c) $10^{1.5} \approx 31.623$
7. (a) New York: 18.2 million; Florida: 14 million
(b) Florida's population will equal New York's population at about the beginning of 2011. At this time both populations will be approximately 18.5 million.

SECTION 5.5 (pp. 444–446)
1. 1.447; Property 2   **3.** 2.197; Property 3
5. 3.010; Property 4   **7.** $\log_2 a + \log_2 b$
9. $\ln 7 + 4 \ln a$ **11.** $\log 6 - \log z$
13. $2 \log x - \log 3$ **15.** $\ln 2 + 7 \ln x - \ln 3 - \ln k$
17. $2 + 2 \log_2 k + 3 \log_2 x$
19. $2 + 3 \log_5 x - 4 \log_5 y$
21. $4 \ln x - 2 \ln y - \frac{3}{2} \ln z$
23. $-1 + 3 \log_4 (x + 2)$
25. $3 \log_5 x - 4 \log_5 (x - 4)$
27. $\frac{1}{2} \log_2 x - 2 \log_2 z$
29. $\frac{1}{3} \ln (2x + 6) - \frac{5}{3} \ln (x + 1)$
31. $\frac{1}{3} \log_2 (x^2 - 1) - \frac{1}{2} \log_2 (1 + x^2)$
33. $\log 6$ **35.** $-\frac{3}{2} \ln 5 = \ln 5^{-3/2}$ **37.** $\log 2$
39. $\log 6$ **41.** $\log_7 5k^2$ **43.** $\ln x^3$ **45.** $\frac{3}{2} \log x = \log x^{3/2}$
47. $\ln \frac{x^3 z^4}{\sqrt{y^3}}$ **49.** $\ln \frac{2}{e}$ **51.** $\ln \frac{x^2 \sqrt{z}}{y^4}$
53. $\log \left(4 \sqrt{x^5}\right)$ **55.** $\log \dfrac{(x^2 - 1)^2(x - 2)^4}{\sqrt{y}}$
57. (a) Yes   **(b)** By Property 2:
$\log 3x + \log 2x = \log (3x \cdot 2x) = \log 6x^2$
59. (a) Yes   **(b)** By Property 3:
$\ln 2x^2 - \ln x = \ln \left(\frac{2x^2}{x}\right) = \ln 2x$
61. (a) Yes   **(b)** By Property 4:
$\ln x^4 - \ln x^2 = 4 \ln x - 2 \ln x = 2 \ln x$
63.

65.

67. $\dfrac{\log 25}{\log 2} \approx 4.644$ **69.** $\dfrac{\log 130}{\log 5} \approx 3.024$
71. $\dfrac{\log 5}{\log 2} + \dfrac{\log 7}{\log 2} \approx 5.129$ **73.** $\sqrt{\dfrac{\log 46}{\log 4}} \approx 1.662$
75. $\dfrac{\log 12/\log 2}{\log 3/\log 2} = \dfrac{\log 12}{\log 3} \approx 2.262$ **77.** 4.714
79. ± 2.035 **81.** $L(x) = \dfrac{3 \ln x}{\ln 10}$; $L(50) \approx 5.097$; yes
83. 10 decibels   **85.** $I = I_0 e^{-kx}$

87. (a) $x = \frac{1}{0.013} \ln \frac{P}{34}$ **(b)** About 8.6; the population is expected to reach 38 million during 2009.

89. $t = \frac{\ln \frac{A}{P}}{r}$

91. $\log (1 \cdot 2^2 \cdot 3^3 \cdot 4^4 \cdot 5^5) = \log 86{,}400{,}000$

SECTION 5.6 (pp. 456–459)

1. (a) About 2   **(b)** $\ln 7.5 \approx 2.015$
3. (a) About 2   **(b)** $5 \log 2.5 \approx 1.990$
5. $\ln 1.25 \approx 0.2231$ **7.** $\log 20 \approx 1.301$
9. $-\frac{5}{6} \ln 0.4 \approx 0.7636$ **11.** $\frac{\ln 0.5}{\ln 0.9} \approx 6.579$
13. $1 + \frac{\log 4}{\log 1.1} \approx 15.55$ **15.** 1   **17.** $\frac{1}{5}$
19. 1, 2   **21.** No solutions
23. $2 + \frac{\log (1/3)}{\log (2/5)} \approx 3.199$ **25.** $\frac{\log 4}{\log 4 - 2 \log 3} \approx -1.710$
27. $\frac{3}{1 - 3 \ln 2} \approx -2.779$ **29.** $\frac{\log (64/3)}{\log 1.4} \approx 9.095$
31. $1980 + \frac{\log (8/5)}{\log 1.015} \approx 2012$ **33.** 5
35. $10^{2/3} \approx 4.642$ **37.** $\frac{1}{2}e^5 \approx 74.207$
39. $\pm\sqrt{50} \approx \pm 7.071$ **41.** 6   **43.** −39
45. 10^{-11} **47.** $e \approx 2.718$ **49.** $2^{2.1} \approx 4.287$
51. $\sqrt{50} \approx 7.071$ **53.** $-\frac{998}{3}$
55. 2 $\left(\text{extraneous} -\frac{5}{3}\right)$
57. $1 + \sqrt{2} \approx 2.414$ $\left(\text{extraneous } 1 - \sqrt{2}\right)$
59. 2 $\left(\text{extraneous} -2\right)$ **61.** 2 $\left(\text{extraneous} -4\right)$
63. 4 $\left(\text{extraneous} -4\right)$ **65.** 0.31   **67.** 1.71
69. 2.10   **71.** 1977   **73.** $\frac{\ln 0.25}{-0.12} \approx 11.55$ ft
75. $\frac{1}{0.014} \ln 50 \approx 279.4$ min   **77.** About 1981
79. (a) $C = 1, a \approx 1.01355$
(b) $P(2010) \approx 1.14$ billion   **(c)** About 2030
81. About 2012   **83. (a)** $T_0 = 32, D = 180, a \approx 0.045$
(b) About 139°F   **(c)** About 1 hr
85. (a) About 16.7°C   **(b)** About 1.09 hr
87. About 2.8 acres   **89.** $100 - 10^{1.16} \approx 85.5\%$
91. (a) $\frac{\ln 1.15}{6} \approx 0.0233$ **(b)** About 1.45 billion per liter
(c) About 2.36%   **93.** 8.25 yr   **95. (a)** $\frac{\ln 1.5}{0.09} \approx 4.505$
(b) \$500 invested at 9% compounded continuously results in \$750 after 4.5 years.   **97.** About 8633 yr
99. $-2 \ln 0.5 \approx 1.39$ min   **101.** About 3.6 hr
103. About 13.5 yr   **105. (a)** 11 milligrams/liter
(b) About 2.11 hr   **107.** $t = \frac{\log (A/P)}{n \log (1 + r/n)}$

5.6 EXTENDED AND DISCOVERY EXERCISE (p. 460)

1. $f(x) = Ce^{x \ln a}$ and $g(x) = e^{x \ln 2}$; that is, $k = \ln a$.

CHECKING BASIC CONCEPTS FOR SECTIONS 5.5 AND 5.6 (p. 460)

1. $2 \log x + 3 \log y - \frac{1}{3} \log z$
3. (a) $\frac{\log (29/5)}{\log (1.4)} \approx 5.224$ **(b)** $\frac{1}{3}$
5. (a) It levels off at 80°F.   **(b)** 17 min

SECTION 5.7 (pp. 466–468)

1. Logarithmic   **3.** Exponential
5. Exponential: $f(x) = 1.2(1.7)^x$
7. Logarithmic: $f(x) = 1.088 + 2.937 \ln x$
9. Logistic: $f(x) = \frac{9.96}{1 + 30.6e^{-1.51x}}$

11. (a) $[25, 75, 5]$ by $[-100, 2100, 200]$

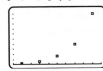

(b) $f(x) = 1.568(1.109)^x$ **(c)** About 6164 deaths per 100,000
13. $f(x) = 9.02 + 1.03 \ln x$
15. (a) $f(x) = 1.4734(0.99986)^x$
(b) Approximately 0.55 kg/m³
17. (a) $f(x) = \frac{4.9955}{1 + 49.7081e^{-0.6998x}}$
(b) About 5 thousand per acre
19. (a) $N(x) = 213.78x - 414{,}673.4$
(b) $[1950, 2010, 10]$ by $[3000, 14000, 1000]$

(c) During 1984
21. (a) $A(w) = 101w^{0.662}$
(b) $[0, 20, 2]$ by $[100, 700, 50]$

(c) About 11.2 lb
23. (a) 3 ft; after 5 years, the tree is 3 feet tall.
(b) $H(x) = \frac{50.2}{1 + 47.4e^{-0.221x}}$
(c) $[0, 45, 5]$ by $[0, 55, 5]$

(d) About 17.4 yr
(e) Interpolation

25. (a) The data are not linear.
$[-2, 32, 5]$ by $[0, 80, 10]$

(b) $f(x) = 12.42(1.066)^x$
(c) $f(39) \approx 150$; fertilizer use increased but at a slower rate than predicted by f.

5.7 EXTENDED AND DISCOVERY EXERCISE (p. 469)

1. (a) $f(x) = 0.0904(0.844)^x$ (regression) or
$f(x) \approx 0.128(0.777)^x$ (trial and error)   **(b)** About 4.6 min
(c) Answers may vary.

CHECKING BASIC CONCEPTS FOR SECTION 5.7 (p. 469)

1. Exponential: $f(x) = 0.5(1.2)^x$
3. Logistic: $f(x) = \dfrac{4.5}{1 + 277e^{-1.4x}}$

CHAPTER 5 REVIEW EXERCISES (pp. 473–476)

1. (a) 8   **(b)** 0   **(c)** -6 **(d)** Undefined
3. (a) 7   **(b)** 1   **(c)** 0   **(d)** $-\frac{9}{2}$
5. (a) 2   **(b)** 1   **(c)** 2
7. (a) $\sqrt{6}$ **(b)** 12
9. $(f \circ g)(x) = \left(\frac{1}{x}\right)^3 - \left(\frac{1}{x}\right)^2 + 3\left(\frac{1}{x}\right) - 2; D = \{x \mid x \neq 0\}$
11. $(f \circ g)(x) = x$; all real numbers
13. $f(x) = x^2 + 3, g(x) = \sqrt{x}$ (answers may vary)
15. Subtract 6 from x and then multiply the result by 10;
$\frac{x}{10} + 6$ and $10(x - 6)$ **17.** f is one-to-one.
19. f is not one-to-one.
21.

x	6	4	3	1
$f^{-1}(x)$	-1	0	4	6

For $f: D = \{-1, 0, 4, 6\}$ and $R = \{1, 3, 4, 6\}$;
for $f^{-1}: D = \{1, 3, 4, 6\}$ and $R = \{-1, 0, 4, 6\}$
23. $f^{-1}(x) = \dfrac{x + 5}{3}$
25. $(f \circ f^{-1})(x) = f\left(\frac{x + 1}{2}\right) = 2\left(\frac{x + 1}{2}\right) - 1 = x$;
$(f^{-1} \circ f)(x) = f^{-1}(2x - 1) = \dfrac{(2x - 1) + 1}{2} = x$
27. 1   **29.** $f^{-1}(x) = x^2 - 1, x \geq 0$; for
$f: D = \{x \mid x \geq -1\}$ and $R = \{y \mid y \geq 0\}$;
for $f^{-1}: D = \{x \mid x \geq 0\}$ and $R = \{y \mid y \geq -1\}$
31. $C = 3, a = 2$
33.

35.

D = all real numbers $D = \{x \mid x > 0\}$
37. $C = 2, a = 2$ **39.** \$1562.71
41. $\ln 19 \approx 2.9444$ **43.** 3   **45.** -21
47. 2   **49.** 1   **51.** 2.631   **53.** $\log 125 \approx 2.097$
55. $10 \ln 5.2 \approx 16.49$ **57.** $-\dfrac{1}{\log 5} \approx -1.431$
59. $10 + \dfrac{\log (29/3)}{\log (0.78)} \approx 0.869$ **61.** $f(x) = 1.5(2)^x$
63. $10^{1.5} \approx 31.62$ **65.** $e^{3.4} \approx 29.96$ **67.** $\log 30x$
69. $\ln y - 2 \ln x$ **71.** $10^{1/4} \approx 1.778$
73. $\dfrac{100,000}{3} \approx 33,333$ **75.** 1 (extraneous -4)

77. The x-intercept is b. If $(0, b)$ is on the graph of f, then
$(b, 0)$ is on the graph of f^{-1}.
79. (a) 11,022/mL   **(b)** About 1.86 hr   **81.** $h(x) = 15x$
83. $x \approx 2.74$; the fish weighs 50 milligrams at about 3 weeks.
85. About 3.8 min   **87.** Logistic: $f(x) = \dfrac{171.4}{1 + 18.4e^{-0.0744x}}$
89. $a \approx 3.50, b \approx 0.74$

[0, 5, 1] by [0, 3, 1]

CHAPTER 5 EXTENDED AND DISCOVERY EXERCISES (pp. 476–477)

1. The data points $(\ln x, \ln y)$ seem to be almost linear.
Using linear regression gives the equation $y = 1.5x - 8.5$.
Since $b = 1.5$ and $a = e^{-8.5} \approx 0.0002$, we get the power
function $f(x) = 0.0002x^{1.5}$.
3. (a) $(T \circ R)(x) = 0.206e^{0.0124x}$
(b) $(T \circ R)(100) \approx 0.7119$; in the year 1900, radiative
forcing due to increased greenhouse gases was responsible
for a 0.7°F increase in average global temperature.
5. (a) $e^{\pi\sqrt{163}} \approx 262,537,412,640,768,743.99999999999925$;
it is not an integer.

CHAPTER 6: Trigonometric Functions

SECTION 6.1 (pp. 488–492)

1. (a)

(b)

(c)

(d)

3. Answers may vary.   **5.**

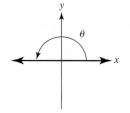

7. Answers may vary.   **9.** Answers may vary.

11. (a) $\frac{1}{4}$ (b) $\frac{1}{12}$ (c) $\frac{1}{6}$ (d) $\frac{1}{8}$

13. $510°, -210°$ (answers may vary)

15. $288°, -432°$ (answers may vary)

17. $\frac{5\pi}{2}, -\frac{3\pi}{2}$ (answers may vary)

19. $\frac{9\pi}{5}, -\frac{11\pi}{5}$ (answers may vary)

21. $125.25°$ **23.** $108.76°$ **25.** $125°18'$

27. $51°21'36''$ **29.** $\alpha = 34.1°, \beta = 124.1°$

31. $\alpha = 4°36'15'', \beta = 94°36'15''$

33. $\alpha = 66°19'25'', \beta = 156°19'25''$

35. $\theta = 2$ radians; $\theta \approx 114.6°$

37. $\theta = 1.3$ radians; $\theta \approx 74.5°$

39. (a) $\frac{\pi}{4}$ (b) $\frac{3\pi}{4}$ (c) $-\frac{2\pi}{3}$ (d) $-\frac{7\pi}{6}$

41. (a) $\frac{37\pi}{180}$ (b) 2.15 (c) -1.61 (d) 4.02

43. (a) $30°$ (b) $12°$ (c) $-300°$ (d) $-210°$

45. (a) $45°$ (b) $25.71°$ (c) $177.62°$ (d) $-143.24°$

47. $s = \frac{2\pi}{3}$ in.   **49.** $\theta = \frac{12}{5}$ radians   **51.** $r = \frac{5}{\pi}$ ft

53. $\frac{\pi}{4}$ m   **55.** π ft   **57.** $\frac{7\pi}{360}$ mi   **59.** 2π in.; $\frac{2\pi}{15}$ in./min

61. 10π in.; $\frac{2\pi}{15}$ in./min   **63.** 1.5π in^2 **65.** 4.5π in^2

67. $\frac{17,161\pi}{3000} \approx 5.72\pi$ cm^2 **69.** $\frac{3\pi}{16} \approx 0.59$ ft^2

71. 240π in^2 **73.** 4292π cm^2

75. 16.25 ft/sec; about 11.1 mi/hr   **77.** $49.8''$/yr

79. (a) $1000\pi \approx 3141.6$ radians/min

(b) $15,000\pi \approx 47,123.89$ in./min, or about 65.4 ft/sec

81. 810 mi   **83.** About 69 ft/sec   **85.** 14.5 in.

87. (a) 2.5 revolutions   (b) $\frac{65\pi}{6} \approx 34$ ft/sec

89. (a) $78,370$ mi/hr   (b) $66,630$ mi/hr

(c) $29,250$ mi/hr   (d) $12,160$ mi/hr

Planets farther from the sun have slower orbital velocities.

91. 137.2 m   **93.** (a) 388.8 m   (b) 881.8 m

6.1 EXTENDED AND DISCOVERY EXERCISES (p. 492)

1. $s = r\theta \left(\frac{\pi}{180°}\right)$, where θ is in degrees. The formula for radian measure is simpler.

SECTION 6.2 (pp. 500–505)

1.

3.

5. $\frac{\sqrt{3}}{2}$ **7.** $\frac{\sqrt{3}}{2}$ **9.** 2 **11.** 1 **13.** 1 **15.** $\frac{1}{\sqrt{2}}$

17. $\sin\theta = \frac{4}{5}, \cos\theta = \frac{3}{5}, \tan\theta = \frac{4}{3}$,

$\csc\theta = \frac{5}{4}, \sec\theta = \frac{5}{3}, \cot\theta = \frac{3}{4}$

19. $\sin\theta = \frac{12}{13}, \cos\theta = \frac{5}{13}, \tan\theta = \frac{12}{5}$,

$\csc\theta = \frac{13}{12}, \sec\theta = \frac{13}{5}, \cot\theta = \frac{5}{12}$

21. $\sin 60° \approx 0.866, \cos 60° = 0.5, \tan 60° \approx 1.732$,

$\csc 60° \approx 1.155, \sec 60° = 2, \cot 60° \approx 0.577$

23. $\sin 25° \approx 0.423, \cos 25° \approx 0.906, \tan 25° \approx 0.466$,

$\csc 25° \approx 2.366, \sec 25° \approx 1.103, \cot 25° \approx 2.145$

25. $\sin 5°35' \approx 0.097, \cos 5°35' \approx 0.995$,

$\tan 5°35' \approx 0.098, \csc 5°35' \approx 10.278$,

$\sec 5°35' \approx 1.005, \cot 5°35' \approx 10.229$

27. $\sin 13°45'30'' \approx 0.238, \cos 13°45'30'' \approx 0.971$,

$\tan 13°45'30'' \approx 0.245, \csc 13°45'30'' \approx 4.205$,

$\sec 13°45'30'' \approx 1.030, \cot 13°45'30'' \approx 4.084$

29. $\sin 1.05° \approx 0.018, \cos 1.05° \approx 1.000$,

$\tan 1.05° \approx 0.018, \csc 1.05° \approx 54.570$,

$\sec 1.05° \approx 1.000, \cot 1.05° \approx 54.561$

31. 3 **33.** $\frac{13}{12}$ **35.** $\frac{24}{7}$ **37.** $a \approx 13.86, b = 8$

39. $b \approx 5.03, c \approx 7.83$ **41.** $a \approx 5.25, b \approx 6.04$

43. $a \approx 16.82, c \approx 23.28$

45. $a = 12, b = 12\sqrt{3}, c = 12\sqrt{6}, d = 12\sqrt{3}$

47. $a = \frac{14\sqrt{3}}{3}, b = \frac{7\sqrt{3}}{3}, c = \frac{14\sqrt{3}}{3}, d = \frac{14\sqrt{6}}{3}$

49. $a \approx 20.78$ **51.** $c \approx 168.98$

53. (a) $\cos 20° \approx 0.9397$ (b) $\sin 50° \approx 0.7660$

55. (a) $\sec 41° \approx 1.3250$ (b) $\csc 27° \approx 2.2027$

57. $1500\tan 37°30' \approx 1151$ ft   **59.** 7.8 ft

61. About 34 ft   **63.** $52,000$ ft   **65.** About 128.2 ft

67. About 194.5 ft   **69.** $19,600$ ft   **71.** 114 ft

73. Barnard's Star: 3.5×10^{13} mi, 5.9 light-years;

Sirius: 5.1×10^{13} mi, 8.6 light-years;

61 Cygni: 6.6×10^{13} mi, 11.1 light-years;

Procyon: 6.7×10^{13} mi, 11.3 light-years

75. Min: 2.9×10^7 mi; max: 4.4×10^7 mi

77. $12,534$ mi

79. (a) About 704 ft **(b)** About 595 ft
(c) Increasing θ decreases r.

X	Y1
0	968
1	860.52
2	774.48
3	704
4	645.18
5	595.31
6	552.49

X=0

81. $d = 625\left(\frac{1}{\cos 54°} - 1\right) \approx 438$ ft   **83.** $A = \frac{\sqrt{3}}{4} s^2$

CHECKING BASIC CONCEPTS FOR SECTIONS 6.1 AND 6.2 (p. 505)

1. (a) $\frac{\pi}{4}$ **(b)** $\frac{5\pi}{12}$ **3.** $s = 2\pi$ in.; $A = 12\pi$ in^2

5. $\sin \theta = \frac{5}{13}$, $\cos \theta = \frac{12}{13}$, $\tan \theta = \frac{5}{12}$, $\csc \theta = \frac{13}{5}$,
 $\sec \theta = \frac{13}{12}$, $\cot \theta = \frac{12}{5}$

SECTION 6.3 (pp. 518–522)

1. (a) 13 **(b)** $\sin \theta = \frac{5}{13}$, $\cos \theta = \frac{12}{13}$

3. (a) 17 **(b)** $\sin \theta = \frac{8}{17}$, $\cos \theta = -\frac{15}{17}$

5. $\sin \theta = \frac{3}{5}$, $\cos \theta = \frac{4}{5}$

7. $\sin \theta = -\frac{2}{\sqrt{5}}$, $\cos \theta = \frac{1}{\sqrt{5}}$

9. $\sin \theta = \frac{2}{\sqrt{5}}$, $\cos \theta = \frac{1}{\sqrt{5}}$

11. $\sin \theta = -\frac{3}{\sqrt{10}}$, $\cos \theta = \frac{1}{\sqrt{10}}$

13. $\sin 45° = \frac{1}{\sqrt{2}}$, $\cos 45° = \frac{1}{\sqrt{2}}$

15. $\sin(-30°) = -\frac{1}{2}$, $\cos(-30°) = \frac{\sqrt{3}}{2}$

17. $\sin 225° = -\frac{1}{\sqrt{2}}$, $\cos 225° = -\frac{1}{\sqrt{2}}$

19. $\sin(-420°) = -\frac{\sqrt{3}}{2}$, $\cos(-420°) = \frac{1}{2}$

21. $\sin \frac{\pi}{3} = \frac{\sqrt{3}}{2}$, $\cos \frac{\pi}{3} = \frac{1}{2}$

23. $\sin\left(-\frac{\pi}{2}\right) = -1$, $\cos\left(-\frac{\pi}{2}\right) = 0$

25. $\sin \frac{7\pi}{6} = -\frac{1}{2}$, $\cos \frac{7\pi}{6} = -\frac{\sqrt{3}}{2}$

27. $\sin\left(-\frac{9\pi}{4}\right) = -\frac{1}{\sqrt{2}}$, $\cos\left(-\frac{9\pi}{4}\right) = \frac{1}{\sqrt{2}}$

29. $\sin 93.2° \approx 0.9984$, $\cos 93.2° \approx -0.0558$

31. $\sin 123°50' \approx 0.8307$, $\cos 123°50' \approx -0.5568$

33. $\sin(-4) \approx 0.7568$, $\cos(-4) \approx -0.6536$

35. $\sin \frac{11\pi}{7} \approx -0.9749$, $\cos \frac{11\pi}{7} \approx 0.2225$

37. $\sin \theta = \frac{3}{5}$, $\cos \theta = \frac{4}{5}$ **39.** $\sin \theta = -\frac{5}{13}$, $\cos \theta = \frac{12}{13}$

41. $\sin \frac{\pi}{2} = 1$, $\cos \frac{\pi}{2} = 0$

43. $\sin \frac{7\pi}{6} = -\frac{1}{2}$, $\cos \frac{7\pi}{6} = -\frac{\sqrt{3}}{2}$

45. $\sin\left(-\frac{3\pi}{4}\right) = -\frac{1}{\sqrt{2}}$, $\cos\left(-\frac{3\pi}{4}\right) = -\frac{1}{\sqrt{2}}$

47. $\sin \frac{5\pi}{2} = 1$, $\cos \frac{5\pi}{2} = 0$

49. $\sin\left(-\frac{\pi}{3}\right) = -\frac{\sqrt{3}}{2}$, $\cos\left(-\frac{\pi}{3}\right) = \frac{1}{2}$

51. $(-1, 0)$ **53.** $(1, 0)$ **55.** $(0, -1)$ **57.** $(0, -1)$

59. $\left(-\frac{1}{\sqrt{2}}, -\frac{1}{\sqrt{2}}\right)$ **61.** $\left(-\frac{1}{\sqrt{2}}, \frac{1}{\sqrt{2}}\right)$ **63.** $\left(\frac{\sqrt{3}}{2}, -\frac{1}{2}\right)$

65. $\left(\frac{1}{2}, -\frac{\sqrt{3}}{2}\right)$ **67.** $\sin 3\pi = 0$, $\cos 3\pi = -1$

69. $\sin(-2\pi) = 0$, $\cos(-2\pi) = 1$

71. $\sin \frac{3\pi}{2} = -1$, $\cos \frac{3\pi}{2} = 0$

73. $\sin\left(-\frac{5\pi}{2}\right) = -1$, $\cos\left(-\frac{5\pi}{2}\right) = 0$

75. $\sin \frac{5\pi}{4} = -\frac{1}{\sqrt{2}}$, $\cos \frac{5\pi}{4} = -\frac{1}{\sqrt{2}}$

77. $\sin\left(-\frac{5\pi}{4}\right) = \frac{1}{\sqrt{2}}$, $\cos\left(-\frac{5\pi}{4}\right) = -\frac{1}{\sqrt{2}}$

79. $\sin \frac{11\pi}{6} = -\frac{1}{2}$, $\cos \frac{11\pi}{6} = \frac{\sqrt{3}}{2}$

81. $\sin\left(-\frac{7\pi}{3}\right) = -\frac{\sqrt{3}}{2}$, $\cos\left(-\frac{7\pi}{3}\right) = \frac{1}{2}$

83. (a) $[-2\pi, 2\pi, \pi/2]$ by $[-4, 4, 1]$

$R = \{y \mid -3 \le y \le 3\}; f\left(\frac{3\pi}{2}\right) = -3$
  (b) $[-2\pi, 2\pi, \pi/2]$ by $[-4, 4, 1]$

$R = \{y \mid -1 \le y \le 1\}; f\left(\frac{3\pi}{2}\right) = 1$

85. (a) $[-2\pi, 2\pi, \pi/2]$ by $[-4, 4, 1]$

$R = \{y \mid -1 \le y \le 3\}; f\left(\frac{3\pi}{2}\right) = 1$
  (b) $[-2\pi, 2\pi, \pi/2]$ by $[-4, 4, 1]$

$R = \{y \mid -2 \le y \le 0\}; f\left(\frac{3\pi}{2}\right) = -2$

87. (a) 3% **(b)** $25,000\left(\dfrac{3}{\sqrt{10,009}}\right) \approx 750$ lb

89. First quarter: $\dfrac{\pi}{2} + 2\pi n$

91. (a) V is sinusoidal and varies between -310 volts and 310 volts. **(b)** $V(1/120) = 0$; after $1/120$ second, the voltage is 0. **(c)** $310/\sqrt{2} \approx 219$ volts

93. (a) 201 ft **(b)** 258 ft **(c)** It is easier to stop going uphill ($\theta > 0$) than downhill ($\theta < 0$).

95. (a) 31.4 ft **(b)** 78.3 ft **(c)** When the speed limit is higher, more land needs to be cleared on the inside of the curve.

97. (a) [0, 13, 1] by [3, 8, 1]   **(b)** [0, 13, 1] by [3, 8, 1]

From the close data fit shown, we may conclude that flying squirrels become active near sunset.

6.3 EXTENDED AND DISCOVERY EXERCISES (p. 522)

1. $[-2\pi, 2\pi, \pi/2]$ by $[-2, 2, 1]$

The translated graph and the graph of g are identical.

SECTION 6.4 (pp. 534–537)

1. $\sin\theta = \dfrac{12}{13}$, $\cos\theta = \dfrac{5}{13}$, $\tan\theta = \dfrac{12}{5}$, $\csc\theta = \dfrac{13}{12}$, $\sec\theta = \dfrac{13}{5}$, $\cot\theta = \dfrac{5}{12}$ **3.** $\sin\theta = -\dfrac{24}{25}$, $\cos\theta = \dfrac{7}{25}$, $\tan\theta = -\dfrac{24}{7}$, $\csc\theta = -\dfrac{25}{24}$, $\sec\theta = \dfrac{25}{7}$, $\cot\theta = -\dfrac{7}{24}$

5. $\sin 90° = 1$, $\cos 90° = 0$, $\tan 90°$ is undefined, $\csc 90° = 1$, $\sec 90°$ is undefined, $\cot 90° = 0$

7. $\sin(-45°) = -\dfrac{1}{\sqrt{2}}$, $\cos(-45°) = \dfrac{1}{\sqrt{2}}$, $\tan(-45°) = -1$, $\csc(-45°) = -\sqrt{2}$, $\sec(-45°) = \sqrt{2}$, $\cot(-45°) = -1$

9. $\sin\pi = 0$, $\cos\pi = -1$, $\tan\pi = 0$, $\csc\pi$ is undefined, $\sec\pi = -1$, $\cot\pi$ is undefined

11. $\sin\left(-\dfrac{\pi}{3}\right) = -\dfrac{\sqrt{3}}{2}$, $\cos\left(-\dfrac{\pi}{3}\right) = \dfrac{1}{2}$, $\tan\left(-\dfrac{\pi}{3}\right) = -\sqrt{3}$, $\csc\left(-\dfrac{\pi}{3}\right) = -\dfrac{2}{\sqrt{3}}$, $\sec\left(-\dfrac{\pi}{3}\right) = 2$, $\cot\left(-\dfrac{\pi}{3}\right) = -\dfrac{1}{\sqrt{3}}$

13. $\sin\left(-\dfrac{\pi}{2}\right) = -1$, $\cos\left(-\dfrac{\pi}{2}\right) = 0$, $\tan\left(-\dfrac{\pi}{2}\right)$ is undefined, $\csc\left(-\dfrac{\pi}{2}\right) = -1$, $\sec\left(-\dfrac{\pi}{2}\right)$ is undefined, $\cot\left(-\dfrac{\pi}{2}\right) = 0$

15. $\sin 360° = 0$, $\cos 360° = 1$, $\tan 360° = 0$, $\csc 360°$ is undefined, $\sec 360° = 1$, $\cot 360°$ is undefined

17. $\sin\dfrac{\pi}{6} = \dfrac{1}{2}$, $\cos\dfrac{\pi}{6} = \dfrac{\sqrt{3}}{2}$, $\tan\dfrac{\pi}{6} = \dfrac{1}{\sqrt{3}}$, $\csc\dfrac{\pi}{6} = 2$, $\sec\dfrac{\pi}{6} = \dfrac{2}{\sqrt{3}}$, $\cot\dfrac{\pi}{6} = \sqrt{3}$

19. $\sin\dfrac{4\pi}{3} = -\dfrac{\sqrt{3}}{2}$, $\cos\dfrac{4\pi}{3} = -\dfrac{1}{2}$, $\tan\dfrac{4\pi}{3} = \sqrt{3}$, $\csc\dfrac{4\pi}{3} = -\dfrac{2}{\sqrt{3}}$, $\sec\dfrac{4\pi}{3} = -2$, $\cot\dfrac{4\pi}{3} = \dfrac{1}{\sqrt{3}}$

21. $\sin(-225°) = \dfrac{1}{\sqrt{2}}$, $\cos(-225°) = -\dfrac{1}{\sqrt{2}}$, $\tan(-225°) = -1$, $\csc(-225°) = \sqrt{2}$, $\sec(-225°) = -\sqrt{2}$, $\cot(-225°) = -1$

23. $\sin\left(-\dfrac{13\pi}{6}\right) = -\dfrac{1}{2}$, $\cos\left(-\dfrac{13\pi}{6}\right) = \dfrac{\sqrt{3}}{2}$, $\tan\left(-\dfrac{13\pi}{6}\right) = -\dfrac{1}{\sqrt{3}}$, $\csc\left(-\dfrac{13\pi}{6}\right) = -2$, $\sec\left(-\dfrac{13\pi}{6}\right) = \dfrac{2}{\sqrt{3}}$, $\cot\left(-\dfrac{13\pi}{6}\right) = -\sqrt{3}$

25. $\sin\theta = \dfrac{4}{\sqrt{17}}$, $\cos\theta = -\dfrac{1}{\sqrt{17}}$, $\tan\theta = -4$, $\csc\theta = \dfrac{\sqrt{17}}{4}$, $\sec\theta = -\sqrt{17}$, $\cot\theta = -\dfrac{1}{4}$
The slope of the line equals $\tan\theta$.

27. $\sin\theta = -\dfrac{6}{\sqrt{37}}$, $\cos\theta = -\dfrac{1}{\sqrt{37}}$, $\tan\theta = 6$, $\csc\theta = -\dfrac{\sqrt{37}}{6}$, $\sec\theta = -\sqrt{37}$, $\cot\theta = \dfrac{1}{6}$
The slope of the line equals $\tan\theta$.

29. $\tan\theta = \dfrac{3}{4}$, $\cot\theta = \dfrac{4}{3}$, $\csc\theta = \dfrac{5}{3}$, $\sec\theta = \dfrac{5}{4}$

31. $\sin\theta = -\dfrac{15}{17}$, $\cos\theta = -\dfrac{8}{17}$, $\tan\theta = \dfrac{15}{8}$, $\cot\theta = \dfrac{8}{15}$

33. $\sin\theta = \dfrac{5}{13}$, $\csc\theta = \dfrac{13}{5}$, $\cot\theta = \dfrac{12}{5}$, $\sec\theta = \dfrac{13}{12}$

35. $\cos\theta = \dfrac{4}{5}$, $\tan\theta = -\dfrac{3}{4}$, $\csc\theta = -\dfrac{5}{3}$, $\sec\theta = \dfrac{5}{4}$, $\cot\theta = -\dfrac{4}{3}$

37. $\sin\theta = -\dfrac{3}{5}$, $\tan\theta = \dfrac{3}{4}$, $\csc\theta = -\dfrac{5}{3}$, $\sec\theta = -\dfrac{5}{4}$, $\cot\theta = \dfrac{4}{3}$

39. $\sin(-7\pi) = 0$, $\cos(-7\pi) = -1$, $\tan(-7\pi) = 0$, $\csc(-7\pi)$ is undefined, $\sec(-7\pi) = -1$, $\cot(-7\pi)$ is undefined

41. $\sin\dfrac{7\pi}{2} = -1$, $\cos\dfrac{7\pi}{2} = 0$, $\tan\dfrac{7\pi}{2}$ is undefined, $\csc\dfrac{7\pi}{2} = -1$, $\sec\dfrac{7\pi}{2}$ is undefined, $\cot\dfrac{7\pi}{2} = 0$

43. $\sin\left(-\dfrac{3\pi}{4}\right) = -\dfrac{1}{\sqrt{2}}$, $\cos\left(-\dfrac{3\pi}{4}\right) = -\dfrac{1}{\sqrt{2}}$, $\tan\left(-\dfrac{3\pi}{4}\right) = 1$, $\csc\left(-\dfrac{3\pi}{4}\right) = -\sqrt{2}$, $\sec\left(-\dfrac{3\pi}{4}\right) = -\sqrt{2}$, $\cot\left(-\dfrac{3\pi}{4}\right) = 1$

45. $\sin \frac{7\pi}{6} = -\frac{1}{2}$, $\cos \frac{7\pi}{6} = -\frac{\sqrt{3}}{2}$, $\tan \frac{7\pi}{6} = \frac{1}{\sqrt{3}}$,

$\csc \frac{7\pi}{6} = -2$, $\sec \frac{7\pi}{6} = -\frac{2}{\sqrt{3}}$, $\cot \frac{7\pi}{6} = \sqrt{3}$

47. $\sin \theta = \frac{1}{\sqrt{2}}$, $\cos \theta = \frac{1}{\sqrt{2}}$, $\tan \theta = 1$, $\csc \theta = \sqrt{2}$,

$\sec \theta = \sqrt{2}$, $\cot \theta = 1$

49. $\sin \theta = -\frac{12}{13}$, $\cos \theta = \frac{5}{13}$, $\tan \theta = -\frac{12}{5}$, $\csc \theta = -\frac{13}{12}$,

$\sec \theta = \frac{13}{5}$, $\cot \theta = -\frac{5}{12}$

51. -1 **53.** 2   **55.** $\frac{2}{\sqrt{3}}$

57. $\sin \frac{9\pi}{2} = 1$, $\cos \frac{9\pi}{2} = 0$, $\tan \frac{9\pi}{2}$ is undefined,

$\csc \frac{9\pi}{2} = 1$, $\sec \frac{9\pi}{2}$ is undefined, $\cot \frac{9\pi}{2} = 0$

59. $\sin (-\pi) = 0$, $\cos (-\pi) = -1$, $\tan (-\pi) = 0$,

$\csc (-\pi)$ is undefined, $\sec (-\pi) = -1$,

$\cot (-\pi)$ is undefined

61. $\sin \frac{4\pi}{3} = -\frac{\sqrt{3}}{2}$, $\cos \frac{4\pi}{3} = -\frac{1}{2}$, $\tan \frac{4\pi}{3} = \sqrt{3}$,

$\csc \frac{4\pi}{3} = -\frac{2}{\sqrt{3}}$, $\sec \frac{4\pi}{3} = -2$, $\cot \frac{4\pi}{3} = \frac{1}{\sqrt{3}}$

63. (a) 0.9984   **(b)** 1.0016
65. (a) 1.4045   **(b)** 0.7120
67. (a) -0.8637 **(b)** -1.1578
69. (a) 0.2225   **(b)** 4.4940
71. Origin symmetry   **73.** Origin symmetry

75. y-axis symmetry

77. D = all real numbers, $R = \{y \mid -1 \le y \le 1\}$,
period $= 2\pi$
79. $D = \left\{t \mid t \ne \pm\frac{\pi}{2}, \pm\frac{3\pi}{2}, \pm\frac{5\pi}{2}, \dots\right\}$,
R = all real numbers, period $= \pi$

81. $D = \left\{t \mid t \ne \pm\frac{\pi}{2}, \pm\frac{3\pi}{2}, \pm\frac{5\pi}{2}, \dots\right\}$,
$R = \{y \mid |y| \ge 1\}$, period $= 2\pi$
83. 0.900   **85.** 0.527
87. (a)

(b) Just after sunrise ($t = 0$), the shadow is very long. As the elevation of the sun increases, the shadow decreases in length until it is 0 when $t = \pi/2$. In the afternoon, the shadow increases in length in the opposite direction until sunset ($t = \pi$).

89. About $17.95''$; $\theta = \frac{57.3 \sin \alpha}{\cos \alpha}$ **91.** About 1300 ft
93. About 41%
95. As θ increases, the values of y_1 and y_2 become closer together. When $\theta = 20°$, the difference is only about 0.02, or 2%.

97. $[0, 20000, 5000]$ by $[0, 4000, 1000]$

(a) About $(7612, 2197)$ **(b)** About 15,223 ft

6.4 EXTENDED AND DISCOVERY EXERCISE (p. 537)
1. Using the small right triangle yields
$\sin \theta = \frac{\text{Opp.}}{\text{Hyp.}} = \frac{\text{Opp.}}{1} = $ Opposite side. Similarly,
$\cos \theta = \frac{\text{Adj.}}{\text{Hyp.}} = \frac{\text{Adj.}}{1} = $ Adjacent side.
Then using the large right triangle yields
$\tan \theta = \frac{\text{Opp.}}{\text{Adj.}} = \frac{\text{Opp.}}{1} = $ Opposite side. Similarly,
$\sec \theta = \frac{\text{Hyp.}}{\text{Adj.}} = \frac{\text{Hyp.}}{1} = $ Hypotenuse.

CHECKING BASIC CONCEPTS FOR SECTIONS 6.3 AND 6.4 (p. 537)
1. $\sin \theta = \frac{6}{\sqrt{85}}$, $\cos \theta = -\frac{7}{\sqrt{85}}$, $\tan \theta = -\frac{6}{7}$,

$\csc \theta = \frac{\sqrt{85}}{6}$, $\sec \theta = -\frac{\sqrt{85}}{7}$, $\cot \theta = -\frac{7}{6}$

3. Sine

Cosine

9. Shift the cosine graph $\frac{\pi}{2}$ units to the left, increase the amplitude to 2, and reflect the graph across the x-axis.

Tangent

11. Shift the cosine graph $\frac{1}{\pi}$ unit to the right, shorten the period to 2, and decrease the amplitude to $\frac{1}{2}$.

SECTION 6.5 (pp. 548–552)

1. Period: 4π; amplitude: 3

3. Period: $\frac{2\pi}{3}$; amplitude: 2

13. Amplitude: 3; period: $\frac{\pi}{2}$; phase shift: $\frac{\pi}{4}$; vertical shift: -4
15. Amplitude: 4; period: 4; phase shift: 1; vertical shift: 6
17. Amplitude: $\frac{2}{3}$; period: $\frac{\pi}{3}$; phase shift: $-\frac{\pi}{2}$; vertical shift: $-\frac{5}{2}$
19. $a = 3, b = 2$
21. Amplitude: 3; period: 4π; phase shift: 0
23. c   **25.** d   **27.** a   **29.** $y = 3 \sin\left(\frac{1}{2}x\right)$

5. Period: 2; amplitude: $\frac{1}{2}$

31. Amplitude: 2; period: 2π; phase shift: 0

33. Amplitude: 1; period: 4π; phase shift: 0

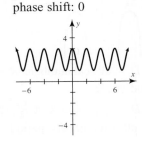

7. Shorten the period of the sine graph to π, increase the amplitude to 3, and shift the graph downward 1 unit.

35. Amplitude: 1; period: 2π; phase shift: 0

37. Amplitude: 1; period: 2; phase shift: 0

39. Amplitude: 1; period: π; phase shift: $-\pi$

41. Amplitude: 1; period: 2π; phase shift: $\frac{\pi}{2}$

43. Amplitude: $\frac{1}{2}$; period: π; phase shift: 0

45. Amplitude: 2; period: π; phase shift: $-\frac{\pi}{4}$

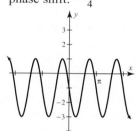

47. Amplitude: 1; period: π; phase shift: $\frac{\pi}{2}$

49. Amplitude = 2; period = π; phase shift = 0

$[-2\pi, 2\pi, \pi/2]$ by $[-4, 4, 1]$

51. Amplitude = $\frac{1}{2}$; period = $\frac{2\pi}{3}$; phase shift = $-\frac{\pi}{3}$

$[-2\pi, 2\pi, \pi/2]$ by $[-4, 4, 1]$

53. Amplitude = 2.5; period = π; phase shift = $-\frac{\pi}{4}$

$[-2\pi, 2\pi, \pi/2]$ by $[-4, 4, 1]$

55. Amplitude = 2; period = 1; phase shift = $-\frac{1}{8}$

$[-2\pi, 2\pi, \pi/2]$ by $[-4, 4, 1]$

57. Period = $\frac{\pi}{2}$; phase shift = 0
Asymptotes: $x = \pm\frac{\pi}{4}, \pm\frac{3\pi}{4}, \pm\frac{5\pi}{4}, \pm\frac{7\pi}{4}$

59. Period = π; phase shift = $\frac{\pi}{2}$
Asymptotes: $x = 0, \pm\pi, \pm2\pi$

61. Period = $\frac{\pi}{2}$; phase shift = 0
Asymptotes: $x = 0, \pm\frac{\pi}{2}, \pm\pi, \pm\frac{3\pi}{2}, \pm2\pi$

63. Period = $\frac{\pi}{2}$; phase shift = $\frac{\pi}{4}$
Asymptotes: $x = \pm\frac{\pi}{4}, \pm\frac{3\pi}{4}, \pm\frac{5\pi}{4}, \pm\frac{7\pi}{4}$

65. Period = 4π; phase shift = 0
Asymptotes: $x = \pm\pi$

67. Period $= 2\pi$; phase shift $= \pi$
Asymptotes: $x = 0, \pm\pi, \pm2\pi$

69. Period $= 6\pi$; phase shift $= \frac{\pi}{6}$
Asymptotes: $x = -\frac{4\pi}{3}, \frac{5\pi}{3}$

71. (a) $40°F, -40°F$ **(b)** Amplitude $= 40$, half the difference between the maximum and minimum monthly average temperatures; period $= 12$ means that the temperature pattern repeats every 12 months. **(c)** The months when the average temperature is $0°F$
73. (a) Amplitude $= 34$; period $= 12$; phase shift $= 4.3$
(b) $f(5) \approx 12.2°F; f(12) \approx -26.4°F$ **(c)** About $0°F$
75. (a) [0, 25, 2] by [0, 80, 10]

(b) $f(x) = 14 \sin\left(\frac{\pi}{6}(x - 4)\right) + 50$
(c) [0, 25, 2] by [0, 80, 10]

77. (a) $f(x) = 17 \cos\left(\frac{\pi}{6}(x - 7)\right) + 75$
(b) Yes. The period of the graph is 12, so, for example, the phase shift could be $c = 7 + 12 = 19$ or $c = 7 - 12 = -5$.
79. (a) About 18.5 hr; June 21 **(b)** About 6 hr; December 21 **(c)** The amplitude represents half the difference in daylight between the longest and shortest day; the period represents one year (answers may vary).

81. (a) Max ≈ 8 in.; min ≈ 0.5 in. **(b)** Amp ≈ 3.75. The amplitude represents half the difference between the maximum and minimum monthly average precipitations.
(c) $f(x) = 3.75 \cos\left(\frac{\pi}{6}x\right) + 4.25$
83. (a) $f(t) = 2 \cos\left(\frac{\pi}{6}(t - 1)\right) + 4$
(b) $f(1) = 6, f(7) = 2$
85. (a) [0, 25, 2] by [60, 90, 5]

(b) $f(x) = 6.5 \sin\left(\frac{\pi}{6}(x - 4)\right) + 78.5$
87. [0, 12, 3] by [0, 50, 10]

At 9:00 A.M. the outdoor temperature is 20°C. The temperature increases to a maximum of 35°C at 3:00 P.M. Then it begins to fall until it reaches 20°C again at 9:00 P.M.
89. (a) [0, 1/100, 1/880] by [−1.5, 1.5, 0.5]

(b) $\frac{1}{440} \approx 0.00227$ sec **(c)** 440 cycles per second
91. (a) $s(t) = 2 \cos(4\pi t)$ **(b)** $s(1) = 2$. The weight is moving neither upward nor downward. At $t = 1$ the motion of the weight is changing from up to down.
93. (a) $s(t) = -3 \cos(2.5\pi t)$ **(b)** $s(1) = 0$. The weight is moving upward.
95. $a = 0.21, b = 55\pi$; $Y_1 = 0.21 \cos(55\pi X)$
[0, 0.05, 0.01] by [−0.3, 0.3, 0.1]

97. $a = 0.14, b = 110\pi$; $Y_1 = 0.14 \cos(110\pi X)$
[0, 0.05, 0.01] by [−0.3, 0.3, 0.1]

99. $y \approx 13.2 \sin (0.524x - 2.18) + 49.7$
[0, 25, 2] by [30, 80, 10]

101. $y \approx 16.9 \sin (0.522x - 2.09) + 75.4$
[0, 25, 2] by [50, 100, 10]

SECTION 6.6 (pp. 562–565)

1. one-to-one   **3.** π **5.** 0   **7.** $\frac{\pi}{2}$ **9.** $-\frac{\pi}{4}$ **11.** $\frac{2\pi}{3}$

13. (a) $\frac{\pi}{2}$, or $90°$ **(b)** Undefined   **(c)** $-\frac{\pi}{3}$, or $-60°$

15. (a) $\frac{\pi}{2}$, or $90°$ **(b)** π, or $180°$ **(c)** $\frac{\pi}{3}$, or $60°$

17. (a) $\frac{\pi}{4}$, or $45°$ **(b)** $-\frac{\pi}{4}$, or $-45°$ **(c)** $\frac{\pi}{3}$, or $60°$

19. (a) Undefined   **(b)** 1.47, or $84.3°$ **(c)** 1.82, or $104.5°$

21. (a) β **(b)** α **(c)** β **23.** 1   **25.** $\frac{3\pi}{4}$ **27.** -3

29. (a) $90°$ **(b)** $90°$ **(c)** $90°$; $\sin^{-1} x + \cos^{-1} x = 90°$ whenever $-1 \le x \le 1$.

31. *Verbal:* Determine the angle (or real number) θ such that $\sin \theta = 2x$ and $-\frac{\pi}{2} \le \theta \le \frac{\pi}{2}$.

Numerical:

Graphical:
$[-1, 1, 1]$ by $[-2, 2, 1]$

33. *Verbal:* Determine the angle (or real number) θ such that $\cos \theta = \frac{1}{2}x$ and $0 \le \theta \le \pi$.

Numerical:

Graphical:
$[-3, 3, 1]$ by $[-1, 4, 1]$

35. $\alpha = \tan^{-1} \frac{7}{24} \approx 16.3°$, $\beta = \tan^{-1} \frac{24}{7} \approx 73.7°$, $c = 25$

37. $\alpha = \sin^{-1} \frac{6}{10} \approx 36.9°$, $\beta = \cos^{-1} \frac{6}{10} \approx 53.1°$, $b = 8$

39. $\beta = 35°$, $b = \frac{5}{\tan 55°} \approx 3.5$, $c = \frac{5}{\sin 55°} \approx 6.1$

41. $\tan \theta = \dfrac{x}{\sqrt{1 - x^2}}$ **43.** $\cos \theta = \dfrac{1}{\sqrt{1 + x^2}}$

45. $\sqrt{1 - u^2}$ **47.** $\dfrac{\sqrt{1 - u^2}}{u}$ **49.** u **51.** $90°$

53. $45°$ **55.** $90°$ **57.** $82.8°$ **59.** $80.5°$ **61.** $68.9°$

63. -0.197 **65.** 1.102   **67.** -0.282 **69.** 1.369, 1.772

71. $\tan^{-1} \frac{50}{85} \approx 30.5°$ **73. (a)** $4.6°$ **(b)** $2.1°$ **(c)** $-2.9°$

75. $\tan^{-1} \frac{4}{10} \approx 21.8°$ **77. (a)** $60°$ **(b)** $120°$

79. $\theta \approx 41.9°$ **81. (a)** 14.9 hr   **(b)** 13.8 hr   **(c)** 14.6 hr

83. $\theta_1 \approx 35.8°$, $\theta_2 \approx 54.2°$

85. Let α and β represent the angles of elevation from the shrub to the shorter and taller buildings, respectively. The distance from the shrub to the shorter building is $100 - x$; thus $\alpha = \arctan \frac{75}{100 - x}$. Similarly, $\beta = \arctan \frac{150}{x}$. Because the angles α, θ, and β form a straight angle, $\theta = \pi - \alpha - \beta$. That is, $\theta = \pi - \arctan \frac{75}{100 - x} - \arctan \frac{150}{x}$.

6.6 EXTENDED AND DISCOVERY EXERCISE (p. 565)

1. The maximum value of θ is $38.7°$ when $x \approx 6.24$.
[0, 50, 10] by [0, 50, 10]

CHECKING BASIC CONCEPTS FOR SECTIONS 6.5 AND 6.6 (p. 566)

1. Amplitude $= 3$; period $= \frac{2\pi}{2} = \pi$; phase shift $= \frac{\pi}{4}$

3. (a) $\sin^{-1} 0 = 0°$ **(b)** $\cos^{-1}(-1) = 180°$
(c) $\tan^{-1}(-1) = -45°$ **(d)** $\sin^{-1} \frac{1}{2} = 30°$
(e) $\tan^{-1} \sqrt{3} = 60°$ **(f)** $\cos^{-1} \frac{1}{2} = 60°$

5. (a) $\sin^{-1} 0.55 \approx 0.582$ **(b)** $\cos^{-1}(-0.35) \approx 1.93$
(c) $\tan^{-1}(-2.9) \approx -1.24$

CHAPTER 6 REVIEW EXERCISES (pp. 569–571)

1. (a)   **(b)**

3. (a) $60°$ (b) $5°$ (c) $-150°$ (d) $-315°$

5. 2π ft **7.** $\frac{1}{2}$ **9.** $\frac{1}{\sqrt{3}}$ **11.** $\sqrt{2}$

13. $\sin\theta = \frac{8}{\sqrt{145}}$, $\cos\theta = \frac{9}{\sqrt{145}}$, $\tan\theta = \frac{8}{9}$,

$\csc\theta = \frac{\sqrt{145}}{8}$, $\sec\theta = \frac{\sqrt{145}}{9}$, $\cot\theta = \frac{9}{8}$

15. $\csc\theta = 3$

17. $\sin 25° \approx 0.423$, $\cos 25° \approx 0.906$, $\tan 25° \approx 0.466$, $\csc 25° \approx 2.366$, $\sec 25° \approx 1.103$, $\cot 25° \approx 2.145$

19. $\sin\theta = -\frac{2}{\sqrt{5}}$, $\cos\theta = \frac{1}{\sqrt{5}}$, $\tan\theta = -2$,

$\csc\theta = -\frac{\sqrt{5}}{2}$, $\sec\theta = \sqrt{5}$, $\cot\theta = -\frac{1}{2}$

21. $\sin\theta = \frac{\sqrt{3}}{2}$, $\cos\theta = -\frac{1}{2}$, $\tan\theta = -\sqrt{3}$,

$\csc\theta = \frac{2}{\sqrt{3}}$, $\sec\theta = -2$, $\cot\theta = -\frac{1}{\sqrt{3}}$

23. -1 **25.** 0

27. $\tan\theta = -\frac{4}{3}$, $\csc\theta = -\frac{5}{4}$, $\sec\theta = \frac{5}{3}$, $\cot\theta = -\frac{3}{4}$

29. Amplitude $= 3$; period $= \pi$; phase shift $= 0$

31. Amplitude $= 3$; period $= 4\pi$; phase shift $= \pi$

33. $a = 2$, $b = 3$

35. Period $= \frac{\pi}{2}$; phase shift $= 0$

37. (a) $-\frac{\pi}{2}$, or $-90°$ (b) $\frac{\pi}{3}$, or $60°$ (c) $\frac{\pi}{4}$, or $45°$

39. (a) -0.64, or $-36.9°$ (b) 1.37, or $78.7°$
(c) 1.45, or $83.1°$

41. $\alpha = \tan^{-1}\frac{5}{3} \approx 59.0°$, $\beta = \tan^{-1}\frac{3}{5} \approx 31.0°$, $c = \sqrt{34}$

43. $30°$ **45.** $78.5°$ **47.** -0.6435

49. (a) $\frac{\pi}{25}$ radian/sec (b) $\pi \approx 3.14$ ft/sec **51.** 89 ft

53. (a) $\sin^{-1}\frac{350}{6000} \approx 3.3°$ (b) $\sin^{-1}\frac{160}{4500} \approx 2.0°$

55. 161 mi

57. (a) $[0, 25, 2]$ by $[0, 70, 10]$ (c) $[0, 25, 2]$ by $[0, 70, 10]$

(b) $f(x) = 26\cos\left(\frac{\pi}{6}(x - 7)\right) + 32$

CHAPTER 6 EXTENDED AND DISCOVERY EXERCISES
(pp. 571–572)

1. (a) $x_Q = d\sin\theta + x_p$ and $y_Q = d\cos\theta + y_p$
(b) Approximately $(233.9, 377.2)$
3. (a) $f(x) = 12.5\cos\left(\frac{\pi}{6}(x - 1)\right) + 61.5$
(b) $[0, 25, 2]$ by $[40, 80, 10]$

Answers may vary.

(c) The high temperatures occur in January in the Southern Hemisphere as opposed to July in the Northern Hemisphere. This affects the phase shift of f.

CHAPTERS 1–6 CUMULATIVE REVIEW EXERCISES
(pp. 572–576)

1. 1.25×10^5; 0.00467

3. (a)   (b)

(c)   (d)

(e)   **(f)**

5. (a) $D = \{x \mid x \neq 2, x \neq -2\}$

(b) $f(-1) = \frac{1}{4}$; $f(2a) = \frac{1}{8(a+1)}$

7. 90   **9.** $y = -\frac{2}{3}x - \frac{5}{3}$

11. (a) $-\frac{4}{5}, \frac{12}{5}$ **(b)** $\ln 14 \approx 2.64$ **(c)** $0, 1, 2$

(d) $\pm 2, \pm \sqrt{2}$ **(e)** $-1, 2$ **(f)** 5   **(g)** $3 \pm \sqrt{7}$

(h) 65,535

13. (a) $\left(-\infty, \frac{9}{5}\right)$ **(b)** $\left(-\frac{2}{3}, \frac{7}{3}\right]$ **(c)** $\left(-\infty, -\frac{3}{2}\right] \cup [3, \infty)$

(d) $[1, 4]$ **(e)** $(-1, 0) \cup (1, \infty)$ **(f)** $\left(-2, -\frac{5}{3}\right]$

15. $\left(\frac{2}{3}, -\frac{1}{3}\right)$ **17.** $\frac{-2 \pm \sqrt{6}}{2}$

19. (a) Increasing: $[-2, 0] \cup [2, \infty)$; decreasing:
$(-\infty, -2] \cup [0, 2]$ **(b)** $-2.8, 0, 2.8$ (approximate)
(c) $(-2, -4), (0, 0), (2, -4)$ **(d)** Local minimum: -4;
local maximum: 0

21. (a) $\frac{5a^2}{2} - 1 + \frac{2}{a^2}$ **(b)** $x - 3 + \frac{4}{x^2 + 1}$ **23.** $\pm 2i, 1$

25. $D = \left\{x \mid x \neq \frac{7}{3}\right\}$; vertical: $x = \frac{7}{3}$; horizontal: $y = \frac{2}{3}$

27. (a) 5   **(b)** Undefined   **(c)** 2   **(d)** 0

29. (a) 8   **(b)** 0   **(c)** $(f - g)(x) = x^2 + 2x$

(d) $(f \circ g)(x) = x^2 - x - 4$

31.

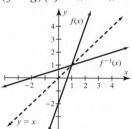

33. $C = \frac{1}{2}, a = 2$

35. (a) 2   **(b)** 4   **(c)** -2 **(d)** $\frac{1}{3}$ **(e)** 2   **(f)** 1

37. 4.395   **39.** $\alpha = 79°26'$; $\beta = 169°26'$

41. $\frac{5\pi}{6}$ **43.** $\frac{\pi}{4} \approx 0.79$ ft   **45.** $\frac{9}{\tan 30°} \approx 15.6$

47. $\sin \theta = -\frac{1}{\sqrt{2}}$, $\cos \theta = -\frac{1}{\sqrt{2}}$, $\tan \theta = 1$,

$\csc \theta = -\sqrt{2}$, $\sec \theta = -\sqrt{2}$, $\cot \theta = 1$

49. $\sin \theta = -\frac{12}{13}$, $\cos \theta = \frac{5}{13}$, $\tan \theta = -\frac{12}{5}$,

$\csc \theta = -\frac{13}{12}$, $\sec \theta = \frac{13}{5}$, $\cot \theta = -\frac{5}{12}$

51. (a)   **(b)**

(c)

53. $\alpha \approx 33.4°$; $\beta \approx 56.6°$; $b \approx 16.7$

55. (a) 16 ft/sec; 48 ft/sec

(b) During the first second, the average speed is
16 ft/sec. During the next second, the average
speed is 48 ft/sec. The object is speeding up.

(c) $32t + 16h$

57. 24 lb

59. (a) $C(t) = 135(t - 1980) + 800$

(b) 1996; the formula gives the correct year.

61. (a) $f^{-1}(x) = \frac{9}{5}x + 32$

(b) f^{-1} converts degrees Celsius to degrees Fahrenheit.

63. 20 ft/sec

65. $f(x) \approx 14.5 \sin \left(\frac{\pi}{6}(x - 4)\right) + 75.5$

CHAPTER 7: Trigonometric Identities and Equations

SECTION 7.1 (pp. 588–591)

1. $\cot \theta = 2$ **3.** $\sec \theta = \frac{7}{2}$ **5.** $\cos \theta = -\frac{1}{4}$

7. $\frac{\sin^2 \theta}{\cos^2 \theta}$ **9.** $\cos^2 \theta$ **11.** $\sin \theta + \cos \theta$

13. $\tan \theta = -\frac{3}{4}$, $\csc \theta = \frac{5}{3}$, $\sec \theta = -\frac{5}{4}$, $\cot \theta = -\frac{4}{3}$

15. $\cos \theta = -\frac{7}{25}$, $\tan \theta = \frac{24}{7}$, $\csc \theta = -\frac{25}{24}$, $\sec \theta = -\frac{25}{7}$

17. $\tan \theta = \frac{60}{11}$, $\cot \theta = \frac{11}{60}$, $\csc \theta = -\frac{61}{60}$, $\sec \theta = -\frac{61}{11}$

19. $\sin \theta = \frac{1}{\sqrt{2}}$, $\cos \theta = -\frac{1}{\sqrt{2}}$, $\tan \theta = -1$, $\cot \theta = -1$

21. 1   **23.** 1   **25.** 1   **27.** $\cos^2 \theta$ **29.** $\tan^2 \theta$

31. $-\tan \theta$ **33.** $\sec \theta$ **35.** $\tan^2 \theta$ **37.** $\cos x$

39. 1   **41.** $\sec x - \csc x$ **43.** Yes   **45.** No

47. Quadrant IV   **49.** Quadrant III   **51.** Quadrant II

53. $\sin\theta = -\frac{\sqrt{3}}{2}$, $\tan\theta = -\sqrt{3}$, $\csc\theta = -\frac{2}{\sqrt{3}}$,

$\sec\theta = 2$, $\cot\theta = -\frac{1}{\sqrt{3}}$

55. $\sin\theta = -\frac{11}{61}$, $\cos\theta = \frac{60}{61}$, $\cot\theta = -\frac{60}{11}$,

$\csc\theta = -\frac{61}{11}$, $\sec\theta = \frac{61}{60}$

57. $\cos\theta = \frac{24}{25}$, $\tan\theta = \frac{7}{24}$, $\cot\theta = \frac{24}{7}$, $\csc\theta = \frac{25}{7}$,

$\sec\theta = \frac{25}{24}$

59. $\cos\theta = -\frac{\sqrt{8}}{3}$, $\tan\theta = \frac{1}{\sqrt{8}}$, $\cot\theta = \sqrt{8}$,

$\csc\theta = -3$, $\sec\theta = -\frac{3}{\sqrt{8}}$

61. $\sin\theta = -\frac{35}{37}$, $\cos\theta = \frac{12}{37}$, $\tan\theta = -\frac{35}{12}$,

$\csc\theta = -\frac{37}{35}$, $\cot\theta = -\frac{12}{35}$

63. $\sin\theta = \frac{3}{7}$, $\cos\theta = -\frac{\sqrt{40}}{7}$, $\tan\theta = -\frac{3}{\sqrt{40}}$,

$\sec\theta = -\frac{7}{\sqrt{40}}$, $\cot\theta = -\frac{\sqrt{40}}{3}$

65. $\sin x = \sqrt{1-\cos^2 x}$; $\tan x = \frac{\sqrt{1-\cos^2 x}}{\cos x}$

67. $\sin x = -\frac{\tan x}{\sqrt{1+\tan^2 x}}$; $\sec x = -\sqrt{1+\tan^2 x}$

69. $\cot x = -\sqrt{\csc^2 x - 1}$; $\cos x = -\frac{\sqrt{\csc^2 x - 1}}{\csc x}$

71. $\cos\theta = \sqrt{1-x^2}$, 0.8586

73. $\sin\theta = -\frac{1}{\sqrt{1+x^2}}$, -0.8899

75. $\tan\theta = \frac{x}{\sqrt{1-x^2}}$ **77.** $-\sin 13°$

79. $-\tan\frac{\pi}{11}$ **81.** $\sec\frac{2\pi}{5}$ **83.** $d = 93,000,000\csc\theta$

85. **(a)** The sum of L and C equals 3.

$[0, 10^{-6}, 10^{-7}]$ by $[-1, 4, 1]$

(b) Let $Y_1 = L(t)$, $Y_2 = C(t)$, and $Y_3 = E(t)$.
$E(t) = 3$ for all inputs

X	Y2	Y3
0	0	3
1E-7	.95646	3
2E-7	2.6061	3
3E-7	2.8451	3
4E-7	1.3688	3
5E-7	.05974	3
6E-7	.58747	3

$Y_3 = Y_1 + Y_2$

(c) $E(t) = L(t) + C(t)$

$= 3\cos^2(6{,}000{,}000t) + 3\sin^2(6{,}000{,}000t)$

$= 3(\cos^2(6{,}000{,}000t) + \sin^2(6{,}000{,}000t))$

$= 3(1) = 3$

87. **(a)** The graph has y-axis symmetry. The monthly high temperatures x months before and x months after July are equal.

$[-6, 6, 1]$ by $[0, 100, 10]$ **(b)** f is even.

(c) $f(-x) = f(x)$

89. **(a)** $P = 16k\cos^2(2\pi t)$

(b) Let $Y_1 = 32(\cos(2\pi X))^2$. Y_1 has a maximum value of 32 when $t = 0, 0.5, 1, 1.5, 2.0$, and Y_1 has a minimum value of 0 when $t = 0.25, 0.75, 1.25, 1.75$. The spring is either stretched or compressed the most when Y_1 is maximum.

$[0, 2, 0.5]$ by $[-1, 40, 8]$

(c) $P = 16k(1 - \sin^2(2\pi t))$

SECTION 7.2 (pp. 597–598)

1. **(a)** $1 - x^2$ **(b)** $\cos^2\theta$

3. **(a)** $x^2 - x$ **(b)** $\sec^2\theta - \sec\theta$

5. **(a)** $(x + 1)(x + 1)$ **(b)** $(\cos\theta + 1)(\cos\theta + 1)$

7. **(a)** $x(x - 2)$ **(b)** $\sec t(\sec t - 2)$

9. **(a)** $\frac{2}{1 - x^2}$ **(b)** $2\csc^2\theta$

11. **(a)** $\frac{x^2 + y^2}{xy}$ **(b)** $\sec t\csc t$

13. **(a)** $y^2 + x^2$ **(b)** 1 **15.** **(a)** x **(b)** $\cos\theta$

17. $\sin\theta$ **19.** $\sin\theta - \sec\theta$ **21.** $2\tan t + \sec^2 t$

23. $\cos^2\theta$ **25.** $(1 - \tan\theta)(1 + \tan\theta)$

27. $(\sec t - 3)(\sec t + 2)$ **29.** $\sec^2\theta(\tan^2\theta + 2)$

31. $\csc^2\theta - \cot^2\theta = 1 + \cot^2\theta - \cot^2\theta = 1$

33. $(1 - \sin t)^2 = 1 - 2\sin t + \sin^2 t$ (FOIL)

35. $\frac{\sin t + \cos t}{\sin t} = \frac{\sin t}{\sin t} + \frac{\cos t}{\sin t} = 1 + \cot t$

37. $\sec^2\theta - 1 = (1 + \tan^2\theta) - 1$

$= \tan^2\theta$

39. $\frac{\tan^2 t}{\sec t} = \frac{\sec^2 t - 1}{\sec t}$

$= \sec t - \frac{1}{\sec t}$

$= \sec t - \cos t$

41. $\cot x + 1 = \frac{\cos x}{\sin x} + \frac{\sin x}{\sin x}$

$= \frac{1}{\sin x}(\cos x + \sin x)$

$= \csc x(\cos x + \sin x)$

43. $\frac{\sec t}{1 + \sec t} = \frac{\sec t}{1 + \sec t} \cdot \frac{\cos t}{\cos t}$

$= \frac{1}{\cos t + 1}$

45. $(\sec t - 1)(\sec t + 1) = \sec^2 t - 1$
$= \tan^2 t$

47. $\dfrac{1 - \sin^2\theta}{\cos\theta} = \dfrac{\cos^2\theta}{\cos\theta} = \cos\theta$

49. $\dfrac{\sec t}{\tan t} - \dfrac{\tan t}{\sec t} = \dfrac{\sec^2 t - \tan^2 t}{\sec t \tan t}$
$= \dfrac{(1 + \tan^2 t) - \tan^2 t}{\sec t \tan t}$
$= \dfrac{1}{\sec t \tan t}$
$= \cos t \cot t$

51. $\dfrac{\cot^2 t}{\csc t + 1} = \dfrac{\csc^2 t - 1}{\csc t + 1}$
$= \dfrac{(\csc t + 1)(\csc t - 1)}{\csc t + 1}$
$= \csc t - 1$

53. $\dfrac{\cot t}{\cot t + 1} = \dfrac{\cot t}{\cot t + 1} \cdot \dfrac{\tan t}{\tan t}$
$= \dfrac{\cot t \tan t}{\cot t \tan t + \tan t}$
$= \dfrac{1}{1 + \tan t}$

55. $\dfrac{1}{1 - \sin t} + \dfrac{1}{1 + \sin t} = \dfrac{(1 + \sin t) + (1 - \sin t)}{(1 - \sin t)(1 + \sin t)}$
$= \dfrac{2}{1 - \sin^2 t}$
$= \dfrac{2}{\cos^2 t}$
$= 2\sec^2 t$

57. $\dfrac{\csc t + \cot t}{\csc t - \cot t} = \dfrac{\csc t + \cot t}{\csc t - \cot t} \cdot \dfrac{\csc t + \cot t}{\csc t + \cot t}$
$= \dfrac{(\csc t + \cot t)^2}{\csc^2 t - \cot^2 t}$
$= \dfrac{(\csc t + \cot t)^2}{\csc^2 t - (\csc^2 t - 1)}$
$= (\csc t + \cot t)^2$

59. $\dfrac{\cos^2 t}{1 - \sin t} = \dfrac{\cos^2 t}{1 - \sin t} \cdot \dfrac{1 + \sin t}{1 + \sin t}$
$= \dfrac{\cos^2 t(1 + \sin t)}{1 - \sin^2 t}$
$= \dfrac{\cos^2 t(1 + \sin t)}{\cos^2 t}$
$= 1 + \sin t$

61. $\dfrac{1}{1 + \sin\theta} = \dfrac{1}{1 + \sin\theta} \cdot \dfrac{1 - \sin\theta}{1 - \sin\theta}$
$= \dfrac{1 - \sin\theta}{1 - \sin^2\theta}$
$= \dfrac{1 - \sin\theta}{\cos^2\theta}$

63. $\sqrt{1 - \sin^2\theta} = \sqrt{\cos^2\theta}$
$= |\cos\theta|$
$= \cos\theta$, where θ is acute

65. $\dfrac{1 + 2\sin x + \sin^2 x}{\cos^2 x} = \dfrac{(1 + \sin x)^2}{1 - \sin^2 x}$
$= \dfrac{(1 + \sin x)(1 + \sin x)}{(1 - \sin x)(1 + \sin x)}$
$= \dfrac{1 + \sin x}{1 - \sin x}$

67. $(1 - \cos^2 x)(1 + \cos^2 x) = \sin^2 x\,(1 + (1 - \sin^2 x))$
$= \sin^2 x\,(2 - \sin^2 x)$
$= 2\sin^2 x - \sin^4 x$

69. $\cot\theta \sin\theta = \dfrac{\cos\theta}{\sin\theta} \cdot \sin\theta$
$= \cos\theta$

71. $(1 - \cos^2\theta)(1 + \tan^2\theta) = \sin^2\theta \sec^2\theta$
$= \dfrac{\sin^2\theta}{\cos^2\theta}$
$= \tan^2\theta$

73. $\cos t(\tan t - \sec t) = \cos t\left(\dfrac{\sin t}{\cos t} - \dfrac{1}{\cos t}\right)$
$= \sin t - 1$

75. $\dfrac{\tan(-\theta)}{\sin(-\theta)} = \dfrac{-\tan\theta}{-\sin\theta}$
$= \dfrac{1}{\cos\theta}$
$= \sec\theta$

77. (a) $W(t) = 5(1 - \sin^2(120\pi t)); W = 0$
(b) Whenever there is a peak or valley on the graph of V, the graph of W intersects the x-axis, which corresponds to a zero of W.
$[0, 1/15, 1/60]$ by $[-30, 30, 10]$

CHECKING BASIC CONCEPTS FOR SECTIONS 7.1 AND 7.2 (p. 598)

1. Quadrant III
3. (a) $\cos^2\theta$ (b) $\tan^2 t$
5. (a) $(1 - \sin^2\theta)(1 + \cot^2\theta) = \cos^2\theta\csc^2\theta$
$= \dfrac{\cos^2\theta}{\sin^2\theta} = \cot^2\theta$
(b) $\dfrac{\cot^2 t}{\csc t} = \dfrac{\csc^2 t - 1}{\csc t}$
$= \dfrac{\csc^2 t}{\csc t} - \dfrac{1}{\csc t}$
$= \csc t - \sin t$

SECTION 7.3 (pp. 609–612)

1. $60°$ 3. $85°$ 5. $65°$ 7. $\dfrac{\pi}{6}$
9. $\dfrac{\pi}{3}$ 11. (a) $90°; \dfrac{\pi}{2}$ (b) $270°; \dfrac{3\pi}{2}$
13. (a) $60°, 240°; \dfrac{\pi}{3}, \dfrac{4\pi}{3}$ (b) $120°, 300°; \dfrac{2\pi}{3}, \dfrac{5\pi}{3}$
15. (a) $60°, 300°; \dfrac{\pi}{3}, \dfrac{5\pi}{3}$ (b) $120°, 240°; \dfrac{2\pi}{3}, \dfrac{4\pi}{3}$
17. (a) No solutions   (b) No solutions
19. (a) $\dfrac{\pi}{6} + 2\pi n, \dfrac{5\pi}{6} + 2\pi n$ (b) $\dfrac{7\pi}{6} + 2\pi n, \dfrac{11\pi}{6} + 2\pi n$
21. (a) $\dfrac{\pi}{4} + \pi n$ (b) $\dfrac{3\pi}{4} + \pi n$
23. (a) $\dfrac{\pi}{3} + 2\pi n, \dfrac{5\pi}{3} + 2\pi n$ (b) $\dfrac{2\pi}{3} + 2\pi n, \dfrac{4\pi}{3} + 2\pi n$
25. $\dfrac{\pi}{4}, \dfrac{5\pi}{4}$ 27. $\dfrac{\pi}{3}, \dfrac{5\pi}{3}$
29. (a) $\dfrac{1}{2}$ (b) $30°, 150°$

31. (a) 0, 1   (b) 0°, 90°, 180°
33. (a) −1, 1   (b) 45°, 135°, 225°, 315°
35. (a) −2, 1   (b) 0°
37. $\frac{\pi}{3}, \frac{2\pi}{3}, \frac{4\pi}{3}, \frac{5\pi}{3}$ **39.** No solutions   **41.** $\frac{\pi}{2}, \frac{3\pi}{2}$
43. $\frac{\pi}{4}, \frac{5\pi}{4}$ **45.** $\frac{\pi}{6}, \frac{5\pi}{6}, \frac{7\pi}{6}, \frac{11\pi}{6}$ **47.** $0, \frac{\pi}{2}, \pi, \frac{3\pi}{2}$
49. $\frac{\pi}{3}, \frac{5\pi}{3}$ **51.** 0   **53.** $\frac{\pi}{4} + \pi n$
55. $\frac{\pi}{6} + 2\pi n, \frac{5\pi}{6} + 2\pi n$
57. $\frac{\pi}{6} + 2\pi n, \frac{5\pi}{6} + 2\pi n, \frac{\pi}{2} + 2\pi n$
59. $\frac{2\pi}{3} + 2\pi n, \frac{4\pi}{3} + 2\pi n, \pi + 2\pi n$
61. $\frac{\pi}{4} + \frac{\pi n}{2}$ **63.** No solutions   **65.** πn **67.** $\frac{\pi}{4} + \pi n$
69. $\frac{\pi n}{2}$ **71.** $2\pi n, \frac{\pi}{2} + 2\pi n$ **73.** $\frac{\pi}{18} + \frac{2\pi}{3} n, \frac{5\pi}{18} + \frac{2\pi}{3} n$
75. $\frac{5\pi}{24} + \frac{\pi}{2} n, \frac{7\pi}{24} + \frac{\pi}{2} n$ **77.** $\frac{\pi}{20} + \frac{\pi}{5} n$
79. $\frac{7\pi}{24} + \frac{\pi}{2} n, \frac{11\pi}{24} + \frac{\pi}{2} n$ **81.** $\frac{3\pi}{16} + \frac{\pi}{2} n, \frac{5\pi}{16} + \frac{\pi}{2} n$
83. $\frac{\pi}{16} + \frac{\pi}{4} n$ **85.** $\frac{\pi}{16} + \frac{\pi}{4} n$
87. $0.456 + \frac{2\pi}{3} n, 1.638 + \frac{2\pi}{3} n$
89. $0.170 + \pi n, 1.401 + \pi n$
91. 1.085, 5.198   **93.** 0.737, 2.404
95. 0.659, 2.483, 3.801, 5.624
97. 0, 4.49   **99.** 0.39, 1.96   **101.** 3.60
103. 30°, 210°; 30° + 180° · n **105.** 1
107. $-\frac{\sqrt{3}}{2} \approx -0.866$ **109.** −1   **111.** 0
113. $\frac{1}{\sqrt{2}} \approx 0.707$ **115.** 60° + 360° · n, 300° + 360° · n
117. About 9.1°
119. 2.6, 10.7; near February 17 and October 22
121. 2.6, 10.7; near February 17 and October 22
123. (a) $f(x) = 122.3 \sin (0.524x - 1.7) + 367$ (answers may vary)
(b) $2.98 \le x \le 9.51$ (approximately). At 50°N latitude, the maximum number of monthly hours of sunshine is greater than or equal to 350 hours roughly from March through September.
125. 0.26
127. (a) [0, 0.01, 0.005] by   [0, 0.01, 0.005] by
        [−0.005, 0.005, 0.001]   [−0.005, 0.005, 0.001]

 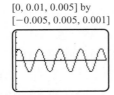

[0, 0.01, 0.005] by [−0.005, 0.005, 0.001]

(b) Maximum: $P \approx 0.004$
(c) No; the maximum of P_1 is 0.003, and the maximum of P_2 is 0.002.

SECTION 7.4 (pp. 621–623)

1. $\frac{\sqrt{6} - \sqrt{2}}{4}$ **3.** $2 - \sqrt{3}$ **5.** $\frac{\sqrt{6} - \sqrt{2}}{4}$
7. $\frac{\sqrt{6} - \sqrt{2}}{4}$ **9.** $\frac{\sqrt{6} + \sqrt{2}}{4}$
11. (a) *Graphical:* Graphs of $y = \sin\left(t + \frac{\pi}{2}\right)$ and $y = \cos t$ are the same.
Verbal: If the sine graph is translated $\frac{\pi}{2}$ units left, it coincides with the cosine graph.
(b) $\sin\left(t + \frac{\pi}{2}\right) = \sin t \cos \frac{\pi}{2} + \cos t \sin \frac{\pi}{2}$
$= \sin t (0) + \cos t (1)$
$= \cos t$
13. (a) *Graphical:* Graphs of $y = \cos (t + \pi)$ and $y = -\cos t$ are the same.
Verbal: If the cosine graph is translated π units left, it coincides with the cosine graph reflected about the x-axis.
(b) $\cos (t + \pi) = \cos t \cos \pi - \sin t \sin \pi$
$= \cos t (-1) - \sin t (0)$
$= -\cos t$
15. (a) *Graphical:* Graphs of $y = \sec\left(t - \frac{\pi}{2}\right)$ and $y = \csc t$ are the same.
Verbal: If the secant graph is translated $\frac{\pi}{2}$ units right, it coincides with the cosecant graph.
(b) $\sec\left(t - \frac{\pi}{2}\right) = \dfrac{1}{\cos\left(t - \frac{\pi}{2}\right)}$
$= \dfrac{1}{\cos t \cos \frac{\pi}{2} + \sin t \sin \frac{\pi}{2}}$
$= \dfrac{1}{\cos t (0) + \sin t(1)}$
$= \dfrac{1}{\sin t}$
$= \csc t$
17. $\sin\left(\frac{\pi}{2} - t\right) = \sin \frac{\pi}{2} \cos t - \cos \frac{\pi}{2} \sin t$
$= (1) \cos t + (0) \sin t$
$= \cos t$
19. $\sec\left(\frac{\pi}{2} - t\right) = \dfrac{1}{\cos\left(\frac{\pi}{2} - t\right)}$
$= \dfrac{1}{\cos \frac{\pi}{2} \cos t + \sin \frac{\pi}{2} \sin t}$
$= \dfrac{1}{(0) \cos t + (1) \sin t}$
$= \dfrac{1}{\sin t}$
$= \csc t$
21. (a) $\frac{56}{65}$ (b) $\frac{33}{65}$ (c) $\frac{56}{33}$ (d) I
23. (a) $-\frac{988}{1037}$ (b) $\frac{315}{1037}$ (c) $-\frac{988}{315}$ (d) IV
25. (a) $-\frac{33}{65}$ (b) $-\frac{56}{65}$ (c) $\frac{33}{56}$ (d) III
27. (a) $-\frac{775}{793}$ (b) $-\frac{168}{793}$ (c) $\frac{775}{168}$ (d) III

29. $\cos\left(t - \frac{\pi}{4}\right) = \cos t \cos\frac{\pi}{4} + \sin t \sin\frac{\pi}{4}$

$= \cos t\left(\frac{\sqrt{2}}{2}\right) + \sin t\left(\frac{\sqrt{2}}{2}\right)$

$= \frac{\sqrt{2}}{2}(\cos t + \sin t)$

31. $\tan\left(t + \frac{\pi}{4}\right) = \dfrac{\tan t + \tan\frac{\pi}{4}}{1 - \tan t \tan\frac{\pi}{4}}$

$= \dfrac{\tan t + 1}{1 - \tan t(1)}$

$= \dfrac{1 + \tan t}{1 - \tan t}$

33. $\dfrac{\cos(x - y)}{\cos(x + y)} = \dfrac{\cos x \cos y + \sin x \sin y}{\cos x \cos y - \sin x \sin y}$

$= \dfrac{\dfrac{\cos x \cos y}{\cos x \cos y} + \dfrac{\sin x \sin y}{\cos x \cos y}}{\dfrac{\cos x \cos y}{\cos x \cos y} - \dfrac{\sin x \sin y}{\cos x \cos y}}$

$= \dfrac{1 + \tan x \tan y}{1 - \tan x \tan y}$

35. $\dfrac{\cos(\alpha - \beta)}{\cos\alpha \sin\beta} = \dfrac{\cos\alpha \cos\beta + \sin\alpha \sin\beta}{\cos\alpha \sin\beta}$

$= \dfrac{\cos\alpha \cos\beta}{\cos\alpha \sin\beta} + \dfrac{\sin\alpha \sin\beta}{\cos\alpha \sin\beta}$

$= \dfrac{\cos\beta}{\sin\beta} + \dfrac{\sin\alpha}{\cos\alpha}$

$= \cot\beta + \tan\alpha$

$= \tan\alpha + \cot\beta$

37. $\sin 2t = \sin(t + t)$

$= \sin t \cos t + \cos t \sin t$

$= 2\sin t \cos t$

39. $\sin(\alpha + \beta) + \sin(\alpha - \beta)$

$= \sin\alpha \cos\beta + \cos\alpha \sin\beta + \sin\alpha \cos\beta$

$\quad - \cos\alpha \sin\beta$

$= \sin\alpha \cos\beta + \sin\alpha \cos\beta$

$= 2\sin\alpha \cos\beta$

41. $\tan(\pi - \theta) = \dfrac{\tan\pi - \tan\theta}{1 + \tan\pi \tan\theta}$

$= \dfrac{0 - \tan\theta}{1 + (0)\tan\theta}$

$= -\tan\theta$

43. $\tan(x - y) - \tan(y - x)$

$= \dfrac{\tan x - \tan y}{1 + \tan x \tan y} - \dfrac{\tan y - \tan x}{1 + \tan y \tan x}$

$= \dfrac{\tan x - \tan y - \tan y + \tan x}{1 + \tan x \tan y}$

$= \dfrac{2(\tan x - \tan y)}{1 + \tan x \tan y}$

45. $\dfrac{\sin(x + y)}{\cos x \cos y} = \dfrac{\sin x \cos y + \cos x \sin y}{\cos x \cos y}$

$= \dfrac{\sin x \cos y}{\cos x \cos y} + \dfrac{\cos x \sin y}{\cos x \cos y}$

$= \dfrac{\sin x}{\cos x} + \dfrac{\sin y}{\cos y}$

$= \tan x + \tan y$

47. $\frac{7}{79}$

49. $\tan\theta = \tan(\beta - \alpha)$

$= \dfrac{\tan\beta - \tan\alpha}{1 + \tan\alpha \tan\beta}$

$= \dfrac{m_2 - m_1}{1 + m_1 m_2}$

51. $\theta = \tan^{-1}\frac{7}{11} \approx 32.5°$

53. (a) $F \approx 409$ lb (b) About 0.81, or 46.2°

55. (a) $[0, 0.02, 0.001]$ by $[-6, 6, 1]$

(b) $a = 5, k \approx 0.9272$

(c) $5\sin(220\pi t + 0.9272)$

$= 5(\sin(220\pi t)\cos(0.9272)$

$\quad + \cos(220\pi t)\sin(0.9272))$

$\approx 5(0.6\sin(220\pi t) + 0.8\cos(220\pi t))$

$= 3\sin(220\pi t) + 4\cos(220\pi t)$

7.4 EXTENDED AND DISCOVERY EXERCISES (p. 623)

1. (a) $[0.15, 1.15, 0.05]$ by $[-0.01, 0.01, 0.001]$

There are 3 beats in 1 second.

(b) $[0.15, 1.15, 0.05]$ by $[-0.01, 0.01, 0.001]$

There are 4 beats in 1 second.

(c) When the frequencies are F_1 and F_2, the rate of beats per second is given by $|F_2 - F_1|$.

3. $[0.2, 1.2, 0.05]$ by $[-0.01, 0.01, 0.001]$

There are 3 beats in 1 second.

CHECKING BASIC CONCEPTS FOR SECTIONS 7.3 AND 7.4 (p. 623)

1. (a) $45°$ (b) $\frac{\pi}{6}$

3. (a) $\frac{3\pi}{4} + \pi n$ (b) $\frac{2\pi}{3} + 2\pi n; \frac{4\pi}{3} + 2\pi n; 2\pi n$

5. $\sin(t - \pi) = \sin t \cos\pi - \cos t \sin\pi$

$= \sin t(-1) - \cos t(0)$

$= -\sin t$

SECTION 7.5 (pp. 636–639)

1. $\frac{1 - \cos 20t}{2}$ **3.** $\frac{1 - \cos 10t}{1 + \cos 10t}$

5. $2 \sin 10x \cos 10x$ **7.** $\frac{1 - \cos 10x}{\sin 10x}$

9. (a) 1   (b) $\frac{\sqrt{3}}{2}$; not equal

11. (a) 1   (b) $-\frac{1}{2}$; not equal

13. (a) 2   (b) Undefined; not equal

15. *Graphical:* Graphs of $y = \tan 2\theta$ and $y = 2 \tan \theta$ are not the same.

Symbolic: $\tan 2\theta = \frac{2 \tan \theta_2}{1 - \tan \theta} \neq 2 \tan \theta$,

unless $\tan \theta = 0$.

17. $\sin 2\theta = \frac{24}{25}$, $\cos 2\theta = \frac{7}{25}$, $\tan 2\theta = \frac{24}{7}$

19. $\sin 2\theta = -\frac{336}{625}$, $\cos 2\theta = -\frac{527}{625}$, $\tan 2\theta = \frac{336}{527}$

21. $\sin 2\theta = -\frac{1320}{3721}$, $\cos 2\theta = \frac{3479}{3721}$, $\tan 2\theta = -\frac{1320}{3479}$

23. $\sin 2\theta = \frac{336}{625}$, $\cos 2\theta = \frac{527}{625}$, $\tan 2\theta = \frac{336}{527}$

25. 0   **27.** $\frac{527}{625}$ **29.** $\frac{828}{2197}$ **31.** $\frac{2x}{x^2 + 1}$ **33.** $\frac{24}{25}$

35. $\sin 2\theta$ **37.** $\frac{1}{2} \sin 2\theta$ **39.** $\cos 4\theta$

41. 1   **43.** $\cot^2 5x$ **45.** $\frac{2 + \sqrt{2}}{4}$ **47.** $\frac{2 + \sqrt{3}}{2 - \sqrt{3}}$

49. (a) $\frac{\sqrt{2 + \sqrt{3}}}{2}$

(b) $-\sqrt{\frac{2 - \sqrt{3}}{2 + \sqrt{3}}}$, or $-\frac{1}{2 + \sqrt{3}}$, or $\sqrt{3} - 2$

51. (a) $\sqrt{\frac{2 - \sqrt{2}}{2 + \sqrt{2}}}$, or $\frac{\sqrt{2}}{2 + \sqrt{2}}$, or $\sqrt{2} - 1$

(b) $-\frac{\sqrt{2 - \sqrt{2}}}{2}$

53. $\sin 30°$ **55.** $\cos 25°$ **57.** $\tan 20°$

59. $\sin \frac{\theta}{2} = \frac{1}{\sqrt{10}}$, $\cos \frac{\theta}{2} = \frac{3}{\sqrt{10}}$, $\tan \frac{\theta}{2} = \frac{1}{3}$

61. $\sin \frac{\theta}{2} = -\frac{1}{\sqrt{26}}$, $\cos \frac{\theta}{2} = \frac{5}{\sqrt{26}}$, $\tan \frac{\theta}{2} = -\frac{1}{5}$

63. $\sin \frac{\theta}{2} = \frac{4}{5}$, $\cos \frac{\theta}{2} = \frac{3}{5}$, $\tan \frac{\theta}{2} = \frac{4}{3}$

65. $4 \sin 2x = 4(2 \sin x \cos x)$
$= 8 \sin x \cos x$

67. $\frac{2 - \sec^2 x}{\sec^2 x} = \frac{2}{\sec^2 x} - 1$
$= 2 \cos^2 x - 1$
$= \cos 2x$

69. $\sec 2x = \frac{1}{\cos 2x}$
$= \frac{1}{1 - 2 \sin^2 x}$

71. $\sin 3\theta = \sin(2\theta + \theta)$
$= \sin 2\theta \cos \theta + \cos 2\theta \sin \theta$
$= (2 \sin \theta \cos \theta) \cos \theta + (1 - 2 \sin^2 \theta) \sin \theta$
$= 2 \sin \theta \cos^2 \theta + \sin \theta - 2 \sin^3 \theta$
$= 2 \sin \theta (1 - \sin^2 \theta) + \sin \theta - 2 \sin^3 \theta$
$= 2 \sin \theta - 2 \sin^3 \theta + \sin \theta - 2 \sin^3 \theta$
$= 3 \sin \theta - 4 \sin^3 \theta$

73. $\sin 4\theta = 2 \sin 2\theta \cos 2\theta$
$= 2(2 \sin \theta \cos \theta) \cos 2\theta$
$= 4 \sin \theta \cos \theta \cos 2\theta$

75. $\frac{\sin 2\theta}{\sin \theta} = \frac{2 \sin \theta \cos \theta}{\sin \theta}$
$= 2 \cos \theta$

77. $2 \cos^2 \frac{\theta}{2} = 2\left(\frac{1 + \cos \theta}{2}\right)$
$= 1 + \cos \theta$

79. $\cos^4 \theta - \sin^4 \theta = (\cos^2 \theta - \sin^2 \theta)(\cos^2 \theta + \sin^2 \theta)$
$= (\cos 2\theta)(1)$
$= \cos 2\theta$

81. $\csc 2t = \frac{1}{\sin 2t}$
$= \frac{1}{2 \sin t \cos t}$
$= \frac{\csc t}{2 \cos t}$

83. $\tan \frac{x}{2} = \frac{\sin \frac{x}{2}}{\cos \frac{x}{2}}$
$= \frac{2 \sin \frac{x}{2} \cos \frac{x}{2}}{2 \cos \frac{x}{2} \cos \frac{x}{2}}$
$= \frac{\sin x}{2 \cos^2 \frac{x}{2}}$
$= \frac{\sin x}{1 + \cos x}$

85. (a) $\frac{1}{2}(\sin 70° - \sin 30°)$ (b) $\frac{1}{2}(\cos 3x + \cos x)$

87. (a) $\frac{1}{2}(\sin 10\theta + \sin 4\theta)$ (b) $\frac{1}{2}(\cos 4x - \cos 12x)$

89. (a) $2 \sin 35° \cos 5°$ (b) $2 \cos 40° \cos 5°$

91. (a) $2 \cos 5\theta \cos \theta$ (b) $2 \sin \frac{11x}{2} \cos \frac{3x}{2}$

93. $0°, 180°$ **95.** $90°, 210°, 270°, 330°$ **97.** $180°$

99. $90°, 270°$ **101.** $\frac{\pi}{12} + \pi n, \frac{11\pi}{12} + \pi n$

103. $\pi n, \frac{2\pi}{3} + 2\pi n, \frac{4\pi}{3} + 2\pi n$ **105.** $\frac{\pi}{3} + 4\pi n, \frac{5\pi}{3} + 4\pi n$

107. $\frac{\pi}{6} + 2\pi n, \frac{5\pi}{6} + 2\pi n, \frac{3\pi}{2} + 2\pi n$

109. $\frac{\pi}{8} + \pi n, \frac{5\pi}{8} + \pi n$, or equivalently, $\frac{\pi}{8} + \frac{\pi n}{2}$

111. $\frac{2\pi}{3} + 4\pi n, \frac{10\pi}{3} + 4\pi n$

113. $\frac{\pi}{8} + \pi n, \frac{5\pi}{8} + \pi n$, or equivalently, $\frac{\pi}{8} + \frac{\pi n}{2}$

115. $\frac{\pi}{12} + \pi n, \frac{5\pi}{12} + \pi n, \frac{3\pi}{4} + \pi n$ **117.** 0.333, 4.379

119. (a) [0, 0.04, 0.01] by [−500, 2500, 500]

(b) $a = -1085$, $k = 240$, $d = 1085$

121. $\frac{1}{720} + \frac{n}{60}$, $\frac{5}{720} + \frac{n}{60}$ sec

123. (a) $d = 600\left(1 - \cos\frac{80°}{2}\right) \approx 140.4$ ft

(b) *Hint:* First show that $r = d + r\cos\frac{\beta}{2}$. See Student's Solutions Manual.

(c) No, since $\cos\frac{\beta}{2} \neq \frac{1}{2}\cos\beta$ in general

125. Let $f(t)$ model the tone for the number 3 and $g(t)$ model the tone for number 4.

(a) $f(t) = \cos(1394\pi t) + \cos(2954\pi t)$
$g(t) = \cos(1540\pi t) + \cos(2418\pi t)$
(b) $f(t) = 2\cos(2174\pi t)\cos(780\pi t)$
$g(t) = 2\cos(1979\pi t)\cos(439\pi t)$

CHECKING BASIC CONCEPTS FOR SECTION 7.5 (p. 639)

1. $\sin 2\theta = -\frac{336}{625}$, $\cos 2\theta = -\frac{527}{625}$

3. $\sin\frac{\theta}{2} = \frac{1}{\sqrt{5}}$, $\cos\frac{\theta}{2} = \frac{2}{\sqrt{5}}$ **5.** 90°, 270°

CHAPTER 7 REVIEW EXERCISES (pp. 642–644)

1. Quadrant II

3. $\tan\theta = -\frac{3}{4}$, $\csc\theta = \frac{5}{3}$, $\sec\theta = -\frac{5}{4}$, $\cot\theta = -\frac{4}{3}$

5. $\sin\theta = -\frac{7}{25}$, $\csc\theta = -\frac{25}{7}$, $\sec\theta = \frac{25}{24}$, $\cot\theta = -\frac{24}{7}$

7. $-\sin 13°$ **9.** $\sec\frac{3\pi}{7}$ **11.** 1   **13.** 1   **15.** $\cos\theta$

17. $\sin\theta \approx 0.7776$, $\cos\theta \approx 0.6288$, $\csc\theta \approx 1.2860$, $\sec\theta \approx 1.5904$, $\cot\theta \approx 0.8086$

19. $\sin\theta \approx 0.8908$, $\tan\theta \approx -1.9604$, $\csc\theta \approx 1.1226$, $\sec\theta \approx -2.2007$, $\cot\theta \approx -0.5101$

21. $(\sin\theta + 1)(\sin\theta + 1)$

23. $(\tan\theta + 3)(\tan\theta - 3)$

25. $(\sec\theta - 1)(\sec\theta + 1) = \sec^2\theta - 1$
$= (1 + \tan^2\theta) - 1$
$= \tan^2\theta$

27. $(1 + \tan t)^2 = 1 + 2\tan t + \tan^2 t$
$= \sec^2 t + 2\tan t$

29. $\sin(x - \pi) = \sin x \cos\pi - \cos x \sin\pi$
$= \sin x(-1) - \cos x(0)$
$= -\sin x$

31. $\sin 8x = \sin(2 \cdot 4x)$
$= 2\sin 4x \cos 4x$

33. $\sec 2x = \frac{1}{\cos 2x}$
$= \frac{1}{2\cos^2 x - 1}$

35. $\cos^4 x \sin^3 x = \cos^4 x \sin^2 x \sin x$
$= \cos^4 x(1 - \cos^2 x)\sin x$
$= (\cos^4 x - \cos^6 x)\sin x$

37. $\sec^4\theta - \tan^4\theta = (\sec^2\theta - \tan^2\theta)(\sec^2\theta + \tan^2\theta)$
$= (1)(1 + \tan^2\theta + \tan^2\theta)$
$= 1 + 2\tan^2\theta$

39. 60°   **41.** $\frac{2\pi}{7}$ **43.** $\frac{\pi}{2}, \frac{3\pi}{2}$

45. (a) 60°, 240°   **(b)** 150°, 330°

47. 60°, 300°   **49.** 90°   **51.** $\frac{\pi}{4}, \frac{5\pi}{4}$

53. $\frac{\pi}{6} + \pi n, -\frac{\pi}{6} + \pi n$; $30° + 180° \cdot n, -30° + 180° \cdot n$

55. $\frac{\pi}{2} + \pi n$; $90° + 180° \cdot n$

57. $-\frac{\sqrt{2 - \sqrt{3}}}{2}$ **59.** $-4.43, 1.13, 4.53$

61. (a) $\sin(\alpha + \beta) = \frac{63}{65}$ **(b)** $\cos(\alpha + \beta) = \frac{16}{65}$,

(c) $\tan(\alpha + \beta) = \frac{63}{16}$ **(d)** Quadrant I

63. $\sin 2\theta = -\frac{24}{25}$, $\cos 2\theta = -\frac{7}{25}$, $\tan 2\theta = \frac{24}{7}$

65. $\sin\frac{\theta}{2} = \sqrt{\frac{3}{8}}$, $\cos\frac{\theta}{2} = \sqrt{\frac{5}{8}}$, $\tan\frac{\theta}{2} = \sqrt{\frac{3}{5}}$ **67.** $\frac{3479}{3721}$

69. $x = 4.7, 8.7$, about April 21 and August 22

71. (a) [0, 0.06, 0.01] by [−0.012, 0.012, 0.002]

(b) $a = 0.01$, $k \approx 0.6435$
(c) $0.01\sin(100\pi t + 0.6435)$
$= 0.01(\sin(100\pi t)\cos(0.6435)$
$\quad + \cos(100\pi t)\sin(0.6435))$
$\approx 0.01(0.8\sin(100\pi t) + 0.6\cos(100\pi t))$
$\approx 0.008\sin(100\pi t) + 0.006\cos(100\pi t)$

73. $W(t) = 7 - 7\sin^2(240\pi t)$; when V is maximum or minimum, $W = 0$.

75. $\theta \approx 30.11°$, or 0.53 radian

CHAPTER 7 EXTENDED AND DISCOVERY EXERCISES (pp. 644–645)

1. (a) i. [0, 0.01, 0.002] by [−0.005, 0.005, 0.001]

ii. [0, 0.01, 0.002] by [−0.005, 0.005, 0.001]

iii. [0, 0.01, 0.002] by [−0.005, 0.005, 0.001]

iv. [0, 0.01, 0.002] by [−0.005, 0.005, 0.001]

v. [0, 0.01, 0.002] by [−0.005, 0.005, 0.001]

(b) The graph approximates a saw-tooth shape.
(c) The maximum pressure of P is approximately 0.00317.
(d) The pure tone is modeled by a smooth graph, whereas the piano tone is modeled by a saw-tooth shape.
3. (a) [0, 1, 0.1] by [−0.15, 0.15, 0.05]

(b) These graphs bound the changing amplitude of A.
[0, 1, 0.1] by [−0.15, 0.15, 0.05]

(c) $t \approx 0.55$ second (answers may vary slightly)

Chapter 8: Further Topics in Trigonometry

SECTION 8.1 (pp. 655–658)

1. No **3.** No **5.** Yes **7.** Yes
9. $\beta = 60°, a \approx 5.5, b \approx 4.9$
11. $\beta = 35°, a \approx 15.1, c \approx 7.4$
13. $\beta_1 \approx 57.1°, \gamma_1 \approx 76.9°, c_1 \approx 8.1$
    $\beta_2 \approx 122.9°, \gamma_2 \approx 11.1°, c_2 \approx 1.6$
15. There are no solutions.
17. $\beta_1 \approx 67°13', \gamma_1 \approx 62°35', c_1 \approx 11.6$
    $\beta_2 \approx 112°48', \gamma_2 \approx 17°0', c_2 \approx 3.8$
19. $\gamma = 93°, a \approx 7.8, c \approx 14.6$
21. $\beta_1 \approx 26.1°, \gamma_1 \approx 133.9°, c_1 \approx 14.7$
    $\beta_2 \approx 153.9°, \gamma_2 \approx 6.1°, c_2 \approx 2.2$
23. $\gamma = 90°, \alpha = 60°, a = 10\sqrt{3} \approx 17.3$
25. $\alpha \approx 52.9°, \beta \approx 25.1°, b \approx 22.4$
27. $\beta = 10°, a \approx 92.2, c \approx 101.9$
29. There are no solutions.   **31.** There are no solutions.
33. $\alpha_1 \approx 60°56', \gamma_1 \approx 72°19', c_1 \approx 6.5$
    $\alpha_2 \approx 119°4', \gamma_2 \approx 14°11', c_2 \approx 1.7$
35. $\gamma = 99°45', a \approx 84.6, b \approx 40.9$ **37.** 3629 ft
39. The calculated distance to the moon changes to about 343,000 km, a difference of about 76,000 km. A small error in measuring lunar angle could result in large errors in calculating the distance to the moon.
41. $d \approx 7.2$ mi   **43.** About 28.8 ft

45. About 118.0 m   **47.** About 3.86 mi
49. About 0.49 mi   **51.** $AB \approx 105.4$ ft
53. (a) 3.57 mi   **(b)** 48°   **55.** 630 ft

SECTION 8.2 (pp. 665–669)

1. (a) SAS   **(b)** Law of cosines
3. (a) SSA   **(b)** Law of sines
5. (a) ASA   **(b)** Law of sines
7. (a) ASA   **(b)** Law of sines
9. 7   **11.** 30°
13. $a \approx 5.4, \beta \approx 40.7°, \gamma \approx 78.3°$
15. $\alpha \approx 22.3°, \beta \approx 108.2°, \gamma \approx 49.5°$
17. $\alpha \approx 33.6°, \beta \approx 50.7°, \gamma \approx 95.7°$
19. $c \approx 28.8, \alpha \approx 116.5°, \beta \approx 28.5°$
21. $\alpha \approx 101.0°, \beta \approx 44.0°, \gamma \approx 34.9°$
Angles do not sum to 180° because of rounding.
23. $a \approx 9.0, \beta \approx 150.9°, \gamma \approx 18.6°$
25. $\alpha \approx 23.1°, \beta \approx 107.2°, \gamma \approx 49.7°$
27. $b \approx 30.7, \alpha \approx 33°26', \gamma \approx 24°24'$
29. $\alpha \approx 45.1°, \beta \approx 63.5°, \gamma \approx 71.5°$
Angles do not sum to 180° because of rounding.
31. No, since $a + b < c$
33. No, since $89° + 112° > 180°$
35. Yes, since we are given ASA and $\alpha + \gamma < 180°$
37. 86.8   **39.** 5.3   **41.** 50.9   **43.** 18.3
45. 2.1   **47.** 18.3   **49.** 66   **51.** 160.4
53. 169 ft   **55.** About 29.8 mi   **57.** 4.4 ft; 7.7 ft
59. A to B: 76°; B to C: 309°
61. (a) $\alpha \approx 75.1°, \beta \approx 65.6°, \gamma \approx 39.4°$
Angles do not sum to 180° because of rounding.
(b) 6299 ft^2
63. $\theta \approx 40.5°$ **65.** 302 mi   **67.** About 1452 ft
69. About 745 mi   **71.** 147.8 ft^2
73. (a) $9\sqrt{3} \approx 15.6$ in^2 **(b)** The results are equal.
75. 149,429 ft^2 **77.** 21,309 ft^2

8.2 EXTENDED AND DISCOVERY EXERCISES (p. 669)

1. About 2000 km

CHECKING BASIC CONCEPTS FOR SECTIONS 8.1 AND 8.2 (p. 669)

1. $\beta = 74°, b \approx 16.6, c \approx 15.3$
3. (a) $b \approx 7.1, \alpha \approx 63.0°, \gamma \approx 66.0°$
(b) $\alpha \approx 110.7°, \beta \approx 37.0°, \gamma \approx 32.3°$

SECTION 8.3 (pp. 681–686)

1. (a) $a_1 \approx 3, a_2 \approx 4$ **(b)** $\|\mathbf{v}\| = 5$
3. (a) $a_1 \approx -5, a_2 \approx -12$ **(b)** $\|\mathbf{v}\| = 13$
5. (a) ↓ $\mathbf{v}$ (20 mi/hr)   **(b)** $\mathbf{v} = \langle 0, -20 \rangle$

(c) $2\mathbf{v} = \langle 0, -40 \rangle$; this represents a 40-mi/hr north wind. $-\frac{1}{2}\mathbf{v} = \langle 0, 10 \rangle$; this represents a 10-mi/hr south wind.

7. (a) 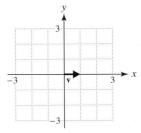 v (5 mi/hr)

(b) $\mathbf{v} = \left\langle \dfrac{5}{\sqrt{2}}, -\dfrac{5}{\sqrt{2}} \right\rangle$, or $\left\langle \dfrac{5}{2}\sqrt{2}, -\dfrac{5}{2}\sqrt{2} \right\rangle$

(c) $2\mathbf{v} = \left\langle \dfrac{10}{\sqrt{2}}, -\dfrac{10}{\sqrt{2}} \right\rangle$, or $\langle 5\sqrt{2}, -5\sqrt{2} \rangle$;

this represents a 10-mi/hr northwest wind.
$-\frac{1}{2}\mathbf{v} = \left\langle -\frac{5}{4}\sqrt{2}, \frac{5}{4}\sqrt{2} \right\rangle$; this represents a
2.5-mi/hr southeast wind.

9. (a) 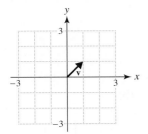 v (30 lb)   **(b)** $\mathbf{v} = \langle 0, 30 \rangle$

(c) $2\mathbf{v} = \langle 0, 60 \rangle$; this represents a 60-lb force upward.
$-\frac{1}{2}\mathbf{v} = \langle 0, -15 \rangle$; this represents a 15-lb force downward.

11. (a) Horizontal = 1, vertical = 1
(b) $\|\mathbf{v}\| = \sqrt{2}$; **v** is not a unit vector.
(c) $\|\mathbf{v}\|$ represents the length of **v**.

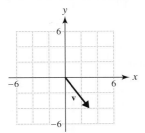

13. (a) Horizontal = 3, vertical = −4
(b) $\|\mathbf{v}\| = 5$; **v** is not a unit vector.
(c) $\|\mathbf{v}\|$ represents the length of **v**.

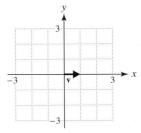

15. (a) Horizontal = 1, vertical = 0
(b) $\|\mathbf{v}\| = 1$; **v** is a unit vector.
(c) $\|\mathbf{v}\|$ represents the length of **v**.

17. (a) Horizontal = 5, vertical = 12
(b) $\|\mathbf{v}\| = 13$; **v** is not a unit vector.
(c) $\|\mathbf{v}\|$ represents the length of **v**.

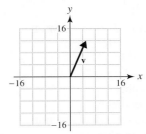

19. 3; 0°   **21.** $\sqrt{2}$; 135°   **23.** 2; 330°
25. 13; 247.4°   **27.** 85; 278.8°   **29.** 29; 133.6°
31. −4, 0   **33.** −1, 1   **35.** 13.5, 18.6
37. 22.8, −25.3   **39.** $\langle 3\sqrt{3}, 3 \rangle$
41. $\left\langle -\dfrac{9\sqrt{2}}{2}, -\dfrac{9\sqrt{2}}{2} \right\rangle$ **43.** $\langle 3.06, 2.57 \rangle$
45. $\langle 4.10, -2.87 \rangle$
47. (a)   **49. (a)**

(b) $\overrightarrow{PQ} = \langle -1, 2 \rangle$ **(b)** $\overrightarrow{PQ} = \langle 2, 4 \rangle$
(c) $\|\overrightarrow{PQ}\| = \sqrt{5}$ **(c)** $\|\overrightarrow{PQ}\| = \sqrt{20}$

51. (a)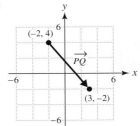

(b) $\overrightarrow{PQ} = \langle 5, -6 \rangle$ **(c)** $\|\overrightarrow{PQ}\| = \sqrt{61}$
53. 94.2 lb   **55.** 24.4 lb
57. $\mathbf{F} = -5\mathbf{i} + 12\mathbf{j}, \|\mathbf{F}\| = 13$
59. $\mathbf{F} = \langle 9, -22 \rangle, \|\mathbf{F}\| = \sqrt{565}$
61. $\mathbf{F} = -\mathbf{i} - 5\mathbf{j}, \|\mathbf{F}\| = \sqrt{26}$
63. (a) $\langle 3, 2 \rangle$ **(b)** $\langle -3, 2 \rangle$
65. (a) $3\mathbf{i} - \mathbf{j}$ **(b)** $\mathbf{i} + 3\mathbf{j}$
67. (a) $\left\langle 0, -\frac{1}{4} \right\rangle$ **(b)** $\left\langle -2\sqrt{2}, \frac{5}{4} \right\rangle$
69. (a) $\dfrac{\sqrt{2}}{2}\mathbf{i} + \dfrac{\sqrt{2}+2}{2}\mathbf{j}$ **(b)** $\dfrac{\sqrt{2}}{2}\mathbf{i} + \dfrac{\sqrt{2}-2}{2}\mathbf{j}$

71. (a) $\langle -4, 16 \rangle$ **(b)** $\langle -12, 0 \rangle$ **(c)** $\langle 8, -8 \rangle$

73. (a) $\langle 8, 0 \rangle$ **(b)** $\langle 0, 16 \rangle$ **(c)** $\langle -4, -8 \rangle$

75. (a) $\langle 0, 12 \rangle$ **(b)** $\langle -16, -4 \rangle$ **(c)** $\langle 8, -4 \rangle$

77. (a) 2 **(b)** $4\mathbf{i}$ **(c)** $7\mathbf{i} + 3\mathbf{j}$

79. (a) $\sqrt{5}$ **(b)** $\langle -2, 4 \rangle$ **(c)** $\langle 7, 4 \rangle$

81. (a) $-21\mathbf{i} + 10\mathbf{j}$ **(b)** $17\mathbf{i} - 10\mathbf{j}$

83. (a) $\langle -13, 24 \rangle$ **(b)** $\langle 11, -23 \rangle$

85. (a) $\langle 43, -2 \rangle$ **(b)** $\langle -41, -1 \rangle$

87. (a) 1 **(b)** $81.9°$ **(c)** Neither

89. (a) 0 **(b)** $90°$ **(c)** Perpendicular

91. (a) 122 **(b)** $0°$ **(c)** Parallel, same direction

93. (a) -4 **(b)** $143.1°$ **(c)** Neither

95. 150 ft-lb **97.** 100,000 ft-lb

99. Work $= 590$ ft-lb, $\|\mathbf{F}\| = \sqrt{500} \approx 22.4$ lb

101. Work $= 27$ ft-lb, $\|\mathbf{F}\| = \sqrt{34} \approx 5.8$ lb

103. 24 **105.** 4

107. $\mathbf{v} = \langle 2, 3 \rangle$, speed $= \sqrt{13} \approx 3.6$ mi/hr

109. $\mathbf{v} \approx \langle -364.6, -35.4 \rangle$, groundspeed ≈ 366.3 mi/hr, bearing $\approx 264.5°$

111. Groundspeed ≈ 431.3 mi/hr, bearing $\approx 159.1°$

113. Airspeed ≈ 149.3 mi/hr, groundspeed ≈ 154.6 mi/hr

115. (a) $\|\mathbf{R}\| = \sqrt{5} \approx 2.2$, $\|\mathbf{A}\| = \sqrt{1.25} \approx 1.1$. About 2.2 inches of rain fell. The area of the opening of the rain gauge is about 1.1 square inches. **(b)** $V = 1.5$; the volume of rain collected in the gauge was 1.5 cubic inches. **(c)** $\mathbf{R}$ and $\mathbf{A}$ must be parallel and point in opposite directions.

117. (a) $\mathbf{c} = \mathbf{a} + \mathbf{b} = \langle 1, 4 \rangle$ **(b)** $\sqrt{17} \approx 4.1$ ft

(c) $3\mathbf{a} + \frac{1}{2}\mathbf{b} = \langle 8, 7 \rangle$

119. (a) $(2, 2)$ **(b)**

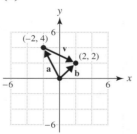

121. $W \approx 297,228$ ft-lb

123. Speed ≈ 180 mi/hr, bearing $\approx 128.2°$

8.3 EXTENDED AND DISCOVERY EXERCISES (pp. 686–687)

1. Blue **3. (a)** Blue **(b)** White

SECTION 8.4 (pp. 694–696)

1. (a)

t	0	1	2	3
x	-1	0	1	2
y	0	2	4	6

(b)

(c) Line segment

3. (a)

t	0	1	2	3
x	2	3	4	5
y	4	1	0	1

(b)

(c) Lower portion of a parabola

5. (a)

t	0	1	2	3
x	3	$\sqrt{8}$	$\sqrt{5}$	0
y	0	1	2	3

(b)

(c) Portion of a circle with radius 3

7. (a)

t	0	1	2	3
x	0	1	2	3
y	3	$\sqrt{8}$	$\sqrt{5}$	0

(b)

(c) Portion of a circle with radius 3

9. $y = \frac{1}{3}x - 1$;
line

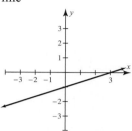

11. $x = 3(y - 1)^2$;
parabola

13. $x = \sqrt{1 - y^2}$; portion
of a circle with radius 1

15. $y = \frac{1}{2}x^2$; portion of
a parabola

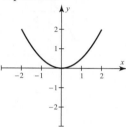

17. $y = x^2 + 4x + 5$;
portion of a parabola

19. $x^2 + y^2 = 9$;
circle with radius 3

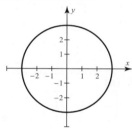

21. $x^2 + y^2 = 4$;
circle with radius 2

23. $x^2 + y^2 = 9$;
circle with radius 3

25.

27.

29.

31.

33. $[0, 6, 1]$ by $[-2, 2, 1]$

35. $[-2, 20, 2]$ by $[-1, 3, 1]$

37. $[-4.7, 4.7, 1]$ by
$[-3.1, 3.1, 1]$

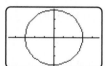

39. $[-1.5, 1.5, 0.5]$ by
$[-1, 1, 0.5]$

41. $[-4.7, 4.7, 1]$ by $[-3.1, 3.1, 1]$

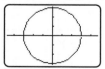

43. $x = t, y = 4 - 2t$ **45.** $x = t, y = 4 - t^2$
47. $x = t^2 + t - 3, y = t$ **49.** $x = 2\cos t, y = 2\sin t$
51. $x = t, y = e^{0.1t^2}$ **53.** $x = t^2 - 2t + 1, y = t$
55. (a) The curve traces
a circle of radius 3 once.
$[-4.7, 4.7, 1]$ by $[-3.1, 3.1, 1]$

(b) The curve traces
a circle of radius 3 twice.
$[-4.7, 4.7, 1]$ by $[-3.1, 3.1, 1]$

57. (a) The curve traces a circle of radius 3 once counterclockwise, starting at $(3, 0)$.
$[-4.7, 4.7, 1]$ by $[-3.1, 3.1, 1]$

(b) The curve traces a circle of radius 3 once clockwise,
starting at $(0, 3)$.
$[-4.7, 4.7, 1]$ by $[-3.1, 3.1, 1]$

59. (a) Circle of radius 1 centered at $(-1, 2)$
$[-4.7, 4.7, 1]$ by $[-3.1, 3.1, 1]$

(b) Circle of radius 1 centered at $(1, 2)$
$[-4.7, 4.7, 1]$ by $[-3.1, 3.1, 1]$

61. F
$[0, 6, 1]$ by $[0, 4, 1]$

63. D
$[0, 6, 1]$ by $[0, 4, 1]$

65. $x_1 = 0, y_1 = 2t; x_2 = t, y_2 = 0; 0 \le t \le 1$
67. $x_1 = \sin t, y_1 = \cos t; x_2 = 0, y_2 = t - 2; 0 \le t \le \pi$
69. Answers may vary.
71. The ball hit at $35°$ travels about 128 feet. The ball hit at $50°$ travels about 134 feet.
73. Yes   **75.** About 285 feet
77. $[-6, 6, 1]$ by $[-4, 4, 1]$

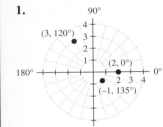

79. $[-6, 6, 1]$ by $[-4, 4, 1]$

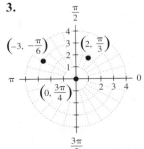

8.4 EXTENDED AND DISCOVERY EXERCISES (pp. 696–697)

1. $F_2 = 100$ **3.** $F_2 = 300$

CHECKING BASIC CONCEPTS FOR SECTIONS 8.3 AND 8.4 (p. 697)

1. (a) $\mathbf{v} = \langle 4, 4 \rangle$ **(b)** $\|\mathbf{v}\| = 4\sqrt{2}$ **(c)** $\langle 0, 0 \rangle$
3. (a) -9 **(b)** $142.1°$

SECTION 8.5 (pp. 706–708)

1.

3.

5. Yes   **7.** No   **9.** Yes

11. $\left(\dfrac{3}{\sqrt{2}}, \dfrac{3}{\sqrt{2}} \right)$, or $\left(\dfrac{3\sqrt{2}}{2}, \dfrac{3\sqrt{2}}{2} \right)$

13. $(0, 10)$ **15.** $(5, 0)$
17. $\left(-\dfrac{3}{2}, -\dfrac{3\sqrt{3}}{2} \right)$ **19.** $(0, -2)$
21. (a) $(3, 90°)$ **(b)** $(-3, -90°)$
23. (a) $(2, 240°)$ **(b)** $(-2, 60°)$
25. (a) $(\sqrt{18}, 315°)$ **(b)** $(-\sqrt{18}, 135°)$
27. $(25, 1.29)$ **29.** $(13, 1.97)$
31.

33. (a)

θ	0°	90°	180°	270°
r	2	2	2	2

(b)

35. (a)

θ	0°	90°	180°	270°
r	0	3	0	-3

(b)

37. (a)

θ	0°	90°	180°	270°
r	2	4	2	0

(b)

39. (a)

θ	0°	90°	180°	270°
r	1	2	3	2

(b)

41.

43.

45.

47.

49.

51.

53.

55.

57.

59. $r = 3 \csc \theta$ **61.** $\theta = \frac{\pi}{4}$ **63.** $r = 3$

65. $r = 2 \cos \theta$ **67.** $x^2 + y^2 = 9$ **69.** $x = 2$

71. $2x + 4y = 3$ **73.** $x^2 + y^2 = x$

75.
$[-9.4, 9.4, 1]$ by $[-6.2, 6.2, 1]$

77.
$[-9.4, 9.4, 1]$ by $[-6.2, 6.2, 1]$

79.
$[-9.4, 9.4, 1]$ by $[-6.2, 6.2, 1]$

81.
$[-9.4, 9.4, 1]$ by $[-6.2, 6.2, 1]$

83. $[-4.7, 4.7, 1]$ by $[-3.1, 3.1, 1]$

85. Use radian mode with $0 \le \theta \le \frac{9\pi}{2}$.
$[-9.4, 9.4, 1]$ by $[-6.2, 6.2, 1]$

87. Let $r_1 = \sqrt{2 \sin (2\theta)}$ and $r_2 = -\sqrt{2 \sin (2\theta)}$.
$[-3, 3, 1]$ by $[-2, 2, 1]$

89. 30°, 150°   **91.** 210°, 330°   **93.** 30°, 150°

95.
$[-30, 30, 5]$ by $[-20, 20, 5]$

97.
$[-3, 3, 1]$ by $[-2, 2, 1]$

99. The radio signal can be received inside the "figure eight." This region is generally in an east–west direction from the two towers, with a maximum distance of 200 miles.
[−300, 300, 100] by [−200, 200, 100]

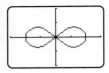

8.5 EXTENDED AND DISCOVERY EXERCISES (p. 708)

1. $y = x \tan (\ln (x^2 + y^2))$

$$\frac{y}{x} = \tan (\ln r^2)$$

$$\tan \theta = \tan (\ln r^2)$$

$$\theta = \ln r^2$$

$$\theta = 2 \ln r$$

$$\frac{\theta}{2} = \ln r$$

$$e^{\theta/2} = r$$

SECTION 8.6 (pp. 717–719)

1.

3.

5. $\sqrt{2}$ **7.** 13 **9.** 6 **11.** $\sqrt{13}$

13. $\sqrt{2} (\cos 135° + i \sin 135°)$

15. $5(\cos 0° + i \sin 0°)$ **17.** $4(\cos 90° + i \sin 90°)$

19. $2(\cos 120° + i \sin 120°)$ **21.** $2(\cos 30° + i \sin 30°)$

23. $2(\cos \pi + i \sin \pi)$ **25.** $\sqrt{8}\left(\cos \frac{3\pi}{4} + i \sin \frac{3\pi}{4}\right)$

27. -5 **29.** $\sqrt{2} + i\sqrt{2}$ **31.** $\sqrt{3} + i$ **33.** 3

35. $z_1 z_2 = \frac{27}{2} + \frac{27\sqrt{3}}{2} i, \frac{z_1}{z_2} = \frac{3\sqrt{3}}{2} + \frac{3}{2} i$

37. $z_1 z_2 = -6, \frac{z_1}{z_2} = 6i$

39. $z_1 z_2 = \frac{\sqrt{3}}{2} - \frac{1}{2} i, \frac{z_1}{z_2} = \frac{1}{2} + \frac{\sqrt{3}}{2} i$ **41.** $8i$ **43.** 1

45. $-\frac{25}{2} + \frac{25\sqrt{3}}{2} i$ **47.** $-2 + 2i$ **49.** $-16\sqrt{3} + 16i$

51. $1 + i\sqrt{3}, -1 - i\sqrt{3}$ **53.** $-1, \frac{1}{2} + \frac{\sqrt{3}}{2} i, \frac{1}{2} - \frac{\sqrt{3}}{2} i$

55. $\frac{\sqrt{2}}{2} + \frac{\sqrt{2}}{2} i, -\frac{\sqrt{2}}{2} - \frac{\sqrt{2}}{2} i$

57. $-2, 1 + i\sqrt{3}, 1 - i\sqrt{3}$

59. $-4i, 2\sqrt{3} + 2i, -2\sqrt{3} + 2i$

61. $\pm 3, \pm 3i$ **63.** Yes **65.** No

67. (a) $Z = 110 + 32i$ **(b)** $\sqrt{13,124} \approx 114.6$ ohms

8.6 EXTENDED AND DISCOVERY EXERCISES (p. 719)

1. (a) $\langle 2 \cos 30°, 2 \sin 30° \rangle$ **(b)** $(2, 30°)$

(c) $2(\cos 30° + i \sin 30°)$

3. (a) $\langle 5 \cos 323.1°, 5 \sin 323.1° \rangle$ **(b)** $(5, 323.1°)$

(c) $5(\cos 323.1° + i \sin 323.1°)$

CHECKING BASIC CONCEPTS FOR SECTIONS 8.5 AND 8.6 (p. 719)

1.

3.

5. $\frac{\sqrt{3}}{2} + \frac{1}{2} i, -\frac{\sqrt{3}}{2} + \frac{1}{2} i, -i$

CHAPTER 8 REVIEW EXERCISES (pp. 722–724)

1. $\gamma = 70°, a = 10.1, b \approx 6.9$

3. $a \approx 5.5, \beta \approx 59.1°, \gamma \approx 78.9°$

5. $\gamma = 115°, a \approx 5.9, c \approx 16.4$

7. $\beta \approx 14.4°, \alpha \approx 145.6°, a \approx 18.2$

9. $c \approx 13.2, \beta \approx 93.6°, \alpha \approx 51.4°$

11. 53.0 **13.** 891.4

15. (a) Horizontal $= 3$, vertical $= 4$ **(b)** $\|\mathbf{v}\| = 5$

(c) $\|\mathbf{v}\|$ represents the length of $\mathbf{v}$.

17. (a)

(b) $\overrightarrow{PQ} = -2\mathbf{i} - 4\mathbf{j}$ **(c)** $\|\overrightarrow{PQ}\| = \sqrt{20}$

19. (a) $2\mathbf{a} = \langle 6, -4 \rangle$ **(b)** $\mathbf{a} - 3\mathbf{b} = \langle 0, -5 \rangle$

(c) $\mathbf{a} \cdot \mathbf{b} = 1$ **(d)** $\theta \approx 78.7°$

21. (a) $2\mathbf{a} = 4\mathbf{i} + 4\mathbf{j}$ **(b)** $\mathbf{a} - 3\mathbf{b} = -\mathbf{i} - \mathbf{j}$
(c) $\mathbf{a} \cdot \mathbf{b} = 4$ **(d)** $\theta = 0°$
23. About 207.1 lb
25.

[−4.7, 4.7, 1] by [−3.1, 3.1, 1]

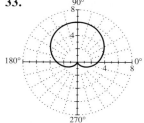

27.

[−4.7, 4.7, 1] by [−3.1, 3.1, 1]

29.

31.

33.

35. Imaginary axis

37. $z_1 z_2 = -8, \frac{z_1}{z_2} = -1 + i\sqrt{3}$
39. $\sqrt{3} + i, -\sqrt{3} - i$ **41.** About 701.6 mi
43. 7204 ft **45.** 15,600 ft²
47. (a) $60\mathbf{i} + 10\mathbf{j}$ **(b)** 60.8 cm **(c)** $80\mathbf{i} + 40\mathbf{j}$
49. About 78.1 ft

CHAPTER 8 EXTENDED AND DISCOVERY EXERCISES
(pp. 724–726)

1. (a) About 56 miles per second
(b) About 87 miles per second
3. (a) 6.02 in. **(b)** About 7470 ft
5. $\|\mathbf{a} - \mathbf{b}\|^2 = \|\mathbf{a}\|^2 + \|\mathbf{b}\|^2 - 2\|\mathbf{a}\| \|\mathbf{b}\| \cos \theta$
$\Rightarrow \|\mathbf{a} - \mathbf{b}\|^2 - \|\mathbf{a}\|^2 - \|\mathbf{b}\|^2 = -2\|\mathbf{a}\| \|\mathbf{b}\| \cos \theta$
$\Rightarrow \left(\sqrt{(a_1 - b_1)^2 + (a_2 - b_2)^2}\right)^2 - \left(\sqrt{a_1^2 + a_2^2}\right)^2$
$- \left(\sqrt{b_1^2 + b_2^2}\right)^2 = -2\|\mathbf{a}\| \|\mathbf{b}\| \cos \theta$
$\Rightarrow (a_1 - b_1)^2 + (a_2 - b_2)^2$
$- (a_1^2 + a_2^2) - (b_1^2 + b_2^2) = -2\|\mathbf{a}\| \|\mathbf{b}\| \cos \theta$
$\Rightarrow a_1^2 - 2a_1 b_1 + b_1^2 + a_2^2 - 2a_2 b_2$
$+ b_2^2 - a_1^2 - a_2^2 - b_1^2 - b_2^2 = -2\|\mathbf{a}\| \|\mathbf{b}\| \cos \theta$
$\Rightarrow -2a_1 b_1 - 2a_2 b_2 = -2\|\mathbf{a}\| \|\mathbf{b}\| \cos \theta$
$\Rightarrow a_1 b_1 + a_2 b_2 = \|\mathbf{a}\| \|\mathbf{b}\| \cos \theta$
$\Rightarrow \mathbf{a} \cdot \mathbf{b} = \|\mathbf{a}\| \|\mathbf{b}\| \cos \theta$

CHAPTERS 1–8 CUMULATIVE REVIEW EXERCISES
(pp. 726–729)

1. 9.12×10^4; 0.006734 **3.** $\sqrt{65}$
5. $\{x \mid x \le 4\}$; 3 **7.** $8x + 4h$
9. x-intercept: −3; y-intercept: −4

11. (a) $-\frac{2}{3}$ **(b)** $-\frac{1}{2}, \frac{11}{2}$ **(c)** $-4, \frac{1}{3}$ **(d)** $-1, 0, 1$
(e) $\pm i, \pm\sqrt{3}$ **(f)** $-\frac{4}{7}, \frac{1}{2}$ **(g)** 8 **(h)** $\frac{\ln(51/2)}{3} \approx 1.08$
(i) 2 **(j)** $\frac{\pi}{6}, \frac{5\pi}{6}$ **(k)** $\frac{\pi}{3} + \frac{\pi}{2}n$
(l) $\frac{\pi}{3} + 2\pi n, \pi + 2\pi n, \frac{5\pi}{3} + 2\pi n$
13. (a) $\left(\frac{13}{5}, \infty\right)$ **(b)** $[-1, 2]$
(c) $(-\infty, -1) \cup (3, \infty)$ **(d)** $(-2, 0) \cup (2, \infty)$
(e) $[0, 1)$ **(f)** $\left[-\frac{8}{3}, \frac{8}{3}\right]$
15. $f(x) = \frac{1}{2}(x + 2)^2 - 3$ **17.** $g(x) = x^2 + 6x + 2$
19. Neither
21. (a) $3x^2 - 6x + 11 + \frac{-20}{x + 2}$ **(b)** $x^2 - x + \frac{-1}{2x - 1}$
23. $f(x) = 6(x - 1)(x + 1)(x - i)(x + i); f(x) = 6x^4 - 6$
25. $D = \left\{x \mid x \ne \frac{2}{3}\right\}$; vertical asymptote: $x = \frac{2}{3}$; horizontal asymptote: $y = -1$
27. $f^{-1}(x) = \frac{x + 2}{3}$ **29.** $C = 5000; a = 2^{2/3}$
31. (a) -3 **(b)** -3 **(c)** $\frac{1}{3}$ **(d)** 3
33. $\log_2 x^2 y^3 \sqrt{z}$ **35.** $\frac{5\pi}{4}$
37. $\sin \theta = \frac{5}{13}, \cos \theta = \frac{12}{13}, \tan \theta = \frac{5}{12}, \csc \theta = \frac{13}{5},$
$\sec \theta = \frac{13}{12}, \cot \theta = \frac{12}{5}$
39. $\sin \theta = -\frac{60}{61}, \cos \theta = \frac{11}{61}, \tan \theta = -\frac{60}{11}, \csc \theta = -\frac{61}{60},$
$\sec \theta = \frac{61}{11}, \cot \theta = -\frac{11}{60}$
41. $\sin \theta = -\frac{7}{25}, \cos \theta = -\frac{24}{25}, \tan \theta = \frac{7}{24}, \csc \theta = -\frac{25}{7},$
$\sec \theta = -\frac{25}{24}, \cot \theta = \frac{24}{7}$
43. $c = 13, \theta \approx 22.6°, \beta \approx 67.4°$
45. $(\cot \theta - 1)^2$ **47.** $\frac{63}{65}$
49. (a) $\beta = 96°, a \approx 7.8, c \approx 12.0$
(b) $\beta \approx 25.4°, \gamma \approx 123.6°, c \approx 9.7$
(c) $\alpha \approx 47.0°, \gamma \approx 77.0°, b \approx 6.8$
(d) $\alpha \approx 46.6°, \beta \approx 57.9°, \gamma \approx 75.5°$

51. (a) 25 **(b)** $\langle -31, 96 \rangle$ **(c)** -323
(d) About $173.6°$
53.

55. (a) 48 ft/sec; 120 ft/sec **(b)** From 1 to 3 seconds the car's average speed is 48 feet per second, and from 4 to 6 seconds the car's average speed is 120 feet per second.
57. 15 by 19 inches
59. 12 by 6 by 4 inches or 7.4 by 3.7 by 10.5 inches
61. (a) $C(t) = 630(t - 1980) + 3600$
(b) About 1992, which is one year from the actual value
63. 2.5 feet **65.** $56.2°$
67. (a) About 247 feet **(b)** About 18,988 ft^2
69. About 249.4 feet

CHAPTER 9: Systems of Equations and Inequalities

SECTION 9.1 (pp. 745–750)

1. 20; the area of a triangle with base 5 and height 8 is 20.
3. 13 **5.** -18 **7.** $\frac{4}{5}$ **9.** $f(x, y) = y + 2x$
11. $f(x, y) = \frac{xy}{1 + x}$ **13.** $x = \frac{4y + 7}{3}$; $y = \frac{3x - 7}{4}$
15. $x = y^2 + 5$; $y = \pm\sqrt{x - 5}$ **17.** $x = 2y$; $y = \frac{x}{2}$
19. $(2, 1)$; linear **21.** $(4, -3)$; nonlinear
23. $(2, 2)$ **25.** $\left(\frac{1}{2}, -2\right)$
27. $x - y = 2$, $2x + 2y = 38$, where x is width and y is height; 10.5 in. wide, 8.5 in. high
29. $x + y = 75$, $4x + 7y = 456$, where x is the number of child tickets and y is the number of adult tickets; 23 child tickets, 52 adult tickets **31.** Consistent with solution $(2, 2)$
33. Inconsistent; no solutions
35.

No solutions; inconsistent

37.

$(2, -1)$; consistent, independent

39.

$(-2, 2)$; consistent, independent

41.

$\{(x, y) \,|\, 2x - y = -4\}$; consistent, dependent

43. $(-2, 1)$ **45.** $\left(\frac{1}{2}, 2\right)$ **47.** $(6, 8)$
49. $\{(x, y) \,|\, 3x - 2y = 5\}$ **51.** No solutions
53. $(4, 3)$ **55.** $(-2, 4), (0, 0)$ **57.** $(2, 4), (4, 2)$
59. $(2, 4), (-2, -4)$ **61.** No real solutions
63. $\{(x, y) \,|\, 2x^2 - y = 5\}$ **65.** $(-2, 0), (2, 0)$
67. $(-2, -2), (0, 0), (2, 2)$ **69.** $l = 7, w = 5$
71. $(14, 6)$; consistent, independent
73. $(1, 3)$; consistent, independent
75. $\{(x, y) \,|\, x + y = 500\}$; consistent, dependent
77. No solutions; inconsistent
79. $\left(-\frac{11}{7}, \frac{12}{7}\right)$; consistent, independent
81. $(2, -4)$ **83.** $\{(x, y) \,|\, 7x - 3y = -17\}$
85. No solutions **87.** $(10, 20)$
89. $(2, 1)$ **91.** $(-5, -4)$ **93.** $(3, 3), (-3, 3)$
95. $(4, 3), (-4, 3), (3, 4), (-3, 4)$
97. $(2, 0), (-2, 0)$ **99.** $(-\sqrt{8}, -\sqrt{8}), (\sqrt{8}, \sqrt{8})$
101. $(6, 2), (-2, -6)$ **103.** $(1, -1)$ **105.** $(1, 2)$
107. $(-1.588, 0.239), (0.164, 1.487), (1.924, -0.351)$
109. $(1.220, 0.429)$ **111.** $(0.714, -0.169)$
113. (a) $x + y = 670, x - y = 96$ **(b)** $(383, 287)$
(c) Consistent; independent
115. $W_1 = \dfrac{300}{1 + \sqrt{3}} \approx 109.8$ lb,

$W_2 = \dfrac{300\sqrt{3}}{\sqrt{6} + \sqrt{2}} \approx 134.5$ lb
117. $r \approx 1.538$ in, $h \approx 6.724$ in.
119. 12 by 12 by 4 in. or 9.10 by 9.10 by 6.96 in.
121. (a) $x + y = 831,000, x - y = 15,000$
(b) & (c) $(423000, 408000)$
123. (a) $x + y = 3000, 0.08x + 0.10y = 264$
(b) $1800 at 8%; $1200 at 10%
125. There are no solutions. If loans totaling $3000 are at 10%, then the interest must be $300.
127. (a) $l = 13$ ft, $w = 7$ ft **(b)** $A = 20w - w^2$
(c) 100 ft^2; a square pen will provide the largest area.
129. Airplane: 520 mi/hr; wind speed: 40 mi/hr
131. 53,077 deaths in WWI; 291,923 deaths in WWII
133. (a) First model: about 245 lb; second model: about 253 lb **(b)** Models agree when $h \approx 65.96$ in.
(c) First model: 7.46 lb; second model: 7.93 lb

135. About 1.77 m² **137.** $S(60, 157.48) \approx 1.6$ m²
139. 0.51 **141.** 32.4
143. Approximately 10,823 watts
145. Approximately 2.54 cords
147. $S(w, h) = 0.0101w^{0.425}h^{0.725}$

SECTION 9.2 (pp. 759–762)

1.

3.

5.

7.

9.

11.

13. c; (2, 3) (answers may vary)
15. d; (−1, −1) (answers may vary)
17. (0, 2)
(answers may vary)

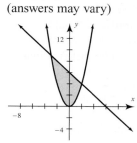

19. (0, 0)
(answers may vary)

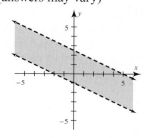

21. (−1, 1) (answers may vary)

23.

25.

27.

29.

31.

33.

35. $x = 300, y = 350$
37. This individual weighs less than recommended for his or her height.
39. $25h - 7w \leq 800, 5h - w \geq 170$

41.

43.

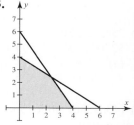

45. Maximum: 65; minimum: 8
47. Maximum: 66; minimum: 3
49. Maximum: 100; minimum: 0
51. $x + y \le 4, x \ge 0, y \ge 0$
53. Minimum: 6
55. Maximum: $z = 56$; minimum: $z = 24$
57. 25 radios, 30 CD players
59. 2.4 units of Brand A, 1.2 units of Brand B
61. $600   **63.** Part X: 9, part Y: 4

CHECKING BASIC CONCEPTS FOR SECTIONS 9.1 AND 9.2 (p. 762)

1. $d(13, 18) = 20$ **3.** $y = \pm\sqrt{z - x^2}$
5.

7. (a) $x + y = 32, 1.9x - y = 0$
(b) Approximately (11, 21); in 1999 about 11 million television sets were sold with stereo sound and about 21 million were sold without stereo sound.

SECTION 9.3 (pp. 768–770)

1. No   **3.** 2   **5.** $(0, 2, -2)$ is not, but $(-1, 3, -2)$ is a solution.   **7.** Both are solutions.
9. $(1, 2, 3)$ **11.** $(1, 0, 1)$ **13.** $\left(\frac{1}{2}, \frac{1}{2}, -2\right)$
15. $\left(\frac{z + 3}{2}, \frac{-z + 3}{2}, z\right)$
17. No solutions   **19.** $\left(-\frac{5}{2}, -2, 4\right)$
21. No solutions   **23.** $\left(-\frac{1}{2}, \frac{1}{2}, -\frac{1}{2}\right)$
25. $\left(\frac{-5z + 18}{13}, \frac{-6z + 19}{13}, z\right)$
27. $\left(8, -11, -\frac{1}{2}\right)$ **29.** $(2, 3, -4)$
31. 120 child, 280 student, and 100 adult tickets
33. No solutions; at least one student was charged incorrectly.
35. (a) $x + y + z = 180$
  $x \quad\quad - z = 25$
  $-x + y + z = 30$
(b) 75°, 55°, 50°
37. $2500 at 5%, $7500 at 7%, $10,000 at 10%
39. (a) $N + P + K = 80$
  $N + P - K = 8$
  $9P - K = 0$
(b) (40, 4, 36); 40 lb of nitrogen, 4 lb of phosphorus, 36 lb of potassium

SECTION 9.4 (pp. 783–787)

1. (a) 3×1 **(b)** 2×3 **(c)** 2×2
3. Dimension: 2×3 **5.** Dimension: 3×4
$$\begin{bmatrix} 5 & -2 & 3 \\ -1 & 3 & -1 \end{bmatrix}$$
$$\begin{bmatrix} -3 & 2 & 1 & -4 \\ 5 & 0 & -1 & 9 \\ 1 & -3 & -6 & -9 \end{bmatrix}$$
7. $3x + 2y = 4$
  $y = 5$
9. $3x + y + 4z = 0$
  $5y + 8z = -1$
  $-7z = 1$
11. (a) Yes   **(b)** No   **(c)** Yes
13. $(5, -1)$ **15.** $(2, 0)$ **17.** $(2, 3, 1)$
19. $(3 - 3z, 1 + 2z, z)$
21. No solutions
23. $\begin{bmatrix} 1 & -2 & 3 & 5 \\ -3 & 5 & 3 & 2 \\ 1 & 2 & 1 & -2 \end{bmatrix}$ **25.** $\begin{bmatrix} 1 & -1 & 1 & 2 \\ 0 & 1 & -1 & 2 \\ 0 & 8 & -1 & 3 \end{bmatrix}$
27. $(-17, 10)$ **29.** $(0, 2, -1)$ **31.** $(-1, 2, 4)$
33. $(3, 2, 1)$ **35.** No solutions
37. $\left(-1 - z, \frac{9 - 5z}{2}, z\right)$
39. $(1, 1, 1)$ **41.** $\left(\frac{1}{2}, -\frac{1}{2}, 4\right)$ **43.** $(-2, 5, -1)$
45. No solutions   **47.** $(-1, -2, 3)$
49. $\left(\frac{3z + 1}{7}, \frac{11z - 15}{7}, z\right)$ **51.** $(12, 3)$
53. $\left(-2, 4, \frac{1}{2}\right)$ **55.** $(4 - 2z, z - 3, z)$
57. No solutions   **59.** $(3, 2)$ **61.** $(-2, 1, 3)$
63. $(-2, 5, 7)$ **65.** No solutions
67. $(-9.226, -9.167, 2.440)$
69. $(5.211, 3.739, -4.655)$
71. $(7.993, 1.609, -0.401)$
73. (a) $F = 0.5714N + 0.4571R - 2014$ **(b)** $5700
75. Pump 1: 12 hours; pumps 2 and 3: 24 hours
77. $(3.53, 1.62, 1.91)$
79. (a) $x + y + z = 5000$
  $x + y - z = 0$
  $0.08x + 0.11y + 0.14z = 595,$
where x is amount invested at 8%, y is amount invested at 11%, and z is amount invested at 14%
(b) $1000 at 8%; $1500 at 11%; $2500 at 14%
81. (a) At intersection A, incoming traffic is equal to $x + 5$. The outgoing traffic is given by $y + 7$. Therefore, $x + 5 = y + 7$. The other equations can be justified in a similar way.
(b) The three equations can be written as
  $x - y = 2$
  $x - z = 3$
  $y - z = 1$
The solution can be written as $\{(z + 3, z + 1, z) | z \ge 0\}$.
(c) There are infinitely many solutions, since some cars could be driving around the block continually.

83. **(a)**
$$\begin{aligned} c &= 3 \\ 4a + 2b + c &= 55 \\ 16a + 4b + c &= 150 \end{aligned}$$

(b) $f(x) = 5.375x^2 + 15.25x + 3$

(c) $[-0.5, 5, 1]$ by $[-10, 175, 25]$

(d) In 2012 predicted sales are $f(8) = 469$ million (answers may vary).

85. **(a)**
$$\begin{aligned} 1990^2 a + 1990b + c &= 11 \\ 2010^2 a + 2010b + c &= 10 \\ 2030^2 a + 2030b + c &= 6 \end{aligned}$$

(b) $f(x) = -0.00375x^2 + 14.95x - 14{,}889.125$

(c) $[1985, 2035, 5]$ by $[5, 12, 1]$

(d) In 2015 the predicted ratio is $f(2015) \approx 9.3$ (answers may vary).

9.4 EXTENDED AND DISCOVERY EXERCISES (p. 787)

1. $(1, -1, 2, 0)$

CHECKING BASIC CONCEPTS FOR SECTIONS 9.3 AND 9.4 (p. 787)

1. **(a)** $(3, 2, -1)$ **(b)** $\left(\frac{4 - 5z}{3}, \frac{5 - z}{3}, z\right)$
(c) No solutions   **3.** $(2, -1, 0)$

SECTION 9.5 (pp. 797–800)

1. **(a)** $a_{12} = 2, a_{21} = 4, a_{32}$ is undefined.   **(b)** 23
3. **(a)** $a_{12} = -1, a_{21} = 3, a_{32} = 0$ **(b)** 13
5. $x = 1, y = 1$ **7.** Not possible
9. **(a)** $a_{12} = 3, b_{32} = 1, b_{22} = 0$ **(b)** -2 **(c)** $x = 3$

11. **(a)** $A + B = \begin{bmatrix} 3 & 3 \\ 3 & 3 \end{bmatrix}$ **(b)** $B + A = \begin{bmatrix} 3 & 3 \\ 3 & 3 \end{bmatrix}$

(c) $A - B = \begin{bmatrix} 5 & -5 \\ -5 & 5 \end{bmatrix}$

13. **(a)** $A + B = \begin{bmatrix} 14 & 9 & -3 \\ 4 & -10 & 14 \\ 4 & 11 & 16 \end{bmatrix}$

(b) $B + A = \begin{bmatrix} 14 & 9 & -3 \\ 4 & -10 & 14 \\ 4 & 11 & 16 \end{bmatrix}$

(c) $A - B = \begin{bmatrix} -8 & -1 & 1 \\ -4 & 4 & -10 \\ -8 & -1 & 4 \end{bmatrix}$

15. **(a)** $A + B = \begin{bmatrix} 1 & -6 \\ 1 & 4 \end{bmatrix}$ **(b)** $3A = \begin{bmatrix} 6 & -18 \\ 9 & 3 \end{bmatrix}$

(c) $2A - 3B = \begin{bmatrix} 7 & -12 \\ 12 & -7 \end{bmatrix}$

17. **(a)** $A + B$ is undefined.

(b) $3A = \begin{bmatrix} 3 & -3 & 0 \\ 3 & 15 & 27 \\ -12 & 24 & -15 \end{bmatrix}$

(c) $2A - 3B$ is undefined.

19. **(a)** $A + B = \begin{bmatrix} 0 & -2 \\ -2 & 2 \\ 9 & -8 \end{bmatrix}$ **(b)** $3A = \begin{bmatrix} -6 & -3 \\ -15 & 3 \\ 6 & -9 \end{bmatrix}$

(c) $2A - 3B = \begin{bmatrix} -10 & 1 \\ -19 & -1 \\ -17 & 9 \end{bmatrix}$

21. $\begin{bmatrix} 0 & 2 \\ 13 & -5 \\ 0 & 1 \end{bmatrix}$ **23.** $\begin{bmatrix} 2 & 6 \\ 11 & -9 \end{bmatrix}$

25. $\begin{bmatrix} 7 & 4 & 7 \\ 4 & 7 & 7 \\ 4 & 7 & 7 \end{bmatrix}$ **27.** $A = \begin{bmatrix} 1 & 2 & 1 \\ 1 & 2 & 1 \\ 1 & 2 & 1 \end{bmatrix}$

29. $B = \begin{bmatrix} -1 & 1 & -1 \\ -1 & 1 & -1 \\ -1 & 1 & -1 \end{bmatrix}, A + B = \begin{bmatrix} 0 & 3 & 0 \\ 0 & 3 & 0 \\ 0 & 3 & 0 \end{bmatrix}$

31. $AB = \begin{bmatrix} -3 & 1 \\ -4 & 6 \end{bmatrix}, BA = \begin{bmatrix} 4 & 2 \\ 5 & -1 \end{bmatrix}$

33. AB and BA are undefined.

35. $AB = \begin{bmatrix} -15 & 22 & -9 \\ -2 & 5 & -3 \\ -32 & 18 & 6 \end{bmatrix}, BA = \begin{bmatrix} 5 & 14 \\ 20 & -9 \end{bmatrix}$

37. AB is undefined. $BA = \begin{bmatrix} -1 & -3 & 19 \\ -1 & 1 & -39 \end{bmatrix}$

39. AB and BA are undefined.

41. $AB = \begin{bmatrix} -1 & 15 & -2 \\ 0 & 1 & 3 \\ -1 & 14 & -5 \end{bmatrix}, BA = \begin{bmatrix} 0 & 6 & 0 \\ 6 & -5 & 9 \\ 0 & 1 & 0 \end{bmatrix}$

43. $AB = \begin{bmatrix} -1 \\ 6 \end{bmatrix}$ BA is undefined.

45. $AB = \begin{bmatrix} -7 & 8 & 7 \\ 18 & -32 & -8 \end{bmatrix}, BA$ is undefined.

47. $AB = \begin{bmatrix} -5 \\ 10 \\ 17 \end{bmatrix}$, BA is undefined.

49. $\begin{bmatrix} -19 & 19 & 11 \\ 21 & -7 & -48 \\ -22 & 23 & -58 \end{bmatrix}$ **51.** $\begin{bmatrix} 83 & 32 & 92 \\ 10 & -63 & -8 \\ 210 & 56 & 93 \end{bmatrix}$

53. They both equal $\begin{bmatrix} 36 & 36 & 8 \\ -15 & -38 & -4 \\ -11 & 13 & 10 \end{bmatrix}$. The distributive property appears to hold for matrices.

55. They both equal $\begin{bmatrix} 50 & 3 & 12 \\ -6 & 55 & 8 \\ 27 & -3 & 29 \end{bmatrix}$. Matrices appear to conform to rules of algebra except that $AB \neq BA$.

57. $B = \begin{bmatrix} 3 & 3 & 3 \\ 3 & 3 & 3 \\ 3 & 3 & 3 \end{bmatrix}$, $B - A = \begin{bmatrix} 3 & 0 & 3 \\ 3 & 0 & 3 \\ 3 & 0 & 3 \end{bmatrix}$

59. $A = \begin{bmatrix} 3 & 3 & 3 & 3 \\ 3 & 0 & 0 & 0 \\ 3 & 3 & 3 & 0 \\ 3 & 0 & 0 & 0 \\ 3 & 0 & 0 & 0 \end{bmatrix}$

61. (a) One possibility is $A = \begin{bmatrix} 3 & 3 & 3 & 3 \\ 0 & 0 & 3 & 0 \\ 0 & 3 & 0 & 0 \\ 3 & 3 & 3 & 3 \end{bmatrix}$.

(b) $B = \begin{bmatrix} 3 & 3 & 3 & 3 \\ 3 & 3 & 3 & 3 \\ 3 & 3 & 3 & 3 \\ 3 & 3 & 3 & 3 \end{bmatrix}$

63. (a) One possibility is $A = \begin{bmatrix} 3 & 0 & 0 & 0 \\ 3 & 0 & 0 & 0 \\ 3 & 0 & 0 & 0 \\ 3 & 3 & 3 & 3 \end{bmatrix}$.

(b) $B = \begin{bmatrix} 3 & 3 & 3 & 3 \\ 3 & 3 & 3 & 3 \\ 3 & 3 & 3 & 3 \\ 3 & 3 & 3 & 3 \end{bmatrix}$

65. (a) $A = \begin{bmatrix} 12 & 4 \\ 8 & 7 \end{bmatrix}$, $B = \begin{bmatrix} 55 \\ 70 \end{bmatrix}$

(b) $AB = \begin{bmatrix} 940 \\ 930 \end{bmatrix}$. Tuition for Student 1 is $940, and tuition for Student 2 is $930.

67. (a) $A = \begin{bmatrix} 10 & 5 \\ 9 & 8 \\ 11 & 3 \end{bmatrix}$, $B = \begin{bmatrix} 60 \\ 70 \end{bmatrix}$

(b) $AB = \begin{bmatrix} 950 \\ 1100 \\ 870 \end{bmatrix}$. Tuition for Student 1 is $950, for Student 2 is $1100, and for Student 3 is $870.

69. $AB = \begin{bmatrix} 350 \\ 230 \end{bmatrix}$. The total cost of order 1 is $350, and the total cost of order 2 is $230.

9.5 EXTENDED AND DISCOVERY EXERCISES (pp. 800–801)

1. Aquamarine is represented by (0.369, 0, 0.067) in CMY.

3. $\begin{bmatrix} R \\ G \\ B \end{bmatrix} = \begin{bmatrix} 1 \\ 1 \\ 1 \end{bmatrix} - \begin{bmatrix} C \\ M \\ Y \end{bmatrix}$

SECTION 9.6 (pp. 810–814)

1. Yes **3.** Yes **5.** No
7. $k = 1$ **9.** $k = 2.5$ **11.** A **13.** A

15. $\begin{bmatrix} 3 & -2 \\ -1 & 1 \end{bmatrix}$ **17.** $\begin{bmatrix} 5 & 2 \\ 3 & 1 \end{bmatrix}$ **19.** $\begin{bmatrix} -\frac{1}{2} & \frac{5}{2} \\ 1 & -4 \end{bmatrix}$

21. $\begin{bmatrix} 0 & 1 & 0 \\ 0 & 0 & 1 \\ 1 & 0 & 0 \end{bmatrix}$ **23.** $\begin{bmatrix} -2 & 1 & -1 \\ -5 & 2 & -1 \\ 3 & -1 & 1 \end{bmatrix}$

25. $\begin{bmatrix} -10 & 3 & -5 \\ 4 & -1 & 2 \\ -3 & 1 & -1 \end{bmatrix}$ **27.** $\begin{bmatrix} -1 & -1 & -1 \\ -\frac{2}{5} & -\frac{1}{5} & -\frac{4}{5} \\ \frac{1}{5} & \frac{3}{5} & \frac{2}{5} \end{bmatrix}$

29. $\begin{bmatrix} -10 & 30 \\ -4 & 10 \end{bmatrix}$ **31.** $\begin{bmatrix} 0.2 & 0 & 0.4 \\ 0.4 & 0 & -0.2 \\ 1.4 & -1 & -1.2 \end{bmatrix}$

33. $\begin{bmatrix} 0.5 & 0.2 & 2.1 \\ 0 & 0.2 & 1.6 \\ 0 & 0 & -1 \end{bmatrix}$ **35.** $\begin{bmatrix} 0.5 & 0.25 & 0.25 \\ 0.25 & 0.5 & 0.25 \\ 0.25 & 0.25 & 0.5 \end{bmatrix}$

37. $\begin{bmatrix} 1.2\overline{6} & 0.2\overline{6} & 0.0\overline{6} & 0.0\overline{6} \\ 0.2\overline{6} & 0.2\overline{6} & 0.0\overline{6} & 0.0\overline{6} \\ 0.0\overline{6} & 0.0\overline{6} & 0.2\overline{6} & 0.2\overline{6} \\ 0.0\overline{6} & 0.0\overline{6} & 0.2\overline{6} & 1.2\overline{6} \end{bmatrix}$

39. $AX = \begin{bmatrix} 2 & -3 \\ -3 & -4 \end{bmatrix} \begin{bmatrix} x \\ y \end{bmatrix} = \begin{bmatrix} 7 \\ 9 \end{bmatrix} = B$

41. $AX = \begin{bmatrix} \frac{1}{2} & -\frac{3}{2} \\ -1 & 2 \end{bmatrix} \begin{bmatrix} x \\ y \end{bmatrix} = \begin{bmatrix} \frac{1}{4} \\ 5 \end{bmatrix} = B$

43. $AX = \begin{bmatrix} 1 & -2 & 1 \\ 0 & 3 & -1 \\ 5 & -4 & -7 \end{bmatrix} \begin{bmatrix} x \\ y \\ z \end{bmatrix} = \begin{bmatrix} 5 \\ 6 \\ 0 \end{bmatrix} = B$

45. $AX = \begin{bmatrix} 4 & -1 & 3 \\ 1 & 2 & 5 \\ 2 & -3 & 0 \end{bmatrix} \begin{bmatrix} x \\ y \\ z \end{bmatrix} = \begin{bmatrix} -2 \\ 11 \\ -1 \end{bmatrix} = B$

47. (a) $AX = \begin{bmatrix} 1 & 2 \\ 1 & 3 \end{bmatrix} \begin{bmatrix} x \\ y \end{bmatrix} = \begin{bmatrix} 3 \\ 6 \end{bmatrix} = B$ **(b)** $X = \begin{bmatrix} -3 \\ 3 \end{bmatrix}$

49. (a) $AX = \begin{bmatrix} -1 & 2 \\ 3 & -5 \end{bmatrix} \begin{bmatrix} x \\ y \end{bmatrix} = \begin{bmatrix} 5 \\ -2 \end{bmatrix} = B$

(b) $X = \begin{bmatrix} 21 \\ 13 \end{bmatrix}$

51. (a) $AX = \begin{bmatrix} 1 & 0 & 1 \\ 2 & 1 & 3 \\ -1 & 1 & 1 \end{bmatrix} \begin{bmatrix} x \\ y \\ z \end{bmatrix} = \begin{bmatrix} -7 \\ -13 \\ -4 \end{bmatrix} = B$

(b) $X = \begin{bmatrix} 5 \\ 13 \\ -12 \end{bmatrix}$

53. (a) $AX = \begin{bmatrix} 1 & 2 & -1 \\ 2 & 5 & 0 \\ -1 & -1 & 2 \end{bmatrix} \begin{bmatrix} x \\ y \\ z \end{bmatrix} = \begin{bmatrix} 2 \\ -1 \\ 0 \end{bmatrix} = B$

(b) $X = \begin{bmatrix} -23 \\ 9 \\ -7 \end{bmatrix}$

55. (a) $AX = \begin{bmatrix} 1.5 & 3.7 \\ -0.4 & -2.1 \end{bmatrix} \begin{bmatrix} x \\ y \end{bmatrix} = \begin{bmatrix} 0.32 \\ 0.36 \end{bmatrix} = B$

(b) $X = \begin{bmatrix} 1.2 \\ -0.4 \end{bmatrix}$

57. (a) $AX = \begin{bmatrix} 0.08 & -0.7 \\ 1.1 & -0.05 \end{bmatrix} \begin{bmatrix} x \\ y \end{bmatrix} = \begin{bmatrix} -0.504 \\ 0.73 \end{bmatrix} = B$

(b) $X = \begin{bmatrix} 0.7 \\ 0.8 \end{bmatrix}$

59. (a) $AX = \begin{bmatrix} 3.1 & 1.9 & -1 \\ 6.3 & 0 & -9.9 \\ -1 & 1.5 & 7 \end{bmatrix} \begin{bmatrix} x \\ y \\ z \end{bmatrix} = \begin{bmatrix} 1.99 \\ -3.78 \\ 5.3 \end{bmatrix} = B$

(b) $X = \begin{bmatrix} 0.5 \\ 0.6 \\ 0.7 \end{bmatrix}$

61. (a) $AX = \begin{bmatrix} 3 & -1 & 1 \\ 5.8 & -2.1 & 0 \\ -1 & 0 & 2.9 \end{bmatrix} \begin{bmatrix} x \\ y \\ z \end{bmatrix} = \begin{bmatrix} 4.9 \\ -3.8 \\ 3.8 \end{bmatrix} = B$

(b) $X \approx \begin{bmatrix} 9.26 \\ 27.39 \\ 4.50 \end{bmatrix}$

63. (a) $(2, 4)$ **(b)** It will translate $(2, 4)$ to the left 2 units and downward 3 units, back to $(0, 1)$;

$A^{-1} = \begin{bmatrix} 1 & 0 & -2 \\ 0 & 1 & -3 \\ 0 & 0 & 1 \end{bmatrix}$ **(c)** I_3

65. $A = \begin{bmatrix} 1 & 0 & -3 \\ 0 & 1 & -5 \\ 0 & 0 & 1 \end{bmatrix}$ and $A^{-1} = \begin{bmatrix} 1 & 0 & 3 \\ 0 & 1 & 5 \\ 0 & 0 & 1 \end{bmatrix}$.

A^{-1} will translate a point 3 units to the right and 5 units upward.

67. (a) $BX = \begin{bmatrix} -2 \\ 0 \\ 1 \end{bmatrix} = Y$ **(b)** $B^{-1}Y = \begin{bmatrix} -\sqrt{2} \\ -\sqrt{2} \\ 1 \end{bmatrix} = X$.

B^{-1} rotates the point represented by Y counterclockwise 45° about the origin.

69. (a) $ABX = \begin{bmatrix} 2 \\ 2 \\ 1 \end{bmatrix} = Y$

(b) The net result of A and B is to translate a point 1 unit to the right and 1 unit upward.

$AB = \begin{bmatrix} 1 & 0 & 1 \\ 0 & 1 & 1 \\ 0 & 0 & 1 \end{bmatrix}$. **(c)** Yes **(d)** Since AB translates a point 1 unit right and 1 unit upward, the inverse of AB would translate a point 1 unit left and 1 unit downward. Therefore

$(AB)^{-1} = \begin{bmatrix} 1 & 0 & -1 \\ 0 & 1 & -1 \\ 0 & 0 & 1 \end{bmatrix}$.

71. A: \$10.99; B: \$12.99; C: \$14.99
73. (a) $a + 1500b + 8c = 122$
$a + 2000b + 5c = 130$
$a + 2200b + 10c = 158$
$a = 30, b = 0.04, c = 4$
(b) \$130,000 **75. (a)** $\left(\frac{17}{12}T, \frac{5}{6}T, T\right)$
(b) Service: 85 units; electrical: 50 units

CHECKING BASIC CONCEPTS FOR SECTIONS 9.5 AND 9.6 (p. 814)

1. (a) $A + B = \begin{bmatrix} 0 & 1 & 3 \\ -1 & 5 & 3 \\ 2 & 1 & 0 \end{bmatrix}$

(b) $2A - B = \begin{bmatrix} 3 & -1 & 0 \\ -2 & -2 & 3 \\ 1 & 8 & 0 \end{bmatrix}$

(c) $AB = \begin{bmatrix} 0 & -1 & 2 \\ 3 & -1 & -1 \\ -1 & 13 & 5 \end{bmatrix}$

3. (a) $AX = \begin{bmatrix} 1 & -2 \\ 2 & 3 \end{bmatrix}\begin{bmatrix} x \\ y \end{bmatrix} = \begin{bmatrix} 13 \\ 5 \end{bmatrix} = B; X = \begin{bmatrix} 7 \\ -3 \end{bmatrix}$

(b) $AX = \begin{bmatrix} 1 & -1 & 1 \\ -1 & 1 & 1 \\ 0 & 1 & -1 \end{bmatrix}\begin{bmatrix} x \\ y \\ z \end{bmatrix} = \begin{bmatrix} 2 \\ 4 \\ -1 \end{bmatrix} = B; X = \begin{bmatrix} 1 \\ 2 \\ 3 \end{bmatrix}$

(c) $AX = \begin{bmatrix} 3.1 & -5.3 \\ -0.1 & 1.8 \end{bmatrix}\begin{bmatrix} x \\ y \end{bmatrix} = \begin{bmatrix} -2.682 \\ 0.787 \end{bmatrix} = B;$

$X = \begin{bmatrix} -0.13 \\ 0.43 \end{bmatrix}$

SECTION 9.7 (pp. 821–822)

1. $\det A = 1 \neq 0$. A is invertible.
3. $\det A = 0$. A is not invertible.
5. $M_{12} = 10, A_{12} = -10$
7. $M_{22} = -15, A_{22} = -15$
9. $\det A = 3 \neq 0$. A^{-1} exists.
11. $\det A = 0$. A^{-1} does not exist.
13. 30   **15.** 0   **17.** -32 **19.** 0
21. 643.4   **23.** -4.484 **25.** $\left(-\frac{13}{9}, \frac{16}{9}\right)$ **27.** $\left(\frac{49}{2}, 19\right)$
29. $(5, -3)$ **31.** $(0.45, 0.67)$ **33.** 7 square units
35. 6.5 square units   **37.** The points are collinear.
39. The points are not collinear.
41. $x + y = 3$ **43.** $2x + y = 5$

9.7 EXTENDED AND DISCOVERY EXERCISES
(pp. 822–823)

1. $(1, 3, 2)$ **3.** $(1, -1, 1)$ **5.** $(-1, 0, 4)$
7. $x^2 + y^2 - 4 = 0$ **9.** $5x^2 + 5y^2 - 15x - 5y = 0$

CHECKING BASIC CONCEPTS FOR SECTION 9.7 (p. 823)

1. $\det A = 19$; A is invertible.

CHAPTER 9 REVIEW EXERCISES (pp. 827–829)

1. $A(3, 6) = 9$ **3.** $(1, -2)$ **5.** $(2, 3)$; consistent
7. $\{(x, y) \mid 2x - 5y = 4\}$; consistent

9. $\left(\frac{3\sqrt{2}}{2}, \frac{1}{2}\right), \left(-\frac{3\sqrt{2}}{2}, \frac{1}{2}\right)$

11.

13. $(2, 2)$ (answers may vary)

15. $(-1, 2, 1)$ **17.** No solutions
19. $(-9, 3)$ **21.** $(-2, 3, 0)$ **23.** $(1, -2, 3)$
25. (a) 5   **(b)** -10

27. (a) $A + 2B = \begin{bmatrix} 7 & 1 \\ -8 & 1 \end{bmatrix}$

(b) $A - B = \begin{bmatrix} -2 & -5 \\ 7 & -2 \end{bmatrix}$ **(c)** $-4A = \begin{bmatrix} -4 & 12 \\ -8 & 4 \end{bmatrix}$

29. $AB = \begin{bmatrix} -2 & -4 \\ 17 & 31 \end{bmatrix}, BA = \begin{bmatrix} 8 & -6 \\ -27 & 21 \end{bmatrix}$

31. $AB = \begin{bmatrix} 3 & 7 \\ -2 & 8 \end{bmatrix}, BA = \begin{bmatrix} 2 & -1 & 3 \\ 2 & 9 & -3 \\ 6 & 12 & 0 \end{bmatrix}$ **33.** Yes

35. (a) $AX = \begin{bmatrix} 1 & -3 \\ 2 & -1 \end{bmatrix}\begin{bmatrix} x \\ y \end{bmatrix} = \begin{bmatrix} 4 \\ 3 \end{bmatrix} = B$

(b) $X = \begin{bmatrix} 1 \\ -1 \end{bmatrix}$

37. (a) $AX = \begin{bmatrix} 12 & 7 & -3 \\ 8 & -11 & 13 \\ -23 & 0 & 9 \end{bmatrix}\begin{bmatrix} x \\ y \\ z \end{bmatrix} = \begin{bmatrix} 14.6 \\ -60.4 \\ -14.6 \end{bmatrix} = B$

(b) $X = \begin{bmatrix} -0.5 \\ 1.7 \\ -2.9 \end{bmatrix}$ **39.** $\begin{bmatrix} -1 & -2 \\ -1 & -1 \end{bmatrix}$

41. $\det A = 25$ **43.** $\det A = -1951 \neq 0$. A is invertible.
45. $l = 11, w = 7$
47. Both methods yield $1200 at 7%, $800 at 9%.
49. A: $11.49; B: $12.99
51. 10.5 square units   **53.** 4500

CHAPTER 9 EXTENDED AND DISCOVERY EXERCISES
(pp. 830–831)

1. (a) $A^T = \begin{bmatrix} 3 & 2 & 4 \\ -3 & 6 & 2 \end{bmatrix}$ **(b)** $A^T = \begin{bmatrix} 0 & 2 & -4 \\ 1 & 5 & 3 \\ -2 & 4 & 9 \end{bmatrix}$

(c) $A^T = \begin{bmatrix} 5 & 1 & 6 & -9 \\ 7 & -7 & 3 & 2 \end{bmatrix}$

3. $f(x) = 2.6314x + 2.2714$
$[-1, 6, 1]$ by $[0, 18, 2]$

5. (a) HELP is coded as the word UBNL.
(b) LETTER is coded as the word CHHPBQ.

CHAPTER 10: Conic Sections

SECTION 10.1 (pp. 840–843)

1.

3.

5.

7.

9.

11. e   **13.** a   **15.** d
17. Vertex: $V(0, 0)$;
focus: $F(0, 4)$;
directrix: $y = -4$

19. Vertex: $V(0, 0)$;
focus: $F(2, 0)$;
directrix: $x = -2$

21. Vertex: $V(0, 0)$;
focus: $F(-1, 0)$;
directrix: $x = 1$

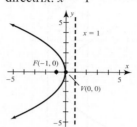

23. Vertex: $V(0, 0)$;
focus: $F(0, -2)$;
directrix: $y = 2$

25. Vertex: $V(0, 0)$;
focus: $F(-1, 0)$;
directrix: $x = 1$

27. $x^2 = 4y$

29. $y^2 = -12x$

31. $x^2 = 3y$ **33.** $y^2 = -8x$ **35.** $y^2 = 4x$
37. $y^2 = -x$ **39.** $y^2 = 4x$ **41.** $x^2 = -12y$
43. $y^2 = -4x$
45.

47.

49. c   **51.** a

53. Vertex: $V(2, -2)$; focus: $F(2, 0)$; directrix; $y = -4$

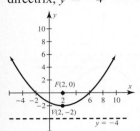

55. Vertex: $V(2, -3)$; focus: $F(1, -3)$; directrix: $x = 3$

57. Vertex: $V(-2, 0)$; focus: $F(-2, -1)$; directrix: $y = 1$

59. $x^2 = 4(y - 1)$

61. $y^2 = 4(x + 1)$

63. $(x + 1)^2 = -8(y - 5)$ **65.** $(y - 3)^2 = -\frac{9}{2}(x + 2)$

67. $(y - 0)^2 = -8\left(x + \frac{5}{4}\right)$ **69.** $(y + 1)^2 = \frac{1}{2}(x + 3)$

71. $\left(x - \frac{3}{2}\right)^2 = 2\left(y - \frac{7}{8}\right)$ **73.** $\left(y + \frac{1}{2}\right)^2 = \frac{5}{4}\left(x + \frac{6}{5}\right)$

75. $[-6, 6, 1]$ by $[-4, 4, 1]$ **77.** $[-9, 9, 1]$ by $[-6, 6, 1]$

Note: If a break in the graph appears near the vertex, it should not be there. It is a result of the low resolution of the graphing calculator screen.

79. $[-6, 6, 1]$ by $[-4, 4, 1]$

81. $\left(\pm\sqrt{2}, 1\right)$ **83.** $\left(-0.1, \pm\sqrt{0.9}\right)$

85. $(0, 0), (3, -1)$ **87.** $p = 3$ ft

89. (a) $y = \frac{32}{11,025}x^2$ **(b)** About 86.1 ft

91. (a) $(25, 0)$ **(b)** 25 million mi   **93.** $k = 6$

SECTION 10.2 (pp. 853–857)

1. Foci: $F\left(0, \pm\sqrt{5}\right)$; vertices: $V(0, \pm 3)$; endpoints of the minor axis: $U(\pm 2, 0)$

3. Foci: $F\left(\pm\sqrt{20}, 0\right)$; vertices: $V(\pm 6, 0)$; endpoints of the minor axis: $U(0, \pm 4)$

5. Foci: $F\left(\pm\sqrt{300}, 0\right)$; vertices: $V(\pm 20, 0)$; endpoints of the minor axis: $U(0, \pm 10)$

7. Foci: $F(0, \pm 4)$; vertices: $V(0, \pm 5)$; endpoints of the minor axis: $U(\pm 3, 0)$

9. b   **11.** c

13. $\frac{x^2}{36} + \frac{y^2}{16} = 1$; vertices: $V(\pm 6, 0)$; endpoints of the minor axis: $U(0, \pm 4)$; foci: $F\left(\pm\sqrt{20}, 0\right)$

15. $\frac{x^2}{4} + \frac{y^2}{16} = 1$; vertices: $V(0, \pm 4)$; endpoints of the minor axis: $U(\pm 2, 0)$; foci: $F\left(0, \pm\sqrt{12}\right)$

17. $\frac{x^2}{25} + \frac{y^2}{9} = 1$ **19.** $\frac{x^2}{5} + \frac{y^2}{9} = 1$

21. $\frac{x^2}{12} + \frac{y^2}{16} = 1$ **23.** $\frac{x^2}{36} + \frac{y^2}{11} = 1$

25. $\frac{x^2}{16} + \frac{y^2}{9} = 1$ **27.** $\frac{x^2}{9} + \frac{y^2}{5} = 1$

29. $\dfrac{(x-2)^2}{4} + \dfrac{(y+1)^2}{3} = 1$ **31.** $\dfrac{(x+3)^2}{2} + \dfrac{(y+4)^2}{9} = 1$

59. $\dfrac{\left(x-\frac{1}{2}\right)^2}{4} + \dfrac{\left(y+\frac{3}{2}\right)^2}{16} = 1$; center: $C\left(\frac{1}{2}, -\frac{3}{2}\right)$;
vertices: $V\left(\frac{1}{2}, \frac{5}{2}\right), V\left(\frac{1}{2}, -\frac{11}{2}\right)$

61. $[-6, 6, 1]$ by $[-4, 4, 1]$ **63.** $[-4.7, 4.7, 1]$ by $[-3.1, 3.1, 1]$

33.

35.

65. $(0, 3), \left(\frac{24}{13}, \frac{15}{13}\right)$

37.

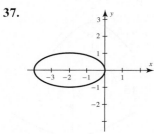

67. Four solutions:
$\left(\pm\sqrt{\frac{20}{3}}, \pm\sqrt{\frac{7}{3}}\right)$

69. $(3, 0), (-3, 0)$

39. d   **41.** c

43. Foci: $F(1, 1 \pm 4)$;
vertices: $V(1, 1 \pm 5)$

45. Foci: $F\left(-4 \pm \sqrt{7}, 2\right)$;
vertices: $V(-4 \pm 4, 2)$

71. $(0, 2)$ **73.** Four solutions: $\left(\pm\sqrt{\frac{4}{3}}, \pm\sqrt{\frac{4}{3}}\right)$

75. $\left(1, \sqrt{8}\right), \left(1, -\sqrt{8}\right)$

77.

79.

47. $\dfrac{(x-2)^2}{5} + \dfrac{(y-1)^2}{9} = 1$ **49.** $\dfrac{x^2}{9} + \dfrac{(y-2)^2}{5} = 1$

51. $\dfrac{(x-2)^2}{16} + \dfrac{(y-4)^2}{4} = 1$

53. $\dfrac{(x+1)^2}{4} + \dfrac{(y-1)^2}{9} = 1$; center: $C(-1, 1)$;
vertices: $V(-1, -2), V(-1, 4)$

55. $\dfrac{(x+1)^2}{1} + \dfrac{(y+1)^2}{4} = 1$; center: $C(-1, -1)$;
vertices: $V(-1, -3), V(-1, 1)$

57. $\dfrac{(x+2)^2}{5} + \dfrac{(y-1)^2}{4} = 1$; center: $C(-2, 1)$;
vertices: $V\left(-2 - \sqrt{5}, 1\right), V\left(-2 + \sqrt{5}, 1\right)$

81.

83.

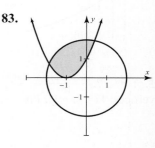

85. $A = 6\pi \approx 18.85 \text{ ft}^2$

87. $A = 20\pi \approx 62.83 \text{ ft}^2$

89. $\dfrac{x^2}{0.387^2} + \dfrac{y^2}{0.379^2} = 1$; sun: $(0.0797, 0)$
$[-0.6, 0.6, 0.1]$ by $[-0.4, 0.4, 0.1]$

91. 6.245 in.   **93.** $A = \pi(64.03)(40) \approx 8046.25 \text{ ft}^2$
95. 348.2 ft   **97.** About 21.65 ft
99. Maximum: 668 mi; minimum: 340 mi

**10.2 EXTENDED AND DISCOVERY EXERCISES
(p. 858)**

1. $\dfrac{x}{2} + \dfrac{y}{3} = 1$; x-int: 2, y-int: 3

3. $\dfrac{x}{2.5} + \dfrac{y}{5} = 1$; x-int: 2.5, y-int: 5

5. x-int: ± 5, y-int: ± 3

**CHECKING BASIC CONCEPTS FOR SECTIONS 10.1
AND 10.2 (p. 858)**

1. Focus: $F\left(\frac{1}{2}, 0\right)$;
directrix: $x = -\frac{1}{2}$

3. Foci: $F(0, \pm 8)$;
vertices: $V(0, \pm 10)$;
endpoints of the
minor axis: $U(\pm 6, 0)$

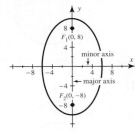

5. 1 ft   **7.** $\dfrac{(x-2)^2}{16} + \dfrac{(y+1)^2}{4} = 1$; center: $C(2, -1)$;
vertices: $V(-2, -1)$, $V(6, -1)$

SECTION 10.3 (pp. 865–867)

1. Asymptotes: $y = \pm \frac{7}{3}x$;
$F\left(\pm \sqrt{58}, 0\right)$; $V(\pm 3, 0)$

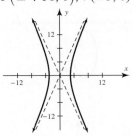

3. Asymptotes: $y = \pm \frac{3}{2}x$;
$F\left(0, \pm \sqrt{52}\right)$; $V(0, \pm 6)$

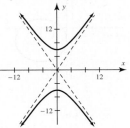

5. Asymptotes: $y = \pm x$;
$F\left(\pm \sqrt{18}, 0\right)$; $V(\pm 3, 0)$

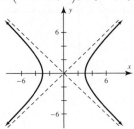

7. Asymptotes: $y = \pm \frac{4}{3}x$;
$F(0, \pm 5)$; $V(0, \pm 4)$

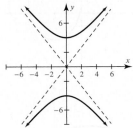

9. d   **11.** a

13. $\dfrac{x^2}{16} - \dfrac{y^2}{9} = 1$

15. $\dfrac{y^2}{36} - \dfrac{x^2}{64} = 1$

17. $\dfrac{y^2}{144} - \dfrac{x^2}{25} = 1$; asymptotes: $y = \pm \frac{12}{5}x$

19. $\dfrac{y^2}{4} - \dfrac{x^2}{21} = 1$; asymptotes: $y = \pm \dfrac{2}{\sqrt{21}}x$

21. $\dfrac{x^2}{9} - \dfrac{y^2}{4} = 1$; asymptotes: $y = \pm \frac{2}{3}x$

23. $\dfrac{x^2}{16} - \dfrac{y^2}{9} = 1$; asymptotes: $y = \pm \frac{3}{4}x$

25. $\dfrac{x^2}{10} - \dfrac{y^2}{9} = 1$; asymptotes: $y = \pm \dfrac{3}{\sqrt{10}}x$

27. Vertices: $V(1 \pm 4, 2)$; foci: $F\left(1 \pm \sqrt{20}, 2\right)$;
asymptotes: $y = \pm \frac{1}{2}(x - 1) + 2$

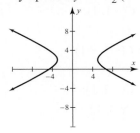

29. Vertices: $V(-2, 2 \pm 6)$; foci: $F\left(-2, 2 \pm \sqrt{40}\right)$;
asymptotes: $y = \pm 3(x + 2) + 2$

31. Vertices: $V(\pm 2, 1)$; foci: $F\left(\pm \sqrt{5}, 1\right)$;
asymptotes: $y = \pm \frac{1}{2}x + 1$

33. b   **35.** c
37. $\frac{(y + 4)^2}{16} - \frac{(x - 4)^2}{4} = 1$; vertices: $V(4, -4 \pm 4)$;
foci: $F\left(4, -4 \pm \sqrt{20}\right)$; asymptotes: $y = \pm 2(x - 4) - 4$
39. Vertices: $V(1 \pm 2, 1)$; foci: $F\left(1 \pm \sqrt{8}, 1\right)$;
asymptotes: $y = \pm(x - 1) + 1$

41. Vertices: $V(1, -1 \pm 4)$; foci: $F(1, -1 \pm 5)$;
asymptotes: $y = \pm \frac{4}{3}(x - 1) - 1$

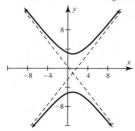

43. $(x - 2)^2 - \frac{(y + 2)^2}{3} = 1$ **45.** $y^2 - \frac{(x + 1)^2}{8} = 1$

47. $\frac{(x - 1)^2}{4} - \frac{(y - 1)^2}{4} = 1$; center: $C(1, 1)$;
vertices: $V(-1, 1), V(3, 1)$
49. $\frac{(y + 4)^2}{2} - \frac{(x - 3)^2}{3} = 1$; center: $C(3, -4)$;
vertices: $V\left(3, -4 - \sqrt{2}\right), V\left(3, -4 + \sqrt{2}\right)$
51. $\frac{(x - 3)^2}{2} - \frac{(y - 0)^2}{1} = 1$; center: $C(3, 0)$;
vertices: $V\left(3 - \sqrt{2}, 0\right), V\left(3 + \sqrt{2}, 0\right)$
53. $\frac{(y + 4)^2}{5} - \frac{(x + 1)^2}{4} = 1$; center: $C(-1, -4)$;
vertices: $V\left(-1, -4 - \sqrt{5}\right), V\left(-1, -4 + \sqrt{5}\right)$
55. $[-15, 15, 5]$ by $[-10, 10, 5]$ **57.** $[-9, 9, 1]$ by $[-6, 6, 1]$

59. Four solutions: $\left(\pm \sqrt{\frac{13}{2}}, \pm \sqrt{\frac{5}{2}}\right)$

61. $(2, 0), (-5.2, 7.2)$ **63.** Four solutions: $\left(\pm 2, \pm \frac{2}{\sqrt{3}}\right)$
65. $\left(\frac{2}{\sqrt{11}}, \frac{6}{\sqrt{11}}\right), \left(-\frac{2}{\sqrt{11}}, -\frac{6}{\sqrt{11}}\right)$
67. (a) Elliptic   **(b)** Its speed should be 4326 m/sec or
greater.   **(c)** If D is larger, then $\frac{k}{\sqrt{D}}$ is smaller, so smaller
values for V satisfy $V > \frac{k}{\sqrt{D}}$.

10.3 EXTENDED AND DISCOVERY EXERCISES (pp. 867–868)
1. (a) $x = \sqrt{y^2 + 2.5 \times 10^{-27}}$; this equation represents
the right half of the hyperbola.   **(b)** About 1.2×10^{-13} m

CHECKING BASIC CONCEPTS FOR SECTION 10.3 (p. 868)
1. $\frac{x^2}{16} - \frac{y^2}{9} = 1$
3. $\frac{(x - 1)^2}{9} - \frac{(y - 3)^2}{4} = 1; F\left(1 \pm \sqrt{13}, 3\right)$

CHAPTER 10 REVIEW EXERCISES (pp. 870–872)
1.

3.

5.

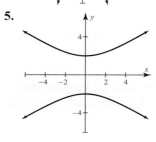

7. d   **9.** a   **11.** e

13. $y^2 = 8x$ **15.** $\frac{x^2}{25} + \frac{y^2}{9} = 1$ **17.** $\frac{y^2}{64} - \frac{x^2}{36} = 1$

19. $F(0, -1)$ **21.** $F\left(\pm\sqrt{21}, 0\right)$

23. $F(\pm 5, 0)$ **25.** Both foci are located at $(3, -1)$.

27. Center: $C(1, -1)$

29.

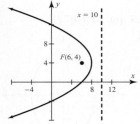

31. $[-5, 5, 1]$ by $[-5, 5, 1]$

Note: If breaks in the graph appear near the vertices, they should not be there. It is a result of the low resolution of the graphing calculator screen.

33. $(y - 0)^2 = -10\left(x + \frac{7}{5}\right)$

35. $\frac{(x + 1)^2}{25} + \frac{(y - 5)^2}{4} = 1$; center: $C(-1, 5)$;
vertices: $V(-6, 5), V(4, 5)$

37. $\frac{(x + 2)^2}{4} - \frac{(y - 3)^2}{1} = 1$; center: $C(-2, 3)$;
vertices: $V(-4, 3), V(0, 3)$

39. Four solutions: $\left(\pm\sqrt{\frac{8}{3}}, \pm\sqrt{\frac{4}{3}}\right)$

41.

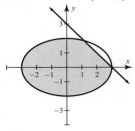

43. (a) Minimum: 4.92 million mi; maximum:
995.08 million mi   **(b)** $2\pi\sqrt{\frac{500^2 + 70^2}{2}} \approx 2243$ million
miles, or 2.243 billion miles   **45.** About 29.05 ft

**CHAPTER 10 EXTENDED AND DISCOVERY EXERCISES
(p. 872)**

1. Neptune: 0.271; Pluto: 9.82

3. Neptune: $\frac{(x - 0.271)^2}{30.10^2} + \frac{y^2}{30.10^2} = 1$;

Pluto: $\frac{(x - 9.82)^2}{39.44^2} + \frac{y^2}{38.20^2} = 1$

5. No. Because Pluto's orbit is so eccentric, there is a period of time when Pluto is not farther from the sun. However, its average distance a from the sun is greater.

CHAPTER 11: Further Topics in Algebra

SECTION 11.1 (pp. 883–886)

1. $a_1 = 3, a_2 = 5, a_3 = 7, a_4 = 9$

3. $a_1 = 4, a_2 = -8, a_3 = 16, a_4 = -32$

5. $a_1 = \frac{1}{2}, a_2 = \frac{2}{5}, a_3 = \frac{3}{10}, a_4 = \frac{4}{17}$

7. $a_1 = -\frac{1}{2}, a_2 = \frac{1}{4}, a_3 = -\frac{1}{8}, a_4 = \frac{1}{16}$

9. $a_1 = \frac{2}{3}, a_2 = -\frac{4}{5}, a_3 = \frac{8}{9}, a_4 = -\frac{16}{17}$

11. $a_1 = 3, a_2 = 8, a_3 = 17, a_4 = 32$

13. 2, 4, 3, 5, 3, 6, 4

15. (a) $a_1 = 1, a_2 = 2, a_3 = 4, a_4 = 8$
(b) $[0, 5, 1]$ by $[0, 9, 1]$

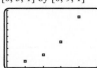

17. (a) $a_1 = -3, a_2 = 0, a_3 = 3, a_4 = 6$
(b) $[0, 5, 1]$ by $[-4, 7, 1]$

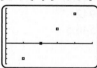

19. (a) $a_1 = 2, a_2 = 5, a_3 = 14, a_4 = 41$
(b) $[0, 5, 1]$ by $[0, 45, 5]$

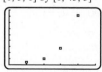

21. (a) $a_1 = 2, a_2 = 5, a_3 = 3, a_4 = -2$
(b) $[0, 5, 1]$ by $[-3, 6, 1]$

23. (a) $a_1 = 2, a_2 = 4, a_3 = 16, a_4 = 256$
(b) $[0, 5, 1]$ by $[0, 300, 50]$

25. (a) $a_1 = 1, a_2 = 3, a_3 = 6, a_4 = 10$
(b) $[0, 5, 1]$ by $[0, 12, 1]$

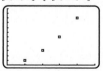

27. (a) $a_1 = 2, a_2 = 3, a_3 = 6, a_4 = 18$
(b) $[0, 5, 1]$ by $[0, 20, 2]$

29. (a)

n	1	2	3	4	5	6	7	8
a_n	1	3	5	7	9	11	13	15

(b) $[0, 10, 1]$ by $[0, 16, 1]$　**(c)** $a_n = 2n - 1$

31. (a)

n	1	2	3	4	5	6	7	8
a_n	7.5	6	4.5	3	1.5	0	-1.5	-3

(b) $[0, 12, 1]$ by $[-4, 8, 1]$　**(c)** $a_n = -1.5n + 9$

33. (a)

n	1	2	3	4	5	6	7	8
a_n	$\frac{1}{2}$	2	$\frac{7}{2}$	5	$\frac{13}{2}$	8	$\frac{19}{2}$	11

(b) $[0, 9, 1]$ by $[0, 12, 1]$　**(c)** $a_n = \frac{3}{2}n - 1$

35. (a)

n	1	2	3	4	5	6	7	8
a_n	8	4	2	1	$\frac{1}{2}$	$\frac{1}{4}$	$\frac{1}{8}$	$\frac{1}{16}$

(b) $[0, 10, 1]$ by $[-1, 9, 1]$　**(c)** $a_n = 8\left(\frac{1}{2}\right)^{n-1}$

37. (a)

n	1	2	3	4	5	6	7	8
a_n	$\frac{3}{4}$	$\frac{3}{2}$	3	6	12	24	48	96

(b) $[0, 10, 1]$ by $[-10, 110, 10]$　**(c)** $a_n = \frac{3}{4}(2)^{n-1}$

39. (a)

n	1	2	3	4	5	6	7	8
a_n	$-\frac{1}{4}$	$-\frac{1}{2}$	-1	-2	-4	-8	-16	-32

(b) $[0, 9, 1]$ by $[-36, 4, 4]$　**(c)** $a_n = -\frac{1}{4}(2)^{n-1}$

41. $a_n = -2n + 7$　　**43.** $a_n = 3n - 8$
45. $a_n = 2n + 1$　　**47.** $a_n = 3n + 5$
49. $a_n = 0.5n - 6.5$　　**51.** $a_n = 2\left(\frac{1}{2}\right)^{n-1}$
53. $a_n = \frac{1}{2}\left(-\frac{1}{4}\right)^{n-1}$　　**55.** $a_n = 8\left(\frac{1}{2}\right)^{n-1}$
57. $a_n = -5(-5)^{n-1}$　　**59.** $a_n = -\frac{1}{2}(2)^{n-1}$
61. No　　**63.** Yes　　**65.** Yes　　**67.** Yes
69. Yes　　**71.** No　　**73.** No　　**75.** No
77. Arithmetic　　**79.** Geometric　　**81.** Neither
83. Arithmetic, $d < 0, d = -1$
85. Geometric, $r < 0, |r| < 1$
87. The insect population density increases rapidly and then levels off near 5000 per acre.

89. (a) $a_n = 0.8a_{n-1}, a_1 = 500$ **(b)** $a_1 = 500$,
$a_2 = 400, a_3 = 320, a_4 = 256, a_5 = 204.8$, and
$a_6 = 163.84$. The population density decreases by 20%
each year.   **(c)** $a_n = 500(0.8)^{n-1}$
91. (a) $a_1 = 8, a_2 = 10.4, a_3 = 8.528$
(b) [0, 21, 1] by [0, 14, 1]

The population density oscillates above and below approximately 9.5.
93. (a) 1, 1, 2, 3, 5, 8, 13, 21, 34, 55, 89, 144
(b) $\frac{a_2}{a_1} = 1, \frac{a_3}{a_2} = 2, \frac{a_4}{a_3} = 1.5, \frac{a_5}{a_4} = \frac{5}{3} \approx 1.6667$,
$\frac{a_6}{a_5} = \frac{8}{5} = 1.6, \frac{a_7}{a_6} = \frac{13}{8} = 1.625, \frac{a_8}{a_7} = \frac{21}{13} \approx 1.6154$,
$\frac{a_9}{a_8} = \frac{34}{21} \approx 1.6190, \frac{a_{10}}{a_9} = \frac{55}{34} \approx 1.6176$,
$\frac{a_{11}}{a_{10}} = \frac{89}{55} \approx 1.6182$, and $\frac{a_{12}}{a_{11}} = \frac{144}{89} \approx 1.6180$.
The ratio appears to approach a number near 1.618.
(c) $n = 2: a_1 \cdot a_3 - a_2^2 = (1)(2) - (1)^2 = 1 = (-1)^2$
$n = 3: a_2 \cdot a_4 - a_3^2 = (1)(3) - (2)^2 = -1 = (-1)^3$
$n = 4: a_3 \cdot a_5 - a_4^2 = (2)(5) - (3)^2 = 1 = (-1)^4$
95. (a) $a_n = 2000n + 28,000$, or
$a_n = 30,000 + 2000(n - 1)$; arithmetic
(b) $b_n = 30,000(1.05)^{n-1}$; geometric
(c) Since $a_{10} = \$48,000 > b_{10} \approx \$46,540$, the first salary
is higher after 10 years. Since $a_{20} = \$68,000 < b_{20}$
$\approx \$75,809$, the second salary is higher after 20 years.
(d) With time, the geometric sequence with $r > 1$
overtakes the arithmetic sequence.
[0, 30, 10] by [0, 150000, 50000]

97. $a_6 \approx 1.414213562, \sqrt{2} \approx 1.414213562$
99. $a_6 = 4.582581971, \sqrt{21} \approx 4.582575695$
101. By definition,
$a_n = a_1 + (n - 1)d_1$ and $b_n = b_1 + (n - 1)d_2$.
Then $c_n = a_n + b_n$
$\qquad = [a_1 + (n - 1)d_1] + [b_1 + (n - 1)d_2]$
$\qquad = (a_1 + b_1) + [(n - 1)d_1 + (n - 1)d_2]$
$\qquad = (a_1 + b_1) + (n - 1)(d_1 + d_2)$
$\qquad = c_1 + (n - 1)d$,
where $c_1 = a_1 + b_1$ and $d = d_1 + d_2$.

SECTION 11.2 (pp. 898–900)

1. 2, 4, 6, 8, 10, 12
3. 2 + 4 + 6 + 8 + 10 + 12; 42
5. $A_5 + A_6 + A_7 + A_8 + A_9$

7. 45   **9.** 25   **11.** 60   **13.** $\frac{71}{20}$
15. 80   **17.** 1275   **19.** 1739   **21.** 4100
23. 31   **25.** 460   **27.** 105   **29.** 1942
31. 948   **33.** 545   **35.** 255   **37.** 546.5
39. 3,145,725
41. 0.625; 0.671875; 0.666015625
43. 5; 42.333333333; 341
45. $\frac{3}{2}$ **47.** $\frac{18}{5}$ **49.** $\frac{10}{11}$
51. $\frac{2}{3} = 0.6 + 0.06 + 0.006 + 0.0006 + \cdots$
53. $\frac{9}{11} = 0.81 + 0.0081 + 0.000081 + \cdots$
55. $\frac{1}{7} = 0.142857 + 0.000000142857 + \cdots$
57. $\frac{8}{9}$ **59.** $\frac{5}{11}$ **61.** 2 + 3 + 4 + 5 = 14
63. 4 + 4 + 4 + 4 + 4 + 4 + 4 + 4 = 32
65. 1 + 8 + 27 + 64 + 125 + 216 + 343 = 784
67. 12 + 20 = 32
69. $\sum_{k=1}^{6} k^4$ **71.** $\sum_{k=1}^{7} \left(\frac{2k}{k+1}\right)$
73. $\sum_{k=1}^{\infty} \left(\frac{1}{k^2}\right)$ **75.** $\sum_{n=1}^{4} (n+5)^3$ **77.** $\sum_{n=1}^{24} (3n+22)$
79. $\sum_{n=1}^{37} (n^2 + 27n + 180)$
81. 540   **83.** 600   **85.** 1395   **87.** 5525
89. 1360   **91.** 290
93. $\sum_{k=1}^{n} k = 1 + 2 + 3 + \cdots + n$ is an arithmetic series
with $a_1 = 1$ and $a_n = n$. Its sum equals
$S_n = n\left(\frac{a_1 + a_n}{2}\right) = n\left(\frac{1 + n}{2}\right) = \frac{n(n + 1)}{2}$.
95. (a) & (b) \$819,000
97. \$91,523.93   **99.** \$62,278.01
101. $S_9 = 9\left(\frac{7 + 15}{2}\right) = 99$ logs
103. (a) $\sum_{k=1}^{n} 0.5(0.5)^{k-1}$ **(b)** Infinitely many filters
105. 2
107. $1 + 1 + \frac{1}{2} + \frac{1}{6} + \frac{1}{24} + \frac{1}{120} + \frac{1}{720} + \frac{1}{5040}$
$\approx 2.718254; e \approx 2.718282$
109. $S_2 = \frac{4}{3} \approx 1.3333, S_4 = \frac{40}{27} \approx 1.4815, S_8 \approx 1.49977$,
$S_{16} \approx 1.49999997; S = 1.5$. As n increases, the partial
sums approach S.
111. $S_1 = 4, S_2 = 3.6, S_3 = 3.64, S_4 = 3.636$,
$S_5 = 3.6364, S_6 = 3.63636; S = \frac{40}{11} = 3.\overline{63}$.
As n increases, the partial sums approach S.

**CHECKING BASIC CONCEPTS FOR SECTIONS 11.1
AND 11.2 (p. 901)**

1. [0, 7, 1] by [−10, 2, 1]

n	u(n)
1	1
2	-1
3	-3
4	-5
5	-7
6	-9
7	-11
u(n) = -2n+3	

3. (a) Arithmetic; $S_{10} = 190$
(b) Geometric; $S_6 = \frac{364}{81} \approx 4.494$
(c) Geometric; $S = \frac{8}{3} \approx 2.667$ **(d)** Geometric; $S = 1$
5. (a) 150   **(b)** 6622

SECTION 11.3 (pp. 909–911)

1. $2^{10} = 1024$
3. $2^5 4^{10} = 33{,}554{,}432$
5. $10^3 \cdot 26^3 = 17{,}576{,}000$
7. $26^3 \cdot 36^3 = 820{,}025{,}856$
9. $3^5 = 243$ **11.** $5^5 = 3125$ **13.** 2   **15.** 24
17. 1,000,000   **19.** $2^{12} = 4096$
21. No; there are 35,152 call letters possible.
23. 24   **25.** 8,000,000   **27.** 720   **29.** 3,628,800
31. 60   **33.** 8   **35.** 210   **37.** 600   **39.** 5040
41. 24   **43.** 2730   **45.** 210   **47.** 30,000
49. 360   **51.** 362,880   **53.** About 6.39×10^{12}
55. 6,400,000,000   **57.** 3   **59.** 20   **61.** 1
63. 28   **65.** 190   **67.** 575,757   **69.** 30
71. 50   **73.** 7920   **75.** 2024
77. $P(n, n - 1) = \frac{n!}{(n - (n - 1))!} = \frac{n!}{1} = n!$ and
$P(n, n) = \frac{n!}{(n - n)!} = \frac{n!}{0!} = \frac{n!}{1} = n!$. For example,
$P(7, 6) = 5040 = P(7, 7)$.

SECTION 11.4 (pp. 915–916)

1. 5   **3.** 1   **5.** 6   **7.** 1   **9.** 10   **11.** 70
13. 1   **15.** 5   **17.** $x^2 + 2xy + y^2$
19. $m^3 + 6m^2 + 12m + 8$
21. $8x^3 - 36x^2 + 54x - 27$
23. $p^6 - 6p^5q + 15p^4q^2 - 20p^3q^3 + 15p^2q^4 - 6pq^5 + q^6$
25. $8m^3 + 36m^2n + 54mn^2 + 27n^3$
27. $1 - 4x^2 + 6x^4 - 4x^6 + x^8$
29. $8p^9 - 36p^6 + 54p^3 - 27$
31. $x^2 + 2xy + y^2$
33. $81x^4 + 108x^3 + 54x^2 + 12x + 1$
35. $32 - 80x + 80x^2 - 40x^3 + 10x^4 - x^5$
37. $x^8 + 8x^6 + 24x^4 + 32x^2 + 16$
39. $256x^4 - 768x^3y + 864x^2y^2 - 432xy^3 + 81y^4$
41. $m^6 + 6m^5n + 15m^4n^2 + 20m^3n^3 + 15m^2n^4 + 6mn^5 + n^6$
43. $8x^9 - 12x^6y^2 + 6x^3y^4 - y^6$
45. $84a^6b^3$ **47.** $70x^4y^4$ **49.** $40x^2y^3$ **51.** $-576xy^5$

CHECKING BASIC CONCEPTS FOR SECTIONS 11.3 AND 11.4 (p. 916)

1. $2^8 = 256$ **3.** $26 \cdot 36^5 = 1{,}572{,}120{,}576$

SECTION 11.5 (pp. 922–923)

1. $3 + 6 + 9 + \cdots + 3n = \frac{3n(n + 1)}{2}$
(i) Show that the statement is true for $n = 1$:
$3(1) = \frac{3(1)(2)}{2}$
$3 = 3$
(ii) Assume that S_k is true:
$3 + 6 + 9 + \cdots + 3k = \frac{3k(k + 1)}{2}$
Show that S_{k+1} is true:
$3 + 6 + \cdots + 3(k + 1) = \frac{3(k + 1)(k + 2)}{2}$
Add $3(k + 1)$ to each side of S_k:
$3 + 6 + 9 + \cdots + 3k + 3(k + 1)$
$= \frac{3k(k + 1)}{2} + 3(k + 1)$
$= \frac{3k(k + 1) + 6(k + 1)}{2}$
$= \frac{(k + 1)(3k + 6)}{2}$
$= \frac{3(k + 1)(k + 2)}{2}$
Since S_k implies S_{k+1}, the statement is true for every positive integer n.
15. 1, 2   **17.** 2, 3, 4
19. $(a^m)^n = a^{mn}$
(i) Show that the statement is true for $n = 1$:
$(a^m)^1 = a^{m \cdot 1}$
$a^m = a^m$
(ii) Assume that S_k is true:
$(a^m)^k = a^{mk}$
Show that S_{k+1} is true:
$(a^m)^{k+1} = a^{m(k+1)}$
Multiply each side of S_k by a^m:
$(a^m)^k \cdot (a^m)^1 = a^{mk} \cdot a^m$
$(a^m)^{k+1} = a^{mk+m}$
$(a^m)^{k+1} = a^{m(k+1)}$
Since S_k implies S_{k+1}, the statement is true for every positive integer n.
31. $P = 3\left(\frac{4}{3}\right)^{n-1}$ **33.** $2^n - 1$

SECTION 11.6 (pp. 932–935)

1. Yes   **3.** No   **5.** Yes   **7.** No   **9.** $\frac{1}{2}$ **11.** $\frac{1}{6}$
13. $\frac{1}{2}$ **15.** $\frac{4}{52} = \frac{1}{13}$ **17.** $\frac{1}{10{,}000}$
19. (a) 0.57, or 57%   **(b)** 0.33, or 33%
21. $\frac{1}{4}$ **23.** $\frac{1}{27}$ **25.** $\frac{1}{36}$
27. $\frac{625}{1296} \approx 0.482$ **29.** $\frac{1}{270{,}725}$
31. $\dfrac{\binom{13}{3} \cdot \binom{13}{2}}{\binom{52}{5}} \approx 0.0086$, or a 0.86% chance
33. $\frac{4}{20} = 0.2$

35. (a)

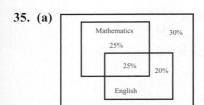

(b) 0.7, or 70%   **(c)** Let M denote the event of needing help with math and E the event of needing help with English. Then $P(M \cup E) = P(M) + P(E) - P(M \cap E)$ $= 0.5 + 0.45 - 0.25 = 0.7.$

37. (a) $\frac{6095}{119,923} \approx 0.051$ **(b)** $\frac{106,232}{119,923} \approx 0.886$

39. $\frac{94}{100,000} = 0.00094$

41. (a) $\frac{158,502}{984,155} \approx 0.161$ **(b)** $\frac{825,653}{984,155} \approx 0.839$

(c) $\frac{79,005}{984,155} \approx 0.080$

43. $\frac{9}{36} = \frac{1}{4}$ **45.** $\frac{12}{52} = \frac{3}{13}$ **47. (a)** 0.7   **(b)** 0.09

49. (a) 0.09   **(b)** 0.12

51. $\frac{1}{1000}$

53. (a) $\frac{22}{50} = 0.44$ **(b)** $\frac{28}{50} = 0.56$ **(c)** $\frac{28}{50} = 0.56$

55. $\frac{3}{51} = \frac{1}{17}$ **57.** $\frac{1}{3}$ **59.** $\frac{19}{27}$ **61.** 40%   **63.** $\frac{1}{3}$

65. (a) $\frac{18}{235}$ **(b)** $\frac{7}{18}$ **(c)** $\frac{7}{235}$

67. (a) $\frac{8}{15}$ **(b)** $\frac{7}{15}$ **(c)** $\frac{2}{5}$ **(d)** $\frac{1}{3}$ **(e)** $\frac{1}{15}$

CHECKING BASIC CONCEPTS FOR SECTIONS 11.5 AND 11.6 (p. 935)

1. $4 + 8 + 12 + \cdots + 4n = 2n(n + 1)$
(i) Show that the statement is true for $n = 1$:
$4(1) = 2(1)(1 + 1)$
$4 = 4$
(ii) Assume that S_k is true:
$4 + 8 + 12 + \cdots + 4k = 2k(k + 1)$
Show that S_{k+1} is true:
$4 + 8 + \cdots + 4(k + 1) = 2(k + 1)(k + 2)$
Add $4(k + 1)$ to each side of S_k:
$4 + 8 + 12 + \cdots + 4k + 4(k + 1)$
$= 2k(k + 1) + 4(k + 1)$
$= 2k^2 + 6k + 4$
$= 2(k + 1)(k + 2)$
Since S_k implies S_{k+1}, the statement is true for every positive integer n.

3. $\frac{1}{16}$ **5.** $\dfrac{\binom{4}{4} \cdot \binom{4}{1}}{\binom{52}{5}} = \dfrac{4}{2,598,960} \approx 0.0000015$

CHAPTER 11: REVIEW EXERCISES (pp. 939–940)

1. $-1, -4, -7, -10$ **3.** $0, 1, 3, 7$
5. $5, 3, 1, 2, 4, 6$

7. (a)

n	1	2	3	4	5	6	7	8
a_n	3	1	-1	-3	-5	-7	-9	-11

(b) $[0, 10, 1]$ by $[-12, 4, 1]$

(c) $a_n = -2n + 5$
9. $a_n = 4n - 15$ **11.** Arithmetic   **13.** Geometric
15. 65   **17.** 90   **19.** 3280   **21.** 3
23. $6 + 11 + 16 + 21 + 26$

25. $\sum_{k=1}^{6} k^3$ **27.** -1155

29. $0.18 + 0.0018 + 0.000018 + \cdots$
31. 120
33. $1 + 3 + 5 + \cdots + (2n - 1) = n^2$
(i) Show that the statement is true for $n = 1$:
$2(1) - 1 = 1^2$
$1 = 1$
(ii) Assume that S_k is true:
$1 + 3 + 5 + \cdots + (2k - 1) = k^2$
Show that S_{k+1} is true:
$1 + 3 + \cdots + (2(k + 1) - 1) = (k + 1)^2$
Add $2k + 1$ to each side of S_k:
$1 + 3 + 5 + \cdots + (2k - 1) + (2k + 1)$
$= k^2 + 2k + 1$
$= (k + 1)^2$
Since S_k implies S_{k+1}, the statement is true for every positive integer n.

35. $\frac{1}{2}$ **37.** 120   **39.** $4^{20} \approx 1.1 \times 10^{12}$ **41.** 6,250,000
43. (a) 4, 3.6, 3.24, 2.916, 2.6244; geometric
(b) $[0, 6, 1]$ by $[0, 6, 1]$ **(c)** $a_n = 4(0.9)^{n-1}$

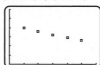

45. 20   **47.** 210   **49.** $\frac{91}{120} \approx 0.758$

51. (a) $\frac{27}{60} = 0.45$ **(b)** $\frac{33}{60} = 0.55$ **(c)** $\frac{13}{60} \approx 0.217$
53. The population density grows slowly initially, then increases rapidly, and finally levels off near 4,000,000 per acre.
$[0, 16, 1]$ by $[0, 5000, 1000]$

CHAPTER 11 EXTENDED AND DISCOVERY EXERCISES (pp. 940–941)

1. (a) $P = \begin{bmatrix} 1 & 0 & 0 \\ \frac{1}{6} & \frac{2}{3} & \frac{1}{6} \\ 0 & 0 & 1 \end{bmatrix}$

(b) The greatest probabilities lie on the main diagonal: $1, \frac{2}{3}, 1$; this means that a mother cell is most likely to produce a daughter cell like itself (answers will vary).

3. The quantity $\frac{a_1 + a_n}{2}$ represents not only the average of a_1 and a_n, but also the average of the terms $a_1, a_2, a_3, \ldots, a_n$. This is true whether n is odd or even. The total sum is equal to n times the average of the terms.

CHAPTERS 1–11 CUMULATIVE REVIEW EXERCISES (pp. 941–946)

1. 3.45×10^4; 0.000152 **3.** $\sqrt{41}$

5. (a) 2; $\sqrt{-a}$ **(b)** $D = \{x \mid x \le 1\}$, or $(-\infty, 1]$

7. 7 **9.** $y = -\frac{6}{5}x - \frac{8}{5}$

11. (a) $\frac{3}{4}; -1; \frac{4}{3}$ **(b)** $f(x) = \frac{3}{4}x - 1$ **(c)** $\frac{4}{3}$

13. (a) $\frac{17}{10}$ **(b)** $-\frac{1}{3}, \frac{5}{2}$ **(c)** $\frac{1 \pm \sqrt{13}}{2}$ **(d)** $-2, -1, 2$

(e) $\pm\sqrt{3}, \pm 1$ **(f)** $\frac{17}{3}$ **(g)** $\frac{\ln(28/3)}{2} \approx 1.117$

(h) $10^{3/2} - 1 \approx 30.623$ **(i)** 6 **(j)** $-\frac{4}{3}, 2$

15. f is continuous.

17. (a) $-2, -1, 1, 2$

(b) $\{x \mid x < -2 \text{ or } -1 < x < 1 \text{ or } x > 2\}$, or $(-\infty, -2) \cup (-1, 1) \cup (2, \infty)$

(c) $\{x \mid -2 \le x \le -1 \text{ or } 1 \le x \le 2\}$, or $[-2, -1] \cup [1, 2]$

19. $\left(\frac{3}{2}, \frac{31}{4}\right)$

21. (a) Incr: $\{x \mid x \le -2 \text{ or } x \ge 1\}$, or $(-\infty, -2] \cup [1, \infty)$; decr: $\{x \mid -2 \le x \le 1\}$, or $[-2, 1]$

(b) $-3.3, 0, 1.8$ **(c)** $(-2, 2), (1, -0.7)$

(d) Local minimum: -0.7, local maximum: 2

23. (a) $3x^2 - 1 + \frac{1}{2x^2}$

(b) $2x^3 - 5x^2 + 5x - 6 + \frac{8}{x+1}$

25. $f(x) = 3(x + 1)(x - 3i)(x + 3i)$; $f(x) = 3x^3 + 3x^2 + 27x + 27$

27. $-1 \pm 2i$ **29.** $(x + 1)^{3/5}$; 8

31. (a) 3 **(b)** -1 **(c)** 1 **(d)** -4

33. $f^{-1}(x) = -\frac{x}{x-1}$ **35.** $f(x) = 2(3)^x$

37. $\$1221.61$

39. (a) $D = \{x \mid -\infty < x < \infty\}$, or $(-\infty, \infty)$; $R = \{x \mid x \ge 0\}$, or $[0, \infty)$

(b) $D = \{x \mid -\infty < x < \infty\}$, or $(-\infty, \infty)$; $R = \{x \mid x > 0\}$, or $(0, \infty)$

(c) $D = \{x \mid x > 0\}$, or $(0, \infty)$; $R = \{x \mid -\infty < x < \infty\}$, or $(-\infty, \infty)$

(d) $D = \{x \mid x \ne 0\}$, or $(-\infty, 0) \cup (0, \infty)$; $R = \{x \mid x \ne 0\}$, or $(-\infty, 0) \cup (0, \infty)$

41. $\log \dfrac{x^2 y^3}{\sqrt[3]{z}}$

43. (a)

(b)

(c)

(d)

45. (a) $\frac{5\pi}{6} + 2\pi n, \frac{7\pi}{6} + 2\pi n$ **(b)** $\frac{\pi}{4} + \frac{\pi}{2}n$

(c) $\pi n, \frac{7\pi}{6} + 2\pi n, \frac{11\pi}{6} + 2\pi n$ **(d)** $\frac{\pi}{2} + \pi n$

47. $5°32'24''$

49. $225°$

51. $\sin\theta = \frac{24}{25}, \cos\theta = -\frac{7}{25}, \tan\theta = -\frac{24}{7},$ $\csc\theta = \frac{25}{24}, \sec\theta = -\frac{25}{7}, \cot\theta = -\frac{7}{24}$

53. $\sin\theta = -\frac{60}{61}, \cos\theta = -\frac{11}{61}, \tan\theta = \frac{60}{11},$ $\csc\theta = -\frac{61}{60}, \sec\theta = -\frac{61}{11}, \cot\theta = \frac{11}{60}$

55. $b = 8; \alpha \approx 36.9°; \beta \approx 53.1°$

57. $1 - \sin^2\theta + \cot^2\theta - \sin^2\theta \cot^2\theta$
$= 1 - \sin^2\theta - \sin^2\theta \cot^2\theta + \cot^2\theta$
$= 1 - \sin^2\theta(1 + \cot^2\theta) + \cot^2\theta$
$= 1 - \sin^2\theta(\csc^2\theta) + \cot^2\theta$
$= 1 - 1 + \cot^2\theta$
$= \cot^2\theta$

59. (a) $b \approx 15.4; c \approx 11.8; \alpha = 107°$

(b) $a_1 \approx 8.5; \alpha_1 \approx 68.9°; \beta_1 \approx 61.1°$ or $a_2 \approx 1.8; \alpha_2 \approx 11.1°; \beta_2 \approx 118.9°$

(c) $a \approx 5.7; \beta \approx 58.6°; \gamma \approx 77.4°$

(d) $\alpha \approx 51.3°; \beta \approx 59.2°; \gamma \approx 69.5°$

61. (a) 5 **(b)** $\langle -26, 56 \rangle$ **(c)** -63 **(d)** About $165.75°$

63.

65. (a) $(-1, 2)$ **(b)** $\{(x, y) \mid 4x - y = -2\}$
(c) Four solutions: $\left(\pm 3, \pm\sqrt{7}\right)$ **(d)** $\left(\frac{5z - 24}{3}, \frac{z + 6}{3}, z\right)$

67. (a)   **(b)**

69. $\begin{bmatrix} -2 & -1 \\ -1.5 & -0.5 \end{bmatrix}$

71. (a) -11 **(b)** 22 **73.** $y^2 = 3x$
75. $\frac{y^2}{25} - \frac{x^2}{144} = 1$ **77.** $a_n = 4n$
79. (a) 950 **(b)** $\frac{2}{9}$ **81.** $175{,}760{,}000$
83. $16x^4 - 32x^3 + 24x^2 - 8x + 1$
85. $\frac{16}{52} = \frac{4}{13}$ **87.** $\frac{8}{20} = \frac{2}{5}$ **89.** 146.5 mi
91. 0.6 hr at 7 mi/hr; 0.7 hr at 9 mi/hr
93. $\frac{20}{9} \approx 2.22$ hr
95. (a) $C(x) = x(405 - 5x)$ **(b)** 25 or 56; the cost is
$7000 when 25 or 56 tickets are purchased.
(c) $8200; the cost is $8200 when 40 or 41 tickets are
purchased.
97. 1.125 ft   **99.** $\frac{49}{77} = \frac{7}{11}$
101. 116.6 ft   **103.** $40.6°$

CHAPTER R: Reference: Basic Concepts from Algebra and Geometry

SECTION R.1 (pp. R-5–R-7)

1. Area: 105 ft^2; perimeter: 44 ft
3. Area: 3500 m^2; perimeter: 270 m
5. Area: $3xy$; perimeter: $6x + 2y$
7. Area: $2W^2$; perimeter: $6W$
9. Area: $W(W + 5)$; perimeter: $4W + 10$
11. 20 cm^2 **13.** 20 in^2 **15.** 3686.5 m^2 **17.** $6x^2$

19. $\frac{5}{2}z^2$ **21.** Circumference: $8\pi \approx 25.1$ m; area:
$16\pi \approx 50.3$ m^2 **23.** Circumference: $38\pi \approx 119.4$ in.;
area: $361\pi \approx 1134.1$ in^2 **25.** Circumference: $4\pi x$; area:
$4\pi x^2$ **27.** $c = 61$ ft; perimeter: 132 ft   **29.** $b = 12$ cm;
perimeter: 30 cm   **31.** $a = \sqrt{51} \approx 7.1$ mm; perimeter:
$17 + \sqrt{51} \approx 24.1$ mm   **33.** Area: 9 ft^2
35. Area: $\frac{11}{2}\sqrt{104} \approx 56.1$ in^2 **37.** Volume: 24 ft^3;
surface area: 52 ft^2 **39.** Volume: 216 in^3; surface area:
240 in^2 **41.** Volume: $6x^3$; surface area: $22x^2$
43. Volume: $6xyz$; surface area: $4xy + 6xz + 12yz$
45. W^3 **47.** Volume: $36\pi \approx 113.1$ ft^3; surface area:
$36\pi \approx 113.1$ ft^2 **49.** Volume: $43.7\pi \approx 137.3$ m^3; surface
area: $41.0\pi \approx 128.8$ m^2 **51.** Volume: $\frac{1}{2}\pi \approx 1.6$ ft^3; side
surface area: $2\pi \approx 6.3$ ft^2; total surface area: $\frac{5}{2}\pi \approx 7.9$ ft^2
53. Volume: $3456\pi \approx 10{,}857.3$ mm^3; side surface area:
$576\pi \approx 1809.6$ mm^2; total surface area: $864\pi \approx 2714.3$ mm^2
55. Volume: 157.1 cm^3; side surface area: 122.7 cm^2
57. Volume: 12.6 ft^3; side surface area: 22.7 ft^2
59. Volume: 43.4 ft^3; side surface area: 57.2 ft^2
61. $x = \frac{20}{3} \approx 6.7$ **63.** $x = \frac{21}{2} = 10.5$

SECTION R.2 (pp. R-12–R-13)

1. No; $2^3 = 8$ and $3^2 = 9$.   **3.** $\frac{1}{7^n}$ **5.** 5^{m-n} **7.** 2^{mk}
9. 5000 **11.** 2^3 **13.** 4^4 **15.** 3^0 **17.** 125 **19.** -16
21. 1 **23.** $\frac{8}{27}$ **25.** $\frac{1}{16}$ **27.** $\frac{1}{64}$ **29.** 16 **31.** $\frac{64}{27}$
33. $\frac{8}{9}$ **35.** $6^{-1} = \frac{1}{6}$ **37.** $6x^3$ **39.** $10^8 = 100{,}000{,}000$
41. $\frac{5}{2}$ **43.** $8a^{-1}b^{-3} = \frac{8}{ab^3}$ **45.** $5^2 = 25$ **47.** $\frac{1}{a^6}$
49. $4x^2$ **51.** $\frac{2b}{3a^2}$ **53.** $\frac{3y^6}{x^7}$ **55.** $\frac{1}{5^3} = \frac{1}{125}$ **57.** $\frac{1}{y^8}$
59. $4^3 y^6 = 64y^6$ **61.** $\frac{4^3}{x^3} = \frac{64}{x^3}$ **63.** $\frac{z^{20}}{2^5 x^5} = \frac{z^{20}}{32x^5}$
65. $2ab$ **67.** $\frac{t^2}{16}$ **69.** $\frac{3a}{2b}$ **71.** $\frac{rs^3}{5t^5}$ **73.** $\frac{y^6}{9x^4}$ **75.** 1
77. $\frac{27t^9}{8}$ **79.** $\frac{n^3}{m^3}$ **81.** $\frac{a^{12}}{81b^4}$ **83.** $\frac{2y^2}{x}$ **85.** $\frac{1}{rt^3}$ **87.** y
89. $\frac{125r^{15}}{t^9}$

SECTION R.3 (pp. R-19–R-20)

1. $8x^3$ **3.** $-3y^7$ **5.** $6x^2 + 8x$ **7.** $3x^2 + 3x$
9. $5x^2 + 13x - 2$ **11.** $10x^2 y + 2y$
13. Degree: 2; leading coefficient: 5   **15.** Degree: 3; leading
coefficient: $-\frac{2}{5}$ **17.** Degree: 5; leading coefficient: 1
19. $3x + 12$ **21.** $3x - 2$ **23.** $-0.5x + 1$
25. $-7x^4 - 2x^2 - \frac{9}{2}$ **27.** $2z^3 + z^2 + 2z - 4$
29. $-7x^3$ **31.** $-19z^5 + 5z^2 - 3z$ **33.** $-z^4 + z^2 + 9$
35. $3x - 7$ **37.** $6x^2 - 5x + 5$ **39.** $3x^4 + 4x^2 - 4$
41. $-3x^4 - 3x - 8$ **43.** $5x^2 - 25x$ **45.** $-15x - 5$
47. $5y + 10$ **49.** $-10x - 18$ **51.** $6y^2 - 18y$
53. $-20x + 4y$ **55.** $y^2 - 2y - 35$ **57.** $-2x^2 - 3x + 9$
59. $-2x^2 + 7x - 6$ **61.** $x^2 - \frac{1}{4}x - \frac{1}{8}$

63. $2x^4 + x^2 - 1$ **65.** $x^2 - xy - 2y^2$
67. $6x^3 - 3x^2 - 3x$ **69.** $-2x^5 + x^3 - 10x$
71. $6x^4 - 12x^3 + 3x^2$ **73.** $x^3 + 3x^2 - x - 3$
75. $6x^4 - 19x^3 + 4x^2 + 19x - 10$
77. $3x^3 - 2x^2 + 6x - 4$ **79.** $x^2 - 49$
81. $9x^2 - 16$ **83.** $4x^2 - 9y^2$ **85.** $x^2 + 8x + 16$
87. $4x^2 + 4x + 1$ **89.** $x^2 - 2x + 1$
91. $4 - 12x + 9x^2$ **93.** $3x^3 - 3x$ **95.** $4 - 25x^4$

SECTION R.4 (pp. R-28–R-30)

1. $5(2x - 3)$ **3.** $x(2x^2 - 5)$ **5.** $4x(2x^2 - x + 4)$
7. $5x^2(x^2 - 3x + 3)$ **9.** $5x(3x^2 + 2x - 6)$
11. $2r^3(3r^2 - 4r + 6)$ **13.** $8x^2y^2(1 - 3y)$
15. $6mn^2(3 - 2mn)$ **17.** $-2a(2a + b - 3b^2)$
19. $(x + 3)(x^2 + 2)$ **21.** $(3x - 2)(2x^2 + 3)$
23. $(z - 5)(z^2 + 1)$ **25.** $y(y + 2)(y^2 - 5)$
27. $(x^2 + 1)(2x - 3)$ **29.** $(x^3 + 2)(2x - 1)$
31. $(2x - y)(a - 3b)$ **33.** $(x + 2)(x + 5)$
35. $(x + 2)(x + 6)$ **37.** $(z - 6)(z + 7)$
39. $(z + 3)(z + 8)$ **41.** $(4x + 3)(6x - 1)$
43. $(2x + 1)(3x - 2)$ **45.** $(1 - x)(1 + 2x)$
47. $(5 - 2x)(4 + 3x)$ **49.** $x(x - 1)(5x + 6)$
51. $3x(x + 3)(x + 1)$ **53.** $2(x - 5)(x - 2)$
55. $10t^2(t + 4)(6t - 1)$ **57.** $2m(m + 3)(2m - 1)$
59. $(x - 5)(x + 5)$ **61.** $(2x - 5)(2x + 5)$
63. $4(3x - 5)(3x + 5)$ **65.** $z^2(8 - 5z)(8 + 5z)$
67. $(2x - y)(2x + y)(4x^2 + y^2)$ **69.** Does not factor
71. $(2 - rt)(2 + rt)$ **73.** $(x - 5)(x + 3)$
75. $-(z + 1)(z + 5)$ **77.** $(x + 1)^2$ **79.** $(2x + 5)^2$
81. $(x - 6)^2$ **83.** $z(3z - 1)^2$ **85.** $y(3y + 5)^2$
87. $(2x - 3y)^2$ **89.** $ab(3a - 2)^2$
91. $(x - 1)(x^2 + x + 1)$ **93.** $(y + z)(y^2 - yz + z^2)$
95. $(2x - 3)(4x^2 + 6x + 9)$ **97.** $x(x + 5)(x^2 - 5x + 25)$
99. $(2r^2 - t)(4r^4 + 2r^2t + t^2)$
101. $10(m^3 - 3n^2)(m^6 + 3m^3n^2 + 9n^4)$
103. $(4x - 5)(4x + 5)$ **105.** $(x - 4)(x^2 + 4x + 16)$
107. $(x + 8)^2$ **109.** $(x - 8)(5x + 2)$
111. $x(x + 2)(x^2 - 2x + 4)$
113. $8(2x + y)(4x^2 - 2xy + y^2)$ **115.** $(x + 1)(3x - 8)$
117. $a(a + 3)(7a - 1)$ **119.** $(x^2 + 3)(2x - 1)$
121. $x^2(x - 5)(2x + 5)$ **123.** $(x^2 + 1)(2x^2 + 3)$
125. $(x^2 + 1)(x + 3)$ **127.** $5(x^2 + 2)(x - 1)$
129. $(a + b)(x - y)$ **131.** $2(3x + 1)^2$
133. $-4x(x - 3)^2$ **135.** $(3x - 2)(9x^2 + 6x + 4)$
137. $-x(x + 2)(x^2 - 2x + 4)$
139. $(x - 1)(x - 2)(x^2 + x + 1)$
141. $(r - 2)(r + 2)(r^2 + 4)$ **143.** $(5x - 2a)(5x + 2a)$
145. $2(x - y)(x + y)(x^2 + y^2)$ **147.** $3x(3x - 1)(x + 1)$
149. $(z - 5)(z + 1)$
151. $3(x + 1)(x^2 - x + 1)(x - 3)(x + 3)$
153. $2(x + 2)^2(x + 4)^3(x + 3)$

155. $2(6x + 1)(8x - 3)^3(x - 2)$
157. $2x(5x - 1)^5(7x - 1)$

SECTION R.5 (pp. R-37–R-39)

1. $2x$ **3.** $x + 5$ **5.** $x + 4$ **7.** $\frac{1}{2x - 1}$ **9.** $-\frac{1}{4}$
11. $\frac{x}{x + 1}$ **13.** $a^2 - ab + b^2$ **15.** $\frac{1}{6}$ **17.** $\frac{2}{3}$ **19.** $\frac{1}{2}$
21. $\frac{4}{5}$ **23.** 1 **25.** $-\frac{1}{7}$ **27.** $\frac{37}{33}$ **29.** $\frac{7}{10}$ **31.** $\frac{55}{84}$
33. $\frac{3}{2x}$ **35.** 1 **37.** $\frac{x}{2x - 5}$ **39.** $x(x + 3)$
41. $\frac{(x - 1)(x + 5)}{2(2x - 3)}$ **43.** $\frac{4}{b^3}$ **45.** $\frac{3a(3a + 1)}{a + 1}$
47. $\frac{x^2}{(x - 5)(x^2 - 1)}$ **49.** $\frac{1}{x^2}$ **51.** $\frac{(x - 2)(x - 1)}{(x + 2)^2}$
53. 1 **55.** $\frac{15}{y^2}$ **57.** $\frac{1}{x}$ **59.** 36 **61.** $10a^3$
63. $z(z - 4)^2$ **65.** $(x - 2)(x - 3)^2$ **67.** $7(x + 1)$
69. $(x + 4)(x - 4)$ **71.** $2(2x + 1)(x - 2)$ **73.** $\frac{7}{x + 1}$
75. $-\frac{1}{x + 1}$ **77.** $\frac{x^2 - x - 1}{x(x + 4)}$ **79.** $\frac{-4x^2 + x + 2}{x^2}$
81. $\frac{x^2 + 5x - 34}{(x - 5)(x - 3)}$ **83.** $\frac{-2(x^2 - 4x + 2)}{(x - 5)(x - 3)}$ **85.** $\frac{x(5x + 16)}{(x - 3)(x + 3)}$
87. $-\frac{1}{2}$ **89.** $\frac{6x^2 - 1}{(x - 5)(3x - 1)}$ **91.** $\frac{(x + 1)^2}{(x - 1)^2(x + 3)}$
93. $\frac{x^2 + 4x - 2}{(x - 4)(x - 1)(x + 2)}$ **95.** $\frac{x^2 + x + 2}{(x - 2)(x + 2)^2}$
97. $\frac{3(x^2 - xy - y)}{(x - y)^2}$ **99.** $\frac{x^3 - x + 2}{(x - 1)(x + 1)}$ **101.** $\frac{9t^2 - 17t + 2}{t(t - 2)(t - 1)}$
103. -3 **105.** $-1, \frac{1}{3}$ **107.** -1 **109.** $-2, 1$ **111.** $\frac{5}{3}$
113. $\frac{x + 1}{x - 1}$ **115.** $\frac{x}{3x - 20}$ **117.** $\frac{2}{3 - x}$
119. $\frac{3(x + 1)}{(x + 3)(2x - 7)}$ **121.** $\frac{4x(x + 5)}{(x - 5)(2x + 5)}$ **123.** $\frac{ab}{2(a + b)}$

SECTION R.6 (pp. R-44–R-45)

1. $-5, 5$ **3.** $-\frac{4}{5}, \frac{4}{5}$ **5.** $-3.32, 3.32$ **7.** 12 **9.** 4.80
11. $\frac{2}{7}$ **13.** $-b$ **15.** 3 **17.** -2 **19.** $\frac{1}{3}$ **21.** b^3
23. 3 **25.** -2.24 **27.** 3 **29.** -4 **31.** 1.71
33. $-x^3$ **35.** $4x^2$ **37.** 3 **39.** -1.48 **41.** $\sqrt{6}$
43. $\sqrt{xy}$ **45.** $\frac{1}{\sqrt[5]{y}}$ **47.** $\sqrt[3]{27^2}$, or $\left(\sqrt[3]{27}\right)^2$; 9
49. $\sqrt[3]{(-1)^4}$, or $\left(\sqrt[3]{-1}\right)^4$; 1 **51.** $\frac{1}{\sqrt[3]{8}}; \frac{1}{2}$
53. $\frac{1}{\sqrt[5]{13^3}}$, or $\frac{1}{(\sqrt[5]{13})^3}$ **55.** 4 **57.** 4 **59.** 2 **61.** 16
63. $2^{7/6} \approx 2.24$ **65.** $\frac{2}{3}$ **67.** $4^{1/6} \approx 1.26$ **69.** $\frac{1}{2}$
71. $-\frac{1}{2}$ **73.** 2 **75.** $2^{3/2} \approx 2.83$ **77.** x^3 **79.** xy^4
81. xy^2 **83.** $\frac{y^2}{x}$ **85.** $y^{13/6}$ **87.** $\frac{x^4}{9}$ **89.** $\frac{y^3}{x}$ **91.** $y^{1/4}$
93. $\frac{1}{a^{2/3}}$ **95.** ab^2 **97.** $\frac{1}{k^2}$ **99.** $b^{3/4}$ **101.** $z^{23/12}$
103. $p^2 + p$ **105.** $x^{5/6} - x$

SECTION R.7 (pp. R-52–R-53)

1. 3 **3.** 10 **5.** 4 **7.** $\frac{3}{5}$ **9.** $\frac{1}{4}$ **11.** $\frac{x}{4}$ **13.** 3
15. 3 **17.** -2 **19.** a **21.** $\frac{x}{2}$ **23.** $2x^2$ **25.** $2x\sqrt[4]{y}$
27. $6x$ **29.** $2x^2yz^3$ **31.** $\frac{3}{2}$ **33.** $5\sqrt{z}$ **35.** $\frac{1}{a}$

37. $10\sqrt{2}$ **39.** $3\sqrt[3]{3}$ **41.** $2\sqrt{2}$ **43.** $-2\sqrt[5]{2}$

45. $2n\sqrt{2n}$ **47.** $2ab^2\sqrt[3]{3b}$ **49.** $-5xy\sqrt[3]{xy^2}$

51. $5\sqrt[3]{5t^2}$ **53.** $\dfrac{3t}{r\sqrt[4]{5r^3}}$ **55.** $\sqrt[6]{3^5}$ **57.** $2\sqrt[12]{2^5}$

59. $x\sqrt[12]{x}$ **61.** $\sqrt[12]{r^{11}t^7}$ **63.** $9\sqrt{3}$ **65.** $2\sqrt{x}-\sqrt{y}$

67. $-5\sqrt[3]{6}$ **69.** $9\sqrt{7}$ **71.** $-2\sqrt{11}$ **73.** $5\sqrt[3]{2}-\sqrt{2}$

75. $-\sqrt[3]{xy}$ **77.** $3\sqrt{x+2}$ **79.** $\dfrac{71\sqrt{2}}{10}$

81. $4\sqrt[3]{b}(5b-1)$ **83.** $20\sqrt{3z}$ **85.** $(3ab-1)\sqrt[4]{ab}$

87. $(n-2)\sqrt[3]{n}$ **89.** 2   **91.** $x-64$ **93.** $ab-c$

95. $x+\sqrt{x}-56$ **97.** $\dfrac{4\sqrt{3}}{3}$ **99.** $\dfrac{\sqrt{5}}{3}$ **101.** $\dfrac{\sqrt{3b}}{6}$

103. $\dfrac{3+\sqrt{2}}{7}$ **105.** $\sqrt{10}-2\sqrt{2}$ **107.** $\sqrt{7}+\sqrt{6}$

109. $\dfrac{z+3\sqrt{z}}{z-9}$ **111.** $\dfrac{a+2\sqrt{ab}+b}{a-b}$

APPENDIX C: Partial Fractions (p. AP-30)

1. $\dfrac{5}{3x}+\dfrac{-10}{3(2x+1)}$ **3.** $\dfrac{6}{5(x+2)}+\dfrac{8}{5(2x-1)}$

5. $\dfrac{5}{6(x+5)}+\dfrac{1}{6(x-1)}$

7. $\dfrac{-2}{x+1}+\dfrac{2}{x+2}+\dfrac{4}{(x+2)^2}$ **9.** $\dfrac{4}{x}+\dfrac{4}{1-x}$

11. $\dfrac{15}{x}+\dfrac{-5}{x+1}+\dfrac{-6}{x-1}$ **13.** $1+\dfrac{-2}{x+1}+\dfrac{1}{(x+1)^2}$

15. $x^3-x^2+\dfrac{-1}{3(2x+1)}+\dfrac{2}{3(x+2)}$

17. $\dfrac{1}{9}+\dfrac{-1}{x}+\dfrac{25}{18(3x+2)}+\dfrac{29}{18(3x-2)}$

19. $\dfrac{-3}{5x^2}+\dfrac{3}{5(x^2+5)}$ **21.** $\dfrac{-2}{7(x+4)}+\dfrac{6x-3}{7(3x^2+1)}$

23. $\dfrac{1}{4x}+\dfrac{-8}{19(2x+1)}+\dfrac{-9x-24}{76(3x^2+4)}$

25. $\dfrac{-1}{x}+\dfrac{2x}{2x^2+1}+\dfrac{2x+3}{(2x^2+1)^2}$

27. $\dfrac{-1}{x+2}+\dfrac{3}{(x^2+4)^2}$

29. $5x^2+\dfrac{3}{x}+\dfrac{-1}{x+3}+\dfrac{2}{x-1}$

APPENDIX D: Rotation of Axes (p. AP-36)

1. Circle or ellipse or a point

3. Hyperbola or two intersecting lines

5. Parabola or one line or two parallel lines

7. 30°   **9.** 60°   **11.** 22.5°

13.

$$\dfrac{x'^2}{12}+\dfrac{y'^2}{4}=1$$

15.

$$\dfrac{x'^2}{9}+\dfrac{y'^2}{4}=1$$

17.

$$\dfrac{x'^2}{4}+\dfrac{y'^2}{2}=1$$

19.

$$\dfrac{x'^2}{5}-\dfrac{3y'^2}{5}=1$$

21.

$$\dfrac{x'^2}{4}+\dfrac{y'^2}{16}=1$$

23.

$$y'^2=x'$$

25.

$$\dfrac{x''^2}{2}-\dfrac{y''^2}{10}=1$$

27.

$$x''^2\approx-8.94y''$$

29.

$$x''^2=-6y''$$

Photo Credits

Index of Applications

INDEX

Trigonometric Functions and Identities

Right Triangle Trigonometry

$$\sin \theta = \frac{\text{side opposite}}{\text{hypotenuse}} \qquad \cos \theta = \frac{\text{side adjacent}}{\text{hypotenuse}}$$

$$\csc \theta = \frac{\text{hypotenuse}}{\text{side opposite}} \qquad \sec \theta = \frac{\text{hypotenuse}}{\text{side adjacent}}$$

$$\tan \theta = \frac{\text{side opposite}}{\text{side adjacent}}$$

$$\cot \theta = \frac{\text{side adjacent}}{\text{side opposite}}$$

Hypotenuse, Side opposite, A Side adjacent C, B, θ

The Unit Circle

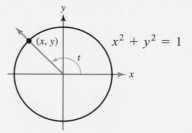

(x, y) t $x^2 + y^2 = 1$

$$\sin t = y \qquad \cos t = x \qquad \tan t = \frac{y}{x}$$

$$\csc t = \frac{1}{y} \qquad \sec t = \frac{1}{x} \qquad \cot t = \frac{x}{y}$$

The Unit Circle and Special Angles

Pythagorean

$$\sin^2 \theta + \cos^2 \theta = 1$$

$$1 + \tan^2 \theta = \sec^2 \theta$$

$$1 + \cot^2 \theta = \csc^2 \theta$$

Sum and Difference

$$\cos (\alpha + \beta) = \cos \alpha \cos \beta - \sin \alpha \sin \beta$$

$$\cos (\alpha - \beta) = \cos \alpha \cos \beta + \sin \alpha \sin \beta$$

$$\sin (\alpha + \beta) = \sin \alpha \cos \beta + \cos \alpha \sin \beta$$

$$\sin (\alpha - \beta) = \sin \alpha \cos \beta - \cos \alpha \sin \beta$$

$$\tan (\alpha + \beta) = \frac{\tan \alpha + \tan \beta}{1 - \tan \alpha \tan \beta}$$

$$\tan (\alpha - \beta) = \frac{\tan \alpha - \tan \beta}{1 + \tan \alpha \tan \beta}$$

Reciprocal

$$\sin \theta = \frac{1}{\csc \theta} \qquad \cos \theta = \frac{1}{\sec \theta} \qquad \tan \theta = \frac{1}{\cot \theta}$$

$$\csc \theta = \frac{1}{\sin \theta} \qquad \sec \theta = \frac{1}{\cos \theta} \qquad \cot \theta = \frac{1}{\tan \theta}$$

Quotient

$$\tan \theta = \frac{\sin \theta}{\cos \theta} \qquad \cot \theta = \frac{\cos \theta}{\sin \theta}$$

Negative-Angle

$$\sin (-\theta) = -\sin \theta \qquad \csc (-\theta) = -\csc \theta$$

$$\cos (-\theta) = \cos \theta \qquad \sec (-\theta) = \sec \theta$$

$$\tan (-\theta) = -\tan \theta \qquad \cot (-\theta) = -\cot \theta$$

Double-Angle

$$\sin 2\theta = 2 \sin \theta \cos \theta$$

$$\cos 2\theta = \cos^2 \theta - \sin^2 \theta$$

$$= 2 \cos^2 \theta - 1$$

$$= 1 - 2 \sin^2 \theta$$

$$\tan 2\theta = \frac{2 \tan \theta}{1 - \tan^2 \theta}$$

Power-Reducing

$$\sin^2 \theta = \frac{1 - \cos 2\theta}{2}$$

$$\cos^2 \theta = \frac{1 + \cos 2\theta}{2}$$

$$\tan^2 \theta = \frac{1 - \cos 2\theta}{1 + \cos 2\theta}$$

Half-Angle

$$\sin \frac{\theta}{2} = \pm \sqrt{\frac{1 - \cos \theta}{2}} \qquad \cos \frac{\theta}{2} = \pm \sqrt{\frac{1 + \cos \theta}{2}}$$

$$\tan \frac{\theta}{2} = \frac{1 - \cos \theta}{\sin \theta} \qquad \tan \frac{\theta}{2} = \frac{\sin \theta}{1 + \cos \theta}$$

Formulas from Geometry

Rectangle

$A = LW$

$P = 2L + 2W$

Triangle

$A = \frac{1}{2}bh$

$P = a + b + c$

Pythagorean Theorem

$c^2 = a^2 + b^2$

Circle

$C = 2\pi r$

$A = \pi r^2$

Equilateral Triangle

$A = \dfrac{\sqrt{3}}{4}s^2$

$P = 3s$

Parallelogram

$A = bh$

$P = 2a + 2b$

Trapezoid

$A = \frac{1}{2}(a + b)h$

$P = a + b + c + d$

Ellipse

$A = \pi ab$

$P \approx 2\pi\sqrt{\dfrac{a^2 + b^2}{2}}$

Sum of the Angles in a Triangle

$A + B + C = 180°$

Sum of the Angles in a Quadrangle

$A + B + C + D = 360°$

Rectangular (Parallelepiped) Box

$V = LWH$

$S = 2LW + 2LH + 2WH$

Cylinder

$V = \pi r^2 h$

$S = 2\pi rh + 2\pi r^2$

Sphere

$V = \frac{4}{3}\pi r^3$

$S = 4\pi r^2$

Cone

$V = \frac{1}{3}\pi r^2 h$

$S = \pi r^2 + \pi r\sqrt{r^2 + h^2}$